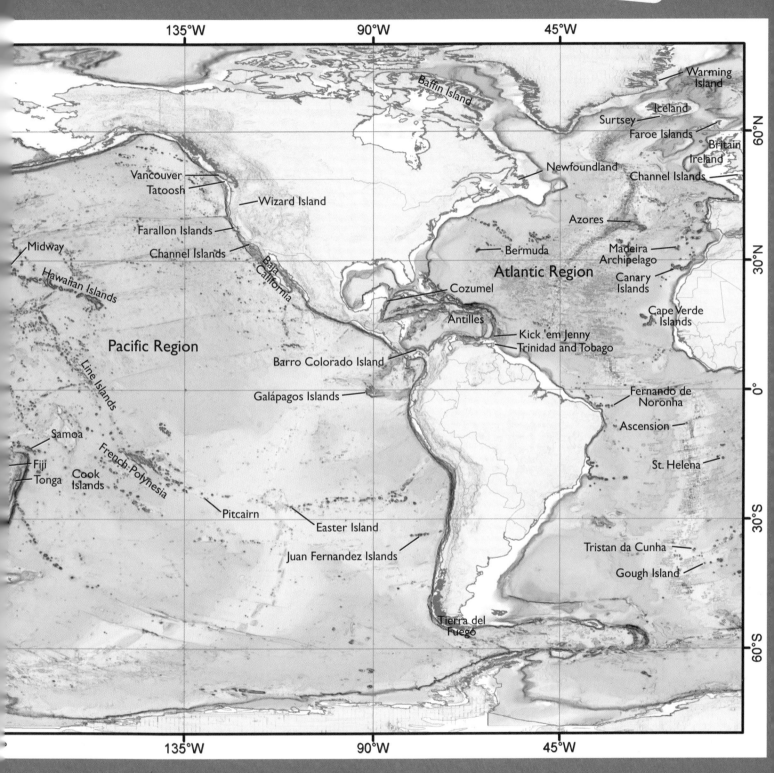

ENCYCLOPEDIA OF ISLANDS

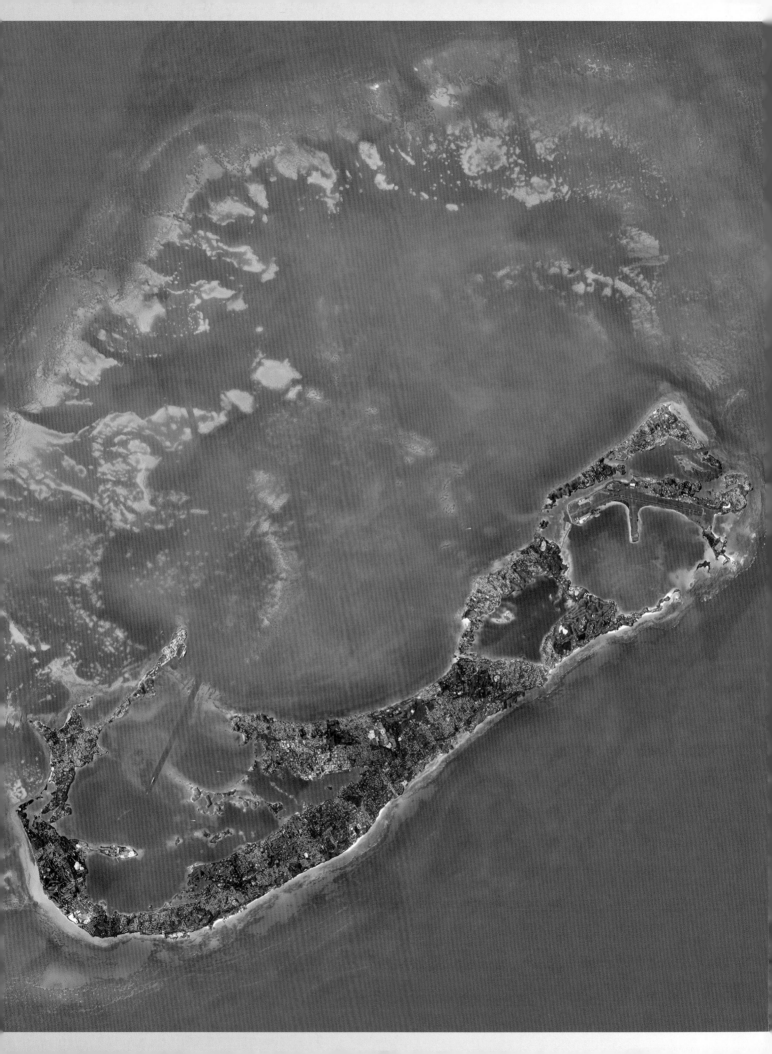

ENCYCLOPEDIA OF ISLANDS

EDITED BY

ROSEMARY G. GILLESPIE
University of California, Berkeley

DAVID A. CLAGUE
Monterey Bay Aquarium Research Institute

UNIVERSITY OF CALIFORNIA PRESS
Berkeley   Los Angeles   London

University of California Press, one of the most distinguished university presses in the United States, enriches lives around the world by advancing scholarship in the humanities, social sciences, and natural sciences. Its activities are supported by the UC Press Foundation and by philanthropic contributions from individuals and institutions. For more information, visit www.ucpress.edu.

Encyclopedias of the Natural World, No. 2

University of California Press
Berkeley and Los Angeles, California

University of California Press, Ltd.
London, England

© 2009 by the Regents of the University of California

Library of Congress Cataloging-in-Publication Data

Encyclopedia of islands / edited by Rosemary G. Gillespie and David A. Clague.
    p. cm. — (Encyclopedias of the natural world ; 2)
  Includes bibliographical references and index.
  ISBN 978-0-520-25649-1 (cloth : alk. paper)
    1. Islands—Encyclopedias. I. Gillespie, Rosemary G., 1957- II. Clague, D. A.
GB471.E53   2009
551.4203—dc22   2008037221

Printed in China
16 15 14 13 12 11 10 09
10 9 8 7 6 5 4 3 2 1

The paper used in this publication meets the minimum requirements of ANSI/NISO Z39.48-1992 (R 1997) (Permanence of Paper).{infcir}

Cover photograph: Mercherchar Island, Palau, courtesy Patrick L. Colin. Insets, from left: Lava from Piton de la Fournaise, Rèunion, © KM KRAFFT/CRI-Nancy-Lorraine, used with permission; Greene's dudleya (*Dudleya greenei*) on Santa Cruz Island, © Kathy deWet-Oleson; male orange dove (*Chrysoenas victor*), endemic to Fiji, courtesy Paddy Ryan; aerial view of Bermuda, courtesy Bermuda Zoological Society (see also frontispiece).

CONTENTS

Contents by Subject Area / xi

Contributors / xv

Guide to the Encyclopedia / xxix

Preface / xxxi

■

Adaptive Radiation / 1
Rosemary G. Gillespie

Anagenesis / 8
Tod F. Stuessy

Antarctic Islands, Biology / 10
Steven L. Chown; Jennifer E. Lee

Antarctic Islands, Geology / 17
John L. Smellie

Antilles, Biology / 20
Charles A. Woods; Florence E. Sergile

Antilles, Geology / 29
Richard E. A. Robertson

Ants / 35
Brian L. Fisher

Archaeology / 41
Marshall I. Weisler

Arctic Islands, Biology / 47
Inger Greve Alsos; Lynn Gillespie;
Yuri M. Marusik

Arctic Islands, Geology / 55
Michael J. Hambrey

Arctic Region / 59
Stephen D. Gurney

Ascension / 61
David M. Wilkinson

Atlantic Region / 63
Andreas Klügel

Atolls / 67
Edward L. Winterer

Azores / 70
Paulo A. V. Borges; Isabel R. Amorim;
Rosalina Gabriel; Regina Cunha; António
Frias Martins; Luís Silva; Ana Costa;
Virgílio Vieira

■

Baffin / 76
Lynda Dredge

Baja California: Offshore Islands / 78
Martin L. Cody

Barrier Islands / 82
Miles O. Hayes

Barro Colorado / 88
Egbert Giles Leigh, Jr.

Beaches / 91
Bruce Richmond

Bermuda / 95
Anne F. Glasspool; Wolfgang Sterrer

Biological Control / 99
Mark Gillespie; Steve Wratten

Bird Disease / 103
David Cameron Duffy

Bird Radiations / 105
Jeffrey Podos; David C. Lahti

Borneo / 111
Swee-Peck Quek

Britain and Ireland / 116
Rosemary G. Gillespie; Mark Williamson

Canary Islands, Biology / 127
Javier Francisco-Ortega; Arnoldo Santos-Guerra; Juan José Bacallado

Canary Islands, Geology / 133
Kaj Hoernle; Juan-Carlos Carracedo

Cape Verde Islands / 143
Maria Cristina Duarte; Maria Manuel Romeiras

Caroline Islands / 148
James R. Hein

Caves, as Islands / 150
David C. Culver; Tanja Pipan

Channel Islands (British Isles) / 154
Edward P. F. Rose

Channel Islands (California), Biology / 155
Aaron Moody

Channel Islands (California), Geology / 161
Janet Hammond Gordon

Cichlid Fish / 165
Ole Seehausen

Climate Change / 169
David A. Burney

Climate on Islands / 171
Thomas A. Schroeder

Cold Seeps / 174
Charles K. Paull

Comoros / 177
D. James Harris; Sara Rocha

Continental Islands / 180
David M. Watson

Convergence / 188
Tadashi Fukami

Cook Islands / 191
Tegan Hoffmann

Coral / 197
Daphne G. Fautin; Robert W. Buddemeier

Cozumel / 203
Alfredo D. Cuarón

Crickets / 206
Daniel Otte; Greg Cowper

Cyprus / 212
Ioannis Panayides

Darwin and Geologic History / 217
James H. Natland

Deforestation / 221
Barry V. Rolett

Dispersal / 224
Isabelle Olivieri

Dodo / 228
J. P. Hume

Drosophila / 232
Patrick M. O'Grady; Karl N. Magnacca; Richard T. LaPoint

Dwarfism / 235
Shai Meiri; Pasquale Raia

Earthquakes / 240
Paul Okubo; David A. Clague

Easter Island / 244
Grant McCall

Ecological Release / 251
Rosemary G. Gillespie

Endemism / 253
Quentin C. Cronk; Diana M. Percy

Ephemeral Islands, Biology / 258
Sofie Vandendriessche; Magda Vincx; Steven Degraer

Ephemeral Islands, Geology / 259
Kazuhiko Kano

Erosion, Coastal / 261
Wayne Stephenson

Eruptions: Laki and Tambora / 263
T. Thordarson

Ethnobotany / 271
W. Arthur Whistler

Exploration and Discovery / 276
Scott M. Fitzpatrick

Extinction / 281
Kevin J. Gaston

Farallon Islands / 287
Phil Capitolo

Faroe Islands / 291
Gina E. Hannon; Simun V. Arge; Anna-Maria Fosaa; Ditlev L. Mahler; Bergur Olsen; Richard H. W. Bradshaw

Fernando de Noronha / 297
R. V. Fodor

Fiji, Biology / 298
Paddy Ryan

Fiji, Geology / 305
Howard Colley

Fish Stocks/Overfishing / 310
Jon Brodziak

Flightlessness / 311
Curtis Ewing

Fossil Birds / 318
Trevor H. Worthy

Founder Effects / 326
Michael C. Whitlock

Fragmentation / 328
Luis Cayuela

Fraser Island / 330
Brad Balukjian

French Polynesia, Biology / 332
Jean-Yves Meyer; Bernard Salvat

French Polynesia, Geology / 338
Alain Bonneville

Freshwater Habitats / 343
Alan P. Covich

Frogs / 347
Rafe M. Brown

Galápagos Finches / 352
Heather Farrington; Kenneth Petren

Galápagos Islands, Biology / 357
Terrence M. Gosliner

Galápagos Islands, Geology / 367
Dennis Geist; Karen Harpp

Gigantism / 372
Pasquale Raia

Global Warming / 376
David R. Lindberg

Granitic Islands / 380
Millard F. Coffin

Great Barrier Reef Islands, Biology / 382
Harold Heatwole

Great Barrier Reef Islands, Geology / 386
Scott G. Smithers

Greek Islands, Biology / 388
Kostas A. Triantis; M. Mylonas

Greek Islands, Geology / 392
Michael D. Higgins

Hawaiian Islands, Biology / 397
Jonathan Price

Hawaiian Islands, Geology / 404
David R. Sherrod

Honeycreepers, Hawaiian / 410
Robert C. Fleischer

Human Impacts, Pre-European / 414
Patrick V. Kirch

Hurricanes and Typhoons / 418
Thomas A. Schroeder

Hydrology / 420
Christian Depraetere; Marc Morell

Hydrothermal Vents / 424
Robert C. Vrijenhoek

Iceland / 428
Sigurdur Steinthorsson

Inbreeding / 436
Leonard Nunney

Indian Region / 437
Frederick A. Frey

Indonesia, Biology / 446
Tigga Kingston

Indonesia, Geology / 454
Robert Hall

Insect Radiations / 460
Diana M. Percy

Inselbergs / 466
Stefan Porembski

Introduced Species / 469
Daniel Simberloff

Invasion Biology / 475
George Roderick; Philippe Vernon

Island Arcs / 481
Richard J. Arculus

Island Biogeography, Theory of / 486
José María Fernández-Palacios

Island Formation / 490
Patrick D. Nunn

Island Rule / 492
Shai Meiri

■

Japan's Islands, Biology / 497
Lázaro M. Echenique-Diaz; Masakado Kawata; Jun Yokoyama

Japan's Islands, Geology / 500
S. Maruyama; S. Yanai; Y. Isozaki; D. Hirata

Juan Fernandez Islands / 507
Simon Haberle

■

Kick 'em Jenny / 510
Jan Lindsay

Kīpuka / 512
Amy G. Vandergast

Komodo Dragons / 513
Tim Jessop

Kon-Tiki / 515
Robert C. Suggs

Krakatau / 517
Robert J. Whittaker

Kurile Islands / 520
Alexander Belousov; Marina Belousova; Thomas P. Miller

■

Lakes, as Islands / 526
Shelley Arnott

Land Crabs on Christmas Island / 532
Peter Green

Landslides / 535
Simon J. Day

Land Snails / 537
Brenden S. Holland

Lava and Ash / 542
Katharine V. Cashman

Lava Tubes / 544
Jim Kauahikaua; Frank Howarth; Ken Hon

Lemurs and Tarsiers / 549
Robert D. Martin

Line Islands / 553
Christopher Charles; Stuart Sandin

Lizard Radiations / 558
Miguel Vences

Lophelia Oases / 564
Sandra Brooke

Lord Howe Island / 568
Carole S. Hickman

■

Macquarie, Biology / 573
Jenny Scott

Macquarie, Geology / 575
Arjan Dijkstra

Madagascar / 577
Steven M. Goodman

Madeira Archipelago / 582
Dora Aguin-Pombo; Miguel A. A. Pinheiro de Carvalho

Makatea Islands / 585
Lucien F. Montaggioni

Maldives / 586
Paul Kench

Mammal Radiations / 588
Lawrence R. Heaney; Steven M. Goodman

Mangrove Islands / 591
Peter Saenger

Marianas, Biology / 593
Gordon H. Rodda

Marianas, Geology / 598
Frank A. Trusdell

Marine Lakes / 603
Michael N Dawson; Laura E. Martin; Lori J. Bell; Sharon Patris

Marine Protected Areas / 607
Alan M. Friedlander

Marshall Islands / 610
Nancy Vander Velde

Mascarene Islands, Biology / 612
Christophe Thébaud; Ben H. Warren; Dominique Strasberg; Anthony Cheke

Mascarene Islands, Geology / 620
Robert A. Duncan

Mediterranean Region / 622
John Wainwright

Metapopulations / 629
Dag Øystein Hjermann

Midway / 631
Elizabeth Flint

Missionaries, Effects of / 633
Alan I. Kaplan; Vincent H. Resh

Moa / 638
Allan J. Baker

Motu / 641
Francis J. Murphy

■

New Caledonia, Biology / 643
Jérôme Murienne

New Caledonia, Geology / 645
Timothy J. Rawling

Newfoundland / 649
Harold Williams

New Guinea, Biology / 652
Allen Allison

New Guinea, Geology / 659
Hugh L. Davies

New Zealand, Biology / 665
Steven A. Trewick; Mary Morgan-Richards

New Zealand, Geology / 673
Hamish Campbell; Charles A. Landis

Nuclear Bomb Testing / 680
Edward L. Winterer

■

Oases / 686
Slaheddine Selmi; Thierry Boulinier

Oceanic Islands / 689
Patrick D. Nunn

Orchids / 696
David L. Roberts; Richard M. Bateman

Organic Falls on the Ocean Floor / 700
Craig R. Smith

Pacific Region / 702
Anthony A. P. Koppers

Palau / 715
Alan R. Olsen

Pantepui / 717
Valentí Rull

Peopling the Pacific / 720
Patrick V. Kirch

Philippines, Biology / 723
Rafe M. Brown; Arvin C. Diesmos

Philippines, Geology / 732
Graciano Yumul, Jr.; Carla Dimalanta; Karlo Queaño; Edanjarlo Marquez

Phosphate Islands / 738
James R. Hein

Pigs and Goats / 741
Elizabeth Matisoo-Smith

Pitcairn / 744
Naomi Kingston; Noeleen Smyth

Plant Disease / 747
Ulla Carlsson-Granér; Lars Ericson; Barbara E. Giles

Plate Tectonics / 752
Roger C. Searle

Pocket Basins and Deep-Sea Speciation / 755
Bruce H. Robison; William M. Hamner

Polynesian Voyaging / 758
Atholl Anderson

Popular Culture, Islands in / 761
Vincent H. Resh; Jonathan P. Resh

Population Genetics, Island Models in / 766
Jeffrey D. Lozier

Prisons and Penal Settlements / 767
Ephraim Cohen

Radiation Zone / 772
Kostas A. Triantis; Robert J. Whittaker

Rafting / 775
Christophe Abegg

Reef Ecology and Conservation / 779
Robert H. Richmond; Willy Kostka; Noah Idechong

Refugia / 785
Angus Davison

Relaxation / 787
Kenneth J. Feeley

Research Stations / 788
Neil Davies

Rodents / 792
David Towns

Rottnest Island / 796
Anne Brearley

Samoa, Biology / 799
A. C. Medeiros

Samoa, Geology / 802
James H. Natland

São Tomé, Príncipe, and Annobon / 808
D. James Harris

Seabirds / 811
Mark J. Rauzon; Sheila Conant

Sea-Level Change / 815
W. H. Berger

Seamounts, Biology / 818
Malcolm Clark

Seamounts, Geology / 821
Paul Wessel

Sexual Selection / 825
Kenneth Y. Kaneshiro; Richard T. LaPoint

Seychelles / 829
Justin Gerlach

Shipwrecks / 833
James Hayward

Silverswords / 835
Bruce G. Baldwin

Sky Islands / 839
John E. McCormack; Huateng Huang; L. Lacey Knowles

Snakes / 843
Gordon H. Rodda

Socotra Archipelago / 846
Kay Van Damme

Solomon Islands, Biology / 851
Orlo C. Steele

Solomon Islands, Geology / 854
Hugh L. Davies

Species–Area Relationship / 857
David A. Spiller; Thomas W. Schoener

Spiders / 861
Miquel A. Arnedo

Spitsbergen / 865
Maria Włodarska-Kowalczuk

Sri Lanka / 866
Colin Groves; Kelum Manamendra-Arachchi

St. Helena / 870
Philip Ashmole; Myrtle Ashmole

Sticklebacks / 873
Michael A. Bell

Succession / 877
Beatrijs Bossuyt

Surf in the Tropics / 879
Graham Symonds; Thomas C. Lippmann

Surtsey / 883
Sturla Fridriksson

Sustainability / 888
R. R. Thaman

Taiwan, Biology / 897
Man-Miao Yang; Kuang-Ying Huang

Taiwan, Geology / 902
Yue-Gau Chen

Tasmania / 904
Alastair M. M. Richardson

Tatoosh / 909
Egbert Giles Leigh, Jr.; Robert T. Paine

Taxon Cycle / 912
Mandy L. Heddle

Tides / 914
Marlene Noble

Tierra del Fuego / 917
Matthew J. James; John M. Woram

Tonga / 918
Donald R. Drake

Tortoises / 921
Charles R. Crumly

Trinidad and Tobago / 926
Christopher K. Starr

Tristan da Cunha and Gough Island / 929
Peter G. Ryan

Tsunamis / 933
Emile A. Okal

Vancouver / 937
Martin L. Cody

Vanuatu / 939
Jérôme Munzinger

Vegetation / 941
Dieter Mueller-Dombois

Vicariance / 947
Michael Heads

Volcanic Islands / 950
John M. Sinton

Voyage of the *Beagle* / 954
Jere H. Lipps

Wallace, Alfred Russel / 962
Elin Claridge

Wallace's Line and Other Biogeographic Boundaries / 967
Jeremy D. Holloway

Warming Island / 971
Kurt M. Cuffey

Whale Falls / 973
Amy Baco

Whales and Whaling / 975
Joe Roman

Wizard Island / 979
David W. Ramsey

Zanzibar / 982
N. D. Burgess; R. A. D. Burgess

Glossary / 987

Index / 1019

CONTENTS BY SUBJECT AREA

GEOGRAPHY

Arctic Region

Atlantic Region

Indian Region

Mediterranean Region

Pacific Region

ISLAND TYPES

Atolls

Barrier Islands

Caves, as Islands

Cold Seeps

Continental Islands

Ephemeral Islands, Biology

Ephemeral Islands, Geology

Granitic Islands

Hydrothermal Vents

Inselbergs

Island Arcs

Kīpuka

Lakes, as Islands

Lava Tubes

Lophelia Oases

Makatea Islands

Mangrove Islands

Marine Lakes

Motu

Oases

Oceanic Islands

Organic Falls on the Ocean Floor

Pantepui

Phosphate Islands

Pocket Basins and Deep-Sea Speciation

Seamounts, Biology

Seamounts, Geology

Sky Islands

Volcanic Islands

Whale Falls

IMPORTANT ISLANDS

Ascension

Azores

Baffin

Baja California: Offshore Islands

Bermuda

Borneo

Britain and Ireland

Cape Verde Islands

Caroline Islands

Channel Islands (British Isles)

Comoros

Cook Islands

Cozumel

Cyprus

Easter Island

Farallon Islands

Faroe Islands

Fernando de Noronha

Fraser Island

Iceland

Juan Fernandez Islands

Krakatau

Kurile Islands

Line Islands

Lord Howe Island

Madagascar

Madeira

Maldives

Marshall Islands

Newfoundland

Palau

Rottnest Island

São Tomé, Príncipe, and Annobon

Seychelles

Socotra Archipelago

Spitsbergen

Sri Lanka

St. Helena

Surtsey

Tasmania

Tonga

Trinidad and Tobago

Tristan da Cunha and Gough Island

Vancouver

Vanuatu

Warming Island

Wizard Island

Zanzibar

GEOLOGY

Antarctic Islands, Geology

Antilles, Geology

Arctic Islands, Geology

Beaches

Canary Islands, Geology

Channel Islands (California), Geology

Fiji, Geology

French Polynesia, Geology

Galápagos Islands, Geology

Great Barrier Reef Islands, Geology

Greek Islands, Geology

Hawaiian Islands, Geology

Indonesia, Geology

Japan's Islands, Geology

Kick 'em Jenny

Macquarie, Geology

Marianas, Geology

Mascarene Islands, Geology

New Caledonia, Geology

New Guinea, Geology

New Zealand, Geology

Philippines, Geology

Samoa, Geology

Solomon Islands, Geology

Taiwan, Geology

GEOLOGIC PROCESSES

Earthquakes

Erosion, Coastal

Eruptions: Laki and Tambora

Island Formation

Landslides

Lava and Ash

Plate Tectonics

BIOGEOGRAPHY

Antarctic Islands, Biology

Antilles, Biology

Arctic Islands, Biology

Canary Islands, Biology

Channel Islands (California), Biology

Fiji, Biology

French Polynesia, Biology

Galápagos Islands, Biology

Great Barrier Reef Islands, Biology

Greek Islands, Biology

Hawaiian Islands, Biology

Indonesia, Biology

Japan's Islands, Biology

Macquarie, Biology

Marianas, Biology

Mascarene Islands, Biology

New Caledonia, Biology

New Guinea, Biology

New Zealand, Biology

Philippines, Biology

Samoa, Biology

Solomon Islands, Biology

Taiwan, Biology

ECOLOGY AND EVOLUTION

Adaptive Radiation

Anagenesis

Convergence

Dispersal

Dwarfism

Ecological Release

Endemism

Flightlessness

Fossil Birds

Founder Effects

Fragmentation

Freshwater Habitats

Gigantism

Inbreeding

Invasion Biology

Island Biogeography, Theory of

Island Rule

Metapopulations

Population Genetics, Island Models in

Radiation Zone

Rafting

Refugia

Relaxation

Sexual Selection

Species-Area Relationship

Succession

Taxon Cycle

Vegetation

Vicariance

Wallace's Line and Other Biogeographic Boundaries

OCEANOGRAPHY AND CLIMATOLOGY

Climate Change

Climate on Islands

Hurricanes and Typhoons

Hydrology

Sea-Level Change

Surf in the Tropics

Tides

Tsunamis

PLANTS AND ANIMALS

Bird Radiations

Cichlid Fish

Coral

Crickets

Dodo

Drosophila

Frogs

Galápagos Finches

Honeycreepers, Hawaiian

Insect Radiations

Komodo Dragons

Land Crabs on Christmas Island

Land Snails

Lemurs and Tarsiers

Lizard Radiations

Mammal Radiations

Moa

Orchids

Reef Ecology and Conservation

Seabirds

Silverswords

Spiders

Sticklebacks

Tortoises

HUMAN IMPACT

Ants

Biological Control

Bird Disease

Deforestation

Ethnobotany

Extinction

Fish Stocks/Overfishing

Global Warming

Introduced Species

Marine Protected Areas

Pigs and Goats

Plant Disease

Popular Culture, Islands in

Research Stations

Rodents

Snakes

Sustainability

Whales and Whaling

HISTORY AND PRE-HISTORY

Archaeology

Barro Colorado Island

Darwin and Geologic History

Exploration and Discovery

Human Impacts, Pre-European

Kon-Tiki

Midway

Missionaries, Effects of

Nuclear Bomb Testing

Peopling the Pacific

Pitcairn

Polynesian Voyaging

Prisons and Penal Settlements

Shipwrecks

Tatoosh

Tierra del Fuego

Voyage of the *Beagle*

Wallace, Alfred Russel

CONTRIBUTORS

CHRISTOPHE ABEGG
German Primate Centre
Padang, Sumatra Barat, Indonesia
Rafting

DORA AGUIN-POMBO
University of Madeira, Portugal
Madeira Archipelago

ALLEN ALLISON
Bishop Museum
Honolulu, Hawaii
New Guinea, Biology

INGER GREVE ALSOS
University Centre of Svalbard
Longyearbyen, Norway
Arctic Islands, Biology

ISABEL R. AMORIM
University of the Azores
Terceira, Portugal
Azores

ATHOLL ANDERSON
Australian National University, Canberra
Polynesian Voyaging

RICHARD J. ARCULUS
Australian National University, Canberra
Island Arcs

SIMUN V. ARGE
National Museum of the Faroe Islands, Tórshavn
Faroe Islands

MIQUEL A. ARNEDO
University of Barcelona, Spain
Spiders

SHELLEY ARNOTT
Queen's University
Kingston, Ontario, Canada
Lakes, as Islands

MYRTLE ASHMOLE
Kidston Mill
Peebles, Scotland, United Kingdom
St. Helena

PHILIP ASHMOLE
Kidston Mill
Peebles, Scotland, United Kingdom
St. Helena

JUAN JOSÉ BACALLADO
Museum of Natural Sciences
Tenerife, Spain
Canary Islands, Biology

AMY BACO
Associated Scientists at Woods Hole
Woods Hole, Massachusetts
Whale Falls

ALLAN J. BAKER
Royal Ontario Museum
Toronto, Ontario, Canada
Moa

BRUCE G. BALDWIN
University of California, Berkeley
Silverswords

BRAD BALUKJIAN
University of California, Berkeley
Fraser Island

RICHARD M. BATEMAN
Royal Botanic Gardens, Kew
Richmond, Surrey, United Kingdom
Orchids

LORI J. BELL
Coral Reef Research Foundation
Koror, Palau
Marine Lakes

MICHAEL A. BELL
Stony Brook University
Stony Brook, New York
Sticklebacks

ALEXANDER BELOUSOV
Institute of Volcanology and Seismology
Petropavlovsk, Russia
Kurile Islands

MARINA BELOUSOVA
Institute of Volcanology and Seismology
Petropavlovsk, Russia
Kurile Islands

W. H. BERGER
University of California, San Diego
Sea-Level Change

ALAIN BONNEVILLE
Institut de Physique du Globe de Paris, France
French Polynesia, Geology

PAULO A. V. BORGES
University of the Azores
Terceira, Portugal
Azores

BEATRIJS BOSSUYT
University of Ghent, Belgium
Succession

THIERRY BOULINIER
Centre d'Écologie Fonctionnelle et Évolutive
Montpellier, France
Oases

RICHARD H. W. BRADSHAW
Liverpool University, United Kingdom
Faroe Islands

ANNE BREARLEY
University of Western Australia
Crawley, Western Australia, Australia
Rottnest Island

JON BRODZIAK
Pacific Islands Fisheries Science Center,
NOAA Fisheries Service
Honolulu, Hawaii
Fish Stocks/Overfishing

SANDRA BROOKE
Marine Conservation Biology Institute
Bellevue, WA
Lophelia Oases

RAFE M. BROWN
University of Kansas, Lawrence
Frogs
Philippines, Biology

ROBERT W. BUDDEMEIER
Kansas Geological Survey
Lawrence, Kansas
Coral

N. D. BURGESS
University of Cambridge
Cambridge, United Kingdom
Zanzibar

R. A. D. BURGESS
Cambridge Regional College
Cambridge, United Kingdom
Zanzibar

DAVID A. BURNEY
National Tropical Botanical Garden
Kalaheo, Hawaii
Climate Change

HAMISH CAMPBELL
GNS Science
Lower Hutt, New Zealand
New Zealand, Geology

PHIL CAPITOLO
Berkeley, California
Farallon Islands

JUAN-CARLOS CARRACEDO
Estación Volcanológica de Canarias
Tenerife, Spain
Canary Islands, Geology

ULLA CARLSSON-GRANÉR
Umeå University, Sweden
Plant Disease

KATHARINE V. CASHMAN
University of Oregon, Eugene
Lava and Ash

LUIS CAYUELA
University of Alcalá
Madrid, Spain
Fragmentation

CHRISTOPHER CHARLES
Scripps Institution of Oceanography
La Jolla, California
Line Islands

ANTHONY CHEKE
Oxford, United Kingdom
Mascarene Islands, Biology

YUE-GAU CHEN
National Taiwan University
Taipei, Taiwan
Taiwan, Geology

STEVEN L. CHOWN
Stellenbosch University
Matieland, South Africa
Antarctic Islands, Biology

DAVID A. CLAGUE
Monterey Bay Aquarium Research Institute
Moss Landing, California
Earthquakes

ELIN CLARIDGE
Richard B. Gump Moorea Field Station
University of California, Berkeley
Moorea, French Polynesia
Wallace, Alfred Russel

MALCOLM CLARK
National Institute of Water and Atmospheric Research
Kilbirnie, Wellington, New Zealand
Seamounts, Biology

MARTIN L. CODY
University of California, Los Angeles
Baja California: Offshore Islands
Vancouver

MILLARD F. COFFIN
University of Southampton, United Kingdom
Granitic Islands

EPHRAIM COHEN
Hebrew University of Jerusalem
Rehovot, Israel
Prisons and Penal Settlements

HOWARD COLLEY
Higher Education Academy
Heslington, York, United Kingdom
Fiji, Geology

SHEILA CONANT
University of Hawaii, Honolulu
Seabirds

ANA COSTA
University of the Azores
Vairão, Portugal
Azores

ALAN P. COVICH
University of Georgia, Athens
Freshwater Habitats

GREG COWPER
Academy of Natural Sciences
Philadelphia, Pennsylvania
Crickets

QUENTIN C. CRONK
University of British Columbia
Vancouver, British Columbia, Canada
Endemism

CHARLES R. CRUMLY
University of California, Berkeley
Tortoises

ALFREDO D. CUARÓN
Servicios Ambientales, Conservación Biológica y Educación
Morelia, Michoacan, Mexico
Cozumel

KURT M. CUFFEY
University of California, Berkeley
Warming Island

DAVID C. CULVER
American University
Washington, DC
Caves, as Islands

REGINA CUNHA
University of the Azores
Varão, Portugal
Azores

HUGH L. DAVIES
University of Papua New Guinea
New Guinea, Geology
Solomon Islands, Geology

NEIL DAVIES
Richard B. Gump Moorea Field Station
University of California, Berkeley
Moorea, French Polynesia
Research Stations

ANGUS DAVISON
University of Nottingham, United Kingdom
Refugia

MICHAEL N DAWSON
University of California, Merced
Marine Lakes

SIMON J. DAY
University College London, United Kingdom
Landslides

STEVEN DEGRAER
Ghent University, Belgium
Ephemeral Islands, Biology

CHRISTIAN DEPRAETERE
Global Islands Network
Grenoble, France
Hydrology

ARVIN C. DIESMOS
National Museum of the Philippines
Manila, Philippines
Philippines, Biology

ARJAN DIJKSTRA
University of Neuchâtel, Switzerland
Macquarie, Geology

CARLA DIMALANTA
University of the Philippines
Diliman, Quezon City, Philippines
Philippines, Geology

DONALD R. DRAKE
University of Hawaii, Manoa
Tonga

LYNDA DREDGE
Geological Survey of Canada
Ottawa, Ontario, Canada
Baffin

MARIA CRISTINA DUARTE
Instituto de Investigação Científica Tropical
Lisbon, Portugal
Cape Verde Islands

DAVID CAMERON DUFFY
University of Hawaii, Manoa
Bird Disease

ROBERT A. DUNCAN
Oregon State University, Corvallis
Mascarene Islands, Geology

LÁZARO M. ECHENIQUE-DIAZ
Tohoku University
Aoba-ku, Sendai, Japan
Japan's Islands, Biology

LARS ERICSON
Umeå University, Sweden
Plant Disease

CURTIS EWING
University of California, Berkeley
Flightlessness

HEATHER FARRINGTON
University of Cincinnati, Ohio
Galápagos Finches

DAPHNE G. FAUTIN
University of Kansas, Lawrence
Coral

KENNETH J. FEELEY
Wake Forest University
Winston-Salem, North Carolina
Relaxation

JOSÉ MARÍA FERNÁNDEZ-PALACIOS
La Laguna University
Tenerife, Spain
Island Biogeography, Theory of

BRIAN L. FISHER
California Academy of Sciences
San Francisco, California
Ants

SCOTT M. FITZPATRICK
North Carolina State University, Raleigh
Exploration and Discovery

ROBERT C. FLEISCHER
Smithsonian Institution
Washington, DC
Honeycreepers, Hawaiian

ELIZABETH FLINT
U.S. Fish and Wildlife Service
Honolulu, Hawaii
Midway

R. V. FODOR
North Carolina State University, Raleigh
Fernando de Noronha

ANNA-MARIA FOSAA
Faroese Museum of Natural History, Tórshavn
Faroe Islands

JAVIER FRANCISCO-ORTEGA
Florida International University, Miami
Canary Islands, Biology

FREDERICK A. FREY
Massachusetts Institute of Technology, Cambridge
Indian Region

STURLA FRIDRIKSSON
Reykjavik, Iceland
Surtsey

ALAN M. FRIEDLANDER
University of Hawaii, Honolulu
Marine Protected Areas

TADASHI FUKAMI
Stanford University
Stanford, California
Convergence

ROSALINA GABRIEL
University of the Azores
Terceira, Portugal
Azores

KEVIN J. GASTON
University of Sheffield, United Kingdom
Extinction

DENNIS GEIST
University of Idaho, Moscow
Galápagos Islands, Geology

JUSTIN GERLACH
Nature Protection Trust of Seychelles
Cambridge, United Kingdom
Seychelles

BARBARA GILES
Umeå University, Sweden
Plant Disease

LYNN GILLESPIE
Canadian Museum of Nature, Ottawa
Arctic Islands, Biology

MARK GILLESPIE
Lincoln University
Christchurch, New Zealand
Biological Control

ROSEMARY G. GILLESPIE
University of California, Berkeley
Adaptive Radiation
Britain and Ireland
Ecological Release

ANNE F. GLASSPOOL
Bermuda Zoological Society
Flatts, Bermuda
Bermuda

STEVEN M. GOODMAN
The Field Museum
Chicago, Illinois
Madagascar
Mammal Radiations

JANET HAMMOND GORDON
Pasadena City College, California
Channel Islands (California), Geology

TERRENCE M. GOSLINER
California Academy of Sciences
San Francisco, California
Galápagos Islands, Biology

PETER GREEN
La Trobe University
Bundoora, Victoria, Australia
Land Crabs on Christmas Island

COLIN GROVES
Australian National University, Canberra
Sri Lanka

STEPHEN D. GURNEY
University of Reading, United Kingdom
Arctic Region

SIMON HABERLE
Australian National University, Canberra
Juan Fernandez Islands

ROBERT HALL
Royal Holloway University of London
Egham, Surrey, United Kingdom
Indonesia, Geology

MICHAEL J. HAMBREY
Aberystwyth University
Ceredigion, Wales, United Kingdom
Arctic Islands, Geology

WILLIAM M. HAMNER
University of California, Los Angeles
Pocket Basins and Deep-Sea Speciation

GINA E. HANNON
Southern Swedish Forest Research Centre
Alnarp, Sweden
Faroe Islands

KAREN HARPP
Colgate University
Hamilton, New York
Galápagos Islands, Geology

D. JAMES HARRIS
University of Porto
Vila do Conde, Portugal
Comoros
São Tomé, Príncipe, and Annobon

MILES O. HAYES
Research Planning, Inc.
Columbia, South Carolina
Barrier Islands

JAMES HAYWARD
University of California, Berkeley
Shipwrecks

MICHAEL HEADS
Ngaio, Wellington, New Zealand
Vicariance

LAWRENCE R. HEANEY
The Field Museum
Chicago, Illinois
Mammal Radiations

HAROLD HEATWOLE
North Carolina State University, Raleigh
Great Barrier Reef Islands, Biology

MANDY L. HEDDLE
Center for Environmental Education
Ahmedabad, India
Taxon Cycle

JAMES R. HEIN
U.S. Geological Survey
Menlo Park, California
Caroline Islands
Phosphate Islands

CAROLE S. HICKMAN
University of California, Berkeley
Lord Howe Island

MICHAEL D. HIGGINS
University of Québec
Chicoutimi, Canada
Greek Islands, Geology

D. HIRATA
Kanagawa Prefectural Museum of Natural History
Odawara, Japan
Japan's Islands, Geology

DAG ØYSTEIN HJERMANN
University of Oslo, Norway
Metapopulations

KAJ HOERNLE
Leibniz Institute of Marine Sciences (IFM–GEOMAR)
Kiel, Germany
Canary Islands, Geology

TEGAN HOFFMANN
T. C. Hoffmann and Associates, LLC
Oakland, California
Cook Islands

BRENDEN S. HOLLAND
University of Hawaii, Manoa
Land Snails

JEREMY D. HOLLOWAY
The Natural History Museum
London, United Kingdom
Wallace's Line and Other Biogeographic Boundaries

KEN HON
University of Hawaii, Hilo
Lava Tubes

FRANK HOWARTH
Hawaii Biological Survey, Bishop Museum
Honolulu, Hawaii
Lava Tubes

HUATENG HUANG
University of Michigan, Ann Arbor
Sky Islands

KUANG-YING HUANG
Yangmingshan National Park
Taipei, Taiwan
Taiwan, Biology

J. P. HUME
The Natural History Museum
London, United Kingdom
Dodo

NOAH IDECHONG
Palau National Congress, Koror
Reef Ecology and Conservation

Y. ISOZAKI
University of Tokyo, Japan
Japan's Islands, Geology

MATTHEW J. JAMES
Sonoma State University
Rohnert Park, California
Tierra del Fuego

TIM JESSOP
Zoos Victoria
Parkville, Victoria, Australia
Komodo Dragons

KENNETH Y. KANESHIRO
University of Hawaii, Manoa
Sexual Selection

KAZUHIKO KANO
Geological Survey of Japan
Tsukuba, Ibaraki, Japan
Ephemeral Islands, Geology

ALAN I. KAPLAN
El Cerrito, California
Missionaries, Effects of

JIM KAUAHIKAUA
Hawaiian Volcano Observatory
U.S. Geological Survey, Hawaii National Park
Lava Tubes

MASAKADO KAWATA
Tohoku University
Aoba-ku, Sendai, Japan
Japan's Islands, Biology

PAUL KENCH
University of Auckland, New Zealand
Maldives

NAOMI KINGSTON
National Parks and Wildlife Service
Dublin, Ireland
Pitcairn

TIGGA KINGSTON
Texas Tech University
Lubbock, Texas
Indonesia, Biology

PATRICK V. KIRCH
University of California, Berkeley
Human Impacts, Pre-European
Peopling the Pacific

ANDREAS KLÜGEL
University of Bremen, Germany
Atlantic Region

L. LACEY KNOWLES
University of Michigan, Ann Arbor
Sky Islands

ANTHONY A. P. KOPPERS
Oregon State University, Corvallis
Pacific Region

WILLY KOSTKA
Micronesia Conservation Trust
Kolonia, Pohnpei
Reef Ecology and Conservation

DAVID C. LAHTI
University of Massachusetts, Amherst
Bird Radiations

CHARLES A. LANDIS
University of Otago
Dunedin, New Zealand
New Zealand, Geology

RICHARD T. LAPOINT
University of California, Berkeley
Drosophila
Sexual Selection

JENNIFER E. LEE
Stellenbosch University
Matieland, South Africa
Antarctic Islands, Biology

EGBERT GILES LEIGH, JR.
Smithsonian Tropical Research Institute
Balboa, Panama
Barro Colorado
Tatoosh

DAVID R. LINDBERG
University of California, Berkeley
Global Warming

JAN LINDSAY
University of Auckland, New Zealand
Kick 'em Jenny

THOMAS C. LIPPMANN
Ohio State University, Columbus
Surf in the Tropics

JERE H. LIPPS
University of California, Berkeley
Voyage of the Beagle

JEFFREY D. LOZIER
University of Illinois, Urbana-Champaign
Population Genetics, Island Models in

KARL N. MAGNACCA
University of California, Berkeley
Drosophila

DITLEV L. MAHLER
National Museum
Copenhagen, Denmark
Faroe Islands

KELUM MANAMENDRA-ARACHCHI
Wildlife Heritage Trust
Colombo, Sri Lanka
Sri Lanka

EDANJARLO MARQUEZ
University of the Philippines, Manila
Philippines, Geology

LAURA E. MARTIN
University of California, Merced
Marine Lakes

ROBERT D. MARTIN
The Field Museum
Chicago, Illinois
Lemurs and Tarsiers

ANTÓNIO FRIAS MARTINS
University of the Azores
Vairão, Portugal
Azores

YURI M. MARUSIK
Institute for Biological Problems of the North
Magadan, Russia
Arctic Islands, Biology

S. MARUYAMA
Tokyo Institute of Technology
Meguro, Tokyo, Japan
Japan's Islands, Geology

ELIZABETH MATISOO-SMITH
University of Auckland, New Zealand
Pigs and Goats

GRANT MCCALL
University of New South Wales
Sydney, Australia
Easter Island

JOHN E. MCCORMACK
University of Michigan, Ann Arbor
Sky Islands

A. C. MEDEIROS
Pacific Island Ecosystems Research Center,
U.S. Geological Survey
Makawao, Hawaii
Samoa, Biology

SHAI MEIRI
Imperial College London
Ascot, United Kingdom
Dwarfism
Island Rule

JEAN-YVES MEYER
Department of Research of the Government of French Polynesia
Papeete, Tahiti, French Polynesia
French Polynesia, Biology

THOMAS P. MILLER
Alaska Volcano Observatory, U.S. Geological Survey
Anchorage, Alaska
Kurile Islands

LUCIEN F. MONTAGGIONI
University of Provence
Marseille, France
Makatea Islands

AARON MOODY
University of North Carolina, Chapel Hill
Channel Islands (California), Biology

MARC MORELL
Institut de Recherche pour le Développement
Fort de France, Martinique
Hydrology

MARY MORGAN-RICHARDS
Massey University
Palmerston North, New Zealand
New Zealand, Biology

DIETER MUELLER-DOMBOIS
University of Hawaii, Manoa
Vegetation

JÉRÔME MUNZINGER
Institut de Recherche pour le Développement
Nouméa, New Caledonia
Vanuatu

JÉRÔME MURIENNE
Harvard University
Cambridge, Massachusetts
New Caledonia, Biology

FRANCIS J. MURPHY
Richard B. Gump Moorea Field Station
University of California, Berkeley
Moorea, French Polynesia
Motu

M. MYLONAS
University of Crete
Irakleio, Greece
Greek Islands, Biology

JAMES H. NATLAND
University of Miami, Florida
Darwin and Geologic History
Samoa, Geology

MARLENE NOBLE
U.S. Geological Survey
Menlo Park, California
Tides

PATRICK D. NUNN
University of the South Pacific
Suva, Fiji
Island Formation
Oceanic Islands

LEONARD NUNNEY
University of California, Riverside
Inbreeding

PATRICK M. O'GRADY
University of California, Berkeley
Drosophila

EMILE A. OKAL
Northwestern University
Evanston, Illinois
Tsunamis

PAUL OKUBO
Hawaiian Volcano Observatory
U.S. Geological Survey, Hawaii National Park
Earthquakes

ISABELLE OLIVIERI
University of Montpellier, France
Dispersal

ALAN R. OLSEN
Belau National Museum
Koror, Palau
Palau

BERGUR OLSEN
Faroese Fisheries Laboratory, Tórshavn
Faroe Islands

DANIEL OTTE
Academy of Natural Sciences
Philadelphia, Pennsylvania
Crickets

ROBERT T. PAINE
University of Washington, Seattle
Tatoosh

IOANNIS PANAYIDES
Cyprus Geological Survey
Lefkosai, Nicosia, Cyprus
Cyprus

SHARON PATRIS
Coral Reef Research Foundation
Koror, Palau
Marine Lakes

CHARLES K. PAULL
Monterey Bay Aquarium Research Institute
Moss Landing, California
Cold Seeps

DIANA M. PERCY
University of British Columbia
Vancouver, British Columbia, Canada
Endemism
Insect Radiations

KENNETH PETREN
University of Cincinnati, Ohio
Galápagos Finches

MIGUEL A. A. PINHEIRO DE CARVALHO
University of Madeira, Portugal
Madeira Archipelago

TANJA PIPAN
Karst Research Institute
Postojna, Slovenia
Caves, as Islands

JEFFREY PODOS
University of Massachusetts, Amherst
Bird Radiations

STEFAN POREMBSKI
University of Rostock, Germany
Inselbergs

JONATHAN PRICE
University of Hawaii, Hilo
Hawaiian Islands, Biology

KARLO QUEAÑO
Department of Environment and Natural Resources
Diliman, Quezon City, Philippines
Philippines, Geology

SWEE-PECK QUEK
Harvard University
Cambridge, Massachusetts
Borneo

PASQUALE RAIA
University of Naples, Italy
Dwarfism
Gigantism

DAVID W. RAMSEY
U.S. Geological Survey
Vancouver, Washington
Wizard Island

MARK J. RAUZON
Marine Endeavours
Oakland, California
Seabirds

TIMOTHY J. RAWLING
University of Melbourne
Melbourne, Victoria, Australia
New Caledonia, Geology

JONATHAN P. RESH
Undaunted Design Co.
Chicago, Illinois
Popular Culture, Islands in

VINCENT H. RESH
University of California, Berkeley
Missionaries, Effects of
Popular Culture, Islands in

ALASTAIR M. M. RICHARDSON
University of Tasmania
Hobart, Tasmania, Australia
Tasmania

BRUCE RICHMOND
U.S. Geological Survey
Santa Cruz, California
Beaches

ROBERT H. RICHMOND
University of Hawaii, Honolulu
Reef Ecology and Conservation

DAVID L. ROBERTS
Royal Botanic Gardens, Kew
Richmond, Surrey, United Kingdom
Orchids

RICHARD E. A. ROBERTSON
University of the West Indies
St. Augustine, Trinidad and Tobago, West Indies
Antilles, Geology

BRUCE H. ROBISON
Monterey Bay Aquarium Research Institute
Moss Landing, California
Pocket Basins and Deep-Sea Speciation

SARA ROCHA
University of Porto
Vila do Conde, Portugal
Comoros

GORDON H. RODDA
U.S. Geological Survey
Fort Collins, Colorado
Marianas, Biology
Snakes

GEORGE RODERICK
University of California, Berkeley
Invasion Biology

BARRY V. ROLETT
University of Hawaii, Honolulu
Deforestation

JOE ROMAN
University of Vermont, Burlington
Whales and Whaling

MARIA MANUEL ROMEIRAS
Instituto de Investigação Científica Tropical
Lisbon, Portugal
Cape Verde Islands

EDWARD P. F. ROSE
University of London
Bucks, United Kingdom
Channel Islands (British Isles)

VALENTÍ RULL
Botanic Institute of Barcelona, Spain
Pantepui

PADDY RYAN
Johnson and Wales University
Denver, Colorado
Fiji, Biology

PETER G. RYAN
University of Cape Town
Rondebosch, South Africa
Tristan da Cunha and Gough Island

PETER SAENGER
Southern Cross University
Lismore, New South Wales, Australia
Mangrove Islands

BERNARD SALVAT
University of Perpignan, France
French Polynesia, Biology

STUART SANDIN
Scripps Institution of Oceanography
La Jolla, California
Line Islands

ARNOLDO SANTOS-GUERRA
Jardín de Aclimatación de La Orotava
Tenerife, Spain
Canary Islands, Biology

THOMAS W. SCHOENER
University of California, Davis
Species–Area Relationship

THOMAS A. SCHROEDER
University of Hawaii, Manoa
Climate on Islands
Hurricanes and Typhoons

JENNY SCOTT
University of Tasmania
Hobart, Tasmania, Australia
Macquarie, Biology

ROGER C. SEARLE
Durham University, United Kingdom
Plate Tectonics

OLE SEEHAUSEN
EAWAG Center for Ecology, Evolution, and Biogeochemistry
Kastanienbaum, Switzerland
Cichlid Fish

SLAHEDDINE SELMI
Faculté des Sciences de Gabès, Tunisia
Oases

FLORENCE E. SERGILE
University of Florida, Gainesville
Antilles, Biology

DAVID R. SHERROD
U.S. Geological Survey
Vancouver, Washington
Hawaiian Islands, Geology

LUÍS SILVA
University of the Azores
Vairão, Portugal
Azores

DANIEL SIMBERLOFF
University of Tennessee, Knoxville
Introduced Species

JOHN M. SINTON
University of Hawaii, Honolulu
Volcanic Islands

JOHN L. SMELLIE
British Antarctic Survey
Cambridge, United Kingdom
Antarctic Islands, Geology

CRAIG R. SMITH
University of Hawaii, Manoa
Organic Falls on the Ocean Floor

SCOTT G. SMITHERS
James Cook University
Townsville, Queensland, Australia
Great Barrier Reef Islands, Geology

NOELEEN SMYTH
National Botanic Gardens
Dublin, Ireland
Pitcairn

DAVID A. SPILLER
University of California, Davis
Species–Area Relationship

CHRISTOPHER K. STARR
University of the West Indies
St. Augustine, Trinidad and Tobago, West Indies
Trinidad and Tobago

ORLO C. STEELE
University of Hawaii, Hilo
Solomon Islands, Biology

SIGURDUR STEINTHORSSON
University of Iceland, Reykjavik
Iceland

WAYNE STEPHENSON
University of Melbourne
Melbourne, Victoria, Australia
Erosion, Coastal

WOLFGANG STERRER
Bermuda Zoological Society
Flatts, Bermuda
Bermuda

DOMINIQUE STRASBERG
University of La Réunion
Saint-Denis, Réunion
Mascarene Islands, Biology

TOD F. STUESSY
University of Vienna, Austria
Anagenesis

ROBERT C. SUGGS
Boise, Idaho
Kon-Tiki

GRAHAM SYMONDS
CSIRO Marine and Atmospheric Research
Wembley, Western Australia, Australia
Surf in the Tropics

R. R. THAMAN
University of the South Pacific
Suva, Fiji
Sustainability

CHRISTOPHE THÉBAUD
Paul Sabatier University
Toulouse, France
Mascarene Islands, Biology

T. THORDARSON
University of Edinburgh, United Kingdom
Eruptions: Laki and Tambora

DAVID TOWNS
New Zealand Department of Conservation
Newton, Auckland, New Zealand
Rodents

STEVEN A. TREWICK
Massey University
Palmerston North, New Zealand
New Zealand, Biology

KOSTAS A. TRIANTIS
Oxford University, United Kingdom
Greek Islands, Biology
Radiation Zone

FRANK A. TRUSDELL
Hawaiian Volcano Observatory
U.S. Geological Survey, Hawaii National Park
Marianas, Geology

KAY VAN DAMME
Ghent University, Belgium
Socotra Archipelago

SOFIE VANDENDRIESSCHE
Instituut voor Landbouw- en Visserijonderzoek
Oostende, Belgium
Ephemeral Islands, Biology

AMY G. VANDERGAST
U.S. Geological Survey
San Diego, California
Kīpuka

NANCY VANDER VELDE
Majuro, Marshall Islands
Marshall Islands

MIGUEL VENCES
Technical University of Braunschweig, Germany
Lizard Radiations

PHILIPPE VERNON
University of Rennes 1
Paimpont, France
Invasion Biology

VIRGILIO VIEIRA
University of the Azores
Ponta Delgada, Portugal
Azores

MAGDA VINCX
Ghent University, Belgium
Ephemeral Islands, Biology

ROBERT C. VRIJENHOEK
Monterey Bay Aquarium Research Institute
Moss Landing, California
Hydrothermal Vents

JOHN WAINWRIGHT
University of Sheffield, United Kingdom
Mediterranean Region

BEN H. WARREN
University of La Réunion
Saint-Denis, Réunion
Mascarene Islands, Biology

DAVID M. WATSON
Charles Sturt University
Albury, New South Wales, Australia
Continental Islands

MARSHALL I. WEISLER
University of Queensland
St. Lucia, Queensland, Australia
Archaeology

PAUL WESSEL
University of Hawaii, Manoa
Seamounts, Geology

W. ARTHUR WHISTLER
University of Hawaii, Manoa
Ethnobotany

MICHAEL C. WHITLOCK
University of British Columbia
Vancouver, British Columbia, Canada
Founder Effects

ROBERT J. WHITTAKER
Oxford University, United Kingdom
Krakatau
Radiation Zone

DAVID M. WILKINSON
Liverpool John Moores University, United Kingdom
Ascension

HAROLD WILLIAMS
Memorial University
St. John's, Newfoundland, Canada
Newfoundland

MARK WILLIAMSON
University of York, United Kingdom
Britain and Ireland

EDWARD L. WINTERER
Scripps Institution of Oceanography
La Jolla, California
Atolls
Nuclear Bomb Testing

MARIA WŁODARSKA-KOWALCZUK
Institute of Oceanology, Polish Academy of Sciences
Sopot, Poland
Spitsbergen

CHARLES A. WOODS
University of Vermont, Island Pond
Antilles, Biology

JOHN M. WORAM
Rockville Centre, New York
Tierra del Fuego

TREVOR H. WORTHY
University of Adelaide
Unley, Adelaide, Australia
Fossil Birds

STEVE WRATTEN
Lincoln University
Christchurch, New Zealand
Biological Control

S. YANAI
Japan Communications, Co. Ltd.
Tokyo, Japan
Japan's Islands, Geology

MAN-MIAO YANG
National Chung Hsing University
Taichung, Taiwan
Taiwan, Biology

JUN YOKOYAMA
Yamagata University, Japan
Japan's Islands, Biology

GRACIANO YUMUL, JR.
University of the Philippines
Diliman, Quezon City, Philippines
Philippines, Geology

GUIDE TO THE ENCYCLOPEDIA

The *Encyclopedia of Islands* is a comprehensive, complete, and authoritative reference dealing with all of the physical and biological aspects of islands and island habitats. Articles are written by researchers and scientific experts and provide a broad overview of the current state of knowledge on these fascinating places. Biologists, ecologists, geologists, climatologists, oceanographers, geographers, and zoologists have contributed reviews intended for students as well as the interested general public.

In order for the reader to easily use this reference, the following summary describes the features, reviews the organization and format of the articles, and is a guide to the many ways to maximize the utility of this *Encyclopedia*.

SUBJECT AREAS

The *Encyclopedia of Islands* includes 236 topics that review the various ways scholars have studied islands. The *Encyclopedia* comprises the following subject areas:
- Geography
- Island Types
- Important Islands
- Geology
- Geologic Processes
- Biogeography
- Ecology and Evolution
- Oceanography and Climatology
- Plants and Animals
- Human Impact
- History and Prehistory

ORGANIZATION

Articles are arranged alphabetically by title. An alphabetical table of contents begins on page v, and another table of contents with articles arranged by subject area begins on page xi.

Article titles have been selected to make it easy to locate information about a particular topic. Each title begins with a key word or phrase, sometimes followed by a descriptive term. For example, "Seamounts, Biology" is the title assigned rather than "Biology of Seamounts," because *seamounts* is the key term and, therefore, more likely to be sought by readers. Articles that might reasonably appear in different places in the *Encyclopedia* are listed under alternative titles—one title appears as the full entry; the alternative title directs the reader to the full entry. For example, the alternative title "Darwin's Finches" refers readers to the entry entitled Galápagos Finches.

ARTICLE FORMAT

The articles in the *Encyclopedia* are all intended for the interested general public. Therefore, each article begins with an introduction that gives the reader a short definition of the topic and its significance. Here is an example of an introduction from the article "Biological Control":

> Biological control (or biocontrol) is the use of natural enemies to suppress pest species populations to less damaging densities. When certain native and introduced invertebrates, plants, pathogens and vertebrates increase in abundance and become pests through human influence or through other causes, economic crop damage and threats to natural resources are likely. Islands are particularly vulnerable to pest outbreaks. With high endemism, low species diversity, small land areas, and a history less affected by forces that develop adaptability compared to continents, islands are more susceptible to the effects of habitat changes and species introductions. The enhancement of the efficacy of the natural enemies of pest organisms is a potentially more environmentally sound and sustainable control option than chemical or mechanical control strategies.

Within most articles, especially the longer ones, major headings help the reader identify important subtopics within each article. The article "Cook Islands" includes the following headings: "Geology and Geomorphology," "Climate and Weather," "Biological Resources," "Settlement, History, and Culture," and "Environmental Issues."

CROSS-REFERENCES

Many of the articles in this *Encyclopedia* concern topics for which articles on related topics are also included. In order to alert readers to these articles of potential interest, cross-references are provided at the conclusion of each article. At the end of "*Drosophila,*" the following text directs readers to other articles that may be of special interest:

SEE ALSO THE FOLLOWING ARTICLES

Flightlessness / Founder Effect / Hawaiian Islands, Biology / Insect Radiations / Sexual Selection

Readers will find additional information relating to *Drosophila* in the articles listed.

BIBLIOGRAPHY

Every article ends with a short list of suggestions for "Further Reading." The sources offer reliable in-depth information and are recommended by the author as the best available publications for more lengthy, detailed, or comprehensive coverage of a topic than can be feasibly presented within the *Encyclopedia*. The citations do not represent all of the sources employed by the contributor in preparing the article. Most of the listed citations are to review articles, recent books, or specialized textbooks, except in rare cases of a classic ground-breaking scientific article or an article dealing with subject matter that is especially new and newsworthy. Thus, the reader interested in delving more deeply into any particular topic may elect to consult these secondary sources. The *Encyclopedia* functions as ingress into a body of research only summarized herein.

GLOSSARY

Almost every topic in the *Encyclopedia* deals with a subject that has specialized scientific vocabulary. An effort was made to avoid the use of scientific jargon, but introducing a topic can be very difficult without using some unfamiliar terminology. Therefore, the contributors were asked to define a selection of terms used commonly in discussion of their topics. All these terms have been collated into a glossary at the back of the volume after the last article. The glossary in this work includes over 900 terms.

INDEX

The last section of the *Encyclopedia of Islands* is a subject index consisting of more than 3,200 entries. This index includes subjects dealt with in each article, scientific names, topics mentioned within individual articles, and subjects that might not have warranted a separate stand-alone article.

ENCYCLOPEDIA WEBSITE

To access a website for the *Encyclopedia of Islands,* please visit:

http://www.ucpress.edu/books/pages/10384.php

This site provides a list of the articles, the contributors, several sample articles, published reviews, and links to a secure website for ordering copies of the *Encyclopedia*. The content of this site will evolve with the addition of new information.

PREFACE

Islands are peaceful and tranquil, and also mysterious and intriguing—paradigm, paradox, paradise, and prison—and as such have inspired scientists, writers, and painters for centuries. Metaphorically, writers have used the idea of "island" over and over—islands in the sky, an island of discontent, an island of sanity, and a tropical island of mystery—*Robinson Crusoe*, *Lord of the Flies*, *Joe vs. the Volcano*. Everybody knows what an island is. More prosaically, scholars generally define an island as an isolated piece of habitat that is surrounded by dramatically different habitat, such as water. A high moist mountain surrounded by a dry desert, a patch of fertile soil in a "sea" of jumbled rocks, or the corpse of a whale at the bottom of the ocean—all of these are also islands. Islands have been studied precisely because isolation, and many biological ideas about isolation, have been so essential in uncovering biological patterns and processes. Darwin's theory of evolution might not have emerged in the way it did had he not experienced the reality and significance of islands.

Scale and context matter. A lake in the middle of a continent may be isolated for a fish, but not so much for a horse; an oceanic island may be isolated for a frog or an insect, but not for a whale; and a rock in the middle of a forest may be isolated for an ant, though not for a pig. The *Encyclopedia of Islands* is meant to pull together all the traditional ideas of islands, as well as the biological and geological concepts that have been erected around islands as isolated locations. Isolation can be achieved in many ways, but most often through some geological process. New volcanoes grow from the sea floor to create islands; global plate tectonics fragment continents, recombine them in different configurations, and fragment them again. This book examines the geologic and other processes that create isolation, and then looks at the biologic consequences of isolation. The extent of biological isolation depends on the ability of organisms to move and disperse as well as their ability to withstand terrain that is inhospitable. In visualizing biological islands, it becomes clear that almost anything can serve as an island—a water-filled tree hole for many invertebrates; a human body for the parasites it contains; a crack in the sidewalk for a weed. Given the astonishingly broad importance of the idea of "island," our purpose is to cover a range of topics in sufficient detail to give the reader a general understanding of guiding principles, and lead them to the vast literature on island science.

Circumstantially, this reference might be thought of as a collection of islands, each article being an island itself. Our hope is that readers of this reference will "travel" from island to island, from topic to topic, and learn the myriad ways that the idea of "island" has contributed to a remarkable scientific voyage whose destination will always lie just over the horizon.

We wish to thank the many busy scientists who took the time and went to the effort to create the individual articles that constitute this volume. Many of our authors helped define the book by suggesting additional topics and proposing other potential authors. We also thank the staff at the University of California Press for their assistance in preparing this volume. In particular, Chuck Crumly originated the idea of an encyclopedia of islands, convinced us we were the right scientists to assemble it, and oversaw the developmental stages. Gail Rice expertly managed the manuscripts and revisions and persisted in extracting articles from even the most recalcitrant of contributors. Without her efforts there would be no *Encyclopedia of Islands*. We finally thank our families—George,

William, and Melrose; Andrea and Gillian—for their patience, understanding, and encouragement as our time and thoughts were diverted to islands around the world and of all types.

Rosemary G. Gillespie
University of California, Berkeley

David A. Clague
Monterey Bay Aquarium Research Institute
Moss Landing, California

A

ADAPTIVE RADIATION

ROSEMARY G. GILLESPIE

University of California, Berkeley

Adaptive radiation is one of the most important outcomes of the process of evolution, and islands are places where it is best observed. The term itself was first used by H. F. Osborn in 1902 in describing parallel adaptations and convergence of species groups on different land masses. Since then, adaptive radiation has been widely recognized and defined in multiple ways to emphasize the contribution of key features thought to underlie the phenomenon, including adaptive change, speciation within a monophyletic lineage, and time. Dolph Schluter, a prominent researcher in the field, defines it as "the evolution of ecological diversity within a rapidly multiplying lineage."

There are radiations that are not adaptive. These are mostly caused by changes in topography that, instead of opening up new habitats, have served simply to isolate a previously more widespread species. Island radiations are likely to include adaptive as well as non-adaptive evolutionary elements. In this article, the focus is on adaptive evolutionary radiations.

FACTORS UNDERLYING ADAPTIVE RADIATION

The most widely recognized "trigger" for adaptive radiation is the opening up of ecological space. This may occur following evolution of a key innovation or when climatological or geological changes lead to the appearance of novel environments. Islands of all sorts provide examples of such dynamic environments, which contribute to the emergence of novel evolutionary changes.

Key Innovations

Key innovations are, for example, adjustments in morphology/physiology that are essential to the origin of new major groups, or features that are necessary, but not sufficient, for a subsequent radiation. It appears that the new trait can enhance the efficiency with which a resource is

FIGURE 1 Adaptive radiation of weevils, genus *Cratopus* (Coleoptera, Curculionidae), on the small island of La Réunion. The southwestern Indian Ocean is a hotspot of terrestrial diversity and endemism. The genus *Cratopus* has undergone intense diversification within the islands of the Mascarenes, including a total of 91 species, of which 43 are present on La Réunion (21 endemic), 36 on Mauritius (28 endemic), and five on Rodrigues (all endemic). Research is currently under way to elucidate the nature of diversification in this group. Photographs by Ben Warren, Christophe Thébaud, Dominique Strasberg, and Antoine Franck, with permission.

utilized and hence can allow species to enter a "new" adaptive zone in which diversification occurs. For example, with the origin of powered flight in birds, new feeding strategies, life history patterns, and habitats became available. Likewise, the origin of jaws in vertebrates allowed rapid diversification of predatory lineages. However, key innovations only set the stage for changes in diversity; they do not, by themselves, cause the change.

Key innovations can occur over and over, such as heterospory in plants. They have been extensively implicated in the adaptive radiation of interacting species. For example, symbioses can be an "evolutionary innovation," allowing the abrupt appearance of evolutionary novelty, providing a possible avenue through which taxonomic partners can enter into a new set of habitats unavailable to one or both of the symbiotic partners alone. The development of toxicity in plants (which can allow them to "escape" the predatory pressure of insects) can lead to a subsequent development of tolerance to the toxin in insects, allowing the insects to radiate onto the plants. Paul Ehrlich and Peter Raven envisioned this as a steplike process in which the major radiations of herbivorous insects and plants have arisen as a consequence of repeated openings of novel adaptive zones that each has presented to the other over evolutionary history. This idea, termed the "escalation/diversification" hypothesis, has been supported by studies of phytophagous beetles. A recent study in butterflies has shown that a key innovation (evolution of the NSP glucosinolate detoxification gene) has allowed butterflies (Pierinae) to radiate onto plants (Brassicales) that had temporarily escaped herbivory from the caterpillars.

Novel Environments

When a new habitat forms close to a source, it is generally colonized by propagules from the adjacent mainland. However, if the new habitat is isolated, as in the case of the formation of islands in the ocean, then colonization may be very slow, to the extent that evolution may contribute to the formation of new species more rapidly than does colonization. It is under such conditions that some of the most remarkable adaptive radiations have occurred (e.g., Fig. 1). Geological history is punctuated with episodes of extinction, presumably induced by catastrophic environmental changes, with each extinction episode setting the stage for the subsequent adaptive radiation of new (related or unrelated) groups. Environmental change has, for example, been implicated in Phanerozoic revolutions and the repeated radiations of ammonoids throughout the geological record, in which each radiation appears to have originated from a few taxa

FIGURE 2 Adaptive radiation of Hawaiian lobeliad plants. The very large and spectacular radiation of plants (125 species in six genera) appears to have occurred together with that of its pollinators, the Hawaiian honeycreepers. Species shown are (A) *Trematolobelia macrostachys*, Poamoho Trail, Ko'olau Mountains, O'ahu; (B) *Clermontia samuelii*, Upper Hana rain forest, East Maui; (C) *Cyanea koolauensis*, Ko'olau Mountains, O'ahu; (D) *Trematolobelia macrostachys*, Wai'anae Mountains, Ka'ala, O'ahu; (E) *Clermontia kakeana*, O'ahu; (F) *Cyanea leptostegia*, Koke'e, Kaua'i; (G) *Brighamia insignis*, Kaua'i (photograph by H. St. John, courtesy of Gerald D. Carr). All photographs by Gerald D. Carr, with permission, except where noted.

that went on to give rise to morphologically diverse lineages. Some of the best examples of recent, or ongoing, adaptive radiations come from isolated islands, including oceanic archipelagoes, continental lakes, and mountaintops. Galápagos island finches, Hawaiian honeycreepers, Madagascar lemurs, and African cichlids are frequently cited as some of the best examples of adaptive radiation. The finches are well known because of their historical importance, being a focus of discussion by Darwin in the *Origin of Species*. Thirteen species of Galápagos finch (Geospizinae, Emberizidae) have diversified on the Galápagos Islands, with each filling a different niche on a given island. The clear similarity among species in this radiation, coupled with the dietary-associated modifications in beak morphologies, played a key role in

the development of Darwin's theory of evolution through natural selection. Research on the finches has continued to illuminate evolutionary principles through the work of Rosemary and Peter Grant.

Hawaiian honeycreepers (Drepaniinae, Emberizidae) also originated from a finchlike ancestor and diversified into approximately 51 species in three tribes, with extraordinary morphological and ecological diversity. However, a large proportion of these species (about 35) are now extinct.

Isolation is a relative phenomenon. Accordingly, islands that are isolated for mammals and lizards are less so for many insects, with mammal radiations known in less isolated islands such as the Philippines (radiation of Old World mice/rats, Murinae), and lizards in the Caribbean (*Anolis* diversification in the Greater Antilles); prior to human transportation, few vertebrates, except for birds and an occasional bat, succeeded in colonizing more remote islands. However, the most remote islands of the

ADAPTIVE RADIATION 3

central Pacific are well known for some of the most spectacular radiations of arthropods and plants.

Examples of Insular Radiation

Some adaptive radiations are perhaps best known because of the number and rate of species that have been produced. One of the most spectacular in this regard is the Hawaiian Drosophilidae, with an estimated 1000 species divided into two major lineages (Hawaiian *Drosophila* and *Scaptomyza*). However, the cricket genus *Laupala* has been documented as having the highest rate of speciation so far recorded in arthropods, with 4.17 species per million years, supporting the argument that divergence in courtship or sexual behavior can drive rapid speciation in animals. Very rapid rates of speciation have also been reported for the African cichlid fish, with an estimated 2.02–2.09 species being formed per million years in Lake Malawi and Lake Victoria. Here again, sexual selection is implicated in the diversification of the group. A recent study, which awaits explanation, is that of Hawaiian bees (*Hylaeus,* Colletidae), which indicates that the 60 species known to occur on the islands originated and radiated on the island of Hawaii less than 700,000 years ago.

Most island adaptive radiations occur within an archipelago setting, and the argument is often made that different islands are necessary to provide sufficient isolation for adaptive radiation to take place. Interesting in this regard is a radiation of small flightless weevils in the genus *Miocalles* (Coleoptera: Curculionidae: Cryptorhynchinae) on the single small island of Rapa in the southern Australs of French Polynesia. Rapa is home to almost half of the 140 species that occur across the western Pacific and Australia. Here, the beetles collectively feed on 24 genera of native plants and show varying degrees of host specificity, thus utilizing almost all genera of native plants found on Rapa.

Some radiations have been studied because they are readily accessible to scrutiny of the process of adaptive radiation, in part because they have occurred very recently. In this context, research of Dolph Schluter has focused on stickleback fish of the deglaciated lakes of coastal British Columbia, Canada. These lakes harbor a number of sibling species of fish, and repeated co-occurrence of pairs of species has been attributed to novel ecological opportunities provided by deglaciation and recolonization. In particular, the threespine stickleback, *Gasterosteus aculeatus,* is a species complex that has diversified in each lake such that no more than two species occur in any one lake. Interestingly, it appears that pairs of species in different lakes have evolved independently of other pairs, and species have diverged as a result of parallel episodes of selection for alternate feeding environments. Research in this system has highlighted the role of divergent natural selection as a mechanism underlying adaptive radiation.

Among plants, the largest known radiation is that of the Hawaiian lobeliads (*Brighamia, Cyanea, Clermontia, Delissia, Lobelia,* and *Trematolobelia*—Campanulaceae), with the more than 100 species now thought to have arisen from a single colonization event (Fig. 2). The radiation exhibits extraordinary diversity in vegetative and flower morphology, with species inhabiting a huge array of habitats. The diversity is considered to have arisen in concert with that of the Hawaiian honeycreepers, the lobeliads displaying a suite of morphological characteristics associated with bird pollination, including deep, tubular, long-lived inflorescences, with an abundance of nectar and no odor. Another spectacular example of adaptive radiation in plants is the Hawaiian silversword alliance, *Argyroxiphium, Dubautia, Wilkesia* (Asteraceae–Madiinae), in which 28 species are known that display a huge diversity in life form, from trees to shrubs, mats, rosettes, cushions, and vines, occurring across habitats from rain forests and bogs to desert-like settings. An analogous radiation of 23 species in the genus *Argyranthemum* (Asteraceae–Anthemideae) has occurred in the Macaronesian islands, although the largest radiation of plants in the Canaries is that of the succulent, rosette-forming species of the genus *Aeonium* (Crassulaceae).

Predisposition to Adaptive Radiation

Are certain taxa predisposed to adaptive radiation? Some have suggested that plant-associated insects are constrained by their narrow host range, which prevents adaptive diversification. However, this argument is not well supported, as multiple insects with specialized host affinities have succeeded in colonizing remote islands and have also undergone some of the most spectacular adaptive radiations. Based on the available information, there is no a priori means of predicting whether or not a species will undergo adaptive radiation upon being provided an ecological opportunity that it can exploit.

At the same time, some lineages do show multiple independent episodes of adaptive radiation; for example, Hawaiian silverswords have a parallel sister radiation of California tarweeds, and one sister group, the shrubby tarweeds (*Deinandra*), has undergone adaptive radiation on the California Channel Islands (Fig. 3). Hawaiian long-jawed spiders (genus *Tetragnatha*) have undergone independent radiations on different archipelagoes of the Pacific, presumably because they are adept at overwater dispersal and readily adapt to insular environments.

In general, although there may be a substantial random element to colonization, successful colonization of very isolated locations requires high dispersal abilities, so representation of taxa within biotas in isolated areas will be skewed toward those with high dispersal abilities. However, subsequent establishment of the initial colonists on remote islands is frequently associated with a dramatic loss of dispersal ability and/or attainment of a more specialized habitat. Indeed, loss of dispersal ability is implicated as a key factor in allowing diversification to proceed.

PROCESS OF ADAPTIVE RADIATION

Ecological Release

Ecological release has been inferred to occur at the outset of adaptive radiation. This sort of release is the expansion of range, habitat, and/or resource usage by an organism when it reaches a community from which competitors, predators, and/or parasites may be lacking. Indeed, regular cycles of ecological and distributional expansion following colonization of islands are well known and have been documented in a number of groups, with some showing subsequent shifts toward specialization, as part of the phenomenon of "taxon cycles."

Adaptive Plasticity

Behavioral and ecological plasticity have recently been implicated as playing a key role in adaptive radiation. Although initially thought to impede adaptive diversification because it allows a single taxon to exploit a broad environmental range without requiring evolutionary shifts, recent work by Mary Jane West-Eberhard indicates that adaptive plasticity (including behavior) can promote evolutionary shifts, in particular when the environment is variable. Accordingly, plasticity in Caribbean *Anolis* lizards may allow species to occupy new habitats in which they otherwise might not survive. Once in these habitats, selection may act to accentuate attributes that allowed them to live in this habitat. Recent studies of threespine stickleback, in which ancestral oceanic species have changed little since colonization and diversification of freshwater species, have lent support to the importance of behavioral plasticity in allowing adaptive radiation to proceed. An interesting example of how natural selection can promote plasticity, potentially leading to species radiation, is found in pitohui birds in New Guinea (Fig. 4).

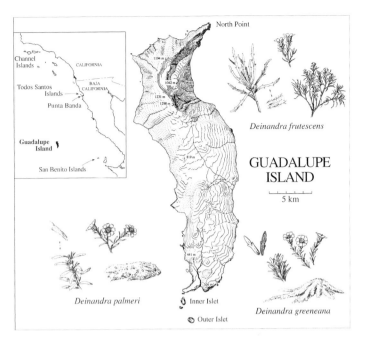

FIGURE 3 Radiation of the shrubby tarweed genus *Deinandra* on Guadalupe Island, Mexico. Guadalupe, the highest and most remote of the California Islands, is home to three endemic taxa of shrubby tarweeds (*Deinandra*). This small radiation in the California Islands parallels diversification of the sister lineage, the Hawaiian silversword alliance (Compositae-Madiinae), indicating the propensity of this group of plants to diversify in isolated settings. From B. G. Baldwin, 2007.

Speciation in Adaptive Radiation

A key feature of adaptive radiation is rapid speciation coupled with phenotypic diversification. This introduces an interesting paradox in that adaptive radiations are simultaneously characterized by minimal genetic diversity (very small numbers of individuals involved in the initial colonization) and very high morphological/ecological/behavioral diversity. Given the circumstances, a number of processes have been implicated as operating together to allow adaptive radiation to occur.

Founder Events

A founder event occurs when a new population is composed of only a few colonists, inevitably carrying only a small sample of the genetic diversity of the parent population. This small population size means that the colony may have reduced genetic variation and a non-random sample of the genes relative to the original population. Many studies have suggested that founder events play a role in adaptive radiation, as taxa within a radiation are generally characterized by small population sizes with ample opportunity for isolation. This could potentially lead to a cascade of genetic changes leading to evolutionary differentiation, an idea first formulated in the "genetic revolution" model,

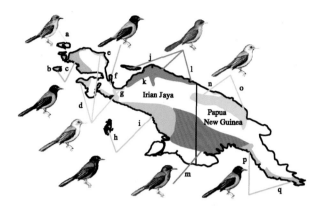

FIGURE 4 Example of morphological plasticity in a single clade showing how natural selection might lead to rapid diversification under strong natural selection. Bird species in the genus *Pitohui*, endemic to the island of Papua New Guinea, are chemically defended by a potent neurotoxic alkaloid in their skin and feathers. It appears that they cannot produce the toxin themselves but rather rely on eating a melyrid beetle and sequestering the beetle toxin. The two most toxic species are the hooded pitohui (*P. dichrous*) and the variable pitohui (*P. kirhocephalus*). *Pitohui kirhocephalus* is considered to be a single species based on the similarity of its members' songs and habits, because of clinal variation between certain races, and because no two races have been found in sympatry; *Pitohui dichrous*, in contrast, shows little geographic variation throughout its range. However, the two species are virtually identical in color pattern in many areas of co-occurrence. It has been suggested that Müllerian mimicry is driving the similar color patterns between the "mimetic" *P. kirhocephalus* phenotype and *P. dichrous*. The map shows *P. kirhocephalus* subspecific ranges and phenotypes; letters indicate the subspecies, in which certain phenotypes (ranges shown in orange and green in b, c, f, h, i, p, q) are thought to be potential mimics of *P. dichrous*. From Dumbacher and Fleischer (2001).

which posits that the reduced levels of heterozygosity following founder events affect the nature of co-adapted gene complexes. However, the precise role of founder events remains unclear: During the bottleneck, (1) a large proportion of alleles is lost, and few new mutations can occur with the population at small size; (2) genetic change will occur through drift, but the effect becomes weaker as the population starts to grow. As a result of these opposing forces, the number of beneficial mutations fixed per generation will change little because of the bottleneck. However, subsequent to a genetic bottleneck, selection can preserve alleles that are initially rare and that would otherwise tend to be lost through stochastic events. Founder events have been implicated in the adaptive radiation of such large groups as Hawaiian *Drosophila* and other insects, but there is little empirical evidence to support their role in species formation among vertebrates.

Divergent Natural Selection

Although adaptive radiation is generally associated with reduced competition, the process through which species are formed through adaptive shifts is generally considered to require competition between similar taxa. The ecological theory of adaptive radiation suggests that speciation and the evolution of morphological and ecological differences are caused by divergent natural selection resulting from interspecific competition coupled with environmental differences. Recent studies with walking stick insects and bacteria have suggested that predation may operate with (or instead of) competition to allow divergent natural selection.

Hybridization and Gene Flow

A traditional argument has been that gene flow among diverging populations, or hybridization between incipient species, acts to slow the process of diversification. However, recent research suggests that divergence between lineages can be increased through moderate levels of gene flow, a phenomenon termed "collective evolution." Likewise, interspecific hybridization may be a possible source of additional genetic variation within species, increasing the size of the gene pool on which selection may act.

Sexual Selection

Sexual selection has been linked to the diversification of species within some of the most rapid adaptive radiations. In particular, although ecological diversification still plays a role, it appears that sexual selection may drive species proliferation in African haplochromine cichlids, Hawaiian *Drosophila* flies, *Laupala* crickets, and Australasian birds of paradise. In each case, female choice is implicated in driving speciation: Males (rather than females) exhibit striking colors (in the case of the cichlids), demonstrate complex mating systems that involve modifications of the mouthparts, wings, and/or forelegs with associated elaborate behaviors and leks which are visited by females (in the flies), and have distinct courtship songs (in the crickets). Likewise, diversification in jumping spiders in the sky islands of the western United States appears to be the product of female preference for greater signal complexity or novelty.

COMMUNITY ASSEMBLY

Communities on an island are generally assembled by the interplay between colonization of species from the same niche in another region (e.g., on a mainland source or another island) or, in cases where isolation means that the number of available colonists are insufficient to fill a community, by adaptive shifts from one niche to allow a taxon to exploit a new niche. Accordingly, during the course of adaptive radiation, speciation appears to play a role similar to that of immigration—although over an extended time period—in adding species to a community.

One striking aspect of adaptive radiation is the role of convergent evolution in similar habitats, and the associated parallel evolution of similar ecological forms, resulting in the production of strikingly predictable communities during the course of diversification. Some of the best examples here are the *Anolis* lizards of the Caribbean, where studies by Jonathan Losos and colleagues have shown that on the islands of the Greater Antilles (Cuba, Hispaniola, Jamaica, and Puerto Rico) multiple species co-occur. Each species can be recognized as an "ecomorph," occupying a characteristic microhabitat (e.g., tree twigs, grass, tree trunks) with corresponding morphological and behavioral attributes named for the part of the habitat they occupy. Most remarkably, similar sets of ecomorphs are found on each island and have generally arisen through convergent evolution on each island, showing that similar communities on different islands evolved independently. Studies have now demonstrated similar patterns of multiple convergences and independent evolution of similar sets of ecomorphs among multiple island settings—for example, spiders in the Hawaiian Islands, cichlid fishes of the African Rift lakes, and Madagascan and Asian ranid frogs.

The phenomenon of repeated evolution of similar forms among species undergoing adaptive radiation has led to research on the molecular basis of such apparently complex changes. Increasingly, these studies are indicating that rather small developmental shifts may lead to large shifts in morphology (Fig. 5). Accordingly, very minor developmental shifts may allow some very striking morphological and ecological shifts, which can readily be lost or gained.

SEE ALSO THE FOLLOWING ARTICLES

Ecological Release / Founder Effects / Radiation Zone / Sexual Selection / Taxon Cycle

FURTHER READING

Baldwin, B. G. 2007. Adaptive radiation of shrubby tarweeds (*Deinandra*) in the California Islands parallels diversification of the Hawaiian silversword alliance (Compositae-Madiinae). *American Journal of Botany* 94: 237–248.

Carlquist, S., B. G. Baldwin, and G. D. Carr, eds. 2003. *Tarweeds and silverswords: evolution of the Madiinae (Asteraceae)*. St. Louis: Missouri Botanical Garden Press.

Dumbacher, J. P., and R. C. Fleischer. 2001. Phylogenetic evidence for colour pattern convergence in toxic pitohuis: Müllerian mimicry in birds? *Proceedings of the Royal Society Biological Sciences* 268: 1971–1976.

FIGURE 5 Diversity of scarab beetle "horns," used as weapons in males for access to mates. Dung beetles (Scarabaeinae) and rhinoceros beetles (Dynastinae) are both common inhabitants of islands, and the dung pats that the beetles use can themselves be considered ephemeral islands of nutrients. Sample taxa illustrated are (A) *Dynastes hercules* (Dynastinae) from the New World tropics, including different subspecies on the islands of the Lesser Antilles; (B) representative of the genus *Golofa* (*G. porteri*) (Dynastinae), a neotropical lineage known for its "Mesoamerican mountaintop" pattern of biogeography; (C) *Allomyrina dichotoma* (Dynastinae) from the Old World tropics, including Japan, Taiwan, and associated islands; (D) *Proagoderus tersidorsis* (Scarabaeinae) from South Africa; (E) *Onthophagus nigriventris* (Scarabaeidae: Scarabaeinae) from the East Africa highlands, and now in Australia and Hawaii. The huge diversity of horns appears to result from subtle changes in the relative activities of different developmental pathways. Photograph montage by Douglas J. Emlen, with permission.

Ehrlich, P. R., and P. H. Raven. 1964 Butterflies and plants: a study in coevolution. *Evolution* 18: 586–608.

Emlen, D. J., Q. Szafran, L. S. Corley, and I. Dworkin. 2006. Insulin signaling and limb-patterning: candidate pathways for the origin and evolutionary diversification of beetle "horns." *Heredity* 97: 179–191.

Givnish, T. J., and K. J. Sytsma. 1997. *Molecular evolution and adaptive radiation*. Cambridge: Cambridge University Press.

Grant, P. R., and B. R. Grant. 2007. *How and why species multiply: the radiation of Darwin's finches*. Princeton, NJ: Princeton University Press.

Losos, J. B. 2009. *Lizards in the evolutionary tree: the ecology of adaptive radiation in Anoles*. Berkeley: University of California Press.

Ricklefs, R. E., and E. Bermingham. 2007. The causes of evolutionary radiations in archipelagoes: passerine birds in the Lesser Antilles. *American Naturalist* 169: 285–297.

Schluter, D. 2000. *The ecology of adaptive radiation*. Oxford: Oxford University Press.

Seehausen, O. 2006. African cichlid fish: a model system in adaptive radiation research. *Proceedings of the Royal Society B—Biological Sciences* 273: 1987–1998.

West-Eberhard, M. J. 2003. *Developmental plasticity and evolution*. New York: Oxford University Press.

ALEUTIAN ISLANDS

SEE PACIFIC REGION

AMSTERDAM

SEE INDIAN REGION

ANAGENESIS

TOD F. STUESSY

University of Vienna, Austria

Anagenesis is a process of gradual speciation whereby only single endemic species within respective genera diverge within oceanic islands (Fig. 1). In this case, the founding immigrant population does not dramatically and rapidly change and split (cladogenesis) into two or more different species after arrival, which is what happens during the better-known pattern of adaptive radiation.

THE PROCESS OF ANAGENETIC SPECIATION

The process of speciation through anagenesis is very different from that via cladogenesis (resulting in adaptive radiation). During anagenesis an immigrant population establishes in a suitable new oceanic island habitat. In the absence of ecological opportunity, such as may occur on an island with low elevation, the pioneer population proliferates but maintains genetic cohesiveness because of the absence of geographical and ecological isolating barriers. Genetic variation begins to accumulate throughout the growing population as a result of mutation and recombination. Through time and genetic drift, enough variation accumulates such that the island population (or metapopulation) appears morphologically and genetically distinct at the specific level from continental relatives. The level of morphological divergence is not great, as selection has not been directional and intense. Despite initial genetic reduction due to founder effect in the immigrant population, over time the level of genetic variation within the island metapopulation may approximate that of the continental progenitor. This is in marked contrast to the genetic consequences of cladogenesis and adaptive radiation, whereby the amount of genetic divergence between species is minimal, accompanied, however, by dramatic morphological differences.

EVIDENCE FOR ANAGENESIS IN ISLAND ARCHIPELAGOES

Evidence from endemic vascular plants of Ullung Island, Korea, provides a good example for the process of anagenesis. Ullung is a single low island 130 km east of South Korea in the Eastern (Japan) Sea, just under 1000 m elevation and with only moderate vegetational zonation. Of the 23 endemic vascular plant taxa found on Ullung Island, most are alone in their respective genera. This suggests that these endemic species have originated by anagenetic speciation. They clearly do not belong to large, adaptively radiated species complexes.

Recent molecular studies in Ullung Island of the endemic and anagenetically derived *Dystaenia takesimana* (Apiaceae) from its Japanese progenitor, *D. ibukiensis*, provide further insight on the genetic components of anagenetic speciation. The advantage of this pair of species is that (1) they are the only two species in the genus, which excludes complex origins (e.g., via hybridization) with other potential relatives; and (2) Ullung island is only 1.8 million years old, much younger than the island of Japan, and therefore it is more likely that *D. takesimana* was derived from *D. ibukiensis* rather than the reverse. A deletion in one of the DNA sequences (*trnL-F*) in the chloroplast genome of the island endemic species, which affects the secondary structure of the molecule, also supports this interpretation. Surprisingly, the levels of genetic variation are similar in both species. This suggests that the colonizing populations may have passed through a genetic bottleneck as a result of the original founder effect, but that subsequently the level of genetic variation built back up to that of the progenitor.

FREQUENT OCCURRENCE OF ANAGENESIS IN ISLAND SYSTEMS

Ullung Island is, no doubt, an exceptional case with very high levels of endemic species having been derived

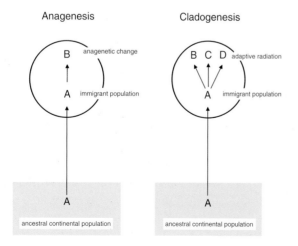

FIGURE 1 Diagrammatic contrast between anagenetic and cladogenetic speciation in oceanic islands.

TABLE 1
Features of Island Systems, Numbers of Endemic Species, and Estimated Levels of Anagenetic vs. Cladogenetic Speciation in Selected Continental and Oceanic Islands/Archipelagoes

Island System	Number of Islands	Size (km²)	Distance from Mainland (km)	Age (million years)	Elevation (m)	Vegetation Heterogeneity[a]	No. Endemic Species	Anagenetic Speciation (%)	Cladogenetic Speciation (%)
Oceanic									
Hawaii	8	16885	3660	5	4250	6	828	7	93
Canary	7	7601	100	21	3710	6	429	16	84
Tristan da Cunha	4	208	2580	18	2060	3	27	33	67
Juan Fernandez	3	100	600	4	1319	5	97	36	64
Cape Verde	12	4033	570	23	2829	2	68	37	63
Galapagos	16	7847	930	5	1707	4	133	43	57
Madeira	3	792	630	14	1862	3	96	48	52
Ogasawara	12	99	800	Tertiary	916	3	118	53	47
St. Helena	1	123	1850	15	826	2	36	53	47
Ullung	1	73	130	2	984	2	33	88	6
Continental									
Taiwan	1	35800	130	5	3950	3	724	29	71
Chatham	1	963	300	80	294	1	37	62	38
Falkland	2	8500	410	Tertiary	705	2	14	71	29

NOTE: From Stuessy et al. 2006.
[a]The higher the value, the greater the vegetation heterogeneity (vegetation zones). For calculations, refer to Stuessy et al. (2006).

by anagenesis (at least 88% of endemic species). It is of interest, therefore, to examine other islands/archipelagoes of the world, of both oceanic and continental origins, to determine relative levels of anagenesis and cladogenesis (Table 1). Results show very different levels of both processes, the lowest level of anagenesis being in the Hawaiian archipelago (7%) and the highest in Ullung Island (88%).

The levels of anagenesis or cladogenesis in different islands/archipelagoes clearly relate to two factors: (1) elevation of the island and (2) habitat heterogeneity. In islands that are high in elevation and with strong and diverse vegetational zones, the highest levels of cladogenesis occur. In contrast, low islands with few vegetation zones show the highest level of anagenetically derived species. Likewise, in continental islands also with high elevation and habitat diversity, higher levels of cladogenesis are also seen (Table 1). These levels do not relate directly to size of island, age, or distance from mainland source areas.

THE EVOLUTIONARY IMPORTANCE OF ANAGENESIS

The importance of anagenesis is that it represents another major model for speciation in oceanic islands. Cladogenetic examples involving adaptive radiation will continue to capture our imagination by virtue of their dramatic natures, but this process does not explain all endemic species in all islands. In fact, for some islands, anagenesis is fundamental for explaining patterns of diversity. We have estimated conservatively that 25% of all island species have originated by anagenesis.

An important related point is that the genetic and specific diversity of species originating from anagenesis or cladogenesis will not only be different but will also vary over time (Fig. 2). Oceanic islands have short existences, often enduring little more than 6 million years before they erode and subside under the sea. Depending upon the stage of island ontogeny, the existing genetic and specific diversity will vary, starting with little variation, peaking in mid-life of the island, and declining as the size and ecological diversity of the island diminishes through time. In making comparisons of genetic variation in populations between islands, therefore, as well as between those of continental regions and islands, we need to keep the island ontogeny firmly in mind. It makes little sense to compare implications of the founder effect, for example, between populations in continental regions and those on two different islands of very different ages. Likewise, it is important to know whether the genetic variation being analyzed occurs in a species having evolved cladogenetically or anagenetically, as this can also help suggest reasons for the observed patterns.

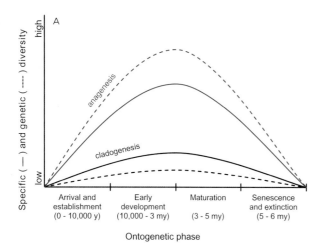

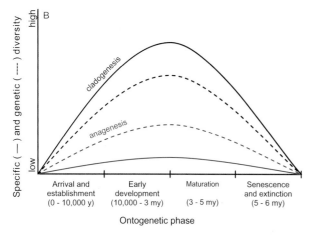

FIGURE 2 Change in genetic and specific diversity during ontogeny of floras of low (A) and high (B) elevation oceanic islands. From Stuessy (2007).

SEE ALSO THE FOLLOWING ARTICLES

Endemism / Founder Effects / Metapopulations / Population Genetics, Island Models in

FURTHER READING

Crawford, D. J., and T. F. Stuessy. 1997. Plant speciation in oceanic islands, in *Evolution and Diversification of Land Plants*. K. Iwatsuki and P. H. Raven, eds. Tokyo: Springer-Verlag, 249–267.

Pfosser, M., G. Jakubowsky, P. M. Schlüter, T. Fer, H. Kato, T. F. Stuessy, and B.-Y. Sun. 2005. Evolution of *Dystaenia takesimana* (Apiaceae) endemic to Ullung Island, Korea. *Plant Systematics and Evolution* 256: 159–170.

Stuessy, T. F. 2007. Evolution of specific and genetic diversity during ontogeny of island floras: the importance of understanding process for interpreting island biogeographic patterns, in *Biogeography in a Changing World*. M. C. Ebach and R. S. Tangney, eds. Boca Raton, FL: CRC Press, 111–123.

Stuessy, T. F., D. J. Crawford, and C. Marticorena. 1990. Patterns of phylogeny in the endemic vascular flora of the Juan Fernandez Islands, Chile. *Systematic Botany* 15: 338–346.

Stuessy, T. F., G. Jakubowsky, R. Salguero Gómez, M. Pfosser, P. M. Schlüter, T. Fer, B.-Y. Sun, and H. Kato. 2006. Anagenetic evolution in island plants. *Journal of Biogeography* 33: 1259–1265.

Stuessy, T. F., and M. Ono, eds. 1998. *Evolution and Speciation of Island Plants*. Cambridge, UK: Cambridge University Press.

Whittaker, R. J., and J. M. Fernández-Palacios. 2007. *Island Biogeography: Ecology, Evolution, and Conservation*, 2nd ed. Oxford: Oxford University Press.

ANTARCTIC ISLANDS, BIOLOGY

STEVEN L. CHOWN AND JENNIFER E. LEE

Stellenbosch University, South Africa

Antarctic islands vary substantially, from small exposed mountain peaks and large dry valleys surrounded by ice to the highly variable Southern Ocean islands, which have a considerable range of sizes and geological histories. Reflecting the diverse locations and origins of Antarctic islands, their biota varies substantially. Some ice-free areas of the continent are devoid of anything except microbial life, while others support bryophytes, lichens, nematodes, arthropods, and, occasionally, breeding seabirds. The Southern Ocean islands range from those virtually covered by glaciers to others that have lush, vegetated landscapes at their lower elevations, riddled by the burrows of breeding seabirds and home to large colonies of penguins, albatrosses, and seals.

ANTARCTIC ISLANDS

Despite its considerable area, only 0.32% of the Antarctic continent is exposed. Thus, land-surfaces are effectively islands surrounded by a sea of ice. These "islands" vary from small nunataks (Fig. 1), no larger in some cases than tens of square meters, to much larger areas such as the McMurdo Dry Valleys (4000 km^2), Bunger Hills (950 km^2), and Vestfold Hills (420 km^2). Along the Antarctic Peninsula, more conventional islands abound. Among the better known of these are those of the Scotia Arc: the South Shetland, South Orkney, and South Sandwich Islands and South Georgia. The latter is typically classified as a sub-Antarctic island, together with several archipelagos

FIGURE 1 Cairn Peak, a nunatak in the Robertskollen nunatak group, Western Dronning Maud Land (Queen Maud Land), Antarctica.

FIGURE 2 An aerial view of the Tafelberg area on sub-Antarctic Marion Island. Take note of the mire and fernbrake vegetation in the foreground and the open fellfield on the higher, exposed areas.

to the east, including the Prince Edward Islands (Fig. 2), Crozet Islands, Kerguelen Islands, Heard and McDonald Islands, and Macquarie Island. These islands all lie in the Antarctic Polar Frontal Zone, on either side of the Antarctic Polar Front. More remote and poorly known Antarctic islands include the extremely isolated Bouvetøya (~54° S, 3° E), Balleny Islands (~67° S, 163° E), and Peter I Øy (~69° S, 90° W). On biogeographic grounds, the Antarctic is typically divided into three major regions: the continental Antarctic (most of the continent, the eastern and southern regions of the Peninsula, and the Balleny Islands), maritime Antarctic (western coastal regions of the Peninsula south to Alexander Island, as well as the South Shetlands, South Orkneys, Bouvetøya, and Peter I Øy), and sub-Antarctic.

Biogeographically, the sub-Antarctic islands form an almost indistinguishable continuum with the other islands of the Southern Ocean. Indeed, in terms of the relationships among their biotas, the New Zealand sub-Antarctic islands (Auckland, Campbell, Snares, Antipodes, Bounty) and the islands of the Tristan da Cunha/Gough and St. Paul/New Amsterdam archipelagoes are not entirely distinguishable from those further to the south and are often treated as part of the Antarctic region. This is true also of the Falkland Islands, though these various treatments are not without controversy. Here, we include these island groups along with those in the sub-Antarctic as Southern Ocean islands (Fig. 3).

Given their occurrence right around the Southern Ocean, the islands differ considerably in their geological histories, past and current glacial extents, current climates, and vegetation. Although some of the islands, such as Prince Edward Island, are entirely volcanic, young (< 500,000 years), and show no signs of glaciation at the height of the last glacial maximum (LGM), others

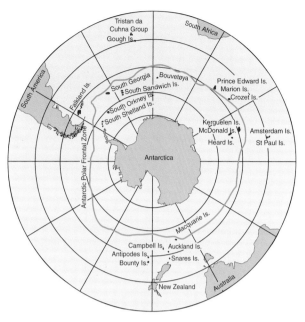

FIGURE 3 Schematic map of the position of the Southern Ocean Islands, which straddle the Antarctic Polar Frontal Zone.

have a more complex geology and history. Macquarie Island constitutes a raised section of seafloor. The Kerguelen Islands (a large archipelago) are still partly glaciated and have a complex history associated with the 100-million-year geological evolution of the large igneous province of the Kerguelen Plateau, of which some parts were subaerial (i.e., above sea level) at least as far back as 93 million years ago and consistently for at least 40 million years. The geological history of the Crozet archipelago remains something of a conundrum, whereas the Falkland Islands have a complex geological history spanning some 2500 million years.

In terms of climates, similar variation can be found, from the temperate, warmer islands, such as Gough Island, the Falklands, and the Auckland Islands, to the north of the Polar Frontal Zone, to the much colder islands south of the zone, such as South Georgia and Heard Island. The islands also differ in the extent to which they are influenced by frontal weather, and in some cases the considerable height of the islands (e.g., highest peaks of 2950 m for South Georgia; 2745 m for Heard Island) means that the climates on the weather and lee sides of the islands are wholly different.

BIOLOGICAL DIVERSITY

The Antarctic continent is depauperate by comparison with other terrestrial regions. Microbes predominate, but, unlike most areas elsewhere on the planet, terrestrial areas of the Antarctic are frequently either devoid of or very poor in other life forms. How the identity and diversity of Antarctic microbial communities compares with other areas globally is not yet certain, largely because similar sampling regimes have not been as comprehensively implemented in many other areas. Algae are the most widespread and diverse primary photosynthetic organisms in the terrestrial Antarctic, with 700–1000 species likely present on the continent. Lichens are represented by approximately 420 species, with the majority known from the maritime Antarctic. The mosses are also more diverse in the maritime than in the continental Antarctic. In total, 113 species are known from the region, of which only 18 have been recorded in continental Antarctica. Liverworts are much less diverse, with approximately 27 species known from the maritime Antarctic and only one from the continental Antarctic. Only two species of vascular plants, the grass *Deschampsia antarctica* and the forb *Colobanthus quitensis,* are known from Antarctica, and specifically the Peninsula area.

Multicellular animal life is absent from several ice-free areas. Moreover, some sites are unusual in that they lack groups of animals that are characteristic of virtually all systems globally (such as nematode worms, springtails, and mites). The continental and maritime Antarctic house approximately 29 and 54 mite species, respectively, and some 30 springtail species. Approximately 70 tardigrade (water bear) species are known from the region, and rotifer richness is in the region of 150. Nematode diversity is likely high but is poorly known because of the lack of comprehensive sampling. It is only on the Antarctic Peninsula that true insects (two species of chironomid midge: *Belgica antarctica* and *Parochlus steinenii*) can be found.

As might be expected, species richness of most groups is substantially higher on the Southern Ocean islands than on the Antarctic Continent. For example, the sub-Antarctic Prince Edward Islands (~340 km^2) are characterized by an indigenous biota comprising 90 moss species, 40 liverworts, 118 lichens, 22 vascular plants, ~63 mites, 11 springtails, and 19 insect species. The warmer Gough Island is home to 57 indigenous vascular plant species and 28 insect species. Islands further to the north, such as the Falklands and the Auckland Islands, are considerably more species rich. For example, East Falkland is home to ~149 indigenous vascular plant and 132 indigenous insect species, and in total 188 indigenous vascular plant and 237 indigenous insect species have been recorded from the Auckland islands. For many islands, higher taxa such as the bryophytes, springtails, and mites remain poorly known, owing to lack of systematic surveys and modern systematic treatments. In consequence, biogeographic conclusions concerning these groups as a whole across the region are likely to be speculative at best.

By contrast, the pelagic, though land-breeding, seabirds and seals are well known. Islands such as Possession Island in the Crozet group and Prince Edward Island in the Prince Edward Islands group are home to 33 and 28 species of seabirds respectively. On Prince Edward Island these include albatrosses (wandering, gray-headed, yellow-nosed, dark-mantled sooty and light-mantled sooty), penguins (king, rockhopper, macaroni, and gentoo), petrels (including the white-chinned, gray, great-winged, Kerguelen, soft-plumaged, Georgian diving, and gray-backed storm), and several other species, including the unusual, nonpelagic lesser sheathbill. The diversity of seals is lower, but on some islands includes breeding southern elephant seal, Antarctic fur seal, and sub-Antarctic fur seal, and vagrant species such as leopard seal.

BIOGEOGRAPHY

Although the Antarctic is typically divided into three biogeographic regions (continental Antarctic, maritime Antarctic, and sub-Antarctic, as just described), more recently an extremely clear distinction between faunas of the Antarctic Peninsula and the remainder of continental Antarctica has been recognized. The zone of division between them has been named the Gressitt Line (roughly south of Alexander Island and the English and Bryan Coasts, north of inland Ellsworth nunatak ranges, and south of the Wakefield Mountains on the east coast of the Peninsula). This distinction is not shown by the bryophytes, perhaps reflecting their greater dispersal ability. Variation in dispersal ability also accounts for differences in the biogeography of groups found on the Southern Ocean islands. For those taxa with limited dispersal ability, biogeographic affinities of the islands tend to be to the nearest continental landmass, whilst in the case of more mobile taxa, distributions tend to be more circum-Antarctic, with mean annual temperature, rather than identity of the nearest landmass, being more significant a correlate of the composition of the assemblages. Nonetheless, some relationships remain enigmatic. For example, although recent phylogeographic work has revealed that islands such as the Prince Edwards were probably colonized by ectemnorhine weevils shortly after they became subaerial (despite the present flightlessness of all members of the group), and it is known that the weevils probably arose on older islands (e.g., Kerguelen, which has been subaerial since the Cretaceous), the closest relatives of the group remain a matter of conjecture. Indeed, the origins of many groups in the Kerguelen Biogeographic Province (or South Indian Ocean Islands) remain controversial.

In other areas of the Antarctic, phylogeographic work has resolved similar questions, demonstrating, for example, that the midge species endemic to the Antarctic Peninsula achieved their distributions as a consequence of vicariance following opening of the Drake Passage ~20–40 million years ago. Similarly, springtails within and those closely related to the genus *Cryptopygus* likely diverged following glaciation of the continent 10–23 million years ago and colonized the sub-Antarctic islands much later on several different occasions (< 2 million years ago). By contrast, mites in the genera *Halozetes* and *Alaskozetes* likely colonized the continent and the sub-Antarctic islands over the last 2–3 million years. Within particular regions, the smaller-scale phylogeographic signal may vary substantially. For example, a relatively straightforward isolation by distance signal is characteristic of the springtail *Gomphiocephalus hodgsoni* in Victoria Land. By contrast, dozens of haplotypes are found in each of several mite and springtail species at Marion Island (e.g., *Cryptopygus antarcticus travei; Eupodes minutus; Halozetes fulvus*), showing no isolation by distance, but rather a complex pattern of relationships relating to glacial and volcanic dissection of the landscape. Reflecting their origins from a small number of founding individuals, virtually all alien species that have been examined to date show minimal phylogeographic structure, and often very large populations are characterized by a single haplotype.

As is the case on other isolated islands, some groups have shown adaptive radiation. For example, the ectemnorhine weevils are restricted to the islands of the South Indian Ocean province and comprise some 36 species. Endemicity in the insects is also high for other archipelagoes such as the Aucklands. By contrast, the vascular plants show much lower levels of endemicity, and this is true also of more mobile taxa such as bryophytes and the pelagic birds and seals.

From the perspective of ecological biogeography, and for better known groups, such as vascular plants, insects, seabirds and, where these are present, landbirds, variation in indigenous and alien species richness across the Southern Ocean islands has been thoroughly investigated. Vascular plant species richness covaries strongly with available energy and with island area, insect richness covaries with available energy, seabird richness with energy at sea (chlorophyll concentration) and energy, and landbird richness with indigenous insect and plant richness. These patterns are in keeping with the kinds of species–energy relationships found elsewhere on both continents and islands. However, their mechanistic basis remains as elusive here as it does elsewhere. That the alien vascular plant and insect species richness (see below) also show strong relationships with energy suggests that extinction, rather than speciation, is a major determinant of richness variation, at least over the short term. Of course, for islands, colonization rates must also be significant, and indeed for the alien species, human visitor frequency to the islands is also a significant covariate of richness variation.

SMALL-SCALE PATTERNS IN DIVERSITY

Over small spatial scales, variation in the presence, richness and abundance of Antarctic plants and animals is determined mainly by water availability, temperature (which also influences water availability), protection from wind, the availability of nutrients, the extent of lateral water movement, and the extent of soil movement and ice formation. Of these, water availability (and the elevated temperatures that drive it) is thought to be most significant on the Antarctic continent and Peninsula, while nutrient availability, soil water movement and temperature are most significant in the sub-Antarctic. At least on the continent, extreme abiotic conditions preclude life in many ice-free areas and, unlike the situation across most of the planet, abiotic rather than biotic stressors exert a controlling influence on life histories. Indeed, a few abiotic and spatial factors together may account for more than 80% of the variation in the abundance of arthropods at a given site. In addition to free water, the availability of nitrogen has a substantial influence on assemblage patterns at some continental Antarctic sites (e.g. nunataks supporting vs. those devoid of seabird colonies have very different assemblages). Nonetheless, the limiting resource in many continental Antarctic systems is carbon. Indeed, at some sites, such as the McMurdo Dry Valleys, it is clear that palaeo deposition of carbon supports current, slow biogeochemical activity. Carbon limitation is highly unusual by comparison with other regions globally.

Among the Southern Ocean islands, plant assemblages vary substantially mostly owing to climate. For example, Heard Island (53° S) has closed vegetation communities only in coastal areas and in some deglaciated valleys. Above approximately 50 m in elevation, vegetation is open, and above about 200 m comprises cryptogams only. From about 300 m the slopes are almost entirely ice-covered. By contrast, Gough Island (40° S) supports trees (*Phylica arborea, Sophora macnabiana*) and tree ferns (*Blechnum palmiforme*) at lower elevations, but above ~300 m it comprises mostly wet heath and moorland vegetation. Similarly, whereas at South Georgia closed vegetation is mostly restricted to the lowlands, a structurally complex flora can be found at the Auckland Islands.

On particular islands, plant assemblage variation is a function of the influence of salt spray and manuring by pelagic species that come ashore to breed; water availability and the extent of lateral water movement; and temperature. For example, on Marion Island at the whole-island level the most pronounced variation is with altitude. High-elevation areas are free of vascular plants and are classified as polar desert. Intermediate elevations are dominated by fellfield vegetation (with the cushion-forming *Azorella selago* (Apiaceae) most abundant). Coastal plains are characterized by mires and fernbrakes (Fig. 2), and where seabird and seal colonies are present, tussock grasslands and herbfield predominate, characterized by nitrophilous or salt-tolerant species. In low-land areas, finer spatial-scale variation is considerable and is most clearly reflected by changes in the species composition of bryophytes, in response especially to the extent of waterlogging and likelihood of lateral movement of water. Similar patterns characterize other islands, though with the absolute distance over which assemblage differences develop and the nature of the high elevation assemblages depending on the latitudinal position and altitudinal extent of the island in question.

Invertebrate abundances and distributions tend to reflect those of the major plant assemblages, mostly because the invertebrates respond to similar environmental conditions. Nonetheless, pronounced differences do occur, such as the much higher richness and habitat specificity of species in the rocky, epilithic biotope dominated by cryptogams, compared with the lower richness and habitat specificity of those of the vegetated biotope that is dominated by vascular plants.

Because temperature and water availability are major determinants of the abundance and distribution of life (and therefore of survival and reproduction at the population level), much attention has been focussed on the survival strategies and life histories of Antarctic and sub-Antarctic organisms. A variety of strategies are adopted by Antarctic organisms to survive low temperatures, and, at least on the Peninsula and in the sub-Antarctic, unpredictable changes in weather. These strategies range from anhydrobiosis in tardigrades and nematodes, to very rapid alterations (within hours) of the freezing point of springtails, to substantial protection against ultraviolet radiation in a variety of bryophytes. In sub-Antarctic insects it appears that moderate freeze tolerance is used as a strategy to overcome the marked and unpredictable short-term variation in temperature so characteristic of these cold islands. Indeed, sub-Antarctic insects are very different from their sub-Arctic counterparts, which tend to be characterized mostly by freezing intolerance and a strategy of lowering their freezing points in anticipation of winter conditions. This asymmetry in strategies seems typical of many groups and may indeed represent one of a variety of ways, at several levels of the biological hierarchy, in which the cold temperate zones of the northern and southern hemispheres differ.

HUMAN IMPACTS

In the terrestrial ecosystems of the Antarctic, human impacts are smaller than they have been elsewhere. Humans first landed on the continent in the Peninsula area around 1821, and on East Antarctica in 1895. Many of the sub-Antarctic islands have equally short human histories (sealing commenced mostly in the early to late nineteenth century). Intriguingly, the number of humans likely to be present on an island annually is strongly related to the area of the island and to its mean annual temperature – people prefer large, warm islands. However, the situation might now be changing because numbers of tourists to the Antarctic, and particularly to the Peninsula region and a few sub-Antarctic islands, are growing almost exponentially.

Early human impacts were mostly restricted to marine systems as a consequence of sealing and whaling, with changes to the terrestrial environment being localized in their extent and nature. Now the situation is quite different, and both the direct local and indirect influences of humans are increasing across the region. For example, invasive alien species have profoundly altered species assemblages and ecosystem functioning on most sub-Antarctic islands, and their direct effects are starting to be felt on the continent itself, often in ways that are not immediately obvious.

At only two sites on the Antarctic continent have reproducing populations of alien multicellular species established themselves outside research stations (the same is not true of microbes): the grasses *Poa pratensis* (Cierva Point, northern Antarctic Peninsula) and *Poa annua* (King George Island). A single individual of an alien grass (*Poa trivialis*) is known from the vicinity of Syowa station (East Antarctica). Two populations of *Lycoriella* midge have established on the continent: one survives in the sewage system at Casey Station (East Antarctica) and one, now eradicated, in the alcohol bond store at Rothera station (Antarctic Peninsula). The origins of these species remain poorly known. On Signy Island (South Orkney Islands), the chironomid midge *Eretmoptera murphyi* and the enchytraeid worm *Christensenidrilus blocki* have become established after accidental introduction during reciprocal transplant experiments investigating plant performance.

Alien species are typical of the large majority of Southern Ocean islands. Many of these species are thought to have been introduced either by sealing and whaling activities in the nineteenth and twentieth centuries or by scientific and logistic operations that have taken place since the early to mid-twentieth century. On some islands, the alien component is substantially larger than the indigenous species richness of the same groups, such as for the vascular plants on Possession Island (Crozet Group) and the insects on Gough Island. Alien species include aggressive transformer species such as feral cats, rabbits, mice, reindeer, and weedy grasses that have had or continue to have substantial impacts on local ecosystems and on several of their constituent species.

The overwhelming influence of abiotic factors on the distribution and abundance of organisms in the Antarctic and on the Southern Ocean islands suggests that assemblages will be highly sensitive to climate change. At least in some parts of the continent, change in temperature and precipitation has been considerable. Thus, over the past 50 years, mean annual temperature at Faraday/Vernadsky station (Western Peninsula) has increased by 0.56 °C per decade, with much of the warming taking place in the winter months. Liquid precipitation is also on the increase. By contrast, temperatures in some Eastern Antarctic locations are declining, with the Amundsen-Scott station (South Pole) experiencing a decline at a rate of −0.17 °C per decade. Elsewhere, little change on the ground has taken place, although generally tropospheric warming is characteristic of the region. Regional variation in local responses to global climate change is also characteristic of the Southern Ocean islands, with some islands, such as Gough Island, showing little more than the average change for the planet in terms of temperature and little alteration in rainfall, while others, such as Kerguelen, Macquarie, and Marion Islands, have shown substantial increases in mean annual temperature and declines in precipitation (~1.2 °C increase in mean annual temperature and a decline of 600 mm in total annual precipitation on Marion Island). Less obvious changes include an increase in the frequency of strong winds, a change in wind direction, and an increase in the number of clear sky evenings.

In response to warming on the Antarctic Peninsula, the two vascular plant species have shown a considerable increase in local abundance and in overall distributional range. Not all species have shown such clear-cut responses. Indeed, the effects of interactions between changes in UV-B radiation, temperature and water availability, and the life histories and physiological responses of different taxa make responses to environmental change complex. Although experimental work has demonstrated increases in abundance of some groups with warming, in others the effects may be in the opposite direction, especially where water stress increases. Moreover, even within higher taxa, responses vary considerably between sites and between studies. Nonetheless, the exposure of new ground as glaciers retreat and the increasing availability of water in some areas suggest that the overall area occupied by multicellular life is likely to increase, at least in the Peninsula region.

In the sub-Antarctic, increases in temperature and declines in water availability are having considerable impacts on arthropod assemblages and on several plant species. Glacial retreat on some islands is exposing new ground, which is often colonized by invasive alien species. For example, pioneer species on glacier forelands on Kerguelen Island include *Poa annua* and *Cerastium fontanum*, species that are invasive throughout the region. On other islands the upper altitudinal distribution of vascular plants has increased substantially over the past half century. The impacts of climate change can also be more subtle. For example, increases in freeze-thaw cycles associated with more frequent clear skies are predicted to have a substantial impact on a keystone caterpillar species on Marion Island (the moth *Pringleophaga marioni*) because of its inability to sustain growth after multiple low-temperature exposures. These impacts will be compounded by the predation of caterpillars by introduced feral house mice, which are showing a population increase due to their positive response to warming, drying conditions. Indeed, mice are substantially altering nutrient cycling and community level changes in plant abundances on Marion and many other Southern Ocean islands. Interactions between climate change and invasive species show further subtleties. For example, it appears that alien springtail species on Marion Island have greater desiccation resistance following acclimation to high temperatures (15 °C), whereas the converse is true of the indigenous species. These differences in the form of phenotypic plasticity account for the positive response to warming and drying in the alien species and the negative response in the indigenous ones in large-scale manipulative field trials. Field surveys have also demonstrated the predominance of alien springtails in low-altitude assemblages on the island and their absence from higher elevations. Laboratory work has in turn demonstrated steeper development rate–temperature relationships in the alien than in the indigenous species, and a greater tolerance of low temperatures and less of a tolerance of high temperatures in the indigenous compared with the alien species. Thus, abiotic controls are likely playing a major role in influencing the way in which climate change affects the interplay between indigenous and alien species on the island.

Interactions between alien and indigenous species are predicted to grow in scope and complexity with climate change because more alien species are likely to establish in warmer climates. Indeed, an increasing level of establishment of alien species is predicted both for the Antarctic Peninsula and for the Southern Ocean islands because of ameliorating climates and increases in human activity (scientific, logistic, and tourist). Although ongoing monitoring has not taken place at many sites, it is clear that at several of the sub-Antarctic islands, alien species continue to establish despite strict management controls. In many instances these introductions have substantially increased food web complexity by adding trophic groups (e.g., parasitoids) that were either scarce or absent from terrestrial systems.

CONSERVATION

Conservation south of 60 °S is implemented through the Antarctic Treaty System (the Antarctic Treaty itself was signed on December 1, 1959, and entered into force on June 23, 1961), specifically via the provisions of the Protocol on Environmental Protection to the Antarctic Treaty, or the Madrid Protocol. The Committee for Environmental Protection is the body responsible for conservation decision making in the region, by consensus among the Treaty Parties, which are then individually responsible for implementing decisions via their national legislation. Although formal conservation planning at a regional scale has been absent from the Antarctic region until recently, new efforts are underway to remedy the situation.

Because the Southern Ocean Islands have sovereignty claims, each island is administered by the claimant nation, and the level of management varies considerably. Nonetheless, the exceptional biological value of these islands is broadly recognized. The islands are governed by five different nations: the United Kingdom, South Africa, Australia, New Zealand, and France. Although the Antarctic Treaty does not apply to the islands, international agreements to which the above states are party, such as the Convention on Biodiversity and the Agreement for the Conservation of Albatrosses and Petrels, apply to the islands. Moreover, most of the islands enjoy a high conservation status. The five New Zealand sub-Antarctic island groups (Snares, Bounty, Antipodes, Auckland, and Campbell Islands), Heard and McDonald Islands, Gough Island, and Macquarie Island are all World Heritage Areas (at the highest IUCN Reserve Status of Category Ia). Macquarie Island is listed as a UNESCO Biosphere Reserve. At a national level, the New Zealand sub-Antarctic islands are all National Nature Reserves. Macquarie Island and Heard and McDonald Islands (Australia) have the highest reservation status, Nature Reserve and Commonwealth Reserve, respectively, under their governing legislations (state and federal). Marion and Prince Edward Island (South Africa) are classified as a Special Nature Reserve under South African legislation, South Georgia has National status (United Kingdom) as a Protected Area, and the Kerguelen and Crozet Islands (France) and Gough Island (United Kingdom) are National Nature Reserves, either as a whole or, for larger islands, in part.

In most cases, the indigenous biotas of the broader Antarctic region are strictly protected, and considerable care is given to their management and the assessment of population trends, especially for seabirds and seals. At least for many of the Southern Ocean islands, strict quarantine procedures are typically in place to prevent the introduction of new alien species, although for some of the islands more needs to be done in terms of their implementation. For the region, more also needs to be done in terms of the documentation of indigenous biotas and their spatial variability so that appropriate conservation management procedures can be developed and implemented. Nonetheless, most Antarctic and Southern Ocean islands enjoy considerable protection, probably more so than many other parts of the world. Whether this situation will change when it becomes financially viable to mine the Antarctic continent or the seabed for fossil fuels or minerals remains to be seen.

SEE ALSO THE FOLLOWING ARTICLES

Antarctic Islands, Geology / Climate Change / Flightlessness / Invasion Biology / Whales and Whaling

FURTHER READING

Bergstrom, D., and S. L. Chown. 1999. Life at the front: history, ecology and change on Southern Ocean islands. *Trends in Ecology and Evolution* 14: 472–477.

Bergstrom, D., P. Convey, and A. H. L. Huiskes, eds. 2006. *Trends in Antarctic terrestrial and limnetic ecosystems.* Berlin: Springer-Verlag.

Chown, S. L., and P. Convey. 2007. Spatial and temporal variability across life's hierarchies in the terrestrial Antarctic. *Philosophical Transactions of the Royal Society of London Series B Biological Sciences* 362: 2307–2331.

de Villiers, M. S., J. Cooper, N. Carmichael, J. P. Glass, G. M. Liddle, E. McIvor, T. Micol, and A. Roberts. 2006. Conservation management at Southern Ocean Islands: towards the development of best-practice guidelines. *Polarforschung* 75: 113–131.

Frenot, Y., S. L. Chown, J. Whinam, P. M. Selkirk, P. Convey, M. Skotnicki, and D. M. Bergstrom. 2005. Biological invasions in the Antarctic: extent, impacts and implications. *Biological Reviews* 80: 45–72.

Riffenburgh, B., ed. 2007. *Encyclopedia of the Antarctic*. New York: Routledge.

Turner, J., J. E Overland, and J. E. Walsh. 2007. An Arctic and Antarctic perspective on recent climate change. *International Journal of Climatology* 27: 277–293.

ANTARCTIC ISLANDS, GEOLOGY

JOHN L. SMELLIE
British Antarctic Survey, Cambridge, United Kingdom

Islands in the Antarctic region (south of 60° S) have an importance beyond a simple curiosity related to their current geographical and climatic isolation. They contain a repository of geology that is representative of much of the geology of Antarctica; they are typically more accessible and often better exposed than elsewhere in Antarctica as a result of recent marine-related stripping of superimposed snow and ice; and some islands may also have acted as persistent ice-free refugia for fugitive plant and animal communities at the end of each interglacial, when the continent itself was swathed in extensive ice sheets. The survival of those refugia may have been critical in subsequently determining evolutionary trends.

DISTRIBUTION AND CLASSIFICATION OF ISLAND TYPES

Islands included in this article form two geographically, geomorphologically, and tectonically distinct groups, comprising: (1) South Orkney Islands, South Shetland Islands, Joinville Island group, James Ross Island group, Seal Nunataks, Anvers and Brabant islands, Adelaide Island, Alexander Island, and Thurston Island; and (2) Peter I Island, Siple Island, Ross Island, Black and White islands, Franklin and Beaufort islands, Coulman Island, Scott Island, and Balleny Islands. The first group is mainly situated close to the Antarctic Peninsula and owes its origins geologically and tectonically to subduction of Pacific Ocean crust (Fig. 1), whereas the second group of islands is geographically more scattered and is composed of several isolated large alkaline volcanoes situated in an intraplate setting (both continental and oceanic) and probably related to impingement of one or more deep thermal anomalies (also called mantle

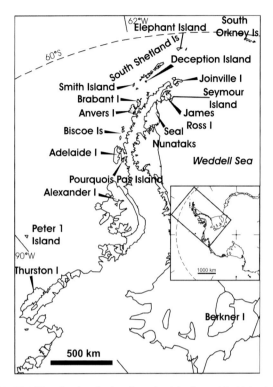

FIGURE 1 Map showing the location of subduction-related islands in Antarctica.

plumes) (Fig. 2). Although some of the islands are surrounded by open sea, in summer, at least, others are joined to mainland Antarctica by permanent thick shelf ice, such as the Seal Nunataks (originally named Seal Islands), Alexander Island, Thurston Island, Siple Island, Ross Island, and Black and White Islands. Others, such as Berkner Island and Roosevelt Island, are wholly enshrouded by the large ice shelves facing the Weddell and Ross seas; although covered by topographically prominent ice domes, the rock surfaces in these are probably well below sea level.

SUBDUCTION-RELATED ISLANDS

By Antarctic standards, the subduction-related islands are comparatively well exposed (e.g., large snow-free peninsulas on King George and Livingston islands in the South Shetland Islands; Deception Island; northern James Ross Island; and Seymour Island). Others are almost wholly encased in snow and ice (e.g., Joinville Island group, Biscoe Islands). They range from mountainous islands with sharp peaks rising to summits over 2000 m above sea level (e.g., South Orkney Islands; Elephant and Clarence Islands group; Smith Island; Brabant Island; and eastern Anvers, Adelaide, and Alexander islands) to subdued low-lying islands typically < ~1000 m in elevation, dominated

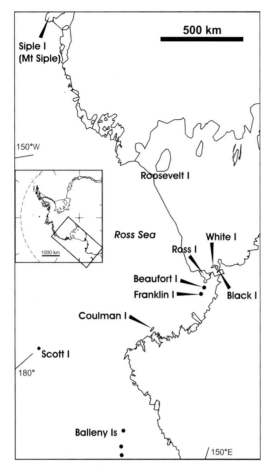

FIGURE 2 Map showing the location of intra-plate islands in Antarctica.

by thick snow and ice domes (e.g., many of the South Shetland Islands; western Anvers, Adelaide, and Alexander islands). Unusually for the region, the South Shetland Islands are rich in raised beaches and raised marine platforms. Similar features have also been described below sea level.

The geology of these islands can be described within a plate tectonic framework of eastward subduction of oceanic crust beneath the Antarctic Peninsula, a process that may have started as early as the mid-Paleozoic but that is mainly represented by rocks < 200 million years old. Thus, the region is dominated by remnants of a near-complete Mesozoic arc–trench system. An alternative view, involving the sequential accretion of suspect terranes, has also been proposed. However, the terrane accretion also took place within a plate tectonic framework that does not greatly detract from the basic description of the principal tectonic elements given here. Subduction ceased northward as a series of sections of a mid-ocean ridge arrived at the Antarctic Peninsula trench, beginning about 50 million years ago off southern Alexander Island and continuing until about 4 million years ago off Anvers Island. The South Shetland trench is the sole surviving remnant. Although this trench is largely quiet seismically, subduction is believed to be occurring there still, but at a very slow rate. Active tectonic processes may have been responsible for the relatively recent rapid uplift of sections of crust that were formerly buried very deeply and are now quite mountainous (e.g., Smith Island, and Elephant and Clarence Islands Group).

Subduction complexes are exposed in islands in the Joinville Island group, South Shetland Islands, South Orkney Islands, and Alexander Island. The oldest, which may be part of the local basement on which the Mesozoic magmatic arc was founded, comprises outcrops of deformed metasedimentary sequences, mainly quartzose sandstones and mudstones (Trinity Peninsula Group). The strata are sparsely exposed on Joinville Island, and they form Laurie Island and part of Powell Island (South Orkney Islands). Although commonly attributed to a subduction complex, the tectonic setting is not well understood. Deposition may have been Permo-Triassic (250–230 million years ago), with accretion probably completed by Early Jurassic time (210–190 million years ago). Further south, quartzofeldspathic metasedimentary and rarer volcanic rocks occupy an outcrop over 300 km in width in Alexander Island (LeMay Group). They are deformed and metamorphosed up to blueschist grade. Sedimentation was probably mainly Early Jurassic–Early Cretaceous (200–100 million years ago), but it may extend back to Permo-Carboniferous times (350–260 million years ago). Other distinctive subduction complex outcrops form Smith Island and the Elephant and Clarence Islands Group (South Shetland Islands) and parts of the South Orkney Islands. The metamorphic outcrops include distinctive blueschists; ages around 200–240 million years have been obtained in the South Orkney Islands and 90–110 million years in the South Shetland Islands. The South Orkney Islands owe their present isolated position, some 600 km northeast of the Antarctic Peninsula, to late Cenozoic faulting associated with Scotia Sea formation during the separation of Antarctica from southern South America.

Undeformed, highly fossiliferous fore-arc basin sedimentary sequences are preserved in Alexander Island, where they unconformably overlie the LeMay Group.

The sediments are known as the Fossil Bluff Group, and they comprise a weakly deformed and essentially unmetamorphosed marine to fluvial sequence up to 7 km in thickness that is Early Jurassic to mid-Cretaceous in age (200–100 million years).

Magmatic rocks, both volcanic and plutonic, are widely exposed all the way from Thurston Island to the South Shetland Islands. They were formed between about 170 and 20 million years ago and comprise mainly basaltic to andesitic lavas and fragmental deposits, and more rarely dacites to rhyolites. The volcanic rocks are hydrothermally altered by a range of compositionally similar coeval plutonic intrusions. Most of the magmatic rocks are related to partial melting caused by subduction of old Pacific Ocean crust, but those on Alexander Island include distinctive magnesium-rich andesites linked to subduction of a segmented ancient mid-ocean ridge after about 80 million years ago, causing anomalously shallow heating and melting in the fore-arc region and the northward migration of the volcanic centers on the island. Volcanic centers in the South Shetland Islands also migrated in a northeasterly direction, but the cause is unknown. The youngest magmatic arc rocks (< 30 million years ago) on King George Island are interbedded with three important glacial sedimentary units. The two older sequences contain a rich diversity of exotic stones that indicate that glaciation was also under way in East Antarctica.

Late Cretaceous to Eocene/Oligocene (90–35 million years ago) back-arc basin sedimentary sequences are widely exposed on Seymour, Snow Hill, and James Ross islands. The basin fill is possibly 5 km thick, and it contains a high-resolution record of both terrestrial and marine climate change. The excellent exposure of highly fossiliferous sediments contains a wide range of invertebrate and vertebrate fossils, which are often very well preserved, but the very soft nature of the strata means the outcrop areas form very low-lying ground.

Extension and rifting of the northwestern margin of the Antarctic Peninsula within the past 20 million years culminated in the formation of a wide marginal basin in Bransfield Strait and eruption of basalt lavas. Deception Island, an active volcano with a large flooded caldera, is related to this process (Fig. 3), as are Bridgeman Island, Penguin Island, and numerous small volcanic centers on Livingston and Greenwich Islands. Three large overlapping shield volcanoes less than 200,000 year in age were also constructed on Anvers and Brabant islands, at the likely southern limit of marginal basin tectonic effects. Topographic expression

FIGURE 3 Aerial view of Deception Island, a largely ice-free active volcano with extensive areas of snow-free ground. The volcano has a flooded interior caused by a major collapse following a very large eruption, probably less than 10,000 years ago. Photograph by John Smellie.

of Cenozoic extension in the Antarctic Peninsula region is widespread and is shown by deep north–south channels separating the islands from the mainland. Outlet glaciers formerly occupied the channels, and buttressing of the large shield volcanoes on Brabant and Anvers islands may have enhanced the glacial erosion.

Back-arc volcanism is well represented in the James Ross Island group, which contains an extensive basaltic volcanic field dominated by a single large volcano (Mt. Haddington). Many of the smaller islands in the group are satellite volcanic centers, and the volcanic field may still be active. There are numerous interbedded glacial sedimentary rocks, some highly fossiliferous. The volcanic rocks also show unequivocal evidence for repeated eruptions beneath an ice sheet, and, together with the sedimentary rocks, they are a major source of paleoenvironmental information.

Possibly as a consequence of mid-ocean ridge collisions with the Antarctic Peninsula trench, gaps were created in the subducting slab and resulted in eruption of post-subduction volcanic rocks, mainly in Alexander Island and Seal Nunataks, from 7 million years ago on. The volcanism is basaltic, related to upwelling of mantle through the gaps in the slab. Surprisingly, the timing of post-subduction eruptions shows no obvious progression mirroring the ridge collisions.

INTRAPLATE ISLANDS

More than 90% of the surfaces on the islands in this group are covered by permanent snow and ice. Almost all are well-formed volcanic cones that contrast with the much more severely eroded subduction-related islands. This contrast is probably due to the very different ages and glacier thermal regimes in the two island groups: almost all of the subduction-related islands are older and are deeply eroded by wet-based ice, whereas the intraplate islands (excepting the oceanic Peter I Island, Scott Island, and Balleny Islands) are less than 7 or 8 million years old and have been affected by a largely nonerosive cold polar regime. The intraplate islands are all volcanoes, typically rising steeply to elevations between 1000 and 3000 m above sea level. A few have snow- and ice-filled calderas, but most simply have small summit craters; in one case (Mt. Erebus) the crater contains a persistently active lava lake that frequently erupts pumiceous bombs with distinctive large (to 10 cm) feldspar crystals. At least one other volcano in the group may be active (Mt. Siple, Siple Island). A few are still undated (Scott Island, Balleny Islands, Beaufort Island). The volcanoes range in composition from basalt to mugearite, trachyte and phonolite, characteristic of an intraplate extensional environment. However, there are differences in tectonic setting. Mt. Siple is probably part of the large continental alkaline volcanic province in Marie Byrd Land, which is causally related to a mantle plume, regional updoming, and progressive fracturing along orthogonal continental crustal fractures. Volcanoes in the western Ross Sea region may be associated with reactivated deep faults related to large-scale Cenozoic Australia–Antarctica plate separation tectonics. Finally, Peter I Island, Scott Island, and Balleny Islands have been linked to eruptions along "leaky" oceanic fracture zones.

SEE ALSO THE FOLLOWING ARTICLES

Antarctic Islands, Biology / Plate Tectonics / Refugia / Volcanic Islands

FURTHER READING

Barker, P. F. 1982. The Cenozoic subduction history of the Pacific Margin of the Antarctic Peninsula: ridge crest–trench interactions. *Journal of the Geological Society of London* 139: 787–801.

LeMasurier, W. E., and J. W. Thomson, eds. 1990. Volcanoes of the Antarctic Plate and Southern Oceans. *American Geophysical Union, Antarctic Research Series* 48: 1–487.

Smellie, J. L., W. C. McIntosh, and R. Esser. 2006. Eruptive environment of volcanism on Brabant Island: evidence for thin wet-based ice in northern Antarctic Peninsula during the late Quaternary. *Palaeogeography, Palaeoclimatology, Palaeoecology* 231: 233–252.

Storey, B. C., and S. W. Garrett. 1985. Crustal growth of the Antarctic Peninsula by accretion, magmatism and extension. *Geological Magazine* 122: 5–14.

Storey, B. C., R. J. Pankhurst, I. L. Millar, I. W. D. Dalziel, and A. M. Grunow. 1991. A new look at the geology of Thurston Island, in *Geological Evolution of Antarctica*. M. R. A. Thomson, J. A. Crame, and J. W. Thomson, eds. Cambridge, UK: Cambridge University Press, 399–403.

Vaughan, A. P. M., and B. C. Storey. 2000. Terrane accretion and collision: a new tectonic model for the Mesozoic development of the Antarctic Peninsula magmatic arc. *Journal of the Geological Society of London* 157: 1243–1256.

ANTILLES, BIOLOGY

CHARLES A. WOODS

University of Vermont, Island Pond

FLORENCE E. SERGILE

University of Florida, Gainesville

The "Antilles" is an archipelago of over 7000 large and small islands, cays, reefs, and exposed offshore banks with long and diverse geological and biological histories. The total human population of the Antilles is 34.5 million. The heterogeneous assemblage of islands stretches in an arc over 3200 km long. The islands originated in a variety of ways (volcanism, uplifted island arcs, exposed and uplifted coral banks, movements of major plates, changes in sea levels) and have had a history of numerous vicariance (separation) events, as pieces of islands as well as whole islands became attached and unattached, submerged and reemerged. This dynamic geological history has resulted in extremely high levels of biodiversity and insular endemism. Because of the unique combination of a well-preserved fossil record of vertebrates in sinkholes and caves (mostly found in areas of karst geology) and the extraordinarily rich record of microfossils found in Dominican amber, the past and present biodiversity of these islands is especially well documented.

BIOGEOGRAPHY OF THE GREATER ANTILLES

The Greater Antilles includes Cuba, Hispaniola (Haiti and the Dominican Republic), Jamaica, and Puerto Rico (including the Virgin Islands) (Fig. 1). Parts of many of these islands have changed position in the past. For example, the southern part of Hispaniola ("south island") was separate from the rest of Hispaniola for much of its early history and was located well west of that island. The mountains of eastern Cuba were likely part of northern Hispaniola. It is also likely that parts of eastern Hispaniola were united with Puerto Rico. The Greater Antilles has extensive areas of limestone, and some have been uplifted into high mountain plateaus. Rainfall falling on these karst areas creates crevices, hollowing out sinkholes and even deep caverns. These features provide additional important habitats that further increase biodiversity. These sinkholes and caves are also places where fossil and semifossil remains of animals are found. Their presence is one reason that there is such an excellent record of the kinds and ages of vertebrates that lived on the islands.

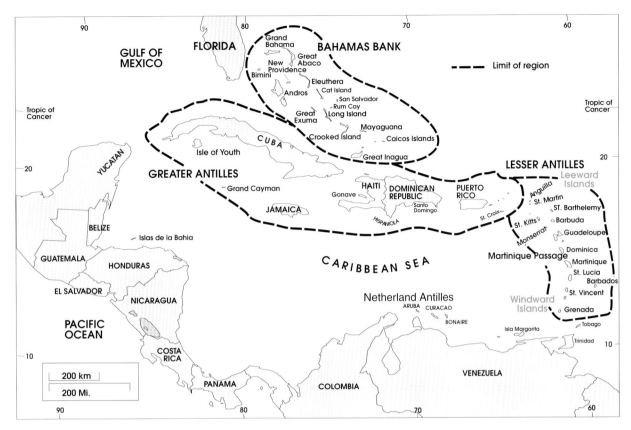

FIGURE 1 Map of the Antilles showing its biological regions.

In the Greater Antilles, many plants and animals are derived from South American ancestors. One hypothesis as to how South American terrestrial vertebrates dispersed to the Greater Antilles is the GAARlandia theory proposed by Ross MacPhee and Manuel Iturralde-Vinent. They believe that about 35 million years ago proto-Cuba, Hispaniola, and Puerto Rico were connected to northwestern mainland South America via the now submerged Aves Ridge, and they give the name GAARlandia (from Greater Antilles + Aves Ridge) to the overall structure. This bridge of emergent islands would have been above sea level for about 2 million years (between 35 and 33 million years ago). Terrestrial mammals such as sloths, monkeys, and rodents would have been able to make their way to GAARlandia at that time, later dispersing to other islands over water or via vicariance events. Rodents, sloths, monkeys, and even the Antillean piculet (*Nesoctites micromegas*), a primitive woodpecker, may have dispersed to the central Greater Antilles via this route. Other plants and animals, such as soricomorphs (including the large solenodons and the much smaller island shrews of the genus *Nesophontes*), most likely originated in North America and were able to disperse to the Greater Antilles via vicariance events when pieces of the proto-Antilles passed close to the southern tip of the North American continent. There is recent evidence for the antiquity and uniqueness of solenodons. Analysis of a sequence of 13,885 base pairs of both nuclear and mitochondrial genes of *Solenodon paradoxus* indicates that solenodons separated from other placental mammals 76 million years ago. The last connection between the proto-Antilles and mainland North America was severed 70–80 million years ago, and it is likely that solenodons became part of the Antillean fauna by vicariance at that time.

Cuba

Cuba is the largest island of the Antilles (110,860 km^2). It is a long and relatively narrow island organized into areas of high mountains surrounded by lowlands. In the past, these lowland areas were flooded by rising sea levels, and Cuba was fragmented into as many as three separate islands. The main island is fringed along the north and south coasts by small islands and cays that are protected from invasive species and are rich in biodiversity. Off the south coast is the Isle of Youth, itself a complex island formed of two parts, which is especially rich in biodiversity. Overall, the Cuban Bank, with its diversity of habitats and smaller islands, contains the most complex and varied biota in the Antilles. Cuba has the largest number of educational and conservation programs, and the status of the islands biota is therefore in better overall shape than any other island in Antilles.

Hispaniola

Hispaniola (76,193 km^2) is located on the tectonically active Caribbean plate and is the most geologically complex of the Antilles. Pico Duarte in the Central Mountains of the Dominican Republic (3175 m) is the highest mountain peak east of the Mississippi River and the Andes. There are a number of high ranges on the island with cool, well-forested peaks (that is, sky islands), and these high mountain ranges often create rain shadows downwind with dry, and even desert, habitats. Along the fault zone where the old north and south islands join sits Lake Enriquillo, which is 45 m below sea level. Hispaniola lacks the numerous offshore cays that in Cuba have protected endemic species from competition from invasive species, but there are several large offshore islands such as Gonave Island near Port-au-Prince and Turtle Island in the north.

Jamaica

Jamaica (11,424 km^2) is much smaller than Cuba and Hispaniola, but the island has high mountain areas (the Blue Mountains rise to 2256 m) and significant habitat diversity, including semi-desert areas. After the Middle Eocene, Jamaica was completely submerged. It was once again fully above sea level in the Middle Miocene, about 16 million years ago, thus making the island biologically the youngest of the Greater Antilles based on current (extant) biota. The submerged platform forming the base of Jamaica is narrow, and as a result, unlike Cuba or Hispaniola, the island has few offshore cays and islands that form disjunct ecosystems. As a consequence, Jamaica has much less biodiversity than either Cuba or Hispaniola. The Seven Rivers vertebrate site in west central Jamaica (15 km south of Montego Bay) is of late Early or early Middle Eocene age. This site has produced the remains of many aquatic vertebrates, including the earliest known fully quadrupedal sea cow (sirenian), *Pezosiren portelli*. The site also produced the remains (right dentary) of a large rhinoceros-like perissodactyl (*Hyrachys* sp.). The specimen is important because it is unlikely that such a terrestrial mammal as a rhinoceros could have dispersed to Jamaica across a broad open-water area, making it likely that Jamaica had a land bridge connection with North America at the time the rhino dispersed to the island.

Puerto Rico

Puerto Rico was part of northern Hispaniola before it separated and moved eastward about 15 million years ago. After that, and before the high mountains of the center of the island arose, the island largely submerged again before emerging for good about 12 million years ago. The submerged platform on which the island rests is very broad, and Puerto Rico has numerous surrounding cays and offshore islands such as Culebra Island, Vieques Island, and the northern Virgin Islands.

THE BAHAMA BANKS

The Bahama Banks has had a dynamic history as a result of changing sea levels during the Pleistocene ice age cycles. This region of over 2000 small islets ("cays") and 700 larger islands has a separate geological origin and is biogeographically distinct from the Greater Antilles. The geological origin of the Bahamas is controversial. The archipelago may have arisen via activity associated with plate tectonics 200 million years ago, or these low-lying islands and cays may be the remnants of a much larger platform that formed in tropical waters. The islands are all less than 61 m above sea level. The flora and fauna are mostly derived from Cuba and Hispaniola, having dispersed there when sea levels were much lower. At the height of the last ice age, when the sea level was over 90 m lower, many islands of the Bahamas became contiguous and formed two mega-islands corresponding to the "Great Bahama Bank" and the "Little Bahama Bank." There is little topographical or ecological diversity in the Bahamas, and as a result there is relatively little diversity of plants and animals. Biogeographically, the Bahamas also include the Turks and Caicos Islands, as well as the Navidad Bank and Silver Bank off the north coast of the Dominican Republic, which are important wintering areas for whales, especially the humpback whale.

LESSER ANTILLES

The Lesser Antilles begins east of the Virgin Islands (which here are considered part of the Puerto Rican Bank). The 25 larger and hundreds of smaller-to-tiny islands lie along the leading edge of the Caribbean plate, which is colliding with the North American plate (subduction). Many of the islands are of volcanic origin, and most of the remaining small cays and islets are of coral origin. Overall, this region of small volcanic and coral islands contains only 6% of the overall landmass of the Antilles. The islands in the northwest are collectively called the Leeward Islands. Within this subregion is the St. Martin Bank, which includes the islands of St. Martin, Anguilla, and St. Barthelemy. At times of low sea levels, they would have formed a single island approximately 5949 km². South of this, but still within the Leeward Islands, is the St. Kitts Bank, including the islands of St. Eustatius, St. Christopher (i.e., St. Kitts), and Nevis. At low sea levels, this bank would have been a single island of approximately 1546 km². The islands of Barbuda and Antigua are located on a bank east of the above, which at times of low sea levels would have been over 4274 km² in area. The Leeward Islands are relatively low-lying and similar to the Virgin Islands. South of these are the much more mountainous islands of Montserrat, Guadeloupe, and Dominica, still considered part of the Leeward Islands by most biogeographers.

South of the Martinique Passage (located between the islands of Dominica and Martinique), a series of islands are isolated from one another by deep-water passages. These islands are collectively known as the Windward Islands and range from Martinique in the north to Grenada in the south. There is biological evidence for excluding Trinidad, Tobago, Margarita, and the Netherlands Antilles from the true Antilles.

MICROFAUNA KNOWN FROM DOMINICAN AMBER

The nature of the biodiversity of small organisms such as ants, bees, and forest insects is poorly understood on many islands and archipelagoes but is fabulously preserved in amber deposits of the Dominican Republic. This record, preserved in fossilized resin of a now extinct algarrobo tree (*Hymenaea protera*), is a record of the broad biodiversity of plants and animals living together in a complex moist tropical forest between 15 and 45 million years ago in the western part of what is now Hispaniola. This tall forest habitat, referred to by George and Roberta Poinar (1999) as the "Amber Forest," was an ancient forest ecosystem of 40-m-tall canopy trees such as the caoba (*Swietenia*), the algarrobo (*Hymenaea*), and the nazareno (*Peltogyne*), as well as a diverse understory layer of shrubs, ferns, and flowers.

The algarrobo trees produced copious quantities of sticky resin in which small organisms were trapped and preserved. The trees formed a complex multistory habitat rich in animal biodiversity attracted by the leaves, flowers, pollen, and fruits of the trees. Ants, bees, wasps, moths, butterflies, birds, bats, rodents, and even small insectivores became trapped in the sticky resin and are now preserved in chunks of amber. The result is one of the best-documented histories of forest biodiversity known from an ancient tropical ecosystem.

The Poinars examined over 3000 pieces of amber, and the most frequently occurring organisms were worker ants (497), winged adult ants (286), gall midges (197), bark lice (173), and stingless bees (156). They also documented the remains of small vertebrates such anoles and geckos, soft ticks and hairs from rodents, the feather of an Antillean piculet (a small primitive woodpecker-like bird), and a few vertebrae and ribs of a tiny island shrew (*Nesophontes*).

What is remarkable is how many Antillean plants and animals documented in Dominican amber, including the algarrobo tree itself, are now extinct or extirpated from the Antilles. For example, of the seven genera and subgenera of bees documented by the Poinars in amber, all are now extinct in the Antilles (whereas relatives survive in tropical Central and South America). The reasons and timing of these extinction events in the Antilles may relate to climate change, because ice ages during the Pleistocene in North America led to drying and cooling in the Antilles, and lowering sea levels changed the size and shape of Antillean islands.

MAMMALIAN FAUNA

There are few surviving examples of the rich mammalian fauna that characterized the Antilles 18,000 years ago. At that time, land areas of the Greater Antilles were much larger than today because of lower sea levels. The mammals that inhabited these islands included a variety of rodents, sloths, primates, and primitive insectivores. The rich diversity of endemic terrestrial mammals was especially true on the large, topographically diverse islands of Hispaniola and Cuba.

Antillean Rodents

The best-known taxa are members of the rodent family Capromyidae. They are known as "jutias" in Cuba and the Dominican Republic, "zagoutis" in Haiti, and "conies" in Jamaica and the Bahamas. We will call them hutias. They reached high levels of diversity in the Cuban archipelago (five genera, three endemic) and on Hispaniola (four endemic genera). They are not known to have occurred in the Lesser Antilles, but one genus (*Geocapromys*) with three subspecies occurred in the Bahamas and dispersed to the Cayman Islands and to very small Little Swan Island. There were at least 55 species, of which 42 have become extinct (76%). The reasons for the extinction of these rodents, many of which were the size of squirrels and large house cats, was likely overhunting by Amerindians, habitat destruction, and predation by introduced dogs, cats, and the mongoose. The largest numbers of surviving hutias occur on Cuba, where 11 species are found in habitats ranging from small isolated offshore islands to mainland Cuban swamps and forested areas. One species (*Capromys pilorides*) is common throughout Cuba, where it is well known and frequently eaten as "bush meat." On Hispaniola, the single surviving species of hutia (*Plagiodontia aedium*) has two extant subspecies, one of which is mainly restricted to southern Hispaniola (*P. a. aedium*) and the other to northern Hispaniola (*P. a. hylaeum*). A closely related hutia (the now extinct *Rhizoplagiodontia lemkei*) was restricted to the far west of the southern peninsula of Haiti in the Massif de la Hotte. It was very abundant and was part of the cluster of endemic plants and animals that characterized the Massif de la Hotte hotspot. On Cuba, a fossil rodent of early Miocene age (*Zazamys veronicae*) has been discovered, which is ancestral to Hispaniolan hutias of the genus *Isolobodon*. This species, like most other capromyid rodents, was a delicacy, and its remains are common in Amerindian kitchen middens. "Conies" of the genus *Geocapromys* still survive in Jamaica and the Bahamas. Thus, it is still possible to observe the last remnants of the once great Antillean radiation of capromyid rodents.

In addition to capromyid rodents, large, heavy-bodied, wide-toothed, hutia-like rodents were present on Jamaica, Hispaniola, and Puerto Rico as well as on two small islands (Anguilla and St. Martin [St. Martin Bank]) in the northernmost Lesser Antilles. Some of these forms, such as the gigantic *Amblyrhiza inundata* from Anguilla and St. Martin, were as large as 200 kg. These giant rodents (family Heptaxodontidae) were known as "twisted-toothed giant hutias." They apparently did not radiate into many species, but they were of exceptional mass.

Hispaniola, Cuba, and Puerto Rico shared a radiation of small, spiny, rat-like rodents. They were smaller than hutias in mass and had cheek teeth similar in morphology to South American spiny rats. Two species are common in cave deposits on Hispaniola, and another two species occurred in Cuba. Three species are known from Puerto Rico. The remains of most of these species are fresh in appearance, indicating that they became extinct in historical times, especially on the high plateaus of southwestern Haiti. Remains are common in sinkholes, in cave deposits, and in Amerindian kitchen middens. They were likely driven to extinction by competition from introduced rats and predation by introduced dogs, cats, and possibly even the mongoose.

The radiation of rodents in the Lesser Antilles was much more limited and reflects the very different origin of this long chain of volcanic islands. There are no capromyids, spiny rats, or (with the exception of Anguilla and St. Martin) giant hutias. Instead, the rodents of these small islands were sigmodontine rodents of the genera *Oryzomys* and *Megalomys* that made their way onto this chain of oceanic islands by overwater dispersal from South America. There are described and undescribed taxa of sigmodontines from Anguilla, Antigua, Barbados, Barbuda, Guadeloupe, Marie Galante, Montserrat, St. Eustatius, St. Kitts, St. Lucia, and St. Vincent. One species of *Oryzomys* is known from Jamaica, which likely dispersed there over water from Central America. All of these forms are now extinct.

Antillean Sloths

The second largest adaptive radiation of land mammals in the Antilles occurred on Hispaniola (four genera and six species) and Cuba (five genera and six species), where sloths filled a number of niches. They ranged in mass from the size of a very large dog to smaller than a house cat (the smallest known sloth). The radiations of sloths on Cuba and Hispaniola were remarkably similar to each other. Sloths are also known to have occurred on Puerto Rico (one named form and one unnamed form). This distribution confirms the close biogeographic affinity of these three islands. Sloths are not known to have occurred on Jamaica. Fragmentary dental remains of what is clearly sloth material are known from Grenada in the Lesser Antilles. No sloths have survived in the Antilles into historical times. Two large, ground-dwelling (i.e., megafaunal) sloths were present on Hispaniola, and they were likely hunted to extinction by early Amerindians. Both are closely related to the giant megalonychid ground sloth from Cuba (*Megalocnus rodens*), estimated to have weighed 270 kg. Another part of the sloth radiation on Hispaniola and Cuba (subfamily Choloepodinae, tribe Acratocnini) are forms similar to the two-toed sloth (*Choloepis*) from Central and South America. These much smaller sloths, such as "Yesterday's Acratocnus" (*Acratocnus ye*) are known from well-forested and savanna upland habitats in the southwest. It is likely that this form hung from branches as two-toed sloths do. Hispaniola also had three species of a smaller sloth group (tribe Cubanocnini). The larger form was the "Hispaniolan Neocnus" (*Neocnus comes*), which was widespread on Hispaniola. Also present on Hispaniola was the "Slow Neocnus" (*Neocnus dousman*) and the "Least Neocnus" (*Neocnus toupiti*). All three forms were excellent climbers and moved easily in the treetops but did not hang under branches (according to sloth expert Jennifer White).

Why and when the extensive radiation of sloths became extinct, in spite of their wide range of size, food habits, and locomotor habits, is not yet resolved. The well-preserved remains of sloths are common in sinkholes along the karst-covered plateaus of the southern peninsula of Haiti. These plateaus were well forested and probably represented the last haunts of sloths on the island. It is likely that they became extinct within the last 2000 years as a result of human activity.

Antillean Primates

Monkeys were first documented in the Antilles as an endemic platyrrhine (*Antillothrix bernensis*) in Hispaniola. Fossil endemic monkeys are also known from Cuba (*Paraloutta varonai*) and Jamaica (*Xenothrix mcgregori*). Fossil evidence of monkeys in the Greater Antilles extends back as far as the Early Miocene. The Hispaniolan monkey is similar to a living squirrel monkey (*Saimiri*). It was originally described as *Saimiri bernensis* but is now considered to be an endemic genus closely related to the Cuban monkey. The extinction of monkeys in the Greater Antilles was again probably the result of overhunting and deforestation by Amerindians.

Living Fossils and "Island Shrews"

The surviving solenodons and very recently extinct (if they are extinct) island shrews are like "living fossils" from an earlier and long-extinct North American radiation. They are distributed on Hispaniola, Cuba (and the nearby Cayman Islands), and Puerto Rico (including adjacent Vieques Island). Presumably, solenodontids dispersed from North America to proto-Cuba between 70 and 80 million years ago in the Late Cretaceous, when the proto-Antilles and North America were last connected. They could have dispersed to what is now northern Hispaniola when eastern Cuba and northern Hispaniola were connected during the Oligocene. There is DNA evidence that *Solenodon paradoxus* and *S. cubanus* diverged 25 million years ago. This is about the time that eastern Cuba and northern Hispaniola began separating. The dispersal of *Solenodon* and *Nesophontes* to southern Haiti and the adjacent southern Dominican Republic could not have occurred until the Late Miocene or Early Pliocene, when the isolated "south island" crunched into true Hispaniola.

SOLENODONS

Solenodons are found on Cuba (*Solenodon cubanus*) and Hispaniola (*Solenodon paradoxus*). Both are now *very* rare and are threatened with extinction. Saving these forms from extinction has been given top priority by the Zoological Society of London's EDGE scheme (Evolutionarily Distinct and Globally Endangered). They are truly evolutionarily distinct, having become isolated from all other placental mammals 76 million years ago. DNA analyses suggest that the two solenodons are so genetically divergent that the Cuban form could be considered as a separate genus (*Atopogale*). Solenodons never radiated into many species (an additional extinct species is known from each island). *Solenodon paradoxus* has a widespread distribution in the mountains and in appropriate lowland karst zones of southern Haiti and much of the Dominican Republic. In Cuba, *Solenodon cubanus* is now restricted to the high karst mountains of eastern Cuba. These living fossils have poor vision and are slow moving. They are easily preyed upon by dogs and the ever more abundant introduced mongoose, so their future survival, even under the best of conservation efforts, is questionable.

ISLAND SHREWS

Endemic "island shrews" of the genus *Nesophontes* (Fig. 2) are part of the other soricomorph radiation in the Greater Antilles. The best current hypothesis is that solenodons and island shrews are closely related to each other. Cuba and Hispaniola each have three species of island shrew. In Hispaniola the remains of all three species of *Nesophontes* are abundant in some cave and sinkhole deposits under barn owl roosting sites. Some bones still have bits of dried tissue on them, and they are often mixed with the fresh-looking remains of introduced rats. The extinction of island shrews in both Cuba and Hispaniola was likely quite recent and may have been caused by competition and predation from black rats and mongooses. It is believed that the final extinction of *Nesophontes* in both Cuba and Hispaniola may have been as recent as the last 60 years. In Puerto Rico, there is just one, much larger species (*Nesophontes edithae*), which may have been an ecomorph of *Solenodon*. Remains are abundant on the island but are not nearly as recent in appearance as remains from Cuba and Hispaniola, and it is likely that this species became extinct before the arrival of Europeans in the Antilles, and before rats and dogs were introduced.

FIGURE 2 Reconstruction of the largest Hispaniolan island shrew (*Nesophontes paramicrus*).

Other Antillean Mammals

Bats are the best known of the other major groups of Antillean mammals. There are 56 known extant species of bats in the Antilles, 28 (50%) of which are endemic. Cuba has the largest and most diverse bat fauna, with 26 living species (plus four extinct and three extirpated species). The next largest assemblage of bats is surprisingly (because it is so much smaller and less diverse than Hispaniola) found on Jamaica, with 21 extant species and three extirpated forms. The much larger island of Hispaniola has 18 extant species and three extinct. In the Lesser Antilles, most major islands have between 10 and 13 extant bat species, with small islands such as Saba (three species) and the Grenadines (four species) having many fewer. Fossil species known from the Lesser Antilles include one extinct form and three locally extinct species (but which still occur in the Greater Antilles). The five most common and widespread species of Antillean bats are *Monophyllus redmani, Brachyphylla cavernarum, Artibeus jamaicensis, Noctilio leporinus,* and *Molossus molossus.*

Carnivores appear to be lacking as native species in the Antilles, although they are ecologically present in the form of feral dogs and cats, and the small Indian mongoose (*Herpestes javanicus*) that was introduced to Trinidad from India in 1870, and has now spread to 29 Antillean islands. The extinct "wild dogs" described from fossils in Cuban cave deposits (*Cubacyon* and *Paracyon*) are most likely remains of deformed domestic dogs. There are also reports of native raccoons in the Antilles. Recent molecular studies of the DNA of raccoons have confirmed that what were previously considered to be endemic island species (*Procyon maynardi* from the Bahamas, *P. minor* from Guadeloupe, and *P. gloveralleni* from Barbados) are conspecific with the North American raccoon *Procyon lotor.*

BIRDS

There are 604 bird species in the Antilles, and 163 of these are endemic (27% endemism), some with very restricted distributions. In the Antilles, there are 36 endemic genera and two endemic families (palmchats [Dulidae] and todies [Todidae]). Within the boundaries of the Antilles, 48 bird species are currently threatened with extinction. Fourteen bird species and subspecies have already become extinct.

Because the Antilles lacked predatory mammals, an unusually large number of birds became flightless during prehistoric times, and some birds filled predatory niches. There was at least one large ground-dwelling predatory owl (*Ornimegalonyx oteroi*), several other large barn owls, and huge birds of prey. There were 15 or 16 parrot-like species endemic to the Antilles, all of which are now extinct. Past climate and sea-level changes during the Pleistocene, as well as the impact of humans, introduced predators (dogs, cats, mongoose), and invasive species have combined to destroy most ground-nesting and flightless species.

The Antilles, especially the Greater Antilles, is an important area for migratory birds. Many North American migrants spend the winter months on these islands. Many individuals return to specific locations each winter. This creates numerous biological challenges, such as competition for limited resources with native species. Some migrants such as black-throated blue warblers have developed winter strategies whereby males and females winter at different elevations and in different habitats. A small percentage of the population of some migrants remains behind each year (e.g., in the case of redstart warblers in Hispaniola and Cuba). There are subspecies of North American migrants that have become endemic breeding forms (e.g., pine warblers, yellow warblers, and white-winged crossbills). One of the more interesting of site-specific North American migrants is the Bicknell's thrush, which nests on specific mountaintops in northern New England and winters in Hispaniola in specific locations, usually primary montane forests above 1000 m elevation. These examples reveal the complex mixture of resident, endemic, migratory, and modified migratory species that makes up the avifauna of the Antilles.

The distribution of Hispaniolan birds reflects the way the island was formed. There are two species of palm tanagers, with the black-crowned palm tanager (*Phaenicophilus palmarum*) being restricted to northern Hispaniola (the old north island). The gray-crowned palm tanager (*Phaenicophilus poliocephalus*) likely arose from a dispersal of a small group of juvenile *P. palmarum* to a western piece of the south island in the Pleistocene. The gray-crowned palm tanager is still restricted to southwestern Haiti west of Jacmel (an old separate part of the south island). There are two species of endemic tody on Hispaniola. Cuba, Jamaica, and Puerto Rico each have only one species. There is also a species pair of high-mountain chat-tanagers (*Calyptophilus*) on Hispaniola. The western chat-tanager (*C. tertius*) is restricted to high mountain areas of the old south island. The eastern chat-tanager (*C. frugivorus*) is restricted to areas of the old north island in Haiti (including Gonave Island) and to the northern part of the Dominican Republic. The morphology and distribution of the western chat-tanager further reflects the past geographical history of the island. The western species can be divided into a far western subspecies (*C. t. tertius*) in the Massif de la Hotte of Haiti and an eastern subspecies (*C. t. selleanus*) in the Massif de la Selle of Haiti and the Sierra de Bahoruco of the Dominican Republic, reflecting the division of the old south island into two separate sub-islands. The above examples of bird species, subspecies, and distributions that mirror the past geological and biological history of Hispaniola is biogeography at its living best.

REPTILES AND AMPHIBIANS

There are 499 native species of reptiles in the Antilles, 468 (94%) of which are endemic. This high level of endemism is one important reason that the Antilles is considered as a world hotspot by many conservation organizations. The principal radiations of reptiles in the Antilles include several species swarms such as in the genus *Anolis* with 154 species, 150 (97%) of which are endemic. Other species swarms include the beautiful dwarf geckos of the genus *Sphaerodactylus* with 86 species, 82 (95%) of which are endemic (Fig. 3), and curly-tailed lizards (*Leiocephalus*) with 23 species, all of which are endemic to the Antilles. Other important adaptive radiations of reptiles in the Antilles include 11 species of large rock iguanas (*Cyclura*) that thrive in dryer areas of the islands, nine species of large boa constrictors of the genus *Epicrates*, many of which are large enough to kill and eat large capromyid rodents. Other Antillean snakes include racers (*Alsophis*) with 13 endemic species and boldly colored snakes of the genus *Tropidophis* with 26 endemic species. There are two species of crocodiles in the Antilles: *Crocodylus acutus* (American crocodile), found in Cuba, the Cayman Islands, Jamaica, Hispaniola, and possibly Martinique, and *Crocodylus rhombifer* (the Cuban crocodile), which is now restricted to the Zapata Swamp and to a large swamp in the central area of the nearby Isle of Youth. The Cuban crocodile has the smallest range of any species of crocodile in the world. The only poisonous snakes in the Antilles are pit vipers (fer-de-lance) of the genus *Bothrops* in Martinique and Guadeloupe (*B. lanceolatus*).

Amphibians are not as diverse as reptiles in the Antilles. There are only 165 species, and all of them are frogs and toads. All but one of these frogs is endemic to the Antilles. Almost all of these frogs are from the single genus *Eleutherodactylus* (139 species). Frogs of this genus lay their eggs on land and hatch into adults without passing through a tadpole stage. Some species can be tiny, such as the Cuban species *Eleutherodactylus iberia,* one of the smallest tetrapods in the world (only 10 mm in length). The largest concentration of *Eleutherodactylus* in the Antilles is on mid-elevation slopes of the Massif de la Hotte of western Haiti, where 19 species occur sympatrically. The Alliance for Zero Extinction (AZE) has designated the Macaya Biosphere Reserve in this area of Haiti as the site in the world with the largest number of critically endangered species, 13 of which are *Eleutherodactylus* frogs. The most famous frog in the Antilles is the coqui of Puerto Rico (*Eleutherodactylus coqui*), whose loud vocalizations in the treetops are recognized as the typical sound of the night by almost all Puerto Ricans. In addition to frogs, the Antilles also has 12 species of endemic toads (Bufonidae), with eight species on Cuba, three on Hispaniola, and one on Puerto Rico. Most islands of the Greater and Lesser Antilles also have breeding populations of the introduced *Bufo marinus.*

The greatest biodiversity of amphibians and reptiles in the Antilles is on Hispaniola (a pattern that is true for almost all Antillean flora and fauna). There are 217 species, 209 of which are endemic (96% endemism). On Hispaniola, there are 64 amphibian species, 62 of which are endemic, and 153 reptiles, 147 of which are endemic. The small frogs of the genus *Eleutherodactylus,* which have very limited abilities to disperse, have 28 species in Haiti's southwestern mountain range (the Massif de la Hotte) and 15 species in the Central Mountains of the Dominican Republic. The two areas share only six species. The Massif de la Hotte has the highest density of frog species anywhere in the Antilles (Fig. 4). There, tall pine trees (*Pinus occidentalis*) form the forest canopy along with mid- and understory layers of other trees and shrubs rich in biodiversity. This relictual pine forest is in an area of tremendous rainfall and is the closest remaining forest in the Antilles structured like the long-lost moist tropical algarrobo forest of the Dominican Republic, the record of which is preserved in Dominican amber.

FIGURE 3 A dwarf gecko (*Sphaerodactylus*) from the Massif de la Hotte hotspot in southwestern Haiti. Courtesy of Charles Woods.

FIGURE 4 Photograph of Pic Macaya (right) and Pic Formon (left) in the core area of the Massif de la Hotte of southwestern Haiti. Courtesy of Charles Woods.

FRESHWATER FISHES

The Antilles have over 160 species of freshwater fishes, but many of these are introduced species or species of marine origin that occupy freshwater and brackish habitats. The true "freshwater" fish fauna of the Antilles includes 71 species in nine families. Most (65) freshwater species are endemic. It is not a surprise that Cuba and Hispaniola, with their complex biomes, have the largest number of species (89% of the total freshwater fish fauna). Jamaica (six species), the Bahamas (five species), and the Cayman Islands (four species) have much smaller freshwater fish faunas than do Cuba (28 species) and Hispaniola (32 species), in spite of these islands being geographically close to one another. Puerto Rico, in spite of being close to Hispaniola and having a broad platform with wide coastal features, totally lacks a *native* freshwater fish fauna. A measure of the complexity of any analysis of freshwater fishes, which are often kept (and sold) as pets in the aquarium trade and are sometimes raised for food (i.e., *Tilapia* in fish farms), is the abundance of 24 well-established introduced freshwater fishes on Puerto Rico. The Lesser Antilles have many fewer freshwater fishes, with one species (*Rivulus cryptocallus*) endemic to Martinique, and a second (*Rivulus ocellatus*) widespread in the Greater Antilles and many islands in the Lesser Antilles. The most successful family of Antillean freshwater fishes is the live-bearing killifishes of the family Poeciliidae (five genera and 46 species).

The complex freshwater fish fauna on the Antilles includes a total of 71 species, including 27 exotic species and five species that are of marine origin. The component of freshwater fishes derived from North America (seven genera, 27 species) is almost equal to the number derived from South America (five genera, 39 species). Almost all of these fishes are endemic to a single island, or at least to a single island group.

BUTTERFLIES AND OTHER INSECTS

There are about 350 butterflies known in the Antilles. In a pan-Antillean sense, the biogeography of butterflies is similar to the pattern observed for other groups. The greatest diversity of butterflies is on Hispaniola. Hispaniola also has the greatest number of species (201), 75 of which are endemic. The multipartite history of Hispaniola increases species numbers and diversity of butterflies just as has been observed in West Indian mammals, birds, reptiles, amphibians, and freshwater fishes. The north island–south island species dichotomy observed in other groups shows up clearly in butterflies too. For example, in the genus *Callisto*, 11 species originated on the old south island, and 22 species are associated with areas that were on the north island.

There are several important questions that are still unresolved in studies on the biology of Antillean butterflies. One is how butterflies dispersed to the islands: specifically whether strong fliers dispersed (flew) over open water whereas generally poor fliers mainly dispersed to the Antilles by vicariance. The large number of endemic species of weak flying butterflies such as satyrids (genus *Calisto*) is in marked contrast with the distribution of strong fliers such as sphingids (hawkmoths) with 47 species, only seven of which are endemic. It is likely that weak-flying groups of butterflies are descendents of ancestors that arrived in the Antilles by vicariance events long ago, whereas strong fliers dispersed on numerous occasions by overwater dispersal.

CONSERVATION BIOLOGY

The Antilles is recognized as one of the world's most diverse and richest biological regions, with an unusually large number of island endemics. For example, of the 13,000 plant species, 6550 are endemic to single islands in the Antilles. This long (3200 km) curving arch of islands has an overall land area of 236,000 km^2. The problem of protecting the biodiversity of the Antilles, however, is reflected in the equilibrium theory of island biogeography as well as in the size and shape of the islands. The slope of species-area curves for most species jumps sharply above a threshold of island size of about 3000 km^2. On islands above this threshold, speciation exceeds immigration, and species proliferation increases. This is one reason that biodiversity is especially rich on Cuba, which has over 6505 plant species, 3224 of which are endemic. The percentage of endemic plant species on Cuba alone is 54% of all of the endemic plants of the Antilles. Thus, as a consequence of the size (105,806 km^2) and shape (numerous offshore archipelagoes) of Cuba, it is geographically and biologically advantaged over smaller islands of the Antilles. Indeed, about 50% of the entire land area of the Antilles is found on Cuba alone. Ninety percent of the area of the Antilles occurs on the four major islands of the archipelago (Cuba, Hispaniola [73,929 km^2], Jamaica [11,190 km^2], and Puerto Rico [9100 km^2]).

But island size alone does not determine the chances of retaining an islands biodiversity. Other important influences are cultural history, population size and distribution, and the percentage of an islands area that is "protected." Once again, Cuba (with a population of 11.5 million and a population density of 103 people per km^2) sets the standard for the Antilles, with 15% of its area being "protected" (about the same percentage of the island that remains forested). Overall, in the Antilles, 12.9% of the area is "protected" (although only 7.1% is designated conserved in IUCN categories I–IV).

Keeping in mind that sustainable biodiversity requires habitat (remember the lesson of species-area curves), what are the prospects for biodiversity in other parts of the Antilles? Island size alone is not the critical indicator, but rather population density and the presence of suitable habitat. For example, in the Dominican Republic, with a population of over 9 million (186 people per km^2), 20% of the land area of the country is "protected" (19 national parks, six scientific reserves, 15 natural preserves, two marine sanctuaries). In adjacent Haiti, with a population of 8.7 million (population density 316 people per km^2), only 1.7% of the land area is protected in three barely functional national parks. The standard for protected land area in the Antilles is the small Lesser Antillean island of Dominica (754 km^2) with a population of 71,540 (a population density of 101 people per km^2 and declining), which protects 21.4% of the island in three national parks. Barbados and Aruba have preserved less that 1% of their land areas, and tiny Anguilla (only 102 km^2) is developing its first national park (Fountain Cavern).

SEE ALSO THE FOLLOWING ARTICLES

Antilles, Geology / Freshwater Habitats / Lizard Radiations / Mammal Radiations / Vicariance

FURTHER READING

Crother, B. I. 1999. *Caribbean amphibians and reptiles*. San Diego, CA: Academic Press.
Fernández, E. 2007. *Hispaniola: a photographic journey through island biodiversity/Bioversidad a través de un recorrido fotografico* [bilingual ed.]. Cambridge, MA: Harvard University Press.
Poinar, G., and R. Poinar. 1999. *The Amber Forest*. Princeton, NJ: Princeton University Press.
Raffaele, H., J. Wiley, O. Garrido, A. Keith, and J. Raffaele. 1998. *Field guide to the birds of the West Indies*. Princeton, NJ: Princeton University Press.
Ricklefs, R., and E. Bermingham. 2007. The West Indies as a laboratory of biogeography and evolution. *Philosophical Transactions of the Royal Society of London B—Biological Sciences.* doi: 10.1098/rstb.2007.2068.
Schwartz, A., and R. W. Henderson. 1991. *Amphibians and reptiles of the West Indies: descriptions, distributions, and natural history*. Gainesville: University of Florida Press.
Sergile, F. E. 2005. *A la découverte des oiseaux d'Haïti*. Port-au-Prince: Société Audubon Haïti.
Smith, D. S., L. D. Miller, and J. Y. Miller. 1994. *The butterflies of the West Indies and South Florida*. Oxford: Oxford University Press.
Woods, C. A. 1989. *Biogeography of the West Indies: past, present and future*. Gainesville, FL: Sandhill Crane Press.
Woods, C. A., and F. E. Sergile. 2001. *Biogeography of the West Indies: patterns and perspectives*, 2nd ed. Boca Raton, FL: CRC Press.

ANTILLES, GEOLOGY

RICHARD E. A. ROBERTSON

University of the West Indies, St. Augustine, Trinidad and Tobago

Located at the eastern edge of the Caribbean plate where it borders with the North and South America plates, the Lesser Antilles is a region of high seismicity, tectonism, and active volcanism that typifies oceanic-island arc volcanism. The island chain forms an 850-km arc that is convex toward the Atlantic and extends from Sombrero in the north to the southern island of Grenada. The arc splits north of St. Lucia into two: an Eocene-to-Miocene eastern arc and a Pliocene-to-Recent western arc. The northeastern islands extending from Marie Galante to Sombrero are characterized by Cenozoic limestones and are called the Limestone Caribbees. The inner arc extending from Grenada to Saba comprises the present day volcanic front and is called the Volcanic Caribbees. These have been mainly active from Eocene to mid-Oligocene (Fig. 1). The entire region is one of active volcanism, which exhibits both explosive and effusive activity and which currently poses a threat to vulnerable island communities throughout the Eastern Caribbean.

STRUCTURE AND TECTONICS

The main structural components of the arc consist of the Atlantic oceanic crust; a fore-arc upon which is built an accretionary wedge (with the island of Barbados as the exposed top); the island arc itself; and the Grenada basin behind the arc. The dominant tectonic process is active

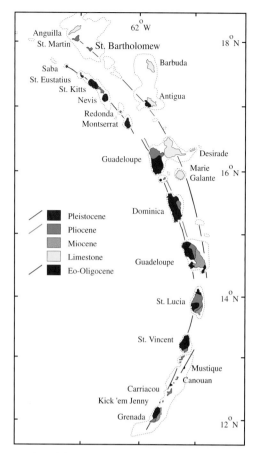

FIGURE 1 Map of the Lesser Antilles island arc, showing the ages of the exposed rocks and the positions of the volcanic front during the Eocene-Oligocene (red line), Pliocene (blue line), and Pleistocene (black line). The dashed line is the 200-m isobath (adapted from Fig. 9.3 of Wadge 1994).

crustal subduction with underthrusting of the Caribbean plate, by the Atlantic Ocean crust of Cretaceous to Jurassic age along the axis of the arc (Fig. 2). Subduction rates are estimated to be 2.2–3.8 cm/year. Plate subduction with volcanism commenced along the arc after cessation of similar activity along the Aves ridge to the west (at about the time of the Oligocene).

The island arc is bounded to the north and northwest by the Greater Antilles, which consists of deformed and metamorphosed sediments and volcanics of Jurassic to Eocene age located to the northwest of the arc. The Puerto Rico trench separates the Greater Antilles from the island arc. To the west, the island arc is bounded by the Venezuela basin and Grenada trough, between which lies the north–south striking Aves ridge, the site of active subduction during the Upper Cretaceous.

Investigations of regional tectonics have indicated that the Caribbean plate is moving eastward relative to the American plate, although the velocity and sense of movement is debated. The suggested direction of relative movement has included east–west, left-lateral strike slip to northeast–southwest oblique convergence. Various estimates exist for the rate of convergence within the region, depending on the period considered and the specific part of the arc examined. All estimates suggest that the rate of convergence in the Lesser Antilles (from as low as 1.3 cm per year to as high as 4 cm per year) is lower than in most arc systems.

The Benioff zone has an average dip of 45° and is at 100- to 120-km depth beneath the active volcanic arc. The configuration of the Benioff zone beneath the Lesser Antilles has been established using the hypocentral locations of earthquakes recorded from 1978 to 1984. The arc is segmented: To the north of Martinique, the Benioff zone dips at 60–50° and trends about 330°. In contrast, the southern segment trends at 20°, dips at 50–45° in the north, and is vertical in the south at Grenada.

Gravity anomalies occur as an arcuate pair associated with the fore-arc (negative anomalies) and the arc massif (positive anomalies). The anomalies are due to either the presence of dense igneous rocks (positive anomalies) or the presence of thick, low-density sedimentary rocks (negative anomalies). Crustal thickness beneath the arc is about 30 km and has typical island arc and oceanic crustal seismic velocities. Seismic reflection data indicate that the arc crust consists of three layers: an upper layer of sedimentary and volcanic rocks, a middle layer of intermediate plutonic rocks, and a lower layer of basic rocks. Gravity anomalies supported by seismic refraction data indicate significant differences between the northern and southern segments of the arc.

EVOLUTION AND STRATIGRAPHY

The Caribbean plate was generated over the Galápagos hotspot 100–75 million years ago and was inserted between the North American and South American plates sometime between the Late Campanian and Late Eocene. A subduction zone and associated volcanic arc developed at its leading edge, and volcanic activity, now represented by the Aves ridge, extended from the Upper Turonian to the Lower Palaeocene. The Lesser Antilles volcanic arc developed in the area where the accretionary prism associated with the Aves ridge developed. The Aves ridge accretionary prism was separated from the active volcanic arc by the development of a back-arc basin, now represented by the Grenada trough, a 3-km-deep basin, which contains a sedimentary pile estimated to vary from 4.2–7 km to 7–12 km thick.

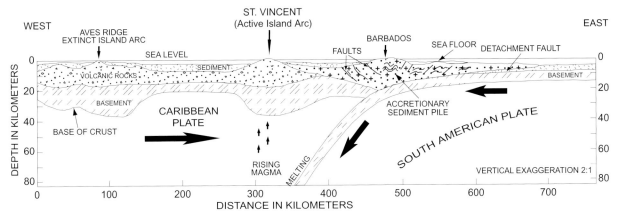

FIGURE 2 Cross-section drawn through the Lesser Antilles at the latitude of St. Vincent, showing the subduction of the South American plate underneath the Caribbean plate (after Westerbrook et al. 1984).

Basement rocks in the Lesser Antilles consist of Cretaceous/Palaeocene island arc rocks that underlie much of the arc from Guadeloupe northward, but much of the evidence for pre-Cenozoic arc rocks is fragmentary and difficult to correlate. Barbados has a unique position in the arc as the top of the accretionary wedge of sediments scraped off the ocean floor during the subduction process. The island is composed mostly of Pleistocene reef limestones that have been rapidly uplifted, but there are two groups of older accretionary rocks exposed on the northeast coast.

The Limestone Caribbees, which extend from Anguilla to Marie Galante, are composed of Cenozoic limestones of varying ages (Eocene in St. Barthelemy; Oligocene on Antigua; Miocene on Anguilla and St. Martin; and Pliocene-Quaternary on Barbuda, Guadeloupe, and Marie Galante). They are underlain by volcanic rocks that represent the older, eastern branch of the arc where volcanism ceased millions of years before. Generally, the rocks exposed on these islands increase in age as one moves northward along this part of the island chain.

The Volcanic Caribbees, which extend from Grenada to Saba, represent the areas where magma produced by the subduction process reaches the surface. The oldest exposed rocks date back to the Eocene and occur in the islands from Martinique southward. The age of the rocks in the islands north of Martinique extend only to the Miocene, and in some cases to the Early Pliocene. In the southernmost part of the arc, Grenada and the Grenadines, sedimentary rocks of Middle Eocene to Middle Miocene age, are abundant.

Evidence for the transition from the eastern Limestone Caribbees front to the current Volcanic Caribbees is not well preserved on the islands. Only in Martinique, where a record of almost continuous migration of volcanic activity from the tholeiitic products in the east (~16 million years ago) to calc-alkaline rocks in the west (~6 million years ago) exists, is the evidence well preserved. In fact, determination of stratigraphic relationships in the arc is fraught with problems given the abundance of volcanic deposits and the essentially point-source evidence for arc evolution provided by the islands.

GEOCHEMISTRY

The Lesser Antilles arc consists of three geochemically and structurally distinct zones. The northern segment, from Saba to Montserrat, contains andesites and minor dacites and belongs to the island-arc tholeiitic magma suite. The islands of this group have low volumes of basalts and rare rhyolites (St. Kitts and St. Eustatius). The central group (Guadeloupe to St. Lucia) contains the most prolific volcanoes in the Quaternary and has total erupted volumes among the largest in the Lesser Antilles. The predominant rock type is again andesites, with some basalts and dacites and rare rhyolites, but the magmas belong to the calc-alkaline magma suite. The southern group extends from St. Vincent to Grenada and consists of predominantly basalts and basaltic andesites with rare andesites. This group includes an alkalic suite of magmas associated with highly undersaturated lavas enriched in incompatible and transition elements.

In addition to major rock types outlined above, three types of plutonic nodules have been found among the strata exposed in the Lesser Antilles. Cognate inclusions are phenocryst clusters and fine-grained-to-porphyritic crystal clots of differing textures to the host magma. Metamorphic xenoliths are cordierite-bearing hornfels and metasediments with relict bedding and cross-stratification. Rare samples of these xenoliths have been found at the Soufrière volcano. Finally, there are plutonic cumulate inclusions and nodules.

Cumulate-textured blocks are a common occurrence in most islands of the Lesser Antilles. The blocks vary in size from 1 cm to several tens of centimeters and are particularly abundant in some areas (e.g., the Soufrière of St. Vincent). The rocks exhibit a wide variety of textures and mineralogies with plagioclase and amphibole being dominant in most islands.

SEISMICITY

The Eastern Caribbean is significantly seismically active (Fig. 3). Tectonic earthquakes associated with the subduction process and volcanic earthquakes associated with the rise of magma are the two types of earthquakes experienced in the region. As noted previously, the hypocenters of the tectonic earthquakes define the shape of the subducting plate or the Wadati-Benioff zone. Energy release from major historical earthquakes indicates a slip rate of 1–5 mm/year, significantly less than the 20 mm/year predicted from global plate tectonic models.

The relatively slow plate convergence rate of 2 cm per year contributes to long intervals between the largest earthquakes generated by the system. These earthquakes occur in the rigid crustal material on either side of the collision boundary and within the descending slab. In the subduction zone environment, earthquakes within the crust are described as shallow, and those occurring in the descending slab are described as occurring at intermediate depth or as being deep.

Annually, there are over 1000 earthquakes recorded by the seismograph networks in the Eastern Caribbean with epicenters located in the Lesser Antilles island arc. The general pattern observed during the more than 50 years that continuous monitoring has been taking place is a broad zone of shallow seismicity with better defined, overlapping bands of intermediate depth and deeper seismicity. In general, the deepest events occur to the west of the arc. Earthquake activity in the Eastern Caribbean is not distributed uniformly throughout the region, and some areas exhibit more intense activity than others do. The zones near Antigua, north of the Paria Peninsula and Gulf of Paria, are the areas where higher levels of seismicity are manifested. The lowest level of seismicity is seen in the area from Grenada to St. Lucia. That pattern has been attributed to a smoothly descending slab or to the accumulation of strain energy, which is yet to reach its limit. The area around Barbados also displays a relatively low level of seismic output, which is considered consistent with its location away from active subduction.

VOLCANISM

Volcanism in the Lesser Antilles dates back as far as the Eocene, and its general nature appears to have remained unchanged throughout. Volcanic centers exhibit a wide range of isotopic and chemical compositions, which reflect the variety and nature of sources and evolutionary processes that led to their genesis. Westward translation of the volcanic arc occurred in the northern islands during the Miocene, but in the south, new volcanic centers formed adjacent to the older ones. Although the nature of volcanism has remained unchanged with time, there has been migration of the center of activity within islands. There are at least three examples of progressive intra-island migration of volcanism during the Plio-Pliestocene period in the Lesser Antilles: St. Kitts (northward), Guadeloupe (southward), and St. Vincent (northward). In each case, migration (at rates of 4 to 10 km/million years), may represent the movement of a single magma source or plume trace along the active front, creating lines of volcanoes with linearly decreasing age.

The largest volcanoes occur in the central part of the arc extending from St. Vincent to Guadeloupe. These have created large islands from overlapping volcanic deposits produced by repeated eruptions of volcanic cen-

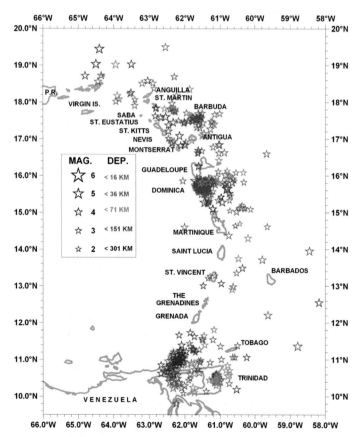

FIGURE 3 Eastern Caribbean earthquakes for the period July 1, 2004, to July 31, 2006. (From Fig. 3 of Seismic Research Centre, 2007).

ters. Large, mature, and complex stratovolcanic centers have been created on these islands usually with a central core of lava domes surrounded by primary and reworked pyroclastic deposits. Determination of stratigraphic relationships is often difficult and has relied in the past heavily on the incidence of dateable charcoal created by the carbonization of wood during pyroclastic flows.

The islands from Montserrat to Saba in the northern segment of the arc are smaller and are comprised of lower-altitude volcanoes that appear incapable of producing the size of eruptions preserved in the rock record in the central islands. One effect of the smaller size of volcanic mountains is the greater interaction with the sea. As such, on St. Kitts and St. Eustatius there are raised limestone platforms created by the uplifting of shallow, submarine shelves during the emplacement of cryptodomes.

VOLCANIC HAZARDS

Pleistocene-to-Recent volcanoes (occurring less than 2 million years ago) occurs in narrow zones (less than 10 km wide), which appear to define three segments: Saba to Montserrat, Guadeloupe to Martinique, and St. Lucia to Grenada. Active volcanism has been characterized both by effusive eruptions producing lava flows and domes and by explosive eruptions producing various types of pyroclastic deposits. The total volumetric volcanic production over the past 0.1 million years is symmetrical about Dominica, which has produced ~40 km^3 of volcanic deposits. This compares with 8 km^3 for Guadeloupe and Martinique and 0–5 km^3 for the islands located to the north and south of these central islands. The mean spacing between active volcanoes during the past 0.1 million years is in the range of 15 to 125 km.

The Lesser Antilles contain 21 live volcanoes distributed among 11 volcanically active islands (Fig. 4). There have been at least 34 eruptions of Lesser Antilles volcanoes (Table 1) during the "historic period" (the past 400 years), and currently about 1 million people live within the areas that can be affected by the direct impacts of eruptions in the future. Volcanic eruptions have killed over 30,000 people in the past, and volcanic hazards are some of the main geologic hazards that threaten the Eastern Caribbean region. All of the islands of the Volcanic Caribbees have at least one volcano that may erupt in the future.

Twenty-one of the historic eruptions have occurred since 1900: nine on land from volcanoes on Guadeloupe, Martinique, St. Vincent, Montserrat, and Dominica, and 12 from the submarine volcano Kick 'em Jenny, ~9 km north of Grenada. These eruptions have all shown a wide variety of eruptive style, magnitude, and impact on the local population. Several eruptions have been phreatic in nature, involv-

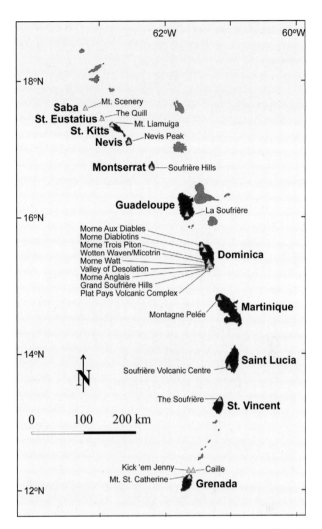

FIGURE 4 The Lesser Antilles region showing the location of the 21 volcanic centers considered active or potentially active. The islands that make up the Volcanic Caribbees are shown in brown and those of the Limestone Caribbees in yellow.

ing the interaction of groundwater with rising magma. One of these was a minor phreatic eruption in Dominica in 1997 that went largely unnoticed; two were much more serious phreatic eruptions in Guadeloupe in 1956 and 1976–1977.

The 1902–1907 magmatic eruption of Montagne Pelée on Martinique was characterized by both effusive dome formation and explosive dome collapse, and led to the total destruction of the town of St. Pierre and the deaths of approximately 30,000 people. A similar eruption occurred from Montagne Pelée several years later, between 1929 and 1932, this time with no reported casualties.

Explosive magmatic eruptions have included the 1902 eruption from the Soufrière in St. Vincent that resulted in the deaths of at least 1500 people. In contrast, the 1971–1972 eruption of this volcano was an effusive magmatic

TABLE 1
Historical Volcanic Eruptions of the Lesser Antilles

Country	Volcano	Date	Type of Eruption
Saba	Mt. Scenery	~1670 (280 ± 80 years BP)[a]	Explosive
Montserrat	Soufrière Hills	1667 ± 40 (285 years BP; $n=10$)[b]	Dome-forming
	Soufrière Hills	1995–present	Dome-forming
Guadeloupe	La Soufrière	1690	Phreatic
	La Soufrière	1797–1798	Phreatic
	La Soufrière	1812	Phreatic
	La Soufrière	1836–1837	Phreatic
	La Soufrière	1956	Phreatic
	La Soufrière	1976–1977	Phreatic
Dominica	Valley of Desolation	1880	Phreatic
	Valley of Desolation	1997	Phreatic
Martinique	Montagne Pelée	1792	Phreatic
	Montagne Pelée	1851–1852	Phreatic
	Montagne Pelée	1902–1905	Explosive and dome-forming
	Montagne Pelée	1929–1932	Explosive and dome-forming
St. Lucia	Soufrière Volcanic Center	1766	Series of phreatic eruptions
St. Vincent	The Soufrière	1718	Explosive magmatic
	The Soufrière	1780	Dome-forming
	The Soufrière	1812–1814	Explosive magmatic
	The Soufrière	1902–1903	Explosive magmatic
	The Soufrière	1971–1972	Dome-forming
	The Soufrière	1979	Phreatomagmatic and dome-forming
Grenada	Kick 'em Jenny	July 24, 1939	Phreatomagmatic
	Kick 'em Jenny	October 5–6, 1943	Submarine
	Kick 'em Jenny	October 30, 1953	Submarine
	Kick 'em Jenny	October 24, 1965	Submarine
	Kick 'em Jenny	May 5–7, 1966	Submarine
	Kick 'em Jenny	August 3–6, 1966	Submarine
	Kick 'em Jenny	July 5, 1972	Submarine
	Kick 'em Jenny	September 6, 1974	Phreatomagmatic
	Kick 'em Jenny	January 14, 1977	Submarine, dome-forming
	Kick 'em Jenny	December 29–30, 1988	Phreatomagmatic, dome-collapsing
	Kick 'em Jenny	March 26–April 5, 1990	Submarine
	Kick 'em Jenny	December 4–6, 2001	Submarine

NOTE: From Lindsay et al. 2005.
[a] There are no written accounts of this eruption, but radiocarbon dates place it well within the historical period for the region.
[b] There are no written accounts of this eruption, but radiocarbon dates place it well within the historical period for the region.

eruption that resulted in the formation of a lava dome confined within the summit crater. The 1979 eruption of the Soufrière was again explosive but was followed by dome growth, and although there was some property damage, no lives were lost.

The 12 submarine eruptions from Kick 'em Jenny are believed to have been dominantly explosive, although in at least one case a lava dome was extruded. The Soufrière Hills volcano on Montserrat has been in active eruption since 1995 and has had a major impact on the island's population. The eruption is characterized by periods of dome growth interspersed with dome collapse and minor explosions. The Soufrière Hills volcano is the only volcano currently erupting in the Eastern Caribbean.

VOLCANO MONITORING IN THE LESSER ANTILLES

The responsibility for monitoring volcanic and seismic activity in the Lesser Antilles is divided between three main organizations. The Seismic Research Centre, which is part of the University of the West Indies (UWI), is based in St. Augustine, Trinidad, and is responsible for monitoring activity in the independent Commonwealth countries of the Lesser Antilles, namely St. Kitts and

Nevis, Dominica, St. Lucia, St. Vincent and the Grenadines, and Grenada. The Institut de Physique du Globe de Paris (IPGP) monitors volcanic activity in Martinique and Guadeloupe, and the Montserrat Volcano Observatory (MVO) operates a monitoring network in Montserrat. In all of the islands of the Lesser Antilles, these agencies work closely with the civil authorities (typically known locally as national disaster preparedness organizations), which represent the respective local governments.

The mainstay of all volcanic monitoring in the Lesser Antilles is the seismograph network. The Seismic Research Centre of UWI maintains 40 seismic stations in the volcanic islands for which they are responsible; these are located near the 18 live volcanoes spread across these countries. The IPGP maintains eight stations in Guadeloupe and eight in Martinique to monitor La Soufrière and Montagne Pelée, respectively. The seismograph network on Montserrat comprises 11 stations, eight maintained by the MVO and three by the Seismic Research Centre. All these stations form part of the regional seismograph network, which includes a further 16 UWI stations on the surrounding non-volcanic islands (Trinidad, Tobago, Barbados, Antigua, Barbuda, St. Martin); nine stations in northeast Venezuela maintained by the Universidad de Oriente, Cumana, Venezuela; and the Fundacion Venezolana de Investigaciones Sismologicas and several French stations in eastern Guadeloupe and southern Martinique.

In addition to seismic monitoring, programs of volcanic gas surveillance and ground deformation monitoring are also maintained in the volcanic islands of the Lesser Antilles.

SEE ALSO THE FOLLOWING ARTICLES

Antilles, Biology / Earthquakes / Island Arcs / Kick 'em Jenny / Lava and Ash

FURTHER READING

Biju-Duval, B., and J.C. Moore. 1984. *Initial report of the Deep Sea Drilling Project*. 78A. Washington, DC: Government Printing Office.

Briden, J.C., D.C. Rex, A.M. Faller, and J.F. Tomblin. 1979. K-Ar geochronology and palaeomagnetism of volcanic rocks in the Lesser Antilles island arc. *Philosophical Transactions of the Royal Society of London* A291: 485–528.

Brown, G.M., J.G. Holland, H. Sigurdsson, and J.F. Tomblin. 1977. Geochemistry of the Lesser Antilles volcanic island arc. *Geochimica et Cosmochimica Acta* 41: 785–801.

Fox, P.J., and B.C. Heezen. 1975. *Geology of the Caribbean crust. The Ocean Basins and Margins* series, volume 3. A. E. M. Nair and F. G. Stehli, eds. New York: Plenum.

Lindsay, J.M., R.E.A. Robertson, J.B. Shepherd, and S. Ali, eds. 2005. *Volcanic hazard atlas of the Lesser Antilles*. Trinidad and Tobago: Seismic Research Unit, University of the West Indies.

Macdonald, R., C.J. Hawkesworth, and E. Heath. 2000. The Lesser Antilles volcanic chain: a study in arc magmatism. *Earth-Science Reviews* 49: 1–76.

Pindell, J.L., and S.F. Barrett. 1990. Geological evolution of the Caribbean region: a plate tectonic perspective, in *The geology of North America, Vol. H, The Caribbean region*. G. Dengo and J.E. Case, eds. Boulder, CO: Geological Society of America, 405–432.

Speed, R.C., and G.K. Westbrook. 1984. *Lesser Antilles and adjacent terranes, Atlas 10, Ocean Margin Drilling Program regional atlas series 28*. Woods Hole, MA: Marine Science International.

Wadge, G. 1994. The Lesser Antilles, in *Caribbean geology: an introduction*. S.K. Donovan and T.A. Jackson, eds. Kingston, Jamaica: UWI Publishers Association.

Westerbrook, G.K., A. Mauffret, A. Munschy, R. Jackson, B. Biju-Dival, A. Mascle, and J.W. Ladd. 1984. Thickness of sediments above acoustic basement, in *Lesser Antilles arc and adjacent terranes, Atlas 10, Ocean Margin Drilling Program regional atlas series*. R.C. Speed and G.K. Westbrook, eds. Woods Hole, MA: Marine Science International.

ANTS

BRIAN L. FISHER

California Academy of Sciences, San Francisco

The rise of ants to ecological dominance has been called one of the great epics in evolution. The same features associated with their ecological success also make them destructive invaders. Islands provide an exceptional model for studying ant dispersal, extinction, and radiation. Ants often reach oceanic islands via accidental "sweepstake routes," leading to a unique cluster of ant species on different islands. The chance dispersal to islands results in high species turnover between islands and within islands over time. The composition of the ant fauna on any particular island can therefore reflect the age, size, and relative isolation of the island. At the same time, the limited land area and biodiversity of islands also increase their vulnerability to incursions by invasive ant species. Increased habitat fragmentation and the accelerated pace of ant species introductions put endemic island ecosystems at increased risk for invasional meltdown.

ISLAND ANTS

Ants are the glue that holds ecosystems together. These social insects dominate almost every terrestrial habitat throughout the world, in terms of both sheer numbers and ecological interactions. This dominance is particularly remarkable because ants constitute only about 1% of all described insect species. Understanding the processes driving the phenomenal success of ants is an active area of research. Current techniques involve the careful analysis of species distributions as well as the historical and geographical factors affecting dispersal and radiation.

Overall, the study of ant ecology is limited by the fact that up to half of the estimated 20,000 species of ants in the world have yet to be described. However, because islands are much smaller in area and harbor less diverse faunas than do continents, an exhaustive inventory of their ant species is feasible. Analysis of this more limited assemblage can then shed light on processes that affect ant composition and dispersal.

In general, the bigger an island, the more diverse its assemblage of ants. This is certainly true for the world's three largest tropical islands: New Guinea, Borneo, and Madagascar. It is no coincidence that these three land masses also feature more endemic ant genera and species than any other islands on Earth.

Chance Dispersal

Ants on oceanic islands often arrive via a so-called "sweepstakes route," a term that aptly describes the rarity of a successful island landing but also the huge potential payoff. Ant species that do manage to establish themselves on a large island can often then radiate to fill many empty ecological niches.

Island ants typically arrive via one of four common dispersal routes. In many ant species, the reproductive form is a winged queen. Newly inseminated queens taking wing to establish a new colony can be blown across the open ocean to distant shores. Alternatively, an entire ant colony can raft to an island inside a rotten log or tree washed out to sea during a storm. Some islands have received new fauna from the mainland or other islands via land bridges exposed during periods of low sea level. Finally, humans have been transporting ants inadvertently wherever they travel, including on island voyages.

An island's geography can be the determining factor in whether or not dispersing ants can land and establish a foothold. Islands in proximity to other sources of ants are easily reached by prevailing winds or ocean currents and will be colonized more often. The older an island, the more time ants will have had to arrive and establish themselves. By the same token, larger islands offer a bigger target to dispersers.

The extreme isolation and relative youth of the Polynesian islands east of Samoa, including Hawaii, have placed these islands among the few places on Earth that lack native ant species. On these islands, ants have had little opportunity to arrive on their own. In fact, Hawaii's native fauna includes no social insects of any kind. Yet today, 50 ant species are established in Hawaii, having

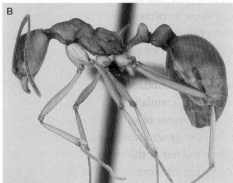

FIGURE 1 Endemic species on islands such as Mauritius and Madagascar are characterized by species with wingless reproductive queens. These colonies reproduce by fission, when the newly inseminated queen walks away from the parent nest with a few workers to start her new colony. (A) *Odontomachus coquereli*; (B) *Aphaenogaster* sp.; (C) *Pristomyrmex bispinosus*.

been brought to the islands by humans over the past century. Among them are some of the world's most widespread and damaging invasive species (see below). These recent arrivals have been devastating the highly endemic arthropod fauna, which never evolved defenses against ants and lacks specialized ant predators.

In contrast, as a very old island long isolated in the southwestern Indian Ocean, Madagascar today harbors an unusually diverse ant fauna. Of its more than 1000 ant species, over 95% are endemic to the island. Madagascar originally formed part of the ancient supercontinent of Gondwana but broke away from Africa approximately 120 million years ago. Its assemblage of native ant species likely evolved only after this breakup. Biologists now believe the extant ant lineages on Madagascar arrived via oceanic dispersal, primarily from Africa, but also from Asia. Although Madagascar is much closer to Africa, ocean and wind currents from the east may explain the connection to Southeast Asia.

On the other side of the world, the Antilles, which arc across the Caribbean in a chain of more than 7000 islands, provide another good example of sweepstakes colonization. Here, islands at increasing distance from the mainland show a corresponding drop in the number of ant genera. Moreover, the larger the size of an island, the greater the number of endemic species it contains; few to no endemics live on Caribbean islands under 1000 km². The exception is Trinidad, a large continental island just 11 km from the mainland of Venezuela. The ant fauna there is an extension of species found in South America and includes 17 genera widespread on the continent but absent from the rest of the archipelago.

The existing ant community, vegetation, and habitat of the island also help shape how hospitable the island will be once an ant makes landfall. Despite the African and Asian origins of Madagascar ants, the ant fauna is quite unique when compared to the faunas of neighboring continents. Several of the ant species that dominate ecosystems in Africa and Asia are absent from Madagascar. Among these are army ants (*Aenictus, Dorylus*) of the forest floor and weaver ants (*Oecophylla*) of the forest canopy. Weaver and army ants, especially of the genus *Dorylus* (driver ants), are major predators of other ants. Their presence influences the structure of ant populations as well as the diversity of ant communities. The absence of such keystone species is a common feature of many other islands (e.g., Cuba, Hispaniola, Fiji) where ants have radiated. These island systems thus constitute a natural experiment for evaluating how these dominant ants affect biological communities and how their absence allows the diversification of some unusual groups.

The characteristics of an individual species may also affect how likely it is to become a pioneer. For example, in most continental species, ants reproduce and disperse through a winged queen. After mating, she flies to a new location to establish a new colony. Winged queens are clearly an advantage for dispersing to an island. Interestingly, winged queens may become a drawback once a species has reached an island, because winged queens are more likely to be blown offshore and into the ocean. This may explain why many endemic ant species of the southwestern Indian Ocean islands have evolved wingless queens (Figs. 1, 2).

FIGURE 2 Endemic species with wingless reproductive queens. (A) *Terataner* sp.; (B) *Mystrium mysticum*; (C) *Cerapachys* sp.

Radiation

Once an ant species establishes on an island, it may undergo adaptive radiation to fill vacant niches. The number of endemic species tends to be greater on older and larger islands, where ants have had more time to evolve and there has been a greater complexity of local habitats to occupy.

In the Caribbean, for example, endemic ants make up a disproportionate share of the ant faunas of both Cuba and Hispaniola. Endemic radiations of a single genus, *Temnothorax*, now constitute more than 25% of the ant fauna in Cuba alone. The group includes species that have become specialists at nesting in habitats such as soil, limestone crevices, or epiphytic plants. The stunning morphological diversity of these species is comparable to the range usually seen in several genera. The absence of army or driver ants on these islands may have encouraged this evolutionary profusion. Likewise, on the island of Fiji, the diversification of the genus *Leptogenys* might have been possible because of the absence of army ants and the relatively low number of other endemic species from similar genera.

On Madagascar, many groups of ants have undergone an equally spectacular radiation. The five most species-rich ant genera on the island, *Camponotus, Hypoponera, Pheidole, Strumigenys,* and *Tetramorium,* all contain over 100 species. Each group exhibits remarkable morphological and niche diversity. Local diversity is also amazing. For example, on the Masoala Peninsula alone, 98 species from these five genera co-occur. As on Hispaniola, the absence of army and weaver ants likely allowed certain lineages to persist and others to radiate and flourish on the island. An example is the diversification of the tribe Cerapachyinae (*Cerapachys* and *Simopone*), which includes an unprecedented, morphologically diverse assemblage of more than 50 species on Madagascar. In fact, certain *Cerapachys* species show morphological similarity to the army ant genus *Aenictus* found in Africa. Whether the absence of army ants led to the diversification of *Cerapachys* in Madagascar, or simply permitted their persistence, remains unclear.

Another remarkable example of ant radiation in Madagascar is illustrated by the two closely related genera of the ant tribe Dacetini. Dacetine ants rely on the trap-like action of their mandibles to capture and subdue live food. Most of the differences between species reflect various methods of seizing prey. With 89 described species in Madagascar, the dacetines are the island's dominant predatory leaf-litter insect. Local diversity, too, is off the charts. For example, 25 species of dacetines have been recorded in an area roughly 1 km^2 on the Masoala Peninsula.

The ecological context in which ant colonists find themselves may also influence whether or not a lineage has the opportunity to radiate. For example, the relative diversity of *Strumigenys* and *Pyramica* in Africa is very different from their diversity on Madagascar. These specialized trap-jaw predators have oddly shaped mandibles and pear-shaped heads, and are often covered in bizarre hairs and strange outgrowths of whitish sponge-like tissue around their waist segments. *Strumigenys* are quite diverse in Madagascar, with 74 species on the island compared to 50 in Africa. *Pyramica*, in contrast, are much more diverse on the continent, with 81 species in Africa and only 15 in Madagascar. Why *Strumigenys* has undergone this diverse island radiation whereas *Pyramica* has not remains somewhat of a mystery. Maybe the first *Pyramica* arrived after *Strumigenys* had already radiated and filled potential niches.

Taxon Cycle

While studying ants in Melanesia, E. O. Wilson observed that species pass through sequential phases of expansion and contraction in distribution. He coined the term "taxon cycle" to describe the phenomenon. On islands, expanding taxa tend to be recent arrivals from the mainland that occupy lowland habitats along the coastlines of islands. By contrast, contracting taxa have reduced and fragmented ranges and tend to occupy interior and montane habitats. These differences suggest that existing species in the contraction phase of the cycle are pushed upslope and into new habitats by competition from more recent arrivals along the coast.

This pattern certainly holds in Madagascar, where endemic and possibly older ant groups are restricted to mountaintops. An example is the genus *Anochetus,* also known as little trap-jaw ants. Within this group, two species thought to have resided on the island for a long period of time are related to taxa in Asia and found only at interior higher elevation habitats. Two other species, related to taxa in Africa, are widespread across the entire lowlands. Together, these data support the idea that *Anochetus* species arriving on Madagascar first settled in marginal coastal habitats before shifting to anterior lowland forest and finally up to montane forest.

Turnover

The composition of ant species can vary considerably across an island's history. The primary forces that impact island biogeography—size, isolation, and habitats—also exert great influence on species turnover through time.

On Hispaniola, distance to source populations has had a dramatic effect on faunal assembly. Studies of Dominican amber indicate that 20 million years ago, during the Miocene, the island's ant fauna was closely related to the continental fauna of Mexico. Of the 38 genera and subgenera found in amber, only 22 persist today on Hispaniola, whereas 15 native genera have colonized the island since. Interestingly, the amber fauna includes army ants no longer present on the islands. This dramatic species turnover reflects the fact that during the Miocene, the Greater Antilles (including Cuba, Jamaica, Hispaniola, and Puerto Rico) were located nearer to the mainland. Lying further from the mainland today, Hispaniola has lost some of its continental taxa. Highly specialized species or those less able to establish themselves on new ground were the most likely to disappear.

Similar turnovers in ant species have probably swept across many islands as shifts in climate, volcanic eruptions, or other geological factors changed their habitats.

Further Research

Islands are natural laboratories for understanding the processes of faunal distribution and diversification. New methods combining species inventories, taxonomic research, and phylogenetic findings are enabling scientists to investigate these processes through the study of island ants. A comprehensive research effort is now in progress in the southwestern Indian Ocean (SWIO) islands. The coralline, volcanic, and Gondwanaland fragments of this region vary widely in age, size, degree of isolation, and habitat type, making them an ideal place to explore how each of these factors affects species diversity. Some of the questions researchers seek to answer include the geographic origins of the ant fauna and whether the estimated ages of endemic groups correlate with the ages of the islands themselves.

INVASIVE ANTS

Although colonization and species turnover are natural island processes, modern-day incursions of invasive ants are a major threat to natural ecosystems. Small size and limited biodiversity make islands inherently vulnerable to new species introductions. Today, habitat fragmentation caused by development, together with species introductions accelerated by global trade, has further increased this vulnerability.

Combating invasive species is of particular importance on smaller islands. In Mauritius, where only a few patches of original forest still remain, invasive ants may have driven the entire lowland ant fauna to extinction. The big-headed ant, *Pheidole megacephala*, has been implicated in the blanket decimation of Hawaii's lowland arthropods. Entomologists in the early twentieth century described in detail how the native beetle fauna was defenseless again the onslaught of the invading big-headed ant. On the smaller, granitic islands of the Seychelles, Christmas Island, and Zanzibar, invasive ants such as *P. megacephala* have already extirpated native ants and are now threatening nesting bird populations.

Larger islands such as Madagascar, where habitats are severely fragmented, can be just as vulnerable to invasion as their smaller counterparts. And although parks and reserves bolster the chances of survival of native species by protecting habitat, they cannot prevent aggressive exotic ants from driving native species locally extinct.

An island's ant fauna may include dozens of invasive species. However, a few bad actors can cause enough damage to destroy an island system. The usual culprits include the yellow crazy ant *Anoplolepis gracilipes*, the white-footed ant *Technomyrmex albipes*, the little fire ant *Wasmannia auropunctata*, the big-headed ant *Phiedole megacephala*, the tropical fire ant *Solenopsis geminata*, and the Argentine ant *Linepithema humile* (Figs. 3, 4). These ants are especially dangerous when they join forces with another insect like mealybugs, a type of sap-sucking plant parasite that is often invasive on islands. In exchange for protection, the mealybugs provide drops of sugary honeydew to the ants. Fueled by these bonus sources of sugar, invasive ant populations can easily multiply out of control on an island system and trigger an invasional meltdown. This scenario has played out on Christmas Island, where such interactions and the concomitant vast numbers of ants (*Anoplolepis gracilipes*) have led to decimation of the native land crab–dominated ecosystem.

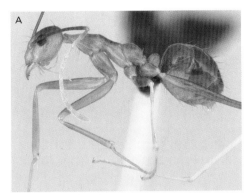
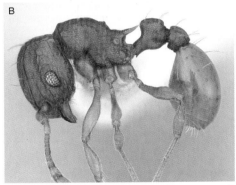

FIGURE 3 The most notorious invasive ants on islands. (A) yellow crazy ant, *Anoplolepis gracilipes*; (B) little fire ant, *Wasmannia auropunctata*; (C) tropical fire ant, *Solenopsis geminata*.

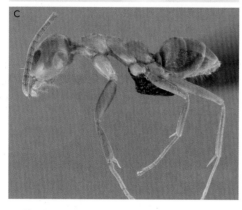

FIGURE 4 The invasive (A) white-footed ant, *Technomyrmex albipes;* (B) big-headed ant, *Phiedole megacephala;* and (C) Argentine ant, *I inepitherna humile.*

Once non-native, invasive ants become established in natural settings, they are difficult, if not impossible, to eradicate. Thus, when preserving an island's native ant species, an ounce of prevention is truly worth a pound of cure.

Impact of Invasives

Worldwide, invading ants have caused impacts that reverberate throughout local ecosystems. In some cases, invasive ants have reduced the abundance and diversity of native ants by more than 90%. Nor are the consequences of ant invasions limited to other ants. The intruders also cause decreases in the diversity of insect herbivores, mammals, lizards, birds, and even plants.

Invasional meltdowns of island ecosystems may be caused in part by the formation of ant supercolonies. The entire population of a newly arrived species may derive from the landing of a single queen or colony. Because all ants of this species are so closely related, they may lose aggression toward others of their own kind, a feature that normally limits colony densities. The resulting supercolonies can attain extremely high densities, can decimate local arthropod communities in the region, and can lead to an oversimplified invertebrate community that fails to provide essential ecosystem services such as nutrient cycling, plant seed dispersal, and a prey base for higher trophic levels.

On Christmas Island, for example, researchers have documented the devastating impact of a supercolony on a local ecosystem. After an accidental introduction, the invasive crazy ant (*Anoplolepis gracilipes*) formed massive supercolonies that tended scale insects. Through direct predation, the supercolonies practically eliminated the red ground crab in the infested area. Without the crab, the principal litter consumer and seed disperser, the habitat changed dramatically. The scale insects killed off many trees and impacted ground-nesting birds.

History

Ants have probably hitched rides with humans since the dawn of history. A recent dig at a Roman bath in Britain uncovered the bodies of 2000-year-old invasive ants. Over the past 500 years, there have been many accounts of ant plagues on different islands. For example, between the sixteenth and eighteenth centuries, several tropical West Indian islands were stricken with a series of ant plagues. Historical documents prove that environmental problems caused by invasive ants ensued just decades after Europeans came to the New World.

At least two different ant species were the culprits of the Caribbean ant plagues. *Solenopsis geminata,* the tropical fire ant, was brought to the islands in the early 1500s; *Pheidole megacephala,* an African ant, came in the late 1700s. The plagues caused widespread crop destruction and may have been accelerated by the arrival of sapsucking insects. The same scenario has been replayed again and again on other islands.

One interesting feature of ant plagues is that they are relatively short lived. The invasive ants are soon either repressed or driven extinct by later invading ants. For example, *Pheidole megacephala* was the dominant ant species on the Atlantic islands of Bermuda for much of the twentieth century. In 1940, however, the Argentine ant (*Linepithema humile*) arrived in the area and quickly outcompeted the earlier champion. *Pheidole megacephala,* however, persists, and ever-shifting battlefronts now criss-cross most of the islands.

An important note is that islands have a long history of cycling through taxa. For example, historical records for the Indian Ocean island of Réunion indicate that the invasive ant *Anoplolepis gracilipes* was already abundant on the island in 1895. With the capacity to attain extremely high densities, this species can decimate resident vertebrates and invertebrate populations. In the ensuing 100 years, however, *A. gracilipes* has become rare. Research suggests that competition with dominant species such as *Pheidole megacephala* and *Solenopsis geminata* may have reduced its foraging efficiency and, therefore, its abundance. Although *A. gracilipes* is less competitive than other invasives, it thrives on a wide range of islands and must possess superior colonizing ability. The fortunes of this species have followed a similar trajectory in the Seychelles.

The modern twist to this phenomenon is the speed with which species turnover now occurs. As planes and ships have multiplied, and transport times have shrunk, pressures on native species have increased apace.

Further Research

The first step in understanding the level of threat posed by invasive ants is to inventory and map the extent of ant invasions. However, an assessment of invasives has yet to be initiated for many island systems. Once the invasives have been catalogued and mapped, a number of critical questions can be addressed. These include evaluating the risk of spread to other regions and habitats, predicting the effects on native fauna and flora, and assessing the impact of climate change on species interactions. The answers can help provide guidelines for conservation policies and control or eradication initiatives such that native island ants can persist.

SEE ALSO THE FOLLOWING ARTICLES

Dispersal / Insect Radiations / Land Crabs / Madagascar / Species–Area Relationship / Taxon Cycle

FURTHER READING

Fisher, B. L. 2005. A model for a global inventory of ants: a case study in Madagascar. *Proceedings of the California Academy of Sciences* 56: 86–97.
Holway, D. A., L. Lach, A. V. Suarez, N. D. Tsutsui, and T. J. Case. 2002. The causes and consequences of ant invasions. *Annual Review of Ecological Systems* 33: 181–233.
Molet, M., C. Peeters, and B. L. Fisher. 2007. Permanent loss of wings in queens of the ant *Odontomachus coquereli* from Madagascar. *Insectes Sociaux* 54: 174–182.
O'Dowd, D. J., P. T. Green, and P. S. Lake. 2003. Invasional 'meltdown' on an oceanic island. *Ecology Letters* 6: 812–817.
Ricklefs, R. E., and E. Bermingham. 2002. The concept of the taxon cycle in biogeography. *Global Ecology and Biogeography* 11: 353–361.
Suarez, A. V., D. A. Holway, and P. S. Ward. 2005. The role of opportunity in the unintentional introduction of invasive ants. *Proceedings of the National Academy of Sciences USA* 102: 17,032–17,035.
Underwood, E. C., and B. L. Fisher. 2006. The role of ants in conservation monitoring: if, when, and how. *Biological Conservation* 132: 166–182.
Wilson, E. O. 1988. The biogeography of the West Indian Ants (Hymenoptera: Formicidae), in *Zoogeography of Caribbean insects*. J. K. Liebherr, ed. Ithaca, NY: Cornell University Press, 214–230.

ARCHAEOLOGY

MARSHALL I. WEISLER

University of Queensland, St. Lucia, Australia

Archaeology is the systematic study of material remains. The diversity of island environments and biota presents fascinating opportunities for studying the evolution of Oceanic societies that developed from ~40,000 to ~800 years ago.

THE FASCINATION OF ISLANDS

The vast expanse of the Pacific Islands hosted the greatest maritime migration in human history. Venturing across uncharted waters, Pacific colonists over millennia developed remarkable voyaging skills; built massive double-hulled sail-rigged canoes up to 10 m long; and carried vital plants, animals, and indeed the mental templates of their community layouts or "transported landscapes," to every speck of land in this watery world known as Oceania. This journey began ~40,000 years ago in the western Pacific and culminated a millennium ago in the settlement of the isolated outposts of Hawai'i, New Zealand, and Easter Island—the final islands colonized before encountering the continental barrier of South America.

Thousands of islands, scattered over more than one-third of the Earth's surface, display a wide range of size and elevation fostering a diversity of geology, landforms, soils, vegetation, climate, marine biota, and degrees of isolation, providing an endless array of "natural experiments" for human colonists. Archaeologists are challenged to understand, for example, how and why founding human groups settling in the Hawai'ian Islands about AD 900 developed into one of the most highly stratified Oceanic societies, yet with more than 30,000 years of occupation, communities in the southwestern Pacific never attained this level of social complexity. How did humans adapt to the remarkable variability of environments found on continental, high volcanic, and raised limestone (*makatea*) islands and on low coral atolls? Why were some societies sustainable, whereas others fell into an endless spiral of environmental degradation, warfare, and ultimate collapse? Oceanic islands, in all their endless variety, provide remarkable opportunities for investigating the historical development of human societies.

NEAR VS. REMOTE OCEANIA

Having traversed the great breadth of Oceania nearly two centuries ago, the great French explorer and naval commander Dumont d'Urville was the first to classify Pacific peoples into three seemingly distinct groups. Melanesians ("dark islanders") occupied islands in the southwest Pacific including New Guinea, the Bismarck Archipelago, the Solomons, New Caledonia, Fiji, and a few others. These are generally dark-skinned people who have inhabited Oceania the longest and speak the greatest diversity of languages. Consisting mainly of low coral atolls scattered along the northern tropical latitudes, Micronesians ("small islanders") are not a homogeneous group linguistically, in terms of the range of human phenotypes, nor do they share a common point of origin. Decades before d'Urville, the famous British commander Capt. James Cook, after visiting Tahiti, New Zealand, and Hawai'i, remarked on the similarities of language, physical appearance, and customs of the "Indians of the South Seas"—the region d'Urville labeled Polynesia or "many islands." Although not without its problems, this tripartite division of Melanesia, Micronesia, and Polynesia has remained in use to this day.

Roger Green, in 1991, developed far more meaningful terms to differentiate among the islands east of Wallace's Line and reflect the biogeographic differences of Oceania, importantly reflecting the constraints and opportunities for human colonists and the subsequent evolution of island societies. Green's Near Oceania, consisting mostly of New Guinea, the Bismarck Archipelago, and the Solomon Islands, has the greatest biodiversity in Oceania and the only Pleistocene human occupation (Fig. 1). Some resident Papuan speakers eventually mixed with the Austronesians who would settle the region much later. The islands are generally larger and more closely spaced than those of Remote Oceania, a fact that had important implications for human colonization and subsequent exploratory voyaging. Geoffrey Irwin refers to this region, sheltered from cyclones, as a "voyaging nursery" where the knowledge of open-water voyaging was learned—an essential skill for traversing the vast distances separating the far-flung islands of Remote Oceania. Here, the biodiversity declines progressively in an easterly direction.

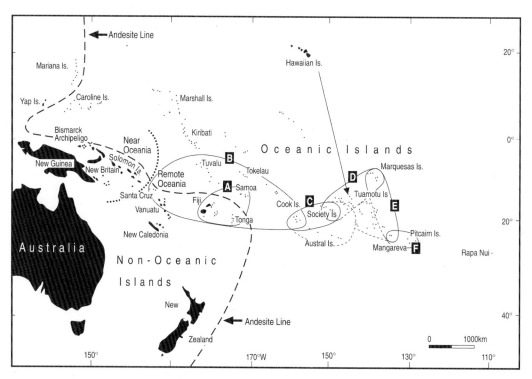

FIGURE 1 Map of the Pacific showing the division between Near and Remote Oceania and trade (interaction) spheres defined by the identification of exotic artifacts, including the movement of adze rock from Hawaii to the Tuamotu Archipelago (adapted from Weisler 1998).

Settled by Austronesian speakers no earlier than about 1500 BC, the islands of Remote Oceania were among the last places settled on Earth. It is little wonder, given that the islands are generally smaller and spaced much farther apart than those in Near Oceania. Human colonists had to develop the voyaging skills required to routinely sail out of sight of land in the quest for new islands. Given the larger distances between landfalls, it was previously thought that once colonized, island societies developed in near isolation. Recent studies of ancient Polynesian trade have shown otherwise and are discussed below after a brief overview of the foundation culture of Remote Oceania, known as Lapita.

LAPITA

Nearly four millennia ago, pottery-using agricultural peoples inhabiting Taiwan, the Philippines, and the immediate islands arcing from Southeast Asia initiated a second wave of Oceanic exploration. Venturing into New Guinea, the Bismarck Archipelago, and the Solomon Islands, these Austronesian speakers encountered small human populations in scattered settlements, unlike the larger agricultural communities from whence they came. The small, kin-based social groups of Near Oceania made some use of marine resources, developing a shell technology for making adzes, fishhooks, arm rings, and beads, but they had a more inland focus—a gathering-hunting adaptation for exploiting brackish to freshwater environments and the forest for nut- and fruit-bearing trees, tubers, rats, reptiles, and introduced marsupials such as the cuscus. Inter-island movement is indicated by the transport of obsidian. This way of life, the dominant pattern for ~30,000 years, was supplanted by "late" arrivals from the west—the Lapita culture.

Jack Golson was the first to realize that highly decorated pottery finds from archaeological sites spanning the Near and Remote Oceania divide represented a "community of culture." This ceramic horizon developed in Near Oceania and was identified by the unique, fine dentate-stamped designs on low-fired, earthenware pottery (Fig. 2). The geographic spread of Lapita covers Near Oceania and the islands east to Tonga and Samoa and is the foundation culture for all of Polynesia. In the west the earliest sites are ~3500 years old and consist of large villages on coastal beach terraces or clusters of stilt-houses built over the shallow lagoons or calm shorelines. The average settlement size was ~5000 m^2, consisting of 15 to 30 dwellings, whereas the largest communities may have had up to 100 houses. These were sophisticated agriculturalists who managed root and tree crops and introduced

FIGURE 2 A Lapita pottery sherd exhibiting characteristic dentate-stamping and lime-filled design (photograph by M.I. Weisler; used with permission of P. Kirch).

pigs, dogs, and chickens from their homeland. There was nearby inter-island and long-distance trade spanning hundreds of kilometers for exchanging pottery, obsidian, chert, volcanic oven stones, and undoubtedly marriage partners as well as perishable foodstuffs, which rarely survive in archaeological contexts. In fact, one of the longest documented movements of any commodity during world prehistory is that of obsidian, traded 4500 km across the breadth of Near Oceania into Remote Oceania, undoubtedly as a series of linked moves.

The Lapita diaspora is also remarkable for its speed, taking perhaps two to three centuries to traverse the 5000 km from New Guinea and the Bismarck Archipelago to Tonga and Samoa in West Polynesia, establishing viable communities en route. The reasons for such rapid migration are clearly not based on population pressure, because density-dependent models require more time between stages of migration. Part of the reason lies in the search for pristine islands with their abundant stocks of seabirds (which could number into the tens of thousands) and unexploited marine shores teaming with fish, molluscs, crabs, and lobsters—or perhaps even new sources of stone for making the ubiquitous adze and other cutting tools. Imagine laying claim to a new, high-grade stone source—a veritable gold mine by today's standards.

Economics aside, there were important social reasons for risking all to found new colonies. In traditional Oceanic societies the first-born is often bestowed with all the property and accumulated family wealth. This, alone, would be a real incentive for junior siblings to split off from the parent community to establish new settlements where they could assume seniority. Archaeologists are now charged with filling in the details of the Lapita horizon and all its variability across this large swath of the Pacific. Using Roger Green's terms of intrusion, innovation, and integration of the Lapita phenomenon, how did these processes play out over the entire spatial and temporal span that this horizon occupied? What was the sequence of island colonization? With rare exception, excavation samples for any one Lapita site are still small, thus limiting our understanding of intra-site variation or archaeology at the "household" level. What was the role of exchange, not only for establishing new colonies, but also throughout the duration of settlement? These are just a few of the questions that will occupy archaeologists in the coming decades.

DATING HUMAN COLONIZATION OF PACIFIC ISLANDS

There is scarcely any Pacific island where archaeologists agree on the timing of human colonization. This is no small problem, as the arrival of humans on previously uninhabited islands "starts the clock" for examining the speed and tempo of adaptations, the rate of human-caused animal extinctions, and transformations in the economic and social fabric of society. What kinds of data are useful for clearly documenting the first presence of people on islands? All archaeologists agree that artefacts, charred and patterned fragments of food remains, subterranean ovens and hearths, and post molds are obvious evidence for human habitation. In addition to radiocarbon chronologies for anchoring island sequences, pollen, sediment, and charcoal depositional events can signal human colonization of pristine islands and subsequent human-induced landscape modification, which often coincides with bird extinctions or reductions, changes in land snail assemblages, and marked alterations in vegetation communities.

Paleoenvironmental estimates for human arrival are often inferred from pollen sequences, developed from sediment cores taken most often in lakes. Colonists cleared land for agricultural production, which resulted in a marked decline in forest taxa, an increase of disturbance indicators (such as open ground ferns), and an influx of charcoal particles. However, these characteristics can also develop in response to climatic change such as drought, or short-term perturbations, including hurricanes, which can influence vegetation dynamics. For example, in Palau, western Micronesia, sediment cores show a decline in forest pollen and increases in ferns and microscopic charcoal particles ~4500 years ago. This is more than 1000 years earlier than the documentation of cultural materials (artifacts, food remains, etc.). In Fiji, charcoal particles from pollen cores were identified at ~4300 years ago, yet there is no reliable indication of human settlement until 2900 years ago. Mangaia, in the southern Cook Islands, shows evidence of habitat disturbance at 2500 years ago, but no in situ cultural remains until ~1000 years later. Pollen sequences are most useful when they include evidence of humanly transported plants such as aroids (e.g., taro, giant swamp taro) and ornamental and medicinal tree and shrub species. Whereas pollen cores are an important source of environmental data—whether they document human disturbance or not—multiple, well-dated sediment cores displaying similar disturbance-influenced environmental sequences and pollen indicative of human-introduced plants are robust data for inferring the presence of people on islands.

More holistic approaches to determining human colonization of islands can be achieved by using a broader array of data. Are the earliest habitation sites located in favorable environmental settings? Using a settlement pattern approach, we would expect to find the earliest sites situated near formerly abundant terrestrial and marine resources, with easy access to the sea. In other words, is the "earliest" archaeological site located in a logical environmental position? Exploitation of abundant and pristine resources should be found at the earliest cultural layers. The lowest, and oldest, cultural layers in early archaeological sites usually have the densest concentrations of food bones with evidence of extinct species, often those of flightless birds. In fact, extensive research has shown that most Pacific islands had one or more species of flightless bird that became extinct after human arrival. This contextual information, of settlement patterns, subsistence, and extinction events, provides a better platform for clearly documenting early sites.

THE MYTH OF ISOLATION: DOCUMENTING ANCIENT TRADE AND EXCHANGE

Once settled, small founding groups on previously uninhabited islands were thought to have diversified and evolved in relative isolation. However, early trade links between parent and daughter communities—documented through the identification of exotic artifacts—were seen as integral to the process of colonization where "lifelines" were maintained until founding colonists were well established. (The term "trade" is used loosely here to mean interaction at some level, but not necessarily the two-way movement of commodities.) The classic example is that of the Pitcairn Group in southeast Polynesia. The four islands of the Pitcairn Group consist of the Pleistocene volcano of Pitcairn (only 4.5 km^2), the raised limestone (or *makatea*) island of Henderson (37 km^2), and the two diminutive atolls of Oeno and Ducie. These islands are characterized by their isolation, small size, lack of reliable freshwater, nutrient-poor soils, and limited reefs (especially in the case of Pitcairn and Henderson). It is little wonder that Peter Bellwood coined the term "mystery islands" for these and nearly two dozen other ecologically marginal islands found throughout Polynesia, for it is these "mystery" islands that have ample evidence of prehistoric habitation, but were found abandoned when rediscovered by Europeans. Why were these islands settled in the first place, and why were they abandoned? The Pitcairn Group provides an illuminating example.

The parent population for the Pitcairn communities was the diverse and well-watered volcanic islands of Mangareva, situated ~400 km west. Extensive archaeological survey and excavations throughout Mangareva and the Pitcairn Group by Weisler in the 1990s detailed a history of trade between Pitcairn and Henderson islands, and on to Mangareva. Traded commodities included fine-grained basalt adzes (Fig. 3) and volcanic glass (a silica-rich rock made into sharp flakes for cutting) from Pitcairn, which were exported to Henderson because it had no sources of volcanic rocks for vital tool making. Mangareva received adzes from Pitcairn and exchanged black-lipped pearlshell (primarily for making fishhooks), which is ubiquitous in its 20-km-wide lagoon. Mangareva also supplied Henderson with volcanic oven stones, planting stock for coconuts, giant swamp taro, and bananas,

FIGURE 3 A typical late prehistoric, fine-grained basalt adze blade from Pitcairn Island. Adze material from Pitcairn Island was traded to Henderson Island (100 km north) and Mangareva (400 km west).

as well as marriage partners and pigs. Trade was essential for the long-term viability of the Henderson community because the island had no sources of these necessary resources. The commodities transferred within this interaction sphere are shown in Fig. 4. The detailed radiocarbon chronology for Henderson settlements documented a period of sustained interaction between Pitcairn Island and Mangareva for nearly six centuries beginning about AD 900. After AD 1450 imported artefacts are no longer found in the Henderson archaeological sites, signaling an

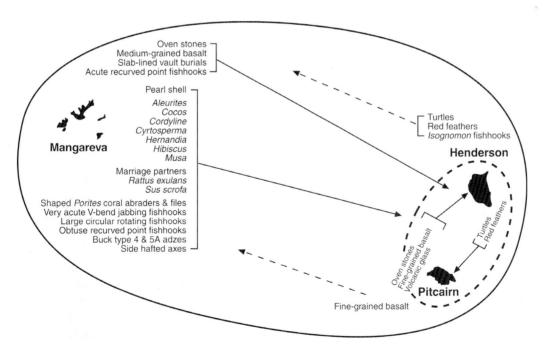

FIGURE 4 The southeast Polynesia interaction sphere between the ecologically marginal Pitcairn Group and the high volcanic islands of Mangareva. This system operated for six centuries beginning about AD 900 (Green and Weisler 2002).

end to external trade. Perhaps one or two human generations later, after the lifeline to Mangareva was severed, the small communities on Henderson died out or relocated to their parent homeland.

During the past two decades archaeologists, often working in concert with geochemists, have identified exotic volcanic adzes, pottery, and obsidian in distant archipelagoes. When exotic artefacts are found in dated layers of archaeological sites, the spatial and temporal dynamics of long-distance interaction can be defined. Fig. 1 shows the limits of Polynesian interaction spheres based on the "sourcing" of exotic adzes. Recent research has also identified a rock from Hawai'i that was fashioned into an adze in the Tuamotu Archipelago, a distance of more than 4000 km, making this one of the longest uninterrupted maritime movements in world prehistory. This interdisciplinary research makes it quite clear that Oceanic communities were far from isolated and, in many cases, had a long history of interaction with distant islands. It is these interactions that archaeologists must define to better understand the external forces that shaped and transformed island societies.

SOCIOPOLITICAL CHANGE

Throughout much of Remote Oceania, surface residential, religious, and fortification architecture are important sources of information for documenting changes in the social organization of island societies. East Polynesia is known for its household architecture in which dry-laid stones were stacked to delimit various functional spaces and to provide support for pole-and-thatch superstructures. Fig. 5 illustrates a substantial L-shaped wall (house foundation), an attached stone terrace (shrine), and a C-shaped shelter (cookhouse) forming a late prehistoric residential complex on Moloka'i, Hawai'ian Islands. A comparative analysis of the size and content of similar residential complexes across the landscape can provide information about the status of the occupants. This was clearly shown in the late prehistoric settlement pattern of Kawela, located along the central south shore of Moloka'i, where extensive excavations of ten residential complexes and a contrastive analysis of the artefacts, food remains, and architectural complexity revealed the status of the occupants.

Religious architecture has a wide range of forms and sizes: small family shrines (Fig. 5); small to large stone platforms, some larger than 2000 m^2; and the well-known statues (*moai*) on Easter Island, an extreme example of a familiar East Polynesian pattern of temple layout (Fig. 6). Monumental architecture serves a multitude of functions within society. Ritualized ideology is codified by a series of temples that were strategically placed to reinforce social

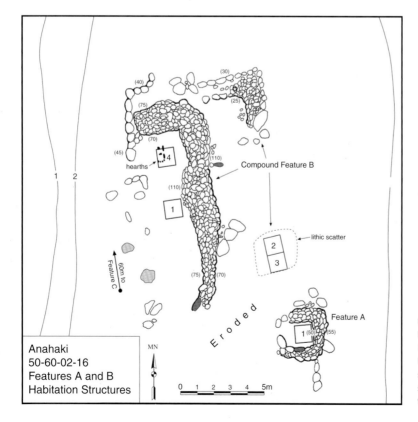

FIGURE 5 An early seventeenth-century late prehistoric residential complex from Moloka'i, Hawaiian Islands. The long wall anchored a pole and thatch house with an internal slab-lined hearth. The small stone terrace, immediately east, was a family shrine. Feature A was a cookhouse with a stone tool-working area just north.

FIGURE 6 Monumental architecture on Easter Island (Rapa Nui). This temple site (*marae*) consists of statues (*moai*), a platform (*ahu*), and a paved court. Photograph courtesy of M. I. Weisler.

order. This can be seen in the distribution of Easter Island temples, whose placement relates to traditional land unit boundaries. The size, number, and position of temples coincide with sociopolitical complexity where human activities were regulated.

Prominent examples of prehistoric defensive architecture are found along the ridges of Babeldoab on the Palau Islands in western Micronesia, on the hilltops of Rapa on the Austral Islands in East Polynesia, and throughout much of New Zealand. Many fortifications consisted of a series of ditches, embankments, and palisades that ensured protection from warring tribes. These sites were built in response to social conflicts that developed, in part, from increasing populations, low soil productivity for crop production, and intertribal competition between late prehistoric social groups. In essence, fortifications symbolize periods of great social upheaval.

CONCLUSIONS

The archaeology of the Pacific Islands provides ample opportunities to investigate many of the important problems in world prehistory, not all of which could be discussed in this short essay. Oceania has some of the earliest examples of crop domestication, the longest transport of commodities, some of the most complex chiefdoms in the world, and certainly the most linguistically diverse regions on Earth. Modern archaeological research is only a few decades old in the Pacific, and many exciting and innovative studies await future generations.

SEE ALSO THE FOLLOWING ARTICLES

Easter Island / Exploration and Discovery / Human Impacts, Pre-European / Peopling the Pacific / Polynesian Voyaging / Wallace's Line

FURTHER READING

Collerson, K. D., and M. I. Weisler. 2007. Stone adze compositions and the extent of ancient Polynesian voyaging and trade. *Science* 317: 1907–1911.
Green, R. C., and M. I. Weisler. 2002. The Mangarevan sequence and dating of the geographic expansion into Southeast Polynesia. *Asian Perspectives* 41: 213–241.
Irwin, G. 1992. *The prehistoric exploration and colonisation of the Pacific.* Cambridge, UK: Cambridge University Press.
Kirch, P. V. 2000. *On the road of the winds: an archaeological history of the Pacific islands before European contact.* Berkeley: University of California Press.
Rainbird, P. 2004. *The archaeology of Micronesia.* New York: Cambridge University Press.
Weisler, M. I., ed. 1997. *Prehistoric long-distance interaction in Oceania: an interdisciplinary approach.* New Zealand Archaeological Association Monograph 21. Auckland: New Zealand Archaeological Association.
Weisler, M. I. 1998. Hard evidence for prehistoric interaction in Polynesia. *Current Anthropology* 39: 521–532.

ARCTIC ISLANDS, BIOLOGY

INGER GREVE ALSOS

University Centre of Svalbard, Longyearbyen, Norway

LYNN GILLESPIE

Canadian Museum of Nature, Ottawa

YURI M. MARUSIK

Institute for Biological Problems of the North, Magadan, Russia

Arctic islands constitute a major part of the arctic land masses. Low temperatures and short summers are strong environmental filters that exclude most organisms from living there. Thus, the diversity of most species groups is lower on arctic islands than on the arctic mainland and more southern latitudes. Arctic species exhibit many different adaptations to cope with these harsh environmental conditions.

EFFECT OF PAST AND PRESENT CLIMATE ON PATTERNS OF BIODIVERSITY AND ENDEMISM

Repeated periods of glaciation during the Pleistocene have strongly influenced the biota of arctic islands. During the Last Glacial Maximum (LGM; about 20,000 years ago), major ice caps wiped out most species in the Canadian Arctic Archipelago (CAA), Greenland, Novaya Zemlya, Severnaya Zemlya, Franz

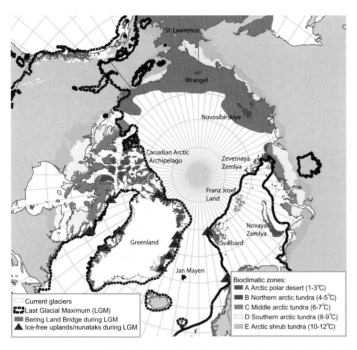

FIGURE 1 Bioclimatic zones (http://www.arcticatlas.org/maps/catalog/index.shtml) and glaciations of Arctic islands.

FIGURE 2 Vascular plant (red bars) and springtail (Collembola, blue bars) diversity on arctic islands. Data compiled from various sources. The bar for total species of vascular plants in Greenland represents 515 species. The bars for springtails have been doubled to visualize them.

Josef Land, and Svalbard (Fig. 1). Some ice-free nunataks and uplands existed, however, and it is debated whether some plants and animals survived the periods of glaciation (glacial persistence or glacial survival hypothesis), or became locally extinct and later recolonized from areas outside the main ice caps (*tabula rasa* hypothesis). Although a few molecular studies have found support for the glacial survival hypothesis, no paleorecords support continuous *in situ* existence of life within the glaciated islands.

In contrast, the islands and areas around the Bering Strait (Beringian islands such as Novosibirskiye Ostrova, Wrangel, St. Lawrence, and Diomede) remained unglaciated throughout the Pleistocene. The lowered sea levels during glacial periods resulted in a large shelf area connecting present-day islands with the Russian and Alaskan mainland (the Bering land bridge). These altering connections to the mainland, and the Beringian islands remaining unglaciated, have strongly influenced both speciation processes and distribution of species on these islands. The number of endemic species is larger on Beringian islands than on other arctic islands and the diversity of vascular plants and springtails on, for example, Wrangel Island is extremely high (Fig. 2).

The current summer temperatures of the warmest month range from 10–12 °C in the arctic shrub tundra zone to 1–3 °C in the arctic polar desert zone. The polar desert and northern Arctic tundra zones are almost exclusively found on arctic islands. Within a geographical region, summer temperature is the environmental variable that best predicts the diversity of species. For example, the number of vascular plants decreases towards the north in the Canadian Arctic Archipelago and from Novaya Zemlya to Franz Josef Land in Russia. However, some exceptions exist. In bryophytes, species diversity depends more on substrate than on temperature, and thus the difference in species numbers between north and south Greenland is low.

Although the total number of species decreases towards the north or with lower temperatures, the proportion of widespread species increases. Of 115 taxa of vascular plants found in the arctic polar desert zone, 91.3% occur in both North America and Eurasia, and only one species is endemic to a region. Similarly, in the small arthropods known as springtails, the proportion of widespread species is highest in previously glaciated, high arctic islands.

ISLAND GROUPS

Svalbard and Jan Mayen

The Svalbard archipelago is situated from 74° to 81° N and from 10° to 30° E. The land area is 61,000 km², but about 60% of this is covered by glaciers. The influence of the warm North Atlantic Current gives a more oceanic

TABLE 1
Estimated Numbers of Different Species Groups

	Canadian Arctic Archipelago	Greenland	Svalbard	Wrangel Island
Size (km²)	1,420,000	2,170,000	61,000	7,600
Ice-free area	1,260,000	410,000	24,000	7,608
Vascular plants	349	515 (32)	165 (4)	417 (23)
Mosses	346	477	288	239
Liverworts	—	135	85	87
Fungi	—	1600	624	—
Lichens	>750	1094	764 (12)	350
Terrestrial mammals	11	8	2 (1)	3 (2)
Marine mammals	7	22	8	12
Birds (regularly nesting)	61	58	38	47 (1–3)
Freshwater fishes	10	3	1	0

NOTE: Data are for Canada (various sources), Greenland (Jensen and Christensen 2003), Svalbard (updated from Elvebakk and Prestrud 1996; Prestrud et al. 2004), and Wrangel Island (Stishov 2004). Estimated numbers of endemic species are given in parentheses. Note that differences in degree of exploration and as well as taxonomical view make the numbers inaccurate.

and warmer climate compared to other islands at this latitude. This is also reflected in the species diversity, which is comparatively high in Svalbard (Table 1).

Svalbard was almost entirely glaciated during the last glacial maximum, and paleorecords show a sparse arctic vegetation subsequent to 10,000 years ago. Although this is one of the most remote arctic archipelagos, molecular analyses of plant species show that it was repeatedly colonized during the Holocene from several source areas (Fig. 3). The main source areas were in northern Russia/Siberia and northeastern Greenland, areas connected to Svalbard by winter sea ice. Thus, sea ice, probably in combination with wind, might be an important dispersal mechanism to arctic islands. Exceptionally warm winds may also carry insects directly from areas such as Russia to Svalbard, as was observed for the nonresident migratory diamondback moth *Plutella xylostella*. The few endemic species or subspecies in Svalbard (i.e., the Svalbard reindeer, the Svalbard aphid, and four plant species) have probably evolved recently from species that immigrated after the LGM.

In contrast to most arctic islands, there are no rodents on the archipelago (except the locally introduced sibling vole). The main herbivores are geese and reindeer. The arctic fox feeds mainly on eggs and chicks of sea birds and

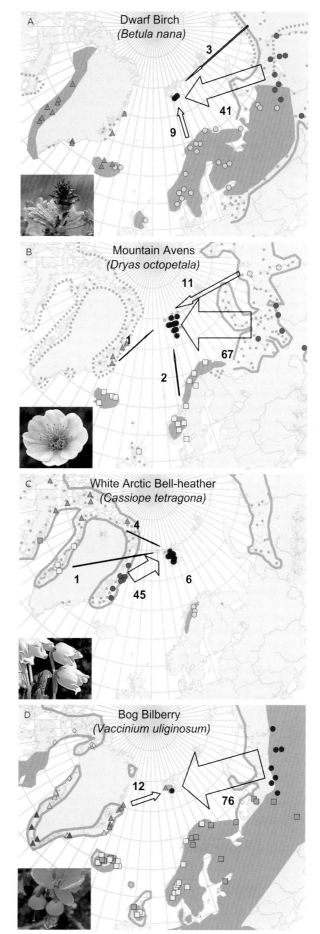

FIGURE 3 Source regions for past colonization of (A) dwarf birch (*Betula nana*), (B) mountain avens (*Dryas octopetala*), (C) white arctic bell-heather (*Cassiope tetragona*), and (D) bog bilberry (*Vaccinium uliginosum*) to Svalbard. Source regions are inferred from genetic data (amplified fragment length polymorphism). Colors represent main genetic groups, and symbols represent sub-groups. Numbers on the arrows are percentages assumed to have arrived from the source region. The geographic distribution of the species is shaded. Reproduced with permission from *Science*.

geese, as well as carcasses of seals and reindeer. There are many seabirds breeding in the archipelago, and these contribute a significant input of nutrients for plant growth. The only resident bird is the Svalbard ptarmigan (*Lagopus mutus hyperboreus*), which is endemic to Svalbard and Franz Josef Land.

Jan Mayen is a small (373 km^2) volcanic island east of Greenland. It has an extremely oceanic climate with mild winters and relatively cold and wet summers. About two-thirds of the 66 vascular plant species found there are circumpolar, whereas the other third is amphi-Atlantic. The only endemic species found are apomictic dandelions (*Taraxacum*). The arctic fox is the only terrestrial mammal on the island. Large seabird colonies are found during the summer, but only the fulmar (*Fulmarus glacialis*) stays during winter.

Greenland

Greenland is the world's largest island. Including the numerous smaller islands along the shore, its total size is 2.17 million km^2. The majority of species are confined to the ice-free margins, which cover only approximately 410,000 km^2. Greenland stretches from 59°45′ N to almost 84° N and spans a vegetation gradient from birch forest in the south to polar desert in the north (Fig. 4).

Considering the large size of Greenland, species diversity is relatively low, and it decreases from south to north as, for example, in vascular plants. Also, there are only a limited number of species that are endemic to Greenland. Of the total 515 vascular plant species, 32 taxa are endemic. However, 15 of the endemics belong to the apomictic hawkweed genus (*Hieracium*), which rapidly evolves new species. Endemic species of algae and three spider species have also been recorded, and a few bird subspecies breed only in Greenland, but they overwinter elsewhere.

The relatively low diversity and endemism found in this large island are probably due to its glacial history. Ice-free areas existed in Greenland throughout the glacial period, but according to climate data derived from ice cores, it was so cold during the LGM that only the most cold-adapted species could have survived there. Thus, it is assumed that the majority of species colonized Greenland during the last 11,500 years. This view is supported by molecular studies of several plant species. A large proportion of Greenland's plants and animals are also found in northwestern Europe, indicating that they arrived from there. Although this distance is long, the Faroe Islands and Iceland form steppingstones along the route. Further, the majority of Greenlandic birds migrate from Europe and could have transported seeds, spores, and even some invertebrates. The majority of spiders and some groups of insects are Nearctic, indicating a high proportion of immigration also from northeastern Canada.

Canadian Arctic Archipelago

The Canadian Arctic Archipelago (CAA) covers an immense area, ~1.42 million km^2, and comprises numerous large and many more smaller islands. It extends about 3000 km from below the Arctic Circle to the northern tip of Ellesmere, and 3000 km east-west from Baffin to Banks Islands. Ice caps occur in the mountainous northern and eastern parts on Axel Heiberg, Ellesmere, Devon, and Baffin Islands (maximum elevation 2615 m), but overall, glaciers cover only about 11% of the archipelago. Thus, the ice-free area of CAA is three times as large as the ice-free area of Greenland and more than 50 times as large as the ice-free area in Svalbard.

Recolonization after the LGM occurred primarily from mainland areas to the south, a distance as short as 1–20 km at several locations. Unglaciated Beringia was also an important source area for many groups, contributing to east-west differences in species composition (e.g., higher diversity of legumes on Banks and Victoria Islands). Glacial refugia on Banks Island provided additional source areas, while postulated refugia on Ellesmere and other islands have yet to be confirmed.

Considering its large land area, species diversity is low on the Canadian Arctic Archipelago. The strong south-to-north decrease in diversity is correlated with summer temperature and distance from the mainland. However, topography and oceanic influences modify this gradient,

FIGURE 4 Low-stature vegetation with prostrate or cushion-formed herbs such as moss champion (*Silene acaulis*) dominate the middle arctic tundra zone. Ammassalik district in southeastern Greenland. Photograph by Inger Greve Alsos.

creating a more complex pattern. Rain shadow effects are responsible for warmer drier summers and the relatively high diversity of the "polar oases" of the Forsheim Peninsula and Lake Hazen area on Ellesmere Island. Cool summers with extensive cloud and fog are responsible for the low vascular plant and arthropod diversity on Ellef Ringnes and nearby islands. Located north of the "shrub line," this barren region lacks woody plants, which are so characteristic of tundra vegetation (e.g., willows, mountain avens, Ericaceae).

No endemic vascular plants, bryophytes, lichens, mammals, or arthropods are known from the CAA, but several species are confined to the Archipelago and Greenland, such as Peary caribou, the alkali grass *Puccinellia bruggemannii*, the moth *Gynaephora groenlandica*, and the wolf spider *Alopecosa exasperans*. Also, at least one undescribed species of spider has been found on Banks Island.

Russian and Beringian Islands

The Russian arctic islands can be divided into five main groups: (1) Novaya Zemlya ("New Land") with adjacent Vaigach, Kolguyev, and some smaller islands; (2) Franz Josef Land; (3) Severnaya Zemlya ("North Land"); (4) Novosibroskiye Ostrova ("New Siberian Islands"); and (5) Wrangel and Gerald Islands. Besides these main groups there many small islands at the Ob', Yenisei, Kheta, Lena, and Kolyma deltas and near Taimyr Peninsula. In addition, there are several arctic islands belonging to Russia and the United States in the Bering Strait and Bering Sea (e.g., St. Lawrence, Yttygran, Arakamchechen, Diomede and King Island).

The most biologically diverse and well studied island in the Russian Arctic is Wrangel Island. It is a remote, relatively small island of about 7,600 km², with the highest elevation above 1000 m. A unique feature of this island is the very limited extent of Pleistocene glaciations combined with the lowered sea level during LGM, making Wrangel a part of the Bering land bridge. This has enabled enrichment of the fauna and flora by very different elements originating in boreal, forest-tundra, tundra, arctic polar desert, and even steppe zones from Asia as well as North America, which has resulted in a species composition on Wrangel Island different from those on all other islands. For vascular plants, 417 species are known, more than for the whole CAA (349 species) and the northwest sector of the Siberian arctic (387 species), and approaching that of Taimyr Peninsula (494) and Greenland (515). Similarly, the diversity of spiders, beetles, and birds is high on Wrangel Island compared to Svalbard and Greenland. The diversity of many insect families and orders is higher than on any other arctic island, including Greenland.

The recurrent connections and disconnections of Wrangel Island also led to speciation in mammals, vascular plants, and some groups of arthropods. The number of endemic species on the island is extraordinarily high for the Arctic in general and for arctic islands particularly. There are 23 endemic vascular plant species, four spider species, 20% of weevils, both of the rodents *Dicrostonyx groenlandicus vinogradovi* and *Lemmus sibiricus portenkoi* (Fig. 5), and at least one bird subspecies *Cepphus grylli tajani*. If the recently (3500 years old) extinct dwarf mammoth is counted, the level of endemism of mammals would be higher. In the late Pleistocene several other ungulates such as Przewalski's horse, woolly rhinoceros, primeval bison (*Bison priscus*), musk ox, woolly mammoth, and reindeer occurred on the island.

The Novosibroskiye Ostrova (New Siberian Islands) consists of two larger groups of islands, Lyakhovsky and Anzhu, and one small group called De-Longa. With its area of about 36,000 km², this region is about five times larger than Wrangel Island, and like Wrangel, it was also unglaciated and connected by the Bering land bridge. However, the archipelago is rather flat, which limits habitat diversity, and the diversity of plants, springtails, birds, and beetles is less than on Wrangel. There are some species on Novosibroskiye Ostrova that do not occur on Wrangel, including a willow grouse species and two goose species. The mammal fauna consist of wolf, wild reindeer, one lemming species, and arctic fox. The fossil mammoth fauna is even richer than on Wrangel Island with additional species such as saiga antelope, cave lion, and voles. Also, the fossil beetle fauna is much richer

FIGURE 5 Portenkoi's lemming, *Lemmus sibiricus portenkoi*, endemic to Wrangel Island, has an important role in the ecosystem. Photograph by Igor Dorogoi.

(about 100 species during the past 200,000 years, or 58 species during last 115,000 years) than the present beetle fauna (about 10 species). In the nineteenth and beginning of the twentieth centuries, digging up and selling ivory from fossil mammoths was a profitable business.

The Severnaya Zemlya Ostrova consist of four large (October Revolution, Bolshevik, Komsomolets, Pioneer) and about 70 smaller islands, covering a total area of about 37,000 km². Although the island group remained partly unglaciated throughout the Pleistocene, it was not connected by the Bering land bridge, and it is also situated rather far north. Thus, species diversity is lower than on most other Russian archipelagoes with, for example, about 78 species of vascular plants. Only 17 of the 32 bird species that have been observed on the islands breed there. Six terrestrial mammals are known there: lemming, arctic fox, wolf, ermine, arctic hare, and reindeer. Four species of beetles have been found on the archipelago, but only on the southernmost island.

The Novaya Zemlya Archipelago consists of two large and several smaller islands, in total about 81,000 km². The number of vascular plants and springtails is similar to that on many other islands that have been previously glaciated. The flora represents a transition between the arctic Europe and Asia but with a separate mountain range element connected to the Urals. Some endemic vascular plants in Novaya Zemlya have been proposed, but they are dubious. There are two lemming species, a local reindeer subspecies, and arctic fox on the archipelago.

Franz Josef Land (16,100 km²) is the northernmost archipelago and consists of almost 200 islands. Glaciers cover 85% of the archipelago. It is the most species-poor arctic archipelago, with only about 50 vascular plant species, about 150 bryophytes, over 300 lichens, one terrestrial mammal (arctic fox), at least 14 springtails, two spiders, and no beetles. Only 14 bird species breed on the island, but almost 30 other bird species have been observed visiting.

Aleutian Islands

The Aleutian and Commander Islands are a 2,100-km long archipelago to the south of the Arctic, separating the Bering Sea from the North Pacific Ocean. Although the U.S. government includes this archipelago in its definition of the Arctic because of the treeless landscape that prevails here, this is caused largely by strong winds rather than low temperatures and short growing season, as is the case in the Arctic. These islands serve as a natural bridge between Old World and New World flora and fauna, although

FIGURE 6 The sedge *Carex bigelowii* ssp. arctisibirica in Svalbard. Most sedges have clonal growth and can become very old; up to 3000-year-old clones have been found in the Arctic. Photograph by Inger Greve Alsos.

physical evidence suggests that this archipelago was under ice during the LGM, so terrestrial species on these islands should be recent colonists (i.e., since the last glaciation, or less than 10,000 years ago). However, the relatively high levels of endemism (for high-latitude organisms) that characterize the Aleutian and Commander Islands suggests that many of these taxa were isolated for longer periods of time, probably in "cryptic" glacial refugia: ice-free areas that harbored multiple taxa through the Quaternary glacial cycles, though so far only evident through the biological record. Moreover, natural selection has resulted in local adaptations to the harsh conditions of the islands, with evidence of traits such as increased body size in some bird species. In the same way, the Alexander archipelago, a chain of over 1000 islands off the southeastern coast of Alaska that is also recently glaciated, though currently covered by evergreen forest and even temperate rain forest, is characterized by a number of monophyletic lineages, which may be attributed to multiple Holocene invasions or the persistence of taxa in refugia during Pleistocene glacial advances.

CHARACTERISTICS OF SPECIES

Plants

The arctic flora ranges from shrub tundra in the south to almost barren polar desert, where no woody plants live and only scattered herbs, bryophytes, and lichens are found. The majority of species are long-lived perennials with relatively low resource allocation to sexual reproduction and high reliance on asexual reproduction for population maintenance, but a high variety of life strategies exists. There are fewer pollinators and lower pollinator activity in the Arctic than in other regions. Plant species adapted for wind pollination are dominant, and self-pollination is common. The growth forms are often prostrate mats, tussocks, rosettes, or cushions, which reduce desiccation and mechanical damage from the strong wind and maximize heat absorption (Fig. 6). In addition to the low temperatures and short growing season, drought places a very significant stress on plant life. The majority of bryophytes and lichen species are well adapted to periods of drought and increase both in abundance and ecosystem importance northwards.

Invertebrates

Invertebrate groups occurring on arctic islands have evolved from species in the boreal biome, where winter temperatures are often lower than in the tundra zone, and they have similar adaptations to cold resistance as boreal species. The main limiting factor for invertebrates is heat deficit in the short (2–3 month), cool growing season, and therefore the main adaptations are directed towards shortening of the life cycle (vivipary, reduction of size) or extension of the life cycle to several years. They survive the winter by producing cryoprotectors, being able to dehydrate, or overwintering as cold-resistant eggs. In addition, behavioral strategies may assist in avoiding low-temperature extremes, for example seeking protected places to avoid winter cold, such as under thick snow cover or close to non freezing water currents.

In arthropods, adaptations to the arctic climate lead to dominance by small-sized groups such as mites, spiders, and springtails, which have relatively high species diversity on arctic islands. Large sized insects, such as large ground beetles and bumblebees, are lacking on most arctic islands, whereas beetles and some other megadiverse groups, such as moths and true bugs, are represented on arctic islands by fewer species than small-sized groups. There is a decrease in species numbers of herbivorous insects (especially among beetles, butterflies, moths, and true bugs) in comparison to predaceous ones.

Mammals and Birds

To survive the harsh winter of the Arctic, mammals and birds have developed morphological, physiological, and behavioral adaptations (Fig. 7). The difference between ambient temperature and body temperature may be as high as 90 °C. Larger arctic animals have developed thicker and denser fur or plumage or a thick fat/blubber layer to keep warm without spending much energy. Smaller animals such as lemmings and voles live mainly under the snow, which acts as a thick insulating layer. Ptarmigans stay in "dock" (under snow) during bad weather conditions to reduce heat loss. The blood circulation system is also adapted to minimize heat loss by countercurrent heat exchange and by slowing down the circulation to extremities. Many arctic mammals have enlarged nasal cavities, and circulatory adjustments in the nose reduce water and heat loss. Some arctic animals, such as the Svalbard reindeer and ptarmigan, store large amounts of fat during the summer and autumn season, which is used to survive the winter (Fig. 8). Most arctic birds species migrate south before the winter period. When they arrive

FIGURE 7 In the Arctic, many animals are white in winter and brown during summer, which gives them a good camouflage. In the High Arctic some animals, such as this arctic hare on Ellesmere Island, remain white year-round, an adaptation to the very short snow-free summer period. Photograph by Lynn Gillespie.

FIGURE 8 Many arctic mammals, such as the Svalbard reindeer, put on large fat reserves during the autumn. Photograph by Inger Greve Alsos.

on the breeding ground in spring, their breeding success is closely linked to the timing of breeding relative to snow melt and peak food production.

HUMANS ON ARCTIC ISLANDS

Humans have long been a part of the arctic environment, intimately connected to the local resources on land and sea. The indigenous peoples harvest natural resources both from the terrestrial (arctic fox, ptarmigan, reindeer, caribou, musk ox) and the marine environment (fish, whale, seal, polar bear; Fig. 9). In Greenland, fishing is the all-dominating trade and accounts for 95% of total exports, but in the hunting districts of the outlying areas the seal and whale catch is of great importance and forms a stable existence for one-fifth of the Greenlandic population. Reindeer herding is of local importance only on few arctic islands, such as in northernmost Norway.

The 15 communities in the CAA are mainly inhabited by Inuit. Most Russian arctic islands are not inhabited except by the staff of small military camps, nature reserve stations and weather stations, but indigenous people live on some islands; for example, some Nenet families live on Novaya Zemlya. Fifty-seven thousand people, predominantly Inuit, live in towns and small settlements on Greenland. In Svalbard there is one Norwegian settlement with around 2000 inhabitants and one Russian settlement with about 500 inhabitants. Industrial activity is found only on a few arctic islands, such as mining activities in Svalbard, on Kolguyev, and, until recently, on Little Cornwallis and Baffin Island. There are huge reserves of oil and gas on arctic islands and the surrounding sea floor, such as the Sverdrup Basin in the Canadian high arctic, and exploration drilling is done in several locations.

FIGURE 9 Inuit hunters skinning a seal in Grise Fiord, Ellesmere Island, Canada. Photograph by Olivier Gilg.

Tourism and research activities are increasing on some arctic islands such as Svalbard and Baffin.

CONSERVATION OF ARCTIC FLORA AND FAUNA

The arctic flora and vegetation are vulnerable to physical disturbance, and vehicle tracks often last for decades. Humans have overexploited many species, such as whales, polar bear, and arctic fox. Although some species and populations have recovered, others are still threatened. Long-range pollution from the industrial part of the world, such as heavy metals, persistent organic pollutants (POPs), and radionuclides, has reached arctic islands, and such pollutants are accumulating in some organisms. Climate change is predicted to be of higher magnitude in the Arctic than in other places in the word. Because the arctic islands represent the "end of land," high arctic species have no further place to migrate if they are outcompeted by more southern species, and they may thus become extinct. Global warming will also open up the northern sea routes both in Canada and Russia and make the arctic oil and gas reserves more accessible, which would potentially lead to increased pollution and disturbance.

Knowledge necessary for conservation is lacking for many islands, species, and ecological processes in the Arctic. For example, the identification and classification of arctic invertebrates, fungi, bryophytes, and microorganisms is limited. Although some monitoring programs exist, information on the status and trends of arctic populations is fragmentary. For proper management in a changing climate, more knowledge is needed about the species found on arctic islands, the ways they interact, and how they respond to the changing physical environment, especially climate.

SEE ALSO THE FOLLOWING ARTICLES

Arctic Islands, Geology / Global Warming / Mammal Radiations / Refugia / Whales and Whaling

FURTHER READING

Aiken, S. G., M. J. Dallwitz, L. L. Consaul, C. L McJannet, L. J. Gillespie, R. L. Boles, G. W. Argus, J. M. Gillett, P. J. Scott, R. Elven, and M. C. LeBlanc. 2007. *Flora of the Canadian Arctic Archipelago*. Ottawa: NRC Press (CD-ROM).
Born, E. W., and J. Böcher. 2001. *The ecology of Greenland*. Nuuk, Greenland: Ministry of Environment and Natural Resources.
Conservation of Arctic Flora and Fauna. 2001. *Arctic flora and fauna. Status and conservation*. Helsinki: Edita.
Chapin, F. S. III, and C. Körner. 1995. *Arctic and alpine biodiversity: pattern, causes and ecosystem consequences*. Berlin: Springer-Verlag.
Danks, H. V. 1981. *Arctic arthopods: a review of the systematics and ecology with particular reference to the North American fauna*. Ottawa: Entomological Society of Canada.

Elvebakk, A., and P. Prestrud. 1996. A catalogue of Svalbard plant, fungi, algae and cyanobacteria. *Norsk Polarinstitutt Skrifter* 198.

Jensen, D. B., and K. D. Christensen. 2003. *The biodiversity of Greenland—a country study. Technical report.* Pinngortitaleriffik: Grønlands Naturinstitut (Greenland Institute of Natural Resources).

Kristinsson, H., E. S. Hansen, and M. Zhurbenko. 2008. *Panarctic lichen checklist.* Conservation of Arctic Flora and Fauna. http://arcticportal.org/en/caff/.

Prestrud, P., H. Strøm, and H. V. Goldman. 2004. A catalogue of the terrestrial and marine animals of Svalbard. *Norsk Polarinstitutt Skrifter* 201.

Stishov, M. S. 2004. [*Wrangel Island—master pattern of nature and nature anomaly.*] Yoshkar-Ola: Mariyski Printing Factory Press (In Russian).

ARCTIC ISLANDS, GEOLOGY

MICHAEL J. HAMBREY

Aberystwyth University, United Kingdom

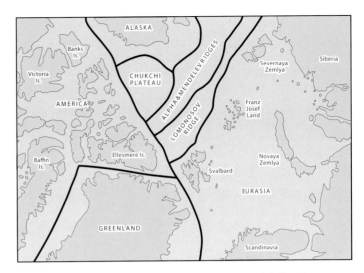

FIGURE 1 Plate tectonic reconstruction of the Arctic as it is believed to have appeared approximately 70 million years ago prior to opening of the Arctic, North Atlantic, and North Pacific oceans. (Reprinted with permission of Cambridge University Press from Dowdeswell and Hambrey 2002: 32, adapted from Worsley and Aga 1986.)

The geological history of the Arctic spans nearly four billion years and includes some of the oldest rocks on Earth. A vast range of sedimentary, igneous, and metamorphic rocks are present, but few were formed in their current position. The geological record for many Arctic islands reflects the drift of fragments of continental crust from a position south of the equator to their current polar position. As a consequence, the rocks record a range of climates from tropical to glacial, as well as a fascinating glimpse of biological evolution from the algae of the Precambrian to the high-order animals and plants of today.

TECTONIC EVOLUTION

By outlining the tectonic components and history of the whole Arctic region, a context is provided for the main phases of geological evolution of the region. The geological attributes of the Arctic islands reflect the disparate nature of individual continental fragments and their movement by plate tectonic processes through time. These processes involved continental breakup, continental collision, and sea floor spreading. Indeed, many parts of the Arctic have rocks that once were formed south of the equator; plate movements have resulted in their slow progression to a northern polar position today. All the continental fragments are believed to have been united as one supercontinent about 70 million years ago (Fig. 1). Since then, the Arctic Ocean basin has opened, along with the North Pacific Ocean, Baffin Bay, and the Norwegian-Greenland Sea. The oldest rocks are Archean to Proterozoic metamorphic rocks that represent stable crystalline shields. In many areas they are overlain by sedimentary rocks. Periodically, the sedimentary strata were intruded by igneous rocks, subjected to metamorphism deep in the crust, and deformed during mountain-building events or "orogenies," when continents collided. The sedimentary strata reflect the climatic regimes and topographic/bathymetric settings under which they formed, including under ice age, temperate, and tropical climates and deep-sea, continental shelf, fluvial, glacial, estuarine, and mountain environments.

This pattern of northward drift has been determined with a reasonable degree of certainty over the past 600 million years, especially for Svalbard and East Greenland. Ongoing tectonic processes are focussed on the continuing opening of the North Atlantic. New oceanic crust continues to form along the Mid-Atlantic Ridge, upon which the volcanic islands of Iceland and Jan Mayen are located, surrounded by deep ocean. New basaltic rocks formed on the ridge continue to push Europe away from Greenland and North America, a process that began about 60 million years ago. In contrast, continental margins are "passive" and thus relatively stable today, lacking significant earthquake activity.

The other main plate tectonic process, continental collision, is not a feature of the Arctic at the present-day. However, the older geological record shows dramatic evidence of this process on several Arctic islands. The "Caledonian Orogeny" of the early Paleozoic Era resulted in a chain of mountains of Himalayan proportions, with evidence of volcanism, metamorphism, folding, strike-slip and thrust faulting, and igneous intrusion on a vast scale.

It will be evident from the above that the islands do not coincide neatly with geological boundaries. Some, such as Svalbard, consist of a number of slivers of crust called terranes, each of which has a distinct geological history. In contrast, some now widely separated regions share a common history. Strike-slip faulting on a scale of hundreds or thousands of kilometers was the process whereby terranes joined or separated.

Although the geology of the Arctic islands is complex, a number of key stages of evolution (summarized in Fig. 2) can be identified, as outlined below.

EARLY CRUSTAL EVOLUTION

The geological foundations of many continental areas are represented by ancient crystalline shields, the Arctic region being no exception. The islands contain fragments of two main shields that were joined together prior to 70 million years ago (Fig.1). The Laurentian shield includes much of North America, including Greenland, and Baltica is the foundation of much of Europe, including parts of Svalbard. Even where these rocks are not exposed, they underlie younger sedimentary cover rocks. The rocks are strongly deformed and metamorphosed, the dominant rock types being gneiss, schist, and igneous rocks, some of which are among the oldest rocks on the planet; rocks in southwest Greenland date back to 3800 million years. Typical shield landscapes are characterized by low rocky hills with intervening boggy areas and lakes that are commonly aligned parallel to the dominant structures such as the metamorphic layering (foliation) or faults.

Million Years	Period	Some key events in Arctic
0	Quaternary	Glacial/interglacial cycles
1.8	Neogene	Arctic acquires current configuration
23.8	Palaeogene	Cool temperate climates; Extensive forests → coal
65	Cretaceous	North Atlantic & Arctic Oceans begin opening
142	Jurassic	Dinosaurs
206	Triassic	
248	Permian	Younger sedimentary basins; Tropical climates
290	Carboniferous	
354	Devonian	Old Red Sandstones deserts and rivers
417	Silurian	
443	Ordovician	Continental collision between Eurasia and mountain building: the Caledonian Orogeny
495	Cambrian	
545	Pre-Cambrian: Proterozoic Eon	Major ice age; Early life forms (algae) preserved
2500	Pre-Cambrian: Archaean Eon	Oldest rocks, 3800m.y., in Greenland
4600		

FIGURE 2 Summary of the geological time scale and key events affecting the Arctic islands. (Reprinted with permission of Cambridge University Press from Dowdeswell and Hambrey 2002: 34.)

EARLY SEDIMENTARY BASINS

The crystalline shield areas are overlain or bordered by zones of sedimentary rocks that range in age from approximately 1000 to 400 million years (late Proterozoic to early Paleozoic). The rocks are variably metamorphosed, some of which were partially melted deep in the crust (Fig. 3). However, those least affected give unique insights to early climates and life. These sedimentary rocks were laid down in rift basins where the crust was being stretched and flooded by the sea, and supplied by detritus from the rift margins. The shallow marine sediments are dominated by limestone, dolomite, mudstone, and sandstone that collectively attain a thickness of 10 to 20 kilometers. Particularly striking are stromatolites, layers of algae that trapped carbonate sediment, forming mounds or columns several meters high. Impressive examples occur in eastern Greenland and northeastern Svalbard. The best known modern examples are found in Shark Bay, Western Australia, but they never attain the spectacular scale of those of Proterozoic age in the Arctic. In contrast to these probably warm-water features, there is abundant evidence of glaciation, in the form of tillites (Fig. 4). These rocks are best known from eastern and northern Greenland and from Svalbard, and represent the Arctic manifestation of a global ice age, when according to some geologists the Earth

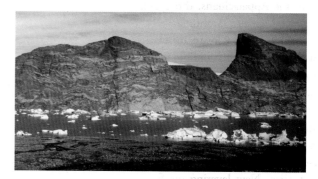

FIGURE 3 Ice-smoothed wall of inner Nordvestfjord, East Greenland, displaying deformed and highly metamorphosed rocks of Proterozoic age called migmatites. Photograph by M. J. Hambrey.

FIGURE 4 Neoproterozoic tillite, indicative of a global ice age, Ella Ø, East Greenland. Photograph by M. J. Hambrey.

was locked in a deep freeze of global proportions. This controversial hypothesis is known as "Snowball Earth," but even if this extreme view is incorrect, the evidence for continental-scale glaciation is unequivocal. It is possible to match strata in these early sedimentary basins across different areas, sometimes even on a bed-by-bed basis, as between eastern Greenland and northeastern Svalbard. Even though the rocks are today separated by not only an ocean but other tectonic terranes, the evidence points to their being formed in the same sedimentary basin. Subsequent plate movements, notably strike-slip faulting, have since separated these two areas.

THE CALEDONIAN OROGENY

An ancient Celtic tribe, the Caledones, in Scotland gives its name to the most important mountain-building phase that affected the Arctic islands. The Caledonian Orogeny was the result of closure of an ancient ocean, now called Iapetus, that separated North America (with Scotland) from Europe (with England and Wales). Closure of the ocean and subduction led to the growth of a mountain range 100–200 km wide that extended from the Appalachians, through the northwestern British Isles, western Scandinavia, eastern Greenland, and eastern Svalbard. Closure of the ocean was accompanied by metamorphism, folding, and thrusting of the predominately shallow marine strata of the early sedimentary basins. Metamorphism ranged from low-grade (low temperature, high pressure) to high-grade (high temperature, high pressure). The greatest pressures and highest temperatures occurred at depths of 30–40 km, permitting the growth of new minerals such as garnet, hornblende, and mica. New layering, or foliation, defined by new minerals, overprinted the original sedimentary structures. Simultaneously, the rocks behaved plastically and were subject to folding on scales ranging from kilometers to millimeters. In the upper few kilometers of the crust, the rocks were only slightly metamorphosed and were more brittle and prone to fracturing. Thrusts allowed translocations of large bodies of rocks (known as nappes) by tens of kilometers. Spectacular examples are present along the walls of eastern Greenland's fjords. Intense deformation took place sporadically over the period from Ordovician to Silurian times (480–420 million years ago). The latter stages of the Caledonian Orogeny were accompanied by extensive intrusion of granite as a result of melting of the upper crust; again these rocks are well exposed in Svalbard and eastern Greenland. The Orogeny ended with reordering of continental fragments along strike-slip faults in a manner analogous to movements along the San Andreas Fault in California today.

The end result of the Caledonian Orogeny was the unification of North America and Europe, a state of affairs that lasted until the modern North Atlantic began to open some 360 million years later. Similar, but smaller scale orogenies affected western Svalbard, northern Greenland, and Ellesmere Island and the Urals, Severnaya Zemlya, and Novaya Zemlya. The vast, high mountain chain was subject to erosion, and as the crust relaxed, intermontane basins developed. The erosion products accumulated in these dry continental basins as red beds: sandstone and conglomerate delivered by flash floods. Ephemeral lakes developed, but some lasted sufficiently long for early fish to flourish. These strata are referred to as the Old Red Sandstones and, broadly speaking, were deposited during the Devonian Period.

YOUNGER SEDIMENTARY BASINS

Devonian events heralded a change throughout the Caledonian fold belt from continental collision through strike-slip motions to crustal extension. Sedimentary basins bounded by faults developed next, not only in the fold belt but in other parts of the High-Arctic. The Canadian Arctic Archipelago, western and eastern Greenland, Svalbard (Fig. 5), and Franz Josef Land have well-developed sedimentary sequences that span all geological periods from Carboniferous to Cretaceous (380–65 million years ago). Active phases of faulting led to pulses of sedimentation in both terrestrial and marine settings. Sediments include mud, sand, limestone, dolomite, and occasional coal and evaporite deposits such as salt. Marine environments were commonly rich in shelly faunas, whereas on land, plants and animals (including dinosaurs later on Svalbard) were abundant at times. Organic matter, especially in muddy sediments, decayed

FIGURE 5 Carboniferous limestone and dolomite exposed in the walls of Billefjorden, Svalbard. Photograph by M. J. Hambrey.

to produce hydrocarbons, with oil and gas migrating into adjacent porous sandstones. Extensive hydrocarbon exploration has taken place in Svalbard, Greenland, and the Canadian Arctic, with limited success, but large reserves of gas and oil are most likely to be found on the surrounding continental shelves. Oil and gas reserves have already been heavily exploited in equivalent rocks on the mainland areas, such as the North Slope of Alaska and Siberia. In Svalbard, localized deposits of Carboniferous coal were extracted until a decade ago. These deposits all carry a strong climatic signal and, in some regions, reflect movements from equatorial through tropical and temperate climatic zones.

Intrusive activity occurred sporadically within these sedimentary basins, reflecting enhanced phases of crustal stretching. Prominent sills of Mesozoic basalt (a basic igneous rock) intrude sedimentary strata in Svalbard and the Canadian Arctic Archipelago, forming resistant cliffs in otherwise relatively soft sediment.

VOLCANISM AND OPENING OF THE NORTH ATLANTIC OCEAN

The supercontinent that embraced Europe and North America was beginning to split apart toward the end of the Cretaceous Period. Separation began in the south with rifting, followed by sea floor spreading and the initiation of the North Atlantic Ocean. Opening of the ocean propagated northward, branching on either side of Greenland to form Baffin Bay and the Norwegian-Greenland Sea. The opening of the ocean was heralded by extensive igneous activity, notably large-scale intrusion of granite derived from melting of the crust, and gabbros and basalts derived from the mantle. In addition, huge volumes of basalt were extruded, forming columnar cliffs and flat-topped plateaus. The most spectacular examples of these volcanic processes occur in eastern Greenland, where glaciers have carved out cross-sections through the intrusions and lavas, producing the most rugged scenery in the Arctic. These igneous events took place in the early Paleogene Period, being dated mainly to ~52–55 million years ago in eastern Greenland and 65–55 million years in western Greenland and on Baffin Island. As new ocean floor was created at the northward-extending Mid-Atlantic Ridge, the continental areas became increasingly separated. Igneous activity ceased in the continental areas and became focused mainly on the submarine ridges. The most obvious manifestation of continental separation is evidenced where the Mid-Atlantic Ridge rises above the ocean, in Iceland and Jan Mayen. Sea floor spreading and active volcanism continues to this day, pushing North America away from Europe at a rate of several millimeters a year.

Centered on the North Pole, the Arctic Ocean formed simultaneously, as a result of spreading from several mid-oceanic ridges, such as the Nansen-Gakkel Ridge. Today the Arctic Ocean consists two main basins, the Amerasia and the Eurasia, underlain by sea floor basalt.

EMERGENCE OF PRESENT-DAY GEOGRAPHY UNDER A TEMPERATE CLIMATE

In many parts of the Arctic, igneous activity was accompanied by uplift, and the Arctic islands configuration began to emerge at this time. Crustal deformation and even localized mountain building (in western Svalbard) took place in the early Paleogene Period, as tectonic plates shuffled into their current positions. On the land areas, temperate climatic conditions prevailed, promoting the growth of extensive forests, comprising both coniferous and broad-leaved trees. Extensive tree fossils are preserved on Axel Heiberg Island and Spitsbergen, and on the latter island vegetation accumulated sufficiently to generate exploitable coal measures, notably at Barentsburg, Longyearbyen, and Svea. These fossil forests clearly indicate that the climate was warmer and wetter than the present day, even though the islands were already in a high latitude.

THE ICE AGE

The succeeding Neogene Period, approximately 10 million years ago, saw most of the principal elements of the Arctic land masses achieving approximately their current configuration. Along with many other parts of the world, the Arctic experienced sharp climatic cooling at this time. Glaciers began to form over Greenland, possibly as early as 11 million years ago, as evidenced by ice-rafted sediments on the ocean floor, delivered to the coast by tidewater glaciers and transported by icebergs offshore. The main growth of the northern hemisphere ice sheets over the Canadian Arctic, Greenland, and northern Eurasia is believed to have taken place in late Neogene/early Quaternary times, approximately 2.5 million years ago. The Arctic islands, especially Greenland, were affected by many glacial periods, but evidence remains only of the last one, named the Wisconsinan in North America, and the Weichselian in northwestern Europe, which spanned the interval 80,000 to 10,000 years ago. Glaciers remain extensively developed on many of the Arctic islands, and reached a recent historical peak around AD 1900, leaving behind extensive glacial and glaciofluvial deposits (Fig. 6).

Beyond the ice limits, most of the Arctic islands are influenced by frozen ground (permafrost), in some places to depths of several hundred meters. Modern coastal and fluvial processes continue to rework sediments, especially those of glacial origin. Braided river plains are a characteristic feature of many parts of the Arctic.

THE FUTURE

Arctic island geology continues to evolve. On the multi-million-year time scale, plate movements will continue to change the geographical configuration of the Arctic, and islands will come and go. Today, in a sense, several Arctic islands are still under "ice age" conditions, with extensive glaciers and permafrost, and the surrounding seas are covered by sea ice in winter. However, under the influence of human-generated atmospheric pollutants, the Arctic region is warming at a rate faster than almost anywhere else on the planet, and major changes are already occurring, including recession of glaciers, melting of permafrost, and thinning and shrinkage of the area covered by sea ice, to name but a few. Belated recognition of global warming, induced by humans, and concerted action to reduce emissions, is too late to halt these changes on a centennial time scale; the best that can be anticipated is a reduction in the rates of change.

SEE ALSO THE FOLLOWING ARTICLES

Arctic Islands, Biology / Climate Change / Continental Islands / Plate Tectonics

FURTHER READING

Dowdeswell, J. A., and M. J. Hambrey. 2002. *Islands of the Arctic*. Cambridge, UK: Cambridge University Press.
Haller, J. 1971. *Geology of the East Greenland Caledonides*. New York: Wiley & Sons.
Harland, W. B. 1997. *The geology of Svalbard*. Memoir no. 17. London and Bath: Geological Society.
Henriksen, N. 2008. *Geological history of Greenland*. Copenhagen: Geological Survey of Denmark and Greenland (GEUS).
Hjelle, A. 1993. *Geology of Svalbard*. Polarhåndbok No. 7. Oslo: Norsk Polarinstitutt.
Escher, A., and S. W. Watt, eds. 1976. *The geology of Greenland*. Copenhagen: The Geological Survey of Greenland.
Worsley, D., and O. J. Aga. 1986. *The geological history of Svalbard*. Stavanger, Norway: Den Norske Stats Oljeselskap A.S.

ARCTIC REGION

STEPHEN D. GURNEY
University of Reading, United Kingdom

The Arctic is that region of the northern hemisphere where the sun, for some time in summer, does not set and, for some time in winter, does not rise. It is a land of contrasts, of the polar night and the polar day, and its southerly limit is the Arctic Circle at a latitude of 66°33′ N. Islands within the Arctic region (see Fig. 1) include the Canadian Arctic islands, Greenland (considered to be the largest island on the planet) and its surrounding islands, Jan Mayen, Bjørnøya, the Svalbard archipelago, the Lofoten Islands, and the islands of the Russian Arctic including Novaya Zemlya, Zemlya Frantsa Iosifa, Severnaya Zemlya, Novosibirskiye Ostrova, and Wrangel Island. The North Pole is not on land; rather, it can be visited only by venturing onto the frozen surface of the Arctic Ocean.

CLIMATIC SETTING

The climate of the islands of the Arctic region varies greatly, although it is typified by a negative heat balance

FIGURE 6 Thompson Glacier with push-moraine and braid-plain on Axel Heiberg Island, Canadian Arctic Archipelago. Photograph by M. J. Hambrey.

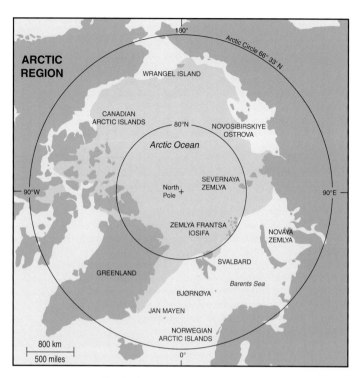

FIGURE 1 The Arctic Region and the islands within it.

or budget, meaning that mean annual air temperatures can be lower than -15 °C and winter temperatures may fall to lower than -50 °C. Winters are generally cold and stormy, and snow cover can last six months or longer. Summers are short and often cloudy, but they may be mild. The proximity of sea ice or the existence of a warm ocean current can result in colder or warmer conditions than would otherwise be expected. All areas experience the "Arctic Night" in winter when the sun does not rise above the horizon for a period of up to several months (depending on latitude). In summer, of course, the "Midnight Sun" creates 24-hour daylight.

GLACIERS

The largest body of ice found on the Arctic islands is the Greenland Ice Sheet; however, it does not lie totally within the Arctic (the southern tip of Greenland is found at 60° N). Although smaller than the East Antarctic and West Antarctic Ice Sheets, it is still vast at over 1.8 million km². The nature of other ice masses on the Arctic islands is generally dictated by the terrain and precipitation regime. For example, Baffin Island supports lowland ice caps, whereas Novaya Zemlya supports mountain ice caps and valley glaciers of the outlet type. In some areas the glaciers flow down to meet the sea and form floating glacier tongues which calve (create) icebergs into the sea. The largest icebergs originate from the outlet glaciers of the Greenland Ice Sheet, and these drift out into the North Atlantic via fjord systems.

PERMAFROST

In areas without glaciers, the cold climate results in the formation and maintenance of permafrost. Permafrost is a thermal condition of the ground whereby it is frozen throughout the year to depths of up to several hundred meters. In the brief Arctic summer, however, the positive air temperatures thaw a thin layer of the surface materials, known as the "active layer." Under certain conditions the summer warmth can lead to extensive thawing of ice lenses at the base of the active layer, which can cause it to become detached, resulting in large slope failures.

GEOMORPHOLOGY

The geomorphology of the Arctic islands is varied but is dominated by glacial processes and landforms where glaciers are present and by permafrost and periglacial processes where they are not. Permafrost landscapes are often dominated by ground-ice landforms, the most important of which are ice wedges. These features are created over many years because the ground cracks open in winter as a result of thermal contraction. The cracks become filled, first with veins and then later with wedges of ice, and these grow over successive winters until they may be as much as 4 m wide at their tops and 6 m deep. At the surface they can be seen through the polygons that they produce, which may be low- or high-centered and can have a surface relief of up to 2 m. The ice wedges themselves are hidden from view by the active layer, which closes the uppermost portion of the ground cracks in summer as it thaws. The slopes in permafrost terrain may be composed of solifluction sheets or lobes, which slowly creep downhill through the growth and subsequent thaw of ice within the active layer on a seasonal basis; such processes can operate on slopes with angles of only a few degrees. Pingos are small ice-cored hills and are another feature unique to permafrost. Some of the best examples are seen in the Arctic islands, for example on Traill, in east Greenland, and in the Svalbard archipelago, where they can attain heights of up to 40 m and diameters of over 500 m. They grow when water under pressure makes its way toward the surface but freezes just beneath it to form a conical ice core. As the ice core grows it forces up the ground above it into a hill, which is the pingo. In lowland tundra areas the source of water under pressure results from permafrost growth into the saturated sediments exposed by lake drainage. In mountainous regions the source is upwelling groundwater that originates on the hillsides high above the valley floor. Rock glaciers also form in the mountainous permafrost environments of the Arctic islands. These take the form of lobes or tongues of rock debris where the rock fragments are cemented together by ice and flow slowly downhill through the slow deformation of the ice.

COASTAL PROCESSES

The coastal processes and geomorphology of the islands of the Arctic region are similar to those found in other areas, except where there is a role played by permafrost or sea ice. Where ground ice contents are high, the seawater has the ability to erode thermally—that is, its relative warmth thaws the ice within the sediments that it contacts—and this can lead to a loss of strength and to enhanced erosion. In a microtidal environment, such as is found in much of the Arctic, this can lead to the undercutting of soft sediment cliffs and to block collapse. In areas with sea ice, coastal morphology can be shaped by the effects of the ice being blown or driven onto the island fringes by winds, tides, or currents, especially where the islands are low-lying.

FURTHER READING

Burn, C. R. 1995. Where does the polar night begin? *The Canadian Geographer* 39: 68–74
Dowdeswell, J. A., and M. J. Hambrey. 2002. *Islands of the Arctic.* Cambridge, UK: Cambridge University Press.
Gurney, S. D. 1998. Aspects of the genesis and geomorphology of pingos: perennial cryogenic mounds. *Progress in Physical Geography* 22: 307–324.
Humlum, O. 2000. The geomorphic significance of rock glaciers: estimates of rock glacier debris volumes and headwall recession rates in west Greenland. *Geomorphology* 35: 41–67.
Lewkowicz, A. G. 2007. Dynamics of active-layer detachment failures, Fosheim peninsula, Ellesmere Island, Nunavut, Canada. *Permafrost and Periglacial Processes* 18: 89–103.
Mackay, J. R. 2000. Thermally induced movements in ice wedge polygons, western Arctic coast: a long term study. *Géographie Physique et Quaternaire* 54: 41–68.

ASCENSION

DAVID M. WILKINSON

Liverpool John Moores University, United Kingdom

Remote oceanic islands such as Ascension have long played an important role in the study of ecology and evolutionary biology. Although historically less important for these subjects than the Galápagos, Ascension was visited by some of the leading nineteenth-century traveling scientists, including Charles Darwin and Joseph Dalton Hooker, and it has a range of endemic species of both scientific and conservation importance.

PHYSICAL GEOGRAPHY AND LANDSCAPE

Ascension is a volcanic island located in the tropical south Atlantic (7°57′ S, 14°22′ W) with an area of about 97 km². Much of its scientific interest comes from its remote location. The nearest point in Africa is 1504 km away, and the island is 2232 km from South America; the closest other island is St. Helena, 1300 km to the southeast. The oldest recorded rocks on the island are approximately 1 million years old. Therefore, the ancestors of all the organisms living on the island either dispersed to it naturally during the last million years or have been brought there by humans in the last few centuries.

Most of the island is very arid, providing a dramatic volcanic landscape dotted with scoria cones—large cones of volcanic material formed by explosive eruptions. When the marine biologist and amateur watercolorist Alister Hardy visited in 1925, he described these cones thus: "although barren they present a great variety of colour: raw sienna, reds, browns, dark and light grays and yellows, while some are almost crimson—all changing tone with the light and shade from passing clouds" (Fig. 1). The highest point on the island is Green Mountain (845 m; declared a national park in 2005). Here there is lush tropical vegetation mainly composed of introduced plant species. It is probably best described as cloud forest—the water supply for this vegetation coming from clouds blown onto the summit from the ocean by the trade wind. The very top of the "mountain" is dominated by bamboo

FIGURE 1 Much of Ascension is an arid volcanic landscape. The photograph shows the volcanic cone of "Sisters Peak." The white marks on the rocks in the foreground are the remains of guano from a former seabird colony (often referred to as "ghost colonies"); such easily accessible colonies failed to survive predation by introduced cats.

FIGURE 2 Green Mountain photographed from the top of Breakneck Valley. The peak is enveloped in cloud, which provides the water to support the cloud forest vegetation on the summit, largely composed of introduced plant species. Much of the northeastern side of the "mountain" is exposed to the full force of the trade winds and is covered in a grass-dominated vegetation—again mainly comprising introduced species.

and has an artificially dug pond, making it one of the very few freshwater habitats on the island (Fig. 2).

At sea level the temperature is relatively constant, usually varying between 27 and 31 °C, whereas at 660 m on Green Mountain, temperatures in the low 20s are more often recorded. Rainfall at sea level is approximately 140 mm per year, although the island was presumably wetter at some point in the past; the crater called the "Devil's Riding School" contains old lake deposits—first recognized as such by Charles Darwin when he visited in 1836 on the return leg of the *Beagle* voyage.

BIOLOGY

As with many remote islands, one of the reasons Ascension is of interest to conservationists and evolutionary biologists is its endemic species—which must have evolved there during the last 1 million years. The flora, prior to human introduction, was very limited, presumably due to a combination of the island's remote location and geologically recent origin. When humans first arrived on the island, in the early sixteenth century, it is thought that there were around 25 native plant species, ten of which were endemic (four of these endemics are now extinct). Today an additional 280-plus introduced plant species grow on the island, threatening the survival of some of the remaining endemics. Many of these species were deliberately introduced during the nineteenth century in an attempt to improve the environment for humans. The only tree in the native flora was *Oldenlandia adscensionis* (this may always have been rare as Darwin failed to notice it when he visited); it was last recorded in 1889.

There is currently one species of endemic bird on the island, the Ascension Frigatebird *Fregata aquila;* however, sub-fossil remains of two extinct endemic birds have been found in cave deposits. There was a flightless rail *Mundia elpenor* (formerly placed in the genus *Atlantisia*), which was still present when humans arrived in the sixteenth century, and a small night heron *Nycticorax olsoni,* whose date of extinction is not known. The island is an important breeding ground, not only for the frigatebird but also for many species of tropical seabird, and it is also famous for the green turtles, *Chelonia mydas,* which nest on its beaches and have formed the subject of extensive scientific research.

There is an equally interesting invertebrate fauna, containing at least 25 endemic terrestrial species; this includes five species of endemic pseudoscorpions. One of these, *Garypus titanius,* confined to Boatswain Bird Island, is the largest known pseudoscorpion in the world. There is also a land crab, *Johngarthia* (*Gecarcinus*) *lagostoma,* which is restricted to Ascension and a few islands off the coast of Brazil. On Ascension it mainly lives above 200 m, where there is more vegetation than in the arid lowlands; however, it has to migrate across these lowlands to the sea to breed. Little is known about the microorganisms on Ascension; a recent study of the soil protozoa recorded 52 species, all of which have wide global distributions.

CURRENT STATUS AND CONSERVATION

The discovery of the island in the early sixteenth century led to the introduction of many non-native species. These included a long list of plant species, which have competed with the limited native flora, and animals such as rats and cats which proved very destructive to seabird populations. During the twentieth century, with the exception of sooty terns, *Sterna fuscata,* all these seabirds have been confined to breeding either on the steep cliffs of the main island or on Boatswain Bird Island off the east coast—to escape predation by introduced cats and rats. Guano deposits show that seabird colonies were much more extensive on the main island in the past (Fig. 1). The successful eradication of feral cats (completed in 2004) has lead to some seabirds starting to breed on the main island again.

Ascension is also famous for the green turtles, *Chelonia mydas,* which nest on its beaches. Recent analyses of historical records suggests that there was a decline in the numbers breeding between 1822 and 1935, during which time there was a commercial harvest of turtles, this harvest stopped in the 1940s. More recently numbers have increased greatly—by an estimated 285% since the 1970s.

Plant conservation has been more limited than seabird conservation, but it has recently included attempts to propagate some of the endemic species—such as the fern *Pteris adscensionis*—under nursery conditions.

SEE ALSO THE FOLLOWING ARTICLES

St. Helena / Tristan da Cunha and Gough Island / Voyage of the *Beagle*

FURTHER READING

Ashmole, P., and M. J. Ashmole. 2000. *St Helena and Ascension Island: a natural history*. Oswestry, UK: Anthony Nelson (current distributor: www.kidstonmill.org.uk.).
Gray, A., T. Pelembe, and S. Stroud. 2005. The conservation of the endemic vascular flora of Ascension Island and threats from alien species. *Oryx* 39: 449–453.
Hartnoll, R. G., T. MacKintosh, and T. J. Pelembe. 2006. *Johngarthia lagostoma* (H. Milne Edwards, 1837) on Ascension Island: a very isolated land crab population. *Crustaceana* 79: 197–215.
Wilkinson, D. M. 2004. The parable of Green Mountain: Ascension Island, ecosystem construction and ecological fitting. *Journal of Biogeography* 31: 1–4.
Wilkinson, D. M., and H. G. Smith. 2006. An initial account of the terrestrial protozoa of Ascension Island. *Acta Protozoologica* 45: 407–413.

ATLANTIC REGION

ANDREAS KLÜGEL

University of Bremen, Germany

The Atlantic is the world's second largest ocean, forming an elongated basin between the Arctic Ocean in the north and Antarctica in the south. It owes its existence to the break-up of the supercontinent Pangaea, which began around 180 million years ago in the North Atlantic and 130 million years ago in the South Atlantic. Today the Atlantic continues to widen at rates of approximately 1.8–3.5 cm per year by seafloor spreading along the Mid-Atlantic Ridge, a submarine mountain range separating the ocean into an eastern and a western basin. The Atlantic harbors a number of islands of mostly volcanic origin that experience a wide range of maritime climates, from polar in the high latitudes to tropical around the equator (Fig. 1). Many of these islands and associated seamounts, for example the Canary Islands, form volcanic chains, which reflect movement of the underlying tectonic plates over a hotspot. Overall, these volcanic chains are less conspicuous and less frequent in the Atlantic than in the Pacific. This article gives an overview of the Atlantic's islands, their ages, and how they were formed; it is orga-

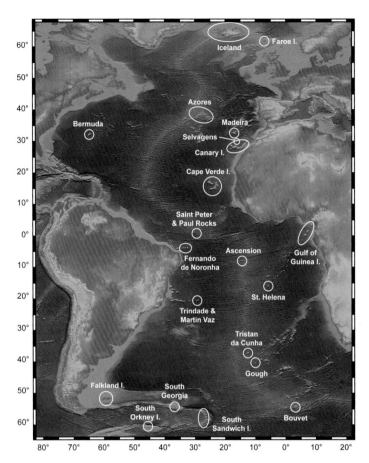

FIGURE 1 Map of the Atlantic with locations of islands covered in this article (GLOBE Task Team and others. 1999. *The Global Land One-kilometer Base Elevation (GLOBE) Digital Elevation Model, Version 1.0*. Boulder, CO: National Oceanic and Atmospheric Administration, National Geophysical Data Center, URL: http://www.ngdc.noaa.gov/mgg/topo/globe.html).

nized by geographic-geologic groups roughly from north to south.

ICELAND, SURTSEY, FAROE ISLANDS

Iceland is a highly active volcanic island, located just where a hotspot meets the Mid-Atlantic Ridge. It is the only place where the west-east spreading of the Eurasian and North American tectonic plates can be witnessed on land. The oldest rocks of Iceland (formed 13 to 16 million years ago) are therefore exposed at its western and eastern ends, and the youngest rocks are found closer to the center. The cause for the hotspot and for the uplift of the Iceland plateau is a deep-seated mantle plume, the track of which is represented by the Greenland-Faroe ridge, which formed during the opening of the northeast Atlantic. The Faroe Islands represent an earlier stage of the Iceland hotspot; they are of volcanic origin and consist predominantly of lava flows between 54 and 58 million years old.

Iceland's volcanically active areas are the Western, Eastern, and Northern Volcanic Zones, which are rift zones crossing the island in north-south to northeast-southwest directions. The current locus of the Iceland plume axis is at the Vatnajökull glacier, where subglacial volcanic eruptions periodically cause large meltwater floods known as jökulhlaups. Between 1963 and 1967, volcanism off southwest Iceland gave birth to Surtsey Island, which is one of the youngest islands in the world and a classic natural reserve for the study of biocolonization. The volcanic activity of Iceland is also reflected by abundant hot springs and geysers.

Iceland is composed primarily of basaltic lavas and pyroclastics erupted from fissures and shield volcanoes. Central volcanoes located within fissure swarms have also produced rhyolitic ash-flow deposits in caldera-forming eruptions. About 11% of Iceland's land area is covered by glaciers, and about half is of Quaternary volcanic origin forming much of the inhospitable and uninhabited central highland. Despite its location immediately south of the Arctic Circle, Iceland's climate is temperate because of the Gulf Stream's moderating influence.

ISLANDS OFF NORTHWESTERN AFRICA (MACARONESIA)

Macaronesia (Greek for "fortunate islands") covers the archipelagoes of the Azores (Portugal), Madeira with Selvagens (Portugal), the Canary Islands (Spain), and Cape Verde, which have pleasant subtropical to tropical climates. All these islands are entirely of volcanic origin and are the expression of several hotspots likely underlain by mantle plumes. Apart from the Azores, the Macaronesia islands and associated seamounts form part of a remarkable volcanic belt off northwestern Africa extending from the Azores-Gibraltar Fracture Zone to the Sierra Leone Rise. The origin of this belt, however, is not yet fully understood.

The Azores archipelago consists of nine major islands rising from a broad submarine plateau; hence, there is a topographic Azores high in addition to the meteorologic one. The archipelago likely owes its existence to a mantle plume and to its special tectonic setting near a triple junction. The westernmost islands Flores and Corvo are located on the North American plate, and the other islands are on the Azores microplate, which is bounded by the North American plate along the Mid-Atlantic Ridge in the west, by the Eurasian plate along the Terceira Rift (a spreading center) in the northeast, and by the African plate along the East Azores Fracture Zone in the south. The oldest island of the archipelago is Santa Maria (more than 8 million years old) and the youngest is Pico, where the archipelago's highest peak (2351 m) is located. Historic volcanic eruptions occurred on São Miguel, Pico, São Jorge, Faial, and Terceira, with São Miguel being the most active island. The last volcanoes to erupt were Capelinhos in the western part of Faial in 1957 and the shallow João de Castro seamount between São Miguel and Terceira in 1997. The volcanic rocks of the Azores span a wide range in composition from alkalic basalts to alkali-rich rhyolites.

The Madeira Archipelago comprises Porto Santo (14.3–11.1 million years old), Madeira (5.3 million to ~6,000 years old), and the three inhabited Desertas Islands (5.1–1.9 million years old). Madeira island is the present locus of a discontinuous hotspot track that can be traced back 67 million years along Seine, Ampère, and Ormonde seamounts to the northeast. The archipelago's highest peak (1862 m) is located within the deeply eroded central highland of Madeira. The islands consist dominantly of basaltic lava flows and pyroclastics, and the elongated and spectacular Desertas Islands show the deeply eroded interior of a volcanic rift zone. At Porto Santo the shallow submarine stage of an emerging seamount is well exposed. Most rocks of the archipelago have compositions ranging from basanite and alkalic basalt to basaltic trachyandesite to trachyte.

The Selvagen Islands are located between the Madeira and Canary Archipelagoes and form the summits of two extinct shield volcanoes. Geochemical studies reveal that they are part of the Canary hotspot track (thus, though the Selvagens belong to Portugal politically, they belong to Spain geologically). The small islands consist predominantly of alkalic basaltic, basanitic, and phonolithic volcanic rocks ranging in age from 29 to 3 million years old. The presence of marine carbonate sediments on Selvagem Grande show that the top of the volcano was eroded and submerged beneath sea level during a volcanic hiatus between 24 and 12 million years ago.

The Canary Islands form a chain of seven major volcanic islands that are all underlain by Jurassic oceanic crust. The ages of the islands' oldest subaerially erupted rocks decrease roughly from 21 million years at Fuerteventura in the east to 1.7 million at La Palma and 1.1 million at El Hierro in the west, consistent with a hotspot origin. The hotspot track can be traced back 68 million years along a chain of seamounts to the northeast. Despite their age progression, all of the Canary Islands except for La Gomera are still volcanically active. Historic eruptions occurred on La Palma (the most active

Canary Island in historic times), Tenerife, and Lanzarote. La Palma features the world's largest erosional crater, in which an uplifted seamount is exposed, and Pico de Teide on Tenerife is the Atlantic's highest peak (3717 m). The Canary Islands comprise almost the entire spectrum of volcanic eruption products including lava flows, fallout, nonwelded and welded ash flows, and surge deposits. The compositional spectrum of the rocks is also extremely large, ranging from tholeiitic basalts to highly silica-undersaturated alkalic basalts to phonolites and peralkaline rhyolites.

The Cape Verde Archipelago is horseshoe shaped and comprises ten main islands along a northern (Barlavento) and southern (Sotavento) group. The islands are situated on the Cape Verde Rise, which represents a hotspot swell. Because of large local uplift in the lithosphere, Jurassic oceanic crust is now exposed on Maio and São Tiago islands. Subaerial volcanic activity on the archipelago took place from around 16 million years ago (Sal island) until the present, and although there is no simple age progression, a crude decrease in age from east to west is recognized, with Santo Antão and Fogo islands showing the greatest amount of recent volcanism. Historic volcanic activity occurred on Fogo only, which also features the highest peak of the archipelago (2829 m) and a huge recent collapse scarp. The islands show a variety of rock types ranging from nephelinites and alkalic basalts to phonolites. Carbonatites (carbonate-rich igneous rocks), which are exceptional in ocean settings, occur on Brava, Fogo, São Tiago, Maio, and São Vicente.

BERMUDA ISLANDS

The Bermuda islands belong to the 1500-km-long Bermuda Rise, an elongated platform built on 100–105-million-year-old oceanic crust that was caused by some type of hotspot. The islands are the surface expression of an extinct shield volcano that had been uplifted and eroded. Submarine intrusive and extrusive activity of this volcano occurred between 45 and 33 million years ago. The present Bermuda islands rise up to 76 m above sea level, forming a 15–100 m-thin carbonate cap on the 4500-m-tall truncated stump of the old volcano. They are predominantly composed of Quaternary carbonate sandstones originally eroded from biogenic, primarily coral reef limestones during the low sea levels of several glaciations. The islands show a central lagoon, but the relation of the present morphology to the volcanic basement is unknown. There is also no evidence for a volcanic caldera often said to be outlined by the islands. The Bermuda platform is the most northerly coral reef habitat in the Atlantic Ocean and shows similarities to Pacific atolls. The climate of the islands is warm-temperate or oceanic with high humidity.

GULF OF GUINEA ISLANDS

The islands of Bioko (or Fernando Po), Príncipe, São Tomé, and Pagalu (or Annobón) are of volcanic origin and are extensions of the Cameroon line, a 1600-km-long chain of intra-plate volcanoes extending from inland Cameroon to Pagalu island. The ages of the oldest exposed rocks decrease from 31 million years on Príncipe to 13 million years on São Tomé to 5 million years on Pagalu, consistent with movement of the African plate over a hotspot. Bioko is an exception from this age progression, being a young volcanic island located close to the continental Cameroon line volcanoes. Although none of the islands have had historic eruptions, São Tomé and Pagalu show signs of recent volcanic activity, whereas Príncipe is more deeply eroded. The volcanic rocks range in composition from alkalic basalts to phonolites and rare trachytes, some of which form morphological domes and plugs. Being situated close to the equator, the islands experience a hot and humid tropical climate.

ISLANDS OFF EASTERN BRAZIL

The Fernando de Noronha archipelago consists of the main island and 20 small islets, which are all of volcanic origin and rise more than 4000 m above the seafloor. The volcanic rocks of Fernando de Noronha are between 12 and 1.7 million years old, and the youngest volcanic activity is represented by a single lava flow on the islet of São José. The islands are part of a 500-km-long hotspot track that can be traced back towards Fortaleza, where 30 million-year-old volcanic rocks are exposed. A seamount to the east of Fernando de Noronha may be volcanically active and represent the current hotspot location. The archipelago is composed mainly of highly alkaline, silica-undersaturated rocks including phonolitic and trachytic plugs and domes and various basalts. The climate is tropical oceanic with well-defined rainy and dry seasons.

Trindade and Martín Vaz are small rugged volcanic islands rising 5500 m above the seafloor. They are the easternmost and youngest expression of a 1200-km-long hotspot track that is well defined by a seamount chain to the west. Trindade is about 10 km² in size, up to 600 m high and ~5 million years old, with the last eruption having produced Vulcão de Paredão 10,000 to 5000 years ago. It is uninhabited except for a permanent Brazilian navy base. The tiny Martín Vaz islets (0.3 km²), located 47 km east of Trindade and the easternmost Brazilian territory, are hard to access and are about 1 million years old. Trindade and

Martín Vaz, together with Fernando de Noronha, represent the most alkaline province among oceanic volcanic islands of the world. The extremely sodium-rich and silica-undersaturated rocks comprise abundant phonolitic domes and plugs and different types of basalts.

The Saint Peter and Paul Rocks are located 870 km northeast of Fernando de Noronha close to the Mid-Atlantic Ridge. They consist of five rugged islets and four rocks reaching less than 20 m above sea level within an area of about 200 by 350 m, and they are the surface expression of a 3800-m-high sigmoidal massif. Only the largest rock, on which a lighthouse was built, shows some low vegetation. The rocks are not of volcanic origin but are composed of peridotite, an olivine-rich rock typical of the Earth's mantle, and are thus a unique surface outcrop of mantle rocks within an ocean. The cause for the uplift of the St. Peter and Paul Rocks massif is still-occurring tectonic activity along the St. Paul Fracture Zone.

ISLANDS IN THE SOUTH CENTRAL ATLANTIC

These volcanic sister islands are British overseas territories and are among the world's most remote islands. Ascension (98 km^2) rises near the Mid-Atlantic Ridge from a depth of 3000 m to a height of 859 m above sea level. It is a hotspot-type island but is not associated with a hotspot track. Subaerial volcanism occurred from approximately 1 million years ago to the present, and much of the island is morphologically young, with many cinder cones; however, no eruptions have been reported since its discovery in 1501. The volcanic rocks range from mildly alkaline basalts to trachytes and some rhyolites comprising abundant pyroclastics and lava flows.

St. Helena (120 km^2) rises from a 4400-m depth to 823 m above sea level and is near the end of a broad and diffuse hotspot track heading northeast toward Cameroon (but not belonging to the Cameroon line hotspot track). Subaerial volcanic activity on St. Helena occurred from 14 to 7 million years ago, and since then erosion has produced tall cliffs. The island is dominated by basaltic to trachytic lava flows, some trachyte intrusions, and subordinate pyroclastics.

Tristan da Cunha and the adjacent uninhabited Nightingale and Inaccessible Islands form the end of a prominent but broad hotspot track that includes Walvis Ridge to the northeast and is the expression of the Tristan/Gough plume. Tristan (98 km^2) is a huge, almost circular volcano rising from a 3700-m depth to 2062 m above sea level; the subaerial rocks are less than 1 million years old. Because of prevailing steep slopes and tall cliffs, a small plateau where the settlement of Edinburgh is located is the only habitable part of the island. It is exactly there that a volcanic eruption occurred in 1961. The Tristan consists predominantly of pyroclastics and lava flows of alkali basaltic through trachytic composition.

Gough Island (or Diego Alvarez island; 65 km^2) is located about 350 km southeast of Tristan da Cunha and is part of the same hotspot chain. It rises to 907 m above sea level and is less than 1 million years old. The compositions of the lavas, pyroclastics, and dikes on the island range from mildly alkaline basalt through trachyte. Except for a weather station, the remote island is uninhabited. Gough is a protected wildlife reserve and harbors one of the Atlantic's least disrupted ecosystems. The climate of Gough and Tristan da Cunha is marine subtropical with only small temperature variations.

FALKLAND ISLANDS

The Falkland Archipelago (Spanish: Islas Malvinas) consists of two main islands, East Falkland and West Falkland, and several hundred smaller ones. It has a total land area of about 12,000 km^2 and rises to 705 m. The islands are located 500 km off southern Argentina on the Patagonian shelf and have a cold, windy, and humid maritime climate. They represent fragments of continental crust resulting from the break-up of Gondwana and the opening of the South Atlantic that began about 130 million years ago. Because of this breakup, the Falkland Islands display similar rock sequences as southern Africa and Queen Maud Land, Antarctica. The crystalline basement consists of 1100–1000-million-year-old gneisses and granitoids, which are overlain by a Devonian through Triassic siliciclastic succession (about 400 to 210 million years old) dominated by quartzite, sandstone, and mudstone. Beginning about 280 million years ago, the rocks came to be strongly deformed and folded, and subsequently intruded by Jurassic mafic dikes (190 million years old), the youngest rocks. The islands' surface was finally modified during the Pleistocene glaciations.

SOUTH GEORGIA, SOUTH SANDWICH ISLANDS, SOUTH ORKNEY ISLANDS

These islands are the subaerial part of the Scotia arc that encircles the Scotia Sea (a back-arc basin) and extends from Tierra del Fuego along the submarine Scotia Ridge, South Georgia, South Sandwich, and South Orkney Islands to the Antarctic Peninsula. They owe their existence to complex tectonic processes related to the breakup of Gondwana and the opening of the South Atlantic. The climate of the islands is harsh, cold, wet, and windy.

The South Georgia Islands lie 1400 km east of the Falkland Islands and are the emergent part of an isolated microcontinental block with similar geology as Tierra del Fuego. The main island (3528 km^2) rises to 2934 m and is largely covered by glaciers, ice caps, and snow fields. The islands largely consist of thick late Jurassic to early Cretaceous (140–110-million-year-old) turbidite sequences of volcaniclastic and siliciclastic sandstones and shales that were strongly deformed and folded around 91–82 million years ago. The oldest rocks are gneisses and schists intruded by middle Jurassic (164-million-year-old) granites and gabbros. In southeastern South Georgia some uplifted 150-million-year-old basaltic ocean crust (an ophiolite) is exposed. The youngest rocks, about 80–100 million years old, are remnants of a former volcanic arc.

The South Sandwich Islands comprise 11 uninhabited volcanic islands with a total area of 310 km² and the highest peak, a stratovolcano, reaching 1370 m in height. They represent a volcanic arc west of the Scotia Trench where the South American plate is being subducted from the east. The islands are volcanically active with rocks created between about 4 million years ago and the present. The compositions of the volcanic rocks is calc-alkaline ranging from basalt to rhyolite.

The South Orkney Islands are part of a continental fragment on the southern Scotia arc. They comprise five main islands about 600 km northeast of the Antarctic Peninsula with a total area of 620 km² and a height of up to 1266 m. The islands consist largely of metamorphosed and folded sedimentary and volcanic rocks of early to middle Jurassic age (205–176 million years old) as well as Triassic to early Cretaceous sediments. The subantarctic islands are barren, uninhabited, and largely covered by ice and snow.

BOUVET ISLAND

Bouvet is a small volcanic island (about 50 km²) rising from near the southernmost end of the Mid-Atlantic Ridge to 780 m above sea level. It is the world's most remote island lying about 1600 km from the nearest land, Queen Maud Land, Antarctica. Bouvet owes its existence to a melting anomaly of the underlying mantle or a small mantle plume. The island forms the top of a large stratovolcano, the maximum age of which is constrained by the age of the underlying oceanic crust of 6 million years. The volcano is considered active, as shown by the formation of a lava shelf on the west coast in the 1950s. The main eruptive center is in the northwest of the island, where fumaroles occur and a caldera may have collapsed. The lava compositions range from basalts to trachyandesites to alkalic rhyolites and obsidian. Most of Bouvet is covered by glaciers, which together with high cliffs and harsh weather conditions make landing extremely difficult. The harsh polar climate at Bouvet restricts the vegetation to litchens and mosses, and the fauna comprises seals, penguins, and seabirds. The island is uninhabited and was designated a nature reserve in 1971.

FURTHER READING

Faure, G. 2001. *Origin of igneous rocks: the isotopic evidence.* Berlin: Springer.
Mitchell-Thomé, R. C. 1970. *Geology of the South Atlantic islands.* Berlin: Borntraeger.
Sigurdsson, H., ed. 2000. *Encyclopedia of volcanoes.* San Diego, CA: Academic Press.
Vogt, P. R., and B. E. Tucholke. 1986. *The geology of North America, vol. M: the western North Atlantic region.* Boulder, CO: Geological Society of America.
The Great Plume Debate Website. http://www.mantleplumes.org/.

ATOLLS

EDWARD L. WINTERER
Scripps Institution of Oceanography, La Jolla, California

Atolls are a special type of coral reef complex formed in tropical seas by a ring of reef coral enclosing a lagoon. The characteristic features of an atoll include a reef rim, from 100 to 500 m across, which is mainly awash at high tide, and flattish islands (motu), which remain a few meters above sea level and on which people may live. The word *atoll* itself comes from the language of the Maldive Islands, a chain of large atolls in the Indian Ocean (Fig. 1). Atolls range in size from a few to as many

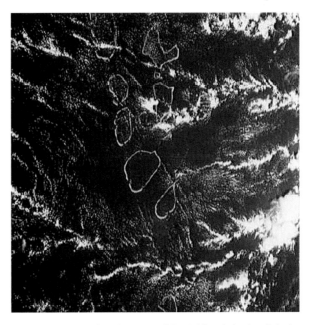

FIGURE 1 Air view of southern part of the Maldive chain of atolls in the (cloudy) Indian Ocean. Taken looking north from 2.4° N, 73.3° E. Kolumadulu Atoll, in center (at 2°25′ N, 73°10′ E), is about 50 km across. NASA photograph STS037-97-46. Image courtesy of the Image Science and Analysis Laboratory, NASA Johnson Space Center.

as 40,000 km², and lagoon depths range from almost nothing to as much as 100 m.

DISTRIBUTION

The minimum water temperature for the approximately 600 atolls around the globe is about 20 °C, a temperature that prevails from about 24° N to 29° S. Reef-building corals are partly dependent on their embedded photosynthesizing algal symbionts for adequate oxygen, and thus grow healthily only in lighted waters less than a few tens of meters deep. Slow-growing, deep-water coral reefs, without algal symbionts, are known to exist at depths down to about 6000 m. Many atoll lagoons are connected directly to the sea by inlet channels, but few channels are wide or deep enough to be navigable by anything but canoes. Atolls with wide, deep channels are favored for tourism and for administrative headquarters. People from other atolls in the region tend to migrate to these centers, and overcrowding is a problem on some. The main source of income on most atolls is copra from coconuts, and tourism is important on a few.

THE REEF RIM

Between the ocean and the lagoon, the reef comprises an outer, wave-resistant algal ridge a few tens to a hundred meters wide at sea level, against which the ocean swell breaks. Seaward, the ridge drops steeply off into deeper, less turbulent water, and corals thrive along these slopes. The outer part of the algal ridge is commonly notched to a depth of 10–20 m by a reticulate network of narrow (1–3 m) grooves and surge channels that funnel water across the reef rim toward the lagoon. Behind the algal ridge is a reef flat, close to low tide level, where a pavement of living and dead coral and coral debris from the size of sand grains to the size of boulders extends lagoonward. Reef debris thrown onto the flat during typhoons lies in patches and windrows 1–2 m high in some places.

Large and small islands, termed motu, surmount the reef flat in places and rise to as much as 4 m above sea level (Fig. 2). They are commonly edged by a steep scarp toward the sea, created by waves beating against the island. It is only on these islands that the people of the atolls can live, where they are safely above the surf. The motu are weathered remnants of reef flats dating from the time of a regionally (in low latitudes) higher stand of sea level, about 2000–4000 years ago. The highstand resulted from the isostatic response to a redistribution of ocean waters after deglaciation of the northern continental ice sheets, beginning about 18,000 years ago. Between motu are shallow channels (hoa) that direct washover from ocean waves toward the lagoon. Shallow lakelike depressions, termed faros, pock reef flats on some atolls. These are the now-drowned remnants of karstic depressions dissolved by meteoric waters during the last lowstand of sea level. In addition to carefully collecting and storing rainwater, atoll islanders depend on a precarious groundwater supply beneath the motu, which accumulates as a freshwater lens floating on the denser saltwater beneath. Seepages of this water are used to water vegetable gardens, and wells tap the lens for potable water.

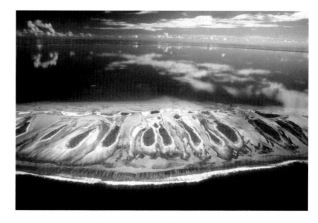

FIGURE 2 Leeward side of Tikehau atoll in the Tuamotu archipelago (15° S, 148° W), showing closely spaced surge channels next to the shore, the reef flat surmounted by wooded motu separated by channels (hoa) leading to the lagoon. The opposite shore of the lagoon is visible in the background as a dark line. Photograph by Pierre Labout, with permission of the Institut de Recherche pour le Développement, Paris.

LAGOONS

Atoll lagoons are mainly of fairly normal ocean salinity, except for completely landlocked lagoons surrounded by relatively high islands. These lagoons may include shallow salt pans and briny lakes. Detailed charts of atoll lagoons commonly show patch reefs, some rising to sea level but others submerged so as to constitute hazards to navigation. Sediment builds up in the lagoon almost entirely during sea-level highstand times, when ocean waves can wash fine sediment across the reef. Shells of organisms living in the lagoon add to this thickness. At lowstand times, rain falling into the perched lagoonal depression may drain to the surrounding sea through erosional channels cut through the reef rim.

ORIGIN OF THE ATOLL FORM

The origin of the ring-around-a-lagoon form has attracted the attention of geologists for two centuries, beginning with Charles Darwin (1842), who never actually visited an atoll during his *Beagle* voyage but made a painstaking study of charts, supplemented by his own observations on other coral-surrounded islands. He showed that tropical islands displayed a series of reef forms, from fringing reefs that cling to the shore, to barrier reefs separated from the island shore by a lagoon, to atolls, with a lagoon but no bedrock island. From his experiences in South America, where his hikes took him to shelly marine deposits high on the Andes slopes, connoting tectonic uplift, he reasoned that there must be a compensating subsidence in the Pacific basin and that the array of reef forms—fringing, barrier, atoll—must be due to regional subsidence of the islands. According to Darwin, the original fringing reef would grow upward, keeping pace with the relative rise of sea level during island subsidence, but as subsidence continued, the original fringing reef would continue to grow upward, and coral growth would be inhibited in the less turbulent and less fertile waters inland from the reef front. The perimeter would keep up with the rising sea level, whereas the more interior parts of the reef would fall further and further behind, resulting in a lagoon, first behind a barrier reef and then within an atoll. (Fig. 3)

Since Darwin's theory was published, in 1842, the reality of island subsidence has been confirmed by the drilling of atolls in several places, starting with a testing expedition by the Royal Society to the Pacific atoll of Funafuti, in the Ellice Islands, in 1896–1898, The drill reached a depth of 340 m, at a level near the Pleistocene–Pliocene boundary (~2 million years old according to Ohde et al. 2002) in dolomitized reefal deposits, a finding that demonstrated rapid subsidence of the island foundations. Later drilling at Bikini and Enewetak (Eniwetok) atolls in the Marshall Islands, as part of the nuclear bomb testing program, reaffirmed the subsidence theory. The history of reef construction at Enewetak goes back some 50 million years, to Eocene times.

Difficulties with Darwin's hypothesis for the origin of the lagoon provoked others (e.g., Daly 1915) to advocate other schemes, mainly in hypothesizing planation (including dissolution) of carbonate banks during lowered sea-level periods followed by construction of a perimeter reef during the following sea-level rise, thus leaving a central lagoon. These hypotheses were disproved by drilling at Bikini, Enewetak, and Mururoa, as part of nuclear bomb testing, which showed no such erosive platform.

THE DISSOLUTION, OR KARST, THEORY

The Darwinian hypothesis for the origin of the lagoon has been replaced gradually by the dissolution, or karst, theory, in which the lagoonal depression is caused by dissolution of the reef carbonate by rainwater during periodic lowstands of sea level in the Pleistocene, when the sea level dropped repeatedly by about 100 m, on a 100,000-year time scale. The freshwater dissolved a hollow in the emergent carbonate bank, a hollow that was later filled with seawater when the sea level rose again. The two most important controls on lagoon depth are rainfall catchment area and average rainfall. The bigger the atoll and the rainier the climate, the deeper the lagoon. Present-day global rainfall patterns (Fig. 4) probably reasonably reflect rain patterns (wet north and south of the equator, with an equatorial dry belt in the eastern Pacific) during drier glacial times. Erosion of the atoll rim, generally by about 8 m, occurs during lowstands, but the rim is rebuilt during sea-level rise, and then still more rim carbonate is added during slow tectonic subsidence at highstand times. At Enewetak atoll, drilling shows that since Eocene times, a flattish carbonate bank built up to a thickness of about 1400 m, keeping pace with tectonic subsidence, but interrupted from time to time by periods of emergence, when soils developed. The drilling and seismic work there also showed that the atoll form itself likely did not develop until Pleistocene times, when sea-level fluctuations were large and glacial intervals long, whereas earlier sea-level fluctuations were of lesser magnitude and duration. Atolls are thus mainly a phenomenon of the Pleistocene.

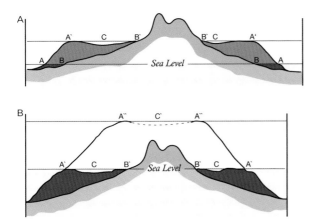

FIGURE 3 The stages in the evolution of an atoll, according to Darwin (1842). (A) A subsiding island is first girt by a fringing reef; then, as subsidence continues, the island is surrounded by a barrier reef, with a lagoon formed because growth rates of "interior" corals do not keep pace with better nourished "exterior" corals. (B) Continued subsidence, with the same growth-rate difference, leads to disappearance of the island and to an atoll form.

THE FUTURE OF ATOLLS

Atolls and their peoples are threatened today both by wave attacks on the fragile motu and by rising sea levels associated with global warming. These combine to reduce atoll relief to ephemeral patches of storm debris on reef flats awash at

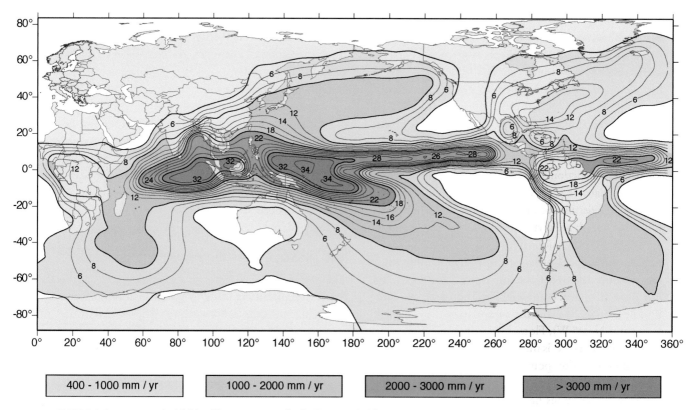

FIGURE 4 Average annual rainfall in millimeters per year for the 17-year period from 1979 to 1995. Interpolated and modified from Xie and Arkin (1997).

high tide. This is to say nothing of the threats to healthy coral growth posed by high water temperatures and possible increasing acidification of the ocean. Very high carbon dioxide levels probably occurred during the Cretaceous, a time of global warmth but also of very healthy populations of calcareous animals and plants. Whatever trends toward major acidification in the Cretaceous world existed were partly offset by dissolution of calcareous sediments on the deep seafloor. Ultimately, the same balancing may happen in the modern ocean, but this will take time. Over the shorter term, Waterworld awaits.

SEE ALSO THE FOLLOWING ARTICLES

Darwin and Geologic History / Makatea Islands / Marshall Islands / Motu / Reef Ecology and Conservation

FURTHER READING

Daly, R. A. 1915. The glacial control-theory of coral reefs. *Proceedings of the American Academy of Arts and Sciences* 51: 155–251.
Darwin, C. 1962. *The structure and distribution of coral reefs.* H. W. Menard, ed. Berkeley: University of California Press.
Dickinson, W. R. 2003. Impact of mid-Holocene hydro-isostatic highstand in regional sea level on habitability of islands in Pacific Oceania. *Journal of Coastal Research* 19: 489–502.
Purdy, E. G., and E. L. Winterer., 2001. Origin of atoll lagoons. *Geological Society of America Bulletin* 113: 837–854.

ATOMIC TESTING

SEE NUCLEAR BOMB TESTING

AZORES

PAULO A. V. BORGES, ISABEL R. AMORIM, AND ROSALINA GABRIEL

University of the Azores, Terceira, Portugal

REGINA CUNHA, ANTÓNIO FRIAS MARTINS, LUÍS SILVA, AND ANA COSTA

University of the Azores, Vairão, Portugal

VIRGÍLIO VIEIRA

University of the Azores, Ponta Delgada, Portugal

The Azores are a remote and geologically recent archipelago consisting of nine volcanic islands located in the North Atlantic Ocean (Fig. 1). Of the 4467 species and subspecies of terrestrial plants and animals known to inhabit this archipelago, 420 are endemics. These islands were discovered in the fifteenth century, and more than 500 years of human settlement have taken their toll on the local fauna and flora. Approximately 70% of the vascular plants and 58% of the arthropods found in the Azores are exotic, many of them invasive, and only 20% of the archipelago's terrestrial realm is protected, which raises serious long-term conservation concerns for the Azorean endemic biota.

GEOLOGICAL SETTING AND ENVIRONMENT

The Azorean archipelago is located in the North Atlantic Ocean, at the junction of the Eurasian, African, and North American plates (Fig. 1). The archipelago consists of nine volcanic islands, aligned on a west-northwest–east-southeast trend, which are divided into three groups: the western group of Corvo and Flores; the central group of Faial, Pico, Graciosa, São Jorge, and Terceira; and the eastern group of São Miguel and Santa Maria (Fig. 1). The largest island is São Miguel (745 km^2), and the smallest is Corvo (17 km^2). Santa Maria is the southern- and easternmost island (37° N, 25° W), Flores is the westernmost (31° W), and Corvo (39°42′ N) is the northernmost island. Pico has the highest elevation point (2351 m above sea level), and Graciosa the lowest (402 m above sea level). Five other islands have elevations near 1000 m above sea level. The three island groups are separated by 1000–2000-m-deep sea channels, except for Faial and Pico islands, between which the channel is, in many parts, only 20 to 50 m deep. The Azores are separated from the most western point of mainland Europe (i.e., Cabo da Roca, Portugal) by 1390 km. Located in the Atlantic Ocean at a mean latitude of 38°30′ N, the Azores enjoy a distinctly oceanic climate. The insignificant variation in the seasonal temperature and the high humidity and precipitation that characterize the archipelago's climate are mostly due to the influence of the Gulf Stream, which transports warm waters and humid air masses and is responsible for the high-pressure systems over the Azores.

Geologically, the Azores comprise a 20–36-million-year-old volcanic plateau; the oldest rocks (composing Santa Maria Island) emerged 8.120 million years ago, whereas the youngest (forming Pico Island) are about 250,000 years old. The geostructural environment of the Azores Plateau, defined by the 2000-m bathymetric contour line, is dominated by the confluence of the American, Eurasian, and African lithospheric plates. Thus the Azores are characterized by high volcanic activity typical of a ridge-hotspot interaction (i.e., a hotspot on a slow-moving plate). As opposed to the Hawaiian islands, which are chronologically arranged, the Azorean islands do not show any correlation between their distances to the hotspot and their individual ages of emergence. The eastern parts of all Azorean islands are geologically the oldest, which is the result of the particular seismovolcanic mechanisms of this archipelago. This tectonic feature is responsible for many volcanic eruptions (e.g., Capelinhos, Faial Island, 1957–1958) and tectonic earthquakes (e.g., Terceira and São Jorge islands, 1980; Faial and Pico islands, 1998).

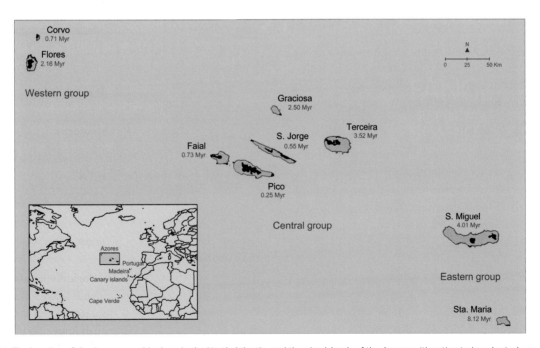

FIGURE 1 The location of the Azorean archipelago in the North Atlantic, and the nine islands of the Azores with estimated geological age. Shaded areas correspond to protected island areas based on recent IUCN classification (almost 20% of the total territory of the archipelago).

As a result of several recent historical lava flows, there is a great concentration of lava tube caves and pits in the Azores. A total of 250 underground cavities, including lava tubes, volcanic pits, pit-caves, and sea-erosion caves, are known to exist on the Azores, creating many kilometers of cave passages, extraordinary geological formations, and unique fauna adapted to caves.

Regarding native plant communities, laurisilva,—a humid evergreen broadleaf laurel forest—was considered in the past to be the predominant vegetation form in the Azores. However, more recent studies have shown the existence of a wide variety of plant communities, including coastal vegetation, wetland vegetation (lakeshore and seashore communities and a variety of bogs), several types of meadows, and different types of native scrub and forest. Moreover, the Azorean laurisilva differs from that found on Madeira and on the Canary Islands, as it includes a single species of Lauraceae, several species of sclerophyllous and microphyllous trees and shrubs, and luxuriant bryophyte communities, covering all available substrata. In contrast to other Macaronesian archipelagoes, the Azores only has one endemic genus of vascular plants (*Azorina*). After human settlement, other types of vegetation cover have become progressively dominant. Presently, they include pastureland, production forest (mostly with *Cryptomeria japonica*), mixed woodland (dominated by nonindigenous taxa), field crops and orchards, vineyards, hedgerows, and gardens.

OVERALL BIODIVERSITY: SPECIES INHABITING THE AZORES

The terrestrial flora and fauna of the archipelago were recently listed (summary in Table 1). It is believed that the Azores, especially the younger islands, are not saturated with species. The islands are probably in a nonequilibrium condition as a consequence of (1) the dispersal difficulties imposed by the isolation of the archipelago, which are much greater than the dispersal abilities of a wide range of taxa; (2) the vicissitudes of the Pleistocene environment; (3) the destructive influence of volcanic activity; and, more recently, (4) the impact of human activities.

BIOGEOGRAPHY

Even factoring out the area of the islands, the native fauna and flora of the Azores is impoverished when compared to the other Macaronesian archipelagoes (Madeira and the Canary Islands). For example, the number of Azorean endemic species is about three times less than

TABLE 1

Number of Currently Known Terrestrial Species and Subspecies in the Fauna and Flora of the Azores

	Total	Endemic
Algicolous fungi	1	0
Lichenicolous fungi	22	0
Lichens	551	12
Bryophyta	438	9
Plantae	947	68
Nematoda	80	2
Annelida	21	0
Mollusca	111	49
Arthropoda	2227	267
Chordata	69	13
Total	4467	420

NOTE: Table based on the catalog of Borges *et al.* (2005).

the number of endemics of the Madeira archipelago (three times smaller but older and nearer to the mainland) and ten times less than the number of endemics of the Canary Islands (three times larger). Given the isolation of the Azores, the ancestors of all the terrestrial endemic species found in the archipelago had to travel over a significant water distance (more than 1200 km) from neighboring Europe and about 800 km from Madeira Island. Additionally, colonization of the Azores has occurred over a short geological period, since the oldest island (Santa Maria) emerged 8.12 million years ago. Accordingly, it is of no surprise that the only indigenous terrestrial vertebrates are bats (two species) and birds (16 species). Nevertheless, the presence of many endemic flightless beetles in the Azores, whose ancestors are believed to have also been flightless, suggests that long-distance ocean dispersal, by air and on the water surface (rafting), must have been an important colonization mechanism. Most storms and prevailing winds come from the West, but the Azorean biota is mainly of Palearctic and Macaronesian origin. A similar situation occurs with the Azorean terrestrial molluscs, which are clearly of Palearctic origin and, at the same time, exhibit Macaronesian relationships in some taxa: *Leptaxis* lives also in Madeira, and *Napaeus* in the Canary Islands. The preferred explanation for this "anomaly" is the large distance to the American continent and the possibility that the paleo-winds may have blown in a different direction from the current prevailing winds. However, a simpler explanation for the biota composition of the Azores is the arrival of colonizers through "sandstorm" dispersal originally coming from the Sahara. Clear exceptions to the Palearctic/Macaronesian origin can only be found in organisms with great dispersal abilities, such as bryophytes, some species of which, found in the Azores, are unequivocally of American origin.

The most general pattern in ecology is the species-area relationship (SAR). Considering only the area above an altitude of 300 m (because native habitats can only be found above that elevation in almost all Azorean islands), a significant relationship is observed in the Azores for indigenous bryophytes, vascular plants, and arthropods, but not for land molluscs (Fig. 2). Arthropods show the steepest slope of the SAR curves, which implies a higher beta-diversity for this group and, consequently, a more heterogeneous species composition among the islands. However, the time factor should be accounted for in the case of endemics; this could explain the absence of SAR for land snails, as small, older islands (e.g., Santa Maria) harbor more endemic species.

Some of the most diverse Azorean genera with endemic species are also diverse in Madeira and the Canary Islands (e.g., the beetles *Trechus* and *Tarphius,* the hemipteran *Cixius,* and the land-snails *Napaeus* and *Plutonia*), thus reinforcing the hypothesis of a Macaronesian interarchipelago dispersal.

EVOLUTION

There are several aspects of the evolutionary history of the Azorean biota that are still not clear. Traditionally, many biologists considered the Azorean and the Macaronesian endemic flora in general to be ancient, consisting of many paleoendemic species, relicts of the vegetation that originally covered most of Western Europe during the Tertiary period. However, there is increasing evidence from molecular data that many of the endemic Macaronesian plant species are the result of in situ evolution after a relatively recent colonization (neoendemics). This theory may also apply to invertebrates (e.g., arthropods and terrestrial molluscs), in which colonization followed by isolation has led to the evolution of a highly original neoendemic fauna. The most diverse genera in the Azores belong to the animal realm (classes Gastropoda and Insecta). Molecular data on insects show that many endemic species belonging to speciose genera are monophyletic in the Azores (e.g., *Tarphius, Trechus, Hipparchia*), thus implying that all species within a particular genus originated by speciation events occurring after the arrival of a single ancestor to the archipelago. In spite of some evidence that evolution has proceeded to a subgeneric level in some land molluscs (e.g., *Macaronapaeus, Atlantoxychilus, Drouetia*), natural arrival to the Azores is an uncommon event, and most of the Azorean endemics are neoendemics. Thus, dispersal limitation may be viewed as one of the main driving forces that has shaped the Azorean native biota. Available data suggest that the "progression rule"

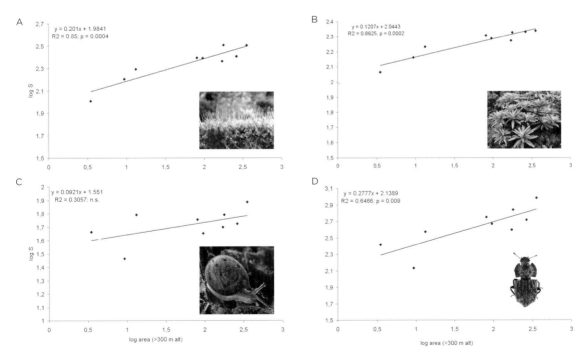

FIGURE 2 Species-area curves for (A) indigenous bryophytes, (B) vascular plants, (C) terrestrial molluscs, and (D) arthropods (see text for further explanations).

(i.e., a nonstochastic pattern of colonization from older to younger islands) applies to the Azores: Santa Maria has generally been the first to be colonized, accompanied by subsequent lineage splitting as individuals disperse to the younger western islands. Future phylogenetic studies, aimed at understanding patterns of dispersal, colonization, and speciation in the Azores, should clarify the presence of more complex patterns within and among island speciation as well as the possibility of back colonization events from younger to older islands.

The high volcanic activity in the Azores is responsible for the formation of new habitats, such as lava tubes and volcanic pits, from which 20 neoendemic troglobitic arthropod species have been described to date. These cave-limited species exhibit different levels of adaptation to the underground environment and therefore constitute an excellent opportunity to investigate ongoing evolutionary processes. Many of the cave species known to occur in the Azores belong to genera that have representatives in the troglobitic fauna of other Macaronesian islands (e.g., the Canary Islands), thus serving as a model for further studies of inter-archipelago speciation.

An interesting example of an island syndrome in the Azores is given by the damselfly *Ischnura hastata* (Insecta, Odonata): There is no evidence of parthenogenetic populations in the New World, so the Azorean populations probably developed parthenogenesis after colonization.

CONSERVATION REMARKS

The relatively high level of endemism in the Azores gives to the archipelago's biota great conservation relevance. Its preservation has also been recognized by the local government through the establishment of protected areas for conservation purposes since the early 1980s. The Azorean Protected Areas Network is currently being reformulated according to IUCN criteria and includes 23 Sites of Community Importance and 15 Special Protected Areas, which are part of the NATURA 2000 network of nature protection areas. Most of this protected area includes the richest sites in endemic arthropods and also in rare European bryophyte species, but the area does not protect all native fauna and flora.

Although expanding, unregulated tourism has not yet raised conservation concerns in the archipelago. Fragmentation and degradation of habitats together with the spread of nonindigenous species are the greatest threats to the terrestrial biodiversity in the Azores. Intentional introduction of many plant species for agriculture, forestry, and aesthetic purposes has had an enormous impact on the current flora of these islands. Many of the imported species "escaped into the wild," and a considerable proportion have become naturalized, causing problems in agriculture and forestry. The impact of these species—in particular, invasive vascular plants, which are disrupting native plant communities with unknown consequences for overall native biodiversity—is of great concern. A negative impact on the indigenous community of phytophagous insects is expected, as well as changes in vegetation structure, difficulties in the regeneration of endemic species, and competition for dispersal agents, leading to a reduction in the frequency and abundance of indigenous plant taxa. Humans are clearly implicated in the establishment of exotic species: 70% of the vascular plants and 58% of the arthropod species and subspecies have been introduced on purpose or as stowaways. Moreover, the density of human population is correlated with the diversity of exotic taxa (vascular plants: $r = 0.86$; $p = 0.003$; arthropods: $r = 0.93$; $p = 0.0002$), and there is a remarkable correlation between the richness of exotic plant species and that of exotic arthropod species ($r = 0.96$; $p < 0.0001$).

Protected areas are strategically important in order to guarantee a successful management of biodiversity conservation policy in the Azores. Progress in the conservation of Azorean biodiversity depends predominantly on long-term studies on the distribution and abundance of focal species and the control of invasive species. This research requires serious commitment from scientists, politicians, and the general public. The definition of genetic units for conservation purposes in the Azores is also extremely important, particularly for widespread endemic species. For some of those endemics that are geographically structured, part of their genetic variability is locally endangered due to threats to specific populations or to the refuge-type distribution. The conservation of the Azorean natural heritage will largely depend on the definition of a global and integrated global strategy focusing on the management of both indigenous and nonindigenous species, and paramount attention needs to be paid to the implementation of a sustainable use of the archipelago's natural resources, including its biodiversity, in a trade-off with human activities and increasing inhabitants' commitment to environmental values.

SEE ALSO THE FOLLOWING ARTICLES

Canary Islands, Biology / Fragmentation / Lava Tubes / Madeira / Species–Area Relationship

FURTHER READING

Borges, P. A. V., and V. K. Brown. 1999. Effect of island geological age on the arthropod species richness of Azorean pastures. *Biological Journal of the Linnean Society* 66: 373–410.

Borges, P. A. V., R. Cunha, R. Gabriel, A. F. Martins, L. Silva, and V. Vieira., eds. 2005. A list of the terrestrial fauna (*Mollusca* and *Arthropoda*) and flora (*Bryophyta, Pteridophyta* and *Spermatophyta*) from the Azores. Direcção Regional do Ambiente and Universidade dos Açores, Horta, Angra do Heroísmo and Ponta Delgada.

Cameron, R. A. D., R. M. T. da Cunha, and A. M. Frias Martins. 2007. Chance and necessity: land snail faunas of São Miguel, Açores, compared with those of Madeira. *Journal of Molluscan Studies,* 73: 11–21.

Gabriel, R., and J. W. Bates. 2005. Bryophyte community composition and habitat specificity in the natural forests of Terceira, Azores. *Plant Ecology* 177: 125–144.

Silva, L., and C. W. Smith. 2004. A characterization of the non-indigenous flora of the Azores Archipelago. *Biological Invasions* 6: 193–204.

B

BAFFIN

LYNDA DREDGE

Geological Survey of Canada, Ottawa

Baffin Island, with an area of more than 500,000 km², is one of the principal islands of the Canadian Arctic archipelago and the world's fifth-largest island (Fig. 1). It lies within the territory of Nunavut, at the eastern portal of the Northwest Passage, the grail of early explorers and a potential shipping route in years to come. Half the island's 11,000 inhabitants live in Iqaluit, the administrative capital for Nunavut; the remainder live in seven other coastal communities. The island is named after William Baffin, a seventeenth-century British explorer, although it was known previously to the Norse as Helluland, the land of flat stones.

HISTORY

Early settlement of the island dates back about 4000 years to the Paleoeskimo Pre-Dorset people, and their successors, the Dorset people. These cultural groups are named after key sites at Cape Dorset, along the south coast of Baffin Island. Both cultures were based on a maritime economy dominated by the hunting of sea mammals, especially walrus, narwhal, and beluga. The Dorset culture is particularly renowned for superb miniature ivory carvings of sea animals, polar bears, humans, and magical beings. About 1000 years ago, as the climate warmed, the Thule people, the ancestors of the present Inuit, migrated eastward into the region, and for the first time dog teams were used to pull sleds. Although seals and bowhead

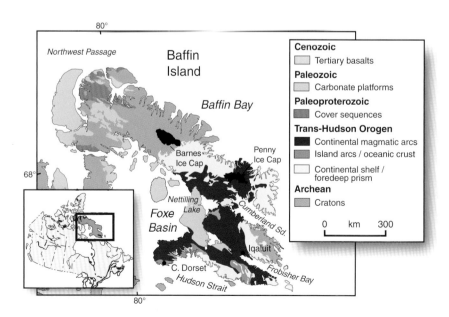

FIGURE 1 Map showing location (inset) and geology of Baffin Island.

whales were the main food sources of the Thule people, all early peoples of Baffin Island had a diversified diet that included sea mammals, caribou, fish, and birds' eggs.

The first definitive European exploration began in 1576 when Martin Frobisher, an English privateer-turned-explorer, sailed into what is now known as Frobisher Bay. There, he found "black ore," which he promoted as gold-bearing. He brought back more than 1200 tons to England in three mining expeditions backed by investors, but all the ore was hornblende and pyrite—fool's gold. Frobisher thus instigated the first mining scam in the New World.

Today, the island's main inhabitants are the Inuit, and their economy is based on government administration, hunting and fishing, mineral resources, and tourism, as well as the carving and printmaking first established at Cape Dorset.

GEOLOGY AND LANDFORMS

Baffin Island is part of the Canadian Shield. The oldest rocks are Archean-age (2.8 billion years ago) granite, gneiss, and metamorphosed volcanic and sedimentary rocks that form part of the Rae Craton, a proto-continent. A large fold belt crossing the central and southern parts of the island consists of a continental margin succession of quartzite, marble, shale, and turbidite wackes of Proterozoic age (2.16–1.90 billion years ago), deposited on the flank of the craton. Associated volcanic rocks resulted from eruptions about 1.93 billion years ago. The Precambrian rocks were deformed by plate convergence and continental accretion during the Trans-Hudson Orogeny 1.8 billion years ago, a major event in the welding together of the Canadian Shield. Later, in the Paleozoic era (450 million years ago), calcareous marine sediments were laid down over the craton. The flat-lying strata of dolostone and limestone that form surface rocks on western parts of the island are remnants of this event. Gold, iron ore, and sapphires from the Precambrian rocks, lead-zinc deposits in the Paleozoic carbonates, and diamonds in younger kimberlite intrusions are the main mineral resources on the island.

In the Cenozoic era, passive-margin plate tectonics associated with the opening of the Atlantic Ocean and the drifting of Greenland caused rifting and uplift along the northern and eastern edge of Baffin Island and shaped the island into what we know today—a plateau tilted downward toward the southwest. The eastern side of the island rises 2100 m out of the waters of Baffin Bay, creating an ice-capped mountainous edge that is deeply dissected by fjords and sounds, some of which penetrate more that 100 km

FIGURE 2 The northern coast. Rock-walled Royal Society Fjord, with snowfields on the highlands and small outlet glaciers descending to tidewater. Photograph GSC 2007-193 by D. A. Hodgson.

inland (Fig. 2). The land slopes gently across a central plateau to the relatively shallow waters of Foxe Basin.

In the last 2 million years, Baffin's landforms have been modified by the great ice sheets that covered substantial parts of the northern hemisphere. During the last glaciation, culminating 20,000 years ago, Laurentide ice from the Canadian mainland crossed to the outermost edges of Baffin Island, scraping and polishing rocks, deepening fjords, and depositing a layer of till (a mixture of glacially eroded debris). Ice began to melt away from the northeast highlands about 8600 years ago and from Foxe Basin about 7800 years ago, shifting one of the glacial dispersal centers onto central Baffin Island. The 700-m thick Barnes Ice Cap (Fig. 3) on the central plateau and the Penny Ice Cap are the last remaining remnants of the once-vast continental ice sheet. The numerous ice fields and small outlet glaciers in the mountains covering the northern island rim are more recent and have formed in

FIGURE 3 The edge of the Barnes Ice Cap, ice-marginal moraine ridges, proglacial meltwater streams, and nearby tundra barrens. Photograph GSC 2007-195 by L. A. Dredge.

the postglacial period. As the ice sheet melted, marine waters inundated the lowlands along Foxe Basin, depositing sand and mud. With glacioisostatic rebound, these deposits now lie between present sea level and an elevation of about 100 m and form the grassy coastal lowlands that are major wildlife habitat.

FLORA AND FAUNA

Most of Baffin Island lies above the Arctic Circle and is thus subjected to midnight sun in summer and polar night during the winter. It experiences a maritime arctic to continental-arctic climate. Mean annual temperatures average about –15 °C; precipitation ranges from 400 mm on the eastern lowlands to less than 200 mm on the upland plateau, which forms a polar desert. Below a shallow layer of soil that thaws seasonally, the ground remains frozen to a depth of about 400 m.

Most of the land is rocky and supports tundra-barrens vegetation, mainly a cover of lichen-heath (Fig. 4). Foxes, wolves, hares, and lemmings are found in this habitat. However, extensive areas of the plateau above an elevation of 550 m were covered with persistent snowfields during the Little Ice Age, between AD 1600 and 1850. These areas remain almost bare of any vegetation, including lichens, and are devoid of wildlife. In contrast, the lowlands around Foxe Basin consist of sedge and grass meadows containing dwarf birch and willow shrubs, interspersed with shallow tundra ponds. These areas are prime habitats for barren-ground caribou and a myriad of migratory birds, including various species of geese, sandpipers, murres, plovers, and gulls. The lowland around Nettilling Lake, known as the Great Plain of the Koudjuak, lies along the Eastern Flyway and is a major nesting area for Canada and brant geese. Whales, particularly the bowhead, beluga, and narwhal, are prevalent in open waters, and walruses, polar bears, and ringed and bearded seals live on the pack ice that surrounds the island for much of the year.

SEE ALSO THE FOLLOWING ARTICLES

Arctic Region / Climate Change / Sea-Level Change

FURTHER READING

Andrews, J. T. 1989. Quaternary geology of the northeastern Canadian Shield. *Geological Survey of Canada, Geology of Canada* 1: 276–318.
Bird, J. B. 1967. *The physiography of Arctic Canada.* Baltimore, MD: Johns Hopkins Press.
Dredge, L. A. 2002. Surficial materials, central Baffin Island. *Geological Survey of Canada, Current Research* 2002-C20.
McGhee, R. 1996. *Ancient people of the Arctic.* Vancouver: University of British Columbia Press.
Porsild, A. E. 1964. Illustrated flora of the Canadian Arctic Archepelago. *National Museum of Canada Bulletin* 146.
Scott, D. J., M. St-Onge, and D. Corrigan. 2003. Geology of the Archaean Rae Craton and Mary River Group and the Paleoproterozoic Piling Group, central Baffin Island, Nunavut. *Geological Survey of Canada, Current Research* 2003-C26.

BAJA CALIFORNIA: OFFSHORE ISLANDS

MARTIN L. CODY

University of California, Los Angeles

The Baja (lower) California Peninsula, in extreme northwest Mexico, stretches more than 1000 km south from the southern (Alta) California border to the Tropic of Cancer, spanning latitudes of 32°30′ to 23° N. Originating about 500 km south of its present position and rifting north some 4 cm per year since the Miocene, the peninsula was sheared from the North American plate by contact with the northwesterly moving Pacific plate. The East Pacific Rise runs up the center of the Sea of Cortés (or Gulf of California), which separates the peninsula from mainland Mexico; it continues north as the classic strike-slip San Andreas Fault. Resulting largely from complex plate-boundary dynamics involving, besides those mentioned, the fragmented remnants of the largely subducted Farallon plate, islands have been generated around the peninsula, west and south in the Pacific Ocean and east in the Sea of Cortés off the peninsula's trailing edge. Many islands had Pleistocene connections to the mainland; others are deep-water islands, some being blocks faulted from the trailing

FIGURE 4 Lichen-covered, glaciated, granitic rocks and tundra barrens of central Baffin Island. Stacked boulders are left where rocks transported within glaciers have been gently deposited as the ice melts. Photograph GSC 2007-196 by L. A. Dredge.

edge of the peninsula, variously tilted and uplifted, and others being the products of seafloor spreading at submarine fracture zones. The larger and more interesting of these islands are discussed here.

ISLANDS TO THE WEST OF THE PENINSULA

Isla Guadalupe

Isla Guadalupe is an oceanic island at latitude 29° N, with an area of 244 km² and an elevation of 1295 m, and distant some 259 km from the nearest point on the Baja California Peninsula. It is the emergent tip of submarine shield volcanoes of plate-boundary origin and Miocene age. Guadalupe had evolved an extremely interesting flora and vertebrate fauna (birds only; no mammals reached the island, and the only herpetological record is an old one of treefrog tadpoles with no further substantiation). At least 20% of the native biota is endemic at genus (e.g., *Baeropsis*, Asteraceae; *Hesperelaea*, Oleaceae), species (e.g., Caracara *Polyborus lutosa*, Falconidae), or subspecies level; snails and insects may be similarly distinct. The endemic shrubby tarweeds *Deinandra* (first studied by Sherwin Carlquist) have radiated from ancestral annuals in a similar fashion to the classical tarweed of the Hawaiian islands. Mere remnants of the biota remained into the twentieth century, and fewer still survive to the present. Around two centuries of residence by goats almost completely denuded the island, such that many endemics (and dependent fauna) became extinct (26 of 154 native species) and much of the remainder nearly so. To add to the debacle, goatherds deliberately hounded to extinction the endemic caracara (perceived as a goat predator) by around 1901. Just three of the original nine endemic bird taxa are still extant: the rock wren *Salpinctes obsoletus guadalupensis*, the house finch *Carpodacus mexicanus amplus*, and the junco *Junco insularis*.

A recently instigated (2003) goat removal program was successful, and it promises a limited rebound among the plants (e.g., with recent rediscoveries of taxa thought extinct, and renewed regeneration in the endemic trees *Pinus radiata binata, Cupressus guadalupensis*). The island is a now a biosphere reserve receiving international attention and a pinniped sanctuary (with hopeful great white sharks *Carcharodon megalodon* in attendance—another tourist attraction!); extant marine mammals include seals, sea lions, elephant seals, and the one remaining (relictual) breeding colony of Guadalupe fur seal *Arctocephalus townsendi*.

Isla de Cedros

At latitude 28° N, Isla de Cedros is the largest island west of Baja California (360 km², elevation 1200 m), 19 km northwest of the tip of the Vizcaino Peninsula, to which it was attached during late Pleistocene times when sea levels were lower. It supports a diverse flora of conspicuous (relictual) northern affinities, including pine forest (*Pinus radiata*) at higher elevations, California juniper (*Juniperus californicus*), and various chaparral shrub taxa (e.g., *Arctostaphylos*) on lower slopes. Around a dozen plant species are endemic at the species level; there are also endemic species of reptiles (horned-lizard *Phrynosoma*, alligator-lizard *Elgaria*, rattlesnake *Crotalus*) and mammals (pocket mouse *Chaetodipus*, packrat *Neotoma*) as well as a plethora of endemic subspecies (especially in other taxa indicative of landbridge status, such as deer and cottontails). The island's birds, unsurprisingly, are not distinct. There is a substantial human presence on the island, with many of the 10,000 souls occupied with inshore fishing, shellfish exploitation, and canneries; recent archaeological excavation shows basically similar occupations in sizable human populations that thrived there 11,000 years ago. Among the usual wide variety of introduced mammals, feral cats and dogs pose the most serious conservation threats.

ISLANDS TO THE SOUTH OF THE PENINSULA

Las Tres Marías

The Tres Marías (Maria Madre, Maria Magdalena, and Maria Cleofas, with the much smaller San Juanito lying to the north) are an island trio at 21°30′ N, with a total area of 245 km² and a maximum elevation of 615 m. They are situated 100 km off Nayarit, western Mexico, and around 350 km southeast of Cabo San Lucas at the tip of the Baja California Peninsula; before rising sea levels partitioned them, they were likely parts of a single large (80 km²) island. They are underlain mostly by uplifted Miocene volcanics of plate-boundary origin, although shallow mainland-island channel depth assures landbridge status. There is extensive and varied habitat on the islands, from tall tropical dry forest and shorter deciduous woodland to low thornscrub, comparable to vegetation on the adjacent mainland (e.g., south of Puerto Vallarta, Jalisco, Mexico).

Some 40 breeding bird species have been recorded, 16 represented by endemic subspecies. All populations are thought to be quite healthy except that of the blue-rumped parrotlet *Forpus cyanopygius insularis*, which may be close to extinction. There is an endemic raccoon (*Procyon insularis*), cottontail *Sylvilagus graysoni*, bat *Myotis findleyi*, and mouse *Peromyscus madrensis*, but the endemic rice rat *Oryzomys nelsoni* is now thought extinct, replaced apparently by ubiquitous *Rattus rattus*. Around a thousand people are resident, most on Maria Madre and involved with a penal colony. Inmates cultivate agave to supply their henniquen mill, an operation that is an obvious threat to the persistence of the native biota.

Las Islas Revillagigedo

Some 480 km just west of south from Cabo San Lucas is Isla Socorro (132 km^2; elevation 1130 m), largest of the oceanic Islas Revillagigedo archipelago and around 600 km from the closest coast in Colima. It was discovered in 1533 by the Spanish explorer Hernando de Grijalva and named after the 53rd viceroy of New Spain. Of the four islands between latitudes 18° and 19°20′ N, the second largest, Clarión (20 km^2; elevation 335 m), lies a further 370 km to the west. Their volcanic origins along the Clarión Fracture Zone were highlighted in August 1952 with the birth of Volcan Bárcena on the third-largest and more northerly Isla San Benedicto (6 km^2); in the ensuing six months, 300 million m^3 of fresh tephra and lava added both height and area to this islet.

Clarión's native vegetation is composed mostly of grasses and scrubby cacti, but on Socorro endemic trees of *Bumelia, Ilex, Guettarda,* and *Psidium* form low and dense woodlands at higher elevations. The native reptiles (other than marine sea turtles) are an endemic snake *Masticophis anthonyi* on the older Clarión and two endemic brush lizards, *Urosaurus clarionensis* on Clarión and *U. auriculatus* on Socorro. These brush lizards are not closely related, apparently representing independent invasions from the mainland. Among the ten species of nesting seabirds, one (Townsend's shearwater, *Puffinus auricularis*) is endemic, persisting on Socorro but thought to be near extinction on Clarión, where feral pigs excavate its burrows. Fifteen species of landbirds are all endemic at generic, species, or subspecies levels, and none are shared among the three larger islands. San Benedicto lost its endemic rock wren *Salpinctes obsoletus exsul* after the volcanic eruption there. Two nonendemic landbirds, both on Socorro, appear to be playing out a recurrent island biogeographic theme: Repeated invasion of an island by the same mainland taxon causes the demise of the resident endemic relative and drives a "taxon cycle." Northern mockingbird *Mimus polyglottos* reached Soccoro sometime between the 1950s and 1978; it has become common since, during which time the endemic *Mimoides graysoni,* the product of an earlier mockingbird invasion, has become rare and restricted. During the same pre-1978 period, mourning dove *Zenaida macroura* reached Socorro, where earlier colonization had produced an endemic species *Zenaida graysoni*. Mourning doves have become abundant on Socorro, whereas the endemic species has declined precipitously and may now be extinct. These might be considered natural extinctions—inevitable faunal turnover—but casual introductions of herbivores and predators via the small naval garrisons on Socorro and Clarión (house mice, cats, and sheep, with pigs and rabbits on Clarión as well) likely accelerate the process.

ISLANDS TO THE EAST OF THE PENINSULA

Islands in the Sea of Cortés have received much attention over the last several decades from biogeographers and ecologists, as well as conservationists, ecotourists, and politicians. Travel to the region is now well served by land, sea, and air, and the spectacular natural beauty and biological wonders of the area are readily available for study, admiration, or exploitation.

The two largest islands, both in the northern gulf between 28°30′ N and 29°30′ N, are Ángel de la Guarda (west) and Tiburón (east). These and the several islands nearby (Partida Norte, Rasa, Salsipuedes, San Lorenzo, San Esteban) constitute the "Midriff," and biotic affinities change sharply from east to west, being mainland Sonoran on Tiburón, intermediate on San Esteban, and peninsular on Ángel and the remaining islands. North of this area the gulf is shallow, filled with sediments from the Colorado River and bordered by broad alluvial fans. From Ángel (area 936 km^2) south to San Lorenzo, the islands are faulted blocks separated from the peninsula by channels of modest width (12–20 km), but with troughs and basins reaching depths of up to 1400 m, some of which are active spreading centers. Tiny Rasa (0.68 km^2) is volcanic and formed of basalts derived from seafloor spreading; centrally located San Esteban is comprised of Miocene volcanics with a similar genesis. Across by the Sonoran coast, Tiburón is both very large (1224 km^2) and a landbridge island with prior mainland connections; it has essentially a mainland flora (298 species) and fauna (reptiles: 29 species; mammals: 14 species; and landbirds: 34 species; totals far richer than those of other Sea of Cortés islands).

From latitude 27° N south to 24° N and closely adjacent to the peninsula is a series of landbridge islands: San Marcos, Coronados, Carmen, Danzante, Santa Cruz, San Diego, San José, and San Francisco, with Monserrat of questionable classification. The troughs and basins, with depths to 2400 m, here lie to the east in the center of the gulf. Northeast of the city and bay of La Paz are Espíritu Santo and Partida Sur, nearly united and likewise landbridge islands across a channel 6 km across and 12 m deep from the peninsula. Carmen, San José, and Espíritu Santo/Partida Sur are the largest of these islands, at 143, 187, and 107 km^2, respectively; almost all are faulted and variously uplifted blocks.

Several remaining islands are deepwater islands and are more interesting because of their increased isolation over time and space. South of the Midriff, Isla San Pedro Mártir (2.9 km²) in the central gulf is a towering (320 m) basement rock, a biosphere reserve, and a celebrated breeding site for seabirds and marine mammals. Further south Isla Tortuga (latitude 27°26′ N, 11.4 km²) sits in water 1200 m deep and, like Rasa, is a product of recent (Holocene) volcanic activity. At latitude 25°39′ N, Santa Catalina is another block-faulted island, 41 km², which in Pleistocene times lay just a few kilometers east from the now-submerged peninsular shelf. Lastly, the southernmost island of Cerralvo is old (Pliocene), large (140 km²), and high (767 m), with no record of past peninsular or mainland connections. It is thus an island on which, a priori, interesting biology might be expected, with expectations fulfilled by numerous endemic plant, lizard, snake, and mammal species, and even a locally endemic lizard genus (*Sator*; Iguanidae).

The prevalent vegetation throughout the Sea of Cortés region is Sonoran Desert scrub, open and xeric with a preponderance of stem-succulent cacti, drought-deciduous shrubs, and short, phreatophytic trees (Fig. 1). Some 80–150 mm of annual precipitation sustain life here; most is winter rainfall except in the south (e.g., near Cerralvo), where summer precipitation increases, and a taller, denser thorn-scrub prevails. With many recent (late Pleistocene) land bridge connections, similar conditions island to mainland, and modest isolation distances, there is very little endemism in the more vagile taxa, namely plants and birds. About 650 plant species are found on the islands; 17 of them endemic, 8 of these cacti (*Echinocereus,*

FIGURE 1 Sonoran Desert scrub vegetation is typical of most of the islands in the Sea of Cortés. Note the prevalence of cacti, such as the giant tree-like cardon cacti (*Pachycereus pringlei*), the sprawling *Stenocereus gummosus* in the left foreground, and the bushy cholla (*Opuntia cholla*) behind it.

Ferocactus, Mammillaria), and are arguably no more than single-island growth form variants. Of more interest are plants whose present island occurrences reflect historical legacies. When Mojave Desert vegetation moved south onto the Baja California Peninsula in cooler and wetter Pleistocene climates, several typical Mojave Desert taxa colonized Ángel de la Guarda: for example, the phreatophyte *Acacia greggii,* the shrub *Gutierrezia microcephala,* the composite forb *Trichoptilum incisum,* and the annual *Plagiobothrys jonesii*. When peninsular relatives retreated north with the onset of warmer, drier times, insular populations remained marooned on that island.

There are, unremarkably, only a few weakly defined subspecies in the island avifauna, composed of typical Sonoran Desert birds that are island-area dependent in terms of species richness and even species identity. The thrasher genus *Toxostoma* has allopatric ecological counterparts across the gulf, the curve-billed thrasher *T. curvirostre* to the east and the gray thrasher *T. cinereum* to the west. Each occupies several gulf islands but no islands are co-occupied. Several seabird species are near-endemic to the gulf in their breeding ranges; one such is Heermann's gull *Larus heermanni,* of which 200,000 congregate on Rasa yearly and defend nesting territories a meter or so across. Following El Niño winters, food availability declines with the higher sea surface temperatures, females lose body weight, and chick survival drops to near zero; similar trends are recorded in other breeding seabirds such as elegant terns, *Sterna elegans,* and brown pelicans, *Pelecanus occidentalis.*

Birds are regarded as good colonists because they fly, but their high metabolic rates and short longevities translate to poor persistence. Mammals also are poor persisters for the same reason, but in addition, they are poor colonists (excepting bats, they do not fly). On the gulf islands mammals are scarce overall, and many species are found only on land bridge islands. This is illustrated by taxa such as shrews (*Notiosorex* on Tiburón), jackrabbits (*Lepus* on Carmen, endemic species on Espíritu Santo), cottontails (endemic *Sylvilagus* species on San José), coyote *Canis latrans,* ringtail *Bassariscus astutus,* and mule deer *Odontocoileus hemionus* on San José as well as Tiburón, and many others.

The packrat *Neotoma lepida* reached one non–land bridge island (Ángel), but the only taxon with a respectable colonization record throughout the gulf islands is the mouse *Peromyscus,* with ten endemic species and numerous subspecies described. However, most taxa are single-island records, and all islands except Tiburón have but a single taxon. Thus, there has been a substantial evolution of island forms, but no adaptive radiation.

From an island biogeography viewpoint, the most interesting animals on Sea of Cortés islands are reptiles. They colonize infrequently but persist well, even on small, unproductive islands; thus, they readily form endemics (one-fourth to one-third are endemic at the species level). The endemic genus *Sator* occurs on three islands: *S. grandaevus* on Cerralvo and *S. angustus* on adjacent San Diego and Santa Cruz. The gecko *Phyllodactylus,* the side-blotched lizard *Uta* (with six described endemic species), and several snake genera have demonstrated an ability to colonize islands over water. The Sonoran (mainland) western diamondback rattlesnake has reached San Pedro Mártir, where it is an island dwarf, and also Tortuga, where it constitutes an endemic species *Crotalus tortugensis*. The endemic San Esteban rattlesnake likewise is a dwarf and also is derived from a Sonoran species, the black-tailed *C. molossus*. All the preceding are one-snake islands, but on Ángel there are two rattlesnake species. Peninsular red rattlesnake *C. ruber* is present but as a dwarf, much smaller than peninsular specimens. On the other hand, the speckled rattlesnake derivative there, from peninsular *C. mitchelli,* is a veritable giant. The two have reversed positions on the island, relative to the peninsula, in a size segregation sequence.

Dramatic body size shifts are evidenced in island chuckawallas *Sauromalus*. Whereas peninsular and mainland chuckawallas *S. obesus* average 300 g in mass, the endemic island derivative *S. hispidus* on Ángel de la Guarda and neighbors is a 1400-g giant, and another endemic, *S. varius* on San Esteban, is truly huge (1800 g). These herbivorous lizards reproduce successfully only in El Niño years, when higher rainfall produces more edible plant material (i.e., a pattern opposite that of seabirds). The island forms show rapid growth rates and occur in high densities; they are thought to evolve gigantism in response to the absence of large predators on islands, and possibly also to a general lack of herbivorous competitors. One historic predator was the Seri Indians, who visited at least the Midriff islands on foraging trips. The giant chuckawallas must have made attractive and easy prey, so much so that the Indians are known to have transplanted the lizards to chuckawalla-free islands to make living food caches, just as the old ocean-going mariners did with goats.

SEE ALSO THE FOLLOWING ARTICLES

Gigantism / Pigs and Goats / Snakes / Taxon Cycle / Vegetation

FURTHER READING

Axelrod, D. 1979. Age and origin of Sonoran Desert vegetation. *Occasional Papers of the California Academy of Sciences* 132: 1–74.
Baldwin, B. A. 2007. Adaptive radiation of shrubby tarweeds (*Deinandra*) in the California Islands parallels diversification of the Hawaiian silversword alliance (Compositae-Madiinae). *American Journal of Botany* 94: 237–248.
Brattstrom, B. H. 1990. Biogeography of the Islas Revillagigedo, Mexico. *Journal of Biogeography* 17: 177–183.
Case, T. J., M. L. Cody, and E. Ezcurra. 2002. *A new island biogeography of the Sea of Cortés*. Oxford: Oxford University Press.
De la Luz, J. L. L., J. P. Rebman, and T. Oberbauer. 2003. On the urgency of conservation on Guadalupe Island, Mexico: is it a lost paradise? *Biodiversity and Conservation* 12: 1073–1082.
Des Lauriers, M. R. 2006. Terminal Pleistocene and early Holocene occupations of Isla de Cedros, Baja California, Mexico. *Journal of Island and Coastal Archaeology*, 1: 255–270.
Jehl, J. R., and K. C. Parkes. 1982. The status of the avifauna of the Revillagigedo Islands, Mexico. *Wilson Bulletin* 94: 1–19.
Mellinck, E. 1993. Biological conservation of Isla de Cedros, Baja California, México: assessing multiple threats. *Biodiversity & Conservation* 2: 62–69.
Shreve, F., and I. L. Wiggins. 1964. *Vegetation and flora of the Sonoran Desert*. Stanford, CA: Stanford University Press.
Stager, K. 1957. The avifauna of the Tres Marias Islands, Mexico. *Auk* 74: 413–432.

BALEARIC ISLANDS

SEE MEDITERRANEAN REGION

BARRIER ISLANDS

MILES O. HAYES

Research Planning, Inc., Columbia, South Carolina

Barrier islands are elongate, shore-parallel accumulations of unconsolidated sediment, some parts of which are situated above the high-tide line (supratidal) most of the time, except during major storms (see example in Fig. 1). They are separated from the mainland by bays, lagoons, estuaries, or wetland complexes and are typically intersected by deep tidal channels called tidal inlets. A large percentage of the major barrier islands of the world occur along the coastlines of the trailing edges of continental plates and of epicontinental seas and lakes (e.g., Caspian and Black Seas). Because they are composed of unconsolidated sediments (primarily sand, with gravel being present in some Arctic regions), they most commonly occur on coastal plain and deltaic shorelines (depositional coasts), where the sediment that makes up the islands was ultimately brought to the shore by rivers and streams. Some barrier islands do occur, primarily as spit forms, on leading edges of continental plates and on some glaciated coasts, but they are a minority coastline type in those areas.

FIGURE 1 Oblique view with infrared film looking southwest at Kiawah Island, South Carolina, which clearly illustrates the drumstick-like configuration of the 13.7-km-long, prograding barrier island. Note the presence of linear ridges of sand vegetated by maritime-forest, which indicate the positions of the backbeach foredunes at earlier stages in the growth of the island. Photograph taken by Dennis K. Hubbard on March 18, 1976. Since that time, this island has been developed for residences and golf courses.

OCCURRENCE AND IMPORTANCE

The largest single chain of barrier islands in the world occurs along the East and Gulf Coasts of North America. Many of the largest of these North American barrier islands are highly developed with human habitation, some with moderate-sized cities, such as Galveston, Texas, and Atlantic City, New Jersey. Because of the dynamic nature of the coastal zone, beach erosion is a serious problem for some of the more developed barrier islands, invoking major engineering efforts to stem the erosion. One commonly applied technique is the addition of massive volumes of sand derived from elsewhere along the eroding shoreline, a process called beach nourishment. The largest of these beach nourishment projects commonly cost many millions of dollars. Some of the islands have been preserved as national and state parks, to which vacationers flock in season. Consequently, barrier islands have a high socioeconomic profile in North America, especially for the large percentage of the population that lives in coastal areas. Barrier islands also occur on the shorelines of northwestern Europe, the Mediterranean and Caspian Seas, West Africa, and elsewhere.

ROLE OF TIDES, WAVES, AND SEA LEVEL

Depositional coasts have characteristic morphology and sediment distribution patterns controlled by the interaction of waves and tides, with the magnitude of the tides being of particular importance. Accordingly, depositional coasts are commonly classified as microtidal (tidal range, or T.R. = 0–2 m), mesotidal (T.R. = 2–4 m), and macrotidal (T.R. = > 4 m). As a generalization, depositional features on microtidal coasts are highly influenced by waves (wave-dominated coasts), whereas those on macrotidal coats are highly influenced by tides (tide-dominated coasts) and those on mesotidal coasts respond to the effects of both waves and tides (mixed-energy coasts). For example, barrier islands do not occur on open-ocean, coastal-plain shorelines with tidal ranges greater than about 4 meters (macrotidal coasts). This is because their primary mechanism of formation, wave action, is not focused long enough at a single level during the tidal cycle to form the island. Furthermore, the strong tidal currents associated with such large tides transport the available sediments to the offshore regions.

Barrier islands exposed to open-ocean waves and tides that are in a progradational mode (i.e., consistently building in an offshore direction) show major differences depending on whether the tides are microtidal or mesotidal. Prograding barrier islands along microtidal shorelines are long and linear, commonly over 25–50 km in length, with a predominance of storm washover features. Those on mesotidal shorelines are stunted, usually less than 16 km in length, with an abundance of large tidal inlets. More tidal inlets are required on mesotidal coasts, because of the large amount of water that moves into and out of the backbarrier regions during a single tidal cycle. During major storms with significant storm surges, microtidal barrier islands are usually washed over and permanent washover fans are formed (e.g., those on the Texas coast), whereas on mesotidal barrier islands, permanent washover fans are not so common, because the system is already adjusted to major influx and outflow of ocean waters during normal spring tides (e.g., those on the Georgia and South Carolina coasts).

The long-term patterns of morphology and sedimentation on most coastal plain and deltaic shorelines have been significantly impacted by the major changes in sea level that occurred during the glacial episodes of the Pleistocene Epoch. During each major glaciation, sea level was lowered significantly, over 100 m during the last glaciation (Wisconsin). When the sea level was lower, major valleys, called lowstand valleys, were carved across the coastal plains and continental shelves. As sea level rose, the valleys were flooded to become major estuaries.

MORPHOLOGY AND STRATIGRAPHY

A major consideration with regard to the morphology and stratigraphy of barrier islands is whether they consistently migrate landward (transgressive) or build in an offshore direction (prograding). The general patterns of

FIGURE 2 View of the southwest end of Cape Romain, South Carolina a few days after the passage of Hurricane Hugo (1989). The white band of sediment, a mixture of sand and shell, is a washover terrace that advanced landward some 10 m during the storm. The dark layer seaward of the washover terrace is exposed, muddy backbarrier sediments. As a result of the landward migration of this transgressive barrier island, a new tidal inlet was created where it intersected the large tidal channel in the foreground.

prograding barrier islands with respect to the effect of tidal range were discussed in the previous section of this article. However, both types of barrier islands, transgressive and prograding, may be present on either microtidal or mesotidal coastlines, depending upon the rate of sea-level change relative to sediment (usually sand) supply at that location (not the tidal range). Diminished, or low, sand supply and rapid sea-level rise both promote the development of landward-migrating islands, and *vice versa* for those that build in an offshore direction. Both prograding and transgressive barrier islands clearly tend to change to some extent over time, with the rates of landward migration and offshore growth being different from place to place. In some parts of the South Carolina coast, for example, the landward-migrating barrier islands may move 3 m or more per year (example in Fig. 2). However, those that build seaward usually grow more slowly, at rates of < 3 m per decade (example in Fig. 1).

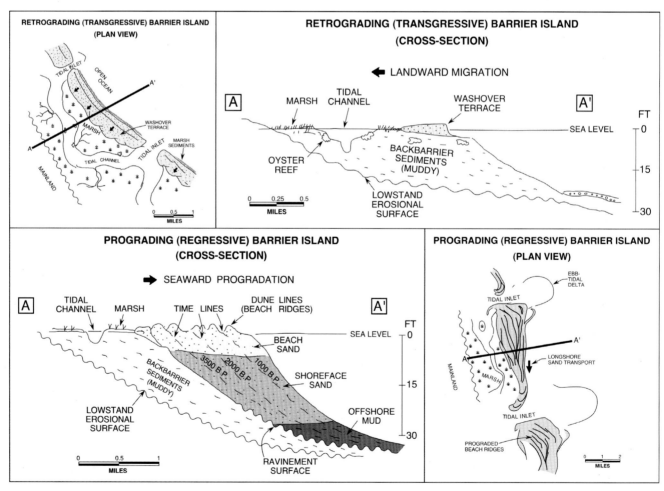

FIGURE 3 Morphology and subsurface three-dimensional configuration (stratigraphy) of prograding and transgressive barrier islands.

As illustrated in Figs. 2 and 3, transgressive (landward-migrating) barrier islands are composed of coalescing washover fans, or a washover terrace, that is overtopped at high tides, usually several times a year. In the process of migration, the entire washover terrace complex moves landward, leaving an eroded nearshore zone in its wake. As a result of this type of migration, in three dimensions the entire complex consists of a relatively thin (< 1–3 m) wedge of sand and shell of the washover terrace, which overlies muddy sediment originally deposited in the lagoons or wetlands landward of the islands (see cross section in Fig. 3). Because of their continual landward migration, these types of islands are, needless to say, impractical sites for human development. The transgressive (landward-migrating) barrier islands in South Carolina are relatively short, 2–8 km on the average, because new inlets are created where the migrating islands intersect tidal channels (see Fig. 2).

Prograding (seaward-building) barrier islands (Figs. 1 and 3) are typically composed of multiple, relatively parallel linear ridges of sand topped by vegetated sand dunes that originally formed as front-line dunes on the back-beach (called foredunes). The most notable changes on these types of islands occur where adjacent tidal inlets migrate into them or when the inlets expand dramatically during hurricane storm surges. As a result of their offshore growth, in three dimensions these types of barrier islands typically consist of a wedge of sand 7–9 m thick that has built over offshore muds (see cross section in Fig. 3). Most of the major developed barrier islands along the east coast of the United States, which typically are greater than 16 km long, are of this type (e.g., Kiawah Island and Hilton Head Island, South Carolina). When human development occurs on these types of islands, buildings are usually secure from all but the most extreme hurricanes if they have been set back an adequate distance from the front-line dunes and tidal inlets. That security will vanish, however, if a major rise of sea level occurs in the near future as a result of global warming.

The morphology of the prograding barrier islands longer than about 11 km takes on a characteristic drumstick appearance, as shown in Fig. 4. This pattern is most common on mesotidal shorelines. Two factors that enhance the development of the drumstick shape are:

1. The occurrence of significant masses of sand in the form of large, wave-built intertidal sand bars (swash bars) that develop along the outer margin of a large lobe of sand deposited on the seaward side of the tidal

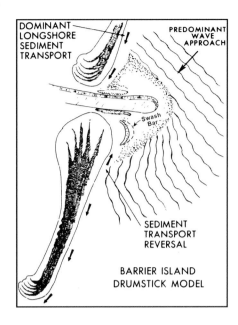

FIGURE 4 Barrier island drumstick model, primarily the result of welding of masses of sand derived from the ebb-tidal delta in the form of large swash bars. A sediment transport reversal resulting from the refraction of the dominant waves around the ebb-tidal delta is also a factor. These types of barrier islands are most common on prograding, mesotidal (tidal range = 2-4 m) shorelines.

inlet by ebb-tidal currents (called the ebb-tidal delta). These huge swash bars eventually move toward shore and weld to the beach (Fig. 4). This welding process builds out the end of the island that faces the direction from which the sediments come, accentuating its drumstick shape (see photographs in Figs. 1 and 5).

2. Refraction of the dominant waves around the ebb-tidal delta, a process that enhances deposition on the same end of the island where the huge swash bars come ashore. The refracted waves create currents that transport sediment in the opposite direction from that on the open coast. This is a relatively minor reversal from the normal longshore transport direction (see model in Fig. 4), but it allows sand to remain in the inlet area and aids in its accumulation on that end of the island.

ROLE OF CLIMATE

Climate plays a significant role in the nature of barrier islands, not so much regarding their morphology and stratigraphy, as was demonstrated for tides and waves, but in the production of sediment types, occurrence of storms, and the types of vegetation present. This is especially true in extreme climates, such as polar regions. For example, the barrier islands on the North Slope of Alaska are eroding away at alarming rates. This erosion continues despite the fact the Arctic Ocean is frozen for many months of the year, producing limited fetch for the waves, hence relatively small waves, and a short season for waves to occur. Even in August, blocks of ice sometimes occur near shore. Composed mostly of gravel, these islands are short (average length < 3 km) and low, with numerous wide inlets. Warming of the Arctic Ocean may be playing a role in the islands' demise, with melting of the permafrost possibly being a major factor.

FIGURE 5 View looking east of a prograding, drumstick-shaped barrier island, Egg Island, Alaska (on the Copper River Delta). Note the recurved spits on the western end of the island, clear evidence of the strong east-to-west longshore transport of sediment along the barrier islands in this area. Also note where several large swash bars have welded to the eastern end of the island (compare with model in Fig. 4). Photograph of this mesotidal barrier island taken at low tide in the spring of 1975.

In the other extreme, barrier islands in hot, arid regions are more prone to have barren sand dunes and extensive sand transport across them. In shallow seas with warm water and high salinities, the sediments of the islands beaches may be cemented with *beachrock*. As the satellite image of the shoreline of Abu Dhabi in Fig. 6 shows, although probably the hottest and driest barrier island chain in the world, these islands have many of the characteristics of mesotidal barrier islands (the tidal range in Abu Dhabi is 2.5 m). Features such as stunted islands, large tidal inlets with huge ebb-tidal deltas, and complex backbarrier regions stand out. However, typical inlet migration and beach erosion patterns are inhibited by the *beachrock*, producing some jagged shorelines cemented in place. Also, the ebb-tidal delta sediments are composed of carbonate oolite sand, another signature of arid tropic regions.

ORIGIN OF BARRIER ISLANDS

The origin of barrier islands has been a matter of conjecture in the geological literature for well over 100 years. The processes of beach accretion between storms and the formation of a foredune landward of the beach, primary components of barrier islands, are well understood, and their common occurrence on depositional shorelines is not surprising. The aspect of barrier islands that is most difficult to understand is the fact that they occur offshore of the mainland separated by a topographically low area, a lagoon or estuarine complex in most cases. Any theory for the origin of these islands must account for their mysterious offshore location.

Numerous hypotheses for the origin of barrier islands have been proposed, including the following three examples.

1. Elongation of sand spits away from some kind of headland, with segmentation of the spits as they grow as a result of the formation of permanent tidal inlets through them during storms, creating individual islands separated from the mainland by an open-water lagoon (as illustrated in Fig. 7).
2. Elevation of an offshore bar or flooding of a line of foredunes along the shore.
3. The transgressive–regressive interfluve hypothesis, which has been well documented on the coasts of North Carolina and South Carolina.

One of the major proponents of the spit elongation hypothesis, John Fisher, cited the northern part of the Outer Banks of North Carolina as one of his examples of barrier islands that have been formed by that mechanism. This mode of origin has been suggested for other areas in the world, for example those islands on the central Texas coast. This clearly is one way barrier islands can form, but not the only one.

There is no doubt that many barrier islands have formed without the aid of spit elongation. Two major ideas have been proposed to account for the elevation and permanence of barrier islands independent of spit elongation. Some observations along the Gulf Coast of the United States illustrate that during the elevated tides and unusually high water levels that accompany hurricanes, a major offshore bar may be formed by the large waves. When the water level recedes, the bar emerges and, under the right circumstances, may survive to become a barrier island. A second idea is that a line of typical foredunes forms along the beach during a stillstand or slowly rising sea level. A relatively abrupt rise in sea level floods the line of foredunes. During this abrupt rise, the foredunes are not completely eroded away by the waves. The remnant of the original dune line becomes an island. Both of these processes seem reasonable, but their documentation is still somewhat in question.

The mode of formation for most of the larger prograding barrier islands on the South Carolina coast, such as Kiawah Island, is clear and well documented in studies by Tom Moslow and D. J. Colquhoun. Four major steps take place in this mode of formation:

1. A narrow, landward-migrating barrier island moved rapidly across what is now the inner continental shelf, leaving behind a thin lag of coarse material on top of an erosion surface across the continental shelf, called the transgressive surface of erosion.
2. The topography over which the shoreline advanced was irregular, and estuarine waters flooded the numerous river valleys formed when the shoreline was further offshore. Isolated, primary transgressive barrier islands, consisting of washover terraces composed of coarse-grained sand and shell, continued to migrate landward on the exposed interfluves between the drowned lowstand valleys.
3. When sea level stopped rising and a relative stillstand occurred about 4500 years ago, shoals developed at the entrances of the estuaries created by the drowning of the valleys, and a longshore sediment transport system was initiated along the face of the stranded barrier islands. Over time, beach ridges began to develop, eventually impinging upon the adjacent estuary entrances. As a well defined inlet throat evolved, a shoal off the entrance (ebb-tidal delta) formed, around which sediment was bypassed, augmenting beach-ridge growth on the adjacent barrier island.
4. As the barrier island matured, and minor fluctuations of sea level occurred, parts of some of the originally prograding beach ridges were eroded as a result of tidal-creek and tidal-inlet migration.

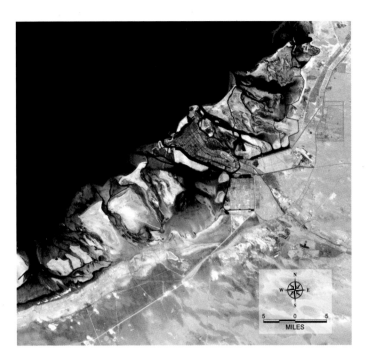

FIGURE 6 Satellite image of the barrier island chain in Abu Dhabi acquired in 2000, courtesy of Earth Science Data Interface (ESDI) at the Global Land Cover Facility. These barrier islands show many of the characteristics of mesotidal barrier islands—short length, large ebb-tidal deltas, and absent flood-tidal deltas (tidal range = 2.4 m this area). The huge ebb-tidal deltas are composed of carbonate sand (with abundant oolites).

The end result of all this was a prograding, drumstick-shaped barrier island, such as the ones illustrated in Figs. 1, 4, and 5. However, this hypothesis does not account for how the transgressive element, the original barrier island, formed.

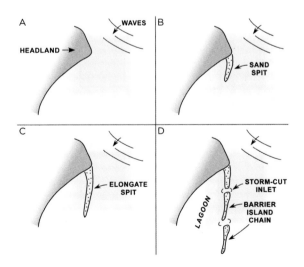

FIGURE 7 The spit elongation hypothesis for the origin of barrier islands, based on ideas presented by Fisher (1967) and others.

SEE ALSO THE FOLLOWING ARTICLES

Beaches / Hurricanes and Typhoons / Sea-Level Change / Tides

FURTHER READING

Curray, J. R. 1964. Transgressions and regressions, in *Papers in marine geology: Shepard commemorative volume*. R. L. Miller, ed. New York: MacMillian and Co, 175–203.

Davis, R. A. Jr., ed. 1994. *Geology of the Holocene barrier island systems*. Berlin: Springer-Verlag.

Fischer, J. J. 1967. Origin of barrier island chain shoreline, Middle Atlantic States. *Geological Society of America Special Paper* 115: 66–67.

Hayes, M. O. 1979. Barrier island morphology as a function of tidal and wave regime, in *Barrier islands, from the Gulf of St. Lawrence to the Gulf of Mexico*. S. Leatherman, ed. New York: Academic Press, 1–27.

Hoyt, J. H. 1967. Barrier island formation. *Geological Society of America Bulletin* 78: 1125–1136.

Pilkey, O. H., and M. E. Fraser. 2003. *A celebration of the world's barrier islands*. New York: Columbia University Press.

BARRO COLORADO

EGBERT GILES LEIGH, JR.

Smithsonian Tropical Research Institute, Balboa, Panama

Barro Colorado is a 1500-ha island situated at 9°9′ N, 79°51′ W in central Panama, first isolated from the surrounding mainland in 1914 by the rising waters of Gatun Lake, after the Chagres River was dammed to form part of the Panama Canal. The island is covered by seasonal tropical forest, half of it old growth, which offers beauty and fascination enough to fill any biologist's lifetime. Barro Colorado's primary claim to the reader's attention is its contribution to our understanding of tropical biology.

HISTORY

The governor of the Canal Zone declared this island a reserve in 1923 in response to two groups. Biologists from Harvard and the American Museum of Natural History wished to preserve this island for their research, whereas Canal Zone entomologists, who were spraying mainland areas to eliminate malaria-carrying mosquitoes, wished to keep the island unsprayed, a standard for measuring the effectiveness of their spraying. Their standard was high. Most mosquitoes on this island breed in water-filled tree holes, which also shelter resident nymphs of 100-mm-long giant damselflies. These nymphs eat most of the wrigglers, keeping Barro Colorado relatively free of mosquitoes.

Barro Colorado's first buildings were erected in 1924. James Zetek, an entomologist, directed the laboratory during its first, impoverished, 30 years. The Smithsonian was given charge of Barro Colorado in 1946. In 1957, Martin Moynihan, a student of animal behavior, was appointed resident naturalist. He set about assembling a staff of resident scientists. In 1966, the Smithsonian Tropical Research Institute, including Barro Colorado and marine laboratories on both coasts, was declared a bureau of the Smithsonian Institution. Moynihan was extraordinarily perceptive and was interested in nearly everything. He distrusted team research, believing that the best ideas come from people pursuing projects they devised themselves. He established a sound intellectual basis for further research at his Institute. Ira Rubinoff succeeded him in 1973, established agreements with the Republic of Panama for the Institute's continued activity, broadened the range of scientific specialties at the Institute, and built the laboratories needed to support this research.

THE CENTRAL PROBLEM OF ANIMAL BEHAVIOR AND ITS ANALOGUES

An early theme of Barro Colorado's research was animal behavior. In the 1930s, C. R. Carpenter was studying howler monkeys, which live in groups of 15 or more, and T. C. Schnierla was studying army ants, whose workers cooperate in organized swarms of up to a half million ants to overcome prey. A central problem of animal behavior is why animals live in groups. An animal's fellow group members are usually its closest competitors for food, shelter, and mates: What keeps competition among a group's members from becoming a cheating contest? This problem has many analogues. One is far older than Plato: How can the advantage of individuals be aligned with the good of their society? Second, a person's body consists of a multitude of cells that normally cooperate for that person's good, yet rogue cell lines—cancers—can spread and kill that person. Why are cancers not much more common? Third, partnerships between members of different species confront similar tensions. For example, each of our cells has mitochondria, organelles that turn sugar into energy. These mitochondria descend from parasitic bacteria that invaded our one-celled ancestors over a billion years ago. Conflicts of interest between cells and their mitochondria sometimes erupt, but why so rarely?

One partnership between species, a classical example of coevolution, profoundly influences this island's ecosystem. In 1980, Allen Herre began working on fig trees, each species of which maintains one or more species of minute wasp as dedicated pollinators. A fig "fruit" is a flowerhead turned outside in, so the flowers line the inside of a perforated ball. One or more pollen-bearing wasps enter each

FIGURE 1 A view of mature forest on Barro Colorado Island. From the left, the drought-sensitive species *Poulsenia armata*; the long-lived species *Anacardium excelsum*, which colonizes large tree-fall gaps; and *Quararibea asterolepis*, a shade-tolerant canopy species. At far right, the midstory tree *Gustavia superba*, with rosettes of very long leaves. Drawing by Daniel Glanz.

"fruit," pollinate its flowers, and lay eggs in about half of them. Each wasp larva grows within a single fig seed and metamorphoses into an adult. In each fruit, hundreds of wasps emerge and mate with each other. Fertilized females fly off in search of trees in "fruit" in which to lay their eggs; they pick up pollen and carry it from tree to tree. In the 1990s, John Nason found that although these wasps live less than 72 hours, they often pollinate trees over 10 km away, maintaining great genetic diversity in their fig species even if it is very rare. To maintain its pollinators, each species of fig must always have some trees in fruit. On Barro Colorado, this year-round fruit supply supports ten species of fig-eating bat. The island's 17 fig species bear different-sized fruit. Larger fruit attracts bigger bats, which carry seeds further from parents and their pests. Seedlings further from adults of their species are more likely to grow to a safe size before pests specialized on their species find them, so these plants need invest less in defenses against their pests, freeing resources for faster growth. Wasps die, however, if their fruit overheats: Large fruits must evaporate much water to keep cool, so large-fruited figs must have reliable access to water. This mutualism, however, has its tensions. In 2005, Charlotte Jander began to study how fig trees make their wasps pollinate their fruit. Another question, as yet unanswered, is how fig trees keep their wasps from laying eggs in more than half their flowers.

Barro Colorado's denizens live by such partnerships. Most species of plants need animals that will carry pollen from one plant to another of its species, even when this species is kept rare by its pests. Likewise, most plants need mycorrhizae, root fungi, to help extract mineral nutrients from the soil. Many need animals to disperse their seeds, and perhaps even bury them, beyond the reach of the parents' pests. Barro Colorado's early emphasis on social behavior encouraged study of the many different ways in which the plants and animals of tropical forest depend on each other, and the ways participants defend cooperative relationships against subversion by cheaters.

DIVERSITY OF ANIMALS AND PLANTS

A second research theme is the great diversity of Barro Colorado's plants and animals. The first step in this study is identifying them. The Smithsonian's Paul Standley first catalogued the island's plants in 1927. In 1935, a Swarthmore professor, Robert Enders, summarized what was known of the island's mammals. In 1952, the American Museum's Eugene Eisenmann published a list of its birds. Such work continues.

How can forests like Barro Colorado's harbor so many kinds of plants and animals? One light in this darkness is the "principle of competitive exclusion": If two species make their living in the same way, in the same place, one will be better and will replace the other. Diversity reflects trade-offs, circumstances where enhancing the aptitude for one task diminishes the aptitude for others. Where trade-offs occur, the threat of competitive exclusion drives adaptive divergence in competing populations. Work on Barro Colorado helped to reveal factors driving the trade-off plants face between growing fast in bright light and surviving in shade. This trade-off allows fast-growing species of "weed tree" that survive only in tree-fall gaps to coexist with shade-tolerant tree species of mature forest (Fig. 1). Insects and disease-causing microbes face trade-offs in ability to handle defenses of different kinds of plant. The most damaging pests specialize on particular species or genera of plants.

Pests play a major role in the lives of this island's trees. From observations that he began when he first came to the island in 1967, Robin Foster found that trees of the canopy species *Tachigali versicolor* flower and fruit once, then die. Only every four years do some *Tachigali* have fruit. Yet a weevil's specialist larvae destroy 20% of their fruit. Did trees that fruited a second time lose their whole second crop to these weevils? Work on this island helped to show how, in some tree species, a parent's pests destroy its nearby young, making room for trees of other species to grow up between a parent and those of its young far enough away to survive.

To document and explain the island's tree diversity, Robin Foster and Stephen Hubbell set up a 50-ha Forest

Dynamics Plot in 1980, mapping, marking, measuring, and identifying every free-standing woody stem over 1 cm diameter. In 1982, this plot had 305 species among 235,000 stems over 1 cm in diameter, and 238 species among its 20,881 trees over 10 cm trunk diameter. The plot was recensused every five years from 1985 on. To document tree diversity, plots like it were established all through the tropics. A 52-ha plot in Lambir, Sarawak, had 1008 species among 32,661 trees over 10 cm trunk diameter. Comparing these plots shed light on the causes of tree diversity. In these plots, trees usually reproduce less well where they are more common, so no one kind of tree can crowd out the others, allowing many species to coexist. The next question: Are these forests so diverse because each species is kept rare by its specialist pests?

HERBIVORES AND THE FOREST

The third research theme is, What keeps the animals from stripping the forest bare? Exploring this theme illustrates the importance of "unity of place," showing how many different kinds of study, done in one small area, can cohere into a unified picture.

Starting in August 1969, Robin Foster measured fruit fall on Barro Colorado every week or two, for two years. In both years, little fruit fell from November through February. In striking contrast to 1969, however, little fruit fell from August through October in 1970. Many mammals starved, the vultures could not keep up with the corpses, and the forest stank. Were populations of vertebrate fruit-eaters limited by seasonal shortage of suitable food? This question dominated the next 15 years of research on Barro Colorado.

The staff scientist Stanley Rand established an environmental monitoring program that maintained records on weather, stream flow, soil moisture, fruit fall, the times of flowering, fruiting and leaf flush in many tree species, and the responses to these events of many animal populations. Year-long studies of the behavior of agoutis, pacas, sloths, coatis, howler monkeys, many kinds of bats, and selected birds suggested that their populations were limited by seasonal shortage of fruit or new leaves. The island's mammals were censused. Knowing their numbers and sizes, one can calculate their food requirement. In December and January, the measured fruit fall was too little to feed the ground-dwelling fruit-eaters. True, these animals do not starve then. Indeed, each species has a different fallback specialty, which is why no species is entirely replaced by its competitors. These specialties, however, barely tide them over to the next fruiting season: The season of fruit shortage is obviously a time of dearth.

Especially where many different scientific specialties are represented, questions beget questions. How do different kinds of plants know when to flower, fruit, or flush new leaves? In particular, what keeps most plants from fruiting during the season of shortage? Observations suggest that some plant species flower only after sensing a clear transition from dry to rainy season: If the dry season is too wet, they will not flower. Others flower in response to dry season rains, if the soil is dry enough before the rain. To test these ideas, the staff scientist S. Joseph Wright watered two 2.25-ha plots through five successive dry seasons. The behavior of nearly all the larger trees was unchanged by watering. Everyone was astonished by this lack of response. Plant physiologists found that most of the island's big trees had reliable access to water all year long. They must time their activities in response to atmospheric conditions, such as humidity or solar radiation. Some trees even flower in response to seasonal changes in day length. There is still much to learn about how different trees choose the times for their activities.

Insects also eat plant leaves. Mature leaves are too tough for most insects to eat, but young leaves are tender and nutritious. In 1977, Foster's student Phyllis Coley began work that eventually showed that young leaves on Barro Colorado are eaten 20 times more rapidly than their temperate-zone counterparts, despite being more poisonous than the latter. To protect their young leaves, some plants shoot them out and toughen them as soon as possible. Others stock slower-growing leaves with a cocktail of poisons, different for each species. Coley found some of these poisons to be of medical interest. Specialist herbivores deal best with the poisons in their host's leaves. These specialists do the most damage. Do they keep their host species rare enough that many of its plants escape the pest's attentions until big enough to survive them? Barro Colorado's forest loses at least 7% of its leaf production—over a half ton dry weight of leaves per hectare per year—to leaf-eating insects. Judging from the numbers, weights, and diets of the island's birds, a third of this eaten foliage feeds insects birds eat. The forest can defend itself from vertebrates fond of vegetable matter without help from jaguars, but it needs help from birds, wasps, and spiders to deal with its insect pests.

THE CONSEQUENCES OF BECOMING AN ISLAND

The fourth research theme is, What are the consequences of becoming an island? Since Barro Colorado became an island, its species diversity has been declining toward a new, lower equilibrium as predicted by the theory of island biogeography. Barro Colorado is now too small to support white-lipped peccaries, which live in herds of a hundred or more. It is too small to guarantee the presence of very young forest. Accordingly, a pygmy squirrel that lived in such forest died out after its forest matured. In 1982, Barro Colorado's 50-ha plot had 238 species among its trees over 10 cm trunk diameter, but it had only 226 in 2000: Did this happen because the plot is on an island? Many of the island's bird species, 35 edge and 30 forest species, died out after their home became an island. Most of the 30 forest species are understory insectivores. Experiments suggest that in most of the extinct understory species, birds cannot fly 100 m over open water, either because bright sunlight dazzles them or because they lack the stamina to fly so far. When these species die out, they cannot return. On the other hand, Barro Colorado, with 74 species of bat, lacks few of the bat species found on the nearby mainland.

Indeed, Barro Colorado's ecosystem is not sufficient unto itself. Some of its insect populations are replaced from the nearby mainland when they die out. Some North American songbirds spend the winter there. Some bats leave the island during certain seasons: No one knows where they go. The island lies athwart the route of many species of butterfly that migrate seasonally within the tropics. Many species of bat and bird, even hummingbirds, regularly fly between the island and the nearby mainland. Many animals, including pumas and ocelots, swim to the mainland and back. This traffic has delayed the onset of inbreeding in many populations, prevented extinctions, and played an essential role in maintaining the integrity of Barro Colorado's ecosystem.

Ironically, by becoming a (reasonably large) island, Barro Colorado became a good place to learn how mainland forest ecosystems are organized. Concentrating research there provided "unity of place": Each project there provided both the empirical background data and the intellectual foundation for further work, and essential context for other, very different projects. These many projects accordingly cohere into a common story of this forest's function. Work there has also helped to answer many other questions: How mutualism evolves, why there are so many kinds of tropical trees, where to look for chemical compounds most likely to be useful for future medicines, and how a large group of mindless army ants following simple rules can form a well-organized swarm of raiders.

SEE ALSO THE FOLLOWING ARTICLES

Climate on Islands / Deforestation / Island Biogeography, Theory of / Vegetation

FURTHER READING

Leigh, E. G., Jr. 1999. *Tropical forest ecology: a view from Barro Colorado Island.* New York: Oxford University Press.
Leigh, E. G., Jr., A. S. Rand, and D. M. Windsor, eds. 1996. *The ecology of a tropical forest: seasonal rhythms and long-term changes.* Washington, DC: Smithsonian Institution Press.
Losos, E. C., and E. G. Leigh, Jr., eds. 2004. *Tropical forest diversity and dynamism.* Chicago: University of Chicago Press.
Robinson, W. D. 1999. Long-term changes in the avifauna of Barro Colorado Island, Panama, a tropical forest isolate. *Conservation Biology* 13: 85–99.
Ziegler, C., and E. G. Leigh, Jr. 2002. *A magic web.* New York: Oxford University Press.

BEACHES

BRUCE RICHMOND

U.S. Geological Survey, Santa Cruz, California

Beaches are shoreline accumulations of loose sand, gravel, or a mixture of the two, that are formed primarily by the action of waves. Beach sediment can be derived from a variety of sources including insular shelves, the adjacent land and upland sources, or other beach locations through alongshore movement of material. Beaches provide critical coastal habitat, such as nesting sites for sea turtles; they act as a buffer protecting adjacent land from storm wave attack; and they are an important cultural and recreational resource. Island beaches are the same as those on the continents, but island beach characteristics typically change over very short distances on account of rapid changes in coastline orientation, exposure to waves, and sediment source.

BEACH MORPHOLOGY

Beaches straddle the boundary between land and water, extending from the low-water line to the supratidal backshore. A typical beach consists of a beach face or foreshore, a berm or crest, and backbeach or backshore (Fig. 1). The foreshore occurs between the low- and high-tide water levels and typically has a concave-upward topography. The base of the beach commonly ends with a steep step or "beach toe." In areas of ample sediment supply and suitable nearshore conditions, wave-formed bars occur in the surf zone. The crest, or berm, of the beach is the boundary between the relatively steep beach face and the flatter backbeach area. The landward limit of the backshore is typically marked by permanent vegetation, dunes, or

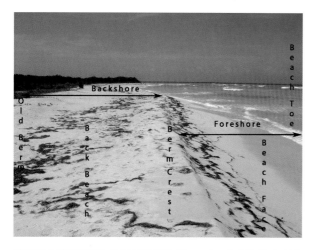

FIGURE 1 Photograph of a beach from the west end of St. Croix, U.S. Virgin Islands, showing the beach toe, foreshore, berm crest, backshore, and an older berm marked by an erosional scarp. The vegetation marks the more stable section of beach. This is a popular sea turtle nesting site. Photograph by the author.

coastal cliffs and indicates the highest reach of the waves during storms.

Beach planform is dependent primarily upon the local physiographic setting. For example, crenulated coastlines typically produce confined pocket beaches, whereas straighter coastlines tend to be bordered by long, gently curving beaches.

Beach types include stream valley with bayhead beach, structural embayment with beach, pocket beach, coastal plain/cuspate foreland beach, delta beach, and, perched beach. In addition to these enclosed beach systems, some islands such as barrier islands and sand cays are for the most part entirely bordered by beaches along their shorelines.

The following are brief descriptions of island beach physiographic settings and their important characteristics (adapted from Richmond 2002).

Stream Valley with Bayhead Beach

Stream valleys, which are typically cut during sea-level lowstands and subsequently drowned during subsequent sea-level rise, are often the sites of bayhead or barrier beach deposition. Where there is an open inlet to the sea, the beach may form a spit, which seasonally opens or closes depending upon stream outflow and wave energy. Sediment composition is typically a mixture of terrigenous and marine-derived material.

Structural Embayment with Beach

Embayments of either fluvial or structural origin can be distinguished by the presence or absence of a major stream valley. Drowned stream valleys may have a structural origin that is further emphasized by stream development, whereas embayments of a strictly structural origin have no pronounced stream development. Structural embayments can be created by a variety of processes including volcano growth characteristics (including coalescing volcanoes), volcano decay (e.g., landslides and crater collapse), faulting, and vertical tectonic displacement.

Pocket Beach

Pocket beaches are common features of rocky coasts worldwide and are essentially smaller (< 1 km long) versions of the drowned stream valley or structural embayments (Fig. 2). Pocket beach sediment can be relatively isolated from adjacent systems, and there may be a marked difference in composition from one beach to the next.

FIGURE 2 Oblique aerial photograph of the beach at Waimea Bay on the north shore of O'ahu, Hawai'i. This pocket beach, composed mostly of carbonate sand, is formed at the mouth of a stream within a structural embayment. Photograph by the author.

Coastal Plain/Cuspate Foreland Beach

Broad coastal plains (hundreds of meters wide, 1+ km long), composed mostly of aggraded beach deposits, are indicative of long-term stability and deposition. Many of the larger coastal plains on islands tend to occur along leeward shores. The convex-seaward plan form of the cuspate foreland (Fig. 3) distinguishes them from beaches within embayments. Cuspate forelands often migrate laterally by erosion on one side accompanied by deposition on the other. Changes in weather patterns can lead to reversals in transport directions, resulting in a very dynamic coastline prone to large changes in shoreline position on both seasonal and decadal scales. For example, the seasonal shifts in monsoon winds in the Maldives results in a significant

FIGURE 3 Oblique aerial photograph of the cuspate foreland shoreline and coastal plain on the southwest coast of Isla de Mona, Puerto Rico. The arrows point to areas of exposed beachrock that denote former shoreline positions. Photograph by the author.

FIGURE 4 Perched beach at Pu'uhonua o Honaunau National Historic Park (PUHO) on the west (Kona) coast of the island of Hawai'i. The beach, which is composed mostly of reef-derived carbonate sand and gravel and minor amounts of volcanic debris, is perched on top of a basalt platform that lies just above mean sea level. View to the north. Photograph by the author.

shift in shoreline position between seasons, and El Niño–Southern Oscillation (ENSO) events commonly increase erosion of leeward shores of Pacific islands through an increase of westerly winds and waves.

Delta Beach

On coastlines where stream deposition is greater than the rate of longshore dispersion of sediment, a seaward-protruding delta beach will form. Delta beaches are usually limited to older islands with well-developed stream systems. Sediment composition is typically mostly terrigenous. Ephemeral deltas may form as a result of a large storm/rainfall event and slowly disappear as sediment is reworked alongshore.

Perched Beach

Along rocky coasts where low terraces are developed, storms can form beaches above the normal wash of the tides (Fig. 4). These perched beaches are active only during storms and other large-wave events. Perched beaches occur on both volcanic islands where basalt benches have formed near sea level and on limestone islands where elevated limestone platforms border the coast.

Sand Cay

Sand cays are small, roughly circular to oval-shaped islands, which occupy the tops of intertidal to shallow subtidal platforms (commonly a reef platform in tropical settings). The entire shoreline of a sand cay is beach deposits. The location of the sand cay on the underlying platform represents some long-term average sediment depocenter. The shape of the cay can change in response to seasonal weather variations and possibly during individual storms.

Barrier Island

Barrier islands are depositional features offshore of a mainland coast with shorelines that are typically composed almost entirely of beach deposits (Fig. 5). Barrier islands are more common in temperate and higher latitude settings where there is a sufficient supply of sediment and a gently sloping continental shelf.

FIGURE 5 Oblique aerial photograph of Cross Island, a barrier island in the Beaufort Sea near Prudhoe Bay off the north coast of Alaska. The entire island is bounded by beaches that are modified by waves during ice-free periods, typically a few months during the summer, and by sea ice the rest of the year. Photograph by the author.

TABLE 1
Approximate Coastline Length, Percent Beach Shoreline, and Island Type for Selected Pacific Islands

Island	Approx. Coastline Length (km)	Approx. % Beach	Island and Reef Type
Hawai'i	492	8	Active volcanic
Maui	255	21	Young volcanic
Moloka'i	170	22	Volcanic (partial fringing reef)
Lana'i	84	34	Volcanic (partial fringing reef)
O'ahu	320	28	Volcanic (partial fringing reef)
Kaua'i	182	43	Volcanic (partial fringing reef)
Kosrae, FSM	90	31	Volcanic (continuous fringing reef)
Pohnpei, FSM	180	1	Volcanic (barrier reef)
Chuuk, FSM	110	30	Volcanic (almost-atoll/barrier)
Majuro, RMI	140	100	Atoll
Yap, FSM	124	9	Mixed (uplifted; fringing reef)
Guam	60	33	Mixed (uplifted; fringing and barrier reef)
Saipan, CNMI	55	31	Mixed (uplifted; fringing and barrier reef)

NOTE: Approximate coastline length and percent of beach shoreline for the main Hawaiian Islands are modified from Moberly and Chamberlain (1964), and the values for the islands in the Federated States of Micronesia (FSM) are interpolated from U.S. Geological Survey topographic maps. CNMI = Commonwealth of the Northern Mariana Islands; RMI = Republic of the Marshall Islands. Modified from Richmond (2002).

BEACH SEDIMENT

Beaches comprising sand-size material are the most common and occur throughout the world. Gravel beaches are much less common and tend to occur where there is a local gravel source, such as near gravel-rich rivers, gravel-bearing cliffs, or coral reef environments. Although mixed sand and gravel beaches exist, in general beach sediment tends to be moderately well-sorted, with mean grain size decreasing in a downdrift direction away from the source area.

The composition of the beach sediment is directly related to the local source material. For example, in tropical environments away from the effects of rivers, reef-derived biogenic carbonate sediment dominates the beach composition. Carbonate components may include fragments of coral, mollusc, foraminifera, coralline algae, and calcareous green alga (*Halimeda*). The amount of carbonate sediment decreases with distance from the reef and/or increasing terrigenous input, mostly from streams. Temperate and high-latitude beaches are typically composed of terrigenous sediment delivered primarily from rivers but with contributions from erosion of the adjacent landscape. In islands formed by continental-type rocks, quartz and feldspar sediments are dominant with smaller amounts of heavy minerals. Young volcanic island beaches often have high heavy mineral content and volcanic clasts.

The percentage of beach that is comprised in island shorelines is related to the type of island and island age (Table 1). For example, in the main Hawaiian Islands, the percent of beach volume increases with island age, with a low percentage on Hawai'i (youngest) and a much higher percentage on Kaua'i (oldest). The increase in beach volume on older islands is due to a combination of well-developed stream systems, which can deliver large amounts of sediment to the coast, and more pronounced reef development, which provides a consistent supply of reef-derived carbonate sediment. The development of large beach systems requires a relatively stable landmass, adequate coastal accommodation space for beach deposition, and a steady and abundant supply of sediment. Although modern beaches are essentially Holocene features, well-developed beach systems require a much longer time frame to evolve. Longer time for development results in increased reef formation and greater land denudation—both of which lead to increased sediment production.

In summary, island beaches are best developed and most stable where there is an abundant source of sediment and sufficient time and space for the sediment to accumulate, the rate of vertical tectonic movement is low (i.e., a stable island), and anthropogenic disturbances are minimal.

SEE ALSO THE FOLLOWING ARTICLES

Atolls / Barrier Islands / Erosion, Coastal / Motu / Surf in the Tropics / Tides

FURTHER READING

Bascom, W. 1980. *Waves and beaches, revised and updated.* New York: Anchor Doubleday.
Komar, P. D. 1998. *Beach processes and Sedimentation,* 2nd Ed. Upper Saddle River, NJ: Prentice Hall.
Moberly, R., and Chamberlain, T. 1964. *Hawaiian beach systems.* Final Report prepared for Department of Planning and Economic Development, State of Hawai'i, Hawai'i Institute of Geophysics HIG-41.
Richmond, B. M. 2002. Overview of Pacific Island carbonate beach systems, in *Carbonate beaches 2000. Proceedings of the First International Symposium on Carbonate Sand Beaches.* L. L. Robbins, O. T. Magoon, and L. Ewing, eds. Reston, VA: ASCE, 218–228.
Robbins, L. L., O. T. Magoon, and L. Ewing. 2002. *Carbonate beaches 2000. Proceedings of the First International Symposium on Carbonate Sand Beaches.* Reston, VA: ASCE.
Short, A. D., ed. 1999. *Handbook of Beach and Shoreface Morphodynamics.* Chichester, UK: John Wiley and Sons.

BERMUDA

ANNE F. GLASSPOOL AND
WOLFGANG STERRER

Bermuda Zoological Society, Flatts

Located at 32°18′ N, 64°46′ W in the Sargasso Sea (northwestern Atlantic Ocean), Bermuda is a small, low-lying oceanic archipelago of solidified wind-blown dunes (aeolianite) lying atop an eroded volcanic platform. Because of the Gulf Stream's influence Bermuda boasts a subtropical climate that supports the world's northernmost coral reef and mangrove systems.

GEOLOGY AND GEOGRAPHY

The Bermuda Seamount originated at least 33 million years ago, probably on top of a much earlier eruptive episode (110 million years ago), as a towering mid-ocean volcano with two side peaks (now Challenger Bank and Argus Bank). Drifting westward on the North American Plate, the extinct volcano eventually eroded down to sea level and, over the Pleistocene (1.8 million to 10,000 years ago), acquired a surface topography of solidified calcareous sand dunes derived from surrounding coral reefs (Fig. 1). Aeolian (wind-blown) dunes accumulated during warmer, interglacial periods of rising sea levels (e.g., 400,000 ago, to 22 m above today's). During cooler, glacial periods of falling sea levels (e.g., 18,000 years ago, to 120 m below today's), dunes stabilized and solidified when rain first dissolved, then recrystallized carbonate sand grains. The youngest limestone formations (called Southampton), therefore, are barely consolidated calcareous sand, whereas the oldest (Walsingham) are largely recrystallized, containing extensive karst features such as solutional caves and cave collapses, most of which are now below sea level. Limestone formations are separated by geosols (fossil soils) made of dissolved limestone, plant debris, and atmospheric dust from as far as the Sahara Desert.

Present-day Bermuda is a low-lying (maximum elevation 79 m), fishhook-shaped chain of four larger islands surrounded by hundreds of islets totaling 53.7 km^2. Together these enclose significant inshore basins and line the southeastern margin of an extensive (750 km^2) but shallow (average depth 10 m) oval lagoon, which is, in fact, formed by the truncated top of the volcanic pinnacle rising from the deep sea. During the Pleistocene, eustatic sea level fluctuations alternately exposed and flooded the platform about every 100,000 years, and the Island's land mass oscillated between two extremes—a contiguous area of nearly 1000 km^2 surrounded by a narrow reef fringe at low sea levels, and a string of a few islets totaling less than today's land area surrounded by an extensive shallow reef lagoon at high sea levels.

CLIMATE

Bermuda's climate is influenced by the Mid-Atlantic (or Bermuda-Azores) High. In the summer (May–October), winds are relatively weak and southeasterly, whereas during the winter (November–March) westerlies predominate, including gales. The tropical storm season is between May and November, with one hurricane approaching Bermuda every year on average and a severe hurricane expected every 4–5 years. The annual total rainfall of 150 cm is evenly distributed through the year, and humidity is uniformly high (70–82%). Mean monthly air temperatures range between 18.5 °C (February) and 29.6 °C (August), and mean sea temperatures between 17.3 °C (February) and 28.0 °C (August). Tides are semidiurnal, with a mean annual range of 0.75 m.

PRECOLONIAL ECOLOGY

Precolonial Bermuda was a low, hilly landscape with freshwater marshes (but no streams), densely wooded with some 15 species of endemic evergreen plants such as the Bermuda cedar (*Juniperus bermudiana*), Bermuda palmetto (*Sabal bermudana*), olivewood bark (*Cassine laneana*), and some 150 native plants. Forests and marshes were populated by a species-poor invertebrate fauna consisting of about 200 native and some 92 endemic species (including the only endemic genus, the land snail

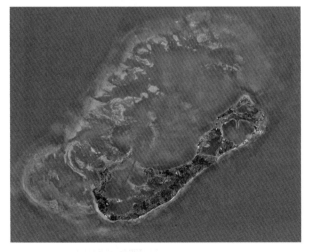

FIGURE 1 Bermuda, an aerial view taken in 1997, showing the land and surrounding reef complex, the most northerly in the world. Photograph courtesy of Bermuda Zoological Society.

Poecilozonites). Bermuda was a breeding ground for at least four species of Atlantic sea turtles, a dozen species of landbirds (including an endemic subspecies of the white-eyed vireo, *Vireo griseus bermudianus*), and at least six species of seabirds including the endemic Bermuda petrel or cahow (*Pterodroma cahow*, Fig. 2), which may have had a nesting colony of a million birds. There were no amphibians or terrestrial mammals, and a small lizard, the endemic Bermuda skink (*Eumeces longirostris*) was the only four-legged land animal.

Thanks to the warm Gulf Stream, which passes halfway between the island and Cape Hatteras, Bermuda has the complete range of tropical marine habitats, from drowned karst caves and brackish ponds to inshore lagoons, sandy and rocky shores, mangroves, sea grass beds, and coral reefs. Bermuda's coral reefs extend over the entire lagoon.

FIGURE 2 The endemic Bermuda petrel or cahow (*Pterodroma cahow*), "rediscovered" after it was believed to have been extinct for nearly 300 years, is now the subject of an intensive translocation initiative. Photograph by Andrew Dobson.

The entire southern coast is paralleled by a string of circular "algal-vermetid" reefs (also called mini-atolls, or "boilers") which, arising from an ancient shoreline, are constructed mainly by sedentary worm snails (Vermetidae) and encrusting red algae. The northern lagoon is littered with "coralgal" reefs (i.e., constructed mainly by corals and coralline algae), which have irregular round or elliptic shapes in the lagoon, but are rather linear at the rim of the platform where they form a near-impenetrable bulwark of breakers.

BIOGEOGRAPHY

Bermuda's native flora and fauna are a subset of southeastern North America and the tropical Caribbean, as might be expected if one accepts the Gulf Stream as the major vector for drifting and rafting colonizers. Of about 60 species of shallow-water reef corals (Scleractinia) known from the Caribbean, for example, 24 species (40%) also occur in Bermuda. The island's fossil record does not predate the Pleistocene, with salient terrestrial records of snails and birds showing significant extinctions and recolonizations, including an endemic tortoise (*Hesperotestudo bermudae*) known only as a single 300,000-year-old fossil. Despite its isolation and great age, Bermuda has very few endemic species in both marine and terrestrial habitats, as highlighted in a comparison with Hawaii (Sterrer, 1998), where marine endemism is five times greater (10.6% vs. 2.1% in Bermuda), and terrestrial/freshwater endemism is 13 times greater (48.3% vs. 3.7% in Bermuda). Most of Bermuda's endemics are found in drowned (anchialine) caves (Fig. 3), where of a total of 86 species recorded (mostly Crustacea), 80 (93%) are endemic (including three endemic genera). The Island's low rate of endemism is best explained by its relative paucity of terrestrial habitats, its down-current position with regard to the Gulf Stream (which maintains genetic continuity with Caribbean biota), and by habitat discontinuity during the Pleistocene, when low sea levels would have favored extensive forests and marshes at the expense of shallow reef environments, with high sea levels reversing this ratio. Karst caves, whose depth exceeded sea level fluctuations, may have been among the few temporally continuous habitats, and thus favored the origin and persistence of endemics.

HUMAN COLONIZATION AND ITS IMPACT

Discovery by the Spanish in 1503 marked a significant turning point in the natural history of Bermuda. Initially,

FIGURE 3 Divers explore one of the many submerged passages in Bermuda's anchialine cave system. Photograph by Christian Lascu.

fears of the treacherous reefs, coupled with haunting nocturnal cries of the cahow, deterred sailors from actually landing on the "Isle of Devils." However, in 1609 the shipwrecking of the *Sea Venture* off Bermuda during a hurricane, followed by the survival of all on board as they sought refuge on the island, inspired a new perspective, and permanent settlement followed.

As word of the Island's bountiful resources spread (notably fish, turtles, and whales), and new settlers experimented with a range of introduced crop species and domestic animals, Bermuda for a time came to be viewed as a mid-Atlantic provision store. Rapid depletion of the natural resources inevitably followed, and this, coupled with a suite of less desirable introductions—including soon-to-be terrestrial invasives (such as the common rat [*Rattus rattus*], American cockroach [*Periplaneta americana*], weevils, and wireweed [*Sida carpinifolia*])—also marked the beginning of widespread and largely irreversible changes to the native biodiversity. By 1684, when it was apparent that many of the early economic speculations (ambergris, pearls, and silk) and cash crops (tobacco and sugar cane) were not viable, Bermudians had to reinvent their survival strategy and embarked upon a 150-year era of shipbuilding, whaling, maritime commerce, and privateering. Fueling this activity was the endemic cedar, which provided food and lumber, inevitably leading to heavy deforestation, which in turn provided the opportunity for the continual influx of exotic species to firmly establish themselves. Agriculture reemerged as an export business in the mid-1800s until the 1930s, and for 200 years after 1783 Bermuda was also a strategic naval and military outpost, most significantly during World War II. Inspired by this rich history and the more than 300 documented shipwrecks, Bermudians have been pioneers in the development of modern archeological methods.

The twentieth century saw Bermuda become transformed into an almost exclusively service-based economy, first through tourism and more recently through international business. Today, Bermuda supports a resident population of 68,000, with 400,000 visitors a year, and the island has been transformed into a largely suburban landscape (Fig. 4). More than 50% of the land mass is considered developed, with 14% covered by artificial surfaces. Gardens, golf courses, and arable land occupy a further 20%. No undisturbed upland valleys remain, 75% of upland coastal habitat has been developed, and remaining upland forest occupies just 39%

FIGURE 4 Most of Bermuda's landscape is now considered suburban, as evidenced in this aerial photograph. Photograph courtesy of Bermuda Government Information Services.

of its former area. Only the coastal habitats remain relatively unchanged (Table 1).

Bermuda's close proximity to North America and easy access to deep water has encouraged a rich history of scientific exploration, including the HMS *Challenger* expedition of the 1870s, and William Beebe's deep-sea bathysphere triumph in the 1930s. Furthermore, inspired by the rediscovery of the cahow in 1951, which was believed to have been extinct for nearly 300 years, the creation of the Nonsuch Island Living Museum is acknowledged as one of the first global examples of an island restoration project. It also

TABLE 1
Change in the Area of Bermuda's Habitats since Colonization

Habitat, in Hectares	AD 1600	%	AD 2000	%
Beach and dune	76	1.5	76	1.4
Mangrove swamp	24	0.5	18	0.3
Salt marsh	4	0.1	1	0.0
Marine pond	17	0.3	17	0.3
Rocky coastal	162	3.2	90	1.7
Upland coastal	1382	26.9	348	6.5
Upland hillside	2303	44.8	903	16.8
Brackish/fresh pond	8	0.1	10	0.2
Peat marsh	119	2.3	45	0.8
Upland valley	921	17.9	0	0.0
Limestone sink	125	2.4	67	1.2
Arable field/pasture	0	0.0	178	3.3
Garden	0	0.0	669	12.5
Golf course	0	0.0	260	4.8
Developed, including hedgerows, walls, wayside, road verges	0	0.0	2689	50.1
Total	5141	100	5371	100

NOTE: Bermuda's total land area has increased since AD 1600 as a result of land reclamation, primarily for the airport.

TABLE 2
Change in Bermuda's Major Terrestrial Taxa since Colonization

Terrestrial Species	Endemic	(of Which Extinct)	Non-Endemic Native	Introduced Naturalised	Total Species	% Aliens
Flowering plants[a]	10	0	150	371	531	70
Ferns, mosses	6	0	15	17	38	45
Molluscs	18	6	6	33	57	58
Insects[b]	44	16	172	703	919	76
Spiders	2	0	5	34	41	83
Amphibians	0	0	0	3	3	100
Reptiles	1	0	0	4	5	80
Birds	4	3	7	9	20	45
Mammals	0	0	0	4	4	100
Total	85	25	355	1178	1618	73

[a]The total for introduced species includes only naturalized, self-propagating species.
[b]The total for introduced species excludes interceptions and isolated records.

marked the real start of the conservation movement in Bermuda, triggered by the devastating effects of the cedar blight, a disease of Bermuda's cedars caused by two accidentally introduced scale insects, the juniper scale (*Carulaspis minima*) and the oyster-shell scale (*Insulaspis pallida*), in the 1940s. This prompted perhaps one of the biggest attempts at biological control ever tried, with the introduction of over 100 species of insect predators, mainly beetles and wasps. With the loss nevertheless of 94% of the island's cedar trees, a huge reforestation effort was initiated with a suite of alien species, including casuarina (*Casuarina equisetifolia*), Brazil pepper (*Schinus terebinthifolius*), and Indian laurel (*Ficus retusa*), resulting in wholesale change of the landscape. Recent island-wide vegetation surveys have revealed that 22 invasive plant species are now a dominant feature of the 33% of Bermuda's land area that remains undeveloped. And of more than 1600 resident terrestrial plant and animal species, only 27% are native (Table 2). Twenty-five endemic species have become extinct, 200 native species have declined significantly, and at least 1200 exotic species have become naturalized (mainly flowering plants, insects, spiders, snails, birds, reptiles, and amphibians).

Bermuda's shallow water marine platform has been less impacted by humans, with only ten native species and one endemic known to have been extirpated. Today, shipping and shoreline development pose the main threats to the inshore marine environment; however, the reefs remain extremely healthy particularly in comparison with the neighboring Caribbean, with coral coverage as high as 75% on some main terrace reefs. With the establishment of the first Marine Protected Areas in 1966, legislative enactment of the whole island as a coral reef preserve in 1972, and the banning of fish traps in 1990, Bermuda has generally been ahead of the game in marine resource management.

SEE ALSO THE FOLLOWING ARTICLES

Caves as Islands / Coral / Hurricanes and Typhoons / Mangrove Islands / Marine Protected Areas / Seamounts, Geology

FURTHER READING

Anderson, C., H. De Silva, J. Furbert, A. Glasspool, L. Rodrigues, W. Sterrer, and J. Ward. 2001. *Bermuda biodiversity country study.* Bermuda Zoological Society.

Curran, H. A., and B. White, eds. 1995. Terrestrial and shallow marine geology of the Bahamas and Bermuda. *The Geological Society of America Special Paper 300.* Boulder, CO: The Geological Society of America.

Iliffe, T. M. 1993. A review of submarine caves and cave biology of Bermuda. *Boletino Sociedad Venezolana de Espeleologia* 27: 39–45. http://www.tamug.edu/cavebiology/Bermuda/BermudaIntro.html

Morris, B., J. Barnes, F. Brown, and J. Markham. 1977. The Bermuda marine environment. *Bermuda Biological Station Special Publication #15.*

Sterrer, W. 1998. How many species are there in Bermuda? *Bulletin of Marine Science* 62: 809–840.

Sterrer, W., A. Glasspool, H. De Silva, and J. Furbert. 2004. Bermuda—an island biodiversity transported, in *The effects of human transport on ecosystems: cars and planes, boats and trains.* J. Davenport and J. Davenport, eds. Dublin: Royal Irish Academy, 118–170.

Thomas, M. 2004. *The natural history of Bermuda.* Bermuda Zoological Society.

Vesey, T. 2002. When disaster struck. *The Bermudian* Sept.: 10–21.

Vogt, P. R., and W.-Y. Jung. 2007. Origin of the Bermuda volcanoes and Bermuda rise: history, observations, models and puzzles, In *Plates, plumes and planetary processes.* G. R. Fowler and D. M. Jurdy, eds. Geological Society of America Special Paper. 430: http://www.mantleplumes.org/P%5E4/P%5E4Chapters/VogtBermudaP4AcceptedMS.pdf

Wingate, D. B. 1997. The pre-colonial and current status of Bermuda's seabird population. *El Pitirre* 10: 36–37.

BIOLOGICAL CONTROL

MARK GILLESPIE AND STEVE WRATTEN
Lincoln University, Christchurch, New Zealand

Biological control (or biocontrol) is the use of natural enemies to suppress pest species populations to less damaging densities. When certain native and introduced invertebrates, plants, pathogens, and vertebrates increase in abundance and become pests through human influence or through other causes, economic crop damage and threats to natural resources are likely. Islands are particularly vulnerable to pest outbreaks. With high endemism, low species diversity, small land areas, and a history less affected by forces that develop adaptability compared to continents, islands are more susceptible to the effects of habitat changes and species introductions. The enhancement of the efficacy of the natural enemies of pest organisms is a potentially more environmentally sound and sustainable control option than chemical or mechanical control strategies.

THE IMPORTANCE OF PESTS

The presence of pest animal and plant species can heavily impact agricultural, forest, and urban ecosystems, costing national economies billions of dollars in management and lost revenue, compromising biosecurity internationally, and threatening natural ecosystems and endangered species. Weeds alone are estimated to cost the New Zealand economy NZ$100 million annually, for example. However, the dairy industry in that country has an annual economic value of NZ$7 billion, so in the short term at least, weeds could be considered to be a "minor" problem.

There are two main causes of pest outbreaks: the structure of agricultural systems and deliberate or accidental introductions of non-native species. In many agricultural systems, the selection of crops for rapid growth at the expense of defense mechanisms such as herbivore resistance and the planting of large uniform monocultures bereft of biodiverse non-crop habitats can disproportionately benefit herbivores, unwanted plants, and pathogens over their natural enemies, such that they reach economically damaging densities. It is estimated that about 37% of global crops are lost to native and introduced pests.

Non-native species, introduced to new areas either deliberately or accidentally, are also one of the biggest threats to indigenous flora and fauna. These adventive species often arrive without their coevolved natural enemies, and many are then able to thrive in an enemy-free environment. They may even change fundamental ecological properties such as water availability or soil chemistry, resulting in a further loss of competitiveness among natives. Before the advent of global transport and commerce, the arrival of introduced species is likely to have been once every 35,000 years in Hawaii, for example. Between 1962 and 1985, there were on average 19–20 immigrant species a year entering that island group, 3.5 of which became pest species.

The Issues for Islands

This article focuses on the issues facing the more remote islands such as those in the Pacific. Although pest origins and control principles are similar in all parts of the world, the characteristics of the more remote islands make them particularly vulnerable to both the damaging impacts of exotic pests and potential nontarget effects of the introduced biocontrol species (the agent).

Islands that have been geographically and evolutionarily isolated for millions of years tend to accommodate large proportions of endemic species, low species diversity, restricted genetic diversity, low natural immigration rates, and narrow home ranges of species, compared with larger land masses. During their geological development, they have not faced the effects of mammalian herbivore browsing and virulent diseases that are routinely encountered on continents. Because of these factors, natural enemies of invasive species are less likely to be present on such islands, and other affected species are less likely to adapt to and compete with new arrivals. Intense human activities on the smaller land areas of islands can also marginalize native species by reducing the availability of refugia. With low adaptability, restricted ranges, and sudden occurrence of hitherto absent competitive or predatory species, islands are at greater risk of losing globally important endemic species than are continents or less fragile islands.

An example of this vulnerability is the arrival of mammals in New Zealand. Before Polynesian and European settlement, the New Zealand archipelago held unique species assemblages that had evolved in isolation for over 80 million years. When humans arrived, they deliberately or inadvertently brought rats, livestock, cats, domestic dogs, mustelids, and plants and irrevocably altered unique habitats. Such rapid changes were unlike any the native flora and fauna had been faced with before, and species such as ground-dwelling, flightless birds with ineffective mammalian predator defense strategies were easy prey. As a result of this and widespread hunting, many of New Zealand's archaic and mammal-niche species were driven to extinction, and the new, enemy-free species thrived.

With increased global travel and current tourism trends toward "island paradises," the numbers of introduced species are reaching high levels, a situation that further increases the need for control. However, the above factors are also applicable to the introduction of natural enemies to control the pests, and examples given later of agents impacting upon nontarget species demonstrate the fragility of islands and the difficulty of achieving sustainable pest control.

The Limitations of Nonbiological Control

The negative effects of pesticides have been documented comprehensively, from adversely affecting nontarget natural enemies and other beneficial organisms to accumulating in soil, water, and even humans. Of particular concern is the capability of many pests to become resistant to pesticides. Many of them can also "create" pests by killing the natural enemies of previously harmless species, allowing them to thrive as "secondary pests."

In many cases, chemical and mechanical strategies such as manual removal are not desirable because of the fragility of island flora and fauna, or are not possible because of the cost and labor requirements of such methods. Concerns about these control methods have led many countries to seek ways to reduce chemical use, and biocontrol is billed as a technique that can help to limit the use of more damaging or inappropriate control methods.

THE PRINCIPLES OF BIOLOGICAL CONTROL

Although biological control occurs naturally everywhere, it is the human manipulation of the processes that forms the techniques we call biocontrol. These were first utilized in China and Yemen thousands of years ago, but they have gained popularity only since the turn of the nineteenth century, alongside advances in scientific knowledge of the ecological processes involved. Since then, many programs have been attempted with the aim of developing a sustainable and cost-effective technology.

The most common targets of biocontrol are insects and weeds, although vertebrates, other invertebrates such as mites and snails, and pathogens are just as problematic and are currently receiving much attention in the scientific literature. The agents, natural enemy organisms that can be used in biological control programs, also vary widely and include invertebrates, vertebrates, and pathogens, with parasitic wasps that attack insect pests being the most common type of program.

Broadly, biocontrol considers that a contribution to solving pest problems rests with conserving natural enemies and/or reconnecting disrupted food webs. Three general applications of this theory have evolved over the last 120 years. Classical biocontrol, the introduction of one or more appropriate indigenous and/or exotic natural enemies to pest-infested areas, is the oldest and most common method. This method also carries the most risk, because, although trials are made to ensure the agent has no nontarget effects, some effects are unpredictable. Augmentation biocontrol, largely a commercial industrial technology and application, aims to increase the abundance of natural enemies already present in low densities or arriving late relative to the pests, through timely releases either inundatively or by innoculation of smaller numbers of natural enemies into the cropping system. Finally, conservation biocontrol, a more recently recognized method, attempts to conserve natural enemies by rectifying negative influences on these beneficial organisms. These influences may include pesticide use and the removal of natural-enemy overwintering sites. This method is often employed as part of an integrated pest management program or following an initial classical introduction of natural enemies in the case of exotic pests.

In successful programs, the agents will continue to provide control after establishment or conservation without the need for continued management and will also persist when pest densities are low, greatly reducing both the effort and cost of the control method. Similarly, both the pest and the introduced specialized enemy ideally decline in abundance over time, reaching densities similar to those in the pest's home range; preferably, a community structure is achieved that is similar to that existing before the pest invasion.

BIOLOGICAL CONTROL ON ISLANDS

The Benefits of Successful Biological Control

When biocontrol works, it is self sustaining, non-polluting, and cost-effective. In Hawaii alone, programs have contributed to the control of over 200 agricultural pest species, saving millions of dollars and tons of pesticides annually. Worldwide, successful introduction programs number over 700 (with more than 200 providing complete control and over 500 giving substantial or partial control). Economic analyses of these programs suggest that benefit-cost ratios for successful programs can exceed 145:1. A well known example is the control of the floating fern, *Salvinia molesta* (Fig. 1), which was introduced to Papua New Guinea from Brazil as an aquarium plant and botanical curiosity. By the 1970s, the fern had established sufficiently to spread over the surface of the Sepik River in a mat up to 1 m thick, preventing navigation, killing submerged vegetation, and isolating villages. In 1978, the introduction of a weevil from the fern's home range led to effective control and rid the river of the fern permanently, delivering control at a moderate cost when all other methods were considered unfeasible.

FIGURE 2 The glassy-winged sharpshooter, *Homalodisca coagulata*, alongside native species of cicadellid. The *Tharra sp.* (middle) is the largest species native to Tahiti, whereas the *Nesophyla sp.* (left), is an average sized native cicadellid in Tahiti. (Courtesy of J. Grandgirard and J. Petit, Gump Station, University of California, Berkeley).

FIGURE 1 A farm pond covered in the aquatic weed *Salvinia molesta*. Such infestations can kill submerged flora and fauna and threaten livelihoods. (Courtesy of Ted D. Center, USDA Agricultural Research Service.)

Success has also recently been seen in French Polynesia, where a bug, the glassy-winged sharpshooter, *Homalodisca coagulata* (Fig. 2), is a major pest of agricultural, ornamental, and native plants, reaching densities up to 1000 times greater than in its home range in the southeastern United States and Mexico. However, biocontrol scientists identified a suitable egg parasitoid from the home range of *H. coagulata*, and in September 2005, 14,000 parasitoids were introduced to 27 sites on Tahiti. By December 2006, a 99% decrease in *H. coagulata* nymph densities was recorded. To date, both pest and natural enemy remain at low densities on every island to which they were introduced.

In both these cases, a high degree of scientific expertise was employed, strict biocontrol standards were adhered to, and measures were taken to ensure that the pest control agent did not impact native species. The selection of the most appropriate natural enemy to introduce is not as straightforward as it may seem however, and past errors have detracted from successes.

The Limitations of Biological Control

Biological control has thus far not been the panacea that the early pioneers had hoped. There have been some attempts that have had disastrous consequences over the years, which some critics claim have led to nontarget effects as serious as species extinctions.

The rosy wolf snail, *Euglandina rosea*, for example, was introduced to Hawaii in the 1950s after the government responded under pressure from residents to control the invasive giant African land snail, *Achatina fulica*, despite incomplete biocontrol trials and the fact that the snails do not share the same home range. Although *A. fulica* did decrease in abundance following the introduction of *E. rosea*, reasons other than the introduction of biocontrol agents have been posited. In contrast, there is ample evidence that *E. rosea* negatively impacted many native snail species, to the extent that they caused extinctions of several endemic species. Despite this evidence, *E. rosea* was subsequently introduced to the Society Islands for the same reason and with the same devastating impact on native *Partula* snails.

In the 1920s, a moth species, *Levuana iridescens*, was heavily damaging coconut production on Fiji. Entomologists searching outside Fiji failed to locate *L. iridescens*, and therefore its natural enemies, and instead selected and imported a parasitic fly from Malaysia that attacked a similar host species. *Levuana iridescens* was brought under control rapidly and has been cited as a leading example of biocontrol. However, recent work argues that *L. iridescens* was endemic to the Fijian archipelago and that the agent introduced for control may have driven it to extinction. The matter is as yet unresolved, but it illustrates the fragility of island species and the risks involved with biocontrol programs.

When programs have drifted from the fundamental principles of biocontrol (e.g., natural enemies should be highly host specific), nontarget effects have often occurred that could have been avoided. Badly conceived projects fail largely because inappropriate natural enemies are selected. Generalist predators are inappropriate, for example, because of their adverse effects on native organisms and lack of "fidelity" to the target pest. In some cases, groups other than skilled technicians have carried out the control with no scientific grounding or government oversight, often in order to sidestep rigorous procedures to achieve "quick fixes." Alternatively, projects were carried out before the implementation of government regulation of control agent importation.

Many nontarget effects are actually also open to controversy because direct evidence linking the exotic natural enemy to the decline in indigenous organisms is often not available. For example, the parasitoid wasp *Pteromalus puparum*, which was introduced to New Zealand in 1933 to control the cabbage white butterfly, *Pieris rapae*, also attacks the endemic Red Admiral butterfly, *Bassaris gonerilla*. However, a self-introduced parasitoid, *Echthromorpha indicatoria*, has been shown to have a greater impact on *B. gonerilla* abundance than does the introduced biocontrol agent.

Although poor planning and technical knowledge may be blamed for many inappropriate biocontrol programs, even the strictest program can suffer from risk and unpredictability. Importantly, though, problematic control attempts have taught lessons and led to the guidelines that most countries now follow when embarking upon a biological control program. Criticism of the adverse affects of classical control has also led to guidelines set down by the Food and Agriculture Organization of the United Nations (FAO). New Zealand and Hawaii, for example, currently have the some of the most rigorous national legal regulations for the importation of potential biological control agents, and this is designed to limit such risks, although they cannot be eliminated.

FUTURE PROSPECTS FOR BIOLOGICAL CONTROL TECHNOLOGY

The risks of nontarget impacts need to be considered alongside the risks of inaction and those of pesticide use on insecticide resistant pests. The risks of biocontrol are usually taken seriously in today's environmentally aware social climate. In the years since strict quarantine screening and risk-benefit analysis were implemented, the safety record of biocontrol science has been particularly strong. For example, although a study in Hawaii found that 83% of parasitoids reared from native moths were biological control agents with parasitism reaching 28% in some species, all of those agents were species released before 1945. This suggests that recent guidelines have been effective.

The future of biological control is, however, hampered by a lack of three things: research funding, continuity of effort, and a long-term emphasis. Pacific islands for example, make up only 2% of the Pacific Ocean; have great differences in flora between them; and have low population densities, a low proportion of commercial agriculture, and a certain reliance on foreign aid. As a result, funding agencies are reluctant to contribute to such apparently low-benefit causes, overlooking the conservation value of these islands and the fact that collaboration with a developed country is important if control and quarantine of new species is to be successful. Indeed, biological control is often the only economically viable technology in such regions.

Nonetheless, although young, the science of biocontrol is growing: Novel biocontrol techniques are being constantly explored, and new targets are being frequently studied. New approaches to thus-far difficult targets, such as feral cats, including sexual transmission of host-specific diseases and immunocontraception as means to reduce the fecundity of noxious vertebrates, are being actively pursued in New Zealand and Australia.

This continued development and further international and interdisciplinary collaboration are vital if biocontrol is to fulfill the potential it promises. Furthermore, in today's scientific and political environment, risks are likely to be minimized, and biological control will develop as a viable, safe, and sustainable solution to pest invasions of islands.

SEE ALSO THE FOLLOWING ARTICLES

Extinction / Fiji, Biology / Hawaiian Islands, Biology / Introduced Species / Invasion Biology / New Zealand, Biology

FURTHER READING

DeBach, P., and D. Rosen. 1991. *Biological control by natural enemies.* Cambridge: Cambridge University Press.
Gurr, G., and S. Wratten. 2000. *Biological control: measures of success.* Dordrecht: Kluwer Academic Publishers.
Hoddle, M. S. 2002. Restoring balance: using exotic species to control invasive exotic species. *Conservation Biology* 18: 38–49.
Julien, M. H., J. K. Scott, W. Orapa, and Q. Paynter. 2007. History, opportunities and challenges for biological control in Australia, New Zealand and the Pacific Islands. *Crop Protection* 26: 255–265.

Messing, R. H., and M. G. Wright. 2006. Biological control of invasive species: solution or pollution? *Frontiers in Ecology and Environment* 4: 132–140.

Van Driesche, R. G., and T. S. Bellows. 1996. *Biological control.* New York: Chapman and Hall.

Wilson, K.-J. 2004. *Flight of the huia: ecology and conservation of New Zealand's frogs, reptiles, birds and mammals.* Christchurch: Canterbury University Press.

BIRD DISEASE

DAVID CAMERON DUFFY

University of Hawai'i, Manoa

Bird diseases have had a profound effect on some island avifauna, but, as with human diseases, our full understanding of their extent and impact has been limited by isolation, the paucity of qualified observers, and the possibility that many of the strongest effects have already occurred. Despite these limitations, research suggests that bird diseases on islands fall into two main, contrasting groups, both the result of human intervention: alien diseases attacking naïve, susceptible populations, with often devastating effects on an island's species and ecosystems, and (more rarely) "emerging diseases," resulting from human contact with endemic avian pathogens on previously isolated islands.

DISEASE SUSCEPTIBILITY IN ISLAND BIRDS

An island's size and isolation affect its avian disease ecology much as they determine its avifauna. Only a few birds reach more isolated islands, representing a subset of the source population's genes, and they may also carry a reduced set of pathogens compared to the source pool of pathogens. With a small host population, birds and pathogens would undergo rapid selection for resistance and reduced mortality. At the community level, small avifaunas would theoretically also reduce the total pool of pathogens and the subsequent risk of inter-specific transfer of diseases between bird species. Unfortunately, the role of disease in organizing natural island bird communities in the absence of humans can probably no longer be explored because avifaunas have been changed so greatly by human activity.

The lack of pathogen challenge may have led to reduced immunologic defenses in island birds, making them more vulnerable to new diseases, but this theory remains controversial. In theory, there are costs to immune defenses, and, in the absence of a challenging disease environment, birds should redirect their resources to other activities such as reproduction. On the other hand, island birds remain in an evolutionary race with those diseases that are present, they may be able to adapt different components of their immune systems to the pathogen community, and changes in their immune systems may be limited by mutation and genetic drift.

It does appear that island birds exhibit reduced immune functions following inbreeding during bottlenecks, when populations are reduced to small numbers, whether caused by anthropogenic damage, reintroduction, or perhaps even during natural colonization. The resulting combination of small populations and weakened immune systems would make populations vulnerable to extinction from disease.

INTRODUCED DISEASES

Acute Catastrophic Effects

The first human contact with any island has frequently been coupled with the introduction of alien pathogens, often with disastrous consequences to native ecosystems. Sailing vessels carried a menagerie of rats, cats, chickens, and mosquitoes, and their long voyages meant they had to stop at isolated islands to take on water and food, leaving ample opportunity for various hosts and vectors of pathogens to disembark. Such introductions and the resulting diseases often had devastating effects for island birds, much as the diseases of sailors themselves had for island peoples.

The best examples of epidemics or epizootics are mosquito-borne avian pox virus and avian malaria (*Plasmodium relictum*) in the Hawaiian Islands. Although a mosquito vector was present from the early 1800s, it was probably not until the late nineteenth or early twentieth centuries that the two diseases devastated the native avifauna. One observer reported "scores" of forest birds "dead or dying" and others transformed with hideous pox lesions that in some cases left them unable to move. Although a few species remained unaffected, most retreated up the slopes of the Hawaiian volcanoes, above the range of the mosquito vector. In subsequent experiments, native birds developed severe pox and malaria, and many died.

Although avian malaria and avian pox had catastrophic effects on Hawaii's bird populations, there is some evidence that an "equilibrium" is forming between pathogen and host. Within a century of the introduction of avian malaria, two species in Hawaii show signs of developing resistance to the disease. Avian pox, initially associated with mortality in red-tailed tropicbirds (*Phaethon rubricauda*) on Midway Island in 1963, continues to affect nestling Laysan albatross (*Pheobastria* [*Diomedea*] *immutabilis*) populations in Hawai'i, but with little mortality, suggesting either a decrease in virulence by the virus or an increase in resistance by the birds.

As alien diseases continue to be introduced to island ecosystems, other significant mortality events for island birds have included avian malaria in New Zealand on yellow-eyed penguins (*Megadyptes antipodes*) and in South Africa on African penguins (*Spheniscus demersus*); avian cholera and Newcastle disease on the guano islands of Peru (various seabird species); avian cholera, coccidiosis, and influenza A off South Africa (various seabird species); Newcastle disease on double-crested cormorants (*Phalacrocorax auritus*) in the Great Lakes and California; a strain of *Salmonella* on an endangered New Zealand bird, the hihi (*Notiomystis cincta*), and avian pox in Galapagos (various species). Avian pox has also infected two endemic land birds in the Canary Islands, Berthelot's pipits (*Anthus berthelotti*) and short-toed larks (*Calandrella rufescens*), exhibiting 28% and 50% frequencies of lesions referable to pox, thus suggesting that pox is a serious threat to the two species.

The route of transmission of disease to a new population is not always direct or obvious. An ornithosis outbreak in the 1930s in the Faeroe Islands and Iceland involving northern fulmars (*Fulmarus glacialis*) may have been triggered by avian scavenging on diseased parrots tossed overboard while being transported to Europe to be sold. The disease spread to islanders through consumption of young birds. The death rate among infected people was 20% (up to 80% in pregnant women), and the situation forced a ban on the traditional harvest of nestlings. In many cases, the causes of mortality events remain unknown or are suspected to be the result of a combination of factors, with disease sometimes playing only a secondary role.

Chronic Effects or Their Absence

Not all new pathogens produce epidemics. Some pathogens may not have the best mode of transmission, or the social structures of bird species may not be conducive to spreading the disease, or bird species may not be phylogenetically vulnerable to a particular pathogen. On the other hand, it is possible that observers arrived only after a pathogen had lost its initial virulence or that the disease had sublethal effects that would not be obvious without study. Some diseases can be hosted in a carrier population of birds that remain unaffected even as they transmit the disease to other species. *Campylobacter jejuni,* a cause of human bacterial enteritis, has been detected in Macaroni penguins (*Eudyptes chrysolophus*) on an island off South Georgia, occurring without apparent negative effects on the birds. Similarly, infectious bursal disease virus, which impairs the immune system in chickens, is present in Adelie penguins (*Pygoscelis adeliae*) in the Antarctic with no apparent clinical effects. *Borrelia burgdorferi,* the infectious agent for Lyme disease, has been found in a wide range of seabirds in both hemispheres, and antibodies to it have been detected in Faeroe Islanders who harvest seabirds, without any evidence of clinical symptoms in either birds or humans.

Some pathogens cause devastation in some geographic areas while occurring without detriment in others. Avian malaria, despite its devastation elsewhere, appears to have no effect on the avifaunas of American Samoa, Bermuda, Moorea, and some South Pacific islands, suggesting the pathogen has been present for long enough for the avifaunas to evolve resistance to it or for any vulnerable species to have already gone extinct.

ENDEMIC DISEASES

A long list of obscure avian pathogens, especially viruses, have been discovered on islands throughout the world, following what were usually only superficial surveys. After so much ship traffic, we cannot now always be sure which diseases are indigenous, just as we have no idea what diseases might have already vanished after exterminating their hosts.

Much research has focused on pathogens that have the potential to affect humans, especially viruses associated with seabird-parasitizing ticks (Acari), the soft-bodied genus *Ornithodoros* in the tropics and the hard-bodied genus *Ixodes* in the temperate and polar regions. Whereas the ticks are known to cause anemia, exanguination, nest desertion, and resulting mass mortality of abandoned young, little to nothing is known of the effect of viruses on seabirds themselves. Apparent human illness associated with tick bites, but perhaps caused by viruses, has been reported from Arabia, the Seychelles, Peru, Morocco, and France, primarily in guano workers, farmers, and researchers. Finally, there is the enigmatic Laysan fever from the northwest Hawaiian Islands that attacked field biologists for a period and then apparently died out.

Mosquito-borne viruses that use birds as hosts have the potential to cause widespread infection among humans, but endemic forms appear rare on islands. Whataroa virus, affecting humans in New Zealand with an influenza-like disease, now appears to be sustained by the introduced song thrush (*Turdus philomelos*), but it is spread by endemic mosquitoes in a human-modified landscape with relatively few native birds. Its original host remains unknown, but unlike many islands, New Zealand had endemic mosquitoes, which could have coevolved with viruses.

Another significant disease because of its pandemic potential is avian influenza. The normal hosts are waterfowl, shorebirds, and seabirds. Sampling for influenza antibodies in Australian seabirds shows variation between different species (wedge-tailed shearwaters *Puffinus pacificus* and black noddies *Anous minutus*) and local breeding islands, suggesting they may host local lineages of influenza.

THE FUTURE

Our understanding of the effects of introduced bird diseases on island avifauna has begun to lead to protective measures such as the removal of alien-host rock doves (*Columba livia*) in Galapagos and vector mosquitoes (*Culex quiquefasciatus*) on Midway Island in the Pacific, the implementation of quarantine systems of varying degrees of competence, the sterilization of boots in the Antarctic, and the institution of bans on importation of high-risk hosts. Despite these efforts, diseases such as Newcastle disease, West Nile, and avian influenza remain ever ready to make an appearance, facilitated by today's rapid forms of transportation and international trade.

The potential for emerging diseases from islands also continues and may even increase as ecotourism brings travelers onto ever more remote islands and then rapidly back to sophisticated medical facilities where their ailments can be diagnosed as something beyond "fever."

The study of bird diseases on islands remains one of the last frontiers for exploratory biology. It is likely to continue to provide unexpected insights into immunology, public health, community ecology, and conservation biology, as additional islands, their avifaunas, and their diseases are examined.

SEE ALSO THE FOLLOWING ARTICLES

Extinction / Honeycreepers, Hawaiian / Inbreeding / Seabirds

FURTHER READING

Clifford, C. M. 1979. Tick-borne viruses of seabirds, in *Arctic and tropical arboviruses*. E. Kurstak, ed. New York: Academic Press, 83–99.
Daszak, P., A. A. Cunningham, and A. D. Hyatt. 2001. Anthropogenic environmental change and the emergence of infectious diseases in wildlife. *Acta Tropica* 78: 103–116.
Matson, K. D. 2006. Are there differences in immune function between continental and insular birds? *Proceedings of the Royal Society B* 273: 2267–2274.
Van Riper, C., S. G. Van Riper, M. L. Goff, and M. Laird. 1986. The epizoootiology and ecological significance of malaria in Hawaiian land birds. *Ecological Monographs* 56: 327–344.
Wikelski, M., J. Foufopoulos, H. Vargas, and H. Snell. 2004. Galápagos birds and diseases: invasive pathogens as threats for island species. *Ecology and Society* 9:5. http://www.ecologyandsociety.org/vol9/iss1/art5/.

BIRD RADIATIONS

JEFFREY PODOS AND DAVID C. LAHTI

University of Massachusetts, Amherst

Bird radiations provide informative illustrations of ecological and evolutionary processes, including those that help to generate biodiversity as ancestral species radiate into multiple descendent species. Adaptive radiation involves initial phases of divergence among populations or incipient species, accompanied by or followed by the evolution of reproductive isolation. In this article, aspects of these two key processes—divergence and the evolution of reproductive isolation—are outlined and examined with specific reference to island bird radiations.

BIRD RADIATIONS ON ISLANDS: ECOLOGICAL AND EVOLUTIONARY INSIGHTS

Island habitats provide biologists valuable if not unique opportunities for the study of ecology, evolution, and animal behavior. Relative to continental habitats, islands tend to express low biological diversity, high abundance of constituent species, streamlined food webs, and environmental fluctuations that can be both marked and unpredictable. The simplified and dynamic profile of island biology has facilitated the study of a wealth of biological patterns and processes, with results gleaned from island studies often extrapolated to more complex and empirically less accessible habitats such as tropical rain forests.

Birds have taken center stage in many field studies of island ecology and evolution, following a tradition set by David Lack in the mid-twentieth century. Bird radiations have been numerous and varied, featuring outcomes such as flightlessness and robust variations in plumage, size, and beak form and function (Table 1). This entry provides a brief overview of some ecological and evolutionary processes that drive island bird radiations and then discusses recent findings from the authors' own studies of two island bird systems: introduced island populations of African weaverbirds, and Darwin's finches of the Galápagos Islands.

TABLE 1
Representative Examples of Bird Radiations on Islands

Bird Assemblage	Islands	Distinctive Features of the Radiation
Vangas (15–22 species) and Malagasy songbirds (9+ species)	Madagascar	The vangas' common ancestor presumably arrived on Madagascar about 25 million years ago and subsequently radiated into a group unusually diverse in both plumage and bill morphology. Another diverse group, Malagasy songbirds (previously placed into three different families of bulbuls, babblers, and warblers), represent a radiation from a single colonizing species 9–17 million years ago. At 587,000 km^2, Madagascar is the smallest single island in the world with a prominent bird radiation.
Moas (14 species)	New Zealand	Beginning about 20 million years ago, these herbivorous and flightless birds evolved in the absence of predators, radiating mostly within the last 10 million years into a variety of body forms and sizes.
Darwin's finches (14–15 species)	Galápagos and Cocos islands	A grassquit or warbler-like ancestral species colonized the Galápagos about 2–3 million years ago, followed by a pattern of island-hopping and speciation; descendant species eventually redistributed themselves throughout the islands without interbreeding. Bill morphology diverged by adaptation to available seed types, song subsequently diverged as a by-product; female preference for both traits is thought to have promoted reproductive isolation upon secondary contact.
Myiarchus flycatchers (five species)	West Indies	A single immigration to Jamaica from Central America about 4 million years ago, led to subsequent island-hopping and divergence. Several island-specific forms are designated as subspecies; thus this is apparently a radiation in process.
Honeycreepers (~52 species)	Hawaiian Islands	Beginning about 4–5 million years ago, the descendants of a single colonizing finch species radiated explosively; highly diverse feeding methods and bill types were faciliated by a broad range of vacant feeding niches.
Golden whistler *Pachycephala pectoralis* (66 forms, perhaps can be grouped into 5 species)	Northern Melanesia and New Guinea	Called the world's "greatest speciator" by Mayr and Diamond (2001), this bird has diverged greatly in color patterns, but species boundaries are difficult to determine. Some forms arose by divergence but others through hybridization (mixing) of existing forms.
Chaffinches *Fringilla coelebs* (four forms, designated as subspecies)	North Atlantic islands	DNA evidence has indicated the history of chaffinch expansion from an ancestral stock that ranges across Europe and Africa, to several nearby island groups: a Portuguese chaffinch population colonized the Azores, birds from the Azores colonized Madeira, and birds from Madeira then colonized the Canary Islands twice.

Island Ecology

COMPETITION AND CHARACTER DISPLACEMENT

Any given habitat supports a diversity of species. Species that share habitats are thought to partition limited sets of resources such as space and nutrients. Shorebirds, for instance, have evolved to forage in distinct shoreline zones. Moreover, resource partitioning appears to be adaptive; sandpipers and plovers, to illustrate, are adept in extracting distinct food sources at particular depths within the sand or mud, by means of distinct bill lengths and shapes. Adaptive partitioning of resources among present-day species implies two historical processes: competition for limited resources, followed by selection favoring the occupation of divergent niches ("character displacement"). These processes are thought to be highly generalized, explaining, for example, how tropical rainforests can support a rich biota. However, these processes have proven notoriously difficult to study in field environments.

A clear demonstration of competition leading to character displacement has been reported by Peter and Rosemary Grant, who have been conducting field studies on Galápagos finches for over three decades. During an initial decade of study, beginning in 1973, the small central island of Daphne Major was found to support a large breeding population of the medium ground finch *Geospiza fortis*. This population featured birds of a wide range of beak sizes feeding on a diverse array of seed types, including the relatively hard seeds of caltrop *Tribulus cistoides*. Observations of feeding finches revealed that only the largest-beaked *G. fortis* were able to crack *Tribulus* seeds, a capability that served these large-beaked birds well during a severe drought in 1977. With exclusive access to this food resource, large-beaked *G. fortis* were able to survive and proliferate. In 1982, however, a breeding population of the large ground finch (*Geospiza magnirostris*) began to take root on Daphne, building slowly and peaking at about

350 individuals in 2003. *G. magnirostris* also feeds on the seeds of *Tribulus*, and thus introduced a competitive threat to large-beaked *G. fortis*. The effects of interspecies competition were realized during an intense drought in 2003, during which time the population of *G. fortis* crashed as a result of widespread starvation. Subsequent analysis revealed that large-beaked *G. fortis* had suffered disproportionately, presumably because of the depletion of the favored *Tribulus* seed resource by *G. magnirostris*. Smaller-beaked *G. fortis* survived and bred in disproportionate numbers, thus leading to an evolutionary reduction in *G. fortis* beak size in the next generation. The character of beak size had thus been displaced through competition from another species, through selection favoring those individuals that could avoid direct interspecies competition.

ISLAND BIOGEOGRAPHY

Given evidence of resource partitioning through competition and displacement, it follows that habitats with more ecological niches might support relatively greater levels of species biodiversity. This idea has received particular attention in the theory of island biogeography, which focuses on "rules" that might govern the diversity and evolution of species on terrestrial islands. A first rule is that large islands are expected to support more species than small islands, because of a greater abundance of ecological niches, resulting not only from their larger areas but also from broader elevational gradients. A tenfold increase in the size of an island is expected to lead to an approximate doubling of species number. A second rule of island biogeography is that more remote islands should be relatively depauperate, because chance immigrations of new species are less likely when mainland source populations are more distant. Larger and more isolated islands are also expected to support greater numbers of endemic species, because of increased opportunities for genetic isolation. Island species diversity can of course be influenced by other factors including the presence or absence of predators, the geological age of islands, the extent and diversity of appropriate habitats, and factors that can affect animal movement such as storm paths and migration routes. In modern times, humans have also imposed profound impacts on island species diversity through activities such as hunting, habitat conversion, and species introductions.

Birds on the islands of the Indian Ocean well illustrate some of the principles of island biogeography, particularly those related to endemism. Of fourteen islands, endemic species occur in much greater frequency on the larger islands. The two smallest islands (< 100 km² area) support no endemics, whereas the three largest islands (> 1000 km²) support over a dozen each. A parallel pattern is observed across the four inhabited Comoro Islands. The largest of these islands, Grande Comore, supports 14 endemic species, whereas the other three islands support only two or three endemics each. The Comoro Islands also highlight the importance of elevation to bird diversity; all of the Grande Comore endemics can be found on its Mount Karthala, the only mountain of the archipelago. The Indian Ocean Islands also illustrate a relationship, albeit more subtle, between endemic species and degrees of island isolation. Very small and distant islands have no endemics at all, as predicted given that they are only seldom visited and that any occasional founding population is likely to go extinct. Of islands with at least one endemic species, the more remote ones tend to have higher proportions of endemics. For instance, Rodrigues, Mauritius, and Réunion make up the Mascarenes, the most remote archipelago in the Indian Ocean. Not surprisingly, none of the birds native to these islands are found anywhere else. The small, isolated island of Rodrigues illustrates both predictions of low total species and a high proportion of endemism. Before it was inhabited by humans, 13 landbird species lived on the island, 12 of which were endemic. Today there are only two native landbird species on the island, both of which are endemic (the Rodrigues fody and the Rodrigues warbler). The sharp decline in species on Rodrigues highlights the often overwhelming influence of human impacts, which usually affect species diversity to at least as great a degree as island biogeographic processes and thus interfere with our ability to detect those processes. The Indian Ocean islands are unusual in that anthropogenic extinctions and introductions apparently have not yet obliterated the effects of island size and distance.

Evolution on Islands

POPULATION DIVERGENCE

Populations that colonize island habitats can undergo rapid evolutionary change, through any number of evolutionary mechanisms. Colonizing populations are small (normally comprising a small fraction of individuals from source populations) and thus normally experience severe population bottlenecks, with subsequent trajectories of evolution altered by founder effects. Founder effects may be sustained in "island hopping" scenarios, because of multiple sequential bottlenecks. Small population sizes also amplify the effects of genetic drift, which can theoretically result in the elimination or the fixation of genetic traits. Evolutionary change may also be facilitated by the fact that island habitats often differ substantially from mainland habitats in resource structure and competition regimes. The genetic profiles of different populations can diverge even as they adapt to similar environmental challenges, because evolutionary changes in phenotypes can be achieved through distinct, parallel genetic modifications.

A well-known, comprehensive illustration of adaptive divergence is found again in the research of the Grants and colleagues. Comparative and phylogenetic analysis of Galápagos finches and sister taxa suggests that the common ancestor of Darwin's finches was relatively small-bodied and small-beaked, akin to present-day warbler finches. The radiation of Darwin's finches toward mostly larger-beaked forms reflects a general trend in island radiations toward larger beak sizes, which may be powered both by "release" from competition in low-diversity environments and by a relative abundance of hard foods in ocean island habitats. The microevolutionary process of adaptation through natural selection has been documented several times for the Daphne finch populations. During the severe drought of 1977 (mentioned previously in the discussion of competition and character displacement), large-beaked members of the *Geospiza fortis* population held a relative survival advantage in being able to crack *Tribulus* seeds, which were available in quantity at that time. The softer, smaller seeds on which smaller-beaked *G. fortis* relied had become depleted, both because of a lack of rain and new seed growth and because of high demand for seeds given the relatively large *G. fortis* population resident on Daphne at the time. As the *G. fortis* population crashed, large-beaked *G. fortis* thus survived to breed in greater numbers, and the subsequent generation exhibited larger beaks than their parental generation.

REPRODUCTIVE ISOLATION

Divergence is a prelude to radiation. In the opinion of many biologists, the currency of radiation—the generation of new species—is achieved only when divergence causes populations to become reproductively isolated from each other. Consider an ancestral population that colonizes multiple island habitats. If offshoot populations remain geographically isolated, they will accumulate, via divergent trajectories of drift or selection, differences in genetic and phenotypic traits. On secondary contact, insufficient reproductive isolation among offshoot populations would result in cross-breeding and reunification of gene pools, thus obstructing speciation. Alternatively, limited cross-breeding success would prevent mixing of gene pools and thus foster continued interpopulation divergence.

In many groups of birds, reproductive isolation is driven in large measure by the divergence of mating signals and mate recognition systems. As mating signals diverge among populations, individuals are less likely to mate erroneously with members of other populations. Hybrid offspring, if viable, may suffer disadvantages in survival and reproduction, thus depressing the fitness of the hybrid's parents. Selection is thus thought to favor individuals that breed within-population, a pressure that in turn favors the divergence of mating signals. In a recent review of this literature, Trevor Price finds that island bird populations show unusually wide divergence from both ancestral and sister species in signals used in mate and species recognition, including color patterns, feather ornaments, and vocal structure.

Reproductive isolation of diverging island birds has been documented by Peter Ryan and colleagues for *Neospiza* buntings of the Tristan da Cunha archipelago. Two species of bunting, *Neospiza acunhae* and *N. wilkinsi*, are recognized on two islands, Nightingale and Inaccessible. On Inaccessible Island, the two species interbreed regularly, as indicated by an abundance of hybrids within a specific habitat (coastal tussock grass) favored by both species. On Nightingale Island, by contrast, reproductive isolation appears to be complete, with the two species forms showing highly distinctive body and beak forms and habitat preferences. Genetic analyses suggest that the diversity of subspecies in this group of birds has emerged through parallel trajectories of within-island interpopulation divergence, with reproductive isolation building up over time within each island, but not yet having reached completion on Inaccessible Island. Divergence among these species appears to have been facilitated not only by habitat segregation but also by the evolutionary divergence of mating signals.

FOCUS EXAMPLE: EVOLUTION OF VILLAGE WEAVERBIRDS ON ISLANDS

The village weaverbird *Ploceus cucullatus*, a species that is common across subsaharan Africa, has been introduced to a number of different islands worldwide (Fig. 1). In the 1700s village weaverbirds were introduced from West Africa to the Caribbean island of Hispaniola (present-day Dominican Republic and Haiti). In the 1880s weaverbirds from South Africa were introduced to Mauritius and Réunion, two islands in the Indian Ocean. In the 1980s the village weaverbird also became established on the Caribbean island of Martinique. The Cape Verde and

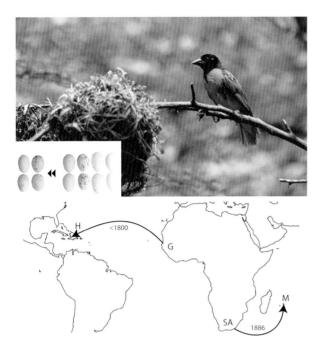

FIGURE 1 Egg color as an example of trait divergence on islands. The village weaver (*Ploceus cucullatus*) has been introduced from two African populations (G = The Gambia, SA = South Africa), to two islands (H = Hispaniola, M = Mauritius). Egg color has evolved in this species in both introduced populations. Variation in egg color and spotting between individuals has decreased, and variation within an individual's eggs has increased. Shown are an example of the range of egg colors in The Gambia (to the right of the paired arrows) and the reduced variation on Hispaniola over 200 years since introduction (to the left of the paired arrows). The pair of eggs in each column are from the same nest. Evolutionary theory predicts this divergence because of the presence of parasitic cuckoos in Africa but not on the islands (see text).

São Tomé Islands, off the coast of Africa, also appear to support small populations.

Evidence suggests that recently established island populations of village weaverbirds are undergoing rapid divergence from their mainland ancestral counterparts. One major axis of divergence hinges upon interactions between weaverbirds and the diederik cuckoo *Chrysococcyx caprius*. Across the African continent, diederik cuckoos lay eggs in the nests of numerous "host" species, including the village weaverbird. Cuckoo chicks normally expel host eggs and nestlings, and unwitting host parents rear parasite chicks until they achieve independence. Given the obvious detriments of brood parasitism, it is not surprising that antiparasite adaptations are common in host species. One such adaptation in village weavers is their ability to identify and eject parasite eggs from their nests, based largely on the visual cues of egg color and spotting. Diederik cuckoos, in response, have evolved eggs that mimic weaver eggs. As a coevolutionary response, weavers have evolved unusually high individual distinctiveness of egg color and spotting, by means of an increased consistency of eggs within clutches, and by means of diverse color patterns among females in a given population. These counteradaptations ensure that cuckoo eggs rarely present a perfect match to a host's eggs, thus helping the weaver to detect and evade the fraud.

In at least two weaver island populations, Mauritius and Hispaniola, there are no diederik cuckoos nor any other brood parasite that mimics weaver eggs. Thus the ancestral traits—wide population variation of egg spotting patterns, individual consistency of egg spotting patterns, and refined egg discrimination abilities—presumably no longer retain their original function. Traits that lose their original functions and attain no new functions are expected to decay through evolution, especially if there is some cost to their maintenance. After one hundred years in the absence of the cuckoo on Mauritius, and after more than two hundred years on Hispaniola, weavers have lost some of their egg spotting, have showed reduced within-bird consistency of egg color and spotting, and express comparatively reduced variation among individuals. As a result of these changes the efficiency of egg rejection has declined as well. Notably, these changes have been more pronounced on Hispaniola, where the weavers had about twice as much time to evolve in the absence of the cuckoo as on Mauritius. Thus in only about 75 and 150 generations, two island populations diverged from their respective source populations.

Other differences have arisen between the island and mainland populations. Egg color of village weavers on the islands of Mauritius and Hispaniola, for instance, are converging on a shade of medium blue-green that is roughly what would be expected if eggshell pigment were to evolve to optimally protect embryos from solar radiation, and ultraviolet (UV) rays in particular. This convergence suggests that in Africa egg colors tend to be determined mostly by the importance of avoiding cuckoo parasitism, whereas in the absence of the cuckoo the egg colors are determined by a weaker source of selection, possibly avoidance of damage from solar radiation. Some divergent behavioral traits might be learned or innovated rather than having evolved. For instance, weavers nest higher in trees on Mauritius than they do in South Africa, which makes sense because hawks are a dangerous predator of nests in South Africa but not on Mauritius. Also, weavers depend on rainfall and the proximity of water in order to breed in Africa, but on Hispaniola they were found to breed in a remarkably arid region, with the lack of nearby water mitigated by the presence of a common juicy cactus fruit ("pitaya," *Stenocereus*).

Further insights into the evolutionary divergence of island populations are being gleaned from genetic analyses. In village weaverbirds, DNA has been found to be more variable in a continental African population than in a similarly sized region of Hispaniola. More specifically, DNA sequences from weaver mitochondria, which are transmitted from mother to offspring, are identical in all Hispaniolan birds sampled, whereas their ancestral population in The Gambia shows a great deal of variation. This result suggests the action of a founder effect on Hispaniola, which might be partly responsible for some of the evolutionary changes observed in egg features, most plausibly the decrease in egg color and spotting pattern variation. On the other hand, variation in nuclear genes, which are inherited by offspring through both the mother and father, have declined to a lesser extent following introduction. This indicates that female-inherited mitochondrial DNA has experienced a stronger or more persistent founder effect than biparental nuclear DNA. The contrast here might be explained by genetic drift or, alternatively, might indicate genetic peculiarities of the species introduction. Given that only males sing and have bright plumage, it is likely that many more males than females were brought to Hispaniola. If so, then perhaps only a few forms of weaver mitochondrial genes were retained across generations, whereas the large number of males ensured that a greater variation in nuclear DNA were maintained in the introduced Hispaniolan population. In summary, these weaver studies illustrate how species introductions can drive ecological and genetic changes that underlie divergence of populations, the raw material for adaptive radiation.

FOCUS EXAMPLE: EVOLUTION OF REPRODUCTIVE ISOLATION IN GALÁPAGOS FINCHES

Reproductive isolation and speciation in Galápagos finches have been facilitated by a diverse array of factors. The dispersed geography of the Galápagos archipelago, which contains some isolated peripheral islands, has caused cross-island immigration events to be relatively rare. Population sizes of these birds tend to be small, and population characteristics are thus subject to unusually rapid change via genetic drift or natural selection. A relative paucity of other terrestrial bird species has allowed Darwin's finches to occupy a broad range of ecological niches. Despite the plethora of evolutionary pressures driving populations apart, however, speciation in this clade remains a fairly rare occurrence, as can be observed in the propensity for diverging populations to undergo fusion via hybridization.

The regular occurrence and viability of hybrids indicates that species barriers in this group of birds are normally maintained through pre-mating rather than post-mating barriers. The question of whether populations in secondary contact will fuse or split thus depends largely on the behavioral processes by which Darwin's finches choose their mates. As in many other songbird groups, male Darwin's finches compete for access to females, who in turn choose mates. Towards this end, females attend to mating cues or signals expressed by males, in part to assess the potential quality of male partners as parents for their offspring. Such abilities result from a process of sexual selection. Females are also presumably under selection to be able to identify males that are of their own population or species, given that hybrid offspring production might be disadvantageous.

A primary cue by which females might recognize preferred mates is the size and shape of the beak. Studies on the responses of territorial males to taxidermic mounts provide indirect support for this idea. Females also appear to assess male identity and quality by attending to male vocal mating signals, or songs, which have been shown to vary across species and populations. Territorial male finches have been shown to respond more aggressively to playback of songs of their own species, population, and locality.

Recent research on vocal mechanics and performance in Darwin's finches and other songbirds indicates a relationship between song structure and beak morphology. The beak and upper vocal tract of songbirds serve as a resonance chamber for sounds produced at the syrinx (the sound-producing organ of birds). As birds modify source frequencies, they often conduct parallel modifications of vocal tract structure, presumably in order to retain the resonance function across a wide range of frequencies. These modifications include (but are not limited to) changes in beak gape. For example, when birds produce high-frequency notes, they will often flare their beaks, thus reducing the volume of the vocal tract and maintaining an effective resonance function for the vocal tract. With respect to variation in beak morphology, it is thought that birds with large beaks should face relatively severe limitations in being able to rapidly adjust beak gape, whereas birds with small beaks should face fewer such limitations. This expectation is based on a predicted trade-off between bite force production

FIGURE 2 Evidence of assortative mating by beak morph in a population of Darwin's finches. The population of medium ground finches (*Geospiza fortis*) at El Garrapatero, Santa Cruz Island, shows distinctive small- and large-beaked morphs, with few birds of intermediate beak size. Breeding season observations reveal that females generally choose as mates males of similar beak size. The panels show representative pairs of breeding finches matched for being small-beaked (A) and large-beaked (B).

and the velocity by which birds can open and close their beaks while singing. The vocal tract constraint hypothesis posits that song structure should evolve as an incidental byproduct of beak evolution. These predictions have been borne out to some extent in field studies of Darwin's finches.

Correlations between vocal signals and visual cues would appear to provide a measure of redundancy for females attempting to assess male identity. This redundancy could potentially facilitate the process of reproductive isolation, by providing females with redundant information they could use to make "correct" mating decisions. This possibility has been raised in a recent study of *G. fortis* at El Garrapatero on Santa Cruz Island. This population features birds of either large beaks or small beaks, with few birds of intermediate beak size. These beak morphs produce distinctive songs, in a manner consistent with a vocal tract constraint hypothesis. Females of this population thus can rely on both visual cues and vocal signals when selecting mates. Indeed, females have been found to choose mates with similar beak sizes to their own (assortative mating, Fig. 2). Persistence of strong assortative mating in the absence of gene flow would presumably lead to further genetic divergence, further song divergence, and strengthened barriers to premating isolation.

SEE ALSO THE FOLLOWING ARTICLES

Convergence / Ecological Release / Flightlessness / Fossil Birds / Founder Effects / Galápagos Finches / Island Biogeography, Theory of / Population Genetics, Island Models in

FURTHER READING

Grant, P. R., and B. R. Grant. 2006. Evolution of character displacement in Darwin's finches. *Science* 313: 224–226.

———. 2008. *How and why species multiply: the radiation of Darwin's Finches.* Princeton, NJ: Princeton University Press.

Huber, S. K., L. F. De Leon, A. P. Hendry, E. Bermingham, and J. Podos. 2007. Reproductive isolation of sympatric morphs in a bimodal population of Darwin's finches. *Proceedings of the Royal Society of London Series B Biological Sciences* 274: 1709–1714.

Lahti, D. C. 2005. Evolution of bird eggs in the absence of cuckoo parasitism. *Proceedings of the National Academy of Sciences of the United States of America* 102: 18057–18062.

Mayr, E., and J. M. Diamond. 2001. *The birds of northern Melanesia: speciation, ecology, and biogeography.* New York: Oxford University Press.

Podos, J. 2001. Correlated evolution of morphology and vocal signal structure in Darwin's finches. *Nature* 409: 185–188.

Price, T. 2008. *Speciation in birds.* Greenwood Village, CO: Roberts & Company Publishers.

Ricklefs, R. E., and E. Bermingham. 2007. The causes of evolutionary radiations in archipelagoes: passerine birds in the Lesser Antilles. *American Naturalist* 169: 285–297.

Ryan, P. G., P. Bloomer, C. L. Moloney, T. J. Grant, and W. Delport. 2007. Ecological speciation in South Atlantic Island finches. *Science* 315: 1420–1423.

Schluter, D. 2000. *The ecology of adaptive radiation.* Oxford: Oxford University Press.

Weiner, J. 1994. *The beak of the finch: a story of evolution in our time.* New York: Knopf.

Yamagishi, S., M. Honda, K. Eguchi, and R. Thorstrom. 2001. Extreme endemic radiation of the Malagasy vangas (Aves: Passeriformes). *Journal of Molecular Evolution* 53: 39–46.

BORNEO

SWEE-PECK QUEK

Harvard University, Cambridge, Massachusetts

Straddling the equator between 7° N and ~4° S at the southeasternmost extremity of the Eurasian continental crust, Borneo, at ~740,000 km², is the third largest island after Greenland and New Guinea. Early explorers in the Indo-Malaysian archipelago encountered a Borneo that was cloaked in dense rain forests from coast to coast and inhabited by headhunting tribes. The original forests of Borneo have been reduced to about half their former extent but still hold an exceptionally rich biota teeming with endemics. Borneo's forests rank among the most diverse of the world's rain forests, and, as recently as 2005, have even yielded a new mammal species. Borneo forms part of the Sundaland Biodiversity Hotspot and is home to two World Heritage Sites: Kinabalu Park and Gunung Mulu National Park (Fig. 1).

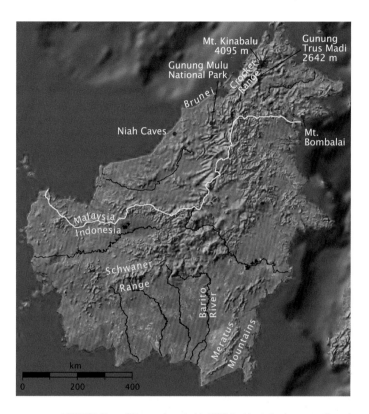

FIGURE 1 Map of Borneo (created in ESRI ArcMap). Features mentioned in text are shown.

GEOGRAPHY

Borneo is under the administration of three nations (Fig. 1). Just under three-quarters of its area is in the Indonesian state of Kalimantan, about 27% lies the Malaysian states of Sabah and Sarawak, and less than 1% is occupied by Brunei. In 2006, the population of Borneo was estimated at 17.7 million, of which 17% were indigenous Dayak. Commercially extracted minerals include diamonds and other precious and semi-precious stones, iron ore, gold, and limestone. Borneo also produces petroleum, coal, and natural gas.

Over half of Borneo lies below 150 m. Geomorphically, it is dominated by a central mountain massif, which peaks in the north at Mount Kinabalu. At 4095 m, it is the highest peak between the Himalayas and New Guinea. The next highest peak on Borneo, Gunung Trus Madi, is a distant 2642 m. From the central mountains, the major rivers of Borneo radiate toward the coast and constitute major routes of transportation and communication.

Borneo experiences a perhumid climate that is warm and wet year-round. Most months experience more than 200 mm of rainfall, and most of the island receives more than 2000 mm annually. As much as 7000 mm of rainfall per year has been recorded in Gunung Mulu National Park. In the lowlands, daily temperatures typically range between 25 °C and 35 °C, and on Mount Kinabalu, below-freezing nighttime temperatures are common. The wet northeast monsoon prevails from November to April, and a milder southwest one from May to October.

TECTONIC AND GEOPHYSICAL HISTORY

Borneo, together with Sumatra, Java, and the Malay Peninsula make up the major land masses of Sundaland, the currently emergent portions of the Sunda shelf (Fig. 2). The Sunda shelf is the southeasternmost extension of the Eurasian continental shelf, a tectonic block that appears to be moving independently of Eurasia. The geophysical history of this block has been dynamic and highly complex. It has experienced changing climates and fluctuating sea levels superimposed upon a longer-term intricate tectonic history of micro-plate movements with their associated subduction zones, island arcs, continental volcanism, and orogenic activity with subsequent erosion. Not surprisingly, the history of Sundaland, and thus of Borneo, is incompletely understood. The following account reflects our current knowledge.

By the late Mesozoic, the rudiments of the Sunda Shelf had formed. The portions that make up Borneo comprise ophiolites (sections of oceanic crust infused with upper mantle), island arc material, and bits of continental crust derived from Gondwana and South China that had accreted onto a Paleozoic core (visible in the Schwaner Range, Fig. 1). This was followed by sediments laid down in the Eocene and Oligocene, forming most of the Malaysian and Bruneian territory, completing the major geologic regions of the island.

The current stability of Borneo belies a much more dynamic past. From 25 to 10 million years ago, Borneo underwent a counterclockwise rotation of ~40°, arriving at its present orientation. Beginning with the onset of rotation, tectonically active boundaries abutted what is now its northern portion. Reconstruction models suggest that this rotation was accompanied initially by increased volcanism, which has all but subsided, leaving only pockets of geothermal activity around Sabah, including Mount Bombalai, Borneo's youngest (and currently inactive) volcano. Subduction at the northern boundaries through the Oligocene to early Miocene gave rise to the Crocker range and, during the middle Miocene, to the central mountains. The Meratus Mountains in the southeast were uplifted sometime during the Neogene. Unusually thick sediments have developed on and around Borneo as a consequence of rapid Neogene erosion of its mountains. Within the past 15 million years, a granitic pluton intruded the Crocker Range, and its subsequently dramatic uplift and exposure produced Mount Kinabalu, whose apex was glaciated throughout much of the Pleistocene. Despite the change in shape experienced by Sundaland during the Cenozoic, Borneo's latitude has remained little changed.

Borneo's identity as an island has a relatively recent history. By the Late Cretaceous, the Sunda Shelf was above sea level, and a large area spanning the Malay Peninsula and western Borneo appears to have remained elevated throughout most of the Cenozoic, with the extent of emergent land in Borneo increasing through the Miocene and Pliocene. During the Pleistocene (and to a lesser extent, the Pliocene), oscillating sea levels, coinciding with Milankovitch-driven glacial-interglacial cycles, caused brief but regular submersions of parts of the Sunda Shelf, creating islands as seen today. The current insular (rather than entirely exposed) configuration of the shelf is the result of high-standing sea levels associated with the current interglacial phase and concomitant perhumid climate in equatorial regions favoring the development of everwet forests. Approximately a dozen such interglacial phases have occurred within the past million years. By contrast, glacial phases produced low-standing sea levels (and thus greater exposure of the Sunda Shelf) and cooler and drier climates, causing the contraction of everwet forests and the expansion of more seasonal forest types. During the Pleistocene, the vegetation of Sundaland, and therefore of Borneo, thus alternated between the everwet forests, as seen at present, and more seasonal forest types similar to those currently found in Thailand.

BIOTA

The biodiversity of the Asian biogeographic province culminates in the Sundaland Biodiversity Hotspot, comprising Borneo, Sumatra, Java, and the Malay Peninsula. Borneo's biota thus has much in common with its Sundaland neighbors, in particular Sumatra, its closest neighbor (~400 km to the west at its closest point) on the Sunda Shelf. Within Sundaland, Borneo's biota is the richest, boasting famous members such as the orangutan, the parasitic flowering plant *Rafflesia* with

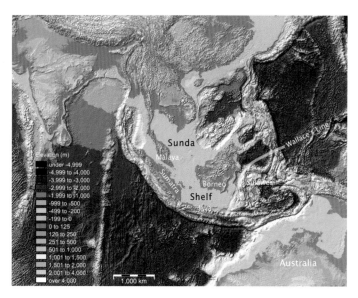

FIGURE 2 The geographic and bathymetry context of Borneo and the Sunda Shelf (created in ESRI ArcMap).

its giant flowers, the Borneo pygmy elephant, the river dolphin, the smallest rhinoceros species (the Sumatran rhino), and the comical, endemic proboscis monkey, all of which are endangered or critically endangered. During the Pleistocene, sea levels were as much as 120 m below present levels, exposing the entire Sunda Shelf (Fig. 2) and facilitating biotic exchanges among what are now (pen)insular lands. The last land connection between Borneo and Sumatra may have been severed as recently 10,000–8000 years ago. Studies of several faunal groups distributed between Borneo and the other Sundaland areas suggest vicariant lineages that diverged from one another at time depths ranging from the Miocene through to the Pliocene and Pleistocene. Despite the opportunities for biotic exchange afforded by frequent and prolonged land bridges, Borneo's biota contains a fair proportion of endemics, ranging from 9% in resident birds to 60% in frogs (Table 1). Invertebrates have been poorly studied in proportion to their diversity, but available data suggest endemism rates of up to 30–40% as suggested by the Lepidoptera and Coleoptera (Table 1). For many groups, Borneo's endemism rate is the highest within Sundaland. Many of the endemics are apparently concentrated in the northern and northwestern regions approximated by Malaysian territory. This high rate of endemism may be the result of (1) repeated isolation of the entire island and of specific habitats due to the cyclical climatic conditions of the Pleistocene, (2) high elevation (Mount Kinabalu in particular) creating persistent "islands" of

TABLE 1
Estimated Species Diversity and Endemism for Select Groups in Borneo

Taxon	Number of Species	Number of Endemic Species	Proportion Endemic
Plants	15,000	6000	40%
Land mammals	223	45	20%
Resident birds	434	39	9%
Snakes	153	31	20%
Frogs	150	90	60%
Lizards	109	37	34%
Turtles	19	2	11%
Freshwater fish	424	179	42%
"Macro" moths	4250	1120	26%[a]
Lepidoptera	8000–10,000	unknown	28–30%[b]
Galerucine beetles (Chrysomelidae)	232	91	39%

NOTE: Some of the figures are likely underestimates of true diversity because of lack of comprehensive study. Estimates for endemism are particularly problematic, because they additionally reflect the extent of research (or lack thereof) on neighboring regions, especially Sumatra.
[a]Endemism rates range from 1% to 34% across some better-studied groups.
[b]Based on projected estimate.

temperate-aspect conditions during the Pleistocene intrusions of perhumid climatic phases, (3) continuity of wet conditions throughout the Pleistocene in some regions (favorable to the persistence of everwet rainforest refugia) despite the cooler, relatively drier climate characteristic of that epoch, (4) extinction elsewhere, or (5) inadequate research elsewhere.

A mere ~120 km to the east of Borneo lies Sulawesi, the flagship of Wallacea (the biogeographic province famed for its intermingling of Oriental and Australian biotas). Although this distance is less than a third of that between Borneo and Sumatra, the biotas of Borneo and Sulawesi bear surprisingly little resemblance to each other given their proximity. This biotic divide, first noted by Alfred Russel Wallace in 1858, has been maintained by a marine trench over 1500 m deep (more than 2000 m deep along most of its length), presenting a formidable barrier to migration even at the lowest Pleistocene sea levels. For some taxa, however, the Wallace Line was a filter rather than a barrier, as indicated by the presence of Australian elements in the biota of Sundaland (e.g., *Phyllocladus* and *Podocarpus* in the Podocarpaceae; *Casuarina;* several genera of Myrtaceae; etc.) and Asian elements east of Wallace's Line. The migrations that did occur between Borneo and Sulawesi may have taken place via the Philippine islands in the north, via the islands to the south, or by wind or water dispersal. Wallace's Line acquires relevance only beginning with the early Miocene. Prior to that, the islands of eastern Indonesia—the stepping stones and mixing zones that now make up Wallacea—were largely not yet in place, separating Sundaland from Australia by a large oceanic expanse offering little opportunity for biotic exchange.

The perhumid climate of Borneo produces evergreen (or everwet) forests, as opposed to the seasonal, monsoon, or savanna forests that also occur in the more seasonal parts of Southeast Asia. The original vegetation cover of Borneo comprises predominantly lowland dipterocarp forest—the most species-rich forest type with a canopy dominated by members of the Dipterocarpaceae. Also found in the lowlands are mangrove forests, peat swamp forests, freshwater swamp (or flooded) forests, and heath forests (associated with nutrient-poor sandy soils). Limestone and ultrabasic soils occurring in scattered pockets on Borneo also support distinct forest types. At higher elevations, the montane forests have a lower canopy dominated by members of Fagaceae and Lauraceae. Subalpine forest and alpine scrub vegetation occur on the high elevations of Mount Kinabalu. The Kinabalu flora contains lineages with affinities to other high mountain floras stretching from Sumatra to New Guinea, and to high-latitude floras in the Sino-Himalayan region and Japan, suggesting that these Kinabalu lineages represent relicts from a time when vegetation associated with a cooler and/or more seasonal climate (such as occurred during the Pleistocene glacial phases or prior to the Miocene) enjoyed a more continuous distribution throughout the region.

HUMAN COLONIZATION

The Niah Caves in Sarawak have yielded the longest record of human occupation in insular Southeast Asia, and archeological excavations there in 1958 unearthed a skull that has proven thus far to be the earliest modern human recorded in eastern Asia, dated at 45,000–39,000 years ago. The skull was deemed to bear morphological affinities to Australo-Melanesians (a diverse group that includes Australian aborigines, Melanesians, and Negrito populations scattered across the Philippines, the Malay Peninsula, and the Andaman Islands), in particular Tasmanians, thus suggesting little relation to the current inhabitants of Borneo.

Around 4500 years ago, an Austronesian-speaking agricultural people expanded out of Taiwan into the northern Philippines to begin a spectacular colonization of islands that spanned more than half the globe. From the Philippines, they expanded into northern Borneo possibly by 4000 years ago (and certainly by 2500 years ago). Within Borneo, the settlers spread throughout the coastal and hinterland regions, differentiating into the various ethnic and linguistic groups that are now indigenous to the island. Migrations from neighboring Austronesian-speaking populations, largely to coastal regions, have also contributed to Borneo's ethnic diversity. Between 2000 and 1500 years ago, a group of Ma'anyan speakers, whose present-day descendents occupy the Barito river region in southern Borneo, arrived in Madagascar to establish an isolated outpost of Austronesian speakers to the west.

The native inhabitants of present-day Borneo can be loosely divided into the interior peoples, often collectively referred to as the Dayak, and the coastal Malay/Muslim peoples. Together, they fall into between ten and 50 ethnic groups, depending on the classification criterion consulted. The interior versus coastal distinction, however, is often challenged because of intermarriage, migration, and religious conversion. The indigenous languages of Borneo, many extinct or endangered, number approximately 50 or more (depending on the criteria used to distinguish languages from dialects) and fall into ten Austronesian linguistic subgroups. The coastal peoples traditionally rely on fishing and trade, whereas the interior peoples (with some exceptions) generally share the custom of living in long communal dwellings and practice a mix of cultivation (fruit and nut trees, rice, and tubers) and hunting and gathering of forest products. One group, the egalitarian Penan, were by contrast predominantly nomadic hunter-gatherers until forced into a more settled lifestyle in recent decades by large-scale forest destruction. A few Penan have resisted settlement and continue to gather and hunt. As with the Penan, the ancestral ways of life for many indigenous groups are vanishing under the pressures of development and deforestation.

Borneo's wealth in gold, diamonds, and other natural products has sustained the interests of polities near and far since early historical times. In the first few centuries AD, contact with India led to the establishment of a Hindu kingdom in Borneo, and Chinese historical records from the sixth to seventh centuries show trading links with Borneo. Records of contact with Java appear by the ninth century, and Islam became established in Brunei possibly by the fifteenth century. Europeans first visited the island in the early sixteenth century, and the seventeenth century onward saw the arrival of Chinese gold miners. In the nineteenth century, the Dutch and British staked out territories, and the subsequent establishment of colonial rule opened the door to western missionaries, explorers, and naturalists. The Bugis (a maritime people historically centered in southern Sulawesi) and Siamese have also ranked among Borneo's trading partners.

The coastal centers of present-day Borneo are also home to a sizeable number of inhabitants of Chinese and Indian descent who migrated for trade and commerce in recent centuries. Recent policy-driven movement of peoples has also brought in ethnic groups from other parts of Indonesia and Malaysia. In the past decades, timber and petroleum mining have also drawn people from around the world to Borneo.

CONSERVATION

Much of Borneo's original forests have been laid bare by logging and have given way to plantations of oil palm and rubber, cultivated land, secondary regrowth, human settlements (especially from transmigration), and commercial activity. An increasing frequency of fires has also destroyed millions of hectares. Less than 50% of original forests remain on Borneo, and deforestation continues at an estimated 3.9% per year. Although 9% of Kalimantan, 8% of Sarawak, and 14% of Sabah have been officially declared as protected, those areas continue to face threats from human encroachment, hunting and poaching, mining, fires, and illegal logging (encouraged by corruption and known to have felled up to half of one of Borneo's national parks). Many species have probably already become extinct in recent decades without ever having been recorded by scientists. However, the interior-most reaches of the island still contain large swaths of virgin rain forests where 52 new species of plants and animals (bearing in mind severe underreporting in arthropods) have been found within the past year alone. A project led by the World Wildlife Fund has been undertaken to conserve this area, occupying nearly one-third of the island, and may offer survival opportunities for many critically endangered species.

SEE ALSO THE FOLLOWING ARTICLES

Deforestation / Indonesia, Geology / New Guinea, Biology / Peopling the Pacific / Wallace's Line

FURTHER READING

Bellwood, P. 1997. *Prehistory of the Indo-Malaysian archipelago*. Honolulu, HI: University of Hawai'i Press.
Butler, R. A. 2006. www.mongabay.com/borneo.html#conservation
Hall, R. 2001. Cenozoic reconstructions of SE Asia and the SW Pacific: changing patterns of land and sea, in *Faunal and floral migrations and evolution in SE Asia-Australasia*. I. Metcalfe, J. M. B. Smith, M. Morwood, and I. Davidson, eds. Lisse: A. A. Balkema Publishers (Swets and Zeitlinger), 35–56.

MacKinnon, K., G. Hatta, H. Halim, and A. Mangalik. 1996. *The ecology of Kalimantan*. Hong Kong: Periplus Editions.
Morley, R. J. 2000. *Origin and evolution of tropical rain forests*. Chichester, UK: John Wiley & Sons.
Muller, K. 1996. *Borneo: journey into the tropical rainforest*. Lincolnwood, IL: Passport Books (NTC Publishing Group).
Rautner, M., M. Hardiono, and R. J. Alfred. 2005. *Borneo: Treasure Island at risk. Status of forest, wildlife and related threats on the Island of Borneo*. Frankfurt am Main, Germany: World Wildlife Fund. Available at http://www.panda.org/about_wwf/where_we_work/asia_pacific/our_solutions/borneo_forests/publications/index.cfm?uNewsID=21037
Taylor, D., P. Saksena, P. G. Sanderson, and K. Kucera. 1999. Environmental change and rain forests on the Sunda shelf of Southeast Asia: drought, fire and the biological cooling of biodiversity hotspots. *Biodiversity and Conservation* 8: 1159–1177.

BOUGAINVILLE

SEE PACIFIC REGION

BRITAIN AND IRELAND

ROSEMARY G. GILLESPIE
University of California, Berkeley

MARK WILLIAMSON
University of York, United Kingdom

Britain and Ireland are a pair of archipelagoes of which the island of Britain, which comprises most of England, Scotland and Wales, is the largest (Fig. 1). The next largest island is Ireland, comprising Northern Ireland, which is part of the United Kingdom (UK), and the Irish Republic. The islands range from 60°51′ N in the north (Muckle Flugga, Shetland Islands) to 49°51′ N (Pednathise Head, Isles of Scilly) in the south, and from 10°35′ W in the west (Tearaght Island, Kerry, Ireland) to 01°45′ E in the east (Lowestoft, Suffolk, England). The total area of the UK is about 243,000 km² (England, about 130,400 km²; Scotland, 78,300 km²; Wales, 20,750 km²; and Northern Ireland, 13,850 km²). The Irish Republic is about 70,280 km². The islands are well known for a long and complex political history, which is replicated by an equally complicated geological history. For example, in the context of islands, more than 1000 smaller islands ring the two largest, including the Isle of Man, the Hebrides, the Orkneys, and many others. Given the diversity and complexity of the islands, the following review is necessarily cursory. The interested reader left wanting more about the biology and geology of Britain and Ireland should refer to the Further Reading at the end of this account.

FIGURE 1 Map of Britain and Ireland showing the main islands mentioned in the text.

CLIMATE AND TOPOGRAPHY

The islands vary from low-lying in eastern England to mountainous in many parts of Wales, Scotland, and Ireland. Important mountain chains include the Pennines, which form a ridge down northern England; the Cambrian Mountains, which stretch across Wales (Snowdon in the northwest rises to 1,085 m); the Sperrin, Antrim, and Mourne Mountains in Northern Ireland; a number of peaks such as Errigal in Donegal, Croagh Patrick in Mayo, and the Twelve Bens in Galway, in the Irish Republic; and several ranges in Scotland, the highest being the Grampian range with Ben Nevis (1,343 m, the highest point in the islands).

The climate is mild, cool-temperate, and oceanic, warmer than one might expect given the latitude, largely as a result of the Gulf Stream, which warms the air currents across the Atlantic and makes the rainfall unpredictable.

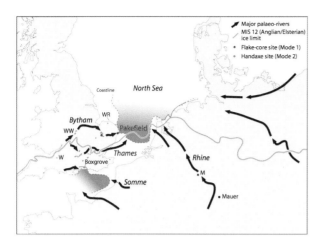

FIGURE 2 A reconstruction of the paleogeography of northwest Europe during the early Middle Pleistocene. These sites provide a record of intermittent early human occupation of ~300,000 years duration. Human remains are known from Mauer and Boxgrove. M = Miesenheim I, W = Westbury-sub-Mendip, WR = West Runton, WW = Waverley Wood. Recent reports from flint artifacts at Pakefield (Suffolk), from an interglacial sequence provide the earliest (~700,000 years old) evidence for human presence north of the Alps. From Parfitt et al. (2005), reprinted by permission from Macmillan Publishers Ltd.

QUATERNARY HISTORY

The ancient geologic history of Great Britain and Ireland is closely related to that of eastern North America and is covered elsewhere in the volume. Since its separation from North America, one of the key geological periods affecting the biology of the region was the Quaternary. This period was dominated by a series of recent ice ages, resulting in U-shaped valleys in highland areas carved by glaciers, and yielding fertile, stony soils in southern Britain.

Britain has been occupied by humans intermittently for at least 500,000 years. During that time sea levels have dropped and returned to near-current levels, repeatedly linking the country to continental Europe. The richest Paleolithic record is found during the first Hoxnian Interglacial, about 400,000 years ago (Fig. 3). The subsequent period, the Lower–Middle Paleolithic transition about 300,000–180,000 years ago, marked the first appearance of the new Levallois flint knapping technique. The population collapsed during the Middle Paleolithic, between 180,000 and 60,000 years ago. Humans recolonized Britain before the height of the most recent glacial of 22,000–13,000 years ago. During the late Devensian, the period of maximum ice advance, polar desert conditions likely precluded human habitation. Recolonization again occurred after the last glacial maximum, between 13,000 and 8,800 years ago.

Before the end of the Devensian, the islands of Britain and Ireland were all part of continental Europe. The North Sea and almost all of those islands were covered with ice. The sea level was about 120 m lower than it is today, and much of the English Channel consisted of low-lying tundra. The Thames, Seine, and Rhine joined and flowed along the (above sea level) English Channel as a wide, slow river, eventually reaching the Atlantic Ocean. Ice sheets (about 1,500 m thick) covered northern Britain, Ireland, and much of the Continent and blocked the North Sea, creating a giant lake, which overflowed, cutting deeper valleys through the "land bridge." Scandinavian and British ice caps subsequently merged during the glacial maximum and impounded a large lake fed by the Rhine. In the end the water poured out through the lowest gap in the hills, with the deepening valleys becoming the Straits of Dover after the Ice Age. As the climate warmed 10,000 years ago, the rivers again flowed out to the North Sea (Fig. 2). Early humans and animals returned to northern France and crossed to England over the still-emergent English Channel. New forests and grasslands covered the lowland, and humans were attracted to the abundant game.

A key factor in shaping biodiversity in Britain was the transition from hunting to husbandry and the spread of economic and cultural elements associated with Neolithic farming across Europe between the seventh and fourth millennia BC, with associated management of woodlands. Although practices differed in England, as compared to

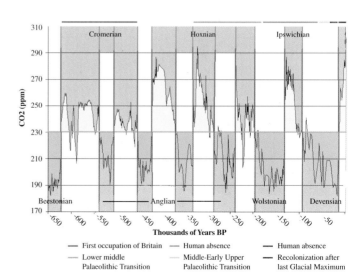

FIGURE 3 Late Pleistocene in Britain: Atmospheric CO_2 (ppm) and glacial cycles (650,000 years ago–present). Glacial periods affecting the UK during the Quaternary. Data from Petit et al. (1999); Raynaud et al. (2000); Thompson (2000).

Scotland, Ireland, and Wales, overall woodland cover in the British Isles declined from an estimated 80% in 5000 BC to 20% in 1000 AD and to only 4% in the early twentieth century, based on data from the Forestry Commission/Countryside Commission (1996); however, recent years show increased cover (currently close to 12%). Given the long history of fragmentation, any species with large area requirements would have been lost long ago, with the remaining species adapted to the fragmented landscape. Initiatives now under way will recreate woodland networks and combat ecological isolation without the need for large-scale expansion of woodland.

Early and middle Holocene events dramatically influenced the landscape and vegetation of today. Even more recent events, many part of written history, underscore that ecosystem changes have been dictated by human land use. There remain no true "natural areas." For example, the late eighteenth century heralded major changes in the Scottish Highlands economy with the creation of extensive sheep farms, setting aside land for recreation. As people left the Highlands (the "Highland Clearances"), sheep, and deer and capercaillie (extinct in Scotland since 1785, subsequently reintroduced), managed for game species, created apparent "natural woodlands" and "native species" that are actually replacements or reintroductions, often from different parts of Britain or Europe. The biota of Ireland also reflects a history of "natural" reintroduction after the last Ice Age, coupled with a long history of human influence. It appears that much of the biota reached Ireland by means of land bridges. A biogeographic mystery is the existence of "Lusitanian" species—species that appear to have reached Ireland directly from the region of Lusitania (northwest Iberia), the best-known examples being the natterjack toads (Ireland's only toad species), the Kerry slug (*Geomalacus maculosus*), and the strawberry tree *Arbutus unedo* (Ericaceae).

CURRENT IMPACTS

Current human impacts in Britain and Ireland appear deceptively muted. A recent study by Jeremy Thomas and colleagues has suggested that this may be because (1) major removal of primary vegetation occurred earlier; (2) climate warming, to date, has enhanced the net capacity of British ecosystems to support certain species; (3) exotics have not colonized British ecosystems with the frequency and damaging impacts found in many communities elsewhere; and (4) targeted conservation measures, including regulation of collecting and hunting, have reversed the former declines of several species. Despite these ameliorating effects, a recent survey showed that 28% of native plant species have decreased in Britain over the past 40 years, and 54% of native bird species and 71% of butterfly species have decreased over the last ~20 years. In 1993, in response to the Rio Convention, the UK Biodiversity Action Plan was established to deal with biodiversity conservation. The unit is charged with describing the UK's biological resources and planning for the protection of these resources. These tasks are well under way (see http://www.ukbap.org.uk/).

FLORA

Repeated glaciation and recolonization has led to a depauperate flora, though, using an estimate of 1363 recognizable (morphologically distinct) species and well over 800 microspecies (often morphologically indistinguishable) and 400 hybrids (many sterile), the islands are estimated to contain about 2800 species of native flowering plants and ferns (see Fig. 4B for patterns of diversity). The number of endemic species is controversial, as it depends on the taxonomic status of microspecies (i.e., apomictically reproducing plants that are genetically identical from one generation to the next, so that each has diagnostic characters as in true species, but differences between such species are much smaller than usual, making them very difficult to identify). Such species are common in the genera *Hieracium, Limonium, Taraxacum, Rubus, Sorbus,* and *Festuca.* If microspecies are accepted as legitimate, there are approximately 470 endemic species including microspecies in the genera *Rubus* (200), *Hieracium* (149), and *Taraxacum* (39), together with 30 endemic subspecies. However, although microspecies may display local endemism, their conservation may be less critical relative to the standard sexually reproducing species because of their lower genetic diversity, together with their often dubious and confusing taxonomy. In *Taraxacum,* for example, there is a hierarchy of groups and species lines that can be drawn at different places in the hierarchy. However, there is some interest in the value of these species in generating biodiversity, for example the formation of multiple *Sorbus* species through hybridization on the island of Arran (see discussion of Arran below).

An analysis of the flora of the UK by Chris Preston and colleagues indicated that the most species-rich location in Britain is in Dorset, around Wareham, where a wide range of species-rich habitats includes heathland, chalk grassland, two lowland rivers with grazing marsh ditches, and a range of coastal plant communities. This site contains 844 species, including 56 rare and scarce species (38 are native to the area, 13 are native elsewhere in Britain but here only as garden escapes, and 5 are archeophytes, species introduced before 1500 AD).

Most of the British plants share affinities across northern Europe. An interesting exception is the several Lusitanian heathers from Ireland, though these may have reached the island in packing around Spanish wine casks.

FAUNA

The fauna of the British Isles has been shaped by a history of change, both natural/climatic and anthropogenic, which has decimated some of the more restricted species while providing conditions ensuring the survival of others. As with the plants, repeated glaciation has resulted in a relatively small fauna, which shows a strong northwest–southeast diversity gradients (Fig. 4A).

Mammals

There is evidence of many more mammals in prehistoric times than currently occur in the islands, their disappearance linked to the Holocene extinction event. For example, the cave lion, or European lion, *Panthera leo spelaea,* is an extinct subspecies of lion known from fossils and prehistoric art. Cave paintings and remains found in ancient campsites indicate that they were hunted by early humans, who may also have contributed to their demise. Similarly, the cave hyena, *Crocuta crocuta spelaea,* an extinct subspecies of spotted hyena native to Eurasia and known from fossils and prehistoric cave art, is thought to have become extinct near the end of the last ice age. A similar time frame and reason (climate change) has been proposed for the demise of the cave bear, *Ursus spelaeus,* which lived in Europe during the Pleistocene and became extinct about 20,000 years ago.

Many well-known British mammals have a much broader range outside Great Britain and Ireland. The red squirrel, *Sciurus vulgaris,* is common through Eurasia, though in Britain and Ireland its numbers have decreased drastically largely because of the introduction of the larger eastern gray squirrel from North America. A number of conservation efforts have been established in the UK, including culling and removal programs for the eastern gray squirrel. Interestingly, recent genetic studies by Marie Hale and colleagues have shown that, although British populations themselves show little genetic structure because of documented translocations within the country, northeastern British *S. vulgaris* populations appear to contain mostly individuals with a recent (last 40 years or so) continental European (mostly Scandinavian) ancestry.

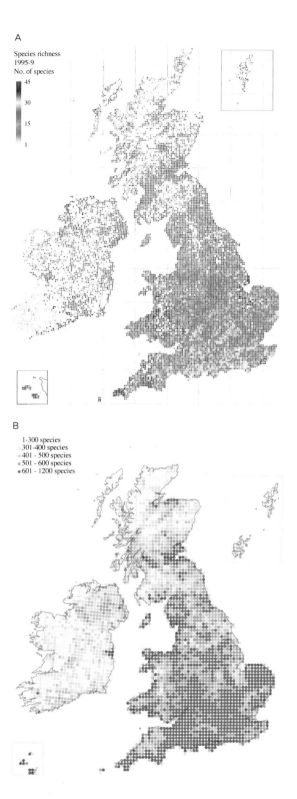

FIGURE 4 Species diversity patterns in butterflies and plants. (A) Number of butterfly species recorded in each 5-km square in the Butterflies for the New Millenium survey. From Asher *et al.* (2001:45), reprinted by permission from Oxford University Press. (B) Number of species of plants recorded in each 10-km square from 1970 onward. From Preston *et al.* (2002:28).

Other well-known British mammals that have broad ranges include the European hedgehog, *Erinaceus europaeus*, which occurs throughout the Palaearctic region, except in the Himalayas and North Africa. Although native to much of Britain, it has only recently (last 30–40 years) been introduced into the Western Isles and Orkneys of Scotland, where it has caused serious declines in ground-nesting shorebirds, and population control is now underway. The European beaver, *Castor fiber*, widely distributed in Europe and closely related to (but heavier than) the North American species, was hunted almost to extinction in Europe, becoming extinct in Great Britain in the sixteenth century. Recently (2001) captive populations have been brought in (confined) to specific sites with reintroduction scheduled for 2009 in Knapdale, Scotland. The common (or hazel) dormouse, *Muscardinus avellanarius*, known for its extended period of hibernation, is native to northern Europe and Asia Minor. The European badger, *Meles meles* (Mustelidae) is native to most of Europe and many parts of Asia. The red fox, *Vulpes vulpes* (Canidae) has a distribution spanning North America and Eurasia.

Since the end of the last ice age, the mammal fauna of the British Isles has changed radically; 21 native species have disappeared including all the large predators such as lynx, wolf, and brown bear (Table 1). Currently, approximately 60 species of terrestrial mammals breed in Britain, although this number is somewhat inexact because of species splitting, the uncertainty of including species that have been recorded in the islands only very infrequently, and whether to include marine mammals such as the bottle-nosed dolphin and porpoise.

In addition, some species are introduced while others have recently gone extinct, or been exterminated. The *Handbook of British Mammals* lists 44 native terrestrial species in Britain, 18 in Ireland, and 14 introduced species in Britain, 13 in Ireland. Other issues affecting taxon designations of British mammals are hybridization and translocation. For example, the native wild cat, *Felis sylvestris*, is still extant, though most are likely hybrids and are confined to northern Scotland. The range of polecat, *Mustela putorius*, has been expanding in Wales and the Midlands, and it has been reintroduced into two areas of Scotland and to Cumbria in England. Populations of the otter, *Lutra lutra*, have also shown recovery over recent years.

Current distributions and population structure are often the result of reintroductions and translocations that have occurred over extended periods and have been particularly common in game and domesticated species. Archaeological records of fur-bearing species in Scotland highlight the presence of foxes, badgers, and other mustelids outside their modern-day geographic range (e.g., foxes on Orkney from perhaps the last few centuries BC to the mid to late first millennium AD, pine marten on Orkney in the Neolithic, and badgers on the Outer Hebrides in the sixth–seventh centuries AD). Although zooarchaeological analysis of the data suggests that the evidence from the Outer Hebrides is indicative of imported products of fur-bearing species, such as skins or "trophies," the evidence from Orkney implies that humans may have purposefully introduced populations of fur-bearing species. Native populations of red deer, *Cervus elaphus*, likely went extinct in the lowlands following deforestation of native woodlands and persecution by the eighteenth century. The species has subsequently been reintroduced, and further translocations through the country are likely to have blurred natural genetic patterns in red deer throughout Great Britain and Ireland. Since the end of the eighteenth century, red deer numbers have generally increased; one of the best-known herds of red deer is that on the island of Rum in Scotland (see below), which are known to have been reintroduced in about 1845.

Domestication itself has played a key role in shaping biodiversity in Great Britain and Ireland. Aurochs are bovids, initially placed by Linnaeus in the same species as

TABLE 1

Mammals That Have Become Extinct in Britain over the Last 15,000 Years, with Likely Dates and Causes of Extinction

Common Name	Species	Date	Cause
Mammoth	*Mammuthus primigenius*	12,500 BP	Climate
Saiga antelope	*Saiga tatarica*	12,400 BP	Climate
Arctic fox	*Alopex lagopus*	12,400 BP	Climate
Lemming	*Lemmus lemmus*	10,500 BP	Climate
Arctic lemming	*Dicrostonyx torquatus*	10,500 BP	Climate
Narrow-headed vole	*Microtus gregalis*	10,500 BP	Climate
Pika	*Ochotona pusilla*	10,000 BP	Climate
Wild horse	*Equus ferus*	9330 BP	Climate
Giant elk	*Megaloceros giganteus*	9225 BP	Climate
Reindeer	*Rangifer tarandus*	8300 BP	Climate
Wolverine	*Gulo gulo*	8000 BP	Hunting
Northern vole	*Microtus oeconomus*	3500 BP	Climate
Elk	*Alces alces*	3400 BP	Hunting
Aurochs	*Bos primigenius*	3250 BP	Hunting
Lynx	*Lynx lynx*	200 AD	Hunting
Brown bear	*Ursus arctos*	500 AD	Hunting
Beaver	*Castor fiber*	1300 AD	Hunting
Wild boar	*Sus scrofa*	1500 AD	Hunting
Wolf	*Canis lupus*	1700 AD	Hunting

NOTE: Climate change affected vegetation, with the Holocene vegetation during colder Younger Dryas times probably unlike the mosaic environment of previous interglacials, thus contributing to the demise of species such as the antelope, elk, and reindeer. Data from Yalden (1999); Yalden and Kitchener (2008). BP = before present.

European domesticated cattle; however, domestication of the aurochs began in the southern Caucasus and northern Mesopotamia from about 6000 BC with substantial changes in morphology and physiology, such that domestic cattle are often considered separate species. Accordingly, when one complete aurochs skeleton was excavated in the nineteenth century, it was described as a new species, *Bos primigenius* Bojanus, 1827. Analysis of bones from aurochs that lived contemporaneously with domesticated cattle showed no genetic contribution to modern breeds. Thus, modern European cattle are now thought to be descended directly from the Near East domestication. The original range of the aurochs was from the British Isles and southern Scandinavia to northern Africa, the Middle East, India, and central Asia. By the thirteenth century AD, the aurochs's range was restricted to eastern Europe; the last aurochs died in 1627.

The sheep, *Ovis aries,* is thought to be one of the first animals domesticated, about 11,000 years ago from wild sheep in Southwest Asia. As sheep moved with humans to more northern climates, humans selected sheep with more wool fiber types preferentially over those with the coarser hair in ancestral wild types. Sheep arrived in England about 9000 years ago. Soay sheep—mentioned in recent history from the islands of St. Kilda, but neither evolved nor domesticated there—are likely one of the earliest breeds. Since 1955 the total number of Soay sheep on the Island of Hirta has been counted each summer, population sizes showing periods of rapid increase to high density, followed by crashes, typical of food-limited populations of vertebrates in northern latitudes.

Other species have been introduced to Great Britain and Ireland over the years for a variety of reasons (Table 2), and several have become serious pests, such as the gray squirrel mentioned above and the brown rat (see discussion of islands below). Another example is the European wild rabbit, *Oryctolagus cuniculus,* native to southern Spain and North Africa, which was introduced into Great Britain and Ireland in the twelfth century by the Normans, who bred captive populations for food and fur. With fewer predators, wild populations of rabbits increased rapidly, and had become abundant by the sixteenth century, with populations having expanded over most of what is now the UK. The myxoma virus was introduced to France in 1952 and had spread to Britain by 1953, initially causing mortality of 99.9%. However, the subsequent history of the virus and rabbits is a fascinating story of coevolution, showing the tendency of a pathogen to evolve toward an optimum level of virulence dictated by the mode of transmission of the

TABLE 2
Mammals Considered to Have Been Introduced to Britain in the Last 15,000 Years

Common Name	Species	Date	Reason
?Dog *	*Canis "familiaris"*	9000 BP	Companion animal
Sheep *	*Ovis "aries"*	5400 BP	Food
Goat	*Capra "hircus"*	5400 BP	Food
?Ox *	*Bos "taurus"*	5400 BP	Food
?Pig **	*Sus scrofa*	5400 BP	Food
Orkney vole	*Microtus arvalis*	5400 BP	Accident
Horse *	*Equus "caballus"*	4000 BP	Transportation
?Harvest mouse	*Micromys minutus*	4000 BP?	Accident
Scilly shrew	*Crocidura suaveolens*	4000 BP?	Accident
House mouse	*Mus domesticus*	3500 BP	Accident
Cat	*Felis "catus"*	2500 BP	Pest control/pet
?Brown hare	*Lepus europaeus*	2500 BP?	Food or cult
Black rat	*Rattus rattus*	200 AD	Accident
Fallow deer	*Dama dama*	1100 AD	Food
Rabbit	*Oryctolagus cuniculus*	1150 AD	Food
Brown rat	*Rattus norvegicus*	1728 AD	Accident
Sika	*Cervus nippon*	1860 AD	Amenity
Gray squirrel	*Sciurus carolinensis*	1876 AD	Amenity
Edible dormouse	*Glis glis*	1902 AD	Amenity
Chinese muntjac	*Muntiacus reevesi*	1922 AD	Amenity
Muskrat	*Ondatra zibethica*	1927 AD	Fur
Red-necked wallaby	*Macropus rufogriseus*	1940 AD	Amenity
Coypu	*Myocastor coypus*	1944 AD	Fur
Chinese water deer	*Hydropotes inermis*	1945 AD	Amenity
American mink	*Mustela vison*	1958 AD	Fur

NOTE: Uncertainties are indicated with ?. For certain species, domestication could have been in Britain, but more likely happened elsewhere. For twentieth-century additions, dates indicate time of escape/establishment, not original importation. Data based on sources listed in Table 1. * indicates no feral populations; ** feral only recently.

pathogen. In Britain the virus is transmitted mechanically by fleas, so successful transmission requires a large amount of virus in the skin over the lesions, but too much virus means that the host will die too quickly for transmission. Accordingly, the virus evolves to intermediate levels of virulence, with a subsequent slower coevolutionary process. At the same time, the rabbits are also evolving greater resistance to the virus. Accordingly, continued evolution of the virus (in a different direction from the rabbits) means that levels of virulence among unselected rabbits are almost as high as that of the original strain of virus. However, the current situation is that the virus no longer has a large impact on rabbit populations in general, mortality being about 20%, which has little effect on population densities.

Birds

The birds in Britain are largely similar to those in Europe, but with fewer breeding species. The number of species listed in Britain in 2006 was 572, with 552 in category A (species that have been recorded in an apparently natural state ≥ 1 time since 1950), 10 in B (species recorded in an apparently natural state ≥ 1 time between 1800 and 1949, but have not been recorded subsequently), and 10 in C (species that, although introduced, now derive from the resulting self-sustaining populations). As with the mammals, the list is somewhat arbitrary, because the A list includes all vagrants that have arrived without direct human introduction. A more reasonable estimate of total numbers of birds is approximately 250, which includes regular breeders, regular migrants, and common casuals in both categories. With its mild winters, Britain has substantial populations of wintering species, particularly ducks, geese, and swans. There are also a number of species, such as the oystercatcher, that are resident but elsewhere are migrants. Britain also receives a number of vagrants from Asia and North America. However, the best-known birds of Great Britain and Ireland are resident year-round (Fig. 5).

Like the mammals, birds of Britain and Ireland have experienced considerable change and movement through history. Approximately 20 bird species have gone extinct in Britain; seven have been reintroduced (barnacle goose, capercaillie, great bustard, and white-tailed eagle) or have naturally recolonized (common crane, great bittern, and osprey). Other species have been supplemented by European stocks when numbers became very low (e.g., red kite). The islands off the coast of the UK are particularly well known for their seabird colonies (see below). For example, the Bass Rock, a volcanic plug 100 m high in the Firth of Forth east of Scotland, is a Site of Special Scientific Interest (SSSI) due to its gannet colony, the largest single rock gannetry in the world.

Reptiles and Amphibians

Britain and Ireland have only seven native species of amphibians (including three species of newt, the great crested being extremely rare) and six land reptiles: the adder, grass snake, smooth snake, slow-worm (legless lizard), sand lizard, and common lizard. Ireland is unusual in having no snakes and also in having the Lusitanian natterjack toad, *Epidalea calamita*. Recent genetic studies suggest that the species survived in north European refugia, as well as in Iberia, since the last glacial maximum around 20,000 years ago, with subsequent local recolonization after the Younger Dryas cooling around 11,000 years ago.

Invertebrates

In 1995 there had been about 22,500 insects species recorded from Britain. Overall, the islands show a strong northwest–southeast diversity gradient (see gradient for butterfly and plant diversity, Fig. 4). An estimated 10.8% of British insects are rare, vulnerable, or endangered; a series of these were recently featured on postage stamps (Fig. 6). Updated lists of rare and endangered species are held by the Joint Nature Conservation Committee (JNCC), the statutory adviser to the Government on UK and international nature conservation. Reasons for endangerment vary; for example, the draining of fens in England is thought to be responsible for the extirpation of the butterfly *Lycaena dispar* (and likely other insects) between 1851 and 1864. Lack of ants is known to have caused the demise of the large blue butterfly, *Maculinea arion*, in 1979, because its larvae are obligate parasites of red ant colonies of one species, *Myrmica sabuleti*, and plant communities were not managed for these red ants; the butterfly has since been reintroduced successfully to appropriately managed sites using a subspecies from Sweden.

Surprisingly, the successive glacial and interglacial periods did not result in widespread extinctions among insects, at least based on evidence from beetles. In the past few decades, intensive investigations of Quaternary insect fossils (mostly beetles) from lake beds in which terrestrial

FIGURE 5 Stamps showing four of Britain's most familiar birds, issued by the British Postal Service in 1966. (A) The small black-headed gull, *Larus ridibundus* (Laridae), breeds in much of Europe and Asia, and is largely migratory, wintering further south. (B) The blue tit, *Cyanistes caeruleus* (Paridae), is a widespread and common resident breeder throughout temperate and subarctic Europe and western Asia in deciduous or mixed woodlands. (C) The European robin, *Erithacus rubecula* (Muscicapidae), occurs through much of Eurasia. Most British robins are resident, though a few migrate to southern Europe during winter; the prefer to live in parks and gardens. (D) The blackbird, *Turdus merula* (Turdidae), is a species of true thrush. It breeds in Europe, Asia, and North Africa and can be either resident or migratory. Stamps reprinted with permission from the Royal Mail.

FIGURE 6 Stamps showing some rare British insects, issued by the British Postal Service in 2008. (A) The stag beetle, *Lucanus cervus*, is Britain's largest beetle, living in damp, decaying timber around London, and threatened through intensive habitat modification. (B) The Adonis blue, *Polyommatus bellargus* (Lycaenidae), is a small butterfly that suffered dramatic decline in the 1950s when much of its habitat of chalky, short grassland was lost. (C) The southern damselfly *Coenagrion mercuriale* is internationally endangered, and the British Isles remain its last stronghold; the population has declined by 30% over the last 40 years. (D) The Purbeck mason wasp, *Pseudepipona herrichii*, is a small wasp restricted to a few heathland sites in the Poole Basin area of Dorset. (E) The red-barbed ant, *Formica rufibarbis*, is one of the UK's rarest native species, restricted to just one heathland site in Surrey, with loss of habitat blamed for its critical status. (F) The barberry carpet moth, *Pareulype berberata*, was already close to extinction in the 1950s due to removal of ancient hedgerows and woodland margins containing it's main food source, the barberry bush. (G) The hazel pot beetle, *Cryptocephalus coryli*, is restricted to small areas in Lincolnshire, Surrey, and Berkshire. Its decline could be due to the reduction of hazel coppicing and widespread removal of birch from heathland. (H) The field cricket, *Gryllus campestris*, was always uncommon, but 20 years ago was reduced to just one site in West Sussex; it is now one of the UK's rarest native species, and through a captive breeding program two smaller colonies have been reintroduced in southern England. (I) The silver-spotted skipper butterfly, *Hesperia comma*, was once widespread in the UK but declined dramatically in the 1950s, although since the 1980s numbers have increased by about 30%. (J) The noble chafer, *Gnorimus nobilis*, is found in old orchards and woodlands, in damp, decaying timber. Rare for over a century, it is now found only in small pockets in the New Forest, Evesham, and Wyre Forest. Stamps reprinted with permission from the Royal Mail.

debris has accumulated, peat bogs, and other such sites in Britain have shown that species remained constant both in their morphology and ecology throughout the Quaternary. For example, Britain has more than 2000 species of beetle represented by fossils that match precisely their present-day equivalents. It appears that the rapidity of glacial advance and retreat has minimized adaptation or regional allopatric speciation, while the mobility of insects allows them to alter their ranges and track climate changes. In many instances, insects isolated on an island (e.g., an oceanic island or equatorial mountaintop), from which escape is impossible, either evolve or become extinct. Thus, long-term species constancy and low rate of extinction in the insects in Britain can be considered a "mainland effect." Moreover, because species were tracking acceptable climates, populations may have continuously split and fused, making directional evolution unlikely and hence allowing species to survive over extended periods.

SOME MAJOR ISLANDS

Orkney and Shetland

Orkney, an archipelago 16 km off the north coast of Scotland, comprises more than 70 islands, about 20 inhabited. It was originally colonized by Neolithic tribes (3500 BC), then by Picts, and finally invaded and settled by Norse in 875. As with the Shetlands, the islands are notable for the absence of trees. The abundance of peat (partially decayed vegetation) indicates that this was not always the case, and, based on the ancient stone settlements, it appears that deforestation occurred shortly after human arrival. Indeed, paleoecological evidence indicates that the Western Isles, Orkney, and Shetland supported woodland during the early Holocene (11,000–5000 years ago). The development of large expanses of mire, subsequent to the arrival of Neolithic people and grazing animals, is therefore probably largely anthropogenic in origin. The pattern of vegetation change on Orkney has been different from that on other island groups, largely because of the more fertile soil on top of the Old Red Sandstone, and consequently different land-use practices. The Shetlands have some of the largest bird colonies in the North Atlantic. Many Arctic birds spend the winter on the Shetlands, including the whooper swan and great northern diver.

Outer Hebrides (or Western Isles)

This area, to the northwest of Scotland in the north Atlantic, includes the islands of St. Kilda, Lewis and Harris, North and South Uist, and Barra. The area is well known for birds (327 species recorded, of which more than 100 breed), and the islands provide a natural flyway for migrating landbirds to and from their Arctic breeding grounds and a refuge for wind-blown vagrants from America and northern Europe. St. Kilda (8.5 km^2), an isolated archipelago 65 km northwest of North Uist, was permanently inhabited for at least 2000 years, although recent evidence documents an earlier Neolithic settlement on the islands. The islanders fed largely on seabirds (eggs and young birds), though they also kept sheep and a few cattle and were able to grow a limited amount of food crops. After several outbreaks of disease and extensive emigration, the remaining islanders were finally evacuated in 1930. Given the small size and isolation of St. Kilda, coupled with a lack of trees, biodiversity is generally low (only 58 species of Lepidoptera compared to 367 recorded from the Western Isles), although there are more than 130 flowering plants (diversity is limited to those that can tolerate the salt spray, strong winds, and acidic peaty soils), and about 160 species each of each of fungi and bryophytes. Many important seabirds breed here, including northern gannets (world's largest colony), Leach's petrel, Atlantic puffins, and northern fulmars; the last great auk, *Pinguinus impennis*, of Britain was killed in St. Kilda in 1840. St. Kilda is also known for its herd of feral Soay sheep (see above). In 1986 the islands became the first place in Scotland to be inscribed as a UNESCO World Heritage Site, for its natural features, including its habitats for rare and endangered species, and large population of seabirds.

Lewis and Harris (single island, approximately 2,180 km^2) lies 39 km west of the Scottish mainland. Lewis was once covered by woodland, but today only small pockets remain on inland cliffs and on islands within lochs, away from fire and sheep. The perimeter of the island is sandy beaches and dunes on the east coast, and rugged Atlantic coastline elsewhere surrounding a central peat-covered plateau. The "machair" between the beach and peat bogs possesses a high seashell content, which confers fertility to the soil and surrounding grassland, allowing growth of rare species of flowers. The island of Lewis is well known for seabirds, a number of birds of prey, and a diversity of waterfowl, including the eider and long-tailed duck. The rivers and freshwater lakes are known for salmon and trout, with Arctic char, eel, sticklebacks, mullet, and flounder also present.

Inner Hebrides

The islands of the Inner Hebrides are diverse in size and isolation, features that have been used to study effects such as the evolution of body size. A study by Thomas White and Jeremy Searle on these islands showed that the well-documented tendency of small mammals to be larger on islands was positively related to isolation and negatively to island size. Skye (1,656 km^2), the second largest island in Scotland after the island of Lewis and Harris, is wet and windy, especially on the exposed coasts. There is a Mesolithic hunter-gatherer site dating to the seventh millennium BC, one of the oldest archaeological sites in Scotland, and additional settlements dating from the Neolithic, Bronze, and Iron Ages. Each of the main peninsulas has an individual flora, and the island is known for a diversity of Arctic and Alpine plants. The primary conservation concerns are the spread of ragwort and bracken as well as invasive species, including Japanese knotweed, rhododendron, New Zealand flatworm, and mink, and over-grazing by sheep and red deer.

The Small Isles comprise Rum, Eigg, Muck, Canna, and Sanday. The Isle of Rum was acquired by the Nature Conservancy Council (NCC) in 1958 primarily so that it could be used as an outdoor laboratory for long-term ecological studies. In particular, the intention was to use Rum for studies of the ecology of red deer. Over the years, research on the red deer on Rum has addressed diverse topics such as aging, long-term population trends, and the inheritance of individual characteristics that can rarely be explored in detail in natural settings (see www.zoo.com.ac.uk/ZOOSTAFF/larg/pages/Rum.html). This Kilmory red deer research project now represents one of the longest and most complete scientific studies of vertebrates in the world. Over a hundred widely cited scientific papers and several books include groundbreaking research in behavioral ecology, population ecology, and evolutionary biology. Canna and Sanday were declared a Site of Special Scientific Interest in 1987 and a Special Protected Area (SPA) in 1997 for their seabird and raptor populations, especially Manx shearwater, shag, and white-tailed eagle. The Highland Ringing Group has studied declines, attributed to the brown rat, in several species of seabirds. A full rat eradication program was initiated in 1997. A census in 2004 showed that many seabird species were still undergoing major declines on Canna, because predation by brown rats continues to be a problem. Eigg, consisting largely of a moorland plateau that reaches 393 meters, is also noted for its diversity of birds, mostly raptors. Muck is known for its seal and porpoise populations.

The islands of Mull (~875 km^2), Islay (~620 km^2), and Jura (~370 km^2) are also included in the Inner Hebrides. These islands, warmed by the Gulf Stream, are home to many bird species, including the white-tailed eagle on Mull, migrating to this island following reintroduction to Rum. Jura is known for its large population of red deer. Marine species include minke whales, porpoises, and dolphins.

Firth of Clyde

The largest island in the Firth of Clyde is the Isle of Arran (430 km^2), which is comprised of rugged granitic mountains in the northern half and undulating moorland in the southern half. The wide diversity of habitat on the island results in a rich array of species. Arran is particularly well known for its actively evolving *Sorbus* tree complex, including two endemic species, *S. arranensis* (Arran whitebeam) and *S. pseudofennica* (Arran service tree), found only in the north of Arran. Both species are of hybrid origin. *Sorbus arranensis* is triploid and, although an evolutionarily dead end in itself, has given rise on multiple occasions to a tetraploid taxon, *S. pseudofennica*. It may be that additional diversity is being created by crosses between *S. pseudofennica* and the widespread species *S. aucuparia*. Conservation programs are being developed that focus on the evolutionary dynamics of this lineage. North of Arran, the island of Bute (120 km^2) lies close (300 m at its closest) to the Scottish mainland. The wide variety of habitat on Bute range from intensively farmed land to open moorland and bog, woodland, and coastal sites; a number of areas have been designated for their ecological value. East of Bute lie the islands of Greater Cumbrae (12 km^2) and Little Cumbrae (3 km^2). A well-known site on Greater Cumbrae is the University Marine Biological Station at Millport, which hosts students for research and classes throughout the academic year. Current research focuses on marine biodiversity, ecology, and conservation, including human impacts, and sustainability. South of Arran, Ailsa Craig (1 km^2) is a 340-m volcanic plug remnant of a past volcano that now serves as a bird sanctuary, including a large gannet colony.

Isle of Man

The Isle of Man (572 km^2) lies in the middle of the northern Irish Sea and is bisected by a central valley. The most northern area is very flat, consisting mainly of glacial deposits; the remainder consists of low mountains (highest elevation 621 m). The island is a British crown dependency but is not part of the UK or the European Union. The island is known for a number of unique species and breeds, including the Manx cat, a breed of cats with a naturally occurring mutation of the spine that shortens the tail to a small stub; the Manx loaghtan, a breed of sheep known for its dark brown wool and supernumerary horns; the Manx robber fly, *Machimus cowini* (Asilidae), originally believed to be endemic to the island but now also known from sand dunes on the east coast of the Republic of Ireland; the Isle of Man cabbage, *Coincya monensis* subsp. *monensis* (Brassicaceae), found in coastal habitats on the island, though also in similar sites on the west of Great Britain; and the Manx Marvel tomato, a variety thought to have originally been bred on the island. The Isle of Man also had Manx breeds of cattle, horse, sheepdog, and pig, all of which are now extinct.

Farne Islands

Located off the coast of Northumberland, the Farne Islands comprise 15–20 or more islands (depending on the tide) about 2.5–7.5 km from the mainland. They were formed as a result of both erosion of the weaker surrounding rock and sea-level rise following the last ice age. A diversity of seabirds reside here including puffins, eiders, and Arctic terns, and also a large colony of gray seals. Saint Cuthbert, one of the first residents, introduced special laws in 676 protecting seabirds nesting on the islands; these are thought to be the earliest bird protection laws in the world. Nearby, Lindisfarne, or Holy Island, is a tidal island, connected to the mainland at low tide. The island is a haven for wintering, migrating, and (on occasion) rare birds, and large parts are protected as a national nature reserve.

Anglesey

Anglesey (714 km^2) is a large, relatively low-lying island only 200 m off the north coast of Wales. The coastline, with its many sandy beaches and dramatic cliffs, has been designated an Area of Outstanding Natural Beauty, with terns, puffins, razorbills, and guillemots; several lakes are of significant ecological interest, and wetland sites are protected. Red squirrels used to be widespread on Anglesey. Starting in the early 1970s, gray squirrels became increasingly common, leading to a parallel decline in the red squirrel population in the 1980s. By 1998 there were only about 40 red squirrels left, confined to a single conifer plantation. Since 1998, eradication programs have led to the removal of most of the gray squirrels, and the red squirrels are recovering well with the help of reintroduction efforts.

Isle of Wight

The Isle of Wight (380 km^2) lies in the English Channel 5 km from the south coast of England. As such, it serves as a microcosm of southeast England, with a diversity of habitats including chalk grasslands, maritime cliffs and slopes, and estuaries. Moreover, its mild maritime climate, coupled with the lack of a number of introduced species such as gray squirrels, has allowed it to serve as a refuge for red squirrels, dormice, bats, water voles, early gentian, skylark, Adonis blue butterfly, and pearl-bordered fritillary.

Isles of Scilly

The Scillies (16 km^2) form an archipelago of low-lying, granite islands, 45 km off the southwest coast of England. They are largely treeless and windswept and possess a maritime influenced climate. Habitats include many white sandy beaches, sand dunes, maritime heaths, and grasslands. Sea-level rise is evident from the field walls beneath the water between some of the main islands; many of the islands were joined into one until quite recently. The islands were a focus of human activity from at least Neolithic times, when they were much larger, but their outstanding prehistoric monuments are the chambered barrows and standing stones of the late Neolithic/early Bronze Age period. Currently, the mild climate makes the islands important for growing cut flowers. The islands were designated as an Area of Outstanding Natural Beauty in 1975 and are well known as a haven for rare birds, partly because of their position, making them a prime site for migratory and vagrant species. Some islands also have large terneries and colonies of seals, and Scilly is known as the only area in Britain to harbor the lesser white-toothed shrew, *Crocidura suaveolens,* presumably introduced from the continent.

CONCLUSION

The islands of Great Britain and Ireland have had a rich and varied history, and the current summary only scratches the surface of the story. The islands have been studied intensively for a very long time, and recent molecular and archaeological work have allowed events to be dated fairly precisely. The overall picture indicates that biodiversity in the islands has been shaped by the glacial history of the region, in concert with an increasingly profound influence of human activities. Humans have not only modified the landscape, but also moved species around, both deliberately (for farming domesticated species, enhancing populations of game or fur-bearing species, etc.) and accidentally (including range expansion of species purposefully introduced, such as rabbits). Recent conservation efforts have allowed some recovery of woodlands, and certain target species, although population growth and climate change make the future uncertain.

SEE ALSO THE FOLLOWING ARTICLES

Channel Islands (British Isles) / Climate Change / Extinction / Introduced Species / Newfoundland / Seabirds

FURTHER READING

Coope, G. R., and A. S. Wilkins. 1994. The response of insect faunas to glacial-interglacial climatic fluctuations. *Philosophical Transactions of the Royal Society of London B* 344: 19–26.
Dudley, S. P., M. Gee, C. Kehoe, T. M. Melling, and the British Ornithologists' Union Records Committee. 2006. The British List: a checklist of birds of Britain (7th edition). *Ibis* 148.3: 526–563.
Hale, M. W., P. W. W. Lurz, and K. Wolff. 2004. Patterns of genetic diversity in the red squirrel (*Sciurus vulgaris* L.): footprints of biogeographic history and artificial introductions. *Conservation Genetics* 5: 167–179.
Harris, S., and D. W. Yalden, eds. 2008. *Mammals of the British Isles: handbook,* 4th ed. Southampton, UK: The Mammal Society.
Preston, C. D., D. A. Pearman, and T. D. Dines, eds. 2002. *New atlas of the British and Irish flora.* Oxford: Oxford University Press.
Read, H. J., M. Frater, and J. Wright. 1999. *Woodland habitats.* London: Routledge.
Red Deer Research Project on the Isle of Rum. http://www.zoo.cam.ac.uk/ZOOSTAFF/larg/pages/Rum.html.
Thomas, J. A, M. G. Telfer, D. B. Roy, C. D. Preston, J. J. D. Greenwood, J. Asher, R. Fox, R. T. Clarke, J. H. Lawton. 2004. Comparative losses of British butterflies, birds, and plants and the global extinction crisis. *Science* 303: 1879–1881.
Vincent, P. 1990. *The biogeography of the British Isles.* London: Routledge.
Woodcock, N. H. 2000. *Geological history of Britain and Ireland.* Oxford: Blackwell Publishing.

REFERENCES

Asher, J., M. Warren, R. Fox, P. Harding, G. Jeffcoate, and S. Jeffcoate. 2001. *The millennium atlas of butterflies in Britain and Ireland.* Oxford: Oxford University Press.
Parfitt, S. A., R. W. Barendregt, M. Breda, I. Candy, M. J. Collins, G. R. Coope, P. Durbidge, M. H. Field, J. R. Lee, A. M. Lister, R. Mutch, K. E. H. Penkman, R. C. Preece, J. Rose, C. B. Stringer, R. Symmons, J. E. Whittaker, J. J. Wymer, and A. J. Stuart. 2005. The earliest record of human activity in Northern Europe. *Nature* 438: 1008–1012.
Petit, J. R., J. Jouzel, D. Raynaud, N. I. Barkov, J. M. Barnola, I. Basile, M. Bender, J. Chappellaz, J. Davis, G. Delaygue, M. Delmotte, V. M. Kotlyakov, M. Legrand, V. M. Lipenkov, C. Lorius, L. Pépin, C. Ritz, E. Saltzman, and M. Stievenard. 1999. Climate and atmospheric history of the past 420,000 years from the Vostok ice core, Antarctica. *Nature* 399: 429–436.
Preston, C. D., D. A. Pearman, and T. D. Dines, eds. 2002. *New atlas of the British and Irish flora.* Oxford: Oxford University Press.
Raynaud, D., J. M. Barnola, J. Chappellaz, T. Blunier, A. Indermuhle, and B. Stauffer. 2000. The ice record of greenhouse gases: a view in the context of future changes. *Quaternary Science Reviews* 19: 9–17.
Thompson, L. G. 2000. Ice core evidence for climate change in the Tropics: implications for our future. *Quaternary Science Reviews* 19: 19–35.
White, T. A., and J. B. Searle. 2007. Factors explaining increased body size in common shrews (*Sorex araneus*) on Scottish islands. *Journal of Biogeography* 34: 356–363.
Yalden, D. W. 1999. *The history of British mammals.* London: T. & A. D. Poyser.
Yalden, D. W., and A. C. Kitchener. 2008. History of the fauna, in *Mammals of the British Isles: handbook,* 4th ed. S. Harris and D. W. Yalden, eds. Southampton, UK: The Mammal Society, 17–31.

CANARY ISLANDS, BIOLOGY

JAVIER FRANCISCO-ORTEGA
Florida International University, Miami

ARNOLDO SANTOS-GUERRA
Jardín de Aclimatación de La Orotava, Tenerife, Spain

JUAN JOSÉ BACALLADO
Museum of Natural Sciences, Tenerife, Spain

FIGURE 1 Located in Parque Nacional Cañadas del Teide (Tenerife Island), Peak Teide (3718 m) is the highest mountain of Spain. Photograph by J. J. Bacallado.

The Canary Islands have an area of 7447 km² and are located close to the northwestern Saharan coast (~28° N latitude and 16° W longitude). They are composed of seven volcanic islands and a few islets. Several environmental factors contribute to the ecological peculiarities of these islands including the heavy influence of the cold Oceanic Canary current and the northeastern trade winds, the altitude (the highest mountain of Spain, Peak Teide [3718 m], is located on Tenerife [Fig. 1]), the occasional dry winds from the Western Sahara, the high elevation anti-trade dry winds from tropical latitudes, and a topography highly dissected by volcanoes, steep cliffs, caves, and deep gorges and gullies. The archipelago has never been connected to the mainland; however, during the last ice age, Lanzarote and Fuerteventura collectively formed one large island. Quaternary sea-level changes have also influenced the biogeography of the archipelago. It is likely that several "sea mountains" located between the Canaries and Madeira/the mainland were exposed during glacial periods and served as stepping stones for dispersal of faunal and floral elements. Fossil beaches located a few meters above the coastline provide further evidence for a changing topography caused by fluctuations in sea level.

BIOGEOGRAPHY AND ECOLOGY

The Canary Islands and four additional volcanic archipelagoes (i.e., the Azores, Madeira, the Selvagens, and Cape Verde) form the Macaronesian Islands. There has been a long debate on whether these archipelagoes should be considered a distinct biogeographical unit. Macaronesia as a biogeographical entity is supported by the many plant genera and species that are endemic to more than one archipelago. In addition, several evolutionary lineages (clades) with considerable numbers of species are shared across some of these archipelagoes. Opponents of a single biogeographical unit contend that major bioclimatological differences among the archipelagoes support the Canary Islands and Madeira floras belonging to the Mediterranean region, whereas the floras of the Azores and Cape Verde are part of the Medioeuropean and Sudano-Zambesian regions, respectively.

The Canarian biota has been the focus of several phytosociological and bioclimatological studies, and five major terrestrial ecosystems, or life zones, can be recognized. The coastal thicket—low elevation arid woodland (*Kleinio-Euphorbietea canariensis*) is present in all the islands at low altitude (0 to 400 m on southern slopes, and a predominant coastal distribution on northern slopes [Fig. 2]). This zone is devoid of large trees and is mostly filled with small shrubs and perennial plants with succulent leaves and stems (e.g., *Euphorbia* spp., *Kleinia neriifolia*, *Ceropegia* spp., *Aeonium* spp., *Plocama pendula*) or coriaceous leaves (e.g., *Rubia fruticosa*, *Cneorum pulverulentum*, *Echium* spp.). Annual rainfall in this zone is below 250 mm.

FIGURE 2 The coastal thicket, low-elevation arid woodland (*Kleinio-Euphorbietea canariensis*) on northern slopes of Teno (Tenerife Island). Foreground: large plant of *Euphorbia canariensis*. Photograph by J. J. Bacallado.

Dry sclerophyllous forests (*Rhamno crenulatae–Oleetea cerasiformis*) occur between 400 and 600 m (on southern slopes) and between coastal areas and 600 m (on northern slopes) on all islands. This plant community receives an average annual rainfall of 400 mm and has strong floristic links to the Mediterranean Thermophile forests, with those on northern slopes being floristically richer than those on southern slopes. Indicator plants for this ecosystem include trees such as *Olea europaea* subsp. *guanchica*, *Dracaena draco*, *Juniperus turbinata* subsp. *canariensis*, *Pistacia atlantica*, *Visnea mocanera*, *Phoenix canariensis* and small shrubs such as *Cheirolophus* spp., *Crambe* spp., *Echium* spp., *Rhamnus crenulata*, and *Sideritis* spp.

The humid evergreen forests (*Pruno hixa–Lauretea novocanariensis*) are restricted to those slopes of the islands that face the northeast trade winds and are located between 600 and 1200 m. These forests, known as "laurel forests" because they are characterized by several tree species (i.e., *Apollonias barbujana*, *Laurus novocanariensis*, *Ocotea foetens*, and *Persea indica*) in the plant family Lauraceae, are usually cloud covered (average annual rainfall 800–1000 mm), with moisture levels enhanced by the extensive condensation of moisture on leaves, a phenomenon locally known as *lluvia horizontal* ("horizontal rain"). The humid evergreen forests are not found on the most easterly islands of Fuerteventura and Lanzarote, although some small pockets were likely present in Fuerteventura prior to the arrival of European settlers.

The fourth life zone is the Canary pine forest (*Chamaecytiso–Pinetea canariensis* alliance *Cisto–Pinion canariensis*). The only species of pine present in the islands, the endemic *Pinus canariensis* is the representative plant element of this zone. This vegetation type is also absent in Lanzarote and Fuerteventura, and has few small natural pockets on the island of La Gomera. This forest occupies northern (1200–2000 m) and southern (600–2300 m) slopes (average rainfall 200–800 mm). A transitional zone, known locally as Fayal-brezal (*Andryalo–Ericetalia arboreae*), occurs between the laurel and pine forests. Two trees, "brezo" (*Erica arborea*) and "faya" (*Morella faya*) are the predominant plant species of this zone.

Finally, a fifth life zone is a high-elevation dry woodland (*Chamaecytiso–Pinetea canariensis* alliance *Spartocytision supranubii*) and is confined to slopes over 2000 m on La Palma and Tenerife (average annual precipitation 400 mm, most as snow). During winter, frosts are common. Currently, this vegetation type has the shrubs *Adenocarpus viscosus* (on La Palma) and *Spartocytisus supranubius* (on Tenerife) as dominant species, and several endemic species such as *Echium wildpretii* (Fig. 3). However, in the past, this was an open forest where the "cedro canario" tree (Canary Island juniper, *Juniperus cedrus*) was an important species. This species was almost driven to extinction because it was extensively used for timber. Today, the Canary Island juniper is almost a memory on these islands and is mostly relegated to inaccessible landscapes on La Palma.

The humid evergreen forest has the highest number of endemic plants, invertebrates, and vertebrates. However, the two ecosystems with the lowest plant species diversity, the coastal thicket, low-elevation arid woodland and the dry, high-elevation open woodland, also possess an endemic flora/fauna adapted to their ecological peculiarities. For instance the endemic beetle *Lepromoris*

FIGURE 3 Plant of *Echium wildpretii*. (Boraginaceae) growing on the high-elevation dry woodland (*Chamecytiso-Pinetea canariensis* alliance *Spartocytision supranubii*) of Tenerife Island, at Parque Nacional Cañadas del Teide. Photograph by J. J. Bacallado.

gibba feeds exclusively on succulent species of *Euphorbia* restricted to the coastal thicket. Likewise, the praying mantis *Pseudoyersinia teydeana* is only found on high-elevation ecosystems of Tenerife. Concerning vertebrates, the coastal ecosystem has an endemic mammal (the Canarian shrew, *Crocidura canariensis*) restricted to Fuerteventura and Lanzarote, and an endemic bird (the Canary Islands stonechat, *Saxicola dacotiae*) confined to Fuerteventura. There are no endemic vertebrates in the dry, high-elevation woodland, although both the "pinzon azul del Teide" (Teide blue chaffinch, *Fringilla teydea*) and the "murcielago orejudo" (long-eared bat, *Plecotus teneriffae*) are found in this ecosystem and in the pine forest.

TERRESTRIAL BIODIVERSITY: PLANTS

The Canary Islands house a veritable treasure trove of botanical diversity with over 1300 species of native vascular plants found in approximately 102 families and 712 genera. It is estimated that there are over 600 endemic species of seed plants, comprising 40% of the native flora. At least 22 genera of seed plants are endemic to the Canaries, with seven belonging to the family Asteraceae. In addition, 15 genera are endemic to the Canaries and at least one of the other Macaronesian archipelagoes. The four genera with the highest number of endemic species are *Aeonium* (26 spp.), *Sideritis* (24 spp.), *Echium* (23 spp.), and the Macaronesian genus *Argyranthemum* (Asteraceae, 19 spp.).

Tenerife is the largest island and has the highest number of endemic species (144 spp.). However, the second largest island, Fuerteventura, which is relatively low and has little rainfall, has only 13 endemic species. Approximately 60 of the native plant species are shared exclusively by the Canaries and at least one of the other Macaronesian archipelagoes, and are not known to occur on the continent. Most of the plant families occurring in the Mediterranean basin are also present in the Canary Islands; however, there are some notable exceptions. The family Fagaceae does not currently have any native species on the islands, despite oaks being one of the most important components of the Mediterranean ecosystems. However, recent paleobotanical data suggest that oaks were part of the natural vegetation of Tenerife until they started declining around 2000 years ago. Likewise, the Ternstroemiaceae is not present in the Mediterranean basin but has one representative in the Canaries and Madeira.

Molecular and biogeographical data reveal that the vast majority of the closest continental relatives of the Canarian plants are found in the Mediterranean region. However there are exceptions, with some Canarian endemics having their closest relatives in southern and eastern Africa (e.g., *Canarina canariensis*, *Ceropegia* spp., *Cicer canariensis*, *Parolinia* spp., *Solanum vespertilio*, and *S. lidii*). Traditionally, a great proportion of the Canarian flora has been considered to belong to ancient lineages that became extinct on the continent following major global climatic changes after the Tertiary. Phylogenetic studies suggest that most of the endemics seem to have arrived and diversified on the islands relatively recently. It is estimated that over 156 independent dispersal events from the mainland have produced the current endemic flora of the archipelago. Based on the available molecular data, a few of the Canarian endemics appear to belong to early evolutionary branches, among them *Bosea yervamora*, *Limonium dendroides*, and the endemic genus *Navaea*.

Many of the endemic groups (e.g., the *Bencomia* alliance, the *Dendrosonchus* alliance, *Argyranthemum* spp., *Echium* spp.), follow a pattern found in other oceanic islands, as they tend to be arborescent and/or to exhibit woodiness as their main growth form. In most cases, these shifts appear to have originated on the Canarian archipelago following dispersal and speciation; the continental ancestral forms of these species appear to have been herbaceous.

Concerning non-vascular plants, approximately only 5% of the non-vascular native flora is endemic to the Canary Islands. There are 1634 native species of fungi (107 endemics), over 1294 of lichens (26 endemics), and 464 of mosses and liverworts (ten endemics).

Some of the endemic species have a broad ethnobotanical use. Two endemic legumes, tagasaste (*Chamaecytisus proliferus* subsp. *palmensis*) and gacia (*Teline stenopetala* var. *stenopetala*), and one endemic buckwheat (*Rumex lunaria*) are cultivated as fodder crops in several regions of the islands. The sap of the endemic palm *Phoenix canariensis* is commonly used to make a syrup on the island of La Gomera. In addition, other endemic/native species are highly valued for timber (i.e., *Pinus canariensis*, *Tamarix canariensis*, *Apollonias barbujana*).

TERRESTRIAL BIODIVERSITY: INVERTEBRATES

There are approximately 7608 species of terrestrial invertebrates in the archipelago, 95% of which are arthropods. The vast majority of them are insects (~77 endemic genera, 5953 native species [Fig. 4A]), and over 40% of the native species are endemic to the archipelago. However, most of the endemic genera have fewer than two species. Indeed the genus with the highest number of endemic species, the weevil *Laparocerus* (102 endemic species), is native to Morocco and the Macaronesian Islands. Gastropods are also important. They include over 241 native species, and 88.8% of them are endemic. Six mollusc genera are endemic to the Canaries. The snail Canarian genus *Hemicycla* has 76 species and is the second most species-rich genus of invertebrates in the archipelago. The Arachnida (pseudo-scorpions, spiders, and mites) are represented by approximately 800 native species, half of which are endemic. Twelve of the genera in the Arachnida are endemic. With over 43 species, the non-endemic genus of spiders *Dysdera* has the highest number of endemics.

Tenerife harbors the highest number of invertebrate endemic genera and of endemic species of beetles. Recently, DNA data have been used to reconstruct the origin and evolution of certain groups of Canarian insects. Most of these studies suggest a recent (Quaternary) origin for these organisms, although it has been suggested that species from the lowlands are older than those occurring at higher elevations, particularly on the humid evergreen forest and pine forest.

Some of the endemic insects display features that are common on other oceanic islands and have lost their dispersal ability. For instance the grasshopper, *Calliphona konigi*, the aforementioned praying mantis *Pseudoyersinia*

FIGURE 4 Representative fauna of the Canary Islands. (A) *Vanessa vulcanica*, a relatively common endemic butterfly of Madeira and the Canary Islands. (B) *Gallotia galloti* subsp. *gallotii*, a lizard endemic to Tenerife Island. (C) *Fringilla teydea* subsp. *teydea*, a finch endemic to Tenerife Island. Photographs by J. J. Bacallado.

teydeana, and endemic species of the beetle genera *Paradromius* and *Broscus* are either wingless or their wings are poorly developed and non-functional.

The islands have an extensive cave system, which is mostly composed of volcanic tubes. This underground environment has one of the most peculiar ecosystems of the Canary Islands, with a highly endemic invertebrate fauna and with roosting sites for bats. This fauna was relatively unknown until 1980, when the first endemic invertebrate (*Callartida anopthalma*, Heteroptera) was discovered in a cave system of El Hierro. Approximately 168 endemic invertebrate species thrive in this ecosystem; 124 of them are terrestrial, and the rest occur in aquatic environments. The vast majority of these invertebrates are insects (~80%), and 27% of them are spiders. A great proportion of these species are blind, lack any body pigmentation, and have large legs and antennae.

TERRESTRIAL BIODIVERSITY: VERTEBRATES

The Canary Islands have a poor vertebrate fauna without native amphibians or freshwater fishes. There are a total of 91 native vertebrate species: 69 of them are birds, 13 are reptiles, and nine are mammals. Endemicity at species levels is 23%. There is one endemic genus, the giant lizard *Gallotia* (six species [Fig. 4C]). The rest of the reptile species are endemic and belong to the gecko genus *Tarentola* (four species) and the skink genus *Chalcides* (three species). Eight of the mammal species are bats; the other native mammal is the aforementioned Canarian shrew, *Crocidura canariensis*. Only one of the species of bats, *Plecottus teneriffae*, is endemic to the archipelago. Concerning birds, there are two endemic pigeons (*Columba bollii*, *Columba junoniae*), one endemic finch (*Fringilla teydea* [Fig. 4B]), one endemic stonechat (*Saxicola dacotiae*), and one endemic leaf warbler (*Phylloscopus canariensis*).

All of the endemic vertebrates are linked to the Mediterranean and European fauna. Molecular data suggest that *Gallotia* belongs to an ancient lineage that diverged in the Tertiary. These endemic reptiles exhibit gigantism and are the largest species of the family Lacertidae. Fossil records provide additional evidence for gigantism and other island syndrome features found in the endemic vertebrates. There are Late Tertiary–Early Quaternary fossils of giant terrestrial tortoises assigned to the genus *Geochelone*, and Holocene fossils of giant rats (two species belonging to the endemic genus *Canariomys*). In addition, there are fossils from a poorly known giant flightless bird from Lanzarote (approximately 6 million years old). Other extinct flightless birds include a passerine (*Emberiza alcoveri*) and a finch (*Carduelis triasi*); it is believed that these two species became extinct relatively recently.

MARINE BIODIVERSITY

The oceanic current systems that occur in the Canaries and the proximity of the archipelago to the Western Sahara coast are the most important environmental factors that influence the marine ecosystems of the archipelago. These factors promote limited isolation between the Canarian organisms and those occurring along the mainland coasts. Therefore, it is not surprising to find low levels of endemicity within the marine groups. The cold oceanic current of the Canaries provides a dispersal avenue for marine organisms from Europe, North Africa, and tropical and subtropical regions of the New World. The northeastern trade winds also have a major influence, and they are major contributors for coastal upwelling from the African coasts. This upwelling reaches a peak during the summer season when nutrient-rich and low-salinity waters move from the bottom of the sea to the surface.

Because of the strong influence of the Canary current and the trade winds, most of the marine biota of the Canaries is not linked to the one found in the tropical waters of Africa. Therefore, the majority of the marine organisms are biogeographically linked to the Atlantic–Mediterranean regions and to the subtropical/tropical waters of the western Atlantic.

The Canarian marine flora has approximately 700 species (including over 23 species of blue-green algae and three species of flowering plants). Approximately 16 of the algal species are endemic to the Canarian archipelago. The majority of the native flora (~391 species) are red algae (Rhodophyta). Over 30% of the native species are also found in tropical and subtropical areas, and several of these species have a disjunct distribution with neotropical areas. The island with the highest number of species is Tenerife (476 spp.). In contrast, the western islands of La Palma and El Hierro have a poor flora with 196 and 189 species, respectively.

The Canary Island marine environments have a rich vertebrate fauna that includes 730 native species of fish. Only three fish species are regarded as endemic; however, it is likely that they have a broader distribution outside the Canarian marine boundaries, as they occur in deep waters. In the last 20 years, there has been an increase in the number of records of tropical and subtropical fishes, perhaps associated with current trends toward a warmer climate and also with the movement of vessels, as the Canarian ports are among the most active ones in this region of the Atlantic.

Most of the native marine fish fauna belongs to groups with a wide tropical and/or Atlantic distribution or occurring in warm and temperate regions of the western Atlantic; only 17 of the fish species found in the Canaries are restricted to the Macaronesian region. Indeed, only 6% of the fish fauna is linked to that of the tropical coasts of West Africa.

The islands are also rich in marine reptiles and mammals. Four species of marine turtles occur on the islands, but only the leatherback turtle (*Dermochelys coriacea*) has been reported to nest on the islands. Twenty-eight cetacean species are known to occur in the archipelago. Most of these species are migrants, although at least three species (i.e., short-finned pilot whales [*Globicephala macrorhynchus*], sperm whales [*Physeter macrocephalus*], and bottlenose dolphins [*Tursiops truncatus*]) form permanent colonies in the archipelago waters.

There are over 2831 species of invertebrates in the marine environments; approximately 1180 of them (42%) are molluscs, and 1100 are arthropods (38%). However, very few of these are endemic to the Canaries.

CONSERVATION BIOLOGY

Since the arrival of the first pre-Hispanic population to the islands (~2500 years ago), the Canarian biota has been extensively modified. It is believed that the pre-Hispanic inhabitants drove to extinction the endemic species of giant rats, at least one species of the giant lizards (*Gallotia goliath*), and two endemic birds, the Canarian quail (*Coturnix gomerae*) and the shearwater (*Puffinus olsoni*). Early settlers introduced goats, dogs, and pigs.

Shortly after the arrival of the first Europeans, land was heavily cleared for both urban and agricultural development. Sugarcane was the main cash crop of the islands during the fifteenth and sixteenth centuries. The need for charcoal and building materials for the sugarcane industry had a very negative effect on the dry and humid evergreen forests, and these ecosystems suffered a severe reduction in size during this period. In addition, the forests were severely exploited for timber, pitch, and torch poles. More recently, the coastal zones have been the main focus of human development. Coastal and low-elevation ecosystems have been the subject of recent and intensive urban/tourism development and road construction. There is no doubt that the flora and fauna of these areas should have the most immediate priority for conservation.

Two species of plants are considered recently extinct in the wild: *Solanum nava* from the evergreen forests of Tenerife and Gran Canaria, and *Kunkelliela psilotoclada* from the dry sclerophyllous forests of western Tenerife. The Gomera endemic *Viola plantaginea* has not been found since it was described in the nineteenth century. Likewise, the recently discovered *Helianthemum cirae* A. Santos, ined. from the Caldera de Taburiente National Park (La Palma) has not been found since 1992. Only one plant of *H. cirae* was originally found in the wild on an area that is heavily grazed. Since the arrival of the Europeans, at least three endemic/native species of vertebrates have vanished from the islands. Native populations of the red kite (*Milvus milvus*) and the Mediterranean monk seal (*Monachus monachus*) have totally disappeared. The endemic oystercatcher, *Haematopus meadewaldoi,* a shorebird restricted to the eastern islands of Lanzarote and Fuerteventura, was last recorded in 1913 and is believed to have gone extinct in the 1940s.

Two non-native game mammals are currently widespread in two national parks—the Barbary sheep (*Ammotragus lervia*) in Caldera de Taburiente National Park and the European mouflon (*Ovis gmelini*) in the Cañadas del Teide National Park (Tenerife). Other introduced mammals include goats, domestic cats, European rabbits, black and brown rats, house mice, the Barbary ground squirrel, two species of shrew (*Crocidura russula* and *Suncus etruscus*), and the Algerian hedgehog (*Atelerix algirus*). There is ample evidence that all of the introduced species have had a detrimental effect on the native flora and fauna. For instance, the recently discovered and critically endangered giant lizard of La Gomera (*Gallotia gomerana*) has been driven to the verge of extinction by feral cats. Likewise, it is well known that introduced rats are the main predators on eggs of the two endemic species of pigeon and that ungulates modify the landscape severely through grazing.

Among vascular plants, it is estimated that over 400 species introduced by humans (approximately 32% of the flora) are currently established and naturalized in the Canarian ecosystems. Among them, the grass *Pennisetum setaceum* represents one of the most immediate threats to the native vegetation of the lowlands. Other "aggressive" invasive plants occurring at low elevation include the prickly pear (*Opuntia* spp., particularly *O. dillenii* and *O. maxima*) and the American aloe (*Agave* spp., particularly *Agave americana*). At higher elevations, the neotropical sunflowers *Ageratina adenophora* and *A. riparia,* and the spiderwort *Tradescantia fluminensis,* are major concerns in the evergreen forests. The California poppy, *Eschscholzia californica,* covers large areas of the open dry pine forests.

Approximately 45% of the Canarian land mass is officially protected in four national parks and 141 regional reserves/parks. The national parks of Cañadas del Teide and of Garajonay belong to the UNESCO World Heritage network. In addition, there are four UNESCO biosphere reserves in La Palma, Gran Canaria, El Hierro, and Lanzarote. There are also three marine reserves, and they cover the northern Lanzarote coasts and its offshore islets, the southeastern sector of El Hierro, and the southwestern coast of La Palma. These reserves aim to develop a sustainable management plan for the local fisheries.

SEE ALSO THE FOLLOWING ARTICLES

Azores / Canary Islands, Geology / Cape Verde Islands / Caves as Islands / Fossil Birds / Madeira

FURTHER READING

Bacallado, J. J., G. Ortega, G. Delgado, and L. Moro. 2006. *La fauna de Canarias*. Santa Cruz de Tenerife and Las Palmas de Gran Canaria, Canary Islands: Gobierno de Canarias, Consejería de Medio Ambiente y Ordenación Territorial, Centro de la Cultura Popular Canaria.

Bramwell, D., and Z. Bramwell. 2001. *Wild flowers of the Canary Islands*, 2nd ed. Madrid: Editorial Rueda.

Brito, A., P. J. Pascual, J. M. Falcón, A. Sancho, and G. González. 2002. *Peces de las Islas Canarias: catálogo comentado e ilustrado*. La Laguna, Canary Islands: Francisco Lemus Editor.

Fernández-Palacios, J. M., and J. L. Martín, eds. 2001. *Naturaleza de las Islas Canarias: ecología y conservación*. Santa Cruz de Tenerife, Canary Islands: Publicaciones Turquesa.

Haroun, R., M. C. Gil-Rodríguez, and W. Wildpret. 2003. *Plantas marinas de las Islas Canarias*. Talavera de la Reina, Spain: Canseco.

Izquierdo, I., J. L. Martín, N. Zurita, and M. Arechavaleta, eds. 2004. *Lista de especies silvestres de Canarias (hongos, plantas y animales terrestres) 2004*. La Laguna, Canary Islands: Consejería de Medio Ambiente y Ordenación Territorial, Gobierno de Canarias.

Juan, C., B. C. Emerson, P. Oromí, and G. M. Hewitt. 2000. Colonization and diversification: towards a phylogeographic synthesis for the Canary Islands. *Trends in Ecology and Evolution* 15: 104–109.

Machado, A. 1998. *Biodiversidad: un paseo por el concepto y las Islas Canarias*. Santa Cruz de Tenerife, Canary Islands: Cabildo Insular de Tenerife.

Martín, A., and J. A. Lorenzo. 2001. *Aves del archipiélago Canario*. La Laguna, Canary Islands: Francisco Lemus Editor.

Moro, L., J. L. Martín, M. J. Garrido, and I. Izquierdo, eds. 2003. *Lista de especies marinas de Canarias (hongos, plantas y animales terrestres) 2003*. La Laguna, Canary Islands: Consejería de Medio Ambiente y Ordenación Territorial, Gobierno de Canarias.

CANARY ISLANDS, GEOLOGY

KAJ HOERNLE

Leibniz Institute for Marine Sciences (IFM-GEOMAR), Kiel, Germany

JUAN-CARLOS CARRACEDO

Estación Volcanológica de Canarias, Tenerife, Spain

The Canary Islands, located between 100 and 500 km from the coast of northwestern Africa (Morocco), consist of seven major volcanic islands forming a rough west-southwest to east-northeast trending archipelago. Together with the Selvagen Islands and a group of seven major seamount complexes (some of which were former Canary Islands) to the northeast, they form the Canary volcanic province. Volcanism in this ~800-km-long and ~400-km-wide volcanic belt (located at 33–27° N and 18–12° W) decreases in age from the northeast (Lars Seamount, 68 million years) to the southwest (Hierro Island, 1 million years) and is interpreted to represent the Canary hotspot track (Fig. 1). The Canary volcanic province is located on Jurassic ocean crust (~150 million years old beneath the western part of the province to ~180 million years old beneath the eastern part of the province), and contains some of the oldest ocean crust preserved in ocean basins.

GEOLOGICAL OVERVIEW OF THE EVOLUTION OF THE ISLANDS

The morphology of the Canary volcanic province show systematic changes from southwest to northeast, reflecting an increase in age (Figs. 1 and 2) and a change in evolutionary stage. As the volcanoes age, they originally go through a constructive phase of evolution in which growth of the edifice through volcanic activity outpaces its

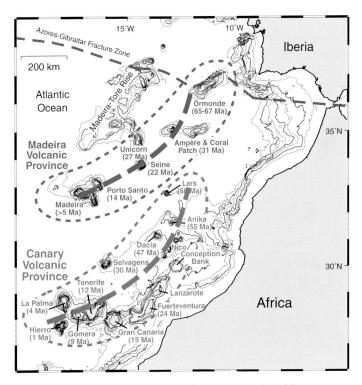

FIGURE 1 Bathymetric map showing the Canary (red) and Madeira (blue) volcanic provinces, including islands and associated seamounts, in the eastern central North Atlantic. Thick dashed lines mark centers of possible hotspot tracks. For clarity, only depth contours above 3500 m are shown. Bathymetric data from Smith and Sandwell (1997); ages and location of the Azores-Gibraltar fracture zone from Geldmacher et al. (2005) and Guillou et al. (1996).

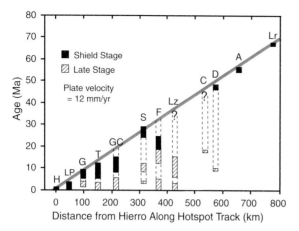

FIGURE 2 Radiometric ages of shield stages and late (rejuvenated or posterosional) stages of magmatic activity on islands and seamounts in the Canary volcanic province versus distance from Hierro. The center of each island is projected onto the proposed curved hotspot track in Fig. 1 along a line perpendicular to the hotspot track. Distances were measured from Hierro along the proposed hotspot track. The regression line calculated for oldest available ages of shield-stage volcanism gives an average rate of plate motion of 12 mm/year. On older islands and seamounts, volcanic units that had low $^{206}Pb/^{204}Pb$ (≤ 19.5) were assigned to the late stage of volcanism, with the exception of the single sample from Lars seamount, which had a slightly lower $^{206}Pb/^{204}Pb$ (19.44). Abbreviations for islands and seamounts: H = Hierro, LP = La Palma, G = Gomera, T = Tenerife, GC = Gran Canaria, S = Selvagen Islands, F = Fuerteventura, Lz = Lanzarote, C = Conception seamount, D = Dacia seamount, A = Anika seamount, and Lr = Lars seamount. References for age data from Guillou et al. (1996) and Geldmacher et al. (2005).

destruction through mass wasting (e.g., landsliding) and erosion. The constructive phase occurs primarily during the shield-building (or shield) cycle of activity, during which eruptive rates are high, and most of the volcanic edifice is formed. Even though mass wasting is an important process during the shield stage, the volcano continues to increase in size, despite short-term setbacks. The constructive phase of island/volcano evolution can extend into the first late (also commonly referred to as post-erosional or rejuvenated) cycle of volcanism, during which volcanic eruptive rates are drastically lower, but magmas can be more evolved (silicious and thus more viscous), contributing to an increase in volcano height. Late cycles of volcanism are generally separated from the shield stage of volcanism by extended periods of volcanic inactivity or drastically reduced activity. During the destructive phase of evolution, mass wasting and erosion outpace volcanic growth, and the volcanoes (islands) decrease in size until they are eroded to sea level. As the plate moves away from the magma source, it cools and subsides, and the now flat-topped volcanic edifices sink beneath sea level, forming guyots. Despite the differences in age of the volcanoes,

all of the islands have had Holocene acitivity except La Gomera and Fuerteventura. The islands of La Palma, Tenerife, and Lanzarote have had historical volcanic activity, and thus youthful volcanic structures can be found across the entire archipelago, even on the oldest islands.

The two youngest and westernmost islands of Hierro (1500 m above sea level, with oldest dated subaerially erupted rocks at 1.2 million years) and La Palma (2426 m; 1.8 million years old) have been the most active within the Holocene. Both volcanic islands are characterized by mafic alkaline volcanism, high eruptive rates, and volcanism along magmatic rift zones, commonly associated with the shield cycle of volcanism on ocean island volcanoes. Both volcanoes (and associated islands) are expected to continue to grow in size in the future.

The three central islands, intermediate in age, were in their shield stage in the Middle to Late Miocene and have had low levels of late or rejuvenated volcanism in the Pliocene and/or Quaternary. Tenerife in the central western part of the archipelago forms the largest island and is also the third largest and highest (more than 7000 m in elevation above the sea floor and 3718 m in elevation above sea level; 11.9 million years old) volcanic structure on Earth after the Hawaiian volcanoes of Mauna Loa and Mauna Kea. The highest peak of Tenerife is formed by the highly differentiated (phonolitic) Teide volcano, nested in a lateral collapse caldera (formed through mass wasting), indicating that this volcano is at the transition from its constructive to destructive phase. Considering the significantly lower eruption rates of the Canaries as compared to the Hawaiian volcanoes, the similarity in size reflects Tenerife's older age (longer life-span), most likely related to the almost-order-of-magnitude slower motion of the plate beneath the Canary Islands (~12 mm/year) as compared to the motion of the plate beneath the Hawaiian Islands. Therefore, it has taken much longer for Tenerife to move away from its magmatic source than has been the case with the Hawaiian volcanoes. Although crudely round in outline, the central volcanoes of La Gomera (1487 m; 9.4 million years old) and Gran Canaria (1950 m; 14.5 million years old) no longer have conical shapes and are characterized by deeply incised canyons, indicating that these two volcanoes are well into their destructive phase of evolution, with erosion and mass wasting outpacing growth through magmatic activity. Erosion has exposed intrusive complexes and dike swarms on both islands. The compositions of volcanic rocks are highly variable, ranging from mafic (transitional tholeiite to melilite nephelinite) to highly evolved (peralkaline rhyolite to trachyte to phonolite), reflecting the mature nature of these volcanoes.

On the two easternmost, oldest, and lowest islands of Fuerteventura (807 m; 20.2 million years old) and Lanzarote (670 m; 15.6 million years old), erosion is clearly the main process shaping the morphology of the islands, even though both islands have had rejuvenated volcanism within the last 150,000 years, and Lanzarote even within historical times. Both islands are realtively flat, with little of the original shield volcano morphology being preserved. Instead, they are characterized by isolated older volcanic sequences or erosional remnants and broad valleys. Surprisingly, the largest historical eruption in the Canary Islands and the second largest basaltic fissure eruption ever recorded (after the 1783 Laki eruption on Iceland) was the 1730–1736 Timanfaya eruption on Lanzarote, which produced ~1 km^3 of primarily tholeiitic basalts.

AGE OF VOLCANISM

The age of volcanism in the Canary Islands is well known from radiometric age dating. More than 600 K/Ar, Ar/Ar, and ^{14}C ages have been obtained by different groups on igneous samples from the islands. There have been a number of problems, however, primarily in dating older volcanic rocks and the uplifted portions of the seamount formations, some of which are clearly affected by alteration. Samples with excess Ar have produced ages that are too old, artificially increasing the duration of the basaltic shield stage on the islands. Ages obtained by newer techniques, such as Ar/Ar step heating, single crystal laser age dating, and the K/Ar unspiked method, and the employment of stringent age-dating requirements (e.g., sampling from well-controlled stratigraphic sections, and performing replicated analyses and systematic comparison of the palaeomagnetic polarities of the samples with the currently accepted geomagnetic reversal timescales) demonstrate that there is a progression of increasing age of the shield stage of volcanism from west to east in the Canary archipelago and from southwest to northeast in the Canary volcanic province, even though most of the Canary volcanoes have had a very long, complicated history, often including multiple cycles of late-stage volcanism (Figs. 1 and 2).

VOLCANIC HAZARDS

Geological hazards are moderate in the Canary Islands compared, for instance, with the Hawaiian Islands, which have a similar area and population but much more frequent and intense volcanic activity and seismicity. Although high magnitude eruptions (plinian) occurred in the geological past of the Canarian archipelago, only moderately explosive activity (strombolian to subplinian) took place in the last 200,000 years. Holocene eruptions, predominantly basaltic fissure eruptions, occurred on all the islands except La Gomera and Fuerteventura. Most of these, however, have been located on La Palma, El Hierro, and Tenerife, with only 10–12 events on Gran Canaria during this period and two on Lanzarote (1730–1736 and 1824). The most recent eruption in the Canary Islands was in 1971, from the Teneguia volcano at the southern tip of La Palma. During the Holocene, phonolitic strombolian to subplinian eruptions were associated with lava dome growth in the Teide volcanic complex on Tenerife, and to a lesser extent on the Cumbre Vieja rift on La Palma.

No casualties have been reported in the 16 eruptions recorded after the colonization of the archipelago at the end of the sixteenth century. Reliable prediction of when future eruptions will occur is not feasible because of the low frequency of events and great variability of inter-eruptive periods, from a few years to several hundred years (e.g., on Cumbre Vieja, the most active volcano in the historic epoch, repose periods varied from 26 to 237 years).

GEOCHEMICAL OVERVIEW OF ISLAND EVOLUTION

The geochemistry of the volcanic rocks from the Canary Islands is well understood in the context of volcano evolution. During each growth cycle (shield and late cycles), highly silica undersaturated rocks (nephelinites and basanites) dominate the early stage, more silica-saturated basaltic melts (alkali basalts and transitional thoeleiites) are produced during the peak of the cycle, and both mafic and evolved alkalic volcanic rocks (alkali basalts–basanites–nephelinites to trachytes–phonolites) are erupted during the waning stages. The almost complete lack of tholeiitic rocks and the low eruption rates during shield-cycle volcanism on the Canary volcanoes is likely to reflect a combination of low sublithospheric upwelling (mass flux) rates of the Canary plume and deep depths of melting beneath the thick Jurassic oceanic lithosphere. Although there are no systematic differences in major and trace element composition, systematic variations in isotopic composition do exist, with shield-stage volcanism being generally characterized by higher Pb and Sr and lower Nd and Hf isotopic ratios as compared to late-stage volcanism. These differences are likely to reflect greater amounts of melting of depleted upper or plume type mantle during the late stages of volcanism, as compared to enriched plume material during the shield stage of volcanism.

GEOLOGY OF THE INDIVIDUAL CANARY ISLANDS

El Hierro

The youngest and smallest of the Canary Islands is formed by three overlapping Quaternary stages of primarily mafic alkaline volcanism from oldest to youngest: Tiñor, El Golfo, and Rift (Fig. 3). A prominent three-branched rift system and related arcuate lateral collapse embayments (landslide scars) between the rifts define the characteristic shape of the island. Four lateral collapses have been recognized on El Hierro. The Tiñor collapse (~0.8-million-year-old) embayment in the northwestern part of the island was rapidly filled by subsequent volcanism, developing a 2000-m-high, 20-km-wide volcano that collapsed (between 130,000 and 20,000 years ago) to form the present El Golfo embayment. The El Julan lateral collapse (occurring more than 158,000 years ago) removed the southwestern flank of the island, whereas the incomplete failure of the southeastern flank (between ~261,000 and 176,000 years ago) generated the San Andrés fault system and the Las Playas slump. The latest eruptive activity of El Hierro, occuring over the last ~145,000 years, forms a conspicuous three-branched rift system, capping most of the island with eruptive vents (at the rift crests) and lava flows (at the flanks), partially filling the respective collapse embayments. The last eruption on the island, from a small vent on the northeastern rift, was dated by ^{14}C at 2500 years ago. An intense seismic crisis, believed to be related to an impending eruption, almost caused the total evacuation of the island in 1793.

La Palma

La Palma, the most active island of the Canaries in the Holocene, is formed by two coalesced volcanoes: the northern circular Taburiente Volcano and the younger north–south elongated Cumbre Vieja volcano to the south (Fig. 3). Two lateral collapses removed much of the southwestern flank of the Taburiente volcano. The uplifted (by about 1000 m) and tilted Pliocene seamount formations are exposed on the floor and walls of the Caldera de Taburiente, formed by the younger gravitational landslide (~0.5 million years ago). The seamount volcanism is composed of basaltic to trachytic pillow lavas and hyaloclastites. Fossils in submarine sequences suggest that the uplifted portion of the seamount stage may be 3–4 million years old, but there is continuing controversy about the validity of these fossil ages.

The oldest subaerial volcanism is separated from the underlying submarine volcanics by an angular unconformity and a 400–600-m-thick sedimentary unit made up of breccias, agglomerates, and sediments. During the Taburiente stage of volcanism, continuous mafic alkaline eruptive activity formed a 3000-m-high shield volcano. Subsequent volcanism migrated southward along the southern Cumbre Nueva rift zone. Continuing southward migration of volcanism ultimately resulted in the extinction of the northern shield ~0.4 million years ago and in the formation in the last 130,000 years of the Cumbre Vieja rift zone, a 20-km-long, 1949-m-high ridge, composed predominantly of mafic alkaline lavas.

Half of the historical (i.e., occurring in the last 500 years) eruptions of the Canary Islands, including the most recent event in 1971, occurred along the Cumbre Vieja rift system (Fig. 3). These eruptions have been characterized by Strombolian activity forming cinder cones and lava flows. The 1971 eruption added several square kilometers of new land to the island, clearly demonstrating that the island is still growing. Phonolites have been extruded in several of the historical eruptions and can form from basanitic parental magmas within 1000–2000 years. The 1585 eruption is famous for the emplacement of giant phonolitic spines (tens of meters high), which, according to an eyewitness account of a monk, rose from the ground like "the devil's horns" at the beginning of the almost exclusively mafic eruption.

La Gomera

La Gomera, presently undergoing a volcanic hiatus, is a heavily eroded, circular (22 km in diameter) volcano (Fig. 4). During the Late Miocene, the mafic alkaline shield volcano (~9–8 million years old) experienced a northward lateral collapse. Continued volcanic activity filled the collapse embayment and spread over the entire island, forming a central volcano with differentiated rocks at its terminus. The remains of a central caldera in this volcano crop out north of Vallehermoso, comprising trachytic and phonolitic lavas and intrusives with the latter forming radial dike swarms and cone sheets. A local unconformity separates the late-cycle Pliocene (occurring primarily 5–4 million years ago) mafic alkaline eruptions from the Miocene shield. Numerous phonolitic domes, some of the most conspicuous and spectacular volcanic features of the island, intrude both the Miocene and Pliocene mafic sequences. Volcanic activity was sparse during the last 4 million years and ended completely on the island

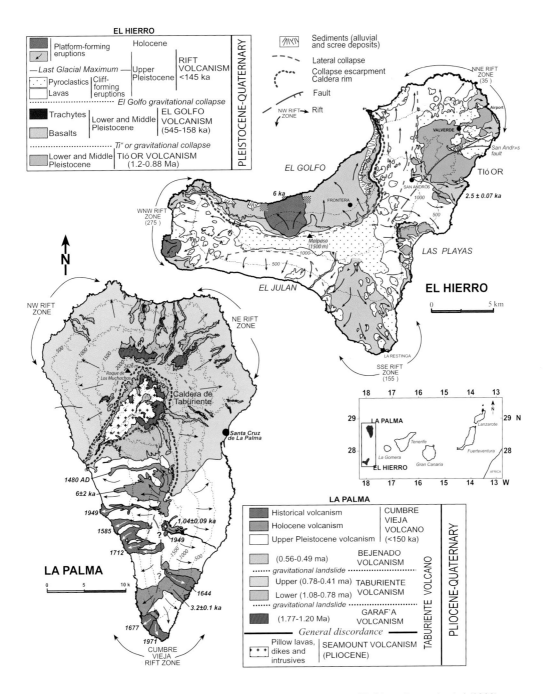

FIGURE 3 Simplified geological map of the western Canaries: El Hierro and La Palma. Modified from Carracedo et al. (2002).

1.9 million years ago, the age of the younger intra-canyon lavas located south of the capital San Sebastián.

Tenerife

The triangular-shaped island of Tenerife is the largest and highest in the Canarian archipelago and has a complex volcanic history (Fig. 4). The oldest visible volcanic rocks on Tenerife correspond to three deeply eroded mafic alkaline massifs: the Central, Anaga, and Teno shield volcanoes. The Central shield, the first to develop, is at present covered by later volcanism, with the exception of some outcrops in the southern flank of the island (the Roque del Conde massif). The shield stage of the island was completed with the growth of two new volcanoes, the

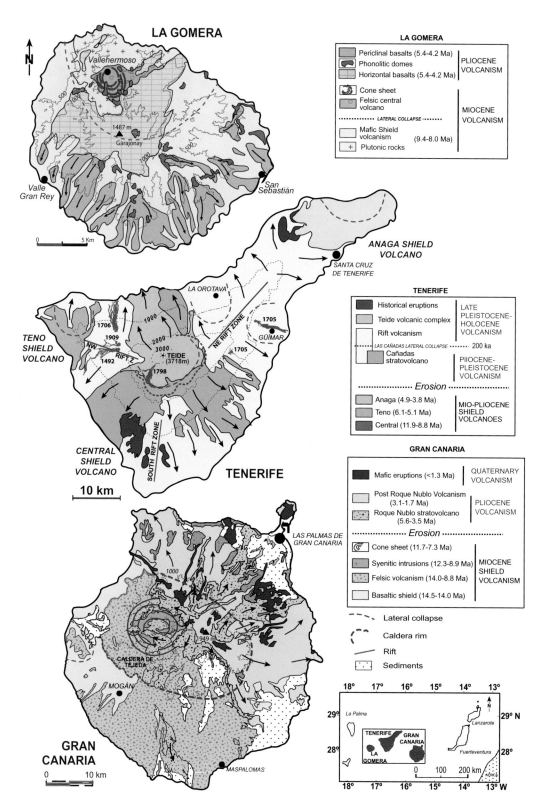

FIGURE 4 Simplified geological map of the central Canaries: La Gomera, Tenerife, and Gran Canaria. Modified from Carracedo et al. (2002).

Teno volcano, overlying the western flank of the Central volcano, and the Anaga volcano, which developed at the northeastern part of the island.

Late-cyle volcanism took place after 4–5 million years of eruptive quiescence and discordantly overlies the Central volcano. In this rejuvenated cycle, a 40-km-wide, 3000-m-high, progressively differentiated central volcano (Las Cañadas volcano) developed with contemporaneous basaltic eruptions along a three-arm rift system on the flanks of the Central volcano. Major explosive eruptions from the central volcano formed trachytic ignimbrites, pyroclastic flows, and ash-fall deposits mantling the southern flank of the island (the Bandas del Sur formation).

About 200,000 years ago, the summit of the Cañadas central volcano collapsed, forming the Caldera de Las Cañadas and the La Guancha–Icod depressions. Activity on the northeastern and northwestern rift zones may have played an important role in activating this gravitational landsliding. This catastrophic event marked the onset of the latest eruptive phase in Tenerife, characterized by continued mafic alkaline activity on the rifts and by the subsequent growth of the phonolitic Teide central volcano, nested within the collapse embayment. Parts of the northeastern rift zone also collapsed to form the La Orotava and Güímar valleys.

Relatively abundant eruptive activity occurred in this latest volcanic phase of Tenerife. At least 42 events took place during the Holocene, the greater part (60%) as mafic alkaline fissure eruptions on the northwestern and northeastern rift zones and 40% as phonolitic eruptions inside the Caldera de Las Cañadas and on the Teide–Pico Viejo volcanoes, particularly as lava domes and coulees. The latest eruption of the Teide volcano (Lavas Negras eruption) is generally believed to be the eruption seen by Christopher Columbus in 1492 on his way to discovering America. The radiocarbon age of 1150 ± 140 years BP, however, indicates that this could not have been the eruption observed by Columbus. The 1492 sighting of Columbus probably corresponds to Montaña Boca Cangrejo, a vent located in the northwestern rift zone with a radiocarbon age of 400 ± 110 years BP. The four most recent eruptions took place in 1705, 1706, 1798, and 1909.

Gran Canaria

The subaerial evolution of Gran Canaria can be divided into three cycles of volcanic activity: a Miocene shield cycle (~14.5–8.8 million years ago) and two late cycles in the Pliocene (~5.6–1.7 million years ago) and Quaternary (less than 1.3 million years ago). The oldest subaerial rocks on Gran Canaria (older than 14.5–14.0 million years), outcropping primarily in the southwest of the island, are alklai basaltic lava flows belonging to the shield stage of volcanism (Fig. 4). The Miocene basalts are overlain by up to more than 1000 m of evolved ignimbrites, lavas, and fallout tephras, respresenting the largely explosive outflow facies of the large elliptical, northwest–southeast trending, 20 by 17–km, 1000-m-deep Caldera de Tejeda. The Tejeda caldera may be unique in the Canary Islands in that it formed by collapse of the summit of the basaltic shield volcano above a large shallow magma chamber (at a 4–5-km depth). The onset of the collapse occurred upon the eruption of the widespread (covering an area of more than 400 km^2), voluminous (~45 km^3) P1 ignimbrite 14 million years ago, zoned from rhyolite to trachyte to basalt (from bottom to top). The outflow facies of the caldera become more silica-undersaturated in their upper sections (changing from predominantly peralkaline rhyolites to phonolites), with most activity having ended by ~9 million years ago. Alkali syenite intrusives and a spectacular trachytic–phonolitic cone sheet swarm of dikes, which erosion has exposed in the central part of the island, provide the last evidence of Miocene magmatic activity lasting to ~7 million years ago.

Late or posterosional volcanism started around 5.6 million years ago with the eruption of low volumes of nephelinite to basanite lavas and pyroclastic rocks from monogenetic centers. The Roque Nublo stratovolcano (~4.2–3.5 million years old) began with thick sequences of mafic lava flows (transitional tholeiite to basanite) filling canyons in the deeply eroded Miocene volcanic edifice. Upsection, the volcanic rocks become more evolved (ranging primarily from hawaiite to trachyte and tephrite to phonolite) and grade into tuffaceous rocks, massive breccia sheets (representing block and ash flows, lahars, and landslides), and volcanic plugs of highly evolved hauyne–phyric phonolites such as above Risco Blanco at the head of Barranco de Tirajana. The upper portions of the Roque Nublo group outcrop primarily in the center of the island, where the well-known Roque Nublo monolith, formed from a large breccia block, forms one of the highest peaks on the island. The Roque Nublo stratocone is likely to have reached an elevation of more than 3000 m and thus may have been similar in height to the present Pico de Teide volcano on Tenerife. Overlying an erosional unconformity in some parts of the island is a package of primarily basanitic to melilite nephelinite lava flows and dikes (~3.1–1.7 million years old), outcropping exclusively on the northeastern half of the island. Quaternary volcanism (occurring in the last 1.3 million years), primarily nephelinitic to basanitic in composition, has been increasing over the last 600,000 years, suggesting that Gran Canaria may be entering a new late cycle of activity. The youngest dated volcanic structure on the island is the Bandama caldera and strombolian cone (~2000 years old).

Fuerteventura

Fuerteventura has the oldest exposed rocks of the Canary Islands and is closest (100 km) to the African coast, situated on up to 10 km of continental rise sediments derived from Africa. Four main lithological units are exposed on Fuerteventura (Fig. 5): (1) Mesozoic oceanic crust and sediments, (2) submarine volcanics and intrusives and sediments, (3) Miocene shield volcanoes, (4) and late or rejuvenated Pliocene-Quaternary volcanic rocks and sediments. The Mesozoic oceanic crust, presumably uplifted as a result of extensive intrusive activity, is exposed along the western coast of the island and comprises a thick (~600-m) sedimentary sequence of Early Jurassic age with rare ocean crust volcanic rocks. Alkalic volcanic rocks with ages near the Cretaceous–Tertiary boundary may be associated with seamount volcanism unrelated to the formation of the Canary Islands.

The submarine volcanic rocks, comprising mafic pillow lavas and hyaloclastites, are unconformably overlain by littoral and shallow-water marine deposits (reefal bioclastic sediments, beach sandstones, and conglomerates), gently tilted to the west and apparently representing the transition from submarine to subaerial activity. A sequence of plutonic and hypabyssal intrusions, including an early syenite–ultramafic formation (alkali pyroxenites, amphibolites, and amphibole gabbros intruded by syenites) and later ijolite–syenite–carbonatite complexes (rarely exposed in intraplate island volcanoes), appear to be associated with the submarine volcanic complex.

Beginning about 20 million years ago, three adjacent basaltic shield volcanoes developed: the Central, Northern, and Southern or Jandía shields, comprising mainly alkali basaltic and trachybasaltic lavas, with minor trachytic differentiates. The extremely dense northeast–southwest dike swarm crossing the Miocene shields may be related to crustal stretching of as much as 30 km. Plutonic rocks (pyroxenites, gabbros, and syenites) and a basaltic-trachybasaltic dike swarm, which represent the hypabyssal roots of the evolving subaerial volcanic complexes, outcrop in the core of both the Central and Northern shields. Later plutonic activity includes concentric intrusions of gabbros and syenites, forming a spectacular ring complex.

Late-cycle volcanism extended from about 5.1 million to 134,000 years ago, the time of the last eruption dated on the island, although cones in the northern part of the island may be younger. Abundant littoral marine fauna developed during Pliocene equatorial climatic phases, and large volumes of aeolian sands derived from the Sahara covered most of the island at the end of the Pliocene.

Lanzarote

Lanzarote is the continuation to the northeast of Fuerteventura; these two islands are not geologically different, *sensu stricto,* because they are only separated by a narrow stretch of sea, which is shallower (less than 40 m) than the lowest sea level at maximum glacials. The oldest rocks in Lanzarote correspond to two main basaltic volcanoes, which developed as independent island volcanoes: the southern Ajaches volcano and the northern Famara volcano (Fig. 5). Isotope geochemistry suggests that both volcanoes represent late cycles of volcanism, and therefore that the shield cycle forming the bulk of this island may not be subaerially exposed. After a period of general quiescence of about 4 million years, a new cycle of rejuvenated activity resumed in the Quaternary with basaltic eruptive centers aligned in the northeast–southwest direction. The Corona volcanism in the northern part of the island (occurring 21,000 ± 6500 years ago) produced one of the largest known lava tubes, 6.8 km long with sections reaching 25 m in diameter. The tube formed as the lavas advanced along a wave-cut platform located 70 m below the present sea level. The last 2 km of the lava tube are at present submerged (−80 m) following sea-level rise after the last glaciation.

Holocene eruptions are probably limited to the historical 1730–1736 and 1824 events. During the 1730–1736 Timanfaya eruption, which lasted for 68 months, over 30 volcanic vents, aligned along a 14-km-long, N 80° E–trending fissure, were formed in five main multi-event eruptive phases. The Timanfaya eruption began with the eruption of basanite and alkali basalt during the first half-year and then erupted tholeiite for the rest of the eruption. In stark contrast to Hawaii, where tholeiites represent the most common rock type during the shield stage of volcanism but are generally absent in the rejuvenated stage, the late-cycle Timanfaya eruption produced the most tholeiitic rocks found thus far in the Canary Islands.

OTHER VOLCANOES OF THE CANARY VOLCANIC PROVINCE

The Selvagen Islands and neighboring large seamounts, located north and northeast of the Canary Islands, are

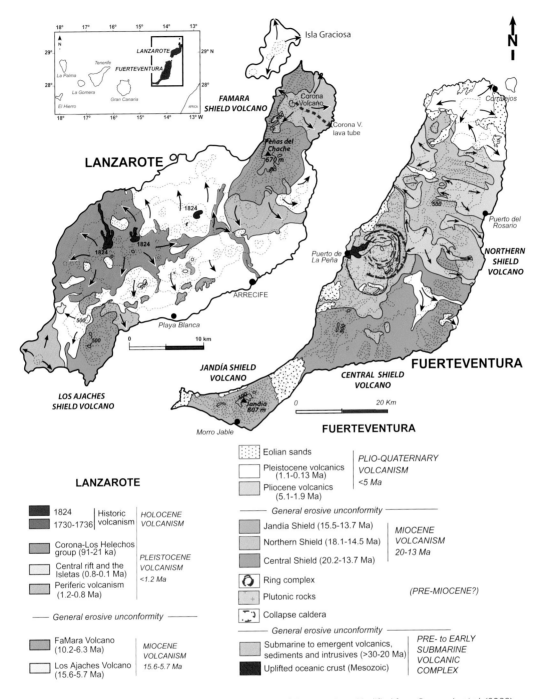

FIGURE 5 Simplified geological map of the eastern Canaries: Lanzarote and Fuerteventura. Modified from Carracedo et al. (2002).

located on the same bathymetric high as the Canary Islands and have similar geochemical (major and trace element and Sr–Nd–Pb–Hf isotopic) compositions to the Canary Island magmatic rocks, but distinct compositions from the Madeira Island rocks to the north (Fig. 1), consistent with their derivation from a similar source as the Canary Islands. Therefore, these volcanic structures are also included in the Canary volcanic province and can provide clues to the earlier history and origin of the Canary Islands.

The Selvagen Islands, located about 200 km north of the Canary Islands, form the summits of two shield volcanoes that ascend from 4000 m water depth, merging at depths between 500 and 1000 m. The main island of Selvagem Grande (163 m; area of 2.4 km²) is located on the summit of the northeast volcano, and a group of small islands, the largest being Selvagem Pequena (49 m; 0.2 km²), is located on the summit of the second volcano. A basanitic to phonolitic intrusive complex (29 million years in age) is exposed on Selvagem Pequena belonging to the shield cycle of activity. Despite its small size, three magmatic stages have been identified on Selvagem Grande: (1) a late Oligocene (occurring 26–24 million years ago) tephritic to phonolitic shield cycle intrusive complex, (2) a Middle to Late Miocene (occurring 12–8 million years ago) mafic alkaline late cycle, and (3) a Pliocene (occurring 3.4 million years ago) mafic alkaline late cycle.

Seven large seamount complexes are located to the northeast of the Canary Islands. Many of the seamounts in the northeastern portion of the volcanic province are guyots with flat tops from which beach cobbles were obtained by dredging, indicating that these guyots were former islands that were eroded to sea level by wave action and then subsequently subsided beneath sea level. Age-dated samples from four seamounts can be placed in shield and late cycles using geochemical criteria (e.g., Pb isotopic composition). When only the oldest ages from the shield-stage volcanic rocks are considered, there is an age progression with ages increasing from southwest to northeast through the Canary volcanic province, consistent with motion of the African plate in this region over a relatively fixed magma source at a rate of ~1.2 mm/year over the last ~70 million years.

MODELS FOR THE ORIGIN OF THE CANARY VOLCANIC PROVINCE

A wide variety of models have been proposed to explain the origin of the Canary Islands and the Canary volcanic province. These include volcanism resulting from decompression melting (1) along a leaky transform fault or propagating fracture (i.e., most likely an extension of the South Atlas fault system), (2) beneath rising lithospheric blocks as a result of tectonic shortening, (3) beneath a suture zone running along the Atlantic margin of northwestern Africa, or (4) of an upwelling mantle plume. The first three sets of models rely upon structures that cut through the lithosphere to cause and to control the location of the volcanism, whereas the origin and location of volcanism in accordance with the fourth model is largely independent of the lithosphere. The curved northeast–southwest alignment of the Canary and Madeira volcanic provinces clearly deviates from the east–west orientation of fracture zones in the East Atlantic crust. Furthermore, there is no evidence for such faults in the Canary or Madeira volcanic provinces or for suture zones in these areas. In addition, movements along structures cutting thick Mesozoic lithosphere would only generate very small volumes of melt as a result of the passive upwelling of normal asthenospheric mantle. Finally, there is no explanation why two roughly parallel curved fracture zones would form and propagate in the same curved direction and at the same average rate for at least 70 million years in this region.

The hotspot or mantle plume model, however, can adequately explain the parallel age progressions of oldest shield-cycle volcanism along curved southwest to northeast paths for the Canary and Madeira volcanic provinces. The age progressions of ~12 mm/year for both volcanic provinces are consistent with the rotation of the African plate around a common Euler pole (at 56° N, 45° W between 0 and 35 million years ago, and at 35° N, 45° W between 35 and 64 million years ago at an angular velocity of ~0.20° ± 0.05°/million years) above relatively fixed but distinct (based on differences in geochemistry for the two volcanic provinces) magmatic sources (i.e., mantle plumes). Strong additional support for the existence of a mantle plume comes from seismic tomography, which has imaged a mantle plume to the core–mantle boundary (at 2900 km) beneath the Canary Islands. The long history of volcanism on individual volcanoes could result through small amounts of melting of plume material flowing in the asthenosphere to the east and northeast beneath the older volcanoes.

SEE ALSO THE FOLLOWING ARTICLES

Canary Islands, Biology / Eruptions / Seamounts, Geology

FURTHER READING

Carracedo, J.C. 1994. The Canary Islands: an example of structural control on the growth of large oceanic-island volcanoes. *Journal of Volcanology and Geothermal Research* 60: 225–242.

Carracedo, J.C., F.J. Pérez Torrado, E. Ancochea, J. Meco, F. Hernán, C.R. Cubas, R. Casillas, E. Rodríguez Badiola, and A. Ahijado. 2002. Cenozoic volcanism II: the Canary Islands, in *The geology of Spain*. W. Gibbons and T. Moreno, eds. London: Geological Society, 439–472.

Carracedo, J.C., E. Rodríguez Badiola, H. Guillou, M. Paterne, S. Scaillet, F.J. Pérez Torrado, R. Paris, and U. Fra-Paleo. 2007. Eruptive and structural history of Teide volcano and rift zones of Tenerife, Canary Islands. *GSA Bulletin* 119: 1027–1051.

Geldmacher, J., K. Hoernle, Pvd. Bogaard, S. Duggen, and R. Werner. 2005. New $^{40}Ar/^{39}Ar$ age and geochemical data from seamounts in the Canary and Madeira volcanic provinces: a contribution to the "Great Plume Debate." *Earth and Planetary Science Letters* 237: 85–101.

Guillou, H., J.C. Carracedo, F. Pérez Torrado, and E. Rodríguez Badiola. 1996. K-Ar ages and magnetic stratigraphy of a hotspot-induced, fast grown oceanic island: El Hierro, Canary Islands: *Journal of Volcanology and Geothermal Research* 73: 141–155.

Hoernle, K., and H.-U. Schmincke. 1993. The role of partial melting in the 15 Ma geochemical evolution of Gran Canaria: a blob model for the Canary hotspot. *Journal of Petrology* 34: 599–626.

Masson, D.G., A.B. Watts, M.J.R. Gee, R. Urgelés, N.C. Mitchell, T.P. Le Bas, and M. Canals. 2002. Slope failures on the flanks of the western Canary Islands. *Earth-Science Reviews* 57: 1–35.

Paris, R., H. Guillou, J. C. Carracedo, and F. J. Pérez-Torrado. 2005. Volcanic and morphological evolution of La Gomera (Canary Islands), based on new K-Ar ages and magnetic stratigraphy: implications for oceanic island evolution. *Geological Society [London] Journal* 162: 501–512.

Schmincke, H. U. 1993. *Geological field guide of Gran Canaria*, 6th. ed. Kiel, Germany: Pluto-Press.

Smith, W. H. F., and D. T. Sandwell. 1997. Global seafloor topography from satellite altimetry and ship depth soundings. *Science* 277: 1956–1962.

CAPE VERDE ISLANDS

MARIA CRISTINA DUARTE AND MARIA MANUEL ROMEIRAS

Instituto de Investigação Científica Tropical, Lisbon, Portugal

Cape Verde's natural heritage is unique. The physical environment of these islands creates a multiplicity of habitats with a great wealth of fauna and flora. Nevertheless, this biodiversity is naturally restricted to the narrow geographical limits of the islands and is extremely vulnerable to disturbances caused by human activities. The threatened island endemics are thus likely to benefit from conservation management programs that are urgently needed if Cape Verde's natural levels of diversity are to be maintained.

PHYSICAL ENVIRONMENT

Geography and Geomorphology

The Cape Verde archipelago is grouped together with the Azores, Madeira, the Selvagens, and the Canary Islands in the Macaronesian region, which is situated in the North Atlantic Ocean, close to the West African coast and the West Mediterranean region. The Cape Verde archipelago consists of ten volcanic islands and several islets situated between 14°45′–17°10′ N and 22°40′–25°20′ W (Fig. 1). It lies 1500 km south of the Canary Islands, and a mere 570 km separate Boavista Island from the African mainland (the coast of Senegal). The archipelago is spread over 58,000 km² of ocean and has about 1050 km of coastline.

The Cape Verde Islands are usually classified in three groups: Northern Islands, Eastern Islands, and Southern Islands (Table 1). However, other classification groups are also considered: the Windward Islands (Santo Antão, São Vicente, Santa Luzia, São Nicolau, Sal, and Boavista), and the Leeward Islands (Maio, Santiago, Fogo, and Brava).

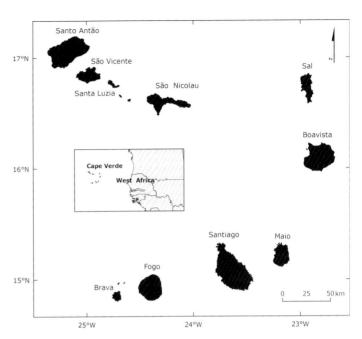

FIGURE 1 Cape Verde Islands: geographic location.

Santiago is the largest island and is home to more than half of Cape Verde's total population (434,625 inhabitants as of the 2000 census), whereas the smallest island—Santa Luzia—is uninhabited.

The northern and southern groups are characterized by high mountains and offer a wide range of habitats in relatively small areas. Slopes can be extraordinarily steep (average gradients of 25° to 41°), with active fluvial erosion. The eastern group is composed of flat islands, and such peaks as do exist reach only a few hundred meters in height and are surrounded by relatively broad extents of plain land, where deposition is dominant. Fogo Island contains Cape Verde's tallest mountain—Pico do Fogo—which rises to 2829 m (Table 1).

TABLE 1
Some Geophysical Features of the Cape Verde Islands

Island Group	Island Names	Area (km²)	Max. Altitude (m)	Population (as of 2000)
Northern	Santo Antão	779	1979	47,170
	São Vicente	227	725	67,163
	Santa Luzia	35	395	0
	São Nicolau	343	1304	13,661
Eastern	Sal	216	406	14,816
	Boavista	620	387	4209
	Maio	269	436	6754
Southern	Santiago	991	1392	236,627
	Fogo	476	2829	37,421
	Brava	64	976	6804

Geology and Soils

All the Cape Verde Islands are volcanic in origin. The archipelago's oceanic basement formed during the Late Jurassic and it is generally assumed that the islands originated with the development of hotspot activity above a mantle plume.

Cape Verde appears to be composed of the oldest rocks in Macaronesia. The subaerial volcanism probably began during the Miocene (the age of the oldest rocks found in the archipelago—on Sal Island—to date is about 25.6 ± 1.1 million years) or pre-Miocene. There is a general progression of the oldest volcanism, which is thought to have been induced by the slow movement of the African tectonic plate. The islands of Santo Antão and Brava thus present the youngest subaerial Tertiary volcanism, which can be dated to 7.57 ± 0.56 million years ago, and 5.9 ± 0.1 million years ago, respectively. Moving east, Santiago Island dates to 10.3 ± 0.6 million years ago, and Maio Island 21.1 ± 6.3. Quaternary igneous activity is concentrated at the western end of the archipelago, whereas Fogo is the only island with recent volcanic activity—the latest eruptions occurred in 1951 and 1995.

Soils are mainly of volcanic origin. They originated from volcanic rocks like basalts, phonolites, trachytes, andesites, tuffs, scorias, and sedimentary rocks—mainly limestone. Despite the small size of the archipelago and the similarity of the parent material, a wide variety of soils are present in Cape Verde and reflect its microclimatic and topographic diversity. They possess good physical and chemical properties, especially on the northeastern slopes of the higher mountains. At lower altitudes, the arid climate means that soils are either incipient or not developed, with low organic matter and nitrogen contents. In the Cape Verde archipelago, active dunes display maximum development on the Eastern Islands—namely Sal and Boavista—whereas there are substantial fossil dunes on Maio Island.

Climate

Cape Verde is included in the African Sahelian arid and semi-arid climatic region, which presents a wet season of one to three months. The archipelago experiences climates ranging from tropical dry to semi-desertic, and potential evaporation exceeds precipitation throughout the year.

The climate is mainly governed by the Azores anticyclone, the Intertropical Convergence Zone (ITCZ), and the mid-Atlantic air mass movements induced by their seasonal changes of location. The annual movement of the ITCZ around the equator, and its migration to $10° \pm 20°$ northern latitudes, brings a temporary southwest monsoon climate to the islands during the months of July to October (the rainy season). However, as a result of the high- and low-pressure zone oscillations, the summer months can be completely dry in some years. Both annual and monthly precipitations are usually low, although daily or storm precipitation can be very high, and the erosion resulting from this pluvial regime can be severe. Annual precipitation ranges from 80–300 mm in the arid coastal zones to 1200–1600 mm in the highlands of the mountain islands.

The Cape Verde archipelago is subject to the northeast trade winds throughout the year, and they especially affect northeastern slopes above 300–400 m in the mountain islands. Another important wind mass is the Harmattan. This dust- or sand-laden, hot, dry wind occasionally blows from the southern Sahara Desert and is more common between November and May. Despite its relatively short duration, it has a devastating effect on agriculture. Mean annual temperatures range from 23–27 °C at sea level to 18–20 °C at high altitude, but temperatures as high as 35–40 °C can occur in inner regions of the arid Eastern Islands.

CAPE VERDE'S BIODIVERSITY

Fauna

On the Cape Verde Islands, the vertebrate terrestrial fauna is poor, and there are few endemic species. Particularly, the terrestrial mammals are only represented in this archipelago by five bats (Chiroptera) and the introduced green monkey (*Cercopithecus aethiops*).

Similar to other Macaronesian Islands, the Cape Verde Islands are home to some endemic birds, which are estimated to include six species and 16 subspecies (e.g., the Cape Verde swift *Apus alexandri*, the Raso lark *Alauda razae*, the Cape Verde warbler *Acrocephalus brevipennis*, and the Cape Verde sparrow *Passer iagoensis*). Moreover, this archipelago offers important breeding habitats for 36 breeding species, nine of which are seabirds. Several ocean-related bird species occur in the region, especially during the winter months (about 130 species of migrants), and come mainly from Palearctic region.

Concerning the terrestrial reptile species, there are 28 taxa, 90% of which are accepted as being endemic in the Cape Verde archipelago. The largest genera are the gekkonid lizards (*Tarentola* spp.; Gekkonidae) and the scincid lizards (*Mabuya* spp.; Scincidae).

In contrast to the poor terrestrial vertebrate diversity, there is an enormous wealth of invertebrate species, in particular arthropods, many of them introduced, whose distribution ranges from the sea to the mountain peaks of the Cape Verde Islands. Although many arthropod species remain undiscovered, and accurate studies are lacking, a considerable number of endemic species are accepted (e.g., 111 species of spiders, 41% endemic; 470 species of Coleoptera, 33% endemic; 251 species of Hymenoptera, 26% endemic; 204 species of Diptera, 26% endemic).

The marine fauna of the Cape Verde archipelago ecosystem is essentially tropical. It includes a large number of species, namely invertebrates (e.g., molluscs such as cone shells and sea slugs; corals; crustaceans such as lobsters, shrimps, and crabs) and vertebrates (such as sharks and other fishes, and mammals such as dolphins and whales). Within the marine realm, invertebrates of interest include corals (e.g., *Porites astreoides*, *P. porites*, *Favia fragum*, and *Sclerastrea radians*), especially between Boavista and Maio Islands, and they represent a unique marine environment that hosts a high concentration of fish and other species.

The marine crustacean group has permanent populations of lobsters (Decapoda) along the coast of the Cape Verde Islands, namely the green lobster *Panulirus regius*, the pink lobster *Palinurus charlestoni*, the brown lobster *Panulirus echinatus* (Palinuridae), and *Scyllarides latus* (Scyllaridae). Because of their high economic value, the lobster populations have been depleted over the past few decades and now are considered to be threatened species. Crustaceans, cephalopods (e.g., *Sepia officinalis*, *Loliolopsis chiroctes*, and *Octopus vulgaris*), and fishes (e.g., *Thunnus albacares*, *Katsuwonus pelamis* [Scombridae]; *Decapterus macarellus*, *D. punctatus*, *Selar crumenophthalmus* [Carangidae]; *Epinephelus guaza*, *Cephalopholis taeniops* [Serranidae]) are common along the coast, and most of them constitute economically important resources.

In the deep water around the Cape Verde Islands, squaliform sharks of the genus *Centrophorus*, *Mustelus mustelus* (Triakidae), are commonly found, together with one of the largest species—the tiger shark (*Galeocerdo cuvier*). In addition, this archipelago is the only area in the North Atlantic outside the West Indies where the existence of humpback whales *Megaptera novaeangliae* (Cetacea, Mysticeti) is recognized. Humpback whales migrate thousands of kilometers each year, from summer feeding grounds at high latitudes to winter breeding grounds at low ones, and they traditionally congregate during the winter months to calve, to nurse their offspring, and presumably to breed near this archipelago.

Five species of marine turtles (Reptilia) have been reported for Cape Verde: the leatherback (*Dermochelys coriacea*), the hawksbill (*Eretmochelys imbricata*), the olive ridley (*Leidochelys olivacea*), the green (*Chelonia mydas*), and the loggerhead (*Caretta caretta*). Every year, from late May to September, more than 3000 loggerhead turtles come ashore on Cape Verde's beaches, particularly Ervatão beach on Boavista Island. This island was recently described as the third most important loggerhead nesting site in the world.

Flora

Cape Verde hosts a high level of endemic plant diversity. It presents a total of 320 lichen species, including a monospecific genus (*Gorgadesia*) and seven species which are accepted as endemic to the archipelago. In the archipelago, there are 36 liverwort species and 110 moss species, six of which are endemic. They can mostly be found on moist rocks, over the soil surface, or as epiphytes. The vascular flora comprises about 750 taxa, including more than a hundred families, of which the Asteraceae, Euphorbiaceae, Fabaceae, Malvaceae, Solanaceae, Cyperaceae, and Poaceae are the best represented.

When they were discovered in the fifteenth century, the Cape Verde Islands were uninhabited. Particularly as a result of human colonization and trading routes during the sixteenth and seventeenth centuries, most of the archipelago's flora is presently composed of exotic naturalized species. Despite the low number of native species (33 pteridophytes and 240 angiosperms), Cape Verde is rich in endemic taxa, including one fern (*Dryopteris gorgonea*) and 85 flowering plants from 42 genera, one of them endemic (*Tornabenea*: family Apiaceae). Among the native flora, annual and perennial herbaceous plants and some shrubs prevail. There are few trees, most of which are threatened: They include the endemics marmulan (*Sideroxylon marginata*) and date palm (*Phoenix atlantica*); the dragon's blood tree (*Dracaena draco*); the fig tree (*Ficus sycomorus* ssp. *gnaphalocarpa*, and *F. sur*); and the tamarisk (*Tamarix senegalensis*), fossilized stalks and roots of which are found in sand dunes on Boavista, S. Vicente, Sal, and Maio.

Hypotheses about the origin of the Macaronesian flora were first formulated about 180 years ago. Recent molecular studies suggest that the origins of the majority of the endemic species are in the western Mediterranean region, though other origins (e.g., North America, Euro-Siberia and northeast and southern Africa) are also reported. Furthermore, the insular woodiness characteristic of some endemic groups, especially those from Madeira, the Canary Islands, and the Cape Verde Islands, is more recent than the herbaceous ones in terms of evolution (e.g., *Echium* [Boraginaceae]; *Sonchus* [Asteraceae]).

Intense speciation processes have led to the large number of endemic species that presently exist in Cape Verde, approximately one-third of which are single-island endemics (e.g., *Conyza schlechtendalii* on S. Nicolau; *Diplotaxis vogelli* on S. Vicente; *D. gracilis* on S. Nicolau; *Echium vulcanorum* on Fogo; *Tornabenea annua* on Santiago; and *T. bischoffii* on Santo Antão).

Climatic factors related to altitude and aspect, as well as to soil characteristics, are responsible for the different plant communities found on Cape Verde. In the arid and semi-arid zones (up to 300–400 m) herbaceous formations like savannah are dominant and can extend up to 700–800 m on leeward slopes or be restricted to coastal zones in areas exposed to trade winds. The native flora in arid and semi-arid zones presents Saharo-Arabian affinities. Humid and sub-humid grasslands and scrub vegetation occur with increasing altitude, and some of the species of native flora display Canarian-Madeiran (and Mediterranean) or Sudano-Zambesian-Sindian affinities.

Unlike along other tropical and subtropical coasts, marine flora and particularly seagrasses (i.e. marine flowering plants) are not common in the Cape Verde archipelago, and only some doubtful records are known. Seaweed flora includes 330 species: 57 green algae (*Chlorophyceae*), 53 brown algae (*Phaeophyceae*), and 220 red algae (*Rhodophyceaea*); this tropical region of the eastern Atlantic is considered one of the poorest in terms of seaweed diversity.

CONSERVING CAPE VERDE'S NATURAL HERITAGE

Cape Verde's biodiversity is of enormous scientific value. Many endemic and native species, some of which are economically valuable, are at risk of extinction and make conserving the archipelago's biodiversity a world concern.

The first Cape Verde *Red List* shows that most of the native fauna are endangered. Some of the main threats are due to small population sizes and restricted distributions; the arid climate, which presents long periods of drought; the introduction of invasive species such as the wild rabbit (*Oryctolagus cuniculus*) and rats (*Rattus rattus* and *R. norvegicus*); and the human impact on natural habitats.

The *Red List* considers a significant number of taxa to be threatened: 59% of terrestrial mollusc species (Actophilia, Stylommatophora: Gastropoda); 64% of arachnid species (Arachnida: Araneida); 64% of beetle species (Insecta: Coleoptera); 25% of terrestrial reptile species (Reptilia); and 47% of bird species. Some of these taxa require urgent conservation measures, as do the small number of Caenogastropoda species (Basommatophora: Gastropoda). Others are already extinct. These include the endemic *Ancylus milleri* (Basommatophora); the three species of the Crustaceae group Decapoda (Natantia) (*Atya sulcatipes, Macrobrachium chevalieri,* and *M. vollenhovenii*), which are considered to have been extinct since 1954; and the giant scincid *Macroscincus coctei* (Reptilia), which has been extinct since the beginning of the twentieth century. Some endemic birds are classified as "critically endangered," including *Buteo bannermani, Milvus fasciicauda* (Accipitridae), and *Ardea purpurea* ssp. *bournei* (Ardeidae), which is restricted to Santiago Island. Furthermore, the emblematic marine turtles (Reptilia) are all threatened species, namely *Dermochelys coriacea* and *Eretmochelys imbricata,* which are classified as critically endangered, and *Leidochelys olivacea, Chelonia myda,* and *Caretta caretta,* which are classified as endangered. The major threat to turtle survival is the degradation of nesting habitats, namely Boavista beaches, as a result of the great increase in tourism in the archipelago.

Concerning the native flora, the prevalence of human activities, especially extensive agriculture, livestock herding (especially goats), and firewood collection, has led to a major destruction of the archipelago's natural habitats. Moreover, the uncontrolled spread of some exotic species like *Lantana camara, Furcraea foetida, Leucaena leucocephala,* and *Prosopis* spp., which are widely used for forestation or land conservation, has led to an invasion of the habitats of the native vegetation and in some cases, as with *Prosopis* in temporary river beds, to the depletion of the already inadequate water resources.

According to the *Red List*, 26% of the angiosperms and 65% of the pteridophytes are threatened; moreover, it is estimated that 30% of the lichens are extinct or threatened, as are 41% of Cape Verde's bryophytes. Overexploitation (e.g., folk medicine, woody fuel, alimentation, fodder, etc.), agricultural practices, and the historical cyclic droughts have a negative effect on the islands' flora. Three species are presently considered extinct (*Fumaria montana, Eulophia guineensis, Nervilia crociformis*), four critically endangered (*Carex antoniensis, C. paniculata* ssp. *hansenii, Conyza schlechtendalii, Ficus sycomorus* ssp. *gnaphalocarpa*), and 13 endangered. About 100 taxa are referred to as medicinal (e.g., *Artemisia gorgonum, Campylanthus glaber, Sarcostemma daltonii, Satureja forbesii*), woody fuel (e.g., *Echium vulcanorum, Periploca chevalieri, Sideroxylon marginata*), fodder (e.g., *Lotus purpureus, Sonchus daltonii, Tornabenea bischoffii*), or are of use for tanning (e.g., *Euphorbia tuckeyana*), and this is responsible for the high rates of threat to the endemic flora (about 54%).

Even though the natural landscape has mainly been altered by human activity, there are still natural communities which remain relatively undisturbed. They include the cliffs at high altitudes, which are rich in endemic species such as *Campylanthus glaber* ssp. *glaber, Conyza feae, Euphorbia tuckeyana, Globularia amygdalifolia, Lavandula rotundifolia, Lobularia canariensis* ssp. *fruticosa, Nauplius daltonii* ssp. *vogelii, Periploca chevalieri, Satureja forbesii, Sideroxylon marginata, Sonchus daltonii,* and *Tolpis farinulosa;* the ancient lava flows on Fogo Island, with *Artemisia gorgonum, Echium vulcanorum, Euphorbia tuckeyana, Erysimum caboverdeanum, Globularia amygdalifolia,* and *Verbascum cystolithicum;* the litoral halophilous communities, with *Arthrocnemum macrostachyum, Suaeda vermiculata,* and *Zygophyllum waterlotii;* and the coastal sand dunes, with *Sporobolus spicatus, Cyperus crassipes, Cistanche brunneri, Lotus brunneri, Limonium brunneri,* and *Zygophyllum waterlotii,* among others.

The Cape Verde authorities have recognized several natural areas that are to be included in the National Network of Protected Areas, and special conservation measures will guarantee the safeguarding of the islands' biological heritage. A large number of protected areas have been established to preserve plant species. Most are at high altitudes, where the mildest climate permits a richer flora and fauna, and where many endemic species—albeit with small population sizes—have their preferential habitats (e.g., Fajã de Cima on São Nicolau Island, which is the main area of occurrence of *Dracaena draco,* and Ribeira da Vinha on São Vicente, which is designed to preserve *Tamarix senegalensis,* which is presently threatened by the spread of *Prosopis*).

Several islets are particularly suited to conservation because they offer important breeding habitats for birds. They include Branco, Raso, Rombo, and Curral Velho, which are home to important colonies of seabird species and their respective breeding areas; Lagoa de Rabil (Boavista), with wintering migrant waders, herons, and terns; some coastal cliffs on Santiago; the central mountain range of São Nicolau; and the volcano crater on Fogo. The archipelago is included in the Endemic Bird Areas of the World and contains 12 "important bird areas," some of which are on the Ramsar List of Wetlands of International Importance. Furthermore, a proposal to classify some continental shelves, coastal areas, dune systems, and lagoons on Sal, Boavista, and Maio Islands as biosphere reserves is being considered in light of the importance of their marine biodiversity, which includes marine turtles, corals, and halieutic resources.

FIGURE 2 Fogo Island. Endemic species of *Echium vulcanorum* from high elevations (about 1800 m) on Pico do Fogo.

The Conservation International considered the "Mediterranean Basin," where the Macaronesian archipelagos were included as one of the "2005 Hotspots."

Finally, a word about the natural and humanmade landscapes—for example, the impressive orography of the mountainous regions, the volcano on Fogo (Fig. 2), the bays and the long sand beaches, and the dunes and the inland salt explorations, each of which possesses a distinctive natural heritage and a particular cultural legacy that results from the combined works of nature and man and gives rise to landscapes with an outstanding value from the point of view of science, conservation, and natural beauty.

SEE ALSO THE FOLLOWING ARTICLES

Atlantic Region / Azores / Canary Islands / Madeira / Volcanic Islands

FURTHER READING

Barbosa, L. A. G. 1968. L'archipel du Cap-Vert, in *Conservation of vegetation in Africa south of the Sahara.* I. Hedberg and O. Hedberg, eds. *Acta Phytogeographica Suecica* 54: 94–97.
Brochmann C., O. H. Rustan, W. Lobin, and N. Kilian. 1997. The endemic vascular plants of the Cape Verde Islands. *Sommerfeltia* 24: 1–356.
Chevalier, A. 1935. Les îles du Cap Vert. Géographie, biogéographie, agriculture. Flore de l'archipel. *Revue de Botanique Appliquée et d'Agriculture Tropicale* 15: 733–1090.
Ferreira, D. B. 1987. La crise climatique actuelle dans l'archipel du Cap Vert. Quelques aspects du problème dans l'île de Santiago. *Finisterra* 43: 113–152.
Hazevoet, C. J. 1995. *The birds of the Cape Verde Islands.* Tring, UK: British Ornithologists' Union (Checklist 13).

Leyens, T., and W. Lobin, eds. 1996. Primeira lista vermelha de Cabo Verde. *Courier Forschungsinstitut Senckenberg* 193: 1–140.

Mitchell-Thomé, R. C. 1987. Some geomorphologic aspects of the Cape Verde archipelago. *Boletim do Museu Municipal do Funchal* 39: 91–115.

CAROLINE ISLANDS

JAMES R. HEIN

U.S. Geological Survey, Menlo Park, California

The Caroline Islands form an archipelago just north of the equator in Micronesia, western Pacific. At various times in the past, the Caroline Islands encompassed all the islands that now comprise the Federated States of Micronesia (FSM), the Republic of Belau (formerly known as Palau), Guam and the southern Mariana Islands, and the southwestern islands of the Republic of the Marshall Islands (Fig. 1). Today, the name Caroline Islands refers to only the islands of the FSM and Belau (sometimes referred to as Palau).

HISTORY OF DISCOVERY

The earliest settlement of Micronesia is thought to have occurred about 2000 years ago, when immigrants from the south arrived on the western mountainous islands. In 1525, Portuguese explorers landed on Yap and Ulithi Islands (FSM) during their search for the Spice Islands (Indonesia). Subsequent Spanish expeditions made the first European contacts with the rest of the Caroline Islands, which were named the Carolinas in 1526 by Toribio Alonso de Salazar. Although visited many times through the years, Spain did not formally occupy the Carolinas until 1886. In 1899, Spain sold the islands to Germany, and then they were taken over by Japan in 1914 and mandated to Japan by the League of Nations in 1920. The United States occupied the islands in 1944–1945, and the United Nations placed them under U.S. administration as Trust Territories in 1947. The FSM gained independence in 1986, as did Belau in 1994; both have Compacts of Free Association with the United States.

GEOGRAPHY

The Caroline Islands have a tropical climate, and crops include taro, yams, breadfruit, coconuts, sugarcane, tapioca, and pepper. Copra and tapioca are important exports, as are handicrafts and fish, mostly dried bonito. The cultures of the FSM are as varied as the islands. Eight different languages are spoken, and nine Micronesian and Polynesian ethnic groups inhabit the FSM. Populations on Nukuoro and Kapingamarangi Atolls are mostly

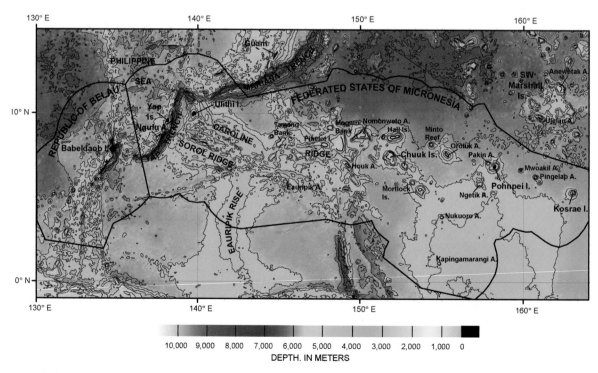

FIGURE 1 Bathymetry of the Caroline Islands, which includes islands in the Federated States of Micronesia (FSM) and the Republic of Belau (Palau); black lines show the 370-km Exclusive Economic Zones of FSM and Belau; I = island, Is = islands, A = atoll; contour interval is 500 m.

Polynesian; those on the remaining islands are mostly Micronesian. Belau is about 70% Micronesian, with the remainder of the population being Filipino, Chinese, or of another Asian ethnicity.

The Caroline Islands include several high islands, which are the erosional remnants of long-inactive volcanoes, now partially submerged. The degree of submergence has determined in part the types of reefs surrounding the islands. For most of the islands, the volcanic edifice subsided completely, and only a limestone platform or coral atoll remains above sea level.

The FSM consists of 607 islands and islets comprising a total land area of 702 km^2. The islands vary from high mountainous islands (Pohnpei, Kosrae, Yap, and part of Chuuk) to low-lying coral atolls. More than half of the population of approximately 110,000 lives in Chuuk State, which consists of seven major island groups, the largest group being Chuuk proper. The Chuuk group is a complex of 98 islands and islets, 14 of which are volcanic, surrounded by a barrier reef that encloses a large lagoon. Pohnpei, the largest island in the FSM, is mountainous (Mt. Totolom, 791 m high, and Mt. Nahna Laud, 798 m), with 342.4 km^2 of land. The impressive ruins at Nan Mandol on islets off southeastern Pohnpei date from about AD 1200. This extensive array of ancient structures was constructed from columnar basalt quarried on the main island.

Belau consists of about 300 islands and islets, with a total land area of 460 km^2; nine of the islands are permanently inhabited. The Belau chain extends for about 240 km. Babeldaob, the largest island in Belau, has a maximum elevation of 287 m; a ring road has recently been built around the island. The Belau archipelago is encircled by a part-barrier, part-fringing reef, except for the northernmost island, Ngcheangel (formerly Kayangel), and the southernmost island, Ngeaur. Ngcheangel constitutes an atoll, and Ngeaur is an uplifted coral platform with a maximum elevation of 56 m.

GEOLOGIC SETTING

The ocean crust on which the Caroline Islands rest can be divided into two broad geologic provinces. The first is a complex array of ancient volcanoes and ridges that make up the 2500-km-long Caroline Ridge (Fig. 1). The second is a volcanic-arc/subduction-zone system comprising the Yap and Belau trenches, the Yap and Belau volcanic arcs, and the Philippine Sea back-arc area.

The eastern third of Caroline Ridge, from Kosrae Island to Chuuk Atoll, consists of isolated atolls and seamounts formed at a hotspot that was active between 12 and 1 million years ago. Central Caroline Ridge, Chuuk Atoll to Tamang Bank, consists of atolls and large submarine carbonate banks. The western third of Caroline Ridge, Tamang Bank to the Yap trench, consists of a large ridge bounded by narrow troughs created by strike-slip faults (southern margin) and normal faults (northern margin). The western two-thirds of Caroline Ridge is generally at less than 2500 m water depth, and it formed about 28 to 24 million years ago. This part of the ridge has been variously interpreted to be a relict volcanic arc, a transform fault along which volcanism occurred, a transform fault combined with a hotspot trace, or an extinct spreading-ridge/transform-fault system.

The Yap arc and trench represent a 34- to 1.8-million-year old subduction zone that is anomalously narrow between the arc and the trench compared to other western Pacific volcanic-arc systems. It has been suggested that subduction ended about 5 million years ago and that back-arc-basin crust was obducted (uplifted) onto the volcanic arc; however, subduction likely occurs at a very slow rate along the Yap trench.

The Belau volcanic arc began forming about 40 million years ago as the Pacific plate was being subducted beneath the Philippine plate. The arc remained volcanically active until about 20 million years ago, when subduction ceased; subduction continued further northeast along the Yap and Mariana trenches. The Belau arc is now part of the Belau-Kyushu Ridge, an extinct volcanic arc located for the most part in the Philippine Sea back-arc basin.

Gold occurs on Maap (also spelled Map) and Gagil-Tomil Islands in the Yap island group, but it has yet to be mined. Copper ore was mined on Gagil-Tomil and Maap Islands during the Japanese occupation. Also during that time, nickel-rich laterite was mined from an open pit on Gagil-Tomil Island.

In Belau, bauxite (aluminum ore) was mined on Babeldaob Island between 1938 and 1944 from two (Ngardmau, Ngeremlengui) of the dozens of prospects, and proven reserves still exist. A gold deposit occurs at Rois Malk on southeast Babeldaob Island, but it has yet to be mined. Locally, the rocks in the same area are highly enriched in organic matter, constituting lignite (coal) deposits. The lignite was briefly mined during World War II, along with interbedded layers of kaolinite (white clay), which was used for ceramics. Phosphate was mined on Ngeaur, Beliliou, and Mecherchar (Eil Malk) Islands, and proven phosphate reserves are still present on those islands as well as on Tobi, Pulo Anna, and Sonsorol islands.

SEE ALSO THE FOLLOWING ARTICLES

Atolls / Coral / Exploration and Discovery / Pacific Region / Seamounts, Geology

FURTHER READING

Hein, J. R., B. R. McIntyre, and D. Z. Piper. 2005. Marine mineral resources of Pacific islands—a review of the Exclusive Economic Zones of islands of U.S. affiliation, excluding the State of Hawaii. *U.S. Geological Survey Circular 1286*.

U.S. Office of Insular Affairs, Federated States of Micronesia, http://www.doi.gov/oia/Islandpages/fsmpage.htm.

U.S. Office of Insular Affairs, Palau, http://www.doi.gov/oia/Islandpages/palaupage.htm.

CAVES, AS ISLANDS

DAVID C. CULVER

American University, Washington, DC

TANJA PIPAN

Karst Research Institute, Postojna, Slovenia

At a very simple level, there is an analogy between real islands and the "virtual islands" of caves. Caves are islands of cavities surrounded by impenetrable rock and connected only by the "ocean" of the surface. Landscapes with caves are island-like at several scales, from the small solution pockets above caves (epikarst) to their subterranean drainage basins. They are island-like in evolutionary time. Many cave species have evolved unique morphological traits—reduced or missing eyes, reduced or missing pigment, elaboration of appendages, and hypertrophy of extrasensory structures. They are also island-like in ecological time, and dispersal among "islands" can occur, especially at smaller landscape scales.

THE PHYSICAL ANALOGY BETWEEN CAVES AND ISLANDS

Like islands, caves are isolated habitats, but the "ocean" for caves is the surface, with the incumbent dangers of predators, sunlight, and environmental variation. For cave-adapted organisms these barriers are as formidable as the open ocean is to land animals.

The reality is, of course, more complicated. In almost all cases, the rock in which caves occur has fractures or small solution tubes that allow subsurface connections between caves. Nevertheless, movement between caves is very restricted. There are certainly areas where caves (and islands) occur as more or less distinct entities with few, if any, subsurface connections. An example is Segeberger Höhle, a cave formed in a salt dome in northern Germany. There are few if any other caves in the salt dome. More commonly, caves occur in relatively dense clusters. For example, in the United States there are over 40,000 reported caves. In cave-rich Slovenia, there are over 8000 caves in a land area of 20,000 km^2. In some places in Slovenia, caves reach a density of more than 5/km^2. At densities such as these, the analogy with islands weakens, or rather, the number of islands becomes very large. Nevertheless, for the terrestrial environment, the physical analogy with islands seems valid: Both caves and islands are surrounded by inhospitable habitat.

For aquatic subterranean environments, the situation is yet more complicated. With regions of soluble rock, sinkholes, springs, and caves (termed karst), a boundary layer between the soil and rock forms that has numerous small water-filled cavities and poorly integrated channels. This "epikarst" both connects caves within a karst region, it also itself serves as an important habitat for subterranean species. Below the water table, there may be further connections via solution tubes and passages (Fig. 1). At the scale of individual caves, what appear to be distinct units are in fact directly connected to each other by passages below the water table and by epikarst. At this scale, the analogy with islands breaks down. For subsurface habitats outside of cave regions, such as the underflow of streams and rivers, the analogy breaks down even further because the habitat is branchlike and linear.

There are two other scales at which the island analogy may be useful—one much smaller and one much larger.

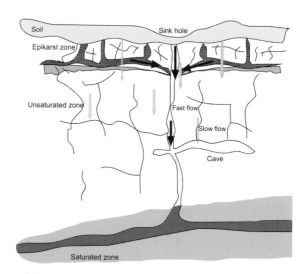

FIGURE 1 Sketch of epikarst and the connections with a cave passage underneath. Drawing by T. Pipan.

The epikarst habitat itself is island-like, albeit at a smaller scale than caves. Connections between solution pockets (islands) in the epikarst are poorly integrated (Fig. 1). Most of what we know about epikarst comes from the study of drips in caves—water exiting epikarst. One of the ways we know epikarst is poorly integrated is that the water chemistry of drips can differ significantly for drips only a few meters apart. So, at a scale of tens to hundreds of meters, rather than tens of kilometers as is the case for caves, epikarst is island-like.

There is also an island analogy at the landscape scale. Groups of caves that are connected by subterranean watercourses form a karst drainage basin, in an analogous way to drainage basins of surface water. Different karst drainages are analogous to islands because movement between these basins is highly restricted. For both aquatic and terrestrial species, caves are associated with contiguous karst regions, which can be quite large. For example, there is a major karst area in the central United States—the Interior Low Plateau—which covers part of Indiana, Kentucky, Alabama, and Tennessee. The most diverse karst area is undoubtedly the Dinaric karst, which ranges from northeast Italy to Montenegro. These karst regions are island-like and are certainly surrounded by inhospitable habitat.

Ultimately, it is the organisms that tell us the extent of isolation of caves and other karst features (like epikarst). If individual species of obligate subterranean species are widespread and ubiquitous among the virtual islands, then the caves are not isolated. We can consider the island analogy in two time frames—evolutionary and ecological.

CAVES AND KARST FEATURES AS ISLANDS IN EVOLUTIONARY TIME

In evolutionary time, populations either actively or passively invade caves. Passive invasion is often connected with relict populations, which occurs during periods of climate change. This is typically associated with extinction of surface populations, and allopatric speciation occurs. Active invasion can occur independently of climate change, especially when cave habitats are favorable habitats in some way, as is the case for lava tubes in young lava fields. In this situation, speciation may be parapatric and the surface populations may not go extinct. Under both of these scenarios, different species in caves are the result of independent invasions of caves. It is possible that speciation can occur by subsurface dispersal and subsequent isolation. Strong clustering of single-cave endemic species in particular regions that cannot be explained by habitat availability led Christman *et al.* to conclude that at least some subsurface dispersal of terrestrial cave species was occurring.

The hallmarks of adaptation include lengthening of appendages, thinning of the exoskeleton of arthropods, and increases in size and complexity of extra-optic sensory structures, which are advantageous traits in an aphotic environment. Eye and pigment loss or reduction may be a direct by-product of adaptation because it may make adaptation possible, or it may be a by-product of selection because eyes and pigments may be reduced as a result of relaxation of selection (i.e., neutral mutation). Whatever the causes, there is a remarkable convergence of morphology of obligate cave-dwelling organisms, and the troglomorphic animal, with long, spindly appendages and no eyes or pigment, is remarkably maladapted to any surface environment. Its fate in a surface environment is almost certainly rapid death, either from predation or UV radiation. This situation parallels evolution of flightlessness and the complex phenomenon of evolution of both gigantism and dwarfism on real islands. These adaptive features tend to restrict island inhabitants to their islands.

Given the reduced opportunities for migration, it is not surprising that endemism is widespread among species limited to subterranean habitats (terrestrial troglobionts and aquatic stygobionts). Among the most extreme cases are epikarstic stygobionts known from a single drip pool. One such case is the copepod *Paramorariopsis irenae* Brancelj, known only from a drip pool in Letuška jama in northeast Slovenia. Epikarst amphipods also often have such highly restricted distributions. *Stygobromus cooperi* Holsinger is only known from drip pools in Silers Cave in eastern West Virginia. At the level of individual caves, examples of endemism abound. Among the approximately 450 troglobiotic species in the eastern United States, over 250 are known from a single cave, making single-cave endemism the rule rather than the exception. A rather exotic example of this is the onychophoran *Speleoperipatus spelaeus* Peck, known only from Pedro Great Cave in Jamaica. Single-cave endemism is less pronounced among the aquatic fauna but is still not uncommon. Three species of stygobiotic fish of the genus *Ituglanis* are each known from a single cave in Brazil.

At the scale of karst basins and karst regions, endemism is rampant. The amphipod *Gammarus minus* occurs in six karst basins in a 1000-km^2 area in southern West Virginia, and the populations in the six basins are isolated from each other and, in fact, have a separate origin from surface-dwelling populations. Among karst regions in the United States, nearly all species are limited to a single region, and in fact there is relatively little overlap even among genera (Fig. 2 and Table 1).

TABLE 1
Proportion of Genera that Overlap between Regions Shown in Figure 2

Region	1	2	3	4	5	6	7	8	9
1	8	0.250	0.500	0.375	0.125	0	0	0	0
2	0.030	67	0.612	0.254	0.149	0.045	0.060	0.030	0.090
3	0.048	0.494	83	0.181	0.121	0.036	0.048	0.024	0.048
4	0.083	0.472	0.417	36	0.194	0.056	0.111	0.028	0.167
5	0.017	0.170	0.170	0.119	59	0.136	0.051	0.017	0.068
6	0	0.231	0.231	0.154	0.615	13	0.154	0	0.077
7	0	0.308	0.308	0.308	0.231	0.154	13	0	0.231
8	0	1.000	1.000	0.500	0.500	0	0	2	0.500
9	0	0.545	0.364	0.545	0.364	0.091	0.273	0.091	11

NOTE: The diagonal is the number of genera of cave-limited species found in the region. Each row contains the proportion of genera in the region associated with that row, and which also are found in the region associated with the column. Regions are (1) Florida Lime Sinks; (2) Appalachians; (3) Interior Low Plateau; (4) Ozarks; (5) Edwards Plateau/Balcones Escarpment; (6) Guadalupe Mountains; (7) Mother Lode; (8) Black Hills; (9) Driftless Area. From Culver et al. 2003.

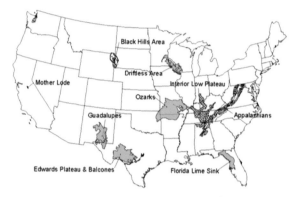

FIGURE 2 Map of major karst regions in the United States. Table 2 shows the number of stygobionts and troglobionts in each. Some additional karst areas are in the western United States, but they have been little studied biologically. From Culver et al. 2003.

CAVES AND KARST AS ISLANDS IN ECOLOGICAL TIME

Caves as Islands for Bats

Unlike stygobionts and troglobionts, bats enter and leave caves either daily or seasonally, depending on the circumstances. Brunet and Medellín, in a study of caves in central Mexico, found that the number of bat species roosting in a cave was strongly affected by ceiling area. The relationship between number of species (S) and area (A) was $S = 0.71 A^{0.31}$, well within the range of values found for fauna on oceanic islands. They suggest that area is also a surrogate for habitat diversity, especially solution pockets in the ceiling that apparently act to retain body heat of roosting bats. The maximum distance between any two caves was 13 km, well within the flight range of bats, so isolation was not a factor.

Drip Pools as Islands

Perhaps the best analogy between caves and islands in ecological time is that of cave pools fed by percolating water, a habitat that bears some superficial similarities to the water trapped in bromeliads, also island-like habitats. These drip pools have a diverse copepod fauna that comes from water dripping from epikarst. But the composition of the copepod fauna is distinct from that of the fauna present in dripping water. The analogy with islands is that epikarst (the source of dripping water) is the mainland, and the pools are islands. A general idea of the dynamics of this system can be seen in Table 2, which shows the relative number of species and individuals in five pools and the associated drips in some Slovenian caves. In general, pool area was about 1 m². The advantages of this system as a model of virtual islands are that (1) it is highly replicated, (2) it is possible to census immigration completely over long periods of time, and (3) given the high rate of migration from epikarst relative to the pool population, the pool fauna is likely to be in equilibrium between migration and extinction rates. For all five pool "islands," the number of species found among migrants was approximately the same as the number of species present in pools. For all but one case, the ratio of resident to migrating individuals was between 3 and 10. These five pools were the most diverse among 35 examined. There were habitat differences among pools, with the most diverse fauna being found in pools with heterogeneous substrates but not necessarily high organic matter, indicating that "island" quality varied.

When species composition is considered, some indirect idea of the extinction can be obtained. On a per-pool basis, a total of 24 species immigrations were recorded (Table 2). Of these, only 11 were present in the associated pool, indicating that most immigrations fail. Furthermore, relatively few species were common in both drips and pools. Other species were good migrators but were rare in pools, and yet other species were common in pools but rare in dripping

TABLE 2
Rates of Migration and Size of Populations in Five Pools, and Associated Drips in Four Slovenian Caves

Cave and Drip Number	Number of Immigrating Species/Year	Number of Immigrating Individuals/Year	Number of Species in Pool	Number of Individuals in Pool
Pivka jama 1	6	158	7	57
Črna jama 1	5	11	4	74
Županova jama 2	5	9	4	93
Županova jama 3	5	18	4	56
Škocjanske jame 4	3	8	6	23

NOTE: Drip numbers correspond to those of Pipan (2005), and data are from Pipan (2003). Immigration data are from a complete sample of a drip through time, but of course more than one drip may feed the pools. Pools themselves were completely sampled prior to the measurement of migration rates.

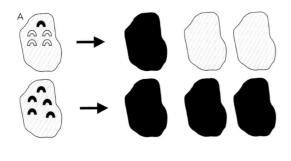
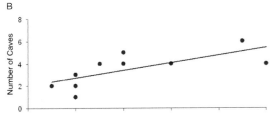

FIGURE 3 Relationship between occupancy of drips and drip pools in a cave and occupancy among caves for stygobiontic copepods. (A) The black semicircles represent drips within a cave with a species present, and the white semicircles represent drips within a cave with a species absent. On the right side, caves with the species are black, and those without a species are gray. Drawing by D. C. Culver. (B) The graph depicts the relationship between number of drips occupied in the Postojna Planina Cave System by a stygobiotic copepod and the number of Slovenian caves in which the species occurs. The solid line is the regression line and accounts for 49% of the variance in number of caves occupied.

water. In this context, the phrase "good migrator" is something of a misnomer because migrating individuals are, in fact, being washed out of their habitat—epikarst. Finally, the question remains as to whether any pool populations are source populations or whether, in the absence of a continuing rain of copepods from the epikarst, all populations would go extinct (i.e., they are sink populations). This is not known, but it is subject to experimental manipulation.

Caves as Archipelagoes

Caves can also be analogous to archipelagoes, with migration among islands (caves) rather than immigration from the mainland (surface environments). Although the analogy is an appealing one, it is difficult to directly or even indirectly measure migration and extinction rates; except in special circumstances, these rates are likely to be too low to be easily measured. We do know that migration between caves can occur, as shown by high degrees of genetic similarity between populations in adjacent and nearby caves. We also know that patterns at small scales, such as individual epikarst drips in a cave, are repeated at larger scales, such as different caves in a karst region (Fig. 3). In this situation, migration seems to be occurring, not only within small regions of epikarst drained by several drips, but also among caves. These kinds of results also indicate that the cave fauna is in a dynamic, changing state, and not simply the result of isolation in caves without further dispersal.

SEE ALSO THE FOLLOWING ARTICLES

Adaptive Radiation / Convergence / Dispersal / Endemism / Lava Tubes / Species–Area Relationship

FURTHER READING

Brunet, A. K., and R. A. Medellin. 2001. The species-area relationship in bat assemblages in tropical caves. *Journal of Mammalogy* 82: 1114–1122.

Christman, M. C., D. C. Culver, M. Madden, and D. White. 2005. Patterns of endemism of the eastern North American cave fauna. *Journal of Biogeography* 32: 1441–1452.

Culver, D. C. 1970. Analysis of simple cave communities. I. Caves as islands. *Evolution* 24: 463–474.

Culver, D. C., M. C. Christman, W. R. Elliott, H. H. Hobbs III, and J. R. Reddell. 2003. The North American obligate cave fauna: regional patterns. *Biodiversity and Conservation* 12: 441–468.

Culver, D. C., and W. B. White, eds. 2005. *Encyclopedia of caves*. Amsterdam: Elsevier.

Gunn, J., ed. 2004. *Encyclopedia of caves and karst*. New York: Fitzroy Dearborn.

Howarth, F. G. 1980. The zoogeography of specialized cave animals: a bioclimatic model. *Evolution* 28: 365–389.

Pipan, T. 2005. *Epikarst—a promising habitat. Copepod fauna, its diversity and ecology: a case study from Slovenia (Europe)*. Ljubljana, Slovenia: ZRC Publishing, Karst Research Institute.

CHAGOS ARCHIPELAGO

SEE INDIAN REGION

CHANNEL ISLANDS (BRITISH ISLES)

EDWARD P. F. ROSE

University of London, United Kingdom

The Channel Islands are British Crown Dependencies situated close to the Normandy coast of France (Fig. 1): Jersey (116 km^2); Guernsey (64 km^2); Alderney (8 km^2); the lesser islands of Sark, Herm, Jethou, and Brecqhou; and adjacent (mostly uninhabited) islets, rocks, and reefs. Their laws and administrative systems are distinct from those of England, and therefore the United Kingdom, having been independently derived from the medieval Duchy of Normandy. Each of the four largest islands is largely self-governing, but Jersey and Guernsey form the centers of two separate bailiwicks with, among other differences, their own postage stamps, banknotes, and coinage.

GEOLOGY

The islands comprise the eroded remnants of part of the Armorican Massif more extensively exposed in Brittany and western Normandy, isolated from France primarily

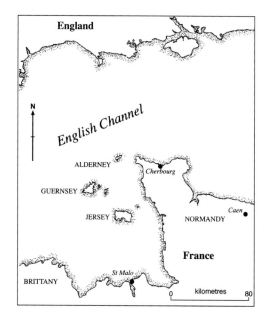

FIGURE 1 Map showing the location of the largest Channel Islands.

by Holocene sea-level rise. Studied intermittently for over 200 years, they represent a classical area of British geology. The northern Massif comprises (1) the Pentevrian, a crystalline basement yielding isotopic ages older than 900 million years (exposed on Guernsey, less certainly on Sark and Alderney), and (2) a series of metamorphic and igneous rocks (Guernsey, Sark, and Alderney), low-grade metamorphic sediments (Jersey and Guernsey), plutonic intrusions (all main islands), volcanic rocks (Jersey), a conglomerate (Jersey), and sandstone (Alderney) all associated with a 700–400-million-year-old cycle of plate tectonic subduction and mountain building commonly known as the Cadomian Orogeny.

Jersey is composed principally of metasediments and overlying volcanics of very late Precambrian age, intruded by end-Precambrian and Ordovician granites and associated igneous rocks and unconformably overlain to the northeast by Cambro-Ordovician conglomerates with subordinate sandstones and mudstones. The metasediments are predominantly fine- to medium-grained turbiditic sandstones; the volcanics are lavas, tuffs, and agglomerates of andesitic and basaltic composition, succeeded by rhyolitic ignimbrites, lavas, and air-fall or water-laid pyroclastics—all affected by low-grade regional metamorphism. Minor intrusions are common, mostly as dykes, and indicate several phases of emplacement.

Guernsey comprises two distinct parts: (1) a southern metamorphic complex of largely Pentevrian basement composed of metasediments, granite gneisses (the most abundant dated to around 2000 million years ago), and quartz-diorite to granodiorite gneisses—all cut by a large number of minor intrusions, mostly dykes, and (2) a northern plutonic igneous complex of Cadomian age, which is largely diorite, flanked to the east by gabbro and to the north and west by granites—mostly emplaced (with relatively few minor intrusions) by around 500 million years ago.

Alderney comprises a Precambrian basement (of gneiss and diorite) to the west, and sandstone (which is possibly coeval with the post-orogenic conglomerates of Jersey) to the east.

Deposition of the conglomerates and sandstone was followed by a time span of around 400 million years that left no sedimentary record. The three largest islands developed a surface topography of low plateaus, now gently inclined and incised by small streams in steep valleys. Their youngest rocks are unconsolidated superficial sedimentary deposits of Quaternary age, which reflect changing climates and relative sea levels from the Middle Pleistocene to the present day. Largely deposits of interglacial or periglacial origin, they comprise raised beaches, blown sands, "head" breccias, peat, or alluvium in coastal or valley areas, and widespread loess.

COAST, CLIMATE, BIOTA, AND HISTORY

Steep cliffs fringe much of the present coast of all the large islands, indented by sandy bays. Tides have a high range (around 8 m), locally exposing extensive rocky wave-cut platforms or sandy beaches at low water.

Climate is maritime, lacking marked variations or extremes. Thus, Jersey averages a mean daily air temperature of 11.5 °C, experiences ground frosts on only 60 days per year, and has snow or sleet on only 12 days per year. The average long-term annual rainfall is 877 mm on Jersey and 790 mm on Guernsey, but actual annual rainfall may vary considerably from the long-term mean. Investigation of groundwater resources, begun when the islands were occupied by German troops during World War II and continued in more recent years by the British Geological Survey, has facilitated an exceptionally detailed appraisal (on Jersey and Guernsey) of the hydrology of island basement aquifers in three phases, over a 65-year time span.

The largest islands are predominantly rural, with intensively worked arable land and grassland used for dairy farming, plus small areas of heathland, semi-natural mixed deciduous woodland, and wetland. Paleobotanical records reveal that from some 10,000 years ago, periglacial tundra was succeeded by a vegetation mosaic in which woodland was important, its composition varying with changes in climate and sea level. The present biota has closest affinity with neighboring France. Apart from birds, wild flora and fauna are relatively restricted in variety, with few mammals or larger reptiles, although the islands host the most northerly occurrence of some southern European plants and contain some unique subspecies of small mammals.

Historically, Jersey at least was accessible to archaic humans by around 200,000 years ago, who were succeeded in turn by Neanderthals and modern humans. Early inhabitants seemingly adopted a hunter-gatherer lifestyle, exploiting resources both terrestrial (e.g., wild ox, pig, red deer, horse, and goat) and marine (e.g., limpets, periwinkles, and winkles). However, all the large islands were extensively cultivated from around 6000 years ago onward, with inhabitants clearing woodland and leaving some notable archaeological remains. There is evidence of cultural contact with both France and England from at least 3000 years ago, although closest links developed with nearby Normandy. Normandy owed allegiance to the English monarch from the Norman Conquest of England in 1066 until it was incorporated within the then–Kingdom of France in 1204. Thereafter, the Channel Islands faced potential attack by the French and were progressively fortified by the English. A few strong castles were developed on Jersey and Guernsey from the thirteenth to seventeenth centuries, more numerous and widespread redoubts and batteries in the eighteenth century, a series of coastal towers around the beginning of the nineteenth century, and headland forts (particularly on Alderney) within the nineteenth century. Massive coastal fortifications constructed during the German occupation of 1940–1945 are additional features of significant tourist interest.

SEE ALSO THE FOLLOWING ARTICLES

Archaeology / Britain and Ireland / Granitic Islands / Hydrology

FURTHER READING

Bishop, A. C., D. H. Keen, S. Salmon, and J. T. Renouf. 2003. *The geology of Jersey, Channel Islands*. Geologists' Association guide no. 41. London: Geologists' Association.

Coysh, V., ed. 1977. *The Channel Islands: a new study*. Newton Abbot, UK: David & Charles.

Gardiner, V. 1998. *The Channel Islands*. World bibliographical series vol. 209. Oxford, UK: Clio Press.

Jones, R. L., D. H. Keen, J. F. Birnie, and P. V. Waton. 1990. *Past landscapes of Jersey: environmental changes during the last ten thousand years*. Jersey: Société Jersiaise.

Power, G. M. 1997. Great Britain: Channel Islands, in *Encyclopedia of European and Asian regional geology*. E. M. Moores and R. W. Fairbridge, eds. London: Chapman & Hall: 276–277.

Roach, R. A., C. G. Topley, M. Brown, A. M. Bland, and R. S. D'Lemos. 1991. *Outline and guide to the geology of Guernsey*. Monograph no. 3. Guernsey: Guernsey Museum.

Robins, N. S., and E. P. F. Rose. 2005. Hydrogeological investigation in the Channel Islands: the important role of German military geologists in World War II. *Quarterly Journal of Engineering Geology and Hydrogeology* 38: 351–362.

Rose, E. P. F. 2005. Specialist maps of the Channel Islands prepared by German military geologists during the Second World War: German expertise deployed on British terrain. *The Cartographic Journal* 42: 111–136.

CHANNEL ISLANDS (CALIFORNIA), BIOLOGY

AARON MOODY
University of North Carolina, Chapel Hill

The California Channel Islands lie near the coast of southwestern California between Point Conception and the U.S.–Mexico border (Fig. 1, Table 1). They contain a rich terrestrial flora and high rates of floral and faunal endemism. The islands also provide significant foraging

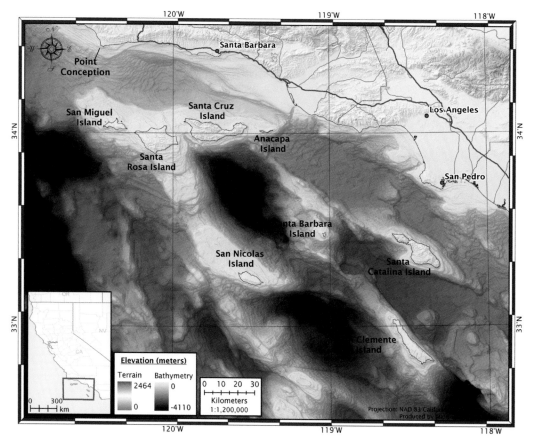

FIGURE 1 The California Channel Islands shown with topography and bathymetry (© Matthew C. Simon). Roads, hydrology, county boundaries, and terrain data were obtained from the California Spatial Information Library (CaSIL) 20080110. Bathymetry was obtained from the National Geophysical Data Center (NGDC) Coastal Relief Model 20080129. The U.S. base map was obtained from Environmental Systems Research Institute (ESRI), ArcGIS 9.2.

and breeding habitat for marine mammals and seabirds, and they harbor rich and productive rocky intertidal and kelp forest communities. Introduced taxa constitute a major component of the terrestrial biota, and the legacies of uncontrolled livestock ranching from the mid-nineteenth to the late twentieth century are readily observed today. The islands are considered to have high conservation value globally.

ISLAND FLORA

General Description

The form and life history of native plants on the California Channel Islands are typical of the vegetation found in Mediterranean-type environments around the world, where cool, wet winters, and hot, dry summers select for drought adaptations. Typical plant traits include shrub and

TABLE 1

Geographic Situation and Floristic Richness on the California Channel Islands

	Area (km²)	Closest Mainland Point (km)	Highest Elevation (m)	Channel Island Endemic Plants	Native Plants	Exotic Plants
San Miguel	37	42	253	18	198	69
Santa Rosa	217	44	484	42	387	98
Santa Cruz	249	30	753	45	480	170
Anacapa	2.9	20	283	22	190	75
Santa Barbara	2.6	61	194	14	88	44
San Nicolas	58	98	277	18	139	131
Santa Catalina	194	32	648	37	421	185
San Clemente	145	79	599	47	272	110

sub-shrub growth form, sclerophyllous leaf physiognomy, drought-deciduous habit, and high energy allocation to root growth. Grassland, coastal sage scrub, and various phases of coastal bluff scrub are the most extensive vegetation types, with significant areas of chaparral on larger islands. Mixed broadleaf woodlands, oak woodlands, and conifer woodlands occur throughout the larger islands on sheltered slopes, canyons, or ridges exposed to frequent fog. Unique and/or rare communities include coastal dune scrub, low baccharis and caliche scrub, riparian woodland, floodplain scrub, coastal bluff scrub, coastal marsh and estuaries, freshwater seeps and springs, and vernal pools. Most of these types occur only on the larger islands. The most obvious legacy of ranching is the expansive replacement of native shrub cover by non-native grasses.

Santa Cruz Island is the largest, most environmentally diverse, and most speciose of the California Channel Islands. Its flora exhibit strong species and community affinities to certain geologic types. Species richness and endemism are highest on steep, north-facing, volcanic sea cliffs. The westernmost island, San Miguel, is also the most windswept, enforcing low vegetative growth forms over the rounded, low-lying plateaus of the island. The smaller islands have low diversity of community types and have been the most greatly transformed by ranching, exotic animals, and non-native plants, but they nevertheless harbor some rare and endemic plants.

Endemic Plants

There are approximately 1000 native plant species, subspecies, or varieties on the islands, representing 29 families. About 100 of these are endemics, of which nearly half are single-island endemics. Plant families and genera showing especially high rates of endemism include the buckwheat (Polygonaceae, *Eriogonum*), heath (Ericaceae, *Arctostaphylos*), madder (Rubiaceae, *Galium*), stonecrop (Crassulaceae, *Dudleya*) (Fig. 2), sunflower (Asteraceae, *Malacothrix* and *Deinandra*), and pea families (Fabaceae, *Lotus*). High rates of single-island endemism at the species or subspecies level within some of these families is suggestive of autochthonous (in situ) diversification, or adaptive radiation, of plants on the islands. Molecular-genetic data for the shrubby tarweeds (*Deinandra*) provide strong evidence of adaptive radiation within this group on Guadalupe Island, approximately 525 km south of the California Channel Islands and 250 km west of Baja California, Mexico.

Many Channel Island endemics are relicts of more widespread historic distributions. *Lyonothamnus*, the only

FIGURE 2 Greene's dudleya (*Dudleya greenei*) on Santa Cruz Island (© Kathy deWet-Oleson).

endemic plant genus on the islands, is a paleorelict from a much larger range that contracted and disappeared from the mainland 5–6 million years ago. *Lyonothamnus floribundus* occurs on Santa Rosa and Santa Cruz Islands, and an endemic subspecies is found on San Clemente Island (*L. floribundus floribundus*). The island oak (*Quercus tomentella*) is a relictual endemic on Guadalupe Island and on five of the California Channel Islands (Santa Rosa, Santa Cruz, Anacapa, Santa Catalina, and San Clemente Islands). The mainland fossils of island oak date from 2 to 10 million years ago. A single relictual stand of Torrey pine (*Pinus torreyana insularis*) occurs on Santa Rosa Island. Torrey pine is among the rarest of all pines, with only one other population of the mainland counterpart (*P. torreyana*) near San Diego, California.

The Catalina cherry (*Prunus ilicifolia* ssp. *lyonii*) was long considered endemic to the California Channel Islands until it was discovered hundreds of kilometers south in the mountains of central Baja California, Mexico. Other common endemic chaparral shrubs with close relatives on the mainland include northern island bush poppy (*Dendromecon rigida harfordii*), mountain mahogany (*Cercocarpus betuloides blancheae*), and big-pod ceanothus (*Ceanothus megacarpus insularis*).

A notable near-endemic on the islands is *Coreopsis gigantea*, a large, charismatic daisy that occurs commonly on the Channel Islands and rarely along a narrow coastal strip from San Luis Obispo County to Los Angeles County.

FAUNA

Although diversity of non-volent terrestrial fauna on the islands is low, there are a number of noteworthy extant and ancient faunal groups.

Paleofauna

During the glacial maximum of the Late Pleistocene (~18,000 years ago), the four northern islands formed a single larger island, Santarosae. The southern islands were all separate during this period, but were much larger than at present. The island mammoth (*Mammuthus exilis*), a dwarf descendent of the Columbian mammoth (*M. columbi*), occupied the islands at this time. *M columbi* probably colonized Santarosae 20,000 to 40,000 years ago, presumably by swimming across the channel from the mainland to Santarosae. Mammoth fossils are sufficiently abundant on Santa Rosa, San Miguel, and Santa Cruz to provide a record of decreasing size through time. *M. exilis* went extinct about 12,000 years ago.

Modern Terrestrial Mammals

The island fox (*Urocyon littoralis*) and the island deer mouse (*Peromyscus maniculatus*) are recognized as endemic species, with distinct subspecies upon each of the islands they inhabit. The fox occurs on the six largest islands. The deer mouse occurs on all eight islands and is larger than mainland deer mouse. The island spotted skunk (*Spilogale gracilis amphiala*) is endemic to Santa Rosa and Santa Cruz Islands.

ISLAND FOX

At 30 to 35 cm and about 2 kg, the island fox is the smallest North American canid (Fig. 3). A descendent of the mainland gray fox, the island fox evolved into a unique species over 10,000 years ago. Although small for a fox, *U. littoralis* is the largest of the Channel Islands' few native mammals. The fox consumes fruits of native and exotic plants, as well as insects, snails, marine invertebrates, ground birds, and the few reptile and rodent species available. The fossil record shows evidence of foxes on Santarosae between 10,400 and 16,000 years ago. The gray fox probably arrived from the mainland by "rafting" on debris propelled by storms or currents during the last glacial maximum, although the possibility that foxes were introduced by people has not been ruled out. Genetic data suggest that island foxes arrived on the southern islands 800–4300 years ago, probably brought by Chumash people, for whom the fox was a sacred animal, and who traded with the Gabrieliño people of the southern islands.

Steep declines in island fox populations coincided with the arrival of the golden eagle, the first known predator of the fox, which was sighted and found nesting on San Miguel and Santa Cruz Islands in the mid-1990s. Golden eagles colonized the islands following local extinction of the largely piscivorous bald eagle. Juvenile feral pigs provided an abundant year-round food source for golden eagles, a terrestrial predator that also reduced fox populations on Santa Cruz, Santa Rosa, and San Miguel Islands. The Santa Catalina Island fox has also declined because of canine distemper virus.

BATS

Bats are often the most diverse mammal group on islands because of their dispersal ability. This is the case on the California Channel Islands. For example, 11 species of bats have been found on Santa Cruz Island, which is also home to a colony of the rare Townsend's big-eared bat (*Corynorhinus townsendii*).

Birds

SEABIRDS

The California Channel Islands provide a major seabird breeding area in the eastern North Pacific. San Miguel, Santa Barbara, and Anacapa Islands represent the southernmost or northernmost breeding habitats for many seabirds, and play a vital role in the conservation of several species. The Channel Islands provide critical breeding habitat for ashy storm petrels (*Oceandroma homochroa*), brown pelicans (*Pelecanus occidentalis*) (Fig. 4), and Xantus's murrelets (*Synthliboramphus hypoleucus*), whose northernmost colonies occur on San Miguel Island. The southernmost breeding colonies of the pigeon guillemots (*Cepphus columba*) occur on Santa Barbara Island. The tufted puffin (*Fratercula cirrhata*) and common murre (*Uria aalge*), both birds of the North Pacific, have disappeared from the Channel Islands, which used to provide their southernmost breeding habitat. Other notable sea-

FIGURE 3 Santa Cruz Island fox (*Urocyon littoralis santacruzae*) on east Santa Cruz Island (© Kathy deWet-Oleson).

FIGURE 4 California brown pelicans (*Pelecanus occidentalis*) on east Anacapa Island (© Kathy deWet-Oleson).

birds that breed on the islands are Cassin's auklet (*Ptychoramphus aleuticus*), Brandt's cormorant (*Phalacrocorax penicillatus*), western gulls (*Larus occidentalis*), and snowy plovers (*Charadrius alexandrius*), which are declining on the Pacific coast.

LANDBIRDS

Island landbirds of particular note include the island scrub jay (*Aphelocoma insularis*), an endemic to Santa Cruz Island, and ten other more widely spread island endemics, including the declining island loggerhead shrike (*Lanius ludovicianus anthonyi*) (an endemic to the northern island group), two sparrows, a warbler, a wren, a flycatcher, a horned lark, a towhee, northern and southern house finches, and a hummingbird. A subspecies of song sparrow (*Melospiza melodia graminea*) endemic to Santa Barbara Island disappeared because of habitat disruption by introduced rabbits, direct predation by feral cats, and a fire that destroyed much of its habitat on the tiny island.

Landbird populations and communities on the islands can change from year to year. Both bald eagles (*Haliaeetus leucocephalus*) and peregrine falcons (*Falco peregrinus anatum*) historically bred on the islands but disappeared because of harassment, shooting, egg thievery, and reproductive failure caused by bioaccumulation of DDT. Both of these species are recovering in response to reintroduction efforts.

Reptiles and Amphibians

The island night lizard (*Xantusia riversiana*) is endemic on Santa Barbara, San Nicolas, and San Clemente Islands. It is the most morphologically distinct of the endemic vertebrates on the Channel Islands, possibly indicative of a prolonged isolation. *X. riversiana* can live up to 20 years, but spends life in territories as small 30 m². Snakes are found on Santa Catalina, Santa Rosa, and Santa Cruz Islands, including the western rattlesnake (*Crotalus viridis*) on Santa Catalina. The island gopher snake (*Pituophis melanoleucus pumilis*) is endemic to the northern islands. Ten to 12 nonendemic snakes and lizards are also found on the islands. These are most diverse on Santa Catalina and Santa Cruz Islands. The Channel Island slender salamander (*Batrachoseps pacificus pacificus*) and the black-bellied salamander (*Batrachoseps nigriventris*) are highly differentiated genetically from their mainland counterparts and between the northern islands, with both cases indicating isolation for as much as 2 million years.

Insects

There are at least 27 endemic moths and butterflies collected to date, including 15 single-island endemics, and a number of endemic Orthoptera, including several relatives of the Jerusalem crickets. Recent work indicates several endemic spiders, including the agelinid genus *Rualena,* which includes several single-island endemics.

Marine Biota

Mixing of the cold California current and the warm California countercurrent promotes productivity and diversity in marine coastal systems around the California Channel Islands. Their situation at the boundary of the Oregonian and the Californian marine biogeographic provinces, their diversity of habitat types, and their exposure to dynamic oceanographic conditions set the stage for abundant and species-rich marine ecosystems around the islands.

KELP FOREST

Giant kelp, *Macrocystis pyrifera,* attaches on the sea floor and grows to the surface where its leafy fronds form extensive canopies. Kelp forests provide forage, cover, and nursery habitat for a diversity of organisms, including warmer-water species such as Garibaldi, moray eels, and spiny lobsters, and colder-water organisms such as black rockfish, and the sunflower star.

MARINE MAMMALS

Many species of marine mammals use the coastal areas of the California Channel Islands and surrounding waters for foraging, breeding, or migratory habitat. Pinnipeds are remarkably rich on the islands, including California sea lions, northern sea lions, northern fur seals, Guadalupe fur seals, harbor seals, and northern elephant seals (Fig. 5). Sea otters, dolphins, porpoises, gray whales, killer whales,

FIGURE 5 Young male northern elephant seals (*Mirounga angustirostrus*) at Cardwell Point, San Miguel Island (© Kathy deWet-Oleson).

humpback whales, and blue whales are regularly observed in the waters surrounding the islands.

ROCKY INTERTIDAL ZONE

The rocky intertidal habitats on the Channel Islands are much better preserved than on the mainland because of relatively low human impact. Rocky substrate and dense algal and plant growth provide habitat for a charismatic assemblage of sessile and vagile marine invertebrates, including numerous nudibrancs, crabs, shellfish, anemones, chitons, limpets, mussels, sea cucumbers, and octopus. From the standpoint of biological diversity or ecological condition, the rocky intertidal habitats on the islands are unparalleled by even the most unspoiled, remote shorelines along California's mainland.

IMPRINT OF INSULARITY

As with islands generally, the biota of the California Channel Islands contains fewer species and exhibits taxonomic bias relative to the adjacent mainland. For example, there are at most four non-volent mammals native to any island. The result is an ecological setting qualitatively different from the mainland, marked, for example, by the absence of large herbivores or predators and by high population densities of certain taxa coincident with the absence of predators and/or competitors.

Gigantism, dwarfism, and flightlessness are noteworthy evolutionary outcomes of resource constraint and competition, or of predator or competitor release. On the California Channel Islands, these artifacts are seen in the extinct (~1870) giant deer mouse (*Peromyscus nesodytes*) from Santarosae; the island scrub jay (*Aphelocoma insularis*), which is larger than its mainland counterpart; the island fox, which is much smaller than its mainland counterpart; and the Late Pleistocene island, or pygmy, mammoth, which was about one-third the size of its mainland ancestor.

IMPRINT OF HUMAN HISTORY

Two femurs from "Arlington Springs Woman" were found on Santa Rosa Island in 1959–1960. Her remains, dated to 13,000 years ago, are among the earliest human remains known from North or South America. Her presence on the island suggests that watercraft were in use along the northwest coast of North America at that time, and lends some support to the coastal migration theory that ancient humans entered North America by boat down the Pacific coast from Alaska. Recent radiocarbon dating of pygmy mammoth fossils from Santa Rosa Island suggests that the last of the mammoths may have been present at the time of human arrival.

Much is known of the human prehistory on the islands from a superb set of cultural resources found on San Miguel Island, including at least 600 archeological sites recording over 11,000 years of indigenous history. Permanent indigenous presence on the islands began about 8000 years ago, dwindled through the Spanish and Mexican eras of control, and ended in the early nineteenth century. Inhabitants of the northern islands were part of the Chumash alliance, and the southern islands were occupied by the Gabrieliño group. These peoples relied heavily on marine resources for their diet and material culture. They may have transferred some terrestrial organisms, such as the island fox, between islands.

Systematic harvesting of marine mammals by Russian, Portuguese, Spanish, Aleut, English, and French hunters began in the late eighteenth century and decimated sea otter (*Enhydra lutris*), northern fur seal (*Callorhinus ursinus*), and northern elephant seal (*Mirounga angustirostris*) populations within a century. Rapid growth of abalone and commercial fishing operations during the second half of the nineteenth century included established encampments at suitable coastal sites. These activities did not resume following World War II, during which the U.S. military co-opted the islands for coastal surveillance.

Intensive and widespread grazing of domestic livestock began in the mid-nineteenth century, and the ranching history of the islands is particularly important to their modern biology. Domestic animals such as sheep, cattle, bison, horses, pigs, goats, cats, and dogs; nuisance species such as rats and mice; and game species including rabbits, wild turkey, California quail, deer, elk, and antelope were all introduced on at least one of the islands, which had previously hosted small and naïve faunal communities. Highly invasive introduced plants include ice plant, fennel, mustard, and several European grasses. Severe impacts have resulted from feral pigs, whose population exploded on Santa Cruz Island upon the eradication of sheep there in the mid-1980s. Feral pigs destroyed native vegetation including rare plants, caused widespread erosion, and disturbed archeological sites.

Modern anthropogenic activity on the islands includes scientific research, ecological restoration, eradication of non-native species, recreational boating and diving, camping and hiking, and legal protection from resource exploitation within the Channel Islands National Park and National Marine Sanctuary. Infrastructure includes research facilities, experiment and restoration sites, unpaved road networks and trails, several small ranch sites, and several small naval installations. Residential and commercial habitation is restricted to the town of Avalon on Santa Catalina Island.

SCIENCE AND CONSERVATION

The Santa Barbara Museum of Natural History curates the official archival collections from the islands and has published five California Channel Island Symposia between 1980 and 2002. The first symposium, in 1967, was published by the Santa Barbara Botanic Garden, where one can see many of the islands' endemic plants in the Garden's island section.

Conservation

Many of the feral and exotic animal species have been removed from the islands. Native vegetation has responded rapidly to the removal of feral sheep on Santa Cruz Island and has recovered over much of San Miguel Island following removal of sheep and burros. Golden eagles are being captured and relocated to northeast California; bald eagles and foxes are being reintroduced, and their populations are recovering. Rat eradication from Anacapa Island has led to significant recovery of the island deer mouse, side-blotched lizard, slender salamander, and Xantus's murrelet, whose population has also recovered on Santa Barbara Island after the removal of feral cats in 1978. Also making a comeback, after protection from hunting, are seals, sea lions, and other pinnipeds that breed, pup, and haul out on isolated coastlines.

In January 1969, an offshore oil platform spilled 200,000 gallons of crude oil into the Santa Barbara Channel, impacting all of the northern Channel Islands and nearby mainland beaches, and killing thousands of seabirds and marine mammals. At about the same time, a steep decline in the breeding success of California brown pelicans was linked to biomagnification of DDT through the marine food web on which the pelican depends. DDT caused eggshell thinning and consequent reproductive failure of the pelican, driving the bird near to extinction and promoting the regulation of DDT. These events mark watershed moments in the history of the environmental movement.

The U.S. National Park Service established the Channel Islands National Park and Channel Islands National Marine Sanctuary in 1980. The Channel Islands National Park includes the four northern islands and Santa Barbara Island. The Channel Islands National Marine Sanctuary extends from mean high tide to six nautical miles offshore around each of these five islands, and within this zone is a network of (at present) nine marine reserves, designated as no-take zones. The Nature Conservancy and the National Park Service have cooperated in overseeing conservation and restoration on the islands.

SEE ALSO THE FOLLOWING ARTICLES

Channel Islands (California), Geology / Introduced Species / Pigs and Goats / Seabirds

FURTHER READING

Garcelon, D. K., and C. A. Schwemm, eds. 2005. *Proceedings of the Sixth California Islands Symposium.* Arcata, CA: Institute for Wildlife Studies. National Park Service Technical Publication CHIS-05-01.

Halvorson, W. L., and G. J. Maender, eds. 1994. *The Fourth California Islands Symposium: update on the status of resources.* Santa Barbara, CA: Santa Barbara Museum of Natural History.

Hochberg, F. G., ed. 1993. *Third California Islands Symposium: recent advances in research on the California Islands.* Santa Barbara, CA: Santa Barbara Museum of Natural History.

Philbrick, R. N., ed. 1967. *Proceedings of the Symposium on the Biology of the California Islands.* Santa Barbara, CA: Santa Barbara Botanic Garden.

Power, D. N., ed. 1980. *The California Islands: proceedings of a multidisciplinary symposium.* Santa Barbara, CA: Santa Barbara Museum of Natural History.

Schoenherr, A. A., C. R. Feldmeth, and M. J. Emerson. 1999. *Natural history of the islands of California.* Berkeley: University of California Press.

CHANNEL ISLANDS (CALIFORNIA), GEOLOGY

JANET HAMMOND GORDON
Pasadena City College, California

The geology of California's Channel Islands records the transition of the North American plate boundary from a subduction zone to a transform boundary that is now the San Andreas fault. This transition was particularly complex at the latitude of the Channel Islands. Geological evidence on the islands indicates that some were rotated and translated by as much as 300 km up the California coast from San Diego. Others are the product of concurrent rifting and volcanism.

REGIONAL SETTING

The Channel Islands reside in the California Continental Borderland, the offshore area between the continental slope and the coast (Fig. 1). The borderland has three sections. The partly onshore Western Transverse Ranges block includes the four northern Channel Islands (Anacapa, Santa Cruz, Santa Rosa, and San Miguel). On the west, the Outer Borderland hosts San Nicolas Island. The Western Transverse Ranges block and the Outer Borderland consist mostly of sedimentary rocks deposited on the continental margin and in the trench of the Mesozoic age subduction zone. The Inner Borderland is floored by Catalina schist of Mesozoic age and includes abundant Miocene-age igneous rocks. Santa Barbara, San Clemente, and Santa Catalina Islands are in the Inner Borderland.

The Catalina schist is an important complex of metamorphic rocks that were oceanic lithosphere and sea-floor sediments before they were subducted and metamorphosed. Blueschist, a gray-blue metamorphic rock, is a distinctive component of the Catalina schist.

NORTHERN CHANNEL ISLANDS

The northern Channel Islands lie on an east–west trending platform aligned with the Santa Monica Mountains onshore to the east. The platform and islands are a thick series of mostly marine sedimentary rocks and volcanic rocks of Late Cretaceous to Middle Miocene age deposited on a basement of Mesozoic metamorphic and plutonic igneous rocks. Paleomagnetic studies show that this island group has undergone a least 90° of clockwise rotation in the last 20 million years.

Mountainous Santa Cruz Island has an area of 250 km^2 with a maximum relief of 750 m. The east–west trending Santa Cruz Island fault separates the island into geologically distinct southern and northern blocks. The oldest rocks are exposed in the southern block where Jurassic plutons intrude older Mesozoic schist. The plutonic rocks include abundant hornblende gabbro plus diorite and tonalite of chemical compositions and ages akin to those of the Peninsular Ranges plutons of southern and Baja California.

The overlying sequence of sedimentary rocks is of Paleocene through Miocene age. It has an exposed thickness of about 3000 m and a regional dip to the south. The Eocene Jolla Vieja formation, a thick-bedded sandstone and conglomerate exposed in a fold in the southern block, is of particular note. The abundant volcanic cobbles in the conglomerate match the distinctive hard rhyolite cobbles in the Eocene Poway conglomerate of the San Diego area and indicate that Jolla Vieja formation was deposited far to the south of the island's present position.

The Miocene San Onofre breccia and Blanca formation document a major change from Eocene conditions. The San Onofre breccia, as it occurs on Santa Cruz Island, contains a mixture of mostly angular pebbles, cobbles, and boulders of blueschist, diorite, quartz, and volcanic rock. In some locations, blueschist fragments make up nearly half of the rock. The overlying Blanca formation is dominantly conglomerate and sandstone

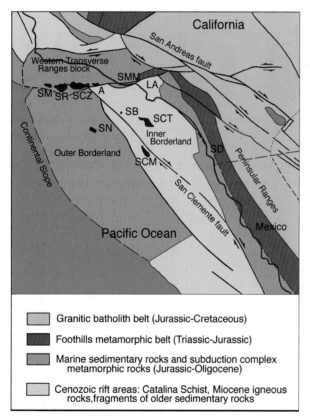

FIGURE 1 California Continental Borderland geologic belts and locations of California Channel Islands. Red line marks today's coastline. SM—San Miguel Island, SR—Santa Rosa Island, SCZ—Santa Cruz Island, A—Anacapa Island, SBI—Santa Barbara Island, SN—San Nicolas Island, SCT—Santa Catalina Island, SCM—San Clemente Island, SD—San Diego, LA—Los Angeles, SMM—Santa Monica Mountains. Adapted from Couch and Suppe (1993).

composed mostly of reworked volcanic ash and cobbles of dacite and rhyolite. Both formations have depositional ages of about 19 million years. Nearby highlands of Catalina schist must have risen by this time to provide blueschist for the San Onofre breccia, and almost simultaneously, volcanoes began shedding debris into the Blanca formation.

The island's northern block is composed principally of Santa Cruz Island volcanics. These north-dipping flows and intrusions range from basalt and andesite to rhyolite. They were probably erupted from a center near the highest peak on the island about 17 million years ago. The Monterey shale overlies the volcanic rocks on the eastern portion of the island. It contains diatoms and volcanic ash deposited in a deep offshore basin.

Small Anacapa Island consists of three elongated segments composed of north-dipping volcanic flows. The rocks are identical in composition and age to the andesite portion of the Santa Cruz Island volcanics just to their west.

The narrow canyons of hilly Santa Rosa Island expose mostly marine sedimentary formations of Eocene to Middle Miocene age. The Eocene South Point formation is a thick-bedded sandstone and conglomerate. Its pebbles and a large proportion of its coarse sand grains match the continental source of the Poway formation. The overlying Sespe and Vaqueros formations also contain Poway-type sand and pebbles. Sediment transport directions for all these formations were generally from south to north. No reasonable sediment source exists to the south of Santa Rosa Island now, but a previous position for the island adjacent to the Peninsular Ranges fits the transport directions.

Miocene rocks include basaltic flows and intrusions with ages of about 19 to 17 million years, and the Monterey shale. The Monterey shale grades upward into the Beechers Bay formation of siltstone, sandstone, and conglomerate made of components from nearby active volcanoes.

San Miguel Island is a 36-km^2 windswept platform of marine sedimentary rock formations of Late Cretaceous to Middle Miocene age plus volcanic rocks erupted about 18 to 17 million years ago. The Late Cretaceous Point Bennett formation is gray feldspar–rich sandstone with few shale and conglomerate interbeds. Two conglomerate layers contain cobbles of black andesite, hard granitic rocks, and quartzite that correlate with the conglomerate of the Rosario group of northern Baja California. The Paleocene to Miocene rocks are similar to those of Santa Rosa and Santa Cruz Islands.

SOUTHERN CHANNEL ISLANDS

San Nicolas Island, located in the Outer Borderland, has geology related to the northern islands. In contrast, Santa Barbara, San Clemente, and Santa Catalina Islands of the Inner Borderland have no rocks with sources in the Peninsular Ranges except for scanty outcrops on Santa Catalina Island.

San Nicolas Island consists mostly of Eocene marine sandstone and siltstone that probably had the same Peninsular Ranges source as the Eocene rocks of the northern Channel Islands. The sediment transport direction and paleomagnetic evidence indicate that although San Nicolas Island was translated away from the Peninsular Ranges, it was not significantly rotated. Miocene volcanic rocks are limited to several small andesite dikes.

Tiny Santa Barbara Island is constructed of submarine basalt and andesite flows, layers of volcanic debris, and associated andesite intrusions ranging in age from 17 to 15 million years. The flows dip gently and are slightly folded.

Elongated San Clemente Island is the emerged portion of an arched block bounded on the northeast by the right-lateral San Clemente fault. The island is principally volcanic rock with minor marine sedimentary rocks of Miocene age. Andesite flows dominated the bulk of the volcanic activity that began about 16 million years ago and ended by 14 million years ago. Pleistocene sedimentary rocks and sediments blanket a significant portion of the island.

About half of the area of Santa Catalina Island is Mesozoic or older Catalina schist. It includes blueschist, greenschist, amphibolite, and serpentinite—all formed by metamorphism during subduction of an oceanic plate. Much of the remainder of the island is Miocene igneous rock including lava flows, volcanic breccias, abundant dikes, and small plutons. Most of the flows are andesite, and the main pluton is hornblende quartz diorite dated at 19 million years. Younger 15-to-14-million-year-old flows and layers of volcanic ash and debris overlie the plutons.

Cretaceous or possibly Paleocene to Eocene sedimentary rocks are exposed only at the east end of the island. These appear to have had sediment sources in the San Diego area. Miocene sedimentary rocks include layers of volcanic sediment plus breccia containing Catalina schist fragments. Remnants of diatom-rich shale and shale made from volcanic ash top the sequence.

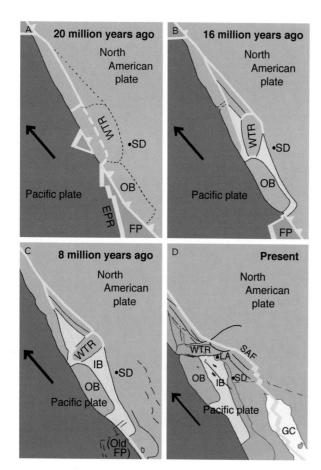

FIGURE 2 Evolution of the North American plate margin. Arrow indicates Pacific plate motion direction. Triangles mark the subduction zone. EPR—East Pacific rise, WTR—Western Transverse Ranges block, FP—Farallon plate, OB—Outer Borderland, IB—Inner Borderland, GC—Gulf of California, SAF—San Andreas fault, SD—San Diego, LA—Los Angeles. (A) Plate configuration at 20 million years ago. (B) Pacific plate pulls old Farallon plate, Western Transverse Ranges, and Outer Borderland blocks from North America at 16 million years ago. (C) Western Transverse Ranges block rotates, and Inner Borderland rift continues to open at 8 million years ago. (D) Plate boundary steps east to the San Andreas fault as the Gulf of California opens, and the borderland is compressed against North America. Adapted from animations by Dr. Tanya Atwater. To view this and related animations please visit http://emvc.geol.ucsb.edu/.

ROTATION, TRANSLATION, AND VOLCANISM

During the Mesozoic and early Cenozoic, the oceanic Farallon plate was subducting beneath the North American plate while the continent moved west toward the spreading center at the East Pacific rise (Fig. 2). During this time, the Cretaceous to Eocene sedimentary rocks exposed on the Channel Islands were being deposited in or near the subduction zone.

Eventually the East Pacific rise intersected with the continent, bringing the Pacific plate in contact with the North American plate. Then, 20 million years ago (Fig. 2A), the East Pacific rise ceased spreading at the latitude of the California Continental Borderland, and a remnant of the Farallon plate become attached to the Pacific plate. As the Pacific plate traveled northeast, it pulled the attached Farallon plate remnant with it. Portions of North American crust overlying the Farallon plate began to break apart (Fig. 2B). This opened major holes or rifts that exposed the underlying and more easily stretched Catalina schist forming the Inner Borderland. Highlands of Catalina schist rose in the rifts and supplied blueschist fragments for Miocene sedimentary formations. At the same time, one piece broken from the North America plate rotated to become the Western Transverse Ranges block; another traveled northwest to become the Outer Borderland (Fig. 2C).

Miocene volcanism began with the opening of the rifts about 19 million years ago. A faulted and thinned crust allowed the very hot underlying mantle to rise and melt by decompression to produce basalt. Some basaltic magma reacted with the Catalina schist and other rocks and evolved to create andesite, dacite, and rhyolite.

About 6 million years ago, the southern part of the plate boundary stepped eastward as the Gulf of California opened and the southern portion of the San Andreas fault became the boundary (Fig. 2D). The Pacific plate was now pushing the attached California Continental Borderland northwestward into the North American plate. This began the folding, faulting, and uplift of the islands that continues today.

SEE ALSO THE FOLLOWING ARTICLES

Channel Islands (California), Biology / Pacific Region / Plate Tectonics

FURTHER READING

Couch, J. K., and J. Suppe. 1993. Late Cenozoic tectonic evolution of the Los Angeles basin and Inner California Borderland: a model for core complex-like crustal extension. *Geological Society of America Bulletin* 105: 1415–1434.

Dibblee, T. W., Jr., and H. E. Ehrenspeck. 1998. General geology of Santa Rosa Island, California, in *Contributions to the geology of the Northern Channel Islands, Southern California*. P. W. Weigand, ed. Bakersfield, CA: American Association of Petroleum Geologists, 49–75.

Gordon, J. H., J. R. Boles, and P. W. Weigand. 2001. Geology of Santa Cruz Island: key to understanding the evolution of the Southern California Borderland, in *Geologic excursions in southwestern California*. G. Dunne and J. Cooper, eds. Pacific Section SEPM, Book 89, 147–185.

Merifield, P. M., D. L. Lamar, and M. L. Stour. 1971. Geology of central San Clemente Island, Calfornia. *Geological Society of America Bulletin* 82: 1989–1994.

Nicholson, C., C. C. Sorlien, T. Atwater, J. C. Crowell, and B. P. Luyendyk. 1994. Microplate capture, rotation of the Western Transverse Ranges, and initiation of the San Andreas transform as a low-angle fault system. *Geology* 22: 491–495.

Vedder, J. G., D. G. Howell, and J. A. Forman. 1979. Miocene strata and their relation to other rocks, Santa Catalina Island, California, in *Cenozoic paleogeography of the western United States*. J. M. Armentrout, M. R. Cole, and H. Terbest Jr., eds. Los Angeles: Pacific Section, Society of Economic Paleontologists and Mineralogists, 239–256.

CICHLID FISH

OLE SEEHAUSEN

EAWAG Center for Ecology, Evolution, and Biogeochemistry, Kastanienbaum, Switzerland

To lake-adapted fish, lakes are what islands in the sea are to continental organisms. Extraordinarily diverse species assemblages have evolved, some during surprisingly short timespans, in many African lakes. These can help us understand the mechanisms behind the phenomenon of adaptive radiation in island biota. The group of organisms that by far exceeds all others in terms of its endemic lacustrine species diversity is cichlid fish, a group of secondary freshwater fishes, members of the large ancient marine radiation of perchlike fish. Altogether, some 1300 to 1500 species of cichlids are endemic to the African Great Lakes.

CICHLID DIVERSITY IN AFRICAN LAKES

The perchlike cichlid fish (Cichlidae) family is one of the most species-rich families of bony fish, widely distributed in Africa, Madagascar, the Middle East, South and Central America, and the coastal plains of the Indian subcontinent. About 1500 species are described, and at least 700 others are known but still undescribed. The family is probably more than 100 million years old, and the deepest phylogenetic splits in the family tree are very old, with each continent having one to several very divergent lineages. Species diversity, however, is very unevenly distributed over the continents: Only three species are known from India, the continent with the oldest lineage; some 25 to 40 from Madagascar; 110 from Central America; 450 from South America; and about 1600 to 1800 from Africa (including seven from the Middle East). Perhaps most remarkably, of the 1600 to 1800 African species, 1300 to 1500 (about 5% of the world's known bony fish species) are endemic to lakes in and around the Great Rift Valley.

Lakes Malawi and Tanganyika are long and deep, and together with the smaller Lake Kivu, Edward, Albert, and Turkana, lie at the bottom of the Rift Valley. Lake Victoria, the biggest of all, is saucer shaped and shallow. Together with Lake Kyoga, it fills the plateau between the western and the eastern branches of the Rift. The dimensions of these large lakes generate ecological conditions that, even though freshwater, are in many regards more comparable to conditions in the sea. Life in the large open waters or at great depths requires adaptations that many river- and small lake–dwelling organisms do not possess. To lake-adapted fish, the different Great Lakes are therefore what islands in the sea are to continental organisms.

Most of the astounding African cichlid species diversity has not evolved in the large and widespread river network (the "continent"), but on the "islands": the lakes. It has all happened in the most recent 5–10% of the evolutionary history of the family within just one phylogenetic lineage. This lineage is often referred to as the East African lineage, or the haplochromines in the wider sense. It originated between 13 and 30 million years ago in East Africa, possibly in a precursor lake to Lake Tanganyika. Lake Tanganyika is the oldest of the Great Lakes (about 12–20 million years old) and is a reservoir of relatively ancient evolutionary branches in the East African cichlid lineage, represented by the tribes Bathybatini, Trematocarini, Lamprologini, Ectodini, Cyprichromini, Limnochromini, Perissodini, Eretmodini, and Haplochromini (the narrow sense haplochromines). Yet its species richness is relatively modest, with current estimates of around 200.

Although much younger than Lake Tanganyika, Lake Malawi (~4–5 million years old) and Lake Victoria (~15,000 years since the latest desiccation) are more species-rich, each having more than 500 endemic species. Hence, speciation in these lakes has been extraordinarily fast. However, with the exception of three and two endemic tilapiine cichlids in Lakes Malawi and Victoria, respectively, these larger species assemblages are composed of representatives of only one of the nine Tanganyikan lineages, the haplochromines in the narrow sense.

CICHLID RADIATIONS' MULTIPLE ANCESTORS

Recent recalibrations of the cichlid molecular clock have confirmed earlier suggestions that the cichlid assemblage of Lake Tanganyika is derived from several distantly related founder lineages. Mitochondrial DNA sequence variation had long been interpreted as evidence that Lakes Victoria and Malawi were each colonized by a single lineage, rendering each radiation strictly monophyletic. This picture has recently changed with the advent of genomic tools, permitting the investigation of a large number of different genetic loci. It now appears that the species diversity observed within these lakes has sprung from genetically diverse hybrid swarms seeded by several founding species that hybridized.

THE GEOGRAPHY OF INTRALACUSTRINE SPECIATION

Whether speciation can occur in the absence of geographical isolation between populations (parapatric or sympatric speciation) is a controversial issue in evolutionary biology. Even though several cichlid lineages colonized each lake, the number of colonizers is one or two orders of magnitude fewer than there are endemic species in any of the lakes. By inference, although most of the diversity of genes almost certainly predates the intralacustrine diversifications, much of the species origination must have occurred within lakes. This has often been taken as support for sympatric speciation. Indeed, there is good evidence that sympatric speciation is possible in cichlids. The best example of this is a species flock of endemic tilapiine cichlids in crater lake Barombi Mbo in Cameroon. Mitochondrial and nuclear genomic evidence taken together suggest that the endemic species have arisen within the lake, which is thought to be too small to allow within-lake geographical isolation. In the larger lakes, however, two models of speciation involving geographical isolation may account for a significant part of the species diversity. The first involves geographical speciation in isolated satellite lakes during lake-level lowstands, followed by secondary sympatry during subsequent lake-level highstands. Although proposed dozens of times, the only strong evidence that this may work is from recent work on Lake Malawi cichlids. Yet the explanatory power of this mechanism appears limited in lakes where many species evolved during periods of insignificant water-level fluctuations. The most extreme such case is Lake Victoria, where most speciation probably occurred after the last lake desiccation 15,000 years ago.

The second model involves microallopatric isolation by ecological barriers, such as unsuitable habitat. There is strong support for such microallopatric speciation from distribution patterns in cichlid groups that are restricted to patchily distributed habitat such as rocky shores. However, given that such habitat specialization does not exist in the riverine cichlid ancestors that first colonized the lakes, it is perhaps unlikely that the beginning of these cichlid adaptive radiations was by microallopatric speciation.

A combined model of sympatric and microallopatric speciation in the evolution of cichlid species flocks is supported by the shape of the evolutionary species–area relationship for African cichlid fish. The number of endemic species has a flat relationship with lake size across several orders of magnitude of lake size variation, but it suddenly increases steeply when lakes become larger than 1000 km^2 in surface area (Fig. 1). It seems that speciation did indeed occur in many lakes without much opportunity for geographical isolation but that many more species are produced when there is opportunity for geographical isolation. The two arms of the curve in Fig. 1 are reminiscent of the relatively flat and steep slopes (z), respectively, of intra- and interprovincial species-area curves in island biogeography. Allopatric speciation is common between provinces, but not within.

Some have argued that perhaps the strongest test of sympatric speciation is whether or not in situ speciation can be detected in the evolution of island biota where islands are too small to afford opportunities for geographical isolation. Whereas work on islands in the sea has only very recently yielded a first strong example, cichlid fish in lakes provide strong support for this scenario. Having said this, it remains to be seen whether cichlid fish speciation in small lakes is best described as truely sympatric or as narrowly parapatric.

MECHANISMS OF SPECIATION

Cichlid fish radiations speak to another major issue in evolutionary biology, too: the relative importance of drift and selection in speciation. On theoretical grounds, waiting times for speciation by genetic drift alone are far too long to explain the speciation rates in some African lakes. Hence, divergent selection must be invoked.

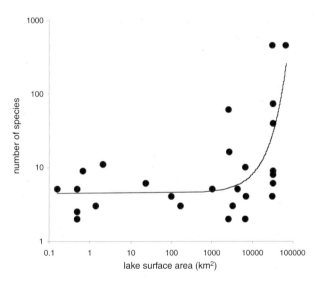

FIGURE 1 The evolutionary species-area relationship for African cichlid radiations, calculated using 29 African cichlid radiations for which lake surface area and number of species were available. Reprinted with permission from *Proceedings of the Royal Society of London* (Seehausen 2006).

Strikingly similar phenotypes have evolved in many of the lakes, including an extraordinary diversity of color and melanin patterns and of feeding types (e.g., pelagic plankton-feeders and their larger mackerel-shaped predators, snail-eaters that crush their prey between specialized oral jaws or in a heavily enlarged pharyngeal apparatus, blunt-snouted algae scrapers with highly specialized brush-like dentition, scale-eaters specialized to tear off scales from other fish, paedophages feeding on eggs and fry of other species, and dwarfs that breed in empty snail shells) (Fig. 2). Interestingly, ecological and morphological diversity of the younger radiations approaches those of the older Tanganyikan flock, whereas the diversity of social, mating, and parental care behaviors is much larger in the latter. Only the Lake Tanganyika flock includes monogamous, harem forming, and polygynous cichlids; substrate-, cave-, and mouthbrooders; and species where both mother and father and even sibs guard the fry and others where parental care is exclusively maternal. In contrast, all species in Lakes Victoria and Malawi are mouthbrooders with maternal care. Adaptive radiation in cichlids seems therefore driven primarily by divergent selection on habitat use, feeding ecology, and color pattern.

ISLAND EFFECTS

Two different island effects explain part of the cichlid phenomenon that so much more diversity has evolved in the geographically small area of the African lakes than in the geographically extensive African river networks. The first is is that cichlid communities of different major lakes do not share any species such that diversity in

FIGURE 2 A representative sample of evolutionary lineages, species, and phenotypes from Lakes Tanganyika, Malawi, and Victoria. (A) Lake Tanganyika (from left to right, top to bottom): pelagic zooplanktivore *Cyprichromis leptosoma* (tribe Cyprichromini), pelagic piscivore *Bathybates leo* (tribe Bathybatini), rock-dwelling algae scraper *Tropheus moorii* (tribe Haplochromini), rock-dwelling insectivore *Lobochilotes labiatus* (tribe Haplochromini), paedophage *Greenwoodichromis bellcrossi* (tribe Limnochromini), generalized insect/plankton-eater *Neolamprologus savoryi* (tribe Lamprologini), scale-eater *Perissodus eccentricus* (tribe Perissodini), snail-crusher *Neolamprologus tretocephalus* (tribe Lamprologini) (all photographs copyright Ad Konings), river- and lake-shore-dwelling generalist *Astatotilapia burtoni* (tribe Haplochromini) (copyright Hans-Joachim Richter). (B) Lake Malawi: river- and lake-dwelling generalist *Astatotilapia calliptera,* one of the seeding species; pelagic zooplanktivore *Diplotaxodon macrops;* pelagic piscivore *Rhamphochromis esox;* rock-dwelling algae scraper *Pseudotropheus tropheops;* sand/rock-dwelling insectivore *Placidochromis milomo;* paedophage *Caprichromis liemi;* generalized insect/plankton-eater *Copadichromis virginalis;* scale-eater *Corematodus taeniatus;* snail-crusher *Chilotilapia rhoadesii* (all tribe Haplochromini) (all photographs copyright Ad Konings). (C) Lake Victoria: generalist *Thoracochromis pharngalis,* one of the seeding species; pelagic zooplanktivore *Yssichromis pyrrhocephalus;* pelagic piscivore *Prognathochromis macrognathus* (photograph by Frans Witte/University of Leiden); rock-dwelling algae-scraper *Neochromis omnicaeruleus;* rock-dwelling insectivore *Paralabidochromis chilotes;* paedophage *Lipochromis microdon;* generalized insect/plankton-eater *Pundamilia nyererei;* scale-eater *Schubotzia eduardiana* (from Lake Edward); snail-crusher *Macropleurodus bicolor* (all tribe Haplochromini).

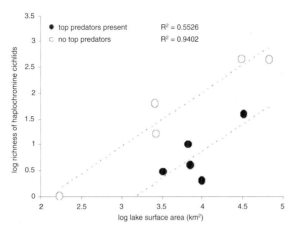

FIGURE 3 The relationship between lake size and species richness of *Astatotilapia*-related haplochromine cichlids in African Great Lakes without and with native top predator communities. Trend lines are estimated as least-squares regressions on a loglog scale (only lakes that hold *Astatotilapia*-related cichlids are included). Reprinted with permission from Nature Publishing Group (Seehausen 2007).

different lakes is completely additive. This must be due to the limited capacity of lake-adapted cichlid species to disperse through river corridors. The other explanation that has often been proposed is evolutionary release in the lakes from competition by other fish taxa. However, the fish fauna of lakes in which cichlids radiated is no less diverse than that of lakes in which cichlids failed to radiate. On the other hand, among the lakes colonized by the *Astatotilapia* group of haplochromine cichlids (which gave rise to the major radiations), those that have naturally occurring top predators do support fewer species of haplochromines (Fig. 3). Another noteworthy pattern is that tilapiine cichlids, which have radiated into small species flocks in several lakes, usually only do so in lakes that are not also colonized by haplochromines (Fig. 4). Hence, top predators and other cichlids may indeed constrain cichlid fish adaptive radiation.

VARYING INTRINSIC PROPENSITIES TO SPECIATION

Release from competition and divergent natural and sexual selection in an ecologically diverse environment alone cannot explain the African lake cichlid phenomenon. Most cichlid lineages consistently failed to speciate even in apparently suitable lakes. The propensity to diversify in lakes appears to have emerged sequentially through multiple episodes of lacustrine radiations in the East African cichlid lineage. It is not fully understood yet what it is that explains the propensity of this lineage to radiate in lakes. A genome amenable to interspecific hybridization and particular genetic architectures of eco-

logical and mate choice traits have been proposed (e.g., major phenotypic effects of individual genes that permit the persistence of large phenotypic variance in the face of gene flow). The exploration of these and other hypotheses is currently a dynamic field of research. A second, albeit smaller, set of lacustrine cichlid radiations has occurred in the Central American cichlid lineage in and around Lake Nicaragua and perhaps in some Mexican lakes. Evolutionists have begun to compare the radiating African and Central American lineages with their relatives that do not show any evidence for diversification in lakes.

EVOLUTION "OUT OF THE ISLANDS"

Even though African rivers support much lower cichlid species richness than do lakes, phylogeographic evidence suggests that many now continentally distributed river cichlid lineages originated in lacustrine radiations. Lamprologines, haplochromines, and perhaps others that probably originated in an ancient precursor lake to Lake Tanganyika are now occupying African rivers, some with continent-wide lineage distributions. At a more recent time scale, much of the haplochromine cichlid diversity in rivers of southern Africa originated in a large paleolake that dried out only very recently. Hence, adaptive radiation in lakes can be an engine of ecological and evolutionary innovation, which subsequently enriches continental biota.

ISLANDS VERSUS LAKES

Future research will hopefully explore the emerging pattern that "intra-island speciation" without geographical isolation is more common in lakes than on islands in the sea and present hypotheses of why this might be so. One reason why cichlids diversify more in lakes than in rivers may possibly be the larger physical dimensionality of lakes. This must increase the dimensionality of total niche space, facilitating speciation along ecological gradients (surface to lake floor, shore to offshore) even in the absence of geographical isolation. Perhaps this would also help explain why more examples of sympatric speciation can be found from lakes than from islands in the sea.

CONSERVATION

African lake cichlid fish unfortunately are a textbook example not just for speciation and adaptive radiation, but for anthropogenically driven extinction as well. About 200 to 250 of the approximately 500 endemic cichlid species of Lake Victoria disappeared in the last two decades of the twentieth century. The causes of this largest mass extinction ever witnessed by scientists are the anthropogenic

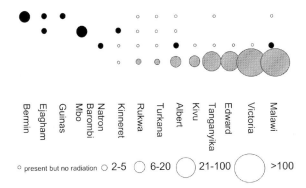

FIGURE 4 Geographical patterns of adaptive radiation imply the possibility of suppression between unrelated cichlid lineages. Two major phylogenetic lineages are responsible for most of the adaptive radiations of cichlid fish in African lakes, the tilapiines (black circles, three genera from top to bottom *Tilapia, Oreochromis, Sarotherodon*) and the *Astatotilapia*-related haplochromines (hatched circles; circle size represents number of species). Both groups are widely distributed across African lakes, and they often co-occur. However, they very rarely radiate in the same lake: In lakes in which haplochromines radiated, tilapiines did most often not, and most tilapiine radiations occurred in lakes from which haplochromines are absent. Reprinted with permission from Nature Publishing Group (Seehausen 2007).

eutrophication of Lake Victoria and the sudden boom in the 1980s of a large top predator, the Nile perch (*Lates niloticus*), introduced to the lake two decades before. The Nile perch depressed abundances of many cichlid species; the increased water turbidity made ecological specialization and behavioral reproductive isolation of coexisting cichlid species ineffective. The two effects together led to a sudden and rapid collapse of species diversity. Lake Victoria can be seen as a model in island conservation biology, where even localized human activities can have devastating effects on species diversity.

SEE ALSO THE FOLLOWING ARTICLES

Adaptive Radiation / Extinction / Freshwater Habitats / Lakes as Islands / Species–Area Relationship / Sticklebacks

FURTHER READING

Genner, M. J., P. Nichols, G. Carvalho, R. L. Robinson, P. W. Shaw, A. Smith, and G. F. Turner. 2007. Evolution of a cichlid fish in a Lake Malawi satellite lake. *Proceedings of the Royal Society of London B* 274: 2249–2257.
Genner, M. J., O. Seehausen, D. H. Lunt, D. A. Joyce, P. W. Shaw, G. R. Carvalho, and G. F. Turner. 2007. Age of cichlids: new dates for ancient lake fish radiations. *Molecular Biology and Evolution* 24: 1269–1282.
Joyce, D. A., D. H. Lunt, R. Bills, G. F. Turner, C. Katongo, N. Duftner, C. Sturmbauer, and O. Seehausen. 2005 An extant cichlid fish radiation emerged in an extinct Pleistocene lake. *Nature* 435: 90–95.
Kocher, T. D. 2004 Adaptive evolution and explosive speciation: the cichlid fish model. *Nature Reviews Genetics* 5: 288–298.
Kwanabe, H., M. Hori, and N. Makoto, eds. 1997. Fish communities in Lake Tanganyika. Kyoto: Kyoto University Press.
Rosenzweig, M. L. 2001. Loss of speciation rate will impoverish future diversity *Proceedings of the National Academy of Sciences* 98: 5404–5410.
Salzburger, W., T. Mack, E. Verheyen, and A. Meyer. 2005. Out of Tanganyika: genesis, explosive speciation, key innovations and phylogeography of the haplochromine cichlid fishes. *BMC Evolutionary Biology* 5: 17.
Schliewen, U., and B. Klee. 2004. Reticulate sympatric speciation in Cameroonian crater lake cichlids. *Frontiers in Zoology* 1: 1–12.
Seehausen, O. 2006. African cichlid fish: a model system in adaptive radiation research. *Proceedings of the Royal Society of London B* 273: 1987–1998.
Seehausen, O. 2007. Chance, historical contingency and ecological determinism jointly determine the rate of adaptive radiation. *Heredity* 99: 361–363.
Stager, J. C., and T. C. Johnson. 2008. The Late Pleistocene desiccation of Lake Victoria and the origin of its endemic biota. *Hydrobiologia* 596: 5–16.

CLIMATE CHANGE

DAVID A. BURNEY

National Tropical Botanical Garden, Kalaheo, Hawaii

Climate changes are well documented for islands throughout the world on many scales, from hundreds of millennia to recent decades. These changes in temperature and moisture have had large effects on many other ecological factors, including sea level, coastal dynamics, biogeography, extinctions, and human culture. Climate changes on islands have varied in severity according to the size of the island, its latitude and elevation range, and the effects of human activities. Climate changes predicted for the near future can be expected to have drastic effects on all islands, even to the point of destroying some island ecosystems, challenging human lifeways and culture, and driving extinction events. Some low islands may disappear entirely if global warming–driven sea-level rise is as great as predicted by some models.

PAST CLIMATE CHANGE

Global Effects on Island Climate

Paleoclimatologists have revealed that global climate is always changing. These scientists have used evidence from sediments, stable isotopes, and microfossils to show that, in the deep-sea sediments, ice cores, and lake deposits they have studied, there is abundant evidence that earth's climate at all latitudes has gone through roughly 20 glacial-interglacial cycles over the last 2 million years. The strongest cycle is approximately 100,000 years in duration, but superimposed on this pattern are other cycles at approximately 20,000 and 40,000 years, all apparently driven by variations in the earth's orbit and tilt (Fig. 1). Complex feedback mechanisms in the earth's many climatic variables have resulted in a series of ice ages, each several tens of millennia in length, interspersed with briefer warm interglacial intervals of 10 to 20 millennia, such as our present Holocene period, which has lasted for about 11,000 years. These global trends have had some drastic effects on islands. For instance, on very large, high islands such as Madagascar and New Guinea, full ice-age conditions drove the tree line on mountains down about 1000 m in elevation, such that large areas of these tropical islands were too cold to support tropical forest over large areas. Islands with very high mountains, such as the Big Island of Hawaii, had alpine glaciers despite their tropical latitude. Under these cooler regimes, some lowland areas were much drier than they are today, with the resultant spread of savanna vegetation. Conversely, during interglacials such as the present Holocene, forest has spread over large areas of many tropical islands.

For islands in particular, these glacial-interglacial vicissitudes of climate may have another drastic consequence: During glacial maximum, the coldest time of an ice age, so much water is locked up in the polar caps and continental glaciers that sea level is lowered by 120 m or more. For an island group such as the Bahamas, this means that today's relatively small islands coalesce into a few quite big islands—at times during the Pleistocene, the Bahamas have been about ten times their present area. This growing together of islands has profound biogeographic effects. The flora and fauna of Maui, Lanai, Kahoolawe, and Molokai in the Hawaiian Islands has more similarity than those of other islands in the chain, likely because they formed one large island during low Pleistocene sea stands.

Local Effects on Island Climate

On more recent time scales, climate has changed on islands primarily through the effects that humans have had on the local ecology. Although difficult to measure, the large-scale deforestation that has occurred since human arrival to remote islands has reduced rainfall, increased runoff, and allowed the soils of these sites to heat up. Famous extreme examples are New Zealand, Madagascar, and Rapa Nui (Easter Island), where much of the original forest cover has been lost, and changes in the local microclimate have hindered reforestation efforts.

A debate has raged in the scientific community regarding the role of "natural" and human-caused climate change in the catastrophic extinctions that have been documented for many of the world's islands. Unique faunas have lost much of their diversity over millennial time scales, challenging scientists to explain this powerful trend. A look at the major variations in global climate, plotted alongside these major extinction events, however, shows that most island extinctions have coincided not with glacial-interglacial cycles but with the advent of humans to these systems (Fig. 1). This pattern holds for the islands reached earliest by humans (Australia and New Guinea, perhaps 50,000 years ago), continues through the Mediterranean and Caribbean islands reached in the mid-Holocene, and culminates with those islands reached last, such as Hawaii, Madagascar, and New Zealand in late prehistoric times and the Mascarenes and Galapagos in the historical period.

Thus, despite the drastic effects of climate change on sea level, coastal dynamics, biogeography, and many other island factors, it has ultimately been humans, not climate, that have changed islands most.

FUTURE CLIMATE CHANGE

One sad irony of modern life is that humans are now showing their capability to modify climate not only locally, but perhaps also on a global scale. There is virtual unanimity among climatologists that global warming is not just a real threat in the future but is in fact happening now. Events

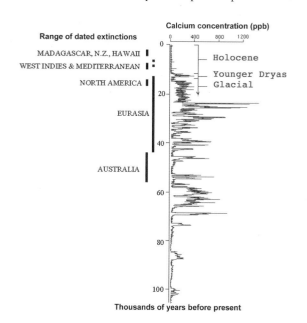

FIGURE 1 A high-resolution calcium concentration record from the GISP2 Greenland ice core indicating the relative amount of atmospheric dust, an index for cool and dry (long horizontal bars) versus wet and warm conditions (much shorter bars). Note that the interglacial-type climate, such as in the present Holocene, is generally warmer and wetter globally than the typical late Pleistocene full-glacial climate, which was cooler and drier in many locations. This indirect measure of the extent of vegetative cover, as well as other Pleistocene climatic indices, shows no correlation with the distribution of "last occurrence" dates for extinct island fauna or continental extinctions (vertical bars). From Burney and Flannery 2005.

detected in the fossil record may now repeat themselves on an accelerated time scale. The well-documented melting of the polar ice caps and alpine glaciers is causing sea level to rise, threatening coastlines with erosion, flooding coastal wetlands, and inundating low atoll islands. Island countries such as the Maldives and many of the islands of Micronesia, where elevations scarcely exceed 2 m, may virtually disappear in coming decades if the model predictions for sea level rise in a CO_2-enriched world prove true. Island nations may soon be faced with the prospect of thousands or even millions of environmental refugees.

Just as in prehistoric times, warming will drive cool-adapted island species higher up the mountains, but many such species may go extinct if the island lacks sufficient area at high elevations to accommodate them. Range shifts in response to climate change on islands today are further complicated by the fact that the mid-elevations and highlands of many islands are already heavily populated and transformed by humans.

Cloud forests and other upper-elevation wet zones, which harbor the richest diversity of endemic species on some islands, may be particularly affected by global climate change. Paleoclimatological data for Maui, for instance, suggest that climatic warming may raise the mean elevation of the base of the cloud bank, thus reducing or completely eliminating these cloud forest habitats on small islands that lack large areas of high elevation.

Another worrisome consequence of future climate change for islands is the predicted increase in extreme weather events. Because of their smaller land area and exposure to marine influences, islands are especially vulnerable to increased frequency and violence of storms, a well-supported prediction from climatological studies of the effects of increased ocean temperature on tropical storm formation and other extreme marine events. Adding further to the climatic uncertainty is the apparent correlation between warmer oceans and the frequency and severity of El Niño–Southern Oscillation (ENSO) events and other large-scale feedback connections between ocean and atmosphere.

The predicted increase in uncertainty in climate patterns may be a special problem for islands, as floods, droughts, and heat waves pose great challenges to the stability of agriculture, fisheries, and other human enterprises on islands, as well as the native biota. Tropical island forests have declined in some areas due to severe droughts. For instance, in the past decade, vast areas of rain forest in Borneo dried out and burned during a drought apparently related to a "Super-ENSO" event.

Increased storminess, wildfires, and rising temperatures have been observed to have synergistic effects on another great challenge to the biota of islands—biological invasions. Many of the alien imports that are driving island species to extinction and interfering with agriculture on islands throughout the world are favored by increased disturbance from hurricanes and other climatic extremes. In this way, even the world's most remote islands share a fate with the large, developed land masses of the planet, because the human-wrought changes to the atmosphere affect the climate everywhere.

SEE ALSO THE FOLLOWING ARTICLES

Climate on Islands / Deforestation / Global Warming / Invasion Biology / Sea-Level Change

FURTHER READING

Burney, D. A., and T. F. Flannery. 2005. Fifty millennia of catastrophic extinctions after human contact. *Trends in Ecology and Evolution* 20: 395–401.

Burney, D. A., R. V. DeCandido, L. P. Burney, F. N. Kostel-Hughes, T. W. Stafford, and H. F. James. 1995. A Holocene record of climate change, fire ecology, and human activity from montane Flat Top Bog, Maui. *Journal of Paleolimnology* 13: 209–217.

Burney, D. A., L. P. Burney, L. R. Godfrey, W. L. Jungers, S. M. Goodman, H. T. Wright, and A. J. T. Jull. 2004. A chronology for late prehistoric Madagascar. *Journal of Human Evolution* 47: 25–63.

Graham, M. H., P. K. Dayton, and J. M. Erlandson. 2003. Ice-ages and ecological transitions on temperate coasts. *Trends in Ecology and Evolution* 18: 33–40

Mayewski, P. A., L. D. Meeker, S. Whitlow, M. S. Twickler, M. C. Morrison, P. Bloomfield, G. C. Bond, R. B. Alley, A. J. Gow, D. A. Meese, P. M. Grootes, M. Ram, K. C. Taylor, and W. Wumkes. 1994. Changes in atmospheric circulation and ocean ice cover over the North Atlantic during the last 41,000 years. *Science* 263: 1747–1751.

Rizvi, H. 2007. Climate change: leaders sound the alarm on island peoples, economies. http://ipsnews.net/news.asp?idnews=37525.

Shackleton, N. J., and N. D. Opdyke. 1973. Oxygen isotope and palaeomagnetic stratigraphy of equatorial Pacific core V28-238: oxygen isotope temperatures and ice volumes on a 10^5 year and 10^6 year scale. *Quaternary Research* 3: 39–55.

CLIMATE ON ISLANDS

THOMAS A. SCHROEDER

University of Hawaii, Manoa

Climate is the average of weather conditions over a long period of time. The period is at least 30 years. The standard weather parameters considered are temperature and precipitation, which are primary controls on distributions of vegetation. In addition to the average conditions, the variance of these conditions is equally important. Year-to-year or decade-to-decade oscillations in climate constitute "variability"; long-term trends constitute climate "change." Climate on islands depends upon a number of geographical and physical conditions.

GEOGRAPHICAL AND PHYSICAL CONTROLS

A primary control on island climate is the climate of the surrounding ocean and atmosphere. Latitude is a major factor. Atmospheric and ocean circulations are organized in planetary structures. Atmospheric features range from the belt of variable weather systems (the polar frontal storm belt) to steady systems such as the trade winds of the subtropics. Equally important are large-scale ocean circulations. Many extremely dry islands ("desert islands") lie in regions of strong oceanic upwelling, regions typified by very stable air and hence limited clouds and precipitation. The Hawaiian Islands lie at the latitude of the great subtropical deserts but are cooled by a steady transport of cool waters by the North Equatorial Current and by the attendant northeast trade winds. High-latitude islands such as Spitsbergen are warmed by northward moving currents.

Continental land masses markedly affect climates of nearby islands. The Australasian (and African) monsoon circulation spans the equator and at least half the longitudinal extent of the tropics. Consequently, islands in the Indian and western Pacific Oceans have monsoonal climates. Many islands in Southeast Asia and the southwestern tropical Pacific comprise what is termed "the Maritime Continent." These islands and archipelagoes have climatic characteristics similar to equatorial Africa or South America.

Horizontal and vertical dimensions and shape matter. Islands range from very low, small atolls such as Canton and Tarawa through high, large islands such as Hawaii and New Guinea. Larger islands actually "produce their own climate (weather)." An example of the influences of size and shape on island climate can be seen in Sulawesi, which consists of several mountainous peninsulas stretching like tentacles. Distinctive climates exist on the peninsulas and in the surrounding bays.

ISLAND EFFECTS

Orographic Effects

If a mountainous island is exposed to prevailing winds, air may be forced up and over mountains (Fig.1). An example is the Koolau range of Oahu, Hawaii, which is long (approximately 50 km), narrow, and up to 1 km high, forming a rampart aligned almost perpendicular to the prevailing winds. If the air rises to its condensation level, clouds and precipitation will develop. These clouds then dissipate as air descends the leeward slope. In this instance, distinct climate zones develop along windward and leeward sides of the range. Gradients of climate parameters can be large (discussed below). The persistent cloud is termed an "orographic cloud."

Moderating factors include distance of the mountains from the sea, intervening terrain, wind velocity, and thickness and stability of the moist layer. The atmosphere is stratified such that most moisture resides in the lowest few kilometers (the "moist layer"). The moist layer is typically deeper in the deep tropics and, in the trade wind belts, is capped by an inversion that restricts cloud thickness. A consequence is that on Oahu's Koolau Range (1 km elevation) maximum annual precipitation (7500 mm) falls at its summit, but Hawaii Island's Mauna Kea (4.2 km elevation) is desert-like at the summit with maximum precipitation on the windward slopes. A second Oahu range (Waianae) lies downwind of Koolau, and the annual maximum rainfall is a "mere" 2000 mm.

Land and Sea Breezes

Water has a much higher heat capacity than soil. Water also conducts and convects heat more efficiently than soil. After sunrise, the temperature of a land surface quickly exceeds that of surrounding oceans. A simple two-dimensional circulation develops in which air rises

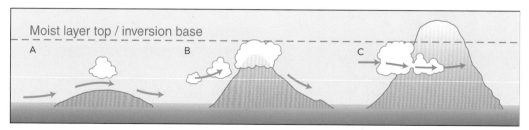

FIGURE 1 Schematic of air flow over and around islands exposed to a steady wind. Moist layer top is indicated. (A) Flow over a low island, such as St. Helena; (B) flow over a high island, which does not reach the moist layer top, such as in the case of Oahu, Hawaii; (C) flow around an island, which penetrates the moist layer, such as in the case of Hawaii (Mauna Loa and Mauna Kea).

over land, moves out to sea, sinks, and returns to land as a "sea" breeze. Upon sunset the process reverses, producing a "land breeze." In regions of weak large-scale winds (such as the Maritime Continent) these breezes dominate the local weather. Clouds and precipitation may form at the leading edge of the sea breeze on shore. At night, clouds and precipitation occur on the leading edge of the land breeze off shore. The sea breeze air and clouds limit daytime temperatures. In regions of strong prevailing winds such as the trade wind belts, the sea breeze is overwhelmed, although the influence of heating may appear in a diurnal variation in wind speed.

Mountain-Valley Breezes

Simple thermal circulations also develop along mountain slopes due to the differing thermal properties of land and the surrounding air. In this instance, air moves upslope during daylight (the "valley breeze") and drains downslope at night (the "mountain breeze"). On mountainous islands, daytime sea breezes, nighttime land breezes, and valley and mountain breezes may interact.

Other Effects

An island can be compared to a rock in the stream. Air encountering the island may rise up and flow over it or be diverted around it. In the latter instance, air accelerates at the corners. Air sinks at the corner to replace the mass accelerating away. The sinking creates local cloud-free regions. Similar accelerations occur at notches in mountain ranges.

SOME CHARACTERISTICS

Temperatures

Over a flat island, average temperature is nearly constant. However, the diurnal range will increase with distance from the coast. Willis Island in the Coral Sea is slightly larger than a supertanker (400 m by 150 m). In trade wind conditions (when winds blow along the maximum length of island), the diurnal range at the coast is 0.7 °C; in the middle of the island, it is 3.0 °C. When winds veer 90 degrees, the difference disappears. On high islands, temperatures decrease with height at rates similar to the free air (~6 °C/km). This "lapse" of temperature is less in cloudy regions and actually may reverse on high mountains on which the inversion intercepts the slope. At higher elevations the rate of decrease is less.

Precipitation

Orographic influences are evident globally. In the South Atlantic, Jamestown (12 m elevation on St. Helena) receives 125 mm of rain annually, while 800 mm falls at 855 m elevation inland (Hutt's Gate). Horizontal gradients can be large. At the summit of Kauai (Mt. Waialeale) annual rainfall varies by 1.8 mm/yr per meter (7 inches over the length of a football field).

In complex terrain, distributions and diurnal cycles become complicated. The island of Hawaii features four distinctive rainfall maxima (each maximum exceeds 2500 mm/year) and three distinctive minima (as low as 250 mm/year) as well as a wide range of diurnal cycles (Fig. 2). These result from interactions of five mountains (three of which penetrate the trade wind inversion layer) and the prevailing trade winds (present 70–80% of all days). One rainfall maximum (the leeward slopes of Mauna Loa) coincides with a different rainy season than the rest of the state.

INTERANNUAL VARIABILITY

Year-to-year climate variability is especially significant in the contiguous tropical belt encompassing the Indian Ocean, Maritime Continent, and Pacific Ocean (and the immediate surroundings). The El Niño–Southern Oscillation (ENSO) is the largest global example of interannual climate variability. The most dramatic variations occur along the equator from Indonesia (Maritime Continent) to South America. Canton Island lies in the equatorial dry

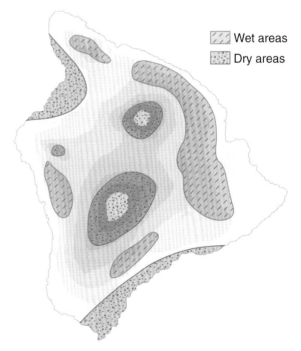

FIGURE 2 Map of the Island of Hawaii. Featured are four local rainfall maxima (slant dashes) and three types of rainfall minima (dotted). Three mountain summits are one type of minimum.

zone (latitude 2° S, longitude 172° W). Equatorial upwelling and subsiding trade winds maintain a dry climate. However, in the El Niño of 1972–1973, the December through February rainfall exceeded 1300 mm, contrasted with 35 mm in the same period of the previous "normal" year. Islands west of the international date line are dry, and those to the east become wet.

CLIMATE CHANGE

Preliminary studies suggest that projected global climate change will cause changes in large-scale circulations of the atmosphere and oceans. Among the possible outcomes are

1. Changes in the characteristics of the Australasian monsoon.
2. More frequent El Niño events.
3. Changes in the depth of the moist layer.
4. Weakening near-equatorial trade winds.

It is too early to tell exactly how individual island climates will change, but consequences could severely impact island flora and fauna.

SEE ALSO THE FOLLOWING ARTICLES

Global Warming / Hurricanes and Typhoons / Spitsbergen / St. Helena / Vegetation

FURTHER READING

Giambelluca. T. W., and T. A. Schroeder. 1998. Climate, in *Atlas of Hawaii, third edition*. S. P. Juvik and J. O. Juvik, eds. Honolulu: University of Hawaii Press, 49–59.
Neal, A. B. 1973. *The meteorology of the Australian trades: part 2: the perturbing effect of Willis Island on the oceanic trade wind flow*. Tech Report, 5(2). Melbourne: Australian Bureau of Meteorology.
Ramage, C. S. 1995. *Forecasters guide to tropical meteorology*. Air Weather Technical Report 240 (updated), Air Weather Service, Scott AFB, IL.

COLD SEEPS

CHARLES K. PAULL

Monterey Bay Aquarium Research Institute, Moss Landing, California

Cold seeps are springs that carry waters at near-ambient temperatures containing dissolved hydrogen sulfide, methane, and other hydrocarbons from subsurface aquifers out onto the sea floor. The sea floor environment surrounding cold seep sites is profoundly altered as a consequence of the chemical and biological reactions that occur when these reduced compounds encounter the oxygenated waters at the ocean floor. Many of the biological communities and geochemical processes that occur around cold seeps are similar to those found at deep-sea hydrothermal vents. Additional heat, however, is not of primary importance to either the biology or the geochemical reactions occurring at cold seep sites.

BIOLOGICAL COMMUNITIES

The most obvious characteristic of cold seeps is the presence of dense communities of distinct organisms, frequently called chemosynthetic biological communities (CBC). In contrast to most deep-sea communities, which ultimately depend on organic matter that was produced by photosynthesis in the sunlight-bathed surface and has settled through the water column to the sea floor, cold seep CBCs are dependent on local organic-matter production fueled by energy derived from dissolved compounds carried within the seeping fluids. Chemoautotrophic microorganisms (Archaea and Eubacteria) use the energy produced during the oxidation of hydrogen sulfide, methane, and other hydrocarbons in the seeping waters to synthesize organic matter. The locally produced organic matter then supports other animals in the surrounding area.

Although the microorganisms that form the basis of the local food chain within cold seep CBCs are sometimes visible as white mats that cover the sea floor, megafaunal organisms such as vesicomyid clams, *Bathymodiolus* mussels, and vestimentiferan tubeworms are the most visibly obvious taxa associated with CBCs. These megafaunal organisms are nourished via a symbiotic relationship with chemoautotrophic microorganisms harbored in their guts. Some species have become so dependent on these microbes that over time their mouths and anuses have become vestigial organs.

The density of non-chemosynthetic organisms also increases in the areas surrounding the cold seep as a consequence of higher biomass associated with the addition of local primary production. These include galatheid crabs, gastropods, and holothurians. Because the CBCs surround discrete local fluid sources, analogies with terrestrial oases are commonly made. Biologists are especially interested in the dispersal and evolution of communities that are specific to these isolated and potentially ephemeral ocean floor environments.

DISTRIBUTION

Deep-diving submersibles have played a critical role in the exploration of cold seep communities. The first discovery of cold seep communities was made in 1984 at the base of the Florida Escarpment in the Gulf of Mexico during *Alvin* dives. This discovery occurred just seven years after CBCs were discovered surrounding mid-ocean ridge hydrothermal vents. Originally, it was presumed that deep-sea CBCs were restricted to active submarine volcanic areas, and thus were extremely rare, but the realization that they also occur on passive margins greatly expanded their potential range. Cold seep communities and environments now appear to be a common occurrence in continental margin settings and occur wherever adequate amounts of reduced fluids make it out onto the ocean floor before being oxidized.

Similar communities also have been discovered at whale falls and over huge clumps of seaweed, both of which constitute massive point-source additions of organic matter onto the sea floor. As these decompose, hydrogen sulfide and even methane may be generated locally, which in turn support CBCs.

To adequately understand the extent and distribution patterns of modern cold seeps, large portions of the sea floor must be surveyed with techniques that allow for the cold seeps' detection. Unfortunately, the detection of cold seeps is still primarily based on making visual observations of the sea floor using submersibles or remotely operated vehicles. To date, only a tiny portion of the sea floor has been visited with these tools, even in the best-studied areas.

In some places, echo-sounder records collected from surface vessels show water column anomalies that are associated with plumes of gas bubbles or oil. Sonar is the only other remote-sensing tool that has been applied to seep detection with any success. The organisms and authigenic mineralization associated with cold seep sites produce a rough sea floor that appears as an area of high reflectivity in sonar data. Unfortunately, sonar data cannot be used to independently map distributions of cold seeps, because many other processes also result in a reflective sea floor.

SEA FLOOR HYDROLOGY

The occurrence of CBCs require that methane, sulfide, or oil be in the immediate subsurface, which usually requires fluid flow. Fluid flow can be stimulated within marine sediments and the oceanic crust by pressure differences associated with tectonic compression (especially along active continental margins), sediment compaction, lateral density and thermal differences, and even hydrologic heads associated with adjacent land masses.

Although seepage of freshwater or minimally altered seawater may be very common, especially in shallow waters, these sites are not associated with obvious changes in the appearance of the sea floor. Seeps where the water lacks methane, hydrogen sulfide, or oil are detected only when they are vigorous enough to produce visually observable plumes of shimmering water or salinity anomalies, or to leave isotope traces that are detectable within the overlying water masses. Thus, the distribution and frequency of slow seepage of freshwater or minimally altered seawater is very uncertain.

Most cold seep sites in deep water are associated with distinct morphologic features that suggest the site is a conduit for focused fluid flow. For example, CBCs commonly occur on such topographic structures as mud volcanoes and diapirs, within pockmark depressions, or at distinct breaks in slope, which commonly are also hydrologic contacts. The occurrence of CBCs has been used to assist in mapping active sea floor faults. CBCs are also common along fresh slump scars. In some cases, the CBCs along fault scars may occur because slumping exposes methane-bearing sediments on or near the sea floor, making the connection to deep-seated faulting unnecessary.

WINDOWS INTO THE SUBSURFACE

The compositions of the fluids that arrive at the sea floor provide information on subsurface conditions. In some areas, the seeping waters carry dissolved salt at concentrations that indicate a connection with underlying evaporite deposits and/or diapirs. In other cases, the volatile, trace element, and isotopic compositions show a connection with deep-seated processes and can indicate sources from subducted slabs at depths of more than tens of kilometers, or even in the mantle.

Elemental and isotopic composition of hydrocarbon gases can be used to determine whether the gas was generated by thermal maturation of organic matter or by microbial processes within the sediments. When the gas source is thermal maturation, there will be distinctly more ethane, propane, and higher hydrocarbons mixed with the methane. Also, the isotopic fractionation associated with the formation of methane by microbial processes is more pronounced; thus, microbial methane typically contains more ^{13}C-depleted carbon. The production of methane by microbial processes occurs within the organic-rich sediments at relatively shallow depths on continental margins worldwide and within relatively young sediments. In contrast, thermogenic gas production takes elevated temperatures and pressures and considerably more time. Thus, the presence of thermogenic methane in the seeping fluids indicates a connection to a relatively deep source.

Although the visible fauna of some cold seep communities are supported by hydrogen sulfide, both methane and hydrogen sulfide may be playing important roles. In more isolated sub-sea-floor environments, where the upward-migrating methane meets sulfate diffusing downward from the overlying seawater, populations of microorganisms engage in anaerobic oxidation of methane (AOM):

$$CH_4 + SO_4^{2-} \rightarrow HCO_3^- + HS^- + H_2O$$

The AOM generates hydrogen sulfide. Although methane is energetically a more favorable compound, it cannot be stored, and chemoautotrophic microorganisms require a continuous supply. In contrast, hydrogen sulfide can be converted into elemental sulfur and stored within the tissue of the organisms. The stored elemental sulfur is like fat because it can be utilized as an energy source during leaner times. Thus, although the CBCs that are visible on the sea floor are supported by sulfide, it is the availability of methane in the immediate subsurface that enables them to exist.

DIAGENETIC PROCESSES

The presence of reduced compounds carried in the seeping fluids also stimulates rapid diagenesis within the sea-floor sediments. One of the most obvious effects is the generation of authigenic carbonate masses in the sediments underlying the cold seep environment. Authigenic pyrite (and associated iron sulfides) and barite are also commonly found in close association with these carbonates.

The precipitation of carbonate is a consequence of AOM, and these carbonates commonly contain a substantial amount of methane-derived carbon. AOM converts methane carbon into HCO_3^-, spiking the porewater-dissolved inorganic carbon pool with ^{13}C-depleted carbon and increasing the alkalinity of the water. This alkalinity increase will stimulate the formation of authigenic carbonates that contain a component of methane-derived carbon.

FIGURE 1 Video images of chemosynthetic biological communities collected using MBARI's ROVs *Ventana* and *Tiburon* in Monterey Bay. (A) Sea floor mats rimmed by clams; (B and C) dense beds of clams; (D) tube worms coming out of an overhang, which also is being occupied by an octopus.

Methane-derived authigenic carbonate samples collected from continental margins exhibit distinct morphologic features and habits that are easily distinguished in hand samples and thin sections. Samples collected from the sea floor come in a variety of forms including pavements, slabs, irregular nodules, and what have been described as chimneys. The samples collected from the sea floor can be divided into two distinct facies based on whether they have coarse- or fine-grained textures. Coarser-grained authigenic carbonates contain multiple styles of methane-derived cements, including aragonite needles filling voids, aragonite as acicular cements, sparry calcite cements of both low- and high-Mg calcite, and occasionally dolomite. Most common samples of methane-derived authigenic carbonates are composed of micrite cements formed within fine-grained clastic sediments. The micrite cements are predominantly composed of low-Mg calcite and occasionally dolomite. Typically these samples have a massive appearance, and there is no evidence of significant primary porosity within the host sediments. Apparently, these are the result of AOM within the subsurface sediments surrounding the seep sites or wherever methane-bearing sediments are exposed near the surface. Such carbonates are common in areas of active seepage and sea-floor erosion. The authigenic sulfides are also a consequence of reactions with the host sediments and the hydrogen sulfide that is either carried in the seeping fluid or produced locally by AOM. The seeping fluids may also carry the barium for barite formation.

The generation of isotopically distinct carbonate cements helps to preserve and allow for the identification of paleo–cold seep environments in the rock record. Fossil cold seep communities and deposits have been found within rock sequences and show that seep-vent-type megafaunal communities have existed since the Paleozoic, and that similar microbial seep communities probably have existed since the early Precambrian.

SEE ALSO THE FOLLOWING ARTICLES

Hydrology / Hydrothermal Vents / Organic Falls on the Ocean Floor / Pocket Basins and Deep-Sea Speciation

FURTHER READING

Campbell, K. A. 2006. Hydrocarbon seep and hydrothermal vent paleoenvironments and paleontology: past developments and future research directions. *Palaeogeography, Paleoclimatology, Paleoecology* 23: 362–407.

Hovland, M., and A. G. Judd. 1988. *Seabed pockmarks and seepages.* London: Graham and Trotman.

Paull, C. K., J. Chanton, A. C. Neumann, J. A. Coston, C. S. Martens, and W. Showers. 1992. Indicators of methane-derived carbonates and chemosynthetic organic carbon deposits: examples from the Florida Escarpment. *Palaios* 7: 361–375.

Roberts, H. H. 2001. Fluid and gas expulsion on the northern Gulf of Mexico continental slope: mud-prone to mineral-prone responses, in *Natural gas hydrates, occurrence, distribution and detection.* C. K. Paull and W. B. Dillon, eds. AGU Geophysical Monograph 124, 145–161.

Sibuet, M., and K. Olu. 1998. Biogeography, biodiversity and fluid dependence of deep-sea cold-seep communities at active and passive margins. *Deep Sea Research II* 45: 517–567.

COMOROS

D. JAMES HARRIS AND SARA ROCHA

University of Porto, Vila do Conde, Portugal

The Comoros are a set of four major oceanic islands—Mayotte (Maore), Anjouan (Ndzuani), Moheli (Mwali), and Grand Comoro (Ngazidja)—and surrounding islets, situated in the middle of the Mozambique Channel, approximately halfway between Madagascar and the East African coast (about 300 km from each) in the southwestern Indian Ocean (Fig. 1). Like most other oceanic island groups, the numbers of species found in the Comoros is much lower than that found in nearby continental systems, but there is a high degree of endemism. The greatest conservation issues for the islands revolve around their high population density, which has led to huge declines in forest cover, followed by the relatively unknown effects of the large number of introduced species.

GEOGRAPHICAL AND GEOLOGICAL BACKGROUND

Grand Comoro is the largest island (1148 km^2), followed by Anjouan (424 km^2), Mayotte (314 km^2), and Moheli (211 km^2). Their origins considerably postdate the separation between Madagascar/India and Africa, which is known to have occurred approximately 165 million years ago, as well as the drift of Madagascar/India relative to Africa, which ended around 121 million years ago. They were formed by progressive hotspot volcanic activity during the last 15 million years, the earliest manifestations of which can be found in the Seychelles plateau. In other words, the Comoros are the youngest result of hotspot activity that earlier produced volcanism in northern Madagascar, the Farquhar and Amirante islands (now coral atolls), and Tertiary basalt and syenite outcrops in the Seychelles islands.

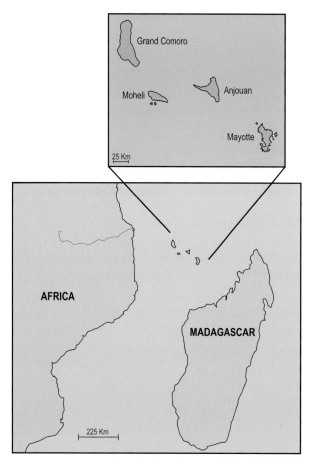

FIGURE 1 Map of the Comoro Islands.

The estimated ages, calculated using the rate of the Somali plate motion and the distance to the present volcanically active hotspot of Karthala volcano on Grand Comoro, are 10–15 million years ago for Mayotte, 11.5 million for Anjouan, and 0.5 million for Grand Comoro, although the ages of the oldest exposed lavas considerably postdate these. Island maximum altitudes are inversely correlated with island age because older islands have suffered erosion for longer periods. Thus, Mount Karthala on Grand Comoro reaches 2361 m, and Mount Ntingui on Anjouan 1595 m, whereas maximum altitudes on Moheli and Mayotte are only 790 m and 660 m, respectively. The Comoros are surrounded by sea depths that vary from 400 m to 3000 m. The lowest historical sea level in the Indian Ocean was −145 m ± 5 m 18,400 years ago, and thus the Comoros are true oceanic islands in the sense that they have never been connected to each other or to Africa or Madagascar.

CLIMATE AND VEGETATION

Generally, the climate is tropical, with a hot wet season between November and April when monsoon winds blowing from the northwest bring strong rains, and a drier cooler season from May to October. On the higher slopes of Mount Karthala on Grand Comoro it is more humid even during the dry season. The north and west flanks of this volcano are the most humid, with the higher parts acting as a rain trap. Average temperatures range between 24 and 28 °C, with an absolute maximum of around 35 °C. On the top of Mount Karthala, temperatures at night can drop almost to freezing. The islands are affected by tropical depressions, and cyclones occur fairly regularly.

The primary natural vegetation across all four islands was forest, with mangrove swamps along some coasts. Biogeographic affinities of Comoroan flora are mostly to Madagascar, and to a lesser extent to the African continent. At least 935 species of plants are registered on these islands, of which approximately 500 species are endemic. Many endemic trees have been identified—*Weinnmania comorensis, Ophiocolea comoriensis,* and the takamaka, *Khaya comorensis*, for example. On Mount Karthala, the higher slopes are very different because of the cooling effects of the altitude and the higher humidity. Here, giant heathers dominate, which can reach 8 m in height. The Karthala white-eye, *Zoosterops mouroniensis,* is an endemic bird found only in this habitat. Fortunately, this higher-altitude forest, found only on Grand Comoro, remains largely intact thanks to its inaccessibility on the steep volcano.

ORIGINS OF COMOROAN FAUNA

After their formation, the Comoro islands were colonized mainly by fauna from Africa and Madagascar, and also eventually from Southeast Asia. Cenozoic dispersal, possibly using islands like the Comoros as steppingstones, seems to be an emerging pattern of colonization of Madagascar from the continent. Later dispersal events from Madagascar to the surrounding oceanic islands including the Comoros, and back to Africa in some cases, subsequently partially obscured this biogeographic signal. East Africa is the other major source for Comoroan taxa. However, estimating origins of island taxa requires an extensive assessment of the diversity within putative areas of origin, and unfortunately, as for many tropical areas where most biodiversity is found, little is known regarding genetic diversity within the fauna of this region. As generators of much of the planet's biodiversity, it is most important that these tropical regions are studied phylogeographically. This means that although Comoroan taxa have diverse affinities, the origins of most taxa remain unknown. However, the reptile fauna of the Comoros has recently been studied and appears to result from three sources: ancient natural colonizations (*Furcifer* chameleons, *Mabuya* skinks, *Phelsuma* day geckos), more recent natural colonizations (*Cryptoblepharus* skinks, *Phelsuma* day geckos), and recent, human-aided dispersals. Multiple colonizations are common in most taxa, and oceanic dispersal, more than archipelago radiations or within-island divergence, has led to species diversity in the region. The emerging biogeographic pattern for the Comoros herpetofauna is that most endemic species have their closest relatives in Madagascar and not in East Africa. This is the case of chameleons, day geckos, *Oplurus* iguanas, skinks of the genera *Cryptoblepharus* and *Amphiglossus,* and *Paroedura* geckos, but it still needs to be tested for other animal groups and for plants.

Although impoverished and unbalanced relative to continental systems, the fauna of the Comoros islands comprises a diverse array of mammals, birds, reptiles, fishes, and invertebrates. Other classes are less abundant: Amphibians, in particular, are represented by only two species, which, apparently, reached the archipelago by natural dispersal from Madagascar. There is a high degree of endemism within the Comoroan fauna, but also a considerable number of widespread and introduced species.

ENDEMISM

Interestingly, despite being the youngest island, Grand Comoro shows the highest number of endemic bird species, with five single-island endemics on Grand Comoro, three on Moheli, one on Anjouan, and two on Mayotte, out of a total of 16 archipelago endemics. This is probably because it is larger and higher than the other islands, and thus provides a greater diversity of habitats. Among the birds, representative examples of radiations within the archipelago include the genera *Turdus, Terpsiphone, Zoosterops, Foudia,* and *Nectarina*. Some species are Comoros endemics, whereas others also occur in other nearby island systems, such as Aldabra and the Mascarenes, but are represented in the Comoros by endemic subspecies. The humid forest is the key habitat for most of the endemic bird species, whereas larger freshwater lakes are important for many visiting migratory birds. Several bird species are threatened, in particular the Scops owls, which include an endemic species on each of Grand Comoro, Moheli, and Anjouan.

The only endemic mammals are fruit bats of the family Pteropodidae. The commonly seen species, *Pteropus seychellensis,* is also found on various other Indian Ocean islands, including the Seychelles, Aldabra, and Mafia off the coast of Tanzania. A second, larger species, *P. livingstonii,* is endemic to Moheli and Anjouan. Although described as abundant in the nineteenth century, surveys in 2001 recorded just 1200 individuals of this species. Another, much smaller, fruit bat species, *Rousettus obliviousus,* was only discovered in 1978 and is endemic to Grand Comoro, Moheli, and Anjouan. Several widespread insectivorous bats species can also be found in the Comoros, some of which include endemic subspecies.

Endemic species of both snakes and lizards occur on the Comoros. Snakes are represented by two typhlopid and three colubrid species—although the typhlopids, or blind snakes, are poorly known—and other species may occur. Of these, three are endemic, one is introduced, and one is also found in Madagascar. None is dangerous to man, and all are rarely encountered. The most widespread and visible lizards are the day geckos of the genus *Phelsuma,* of which several endemic species can be found on all four islands along with two introduced species, *P. dubia* and *P. laticauda*. Two endemic chameleons (*Furcifer*) also occur, as well as skinks (*Mabuya, Amphiglossus,* and *Cryptoblepharus*), an iguanid (*Oplurus*), and other geckos (*Hemidactylus, Ebenavia, Paroedura,* and *Geckolepis*). Two frog species have been reported, both of which were historically presumed to be introduced, but recent molecular analyses suggest they are endemic forms. There are few freshwater fish, with only 17 certain species and several unconfirmed introductions. They are poorly studied, but none appear to be Comoroan endemics. Similarly, studies of invertebrates are limited even though the islands contain several striking endemic radiations, such as the longhorn beetles *Sternotomis,* with an endemic species on each island.

RECENT INTRODUCTIONS

Introduced species can devastate island ecosystems, both through direct competition with native species and also through the introduction of parasites and diseases. Sadly, this all-too-familiar scenario is clearly occurring in the Comoros. As well as typical domestic mammals, such as cows, sheep, dogs, and cats, early settlers brought mice and rats. In a vain attempt to control the rats, lesser Indian civets were then introduced, which are instead decimating the endemic snake populations. The hedgehog-like tenrec from Madagascar was probably introduced for food. These are thought to damage gecko populations in the Seychelles, where they were also introduced, and are probably equally destructive in the Comoros. Unfortunately, the rate of new introductions continues unabated. Mongooses—another potentially devastating predator for many small birds, mammals, and reptiles—were introduced to Grand Comoro apparently during the 1960s or 1970s. Of the reptiles, several house geckos are recent introductions, as well as a blind snake and an agamid. Lemurs were also introduced from Madagascar. Various bird species have been introduced, including chickens, pigeons, and sparrows. Some freshwater fish, such as the very common guppies on Mayotte, are known to be introduced, whereas the status of others remains unclear. The status of many invertebrate species is similarly unknown.

HUMAN HISTORY, GEOGRAPHY, AND INFLUENCE

The first-known human inhabitants of the Comoro islands were Polynesians and Melanesians, around the sixth century BC, although settlement may have occurred earlier. Since then, diverse groups have arrived, from the coast of Africa, the Persian Gulf, and Madagascar. Portuguese explorers visited in 1505, and France established colonial rule in 1841. In 1975 the Comoroan parliament declared independence. In the following referendum, the population of Mayotte voted against independence, and currently Mayotte remains under French political control, whereas the other islands form the Union of the Comoros. The current population (2005 estimate: 798,000) means an average human density of over 275/km^2, although values are higher on Anjouan and lower on Moheli. Agriculture, including fishing and forestry, is the leading sector of the economy, and this way of life, combined with high population density, means that the Comoros are facing an environmental crisis. Between 1973 and 1983, forests declined by 73% on Anjouan, and across the islands approximately half the forest was lost. The forest has been predominantly replaced with fields and plantations of banana, coconuts, and ylang-ylang. Extensive felling of forest to develop fields has led to much greater levels of soil erosion on the steep volcanic landscape. Freshwater supplies have also been seriously influenced by this, with the number of perennial rivers being greatly reduced. Opening of roads into forested regions intensifies this process. The effect of this on endemic plants and invertebrates is unknown, but it is probably catastrophic. Few endemic birds survive in the non-forest areas, although the sunbirds (*Nectarina*) are better adapted to deal with the changes, flourishing on the introduced exotic flowers. Similarly, many of the reptiles appear relatively capable of surviving in secondary forest areas. For most of these, introduced species, such as rats, mice, and civets, pose a possibly more serious threat.

SEE ALSO THE FOLLOWING ARTICLES

Deforestation / Introduced Species / Lizard Radiations / Madagascar / Oceanic Islands

FURTHER READINGS

Goodman, S. M., and J. P. Benstead, eds. 2003. *The natural history of Madagascar*. Chicago: University of Chicago Press.
Louette, M., D. Meirte, and R. Jocque. 2004. *La faune terrestre de l'archipelago des Comores*. Studies in Afrotropical Zoology 293. Tervuren: MRAC.

CONTINENTAL ISLANDS

DAVID M. WATSON

Charles Sturt University, Albury, Australia

As evidenced by the breadth of systems covered in this volume, the term "island" can be applied to a range of ecosystems extending well beyond conventional notions of land masses surrounded by water. Be they caves, rock outcrops, mountaintops, or oases, these systems share many biotic and abiotic features with their oceanic counterparts. One of the primary distinctions between different island systems is whether they were formed de novo, or instead formed by fragmentation. In the former case, the islands have never been in contact with the source of colonists and have abundant "empty" ecological niche space. Fragments are fundamentally different. In these patches, the ecological space will initially be filled as a consequence of connection to the source of colonists prior to insularization. In this article, continental islands are considered to be those formed through the process of fragmentation, and the resulting characteristics of such islands (and how they differ from islands formed de novo) are discussed.

FRAGMENTS VERSUS DE NOVO ISLANDS

Rather than being a trivial technicality or a question of semantics, establishing the historic origin of an island is crucial to formulating expectations of the ecological patterns represented. Many features referred to as islands are actually fragments of previously more widespread ecosystems. Islands formed de novo (such as oceanic islands) are habitat patches that were formed more recently than their surrounding habitat—they are relatively new features that are initially unoccupied, becoming colonized in a variably predictable fashion by a sequence of organisms (Fig. 1A). In contrast, continental islands are patches defined by continental rifting (e.g., causing a fragment to break away from a continent) or by the imposition of a new or modified matrix (e.g., isolating a habitat island

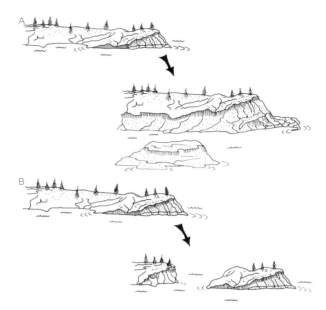

FIGURE 1 Schematic representation of the formation of a generic island de novo and by fragmentation. (A) In this example, lowering sea or lake levels expose a new habitat patch—a small island without any preexisting terrestrial biota. As with all islands, the biota gradually develops from two classes of organism: inter-patch dispersers coming from similar habitat elsewhere and matrix-derived species coming from the surrounding habitat. (B) Here, the fragment becomes isolated by subsidence, with the matrix habitat (in this case, water) surrounding the habitat patch. The fragment biota is initially dominated by relictual organisms (represented by trees in this example), with matrix-derived and inter-patch disperser taxa adding to the fragment biota. Illustrations by Margaret J. Watson.

and diminishing movement rates of organisms into and out of the habitat patch) (Fig. 1B).

Three features of this distinction are important. First, note the complete lack of reference to water—islands may occur on land, in water: anywhere. Thus, a salt lake can be considered analogous to an oceanic island (formed de novo), whereas an oxbow lake qualifies as a fragment; most caves are formed de novo, whereas icebergs are fragments. Second, there is no reference to temporal or spatial scale: Tasmania can be considered a fragment of the Australian continent isolated by rising sea levels around 14,000 years ago, just as a tidepool is a fragment lasting for less than a day. Finally, this functional distinction demonstrates that the key difference between islands formed de novo and fragments lies in the sequence of events that defines them. For islands formed de novo, the patch itself is the new feature, whereas for fragments the habitat comprising the patch is preexisting, and patch boundaries are determined by the imposition of a new or modified matrix habitat. Spanning a wide range of ecosystems ranging in extent from centimeters to thousands of kilometers, and time periods from hours to millions of years, this distinc-

tion provides a useful framework to arrange our current understanding of how inherently patchy ecosystems function and to identify which factors best explain patterns of diversity and distribution within them.

A FUNCTIONAL CLASSIFICATION OF ISLANDS

Islands that have been formed de novo as well as those formed through fragmentation can be further divided into old systems (older than 1000 years) and young landscapes (younger than 100 years), allowing for a surprising number of generalities. These four classes can be further subdivided using a convenient third factor relating to the contrast between the patch (de novo island or fragment) and the surrounding habitat or matrix. Phase differences (i.e., land-locked water bodies, patches of land surrounded by water [Fig. 2]) are deemed high contrast, whereas within-phase differences (e.g., a forest patch surrounded by grassland, rock outcrops within heathland) are low contrast. Using this three-factor approach and some objective thresholds, most patchy landscapes can be readily assigned to one of eight classes (Table 1).

The first step in applying this classification is to establish whether or not a particular patch is a de novo island or a fragment. Although usually simple, it is not always straightforward, and historic information may be equivocal or simply unavailable. Once the origin of the patch is established, patch age and matrix contrast determine which of eight classes it represents. As summarized in Table 1, this classification groups together a wide range of ecosystems, often studied using divergent methods and typically not regarded as islands or fragments.

FIGURE 2 A small patch of lowland rain forest in Gatun Lake, Panama, created by erosion from wave action associated with passing ships. Although referred to informally as an island, this patch is actually a habitat fragment, isolated from similar habitat by a high-contrast matrix. This photograph also illustrates the species-area relationship: Of the 250 plus species of birds recorded from the adjacent Barro Colorado Island (a 1500-ha fragment isolated in 1914), this tiny fragment supports a single species—the black vulture *Coragyps atratus*.

TABLE 1
Functional Classification of Patchy Landscapes

Fragments		Islands	
Low-Contrast	High-Contrast	Low-Contrast	High-Contrast
Young (< ~100 years)			
Forest fragment	Floating raft/mat	Boulder	Artificial dam, pond
Internal fragment/clearing	Flooded tree	City, town	Artificial island/oil rig
Refugium from fire	Hydroelectric island	Lava flow	Mine shaft
Seagrass fragment	Oxbow lake	Plantation, orchard	Sand shoal
	Pack-ice/iceberg	Tree-fall gaps	Volcanic island
	Tidal pool	Vacant lot	
Old (> ~1000 years)			
Cliff-face fragment	Continent	Alpine grassland	Cave, cavern
Lava-flow refugium	Inter-river valley	Bog, fen, pothole	Desert oasis
Mesa/tepui	Land-bridge island	Canyon/gorge	Geothermal vent/seamount
Montane remnant	Riverine/lacustrine island	Inselberg, kopje	Hot spring
Riparian fragment		Landslide	Mound spring
		Morraine field	Oceanic island
		Mound-reef	Salt lake

NOTE: By using age, origin, and matrix contrast to define eight forms of patch, this approach holds across a wide range of spatial and temporal scales, from continents to ephemeral ponds.

Using this approach, physically similar systems are revealed to be functionally divergent. Thus, inselbergs and caves—two ecosystems that have received considerable attention from researchers—are de novo islands, whereas tepuis, mesas, and most other montane patches are functionally fragments. One implication of this historic difference is manifested in the origin of organisms occurring within these patches, with an additional class of organism being associated with habitat fragments. Known as relictual taxa, these are organisms that were living in the habitat prior to fragmentation and became isolated as the matrix surrounded the patch. A pertinent example comes from the cloud forests of Mesoamerica, which are home to many groups of moisture-dependent bolitoglossine salamanders (Fig. 3). These animals were originally more widespread, but, just like fish in an oxbow lake, these habitat specialists became trapped by the surrounding inhospitable habitat. Although they need not be restricted to the fragment and may occasionally or regularly move to other patches, relictual taxa are typically lower-vagility organisms that either persist or become locally extinct after fragmentation.

Two other classes of organism occur on both de novo islands and fragments, albeit in differing proportions. Matrix-derived taxa are those organisms originally associated with the matrix habitat, secondarily occupying the habitat patches. Although this may be short term or opportunistic (e.g., grassland birds nesting in forest fragments, migrating raptors using inselbergs as staging areas), it becomes most evident over evolutionary time, as the organisms adapt to the new habitat. Examples of matrix-derived species include many cave-dwelling arthropods, which exhibit a suite of adaptations that confer advantages in the new habitat. The final class of organism—inter-patch dispersers—consists of medium- to high-vagility taxa that are associated primarily with the patchy habitat, but which can readily move between patches. Thus, swiftlets and rock wallabies are inter-patch dispersers, being dependent on specific habitat patches (caves and rocky massifs, respectively) but being variously able to move between nearby patches across the intervening matrix.

FIGURE 3 A minute bolitoglossine salamander in the genus *Thorius* from a humid pine-oak forest fragment in southern Mexico. These poorly known species illustrate how habitat fragmentation can promote diversification, with populations having become separated as the cold-adapted forests retreated upslope when the region became warmer and drier after the last glacial period ended approximately 40,000 years ago. The evolutionary consequences of vicariance, which affect many groups in this region, are most obvious with low-vagility species like salamanders, with the two mountain ranges separated by the Oaxaca Valley supporting completely complementary assemblages.

CONTINENTAL ISLANDS: FRAGMENTS SURROUNDED BY WATER

A special class of fragment that illustrates some of the key differences between de novo islands and fragments consists of what is known as continental "islands": large areas of continental shelf surrounded by water that were isolated from adjacent continents, either by rising sea levels (e.g., Sicily, Great Britain, Tasmania) or by rifting over millions of years (New Caledonia, Madagascar, New Zealand). Unlike the biota of oceanic islands that formed de novo, where species numbers reflect a dynamic balance between immigration, speciation, and local extinction, continental islands did not start with vacant ecological space. Rather, they carry with them the descendants of those organisms initially present on the land mass prior to fragmentation, with diversity patterns subsequently changing through the process of relaxation. Thus, they start with a "full" complement of organisms and gradually lose taxa over time, unlike oceanic islands, which start out empty and progressively gain taxa. During this period, taxa may become extinct in the remainder of their range but persist in isolation. Over time, this process yields an increasing proportion of species, genera, families, and even orders that are restricted to continental islands—taxa known as paleoendemics (as compared with neoendemics—that is, the comparatively recently formed species that diverged after colonizing oceanic islands).

Continental islands, and the differences between paleoendemics and neoendemics, are exemplified by New Zealand, which became isolated from Gondwanaland around 80 million years ago. New Zealand contains a large number of relictual forms found nowhere else: Considering birds alone, there are the Zealand wrens (Acanthisittidae [Fig. 4A]), widely considered to represent the sister taxon to the diverse passerine order; kakas; keas (Fig. 4B); and kakapos (Nestoridae), which are con-

FIGURE 4 Two examples of relictual species from families wholly restricted to a continental island: New Zealand. (A) A rockwren *Xenicus gilviventris* on the South Island of New Zealand, one of two surviving representatives of the New Zealand wren family (Acanthisittidae). Numerous phylogenetic studies have suggested that these small, almost tail-less birds are sister taxa to the entire Passerine order. (B) A kea *Nestor notabilis* in the alpine grasslands of New Zealand. These intelligent birds and their close relatives in the family Nestoridae are sister to all other parrots.

sidered to be the basal clade in the parrot order; there are also two other endemic families, which include the wattlebirds (Calleatidae) and kiwis (Apterygidae). Even more striking are the two species of tuatara (Fig. 5), the sole living representatives of the order Rhyncocephalia, which became extinct elsewhere in the Jurassic. This diverse assemblage of paleoendemics is complemented by neoendemics, many of which are derived from colonists from nearby Australia. This second source of endemics is exemplified by the takahe (Fig. 6), a large flightless gallinule, which eats a restricted diet of tussock grasses and herbs. In the absence of any terrestrial mammals, a variety of novel morphotypes evolved within birds, including a radiation of gregarious browsers and grazers (moas), a role occupied by mammals elsewhere. Rather than being exceptional, these examples are representative of patterns seen repeatedly on continental islands: aberrant forms evolving in isolation, and relictual taxa surviving long after becoming extinct elsewhere. Despite being frequently labeled islands, these systems and the patterns of diversity associated with them show many parallels with other fragmented landscapes, allowing insight into the long-term consequences of habitat fragmentation.

MONTANE REMNANTS AS CONTINENTAL ISLANDS

One class of patchy ecosystem that has received disproportionate attention is montane remnants, and they serve as instructive examples to evaluate in more depth. Mountaintops often contain vegetation types and specific habitats that are altitudinally restricted—found only under specific climatic conditions associated with high elevations; such habitats include cloud forests, alpine grasslands, herbfields, and paramo. Despite being poetically referred to as "sky islands" by some researchers, these vegetation types are often relictual fragments of previously more widespread habitats, which have been isolated by changing climate (Fig. 7).

These habitat fragments are similar in many ways to the woodland and forest fragments isolated by agricultural development, but they differ dramatically in age. Most anthropogenically created fragmented landscapes were formed by broad-scale clearing 150 to 80 years ago, as the regions of North America, Australasia, Latin America, and Southeast Asia were developed. Montane remnants worldwide were defined much earlier, as regional climates became warmer and drier—coinciding in most cases with

FIGURE 5 A tuatara *Sphenodon punctatus* on Kapiti Island, New Zealand. These aberrant reptiles are the sole living representatives from the ancient Rhynchocephalian order, widespread in the fossil record prior to the Jurassic.

FIGURE 6 A takahe *Porphyrio mantelli* grazing on Kapiti Island, New Zealand. This flightless rail is a neoendemic, diverging from a gallinule ancestor that colonized New Zealand from Australia or elsewhere in the Pacific. Both the takahe and the tuatara are now extinct on the main two islands of New Zealand and only survive on small offshore islands where predators either were not introduced or were eradicated.

FIGURE 7 The central Oaxaca Valley near Yagul, with arid thornscrub in the foreground, agricultural fields in the middle distance, and mountain ranges in the background. The vegetation growing atop these ranges represents fragments of the humid pine-oak forest originally widespread throughout Mesoamerica, attaining their present configuration approximately 12,000 years ago. Disturbance associated with agriculture has largely been restricted to the lowlands and has occurred continuously for more than eight millenia (the wild ancestors of corn originate in this valley).

the last interglacial period in the late Pleistocene. Hence, the fragments in many of these landscapes are 10,000 to 30,000 years old; although subject to further and ongoing disturbance, including additional clearing, these landscapes still retain the imprint of this initial shift.

Although most habitat change associated with anthropogenic activities (i.e., clearing, agricultural intensification, development) occurs over periods of years, vegetation change mediated by regional climatic shifts is often considered to take tens of thousands of years. Recent information, however, is dispelling this assumption. A growing body of research has examined changes in plant distributions associated with recent climate change, demonstrating that even subtle changes in mean temperature and rainfall can effect dramatic shifts in plant occurrence. These data are reinforced by studies of fluctuations in the northern extent of boreal forest during the Pleistocene, as forest boundaries changed rapidly in response to drought associated with postglacial warming. Hence, the creation of elevational remnants may have occurred far more rapidly than is frequently assumed, and many regions may have been transformed from continuous habitat into mosaics of fragments within several thousand years or less.

Compared with anthropogenically fragmented landscapes that have been the focus of several thousand separate ecological investigations, these montane remnants are less thoroughly understood: Determinants of diversity have been studied in less than 50 landscapes. Many studies of montane assemblages were conducted in response to MacArthur and Wilson's equilibrium theory of island biogeography, to find out whether similar processes operated in these continental ecosystems. Having established presence/absence data for selected organism groups, the researchers then related these richness estimates to fragment area, distance to nearest continuous habitat patch (or "mainland"), and a range of other biogeographic metrics.

Unlike anthropogenically created fragments, which are typically embedded within a developed matrix to which access is straightforward, many montane remnants occur in remote and inaccessible regions. These logistical issues are compounded by a pronounced taxonomic impediment. Far fewer researchers have visited many of these areas, and, given the pronounced environmental heterogeneity in these mountainous landscapes, species often exhibit highly restricted distributions, with small-scale neoendemics being well represented. Thus, there are many undescribed or poorly known species, and working out the identities of the species present can pose very real challenges. Moreover, the repeated visits and thorough sampling required to generate accurate richness estimates are often not possible, thus reducing the accuracy of analyses and limiting a researcher's ability to explain recorded patterns of occurrence.

CASE STUDY 1: MAMMALS IN GREAT BASIN BOREAL FOREST FRAGMENTS

The implications of these logistical and procedural issues are exemplified by studies of mammals living in boreal forest fragments in the Great Basin of western North America, isolated around 8000 years ago when this forest type retreated northward, leaving relictual stands on high mountain ranges throughout the region. This landscape is the best studied of all montane remnants, with multiple studies having been conducted of resident birds, mammals, butterflies, and plants. Noted ecologist James H. Brown studied 15 cold-adapted subalpine mammal species (excluding large carnivores, ungulates, and bats) in these forests from 1968 to 1970, complementing published records from the 1930s, 1940s, and 1950s with his own observations to build faunal lists for 17 forest fragments. His analysis demonstrated that species richness did not reflect a balance between colonization and extinction expected under an equilibrium model. Rather, the mammalian fauna of these forest fragments was extinction driven, with larger patches retaining greater numbers of species and greater proportions of larger species with more strict habitat and dietary needs. This study was instrumental in developing the nonequilibrium or relaxation biogeographic model, which eventually became a widely accepted paradigm.

Some 25 years later, mammalogist Tim Lawlor returned to these forests, spending at least three days in each forest fragment and confirming species identities with voucher specimens (essential for morphologically conservative groups such as chipmunks and shrews). He added 25 new records to Brown's dataset, involving 13 of the original 15 subalpine species. Given the distance between mountains and the arid intervening habitat, it is unlikely that these additional records represent recent colonists. Although Brown acknowledged that his data matrix was likely incomplete, he suggested that additional data would have little qualitative effect on the findings. Surprisingly, Lawlor's reanalysis revealed a much weaker association between fragment area and species richness, with similar weak effects for individual species. Isolation from various "mainlands" continued to have little effect on species richness, with patch-restricted relictual taxa occurring in lower proportions than inter-patch dispersers.

Rather than being exceptional, this example is likely to be representative of many other landscapes, most of which have not been studied more than once. In addition to issues of inventory completeness, many historic studies await reanalysis. As described earlier, most studies of elevational fragments were conducted in the 1970s and 1980s, using the analytical tools available to ecologists at the time. Many of these techniques now seem rudimentary, especially when compared with the nestedness analyses and spatially explicit multivariate approaches routinely used to infer determinants of diversity in anthropogenically fragmented landscapes today.

CASE STUDY 2: BIRDS IN HUMID PINE-OAK FORESTS IN SOUTHERN MEXICO

One of the few studies to have applied these modern analytical tools to studying ecological patterns of diversity in montane remnants was a study of resident birds in humid pine–oak forests in southern Mexico conducted by the author. Dominated by various species of pine and oak with associated alder, fir, and other palearctic plants, these temperate forests are restricted to elevations of 2500 m or higher. Originally widespread throughout Mesoamerica (from Texas to Colombia), these cold-adapted forests retreated upslope during the last interglacial period beginning approximately 30,000 years ago, attaining their present configuration approximately 12,000 years ago. The Oaxaca Valley occurs between the two dominant ranges of the region (the Sierra Madre Oriental and Occidental) and has been the focus of comprehensive archaeological research including detailed paleoclimate and vegetation reconstructions for the past 40,000 years.

Exhaustive bird surveys were conducted in 17 patches of humid pine–oak forest, ranging in size from 2 to 160,000 ha. To minimize the issues associated with incomplete inventories, particular attention was paid to the methodology used to survey resident bird assemblages. Because no other study had attempted to conduct inventories for patches spanning five orders of magnitude, a new sampling approach was devised, which applied results-based stopping rules to sample forests at the patch scale to the same degree of completeness (known as the "standardized search"). These field data were complemented with historic data from unpublished records and with specimen data, yielding comparable inventories of consistently high completeness. These data were analyzed using a range of methods and compared qualitatively with findings from comparable studies of other montane remnants.

The most striking finding from this work was the consistent importance of patch-scale factors in determining species richness. Whereas isolation (measured as distance from the nearest forest larger than 100,000 ha) and other landscape-scale factors had little effect (either on overall richness or occurrence patterns of 60 resident species), fragment area, vegetation complexity, and elevational range all emerged as key drivers of diversity. All three predictors were closely interrelated, however, such that larger forests typically contained a higher number of habitats and spanned greater elevational ranges. Patterns of richness among the 17 fragments were found to be highly nested, with the arrangement of subsets being closely related to patch area. These findings echo previous results from other montane remnants elsewhere, with patch-scale factors consistently overwhelming landscape-scale factors. This suggests that much of the isolation effect described in younger systems or patches surrounded by high-contrast matrices is transitory, with fragment size and other patch-scale factors being more crucial to retaining species over thousands of generations. Examining occurrence patterns of individual species, 27 of the 30 habitat specialists were found only in forests larger than 3000 ha, with eight especially sensitive species being found only in montane remnants larger than 30,000 ha (Fig. 8).

THE ROLE OF TEMPORAL SCALE

Comparing findings from these old fragmented systems with those from other classes of patchy landscape, time since formation emerges as critical. Indeed, after thousands of years, continental islands converge in many community-scale properties. For those patches in low-contrast matrices (including montane remnants and inselbergs),

FIGURE 8 A red warbler *Ergaticus ruber*, an elusive species of wood warbler endemic to the humid pine-oak forests of southern Mexico. Along with 26 of the 30 other habitat specialists evaluated, this species was restricted to forest fragments larger than 3000 ha, exemplifying the large areas needed for biota associated with montane remnants to persist over the long term. All photographs by David M. Watson.

diversity stabilizes over time, and the biota becomes distinct from the interstitial matrix. De novo islands become dominated by matrix-derived taxa, whereas relictual taxa continue to dominate fragments, with inter-patch dispersers showing indistinguishable patterns of occurrence. Hence, despite different origins and initial differences, the biotas of older islands, whether formed de novo or by fragmentation, exhibit broad similarity, explaining why they are so frequently studied together.

Over evolutionary time, the process of habitat fragmentation can lead to vicariant speciation—that is, isolated populations of organisms diverging and acquiring novel traits. At the global scale, tectonic movement driving the breakup of Gondwana and Laurasia exemplifies this process, but, so long as patch size and quality are sufficient to allow populations to persist, allospecies can result from fragmentation at much smaller scales. The proportion of these endemics is generally greater in high-contrast systems, but populations isolated by changing courses of rivers, intrusion of lava flows, or shifting dunefields in terrestrial systems can also go on to form distinct species.

Because fragmentation necessarily leads to smaller populations in the short term, it can initiate a sequence of demographic processes culminating in local or complete extinction. Aside from smaller population sizes, fragments have higher proportions of edge habitat, attracting early successional species that can eventually replace the relictual taxa. This explains why patch size consistently emerges as so important in explaining distribution patterns in these older landscapes, and, as revealed by the findings from Oaxaca, why these thresholds can be surprisingly high.

LESSONS FROM THE PAST: IMPLICATIONS FOR RECENT FRAGMENTS

Despite the concerted efforts of ecologists to derive generalized principles that explain the short- to medium-term effects of habitat fragmentation on biodiversity, overall trends are obscured by site- and taxon-specific factors. Following fragmentation of any previously continuous habitat, a variable proportion of relict species becomes locally extinct, initially because of loss of habitat but subsequently because of demographic and stochastic processes. Many studies have focused on these processes, identifying ecological traits of those species especially prone to extinction and patch attributes that may minimize these extinctions. Initial diversity losses can be offset by an influx of matrix-derived species, occasionally resulting in fragmentation having a positive effect on diversity.

These adjustments in community composition can take many generations to occur. Most studies, however, focus on habitats dominated by woody plants (i.e., forests and woodlands), ecosystems defined by organisms with lifespans that generally exceed the time since initial disturbance. Accordingly, net effects of recent fragmentation on diversity patterns can rarely be measured with confidence, and some theoretical estimates suggest that hundreds to thousands of years are required before communities adjust. These concerns notwithstanding, most research carried out in anthropogenically fragmented systems does not focus on long-term effects. Rather, most studies are motivated by site-specific, application-oriented goals, generating valuable data on the short-term responses of communities to fragmentation.

In contrast to the highly divergent consequences of habitat fragmentation in the short term, studies of older systems reveal broad similarities in the long-term effects of fragmentation. Biotas remain composed primarily of relict taxa, with diversity patterns being best explained by area, age, and other patch-scale variables, and landscape-scale factors such as isolation consistently have little influence. Hence, despite comprising less than 50 studies, inferences generated from studying these older, high-elevation fragments yield valuable insight into the future awaiting anthropogenically altered landscapes. Moreover, the close interrelationships between the patch-scale variables in studies of older landscapes suggest that much of the debate between the influence of habitat versus area is needless, with these and other patch-scale factors interacting and combining in their influence on biodiversity occurrence over the long term.

SEE ALSO THE FOLLOWING ARTICLES

Caves, as Islands / Fragmentation / Inselbergs / Pantepui / Relaxation / Sky Islands / Vicariance

FURTHER READING

Brown, J. H. 1971. Mammals on mountaintops: nonequilibrium insular biogeography. *The American Naturalist* 105: 467–478.
Gillespie, R. G., and G. K. Roderick. 2002. Arthropods on islands: evolution and conservation. *Annual Review of Entomology* 47: 595–632.
Larson, D. W., U. Matthes, and P. E. Kelly. 2000. *Cliff ecology: pattern and process in cliff ecosystems.* Cambridge: Cambridge University Press.
Lawlor, T. E. 1998. Biogeography of Great Basin mammals: paradigm lost? *Journal of Mammalogy* 79: 1111–1130.
Porembski, S., and W. Barthlott. 2000. *Inselbergs: biotic diversity of isolated rock outcrops in tropical and temperate regions.* New York: Springer-Verlag.
Watson, D. M. 2002. A conceptual framework for the study of species composition in islands, fragments and other patchy habitats. *Journal of Biogeography* 29: 823–834.
Watson, D. M. 2003. Long-term consequences of habitat fragmentation: highland birds in Oaxaca, Mexico. *Biological Conservation* 111: 283–303.
Worthy, T. H., and R. N. Holdaway. 2002. *Prehistoric life of New Zealand: the lost world of the Moa.* Bloomington: Indiana University Press.

CONVERGENCE

TADASHI FUKAMI

Stanford University, California

In ecology and evolutionary biology, the term "convergence" refers to intriguing resemblance in the characteristics of distantly related organisms or in the structure of independently developed communities. These phenomena are thought to be caused by similarity in the environmental conditions affecting the organisms or communities. Convergence is observed on both islands and continents, but it can be more striking and more easily identified on islands, particularly when they are remotely located from other land masses.

CONVERGENCE IN ORGANISMS

Different species living under similar environmental conditions sometimes show close resemblance to one another in appearance, habit, and lifestyle, even if they are distantly related and geographically isolated. The resemblance is not because the species share a recent common ancestor, but because they have evolved, independently of one another, to acquire similar adaptations to similar environments. This phenomenon is called convergent evolution. A classic example involves plants in Mediterranean shrublands. The Mediterranean climates occur in several regions including California, Chile, Italy, and South Africa. The leaf morphology and growth form of many plant species found in these regions closely resemble one another even though they have evolved independently. Another well-known example of convergent evolution is seen between marsupial mammals in Australia and placental mammals in other continents (Fig. 1). An equally interesting example is convergent evolution of the web-building behavior of Hawaiian spiders (Fig. 2).

CONVERGENCE IN COMMUNITIES

When many pairs of species are convergent (Fig. 1), it is tempting to suggest that the overall composition of the ecological communities is also convergent. Implicit in this idea is that it is not only adaptation to the physical envi-

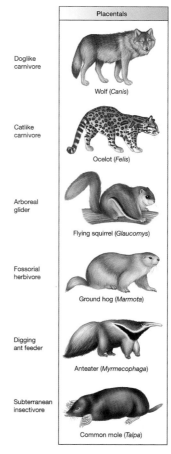

FIGURE 1 Convergent evolution of placental and marsupial mammals. The pairs of species are similar in appearance and habit, and usually, though not always, in lifestyle. Adapted with permission from Begon et al. 2006.

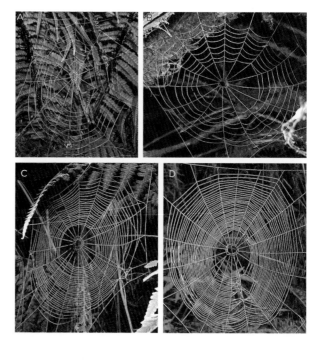

FIGURE 2 Convergence of web-building behaviors among species of *Tetragnatha* on three different islands in the Hawaiian archipelago (Maui, Oahu, and Hawaii). Similar webs are spun by *T. eurychasma* (Maui) (A) and *T.* sp. "emerald ovoid" (Oahu) (B), and by *T. stelarobusta* (Maui) (C) and *T. hawaiensis* (Hawaii) (D). Photographs by Todd A. Blackledge and Rosemary G. Gillespie.

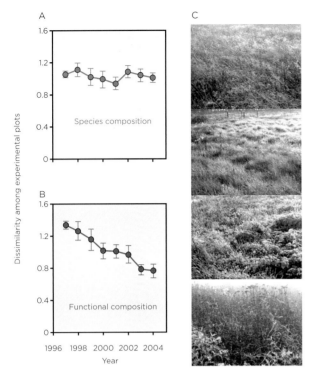

FIGURE 3 Changes in dissimilarity among plots in plant community composition in an experiment in which different species of grassland plants were sown in different plots in 1996. This site was formerly agricultural land and had bare ground at the start of the experiment. Dissimilarity between plots decreased over time in functional composition, indicating community convergence (B). However, these communities did not converge in species composition (A), as apparent in the photographs of some of the plots taken in 2004 (C). Adapted with permission from Fukami et al., 2005. Photograph (C) by Tadashi Fukami.

Australia shows similar relationships to the diversity in foliage height, an index of bird habitat complexity.

We have so far considered adaptation, an evolutionary process, as the mechanism of convergence, but communities can also converge without adaptation. For example, in a study conducted at a Dutch grassland site, experimental plots in which different plant species were sown converged in terms of the types of species that remained in the plots after nine years, although the communities did not converge in the specific identities of species (Fig. 3). Nine years is probably too short for adaptation to play a significant role in these plants. Rather, it is the ecological "sorting" of species involving immigration, interaction, and extinction that caused convergence. Ecological sorting is conceptually similar to evolutionary convergence, in which functionally similar organisms emerge in different communities even though they are not the same species (Fig. 4). Ecological sorting generally occurs over much shorter time than evolutionary sorting, though the two may operate simultaneously.

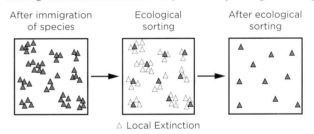

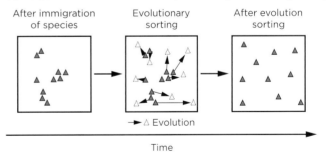

FIGURE 4 Schematic representation of hypotheses regarding how ecological sorting (A) and evolutionary sorting (B) lead to convergence in the structure of biological communities. The square area represents functional space, and each triangle represents a species. Ecological sorting, involving species interactions and consequent local extinctions from the communities, and evolutionary sorting, involving co-evolution within the communities, can both cause an even distribution of species in functional space. If the same sorting occurs independently in different localities (or on different islands), the independently developed communities will converge in functional composition. Adapted with permission from Wilson 1999.

ronment (abiotic environment), but also to other species (biotic environment), that causes convergence. According to this idea, species in a given location co-evolve in such a way that they minimize negative effects (e.g., competition and predation) and maximize positive effects (e.g., predation and mutualism) of other species. If the physical environment dictates the set of species that communities will eventually contain as a result of such co-evolution, then one would expect convergence between communities sharing similar environmental conditions. Although this idea is difficult to demonstrate empirically, there are a limited number of well-described examples from islands, including Hawaiian *Tetragnatha* spiders (Fig. 2) and Caribbean *Anolis* lizards (Fig. 5).

Convergence may also be observed in species diversity. Different locations, despite having independent evolutionary histories, may converge in diversity if these locations share the same physical environment. For example, the diversity of bird species in eastern North America and

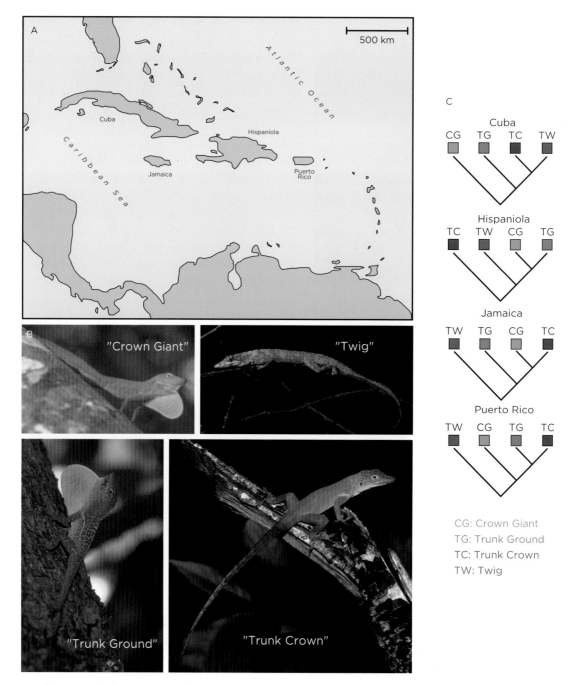

FIGURE 5 Convergence in the composition of the *Anolis* lizard community on the four islands of the Greater Antilles. All islands (A) have the same set of four ecomorphs that occupy different ecological niches (B). But the same ecomorphs found on different islands are different species (the species shown in (B) are from Jamaica), and the evolutionary relationships of these species indicate that different ecomorphs have evolved on these islands independently (C). Adapted with permission from Losos *et al.* 1998, and Losos *et al.* 2004. Photographs (B) by Jonathan B. Losos.

CONVERGENCE ON ISLANDS VERSUS CONTINENTS

Although the concept of ecological and evolutionary sorting is applicable to both islands and continents, islands sometimes provide clearer examples. This is because islands, particularly remote ones, often lack some of the major phylgenetic groups of organisms that are common on continents. On these islands, the ecological niches that would be occupied by these organisms were left empty and thus available for other, phylogenetically unusual (by continental standards) organisms to evolve to occupy, resulting in clearly recognizable convergent evolution. For example, when woodpeckers are absent, as is often the case on isolated islands, other species may evolve to fill the woodpeckers' niche and consequently become convergent with one another.

Islands, particularly when remote and small, provide illustrative examples of convergence at the community level as well. Besides the empty niches that promote convergent evolution, islands are characterized by clear boundaries (shorelines) within which the members of the biota—at least terrestrial ones—are confined to live and interact with other members. Species interactions in such confined space may cause evolutionary and ecological sorting to operate more strongly than on continents, leading to strong convergence. Moreover, when multiple similar islands exist within an archipelago, these islands are essentially replicated ecosystems. When these islands are found to have the same set of ecomorphs that occupy different ecological niches, community convergence is indicated. Furthermore, when a given ecomorph on one island is found to have evolved independently of the same ecomorph on other islands, evolutionary sorting is indicated as a cause of the convergence. Hawaiian *Tetragnatha* spiders (Fig. 2) and Caribbean *Anolis* lizards (Fig. 5) are well-described cases of evolutionary community convergence.

SEE ALSO THE FOLLOWING ARTICLES

Adaptive Radiation / Lizard Radiations / Spiders / Succession

FURTHER READING

Cody, M. L., and H. A. Mooney. 1978. Convergence versus nonconvergence in Mediterranean-climate ecosystems. *Annual Review of Ecology and Systematics* 9: 265–321.

Fukami, T., T. M. Bezemer, S. R. Mortimer, and W. H. van der Putten. 2005. Species divergence and trait convergence in experimental plant community assembly. *Ecology Letters* 8: 1283–1290.

Gillespie, R. G. 2005. The ecology and evolution of Hawaiian spider communities. *American Scientist* 93: 122–131.

Losos, J. B., R. E. Glor, J. J. Kolbe, and K. Nicholson. 2006. Adaptation, speciation, and convergence: a hierarchical analysis of adaptive radiation in Caribbean *Anolis* lizards. *Annals of the Missouri Botanical Garden* 93: 24–33.

Schluter, D., and R. E. Ricklefs. 1993. Convergence and the regional component of species diversity, in *Species diversity in ecological communities: historical and geographical perspectives*. R. E. Ricklefs and D. Schluter, eds. Chicago: University of Chicago Press, 230–240.

Wilson, J. B. 1999. Assembly rules in plant communities, in *Ecological assembly rules: perspectives, advances, retreats*. E. Weiher and P. Keddy, eds. Cambridge: Cambridge University Press, 130–164.

COOK ISLANDS

TEGAN HOFFMANN

T. C. Hoffmann and Associates, LLC, Oakland, California

Located in the heart of the tropical South Pacific between the Society Islands to the east, Tonga and Samoa to the west, and New Zealand about 3500 km to the southwest, the Cook Islands are part of Polynesia. The nation's 15 islands range from the towering, volcanic Rarotonga, the country's largest island and capital, to the northern solitary atolls. Divided into a Northern Group of six low-lying, sparsely populated atolls and, 1000 km away, a Southern Group of nine fertile, volcanic, and elevated islands, where most of the population (estimated at 21,000 in 2007) lives, the Cook Islands lie south of the equator and slightly east of the International Date Line. Although the Cooks' exclusive economic zone (EEZ) covers a total of 1.8 million km^2 of ocean, their total land area is only 240 km^2 (Table 1). The Cook Islanders' language and culture closely resemble those of other Polynesian islands, including Tahiti and Hawai'i.

GEOLOGY AND GEOMORPHOLOGY

The Cook Islands, which rest on the Pacific plate and move northwest at about 10 cm per year, are composed of many types of oceanic islands. The Northern Group contains mostly classic Pacific atolls—older than the Southern Group, which is composed of low-lying, coral atolls, which are groups of small outer reefs and islets encircling a lagoon. Except for the Penrhyn atoll, all the islands in the Northern Group are located at the margin of the Manihiki Plateau, an area of shallow water and thick crust that formed about 110 million years ago.

The ages of the Cooks do not fit within a single hotspot model. Pleistocene volcanism on Aitutaki may have originated from the same hotspot that formed Rarotonga, because the two islands lie within a 300-km diameter of volcanism that defines hotspots. Alternately, Pleistocene volcanism may have been linked to the crustal loading/flexure

TABLE 1
Cook Islands

Island	Area (km²)	Elevation (m)	Geography
Northern Group			
Manihiki	5.2	—	Atoll of 40 islets around a lagoon
Nassau	1.2	—	Only one in group without a lagoon
Tongareva/Penrhyn	9.8	—	Largest atoll of group, ringed by coral
Pukapuka	7	4	Remote atoll of 3 islets and sandbank
Rakahanga	4	—	Rectangular atoll of 7 islets
Suwarrow	0.4	3	atoll surrounding lagoon
Southern Group			
Aitutaki	106	119	3 volcanic, 12 coral islets
Atiu	27	72	Raised volcanic island, ringed by makatea
Mangaia	71	169	Raised volcanic island, ringed by makatea
Manuae	22	5	Uninhabited atoll, 2 islets
Mauke	18.4	30	Raised atoll, ringed by makatea
Mitiaro	22.3	—	Volcanic origin, ringed by makatea
Palmerston	2.6	6	Only true atoll in Southern Group
Rarotonga	67	650	Youngest of Cooks, ringed by lagoon
Takutea	1.3	5	Uninhabited makatea; wildlife sanctuary

that formed Rarotonga. The volcanic edifices of the Northern Group are thought to be the same age as the rest of the plateau, although they have not been age dated (Fig. 1).

The Southern Group, which constitutes nearly 90% of the land area of the Cooks, is actually a continuation of the Austral Islands chain of French Polynesia, lying along the same northeast–southwest fracture of the Earth's crust. This chain was created by molten lava that escaped through the Earth's crust. The Southern Group forms two linear chains that converge to the southeast on the volcanically active Macdonald Seamount, a supposed hotspot volcano. The eastern chain of the Southern Group consists of Aitutaki, Manuae, Takutea, Atiu, Mitiaro, and Mauke, whereas the western chain includes Palmerston, Rarotonga, Mangaia, and seamounts to the southeast.

Great geological variation exists among the Southern Group, where five types of islands are present. Atiu, Mauke, Mitiaro, and Mangaia, classified as makatea islands (the name given to the coastal raised reef), formed as coral reefs gradually encircled volcanic islands. The central volcanic core of these makatea islands range in age from about 20–17 million years ago for Mangaia, 12 million years for Mitiaro, 10–7 million years for Atiu, and 6 million years for Mauke. Tertiary reef limestone of unknown age covers the volcanic core. As the Pacific plate moved on, the volcanic cones sank over time. The four makatea islands' remarkable topography shows signs of karstification and cave development. In the last 100,000 years, young coral reefs developed around these islands; the remains of the older volcanic rocks can be identified in the central parts of the islands as swampy depressions formed by karst erosion in the middle of the limestone plateaus. These depressions are very significant for the supply of fresh water they provide.

The main island of Rarotonga, the youngest of the group, is a high volcanic island surrounded by narrow fringing reefs from the Holocene era that lack substantial raised reef deposits. Rarotonga formed about 2.3 million years ago, when the sea floor buckled in the area nearby. The crustal loading from the formation of Rarotonga uplifted Mauke, Mitiaro, Atiu, Manuae, and Takutea, raising these islands above sea level and exposing the fringing reefs, which eventually became the rocky coastal areas known as makatea. Rarotonga's exposed coral reefs, which became fossilized coral limestone, created the rugged coastline and plateaus, whereas the interior is composed of deeply incised volcanic peaks.

The triangular-shaped Aitutaki is unique in that it is considered an "almost atoll," or an island in a transitional stage of development (Fig. 2). An atoll until volcanic activity formed the main island, it rises 4000 m from the ocean floor and consists of a remnant volcanic cone surrounded by fringing and barrier reefs that surround a shallow lagoon.

Takutea and Nassau are reef islands, or sand and gravel cays that emerged on top of small reef platforms. The wildlife sanctuary of Takutea, a low-lying carbonate island without a lagoon and with a single small reef top, may be a remnant of an atoll annular rim that was destroyed by a submarine slide.

CLIMATE AND WEATHER

The warm, tropical climate of the Cook Islands resembles that of the Hawai'ian Islands, which lie about the same distance north of the equator as the Cooks are south, but with the seasons reversed: the wettest, hottest months occur between December and March, with the cyclone season starting in November, whereas the cooler months are June through September. Because of the longitudinal spread of the Cooks between 8 and 23° S, the climate is not uniform on each island.

The greatest factor moderating the climate is the presence of trade winds. The Northern Group lies in the path of the persistent trade wind belt of the South Pacific. The Southern Group falls within the subtropical high-pressure zone of the South Pacific, which creates a semi-permanent anticyclone circulation east of the Cook Islands.

Weather patterns diverge greatly even within individual islands. Rarotonga's interior, for example, experiences heavy rain much of the year, whereas its coast remains sunny. Generally, the larger, high islands have wet summits but somewhat drier leeward sides.

The Cooks' variable rainfall, which averages about 350 cm/year in the Southern Cooks and 115 cm/year in the Northern Cooks, has also led to unpredictable, and often insufficient, water supplies. The quality of groundwater depends upon the elevation of the water table above sea level, but the evaluation has not been systematically determined throughout the islands. Groundwater is better from the aquifers in thick volcanic rocks at higher elevations than from low-lying makateas.

BIOLOGICAL RESOURCES

Despite the great range of biomes and habitats, biological diversity for both terrestrial and marine ecosystems is relatively low throughout the entire Cook Islands. Generally speaking, the Cooks' different geological island types contain varying degrees of natural biodiversity. The number and abundance of terrestrial species is greatest on the highest islands, followed by the atoll islands, then the sand cays, with the least on the uplifted islands. In the Northern Group, infertile, calcareous sands and soils have limited the evolution of terrestrial flora and fauna. The Southern Group, as a whole, with its more fertile soils derived from basaltic flows, boasts more varied terrain and habitats, and thus greater diversity. Rarotonga, followed by Mangaia, Atiu, Mauke, and Mitiaro, contain the richest terrestrial flora and fauna. The variety of vegetation found throughout the Cooks reflects their ecological diversity. Of the 538 known angiosperm species in the Southern Group, 4% are endemic, and 130 plant species are native. The endemic Mitiaro fan palm (Iniao) is found nowhere else in the world. There are many biogeographic affinities with islands to the west (Tonga), east (Society Islands), and north, (Hawai'i). For example the Hawai'ian *Tetramolopium* has 11 species that are differentiated by a combination of morphological traits and adaptations to different ecological zones, and derivation of a species that originated after long-distance dispersal to the Cook Islands. The Cooks' biomes range from the semi-deciduous forests on Pukapuka to the disturbed, lowland

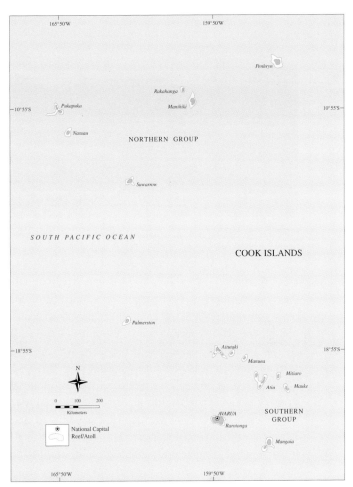

FIGURE 1 Map of the Cook Islands.

rain forests on Mauke and the montane rain forest in central Rarotonga, which may be representative of the original mountain forests of the Cook Islands (Fig. 3). The islands' habitats also include ample beach forests, particularly on the atoll and reef islets; extensive and common scrub, especially on Rarotonga and in frequently burned

FIGURE 2 Aitutaki Motu, one of the small islands of Aitutaki, the "almost atoll." (Photograph courtesy of Tegan Churcher Hoffmann)

FIGURE 3 The forests and hills of Rarotonga. (Photograph courtesy of Tegan Churcher Hoffmann)

areas; grassland on the drier islands; extensive and important seabird rookeries; a permanent lake with endemic eels in the center of Mitiaro, and Lake Tiriara on Mangaia; and, a mountain stream on Rarotonga. Rarotonga has one of the last pristine native forests in Polynesia, which grows the endemic Cook Islands' tree *Homalium acuminatum*. These forests are some of the best remaining examples of primary montane rain and cloud forest in eastern Polynesia. An estimated 60% of the total land area in the Cook Islands is still covered by forest. This is relatively high forest coverage compared to other developing and Pacific Island nations.

In contrast to the Cooks' ample native vegetation, the nation's only native mammal is the Pacific fruit bat (*Pteropus tonganus*), which resides on Rarotonga and Mangaia. The Cook Islands represent its easternmost range, as it is also found in Fiji, Samoa, and Tonga. The herpetofauna of the Cook Islands consists mainly of species found throughout the tropical Pacific and generally includes species transported by humans. However, there is one species, *Emoia trossula*, which has a restricted range (Cooks, Fiji, Tonga) within central Polynesia. The Cooks have six endemic landbirds, four of which are on the IUCN Red List of Endangered Species. The Rarotonga flycatcher is listed as "endangered" (and is conservation maintained), whereas the other four are listed as "vulnerable." The blue lorikeet is a "vulnerable" species with a decreasing native range in French Polynesia; it presently flourishes on Aitutaki. Additional species endemic to the Southern Cooks include the fruit dove, starling, Rarotonga flycatcher, Atiu swiftlet (which breed in caves), warbler subspecies, and Tahitian lorikeet. The Northern Cooks, along with the Phoenix and Line Islands, contain the principal breeding areas of seabirds (and probably sea turtles) for the Central Pacific; rookeries house thousands to millions of birds.

The marine environment harbors a far greater diversity than the Cooks' terrestrial habitats, though, as in the land environment, relatively little development of new species has occurred (Table 2). The Cooks' surrounding waters have abundant, diverse marine species. Overall, the waters surrounding all of the 15 islands boast much greater biological diversity than their terrestrial habitats. The lagoonal and nearshore marine biodiversity is greatest off the atolls of Palmerston and Manuae, and the almost-atoll of Aitutaki. However, in terms of marine biodiversity, the Cook Islands are at the lower end of a west-east gradient of marine diversity in the Pacific. Habitats include freshwater marshes on Mangaia, Rarotonga, Mauke, Mitiaro, and Atiu; tidal salt marshes in Ngatangiia Harbour, Rarotonga; extensive sea turtle nesting areas; algal beds in lagoon bottoms and reef flats; animals in sediments; algal reefs; windward and leeward atoll reefs on Manuae and Palmerston; barrier reefs on Aitutaki; fringing reefs; lagoon reefs; beach; and open lagoons on Aitutaki and Palmerston and a closed lagoon on Manuae.

Several endemics are rare or survive in restricted habitats, such as Rarotonga Garnotia-Grass, Te Manga Cyrtandra, and the Mitiaro fan palm (Iniao). The Cooks have no mangroves or seagrasses, however. Large mammals, including

TABLE 2
Marine Biodiversity in the Cook Islands

Category	Numbers of Species	Native	Comment
Seaweeds	32	Native	—Probably 150
	1	Introduced	—to Aitutaki; failed to establish
Flowering plants	0	N/A	—No mangroves, no sea grasses
Mammals	7	Native	—Positive sightings; several others likely
Reptiles	3	Native	—Two marine turtles and the yellow-bellied sea-snake
Fish—bony	552	Native	—Includes 85 deep-bottom or pelagic species
Sharks and rays	23	Native	—3 rays
Shellfish	390	Native	—304 gastropods, 74 bivalves, 12 others
	4	Introduced	—Button trochus and 3 species of *Tridacna* giant clam
Crustacea	100	Native	—Includes 35 shrimps
Echinoderms	50	Native	—Includes 20 sea cucumbers
Worms	51	Native	—Covers "wormlike" animals; underrepresented

SOURCE: Natural Heritage Project.

sharks and humpback whales, visit Rarotonga and other islands every year in August and September; more than a dozen whales travel from Antarctica to mate and calf. Fifteen species of sea cucumbers inhabit the waters around Rarotonga and Aitutaki, as do over 200 species of crabs and the bright blue starfish. The lagoons are also biologically productive; Manihiki's lagoon has abundant blacklip pearl oysters, a situation that renders pearl-shell farming an important commercial activity.

The greatest of the Cook Islands' biological treasures are its coral reefs. Similar to other reefs throughout the South Pacific, the Cooks contain three different types of reefs. Fringing reefs—narrow, near-horizontal reef flats—border platform islands and are found around Rarotonga, Aitutaki, and the makatea islands. They are probably Holocene in age, although Pleistocene reef-flat remnants have been identified at Rarotonga, which is completely surrounded by a fringing reef. Aitutaki still has remnants of a subsided volcano in the center. Barrier reefs characterize the atolls (as well as Aitutaki), where they form annular rims that enclose their respective lagoons. Lagoon patch reefs, which rise from the lagoon floor, are isolated, steep-sided reefs and are found in Suwarrow, Palmerston, Manuae, Manihiki, and Penrhyn. Fifty-seven species of scleractinian, hard corals, have been identified in the Cooks.

SETTLEMENT, HISTORY, AND CULTURE

Between approximately AD 500 and 800, Polynesians from Tonga, Samoa, Tahiti, the Marquesas, and the Society Islands left their overpopulated islands and migrated to the uninhabited Cook Islands, where they lived a communal, subsistence-based lifestyle.

The Cooks remained culturally isolated until the first great age of exploration, when European ships began to chart new trading routes around the world and colonize unfamiliar territories. Perhaps the most famous of the Cooks' early European guests was British Captain James Cook, who visited many of the islands in 1773 and 1779 and called them the Hervey Islands. A Russian cartographer named the Southern Group the Cook Islands in 1824. Later, traders in search of timber, pearls, and coconut visited the islands and then carried their bountiful resources around the world.

The Cook Islands became a British protectorate in 1888. New Zealand annexed the Cooks in 1901, but the islands did not become self-governing in free association with that country until 1965. In contrast to other Pacific islands, the Cooks enjoy universal suffrage, democratic government, and an independent Parliament.

As with other South Pacific island nations, the Cooks' geographic isolation, lack of natural resources, unpredictable climatic events, and weak infrastructure hamper its economic well-being. The domestic market is small and independently unsustainable. Tourism, based in Rarotonga, leads the Cooks' economy and accounts for nearly 40% of the nation's GDP, which in 2005 reached US$183.2 million. Agricultural exports of crops such as copra and citrus fruit are another important part of the islands' economic base.

The reefs, reef flats, and lagoons also play an important role in the Cook Islands' economy, supplying almost 90% of the protein needs for Northern island inhabitants and about 60% for the Southern population. Over the past five years, marine commodities have accounted for more than 70% of the total value of exports. Cultured black pearls from the remote northern atolls of Manihiki and Penrhyn are the most valuable marine export. Other exported marine goods included trochus, ornamental aquarium fish, and fresh fish. The sale of handicrafts and offshore banking, which started in the 1980s, contribute to the economy as well.

ENVIRONMENTAL ISSUES

Although native forests originally covered many of the islands, human activities have significantly affected these areas, particularly in the lowlands, where the planting of coconut palms for copra has supplanted native forest. Furthermore, early Polynesian and European contact introduced alien plant species that are now prevalent throughout the islands. Taro and yam are non-native subsistence crops. Other species, including rats, pigs, dogs, goats, horses, and cattle have all been introduced, with varying (but most often negative) effects on endemic and native species. Invasive species have also changed the islands' make-up. Although English settlers introduced food and ornamental plant types, they also imported plants to improve the soil for agricultural production. Such plant species include mimosa, Desmodium, kudzu (*Pueraria montana* var. *lobata*), the Brazilian lucern (*Stylosanthes guianensis*), and others. The commonly seen mynah bird, a species introduced to Rarotonga in 1906 to control coconut stick insects, has caused great damage to fruit trees and has driven native birds into the mountains. Rats, originally castaways on trading vessels, similarly interfere with the nesting of the endemic Mangaia kingfisher.

Close to 20% of the Pacific's coral reefs have been destroyed, and almost two-thirds are under threat. The degradation of coral reefs, although far from the only environmental problem on the islands, is typical of many of the environmental concerns that plague the Cooks. On Rarotonga, the coral reefs not only have deep cultural and spiritual significance, but also provide food to the human population (Fig. 4). The reefs also promote tourism, the Cooks' economic backbone. The decline of coral reef health results in part from natural factors, including bleaching from El Niño-Southern Oscillation events (which can lead to raised sea temperatures, although many corals seem to be heat tolerant). Human activities, overall, have exerted a far more negative impact, however. Crown-of-thorns starfish (COTS) outbreaks and algal overgrowth, for example, can destroy a coral reef. These are generally caused by increased nutrient levels that result from agricultural runoff. The state of the coral reefs, compounded with other declining species and a polluted land and marine environment, reveals many of the issues that the Cooks must address in order to remain environmentally sustainable.

The Major Issues

Solid waste management is a major problem in the Cook Islands. All 12 inhabited islands lack proper solid and liquid waste disposal systems. As a result, much of the islands' "marginal" land, or that considered undesirable for agriculture or residential use, is seen a potential solid waste disposal site, to the detriment of the ecologically sensitive wetland or foreshore areas.

Unsustainable land use practices, from real estate development to agriculture and tourism, have also negatively affected delicate ecosystems. Coastal tourist development has altered the foreshore areas and the natural drainage systems of the wetland areas. On Atiu, Mangaia, and Mauke, the extensive cultivation of the sloping lands and burning of fernland escarpments produce much of the soil that blocks the underground water outlets going through the makatea. Consequently, the land has become more difficult to cultivate. Pineapple plantations have destroyed sloping lands as well, and the extensive mining of beach sand causes coastal retreat on Rarotonga.

Human activities have also degraded coastal marine resources. Overfishing generally poses few problems, except on Rarotonga and in the Aitukai lagoon, where uncontrolled subsistence fishing puts stress on the environment. Ornamental fishing, however, which entails breaking coral to remove fish and using chemicals to chase them out, goes largely unmonitored and is very destructive.

Terrestrial runoff affects both the land and the sea. Pollutants from red soil, sewage seepage, agricultural pesticides, and fertilizers have severely degraded both the Aitutaki lagoon and the Rarotonga lagoon, which conservation reports have highlighted since 1975. Increased organic and nutrient levels in the ocean arising from these sources alter the delicate ecosystem and can lead to the death of coral reefs.

A Growing Awareness

Despite the issues that still need to be effectively addressed, the Cook Islands have initiated environmental awareness and conservation projects in many areas, including the protection of biologically important land and waters.

Ra'ui marine reserves have allowed for the rejuvenation of natural resources. In the Ra'ui marine reserve (Fig. 5), a traditional conservation system, village chiefs manage protected areas. Thirteen marine protected areas exist around Rarotonga and account for 8% of the reef circum-

FIGURE 4 Women gleaning for shellfish, Rarotonga. (Photograph courtesy of Tegan Churcher Hoffmann)

FIGURE 5 The Ra'ui, traditional system of management and MPA reserve, Rarotonga. (Photograph courtesy of Tegan Churcher Hoffmann)

ference around the island. Monitoring in 2002 showed that fish stocks have recovered in these protected areas. But poor sewage treatment, which has resulted in high nutrient levels in the lagoon and prevented the recovery of corals in these protected areas, still poses a problem. However, the Cooks' government is drafting regulations to ensure that effective sewage treatment is implemented for commercial businesses and homes.

The Cooks have other protected areas as well. The unpopulated Takutea has been a wildlife sanctuary since 1903; traditional Atiu leaders manage the island and control burgeoning ecotourism. The island is the most important seabird breeding site in the Southern Group, with Pacific-wide and global significance as a red-tailed tropicbird site. The island also supports the brown booby, red-footed booby, great frigatebird, and masked booby—four species that breed nowhere else in the region. Takutea is also a significant turtle nesting site. The Cooks' largest national park, Suwarrow Atoll National Park, established in 1978, protects the atoll's marine resources and seabird breeding sites. On Rarotonga, the Takitumu Conservation Area protects upland forest and the rare Rarotonga flycatcher. In 2001, the Cooks established a 2-million-km^2-wide whale sanctuary around the islands.

In addition to designating conservation areas, the Cook Islands have also signed on to many international conventions dealing with deforestation, vegetation, marine resources, biosafety, animals, and other sensitive environmental issues. However, many needs exist in order to improve the Cooks' environment: improved utilization of offshore marine resources and the development of mariculture; implementation of better policies and legislation for the sustainable management and conservation of natural resources; improved public awareness about existing environmental regulation; and better planned infrastructure development at the foreshore, for example.

SEE ALSO THE FOLLOWING ARTICLES

Fish Stocks/Overfishing / Marine Protected Areas / Peopling the Pacific / Reef Ecology and Conservation / Sustainability / Tonga

FURTHER READING

Hein, J. R., S. C. Gray, and B. M. Richmond. 1997. Geology and hydrology of the Cook Islands, in *Geology and hydrology of carbonate islands*. H. Vacher and T. Quinn, eds. Amsterdam: Elsevier Science B.V., 503–535.
Keller, N., and T. Wheeler. 1998. *Rarotonga & the Cook Islands*. Oakland, CA: Lonely Planet Publications.
Stoddart, D. R., and C. S. G. Pillai. 1972. Coral reefs and reef corals in the Cook Islands, South Pacific, in *Oceanography of the South Pacific 1972*. R. Fraser, ed. New Zealand: National Commission for Unesco, 475–483.
Tangianau, T. 2004. *Cook Islands national environment strategic action framework 2005–2009*. Cook Islands: National Environment Service.
Thompson, G. M., J. Malpas, and I. E. Smith. 1998. Volcanic geology of Rarotonga, southern Pacific Ocean. *New Zealand Journal of Geology and Geophysics* 41: 95–104.

CORAL

DAPHNE G. FAUTIN
University of Kansas, Lawrence

ROBERT W. BUDDEMEIER
Kansas Geological Survey, Lawrence

The term "coral" is neither scientific nor precise. This article focuses on those cnidarian polyps capable of secreting skeletons that contribute to formation of reefs, which can produce or play a role in producing islands. These skeletons consist of the mineral calcium carbonate, deposited in the crystal form aragonite. These corals are among the few animals that shape the environment in a profound and large-scale manner—that are biogeomorphic agents.

DEFINITIONS OF "CORAL"

The many meanings of the word "coral" occupy a full page of the *Oxford English Dictionary*. The first entry refers to the skeleton made by some animals of the phylum Cnidaria, and the second refers to the animals themselves. The subsequent five *OED* entries relate to objects that are precious or bright red, which is one of the colors of the skeleton of one species of cnidarian, the so-called precious coral (*Corallium rubrum*). Actually, its skeleton comes in shades ranging from that deep red through pink to white. The *OED* taxonomy of corals is based on skeleton color—its definitions refer also to white, black, blue, and yellow coral.

In zoological taxonomy, "corals" fall into classes Anthozoa and Hydrozoa of phylum Cnidaria (Fig. 1). Animals currently termed "corals" belong to both subclasses of Anthozoa, Octocorallia and Hexacorallia, but only some members of both are considered "corals"—precisely which depends on how the term is used. Among hydrozoans, "corals" belong to the order Anthoathecata. Fig. 1 shows the current taxonomic arrangement of Cnidaria and which of its members contribute to building coral reefs.

The corals discussed here have skeletons of various colors; they live in shallow water of the tropics and subtropics, and secrete substantial skeletons. Most of these

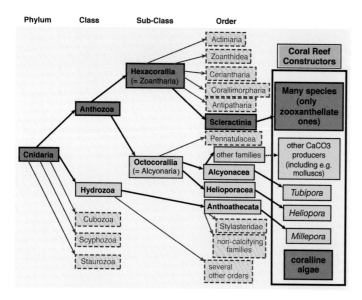

FIGURE 1 Taxonomic hierarchy of phylum Cnidaria, indicating groups containing members that contribute significantly to building coral reefs. Major contributors of calcium carbonate to reefs are indicated in orange, minor ones in yellow. Taxa in boxes bordered with solid lines contain at least some members that contribute significantly to coral reef formation; those in boxes bordered with dashed lines do not. Boxes offset to the right under "orders" are taxa of lower rank. The phylum currently contains as many as five classes, three of which (Anthozoa, Hydrozoa, and Scyphozoa) are of long standing. Cubozoa, which is now widely accepted, and the recently proposed Staurozoa, contain animals that had been considered to constitute orders of Scyphozoa.

animals are colonial, most live attached to a solid surface, and most belong to hexacorallian order Scleractinia. Their restriction to shallow water is due to the possession by these animals of zooxanthellae. Fig. 2 illustrates the anatomy of a scleractinian coral and the skeleton it produces. Zooxanthellae fix carbon through photosynthesis; some of the resulting carbohydrates are leaked to the host animal, which uses them for energy. For reasons that are not entirely understood, the skeletons of zooxanthellate corals are typically denser and more massive than those of azooxanthellate corals. The non-scleractinian cnidarians that are important contributors to some reefs are also zooxanthellate. Among these are members of the hydrozoan genus *Millepora,* the so-called fire corals, the skeleton of which is white, like those of scleractinians. The other two zooxanthellate non-scleractinians that can be important contributors to reef formation are members of Octocorallia: *Heliopora* has a blue skeleton (thus, its common name "blue coral"), and *Tubipora* (the organ-pipe coral) has a red skeleton.

Nearly half the approximately 1600 species of scleractinian corals lack zooxanthellae. Some live in deep water and some are solitary (as opposed to colonial). Most have fragile skeletons by comparison with those of zooxanthellate scleractinians, and few leave large deposits of calcium carbonate after death. Because the deep corals do not contribute to building reefs associated with islands, we will not deal with them. "Black corals," which constitute hexacorallian order Antipatharia, typically live in deep water and have skeletons largely organic in composition and forming a core surrounded by living tissue. These attributes are also characteristic of many octocorals of the order Alcyonacea. Among members of Alcyonacea are the sea fans, sea whips, and "soft corals." The last is an ambiguous term without taxonomic meaning. Polyps of hexacorallian order Corallimorpharia are occasionally erroneously referred to as "soft corals" because they are virtually identical to "hard corals" (scleractinians) except for lacking a skeleton. Most members of Alcyonacea, including the so-called soft corals, possess calcium carbonate sclerites, which may be fused to various degrees. In *Corallium rubrum,* the polyps surround a core of fused sclerites, whereas in *Tubipora,* the skeleton of fused sclerites surrounds the polyps. Many alcyonacean species live on coral reefs; they are important to the ecology particularly of Caribbean reefs. However, except for *Tubipora,* they do not contribute importantly to reef structure, although their sclerites may fill cavities in the reef after the colony dies. Complicating an already ambiguous terminology, many recent scientific publications dealing with alcyonaceans refer to them simply as "corals."

At least some members of two azooxanthellate, colonial families of Hydrozoa form calcified skeletons and so have been termed "corals." Members of Stylasteridae live from shallow to moderate depths, and from polar seas to the tropics. Their skeletons are commonly pink to purple in color, are not robust, and do not contribute much calcium carbonate to reefs on which they occur. A very few species of Hydractiniidae calcify; they live on shells of molluscs, commonly those inhabited by hermit crabs, and are therefore small.

The adjective "coralline" typically applies to a crust of calcium carbonate, especially one reddish in color. It commonly refers to species of calcifying algae, either erect or encrusting, that are common at both temperate and tropical latitudes, on both inorganic and organic substrates.

The term "coral" is also used loosely in geology and engineering to refer to calcium carbonate rock (either solid limestone or sand and gravel) originally formed by corals and/or coralline algae.

CORAL BIOLOGY

In addition to consuming photosynthate, zooxanthellate corals also capture animal prey using nematocysts. Upon being triggered by an appropriate chemical and/or mechanical stimulus, the tubule coiled within the nematocyst everts (turns inside out). The tubule may wrap around or penetrate the prey, depending on the type of nematocyst. Some deliver a toxin—nematocysts are the source of jellyfish stings.

All members of a scleractinian coral colony are morphologically identical, although the terminal polyp of a branch of a coral of the genus *Acropora* (Fig. 3) is larger and physiologically unlike the others, and polyps of some species that come into contact with members of other scleractinian species may grow long tentacles specialized for aggression. In hydrocorals, polyps are dimorphic, with some serving to protect the colony (the cause of the sting that is the source of the name "fire coral") and the others to perform functions such as catching food; gametes are created by small medusae (jellyfish) that have a brief free-living existence in some taxa but are not released in others. Some octocoral colonies have dimorphic polyps (one type devoted to circulating water and food through the colony), but most are monomorphic. In some coral species, members of a colony produce gametes of one sex or the other, and in some, both eggs and sperm are produced.

A coral colony ultimately arises from a single individual polyp, produced through sexual reproduction. Typically, fertilization occurs in the water, where eggs and sperm drift after having been released from parent polyps (or medusae in members of Hydrozoa). The resulting zygote (fertilized egg) develops into a type of larva termed a planula, which matures as it drifts, for days or even weeks. Alternatively, in some corals, eggs are fertilized within the female parent by sperm that are brought in though the mouth, and the larva completes some or all of its development within its mother.

When mature, the scleractinian larva attaches to a suitable substratum, where it secretes a calcium carbonate skeleton around itself. Colony formation is by means of either budding or dividing, depending on the species; these processes produce genetically identical clonemates that remain attached to one another. A piece of a colony that has broken off may manage to attach to the substratum before the living tissue is abraded from it, so even physically separate colonies may be genetically identical.

Basic scleractinian colony form is genetically determined. Massive types form roughly hemispherical lumps, foliose ones form sheets that are typically curved

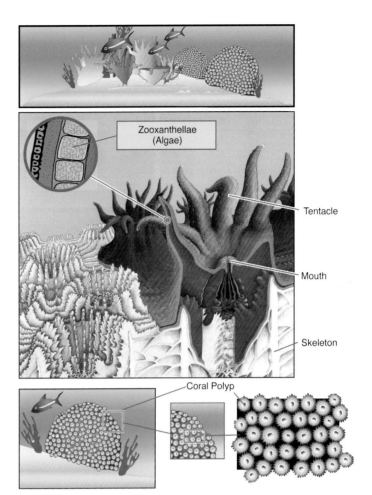

FIGURE 2 A diagrammatic colonial reef-forming scleractinian coral, in distant and close-up views. Living tissue, of the polyp and connecting polyps, covers and secretes the calcium carbonate skeleton. Zooxanthellae live within cells of the polyp and connecting tissue. After the organism dies, the skeleton remains. Adapted with permission from Veron (1986).

and may have polyps only on one surface, and branched ones are treelike. Colony form of particularly the last can be remarkably plastic, with light availability and water action being perhaps its two most powerful determinants. Within a species, with increasing depth, branches are shorter and flatter, until the colony may assume a platelike form. This is presumably due to the fact that at depth nearly all light (essential for the zooxanthellae) comes directly from the sea surface, whereas in shallow water, scattered or reflected light can be important. Within a species, branches tend to be thicker where water motion is greater, and consistently directional water flow can be reflected in colony orientation and polyp position (the extreme in colonial cnidarians is sea fans, which branch in one plane, orienting the colony perpendicular to water flow).

Such variability can make identifying corals difficult. Colonies that experience various microenvironmental regimes may manifest morphological differences so great that separate colonies with features of the extremes might be identified as belonging to different species. However, some coral species are not plastic at all, as evidenced by colonies not changing morphology when transplanted from deep to shallow water and vice-versa. Further complicating the problem of identifying scleractinians is the existence of interspecific and even intergeneric hybrids.

CORAL REEFS

The term "coral reef" is no easier to define than "coral." Among its definitions are geological ones, many of which have to do with net accumulation of calcium carbonate, and biological ones, many of which have to do with complex interspecific interactions. Because our context is islands, our concern is with reefs in which mainly scleractinian corals build the framework. Because those framework producers possess zooxanthellae, they occur only in shallow water. This limitation is consistent with the original definition of "reef," which was a hazard to shipping. Organisms that form reefs (in our sense) (Fig. 4) are confined to water of temperatures typically found in the tropics and in subtropical areas into which waters of tropical origin flow. Water in parts of the tropics where upwelling occurs may not be sufficiently warm for reef development.

Reef-forming corals require firm substrata, as opposed to soft sediments, which prevents their development in some otherwise suitable areas. As larvae and small colonies, they are not good competitors for space (by contrast with their excellent competitive abilities once mature), so reefs typically begin on relatively barren ground, such as

FIGURE 3 Part of a colony of the most speciose genus of reef-forming coral, *Acropora*. Photograph by Daphne Fautin.

FIGURE 4 Scleractinian corals form a reef in the Indo-Pacific. Photograph by Daphne Fautin.

fresh lava flows, land newly inundated by sea-level rise, or surfaces freshly exposed by storms. The classic sequence of island formation involving corals is a reef developing around an island formed by a mid-oceanic volcano. Once volcanic activity ceases, the island gradually sinks and/or erodes, while the flourishing reef expands as rubble from the reef accumulates and is consolidated (this rubble includes skeletons of corals that have died or broken from living colonies, and skeletons of associated organisms such as molluscs and calcifying algae). The original land ultimately disappears, and a ring of reef is left around where the volcano had been—an atoll, which can support newly forming coral islands. This geomorphic sequence, hypothesized by Charles Darwin in the late nineteenth century, was opposed by James Dana, but Darwin's hypothesis was vindicated when drilling on Enewetak atoll in the late 1940s revealed a very thick layer of carbonates, largely scleractinian in origin, capping a basement of basalt.

Other types of reefs (e.g., barrier) occur around larger land masses. For many years, posters produced by the Queensland Tourist and Travel Corporation advertised the Australian Great Barrier Reef as "the world's largest living thing." But it is far from a single entity, and even atolls are not solid rings. Both circularly and linearly arrayed reefs are, in most instances, complexes of individual reefs and sometimes small coral islets separated by channels where corals are sparse or absent. Such islets may be called keys (cays, quays).

There are about ten times more reef-forming species of corals in the Indo-Pacific than in the Atlantic and Caribbean. The area of highest species diversity of reef-forming corals is centered around eastern Indonesia and New Guinea. Organisms of many other taxa inhabit the structure formed by the corals (and contributed to by members of other phyla such as Mollusca, Bryozoa, and Porifera), with many taking refuge in the spaces made by corals, some attaching to dead coral skeletons, and some eating corals or organisms associated with corals. Coral reefs thrive in nutrient-poor ocean waters, so reefs can be patches of high biodiversity in the midst of areas with otherwise low marine biodiversity.

EMERGENT AND DROWNED CORAL REEFS

Sea level has risen and fallen through time; it has been unusually stable during the past five to six millennia, during which human civilization has developed. Corals can grow up to, and even slightly above, mean sea level, as long as the reef organisms can be nourished and stay moist. Although individual corals can extend on the order of 10 to greater than 100 mm per year, reefs as a whole grow more slowly—as much as 14 mm per year, but 10 mm per year is more typical in recent geological time. Corals can grow at considerable depth (100 m or more) if the water is clear enough to transmit adequate light, although growth at depth is less than it is in shallower water.

If sea level falls far enough below the uppermost part of a coral reef, the exposed organisms die, and an island results if the sea does not rise or erode the exposed structure. The surface is essentially flat where the reef had grown up to the surface of the sea, but wave action can pile sand and rubble onto the platform, protecting and further developing the island structure. Gradually, soil develops by decay of organic matter, including drifting materials (ranging from microbes to logs and carcasses); in this soil, terrestrial plants can establish (from propagules that arrive by sea and air). Freshwater, of course, is essential for a terrestrial biota to develop. Rainfall can be trapped in spaces in the porous subsurface structure. Carbonate island groundwater bodies were originally envisioned as simple lenses floating on seawater that filled the spaces in the now-dead reef (they float because freshwater is less dense than seawater); they are now known to be highly irregular in form. Island groundwater characteristics depend on many factors in addition to rainfall such as the nature of the antecedent reef, how firmly the spaces within the reef structure were filled by detritus and carbonate cements, and pumping caused by tide rise and fall.

If sea level rises, new reefs can establish on freshly inundated substratum, if any is available, and the existing reef can extend upward through growth of calcifying marine organisms and infilling with carbonate sediments. The land that forms the island is unlikely to rise. If sea level rises faster than calcifiers can extend the reef upward, the average depth of the reef will increase, and growth will slow because of the attenuation of light. Ultimately, the reef may become so deep that zooxanthellae cannot live, the reef-building organisms are eliminated, and the reef is considered drowned. Sea level must rise more than 10 m in a very few centuries to drown a healthy reef that was initially near the surface. Even if sea level rises more slowly than corals can grow, growth will slow on the lower part of the reef as water deepens over it.

Plate tectonics also plays a role in this dynamic. Mid-ocean volcanoes are often associated with hotspots that erupt through the Earth's crust; chains of islands are the result of a hotspot over which the Earth's crust has moved. For example, the Hawai'ian chain is on a plate moving to the northwest, which explains the current volcanic activity, and youngest island, at the southeast end of the chain. To the northwest of the main chain of seven Hawai'ian islands inhabited by humans are the Northwestern (Leeward) Islands, all of which are near sea level and some of which are mainly under water; northwest of them are the Emperor seamounts, the tops of which are successively deeper as the plate carrying them has moved northwest and the former islands have continued to sink. Remnants of coral reefs remain on some of these seamounts (the flat-topped ones are called guyots)—evidence of their having once been in shallower, warmer waters. The "Darwin point" is the name given to the place where the water becomes too cold and deep for upward coral growth to compensate for seamount sinking, so that reef structure can no longer be maintained.

THE FUTURE OF CORALS AND CORAL REEFS

During most (but not all) of geological history, a feature of shallow marine seas has been reefs formed by animals, typically with mineralizing plant associates. However, not all these reefs were built by corals. During the Middle and Late Cretaceous of the Mesozoic era, most reefal structures were built by rudist bivalves, although scleractinians preceded them (from the Jurassic, if not mid-Triassic) and succeeded them (from the Eocene or so until now). Early in the Paleozoic era, there were archeocyathids (invertebrate animals of uncertain affinity but possibly sponges), and after some time when reefs apparently were absent, there were stony sponges and tabulate corals (cnidarians not clearly related to any currently extant); then, following another reef-free stretch of time, reefs were formed by bryozoans, sponges, and some organisms of unknown affinity. The causes of the demise of most such reefal systems and the shifts in reef-formers are unclear.

Although some individual reefs may be flourishing, as an ecosystem, coral reefs are currently in strife. In 2004, it was estimated that perhaps as little as 30% of the world's coral reefs were thriving and in no imminent danger: As much as 20% had been effectively destroyed, 24% were under imminent risk of collapse because of human pressures, and a further 26% were threatened in the longer term.

Coral reefs are resilient in the face of acute stress, and they persist in the face of chronic stress, but reefs appear unable to cope with both types of stress concurrently—and both are increasing as a result of human activities. For example, the reefs on the north coast of Jamaica have not recovered in the nearly 30 years since Hurricane Allen, although it is likely they had been struck by previous storms. Instead of coral reefs, that area now has large stands of macroalgae, which were essentially absent before the storm. The change has been explained by the inability of coral larvae to establish on the surfaces where reefs had grown because the macroalgae, which previously would have been eaten by herbivorous fishes, could grow thanks to overfishing by humans. Human population growth has also been a factor. As land has been converted from forests to agricultural or urban uses, runoff of sediment has increased, smothering small corals, and runoff of nutrients has increased, encouraging growth of macroalgae.

The zooxanthellate scleractinians (and other corals important to reefs in particular areas) that are the dominant framework builders of reefs are themselves under stress. The coral–zooxanthellae symbiosis has narrow tolerances: It can break down as a result of high or low temperatures, light levels, or salinity, or as a result of other chemical changes. When environmental conditions exceed its tolerances, the coral host expels its algal associates, or the zooxanthellae leave of their own volition (it is uncertain which), a process called bleaching because the normally pigmented corals appear ghostly white. Although some bleaching is natural, in recent years bleaching episodes have become more widespread and more frequent; many have been correlated with episodes of anomalously high seawater temperature, often accompanied by high light levels and calm water. Bleaching does not mean death; corals may be able to reacquire zooxanthellae from the environment or by multiplication of algal cells remaining after the bleaching. However, bleaching reduces the coral's viability and, if repeated or prolonged, is likely to be lethal.

It has been suggested that although global warming might kill corals in the shallow water of the equatorial tropics, it could extend the area where water is warm enough for reef development. The projected effect is geographically limited (Fig. 5); there is not much otherwise suitable shallow-water habitat in regions immediately adjacent to present coral reef habitat. More important is the fact that with increasing latitude (and increasing depth), the degree of saturation of calcium carbonate in seawater decreases. This means that carbonate and calcium ions—the essential building blocks of the coral skeleton—are in diminished supply. Thus, although the temperature of water might become more favorable for reef formation at higher latitudes, reefs are less likely to form there because of water chemistry. This limitation will become stronger concurrently with global warming because the rise in the concentration of carbon dioxide in the atmosphere, one factor in driving up temperature, also occurs in surface seawater because of the rapid equilibration of the partial pressure of gases between the two. Adding carbon dioxide to water lowers its pH, and, as a result, the calcium carbonate concentration state. This makes calcification more energetically costly, so coral skeletons are likely to grow more slowly, become thinner, or both, reducing the animals' ability to survive, compete for space, and repair damage.

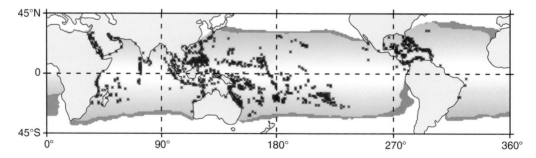

FIGURE 5 Global distribution of coral reefs and potential for their distribution with global warming. Xs indicate where coral reefs currently occur; green indicates areas where the water is a suitable temperature for coral reefs, and blue indicates where water would be within the appropriate range if temperatures were to rise 2 °C. Adapted with permission from Buddemeier et al. (2004).

The concentration of calcium carbonate in surface waters of the equatorial tropics is now 80% of what it was in preindustrial times, and within 50 years, the calcium carbonate saturation state almost everywhere will be lower than that experienced by nearly all the world's reefs as recently as 100 years ago. Accompanying this chemical change will be a temperature increase of 2 °C or more and continued dramatic increases in human population and development on the land adjacent to reefs.

SEE ALSO THE FOLLOWING ARTICLES

Atolls / Darwin and Geologic History / Global Warming / Hydrology / Seamounts, Geology

FURTHER READING

Buddemeier, R. W., J. A. Kleypas, and R. B. Aronson. 2004. *Coral reefs and global climate change.* Arlington, VA: Pew Center on Global Climate Change. [Available online at http://www.pewclimate.org/docUploads/Coral_Reefs.pdf.

Fabricius, K., and P. Alderslade. 2001. *Soft corals and sea fans.* Townsville: Australian Institute of Marine Science.

Stanley, G. D., Jr., ed. 2001. *The history and sedimentology of ancient reef systems.* New York: Kluwer Academic/Plenum Publishers.

Veron, J. E. N. 2000. *Corals of the world* [three volumes]. Townsville: Australian Institute of Marine Science.

Veron, J. E. N. 1986. *Corals of Australia and the Indo-Pacific.* North Ryde, Australia: Angus Robertson Publishers.

Wallace, C. C., and B. L. Willis. 1994. Systematics of the coral genus *Acropora*: implications of new biological findings for species concepts. *Annual Review of Ecology and Systematics* 25: 237–262.

Wells, S., and N. Hanna. 1992. *The Greenpeace book of coral reefs.* New York: Sterling Publishing Co., Inc.

CORALLINE ISLANDS

SEE ATOLLS

CORSICA

SEE MEDITERRANEAN REGION

COZUMEL

ALFREDO D. CUARÓN

Servicios Ambientales, Conservación Biológica y Educación, Morelia, Mexico

Cozumel (~478 km²) is the largest island in the Mexican Caribbean Sea, a site of historical and cultural significance, a popular international resort, and an area of global biodiversity conservation priority. It is located 17.5 km off the northeast coast of the Yucatán Peninsula (20°16′18.2″ to 20°35′32.8″ N; 86°43′23.3″ to 87°01′31.1″ W). It is an oceanic island of coralline origin, which is separated from the mainland by the 400-m-deep Cozumel Channel. Maximum altitude is just over 10 m. The weather is humid tropical, with mean annual temperature of 25.5 °C and mean annual rainfall of 1505 mm.

HISTORY AND ARCHAEOLOGY

The name Cozumel derives from the Maya *Cuzamil* (*cuzam,* swallow, and *luumil,* land), meaning land of swallows. Cozumel was occupied by the ancient Maya at least since 300 BC. It was an important post in the maritime commercial routes of the Maya. It was also an important pilgrimage site, which was visited to adore the moon goddess Ixchel, patroness of childbirth, weaving, sorcery, and medicine. Juan de Grijalva arrived to the island in 1518, making it one of the first sites where the Spaniards landed in what is today Mexico. Hernán Cortés and his crew disembarked in 1519, and from there they continued their journey in the conquest of Mexico. The population declined subsequently, mainly as a result of disease. With the exception of brief British and French pirate incursions or occasional fishermen camps,

FIGURE 1 Cozumel Island, Mexico (NASA Landsat image, April 17, 2001).

the island remained virtually uninhabited for most of the next three centuries. People fleeing from the Yucatán Caste War recolonized Cozumel in 1847.

San Gervasio is the main Maya archaeological site on the island. There are remains of *sacbé,* Maya roads, which connected San Gervasio with other areas on the island. Maya buildings distributed along the eastern coast are believed to have been used to aid navigation. In addition, there are numerous scattered remains of nonceremonial structures throughout the island.

TABLE 1
Endemic Animals (Species and Subspecies) from Cozumel Island

Class/Family	Species or Subspecies	Common Name	Size[a]	Threat Category	
				IUCN[b]	NOM[c]
Crustaceans					
Agostocaridiae[d]	*Agostocaris bozanici*	Cave shrimp		NI	NI
Alpheidae[d]	*Yagerocaris cozumel*	Cave shrimp		NI	NI
Hadziidae[e]	*Bahadzia setodactylus*	Cave amphipod		NI	NI
Fish					
Batrachoididae[f]	*Sanopus splendidus*	Splendid toadfish		VU	NI
Reptiles					
Teiidae	*Aspidoscelis cozumela*	Cozumel whiptail lizard		NI	NI
Mammals					
Didelphidae	*Didelphis marsupialis cozumela*	Cozumel opposum		NI	NI
Muridae	*Peromyscus leucopus cozumelae*	Cozumel white-footed mouse	G	NI	A
	Reithrodontomys spectabilis	Cozumel harvest mouse	G	EN	A
	Oryzomys couesi cozumelae	Cozumel rice rat	G	NI	A
Procyonidae	*Nasua nelsoni*	Dwarf coati	D	EN	A
	Procyon pygmaeus	Pygmy raccoon	D	EN	P
Tayassuidae	*Pecari tajacu nanus*	Pygmy collared peccary	D	NI	P
Birds					
Cracidae	*Crax rubra griscomi*	Cozumel curassow	D	NT	P
Accipitridae	*Buteo magnirostris gracilis*	Cozumel roadside hawk		NI	NI
Trochilidae	*Chlorostilbon forficatus*	Cozumel emerald		NI	NI
Picidae	*Melanerpes aurifrons leei*	Cozumel golden-fronted woodpecker		NI	NI
	Melanerpes pygmaeus pygmaeus	Cozumel woodpecker	D	NI	NI
Tyrannidae	*Attila spadiceus cozumelae*	Cozumel bright-rumped attila		NI	Pr
	Myiarchus yucatanensis lanyoni	Cozumel flycatcher		NI	NI
	Myiarchus tyrannulus cozumelae	Cozumel brown-crested flycatcher		NI	NI
Vireonidae	*Vireo bairdi*	Cozumel vireo		NI	Pr
	Cyclarhis gujanensis insularis	Cozumel rufous-browed peppershrike		NI	Pr
Troglodytidae	*Troglodytes beani* (=*Troglodytes aedon beani*)	Cozumel house wren		NI	Pr
Sylviidae	*Polioptila caerulea cozumelae*	Cozumel blue-gray gnatcatcher		NI	NI
Mimidae	*Toxostoma guttatum*	Cozumel thrasher	D	CR	P
	Melanoptila glabrirostris cozumelana	Cozumel black catbird		NT	NI
Parulidae	*Dendroica petechia rufivertex*	Cozumel yellow warbler		NI	NI
Thraupidae	*Spindalis zena benedicti*	Cozumel western spindalis		NI	NI
	Piranga roseogularis cozumelae	Cozumel rose-throated tanager		NI	NI
Emberizidae	*Tiaris olivaceus intermedius*	Cozumel yellow-faced grassquit		NI	NI
Cardinalidae	*Cardinalis cardinalis saturata*	Cozumel northern cardinal		NI	NI

[a] Refers to known or suspected dwarfism (D) or gigantism (G).
[b] Threat category according to the World Conservation Union (IUCN). NI: Not included; CR: Critically Endangered; NT: Near Threatened.
[c] Official Mexican List of Threatened Species (NOM). NI: Not included; P: [in danger of extinction]; A: [threatened]; Pr: [subject to special protection].
[d] Class Eumalacostraca.
[e] Class Malacostraca.
[f] Class Actinopterygii.

POPULATION AND ECONOMY

Cozumel is a municipality of the state of Quintana Roo. Population (2000) is 60,091, with an annual growth rate of 3%, mainly as a result of immigration. The main town on the island is San Miguel de Cozumel (or simply Cozumel), where the majority of the population lives. The only other town on the island is El Cedral.

During the first half of the twentieth century, the main economic activities were agriculture (coconut palm oil, pineapple, henequén), chicle extraction, and fisheries. Since 1960, Cozumel has experienced an economic boom, mainly as a result of a thriving tourism industry attracted by the natural beauty of the island. As a consequence, the aforesaid economic activities lost their past importance. Cozumel is the main port for cruise ships in the Caribbean, receiving some 3 million ship visitors a year. It is also a popular destination for scuba diving. Large numbers of tourists arrive from the adjacent mainland for day visits. The international airport receives flights from numerous cities in North America and Europe.

BIOLOGY AND CONSERVATION

Cozumel is part of the vast and remarkable Mesoamerican Reef System, second in magnitude only to the Australian Great Barrier Reef. In addition, it has the only coralline algal microatolls, or algal ridge systems, in the western Caribbean.

There are significant coastal lagoons on all flanks of the island, but especially in the north and south (Fig. 1). Because of its karst character, there are no rivers or other superficial bodies of freshwater on the island, with the exception of cenotes and seasonal ponds. These water bodies are crucial for wildlife. Most water for domestic consumption is extracted from underground sources in the central area of the island.

In 2000, roughly 90% of the island was covered with natural vegetation. The dominant vegetation on the island is semi-evergreen tropical forest, also with important areas of subdeciduous tropical forest, *chit* palm forest, halophilus sand dune vegetation, mangroves, and other wetlands. Despite representing only 10% of the area of the state of Quintana Roo, it is estimated that the island holds 40% of the total flora of the state. Cozumel is an important center of species endemism. It is the Mexican island with the largest number of endemic animal taxa (species and subspecies), with at least 31 (Table 1). Endemic crustaceans are found only in a few cenotes, ecological islands within the island. Several of the endemic vertebrates present dwarfism or gigantism. Many of the endemic vertebrates and other native species, although not necessarily included on international or national lists of threatened species, are critically endangered. One rodent (*Peromyscus leucopus cozumelae*) is extinct, and several of the endemic vertebrates (e.g., *Toxostoma guttatum, Nasua nelsoni, Procyon pygmaeus, Reithrodontomys spectabilis*) are on the verge of extinction. The main threat is introduced species (*Boa constrictor*, feral dogs and cats, and house mice and rats). Another threat is the expansion and broadening of the road system, which has fragmented natural vegetation on the island into southwestern and northeastern segments and has affected critical wildlife habitats. The introduction of continental generic counterparts of the endemic vertebrates, usually for pets, is also a problem because they could become sources of genetic introgression. Other threats are parasite and disease spillover from exotic animals, capturing of live wildlife for pets, and to a lesser degree hunting.

Hurricanes are the main natural disturbance. During the twentieth century, at least one hurricane hit the island every decade. The most recent ones were Gilbert (1988; Level 5 in the Saffir-Simpson scale), Roxanne (1995; Level 3), Emily (2005; Level 4) and Wilma (2005; Level 4). There are synergistic effects of natural and anthropogenic disturbance affecting the biota and society of the island.

There are currently two protected areas: Parque Nacional Arrecifes de Cozumel (11,988 ha), which is also a Ramsar site, and Refugio Estatal de Flora y Fauna de Laguna Colombia (1114 ha). There are initiatives to create new protected areas, but vast areas of great conservation importance still remain unprotected.

SEE ALSO THE FOLLOWING ARTICLES

Archaeology / Hurricanes and Typhoons / Introduced Species / Oceanic Islands / Reef Ecology and Conservation / Vegetation

FURTHER READING

Antochiw, M., and A.C. Dachary. 1991. *Historia de Cozumel*. Mexico City: Consejo Nacional para la Cultura y las Artes.

Cuarón, A.D., M.A. Martínez-Morales, K.W. McFadden, D. Valenzuela, and M.E. Gompper. 2004. The status of dwarf carnivores on Cozumel Island, Mexico. *Biodiversity and Conservation* 13: 317–331.

Freidel, D.A., and J.A. Sabloff. 1984. *Cozumel: late Maya settlement patterns*. New York: Academic Press.

IUCN (World Conservation Union). 2007. 2007 IUCN Red List of Threatened Species. http://www.iucnredlist.org/

Romero-Nájera, I., A.D. Cuarón, and C. González-Baca. 2007. Distribution, abundance, and habitat use of introduced *Boa constrictor* threatening the native biota of Cozumel Island, Mexico. *Biodiversity and Conservation* 16: 1183–1195.

Schmitter-Soto, J. J., F. A. Comín, E. Escobar-Briones, J. Herrera-Silveira, J. Alcocer, E. Suárez-Morales, M. Elías-Gutiérrez, V. Díaz-Arce, L. E. Marín, and B. Steinich. 2002. Hydrogeochemical and biological characteristics of cenotes in the Yucatan Peninsula (SE Mexico). *Hydrobiologia* 467: 215–228.

SEMARNAT (Secretaría de Medio Ambiente y Recursos Naturales). 2002. NOM-059-ECOL-2001, Protección Ambiental, especies nativas de México de flora y fauna silvestres—categorías de riesgo y especificaciones para su inclusión, exclusión o cambio—lista de especies en riesgo. *Diario Oficial de la Federación*, (6 de marzo): 95–190.

Steneck, R. S., P. A. Kramer, and R. M. Loreto. 2003. The Caribbean's western-most algal ridges in Cozumel, Mexico. *Coral Reefs* 22: 27–28.

Téllez, O., E. Cabrera, E. Linares, and R. Bye. 1993. *Las plantas de Cozumel: guía botánico-turística de la Isla Cozumel, Quintana Roo.* México City: Instituto de Biología, Universidad Nacional Autónoma de México.

Vega, R., E. Vázquez-Domínguez, A. Mejía-Puente, and A. D. Cuarón. 2007. Unexpected high levels of genetic variability and the population structure of an island endemic rodent (*Oryzomys couesi cozumelae*). *Biological Conservation* 137: 210–222.

CRABS

SEE LAND CRABS ON CHRISTMAS ISLAND

CRICKETS

DANIEL OTTE AND GREG COWPER

Academy of Natural Sciences, Philadelphia, Pennsylvania

FIGURE 1 *Prognathogryllus alternatus* lives in the foliage of the forests above Honolulu. Neighboring males alternate with one another in producing their chirps. This is a member of the tree cricket subfamily Oecanthinae. Photograph by David H. Funk.

FIGURE 2 *Trigonidium varians* from the forests of Kilauea. This is a member of the largest group of Hawaiian crickets. Many of the species hide in foliage and under the bark of a variety of trees during the day. Photograph by David H. Funk.

Virtually all oceanic islands within the tropics and subtropics are occupied by crickets, both those that colonized the islands on their own and those that appear to have been carried there by humans. Islands in the temperate zone are not occupied, probably because crickets are adapted mainly to warmer areas. The diversity of crickets on some islands, as measured by the number of species per unit area, is extraordinary, with Hawaii being more than 2000 times as rich, having at least twice as many cricket species as the continental United States and Canada combined.

RADIATION ON ISLANDS AND CHANCE

Which cricket groups radiate on an island system depends on which groups happen to get there. Hawaii and Galápagos illustrate the fact that it is not always the best colonizers that come to dominate an island, but crickets that arrive randomly through wind or ocean currents.

PROPERTIES OF CRICKETS AND COLONIZING ABILITIES

Crickets are much better at colonizing distant islands than are any other Orthoptera (the jumping insects). Why is this so? Although grasshoppers are some of the most mobile of insects and can fly great distances (the desert locust, *Schistocerca gregaria*, flew across the Atlantic in 1988), they are not nearly as successful in colonizing distant islands as crickets are. This can be seen principally in the distribution of both crickets and grasshoppers in the Pacific. No native grasshopper species occur in Hawaii (three species have been introduced by humans). Possibly adaptation to the new situation is the major barrier, rather than the act of getting there. A graduate student working on birds on Midway Island (part of the northwestern Hawaiian chain) reported a large number of locusts that landed on the island in 1991. The population did not survive because of bird predation and the absence of suitable habitat.

In crickets, dispersal over water is achieved by flying when distances are not large, and it may be aided by favorable winds. But dispersal by rafting, mainly through the laying of eggs into materials that become flotsam and are carried about on ocean currents, is probably the most common mode of dispersal. Rafting of adults themselves may possibly take place over shorter distances. Egg-laying habits therefore predispose certain groups for successful colonization. Long-distance migration is most likely to happen in species that lay their eggs into vegetation that gets washed into the ocean. Crickets that lay their eggs into soil are poorly represented on distant islands.

Successful colonization requires a good match between original and new habitat, which may be frequent within island archipelagoes but less likely between distant places. Species that live along shorelines are more likely to find a good habitat match (because shorelines are quite similar everywhere) than are species living away from the shoreline.

Because crickets are omnivores, they have no specific food requirements in the new habitat, perhaps making it easier to colonize distant habitats. In some instances, the inability of colonists to survive results in dead insects upon which the crickets can feed. A case in point is the cricket *Pteronemobius krakatua* Otte and Cowper, which resides on the lava flows of Anak Krakatau. Krakatau became a study in colonization after its volcanic destruction and sterilization in 1883, and the island of Anak emerged from its benthic remains in 1930. In the mid-1980s Anak Krakatau was still 95% barren when New and Thornton discovered a population of flightless scavenging crickets living on the ash-covered lava. This cricket had successfully become established as the lava's dominant arthropod, with the lava's surface becoming a smorgasbord of dead and dying windborne arthropods (see also the discussion of *Caconemobius* in Hawaii to follow).

EXAMPLES OF COLONIZATION AND ADAPTIVE RADIATION

The approximately 240 known Hawaiian endemic crickets are probably derived from four original colonizing species (two Nemobiinae, one Trigonidiinae, and one Oecanthinae), yet they range in their ecological niches from tree-dwelling, as found in species of *Prognathogryllus* (Fig. 1) and *Trigonidium* (Fig. 2); to species of *Leptogryllus* that live on low foliage and *Thaumatogryllus* (Fig. 3) that live on trunks of trees; to ground-dwelling species in the genus *Laupala* (Fig. 4); and finally, to species of *Caconemobius* (Fig. 5) that live in the dark zones of lava tubes.

Tree crickets (Oecanthinae) are common on continents and relatively rare on islands. On continents they inhabit trees, shrubs, and weeds; most species fly (probably not long distances); and eggs are laid into stems. Under the

FIGURE 3 *Thaumatogryllus mauiensis* lives on trunks of both living and dead trees, where it hides during the daytime. At night, it forages widely in surrounding foliage. This species has lost its acoustical apparatus and probably signals with pheromones only. It is a flightless member of the tree cricket subfamily. Some *Thaumatogryllus* have come to inhabit the dark zone of lava caves. In so doing, they have become pale and blind. Photograph by David H. Funk.

FIGURE 4 *Laupala vespertina*. All members of the genus are very similar to one another but differ strikingly in their songs. The species live near the ground in thick, moist vegetation, but climb up onto foliage at night to feed. Photograph by David H. Funk.

FIGURE 5 *Caconemobius uuku* is known from the caves near Hilo. It lives in the dark zone of lava caves. Roots from the forest above often emerge from the cave roof, then descend and continue growing into the cave floor, providing a food source. Photograph by David H. Funk.

right conditions, stems with eggs could be carried to the ocean in streams. After crossing the ocean, the hatched nymphs must find suitable habitat beyond the shoreline. Because tree crickets do not extend eastward beyond the Solomon Islands, it is reasonable to surmise that the latter barrier might be formidable. The tree crickets common along the coasts of Fiji and New Caledonia were probably transported by humans from Australia.

In Hawaii, tree crickets represent one of the most expansive and dramatic radiations known in insects. Their absence from most Pacific islands suggests that eggs of an ancestor, probably similar to the present-day *Xabea* (Asia) or *Neoxabea* (Americas), rafted to Hawaii from North or South America and exploited a variety of habitats that are normally occupied by different taxa on continents—trees, bushes, weeds, tree stems, bark, and even caves.

Once established on the Hawaiian chain, the original colonist radiated into a variety of forms and habitats. In one lineage, the forewings were reduced, and acoustical communication was lost along with the sound-producing and -receiving mechanisms, and in several lineages all traces of wings were lost. The presence of one species (the largest member of the subfamily, *Thaumatogryllus conanti*) on Nihoa suggests that the oecanthines may have reached the Hawaiian chain long before the large islands of the chain emerged and that they colonized new islands (the present large Hawaiian islands) as they emerged. Subsequently, species went extinct as islands eroded away. But rafting of eggs from the present high islands back to Nihoa cannot be ruled out.

The ground cricket genus *Caconemobius* appears capable of crossing water barriers quite readily. Species live along shorelines (in the wave splash zone) in northern Australia and in Samoa. On the latter island, they live in mangroves and sometimes hide inside empty coconut shells. Because coconuts travel widely around the oceans, it would be relatively easy for *Caconemobius* to be transported with currents. After becoming established on new islands, these crickets have given rise to several new species adapted principally to two new habitats: the surface of new lava flows, and subterranean lava tubes (left after lava has stopped flowing). Because both habitats reach the ocean shore, the ancestral shore species could easily make the transition into the interior of islands. Independently, on different islands, *Caconemobius* has entered lava tubes and crevices, feeding either on roots penetrating from the forest above or on the remains of dead insects, and becoming pale, long-legged, and virtually blind.

The flightless ground cricket *Thetella tarnis*, from along the shore of northeastern Australia, lives among stones, rocks, and coral, and in mangrove thickets. When disturbed, it readily jumps away by hopping on the water. The same species occurs along the shores of Fiji and Hawaii and extends eastward into the Indian Ocean, at least to the Maldive Islands. It, too, is a likely candidate for dispersal by eggs, nymphs, or adults moving about on ocean currents.

EXTRAORDINARY MIGRANTS AND OCCUPATION OF ISLANDS

The following eight species can be considered the most widespread crickets in the world (all inhabit at least some Pacific islands):

Gryllodes sigillatus (Gryllinae) was probably originally from Asia but is now found throughout the tropics and subtropics—on continents and on almost all tropical islands inhabited by humans. It is usually flightless and is normally found near human habitation, usually in crevices in walls and sidewalks. It is almost certainly carried by human traffic. But once established on an island, it spreads very quickly, especially on volcanic islands. In Hawaii, it spreads onto new lava flows and even lives on Lehua Island, the crescent remains of a cinder cone that juts out of the Pacific Ocean waves (off the shore of Niihau Island in the Hawaiian chain).

Acheta domesticus (Gryllinae), the house cricket, is not quite as cosmopolitan as *G. sigillatus*. Its native habitat is Sub-Saharan Africa, where it can be found with 5 or 6 close relatives. It has been carried by humans throughout the world. In the Pacific, it is known only from one small corner of the Hawaiian island of Kauai in dry, sandy habitat that looks very much like the Sub-Sahara. It was almost certainly carried to Hawaii by humans. It is the cricket familiar to hobbyists sold at pet stores as a feeder insect for reptiles and invertebrates.

Gryllus bimaculatus (Gryllinae) has an extraordinary distribution, ranging from the south to the north of Africa, through southern Asia to Singapore on the tip of the Malaysian Peninsula. It is a strong flyer, readily colonizing remote oases in the Sahara. It has also managed to colonize Hawaii, although probably through human transport. It was first noticed in Kauai in 1985, remaining in the port area before spreading the following year to all parts of the island, including to the highest mountain plateau, and the next year to Oahu, this time probably by flying.

Teleogryllus oceanicus (Gryllinae) occurs throughout the central and western Pacific Ocean and northern Australia. It is capable of flight, and in Australia it commonly flies to lights. Humans were probably responsible for it being carried to Hawaii, where it lives mainly in agricultural areas and does not fly to lights, suggesting that flying has been selected against.

Thetella tarnis (Nemobiinae) inhabits shores of continents and islands throughout the western Pacific, and reaches into the Indian Ocean as far as the Maldive Islands. It is a flightless species, which probably migrates by ocean currents in any of its life stages from egg to adult (see the "Colonizing Abilities" section above).

Metioche vittaticolle and *Trigonidomorpha sjostedti* are small flying species that are commonly found associated with human agriculture throughout the western half of the Pacific. Whether these insects were carried naturally by wind or current, or more recently by humans, is not known.

DIVERSITY PATTERNS

Cricket species diversity is generally higher on older and larger islands, but changes in the structure of islands and in the habitat preferences of particular groups can bring about some unexpected results. In the Hawaiian chain, islands keep emerging from the sea floor in a southeasterly direction. They grow in size for a time and then are eroded away to leave atolls with virtually no land above water.

In tree crickets the highest diversity is found on older islands (discounting the small islands that are now mostly rock fragments). In contrast, ground cricket (*Caconemobius*) diversity is highest on the youngest island. This is a result of the fact that only the youngest islands possess the variety of habitats used by *Caconemobius*. As islands age, *Caconemobius* habitats disappear (lava tubes and crevices are filled with erosional debris [Howarth, personal communication]).

DIVERSITY LEVELS AND FAUNAL BALANCE

When islands are very far from source areas, a peculiar pattern of diversity results. The most conspicuous feature of Hawaii's cricket diversity is not the overall diversity, which is low by tropical forest standards, but the fact that it is genus-poor but species-rich because of its great distance from major source areas. The distance leads to few lineages contributing to the final makeup of a community, and lineages that do contribute carry only a small fraction of the original population's genetic variation. The result of this "founder effect" is that new populations may be distinctly different, and growth of distant isolated faunas is mainly through speciation. In islands closer to continents, a larger number of colonists contribute to the final biota and result in a lower species-to-genus ratio and a greater variety of taxa at higher levels. Caribbean islands, close to large source areas and to each other, are poised to gain taxa in this way primarily through immigration.

ADAPTATION TO NEW HABITATS

Hawaiian habitats commonly contain several species, with each producing a characteristic, species-specific mating call. As many as five close relatives may share the same habitat and sing at the same time, with the songs sometimes being the only characteristic by which the species can be distinguished. This is the case in closely related species of *Laupala* (Trigonidiinae) that are both morphologically and ecologically identical.

The mechanism by which the songs of related species become different is in dispute. One of the most likely models postulates that an ancestral species is composed of a number of populations, and the different populations come to reside in different mixes of species. The song of each population then evolves to become distinguishable from the other species in its mix. Eventually, these populations can diverge to the point where they are not genetically compatible with sister populations, at which point they have become different species. But it is also possible that differences in song *alone,* acquired in isolation, prevent males and females from mating when they come together again, even though they have not yet acquired physiological or developmental incompatibilities.

TEMPORAL SEPARATION AND SPECIATION

Temporal separation of closely related crickets is well known in temperate continental cricket faunas. But, unexpectedly, seasonal separation of crickets in Hawaii, a region of extremely low seasonal fluctuations (at least in forested regions), is quite striking in some species pairs. Numerous species with nearly identical songs may now be separated seasonally, but, because their songs are so similar, they have remained undetected.

SPECIATION AND SPECIATION RATE

Recent work by Mendelson and Shaw has shown that speciation rates in the forest-inhabiting Hawaiian *Laupala* crickets may be extraordinarily high, the highest so far recorded in arthropods. They conclude that "speciation on Hawaii is both explosive and ongoing."

Isolation by geographic barriers or distance has a larger effect on species with low mobility. It is probable that most speciation in Hawaiian crickets takes place *within* islands and probably on the same volcano. By the arguments of Wright and Endler and others, population differentiation can take place with only minor barriers to migration, or even without barriers, especially if species are fractionated into small local breeding groups or if species are distributed one-dimensionally along ridges or streams. As islands age, there is an increase in spatial heterogeneity, which increases the opportunities for differential and disruptive selection, but differentiation among isolates or near-isolates may also come from the combined effects of low effective population size, low migration, and local breeding.

OCCUPATION OF CAVES

Although crickets are found in caves on continents, this habitat is usually restricted to one subfamily—the Phalangopsinae (but also in some Nemobiinae in Australia). But distant islands are not easily reached by this group, and perhaps because of their absence, other cricket groups have exploited this niche and adapted to caves.

Two outstanding examples are the occupation of deep caves by Oecanthinae (*Thaumatogryllus*) and Nemobiinae (*Caconemobius*) in Hawaii, groups not known to inhabit caves outside Hawaii. Cave dwelling probably evolved repeatedly in Hawaii—every time a new island emerged from the ocean and provided an abundance of lava caves, species normally living on the surface could exploit the new habitat.

The subfamily Gryllinae is never associated with caves, but in Galápagos two species in the genus *Anurogryllus* have independently gone underground and become at least partially blind (reduced eyes). On continents, *Anurogryllus* species all have large eyes; they make burrows, but emerge at night and feed and mate at the surface.

Cave habitation is very common in Caribbean islands where the ground is often composed of a network of cavities in limestone rocks. All Caribbean cave crickets belong to the Phalangopsinae, which live in similar habitats on the neighboring continents.

LOSS OF DISPERSAL

On islands, a higher proportion of species are flightless than on continents, but different island complexes can be quite different in this regard. The extreme example is the Hawaiian chain, where all native species are flightless, although many retain their forewings, perhaps chiefly because they are used to produce the song. Some species have become mute, with reproductive behavior becoming mediated principally by pheromones. In such species, forewings are greatly reduced and are just large enough to cover pheromonal glands. In other species, only small wing pads remain, and the glandular openings are uncovered.

Janzen's 1967 hypothesis on how tropical (mainly equatorial forest) climates affect dispersal powers of animals—that in areas with little climatic variation, animals are no longer selected to cope with large climatic fluctuations and hence acquire narrower climatic tolerances—may explain much about speciation in Hawaiian crickets.

A comparison of flightless endemic Hawaiian crickets with flightless introduced crickets shows this to be true, with endemic species showing extreme vulnerability to humidity and temperature changes in the laboratory. Continental crickets are much easier to maintain because they are more tolerant of temperature and humidity fluctuations. The higher tolerance of continental crickets can also be seen in the much greater ecological amplitude of the introduced compared to the endemic crickets. *Gryllodes sigillatus* and *Gryllus bimaculatus*, for example, are found on both wet and dry sides of the islands.

LOSS OF ACOUSTIC COMMUNICATION

Loss of acoustical communication is common in crickets, both on continents and islands. On islands, notable cases of loss are (1) reduction of the forewings to small pads in the Hawaiian genera *Leptogryllus* and *Thaumatogryllus*, possibly as a result of acquiring habitats where acoustical signaling is no longer efficacious or is even detrimental (as on the ground, under bark, in caves); (2) complete loss of wings in a species of *Nemobiopsis* on Hispaniola and in a shore-inhabiting *Caconemobius*, perhaps because the noise of the surf makes sound transmission difficult; and (3) the complete loss of singing in the Podoscirtine genus *Insulascirtus* on Lord Howe and Norfolk Islands and in Cuban cave-inhabiting crickets, especially those that are now restricted solely to this dark zone.

ECOLOGICAL ISOLATION

Coexisting species usually differ in their ecological requirements if they are closely related. This separation is often thought to be essential if species are to coexist in space and time. At least some of the Hawaiian crickets seem to violate this rule. For the most part, closely related species appear to have virtually identical ecologies, and differ only in their songs. Doubtless some subtle statistical differences will be found in virtually all cases in which a diligent search is made, but the overlapping resource requirements between species seems great enough to cast doubt that the differences are significant. In *Laupala*, comingling species often show no seasonal, daily, microhabitat, or foraging differences, although slight morphological differences are sometimes evident. Consequently, we are inclined to believe that ecological differences in this genus arise after the species have come into contact, perhaps long after in some cases.

ECOLOGICAL RELEASE

Among Hawaiian crickets, the two species of *Prognathogryllus* that show the greatest amount of within-population variability are isolated from close relatives on Maui and Hawaii. *Prognathogryllus waikemoi* on Maui is the most variable cricket species ever recorded, with songs as variable as the songs of an entire genus. *P. mauka* on the Big Island is also highly variable. Have these species lost a high degree of song specificity because there are no other competing species with which they might be confused?

Another possible example of ecological release can be seen in a podoscirtine cricket on the Pacific island of Pohnpei. This species occupies a huge range of habitats normally occupied by very different subfamilies. They live and sing up in the trees, but they also may be found in rock crevices near flowing water. Up in the trees, they are like typical podoscirtines, but near the water, they behave like phalangopsines.

GIGANTISM

Three examples in crickets illustrate the possibility that island ecology sometimes favors the evolution of large crickets. The largest species of *Gryllus* (field crickets) in the world is *Gryllus alexanderi* from a very small island in the Pacific. The largest tree cricket (Oecanthinae) is the now wingless *Thaumatogryllus conanti* known only from the small island of Nihoa, where it lives among rocks in a seabird colony. The largest cave cricket (Phalangopsinae: *Amphiacusta*) is an undescribed species from Henry Morgan Cave on North Andros Island, Bahamas.

As an aside, at 71 grams, the heaviest adult insect in the world is an orthopteran, *Deinacrida heteracantha,* a giant weta endemic to New Zealand's Little Barrier Island.

MORPHOLOGICAL EVOLUTION

We have noted striking morphological differences that occurred within a group that adapted to different habitats. Two other points need to be emphasized. First, in contrast to the usual rapid evolution of genitalia, reproductive structures diverge only in a minor way among closely related island crickets, and in some cases virtually not at all. At the same time, the more visible aspects of morphology change dramatically in some species. The degree of divergence in overall appearance varies considerably according to the kind of habitat that Hawaiian crickets occupy. Species that hide in leaf litter near the ground or under bark and in other dark crevices (species that rely on hiding less than on camouflage) show little divergence between species (*Trigonidium, Laupala, Caconemobius*). Those that live in foliage and rely on camouflage (*Prognathogryllus, Leptogryllus, Thaumatogryllus*) are often strikingly different in appearance. Variability in appearance is probably the result of two kinds of selection pressures: (1) Species acquire different appearances when they use visual signals and cues in their interactions with conspecifics (no Hawaiian examples are known), or (2) visually hunting predators select for background matching in their prey. If there is competition for escape space (camouflage space) among coexisting species, then a variety of appearances can evolve; this appears to be the basis of virtually all diversity in Hawaiian crickets.

THREATS TO ISLAND CRICKETS

Ants, none native to the islands, and *Pheidole megacephala* in particular, have been linked to huge drops in *Laupala* cricket populations on Kauai. But the principle reason for declines in native species on islands must be the destruction of native habitats. In the early 1980s, the forest in northern Kohala on Hawaii was rich in native vegetation and had at least eight cricket species. By 1990 cattle had entered the forest, completely destroying the understory, and the crickets disappeared. Surprisingly, two hurricanes had no visible effect on the understory cricket species, although they of course removed habitat for tree-inhabiting species.

CONCLUSION

This article considers only islands surrounded by water. Numerous habitat islands within land masses occur and are worthy of study, but less is known about them at this point. The life of crickets on islands is both ancient and ongoing. In the Hawaiian chain, crickets probably evolved on islands now long subsided beneath the ocean. It is a saga of the intricate interplay between the history of populations and the history of islands. A multi-causal mix of geography and climate, biology and chance, and adaptation promotes cricket colonization and diversification on islands. Our understanding of this subject will vastly increase as further archipelagoes and habitats on islands are investigated.

SEE ALSO THE FOLLOWING ARTICLES

Caves, as Islands / Ecological Release / Gigantism / Hawaiian Islands, Biology / Insect Radiations / Rafting

FURTHER READING

Janzen, D. H. 1967. Why mountain passes are higher in the tropics. *American Naturalist* 101: 233–249.
Mendelson, T. C., and K. L. Shaw. 2005. Sexual behavior: rapid speciation in an arthropod. *Nature* 433: 375–376.
New, T. R., and I. W. B. Thornton. 1988. A pre-vegetation population of cricket subsisting on allochthonous aeolian debris on Anak Krakatau. *Philosophical Transactions of the Royal Society of London B* 322: 481–485.
Otte, D. 1994. *The crickets of Hawaii: origin, systematics and evolution.* Orthopterists' Society Publication. Philadelphia: Academy of Natural Sciences.
Otte, D., and G. Cowper. 2007. New cricket species from the Fiji Islands (Orthoptera: Gryllidae). *Proceedings of the Academy of Natural Sciences of Philadelphia* 156: 217–303.

CROZET

SEE INDIAN REGION

CYPRUS

IOANNIS PANAYIDES

Cyprus Geological Survey, Lefkosia, Nicosia

Cyprus is one of the largest Mediterranean islands, with a surface area of 9251 km². It lies in the northeastern corner of the Mediterranean Sea approximately centered on latitude 35° N and longitude 33° E. Two mountain ranges dominate the topography of the island, the Troodos Range in the central region and the Pentadaktylos Range in the north. Between the two ranges lies the Mesaoria Plain, which, together with the narrow alluvial plains along the coast, makes up the bulk of the arable land. Most of the rivers, which flow only in the winter, spring out of the Troodos Mountains, with only one of them having its source in Pentadaktylos.

CLIMATE

Cyprus has a typical Mediterranean climate with mild winters; long, hot, dry summers; and short autumn and spring seasons. Average maximum temperatures in July and August range between 36 °C in the central plain and 27 °C in the Troodos mountains, whereas in January the average minimum temperatures are 5 °C and 0 °C, respectively. Sunshine is abundant during the whole year, with an average duration of sunshine of 11.5 hours per day in summer and 5.5 hours in winter. Because of the arid climate, evapotranspiration is high, corresponding to 80% of the annual rainfall.

Average annual rainfall is about 500 mm and ranges from 300 mm in the central plain and southeastern parts of the island up to 1100 mm at the top of the Troodos Range and 550 mm at the top of Pentadaktylos. Annual rainfall varies both spatially and temporally and can take the form of two- and even three-year-long droughts.

GEOLOGY

Three geological terranes—Pentadaktylos, Troodos, and Mamonia—overlain by the autochthonous sedimentary rocks give rise to four geomorphologic regions—the Pentadaktylos Range, the Troodos Range, the Mamonia Terrane, and the Mesaoria Plain—whose morphological characteristics reflect, to a large extent, the underlying geology (Fig. 1).

The Pentadaktylos Range

The Pentadaktylos Range is the northernmost of the two mountainous units of the island and forms a narrow, steep-sided chain of mountains varying in altitude from 700 to 1024 m. To the north, this very rugged range is separated from the sea by a narrow terraced coastal plain. In contrast, it is flanked to the south by the broad lowlands of the Mesaoria Plain.

The oldest rocks are Permian, very durable, fine-grained, and compact limestones (Kantara Formation) and mostly occur in the eastern part of the range as variously sized blocks (olistoliths) in the younger sediments. However, the three principal limestone lithologies of the range are the Dhikomo (Triassic), Sykhari (Triassic), and Hilarion (Lower Jurassic–Lower Cretaceous) Formations. They are allochthonous and form a series of thickly bedded limestones that have been thrust southward or partly imbricated with the autochthonous younger sediments, namely those of the Lapithos, Kalogrea–Ardana, and Kythrea Formations (Fig. 2).

The oldest autochthonus unit in the Pentadaktylos Range is the Lapithos Formation. It is of Campanian to Eocene age and consists of pelagic marls and chalks with cherts, which occur as faulted and schistose beds containing contemporaneous lava horizons. The Lapithos Formation is followed in upward succession by the Kalogrea–Ardana Flysch, of Upper Eocene age. The Kalogrea–Ardana

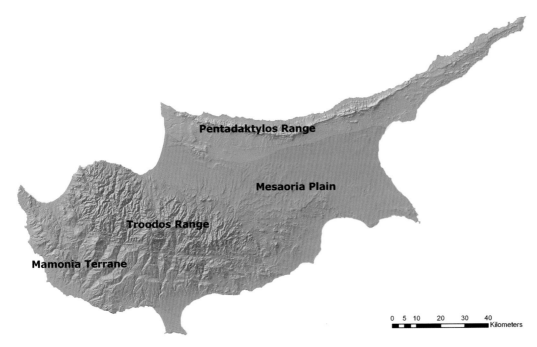

FIGURE 1 The four geomorphologic regions of Cyprus.

Flysch is uncomformably overlain by the Kythrea Formation or Kythrea Flysch, outcropping along both sides of the range as a thick and much-folded sequence of sandstones, siltstones, and marls, of mid-Miocene age.

The Troodos Range

The Troodos Range is the main orographic feature of the island of Cyprus. It covers an area of about 3200 km^2, and its highest peak, Olympus, has an elevation of 1951 m. The Troodos Range, or the Troodos ophiolite, was formed in the Upper Cretaceous by sea floor spreading at a constructive plate boundary. The ophiolite sequence is exposed in two separate areas within the Troodos Range, in the central Troodos and in the Lemesos Forest in the southeastern part of the range.

The ophiolite is characterized by the following sequence of rocks: the ultramafic complex, the plutonic complex, the sheeted-dyke complex, the extrusive sequence, and the chemical sediments.

The ultramafic complex is mainly composed of harzburgite and dunite with 50–80% of the original minerals altered to serpentine, and with serpentinite where the alteration is almost complete. Dunite bodies that vary widely in shape and size occur sporadically throughout the harzburgite. The harzburgites are, in fact, residual upper mantle material.

The plutonic sequence constitutes an assemblage of dunites, werhlites, pyroxenites, gabbros, and plagiogranites, the products of fractional crystallization and magmatic differentiation of a basaltic magma.

The sheeted-dyke complex crops out over most of the area of the massif in a belt around the plutonic core. It consists of a swarm of sub-parallel basaltic dykes (Fig. 3). They represent the infilling of fissures that open along divergent boundaries as two adjoining plates move away from each other. The dykes pass upward into the extrusive section through a transitional zone of dykes with lava screens that is traditionally known as the Basal Group.

The extrusive sequence consists of pillowed and non-pillowed lava flows. The sequence is divided into the upper

FIGURE 2 A block of Hilarion limestone thrusted southward on the autochthonous younger sediments. View from the west.

FIGURE 3 The sheeted-dyke complex consists of a swarm of sub-parallel basaltic dykes. They represent the infilling of fissures that open along divergent boundaries as two adjoining plates move away from each other.

and lower pillow lavas. The lower pillow lavas are mainly basalts and andesites with abundant celadonite and chalcedony. Dykes, sills, and flows form between 30 and 60% of the sequence. The upper pillow lavas are pinkish- or reddish-colored, dyke-free, olivine-bearing basalts with ultrabasic varieties (picrites) at the top of the unit. Pillow lavas make up the bulk of both lava units.

The chemical sediments of the Perapedhi Formation are a sequence of umbers and radiolarian shales, which fill hollows on the lava unit surface, and which were the first sediments to have been deposited on the ophiolitic rocks as a result of hydrothermal activity. Micropaleontological studies of the radiolarian assemblage from these sediments indicate a Turonian to Santonian age, which is consistent with determined radiometric ages for the ophiolite.

Directly associated with the Troodos ophiolite are massive sulfide, chromite, and asbestos mineral deposits. These ore deposits were formed in different stratigraphic units of the ophiolite (lavas, dunite, and harzburgite, respectively) and became exposed as a result of its uplift.

The Mamonia Terrane

The Mamonia Terrane, or Mamonia Complex, consists of a suite of igneous, sedimentary, and some metamorphic rocks, ranging in age from the Middle Triassic to the Upper Cretaceous. These rocks are regarded to be wholly allochthonous and are extensively exposed in western and southwestern Cyprus. Deformation within these rocks is quite intense, as they have been severely faulted and folded. Despite the tectonism, a formal stratigraphy has been recognized that divides the complex into two major groups: the predominantly sedimentary Agios Fotios Group and the mainly igneous Diarizos Group.

The Ayios Fotios Group is a series of sediments that represents a continental margin sequence, comprising Late Triassic to Middle Cretaceous quartz sandstones, calcareous to siliceous lutites, and hemi-pelagic sediments. The Diarizos Group is primarily a pillow lava assemblage with intercalated shallow- to deep-water sediments including pyroclastic tuffs and agglomerates. Large blocks of reefal limestone are closely associated with the volcanic rocks (Fig. 4). Other rock types associated with the Diarizos Group lavas are the serpentinites.

The Mesaoria Plain

The Mesaoria plain is a topographically low, rather flat area, which occupies the central part of the island between the Troodos Range to the south and the Pentadaktylos Range to the north. In its northern part, the folded Kythrea Flysch forms a characteristic hummocky topography, whereas in the southern part, a generally undeformed sequence of the autochthonous sedimentary succession rises towards the Troodos Mountains.

The base of the autochthonous sedimentary succession is marked by the Campanian to mid-Maastrichtian in age Kannaviou Formation, a thick sequence of volcaniclastic sandstones, siltstones, and bentonitic clays. The Kannaviou Formation is topped by the Lefkara Formation, of Paleocene to Miocene age, and consists of typical pelagic marl and white chalk, with or without chert. Where the formation is fully developed, it is represented by four members: lower marls, chalks with layers of chert, massive chalks, and upper marls.

The Lefkara Formation is succeeded by the Pakhna Formation (Miocene), which consists of yellowish chalks, marls, and arenites. Sedimentation in the Pakhna Forma-

FIGURE 4 Exotic blocks of reefal limestone in close association with volcanic rocks (Mamonia Complex).

tion began and terminated in a shallow-water environment with the development of reefal limestones, Tera Member, at the base and Koronia Member at the top of the formation (Fig. 5).

Toward the top of the Pachna Formation there is a sequence of evaporites deposited during a Messinian marine regression. These rocks, known as the Kalavaso Formation, are characterized by gypsum and gypsiferous marls, which occur as irregular bodies or lenses of highly variable thickness marking the top of a mega-regressive sedimentary sequence and overlying all the geological terranes of Cyprus.

A basal Pliocene transgression initiated the deposition of the Nicosia Formation, a sequence of calcareous featureless clayey silts. The silts grade into fine sands and are topped by thin conglomerates indicating the final shallowing of the depositional basin. At the top of the Nicosia Formation is a series of near-horizontal reddish fluvial muds and silts with some conglomerates (Apalos Formation).

The youngest (Pleistocene) sedimentary sequence on the island is the Fanglomerate, a very coarse, widespread alluvial fan deposit with an almost exclusively ophiolite-derived clast lithology, marking the very rapid uplift and erosion of the Troodos massif at this time.

THE GEOLOGICAL EVOLUTION OF CYPRUS

The genesis of Cyprus is intimately associated with the genesis of the Troodos ophiolite in a deep ocean and its subsequent rise to its present impressive elevation through a series of complicated processes that lasted from the Turonian to the Pleistocene. In the Upper Cretaceous, the area of Cyprus was part of the Tethys Ocean and the Troodos ophiolite was formed through sea floor spreading above an intraoceanic subduction zone, where

FIGURE 5 Miocene reefal limestone (Tera Member), with at least five uplifted paleoshorelines.

older crust of the Tethys Ocean and part of the African plate moved northward beneath similar crust attached to the Eurasian plate. As a result of this plate motion, the Tethys was almost completely closed, while new Turonian oceanic crust formed above this subduction zone. Not all of the old Tethyan crust was subducted. A portion of the overlying sediments, intraoceanic volcanic islands, and their fringing reefs were scraped off and accreted onto the edge of the new ophiolitic crust. These rocks form the Mamonia Terrane to the west and southwest of the Troodos massif and are highly tectonized and dismembered.

The mid-Maastrichtian collision of the Arabian promontory with the trench of the subduction zone over which Troodos was being formed resulted in the cessation of subduction and ophiolite generation; the detachment of the Troodos Terrane and its counterclockwise rotation by some 90°; and the emplacement of the Mamonia Complex onto the leading edge of the Troodos Terrane and the amalgamation of the two terranes into a single unit.

Subduction and plate readjustments moved further north, although their southernmost repercussions were felt in the area that the Pentadaktylos Terrane would finally dock. South of this region, tectonic quiescence prevailed with marine sedimentation in progressively shallower seas, with the Troodos massif first appearing above sea level in the Middle Miocene.

At the end of the Miocene and in the northernmost part of the region, which constitutes Cyprus today, a series of allochthonous limestones (Pentadaktylos Terrane) were thrust southward onto the edge of the Troodos Terrane, folding and thrusting any younger sediments in their path. East of Cyprus, the Tethys Ocean was closed, and the Mediterranean coastline acquired much of its present shape. The African plate, however, continued its northward movement, reestablishing plate boundaries and relative plate motions to accommodate for this movement. A new subduction zone developed to the south and west of Cyprus with the two plates sliding past each other further east. For this geometry to work, the African plate and the Arabian continent moved into the underbelly of Turkey, forcing a westward movement expressed as the two Anatolian strike-slip faults. In this way, subduction south of Cyprus became possible. At this time, a fragment of much lighter and water-rich continental crust, the Eratosthenes seamount, approached the new subduction zone south of Cyprus, losing much of its water, which migrated upward, helping to serpentinize the ophiolitic ultramafic mantle. Serpentinite, being lighter, moved upward and therefore helped uplift the Troodos above sea level. It was not, however, the only uplifting agent, for much thicker continental crust was being subducted during the Pliocene and Pleistocene. Being light, it exerted significant buoyancy over the ocean fragment that was to become Cyprus, thus gradually causing it to rise. The rate of uplift, however, was not constant. During periods of rapid uplift, vigorous erosion occurred, and the rivers down-cut the topography, leaving behind remnants of their former beds at different elevations, a feature of many streams in Cyprus. Similarly, uplifted paleoshorelines are found on the coastal zone of the island (Fig. 5).

SEE ALSO THE FOLLOWING ARTICLES

Mediterranean Region / Plate Tectonics

FURTHER READING

Cyprus Geological Survey. 1995. *Geological map of Cyprus*. Scale 1:250,000.
Ducloz, C. 1972. The geology of the Bellapais-Kythrea area of the Central Kyrenia Range. *Bulletin no 6 of the Geological Survey Department, Cyprus*.
Lord, A. R., I. Panayides, E. Urquhart, and C. Xenophontos. 2000. A biochronostratigraphical framework for the Late Cretaceous-Recent circum-Troodos sedimentary sequence, Cyprus, in *Proceedings, Third International Conference on the Geology of the Eastern Mediterranean*. I. Panayides, C. Xenophontos, and J. Malpas, eds. Cyprus Geological Survey, 289–297.
Malpas, J., J. T. Galon, and G. Squires. 1993. The development of a Late Cretaceous microplate suture zone in SW Cyprus, in *Magmatic processes and plate tectonics*. H. M. Prichard, T. Alabaster, N. B. W. Harris, and R. Neary-Christopher, eds. Geological Society Special Publications 76: 177–195.
Robertson, A. H. F. 2000. Tectonic evolution of Cyprus in its easternmost Mediterranean region, in *Proceedings, Third International Conference on the Geology of the Eastern Mediterranean*. I. Panayides, C. Xenophontos, and J. Malpas, eds. Cyprus Geological Survey, 11–44.
Robertson, A. H. F., and C. Xenophontos. 1993. Development of concepts concerning the Troodos ophiolite and adjacents units in Cyprus, in *Magmatic processes and plate tectonics*. H. M. Prichard, T. Alabaster, N. B. W. Harris, and R. Neary-Christopher, eds. Geological Society Special Publications 76: 85–119.

DARWIN AND GEOLOGIC HISTORY

JAMES H. NATLAND
University of Miami, Florida

Before Charles Darwin (1809–1882; Fig. 1) was a biologist, he was a geologist. He is remembered for developing the theory of evolution, but the first independent scientific work he ever did was on the geology of a volcanic island, the one he called St. Jago (Santiago), in the Cape Verde archipelago. This was the first landing on the five-year round-the-world voyage of HMS *Beagle,* the celebrated expedition on which he served throughout as social companion to Robert FitzRoy (1805–1865), the ship's master and commander, and also as naturalist. Of the 15 field notebooks Darwin compiled about his shore excursions during the *Beagle* voyage, most are on geology, and none of them have been published (although facsimiles are now online; see http://darwin-online.org.uk/cwcd_overview.jpg). On Darwin's return to England in 1835, he used these as a basis for composition of three geological monographs, one each on the geology of coral islands, of volcanic islands, and of South America, a project that took more than ten years to complete.

Darwin's first major scientific paper linked descriptions of effects of the great Chilean earthquake of 1835, which he experienced while onshore (finding that it induced giddiness while standing) and which virtually destroyed the city of Concepción, with geological evidence for uplift in the Andes, which he documented by inland excursions. Largely on the strength of this paper, he was elected to

FIGURE 1 Charles Darwin as a young man. Portrait painted by George Richmond in the late 1830s.

the Royal Society of London in 1839 when he was not yet 30 years old. Before the *Origin of Species* was published, he was awarded the Wollaston Medal, the highest honor of the Geological Society, in 1859.

DARWIN'S GEOLOGICAL BACKGROUND

At the outset of the *Beagle* voyage, Darwin's training to do geological work was scant. He was related to the Wedgwoods of ceramic pottery fame, marrying cousin Emma, and thus early on was acquainted with industrial uses and sources of clays. As a boy, he avidly pursued both collecting of beetles and, with brother Erasmus, experiments in chemistry. At the University of Edinburgh, where he started his collegiate career intending to enter the field of medicine, he heard lectures on geology from Robert Jameson (1774–1854), the leading British proponent of the Neptunist doctrines of Abraham Gottlob Werner (1749–1817), but was unimpressed. Being repelled by medicine, after two years he was studying for the ministry at the University of Cambridge where he found as mentor John Stevens Henslow (1796–1861), an ordained minister who held a chair in botany but was also a mineralogist who had done early and important geological work in Wales. Darwin's interaction with Henslow was barely academic, being instead nearly daily conversations during walks and field excursions, and covering all aspects of natural history, which became Darwin's passion. He vowed to become a naturalist, but like Henslow, saw the ministry as the only likely means for him to carry out his avocation.

After completing his formal studies, Darwin at first planned to make an excursion to the volcanic Canary islands, being inspired by the writings of volcanologist Leopold von Buch (1774–1853), another Wernerian. In preparation, he spent some weeks in the field in Wales learning principles of geological mapping and the rudiments of structural geology with Adam Sedgwick (1785–1873), another Cambridge professor (and cleric) who was then just beginning his famous work on the Cambrian System. Then, the opportunity to participate on the *Beagle* voyage suddenly arose, and he was soon off to explore the world; all plans for taking the cloth evaporated. His new friend FitzRoy gave him a copy of Volume I of Charles

SIGNAL POST HILL
A–Ancient volcanic rock
B–Calcareous stratum
C–Upper basaltic lava

FIGURE 2 Darwin's sketch of geological formations including a limestone layer (B) downbowed by the weight of a volcanic cone at St. Jago, Cape Verde Islands. From Darwin (1987, "Geological Observations on Volcanic Islands," p. 7).

Lyell's *Principles of Geology,* which was just then published, to include in his small shipboard library. Although Henslow cautioned him about some of Lyell's generalizations, the book literally transformed Darwin's thinking, serving at once to organize thousands of observations of geological processes under one overarching theme and to provide Darwin with a program for his research during the voyage.

The theme soon came to be known as the doctrine of Uniformitarianism. As Lyell (1797–1875) later expressed it in a letter to Roderick Murchison (1792–1871), "no causes whatever from the earliest time to which we can look back, to the present, ever acted but those now acting; and they never acted with different degrees of energy from that which they now exert." Processes of addition to land by volcanism and subtraction from land by erosion, uplift, and subsidence are intrinsically gradual; they occur in different measure in different parts of the globe; and they take a great deal of time—eons, as it turns out. In the historical introduction to his work, Lyell set up James Hutton (1726–1797), the Scottish geologist who first perceived from stratigraphy the immensity of geological time and the importance of great heat in the Earth's interior, as the first hero of modern (that is, British) geology, and he debunked Werner.

DARWIN ON HMS *BEAGLE*

Darwin imbibed this while experiencing bouts of seasickness during the first stormy transit to Tenerife in the Canary Islands, where no one was allowed to debark for fear of cholera, and then through balmier days down the coast of Africa to the Cape Verdes. Darwin collected Saharan dust en route. At St. Jago, he began to "geologize," as he would term it in a letter to his father, in the full spirit of Charles Lyell. As the ship lowered its anchor, he instantly saw a full illustration and vindication of Lyell's synopsis of Earth processes. A flat bed of limestone with volcanic rock underneath cuts across the face of the bluffs adjoining the anchorage, and above this is perched a small volcanic cone that in turn bows down the rock immediately beneath it (Fig. 2). The first lava beds had sunk below sea level and accumulated a limey sedimentary deposit laden with shells of large shallow-water organisms, and then both were uplifted before the volcanic cone erupted. A day of wandering the beaches and outcrops convinced Darwin that he could, after all, compile more than a mishmash of facts and write an intelligent book about the geological observations he could make during the remainder of the voyage. This he did three times over. During the voyage, Darwin would receive by mail the successive Volumes II and III of Lyell's *Principles,* and each one added to the breadth of his agenda and the scope of his thinking.

In all, Darwin investigated volcanic islands in three oceans: the Cape Verdes (one month), Fernando Nerhona (hours), St. Helena (six days), and Ascension (a day) in the Atlantic; the Galápagos (51 days) and Tahiti (11 days) in the Pacific; and Mauritius (11 days) in the Indian Ocean. He visited one atoll, Keeling (12 days), in the Indian Ocean, and one nonvolcanic islet in the central Atlantic, St. Paul's rocks (one day). The latter is an unusual exposure of sheared alkali peridotites that today is almost entirely awash at high tide. His aggregate of field time on islands thus was 122 days, or about four months, during the five-year voyage. He was singularly alone, being cut off from most scientific discourse except by infrequent arrivals of the mail. His sounding boards were shipmates, not scientists. He did have a great deal of time to read his small library and think, and the thinking at times was without restraint.

THE THEORY OF CORAL REEFS

Darwin never even saw an atoll before conjecturing, on a beach in Chile, with the final chapter of Volume II of Lyell's *Principles* in mind, that, in contrast to all the evidence for uplift he was observing in South America, the oceans must be the most important regions of subsidence. Thus atolls were not, as Lyell argued, reefs rimming the outlines of craters, but the marks of greatly subsided volcanic islands much larger than mere volcanic craters, and kept at sea level by the upward growth of coral. Months later, the work at Keeling thus consisted of confirming by soundings the shallow depths across the lagoon and the rapid dropoff in water depth outside the lagoon (more steep than any volcanic slope) with increasing distance from the surf crashing on the reef. Beyond his description of Keeling, Darwin's monograph on coral reefs of 1842 consisted largely of a global survey of reefs and reef types, these being his now classic fringing reef, barrier reef, and atoll lagoon, the three in his theory representing successive stages of subsidence of an interior volcanic island, its disappearance beneath the sea, and the corresponding upward growth of the coral (Fig. 3). The linchpin of the theory was Darwin's summary of the rapid rate of growth of coral in tropical waters along with the limitation of the environment of coral growth to within or just below the photic zone, some few tens of meters at most. Intact coral rock recovered by the dredge or during sounding at greater depths must, in Darwin's conception, have subsided. The great size of some atolls thus implied disappearance beneath the waves of volcanic summits that must once have reached thousands of meters (or, as he would have said, feet) above sea level.

Darwin's theory of coral reefs was both hailed and assailed over the next hundred years, and was not finally proven until the atoll at Enewetak in the Marshall Islands was drilled in the 1950s. There, over 1720 m of coralline rock were cored before the core bit reached lava of Eocene age that had erupted subaerially. By then, deeply submerged drowned atolls had been discovered by acoustic sounding in the far western Pacific, and oceanographic expeditions were conducted to survey and sample their summits. Recovery of shallow-water reef materials of Mesozoic age confirmed

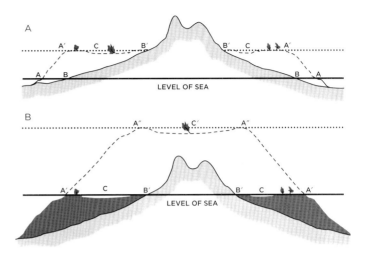

FIGURE 3 Darwin's cross-sectional diagrams of how a coral reef grows about a subsiding volcanic island. In (A), the coral first attaches itself as a fringing reef, on the shoreline (AB), and then as the island subsides (shoreline goes up in the diagram), the reef widens to form a barrier reef (A'B') with an interior lagoon (C). In (B), as the island subsides further, the barrier reef (A'B' again) narrows somewhat but eventually overtops the remnant volcano to form an atoll (A"A") with a full atoll lagoon (C') and investigators in a rowboat.

the former proximity of the summits of these volcanoes, which are termed guyots, to sea level. Profiler records and deep-sea drilling of guyots with thick lagoonal deposits into volcanic basement confirmed total subsidences of 4–5 km. The great subsidences established the ocean floor as the most dynamic part of Earth's surface, and this realization was instrumental in the development of the theories of sea floor spreading and plate tectonics.

VOLCANIC ISLANDS

Today, Darwin's monograph on the volcanic islands reads largely as a compilation of facts that, he thought, "should be of interest to geologists." He describes features of rocks, of textures, of materials in veins, vugs, and amygdales, of lava flows, of scoria, of cones, of dikes, of craters, of the shapes of prominent outcrops, of reduction of volcanic rock by erosion, of rare finds of fossils, and many other things besides. The proper way to perceive this, however, is that the work is almost entirely a pioneering venture into things and places never before discussed as geology. Darwin's foremost predecessor as naturalist aboard a vessel of any nationality was probably Sir Joseph Banks, who sailed on the first of Cook's voyages in the 1770s, in days when the discipline of geology barely existed. The term "geology" itself had not even been invented. Thus, Banks's descriptions of Tahiti as a volcanic island are virtually nil.

For volcanological guidance, Darwin thus only had accounts chiefly of the inactive Auvergne province in France, the active volcanoes of Vesuvius and Etna (both of these being well described, for example, by Lyell), and von Buch's descriptions of the Canaries. But what anyone could say about almost every other volcanic island that Darwin visited was based on skimpy and itinerant accounts by untrained observers who were mainly interested in other things. Thus, in Lyell's otherwise encyclopedic treatment of global geology, which certainly treats of island arcs, there is no hint that the vast regions of the Earth covered by oceans are essentially volcanic in origin, and no discussion of the geology of any volcanic island group of the type Darwin visited other than two near Europe: the Azores and the Canaries. One can fairly say that Darwin was certainly the first geologist to state that, with rare exceptions, the "innumerable islands scattered through the Pacific, Indian and Atlantic Oceans, were composed either of volcanic or of modern coral rock" (Darwin 1844, p. 125).

Darwin also recognized the global tendency, whether on continents or in the ocean basins, of volcanoes to form chains—some of them straight, some of them curved—and clusters where, such as at the Galápagos, historic eruptions have ensued from separate systems of fissures. He associated active volcanoes with regions of uplift, and coral atolls with subsidence. Finally, on the basis of hand specimen observations from the islands he visited, and from published descriptions (but before thin sections were invented), he conjectured on flow orientation of feldspar; that density sorting of crystals from liquids is responsible for the separation of trachyte from an ultimately basaltic precursor; and that such sorting also explains some of the differences between plutonic and eruptive igneous rocks.

THE IMPORTANCE OF ISLAND BIOGEOGRAPHY

During Darwin's field excursions on volcanic islands, he did not merely jot down notes on geology and sample rocks. He observed organisms and sampled those as well. Aboard the *Beagle,* he would annotate and package his specimens, sometimes doing chemical or goniometrical tests on rocks and minerals (here, his boyhood chemical experiments came in handy) and preserving the biological ones by curing and immersion in distilled spirits. The link between Darwin's geology and his biology is not always obvious, but he necessarily carried out both aspects of his work simultaneously. Volume II of Lyell's *Principles* discourses on the distribution of organisms on the present globe and during the past, and presents a view that today would be termed "special creation"—in each place, its own group of divinely created creatures, now and in the past. Darwin puzzled over the applicability of this, especially when noting strong similarities between plants, birds, and other organisms of the Galápagos Islands with those of South America, and considering that distinct species of turtles were endemic to individual islands. He did not originally catch on that the birds he collected on the different islands in the Galápagos were particularly distinct; that was left to John Gould (1804–1881), the ornithologist who examined Darwin's specimens in England. Gould found individual species of wrens, mockingbirds, and finches, but only the mockingbirds had island-of-origin tags. The island-specificity of the finches and their particularly high susceptibility to evolutionary environmental pressures have only become evident with modern studies. Nevertheless, as Darwin noted in his diary at the time, he immediately saw the pertinence of Gould's results to causes for the diversity of species.

Darwin's most recent biographer, David Quammen, in *The Reluctant Mr. Darwin,* has strongly argued for the importance of island biogeography in undergirding Darwin's theory. Implicit in Darwin's work, and in that of his co-discoverer of natural selection, Alfred Russel Wallace (1823–1914), is that islands divide species; indeed, they divide ecosystems, and when such divisions exist, common species on either side diversify and eventually become separate under the pressure of natural selection. The Galápagos are important to Darwin's theory because, of all the volcanic islands he visited, they are the only ones that he appreciated as a group, with small bands of water between the volcanoes. They were also the only ones without significant human intervention, meaning primarily that most of the islands had few or no introduced species.

Formally, a division does not have to be between islands; it could be in the directions of drainage, or on the largest scale, the physical isolation of continents. The logic of Lyellian geology is that it provides for constant, incessant transformation of land surfaces, of landscapes, and therefore of ecosystems. Now that we know about plate tectonics, the message is even clearer. The continents themselves move from frigid to tropical climes in the time it takes to build and erode a mountain range. Organisms have changed accordingly, and in pace with geological change. A huge catastrophe, such as the bolide impact at the end of the Cretaceous, changes everything everywhere and all at once; thence come mass extinctions and a world of fewer organisms struggling to re-establish themselves. Whole lines of major organisms, such as the dinosaurs, disappear, and others, such as the mammals, become dominant. The whole Earth is thus an island.

On a scale accessible to studies of modern organisms, drainages of water, even of small streams, are divisions, but over time, with regions uplifting and subsiding, the millions of small divisions change. Climate changes, soils change, rock subject to erosion changes, volcanic ash buries forests—all sorts of things happen. This is the milieu of constant pressure on all living organisms that forces adaptive change and creates new species. At its core, all this change, including that of the organisms, is geological in origin. Darwin, accepting Lyell's interpretation of geology, saw the fundamental contradiction that underlay Lyell's interpretation of the fossil record. In the end, he was truer to Lyell's vision than was Lyell himself. The Galápagos Islands are a microcosm of island-by-island divisions between ecosystems. They caught Darwin's attention, they underscored the difficulties with Lyell's hypothesis of organic change, and they helped Darwin to formulate a new theory. The Galápagos Islands guide evolutionary studies to this day.

SEE ALSO THE FOLLOWING ARTICLES

Cape Verde Islands / Coral / Galápagos Islands, Geology / Plate Tectonics / Volcanic Islands / Wallace, Alfred Russel

FURTHER READING

Browne, J. 1995. *Charles Darwin, voyaging.* Princeton, NJ: Princeton University Press.
Darwin, C. 1987. The structure and distribution of coral reefs: the geology of the voyage of HMS *Beagle*, part I, in *The works of Charles Darwin*, vol. 7. P. H. Barrett and R. B. Freeman, eds. New York: New York University Press.
Darwin, C. 1987. Geological observations on volcanic islands: the geology of the voyage of HMS *Beagle*, part II, in *The works of Charles Darwin*, vol. 8. P. H. Barrett and R. B. Freeman, eds. New York: New York University Press.
Darwin, C. 2003. The voyage of the *Beagle*, chapter 11; The origin of species, chapter 12. Everyman's Library Edition. New York: Alfred A. Knopf, 804–853.
Herbert, S. 2005. *Charles Darwin, geologist.* Ithaca, NY: Cornell University Press.
Lyell, C. 1990. *Principles of geology,* vol. I, II, and III. Facsimile of the First Edition, 1830–1833. Chicago: University of Chicago Press.
Quammen, D. 2006. *The reluctant Mr. Darwin.* New York: W. W. Norton and Co.
Weiner, J. 1995. *The beak of the finch: a story of evolution in our time.* New York: Knopf Publishing Group.

DARWIN'S FINCHES

SEE GALAPAGOS FINCHES

DECEPTION

SEE ANTARCTIC ISLANDS

DEFORESTATION

BARRY V. ROLETT

University of Hawaii, Honolulu

Deforestation is the loss of trees involving a vegetational succession from forest cover to some other kind of landscape. It causes reduced biodiversity and, under certain conditions, promotes the development of wastelands. The process of deforestation is intertwined with human history as well as with the influence of climate and other environmental factors. Islands are particularly vulnerable to deforestation because their ecosystems tend to be fragile and susceptible to rapid change.

PAST AND PRESENT EVIDENCE FOR DEFORESTATION

Historical records offer abundant evidence for deforestation, especially in relation to the loss of trees through the development of agriculture and timber industries. Records for the Philippines show, for example, that primary-growth forest cover decreased from about 70% to 7% of the total land area over the past 100 years. For islands without such records, or where deforestation occurred in the distant past, there are a number of lines of evidence that can help to identify the occurrence and extent of deforestation. Pollen profiles reconstructed from sediment cores are particularly revealing. Typical pollen profiles indicating deforestation show declining frequencies of tree pollen with simultaneous increases in charcoal and the pollen of grasses. The frequency and size of charcoal particles in the sediments may indicate the role of fire in the loss of tree cover.

Large areas of habitually burned grasses or ferns are another common indicator of deforested landscapes (Fig. 1). Grass- or fern-dominated vegetation forms after burning and is maintained by fire that inhibits the growth of woody competitors. This is the case in New Zealand, where most of the lowlands became deforested following the arrival of Polynesian settlers, and forests were replaced by vast areas dominated by ferns.

Geomorphology can also serve as a marker for deforestation, especially on islands with steep terrain. By removing the tree canopy, deforestation exposes ground surfaces to the direct impact of rainfall. This promotes erosion, which can be accentuated by various factors including slope gradient, rainfall levels, severe weather, and the type of groundcover. In extreme cases, massive erosion results in completely denuded landscapes scarred by deep gullies.

FIGURE 1 View of Hanamiai Valley, Tahuata, Marquesas Islands (French Polynesia). *Miscanthus* grasslands, a form of vegetation maintained by fire, are visible as light-colored areas on deforested ridges leading to the central mountain range. The valley floor is an example of forest replacement, where native trees have been replaced by economically important introduced species of trees. Photograph by B. Rolett, July 1998.

Sediments eroded from slopes accumulate at the base of these slopes and may also be transported by water to valley floors, coastal lowlands, and inshore marine environments. Many islands display physical evidence for erosion and sedimentation, and in certain tropical settings deposits up to 5 m deep have accumulated in less than 1000 years.

ENVIRONMENTAL PREDICTORS OF ISLAND DEFORESTATION

Comparative studies of islands help to identify environmental factors contributing to deforestation. Small, dry, cool, remote islands, with low topography, low inputs of soil-enriching continental dust, and little or no volcanic ash fallout are the most vulnerable to deforestation. Environmental factors aiding the retention of forest cover include high rainfall, large island size, and high levels of volcanic ash. Dry islands are predisposed to deforestation because rainfall is a primary determinant of plant growth rate. Low rainfall also increases forest vulnerability to fire. By contrast, high rainfall favors plant growth and decreases the chance of deforestation.

Large, high islands generally retain more forest than do small, low islands. This is particularly true on islands with tracts of mountainous land that are inaccessible to humans and their fires. Large islands may also have greater habitat and tree species diversity, increasing the likelihood that some species will survive. Furthermore, large islands create their own rain (orographic rainfall) when moist air blowing from the ocean hits mountains and rises until it condenses. Geological age is relevant because soil nutrients become lost from older formations with time, especially by rain leaching. Lost nutrients can be restored by volcanic ash and continental dust fallout. Islands receiving little or no aerial dust or ash are thus more vulnerable to deforestation than are islands closer to volcanoes ejecting tephra and ash. Guano from nesting birds enriches soils on islands where birds are or were common, but it is difficult to quantify the importance of this factor.

THE HUMAN ELEMENT

Although geography is important, cultural practices also have guided the contrasting ecological histories of different islands. Human populations play a direct role in deforestation by clearing land for gardens, cutting trees to obtain timber and firewood, and so on. The importance of humans in these activities varies significantly over time and from one island to another depending upon factors including population size and the scale of the economy. On many forested islands, native species of trees have been replaced by economically valuable introduced species. This process of forest replacement can transform the species composition of the forest without greatly altering the ecology. A classic example is Tahiti, where Polynesian settlers introduced many different species of fruit trees to thoroughly transform the character, but not the ecology, of the island's lowland forests. Captain Cook and other early European explorers mistook these bountiful anthropogenic forests for unaltered environments, and this view contributed to the European idealized image of Tahitians as a people living in harmony with nature.

FIRE

Because most fires are set by people, human colonization introduces fire as a powerful force in island deforestation. It is particularly significant in the deforestation of dry islands. Traditional agricultural systems often rely on fire for clearing fields, and on dry islands there is a high risk of fires spreading from gardens and burning out of control. When this occurs repeatedly, forests are replaced by savannahs of grasses or ferns, which can survive repeated firing. On dry islands throughout the eastern Pacific, fire-associated grasslands unsuitable for cultivation cover large tracts of land.

THE ROLE OF INTRODUCED ANIMALS IN DEFORESTING ISLANDS

Most islands have impoverished native floras and faunas as a consequence of their isolation. When this isolation is broken by the human introduction of alien species, the environmental consequences are highly disruptive. Introduced grazing animals, especially goats and sheep, are particularly destructive. They inhibit forest regeneration by grazing on young saplings, while also exacerbating soil erosion through trampling the ground with their hard hooves. Goats and sheep were introduced (together with agriculture) to the Mediterranean during the Neolithic era, and they are clearly implicated in the deforestation of this region. Yet because people also played a direct role in deforesting the Mediterranean islands, it is difficult to gauge the relative importance of grazing animals, as opposed to humans, in this process.

There are a number of examples in which sheep and goats were left alone on uninhabited islands, and in all cases the outcome was an environmental disaster. Two of these islands are in the Marquesas, a South Pacific archipelago of ten high volcanic islands known for rugged, verdant landscapes. When these islands were first discovered, by Polynesians around 1200 years ago, terrestrial mammals were absent and had never existed in the Marquesas. Seafaring Polynesians introduced pigs, dogs, rats, and chickens, in addition to numerous plants, including root crops, fruit trees, and other cultigens. Human colonization and the introduction of alien species led to the mass extinction of native landbirds, an event that surely reflects broad environmental shifts during the first centuries of Polynesian settlement. Nevertheless, Marquesans developed a flourishing culture, and they achieved sustainable growth for more than 1000 years, until the dawn of European contact.

When Europeans and Americans began to frequent the Marquesas during the late eighteenth century, they initiated a new set of experiments involving the introduction of terrestrial animals. Ship captains released breeding pairs of sheep and goats in various locations across the Marquesas. Goats quickly became feralized throughout the archipelago, whereas herds of sheep sprang up on the uninhabited islands of Eiao and Mohotane. In contrast to the large valleys, deep bays, and accessible coasts of the principal islands, Eiao and Mohotane are dominated by dry plateaus, and their coastlines consist mainly of cliffs. The seamen who introduced these animals could hardly have anticipated the long-term environmental consequences. Both islands supported dryland forests and perennial streams when sheep and goats were introduced. Marquesan forests typically have a canopy of tall trees rising above a dense understory. Today on Mohotane, the stands of tall trees are dwindling, and grazing has completely eliminated the understory (Fig. 2). As the surviving trees reach the end of their life cycle, the forest is vanishing, and as this occurs, the ground is exposed to the direct impact of rainfall. Heavy rains and runoff have produced deep gullies cutting down to lateritic clays and lava bedrock.

The situation is similar but worse on Eiao, another island in the Marquesas (Figs. 3, 4). Eiao's plateau is a red

FIGURE 2 Stands of surviving trees on Mohotane, Marquesas Islands (French Polynesia). Grazing by sheep, introduced less than 100 years ago, has eliminated the understory and has largely deforested the island. Photograph by B. Rolett, April 1987.

FIGURE 3 View of Eiao, Marquesas Islands (French Polynesia), deforested by sheep and goats introduced less than 200 years ago. One of the few surviving trees (at right) is near the end of its life cycle. Photograph by B. Rolett, December 1999.

FIGURE 4 Close-up of gullies produced by erosion and runoff on the plateau of Eiao, Marquesas Islands (French Polynesia), as a result of deforestation. Photograph by B. Rolett, December 1999.

clay desert, and the only surviving trees live in sheltered valleys that trap soils washing off the surrounding slopes. The dense understory in these places denies access to the sheep, preserving small patches of vegetation on an otherwise denuded island. However, Eiao can be contrasted with its close neighbor Hatuta'a, which has an equally fragile landscape. Not only is Hatuta'a fully forested, but it also is considered one of the least disturbed tropical dry forests in Polynesia. Although it is located only a few miles from Eiao, Hatuta'a has largely escaped the attention of humans. Polynesians visited but never settled the island, which has no bays or good landings. The Pacific rat is the only introduced animal from any period in history, and the forests have survived nearly unaltered. All evidence suggests that Eiao's original vegetation, now entirely lost, was similar to the forests still present on neighboring Hatuta'a. The harshly contrasting environmental histories of Eiao and Hatuta'a are mainly attributable to the introduction of sheep and goats. This comparison illustrates the inherent fragility of island environments, while also highlighting the dangerous role of grazing animals in deforestation.

AVOIDING DEFORESTATION

Although islands are susceptible to rapid environmental change after the breaching of their natural isolation, different islands have experienced varying rates and degrees of deforestation. The different outcomes can be explained by a combination of environmental and cultural factors. Forest retention is usually greater on islands where people developed alternatives to fire-intensive forms of agriculture. The most common fire-intensive form of agriculture is shifting cultivation, in which forested land is cleared by burning. In shifting cultivation, gardens are planted for a series of cropping cycles, and then the fields are allowed to lie fallow before they are cleared and burned again to prepare them for a new cropping cycle. One cultural response that favors the retention of forests is the development of irrigated agriculture as an alternative to shifting cultivation. Irrigated fields (with crops such as rice or taro) generally produce higher yields than rainfed gardens do, so irrigated agriculture conserves forests by limiting the amount of arable land needed for food production. In addition, irrigated agriculture does not require the use of fire. This is an advantage on dry islands, where fires can burn out of control. The development of arboriculture also favors the retention of forests. People who rely on arboriculture are not forced to cut trees in order to expand their subsistence economy, because the forest itself is a primary source of food. Instead, in arboriculture people transform forests without destroying them, by replacing native trees with useful species such as breadfruit, chestnuts, and olives. Other cultural responses by people living on islands also favor the retention of forests. These include controlling human population growth, conserving resources, and limiting the introduction of non-native species, especially grazing animals.

SEE ALSO THE FOLLOWING ARTICLES

Climate on Islands / Human Impacts, Pre-European / Lava and Ash / Pigs and Goats / Sustainability / Vegetation

FURTHER READING

Diamond, J. 2005. *Collapse: how societies choose to fail or succeed.* New York: Viking Penguin.
Rolett, B. V. 2008. Avoiding collapse: pre-European sustainability on Pacific Islands. *Quaternary International* 184: 4–10.
Rolett, B. V., and J. Diamond. 2004. Environmental predictors of pre-European deforestation on Pacific Islands. *Nature* 431: 443–446.

DISEASE

SEE BIRD DISEASE, PLANT DISEASE

DISPERSAL

ISABELLE OLIVIERI

University of Montpellier, France

Dispersal (or migration) can be defined as any movement of individuals with potential consequences for gene flow across space. Dispersal can be extended to include movement through time (i.e., dormancy). Dispersal is a fundamental life history trait, with multiple demographic and genetic consequences. In particular, it determines the amount of local adaptation and the likelihood of various speciation mechanisms. It is also an evolving trait.

EVOLUTIONARY CAUSES OF DISPERSAL

Overall, the different causes for dispersal evolution can be classified into three broad classes of explanation. First, the evolution of dispersal may be understood in the light of habitat selection theory, whereby behaviors at departure or arrival help individuals to preferentially establish in certain (supposedly better) types of habitat. In particular, by allowing escape from crowding or inbreeding, dispersal, whether or not triggered by density or local relatedness, may help individuals to reach sites of better quality. Second, dispersal can be viewed as an altruistic trait and its evolution understood in the light of kin selection theory. Third, in fluctuating environments, dispersal can be seen as a bet-hedging strategy. Dispersal may entail different types of costs: cost of producing the dispersal structure (Fig. 1), tradeoff between dispersal ability and other fitness components (e.g., smaller seeds could be better dispersed but at a possible cost of competitive ability), tradeoff between dispersal structures and fecundity, increased predation risk and loss of energy while traveling, decreased local adaptation in the new population, and risk of landing in an unsuitable habitat.

Although most models on the evolution of dispersal consider this trait as a fixed strategy (either pure or mixed, but only genotype-dependent), it is clear that, in many species, dispersal is a plastic behavior, better described by a norm of reaction, whereby dispersal strategy depends on the local environment, social context, maternal effects, and individual condition. Therefore, these models usefully indicate the direction of selection under various ecological scenarios but are unlikely to correctly predict the outcome of evolution. There are, however, a handful of evolutionary models of condition-dependent dispersal.

Dispersal as a Habitat Selection Trait

It is clear that directed or condition-dependent dispersal allows individuals to reach more suitable habitats. However, even random dispersal can also be considered as a habitat selection trait, if the local conditions experienced by migrants are on average better than those experienced by residents. This happens in particular when there are local extinctions, such that dispersal allows reaching low-density (more favorable) patches created by such extinctions. Thus, when the probability of local extinctions increases, the optimal dispersal rate increases as well. However, when local extinctions are very frequent, the total metapopulation size is lower, such that the number of colonizers is low enough for within-patch competition to be decreased. Under these conditions, it no longer pays to disperse to escape competition, so there is an optimal extinction rate, above which increased philopatry is selected for. Dispersal is also advantageous in populations with cyclic or chaotic dynamics. Conversely, spatial variation of local conditions selects against dispersal. Similarly, positive autocorrelation of habitat quality in time decreases the evolutionarily stable dispersal rate. When local conditions vary in both space and time, it is in theory possible for a polymorphism of dispersal rates to be maintained. Such polymorphism can emerge from a monomorphic population, and the low-dispersal phenotype is then expected to be found mostly in good habitat patches, whereas the high-dispersal phenotype should be distributed among habitats according to their frequencies. Such a pattern could explain the observation that specialist species of spiders tend to disperse less than do generalist ones.

Furthermore, even in a stable environment, inbreeding depression (under the form of drift load [i.e., the random fixation of deleterious mutations in patches of small size] and its counterpart, heterosis) could be sufficient to promote dispersal as a habitat selection mechanism. Inbreeding avoidance has been put forward as a mechanism to explain dispersal, especially when restricted to one sex, in particular in mammals. However, when the total metapopulation size is very small, heterosis vanishes, so the advantage of dispersal as a habitat selection strategy decreases as well.

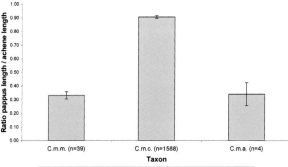

FIGURE 1 Variation in fruit morphology in *Centaurea maculosa* subspecies. C.m.m.—*Centaurea maculosa maculosa*; C.m.c.—*C. maculosa corymbosa*; C.m.a.—*C. maculosa albida*. A 95% confidence interval is indicated for each taxon. Photograph and data provided by Hélène Fréville and Agnès Mignot.

Dispersal as an Altruistic Trait

Conversely, in a stable population, dispersal can evolve as a sib-competition avoidance mechanism (through a kin selection phenomenon), as long as relatedness among individuals is spatially structured. In such populations, short-distance dispersal allows dispersers to experience competition with non-kin rather than with kin. Furthermore, when dispersal is costly, and assuming that all individuals of the same genotype have the same propensity to disperse, there are necessarily fewer immigrants than there are emigrants. As resources freed by an emigrant are shared among all its neighbors, the inclusive fitness benefit will depend on its relatedness with other competitors in the natal patch compared to its relatedness with the occupants of its new patch. Thus, an emigrant, although paying the cost of dispersal, increases its inclusive fitness by increasing the chance that its kin survive and reproduce. Dispersal can thus be viewed as an altruistic trait, whereby emigrants sacrifice their own reproduction to enhance that of their kin. Under such a mechanism, depending on the genetic determinism of dispersal (e.g., maternally inherited or determined by the genotype of the dispersing organism), the dispersal rate that will evolve will be larger or smaller.

Dispersal as a Bet-Hedging Strategy

Dispersal has been suggested to evolve as a bet-hedging strategy—that is, as a strategy maximizing the geometric mean fitness through a reduction of its variance. Thus, one can often read that dispersal evolves "to avoid local extinctions." In fact, this can occur only when the number of patches (or islands) is finite. Otherwise, it can be shown that the geometric mean fitness is not affected by dispersal.

Clearly, there must be interactions between the three above mechanisms. For instance, it is possible to incorporate both local extinctions and kin selection in the same model. One then finds a simple expression for the evolutionarily stable (ES) dispersal rate, as a function of the local extinction probability, the cost of dispersal (mortality caused by emigration), the average relatedness among individuals born in the same patch, and the probability of common origin of immigrants.

DISPERSAL AND OTHER TRAITS

Dispersal coevolves with other life history traits, such as dormancy rates, reproductive effort, and longevity. Fecundity modifies selection on dispersal differently according to the main mechanism involved in dispersal (e.g., kin selection or competition avoidance). Conversely, selection on fecundity and reproductive effort depends on dispersal, and thus these traits are expected to jointly evolve with dispersal. It has also long been recognized that dispersal and dormancy antagonistically coevolve: Dormancy generally reduces the evolutionarily stable level of seed dispersal. Some models, however, predict that in the presence of positive temporal correlations of habitat quality, dormancy instead increases dispersal. Finally, in a spatially heterogeneous environment, selection on dispersal depends on local adaptation, and vice versa. There are only few models on the joint evolution of dispersal and local adaptation.

Apart from their larger propensity to disperse, dispersing individuals are usually not a random sample of the original population. For instance, in insects with wing dimorphism, insects with larger wings often have a higher fecundity than do insects with smaller wings. Similarly, in plants with a seed dimorphism for dispersal, seeds without a dispersal structure (pappus) are sometimes smaller than seeds with a pappus (Fig. 2). Whether this is the result of a coevolution with dispersal traits or whether it is a mere consequence of pleiotropic effects is not always known.

VARIOUS STAGES OF DISPERSAL

Even from the strict point of view of dispersal in space, organisms can vary in the stage being dispersed (e.g., gametes or zygotes, juveniles or adults, males or females). When several life forms disperse, antagonism arises, which can lead to, for example, a single dispersing sex. In species in which males and females have the same ploidy level, with dispersal of unfertilized females and equal costs of dispersal of males and females, the kin competition effects of local mate competition in males and local resource competition in females balance each other, such that the optimal dispersal rates of males and females are predicted to be equal. Differential migration between sexes is expected under any deviation from the above assumptions (e.g., in the presence of inbreeding depression, when females are mated prior to dispersal, or when females are diploid and males haploid). Haplo-diploidy or dispersal of fertilized females in diploids creates asymmetry in the kin competition parameters of females and males, such that female dispersal is favored over male dispersal. Pollen and seed dispersal in diploid plant species is not rigorously equivalent to dispersal of males and females in these cases, as pollen is haploid and seeds are zygotes rather than mated females. Still, kin competition favors among-patch seed dispersal over among-patch pollen dispersal, with this trend being generally conserved with inbreeding load. This result stems from

EVOLUTION OF DISPERSAL ON ISLANDS AND FRAGMENTATION

Following introductions on islands, many insect species have evolved a reduction or a loss of wings. Similarly, island birds often show reduced flight ability or willingness to fly. There is some evidence that the ability to disperse seeds by wind decreases with the number of years since colonization of islands. Darwin was the first to suggest the "wind hypothesis" to explain the early-on documented loss of dispersal ability on islands: Wings would allow wind to blow organisms away into unfavorable water masses, so the loss of wings would therefore be favored by natural selection on small, isolated islands. Thus, on islands, there should be selection against long-distance dispersal.

However, in order to colonize an island, one needs to have dispersal mechanisms in the first place. Thus, it is likely that, besides selection against dispersal following colonization, there is selection for dispersal at foundation—what has been called a "metapopulation effect." There is some evidence that colonizers are indeed a non-random sample of mainland species.

Are such antagonistic forces between the need to arrive and the cost of dispersal sufficient to understand the evolution of dispersal on islands? Most likely, they are not. First, Darwin's "wind hypothesis" has received little support. Instead, the high proportion of flightless insects in Antarctica seems associated with low temperatures. Second, the other forces involved in the evolution of dispersal on the mainland are also likely to occur on islands. For instance, even on islands, short-distance dispersal can evolve as a way to avoid sib-competition and inbreeding. Furthermore, such short-distance dispersal can allow the colonizing of vacant sites within the island once it has been reached. Conversely, increased fragmentation of mainland habitats results in species occupying more and more isolated fragments, which can each be viewed as an island.

Landscape fragmentation is a complex process, affecting the number of habitat fragments in the landscape, their size, the distance separating them, and the nature of the sub-optimal habitat between. Understanding the evolutionary consequences of habitat fragmentation for the evolution of dispersal requires taking into account all these dimensions. Models often mimic increasing fragmentation by increasing dispersal cost. When local population sizes in remnant fragments are large, and mortality during dispersal is low, increasing such mortality results in the evolution of lower dispersal rates. Yet, when the landscape is initially highly fragmented with small habitat fragments and high dispersal mortality, a further increase

FIGURE 2 (A) Slender thistles of *Carduus pycnocephalus* (Asteraceae). (B) Flowerhead of *C. pycnocephalus*. (C) Achene dimorphism for dispersal in *C. pycnocephalus*. Each flowerhead contains at maturity ten to 20 achenes with a pappus, and one to five achenes without a pappus. Genetic variation exists for the proportion of the two types of achenes. Photographs provided by Pierre-Henri Gouyon and Isabelle Olivieri.

relatedness between seeds being higher than relatedness between pollen grains. There are some examples where wind pollination has evolved from animal pollination following colonization of islands, but the mechanism underlying such evolution, which results in increased pollen dispersal on islands, is unclear.

in dispersal mortality unexpectedly selects for increasing mobility. Decreasing patch size and holding dispersal mortality constant also selects for increasing mobility. Concurrent changes in dispersal mortality and fragment size thus result in complex patterns of dispersal evolution as fragmentation proceeds through complex interactions between kin competition intensity and extinction-recolonization dynamics. Nevertheless, in wolf spiders, ballooning propensities were found to decrease both with decreasing salt marsh size and connectivity. Spatially explicit simulations have explored how spatial variation in the cost of dispersal would affect spatial distribution of dispersal genotypes. Isolated habitat fragments are expected to contain more dispersive genotypes than would well-connected fragments when they had just been recolonized, but the reverse trend holds for fragments colonized for a longer time. Similarly, species and/or genotypes with a larger propensity or ability to disperse are more likely to colonize new islands, or "Darwinian islands," whereas "fragment islands" are increasingly isolated one from another, and only those philopatric species or genotypes are likely found on these fragments. Along the same lines, dispersal has also been found to decrease during ecological succession, whereas it increases during biological invasions or range expansions.

SEE ALSO THE FOLLOWING ARTICLES

Flightlessness / Fragmentation / Inbreeding / Invasion Biology / Metapopulations / Succession

FURTHER READING

Bowler, D. E., and T. G. Benton. 2005. Causes and consequences of animal dispersal strategies: relating individual behaviour to spatial dynamics. *Biological Reviews* 80: 205–225.
Bullock, J. M., R. E. Kenward, and R. S. Hails, eds. 2002. *Dispersal ecology*. Oxford, UK: Blackwell Publishing.
Clobert, J., E. Danchin, A. A. Dhondt, and J. D. Nichols, eds. 2001. *Dispersal*. Oxford: Oxford University Press.
Dingle, H. 1996. *Migration: the biology of life on the move*. New York: Oxford University Press.
Gillespie, R. G., and G. K. Roderick. 2002. Arthropods on islands: colonization, speciation, and conservation. *Annual Review of Ecology and Systematics* 47: 595–632.
Hanski, I., and O. E. Gaggiotti, eds. 2004. *Ecology, genetics, and evolution of metapopulations*. Amsterdam: Academic Press.
Olivieri, I., and P.-H. Gouyon. 1997. Evolution of migration rate and other traits: the metapopulation effect, in *Metapopulation biology: ecology, genetics, and evolution*. I. Hanski and M. E. Gilpin, eds. San Diego, CA: Academic Press, 293–323.
Ronce, O. 2007. How does it feel to be like a rolling stone? Ten questions about dispersal evolution. *Annual Review of Ecology, Evolution and Systematics* 38: 231–253.
Silvertown, J., and J. Antonovics, eds. 2001. *Integrating ecology and evolution in a spatial context*. Oxford: The British Ecological Society and Blackwell Science.

DODO

J. P. HUME

Natural History Museum, London, United Kingdom

The dodo, *Raphus cucullatus* (family Columbidae), has become one of the most famous birds in the world, a true icon of extinction, with probably more written about it than any other species, yet we know practically nothing about the bird in life. Contemporary accounts and illustrations are often contradictory, plagiarized from earlier sources or simply manufactured from pure imagination. This has resulted in a wealth of scientific myths and misconceptions based on totally inadequate source material.

DISCOVERY OF THE MASCARENES

The volcanic and isolated Mascarenes Islands, comprising Mauritius, Réunion, and Rodrigues, are situated in the western Indian Ocean. Mauritius, the home of the dodo, lies 829 km east of Madagascar, the nearest large land mass. Arab traders probably discovered the Mascarene Islands as early as the thirteenth century, followed by the Portuguese in the early sixteenth century, but neither the Arabs nor Portuguese, as far as it is known, settled there. Following the acquisition of Mauritius by the Dutch East India Company (VOC) in 1598, the islands were used as ports of call for provisioning ships. For a short period thereafter, the Dutch and other European nations recorded in ships' logs and journals vague and inadequate references to the original fauna and flora, including the dodo. These early accounts are invaluable in determining the island's original ecological composition, because by the end of the seventeenth century, Mauritius had been altered beyond recognition due to the ravages of humans and their commensal animals. It was during this century of human occupation on Mauritius that the dodo became extinct.

WRITTEN EVIDENCE

It was standard practice for VOC fleets to record in ships' logs and journals all details concerning their voyages, including shipping routes and safe harbors for ship refurbishment. Upon the return of the fleets, the journals became important source material for future VOC voyages, artists, scientists, and book publishers. It was these publications, often expanded and illustrated long after the voyage itself, that have become the source material for scientific study. Except in the case of the skilled artist

Joris Joostensz Laerle, who accompanied the Gelderland fleet to Mauritius in 1601–1603, most observers were not trained naturalists. The observations of contemporary voyagers, therefore, are based primarily on the culinary aspect of the fauna and only secondarily on its appearance or habits. Researchers have used these drawings and accounts as a basis for determining the ecology and morphology of species now extinct, notably the dodo, and as a result, a continuous series of misinterpretations has been made. Despite the wealth of material that has been written based on the contemporary accounts, very little is reliable and based on actual observation.

It was during the voyage of Admiral Jacob Cornelisz van Neck in 1598 that Mauritius was claimed for the Netherlands, although van Neck never actually visited the island. The discovery was made by Vice Admiral Wybrandt Warwijck, who had been separated from van Neck during a storm. While anchored in Vieux Grand Port, southeast Mauritius, Warwijck sent a reconnaissance party on shore including ship's mate Heyndrick Dircksz Jolinck, who saw and described the dodo for the first time. Upon the return of the fleet to the Netherlands in 1599–1600, the dodo was mentioned in a small publication entitled "A True Report," which also gave an account of the voyage. Enlarged and expanded editions were published in 1600 and 1601, which included a copper engraving, illustrating not only Dutch activities on shore but also, for the first time, the dodo and other animals (Fig. 1). Van Neck's account has been plagiarized more than any other.

For the next few decades, Dutch fleets called at Mauritius, either on the way to or from the East Indies. Only a few accounts mentioned the fauna; the last detailed description was made in 1631. Despite visiting Mauritius, the Englishmen Peter Mundy in 1628 and Thomas Herbert in 1629 made further observations, but these were from captive dodos in Surat, India. The Frenchman Francois Cauche visited Mauritius in 1638 but did not publish his account until 1651. Furthermore, his account is untrustworthy. He is the only observer to record the call, nest, and egg of the dodo, but he appears to have mingled these descriptions with that of the cassowary *Casuarius casuarius*. In 1662, ship-wrecked Dutch sailor Volkert Evertszen saw living dodos on an islet off Mauritius and managed to catch some after a chase. Two Dutch governors of Mauritius, Commander Hubert Gerritsz Hugo between 1673 and 1677 and Opperhoofd Isaac Joan Lamotius between 1677 and 1692, recorded dodos. Unfortunately, they gave no descriptions other than the Dutch name "Dodaersen," which has resulted in some confusion, as another species of flightless bird, the red rail *Aphanapteryx bonasia*, which survived until around 1700, was similarly named. It may have been the dodo or the rail to which they referred. According to Lamotius, who mentions the dodo for the last time in 1688, the dodo had become extremely rare by this time and presumably died out shortly after.

PICTORIAL EVIDENCE

Very few images of dodos derived from ships' journals exist, and only one observer, the aforementioned Joris Joostensz Laerle, drew a dodo on Mauritian soil. How many of the other illustrations were based on live birds is not known, as most were plagiarized from other sources or derived by artists from mariners' verbal accounts to illustrate popular books. These attempts at illustration, although charming in their own way, leave much to be desired and can only be used as rough interpretations of the living bird. Furthermore, some illustrations were made from memory and not published until many years later, so it is inevitable that inconsistencies exist. Living or stuffed dodos made the journey to the Far East and Europe, particularly the Netherlands, and these specimens formed the basis for numerous paintings, all differing to varying degrees in posture and coloration. Despite the lack of scientific credibility, some authorities continue to erect new species based on this illustrative and written evidence. The existence of the supposed white dodo from Réunion was based entirely on four white dodo paintings and contemporary accounts. However, recent work has shown that the accounts are referable to a white ibis *Threskiornis solitaria*, not a dodo, and the paintings are based on an albinistic dodo from Mauritius.

The most famous and prolific dodo artist was Roelandt Savery (1576–1639), who drew and painted the bird at least ten times, but nearly always in the same pose. It is

FIGURE 1 Dutch activities on Mauritius in 1598. This image was published in the Netherlands in 1601, three years after the Dutch discovery of Mauritius, and included the first illustration of the dodo (center left). Reprinted from Strickland and Melville, 1848.

not known whether the model was a live or stuffed bird, but variations in his renditions imply that some were based on memory and others created purely for artistic composition. Much scientific speculation has been founded on these variations, including seasonal fat and thin cycles, sexual dimorphism, and age, without taking into account the abilities or intentions of the artists involved. Therefore, it is impossible to obtain any scientific credence from them. This is certainly not the case with a dodo portrayal by Ustad Mansur from around 1625. This specimen was part of a menagerie kept by Emperor Jahingir in Surat, and is illustrated among other accurately portrayed species of birds. It is almost certainly the most reliable colored rendition of the dodo that has survived.

TRANSPORTATION OF SPECIMENS, LIVING OR DEAD

Because of the minor variations in dodo illustrations, it has been postulated by some authorities that each image represents a different individual. Based on this assumption, at least 17 dodos must have been transported to Europe. However, physical and documentary evidence suggests that as few as two or three dodos made the journey alive to Europe, and a similar number survived the journey east. One specimen, the only dodo to have unequivocally reached Europe alive, was exhibited in a shop somewhere in London in 1638. This individual may have been the one that ended up in the Ashmolean Museum (now University Museum), Oxford. Despite popular belief, this unique stuffed specimen was not thrown onto a fire to be burned. After examining the disintegrating specimen in 1755, the trustees could save only the head and one foot, disposing of the rest. The head, which still retains soft tissue, and the bony core of the foot still reside in Oxford today, whereas all other dodo skin remnants have long since decayed.

EXTINCTION

Inferences made from the few accounts of dodos on Mauritius indicate that the birds disappeared on the mainland concomitant with ever-increasing encroachment by humans and their commensal animals. Hunting was probably restricted to the coastal areas and extremely limited because of the small human population; therefore, it was almost certainly competition and predation by introduced animals, such as rats, monkeys, pigs, goats, and deer, that were responsible for the dodo's demise. Although still a matter of debate, dodos may have survived until at least 1688, but they had probably ceased to breed long before, with the last aged survivors hanging on in just a few remote places.

RACE TO FIND THE FIRST FOSSIL EVIDENCE

Until the discovery of dodo bones in 1865, virtually no physical remains existed, leading some authorities to doubt that the dodo had ever existed. However, the stuffed Oxford remnants, a foot in London and a skull in Copenhagen, still survived. The British elements formed the basis for the first anatomical study by Hugh Strickland and Alexander Melville in 1848, after which scientific interest in procuring dodo specimens intensified—in particular, the need to discover fossil material on Mauritius.

George Clark, Master of the Diocesan School at Mahebourg, Mauritius, spent some years searching the island hoping to discover dodo fossil material. At the same time, Harry Higginson, a railway engineer, was constructing a railway embankment alongside a marsh called the Mare aux Songes, near the spot where the Dutch first landed. In October 1865, Higginson noted that the laborers were stockpiling bones from the marsh. Informing Clark of the discovery, preliminary excavations revealed the first dodo bones, after which Clark immediately monopolized the site. He sent the first consignment of material in late October to Richard Owen, comparative anatomist at the then British Museum (now the Natural History Museum) and was paid £100 for 100 bones. Owen wasted no time in formally describing and illustrating the dodo's anatomy the following year (Fig. 2),

FIGURE 2 Richard Owen's reconstruction of the dodo in 1869. After illustrating a stout, squat dodo in 1866, Owen, upon obtaining more fossil material, had the bird redrawn in 1869, producing a much more accurate reconstruction. Reprinted from Hume, 2006.

amidst a commotion of high-profile public lectures and engagements.

The Mare aux Songes marsh was reworked more intensively in 1889 by Théodore Sauzier and again by Paul Carié in the early 1900s, resulting in the retrieval of many more dodo bones. Such was the abundance of dodo material collected from the marsh—albeit a composite of many different individuals—that almost all dodo remains held in museum collections today are derived from this one site. Around 1904, Louis Etienne Thirioux, a hairdresser by trade, discovered a complete dodo in a cave near Le Pouce Mountain. It can still be seen in the Mauritius Institute, Port Louis. Until 2007, this was by far the most important fossil dodo discovery made.

AFFINITIES

The affinities of the dodo were explored by numerous authors, and it was often preposterously placed within a large assortment of bird orders—for example, a miniature ostrich, a rail, or even a vulture. After examining a skull in Copenhagen, Professor J. T. Rheinhardt proposed that the dodo was related to Columbiformes (pigeons and doves). This assertion was initially met with ridicule, but after Strickland and Melville confirmed his theory by examination of the Oxford dodo head, the idea became universally accepted. DNA studies have now concluded that the dodo and closely related solitaire *Pezophaps solitaria* of Rodrigues are a sister clade nested within the family Columbidae and derived from the same ancestor as the southeast Asian Nicobar pigeon *Caleonas nicobarica*.

RECENT DISCOVERIES

In 2005, a Dutch expedition discovered fresh fossil material at the Mare aux Songes. This resulted in a full scale excavation in 2006 and 2007, which revealed thousands of bones beneath a layer of rubble, put in place by the British Army during the 1940s to prevent malaria. The fossil layer also contained seeds, tree trunks and branches, leaves, insects, land snails, and even fungi, deposited long before humans arrived on the island. The fossil remains are dominated by extinct giant tortoises *Cylindraspis* sp., but they also include snakes, lizards, owls, hawks, rails, parrots, pigeons, and songbirds. The flora comprised palms, canopy trees such as tambalacoque *Sideroxylon grandiflorum* and ebony *Diospyros* sp., and a host of smaller plant species, enabling scientists to reconstruct the dodo's habitat in a pristine state.

In 2007, more important discoveries were made. A complete but degraded dodo skeleton, "Dodo Fred," was discovered in a cave in the highlands, with further fossil discoveries made in lowland caves, which increased not

FIGURE 3 A reconstruction of the Mare aux Songes before the arrival of humans. The dodo (bottom center) stands amidst giant tortoises, red rails, parrots, pigeons, and other animals surrounded by a palm-rich forest. Today, less than 2% of native forest exists, and all but four animal species included in the illustration are extinct. From an original painting by the author.

only the known distribution of the dodo, but also the chances of obtaining good quality DNA.

As a result of this new physical evidence, it is now possible to make scientifically valid conclusions about the dodo's ecology. The dodo was found close to the coast as well as in the mountains, occupying dry and wet forest zones, and feeding on fallen fruits and seeds. Judging by the number of individuals preserved, it was abundant in the lowlands at least, and it coexisted with vast numbers of giant tortoises and other animals. The seasonal lake/marsh environment at the Mare aux Songes not only abounded with fruiting trees and shrubs but probably acted as an oasis on the otherwise dry leeward side of the island, thus attracting numerous birds (Fig. 3). However, there are still uncertainties about the dodo's morphology, and no doubt, more interpretation will be made from the few images and contemporary accounts. What is certain is that the dodo was unable to cope with the rapid changes brought about by anthropogenic agencies and died out less than a century after being discovered.

SEE ALSO THE FOLLOWING ARTICLES

Extinction / Flightlessness / Fossil Birds / Mascarene Islands, Biology

FURTHER READING

Cheke, A., and J. P. Hume. 2008. *Lost land of the dodo*. London: A & C Black Publishing Ltd.
Fuller, E. 2002. *Dodo: from extinction to icon*. London: Harper Collins.
Hachisuka, M. 1953. *The dodo and kindred birds, or the extinct birds of the Mascarene Islands*. London: H. F. & G. Witherby.
Hume, J. P. 2003. The journal of the flagship Gelderland—dodo and other birds on Mauritius 1601. *Archives of Natural History* 30: 13–27.

Hume, J. P. 2006. The history of the dodo *Raphus cucullatus* and the penguin of Mauritius. *Historical Biology* 18: 65–89.

Hume, J. P., and A. S. Cheke. 2004. The white dodo of Réunion Island: unravelling a scientific and historical myth. *Archives of Natural History* 31: 57–79.

Moree, P. 1998. *A concise history of Dutch Mauritius, 1598–1710*. London: Kegan Paul International.

Mourer-Chauviré, C., R. Bour., S. Ribes., and F. Moutou. 1999. The avifauna of Réunion Island (Mascarene Islands) at the time of the arrival of the first Europeans. *Smithsonian Contributions to Paleobiology* 89: 1–38.

Owen, R. 1866. *Memoir of the dodo* (*Didus ineptus, Linn.*). London: Taylor and Francis.

Strickland, H. E., and A. G. Melville. 1848. *The dodo and its kindred*. London: Reeve, Benham & Reeve.

DROSOPHILA

PATRICK M. O'GRADY, KARL N. MAGNACCA, AND RICHARD T. LAPOINT

University of California, Berkeley

The genus *Drosophila* provides excellent opportunities to study evolution on island systems. The endemic Hawaiian *Drosophila* are a classic example of adaptive radiation and rapid speciation in nature evolving in situ over the course of the past 25 million years. Other groups of *Drosophila*, found on true islands or in island-like systems (e.g., the Madrean Archipelago), are invaluable tools to understanding evolutionary biology and have served as theoretical and empirical model systems for over 50 years.

HAWAIIAN *DROSOPHILA*

The endemic Hawaiian Drosophilidae, with an estimated 1000 species, consists of two major lineages, the Hawaiian *Drosophila* and the genus *Scaptomyza*. The high degree of species diversity in Hawai'i is extraordinary, with about one-sixth of the world's known Drosophilidae being endemic to this small archipelago. Phylogenetic analyses indicate that the family colonized the Hawaiian Islands only once, roughly 25 million years ago. The genus *Scaptomyza*, which also contains a large number of mainland taxa, seems to have escaped from Hawai'i and undergone subsequent diversification on the mainland (see below).

There are currently 411 described species of Hawaiian *Drosophila*, with an additional ~150 awaiting description. These species have been divided into eight species groups (*picture wing, nudidrosophila, ateledrosophila, antopocerus, modified tarsus, modified mouthpart, haleakalae, rustica*) based largely upon sexually dimorphic characters possessed by males and thought to be used mainly in courtship and mating. These characters range from elaborately pigmented wings to elongate setae, cilia, and bristles on the forelegs to unique structures on the mouthparts and forelegs (Fig. 1). The wide variety of secondary sexual characters possessed by males of the various species groups suggests that sexual selection may have played an important role in the diversification of this group.

Approximately 85% of Hawaiian *Drosophila* are single-island endemics, possibly owing to their relatively poor flight abilities and low tolerance for desiccation. These physiological constraints, coupled with the unique geological history of the Hawaiian Islands, have led to a spatially distributed pattern of diversification referred to as the progression rule, where older species are found on older islands, and younger species are found on younger islands (Fig. 1).

One explanation for the large numbers of drosophilid species in the Hawaiian Islands involves adaptation to so-called empty niches. This atmosphere of reduced competition allowed these species to experiment with novel life history strategies, and thus to diversify. Several adaptations unique to the Hawaiian Drosophilidae seem to bear this out. For example, the small group of ~15 species placed in the *Scaptomyza* subgenus *Titanochaeta* have specialized to oviposit in spider egg sacs. Larvae develop and parasitize the spider eggs while they are being guarded by adult spiders. Sixty-seven percent of Hawaiian *Drosophila* for which data is available breed on only a single host plant family, whereas 79% are specific to a single host substrate, such as leaves, bark, or sap flux (Magnacca et al., 2008).

Hawaiian *Drosophila* have served as a model system to address a number of evolutionary phenomena, including how founder events and mating asymmetries can drive species formation (Fig. 2). Throughout their evolutionary history, Hawaiian *Drosophila* have repeatedly undergone founder events, either as they colonize new islands or when populations are subdivided (e.g., by lava flows or erosional processes). Hampton Carson utilized Hawaiian *Drosophila* to illustrate his founder flush theory, a type of founder effect speciation that proposes a reduction in intraspecific competition and an increase in population size following a colonizing event. Once the population size becomes large again, selection is reasserted, and the population may constitute a new species (Fig. 2).

Alan Templeton also used Hawaiian *Drosophila* to explain his transilience founder effect theory. Templeton

Kaneshiro extended founder effect theories to include the complex mating behaviors and secondary sexual characteristics exhibited by Hawaiian *Drosophila*. In most populations, females are very choosy in selecting males with which to mate. However, small, colonizing populations are initially subject to founder effects, in which the scarcity of males forces females to be less selective when choosing a mate: If they are too choosy, they will not find a mate. Additionally, males from this small population display highly variable mating behaviors because of reduced intraspecific competition and relaxed selection in what are normally highly selected behaviors (Fig. 2). Choosy females from the larger source population will not mate with the males of the founder population, although females of this founder population will accept males from the larger source population.

CARIBBEAN *DROSOPHILA*

Patterns of in situ species formation are not as clear in the Caribbean islands because many widely distributed species are also found in mainland North and South America. Historically, this has led to a situation in which gene flow from closely related mainland populations can act to homogenize any unique genetic differences that may accumulate in island populations. In spite of the close proximity of mainland ancestors, some Caribbean *Drosophila* have become genetically distinct from these widespread ancestral species. The *repleta* and *cardini* groups highlight two factors, ecological specialization and morphological adaptation, that may act to drive the formation of new species in the presence of gene flow from the mainland.

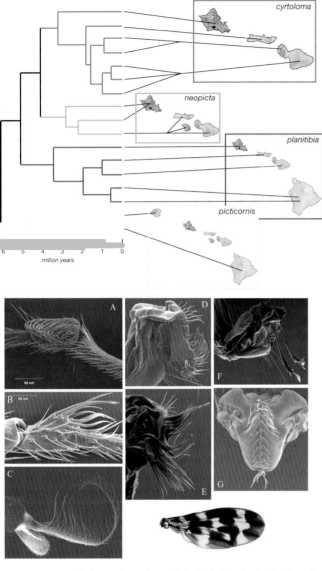

FIGURE 1 Phylogenetic and morphological diversity in the Hawaiian Drosophilidae. Phylogeographic patterns in the *planitibia* species group of Hawaiian *Drosophila* showing four independent examples of the progression rule in the *cyrtoloma* (blue), *neopicta* (green), *planitibia* (red), and *picticornis* (yellow) subgroups. Secondary sexual modifications on the forelegs of (A) *D. waddingtoni* and (B) *D. quasiexpansa;* the arista of (C) *D. tanythrix;* mouthparts of (D) *D. xenophaga,* (E) *D. hystricosa,* (F) *D. adventitia,* and (G) *D. freycinetae;* and the wing of *D. grimshawi*.

discussed how, after a bottleneck derived from a founder event, a highly outbred colonizing population would experience a shuffling between epistatic loci that were adapted to work best in certain combinations (Fig. 2). This would cause an instant shift in adaptive peaks to ones optimal for these randomly recombined loci. Although he concludes that this is likely a rare event, the ecological and genetic nature of the Hawaiian *Drosophila* lend them to this sort of speciation.

	Carson's Founder-Flush	Templeton's Genetic Transilience	Kaneshiro's Mating Asymmetry
ancestral population	outcrossed, coadapted and polymorphic	outcrossed, coadapted and polymorphic	outcrossed, coadapted and polymorphic
	founder event causes drastic reduction in effective population size		
drift	disrupts coadapted gene complexes	disrupts coadapted gene complexes (major genes)	reduces female selectivity on male courtship display
recombination	yes	yes	yes
pleiotropic balance	altered	altered	altered
genetic variation	carries over through founding event	carries over through founding event	carries over at most loci, variation in behavior loci may increase in males and decrease in females
	flush generates a period of rapid population growth		
selection	natural; relaxed until after flush; external and environmental	natural; strong during flush but with high variation at modifier loci; entirely genetic	sexual; relaxed until after flush; females accept wider range of male displays
genetic response	polygenic, but most loci not affected	only a few major genes and their modifiers affected	only loci involved in mating display and acceptance affected
behavioral response	maybe	maybe	shift in mating patterns may become fixed

FIGURE 2 Comparison of founder flush, genetic transillience, and asymmetrical mating hypotheses.

Ecological adaptation may be an important component of species formation. The *Drosophila repleta* group contains about 100 described species, the majority of which are endemic to the New World and have diversified on various species of necrotic cacti. One group, the *mayaguana* triad, is restricted to the Caribbean and includes the widespread species *D. mayaguana,* and the more narrowly restricted taxa *D. straubae* and *D. parisiena.* Although few morphological differences distinguish these species, they do have unique polytene chromosome inversions and ecological associations (different host cacti) that may have acted as a reproductive barrier to generate distinct species.

Caribbean island *Drosophila* have also been used as models to understand morphological variation and phenotypic plasticity. The *Drosophila cardini* group consists of 16 neotropical species in two subgroups, *cardini* and *dunni*. The *dunni* subgroup is entirely Caribbean in distribution; each species is endemic to a specific island in the Greater and Lesser Antilles. These species display a cline of abdominal pigmentation, with more lightly pigmented species being found in the northwest, and darker species in the southeast (Fig. 3). This pattern may be due to island endemics being more geographically isolated and having smaller population sizes, which can cause traits (e.g., color patterns) to become rapidly fixed in the population. However, studies on the genetic basis of this color variation suggest that genetic control of abdominal pigmentation is highly malleable, with similarities in coloration not necessarily reflecting relatedness.

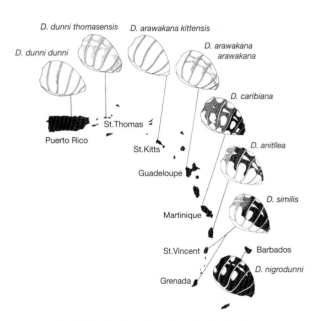

FIGURE 3 Spatial distribution of species and abdominal pigmentation patterns in the *Drosophila cardini* group.

SCAPTOMYZA AND ISLAND COLONIZATION

The cosmopolitan genus *Scaptomyza* (Diptera: Drosophilidae), with nearly 300 described species placed in 20 subgenera constitutes a major radiation within the family Drosophilidae. About 140 described species are endemic to Hawai'i, and an additional ~250 from this archipelago await description. Phylogenetic analyses strongly support the monophyly of *Scaptomyza* and its placement as the sister group to the endemic Hawaiian *Drosophila*. Unlike the Hawaiian *Drosophila,* however, *Scaptomyza* males generally lack secondary sexual characteristics and do not perform elaborate courtship displays. Instead, they have complex genitalia that can be quite different, even between closely related taxa, and may act to reinforce a "lock and key" form of species recognition.

Biogeographic analyses suggest that the genus *Scaptomyza* originated in the Hawaiian Archipelago and has subsequently "escaped" from the islands and diversified on the continent. This is a particularly interesting biogeographic pattern, because no other known group of organisms has colonized a continent from Hawai'i. The genus *Scaptomyza,* perhaps because of its history of radiation on and migration from Hawai'i, is extremely successful as a colonist of remote island archipelagos. Of the 20 currently recognized subgenera, over half possess successful island radiations: Seven (*Alloscaptomyza, Elmomyza, Engiscaptomyza, Exalloscaptomyza, Grimshawomyia, Tantalia, Titanochaeta*) are endemic to Hawai'i, two (*Bunostoma, Rosenwaldia*) are known from the Pacific including Hawai'i, one (*Boninoscaptomyza*) is endemic to the Ogasawara Islands south of Japan, and three (*Lauxanomyza, Macroscaptomyza, Trogloscaptomyza*) are endemic to the South Atlantic islands of St. Helena and Tristan da Cunha. The South Atlantic islands Tristan da Cunha and St. Helena also host representatives of two widespread subgenera, *Scaptomyza* (2 spp.) and *Parascaptomyza* (6 spp.), for a total of eight endemic species. Other species in the subgenus *Parascaptomyza* are island endemics from such disjunct islands as the Azores, Canaries, Marquesas, Cape Verde, and Java.

Many insect species lose the ability to fly when they become fully adapted to island life. One species of brachypterous *Scaptomyza, S. altissima,* is known from Tristan da Cunha and Gough Islands in the South Atlantic. Outside of this example, the loss of flight in this family is rare, although the endemic Hawaiian *Drosophila* are significantly less mobile than their mainland counterparts.

ISLAND-LIKE SYSTEMS

The Madrean Archipelago is a system of mountain islands distributed across the southwestern United States and northern Mexico. During the middle Pliocene, this area was characterized by a mild, warm, semiarid climate. Rapid desertification as a result of the uplift of the Sierra Nevada and the coincident elevation in temperature over the past ~2.5 million years have led to the break-up of these plant and animal communities and their restriction to higher elevations. Present-day mountaintops in this region are characterized by cooler, more mesic habitats than are the surrounding arid deserts. Several groups of mycophagous species in the *macroptera, rubrifrons, quinaria,* and *testacea* species groups have formed in association with these island systems because of their dependence on the fungi that are geographically restricted to these areas.

In *Drosophila innubila,* paleoclimatic data suggest that populations may have been panmictic until 18,000 years ago, when they became fractured into their present-day distributions. Significant genetic differentiation between populations suggests that gene flow is restricted among these sky islands, and the high geographic variation in mitochondrial DNA may be due to an association with the male-killing *Wolbachia* endosymbiont, which tends to accentuate the effects of genetic drift.

PARALLEL EVOLUTION IN ISLAND *DROSOPHILA*

Island endemic species often evolve unique morphological structures, behaviors, or ecological associations. Some adaptations may generate large radiations, such as the Hawaiian *Drosophila,* whereas others, like the spider predators in *Scaptomyza* (*Titanochaeta*), are represented by only a few species. An association between drosophilids and land crabs, one of the most interesting ecological adaptations in this family, has occurred independently on three different island systems. Although the majority of species in the *repleta* group are cactophilic, the West Indies endemic *D. carcinophila* has been reared from the crab species *Gecarcinus ruricola*. Likewise, although the entire *quinaria* group is mycophagous, one representative, *D. endobranchia,* has also been recorded from *G. ruricola* in the Cayman Islands and Cuba. Similarly, within the predominantly mycophagous *Lissocephala, L. powelli* from Christmas Island has been recorded from several crab genera, including *Gecarcoidea, Birgus, Geograpsus,* and *Cardisoma*. All three crab fly species have independently evolved the ability to utilize the external nephric groove or branchial chamber of these land crabs. These structures are involved in the excretion of nitrogenous and food wastes, respectively. The nephric groove and its associated green gland host a unique assemblage of microorganisms that are sufficient to support the complete development of *D. carcinophila* and at least one instar of *D. endobranchia* and *L. powelli*. Adaptation to this lifestyle from either a cactus or fungal ancestor most likely involves a suite of changes in oviposition preference, olfactory and gustatory receptors, and larval behavior. Recent genome sequencing has facilitated the understanding the genetics of host-plant association in the genus *Drosophila,* and similar studies will shed light on this fascinating ecological adaptation.

SEE ALSO THE FOLLOWING ARTICLES

Flightlessness / Founder Effect / Hawaiian Islands, Biology / Insect Radiations / Sexual Selection

FURTHER READING

Brisson, J. A., A. Wilder, and H. Hollocher. 2006. Phylogenetic analysis of the *cardini* group of *Drosophila* with respect to changes in pigmentation. Evolution 60: 1228–1241.

Carson, H. L. 1968. The population flush and its genetic consequences, in: *Population biology and evolution*. R. C. Lewontin, ed. New York: Syracuse University Press, 123–137.

Carson, H. L., and A. R. Templeton. 1984. Genetic revolutions in relation to speciation phenomena: the founding of new populations. *Annual Review of Ecology and Systematics* 15: 97–131.

Craddock, E. M. 2000. Speciation processes in the adaptive radiation of Hawaiian plants and animals. *Evolutionary Biology* 31: 1–53.

Kaneshiro, K. Y., and C. R. B. Boake. 1987. Sexual selection and speciation: issues raised by Hawaiian *Drosophila*. *Trends in Ecology and Evolution* 2: 207–212.

Magnacca, K. N., D. Foote, and P. M. O'Grady. Breeding ecology of the endemic Hawaiian Drosophilidae. *Zootaxa* in press.

Markow, T., and P. M. O'Grady. 2005. Evolutionary genetics of reproductive behavior in *Drosophila*: connecting the dots. *Annual Review of Genetics* 39: 263–291.

O'Grady, P. M. 2002. Species to genera: phylogenetic inference in the Hawaiian Drosophilidae, in *Molecular systematics and evolution: theory and practice*. R. DeSalle, G. Giribet, and W. Wheeler, eds. Berlin: Birkhauser Verlag, 17–30.

DWARFISM

SHAI MEIRI

Imperial College London, United Kingdom

PASQUALE RAIA

University of Naples, Italy

Insular dwarfism is a tendency of many island animals to evolve a smaller size than their ancestors on the near mainland. Dwarfism may be relatively minor but can sometimes be drastic, with some Mediterranean island elephants reduced in weight by up to 99%. The main selection pressures thought to promote dwarfism involve responses to resource limitation, small prey size, and enhanced reproductive output (in mammals). Dwarfism may more easily be attained on islands that lack, or have fewer, competitors and predators. Whereas members of some clades (e.g., ungulates) dwarf regularly on islands, others have a much wider range of size responses to insularity, often showing either dwarfing or gigantism on islands of the same archipelago that differ in their community composition. Although insular communities are sometimes perceived as having more dwarf species than would be expected by chance, this has not been quantitatively shown.

INSULAR DWARFISM IN EXTANT LINEAGES

Islands the world over have populations and species that are markedly smaller-bodied than their close relatives on the mainland. Most of the well-documented cases of dwarfism are of mammals, in which dwarfism is known to have occurred in a large number of species. Among primates, dwarfism has been demonstrated in Southeast Asian macaques (*Macaca nemestrina*, *M. fascicularis*), gibbons, and proboscis monkeys. Among even-toed ungulates, dwarfism is known to have occurred in some populations of Japanese pigs, deer from the southeastern United States, and north Pacific populations of reindeer. Dwarf carnivores include some populations of Southeast Asian civets, mongooses, and leopard cats. Dwarfism also occurs often (but not always) among raccoons, which have many dwarf populations and species inhabiting islands in the Caribbean, in the Florida Keys, and on Cozumel Island. Cozumel is also home to dwarf coatis and gray foxes, which, on their part, also show remarkable dwarfism (between 5 and 40% in different populations) on the California Channel Islands. Some bats also seem to dwarf on islands, and Hawaiian *Lasiurus cinereus* is almost 50% smaller than its North American conspecifics. Dwarfism has also been demonstrated in Japanese moles; in Southeast Asian, eastern Pacific, and Channel Island shrews among insectivores; and in some populations of tree shrews. Many rodents, a clade often noted for its tendency toward insular gigantism, actually have dwarf insular populations. Examples include some populations of Southeast Asian squirrels of the genus *Callosciurus,* mice of the genera *Maxomys* and *Sundamys,* western American island heteromyids of the genus *Chaetodipus,* pack rats (*Neotoma*), Senegalese islands *Mastomys* rats, Minorca dormice (*Eliomys quercinus*), and others. Panamanian sloths and some Australian marsupials (e.g., *Parantechinus apicalis*) also have populations of insular dwarfs.

Other vertebrates for which insular dwarfism has been shown include birds, such as many insular endemic ducks, streak-backed orioles (*Icterus pustulatus*) on Islas Marias off the Pacific Coast of Mexico, and South Pacific rails. Insular dwarfism has also been observed in quite a few reptiles: Caribbean boa constrictors, European grass snakes on an island off the Swedish coast, four-lined rat-snakes on small southern Japanese islands, Australian tiger snakes, *Anolis luteogularis* on islands off Cuba, *Egernia cunninghami* on islands off southern Australia, and even Komodo dragons on small islands near Flores and Komodo.

At a larger scale, the smallest species in some clades are found on islands: The smallest snake (the Lesser Antillean threadsnake *Leptotyphlops bilineata*), the smallest constrictor (*Tropidophis pardalis* of Cuba), and many of the smallest lizards (Caribbean members of the genus *Sphaerodactylus* and Madagascar chameleons of the genus *Brookesia*) are insular endemics. The smallest representatives of some bird orders are also insular: The smallest bird is often considered to be the bee hummingbird of Cuba; the smallest swift (*Collocalia troglodytes*) inhabits the Philippines; the smallest member of the order Coraciiformes, *Todus mexicanusi,* is endemic to Puerto Rico; and the world's smallest parrot (*Micropsitta pusio*) inhabits Papua New Guinea.

It is far from clear, however, if these examples generalize to a common pattern whereby dwarfism is exceedingly common on islands. Reptiles, terrestrial mammals, and especially birds and bats are extremely diverse on islands. Over evolutionary time, it is reasonable that at least some lineages, especially those with many insular species, will evolve diverse body sizes. Thus, the fact that the smallest member of many groups is an insular endemic can be expected to some extent to have occurred by chance alone. Whether more of the smallest members of different clades are insular than would have been predicted by chance has not yet been properly studied.

An appropriate statistical test of this pattern will likely involve randomization tests, in which a given number of species equivalent to the known number of insular endemics are drawn from a global species pool, and their masses compared to those actually observed. As things stand now, however, whether islands really harbor an unusual number of dwarfs is not clear.

INSULAR DWARFISM IN EXTINCT LINEAGES

Dwarf species are common in extinct faunas. Among mammals, insular dwarfs arose repeatedly in elephants, deer, and hippopotamuses. In Mediterranean islands, extinct dwarf elephants are known from Sicily, Favignana, Malta, Crete, Cyclades (Naxos and Delos), Dodecanese

(Rhodes and Tilos), and Cyprus. They also occur in Japan. Diminutive mammoths lived in Sardinia and Crete but are also known to have inhabited the Channel Islands off the California coast. A small, but perhaps not really dwarf, mammoth survived the end-Pleistocene extinction of the continental lineage on Wrangel Island, north of the Bering Strait. Remains of this species have been ^{14}C-dated at some 4000 years before present. A very similar age estimate was calculated for the Tilos elephant. Another fossil proboscidean genus, *Stegodon,* went dwarf several times in the Indonesian Archipelago, on Flores, Java, and Sulawesi. Elephants showed perhaps the greatest degree of dwarfism: *Elephas falconeri* from the middle Pleistocene of Malta and Sicily (Fig. 1) weighed approximately 100 kg, or just 1% the weight of its putative European ancestor, *Elephas antiquus.*

Dwarf fossil deer occur in Crete, Karpathos, Sicily, Malta, Capri, Sardinia, Corsica, Jersey, and Japan. Pigmy hippopotamuses lived in Malta, Sicily, Cyprus, Crete (two species), and Madagascar (three species). The Malagasy *Hippopotamus lemerlei* was still living some 1000 years ago. Dwarf bovid occurred in Elba Island (off of Italy's west coast) and Sicily. A dwarf buffalo, *Bubalus cebuensis,* from Cebu Island (Philippines) has been recently described. Enigmatic small goats of the genus *Myotragus* lived in the Balearic archipelago. The descent of *Myotragus* is unclear; hence, its degree of dwarfism cannot be calculated. A similar case could be put forward for the Sardinian *Nesogoral.*

Carnivores were much rarer on islands, and it is claimed that they defy the island rule. Consequently, dwarf extinct carnivores are exceedingly rare. A possible exception is the Sardinian *Cynotherium sardous,* a small canid whose ancestry is unknown.

Birds are common in island fauna; especially in oceanic islands that they reached by flight. There, they often come to occupy ecological niches whose continental equivalents are usually taken up by mammals. This niche shift entails flightlessness. Thereby, freed by the weight constraint imposed by flight, island birds often became gigantic. As a consequence, dwarfism is much rarer. A supposed exception is a rail species on the Balearic Islands, for a smaller-sized relative of extant water rail *Rallus aquaticus* lived on Ibiza and possibly on Formentera.

Dwarf reptiles occurred in several clades. Small crocodiles inhabited South Pacific islands. The 2-m-long crocodile *Mekosuchus inexpectatus* lived in New Caledonia. The congeneric *M. kalpokasi* lived on Vaté Island, Vanuatu. A further mekosuchid, *Volia athollandersoni,* settled on Viti Levu, Fiji Archipelago. Although these species are comparatively small by crocodilian standards, the group they belong to, *Mekosuchinae,* includes other small species that once inhabited mainland Australia; hence, their dwarfism, although probable, has not been definitively proven.

Some claims have even been made for insular dwarfism in sauropod dinosaurs, although none of these has convincingly showed that the small sauropods in question were actually smaller than any putative mainland ancestor, or, indeed, that they were even truly insular. These include the titanosaurid *Magyarosaurus dacus* and the basal macronarian *Europasaurus holgeri.*

Albeit described quite recently, the most famous case of dwarfism is the Flores Man, *Homo floreisensis*. These little hominids (nicknamed "hobbits") stood some 1-m tall and lived at the very end of the Pleistocene, some 18,000 years ago. Although some researchers argue that *H. floreisensis* is not a real species but represents pathological conditions of modern humans, the notion of dwarfism in this possible *Homo erectus* descendant holds out. It is the smallest and least-brained of all hominids, yet it was capable of producing stone tools and surviving in a restricted habitat along with a dwarf *Stegodon* until a few millennia ago. *H. floreisensis* wrist bones are quite apelike, supporting the theory that it is a primitive species. Some believe, however, that *H. floreisensis* was actually a microcephalic modern human (or that it was a modern human with a defective thyroid), based on paleopathologic investigations. Evidence for microcephalia, however, is contradicted by computed tomography (CT) analysis of a skull, and the debate goes on.

SELECTION PRESSURES PROMOTING DWARFISM

Dwarfism is thought to result from a combination of several selection pressures. Chief among these is resource limitation. Because islands are small, they have lower

FIGURE 1 Comparison of molars of the woolly mammoth (left) and the diminutive Sicilian *Elephas falconeri* (right). Both teeth belonged to full-grown individuals.

absolute resource abundance than do continents (even though local abundance may not be lower, and, through density compensation, it may even be higher on islands). The notion, therefore, is that animals will benefit because growing smaller will lower their overall energetic needs. Although this may be plausible for very large animals on very small islands, it probably does not account for some of the more drastic cases of insular dwarfism. Louise Roth famously calculated that the island of Sicily, at 26,000 km^2, could easily have supported a healthy population of even 15-ton elephants; thus, overall resource abundance could not explain the evolution of 100-kg dwarfs.

On very small oceanic islands, low overall food abundance may explain some cases of dwarfism, often accompanied by a great reduction in the relative mass of the energy-hungry flight muscles in birds—leading to insular flightlessness. Small individuals, however, often lose out when competing with their larger conspecifics for food or mates. Although dwarfism may be for the good of the population, individuals bearing mutations that make them grow larger may be more successful than their dwarf kin. Dwarfism is therefore not an evolutionarily stable strategy.

A more general explanation for insular dwarfism, however, likely involves low local food abundance (calories per square meter, rather than overall caloric value of food on the entire island). Local food shortage is well known to lead to size reduction on continents as well as islands. For predators preying on small prey, dwarfism may be beneficial if it enhances their efficiency in catching such prey.

Two of the most widely invoked explanations for insular dwarfism, reduced competition and predation pressures on islands, cannot in themselves explain the evolution of small body size. Although it is true that on islands a smaller competitor will often be absent, *allowing* its larger competitor to grow smaller, there must be some adaptive advantage to actually achieve this. Thus, although the absence of competitors can permit dwarfism, it is insufficient in itself to explain it. A similar argument holds for predation. Although it may well be true that large size is an adaptation to predator avoidance, the absence of predators can allow dwarfism to evolve, but large size must be a costly adaptation, or small size a highly advantageous one, for dwarfism to evolve.

Two interesting theories predict dwarfism as being favorable for large mammals as a whole. The first suggests that there is a single, optimal body size for mammals. Under this scenario, this size is ecologically or physiologically optimal, allowing animals to control the most resources or to allocate the greatest part of their available energy toward offspring production. Smaller and larger animals are, according to these theories, prevented from evolving such size by competition pressures. On competitor-free islands, large mammals are thus expected to evolve to a smaller size. These theories, however, have some serious theoretical shortcomings and have failed nearly all attempts at empirical verification.

Another interesting theory relies on the tendency of smaller mammals to start breeding earlier and to have larger and more frequent litters. Usually this entails higher juvenile mortality, yet, on islands, two major sources of mortality in young individuals—predation and competition—are reduced, and large mammals can therefore dwarf to enhance their reproductive success. In groups where large size is associated with larger clutches and litters (e.g., reptiles), this mechanism is unlikely to drive dwarfism.

COMPARATIVE STUDIES AND THE EVOLUTION OF DWARFISM

Comparative studies of size evolution on islands usually examine both dwarfism and gigantism, under the pretext of the "island rule," rather than restricting their discussion to dwarfism. Some exceptions are studies that address groups in which dwarfing predominates, almost invariably fossil elephants and artiodactyls, which mostly support theories emphasizing the effect of competition. Comparative studies of groups of modern animals usually try to establish either which groups tend towards dwarfism or what circumstances lead to dwarfism. The former line of investigation seems to indicate that dwarfism is the general response to insularity in elephants, deer, and hippopotamuses, as well as in mongooses and civets. Evidence regarding the predominance of dwarfism among primates and leporids (rabbits and hares) is more contentious.

A perhaps more interesting line of study tries to establish what conditions promote dwarfism, rather than gigantism or no size change. It seems that many species can rapidly evolve either large or small body sizes on islands in response to the biotic characteristics of the island they find themselves on. Some studies indicate that abiotic conditions can determine the course of evolution, with dwarfism in birds evolving in more equatorial islands in agreement with Bergmann's Rule. Aspects such as poor soils may also promote dwarfism through their influence on the amount of available food. Studies of insular populations of snakes, predatory lizards, and carnivorous mammals have repeatedly identified the size and abundance of available prey as a key determinant of body size evolution. Insular populations grow large if abundant and/or large-bodied prey exists on the island, but small size will evolve if prey is scarce or small.

DWARFISM OR NOT GROWING LARGER?

Showing that individuals in insular populations are smaller than their relatives on the mainland does not necessarily mean the insular population is dwarfed. Dwarfism means that the island was colonized by mainland individuals, and evolution of smaller size followed. Two other plausible explanations are either (1) that the mainland was colonized by insular individuals, and the mainland population subsequently evolved large size, or (2) that members of the insular population remained the same size or even grew larger since colonization, but the mainland population evolved an even larger size since. The second scenario was suggested as early as 1909 by Charles Deperet, who is often acknowledged as defining what is nowadays commonly referred to as Cope's Rule. Under this rule, animal lineages increase in size over their evolutionary history. Deperet believed, therefore, that insular elephants were not dwarfs, because this would have negated his rule; therefore, they must have simply "failed" to evolve to a larger size. The evolutionary forces affecting body size, however, are likely to be similar, whether or not an insular population is technically dwarfed or is simply comprised of small-sized "failed giants."

FUTURE PROSPECTS

Although well known and intensively studied, dwarfism has not been thoroughly explained. First, recent evidence has demonstrated that it is highly clade specific (that is, some groups of phylogenetically related animals are much more likely to experience dwarfism than others). Second, dwarfism is a product of (decreased) growth. Yet growth reduction could result from a shorter total growth period (of the insular species as compared to its mainland ancestor), a lower growth rate, or delayed growth onset. These shifts in timing of ontogenetic events are parts of the evolutionary phenomenon called heterochrony. Surprisingly, heterochrony has been little investigated in relation to dwarfism, although some evidence suggests they are related. For instance, a species of dwarf elephant has been said to be paedomorphic. Altered growth rates have also been reported in a Jurassic "dwarf" dinosaur. An explicit test of heterochrony would help discriminate among competing explanations advanced for dwarfism. For example, life history–based theories predict that dwarfism results from reduced somatic development (allowing resources to be reallocated toward reproduction). Theories focusing on reduced caloric intake, on the other hand, predict that animals would dwarf without paedomorphosis.

SEE ALSO THE FOLLOWING ARTICLES

Gigantism / Island Rule / Rodents / Snakes / Tortoises

FURTHER READING

Case, T. J. 1978. A general explanation for insular body size trends in terrestrial vertebrates. *Ecology* 59: 1–18.
Clegg, S. M., and I. P. F. 2002. The "island rule" in birds: medium body size and its ecological explanation. *Proceedings of the Royal Society B.* 269: 1359–1365.
Gordon, K. R. 1986. Insular evolutionary body size trends in Ursus. *Journal of Mammalogy* 67: 395–399.
Heaney, L. R. 1978. Island area and body size of insular mammals: evidence from the tri-colored squirrel (*Callosciurus prevosti*) of Southeast Asia. *Evolution* 32: 29–44.
Lomolino, M. V. 1985. Body size of mammals on islands: the island rule reexamined. *American Naturalist* 125: 310–316.
McNab, B. K. 1994. Energy conservation and the evolution of flightlessness in birds. *The American Naturalist* 144: 628–642.
Meiri, S. 2007. Size evolution in island lizards. *Global Ecology and Biogeography* 16: 702–708.
Meiri, S., N. Cooper, and A. Purvis. 2008. The island rule: made to be broken? *Proceedings of the Royal Society B.* 275: 141–148.
Millien, V., and J. Damuth. 2004. Climate change and size evolution in an island rodent species: new perspectives on the island rule. *Evolution* 58: 1353–1360.
Raia, P., C. Barbera, and M. Conte. 2003. The fast life of a dwarfed giant. *Evolutionary Ecology* 17: 293–312.
Raia, P., and S. Meiri. 2006. The island rule in large mammals: paleontology meets ecology. *Evolution* 60: 1731–1742.

EARTHQUAKES

PAUL OKUBO
U.S. Geological Survey, Hawaiian Volcano Observatory

DAVID A. CLAGUE
Monterey Bay Aquarium Research Institute,
Moss Landing, California

An earthquake is a sudden movement in the Earth, occurring sufficiently quickly that the movement generates seismic waves. Basic descriptors of an earthquake are origin time, location or hypocenter, and magnitude, which are inferred by analyzing the times and amplitudes of seismic wave arrivals at recording stations equipped with seismographs. Although the vast majority of earthquakes recorded worldwide are small and are detected only by sensitive instruments that gauge the radiated seismic waves, large earthquakes can have widespread and devastating effects. Large earthquakes are also capable of triggering tsunamis when they occur at shallow depths in submarine or near-coastal regions. Earthquakes occur worldwide, but many are located on and near islands because many islands are active volcanoes or are near tectonic plate boundaries. Emergency response on islands is more difficult because of their isolation, making appropriate land use planning, adoption and enforcement of appropriate building codes, and emergency planning more critical there than elsewhere.

CAUSES OF EARTHQUAKES

The location of an earthquake is fundamental to understanding its cause. Globally, the vast majority of earthquakes are located on the margins of the Pacific basin—delineating what has become known as the Pacific "Ring of Fire"—with close associations in some areas with active volcanism. Elsewhere, the distributions of earthquakes help outline the Earth's tectonic plates that move atop the underlying asthenosphere. The global distribution of earthquakes, showing the Ring of Fire and otherwise outlining tectonic plates, is illustrated in Fig. 1.

Different cultures developed their own explanations for the cause of earthquakes that affected them. The legends include giant catfish beneath Japan, elephants standing on top of a turtle on top of a cobra beneath India, subterranean winds beneath Greece, and numerous supernatural beings who shake the Earth, sometimes at whim, sometimes in anger.

Studies of the San Francisco, California, earthquake of 1906 provided a strong scientific basis for the relationship between earthquakes and faults. From this work, the "elastic rebound theory" for earthquakes and faults was proposed, wherein the Earth's crust around a locked fault deforms and stores strain energy until the fault abruptly ruptures in an earthquake. Since 1906, earthquakes, faults, and faulting processes have been studied extensively to further develop basic earthquake science. Although earthquake models have been generalized to account for other geological settings beyond the San Andreas fault in northern California, the basic ideas of elastic rebound theory still apply.

Many islands along the Ring of Fire, as well as islands like Hawai'i in intraplate regions, are also active volcanoes. In volcanic settings, earthquakes occur in response to magma movement. In some cases, magma movement results in changes in stress that subsequently lead to earthquakes; these are referred to as volcano-tectonic earthquakes. In other cases, magma movement creates oscillations within the magma transport system. Sustained oscillations are

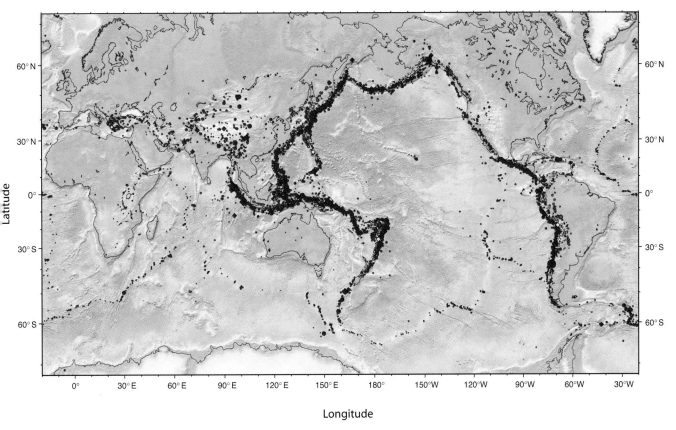

FIGURE 1 Global map of earthquake epicenters, 1900–June 2008, adapted from the Centennial Catalog (Engdahl and Villasenor 2002) to show earthquakes with magnitudes larger than 7.5 since the year 2000 as diamond symbols. Symbol colors indicate earthquake depths: red, 0–70 km deep; yellow, 70–350 km deep; blue, deeper than 350 km. In addition, earthquakes of magnitudes 8 or greater are indicated as symbols outlined with a heavy black line, and the December 2004 M9.0 Sumatran and May 2008 M7.9 Sichuan earthquakes are shown as stars. The reporting threshold for completeness of the Centennial Catalog since 1964 is considered M5.5.

referred to as volcanic tremors, whereas discrete events are referred to as volcanic earthquakes. It is common for volcanic tremors or earthquakes to feature a dominant frequency of oscillation. In these cases, they are referred to as harmonic tremors and long-period, or LP, earthquakes.

EARTHQUAKE EFFECTS

The defining effects of an earthquake are seismic waves radiated from the earthquake focus or hypocenter. For small earthquakes or earthquakes in remote areas, the registration of seismic waves on seismographs can often be the only evidence of an earthquake's occurrence. The two principal types are body waves and surface waves. There are two types of body waves: P-waves and S-waves. There are also two types of surface waves: Love waves and Rayleigh waves. The seismic waves produce earthquake shaking. Current practice includes compiling records of earthquake effects according to the Modified Mercalli Intensity scale. (Popular discussions of earthquakes often mention the Richter scale, which is an estimate of the earthquake's magnitude, or total amount of energy released. The Richter magnitude scale was developed to quantify earthquake sizes in southern California, by recording earthquake waves and comparing their respective amplitudes on a collection of standard seismic instruments; since then, magnitudes have been generalized to allow estimation from any calibrated seismic recording system. The Mercalli scale, abbreviated MMI, on the other hand, is an assessment of the quake's intensity, or effect at a given location.)

Seismic body waves travel through the Earth. The first-arriving seismic wave is the primary wave, or P-wave. The P-wave motions are longitudinal or compressional, or traveling in the direction of wave propagation. The motions of the later-arriving secondary wave, or S-wave, are transverse, or perpendicular to the direction of wave propagation. S-waves are also referred to as shear waves. S-wave amplitudes are several times larger than P-wave amplitudes. Surface waves propagate at and near the Earth's surface. They travel more slowly than do body waves, with Love waves being faster than Rayleigh waves. At greater distances from the earthquake source, surface waves are larger in amplitude and have longer periods of oscillation than body waves. The different types of seismic waves are illustrated in Figs. 2 and 3, with examples from locally and teleseismically recorded earthquakes.

In addition to seismic waves, other earthquake effects include surface faulting, ground failure, and tsunamis. Close enough to a large, shallow earthquake, surface faulting can be observed as relative offsets across the two sides of the causative fault. Seismic shaking can lead to other styles of ground failure. Earthquakes can trigger landslides on steep slopes covered by unconsolidated or weak materials. Strong seismic shaking can also lead to an effect called liquefaction, which is a dynamically induced loss of soil strength or cohesion such that soils temporarily behave as though they were fluid rather than solid. Tsunamis are generated when the earthquake faulting itself, or landslides triggered by earthquakes, vertically displaces the oceanic water column. The December 26, 2004, M9 Sumatran earthquake triggered a tsunami with widespread devastating effects across the Indian Ocean. Consequences of ground shaking and subsidence are shown for several large earthquakes in Figs. 4–7.

EARTHQUAKE HAZARDS AND RISKS

Because of their historical impact, there has been considerable interest in understanding earthquakes. Data and technology have complemented history and experience, and earthquake science encompasses a wide range of observational, theo-

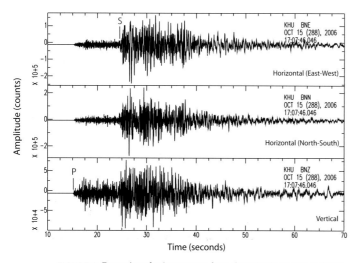

FIGURE 2 Examples of seismograms from the seismic station, Kahuku, Hawai'i, from the October 15, 2006, M6.7 Kiholo Bay, Hawai'i, earthquake recorded locally (earthquake to station distance of 77 km) showing the different types of seismic waves described in the text.

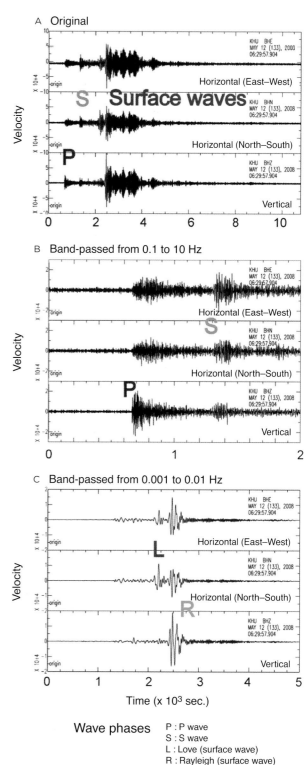

FIGURE 3 Examples of seismograms from the seismic station, Kahuku, Hawai'i, from the May 12, 2008, M7.9 Sichuan, China, earthquake recorded teleseismically (earthquake to station distance of 9930 km) showing the different types of seismic waves described in the text.

242 EARTHQUAKES

FIGURE 4 A house that collapsed during the 1989 Kalapana M6.1 earthquake. Photograph by J. D. Griggs, U.S. Geological Survey.

FIGURE 5 Road damage along the rim of Kilauea caldera caused by the 1983 Kaoiki M6.6 earthquake. Photograph by J. T. Takahashi, U.S. Geological Survey.

FIGURE 6 Damage to the bridge and highway at Paauilo, Hawai'i, following the Oct 15, 2006, M6.7 Kiholo Bay earthquake. Photograph by A. M. Preister, U.S. Geological Survey.

retical, and computational investigation. Although predicting earthquakes remains an arguably worthy goal, it is also quite important to understand and correspondingly mitigate against the hazards and risks that earthquakes pose.

FIGURE 7 Coastal subsidence on the south flank of Kilauea Volcano caused by the 1975 Kalapana M7.2 earthquake. Photograph by P. Lipman, U.S. Geological Survey. This earthquake also triggered a small tsunami.

Seismic hazards across the United States are now consistently presented as the product of probabilistic seismic hazard modeling. Details of earthquake history, seismic wave propagation, near-surface geology, and tectonic plate motion and fault displacement rates are input into computer calculations to determine exposure to strong earthquake shaking. The results are typically shown as maps that contour the levels of strong ground shaking with a specified probability of being exceeded in a specified timeframe, as shown in Fig. 8 for the Hawaiian Islands. A principal application of this modeling lies in the development and adoption of building codes.

With or without seismic hazard modeling, islands exposed to earthquake hazards face special challenges in developing response plans to mitigate the effects of earthquakes and in executing an effective response once an earthquake occurs. Critical infrastructure and facilities, especially on smaller islands, represent single points of failure, and, should they fail or be compromised, response and recovery are susceptible to life-threatening and costly delays. These facilities include hospitals, fire, and police facilities; port and airport facilities; roads and bridges; power generation, water, and other utilities; stores; and fuel depots within the affected areas. Land use planning and adoption and enforcement of appropriate building codes are the first lines of defense, but detailed response and recovery planning based on prior experience and understanding of the geologic record and environment will prove invaluable when disaster strikes.

SEE ALSO THE FOLLOWING ARTICLES

Landslides / Plate Tectonics / Tsunamis

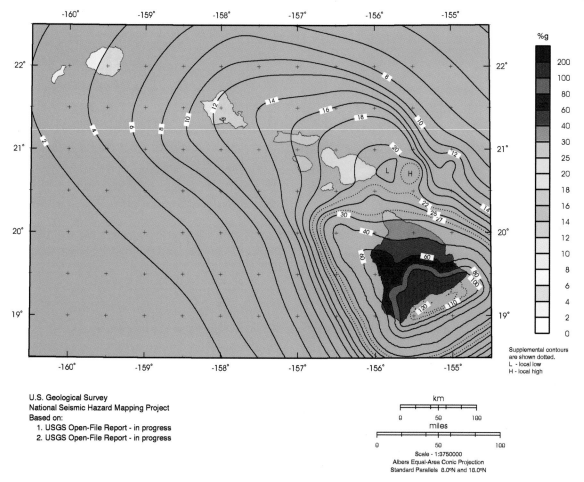

FIGURE 8 Probabilistic earthquake hazards map of the Hawaiian Islands from Klein *et al.* (2001) showing peak horizontal acceleration as a percentage of gravitational acceleration with 10% probability of exceedence in 50 years. The largest peak accelerations are most likely beneath the south flank of Hawai'i and rapidly decrease for the islands to the northwest.

FURTHER READING

Bolt, B. A. 2006. *Earthquakes.* New York: W. H. Freeman and Company.

Buchanan-Banks, J. M. Structural damage and ground failures from the November 16, 1983 Kaoiki earthquake, Island of Hawaii, in *Volcanism in Hawaii,* U. S. Geological Survey Professional Paper 1350, 1187–1220.

Engdahl, E. R., and A. Villaseñor. 2002. Global seismicity: 1900–1999, in *International handbook of earthquake and engineering seismology,* part A, chapter 41. W. H. K. Lee, H. Kanamori, P. C. Jennings, and C. Kisslinger, eds. San Diego, CA: Academic Press, 665–690.

Hough, S. E. 2002. *Earthshaking science: what we know, and don't know, about earthquakes.* Princeton, NJ: Princeton University Press.

Klein, F. W., A. D. Frankel, C. S. Mueller, R. L. Wesson, and P. G. Okubo. 2000. *Seismic-hazard maps for Hawaii.* U. S. Geological Survey, Geologic Investigations Series I-2724.

Klein, F. W., A. D. Frankel, C. S. Mueller, R. L. Wesson, and P. G. Okubo. 2001. Seismic hazard in Hawaii: high rate of large earthquakes and probabilistic ground motion maps. *Bulletin of the Seismological Society of America* 91: 479–498.

Klein, F. W., R. Y. Koyanagi, J. S. Nakata, and W. R. Tanigawa. 1987. The seismicity of Kilauea's magma system, in *Volcanism in Hawaii.* U. S. Geological Survey Professional Paper 1350, 1019–1185.

Lee, W. H. K., H. Kanamori, P. Jennings, and C. Kisslinger, eds. 2002. *International handbook of earthquake and engineering seismology,* part A. San Diego, CA: Academic Press.

Scholz, C. H. 2002. *The mechanics of earthquakes and faulting.* Cambridge: Cambridge University Press.

Stein, S., and M. Wysession. 2003. *An introduction to seismology, earthquakes, and Earth structure.* Malden, MA: Blackwell Publishing.

Zobin, V. M. 2003. *Introduction to volcanic seismology.* Amsterdam: Elsevier.

EASTER ISLAND

GRANT McCALL
University of New South Wales, Sydney, Australia

Rapanui (Easter Island) is the world's most remote continuously inhabited place, with a unique environment, astonishing traditional evolution, and tragic, but ultimately triumphant, modern history of contemporary development of its vigorous and adaptable population.

A NOTE ON NAMES

"Easter Island," after the Christian holy day on which Europeans first arrived there, is the international name, translated into the various world languages, for the place. "Rapanui" is the name and spelling used mostly by the Islanders themselves for their language, themselves, and their island. "Rapa Nui" is the rendering used in official Chilean sources, as the island is an autonomous region of that South American country. This article uses the locally favored spelling "Rapanui." "Rapanui" was the name that other Pacific Islanders gave to the island in the early 1860s, as they thought that the place was related to the (then) better-known Rapa in the Austral group. In common with other Pacific islands, it is likely that traditionally the Rapanui did not have a single name for their island—just a large number of designations for particular parts of their homeland.

PHYSICAL DESCRIPTION

Rapanui lies at Latitude 27°13 S and Longitude 109°37 W, with an area of 166 km², its highest point being Mount Terevaka at 511 m. The nearest inhabited island is 2011-km-distant Pitcairn, giving Rapanui the top score in the isolation index of 149 (from http://islands.unep.ch). The nearest continental point is the Chilean port of Caldera, some 3568 km away. The local time is set artificially by the governing Chilean administration at –6 hours GMT.

Rapanui resulted from lava flows from three principal volcanoes (in order of appearance): Poike, Rano Kau, and Terevaka. This gives Rapanui its distinctive triangular shape (Fig. 1).

As is the case with Pacific high islands, the leeward side differs in significant ways from the windward one. Leeward (roughly west) the soils are thin and the area rocky, but fishing points are more prevalent. The windier windward side has deeper soils, and the marine resources were exploited more along the shore than in the open sea. This windward/leeward contrast was represented culturally, and there is a still-visible line of marker stones running through the ideological center of the island, which shows the ancient physical and cultural division.

The climate is subtropical with varying cycles of rain (as much as 1250 mm per annum) and drought. The soil is porous and although there is little surface water apart from three large freshwater calderas, there is an abundant aquifer with wells used since ancient times, including even underground outlets into the sea along the shore.

FLORA AND FAUNA

Pollen evidence indicates that, prior to Polynesian colonization, Rapanui was covered with woodland and scrub. Some of the now extinct trees include *Paschalococos disperta*, a palm that appears to have been similar to, but distinct from, the Chilean *Jubaea* palm. It appears that this species grew to large size, indicating that the original forest, at least locally, grew to some stature. Other tree species included *Sophora toromiro*, the only species of tree known on the island in historic times, though now extinct in the wild. Shrubs included the hau hau (*Triumfetta semitriloba*), which is still present, and *Coprosoma* spp., now extinct. Most of the grasslands are now covered by introduced species and the native *Cynodon dactilon*, although several ferns are indigenous to Rapanui. The pollen record also indicates an interesting history of fluctuation in the vegetation, likely due to climatic changes. The indigenous fauna once included about 30 species of seabirds, of which 8–10 no longer breed on the main island and another 13–16 no longer breed even on any of the offshore islets. Landbirds were represented by at least six species in four families, although only the rail has so far been identified. There were also a number of marine mammals, and several indigenous insects, including *Asymphorodes trichogramma* (Cosmopterygidae, endemic to Rapanui and the Marquesas), and land snails.

The arrival of humans and their constant companion, the Polynesian rat (*Rattus exulans*), drove a number of bird species to extinction, eventually terminating the native palm, the shells of the nuts of which are found still in ancient nests of those ravenous animals. Although people always have been able to gather shoreline resources to feed themselves, there is no fringing reef, so the Rapanui people fished using canoes that, together with the trees from which they were built, had all but disappeared (but not the stories associated with them) when the first Europeans arrived.

FIGURE 1 NASA satellite photograph of Rapanui, taken March 27, 2005 (http://visibleearth.nasa.gov/view_rec.php?id=17133).

Most of the remaining native flora and fauna have vanished, largely because of the intense exploitation of the island as a sheep ranch for just over a century (1872–1973). This pattern of development, using mainly Australian-sourced trees and grasses, has given the island the barren appearance most often portrayed in print and documentary today (Fig. 2). Not only are the introduced Australian plants aggressive, but also the sheep ranch operators annually fired parts of the entire island one after another to help propagation and growth of the Australian plants, obliterating local species.

FIGURE 2 The desolate appearance of Rapanui is due to a century of sheep ranch exploitation and the replacement of native flora by imported Australian grasses and trees suitable for livestock. Photograph courtesy of Rapanui/O. I. R. Andreassen.

TRADITIONAL HISTORY

As with most Pacific Islanders, Rapanui trace their origins to a place other than the one they inhabit. In the Rapanui case, it is called Maraerenga and described as being some distance to the west. Variously, the culture hero and leader, Hotu Matu'a, had lost a battle or his land was flooding: He had to leave the homeland, and he took his people with him, eventually landing on Rapanui where his sons became the founding ancestors of the clans reported to researchers in the last century or so.

Linguistic, physical-anthropological, and cultural similarities suggest that the people who became the Rapanui did migrate in one or more waves from the west, perhaps from Mangareva or somewhere in the Marquesas group, about 1500 years ago. They brought with them a Polynesian menu of root crops, rats, and chickens and a belief system in a pantheon of creator gods who continued their occasional interventions in human lives. They brought with them also a Polynesian social organization in which descent groups (extended families) traced their ancestry to a pivotal figure, from whom they might take their collective name and identity. Although the contemporary naming system is derived from Chile, with its Spanish standards, in ancient times, Rapanui were known by one name, usually indicative of an incident in family history or another culturally significant event.

Their landing place most likely was on the western or leeward side, where living looked good with a rocky, shellfish-rich shoreline. From that point, they spread around the island, establishing the distinctive settlement pattern: an altar with a place before it for discussion and perhaps worship, and certainly an open public place known in many parts of eastern Polynesia as a *marae*. The word has been lost on Rapanui, but not the concept. "Marae" survives in the name of the origin place (see above) of the islanders: Maraerenga.

Instructive of the possible landing place is Ahu Tepeu on the coast, which has produced the earliest settlement date so far: It is aligned perfectly over a few kilometers with perhaps the last temple platform (*ahu*) to be completed, Ahu Akivi, one of the most photographed constructions on the island. Akivi is one of the few ahu located inland.

There is little evidence that after settlement the Rapanui had contact with other places, although around 1000 years ago either the Rapanui or other Polynesians made at least one landfall in South America and there made a remarkable discovery: the sweet potato. This nutritious and fast-growing root crop became the staple food for Rapanui and, eventually, for the rest of the Pacific Islands. More astonishing is that the name of this crucial cultivar, *kumara,* traveled with the plant and its tending. *Kumara* is the Quechua name for the sweet potato and is the word used in various forms throughout many of the Pacific Islands. Along with the sweet potato (*Ipomoea batatas*), the Polynesians brought back from Peru what is called today the Polynesian bottle gourd (*Lagenaria siceraria*), used for storage, and sometimes for cooking.

Recent work in Chile suggests that the distinctive Polynesian chicken (*Gallus gallus*) was brought to South America in pre-Columbian times, indicating perhaps more than one contact voyage.

THE ARCHAEOLOGICAL PAST

The arrival of the sweet potato to Rapanui revolutionized life there, as it did elsewhere: More people could be fed well with fewer resources devoted to planting and fishing to live. The result on Rapanui was a florescence, a burst of creative energy directed to megalithic architecture with commemorative figures of ever-increasing size (weighing as much as 60 tons in the end) being carved and transported over the island's surface.

The figures, called *moai*, were stylized busts intended to represent important ancestors. Each had an ancestor's name, and more than a thousand were carved between about 1200 and 700 years ago. The usual speculation is that the figures became larger and larger, although their style did not change, over the production period.

All of the figures were carved at one quarry, Rano Raraku, on the island's southeastern side. First, an outline was pecked into the relatively soft volcanic tuff, using *toki*, or stone hand adzes, of harder material found at various points around the island. Then, the figures were lowered into prepared pits at the base of Rano Raraku, where details were added and smoothing took place.

The next stage was transport, using levers of native palm (now extinct) and rope made from palm and other vegetable fibers. The details of this transport over about half a dozen or so still-visible paths are obscure, but many schemes have been proposed. The fact is that they were indeed moved over as much as 20 km; there are 42 moai en route from Rano Raraku to their prepared destinations. When the moai reached their prepared destinations, they were raised by levers and ropes, with stones being placed under the tonnage of the slowly rising figure until it was erect on its reinforced space.

It was at that point that *aringaora* took place: "Eyes" were placed in the sockets so that the ancestors indeed did look out over their descendants. All the moai face inland, overlooking the open space where people lived and transacted their lives. These eyes were made from white coral, with an obsidian disk in the center as a pupil. All the moai with eye sockets have a haughty appearance because the heads tilt back slightly, which is necessary so that the eye coral and obsidian would stay in place.

As the size of the figures grew, so did the number carved, transported, and erected on a group's ahu, or ceremonial space. Some ahu have no figures, and their design suggests that they were not constructed with that idea in mind. On the other hand, the recently restored Ahu Tongariki has 13 figures, the largest count (Fig. 3).

The moai were colored with various local earth pigments, traces of which can be found today: They probably required periodic refurbishment, and that was part of the veneration. As well, the ahu themselves were constructed by posing stones with similar colors and shapes in patterns, which, again, may be seen today by the trained eye. Some figures also had huge red scoria "topknots," *pukao*, mounted on them using the same piling stone technology. These pukao all were carved at one quarry, Puna Pao, and then transported to their waiting moai.

FIGURE 3 Moai at the restored Ahu Tongariki. Photograph courtesy of Rapanui/O. I. R. Andreassen.

Around 700 years ago, the carving, transport, and erection complex ceased, seemingly suddenly. There are moai in various stages of production and transport, and the period dubbed by one archaeologist *huri moai*, the toppling of the moai, started. Moai were tipped over as carefully as they had been erected, often with a stone placed to exactly break the figure at the neck. Many remain in that state today, with only restored ceremonial centers in various places showing the material culture at its stage of greatest elaboration. The usual explanation is that the figures were destroyed as a result of feuding.

A feature of Rapanui archaeology is that there are no obvious weapons of war found in the record prior to 700 years ago. Stone, including obsidian, was shaped for agriculture, fishing, and even human hair removal, but not as projectile points. No skeletons studied older than 700 years show evidence of trauma from battle. After 700 years ago, however, such weaponry and trauma abound, with scatterings of projectile points, some of them embedded in skeletons.

Another element bolstering the claim of original islandwide peace on Rapanui is that products from one part of the island were distributed around the place. The moai came from one part of Rapanui, as did the stone used in carving and construction. Fish and agricultural products were consumed evenly, although not produced everywhere. The culturally important obsidian came mainly from one location, Maunga Orito, but was found as shaped objects everywhere. Peace must have prevailed to have allowed this trade and distribution.

THE FALL OF THE MOAI

It is unclear why the moai complex ceased so seemingly abruptly and equally unclear how conflict came to characterize late Rapanui history. The evidence of an abrupt cessation of the moai carving and transport usually is of two sorts: the state of the moai and pukao quarries and the presence of more than three dozen moai in the course of transport along the ancient roads.

The quarry at Rano Raraku featured moai in various stages of completion, and for earlier visitors, even in the twentieth century, stone tools scattered about the work place. For that reason, as noted above, one can follow the stages of moai carving. Similarly, the pukao quarry at Punapau shows that production was interrupted. Around the island are 42 moai scattered along ancient roads, bound for their platforms. The receiving platforms themselves are in various stages of preparation.

The usual date for the cessation of the moai carving and transport complex is the fourteenth century AD. The usual explanation is that the Rapanui just ran out of resources. They (as with all humans, the story goes) allowed their population to outgrow their resources and squandered what they did have on dysfunctional, vainglorious cultural practices, such as the moai carving and transport complex. Such accounts use Rapanui as a metaphor for planet Earth, with a warning that if humankind does not control its population and better steward its natural resources, it will go the way of those remote Polynesians: collapse of civilization and descent into conflict and chaos.

Certainly, it is likely that the Rapanui population had increased from its small founding figure to a larger one by 700 years ago, and that would have put strain on the remote subtropical environment. Equally, the obvious devotion to the moai carving and transport complex would have taken up considerable resources (and tons of sweet potatoes), even if such activity did ensure peace and keep envy and war under control. But is that the whole story?

Around the same time, the Norse were closing down their distant settlements in North America and Greenland. The Black Plague was ravaging Europe and throwing it into confusion, marginal islands (such as Pitcairn) were being abandoned, and there were other shifts in cultural history worldwide.

There is global evidence of what Patrick Nunn has called "1300 Event": the "Little Ice Age," a global climate change that affected the entire planet, irrespective of the conduct of local human populations. The 1300 Event brought drought and changed weather conditions that affected navigation and famine.

If the 1300 Event affected the rest of the world, most likely it affected Rapanui, and such a severe climate change would have left the Rapanui starving and in a state of considerable disarray—so much disarray that they would have not had the resources to continue their moai carving and transport complex, whatever its islandwide peace benefits might have been. As in Europe, people would have attributed this violent climate change and its consequences to supernatural punishment for immoral human conduct.

Along with the cessation of the moai complex around 700 years ago, there was a decline in pollen production, detected in the freshwater lakes on Rapanui. This usually is attributed to human intervention. Equally, the Little Ice Age could have caused plants to produce less pollen in drought. A further contributor is the increasing presence of rats brought by Polynesians as passengers on their canoes, as mentioned above in the context of Polynesian settlement of Rapanui. As the human population increased, so did the rat one.

THE BIRDMAN CULT

At about the time that the moai carving and transport complex declined, another complex of belief and practice arose which we might call (as did the nineteenth century Rapanui), the *Orongo,* meaning "the call" or "news," as it does in other Polynesian languages.

Perhaps because the climate became so unpredictable, the one natural event connected with an annual cycle that continued was the arrival of the *Manutara* or sooty tern, formerly *Sterna fuscata,* now classified as *Onychoprion fuscata.* The annual migration, bringing news (*orongo*) of the austral spring in August/September, became a crucial calendrical cultural event.

At a place that came to be called Orongo, the Rapanui built a ceremonial village and associated sites out of rock overhangs with almost free-standing structures, using dry masonry technique with the flat slatelike rocks found at the site. As with the moai complex focus on Rano Raraku, the one site of Orongo served as a focus for the entire Rapanui population. The overt purpose of the yearly event was to result in the selection of a high chief of the island through an ordeal.

The Orongo observance probably took up two or more months. First, there was a meeting of representatives, or perhaps of the entire population of the island itself, on the plain of Mataveri below Rano Kau, the southwest vertex of the triangular island. Clan representatives would then ascend the mountain and lodge on a promontory until the first Manutara was sighted. At that moment, the competitors would scramble down the sharp, steep seaward side of Rano Kau and swim through shark-infested waters and strong current to the rookeries on the islet of Motunui. The first contestant, to find a Manutara egg, call out his discovery, and make it back across the channel and up the cliffs to Orongo would become the *Tangatamanu,* or "Birdman," the paramount leader (*Ariki Mau*) of Rapanui for the year. This dramatic event features in a 13-minute sequence in the 1994 Hollywood film *Rapa Nui* and is worth viewing for its accurate detail. Each Birdman had his image, with a sooty tern mask, carved in the rocks at Orongo. There are 243 such images.

Although the moai carving and transport complex and the Orongo represent very different activities and are founded on different principles, there are links between them. At Orongo, there was a moai carved from local—not Rano Raraku—material set in one of the structures (it was taken to the British Museum in 1868). There also was an ahu at Orongo, from the center of which one can look eastward to Puakatiki (the top of the Poike Peninsula) at the northeast corner of the island and westward the three islets off Rano Kau. These points define a line that oral tradition collected from the end of the nineteenth century designates as the center line of the island, dividing the northern clans (Tu'uaro) from the southern ones (Hotuiti). Modern Rapanui culture guardians actually have walked this line in their quest to achieve land justice from the current Chilean government: It is visible with boundary stones today. This is the boundary line, mentioned previously, that serves as the division between the windward and leeward sides of the island. As a further link between the older moai tradition and the later practice, the Birdman had to spend his year in power in a boat-shaped house on the slope of Rano Raraku.

The Tangatamanu and another new figure, Makemake, are characteristic of the last phase of pre-European history on Rapanui. Makemake was a phallic figure with his nose and eyes clearly being a phallus and testicles, respectively. Makemake was carved on stone around the island, including Orongo. Makemake seems to be later than other carvings because his image is found on overturned moai, on ruined ahu, and in caves that were the sites of refuge and habitation in the later period.

RONGORONGO: AN INDIGENOUS WRITING SYSTEM

In addition to the moai, another great interest people have in Rapanui is Rongorongo, an indigenously developed writing system, still undeciphered in spite of many efforts to do so since its first European study by Tepano Jaussen, Catholic bishop of Tahiti. In 1868 Bishop Jaussen received from Rapanui a gift of a rope tress of human hair wrapped around a piece of wood that turned out to be covered with unfamiliar glyphs. The regularity and order of these markings convinced Jaussen that they were a writing system of some kind, and he asked the priest on Rapanui, Father Hippolyte Roussel, to send any more such tablets to him, which the obedient priest did: three more pieces, one burnt slightly on the edge.

Since then, 18 or so such examples of Rongorongo, carved on wooden objects (often made from *Saphora toromiro*), have turned up from various sources, as have some amusing frauds, given the rarity and value of these artifacts. No Rapanui ever has been able convincingly to read the Rongorongo, although many have been asked and have tried. One Rapanui, known to the author, frequently entertained foreign visitors by writing and reading Rongorongo, but it was a syllabary that this clever man had developed himself based on the original signs.

The first outsider to report Rongorongo was the lay brother Eugène Eyraud, who stayed on Rapanui for nine months in 1863–1864. In his account, he noted that he saw "thousands" of Rongorongo in the dwelling places of the Rapanui: They were that common. Nevertheless, there is no evidence that a Rongorongo object ever was traded during the extensive contact period of 1722 to 1863. Plenty of other and varied pieces were part of the souvenir trade during those 141 years, but no Rongorongo.

So what happened to those thousands of Rongorongo seen by Eyraud? Those of the Rapanui who could read them had been killed or captured a few years earlier by slave raiders.

So, from where did all the others in public and private collections spring? One explanation could be that the Rongorongo script was secularized and became part of the trade in artifacts. Lacking in images of the original inscriptions, however, the forgers who made pieces for subsequent visitors produced them from memory or by copying and reorganizing published examples. The glyphs were applied, but in no particular order, because those masters who could read the script had died.

VISITORS AND TRADE

It is at this point that most people lose interest in Rapanui, the focus for most researchers and amateurs alike being the ancient past of megalithic moai, untranslated Rongorongo, and the peril-ridden Birdman ordeal. Tourist visits to Rapanui include little about what happened on the island after European "discovery" in 1722.

Jakob Roggeveen, licensed to trade in the West Indies, explored the East Indies, including Rapanui, where he arrived in his three ships on Easter Sunday, 1722. Many others—famous, infamous, and just passing—followed. In summary, about 100 ships stopped for varying periods of time at Rapanui between 1722 and the fatal slave raids of 1861–1862. Scholars have gone through the accounts of those who tarried more than a few hours to try to extract any information that would shed light on the ancient, spectacular past. The actual visits themselves as historical events or indicators of Rapanaui personality are rarely studied.

No artifacts were collected during the Dutch 1722 visit and the Spanish one in 1770. In 1774, however, during James Cook's short stay during his second voyage, he and his crew were besieged by islanders pushing their imaginative carvings for trade, as they did with all subsequent visitors. Indeed, the excellent carvings of Rapanui craftsmen adorn the museums and homes of all who have visited the place for any length of time.

For this period, it is enough to say that the islanders developed a detailed series of methods of dealing with outsiders, including a lucrative artifact trade from Cook's time on. There were incidents of kindness and brutality from both Rapanui and visitor. What is clear is that the Rapanui managed to produce an extensive corpus—both numerically and stylistically—of goods that are scattered in collections around (mostly) the European world. The variety and quality of goods traded for such a remote place is astonishing, so eager were the islanders to acquire outsider goods and even services, such as getting a sailor-style haircut!

All of this came to an end in December 1861 when eight ships encircled Rapanui on a hunt for labor for the plantations and middle-class homes of Peru. The raiding was vicious and determined, with about 1500 Rapanui being taken away. Popularly, most authors—copying from others—claim that the Rapanui and other Islanders were set to work in the Peruvian guano islands, although there is no reliable evidence that this was their destination. Instead, it was as indentured agricultural laborers and servants that most who survived were engaged. The trade was short and sharp, with only a few stragglers turning up by March 1862, scarcely four months after it had begun.

These events came just in time for a major smallpox epidemic in Peru, which took scores of people: Of the 1500 Rapanui to be removed from their island, about a dozen returned, some of them carrying the disease. By the time that resident missionaries and traders, mostly from French institutions, arrived in 1866, the Rapanui were dying from disease (smallpox, tuberculosis, flu) and by their own hand with desperate suicides.

Conflict broke out between the French trader Jean-Baptiste Onesime Dutrou-Bornier (arriving in 1868), and Father Roussel, with the former establishing Rapanui as an agricultural station producing European foodstuffs and the latter taking himself and 252 Islanders to Mangareva, both events in 1871.

The population reached its low point in 1877; a Chilean overseer (the French entrepreneur had died) told a visiting party from France that there were 111 people on the island, including himself.

Visitors and trade continued until 1888, when Chile, seeking to boost its stocks as an important nation on the world stage, annexed Rapanui; thus began the South American attempt to sever of Easter Island from its Polynesian neighbors. The Chilean administration set about building up the sheep ranch that became the island's principal activity and quashing any indigenous opposition to metropolitan rule: In 1899, the Rapanui elected king, Riro, was assassinated, and the flag that had been developed in the nineteenth century was torn down and prohibited.

Outsider visits continued for curiosity and study, culminating in the extensive expedition in 1955–1956 led by Thor Heyerdahl, which opened the island to more, although limited, tourism, and the islanders to educational opportunities in Chile.

MODERN RAPANUI

If there is scant interest in the contact history of Rapanui (relative to the preoccupation with the archaeological past), there is even less in the 4000 or so Rapanui in the world today. The tourism and Islander education that started with the landmark Heyerdahl expedition has continued to increase and develop, along with a number of researchers, short- and long-term. In 1966, bowing to Islander pressure, the Chilean government permitted the Rapanui to become full citizens of Chile, with rights of travel and participation in the national life that had been denied them since the annexation in 1888. This trend of liberalization continues today, and in recent years, the island's growing freedom has resulted in its having an autonomous status within the Chilean republic.

Rapanui, whose ancestors braved hazardous and desperate voyages of escape in the 1940s and 1950s from the island to which the Chilean colonial authorities confined them, today are working and living in many countries, including Tahiti, the United States, France, Germany, and elsewhere. Perhaps a quarter of the Rapanui population at any given time lives off the island, studying, working, and making a life elsewhere. Most expatriates say they eventually want to return to Rapanui, and some have done so.

With the increasingly easy contact between Rapanui and the rest of the world, people from many places, notably Chile, have come to live on the island. Every year

40,000 or so people go there to marvel at the megaliths. About a thousand Chileans live on Rapanui, employed by the extensive government bureaucracy and taking advantage of the relative prosperity of the island. This has resulted in considerable intermarriage and a blurring of ethnic lines: Over the last four decades of freedom, there has ceased to be a sharp distinction between Chilean and Rapanui; the language of the island is heard rarely in public, and increasingly, Chilean Spanish is the home and school language as well, although official policy is to promote the indigenous culture.

A sign of the vibrancy of Rapanui culture, in spite of the language decline, is the annual "Easter Island Week" (*Tapati Rapanui*) around the end of January or beginning of February; in this festival, nearly every Islander is either on stage or in the enthusiastic audience, and anyone on Rapanui at the time is invited without charge. The celebration is a delight for tourists, but it is not intended for that outsider audience. Instead, it is a cultural production demonstrating Rapanui values and ideals.

The dominant industry for Rapanui is tourism: People will always want to visit the place. The Islanders are excellent guides (since 1722!) and are pleased that people make such sacrifices to visit the place where they live. The rich volcanic soil provides European and some Polynesian crops in sufficient abundance to subsist on a traditional diet, however, and some few Rapanui choose to do just that, leading a simple country life (with selected European artifacts such as clothing), in contrast to their town-dwelling urbanite counterparts, who live by and around the tourist services.

In song and story Rapanui is often called *Tepito ote henua;* it was so called in the 1888 Chilean annexation treaty. In the Rapanui language this name means "the navel [i.e., center] of the Earth," and in some sense all of us think of the place where we live as being the center of our world at least, but another meaning flavors the place for the Rapanui: "the end of the world": the most remote place, the last place, even the isolated place. Some rueful Rapanui readily give that sort of translation to the place where they live, work, entertain visitors, and seek to end their days.

SEE ALSO THE FOLLOWING ARTICLES

Archaeology / Exploration and Discovery / Human Impacts, Pre-European / Missionaries, Effects of / Peopling the Pacific

FURTHER READING

Fischer, S. R. 2005. *Island at the end of the world: the turbulent history of Easter Island.* London: Reaktion Books.
Flenley, J., and P. Bahn. 2002. *The enigmas of Easter Island.* Oxford: Oxford University Press.
Haun, B. 2008. *Inventing 'Easter Island.'* Toronto: University of Toronto Press.
Heyerdahl, T. 1976. *The art of Easter Island.* London: George Allen & Unwin.
McCall, G. 1994. *Rapanui: tradition and survival on Easter Island,* 2nd ed. Honolulu: University of Hawai'i Press.
Metraux, A. 1940/1971. *The ethnology of Easter Island.* Bulletin 160. Honolulu, HI: Bishop Museum.
Nunn, P. 2007. *Climate, environment and society in the Pacific during the last millennium.* Developments in Earth and Environmental Sciences 6. New York: Elsevier.
Orliac, C., and M. Orliac. 1995. *The silent gods: mysteries of Easter Island.* London: Thames & Hudson.
Van Tilburg, J. 1994. *Easter Island: archaeology, ecology and culture.* London: British Museum Press.
Vargas, P., C. Cristino, and R. Izaurieta. 2007. *1000 años en Rapa Nui: argeología del asentamiento.* Santiago de Chile: Editoria Universitaria.

ECOLOGICAL RELEASE

ROSEMARY G. GILLESPIE

University of California, Berkeley

Ecological release is the expansion of range, habitat, and/or resource usage by an organism when it arrives in a community from which some members are lacking (Fig. 1); effectively, it indicates the difference between the fundamental and realized niche of a species. The phenomenon has frequently been described for species upon initial colonization of islands, where communities may have lower species diversity, or where the existing predators/parasites are not yet capable of exploiting the newly arriving species. For example, Puercos Island (Pearl Archipelago off Panama) has far fewer species of resident birds compared to the mainland, though the density of individuals is generally slightly higher. Here, island colonists have expanded their habitat (wider ranges of vertical foraging, Table 1); density compensation varies between islands depending on the relative effects of (i) replacement of missing mainland species by island colonists less well adapted to the vacated habitat, tending to lower densities; and (ii) relatively fewer large species on islands, tending to increase population densities.

HISTORIC EVIDENCE AND ADAPTIVE RADIATION

A species radiation almost invariably starts when a population moves into a novel and unoccupied set of niches. In the absence of competitors or predators, the popula-

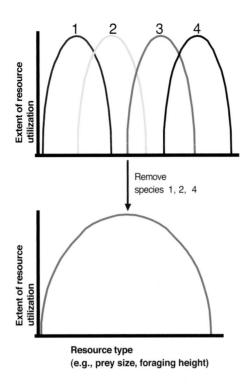

FIGURE 1 Diagram illustrating how ecological release might occur on an island. For species (3), the diagram on the top indicates the realized niche in its original habitat. When (3) colonizes a new area (e.g., an island) from which species (1), (2), and (4) are lacking, it can expand its range to include not only its own niche but also the niches of species (1), (2), and (4), as shown in the diagram below.

tion may undergo ecological release and expand its range, adopting a more generalized habit, the degree to which expansion occurs being a function of preadaptation to that environment. During this period of release, differentiation may occur rapidly as a consequence of diversifying selection (natural or sexual) in the novel and expanded environment.

We can differentiate between two major types of novel niches that a species might colonize at the outset of a radiation: First, new habitat may become available, as in the formation of new (and isolated) islands: Here, colonization is a rare event (because of the isolation), and new arrivals are presented with the opportunity to expand their ecological range (undergo ecological release), which may then allow populations to start diverging. The rate of differentiation will depend on the extent of gene flow between diverging populations and/or the strength of disruptive local selection. Such geographic events may be the most common mechanisms in the initiation of species diversification. Alternatively, species may enter a new "adaptive zone," either because it was preadapted for a niche(s) that becomes available or because it evolves a "key innovation," enabling it to use resources from which it was previously barred. The evolution of key characters or ecological shifts has repeatedly been invoked to explain the initiation of species radiations. Under such conditions a population of a given species may escape the constraints of its current ecological dimension and expand its range across the new dimension (for example, insects may develop resistance to a toxic chemical in potential host plants, allowing them access to an entire suite of new hosts).

After a variable amount of time, the expanded ecological range of the taxon in the new environment will tend to decrease. Reasons for this include (i) enhanced ability of predators or pathogens to exploit the new colonist, leading to reduced numbers in part or all of its range; (ii) competitive displacement from part of its original range by new colonists (close relatives or otherwise); and/or (iii) adaptation of a taxon to different specific local environmental conditions, which may allow evolutionary divergence between populations (and may ultimately lead to adaptive radiation). These tendencies are not mutually exclusive (for example, selection as a result of predation pressure may lead to greater ecological specialization).

Evidence for historic ecological release can often be found in examining phylogenetic histories of lineages that have undergone adaptive radiation: For many of these, the most extensive ecological range expansion has occurred at the outset of the radiation. For example, planthoppers in the genus *Nesosydne* (Delphacidae) are represented by a radiation of approximately 80 species in the Hawaiian Islands. Phylogenetic analysis shows that the lineage diverged from an ancestral group restricted to monocotyledonous plants (especially grasses and sedges) to feed on a total of 28 plant families, mostly dicotyledonous plants, in the Hawaiian Islands.

TABLE 1
Forage Layers of Birds on Puercos

	Puercos	Mainland
Antshrike	3.95	1.24
Wren	3.01	2.55
Yellow-Green Vireo	2.09	1.78
Bananaquit	2.00	1.33

NOTE: On the island of Puercos, off Panama, there are very few antbirds (foliage gleaners) compared to the mainland. Birds on the island have undergone ecological release in habitat and foraging range (height). The table lists the number of layers in which birds were found to forage. From Macarthur et al. (1972).

ISLAND COLONIZATION AND TAXON CYCLES

Early studies on land birds showed an increased population density and greater habitat breadth, together with increased diversity of feeding strategies, in birds that had colonized Bermuda, Puerto Rico, and other islands in the Caribbean, as compared to their mainland counterparts. Different studies of ecological release have shown that an island population may occupy a wider range of elevations, more vegetation types, or forage more broadly as compared to the mainland population. However, considerable differences exist between species in their colonization ability and tendency to undergo ecological release. In some groups, ecological release appears to be the first step in a cycle of distributional change following colonization of islands. E. O. Wilson coined the term "taxon cycle" to describe such a progression in Melanesian ants: The ants initially expand their ecological range upon colonization of the island and occur as widespread, dispersive populations ("Stage I"). These give rise to many more restricted and specialized populations ("Stage II") or species ("Stage III"). Stage I species reflect ecological release; they are "generalist" and dominate in marginal habitats. At Stage II the ants have moved into the forest and specialize on specific substrates. Stage III species show development of the process of becoming local endemics. The concept of the taxon cycle has been adapted for carabid beetles in the related "taxon pulse" hypothesis.

A similar cycle of distributional change has been described for Caribbean birds. Most birds appear to have colonized the islands through a similar process, from ecological release in marginal habitats to more specialized inner forest and montane species. Also, it appears that habitat shifts can intermittently occur in reverse (i.e., from specialized species), associated with colonization of new islands, thus restarting the taxon cycle. As with the broader phenomenon of ecological release, one of the primary drivers of taxon cycles may be coevolutionary responses of predators and prey or of pathogens and hosts: Upon arrival on an island, colonists escape their predators or parasites and expand their ecological amplitude. Over evolutionary time, resident predators or pathogens may adapt to, and hence exploit, the new colonists, which may therefore undergo selection for greater ecological specialization.

INVASIONS

When invasive species first colonize a new area, the rates of population growth and range expansion can vary considerably, with some species showing an immediate and rapid rate of local population growth and range expansion, whereas others show a long lag time between initial introduction and population growth. Explanations for such a lag time are not well understood. However, once populations of an invasive species start to expand, they frequently undergo ecological release, and this phenomenon has now been well documented. Indeed, successful establishment of introduced species is often attributed to ecological release from competitors or predators/parasites in their native range. A related concept is that of "enemy release," in which invasive species gain an advantage over native species by escaping many of their natural enemies upon colonization of a new area. Escape from predators and parasites may also explain the frequently observed increased size and vigor of some invasive plants.

SEE ALSO THE FOLLOWING ARTICLES

Adaptive Radiation / Invasion Biology / Taxon Cycle

FURTHER READING

Diamond, J. M. 1970. Ecological consequences of island colonization by southwest Pacific birds. I. Types of niche shifts. *Proceedings of the National Academy of Sciences of the United States of America* 67: 529–536.

Losos, J. B., and K. de Queiroz. 1997. Evolutionary consequences of ecological release in Caribbean *Anolis* lizards. *Biological Journal of the Linnean Society* 61: 459–483.

Macarthur, R. H., J. Diamond, and J. R. Karr. 1972. Density compensation in island faunas. *Ecology* 53: 330–342.

Ricklefs, R. E., and G. W. Cox. 1978. Stage of the taxon cycle, habitat distribution and population density in the avifauna of the West Indies. *American Naturalist* 112: 875–895.

ELBA

SEE MEDITERRANEAN REGION

ELEPHANT

SEE ANTARCTIC ISLANDS

ENDEMISM

QUENTIN C. CRONK AND DIANA M. PERCY
University of British Columbia, Vancouver, Canada

Island organisms are often not found elsewhere and are taxa distinct from their closest relatives on continental landmasses. This phenomenon is called island endemism. Island endemics may be relicts (paleoendemics) or may be of more recent origin (neoendemics). The evolutionary processes leading to endemism often involve evolutionary radiation (i.e., the rapid evolution of many, often ecologically separated, species from the original single founding species).

PATTERNS OF ENDEMISM ON ISLANDS

The term endemic (*endémique*), as used for geographically restricted organisms, dates to the Swiss botanist and pioneer biogeographer Augustin Pyramus de Candolle in his 1820 essay "Géographie botanique" that appeared in Cuvier's *Dictionnaire des Sciences Naturelles*. Darwin uses the term "endemic" no less than 21 times in *On the Origin of Species* (1859), especially in connection with oceanic islands. In the *Origin,* Darwin is at pains to stress the contrast between the low species diversity on oceanic islands and the high endemism. As he puts it, "Although in oceanic islands the number of kinds of inhabitants is scanty, the proportion of endemic species (i.e., those found nowhere else in the world) is often extremely large."

The main driver of island endemism is geographical isolation, restricting migration to and from islands. Isolation (along with the small size of islands) is also responsible for the low species numbers on islands, so the linkage between these two phenomena noted by Darwin is not unexpected. The colonization history and extent of geographical isolation explains much of the pattern of island endemism.

Some islands are fragments of continental plates that have become detached and isolated. Because of the slow rate of movement of plate fragments into isolated geographical positions, this fragmentation is usually geologically ancient. Such fragments may have large numbers of endemics, often only distantly related to continental species, elsewhere, in some cases, because of extinction of close relatives over time. An example is provided by the island of New Caledonia. A remarkable New Caledonian endemic is the flowering plant *Amborella trichopoda,* which occurs as an understory shrub in the forest. *Amborella* is the sister group to all other flowering plants (i.e., the evolutionary divergence about 130 million years ago between *Amborella* and the other flowering plants is the earliest evolutionary event for extant flowering plants). It is therefore unlikely for *Amborella* to have evolved recently from a close relative elsewhere, as it has no close relatives and may therefore be described as a paleoendemic. New Caledonia also has taxonomically isolated animal endemics. The near-flightless, chicken-sized kagu bird (*Rhynochetos jubatus*) is the only member of its family Rhynochetidae, suggesting another example of paleoendemism.

A second type of island is the volcanic oceanic island. These are often extremely isolated and geologically recent in origin, as they rise up in response to eruptions from volcanic hotspots in the middle of oceans. They are not generally very stable; initially, the hotspot activity may continue to cause terrestrial disturbance, but eventually either it will cease or plate movement will carry the island away from the hotspot. Then, the island, no longer augmented by volcanism, will sink under its own weight into the ocean crust. Oceanic islands may therefore disappear altogether after a relatively short time: a few tens of millions of years at most. Endemics on such islands are the result of long-distance dispersal from continental source populations, and, as a result, they are often closely related to continental species and will be relatively recently diverged from these continental relatives. An example is provided by the Hawaiian violets (*Viola* spp.), a group of ten species related to the boreal *Viola langsdorfii* of Siberia and Alaska. Migratory, arctic-breeding birds are possible dispersal agents for boreal plants to islands of the tropical Pacific. Of the ~250 migratory birds wintering in the Hawaiian islands, some 50 species breed in the arctic, from which place they may have transported seeds to the Hawaiian Islands.

Under certain circumstances (e.g., the "island conveyor belt" effect), endemic lineages may outlive the oceanic islands on which they originally occurred. If oceanic crust is moving over a long-lived hotspot, a chain of islands may form, and endemic lineages may migrate from older to younger islands as they become available for colonization, escaping the older islands before they sink beneath the ocean to become coral atolls or seamounts. This type of sequential colonization along an island chain is well documented from the Hawaiian Islands for a large number of species including *Orsonwelles* spiders and *Psychotria* understory shrubs. That many endemic lineages have species on multiple islands, apparently following the "progression rule" with younger species on younger islands, provides evidence of how endemic lineages may "escape" the disappearance of the original island colonized.

A third type of island is the continental shelf island. The proximity to continental sources of biodiversity, and therefore the frequent occurrences of colonization and back-colonization, results in high species diversity but low levels of endemism. Many such islands have been connected and reconnected to the continental landmass multiple times during the Pleistocene as a result of eustatic and isostatic changes in sea level during glacial and interglacial periods. In cold regions, such islands may be near enough to continental ice sheets during glacial periods to have suffered extinction of biota, which is another factor leading to a paucity of endemics, as subsequent colonization would be too recent for the evolution of endemic species. However, island endemics do evolve, even under such apparently unpromising circumstances. The island of Lundy in the Bristol Channel, only ~15 km off the coast of Britain, has an endemic plant, the Lundy cabbage (*Coincya wrightii*). Even when isolation has been very recent, endemic taxa may occur if we take into consideration differentiation below the species level. For instance, the house mouse (*Mus musculus*) was introduced to the Welsh island of Skokholm around 1890, but the Skokholm mouse is now extremely distinctive compared to other populations of house mouse, both in overall size and in skeletal characteristics. When speciation is extremely abrupt, as in the case of allopolyploid speciation in plants, endemic species can form on islands colonized only a few thousand years ago. Thus, the recently glaciated Scottish island of Arran has three endemic tree species in the genus *Sorbus* (*S. arranensis*, *S. pseudomeincichii*, and *S. pseudofennica*), all allopolyploid derivatives of a series of rowan (*S. aucuparia*) and whitebeam (*S. rupicola*) hybridizations. This is despite Arran having being colonized by trees in the postglacial period only a few thousands of years ago.

HISTORY OF ENDEMICS ON ISLANDS

Endemism on islands may be promoted either by (1) post-colonization differentiation of the island lineage relative to the founding continental lineage (neoendemism), or (2) changes (such as extinction) that occur in the continental lineage (paleoendemism). Both contribute to levels of island endemism but do not typically contribute equally to levels of speciation on islands. Speciation (especially on islands) may occur rapidly, and the majority of these recent endemics will be due to post-colonization speciation (e.g., neoendemic radiations), whereas island relicts of older continental lineages do not appear to contribute many additional neoendemics. Thus, deeply divergent island endemics are known as relicts, or paleoendemics. In contrast, neoendemics are island endemics that have diverged recently, often with sibling taxa on the same or adjacent islands. Island endemic taxa vary greatly in their depth of divergence from the related continental taxa they represent: from Skokholm mice, representing a distinctive population but not an endemic species (a potential neoendemic), to endemic families such as the monotypic *Amborella* in New Caledonia (a paleoendemic).

The best way to assess this depth of divergence is through the "sister-group" relationship, which can be fairly reliably determined using molecular phylogenetics (Fig. 1). In addition, advances in molecular dating increasingly allow reasonably accurate estimates of the divergence date. In the

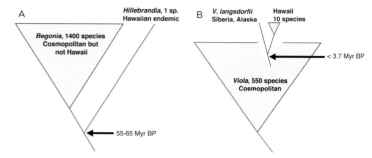

FIGURE 1 Phylogenetic summary diagrams to illustrate the relationships of two Hawaiian endemic taxa, *Viola* section Nosphinium (about ten spp.) and *Hillebrandia* (one sp.). *Viola* shows a phylogenetic pattern consistent with neoendemism and recent colonization, whereas *Hillebrandia* shows a pattern consistent with relict endemism (paleoendemism).

case of *Amborella*, its sister group remarkably appears to be all 250,000 species of flowering plants, and one estimate places the divergence between this endemic and its closest relative at 130 million years ago. This is unlikely to represent the date of colonization of New Caledonia by *Amborella*, as a large amount of extinction has occurred in the flowering plant lineages that were present 130 million years ago. This extinction also makes it difficult to assess how much evolutionary divergence has occurred in situ on New Caledonia since *Amborella* became an island endemic.

Another example of paleoendemism is provided by *Hillebrandia*, a genus endemic to the Hawaiian Islands. Its sister group is the entirety of the large pantropical genus *Begonia*, from which it probably diverged some 50 million years ago. This is older than the oldest of the Hawaiian Islands (29 million years), and it has been suggested that *Hillebrandia* may have colonized an island now completely submerged and then maintained itself in the Hawaiian flora by successive colonization of younger islands as they became available (the "progression rule" mentioned earlier). With the divergence being so ancient, it is difficult to determine how much of the differentiation of *Hillebrandia* from *Begonia* (such as differences in perianth, ovary, fruit dehiscence, and pollen) is allochthonous (off-island, pre-colonization differentiation) and how much is autochthonous (on-island, post-colonization divergence).

A good example of neoendemism is provided by the Hawaiian violets, *Viola* section *Nosphinium*, a small group of species characterized by woodiness, a trait highly unusual in the genus, and by their possession of distinctive few-flowered, cymose inflorescences. The sister group of the Hawaiian violets belongs to the *Viola langsdorfii* species complex from Siberia and Alaska (mentioned earlier). As *V. langsdorfii* is not untypical of the genus as a whole, it is evident that the unique features of section

Nosphinium have evolved rapidly on the Hawaiian Islands (probably within the last 3 million years).

On rare occasions, there may be fossil evidence to assist in the interpretation of endemic status, although oceanic islands are usually actively eroding and thus are taphonomically unpromising for fossil deposition. However, 9.5-million-year-old pollen of the genus *Trochetiopsis* from the 14.5-million-year-old island of St. Helena helped to establish the paleoendemic nature of this endemic genus.

There is no absolute dividing line between paleoendemics and neoendemics, and many islands have a gradation, or "relictual series," between the two. Islands that are very young (but not part of a hotspot chain of islands), or which have been recently denuded by glaciation, have only neoendemics.

PROCESSES OF ENDEMISM ON ISLANDS

The degree of island isolation and habitat richness, when combined with patterns of colonization, results in sharp increases in endemism on more remote islands and on islands with greater diversity of habitats (Tables 1 and 2).

At the moment of colonization of an island that is remote enough to cut off gene flow from the ancestral population, divergence will start with genetic drift. This is inevitable, given small island populations and a small founding population (which may be a single gravid female in animals or a single diaspore in plants), representing a single lineage of a more variable continental population. Levels of endemism are lower on islands that are closer to source populations and in lineages that are highly dispersible (such as the heron family, Aredeidae), as there is greater likelihood of further colonization and gene exchange. Low levels of endemism in the highly dispersible herons can be contrasted with higher endemism in small, poorly dispersible bird families (such as the white-eyes, Zosteropidae) (Table 1).

Furthermore, natural selection, as a consequence of island environments being different from those of continental source areas, is also likely to promote differentiation and speciation. However, the remoteness that cuts off gene flow will also be responsible for a decrease in potential colonists and numbers of species. Thus, high levels of endemism are often related to low absolute species numbers (Table 2). However, habitat richness is important in

TABLE 1
Endemism and Dispersal Ability of Selected Birds in Pacific Islands

Bird Family	Island Groups Occupied	Total Species	Percentage Endemic
Ardeidae (herons)	26	9	0
Anatidae	17	11	27
Accipitridae	9	19	47
Rallidae	22	23	57
Columbidae	26	62	55
Psittacidae	16	34	59
Apodidae	16	7	43
Alcedinidae	15	19	58
Pachycephalidae	21	68	79
Muscicapidae	19	32	66
Sturnidae	15	17	63
Meliphagidae	14	37	84
Zosteropidae (white-eyes)	10	28	89

NOTE: Herons (Ardeidae) are widespread in the Pacific, being strong dispersers, and none are endemic. This contrasts with white-eyes (Zosteropidae), which are poor dispersers with a high degree of endemism.
SOURCE: Adler (1992). Pelagic and migrant species are excluded.

TABLE 2
Bird Endemism in the Pacific, Showing the Effect of Habitat-Richness and Isolation

Archipelago	Total Spp.	Endemic Spp.	Percent Endemic
Solomon Islands	159	59	37
Hawaiian Islands	53	49	92
Bismarck archipelago	168	48	29
Fiji	57	22	39
New Caledonia group	75	20	27
Caroline Islands	33	17	52
Marquesas Islands	13	11	37
Samoa group	33	10	30
Palau (Belau) group	32	10	31
Mariana Islands	21	9	43
Society Islands	18	8	44
Vanuatu (New Hebrides) group	56	7	12
Cook Islands	12	6	50
Austral and Pitcairn Islands	9	5	56
Tuamotu archipelago	10	4	40
Santa Cruz Islands	33	3	9
Marshall, Gilbert, and Ellice Islands supergroup[a]	9	3	33
Tonga group	21	2	9
Rotuma	10	1	10
Niue	11	0	0
Wallis and Futuna group	11	0	0

NOTE: The habitat-rich Solomon Islands and the Hawaiian Islands have the most endemics, whereas the habitat-poor low islands of the Pacific, such as those in the Wallis and Futuna group, have the fewest. Although the Solomon Islands (neighboring Australasia and New Guinea) have the largest number of endemics, this group has a lower percentage of endemics when compared to the very isolated Hawaiian Islands.
SOURCE: Adler, (1992).
[a] "Marshall, Gilbert and Ellice Islands etc" here comprise the low islands of Micronesia including the Marshall Islands, Nauru, Banaba (Ocean Island), Tungaru (Gilbert Islands), Tuvalu group (Ellice Islands), Phoenix Island, Tokelau group, Line Islands and the Wake group. Pelagic and migrant species are excluded.

determining species numbers, and therefore more remote, habitat-rich, high-altitude islands have more species than do habitat-poor, low islands. On some islands, endemism seems disproportionately concentrated at higher altitudes; however, many of the lower elevations on these islands have been subject to habitat disturbance and invasive species, so the levels of endemism may be artificially skewed. On remote islands, the species assemblage present will not only be small but also "disharmonic" (i.e., not a representative selection of all taxonomic and ecological groups, but a poorly sampled or biased one). The result of this appears to be to lessen competition and to provide new ecological opportunities for lineages that may be ecologically conservative in continental source areas. As a result, a single colonization event will often lead to multiple endemic species developing on islands because of evolutionary radiation. As this is ecologically driven, such radiations are usually assumed to be adaptive—although this is rarely formally shown.

The *Tetragnatha* spiders in the Hawaiian Islands provide an example. This endemic lineage has evolved species that are representative of several distinctive "ecomorphs." Remarkably, these ecomorphs have evolved independently a number of times, suggesting that endemic species radiations are highly predictable rather than stochastic. On the Galápagos archipelago, local endemic speciation of the endemic prickly pears (*Opuntia*) appears to have been influenced by the presence or absence of the main endemic herbivores on individual islands. The herbivores capable of eating *Opuntia* pads are the giant tortoise (*Geochelone*) and the land iguana (*Conolophus*) (Figs. 2 and 3). Galápagos *Opuntia* species are usually stoutly trunked (Fig. 4), but on the islands of Marchena and Genovesa, which were never colonized by tortoises, there are small, sprawling, trunkless, and less spiny *Opuntia* ecomorphs.

Climatic as well as biotic factors may have an effect on island endemic speciation. As high oceanic islands are often

FIGURE 3 Photograph of a wild Galápagos giant tortoise (*Geochelone elephantopus*) from the island of Santa Cruz. Along with the land iguana, the giant tortoise is the only other herbivore capable of eating Galápagos *Opuntia* to any great extent.

FIGURE 4 Photograph of the tree form of Galápagos prickly pear or *Opuntia* (*O. echios*) from the island of Santa Cruz.

FIGURE 2 Land iguana (*Conolophus subcristatus*), one of the main herbivores predating *Opuntia* in the Galápagos.

formed from young volcanoes with steep, even precipitous, slopes, the temperature lapse rate with altitude will often ensure that numerous temperature and humidity regimes are found in a single island and that altitudinally based speciation will be common. It has also been suggested that the relatively temperate oceanic climate enjoyed by many islands will provide buffering against extinction driven by glacial-interglacial cycles, thus promoting survival of lineages that may become relict endemics because of the extinction of continental relatives.

SEE ALSO THE FOLLOWING ARTICLES

Adaptive Radiation / Dispersal / Founder Effects / Galápagos Islands, Biology / Hawaiian Islands, Biology / New Caledonia, Biology

FURTHER READING

Adler, G. H. 1992. Endemism in birds of tropical Pacific islands. *Evolutionary Ecology* 6: 296–306.

Baldwin, B. G., and B. L. Wessa. 2000. Origin and relationships of the tarweed-silversword lineage (Compositae-Madiinae). *American Journal of Botany* 87: 1890–1908.

Ballard, H. E., and K. J. Sytsma. 2000. Evolution and biogeography of the woody Hawaiian violets (*Viola*, Violaceae): Arctic origins, herbaceous ancestry and bird dispersal. *Evolution* 54: 1521–1532.

Clement, W. L., M. C. Tebbitt, L. L. Forrest, J. E. Blair, L. Brouillet, T. Eriksson, and S. M. Swensen. 2004. Phylogenetic position and biogeography of *Hillebrandia sandwicensis* (Begoniaceae): a rare Hawaiian relict. *American Journal of Botany* 91: 905–917.

Cronk, Q. C. B. 1992. Relict floras of Atlantic islands: patterns assessed. *Biological Journal of the Linnean Society* 46: 91–103.

Cronk, Q. C. B. 1990. The history of the endemic flora of St Helena: Late Miocene *Trochetiopsis*-like pollen from St Helena and the origin of *Trochetiopsis*. *New Phytologist* 114: 159–165.

Gillespie, R. G., and G. K. Roderick. 2002. Arthropods on islands: colonization, speciation, and conservation. *Annual Review of Entomology* 47: 595–632.

Jansson, R. 2003. Global patterns in endemism explained by past climatic change. *Proceedings of the Royal Society of London Series B: Biological Sciences* 270: 583–590.

EPHEMERAL ISLANDS, BIOLOGY

SOFIE VANDENDRIESSCHE

Instituut voor Landbouw- en Visserijonderzoek, Oostende, Belgium

MAGDA VINCX AND STEVEN DEGRAER

Ghent University, Belgium

Next to islands constituted of real land masses, the surface of oceans and seas is littered with ephemeral floating objects of various sizes and materials, which form isolated and distinct habitats within the quite uniform marine surface layer. Because of the provision of shelter, surface for attachment, and in many cases also additional food sources, these floating objects attract a wide variety of invertebrates, fishes, and birds. The association behavior of the encountered species and their use of the transient resources offered by floating objects have a considerable impact on the composition, biogeography, and ecology of the marine fauna.

LIFE IN THE MARINE NEUSTON

The neuston, comprising all plants and animals inhabiting the surface layer of oceans and seas, is subject to a harsh life in the proximity of the air–water interface. Neustonic organisms are exposed to high levels of ultraviolet and infrared radiation, to dramatic changes in salinity and temperature, to strong wave action, to high concentrations of contaminants, and to a high predation pressure. On the other hand, the neustonic zone is characterized by a high input of nutrients (e.g., aerial precipitation of terrigenic matter and the accumulation of remains and excreta of marine organisms). Consequently, many organisms have developed adaptation strategies to withstand the high amount of stress and to be able to profit from the enhanced productivity. These adaptations include a specific pigmentation to avoid predation, the development of air sacs and bladders to promote buoyancy, and the development of a tendency to mimic and cling to floating objects, both natural (e.g., floating seaweeds [Figs. 1–2], seeds, logs, seagrass, volcanic pumice, and tar from natural seeps) and humanmade (e.g., tar pellets, plastic, rubber, nylon, and glass).

FLOATING OBJECTS AS EPHEMERAL ISLANDS

Although floating objects are ephemeral by nature, they are important sources of small-scale patchiness that significantly influence the faunal species composition of the neuston. The motives of invertebrates, fishes, and birds for associating with floating objects are species-specific, and many authors have already hypothesized on the advantages of this association behavior. The most common motives are probably the availability of a surface for attachment, the provision of shelter from predators, and the presence of a food source (the substrate itself or the associated fauna). Other possible reasons are the functioning of floating objects

FIGURE 1 Multi-species floating seaweed clump off the Belgian coast (North Sea). Photograph by Sofie Vandendriessche.

FIGURE 2 Floating algal mats, south of Chile-Chaitén. Photograph courtesy of Martin Thiel and Iván Hinojosa, Universidad Católica del Norte, Coquimbo, Chile.

as a substitution of the seabed or the littoral zone, as a spawning substrate and nursery area, as a meeting point for the formation and maintenance of schools, or as a cleaning station.

The suitability of a floating object as a temporary habitat differs considerably depending on size, buoyancy, longevity, and value as a food source. Plastic micro-litter, plant seeds, and volcanic pumice are generally small (mm to cm), whereas large items such as trees and whale carcasses can exceed several meters in length or diameter. The persistency of a raft largely depends on the composition and degradability of the raft, resulting in periods of buoyancy of hours up to months. Especially floating seaweeds such as the permanently floating *Sargassum* and uprooted coastal seaweeds score high in buoyancy, longevity, and food value, and form a relatively diverse habitat that can travel considerable distances.

THE APPLICABILITY OF THE TERM "ISLAND" TO FLOATING OBJECTS

Floating seaweeds are the most thoroughly investigated floating objects and have long been termed "biotic islands" or "floating islands" because certain parallels can be drawn with classic islands, such as the observed species turnover during drift and the correlation between the size of the raft and species richness in some studies (and between species richness and the distance to the site of origin, if known). However, ephemeral islands seem to be very complex. Ephemeral floating seaweed clumps, for example, take a part of their fauna with them at the moment of detachment and are colonized by both shore fauna and fauna from the surrounding water column. Furthermore, this type of island is subject to beaching, wave action, fragmentation, and coalescence. Consequently, the origin of the associated species community is hard to reconstruct, and the driving forces of variation within this habitat are often obscured.

Another consequence of the dynamics and passive movement of floating objects is the possibility of rafting, which can greatly enhance species dispersal over distances beyond their swimming abilities. The rafting process thereby increases the connectivity and gene flow between populations, whereas isolation and speciation are central themes in the classic concept of island ecology. Recently, persistent humanmade debris has become very abundant and is now considered to be a threat with regard to the introduction of non-native and possibly even invasive species.

SEE ALSO THE FOLLOWING ARTICLES

Dispersal / Ephemeral Islands, Geology / Rafting

FURTHER READING

Ingólfsson, A. 1995. Floating clumps of seaweed around Iceland: natural microcosms and a means of dispersal for shore fauna. *Marine Biology* 122: 13–21.
Thiel, M., and L. Gutow. 2005. The ecology of rafting in the marine environment I. The floating substrata. *Oceanography and Marine Biology: An Annual Review* 42: 181–264.
Thiel, M., and L. Gutow. 2005. The ecology of rafting in the marine environment II. The rafting organisms and community. *Oceanography and Marine Biology: An Annual Review* 43: 279–418.
Vandendriessche, S., M. Messiaen, S. O'Flynn, M. Vincx, and S. Degraer. 2007. Hiding and feeding in floating seaweed: floating seaweed clumps as possible refuges or feeding grounds for fishes. *Estuarine, Coastal and Shelf Science* 71: 691–703
Zaitsev, Y. P. 1970. *Marine neustonology* (in Russian). Kiev: Naukova Dumka Publishing House.

EPHEMERAL ISLANDS, GEOLOGY

KAZUHIKO KANO
Geological Survey of Japan, Tsukuba

The term "ephemeral island" has been informally applied to a piece of land that is completely surrounded by water or that emerged out of water and is fated to disappear on a time scale of hours to years after its appearance. This term has also been applied to pumice rafts and other floating objects. Ephemeral islands have the potential to temporarily provide organisms with a shelter, a surface for attachment or spawning, and a food source.

BIRTH AND FATE

Ephemeral islands are diverse in birth and fate. They appear and disappear during dynamic surface processes including floods, volcanic eruptions, and even activities of organisms.

Mounds or rocks in valleys, lowlands, and shallow waters are, for example, isolated or submerged by floods or suddenly built dams as the water level rises, but they again become a part of the landmass after the water drains. Debris avalanches, landslides, pyroclastic density currents, and lava flows carry huge blocks and other materials to build dams and mounds that are left behind the heads. Floods also carry a large amount of sediment, which accumulates to form point bars, deltas, and other low-rise sedimentary features in river channels or river mouths. Such low-rise sedimentary features, however, could be channeled by subsequent flood flows, and some areas may remain isolated by meandering water streams. Barrier islands, atolls, barrier reef islands, and other marine low-rises may also be channeled by tsunamis, storms, or high tides in association with strong currents and waves or winds, especially during a period of global sea-level rise or warming. The remnant parts are isolated by resultant channels but may be removed by subsequent catastrophic processes.

Where volcanoes grow to rise slowly above the sea or lake water, explosive interactions of magma and water likely recur to accumulate loose materials around the vents, whereas wave action easily erodes loose volcanic materials to keep the elevations low or reduce them down to a level below the water surface. Volcanoes of this sort are likely to repeatedly emerge and submerge on a time scale of hours to years, as typically observed in island arcs and seamounts.

FLOATING ISLANDS

Except for artificial islands like floating aggregates of water plants on Lake Titicaca, one of the most prominent floating islands is a pumice raft. Pumice is vesicular and mostly buoyant in water if the vesicles are filled with air, steam, or other gas. Pumice fragments are produced by explosive or quench fragmentation of highly vesicular magma and are initially hot, being filled mainly with steam. Explosively projected into the air, hot pumice fragments may fall or flow onto the water surface, and the majority likely remain buoyant and form aggregates for a period of time. When pumice fragments are released

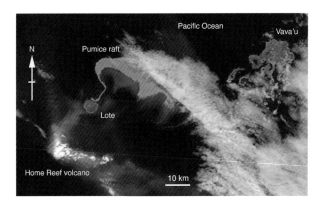

FIGURE 1 Pumice raft drifting 25–55 km away from the source Home Reef volcano along the Tonga Islands in the South Pacific. A large amount of pumice was released from the underwater vent during its growth out of water. This photograph shows a satellite image from August 10, 2006, provided by NASA at http://earthobservatory.nasa.gov/NaturalHazards/view.php?id=17605.

under water from explosive eruption plumes or quench-fragmentation of lava domes, they ascend in water so long as they remain hot. Relatively small fragments of pumice cool rapidly, absorb water, and start sinking to depths. Many large fragments of pumice, however, may reach the water surface and are concentrated to form a raft around the point of upwelling. Pumice rafts produced in this manner may reach a scale of tens of kilometers, drift away from the source with ocean currents and winds, and potentially travel over tens of thousands of kilometers for years (Fig. 1). They are fouled by a variety of organisms during their long journey and carry the organisms to remote places, consequently playing an important role in faunal dispersal.

SEE ALSO THE FOLLOWING ARTICLES

Atolls / Barrier Islands / Ephemeral Islands, Biology / Landslides / Lava and Ash

FURTHER READING

Brian, S. E., A. Cook, J. P. Evans, P. W. Colls, M. G. Wells, M. G. Lawrence, J. S. Jell, A. Greig, and R. Leslie. 2004. Pumice rafting and faunal dispersion during 2001–2002 in the Southwest Pacific: record of a dacitic submarine explosive eruption from Tonga. *Journal of Volcanology and Geothermal Research* 227: 135–154.

Kano, K. 2003. Subaqueous pumice eruptions and their products: a review, in *Explosive subaqueous volcanism, geophysical monograph 140*. J. D. L. White, J. L. Smellie, and D. A. Clague, eds. Washington, DC: American Geophysical Union, 213–229.

White, J. D. L. 1996. Pre-emergent construction of a lacustrine basaltic volcano, Pahvant Butte, Utah (USA). *Bulletin of Volcanology* 58: 249–262.

EQUILIBRIUM

SEE ISLAND BIOGEOGRAPHY, THEORY OF

EROSION, COASTAL

WAYNE STEPHENSON
University of Melbourne, Australia

Coastal erosion is the landward retreat of a datum such as mean high water or a geomorphic feature such as a cliff or beach. The landward retreat is persistent through time as distinct from cyclic fluctuations in the position of the shoreline associated with storm erosion and recovery, typical of beaches. Coastal erosion is also widespread, with 70% of the world's beaches said to be eroding, and 80% of the world's total shoreline composed of eroding cliff and rock morphologies. Islands are particularly susceptible because of their geomorphic settings and limited sediment supply. Although coastal erosion is typically viewed as a hazard and a significant management issue, it must not be forgotten that in many instances, it is a natural and necessary geomorphic process representing the adjustment of the shoreline toward a new equilibrium condition, often in response to post-glacial sea-level rise. For example, erosion of cliffed coasts can provide sediment to beaches. On some islands, cliff erosion is likely to be an important source of sediment compared to continental coasts where rivers and continental shelves are often dominant sources of sediment.

CAUSE

There are three fundamental causes of coastal erosion applicable to all coastal land forms: a relative rise in sea level, a deficiency in the local sediment budget, and a change in wave patterns. However, a rise in sea level does not automatically result in coastal erosion, and in some instances, shorelines can respond by accreting if there is an oversupply of sediment eroded from elsewhere or transported onshore from the continental shelf. Sea-level rise since the last glacial maximum is by far the most important cause of coastal erosion, and many shorelines are still responding to this rise. In addition to eustatic (changes in ocean water volume) causes of sea-level rise, isostatic processes can also cause a relative rise in sea level through subsidence of land masses. Isostatic processes result from the addition or removal of mass from the Earth's crust (e.g., ice sheets, sediment, or water), which causes the crust to sink under loading and to rebound after de-loading. Tectonic processes may also cause subsidence of land masses, giving the appearance of sea-level rising.

Sediment budgeting offers a useful management approach to coastal erosion by accounting for sources and losses of sediment from a defined length of coast known as a littoral cell. A negative sediment budget (where a coastal cell looses sediment over time) equates to coastal erosion and allows the cause to be identified and the amount of sediment being lost from a system to be quantified. Determining the size of the deficit in a sediment budget allows for better assessment of appropriate mitigation techniques. Increases in wave heights have been observed in the Northwest Pacific and the North Atlantic over recent decades, and these have been linked to beach erosion. Increases in wave height and frequencies of storms are expected to increase the incidence of coastal erosion under climate change scenarios.

In addition to natural changes in sea level, sediment budgets, and wave patterns, human activities can also cause or exacerbate coastal erosion. Sea-level rise from global warming will contribute to coastal erosion, although evidence for an acceleration in sea-level rise (which would be evidence for anthropogenic forcing) above natural noise is equivocal. Development activities can remove, slow, or stop sediment entry and transport to and within the coastal system, shifting neutral or positive sediment budgets into negative ones. Dams on, or extraction of sediment from, rivers can cause coastal erosion by denying sediment supply to the coast. Sediment extraction for dunes, beaches, reefs, or the nearshore may also lead to coastal erosion. On Pacific islands a scarcity of sand for construction has often meant that sand is removed from beaches, causing coastal erosion. Seawalls built to stop cliff erosion also prevent sediment delivery to the coast, and where cliffs are important local sources of sediment for beaches, coastal erosion can result. Beach loss is a widely recognized consequence following seawall construction, as wave energy is reflected, causing scour in front of the wall rather than dissipating wave energy as a beach would do. Structures perpendicular to the shoreline such as a groyne, built to intercept longshore sediment transport, may well accumulate sediment on one side (the updrift side) but also cause erosion on the downdrift side or further along the shoreline, by preventing sediment transport. Thus, many actions taken to prevent coastal erosion can make the situation worse or can simply transfer the problem elsewhere along the coast. Breakwaters built for harbors can also inadvertently act as groyne and produce the same effects. Causeways can change wave patterns and tidal currents around islands and can consequently alter sediment transport, depriving some shorelines of sediments. Removal of mangroves for wood fuel or shrimp farming reduces the deposition of sediment and the overall ability of the shoreline to attenuate wave energy, making coastal erosion more likely.

Coastal erosion as a hazard only eventuates when human infrastructure occupies a zone likely to be eroded. That is, a hazard is a social construction, not a natural occurrence, and the commonly used term "natural hazard" is a misnomer. It follows then that coastal and hazard management should be more concerned with managing peoples' activities, rather than the physical coast. On many islands, however, people have a long history of occupying the coastal zone, often out of necessity, and often very successfully based on traditional modes of production and tenure. Many coastal erosion problems have arisen for these communities because of intensified, and hence less resilient, landuse, resulting from economic development.

MEASUREMENT OF COASTAL EROSION

Coastal erosion can be measured using a wide range of techniques, from simple regular surveying of cross shore transects to aircraft using lasers known as LiDAR (light detection and ranging). When instrumented measurement is not possible, historical analysis from maps, charts, and aerial photographs can be used. Regardless of the method used, knowing the rate of shoreline retreat and/or the volume of sediment being lost from a coastal cell is critical for determining appropriate responses. Furthermore, the analysis needs to be over a sufficiently long time scale to remove short-term fluctuations associated with storm erosion and recovery in the case of sedimentary coasts, or noise associated with episodic failures common on cliffed coasts. There are few published data on rates of coastal erosion on islands, and more data will be needed to assess appropriate responses for the future. Published rates of coastal erosion on islands vary from 1 to 4 m/year.

IMPACTS

An obvious impact of coastal erosion is the permanent loss of land, and on islands, this is a significant issue because land area is already small. Erosion of land can lead to loss of economic activity, as in the case of beaches, which serve as tourist amenities. Ecological function can also be lost as erosion progressively removes mangroves or salt marshes; this is a significant issue when the coastal erosion is caused by human activity. Community vulnerability is increased, as coastal erosion reduces the ability of shorelines to dissipate wave energy. Vulnerability may also increase when artificial defenses (e.g., seawalls) are lost or damaged.

MITIGATION

Approaches to erosion mitigation generally take one of two forms: retreat or defend. In the case of retreat, assets or activities can be abandoned or relocated to a safer position further inland from the shore. Defensive mitigation involves building a wide range of structures, such as seawalls, breakwaters, groynes, and beach renourishment systems, where sediment is placed to recreate the natural buffer provided by beaches and dunes. Beach renourishment has become a popular and widely used method of erosion mitigation and is a useful approach because it addresses one of the three fundamental causes of erosion—a deficit in the sediment budget. However, as with any response, it is not without problems, such as high costs, lack of suitable sediment, and ecological impacts. For example, sea turtles are known to be impacted when renourished sand is different from the native sand, in turn affecting nesting success. Alternatively, well-designed beach renourishment can improve nesting success. In situations where there is economic dependence on beach tourism, the cost of renourishment will probably be justifiable. Defensive structures are also expensive to build and maintain and often have detrimental impacts on shorelines, such as beach loss following seawall construction or downdrift erosion after a groyne is built. However, for high value assets, defensive structures are often necessary, and beach loss is accepted. There is no single response that is suitable for all coastal erosion problems, and each case needs to be assessed on its own merits.

In situations where development has not occurred or is proposed, careful assessment of future erosion trends is necessary to avoid creating a new hazard. In such cases, planning schemes and setback zoning can be used to avoid creating erosion hazards through inappropriate development. Predicting the position of future shorelines as sea levels rise is another approach, and a wide variety of models are available for predicting coastal erosion of different shore types. The most common one for sandy beaches where sea level is rising is the Bruun Rule (although it is not a rule). This deterministic model has been criticized and is unlikely to have wide application on islands because the assumptions that underpin the model will most likely not hold. Mitigation of coastal erosion for island states or communities is a different and arguably more difficult challenge compared to continental shorelines. Beach renourishment can be severely limited by a lack of suitable sediment sources, although barrier islands are an obvious exception. Relocation or retreat is often impractical given that space may not be available. In addition, small island communities often lack sufficient resources for erosion mitigation, particularly in the case of developing island states. Well-resourced islands such as Singapore have undertaken substantive coastal protection works that cannot be matched by developing island states. A common problem is the view that shorelines should be rendered stable by erosion mitigation techniques, but this view is at odds with an environment that is naturally dynamic. Mitigation techniques also need to be dynamic, thus providing community resilience against climate change and sea-level rise in the future. Shoreline structures can be modified to return dynamic behavior or built in such a way as to allow dynamic behavior. For example, groynes can be constructed so as to be permeable, to allow some sediment to continue to move alongshore. Restoration of ecological function in mangroves, salt marshes, and coral reefs also provides protection from sea-level rise, cyclones, and tsunami, and hence coastal erosion. Such methods may be a more sustainable approach to erosion management on islands than would be engineered responses.

CONCLUSIONS

Coastal erosion will continue to be a major environmental issue for island states and communities into the future, and in the face of projected climate change it is likely to become worse. Alleviation will come from adaptive strategies utilizing a wide range of mitigation techniques.

SEE ALSO THE FOLLOWING ARTICLES

Beaches / Global Warming / Hurricanes and Typhoons / Sea-Level Change / Tides / Tsunamis

FURTHER READING

Bird, E. C. F. 1996. *Beach management*. Chichester, UK: Wiley and Sons.
Cooper, J. A. G., and O. H. Pilkey. 2004. Sea-level rise and shoreline retreat: time to abandon the Bruun Rule. *Global and Planetary Change* 43: 157–171.
Gilman, E., and J. Ellison. 2007. Efficacy of alternative low-cost approaches to mangrove restoration, American Samoa. *Estuaries and Coasts* 30: 641–651.
Komar, P. D. 1996. The budget of littoral sediments: concepts and applications. *Shore and Beach* 64: 18–26.
Leatherman, S. P. 1996. Shoreline stabilization approaches in response to sea level rise: US experience and implications for Pacific Island and Asian nations. *Water Air and Soil Pollution* 92: 149–157.
Masselink, G., and M. G. Hughes. 2003. *Introduction to coastal processes and geomorphology*. London: Arnold.
Nordstrom, K. F. 2000. *Beaches and dunes of developed coasts*. Cambridge: Cambridge University Press.
Nordstrom, K. F., R. Lampe, and N. L. Jackson. 2007. Increasing the dynamism of coastal landforms by modifying shore protection methods: examples from eastern German Baltic Sea coast. *Environmental Conservation* 34: 205–214.
Viles, H., and T. Spencer. 1995. *Coastal problems: geomorphology ecology and society at the coast*. London: Arnold.

ERUPTIONS: LAKI AND TAMBORA

T. THORDARSON
University of Edinburgh, United Kingdom

The 1783–1784 Laki (Iceland) and 1815 Tambora (Indonesia) eruptions both took place at volcanoes situated on islands and are the two largest volcanic eruptions on Earth in the last 250 years. They had significant environmental and climatic impacts, despite contrasting eruption style and magma composition (Table 1).

LAKI EVENT (1783–1784)

Eruption History

The eight-month long Laki eruption (June 8, 1783–February 7, 1784) in Iceland is the second largest flood lava event in historic time (Fig. 1), after its neighbor the 934–940 Eldgjá eruption. We know more about the Laki eruption than any other of its kind because it is described in many contemporary accounts. However, none are as spectacular as the treatise, *A Complete Description of the Síða Fires*, written by the Reverend Jón Steingrímsson in 1788. This chronicle contains vivid and accurate descriptions of eruptive events on a day-to-day basis and its consequences for the local community.

In 1783, the people of southern Iceland had enjoyed a favorable spring and were looking forward to summer. However, their destiny was about to change. Weak earthquakes in the Skaftártunga district in mid-May were the first sign, and their intensity increased steadily. By June 1 they were felt up to 100 km away from the source and kept escalating until June 8, when a dark volcanic cloud spread over the district, blanketing the ground with ash. The Great Laki eruption had begun (Fig. 2).

Later that day, more than 1000-m-high lava columns rose from a new volcanic fissure in the highlands to the north. The volcanic fumes made the sun appear red as

TABLE 1
Some Basic Facts on the 1783–1784 Laki and 1815 Tambora Eruptions

Eruption	Duration (Days)	Eruption Episodes	Column Height (km)	Magma Volume (km^3)	SO$_2$ Output (Million Tons)
Laki	247	10	7–13 km	15	122
Tambora	7	4	> 40 km	33	54–60

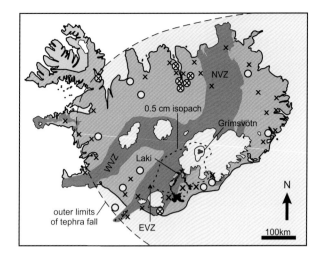

FIGURE 1 Position of the Laki fissures (red) and lava flow (black) in relation to the active volcanic zones (orange) in Iceland. Also shown are the 0.5 cm isopach (broken blue line) of the Laki tephra fall as well as the outer limit of the area (shaded) that was affected by fallout of very fine ash, which covered ~750,000 km² including the bulk of the land surface in Iceland (~100,000 km²). Open circles show locations where fall of fine ash was reported in Iceland.

The green-colored parts of Iceland show where > 60% of the grazing livestock was killed. Crosses indicate locations where deaths are consistent with fluorine poisoning, and large crosses indicate where livestock died within 2–14 days of the onset of the Laki eruption. Modified after Thordarson and Self (2003).

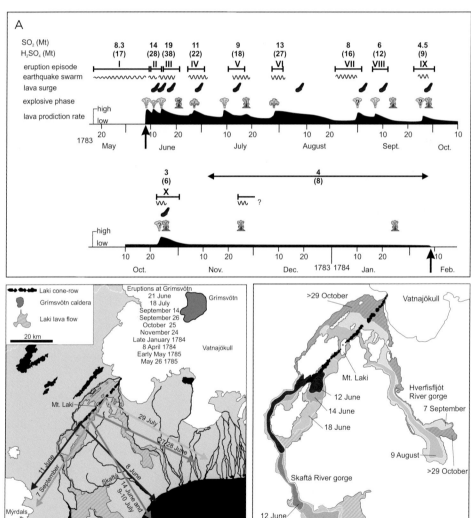

FIGURE 2 Diagrammatic presentation of the sequence of events during the Laki eruption: (A) the timing and duration of the 10 eruption episodes (labeled I–X) as well as the occurrence of key events. Arrows show onset and termination of Laki eruption. Horizontal bars indicate episode duration and numbers the estimated atmospheric SO_2 and H_2SO_4 mass loadings by each episode. Also shown are earthquake swarms (wiggly lines), major lava surges (red symbol), magma discharge with time (blue), and explosive phase at Laki fissures (gray eruption clouds) and Grímsvötn (green eruption clouds). (B) Dispersal directions of main plumes and tephra falls from Laki fissures and timing of eruptions at Grímsvötn volcano in 1783–1785. (C) Growth of the Laki lava flow field showing the position of the flow front at given dates. Modified after Thordarson and Self (1993) and Thordarson et al. (2003).

blood and deprived of its natural shine. The accompanying rainfall was salty and sulfur-smelling, causing smarting of the eyes and skin.

On June 11 lava issuing from the fissures had dammed up the Skaftá River, and the following day a large lava stream surged out of the Skaftá River gorge with incredibly loud cracking noise and rumbles. Recurring tephra falls occurred through June and July, and lava surged from the gorge for the next 45 days. By the time it stopped, lava had filled the gorge, which was up to 100 m deep before the eruption, and covered 350 km^2 of land, burying cultivated areas in front of the gorge and 17 farms.

The Laki fissures were not done. On July 29 heavy tephra fall came over eastern part of the Fire districts, forcing many farmers off their land. On August 3 the Hverfisfljót River dried up, and shortly thereafter lava surged from its gorge, threatening to envelop the Sí a district completely. In the fall, the Fire districts were showered by intermittent tephra falls, and lava flowed from the Hverfisfljót River gorge until end of October, adding 250 km^2 to the lava flow field and destroying four more farms. The Laki fissures continued to produce lava until February 7, 1784.

As if one erupting volcano were not enough! In the summer of 1783 "sandy and muddy ash" fell in the Fire districts from another *eldgjá* (= fire fissure). This was one of many concurrent subglacial eruptions at the Grímsvötn volcano that occurred intermittently until May 1785.

The Laki eruption emerged from a 27-km-long fissure, now delineated by more than 140 cones, and issued ~15 km^3 of magma when calculated as dense rock equivalent (DRE). Ten eruptive episodes occurred during the first five months of activity, each featuring 2–4-day explosive eruptions followed by a longer period of lava effusions. The explosive activity included > 13-km-high eruption columns and produced tephra fallout over ~750,000 km^2. Where the tephra blanket is thicker than 0.5 cm it covers ~7,200 km^2. The volume of tephra (calculated on basis of vesicle-free magma) is 0.4 km^3 or more than double the volume produced by the 1980 Mt. St. Helens eruption in the northwestern United States. In the end, lava covered ~600 km^2 and had a volume of ~14.7 km^3. The eruption emitted about 120 million tons of sulfur dioxide, with ~80% released into the upper troposphere/lower stratosphere (i.e., the westerly jet stream), where it reacted with atmospheric moisture to produce between 165 and 200 million tons of sulfuric acid (H_2SO_4) aerosols. The discharge by each of the Laki eruption episodes was half to equal the total SO_2-discharge of the 1991 eruption at Mt. Pinatubo in the Philippines. The eruption maintained a sulfuric aerosol veil that hung over the Northern Hemisphere for more than 5 months.

Effects and Consequences of Laki

The year 1783 is known in Europe as *Annus Mirabilis* (The Year of Awe), primarily because of the extraordinary state of the atmosphere that caused great public concern. The culprit was the Laki eruption in Iceland, and its consequences were felt across the Northern Hemisphere and affected both the environment and climate.

ENVIRONMENTAL EFFECTS

The direct environmental impact of Laki was greatest in Iceland. The lava flows destroyed valuable land and ruined more than 20 farms, but no humans were killed directly by lava or tephra fall. However, the damaging effects of the sulfuric haze and fallout of very fine ash was noticed everywhere in the country, and it seriously affected vegetation, animals, and people. On the first day of the eruption the plumes carried ash and acid rainfall over the districts closest to Laki, such that fallout burned holes in dock leaves and caused wounds on skin of animals and humans. Elsewhere, the haze was accompanied by a sulfurous smell and fallout of burning (acid) rain, along with black, fine ash and white dust (sulfuric precipitates?) that stained metal objects. People complained of weakness, shortness of breath, irritation in the eyes, and throbbing of the heart. Most of the birch trees, shrubs, and mosses were killed; these plants disappeared from parts of Iceland for 3 to 10 years after the eruption and never returning to some areas. The grass in cultivated fields withered down to the roots, and grass growth was stunted. The grazing livestock suffered chronic fluorosis, resulting in softening and deformation of bones and joints, dental lesions and outgrowth on the molars (known as *gaddur*, "spike," in Iceland). Within a year more than 60% of the grazing livestock had died, which, along with an unusually cold summer and winter (1783–1784), resulted in the disastrous "Haze Famine," killing 20% of the country's population (~10,000 individuals).

The effects of Laki extended well beyond the shores of Iceland. Its eruption columns pumped 100 million metric tons of sulfur dioxide into the westerly jet stream, which governs the atmospheric circulation above Iceland. These sulfur-rich plumes dispersed 165–200 million tons of H_2SO_4 aerosols eastward over the Eurasian continent and across the Arctic to North America, and they eventually covered the Northern Hemisphere from the latitude of 30° N to the North Pole. About four-fifths of the aerosols were removed from the atmosphere in the summer and fall of 1783 by subsiding air masses within high-pressure systems (Fig. 3). This removal formed the infamous "dry fog" that hung over the Northern Hemisphere for more

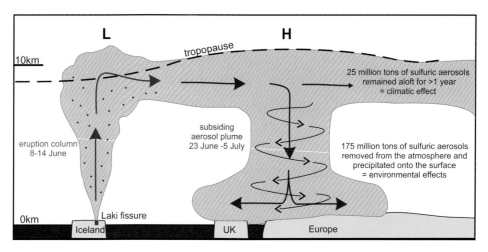

FIGURE 3 Schematic cross section from Iceland to Eurasia depicting the dispersal and development of the Laki plumes over Europe in the first 3 weeks of eruption. Plumes produced by explosive activity at the Laki fissures carried ash and SO₂ and other gases to altitudes of 9–15 km (red arrows) and then dispersed eastward over Europe by the polar jet stream. Because of convergence of airflow at the tropopause, the Laki aerosol cloud was sucked into a large quasi-stationary high-pressure cell (H) located over Europe at the time and reintroduced into the lower atmosphere by the subsiding air masses (blue arrows), spreading in a spiral-like fashion across the continent (black arrows). In total 175 million tons of sulfuric aerosols were removed from the upper atmosphere in summer and fall 1783, producing a widespread volcanic pollution and acid precipitation (equal to ~1000 kg/km² of H₂SO₄) responsible for the observed environmental impact at the surface. The 25 million tons that remained aloft were the main contributor to the climatic effects of the eruption. Modified after Thordarson and Self (2003).

than five months and produced severe air pollution on a continental scale. The concentrations of sulfuric aerosols ($\geq$ 1000 mg m⁻³) within the dry fog clearly passed critical thresholds for human health and were responsible for severe respiratory dysfunction that affected many people. This volcanic pollution may have increased the death rate among humans in England and France by up to 25% (~25,000 individuals). The dry fog also resulted in dry precipitation onto the land, amounting to > 1000 kg of sulfuric acid per square kilometer in summer and fall 1783. This precipitation caused considerable damage to vegetation and crops across Europe. In June, trees in Holland dropped their canopy such that they had the appearance they would ordinarily have in late fall; tree growth in Scandinavia and Alaska was stunted. The Laki haze may also be the sole cause for the late eighteenth-century demise of the Inuit population in western Alaska.

CLIMATIC EFFECTS

The weather in the summer of 1783 was unusual and extreme across the whole Northern Hemisphere. July and August 1783 were dry and hot in southwestern, western, and northwestern Europe but fair in central Europe, as indicated by record grape harvest in Hungary. However, it was unstable and cold in eastern Europe and Siberia, with significant mid-summer snowfall in Rezeszow, Poland, and Moscow, Russia. Unusually intense thunder- and hailstorms were frequent across the Eurasian mainland. In early July, the regions at the foothill of the Altai Mountains experienced harsh overnight frost. Severe drought was reported from North Africa (Egypt), which resulted in death or displacement of one-sixth of Egypt's population. In India, drought resulted in the Chalisa famine. Severe drought is reported from the Yangtze region, and in general the summer was extremely cold all over China. In Japan, the summer of 1783 is singled out as particularly calamitous time because of widespread failure of the rice harvest caused by persistent late summer wet and cold. This weather pattern is attributed to persistent northeasterly winds, induced by blocking of the jet stream by stationary anticyclones off the east coast of Japan. It resulted in the most severe famine in the nation's history, with an estimated death toll in excess of one million people. North America had a fairly normal summer, except that unseasonably cold weather may have affected western and northern Alaska.

The following winter (i.e., 1783–1784) was bitter cold and unusually long in many regions across the Northern Hemisphere. It is still one of the most severe winters on record in both Europe and North America. The straits of Denmark could not be crossed by boats in January because of ice cover, and the Jutland peninsula was still covered by ~1 m of snow in mid-April. Winter lasted well into May, and further south, in Hamburg, Germany, it lasted past mid-March. In Amsterdam people drove wagons on ice across the Markersee and traveled on ice between villages along the North Sea coast, a distance of ~25 km. Reports from Paris describe a long-lasting freeze in January and February with persistent temperatures of −4 °C. The ice and snow hindered commuter travel, causing a severe shortage of firewood in the city. Similar news on shortage of merchandise came from Vienna, because the Danube River was completely frozen over. The arrival of spring thaw raised the water of major rivers in central and south Europe such that severe floods caused damage from Prague, now in the Czech Republic, to Seville, Spain.

The winter appears to have been normal across eastern Asia, Japan, and western North America. However, in the eastern United States, winter commenced early and lasted well into spring (i.e., April to May). It caused the longest known closure by ice in the harbors and channels of Chesapeake Bay, and the Mississippi River was filled up with ice fragments at New Orleans between February 13 and 19, 1784.

Eighteenth-century temperature records from over Europe and North America suggest that the annual cooling that followed the Laki eruption was about 1.3 °C and lasted for two to three years. The Laki haze would have significantly increased the planetary albedo over the Northern Hemisphere, and this negative departure in the annual mean surface temperature following the Laki eruption appears to be consistent with a direct offset of radiative thermal balance in the Northern Hemisphere atmosphere. However, ice core evidence indicate that the bulk of the Laki haze had been removed from the atmosphere by summer 1784. Therefore, low annual temperatures in the years 1785 and 1786 cannot be a directly attributed to radiative forcing caused by the Laki haze. Nonetheless, back-to-back occurrence of cold years in Europe and North America implies a common source, and close temporal association with Laki indicates a connection. Did the Laki aerosol cloud stimulate changes to the atmospheric circulation? Unusually low frequency of progressive (westerly) weather over the British Isles and Central Europe, weaker summer monsoons, and stagnation of the polar front of the northeast coast of Japan support such a notion because they indicate change to dominantly mixed or meridional circulation in the Northern Hemisphere in late 1783 and in 1784.

TAMBORA EVENT (1815)

Eruption History

The April 1815 Tambora eruption was the largest explosive event on Earth in the last 250 years. Mt. Tambora is located on Sanggar Peninsula on Sumbawa island and is one of 28 volcanoes on the Lesser Sunda island chain in the Indonesian archipelago (Fig. 4). Tambora now rises 2850 m above sea level and is crowned by a 6-km-wide

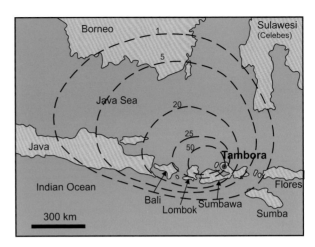

FIGURE 4 Map of the central part of the Indonesian archipelago. Also shown are the location of Tambora volcano on the Sanggar Peninsula of Sumbawa island and isopach contours (in cm) of the ash fallout from the 1815 eruption. Modified after Self et al. (1984), published with permission from S. Self.

and ~700-m-deep caldera, formed in the 1815 eruption. Before that event, the volcano was at least 4000 m high.

Mt. Tambora had not been active in living memory before unrest commenced in 1812. A dark eruption cloud appeared at the summit of Tambora and remained for the next three years, growing darker and larger with time. Rumbling noise of increasing intensity caused such concern among Sumbawa residents that they requested formal investigation. Explosive activity and ash fall in 1814 was noted by an expedition sailing by Sanggar on the way to Makassar (Sulawesi).

In the late afternoon on April 5, 1815, the main eruption began with a violent 2–3-hour-long event that produced ash fall as far as central Java (i.e., ~800 km). The explosions, heard across the Indonesian archipelago up to distances of 1400 km, resembled cannon fire, such that soldiers in Yogyakarta on Java searched the land and sea for possible invaders. Explosive activity of lower intensity continued over the next four days. Around 7 PM on April 10, the 1815 Tambora eruption intensified and entered its climactic phase (Fig. 5). According to eyewitnesses in Sanggar village (~30 km east), this phase began with three columns of flame bursting up from the summit, rising to great heights before uniting in a single plume. Shortly thereafter, the slopes of the volcano appeared like a body of liquid fire, spreading in all directions. Within the hour, nut- to fist-sized stones (pumice?) began to rain down at Sanggar, and between 9 and 10 PM, intense ash fall commenced. Soon after a violent "whirlwind" descended on the village and blew down almost every house. It was much more devastating closer to the volcano; uprooting the largest trees and throwing them along with houses,

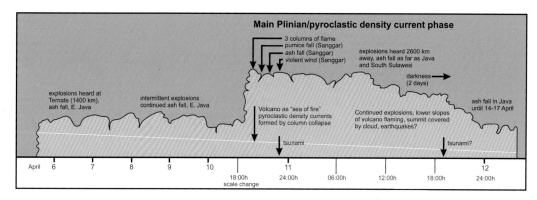

FIGURE 5 Chronology of the 1815 Tambora eruption deduced from contemporary reports. After Self *et al.* (1984), published with permission from S. Self.

humans, and livestock into the air and out to sea. At this time sea level rose suddenly by ~4 m and swept the coastal areas with disastrous consequences. This elevated state of activity appears to have continued through the night and into the evening of April 11, because reports indicate the presence of a sustained plume and flaming slopes as well as explosions at Tambora that were heard across the Indonesian archipelago (~2000 km). The ash plume spread out over a vast area extending from Bima on Sumbawa (~90 km east and upwind from the volcano) to Makassar on Sulawesi (380 km north) and western Java (~1000 km west). It blocked out the sun and caused complete darkness in the region until April 12, when the eruption intensity dropped. Explosions ceased on April 15, but minor ash fall was reported as late as the 17th.

These accounts suggest that the paroxysmal phase started as a very powerful Plinian eruption, producing ≥ 40-km-high columns and widespread pumice and ash falls, which were dispersed disproportionately westward by weak (~6 m s^{-1}) southeasterly monsoonal winds. It also implies that pyroclastic density currents (i.e., surges and/or flows) of significant volume formed within the first hour as indicated by the tsunami in Sanggar around 10 PM. The eruption appears to have maintained this level of intensity for ~24 hours. During that time it produced pumice and ash fall that spread over > 980,000 km^2 and at least seven major pyroclastic density currents that covered the flank of the volcano with up to 25-m-thick pyroclastic surge and flow deposits. It has been suggested that deposition of pyroclastic fall and density currents was more or less synchronous throughout the paroxysmal phase. A sustained but partially collapsing eruption column, augmented by co-ignimbrite ash clouds rising from the pyroclastic density currents and recycled back into the main eruption column, fed an umbrella cloud that produced the extensive 1815 ash fall deposit. This fall deposit, which was 1–2 cm thick at 1000 km distance, has an estimated volume of ~100 km^3, equivalent to ~27 km^3 of vesicle-free lava. To this we must add the volume of ignimbrite (~5.5 km^3) and early (April 5–10) fall deposits (~0.6 km^3), making the total volume of magma expelled by Tambora ~33 km^3, which corresponds to ~8 × 10^{13} kg. Using this mass along with the measured pre- and post-eruption sulfur contents of the magma, the 1815 Tambora eruption is estimated to have released ~56 million tons of SO$_2$, which yielded a stratospheric sulfuric aerosol loading of ~100 million tons.

Effects and Consequences of Tambora

In Europe, the eastern United States, and Canada the year 1816 was known as "the year without a summer" because of unseasonably cold weather and widespread crop failure. The 1815 Tambora eruption is now recognized as the main cause of this unseasonable and grisly summer. The effects of Tambora were not limited to these regions but felt across the globe, and nowhere more strongly than in Indonesia, where the environmental impact proved to be devastating for tens of thousands of its people.

ENVIRONMENTAL EFFECTS

Direct environmental effects of the Tambora eruption were confined to Indonesia and were catastrophic for Sumbawa and the neighboring islands. The princedoms of Pekat and Tambora on Sanggar Peninsula were completely wiped out and their inhabitants killed by pyroclastic density currents, ash falls, or tsunamis. Many people in the village of Sanggar met the same fate, and survivors fled to the princedom of Bima on east Sumbawa. The death toll was between 10,000 and 11,000 people. In other parts of the island, ash fall from the eruption polluted drinking water and destroyed crops, causing immense hardship and starvation. The ensuing famine on

Sumbawa claimed an additional 37,000–38,000 lives and > 75% of the livestock. Another 36,000 individuals fled to other islands. Thus, Sumbawa lost ~50% of its inhabitants because of the 1815 eruption, and it took 20–30 years for the island to recover fully from this disaster. On the neighboring islands (Lombok, Bali, and South Sulawesi), the ash fall caused widespread crop failure, roof collapses, and polluted drinking water. The majority of the fatalities on these islands resulted from the famine induced by the failure of the rice crop, killing between ~35,000 and ~70,000 people. Nonetheless, Lombok and Bali recovered much faster than Sumbawa, and within 12 years rice farming and business were in full swing. At the time, the Tambora eruption was a natural disaster of unprecedented magnitude that had dire economic consequences and killed between 83,000 and 118,000 people, making it the deadliest eruption in historic times.

CLIMATIC EFFECTS

Tambora's sulfuric aerosol cloud had a global distribution, as testified by ice-core acidity peaks in both hemispheres, and it was present in the atmosphere for 2–3 years, producing an estimated cooling of 1.0–1.5 °C. Also, climate records indicate that the entire decade 1810–20 was unusually cold across the globe, setting in abruptly around 1810 and further enhanced after 1815. It is noteworthy that ice core records show a strong volcanic acidity signal in 1809, attributed to a major equatorial explosive eruption from an unknown source. Thus, it is possible that the earlier eruption initiated cooling that was enhanced by the Tambora eruption.

Contemporary accounts and various types of climate proxy records contain abundant valuable information on the state of the atmosphere and weather conditions in the years following the 1815 Tambora eruption. Both of these data sets, however, are strongly biased toward the Northern Hemisphere. Some key observations are summarized below.

In Indonesia the period 1813 to 1819 was unusually dry, with the most severe drought conditions in 1817. In Northern India and neighboring countries the regularity of the seasons was suddenly disrupted in summer of 1815, a condition that persisted until 1819. The cold seasons were unusually severe and wet, whereas the hot seasons were marked by excessive drought. It has been suggested that the cholera epidemic that broke out in the region in mid-1816, and had very high mortality rates, was caused by these oppressive weather conditions. In southeastern India (Madras, now Chennai) rainfall was well below normal in 1815 but above the norm in 1816. Also, in late April 1815 and again in summer 1816, the temperature in Madras dropped below freezing, which is almost unheard of in this region. In the lowlands of eastern China it was abnormally cold from autumn 1815 to summer 1817. Extremely cold and dry weather in the winter of 1815–1816 killed > 50% of the trees on the tropical plains of Hainan, whereas in the north, wet and cold weather destroyed vegetation and crops. Icy and snowy conditions are reported from southeast China and Taiwan. In central Japan, the winter 1815–1816 was severe, with long-lasting cold and heavy snow, whereas the summer of 1816 was hot and dry. Normal weather prevailed in northern Japan and across the Aleutian Islands.

In western North America the weather was normal during the decade 1811–1820, including 1816. In contrast, eastern North America experienced anomalously cold conditions in the period 1815–1818, extending from Hudson Bay to the Atlantic coast. In Ontario, the warm summer of 1815 was terminated early and abruptly by cold and wet stormy weather. The winter 1815–1816 was very cold and dry, and spring arrived 2–3 weeks later than normal. The summer of 1816 was bitterly cold in the region from Hudson Bay to Labrador, with northerly winds bringing temperatures down to near-Arctic condition (i.e., up to 4 °C colder than normal). The region to the south of Hudson Bay experienced unusually strong north and northwest winds and cold conditions. Three major outbreaks of cold Arctic air occurred during the summer of 1816: one on June 5–10 that brought killing frost and snow to the St. Lawrence Valley, New England, and New York State and frost to New Jersey and Pittsburgh; another on July 5–9 with frost in New England and unusually cold temperatures as far south as Philadelphia; and the third on August 20–22 with frost as far south as Philadelphia. The winter of 1816–1817 was again very cold and heavy with snow. Interestingly, the summer of 1816 in southeast Newfoundland was warmer than normal because of persistent southwest winds, whereas conditions in Greenland were wet, cold, and gloomy.

The 1810–1820 decade was one of the coldest over Europe since 1750. Also, in 1816 all seasons were colder than normal, and the departure from the norm (up to −2 °C) was strongest in the spring and summer and caused widespread crop failures across the region. Russia was an exception, with all seasons warmer than today. Nothing unusual is evident in weather or climate proxy records from the Mediterranean region. Data available from the Southern Hemisphere is sparse and nonconclusive but appears to indicate statistically significant departure from normal weather conditions in the wake of Tambora.

The evidence outlined above indicate that during the summer of 1816, the upper troposphere circulation in the Northern Hemisphere was typified by persistent large-scale meridional flow patterns, pushing troughs of cold air further south and ridges of warm air further north than in normal years. The ridges frequently developed into blocking highs, steering surface depressions along more southerly tracks and dragging cold northerly air southward in the process. This circulation pattern implies stronger meridional temperature gradient in mid-latitudes compared to a normal summer season.

LAKI VERSUS TAMBORA

It is fascinating to compare and contrast Laki and Tambora, the two most climatically significant eruptions in recent history, because the difference is so striking in almost every respect. Their geologic and geographic settings are vastly different. The Laki fissures are on a divergent plate boundary that sits right on top of a mantle plume. Geographically it is practically in the Arctic (~64°N). On the other hand, Tambora volcano is part of a volcanic arc formed above a convergent plate boundary and is situated just south of the equator (~8° S). The difference in geographic position explains why the effects of Tambora were of global extent while those of Laki were apparently confined to one hemisphere. They are also characterized by starkly contrasting eruption styles. Laki is classified as a flood lava event (i.e., an effusive eruption) because > 95% of the magma was erupted as lava. Tambora is on the opposite end of the spectrum, a classic example of an explosive eruption, because all of its magma was disintegrated in the conduit and erupted as pyroclasts. It is the most violent terrestrial eruption in the last 250 years. The magma volume emitted by Tambora (~33 km^3) is more than double that produced by Laki (~15 km^3). Yet the amount of sulfur dioxide released into the atmosphere by Laki is close to twice that of Tambora because the sulfur content of its magma was greater by a factor of ~2.5. In other words, the Laki magma (basalt with 49% SiO_2 and 14.5% FeO) had the capacity to carry more sulfur to the surface than the Tambora magma (trachyandesite with 56% SiO_2 and 4.8% FeO). The dissimilarities go on. Tambora was much more explosive than Laki and produced much higher eruption columns. Therefore, it injected its load at much greater altitude (i.e., at ~30–40 km versus 9–13 km), which resulted in greater distribution and longer residence of the sulfuric aerosol cloud in the atmosphere and stronger climatic effects. However, the emissions from Tambora were almost instantaneous, whereas at Laki the sulfur loading occurred in ten eruption episodes over a five-month period. These three factors—higher sulfur content, lower eruption columns, and prolonged sulfur emissions—explain the more pronounced and widespread volcanic surface pollution brought on by Laki and caused havoc in Europe during the summer and fall 1783. The estimated death toll of Tambora ranges from 83,000 to 118,000 people, whereas the same figure for Laki is between 10,000 and 35,000 individuals. These differences highlight the importance of considering the nature of the eruption when evaluating its potential atmospheric effects and socioeconomic impacts.

In light of the contrasts between Laki and Tambora, it is rather remarkable that the most striking similarity between the two events is their impact on the climate. In both instances, the radiative forcing (i.e., reduction of the surface solar energy flux) imposed by the sulfuric aerosol clouds did not produce a uniformly distributed negative temperature anomaly. A straightforward explanation is that both eruptions influenced the long-wave flow pattern and brought on mixed or meridian circulation. Laki did so in the seasons immediately following the eruption (i.e., summer 1783 to summer 1784) and Tambora in summer of 1816, more than a year after the eruption. In this context it is important to note the northerly location of the Laki fissures and the timing of the eruption (June onward) as well as the southerly position of Tambora and the timing of the climatic effects.

The sulfuric aerosol burden from Laki was concentrated in the Arctic. It may have caused strong disruption to the thermal balance in the region over two summer seasons or at a time when the incoming radiative flux was at its peak. The net effect of this perturbation would be heating of the upper Arctic atmosphere and subsequent reduction of the equator–pole thermal gradient. The sulfuric aerosol burden from Tambora, on the other hand, was initially concentrated above the equatorial regions (i.e., in 1815) and not effectively dispersed toward the poles until the following year (i.e., 1816), when it also sank to lower atmospheric levels. The net effect of this transport was similar to that of Laki: heating of the upper Arctic atmosphere and moderation of the equator–pole thermal gradient. In both cases, the consequence of this excess heating is a weaker westerly jet stream and development of mixed or meridional circulation. Thus, when assessing climatic impact of volcanic eruptions, we have to consider both the radiative effects and modifications of atmospheric circulation. If the eruptions are the primary cause of the observed change in atmospheric circulations, the number of their victims increases significantly—in the case of Laki, to more than one million people.

SEE ALSO THE FOLLOWING ARTICLES

Climate Change / Iceland / Indonesia, Geology / Krakatau / Lava and Ash / Volcanic Islands

FURTHER READING

de Jong Boers, B. 1995. Mount Tambora in 1815: a volcanic eruption in Indonesia and its aftermath. *Indonesia* 60: 37–59.

Harington, C. R., ed. 1992. *The year without a summer, world climate in 1816.* Ottawa: Canadian Museum of Nature.

Oppenheimer, C. 2003. Climatic, environmental and human consequences of the largest known historic eruption: Tambora volcano (Indonesia) 1815. *Progress in Physical Geography* 27: 230–259.

Self, S., M. R. Rampino, M. S. Newton, and J. A. Wolff. 1984. Volcanological study of the great Tambora eruption of 1815. *Geology* 12: 659–663.

Sigurdsson, H., and S. Carey. 1989. Plinian and co-ignimbrite tephra fall from the 1815 eruption of Tambora volcano. *Bulletin of Volcanology* 51: 243–270.

Stothers, R. B. 1984. The great Tambora eruption in 1815 and its aftermath. *Science* 224: 1191–1198.

Thordarson, T., and S. Self. 1993. The Laki (Skaftar Fires) and Grimsvotn eruptions in 1783–1785. *Bulletin of Volcanology* 55: 233–263.

Thordarson, T., and S. Self. 2003. Atmospheric and environmental effects of the 1783–1784 Laki eruption: a review and reassessment. *Journal of Geophysical Research* 108(D1), 4011, doi:10.1029/2001JD002042.

Thordarson, T., G. Larsen, S. Steinthorsson, and S. Self. 2003. 1783–85 AD Laki-Grímsvötn eruptions II: appraisal based on contemporary accounts. *Jökull* 51: 11–48.

ETHNOBOTANY

W. ARTHUR WHISTLER

University of Hawaii, Honolulu

Ethnobotany is the study of the relationship between plants and people, but in practice it is usually applied to the use of plants in the developing world (i.e., in cultures other than mainstream "First World" countries). It includes the study of how plants are used and managed as food, medicine, housing materials, cordage, textiles, cosmetics, dyes, and other artifacts and practices that are a part of all cultures. It is considered to be part of ethnobiology, which also includes ethnozoology, but because the majority of most cultures' needs are met by plants, ethnobotany is the more commonly used term for the relationship between people and the biotic world.

TYPES OF ISLANDS

The Pacific Ocean is the largest natural feature on earth, and the majority of the world's islands (particularly the oceanic islands) are located in its tropical zone. Because of this, Oceania, as the area is called, is the main focus here. Most of its islands, ranging over halfway around the world, from Madagascar (which was originally settled by voyagers from Indo-Malaysia) to Easter Island and Hawai'i, share a common culture, assemblage of useful plants, and ethnobotany. Oceanic islands elsewhere include parts of the Caribbean, the mid-Atlantic islands (such as the Azores and Iceland), and the Indian Ocean islands, but these are few in number compared to the islands of Oceania.

Because the ethnobotany of islands requires inhabitants, uninhabited islands (or, more precisely, islands that do not have inhabitants or foraging visitors) are not included in the topic. Inhabited (past or present) continental islands, such as the Channel Islands off the coast of Southern California, the Aleutians extending into the Pacific from Alaska, and the Mediterranean islands are often inhabited by the same ethnic groups as the adjacent mainland areas. The flora of continental islands can be very similar to that of the adjacent mainland flora and depends upon distance from the mainland, size of the islands, and time since they have been separated from the mainland (or *whether* they have been separated). When the islands are large, floristically similar to the adjacent mainland, and inhabited by the same ethnic group present on the mainland, the ethnobotany of the two areas is often very similar, and the islands' ethnobotany is best treated as part of that of the mainland. Consequently, the most relevant islands to the topic of ethnobotany of islands are the inhabited oceanic islands. Also relevant, however, are continental archipelagoes, such as the islands of Indo-Malaysia, that may be far from nearby continents, from which they may have been separated for millions of years. The islands of the Caribbean make up a special case. The original inhabitants of this area, called the Caribs, were virtually exterminated shortly after contact with the Europeans. Most of their original ethnobotany, which developed over thousands of years during their presence in the islands, was irretrievably lost with their demise. In its place was a new ethnobotany largely based upon plants introduced by Europeans and Africans, who replaced the original inhabitants.

BIODIVERSITY OF ISLANDS

The number of native species of plants on an island is determined by several factors, some of which were elucidated in the classic island biogeography work of Robert MacArthur and Edward Wilson. Their theory predicts that, all other things being equal, distant islands will have relatively fewer species than those close to a mainland biota source (i.e., a source of plant dispersal propagules).

Also important are the number of habitats, the elevation, and the age of the island. The theory of MacArthur and Wilson generally deals with native species, but it can also apply to non-native species, which constitute a category far more important to island ethnobotany than are native species.

CATEGORIES OF ETHNOBOTANICAL PLANTS

Island ethnobotanical plants, like the plants of any place, can be divided into two groups: native species and alien (introduced or non-native) species. Native species are those that arrive at the islands by natural means (i.e., without the assistance of humans). The vast majority of them were present when the first colonists arrived. The number of useful native plants on an island is partly determined by the size of the native flora, and it is intuitive that, in general, the larger the native flora, the more native species that may be useful.

Alien plant species, which constitute the most important category of ethnobotanical plants, can be subdivided into two categories based upon method of long-distance dispersal: accidental introductions and intentional introductions. Accidental introductions were unintentionally brought to the islands and are mainly "weeds" with very little use (in fact, they are often harmful). Intentional introductions are plant species that were purposely introduced to an area. Islanders selected the plants (and animals) that were useful to them as food, cordage, and so forth, and transported these species across the ocean to their new islands. In Oceania, this includes plants such as breadfruit (*Artocarpus altilis*), taro (*Colocasia esculenta*), bananas (*Musa* hybrids), and yams (*Dioscorea* spp.). These plants are often called "canoe plants," or the less frequently used term, "portmanteau plants." Intentional introductions compose by far the most important category of ethnobotanical plants on islands. They have often become part of a common "transported landscape" of useful plants.

The time of introduction of useful plants is significant to ethnobotany. Plants that were introduced by the original inhabitants are called "aboriginal introductions" (or more specifically, Polynesian introductions for Polynesia, Micronesian introductions for Micronesia, etc.), and those introduced by more recent arrivals (usually Europeans) are often termed "modern" or "recent" introductions. Most significant to island ethnobotany are the aboriginal introductions, because these are the traditional plants used by the culture, as opposed to the modern introductions, such as mango in much of the tropics, the uses of which are modern rather than traditional. It is interesting to note that the number of canoe plants in Polynesia also adheres to the MacArthur predictions of decreasing numbers of species with increasing distance from the source area, in this case, Indo-Malaysia (insular Southeast Asia and western Melanesia). In Tonga, the Polynesian area closest to Indo-Malaysia (and where the original Polynesians first settled), about 72 ethnobotanical species were cultivated or harvested, whereas on the islands farthest away, Hawai'i, less than half that number were.

When islands are discovered and colonized, the new inhabitants will usually try to bring their useful plants along with them (the canoe plants), because these species are tried and true. One of the main reasons for this is that these plants were selected, used, and often bred for their usefulness. For example, bananas, which originated in New Guinea as a result of natural hybridizations between *Musa acuminata* and *Musa balbesiana*, were selected over millennia so that the most prolific or best tasting varieties were developed, and these were the ones most likely to be carried with people in their colonization of distant islands.

THE USEFUL PLANTS OF OCEANIA

A listing of over 60 recognized uses for plants in Kiribati (formerly called the Gilbert Islands) in Micronesia was compiled by Randy Thaman, but these can apply to ethnobotany in general. The most important of the uses of plants listed include food; cordage; clothing and textiles (including mats and baskets); timber and materials for housing, boats, and tools; dyes; decoration; and medicines.

Food Plants

The vast majority of the food plants of islands are intentional introductions. In Hawai'i, which is one of the most remote parts of Polynesia, virtually no native edible plants were worthy of cultivation or harvesting. Virtually all of the plant diet of the ancient Hawaiians and the rest of the Polynesians was supplied by their canoe plants. The major exception to this was New Zealand, where the roots of a native fern (*Pteridium esculentum*) and a native lily (*Cordyline australis*) were a major part of the diet of the original Maoris. The major edible canoe plants of the Polynesians were taro (*Colocasia esculenta*), giant taro (*Alocasia macrorrhyos*), coconut (*Cocos nucifera*), sweet potato (*Ipomoea batatas*), three species of yams (*Dioscorea alata*, *D. pentaphylla*, and *D. bulbifera*), bananas (*Musa* x *paradisiaca*), breadfruit (*Artocarpus altilis*), Malay apple (*Syzygium malaccense*), Otaheite apple (*Spondias dulcis*), sugarcane (*Saccharum officinarum*), ti plant (*Cordyline fruticosa*), and Polynesian arrowroot (*Tacca leontopetaloides*). Dozens of other food plants were utilized in Indo-Malaysia, most of them native species that were brought into cultivation. Little is known about which plants were utilized as food by the original inhabitants of the Caribbean, as the new inhabitants from Europe and Africa brought their own food plants (e.g., cassava) to supplement the major ones already present.

Cordage Plants

The most important cordage plants in the Pacific Islands were coconut (coir from the husk), beach hibiscus (*Hibiscus tiliaceus*), and *Pipturus argenteus*. These species are likely to have been introduced to the area, although there is evidence of a variety of coconut being native as far east as the Society Islands. In fact, the islanders (most likely Marquesans) probably introduced the coconut to South America, where, at the time of Columbus's discovery of the New World, it was restricted to the west coast of the continent (and absent from the beaches of the Caribbean when Columbus arrived). To make cordage (sennit), coconut husks are soaked and pounded to separate the useful fibers from the useless interstitial material. These fibers are then twisted and braided into coir (Fig. 1). Beach hibiscus is native over part of its pre-European range, but it was probably introduced to many islands, especially to atolls and to eastern Polynesia. Its inner bark was removed from the stem, soaked, scraped, and dried, after which it was split into the desired width. *Pipturus argenteus* may also have been introduced over much of its range, including to atolls, but is it is probably

FIGURE 2 Making a screwpine mat in Tonga.

a native secondary forest species on high islands. Its fibers were extracted in a fashion similar to those of beach hibiscus and were used to fashion fishing lines. One exception to the almost exclusive use of alien or widespread native plants for cordage in Oceania is *Touchardia latifolia* (Urticaceae), which is endemic to Hawai'i. Its bark fibers are one of the best in Oceania. Another example, from the continental islands of New Zealand, is New Zealand flax (*Phorium* spp.). The Maoris who settled the archipelago nearly a millennium ago probably brought their own cordage plants, but these did not do well in New Zealand's temperate climate and were entirely replaced by the native flax.

Material Plants

Material plants in Oceania are a mixture of native and canoe plants. The most important plants for making mats and baskets are screwpine (*Pandanus tectorius*) and coconut, both of which are native eastward at least to the Society Islands (but more useful cultivars of both were also introduced by the early voyagers). Mats form an integral part of tropical island cultures, because they are often used to cover the floors of houses or function as prestige items (e.g., "fine mats"). The making of floor and fine mats was an important and time-consuming occupation for island women (Fig. 2). The most useful plants for thatch (which is today largely replaced by corrugated iron roofs) were screwpines, coconuts, and sugarcane, with the last species being an aboriginal introduction

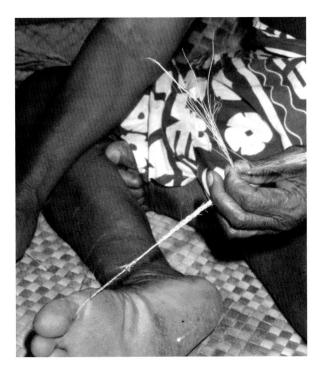

FIGURE 1 Making sennit from coconut husk in Tokelau.

to the area. In Indo-Malaysia, other species, such as native palms, were widely used to make thatch. The most useful thatch plant in Samoa is the sago palm *Metroxylon warburgii*, which is native to Melanesia. The most useful plants for textiles and clothing were paper mulberry (*Broussonetia papyrifera*) and ti (*Cordyline fruticosa*), both of them aboriginal introductions. Tapa cloth is made by stripping the bark from the stems of paper mulberry, scraping off the outer bark, and pounding the material (Fig. 3) until it flattens out to several times its original width. These wide strips are then felted together by further pounding, or pasted using an adhesive made from plant substances, such as the starch from tuber of the Polynesian arrowroot (*Tacca leontopetaloides*). The resulting tapa cloth, white in color, is then usually painted, typically with brown or black dyes. Making tapa cloth from paper mulberry is a dying art in Polynesia, where it is done extensively now only in Tonga. Ti leaves were tied together to make skirts or kilts that were used as the main article of clothing throughout most of Polynesia. In New Zealand, the local flax (*Phorium* spp.) served for making most of the clothing and baskets of the Maoris, as well as the cordage mentioned above.

FIGURE 3 Pounding paper mulberry bark to make bark cloth (tapa) in Samoa.

Timber Plants

In contrast to edible plants, most plants used for timber or wood are native species. The two most important uses for timber are for houses and canoes. This applies equally well to both islands and continental areas, although the floras of the two areas are often quite different. There is, however, a major difference between the species used for timber on atolls and high islands. High islands have many more species, and it is predictable that some of them would be useful for timber. Atolls have only a few tree species present, but some of these are serviceable as timber. The most important of the widespread littoral timber species are Alexandrian laurel (*Calophyllum inophyllum*), Chinese lantern tree (*Hernandia nymphaeifolia*), and Pacific rosewood (*Cordia subcordata*). *Calophyllum* has a hard wood used for houses and boats; the wood of *Hernandia* is soft and is favored for making canoes; and the wood of *Cordia* is fine-grained and used for making plank canoes, paddles, and handicrafts. The most valuable timber tree in western Polynesia is *Intsia bijuga*, which is found on atolls and high islands from the Indian Ocean to Samoa. Hawai'i has its endemic koa tree (*Acacia koa*), which is now very highly prized for its beautifully grained wood. The beach hibiscus (*Hibiscus tiliaceus*) is also important on islands throughout its wide range, especially for tools, tool handles, and light construction. Carved wooden bowls were also important in Oceania, especially for the mixing and serving of kava (*Piper methysticum*), traditionally a very important ceremonial and medicinal plant native to Melanesia, but spread widely by aboriginal peoples in Oceania.

Dye Plants

Many plants are used for dyes on islands, some of them canoe plants and some native species. The best-known dye plants belong to the former category. The most widely used dye plants in Oceania are the Indian mulberry (*Morinda citrifolia*), turmeric (*Curcuma longa*), mangroves (e.g., *Bruguiera gymnorrhiza*), and candlenut (*Aleurites moluccana*). The scrapings of these plants were mixed with water for water-soluble dyes or with oils, such as coconut oil, for water insoluble dyes. The bark and roots of Indian mulberry were used to make both a red and a yellow dye, the different colors depending upon whether or not mordents (usually crushed coral) were used. Turmeric was processed by cooking the rhizome scrapings to extract the yellow dye, which was used for a variety of purposes, including body paint. The bark of mangroves was scraped and made into a dark brown dye used to stain tapa cloth and other materials (Fig. 4). The most common black dye in Oceania was obtained from burnt candlenut kernels. The nuts were roasted or baked, and the hard shell cracked to extract the nut. The nuts were then burned

FIGURE 4 Painting Tongan tapa cloth with a bark dye.

in a partially closed container to produce a sooty smoke that collected on the inside of the container. The jet-black soot was then collected and mixed with coconut oil to be used as a dye for tapa cloth and other textiles, as well as being the main ink used for Polynesian tattoos.

Plants for Ornamentation

A number of plants were used for decoration of habitations and for personal adornment. The majority of these have showy flowers. Most native species have relatively small flowers, so plants selected for their large colorful flowers were among the canoe plants. As with native species, the number of ornamental plants decreases with distance from their Indo-Malaysian source, but some are native to Melanesia and Polynesia. The best known of the latter type are the Tahitian gardenia (*Gardenia taitensis*), which is native to Melanesia, and *Fagraea berteroana,* which is native from Melanesia to Polynesia. The popular perfume plant (*Cananga odorata*) has large, strongly scented, attractive flowers, but it is not clear whether it is native or aboriginally introduced to Polynesia and Melanesia. Also important throughout Oceania is the red hibiscus, a sterile hybrid that has now been joined by numerous other recently introduced hybrids having many colors.

Plants with attractive form, flowers, or leaves are used as ornamentals around houses, but the leaves and flowers are also used for personal adornment. They are worn singly over the ear or are fashioned into neck or head leis. Fragrant flowers or other parts of plants, such as the wood of native sandalwood, are crushed or scraped and mixed with coconut oil that is commonly applied to the body like an oily perfume. Other plant parts are sometimes used for decoration, but are not as popular as flowers because they usually lack scent or attractive colors.

Medicinal Plants

Medicinal plants are used on islands, as throughout the world, for treating ailments. Medicinal treatment was often in the hands of specialist healers (Fig. 5), but some medicines and treatments were used by lay persons and were termed "folk medicine." Many of the plants used as medicines are native species, but many others are canoe plants. Plants that proved effective, or that were believed to be effective, would most likely be carried in the migrations of the islanders. The most important plant in this regard is the Indian mulberry (*Morinda citrifolia*), also known today as noni. It was formerly thought to be a canoe plant, but it has now been determined to have been in eastern Polynesia (but probably not Hawai'i) prior to the arrival of the Polynesians. It is used throughout its

FIGURE 5 A Samoan healer and her medicinal equipment.

range for treating a number of ailments, especially "ghost sickness" (believed to be caused by the actions of malevolent spirits) and wounds and sores. The leaves were the major part used for these purposes, rather than the fruit, which has recently been developed by companies in the West into a controversial and unproven panacea.

Plants for Other Uses

Two other uses should be mentioned. Several plant species were used as fish poisons across the tropical Old World islands, mostly to catch small fish in tide pools. In Oceania, these include the fish-poison tree (*Barringtonia asiatica*), derris (*Derris trifoliata*), and tephrosia (*Tephrosia purpurea*). The last two are probably native species but may have been carried to some islands in the eastern part of their range. The roasted and skewered seeds of the candlenut tree (*Aleurites moluccana*) were used throughout the Old World tropics islands as a source of illumination.

CONCLUSIONS

In summary, the ethnobotany of continental islands (which are associated with and often derived from the adjacent continental areas) is often relatively indistinguishable from that on the adjacent mainland, particularly if the islands are large, close, and inhabited by the same ethnic group. The ethnobotany of inhabited oceanic islands, however, is often quite different from that of the distant continents and continental islands because of the depauperate nature of such islands' flora. On the low oceanic islands called atolls, the flora might consist of a mere 35 or fewer species. Consequently, the ethnobotany of oceanic islands is distinct from continental ethnobotany. The plant species are different, as is how they are prepared and utilized. Only a few native species on newly discovered islands could be utilized by the colonists for anything other than timber and medicine. This problem was partly overcome by the intentional transport of a suite of plants—canoe plants—during the islanders' oceanic voyages. Food plants are nearly all canoe plants. This transportation of the useful plants out of the main source area of the tropical Pacific and Indian Oceans (Indo-Malaysia) accounts for the similarity of ethnobotany from Easter Island and Hawai'i in the east to Madagascar (which was discovered and colonized by Malays) in the west. In addition to Oceania, the Caribbean probably had a unique ethnobotany, but this was almost totally lost when the original inhabitants of the islands (the Caribs) rapidly disappeared in the European Era (after 1492).

SEE ALSO THE FOLLOWING ARTICLES

Endemism / Human Impacts, Pre-European / Introduced Species / Oceanic Islands / Peopling the Pacific / Sustainability

FURTHER READING

Cox. P. A., and S. A. Banack, eds. 1991. *Islands, plants, and Polynesians.* Portland, OR: Dioscorides Press.
Krauss, B. H. 1993. *Plants in Hawaiian culture.* Honolulu: University of Hawai'i Press.
Parkes, A. 1997. Environmental change and the impact of Polynesian colonization, in *Historical ecology in the Pacific Islands: prehistoric environmental and landscape change.* P. V. Kirch and T. L. Hunt, eds. New Haven, CT: Yale University Press, 167–199.
Thaman, R. R. 1990. Kiribati agroforestry: trees, people, and the atoll environment. *Atoll Research Bulletin* 333: 1–29.
Whistler, W. A. 1991. Polynesian plant introductions, in *Islands, plants, and Polynesians.* P. A. Cox and S. Banack, eds. Portland, OR: Dioscorides Press, 25–66.

EXPLORATION AND DISCOVERY

SCOTT M. FITZPATRICK

North Carolina State University, Raleigh

One of the greatest achievements in human history was the construction of seaworthy craft and navigational aids that allowed people to move across the world's seas and oceans and to eventually colonize islands. Archaeological evidence indicates that hominids have been able to cross water gaps since the Pleistocene (at least by 750,000–800,000 years ago) and that *seafaring* (rather than *seagoing,* which as the archaeologist Cyprian Broodbank has noted, implies tentative or "culturally enhanced floating") has been practiced for at least 40,000 years. It was not until much later, during the Middle to Late Holocene, that humans began voyaging across much longer distances.

SEAFARING AND THE ARCHAEOLOGICAL RECORD

A major issue when trying to determine how and when peoples sailed across aquatic realms is the paucity of material remains preserved in the archaeological record. Because boats are made from organic materials that generally do not preserve well, particularly in the hot and humid tropics, researchers must often rely on indirect evidence such as artifacts, rock art, or historical records as indicators of when peoples had crossed seas and oceans and reached islands in the past.

Despite some preservation issues, there is still an abundance of evidence that humans were able to explore the world's island regions. Through time, they became increasingly sophisticated at discovering islands using a variety of boat manufacturing and navigational innovations. Although it is becoming clear that most islands in the world were colonized purposefully, not accidentally as was once thought, there are still many lingering questions about how, when, and why humans first settled islands, particularly in the Pacific, Mediterranean, and Caribbean.

EARLY SEA-CROSSINGS IN THE PACIFIC

Despite the generally poor preservation of organic remains, archaeologists have been able to infer that the earliest colonization of an island probably occurred on Flores in the Indonesian archipelago. Stone artifacts, reminiscent of those made by *Homo erectus,* have been found there that date between 800,000 and 900,000 years ago. Flores was never connected to other land masses, even during periods of lower sea level, suggesting that *H. erectus* was able to cross over 20 km of open ocean using some type of watercraft, even if only a simple raft made of bamboo (Fig. 1). That this task was accomplished so early is, in part, probably related to the importance that coastal migration routes and resources began to play in the lives of highly mobile hunters and gatherers.

However, it was not until much later that humans began colonizing other parts of Oceania from the South-

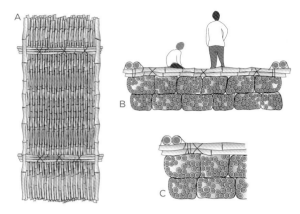

FIGURE 1 Hypothetical raft made from bundled bamboo trunks: (A) plan view, (B) side view, (C) cross-section of raft construction. Drawings by Leslie Hazell.

east Asian mainland. The Indo-Pacific region is notable in this regard because it holds the earliest indirect evidence for seafaring. In addition, the colonization of the Pacific Islands represents one of the most extensive and rapid migrations of peoples prehistorically and displays an astonishing array of sophisticated maritime technologies that had few equals in the ancient world.

Archaeological evidence from numerous Pleistocene-aged sites suggest that Australia was settled by at least 50,000 years ago, New Guinea by around 35,000–40,000 years ago, and the Solomon Islands by approximately 30,000–35,000 years ago. Human remains recovered from the Minatogawa Fissure and Yamashita-cho sites in Okinawa that date between 32,000 and 15,000 years ago, and obsidian transported from Kozu Island to mainland Japan 20,000–25,000 years ago, also indicate early periods of seafaring to islands. Based on a plethora of chronometric dates, artifacts, and other remnants of occupation, it is clear that humans in the western Pacific Rim were beginning to construct more advanced types of watercraft and were honing the seafaring and navigational skills that allowed them to travel greater distances over open ocean. One notable find from the site of Kuahuqiao in the Lower Yangzi River delta of eastern China is a remarkably well-preserved dugout canoe dating to 8000 years ago. This is the earliest direct evidence for watercraft found thus far along the Pacific Rim. Although it is unknown whether this canoe was outfitted with an outrigger (Fig. 2) that would have enabled sailors to harness the wind and venture further, this was surely the beginning of efforts to do so.

Despite a trend toward mastery of the sea over a period of tens of thousands of years, it is likely that humans in the Asian-Pacific region only began constructing watercraft capable for long-distance (200-km+) voyaging some-

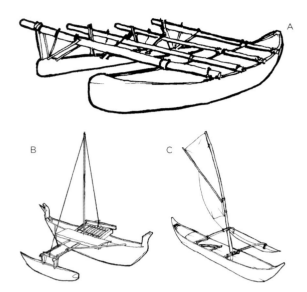

FIGURE 2 Examples of outrigger sailing canoes recorded historically in the Pacific: (A) Samoan dugout canoe with outrigger; four-boom *soatau* (after photograph taken in 1925), (B) Yapese sailing canoe from western Micronesia (without lee platform) (after model held in Cologne Museum), (C) Bonin Islands (Japan) outrigger canoe (after model held Science Museum, South Kensington, London). Drawings by Leslie Hazell, from Haddon & Hornell, 1936-1938.

time during the Middle Holocene, after which humans began locating and settling more remote islands. During the Austronesian expansion around 5000 years ago, some archaeologists theorize that peoples moved southward from Taiwan through the Philippines and into island Southeast Asia and Melanesia. At around 3300–3500 years ago, there was an explosion of seafaring activity as peoples moved eastward from islands fringing the northern part of New Guinea ("Near Oceania," where most islands are intervisible or can be seen at points midway between islands) and into "Remote Oceania," including western Micronesia and West Polynesia. It is during this time that we see "Lapita" groups (the ancestors of Polynesian and most Micronesian peoples), sailing hundreds of kilometers to islands such as Vanuatu, Fiji, Tonga, and Samoa. Lapita peoples were skilled voyagers who were able to manufacture durable, stable, and maneuverable outrigger canoes outfitted with sails that could effectively harness the wind. Similar technologies also allowed Austronesian speakers to cross the Indian Ocean and settle Madagascar around 2000 years ago.

EARLY SEA-CROSSINGS IN THE MEDITERRANEAN

The second earliest locus of seafaring and exploration of islands, after islands in the south/southwest Pacific, is the Mediterranean. Although there is some debate about the timing of sea crossings here, it is generally thought that peoples first crossed the narrow strait of Messina to reach Sicily from the Italian mainland around 30,000 years ago as evidenced by Aurignacian (early modern human) deposits at the site of Fontana Nuova. Around 10,000–15,000 years later, humans crossed the Tyrrhenian Sea, reaching the island of Sardinia. A single human bone found at Corbeddu cave and Upper Paleolithic stone tools recovered from Acquas confirm the presence of humans. Interestingly, there never seems to have been a sustained human occupation on either Sicily or Sardinia after these initial occupations until thousands of years later. It is possible that both were exploratory and temporary endeavors, perhaps fueled by poor environmental conditions on the mainland, which made these islands attractive refuges.

The first good evidence for seafaring in the Mediterranean occurs at around 11,000 BC, when peoples from the Greek mainland traveled 120 km or so (probably in a steppingstone fashion) to the island of Melos to obtain obsidian, flakes of which are found in deposits at Franchthi Cave. It is unknown what types of watercraft were used, but some form of reed bundle raft could have ferried passengers and cargo quite efficiently over shorter distances. Obsidian found in subsequent deposits at Franchthi Cave indicate that repeated trips were made to extract this highly valuable resource. The discovery of lithics, beads, firepits, and faunal remains at the Aetokremnos rockshelter on Cyprus dating back to over 12,000 years ago would have involved a shorter trip than Melos (65–70 km from the Turkish coast, or 100 km from present-day Syria), but a longer overall sea crossing. What these two cases suggest is that peoples in the northern and eastern parts of the Mediterranean were becoming more highly adept at seafaring, scouting out new islands, and setting the stage for more extensive maritime movements across even further distances.

In all instances, it is unknown what types of watercraft were being used to cross over to these islands. Some of the earliest examples of ancient watercraft have been discovered in northern Europe and North Africa. These are of the logboat variety, the earliest of which dates to around 8000 BC in The Netherlands, providing direct evidence that seafaring technology was moving beyond simpler rafts to more sophisticated forms of water transport. By 2600 BC, plank boats fastened with rope lashings were being used in Egypt; those found were associated with the funerary boat of Pharaoh Cheops. A few hundred years later, similar plank boats occurred in Britain, ranging in age from 2000 to 800 BC, and in Demark around 300 BC. Many more examples exist of seafaring technology that allowed explorers to communicate and interact with other people and to initiate and maintain exchange relationships throughout the islands of the Mediterranean and elsewhere in the Old World, although they are typically biased toward log and plank canoes much later in time. The pattern in the Mediterranean, similar to what is seen in virtually every other part of the world, is that successful colonization of islands by humans required the transport of plants and animals to create an agricultural base conducive to sustaining long-term settlements.

THE PRE-COLUMBIAN CARIBBEAN

The earliest colonization of the Caribbean dates back to around 6000 years ago when Lithic/Archaic peoples first arrived in Cuba and Hispaniola, probably originating from somewhere in Mesoamerica. The island of Trinidad was settled as early or even earlier by different Archaic groups from South America, but it was in close proximity to the mainland during periods of lower sea-level, making the island's colonization history different and peripheral from that of the oceanic islands. Archaic age peoples also settled islands in the Lesser Antilles beginning around 4500 years ago, having migrated from South America and settling on a number of islands; the best evidence comes from islands in the northern Antilles (e.g., Antigua). The most extensive migration of peoples into the Antillean chain of islands occurred around 2500 years ago, when ceramic-making horticulturalists known as "Saladoid" began a northward movement from South America into Puerto Rico, the Virgin Islands, and the Lesser Antilles. A thousand years later, peoples had colonized virtually every major island, including Jamaica and the Bahamian archipelago. The Cayman Islands remain the lone exception, having remained free from human contact until the historic period. Similar to the Pacific and Mediterranean, most of the Caribbean Islands were settled relatively late by peoples who were agriculturalists and had gained seagoing and seafaring experience along the coastal margins of mainlands.

Although it is generally agreed that the Lithic/Archaic hunter-gatherers who settled the Greater Antilles originated from Mesoamerica and that South America was the departure point for the other major migrations thereafter, it is debated how people actually moved through the islands. Further explanations suggested that peoples ventured northward from South America through the islands in a steppingstone fashion. Most of the islands in the Lesser Antilles are visible from one to another, especially during clear weather, or can be seen from a point midway between two islands.

However, recent research suggests that migrants may instead have reached the northern islands, such as Puerto Rico, first and then moved east and south back toward the South American mainland. Computer simulations of seafaring have shown, for example, that a direct route from South America to Puerto Rico would have been much quicker and less hazardous. By doing so, seafarers could have avoided dangerous crosscurrents between islands, and the "bottleneck" effect in which fast-moving water in the Atlantic is pushed through narrow inter-island passes by the northeast trade winds. Heavy squalls can also develop as inclement weather passes over islands, with high peaks making sea travel more difficult.

Supportive evidence for a more direct colonization strategy that bypassed the Lesser Antilles comes from an analysis of hundreds of radiocarbon dates from islands throughout the Caribbean. Results suggest that the older Saladoid-era occupations, around 2500 years ago, were in the northern Antilles and are progressively younger as one moves south. Although more archaeological research is necessary to support either hypothesis, it is clear that the peoples who settled the Antilles were perfectly capable of building watercraft that could transport people, plants, animals, and important resources such as stone between mainlands and the Antilles. Current evidence based on the presence of stylistically similar artifacts suggests that interaction was frequent, although local differences do exist. It is important to note that native groups in the Caribbean never developed sophisticated watercraft that took advantage of sails or outriggers. Based on reports from historical chroniclers, they instead seemed to have relied upon dugout canoes. A few examples of these have been recovered from archaeological sites in the Bahamas and coastal South America and are similar to ones still used today (Fig. 3).

FIGURE 3 Generic example of a simple dugout canoe, one of the earliest types of watercraft found in the archaeological record. Drawing by Leslie Hazell.

SEAFARING TECHNOLOGIES AND NAVIGATIONAL TECHNIQUES

For decades, anthropologists and archaeologists have investigated the techniques that peoples used for constructing boats, sailing, and then finding (and returning from) islands. Early computer simulations of seafaring in the Pacific suggested that voyagers may have first sailed against, across, and then down the wind. This strategy would be employed by leaving a given point and then sailing against the wind on a generally eastward track. By using the rising and falling of stars across the sky as an astral compass, they could determine their latitude and sail across the ocean until land was found. If provisions had diminished before any land was sighted, they could simply sail with the wind at their back toward their origination point without having to tack or shunt against prevailing winds on the return voyage. However, some researchers now suggest that periods of wind reversal, as can occur during El Niño–Southern Oscillation (ENSO) events, may have been influential in stimulating migratory movements of people, particularly in East Polynesia.

The use of the star structure or "sidereal" compass was a well-developed technique, especially in Micronesia, where traditional navigators still practice these skills today. In essence, the compass involves tracking the rising and falling stars observed along a north–south axis. Specific stars can then be followed as they move across the sky and a "reciprocal" star, located at 180° on the opposite side of the axis, which is used as a reference point. This was a critical technique, for it allowed navigators to keep track of a boat's heading at night.

In addition to the astral compass, traditional sailors in the Pacific used other tactics such as "island looking," in which the locations of islands were learned in relation to others; "sea knowing," in which the lanes between islands and reefs were memorized; and "dragging," or estimating how far one has traveled by using another island as a point of reference instead of the one that is being targeted. Pacific Islanders also recognized aimers or "living seamarks" such as whales, which could help a navigator regain his course if he were lost or if other techniques failed. The Marshallese, whose survival in an island chain composed only of coral atolls was largely dependent on their ability to navigate between small islets, also developed a unique navigational chart made of sticks and shells called *rebbelib* that provided a relative means for estimating distance and position between islands.

It is well known that Pacific Islanders attempting to discover or relocate to remote islands also knew to look for tell-tale signs of land when sailing across the ocean. For example, traditional navigators recognized that clouds formed over larger islands—this can be attested to by the traditional Maori name for New Zealand, *Aotearoa,* which translates to "the Land of the Long White Cloud." Most of the 25,000 or so islands in the Pacific are small coral atolls, and the reflection of lagoons or dense forests in the clouds also gave hint that land was near. In addition, the presence of seabirds that normally roost on land (e.g., terns and frigates), floating debris (flotsam), fires, and volcanism, were also recognized as indicators that land might not be far off in the distance. Well-experienced sailors also recognized changes in wave and swell patterns as they deflected off an island.

In an attempt to understand traditional boat-building and navigational techniques, a number of researchers worldwide have conducted experiments following the same native techniques as that were recorded by history chroniclers, explorers, and anthropologists. Experimental voyages using replica watercraft, including that of the famous *Hokule'a,* which sailed from Hawai'i to Tahiti and back in 1976 (and has since made numerous other voyages), have not only demonstrated that the Pacific Islands were settled purposefully (not accidentally as some researchers had argued), but also that these islanders had continually refined their tradition of seafaring over thousands of years which enabled them to reach islands scattered among the world's largest ocean. What this and other similar studies have shown is that ancient sailors became extremely adept at building seaworthy craft, and finding and (re)locating to islands. This did not always ensure a successful voyage to their intended destination, however. Unpredictable winds, currents, storms, and other climatic events could have easily altered a craft's course, damaged it beyond repair, and prevented people from landing safely. A number of these cases historically suggest this happened fairly frequently, even among well-trained sailors.

What remains clear is that advances in watercraft construction over time allowed prehistoric peoples to sail farther and faster in search of new land. The single outrigger canoes such as those preferred in Micronesia (*prao*) were extremely fast and maneuverable. The larger double-hulled outriggers found in East Polynesia, constructed of two equally sized hulls, could easily carry dozens of people and supplies to successfully colonize even the most remote islands, allowing humans to explore and settle new lands.

EUROPEAN EXPLORATION

Thousands of years after most Pacific and Caribbean islands had been settled by humans, Europeans began venturing out into the world's largest seas and oceans. Irish monks, and later the Norse, settled Iceland in the eighth and ninth centuries AD; Columbus reached the Caribbean in AD 1492; and Magellan was the first European to cross the tip of South America through the straits that bear his name in AD 1520. This "Age of Discovery" by major European powers, particularly the Portuguese, Spanish, French, British, and Dutch, was precipitated by a number of technological achievements in shipbuilding, map-making, and navigation during the fifteenth and sixteenth centuries. These advancements led other well-known explorers such as Cook, Bougainville, la Perouse, and Vespucci to cross vast swaths of ocean with huge amounts of cargo, animals, plants, people, and weaponry.

As European explorers began vying for power and control of new lands throughout the Americas and the Pacific, they encountered a rich mosaic of native peoples who had already occupied many of these islands for centuries or even millennia. Racial prejudices and the endless search for gold, silver, spices, and other commodities fueled rapid expansion of colonial influence and the spread of new plants, animals, and diseases; this led to what the historian Alfred Crosby called the "Columbian Exchange"—in essence, the transfer of biota between the Old and New Worlds. Many Europeans, believing themselves superior to the native peoples they came into contact with, thought that such peoples were incapable of long-distance voyaging. This was reinforced, in part, by the fact that some societies had lost their knowledge of building long-distance sailing vessels because they were no longer needed. Despite these ethnocentric attitudes, many Pacific Islanders demonstrated to Europeans that they could in fact sail between islands using larger outrigger canoes, and could navigate using the stars and other time-tested techniques.

One of the major challenges that ancient navigators and early explorers faced, however, was the inability to estimate longitude. The invention of the marine chronometer by John Harrison in the mid-1700s eventually allowed sailors to determine their east–west position and more accurately map island and coastal regions. In effect, the development of this important tool was greatly influenced by the desire by Europeans to carve out and lay claim to new territories and establish ports to replenish supplies and trade goods.

CONCLUSIONS

The first colonization of an island by *H. erectus* during the Pleistocene set the stage for humans to harness the power of the world's seas and oceans. By the time of European contact in the Caribbean and Pacific, many islanders were using watercraft as a mechanism of travel, often seeing longer trips such as those in East Polynesia as tests of endurance rather than exceptionally difficult navigational exercises.

But there were many others who had reached islands and had eventually lost their knowledge or consciously discontinued seafaring over time. For some islanders, this may have been the result of having inhabited islands, particularly larger volcanic ones, with comparatively resource-rich environments. As such, they may have no longer needed or felt compelled to venture outside their island domain or to continue these traditions, having had all of the necessary things they needed to survive over the long term. In other cases, the remoteness of islands such as Rapa Nui and Hawai'i created more extreme isolation for peoples that may have made it difficult, if not impossible, to leave once they arrived.

Overall, archaeological, ethnographic, and historical evidence suggests that seafaring was primarily a male-dominated activity that required years of specialized training and practice. In the Pacific, it was also common for seafarers to recite prayers and chants and to conduct special rituals to ensure their protection during a trip or to help recount certain navigational exercises. Sailors, especially navigators, also typically had to abide by a number of taboos, invoke the spirits, and employ protective magic to ensure a successful voyage. Not doing so would endanger the lives of crewmembers and unnecessarily risk the impending trip.

During the "Age of Exploration," Scandinavians and western European powers ventured out into all of the world's major oceans, coming into contact with native peoples on hundreds of different islands. Although it was initially believed by many explorers in the Pacific, for example, that islanders could not have reached islands purposefully, the demonstration of traditional navigational techniques has altered this perception. Nonetheless, the desire for various commodities such as spices and precious metals led Europeans to vie for new island territories, conquer native groups, and exploit resources, quickly leading to significant cultural and economic upheavals that dramatically changed the course of human history.

SEE ALSO THE FOLLOWING ARTICLES

Archaeology / Human Impacts, Pre-European / Peopling the Pacific / Polynesian Voyaging

FURTHER READING

Broodbank, C. 2006. The origins and early development of Mediterranean maritime activity. *Journal of Mediterranean Archaeology* 19: 199–230.
D'Arcy, P. 2006. *The people of the sea: environment, identity, and history in Oceania*. Honolulu: University of Hawaii Press.
Finney, B. 1994. *Voyage of rediscovery: a cultural odyssey through Polynesia*. Berkeley: University of California Press.
Irwin, G. 1992. *Prehistoric exploration and colonisation of the Pacific*. Cambridge: Cambridge University Press.
Kirch, P. V. 2002. *On the road of the winds: an archaeological history of the Pacific Islands before European contact*. Berkeley: University of California Press.
Lewis, L. 1994. *We, the navigators: the ancient art of landfinding in the Pacific*. Honolulu: University of Hawaii Press.

EXTINCTION

KEVIN J. GASTON
University of Sheffield, United Kingdom

In varied combinations, three processes shape the size and composition of the species assemblages occurring on individual islands. Gains in species numbers arise through speciation and the immigration of individuals from elsewhere (other islands or mainland areas). Losses in species numbers take place through extinction. Global extinction occurs when the last individual of a species dies, although one might qualify this by distinguishing between the outright loss of all individuals and extinction in the wild (with individuals remaining in cultivation or captivity). Local extinction, or extirpation, occurs when the last individual of a given population dies, such as the population on one island of a more widespread species.

THE BIG PICTURE

Most of the species that have existed on islands are now extinct. Although data specifically for island species are scant, overall the fossil record suggests that from the appearance of the first individual of a species to the demise of the last takes time on the order of 5–10 million years on average. Some biases in the record will lead this to be an underestimate, others to it being an overestimate. Regardless, the distribution of persistence times is undoubtedly highly skewed, such that the majority of species survive for much shorter periods. This seems likely to be particularly true of island species.

Although species have been going extinct on islands since they first became established on them, the susceptibility of island species to extinction is particularly evidenced by the large numbers that have been lost as a direct or indirect consequence of human activities. In 1994, Pimm and others determined that on the Hawaiian Islands alone, of an original avifauna estimated to comprise roughly 125–145 species, 11 survive in numbers that suggest that they are likely to persist, 12 are endangered, 12 occur in numbers that mean they are unlikely to survive for much longer, and 90–110 species have been driven extinct, mostly after Polynesian but before European colonization. More generally, far greater numbers and higher proportions of human-caused extinctions, and of the subset comprising recent recorded extinctions, have been on islands rather than on continents. Estimates suggest that more than 90% of recent bird species extinctions and approaching 75% of recent mammal species extinctions have been from islands. Likewise, more than half of insular bird species and a third of insular terrestrial mammal species have gone extinct since the arrival of humans on those islands. Indeed, few islands have escaped extinctions as a consequence of human activities, and evidence of recent extinctions on islands that have resulted from other factors is scant.

Although they may interact in potentially complex ways, extinctions on islands can be viewed as resulting from the interplay between two sets of factors. The first are extrinsic features of the environment in which a species occurs, and the second are intrinsic traits that the species exhibits.

EXTRINSIC FACTORS

Four broad classes of extrinsic factors can give rise, or contribute, to extinction on islands: habitat change (loss, degradation, and fragmentation); natural enemies (predation and disease); competition; and cascades of extinction. Frequently, the ultimate fate of a particular species is determined by the combined action of two or more of these factors, rather than necessarily by individual factors in isolation.

Habitat Change

Habitat loss and degradation have doubtless been among the foremost factors that have led to the extinction of species on islands. The processes that drive such change vary greatly, and may be rare and unexpected events, a regular occurrence with a long or a short return time, and ephemeral or long-term. They include such diverse influences as the cyclic variation in the parameters of the earth's orbit (and the resultant Milankovitch cycles in climate), large meteors, volcanic eruptions, tsunamis, cyclones, fire, deforestation, agriculture, and urbanization.

Regardless of whether habitat changes are natural or are the result of human activities, island species seem likely to be prone to extinction as a consequence because of the high likelihood that these changes will influence the majority of individuals, and the restricted opportunities for individuals to move to other areas to avoid them. Changes in the three-dimensional complexity of vegetation on islands may be disproportionately problematic, because this profoundly shapes environmental conditions and associated species assemblages. Forest clearance by humans, for exploitation and agriculture, has been marked on many islands, with the remnants often comprising just a few percent of the original cover, confined to protected areas, and containing many species that are found nowhere else. A stark example is provided by Singapore. Brook and others estimate that since the British first established a presence on this tropical island in 1819, more than 95% of the original vegetation has been entirely cleared, and among the relatively well-studied taxonomic groups 28% of recorded species and possibly as many as 73% of all species (recorded and unrecorded) are estimated to have been lost.

The fragmentation of habitat often accompanies its loss and degradation. As a consequence of previously more continuous areas being reduced to progressively smaller and more dispersed patches, the environmental conditions within these remnants also tend commonly to change. In part this is a consequence of greater edge-to-core ratios, and in part it can be because of the effects of the amount of habitat on prevailing climates. Particularly for species specializing on the habitat type (individuals or gametes of which may be unable or unwilling to cross intervening areas), fragmentation may thus cause reductions in overall population size that are disproportionate to the absolute areal losses of habitat. Moreover, small patches are less likely to sustain local populations at a sufficient size to be viable, giving rise to local extinctions and increasing the likelihood of overall extinction. In general, this fragmentation is regarded as likely to exacerbate the difficulties that species face as a consequence of global climate change, because it limits the ease with which their distributions can respond through spatial shifts to track suitable environments. This is unlikely to be a major issue on small islands, where the possibilities for spatial shifts are by definition limited to nonexistent, and where species extinctions may thus be an inevitable consequence of marked climate changes. On islands that are sufficiently large that different areas experience markedly different environmental conditions, habitat fragmentation may certainly serve to limit spatial responses to climate change and, as a consequence, increase associated extinction rates. On both small and large islands, empirical evidence for recent systematic climate change is accruing rapidly.

Natural Enemies

Many of the best-documented examples of extinctions of species from islands concern increases in mortality or reductions of natality that have been caused by predation. Principally, these have resulted either from direct overexploitation by humans or from the impacts of predatory species that have been intentionally or accidentally introduced by humans. One can envisage similar consequences from the natural arrival of a previously absent predatory species. In many instances, the effects of predators on islands have been exacerbated by the naïveté of island species to predatory species, especially predatory mammals, which has arisen evolutionarily as a consequence of the predators' complete absence from the more isolated islands.

Species losses from islands have in the extreme resulted from the activities of just a single individual of a predatory species. Individual domestic cats drove the Stephens Island wren *Xenicus lyalli* extinct from Stephens Island, and extirpated the Angel de la Guarda deer mouse *Peromyscus guardia* from Estanque Island. More often, the losses have resulted from the activities of multiple individuals of a single predatory species. For example, nearly three-quarters of the *Partula* snail species native to the six Society Islands are extinct as a result of the introduction of the predatory wolf snail *Euglandina rosea* as a biological control agent for another snail species. However, as has been shown for birds on oceanic islands, the probability that a species has been lost may more typically be a positive function of the number of different predatory species introduced to islands. This is because a greater diversity of predator species increases overall predator pressure and also increases the breadth of predatory behaviors (species targeted, habitats exploited, timing of hunting), reducing the availability of enemy-free space. In this latter vein, generalist predators can be particularly problematic, both because they can exert predation pressure on a wider range of organisms and because they can maintain relatively high abundances even when one of their prey species has been driven to low numbers, often maintaining high predation pressure on that scarce species.

Although attention has, quite reasonably, principally focused on the lethal effects of predators on prey species, it is becoming apparent that their sublethal effects may also be important. By altering prey behaviors, including foraging patterns and use of different habitats, predators can have profound population-level effects through the consequences for reproductive success and adult and juvenile survival. This might particularly be the case on smaller islands, where potential prey individuals may repeatedly encounter predators, even when the latter occur in extremely low numbers.

The significance of introduced predators on island extinctions has been demonstrated not simply by the numbers of documented extinctions that have been attributed to their effects. It has also been shown by the successful introduction of potential prey species onto islands from which those predators are absent, and the successful reintroduction or population recovery of such species when predator numbers are reduced or eliminated.

Diseases can rapidly eliminate large numbers of individuals of susceptible species and so have the potential to generate extinction. Again, this may particularly be the case on islands, which are likely to be disproportionately disease-free because hosts are too small in number to maintain persistent disease populations, and where species may often have little resistance to parasites because they have had no previous contact with them. For example, many of the species of birds of the Hawaiian Islands that have gone extinct in the last two centuries, or are currently highly threatened, seem likely to have become so at least in part because of mortality resulting from the spread of avian pox and malaria by introduced mosquitoes. Given evidence that the development of avian malaria within hosts is temperature dependent, this problem may be exacerbated by projected climate changes.

Although predators are more likely to give rise to extinctions if they are generalists, diseases are more likely to do so if they repeatedly spill over from another reservoir species, thereby overcoming the limitation of populations of individual island hosts often being too small to sustain parasite populations in isolation. Even if they do not result directly in extinctions, and do not necessarily themselves persist, both specialist predators and parasites may nonetheless reduce the numbers of their victims to the point where other threatening processes deliver the final blow.

Competition

Habitat change will tend to favor some species at the expense of others. However, exclusion of a species by one or more others may also occur as a consequence of interspecific competition. Such effects on islands seem to occur between native and introduced plant species, with the latter transforming vegetation patterns as a consequence of their effects on the availability of such factors as light, water, and nutrients. However, strong competitive interactions, leading at least to significant declines in native species abundances, have also been documented between native and introduced animal species as a consequence of their effects on the availability of food resources, nesting sites, and refuges from predators.

Island species may be prone to being outcompeted, particularly by species introduced intentionally or accidentally by human agency, because the lower species richness of islands compared with mainland areas means that by contrast with the species being introduced they will often have experienced lower competition in the past.

Cascades of Extinction

Extinctions may not just result from the addition of species interactions, such as a predator, a parasite, or a competitor. They may also follow from the loss of interactions. Most obviously this will take place where one species provides a critical resource (e.g., it is a sole host) or function (e.g., as an agent of dispersal or pollination) for others, but is itself lost to extinction. The circumstances may, however, also be more complex, with the loss of one species resulting not just in the loss of some interactions but in a shift in the balance of others, and leading to further extinctions as a consequence. Such cascades of extinction seem likely to be particularly prevalent on islands, where the possibilities are much reduced for switching between resources, or for other species fulfilling critical functions that would otherwise be lost. They may be more frequent where a critical resource or function was provided by a vertebrate species, given the high levels of extinction that these have experienced on islands.

INTRINSIC FACTORS

There is evidence of taxonomic selectivity or taxonomic clumping of extinctions, with species in particular groups having a higher probability of extinction in a given geological period. In general, this effect tends to be quite small, with the higher taxon to which a species belongs having limited predictive capacity as to the probability of extinction. However, there is some suggestion that the particular threats posed by human activities might heighten the degree of such selectivity, resulting in more marked tendencies for species in some taxonomic groups disproportionately to have experienced recent extinction or to have high likelihoods of extinction in the near future.

Taxonomic clumping of extinctions could reflect extrinsic factors associated with the particular regions and environments that different groups inhabit. It may also reflect intrinsic factors that make some species more vulnerable to extinction than others, with the relationship between the intrinsic characteristics of species and the likelihood of extinction depending fundamentally on the specific extrinsic factors that are posing the threat to continued persistence. Indeed, within taxonomic groups there is good evidence that differences between species in intrinsic factors can contribute significantly to their persistence times, resulting in an element of trait selectivity in extinctions.

Rarity and Restriction

Island species typically have small numbers of individuals, and the area over which they are distributed is very restricted; many are endemic to single islands. The potential for substantial absolute variation in abundance and range size over their life times, from speciation to extinction, is much reduced compared to that of most continental species. They thus contribute disproportionately to the left-hand tails of species–abundance and species–range size distributions (the frequency distributions of species with different abundances or range sizes), which are almost invariably strongly right skewed, and as such may play a key role in shaping those distributions.

This rarity and restriction is one of the principal reasons that island species populations are prone to extinction. Indeed, the strongest correlates of the risk of extinction faced by a species are often its abundance and range size. There are several reasons why rarity may increase the risk of extinction. First, small populations are more readily driven to extinction by more deterministic extrinsic processes, such as habitat loss and degradation, predator and disease pressure, competition, and cascading losses of interactions. Second, small populations can suffer from demographic stochasticity, in which by chance during the same time interval all individuals may die, all may fail to breed, or sex ratios may become highly skewed. Third, small populations may suffer inbreeding because of an unavoidable increase in mating between close relatives, reducing reproductive success in populations of species that naturally outbreed. Island populations may tend inherently to be more inbred, particularly if they experienced genetic bottlenecks when they were founded, and have lower overall levels of genetic variation than mainland populations. Finally, at low densities population growth rates may decline because the probability of individuals exchanging gametes becomes unduly small (Allee effects).

Island populations of species may occur at higher densities than their mainland counterparts, as a consequence of density compensation resulting from competitive release. Nonetheless, they tend to experience a form of "double jeopardy" because not only are the numbers of individuals typically small but so too are geographic range sizes. A small range size tends additionally to increase the risk of extinction because there is a greater likelihood that adverse conditions will affect the distribution as a whole, and thus that a species will simultaneously be in decline or undergo local extinction in all of the areas in which it previously occurred.

There is some evidence that as populations decline, they can enter an extinction vortex, in which mutual reinforcement between threatening processes can occur, driving numbers down under the combined influences of such factors as environmental changes, demographic stochasticity, inbreeding, and reproductive failure. As numbers get lower, declines give rise to further declines, at higher rates, because stochastic factors assume progressively greater importance. The small size of many island populations would suggest that, on average, they are always closer to entering such vortices than are mainland populations.

The powerful influence of abundance and range size on the likelihood of extinction is evidenced by a greater likelihood of populations being lost from smaller islands, and a broad decline in the proportion of species that have become extinct or are threatened with extinction as island size increases.

Niche Breadth

In general, the likelihood of extinction may be influenced by the breadth of conditions or resources exploited by a species, because the narrower these are, the less likely that the species will be able to persist in a changing world. There are many more specialist species than generalist ones, which may serve in part to explain why, overall, such a high proportion of species have rather short persistence times. Island species may tend to be more generalist than their mainland counterparts, because of competitive release and the low abundance of individual resource types. However, the small extent of most islands, and thus the low diversity of resource types available, also greatly restricts the opportunities for generalism. The balance of these two processes is unclear, as is whether the degree of specialism significantly influences the persistence times of most island species. Of the extrinsic factors driving extinctions, specialist species would be expected to be disproportionately threatened by habitat loss and degradation.

Body Size, Reproductive Strategy, and Dispersal

Large body size tends to be associated with low rates of population increase and occurrence at low densities. This means that large-bodied species tend to have a disproportionate risk of extinction in the face of persistent threatening processes that perturb the balance between fecundity and longevity. They may also be more resilient in the face of ephemeral threatening processes, because they are better buffered against environmental extremes, and because their longer life span enables them to persist through periods of low or no reproductive success. However, populations will take a long time to recover in size, meaning that they continue to experience the risks of low numbers. In addition, the intrinsic risks associated with large body size may often be exacerbated because, at least in many trees and vertebrates, large size has tended to increase the likelihood of species being directly exploited by humans.

The influence of body size on extinction risk is perhaps reduced on islands, where very large-bodied species are less likely to occur because the available area is often too small to support viable populations; the maximum size of the species that are present tends to increase with the extent of a land mass. Nonetheless, the largest-bodied species in particular higher taxa have often become extinct from islands in recent times, or are presently highly threatened (e.g., moas in New Zealand, rhinoceros species on islands of Southeast Asia). Although the largest representatives of some insect taxa have occurred disproportionately on oceanic islands, likely because of lower predation pressures through evolutionary time, frequently these are presently at high risk of extinction or have not been observed in the wild for substantial periods (e.g., the St. Helena giant earwig *Labidura herculeana,* the New Zealand giant weta *Deinacrida* spp., the Lord Howe Island stick insect *Dryococelus australis,* and the Fijian longhorn beetles *Xixuthrus heros* and *X. terribilis*).

Beyond the simple correlates of body size, other facets of reproductive strategies may also influence the likelihood of a species becoming extinct. These include (1) asexual reproduction, which reduces the rate at which populations adapt to changing environmental conditions but eases some of the other pressures of low numbers; (2) social facilitation (e.g., breeding in large aggregations to reduce the per capita effects of predation), which makes populations very vulnerable when their sizes are small; and (3) the presence of closely related species, which in some cases facilitates hybridization when population size of both or, more commonly, one of the species is small.

Whether as a response to resource availability, reduced predation pressure, or the isolation of populations, island species in many higher taxa have reduced dispersal abilities. This has proven particularly injurious when predation pressure is increased through the introduction of alien species. For example, human colonization of the Pacific islands led to the extinction of many hundreds of species of rails (Rallidae), a group whose species are especially prone to becoming flightless on isolated islands.

DEBTS, GHOSTS, AND FILTERS

Populations can persist for some time after they have lost long-term viability, or even any potential to reproduce, whatever the combination of extrinsic or intrinsic factors that have brought them to this point. This is particularly so where individuals are long-lived. Perhaps most crucially, such an "extinction debt" or body of "latent extinctions" means that the immediate effects of habitat destruction may underestimate the overall long-term consequences. The extent of this underestimation may often be indicated by the numbers of extant species that are regarded as highly threatened and as having a high probability of extinction in the near future. A high proportion of the native species on many islands can often be characterized as such.

Extinctions may leave behind them the "ghosts of interactions past," sometimes expressed as components of extinction debt. These may, for example, take the form of species without pollination or dispersal agents, and species with behaviors to cope with absent competitors or predators. The frequency of such ghosts on islands is unclear, although, given the high levels of extinction that have been experienced, one would predict that they may be widespread.

Another consequence of past extinctions is a shift in the vulnerability of assemblages to further species extinction events. That is, they may be more resilient if they have faced similar challenges in the past, because these challenges act as "extinction filters," already purging assemblages of the most vulnerable species. This can result in some initially counterintuitive patterns of threat.

For example, the proportion of recently extinct or presently threatened species may be found to decline with increases in the duration of human occupation of islands. This is not because a longer period of occupation reduces the risks posed by humans, but rather because the less vulnerable taxa are those that remain.

CONCLUSION

Extinction is the fate of all species. Nonetheless, there is growing evidence that, after initial marked increases in diversity, rates of speciation and extinction were on average broadly matched for much of the history of life. Island assemblages provided the earliest, and most dramatic, clues that human activities had the potential to disrupt that trend on a scale that had only previously been seen during geological episodes of mass extinction. None of the drivers of extinction in island assemblages are, however, unique to these environments.

SEE ALSO THE FOLLOWING ARTICLES

Bird Disease / Deforestation / Fragmentation / Inbreeding / Introduced Species

FURTHER READING

Baillie, J. E. M., C. Hilton-Taylor, and S. N. Stuart, eds. 2004. *2004 IUCN Red List of threatened species. A global species assessment.* Gland, Switzerland, and Cambridge, UK: IUCN.
Brook, B. W., N. S. Sodhi, and P. K. L. Ng. 2003. Catastrophic extinctions follow deforestation in Singapore. *Nature* 424: 420–423.
Caughley, G., and A. Gunn. 1996. *Conservation biology in theory and practice.* Oxford: Blackwell Science.
Gaston, K. J. 2003. *The structure and dynamics of geographic ranges.* Oxford: Oxford University Press.
Lawton, J. H., and R. M. May, eds. 1995. *Extinction rates.* Oxford: Oxford University Press.
MacPhee, R. D. E., ed. 1999. *Extinctions in near time: causes, contexts, and consequences.* New York: Kluwer/Plenum.
Pimm, S. L., M. P. Moulton, L. J. Justice, N. J. Collar, D. M. J. S. Bowman, and W. J. Bond. 1994. Bird extinctions in the Central Pacific. *Philosophical Transactions of the Royal Society of London B* 344: 27–33.
Steadman, D. W. 2006. *Extinction and biogeography of tropical Pacific birds.* Chicago: University of Chicago Press.

F

FALKLAND ISLANDS (ISLAS MALVINAS)
SEE ATLANTIC REGION

FARALLON ISLANDS

PHIL CAPITOLO

Berkeley, California

The Farallon Islands (Farallones) consist of several sparsely vegetated rocks about 47 km west of San Francisco, California. The South Farallones include Southeast Farallon Island (SEFI), West End Island (WEI), and several adjacent seastacks, totaling less than 50 hectares. The North Farallones are four steep islets about 11 km northwest of SEFI, totaling less than 10 hectares. Middle Farallon is a small, wave-washed rock 3.5 km northwest of SEFI (Fig. 1). The Farallones are perched on the edge of the continental shelf in waters about 60 m deep; less than 10 km to the southwest the ocean floor drops abruptly to depths of 3000–4000 m. The Farallones represent the highest points of an underwater granitic ridge that appears similar geologically to parts of the southern Sierra Nevada mountain range. Within the political boundaries of San Francisco, the islands nevertheless continue to drift slowly northwest along the margin of the Pacific tectonic plate.

ESTABLISHMENT OF NATIONAL WILDLIFE REFUGE

The South Farallones alone support the largest single assemblage of breeding seabirds along the California

FIGURE 1 Aerial view of the Farallon Islands, May 30, 2007. PRBO and USFWS biologists operate a research station from the Victorian-era buildings on Southeast Farallon Island. Middle and North Farallones appear in the background.

coast, with hundreds of thousands of common murres; tens of thousands of Brandt's cormorants, western gulls, and Cassin's auklets; and hundreds of Leach's and ashy storm-petrels, double-crested and pelagic cormorants, pigeon guillemots, rhinoceros auklets, and tufted puffins. Five species of pinnipeds also breed at the Farallones, including the Steller sea lion, a species federally listed as threatened. To protect these and other natural resources, the North and South Farallones received National Wildlife Refuge status in 1909 and 1969, respectively. Additionally, the North Farallones, Middle Farallon, and WEI are congressionally designated Wilderness Areas. Today, the Farallon National Wildlife Refuge is managed by the U. S. Fish and Wildlife Service (USFWS) through a cooperative agreement with PRBO Conservation Science (PRBO, formerly the Point Reyes Bird Observatory).

PAST HUMAN OCCUPATION OF THE FARALLONES

The Farallones were known to Native Americans of the central coast of California as Island of the Dead, and believed to be the residing place of spirits after death. However, there is no evidence that local Native Americans visited the islands. The first known landing was made by British explorers in 1579 to collect seals and seabirds for food during an expedition led by Sir Francis Drake. After Drake's visit, Spanish explorers steered offshore of present-day San Francisco to avoid the hazards of the Farallones and other possible submerged rocks; thus, the discovery of San Francisco Bay did not occur until the land expedition of Gaspar de Portola in 1769. During this time of Spanish exploration, the islands were generally referred to in the accounts of explorers as *farallones,* Spanish for "rocks rising from the sea."

In search of sea otter pelts to trade with Russians in Alaska, the crew of the Boston vessel *O'Cain* made the second known landing on the Farallones in February 1807. Finding instead vast numbers of seals, crews of several Boston ships took over 100,000 northern fur seal pelts by 1812. The Russian American Fur Company then occupied the South Farallones beginning in 1812, after having established a settlement to the north at Fort Ross on the California mainland. The company employed Native Americans from Alaska and California and hunted seals, sea lions, and seabirds to provide food and other products for Russian communities in both regions. More than 1000 fur seals were taken each year initially, but by 1838 the company had taken all remaining animals and abandoned the Farallones. Following their extirpation, northern fur seals did not breed at the South Farallones until 1996; increasing numbers of fur seal pups have subsequently been noted, numbering near 100 by 2006.

Populations of seabird species also were heavily impacted by human disturbance in the 1800s, most notably when commercial take of common murre eggs began in 1849. The size of murre eggs (about twice the volume of a chicken egg), their palatability, and their availability in dense concentrations on the Farallones (common murres can breed in densities of 20 pairs/m^2) made them a valued commodity in San Francisco, where domestic poultry production was insufficient for the growing human population during the Gold Rush. In 1851, a group of eggers claimed ownership of the islands and exclusive rights to egging operations, forming the Farallon (or Pacific) Egg Company. Around the same time, the Lighthouse Board (within the U.S. Department of Treasury) commissioned the construction of a lighthouse and light keepers' residences on SEFI, the lighthouse becoming operational in 1855. Years of conflict between the egg company and light keepers over egging opportunities ensued until a U.S. Marshal evicted the egg company in 1881. Large-scale egging by light keepers and fishermen continued until the Lighthouse Board, following recommendations from the California Academy of Sciences and American Ornithologists Union, prohibited egging in 1896. Some illegal egging continued at least until 1904. In total, it is estimated that more than 14 million murre eggs were harvested during the second half of the nineteenth century. Annual harvest estimates, and likely population decline during the first half of the century, indicate that the murre population at the Farallones was likely more than 1,000,000 breeding birds prior to the arrival of sealers and eggers. Fewer than 20,000 murres were estimated in 1911.

Human occupation of SEFI peaked in the first half of the twentieth century. The U.S. Weather Bureau built a station in 1902, and the U.S. Navy arrived just a few years later, replacing the weather station and also operating a radio compass station. Including Navy personnel, light keepers, and their families, the island's human population reached 78 permanent residents in 1942. After World War II and with developments in navigational techniques, Navy presence was no longer needed. Light keepers remained as employees of the U.S. Coast Guard, which had assumed duties when the Lighthouse Service was dissolved in 1939. Numerous buildings were razed, and by 1965 the Coast Guard no longer allowed keepers' families to reside on SEFI. PRBO was formed in 1965, and their biologists made their first visits to the Farallones in 1967 to conduct initial studies of the impacts of human activities on the islands' ecology. With impending lighthouse automation and removal of Coast Guard personnel, efforts increased to expand the Refuge to encompass all islands and to establish a biological field station on SEFI.

Since April 1968 (and under a cooperative agreement with USFWS since 1972), SEFI has been staffed by biologists from PRBO each day of each year. Utilizing two remaining Victorian-era buildings (built in 1878 and 1880) to house personnel and equipment (Fig. 1), PRBO and USFWS conduct research and conservation efforts and also have facilitated work by many researchers from universities, government agencies, and other organizations. SEFI is closed to the general public to protect sensitive resources and because accessing the island is not routine (Fig. 2). Resident biologists are allowed only limited access to WEI (separated from SEFI by the Jordan

Channel) and certain areas of SEFI. Visitors must obtain special-use permits from USFWS, and boats and aircraft are required to maintain minimum distances from the islands. The waters surrounding the Farallones are within the Gulf of the Farallones National Marine Sanctuary (GFNMS), managed by the National Oceanic and Atmospheric Administration.

FIGURE 2 A derrick at East Landing is used to lift biologists, visitors, and supplies onto Southeast Farallon Island.

ECOLOGY OF THE FARALLONES

South Farallones

Steep talus slopes and a broad marine terrace on its south and west shore characterize the landscape of SEFI (Fig. 1), where most research and monitoring activities have occurred. Ninety species of plants have been documented at the South Farallones, 65 of which are nonnative. Maritime goldfields, a low-lying, endemic annual, is widespread and temporarily colors the marine terrace green in winter and yellow in spring. New Zealand spinach and other invasive plants that threaten habitat of crevice- and burrow-nesting seabirds are targets of control efforts. European rabbits, introduced to the islands for food in the 1880s and abundant until eradicated in 1974, reduced native vegetation and competed with seabirds for nesting cavities. Since the eradication of rabbits, native vegetation has been slowly reclaiming the islands, and rhinoceros auklets, absent since rabbits were established, have recolonized the South Farallones.

Each spring and summer since 1971, the breeding biology of Farallon seabirds has been studied in detail, contributing vastly to the understanding of the life history strategies of these species. One shorebird species, the black oystercatcher, breeds on the Farallones and also is studied. The seabird breeding community consists of crevice or burrow nesters (ashy and Leach's storm-petrels, pigeon guillemots, Cassin's and rhinoceros auklets, and tufted puffins), and surface nesters (Brandt's, double-crested, and pelagic cormorants, western gulls, and common murres). For several species, banding of chicks and subsequent resightings of them as adults has resulted in detailed information on philopatry, age at first breeding, and estimates of survival and lifetime reproductive success. Diet composition data, studied by identifying fish and zooplankton species brought to chicks or found in pellets or regurgitation samples, have provided insight into annual and long-term variability in prey resources as affected by oceanographic conditions. Driven by strong northwesterly winds that push surface waters offshore in spring, coastal upwelling of cold, nutrient-rich waters results in abundant prey resources in the California Current, supporting the great diversity and number of seabirds that breed at the Farallones. El Niño–Southern Oscillation (ENSO) events occur periodically, reducing upwelling and resulting in reduced seabird breeding effort or success for several species. PRBO's long-term datasets on seabird survival, reproduction, and diet are among the most extensive in the Northern Hemisphere and are now being used to assess impacts of climate change and fisheries on marine ecosystem health.

For most species, breeding population sizes also are estimated annually. The breeding colonies of ashy storm-petrel, Brandt's cormorant, and western gull are the largest throughout their ranges. After low numbers following commercial egging, the common murre population had increased to more than 200,000 breeding birds by 2006, despite a steep decline in the mid-1980s due to mortality in oil spills and gill nets, which led to regulations by the California Department of Fish and Game restricting the use of gill nets. Populations of double-crested cormorants and tufted puffins appear stable but remain well below historic levels. The incomplete recovery of these two species may be related to loss of Pacific sardines as a prey item, caused by oceanic changes and overfishing by humans in the 1930s and 1940s. Sardines began reappearing in waters around the Farallones in the early 1990s, but as yet they account for only a small percentage of prey items delivered to murre and rhinoceros auklet chicks.

Ashy storm-petrels must contend with the threat of predation by western gulls, but they also have been impacted by predation by wintering burrowing owls. Several migrant owls reach the South Farallones each fall and find an abundant prey source in house mice (introduced probably in the nineteenth century). After the mouse population crashes with winter rains, the owls remain and switch to storm-petrels as prey. Some owls also die from starvation. In the absence of house mice, it is believed that burrowing owls would continue to migrate to more suitable mainland wintering habitat, both storm-petrel predation and owl starvation would be reduced, and native flora would more quickly become reestablished. Thus, eradication of house mice is now being planned.

Many migrant raptors, shorebirds, waterbirds, and landbirds arrive to the islands in fall, with a smaller passage in spring. Through 2008, more than 416 species of birds had been documented on the Farallones and surrounding waters, including many off-course migrants that breed in eastern North America and Siberia. Landbirds congregate in the four planted trees on SEFI (three Monterey cypress on the lee sides of the houses, and a low, sprawling Monterey pine near the east end of the island), allowing them to be banded, measured, and studied through strategic placement of mist-nets. Arrivals of migrants are greatly affected by weather conditions, but standardized long-term data collection allows analyses of population trends of western species that can be used to affect habitat conservation on both breeding and wintering grounds.

Standardized censuses of other migratory organisms also are conducted in fall. Of five species of bats noted at SEFI, the highly migratory hoary bat is most commonly sighted, and only on SEFI has hoary bat copulation been witnessed. Blue, humpback, and several other species of whales are identified, often by their spouts, as they migrate and forage along the continental shelf break. Gray whales migrate over the shelf, and one to three whales typically reside near SEFI each spring and summer, forgoing the remainder of their northward migration. Censusing from SEFI indicates that most whales and other cetaceans have increased since nearby whaling was curtailed in 1968. White shark sightings peak with maximum immigration and emigration of their preferred prey, immature northern elephant seals. Research at the South Farallones was instrumental in establishing 1994 legislation protecting white sharks in California waters, and PRBO biologists documented the first instance of a killer whale preying on a white shark in 1997. Among nonmigratory organisms, GFNMS biologists have monitored invertebrates and algae in rocky intertidal habitats since 1992. More than 200 taxa have been found; red turf and coralline algae predominate. Less well studied are endemic taxa of camel cricket, California arboreal salamander, and trapdoor spider that may reflect an earlier time when the Farallones were connected to the mainland.

The biology of the northern elephant seal is the focus of research in winter. Cows begin to arrive in December to give birth and breed with adult males that have already established territories. Like the northern fur seal, elephant seals were extirpated in the 1800s. They were resighted at the South Farallones in 1959 and began pupping again in 1972. The number of pups per year increased annually to a peak of 475 in 1983, then declined, and has been mostly stable since 1999 at fewer than 200. Not agile on land, elephant seals are easily tagged to enable studies of survival and reproductive success by age class, as well as immigration and emigration patterns. California sea lions are currently the most abundant pinniped on the South Farallones, where they reach their northern breeding limit. However, only small numbers of pups (<50), and fewer Steller sea lion and harbor seal pups, are born annually in spring and summer. The rapidly growing fur seal population may limit nesting space for seabirds in the future, providing insight into the dynamics of these populations prior to the 1800s. The Guadalupe fur seal and the sea otter are occasionally seen in fall.

North Farallones

Several thousand common murres breed on each of the four islets of the North Farallones with small numbers of nesting Brandt's cormorants on certain islets. These populations are surveyed annually with aerial photographs (also at the South Farallones). The only known on-site biological survey was conducted in fall of 1994 when biologists from Humboldt State University and USFWS landed on the West, East, and South islets. On-site surveys are rarely permitted because of the steep terrain of the islets, frequent rough seas, and the need to prevent disturbance to surface-nesting seabirds and hauled-out pinnipeds. The guano-covered islets are largely devoid of vegetation, with only a small amount of habitat for crevice- and burrow-nesting species. Breeding pigeon guillemots, which may outcompete smaller species for crevice habitat, were found on each surveyed islet, and a small breeding colony of Cassin's auklets was found on the West Islet. Wooden boards on the West Islet were evidence of prior human activity, likely from egg-harvesting activities in the nineteenth century.

SEE ALSO THE FOLLOWING ARTICLES

Granitic Islands / Pacific Region / Seabirds

FURTHER READING

Ainley, D. G., and R. J. Boekelheide, eds. 1990. *Seabirds of the Farallon Islands*. Stanford, CA: Stanford University Press.

DeSante, D. F., and D. G. Ainley. 1980. The avifauna of the South Farallon Islands, California. *Studies in Avian Biology* 4.

Klimley, A. P., and D. G. Ainley, eds. 1996. *Great white sharks: The biology of Charcharodon carcharias.* San Diego, CA: Academic Press.

Pyle, P., D. J. Long, J. Shonewald, R. E. Jones, and J. Roletto. 2001. Historical and recent colonization of the South Farallon Islands, California, by northern fur seals (*Callorhinus ursinus*). *Marine Mammal Science* 17: 397–402.

White, P. 1995. *The Farallon Islands.* San Francisco, CA: Scottwall Associates.

FAROE ISLANDS

GINA E. HANNON
Southern Swedish Forest Research Centre, Alnarp, Sweden

SIMUN V. ARGE
National Museum of the Faroe Islands, Tórshavn

ANNA-MARIA FOSAA
Faroese Museum of Natural History, Tórshavn

DITLEV L. MAHLER
National Museum, Copenhagen, Denmark

BERGUR OLSEN
Faroese Fisheries Laboratory, Tórshavn

RICHARD H. W. BRADSHAW
Liverpool University, United Kingdom

FIGURE 1 The position of the Faroe Islands in the North Atlantic, showing the geology and the localities mentioned in the text.

The Faroe (or Faeroe) Islands, 61° 24′–62°24′ N and 6°15′–7°41′ W, are a series of small islands oriented in a northwest-southeast direction situated in the North Atlantic between Iceland, Norway, and Scotland (Fig. 1). They consist mainly of steep mountainous plateaus with narrow terraces, gorges, and cirque valleys, divided by narrow fjords and sounds. Most of the outer coastal reaches are vertical cliffs several hundred meters in height, which are extremely exposed and subject to erosion by breakers. The islands are slightly tilted, with the eastern parts experiencing some subsidence after the land uplift following the last glaciation, as evidenced by the presence of submerged bogs.

OCEANOGRAPHY

The Faroe Islands are positioned on the ridge that stretches between Scotland and Iceland and further to Greenland. This is the region where Atlantic water enters the Nordic Seas, with main flows on both sides of the Faroes. The water on the Faroe shelf circulates clockwise (anticyclonic), and a persistent tidal front separates the shelf water from the surrounding ocean. This current system provides the basis for a small (8,000 km²) and uniform coastal ecosystem that is surrounded by an oceanic environment. Within this ecosystem there appears to be a trophic relationship between plankton, fish, and seabirds, with marked interannual variability driven by changes in the physical conditions.

PHYSICAL BACKGROUND

The islands are part of the North Atlantic basalt area, which was formed during a period of intense volcanic activity in the Tertiary ~60 million years ago. They are the remnants of large, low-relief flood basalt lavas, built up in near horizontal layers mainly from rift volcanism, which were laid down as the northwestern European and North American continents began to drift apart. The width of the island group from north to south is approximately 113 km, and it is approximately 75 km from east to west. They cover a total land area of almost 1400 km², dominated by steep cliffs in the outer coastal regions on

FIGURE 2 The western cliffs on the island of Skúvoy. Stóra Dímun and Lítla Dímun are in the background. Photograph by Bergur Olsen.

the north and west (Fig. 2). The only habitation areas are close to the coast, where there are flat or sloping areas that can support cultivation. The highest mountain is 882 m above sea level. The basalt sequences can be separated into three distinct volcanic phases called the Upper, Middle, and Lower basalt series (Fig. 1) with intervening layers of volcanic ashes, tuff, slate, and basaltic sandstone and coal-bearing sequences. The latter contain rich pollen and spore floras, including Polypodiaceae (ferns), Lycopodiaceae (club mosses), Sphagnaceae (bog mosses), Ginkgoaceae (maidenhair tree), Taxodiaceae (bald cypress family), Cupressaceae (cypress family), Palmae (palm family), and *Pinus haploxylon* (pine trees).

The U-shaped valleys, fjords, and sounds that separate the basalt plateaus are aligned northwest to southeast and have been cut into by the repeated glaciations of the Quaternary Period. The Faroe Islands had their own ice cap during the last glaciation, with valley glaciers radiating out from a few central areas to a maximum height of 700 m above sea level in the central region of Eysturoy, leaving some of the higher mountain peaks ice-free. One interglacial deposit of clay/gyttja can be seen exposed between two moraines a few meters above sea level in a coastal cliff sequence at Klaksvík on the northern island of Borðoy (Fig. 1). This unit contains *Picea* (spruce) and/or *Larix* (larch) wood, together with a flora distinctly different from the present day flora of the islands, including *Buxus* (boxwood) pollen. The unit yielded an infinite radiocarbon age but has now been identified as being of Eemian interglacial age by the presence of the 5e-Midt/RHY tephra. This tephra has been located in several marine cores in the North Atlantic and has an age of ~124,000 years ago.

CLIMATE

The climate of the islands is strongly maritime with mild winters (mean January temperature of 3.4 °C) and cool summers (mean August temperature of 10.5 °C). Higher altitudes have an arctic climate. Mean annual temperature at Tórshavn, close to sea level, is 6.5 °C, whereas at high altitudes the mean annual temperature can be as low as 1.7 °C. The annual rainfall varies widely both regionally and with altitude. The range is from 800 mm per year on the western island of Mykines to 1300 mm per year toward the center of the islands and increasing to 3300 mm in the northern, higher relief areas. May and June are the driest months, and October and November are the wettest.

FAUNA

The number of breeding bird species is low (around 60). Seabirds breed in large numbers, and the Faroes are home to a substantial proportion of the northeastern Atlantic populations. Twenty-one seabird species breed regularly, and the total populations are estimated to be about 1.7 million pairs. Fulmars (*Fulmarus glacialis*) and puffins (*Fratercula arctica*) are the most numerous breeders, followed by European storm petrels (*Hydrobates pelagicus*) and kittiwakes (*Rissa tridactyla*). The most important breeding habitats are the large seabird breeding colonies found on steep west- and north-facing cliffs (Fig. 3), grass-covered slopes, and boulder screes. Eighteen sites are considered of international importance for their cliff-breeding seabird populations, especially guillemots (*Uria aalge*), puffins, and kittiwakes. European storm petrel numbers are of major international importance, with the islands probably holding about 40% of the world population. In summer, the seabirds mainly feed within 60 km from land, at depths less than 150 m, but in the winter most of them migrate away from the islands, particularly to Norway, the North Sea, and the United Kingdom. Because of the geographical isolation of the islands, there are some endemic subspecies of birds. Among the seabirds, the eider *Somateria mollissima faeroeensis* and the black guillemot *Cepphus grylle faeroeensis* are both endemic for the islands.

The number of fish species recorded in Faroese waters is 235. Because of the circulation pattern with retention of the shelf water, the Faroe Plateau contains self-sustained

FIGURE 3 Guillemots with chicks on the bird cliffs. Photograph by Bergur Olsen.

populations of demersal fish, mainly saithe (*Pollachius virens*), cod (*Gadus morhua*), and haddock (*Melanogrammus aeglefinus*). These support the fisheries that are important to the Faroese economy. Sand-eels (*Ammodytes* sp.) are also found on the Faroe Plateau, and they play an important role in the ecosystem as prey for fish and seabirds. The resident nature of the shelf water gives a distinctive character to the phytoplankton and the zooplankton community composition, which is the basis for production in the higher trophic levels within the ecosystem. Five species of fish are found in Faroese freshwater lakes. These include brown trout (*Salmo trutta*), Arctic char (*Salvelinus alpinus faroensis*), eel (*Anguilla anguilla*), three-spined stickleback (*Gasterosteus aculeatus*), and flounder (*Platichthys flesus*). It may be that Atlantic salmon (*Salmo salar*) once also occupied Faroese fresh waters.

More than 1200 species of invertebrates are recorded in the Faroe Islands, most of them from the classes Insecta, Crustacea, and Arachnida. The terrestrial mammals on the Faroes have all been introduced by humans. Sheep (*Ovis aries*), cattle (*Bos taurus*), and horses (*Equus caballus*) were brought in as domestic livestock by the early settlers, and the mountain hare (*Lepus timidus*) was introduced in 1855 as game. The brown rat (*Rattus norvegicus*) and the house mouse (*Mus musculus*) were introduced unintentionally and are found on only some of the islands. In the sea, the gray seal (*Halichoerus grypus*) breeds in the Faroe Islands, and other seal species are rare. Whales use the Faroes waters as a feeding area. Most notable is the pilot whale (*Globicephala melas*), which feeds primarily on squid. The bottlenose dolphin (*Tursiops truncatus*), harbor porpoise (*Phocoena phocoena*), and white-sided dolphin (*Lagenorhynchus acutus*) are also common in Faroese waters.

VEGETATION AND FLORA

Soils formed during the last 10,000 years or so have developed from the parent basalt materials and are strongly acidic and more or less continuously wet or moist. As the islands are sparsely vegetated and almost treeless, the vegetation has a tundra-like appearance. Cultivated land comprises about 6% of the lowlands, mainly near the coast (Fig. 4). Crops, mainly barley (*Hordeum vulgare* and *Hordeum distichon*), have been grown in the past. Hay is almost the only crop that remains important today which is used as winter food for sheep.

About 70% of the land area of the islands is more than 200 m above sea level, and the vertical vegetation zonation from this point upward is low alpine, grading into an arctic vegetation zone on the mountain tops. At high elevations, grass vegetation is found on both south- and north-facing slopes. The alpine area is dominated by *Racomitrium* heaths, some snow-bed vegetation in areas with late-lying snow, and fell-field vegetation. Common species in this area are *Salix herbacea* (dwarf willow) (Fig. 5), *Carex bigelowii*, *Bistorta vivipara*, and *Racomitrium lanuginosum*. The predominant vegetation on north-facing slopes is grassland from sea level up to the mountain tops.

The lowland island vegetation (less than 200 m above sea level) is usually classified in the temperate vegetation zone. The dominant species in the lowland grassland are *Nardus stricta* and *Anthoxanthum odoratum*, together with *Agrostis capillaris* and *Agrostis canina*, which is common from low to high altitudes. The vegetation is heavily affected by sheep, the most common domestic animals,

FIGURE 4 Viðareiði on the island of Viðoy. Photograph by Gina Hannon.

FIGURE 5 *Salix herbacea*. Photograph by Dánjal Jespersen.

which are allowed unrestricted grazing in uncultivated areas throughout the year and in the infields during the winter months.

Heathland vegetation is found only on the larger islands on warm south- and west-facing slopes up to 200 m above sea level, where this vegetation type disappears. The heathland is very mixed with many grasses, herbs, and mosses. The dominant plant species are *Calluna vulgaris, Empetrum hermaphroditum,* and *Vaccinium myrtillus,* with *Erica cinerea* and *Vaccinium uliginosum.* Of the herb species that are found in the heathland, *Potentilla erecta, Hypericum pulchrum,* and *Polygala serpyllifolia* are worthy of note. The limiting factor for the growth of heathers is probably too little sun, but the heavy sheep grazing and the northern limit of *Calluna* also could have a negative effect on its distribution.

Three types of mires are found in the islands: topogenic mires, which are overgrown lakes; soligenic mires, which are found on hills and slopes; and ombrogenic mires, which are found in valleys and called "blanket mires." Common species in these wetland areas are *Juncus squarrosus, Scirpus cespitosus,* and sedge species such as *Carex nigra* and *Carex panicea* as well as *Eleocharis* species. Common herbs found are *Pinguicula vulgaris* and *Saxifraga stellaris.* There are a great many lakes, which support several genera of submerged water plants including *Isoetes* and *Potamogeton* along with algae such as *Chara* and phytoplankton. Species richness is poor, most likely because of the low nutrient content and the substrate, which is often stony.

The coastal cliffs are covered in algae at and below water level, as well as crustose lichens such as *Verrucaria,* foliose lichens *Xanthoria,* and species of *Ramalina* higher up. *Zostera marina* is recorded in one of the fjords, on the southernmost island, Suðuroy. Where coarse-textured soil has built up in crevices formed by weathering, some species that can tolerate constant salt spray survive. In more protected localities of the inner fjords, where the saltwater has difficulty disappearing, salt marshes are found around the mouths of small streams. Sand dunes are found only on Sandoy, which is the island with the lowest relief. Behind the sand dunes away from the sea, the species diversity is higher, with species such as *Dactylorhiza purpurella, Coeloglossum viride,* and *Ligusticum scoticum.*

The few areas of trees on the islands today have been planted, yet a significant proportion of the islands lie within a climatically suitable zone for tree growth. The first forestry plantation was established at Tórshavn in 1885, but it was only in the mid-1900s that tree-planting became mildly successful in sheltered localities. The only locality to have shown positive presence of trees on the islands during the Holocene is the Viking site of Argisbrekka on the northern island of Eysturoy (Fig. 1), where a thick layer of wood peat dating from ~4250 calendar years ago (2300 BC) contained a network of tree roots and stumps of *Betula pubescens* (tree birch) (Fig. 6). Sub-fossil *Betula, Salix* (willow) and *Juniperus* (juniper) found buried in peat profiles from several islands show that there was at least partial woody vegetation cover until humans and their grazing animals arrived.

FIGURE 6 Sub-fossil tree birch (*Betula pubescens*) roots at Argisbrekka. Photograph by Ditlev Mahler.

PALEOECOLOGY AND HUMAN SETTLEMENT

The onset of deglaciation is not definitely known from the Faroe Islands. The oldest dated lacustrine sediments of 11,360–11,180 calendar years ago come from Hoydalar on the island of Streymoy. The vegetation record begins ~10,980 years ago, with a succession from sparsely vegetated tundra to *Betula* and *Salix* shrub tundra. This is followed by the development of grassland with tall herbs and *Juniperus,* which after the mid-Holocene is largely replaced by *Calluna* heathland. The first people to settle permanently on the islands left a distinctive mark in the paleoecological record, including pollen evidence for the cultivation of cereal crops, cultural macrofossil assemblages, and charcoal fragment as seen at Tjørnuvík on the island of Streymoy (Figs. 1 and 7).

The transformation of the flora of this fragile ecosystem is best expressed by the large number of ruderal, post-settlement plants recorded as plant macrofossils. Many plant macrofossil taxa occurred either just before or at the horizon in which the cultivated crops were recorded and can be associated with human impact. *Montia fontana, Stellaria media, Sagina procumbens, Lychnis floscuculi, Cardamine flexuosa, Linum catharticum, Galaeopsis speciosa, Plantago* sp., *Caltha palustris, Ranunculus repens, Chenopodium album,* and *Rumex* cf. *longifolius* are continually recorded after settlement. Domestic animals were introduced to the islands with humans, and they have contributed to the loss of woody plants and the spread of grassland. A similar situation has been proposed for Iceland, where sheep continue to have a regional influence on the vegetation.

Accelerator-based mass spectrometer (AMS) radiocarbon dates from Tjørnuvík have revealed that the first crops were cultivated in the mid-700s (AD). Comparable dates for cultivation have been recorded from three other localities on the islands: Korkadalur on the western island of Mykines, Heimavatn on the northern end of the island of Eysturoy, and at Hov on the island of Suðuroy (Fig. 1). These results strongly suggest that there was anthropogenic contact and disturbance prior to the formal settlement that left firm archaeological remains, creating a similar debate as exists over the settlement of Iceland. The introduction of sheep and cattle could have been one result of these early contacts. This issue could be further researched by joint paleoecological-archaeological ventures.

The vegetation changes during the latter part of the Holocene reflect intimate interactions between cultural and environmental development. Settlement could well have been the most significant disturbance factor on Faroese vegetation over the last 10,000 years. Pollen records of primarily nonforested landscapes can, however, be difficult to interpret. It is unclear whether the treeless grasslands and heathlands that characterize the islands today are a cultural product, as has been demonstrated elsewhere in northern Europe, or have developed as a result of recent climatic change. Settlement may well have accelerated a degradative process that was initiated by a

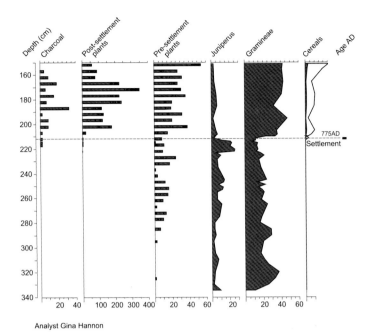

FIGURE 7 Summary pollen percentages (continuous curves), plant macrofossil and charcoal concentration/30 mL (bars) from Tjørnuvík illustrating the effects of first human settlement. The line represents the first cereal pollen. The hollowed graph beside cereal pollen represents exaggeration of the scale by a factor of 10. Redrawn from Hannon and Bradshaw (2000).

changing climate. The general sparseness of species on the Faroes, compared with mainland Europe, is attributable to their isolation and may also contribute to the current absence of a climatic tree line.

ARCHAEOLOGY

The traditional view of settlement has held that the Faroe Islands were colonized from the east. Irish monks arrived on the Faroes about AD 700, and they in turn were driven out by Norwegian settlers about AD 825. The idea of such an early Irish settlement is based on a literary source: Dicuil's *De mensura orbis terrae,* written about AD 825. It contains a brief description of sea voyages by Irish clergy to some uninhabited islands, generally accepted to be the Faroes. The description of the islands is vague, and although earlier visits cannot be excluded, they have left no archaeological traces to date. The earliest archaeological evidence is from settlements during the ninth century, based on an early Norse culture with close contacts to neighbors: the Irish-Scottish area to the south and Scandinavia to the east. During this period Orkney, Shetland, Caithness, and the Hebrides were dominated by a Norse population. Contacts with the south are also reflected in the Faroese vocabulary, which is Norse but contains, among others, words such as *ærgi,* meaning shieling site (huts used by herders during the summer), borrowed from the Gaelic word *áirig*.

The Norse farmers who settled the islands established farms along the coasts and fjords, trying to follow the same way of living that had been developed in their homelands. The enclosed infields that surrounded the farm sites were used for growing grain and hay; in addition, there were the outfields on which the farms relied for their pasture, turf cutting, and other resources. The settlements were permanent, and their location was largely determined by the topography of the islands and the limited lowland areas suitable for cultivation. Settlements were chiefly located along the coasts in the bays.

In contrast to Norway and Iceland, sheep were common and cattle rare. The earliest economy apparently was based on the exploitation of a much larger variety of resources than later in the Middle Ages. For example, bones of domestic animals were only a minor component compared to bird, fish, and shellfish remains in the archaeofauna from recent investigations at Sandur on the island of Sandoy (Fig. 1) covering the ninth through the thirteenth centuries. Pig farming was also a considerable part of the economy from the settlement into the Middle Ages, in contrast to known sites in contemporary Iceland and Greenland.

Viking remains have been excavated at a number of farm complexes and farmsteads. The site at Kvívík on the island of Streymoy (Figs. 1 and 8) is a classic example of a Faroese Viking farm, with longhouse 20 m in length and ~5 m wide. Parallel to the dwelling there was another building, an outhouse, which was later turned into a smaller building with a secondary use as a byre (cattle barn), with stalls along each side and a drainage trench running down the middle. As in other farmsteads from this period, the buildings were basically constructed of wood, protected by wide outer walls of stone and earth. The homeland tradition of building in wood was continued, but adapted to local conditions, as timber was scarce.

The artefacts recovered from excavations provide a glimpse of everyday-life in the Faroes during the Viking and Middle Ages. They also inform about contacts with other people and the outside world and help with dating of the sites. Household equipment was both locally made and imported. The most essential imported artefacts included minerals and wood. Metal objects and iron slag have been found, but decorative objects and jewelry are rather rare. Toys and gaming pieces give evidence of leisure activities for both children and adults. Local materials used included tufa for spindle whorls and lamps, and the oldest evidence of local pottery production dates to the tenth century.

Excavations of the outfield site at Argisbrekka on Eysturoy (Fig. 1), of smaller ruins with walls made of turf, have revealed that the Landnám settlers brought with them a farming system very similar to the Celtic and Norwegian shieling practice. The Viking Period shieling economy was replaced in the eleventh and twelfth centuries by more intensive use of the outfield regions for sheep herding, which by then had become the centerpiece of Faroese economy. The practice of milking sheep is an interesting custom that is not recorded after AD 1300. Zooarchaeological investigations have shown that pig farming was also a part of the Faroese economy during the Viking and Middle Ages but also ceased after this time.

Little is known about religious beliefs in the Faroes before the introduction of Christianity AD ~1000. Only two burial sites have been documented archaeologically, dated to the tenth century: one in Tjørnuvík, the other in Sandur. At both sites 12 burials were revealed, where the dead had been buried together with various personal belongings. Both the Catholic Church and the Norwegian crown had strong interests in the Norse settlements in the Atlantic. By the early Middle Ages the Church and the crown strengthened control over the North Atlantic region as a whole, and the Faroes became linked with the broader economic and cultural network of the North Sea and North Atlantic region.

FIGURE 8 Viking farm at Kvívík. Photograph by Simun Arge.

SEE ALSO THE FOLLOWING ARTICLES

Archaeology / Atlantic Region / Iceland / Seabirds / Vegetation

FURTHER READING

Arge, S.V., G. Sveinbjarnardóttir, K.J. Edwards, and P.C. Buckland. 2005. Viking and medieval settlement in the Faroes: people, place and environment. *Human Ecology* 33: 597–620.

Fosaa, A. 2003. Mountain vegetation in the Faroe Islands in a climate change perspective. Ph.D. thesis, Lund University, Sweden.

Gaard, E., B. Hansen, B. Olsen, and J. Reinert. 2002. Ecological features and recent trends in the physical environment, plankton, fish stocks, and seabirds in the Faroe shelf ecosystem, in *Large Marine Ecosystems of the North Atlantic.* K. Sherman and H. R. Skjoldal, eds. Amsterdam: Elsevier Science, 245–265.

Guttesen, R., ed. 1996. *The Faeroe Islands topographic atlas.* Copenhagen: C. A. Reitzels Forlag.

Hannon, G. E. and R. H. W. Bradshaw. 2000. Holocene vegetation dynamics and impact of human settlement on the Faroe Islands. *Quaternary Research* 54: 404–413.

Hansen, K., and J. Johansen. 1982. Flora and the vegetation of the Faeroe Islands, in *The physical environment of the Faeroe Islands.* G. K. Rutherford, ed. Monographie Biologicae 46. The Hague: Dr W. Junk Publishers, 35–52.

Hansen, S. S. 1990. Toftanes. A Faroese Viking Age farm from the 9th-10th century. *Acta Archaeologica* 61: 44–53.

Jensen, J.-K., D. Bloch, and B. Olsen. 2005. *List of birds seen in the Faroe Islands.* Tórshavn: Føroya Náttúrugripasavn.

Jóhansen, J. 1985. *Studies in the vegetational history of the Faroe and Shetlands Islands.* Tórshavn: Føroya Fró skaparfelag.

Mahler, D. L. 2007. *Sæteren ved Argisbrekka. Economic development during the Viking Age and Early Middle Ages on the Faroe Islands.* Annales Societatis Scientiarum Færoensis Supplementum 47. Tórshavn: Faroe University Press.

FERNANDO DE NORONHA

R. V. FODOR

North Carolina State University, Raleigh

Fernando de Noronha is an east–west trending volcanic archipelago located in the equatorial South Atlantic Ocean 350 km from Brazil (Fig. 1). It is composed of 21 islands and islets for a total area of 26 km². The largest island, 17 km² in area and named Fernando de Noronha, is about 10 km long and 3.5 km at its widest point. Of the 20 other islands and islets, Rata is the next largest, at about 6.8 km². The remaining land comprises a total area of about 2 km². The approximately 3000 inhabitants of Fernando de Noronha are Brazilian citizens, and many serve in island tourism and wildlife protection.

HISTORY

Fernando de Noronha was recorded on a nautical chart dated AD 1500, although its discovery is credited to navigator Amerigo Vespucci in 1503. The Portuguese lord Fernão de Loronha financed Vespucci's expedition, and the island chain was named in his honor in 1504. Afterwards, for part of the seventeenth century,

FIGURE 1 Map showing the location (inset) of Fernando de Noronha relative to Brazil, and outlining the main island (also called Fernando de Noronha), and Rata, the second largest island of the archipelago, and several small islands.

the Dutch occupied Fernando de Noronha and called it Pavonia. Then, in the eighteenth century, the French occupied the island, using the name Ile Delphine. By 1737, however, the Portuguese had reclaimed Fernando de Noronha. Because they believed it to be vulnerable to invasions—as Fernando de Noronha was located along major navigation routes of the times—the Portuguese fortified it. Most of the construction to create roads and buildings, including ten forts, was accomplished by prisoners. To keep them from escaping on rafts, the tree vegetation was greatly destroyed during that period of occupation.

In 1839, naturalist Charles Darwin visited Fernando de Noronha with his scientific crew during the *Beagle* expedition. The biodiversity and rich near-shore marine life that Darwin noted and recorded accounts for why Brazil eventually created in 1988 a national marine preserve and a state environmental protected area on the island. UNESCO certified Fernando de Noronha in 2001 as a "Site of the Natural World Patrimony."

GEOGRAPHY AND GEOLOGY

The archipelago, located below the equator at 3°54′ S and 32°25′ W, is physically situated on a volcanic structure estimated to have a base about 70 km in diameter at an approximately 4000 m depth. Accordingly, the islands represent the tops of this huge volcanic platform rising from the ocean floor.

As a result of erosion, the central part of the main island has a plateau from 30 to 45 m in elevation, which probably represents an ancient sea level. There are two other plateaus on the main island, 150 to 200 m high, plus hills and scattered rocky peaks, the highest of which is 321 m and forms Pico Mountain. The other islands and islets vary in relief from rocky points and cliffs to flat and rolling surfaces, depending on the makeup of their geologic material.

The volcanic rocks of Fernando de Noronha are geologically young, less than 12.3 million years. They represent compositional types that are uncommon within the spectrum of igneous rocks, mainly because of their relatively high abundances of alkali elements (e.g., total Na_2O+K_2O ranges from 4 to 17 wt%) and because many have relatively low SiO_2 abundances (37 to 43 wt%). Mapping by the Brazilian geologist Fernando de Almeida in the 1950s identified two main volcanic groups. The Remedios formation is the older, being 12.3 to 8 million years in age. It consists of phonolite and trachyte composition pyroclastic deposits intruded by lamprophyre dikes, and by phonolite and trachyte plugs and domes. These latter features form some of the spires of the islands. The Remedios formation also includes occurrences of alkalic basalt and andesite lavas.

The second volcanic group, Quixaba formation, is made of lavas that range from 4.2 to 1.5 million years in age and is separated from the older Remedios formation by an erosional surface. The Quixaba formation lavas are largely olivine and pyroxene melanephelinites, nepheline basalt, and basanite. One basanite lava contains fragments, or xenoliths, of rock carried up from the mantle. The Quixaba formation blankets most of the main island and the second largest, Rata, and the thickness of this formation can reach 100 m.

Geologically young sedimentary rocks are present as calcarenites, rock made of cemented shell and coral fragments, which largely represent ancient dunes. In places, calcarenites reach 50 m in thickness and are sometimes interbedded with pebble conglomerates, suggesting that both wind-driven and shallow marine environments shaped the islands' sedimentary history. Radiocarbon dates for these sediments range from 42,000 to 7500 years.

FAUNA AND FLORA

The Fernando de Noronha archipelago supports migratory and resident birds and is home to the largest breeding colonies of all tropical South Atlantic islands. Among the bird species are noddies, terns, and boobies. Rotifer dolphins swim in the bays and are protected by Brazilian law. Sea turtles use Fernando de Noronha beaches for spawning and are protected in the national marine reserve. Vegetation consists of endemic trees in relatively high areas, and of bushes occupying lower, flat areas. Each can become covered by native species of climbing plants. Cashew trees are also present, but only because they were brought to Fernando de Noronha to provide food. The only insular mangrove growth in the South Atlantic is an area along the southern part of the main island.

SEE ALSO THE FOLLOWING ARTICLES

Atlantic Region / Exploration and Discovery / Marine Protected Areas / Volcanic Islands / Voyage of the *Beagle*

FURTHER READING

de Almeida, F. F. M. 1955. Geologia e petrologia do arquipelago de Fernando de Noronha. Divisao de Geologia, Departmento Nactional Production Minerals Monografia 13.

de Almeida, F. F. M. 2000. The Fernando de Noronha archipelago, in *Sitios geologicos e paleontologicos do Brasil*. D. Schobbernhaus, D. A. Campos, E. T. Queiroz, M. Winge, and M. Berbert-Born, eds. Available online at http://www.unb.br/ig/sigep/sitio066/sitio066.htm.

Gerlach, D. C., J. C. Stormer, and P. A. Mueller. 1987. Isotopic geochemistry of Fernando de Noronha. *Earth and Planetary Science Letters* 85: 129–144.

Gunn, B. M., and N. D. Watkins. 1976. Geochemistry of the Cape Verde Islands and Fernando de Noronha. *Geological Society of America Bulletin* 87: 1089–1100.

Ulbrich, M. N. C. 1993. Petrography of alkaline volcanic-subvolcanic rocks from the Brazilian Fernando de Noronha archipelago, southern Atlantic Ocean. *Boletim Instituto de Geociencias-Universidade do Sao Paulo, Serie Cientifica* 24: 77–94.

Ulbrich, M. N. C., V. Maringolo, and E. Ruberti. 1994. The geochemistry of alkaline volcanic-subvolcanic rocs from the Brazilian Fernando de Noronha archipelago, southern Atlantic Ocean. *Geochimica Brasil* 8: 21–39.

Weaver, B. L. 1990. Geochemistry of highly-undersaturated ocean island basalt suites from the South Atlantic Ocean: Fernando de Noronha and Trindade Islands. *Contributions to Mineralogy and Petrology* 105: 502–515.

FIELD STATIONS

SEE RESEARCH STATIONS

FIJI, BIOLOGY

PADDY RYAN

Johnson & Wales University, Denver, Colorado

The Fijian archipelago is centrally located amid the island groups of the southwestern Pacific. Tonga and Samoa are to the east, Vanuatu and the Solomon Islands to the northwest, New Caledonia to the southwest, more distant New Zealand to the south, and Australia to the west. The group consists of between 300 and 844 islands, depending on the authority and on the definition of "island." There are four major islands: Vitilevu (10,388 km^2), Vanualevu (5535 km^2), Taveuni (434 km^2), and Kadavu (408 km^2). Total land area is approximately 18,300 km^2, which occupies an ocean area of around 650,000 km^2.

CLIMATE AND TOPOGRAPHY

The climate is tropical maritime and is thus without great extremes of temperature. Mean monthly temperature ranges from 22 °C in July to 26 °C in January. The predominant winds are the southeast trades. Windward sides of the islands experience increased cloudiness and rainfall. There is a significant rainshadow effect, so lee sides tend to be considerably drier. "Wet zone" areas of the larger islands average around 305–345 cm of rain annually, whereas the "dry zone" areas average 165–229 cm. Tropical cyclones occur relatively rarely, with only 10–15 per decade and only two to four causing severe damage. The El Niño weather pattern occasionally brings periods of drought and high ocean temperatures.

The combination of weather and topography has led to a variety of terrestrial habitats, ranging from beach and mangrove to cloud forest.

The complex arrangement of islands, large rivers, upwellings, and deep oceanic trenches has led to a wide diversity of aquatic habitats. Fiji possesses one of the largest barrier reefs in the world (the Great Sea Reef north of Vanualevu) and a spectacular array of soft corals and gorgonians.

PLANTS

Fiji's terrestrial environments can be loosely divided into wet zone vegetation (including lowland rainforest, montane rainforest, and cloud forest), dry zone vegetation (Sclerophyllous forests and scrublands brought about by repeated burning), and coastal vegetation (littoral, strand, and mangrove)

A. C. Smith has written the major work on Fiji's vascular plants, *Flora Vitiensis Nova: A New Flora of Fiji (Spermatophytes Only)*, in which 2295 species are described in 3189 pages. The vascular flora of Fiji is closely related to the Indo-Malesian floristic realm, with around 90% of the plant genera being in common with New Guinea. The remaining 10% has affinities with Australia, New Zealand, New Caledonia, and French Polynesia. A substantial proportion of Fiji's seed plants are dispersed by frugivorous bats and birds. There are approximately 2600 vascular plant species in Fiji, of which 1600 are native species. Of the 303 fern species, 88 (29%) are endemic. There is similar level of endemism among the native seed plants (25%). There is one endemic family, Degeneriaceae, and only six of the approximately 460 genera are endemic. It is possible that *Degeneria* and *Podocarpus* arrived to Fiji via the Tongan Island of Eua, a Gondwanan fragment that may have come into contact with Vitilevu around 5 million years ago. Some authorities consider *Degeneria* a Malesian relict species. There are 11 gymnosperm species, several of which (*Podocarpus affinis, Acmopyle sahniana,* and *Dacrydium nausoriense*) are considered endangered.

Among the dicotyledons, the genus *Psychotria* (family Rubiaceae) shows the highest endemism with 72 of 76 species (95%). The palms are the only other floristic group to rival this level of endemism, with 24 out of 31 having this status (77%). Several species are highly restricted in distribution and possibly at risk of extinction. The palms of Fiji are ably described and illustrated in Dick Watling's *Palms of the Fiji Islands*. There are currently 164 native orchid species known, of which 51 (31%) are considered endemic.

Forest trees tend to flower during January to March, whereas fruit production is relatively constant from June to September. Fruit bats and frugivorous birds (primarily doves and pigeons) depend upon a year-round supply of fruit, but their breeding seasons coincide with peak fruit abundance.

ANIMALS

Fiji is close enough to other island groups for regular dispersal to occur and has probably had some areas above the water for 40–50 million years. This, together with a wide range of habitats, has led to considerable diversity. Some of Fiji's animals appear to have quite anomalous distributions (a fly species with nearest relatives in Madagascar for instance). With more taxonomic study, it is likely that these apparent relationships will be resolved. Knowledge of the terrestrial invertebrate fauna is still limited, but the Fijian Arthropod Survey run by the Bishop Museum has already vastly expanded our information.

Terrestrial vertebrates are reasonably well known, although it is possible new lizard species may yet be found. Several new freshwater fish species have been described in recent years. There is limited endemism among the marine organisms, and, as with terrestrial ecosystems, there is much to be studied.

Molluscs

Many of Fiji's native land molluscs are endemic. Fifty-eight species of terrestrial molluscs have been recorded from Vitilevu. This is an area that urgently needs more research. There is also significant diversity among the freshwater gastropods with 39 species reported: three are endemic, and one is an endemic monotypic genus (*Fijidoma*).

Arthropods

As is the case with many other developing countries, Fiji's terrestrial arthropod fauna is only patchily known. The arthropod equivalents of the charismatic megafauna (the butterflies, moths, and some of the beetles) have been reasonably well documented. There are over 400 species of macrolepidoptera and seven endemic genera.

Two families, five genera, and 17 species of mayflies have been collected from Fiji. Sixteen species belong to the family Baetidae with one species in the family Caenidae. Mayflies are poor at long-distance dispersal, so their presence in Fiji offers a useful opportunity to provide insight into colonization mechanisms. The most likely explanation is the "Melanesian arc": two chains of islands that converged at their western ends to form what is now New Guinea. With more islands then, shorter dispersal distances may have provided mayflies access to Fiji.

Fijian damselflies have attracted the attention of evolutionary biologists since the diversity of *Melanonesobasis* and *Nesobasis* was revealed by Donnelly. There are currently six described *Melanesobasis* and 24 *Nesobasis*. Several new *Nesobasis* are awaiting description. None of the *Nesobasis* species from Vitilevu is found on Vanualevu, which suggests limited dispersal ability. There is a suggestion of a role reversal amongst the *Nesobasis* damselflies, with the females patrolling and defending territories.

The cicadas in Fiji have been extensively studied. There are 15 species, of which 14 (93%) are endemic. One genus, *Fijipsalta*, is endemic. *Tibicen knowlesi* (*nanai* in Fijian) has an eight-year life cycle. Fijians in the Sigatoka valley collect them as they emerge from their nymphal shells, tie them into strings, and then cook and eat them.

Fiji has a high diversity of ants. There are currently 32 genera and 91 endemic species (66% endemism). A number of genera, notably *Pheidole, Camponotus, Cerapachys,* and *Strumigenys* have undergone significant radiation. One endemic genus, *Poecilomyrma*, appears to have diversified into three distinct species. The genus *Lordomyrma* is represented by six endemic species out of the 16 known worldwide. As with much of the biota, the ants appear to have been derived from the New Guinea region.

There are more beetles (1284) recorded from Fiji than any other arthropod order. Of these, around 140 are longhorn beetles (Cerambycidae) including three extremely large *Xixuthurus* species. The largest of these is *X. heros*, which may be the world's second-largest beetle, with some specimens reaching 15-cm body length. The other species are *X. ganglbaueri* and *X. terribilis*. All three species share the ability to force air through narrow apertures at the bases of the wing covers to produce a hissing sound. This presumably deters potential predators. The larvae are enormous, reputedly growing to 20 cm in length and producing tunnels 5 cm in diameter in the wood in which they live and on which they feed. Village Fijians consider these larvae delicacies but probably encounter them too infrequently for this food source to have a major impact on their population. Other large cerambycids include five species of *Olethrius,* including *O. tyrannus,* the most commonly seen longhorn. Both *Xixuthurus* and *Olethrius* are found in New Guinea, the possible source of these genera, which have undergone minor radiation in Fiji.

Jewel beetles, family Buprestidae, are also well represented with 45 species. There are around 250 weevil species, family Curculionidae.

A recent checklist records 911 species of Lepidoptera; of these, 44 are butterflies. There are several endemic species and subspecies, the largest of which is the endemic Fiji swallowtail *Papilio schmeltzi*. The crow butterflies (so named because of their dark crowlike coloration) are well represented with four species. There has been a minor radiation in the noctuid genus *Lophocoleus*, which has six species, and several other genera show similar local diversity.

The two-winged flies, family Diptera, have been the focus of recent studies. Currently 596 species from 61 families have been reported. Some have unusual affinities. Mythicomyiid bee flies from the high rain forests of Taveuni are otherwise known from the arid cordillera in North and South America, and a simuliid black fly from Taveuni belongs to a South American subgenus. This may reflect our lack of data on dipteran evolution or the relative importance of characters when assigning taxonomic status.

Fifty-seven millipede species have been recorded from Fiji. On an area basis, this is substantial; much larger New Zealand has a similar number, although many more species remain undescribed. Some of the Fijian rain forest millipedes *Salpidobolus* spp. reach large sizes: 20 cm is not uncommon. It is likely that there has been substantial radiation of these giant animals in Fiji. The arachnids are also well represented: 308 species have been recorded, the majority (171) of which are mites. Of the 127 spiders, 38 are jumping spiders (family Salticidae), and many are endemic. One, currently undescribed, mygalomorph has been collected. Other arachnid orders include Schizomida, Amblypygida, the tail-less whip scorpions, also with one species, Thelyphonida; the vinegaroons with one species; the Pseudoscorpionida with three species; and the Scorpionida with four species. The presence of scorpions concerns the tourist industry, but these creatures are rarely encountered and lack a dangerous sting.

Little endemism is apparent in marine or terrestrial crustacea. Ten grapsid and four geocarcinid land crabs are known from Fiji. Among the freshwater shrimp, 14 atyid species have been recorded, five of which are endemic. There are ten native palaemonids. It is likely that more endemic freshwater shrimp and new land crab records will be reported with more research.

Fishes

Fiji has around 1200 marine species in 162 families. Because of the relatively close proximity to other island groups, few are endemic. Eighty-nine freshwater fish species from 26 families are known, with gobies being the best represented. Recent research has added six new species of freshwater goby. Seven sicydiine gobies are recognized by Jenkins *et al.*, three of which are new endemic species; a fourth, in a new genus, awaits description.

The freshwater eels, *Anguilla marmorata* and *A. obscura*, are the most widely distributed freshwater fish; together with "jungle perch" *Kuhlia* species they provide an important source of protein to inland villagers. An ophichthid eel, *Yirrkala gjellerupi*, previously only known from the holotype collected in New Guinea and formally described in 1916, was collected from a stream near Suva. More widespread collecting is likely to reveal other such gems.

Most "freshwater" fish in Fiji spend at least some part of their lives in the sea. This is true of the eels, the jungle perch, and many of the gobies. The sicydiine gobies are amphidromous—that is, they lay eggs in freshwater. The free embryos drift to the sea, where the larvae complete a planktonic phase before returning upriver. In Fiji these youngsters are referred to generically as *cigani* or *tanene*. As in many Pacific Island cultures, this upstream migration is harvested and eaten.

For many years after initial European colonization, Fiji was known as the "shark attack capital of the world." Early reports note that missing fingers and toes from shark attacks were common among inland people and that deaths were frequent. The Reverend David Caygill wrote in 1839, "It abounds with shellfish and ground-sharks . . . Ground-sharks are so numerous in every part of the river that the natives who bathe in it are in danger of being killed or maimed by their bites." The shark in question, the bull shark *Carcharhinus leucas*, is still found in the larger rivers, but attacks have not been recorded in recent years.

There are few endemic marine fish species; these include the deep-water species *Parmops echinites*, *Thamnacous fijiensis*, and *Plectranthias fijiensis*. Shallow water species include the goby *Bryaninops dianneae* and Carlson's damsel *Neoglyphidodon carlsoni*.

Amphibia

There are two endemic frogs: *Platymantis vitiensis*, the Fiji tree frog (Fig. 1), and *P. vitiana*, the Fiji ground frog. These species represent the easternmost penetration of frogs into the South Pacific. The ground frog, which used to be common on Vitilevu, is now rare on Vitilevu and uncommon on Vanualevu, possibly because of predation by the introduced mongoose *Herpestes auropunctatus*. On Viwa Island, in the Rewa delta, ground frogs are frequently found on the beach and may take refuge in the sea when threatened. One captive specimen had eaten a crab. Ground frog females reach 105 mm in body length, and males up to 70 mm.

FIGURE 1 The endemic Fijian treefrog *Platymantis vitiensis* is still reasonably common on the main islands.

The tree frog is widely distributed on Vitilevu and Vanualevu, as well as on a few smaller offshore islands. It seems relatively secure as long as its rain forest habitat remains intact. Females are considerably larger than males and can reach 51 mm body length and weigh up to 10.8 g. The tree frog is highly unusual, as both sexes produce a mating call.

Both frog species are nocturnal and lay eggs, which develop directly without a tadpole stage, hatching as miniature adults. Ground frogs lay around 40 eggs at a time, but the clutches are rarely seen. Tree frog clutches ranged from 5 to 29 eggs with a mean of around 17. Egg laying in both species occurs during the wetter, more humid months, probably December to March.

Reptiles

Terrestrially, Fiji has three iguanas, four snakes, ten geckos, and 12 skinks. Thirteen out of 29 (45%) are endemic. There are nine marine species: five turtles and four sea snakes.

FIGURE 2 The Fijian banded iguana *Brachylophus bulabula* male (right) and female (left). The introduced mongoose remains a threat to their buried egg clutches.

The arboreal iguanas, *Brachylophus vitiensis*, the crested iguana; *B. bulabula*, the banded iguana (Fig. 2); and *B. fasciatus*, currently known from the Lau Group, are fascinating from a biogeographical viewpoint, because their nearest relatives are in Central America. The inescapable conclusion is that the *Brachylophus* ancestors rafted into the South Pacific from this region. Extinct regional species include a large Tongan land iguana, *Brachylophus gibbonsi*, and a much bigger Fijian species, *Laptiguana impensa*, which may have weighed 20 kg. The crested iguana appears to be endangered by habitat loss and exotic predators such as cats.

Of the four terrestrial snakes, one is unequivocally endemic. *Ogmodon vitianus* is the only member of its genus. This elapid snake is a burrower and has rarely been seen or collected. Known in Fijian as *bolo*, there seems no reason to invent an English name: "Fijian burrowing snake" seems so mundane. The bolo is restricted to the interior of Vitilevu. It feeds on worms and other soft-bodied invertebrates, subduing them with its mildly neurotoxic venom. Captive specimens have been extremely docile. As this encyclopedia goes to press, a new burrowing endemic snake has been discovered in south Taveuni. Yet to be described, it is almost certainly a new genus.

The most common terrestrial snake is *Candoia bibroni*, the Pacific boa. Although it can grow to 3 m, 1.5 m is more common in Fiji. They feed on terrestrial vertebrates. Females give birth to 20 to 30 young. The boa is widely distributed, both within the Pacific and Fiji.

The sea snakes include two species of *Laticauda*: *L. colubrina*, the yellow-lipped sea krait, and *L. laticaudata*, the banded sea krait. The black-headed sea snake *Hydrophis coggeri* looks similar to the *Laticauda* sea snakes to the nonspecialist. Unlike the sea kraits, *H. coggeri* is totally marine. They are occasionally seen in the shallow seagrass beds of Laucala Bay.

Of the ten geckos, six species were almost certainly introduced by humans. *Nactus pelagicus*, the Pacific slender-toed gecko, and *Gehyra vorax*, the giant forest gecko, probably arrived by long-distance transoceanic dispersal.

The remaining geckos—*Lepidodactylus gardneri*, the Rotuman forest gecko, and *L. manni*, Mann's forest gecko—are endemic. The Rotuman forest gecko is restricted to the island of Rotuma. Mann's forest gecko is currently known on Vitilevu, Vanualevu, Ovalau, and Kadavu.

The skinks are dominated by the genus *Emoia*: nine of the 12 Fijian species belong to this genus, and four are endemic. The endemic species are *E. campbelli*, the montane skink, which is currently only known from Monasavu; *E. concolor*, the green tree skink, which is found throughout the group; *E. parkeri*, the Fijian copper-headed skink, from Vitilevu, Kadavu, Ovalau, and Taveuni; and *E. mokosariniveikau*, the turquoise tree skink, from Vanualevu, although specimens similar to this species have been collected in Vitilevu as well.

The final skink species (but probably not the last that will be found in Fiji) is *Leiolopisma alazon*, the Lauan ground skink. This genus is associated with New Zealand, so its presence in Fiji is a surprise. It presumably arrived from New Zealand via the Tonga–Kermadec Arc when parts of it were above water.

The remaining reptiles are all sea turtles. Two of these turtles—*Eretmochelys imbricata*, the hawksbill turtle, and *Chelonia mydas*, the green turtle—breed in Fiji. The others—*Caretta caretta*, the loggerhead turtle; *Lepidochelys olivacea*, the olive ridley turtle; and *Dermochelys coriacea*, the leatherback turtle—are occasional visitors. Clare Morrison's *A Field Guide to the Herpetofauna of Fiji* provides additional information on these and other species.

Birds

There are currently 87 species of breeding bird in Fiji. Fifty-seven of these are land birds, and 19 are seabirds; 11 of these species have been introduced. Twenty-seven (46%) of Fijian bird species are endemic. The Fijian birds have been extensively covered by Dick Watling in his book *Guide to the Birds of Fiji & Western Polynesia*. In 2006, BirdLife International published the useful *Important Bird Areas in Fiji*, edited by Vilikesa Masibalavu and Guy Dutson. This gives up-to-date conservation and distribution information. As there are too many species to discuss in an article of this length, only endemic species and some others have been selected.

A number of Fijian birds have become internationally known. The Fijian shining parrots—*Prosopeia personata,* the masked shining parrot; *P. tabuensis,* the red shining parrot; and *P. splendens,* the crimson shining parrot—are good examples. Within Fiji, all the shining parrots are now nominally protected, but the crimson shining parrot is still sought after as a pet. The masked shining parrot is restricted to Vitilevu; the red shining parrot lives in Vanualevu and Taveuni; and the crimson shining parrot, once considered a subspecies of the red shining parrot, is found only in Kadavu. The red-throated lorikeet *Charmosyna amabilis,* another Fijian endemic, originally found in Vitilevu, Taveuni, and Ovalau, is considered critically endangered and may well be extinct. The collared lory *Phigys solitarius,* also endemic, is widespread in the group and is commonly seen. A sixth parrot, *Vini australis,* the blue-crowned lory, occurs in the eastern islands of Fiji and in parts of Tonga and Samoa.

Doves in the genus *Chrysoenas* have undergone a minor radiation in Fiji. There are three endemic species: *C. layardi,* the whistling dove found on Kadavu and Ono; *C. luteovirens,* the golden dove endemic to Vitilevu; and *C. victor,* the stunning orange dove (Fig. 3) found on Vanualevu, Taveuni, and nearby islands. Two *Ptilinopus* species—*P. porphyraceus,* the purple-crowned fruit dove, and *P. perousii,* the many-colored fruit dove—are also widely distributed in the region. Other doves include the endemic *Ducula latrans,* the barking pigeon, found in mature forest on the major islands. The regional endemic *Gallicolumba stairi,* which rejoices in two diametrically opposed common names (the friendly ground dove and the shy ground dove), is considered vulnerable.

FIGURE 3 The spectacular endemic orange dove *Chrysoenas victor.* Females are much less brightly colored.

Another endangered endemic is *Trichocichla rufa,* the long-legged warbler, also called the long-legged thicket bird. It was first collected in 1890, with four more added to the total by 1894. It was not seen again until 1974 when a Vanualevu subspecies was discovered (which has not been seen since). In 2003, scientists from BirdLife International discovered a small population of 12 pairs in Vitilevu. Further study revealed a total population of between 50 and 250 mature birds. This bird is now protected by law. The Fiji bush-warbler *Cettia ruficapilla,* another endemic, lives in a variety of habitats on the four main islands.

Of Fiji's two endemic parrotfinch species, one, *Erythrura pealii,* the Fiji parrotfinch, will be seen by even the most ornithologically inept visitor. These green finches with red heads can often be seen in small flocks industriously gleaning grass areas for seeds. The other species, *E. kleinschmidti,* the pink-billed parrotfinch, is restricted to forest on Vitilevu. Although it will eat flower buds and fruit, it is primarily an insectivore. The IUCN accords it vulnerable status.

The honeyeaters are well represented in Fiji. There are five species, four of which are endemic. One, *Myzomela chermesina,* is found only in Rotuma; another, *Xanthotis provocator,* the Kadavu honeyeater, is restricted to that island. The giant forest honeyeater *Gymnomyza viridis* is more often heard than seen. Also called the yodeling honeyeater, it produces a range of calls, the most notable being a ringing, liquid yodeling. *G. viridis* is found in forest on Vitilevu, Vanualevu, and Taveuni. The orange-breasted myzomela, *Myzomela jugularis,* is widely distributed in Fiji and may often be seen hovering in front of flowers while drinking nectar like an inefficient hummingbird. The final species, *Foulehaio carunculata,* the wattled honeyeater (*kikau* in Fijian), is found in Fiji, Samoa, and Tonga. It often visits suburban gardens, where it appears to spend much of its time chasing off rivals. The Fijian name *kikau* is onomatopoeic.

Fantails, family Dicruridae, are widely distributed through southern Asia and Australasia. Two species occur in Fiji: *Rhipidura spilodera,* the streaked fantail, also known from Vanuatu and New Caledonia, and *R. personata,* a Kadavu island endemic. One of Fiji's most enigmatic birds is *Lamprolia victoriae,* the silktail. For many years, its systematic position was debated, with some authorities even suggesting it was related to the New Guinean birds-of-paradise. It is now considered to be in the family Monarchidae (monarch flycatchers). There are two main silktail populations, one in Taveuni and the other on the Natewa Peninsula of Vanualevu.

Male blue-crested flycatchers, *Myiagra azureocapilla*, are highly distinctive. Their slaty blue head makes them easy to identify. Blue-crested flycatchers live in forest on the three main islands. The closely related *M. vanikorensis*, the Vanikoro flycatcher, also known from the Solomon Islands, is ubiquitous in Fiji. They are also referred to as broadbills.

The other Fijian flycatchers are *Mayrornis lessoni*, the slaty monarch; *M. versicolor*, the Ogea monarch; *Clytorhynchus vitiensis*, the Fiji shrike-bill; and *C. nigrogularis*, the black-throated shrikebill. The slaty monarch is found throughout the Fiji group whereas the Ogea monarch is restricted to that island. Despite the name, the Fiji shrikebill is found in Samoa and Tonga as well as in Fiji. The black-throated shrikebill is a Fijian endemic. Both shrikebill species actively search for insects, ripping off bark, moving dead leaves, and probing rotten wood.

There are two white-eye species. One, *Zosterops lateralis*, is common in Australia, New Zealand, and New Caledonia, whereas the other, *Z. explorator*, the Fiji white-eye, is endemic. Both species feed on insects, fruit, and nectar.

The endemic Fiji woodswallow *Artamus mentalis* is a group breeder. Members of the commune help build the nest and share in the incubation and feeding of the chicks. Foraging birds often perch in high positions that offer a good all-around view of their surroundings and from which they dive to catch prey. They can often be seen on telephone wires. They fearlessly attack birds of prey.

There are four species of raptor currently found in Fiji: *Circus approximans*, the Pacific harrier; *Falco peregrinus*, the peregrine falcon; and the endemic *Accipiter rufitorques*, the Fiji goshawk. The barn owl *Tyto alba* is also present, but *T. longimembris*, the eastern grass owl, is considered extinct in Fiji. The Fiji peregrine is endangered, and studies have shown that members of the local population are practically clones of each other. The Fiji peregrine will take fruit bats and seabirds, often well beyond the reef.

The only endemic seabird is *Pterodroma macgillivrayi*, the Fiji petrel. Originally known from a single fledgling collected in 1855 on Gau, it was dramatically rediscovered by Dick Watling in 1983 when a specimen flew out of the night (attracted to a light set up for the purpose) and hit him in the head.

There have been a number of bird extinctions since the arrival of humans in Fiji; they are discussed below.

Mammals

The only native land mammals are bats. There are six species. The big fruit bats are the flying foxes *Pteropus tonganus* and *P. samoensis*. Both species are reasonably common and are still occasionally hunted for food. They typically roost communally in tall forest trees, and colonies can be heard and smelled from some distance. In Fiji, the fruit bats tend to be primarily nocturnal, emerging at dusk to make long-distance flights to fruit trees and returning at dawn. In places where large diurnal birds of prey are rare or absent, they may fly during the day. The Fiji monkey-faced bat *Mirimiri acrodonta* is currently known only from Taveuni, but it may be found on other islands. The Fijian blossom bat, *Notopterus mcdonaldi*, also found in Vanuatu, is the only Fijian macrochiropteran to possess a long tail. Originally only known from Vitilevu, it has recently been found on Vanualevu and Taveuni.

The other two species are both insectivorous. The Polynesian sheath-tail bat *Emballonura semicaudata* is still quite common, but its numbers have declined on Vitilevu over the last 60 years. It roosts communally in caves, which makes it very vulnerable to disturbance. The Fijian mastiff bat *Chaerephon bregullae* is found only in Vanuatu and Fiji. A number of colonies, including a large maternity colony, have been located, but there are concerns that numbers are declining.

HUMAN IMPACTS AND CONSERVATION

Considerable information is now available about extinctions associated with early human colonization of the Pacific Islands. When humans first colonized Fiji around 3500 years ago, they encountered a very diverse vertebrate fauna. We know that there was a giant pigeon, *Natunaornis gigoura*, on Vitilevu, which approached the size of the dodo. There were also several species of megapode, one of which, *Megavitiornis altirostris*, the giant Fiji megapode, has been described. The giant Fiji ground frog *Platymantis megabotoniviti* was twice the size of the extant *P. vitiana*. The giant Fiji land iguana has already been mentioned. There was also a mekosuchine crocodile, *Volia athollandersoni*, and a horned tortoise, *Meiolania* sp., which has yet to be described. Although it is possible that Polynesian rat *Rattus exulans* introductions had an impact on these species, it is far more likely they were simply eaten to extinction.

Since European colonization in the early 1800s, there have been further extinctions. Currently, these are believed to be the wandering whistling duck *Dendrocygna arcuata*, which is secure elsewhere in its range, and the grass owl *Tyto capenis*, which is also found elsewhere. The only endemic post-European extinction is that of the barred-wing rail *Nesoclopeus poecilopterus*, which was recorded from wetlands on Ovalau and Vitilevu. BirdLife International estimates that the population of red-throated lorikeet *Charmosyna amabilis*, another Fijian endemic, may be less than 50, and there are no breeding colonies in captivity.

There are 11 introduced birds which have successfully established. The two mynahs *Acridotheres tristis*, the common mynah, and *A. fuscus*, the jungle mynah, are widespread. The red-vented bulbul *Pycnonotus cafer* is only common on Vitilevu. One of the more unusual introductions was that of the Java sparrow *Padda oryzivora*, locally common in areas near Suva, Savusavu, and parts of Taveuni. Ironically, it is considered vulnerable on the IUCN list of threatened species in its home range in Indonesia.

The mongoose *Herpestes auropunctatus* was introduced in 1883 to control rats in sugar cane fields. It did a bad job of doing this but was very effective at eliminating the ground frog and reducing the number of several bird and lizard species wherever it occurred. Currently, several biologically important islands are mongoose-free, and it is hoped that they will stay that way.

Other introductions include the usual mammals—rats (*R. rattus* and *R. norvegicus*), mice, cats, dogs, and goats—as well as the unusual (fallow deer on the island of Wakaya). On several islands in the group, goats represent a conservation threat because they remove vegetation. Rats may be impacting banded and crested iguanas by predating both eggs and youngsters.

The giant toad *Bufo marinus* was introduced in 1936, as in so many other places, to control agricultural pests. The giant toad did not succeed at this, but it may have competed directly with the ground frog. It is now found on all of the main islands and many of the smaller ones.

Ramphytophlops braminus, the flowerpot snake, has been recorded from the Suva area. It was probably accidentally introduced in soil from imported plants. It grows to 170 mm and looks vaguely wormlike, as a result of convergent evolution stemming from a common burrowing lifestyle.

In freshwater areas, introduced species such as tilapia, *Oreochromis mossambica*, and mosquito fish, *Gambusia affinis*, may compete with native fish. More information on such interactions is urgently needed.

Although introduced organisms remain a major conservation threat, habitat modification may pose a greater long-term danger. Fiji urgently needs more conservation real estate that is properly policed and supported by the local communities. This politically troubled island nation has some biological gems that are of global significance.

SEE ALSO THE FOLLOWING ARTICLES

Biological Control / Fiji, Geology / Frogs / Insect Radiations / Snakes

FURTHER READING

Fiji Arthropod Survey Website. http://hbs.bishopmusuem.org/fiji/
Heads, M. 2006. Seed plants of Fiji: an ecological analysis. *Biological Journal of the Linnean Society* 89: 407–431.
Masibalavu, V., and G. Dutson. 2006. *Important bird areas in Fiji: conserving Fiji's natural heritage*. Suva, Fiji: Birdlife International Pacific Partnership Secretariat.
Morrison, C. 2003. *A field guide to the herpetofauna of Fiji*. Suva, Fiji: Institute of Applied Sciences, University of the South Pacific.
NatureFiji Website. http://www.naturefiji.org/members2.php
Ryan, P. A. 2001. *Fiji's natural heritage*. Auckland, NZ: Exisle Publishing.
Ryan Photographic Website. http://www.ryanphotographic.com
Smith, A. C. 1979–1996. *Flora vitiensis nova: a new flora of Fiji*, vols. 1–6. Lawai, Kauai: Pacific Tropical Botanical Garden.
Watling, D. 2001. *Guide to the birds of Fiji & western Polynesia*. Suva, Fiji: Environment Consultants Fiji.
Watling, D. 2005. *Palms of the Fiji islands*. Suva, Fiji: Environment Consultants Fiji.

FIJI, GEOLOGY

HOWARD COLLEY

Higher Education Academy, York, United Kingdom

The Fiji Archipelago, which is located about 2400 km northeast of Australia, consists of over 300 islands and is dominated by two major islands, Viti Levu and Vanua Levu. Fiji is very unusual for an archipelago in a deep-oceanic setting. Compared to the more typical adjacent island arcs of Vanuatu to the west and Tonga to the east, Fiji has much larger islands; a much more varied sequence of rocks, including plutonic bodies; an abundance of mineral deposits; and a crustal thickness (up to 30 km) that is comparable to that of continental margins. The early history of Fiji is recorded in rocks deposited and erupted below sea level, and these include pillow lavas and limestones of late Eocene age (40–36 million years ago); the youngest rocks are air-fall volcanic ashes erupted on Taveuni in historic times (about AD 300–500). The geology of Fiji is dominated by volcanic rocks and their sedimentary erosion products (Table 1).

TABLE 1
Summary of Principal Stratigraphic Units in Fiji

Unit	Location	Age	Remarks
Deltaic sediments	SE Viti Levu	Holocene	Rewa delta sediments
Taveuni group	Taveuni	Quaternary–Recent	Oceanic-type basaltic rocks
Kadavu group	Kadvau	Pliocene	Volcanic rocks of andesitic composition
Bua and Koro groups	W Vanua Levu	Pliocene	Oceanic-type basaltic rocks
Natewa, Nararo, and Monkey Face groups	Vanua Levu	Mio-Pliocene	Volcanic rocks of basaltic to andesitic composition
Udu volcanic group	NE Vanua Levu	Mio-Pliocene	Volcanic rocks of dacitic to rhyolitic composition
Ba volcanic group	N Viti Levu	Pliocene	Volcanic rocks of basaltic to shoshonitic composition
Koroimavua volcanic group	NE Viti Levu	Pliocene	Volcanic rocks of calcalkaline to shoshonitic composition
Medrausucu volcanic group	SE Viti Levu	Late Miocene	Volcanic rocks of calcalkaline composition
Nadi, Navosa, Ra, and Tuva sedimentary groups	Viti Levu	Miocene	Conglomerates, breccias, sandstones, and marls
Colo plutonic suite	S Viti Levu	Upper Miocene	Granitoid and gabbroic stocks
Wainimala group	S Viti Levu	Lower Miocene	Mixed arc-type volcanic rocks and derived sediments
Yavuna group	SW Viti Levu	Eocene–Oligocene	Lavas, gabbros, and limestones

REGIONAL SETTING

At the present time, Fiji lies between the opposing eastward-subducting New Hebrides (Vanuatu) and westward-subducting Tonga–Kermadec convergence zones (Fig. 1).

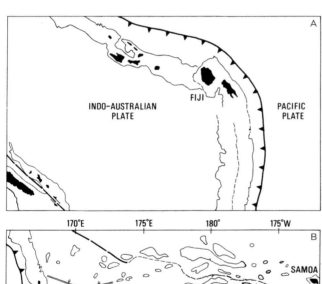

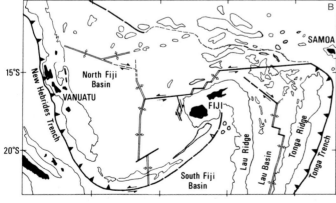

A complex system of transform faults and back-arc basins (e.g., Lau basin) is located between these convergence zones, and this system includes the Fiji platform, a shallow marine area that forms the basement for the Fiji Archipelago. Although the rocks of Fiji are predominantly of island-arc origin, the islands are no longer closely associated with a subduction zone because of extensional movement along the transform system. The early history of Fiji appears to be associated with subduction of the Pacific tectonic plate westward beneath the Indo-Australian plate, creating what is commonly referred to as the Vitiaz Arc, an Outer Melanesian arc system encompassing parts of Tonga, Fiji, Vanuatu, and the Solomon islands that began to form around 40 million years ago. Around 10–12 million years ago this arrangement was disrupted at its northern and western end, possibly by the arrival, through plate spreading, of an exceptionally thick piece of oceanic crust (the Ontong Java plateau) in the subduction system. An alternative explanation suggests that rapid avalanching of cold, dense subduction slab material across the upper-to-lower mantle boundary perturbed convection flow in the mantle. Beginning around 8 million years ago, a new subduction zone was started to the west of Vanuatu, with subduction toward the east of the Indo-Australian plate beneath the Pacific plate. This rever-

FIGURE 1 (A) Configuration of the Vitiaz arc (approximately 40–10 million years ago) with subduction of the Pacific plate westward beneath the Australian plate. (B) Present-day configuration of the Outer Melanesian arcs showing isolation of Fiji from modern subduction zones and development of back-arc basins (e.g., Lau basin) from spreading centers and transform faulting.

sal of polarity was accompanied by development of extensional tectonic features, which included the North Fiji basin and the transform system. This extension continued with the initial formation of the Lau back-arc basin around 3 to 5 million years ago. With extension, Fiji has become progressively more divorced from subduction systems, and the Fiji platform has undergone around 135° of counterclockwise rotation during the last 10 million years.

PRIMITIVE ARC STAGE (40–30 MILLION YEARS AGO)

Rocks characteristic of very young or primitive island arcs are found principally in the western part of Viti Levu (Yavuna group), the main island in Fiji (Fig. 2). The rocks are of Upper Eocene to Lower Oligocene age (40–30 million years old) and include basaltic lavas and breccias and their sedimentary derivatives, small intrusions of gabbro, and numerous small outcrops of limestone. These rocks are intruded by the Yavuna Stock, a granitoid body that is itself intruded by a swarm of diabase dikes. Basaltic lavas commonly occur in pillow form, indicating submarine eruption, and include high-magnesium compositional varieties; elsewhere in the Pacific, these are typically found in primitive arc sequences (e.g., Bonin arc). The limestones are generally massive or crudely bedded and dominated by coralline algal debris, and they contain a variety of foraminifera. Blocks of basalt are common within the limestone, and this strongly suggests that the limestones are contemporaneous with the basaltic rocks. Fossil ages derived from the study of foraminifera range from the Upper Eocene to the Lower Oligocene. Radiometric dates of the granitoid fraction of the Yavuna stock cluster around 35 million years ago, also indicating an age around the Eocene–Oligocene boundary. Elsewhere in Fiji, there are more tentative identifications of Upper Eocene–Lower Oligocene rocks in the Mamanuca and Yasawa island chains west of Viti Levu. Further away, there are very small outcrops of rocks of similar age on 'Eua in Tonga, and on Maewo (in Vanuatu) limestone blocks of this age occur in younger rock sequences. These occurrences, along with the far more voluminous outcrop of the Yavuna group, are taken to represent the earliest evidence of the formation of the Vitiaz arc.

EARLY TO MATURE ARC STAGE (30–7 MILLION YEARS AGO)

With the establishment of the Vitiaz arc, there was a period of prolonged but intermittent volcanic activity centered on Viti Levu, with eruption of predominantly arc-related basaltic, andesitic, and dacitic rocks forming the Wainimala

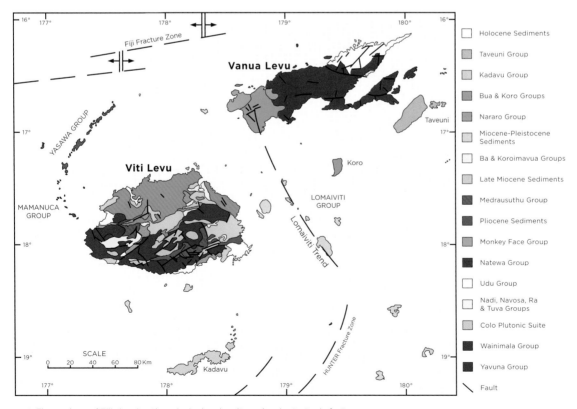

FIGURE 2 The geology of Fiji showing the principal rock units and major tectonic features.

group stratigraphic unit (Fig. 2). Much of this sequence was erupted below sea level and has been reworked by marine processes; it consists of volcaniclastic rocks. These rocks represent dispersal aprons of major volcanic edifices, but the original centers of eruption are difficult to recognize. In western Viti Levu, there is also a deeper-water succession of thinly bedded rocks, which includes deeper-water limestones, underwater debris avalanches (turbidites), and pale-colored volcanic horizons, taken to represent ash fall from island volcanoes. It is thought that the succession represents a fore-arc sequence deposited during a period of diminishing volcanism. This Vitiaz arc succession in Fiji is very poorly dated, but limestone bodies occur in what appear to be younger parts of the sequence in northeast and southwest Viti Levu and give Miocene (19–14-million-year-old) ages. Similar rocks also extend into the Mamanuca and Yasawa island chains to the west of Viti Levu.

Toward the end of the mature arc stage in the late Miocene, Fiji went through a period of geologic activity that is most unusual for an arc in a deep-oceanic setting well away from a continental land mass. Crustal flexuring, referred to as the Colo orogeny, created longitudinal folds more or less parallel to the east–west Vitiaz arc axis in Fiji. Elongate plutonic bodies in the cores of folds are mainly composed of granitoid and gabbroic rocks, and numerous radiometric dates indicate a late Miocene age (12–7 million years ago). The relationship of the Colo plutonic bodies and folds is enigmatic, and further study is required to determine whether the folds are caused by significant crustal shortening or whether they are drapes over the intruded plutonic bodies. In the Yasawa chain, there is a better understanding of folding, which has led to thrust faulting, and this compression can be dated to around 7 million years ago, an age that postdates most of the Colo intrusions. Given that numerous paleomagnetic studies have shown that the Fiji platform was undergoing counterclockwise rotation at this time, crustal shortening is to be expected. Aeromagnetic and gravity surveys indicate that much of southern Viti Levu may be underlain by the plutonic belt and that the volume of plutonic activity is very unusual for an oceanic arc system such as the Vitiaz arc. Following the Colo intrusive period, the first substantial stable landmass was established on the Fiji platform, possibly partly as a consequence of rapid isostatic uplift associated with the plutonism.

ARC-RIFTING STAGE (7–3 MILLION YEARS AGO)

Following the close of the Colo intrusive period, there was a period of widespread, voluminous, and compositionally varied volcanism in Fiji between 5 and 3 million years ago (Fig. 2). Volcanic centers were established in northern and southeastern Viti Levu, and volcanic activity created the islands of Vanua Levu, Ovalau, Gau, Beqa, and Kadavu, and on the eastern side of Fiji, the Lau chain. Much of present-day Fiji was created during this period, which coincides with the onset and continuation of extensional tectonic conditions. Many of the volcanic centers are still recognizable, and more than 30 have been identified. The most studied and best known is the Tavua caldera in northern Viti Levu, which is host to the Emperor gold mine at Vatukoula. This and other centers in northern Viti Levu are dominated by the Ba- and Koroimavua-group potassium-rich basaltic rocks with minor andesitic rocks. The oldest rocks in the Tavua caldera are pillow lavas, and the full succession shows a progression from submarine eruptions to subaerial effusions dominated by fluid lava flows with only minor explosive activity. In southeastern Viti Levu, andesitic volcanic rocks of the Medrausucu group form another large caldera. There was considerable explosive activity, and the sequence is dominated by pyroclastic rocks and their sedimentary derivatives. Basaltic and andesitic volcanism also occurred on Gau and Ovalau, on other islands of the Lomaiviti group, and in the Lau chain. In Vanua Levu, much of the island is composed of submarine volcanic rocks of basaltic composition and their sedimentary derivatives, which are assigned to the Natewa group (Fig. 2). In northeastern Vanua Levu, silica-rich dacitic and rhyolitic rocks of the Udu group are predominantly of shallow submarine explosive origin. Submarine basaltic volcanism also formed most of the Yasawa island chain during this period. Many radiometric dates from islands across Fiji testify to an intense and widespread period of volcanism between 7 and 3 million years ago. Sedimentary rocks of this period are largely derived from volcanic source rocks—for example, the Nadi, Navosa, Ra, and Tuva sedimentary groups—and have commonly been deposited in small sedimentary basins created by extensional tectonic forces. Commonly, the basins are floored by conglomerates and breccias derived at a short distance from the underlying volcanic terrain, and these pass upward into sandstones and marls.

INTRAPLATE STAGE (3 MILLION YEARS AGO TO PRESENT)

Around 3 million years ago, there was a significant change in the composition of volcanic rocks. Alkali olivine basalts of ocean-island composition were erupted, and globally such rocks typify volcanic activity in intraplate (e.g., Hawaiian islands), rather than arc tectonic settings. This is consistent with Fiji becoming progressively more divorced from arc-subduction settings during this period (Fig. 1). The activity is principally located at the Seatura caldera in western Vanua Levu (the Bua group) and on the island of Taveuni. More isolated activity occurs elsewhere in Vanua Levu, Koro, and a few islands in the Lau chain. Radiometric dates indicate an age of 3 million years for Seatura caldera, 2–1 million for Koro, and less than 750,000 for Taveuni, with some deposits being as recent as AD 300–500. The only islands that do not fit this pattern are in the Kadavu chain, at the southern end of the Fiji platform. The chain is wholly volcanic but composed of arc-type andesitic rocks ranging in age from 3.5 million to 360,000 years old. It seems as though the Kadavu volcanism may be related to a complex mix of subduction and strike-slip movement along the southern termination of the New Hebrides trench system (Fig. 1).

MINERALIZATION IN FIJI

The association of Fiji for much of its geological history with subduction along the Pacific and Indo-Australian plate boundary raises the potential for the islands to host hydrothermal mineralization typical of subduction zones: porphyry copper–gold deposits, epithermal gold deposits, and polymetallic massive sulfide bodies. In recent decades some of the world's major ore deposits have been exploited within the Melanesian island arcs, and these include gold and copper mines at Lihir, Ok Tedi, Porgera, and Bougainville in Papua New Guinea and the Emperor mine in Fiji.

Mineral prospects are comparatively rare in the early to mature arc stages in Fiji, with minor occurrences of copper–lead–zinc massive sulfide bodies and manganese deposits associated with submarine volcanic sequences. There are also small copper–lead–zinc mineral vein systems and metasomatic replacement (skarn) bodies with similar metals associated with the granitoid Colo intrusions. Around 75% of the mineral prospects in Fiji occur in the arc-rifting stage between 7 and 3 million years ago. Given that hydrothermal mineralization depends on the provision of heat, abundant water, and a good system of fractures to allow fluid transport, it is perhaps not surprising to find a wealth of deposits in the arc-rifting stage, with extensional tectonic forces creating major fracture systems and the Colo plutonism followed by voluminous volcanism testifying to hot crustal conditions. Much of the early volcanism in the arc-rifting stage was submarine, so seawater would have had access to the explosive and fragmental volcanic successions. Intense mineral exploration in recent decades has located numerous prospects, but so far significant exploitation has been restricted to the Emperor mine. The Emperor mine at Vatukoula in northern Viti Levu is a gold–silver–tellurium epithermal body located within the Tavua caldera. The deposit was formed by hydrothermal fluids during the waning stages of volcanic activity. Since mining began in 1933, the mine has produced around 5 million ounces of gold, making it a world-class producer; however, in late 2006, the mine was closed, and its future remains uncertain. The other major prospect, the Namosi porphyry copper–gold deposit, is found in the eroded Namosi caldera in southwest Viti Levu. Copper minerals with very minor gold are disseminated through vein stockworks associated with small quartz-bearing dioritic porphyry intrusions. Reserves are significant, but at present they are not economically viable for mining. Many of the other volcanic centers formed in the arc-rifting stage show evidence of hydrothermal activity, with a widespread occurrence of pyrite and propylitic (calcite–chlorite–epidote) alteration of volcanic rocks. However, exploration thus far has revealed only minor-to-subeconomic base- and precious-metal mineralization in some of the centers, such as the intermittently mined Mt. Kasi deposit in western Vanua Levu and the Sabeto area of western Viti Levu. The one other significant deposit in the arc-rifting stage is the copper–zinc–lead massive sulfide deposit on the Udu Peninsula in northeast Vanua Levu. The deposit occurs in submarine explosive dacite–rhyolite rocks and has many similarities to mineral deposits in the Kuroko district of Japan. The deposit was mined briefly in 1968, and although there has been regular further exploration, thus far minable quantities of ore have not been located.

SEE ALSO THE FOLLOWING ARTICLES

Fiji, Biology / Island Arcs / Plate Tectonics / Tonga / Vanuatu

FURTHER READING

Colley, H., and D. J. Flint. 1995. *Metallic mineral deposits of Fiji*. Memoir 4, Mineral Resources Department, Government of Fiji.

Cronin, S. J., and V. E. Neall. 2001. Holocene volcanic geology, volcanic hazard, and risk on Taveuni, Fiji. *New Zealand Journal of Geology and Geophysics* 44: 417–437.

Hathway, B. 1995. Deposition and diagenesis of Miocene arc-fringing platform and debris-apron carbonates, southwestern Viti Levu, Fiji. *Sedimentary Geology* 94: 187–208.

Ollier, C. D., and J. P. Terry. 1999. Volcanic geomorphology of northern Viti Levu, Fiji. *Australian Journal of Earth Sciences* 46: 515–522.

Pals, D. W., and P. G. Spry. 2003. Telluride mineralogy of the low-sulfidation epithermal Emperor gold deposit, Vatukoula, Fiji. *Mineralogy and Petrology* 79: 285–307.

Pysklywec, R. N., J. X. Mitrovica, and M. Ishii. 2003. Mantle avalanche as a driving force for tectonic reorganization in the Southwest Pacific. *Earth and Planetary Science Letters* 209: 29–38.

Taylor, G. K., J. Gascoyne, and H. Colley. 2000. Rapid rotation of the Fijian islands: palaeomagnetic evidence and tectonic implications. *Journal of Geophysical Research* 105: 5771–5782.

Wharton, M., B. Hathway, and H. Colley. Volcanism associated with extension in an Oligocene-Miocene arc, southwestern Viti Levu, Fiji. *Special Publication Geological Society of London* 81: 95–114.

FISH STOCKS/OVERFISHING

JON BRODZIAK

Pacific Islands Fisheries Science Center, Honolulu, Hawaii

Fish stocks are renewable resources that have been harvested throughout human history. After vast increases in fishery catches during the twentieth century, overfishing is now a worldwide problem. Fishery harvests by island nations currently account for about 20% of worldwide capture production (Fig. 1), with Japan (26%), Indonesia (22%), the Philippines (10%), and Iceland (10%) accounting for about two-thirds of island catches (Fig. 2). Substantial fisheries occur within island ecosystems, which provide forage areas and spawning habitats for diverse species. Island ecosystems are structured by the width of their continental shelf, or lack thereof, and by their relative latitude (tropical, temperate, or boreal) and associated water temperature. Island fish stocks include insular species, which reside within island ecosystems throughout their lifetimes, and highly migratory species that are temporary residents.

OVERFISHED AND OVERFISHING

Like a natural bank account, a fish stock produces a harvestable surplus each year that depends on its intrinsic growth rate and adult biomass, as well as on environmental conditions. A fish stock is overfished when its adult biomass is lower than what is needed for producing fishery yields, similar to a savings account with a low balance.

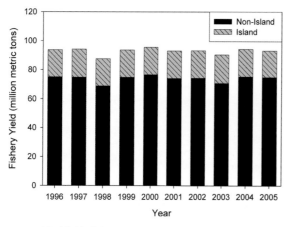

FIGURE 1 Worldwide fish capture production by non-island and island nations during 1996–2005. Source: FAO (2006), Table A2. Capture production by countries or areas.

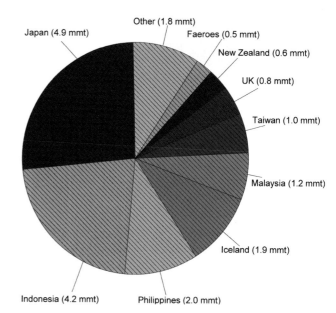

FIGURE 2 Annual average fishery yields by island nations ranked by percentage of total island nation fish capture production during 1996–2005: Japan (26%), Indonesia (22%), Philippines (10%), Iceland (10%), Malaysia (7%), Taiwan (6%), United Kingdom (4%), New Zealand (3%), Faeroes (2%), and other islands (10%).

A stock is said to be experiencing overfishing when the harvest rate exceeds the rate of biomass production, similar to withdrawing more funds than the amount earned through interest. Overfishing can occur because fish are either harvested at too small a size (growth overfishing) or harvested before they have an opportunity to reproduce (recruitment overfishing).

INSULAR FISHES

Insular island fish include a diversity of demersal and pelagic species ranging from groupers to squids to reef sharks to sea urchins. In the tropics, island ecosystems support diverse fish assemblages on coral reefs in nearshore waters. Tropical fish stocks tend to have higher intrinsic growth rates, because of higher energy flux per unit area, and lower carrying capacities, because of limited shelf waters, than do similar species in temperate or boreal ecosystems. Tropical fish are harvested for subsistence and are increasingly being captured for live-fish export markets. As a result, many insular stocks are at risk of being overfished or are experiencing overfishing (Fig. 3). Overfishing of reef fishes is a particularly acute problem. When herbivorous reef fishes are overfished, the abundance of algae increases, which can overgrow and kill corals unless grazed. Marine protected areas where no fishing is allowed can help to conserve fish stocks and coral reefs and have been successful in some island areas (e.g., Palau).

FIGURE 3 Coral reef species at risk of being overfished or experiencing overfishing include the giant humphead wrasse, Cheilinus undulatus (A); the bumphead parrotfish, Bolbometopon muricatum (B); the blacksaddle coral grouper, Plectropomus laevis (C); and the gray reef shark, Carcharhinus amblyrhynchos (D). Photographs by Julia Leung.

Many fish stocks in temperate and boreal island ecosystems have experienced heavier fishing pressure from industrialized fisheries than have tropical stocks. Temperate and boreal island fishes, which often occur within productive continental shelf regions, tend to have lower intrinsic growth rates but higher carrying capacities than tropical stocks. As a result, insular fisheries in temperate and boreal islands often produce more catch. Unfortunately, some of the more abundant resources have been heavily overfished (e.g., Northern cod [*Gadus morhua*] off Newfoundland), whereas other cod stocks are now being harvested at reduced rates to sustain productive fisheries (e.g., Iceland and the United Kingdom).

HIGHLY MIGRATORY FISHES

Highly migratory fishes that transit through island ecosystems include tunas and billfishes as well as salmon. Tropical islands in the Atlantic and Pacific have important tuna fisheries, and currently most of these stocks are fully exploited. The largest tuna species, the highly migratory bluefin tuna (*Thunnus thynnus*), is heavily prized in Japan and other Asian nations. As a result of growing demand, bluefin is overfished in the Atlantic and is in danger of collapse. Although most tuna stocks in the Atlantic and the Pacific are being fully exploited or are experiencing overfishing, only about one-tenth of the tuna catch worldwide comes from artisanal island fisheries, with the vast majority coming from industrialized fleets operating on the high seas outside of the islands' territorial waters. Thus, the problem of overfishing is more complex for highly migratory fishery resources. The long-term sustainability of highly migratory fish stocks will require international cooperation through regional fishery management organizations.

SEE ALSO THE FOLLOWING ARTICLES

Coral / Iceland / Marine Protected Areas / Newfoundland / Palau

FURTHER READING

Allison, G., J. Lubchenco, and M. Carr. 1998. Marine reserves are necessary but not sufficient for marine conservation. *Ecological Applications Supplement* 8: S79–S92.
FAO. 2006. *Yearbooks of Fishery Statistics*. Rome: Food and Agricultural Organization of the United Nations. ftp://ftp.fao.org/fi/stat/summary/default.htm.
Myers, R., J. Hutchings, and N. Barrowman. 1997. Why do fish stocks collapse: the example of cod in Atlantic Canada. *Ecological Applications* 7: 91–106.
Pikitch, E., D. Huppert, and M. Sissenwine, eds. 1997. Global trends in fisheries management. American Fisheries Society Symposium 20. American Fisheries Society, Bethesda, MD.
Sudovoy, Y. 2005. Trouble on the reef: the imperative for managing vulnerable and valuable fisheries. *Fish and Fisheries* 6: 167–185.

FLIGHTLESSNESS

CURTIS EWING

University of California, Berkeley

The evolution of flightless species from ancestors capable of flight as a phenomenon linked to islands was formally proposed by Charles Darwin in *On the Origin of Species* for beetles from the island of Madeira. Subsequently, it has been shown that a statistically significant correlation exists between flightlessness and island habitats among certain birds, although no significant correlation has been proven among insects.

EVOLUTION OF FLIGHTLESSNESS

In an 1855 letter to Joseph Hooker, Darwin proposed an evolutionary mechanism to explain the observation that 36% of the beetle species in the Madeira Islands lacked or possessed reduced wings, saying "powers of flight would be injurious to insects inhabiting a confined locality and expose them to be blown to the sea; to test this, I find that the insects inhabiting the Dezerta Grande, a quite small islet, would be still more exposed to this danger, and here the proportion of apterous insects is even considerably greater than on Madeira proper."

Birds, insects, and bats are the only organisms capable of sustained powered flight and, as such, are the only groups with the potential to evolve flight loss. Numerous species of birds and insects have evolved flightlessness, but there are no examples of flightless bats, either extant or in the fossil record. Determining whether flightlessness has evolved more frequently on islands as compared to continental areas requires knowledge of (1) the number of species in each area, (2) phylogenetic relationships among species and higher taxonomic groups, (3) behavior, and (4) a hypothesis of the ancestral condition. For extant birds, this information is largely available, although among the much more numerous insect species, relatively little data is available for analysis. Possible mechanisms producing flightless species in addition to Darwin's hypothesis are (1) escape from predators, (2) altitude effects (the mountain effect), (3) minimal environmental variability through time, (4) geophily, and (5) patchy ecological zones forming "islands within islands." None of these factors would likely select against flight directly as Darwin's hypothesis requires; rather, they are conditions under which the ability to fly loses much of its value, such that relaxed selection can occur. All of these factors often coincide on islands, which may produce an increased rate in the evolution of flightlessness and allow flightless lineages to survive and proliferate. Many islands have relatively constant temperatures because of the buffering effect of the ocean, and this effect is more pronounced as latitude decreases. Prevailing trade winds and orographic rain combine to produce narrow and persistent ecological zones that can be surrounded by extreme topography, forming islands within islands, which may be small enough for a within-island variant of Darwin's hypothesis to apply. Many of the more isolated islands have few large predators, and dominant arthropod predators such as ants and other social Hymenoptera are absent as well. Tropical rain and cloud forests often have thick, actively decaying litter layers that are rich in arthropod numbers and species. Lastly, nearly any comparison between insect groups native to insular island habitats will be confounded by elevational effects. The majority of extant insect species on insular islands are found at high elevations because of disturbance at low elevations, and low islands are likely both to be small and to have little, if any, undisturbed habitat.

Flightlessness among birds has evolved primarily on volcanic oceanic islands, although Australia and islands derived from Gondwana (including Madagascar, New Caledonia, New Zealand, Papua New Guinea, and other smaller islands) have also evolved high numbers of flightless bird lineages.

FLIGHTLESS BIRDS

The loss of flight is clearly correlated with islands among birds, and the shift to flightlessness often is in conjunction with other morphological and behavioral innovations. Many oceanic islands have few, if any, native mammalian predators, a situation that allows already flightless birds to survive and diversify and allows for the evolution of flightlessness from volant ancestors. Flightlessness has evolved in 25 bird families, with a total of at least 60 transitions proposed for extant and fossil species (Table 1). The vast majority of flightless birds are found in the Southern Hemisphere, reflecting the distribution of mammalian predators and ratites, and the large number of islands in the South Pacific. The relationships among the basic groups and the timing of their divergence help explain the evolution and distribution of flightless birds. Flightlessness has evolved once among the Paleognathae (ratites and tinamous) and more than 60 times among the Neognathae (virtually all other extant species), more than 30 times among rails, and more than 30 times among other groups.

The evolution of flightlessness in birds is considered a classic case of heterochrony. Reduction or loss of the keel of the breast bone; reduction of wing bones; soft, non-aerodynamic feathers; and enlargement of the legs characterize many flightless species. Many flightless species are considered classic examples of neoteny, a specific type of heterochrony in which juvenile characteristics are retained by adults. The feathers of flightless species often lack barbules (which hold flight feathers in a plane) and have a soft, downy appearance.

Flightlessness evolved once in the common ancestor of all ratites on a continent, but islands provided a refuge, allowing species to survive and differentiate after the rise of large predacious mammals. The ancestral continental species achieved release from predation through their large size and strong legs, which function in escape and as weapons. Australia, New Zealand, and New Guinea are isolated and have historically been free of large terrestrial mammals, and the majority of ratite species are known from these areas. The kiwi of New Zealand, which lives in burrows, is the smallest ratite at 1–4 kg, and it evolved from larger ancestors (emus and cassowaries) in the absence of predators.

Flightlessness among the Neognathae has evolved at least 60 times in 18 families (15 extant, three extinct). Only

TABLE 1
Flightless Birds

Locality	Family	Name
Caribbean		
Bermuda	Anatidae	Bermuda Islands flightless duck (*Anas pachyscelus*)[n]
Bermuda	Gruidae	Flightless crane (*Grus latipes*)[n]
Cuba	Gruidae	Cuban flightless crane (*Grus cubensis*)[n]
Cuba	Rallidae	Zapata rail (*Cyanolimnas cerverai*)[y]
Bahamas	Rallidae	*Rallus* undet. sp.[n]
Barbados	Rallidae	Barbados rail (Rallidae gen. et sp. indet.)[n]
Bermuda	Rallidae	Lesser Bermuda crake (*Porzana piercei*)[n]
Bermuda	Rallidae	Lesser Bermuda rail (*Rallus ibycus*)[n]
Cuba	Rallidae	Cuban cave rail (*Nesotrochis picapicensis*)[n]
Haiti	Rallidae	Haitian cave rail (*Nesotrochis steganinos*)[n]
Puerto Rico/Virgin Is.	Rallidae	Antillean cave rail (*Nesotrochis debooyi*)[n]
Cuba	Strigidae	Cuban giant owl (*Ornimegalonyx oteroi*)[n]
Jamaica	Threskiornithidae	Club-winged ibis (*Xenicibis xympithecus*)[n]
Fernandina/Isabela	Phalacrocoracidae	Flightless cormorant (*Phalacrocorax harrisi*)[n]
Hawaii		
Kauai	Anatidae	Turtle-jawed moa-nalo (*Chelychelynechen quassus*)[n]
Maui	Anatidae	Maui moa-nalo (*Ptaiochen pau*)[n]
Maui Nui	Anatidae	Maui Nui moa-nalo (*Thambetochen chauliodous*)[n]
Oahu	Anatidae	Oahu moa-nalo (*Thambetochen xanion*)[n]
Various islands	Anatidae	*Branta*, 4–11 spp. (see Olson and James, 1991)[n]
Big Island	Rallidae	Hawaiian spotted crake (*Porzana sandwichensis*)[n]
Big Island	Rallidae	Small Hawaii rail (*Porzana* sp.)[n]
Big Island	Rallidae	Large Hawaii rail (*Porzana* sp.)[n]
Kauai	Rallidae	Medium Kauai rail (*Porzana* sp.)[n]
Kauai	Rallidae	Large Kauai rail (*Porzana* sp.)[n]
Laysan	Rallidae	Laysan crake (*Porzana palmeri*)[n]
Maui	Rallidae	Smallest Maui rail (*Porzana keplerorum*)[n]
Maui	Rallidae	Medium Maui rail (*Porzana* sp.)[n]
Maui	Rallidae	Large Maui rail (*Porzana severnsi*)[n]
Molokai	Rallidae	Very small Molokai rail (*Porzana menehune*)[n]
Oahu	Rallidae	Small Oahu rail (*Porzana ziegleri*)[n]
Oahu	Rallidae	Medium-large Oahu rail (*Porzana ralphorum*)[n]
Maui	Threskiornithidae	Maui flightless ibis (*Apteribis brevis*)[n]
Molokai	Threskiornithidae	Molokai flightless ibis (*Apteribis glenos*)[n]
Indian Ocean		
Amsterdam Island	Anatidae	Amsterdam Island flightless duck (*Anas marecula*)[n]
Mauritius	Columbidae	Dodo (*Raphus cucullatus*)[n]
Rodrigues	Columbidae	Rodrigues solitaire (*Pezophaps solitaria*)[n]
Mauritius	Psittacidae	Broad-billed parrot (*Lophopsittacus mauritianus*)[n]
Mauritius	Rallidae	Red rail (*Aphanapteryx bonasia*)[n]
Rodrigues	Rallidae	Rodrigues rail (*Aphanapteryx leguati*)[n]
Réunion	Threskiornithidae	Réunion sacred ibis (*Threskiornis solitarius*)[n]
Oceania		
New Zealand	Acanthisittidae	Lyall's wren (*Traversia lyalli*)[n]
New Zealand	Acanthisittidae	Stout-legged wren (*Pachyplichas yaldwyni*)[n]
New Zealand	Acanthisittidae	Long-billed wren (*Dendroscansor decurvirostris*)[n]
New Zealand	Aegothelidae	Owlet nightjar (*Aegotheles novaezelandiae*)[n]
New Zealand	Anatidae	Campbell Island teal (*Anas nesiotis*)[y]
New Zealand	Anatidae	Auckland Islands teal (*Anas aucklandica*)[y]
Rota, Mariana Is.	Anatidae	Undescribed sp.[n]
New Zealand	Anatidae	South Island goose (*Cnemiornis calcitrans*)[n]
New Zealand	Anatidae	North Island goose (*Cnemiornis gracilis*)[n]
New Zealand	Apterygidae	Kiwi, 4 spp. (see Worthy and Holdaway, 2002)[y,p]

(*continued*)

TABLE 1
(*Continued*)

Locality	Family	Name
New Zealand	Aptornithidae	North Island adzebill (*Aptornis otidiformis*)[n]
New Zealand	Aptornithidae	South Island adzebill (*Aptornis defossor*)[n]
Viti Levu, Fiji	Columbidae	Viti Levu giant pigeon (*Natunaornis gigoura*)[n]
New Zealand	Dinornithidae	Moa, 11 spp. (see Worthy and Holdaway, 2002)[n, p]
New Caledonia	Megapodiidae	Giant megapode (*Sylviornis neocaledoniae*)[n]
Viti Levu, Fiji	Megapodiidae	Viti Levu scrubfowl (*Megapodius amissus*)[n]
Viti Levu, Fiji	Megapodiidae	Noble megapode (*Megavitiornis altirostris*)[n]
New Zealand	Psittacidae	Kakapo (*Strigops habroptila*)[y]
Buton/Sulawesi	Rallidae	Snoring rail (*Aramidopsis plateni*)[y]
Tasmania	Rallidae	Tasmanian native hen (*Gallinula mortierii*)[y]
Philippines	Rallidae	Calayan rail (*Gallirallus calayanensis*)[y]
New Zealand	Rallidae	South Island takahē (*Porphyrio hochstetteri*)[y]
New Zealand	Rallidae	Weka (*Gallirallus australis*)[y]
New Zealand	Rallidae	13 additional spp. (see Steadman, 2006)[n]
Numerous islands	Rallidae	Various genera, 30–200+ spp. (see Steadman, 2006; Curnutt and Pimm, 2001)[y, n]
New Caledonia	Rhynochetidae	Kagu (*Rhynochetos jubatus*)[y]
Atlantic Ocean		
Falkland Islands	Anatidae	Flightless steamer duck (*Tachyeres brachypterus*)[y]
Tenerife, Canary Is.	Emberizidae	Long-legged bunting (*Emberiza alcoveri*)[n]
Gough Island	Rallidae	Gough Island moorhen (*Gallinula nesiotis comeri*)[y]
Inaccessible Island	Rallidae	Inaccessible Island rail (*Atlantisia rogersi*)[y]
Ascension Island	Rallidae	Ascension flightless crake (*Mundia elpenor*)[n]
St. Helena	Rallidae	St. Helena swamphen (*Aphanocrex podarces*)[n]
St. Helena	Rallidae	St. Helena crake (*Porzana astrictocarpus*)[n]
Tristan de Cunha	Rallidae	Tristan moorhen (*Gallinula nesiotis*)[n]
St. Helena	Upupidae	St. Helena giant hoopoe (*Upupa antaios*)[n]

NOTE: y = extant, n = extinct, p = primitively flightless (e.g., flightless condition not an island phenomenon).

four of the 60 transitions are associated with continents. The remaining 56 transitions are generally among birds that tend to feed on the ground, primarily the rails. There are only two known instances where flightlessness has evolved among the order Passeriformes (perching birds): the long-legged bunting of Tenerife, Canary Islands, and three flightless wrens native to New Zealand. In most cases, the loss of flight evolved in situ on the island from a volant ancestor.

Many flightless bird species historically endemic to islands have been driven extinct through direct human and other mammalian (rat, cat, dog, and ferret) predation. In the Pacific, the first wave of extinctions occurred during the major expansion of Polynesian cultures between 1500 BC and AD 1000. The second is ongoing and began with European voyages in the seventeenth century. David Steadman estimated that between 442 and 1576 rail species were driven extinct during the past 3500 years in the South Pacific, with the majority being flightless. Others have produced an estimate of approximately 200 species of rails (Fig. 1).

It is not surprising that the majority of flightless birds associated with islands are known from the Pacific. The majority of the world's islands are found there, and their geologic history and geographic isolation largely exclude large predators, providing the greatest opportunity to evolve flightlessness. Flightless birds are also known from islands in the Caribbean, Atlantic Ocean, and Indian Ocean.

FIGURE 1 The Tasmanian native hen (*Gallinula mortierii*) is one of the only flightless rail species not considered endangered. Photograph by Heather Wallace (http://www.flickr.com/photos/heatherw/).

New Zealand

Historically, New Zealand was home to the greatest diversity of flightless birds. The islands of New Zealand are primarily continental in origin, and they separated from Australia and Antarctica approximately 85 million years ago. The New Zealand biota lacked both eutherian and marsupial mammalian predators, and this release led to the evolution of many unique forms.

The extinct moa were large (25 kg) to gigantic (250 kg) ratites unique to New Zealand. As many as 64 species and 20 genera have been historically recognized, with recent reanalysis of fossils reducing the number to 11 species in six genera. Archeological evidence suggests the moa were driven to extinction by AD 1400, approximately 100 years after Polynesian colonization. This "blitzkrieg" extinction pattern, where large numbers of edible terrestrial species were driven quickly to extinction soon after human colonization, was played out on nearly all inhabited Polynesian islands over the past 3500 years.

Of all ratites, the evolution of the kiwi is most closely tied to the island phenomenon (Fig. 2). Regardless of the origin of flightlessness among kiwi, the small size and behavior of kiwi is a direct result of their diversifying in the absence of mammalian predators, which now threaten all of the four surviving species. They are unique among the ratites for their small size, nocturnal behavior, narrow beak with apical nostrils adapted for feeding primarily on insects and other invertebrates, and burrowing behavior. Primary current threats are predation of juveniles (and to a lesser extent adults and eggs) by stoats, cats, and other introduced predators.

The other flightless birds of New Zealand include the kakapo (*Strigops habroptila*) (Fig. 3), the world's heaviest parrot and one of only two flightless parrots known; the flightless wrens, placed in their own endemic lineage (Oligomyodi: Acanthisittidae); the extinct adzebills, primitive flightless Gruiformes remarkable for their large size (estimated at almost a meter tall and over 9 kg in weight) and robust, adze-shaped bill; and the teals from Campbell Island (*Anas nesiotis*) and the Auckland Islands (*A. aucklandica*), relatively small nocturnal ducks from subantarctic islands south of New Zealand that feed primarily on terrestrial and marine invertebrates. New Zealand and its surrounding islands were also home to 13 flightless rails derived from six or more independent volant colonists.

FIGURE 3 The kakapo (*Strigops habroptila*) is native to New Zealand and is the only extant flightless parrot. One other putatively flightless parrot, the broad-billed parrot of Mauritius, was driven to extinction in the 1600s.

Hawaiian Islands

The Hawaiian Islands were also home to a diverse assemblage of flightless birds, which, in contrast to the continental origins of much of the New Zealand avifauna, was entirely descendent from volant ancestors. The extreme isolation and ephemeral nature of each volcanic island in the Hawaiian archipelago renders colonization by flightless birds unlikely in the extreme. Evolving in the absence of predation and competition from other terrestrial vertebrates, the flightless birds of Hawaii no doubt had a strong impact on the ecosystem before the arrival of humans. Subfossils of flightless birds have been collected and described representing four species of giant dabbling ducks, two to three species of ibis, seven or eight species of geese/shelducks, and 11 or 12 species of rails. Interestingly, molecular data suggest that each group is the result of multiple introductions and convergent evolution to flightlessness. Many of the subfossils are too fragmentary to definitively determine species limits and could represent as many as 22 more species, some of which are assumed to represent undescribed species of flightless rails and possibly additional ducks and geese. All but one of the flightless birds endemic to Hawaii were driven to extinction

FIGURE 2 The great spotted kiwi (*Apteryx haastii*) is native to the South Island of New Zealand and is the largest species of kiwi.

after the arrival of Polynesian colonists, probably through a combination of habitat destruction, predation by introduced rats, dogs, and pigs, and over-harvesting for food. Only the Hawaiian rail, *Porzana sandwichensis*, remained when James Cook visited the islands.

Indian Ocean Islands

Arguably the most famous of all extinct and flightless birds, the dodo (*Raphus cucullatus*) was a giant (10–20 kg) dove native to Mauritius in the Mascarene archipelago (Fig. 4). The dodo had short wings and a large beak, and was covered with gray and white feathers with soft barbules that superficially resembled down, and a tuft of soft white tail feathers. It was discovered in the first decade of the sixteenth century and was extinct by the late seventeenth century. Individuals were extremely naïve and had

FIGURE 4 The dodo (*Raphus cucullatus*) was a large, flightless pigeon endemic to the island of Mauritius. This species retained many juvenile characteristics into adulthood.

no fear of humans or introduced predators, so they were quickly overharvested for meat and were preyed upon by introduced dogs, cat, rats, and pigs.

The Rodrigues solitaire (*Pezophaps solitaria*) was the sister species of the dodo, named for its solitary nesting habits. As with the dodo, it was a relatively heavy-bodied bird with very short wings, but it was apparently less specialized than the dodo, having more typical plumage, beak, and tail morphology. It was last recorded in the early to mid-eighteenth century and succumbed to the same pressures as the dodo.

Other flightless birds of the Mascarenes include the Réunion sacred ibis (*Threskiornis solitarius*), the broad-billed parrot (*Lophopsittacus mauritianus*), and two species of flightless rails—the red rail (*Aphanapteryx bonasia*) of Mauritius and the Rodrigues rail (*Aphanapteryx leguati*). All were driven to extinction by the mid-eighteenth century.

Caribbean Flightless Birds

The Caribbean islands have relatively few flightless species, probably because of their close proximity to the American continents. In contrast to Oceania, colonization would have been relatively easy, and immigrants would have been less isolated from continental source populations. All but one of the flightless Caribbean birds was extinct at the time of Western contact, with all 13 species known only from fossil remains. One of the most interesting was the Cuban giant owl, the largest known owl species, standing over a meter tall and weighing approximately 10 kg, with highly reduced wings and very long and powerful legs.

Atlantic Islands

Compared to the South Pacific, the Atlantic Ocean has few isolated islands. Flightless bird species are known from five islands in the Southern Hemisphere: the Ascension Islands had a flightless crake (*Mundia elpenor*), driven to extinction in the late seventeenth century; Tristan da Cunha had the Tristan moorhen (*Gallinula nesiotis*), extinct since the late nineteenth century; and St. Helena three species, the St. Helena swamphen (*Aphanocrex podarces*) and crake (*Porzana astrictocarpus*)—both extinct since the early sixteenth century—and the St. Helena giant hoopoe (*Upupa antaios*) (Fig. 5).

There are two extant flightless rails in the South Atlantic, the Inaccessible Island rail (*Atlantisia rogersi*), which is the world's smallest extant flightless bird, and the nearly flightless Gough Island moorhen (*Gallinula nesiotis comeri*). The Inaccessible and Gough Islands are world heritage sites and have remained predator free, allowing these species to survive to the present. The Falkland

FIGURE 5 The giant hoopoe (*Upupa antaios*) was native to St. Helena Island and is known only from fossil remains. It is speculated that it fed on, among other things, the (possibly extinct) giant flightless St. Helena earwig (*Labidura herculeana*).

Islands are home to a flightless steamer duck (*Tachyeres brachypterus*), which has two close relatives endemic to the shores of South America. Lastly, on Tenerife in the Canary Islands, fossils were discovered of a flightless passerine, the long-legged bunting (*Emberiza alcoveri*).

South Pacific Islands

The South Pacific islands combine all of the geologic features ideally suited to the evolution of flightlessness: They are isolated, numerous, and predator free. The islands are relatively ephemeral, so flightlessness needs to evolve much faster than on continental islands such as New Zealand. The rails appear to be preadapted for the rapid evolution of flightlessness, and in the South Pacific, flightless rails were present in the greatest numbers. Steadman estimates that each island could have had one to three species, and reached an estimate of over 1000 total rail species on these islands. Most of the species described from fossils have been flightless, so it is assumed that most of the estimated species would be as well. New species of flightless rails continue to be described from South Pacific islands, and if estimates of extinct rail taxa are correct, many more new species descriptions can be expected in the future. Although the geology and geography of the South Pacific islands has resulted in the evolution of numerous flightless rails, only a few scattered flightless species are known from other groups.

FLIGHTLESS INSECTS

Flightlessness among insects is correlated with altitude; latitude; cold temperatures; and persistent environments such as late successional forests, leaf litter, and caves, and it does not seem to be correlated with areas with persistent environmental factors such as moisture and temperature. No correlation exists between the number of flightless species and islands per se, when comparing continental and island faunas of similar latitude, although windy and persistent habitat islands, such as high-altitude ridges, mountaintops, and exposed alpine bogs, may select for flightlessness in concordance with Darwin's hypothesis mentioned above. However, to address this question, we must determine whether flightlessness evolves at a greater rate on islands than on continents; this is a much more difficult question to answer.

The island of Madeira (the inspiration for Darwin's hypothesis), as well as subantarctic islands, are known to have a disproportionally high number of flightless insect species. 200 or more of the 550 beetle species present on Madeira were flightless, although the number of loss events this represents remains unknown. On subantarctic islands south of New Zealand, nearly all beetle species and approximately three-quarters of all Lepidoptera have reduced wings, but this has been considered a latitudinal effect, rather than an island effect. Antarctica also has a high percentage of flightless insect species, with only one volant species known. It can be argued that Antarctica has low biodiversity and reduced predator–prey interactions and that it is island-like in ways important to the evolution of flightlessness.

The Hawaiian Islands provide a large number of the environmental and biotic factors linked with the evolution of flightlessness. There are over 100 different biotypes present, with many being small, topographically isolated, and environmentally homogeneous over time. There are also large and relatively homogeneous areas of wet and mesic forest present on the slopes of all the high islands and alpine areas on Maui and Hawaii Island. The fauna is disharmonic (a large number of species in a few higher taxa), with some lineages being numerically and ecologically diverse. The native insects evolved in the absence of ants and aggressive social wasps such as *Vespula*, providing release from predation. The top predators would have been birds, spiders, and insect predators such as carabid beetles and nabid bugs. A number of insect groups in Hawaii have been studied extensively, and the approximate number of species is known, phylogenetic hypotheses exist for some lineages, and the flight ability has been evaluated for a few of these. Several insect lineages have a relatively large number of flightless species or estimated transitions to flightlessness. The loss of flight among these lineages appears to be correlated with their small home ranges and high altitude (Fig. 6).

New Zealand also is home to a perceived concentration of flightless insect species, although evaluation of the number of transitions to flightlessness is currently not

FIGURE 6 *Gonioryctus monticola* (Coleoptera: Nitidulidae) from Oahu (Hawaiian Islands), which is associated with decaying tree fern fronds (*Cibotium* sp.). Flightlessness has evolved at least twice within this endemic Hawaiian genus. Photograph by the author.

possible. Nearly all Orthoptera endemic to New Zealand are flightless, as are many beetles, moths, stoneflies, wasps, and earwigs. Flightlessness among the New Zealand insect fauna appears to be an ancestral trait in most cases, with only a few endemic radiations exhibiting a mix of volant and flightless species.

The most striking flightless species in New Zealand are the weta (Fig. 7), crickets placed in the family Anostostomatidae (not to be confused with cave weta, flightless crickets endemic to New Zealand in the family Rhaphidophoridae). All of the New Zealand endemic weta are primitively wingless, and the flight capability of the common ancestors remains obscure.

FIGURE 7 Weta from New Zealand. (A) A male Wellington tree weta (*Hemideina crassidens*). Note the elongated head and enlarged mandibles characteristic of the males of this species. (B) The bluff weta (*Deinacrida elegans*) is associated with cracks and fissures on nearly vertical rocky bluffs. Photographs by Rod Morris.

A survey of the Orthopteroidea of the Galápagos Islands identified 24 flightless species in 15 lineages. Nine of the lineages are primitively flightless, and six are believed to have lost the ability to fly on the islands. The beetle fauna has a similarly high proportion of flightless species, with nine genera of primitively flightless species and 14 genera that evolved flightlessness after colonization. Flightlessness among the Galápagos beetle fauna is correlated with the arid habitats that predominate there.

CONCLUSIONS

Although no statistically significant correlation has been demonstrated between insects and islands per se, in certain insect families, all or many of the known flightless species have evolved flightlessness in an island setting. A number of biotic and abiotic factors are correlated with flightlessness, whereas the risk of "being blown out to sea," as proposed by Darwin, may be important only on very small and exposed islands. A genetic predisposition for flightlessness, coupled with variation upon which selection can operate and with release from predation, is a prerequisite for the evolution of flightlessness. In stable environments, the cost of maintaining flight musculature, wings, and supporting structures can be higher than the benefit derived by dispersal. The rails are a group that clearly shows a tendency toward flightlessness on islands around the world, but flightlessness among insects endemic to islands occurs in different groups on different islands.

SEE ALSO THE FOLLOWING ARTICLES

Dodo / Fossil Birds / Insect Radiations / Madeira / Moa / New Zealand, Biology

FURTHER READING

Curnutt, J., and S. Pimm. 2001. How many bird species in Hawai'i and the central Pacific before first contact?, in *Evolution, ecology, conservation, and management of Hawaiian birds: a vanishing avifauna*. M. J. Scott, S. Conant, and C. van Ripper III, eds. Lawrence, KS: Cooper Ornithological Society, 15–30.

Haddrath, O., and A. J. Baker. 2001. Complete mitochondrial DNA genome sequences of extinct birds: ratite phylogenetics and the vicariance biogeographic hypothesis. *Proceedings of the Royal Society London B* 268: 939–945.

Olson, S. L., and H. F. James. 1991. Description of thirty-two new species of birds from the Hawaiian Islands: part I. Non-passeriformes. Washington, DC: American Ornithologists' Union.

Peck, S. B. 2006. *The beetles of the Galapagos Islands, Ecuador: evolution, ecology, and diversity* (*Insecta: Coleoptera*). Ottawa, Canada: NRC Research Press.

Roff, D. A. 1994. The evolution of flightlessness: is history important? *Evolutionary Ecology* 8: 639–657.

Steadman, D. W. 2006. *Extinction and biogeography of tropical Pacific birds*. Chicago: University of Chicago Press.

Trewick, S. A., and M. Morgan-Richards. 2004. Phylogenetics of New Zealand's tree, giant and tusked weta (Orthoptera: Anostostomatidae): evidence from mitochondrial DNA. *Journal of Orthoptera Research* 13: 185–196.

Worthy, T. H., and R. N. Holdaway. 2002. *The lost world of the moa*. Bloomington: Indiana University Press.

FOSSIL BIRDS

TREVOR H. WORTHY

University of Adelaide, Australia

Fossil birds are extinct avian species for which there is no historical record. Such taxa may be known from archaeological sites, ranging from a few hundreds of years old (e.g., many islands in Oceania) to a few thousands of years old (e.g., sites in the Mediterranean). Most often they derive from fossil deposits accumulated either by predators, or in pitfall traps in caves, or in water-laid deposits such as in lake and fluvial deposits.

PATTERNS OF DIVERSITY ON ISLANDS

Birds are found on most islands of the world, and if there is terrestrial vegetation, then landbirds may also be present. Seabirds often nest on islands, forming very large colonies, but it is landbirds that this article primarily addresses. The diversity of landbirds on islands is related to island size and distance from source faunas on continents, with small, remote islands having the fewest taxa. Islands are renowned for having produced highly differentiated endemic taxa with no continental equivalents, of which the dodo is archetypical. Flightless taxa occur far more frequently on islands than on continents, and island taxa tend to evolve extremes of size and can be either very much bigger (e.g., rails) or smaller (e.g., teal ducks) than related mainland taxa. By the mid-twentieth century most islands of the world had been explored and their extant avifaunas documented. Many islands or archipelagos had endemic taxa, but generally diversity was perceived as relatively low and derived often from the same few groups (e.g., rails, pigeons).

However, almost from the outset of the great period of scientific exploration in the early nineteenth century and through the period of paradigmatic change as the theory of evolution gained acceptance, fossils from islands revealed a hidden and unexpected diversity of life. Perhaps strangest of all was the radiation of giant birds termed moa (Aves: Dinornithiformes) found in New Zealand (see below). But similar discoveries on Madagascar and on the Mascarenes challenged biologists' understanding of the world and, in particular, raised the specter of extinction as a process in faunal evolution.

DEFINING A FOSSIL

Fossils are the preserved remains of an animal or its traces. For fossil birds this usually means fossil bones, but rare fossil feathers and footprints are known. Fossil bones may be a few hundred to many millions of years old. Some researchers use the term "subfossils" to delimit younger or unmineralized bones, but because bones can be mineralized in just a few thousand years or remain unmineralized for many millions of years, the term has little meaning in terms of paleontology and evolution. It is more informative to use age qualifiers; for example, Holocene fossils are those of the last 10,000 years. Most fossils referred to hereafter are Holocene fossils, but a few are Late Pleistocene in age (100,000–10,000 years ago). Younger fossils, particularly if from cold environments, may contain organic material (DNA, etc.), providing an increasingly important resource for evolutionary studies.

FACTORS PREDICTING THE PRESENCE OF FOSSIL BIRDS ON ISLANDS

Fossils reveal that there are virtually no islands where the impact of recent extinction, in all cases mediated by humans, has not significantly reduced the diversity of the original avifaunas. In very many instances more than half of original avifaunas have been lost. Most such extinct taxa were land and freshwater species. Tropical and subtropical islands throughout the world have been most affected. In these islands, today the extant avifauna is often but a remnant of the original and is not in evolutionary equilibrium, and species exist now separated from coevolved taxa. Often, adaptations and behaviors in surviving species make sense only in the context of the original biota.

Most island fossil birds date from within the Quaternary; that is, they are less than 2 million years old. It is the nature of islands themselves that limits the age of fossil deposits to a great extent. There are very few islands or archipelagoes known to be older than the Miocene, or greater than 25 million years old. Notable among these are Madagascar, New Caledonia, and New Zealand (discussed further below). In each case, these are continental fragments, separated by tectonic forces millions of years ago from larger continental blocks, and their biota has an original vicariant component. In all cases, the original faunas have been much modified by dispersal and extinction throughout time to the present. Among truly oceanic islands, among the oldest is Vitilevu in Fiji, which has supported land perhaps since the Middle Miocene, 15 million years ago. But most islands (e.g., many hundreds in the Pacific Ocean) are very much younger, with duration of emergent land measured in just hundreds of thousands to a few millions of years. Most of these are either volcanic in origin or formed as coral atolls. The age of islands constrains fossil faunas in two ways. Firstly, the age of an island obviously limits the potential age of fossil faunas, so only the continental fragments noted above are likely to have fossil faunas aged in the millions of years. Secondly, the age of the faunas themselves is roughly correlated with their uniqueness, so islands millions of years old are more likely to have had unique taxa evolve on them. However, it is important to note that fossil faunas only a few thousands of years old are important on any island, because it is these that reveal the taxa that have recently become extinct. As reiterated throughout this contribution, those that went extinct were more likely to be the unique, formerly endemic, taxa to those islands.

Fossils have revealed many endemic taxa on islands, so what characteristics of islands facilitate evolution of such taxa? Firstly, the island needs to have been isolated sufficiently from other land masses to prevent regular genetic exchange. This fact is especially relevant in relation to recent sea level changes correlated with glacial–interglacial cycles. Both the most recent glacial maximum about 26–14 thousand years ago and the penultimate glacial period saw sea levels drop perhaps 120 m globally. Thus islands separated from continental blocks or other islands by water less than 120 m deep were connected as recently as 14,000 years ago. Species-level evolution of birds takes longer than this, so avifaunas on such islands are not distinct from those on nearby mainlands. New Zealand, an archipelago of over 800 islands greater than 1 ha in size, actually comprises two main groups (New Zealand mainland and associated islets, and the Chathams group), with the Kermadecs to the north, and a few subantarctic isolates to the south. These are the evolutionary stages—not New Zealand, as in the geopolitical entity. Evolution of the avifauna on the New Zealand mainland had parallels on the Chathams: the two honeyeaters on the mainland had two sister taxa on the Chathams, as did the large parrot the kaka, pigeons, and waterfowl (e.g., the shelducks). Thus, diversity within an archipelago is often by allopatric speciation on its component islands. In conclusion, the potential for fossil faunas to reveal greater diversity in the avifauna of an archipelago is related to the isolation of the constituent islands at low sea levels.

Adequate time is a prerequisite for fossil bird species to have evolved. Insular genera and species can evolve in birds in a surprisingly short period of time. Henderson Island, in the Pitcairn Group in the far eastern Pacific, one of the most remote in Oceania and small (37 km^2), is only about 400,000 years old, yet it has/had among columbid pigeons an endemic species in each of *Gallicolumba, Ducula, Ptilinopus,* and an endemic genus. The longer the island or archipelago has provided a terrestrial habitat, the more unique its fauna is likely to be. New Zealand, separated from nearby continents for 60–80 million years, is old and has one of the most distinct island avifaunas.

Complexity of habitats, related to size and topography of islands, and the availability of freshwater all restrict potential evolution. Small, poorly vegetated islands in the subantarctic can support only a limited terrestrial avifauna. In contrast, sea birds use islands solely to breed on, so complexity of terrestrial ecosystems is a poor predictor of their diversity. Amsterdam Island in southern Indian Ocean is a small windswept island clothed in grasses and perhaps a few shrubs. One nonmarine species is known from there: the fossil teal *Anas marecula,* but at least 20 seabirds did, and most continue to, breed on this island.

FOSSILS THAT REVEAL AVIAN EVOLUTION ON THE MOST REMOTE ISLANDS

Nineteenth-century naturalists were amazed to find endemic landbirds on the most remote of islands. But modern scientists have been amazed to learn via fossil discoveries that the extant endemic inhabitants of remote islands are but a few of those living just a few thousand years ago. The size of the island and its isolation seem to have been no barrier to speciation by birds. Ascension Island, in the mid-Atlantic Ocean, is small and very remote, yet fossils have revealed it had an endemic heron *Nycticorax olsoni* and a rail *Mundia elpenor,* both now extinct.

The Pacific Ocean provides numerous examples of remote islands, but unfortunately fossil avifaunas are as yet known from comparatively few of them. Nevertheless, where there is a fossil record, the most remote islands in Oceania have extinct non-marine species, as with Henderson Island (see above) and Easter Island. The comprehensive treatment of fossil birds of the Pacific by David Steadman (2006) is a landmark work and essential reading for anyone interested in Pacific avifaunas. This book summarizes Steadman's research on island fossil avifaunas over the last 30 years. The outstanding sobering fact arising from this work is that on every island with an adequate fossil record, at least 50% of the original avifauna extant before humans arrived is now extinct. Thus extant diversity drastically underestimates diversity in the faunal evolutionary unit; that is, the basis of the theory of island biogeography wherein diversity might be predicted was found to be seriously flawed. Islands once thought to have depauperate avifaunas were shown to have had considerably richer ones. In many cases, unique endemics in the extant faunas were shown to be relics of far wider distributions.

Analyses of fossils reiterate that just a few groups of birds are responsible for colonizing and evolving taxa on many remote islands. Rails, particularly of the genus *Gallirallus,* have spread eastwards across the Pacific and have colonized most islands with the notable exception of the Hawaiian chain. Having colonized an island, they had a major tendency to evolve into flightless species. Steadman estimates that perhaps up to 2000 flightless species existed at the onset of human occupation of the Pacific. Other prominent groups are the crakes in *Porzana,* which also evolved numerous distinct taxa. Across the equatorial Pacific, a diversity of pigeons (notably in *Ducula, Gallicolumba, Caloenas, Didunculus, Ptilinopus*) and parrots (*Cyanoramphus* derivatives, and lories, e.g., *Vini*), spread and evolved distinct taxa. The further one gets from the source faunas, the more depauperate the resultant faunas are. So in the Pacific, where the spread has generally been from the west, generic and specific diversity is reduced towards the east. Hawaii is unusual among the islands of Polynesia in that its fauna derives from America to the east. The Mascarenes, in the Indian Ocean is far from the Pacific, but there too, pigeons, rails, and parrots dominate the endemic fossil taxa.

EVOLUTION OF AVIFAUNAS ON ISLANDS WITH TERRESTRIAL MAMMALS

Globally, there are a few islands where endemic insular faunas include terrestrial mammals. These include some in the Mediterranean, Madagascar, and some West Indian islands. Fossil birds have been found on all, but as detailed below, they tend to lack the diversity found on mammal-free islands. Nevertheless, unique and interesting radiations occur.

Mediterranean Islands

All avifaunas on Mediterranean islands evolved in the presence of terrestrial mammals, including both vegetarians and predators. Fossil birds, in revealing that there have been few recent changes to avian diversity in the region, show the limiting nature of terrestrial mammals on avifaunal evolution. The western Balearic Islands (Mallorca, Menorca, Ibiza, Formentera) have an insular history since the Miocene, and are famed for the unique caprine *Myotragus* (a goat relative) and endemic leporines (rabbits). Novel birds, however, are restricted to a single flightless rail *Rallus eivissensis* and an undescribed goose (*Anser* sp.) endemic to Ibiza and Formentera. Islands in more eastern parts of the Mediterranean often are millions of years old, (e.g., Corsica, Crete, Malta, or Sardinia) and had endemic mammals including elephantids, hippopotami, deer, and canids. Insular novelties in their avifaunas are few, and restricted mainly to raptors. Several islands had endemic owls, such as Sardinia (*Bubo insularis*), Corsica and Sardinia (*Athene angelis*), Crete (*A. cretensis*), and Malta (*Tyto melitensis*). Corsica and Sardinia had the eagle *Aquila nipaloides*. These raptors coevolved with the insular radiations of rodents, leporines, and larger mammals. Malta and Sicily are exceptional in that fossil birds included large, perhaps flightless, swans *Cygnus falconeri* and *C. equitum,* a crane *Grus melitensis*, and a pigeon *Columba melitensis*.

West Indies, Cuba

The West Indies are paleontologically most famous for their endemic radiations of mammals, particularly giant sloths, insectivores, and rodents. In Cuba, at least 11 genera of mammal were present in the Late Pleistocene, of which at least 82% are now extinct, and most were rodents, with one species up to 200 kg in weight. As in the Eastern Mediterranean, a unique radiation of mammals allowed a unique avifauna to evolve. For example, Cuba has a most unusual assemblage of fossil birds in which predators dominate. There were eight owls (in *Pulsatrix, Ornimegalonyx, Gymnoglaux, Bubo, Tyto*), two falcons, three accipitrids including the giant hawk *Gigantohierax suarezi*, the condor-like *Teratornis olsoni*, a vulture, a stork, and a nightjar; few other fossil species have been described. No doubt, it is the radiation of rodents that enabled the coevolution of this suite of avian predators.

In other West Indian islands that also had endemic rodent radiations, the less well developed fossil avifaunas also notably include predators; for example, Hispaniola had an endemic species each of an accipitrid, an owl, and a falcon.

Madagascar

Madagascar is well known for its spectacular radiation of lemurs, with many extinct taxa, including all of the largest species. However, nearly synonymous with Madagascar, when thinking of fossil birds, are the elephant birds. They are ratites, related most closely to ostriches and assumed to have been stranded on Madagascar when it separated from the supercontinent Gondwana. The largest of these *Aepyornis maximus* is often cited as the largest bird that ever lived, at over 400 kg weight. This is probably an overestimate, with a mass of 200–300 kg likely for larger individuals, but nevertheless this was a giant bird. Only the largest of the mihirung birds *Dromornis stirtoni* of Australia was heavier, and the giant moas, *Dinornis* species, of New Zealand, taller. The eggs of *A. maximus* are the largest known. They were robust and often laid in sand dunes, and many examples are known. Two have even floated across the Indian Ocean, washed up on Australian beaches, been incorporated in Holocene dunes, and found whole!

Up to eight species of elephant bird have been listed in the past. But when consideration is taken that size differences can be explained by sexual dimorphism and temporal variation, qualitative differences support only three distinct species. There are two *Aepyornis* species: *Aepyornis maximus* (includes *A. medius* as its smaller sex) and *A. hildebrandti* (includes *A. gracilis*). The third species is the more gracile *Mullerornis agilis* (includes *M. betsilei* and *M. rudis,* named on unusually small specimens, and *M. grandis* a larger Pleistocene variant). As here defined, the size range of these three elephant birds is less than that of some of the moa species, where much larger samples and well-dated sequences are available.

Other than elephant birds, Madagascar has relatively few other fossil birds. Prominent in fossil collections are bones of the Malagasy shelduck *Alopochen sirabensis* and Malagasy sheldgoose *Centrornis majori*. Other than anatids and elephant birds, most notable are the giant ground cuckoos in the genus *Coua*, with three extinct taxa, a single fossil rail, a plover, and a ground roller. A single fossil eagle *Stephanoaetus mahery* completes the list. This relatively sparse list of fossil birds evolved in the presence of a diverse mammalian fauna, with hippopotami, many kinds of lemurs, including largely terrestrial ones, and predators such as the fossa (*Cryptoprocta ferox*) and its larger extinct relative, the giant fossa (*C. spelea*). This suite of terrestrial mammals probably did not allow the evolution of a diverse array of truly terrestrial birds as seen elsewhere, such as New Zealand.

EVOLUTION OF AVIFAUNAS ON ISLANDS WITH NO TERRESTRIAL MAMMALS

On islands where there were no mammals, birds were free to evolve forms to fill all niches that would otherwise have been occupied by mammals. Particularly, they evolved numerous ground-dwelling forms, many actually or facultatively flightless. And if several distinct islands were available in an archipelago, a single colonizing ancestor could generate a number of distinct species. Obviously such taxa are highly vulnerable to predation by mammals (humans, rats, cats, etc.), so the impact of human arrival in their islands has been catastrophic, resulting in thousands of extinctions. Thus, islands or archipelagoes developed the most distinctive avifaunas when there was significant isolation, no mammals were present, several millions years of time was available for evolution, and there was habitat complexity and/or multiple islands. New Zealand and Hawaii are the best examples of archipelagoes where all these factors exist, and accordingly birds radiated in unique ways and dominated the biota. But the far-flung individual islands of Vitilevu, New Caledonia, and the Mascarenes demonstrate the diversity of species that may evolve given similar conditions.

The Mascarenes are covered elsewhere in this volume, but these islands are special, as Europeans discovered them and recorded the biota in the fifteenth century, making the dodo famous in the process. As a result, the fossil birds of the Mascarenes mostly have a historic record, albeit a sketchy one. Best known were the giant flightless pigeons (the dodo and the solitaire), but flightless species of ibis and unique rails, waterfowl, parrots, and owls were also present.

New Zealand

New Zealand is located some 1400 km east of Australia and is the emergent part of a continental fragment "Zealandia" stretching from New Caledonia in the subtropics to Macquarie Island in the subantarctic, and Chatham Islands 800 km east of South Island. Separation of Zealandia from the Australian part of Gondwana occurred 80–60 million years ago, as the Tasman Sea opened like a zipper from the south. At separation, New Zealand carried a biota, as revealed by fossils, including dinosaurs, and some members of the Recent biota have evolved from such ancestors. Many kinds of invertebrates, both land and freshwater, have relatives on Gondwanan fragments and are most unlikely dispersers. Some plants, such as kauri (*Agathis*), have ancient genetic separation from nearest Australian relatives and have occupied Zealandia since its separation. Similarly, the New Zealand vertebrates such as *Sphenodon* the tuatara, leiopelmatid frogs, and a small flightless terrestrial mammal of Miocene age are best explained as derived from ancestors vicariantly separated on New Zealand. Among birds, perhaps only the largest birds (the moas) and the smallest (the New Zealand wrens, Acanthisittidae) have a Gondwanan origin. The rest are the result of dispersal events that occurred at scattered intervals through the Tertiary, mainly from Australia.

New Zealand has an excellent fossil record of penguins extending from the Paleocene (61 million years ago) to the present, but the only Tertiary record of terrestrial birds is from the Early Miocene (19–16 million years ago) St. Bathans Fauna of Otago. Now about 30 species are known, with waterfowl dominating (8 species), including moa ancestors, rails (2 spp.), a large gruiform, parrots (3 spp.), waders (several spp.), a gull, an eagle, a swiftlet, an owlet nightjar, and several passerines. This fauna is extremely important in that it provides a valuable insight into the evolution of the Recent or Quaternary fauna. But it is the Holocene fauna, by virtue of its age, that is the comparable entity to other island fossil faunas.

The Holocene terrestrial and freshwater fauna of New Zealand lacked mammals other than three bats. It also lacked snakes of any kind, and lizards were restricted to geckoes and skinks, so other than the two aforementioned ancient relict taxa—the tuatara and leiopelmatid frogs—New Zealand was truly a land of birds. However, it is relatively small, about 270,000 km^2, with the main islands spanning about 13 degrees of latitude (34–47° S), so it was affected markedly by the Pleistocene glacial periods. Undoubtedly, the cool times of the glaciations reduced ecological diversity and thus restricted avian diversity.

Fossils reveal that the fauna that greeted Polynesian discoverers at first encounter, in the mid-thirteenth century (just 750 years ago) was one of high generic diversity but low species diversity within genera. But what it lacked in numbers, this fauna made up for in uniqueness.

New Zealand has an extraordinarily rich late Pleistocene and Holocene fossil record derived from deposits in caves, swamps and former lake beds, and sand dunes. The prolific record results from the absence of mammalian scavengers and the recentness of human impact on the fauna. The prehuman period is within the last millennium, in contrast to most places in the world. All known Quaternary species have a Holocene record, and most appear in Polynesian middens; thus the original avifauna at human arrival is known better for New Zealand than for any other large island on Earth.

In a recent analysis of the greater New Zealand archipelago (Kermadecs, New Zealand and nearshore islands, Chathams, and the subantarctic Antipodes, Bounty, Snares, Campbell, and Auckland islands), some 214 original (at first human contact) breeding birds were listed in 110 genera and 46 families, of which 113 (87%) were endemic species to the area. One hundred thirty species (61%) were land and freshwater birds. About 56 species, or 26% of the original fauna, are now extinct, though extinction most affected the land and freshwater component (40%): extinct taxa from the core archipelago and Chatham islands are listed in Table 1. Of the total, 132 species bred on the core of the archipelago (North and South Islands and associated nearshore islands), 62% of which were endemic to the core area. However, both North and South Islands each had a few endemic species of their own. Presently about 50% of the original breeding avifauna of each of the North and South Islands is locally or globally extinct there.

Chatham Islands lie some 800 km east of South Island of New Zealand. Although part of continental Zealandia, the islands were submerged during the Pliocene and came emergent only about one million years ago. The avifauna is therefore derived entirely by long distance over water dispersal, and is no older than this. As is typical of its biota generally, the Chatham avifauna lacks all of the special endemics of New Zealand, such as the moas, kiwis (Apterygidae), adzebills (Aptornithidae), and the passerines New Zealand wrens (Acanthisittidae), wattle-birds (Callaeidae), and piopio (Turnagridae) (Table 1). But it did have endemic flightless rails.

The most unique aspect of the New Zealand avifauna is its ratites. The kiwis (*Apteryx,* five species) and moas (six genera, 10 species) are iconic New Zealand taxa (Fig. 1). The history of taxonomic endeavor with regard to the

FIGURE 1 Images of the moa from New Zealand. (A) Skull of the South Island giant moa *Dinornis robustus;* (B) skeleton of North Island giant moa *D. novaezealandiae;* (C) skull of *Pachyornis elephantopus*, heavy-footed moa; (D) *Euryapteryx gravis,* stout-legged moa. Images by Rod Morris.

moas, spanning over 150 years, is fascinating. Diversity was originally overinterpreted, and up to 38 species were recognized in one list: no less than 64 taxa have been erected in the group. Moas are the only bird in the world for which no skeletal evidence of a wing remains. Ranging from 15 to 250 kg in mass, these were the browsers of New Zealand. The next most speciose group was the acanthi-sittids—now recognized as the sister group to all other passerines. Just two species survive, but formerly seven species in five genera existed. Four of these were flightless, and a fifth was a very weak flier. Only one other flightless passerine is known in the world: the finch *Emberiza alcoveri* from mammal-free Tenerife in the Canary Islands.

The next most speciose group of fossil birds are waterfowl, with eight species (in seven genera) lost from the core archipelago. Most have extant sister taxa in Australia. *Cnemiornis* (Fig. 2), a giant flightless goose equivalent to

TABLE 1
Extinct Birds of Holocene Age from New Zealand and the Chatham Islands

Family	Extinct Species of Core NZ	Common Names	Extinct Species of the Chathams
Apterygidae	*Apteryx* undescribed species	Kiwi	
Dinornithidae	*Dinornis novaezealandiae* (NI)	NI giant moa	
	Dinornis robustus (SI)	SI giant moa	
Emeidae	*Anomalopteryx didiformis*	Little bush moa	
	Emeus crassus (SI)	Eastern moa	
	Euryapteryx curtus (NI)	Coastal moa	
	Euryapteryx gravis	Stout-legged moa	
	Megalapteryx didinus (SI)	Upland moa	
	Pachyornis geranoides (NI)	Mappin's moa	
	Pachyornis elephantopus (SI)	Heavy-footed moa	
	Pachyornis australis (SI)	Crested moa	
Anatidae	*Cygnus atratus*[a]	Black swan	*Cygnus atratus*
	Cnemiornis gracilis (NI)	NI goose	
	Cnemiornis calcitrans (SI)	SI goose	
	Chenonetta finschi	Finsch's duck	
	Malacorhynchus scarletti	Scarlett's pink-eared duck	*Malacorhynchus scarletti*
	Mergus australis[b]	NZ merganser	*Mergus australis*
	Oxyura vantetsi	NZ blue-billed duck	
	Biziura delautouri	NZ musk duck	
		shelduck	*Tadorna* undescribed species
		Chatham duck	*Pachyanas chathamica*
Spheniscidae		Crested penguin	*Eudyptes* undescribed species
Procellariidae	*Puffinus spelaeus*	petrels	*Pterodroma* undescribed species
Phasianidae	*Coturnix novaezealandiae*†	NZ quail	
Aptornithidae	*Aptornis otidiformis* (NI)	NI adzebill	
	Aptornis defossor (SI)	SI adzebill	
Rallidae	*Capellirallus karamu* (NI)	Snipe rail/Chatham rail	*Cabalus modestus*[b]
	Fulica prisca	NZ/Chatham coots	*Fulica chathamensis*
	Gallinula hodgenorum	Hodgens' rail	
	Porphyrio mantelli (NI)	NI takahe	
		Hawkins' rail	*Diaphorapteryx hawkinsi*[b]
		Dieffenbach's rail	*Gallirallus dieffenbachii*
Ardeidae	*Ixobrychus novaezealandiae*[b]	NZ little bittern	*Ixobrychus novaezealandiae*[b]
Accipitridae	*Circus teauteensis*	Eyles's harrier	
	Harpagornis moorei (SI)	Haast's eagle	
Scolopacidae	*Coenocorypha barrierensis* (NI)[b]	NI/Chatham Is snipe	*Coenocorypha chathamica*
	Coenocorypha iredalei (SI)[b]	SI snipe	
Psittacidae			*Nestor* undescribed species
Strigidae	*Sceloglaux albifacies*[b]	Laughing owl	
Aegothelidae	*Aegotheles novaezealandiae*	NZ owlet nightjar	
Acanthisittidae	*Dendroscansor decurvirostris*	Long-billed wren	
	Pachyplichas jagmi (NI)	NI stout-legged wren	
	Pachyplichas yaldwyni (SI)	SI stout-legged wren	
	Traversia lyalli[b]	Lyall's wren	
	Xenicus longipes[b]	Bush wren	
Sylviidae		Chatham Is fernbird	*Bowdleria rufescens*[b]
Meliphagidae		Chatham Is bellbird	*Anthornis melanocephala*[b]
Callaeidae	*Callaeas cinerea* (SI)[b]	SI kokako	
	Heteralocha acutirostris (NI)[b]	Huia	
Turnagridae	*Turnagra tanagra* (NI)[b]	NI piopio	
	Turnagra capensis (SI)[b]	SI piopio	
Corvidae	*Corvus antipodum pycrafti* (NI)	NI/Chatham crow	*Corvus moriorum*
	Corvus antipodum antipodum (SI)	SI crow	

NOTE: Birds are those known from the core archipelago of New Zealand (North and South Islands and near-shore islets) and the Chatham Islands. Taxa endemic to North Island are marked (NI) and those endemic to South Island (SI).

[a]*Cygnus atratus* is an extant species with its principal range in Australia. Its fossils are widespread in New Zealand and Chatham Islands, but it was extinct in both places at European arrival. European introductions in the 1860s saw its eventual re-establishment throughout its former range.

[b]Extinct species that are known historically from the listed region: most were restricted to small areas of their former range at European arrival.

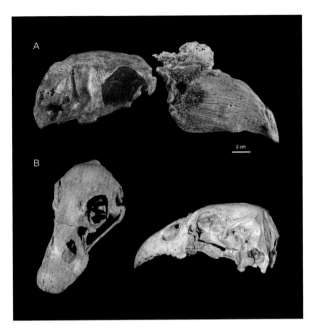

FIGURE 2 (A) Image of a cranium and premaxilla of the New Caledonian *Sylviornis neocaledoniae*. Note the bizarre bony protrusion of the premaxilla, indicating a casque extended back over the cranium. Photograph by Cécile Mourer-Chauviré. (B) Left: Skull of the New Zealand South Island goose *Cnemiornis calcitrans*. Photograph by Rod Morris. (B) Right: Skull of the Hawaiian toothed moa-nalo *Thambetochen chauliodous*. Photograph by Helen James. Scale bar = 2 cm.

the Hawaiian moa-nalo in size, is the sister species to the volant Cape Barren goose.

Gruiforms (Aptornithidae and Rallidae) are the last significant group of fossil birds in the New Zealand region. Aptornithidae contains two highly derived species, both large and flightless, and of uncertain relationships, although probably related to rails. All the fossil rallids (except the coots (*Fulica* spp.) were flightless. Only two flightless species of the original rail fauna survive: the South Island takahe (*Porphyrio hochstetteri*) is critically endangered, and the weka (*Gallirallus australis*) has populations in serious decline.

Compared to West Indian or Mediterranean faunas, New Zealand had few avian predators: a single eagle, one hawk, one falcon (extant), and two owls. The eagle was, however, the largest in the world and was capable of killing large moa with its 25-cm claw span. It clearly coevolved with moas and hunted them in more open areas of the subalpine region and shrub forest mosaics of lowlands in the South Island.

New Caledonia

New Caledonia, specifically Grande Terre, is a large geologically complex island at the northern extremity of Zealandia. Like New Zealand, it is generally considered to have a biota in part derived from vicariant separation from Gondwana perhaps 60 million years ago. Highly endemic faunas and floras attest to great longevity of the biota. Several relationships with the New Zealand biota (e.g., hoplodactyline geckos, bulimulid molluscs *Placostylus*) suggest direct links in the past. New Caledonia lacked frogs, but it has a most speciose radiation of geckoes and skinks. It has some unique birds (e.g., the kagu), but until the 1970s no fossil terrestrial vertebrates were known. As humans only arrived 3200 years ago, some extinct taxa were predicted, and through the 1970s and 1980s a strange assemblage was uncovered. A meiolanid turtle and a small terrestrial crocodile were found to be associated with a meter-high flightless megapode *Sylviornis*. This species had a strange bony casque developed off its deep narrow beak that extended up and over its skull (Fig. 2). So strange is its skull that *Sylviornis* has recently been placed in its own family. New Caledonia has also lost other, more typical megapodes, including *Megapodius molistructor;* a giant gallinule, *Porphyrio kukwiedei,* similar to the takahe of New Zealand; a larger kagu, *Rhynochetos orarius;* a *Coenocorypha* snipe; and at least two ground-frequenting pigeons, *Caloenas canacorum* and *Gallicolumba longitarsus*. Fossil avian predators are also known, with two *Accipiter* species (*A. efficax, A. quartus*) slightly different from extant taxa, and notably a large owl *Tyto? letocarti*.

Fiji

On the western margin of Melanesia, the Fiji Islands comprise numerous islands in two main clusters. In the west, Vitilevu, Vanualevu, Taveuni, and Kadavu are old islands with volcanic cores and, in the case of Vitilevu, known to have emergent land back to the middle Miocene. The smaller surrounding islands, especially in the Lau Group to the east, are much smaller and generally are coral atolls, though a few are volcanic; all are geologically very much younger than the main islands.

Vitilevu has recently been found to have a fossil fauna with a megafaunal component. Like New Zealand and New Caledonia, it too had no land mammals, but like New Caledonia it had a terrestrial crocodilian (*Volia athollandersoni*), a meiolaniid tortoise, and uniquely, a giant iguana (*Lapitiguana impensa*). It lacked ratites and giant waterfowl, but it had two amazing fossil birds: *Megavitiornis altirostris,* a giant flightless megapode just slightly smaller than *Sylviornis* from New Caledonia, and *Natunaornis gigoura,* a giant flightless pigeon rivaling the dodo in size. Associated with these two giant birds, was a more typical, though flightless, megapode *Megapodius amissus;* a long-billed flightless rail, *Vitirallus watlingi;* an undescribed gallinule; a snipe (*Coenocorypha miratropica*) belonging to the New Zealand–centered radiation of forest snipe, now known to have a species in New Caledonia as well; a large fruit pigeon,

Ducula lakeba; and an undescribed teal, *Anas* sp. To these fossil species one further species is consigned to history, with the historic extinction of *Gallirallus poecilopterus*. Younger and smaller islands in the Lau Group have fewer fossil birds: some undescribed *Gallirallus*-derived rails; another megapode, *M. alimentum;* and the pigeon *Ducula lakeba*.

Hawaii

When Europeans discovered Hawaii, it had a remarkable avifauna. The nene, *Branta sandvicensis,* was one of the endemic taxa encountered, but little was it realized that this goose was the last survivor of a spectacular radiation of gooselike waterfowl. *Branta,* it turns out, had spawned three fossil taxa in addition to the nene. Most amazing, however, is a group of astounding large, flightless "waterfowl," the moa-nalos, derived from dabbling duck ancestors (Table 2, Fig. 2). On Hawaii the browsing/grazing ratites of other islands were replaced by waterfowl.

The Hawaiian chain is a linear set of islands that increase in age northwest away from the active volcanic and youngest island, Hawaii. It is thus a perfect staging ground for insular radiation of species with more time for divergence on the oldest islands. Thus the oldest island with a fossil record, Kauai, has the most modified and more kinds of waterfowl than the other islands. *Chelychelynechen* was the most nonducklike duck ever, and with a bill like its namesake (turtles) was clearly a grazer, as were most of the other giant "ducks."

The archipelago had other spectacular radiations. Each of the main islands had its own distinct flightless ibis and a pair of crakes foraging the forest floors. In the absence of *Gallirallus*-type rails, the crakes evolved a large and a small species on each island. Most spectacular of all, however, were the radiations of mohoid "honeyeaters" and drepanidine fringillid finches. Of the finches alone, about 30 fossil species have been identified, with 18 named in 11 genera. The demise of this whole fossil avifauna is related to human arrival with commensal *Rattus exulans* about 800–1000 years ago. Historically, another 11 taxa have gone extinct, and many are perilously close to the same end. Predictably for an island lacking terrestrial mammal predators, there was also a radiation of avian predators. A Hawaiian genus of large owls had evolved a distinct species on each main island except Hawaii, and these were major accumulators of the bones of the finches and honeyeaters and other extinct birds. Also present were an endemic harrier and an eagle inseparable from the bald eagle *Haliaeetus leucocephalus* or white-tailed eagle *H. albicilla,* doubtless supported by the population of large flightless anatids.

SEE ALSO THE FOLLOWING ARTICLES

Bird Radiations / Dodo / Extinction / Flightlessness / Moa

FURTHER READING

James, H. F., and S. L. Olson. 1991. Descriptions of thirty-two new species of birds from the Hawaiian Islands. Part 2. Passeriformes. *Ornithological Monographs* 46: 1–88.
Olson, S. L., and H. F. James. 1991. Descriptions of thirty-two new species of birds from the Hawaiian Islands. Part 1. Non-Passeriformes. *Ornithological Monographs* 45: 1–88.
Steadman, D. W. 2006. *Extinction and biogeography of tropical Pacific birds*. Chicago and London: University of Chicago Press.
Worthy, T. H., and R. N. Holdaway. 2002. *The lost world of the moa: prehistoric life of New Zealand*. Bloomington: Indiana University Press.

TABLE 2
Hawaii's Spectacular Radiation of Fossil Waterfowl (Anatidae)

Subfamily	Taxon	Kauai	Oahu	Molokai	Maui	Hawaii
Anserinae	*Branta hylobadistes*	—	—	—	F	—
	Branta sp., aff B. *hylobadistes*	F	F	—	—	—
	Geochen rhuax	—	—	—	—	F
	Anserine sp., very large	—	—	—	—	F
Tadorninae	Shelduck, ? *Tadorna* sp.	F	—	—	—	—
Anatinae	Duck, ? *Anas* sp.	F	—	—	—	—
	Chelychelynechen quassus	F	—	—	—	—
	Thambetochen chauliodous	—	—	F	F	—
	Thambetochen xanion	—	F	—	—	—
	"Supernumerary Oahu goose"	—	F	—	—	—
	Ptaiochen pau	—	—	—	F	—

FOUNDER EFFECTS

MICHAEL C. WHITLOCK

University of British Columbia, Vancouver, Canada

Because new populations may be founded by a small number of individuals, by chance the new population may be genetically different from its parent population. A population founded by a small group of individuals is likely to have somewhat less genetic variance and different mean phenotypic values from its source population, which may cause the new population to evolve in different ways. Collectively, these changes in the genetic and phenotypic properties of the population are called a founder effect. If the differences between the founding and source populations are great enough, the founding event potentially contributes to the evolution of reproductive isolation, although evidence for this is rare.

GENETIC CHANGES IN FOUNDING POPULATIONS

When an island is colonized by a new species, the founding group is likely to be limited to a small number of individuals. As a result, the new population may, by chance, be different from the source population from which the individuals came. In particular, these founding individuals carry a small sample of the genetic diversity of the larger, source population, and by random sampling alone the new population may differ from the old in the number of alleles at each genetic locus and the frequency of those alleles. If the source population was genetically variable for important traits, the newly founded population may—as a result of this genetic drift during founding—have different average values for important traits. These changes by genetic drift in the means and genetic variances of newly founded populations are called founder effects. (The properties of the population may also change as a result of selection in response to a different environment on the island; these changes, while important, are not traditionally viewed as part of the founder effect.)

Founder effects can cause at least three types of population genetic change. First, some alleles, particularly rare alleles, may not be present in the small population at all. As a result, the genetic diversity of the new population is likely to be lower than its source population. Second, random changes in allele frequency on average result in lower levels of genetic variance, meaning that the population will on average respond more slowly to natural selection. (This is only an average effect, expected to be a reduction in genetic variance proportional to one over twice the founding population size. Genetic variance for some traits will by chance increase in founding populations.) Finally, phenotypic traits will on average change somewhat in newly founded populations. Genetic drift does not directly change the mean values of traits *averaged across multiple founding events,* but for any given founding event the mean value of a given trait is likely to change somewhat as a result of random changes in the allele frequencies of genes affecting the traits.

All of these effects are more pronounced if the size of the founding population is smaller, drift being proportional to the inverse of the population size. None of these factors is likely to be quantitatively important except in a very small founding population, say, of two to twenty individuals. The effects become more pronounced in populations that are unable to grow to a large population size quickly, because if the small population persists, enhanced genetic drift will continue. Effects of drift are lessened, however, if dispersal from the source population to the island continues, because this migration tends to reduce the genetic differentiation between the populations.

Periods of small population size, such as during population founding, are disproportionately important in determining the effects of genetic drift on a population, for two reasons. First, the amount of genetic drift is approximately proportional to the harmonic mean of population size over time, and the harmonic mean is strongly affected by small values. Second, weak natural selection can dominate allele frequency change in large populations, but drift can overwhelm selection in a small population.

FOUNDER EFFECT SPECIATION

The potential importance of founder effects is that they may set the stage for a new evolutionary trajectory. If the mean or variance of the population shifts by genetic drift during founding, then a population may evolve in a different direction in the island compared to the source population, solely because of these initial genetic changes. If certain genetic combinations work better together than others, then shifting the frequencies of some alleles can change the nature of selection on other loci. If such a cascade of changes occurred, then the resulting population may have genotypes that are incapable of breeding well with the ancestral genotype. Some evolutionary biologists have argued that such changes can result in more rapid evolution of reproductive isolation from the source population and as a result be a major contributor to speciation. This idea, called founder effect speciation, originated with Ernst Mayr (1963) and has been developed by Hampton Carson and Alan Templeton, among others.

There have been many attempts to mimic the conditions for founder effect speciation in laboratory experiments. Although some experimental populations develop a small level of reproductive isolation, the effect is rare and of limited magnitude. The majority of evolutionary biologists believe that the environmental differences between island populations and their progenitors are more likely to be important in developing reproductive isolation than are founder effects.

SEE ALSO THE FOLLOWING ARTICLES

Inbreeding / Population Genetics, Island Models in

FURTHER READING

Coyne, J. A., and H. A. Orr. 2004. *Speciation*. Sunderland, MA: Sinauer.
Mayr, E. 1963. *Animal species and evolution*. Boston: Belknap.
Templeton, A. R. 1981. Mechanisms of speciation: a population genetics approach. *Annual Review of Ecology and Systematics* 12: 23–48.
Whitlock, M. C., and K. Fowler. 1999. The changes in genetic and environmental variance with inbreeding in *Drosophila melanogaster*. *Genetics* 152: 345–353.

FRAGMENTATION

LUIS CAYUELA

University of Alcalá, Madrid, Spain

Fragmentation is a process that occurs when originally extensive and continuous habitats are broken into smaller areas and separated by other habitat or land use types that disrupt the continuity of the original habitat. At the landscape level, this process generates patches of a certain habitat type that somewhat resemble islands embedded within a matrix of distinct habitats. Fragmentation of natural habitats can influence the entire suite of ecological processes, from individual behavior through population dynamics to ecosystem fluxes. Although particular attributes of the environment determine the existence of natural fragments, most recent habitat fragmentation at the landscape level is human-caused.

THE FRAGMENTATION PROCESS

In terrestrial ecosystems, fragmentation typically begins with gap formation or perforation of the vegetative matrix as humans colonize a landscape or begin extracting resources there (Fig. 1, Step 1). For a while, the matrix remains as natural vegetation, and species composition and abundance patterns may be little affected (Fig. 1, Step 2). But as the gaps get larger and more numerous, they eventually become the matrix, and the connectivity of the original vegetation is disrupted (Fig. 1, Step 3).

The process of habitat fragmentation has two intrinsic components: (1) reduction of the area of the original habitat; and (2) change of the spatial configuration of what remains (henceforth referred to as "fragments," "patches," or "remnants"). The latter can be described through the

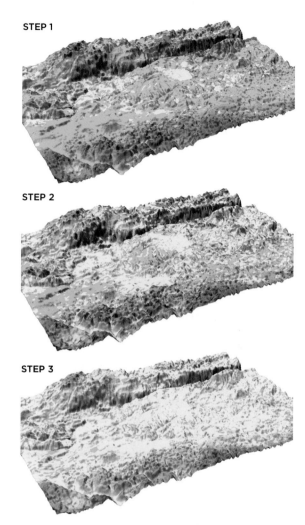

FIGURE 1 Illustration of how forest fragmentation proceeds in a real case study in the Highlands of Chiapas, southern Mexico, during the period 1975–2000. The fragmentation process typically begins with gap formation in the vegetative matrix (in green, Step 1). For a while, the matrix remains as natural vegetation despite the continuing progress of fragmentation (Step 2). As the gaps get larger and more numerous, they become the matrix, and the connectivity of the original vegetation is disrupted (Step 3).

pattern of fragmentation (e.g., size, shape, number, and distribution of habitat patches). Two landscapes can have the same amount of habitat but two completely different patterns of fragmentation. A considerable body of literature exists on how to describe the extent and pattern of habitat fragmentation. However, a situational definition should include some measure of the pattern of fragmentation to place it in context.

Fragmentation is also related to the process of insularization (formation of islands that are isolated from each other and, occasionally, from the mainland). This has brought about the application of island biogeography theory to the study of habitat fragmentation. However, important differences exist between these two processes, in both a temporal and spatial context; thus, caution must be taken when relying on these analogies to investigate the biological consequences of fragmentation.

BIOLOGICAL CONSEQUENCES OF FRAGMENTATION

Any land-use change can potentially result in habitat fragmentation. The most immediate effect is the elimination of species that occurred solely in the portions of the landscape that are destroyed. The habitat that remains is then broken into remnants that are isolated to varying degrees. The time since isolation, the distance between adjacent remnants, and the degree of connectivity between the remnants are all important determinants of the biotic response to fragmentation at any scale.

Fragmentation can affect plant and animal populations at several levels. At the landscape level, fragmentation causes once-continuous populations to break up into smaller subpopulations occupying the remaining habitat patches. Scientists believe that these subpopulations may act as metapopulations (collections of small populations occupying a number of habitat patches). Individuals occasionally move among patches, and populations can go extinct in individual patches as a result of chance events. However, the vacant patches may eventually be colonized and occupied again in the future. If colonization rates of vacant patches are higher than extinction rates, the metapopulation will persist. As fragmentation of the landscape proceeds, patches that are far from other patches will not exchange individuals with the other patches, and the small population remaining in the patch will eventually go extinct. As patches become more and more isolated from each other, the colonization rates will decline to the point that extinction rates exceed colonization rates, and the whole metapopulation will go extinct. Fragmentation thresholds may indeed reflect the point (quantity of habitat loss) where extinction overrides colonization, and species richness collapses at the community-landscape level.

At the fragment level, several factors affect a fragment's value as plant and wildlife habitat. In general, larger fragments are likely to support more species. This principle is supported by theoretical species–area relationships drawn from the field of island biogeography. Individual fragments are also affected by their surroundings. At the forest edge, wind and sunlight result in drier conditions than are found in the interior of the forest patch. Forest edges are also more accessible to predators and parasites that may occur in adjacent fields or developed areas. For example, house cats, which kill small birds, are often more common in forest edges adjacent to residential developments. Cowbirds, which are nest parasites, are also more common in forests adjacent to the open fields where they feed. Cowbirds lay their eggs in the nests of other birds. The host birds will care for the cowbird eggs. When the eggs hatch, the larger cowbird nestlings will outcompete the host nestlings for food and may even push the host nestlings out of the nest. Some plant and animal species ("interior species") are not able to tolerate the drier conditions or the predators and parasites that occur at the habitat edge. These species occur only in the core habitat of remnant patches.

Although most models predict negative effects of habitat fragmentation on biodiversity, empirical evidence to date suggests that the effects of fragmentation can sometimes be positive. For instance, some species do show positive edge effects. For a given amount of habitat, more fragmented landscapes contain more edge; therefore, positive edge effects could be responsible for positive effects of fragmentation on abundance or distribution of some species. Habitat fragmentation can also increase habitat complementation for species that require different kinds of habitats in different stages of their life cycles (e.g., some insects and amphibians), which has a positive effect on biodiversity. In addition, fragmentation, and subsequent habitat isolation, can be a trigger for diversification for many taxa. In the Hawaiian Islands, for example, genetic differences have been documented between habitat patches in forests naturally fragmented by lava (commonly known as kīpuka). This may allow sufficient isolation to initiate diversification.

THE THEORY OF ISLAND BIOGEOGRAPHY

The current concept of habitat fragmentation emerged from the theory of island biogeography. The two predictor variables in this theory are island size and island isolation, or distance of the island from the mainland. The theory of island biogeography essentially states that, as the size of an island increases, so does the number of species it contains. Some of the most plausible explanations to describe this phenomenon are that (1) as area increases, so does the diversity of physical habitats and resources, which in turn supports a larger number of species; (2) as area increases, the size of populations increases, thus reducing the probability of extinction; and (3) for a group of islands, or archipelago, populations within the islands are assumed to work as a metapopulation system. These subpopulations experience an equilibrium between extinction and colonization of species from other islands. Larger islands are less subjected to extinction because they can hold larger populations of their species, which in turn result in a positive imbalance between these two processes. An additional explanation is that only larger islands are likely to contain enough habitat to support species such as large mammals. Consequently, such species can become extinct as the area is reduced.

ISSUES OF SCALE AND THE SPATIAL CONCEPT OF FRAGMENTATION

The parallelism between habitat fragments and islands can be, however, deceptive. When the theory of island biogeography was conceptually extended from island archipelagoes to terrestrial systems of habitat patches, the concept of isolation changed; isolation was then seen as the result of habitat loss, which is enough to explain by itself an important loss of biodiversity.

In temporal terms, there are also important differences. Because island formation is generally a natural geological process, islands may need hundreds and thousands of years to experience the insularization and area effects upon their biota, whereas habitat fragments have generally been created in a much shorter timeframe and have not always had enough time to experience such effects. The temporal scale of investigation may therefore have a strong influence on the results of a study, with short-term crowding effects eventually giving way to long-term extinction debts.

Different organisms and ecological systems also experience the degree of fragmentation of a particular environment in variable, and even contradictory, ways. Species within the same landscape can respond to fragmentation in four distinct ways. First, a species can thrive in the matrix of human land uses; a number of weedy species worldwide fit this description. Second, a species can maintain viable populations within individual habitat fragments, but not in the matrix; this is an option only for species with small home ranges or otherwise modest requirements, such as many plants and invertebrates. Third, a species can be highly mobile and disperse through the matrix; this is the case for many birds. Fourth, a species may require a larger amount of habitat but may be not capable of thriving in or dispersing through the matrix, being thus bound for eventual extinction.

Additionally, the spatial concept of fragmentation often implies that habitat remnants are isolated by areas that function as hostile environments to the organisms within the remnants. There are many instances, however, where this concept of habitat distribution is not applicable. In Central America and other tropical regions of the world, for example, traditional slash-and-burn agriculture has led to a matrix partly dominated by secondary vegetation. The variegated nature of such landscapes must be utilized differently by different taxa. At one extreme, there are native species that grow successfully over the whole range of modifications; for these species, potential habitat forms a continuum across the landscape. The other extreme results from intolerance to most forms of interference; these species exist in a truly fragmented landscape, restricted to remnants in better conditions. The majority of species appear to fall somewhere between these two tolerances. For them, the landscape is a constantly shifting mosaic of habitats of varying suitability.

SEE ALSO THE FOLLOWING ARTICLES

Deforestation / Island Biogeography, Theory of / Metapopulations / Species–Area Relationship

FURTHER READING

Fahrig, L. 2003. Effects of habitat fragmentation on biodiversity. *Annual Review of Ecology and Systematics* 34: 487–515.
Forman, R. T. T., and M. Godron. 1986. *Landscape ecology*. New York: Wiley.
Gustafson, E. J. 1998. Quantifying landscape spatial pattern: what is the state of the art? *Ecosystems* 1: 143–156.
Harris, L. D. 1984. *The fragmented forest: island biogeography theory and the preservation of biotic diversity*. Chicago: The University of Chicago Press.
Lord, J. M., and D. A. Norton, D. A. 1990. Scale and the spatial concept of fragmentation. *Conservation Biology* 4: 197–201.
Newton, A. C., ed. 2007. *Biodiversity loss and conservation in fragmented forest landscapes: the forests of montane Mexico and temperate South America*. UK: CABI Publishing.
Noss, R., B. Csuti, and M. J. Groom. 2006. Habitat fragmentation, in *Principles of conservation biology*, 3rd ed. M. J. Groom, G. K. Meffe, and C. R. Carroll, eds.. Sunderland, MA: Sinauer Associates, Inc, 213–252.
Pimm, S. L. 1998. The forest fragment classic. *Nature* 393: 23–24.
Saunders, D. A., R. J. Hobbs, and C. R. Margules. 1991. Biological consequences of ecosystem fragmentation: A review. *Conservation Biology* 5: 18–32.

FRASER ISLAND

BRAD BALUKJIAN

University of California, Berkeley

Fraser Island, the world's largest sand island, is located off the eastern coast of Australia in the state of Queensland and is both a national park and, since 1992, a World

Heritage Area. Seven vegetation zones and a multitude of freshwater lakes, including half of the world's perched lakes, characterize the island's landscape. Although recently threatened by human impact, the island now has a comprehensive management plan that promises to preserve its stunning natural beauty for years to come.

GEOLOGY AND FORMATION

Fraser Island is a land-bridge island that covers 1653 km² of land area and is 123 km long by 25 km wide at its widest point. It is located at 25°12′ S and 153°6′ E along Australia's eastern Pacific coast, 15 km off the mainland (Fig. 1). Despite being composed of sand, the island is not actually a barrier island; rather, it is an erosional remnant shaped by current wave patterns and sediment transport. Until sea-level rise 6000 years ago, Fraser was connected to the mainland and was part of a Pleistocene dune system that formed from continued deposition by southeast trade winds. The oldest sand found on the island is 700,000 years old, and the island's huge sand deposits provide a continuous record of climatic and sea-level changes. Years of erosion have exposed spectacular shades of yellow, red, and brown sand along the island's sea cliffs, the colors resulting from iron-rich minerals staining the sand.

BIOLOGY AND ENVIRONMENT

Flora

Because Fraser's life as an island has been relatively short (it was separated from the mainland only 6000 years ago), its biota is very similar to that of the nearby mainland. Its proximity to the mainland results in gene flow with the biota there and largely precludes the formation of endemic species (but see below). Nonetheless, Fraser is a large island with much diversity. Seven vegetation zones are found there—rain forest, blackbutt forest, scribbly gum/wallum banksia forest, *Melaleuca* wet forest, coastal shrub, *Callitris* forest, and mangrove/salt marsh. Brush box (*Tristania conferta*), carrol (*Blackhousia myrtifolia*), and kauri pine (*Agathis robusta*) are common species in the rain forest. Internationally threatened plants include stinking cryptocarya (*Cryptocarya foetida*) and *Acacia baueri baueri*.

Fauna

Fraser is host to a diverse fauna. More than 350 bird species have been recorded, including such rare species as the ground parrot (*Pezoporus wallicus*) and the red goshawk (*Erythrotriorchis radiatus*). There are also 48 native terrestrial mammals, including possums, kangaroos, bats, hares, rodents, bandicoots, and one of the purest genetic strains of dingo in Australia. There are also introduced mammals such as horses, pigs, and cats.

Several species of turtle, lizard, and snake are found on Fraser, including the endemic Fraser Island skink (*Coggeria naufragus*), an odd sand-dwelling creature with reduced limbs and the sole member of this monotypic genus. As on many other islands, the invertebrate fauna of Fraser is the least known and understood, although a 2004 study documented 254 ant species alone.

Lakes

A large number of lakes are found on Fraser Island. They come in three main types: perched, barrage, and window. Perched lakes form when organic debris hardens in a depression and collects water; the world's largest perched lake, Boomanjin, is located on the island. Barrage lakes form when sand dunes dam a waterway, and window lakes

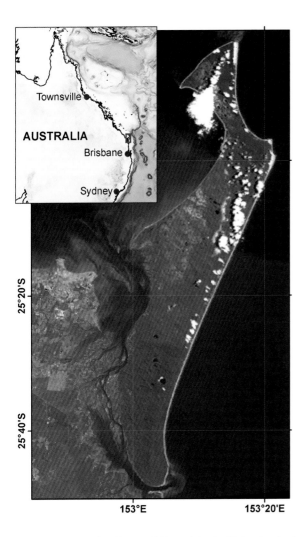

FIGURE 1 NASA Landsat image of Fraser Island with locator inset. Image modified by Jenny Paduan and David Clague, MBARI.

form where the land dips below the water table and fills with groundwater. Fraser's lakes are generally acidic and nutrient-poor, but they are popular sites for recreation. Among the organisms adapted to these acidic conditions are the wallum froglet (*Crinia tinnula*), the freshwater shrimp (*Caridina indistincta*), and the Fraser Island short-neck turtle (*Emydura macquarii nigra*).

HUMAN IMPACT, CONSERVATION, AND THE FUTURE

For thousands of years prior to European contact (the best estimates put human settlement at 5000 years ago), local Aboriginals lived on the island; shell middens, fish traps, and scarred trees are now all that remain of their once vibrant communities. Captain James Cook sighted the island in 1770, marking its initial European discovery, but the early European visitors did not find the island desirable. Matthew Flinders, an English explorer, remarked of the island in 1802, "Nothing can be imagined more barren." The island took its name from James and Eliza Fraser, who were shipwrecked there in 1837.

Horses, sheep, and cattle were introduced with European settlers in the 1840s, and logging of Kauri pines commenced in 1863 and continued until 1992, when the island was designated a World Heritage Area. Mining for valuable minerals began in 1949 but no longer occurs today.

Years of logging, mining, and more recently tourism (the island gets 300,000 visitors per year) have altered vegetation patterns and disrupted local wildlife, but local authorities have taken notice. A comprehensive management plan running through the year 2010 addresses important issues such as fire management, invasive species eradication, and educational programs featuring the indigenous Aboriginal culture. With almost all of the island's land area falling within Great Sandy National Park, the island appears poised for a bright future.

SEE ALSO THE FOLLOWING ARTICLES

Freshwater Habitats / Human Impacts, Pre-European / Sea-Level Change / Vegetation

FURTHER READING

Sinclair, J. 1977. *Discovering Fraser Island.* Surry Hills, New South Wales: Pacific Maps.
Southwell, L. 1975. *Incredible Fraser Island.* Melbourne: Australian Conservation Foundation.
State of Queensland (Environmental Protection Agency). Fraser Island Web site. Brisbane: Queensland Government. http://www.epa.qld.gov.au/projects/park/index.cgi?parkid=1.
State of Queensland (Environmental Protection Agency). 2005. *Great Sandy Region management plan 1994–2010.* Brisbane: Queensland Government. http://epa.qld.gov.au/parks_and_forests/managing_parks_and_forests/management_plans_and_strategies/great_sandy_region/.
Williams, F. 1982. *Written in sand: a history of Fraser Island.* Milton, Queensland: Jacaranda Press.

FRENCH POLYNESIA, BIOLOGY

JEAN-YVES MEYER
Department of Research, Government of French Polynesia, Papeete, Tahiti

BERNARD SALVAT
University of Perpignan, France

Despite their small size and remoteness, the numerous tropical and subtropical oceanic islands of French Polynesia (South Pacific) display a rich array of natural marine and terrestrial ecosystems and habitats. The terrestrial biota provide striking examples of plant and animal speciation and adaptive radiation, high levels of endemism, and a huge number of threatened and extinct species. The large variety of coral reef formations, from open atolls to completely closed lagoons, allows a high diversification of the marine biota. These unique biota are highly susceptible to human impact, particularly resulting from habitat destruction, biological invasion of introduced species, and potentially by global climate change and sea-level rise. The conservation of marine and terrestrial biodiversity is paramount in French Polynesia, not only for its ecological and cultural relevance but also for sustainable socioeconomic development.

A GALAXY OF SMALL REMOTE ISLANDS

French Polynesia, a French Overseas Territory (and an Overseas Country since its political status change in 2004, which gave it increased legislative powers and international autonomy), is located in the eastern corner of the South Pacific. Its maritime territory extends from 5° to 30° S and from 130° to 160° W. The islands themselves lie between 7° and 28° S and 134° and 155° W. It belongs to the Polynesian biogeographic province and to the floristic subregion of southeastern (or eastern) Polynesia, which includes the Cook Islands, the Pitcairn Islands, and Rapa Nui. French Polynesia comprises 120 tropical and subtropical oceanic islands and islets divided into five distinct archipelagoes, namely the Australs, the Societies, the Tuamotu, the Gambiers, and the Marquesas. With the exception of the Tuamotu atolls which result from high volcanic islands formed at the East Pacific ridge after drift (~10 cm/year) and subsequent subsidence, all other island chains, oriented in a southeast to northwest direction, originated from the activity of volcanic hot spots.

These oceanic islands, aged between 30,000 and 60 million years old, are scattered over an ocean area as large as Europe, with an exclusive economic zone of about 5 million km². They are located more than 5000 km away from the nearest continental areas (South America, Southeast Asia, Australia), making French Polynesia one of the most isolated archipelagoes in the world (Fig. 1). However, the presence of many other island archipelagoes located to the west (Cook, Samoa, Tonga) and the northwest (Kiribati, Tokelau, Tuvalu) might have served as "stepping stones" for plant and animal species colonization.

A High Diversity of Natural Ecosystems and Habitats

The French Polynesian islands include 33 high volcanic islands, 81 atolls, and six raised atolls, forming a total land area of only 3520 km² (Table 1), and 15,047 km² of reefs and lagoons. Tahiti is the largest island in French Polynesia with an area of 1045 km², and it has the highest summit of all South Pacific islands (Mt. Orohena, which reaches 2241 m in elevation). The islands are characterized by a large variety of geomorphological types (Table 2), ranging from young volcanic islands (e.g., the volcanic cone of Mehetia in the Societies is 30,000 years old; that of Tahiti is between 0.3 and 1.3 million years old) to barrier-reef old volcanic islands called "almost atolls" (e.g., Bora Bora, Maupiti, and Maiao, which are part of the Societies and are between 2 and 4 million years old), carbonate atolls (e.g., Rangiroa in the Tuamotu, with a lagoon area of 1717 km²), coral islets,

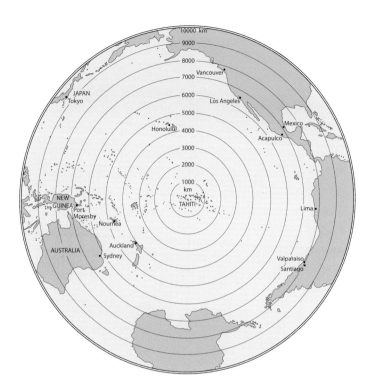

FIGURE 1 Location and isolation of French Polynesia in the South Pacific.

motu and sand cays (e.g., Motu One in the Marquesas), elevated (or raised) coral limestone islands (e.g., Makatea, Niau in the Tuamotu), and composite (volcanic and limestone) islands improperly called "makatea" (Rurutu, with limestone cliffs called "mato," in the Australs).

Climate is tropical oceanic, but some islands are more equatorial (the Marquesas) and others more subtropical (the Gambiers and Australs, especially Rapa, with an absolute minimum recorded at 8.5 °C).

The geological and edaphic nature of the substrate (calcareous or volcanic), the rainfall and temperature gra-

TABLE 1
Geographic and Climatic Characteristics of the Five Archipelagoes of French Polynesia

Archipelago	Latitude, Longitude	Age (Millions of Years)	Mean Annual Rainfall (mm/yr) at Sea Level	Mean Annual Temperature (°C) at Sea Level (min.- max. Values)	Climate Type
Australs	21–28° S, 144–155° W	1.6–12.1	1660–2560	20.6–23.5 (15.3–26.4)	Subtropical to temperate
Gambier	21–24° S, 134–137° W	5.7–6.3	1990	23.7 (18.8–26.5)	Subtropical to temperate
Marquesas	7–10° S, 138–140° W	1.3–5.5	1087–1798	26.4–26.8 (22–31)	Wet tropical
Society	15–18° S, 148–154° W	0.03–6.8	1690–3500	25.8–27 (20.8–31.1)	Wet tropical
Tuamotu	14–24° S, 134–148° W	9.5–60	1300–1900	24.7–28.3 (20.5–31.3)	Wet tropical

TABLE 2
Geomorphological Diversity of French Polynesian Islands

Archipelago	High Volcanic Islands and Rocky Islets	Atolls and Sandy Islets	Raised Atolls and Composite Islands	Total Number of Islands	Land Area (km^2)	Highest Summit (m)
Australs	4	1	2	7	148	650
Gambier	9	13	0	22	40	445
Marquesas	11	2	0	13	1050	1276
Society	9	5	0	14	1598	2241
Tuamotu	0	60	4	64	683	90
Total French Polynesia	33	81	6	120	3519	2241

dients (the mean annual rainfall varies between less than 1500 mm/year and 10,000 mm per year; the temperature decreases by 0.6 °C per 100 m of altitude), and the exposure to the dominant southeast tradewinds (leeward dry coast and windward wet coast) have resulted in a high diversity of natural terrestrial habitats. The mountainous relief of Tahiti (with ten summits above 1500 m including three peaks above 2000 m) impressed Charles Darwin, who wrote in 1852, after coming from the Andes, that "in the Cordillera, I have seen mountains on a far grander scale but for abruptness, nothing all comparable with this." The deep V-shaped valleys separated by knife-edge ridges with nearly vertical cliffs provide unique microclimates and marked isolation, which have facilitated cases of speciation (Fig. 2). Other unique mountainous landscapes in French Polynesia include the phonolitic sugar loaf peaks of Ua Pou (Marquesas), the dissected sea cliffs called the "rocky needles" of Fatu Hiva (Marquesas), the trachytic high-elevation plateaus of Temehani in Raiatea (Society) and the 100-m high limestone sea cliffs of Makatea (Tuamotu).

FIGURE 2 Deep valleys separated by knife-edge ridges have facilitated speciation events in the high volcanic islands of French Polynesia. The highest mountain peaks of Tahiti (here, Mt. Aorai, at 2066 m) harbor pristine montane cloud forests and subalpine vegetation.

The reefs of French Polynesia also have high geomorphological diversity, with some of the highest diversity of coral reef formations and atolls in the world. Reef morphology ranges from fringing reefs to barrier reefs, low-lying atolls (Fig. 3), raised or limestone atolls, and reef banks. Some atolls are open with one pass allowing large exchange of waters between the ocean and the lagoon (e.g., Hao), two passes (e.g., Rangiroa), or three passes (e.g., Amanu), whereas others are closed but feature many channels of communication (locally called "hoa") on the reef flat (e.g., Reao) or only a few such channels (e.g., Takapoto). Other have a completely closed lagoon (e.g., Taiaro) or a filled-in lagoon (e.g., Nukutavake). The combined Tuamotu–Gambier archipelago comprises 77 atolls and raised atolls, the most numerous in an archipelago worldwide, and includes Rangiroa, which is the second largest atoll in the world with a lagoon that is about 30 by 80 km wide.

A UNIQUE TERRESTRIAL FLORA AND FAUNA

Because of their strong geographic isolation, relatively young geological age, and small size, which together imply a lack of topographic and habitat diversity on most of the islands, the islands of French Polynesia's native terrestrial fauna and flora are impoverished in terms of species numbers. However, this geographic isolation, along with the islands' habitat complexity, has resulted in high species endemism.

French Polynesia belongs to the Polynesia–Micronesia terrestrial biodiversity hotspot, one of the 34 key biodiversity areas in the world. It possesses four "endemic bird areas" (namely, the Societies, the Tuamotu, the Marquesas, and Rimatara in the Australs), according to BirdLife International, and one "centre of plant diversity" located in the Marquesas, according to the World Conservation Union (IUCN) and the World Wildlife Fund (WWF).

The terrestrial native vascular flora comprises about 890 native species, including about 550 endemics (62% endemism, up to 74% for the flowering plants only) and

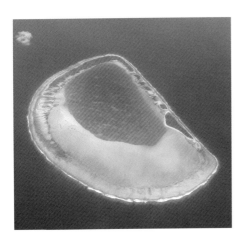

FIGURE 3 Reitoru, one of the 77 atolls of the combined Tuamotu-Gambier archipelago.

11 endemic genera. Among the most speciose genera are *Psychotria* (Rubiaceae) with about 35 endemic species, *Cyrtandra* (Gesneriaceae) with about 30 endemic species, and *Myrsine* (Myrsinaceae) and *Glochidion* (Euphorbiaceae) with more than 20 endemic species. The steep elevation and rainfall gradients have a strong influence on vegetation composition and structure. Eight native vegetation types are recognized (Table 3): coastal or strand vegetation, para-littoral forests, wetlands (from low-elevation brackish marshes to mid-elevation riparian forests and high-elevation freshwater lakes and bogs), low- to mid-elevation dry forests, low- to mid-elevation mesic forests, mid-elevation and valley rainforests, high elevation cloud forests, and subalpine vegetation above 1800 m.

Considering the fauna, the native land snails comprise about 320 described species, nearly all of them endemic to French Polynesia. Among the most speciose taxa are the colorful tree snails of the family Partulidae (*Partula* spp. and *Samoana* spp.) with ~80 species described (all of them endemic except one native widespread species), and the small ground-dwelling snails Endodontidae with ~70 species (100% endemism).

The avifauna comprises 36 species including 30 endemic to eastern Polynesia. With an additional 27 nesting seabirds and 20 migratory birds, the avifauna of French Polynesia is one of the highest among all tropical islands. There is only one southeastern Polynesian endemic bird genera: the monarchs or flycatchers *Pomarea*, with one extinct species in Maupiti, one critically endangered species in Tahiti, one endangered species in Rarotonga (Cook Islands), and four extinct and three threatened species in the Marquesas. One extant species of sandpipers *Prosobonia* (or *Aechmorhynchus*), a genus also found in Kiritimati Islands (Christmas Island), is found on a small number of atolls in the Tuamotu (Morane, Tenararo, Reitoru, Tahanea). Other common endemic land birds restricted to forested areas are fruit doves (*Ptilinopus*), warblers (*Acrocephalus*), kingfishers (*Halcyon* or *Todiramphus*), swiftlets (*Aerodramus* or *Collocalia*) and lorikeets (*Vini*), with surviving species on a few atolls of the Society and Tuamotu Islands (*V. peruviana*), in a few high volcanic islands of the Marquesas (*V. ultramarina*), and on the makatea island of Rimatara in the Australs (*V. kuhlii*).

The entomofauna is less well known, with only 500 native species described. Some groups have undergone adaptive radiation resulting in 100% endemism: Such cases include the small weevil *Miocalles* (Coleoptera: Curculionidae) with 67 endemic species on Rapa, *Mecyclothorax* (Coleoptera: Carabeidae) with 70 endemic species in Tahiti, *Rhyncogonus* with 35 endemic species in French Polynesia, and black flies *Simulium* with 38 endemic species (29 in the Societies, nine in the Marquesas). Many new species of arthropods have been discovered in the islands over the last few years, including long-jawed spiders, black flies, mirid bugs, and most recently four new species of water-skating flies *Campsicnemus* (Dolichopidae). Freshwater ecosystems contain 37 native fishes (including 14 endemics) and 18 decapods (including three endemics), with the highest endemism rate being found in the Marquesas (64%).

The most remote southeastern island, Rapa in the Australs, has an exceptional biodiversity. It is home to ~100 endemic land snails falling within 16 endemic gen-

TABLE 3
Main Natural Vegetation Types and Plant Formations in French Polynesia

	Vegetation Type	Plant Formations
Azonal	Coastal or littoral vegetation	Coastal vegetation on sandy or rocky beaches
	Para-littoral forests	Forests on atolls and raised limestone plateaus
	Wetlands	Vegetation of marshes, lakes and bogs, submangrove, riparian forests
Zonal	Xerophilous (< 1500 mm/an)	Low- to mid-elevation dry to semi-dry forests
	Mesophilous (1500–3000 mm/an)	Low- to mid-elevation moist or mesic forests
	Hygrophilous (3000 mm/an)	Mid- to upper-elevation wet forests (including valley forests)
	Ombrophilous (> 3000 mm/an)	High-elevation (montane) rainforests or cloud forests
	Subalpine	Summit and ridge shrublands

era, 74 endemic moth species (Microlepidoptera) falling within seven endemic genera and one endemic family (Lathroteridae), ~60 island endemic plants falling within three endemic monotypic genera (*Apostates, Metatrophis,* and *Pacifigeron*), two endemic freshwater fishes, and one endemic bird (*Ptilinopus huttoni*), on a land surface of only 40 km².

Among marine taxa, a total of 1024 fish species, 1500 molluscs, 978 crustaceans, 176 coral species, 977 crustaceans (including ~500 crabs), and 425 algae have been reported from French Polynesia. Many species were discovered recently during extensive marine surveys, with new island records and new species to be described. Four marine reptiles (three species of sea turtles and one sea snake) and 16 marine mammals (two whales and 12 dolphins) have been observed. The endemism of the lagoon and marine fauna is relatively poor; the highest rates are found for molluscs in the Marquesas (up to 20%) and in the Australs (10%). The relatively low diversity of marine and reef organisms is correlated with regional diversity, decreasing from west to east. The combined archipelagoes of the Societies, the Tuamotu, and the Marquesas are considered as one of the 43 marine priority ecoregions of the WWF "Global 200 Ecoregions" for their coral diversity and vulnerability.

A TERRESTRIAL BIOTA WITH DIVERSE BIOGEOGRAPHIC AFFINITIES

French Polynesia is characterized by an attenuated Indo-Malesian and Austro-Melanesian flora with a very few New Zealand taxa (e.g., *Myoporum* found in the Australs, *Corokia* and *Hebe* only in Rapa), and a few American components (e.g., *Fuchsia* in Tahiti, *Plakothira* in the Marquesas, *Gouania* in Mangareva). In addition, the Marquesas show some floristic affinities with Fiji (e.g., *Trimenia*), and Hawaii (e.g., *Cheirodendron*). The Australs have also taxa otherwise found only in Hawaii (e.g., *Charpentiera* and *Nesoluma polynesicum*). The small sedge *Oreobolus furcatus* (Cyperaceae), commonly found in the montane bogs of Hawaii, is restricted to a single population located on the highest peak of Tahiti, Mt. Orohena, at a 2240-m elevation. Finally, another principal interest of the French Polynesian flora is that it lies at the easternmost limit of the range of a very large number of genera in the Malesia and Pacific Ocean islands, such as *Alyxia* (Apocynaceae), *Ascarina* (Chloranthaceae), *Cyrtandra* (Gesneriaceae), *Fagraea* and *Geniostoma* (Loganiaceae), *Meryta* (Araliaceae), *Metrosideros* (Myrtaceae), *Planchonella* (Pittosporaceaea), and *Planchonella* (Sapotaceae).

In the same way, among animals, some groups have come entirely from the west (e.g., *Rhyncogonus* weevils, *Inseliellum* black flies, and *Partula* tree snails), whereas other show affinities across the eastern archipelagoes of Polynesia and Hawaii (e.g., Monarch flycatchers, thomisid crab spiders). Some taxa show disjunct boundaries between archipelagoes where lineages from the west meet those from the north/east (e.g., crab spiders likely derived from an American lineage occur in Hawaii, the Marquesas, and the Societies, whereas the Australs are occupied by one from Australasia). Other groups may have colonized each archipelago independently from different mainland sources (e.g., *Tetragnatha* spiders).

RECENT HISTORY AND CURRENT THREATS

The main threats are habitat destruction and fragmentation caused by agriculture; land clearing; housing development and urbanization; shoreline construction and coastal reclamation; infrastructure such as dams, roads, golf courses, and tourism resorts; extraction of coral sand and rocks from the lagoons and reef areas; fires; industrial or domestic pollution (of soil, rivers, and lagoons); overexploitation of natural resources (e.g., of the green sea turtle *Chelonia mydas,* the coconut crab *Birgus latro*); overfishing; intensive harvesting of black-lipped oysters for black pearl production; mining (phosphates on the raised atoll of Makatea in the Tuamotu between 1917 and 1966); nuclear testing (atmospheric tests between 1966 and 1974 and underground tests between 1975 and 1996 on the two atolls of Fangataufa and Morurua in the Tuamotu); biological invasion of accidentally or intentionally introduced species; and global warming (which leads to bleaching of corals, regression of the subalpine flora). French Polynesia is among the Pacific countries that will suffer most from sea-level rise because of its low-lying relief and coastal geomorphology. Coral bleaching because of sea water temperature rise has become more frequent over the last decades with major events in 1991, 1994, and 1998.

Several spectacular and ecologically disastrous biological invasions are well documented in French Polynesia (the invasive miconia tree *Miconia calvescens;* predatory animals such as the black or ship rat *Rattus rattus;* the carnivorous snail *Euglandina rosea;* the swamp harrier *Circus approximans;* aggressive birds such as the common myna *Acridotheres tristis* and the red-vented bulbul *Pycnonotus cafer;* and insects such as fruit flies *Bactrocera* spp.; tramp ants *Pheidole megacephala, Solenopsis geminata, Anoplolepis gracilipes;* and sand flies). In the marine ecosystem, some mollusc species have been introduced for the commercial value of their nacraeous tets (*Trochus niloticus* and *Turbo marmoratus*) but without any apparent disturbances on reef communities. Some native species such as the marine algae *Sargassum mangarevense* and *Turbinaria ornata* have also become invasive; these examples were dispersed from the Society Islands to the Tuamotu atolls. The crown-of-thorns starfish *Acanthaster planci*, which feeds on corals, had periodic demographic outbreaks in the 1970s and the 1980s.

CONSERVATION AND SUSTAINABLE DEVELOPMENT

Two major waves of native species extinctions have occurred in the past. The first was related to overhunting, fire, and forest clearance since Polynesian colonization (1500 years ago) and was associated with introduced animals (Pacific rats, domestic dogs, chickens, and pigs), which drove many endemic birds to extinction, especially flightless rails (*Gallirallus*), swamphens (*Porphyrio*), ground doves (*Gallicolumba*), cuckoo doves (*Macropygia*), lorikeets (*Vini*), sandpipers (*Prosobonia*), and coastal palms and flying foxes (*Pteropus*). The European colonization period, starting ~250 years ago, has led to additional habitat destruction and alteration; to overexploitation and overharvesting (e.g., sandalwood *Santalum*); and to the introduction of grazing mammals (goats, sheep, horses, cattle), predators (black rats, cats, ants, swamp harriers), and aggressive competitors (invasive plants, common mynas). These impacts caused the extinction of many endemic birds (rails, sandpipers, the parakeet *Cyanoramphus*, the fruit doves *Ptilinopus*), plants (the daisy tree *Fitchia*, the coastal legume tree *Sesbania*), and land snails (Endodontidae). During the last 30 years, French Polynesia has been facing a third and "new" wave of extinction: About 60 tree snails (*Partula*) disappeared in the Society Islands following the introduction of the carnivorous snail *Euglandina rosea* (intentionally introduced in the 1970s to control the giant African snail *Achatina fulica*), and four flycatchers (*Pomarea*) vanished because of recent incursions of black rats. Between 40 and 50 endemic plants are directly threatened by the massive invasion of the native rain forests and cloud forests by the invasive miconia tree *Miconia calvescens* in Tahiti.

As a result, French Polynesia has one of the highest number of extinct species worldwide (50 documented species), and about 50 other species are on the brink of extinction (conservation status CR according to IUCN, Table 4). This is an underestimation as many species considered data deficient (DD) are either extinct or critically endangered based on recent and extensive field surveys.

TABLE 4
Status of Threatened Species in French Polynesia, by IUCN Category

IUCN Categories	EX and EW	CR	EN	VU	DD and NT	Total
Vascular plants	6	26	4	17	34	87
Birds	11	5	9	16	3	44
Land molluscs (gastropods)	33	15	2	0	0	50
Mammals (cetaceans)	0	0	1	2	5	8
Reptiles (sea turtles)	0	1	1	0	0	2
Total	50	47	17	35	42	191

NOTE: IUCN categories: EX = extinct; EW = extinct in the wild; CR = critically endangered; EN = endangered; VU = vulnerable; DD = data deficient; NT = near threatened. From the 2007 IUCN Red List of Threatened Species, http://www.iucnredlist.org/.

Rapid increases in transportation, which have led to the exchange of people, goods, and materials between neighboring continents and islands in the Pacific region, are leading to more species introductions, thus enhancing the risk of further invasions (e.g., the little fire ant *Wasmannia auropunctata*, the glassy-winged sharpshooter *Homalodisca vitripennis*, snakes, lizards, frogs, etc.). Rapid population growth in French Polynesia (the population has increased fivefold since the end of World War II and has doubled in the last 30 years from 110,000 in 1975 to 256,000 in 2006), combined with the fact that 70% of its inhabitants live on the island of Tahiti (170,000 in 2005), is leading to strong human pressures on natural resources and is increasing disturbance in native habitats (deforestation, fires, pollution, etc.).

The current situation of park and nature reserves in French Polynesia is critical. Natural protected areas are found on only nine of the 120 islands (Table 5), seven of which are uninhabited, small, high, volcanic islands (including Eiao, Hatutu, and Mohotani in the Marquesas) or uninhabited small atolls (Motu One in the Marquesas, Scilly and Bellinghausen in the Societies, Taiaro in the Tuamotu). The total protected area is ~8200 ha (i.e., only 2% of the total land area of French Polynesia), and it does not include the most ecologically important habitats. All of the protected zones are characterized by a lack of management, with no monitoring or park guards, except for on the atoll of Taiaro, which has been a UNESCO man and biosphere reserve since 1977. This reserve was recently enlarged to include six other Tuamotu atolls in 2006 (Fakarava, Aratika, Kauhei, Raraka, Niau, and Toau), to form the Fakarava Biosphere Reserve. In the absence of humans, the main threat to these protected areas remains the invasion by

TABLE 5
List and Characteristics of the Nine Protected Areas in French Polynesia

Island Name (Archipelago)	Protected Area Type (IUCN Category)	Protected Since	Land Area (ha)	Marine Area (ha)	Elevation Range (m)	Vegetation Type
Taiaro Atoll (Tuamotu)	Natural Reserve, Biosphere Reserve since 1977	1972	340	920	0–10	Coastal vegetation and forest
Mohotani Island (Marquesas)	Natural Reserve, Habitat and Species Management Area since 2000 (IV)	1971	900	0	0–520	Para-littoral forest, dry and mesic forests
Eiao Island (Marquesas)	Natural Reserve, Habitat and Species Management Area since 2000 (IV)	1971	4000	0	0–577	Para-littoral forest, dry and mesic forests
Motu One Sand Islet (Marquesas)	Natural Reserve, Habitat and Species Management Area since 2000 (IV)	1971	50	0	0–10	Coastal vegetation
Hatutu (Hatutaa) Island (Marquesas)	Natural Reserve, Habitat and Species Management Area since 2000 (IV)	1971	750	0	0–420	Para-littoral forest
Scilly (Manuae) Atoll (Society)	Natural Reserve	1971 (lagoon), 1992 (atoll)	900	10 000	0–10	Atoll coastal vegetation and forests
Bellinghausen (Motu One) Atoll (Society)	Natural Reserve	1971 (lagoon), 1992 (atoll)	280	900	0–10	Atoll coastal vegetation and forests
Te Faaiti (Tahiti, Society)	Natural Park (II)	1989	750	0	75–2110	Mesic forests and rainforests
Vaikivi (Ua Huka, Marquesas)	Natural Park and Reserve (II and Ia)	1997	240	0	400–884	Rainforests and cloud forests

alien plants and animal species (the miconia tree in the Te Faaiti natural park, feral sheep on Mohotani, feral sheep and pigs on Eiao).

The conservation of marine and terrestrial biodiversity is paramount in French Polynesia, not only for its ecological relevance but also for sustainable socioeconomic development. The tourism industry (about 250,000 tourists per year), black pearl harvesting (black-lipped oyster *Pinctada margaritifera*) (7 tons of pearls exports in 2006), tuna fishing (about 10,000 tons per year), the production of coconut oil (*Cocos nucifera*), and the harvesting of noni fruits (*Morinda citrifolia*) and vanilla beans (*Vanilla tahitensis*) are the main economic activities and exports. Healthy reefs and lagoon waters are also crucial for human health. The bloom of the toxic dinoflagellate *Gambierdiscus toxicus* on the inshore reefs is correlated with severe natural disturbances such as cyclones but also with human disturbances such as airport construction on reefs (e.g., on the island of Raivavae in the Australs). The importance of traditional Polynesian knowledge should be taken in account because many species have a strong cultural value (e.g., medicinal and ritual plants, plant cultigens or cultivars, legendary animals) and because reef and lagoon environments and resources are so intimately linked with Polynesian life. Thus, future environmental management should draw on both modern and traditional conservation systems and should involve local island communities.

SEE ALSO THE FOLLOWING ARTICLES

Extinction / French Polynesia, Geology / Invasion Biology / Reef Ecology and Conservation

FURTHER READING

Gargominy, O., ed. 2003. *Biodiversité et Conservation dans les Collectivités françaises d'Outre-mer*. Paris: Comité français UICN.
Meyer, J.-Y. 2004. Threat of invasive alien plants to native flora and forest vegetation of eastern Polynesia. *Pacific Science* 58: 357–375.
Salvat, B., *et al.* 2008. Le suivi de l'etat des récifs coralliens de Polynésie française et leur récente évolution. *Revue d'Écologie (Terre et Vie)* 63: 145–177.

FRENCH POLYNESIA, GEOLOGY

ALAIN BONNEVILLE

Institut de Physique du Globe de Paris, France

French Polynesia is located in the south-central part of the Pacific Ocean, between 5° and 30° S and 130° and 160° W. It comprises 118 islands representing an area of 16,000 km² above sea level and having more than 5 million km² of water within the limits of its huge exclusive economic zone. All of French Polynesia's islands are basaltic shield volcanoes that represent only the emerged parts of important submarine mountain chains created by volcanic activity beginning 40 million years ago.

AN EXCEPTIONAL CONCENTRATION OF VOLCANISM

The sea floor of French Polynesia was formed between 25 and 85 million years ago by sea-floor spreading along the ancient Pacific-Farallon ridge. The region has been greatly modified since its creation by mid-plate volcanic activity and associated flexural compensation, and by mass wasting. French Polynesia contains five major volcanic chains (Fig. 1), each attributed (sometimes with great difficulty) to the drift of the Pacific plate over hotspots. The Austral, Society, Pitcairn–Gambier, and Tuamotu island chains trend N120°, the same direction as present Pacific plate motion. The Marquesas Islands have a 140° northern strike. The region is crossed by the Austral and Marquesas fracture zones, with 70° northern strikes. These fracture zones are the main tectonic features of the sea floor and are related to the Farallon ridge. The Austral, Society, Gambier, and Marquesas island chains are composed mainly of high islands, whereas the Tuamotu archipelago is composed entirely of atolls. Altogether, French Polynesia includes 34 islands and 84 atolls.

The islands rise approximately 4500 m above the sea floor, and their maximum elevation above sea level, at Tahiti, is 2200 m. To a first approximation, except for the Tuamotu archipelago, island ages and morphologies are consistent with the hotspot theory, which posits that the age of each island within a chain increases from southeast to northwest, based on radiometric dating of volcanism from the present to about 16 million years ago. The Tuamotu Islands are probably older than 40 million years and rise above a broad plateau. Their origin is uncertain but has been attributed to an aborted ridge, a hotspot, or an emerged submarine chain that originated close to the East Pacific rise. French Polynesia offers all the stages of classical evolution of intraplate volcanoes from the early seamount stage to the ultimate guyot stage. Between these two end-members, one can find all the stages of an elevated island subjected to reef building, tropical erosion, and sea-floor subsidence. This leads to a wide variety of landscapes, which give French Polynesia its unique beauty (Fig. 2).

The age, petrology, and geochemistry of the French Polynesian high islands are relatively well known. All islands and seamounts that have been dredged are mainly composed of oceanic tholeitic and alkali basalts, although some differentiated alkaline lavas and a few outcrops of plutonic rocks can also be found. Oceanic intraplate basalts (OIB) provide insights on the chemical composition and evolution of the Earth's mantle, and French Polynesia is particularly important because its islands present by far the greatest variability in incompatible trace elements and radiogenic isotopes known on Earth. This chemical variation, including that of mid-oceanic ridge basalts (MORB), coupled with geophysical observation, is the key to understanding the dynamics of intraplate volcanism and mantle plumes in this region. Variability can also be observed at the scale of a single volcano, such as Rurutu in the Austral archipelago, where different magmatic phases have different ages and geochemical signatures, emphasizing the importance of structural control, either crustal or lithospheric, in the location of the volcanic activity. A pronounced positive gravity anomaly is also associated with each island. Such anomalies are usually related to shield volcanoes and reflect deep-seated, solidified magma chambers or feeding systems. Field evi-

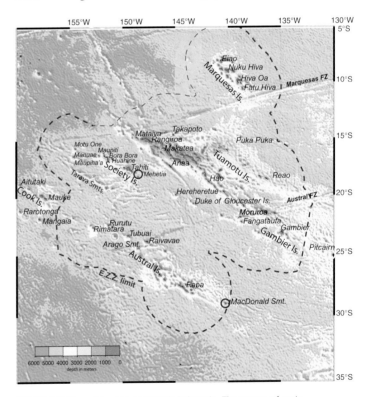

FIGURE 1 Bathymetric map of French Polynesia. The names of main islands and archipelagos are indicated. Polynesian archipelagoes are emerged parts of large submarine mountain chains rising ~7000 m over the surrounding abyssal plains. These chains are composed of high islands, atolls, seamounts, and guyots, with the exception of Tuamotu where the limestone cover, of coral reef origin and more than 2000-m thick, has formed a large plateau. Note the direction of elongation of the chains along the direction of Pacific plate motion. Red circles represent active hotspot locations (recorded by instruments in the last decades), and yellow circles represent the supposed present-day hotspot locations (younger than 200,000 years). EEZ = exclusive economic zone; Is = island; Smt = seamount; FZ = fracture-zone.

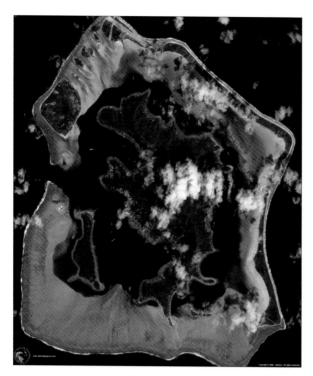

FIGURE 2 Bora Bora viewed from space. It is a typical volcanic island entering the last stage of its life. The basaltic part of the island (the rest of the volcano) in the center is deeply eroded, and the reef barrier has evolved in an almost continuous ring. Ultimately, the mountain will disappear through subsidence and erosion, giving birth to an atoll. IKONOS satellite image courtesy of GeoEye. Copyright 2008. All rights reserved.

dence of such chambers can only be found in Tahiti, where gabbros outcrop in the central caldera.

French Polynesia is located on a broad, anomalously shallow area of sea floor, called the South Pacific Superswell, which is characterized by depth anomalies extended over thousands of kilometers with a maximum amplitude of about 1 km. Other hotspot volcanic chains far from spreading ridges are associated with topographic swells that are typically several hundred kilometers in lateral extent and up to a kilometer in height.

Multibeam bathymetry around the Society and Austral Islands shows evidence of 36 submarine landslides. This inventory shows an evolution of the landslide type with the age of oceanic islands. Submarine active volcanoes are subject to superficial landslides of fragmental material whereas young islands exhibit marks of giant lateral collapses that produced debris avalanches during the period of volcanic activity.

French Polynesia has an excess of mid-plate volcanism, the concentration of which is three to four times the average estimate for normal Pacific lithosphere. This volcanic activity has an apparent rhythm in both space and time. It is increasingly evident that these short-lived and closely spaced hotspots are consistent with a model wherein each hotspot can sample a smaller volume of the large and very heterogeneous mantle plume responsible for the superswell. It has been proposed that this small-scale convection (100 km in surface expression compared to the 2000 km of the superswell) could be due to secondary instabilities developing at the surface of the larger plume in the transition zone of the Earth's mantle.

THE ARCHIPELAGOES

Tuamotu

Oriented N120°, the Tuamotu Archipelago consists of about 60 atolls along two subparallel chains of 1200-km length and 400-km width. Its northern and southern ends are located at the Marquesas fracture zone and the Austral fracture zone, respectively. The age of the sea floor ranges from about 30 million years in the southeast to 65 million years in the northwest. The atolls probably formed atop subsided volcanoes.

The series of atolls are superimposed on an oceanic plateau trending N115°, the predicted direction of absolute motion for the Pacific plate. The Tuamotu may be the southern continuation of the Line Islands. Their combined origin remains unresolved: The chronology of the Line Islands implies the existence of several hotspots, and at least two would have been required for the formation of the Tuamotu Islands. This archipelago has been dated only at its northwestern end, near Mataiva atoll, where K/Ar and Ar/Ar total fusion ages on whole rock dredged samples yielded a minimum age of 47.4 million years.

East of the Tuamotu lies a chain of linear volcanic ridges and sea mounts along a 2600-km corridor, 50–75 km wide, between Puka Puka Island and the western flank of the East Pacific rise. These ridges are closely aligned with the cross-grain gravity lineations. Radiometric ages recorded on the largest ridges indicate nearly simultaneous eruption of volcanoes over a more than 2000-km distance, an observation difficult to reconcile with hotspot theory.

Society

Between the Marquesas and Austral fracture zones and west of Tuamotu plateau, the Society island chain (Fig. 3) is aligned in the direction of drift of the Pacific plate, although some local control by structural discontinuities of the oceanic crust, formed at the Pacific–Farallon ridge, is evident. The age of the oceanic crust under the chain of the Society Islands ranges from 65 million years in the south to

90 million years some 750 km away in the north. The chain is composed of five atolls and nine high islands and extends over at least 500 km and 5 million years between Mehetia, the present hotspot location, and Maupiti, the oldest island (about 4.3 million years old), and it probably continues on further to a northwestern group of atolls, Manuae–Maupihaa–Motu One, for which the age is undetermined. The geochronology, geochemistry, volcanology, and geology of the Society Islands have been reviewed recently by several authors. The aerial activity of these volcanoes can be described by three main sequential stages: the construction of a shield volcano, the formation of a caldera, and the occurrence of post-caldera volcanism.

Tahiti is located at the southeastern end of the Society archipelago. Tahiti is made up of two coalesced eruptive systems (Tahiti Nui and Tahiti Iti) aligned in the direction of plate motion (N120°). The volcanic evolution is marked by the concentration of eruptions through a main east–west rift zone which was responsible for the lateral collapse of the northern slope of the main shield, around 0.87 million years ago; a huge southern landslide has also been identified (Fig. 4). Geological, geomorphological, and geochronological data show that volcanic activity was first concentrated to the north, constructing a second shield within the U-shaped depression. The current hotspot location, around Mehetia, is 50 km farther southeast, as indicated by volcano-seismic crises linked to the growth of several seamounts; this submarine region has been intensely studied over the past 20 years. The last volcanoseismic swarms of 1981–1983 and 1985 were recorded by the seismic network of Tahiti and interpreted as eruptive events.

Another volcanic alignment, the Tarava seamount chain recently mapped and dated at 43 million years, occurs just to the south of the Society chain, with a slightly oblique orientation compared to the Society chain orientation.

Gambier

The Duke of Gloucester–Moruroa–Gambier–Pitcairn volcanic islands chain extends over 1650 km south of the Tuamotu plateau, is oriented N115°, and was built on oceanic crust more than 30 million years old. Available ages are, for Moruroa, 11.8–10.3 million years; Gambier, 7.3–5.70 million years; and Pitcairn, 0.45 million years and zero in the active hotspot area located 70 to 100 km southeast of Pitcairn are compatible with a hotspot origin. The average migration rate of volcanism is estimated at about 11 cm·yr^{-1}, consistent with the average Pacific plate velocity for this period of time. Although there are no dates for the Duke of Gloucester and Hereheretue

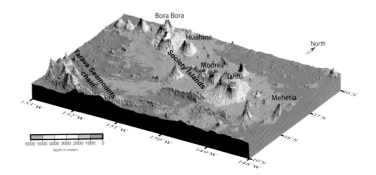

FIGURE 3 Three-dimensional view of the Society and Tarava seamount chains based on complete multibeam bathymetric coverage of the region (see text).

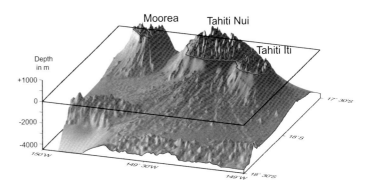

FIGURE 4 Recent detailed multibeam surveys have allowed for the building of detailed bathymetric maps around islands. In some cases, landslides are evident, such as this spectacular one dated at ~800,000 years ago on the submarine western flank of Tahiti.

atolls, it is currently thought that these edifices belong to the same alignment. As is the case for the Austral and the Society chains, preexisting crustal discontinuities influence the locus of volcanism.

Fangataufa and Moruroa, where French nuclear bombs were tested between 1970 and 1997, have been extensively drilled, so considerable data on the properties and geology of these islands exists, and it has been made publicly available in recent years.

Marquesas

To the north, the Marquesas form a 250-km wide by 400-km long linear volcanic chain built on sea floor approximately 55 million years old and more than 4200 m deep. The length of this island chain is only 350 km, but it comprises a dozen islands and as many shallow seamounts. The Marquesas fracture zone is located southeast of the chain. The Marquesas hotspot's volcanic activity was relatively brief, from about 6 million years at Eiao to 1 million at Fatu Hiva. To date, no active hotspot responsible for the Marquesas' volcanism has been identified, but a basalt sample dredged 50 km southeast of Fatu Iva yielded an age of about 0.5 million years. Seismic reflection, bathymetry, and gravity data suggest that the Marquesas fracture zone probably marks the present location of the Marquesas mantle plume.

Like most hotspots on mature lithosphere, the Marquesas are surrounded by a wide shallow region. Most of the seafloor depth anomaly is explained by crustal underplating identified by seismic refraction experiments. The anomalously thick crustal root is a buoyant load that flexes the elastic plate upward from below, creating an important sea floor depth anomaly under the volcanic chain. It is generally accepted that the Marquesas are a classic hotspot chain reflecting the drift of the Pacific plate over a fixed mantle plume. Unlike the other hotspot tracks, however, the orientation of this volcanic chain differs by 20–30° from that of Pacific plate motion (N115°). Because linear volcanic chains supposedly mark the motion of the plate with respect to stationary mantle plumes, the deviation in the trend of the Marquesas chain is quite odd. Scientists often interpret it as an example of a breakdown in the classical hotspot model or argue that Marquesas volcanism could be controlled by a crustal weakness created by the Marquesas fracture zone.

Austral

The Austral volcanic chain, located in southwestern French Polynesia, extends to the northwest for about 1200 km, from Macdonald seamount, an active submarine volcano, to the island of Rimatara. The chain is composed of 11 islands and two atolls with little area above sea level (the largest is 70 km^2). Although oriented roughly in the direction of present Pacific plate motion (11 cm·yr^{-1} along a N115° direction), the pattern of both the aerial and the submarine volcanoes is rather complex. Two recent geophysical cruises have revealed the complexity of the overlapping volcanism and the likely multiplicity of hotspot tracks. The Austral chain is often associated with the Cook Islands because most of islands were formed by the Austral hotspots. The two chains will be so treated here, leading to an alignment of islands longer than 2200 km.

Morphology and geometry of the island groups suggest the existence of two distinct volcanic alignments. The Aitutaki–Mauke islands group, Rimatara, Rurutu, Tubuai, Raivavae, and the Président Thiers bank, form the northeast alignment. Rarotonga and Mangaia Islands, Neilson bank, Rapa, Marotiri, and Macdonald seamount—the only known active volcano of the chain (named after petrologist Gordon Macdonald from Hawaii)—form the southwest branch. Recent eruptions at Macdonald sea mount have been seismically detected (T waves), and in 1987 and 1989, pumice emission was observed from research vessels. Because its summit reaches 40 m under sea level, one of the next eruptions could transform this seamount into an island!

The age of the oceanic crust along the chain ranges from around 35 million to 80 million years. Several good K/Ar or Ar/Ar ages have been measured for almost all the islands and for seamounts in the northern Austral region. In the north of the Cook–Austral region, Rarotonga has an age of 1.1 million years. At Aitutaki, a 1.2-million-year age coexists with an 8.5-million-year stage. At Rurutu, two different volcanic stages have been identified, an old one at 12 million years, compatible with the progression in ages along the northeastern volcanic alignment, and a young one at 1.1 million years.

The initial construction stage of Rurutu can be linked to the magmatic source that formed Tubuai, as it has the same petrologic and geochemical characteristics, and the distance between the two islands is compatible with absolute Pacific plate motion. But the cause of the later 1.1-million-year-old volcanic event on Rurutu must be sought on Arago seamount (named after the French Navy ship that discovered it in 1993), 130 km southeast of Rurutu (Fig. 5). Numerous cones exist between Rurutu and Arago in this 4500-m-deep basin, but no clear crustal swell seems to be associated with this axis.

FIGURE 5 Three-dimensional view of sea floor in vicinity of Arago seamount, the most recently discovered hotspot in French Polynesia (1999). A dredging operation on the illustrated flank (A) collected basalts that have been dated at 200,000 years.

To summarize, Arago seamount is the most recent surface expression of the hotspot responsible for the recent volcanic activity at Rurutu and probably for other volcanoes in the Cook Islands chain. The only possible track for the Macdonald hotspot is along a southeastern path and it could not have supplied the magma for the northern Austral Islands. For these latter islands we must propose an extinct magmatic source close to the Austral fracture zone.

SEE ALSO THE FOLLOWING ARTICLES

Atolls / French Polynesia, Biology / Line Islands / Seamounts, Geology / Volcanic Islands

FURTHER READING

Bonneville, A., L. Dosso, and A. Hildenbrand. 2006. Temporal evolution and geochemical variability of the South-Pacific Superplume activity. *Earth and Planetary Science Letters* 244: 251–269.

Hékinian, R., P. Stoffers, and J.-L. Cheminée. 2004. *Oceanic Hotspots*. Berlin: Springer-Verlag.

McNutt, M. K. 1998. Superswells. *Reviews of Geophysics* 362: 211–244.

White, W. M., and R. A. Duncan. 1996. Geochemistry and geochronology of the Society Islands: new evidence for deep mantle recycling, in *Earth processes: reading the isotopic code*. A. Basu and S. Hart, eds. Geophysical Monograph 95. Washington, DC: American Geophysical Union, 183–206.

FRESHWATER HABITATS

ALAN P. COVICH

University of Georgia, Athens

The remoteness of many islands strongly limits initial dispersal and colonization of insular springs, rivers, and lakes. Colonization of insular freshwaters generally results in some broadly predictable relationships among changes in species richness and the sizes of islands, their locations relative to continental and other island sources of species, and their climate and age. Geologic age, composition of volcanic or sedimentary rocks, and distributions of rainfall all interact to determine the rates of development of drainage networks and availability of aquatic habitats for colonization. Ultimately, the island is completely eroded, its aquatic habitats are filled with sediments, and it is covered by the sea.

GENERAL PATTERNS OF SPECIES RICHNESS

The developmental phases and patterns of succession of freshwater plant and animal communities remain only partially predictable. High numbers of endemic (unique) freshwater species generally are limited to older, remote islands and result from some of same patterns often associated with the diversity of terrestrial species. These patterns are based on dynamic changes in the physical template that modify the diversity of inland aquatic habitats. The need to understand the dynamics of these natural assemblages is increasing as non-native species are rapidly being introduced to many insular freshwater bodies. Some of these non-native aquatic species can alter environmental conditions and affect public health. Introduced mosquitoes and snails are often vectors of diseases that infect native species as well as humans. As global trade increases, especially in inland aquaculture and aquarium commerce, there are many more freshwater species being moved around the globe and invading insular freshwaters.

Total insular freshwater species diversity includes species from five sources: (1) continents; (2) other islands; (3) marine areas (the species slowly adapt to freshwater); (4) endemics evolving in situ; and (5) intentional or accidental human introductions. This total varies with island location, size, age, and climate, especially rainfall. Concentrations of endemic freshwater species generally are found on wet, older, and remote islands. The distributions of insular endemic species appear to reflect the same dynamic patterns often associated with the diversity of insular terrestrial species and continental inland waters. However, specific connections among the various abiotic (such as spatial isolation, drought frequencies, or hurricanes) and biotic variables (such as interspecific competition and predation) that affect speciation are unclear. These connections are still in relatively early stages of study among different insular waters.

GEOLOGY AND EVOLUTION OF INSULAR FRESHWATER COMMUNITIES

Considering the various specific factors influencing freshwater species diversity and community composition, the geological age of an island is of primary importance and, for islands of similar size, shows a strong positive relationship with species richness. Geological processes influence the biotic diversity of insular freshwater communities through initially determining different types and origins of islands. Then the erosion of the island is eventually followed by subsidence and sea-level changes.

Large-scale movements of positions of islands result from continental drift and plate tectonics. These changes result in islands of many different ages and variable locations that influence species colonization. Age and location affect hydrology and availability of diverse freshwater habitats. The ways in which high volcanic mountains intercept clouds and result in wet windward regions and dry leeward regions provide distinct types of habitats of different depths and permanence that influence species richness. Many low-lying islands (such as atolls and old volcanically originated islands) have relatively small drainage basins and fewer types of freshwater habitats. They do not capture sufficient precipitation to produce surface runoff, and they often lack permanent freshwater habitats except perhaps for springs derived from groundwater storage. These types of intermittent freshwaters on low, flat islands are similar to isolated vernal pools on continents, in the sense of having specific types of plants and animals that are well adapted to survive periods of dryness. For speciation to occur, the relevant evolutionary and ecological timescales vary depending the life cycles and generation times of the many types of aquatic species that colonize these habitats.

In general, islands and freshwater communities have a finite existence. Volcanic islands can form rapidly, but then they begin to erode and subside over millions of years. As they originate, develop, erode, and ultimately subside, freshwater species richness initially increases on oceanic islands and then declines. On high volcanic islands, the rate of erosion of mountains and the formation of river drainage networks vary with geology and climate. Although the initial development of new aquatic habitats (springs, rivers, and lakes) and their hydrologic connections enhance total species richness, the total number of species later declines. Accumulations of eroded sediments fill lakes and rivers until they become intermittent or completely dried out as the entire hydrologic cycle changes over time, and freshwater habitats disappear.

As a result of island hydrogeological development, the evolution of endemic freshwater species on islands is expected to result in different levels of species richness depending on island locations and histories. The geologically younger (perhaps less than 5000 or 10,000 years of age) islands that are isolated by long distances from sources of colonizing species are expected to have dispersal-limited communities. These communities would have few endemics and relatively low total species richness in a relatively simple array of alternative freshwater habitats. Younger islands closer to sources of colonization would be expected to have an intermediate level of total species richness depending on their proximity to single or multiple sources of species, currents that transport floating mats of propagules, and rates of weathering. Older (perhaps 10,000 years of age or more), isolated islands would be expected to have the highest diversity, whereas the oldest might experience a decline in the number of endemic species. Islands that were originally derived from continents have a different dynamic process because they began with high species richness from their continental origins. Such islands can increase their total diversity over time if sufficiently isolated, especially from speciation among those fishes and other vertebrates that are not well adapted for oceanic dispersal and colonization. In general, the dominant species that characterize insular rivers have similar distributions in coastal rivers on continents. These communities include a relatively small number of species well adapted for variable salinities in estuaries.

ADAPTATIONS FOR PASSIVE AND ACTIVE DISPERSAL

Passive dispersal is well developed among many widespread freshwater species, especially certain species of algae, seed plants, and invertebrates. Such species have small, lightweight propagules that are well adapted for long-distance dispersal by strong wind storms or by attachment to rafts of floating wood and pumice across oceans. Egg masses and dormant individuals can also be carried by other larger organisms, especially birds and fishes. These resistant stages and cysts remain viable for long periods and are highly effective in moving from continents and among islands. The durability and transportability of eggs, cysts, ephippia, gemules, and other resistant propagules are well documented from continental studies of birds' feet and intestinal pathways. Another type of passive dispersal occurs among some invertebrates and vertebrates with larvae that drift long distances in oceanic currents and colonize insular freshwaters. Consequently, factors such as directions of air and oceanic currents can greatly influence colonization of island freshwaters by species of fishes, decapods, gastropods, insects, and plants. Extremes in prevailing winds and the frequency of storm events can affect active and passive dispersal. Aquatic insects and other "airborne" species can colonize island streams by dispersing from a series of stepping-stone sources. Insular chains with freshwater provide sources for species dispersal by wind, larval drifting, or rafting.

Some species can disperse through sub-surface hydrologic connections. Groundwater flow paths can provide some potential dispersal routes among meta-populations of small species that occur in caves and aquifers. The number of freshwater species that rely on active dispersal by swimming or flying to reach islands is relatively low. Many species use active dispersal within islands along persistent migratory routes along rivers, which are especially critical for many of the marine-derived species. During the early stages of colonization, these initial riverine species are often derived from marine taxa of fishes, crustaceans, and gastropods. Most of these species require riverine migrations along corridors back to marine or estuarine habitats to complete their life cycle. Other species form completely freshwater communities as they continue to adapt to lower and variable salinities over long periods of evolution. Variability of salinities in these coastal rivers can be greatly affected by changes in sea level, storm surges, and saltwater intrusion into groundwater. Effects of high winds and oceanic waves significantly alter estuaries and riverine migrations.

DISTURBANCES AND SPECIES RICHNESS

Island location strongly affects the frequency and intensity of severe storms (hurricanes, typhoons, cyclones). These storms result in extremely intensive rainfall that erodes steep terrain, creates landslides, and floods river valleys, altering channel structure and substrata within river drainage networks. Severe winds remove riparian forest cover resulting in increased sunlight and warmer stream temperatures.

Although some islands experience relatively few tropical storms in any decade, over longer ecological and evolutionary timescales, the effects of these storms on freshwater populations can be important. For example, direct strikes by hurricanes on the Hawaiian Islands are rare relative to Guam and many other Pacific islands. However, the four major hurricanes that impacted Hawaii since 1950 likely had important effects on river discharges and species distributions.

Patterns of severe storms are known to reflect climatic variations in many parts of the world: El Niño–Southern Oscillation (ENSO) and Tropical Pacific Decadal Variability (TPDV). ENSO effects vary at timescales of two to five years, whereas TPDV patterns persist for 10 to 30 years, with occasional shorter reversals. The connections between (1) rainfall and (2) streamflow relative to the strength of ENSO and other climatic variables demonstrate that stream flows are significantly lower than average during and after El Niño periods and higher than average during La Niña periods across the main Hawaiian Islands. The El Niño drought of 1997–1998 resulted in major reductions in rainfall and stream flow from Hawaii to New Caledonia. The number and duration of these droughts have likely had important effects on how insular freshwater species adapt to highly variable habitats.

COMPARATIVE EXAMPLES OF INSULAR SPECIES

To illustrate the effects of different ages and isolation on species richness, several examples provide contrasts among well-studied islands. Ecologists and conservation biologists are especially interested in freshwater diversity "hotspots" on islands of different ages. For example, the "continental islands" (e.g., Trinidad, Madagascar, the Seychelles, Tasmania, and New Caledonia) were geologically derived from mainland sources at different times. They are relatively old and high in species richness and are isolated from their related continental species. Some taxa speciated to form diverse "radiations" of species. For example, Madagascar is a large "continental island" with exceptionally diverse assemblages among freshwater fishes, crayfishes, crabs, and insects. These endemic species have evolved after separation from Africa and India. On New Caledonia, five genera with 54 endemic species of small hydrobid snails occur in widely distributed springs. The highest numbers of endemic species are found on the wettest portions of the islands where, apparently, the spring flows have persisted for long periods. On Tasmania, an endemic species of decapods is one of the largest freshwater crustaceans, *Astacopsis gouldi*. This colorful crayfish species reaches a weight of 5 kg and a length of 80 cm; it is thought to have a 40-year lifespan.

Volcanic islands also provide examples of radiations of aquatic insects and other widely dispersed species. For example, blackflies of the genus *Simulium* provide insights into the pulsed radiations of speciation over varied geologic time periods. As high-elevation, volcanic islands eroded to form their complex drainage basins, numerous habitats and particular ecological niches developed for filter-feeding larvae (e.g., steep waterfall cascades, as well as rivers and streams of different sizes and flows). These filter-feeding blackflies evolved to occupy numerous niches throughout the Cook Islands, Marquesas Islands, and Society Islands. The highest species richness occurs among those on Tahiti, currently the youngest and largest of these islands.

Lakes on recently formed oceanic islands also provide additional types of routes for colonization of freshwater. For example, Lake Wisdom is a caldera lake on a newly formed volcanic island, Long Island. It is younger than nearby Lake Dakataua, another caldera in West New Britain, Papua New Guinea. More species of aquatic plants and aquatic invertebrates are found in the older lake, but there are similar species in both lakes that were likely transported by birds. Besides passive dispersal by birds, these lakes are associated with springs and small streams that flow from the deep volcanic calderas. These streams provide corridors for movements among molluscan and other species that are marine-derived groups. They invade "leaky" volcanic lakes by first adapting to variable salinity in these connecting streams.

Some islands have ancient lakes or persistent springs and rivers with different types of radiations of species. For example, Lake Poso in the Central Sulawesi Islands of Indonesia contains 16 endemic species of hydrobid snails and two endemic fishes (*Adrianichtys kruyti* and *Xenopoecilus poptae*). With a maximum depth of 450 m, this deep lake is home to an interesting assemblage that rivals some of the ancient continental lakes in terms of endemic species.

Because chains of volcanic islands can originate in a temporal sequence (e.g., the Hawaiian Islands), these spatially isolated serial locations can be useful "laboratories" for ecologists and others interested in evolutionary biology. Relatively young islands can be colonized by species from older islands or from continental sources over time. This serial dispersal within an archipelago increases biological diversity by extending the length of time for new species to evolve and for sequential land masses to accumulate species from the nearby island sources as well as from continents. The relationships between the age of a single island and its biodiversity are complexly defined by the potential for multiple sources of colonizing species. Consequently, an historical perspective is needed to understand species relationships among islands of different sizes, areas of catchments, drainage network connectivity, rainfall, and persistent habitats at varied distances from multiple sources for species dispersal.

CONSEQUENCES OF INVASIVE SPECIES

Once non-native species are introduced to insular freshwaters, they can become extremely abundant, die out, or vary in abundance without causing declines in native species or creating new environmental conditions. Once established, some non-native species rapidly dominate insular freshwaters and become invasive pests. Consequently, there is a need to understand both natural dispersal and control of introduced species in critical freshwater ecosystems. The negative impacts of invasive species often depend on how rapidly they spread after their initial colonization in one lake or one coastal river.

Insular freshwaters can be more vulnerable to invasive species if competition with native species does not occur or if native consumers of these new species are lacking. For example, species of fishes, crayfishes, and snails are being actively transported to islands for aquaculture or the aquarium trade throughout the world. Aquacultured fishes, such as *Tilapia mozambica* and *T. nilotica,* and river shrimps such as *Macrobrachium rosenbergii,* escape production ponds and disperse into rivers and lakes. Another example of a widely distributed species is *Procambarus clarkii,* the "red-swamp" crayfish native to the south-central United States and northeastern Mexico. This crayfish has been introduced for aquaculture to insular freshwaters in the Azores, Hispaniola, Puerto Rico, Japan, Taiwan, the Philippines, Sri Lanka, Hawaii, and many other locations. Their omnivorous feeding, tolerance of variable salinities, dissolved oxygen concentrations, warm water temperatures, and high mobility allow them to spread quickly from original places of introduction. Adults consume small fishes, snails, aquatic plants, and many other foods that can lead to declines among native species. Today, these crayfish are being sold in pet stores as miniature "red lobsters." A wide diversity of freshwater fishes, frogs, turtles, and plants are often dispersed through pet stores because aquarists often release them into natural waterways and do not understand the importance of protecting native species. Eventually, simplification of food webs can occur, followed by instability in community structure and shifts in ecosystem processes.

Potentially positive examples are small fishes, such as *Gambusia affinis,* that were intentionally introduced to control mosquito larvae, vectors for many human diseases. They survive in highly variable habitats with wide ranges of salinity and dissolved oxygen. However, they too can affect native species of freshwater fishes and invertebrates that also feed on mosquito larvae or other potential pests.

Recent examples of wide dispersal of freshwater species on islands include several species of mosquitoes (e.g., *Aedes albopictus*) and freshwater snails that are associated with global commerce in areas such as used tires and the aquarium trade. *Melanoides tuberculata* and *Thiara granifera* are snails that parthenogenetically reproduce, bear live young, reach very high densities, and rapidly spread to many rivers and lakes. These thiarid snails serve as hosts for *Paragonimus westermani,* a lung fluke that causes paragonimiasis in humans. Other pulmonate snails (*Biomphalaria glabrata*) are also widely distributed on tropical islands and serve as hosts for liver flukes (*Schistosoma mansoni*). These snail vectors transmit schistosomiasis. In some shallow waters, *Thiara* and *Melanoides* can perhaps outcompete *Biomphalaria*. Another biological control agent, *Marisa cornuarietis,* has been introduced to tropical islands as an omnivorous ampullarid snail that is effective at consuming eggs and juveniles of vector snails such as *Biomphalaria* as well as aquatic plants. This giant "ram's horn" snail is native to South America and now occurs on many islands in the Caribbean. Another ampullarid species, the "apple snail" (*Pomacea canaliculata*), has both lungs and gills. It is well adapted to obtain either dissolved oxygen or atmospheric oxygen in highly variable and productive shallow waters. *Pomacea* is also very well adapted to survive long dry periods by burrowing into the mud and remaining dormant (for more than a year in moist sediments). It recovers quickly after drought and reproduces rapidly. Although it is naturally dispersed by migratory birds, is it even more widely dispersed by aquaculturists. Because of its biological adaptations for dispersal, *Pomacea* is now a pest in rice and poi fields on many islands (Hawaii, Taiwan, Philippines, Indonesia, Malaysia, Thailand, and others). Studies to control its spread provide insights on the strength of these adaptations for dispersal and provide lessons on the importance of understanding each species's biology before introducing it, no matter how good the initial intentions may be.

CONCLUSIONS

Information on colonization of insular freshwaters is beginning to document the importance of speciation and extinction as critical ecological and evolutionary processes. Understanding how different freshwater species assemble and evolve on islands of different ages, locations, and types is relevant to better management of natural communities. Adaptations for active and passive dispersal vary greatly among different groups of plants and animals. Their persistence over time and ability to recover from natural and human-generated disturbances relies on how well adapted they are for dispersal and recolonization. Over time, it will be interesting to determine whether insular freshwater communities are equilibrial assemblages that are saturated with species or whether they can continue to increase in species richness as changes in dominance occur and speciation continues.

It is unclear how many insular freshwater species have gone extinct as a result of the spread of invasive species such as disease-transmitting vectors (mosquitoes and snails). The role of native predators and their dispersal abilities relative to their prey species is of particular interest to ecologists and public health researchers. If proliferation of invasive species leads to declines in species richness and increased spread of waterborne diseases, then the costs for control measures will greatly accelerate. Higher frequency of both intentional and non-intentional introductions of invasive species will require further analysis of how the dispersal of these species compares with those naturally dispersed.

SEE ALSO THE FOLLOWING ARTICLES

Cichlid Fish / Dispersal / Hydrology / Lakes as Islands / Sticklebacks

FURTHER READING

Alvarez, M., and I. Pardo. 2007. Do temporary streams of Mediterranean islands have a distinct macroinvertebrate community? The case of Majorca. *Archiv für Hydrobiologie* 168: 55–70.

Ball, E., and J. Glucksman. 1980. A limnological survey of Lake Dakataua, a large caldera lake on West New Britain, Papua New Guinea, with comparisons to Lake Wisdom, a younger nearby caldera lake. *Freshwater Biology* 10: 73–84.

Benstead, J. B., P. H. De Rham, J.-L. Gattolliat, F.-M. Bibon, P. V. Loiselle, M. Sartori, J. S. Sparks, and M. L. Stiassny. 2003. Conserving Madagascar's freshwater biodiversity. *BioScience* 53: 1101–1111.

Christy, M. T., J. A. Savidge, and G. J. Rodda. 2007. Multiple pathways for invasion of anurans on a Pacific island. *Diversity and Distributions* 13: 598–607.

Covich, A. P. 2006. Dispersal-limited biodiversity of tropical insular streams. *Polish Journal of Ecology* 54: 523–547.

Covich, A. P., T. A. Crowl, C. L. Hein, M. J. Townsend, and W. H. McDowell. 2009 Importance of geomorphic barriers to predator-prey interactions in river networks. *Freshwater Biology* 54: 450–465.

Craig, D. A. 2003. Geomorphology, development of running water habitats, and evolution of black flies on Polynesian islands. *BioScience* 53: 1079–1093.

Jones, J. P. G., F. B. Andreiahajaina, N. J. Hockley, K. A. Crandall, and O. R. Ravoahangimalala. 2007. The ecology and conservation status of Madagascar's endemic freshwater crayfish (Parastacidae; *Astacoides*). *Freshwater Biology* 52: 1820–1833.

McDowall, R. M. 2004. Ancestry and amphidromy in island freshwater fish faunas. *Fish and Fisheries* 5: 75–85.

FROGS

RAFE M. BROWN

University of Kansas, Lawrence

Frogs and toads (collectively termed anurans), together with salamanders and caecilians (eel-like amphibians), constitute the living members of the clade Lissamphibia, which contains all modern amphibians. Amphibians on islands are of intense interest to evolutionary biologists because of the perceived improbability of amphibians naturally crossing saltwater barriers, combined with the undeniable fact that some frogs seem to be relatively proficient at accomplishing this difficult feat. Frogs, in particular, provide an opportunity for evolutionary biologists to address a variety of evolutionary questions within a well-defined historical perspective.

MYSTERIOUS ISLAND FROGS

The mysterious distribution of frogs on islands has piqued the curiosity of explorers, herpetologists, biogeographers, and systematists for over two centuries. The basic questions of how and when frogs arrived on oceanic islands has dominated most of this literature and supplied biogeographers with a wealth of fantastic dispersal or mysterious land-bridge scenarios to ponder and debate. Aside from the questionable plausibility of some of the competing hypotheses is the simple fact that numerous island archipelagos have conspicuously species-rich frog faunas. Some celebrated isolated islands (e.g., Seychelles, Borneo, New Zealand) support phylogenetically deep frog lineages that are often thought of as evolutionary relicts. Other archipelagoes possess spectacularly diverse radiations—very large groups of closely related species. Both of these patterns are testaments to the fact that some frog lineages have had long and complex histories in some of the Earth's most isolated islands and island archipelagoes.

EARLY EXPLORATIONS

The first publications on island frogs were the descriptive works of early Europeans (notably G. A. Boulenger at the British Museum, but also including many others). This "Age of Discovery" marked the introduction of the Western world to many rare, morphologically bizarre, and taxonomically confusing island anurans.

Some of the first comprehensive works by amphibian biologists highlighted the seemingly aberrant distributions of both New and Old World island taxa. G. K. Noble's classic *The Biology of the Amphibia* (published in 1931 by McGraw-Hill) made note of the tailed frogs of New Zealand (family Leiopelmatidae); the flat-headed frogs (family Bombinatoridae) of the Philippines; the Seychelles frogs (Sooglossidae) of the Seychelles islands; the ground frogs, toadlets, and water frogs (Limnodynastidae and Myobatrachidae) of New Guinea and Australia; and the many odd true frogs (Ranidae), toads (Bufonidae), horned frogs, and dwarf litter frogs (Megophryidae) of the Natuna Islands, Borneo, the Philippines, and the Indo-Australian archipelago. Additionally, Noble highlighted the diversity of the various frogs of the families Leptodactylidae and Eleutherodactylidae in the West Indies and the few species the tree frogs (Hylidae) from eastern Indonesia, New Guinea, and Australia. Finally, Noble discussed extensively the wide morphological and taxonomic diversity within Melanesian forest frogs (ceratobatrachine ranids) of Southeast Asia and the islands of the southwestern Pacific, the Madagascar frogs (Mantellidae) of Madagascar and the Comoros Islands, and the narrow-mouth frogs (Microhylidae) of eastern Indonesia and New Guinea.

Building on descriptive and synthetic work of the past, herpetologists of the twentieth century made major advances toward understanding global patterns of frog-species richness on islands. However, it is clear that anuran diversity on several key island archipelagoes is still very poorly known and that much basic descriptive work remains to be carried out.

SPECTACULAR FROG RADIATIONS AND BIZARRE INSULAR SPECIES

No discussion of anuran diversity on islands would be complete without treatment of the extraordinary diversity of several insular radiations. Our understanding of frog diversity in several archipelagoes has progressed with much recent work and a growing sense on the part of herpetologists that the frog fauna of the Earth is still far from well known. Consequently, arriving at comprehensive summaries of anuran species diversity in several island nations has become a major scientific and conservation priority.

One such hotspot of anuran diversity is the endemic frog fauna of Madagascar. In addition to a few species in the family Ranidae, Madagascar is home to high levels of species diversity in the frog families Hyperoliidae (11 species), Microhylidae (60 species), and Mantellidae (163 species), and has an approximate total of 235 species of Malagasy mantellids (Fig. 1). All are endemic to the island and possess a spectacular array of morphological and ecological specializations, suggesting multiple independent origins of many diverse combinations of ecology and morphology into specific types of frogs ("ecomorphs") that mirror those found on the nearby African mainland.

A curious radiation of amphibians of the Seychelles Islands has puzzled herpetologists for two centuries. The endemic amphibians of the Seychelles are few in number, but all are phylogenetically unique. The Seychelles archipelago is home to only 12 species of amphibians but possesses an endemic frog family (Sooglossidae) that consists of four species in three endemic genera. (The caecilians of the Seychelles are also noteworthy; although there are only six species in the archipelago, the level of endemism is unprecedented, with three endemic genera [*Grandisonia*, *Hypogeophis*, and *Praslinia*: 14% of the world's currently recognized caecilian genera].) Interestingly, a single species related to Sooglossidae has been discovered recently in the Western Ghats of India. This unusual burrowing frog may be the sole extant species descendant from a lineage that survived the rafting of the Indian subcontinent

FIGURE 1 Selected exemplars of island frogs from around the world: (A) *Philautus poppiae* (family Rhacophoridae) from Sri Lanka (photograph courtesy of M. Meegaskumbura); (B) *Ceratobatrachus guentheri* (family Ranidae) from the Solomon Islands (photograph courtesy of R.M. Brown); (C) *Scaphiophryne spinosa* (family Microhylidae) from Madagascar (photograph courtesy of F. Glaw and M. Vences); (D) *Barbourula busuangensis* (family Bombinatoridae) from Palawan Island, Philippines (Photograph courtesy of R.M. Brown); (E) an undescribed species of larviparous (live tadpole laying) fanged frog, genus *Limnonectes* (family Ranidae) with its recently deposited clutch of fully formed tadpoles, from Sulawesi (photo: R.M. Brown); (F) *Mantella crocea* from Madagascar (photograph courtesy of F. Glaw and M. Vences).

and its subsequent collision with southern Asia. Other island frog radiations are equivalent in size and content to those found on Madagascar, and others are larger still. Frogs of the family Microhylidae endemic to the island of New Guinea (Papua New Guinea and Indonesian Papua [formerly Irian Jaya]) number somewhere between 190 and 200 species, and an additional approximately 200 species currently await description. As with the mantellid frogs of Madagascar, New Guinea microhylids have undergone substantive diversification; microhylids now occupy all major frog niches on New Guinea, from hefty and rotund burrowing ground frogs to delicate tree frogs. New Guinea also seems to be the source for lineages of microhylid frogs that have dispersed into the small island archipelagoes of eastern Indonesia and given rise to approximately 15 endemic species in the island archipelagos of Wallacea (eastern Indonesia).

Other large islands, notably Sri Lanka and Borneo, possess high species diversity and similarly high levels of endemism. Sri Lanka in particular has been widely cited as a global hotspot of frog biodiversity. With more than 105 species of frogs (Fig. 1), and approximately 65 endemic species in a single family (Rhacophoridae, genus *Philautus*), Sri Lanka boasts very high species richness contained within a single small, isolated landmass. Borneo has 145 species of frogs, 67% of which (97 species) are endemic to this single large island.

In several other island systems, extremely high species diversity can be spread across archipelagos composed of hundreds to thousands of small islands. The world's premier cases of this phenomenon are the frog faunas of the West Indian islands and those of the Indo-Australian archipelagos (Malaysia, the Philippines, and Indonesia). In the former case, more than 170 species of frogs of the genus *Eleutherodactylus* appear to have arisen from a lineage that arrived in the region by over-water dispersal and subsequently diverged via a combination of dispersal and vicariance. In the case of the Indo-Australian archipelagoes, more than 510 species are shared between the primarily island nations of Malaysia, the Philippines, Brunei, and Indonesia. Several groups contribute most heavily to these high levels of species richness. One exemplar is the Southeast Asian and Melanesian forest frogs (considered by some to belong to their own family—Ceratobatrachidae). The approximately 85 species are distributed throughout the islands of the Philippines, eastern Indonesia, Palau, New Guinea, and the Solomon Islands, and the Bismarck, Admiralty, and Fijian archipelagoes. This diverse group has a full complement of ecomorphs (ground frogs, water frogs, tree frogs, dwarf species, and giant species [Fig. 2]), and all are direct developers (i.e., they lay eggs on land, and the hatchlings emerge from their eggs as tiny, fully formed froglets, not tadpoles). Biologists have even hypothesized that direct development may be a crucial evolutionary step that allowed frogs to colonize and persist on islands with little or no standing freshwater.

OVER-WATER DISPERSAL

Until very recently, herpetologists continually debated the origins of insular frog faunas, with arguments centered around three general ideas: that island faunas may have originated from faunal dispersion over land bridges, from break-ups of land masses (vicariance), or from improbable dispersal over saltwater. The desiccation-susceptible physiology of amphibians and their delicate, permeable skin convinced many researchers that amphibians were simply incapable of dispersal over saltwater, especially over long distances. Even the notion of rafting on logs or vegetation seemed implausible, given the near-universal assumption that amphibians should be intolerant of aridity and any exposure to saltwater. In the minds of many,

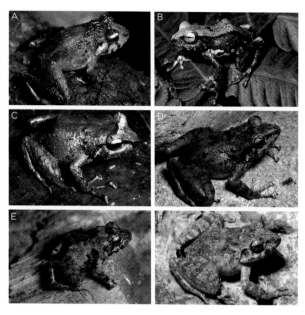

FIGURE 2 Closely related species representing different "ecomorphs" from a radiation of insular forest frogs from the Philippines: (A) *Platymantis dorsalis*, a medium-sized (3-4 g) forest floor species; (B) *Platymantis isarog*, a small (1-2 g) montane shrub frog; (C) *Platymantis banahao*, a large-bodied (4-7 g) tree canopy frog; (D) *Platymantis insulatus*, a large (3-5 g) limestone specialist; (E) *Platymantis pygmaeus*, a tiny (0.5-1.0 g) miniature leaf litter species; and (F) *Platymantis speleaus*, a large-bodied (8-15 g) limestone cave species. Photographs courtesy of R.M. Brown.

this left only a single, alternate possibility: namely, that amphibians must have arrived on today's islands when they were previously connected to continents. These first land-bridge hypotheses included scenarios of land mass vicariance and sea level–mediated ephemeral dry land connections between adjacent land masses. The first of these stems from the fact that many of today's islands were part of former continents and gradually broke off and drifted to their current position (carrying their frog stowaways in the process). The second land-bridge hypothesis suggested that dry land connections were formed by falling Pleistocene sea levels, which would have connected nearby landmasses if separated by sufficiently shallow seas, thus allowing for frogs to cross from one island to another before sea levels rose again.

The advent of new molecular tools for estimating phylogenetic relationships among organisms has provided anuran biologists with powerful tools for addressing these and other related questions. Numerous recent studies have demonstrated that dispersal via a dry land connection can often be rejected in favor of a simple hypothesis of over-water dispersal. Anuran biologists now accept (to varying degrees) the possibility that frogs have often dispersed over marine barriers. The assumption is, of course, that frogs frequently raft on mats of vegetation washed down rivers and into the ocean following heavy rains. Other studies have pointed to the possibility of freshwater "lenses" or channels that often form on top of saltwater following heavy storms. It is assumed that a combination of these phenomena could easily result in a much more hospitable temporary environment that might allow for frog dispersal over saltwater. However implausible this alternative remains, it is clear from many phylogenetic studies that frogs have successfully dispersed over saltwater in numerous ancient and more recent times.

One point to note is that most small oceanic islands that are far from mainland sources have no native amphibians (e.g., Canaries, Capverdian islands, Galápagos, Hawaii, Mascarenes). A few somewhat less-isolated islands such as the Gulf of Guinea islands, the Comoros, Palau, and Fiji provide exceptions and further testament to the astounding dispersal capabilities of some lineages of amphibians.

PROSPECTS AND CHALLENGES FOR THE FUTURE

Current prospects for the study of island frogs are quite promising, especially given the rapid pace of collection of DNA sequence data and promising new analytical and computational tools for estimating phylogenies based on large amounts of DNA sequence data. Interest in amphibian systematics is burgeoning. At least two independent research groups are working toward comprehensive phylogenetic analyses and a "tree of life" for all amphibians. All of these efforts are greatly improving our understanding of patterns of diversity, origins, and processes of diversification in island amphibians.

The future of the study of evolutionary biology of insular frog communities also has several major challenges. First, biologists and policymakers must work together to overcome logistical and bureaucratic obstacles to field research in several biodiversity-rich island nations. At present, several large islands are so poorly studied that even a gross understanding of patterns of species richness is prevented by a near complete lack of biodiversity survey work. For example, nowhere is this situation more serious than on the Indonesian portions of the large islands of Borneo (Kalimantan) and New Guinea (Papua). Most of what we know about diversity of frog communities on Borneo comes from the Malaysian states of Sabah and Sarawak (north Borneo), and yet the much larger Indonesian (southern) portion of Borneo is largely unstudied. Similarly, our information on the frog diversity of New Guinea is dominated by a wealth of studies from the eastern half of the island (Papua New Guinea), whereas the western half of the island (Indonesian

Papua) remains very poorly known. In both of these instances, bureaucratic and political obstacles to research in Indonesia have prevented (and continue to prevent) comprehensive assessments of the biodiversity of these islands.

Second, a global conservation crisis of habitat destruction, climate change, and emerging infectious disease (chytridiomycosis) currently threatens more than one-third of the world's frog and toad species. Amphibians on islands are increasingly susceptible to environmental and atmospheric perturbations and may be at increased risk of extinction because populations are small and genetically homogenous. The ticking clock of this extinction crisis makes the study of insular frog faunas increasingly important if we are to understand the evolution of frog communities on islands. Overcoming these and other challenges will require cooperation between biologists, policymakers, and governments on a scale not yet realized.

SEE ALSO THE FOLLOWING ARTICLES

Borneo / Fiji, Biology / Madagascar / New Guinea, Biology / Philippines, Biology / Seychelles / Solomon Islands, Biology / Sri Lanka

FURTHER READING

Duellman, W. E., ed. 1999. *Patterns of distribution of amphibians: a global perspective*. Baltimore, MD: John Hopkins University Press.

Duellman, W. E., and L. S. Trueb. 1986. *Biology of amphibians*. New York: McGraw-Hill.

Hutchins, M., W. E. Duellman, and N. Schlager, eds. 2003. *Grzimek's animal life encyclopedia*, 2nd ed.: *volume 6, amphibians*. Farmington Hills, MI: Gale Group.

Noble, G. K. 1931. *The biology of the amphibia*. New York: McGraw-Hill.

Savage, J. M. 1973. The geographic distribution of frogs: patterns and predictions, in *Evolutionary biology of the anurans: contemporary research on major problems*. J. L. Vial, ed. Columbia: University of Missouri Press, 351–445.

Stebbins, R. C., and N. W. Cohen. 1995. *A natural history of amphibians*. Princeton, NJ: Princeton University Press.

GALÁPAGOS FINCHES

**HEATHER FARRINGTON
AND KENNETH PETREN**
University of Cincinnati, Ohio

The Galápagos finches, also known as Darwin's finches, are a group of 15 passerine bird species, 14 of which are endemic to the Galápagos Islands of the equatorial Pacific. The fifteenth species is endemic to Cocos Island off the west coast of Costa Rica. Darwin's finches are a textbook example of adaptive radiation, and a valuable model for understanding natural selection in a variable environment. Recent advances in genetic technology have answered many questions regarding the evolutionary history of these birds, yet they continue to captivate and puzzle researchers today.

BACKGROUND

The Galápagos finches were first recorded and collected by scientist Charles Darwin and his shipmates during the famous worldwide voyage of the HMS *Beagle* (1832–1837). Darwin was preoccupied with fossils for much of the voyage until reaching the Galápagos, where the mockingbirds, not the finches, embodied his ideas about how differences in form can arise in isolation. It was not until long after returning to England that John Gould examined Darwin's collection and concluded that several of the plain-colored finches were species unique to the Galápagos. They looked very different from each other, yet they all shared several traits distinct from other passerines, making it clear that they descended from a recent common ancestor.

Darwin's finches are best known for their incredible diversity of beak shapes, which evolved for adaptation to different environmental niches (Figs. 1–2). There are three major groups of Darwin's finches: (1) The ground finches are seed and/or cactus specialists that have broad, deep, triangular beaks specialized for crushing; (2) the tree finches primarily eat concealed insects and have slightly curved beaks to help them grasp and tear vegetation; and (3) the warbler and Cocos Island finches glean surface insects and sip nectar with their long, narrow beaks specialized for probing vegetation. The beak morphology of each species is one of the greatest determinants of foraging behavior. For the ground finches, in particular, the size and shape (width, depth, and length) of the beak constrain each species to seeds or fruits of a particular size or shell hardness. The evolution of the various beak morphologies is thought to be driven by niche specialization to reduce resource competition among species.

In the Galápagos archipelago, various combinations of Darwin's finch species are found in sympatry on each island. Generally, large, high-elevation islands have the most finch species, whereas small, low islands have fewer. This is most likely due to the wider range of habitats that can be supported on a larger island with an elevation gradient. In the Galápagos, low areas tend to be very arid and to have little vegetation, whereas moist forested areas can be found at higher elevations.

EVOLUTION

Darwin's finches are a textbook example of adaptive radiation. Fifteen species of finches evolved from a single group of ancestors that colonized the Galápagos Islands approximately 2–3 million years ago from Central or South America. Several characteristics of this archipelago make it an ideal place for adaptive radiation. First,

FIGURE 1 Darwin's finches illustrating differences in beak size and shape. (A) The large ground finch (*Geospiza magnirostris*) has an exceptionally deep beak for crushing large seeds. (B) The cactus ground finch (*G. scandens*) uses its elongated beak to eat pollen from cactus flowers. Compare with Fig. 2.

FIGURE 2 (A) The warbler finch (*Certhidea olivacea*) is a small bird with a slender beak for gleaning insects from the surface of vegetation and for sipping nectar. (B) The woodpecker finch (*Cactospiza pallida*) uses its long, narrow beak to pull insect larvae from holes; it also uses small sticks as tools to help probe holes.

the isolation of the Galápagos has resulted in few colonizations by mainland organisms. Species-depauperate islands provide ecological opportunities for a colonist to change and adapt to unused niches. The islands within the archipelago also exist in varying degrees of isolation. Remote islands receive fewer immigrants, which allows for local adaptation and for the accumulation of differences between populations. Second, most of the islands are very small (less than 150 km^2) and therefore have relatively small finch populations. These small populations are greatly affected by different selective regimes, stochastic environmental events, and genetic drift, all of which can cause genetic and behavioral differences to accumulate between populations. Environmental conditions on small Galápagos Islands are especially harsh and variable, which results in periods of strong natural selection and rapid evolution of adaptive changes. Natural selection on beak size and shape has been documented repeatedly by Peter and Rosemary Grant, who have studied Darwin's finches with their students and colleagues for more than three decades, and who played an important role in much of the research summarized here.

The adaptive radiation of Darwin's finches presents a unique opportunity to study the relatively recent evolutionary history of a group of closely related species. Knowing that all 15 of these finch species were derived from a recent common ancestor, researchers can reconstruct the sequence of evolutionary events that gave rise to the species we see today. Studying evolution in an isolated island system like the Galápagos can help us better understand the process of evolution in mainland systems. Several phylogenetic studies have attempted to resolve the evolutionary history of this group, including the identity of the original founding species, using various genetic markers (Fig. 3). These studies have generated a number of interesting conclusions regarding the evolutionary history of this radiation.

According to genetic data, the closest living avian taxa to Darwin's finches are the grassquits (Emberizidae: *Tiaris*) and their relatives of mainland Central and South America. Grassquits look much like small ground finches, yet Darwin's finches share a wide variety of different traits such as molting patterns, nest construction, song, and color patterns with many different birds from the mainland and Caribbean.

The most notable aspect of the evolutionary history of Darwin's finches derived from phylogenetic studies is the incongruence between morphological and genetic divergence. Surprisingly, differences in morphology do not correlate well with genetic differences in this group.

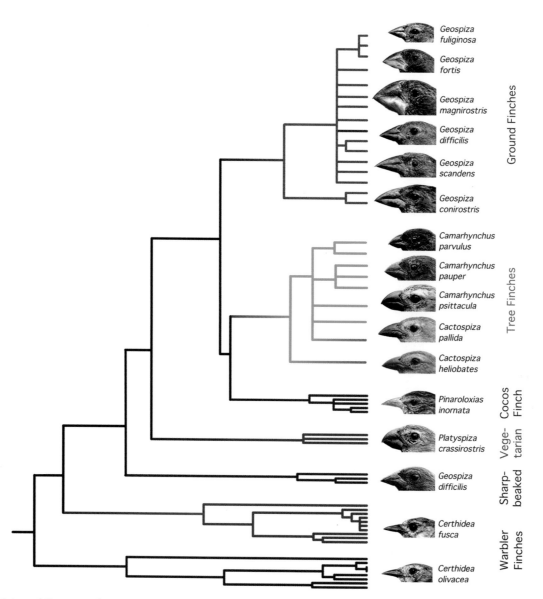

FIGURE 3 An evolutionary tree for Darwin's finches based on recent genetic sequence comparisons. The common ancestor is hypothesized to be on the extreme left. Bifurcations accumulate over time until we reach all current species on the right. Branch lengths are short and do not distinguish the sequence of evolution among most of the very different looking ground finches (orange) and tree finches (light green). In contrast, the very similar warbler finches and some populations of the sharp-beaked ground finches are very distantly related, as reflected by the long branch lengths (blue, red) connecting them.

For example, the warbler finch genus (*Certhidea*) originally contained only one species based on studies of morphological variation. However, genetic comparisons later revealed that there were two genetically distinct species within this genus. The two *Certhidea* species are nearly identical in morphology but have proven to be the most distantly related finches in the radiation. Similarly, genetic relationships among populations of sharp-beaked ground finches (*Geospiza difficilis*) are more distant than relationships among several other species. In contrast, the morphological characteristics among species in the tree and ground finch groups vary dramatically, but there is limited genetic differentiation within these groups, impeding resolution of their evolutionary history. In fact, species relationships among the tree and ground finches are hard to resolve using any type of molecular marker. The interpretation of these patterns is challenging, but the factors that obscure evolutionary history may also play a role in driving rapid speciation.

Hybridization occurs on a rare but regular basis among several species of Darwin's finches. The exchange of genetic material between species may partly explain why evolutionary relationships are hard to recover using genetic markers. Hybridization is also a potential source of new genetic material, which may provide a greater response to natural selection pressures. Cultural inheritance of song can lead to hybridization and introgression among species. Most interspecies pairings occur as a result of mis-imprinting, where a male in the nest learns the song of another species rather than the parent. The mis-imprinted male typically attracts mates from a different species. Hybrids produced by interspecies pairings are often viable and backcross with the species that shares the paternal song. Hybridization resulting in the exchange of genetic variation may be especially important for the evolution and persistence of small island populations that are commonly found in the Galápagos.

ENVIRONMENTAL VARIATION

The intense competition for food resources in the hot, arid environments of the Galápagos drives morphological differentiation among species. However, conditions are not always unfavorable in the Galápagos. El Niño–Southern Oscillation (ENSO) cycles cause drastic changes in island conditions (Fig. 4). Heavy rainfall and cooler temperatures stimulate excessive vegetative growth, which in turn leads to high reproductive output for the finches during times of food abundance. Small, low-elevation islands are the most affected by El Niño events because of the explosive growth of vegetation on usually barren islands. Unfortunately, El Niño events are usually followed by La Niña events, which bring unusually dry conditions on land. These times are marked by harsh selection, in which only the finches that are the most efficient and successful foragers survive. These "boom and bust" cycles keep selective pressures high and reinforce the morphological differences in beak morphology. Interestingly, the marine realm cycles in the opposite way, as warm El Niño conditions cause reduced upwelling and the movement of fish out of local waters. Boom times for finches correspond to dire times for seabirds and other organisms that rely on fish and marine resources.

POPULATION STRUCTURE AND SPECIATION

A common theory in evolutionary biology is that population divergence, which ultimately leads to speciation, must occur in isolation. Early hypotheses of the Darwin's finch radiation assumed that migration between islands in the Galápagos was rare enough to facilitate allopatric speciation. Isolated islands were colonized by individuals from other islands, the new isolated population would diverge significantly from the original population, and when the species came into contact again through a

FIGURE 4 (A) Daphne Major during the "dry" season of a dry period, in 2004. Finches forgo breeding when little or no rain falls and they must survive on a depleting reserve of available seeds. (B) The very same part of Daphne Major during the strong 1998 ENSO event. El Niño years bring rain and a spectacular flush of vegetation, insects, and finch breeding. Based on Petren et al., 2005; photographs by K. Petren, P. R. Grant, and H. Vargas.

colonization event, they would not interbreed because of strong morphological, behavioral or genetic differences.

Recent studies have shown that the finch populations in the Galápagos have a metapopulation structure. Genetic comparisons have revealed low levels of migration between populations throughout the Galápagos that are high enough to constrain local adaptation and prevent speciation under most conditions. The amount of gene flow between populations is constrained by the degree of isolation of each island; thus, we would expect more genetic differentiation to occur among island populations of species on remote islands with less immigration. The morphological differences observed among island populations suggest that, in many instances, natural selection has been strong enough to overcome the homogenizing effects of gene flow. Divergence between partially isolated island populations may represent the earliest stages of speciation, whereas competition for food in sympatry may complete the process and maintain species differences, but the exact sequence of events that led to the adaptive radiation of Darwin's finches still remains largely a mystery.

GENETIC BASIS OF BEAK MORPHOLOGY

Several recent studies have made significant strides toward identifying the genomic basis for the differences in beak morphology among Darwin's finches. Microarray analysis of beak tissues in developing chicks implicated BMP4 (bone morphogenic protein 4) and CaM (calmodulin) as major players in determining beak dimensions. Increased expression of BMP4 occurred in species with wider and deeper beaks, and increased expression of calmodulin occurred in finch species that have longer beaks. However, the exact genetic differences that control these expression differences, which were the targets of natural selection, remain unknown. Identification of the actual genetic mutations that underlie the morphological differences among Darwin's finches would offer incredibly valuable insight into the evolutionary processes of this adaptive radiation.

CONSERVATION

According to historic records, there have been no species extinctions within the Darwin's finch radiation. This contrasts markedly with many other endemic island birds. For instance, the Hawaiian honeycreepers are perhaps a more spectacular adaptive radiation, but more than half of these species are threatened or extinct. Although no Darwin's finches have been lost to extinction, about two dozen island populations have disappeared. More research is needed to determine if the observed population extinctions are a normal part of population dynamics or an early warning of demise and a cause to be concerned for the future of Darwin's finches.

One species of Darwin's finch is of particular concern to conservationists. The mangrove finch (*Cactospiza heliobates*) is listed as critically endangered on the IUCN red list. This species is limited to mangrove swamps on the coasts of Fernandina and Isabela Islands. The population on Fernandina has gone extinct within the last 100 years, whereas the Isabela population has been reduced to an estimated 60 breeding pairs. Efforts are currently under way in the Galápagos to develop a conservation plan to preserve this species and its habitat.

Several factors threaten not only the Galápagos avifauna, but other island species throughout the world as well. The most obvious threat to island species is human disturbance. Human predation and destruction of habitat and resources used by endemic species can drastically reduce population sizes and increase the probability of extinction. Human activities also increase the rate of species introductions (Fig. 5). Introduced animals and plants have drastically altered the vegetative landscape of islands, including many uninhabited Galápagos Islands. The Galápagos have experienced less human impact than many other archipelagoes because of the generally inhospitable terrain and dearth of water. This is changing rapidly as the population has now passed 20,000 residents, and the number of visitors has increased dramatically to more than 130,000 per year. In Galápagos, humans, goats, cats, rats, dogs, and introduced insects such as fire ants and parasitic flies are the major threats. Foreign diseases such as avian malaria and poxviruses have also recently been detected; most likely, they were carried in by introduced mosquitoes and domestic poultry, respectively. Additionally, global warming may cause

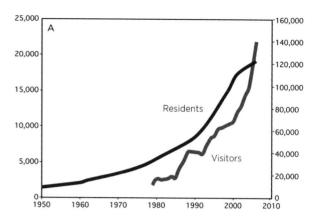

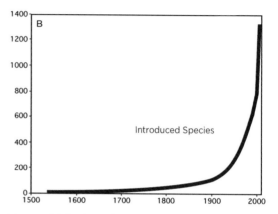

FIGURE 5 (A) Numbers of Galápagos residents (left scale) and tourists (right scale) have increased dramatically over the last 30 years. (B) With the increase in human visitors comes the inevitable increase in the number of introduced species. Source: G. Watkins and F. Cruz, the Charles Darwin Foundation, 2007.

changes in climate and sea level, significantly changing the Galápagos landscape.

SEE ALSO THE FOLLOWING ARTICLES

Bird Radiations / Climate on Islands / Galápagos Islands, Biology / Introduced Species / Metapopulations / Voyage of the *Beagle*

FURTHER READING

Darwin, C. 1859. *On the origin of species by means of natural selection.* London: John Murray, London.

Darwin, C. R. 1842. *Journal of researches into the geology and natural history of the various countries visited during the voyage of H. M. S. 'Beagle' under the command of Captain FitzRoy, R. N. from 1832 to 1836.* London: Henry Colborn.

Grant, P. R. 1999. *Ecology and evolution of Darwin's finches.* Princeton, NJ: Princeton University Press.

Grant, P. R., B. R. Grant, and A. Abzhanov. 2006. A developing paradigm for the development of bird beaks. *Biological Journal of the Linnean Society* 88: 17–22.

Lack, D. 1947. *Darwin's finches.* Cambridge: Cambridge University Press.

Petren, K., P. R. Grant, B. R. Grant, and L. F. Keller. 2005. Comparative landscape genetics and the adaptive radiation of Darwin's finches: the role of peripheral isolation. *Molecular Ecology* 14: 2943–2957.

Weiner, J. 1994. *The beak of the finch: a story of evolution in our time.* New York: Knopf.

GALÁPAGOS ISLANDS, BIOLOGY

TERRENCE M. GOSLINER
California Academy of Sciences, San Francisco

The Galápagos Islands are the classic example that comes to mind when most people think of islands and evolution. From the time of Darwin to the present, the Galápagos have provided remarkable examples of the process of evolutionary change through natural selection. This evolutionary history begins with the formation of new land and continues with its subsequent colonization. From there, new environmental conditions shape the processes of natural selection and adaptation that ultimately result in the appearance of new forms of life that develop novel lifestyles.

THE PROCESS OF COLONIZATION

The oldest rocks of the extant archipelago are about 4–5 million years in age and are found on the southeastern islands of Santa Fe, Española, and Floreana. Recent molecular studies suggest that, based on timing of genetic and molecular clock data, many Galápagos organisms appear to have been separated from presumed mainland relatives for about 10 million years or longer. Thus, the organisms appear to be older than the rocks on which they live! This paradox has been more recently resolved by recent studies that document submerged seamounts off the coast of Central and South America that have rocks 5–17 million years in age and that are part of the Galápagos archipelago. The prevailing view of evolutionary biologists suggests that newly emergent islands situated at the volcanic hotspot on the Nazca plate, near its junction with the Cocos and Pacific plates, provided virgin land initially devoid of life. This provided brand new habitat, rich in minerals and nutrients, to be colonized by newly arriving marine and terrestrial organisms. The predominant view is that the Galápagos were then colonized by organisms from other areas, especially from the mainland Americas. Not all agree that colonization is recent or is largely from the Americas, and some have suggested that the Galápagos Islands and their biota may be part of a persistent ridge dating back to the Cretaceous that migrated across the eastern Pacific over a much longer period of time. Many subsequent workers believe this latter point of view to be unnecessarily complicated and less consistent with both molecular clock and recent geological data.

The process of colonization is complex, and different groups of organisms have employed different mechanisms of dispersal to arrive in the Galápagos. The vast majority of marine species have planktonic larvae that are dispersed widely by oceanic currents. The prevailing Humboldt and Panama currents join near the equator, and then together as the South Equatorial Current bathe the archipelago with larvae-rich waters. In the early part of the calendar year, from January to May, the Panamic and South Equatorial currents dominate, bringing tropical warm waters from the coast of Central America to the Galápagos. In June, the situation is reversed, with the dominance of the Humboldt Current bringing cooler waters to the islands. As a direct result of the prevailing current patterns, most Galápagos marine species are known to also occur in the tropical eastern Pacific. The impact of these current patterns is also reflected in several of the oceanic species. The ancestors of several colder water marine species such as penguins and fur seals have followed the colder Humboldt Current to colonize the Galápagos. Other species, such as sea lions, have traveled from Baja California following the Panama Current to make their way to the Galápagos. Thus, the influences of both of these current regimes have manifested themselves in seeding the unique biota of the archipelago (Fig. 1). The far northern Galápagos Islands of Darwin and Wolf are less influenced by the

FIGURE 1 Plant species capable of ocean flotation and of diverse origins. (A) Red mangrove, *Rhizophora mangle*, Punta Espinosa, Isla Fernandina; (B) Galápagos cotton, *Gossypium davidsonii*, Sierra Negra, Isla Isabela; (C) sea purslane, *Sesuvium portulacastrum*, Cerro Dragon, Isla Santa Cruz; (D) beach morning glory, *Ipomaea pescaprae*, Puerto Egas, Isla Santiago; (E) beach bean, *Canvalia maritima*, Puerto Egas, Isla Santiago.

Humboldt Current. As a result, their marine biota is far more tropical. Many tropical marine species found in the waters of these two islands are found nowhere else in the Galápagos.

In addition to marine organisms arriving in the Galápagos by following oceanic currents, many terrestrial organisms have floated or flown to the islands (Fig. 2). It is hypothesized that many of these organisms were able to raft to the islands on floating masses of vegetation that had washed down South America's rivers. Some widespread plants, especially those found in the coastal strand, have seeds that are well adapted for flotation and maintenance of viability in seawater. These include plants such as the red mangrove, *Rhizophora mangle;* the sea-purslane, *Sesuvium portulacastrum;* the beach morning glory, *Ipomaea pes-caprae;* and the beach bean, *Canvalia maritima.* It has also been suggested that the ancestors of flightless weevils of the genus *Galapaganus,* Galápagos iguanas, and tortoises made their way to the islands by means of flotation.

In addition to rafting, many species can be dispersed by wind over long distances, especially plants with small spores or seeds, and small spiders and insects. Galápagos species that likely were brought to the islands by wind dispersal include most ferns, fungi, some spiders, and many winged insects that have made their way to the archipelago.

Many bird species flew to the Galápagos. Some of these, such as the waved albatross, travel vast distances annually and return to the islands for mating and nesting. Many other birds, especially landbirds, were blown off course, ended up in the archipelago, and became established in the islands. Still other species appear to have arrived by attachment to soil or mud on the feet of birds.

The Galápagos biota, like that of many oceanic islands, is striking in what organisms have not been successful in colonizing the archipelago. These include freshwater fish and amphibians.

Duncan Porter suggested the mechanisms for 378 species of naturally occurring Galápagos plants (Table 1). From these studies, it is evident that bird dispersal has been the most important mechanism of bringing plant species to the islands. Not all elements of the flora are uniformly derived by these mechanisms. For example, all but one of the ferns and their relatives, which have small spores, were likely brought to the islands by wind dispersal.

SISTER-GROUP RELATIONSHIPS OF THE GALÁPAGOS BIOTA

Darwin was the first to recognize the similarities of the Galapagos biota to species occurring in South America. More recent detailed studies suggested that 99% of the

FIGURE 2 Animal species capable of ocean flotation and of diverse origins. (A) Greater flamingo, *Phoenicopterus ruber,* Puerto Moreno, Isla Isabela; (B) Galápagos sea lion, *Zalophus wollebecki,* Isla San Cristóbal; (C) Galápagos fur seal, *Arctocephalus galapagoensis,* Puerto Egas, Isla Santiago.

TABLE 1
Likely Dispersal Mechanisms of Ancestors of Galápagos Plant Species

Bird Dispersal		Wind Dispersal		Oceanic Currents	
225 species	60%	118 species	31%	35 species	9%

NOTE: From Porter (1976).

non-endemic, indigenous species of plants are also found in South America. The remaining 1% occur in Mexico and Central America. Porter further suggested that the endemic plants also have their closest known relatives in Andean South America. This pattern has also been suggested for a wide range of other Galápagos taxa.

With the advent of molecular systematics and modern phylogenetic methods, we have new, powerful tools to test presumed evolutionary relationships and possible ancestors and/or sister taxa of Galápagos endemic species (Fig. 3). Sister taxa are those that are most closely related to another taxon of interest. Most of these studies support Porter's hypothesis. It has been demonstrated that the closest living relative of the Galápagos tortoises, *Geochelone nigra*, is the Chaco tortoise, *Geochelone chilensis*, which is found in southern South America. The sister taxon of the eastern Galápagos lava lizards is a species found in coastal Ecuador and Peru. The Galápagos marine iguana and land iguanas are most closely related to Central and South American iguanas of the genus *Ctenosuara*. The sister taxon of the Galápagos endemic sunflower genus, *Scalesia*, is the Andean endemic genus *Pappobolus*. Another Galápagos endemic genus of Asteraceae is *Lecocarpus*. It is the sister taxon of the South American endemic genus *Acanthospermum*. The Galápagos have two endemic genera of cactus, *Brachycereus* and *Jasminocereus*. They are sister taxa to each other and, in turn, collectively sister taxon to the Ecuadorian cactus genus *Armatocereus*. The sister species of the Galápagos penguin, *Spheniscus mendiculus*, is the Humboldt penguin, *Spheniscus humboldti*, which is known from Peru to Chile. The closest relatives of all of these Galápagos endemics are found in South America.

A few other taxa have sister-group relationships with taxa that are not strictly South American. The Galápagos cotton, *Gossypium klotzschianmum*, is the sister species to *G. davidsonii*, which is known only from southern Baja California, whereas the other Galápagos endemic cotton, *G. darwinii*, is most closely related to *G. barbadense*, which is also found along the mainland coast of South America. Similarly, the Galápagos sea lion, *Zalophus wollebecki*, is the sister species to the California sea lion, *Z. californianus*, which is found as far south as Baja California, Mexico.

Molecular phylogenetic studies also suggest that the Galápagos mockingbirds, *Nesomimus* spp., are most closely related to a clade of mockingbirds, *Mimus* spp., that are commonly found in Mexico south to Central America and the Caribbean. Similarly, the closest relatives of Darwin's finches (Fig. 4) are found in the Caribbean. Other species such as the greater flamingo, *Phoenicopterus*

FIGURE 3 Endemic Galápagos animals. (A) Galápagos mockingbird, *Nesomimus parvulus*, Puerto Moreno, Isla Isabela; (B) Santa Cruz giant tortoise, *Geocholone nigra* ssp. *porteri*, highlands, Isla Santa Cruz; (C) Hood mockingbird, *Nesomimus macdonandi*, Gardiner Bay, Isla Española; (D) flightless cormorant, *Phalacrocorax harrisi*, Punta Espiñosa, Isla Fernandina; (E) swallow-tailed gull, *Creagrus furcatus*, Darwin Bay, Isla Genovesa; (F) Galápagos penguin.

ruber, and the Galápagos pintail duck, *Anas bahamensis galapagensis*, have been derived from Caribbean and Central American distributions or relationships.

Some Galápagos species also have their closest relatives in subantarctic regions. The closest relative of the Galápagos fur seal, *Arctocephalus galapagoensis*, is the South American fur seal, *Arctocephalus australis*.

ENDEMISM

One of the major features of remote islands is the presence of species that are unique (endemic) to different island archipelagoes or to individual islands. It is generally believed that the percentage of endemic species is a reflection of the degree of isolation from colonizing sources and the age of the archipelago or island.

The Galápagos biota is known for its relatively high degrees of endemism, but certainly has lower percentage of endemics compared to the Hawaiian Islands and Madagascar. For the marine fauna, the percentage of endemic shore fishes in the Galápagos has been estimated to be 14–16%. For marine invertebrates, levels of endemism range between 7% and 40%. These differences may

FIGURE 4 Darwin's finches. (A) Vegetarian finch, *Camarhynchis crassirostris*, Isla Pinta; (B) warbler finch, *Certhidia olivacea*, Punta Suarez, Isla Española; (C) sharp-billed finch, *Geospiza difficilis*, Prince Phillip's Steps, Isla Genovesa; (D) small ground finch, *Geospiza fuliginosa*, Puerto Baquerizo Moreno, Isla San Cristóbal; (E) large ground finch, *Geospiza magnirostris*, Darwin Bay, Isla Genovesa; (F) cactus finch, *Geospiza scandens*, Academy Bay, Isla Santa Cruz.

be accounted for partially by differences in life-history strategies. Amphipods (with 40% endemism) have direct development, whereas most other taxa have a high percentage of species with planktonic larvae. In all cases, the largest portion of the Galápagos marine invertebrates is the Panamic component, consisting of species distributed in the Galápagos and elsewhere in the tropical eastern Pacific, rather than species restricted to the archipelago.

In contrast to marine biotas, the terrestrial biota, as a whole, has higher percentages of endemic taxa (Table 2). In all cases, except for seabirds, the percentage of endemic terrestrial organisms is higher than that found for marine species. Seabirds and landbirds have very different rates of endemism owing to different dispersal rates. Within the flowering plants studied by Porter, the ferns and their close relatives (pteridophytes), which have small spores capable of long-distance wind dispersal, have only about 9% endemism. In contrast, the angiosperms, very few species of which have seeds that are capable of long-distance dispersal, have 53% endemism. There are many endemic genera of plants and animals, including four genera of Asteraceae (Fig. 5). Similarly seabirds have much more effective dispersal capabilities and have only 26% endemism whereas landbirds, with reduced dispersal capabilities, have 83%.

ADAPTATIONS

Another aspect of the biology of island biotas is the unique adaptations that are found in island organisms.

TABLE 2
Indigenous Terrestrial Biota of the Galápagos Islands

Taxon	# of Species	# of Endemic Species	% Endemic Species
Plants	522	236	45%
Reptiles	20	20	100%
Mammals	4	2	50%
Seabirds	19	5	26%
Land birds	29	24	83%
Insects	1600	900	56%

The organisms inhabiting the Galápagos may occupy the same ecological niche as their mainland counterparts or may exhibit major modifications. Perhaps the most remarkable example of unusual adaptation is the Galápagos marine iguana, *Amblyrhynchus cristatus,* the only iguanid that has the capability of diving in marine waters to feed upon marine algae. The Galápagos endemic swallow-tailed gull, *Creagrus furcatus,* is the world's only nocturnally active gull.

Additionally, the closest relatives of the Galápagos endemic genus *Scalesia* are members of the South American genus *Pappobolus*. In South America, *Pappobolus* species are shrubs of dry arid slopes. Although most species of *Scalesia* in the Galápagos are also shrubs of arid portions of the islands, one species, *Scalesia pedunculata,* is a tree dominating the moist forests of the Galápagos highlands.

FIGURE 5 Endemic genera of Galápagos Asteraceae. (A) *Lecocarpus darwinii*, Cerro Colorado, Isla San Cristóbal; (B) *Lecocarpus pinnatifidus*, Punta Cormorant, Isla Floreana; (C) *Macraea laricifolia*, Tagus Cove, Isla Isabela; (D) *Darwiniothamnus tenuifolius* var. *glandulosus*, Puerto Moreno, Isla Isabela.

Loss of Dispersal

Many species that become established on islands lose the original dispersal mechanisms that were present in their ancestors. To a species living on a small island, having your seeds or juveniles disperse means you are likely to have them moved out to sea, away from suitable habitat. Seeds often lose awns or other attachment mechanisms and often become larger that their mainland counterparts. Similarly, wings in flying species often become reduced or entirely lost. The flightless cormorant, *Phalacrocorax harrisi*, has stubby wings and is no longer capable of flying. Flightlessness has evolved in many Galápagos insects, including four species of flightless grasshopper in the genus *Halmenus* and cactus weevils of the genus *Gerstaekeria*. Two of the three endemic Galápagos species of the carabid beetle genus *Calosoma* are flightless. The 15 species of weevils of the genus *Galpaganus* are endemic to South America and the Galápagos. The ten Galápagos endemics are all flightless.

Gigantism

On islands, because of the presence of open niches not occupied by competitors, species often are much larger than their ancestral counterparts. The best known case of gigantism is obviously the giant Galápagos tortoises, *Geochelone nigra*, which are much larger than their closest living relative from mainland South America. The ancestors of the Galápagos cacti of the genus *Opuntia* were probably moderately sized species from mainland South America. Treelike forms have evolved on several Galápagos Islands, especially *O. echios gigantea* on Isla Santa Cruz and *O. echios barringtonensis* on Isla Santa Fe.

Dwarfism

Some island species may also become much smaller. The Galápagos penguin is an example of a species that is smaller than its mainland closest relative, the Humboldt penguin, and other close relatives such as the Magellanic and African penguins. It is the second smallest penguin in the world. Species often become smaller in the presence of harsher conditions or more limited nutritional resources.

LOSS OF DEFENSE MECHANISMS

In the absence of historical predators, island species often lose their defense mechanisms. One of the best examples in the Galápagos is the reduction of stiff spines in cactus species where large herbivores such as tortoises were never present. On the islands of Genovesa, Marchena, Darwin, and Wolf, which always lacked tortoises, *Opuntia helleri* has soft rather than rigid spines. Production of defense mechanisms generally has some energetic cost. If those defenses are no longer needed, then species that have lost them will have greater reproductive success.

ADAPTIVE RADIATION

Adaptive radiation, where relatively few ancestral species undergo explosive speciation, is a phenomenon characteristic of most remote islands around the globe. Certain pioneer species have been able to undergo ecological and phenotypic diversification. Certainly, the classic example of insular adaptive radiation is the case of Darwin's finches in the Galápagos Islands. Many popular biology texts credit Darwin with recognizing the significance of this discovery. Darwin was unaware that the finches actually represented closely related species when he visited the islands because they were so different in their appearances. The ornithologist John Gould pointed this out to Darwin upon his return to England. Only after detailed study by David Lack, Bob Bowman, and Peter and Rosemary Grant have the importance and full extent of adaptive radiation and specialization of Darwin's finches become more fully understood.

Another example of adaptive radiation is the differentiation in carapace shape in the various subspecies of Galápagos tortoises. Populations from lower-elevation, more arid islands generally have a "saddleback" carapace, whereas those from higher-elevation, moister islands have a dome-shaped carapace. The saddleback shape permits greater extension of the head and neck to reach vegetation in lowland arid habitats. The populations inhabiting different islands or different volcanoes on the largest islands have all been considered as subspecies.

In contrast, within the Iguanidae (Figs. 6–8), the lava lizards of the genus *Microlophus* are considered to represent seven distinct species. Recent studies have shown that the radiation is likely the result of two distinct colonization events by two different *Microlophus* ancestors from South America.

FIGURE 6 Galápagos iguanid lizards. (A) Hood Island lava lizard, female, *Microlophus dilanonis*, Gardiner Bay, Isla Española; (B) Floreana lava lizard, male, *Microlophus grayi*, Punta Cormorant, Isla Floreana; (C) Marchena lava lizard, male, *Microlophus habellii*, Isla Marchena. See also Figs. 7 and 8.

FIGURE 7 Pinta lava lizards, male (A) and female (B), showing sexual dimorphism, *Microlophus pacificus*, Isla Pinta.

It appears that the sunflower genus *Scalesia* (Figs. 9 and 10), which is endemic to the Galápagos, arose from a single South American species of *Pappobolus*. From this ancestral species, 15 species and five subspecies evolved. Most of the 15 *Scalesia* species have only disk flowers, but one species, *S. affinis*, also has sterile ray florets. Experimental evidence has shown that presence of ray florets in *S. affinis* increases pollination success. The remaining species with similar floral structure exhibit major variation in leaf shape and are found in a wide variety of distinct locations, mostly in arid lowlands.

Members of the cactus genus *Opuntia* (Figs. 11 and 12) have undergone considerable variation in growth form and differentiation within the Galápagos. They have radiated into six distinct species and nine subspecies or varieties. Some species are low-growing forms whereas others may form tall trees more than 12 m high. The various species and subspecies have a strong geographical component to their distributions, and several of the forms are restricted to lowlands or to higher elevations on some islands.

Moths of the genus *Galagete* have radiated into 12 species in the Galápagos Islands. Molecular clock data suggest that the radiation began about 3 million years

FIGURE 8 More Galápagos iguanid lizards. (A) Galápagos land iguana, *Conolophus subcristatus,* Isla Seymour Norte; (B) Santa Fe land iguana, *Conolophus pallidus,* Isla Santa Fe; (C) marine iguana, *Amblyrhynchus cristatus,* Punta Suarez, Isla Española.

ago and required 1.8 million years to reach its present diversity.

Another insect radiation is that of the weevil genus *Galapaganus*. Here, ten endemic species evolved from the original ancestor. This radiation, based on molecular clock data, suggests that the split within the Galápagos endemics occurred about 7.2 million years ago, well before the current islands were formed. This fact suggests that the radiation began on islands that are now below the ocean's

FIGURE 9 Species of *Scalesia.* (A) *Scalesia incisa,* Cerro Brujo, Isla San Cristóbal; (B) *Scalesia villosa,* Punta cormorant, Isla Floreana; (C) *Scalesia baurii baurii,* Isla Pinta; (D) *Scalesia stewartii,* Isla Bartolome. See also Fig. 10.

surface, and it is consistent with other ages for marine iguanas, geckos, and lava lizards that also predate the age of the present islands.

The most extensive explosive speciation in the Galápagos is in the land snail genus *Bulimulus*. Various workers have suggested that at least 63–71 species have evolved from a single ancestor. These species have been remarkably successful and are found in all vegetative zones on the islands except those along the immediate coast zone. Thus, they inhabit the most arid and mesic habitats from sea level to the highest altitudes in the archipelago. In phylogenetic studies, the oldest islands have been shown to have snail species represented by the deepest nodes of the tree. The

FIGURE 10 More species of *Scalesia.* (A) *Scalesia helleri helleri,* Isla Santa Fe; (B) *Scalesia pedunculata,* Los Gemelos, Isla Santa Cruz; (C) *Scalesia affinis,* Sierra Negra, Isla Isabela; (D) *Scalesia pedunculata,* Los Gemelos, Isla Santa Cruz.

FIGURE 11 *Opuntia* cactus species. (A) *Opuntia echios* var. *gigantea*, Academy Bay, Isla Santa Cruz; (B) *Opuntia echios* var. *barringtonensis*, Isla Santa Fe; (C) *Opuntia galapageia* var. *galapageia*, Isla Pinta; (D) *Opuntia megasperma* var. *megasperma*, Post Office Bay, Isla Floreana. See also Fig. 12.

snails found on Isla Santa Cruz represent five independent lineages, suggesting multiple colonization events. The same pattern is evident in snails from Santiago Island. Older islands with greater habitat diversity have the greatest number of distinct species.

CONSERVATION BIOLOGY

The unique biota of the Galápagos Islands has faced serious threat from human activities since the islands were first discovered. The slaughter of giant tortoises by seafarers led to the decimation of populations by the beginning of the twentieth century. Today, three of the tortoise subspecies are extinct: those from the islands of Floreana, Fernandina, and Santa Fe. The Pinta subspecies has only one known extant individual, Lonesome George, in captivity at the Charles Darwin Research Station on Isla Santa Cruz. More recently, with immigration to the islands, a large and growing resident human population of 27,000, and more than 100,000 tourists visiting the islands annually, environmental problems in the islands have been greatly exacerbated.

Extinction

Human-induced extinction is the most dire of the potential environmental problems that plague the Galápagos Islands. As mentioned previously, most of the giant tortoise populations have been seriously reduced. Three subspecies are entirely extinct and have not been seen in more than 100 years. The Pinta tortoise population has been reduced to a single male, and attempts to have him breed have thus proven unsuccessful. Recent genetic studies have indicated that Lonesome George is actually more closely related to tortoises from Española than to those from Wolf Volcano on Isabela, with whom the attempted matings have been focused. Even more recently, a male tortoise with more than 50% Pinta genes was discovered in the population of tortoises from Wolf Volcano. Perhaps, these additional data will inform a strategy that will yield greater success.

The Charles Darwin Research Station has conducted an extensive tortoise-rearing program for more than 40 years. The population of tortoises from Española Island was down to 12 females and two males when the program began. Today, more than 100 tortoises have been returned to the island from successfully captive-bred tor-

FIGURE 12 More *Opuntia* cactus species. (A) *Opuntia helleri*, Darwin Bay, Isla Genovesa; (B) flower, *Opuntia echios* var. *gigantea*, Academy Bay, Isla Santa Cruz; (C) *Opuntia insularis*, Tagus Cove, Isla Isabela.

toises. Other programs and breeding centers are located on the islands of Isabela and San Cristóbal.

During U.S. military activities during World War II, the land iguana population of Isla Baltra was entirely extirpated. Fortunately, scientists during the 1930s had undertaken a transplantation experiment of iguanas from Baltra to Seymour Norte. Individuals from the well-established Seymour Norte population are now being bred at the Charles Darwin Research Station and are being reintroduced to Baltra.

Until recently, five species of flowering plant were considered to be extinct in the Galápagos. Two of them, the Floreana flax *Linum cratericola* and the Santiago daisy *Scalesia atractyloides atractyloides* have been recently rediscovered and are part of ongoing conservation efforts. The amaranth *Blutaphon rigidum* and the gourd *Sicyos villosa* are considered extinct. Another 20 Galápagos plants are critically endangered. Most of these have suffered from critical habitat loss and grazing by feral goats and other introduced herbivores. Remaining critical habitat has been fenced, and feral mammal eradication programs have been highly successful.

Several of the 80 land snail species of *Bulimulus* are critically endangered (*Bulimulus achatellinus*, *B. adelphus*, *B. andserseni*, *B. chemnitzioides*, *B. curtus*, *B. deridderi*, *B. duncanensis*, *B. eos*, *B. galapaganus*, *B. habeli*, *B. lycodus*, *B. saeronius*, and *B. tanneri*). This endangerment is largely due to habitat loss and the impact of non-indigenous species.

One of Darwin's finches, Darwin's large ground finch, *Geospiza magnirostris magnirostris,* is extinct, and the mangrove finch *Camarhynchus heliobates* is critically endangered, with only about 50 breeding pairs remaining in on Isla Isabela. The Galápagos petrel, *Pterodroma phaeopygia*, is also critically endangered.

Three species of Galápagos endemic rodents are extinct: Darwin's Galápagos mouse (*Nesooryzomys darwini*), the Indefatigable (Santa Cruz) Galápagos mouse, (*Nesooryzomys indefessus*), and the San Cristóbal rice rat (*Orysomys galapagoenis galapagoensis*). Probably all were extirpated by competition from introduced rats.

Nonindigenous Species

One of the most serious consequences of increasing human population is the continuing introduction of non-native species that fiercely impact native species. Goats were introduced over several centuries by mariners visiting the islands, to ensure a fresh source of meat. They proliferated and have decimated the vegetation on several islands. Their reproductive potential is legendary. Three goats introduced on Pinta Island went through exponential population growth and by 1971 had reached a population size of 30,000–40,000 goats. Goats have now been entirely eradicated from Española Island, Pinta Island, Santiago Island, and the northern portion of Isabela Island. Similar introductions of rats, donkeys, pigs, cats, and other domestic animals have also had severe impacts on vulnerable native species.

What has changed in the equation is the number of species that have been introduced recently with increased human presence on the islands. There are now at least 1321 introduced species in Galápagos (compared to 112 in 1990), including 748 nonindigenous plants. Insect species such as the fire ant (*Wasmannia auropunctata*), a wasp (*Polistes veriscolor*), and the cottony cushion scale (*Icerya purchasi*) have been especially problematic. Major invasive plant pests include the guava (*Psidium guajava*), the Barbados cedar (*Cedrela odorata*), red quinine trees (*Cinchona pubescens*), and several species of blackberries (*Rubus* spp.). Not only are eradication programs under way by the Charles Darwin Research Station and the Galápagos National Park, but also strict quarantine procedures have been introduced to stem the tide of introductions.

Global Climate Change

As stated earlier, the biota of the Galápagos is dependent on a series of seasonal climatic shifts that bring needed moisture to largely arid regions. The alternation of cold and warm oceanic currents has also provided the context for much of the adaptations found in Galápagos organisms. Human-induced climate change presents particular challenges for the biota of the archipelago. Although El Niño events are part of the natural variation in the Galápagos climate, the frequency and severity of these events has caused devastation to the population of many Galápagos species. The El Niño of 1982–1983 was especially severe. Elevated ocean temperatures and reduced upwelling had severe impacts on physiological tolerances of many species and diminished overall oceanic nutrient levels and productivity. Several endemic species such as the giant sea star (*Luidia superba*), the cup coral (*Tubastrea tagusensis*), the swallow damselfish (*Azurina eupalama*), and the Galápagos barnacle blenny (*Acanthemblemaria castroi*) disappeared after the El Niño event. The blenny was rediscovered several years later, and the cup coral was rediscovered only in 2004. The fates of the giant sea star and the damselfish remain unknown. This El Niño also had extreme consequences to many species dependent on marine resources, such as seabird and marine iguana populations. Marine iguanas have been abundant historically at many sites in the western Galápagos on the islands of Isabela and Fernandina, where the Cromwell Current brings upwelled, nutrient-rich water to these western portions of the archipelago. Populations of marine iguanas in these areas, such as at Punta Espinosa, are now dramatically reduced from historical levels (Fig. 13). Increased anthropogenic global warming poses a serious threat to the survival of the unique biota of the islands.

FIGURE 13 Historical changes in marine iguana density at Punta Espinosa, Isla Fernandina: (A) 1905; (B) 2005.

Nonsustainable Fishing

With increased migration to the Galápagos Islands, many new immigrants to the islands work in the fishing industry. Fisheries such as the shark fishery, long-line fishing, and the sea cucumber (pepino) fishery have had significant impacts on the Galápagos marine biota. Shark finning to provide fins for shark-fin soup and the harvest of sea cucumbers for Asian markets have decimated shark and sea cucumber populations. Several species of sea cucumbers, especially *Stichopus fuscus,* that had been abundant along the western islands are now exceedingly rare. Collection of sea horses and the endemic black coral *Antipathes galapapagensis* for jewelry are also problematic.

The elimination of sea cucumbers is likely related to other changes in the marine environment beyond harvesting by humans. In 2005, we observed the presence of urchin barrens around Fernandina Island that had not been present in many years of earlier observation. The increase in sea urchins had caused the virtual elimination of most benthic algae in near-shore regions, leaving only encrusting coralline algae. This is likely to have adverse impacts on species such as the marine iguana that are dependent on macroalgae as a primary food source.

The problem of fisheries in the Galápagos has been exacerbated by political instability and civil conflict. Existing fishing seasons and quotas have been ignored or suspended. Political protest, violence, and seizure of governmental and conservation facilities such as the Charles Darwin Research Station have caused government authorities to back away from enforcement of existing regulations and sound management practices. These episodic events have undermined the development of an integrated management plan for Galápagos National Park.

Although the Galápagos still have much of their native biota largely intact, human population pressure and lack of political will have created great concern in the international scientific community. Recently, in mid-2007, the United Nations Educational, Scientific and Cultural Organization (UNESCO) placed the Galápagos on a list of endangered World Heritage Sites. It is hoped this action will place greater attention on the situation and bring international pressure to bear on the Ecuadorian government for responsible management. There are recent changes that are encouraging. Late in 2007, Eliecer Cruz, a strict conservationist, was appointed by Ecuadorian President Correa as Governor of the Galápagos in order to rescue the Galápagos from its current UNESCO status. Hopefully, this action will create the necessary stimulus to fundamentally change how the Galápagos are managed and to ensure the long-term preservation of the unique biotic resources that are part of the world's biodiversity heritage.

SEE ALSO THE FOLLOWING ARTICLES

Dispersal / Dwarfism / Endemism / Galápagos Islands, Geology / Gigantism / Insect Radiations / Tortoises

FURTHER READING

Carlquist, S. 1965. *Island life: a natural history of the islands of the world.* New York: American Museum of Natural History.
Darwin, C. 1845. *The voyage of the* Beagle. London: John Murray.
Grehan, J. 2001. Biogeography and evolution of the Galapagos: integration of the biological and geological evidence. *Biological Journal of the Linnean Society* 74: 267–287.
Jackson, M. H. 1993. *Galapagos: a natural history.* Calgary: University of Calgary Press.

Kizirian, D., A. Trager, M. Donnelly, and J. Wright. 2004. Evolution of Galapagos Island lava lizards (Iguania: Tropiduridae: *Microlophus*). *Molecular Phylogenetics and Evolution* 32: 761–769.

Parent, C., and B. Crespi. 2006. Sequential colonization and diversification of Galápagos endemic land snail genus *Bulimulus* (Gastropoda, Stylomatophora). *Evolution* 60: 2311–2328.

Peck, S. 2006. *The beetles of the Galápagos Islands, Ecuador: evolution, ecology and diversity* (*Insecta: Coleoptera*). Ottawa: National Research Council of Canada Research Press.

Porter, D. 1976. Geography and dispersal of Galapagos Islands vascular plants. *Science* 264: 745–746.

GALÁPAGOS ISLANDS, GEOLOGY

DENNIS GEIST
University of Idaho, Moscow

KAREN HARPP
Colgate University, Hamilton, New York

The Galápagos Islands are renowned for their flora and fauna, made famous by Charles Darwin's visit to the archipelago in 1835. Darwin's visit provided an essential foundation for his seminal work, *The Origin of Species*, and the Galápagos remain an important location for the study of evolutionary biology. Even though they are less familiar than his biological discoveries, many of Darwin's most astute observations in the Galápagos were related to the geology of the archipelago. Over the past half century, detailed investigations into the volcanic origins of the islands have revealed that the unique geology of the islands is inextricably intertwined with its biological processes, and many would contend that it is one of the world's foremost natural laboratories for the study of island volcanism and its tectonic origins.

TECTONIC SETTING

The Galápagos Islands are believed to be the surficial manifestation of a mantle plume similar to the one that is responsible for the formation of the Hawaiian Islands. As a column of hot plastic rock rises from deep in the Earth's mantle, it undergoes partial melting near the surface. Molten rock, or magma, being more buoyant than the surrounding rock, separates from the plume and works its way to the surface to supply the active volcanoes of the archipelago with magma. A mantle plume origin for the Galápagos was originally hypothesized in the 1970s on the basis of the pattern of volcanism and the depth of the sea floor surrounding the archipelago, and the plume hypothesis has been supported by a number of geochemical studies since that time. A recent seismic experiment has imaged the Galápagos plume beneath Fernandina and Cerro Azul volcanoes (Fig. 1), across the entire extent of the upper mantle.

The Galápagos plume, like other mantle plumes around the globe, moves slowly compared to the motion of the overlying, rigid plate that forms the crust of the ocean floor. Consequently, as the plate moves over the plume, active volcanoes are carried away from the magma source, and they go extinct. The older volcano then is replaced by a new, active volcanic center. This conveyor belt–like process results in the formation of a chain of volcanoes with the youngest, most active centers being located above the mantle plume and the progressively older, inactive volcanoes arranged in a line extending from the plume, parallel to plate motion. The Galápagos are situated on the Nazca plate, which is moving eastward relative to the plume; recent estimates of the velocity of the Nazca plate range from 22 to 59 km/million years. The Carnegie ridge (Fig. 1) is the ancient trace of the Galápagos plume and is made up of eroded, submerged volcanic islands that were once located where the archipelago is today.

There is an additional tectonic complexity in the Galápagos region that sets it apart from most other ocean island systems. The deep root of the Galápagos plume is centered only about 200 km south of the Galápagos Spreading Center, an east–west trending chain of volcanoes and

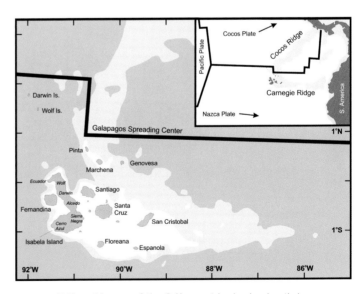

FIGURE 1 Bathymetric map of the Galápagos Islands, showing their proximity to the Galápagos spreading center, which is the boundary between the Nazca and Cocos plates. The six volcanoes of Isabela Island are labeled in italic. Inset shows a regional view, including the Cocos and Carnegie ridges, which are the tracks of the Galápagos hotspot.

fractures responsible for the formation of the ocean crust that constitutes the Nazca and Cocos plates. As the crust forms at the ridge, it spreads away from the ridge crest and is quickly (within only a few million years) carried over the Galápagos mantle plume. Prior to 8 million years ago, the Galápagos Spreading Center was located above the Galápagos plume. At that time, volcanic material produced by the plume was deposited on both the Nazca and Cocos plates simultaneously, resulting in the formation of a second, northeast-trending chain of volcanoes, now recognized as the submarine Cocos ridge (Fig. 1). Since that time, the Galápagos Spreading Center has drifted to the north relative to the plume, and islands no longer are being produced on the plate boundary, but exclusively on the Nazca plate.

The mantle plume hypothesis for the Galápagos Islands predicts a systematic age progression for the formation of the islands, with the youngest lying at the western edge of the archipelago, and increasingly older structures existing to the east. This pattern has been broadly documented using radiometric dating techniques (Fig. 2). Notable exceptions to the age progression remain enigmatic in terms of our understanding of this volcanic system. First, most of the islands have rocks that are younger than would be predicted had they been generated by a single plume source beneath the Nazca plate (i.e., the data lie to the right of the gray field in Fig. 2). Second, the duration of volcanism at individual islands can be millions of years (depicted as vertical arrays on Fig. 2). These deviations from the simplest mantle plume model result in a distribution of volcanoes in the Galápagos that is anomalous, with young volcanoes spread out both in a north–south direction (perpendicular to plate motion) and in an east–west direction (parallel to plate motion).

Many of the deviations from the predicted linear island trend are attributed to the thin lithosphere that underlies the archipelago, which is in turn a reflection of the young age of the ocean crust in the region. The generation of magma by melting occurs mainly by decompression of solid rock, which happens when mantle flows upward as a result of the thermally buoyant plume or plate driving forces (Fig. 3). The solidus is the temperature at which a mantle rock begins to melt, and the farther above the solidus the mantle gets in terms of its pressure and temperature, the more the rock melts. When the ascending mantle reaches the overlying plate, or lithosphere, upwelling and therefore melting are abated. Beneath a

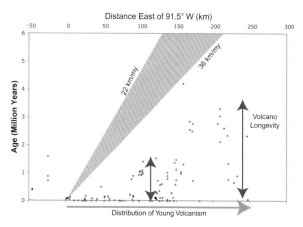

FIGURE 2 A compilation of the ages of volcanic rocks from the Galápagos Islands compared to the eastward component of their distance from Fernandina Island. Fernandina is used as a reference because it is thought to overlie the Galápagos mantle plume. The gray field shows the predicted pattern of ages had all magmas been erupted directly above the mantle plume, for two different velocities that have been estimated for the Nazca plate. Two notable patterns emerge for the Galápagos: The volcanoes are active for millions of years (i.e., they exhibit a wide vertical spread on this diagram), and the volcanoes erupt even when they are more than 200 km away from the mantle plume. Both of these features are attributed to the relatively thin lithosphere in the near-ridge environment and to the anomalously hot mantle from plume-ridge interaction.

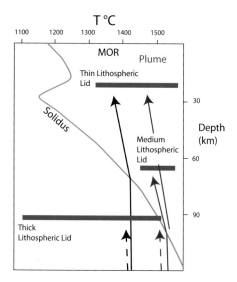

FIGURE 3 Diagram showing the different conditions of melting of the Galápagos mantle. Melt is generated via decompression, as thermally buoyant mantle ascends toward the surface, a process indicated by the red and black arrows on this diagram. Mantle rock begins to melt when it crosses the conditions defined by the solidus (the green line). Upwelling and therefore melting cease when the mantle reaches the overlying plate, or lithosphere. The effects of different lithospheric thicknesses are represented in the diagram by the three horizontal blue lines. Beneath a mid-ocean ridge such as the Galápagos spreading center, the lithosphere is thin, and mantle can rise to shallow depths, producing significant extents of melting. Where the lithosphere is thicker, as it is beneath the main Galápagos archipelago, less melting occurs because of limited ascent by the mantle.

mid-ocean ridge (MOR) such as the Galápagos Spreading Center, even relatively cold mantle melts substantially, because the mantle ascends to shallow depths beneath the thin lithosphere. Beneath thicker lithosphere, ascending mantle cannot experience as much melting, and even hot plumes may not produce much magma.

The Galápagos lie in an area affected by both a MOR and a plume, where the lithosphere thickens with distance from the Galápagos Spreading Center. Consequently, a wide range of these melting conditions exists in the Galápagos region, which may provide an explanation for the wide range in both volcanic activity and eruptive products across the archipelago. Furthermore, tectonic stresses in the lithosphere caused by plate tectonic interactions between the Galápagos Spreading Center and the plume cause additional volcanism where it might not otherwise occur, in the region nearest the mid-ocean ridge.

REGIONAL GEOLOGY OF THE GALÁPAGOS ISLANDS

The Galápagos archipelago comprises four distinct volcanic subprovinces, in which the age and style of volcanism differs, as well as the compositions of the magmas. The western subprovince is made up of the young, historically active shield volcanoes of Fernandina and Isabela Islands (Fig. 1). In contrast, the central islands are older and are mostly inactive. To the east, on the Carnegie ridge, there are drowned islands, effectively a set of proto–Galápagos Islands that have sunk beneath the sea as they were carried away from the plume. The fourth group is the northern islands, scattered edifices that lie between the main Galápagos platform and the Galápagos Spreading Center (Fig. 1).

The Western Subprovince

The seven great shields of the western Galápagos are some of the most active volcanoes on the planet, and this is the type locality of the "Galápagos shield" morphological classification of volcano, for which there are three essential characteristics (Fig. 4A). First, each of these volcanoes is incised by a caldera, which is wide and deep relative to the size of the edifice. Second, Galápagos shields have circumferential fissures close to the caldera rim, but the fissures are radial on the lower flanks. Although there are concentrations of fissures in some sectors of each volcano, linear rift zones like those that characterize Hawaiian shields are absent. Third, Galápagos shields have profiles distinct from those of other ocean island volcanoes around the globe, with gentle lower flanks (less than 4°), steep upper slopes (up to more than 30°), and summit plateaus.

FIGURE 4 Profiles comparing (A) a prototypical western Galápagos shield volcano, Cerro Azul, to (B) a central shield volcano, Santa Cruz. (C) A panorama of the caldera of Fernandina volcano. (D) Wolf Island in the northern archipelago. There are three principal differences between the western and central morphotypes. First, the western shields have circumferential fissures cutting their upper slopes and radial fissures on the lower slopes, whereas the central volcanoes tend to have single rift zones, in the case of Santa Cruz, oriented west-northwest. Second, the western shields have steep upper slopes (up to more than 30°) and more subdued lower slopes (less than 4°); this morphology develops because short, stubby lava flows erupt from the upper reaches, and more voluminous, extensive lava flows erupt on the lower flanks, as is apparent in this photograph. Third, the western volcanoes have prominent calderas, whereas they are lacking in the central volcanoes. Fernandina's caldera is 800 m deep and 4 by 6 km wide. Note that the northern islands have a complete spectrum of morphologies, including some with calderas (Genovesa, Marchena, and Wolf Islands).

The characteristic morphology of the western Galápagos shields is caused by the distribution of fissures on the volcanoes. Higher-elevation circumferential fissures produce low-volume lava flows (Fig. 4A), owing to the higher driving pressure necessary to get the magma up through the low-density summit carapace. These low-volume lavas accumulate near the vents, and this part of the volcano grows upward. In contrast, magmas erupted from the low-elevation radial vents are intruded laterally where they require less driving pressure to erupt. The large-volume, aerially extensive lava flows constitute the shallowly sloping lower flanks. The western Galápagos shields are important analogues for volcanoes on Venus and Mars, some of which have similar morphologies.

The western Galápagos shields are constructed almost entirely of lava, with pyroclastic rocks being mostly restricted to near-vent environments. The 'a'ā lava morphology dominates over pāhoehoe, especially on the middle and lower flanks of the volcanoes.

Several of the shields (Wolf, Sierra Negra, and Fernandina) are built of tholeiitic basalts with monotonous compositions. In contrast, Alcedo has erupted a complete tholeiitic differentiation series, including a Plinian deposit of rhyolitic pumice, and Darwin volcano has erupted icelandites. Cerro Azul and Ecuador lack evolved rocks but have erupted primitive basalts with abundant olivine.

It has been estimated that the western shield volcanoes emerged no more than 500,000 years ago (central Isabela), and most recently less than 50,000 years ago (Fernandina). Recent oceanographic surveys indicate that the western volcanoes are built on a platform comprising lavas produced by the Galápagos plume, which is constructed from a series of stacked, sheet-like terraces, each one hundreds of meters thick. Unfortunately, there is no volcano in a nascent stage of growth in the western part of the archipelago, so we know no details about how Galápagos volcanoes are born and grow before they emerge from the submarine environment.

The Central Subprovince

The volcanoes east of Isabela are considerably older, ranging up to more than 3 million years old. They are completely different morphologically from the western shield volcanoes, with subdued profiles, no calderas, and strongly oriented rift zones (Fig. 4B). Most of these islands are cut by pronounced faults. One possibility is that late-stage volcanism modifies the volcanoes' shape, as the volcano is carried away from the plume, and that a western Galápagos edifice underlies each of these volcanoes. There is little evidence to support this speculation, however; the central volcanoes simply may have been different from the western shields from the time of their emergence above the sea.

The compositions of the central Galápagos lavas differ strongly from those of the western subprovince. On each island, the lavas exhibit diverse compositions, ranging from being more primitive to more evolved than those produced in the western archipelago. For example, Floreana's lavas are olivine-rich and alkaline, and they bear abundant crustal and ultramafic xenoliths. Pinzon and Santiago have complete differentiation series, with rocks as evolved as rhyodacite and trachyte. Santiago, Santa Cruz, and San Cristobal each have lavas that range from enriched alkaline series to depleted tholeiitic rocks similar to lavas erupted from the Galápagos Spreading Center. This compositional diversity is attributed to strongly varying conditions in the part of the mantle that is melting beneath the central archipelago, including temperatures and pressures of melt extraction. Also, these volcanoes lack long-lived magma chambers, so they are not homogenized during ascent, a process that preserves compositional diversity.

Drowned Islands

The hotspot origin of the Galápagos predicts that older islands should be carried eastward on the tectonic plate and replaced by younger volcanic centers, forming a linear chain. This phenomenon is responsible for the Carnegie ridge, which extends toward the South American continent from the Galápagos Islands (Fig. 1). More important evidence was uncovered in 1990, when researchers mapping the Carnegie ridge discovered that several seamounts that make up the ridge have wave-cut terraces and bear rounded beach cobbles, signs that indicate that the volcanoes are drowned islands. This is a classic example of a well-known phenomenon: As the lithosphere moves away from a hotspot, it contracts thermally and the island undergoes erosion, reducing its elevation gradually over time. Thus, the seamounts on the Carnegie ridge were originally Galápagos Islands, but they have sunk below sea level.

The drowned islands have ages consistent with having been formed at the center of the hotspot ~10–12 million years ago, where the large western shield volcanoes of the Galápagos are located today. Similar discoveries were made along the Cocos ridge, on the Cocos plate, extending the timescale for the presence of islands in the Galápagos to nearly 20 million years. Consequently, it is clear that the present-day Galápagos region has been populated by active volcanoes for at least 20 million years. It is possible that there could have been islands for a longer time, but the evidence has been obscured by both subduction and the closing of the Isthmus of Panama. An important implication of the discovery of sunken islands is that there is now a significantly longer timeline for evolution of the unique life forms that populate the archipelago. The endemic organisms may have evolved over tens of millions of years, not simply in the past 3 million years, which is the age of the oldest-known island rock.

Northern Galápagos Province

One of the more enigmatic aspects of the Galápagos Islands is the set of five small volcanoes (Wolf, Darwin, Pinta, Marchena, and Genovesa) and many seamounts that define the northern perimeter of the archipelago (Fig. 1). In particular, these volcanoes form distinct northwest- and east-northeast-trending lineaments. The existence of these islands is inconsistent with the mantle plume model in that they do not lie downstream from the hotspot center in the direction of plate motion. Instead, they occupy the sea floor between the central Galápagos and the Galápagos Spreading Center.

The northern Galápagos volcanoes are hypothesized to be the result of regional stresses generated by the interaction between the mantle plume and the Galápagos Spreading Center, as well as by abnormally high mantle temperatures. The major offset in the Galápagos spreading center at 91° W (Fig. 1) generates a local extension that enables magma from the upper mantle and the hotspot to migrate to the surface along zones of weakness, resulting in the formation of small volcanic centers, some of which grow large enough to become islands. This model predicts that the islands and seamounts of the Northern Galápagos Province are all younger than the formation age of the transform fault, which is less than 3 million years old. Radiometric age determinations confirm this prediction, in that all of the volcanic centers that have been dated in the northern Galápagos appear to be 1 million years old or younger. Genovesa Island, for instance, has only been exposed above sea level for approximately the last 300,000 years, considerably less time than originally believed given its easterly position downstream of the hotspot center.

MAGMA GENESIS AND EVOLUTION

The emergence of the Galápagos Islands is due to crustal growth by the addition of basaltic magma from the mantle to the preexisting Nazca plate crust; although several islands are small fault blocks, there is no evidence of large-scale tectonic uplift anywhere in the archipelago. The two primary sources of Galápagos magmas are thought to be a deeply rooted mantle plume and the shallow asthenosphere, both part of the Earth's mantle. Seismic studies indicate that the mantle beneath the western Galápagos is unusually hot, at least to 600 km depth. The anomalously hot mantle is thought to result from a buoyant upwelling rooted in the deep mantle, perhaps as deep as the core–mantle boundary. As the buoyant plume reaches the base of the lithosphere, it begins to melt via decompression melting (Fig. 3). The main mechanism of magma generation at the Galápagos Spreading Center is also by decompression, but the driving force is different. There, plate spreading from global-scale plate tectonics causes the shallow mantle to well up and melt at even shallower depths (Fig. 3). Because both plume and plate-spreading forces are important in the Galápagos, and because there is a large change in the thickness of the lithosphere across the archipelago, Galápagos magmas are derived from widely ranging temperatures and pressures (Fig. 3).

Chemical and isotopic tracers indicate that there is extensive contribution of the Galápagos plume to the mantle beneath the Galápagos Spreading Center. Likewise, the unusual compositions of magmas from some of the central and northern Galápagos Islands is attributed to contributions from the shallow mantle, which does not ordinarily provide much material to magmas produced by mantle plumes. Obviously, the dual nature of both Galápagos Island and Galápagos Spreading Center compositions is due to the unusual proximity of the two igneous provinces. We do not know at what depth the two-way exchange of mantle material occurs, but recent seismic studies strongly suggest mixing in the upper mantle.

After the magma segregates from its mantle source, it cools and differentiates in magma bodies within the lithosphere. Recent deformation studies show that magma is stored and cools in shallow chambers beneath the western volcanoes. At Fernandina and Sierra Negra volcanoes, the tops of the magma chambers are only 2 km beneath the caldera floor, perhaps explaining why these are the two most active volcanoes in the archipelago.

GEOLOGY AND BIODIVERSITY OF THE GALÁPAGOS

The biology and geology of the Galápagos Islands are related in many complex ways, some of which have only been realized in the past few decades. Island biodiversity is known to correlate to island size in nearly every archipelago, and this holds true for the Galápagos as well. Recent studies have used the Galápagos' more than 120 islands to test relationships between biodiversity and island age and geologic substrate, but no such relationship has yet been found. Even islands less than 1000 years old have the number of vertebrate and plant species that their area would predict, and there is no difference between islands made of tuff, lava, or uplifted submarine rocks. These observations suggest that island colonization is relatively rapid and that the archipelago's pioneer organisms are well adapted to a variety of environmental conditions.

A unique relationship between geology and one of the archipelago's most famous inhabitants, the Galápagos tortoise, has been discovered recently. The subspecies of the giant tortoise found on Alcedo volcano has limited genetic diversity compared to other subspecies of Galápagos tortoises. Approximately 90,000 years ago, the Alcedo line experienced a genetic bottleneck. The timing of the genetic bottleneck corresponds almost perfectly with a voluminous explosive eruption of Alcedo. In fact, genetic studies suggest that the eruption killed all but one female of Alcedo's tortoise population, from which all subsequent tortoises on that volcano have evolved.

The Galápagos are also home to the planet's only reptile that feeds at sea, the marine iguana, as well as its larger cousin, the land iguana. These two organisms can be traced genetically back to a single species of iguana, native to the South American continent. DNA studies revealed that the two types of iguanas must have split from a common ancestor at least 15 and possibly 20 million years ago. When only the present-day islands are considered, it appears as though dry land had been available in the Galápagos Islands only for the past 3 million years (Fig. 2); this abbreviated time-span posed a problem for explaining the evolution of the land and marine iguanas in the islands, as well as several additional evolutionary developments in the archipelago. The drowned islands downstream along the Cocos and Carnegie ridges, however, extend the time over which dry land has been available in the region back to nearly 20 million years (and possibly more), a key step toward understanding the interrelationship between biology and geology in ocean island systems.

In a similar example of the interplay between geology and biology in island chains, the unexpectedly young ages of the northern Galápagos Islands (all less than 1 million years) provide a potential explanation for the surprising colonization patterns of these islands. Specifically, some of these islands lack a number of key species that are abundant in the main archipelago, such as the lava lizard and gecko, possibly one manifestation of their unexpectedly young formation ages.

SEE ALSO THE FOLLOWING ARTICLES

Darwin and Geologic History / Galápagos Islands, Biology / Hawaiian Islands, Geology / Plate Tectonics / Volcanic Islands

FURTHER READING

Christie, D. M., R. A. Duncan, A. R. McBirney, M. A. Richards, W. M. White, and K. S. Harpp. 1992. Drowned islands downstream from the Galápagos hotspot imply extended speciation times. *Nature* 355: 246–248.

Geist, D., D. J. Fornari, M. D. Kurz, K. S. Harpp, S. Adam Soule, M. R. Perfit, and A. M. Koleszar. 2006. Submarine Fernandina: magmatism at the leading edge of the Galápagos hot spot. *Geochemistry, Geophysics, Geosystems* 7: Q12007, doi:10.1029/2006GC001290.

Geist, D., T. R. Naumann, and P. L. Larson. 1998. Evolution of Galápagos magmas: mantle and crustal fractionation without assimilation. *Journal of Petrology* 39: 953–971.

Harpp, K. S., and W. M. White. 2001. Tracing a mantle plume; isotopic and trace element variations of Galápagos seamounts. *Geochemistry, Geophysics, Geosystems* 2: 2000GC000137.

Harpp, K. S., K. R. Wirth, and D. J. Korich. 2002. Northern Galápagos Province: hotspot-induced, near-ridge volcanism at Genovesa Island. *Geology* 30: 399–402.

Werner, R., K. Hoernle, P. v.d. Bogaard, C. Ranero, R. von Huene, and D. Korich. 1999. Drowned 14 m.y.-old Galápagos archipelago off the coast of Costa Rica: implications for tectonic and evolutionary models. *Geology* 27: 499–502.

White, W. M., A. R. McBirney, and R. A. Duncan. 1993. Petrology and geochemistry of the Galápagos Islands: portrait of a pathological mantle plume. *Journal of Geophysical Research* 98: 19,533–19,563.

Wilson, D., and R. Hey. 1995. History of rift propagation and magnetization intensity for the Cocos-Nazca spreading center. *Journal of Geophysical Research* 100: 10,041–10,056.

GIGANTISM

PASQUALE RAIA

University of Naples, Italy

Gigantism is the development of large body size in a species that colonizes an island and remains confined there. In the literature, any island population (or species, if there are no mainland conspecifics alive) is considered gigantic when its average body size exceeds the body size limits of members of the original, mainland population (if living) or of parental species.

DETERMINING WHEN GIGANTISM IS REAL

In discussing island gigantism, we must first be sure that the size change was associated with island colonization. To this end, the first step is to compare an island species's body size to that of its living mainland relative. Although this is straightforward when mainland conspecifics are living, it is much more difficult when the most closely related mainland population or species is unknown or extinct. In the latter case, the species' phylogenetic relationships must be ascertained before any body size comparison can be performed. For instance, the komodo dragon *Varanus komodoensis* is a famed giant reptile that reaches some 3 m in length and can weigh up to 70 kg. Although this monitor lizard has often been regarded as an island giant, recent phylogenetic analyses reveal that it belongs to a clade of similarly large-sized monitor lizards. Hence, there is no island-related size increase in the komodo dragon ancestry. As a consequence, *V. komodoensis* cannot be considered a case of gigantism. Similar reasoning may apply to giant island tortoises, such as the famous *Geochelone* species of Galápagos, Aldabra atoll (Seychelles), and Madagascar.

Some of the most spectacular examples of gigantism come from extinct species, whose phylogenies are often uncertain. Usually, this prevents measuring the magnitude of size change, but it often also means that we cannot be sure whether a fossil island species was indeed giant, dwarf, or neither. Even more problematic are some cases of adaptative radiation on islands. For instance, New Zealand's moas form a paraphyletic clade, including large (genera *Dinornis* and *Pachyornis*) flightless birds that have evolved in situ from smaller relatives, so the principle of comparison to mainland relatives is inapplicable. Although gigantism in moas is proven, its meaning should be recast in terms of their intra-island radiation. In conclusion, without reliable phylogenetic scenarios in hand, defining gigantism as an island phenomenon is uncertain.

GENERAL SURVEY OF GIGANTISM ACROSS TAXA AND ISLANDS

Gigantism is commonplace among smaller representatives of vertebrate clades. Nonetheless, there are many exceptions and, apparently, some important phylogenetic constraints. Some groups tend to present gigantism more often than other, similarly sized groups of closely related animals. For instance, among squamate reptiles, gigantism occurs iteratively between teiid lizards, skinks, geckos, and iguanas, whereas it is much rarer among varanids. In the Canary Islands, species of the lacertid genus *Gallotia*, both living (*G. simonyi*) and extinct (*G. goliath* and *G. gigantea*), have grown to very large sizes. The living *G. simonyi* form El Hierro is some 0.6 m long. The whiptail *Cnemidophorus tigris* in the Sea of Cortez islands often appears to be larger than on the mainland. A giant skink (*Leiolopisma mauritiana*) occurred in Mauritius where it went extinct during the seventeenth century. A sister species of *L. mauritanica* still unnamed has been reported from Réunion Island.

In Cape Verde, two giant skinks were once living. One, *Macroscincus coctei* (with a snout–vent length up to 0.32 m), went extinct in the twentieth century. *Mabuya vaillanti* (with up to a 0.24 m snout–vent length), however, is still living, along with four other congenerics. In New Caledonia lives the large, recently rediscovered *Phoboscincus bocourti*, a skink up to 0.5 m in body length. New Caledonia also hosts the giant *Rhacodactylus leachianus*, which is one of the world's largest geckos at some 0.3–0.4 m in body length. Large, now extinct gekkos occurred on the island of Rodrigues (*Phelsuma gigas*) and in New Zealand (*Hoplodactylus delcourti*). The latter was the largest gecko ever, reaching a total body length of 0.6 m. It is believed that *H. delcourti* is the kawekaweau of Maori oral tradition. Only a bit smaller are giant island chuckwallas *Sauromalus varius* and *S. hispidus*, living on Gulf of California islands.

Snakes are either gigantic or dwarf depending on the size of their prey; hence, the distribution of island snakes' body size is bimodal, with both dwarfism and gigantism being common. Extreme cases of gigantism in snakes are common among colubrids. In the Japanese archipelago, large-bodied forms of *Elaphe quadrivirgata* occur on Tadanae-jima, Kozo-shima, and Nii-jima Islands. Gigantic *Elaphe climacophora* occurs on Kammuri-jima Island, and a giant *Dinodon rufozonatus* occurs on Nakanokamishima. On Cerralvo Island lives a gigantic variant of *Rhinocheilus lecontei*. *Thamnophis sirtalis* has grown gigantic on East Sister, West Sister, and Kelly's Middle Islands, in the Lake Erie archipelago. *Uromacer frenatus* is gigantic on Ile de la Gonâve. The tiger snake *Notechis ater* is a giant elapid living in Chappel, Kangaroo, and Reevesby Islands. Among Viperidae, *Crotalus mitchelli* is gigantic on Angel de la Guarda Island in the Gulf of California.

Islands are surrounded by water; hence, birds are much more common than other vertebrates by their greater dispersability. On islands, they happen to fill the niches large mammals usually leave vacant, thereby developing an astonishing diversity of shapes and sizes. The most interesting forms are perhaps the flightless giants (now extinct) that lived in the past centuries on islands all over the world. On New Zealand, 11 to 13 species of wingless, highly sexually dimorphic, and rapidly growing Moas did occur. The largest, *Dinornis novaezealandiae*, stood more than 2 m and weighed 250 kg. The largest emeid moa, *Pachyornis elephantopus*, weighed about 80 kg. Most other moas are much smaller, although certainly large if compared to moas closest relatives, the living ratites, such as cassowary, kiwi, and emu. Also in Madagascar lived giant ratites, the elephant birds *Aepyornis* and *Mullerornis*.

Aepyornis maximus (Fig. 1) is the largest bird ever. At 1 ton in body weight, it exceeded in size even the

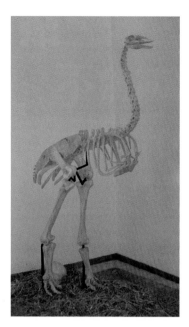

FIGURE 1 The Malagasy elephant bird *Aepyornis maximus*. The skeleton is 248 cm tall. Photograph courtesy of Zoic S.r.l.

immense terror birds, *Phorusracidae,* and the Eocene gruiform *Diatryma/Gastornis.* Bones of *Aepyornis* have been mistaken for remains of the mythological bird *Rukh* cited in the *Thousand and One Nights* collection and in Marco Polo's *Million.* Flightless waterfowl, known as moa-nalos, lived on the Hawaiian Islands. Moa-nalos occurred in four different species: *Chelychelynechen quassus* on Kauai, *Ptaiochen pau* on Maui, *Thambetochen xanion* on Oahu, and *T. chauliodous* on Maui, Molokai, and Lanai. A giant galliform, *Sylviornis,* lived in New Caledonia. *Sylviornis neocaledoniae* was once believed to be a megapode. True giant megapodes are known from the Fiji Islands, as is a giant pigeon, *Natunaornis gigoura,* which stood some 0.80 m tall. Two giant pigeons lived on the Mascarenes. The Rodrigues solitaire *Pezophaps solitaria* and the world-famous dodo *Raphus cucullatus* lived on Mauritius. The dodo and Rodrigues solitaire are sister taxa whose closest living relative is the Nicobar pigeon *Caloenas nicobarica.* Rails are commonly flightless and are much less commonly giant on islands. However most fossil species await a formal description, and additional cases of gigantism will certainly arise. The takahe *Porphyrio hochstetteri* is a flightless giant rail of New Zealand. A giant rail, *Diaphorapteryx hawkinsi,* once inhabited the Chatham Islands, whereas another, *Nesoclopeus poicilopterus,* lived on the Fiji islands. A giant coot, *Fulica chatamensis prisca,* once occurred in New Zealand, and the related subspecies *F. c. chathamensis* still occurs in the Chatham Islands. Other giant coots are claimed to have inhabited Mauritius and Réunion Islands. A giant seabird, Pallas's cormorant *Phalacrocorax perspicillatus,* lived on Bering and other Commander islands. Psittacinae (parrots and macaws) show large forms on New Zealand (e.g., the kakapo *Strigops habroptilus,* which could weight more than 3 kg) and on Mauritius (e.g., *Lophopsittacus mauritianus,* now extinct).

Many birds of prey grew gigantic on islands. The largest was Haast's eagle *Harpagornis moorei* of New Zealand, in which females approached 14 kg in body weight. The species was a sister taxon to the extant and broadly distributed little eagle, *Hieraaetus morphnoides,* and to the booted eagle, *H. pennatus.* Still in New Zealand lived the giant Eyles' Harrier *Circus eylesi.* Another giant eagle was *Aquila chrysaetos szimurgh* from Crete, whose subspecific name refers to the bird *sîmurgh* of Persian mythology. Giant buteonine hawks occurred in Gargano paleoisland (southern Italy) during the Mio-Pliocene (*Garganoaetus freudenthali*), and on Cuba during the Pleistocene (*Titanohierax borrasi*). Congenerics of the Cuban *Titanohierax* occurred on Bahama, Hispaniola, and Grand Cayman Islands. Sea eagles are represented by the giant form of the white-tailed eagle *Haliaeetus albicilla* on Eivissa (Ibiza, Balearic Islands). Large strigiformes (owls) occurred on Cuba (*Ornimegalonyx oteroi,* claimed to have been 2–4 times larger than the extant eagle owl *Bubo bubo*). Another large owl was *Tyto riveroi,* also from Cuba. On Gargano, remains of a barn owl *Tyto gigantea* challenge in size that of *T. riveroi.* Passerines were believed not to follow the Island Rule, although recent analyses suggest the contrary. Yet the most important morphological modification in passerines is probably the acquisition of a large bill, instead of large body size. Large island variants include, among others, the Capricorn silvereye *Zosterops lateralis chlorocephalus* on Heron Island, black bulbuls of the genus *Hypsipetes* on Comoro and Mascarene islands, the extinct drepanid *Hemignathus* from the islands of Hawaii, the extinct flightless long-legged bunting *Emberiza alcoveri* from the Canary Islands, and the superb fairy wren *Malurus cyaneus* on Kangaroo Island.

Among mammals, gigantic island forms are common in rodents. It has been estimated that 85% of rodent species show insular gigantism. Lagomorphs often show the opposite trend (i.e., dwarfism). An example of gigantism in rodents is the Japanese field mouse *Apodemus speciosus* on the Japanese archipelago. The wood mouse *A. sylvaticus* shows gigantism in the western Mediterranean and British islands, and on

Iceland. The yellow-necked mouse *A. flavicollis* is gigantic on several European islands. A large caviomorph, *Tainotherium valei,* which shows adaptation to arboreal life in its femur, occurred on Puerto Rico. A truly gigantic caviomorph, the 200 kg weighting *Amblyrhiza inundata,* lived on the Anguilla Bank, West Indies. Giant black rats *Rattus rattus* live on Congreso Island, Chafarinas archipelago. The Pacific rat *R. exulans* shows repeated cases of size increase on Pacific Islands. A giant rat *Papagomys armandvillei* occurs on Flores Island. On Tenerife (Canary Islands) lived still another giant rat, *Canariomys bravoi.* Wood rats of the genus *Neotoma* are significantly larger on Baja California islands. Three different lineages of the murine rodend *Microtia* developed gigantism on the Gargano paleoisland. Giant dormice occurred in Sicily and Malta (*Leithia melitensis*) Minorca (*Hypnomys mahonensis*) and Majorca (*H. morphaeus*) during the Pleistocene. Giant shrews lived in Sardinia (*Nesiotites similis*) and the Balearic Islands (*N. hidalgo*).

Primates are moderately large by mammalian standards, notwithstanding, cases of gigantism are apparent in Malagasy lemurs. The bizarre sloth lemurs were indriids morphologically convergent with true tree sloths. The largest species, *Archaeoindris fontoynontii,* was estimated to weigh as much as a large male *Gorilla* (i.e., some 200 kg), second in size only to *Gigantopithecus* among primates. A baboon-sized lemur, *Archaeolemur,* was among the most common lemurs in Madagascar. The comparison between *Archaeolemur* and baboons is more than just in their body sizes, as this extinct lemur had large front teeth and little hands and feet, much like true baboons and chimpanzees. Other large extinct lemurs are *Palaeopropithecus, Babakotia* (sloth lemurs; Babakoto is the Malagasy name for the living Indri), *Megaladapis,* and *Hadropithecus.*

Carnivores are usually rare in island biota because of their limited ability to disperse overseas. Among insular carnivores, there is no evidence for the Island Rule, although gigantism is relatively common among mustelids. On British Islands, giant variants of least weasel *Mustela nivalis,* stoat *Mustela erminea,* and European badger *Meles meles* do occur. The American marten *Martes americana* shows gigantism on Moresby and Vancouver Islands. Among Viverridae, the Asian palm civet *Paradoxurus hermaphroditus* is gigantic on Java. As a notable exception to the Island Rule, the bear *Ursus arctos middendorffi* is gigantic on Kodiak Island. There is only one case of gigantism among ungulates: the moose-sized red deer of Crete, *Cervus major,* which lived during the Late Pleistocene.

Both living and recently extinct arthropods exhibit island gigantism to some degree. In insular insects, there is some convergence on other flying creatures, birds. As with island birds, there are island insects that grew gigantic and went flightless. These include the 10-cm-long Madagascar hissing cockroach *Gromphadorhina portentosa,* the cricket *Thaumatogryllus conanti* on Nihoa island (the other three wingless species belong to the same genus that inhabits the Hawaiian archipelago), and the Lord Howe Island phasmid (stick insect) *Dryococelus australis,* which today is restricted to a small islet near Lord Howe. In St. Helena's gumwood forest lived a giant dermapteran, the 8-cm-long *Labidura herculeana,* which is a sister taxon of the cosmopolitan *Labidura riparia.* The most famous examples of island gigantism are, perhaps, New Zealand's wetas. These orthopterans of the genus *Deinacrida* can weigh up to 70 g.

Giant sheet web spiders (genus *Orsonwelles* of the family Linyphiidae) occur in the Hawaii Islands (where they are still producing new species). Other gigantic island forms are less thoroughly studied or less notorious, but are certainly numerous. Good examples are the New Zealand Kauri land snail (genus *Paryphanta*), anurans such as São Tomé *Ptychadena newtoni,* and even plants such as the New Zealand ross lily *Bulbinella rossii* and the Campbell Island daisy *Pleurophyllum speciosum.*

WHY GIGANTISM OCCURS

Some authors have suggested that gigantism occurs because it increases digestion performance in small herbivores. Others maintain the idea that gigantism is a consequence of territorial behavior in some species (larger individuals are better off at defending their own territories). Another idea that has received much attention is the existence of a body size optimum that island mammals would approach. According to this hypothesis, species smaller than the optimum will increase in size, and those larger will decrease, thereby producing the Island Rule.

Much conflicting evidence has been published on all of these theories. They are not mutually exclusive. In fact, they share the basic tenet that species experience ecological release once they colonize the island because of reduced diversity, and hence competition. A major contributor to this reduced diversity is the extreme underrepresentation of mainland carnivores. Hence, both interspecific competition and predation are reduced for insular herbivores. In small species, low competition pressure and reduced predation result in reduced mortality and, consequently, in increased densities. At high densities, large body size is an effective means to succeed in intraspecific contests, with selection producing gigantism. There is some empirical evidence for the effects of reduced island diversity on gigantism: High densities and stable populations have been reported on islands for rodents, lizards, and passerine birds. Many studies have reported that wood mice size trends are influenced by competition suffered from large species. Others have found significant effects produced by the absence of predators.

It seems well established that most island species experience increased levels of intraspecific competition. This latter point inspires yet another class of explanations based on the evolution of life histories. It has been proposed that, at high level of intraspecific competition, selection on large mammals favors individuals that reproduce at smaller body size, whereas selection on smaller species favors larger individuals capable of winning intense intraspecific contests for mates. Taken together, these two tendencies would explain the Island Rule. It is clear, again, that we cannot separate the effect (gigantism) from its primary cause (reduced island diversity). Because almost all theories proposed so far share the same basic causal mechanism, they could hardly rule each other out.

Size changes in predators can not be explained by the above reasoning. They seem not to be affected by competition. On the contrary, a significant effect of prey size and abundance has been reported in carnivorous lizards, snakes, carnivore mammals, and birds of prey.

A special mention should be devoted to flightless birds. In fact, flightlessness is often linked to gigantism. It should be considered that flight imposes strong physical constraints on a bird's weight. Hence, gigantism in some island birds could be a mere side effect of flightlessness.

SEE ALSO THE FOLLOWING ARTICLES

Dwarfism / Flightlessness / Island Rule / Komodo Dragons / Moa / Snakes / Tortoises

FURTHER READING

Adler, G. H., and R. Levins. 1994. The island syndrome in rodent populations. *The Quarterly Review of Biology* 69: 473–490.
Gould, G. C., and B. J. MacFadden. 2004. Gigantism, dwarfism, and Cope's Rule: "nothing in evolution makes sense without a phylogeny." *Bulletin of the American Museum of Natural History* 285: 219–237.
Lomolino, M. V. 2005. Body size evolution in insular vertebrates: generality of the island rule. *Journal of Biogeography* 32: 1683–1699.
Raia, P., and S. Meiri. 2006. The Island Rule in large mammals: paleontology meets ecology. *Evolution* 60: 1731–1742.
Whittaker, R. J. 1998. *Island biogeography: ecology, evolution and conservation.* Oxford: Oxford University Press.

GLOBAL WARMING

DAVID R. LINDBERG

University of California, Berkeley

Islands and their biodiversity are especially vulnerable to global warming because there are fewer options on islands than on larger continental land masses. Island geomorphology and geographical position moderate the effects of global warming and determine the relative importance of sea-level rise, changes in temperature regimes, ocean circulation patterns, and storm tracks. Effects on island ecosystems will likely be severe and will translate into severe social and economic impacts for islanders.

CLIMATE DRIVERS

The Earth's climate changes on a multitude of scales, including seasonal, decadal, centurial, and multi-millennial periods. Other than human activities, drivers of climate change include (1) changes in the shape of the Earth's orbit, the tilt of the Earth's axis, and the amount of wobble around the axis—collectively described by the Milankovitch Theory; (2) changes in solar intensity; (3) changes in the magnitude, direction, and temperature of ocean currents; and (4) volcanoes that release particulates and carbon dioxide (CO_2) into the atmosphere. These climate drivers do not act independently, and a change in one may interact through feedback loops, thereby intensifying or diminishing the initial climate response. However, there is no doubt that greenhouse gases, especially CO_2, from anthropogenic sources are also driving today's warming of the Earth.

Greenhouse gases increase global temperature by trapping infrared radiation in the atmosphere, thereby warming the Earth's surface. The increase in CO_2 and other greenhouse gases has resulted primarily from the burning of fossil fuels, especially since the Industrial Revolution began in the late eighteenth century, and more recently from large-scale deforestations.

ISLANDS

Island geomorphology and geographical position will mediate the biological effects of global warming. One of the most basic distinctions between different types of islands is that between continental and oceanic islands. Continental islands arise from the continental shelf. They typically occur in waters less than 180 m deep and are in close proximity to large land masses (Fig. 1). Oceanic

FIGURE 1 Southeast Farallon Island, a granitic, continental island lying 48 km west of the Golden Gate, San Francisco, California. Photograph by the author.

FIGURE 2 Isla de Guadalupe, a basaltic oceanic island lying 260 km west of Bahia del Rosario, Baja California Norte, Mexico. Photograph by the author.

islands arise from deeper waters (more than 4000 m); are formed by volcanic activity associated with active margins, hotspots, and mid-ocean ridges; and are not in close proximity to large land masses (Fig. 2).

Tropical oceanic islands can be further divided into high islands and low islands. High islands are young volcanic islands. Terrestrial habitats on high islands are better differentiated and more numerous than are those on low islands, and this contributes directly to greater species richness and endemism on high islands. Coral fringing reefs form around the periphery of high islands, and over time catastrophic volcanism, erosion, and subsidence remove the volcanic core, leaving a shallow lagoon surrounded by reefs. These eroded low islands are known as atolls and are often barely above sea level.

The geographical position of an island also mediates the effects of global warming. The eastern side of ocean basins tends to be cooler because of large-scale atmospheric and oceanic patterns. This is especially true for continental islands, where eastern boundary currents and upwelling associated with wind patterns surrounds continental islands in cooler waters. In the tropics, the cooler waters found in the eastern basins also reduce the risk of tropical storms, cyclones, and hurricanes.

SEA LEVEL

Rising sea level is of particular concern for low islands, many of which lie only a few meters above sea level. Since 1880 sea level has risen by approximately 20 cm based on tidal gauge records from around the world. Sea-level rise per se is not a threat to near-shore marine communities. Over the last 18,000 years, sea level has risen approximately 120 m with little evidence of mass extinctions in either tropical or temperate near-shore habitats. In the tropics, coral growth keeps pace with sea-level changes and may actually be stimulated by slowly rising sea levels. However, if coral growth rates are reduced by environmental stresses such as temperature increases, then growth rates may not be able to compensate for rising sea levels. This outcome would be particularly devastating for reef crest communities, which, like the intertidal zone of temperate continental islands, occupy the interface between ocean and terrestrial conditions.

Coastal platforms occur around the periphery of high islands. These platforms are the products of past higher sea levels and uplift, and in the tropics they are backed by the mountainous core of the island, leaving few options for inhabitants in the face of rising sea levels (Fig. 3). Platforms on both oceanic and continental islands also provide rookery and breeding space for a large diversity of sea and shore birds and marine mammals. In temperate settings, these terraces are backed by cliffs, leaving few alternatives for the animals that require space for feeding, breeding, and hauling out.

Mangroves and wetlands also occupy the interface between the ocean and the land and provide important nurseries for fish and invertebrate species and habitat for

FIGURE 3 Most human activity on high islands is concentrated on the narrow coastal planes that surround the mountainous interiors. Paopao, Moorea, French Polynesia. Photograph by the author.

waterfowl and other vertebrates. They also serve as barriers to erosion and buffers to storm surges associated with hurricanes, and they are important filters for the removal of anthropogenic pollution. Although mangroves are particularly abundant on tropical continental islands, they also occur on oceanic high and low islands. Sea-level changes have several effects on mangroves, including changes to the sediment balance and increased saltwater intrusion, that can change species composition. During past sea-level changes, sediment input has kept pace with changes in sea level. However, changes in land use immediately adjacent to mangroves and wetlands reduce the sediment input needed to maintain these habitats, especially if the rate of sea-level rise is accelerated.

Although fully marine taxa are buffered from changes associated with sea-level rise, terrestrial species, especially those found on low islands, are particularly vulnerable. For example, the Bahamas have many species of endemic birds, reptiles, and terrestrial gastropods and over 121 endemic plants and trees. The maximum elevation of the islands of the Bahamas Bank is only 40 m, and most are much lower. Similarly, the 30 endemic species of terrestrial plants and animals in the Florida Keys are also at grave risk. The average elevation of the Keys is 3 m, and any rise in sea level will have substantial consequences.

TEMPERATURE

The distribution of plants and animals on islands is regulated by temperature, and the differences between low and high islands are important here. On continents, terrestrial faunas and floras can compensate for temperature changes by latitudinal and altitudinal shifting. During global warming, a temperate species can shift its distribution northward. In mountainous regions, the same compensation is obtained by moving to higher elevations. For species on islands, altitudinal shifting is the only option, and then only on high islands with their greater range of microclimate variation.

Marine species compensate by latitudinal and depth shifting; the former is better documented and thought to be more common. The distinction between high and low islands is not critical for reef species, but it can matter for species associated with basalt rather than calcium carbonate substrates. More importantly, oceanic islands are not randomly distributed across ocean basins but reflect the larger, underlying processes of plate tectonics, and amenable islands may not lie along a latitudinal gradient, leaving nearshore species with nowhere to go.

Many temperate continental islands provide important breeding rookeries for seabirds and pinnipeds (Fig. 4). Global warming will drive these species poleward, concen-

FIGURE 4 Seabirds and pinnipeds require similar habitats for breeding and roosting/hauling out. Rising sea level will reduce the size of these platforms, increasing interactions between birds and pinnipeds. Flighted seabirds have the option to use precipitous cliffs behind the platforms; flightless seabirds do not. Iquique, Chile. Photograph by the author.

trating more individuals on the remaining islands, where competition for reproductive space between breeding sea and shore birds and pinnipeds is intensified; flightless birds such as penguins are especially vulnerable. Larger concentrations of less-dispersed individuals also are more susceptible to disease.

Most marine organisms live within narrow temperature regimes, and the temperature tolerances of larvae and adults of the same species may vary substantially. Therefore, even short-term temperature increases can have dramatic effects on marine populations. Nowhere has this been better documented than in the case of the impact of temperature increases on corals. It is estimated that short-term high temperatures have contributed to massive declines of coral reefs throughout the tropics. When corals are stressed by high temperatures, they shed their symbiotic algae (zooxanthellae), causing the corals to whiten, a process referred to as coral bleaching (Fig. 5). Temperature rises of as little as 1 °C can cause coral bleaching. Corals without zooxanthellae are less able to cope with physiological stress; many die, and in some cases, entire reefs can be lost. Examples include the elevated temperatures associated with the 1997–1998 El Niño–Southern Oscillation (ENSO) event, which produced the most geographically widespread coral bleaching ever documented, and the 1994 coral bleaching event in American Samoa, which killed over 90% of living corals. Coral bleaching has also devastated the Caribbean, and it is estimated that 80% of the reef corals there have been lost during the last 30 years.

FIGURE 5 Bleached coral among healthy colonies. Subic Bay, Luzon, Philippines. Photograph by the author.

The loss of corals and the reefs they form has major implications beyond the coral species themselves. Coral reefs are three-dimensional structures that form and support one of the most diverse and biologically complex ecosystems on Earth. Thus, the loss of reefs has far-reaching and significant impacts on tens of thousands of organisms that depend on the physical structure of the reef for their survival. Coral reefs also form offshore barriers that protect coastlines by dispersing wave energy over the reef crest and throughout the reef flat. The loss of coral reefs therefore leads to increased storm damage, erosion, and flooding on low islands and on the marine terraces of high islands.

OCEAN ACIDIFICATION

Global warming does not produce ocean acidification, but like global warming it is driven by increasing levels of atmospheric CO_2. When atmospheric CO_2 increases, more CO_2 becomes dissolved in the world's oceans, causing pH to drop and therefore making seawater more acidic. Over the last 30 years, the pH of the world's oceans has dropped about 0.1 pH units, and it is estimated that pH will drop another 0.3 to 0.5 pH units by 2100. Although ocean acidification is a relatively new issue associated with climate change, it was first predicted over 50 years ago.

The chemistry of ocean acidification is well understood, but the consequences for marine biodiversity and ecosystems are largely unknown. Organisms that form calcareous skeletons are most likely to be affected by ocean acidification, including corals, echinoderms, molluscs, and crustaceans. In addition to these invertebrate groups, calcareous skeletons are also produced by several important planktonic groups, including coccolithophores and foraminifera, as well as calcareous algae that are found throughout tropical, temperate, and polar seas. Recent studies also suggest that larval and juvenile fish may be affected by ocean acidification. All of these groups are integral parts of island marine ecosystems.

CURRENTS

Islands depend on currents for their climate, dissolved gases, nutrients, and transport of marine larvae and terrestrial species. Prior to the 1960s, the effects of climate change on oceanic circulation were rarely considered because the immense size of currents was thought to buffer them; changes were thought to take place over geological time, not on centurial or millennial timescales. However, studies of ocean sediments and models suggested that ocean currents could shift over thousands of years, and in the 1980s, the Greenland ice cores revealed that North Atlantic circulation could change on centurial timescales. Today, ocean current systems are viewed as relatively fragile systems that are susceptible to global warming and are capable of relatively rapid change.

Changes in current patterns can have catastrophic effects on island ecosystems that depend on local and global circulation patterns for the retention and recruitment of their fauna and flora. Populations on isolated islands are often maintained by the retention of larvae in gyres that form in the lee of islands in strong unidirectional currents. Other populations are maintained by currents that provide larvae from distant spawning sources to local island recruitment sinks. Changes in current patterns also produce unique ecosystems on islands because of the fortuitous nature of the assembly of both their marine and terrestrial faunas and floras. In many cases source areas are congruent with today's circulation patterns; however, in other cases the pattern is more complex and difficult to decipher in light of contemporary current patterns. Changes in ocean currents can also move food sources farther from islands, causing reproductive failures in island seabird colonies.

STORM TRACKS

Severe weather events are likely to be stronger and more frequent because of global warming. Global warming will also shift storm tracks, increasing rainfall in some regions while producing droughts in others. These changes will affect island terrestrial and stream communities. Increased rainfall will also increase flooding, with concurrent increases in sediment transport, which impacts intertidal and nearshore communities, especially on continental islands (Fig. 6). Increased sediment input also damages corals. El Niño

FIGURE 6 Sediment plume from the Santa Clara River engulfs Anacapa Island in the California Bight. Also note local plumes on the north and west shores of Santa Rosa and Santa Cruz Islands. Quasi true color image from Terra-Modis satellite pass of January 30, 2005, courtesy of Dr. M. Kahru, Scripps Institution of Oceanography, San Diego, California.

events are also expected to become more frequent because of global warming. El Niño events affect precipitation patterns and storm intensity and frequency, and they increase the danger of cyclones in the Pacific. Lastly, changes in storm tracks, like changes in current patterns, can alter the source area of wind-dispersed animals and plants.

GLOBAL WARMING AND ISLANDERS

Global warming will affect every aspect of human life, including medicine and disease, agriculture, biomes and habitats; it will affect every culture and society. The impact and magnitude of these changes are virtually unknown in recent human history, and they will have tremendous impact on global economic, political, and social structures. People who live on islands are especially at risk. Sea-level rise will impact tourism, freshwater supplies, agriculture, and aquaculture. In the worst-case scenario, low islands and perhaps entire atoll countries will completely disappear. Shifts in rainfall regimes will bring droughts to some islands and floods to others. Hurricanes and cyclones are likely to increase in intensity and frequency, and their accompanying storm surges combined with sea-level rise will increase erosion, coastal flooding, and seawater intrusion. Changes in water temperatures and current regimes will have significant impacts on many marine organisms that form important island fisheries. In many cases, these socioeconomic costs will bankrupt small island nations.

These impacts are already being felt, and future impacts on islanders are likely to be even more severe. In 1998, uninhabited islands in the Pacific atoll nation of Kiribati were inundated, and the atoll nation of Vanuatu has evacuated some low islands as a precaution. Sea-level rise, coastal erosion, cyclone activity, and coastal flooding are displacing the inhabitants of the islands of the Sunderbans in the Bay of Bengal. Lohachara Island, which at one time was populated by 10,000 people, has disappeared. In Indonesia, 42 million people live less than 10 m above sea level. Here, the most vulnerable island is Java, with a population of over 110 million, many of whom live near the threatened cities of Jakarta, Surabaya, and Semarang. Mass migrations off islands are likely, and in the Pacific, New Zealand and Australia are already planning for refugees from global warming. The costs of global warming are estimated in the billions of dollars, but the real cost may be in the loss of the varied societies and cultures that have evolved on the world's islands.

SEE ALSO THE FOLLOWING ARTICLES

Hurricanes and Typhoons / Reef Ecology and Conservation / Sea-Level Change / Surf in the Tropics / Warming Island

FURTHER READING

Caldeira, K., and M. E. Wickett. 2003. Anthropogenic carbon and ocean pH. *Nature* 425: 365–368.
Gore, A. 2006. *An inconvenient truth: the planetary emergency of global warming and what we can do about it.* New York: Rodale.
Hoegh-Guldberg, O., P. J. Mumby, A. J. Hooten, R. S. Steneck, P. Greenfield, E. Gomez, C. D. Harvell, P. F. Sale, A. J. Edwards, K. Caldeira, N. Knowlton, C. M. Eakin, R. Iglesias-Prieto, N. Muthiga, R. H. Bradbury, A. Dubi, and M. E. Hatziolos. 2004. Coral reefs under rapid climate change and ocean acidification. *Science* 318: 1737–1742.
MacArthur, R. H., and E. O. Wilson. 1967. *The theory of island biogeography.* Princeton: Princeton University Press.
Mooney, C. 2007. *Storm world: hurricanes, politics, and the battle over global warming.* Orlando: Harcourt.
Sekercioglu, C. H, S. H. Schneider, J. P. Fay, and S. R. Loarie. 2008. Climate change, elevational range shifts, and bird extinctions. *Conservation Biology* 22: 140–150.
United Nations Environment Programme, Islands Web site (http://islands.unep.ch/isldir.htm).

GOUGH ISLAND

SEE TRISTAN DA CUNHA AND GOUGH ISLAND

GRANITIC ISLANDS

MILLARD F. COFFIN

University of Southampton, United Kingdom

Granitic islands are extensions or fragments of continents that are isolated from major continental land masses by the sea or by lakes and rivers. Such islands are common along the margins of continents as well as in lakes and rivers, and they are relatively rare in mid-ocean settings. Variations in continental topography, changing sea and freshwater levels, plate tectonic processes, isostasy, and resistance to erosion account for the existence of granitic islands.

COMPOSITION AND DENSITY

Granite, composed mainly of quartz, feldspar, and mica, is an intrusive igneous rock that typically solidifies from its parent magma at depths of 2 km or more within the Earth's crust (Fig. 1). Granite is a dominant rock in continental crust, the upper part of which has an average density of 2.7 g/cm^3. All granitic islands belong in the category of continental crust, in contrast to oceanic islands. Both oceanic islands and the oceanic crust that encompasses them are dominated by basaltic rock, which has an average density of 2.9 g/cm^3. Aside from geodynamic forces, the Earth's ability to equilibrate isostatically means that these different densities, and hence buoyancies, are the primary control on continental crust, including granitic islands, which lie mostly above sea level, and oceanic crust, which lies overwhelmingly below sea level.

FIGURE 1 View from the summit of Cadillac Mountain (467 m), Mount Desert Island, Maine, toward granitic islands to the south. Granites forming the mountain are Devonian in age, ~365 million years old. Photograph by Millard F. Coffin.

CRUSTAL SETTING, SUBSIDENCE, AND STRUCTURE

Most granitic islands lie along continental margins or within lakes and rivers. Acting in conjunction, variations in continental topography, as well as temporal variations in global sea level and in local lake and river water levels, commonly isolate portions of continental crust, the result of which is granitic islands. Relatively rarely, the plate tectonic processes of divergence and transform motion may isolate slivers of continental crust within ocean basins. In some cases these detached slivers, or microcontinents, remain above sea level, also resulting in granitic islands.

In contrast to the relatively young (< 150 million years old) oceanic crust, continental crust is commonly much older and therefore does not exhibit the typical oceanic crustal/lithospheric subsidence on the order of a few kilometers, resulting from significant cooling following its solidification. Furthermore, granitic islands represent exhumed crust, meaning that 2 km or more of overlying continental crust has been removed sometime during their history, in contrast to new magmatism that produces oceanic islands. These two contrasting aspects of granitic and oceanic islands mean that the former can assume myriad morphologies and structures and can maintain a relatively constant position relative to global average sea level, whereas the latter are typically individual volcanoes, or clusters or linear ridges of volcanoes, which evolve and subside from subaerial volcanoes to atolls to submarine seamounts and guyots.

EROSION AND LIFE

Granite and other igneous rocks comprising continental crust, although relatively enduring, are gradually worn away by the processes of mechanical and chemical erosion. In particular, granitic islands are continually subjected to wave action at their shorelines, which gradually retreat, diminishing the size of the islands. Eventually, granitic islands may erode to sea level and disappear. As with the continents, however, isostasy results in uplift of continental crust (of which granitic islands are a part) as it is eroded downward, thereby counterbalancing the effects of erosion and sustaining granitic islands. Such islands can exist for tens to hundreds of millions of years or more, in sharp contrast to oceanic islands, which have a maximum lifetime of a few tens of millions of years.

Because granitic islands have such long life spans, they can be incubators for unique, endemic flora and fauna if they remain relatively isolated from continents. Although isolated oceanic island chains have the same potential, their generally shorter life spans typically result in less diverse flora and fauna.

TYPE EXAMPLES

The two major types of granitic islands (i.e., those that lie within or are extensions of major continental land masses and those that are isolated from such land masses by oceanic crust) may be considered in slightly more detail by examining an example of each—the Falkland Islands (Islas Malvinas) (Fig. 2) and the Seychelles (Fig. 3), respectively. Although both examples are situated far from major continental land masses, the Falkland Islands lie

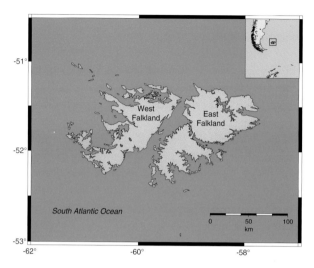

FIGURE 2 Map of the Falkland Islands (Islas Malvinas), South Atlantic Ocean. Inset shows location relative to South America and Antarctica. From Millard F. Coffin.

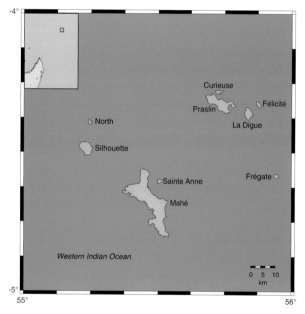

FIGURE 3 Map of the Seychelles, Western Indian Ocean. Inset shows location relative to Madagascar. From Millard F. Coffin.

on a prolongation of South American continental crust, whereas the Seychelles are surrounded by oceanic crust.

The Falklands, lying at ~52° S in the South Atlantic, encompass more than 775 islands and an area of more than 12,000 km², with a maximum relief of 705 m. Of all the islands, only five have an area greater than 10 km². They lie approximately 500 km east of southern Argentina and Tierra del Fuego. They comprise the major subaerial portion of the great submarine Falkland Plateau, continental crust that is flanked by oceanic crust to the north, east, and south mostly created by sea floor spreading between South America and Africa.

The Seychelles, near the equator in the Indian Ocean, consist of 115 islands, of which 32 are predominantly granite, and encompass an area of approximately 450 km². Only the islands situated on the continental Seychelles Bank are known to be granitic; other Seychelles islands—Aldabra, Cosmoledo, Farquhar, Amirante, Plate, and Coetivy—may have originated as oceanic islands. The granitic Seychelles have a maximum relief of 905 m. The Seychelles Bank was once joined with Madagascar and India, but it separated from the former ~84 million years ago, and from the latter ~65 million years ago by plate divergence. The continental Seychelles Bank is now surrounded entirely by oceanic crust.

SEE ALSO THE FOLLOWING ARTICLES

Continental Islands / Oceanic Islands / Plate Tectonics / Seamounts, Geology / Seychelles

FURTHER READING

Baker, B. H., and J. A. Miller. 1963. Geology and geochronology of the Seychelles Islands and structure of the floor of the Arabian Sea. *Nature* 199: 346–348.
Clark, R., E. J. Edwards, S. Luxton, T. Shipp, and P. Wilson. 1995. Geology of the Falkland Islands. *Geology Today* 11: 217–223.
Gilman, R. A., C. A. Chapman, T. V. Lovell, and H. W. Borns Jr. 1988. *The geology of Mount Desert Island: a visitor's guide to the geology of Acadia National Park*. Augusta: Maine Geological Survey.

GREAT BARRIER REEF ISLANDS, BIOLOGY

HAROLD HEATWOLE

North Carolina State University, Raleigh

The biota of the islands of the Great Barrier Reef of Australia varies enormously from island to island owing to the great diversity of geological origins and history of the islands, their differences in isolation from each other and from the mainland, their great latitudinal extent, the extent of disturbance by humans, their diversity of soil, and the variety of weather and patterns of sea currents that impinge upon them. The islands and their resident life are in a continual state of flux.

THE ISLANDS

There are about a thousand islands on the Great Barrier Reef of Australia, ranging from tiny, bare sand bars of only a few meters in diameter and scarcely emergent at high tide to large rocky islands of over 39,000 hectares, rising to elevations of more than 1000 m. They fall into several categories in terms of their origins and geology and include rocky continental islands, sand cays, coral cays, mangrove cays, and low wooded isles. Each of these insular types has its own distinctive biota.

The continental islands were once elevated parts of the Australian mainland, but as glaciers melted and sea level rose during the waning of the last great ice age, low-lying areas were flooded, and former hilltops and mountains, still bearing their resident plants and animals, became isolated as islands. Accordingly, some of their animals and plants represent populations isolated at the time they became separated from the mainland as islands. By contrast, the cays (Fig. 1) have never been connected to any mainland but arose, newly formed, from the living reef itself and as a result their terrestrial life had to reach them by some means of overwater dispersal.

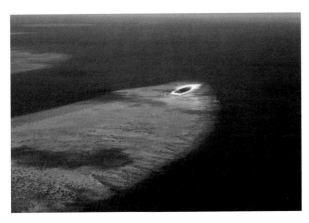

FIGURE 1 Frigate Cay, a small, vegetated coral cay in the Swain Reefs, southern Great Barrier Reef, Australia. Photograph by Harold Heatwole.

BIOTA IN RELATION TO ISLAND TYPE

The differences in origins, types of substrate, and sources of terrestrial life lead to diversity in the kinds of animals and plants on the islands of the Great Barrier Reef. An example is provided by the reptiles. The continental islands, especially the larger ones, have a variety of species of skinks, dragons, snakes, and sometimes even goannas (monitor lizards, family Varanidae), which are the same species as those on the adjacent Australian mainland; most are restricted to the Australian region. By contrast, the cays, if they have terrestrial reptiles at all, have one or a few species of gecko or skink, some of which may have been introduced by humans.

Mammals are not prevalent on barrier reef islands, and when they are extant, it is only on the larger, continental ones; even there, they are only a subset of the smaller species occurring on the adjacent mainland. Landbirds have a similar pattern. Many species have been recorded as single individuals briefly stopping at Barrier Reef islands without nesting or establishing populations. Those that do nest on islands do so mainly on the larger continental ones, but a few species have established populations on forested sand cays. Notable among these are the buff-banded rail (*Gallirallus philippensis*), the bar-shouldered dove (*Geopelia humeralis*), and the insular silvereye (*Zosterops lateralis chlorocephala*).

The invertebrate fauna of Barrier Reef islands is incompletely known, but it is clear that the continental islands harbor a far greater number of species than do coral cays and that forested coral cays have a greater number of species than do those with only low vegetation. Even bare, unvegetated cays, however, often have some invertebrates such as scavenging beetles and flies, and parasitic bird ticks.

BARRIER REEF ISLANDS AS REPRODUCTIVE HAVENS

The eggs of sea turtles and ground-nesting birds are vulnerable to predation on the mainland by goannas, foxes, and other predators that are lacking on most islands. Although nesting by sea turtles and seabirds does occur on mainland beaches, the sand cays of the Great Barrier Reef are their principal sites. Both groups are highly mobile and can reach remote places where they aggregate in prodigious numbers during the breeding season. Raine Island in the northern Great Barrier Reef holds the world record of 11,000 nesting female turtles on the beach in one night. Seabirds often are represented, even on bare sand cays (Fig. 2), by hundreds of pairs.

FIGURE 2 Masked booby (*Sula dactylatra*) and a few brown boobies (*Sula leucogaster*) nesting on a bare sand cay (Gannet Cay) in the Swain Reefs, Great Barrier Reef, Australia. Photograph by Harold Heatwole.

THE EQUILIBRIUM THEORY

Robert H. MacArthur and Edward O. Wilson proposed that islands often are not "saturated"; that is, they do not contain as many species as is ecologically possible, but rather they have a lower number of species representing an equilibrium between the immigration of new species to the island and the extinction of species already present. They further proposed that the number of species present would be affected by the size of islands and their distance from a mainland source of species, with larger islands, everything else being equal, having more species than would smaller ones and islands close to the mainland having a greater number of species than would more remote ones.

This basic model stimulated research, and its predictions were validated; the number of species on islands does generally increase with island size and decrease with distance from a mainland. Also, the implied prediction of measurable rates of turnover of species of plants and animals on islands was upheld for Barrier Reef islands; species come and go, but their total number often remains approximately the same over time as long as the island itself does not change. Various modifications, however, were found necessary, some of them already recognized by MacArthur and Wilson. For example, the presence of stepping-stone islands could influence rates of immigration, and hence the number of species at equilibrium. For some islands on the Great Barrier Reef, the height of the island contributes more to ecological diversity than does insular area, and hence the former is a more important determinant of number of species than is the latter. There is a "small-island effect," in that over the lower range of size of the very smallest islands, there is no change in number of species with increasing insular size. An important aspect not covered by the original theory is that some islands change in size and in other aspects of their ecology, and this influences the numbers, kinds, and persistence of organisms resident on them (see following section). Insular biotic communities are highly dynamic, but in an orderly way.

DYNAMICS OF GREAT BARRIER REEF ISLANDS

Continental islands retain their covering of vegetation at the time they become isolated and thus resemble the mainland. Coral cays start off as bare sand or rubble and must receive their flora via overwater dispersal. Initially, cays are too small for any but the hardiest of plants to survive because of scarcity of freshwater. As cays accrete, however, they reach sufficient size that they can retain rain as a freshwater lens in the sand, and they become suitable for additional species of plants. Plants arrive via buoyant, seawater-resistant propagules, or as sticky or hooked seeds or fruits attached to seabirds; few insular plants are wind dispersed. The first plants to become established on a newly formed cay are usually sea-dispersed pioneer species of grasses and herbs that are salt resistant and that have low requirements for organic matter. Often, they are trailing vines or creeping grasses that can persist in shifting sands; if one part is covered by blowing sand, other parts may still remain above the surface, and if parts are uprooted, other parts remain rooted and able to supply water from the soil. Another important pioneer is octopus bush (*Argusia argentea*), a shrub whose seeds cannot germinate unless subjected to saltwater prior to exposure to freshwater. Mature seeds are carried by sea currents, sometimes for hundreds or even thousands of kilometers, there to be cast ashore by the surf, where they germinate on the strand once soaked by rain. Because this species cannot spread on land even a few meters from the parent bush because of lack of treatment with seawater, they grow only on the upper beach. In time, they often form a shrub ring around a young cay. This ring protects the interior of the cay from wind and salt spray. This ring, and the other pioneer plants, provide shade, lower surface temperatures, and contribute organic matter to the soil. The new conditions permit establishment of plants that formerly could not have survived on the cay. These new arrivals are primarily bird dispersed, and they form a denser cover of vegetation—an herb flat that ameliorates conditions still further. These plants are followed by shrubs and trees that alter the environment even more by providing deeper shading and by contributing additional organic matter. At first, these trees and shrubs are scattered as open woodlands, but in time dense forests, especially of the pisonia tree (*Pisonia grandis*), develop. Thus, there is a progression from bare sand through herb flat (with shrub ring) to open woodland to pisonia forest. These vegetational changes are reflected in the insular fauna.

Some seabirds will nest in open areas on bare substrate and are sometimes found on young cays with sparse or no vegetation. Examples are the black-naped tern (*Sterna sumatrana*) and the masked booby (*Sula dactylatra*). Still others, such as the brown booby (*Sula leucogaster*), nest on the ground, usually where there is at least some low vegetation; others such as the bridled tern (*Sterna anaethetus*) need cover and often place their nests under rather dense low vegetation. The common noddy (*Anous stolidus*) tends to nest in low bushes, such as octopus bush. Finally, species such as the black noddy

(*Anous minutus*) require leafy trees for nesting. As a cay develops its vegetation, it attracts progressively different species of seabirds. All seabird species, regardless of nesting habits, have a profound effect on the ecology of coral cays; they add guano and their dead bodies as organic matter. There is an intimate relationship between the pisonia tree and seabirds. This tree requires high levels of nitrogen in the soil. It also has sticky seeds that attach to birds' feathers and are thereby carried to new islands where ground-nesting seabirds have previously enriched the soil with their guano. The seeds germinate there, and eventually forests are formed that make the interior of the island unsuitable for most ground-nesting seabirds but which supply essential nesting sites for tree-nesters. Some insular shrubs and trees (for example, figs) produce fruits that allow establishment of landbirds, such as the silvereye.

A mature cay is zoned with remnants of previous stages in concentric circles. The beach remains bare toward the tide line, but on the upper beach there is often a band of pioneer species, followed by a shrub ring, inside of which occurs open woodland and/or mature pisonia forest.

The progression of animal life follows that of the vegetation, but often in unexpected ways. The first animals to become established on new cays are mainly scavengers, particularly flies (and some beetles). This is because in early stages there are no plants upon which herbivores can feed, but there is a rich source of food for scavengers in the form of washed-up marine carrion or of guano or carrion from seabirds. The latter are "transfer" species, in that they feed in the sea and deposit food for scavengers on land. After scavengers have become established, predators, such as centipedes, in turn become established and feed upon the scavengers. Finally, once vegetation is present, herbivores and nectar feeders, and subsequently predators upon these, become established.

The progression of life on a coral cay is not always as orderly as outlined above because there are destructive forces that interrupt, or reverse, the sequence. One of the most prevalent is erosion by waves cutting into, or eliminating, the peripheral pioneer species and shrub ring on one side, or even some of the forest. Some eroding cays gradually diminish in size, but for others the eroded sand is redeposited on the opposite side, and the cay "creeps" across the reef, eliminating the established vegetation on the eroding side and opening up new habitat for pioneer vegetation on the prograding side. Thus, erosion alters the original arrangement of concentric zones of vegetation, moving them off center into a lopsided pattern. Sometimes cays move so rapidly over the reef that their outlines from aerial photographs taken 20 years apart do not overlap. On such rapidly moving cays, all vegetation may be eliminated, returning the cay to its original bare condition. In extreme cases, a cay may move over the edge of its reef into deep water and disappear entirely.

Another destructive force is washover by the sea during storms, with resultant loss of vegetation, or even a return of the cay to a completely bare state. Nesting sea turtles uproot plants when excavating nests, thereby exposing the roots and killing most plants. The pioneer vines and creeping plants persist, however, because they are sustained by intact roots from parts of the plant beyond the turtle's nesting site. In this way, turtles can maintain part or all of an island in an early state of development. Nesting seabirds in large numbers trample plants to the extent that low vegetation is destroyed, and a cay may be returned to the bare state. Once humans inhabit an island, they cause enormous changes through the introduction of species, intentional or otherwise, which thereby influences the immigration rate, or by altering the habitat.

Some of these changes, especially those of later stages, are slow and occur over the span of many years, whereas others, particularly those of early stages, may occur rapidly, and differences may be noted over a period of only a few weeks or months; some destructive ones, like overwash and erosion during storms, may occur in only hours.

TROPHIC STRUCTURE

Trophic structure is the relative apportionment of species into categories of diet, such as herbivores, nectar feeders, predators, scavengers, and parasites. A controversial issue in ecology is whether trophic structure remains stable during the dynamic processes of immigration, extinction, and species turnover.

A study of One Tree Island, a coral rubble cay near the southern edge of the Great Barrier Reef and perhaps the most intensively studied cay in the world, allows an answer to this question. Its populations of plants and animals, from soil microarthropods to vertebrates, were monitored over several years. Among the insects, there was a high immigration rate, a high extinction rate, and accordingly a high turnover of species. Of the total number of 396 species recorded, only 26 (6.6%) were permanent, with the rest either first immigrating and establishing during the study, declining during the study and then becoming locally extinct, or both. A drought reduced the low vegetation of an extensive herb flat to a lower biomass of predominantly pioneer species. Insect populations crashed, and the number of species was greatly reduced. Following recovery from the drought, the number of species increased greatly, and the population density of insects changed markedly, overshooting the equilibrium number before returning to it. During such major changes, the ratio of the numbers in the different dietary (trophic) categories remained remarkably stable. Thus, it appears that although which particular species immigrate at any one time is a matter of chance, whether immigrants can become established and persist seems to be influenced by the proportion of kinds of food habits already represented, and a relatively fixed ratio is maintained. If there are too many herbivores, it is more difficult for an immigrating species of herbivore to become established, or one already there is more likely to become extinct, thereby keeping ratios stable. This stability of trophic structure, even in the face of greatly fluctuating population numbers of species and population densities, applies only to a particular stage in the development of islands. When different stages of cays are compared, they differ in trophic structure. Bare cays have a high proportion of scavenging species, for example, with few predators and no herbivores. A forested cay, by contrast, has a high proportion of herbivores, a considerable number of predators, and relatively fewer scavengers. Intermediate stages show transitional trophic structures. Thus, for a given stage of development of a cay, its trophic structure remains relatively stable, but as its major base of resources changes so does its trophic structure. The islands of the Great Barrier Reef are highly dynamic; that dynamism combines the occurrence of chance events whose influence is regulated by orderly processes.

SEE ALSO THE FOLLOWING ARTICLES

Erosion, Coastal / Great Barrier Reef Islands, Geology / Island Biogeography, Theory of / Seabirds / Vegetation

FURTHER READING

Flood, P. G., and H. Heatwole. 1986. Coral cay instability and species-turnover of plants at Swain Reefs, southern Great Barrier Reef, Australia. *Journal of Coastal Research* 2: 479–496.

Heatwole, H. 1971. Marine-dependent terrestrial biotic communities on some cays in the Coral Sea. *Ecology* 52: 363–366.

Heatwole, H. 1984. Terrestrial vegetation of the coral cays, Capricornia Section, Great Barrier Reef Marine Park, in *The Capricornia Section of the Great Barrier Reef: past, present and future*. W. T. Ward and P. Saenger, eds. Brisbane: The Royal Society of Queensland and Australian Coral Reef Society, 87–139.

Heatwole, H. 1991. Factors affecting the number of species of plants on islands of the Great Barrier Reef, Australia. *Journal of Biogeography* 18: 213–221.

Heatwole, H., T. Done, and E. Cameron. 1981. Community ecology of a coral cay: a study of One-Tree Island, Great Barrier Reef. *Monographiae Biologicae* 43: 1–379.

Heatwole, H., and T. A. Walker. 1989. Dispersal of alien plants to coral cays. *Ecology* 70: 501–509.

MacArthur, R. H., and E. O. Wilson. 1967. *The theory of island biogeography*. Princeton, NJ: Princeton University Press.

GREAT BARRIER REEF ISLANDS, GEOLOGY

SCOTT G. SMITHERS

James Cook University, Townsville, Australia

Around 1000 islands occur on Australia's Great Barrier Reef (GBR), with the exact number varying depending on where the boundaries of the GBR are drawn and how islands are defined (e.g., when is a sand bar an unvegetated cay?). The numbers cited below are for the Great Barrier Reef Marine Park (GBRMP). Two main island types occur: continental high islands and low reef islands. Continental high islands are continental outcrops that were separated from the mainland as rising seas flooded the continental shelf after the last ice age. The low reef islands are accumulations of reef-derived carbonate sediments deposited on reef flats formed after the mid-Holocene.

CONTINENTAL HIGH ISLANDS: DESCRIPTION AND DISTRIBUTION

High islands exhibit a range of topography and morphology similar to that observed on the mainland coast, which largely reflects lithology and structure. Some are relatively low and gently sloping (e.g., Stone Island [20°03′ S, 149°15′ E]), but others (e.g., Gloucester Island [20°01′ S, 148°27′ E]) are high and rugged and drop steeply to the sea. They also vary in size, ranging from small rocky outcrops to large islands like Hinchinbrook Island (18°21′ S, 146°14′ E; 390 km^2). Big islands often lack fringing reefs, possibly because of their larger catchments and runoff volumes. More than 50% of the high islands with reefs are located between 20° S and 22° S (Figs. 1A, 2A).

LOW REEF ISLANDS: DESCRIPTION, CLASSIFICATION, AND DISTRIBUTION

Around 300 reef islands occur in the GBRMP. They form where the deposition of reefal carbonate sediments is concentrated by the centripetal action of waves refracted around a reef. Because reef configuration controls wave refraction patterns, reef and reef island shape are often strongly related; oval reefs usually support oval islands, and linear islands commonly develop on elongated reef platforms. Reef islands occur on both very small and very large reefs, suggesting platform size has less influence on their formation, although the competency of waves to transport sediments to a nodal point may be compromised by energy dissipation across wider reef flats. GBR

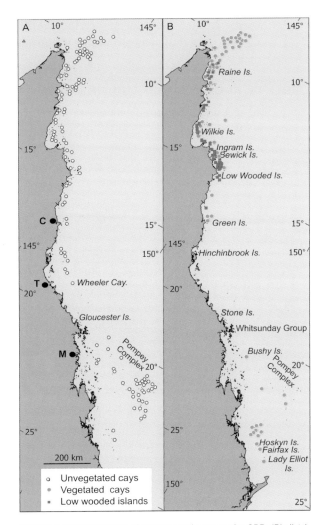

FIGURE 1 (A) Distribution of unvegetated cays on the GBR; (B) distribution of vegetated cays and low wooded islands on the GBR.

reef flats, and thus reef islands, are less than 6000 years old.

Reef islands are usually classified according to shape (linear or compact), sediment type (coarse shingle or sand), and vegetation cover. GBR reef islands dominated by coarse sediments (shingle, rubble) typically develop from linear banks episodically deposited parallel to the windward reef front during storms, or spits trailing lagoonward from these features. Shingle cays may become cemented and relatively stable, but their exposed position inhibits plant colonization.

Sand cays typically accumulate near the leeward reef margin. In the GBRMP, there are 213 unvegetated and 43 vegetated cays (Figs. 1A and B, 2B), although these numbers fluctuate. Unvegetated sand cays are the most dynamic, with several disappearing in recorded history (e.g., Pixie Cay, Ellis Island) and active migration being documented at others (e.g., Wheeler [18°47′ S, 147°32′ E]).

FIGURE 2 (A) Continental high islands of the Whitsunday Islands Group; (B) Ingram Island, a vegetated sand cay with a densely vegetated older central core and lower, younger, and less densely vegetated peripheral areas; (C) Wilkie Island, a low wooded island composed of shingle cay to windward (lower foreground), vegetated sand cay (upper center), and intervening mangroves.

Cays sufficiently large (a width of 120 m has been calculated) and stable to maintain a freshwater lens may support mature vegetation and may develop soils. Phosphatic cay sandstone forms where bird droppings indurate sediments (e.g., Raine Island [11°36′ S, 144°01′ E]), Lady Elliot Island [24°12′ S, 152°45′ E]), which, together with the lithification of beach sediments to form beachrock, can confer additional stability. Stable cays may rise more than 3 m above high water, although recent deposits are generally lower. Both windward shingle and leeward sand cays occur on some reefs (e.g., Fairfax [23°52′ S, 152°22′ E)] and Hoskyn [23°48′ S, 152°18′ E] Islands).

Low wooded islands are a particularly complex type of reef island described in detail from the GBR (Fig. 2C); they comprise windward shingle islands, a leeward sand cay, and an intervening reef top colonized by mangroves (e.g., Bewick Island [14°26′ S, 144°49′ S]), Low Wooded Island [15°05′ S, 145°23′ E]).

Reef island distribution on the GBR reflects regional patterns in relative sea-level history, substrate configuration, and depth. The 44 low wooded islands occur only on the inner shelf north of Cairns (Fig. 1B). The vegetated cays are generally concentrated in the far north and south, with none in the 600 km between Bushy Island (20°57′ S, 150°05′ E) east of Mackay and Green Island (16°46′ S, 148°58′ E) near Cairns (Fig. 1B). Unvegetated sand cays exist throughout the GBR, but like vegetated cays, they are least common in the central GBR. None occur between Wheeler Cay east of Townsville and the northern Pompeys, some 315 km south (Fig. 1A). This absence reflects a combination of deeper reef foundations, a less complete outer barrier, and higher tidal ranges, all of which hinder the development of stable reef islands.

SEE ALSO THE FOLLOWING ARTICLES

Atolls / Barrier Islands / Great Barrier Reef Islands, Biology / Mangrove Islands

FURTHER READING

Flood, P. G. 1986. Sensitivity of coral cays to climatic variations, southern Great Barrier Reef, Australia. *Coral Reefs* 5: 13–18.
Hopley, D. 1997. Geology of reef islands of the Great Barrier Reef, Australia, in *Geology and hydrogeology of Carbonate Islands*. H. L. Vacher and T. Quinn, eds. London: Elsevier Science, 835–866.
Hopley, D., S.G. Smithers, and K.E. Parnell. 2007. *Geomorphology of the Great Barrier Reef: diversity, development and change.* Cambridge: Cambridge University Press.
Stoddart, D. R., R. F. McLean, and D. Hopley. 1978. Geomorphology of reef islands, northern Great Barrier Reef. *Philosophical Transactions of the Royal Society of London Series B* 284: 39–61.

GREEK ISLANDS, BIOLOGY

KOSTAS A. TRIANTIS

Oxford University, United Kingdom

M. MYLONAS

University of Crete, Irakleio, Greece

The Greek islands are heterogeneous in mode and time of formation. They are the only islands with floral and faunal elements originating from three different geographical regions, namely Europe, Asia, and Africa. Additionally, these islands have faced the intensive influence of humans for more than 8000 years.

GREEK ISLANDS IN SPACE AND TIME

Almost one-quarter of the geographical area of Greece consists of islands, which are divided into two major groups: the Aegean to the east of continental Greece and the Ionian to the west. An amazing number of islands and islets (approximately 7582) constitute what is known today as the Aegean archipelago. The southern limit of this group is marked by the South Aegean island arc, which runs from the Peloponnesos Peninsula, through Crete and Rhodes, to southwestern Turkey (Fig. 1). Of these, more than 100 are inhabited. In the Ionian Sea, there are close to 300 islands, of which more than 20 are inhabited.

The geotectonic evolution of the Greek Islands has had a major contribution in shaping the biogeographic patterns of all recent taxa of these areas. The islands have been formed under the influence of three major forces: tectonism, volcanism, and eustatism. Throughout their history, repeated cycles of connection and isolation from neighboring mainlands and insular areas have occurred. These cycles were imposed by the westward movement of the Anatolian plate, the northward movement of the African plate and its subduction under the Eurasian, the Messinian Salinity

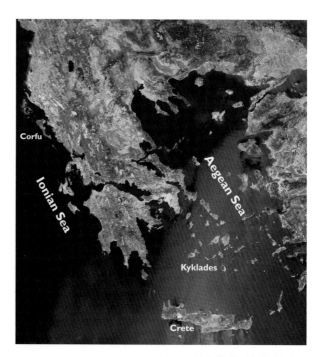

FIGURE 1 Map of the Greek area with the Aegean and Ionian island groups.

Crisis, and the sea-level fluctuations during glacial and interglacial periods. Even for those islands that have experienced long-term isolation, fluctuations in size, submergence, and uplift were very common. Crete, for example, has been crushed, folded, pushed, and shattered, producing some of the biggest fault-scarp cliffs in Europe.

In summary, four main stages in the paleogeographic evolution of the Aegean can be distinguished. During the first stage (23–12 million years ago), a continuous land mass (known as Ägäis) was present. In the second stage (12–5 million years ago), a slow sea transgression occurred, with the formation of the mid-Aegean barrier. During the third stage (5–2 million years ago), there was extensive fragmentation and a widening of the Aegean Sea. Finally, the fourth stage (during the Pleistocene), involved mainly orogenetic and eustatic sea-level changes. Intensive volcanic activity has also contributed to the formation of several islands, a few of which are purely volcanic (e.g., Nisyros Island). On the other hand, the geological evolution of the islands of the Ionian has been quite simple, with most islands becoming isolated from the mainland during the Pleistocene or even more recently (Fig. 2).

Climate in the Aegean and Ionian regions did not exhibit variation analogous to the regions' geology, but it has followed the profound global changes of the last few million years. From the late Miocene (12.5–5 million years ago) until the early Pleistocene, the climate in the eastern Mediterranean was humid and warm with rainy summers (i.e., tropical-subtropical, resulting in extensive forest vegetation). Subsequently, during glacial and interglacial periods of the Pleistocene, fluctuations from humid to dry Mediterranean climate occurred, with deciduous oak forests dominating during glacial periods and maquis, conifers, and phrygana (Mediterranean-type shrublands) dominating during interglacials.

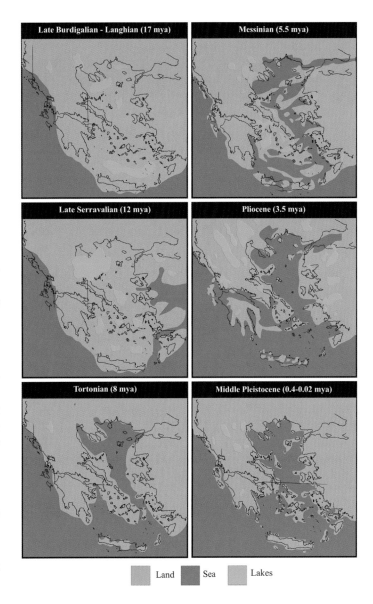

FIGURE 2 The geological evolution of the Greek area from the Miocene era (17 million years ago) to present.

BIOGEOGRAPHIC PATTERNS

Over the last 25 years, numerous ecological, biogeographical, geological, and evolutionary studies have greatly enhanced our knowledge of the biodiversity patterns of the Greek Islands. However, for a significant number of taxa, more thorough studies are needed. Distribution patterns in the Aegean Islands are far more complex than those of the Ionian, because of their larger numbers and greater topographic, paleogeographic, and environmental complexity.

Many distributional patterns in the Greek Islands correlate to some extent with paleo-events. Some can be traced back to the first two phases of the evolution of the Aegean (Fig. 2). During the Middle Miocene, a wide array of taxa invaded the area before the breakup of Ägäis. Their origin was mainly eastern, and they were adapted to humid and warm climates and to the respective vegetation. Mammals constituting the Pikermi-type fauna, such as small horses (*Hipparion*), saber-toothed tigers, monkeys, rhinoceroses, hyenas, and mastodons, as well as several reptiles, are the main representatives of this phase. They gradually became extinct as a result of geological and climatic changes in the area. Among invertebrates and plants, a few taxa of the aforementioned category can still be found; they either display relict distributional patterns, such as those of the scorpion *Iurus dufoureius,* the trees *Zelkova abelicea* and *Phoenix theophrasti* (the Cretan palm tree), and the land snail *Helicodonta gyria,* or are strongly differentiated in many taxa with well-defined distributions.

The biogeographic patterns associated with events of the second evolutionary phase of the Aegean include taxa that are of either European (e.g., the lizard genus *Podarcis*) or Asiatic (e.g., the amphibian genus *Lyciasalamandra* and the land snail *Assyriella*) origin. Irrespective of their origin, these taxa have not overcome the mid-Aegean barrier, which divides the central Aegean islands (Kyklades) from those lying close to the coast of Asia Minor.

Nevertheless, most of the present distributional patterns of taxa in the Greek Islands are directly related to the Pleistocene glacial cycles. The climatic fluctuations of the Pleistocene repeatedly caused invasions and retreats of species. Several endemic genera of the Aegean area either went extinct during glaciations or (at least some of them) were able to survive by occupying extreme habitats such as caves and the subalpine mountainous areas. Characteristic is the case of mammals, because only one endemic species from the Pleistocene fauna has survived: the Cretan white-toothed shrew (*Crocidura zimmermanni*). Most of the typical cave taxa belong to genera whose distributional patterns can be traced back to the middle Miocene.

The biota of the Ionian Islands is very similar to those of the adjacent mainland, although few endemic taxa can still be found, most of which live on the larger and more heterogeneous islands (such as Kefalonia). Additionally, the fauna and flora of the Ionian Islands are far more "harmonic," without profound gaps in their taxonomic composition (such as that shown by amphibians on most Aegean islands and especially Crete).

SPECIES DIVERSITY, ENDEMISM, AND DIVERSIFICATION

The Aegean is characterized by a great overall species diversity and, in many cases, by an exceptionally high percentage of endemism (Table 1). Those animal groups that are known to be poor dispersers, such as land snails, terrestrial isopods, darkling beetles, and some plant taxa, exhibit a level of endemism between 30% and 60% (Table 1). The vast majority of endemic species are not restricted to single islands or small island groups. Most are endemic to larger groups within the Aegean (i.e., the Kyklades), or to the whole Aegean area. Crete, with its satellite islets, is the exception, with a significantly high endemism for many taxa (i.e., ~50% endemism for land snails, ~30% for terrestrial isopods), which is due to its long-term isolation and topographical heterogeneity.

Despite the long-term isolation of many parts of the Aegean and the high degree of endemism, differentiation

TABLE 1
Number of Species Distributed on the Aegean Islands and Respective Percentage of Endemism

Taxon	Species Richness	Endemism
Plants	~3500	10%
Birds	~300	0%
Reptiles and amphibians	60	12%
Land snails	~400	35–45%
Isopods	124	60%
Tenebrionid beetles	~190	>35%
Ants (Formicidae)	128	11%
Centipedes	80	9%
Butterflies	120	8.5%
Lepidoptera (apart from Hesperioidea and Papilionoidea)	1452	3%
Odonata	59	3%

does not reach genus level, a pattern consistent with other less isolated and/or relatively younger archipelagoes. The differentiation of various organisms observed today in the terrestrial ecosystems of the Greek Islands can be categorized largely as either geographical within species or differentiation of closely related taxa with vicariant distributional pattern and doubtful genetic isolation. This is why hybridization is intensive in the contact zones of these taxa. In the case of the Cretan flora, despite the long isolation of the island (more than 5 million years), its complex geological history, and its great topographical heterogeneity, a significant number of its endemic species seem to be paleoendemics that differentiated before the splitting of the Aegean land mass, and have since been conserved in restricted habitats (e.g., the monotypic endemic plant genus *Petromarula* in Crete). In general, it can be argued that all the highly diversified taxa of the Aegean belong to genera that have been present in the region for some 5 to 12 million years (e.g., *Nigella* [plants], *Mastus* and *Albinaria* [land snails], *Dendarus* [beetles], *Schizidium* [terrestrial isopods], *Podarcis* [reptiles], *Rana* [amphibians]). These genera exhibit large numbers of species and high percentages of endemism.

Another striking characteristic of Greek Islands, especially the Aegean ones, is the presence of a vast number of small islets. More than 90% of the Aegean islands are smaller than 10 km^2. Most have been formed quite recently on a geological scale and still "behave" as parts of a continuous land mass, even for taxa with reduced dispersal abilities: species numbers are high, extinctions are marginal, net effects of island size are limited, and there is a significant effect of environmental heterogeneity on species richness. These islets in many cases serve as refugia for endangered species, as in the case of the land snail *Helix godetiana*. In terms of plants, several species are considered "islet specialists."

On the basis of biotic processes such as invasion, differentiation, and extinction, we can distinguish three areas in the Aegean archipelago. The first includes the surrounding mainland and the offshore islands. This zone has faced all the waves of biotic invasions and extinctions. A mixture of relict species, new invaders, and microdifferentiated forms composes the richer and more balanced ecosystems of the Aegean. The second area includes the islands of the central Kyklades, western and central Crete, Ikaria, and Karpathos. Very few of the Pleistocene invaders succeeded in establishing on these archipelagoes. On the contrary, many Pliocene elements have survived to the present day. The ecosystems of the aforementioned islands are more robust and more resistant to human alterations. The third zone includes the most isolated areas of the Aegean (southeastern Kyklades, the more remote islets in the southeastern Aegean, and eastern Crete). Their ecosystems appear to be very poor, with low biodiversity and disharmonic faunas. However, the percentage of Aegean endemics is very high, and their microdifferentiation is distinctive. Relict species are very rare.

HUMAN PRESENCE, HABITAT MODIFICATION, AND ADAPTATION

The Greek islands have undergone intensive human influence for more than 8000 years. The area was the birthplace of many ancient civilizations (e.g., the Minoan of Crete, the Cycladic of the Kyklades, and the Mycenean of the Peloponnesos). The continuous and intense presence of humans in the area led to major alterations of the environment predating the last 2000 years. The first human societies transformed the natural environment through hunting, cultivation, introduction of domestic animals, manipulation of the frequency of fires, and cutting of trees. Thus, the landscape we see now has been shaped almost completely by human intervention. Nevertheless, this intervention was relatively modest until the middle of the last century, and many practices, especially agricultural ones, enhance certain components of biodiversity.

Most recent taxa of the Greek Islands exhibit effective adaptations to the environmental changes and to human impacts. Although human settlements can be disastrous for fragile ecosystems, such as those of oceanic islands, Mediterranean and especially Greek islands are different. Plants for example, have had ample evolutionary experience of repeated fires and browsing and have evolved associated adaptations to avoid or recover from such events. These adaptations have resulted from the establishment of Mediterranean-type ecosystems (MTE) on almost all the islands during the last 2 million years. By the time of the arrival of humans, whose impacts bear analogies with natural temporal changes in MTEs, many species were already "adapted." These are the species that dominate today on most islands.

CURRENT STATUS AND CONSERVATION

Despite these biological features, the scale and intensity of modern human intervention pose serious threats to many components of biodiversity. With the ongoing depletion of water resources, mainly as a result of development driven by tourism, among the most vulnerable ecosystems are those related to inland waters. Hundreds of streams and small wetlands are scattered all over the islands of the Aegean Sea, with small estuaries of seasonal streams being the most common. The insular inland waters act as refuges for most hydrophilic species and host high levels of diversity within very restricted areas, but they are vulnerable to human intrusion and encroachment: They are drained, deprived of their crucial freshwater inputs, overpumped, overgrazed, dumped, split by roads, polluted by sewage, filled with rubble, cultivated, or turned into airports. The reduction of wetlands on the Aegean Islands will drive to extinction many species dependent on them but will also result in a significant reduction of the stopover sites of millions of migrant birds.

A number of conservation plans have been initiated with the recognition of the need to protect and conserve the biodiversity of the Greek Islands. Thus, 96 sites, 82 in the Aegean and 14 in the Ionian Islands, have been included in the NATURA 2000 network of protected areas. Besides the NATURA sites, several other types of sites, such as the National Park of Samaria Gorge (Crete), the National Marine Park of Alonnissos (northern Sporades) and the Protected Natural Monument of the Petrified Forest of Lesvos (Lesvos Island), as well as other types of specially protected areas (aesthetic forests and "important bird areas"), can also be found throughout the Greek Islands. As far as wetlands are concerned, approximately half are under some type of protected status. Nevertheless, progress in the conservation of the Greek Islands' biodiversity depends predominantly on long-term studies in the area and a more organized and targeted research agenda.

CONCLUDING REMARKS

The Greek Islands collectively constitute one of the most outstanding laboratories of nature. The intensiveness and frequency of the environmental dynamics (geological and climatic), both in the past and at present, drastically diminishes the time available for immigration, extinction, and differentiation to establish conditions of equilibrium. In addition, they provide an excellent case study for the dynamics of the interplay between human activities and biodiversity. At the same time, as a result of being a favorite destination for millions of tourists each year during the last several decades, their biota is under serious threat. There is an urgent need for intensification of conservation efforts so that their value, both scientific and cultural, can be preserved.

SEE ALSO THE FOLLOWING ARTICLES

Endemism / Greek Islands, Geology / Refugia / Relaxation / Vicariance

FURTHER READING

Blondel, J., and J. Aronson. 1999. *Biology and wildlife of the Mediterranean region*. Oxford: Oxford University Press.

Poulakakis, N., A. Parmakelis, P. Lymberakis, M. Mylonas, E. Zouros, D. S. Reese, S. Glaberman, and A. Caccone. 2006. Ancient DNA forces reconsideration of evolutionary history of Mediterranean pygmy elephantids. *Biology Letters* 2: 451–454.

Sfenthourakis, S., S. Giokas, and E. Tzanatos. 2004. From sampling stations to archipelagos: investigating aspects of the assemblage of insular biota. *Global Ecology and Biogeography* 13: 23–35.

Stamou, G. P. 1998. *Arthropods of Mediterranean-type ecosystems*. Berlin: Springer-Verlag.

Thompson, J. D. 2005. *Plant evolution in the Mediterranean*. Oxford: University of Oxford Press.

GREEK ISLANDS, GEOLOGY

MICHAEL D. HIGGINS

University of Québec, Chicoutimi, Canada

The geological diversity of the Greek islands reflects long and complex interactions between the Eurasian, Mediterranean (African), and Anatolian tectonic plates. The Mediterranean climate and common paucity of soil have augmented the influence of geology on the cultural development of these islands for the last 5000 years: The nature of the bedrock and water supply has controlled agriculture; exploitation of marble and metals have been important economic activities; volcanic eruptions and earthquakes have directly influenced the lives of the inhabitants. In turn, study of the geology of the islands has contributed much to our knowledge of geological processes elsewhere.

INTRODUCTION TO THE GEOLOGY OF GREECE

Two hundred million years ago the Tethys Ocean lay between Eurasia and Africa, opening out to the east. Since that time, continental fragments have spalled off Africa and been propelled toward Europe by the creation of oceanic tectonic plate material to the south and its consumption in a subduction zone to the north. When these mini-continents collided with Europe, they made mountain ranges—for example, Italy's collision created the Alps. During these collisions some rocks were forced deep into the Earth, where the action of temperature and pressure metamorphosed the rock, changing its mineralogy and appearance. Greece has been the locus of many such collisions, which have contributed to its complex geology.

At the present time the Mediterranean tectonic plate is being subducted beneath the Aegean Sea. Melting of the plate produces molten rock, which rises to the surface as the South Aegean volcanic arc (Fig. 1). The region north of the arc is expanding, opening up tectonic valleys, such as the Gulf of Corinth, and forcing Crete southward. In addition, the Anatolian plate is pushing eastward, separated from the Eurasian plate by the North Anatolian fault and its extension, the North Aegean fault zone. Movements along these plate boundaries occur during earthquakes, and

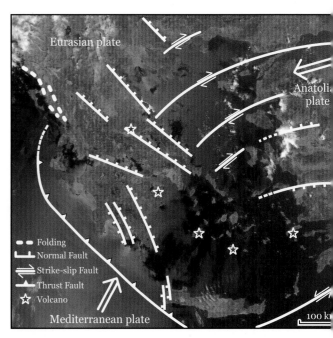

FIGURE 1 Plate tectonics of the Aegean region overlaid on a MODIS satellite image. The Anatolian plate is moving west into the Aegean region. The Mediterranean (African) plate is subducted along a major thrust fault and melts at a depth of 100 km to feed the active volcanoes.

Greece is the most seismically active part of Europe. Fault movements also produce vertical changes in the height of the land, commonly observed by local changes in sea level. Finally, faults provide channels for surface water to descend deep in the Earth and rise as hot springs.

The most common rock in the region is limestone or its metamorphosed equivalent, marble (Fig. 2). Erosion of these rocks produces a special landscape with closed basins, springs, and caves. Early agriculture was enabled by the perennial springs, and caves were important as shelter and for religious purposes. In some areas subsurface water evaporates before it reaches the surface, cementing beach sand to make "beach rock."

It should be remembered that we live in geologically unusual times: Just 20,000 years ago, much of the Northern Hemisphere was covered by ice, and the sea was 120 m below its current level. Most of the Greek Islands were connected to the mainland, and there were vast expanses of shallow sea, with abundant molluscs. The shells of the molluscs were worn down to sand that formed dunes, which were rapidly cemented and transformed into a porous limestone. This useful building material is locally called Poros or Panchina and has been much used in the region for rough construction.

ISLANDS OF THE SARONIC GULF

The Saronic and Corinthian Gulfs are broad, partly flooded valleys produced by almost north–south extension of the crust. The oldest rocks are hard, gray limestones (250–65 million years old) that were deposited in shallow seas to the south. These rocks are well exposed on Salamis, the island closest to Athens. They are also seen on Aegina, the largest island of the group, but only in a small area. Volcanic eruptions started 4 million years ago and covered the southern and eastern parts of the island with lavas and tuffs. After volcanism ceased, the northern part of island was submerged, and marls were deposited. Volcanic activity has continued recently on Methana, a peninsula 10 km to the south, and on the island of Poros, close to the Peloponnese.

EVVOIA

Evvoia (Euboea) is a long island that runs parallel to the Greek mainland, separated from it by a strait that narrows to only 80 m at Khalkis. Here, tidal movements in the North and South Evvoikos gulfs interact to produce chaotic currents that reverse 6 to 14 times a day. The oldest rocks on the island are schists and marbles. The marble in the south of the island was exploited extensively by the Romans, especially for columns—it is now called Cipollino (Italian for onion) because layers rich in muscovite and chlorite give the

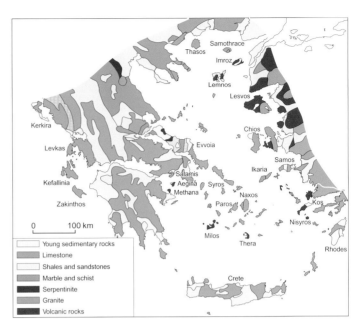

FIGURE 2 Simplified geology of the Aegean region and the Greek Islands.

appearance of an onion (Fig. 3). Variegated colored marble (Fior di Pesca) was also exploited in antiquity near Eretria.

Closure of small ocean basins thrust parts of the ocean floor over the metamorphic rocks. The whole package was then uplifted and weathered under tropical conditions to produce iron- and aluminum-rich "soils" called laterites and bauxites. The former were exploited in antiquity as a source of iron and more recently as a nickel ore. Finally, parts of the island sank down to form swampy basins. Low-grade coal, lignite, formed here and has been exploited for power generation.

IONIAN ISLANDS

Kerkira (Corfu) lies close to the mainland and was indeed connected 8000 years ago when sea level was lower.

FIGURE 3 Partly finished Roman columns 5 m long, from the Cipollino marble quarries on southern Evvoia.

However, the western coast of the island follows a fault, and the sea-floor drops rapidly to over 1000 m. The oldest rocks are hard, gray limestones (250–145 million years old), which crop out in the north and make the highest hill. Further south, the rocks are younger and softer and have developed thick, red soils. Paleolithic implements have been found in this soil, testifying to the long occupation of this fertile island.

The southern Ionian islands (Levkas, Ithaca, Kefallinia, and Zakinthos) lie close to the western edge of a tectonic plate, which is why earthquakes are so common (Fig. 1). Over the long term, such activity has produced strong relief, which is expressed as hills, islands, and lakes. All the islands are dominated by limestone (210–36 million years old), which has been cut by many faults during compression of the region. On Kefallinia this combination has led to an extensive system of sinkholes, caves, and springs. Near Argostoli, on the west coast, there is a very unusual phenomenon: The sea drains into a sinkhole (katavothre) and reappears, mixed with freshwater, on the other side of the island. The process is driven by density differences between seawater and freshwater.

The oldest rocks on Zakinthos (Zante) resemble those of the islands further north but have been overlain by younger rocks. These include gypsum that was formed when the Mediterranean almost completely dried up 6 million years ago. There are natural pools of bitumen (pitch) in the southern part of the island, which formed by seepage of petroleum and evaporation of the more volatile components. Bitumen was used extensively in antiquity for waterproofing ships and jars, as well as for medical purposes. However, there are no significant oil deposits in this region.

CYCLADES

The Cyclades are part of a band of complex metamorphic rocks that stretches north to Attica and Evvoia. Marble and schist dominate, but there are traces of less common minerals and rocks: The blue/mauve mineral glaucophane is widespread and was used as a pigment, jadeite from Syros is a form of jade that may have been used in Neolithic times to make axe heads, and corundum (emery) from Naxos was used to shape and polish marble. White marble was exploited in antiquity from Naxos and Paros; the translucent nature of that from the latter was particularly prized. Granite was intruded into the metamorphic rocks and is abundant on Naxos, Mykonos, and the sacred island of Delos.

Milos and its surrounding islands are dominated by volcanic rocks but have a foundation similar to their

FIGURE 4 The cliffs of the Thera caldera at Oia. Gray lavas at the base of the cliff are covered by red agglomerates. The pale tuff at the top of the cliff is from the Minoan eruption in 1640 BC.

neighbors. Volcanism started 4 million years ago with the eruption of tuffs and lavas and has been expressed most recently by swarms of phreatic explosions, the latest about 2000 years ago. In Paleolithic to Neolithic times, the natural volcanic glass obsidian was exploited for the production of blades. Two domes of obsidian were used, both from the area north of the Bay of Milos. More recently, volcanic rocks have been exploited to make perlite. Melos is also a major producer of the clay bentonite, which is formed by hydrothermal alteration of volcanic rocks.

The most famous and spectacular volcano in the Aegean is on the island of Thera (Santorini, Fig. 4). The volcano was built on a foundation of marble and schist, now exposed on the hills around Ancient Thera. Volcanism started 1.5 million years ago in the southern part of the island, but the main phase only dates from 200,000 years ago. There have been many major eruptions, which are exemplified by the Minoan eruption about 3600 years ago. This started with the rapid eruption of 35 km^3 of volcanic ash, which buried a Bronze Age town in the southern part of the island near Akrotiri. The volcanic summit then collapsed, leading to the formation of a caldera that now makes up the northern part of the Bay of Thera. Construction of a new volcano started shortly afterward with the eruption of lavas in the center of the bay. Volcanic activity continues on the Kameni Islands, which last erupted in 1950. The Colombo Bank underwater volcano, located only 10 km northeast of Thera, erupted in 1650. It will probably make a new volcanic island in a few thousand years.

CRETE

Crete is part of the Hellenic Arc, a series of islands and shallow water that extends from the Peloponnese to Turkey. It formed in response to the subduction of the African plate beneath the Aegean. The plate boundary is immediately to the south, which accounts for the frequency of earthquakes. The lowest rocks exposed on the island are limestones (250–210 million years old), which have been partly recrystallized. During crustal compression almost horizontal faulting has emplaced limestones and other rocks of similar age on top. About 12 million years ago, subduction started to the south and, in response, the Aegean sea to the north expanded. Crete was faulted into many blocks, which moved independently. Some blocks became the mountains, whereas others dropped down, leaving troughs that became filled with sedimentary rocks. These large, and commonly rapid, movements continue to this day: The extreme relief of the Samaria Gorge in western Crete was produced by erosion in response to rapid uplift during the last few thousand years. More recently the harbor at Phalasarna was uplifted by 7 m, possibly during a single earthquake in the fifth century.

ISLANDS OF THE NORTHERN AEGEAN

Thasos is almost completely made up of schist, gneiss, and marble and is an extension of the Rhodope metamorphic massif on the mainland 8 km to the north. The western part of the island has many small metallic mineral deposits. The oldest mines were for red ochre, hydrated iron oxide, which was exploited in Paleolithic times for cult purposes. Indeed, these underground mines were some of the largest in Europe at that time. From the ninth century BC, the same ore was used to make iron metal. There were also significant silver and gold mines, some of which were reopened in the nineteenth century for antimony and zinc. Thasos was also well known in antiquity for white marble.

Samothrace lies to the north of the North Aegean trough, an important plate boundary fault. The island itself is a horst, a block of rock uplifted along faults to the north and south. The oldest rocks are parts of the ocean floor, formed about 150 million years ago. Volcanism 45 million years ago was followed 20 million years later by more volcanism and the emplacement of granite that now makes up some of the highest parts of the island.

Lemnos and Imroz (a Turkish island) lie on the south side of the North Aegean trough. The sea around here is shallow, and indeed both islands were connected to the mainland 20,000 years ago. The oldest rocks on Lemnos are sandstones and marls that were shed from a rising mountain range about 45 million years ago. Similar rocks occur in the Meteora region of central Greece. Much of Lemnos and Imroz are covered by volcanic rocks that were erupted about 20 million years ago. Similar rocks also occur on Lesvos and the Turkish mainland. Although Lemnos is associated with the blacksmith god Hephaestus, there is no evidence of recent volcanic activity. One of the chief products of Lemnos from antiquity onward was Lemnian earth, a medicament. The nature of the earth is not entirely clear: It may have been ochre deposited from springs or a mixture of clay and alum.

EASTERN SPORADES

Lesvos (Mytilene) is a large island close to the Turkish coast. The eastern part of the island is composed of metamorphic rocks—schist and marble. Further west, there is a wide band of serpentinite, part of a section of ocean floor that was thrust up during continental collision. Such rocks do not produce good soils but have been exploited for magnesite. The western part of the island is covered by volcanic rocks, lavas, and tuffs, which are mostly 16–18 million years old. They are part of a much larger volcanic province that extends about 150 km to the east. Fossil pine and sequoia trees have been preserved in volcanic ashes in the western part of the island.

Chios also lies close to the mainland, but its geology is quite different from that of Lesvos except for minor volcanic activity. The oldest rocks are sandstones and shale shed from a mountain range earlier than 250 million years ago. Later, these rocks were overlaid by the limestones that dominate the center of the island. In antiquity the island produced a marble called Marmo Chium or Portasanta, which is salmon pink with red and white inclusions. The rock is a metamorphosed limestone breccia.

Samos is dominated by marble and schist, which are an extension of a metamorphic massif to the east. There are two basins where the younger rocks have been deposited. The eastern basin is well known for fossils, deposited 7 to 9 million years ago around small lakes when Samos was connected to the mainland. The remains include lions, mastodons, rhinos, gazelles, and *Samotherium,* an ancestral giraffe unique to Samos. In antiquity the island was known for its engineering works, including a 1000-m-long tunnel cut through a hill to transport spring water to the city.

DODECANESE ISLANDS

Patmos is a small island built on marble but now largely covered by volcanic rocks 6–7 million years old. These are well exposed in a sacred rock shelter where St. John wrote the Book of Revelation. Marble and schist continue south to Kos but have been largely covered by limestone on the islands of Lipsos and Kalymnos.

The highlands of Kos are made of marble and schist but also contain the earliest volcanic rocks erupted about

FIGURE 5 Recent phreatic explosion craters on Nisyros. These have partly destroyed young volcanic domes visible to the left.

10 million years ago. Volcanism restarted 3 million years ago with the eruption of two volcanic domes (short, thick flows) in the west. Major eruptions 555,000 and 145,000 years ago produced tuffs that covered most of the island and also created calderas, which underlie the sea between Kos and Nisyros. Cold springs at the Asklepieion deposited terraces of travertine, which probably initially attracted attention to the site. Later on, the travertine was quarried to construct the temple and ancient "health center."

Nisyros is the easternmost volcano of the active South Aegean volcanic arc. Volcanic activity started about 200,000 years ago and has continued until recent times. The island is now a single simple volcano with a large crater partly occupied by young volcanic domes. The last volcanic activity was a series of phreatic explosions in 1871–1873 (Fig. 5). Deep wells have drilled for exploitation of geothermal power, but this resource is yet to be exploited.

Rhodes lies just northwest of a major tectonic plate boundary, which accounts for the frequency of earthquakes. One of the most notorious occurred in 226 BC, when it toppled the Colossus of Rhodes, a 30-m high statue of the sun god Helios, one of the seven wonders of the ancient world. The early geological history of Rhodes resembles that of Crete and much of the Peloponnese: Cherty limestones were deposited 210–65 million years ago in shallow water to the south. Later on, overall compression of the crust raised mountains that were eroded. Finally, basins developed and were filled with marls, which make fertile soils. The oldest limestones are resistant to erosion and form the highest point on the island.

SEE ALSO THE FOLLOWING ARTICLES

Earthquakes / Eruptions / Greek Islands, Biology / Mediterranean Region

FURTHER READING

Fassoulas, C. G. 2000. *Field guide to the geology of Crete*. Natural History Museum of Crete.
Friedrich, W. L. 2000. *Fire in the sea: the Santorini volcano: natural history and the legend of Atlantis*. Cambridge: Cambridge University Press.
Higgins, M. D., and R. Higgins. 1996. A *Geological companion to Greece and the Aegean*. Ithaca, NY: Cornell University Press.
Institute of Geology and Mineral Exploration (IGME), Athens, Greece. www.igme.gr
Jacobshagen, V. 1986. *Geologie von Griechenland*. Beiträäge zur regionalen Geologie der Erde, Bd. 19. Berlin: Gebruder Borntraeger.
Pe-Piper, G., and D. J. W. Piper. 2002. *The igneous rocks of Greece: the anatomy of an orogen*. Berlin: Gebruder Borntraeger.
Perissoratis, C., and N. Conispoliatis. 2003. The impacts of sea-level changes during latest Pleistocene and Holocene times on the morphology of the Ionian and Aegean Seas (SE Alpine Europe). *Marine Geology* 196: 145–156.

GREENLAND

SEE ARCTIC REGION

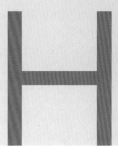

HAWAIIAN ISLANDS, BIOLOGY

JONATHAN PRICE

University of Hawaii, Hilo

The biology of the Hawaiian Islands consists of organisms evolving and interacting in an archipelago characterized by extreme isolation, a distinctive geologic history, and a diverse physical environment. The Hawaiian biota exhibits classic examples of the evolution of endemism, the emergence of ecologic traits typical of islands, and the problems associated with human occupancy and invasive species.

PHYSICAL ENVIRONMENT AND HISTORY

Lying just inside the tropics in the central Pacific, the Hawaiian archipelago is among the most isolated archipelagoes on Earth. The closest point of continental land is the west coast of North America, nearly 4000 km away. The nearest islands of comparable size and with similar environments, the Marquesas, are equally distant. Even the closest tiny atolls are over 1000 km away. These oceanic islands have never been connected to or in proximity to continental land masses. The geologic history of the Hawaiian Islands has ensured their continuous isolation from the time they originally formed to the present.

The Hawaiian Islands consist of a series of volcanoes that emerged from the sea in sequence, with the youngest island, Hawaii, in the southeast and progressively older volcanoes lying to the northwest, extending out to Midway and Kure atolls. After islands form as large land masses, subsidence, or the sinking down of volcanic masses under their tremendous weight, causes them to shrink quickly. Erosion is another process shaping the environment available to organisms. When volcanism ceases, erosive processes gradually reduce islands from large volcanic shields, to mature islands with deeply eroded valleys, to small pinnacles of rock, and ultimately to atolls with no volcanic rock above the sea surface. The duration of these life stages varies among islands, with clear implications for the long-term evolution of organisms. For example, by the time Kauai emerged about five million years ago, the islands that preceded it had largely diminished to small, distantly spaced islands.

The Hawaiian Islands exhibit an enormous range of climates for so small a land area. Within a short distance on a single island, one may encounter arid lowlands, extremely rainy mountain slopes, and cold, dry alpine climates. Although the Hawaiian Islands are situated in the tropics, the presence of tall mountains there results in tremendous variation in average annual temperatures. At the lowest elevations, warm temperatures may persist year-round, while at the highest elevations on Maui, which rises to over 3000 m, and Hawaii, which rises to over 4000 m, freezing temperatures may occur throughout the year.

Variation in annual precipitation is produced by the very frequent influence of trade winds. These winds bring moisture-laden clouds from the northeast, which rise against mountain slopes and produce large amounts of rain. Windward slopes and mountaintops receive over 2500 mm of rain per year, with Mount Waialeale on Kauai receiving an average of over 10,000 mm of rain per year. An inversion layer keeps most clouds and rainfall below an elevation of about 2000 m most of the time; consequently a drier climate is present on the highest mountains on the islands of Maui and Hawaii, which rise above the inversion layer. Additionally, a strong rain shadow produces very dry climates at lower elevations on the leeward (southwestern) sides of all of the major islands. The net result is that most islands contain a wide range of moisture regimes.

The history and climate of the Hawaiian Islands generate numerous landforms and variable soils. As erosion ensues, rugged lava fields give way to complex landscapes that contain cliffs, ridge tops, and valleys with running streams. Porous volcanic rock weathers into viable soil, gradually releasing nutrients. The youngest soils are deficient in important nutrients such as nitrogen and phosphorus, with somewhat older soils (between 20,000 and 500,000 years old) having much greater concentrations. Older soils, however, are subjected to leaching in wet climate zones and lose nutrients over time. This produces a clear pattern in terrestrial vegetation in which growth rate, forest height, and total biomass all track concentrations of soil nutrients. These processes combine to produce a diverse set of terrestrial environments.

COLONIZATION AND EVOLUTION

Being isolated throughout their history, the Hawaiian Islands have been entirely dependent on long-distance dispersal to populate its ecosystems with organisms. The difficulties of dispersing such distances and becoming established have greatly restricted the kinds of organisms that colonized the archipelago. Despite their tropical setting, the Hawaiian Islands lack many groups of plants and animals that are otherwise important components of tropical communities. For example, ants, termites, and members of the philodendron, bromeliad, and ginger plant families all failed to colonize the Hawaiian Islands. Except for a species of bat and a species of monk seal, no mammals have colonized the Hawaiian Islands. Terrestrial reptiles and amphibians entirely failed to colonize.

Instead, organisms with well-developed dispersal mechanisms were much more likely to colonize. Plants that were predisposed to arrive had saltwater-tolerant seeds that could float, tiny seeds that could blow long distances, or small seeds that could be transported by birds either attached to the outside or carried internally after ingestion of fruits. Arthropods that could fly or were small enough to be suspended in wind currents were similarly predisposed to colonize. Nearshore marine organisms with adults or larvae that can survive in the open ocean long enough to reach the Hawaiian Islands also had an advantage in colonization.

Colonists arrived in Hawaii from throughout the Pacific region. The closest relatives of different groups of Hawaiian plants may be found in the Arctic, the desert southwest of North America, Australia, New Zealand, and Mainland Asia. Birds and arthropods share these diverse origins. The progenitors of the Hawaiian biota were therefore limited in number yet diverse in geographic origin. This disharmonic biota is markedly different from those of any other biogeographic region.

The timing of colonization is constrained by the ages of the islands and the dynamic nature of their size and configuration. High islands have been continuously available since about 30 million years ago. Over that period, however, the number and size of islands have fluctuated greatly. Genetic analyses indicate that much of the biota arrived into the archipelago around the time that Kauai formed; few of the organisms that had arrived on older islands were able to disperse to Kauai, both because those islands were largely deteriorated and because they were distantly spaced relative to today's islands. Those organisms whose origins in Hawaii do predate Kauai appear to have arrived during earlier times when larger islands were available.

Upon arrival colonists encountered a diverse physical environment and novel communities of organisms. Facilitated by isolation, which prevented gene flow with ancestral populations, a majority of colonists evolved endemic species, producing very high rates of endemicity for some taxonomic groups (Table 1). These endemics exhibit adaptations to the ecological conditions of the islands, permitting them to occupy habitats quite unlike those of their ancestors. For example, some Hawaiian violet species (Violaceae) live in warm, moderately dry lowland areas, despite being descended from Arctic ancestors. True bugs in the genus *Nysius* (Lygaeidae) have adapted to frigid environments at the summit of Hawaii's highest mountain, Mauna Kea, despite being related to species of tropical lowland origin.

Other adaptations have helped species become more competitive or fill ecological niches that the original colonists could not fill. For example, numerous plants, including violets (Violaceae), beggar's ticks in the genus *Bidens* (Asteraceae), and members of the carnation family (Caryophyllaceae) have evolved woody anatomy from weedy herbaceous ancestors. Birds, especially Hawaiian honeycreepers, adapted feeding behaviors quite different from those of their ancestors, including specialization in floral nectar, eating seeds of particular species of plants, and foraging for insects in very specialized ways with uniquely shaped bills. In one noteworthy insect group, the *Eupithecia* moths, caterpillars have shifted from being herbivores into being ambush predators. Other organisms experienced a release from predation and other pressures experienced by ancestors in their homelands. Birds, including ducks, geese, rails, and ibises, evolved flightless-

TABLE 1
Numbers of Species for Selected Taxa in the Hawaiian Islands

Taxon	Total Natives	Endemic	Endemic (%)	Naturalized	Naturalized (%)	Total Species
Flowering plants	1006	905	90	1101	52	2107
Other plants[a]	719	241	34	47	6	766
Insects	5818	5462	94	2609	31	8427
Other arthropods[b]	456	366	80	735	62	1191
Molluscs	1243	962	77	96	7	1339
Crustaceans	1106	68	6	73	6	1179
Fishes	1143	149	13	73	6	1216
Amphibians	0	0	0	8	100	8
Reptiles[c]	4	0	0	23	85	27
Birds[d]	241	63	26	53	18	294
Mammals	25	2	8	19	43	44

NOTE: Adapted from Eldredge and Evenhuis (2002).
[a] Includes ferns, fern allies, and bryophytes.
[b] Includes arachnids, isopods, centipedes, and millipedes.
[c] Includes marine reptiles (sea turtles).
[d] Includes seabirds and migratory birds.

ness in the absence of mammalian predators. Numerous independent groups of insects also lost the ability to fly.

In many instances, a single ancestral species diversified into numerous endemic descendant species. Speciation occurred in some groups as species moved from older to younger islands, with isolation creating species unique to each given island (a process called the progression rule). In other groups, species differentiated within a given island, resulting in dramatically different morphology and ecology from one another, being adapted to diverse habitats and occupying different ecological niches. In these cases of adaptive radiation, a handful of colonists have produced large portions of the biota representing the full range of adaptations (Table 2). Prominent examples among the plants include the Hawaiian lobelioids (with a total of 125 species in the genera *Brighamia, Clermontia, Cyanea, Delissea, Lobelia,*

TABLE 2
Notable Adaptive Radiations in the Hawaiian Islands

Group	Taxonomic Group	Number of Species	Notable Characteristics
Lobelioidae	Plants (Campanulaceae)	125+	Largest radiation of plant species, includes 6 genera; considerable variation in floral morphology and pollination syndrome
Silversword alliance	Plants (Asteraceae)	30	Includes 3 endemic genera; extreme variation in physiognomy and habitat preference
Schiedea	Plants (Caryophyllaceae)	32	Considerable variation in floral morphology, breeding system, and physiognomy
Hawaiian drosophilids	Insects (Drosophilidae)	1000?	Largest radiation in Hawaii with over 400 described species and many more undescribed; larvae feed on fruit, bark, fungi, and other tissue
Hawaiian gryllid crickets	Insects (Gryllidae)	169+	Species differentiated by distinctive songs; genetic studies suggest extremely rapid speciation rate
Tetragnatha spiders	Arachnids (Tetragnathidae)	29	Largest radiation of arachnids; species with a variety cryptic colorations adapted for different habitats
Achatinelline snails	Molluscs (Achatinellidae)		Largest radiation of terrestrial snails; species exhibit a variety of shell patterns and shapes
Hawaiian honeycreepers	Birds (Fringillidae)	50+	Largest radiation of bird species; extreme variation in bill morphology, feeding behavior, and plumage
Hawaiian honeyeaters	Birds (family?)	7	Unique radiation of nectar-feeding birds; previously considered to be true honeyeaters (Meliphagidae), but now family is uncertain; all species now extinct
Moa nalos	Birds (Anatidae)	4	Originally derived from dabbling ducks, evolved into several large flightless species; all species now extinct

and *Trematolobelia*, in the Campanulaceae), the Hawaiian endemic mints (with a total of 57 species in the genera *Haplostachys, Phyllostegia,* and *Stenogyne* in the Lamiaceae), and the silversword alliance (with a total of 30 species in the genera *Argyroxiphium, Dubautia,* and *Wilkesia* in the Asteraceae). Along with other groups, these havediversified into multiple endemic genera with endemic species specializing in remarkably disparate habitats. Among the birds, the Hawaiian honeycreepers (numerous genera in the Fringillidae) represent the premier adaptive radiation, with over 50 highly specialized and varied species derived from a single finchlike ancestor. Among arthropods, the Hawaiian drosophilid fruit flies have generated the largest number of species from a single ancestor, perhaps as many as 1000. Other noteworthy adaptive radiations of arthropods include crickets (Gryllidae), *Tetragnatha* spiders (Tetragnathidae), and numerous groups of true bugs and moths. In addition, a diverse array of land snails (in the families Achatinellidae and Succinidae) has evolved from a small number of colonists.

NATIVE ECOSYSTEMS

In the Hawaiian Islands, the diverse physical environment and the evolution of specialized endemic organisms have promoted the development of highly unique native ecosystems. Following are general descriptions of the major types of native ecosystems (examples can be seen in Figs. 1–8).

Nearshore Marine Communities

Unlike the open ocean or deep-sea habitats, shallow habitats near the shore such as coral reefs are isolated from similar environments elsewhere and therefore make up distinct communities in Hawaii. This isolation has restricted dispersal such that Hawaii harbors only a fraction of the

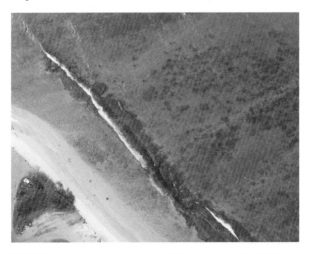

FIGURE 1 A shallow-water coral reef off the leeward coast of Oahu, an example of a Hawaiian native ecosystem.

FIGURE 2 An example of a coastal community, including cliff, beach, and pond, on the north coast of Molokai near Olokui.

number of coral reef organisms found in the Indonesian region, a major source area for colonists. Nevertheless the isolation is sufficient that about 20% of coral reef organisms are endemic to the Hawaiian Islands, a comparatively high rate of endemism. The endemic Hawaiian monk seal can be found throughout the Hawaiian archipelago. Sea turtles can also be found in these waters, including the endangered hawksbill turtle.

Coastal Communities

The coast of the Hawaiian Islands is highly varied, including arid to very wet climatic conditions and containing a variety of landforms. Sea cliffs occur particularly on windward aspects, and often fall within somewhat moist climates. Molokai and Kauai have especially tall and extensive sea cliffs, supporting a rich vegetation of ferns, sedges, succulents, and shrubs. Beaches and sand dunes occur primarily on leeward sides of islands, but also in protected locations on windward sides. These may be calcareous or (particularly on the Island of Hawaii) black volcanic sand and support a vegetation of sprawling plants such as beach morning glory (*Ipomoeia pes-caprae*) and ohai (*Sesbania tomentosa*). Anchialine ponds occur in primarily leeward areas of the island of Hawaii, where fresh groundwater mixes with seawater in ponds that are connected to the ocean only underground through porous lava. Here various species of tiny native shrimp, including opaeula (*Halocaridina rubra*), thrive in the absence of predatory fish.

Dry Forest and Shrubland

The leeward slopes of the Hawaiian Islands typically receive less than about 1250 mm of rain annually. Here a summer dry season presents a challenge to plants. In the very driest areas, which receive less than 500 mm of rain per year, as well as on steep slopes and areas with poor soil development, a sparse vegetation of shrubs and grasses existed, although little is known about vegetation in these areas because they have been strongly modified by human activity. More moderately dry areas receiving more than 500 mm support vegetation consisting of drought-adapted trees. Trees such as lama (*Diospyros sandwicensis*) have small thick leaves that persist through the dry season. A small number of trees, including wiliwili (*Erythrina sandwicensis*), have larger leaves that are shed during drought months, when they flower profusely.

FIGURE 4 A diverse mesic forest at Waimea canyon on Kauai.

FIGURE 3 An example of a lowland dry forest of Wiliwili (*Erythrina sandwicensis*), taken at Ahihi-Kinau on Maui.

Mesic Forests

Areas receiving moderate amounts of rainfall spread throughout the year are typically referred to as mesic. Here conditions are favorable for numerous tree species, and consequently mesic forests are the most diverse, particularly on the older islands of Kauai and Oahu. While ohia trees (*Metrosideros polymorpha*) often dominate in these and other plant communities, in some cases there is no clear dominant tree species. Higher-elevation mesic forests, especially on Maui and Hawaii, may be dominated by very large koa trees (*Acacia koa*). Such forests provide habitat for the endangered akiapolaau (*Hemignathus munroi*), a honeycreeper with a peculiar bill that specializes in digging arthropods out of koa branches.

Wet Forests and Bogs

Plant communities in the wettest areas (receiving over 2500 mm of rain per year) typically consist of wet forest. These forests are usually dominated by the ohia tree (*Metrosideros polymorpha*) and often contain a rich assortment of ferns including hapuu tree ferns (*Cibotium* spp.) and the mat-forming uluhe (*Dicranopteris linearis*). A diverse assortment of epiphytic plants often forms a dense cover on the branches of trees. In poorly drained areas, and particularly on the wettest mountain summits on Kauai and West Maui, bogs occur (although technically these are fens) with continually saturated ground supporting an interesting community of tussock sedges, dwarf shrubs, and showy species such as greenswords (closely related to silverswords) and lobelias.

FIGURE 5 An example of wet forest on the island of Hawaii.

Subalpine and Alpine Communities

Above 2000 m elevation, the trade wind inversion permits little cloud formation or rainfall. In addition, temperatures often drop to below freezing, particularly at night. Consequently, vegetation is adapted to cold, dry conditions. Where warmer and moister habitats directly abut this zone, a treeline often demarcates the upper limit of forest vegetation. Immediately above the treeline, vegetation is stunted and takes the form of a dense shrubland, although trees may be found in gulches, which retain moisture and are protected from colder temperatures. Shrubs adapted to cold, dry conditions, including pukiawe (*Leptecophylla tameiameiae*), hinahina (*Geranium cuneatum*), and aʻaliʻi (*Dodonaea viscosa*), form dense thickets. On the slopes of Mauna Kea, forests of mamane (*Sophora chrysophylla*) support populations of palila (*Loxioides bailleui*), a specialist honeycreeper that feeds on mamane seed pods. At higher elevations, vegetation becomes sparser as conditions become more extreme, and species such as the silversword (*Argyroxiphium sandwicense*) thrive in an otherwise barren landscape. At the highest elevation in the islands, the summit of Mauna Kea is essentially devoid of vegetation, yet it supports populations of the very unique wekiu bug (*Nysius wekiuicola*), which tolerates subfreezing temperatures and feeds on insects that are blown up from lower elevations and stunned by the cold.

FIGURE 6 An alpine community with silversword (*Argyroxiphium sandwicense*), at Haleakala on Hawaii.

Freshwater Habitats

Many perennial streams run particularly along windward slopes, although large watersheds such as that of Waimea Canyon on Kauai begin in wet summit regions, then drain primarily through drier leeward areas. Hawaiian streams are home to numerous types of endemic arthropods, especially conspicuous dragonflies (*Anax* spp.) and damselflies (*Megalagrion* spp.). Oopu (five native species of fish, all but one of which are gobies) spend much of their lives in fresh water and can climb up rocks and even waterfalls against fast-moving stream currents. Larger streams drain into estuaries and small bays, which provide protection and serve as nurseries for many fish species.

FIGURE 7 A lava pioneer community with ohia tree, at Kilauea Iki on Hawaii.

Lava Pioneer Communities

On the island of Hawaii, volcanic eruptions are frequent enough that primary succession occurs across large areas recently covered by lava. Young lava flows tend to support little vegetation, even in areas with very wet climate, because the porous nature of the substrate retains little water and has few available nutrients. The first pioneer species to colonize lava include lichens, ferns (especially ae, *Polypodium pellucidum*), and the nearly ubiquitous ohia. Over time, ohia trees may grow larger, and an increasing number of species may add to the community until a fully formed forest develops (in wet regions) within about 500 years. Young lava flows also may flow around an area of older substrate, leaving an island-like patch of forest surrounded by barren vegetation. These patches, known as kīpuka, behave like miniature islands where certain species (especially *Drosophila* fruit flies) may become isolated and even begin to differentiate from nearby relatives as a result of this natural fragmentation.

FIGURE 8 A freshwater stream and waterfall at Kopiliula on Maui.

HUMAN IMPACTS

The arrival of humans initiated prodigious changes to Hawaiian native ecosystems. The first people to arrive, over 1000 years ago, Polynesians brought a sophisticated island culture, numerous crop and utility plants, and animals, including Polynesian pigs, chickens, dogs, and (unintentionally) the Polynesian rat. These first Hawaiians cleared land for crops, using fire extensively in lowland and dry areas, and likely hunted larger flightless bird species. Rats fed widely on seeds, bird eggs, and nestlings, and probably on arthropods and tree snails. This combination of stresses led to the extinction of approximately half of all endemic bird species, as evidenced by an extensive sub-fossil record. Some plant, arthropod, and snail species also may have become extinct or at least greatly reduced during this period.

Western contact, beginning with Captain Cook in 1778, heralded an acceleration of losses to the Hawaiian biota. Goats, cattle, and two new species of rats were all introduced within decades of European contact. Extensive cutting of sandalwood for export to Asian markets, clearing of more land for plantation agriculture, and the introduction of numerous plant species from around the world (many of which became invasive) greatly increased the rate of habitat loss and degradation. More recently, introduced arthropod pests, game animals such as deer and mouflon sheep, and escaped domestic animals such as cats and large European pigs have further contributed to ecosystem disruption. Large mammals damage native vegetation, which is not adapted to such disturbance, facilitating invasions of invasive plant species that are better adapted. Introduced cats and mongooses prey on native birds, especially their young. Feral European pigs create habitat for introduced mosquitoes, which spread devastating diseases to native forest birds, which have no resistance. Rats continue to degrade vegetation by stripping bark and eating the seeds of endemic plants that are not adapted to such predation. Many more bird species have gone extinct since Western contact, as well as nearly 10% of endemic plant species, and unknown numbers of arthropod and tree snail species.

CONSERVATION

As a result of human activities, a large proportion of the remaining Hawaiian biota is endangered. Consequently, a number of efforts have been employed to stem the loss of species. First, large areas have been set aside for conservation in the Hawaiian Islands, including national parks and federal, state and private nature reserves. These include many terrestrial habitats as well as the marine and coastal habitats of the Northwest Hawaiian Islands National Marine Sanctuary, the largest marine protected area in the world.

Strong measures have been taken to reduce and reverse the damage done by invasive species. By fencing extensive areas and removing large mammals, native vegetation has recovered to some degree, although full recovery may take time and require additional measures. Invasive plant species have been controlled where possible, although new species are becoming established every year. In some cases biocontrol has been used, although with mixed results. Control of rats, cats, mongooses, and ants has been effective only at very local scales.

Some birds have been bred in captivity, although only the puaiohi, or small Kauai thrush (*Myadestes palmeri*), and the nene (*Branta sandwicensis*) have been released with any degree of success. Similarly, numerous plant species have been propagated in nurseries, with many being outplanted in safe locations. Even some land snails have been raised successfully in captivity. However, it is not clear to what degree any of these will establish stable, self-sustaining populations in the long term.

Protection of important nesting sites has become important to the reproduction of sea turtles, and likewise marine protected areas may bolster reproduction of various fish populations.

Despite continued and even very recent extinctions, the combined approach of establishing natural areas and engaging in active management holds promise for long-term conservation. Cooperation among government agencies, nonprofit conservation organizations, and private landowners makes possible coordinated, landscape-scale conservation efforts needed to save what remains of the unique Hawaiian biota.

SEE ALSO THE FOLLOWING ARTICLES

Crickets / *Drosophila* / Hawaiian Islands, Geology / Honeycreepers, Hawaiian / Invasion Biology / Kīpuka / Silverswords

FURTHER READING

Cuddihy, L. W. and C. P. Stone. 1990. *Alteration of native Hawaiian vegetation: effects of humans, their activities and introductions.* Honolulu: University of Hawaii Press.
Culliney, J. L. 2006. *Islands in a far sea: nature and man in Hawaii.* Honolulu: University of Hawaii Press.
Eldredge, L. G., and N. L. Evenhuis. 2002. Numbers of Hawaiian species for 2000. *Bishop Museum Occasional Papers* 68: 71–78.
Stone, C. P., C. W. Smith, and J. T. Tunison, eds. 1992. *Alien plant invasions in native ecosystems of Hawaii.* Honolulu: Cooperative National Park Resources Studies Unit, University of Hawaii.
Price, J. P., and D. A. Clague. 2002. How old is the Hawaiian biota? Geology and phylogeny suggest recent divergence. *Proceedings of the Royal Society of London Series B* 269: 2429–2435.
Price, J. P., and W. L. Wagner. 2004. Speciation in Hawaiian angiosperm lineages: cause, consequence, and mode. *Evolution* 58: 2185–2200.
Vitousek, P. M. 1995. The Hawaiian Islands as a model system for ecosystem studies. *Pacific Science* 49: 2–16.
Wagner, W. L., and V. A. Funk, eds. 1995. *Hawaiian biogeography: evolution on a hot spot archipelago.* Washington, DC: Smithsonian Institution Press.
Ziegler, A. C. 2002. *Hawaiian natural history, ecology, and evolution.* Honolulu: University of Hawaii Press.

HAWAIIAN ISLANDS, GEOLOGY

DAVID R. SHERROD

U.S. Geological Survey, Vancouver, Washington

The Hawaiian Islands described here are the eight principal islands of the Hawaiian Ridge, which, including a series of atolls, extends 2600 km to Kure island (Fig. 1). The ridge is the southeastern part of the Hawaiian–Emperor volcanic chain, the balance of which comprises submarine seamounts that reach to Kamchatka. The chain is convincing evidence for a hotspot melting anomaly deep in the Earth's mantle, even as the stability and depth of some hotspots are being reexamined by new scientific investigations.

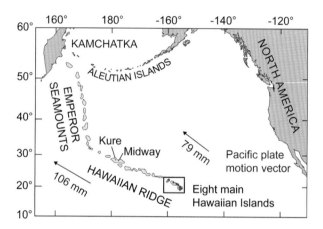

FIGURE 1 Hawaiian Islands and Hawaiian–Emperor volcanic chain, most of which consists of submarine seamounts. Vectors indicate Pacific plate motion, in millimeters per year, relative to presumed fixed mantle hotspot.

HAWAIIAN VOLCANOES BUILD THROUGH A SEQUENCE OF STAGES

The main Hawaiian Islands are built of 15 emergent volcanoes. Two other volcanoes, although now submerged, were once emergent, and a third, newly born about 250,000 years ago, is still building toward sea level.

An idealized Hawaiian volcano passes through four eruptive stages: preshield, shield, postshield, and rejuvenated. These stages likely reflect variations in the amount and rate of heat supplied to the lithosphere as the Pacific plate overrides the Hawaiian hotspot. Volcanic extinction ensues as a volcano moves away from the hotspot.

At inception a Hawaiian volcano is in its preshield stage. Eruptive products, chemically all basalt, are slightly richer in the alkali elements (for example, sodium and potassium) than the typical shield-stage basaltic lava flows. The alkalic character of preshield lava is a consequence of a nascent magma-transport system and less extensive melting at the periphery of the mantle plume fed by the hotspot. Such lava has been dredged from the archetypal preshield-stage volcano, Lōʻihi, youngest of the Hawaiian volcanoes. Lōʻihi's summit lies submerged about 980 m beneath sea level, 30 km south of the Island of Hawaiʻi (Fig. 2). Similar volcanic rocks have been recovered from Kīlauea volcano's south flank by remotely operated and manned submersibles. Recently obtained samples from Hualālai volcano's northwest rift zone, also on the Island of Hawaiʻi, may indicate that the preshield stage is still exposed there. Elsewhere along the chain, these early strata are buried by lava flows of shield-stage volcanism.

The shield stage is the most productive volcanically, and each Hawaiian volcano erupts an estimated 80–95 percent of its ultimate volume during this stage. Shield-stage volcanism marks the time when a volcano is near or above the hotspot and its magma supply system is robust. The degree of mantle melting increases by factor of 2–2.5 compared to the melting that drives preshield-stage magmatism. Magma ascending from at least 60–70 km depth (deepest earthquakes) is stored in reservoirs at the base of the oceanic crust (approximately 20 km depth) and then, at shallow level, in a nexus of sheetlike intrusions and more equant reservoirs 3–7 km beneath the volcano's summit. Some magma erupts at a volcano's summit, but much is shunted into rift zone dikes to feed eruptions and intrusions downslope (Fig. 3). The pāhoehoe and ʻaʻā lava flows of the ongoing eruption that began in 1983 along Kīlauea's east rift zone, 20 km from the volcano's summit, are characteristic of shield-stage volcanism in both style and composition.

Rift zones are prominent topographic features of many Hawaiian shields. Two or three rift zones are typical at an individual volcano. Some of the volcanoes, such as Kauaʻi, are nearly equant in plain view because their three rift zones are equally active (Fig. 2). The presence of an adjacent volcano, however, changes the regional stress regime to favor extension along only two of the rift zones, leading to a substantially elongate volcano. The growth of East Maui has been almost entirely along Haleakalā's east rift zone and its offshore continuation, the Hāna Ridge (Fig. 2); in contrast, the volcano's southwest and north rift zones are stunted. Similarly, Kīlauea's east rift zone and offshore Puna Ridge have been the eruptive loci for most of that volcano's lava flows. Seismic and gravity data indicate that the rift zone dikes of shield-stage volcanism have roots as deep as the base of the volcano and possibly penetrate into the underlying oceanic crust (Fig. 3).

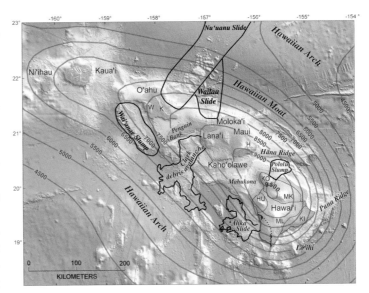

FIGURE 2 Eight main Hawaiian Islands. Shown named are submarine landslides discussed in text. Red lines are contours showing depth, in meters below sea level, to top of oceanic crust (Watts and ten Brink 1989), which is depressed because of the load of young volcanoes and high heat flow in area of Hawaiian hotspot. Some volcanoes indicated by initials: W, Waiʻanae; K, Koʻolau (on Oʻahu); H, Haleakalā (on Maui); KO, Kohala; HU, Hualālai; MK, Mauna Kea; ML, Mauna Loa; KI, Kīlauea (on Hawaiʻi).

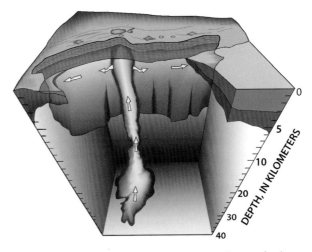

FIGURE 3 Zone of magma throughput into summit area of volcano and downrift transport into rift zone dike system during shield-stage volcanism (generalized from Tilling et al. 1987).

Duration of Shield Stage

Hawaiian volcanoes require at least 0.6 million years to grow from the ocean floor to nearly full size by late in their shield-building stage. They breach the ocean surface about midway through this period. But the entire shield stage persists for 1 million years or more, judging by the 0.6–0.8-million-year span of late-shield ages from subaerially exposed lava on Wai'anae volcano (O'ahu) and West Maui. Stratigraphic accumulation rates on the order of 7.8–8.6 (± 3.1) m per 1000 years are indicated by ages from a deep drill hole into Mauna Kea's shield-stage lava near Hilo. A similar rate, about 6 m per 1000 years, was derived from drill core data on Kīlauea's east rift zone. Volumetrically the magma supply rate at the currently active Kīlauea is in the range 0.09 to 0.11 km^3 per year (calculated as vesicle-free, dense-rock-equivalent magma), on the basis of geodetic and lava effusion-rate data from eruptions in the twentieth century.

Compositional Variation

Of 1000 whole-rock analyses from Mauna Loa and Kīlauea, 99 percent contain between 47% and 54% SiO$_2$—the tholeiitic basalt that builds the islands (Fig. 4). The few analytical outliers are typically only slightly more or less silicic. Early and late-shield strata extend the silica range as alkali basalt and even hawaiite lava flows are sparsely interlayered with tholeiite at some volcanoes. The notable but unusual example of more highly fractionated shield-stage lava comes from Wai'anae volcano, O'ahu. There, the chemical trend includes the Mount Kuwale rhyodacite flow, the most silicic lava in the islands (68 percent SiO$_2$).

Intervolcano compositional differences result mainly from variations in the part of the mantle plume sampled by magmatism and the zoning of sources within it. These distinctions are tracked most successfully by the trace element variation between volcanoes, notably in the radiogenic isotopes of Pb, Sr, and Nd. These data add geochemical significance to the spatial concept of "Loa" and "Kea" trends, in which the volcanoes from O'ahu southeast to the Island of Hawai'i fall into one of two geographic alignments and geochemical groupings. The trend names come from Mauna Loa and Mauna Kea, prominent volcanoes that fix the southeast position of the alignments.

VOLCANO BIRTH PROGRESSES SOUTHEAST

The age of Hawaiian shield-stage volcanism is successively younger from northwest to southeast, an observation made first by the early Polynesian settlers 1200 or more years ago and preserved in their oral history. In the plate tectonic paradigm, this age progression results from the northwestward movement of the Pacific plate over the Hawaiian hotspot at the rate of about 10 cm per year. Radiometric ages from the Ni'ihau shield are as old as about 6.3–5.5 million years. Similar ages have been obtained from Kaua'i (Table 1). Lava from O'ahu dates back to about 3.9 million years from Wai'anae volcano and 3.2 million years from the younger Ko'olau volcano. Ages from Moloka'i and Maui are as old as about 2 million years. All exposed lava flows on the Island of Hawai'i are younger than 0.6 million years.

VOLCANISM COMMONLY PERSISTS INTO THE POSTSHIELD STAGE

As Pacific plate motion rafts Hawaiian volcanoes away from the hotspot, volcanism wanes gradually, passing from the shield stage into the postshield stage. The shallow magma reservoirs (3–7 km depth) of the shield stage volcanoes cannot be sustained as magmatic supply lessens, but smaller reservoirs at 20–30 km depth persist. The rate of extrusion, as measured by upward stratigraphic accumulation, diminishes by a factor of ten late in the shield stage. The composition of erupted lava also changes,

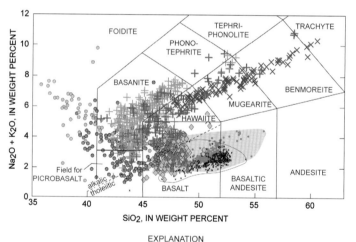

FIGURE 4 Silica vs. total alkalies variation diagram for representative chemical analyses of Hawaiian lava. Rock classification grid from LeBas et al. (1986) and dashed boundary separating tholeiitic and alkalic basalt from Macdonald and Katsura (1964). Stratigraphic formation names shown parenthetically in explanation box. Hawaiian volcanoes evolve compositionally from shield to transitional to post-shield as their magmatic systems become less robust. Rejuvenated-stage lava, where present, follows after a repose. Hawaiian lava data available in electronic format (Sherrod et al. 2007).

TABLE 1
Summary of Volume, Eruptive Age, and Volcanic Stages of Volcanoes on the Main Hawaiian Islands

Island	Volcano	Volume, km³	Stages Present[a]	Oldest Radiometric Age[b]	Youngest Volcanism
Niʻihau	Niʻihau	21.7	S, (PS), R	6.3; 5.54	0.35
Kauaʻi	Kauaʻi	57.6	S, (PS), R	5.77	0.375
Oʻahu	Waiʻanae	53.5	S, PS	3.93	2.89; 2.77
	Koʻolau	34.1	S, R	3.19	0.1
Molokaʻi	West Molokaʻi	30.3	S, PS	1.99	1.73
	East Molokaʻi	23.9	S, PS, R	1.75	0.35
Lānaʻi	Lānaʻi	21.1	S	1.5	1.24
Kahoʻolawe	Kahoʻolawe	26.3	S, (PS)	1.25	0.9
Maui	West Maui	9	S, PS, R	2.15	0.4
	Haleakalā	69.8	S, PS	1.1	ca. AD 1600
Hawaiʻi	Kohala	36.4	S, PS	0.6; 0.46	120 ka, perhaps as young as 80 ka
	Mauna Kea	41.9	S, PS	260 ka in outcrop; 330 ka in drill core	4.6 ka, calibrated age
	Hualālai	14.2	S, PS	114–92 ka	AD 1801
	Mauna Loa	74.0	S	100–200 ka	AD 1984
	Kīlauea	31.6	S	70 ka in outcrop; 220 ka from submarine exposures)	Ongoing continuously since AD 1983

NOTE: Volumetric data from Robinson and Eakins (2006). Stage distribution and age data compiled from Sherrod et al. (2007).
[a]Stages: S, shield; PS, postshield (shown parenthetically where volumetrically minor or not mapped separately); R, rejuvenated.
[b]Radiometric ages in millions of years unless specified; ka = thousands of years.

becoming more alkalic as the degree of melting diminishes (Fig. 4). Of the volcanoes old enough to have seen the transition, only two, Koʻolau and Lānaʻi, lack rocks of postshield composition. Eight have postshield strata sufficiently distinct and widespread to map separately.

The transition to postshield volcanism is brief (less than 0.1–0.2 million years) and commonly too short to measure confidently by radiometric dating. Postshield-stage volcanism generally lasts for another 0.1–0.2 million years, although its endurance is variable. At Mauna Kea it has been ongoing for 0.3 million years. Exceptional is the 1-million-year duration at Haleakalā, where eruptive products of the Kula and Hāna Volcanics have coated the volcano with a thickness as great as 1 km. The occurrence of postshield volcanism at Hawaiian volcanoes may be explained by the thermal structure imposed on the overriding lithospheric plate, as determined from numeric modeling of lithospheric and mantle heat transport, viscosity, and kinematics (Fig. 5), but other, still poorly defined factors must play into the final explanation to account for the widely varying durations mentioned.

REJUVENATED-STAGE VOLCANISM FOLLOWS QUIESCENCE

Five Hawaiian volcanoes have seen rejuvenated-stage volcanism following quiescent periods that ranged from 0.5 to 2.0 million years in duration. The prominent volcanic landforms on Oʻahu, such as Punchbowl, Diamond Head, and Hanauma crater, are products of the rejuvenated-stage Honolulu Volcanics, which began about 0.8 million years ago and have persisted sporadically until as recently

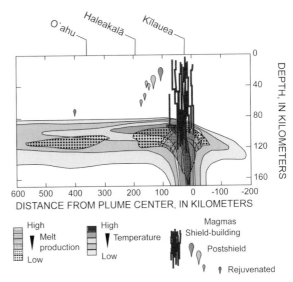

FIGURE 5 Thermal structure of lithosphere beneath Hawaiian Islands, derived from numeric modeling of heat transport, viscosity, and kinematics needed to account for ocean-floor rebound, lateral extent of Hawaiian Arch, and melt production of Hawaiian Ridge. Motion of Pacific plate is from right to left, which leads to asymmetric isotherms leeward of the hotspot. Modified from Ribe and Christenson (1999).

as about 0.1 million years ago. The rejuvenated stage can be brief—only one or two eruptive episodes—or notably durable. That on Niʻihau lasted from 2.2 to 0.4 million years ago; on Kauaʻi, it has been ongoing since 3 million years ago.

The causal mechanisms of rejuvenated-stage volcanism remain difficult to explain. Rejuvenation is likely to be in some way related to decompression melting, either as the lithosphere rebounds from the zone of depression beneath the largest young volcanoes or when hot mantle, dragged initially downward in response to plate motion, rises naturally by its lower density until pressure and temperature are suitable for melt production (Fig. 5). Some models postulate that enhanced crustal fracturing, as might result from the lithospheric rebound, enables magma to percolate upward more easily.

The combination of shield, postshield, and rejuvenated-stage volcanism accounts for the wide range in the age of volcanism along the island chain, even though the age of Hawaiian shields is progressively younger to the southeast. For example, almost every island from Niʻihau to Hawaiʻi had an eruption in the time between 0.3 and 0.4 million years ago, even though only the Island of Hawaiʻi was then hosting volcanoes in their shield stage.

GROWTH OF THE ISLANDS DEPRESSES THE CRUST

The massive outpouring of lava flows from Hawaiian volcanoes weighs upon the oceanic crust, depressing it along an axial Hawaiian Moat (Fig. 2). The periphery of subsidence is marked by the surrounding Hawaiian Arch. The ocean floor, at about 4.5–5.0 km depth adjacent to the Hawaiian Ridge, has subsided to depths as great as 9.5 km along the moat (Fig. 2). Subsidence is ongoing throughout the shield stage and probably into postshield time. The shoreline of Hawaiʻi Island today, loaded down by its five exposed shields, is subsiding at 2–3 mm per year as measured by tide gauges, after correcting for global sea-level rise. The rate in the center of the island is probably twice as great.

One consequence of subsidence is the drowning of coral reefs that drape the submarine flanks of the actively subsiding volcanoes. At least six reefs northwest of Hawaiʻi Island form a stairstep configuration, the oldest being deepest. Drowned reefs on Maui are tilted toward nearby Hawaiʻi Island, where downwarping is greatest.

Large islands built by several volcanoes become a sequence of small islands once volcanic upbuilding ceases. The four islands of Molokaʻi, Lānaʻi, Kahoʻolawe, and Maui were once a single subaerial landmass—Maui Nui—that encompassed about 14,600 km^2, larger by 50% than today's Big Island of Hawaiʻi. Included in this landmass was Penguin Bank, now a broad shoal west of Molokaʻi but originally a volcano that completed its shield-building stage about 2.2 million years ago, when it was briefly connected to Oʻahu. The islands of Maui, Lānaʻi, and Molokaʻi were connected as recently as 18,000 years ago—a connection in this case owing to glacioeustatic sea-level lowering but nonetheless pointing out the shallow ocean depth across the submerged land bridges.

In its heyday, Maui Nui, or "Greater Maui," encompassed extensive lowlands dotted by upland areas. Plants and animals, including flightless or slowly dispersing species, could disperse readily across this terrane; consequently the biologic communities among the Maui Nui islands are more similar than would be found on islands always isolated from each other. Maui Nui's large area increased the chance that airborne or waterborne species would make landfall. Its diverse environments provided a rich storehouse of plants and animals for the subsequent colonization of Hawaiʻi, the youngest island in the chain.

DISSECTION BY LARGE LANDSLIDES MAY OCCUR ANY TIME

Large landslides ring the submarine flanks and seafloor adjacent to Hawaiian volcanoes, a significant discovery of the past 30 years. Recognized first at the emergent volcanoes, these landslides can occur at any stage of volcanic growth or quiescence. For example, submarine Lōʻihi is already gutted by slope failures that have sculpted it into a narrow ridge.

The larger landslides may reach onshore, their headwalls defined by normal faults. These faults may have prominent topographic expression, as seen on the west side of Mauna Loa, for example (Fig. 2). Some faults may result from perturbations in the local stress field after large landslides have removed buttressing slopes. Examples include faults along the crest of Lānaʻi and Kohala, which resulted from, respectively, the Clark debris avalanche about 0.8 million years ago and the Pololū Slump in the past 0.1 million years.

TSUNAMIGENIC DEPOSITS

Disagreement still surrounds the origin of poorly sorted, coralline-bearing sedimentary breccia found at widely ranging altitudes as high as 170 m on the leeward sides of Kohala, West Maui, Lānaʻi, and East Molokaʻi volcanoes. The most widely accepted hypothesis explains these deposits as the consequence of catastrophic, giant waves (megatsunami) generated by prehistoric submarine landslides. The interpretation stems partly from the landward fining of the Lānaʻi deposits and landward fining in the carbonate-clast component of the Molokaʻi deposits. The Lānaʻi deposits were specifically attributed to the ʻĀlika 2 Slide, a slope failure occurring about 125,000 years ago from the west side of Hawaiʻi Island (Fig. 2).

The breccia deposits were originally interpreted as ancient shorelines and attributed to glacioeustatic marine high sea level stands, an explanation that requires substantial uplift of Lānaʻi and Molokaʻi to account for their vertical positioning. Recent estimates for uplift of Oʻahu suggest rates of 0.020–0.024 m per 1000 years for the past 400,000 years. The result has been to expose calcareous reef rock and marine sediment, the heart of the cement industry in the Hawaiian Islands. The absence of these emerged reefs and lagoonal limestone beds elsewhere in the Hawaiian Islands suggests that neither Lānaʻi nor Molokaʻi have seen much uplift. Although rates are imprecisely defined for Lānaʻi, during the past 30,000 years that island has been relatively stable, with uplift or subsidence bracketed between +0.1 and −0.4 m per 1,000 years, on the basis of the depositional character of carbonate deposits on submerged terraces adjacent to the island.

Compelling evidence in favor of the giant-wave hypothesis comes from deposits on Kohala volcano, Island of Hawaiʻi, where the question of uplift is made moot by the ongoing subsidence that has characterized Hawaiʻi Island since its emergence. The calcareous breccia of Kohala, found today at altitudes ranging from sea level to 100 m, must have been deposited originally at altitude 350 to 390 m higher, after correcting for modern rates of subsidence and the age of the deposits.

MAUNA KEA—THE ONLY GLACIATED VOLCANO IN THE CHAIN

Mauna Kea, on the Island of Hawaiʻi, is the only volcano known to have undergone glaciation in the past 200,000 years. Adjacent Mauna Loa had sufficient area above the ice equilibrium-line altitude to maintain an ice cap during the last glaciation 25,000–15,000 years ago, but till and outwash deposits, if they formed, must be buried now by younger lava flows throughout the summit region and down to at least 2,000 m altitude. That altitude is too low to expect depositional traces, if the mapped till and outwash from Mauna Kea are useful benchmarks. Haleakalā (East Maui) lacked sufficient high-altitude terrain to accumulate glaciers in the past 200,000 years, and no deposits have ever been found there.

Mauna Kea's three known glaciations have scoured lava flows at higher altitudes, leaving striated bedrock surfaces that lead down to the moraines and outwash left by the glaciers. Some lava flows in the summit area have ice-contact features such as steep margins, pillow basalt, glassy faces, and palagonitized zones.

The estimated age and duration of Mauna Kea's three glaciations are drawn from the ages of bracketing, dated lava flows, and worldwide correlations using oxygen isotope data to match glacial maxima. The oldest recognized glaciation, the Pōhakuloa (using the Mauna Kea terminology), corresponds to marine oxygen isotope stage 6 and likely occurred sometime between about 180,000 and 130,000 years ago. The Waihū glaciation is thought to have occurred during oxygen isotope stage 4, or roughly 80,000–60,000 years ago. The youngest, the Mākanaka, was under way by about 40,000 years ago. It had ended by 13,000 years ago, the time when a small summit depression became an ice-free lake capable of accumulating sediment.

PERMEABLE VOLCANIC GEOLOGY PUTS GROUNDWATER MOSTLY AT SEA LEVEL

Rain and snow percolate into the highly permeable lava flows of Hawaiian volcanoes. The resulting groundwater may become perched on impermeable ashy beds and zones of secondary mineralization or hydrothermal alteration, but most penetrates to sea level within the island edifice. There the freshwater floats upon the saltwater owing to its lower specific gravity, depressing the interface between them and, in compensation, mounding slightly. The resulting lens-shaped body of freshwater thickens slightly inland, albeit to no more than a few meters above sea level in the center of each island. The slope of the water table is nearly flat, sloping downward roughly 0.3–0.4 m per km toward the coast. Rift zones, with an internal structure rich in vertical dikes, possess somewhat different hydrologic character than a volcano's flanks because the dikes can act as dams to retard the lateral migration of groundwater.

Groundwater in the Hawaiian Islands is derived chiefly from the freshwater lens beneath each island. The resource is fragile, because overpumping allows underlying brackish water to intrude, destroying the quality of the freshwater. Upslope water development typically is limited to elevations where drilling and recovery are economically feasible, currently about 400–600 m above sea level. Potable water at higher altitude comes chiefly from rain catchment systems or by diverting streams into canal systems built in the late nineteenth and early twentieth centuries to bring water from the windward, wetter sides of the higher islands for irrigation purposes.

SEE ALSO THE FOLLOWING ARTICLES

Hawaiian Islands, Biology / Landslides / Lava Tubes / Oceanic Islands / Plate Tectonics / Tsunamis / Volcanic Islands

FURTHER READING

Clague, D. A., and G. B. Dalrymple. 1987. *The Hawaiian–Emperor volcanic chain*. Part I, *Geologic evolution*. U. S. Geological Survey Professional Paper 1350, vol. 1, 5–54.

Eakins, B. W., J. E. Robinson, T. Kanamatsu, J. Naka, J. R. Smith, E. Takahashi, and D. A. Clague. 2003. *Hawaii's volcanoes revealed*. U. S. Geological Survey Geologic Investigations Series Map I–2809, scale about 1:850,000. [available online at http://geopubs.wr.usgs.gov/i-map/i2809].

Gingerich, S. B., and D. S. Oki. 2000. *Ground water in Hawaii*. U. S. Geological Survey Fact Sheet 126-00. http://hi.water.usgs.gov/publications/pubs/fs/fs126-00.pdf.

Moore, J. G., and D. A. Clague. 1992. Volcano growth and evolution of the island of Hawaii. *Geological Society of America Bulletin* 104(11): 1471–1484.

Moore, J. G., D. A. Clague, R. T. Holcomb, P. W. Lipman, W. R. Normark, and M. E. Torresan. 1989. Prodigious submarine landslides on the Hawaiian Ridge. *Journal of Geophysical Research* 94(B12): 17465–17484.

Owen, S., P. Segall, M. Lisowski, A. Miklius, R. Denlinger, and M. Sako. 2000. Rapid deformation of Kilauea Volcano: Global Positioning System measurements between 1990 and 1996. *Journal of Geophysical Research* 105(B8): 18983–18998.

Price, J. P., and D. Elliott-Fisk. 2004. Topographic history of the Maui Nui complex, Hawaii, and its implications for biogeography. *Pacific Science* 58(1): 27–45.

Sherrod, D. R., J. M. Sinton, S. E. Watkins, and K. M. Brunt. 2007. *Geologic map of the State of Hawai'i*. U. S. Geological Survey Open-File Report 2007-1089. http://pubs.usgs.gov/of/2007/1089.

Tilling, R. I., C. Heliker, and T. L. Wright. 1987. *Eruptions of Hawaiian volcanoes: past, present, and future*. U. S. Geological Survey General Interest Publication.

REFERENCES

Le Bas, M. J., R. W. Le Maitre, A. Streckeisen, and B. Zanettin. 1986. A chemical classification of volcanic rocks based on the total alkali-silica diagram. *Journal of Petrology* 27: 745–750.

Macdonald, G. A., and T. Katsura. 1964. Chemical composition of Hawaiian lavas. *Journal of Petrology* 5: 82–113.

Ribe, N. M., and U. R. Christenson. 1999. The dynamical origin of Hawaiian volcanism. *Earth and Planetary Science Letters* 171: 517–531.

Robinson, J. E., and B. W. Eakins. 2006. Calculated volumes of individual shield volcanoes at the young end of the Hawaiian Ridge. *Journal of Volcanology and Geothermal Research* 151: 309–317.

Watts, A. B., and U. S. ten Brink. 1989. Crustal structure, flexure, and subsidence of the Hawaiian Islands. *Journal of Geophysical Research* 94: 10473–10500.

HAZARDS

SEE EARTHQUAKES; ERUPTIONS; HURRICANES AND TYPHOONS; LANDSLIDE; TSUNAMIS

HONEYCREEPERS, HAWAIIAN

ROBERT C. FLEISCHER

Smithsonian Institution, Washington, DC

The Hawaiian honeycreepers, presently classified in the subfamily Drepanidinae (also called drepanidines or Hawaiian finches), are a morphologically and ecologically diverse group of more than 56 species of cardueline finches endemic to the Hawaiian Islands. Molecular data and rate calibrations based on geological age of the islands suggest that they radiated from a single colonizing ancestral cardueline species beginning as little as 3–4 million years ago. The Hawaiian honeycreepers are a highly endangered avian group, with about 70% of the species recently extinct (i.e., recent Holocene), and at least ten of the remaining 17 species currently considered endangered.

TAXONOMY AND PALEONTOLOGY

Specimens of Hawaiian honeycreepers were originally described by early ornithologists as members of many

FIGURE 1 Photograph of museum specimens showing the diversity of extant and recently extinct Hawaiian honeycreepers. Note the diversity in bill shapes, body size, and plumage pattern and color. This represents only a sample of the more than 55 species and the total diversity of the group. Clockwise, from top: Hawaii akepa (*Loxops coccineus*), akiapolaau (*Hemignathus wilsoni*), akohekohe (*Palmeria dolei*), greater koa finch (*Rhodacanthis palmeri*), Nihoa finch (*Telespiza ultima*), apapane (*Himatione sanguinea*), Hawaii akialoa (*Hemignathus obscunis*), palila (*Loxoides balleuxi*), and iiwi (*Vestiaria coccinea*). Photograph by John Steiner.

FIGURE 2 Photographs of several species of living Hawaiian honeycreepers. (A) Nihoa finch (*Telespiza ultima*); (B) akiapolaau (*Hemignathus wilsoni*); (C) palila (*Loxioides balleuxi*); (D) Kauai amakihi (*Loxops stejnegeri*); (E) iiwi (*Vestiaria coccinea*); (F) Hawaii creeper (*Loxops mana*); (G) Hawaii akepa (*Loxops coccineus*); (H) Maui creeper (*Paroreomyza montana*). Photographs by Jack Jeffreys.

different songbird families because of their great diversity in morphology and plumage (Figs. 1 and 2). However, ornithologists studying Hawaiian birds around 1900, such as Hans Gadow, Walter Rothschild, and R. C. L. Perkins, proposed that the diverse species were actually all closely related and belonged to a single family, the Drepanididae. They identified a number of plumage and anatomical characters that united the species, and also noted that most have a distinctive "canvas" odor. More recently, exhaustive osteological and molecular analyses (see below) have largely confirmed the monophyly (descent from a single ancestral species) of the honeycreepers and concluded that the radiation arose from within the cardueline finches (a cosmopolitan group that includes birds such as rosefinches, goldfinches, hawfinches, canaries, and crossbills). Thus, the Hawaiian honeycreepers represent one of the premier examples of insular adaptive radiation within birds.

There is a very limited ancient fossil record for birds in the Hawaiian Islands, but a very rich one from the Holocene (about 10,000 years to present). In fact, the discovery and study of these Holocene fossil birds by Smithsonian scientists Storrs Olson and Helen James revealed that massive extinctions of birds had occurred in the Hawaiian Islands subsequent to their colonization by humans some 1000–1500 years ago. These extinctions included at least 23 species of Hawaiian honeycreepers known only from subfossils. Some of the species had unique morphologies not represented among the remaining extant species. The subfossils are found in old lava tubes, limestone caves, and sand dunes, and many have yielded DNA suitable for sequencing.

Taxonomists have traditionally divided the honeycreeper species into two or three taxonomic subfamilies or tribes, corresponding to whether they are (a) mostly greenish or black and red, or (b) have finchlike bills (Psittirostrini) or warbler-like or long bills (Hemignathini), or are mostly nectar-feeding (Drepanidini). Based on recent osteological studies by Helen James, the 56–60 known species can be placed in 22 genera, but generic-level taxonomies by other authors differ. Of these 56–60 species, 33–37 species were known historically (extant or known from museum specimens collected since the eighteenth century). Several species have only recently become extinct. For example, the poouli (*Melamprosops phaeosoma*) was the most recently discovered Hawaiian honeycreeper. It was found in small numbers on Maui in 1973, and two juvenile specimens were collected by a group of university students. Paleontological studies indicate that it was once one of the most common honeycreepers on Maui at low elevations. Its numbers declined precipitously over the three decades since its discovery, and it was considered extinct in the wild by 2005. The rapid decline and loss of this species serves as a cautionary tale that drastic measures and quick action may often be necessary to keep species from extinction.

EVOLUTIONARY RELATIONSHIPS BASED ON MORPHOLOGY AND MOLECULES

The phylogeny or evolutionary tree of the Hawaiian honeycreepers is difficult to reconstruct, as in many other adaptive radiations, probably because of a relatively low number of defining morphological characters, considerable evidence for convergent or parallel evolution (homoplasy), and an apparently rapid radiation that likely obscures resolution of many branches of the tree. The Hawaiian honeycreepers represent perhaps the most extreme radiation within a single subfamily or family of birds, with forms that have bills of almost any type known from songbirds (thick finch-like, thin warbler-like, parrot-like, long and woodhewer-like, long and decurved to probe flowers for nectar, etc.), and variable plumages that can contain drab browns, blacks, grays, greens, yellows, and bright reds in many patterns (Fig. 1), but never blue hues or iridescence. One species, the akiapolaau (*Hemignathus wilsoni*) has a "dual-purpose" woodpecker bill (Figs. 1 and 2), with a short, straight, robust mandible for hammering holes in bark and a longer, decurved maxilla for probing insects from the hole it makes. Most of the honeycreeper species that take nectar have evolved tongues folded into tubes to facilitate nectar flow, convergent to those found in many other nectar-feeding birds in different families or orders.

The rapidly expanding field of molecular genetics offers powerful methods of DNA amplification via the polymerase chain reaction (PCR) and DNA sequencing to help unravel the evolutionary relationships of the Hawaiian honeycreepers and to offer a means of estimating the timing of their origin and radiation. Evolutionary trees have been constructed using DNA sequences from most of the extant species, and the power of PCR, applied carefully and with proper controls, has allowed the amplification and sequencing of "ancient" DNA from many extinct or endangered honeycreeper species only available as museum specimens or even subfossil bones (Fig. 3). However, even trees based on a large amount of DNA sequence data do not provide strong statistical support to resolve the branching order, and the pattern of the trees suggests a very rapid initial divergence into different morphological types of honeycreepers.

The main Hawaiian Islands follow a progression of ages, with the oldest in the northwest (Kauai at about 5 million years) and the youngest in the southeast (Hawaii at less than 1 million years). These island ages put an age limit on each species, and, with certain assumptions,

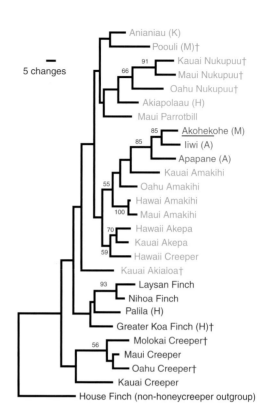

FIGURE 3 A phylogenetic or evolutionary tree based on mitochondrial DNA sequences showing the relationships of a sampling of extant and extinct Hawaiian honeycreepers. The numbers along a branch are the support values from a statistical bootstrap analysis; values over 70 indicate good support for the node. Letters in parentheses indicate islands on which the species exists (when not noted in the name of the species; A = all islands, K = Kauai, M = Maui, H = Hawaii). A † indicates a presumed extinct species.

allow the estimation of rates of DNA sequence evolution. These rates, in turn, can be used to estimate the timing of species formation and even the date when the original finch species colonized the islands. Such calculations indicate that the Hawaiian honeycreepers probably began their radiation about 3 to 4 million years ago and that the original colonization must have occurred when Kauai was the youngest (and largest) island. The timing of the radiation (when most of the morphological types evolved) correlates with the formation of Oahu, the next island that formed after Kauai. This suggests that the formation of the new and uninhabited Oahu may have been a major factor initiating or promoting the adaptive radiation.

ECOLOGY AND BEHAVIOR

The Hawaiian honeycreepers show tremendous diversity in life history characteristics, feeding mode, and other behavioral and ecological traits. They radiated, probably from an ancestral finch-billed granivore, to fit into almost every possible ecological niche of songbirds. There were as many as 19 species of ground-feeding granivores (finches or grosbeaks), at least one specialist frugivore, and at least five flower-probing nectarivores (with a few other species more facultative in their nectar feeding). Among the many insectivorous species, some glean prey from foliage, while others probe bark. Several species, including the aforementioned akiapolaau and the Maui parrotbill (*Pseudonestor xanthophrys*), puncture or tear off bark to find insects within. At least one species, the aforementioned recently extinct poouli, appeared to be a snail specialist but also took other invertebrates, and it had a spatulate tongue well shaped for removing snails from shells or insect prey from crevices. The akepas in the genus *Loxops* have slightly crossed bill tips, which they use like forceps to separate bracts on the ohia tree (*Metrosideros polymorpha*) in search of insect prey. The "creepers" include two related genera (*Oreomystis* and *Paroreomyza*) and a third species on Hawaii Island (*Loxops mana,* formerly *Oreomystis mana*) that is apparently convergent in its "creeping" behavior to the other two genera and is more closely related to the akepas based on osteology and DNA sequences. Two species in the genus *Vangulifer,* known only from subfossil remains, had a shovel-like bill with a rounded tip, unlike any other known songbird. Upon what and how these extinct species fed can only be speculated, but they may have been aerial flycatchers. Whereas some species are highly specialized in feeding mode or prey type, others feed more generally. Some researchers have suggested that these more generalized species are usually less likely to be endangered with extinction than the more specialized forms.

Hawaiian honeycreepers show mostly typical songbird nesting behavior, with most species constructing cup-shaped nests but with a few species known also to nest in cavities. Clutch sizes range from 1 to 4 eggs, and there is considerable variation in other breeding characteristics such as incubation times, growth rates, and egg and nestling survival. Honeycreepers appear to be more highly impacted by introduced mammalian nest predators than are introduced birds, probably because they evolved for millions of years in isolation from such predation pressure.

Hawaiian honeycreepers are like most other songbirds and have a socially monogamous mating system. In the case of the palila (*Loxioides balleuxi*), an endangered, finch-billed species that lives on the island of Hawaii, this social monogamy is matched by genetic monogamy, as a DNA fingerprinting analysis of parentage revealed no cases of extra-pair fertilization. A similar DNA analysis of parentage conducted in the Pearl and Hermes Reef population of the endangered Laysan finch (*Telespiza cantans*) revealed that about 12% of the offspring were the result of extra-pair mating. Both of these values are on the low end of the extra-pair mating range for songbirds, but within the range found for other cardueline finches such as American goldfinches (*Carduelis tristis*), house finches (*Carpodacus mexicanus*), and serins (*Serinus serinus*), and for tropical, sedentary songbirds. Why carduelines and drepanidines, or tropical songbirds, may have lower extra-pair mating levels than other songbirds is not known.

The Hawaiian honeycreepers also vary considerably in vocalizations. The finch-billed forms sing more melodious songs reminiscent of the songs of canaries and other cardueline finches. Most of the insectivorous honeycreepers sing relatively simple songs, such as the flat-toned trill or rattle of the amakihi or the descending trill of the Kauai creeper (*Oreomystis bairdi*), and that are in some ways more convergent to songs of warblers or sparrows. And the nectarivorous honeycreepers (such as the iiwi, *Vestiaria coccinea,* and the apapane, *Himatione sanguinea*) have amazingly versatile and varied songs, with complex series of pure tones, buzzes, clicks, and trills. The apapane has a particularly varied song, with repertoires containing literally dozens of distinct notes, while the iiwi has one of the most divergent and unusual songs, with highly discordant and "creaky" tones. These variable songs are used for a variety of social purposes: for maintaining contact while in feeding flocks, and to effect agonistic and mating interactions.

CONSERVATION AND DISEASE

The Hawaiian honeycreepers have been subjected to a massive Holocene extinction, with nearly 40 of the original 56–60 species disappearing over the past few hundred years. Many of these species disappeared without being discovered by Western naturalists and are known only from subfossil bones. Studies by paleontologists Helen James and Storrs Olson showed that most of these subfossil species disappeared from the fossil record during the period of Polynesian residence in the Hawaiian Islands. This suggests that direct (e.g., hunting) or indirect (e.g., habitat modification, or introduction of predators such as the Polynesian rat, or perhaps disease from chickens) impacts from the Polynesians caused these extinctions.

Of the 17 Hawaiian honeycreeper species that are believed extant, ten species are currently or will soon be listed as endangered under the U.S. Endangered Species Act (note that an additional six species are on the list, but the consensus of most biologists working in Hawaii is that these are actually extinct). Because of this high level of endangerment of the honeycreepers, and 16 other types of Hawaiian birds, Hawaii has been given the moniker of the "capital of endangered species" in the United States. The primary threats to these extant species include habitat loss and degradation (often from introduced pigs and ungulates), introduced mammalian predators (rats, cats, mongooses), invasive diseases vectored by introduced mosquitoes, and possibly competition from introduced birds and insects.

Hawaiian honeycreepers appear to be particularly susceptible to introduced mosquito-vectored diseases such as avian poxvirus and avian malaria (*Plasmodium relictum*). Infection prevalences and parasitemias are particularly high, and mortality is, unfortunately, a common result of infection. Several species of Hawaiian honeycreepers were infected with introduced malaria in a series of controlled aviary experiments by Richard Warner, Charles van Riper, Carter Atkinson, and their colleagues. One species, the amakihi (*Loxops virens*), consistently had lower mortality rates, but for the remaining species tested, such as the iiwi (*Vestiaria coccinea*) and Laysan finch, nearly all individuals succumbed. Introduced birds generally show no mortality and few symptoms when infected with this same malarial strain.

As might be expected, amakihi are the only Hawaiian honeycreeper species that regularly breed at lower elevations where *Culex* mosquitoes and malaria are common, whereas iiwi and most other honeycreeper species are almost nonexistent at these elevations. Genetic structure data suggest that malaria-resistant amakihi survived in small pockets at low elevations and subsequently spread out to fill available habitat, rather than spreading from high to low elevations. Molecular evidence also indicates that the type of avian malaria introduced to Hawaii is widespread across most of the world, particularly on oceanic islands, and likely originated in Africa. Both poxvirus and malaria impact Hawaiian honeycreeper survival, and, along with a variety of additional stressors, continue to challenge this unique and varied taxon with threat of extinction.

SEE ALSO THE FOLLOWING ARTICLES

Bird Disease / Bird Radiations / Fossil Birds / Galápagos Finches / Hawaiian Islands, Biology

FURTHER READING

Beadell, J. S., F. Ishtiaq, R. Covas, M. Melo, B. H. Warren, C. T. Atkinson, T. Bensch, G. R. Graves, Y. V. Jhala, M. A. Peirce, A. R. Rahmani, D. M. Fonseca, and R. C. Fleischer. 2006. Global phylogeographic limits of Hawaii's avian malaria. *Proceedings of the Royal Society, B* 273: 2935–2944.

Fleischer, R. C., C. E. McIntosh, and C. L. Tarr. 1998. Evolution on a volcanic conveyor belt: using phylogeographic reconstructions and K-Ar based ages of the Hawaiian Islands to estimate molecular evolutionary rates. *Molecular Ecology* 7: 533–545.

Fleischer, R. C., C. L. Tarr, H. F. James, B. Slikas, and C. E. McIntosh. 2001. Phylogenetic placement of the po'o-uli *Melamprosops phaeosoma* based on mitochondrial DNA sequence and osteological characters. *Studies in Avian Biology* 22: 98–103.

Foster, J. T., B. L. Woodworth, L. E. Eggert, P. J. Hart, D. Palmer, D. C. Duffy, and R. C. Fleischer. 2007. Genetic structure and evolved malaria resistance in Hawaiian honeycreepers. *Molecular Ecology* 16: 4738–4746.

James, H. F. 2004. The osteology and phylogeny of the Hawaiian finch radiation (Fringillidae: Drepanidini), including extinct taxa. *Zoological Journal of the Linnean Society* 141: 207–255.

James, H. F., and S. L. Olson. 1991. Descriptions of thirty-two new species of birds from the Hawaiian Islands: Part 2. Passeriformes. *Ornithological Monographs* 46: 1–88.

Jarvi, S. I., C. T. Atkinson, and R. C. Fleischer. 2001. Immunogenetics and resistance to avian malaria (*Plasmodium relictum*) in Hawaiian honeycreepers. *Studies in Avian Biology* 22: 254–263.

Pratt, H. D. 2005. *The Hawaiian honeycreepers*. Oxford: Oxford University Press.

Scott, J. M., S. Mountainspring, F. L. Ramsey, and C. B. Kepler. 1986. *Forest bird communities of the Hawaiian Islands: their dynamics, ecology and conservation*. Studies in Avian Biology 9. Lawrence, KS: Allen Press.

van Riper, C. III, S. G. van Riper, M. L. Goff, and M. Laird. 1986. The epizootiology and ecological significance of malaria in Hawaiian land birds. *Ecological Monographs* 56: 327–344.

Warner, R. E. 1968. The role of introduced diseases in the extinction of the endemic Hawaiian avifauna. *Condor* 70: 101–120.

HUMAN IMPACTS, PRE-EUROPEAN

PATRICK V. KIRCH

University of California, Berkeley

Islands, and especially truly oceanic islands (such as those situated on the Pacific Plate), offer numerous historical "experiments" of interactions between previously isolated and often vulnerable ecosystems and colonizing human populations. Because of isolation, biotic disharmony, and lack of competition, island biotas are characterized by high species-level endemism and vulnerability to invasive taxa. Pioneering human populations, such as those of the Lapita and later Polynesian groups, introduced a portmanteau biota along with cultural concepts of land use and ecosystem management. Over time spans ranging from greater than 40,000 to less than 1,000 years, these invading human populations and their "transported landscapes" irreversibly altered these fragile and previously isolated island ecosystems.

TAXONOMY OF HUMAN IMPACTS

Indigenous human populations on islands throughout the Pacific, Indian, and Caribbean Oceans altered their environments in ways that can be characterized as both *direct* and *indirect,* with various subcategories, as indicated in Fig. 1.

Direct impacts are those that result from a variety of consciously directed human actions in the ecosystem, including hunting and gathering of a variety of plant and animal resources, forest clearance and vegetation modification (through both clearing and the use of fire), introduction and planting of agricultural crops and other useful plants, and a wide array of often permanent physical modifications of the landscape, such as terracing slopes, building fish ponds, and digging canals and drains. Such physical manipulation of landscapes often occurred some time after initial island colonization, when human population density had reached relatively high levels, and efforts were made to intensify food production.

Indirect impacts derive in the first instance from the array of portmanteau biota that typically accompanies the movement of a human population into virgin territory. Human bodies themselves may carry disease pathogens and ectoparasites, and along with them go domestic animals, crop and other useful plants, weeds, vermin, and other synanthropic species. This array of synanthropic plants and animals often included species that were highly competitive with the vulnerable native biota of remote oceanic ecosystems. A prime example is the Pacific rat (*Rattus exulans*), which was spread by humans to virtually every island group in the Pacific, and which has been implicated in ecological changes on islands ranging from Rapa Nui (Easter Island) to Hawai'i.

TRANSPORTED LANDSCAPES

Botanist Edgar Anderson originally introduced the concept of "man's transported landscapes" to refer to the fact that when people move into a new area, they carry with them not just physical artifacts, but typically a host of plants and animals, including crops, weeds, and vermin. In addition, they bring cognitive models of how a landscape should be managed, how it should look, and how it should be manipulated. Historian Andrew Crosby extended the notion of transported landscapes with his analysis of the imperial expansion of the West ("ecological imperialism"), and coined the term "portmanteau biota" to refer to the assemblage of plants and animals that accompany expanding human populations. These concepts apply equally well to prehistoric peoples who migrated from continents and larger island masses out into truly oceanic islands.

In the case of the early Polynesians, for example, the portmanteau biota carried on double-hulled voyaging canoes and introduced to islands included pigs, dog, and chickens as purposeful introductions; a small rat (*Rattus exulans*) and several species of gecko and skink (*Lepidodactylus lugubris, Gehyra mutilata, Cryptoblepharus poecilopleurus,* and other species) as probable inadvertent "stowaways"; at least 27 species of crop, narcotic, medicinal, or otherwise useful plants; an unknown number of weed species; and several

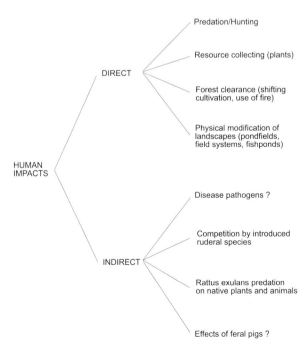

FIGURE 1 A taxonomy of human impacts to island environments.

kinds of terrestrial gastropods (including *Alopeas gracile, Lamellidea pusilla,* and *Gastrocopta pediculus*) the last of these probably adhering to crop plants or existing in associated soil. Some of these newly introduced species quickly escaped beyond the confines of human settlements and began to compete with or have impacts on the native biota. For example, rapid population increases in the Pacific rat *R. exulans* have been invoked as a likely cause for rapid dryland forest reduction in the Hawaiian Islands, and possibly also on Easter Island (Rapa Nui), through the rats' consumption of seeds and young shoots.

The early Polynesians also brought with them cognitive models of land use and management. These included the use of swidden or shifting cultivation ("slash-and-burn") agriculture, in which natural forest vegetation is cut, dried, and fired prior to planting, as well as the planting of taro and other aroids in naturally swampy terrain, or in terraces constructed along stream banks. These cognitive models also included concepts of land ownership and division, with permanent house sites and garden lands forming hereditary estates held by extended family groups. Applying these land-use concepts, the Polynesians transformed island after island into highly managed, anthropogenic landscapes. Recognizing that society and nature become inextricably interconnected in such complex landscapes, following C. Barton we may call them *socioecosystems.*

HUMAN TRANSFORMATION OF ISLAND ECOSYSTEMS

Pre-European transformation of island ecosystems by indigenous human populations has now been well documented through archaeological and associated paleoecological research on many islands in the Pacific, the Caribbean, and the Mediterranean, and on the large island of Madagascar. This article draws mainly on Pacific island examples, but the same fundamental processes can be found on islands throughout the world.

Deforestation was a frequent consequence of pre-European land-use practices, including shifting cultivation, and the use of fire. On many islands in the humid tropics lacking active volcanoes, natural fires appear to have been largely absent prior to human arrival. This is indicated by the absence of microscopic charcoal particles in sediment cores for layers predating human occupation, and by sharp influxes of such charcoal particles following the arrival of humans. Pollen grains extracted from the same sediment cores (on such islands as New Caledonia, Fiji, Atiu, Mangaia, Oʻahu, and Rapa Nui) typically display radical changes in their floristic composition, likewise correlating with the period following initial human settlement. Native arboreal taxa show significant declines whereas pyrophytic or fire-adapted plants (such as *Dicranopteris* ferns and *Pandanus*) increase. The extent and degree of deforestation appears to have varied depending on a number of factors, including island age and consequent nutrient depletion of soils, climate and frequency of drought, human land-use practices, and the size and density of human populations. Some younger volcanic islands such as Tahiti or Rarotonga seem to have been fairly resilient to intensive land use, whereas others (particularly geologically older islands) such as Mangaia, Mangareva, and Rapa Nui were extensively deforested. On the large, near-continental islands of New Zealand, pollen analysis has demonstrated that vast areas were deforested following Polynesian arrival around AD 1200. In part, this may have been due to burning to drive and hunt the large flightless moa birds, and in later times to encourage the growth of indigenous bracken fern, an important food resource.

In part because of deforestation and habitat destruction, and in part because of the direct pressure of hunting for food and feathers (and in some cases also because of predation by rats on ground-nesting eggs and chicks), island bird populations frequently underwent substantial declines, leading to the extirpation and even extinction of many species. Some island birds evolved flightlessness (including flightless ducks and ibises in Hawaiʻi, flightless megapodes in New Caledonia, and the famous moa of New Zealand), and these taxa were especially vulnerable to predation and extinction. Archaeological and paleontological evidence for such major impact on avifauna is now well documented for many Pacific islands. One of the most dramatic cases is New Zealand, where at least 13 species of moa (in the genus *Dinornis* and other taxa), the largest standing up to 3 m tall with neck extended, were driven to extinction in probably less than 200 years after Polynesian colonization. But many other islands, such as Mangareva, Rapa Nui, and Tahuata, show dramatic decreases in bird populations such those of nesting seabirds (petrels, shearwaters, terns) that were evidently abundant in large numbers prior to human arrival but became rare or extirpated thereafter.

Human impacts on native biota extended well beyond birds. On some islands, particularly where the areas of exploitable reef or lagoon were limited, archaeologists have detected evidence for resource depression in marine fauna. On Mangaia Island, for example, this is demonstrated both in significantly diminished size distributions in shellfish (*Turbo* sp.) and in the fish. In Hawai'i, similar impacts have been detected for late prehistoric harvesting pressure on the prized limpet species *Cellana exarata*. On other islands, however, such as Mangareva, marine resources appear to have been more resilient and able to withstand continued pressure of human exploitation without measurable impacts.

The degree to which pre-European human populations irreversibly transformed island ecosystems depended on a number of factors. As noted above, island age and nutrient status was clearly one factor influencing the extent of deforestation. Another important factor was the overall size and density of the human population. In late prehistory, many islands in Remote Oceania saw their human populations attain density levels of 200 or more persons per square kilometer. Under such high population levels, intensification of production systems and conversion of island landscapes to highly managed socioecosystems was inevitable. In some cases, trophic competition between humans and domestic animals (especially pigs) became such that the latter were eliminated (this is now documented for Tikopia, Mangaia, and Mangareva).

IMPLICATIONS FOR BIOGEOGRAPHY

As recently as the 1960s, many naturalists and anthropologists alike assumed that the pre-European populations of the world's islands had had relatively little impact on these insular ecosystems. This viewpoint was a holdover of the eighteenth-century Rousseauian myth of the "noble savage" and can no longer be sustained. Every island that was settled for any length of time in prehistory, and which has now been subjected to archaeological and paleoecological study, has been shown to have had some degree of irreversible impact. In many cases the effect on native biota, through deforestation, species loss, and resource depression, was substantial. For biogeographical studies, this means that the historically recorded data on biodiversity, taxonomic richness, and so on cannot be taken as representative of the truly natural or pristine conditions on islands. Rather, it is essential that paleoecological and archaeological data be used to establish the actual biotic conditions on particular islands as a baseline against which both pre-European and post-European human impacts can be assessed.

INDIGENOUS MANAGEMENT OF LANDSCAPES

The fact that pre-European populations extensively modified island ecosystems through their mix of direct and indirect impacts has now been empirically documented for many islands. The historical record of human impacts should not, however, be misconstrued to mean that indigenous populations were some sort of "eco-vandals" intent on destroying their environments, even though in some cases the consequences of cumulative human actions were indeed calamitous (as on Rapa Nui or Mangaia). To the contrary, island peoples struggled to bring their necessary exploitation of island resources into balance with human population numbers and developed various cultural approaches to land and resource management.

On many Pacific islands, pre-European populations developed intensified methods of agricultural production, especially terracing and irrigation, which allowed some landscapes to be put into near-continuous cropping and food production. In the Hawaiian Islands, for example, alluvial valley soils and colluvial slopes were sculpted into systems of pondfields, with stone-faced embankments, which were flooded with water diverted from streams through stone-lined canals. These pondfield complexes were used to cultivate taro (*Colocasia esculenta*), which responds to such irrigation with high yields. In New Caledonia taro irrigation was also practiced, with earth-banked terraces extending up even very steep hillsides. In addition, the New Caledonians developed intensive yam (*Dioscorea* spp.) culture, planting these xerophytic crops in extensive linear mounds. On other islands, such as Tikopia and Kosrae, and in the Marquesas, land management practices emphasized arboriculture or tree cropping, especially of breadfruit (*Artocarpus altilis*). The Tikopia system of orchard gardening is especially remarkable for the way that it artificially mimics the multi-storey structure of a tropical rainforest. Tikopia appears to represent a true case of long-term sustainability of an indigenous economic production system.

Resource management also extended to reef and lagoon resources. Many island cultures imposed regular prohibitions (*tapu, rahui*) on the taking of certain fish or shellfish resources, in order to allow stocks to replenish. In the Hawaiian Islands, true pond aquaculture was developed for the raising of mullet (*Mugil cephalis*) and milkfish (*Chanos chanos*) in large ponds whose rock walls extended out onto shallow reef flats. More than 400 of these ponds have been archaeologically identified, and they probably greatly augmented pre-European subsistence production.

SEE ALSO THE FOLLOWING ARTICLES

Deforestation / Easter Island / Extinction / Flightlessness / Introduced Species / Peopling the Pacific

FURTHER READING

Dodson, J., ed. 1992. *The naive lands: prehistory and environmental change in Australia and the Southwest Pacific.* Melbourne: Longman Chesire.

Fosberg, R., 1963. *Man's place in the island ecosystem: a symposium.* Honolulu: Bishop Museum Press.

Kirch, P. V. 1997. Microcosmic histories: island perspectives on 'global change.' *American Anthropologist* 99: 30–42.

Kirch, P. V. 2007. Three islands and an archipelago: reciprocal interactions between humans and island ecosystems in Polynesia. *Earth and Environmental Science Transactions of the Royal Society of Edinburgh* 98: 85–99.

Kirch, P. V., and T. L. Hunt, eds. 1997. *Historical ecology in the Pacific Islands: prehistoric environmental and landscape change.* New Haven, CT: Yale University Press.

Rolett, B., and J. Diamond. 2004. Environmental predictors of pre-European deforestation on Pacific Islands. *Nature* 431: 443–446.

Steadman, D. W. 2006. *Extinction and biogeography of tropical Pacific birds.* Chicago: University of Chicago Press.

HURRICANES AND TYPHOONS

THOMAS A. SCHROEDER
University of Hawaii, Manoa

"Hurricane" and "typhoon" are regional terms for intense tropical cyclones. Tropical cyclones are warm-core vortices that develop over tropical oceans. Tropical cyclones are more intense and compact than the extratropical cyclones of middle latitudes (Fig. 1). They impact islands through damaging winds, heavy rains, and coastal flooding.

INTENSITY CLASSIFICATION

Tropical cyclone intensity is determined from either sea-level pressure in the storm center or maximum sustained winds anywhere in the vortex. Intensity class terminol-

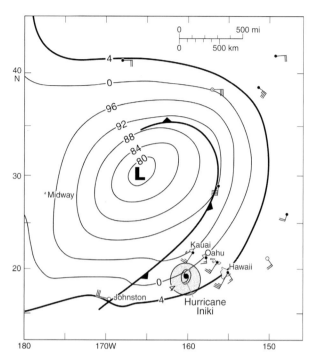

FIGURE 1 Surface weather map for January 8, 1980. Superimposed on the 1980 map is Hurricane Iniki (September 11, 1992). The area of Iniki within the 1004 mb (hPa) isobar is stippled. The equivalent isobar for the 1980 winter storm is in bold. Note the differences in scale of the two systems. Wind symbols (feathered) represent winds reported on January 8, 1980. Meteorological convention is to plot isobars as two digits, where 96 = 996 mb and 04 = 1004 mb.

ogy is basin- and/or forecast-agency dependent. The United States Navy/Air Force Joint Typhoon Warning Center, which issues forecasts for U.S. interests in the western North Pacific Ocean, classifies systems as depressions (winds less than 17 m/s), tropical storms (winds between 17 and 32 m/s) and typhoons (winds greater than 32 m/s). The Japanese Meteorological Agency "typhoon" includes U.S.-defined "tropical storm" and "typhoon."

Sustained winds may be one-minute average (U.S. practice) or ten-minute average. This article uses U.S. Atlantic basin ("hurricane" instead of "typhoon") classifications.

STRUCTURE

A hurricane consists of three regions (Fig. 2):

1. A relatively calm and clear central "eye." The winds are relatively light and speeds increase linearly with increasing radius. Strong sinking motion limits cloud formation. There may be some low clouds and thin, high overcast. Development of an eye is considered to indicate intensification from "tropical storm" to

"hurricane" intensity. Typical eye diameters range from 20 km to 60 km.
2. A "core" or "eyewall." Winds attain maximum intensity, and a solid band of heavy rains normally completely encircles the eye. Smaller intense vortices may exist within the eyewall. Typical eyewall thickness is 20 km to 60 km.
3. An outer region. Winds gradually diminish, and rain squalls ("rain bands") are common. The outer region may extend 500 km from the storm center.

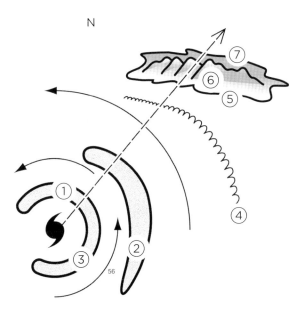

FIGURE 2 Schematic of a hurricane approaching a mountainous island. Featured are (1) eye wall (stippling indicates radar signature), (2) rain band (also stippled), (3) location of maximum sustained winds, (4) waves propagating ahead of the storm, (5) point of expected peak coastal inundation, (6) enhanced rainfall, and (7) likely downslope winds.

The hurricane just described is circularly symmetric. Actual storms are asymmetric. The strongest asymmetry arises from storm motion, which augments the winds in the right-hand sector (relative to storm motion).

FREQUENCIES, FORMATION (GENESIS), AND VARIABILITY

Tropical cyclones are rare events. Annually about 85 storms of all intensity classes occur. One to two tropical cyclones are present daily over an area comprising 50% of the Earth's surface. Storm frequency varies among ocean basins. The most active, in order, are the western North Pacific, the eastern North Pacific, the Southwest Pacific and Australia, the Southwest Indian, and the North Atlantic.

Storm genesis is poorly observed, and the mechanism is debated. Scientists agree upon the fundamental conditions for hurricane formation. These include a minimum sea-surface temperature, a minimum depth of the oceanic mixed layer (a measure of heat content), critical thermodynamic structures in the atmosphere, critical vertical shear of the horizontal winds, and a preexisting seedling disturbance. The necessity that each condition be met explains the rarity of hurricane development.

Annual global storm totals have been constant, but location and frequency of genesis within and among basins varies. One controlling factor is the El Niño–Southern Oscillation (ENSO), which in its warm phase shifts Northwest Pacific typhoon formation zones and inhibits Atlantic hurricanes. Theory suggests that current global warming should lead to a detectable increase in hurricane intensity.

IMPACTS UPON ISLANDS

Hurricanes bring damaging winds, coastal flooding, and torrential rains. The degree of impact varies with the size and topography of the island, with offshore and near shore bathymetry, and with the path of the storm relative to the island.

Hurricane winds destroy vegetation and infrastructure (Fig. 3). These winds extend upward for 3 km. Thus, even on large, high islands, extreme winds penetrate well inland. Storms have easily transited Hispaniola, Jamaica, and New Caledonia with minimal weakening. In steep topography,

FIGURE 3 Hurricane winds battering palm trees on Key West, Florida. Photograph taken at the Weather Forecast Office in Key West, Florida, during the height of Hurricane Charley on Friday, August 13, 2004. Photograph courtesy of the National Weather Service, Key West, Florida.

enhanced downslope winds may devastate supposedly sheltered lee sides (e.g., Hurricane Iniki, 1992).

Embedded in the eyewall and outlying rain bands are intense tornadic vortices. Eyewall vortices were prominent in both Hurricanes Andrew and Iniki in 1992. Rainband tornadoes were observed in the Florida Keys in 1972 during the passage of Hurricane Agnes well to the west.

Coastal flooding is caused by a dome of water (the "storm surge") which grows at the center and right flank of a moving storm. The low pressure in the eye causes an "inverted barometer" which raises the sea level. Additionally, the winds in the right front sector pile water ahead of the storm. Water may rise even higher, depending upon the nearshore bathymetry. The U.S. Gulf Coast, with its broad shelf, favors large storm surges. In 1900 the Galveston Hurricane killed over 6000, all by storm surge. The Hawaiian Islands feature fringing reefs and no significant shelf, but they have nevertheless experienced significant flooding. Hurricane Iniki caused an extreme high-water level of 9.1 m through a combination of surge, high tide during landfall, and run-up.

On high islands coastal inundation is limited to narrow coastal strips. However, low-lying atolls and barrier islands may be completely inundated. Wave action damages reefs and may drastically alter local bathymetry. In one notable instance, Hurricane Bebe (1972) created a completely new debris rampart (the Bebe Rampart) 35 km long, 35 m thick, and up to 3.5 m high along the east side of Funafuti Atoll, Tuvalu.

Hurricanes are prolific rain producers. Precipitation accumulations vary with island topography and speed of storm passage. Most world record rains for periods from one to 15 days have fallen at Réunion Island in the Southwest Indian Ocean. Réunion is a high island featuring a 2.5-km shield volcano. Records range from 1869 mm for 24 hours to 6433 mm for 15 days. The latter occurred as a tropical cyclone (Hyacinthe, 1980) executed two complete loops around the island.

The combination of surge and torrential rains may produce extreme flooding of coastal river deltas and barrier islands. Stream waters running down to the sea encounter elevated sea level. This combination has been especially deadly in the northern Bay of Bengal.

SEE ALSO THE FOLLOWING ARTICLES

Barrier Islands / Climate Change / Climate on Islands / Global Warming / Surf in the Tropics

FURTHER READING

Anthes, R. A., R. W. Corell, G. Holland, J. W. Hurrell, M. C. McCracken, and K. E. Trenberth. 2006. Hurricanes and global warming: potential linkages and consequences. *Bulletin of the American Meteorological Society* 87: 623–628.
Elsberry, R. L., ed. 1995. Global perspectives on tropical cyclones. Report No. TCP-38. World Meteorological Organization.
Emmanuel, K. E. 2005. *Divine wind: the history and science of hurricanes.* New York: Oxford University Press.
Fletcher, C. H., B. M. Richmond, G. M. Barnes, and T. A. Schroeder. 1995. Marine flooding on the coast of Kaua'i during Hurricane Iniki: hindcasting inundation components and delineating washover. *Journal of Coastal Research* 11: 188–204.
Larson, E. 1999. *Isaac's storm.* New York: Crown Publishers.
Maragos, J. E., G. B. K. Baines, and P. J. Beveridge. 1973. Tropical Cyclone Bebe creates a new land formation on Funafuti Atoll. *Science* 181: 1161–1164.
Pielke, R. A., Jr., C. Landsea, M. Mayfield, J. Laver, and R. Pasch. 2005. Hurricanes and global warming. *Bulletin of the American Meteorological Society* 86: 1571–1575.
Simpson, R. H., ed. 2003. *Hurricane! Coping with disaster.* Washington, DC: American Geophysical Union.
Simpson, R. H., and H. Riehl. 1981. *The hurricane and its impact.* Baton Rouge, LA: LSU Press.

HYDROLOGY

CHRISTIAN DEPRAETERE
Global Islands Network, Grenoble, France

MARC MORELL
Institut de Recherche pour le Développement, Fort de France, Martinique

Freshwater is often a critical resource on islands. Island hydrology takes into account the specific effects of the surrounding ocean on the physical processes and water resources budget of islands, and the specific approaches and considerations required to understand and manage freshwater on islands.

UNIQUE CHARACTERISTICS OF ISLAND HYDROLOGY

Islands are special compared to continents when it comes to water resources. This is a consequence of the defining characteristics of islands: their limited size and remoteness from large sources of freshwater. In general, groundwater is the main freshwater resource, although its predominance depends on the area and relief of the island: the smaller and the lower the island, the greater the relative importance of groundwater compared to other sources.

Island hydrological contexts are also as diverse as their continental counterparts. A global survey of islands shows that they can be found in all climatic zones and correspond to a wide range of geological and ecological settings. Nevertheless, it is a geographic fact that they are more directly exposed to marine atmospheric advection, with humid air constantly passing over the island, and undergo less seasonal variation in temperature than larger land masses. Although islands also have great geological diversity, a majority of them are composed of volcanic rock, coralline limestone, or alluvial deposits as direct consequence of specific oceanic, coastal, and tectonic processes.

WATER RESOURCES ON LOW-LYING AND MOUNTAINOUS TROPICAL ISLANDS

The tropical zone includes the contrasted cases of flat low-lying coral atolls and motu (the Tuamotus, the Bahamas, the Maldives) and steep mountainous volcanic islands (Hawaii, the Cape Verde archipelago, the Comoros). On coral islands, all the precipitation immediately percolates and gets stored inside the porous carbonate rocks. Pumping is the only way to get water to the surface. The soils and geological basement of volcanic islands also have high infiltration rates, but during large storms and cyclones the heavy rainfall produces rapid surface runoff on steep slopes, which generates devastating flash floods in the foothills and on coastal plains. In both island types, the use of surface water is limited if not impossible. Both coral and volcanic tropical islands are therefore mostly dependent on their groundwater reserves.

Exploiting this groundwater resource is particularly complicated in the island context. The first difficulty comes from rainfall representing the input variable into the island hydrologic system. The direct advection of humid tropical air masses over mountains produces very complex rain patterns and strong orographic effects that make precise monitoring and estimation of rainfall difficult at the island scale.

Another major drawback for coastal groundwater lies in the fragile equilibrium of the freshwater lens with the surrounding seawater. This body of water is also called a "Ghyben–Herzberg lens," after the two scientists who described how rainwater that percolates into the porous rock and sand floats on the saltwater beneath, depressing it into a profile in the shape of a lens. The form of the lens depends on the hydrostatic equilibrium between the density of saltwater (e_s) and freshwater (e_f):

$$H_b = H_a\, e_f / (e_s - e_f)$$

where H_b and H_a are respectively the depths of freshwater below and above sea level.

The difference of density between saltwater ($e_s = 1.025$) and freshwater ($e_f = 1.0$) is directly responsible for the thickness of the water table at a specific point. For example, if the top stands at 1 m above sea level, then the bottom would be 40 m below sea level. The main challenge is to ensure that pumping from this floating lens will not contaminate it with underlying brackish water. When excessive pumping creates a saltwater wedge, recovery is a slow process with high impact on the island's economy and people. Island groundwater is a fragile resource that is difficult to estimate and shows limited resilience.

AN EXAMPLE OF CONTRASTING ISLANDS WITHIN A TROPICAL ARCHIPELAGO

The archipelago of Guadeloupe in the French Caribbean may be taken as a case study of contrasting hydrological regimes in tropical islands (Fig. 1). On the eastern, windward side are several flat, low-lying islands made of limestone. Consequently, the annual rainfall is below 1500 mm/yr due to the lack of orographic uplift, and the drainage network is poor with no perennial rivers. These properties are typical of dry islands with their water resources mainly stored in underground lenses. Conversely, the western, downwind island is volcanic and features the large mountain of Basse-Terre, which is the major hydrological feature of the whole archipelago. The windward eastern slopes of Basse-Terre experience a drastic rainfall gradient from 2500 mm at sea level up to 11,000 mm at 1400 meters elevation, while the coast on the leeward side receives less than 1400 mm. Although a tropical island represents a small surface area, it usually presents extreme hydroclimatic contrasts.

OVERVIEW OF THE HYDROLOGY OF ISLANDS ACROSS ALL CLIMATIC ZONES

Outside the humid tropical zone, islands face the whole spectrum of hydrological situations depending on the climate. A simple hydroclimatic classification of all the world's islands larger than 0.1 km² is given in Table 1.

In polar regions, all precipitation occurs as snow, permafrost makes the soil impermeable, and no runoff occurs. At subpolar latitudes, melting snow and perma-

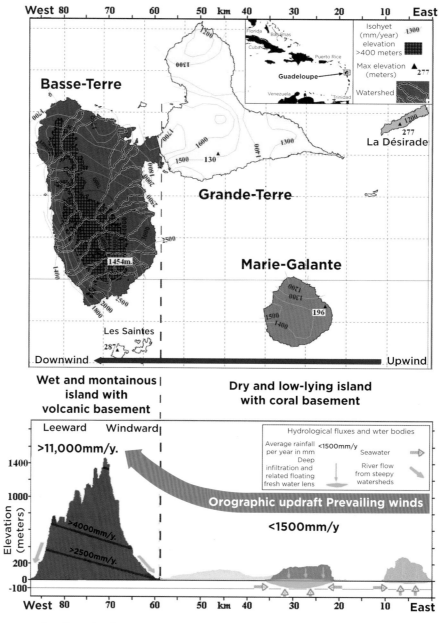

FIGURE 1 Map and cross-section illustrating the water resources and some hydrological processes on the Guadeloupe archipelago (French Caribbean).

frost during the short summer produce a little runoff and myriad pools and puddles on the surface. With cooler average temperatures, precipitation is a mixture of snow and rain. River flows persist through the summer with the melting of snow and ice from glaciers, as is the case on Iceland and Jan Mayen.

In temperate climates, islands may experience seasonal or occasional droughts that may significantly impact water resources and consequently affect or limit agricultural or tourist activities.

At tropical latitudes, islands in dry and hot conditions experience water shortages for economic or even domestic uses, as is the case for Socotra and many other islands in the Red Sea and the Persian Gulf. Scarce rainfall is caught with rooftop catchments and any other waterproof surfaces. Where funds permit, it is possible to use

TABLE 1
Hydroclimatic Classification of All Islands of the World Larger Than 0.1 km²

	Polar $T < -10\ °C$	Cold $-10\ °C < T < -0\ °C$	Cool $0\ °C < T < 10\ °C$	Temperate $10°\ C < T < 20°\ C$	Hot $T > 20\ °C$	Total
Dry $P < 400$ mm	Wrangel (RU) 6979	Spitsbergen (NO) 2366	Isla Porcia (CL) 108	Lanzarote (ES) 319	Socotra (YE) 2675	12447
Moderate 400 mm $< P < 1000$ mm	Peter the Great (AQ) 520	Disco (GL) 7459	Jan Mayen (NO) 10974	Sicily (IT) 2730	Bahamas (BS) 2773	24456
Sub-humid 1000 mm $< P < 1500$ mm	Balleny Islands (AQ) 78	South Sandwich (UK) 174	Iceland (IS) 7927	Tasmania (AU) 3241	Puerto-Rico (PR) 5287	16707
Humid 1500 mm $< P < 2000$ mm	White Island (AQ) 26	0	Kodiak (US) 4318	Honshu (JP) 1307	Hainan (CN) 5877	11528
Very humid $P > 2000$ mm	0	0	Vancouver (CA) 5796	Chiloe (CL) 1399	Java (ID) 13975	21170
Total	7603	9999	29123	8996	30587	86308

NOTE: The number of islands is given for each class of temperature (*T*) and precipitation (*P*). For each class, the name of an island is given as an example along with its ISO country code: AQ = Antarctica, AU = Australia, BS = Bahamas, CA = Canada, CL = Chile, CN = China, ES = Spain, GL = Greenland, ID = Indonesia, IS = Iceland, IT = Italy, JP = Japan, NO = Norway, RU = Russia, PR = Puerto-Rico, UK = United-Kingdom, US = United-States, YE = Yemen.

high-cost alternatives such as desalinization of seawater, piping or barging freshwater from the mainland.

THE LIMITATIONS PLACED ON ISLANDS BY WATER RESOURCES

The UNESCO report on *Hydrology and Water Resources of Small Islands* (1991) is a major milestone in the formal recognition of the specificity of islands in relation to water. The report concludes that the problem of water resources on islands is "usually very serious." The carrying capacity of an island is often determined by available water and is highly vulnerable to the forces of nature. Scientists have highlighted in the International Panel on Climate Change (IPCC) reports that expected climate change and sea level rise will have far-reaching consequences on coastal aquifers. For instance, the atoll groundwater of Kiribati is so reduced by El Niño–Southern Oscillation (ENSO)-related droughts that domestic water wells become too salty to drink. The case of the highly populated, low-lying islands of the Maldives (max elevation 2.3 m!) illustrates the complex problem of sanitation and the thin water lens. With 66,000 inhabitants living on 1.77 km², the islet of Male is exclusively urban, and, despite a great deal of effort by the authorities, preserving the shallow lens from pollutants and sewage is extremely difficult.

There is no doubt that islands more than elsewhere are limited in their development by hydrological issues, both in terms of freshwater quantity and quality. This is true both for insular microstates and for any other island ruled by or dependent on continental countries. The challenge of sustainable development in the insular context is therefore largely dependent on the capacity of local people to manage this crucial and fragile resource for the dual benefit of ecology and economy.

ISLANDS AS BELLWETHERS OF WATER ISSUES

A political leader has recently declared that "Islands are the bellwethers of international environmental policy" at the Global Islands Partnership meeting (GLISPA meeting, 2007). This statement is particularly true for water issues. It is emblematically exemplified in the film *The Naked Island* by Kaneto Shindo (1961, Japan). The action takes place on a small and steep island on which the everyday life of a poor peasant family focuses on collecting water from the neighboring mainland to water their garden on their dry patch of land. Far from the cliché of the Garden of Eden, with water portrayed as the film's heroine, it stresses the hardship of island conditions when hydrologic resources are limited.

SEE ALSO THE FOLLOWING ARTICLES

Climate on Islands / Freshwater Habitats / Maldives / Socotra / Sustainability

FURTHER READING

Baldacchino, G. 2007. *A world of islands: An island studies reader*. Malta: Agenda, 2007.

Falkland, A. C. 1991. *Hydrology and Water Resources of Small Islands: A Practical Guide*. IHP Studies and Reports in Hydrology No. 49. Paris: UNESCO.

Michel, J. A. 2007, Keynote address on the opening of the First Global Island Partnership Strategy Meeting. http://www.cbd.int/island/glispa.shtml.
Shindo, K. 1960. *Hadaka no shima*. (released in English [1962] as *The Naked Island*). BAFTA Awards from Moscow International Film Festival (1961) and BAFTA (1963).

HYDROTHERMAL VENTS

ROBERT C. VRIJENHOEK

Monterey Bay Aquarium Research Institute, Moss Landing, California

Deep-sea hydrothermal vents are submarine hot springs located along the global mid-ocean ridge system, in back-arc basins, and on volcanic seamounts. The vents support lush animal communities fueled by reduced sulfur compounds and methane, rather than sunlight. Chemosynthetic microbes are the primary producers at vents, and they, in turn, are grazed and filtered by a variety of animals or hosted as symbionts by others. Dependence on geochemical energy restricts vent-endemic animals to small island-like habitats scattered throughout the world's oceans. Studies of larval development and ocean circulation coupled with population genetic analyses have revealed a range of physical and biological factors that either facilitate or impede the dispersal of vent animals among these islands. Although some ancient taxa occur at vents, phylogenetic analyses of the dominant vent organisms suggest that these chemosynthetic deep-sea habitats are not stable refugia for living fossils. Instead, species turnover may not differ much from that found in other marine environments.

VENT ISLANDS

Since Darwin's epic voyage on the HMS *Beagle*, the Galápagos Islands have played a seminal role in the growth of our ideas about island biogeography and organic evolution. It is fitting, therefore, that deep-sea hydrothermal vents were discovered along the Galápagos Spreading Center. In 1977, geologists J. Corliss and R. Ballard reported dense aggregations of meter-long tubeworms, giant clams, mussels, crabs, and unusual eel-like fish thriving around submarine hot springs laden with toxic volcanic gases. Hydrothermal vents have since been found along most explored portions of the global mid-ocean ridge system, a 55,000-km-long mountain chain that circles the globe (Fig. 1). Vent communities also occur in back-arc spreading centers and on hydrothermally active seamounts. The discovery of chemosynthetic communities flourishing in darkness, extreme temperatures, and immense pressures of the deep sea changed our views about the limits to life on Earth and opened our minds to the potential for life in extraterrestrial environments.

Most vent animals depend entirely on chemosynthetic primary productivity, constraining them to live around submarine hot springs. Consequently, these inhabitants are sus-

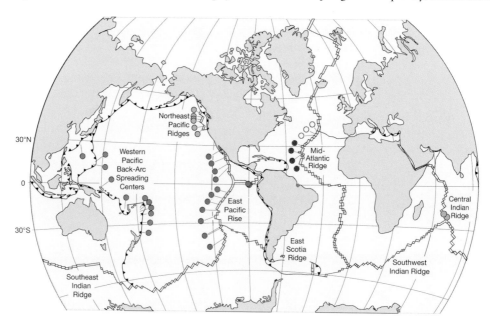

FIGURE 1 Global distribution of well known hydrothermal vent communities. Colors represent biogeographic provinces: dark blue, East Pacific Rise; green, northeast Pacific; pink, western Pacific; red, Mid-Atlantic Ridge; yellow, Azores Plateau; orange, Central Indian Ridge. Spreading centers are shown with double lines, and areas of subduction are marked with arrowheads that point in the direction of subduction. The apparent absence of vents at higher northern and southern latitudes reflects limited exploration in these remote regions.

ceptible to sporadic volcanic eruptions and tectonic events that extinguish old vents and create new ones. The spacing between active vents and the tempo of habitat turnover are linked to the spreading rates of ridge segments. Large transform faults with extended fracture zones occur at frequent intervals along slow-spreading ridge systems like the Mid-Atlantic Ridge. Without active hydrothermal venting, these large offsets are expected to disrupt animal dispersal along a ridge system, particularly if propagules are transported along rift valleys formed between mountainous walls of the ridge axis (Fig. 2). On the other hand, buoyant larvae produced by some vent animals may disperse in hydrothermal plumes that rise above the axial walls. Occasional megaplumes associated with volcanic events are hypothesized to aid long-distance dispersal of these organisms, and the frequency of such events is greater along fast-spreading axes.

Discrete vent fields are typically composed of multiple chimneys and mineralized structures that emit waters as hot as 400 °C. These focused hot vents are often flanked with diffuse flows at much lower temperatures. A vent field can stretch for a few hundred meters along the ridge axis and persist for a few decades before the subterranean plumbing is clogged or the field is obliterated by a lava flow. Along a medium-rate spreading center like the Endeavour Segment of the Juan de Fuca Ridge (Fig. 2A), gaps between vent fields may be a few kilometers long, and gaps between adjacent segments may extend from tens to hundreds of kilometers (Fig. 2B). Lateral offsets like the Blanco Transform Fault displace contiguous ridge segments, disrupting along-axis water currents, and consequently impeding animal dispersal. Intervening seamounts, oceanic microplates, and inflated bathymetry can also displace ridge segments and disrupt currents that contribute to along-axis dispersal.

DISPERSAL BARRIERS

Depending on their unique life histories and modes of dispersal, various vent species are differentially affected by disruptions in the ridge system. Most of the annelids and mollusks are sessile or sedentary as adults and disperse as larvae. The mussel *Bathymodiolus thermophilus* produces numerous small eggs (~50 μm) that hatch into swimming planktotrophic larvae, but the altitude they achieve in the water column is not known. In contrast, the palm worm *Alvinella pompejana* produces large eggs (~200 μm) that hatch into lecithotrophic larvae, which are capable of arresting development in cold abyssal waters and continuing development upon encountering warm waters. The giant tubeworm *Riftia pachyptila* produces intermediate sized eggs (80–100 μm) with sufficient yolk to live for about one month in cold

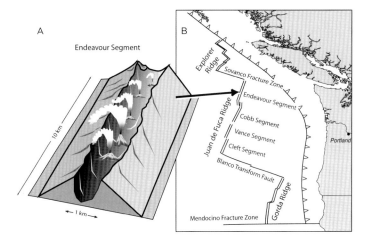

FIGURE 2 Hierarchical structure of vent fields in the northeastern Pacific. (A) Portion of the Endeavour segment (modified from Thomson et al., 2003) has a series of vent fields portrayed as chimneys. (B) Contiguous segments along the Juan de Fuca Ridge are separated by transform faults from the Explorer Ridge and Gorda Ridge. These northeastern Pacific ridge systems are separated from the East Pacific Rise (Fig. 1) by subduction below the North American Plate.

abyssal waters and disperse about 100 km. Vent shrimp and crabs are mobile as adults, but actively swimming larvae and juveniles may feed and grow in the photic zone where they can potentially disperse great distances.

Biological and geographical limits to dispersal are often reflected in patterns of gene flow and population subdivision of a species. For example, large ridge offsets such as the 450-km-long Blanco Transform Fault (Fig. 2) can restrict dispersal between the flanking ridge axes. This fault effectively isolates sister species of limpets *Lepetodrilus fucensis* and *L. gordensis,* which live respectively on the Juan de Fuca and Gorda ridge systems. Northern and southern populations of the tubeworm *Ridgeia piscesae* are not similarly isolated across this region, but gene flow is about six times greater in a southerly direction than in the reverse. Apparently, deep-ocean currents that flow in a southeasterly direction along Blanco Transform Fault force this unidirectional pattern of dispersal. Genetic studies of vent-endemic animals living along the East Pacific Rise and Gálapagos Rift also reveal physical oceanographic barriers to dispersal in some species, complete isolation of others, and no discernable effects on yet others. Various species with differing larval life histories (e.g., planktotrophy vs. lecithotrophy) and buoyancy will spend more or less time and rise to differing heights in the water column. These factors will expose larvae to different current patterns that might facilitate the retention of some species within the axial valleys and subject other species to cross-axis currents that sweep them away from the ridge.

BIOGEOGRAPHIC REGIONS

Hydrothermal vent fauna can be subdivided into six biogeographic provinces associated with various ocean basins and discontinuities in the mid-ocean ridge system (Fig. 1). The northeastern and southeastern Pacific ridge systems (the Explorer, Juan de Fuca, and Gorda ridges versus the East Pacific Rise) were once connected, but subduction under the North American plate eliminated the interconnecting Farallon Ridge about 25 million years ago. Today, these northeastern and southeastern Pacific ridge systems support sister-species pairs of annelids and gastropods that diverged following this vicariant event. Closure of the Isthmus of Panama severed connections between eastern Pacific and Atlantic vent fauna (~10 million years ago), and closure of the Tethys Sea severed connections between Atlantic and Indian Ocean vent fauna (~60 million years ago). Elevation of the Azores Plateau (~20 million years ago) may have separated northern and southern fauna along the Mid-Atlantic Ridge. The relevant timing of these events will likely vary among taxa, as various species differ in the altitudes they achieve when dispersing. Thus, some species, such as vent shrimp, may transcend these boundaries among these provinces, whereas others will not. Chemosynthetic fauna are also found at hydrocarbon seeps and at sites of organic deposition (e.g., fjords, submarine canyons, wood and whale falls). Some researchers hypothesize that these habitats may serve as stepping-stones for the modern-day dispersal of some animals between widely separated vent systems (C. Smith, this volume).

EVOLUTION OF VENT FAUNA

Deep-sea hydrothermal vents were once thought to be refuges for ancient relics, unaffected by catastrophic events that led to global mass extinctions in shallow marine environments at the close of the Paleozoic and Mesozoic eras. Nearly 500 new species of vent-endemic animals have been described since the discovery of vents in 1977. Certain stalked barnacles appear to have occupied vent habitats since the early to middle Mesozoic and are considered living fossils. The great diversity of new genera, families, and putatively new phyla was hypothesized to be a consequence of the stability and longevity of these chemosynthetic environments. Nonetheless, molecular phylogenetic studies reveal that some of the common families of vent invertebrates, such as bresiliid shrimp and bathymodiolin mussels, are products of recent evolutionary radiations following the K-T mass extinction (less than 65 million years ago). Molecular and fossil evidence suggests that chemosynthetic clams and vestimentiferan tubeworms may be somewhat older, having radiated during the late Cretaceous (less than 90 million years ago). Some primitive gastropod families may be older yet (~100 million years), but they are not exclusive to chemosynthetic environments. Ancient deposits from Silurian (~400-million-year-old) hydrothermal vents contain fossils of molluscan groups and brachiopods that are not found in modern vent communities. Thus, deep-sea vents, too, suffered from mass extinction events that revised animal diversity in the photic zone at the close of the Paleozoic.

Although hydrothermal vents have existed since the early eras of our planet, some modern vent taxa may have first diversified in sulfidic cold seeps or in organic deposits, and then subsequently radiated at vents. Molecular evidence suggests that the oldest evolutionary lineages of vestimentiferan tubeworms diversified first in seeps. Vesicomyid clams also radiated first in seeps and then generated a few species capable of exploiting hydrothermal vents. In contrast, bresiliid shrimp and alvinellid polychaetes appear to have radiated recently in vent environments. It remains uncertain whether the most ancient lineages of bathymodiolin mussels first occupied cold seeps, wood falls, or hydrothermal seamounts. All these scenarios are equally probable based on molecular phylogenetic analyses, but the mussels have invaded progressively deeper habitats and have diversified in their use of sulphur- and methane-oxidizing endosymbionts.

THE FUTURE FOR VENTS

Deep-sea hydrothermal vents are not eternally stable refuges, immune to forces affecting Earth's surface. All the animals found at vents are anaerobes, making their livings by eating microbes or hosting them as symbionts. In either case, these primary producers and their hosts must live in a narrow redox zone that provides reduced volcanic gases on one side and oxygen on the other. Oxygen is a byproduct of photosynthesis at the planet's surface, and it is delivered to ocean depths by circulation patterns driven by the heating of surface waters near the equator and cooling near the poles. Periods of intense global warming during the late Cretaceous (~95 million years ago) and associated releases of greenhouse gases during the early Tertiary (~55 million years ago) altered deep-ocean circulation and created anoxic basins that led to extinctions. Vent animals, as well, should be vulnerable to events that increase the boundaries of narrow redox zones that support them. Present-day accumulations of greenhouse gases and global warming, although small compared to catastrophic events during the late Cretaceous, may nevertheless threaten deep-ocean circulation regionally and thereby affect vents similarly.

Mining poses another threat to vent communities. The Papua New Guinea government has granted commercial leases to mine high-grade gold and copper deposits from extensive sulfide mounds at Manus Basin vents. The animals living on these mounds will surely be disturbed, but it not known whether such local disturbances will have greater impacts than the submarine volcanic events that regularly disrupt this region. Finally, scientific visits to vents with manned and robotic submersibles may carry hitchhiking animals, microbes, and potentially even diseases between vents. Thus, the same anthropogenic factors that affect surface islands worldwide (exploitation, habitat destruction, invasive species, and diseases) will also affect deep-sea hydrothermal events. Consequently, two nations have created deep-sea marine protected areas (MPAs). The Lucky Strike and Menez Gwen vent fields were designated the first deep-sea MPAs by the Azores government in June 2002. Canada designated the Endeavour vent field as an MPA in March 2003. Other vent fields that lie in national waters may eventually gain similar status, but the protection of vents that lie outside of exclusive economic zones will require international cooperation.

SEE ALSO THE FOLLOWING ARTICLES

Cold Seeps / Dispersal / Refugia / Vicariance / Whale Falls

FURTHER READING

Corliss, J. B., and R. D. Ballard. 1977. Oasis of life in the cold abyss. *National Geographic Magazine* 152: 441–453.

Little, C. T. S., and R. C. Vrijenhoek. 2003. Are hydrothermal vent animals living fossils? *Trends in Ecology and Evolution* 18: 582–588.

Thomson, R. E., S. F. Mihály, A. B. Rabinovich, R. E. McDuff, S. R. Veirs, and F. R. Stahr. 2003. Constrained circulation at Endeavour Ridge facilitates colonization by vent larvae. *Nature* 24: 545–549.

Tunnicliffe, V., and M. R. Fowler. 1996. Influence of sea-floor spreading on the global hydrothermal vent fauna. *Nature* 379: 531–533.

Tyler, P. A., and C. M. Young. 1999. Reproduction and dispersal at vents and cold seeps. *Journal of the Marine Biological Association of the United Kingdom* 79: 193–208.

Van Dover, C. L. 2000. *The ecology of deep-sea hydrothermal vents.* Princeton, NJ: Princeton University Press.

Van Dover, C. L., C. R. German, K. G. Speer, L. M. Parson, and R. C. Vrijenhoek. 2002. Evolution and biogeography of deep-sea vent and seep invertebrates. *Science* 295: 1253–1257.

ICELAND

SIGURDUR STEINTHORSSON
University of Iceland, Reykjavik

Iceland is the largest volcanic island (103,000 km²) in the Atlantic Ocean. It is one of the most volcanically active places on Earth, with more than 20 eruptions per century, and owing to its northerly location in the middle of the sea, the forces of erosion are very active as well. For these reasons, Iceland is a "natural laboratory" in which a continuous tug-of-war exists between constructive and destructive processes that can be studied in real time as well as in the geological record.

TECTONICS

Tectonic Setting

In terms of global tectonics, Iceland is a hotspot located near a constructive plate boundary. It is the largest mass of land found on the Mid-Atlantic Ridge, and the transverse ridge crossing Iceland from Greenland to Scotland is among the most substantial of aseismic ridges in the oceans (Fig. 1). A progression in age exists from the active central rift of Iceland to the Tertiary basalts of eastern Greenland (65 million years old) and Britain (45–50 million years old). The high-rising Iceland Plateau reflects the buoyancy of a more-than-400-km-deep plume of ascending hot upper mantle presently centered below southeastern Iceland, and the transverse ridge, with over 20-km thickness of basaltic crust, is the passive trail of the hotspot, which has been active since before the opening of the North Atlantic. The V-shaped ridges southwest of Iceland find their explanation in magma pulses travelling ~20 cm/year southward along the ridge.

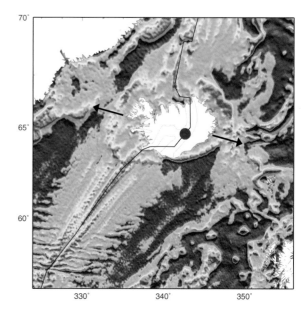

FIGURE 1 Satellite gravity map of the North Atlantic region reflects the submarine topography (Smith and Sandwell, 1995). The black line is the active plate boundary, and the red dot is the assumed center of the Iceland mantle plume. Vectors show the plate movements relative to the spreading center, 1 cm/a in each direction, striking 104°. (Modified by I. Þ. Bjarnason.)

Tectonic Evolution

The Mid-Atlantic Ridge crosses Iceland from southwest to northeast, at which latitude the rate and direction of spreading is 1.95 cm/year and N 104° E, respectively. Relative to the adjacent oceanic Reykjanes and Kolbeinsey Ridges, the plate boundary in Iceland is shifted some 100 km to the east by a set of transform faults: the Tjörnes Fracture Zone (TFZ) in the north and the South Iceland Seismic Zone (SISZ) in the south (Fig. 2). During the geological history of Iceland, the present configuration of volcanic zones has come about in a series of ridge jumps and rift propagations, caused by the gradual west-northwest drift

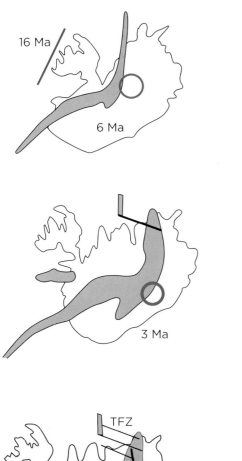

FIGURE 2 Summary of tectonic evolution with reference to the Iceland mantle plume (red circle). About 16 million years ago, the spreading center jumped from west of present Iceland toward the east. The rift zone as shown in the top diagram was active until about 6 million years ago when it again jumped east (center). The Southern Volcanic Zone (SVZ) has been propagating southwestward for the past 3 million years. Present configuration (bottom): SNA: Snæfellsnes; ORA: Öræfajökull; SISZ: South Iceland Seismic Zone; TFZ: Tjörnes Fracture Zone. The different colors denote petrological affinity.

of the North Atlantic plate system relative to the Iceland mantle plume. The three flank zones termed, respectively, the Snæfellsnes, Southern, and Öræfajökull volcanic zones, are all distinguished by volcanics of alkalic tendency resting unconformably upon much older basement, as opposed to the tholeiitic rocks of the rift zones. Each of the three has its own distinct tectonic origin. The Snæfellsnes Zone is a volcanic outlier in the North American plate, representing a dying remnant of an earlier configuration. Prior to the relocation of the rift system 6 million years ago, it corresponded to the Mid-Iceland volcanic zone. Conversely, in southern Iceland, the rift zone is propagating southwestward into a 10-million-year-old plate; the Surtsey eruption of 1963–1967 represents the southernmost site of activity yet. Finally, the Öræfajökull zone may represent the opening up of old crust above the mantle plume.

Kinematics and Crustal Structure

Present-day crustal movements in Iceland are monitored by a range of measurements, including seismics, satellite radar interferometry (SAR), measurements of ground tilt, and measurements of strain in boreholes. A network of seismometers has delineated the plate boundary, relative movement on active faults, and the fact that spreading is sporadic (episodic) along the plate boundary. Tilt measurements (and, more recently, SAR) show that the Icelandic crust is extremely labile, responding almost instantaneously to changes in glacial loading. Recently, the monitoring effort has shown some success in short-term prediction of earthquakes and volcanic eruptions.

Seismics indicate that the basaltic crust in eastern Iceland is 27–33 km thick, and in western Iceland 20–26 km. The upper crust, down to 6–9 km, is characterized by rapid linear increase in P-wave velocity with depth, whereas in the lower crust the velocity changes little with depth. The reason for the difference in thickness may be the west-northwest drift of the plate system, such that the eastern plate is all but stationary above the mantle plume whereas the western plate moves almost 2 cm/year relative to it. Clearly, and in view of the structure of ophiolites, the uppermost crust is composed mostly of subaerial lava flows; dike density increases with depth, and at a certain level gabbros dominate.

Perhaps the most exceptional feature of Iceland's geology, as compared with "classical geology," is embodied in a kinematic model of crustal accretion (Fig. 3). In contrast with the customary idea of horizontal layers accumulating one upon the other—in the fashion of a layer cake—the accumulation of Iceland's basalt succession takes place more in the fashion of a conveyor belt: The source is almost exclusively in the rift zone, from where the formations drift away as new rocks are formed. Both the tilt of the succession toward the rift zone and the seismic layering of the crust are formed in the rift zone as an integral part of the accreting process. Regional synclines in Iceland, therefore, dip toward the rift zone from which they originated, whereas regional anticlines result from rift jumps.

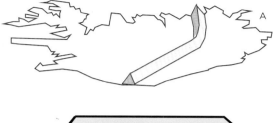

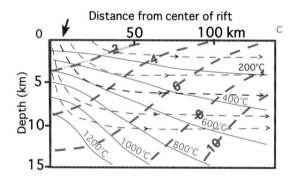
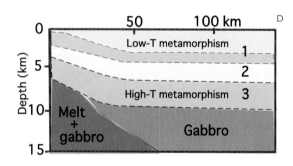

FIGURE 3 Kinematics of crustal accretion. (A) If the 43,000 km³ of volcanics that erupt in Iceland in 1 million years simply piled up along the rift zone, a 300-km-long ridge would result, 20 km wide at the base and 13.6 km high—an unrealistic scenario because the surface remains at constant elevation while the ground sags below new formations. (B) Schematic block diagram of northeastern Iceland showing volcanics formed in the last 2 million years (dark gray) and in the 2 million years prior to that (lighter gray). Blue curves are isochrones showing the surface 2 million years ago, 4 million years ago, and so forth. Black lines are dikes. (C) Kinematic model of the upper crust, quantifying the section in (B): Isochrones (surface at 2 million years ago, four million years ago, etc.), blue; calculated isotherms, red; material trajectories (describing movement of rock erupted in the rift zone), black. Following one such trajectory (arrow): after sinking and heating up for 2 million years, the rock is 400 °C. A maximum of 600 °C is reached in 3 million years, at which time the rock starts cooling as it drifts out of the rift zone (400 °C at 6 million years). (D) As the rocks heat up, largely irreversible recrystallization (progressive metamorphism) takes place. Upon cooling, the mineral assemblage of the maximum temperature is retained, resulting in horizontal metamorphic layering. Rocks forming the upper crust (seismic layers 1, 2, and 3) are volcanic, and the gabbros of the lower crust plutonic.

boundary, is frequently the site of a volcanic center, characterized by evolved rock compositions in addition to basalts and by a high-temperature geothermal system. The volcanic center together with the fissure swarm transecting it form a comagmatic entity termed a volcanic system. The relationship between the plate boundary and volcanic systems has, in southwestern Iceland, been shown seismically: Earthquakes defining the plate boundary below about a 3-km depth cluster along a vertical plane striking N 80° E, whereas shallow earthquakes originate along the strike of the fissure swarms, at about N 40° E. A similar relationship exists for the northeastern branch of the rift.

STRATIGRAPHY AND GEOLOGICAL HISTORY

Overview

The predominantly volcanic rocks of Iceland, which range back about 16 million years in age, are conventionally divided into four stratigraphic formations or series. This division is based on climatic evidence from interlava sediment or volcanic breccias and on paleomagnetic reversal patterns supported by radiometric age datings. The categories are as follows:

Postglacial (Holocene) lavas and detrital rocks of the last 9000–13,000 years. This formation (25,000 km²) occupies, in addition to the regolith, the presently active volcanic areas of Iceland including the median rift zone—the subaerial Mid-Atlantic ridge—traversing the country from southwest to northeast.

Upper Pleistocene "palagonite formation," dating back 0.7 million years, corresponding to the present

The model also explains the decrease in dip and lava thickness with stratigraphic height, the quasi-horizontal zoning of amygdules, and the seismic layering in the upper crust.

Volcanic Systems

Crustal spreading and volcanism in the axial rift zone takes place on discrete fissure swarms straddling the plate boundary in an en-echelon array (Fig. 4). The swarms may be 10–15 km wide and up to 100 km long. They are characterized by extensional tectonic features such as open fissures, graben structures, and crater rows at the surface, and dikes and normal faults at deeper levels. The central, most active part of each system, overlying the plate normal geomagnetic epoch, Brunhes. This formation coincides spatially more or less with the postglacial formations.

Plio-Pleistocene "gray basalt formation," dating back 0.7–3.1 million years ago and including the Matuyama epoch and the Gauss epoch down to the Mammoth event. These rocks occupy a broad belt (25,000 km^2) on either side of the median rift zone.

Tertiary "plateau basalts," older than 3.1 million years. This series covers about half of Iceland (50,000 km^2) in the east, west, and north.

Tertiary formation

The Tertiary plateau basalts, well exposed due to glacial erosion, are made up of regular sequences of flat-lying subaerial lava flows, 5–15 m thick and separated by minor clastic interbeds of volcanic origin. The sequences dip toward the respective rift zone in which they formed; where rift jumps have occurred, the resulting unconformity lies at the crest of an anticline where the beds dip toward the old and the new rift, respectively. The lava pile is composed of elongated lenses, each representing the products of a volcanic system. Along each lens stretches a dike swarm, in the core of which a volcanic center may have developed, characterized by evolved rocks, intrusions of gabbro and granophyre, and hydrothermal alteration.

Mapping the apparently monotonous Tertiary flood basalts calls for special techniques for correlation and age determinations. In the last half century, several "mapping campaigns" employing these new methods have been launched in all Tertiary regions in Iceland. The longest continuous succession studied so far is 8.5 km thick, spanning some 10 million years and encompassing 700 lava flows. The build-up rate varied between 360 and 2600 m per million years, whereas, on average, the eruption rate was one flow per 14,000 years—the common range in the Icelandic Tertiary is 6000–10,000 years between adjacent flows. This means that only exceptionally large lava flows, having spread far out of the volcanic zone, are represented in the exposed sections.

In the flood basalts, three lithologic types can be distinguished in the field for stratigraphic mapping: olivine tholeiite, tholeiite, and feldspar-porphyritic basalt. Apart from the phenocryst content, different appearance of weathered surfaces and diagnostic mineral assemblages in amygdules enable the distinction between rock types. Regional-scale paleomagnetic mapping of lava groups is standard in stratigraphic correlations of basalt successions. The first radiometric datings (1968) of Tertiary samples

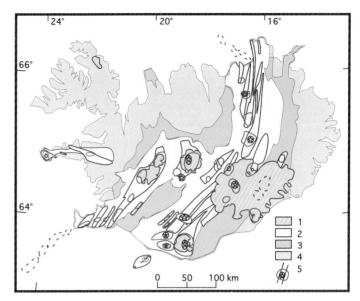

FIGURE 4 Simplified geological map of Iceland showing bedrock and glaciers. (1) ice sheet; (2) Upper Pleistocene and Holocene; (3) Plio-Pleistocene; (4) Tertiary; (5) volcanic system (yellow) with a volcanic center and a caldera.

from Iceland showed the oldest exposed rocks to be about 16 million years old, in sharp contrast to the ~60-million-year-old age assumed on the basis of Iceland being part of an old east-west igneous province extending from eastern Greenland to Britain. Since then, several hundred such age determinations exist of Tertiary basalt successions as well as of silicic volcanic centers within the lava pile. Sedimentary horizons of terrestrial origin, lacustrine or fluvial, occur interspersed with lava flows in the Tertiary basalt succession. These include beds of lignite, which yield a more or less continuous paleobotanical record ranging back 16 million years. By comparison with modern analogues, the mean annual temperature in the North Atlantic region was about 5 °C warmer than now (4–6 °C) some 15 million years ago. A maximum was reached around 12 million years ago (13.5 ± 1 °C, similar to the present southeastern United States) with gradual cooling to 7.4 ± 2 °C at 6 million years ago. The first signs of glaciation appear at about 3.1 million years ago.

Plio-Pleistocene Formation

The boundary between the Tertiary and Plio-Pleistocene is somewhat arbitrarily fixed at the base of the Mammoth geomagnetic event, 3.1 million years ago. A continuously cooling climate in the upper Tertiary led to the appearance of the first tillite horizons interstratified with basalt at about this time. The sedimentary interbeds become thicker and coarser than before; subglacial pillow lavas and hyaloclastites interchanged with thick subaerial lava flows. At least ten glacial/interglacial cycles are known from the Upper Pliocene, as well as at least that many from the Pleistocene. The most complete single section from this period is found on the Tjörnes Peninsula in northern Iceland, where the Plio-Pleistocene sequence is 600 m thick, 400 m of which are shell-bearing sediment of mainly marine and estuarine facies, but in the upper half, also of glacial origin, at least six tillite horizons have been identified. With the onset of glaciation, the topography of Iceland started undergoing radical change from the flat-lying Tertiary lava plain: Elongated ridges and isolated mounds of hyaloclastite piled up in subglacial eruptions, whereas the glaciers sculpted out the fjord landscape now typifying the Tertiary.

Upper Pleistocene Formation

This series encompasses the Brunhes geomagnetic epoch—the last 700,000 years, up to the Holocene. In space it coincides essentially with the neovolcanic zones where its volcanics form two distinct facies: subglacial and subaerial. Within volcanic zones, accumulation proceeded at a much higher rate than did denudation, but in some places groups of lava flows can be seen alternating with tillite beds and hyaloclastite rocks giving information about several interglacial/glacial cycles. Outside the volcanic zones, glacial erosion became very effective. In western Iceland, 800–1000-m-deep valleys were carved out in less than 1.8 million years, and on the Snæfellsnes Peninsula this took place in less than 1 million years.

SUBGLACIAL VOLCANICS

The topography most typical for the Upper Pleistocene are ridges and mounds of hyaloclastite, often with a core of pillow basalts. These are formed in subglacial eruptions, the ridges on long volcanic fissures, the mounds on short fissures or circular craters. Two eruptions of this kind have been followed closely by scientists in Iceland: the submarine Surtsey eruption of 1963–1967 and the subglacial 1996 Gjalp eruption. With decreasing water pressure as the crater builds up below water or ice, pillow lava forms first, followed by tuff, and finally by lava flows as the crater becomes subaerial. By no means will all volcanoes show all three facies: pillow lava at the base and/or subaerial lava on top are often missing, signifying, respectively, that the overlying glacier was too thin or that the eruption failed to melt the overlying ice. At Surtsey it emerged that palagonitiztion—the lithification process transforming the pile of hyaloclastite tuff into rock—takes place between 80 and 150 °C and is completed shortly after the eruption.

Many examples are known of intermediate or silicic volcanics having erupted beneath the Pleistocene ice sheets. The silicic ones are characteristically steep-sided domes consisting of distinctive rock varieties. Among these are lobes of glassy rhyolite encrusted with a thick layer of pitchstone resembling huge pillows. The lobes are embedded in a granulated glass matrix evidently derived from the crust of the lobes themselves.

INTERGLACIAL VOLCANICS

During the interglacials, lava flows and tephra issued from the volcanoes. The lavas tend to have a different aspect from those of the Tertiary, being gray as opposed to black in color and often doleritic. The slaggy surfaces have been scraped off by glaciers, exposing the coarser-grained interiors.

Holocene

Evidence of the Pleistocene-Holocene transition is seen in raised beaches (up to 100 m, but generally 40–60 m), erosional platforms at –30 m, moraines signifying oscillatory retreat of the glaciers, sediments in lakes yielding a continuous record back to 13,000 years ago, and profuse and singular volcanism. The final glacial advance (Younger Dryas) at 10,000 years ago followed an interstadial (Alleröd) at 11,000. The rapid eustatic rise of sea level caused the lowlands to be inundated, but because of the low viscosity of the underlying mantle (10^{18} poise), the land rose isostatically in the span of 500–1000 years, with the formation of erosional platforms at about 30 m below present sea level. The isostatic recovery caused rapid release of pressure in the upper mantle, resulting in very intense volcanism, probably more than 30 times that of present day. Two types of volcanoes appear to be confined to this time (~13,000 to 6000 years ago): small picritic lava shields and large (up to 50 km³) shields of olivine tholeiite. The table mountains (tuyas) are probably subglacial equivalents of the latter, formed near the end of the glaciation.

Postglacial formations comprise lava flows and pyroclastics, loess, unconsolidated marine clays, fluvioglacial and fluvial outwash and soil formed after the deglaciation of the land area.

VOLCANICS

Postglacial volcanism has been confined to the same areas as that of the Upper Pleistocene, with 30 active volcanic systems in the rift zones and the three off-rift zones. Total production in the Holocene is estimated at 400–500 km³, with postglacial lavas covering some 12,000 km². Basaltic lavas predominate (~90%), but the high proportion of intermediate and silicic rocks is unusual for the oceanic rift system—the evolved rocks are confined to volcanic centers. Almost all types of volcanoes known on Earth are found in Iceland (see Volcanoes, this volume).

REGOLITH

The regolith in Iceland is mainly the till of the Weichselian glacial, in addition to fluvioglacial outwash. The latter covers about 5000 km², and much of the southern coast is sandur plains, up to 200 m thick, formed largely by glacial floods (Icelandic: *jökulhlaup*): sediment-charged meltwater from subglacial eruptions. The "organic soil" is mostly peat, rich in mineral matter (often 40–60%) because of frequent tephra falls and deposition of eolian material.

Tephrochronology Numerous eruptions have left tephra layers in the soil, some of which cover most of Iceland (and are found on the sea floor, in northwestern Europe, and in the Greenland ice sheet). These layers, particularly the silicic (light-colored) ones, make useful marker horizons in the soil or sediment, as well as being records of the eruption history of the various volcanoes. Dr. S. Thorarinsson, founder of this branch of geology, worked out the eruption history of Hekla and some other volcanoes, and with improved analytical techniques, most basaltic (black) layers can now be relegated to their respective volcanic system of origin.

Climatic Variations Climatic variations in the Holocene can be traced from biota (including pollen and diatoms) in bogs and lake deposits, and from types of sediment. This work is greatly facilitated by tephrochronological marker horizons. In the early Holocene, 10,000–9000 years ago, *Betula* (birch) appeared in northern Iceland, from which it spread rapidly to the south 9000 years ago, possibly indicating that birch survived the glaciation in some nunataks. This first *Betula* maximum in pollen diagrams corresponds to the Boreal and Lower Atlantic in continental Europe, with annual temperatures some 2 °C higher than present and with lower precipitation. A *Betula* minimum, corresponding to the wet (but still warm) Atlantic, started at about 6500 years ago with bogs replacing the birch forests. During the second *Betula* maximum at 5000–2500 years ago (sub-Boreal), brushes covered half the country, with temperatures 2–3 °C higher than now, low precipitation, and mild winters. At about 2500 years ago, the climate deteriorated, and the birch gave way to bogs. A third *Betula* maximum, corresponding to the Neo-Atlantic, beginning about 1650 years ago, was interrupted by the settlement of Iceland around AD 870, followed by the "Little Ice Age" between AD 1400 and 1900.

PETROLOGY AND CHEMISTRY

Igneous rocks are classified on basis of chemistry (basaltic-intermediate-silicic) and crystallinity (rate of cooling: e.g., glassy hyaloclastite–microcrystalline basaltic lava–coarse-grained gabbro). Hyaloclastites are particularly common in Iceland as a result of subglacial volcanism. Plutonic rocks, gabbro, and granophyre are found in eroded central volcanoes and as xenoliths in volcanic rocks. Increasing alteration and metamorphism with depth is seen in the amygdule-zoning in eastern Iceland and elsewhere (Figs. 3–4), and similar sequences obtain in drill holes into geothermal systems and around intrusions in central volcanoes.

The proportion of evolved rocks in the Tertiary, predominantly rhyolites and dacites (9%), is surprisingly high in Iceland considering its tectonic setting at a spreading center. In this respect Iceland is entirely different from, for instance, the Hawaiian islands, where such rocks hardly occur at all. Intermediate rocks are less common (3%; basalt 80–85%) than silicic, and a Daly gap is frequently present in individual volcanoes. In a typical Tertiary section, volcanogenic sediments amount to some 5–10%; in the Quaternary areas, however, they form a much greater part of the succession owing to the influence of glaciation on both volcanism and denudation. The silicic rocks are confined to volcanic centers, of which about 65 are known in the Tertiary formation and around 30 in the Quaternary and Holocene areas. Investigations have shown that (1) the rift zones give rise to tholeiitic basalts and their derivatives; (2) the off-rift Snæfellsnes, Southern, and Öræfajökull volcanic zones produce alkaline to transitional-alkaline rocks; (3) there is a systematic variation along the rift zone and the adjacent oceanic ridges in various petrochemical properties, including the range of basalt compositions; (4) the volcanic productivity varies systematically along the rift zone, with a maximum in central Iceland (Fig. 5); and (5) each volcanic system shows a range in compositions, from "primitive oceanic tholeiite" to quartz tholeiite, and even to silicic and minor intermediate rocks. Isotope geochemistry indicates that anatexis of hydrated basalt plays an important role in magmatic evolution in rift-zone volcanoes, whereas in the off-rift areas, crystal fractionation is the dominant process. The compositional variation *along* the rift zones has by different authors been ascribed to heterogeneities in

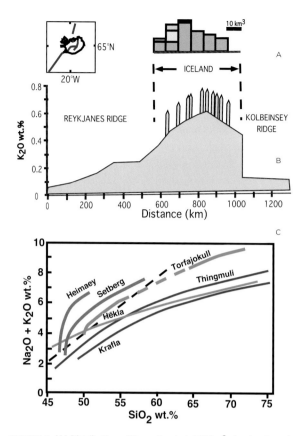

FIGURE 5 (A) Distribution of the estimated 420 km³ of volcanic rocks erupted in Iceland in the Holocene. Dark blue, tholeiites; light blue, alkalic and transitional basalts; yellow, evolved rocks. (B) K₂O-concentration in tholeiitic basalts along the rift zone of the Iceland region (red trace in inset). Note the increase in compositional range toward central Iceland. Yellow arrows are volcanic centers with more evolved rocks. (C) TAS diagram (total alkalis vs. silica) for six volcanic centers in Iceland. Blue, tholeiitic; green, transitional; red, alkalic. Dashed black is the "Hawaii Division Line" separating alkalic and tholeiitic compositions. (Redrawn based on S. P. Jakobsson, in Saemundsson 1979).

the upper-mantle source, to crystal fractionation, or to anatexis and magma mixing in the crust. It now appears that all processes are at work.

VOLCANOES

Three main types of volcanoes have been active in the Holocene: lava shields (shield volcanoes), crater rows, and central volcanoes. Lava shields are typical for the rift-zone volcanism in the early Holocene (before 6000 years ago). Their age, and their similarity to the subglacial table mountains both with regard to petrology and size, suggests that both are related to the isostatic rebound at the close of the Weichselian glaciation about 10,000 years ago. Similar structures exist from earlier glacials and interglacials. The Holocene shields, about 30 in number, are large and small; the apparent volume of the largest ones, Skjaldbreidur and Trölladyngja, is 15–20 km³, but gravity measurements indicate hidden lava flows at depth, increasing their volumes up to 40 km³.

A volcanic center, together with the fissure swarm transecting it, constitutes a volcanic system. Some 24 such volcanic systems have been active in the Holocene, 18 since Iceland's settlement around AD 870. During that time, between 30 and 40 sites have erupted, and during the last 300 years an eruption has started on the average every fifth year.

Based on geological mapping of extinct volcanic centers, it is estimated that their lifespan may be 0.3 to more than 1 million years. The various stages of their evolution can be seen in Iceland, from the "primitive" fissure swarms on the westernmost Reykjanes Peninsula to "mature" centers like Krafla or Askja. The relationship, shown in Fig. 4, indicates that magma enters the upper crust at the intersections between the fissure swarms and the plate boundary. At the depth where isostatic equilibrium is reached, magma pools to gradually form a magma chamber. During spreading episodes, magma empties from the magma chamber into the adjoining fissure swarm, as seen in the Krafla Fires of 1974–1985.

As the rift-zone volcanic centers develop, silicic magma starts forming through the remelting of hydrated basalt. The magma chamber, as it evolves chemically, rises in the crust, and finally the roof caves in, forming a caldera. When the crust is relatively thin, as in the rift zones, it is unable to support lofty edifices, and consequently the volcanic centers never take the form of "classical" central volcanoes, unlike, for example, Snæfellsjökull (1446 m) and Öræfajökull (2111 m), which stand on thick off-rift crust. The largest eruptions in historical time were Eldgja in AD 934 (30-km-long fissure) and Laki in 1783 (25-km-long crater row), each producing 15–20 km³ of basalt. The Laki Fires released huge amounts of sulfur into the atmosphere, affecting climate in the Northern Hemisphere for two years and causing a famine in Iceland that killed over a fifth of the population. Both fissures are fed by a volcanic center, Eldgja by the subglacial Katla volcano and Laki by the subglacial Grimsvötn volcano, the most active in Iceland. Subglacial volcanoes that have given rise to *jökulhlaups* (debris-laden meltwater floods) in historical time are Öræfajökull in 1362 and 1727, Grimsvötn every decade or so, Katla about 20 times (most recently in 1918), and Eyjafjallajökull in 1821.

GLACIERS

Glacier ice (Icelandic: *jökull*) covers about 11% of Iceland. Almost all forms of glaciers are represented, from cirque glaciers to extensive plateau ice caps. By far the largest is the Vatnajökull ice sheet (8300 km^2), the world's largest ice mass after Antarctica and Greenland. Other ice sheets larger than 500 km^2 are Myrdalsjökull (596 km^2), Langjökull (953 km^2), and Hofsjökull (925 km^2), but many high mountains are capped with small glaciers. In the rugged northwest, and especially at the Tröllaskagi Peninsula in north-central Iceland where many mountains reach 1300–1500 m in elevation, there are a great number of "alpine" cirque and valley glaciers. All glaciers in Iceland are of the temperate type (ice in thermal equilibrium with water) and are therefore highly responsive to climatic fluctuations. In the first centuries of settlement in Iceland (870 to the thirteenth century), during the last part of the Medieval Climatic Optimum, the glaciers were much smaller than now. After that, with a cooling climate that culminated in the Little Ice Age, the glaciers grew to their maximum size in postglacial time around 1890. Since then, they have been receding at an ever-increasing rate, and some of the smaller ones have already disappeared.

Glaciers in Iceland are very dynamically active. The most active glacier outlets flow southward from the high plateaus of Myrdalsjökull and Vatnajökull. Only one quarter of the accumulation (annual precipitation exceeds 4000 mm) is melted within the accumulation areas, so enormous amounts of ice are transported by the outlets down to the ablation areas. Surface velocities of these outlet glaciers commonly exceed 1 m/day. Most of the broad-lobed outlets of Vatnajökull are liable to periodic surges (catastrophic advances) with return periods of several decades. The surge in Brúarjökull of 1963–1964 involved 40% of the area of Vatnajökull, with the ice front advancing 8 km at velocities up to 4–5 m/hour. Surges are also typical for outlets of Hofsjökull and Langjökull.

GEOTHERMICS

At present, over 50% of Iceland's energy consumption is geothermal, mostly in the form of space heating (homes and greenhouses), but also for industry and the production of electricity (26%, the rest hydroelectric). The regional heat flow in Iceland falls within the heat flow anomaly of the Mid-Atlantic Ridge crest and varies between about 80 and 300 mW/m^2. Low- and high-temperature thermal areas are distinguished empirically by the geothermal gradient in the uppermost 1 km of the crust. The low-temperature areas are characterized by a gradient of less than 150 °C/km, by a relatively low degree of thermal metamorphism, and by hot springs and geysers. The water is exclusively percolating groundwater, usually mildly alkaline. Hot springs are found in more than 300 localities spread all over the country, although there are very few in the east and southeast. The largest spring, Deildartunguhver in western Iceland, has a discharge of 180 L/s of 100 °C water.

The 22 known high-temperature areas are part of active volcanic systems in the rift zones. Of these, five have so far been harnessed for energy production. The geothermal gradient in the uppermost 1 km exceeds 200 °C/km. Steam vents in the high-temperature areas discharge carbon dioxide, hydrogen sulfide, and hydrogen. Their fluid is derived almost exclusively from percolating groundwater, whereas the heat and the volcanic volatiles are probably derived from cooling intrusions at relatively shallow depths. The mineral content of the high-temperature water depends systematically on the temperature. Large deposits of silica sinter have formed in two high-temperature areas, including that of the Great Geysir in southern Iceland. On the Reykjanes Peninsula, some of the geothermal systems are saline due to infiltration of seawater. The well-known Blue Lagoon is formed by effluent water from the Svartsengi geothermal power station.

EARTHQUAKES

Earthquakes in Iceland are primarily caused by (1) tectonic movements at the plate boundary and (2) volcanic activity. Most large earthquakes ($M > 6$) occur within two transform zones, the SISZ in the south and the TFZ in the north. The SISZ is marked by a 10–15-km-wide, easterly trending epicentral belt. The large earthquakes occur by faulting on north-south striking right-lateral faults. The left-lateral transform motion along the zone thus appears to be taken up by slip on numerous parallel faults and counter-clockwise rotation of the blocks between them (bookshelf tectonics). Within the Tjörnes Fracture Zone, at least three parallel, northwest-trending seismic belts have been identified.

The seismicity of the volcanic zones is characterized by spatial clustering of epicenters. Most clusters coincide with central volcanoes. Rifting structures such as fissure swarms and normal faults are mostly aseismic except during episodes of rifting and magmatism. Several volcanoes exhibit persistent, low-magnitude seismicity. In the Hengill volcano in southwestern Iceland, seismicity is interpreted as the result of extensional failure and heat extraction from a cooling magma chamber.

RATES OF GEOLOGICAL PROCESSES

Iceland's volume above sea level is about 51,500 km^3 (mean altitude 500 m, area 103,000 km^2). The volcanic productivity in the Holocene has been estimated at 420 km^3/10,000 years, which equals 42,000 km^3/million years. At that rate, the whole volume of Iceland, 16 million years old, would be produced in 1.2 million years. This means that powerful "sinks" are at work counterbalancing the high productivity: (1) thermal contraction in the cooling crust as it drifts away from the rift zone, bringing rocks older than ~16 million years old below sea level; (2) isostatic sagging in the rift zones beneath younger volcanics (Fig. 3); and (3) erosion by glaciers, water, and wind. Measured sagging of the surface in the rift zones, such as the Thingvellir graben, is 0.4–1 cm/year (4–10 km/million years), whereas the westward tilt of the Tertiary succession in eastern Iceland indicates an average rate of sinking in the center of the rift zone amounting to 2.7 km/million years. The Pleistocene glaciers carved the fjord topography in the Tertiary lava pile, the weight loss being partly counterbalanced by isostatic uplift. The glacial rivers carry about 0.025 km^3/year (25,000 km^3/million years) of sediment out to sea, where it is temporarily deposited on the insular shelf before continuing down to the abyssal plain.

SEE ALSO THE FOLLOWING ARTICLES

Atlantic Region / Earthquakes / Eruptions: Laki and Tambora / Faroe Islands / Surtsey / Volcanic Islands

FURTHER READING

Jacoby, W. R., and M. T. Gudmundsson, eds. 2007. Hotspot Iceland. *Journal of Geodynamics*, Special Issue, 43.
Saemundsson, K., ed. 1979. Geology of Iceland. *Jökull*, Special Issue, 29.
Sigmundsson, F., L.A. Simonarson, O. Sigmarsson, and O. Ingolfsson, eds. 2008. The dynamic geology of Iceland. *Jö Kull*, Special Issue, 58.
Steinthorsson, S., and S. Thorarinsson. 1997. Iceland, in *Encyclopedia of European and Asian regional geology*. E. M. Moores and R. W. Fairbridge, eds. New York: Chapman and Hall.
Thordarson, Th., and A. Hoskuldsson. 2002. Iceland. *Classic geology in Europe 3*. Hertfordshire, UK. Terra Publishing.

INBREEDING

LEONARD NUNNEY

University of California, Riverside

Inbreeding is a process central to understanding the genetics of populations. Although usually thought of as the mating of close relatives, more generally inbreeding is the local build-up of genetic similarity due to common ancestry. It can cause genetic differentiation among island populations of the same species and the loss of genetic variability within populations. It can also reduce population survival over the short term by inbreeding depression and over the longer term by compromising the ability of the population to adapt to environmental change.

WRIGHT'S THEORY

The theory of inbreeding was developed by Sewell Wright between about 1920 and 1950. Inbreeding is a simple process, but the genetic consequences are not, and Wright's complex theory can be difficult to understand. A good starting point for understanding inbreeding is "identity by descent." Two gene copies are identical by descent (IBD) if they both originated from the same ancestral gene copy. Using this idea, the inbreeding coefficient (F) of a diploid individual is the probability that its two copies (one maternal and one paternal) of any gene are IBD. For example, an offspring of full siblings has a 25% chance of carrying two copies of a gene that are IBD because they were originally present as a single copy in one of its two grandparents. However, if we go back far enough in time (a period called the coalescence time), all gene copies currently segregating in any isolated population (or species) are IBD. This apparent paradox is resolved by recognizing that inbreeding is a relative measure: inbreeding at one level must be interpreted relative to a higher level. The relativity is made explicit in the notation of Wright's hierarchical inbreeding coefficients (e.g., F_{ST} refers to the IBD of gene copies in subpopulations (S) relative to the total (T) population).

To illustrate this point, consider a simple spatial hierarchy of an archipelago (T) of many islands (S), each supporting a population of many individuals (I) of a plant species, and that inter-island movement (via seed and/or pollen) is infrequent. To make the example more interesting, it is assumed that the plants occasionally self-pollinate. Thus, the two gene copies from a randomly chosen individual are, on average, inbred relative to randomly chosen gene copies within the island ($F_{IS} > 0$), a result driven by the presence of the inbred selfed individuals. Similarly, gene copies randomly chosen from any one island are inbred relative to gene copies chosen randomly from the whole archipelago ($F_{ST} > 0$). This island-level inbreeding arises because, on average, gene copies from the same island share more recent common ancestry (i.e., a shorter coalescence time) than those from the larger spatial unit, the archipelago.

GENETIC CHANGES UNDER INBREEDING

A common misconception is that inbred populations generally have a deficit of heterozygotes relative to Hardy-Weinberg (H-W) expectations. In fact, an island population, inbred because of its isolation relative to other populations of the species, is expected to exhibit H-W ratios.

Any relative decrease in heterozygotes compared to H-W ratios would be due to processes internal to the population such as selfing or sibling mating. However, inbreeding does decrease the absolute level of heterozygosity in a population, because genetic variability decreases as a result of the random loss of alleles by genetic drift. This loss also causes initially identical populations to become genetically different at a rate proportional to $1/N_e$, where N_e is the effective size of each population. As a rough guide, significant genetic divergence takes about N_e generations.

For neutral alleles, this random loss affects large and small populations equally (but at different rates); however, as N_e decreases, more genetic variation becomes effectively neutral. The result is that in small populations, deleterious alleles can increase in frequency through random sampling, lowering the average fitness of the population and thereby inducing population-wide inbreeding depression. Thus, in the extreme case of the establishment of a population by few individuals, the founder effect typically includes a dramatic increase in the frequency of some previously rare deleterious alleles (e.g., the high frequency of porphyria variegata in the Afrikaner population of South Africa).

There can also be a positive effect on fitness of such a founding event. Although a few highly deleterious alleles can, by chance, increase in frequency, many more disadvantageous alleles are lost through the same sampling process. Thus, provided that the population can survive the initial founding and grow, selection can reduce the frequency of the remaining deleterious alleles. The net result is that, for a significant period of time, the population can have less of a fitness loss due to disadvantageous alleles than would the parent population (although eventually, new mutations would restore the typical balance). This is one form of a process called purging. Another form of purging is restricted to managed populations and involves mating close relatives within a population to create inbred individuals (i.e., creating a population with more homozygotes that expected under Hardy-Weinberg ratios), allowing selection to more effectively reduce the frequency of deleterious recessives.

INBREEDING DEPRESSION AND EXTINCTION RISK

Inbred populations typically have elevated extinction rates, over the short term because of the negative effects of inbreeding depression and over the long term because the loss of genetic variability resulting from inbreeding reduces adaptive potential, thus increasing susceptibility to environmental change.

Island populations are particularly vulnerable to inbreeding and its negative fitness consequences for two reasons. First, they are very often established by a few immigrants, resulting in a strong founder effect. Second, even after initial growth, island populations tend to be small and isolated, such that additional inbreeding can result in a further loss of genetic variation. These potentially negative effects on fitness may be somewhat offset by the purging of deleterious alleles noted above.

Island populations are especially vulnerable to biotic change such as the arrival of a novel competitor, predator, or pathogen. Examples of the extinction of island endemics following the invasion of such species are well documented; however, when native populations persist, genetic management to reduce inbreeding and promote adaptation to the changing environment can be beneficial. For example, where island populations have become fragmented through habitat loss, inbreeding within the fragments can be minimized by the exchange of migrants among the fragments. On a larger scale, if island populations are not too different, inter-island exchange may be a viable strategy for increasing genetic variability.

SEE ALSO THE FOLLOWING ARTICLES

Extinction / Founder Effects / Invasion Biology / Population Genetics, Island Models in

FURTHER READING

Frankham, R. 1998. Inbreeding and extinction: island populations. *Conservation Biology* 12: 665–675.
Frankham, R. 2005. Genetics and extinction. *Biological Conservation* 126: 131–140.
Hartl, D. L., and A. G. Clark. 2007. Inbreeding, population subdivision and migration, in *Principles of population genetics*. Sunderland, MA: Sinauer Associates, Chapter 6.
Jamieson, I. G. 2007. Has the debate over genetics and extinction of island endemics truly been resolved? *Animal Conservation* 10: 139–144.
Nunney, L. 2001. Population structure, in *Evolutionary ecology: concepts and case studies*. C.R. Fox, D. Roff, and D. Fairbairn, eds. Oxford: Oxford University Press, 70–83.
Willi, Y., J. Van Buskirk, and A. A. Hoffman. 2006. Limits to the adaptive potential of small populations. *Annual Review of Ecology, Evolution, and Systematics* 37: 433–458.

INDIAN REGION

FREDERICK A. FREY

Massachusetts Institute of Technology, Cambridge

Like the submarine igneous crust of the Indian Ocean, most Indian Ocean islands are predominantly formed of volcanic rock, usually basalt; however, others, such as Sri

Lanka, Madagascar, Zanzibar, and the Seychelles archipelago, are predominantly formed of continental crustal rocks, such as granite, sediment, and their metamorphosed equivalents. This article summarizes major geologic and geographic features of many but not all islands within the Indian Ocean basin and discusses the plate tectonic implications of these islands.

GEOLOGIC SETTINGS FOR TERRESTRIAL VOLCANISM

There are three types of geologic settings that lead to basaltic volcanism on Earth.

Regions of Plate Divergence

Volcanoes form when plates move apart because the melting temperature of deep Earth material, the mantle, decreases with decreasing pressure; therefore, where plates move apart, the mantle rocks ascend and partially melt, creating basaltic magmas that erupt because they have lower density than their surroundings. Such magmas form the submarine volcanic crust of the Indian Ocean along the major spreading ridges within the Indian Ocean (i.e., the Southeast, Southwest, and Central Indian ridges) (Fig. 1).

Regions of Plate Convergence

Basaltic volcanoes also form when two plates collide and the denser plate, typically an oceanic plate, subducts beneath the less dense plate, typically a continental plate. Prior to subduction, the subducting oceanic plate interacts with overlying seawater and hydrated minerals form within the oceanic plate; as this plate is subducted to depths of 100 to 200 km, these hydrated phases become unstable and release water into the overlying mantle. Because water, like decreasing pressure, lowers the melting temperature of mantle rocks, basaltic magmas are created in subduction zones and ascend to form the arc volcanoes that are characteristic of convergent zones (e.g., the well-known "Ring of Fire" that surrounds the Pacific Ocean). Arc volcanoes form the eastern boundary of the Indian Ocean (e.g., the islands forming Indonesia).

Intraplate Volcanism

Most of the volcanic islands in the Indian Ocean are distant from plate boundaries; they reflect intraplate volcanism. Often they are referred to as hotspot volcanoes because they apparently form above buoyant mantle, referred to as mantle plumes, which ascend and partially melt because the hotspot (plume) material is hotter and less dense than

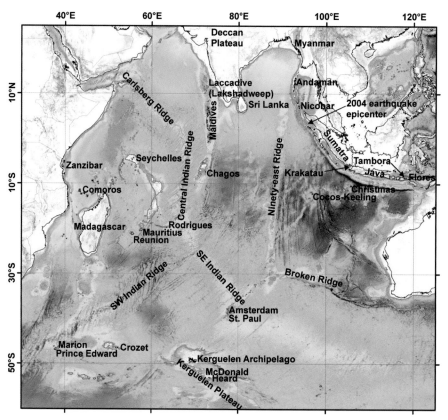

FIGURE 1 Map of the Indian Ocean showing islands and submarine ridges discussed in text. Satellite altimetry data: Geoware GMT Companion CD-R Vol. 1, Version 1.9, June 2006. Image prepared by Jenny Paduan and David Clague, MBARI.

the surrounding mantle. In some regions these hotspots create volcanism for tens of millions of years, and they are relatively fixed in location; consequently, a linear chain of volcanoes forms as an oceanic plate migrates over the hotspot. The Indian Ocean contains two well-defined hotspot tracks, the Ninetyeast ridge in the eastern Indian Ocean and the Laccadive-Maldive-Chagos ridge in the central Indian Ocean.

THE EASTERN INDIAN OCEAN MARGIN: PLATE CONVERGENCE VOLCANISM

Indonesian Volcanic Arc

The Indonesian volcanic arc includes many islands extending over 6000 km from northern Sumatra to the Banda Sea. From northwest to southeast, the arc can be divided into Sumatra; the Sunda arc, extending from Java to Flores; and the Banda arc. Along the Indonesia volcanic arc, there are more than 500 volcanoes, and more than 100 have had Holocene eruptions. These volcanoes are a consequence of eastward subduction of the Indian Ocean plate beneath continental rocks, such as those exposed in Sumatra. Volcanic rocks as old as Paleocene (~55 to 65 million years ago) are present along the Indonesian arc, but volcanism in the Indonesian region is most famous because two of the Earth's largest and most explosive historic volcanic eruptions occurred at Tambora (1815) and Krakatau (1883); more recently, a major submarine earthquake west of Sumatra caused the devastating tsunami on December 26, 2004.

Andaman and Nicobar Islands

The Andaman and Nicobar Islands are in the northeastern Indian Ocean, several hundred kilometers north of Sumatra and south of Myanmar; they are a Union Territory of India. There are ~258 Andaman islands (6408 km^2) and 61 Nicobar islands (1841 km^2); of these, only 37 are inhabited. More than 6000 people were killed by the 10–15-m tsunami arising from the 2004 submarine earthquake off the west coast of Sumatra. In the vicinity of these islands, the convergence of the Indian Ocean plate with southeast Asia is highly oblique, in contrast to the perpendicular convergence at Java. As is typical of convergent zones, there is a deep submarine trench west of these islands and seismic evidence that the Indian Ocean plate has subducted to at least 250 km below the islands; however, the oblique subduction zone geometry has not led to an active volcanic arc. Two outlying Andaman islands, 100 km to the east of the main group, are described as stratovolcanoes; Narcondam Island is an extinct volcano, but Barren Island is an active volcano with eruptions occurring in the 1990s. The majority of Andaman and Nicobar islands, however, are composed of metamorphosed sedimentary rocks (i.e., sandstones, shales, and limestones), which have been related to the Arakan Yoma mountain range that dominates western Myanmar.

THE EASTERN INDIAN OCEAN ISLANDS: INTRAPLATE VOLCANOES

Christmas and Cocos-Keeling Islands

Christmas and Cocos-Keeling Islands, territories of Australia, are several hundred kilometers west of Java. They originally formed as large volcanoes rising several kilometers above the sea floor. Christmas Island, with a population of 1500 and 135 km^2 in area, is composed of interbedded Tertiary volcanic rocks and carbonate sediments with overlying economic deposits of phosphate derived from guano. There are two volcanic sequences of alkalic basalt; the youngest and most voluminous erupted in the Lower Miocene. Subsequent to volcanism, the island was capped by a coral atoll and repeatedly uplifted, as documented by a terraced coastline.

The Cocos-Keeling Islands comprise 27 small coral islands (14 km^2), presumed to be constructed on old submerged volcanoes, which form two distinct coral atolls 24 km apart. Two islands are inhabited by a population of ~620.

Kerguelen Plateau, Broken Ridge, Ninety East Ridge

The Kerguelen Plateau, the Broken Ridge, and the Ninety East Ridge are submarine volcanic structures on the eastern Indian Ocean sea floor. They are broadly consistent with a simple end-member model for intraplate volcanism arising from a hotspot represented by a mantle plume (Fig. 2). In this idealized model, a large igneous province (LIP) is created over a brief time interval, perhaps less than 1 million years, as the large plume head partially melts upon decompression; subsequently, an age-progressive and linear hotspot track (i.e., a line of volcanoes) forms as the oceanic plate migrates for many million years over the less voluminous hotspot tail. The geologic evolution of volcanoes forming a hotspot track as a migrating oceanic plate, such as the Indian plate, overrides a hotspot is as follows: (1) submarine eruptions construct a volcano that eventually emerges above sea level to form an island; (2) subsequently, volcanism wanes as the plate migrates away from the source of volcanism (i.e., the hotspot); (3) erosional processes (e.g., landsliding) become more important than volcanic construction processes; (4) the volcanic island subsides below sea level.

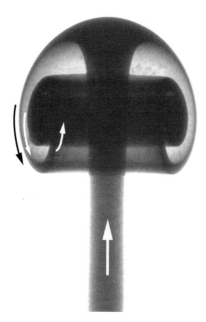

FIGURE 2 Simple plume head/tail model for hotspot volcanism. Photograph of a laboratory model of a starting thermal plume. The red fluid is buoyant hot material derived from a thermal boundary layer; the yellow fluid is cooler material entrained during ascent (modified, with permission, from Campbell 2005). Within the Earth, the buoyancy of mantle plumes are controlled by their composition as well as by their higher temperature; in this case, the plume head and tail structure is likely to be much more complex.

In this hypothesis, the Kerguelen Plateau/Broken Ridge LIP represents volcanism derived from the plume head, and the Ninety East Ridge, a long (more than 5000-km), age-progressive, and linear chain of volcanoes, represents volcanism derived from the plume tail. The Kerguelen Plateau and Broken Ridge were contiguous prior to their separation at ~40 million years ago by the Southeast Indian ridge. Both structures are now submarine, at water depths of 2 to 3 km, with plateaus rising several kilometers above the surrounding deep ocean floor. However, basalt samples recovered from Kerguelen Plateau and Broken Ridge by the Deep Sea Drilling Project and the Ocean Drilling Program, both of which utilized the *JOIDES Resolution* drilling ship, were erupted above sea level (i.e., they formed islands of unknown size). Subsequently, like the Hawaiian volcanoes that form the submarine Hawaiian Ridge, these islands subsided below sea level as volcanism ceased. Depending upon subsidence rates, the emergent parts of the Kerguelen Plateau may have had a role in paleobiogeography by serving as a land bridge between Antarctica and India/Madagascar. Also, basaltic drill cores from the age-progressive Ninety East Ridge (i.e., ~42 million-year-old volcanism in the south to ~77 million-year-old volcanism in the north) show that the Ninety East Ridge formed as a series of islands that emerged above sea level and subsequently subsided as the Indian plate migrated northward over what is known as the Kerguelen hotspot.

Where is the current volcanism created by the Kerguelen hotspot? Three possibilities are recently active volcanic islands constructed on the Kerguelen Plateau, the Kerguelen Archipelago, Heard Island, and the McDonald Islands.

Kerguelen Archipelago

The Kerguelen Archipelago is part of Terres Australes et Antarctique Françaises (TAAF) and is occupied at Port aux Français by a rotating group of ~100 scientists who study the geology, flora, and fauna of the islands. The archipelago consists of a major island (Grande Terre, 6500 km^2) and many nearby, much smaller islands. There is a glacier (Cook) in the western highlands of Grande Terre, and the land surface was extensively glaciated during the Pleistocene; consequently, the coastal morphology is dominated by fjords. These fjords expose steep, ~1 km sections of nearly flat-lying, tholeiitic to transitional flood basalt flows erupted from ~30 to 24 million years ago. These flood basalts, erupted from unexposed fissures, cover 85% of the land surface. In some areas, especially in the west, plutonic rocks, gabbro, syenite, and granite are exposed; such occurrences of slowly cooled (i.e., intrusive) rocks are uncommon in oceanic islands. More recent volcanism, occurring 1 to 0.1 million years ago, in the archipelago erupted from localized vents and created steep-sided volcanoes formed by alkalic lavas, ranging from basanite to trachyte; an example is Mount Ross (1850 m), the highest point in the archipelago. The most recent volcanism (26,000 years ago) is a trachytic ignimbrite. Overall, there has been a long-term trend for archipelago volcanoes to change from tholeiitic and transitional basalt, erupted from 30 to 25 million years ago, to alkalic basalt and more differentiated lavas, such as phonolite and trachyte, erupted from 24 million years ago to the present. Some of these alkalic lavas contain abundant xenoliths of lower crust (granulites) and mantle (peridotite) rocks.

Heard Island

Heard Island, a territory of Australia with no permanent residents, lies 450 km southeast of the Kerguelen Archipelago. It is a world heritage site. The island, largely (more than 70%) covered by glaciers, has two distinct parts: (1) a near circular (20 by 25–km) main island dominated by Big Ben, a steep-sided stratovolcano formed of alkalic basalt and trachyte (Fig. 3). The historically active (2007 eruption) Mawson Peak of Big Ben, with a summit elevation of

FIGURE 3 Big Ben volcano, the largest volcano on Heard Island in the Central Kerguelen plateau; Compton glacier is in the foreground. Photograph courtesy of Dr. W. Powell (GEMOC ARC National Key Centre, Macquarie University, Sydney, Australia).

2745 m, is constructed within a 5–6-m wide caldera that is breached on the southwest flank of the volcano. Like the alkalic basalts in the Kerguelen Archipelago, xenoliths of mantle rock (peridotite) are found in alkalic basalt flows and boulders at Heard Island. The Laurens Peninsula is connected to the main island by a narrow, low isthmus; Mt. Dixon is the principal feature, and the oldest rocks outcrop on this peninsula. They are limestone deposited 45–50 million years ago that are intruded by thin (less than 2-m) sills of basalt.

McDonald Islands

The McDonald Islands, also a territory of Australia and a world heritage site, are a group of islets that lie 40 km west of the Laurens Peninsula. The largest, McDonald Island (1 km^2), is volcanically very active, erupting phonolite lava (2005 eruption); this island doubled in size from 2000 to 2001, destroying all vegetation.

In summary, the current location of the Kerguelen hotspot is uncertain; however, the region occupied by Heard and McDonald Islands is the location of the most recent volcanism.

Amsterdam and St. Paul Islands

Amsterdam and St. Paul Islands are young (less than 1 million years old) tholeiitic basalt volcanoes constructed on the submarine Amsterdam St. Paul (ASP) plateau. They are close—60 and 100 km, respectively—to the actively spreading Southeast Indian ridge (SEIR). Both islands are French (TAAF) territories, and a small scientific base is maintained on Amsterdam Island. This island, with an area of 55 km^2 and a peak elevation of 881 m, is a subcircular (10 by 7–km) volcanic cone that has steep erosional cliffs on the west flank; they perhaps reflect a major landslide collapse. Amsterdam lavas were first inferred to be 0.4 to 0.2 million years in age, which is considerably older than St. Paul Island lavas, but younger, 0.01 to 0.2–million-year-old ages have been recently reported. For Amsterdam Island lavas, no historical eruptions are known, but in 2000 recent volcanism was discovered at Boomerang sea mount, only 18 km northeast of Amsterdam Island.

St. Paul Island, 100 km south of Amsterdam Island, is a small cone, 8.4 km^2 in area and 264 m in elevation, whose northwest sector has collapsed enabling seawater access to the 1.6-m diameter crater. St. Paul Island has been inferred to be younger (by 0.4 to 0 million years) than Amsterdam Island.

Although basalts from these two islands on the ASP plateau are geochemically distinct, they share some geochemical characteristics with basalt erupted along the Ninety East Ridge. This observation, coupled with a series of sea mounts extending to the northeast from the ASP plateau, suggests that the hotspot forming the ASP plateau and its islands may have also played a role in forming the NER.

In summary, volcanism in these islands and a sea mount on the ASP plateau represent recent hotspot-derived volcanism located near a spreading ridge axis, much like the Galápagos Islands and unlike volcanism on the Kerguelen archipelago, Heard Island, and the McDonald Islands, which are many hundreds of kilometers distant from the SEIR.

THE DECCAN PLATEAU, THE LACCADIVE-MALDIVE-CHAGOS RIDGES, AND THE MASCARENE ISLANDS IN THE CENTRAL INDIAN OCEAN: PLUME HEAD AND TAIL VOLCANISM

The Deccan plateau, a continental flood basalt covering much of northwestern India, is a large igneous province derived from a plume head, and the Laccadive–Maldive–Chagos ridges and Mascarene Islands form a hotspot track, which ends at the currently active volcano, Piton de la Fournaise, on Réunion Island. Like the Ninety East Ridge, the oldest part of the Réunion to Deccan hotspot track strikes south–north (i.e., the Laccadive–Maldive–Chagos Ridge), reflecting northward migration of the Indian Plate over the Réunion hotspot. However, the younger part of the track strikes southwest–northeast because it was separated from the south–north segment of the hotspot track by post-Eocene spreading on the relatively young Central Indian Ridge. Age determinations of volcanic rocks recovered from ten geographically and temporally distinct locations along the Réunion to Deccan hotspot track are in excellent agreement with the volcanism originating from a near stationary Réunion hotspot, as the Indian plate migrated northward with subsequent southwest–northeast spreading on the Central Indian ridge.

Deccan Plateau

The Deccan Plateau (often referred to as Deccan Traps) is a well-known continental flood basalt, a type of large igneous province because of its very large volume, 1–10 million km^3, of tholeiitic basalt, which erupted over a relatively narrow time interval of less than 500,000 years 65.5 million years ago, an age that is indistinguishable from the massive extinction event that occurred at the Cretaceous–Tertiary boundary.

Extending southward from the Deccan Plateau, the north–south-trending Laccadive–Maldive–Chagos submarine ridges, which mark the western boundary of the Central Indian basin, extend for more than 2500 km between 14° N and 9° S. Although their upper surfaces are carbonate bank and reef deposits up to 2 km thick, seismic data show that these ridges form a continuous volcanic structure. Sea floor drilling and coring by the *JOIDES Resolution* on the northern margin of the Maldive Ridge and the northern edge of Chagos Ridge recovered tholeiitic basalt that erupted subaerially or at shallow marine depths; these basalts are from ancient, submerged islands underlying carbonate sediments and reef deposits.

Laccadive Islands

The Laccadive Islands, now known as Lakshadweep, are the subaerial portion of the Laccadive Ridge; they are a Union Territory of India in the Arabian Sea, 220–440 km from the southwest (Malabar) coast of India. The archipelago consists of 36 islands, largely atolls and reefs with a land area of 32 km^2. The western rim of the archipelago, exposed to the southwest monsoon, is mostly submerged coral reef, whereas the relatively protected eastern rim includes ten inhabited islands with a combined population of 60,000. These low-lying coral atolls overlie extinct, submerged volcanoes that form the northern end of the Réunion to Deccan hotspot track; these volcanoes are inferred to have formed subsequent to the Deccan Plateau (i.e., less than 65.5 million years ago); however, they have not been sampled by drilling.

Maldive Islands

The Maldive Islands, 595 km south of India, consist of 26 atolls comprising ~1200 islands (298 km^2) extending from 7°10′ N to 0°45′ S. A population of 300,000 inhabitants on ~200 islands form the Republic of the Maldives. These islands are exclusively formed of coral reef deposits with a maximum land elevation of 2.3 m, average of 1.2 m; there is much concern about the future effects of a rising sea level, which has already risen ~20 cm over the last century. Like the Laccadive Islands to the north, the coral atolls of the Maldives were constructed on ancient submerged volcanoes. These volcanoes were sampled by the Ocean Drilling Program at 5°05′ N, 73°50′ E on the eastern margin of the submarine Maldive ridge. Based on a hotspot track model, an age of 55 million years was predicted for the basaltic core recovered from beneath 211 m of shallow water limestone (i.e., reef deposits). Subsequent dating of the recovered tholeiitic basalt using the K-Ar radiogenic decay system determined an eruption age of 57 million years ago, an age that is in excellent agreement with that predicted by the hotspot model.

Chagos Archipelago

The Chagos Archipelago, a British Indian Ocean territory, 456 km south of the southernmost Maldive Islands, is the youngest expression of the submarine Laccadive–Maldives–Chagos volcanic ridge. This archipelago, extending from 04°54′ S to 07°39′ S, consists of seven coral atolls and many islets with a land area of 63 km^2 and a population of 3500. The largest island, Diego Garcia (27 km^2), is a naval and air military base jointly operated by the United States and United Kingdom. Construction of this base required forced evacuation of the local inhabitants to the Seychelles and Mauritius. The consequent human-rights controversy coupled with the strategic military location (i.e., close to volatile Middle Eastern countries) has led to Diego Garcia becoming the most well-known coral atoll in the Indian Ocean. Drilling of the sea floor by the Ocean Drilling Program on the northern margin of Chagos Bank at 4°12′ S recovered tholeiitic basalt below 107 m of sediment. K-Ar dating of this basalt yielded an age of 49 million years, again in excellent agreement with the age range 45–50 million years, predicted by a simple hotspot track model.

Mascarene Islands

The Mascarene Islands in the western Indian Ocean include the volcanic islands of Réunion, Mauritius, and Rodrigues. Reunion and Mauritius define the less-than-10-million-year southwest–northeast trend of the Réunion to Deccan hotspot track that began with formation of the Deccan Plateau at 65.5 million years.

Réunion Island

Réunion Island, 750 km east of Madagascar with an area of 2512 km² and a population of 793,000, is an overseas department of France, and as one of the 26 regions of France, its inhabitants have the same status as citizens of European France. The island, rising 7 km above the surrounding sea floor, consists of two volcanoes, Piton des Neiges and Piton de la Fournaise; the latter is the current volcanic expression of the hotspot that created the Deccan plateau 65.5 million years ago. These two volcanoes at Réunion were simultaneously active, but Piton des Neiges in the central part of the island, with a peak elevation of 3070 m, is older; the earliest volcanism, basalt to hawaiite, began ~2 million years ago and continued until at least 30,000 years ago. Piton des Neiges now appears to be extinct. The surface of the central part of Piton des Neiges is spectacularly rugged because several deeply dissected gorges broaden upstream into large erosional cirques. These cirques provide windows into the early igneous history of the volcano by exposing a basement complex that includes breccias and intrusive rocks, such as layered gabbros.

Piton de la Fournaise, 2631 m at its peak elevation, occupies the eastern part of the island and is crowned by a concentric series of caldera collapse structures. The oldest lavas, dominantly alkalic basalt, erupted ~0.5 million years ago, and volcanism continues today with more than 150 eruptions in the last 300 years; the most recent occurred in April 2007 (Fig. 4). Although the oldest lavas were alkalic basalt, tholeiitic basalt is the most abundant lava type. Piton de la Fournaise and the much smaller McDonald Island on the Kerguelen Plateau are currently the most active volcanoes in the Indian Ocean. The eruptions and lavas at Piton de la Fournaise have been intensively studied, and a volcano observatory, 15 km from the summit, was constructed in 1979 and is operated by the Institut de Physique du Globe in Paris.

Mauritius

Mauritius is an island nation, the Republic of Mauritius, located 1000 km east of Madagascar and 220 km northeast of Réunion Island; it has an area of 2040 km² and a population of 1.25 million. The volcanism that created Mauritius is divided into three stages; the older series, erupted from 7.8 to 5.4 million years ago, is the erosional remnant of a large subaerial shield volcano with a current maximum elevation of 828 m. The oldest lavas of the older series are alkalic olivine basalt; some of these lavas contain xenoliths of cumulate rocks such as dunite, anorthosite, and syenite. Such samples of cumulate rocks provide insight into the

FIGURE 4 Lava from Piton de la Fournaise entering the sea (©KM KRAFFT/CRI-Nancy-Lorraine, with permission).

partial crystallization and resulting mineral segregation processes that control magma evolution during magma ascent. In fact, coarse-grained xenoliths are abundant at each of the Mascarene islands, principally ultramafic and gabbro xenoliths in Réunion lavas and gabbros in Rodrigues lavas. The younger lavas of the older series, aged between 6.8 and 5.5 million years, include more differentiated lavas such as hawaiite and trachyte.

After a 2-million-year hiatus in volcanism, the intermediate series lavas, dominantly alkalic basalt, erupted from 3.5 to 1.9 million years ago. Finally, from 0.7 to 0.3 million years ago, the younger series lavas erupted and covered most of the island with alkalic olivine basalt. In summary, the lava compositions erupted at Réunion and Mauritius Islands are quite different, with predominantly alkalic basalt lavas at Mauritius and predominantly tholeiitic basalt at Réunion. Nevertheless, the radiogenic isotopic ratios of Sr, Nd, Hf, and Pb are similar in the older series (i.e., shield-building lavas on Mauritius and the shield lavas on Réunion Island). This similarity in isotopic ratios and, even more importantly, the inferred 7–8 million year age for construction of the Mauritius shield are consistent with the interpretation that Mauritius belongs to the Deccan to Réunion hotspot track (i.e., the Mauritius shield is 7–8-million-year-old volcanism resulting from the hotspot now underlying the active Piton de la Fournaise volcano on Réunion Island).

Rodrigues Island

Rodrigues Island, the easternmost Mascarene Island, 560 km east of Mauritius, has a population of 40,000; it is part of the Republic of Mauritius. Located on the eastern end of submarine Rodrigues Ridge, a 600-km-long east–west structure, the volcanic island surrounded by a coral reef is 109 km^2 with a peak elevation of 396 m. The submarine Rodrigues Ridge has been sampled by dredging and consists of basalt and trachyte erupted 7–11 million years ago. It has been suggested that the Rodrigues Ridge represents channelized flow of hotspot-related mantle toward the Central Indian Ridge. Rodrigues Island consists predominantly of alkalic basalt erupted ~1.5 million years ago; consequently, unlike lavas erupted at Réunion and Mauritius, Rodrigues Island with its relatively young age is not on the Deccan to Réunion hotspot track.

THE WESTERN INDIAN OCEAN ISLANDS: INTRAPLATE VOLCANOES

Crozet Archipelago

The Crozet Archipelago consists of five main islands located south of the Southwest Indian Ridge (SWIR). They are a French (TAAF) territory and a National Park. There are no permanent residents, but as at Kerguelen and Amsterdam Islands, a rotating group of scientists occupy a permanent base, Port Alfred, on Ile de la Possession. These islands, on the large (4500-km^2) west-northwest–east-southeast–trending submarine Crozet plateau, are subdivided into a western group of three relatively small islands—Ile aux Cochons, a 67 km^2 stratovolcano, and two smaller reef islands, Ile des Pingouins (3 km^2) and Ilots des Apotres (2 km^2)—and an eastern group, 100 km to the east of two larger islands: Ile de la Possession (150 km^2) and Ile de l'Est (130 km^2). These eastern islands, with maximum elevation of ~1 km, are eroded stratovolcanoes, predominantly composed of alkali basalt and associated cumulate (e.g., ankaramite) and differentiated lavas (e.g., phonolite). Ile de l'Est is deeply dissected by 0.5- to 1-km, glacially carved U-shaped valleys, whereas Ile de la Possession has a more subdued surface. The main volcano on Ile de la Possession erupted lavas from ~8 to 0.5 million years ago, but very young appearing, more-than-100-m-high scoria cones, probably Holocene in age, are common on the surface; hot springs are also present.

Marion and Prince Edward Islands

Marion and Prince Edward Islands are South African territories, on the Antarctic plate, 160 to 200 km south of the actively spreading Southwest Indian ridge. The pronounced, north–south trending submarine Madagascar plateau, extending from the southern tip of Madagascar Island to the SWIR, is inferred to be a hotspot track formed as the Indian plate migrated northward relative to the Marion hotspot. The presumed volcanic rocks forming this submarine plateau have not been sampled.

Marion Island (290 km^2) hosts a research station. The island has a low dome-like profile, has a 1230-m summit elevation, and is formed of two basaltic shield volcanoes overlain by ~150 cinder cones. The interior highlands, including a glacier, rise from a coastal plain via a 200–400 m escarpment. The lavas, predominantly alkalic basalt and hawaiite, are divided into two age groups: (1) an older, glaciated gray lava series erupted less than 500,000 years ago (i.e., Pleistocene-age lavas) and (2) a younger black lava series that postdates glaciation (i.e., Holocene-age lavas). Unvegetated lava flows indicate recent volcanism; the first historical eruption occurred in 1980, and a recent eruption occurred in 2004.

The smaller (45-km^2 and 72-m-maximum-elevation) and uninhabited Prince Edward Island, 22 km northeast of Marion Island, also features central highlands bounded by a prominent escarpment. It is a remnant of a shield volcano; Holocene volcanism is reflected by numerous scoria and tuff cones. Apparently, Marion and Prince Edward Islands experienced contemporary volcanism.

Comoros Archipelago

The Comoros Archipelago consists of four islands: Grand Comoro (Ngazidja), Moheli (Mwali), Anjouan (Nzawani), and Mayotte (Maore). The first three islands form the Union of the Comoros, with a population of 676,000; the island of Mayotte is administered by France but claimed by the Union of Comoros. The four islands, midway between Madagascar and northern Mozambique, are the summits of a northwest to southeast trending volcanic ridge that crosses the Mozambique Channel northwest of Madagascar. Grand Comoro, at the northwest end of the archipelago, is the largest (1013 km^2) of the four islands; it consists of two undissected, coalescing shield volcanoes made up largely of alkalic basalt. These shields are La Grille (1087 m), active in the Holocene, and Karthala (2361 m), which has a 3 by 4–km summit caldera; it last erupted in 2006. To the southwest, Moheli and Anjouan Islands are deeply eroded alkalic basalt shields, modified by extensive faulting. At the southwest extremity of the archipelago, Mayotte Island is a subdued (660 m maximum elevation), intensively eroded volcano with an embayed shoreline reflecting subsidence of the shield; the island has a continuous barrier reef reaching 12 km in diameter. The southeast to northwest age progression, obvious from geomorphology and confirmed by K-Ar dating, appears to reflect a hotspot track formed by southeast migration of the Somali plate over a Comoros hotspot. However, unlike the Hawaiian and Réunion shields, which are predominantly composed of tholeiitic basalt, the Comoros volcanoes are composed of alkalic basalt. A distinctive feature of these alkalic lavas is the abundance of xenoliths (i.e., exotic blocks of coarse-grained rocks, predominantly gabbros, dunites, and peridotite). These xenoliths provide insights into the geologic processes occurring at depth within the volcanoes.

INDIAN OCEAN ISLANDS PREDOMINANTLY COMPOSED OF CONTINENTAL CRUST

Sri Lanka

Sri Lanka is a heavily populated (~20 million), large island (65,610 km^2), less than 130 km south of India. At some time there may have been a direct land-bridge connection between Sri Lanka and southern India known as Ramu's Bridge in Hindu mythology. More than 90% of the Sri Lankan land surface is metamorphic Precambrian rock (as old as 2×10^9 years); these rocks are ancient granites and sediments that were strongly recrystallized (i.e., metamorphosed at high pressures and temperatures to form granulites). The Precambrian rocks of Sri Lanka were once part of the ancient supercontinent of Gondwana, originally composed of Antarctica, Australia, India, and Africa, before it began to break apart 200 million years ago. Ongoing research is investigating similarities between the Precambrian rocks of Madagascar, Antarctica, and Southern India.

Madagascar

Madagascar, which has a population of 17 million people, is one of the largest islands on Earth and is the largest island (587,000 km^2) in the Indian Ocean. The Mozambique Channel separates Madagascar from Africa by 250–400 km. As with Sri Lanka, more than two-thirds of Madagascar's land surface is Precambrian rock, as old as $3.2 \times 10 9$ years. Also like Sri Lanka, now far to the northwest, Madagascar was part of Gondwana and was probably contiguous with southern India prior to breakup of Gondwana. Madagascar was created as an island by two major rifting events during Gondwana's break-up: The first was the separation of Madagascar and India from Africa ~180 million years ago; the second resulted in the separation of India and Madagascar 88 million years ago—an event that was accompanied by extensive basaltic (i.e., magma derived from mantle rocks) volcanism. Although much less voluminous than the old continental crustal rocks, this volcanism 88 million years ago is important in the context of intraplate volcanism because the volcanic rocks occurring on the east and west coasts of Madagascar are an extension of the Madagascar submarine plateau that extends from the southern tip of the island to the Southwest Indian ridge. The Madagascar basaltic lavas are inferred to be early volcanism related to the Marion mantle plume, which is currently creating Marion Island.

Zanzibar Archipelago

The Zanzibar Archipelago, only 40 km east of mainland Africa, consists of two main islands, Pemba and Unguja, and several surrounding islets; Pemba is larger and older than Unguja. In contrast to Sri Lanka and Madagascar, these islands are relatively young, Miocene to recent. They formed as a river delta, their maximum elevations are 120 m, and they consist of sedimentary rocks, such as limestone, sand, silt, and clay.

Seychelles Archipelago

The Seychelles Archipelago (455 km^2) consists of ~115 islands, population of ~80,000, covering a large area (1.4 $\times 10^6$ km^2) of the western Indian Ocean north of Madagascar. Surprisingly for islands distant from continental margins—the Seychelles are 480 to 1600 km from the east coast of Africa—~40 Seychelles islands are predominantly composed of Precambrian continental crustal rocks. For example, undeformed and unmetamorphosed granitic rocks formed 750 million years ago are abundant on the major island of Mahé. Further to the south, ~75 smaller Seychelles islands are composed of sedimentary carbonates and coral. However, northwest of Mahé, two islands, Silhouette and Isle du Nord, consist of syenite and related rocks that are ~63 million years old, an age close to that of the Cretaceous–Tertiary boundary. This age is also close to the eruption age (65.5 million years ago) of the Deccan Plateau, a LIP formed when tremendous volumes of basaltic magma (~8 $\times 10^6$ km^3) erupted in northwestern India over less than 1 million years. Moreover, basaltic dikes on Praslin Island, northeast of Mahé, have geochemical characteristics very similar to a distinctive geochemical group of Deccan Plateau basalts. Consequently, it is inferred that the Seychelles granitic islands were previously located at the northwest margin of India, along with Madagascar, and that they were subsequently transported to their present location, far south of India, by sea floor spreading.

In summary, Sri Lanka, Madagascar, and the Seychelles contain ancient (i.e., Precambrian) continental crust that formed well before formation of the Indian Ocean basin, which was created when Gondwana, a large supercontinent composed of Africa, India, Australia, and Antarctica, broke apart between ~200 and 132 million years ago. Therefore, these islands are continental fragments that were stranded within the Indian Ocean after the break-up of the Gondwana supercontinent.

FURTHER READING

Campbell, I. H. 2005. Large igneous provinces and the mantle plume hypothesis. *Elements* 1: 271–275.
Global Volcanism Programs. http://www.volcano.si.edu/
The Great Plume Debate. http://www.mantleplumes.org/
Nairn, A. E. M., and F. G. Stehli, eds. 1982. *The ocean basis and margins, volume 6, the Indian Ocean.* New York: Plenum.
Yoshida, M., B. F. Windley, and S. Dasgupta, eds. 2003. *Proterozoic East Gondwana: supercontinent assembly and breakup.* Geological Society, London, Special Publications 206.

INDONESIA, BIOLOGY

TIGGA KINGSTON
Texas Tech University, Lubbock

Indonesia is one of the most biologically rich countries in the world, yet dramatic land-use changes in recent decades place much of this biodiversity in peril. The high levels of species richness and endemism are in large part attributable to a complex geological history that has both generated a profusion of island speciation centers and brought together two very different biological realms.

BIOGEOGRAPHY OF INDONESIA

Geographic Setting

Straddling the equator from 6° N to 11° S, the Indonesian archipelago comprises over 18,000 islands that extend from Sumatra in the Paleotropical realm to New Guinea and the Aru Islands in the Notogean (Australian) realm. Between these two extremes lies a transitional region, Wallacea, with biological affinities to both realms and a high incidence of endemic species. Fluctuating sea levels and tectonic movements throughout the Cenozoic facilitated connections among many of the island groups, but others, particularly those in the deep seas of the Wallacea region, have been isolated for millions of years, limiting overall diversity but resulting in a profusion of endemic species. As mountain systems arose, many were periodically isolated by rising sea levels, creating additional opportunities for speciation.

Biogeographic History

In 1854, Alfred Russel Wallace set sail from Singapore to explore the biology of the Malay Archipelago (now Malaysia and Indonesia) (Fig. 1). He spent eight years crisscrossing between the islands and collected over 125,000 mammal, bird, and butterfly specimens. His interpretations of the distributions of the species he collected were published in his book *The Malay Archipelago* in 1869, laying the foundation for much of modern biogeography. One of his most intriguing observations was the apparent lack of relationship between the distances among some islands and their faunal affinities. On Borneo he found a diversity of Old World monkeys (Cercopithecidae), wild cats (Felidae), deer (Cervidae), and civets (Viverridae) that were largely absent from Sulawesi (Celebes), some 200 km away across the Makassar Strait. Moreover, Sulawesi lacked the overall diversity of Borneo, but had numerous endemic species. Some endemics had clear Asian affinities: the tusked babirusa (*Babyrousa babyrussa*), a member of the pig family; the dwarf buffaloes or anoas (*Bubalus depressicornis* and *B. quarlesi*); and the endemic macaques (*Macaca*) and tarsiers (*Tarsius*). Others, such as the marsupial cuscus species (*Strigocuscus pelengensis* and *S. celebensis*) and the maleo (*Macrocephalon maleo*), a large turkey-like bird (Megapodiidae) that uses geothermal heat to incubate its eggs, had clear Australasian affinities.

Conversely, despite far greater intervening distances, the Borneo fauna was very like that of Sumatra, Java, and the Malay Peninsula, notable for the shared presence of gibbons (Hylobatidae), rhinoceros (Rhinocerotidae), and elephants (Elephantidae). A similar disjuncture occurred between Lombok and Bali in the Lesser Sundas. The separation provided by the Lombok Strait was as little as 6 km in some places, yet many families of birds that were common on Bali and its western neighbors were greatly reduced or absent east of Lombok (e.g., the barbets [Megalaimidae], bulbuls [Pycnonotidae], and woodpeckers [Picidae]), whereas Lombok seemed to mark the western limit of Australian birds such as the honeyeaters (Meliphagidae), cockatoos (Cacatuidae), and brush-turkeys (Megapodiidae).

The Wallace Line

To delineate the biogeographical divisions he observed, Wallace described a line following the deep water of the Lombok Strait up through the Makassar Strait between Borneo and Sulawesi. Today, the Wallace Line is best viewed as the eastern edge of a region dominated by Asian species and the western limit of Australian species, and Huxley's modification of this line marked the boundary of the Oriental region of the Paleotropical Realm. Lydekker's Line delineates the boundary of the Australian region, part of the Notogean (or Australian) Realm, and represents the eastern limit of Asian species and the western edge of dominance by Australian species (Fig. 2). The area between these lines is now generally accepted as a biogeographic region in its own right, Wallacea. Named for Wallace, the region includes Sulawesi, the Lesser Sunda Islands (except Bali), and the Moluccas (excluding the Aru Islands). Although some view this as a just transition zone between the Oriental and Australian regions, others argue that the high degree of endemicity in most taxonomic groups confers region status.

Biogeographic divisions based on flora differ slightly in that the Australian kingdom does not include New Guinea, and as a consequence all the flora of Indonesia falls into the Malesian subkingdom of the Paleotropical kingdom. The Malesian subkingdom is further split into subdivisions that, again, differ slightly from the faunal regions. The western subdivision includes Sumatra and Borneo (with the Malay Peninsula and the Philippines); Java and the Lesser Sundas are in the southern division; the eastern division encompasses Sulawesi, the Moluccas, and New Guinea.

Tectonic Origins and Pleistocene Sea Levels

Wallace correctly inferred that Sumatra, Borneo, Java, and the Malay Peninsula, separated by wide expanses of uniformly shallow seas (Fig. 2), were at one point part of a single land mass. His boundary traces the edge of the Sunda Shelf, the continental shelf extension of the Southeast Asian mainland. Similarly, Lydekker's Line traces the Sahul Shelf underlying New Guinea, Australia, and the Aru Islands (shallow, smooth water in Fig. 2). Wallace speculated that the similarities among species on islands within these areas might be explained by periods of low sea level, connecting the islands and promoting biotic exchange. However, it was not until the field of plate tectonics came to the fore in the 1970s that a mechanistic explanation for the complex mix of Australian and Oriental faunas could be proffered. Although there are limits to the extent to which geology alone can explain

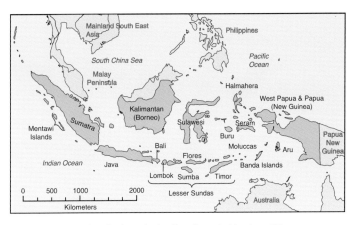

FIGURE 1 Geography of Indonesia detailing the main islands and those discussed in the text. Areas shaded dark green are part of Indonesia; light green areas are other Southeast Asian countries.

biogeography, it does provide a general explanation for the broader distribution patterns seen in Indonesia.

Prior to the Cenozoic, the various land masses that now constitute Sundaland were last in contact with the northern Australia–New Guinea continental margin about 200 million years ago, when they formed part of the great southern continent Gondwana. Continental slivers successively rifted from Gondwana during the Mesozoic, drifting north to eventually collide with Asia and join to form the Sundaland core. Ocean basins created in the process kept the Oriental biota far removed from that of Australia, and it was not until approximately 50 million years ago that tectonic movements split Australia and New Guinea from Gondwana, and the Australia plate began to rapidly drift northward. It brought with it very different flora and fauna from that seen in Asia, with Gondwanan groups that are considered diagnostic for the Notogean Realm including four of the six orders of mar-

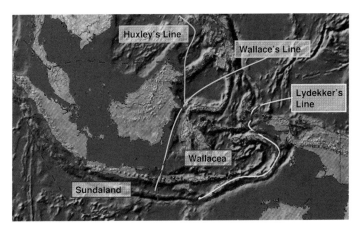

FIGURE 2 Shaded relief map of Indonesia and surrounding Southeast Asian countries to illustrate topography and bathymetry, key faunal boundary lines, and the Sundaland and Wallacea biodiversity hotspots (dashed red lines).

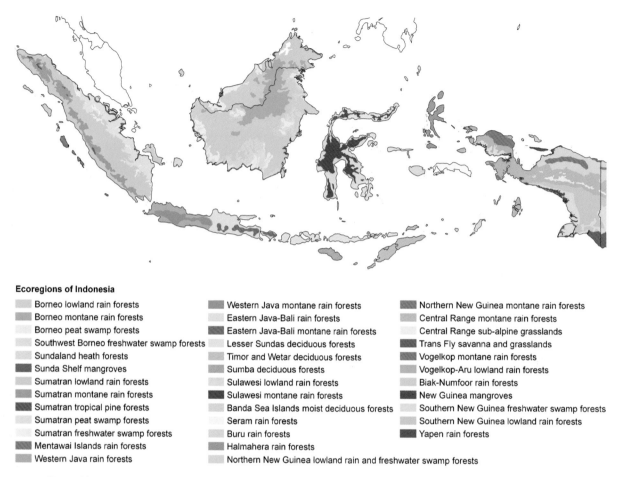

FIGURE 3 Terrestrial Ecoregions of Indonesia. Constructed from the Terrestrial Ecoregion GIS Database of Olson *et al.* (2001).

supials, monotremes (egg-laying mammals such as the platypus [*Ornithorhynchus anatinus*]), ratites (flightless birds such as the cassowary [*Casuarius*]), lungfish, snake-necked turtles (Chelydae), and tree frogs (Hylidae).

In the last 30 million years there has been progressive closure of the marine gap between the Oriental and Australian faunas as the Australian plate collided with the Eurasian plate. The collision also led to the emergence of many of the Wallacea islands, as well as to the uplift of the mountains of New Guinea. However, connections that might have permitted direct migrations between the two faunas appear to have been limited; tectonic movements also opened up deep ocean basins in the Wallacea region, and sea-level fluctuations were such that many of the islands that, as we recognize today, would not have been exposed (and in some cases would not have originated) until approximately 5 million years ago. In contrast, exposure of the Sunda Shelf would have linked Borneo, Java, Sumatra, the Malay Peninsula, and intervening small islands to create a single geographical unit, Sundaland, which was largely emergent or intermittently transected by a very shallow sea throughout the Cenozoic, providing a biogeographic connection from Borneo through to Asia. Cenozoic connections between Australia and New Guinea were limited, as the land masses were not consistently emergent before the last 5 million years.

The most recent opportunity for land connections among the Indonesian islands was during the last glacial maximum, approximately 18,000 years ago. Sea levels were about 120 m lower than today, exposing much of the Sunda Shelf. Although there is some debate about the nature of the exposed vegetation (whether forest and mangrove swamp or savannah), Borneo, Java, Sumatra, and the Malay Peninsula would have been connected. Similar exposure of the Sahul Shelf would have connected New Guinea with Australia and the Aru Islands. The point at which connections between islands were broken coming out of the last ice age varied with the depth of the surrounding sea floor, but by 7000 years ago, sea levels would have risen enough to present the Indonesian archipelago configuration much as we recognize it today.

TABLE 1
Biodiversity of Indonesia

	Number of species		Projected species loss by 2100	
	Recorded (% world total)	Endemic (% Indonesian total)	Indonesian extinctions (min.–max.)	Global extinctions (min.–max.)
Vascular plants	29,375 (11)	17,500 (60)	5797–28,057	3454–16,715
Mammals	667 (14)	216 (32)	161–537	97–323
Birds	1604 (16)	443 (28)	231–624	149–402
Reptiles	749 (9)	227 (32)	32–407	13–167
Amphibians	347 (6)	159 (12)	12–251	10–219

NOTE: Table shows number of species recorded, endemic species, and projected species losses by 2100 for five major components of biodiversity.

Terrestrial Ecoregions

The biotic composition of an individual island also reflects the size of the island and the diversity of habitats that occur on it. Moreover, the extent to which particular groups of species are able to overcome dispersal barriers varies significantly with the ecology of the group or even of the species. Nonetheless, despite the complex interplay of geology, sea levels, and ecology, geographically and biotically distinct assemblages of natural communities have evolved and can be grouped as ecoregions.

Thirty-seven ecoregions are described for Indonesia (Fig. 3), and their distribution reflects the synergy between biogeographic processes. Larger islands with complex topography and geological history typically support several ecoregions. New Guinea is bisected by a very high mountain cordillera (peaks reaching 4000 m) separating the island into distinct north and south regions and supporting several montane habitats. Thus, there are 11 ecoregions in the western half of the island alone. At the same time, the distribution of ecoregions on different islands, particularly those of Wallacea, frequently reflects land connections or low sea levels during the Pleistocene. Despite the proximity of the islands of the Lesser Sundas, they constitute three ecoregions. The Lesser Sundas deciduous forests include Lombok, Sumbawa, and Moya (which were merged as Greater Sumbawa) and Komodo, Flores, and Lembata (Greater Flores). In the outer Banda arc, Sumba was separated from Greater Flores by some 50 km of deeper water and supports its own ecoregion, as do the islands that made up Greater Timor (Roti, Semau, and Timor) and the island of Wetar. In the Moluccas, the Halmahera Islands were most likely connected during the Pleistocene, and they now compose a single ecoregion (Halmahera rainforests). Similarly, Ambon and Seram both fall within the Seram rainforests, but neighboring Buru, which has always been isolated by deep seas, has its own ecoregion.

BIODIVERSITY OF INDONESIA: TRENDS AND KEY FEATURES

Home to nearly 12% of the world's land vertebrates and 10% of the world's vascular plants, Indonesia is one of the most biologically rich countries in the world. Critically, much of the country's flora and fauna is found nowhere else; endemic species account for 45% of the amphibians, 32% of the mammals, 28% of the birds, and a staggering 60% of the vascular plants. Tragically, only about 50% of the original natural forest area remains, and based on current deforestation rates, as little as 11% may survive to the end of this century. As a consequence, even the minimum projected species losses for some groups are of major concern, with the global extinction of 97 of the 667 mammal species and 149 of the 1604 bird species anticipated to occur by 2100 (Table 1).

Indonesia has consequently been considered a "megadiversity country" since the introduction of the concept by Russell Mittermeier in the late 1980s. It encompasses the Sundaland and Wallacea biodiversity hotspots, apolitical areas delineated by high levels of vascular plant endemism (more than 1500 species) and a high degree of habitat threat (70% loss of original vegetation), and prioritized further by considerations of endemism in birds, mammals, reptiles, and amphibians. In the east, about half the New Guinea Tropical Wilderness Area falls within Indonesia (Fig. 2, Table 2).

Sundaland Biodiversity Hotspot

Biotic exchange during periods of exposure of the Sunda Shelf has resulted in major floral and faunal similarities across the hotspot, particularly in the lowland forests, which are dominated by the family Dipterocarpaceae. Nonetheless both species richness and endemism are high in all groups (Table 2), with particularly high levels of endemism in the montane ecoregions.

TABLE 2
Geography and Diversity of the Sundaland and Wallacea Biodiversity Hotspots and New Guinea Wilderness Area

	Sundaland biodiversity hotspot			Wallacea biodiversity hotspot			New Guinea wilderness area		
Geography									
Area (km^2)	1,501,063			338,494			828,818		
Vegetation remaining (%)	6.7			15.0			70.0		
Protected area (%)^a	5.2			5.8			11.0		
Population density (people/km^2)	153			81			5.1		
Terrestrial ecoregions	16			9			17^b		
Species richness	Recorded		Endemic (%)	Recorded		Endemic (%)	Recorded		Endemic (%)
Vascular plants	25,000		15,000 (60)	10,000		1500 (15)	17,000		10,200 (60)
Mammals	380		172 (45)	222		127 (57)	233		146 (63)
Birds	769		142 (18)	647		262 (40)	650		334 (58)
Reptiles	452		243 (54)	222		99 (45)	275		159 (58)
Amphibians	244		196 (80)	48		33 (69)	237		215 (91)

NOTE: Figures include contributions from Malaysia and Brunei to the Sundaland Hotspot and from Papua New Guinea to the New Guinea Wilderness Area.
^{a}Protected areas in IUCN categories I–IV are those managed primarily for science, conservation, or ecosystem function and afford the highest levels of protection.
^{b}12 ecoregions in Indonesian New Guinea.

The hotspot is perhaps best known for its many charismatic but threatened mammals, including two species of orangutan (*Pongo pygmaeus* and *P. abelii*); two critically endangered rhinoceros (*Rhinoceros sondaicus* and *Dicerorhinus sumatrensis*); the Sumatran tiger (*Panthera tigris sumatrae*); two subspecies of Asian elephants (*Elephas maximus sumatrensis* and *E. m. borneensis*); the vulnerable Malayan tapir (*Tapirus indicus*), and the endangered proboscis monkey (*Nasalis larvatus*). Sundaland is also noteworthy for several large endemic threatened reptiles such as the endangered false gharial (*Tomistoma schlegelii*), a freshwater crocodilian species found mostly in Sumatra and Borneo, and two species of critically endangered large river terrapins: the mangrove terrapin (*Batagur baska*) and the painted terrapin (*Callagur borneoensis*).

SUMATRA

Species richness on Sumatra is comparable with the larger islands of Borneo and New Guinea, with approximately 10,000 plant species, 201 mammal species, and 465 breeding bird species. The diverse lowland fauna has much in common with other parts of Sundaland, but the proximity and past connections to peninsular Malaysia mean that a number of Asian species, including 22 mammals, are found that do not occur on other Indonesia islands (e.g., the siamang [*Symphalangus syndactylus*], the great hornbill [*Buceros bicornis*], the monotypic fire-tufted barbet [*Psilopogon pyrolophus*]). In contrast, the more isolated montane forests support eight endemic bird species, such as the threatened Sumatran cochoa (*Cochoa beccarii*) and Sumatran ground-cuckoo (*Carpococcyx viridis*), and seven endemic mammals, including Thomas's leaf-monkey (*Presbytis thomasi*) and the critically endangered Sumatran rabbit (*Nesolagus netscheri*). The drier montane forests are dominated by the Sumatran pine (*Pinus merkusii*), which forms the only coniferous ecoregion in Indonesia. The forests of both Sumatra and Borneo are also noteworthy for 16 species of *Rafflesia*, a genus of parasitic plants that derive energy and nutrients from the tissues of *Tretrastigma* vines. Among their number is the famous *R. arnoldii*, which produces the largest single flower (1 m in diameter) in the world.

Most of the ecoregions have been severely reduced in extent, and total loss of the lowland forests is predicted if current exploitation rates continue. Sumatra is the last refuge for the critically endangered Sumatran orangutan, with less than 3500 individuals in the wild, and it also still contains small populations of the Sumatran rhino (*D. sumatrensis*), the Sumatran tiger, the elephant *E. m. sumatrensis*, and the false gharial.

JAVA AND BALI

Java comprises an island arc of volcanoes that have coalesced to form a large single island, which has probably been consistently emergent only since the Early Pliocene (5 million years ago). It is smaller, more isolated, and less species-rich than are Borneo and Sumatra, with the moister evergreen forests of western Java supporting greater species richness than do the drier forests of eastern Java and Bali. These drier climates may have facilitated the persistence of about 30 seasonal forest birds from mainland Asia (e.g., the green peafowl [*Pavo muticus*], the lineated barbet [*Megalaima lineate*]), which are presumed to have gone extinct on Sumatra and Borneo with the reestablishment of rainforest after the last glacial maximum.

Java is one of the most densely populated islands in the world, and little of its original habitat remains. Each of the four ecoregions harbors critically endangered animals: the western Java rainforests, reduced to less than 5% of their original extent, have the last viable population (less than 40 to 50 individuals) of Javan Rhinoceros (*R. sondaicus*); the western Java montane rainforests (less than 20% of which remain) are home to the most endangered primates in Indonesia, the Javan leaf monkey (*Presbytis comata*) and the Javan gibbon (*Hylobates moloch*); and the eastern Java-Bali rainforest on the island of Bawean is home to an endangered deer (*Axis kuhlii*). Also of concern are the endangered Javan warty pig (*Sus verrucosus*) and several Javan subspecies of carnivore (e.g., the yellow-throated marten [*Martes flavigula robinsoni*] and the leopard *Panthera pardus melas*). The Javan and Balinese subspecies of tigers (*Panthera tigris sondaica* and *P. t. balica*) were last recorded on the islands in 1976 and in the late 1930s, respectively, and both are believed extinct.

Several native bird species are equally imperiled, most famously the Bali starling (*Leucopsar rothschildi*), a critically endangered Bali endemic reduced to just six wild individuals by 2001 by trapping for the illegal cagebird trade. There have been no records of the critically endangered Javanese lapwing (*Vanellus macropterus*) since 1940, and the endangered Javan hawk eagle (*Spizaetus bartelsi*) is confined to the remaining patches of forest.

KRAKATAU

The Krakatau Islands lie in the Sunda Strait between Java and Sumatra and are famous for the cataclysmic eruption of the Krakatau volcano on the main island (Krakatau or Rakata) in 1883. No life was left on the island, and patterns of recolonization of flora and fauna have been central tests of MacArthur and Wilson's Equilibrium Theory of Island Biogeography. Colonization rates of resident land birds to 1933 seemed to suggest that the number of species on the island reflected a dynamic equilibrium between immigration and extinction processes, as MacArthur and Wilson proposed. However, more recent studies, based on longer time series, suggest that persistence probabilities for bird and butterfly species are strongly influenced by successional changes in plant communities.

MENTAWAI ISLANDS

The four Mentawai islands have been separated by an oceanic trench from the west coast of Sumatra for more than 0.5 million years. They support a distinct ecoregion characterized by no less than 17 endemic mammals, including four threatened primates: Mentawai gibbon (*Hylobates klossii*); Mentawai macaque (*Macaca pagensis*); Mentawai leaf-monkey (*Presbytis potenziani*); and the sole representative of an endemic genus, the snub-nosed monkey (*Simias concolor*).

INDONESIAN BORNEO—KALIMANTAN

Kalimantan is the Indonesian region of Borneo and occupies the central and southern two-thirds of the island. Borneo is the third largest island in the world and supports the highest floral diversity in Malesia (more than 15,000 plant species, with 59 endemic genera). It is at the center of dipterocarp diversity with over 265 species, of which at least 155 are endemic, and it further boasts over 3000 tree species and 2000 orchids. Despite a large Sundaland element to the lowland fauna, endemism on Borneo is high for most groups. There are 39 (out of 350) endemic bird species and at least 44 (out of 222) mammals, including the endangered proboscis monkey, the Bornean orangutan (*Pongo pygmaeus*), and the Borneo pygmy elephant (*Elephas maximus borneensis*). There are in the region of 100 endemic amphibians, 47 lizards and 41 snakes, and the island is the only home for the monotypic Lanthanotidae, represented by the very rare Bornean earless monitor lizard (*Lanthanotus borneensis*).

Since 1994, over 400 new species have been discovered on Borneo, and Borneo and Sumatran populations of the clouded leopard have been shown to have diverged from mainland Asian populations about 1.4 million years ago, elevating them to specific status (*Neofelis diardii*). The clouded leopard is the largest predator on Borneo, but in the absence of tigers and leopards, there are at least 26 species of small- to medium-sized carnivores.

The highlands of Borneo are the most extensive, and are among the oldest (with a 20-milion-year history), in Indonesia, and a mix of Asian and Australian plant families has produced the most species-rich montane forests in the Indo-Pacific region, supporting some 150 mammals and 250 birds. In 2007, representatives of the three Bornean governments of Malaysia, Brunei Darussalam, and Indonesia signed the Declaration on the Heart of Borneo Initiative. This historic initiative is intended to conserve and sustainably manage approximately 220,000 km² of rainforest, primarily encompassing the transboundary highlands of Indonesia and Malaysia.

Wallacea Biodiversity Hotspot

Wallacea comprises the large island of Sulawesi and thousands of smaller islands grouped into two archipelagoes, the Lesser Sundas (Nusa Tenggara) and the Moluccas (Maluku). Although the total species richness is generally less than in Sundaland, this is in part due to the much smaller total land area of the region. For most of the major taxonomic groups, richness is actually 1.5–3 times greater than would be expected for an equivalent area on Sundaland. The exception is the amphibians, for which the deep seas appear to have represented a major barrier to dispersal, and only 48 species have been described to date. Levels of endemism are high for all groups, ranging from 15% of vascular plants to 69% of amphibians (Table 2). Although the invertebrate fauna of Wallacea is generally poorly known, endemism is known to be high in the more conspicuous groups such as the huge bird-wing butterflies (*Ornithoptera*) (40 endemic species of 80 recorded) and the tiger beetles (Cicindelidae) (79 of 109).

Endemism at the level of the island is also high, and it seems that the many opportunities for allopatric speciation provided by the thousands of islands have led to rapid radiations in some groups. The avifauna provides a particularly good illustration, with about 25% of the world's accipiters (Accipitridae), 19% of pigeons and doves (Columbidae), 28% of kingfishers (Alcedinidae), and 37% of Megapodiidae. Other groups, such as the Phasianidae and Anseriformes, are greatly underrepresented.

LESSER SUNDA ISLANDS

The Lesser Sundas are an inner volcanic island arc and are unusual in Indonesia because a much drier and more seasonal climate has resulted in deciduous or "monsoon" forests, dominated by the legume *Pterocarpus indicus*. Forest subtypes range from dry thorn to moist deciduous, but human activity has degraded many areas to savanna. The driest islands of Komodo, Padar, Rinca, and Flores are home to the largest lizard in the world, the vulnerable Komodo dragon (*Varanus komodoensis*). Males reach 2.8 m in length and can weigh 50 kg. The only turtle found in Wallacea is also in the Lesser Sundas. McCord's sideneck turtle (*Chelodina mccordi*) represents the westernmost extent of the Chelidae, the dominant turtle family in Australia and New Guinea. It is critically endangered and limited to three populations on the tiny island of Roti where just 70 km² of suitable habitat remain.

SULAWESI

Geologically, Sulawesi is a complex result of convergence of the Australian, Pacific, and Eurasian plates, bringing together five separate paleo-islands. Critical to the biogeography of the region are the timing and positioning of emergent fragments relative to the depth and extent of separating straits and oceans. Parts of southeastern Sulawesi were likely emergent about 20 million years ago, but with few other islands above water, connections to either Sundaland or Australia would have been limited. It was not until 10–5 million years ago that substantial land is known for Sulawesi, but at this point the Makassar Strait would have been fairly wide, limiting colonization from Borneo. This would have likely been the best chance (about 10 million years ago) for land animals and plants to cross Wallacea, as seas were relatively narrow prior to the opening of several deep ocean basins. These deep seas further prevented land connections during the low sea levels of the Pleistocene (with the exceptions of links to the Sula Islands and Tukang Besi Islands), although the Makassar Strait would have been narrower.

Thus, opportunities for "easy" colonization of Sulawesi have been limited, and it provides a good example of an unbalanced endemic fauna, depauperate in species such as large herbivores and large carnivores, which have poor dispersal abilities. Although 62% of the mammals are endemic, there is only a single native carnivore, the endemic Sulawesi palm civet (*Macrogalidia musschenbroekii*), and the only large herbivores are the two anoa species and a deer (*Cervus timorensis*). Similarly, although primate diversity appears relatively high with 12 species, seven are endemic macaques with allopatric distributions, and the rest are tarsiers. Thus, at a single locality, there will typically be one species of macaque and one (or two) tarsier(s). The macaque distributions coincide with the distributions of genetically differentiated populations of toads (*Bufo celebensis*), grasshoppers (*Chitaura*), fanged frogs (*Limnonectes*), and flying lizards (*Draco lineatus*), suggesting that periodic historic barriers to dispersal across the island (e.g., oceanic inundation, large rivers, soil or vegetation shifts) have promoted speciation and created centers of genetic endemism.

MOLUCCAS

Collectively the Moluccas are distinguished by their more even mix of Asian and Australian faunas and by their extraordinary degree of bird endemism, which is greater than anywhere else in Asia. There are 21 species endemic to the Banda Sea Islands moist deciduous forests (225 species), 16 in the Seram deciduous forests (213 species), ten in the Buru rainforests (173 species); the Halmahera rainforests include four endemic monotypic genera (*Habroptila*, *Melitorgrais*, *Lycocorax*, and *Semioptera*) among the 26 endemics (223 species). Many islands are small and uninhabitable but are critical breeding sites for seabirds such as frigatebirds (Fregatidae), tropicbirds (Phaethontidae), boobies (*Sula*), terns (*Sterna*), and smaller species. Typically, the mammal fauna is unbalanced and depauperate, but with both Asian and Australasian affinities, and each ecoregion has several endemic species.

New Guinea Tropical Wilderness Area

New Guinea is the second largest island in the world and is one of the Earth's three remaining major tropical wilderness areas—regions characterized by extent (greater than 10,000 km^2), non-urban population density (fewer than 5 people per km^2), and amount of intact habitat (greater than 70%) (Table 2). The eastern half of the island is an independent nation, Papua New Guinea, but two Indonesian provinces, West Papua and Papua (formerly Irian Jaya), cover the western half. Although much of the forest remains intact, it was also little explored until recent expeditions by Conservation International and the Indonesian Institute of Science uncovered dozens of new species, including 20 new frog species and a giant *Mallomys* rat weighing 1.4 kg. It is likely that all measures of diversity for the area are greatly underestimated.

Levels of diversity and endemism are comparable to Sundaland and Wallacea (Table 1), but, bounded by Lydekker's Line, the flora and fauna of New Guinea are predominantly Australian. Three of the four species of extant monotremes (all echidnas in the family Tachyglossidae) are present, and the genus *Zaglossus* is endemic to New Guinea. The mammalian fauna is otherwise dominated by marsupials, with rats and bats being the only native placental mammals. There are no living large carnivores, although there is fossil evidence of the thylacine, but 16 species of the small- to mid-sized carnivorous Dasyuridae (quolls, marsupial shrews, dasyures, dunnarts, and a planigale) are present. Of the 22 Macropodidae (kangaroos and wallabies), it is the ten endemic tree kangaroos (*Dendrogalus*) and six forest wallabies (*Dorcopsis* and *Dorcopsulus*) that distinguish New Guinea and the surrounding islands, although there are also five species of the smallest macropods, the pademelons (*Thylogale*), mainly in eastern New Guinea (Papua New Guinea). Cuscus (Phalangeridae) evolved on New Guinea and abound with at least 13 species, and there are many small herbivorous species in the families Acrobatidae, Burramyidae, Petauridae, and Pseudocheiridae.

Although the avifauna is largely Australian, there are a number of Asian elements, some of which do not extend into Australia, namely the tree swifts (Hemiprocnidae), shrikes (Laniidae), and sandpipers (Scolopacidae). Australian families exhibiting high diversity in the ancient forests of New Guinea include the famous birds of paradise (Paradisaeidae), bower birds (Ptilonorhynchidae), and honeyeaters (Meliphagidae). New Guinea is also home to all three species of large flightless cassowaries (*Casuarius*). Gondwanan floral elements include the conifers *Podocarpus* and the rainforest emergents *Araucaria* and *Agathis*, as well as tree ferns and several species of *Eucalyptus*.

SEE ALSO THE FOLLOWING ARTICLES

Borneo / Indonesia, Geology / Island Biogeography, Theory of / Krakatau / Wallace's Line

FURTHER READING

Conservation International's Biodiversity Hotspots website http://www.biodiversityhotspots.org/Pages/default.aspx

MacKinnon, K., G. Hatta, H. Halim, and A. Mangalik. 1997. *The ecology of Kalimantan, Indonesian Borneo*. The Ecology of Indonesia Series, volume III. Hong Kong: Periplus Editions.

Marshall, A. J., and B. M. Beehler. 2007. *The ecology of Indonesian Papua: part one*. The Ecology of Indonesia Series, volume VI. Hong Kong: Periplus Editions.

Metcalf, I., J. M. B. Smith, M. Morwood, and L. D. Davidson, eds. 2001. *Faunal and floral migrations and evolution in S E Asia–Australasia*. Lisse, Netherlands: A. A. Balkema (Swets & Zeitlinger Publishers).

Monk, K. A., Y. de Fretes, and G. Reksodiharjo-Lilley. 1997. *The ecology of Nusa Tenggara and Maluku*. The Ecology of Indonesia Series, volume V. Hong Kong: Periplus Editions.

Olson, D. M., E. Dinerstein, E. D. Wikramanayake, N. D. Burgess, G. V. N. Powell, E. C. Underwood, J. A. D'amico, I. Itoua, H. E. Strand, J. C. Morrison, C. J. Loucks, T. F. Allnutt, T. H. Ricketts, Y. Kura, J. F. Lamoreux, W. W. Wettengel, P. Hedao, and K. R. Kassem. 2001. Terrestrial ecoregions of the world: a new map of life on Earth. *BioScience* 51: 933–938. http://www.worldwildlife.org/science/ecoregions/terrestrial.cfm

Sodhi, N., and B. W. Brook. 2006. *Southeast Asian biodiversity in crisis*. Cambridge: Cambridge University Press.

Whitten, T. J., S. J. Damanik, J. Anwar, and N. Hisyan. 2000. *The ecology of Sumatra*. The Ecology of Indonesia Series, volume I. Hong Kong: Periplus Editions.

Whitten, T. J., G. S. Henderson, and M. Mustafa. 2002. *The ecology of Sulawesi*. The Ecology of Indonesia Series, volume IV. Hong Kong: Periplus Editions.

Whitten, T. J., R. E. Soeriaatmadja, and S. A. Affif. 1997. *The ecology of Java and Bali*. The Ecology of Indonesia Series, volume II. Hong Kong: Periplus Editions.

INDONESIA, GEOLOGY

ROBERT HALL

Royal Holloway University of London, United Kingdom

Indonesia is a geologically complex region situated at the southeastern edge of the Eurasian continent. It is bordered by tectonically active zones characterized by intense seismicity and volcanism resulting from subduction. Western Indonesia is largely underlain by continental crust, but in eastern Indonesia there is more arc and ophiolitic crust, and several young ocean basins. The Indonesian archipelago formed over the past 300 million years by reassembly of fragments rifted from the Gondwana supercontinent that arrived at the Eurasian subduction margin. The present-day geology of Indonesia is broadly the result of Cenozoic subduction and collision at this margin.

PRESENT-DAY TECTONIC SETTING

Indonesia is an immense archipelago of more than 18,000 islands extending over 5000 km from east to west between 95° and 141° E, and crossing the equator from 6° N to 11° S (Figs. 1 and 2). It is situated at the boundaries of three major plates: Eurasia, India-Australia, and Pacific-Philippine Sea (Fig. 1). In western Indonesia, the boundary between the Eurasian and Indian plates is the Sunda Trench. Parallel to this in Sumatra is the right-lateral strike-slip Sumatran Fault, which results from the partitioning of oblique plate convergence into normal convergence at the trench and trench-parallel movement further north. Most active deformation in Sumatra occurs between the trench and the Sumatran fault. In contrast, east of Java, active deformation occurs within a complex suture zone up to 2000 km wide, including several small plates and multiple subduction zones; plate boundaries (Fig. 1) are trenches and another major strike-slip zone, the left-lateral Sorong Fault, which runs from New Guinea into Sulawesi. Global Positioning System (GPS) measurements indicate very high rates of relative motions, typically more than several centimeters per year, between tectonic blocks in Indonesia.

Volcanism and Seismicity

The subduction zones are mainly well defined by seismicity extending to depths of about 600 km (Fig. 3) and by volcanoes (Fig. 1). There are at least 95 volcanoes in Indonesia that have erupted since 1500, and most are situated between 100 and 120 km above descending lithospheric slabs. Thirty-two have records of very large eruptions with a volcanic explosivity index (VEI) of greater than 4; 19 have erupted in the last 200 years, including Tambora in 1815 (VEI = 7) and Krakatau in 1883 (VEI = 6). Tambora, on the island of Sumbawa, is known for its impact

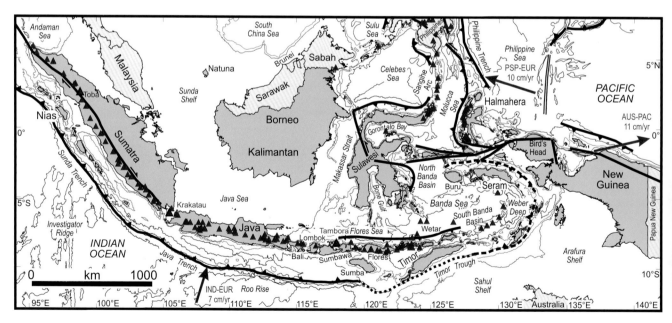

FIGURE 1 Geography of Indonesia and surrounding regions showing present-day tectonic boundaries and volcanic activity. Indonesia is shaded green, and neighboring countries are shaded in pale gray. Bathymetric contours are at 200 m, 1000 m, 3000 m, 5000 m, and 6000 m. The location of the three most famous explosive eruptions known from Indonesia are shown in red text. Red arrows show plate convergence vectors for the Indian plate (IND-EUR) and the Philippine Sea plate (PSP-EUR) relative to Eurasia, and for the Australian plate relative to the Pacific plate (AUS-PAC). There is little thrusting at the Timor trough. The Seram trough and Flores-Wetar thrusts are the sites of active thrusting.

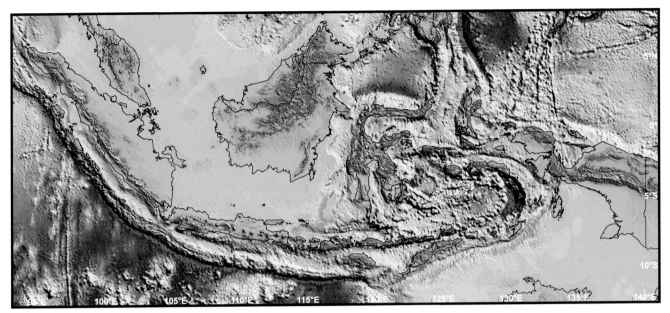

FIGURE 2 Digital elevation model showing topography and bathymetry of the Indonesian region. Compare to Fig. 1 for tectonic and geographic features.

on global climate, and its 1815 eruption resulted in the Northern Hemisphere's 1816 "year without a summer," when crops failed, causing famine and population movements. The eruption of Toba on Sumatra 74,000 years ago was even bigger (estimated VEI of 8) and is the largest eruption known on Earth in the last 2 million years.

Sundaland

The interior of Indonesia (Fig. 2), particularly the Java Sea, Sunda Shelf, and surrounding emergent, but topographically low, areas of Sumatra and Kalimantan (Indonesian Borneo), is largely free of seismicity and volcanism (Figs. 1 and 3). This tectonically quiet region forms part of the continental core of the region known as Sundaland (Fig. 4).

Sundaland extends north to the Thai-Malay Peninsula and Indochina, and formed an exposed landmass during the Pleistocene. Most of the Sunda Shelf is shallow, with water depths considerably less than 200 m (Fig. 2), and its lack of relief has led to the misconception that it is a

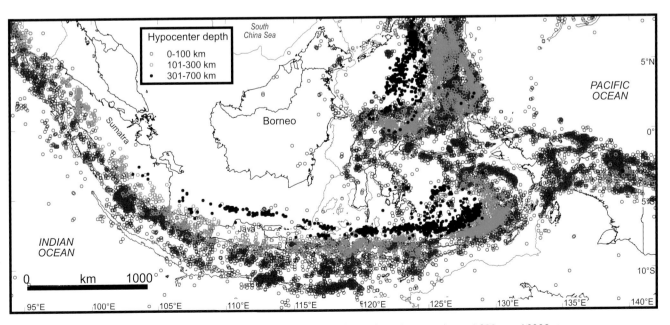

FIGURE 3 Seismicity in the Indonesian region between 1964 and 2000. Bathymetric contours are shown at 200 m and 6000 m.

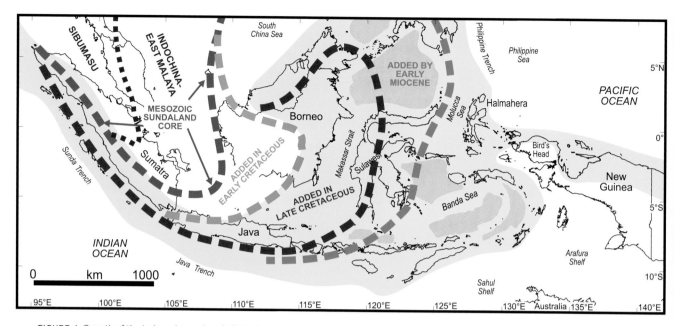

FIGURE 4 Growth of the Indonesian region. Collision between the Sibumasu and East Malaya-Indochina blocks occurred in the Triassic. Additional crust has been added to this Sundaland core, largely by later collisions of continental blocks. The present-day zone of active deformation is shaded yellow. Gray areas within this complex plate-boundary zone are areas underlain by Cenozoic ocean crust.

stable area. Sundaland is often described as a shield or craton, but geological observations, heat flow, and seismic tomography show that this is not the case. There has been significant deformation during the Cenozoic with the formation of deep sedimentary basins and the localized but widespread elevation of mountains. Unlike well-known shields or cratons (e.g., Baltic or Canadian), Sundaland is not underlain by a thick, cold lithosphere formed in the Precambrian. Its interior has high surface heat flow values, typically greater than 80 mW/m^2. At the Indonesian margins, high heat flows reflect subduction-related magmatism, but the hot interior of Sundaland appears to be the consequence of high upper-crustal heat flow from radiogenic granites and their erosional products, the insulation effects of sediments, and a high mantle heat flow. P- and S-wave seismic tomography show that Sundaland is an area of low velocities in the lithosphere and underlying asthenosphere, in contrast to the colder and thicker Indian and Australian continental lithosphere to the northwest and southeast. Such low mantle velocities are commonly interpreted in terms of elevated temperature, and this is consistent with regional high heat flow, but they may also partly reflect a different composition or elevated volatile contents.

Consequences of Long Subduction History

The upper mantle velocities and heat flow observations suggest the region is underlain by a thin and weak lithosphere. In contrast, in the lower mantle beneath Indonesia, there is a high velocity anomaly, suggesting the accumulation of subducted lithosphere.

These features are the consequence of long-term subduction at the Indonesian margins. The active margins are the site of magmatism, heating, and weakening, but the region of weak lithosphere extends many hundreds of kilometers from the volcanic margins. The character of the lithosphere has been a major influence on the development of Indonesia, combined with repeated collisions at the subduction margins that have led to continental growth.

PRE-CENOZOIC HISTORY OF INDONESIA

Western Indonesia, notably the islands of Sumatra and Borneo, contains most of the oldest rocks in Indonesia.

Sumatra: Basement

Sumatra represents the geological continuation of the Malay Peninsula and contains the most extensive outcrops of Paleozoic and Mesozoic rocks. The oldest rocks at the surface are Carboniferous sediments, but possible Devonian rocks have been reported from petroleum boreholes in the Malacca Straits, and granites from boreholes in central Sumatra have been dated as Silurian. Xenoliths in dykes, granite clasts in sediments, and various high-grade metamorphic rocks from different parts of Sumatra suggest a pre-Carboniferous crystalline basement similar to that beneath the Malay Peninsula, which is Proterozoic at depth.

Sumatra: Cathaysian and Gondwana Blocks

In western Sumatra there are Paleozoic sediments that range in age from Carboniferous to Triassic, and Permian volcanic rocks with Cathaysian affinities. These are interpreted to form part of an Indochina–East Malaya block (Fig. 4) that separated from Gondwana in the Devonian and by the Carboniferous was in warm tropical low latitudes where a distinctive flora developed.

In contrast, in eastern Sumatra, Carboniferous sediments include pebbly mudstones interpreted as diamictites that formed in a glacio-marine setting. These indicate cool Gondwana affinities. The Carboniferous rocks and associated Permian and Triassic sediments belong to the Sinoburmalaya or Sibumasu block (Fig. 5), which was at high southern latitudes during the Carboniferous, separated from Gondwana in the Permian, and collided with the Indochina–East Malaya block, already amalgamated with the South and North China blocks, in the Triassic.

Sumatra: Triassic Collision

The collision of the Sibumasu and Indochina–East Malaya blocks was the first stage in the geological development of Indonesia. The widespread Permian and Triassic granites of the Thai-Malay tin belt extend into western Indonesia and are the products of subduction and post-collisional magmatism associated with this event.

Sumatra: Mesozoic

The Mesozoic sedimentary record is very limited but suggests that much of Sundaland, including most of its Indonesian margin, was emergent. During the Mesozoic, there is interpreted to have been reorganization of Sumatran crustal blocks, possibly by strike-slip faulting at an active margin. Isotopic dating in Sumatra indicates that there were several episodes of granite magmatism, interpreted to have occurred at an Andean-type margin, during the Jurassic and Cretaceous. Marine sedimentary rocks were deposited in an intra-oceanic arc that collided with the Sumatran margin in the Middle Cretaceous. The collision added arc and ophiolitic rocks to the southern margin of Sumatra.

Borneo: Mesozoic Collisions

Southwestern Borneo (Fig. 4) may be the eastern part of Triassic Sundaland, or it could be a continental block added in the Early Cretaceous, at a suture that runs south from the Natuna Islands. The Paleozoic is represented mainly by Carboniferous to Permian metamorphic rocks, although Devonian limestones have been found as river boulders in eastern Kalimantan. Cretaceous granitoid plutons, associated with volcanic rocks, intrude the metamorphic rocks in the Schwaner Mountains of southwestern Borneo. To the north, the northwestern Kalimantan domain, or Kuching zone, includes fossiliferous Carboniferous limestones, Permo-Triassic granites, Triassic marine shales, ammonite-bearing Jurassic sediments, and Cretaceous melanges. In Sarawak, Triassic floras suggest Cathaysian affinities and correlations with Indochina. The Kuching zone may mark a subduction margin continuing south from East Asia, at which ophiolitic, island arc, and microcontinental crustal fragments collided and were deformed during the Mesozoic.

Sumatra, Java, Kalimantan, Sulawesi: Cretaceous Active Margin

A Cretaceous active margin is interpreted to have run the length of Sumatra into western Java and then continued northeast through southeastern Borneo and into western Sulawesi, as suggested by the distribution of Cretaceous high pressure–low temperature subduction-related metamorphic rocks in central Java, the Meratus Mountains of southeastern Kalimantan and western Sulawesi. Western Sulawesi and eastern Java (Fig. 4) are underlain in part by Archean continental crust, and geochemistry and zircon dating indicates derivation of this crust from the west Australian margin. Subduction ceased in the Late Cretaceous following collision of this block with Sundaland.

Sundaland: Cretaceous Granites

Cretaceous granites are widespread in the Schwaner Mountains, in western Sarawak, on the Sunda Shelf, and on the Thai-Malay Peninsula. They have been interpreted as the product of Andean-type magmatism at active margins but are spread over a large area and are far from any likely subduction zones. They probably represent post-collisional magmatism following Cretaceous addition of continental fragments in Borneo, eastern Java, and western Sulawesi.

CENOZOIC HISTORY OF INDONESIA

Cenozoic rocks cover most of Indonesia. They were deposited in sedimentary basins in and around Sundaland. There are products of volcanic activity at subduction margins, and there are ophiolites, arc rocks, and Australian continental crust added during collision.

Little is known of the Late Cretaceous and Paleocene because of the paucity of rocks of this age. From Sumatra to Sulawesi, the southern part of Sundaland was probably mostly emergent during the Late Cretaceous and Early Cenozoic, and there was widespread erosion; the oldest Cenozoic rocks typically rest unconformably on Cretaceous or older rocks. There is little evidence of subduction, although there was minor volcanic activity in southern Sumatra and Sulawesi. At the beginning of the Cenozoic, there were probably passive margins around most of Indonesia.

Eocene Subduction Initiation

India moved north during the Cretaceous to collide with Asia in the Early Cenozoic but passed to the west of Sumatra. Australia began to move rapidly northward from about 45 million years ago, in the Eocene. At this time, northward subduction resumed beneath Indonesia, producing widespread volcanic rocks at the active margin. The Sunda arc stretched from Sumatra, through Java and the north arm of Sulawesi, and then continued into the western Pacific. From the Eocene to the Early Miocene, the Halmahera arc was active in the western Pacific, far north of Australia, above a north-dipping subduction zone.

Also in the Eocene, southward subduction of the proto–South China Sea began on the northern side of Sundaland. Sediment carried north from southwestern Borneo was deposited in deep marine fans at this active margin. Early Cenozoic volcanic activity in Kalimantan is not well dated or characterized but appears to be related to this subduction.

Eocene Rifting

In the interior of Sundaland, widespread rifting began at a similar time as subduction and led to the formation of numerous sedimentary basins. These basins, some more than 10-km deep, are filled with Cenozoic sediments and are rich in hydrocarbons. The largest of these are in Sumatra, offshore Java, and eastern Kalimantan. In southeastern Sundaland, Eocene rifting led to the separation of Borneo and western Sulawesi, forming the Makassar Straits (Fig. 2), and by the Oligocene, much of eastern Kalimantan and the straits was an extensive area of deep water. In western Sulawesi, shallow water deposition continued, and there are extensive platform limestones in the south arm. The southern part of the straits is relatively shallow (about 1 km) and underlain by continental crust. The northern straits are connected to the oceanic Celebes Sea, but it is not known whether they are underlain by oceanic or stretched continental crust because there is up to 14 km of sediment below the 2.5-km-deep sea floor. The Makassar Straits are today a major passageway for water and heat from the Pacific to the Indian Ocean and have been an important influence on biogeography. The Wallace Line, marking an important boundary between Asian and Australasian faunas, follows the Makassar Straits south to pass between the islands of Bali and Lombok.

Miocene: Continental Collisions

At the beginning of the Miocene, ophiolites were emplaced by collision between the Australian and the Sundaland continents in Sulawesi, and between the Australian continent and the Halmahera arc much further to the east in the Pacific. The ophiolites of Sulawesi are remnants of fore-arc and oceanic crust between Sundaland and the Australian plate, whereas those of the North Moluccas are fragments of Philippine Sea plate arcs. Later in the Early Miocene, there was collision in north Borneo with the extended passive continental margin of South China. These collisions led to mountain building in eastern Sulawesi and in Borneo. The first Australian continental crust was added in eastern Indonesia (Fig. 5), but as northward movement of Australia continued, there was a change in eastern Indonesia to extension, complicated by minor collisions as microcontinental blocks moved along strike-slip faults. In the Sunda arc, volcanic activity declined in Java for a few million years before a late Miocene increase from Sumatra to the Banda arc.

Neogene: Java-Sumatra

In the Java–Sulawesi sector of the Sunda arc, volcanism greatly diminished during the Early and Middle Miocene, although northward subduction continued. The decline in magmatism resulted from Australian collision in eastern Indonesia, causing rotation of Borneo and Java. Northward movement of the subduction zone prevented replenishment of the upper mantle source until rotation ceased in the late Middle Miocene. Then, about 10 million years ago, volcanic activity resumed in abundance along the Sunda arc from Java eastward. Since the Late Miocene, there has been thrusting and contractional deformation in Sumatra and Java related to arrival of buoyant features at the trench, or increased coupling between the overriding and downgoing plates. Both islands have been elevated above sea level in the last few million years.

Neogene: Borneo

The rise of mountains on Borneo increased the output of sediment to circum-Borneo sedimentary basins. In eastern Kalimantan, thick Miocene to recent sediments filled accommodation space created during Eocene rifting. Most was derived from erosion of the Borneo highlands and inversion of older parts of the basin margins to the north and west, which began in the Early Miocene.

Sedimentation continues today in the Mahakam delta and in the offshore deepwater Makassar Straits.

In parts of central Kalimantan, there was some Miocene magmatism and associated gold mineralization, but volcanic activity largely ceased in Kalimantan after collision. Minor Plio-Pleistocene basaltic magmatism in Borneo may reflect a deep cause such as lithospheric delamination after Miocene collisional thickening.

Neogene: Sulawesi

Sulawesi is inadequately understood and has a complex history still to be unraveled. In eastern Sulawesi, collision initially resulted in thrusting of ophiolitic and Australian continental rocks. However, contractional deformation was followed in the Middle Miocene by new extension. There was Miocene core complex metamorphism in north Sulawesi, extensional magmatism in south Sulawesi, and formation of the deep Gorontalo Bay and Bone Gulf basins between the arms of Sulawesi.

Compressional deformation began in the Pliocene, partly as result of the collision of the Banggai-Sula microcontinent in east Sulawesi, which caused contraction and uplift. Geological mapping, paleomagnetic investigations, and GPS observations indicate complex Neogene deformation in Sulawesi, including extension, block rotations, and strike-slip faulting. There are rapidly exhumed upper mantle and lower crustal rocks, and young granites, near to the prominent Palu-Koro strike-slip fault (Fig. 1). During the Pliocene, coarse clastic sedimentation predominated across most of Sulawesi as mountains rose. The western Sulawesi fold-thrust belt has now propagated west into the Makassar Straits. At present, there is southward subduction of the Celebes Sea beneath the north arm of Sulawesi and subduction on the east side of the north arm of the Molucca Sea toward the west (Fig. 1).

Neogene: Banda

The Banda arc is the horseshoe-shaped arc that extends from Flores to Buru, including Timor and Seram, with islands forming an outer non-volcanic arc and an inner volcanic arc. It is an unusual region of young extension that developed within the Australian-Sundaland collision zone and formed by subduction of an oceanic embayment within the northward-moving Australian plate.

In the Middle Miocene, Jurassic ocean lithosphere of the Banda embayment began to subduct at the Java Trench. The subduction hinge rolled back rapidly to the south and east, inducing massive extension in the upper plate. Extension began in Sulawesi in the Middle Miocene. As the hinge rolled back into the Banda embayment, it led to formation of the Neogene Banda volcanic arc and the opening of the North Banda Sea, the Flores Sea, and later the South Banda Sea. About 3–4 million years ago, the volcanic arc collided with the Australian margin in Timor, causing thrusting. Remnants of the Asian margin and Paleogene Sunda arc are found in the uppermost nappes of Timor and other Banda islands.

After collision, convergence and volcanic activity ceased in the Timor sector, although volcanic activity continued to the west and east. New plate boundaries developed north of the arc between Flores and Wetar (Fig. 1), and to the north of the South Banda Sea, associated with subduction polarity reversal. The Banda region is now contracting. During the last 3 million years, there have been significant shortening and probable intra-continental subduction at the southern margins of the Bird's Head microcontinent south of the Seram trough (Fig. 1).

Neogene: North Moluccas

In eastern Indonesia, the Halmahera and Sangihe arcs (Fig. 1) are the only arcs on Earth currently colliding. Both of the currently active volcanic arcs formed during the Neogene. The Sangihe arc is constructed on Eocene oceanic crust and initially formed at the Sundaland margin in the Early Cenozoic. The modern Halmahera arc is built upon older arcs, of which the oldest is a Mesozoic intra-oceanic arc that formed in the Pacific. Early Miocene arc–Australian continent collision terminated northward subduction, and the north Australian plate boundary became a major left-lateral strike-slip zone in New Guinea. Volcanism ceased, and there was widespread deposition of shallow marine limestones. Arc terranes were moved westward within the Sorong fault zone. At the western end of the fault system, there was subduction beneath the Sangihe arc and collision in Sulawesi of continental fragments sliced from New Guinea.

Initiation of east-directed Halmahera subduction probably resulted from locking of strands of the left-lateral Sorong fault zone at its western end in Sulawesi. The present-day Molucca Sea double subduction system was initiated in the Middle Miocene, and volcanism began in the Halmahera arc about 11 million years ago. The Molucca Sea has since been eliminated by subduction at both its eastern and western sides. The two arcs first came into contact about 3 million years ago, and this contact was followed by repeated thrusting of the Halmahera fore-arc and back-arc toward the active volcanic arc. Collision has formed a central Molucca Sea (Fig. 1) melange wedge, including ophiolite slices from the basement of the Sangihe arc. There are small fragments of continental crust between splays of the Sorong fault.

Neogene: New Guinea

In New Guinea, there was rifting from the Late Triassic onward to form a Mesozoic northern passive margin of the Australian continent, on which there was widespread carbonate deposition during the Cenozoic. To the north of the passive margin were a number of small oceanic basins and arcs developed above subduction zones; the region was probably as complex as the western Pacific today. At the beginning of the Miocene, the arc–Australian continent collision began emplacement of arc and ophiolite terranes, which then moved west in a complex strike-slip zone. The Halmahera arc was one of these. Today, northern New Guinea is underlain by these arc and ophiolitic rocks, fragmented by faulting. In the Pliocene, subduction probably began at the New Guinea Trench (Fig. 1), as there is now a poorly defined slab dipping south that has reached depths of about 150 km. There was isolated, but important, magmatism associated with world-class copper and gold mineralization including the Grasberg and Ertsberg complexes. The rise of the New Guinea main ranges accelerated, and mountains reached their present elevations with peaks more than 5-km high (Fig. 2) capped by glaciers.

SEE ALSO THE FOLLOWING ARTICLES

Earthquakes / Eruptions: Laki and Tambora / Indonesia, Biology / Island Arcs / New Guinea, Geology / Philippines, Geology

FURTHER READING

Barber, A. J., M. J. Crow, and J. S. Milsom, eds. 2005. *Sumatra: geology, resources and tectonic evolution*. Geological Society London Memoir 31.

Bijwaard, H., W. Spakman, and E. R. Engdahl. 1998. Closing the gap between regional and global travel time tomography. *Journal of Geophysical Research* 103: 30,055–30,078.

Bock, Y., L. Prawirodirdjo, J. F. Genrich, C. W. Stevens, R. McCaffrey, C. Subarya, S. S. O. Puntodewo, and E. Calais. 2003. Crustal motion in Indonesia from global positioning system measurements. *Journal of Geophysical Research* 108: doi:10.1029/2001JB000324.

Hall, R. 2002. Cenozoic geological and plate tectonic evolution of SE Asia and the SW Pacific: computer-based reconstructions, model and animations. *Journal of Asian Earth Sciences* 20: 353–434.

Hall, R., and D. J. Blundell, eds. 1996. *Tectonic evolution of SE Asia*. Geological Society of London Special Publication 106.

Hall, R., and C. K. Morley. 2004. Sundaland basins, in *Continent-ocean interactions within the East Asian marginal seas*. P. Clift, P. Wang, W. Kuhnt, and D. E. Hayes, eds. American Geophysical Union, Geophysical Monograph 149, 55–85.

Hamilton, W. 1979. *Tectonics of the Indonesian region*. US Geological Survey Professional Paper 1078.

Hutchison, C. S. 1989. *Geological evolution of South-East Asia*. Oxford Monographs on Geology and Geophysics, Oxford, UK: Clarendon Press.

Metcalfe, I. 1996. Pre-Cretaceous evolution of SE Asian terranes, in *Tectonic evolution of SE Asia*. R. Hall, R. Blundell, and D. J. Blundell, eds. Geological Society of London Special Publication 106, 97–122.

van Bemmelen, R.W. 1949. *The geology of Indonesia*. The Hague, Netherlands: Government Printing Office, Nijhoff.

INSECT RADIATIONS

DIANA M. PERCY

University of British Columbia, Vancouver, Canada

Insect radiations on islands are the evolutionary product of diversification within an insect lineage on an island or a series of islands forming an archipelago. Radiations, by definition, represent in situ diversification and are often characterized by evolutionarily novel adaptations. Island radiations are one of the most important natural phenomena for evolutionary biologists, and, like other animal and plant groups, insects on islands have undergone radiations that range from a modest diversification of species to explosive radiations over a short period of time. Insect radiations vary not only in the numbers of species but in the rates of speciation and in the diversity of adaptive traits that evolve.

COLONIZATION, ESTABLISHMENT, AND DIVERSIFICATION

Island radiations are by definition the product of in situ diversification and the evolution of multiple species from a founding ancestor, producing a lineage of closely related island endemics. All islands represent habitable patches surrounded by uninhabitable environments. These can be terrestrial habitat "islands" such as shifting sand dunes in Namibia, where a lineage of *Scarabaeus* dung beetles has radiated into 12 endemic species, or "sky islands" in the Rocky Mountains, where *Melanoplus* grasshoppers have radiated into 37 species after isolation in multiple glacial refugia. More commonly, islands represent terrestrial habitats surrounded by water.

Insect radiations on islands necessarily begin with a stepwise process involving dispersal from a source population, establishment of a viable population upon colonization, and diversification into multiple species within an island or between islands in an archipelago. In many cases, the distances between source populations on continental land masses and remote oceanic islands (islands that are formed *de novo* and are not terrestrial fragments separated from a once-larger land mass) are great enough that active dispersal by insects (e.g., by flying) is ruled out and passive dispersal methods (e.g., by wind currents, attachment to migrating birds, or rafting with tidal debris) are considered more plausible means for insects to have traveled the distance. Four endemic flightless insect genera, including large flightless beetles, cave crickets, and cockroaches, have dispersed from New Zealand to the Chatham Islands (~800 km) relatively recently (2–6 million years ago), precluding isolation by vicariance. The considerable impediments to such long-distance dispersal events are one of the primary filters that limit which and how many of the taxa in any particular source area will be found on an island. In addition to such dispersal challenges, for many insects the establishment of a viable population in the novel island environment is an equally formidable filter, particularly for habitat specialists (e.g., monophagous herbivores).

These barriers to successful island colonization are the reason why islands are typically lineage-poor, with limited overall representation of the continental biota. Conversely, in situ species radiations can result in unusual species richness within select lineages that manage to establish successfully. Founding species may be at an advantage where colonization barriers have reduced the number of competitors, predators, and parasites; this advantage may promote speciation and radiation in successful colonists. Radiations that are thought to have been initiated under such conditions fall under the "escape and radiate" hypothesis. Conversely, competition between newly formed taxa within a lineage may promote speciation by leading to character displacement that reinforces reproductive isolation in sympatric island taxa (e.g., *Aphanarthrum* bark beetles, *Drosophila*, and *Laupala* crickets).

The more remote and isolated islands are from a source biota, the more asymmetrical or disharmonic the biotas are likely to be. For instance, the Kurile archipelago is an extensive chain of volcanic continental rim islands that are only a few hundred kilometers from continental Asia. These islands have a broad representation of the continental biota and little endemism (<10%). In contrast, the Hawaiian Islands in the Pacific Ocean are more than 3200 km from a continental land mass and have extremely high levels of endemism (>90% in many groups), but the Hawaiian biota has a highly asymmetric representation of its source biotas. On islands where species lineages have radiated and there are numerous multiplications of species from a few ancestral colonists, levels of endemism dramatically increase.

Theoretically, any number of species greater than two is eligible to be considered a radiation, but in practice "radiation" usually refers to higher numbers of derived species. Radiations may be relatively modest in species numbers but notable for their adaptive shifts. The Galápagos finches, which attracted the attention of Charles Darwin, have only 14 species but are considered a classic example of adaptive radiation on the basis of ecologically driven morphological changes between the different species. Insect radiations take many forms, from impressive explosions in the number of species (a primary example being the Hawaiian drosophilid flies, with approximately 1000 extant species derived from a single colonization event) to radiations that are more modest in species number but exhibit dramatic adaptive shifts. A relatively modest radiation of *Caconemobius* ground crickets in the Hawaiian Islands has only nine endemic species, but there have been substantial adaptive shifts in this lineage from inhabiting rocky coastal areas to inhabiting subterranean lava tubes. The derived subterranean species exhibit adaptive morphological changes associated with the shift in habitat, including eye reduction and loss of pigmentation.

Many island radiations combine both the evolution of different adaptive traits and diversification into large numbers of species. There are also nonadaptive radiations, which refer to diversification occurring within a narrow or similar ecological range, such as speciation by specialist herbivores on the same host plant but on different islands (i.e., geographic rather than ecological separation) or nonspecialist, dietary generalists speciating within an island (e.g., *Rhyncogonus* weevils on the island of Rapa). Diversification in the absence of adaptive shifts may be precipitated by other conditions such as escape from a restraint that served to limit diversification on continents (e.g., predators and competitors), resulting in the "escape and radiate" phenomenon already mentioned.

Interpretations of radiation processes are necessarily influenced by the number of species locally extant at a given time, which is determined by a number of factors, including the age of the lineage and rates of extinction. On islands, lineages may have undergone relatively minimal extinction compared to continents, because islands often have stable, temperate climates compared to continents, and in addition, in the case of recent radiations on young islands, there has been less time to the present for extinctions to have occurred. Where much of the diversity in recent radiations survives, evolutionary biologists are presented with some of the clearest examples of speciation processes. In contrast, the process by which an insect lineage dispersed and first established on an island will necessarily be more speculative, because these events, and the ancestral forms and habits of those initial colonists, are frequently no longer apparent from the derived species' morphology and preferred island habitats observed today. In some cases, such as the Hawaiian tree and swordtail crickets, the extant species have diverged from their original founder lineage to such a degree that island groups such as these are often taxonomically misplaced by specialists.

CONSTRAINTS AND PROMOTERS: ISLANDS AND ARCHIPELAGOES

Insect radiations on islands are fairly common, but they are not ubiquitous. Not every insect lineage that successfully colonizes an island or archipelago undergoes a radiation. Finding explanations for what drives or constrains radiations is one of the most challenging aspects of island biology. Potential explanations include morphological or genetic constraints on diversification, and available ecological niches favoring one group over another. An example of a radiation occurring in one insect group but not in a related group can be found in the Canary Islands, which have been colonized by several related lineages of specialist herbivores (psyllids). Diversification in the different psyllid lineages appears to have been limited or promoted by the presence and abundance of familiar host plant species. In contrast, in several Hawaiian herbivore groups, host range has dramatically expanded after colonization. In the case of *Nesosydne* planthoppers, a Hawaiian lineage with more than 80 species derived from continental ancestors typically specialized on monocotyledonous plants, colonization of the Hawaiian Islands has resulted in diversification on many novel (i.e., previously unused) host plant groups, mostly in dicotyledonous plant families. Similarly, the Hawaiian endemic plant bug genus *Sarona*, with at least 40 species, also has a greatly expanded host plant range (even though individual species remain specialists) as compared to continental sister groups, including several novel host plant families. Because the Canary Islands are less isolated from a continental source than the Hawaiian Islands, the presence of multiple colonizing lineages in the Canary Islands may serve to increase competition and maintain ancestral host preferences.

In these herbivorous insect groups, radiation is in large part driven by ecological adaptations to feeding on particular host plants. Macromorphological changes are typically not required in shifting to different hosts, and there is little morphological variation between these species. In contrast, Hawaiian drosophilid flies, in addition to marked ecological habitat specialization, exhibit considerable morphological and behavioral diversity, reflecting the development of divergent secondary sexual characteristics correlated with male mating behavior. In this case, separating out the roles of different processes (e.g., ecological, behavioral) becomes more complex, in particular determining which processes act as a primary promoter rather than secondarily reinforcing speciation.

An important factor in determining the scale of insect radiations is the type of island colonized. Islands come in many forms, from large single- or two-island groups such as Madagascar, New Caledonia, and New Zealand, which share some geological history with other land masses, and are markedly old compared to islands formed de novo from volcanic activity; to small isolated singleton islands of oceanic origin such as the Atlantic island of St. Helena; through to chains of islands sharing a common volcanic hotspot origin such as the Hawaiian Islands. Many of the most impressive examples of species radiations are to be found in these hotspot archipelagoes, such as the Pacific Hawaiian and Galápagos Islands, and the Atlantic Canary Islands. However, a number of insect lineages have also radiated on isolated singleton islands that have no close neighboring islands. Single-island radiations have been less well studied. For instance, there are more than 70 endemic curculionid weevils on the island of St. Helena (an island that is more than 1900 km from the nearest continent and 1100 km from Ascension Island), but there has been no comprehensive systematic study undertaken to assess the number of independent colonizations and therefore the number of individual lineages represented by this diversity. Similarly, on the isolated island of Rapa (part of the Austral Islands in southern French Polynesia, but the most remote of these islands), the weevil genus *Miocalles* is represented by more than 60 species on this small island that is only 40 km^2. As with the St. Helena weevils, the number of independent colonizations has not been established. In New Zealand, at least two radiations of jumping plant lice (comprising more than 80 species) are associated with two highly diverse host plant genera in the Asteraceae. A radiation of Helictopleurini dung beetles in Madagascar is thought to have been promoted by a parallel radiation in lemurs, whose dung the beetles specialize on. These and other insect lineages that diversify within an island may do so by ecological shifts in sympatry or micro-allopatric shifts. For instance, at least 50% of the speciation events in the Hawaiian plant bug genus *Sarona* are estimated to have occurred in sympatry within islands via host plant shifts. In contrast, the monophyletic lineage of *Rhyncogonus* weevils on Rapa are not habitat specialists and therefore may have diversified via small-scale geographic isolation (micro-allopatry) (Fig. 1). Other Hawaiian herbivorous groups, such as the endemic Hawaiian seed bug genus *Neseis*, have speciated primarily by colonizing new islands, often without involving a host plant change. In the case of the *Neseis* seed bugs, an archipelago with multiple islands as geographic isolates has been a necessary promoter for speciation and radiation.

Arguably, it is chains of islands that have produced the most dramatic radiations of insects. For instance, both drosophilid flies and crickets in the Hawaiian Islands have

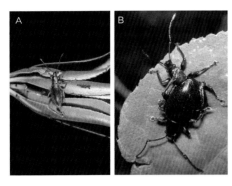

FIGURE 1 *Rhyncogonus gracilis*, endemic to the island of Rapa, has radiated in the Austral Islands; this species is not a habitat specialist. Photograph by Ronald Englund.

lineages that have radiated into more than 100 species, and the Hawaiian drosophilid radiation is unchallenged for species diversity, with approximately 1000 extant species arising from a single founder event, including around 25% of the world's species of *Drosophila*. In terms of rapidity of radiation, the Hawaiian *Laupala* cricket radiation is proposed to have the fastest speciation rate in an insect radiation (4.17 species per million years). Among animals this rate is exceeded only by speciation rates in African cichlid fishes. The Hawaiian *Laupala* crickets are ecologically and morphologically cryptic; they are all forest dwellers with generalist diets, and the primary differences between species are their acoustic mating signals, suggesting that this radiation is driven by sexual selection rather than ecologically adaptive shifts.

One of the signature patterns of archipelago radiations emphasizes the role of multiple island chains. In both Hawaiian crickets and flies, as with many archipelago radiations, there is a pervasive pattern of single-island endemics. However, as with the herbivorous bug groups referred to earlier, there are contrasting modes of speciation. Among the flies, diversification is primarily via inter-island colonization, such that sister taxa typically occur on different islands, whereas among the crickets, diversity is primarily a product of secondary radiations within islands, such that sister taxa are likely to be found on the same island.

Island chains may be particularly productive in terms of species radiations for several reasons. The first is that the additional land area of multiple islands increases both the quantity and usually also the diversity of habitats, particularly if volcanic activity on some islands dramatically increases the elevation available for colonization. Pacific blackfly diversity is greatly increased on younger islands such as Tahiti in French Polynesia, where there are 29 species, because the newly uplifted topography creates the habitats necessary (steep slopes) for a radiation of waterfall-inhabiting species.

Secondly, and perhaps more importantly, each island and its habitats are discretely circumscribed and isolated from other islands by the intervening marine environment. Barriers to gene flow between islands are therefore both persistent (barring dramatic changes in sea level) and repeated with the colonization of each island. This important feature of island chains clearly serves to promote radiations, because each individual island potentially becomes a relay point for a diversifying lineage, and each new island colonized can serve to continue and expand a radiation as well as serving as a refugium if a lineage suffers extinction elsewhere in the archipelago. This is most clearly seen in hotspot archipelagoes, where the successively formed islands are frequently in relatively close proximity to the preceding and following islands and can act as steppingstones for a radiating lineage.

The isolation and uniqueness of individual island lineages is promoted by the frequent loss of dispersal capabilities in island taxa (e.g., marked reduction in wing size and flightlessness). Loss of dispersal capability can dramatically limit inter-island movement, virtually ensuring that populations on different islands become rapidly genetically isolated. The partial or complete loss of forewings in Hawaiian crickets is a classic example. The resulting genealogical patterns, through a combination of intra-island diversification and single-island endemism, are monophyletic island lineages. Each individual island lineage may be characterized by novel adaptations determined by the environmental character of that island. Islands in an archipelago that have similar environmental variables (e.g., altitude, climate, geomorphology, soil, and vegetation zones) may produce independent lineages where species evolve convergent adaptive traits on each island. The species in these lineages may look morphologically and/or ecologically similar but have evolved independently on each island. *Oliarus* planthoppers, with more than 70 species in the Hawaiian Islands, have independently colonized subterranean lava tubes on different islands, but these cave-dwelling species are remarkably morphologically convergent as a result of parallel adaptations to the cavernous habitat. Diversity in a radiation is therefore promoted by both evolution within islands and dispersal between islands. Novel species arising in isolation on one island can "seed" this diversity "forward" onto other, more recently formed islands or "backward" to older islands. In these situations, the presence or absence of genetic or reproductive incompatibility between species that evolved in isolation on different islands will determine to what extent, if any, introgression between different islands will occur after species formation.

Furthermore, the characteristics and extent of any island radiation is influenced by the geological history of the island(s). In an archipelago for instance, whether ancient radiations are able to maintain a continuum by populating new islands as they emerge will depend on whether the replacement rate of old submerged islands by new emergent islands is sufficiently regular in space and time. Molecular dating methods have revealed several insect lineages to be considerably older than the age of the islands inhabited by extant taxa (e.g., Hawaiian drosophilid flies, Galápagos flightless weevils). The likely explanation in these cases is that diversification of the lineage began earlier and outside the current geographic range, and in most cases it is proposed that this earlier diversification took place on older islands now submerged under the sea. Studying insect radiations in archipelagoes with a progressive age range of islands (e.g., the Hawaiian Islands with several islands ranging in age from ~0.5-~5 million years) can reveal how an insect lineage responds to increasing island age and the formation of successively newer islands.

INTERPRETING ISLAND RADIATIONS: THE RULE OF PROGRESSION

Much of our current understanding of island radiations derives from the power of modern molecular phylogenetics. A common pattern found in several archipelagoes is that ancestral species tend to occur on older islands and more recently derived branches of an insect radiation are found on younger islands (Fig. 2). This pattern occurs frequently enough to have become known as the "rule of progression" because the diversification of an insect lineage progresses in step with the age of the islands. There are exceptions to the rule of progression, and these are found even in lineages that show a general progression pattern. A radiation that in large part adheres to the rule of progression may also exhibit retrogressive patterns, such as back-colonizations from younger to older islands, and branches of a radiation that undergo recent secondary radiations within older islands. In this respect, the geological history of individual islands can dramatically influence patterns of island radiations. For instance, new environments on older islands can be created by geological disturbances such as volcanic activity and landslides. Islands of volcanic origin can remain highly volatile, with eruptions and landslides altering the landscape and ecology of large areas at a time. Such events may cause extinctions, but conversely they can also promote new diversification. Volcanically altered landscapes can result

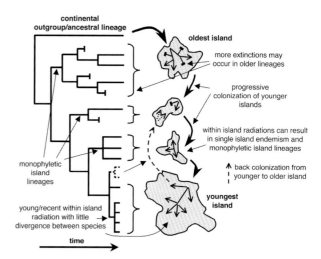

FIGURE 2 This hypothetical phylogeny of island colonization and radiation illustrates how phylogenetic analyses can help interpret patterns of radiation. Illustrated is the colonization and radiation of a hotspot archipelago by a lineage dispersing from a continent. The data depicted here are clocklike, and therefore, shorter branches on the youngest islands indicate more recent colonization and speciation events. The overall branching pattern of individual island lineages within the phylogeny follows the "rule of progression" with successional colonization from older to younger islands. Also illustrated are secondary within-island radiations leading to monophyletic island lineages and species that are single-island endemics. Larger islands with more habitat types result in more intra-island speciation events, but over time, those lineages on older islands may incur more extinction events. Finally, a retrogressive back-colonization from the youngest island to an older island is illustrated.

in geographic range expansions in dominant plant species (*Pinus* in the Canary Islands; *Metrosideros* in the Hawaiian Islands) with similar range expansions in the associated insect fauna. In addition, insect lineages that have already radiated above ground may diversify further in the unique habitats provided by subterranean lava tubes (e.g., crickets and planthoppers).

Insect diversity on isolated singleton islands may be much more severely impacted by the loss of habitat from catastrophic natural phenomena and introduced species, because there are no nearby islands to replenish the lost diversity. On the island of St. Helena in the South Atlantic a number of species (including ground beetles, weevils, earwigs, and grasshoppers) have become endangered or extinct, attributed to ecological changes and habitat destruction. With older islands such as St. Helena (14.5 million years old), it is possible that some species that have become extinct over time may have provided evidence of historic radiations. Unlike the relatively good fossil record for extinct island birds, there is little fossil evidence with which to assess extinct insect faunas on islands, and it is possible that many species are lost before they are ever recorded as having been present.

The recolonization of continental land masses from lineages that have evolved on islands is a less common biogeographic pattern, and most of the known cases are from less remote islands/archipelagoes back to continents, such as *Aphanarthrum* bark beetles from the Canary Islands to North Africa, a distance of 100–500 km. Examples of continental recolonization from more remote oceanic islands are much rarer, but one possible example is the drosophilid genus *Scaptomyza*. This genus appears to have originated in the Hawaiian Islands and subsequently spread to the rest of the world. The loss of dispersal capability (e.g., wing reduction) common in many island lineages likely contributes to the asymmetrical rates of colonization between islands and continents.

A number of other ways in which phylogenetic analyses can reveal patterns and processes of radiation include identifying outgroups and thus determining ancestral source biotas; the number and direction of colonization events; and using sister species pair comparisons to elucidate speciation mechanisms. Clarification of these processes can greatly facilitate our interpretation of evolutionary processes. Molecular data is also critical in revealing radiations among cryptic species such as the *Laupala* crickets. Increasingly, the functional genes involved in radiations are being studied. Radiations that are characterized by multiple adaptive shifts suggests complex evolutionary processes, but adaptively driven radiations can happen rapidly and may involve changes in only one or few genetic loci. Conversely, relatively simple shifts in mate recognition systems that have promoted radiation in Hawaiian crickets may be controlled by multiple genes of small effect.

THE ROLE OF REPRODUCTIVE BEHAVIOR

As a potentially important process in insect radiations, the role of reproductive behavior is less well known and characterized than either geographic or ecological processes. However, in at least three island lineages (crickets, planthoppers, and drosophilid flies in the Hawaiian Islands), reproductive behavior has been shown to be important to interspecific isolation and intraspecific recognition and even a primary factor driving species radiations. In the Hawaiian swordtail cricket genus *Laupala*, the rapid explosion of speciation in this lineage is thought to be driven primarily by divergence in mate recognition systems. Multiple signals are involved in mate recognition in this group, including acoustic signals and, at close range, cuticular hydrocarbons. In contrast to ecologically driven radiations, the *Laupala* cricket species remain ecologically and morphologically similar. Among Hawaiian *Drosophila* several different modes of mate recognition are thought to contribute to species isolation, including acoustic, visual, chemical, and tactile factors. In New Zealand, a radiation of *Kikihia* cicadas (~30 species) includes several morphologically cryptic species that can be differentiated by their acoustic signals (Fig. 3).

FIGURE 3 Two morphologically cryptic *Kikihia* cicada species that are part of a radiation in New Zealand (predominant background color of both species varies from green to yellow-green, as does overall degree of darkness). These species are difficult to distinguish morphologically, but they have well-differentiated songs (scale bar of each oscillogram indicates one second). Photograph by David Marshall.

Whether these behaviors contributed to species isolation and radiation on islands, or developed as important species reinforcement mechanisms post isolation, is not well known. All the Hawaiian crickets exhibit classic island flightlessness, and there is complete loss of forewings in some species that no longer produce sound, but many species have retained the forewings for the purposes of sound production. In Hawaiian drosophilids, morphological characteristics, courtship behavior, and acoustic songs combine to form species-specific mate recognition systems, but in this system there have also been numerous adaptive shifts to different habitats and different larval feeding substrates. Thus, in this group both sexual selection and ecological adaptations are likely to be important in promoting speciation and radiation.

SEE ALSO THE FOLLOWING ARTICLES

Adaptive Radiation / Crickets / *Drosophila* / Endemism / Extinction / Flightlessness / Sexual Selection / Species–Area Relationship

FURTHER READING

Emerson, B. C. 2002. Evolution on oceanic islands: molecular phylogenetic approaches to understanding pattern and process. *Molecular Ecology* 11: 951–966.

Gillespie, R. G., and G. K. Roderick. 2002. Arthropods on islands: colonization, speciation, and conservation. *Annual Review of Entomology* 47: 595–632.

Juan, C., B. C. Emerson, P. Oromi, and G. M. Hewitt. 2000. Colonization and diversification: towards a phylogeographic synthesis for the Canary Islands. *Trends in Ecology and Evolution* 15: 104–109.

Roderick, G. K., and R. G. Gillespie. 1998. Speciation and phylogeography of Hawaiian terrestrial arthropods. *Molecular Ecology* 7: 519–531.

Roderick, G. K., and R. G. Gillespie. 2003. Island biogeography and evolution, in *Encyclopedia of Insects*. V. Resh and R. Cardé, eds. San Diego: Academic Press, 602–604.

Wagner, W. L., and V. A. Funk. 1995. *Hawaiian biogeography: evolution on a hot spot archipelago*. Washington, DC: Smithsonian Institution Press.

INSELBERGS

STEFAN POREMBSKI
University of Rostock, Germany

Inselbergs are isolated rock outcrops that frequently consist of granites and gneisses and form old landscape elements that are widespread on crystalline continental shields. The environmental conditions on inselbergs are extreme both edaphically (because they lack soil) and microclimatically (because they are exposed to intense irradiation and high temperatures), and their vegetation is thus demarcated against the surroundings. Widespread are desiccation-tolerant lichens, mosses, ferns, and angiosperms.

ENVIRONMENTAL CONDITIONS

Patchily distributed habitats that are dominated by a hard, stony surface are known as rock outcrops when they protrude above the surroundings. This definition encompasses a broad range of landforms such as the tepuis of the Guayana shield, the tower karst mountains along the Guilin Li River in China, and inselbergs such as the Pao de Açúcar of Rio de Janeiro (Fig. 1).

Inselbergs embody a spectacular example of island-like rock outcrops that consist of freely exposed slopes. Typically consisting of Precambrian granites and gneisses, they form prominent landscape features in all vegetational and climatic zones. The term "inselberg" (from German *Insel*, island, and *Berg*, mountain) was coined by the German geologist Wilhelm Bornhardt, who noticed their distinctive character as forming island-like habitats. In his honor, a particular type of inselberg that is characterized by a dome-shaped appearance with steeply precipitous flanks is named a "Bornhardt."

Inselbergs are particularly frequent on the old crystalline continental shields. Because of their resistance to erosional processes, they possess a considerable age, frequently surpassing millions of years. The environmental conditions on inselbergs are extreme both edaphically (i.e., in terms of soil and moisture conditions, because they lack soil and nutrients are scarce) and microclimatically (i.e., in terms of atmospheric conditions, because they undergo intense irradiation and temperatures regularly exceeding 60 °C). Even when situated in rainforests they form "micro-environmental deserts." Despite the general lack of moisture, several locally restricted habitat types occur that are seasonally wet. Most prominent are seasonally water-filled depressions ("rock pools"), which may carry water for several consecutive weeks or months. Wet conditions are also present where water seeps continuously over longer periods at the feet of steep rocky slopes. However, unpredictable periods of drought may cause the drying up of wet sites even during the rainy season, thus triggering local extinction events.

With regard to absolute height and surface area, inselbergs cover a vast spectrum. Absolute height ranges from a few meters ("shield inselbergs") to several hundred meters, and surface area reaches from small outcrops cov-

FIGURE 1 Typical examples of steep-flanked inselbergs in a tropical landscape (near Pancas, Espirito Santo, Brazil).

ering several square meters to large domes extending over square kilometers. The degree of geographic isolation of inselbergs varies considerably. They can occur as hills isolated over hundreds of kilometers, or they can form dense clusters with individual outcrops occurring at distances of only a few kilometers.

Throughout the world, cave or rock paintings dating back to prehistoric times are impressive testimonies of the close links of humans to inselbergs. They thus possess an immense cultural importance. However, throughout the tropical and temperate zones, negative human impacts on inselbergs have increased dramatically over the past decades. Of particular importance have been fire, quarrying, and tourism, all of which have led to the degradation of numerous inselbergs, in particular those located near human settlements.

PLANT AND ANIMAL LIFE

Habitat Types

The extent of the expression of the island-like character of inselbergs (i.e., ecological isolation) depends on the surrounding vegetation types, with forests emphasizing particularly profoundly the floristic differentiation between rock outcrops and their surrounding matrix. Based on plant species composition and physiognomic criteria, a limited set of typical plant communities and habitat types can be distinguished. The most important ones are described concisely as follows:

CRYPTOGAMIC CRUSTS

Exposed rocks are covered by cyanobacterial lichens (frequently *Peltula* spp.) and cyanobacteria (e.g., *Stigonema* spp. and *Scytonema* spp.) which are responsible for the often dark coloration of inselbergs. Individual microhabitats such as exposed rocky slopes, boulders, and drainage channels are differentiated floristically. Lichens and cyanobacteria may form dense epilithic surface layers, but they also occur endolithically. Cryptogamic communities on inselbergs have very close relationships to those on other rock surfaces. Particularly well developed affinities exist to the Tintenstrich communities that are ubiquitous on both anthropogenic and natural rocks in temperate and tropical regions.

EPILITHIC VASCULAR PLANTS

This group comprises higher plant species that grow directly on open rock. Frequently, these epilithic species are succulents or xerophytes. Prominent examples are provided by numerous orchids (e.g., in the neotropical genera *Cyrtopodium* and *Laelia* and the aroid genus *Anthurium,* which possess water-storing stems or leaves). Bromeliaceae, too, are very rich in epilithic species, with Brazil being their center of diversity.

MONOCOTYLEDONOUS MATS

On both level and inclined open rocky slopes, dense stands of long-lived (i.e., hundreds of years) monocotyledons occur. These epilithic species are of matlike appearance and are firmly attached to the rock by dense wiry roots. Frequently, monocotyledonous mats occur as isolated patches surrounded by open rock, but large, continuous expanses of mats can also be found. Most typical are Bromeliaceae, Cyperaceae, and Velloziaceae. The last two families contain tree-like, woody-stemmed species that are desiccation tolerant. On South American inselbergs, both tank-forming (e.g., *Alcantarea* spp., *Vriesea* spp.) and xerophytic (e.g., *Dyckia* spp., *Encholirium* spp.) Bromeliaceae occur from sea level up to high altitudes, as is the case in southeastern Brazil. In addition, the bromeliad genus *Tillandsia* occurs with mat-forming species throughout many parts of that country. In tropical Africa and Madagascar, desiccation-tolerant Cyperaceae and Velloziaceae are widespread. The Cyperaceae *Afrotrilepis pilosa* and *Microdracoides squamosus* occur as mat-formers in West Africa (Fig. 2). In southern and eastern Africa and on Madagascar, they are replaced by the Cyperaceae genus *Coleochloa* and by numerous species of the genus *Xerophyta* (Velloziaceae). Remarkably, the woody fibrous stems of mat-forming Cyperaceae and Velloziaceae are occasionally colonized by specific epiphytic orchids. Among the orchids, the genera *Polystachya* (tropical Africa), *Constantia* (Brazil), and *Pseudolaelia* (Brazil) comprise species that are restricted to monocotyledonous mats.

FIGURE 2 Specialized mat-forming monocotyledonous plants (here *Microdracoides squamosus*, Cyperaceae, Cameroon) colonize freely exposed rocky slopes. In being desiccation-tolerant, they are well adapted to survive periods of prolonged droughts.

ROCK POOLS

Usually on level parts, seasonally water-filled rock pools covering a wide range of sizes, forms, and depths occur (Fig. 3). They are products of natural solution processes, have a considerable age, and are typically irregularly shaped depressions of variable depth, covering up to several square meters. They form an unreliable habitat for higher plants because they may dry out even in the rainy season during rainless periods. Typically, epilithic and endolithic cyanobacteria and lichens form a dense cover on open rock walls. Cover by vascular plants is generally very sparse. Widespread are plants that are otherwise

FIGURE 3 Seasonally water-filled rock pools (in Australia called gnammas) provide establishment and growth sites for aquatic plants (e.g., *Dopatrium longidens*, Scrophulariaceae, Ivory Coast). In a sense, they form "islands on islands."

colonizers of marshy ground and ponds. The number of species restricted to rock pools is relatively low. Prominent examples occur within Scrophulariaceae (e.g., in the genera *Amphianthus*, *Dopatrium*, *Glossostigma*, and *Lindernia*) and in the fern genus *Isoetes* (with highly specialized species in the southeastern United States, tropical Africa, and southwestern Australia).

EPHEMERAL FLUSH VEGETATION

Located at the foot of rocky slopes or along the downslope fringes of monocotyledonous mats, this plant community depends on seepage water that is only available during the rainy season. The basic matrix is formed by Poaceae and Cyperaceae with numerous, mostly diminutive annuals imbedded within. Typically the substrate is very shallow with the lowest values occurring toward the transition to the open rock. Nutrient availability is restricted, a situation reflected in the floristic composition of the ephemeral flush vegetation with Lentibulariaceae, Eriocaulaceae, and Xyridaceae being usually well represented. In general, carnivorous plant species (e.g. *Drosera* spp., *Genlisea* spp., *Utricularia* spp.) are widespread and have a center of diversity here.

Plant Adaptive Traits

Inselbergs form centers of diversity for certain plant functional types that are well adapted for survival under environmentally extreme conditions. In particular, water scarcity and low nutrient availability have had a deep impact on the floristic composition of inselbergs. The microclimatical and edaphic dryness of inselbergs is reflected in the presence of numerous drought-adapted plants. Among them, succulents and desiccation-tolerant vascular plants are particularly prominent.

SUCCULENTS

On exposed rocky slopes of tropical inselbergs, succulents occur as perennial lithophytes. In the paleotropics, inselbergs in East Africa and Madagascar are particularly rich in succulents. Here, Aloaceae (*Aloe* spp.), Apocynaceae (e.g., *Pachypodium*), Crassulaceae (e.g., *Kalanchoe*), and Euphorbiaceae (e.g., *Euphorbia*) comprise numerous endemics. On neotropical inselbergs, Bromeliaceae (e.g., *Encholirium*), Cactaceae (e.g., *Coleocephalocereus*), and Orchidaceae (e.g., *Cyrtopodium*) occur with succulent species. Pachycaulous and caudiciformous species are widespread on tropical inselbergs. These plants possess fat water-storing trunks or a subterranean caudex. Comparatively rare are annual leaf succulents that occur on inselbergs in both tropical and temperate regions. Examples are *Cyanotis lanata* (tropical Africa), *Sedum smallii* (southeastern United States), and the genera *Crassula* and *Calandrinia*, which occur on Australian inselbergs.

DESICCATION-TOLERANT VASCULAR PLANTS

Only ~300 species of vascular plants can be classified as being absolutely desiccation tolerant and are known as the so-called *resurrection* plants. Desiccation-tolerant vascular plants are well adapted to withstand long periods of drought by resting in a state of dormancy. Most resurrection plants lose their chlorophyll and other photosynthetic pigments (i.e., they are poikilochlorophyllous, in contrast to relatively few desiccation-tolerant plants that keep their photosynthetic pigments during the process of desiccation and which are thus homoiochlorophyllous). Monocots outnumber dicots among desiccation-tolerant vascular plants, with Velloziaceae, Cyperaceae, and Poaceae being particularly important. Desiccation-tolerant arborescent monocots are unique with regard to the possession of certain morphological and anatomical features. Their fibrous stems consist mainly of adventitious roots and old persistent leaf bases and may attain a height of several meters. Surprisingly, the adventitious roots possess a velamen radicum that might be of functional importance for the rapid capillary uptake of rain water. On inselbergs desiccation-tolerant vascular plants are mainly found as mat-formers, but they also occur in shallow depressions and even in seasonally water-filled rock pools. A striking and highly specialized desiccation-tolerant species in rock pools is the Scrophulariaceae *Lindernia intrepidus* (syn. *Chamaegigas intrepidus*), an endemic to Namibia.

Geographic Patterns of Species Richness

Floristically, inselbergs in different geographical regions are clearly distinct. Based on comparative floristic data, three hotspots of inselberg plant diversity can be identified, which are rich in both species and endemics: (1) southeastern Brazil, (2) Madagascar, and (3) southwestern Australia. It has to be emphasized, however, that for several tropical regions (e.g., Angola, India), our knowledge of inselbergs is still sparse.

SOUTHEASTERN BRAZIL

That the forest vegetation of the Mata Atlântica is rich in species and endemics is well known. However, it is frequently overlooked that in this region (i.e., in particular parts of the Brazilian federal states of Rio de Janeiro, Minas Gerais, and Bahia) rock outcrops not only form dominant landscape elements but also support large numbers of endemics. Remarkably high is the beta diversity (i.e., the degree of floristic differentiation over small distances) in southeastern Brazil, with considerable species turnover between individual outcrops. Their vegetation is extremely rich in drought-resistant perennial species, whereas annuals are relatively rare. Prominent examples are xerophytic and succulent bromeliads (e.g., *Encholirium*, *Orthophytum*, *Pitcairnia*, *Vriesea*, *Tillandsia*), cacti (e.g., *Coleocephalocereus*, *Melocactus*), and orchids (e.g., *Cyrtopodium*, *Laelia*). Moreover, resurrection plants occur abundantly and belong to genera such as *Vellozia* and *Trilepis* and to the fern genera *Anemia*, *Doryopteris*, and *Selaginella*.

MADAGASCAR

Madagascan inselbergs are particularly frequent on the central plateau, where they are famous for their richness in succulent plants. Moreover, desiccation-tolerant vascular plants (e.g., *Coleochloa*, *Myrothamnus*, *Selaginella*, *Xerophyta*) occur profusely. The preliminary data available for this region show that the rock outcrop flora contains an extraordinarily high percentage of endemics. A considerable number of genera (e.g., *Aloe*, *Cynanchum*, *Euphorbia*, *Kalanchoe*, *Pachypodium*, *Senecio*) have obviously radiated on Madagascan inselbergs. Remarkably, many species show a high degree of morphological differentiation over short distances, making the limitation of taxa difficult (e.g., within *Euphorbia* and *Xerophyta*).

SOUTHWESTERN AUSTRALIA

In this region, inselbergs occur along steep climatic gradients from winter rainfall climates to inland deserts. Species richness and endemism decline with increasing aridity. Particularly striking is the richness in annuals, with Asteraceae, Stylidiaceae, Poaceae, and Amaranthaceae being particularly speciose. Resurrection plants are mainly represented by the monocotyledonous genus *Borya*. Apart from a few tiny, short-lived leaf succulents (*Calandrinia* spp., *Crassula* spp.) succulents are absent. Very rich in species are terrestrial *Utricularia* species, which are typical components of seasonally wet vegetation types.

Faunistic Aspects

Inselbergs provide structural niche components that can be used as nest sites, for shade, or for water supply. Special attention among the inselberg specialists is merited by rock lizards, hyraxes, klipspringers, or rock wallabies. Moreover, inselbergs are visited by a large number of animals that use rock outcrops as part of their range.

SEE ALSO THE FOLLOWING ARTICLES

Madagascar / Orchids / Pantepui / Vegetation

FURTHER READING

Barthlott, W., S. Porembski, R. Seine, and I. Theisen. 2007. *The curious world of carnivorous plants: a comprehensive guide to their biology and cultivation*. Portland, OR: Timber Press.

Hopper, S. D., A. P. Brown, and N. G. Marchant. 1997. Plants of western Australian granite outcrops. *Journal of the Royal Society of Western Australia* 80: 141–158.

Hunter, J. T. 2003. Persistence on inselbergs: the role of obligate seeders and respouters. *Journal of Biogeography* 30: 497–510.

Kruckeberg, A. R. 2002. *Geology and plant life: the effects of landforms and rock types on plants*. Seattle: University of Washington Press.

Parmentier, I., T. Stévart, and O. J. Hardy. 2005. The inselberg flora of Atlantic Central Africa: I. determinants of species assemblages. *Journal of Biogeography* 32: 685–696.

Porembski, S., and W. Barthlott, eds. 2000. *Inselbergs—biotic diversity of isolated rock outcrops in tropical and temperate regions*. Ecological Studies, volume 146. Berlin: Springer–Verlag.

Porembski, S., and W. Barthlott. 2000. Granitic and gneissic outcrops (inselbergs) as centers of diversity for desiccation-tolerant vascular plants. *Plant Ecology* 151: 19–28.

Sarthou, C., S. Samadi, and M.-C. Boisselier-Dubayle. 2001. Genetic structure of the saxicole *Pitcairnia geykesii* (Bromeliaceae) on inselbergs in French Guiana. *American Journal of Botany* 88: 861–868.

Twidale, C. R., and J. R. Vidal Romani. 2005. *Landforms and geology of granite terrains*. Leiden, Netherlands: Balkema.

INTRODUCED SPECIES

DANIEL SIMBERLOFF

University of Tennessee, Knoxville

Introduced species are those brought purposefully or inadvertently by humans to new regions, whereas a biological invasion is the establishment and spread of an introduced species into one or more habitats in its new home. Biological invasions are particularly pronounced on islands; worldwide, ~1.6 times as many mammal species and three times as many bird species have been introduced successfully to islands than to continents. Introduced carnivores eliminated many island bird species and subspecies, as well as reptiles and amphibians. Introduced grazers such as sheep, rabbits, and reindeer have eliminated many endemic island plants, and rooting by introduced pigs has massively eroded mountainous islands. Introduced plants have transformed island forests into grassland.

GLOBAL PATTERNS

For birds, 129 species are believed to have gone extinct worldwide since AD 1500; of these, 119 were island endemics. Of these 129 extinctions, scientists implicate a cause for 93 species; of these 93, 78 (84%) were eliminated wholly or partly by introduced species (primarily carnivores, but also herbivores that devastated habitats and birds that vectored introduced diseases). Even these figures do not convey the full impact of introduced species on island birds. In New Zealand, for example, 24 endemic birds were eliminated after the arrival of ancestors of the Maori (probably AD ~1250–1300) but before AD 1400, and some of these may have fallen prey to Pacific rats (*Rattus exulans*), which arrived with the earliest settlers. Island birds are not unique in their susceptibility to invaders. Even a group such as the freshwater fishes, relatively poorly represented on islands because of dispersal difficulties, has been devastated. Of 69 extinctions since AD 1500 (not counting Lake Victoria cichlids, or one extinction on Madagascar), 21 have been on islands; of these, causes are suspected for 20, and in each instance but one, introduced species are implicated. Insufficient data exist for similar analyses on invertebrate extinctions, but introduced species have had catastrophic impacts on a few well-studied groups on islands. For instance, all 144 known extinctions of land snails since AD 1500 were on islands. Causes are not known for many, but ~50 were caused at least partly by the rosy wolf snail (*Euglandina rosea*; Fig. 1), a predatory snail introduced to islands worldwide, but especially in the Pacific, in a failed attempt to control the introduced giant African snail *Achatina fulica*. Predation by introduced rats has also contributed to many island snail extinctions.

Native island plants are at risk from introduced species, just as native animals are. For example, on Phillip Island in the Norfolk group (Australia), 13 indigenous plant species (including two endemics) were eliminated within 140 years of the establishment of pigs, goats, and rabbits, primarily because of grazing by the last of these. Similarly, on Laysan Island (Hawaii), rabbits introduced in 1903 eliminated 26 plant species in just two decades. On St. Helena, beginning with Portuguese discovery in 1502, goats, pigs, cattle, rabbits, horses, donkeys, rats, and mice arrived, with kilometer-long goat herds numbering in the thousands wreaking particular havoc on vegetation. Seven of the 46 endemic plant species known to have been present in 1502 have since been extinguished (plus an unknown number that vanished without records or collections), primarily by these introduced animals. The exotic animals were abetted by introduced plants, far more adapted to the ravages of goats than were the native plants. Today, introduced plant species (at least 77) outnumber native species (53) and dominate most of the island. Even these statistics understate the effect of the introduced species: Many of the native plant species consist of just a few individuals, and two species have now each been reduced to one plant.

For some islands, many impacts of introduced species have been documented. This is especially true of the Hawaiian Islands, New Zealand, the Mascarenes, Macaronesia, the Antilles, and St. Helena. However, all islands have probably been modified by introduced species, whereas only a small fraction of these possible impacts have been studied.

TYPES OF IMPACTS

Impacts of introduced species that lead to endangerment or extinction are varied. Some of the most dramatic and obvious ones, such as an introduced mongoose (Fig. 2) eating a native ground-nesting bird, may not be as consequential as subtler impacts that indirectly affect an entire community, as when introduced plants change nitrogen availability or fire frequency.

A subtle but potentially drastic impact is the fertilization of parts of the island of Hawaii by the shrub *Morella (Myrica) faya*, native to Macaronesia and introduced a century ago. Hawaii is a young, volcanic island, so soil is nitrogen-poor, and the native plants have adapted to this soil regime, which

FIGURE 1 Rosy wolf snail, *Euglandina rosea*. Photograph courtesy of Jack Jeffrey Photography.

most introduced plants cannot tolerate. A nitrogen-fixer, *M. faya* quadruples the input of fixed nitrogen, which in turn stimulates invasions, and ultimately dominance, by a number of other introduced plants. Changing the plant community, in turn, affects many associated animal species. On many Pacific islands, introduced grasses (e.g., in Hawaii, *Melinus minutiflora* from Africa and *Schizachyrium condensatum* from North America) foster more frequent and intense fires that kill most native trees and shrubs, changing diverse, native-dominated woodlands into low-diversity grasslands dominated by exotics.

Introduced herbivores, by removing the dominant plant species, can also affect entire island communities. The goats of St. Helena, noted above, are an example. Similarly, on the Antarctic island of South Georgia, reindeer (*Rangifer tarandus*), introduced from Norway by whalers in the early twentieth century, destroyed great areas of the native tussock grass *Poa flabellata* and also grazed heavily on large lichens. The reindeer have spurred a massive invasion by introduced *Poa annua*, negatively affecting seabird colonies. In addition, the introduced grass has led to a decline in body size of a native beetle that cannot digest it. Impacts on other species are likely but unstudied. North American beavers (*Castor canadensis*), introduced in 1946 to Isla Grande (Tierra del Fuego), have spread to several other islands, and they have changed many closed southern beech forests to grass- and sedge-dominated meadows.

By virtue of gnawing on seeds, Pacific rats now appear to have been at least partly responsible for massive deforestation of endemic palm forests on Oahu and Easter Island, an event that had been attributed wholly to humans. In addition to this impact by herbivory, rats can affect entire island ecosystems by changing soil fertility. On small offshore islands of New Zealand invaded by black rats (*Rattus rattus*) or Norway rats (*R. norvegicus*), predation of seabirds led the latter to abandon these islands, thus disrupting transport of nutrients to the islands, thereby lowering soil fertility. This change, in turn, triggered a cascade of effects on belowground organisms, which in turn affected many ecosystem processes and features, both above and below ground.

Many other impacts of introduced species, including some of the most noteworthy island cases, are seen primarily as particular species affecting particular other species, rather than entire ecosystems. Determining which natives an introduced species affects is not always simple, as many impacts on one species can be propagated to others (for example, when reindeer browsing on a native grass on South Georgia ultimately affect a native beetle).

Predation is foremost among these direct species-level impacts. The hecatomb attributed to the rosy wolf snail has already been depicted. On Guam, the brown tree snake (*Boiga irregularis*), introduced in cargo from the Admiralty Islands just after World War II, has eliminated all but one of the native forest bird species and subspecies. Feral populations of introduced housecats have devastated seabird colonies on Ascension Island, Kerguelen Island, Marion Island, and others. On Kerguelen alone, feral cats are estimated to kill ~1.3 million birds each year. The small Indian mongoose, *Herpestes auropunctatus* (Fig. 2), was introduced in the late nineteenth and early twentieth centuries to the West Indies, Hawaiian Islands, Fiji, Mauritius, Okinawa, and elsewhere, primarily to control rats in agriculture, but also, on the Adriatic islands, to control snakes. It is almost certainly wholly responsible for the global extinction of a rail in Fiji and a petrel on Jamaica, and it contributed to several other avian extinctions on islands, interacting with introduced rats, cats, dogs, and pigs, as well as anthropogenic habitat destruction. *H. auropunctatus* has eliminated several birds that persist on nearby mongoose-free islands. The mongoose is also believed responsible for extinction of four endemic Hispaniolan mammals and possibly several West Indian snake species; it is likely to have caused extirpation of other West Indian snakes on those islands to which it was introduced. Mongooses were almost surely the key cause of extirpation of lizards on particular islands in the West Indies and Fiji, as evidenced by the persistence of these species only on mongoose-free islands. The mongoose similarly extirpated a frog species from three Caribbean islands. In many instances, the exact role of each introduced predator is uncertain, but a battery of them have

FIGURE 2 Small Indian mongoose, *Herpestes auropunctatus*. Photograph courtesy of Jack Jeffrey Photography.

driven a species to the brink of extinction or beyond. For example, the New Zealand kakapo (*Strigops habrotilus*), a large, flightless parrot, now exists only on a few small, predator-free islands to which it was translocated, having been eliminated on all main islands by introduced Pacific rats (*R. exulans*), Norway and black rats, cats, dogs, pigs, stoats (*Mustela erminea*), weasels (*Mustela nivalis*), and ferrets (*Mustela putorius furo*).

Another endangered New Zealand bird, the kokako (*Callaeas cinerea wilsoni*), exemplifies competition for food with introduced species, in this case brush-tailed possums (*Trichosurus vulpecula*), red deer (*Cervus elaphus*), and goats (*Capra hircus*). Two palearctic introduced yellowjacket wasps, *Vespula vulgaris* and *V. germanica*, also compete with threatened native New Zealand birds, especially another endemic parrot, the kaka (*Nestor meridionalis*), in southern beech forests. The competitive effect of *V. vulgaris* is greatly exacerbated by a native scale insect that exudes a dark, sweet "honeydew" on beech trunks. Feeding on this honeydew, *V. vulgaris* reaches densities of 360 wasps/m^2, turning the trunks yellow and harvesting 8.1 kg of honeydew/ha, at least equal to consumption by all birds together. Competition with an invader for food is strongly implicated in the decline of an island native in other cases. Further examples include the ongoing replacement of the European red squirrel (*Sciurus vulgaris*) by the North American gray squirrel (*S. carolinensis*) in Britain and the impact of the introduced house gecko, *Hemidactylus frenatus*, on resident native gecko species on several Pacific islands. Competition can also be for resources other than food: The endangered Puerto Rican parrot (*Amazona vittata*) competes for nest sites with an introduced bird, the pearly-eyed thrasher (*Margarops fuscatus*), as well as introduced honeybees (*Apis mellifera*).

Herbivory by introduced species can affect entire ecosystems, as has been noted above for goats on St. Helena and reindeer on South Georgia. However, many introduced herbivores devastate particular plant species that are not key players in structuring an entire ecosystem. For example, 17 plant species and subspecies are threatened in the California Channel Islands by herbivory by introduced herbivores (primarily goats, pigs, and sheep); most are small plants unlikely to have been community dominants.

Introduced parasites and diseases have frequently devastated island species, contributing to the extinction of many. In the Hawaiian Islands, avian pox and malaria arrived with the extensive introduction of exotic songbirds in the late nineteenth and early twentieth centuries. These diseases have contributed to the decline and extinction of many Hawaiian bird species. Their spread was fostered by introduced mosquitoes, particularly the tropical species *Culex quinquefasciatus*, which gradually adapted to cooler temperatures and thus advanced higher up mountains, restricting surviving native birds to ever-smaller, higher refugia.

A particularly subtle impact of some introduced species is hybridization with native species, which can lead to a sort of genetic extinction even though no lineage is terminated. For instance, both the New Zealand gray duck (*Anas superciliosa superciliosa*) and the Hawaiian duck (*A. wyvilliana*) are endangered partly because they hybridize with the introduced North American mallard (*A. platyrhynchos*). Likewise, the Seychelles turtledove (*Streptopelia picturata rostrata*) has hybridized so extensively with *S. p. picturata*, introduced from Madagascar, that the Seychelles population now consists of a hybrid swarm more similar on average to the invader than to the native.

Sometimes two or more introduced species exacerbate one another's impact on natives, a phenomenon known as invasional meltdown. On Christmas Island, the introduced yellow crazy ant (*Anoplolepis gracilipes*) had long been present but quite innocuous, until the late 1980s. Then, introduction of a scale insect, plus outbreaks of a native scale insect, led to massive production of honeydew. The ants protect the scales and harvest the honeydew, and populations of both scales and ants exploded. The ants then devastated populations of the native red land crab (*Gecarcoidea natalis*), which in turn caused greatly increased growth of ground cover plants, seeds and seedlings of which had previously been harvested by crabs. The invasion of nitrogen-fixing *Morella faya* in Hawaii, described previously, is part of a meltdown in which the seeds of the plant are primarily dispersed by the introduced Japanese white-eye (*Zosterops japonicus*) as well as introduced rats and feral pigs, whereas greatly increased densities of introduced earthworms under *Morella* trees enhance the rate at which nitrogen-rich litter is buried, thereby aiding the invasion of previously nitrogen-limited introduced plants. Often, introduced birds exacerbate invasions of exotic plants by dispersing their seeds. The red-whiskered bulbul (*Pycnonotus jocosus*), introduced to the Mascarenes, has exacerbated invasions by several alien plants, even into previously undisturbed areas. On Isla Grande, beaver-altered sites have higher levels of soil nitrogen, which favors invasion by a number of introduced plants.

Sometimes the impacts of an introduced species initiate a chain reaction that ends up indirectly affecting native species in surprising ways. In Great Britain, caterpillars of the native large blue butterfly (*Maculinea arion*) matured underground in nests of the ant *Myrmica sabuleti*. This ant cannot nest in overgrown habitats, and during the course of centuries, reduced livestock grazing and changing land-use patterns left rabbits (*Oryctolagus cuniculus*), introduced from continental Europe in AD ~1150, as the main species maintaining suitable habitat for the ant. The South American myxoma virus, introduced to France in 1952 to control the rabbit, quickly spread to Great Britain and devastated rabbit populations. Habitats became overgrown, ant populations plummeted, and the butterfly disappeared.

ARE ISLANDS MORE VULNERABLE THAN MAINLAND AREAS TO INTRODUCED SPECIES?

Conventional wisdom is that introduced species are more likely to survive on islands than on mainland areas and to have greater impacts there, because island ecosystems are particularly fragile, and island species weak. Island species have been termed evolutionary "backwaters and dead ends," whereas island ecosystems are seen as presenting less "biotic resistance" to invaders than do continental ecosystems. Probably the most characteristic doomed island species, in the mind of the public, is the hapless, extinct Mauritius dodo.

However, this view of generic weakness of island species and communities is suspect for both a theoretical reason and an empirical one. The theoretical reason is that island species, at least endemic ones, have evolved on their respective islands, and one would expect natural selection to have adapted them to these islands better than a species arriving from elsewhere. The empirical reason is that, for every sort of impact described in the previous section, it is possible to cite similar impacts of species introduced to mainland communities. There are even island species that have wrought havoc with continental communities, such as the New Zealand mud snail *Potamopyrgus antipodarum*, introduced to the Greater Yellowstone region of North America.

The percentage of introduced species is often higher on islands than on the mainland, a fact adduced as evidence of island susceptibility to invasion. For instance, ~35% of Hawaii's insect species are introduced, whereas only ~3% of the insect species of the continental United States are introduced. However, in fact there are as many introduced insect species in the continental United States as in Hawaii: ~3000. The percentage is greater in Hawaii than in the continental United States simply because the latter has ~90,000 native species, and the former only ~6000. The other difficulty in interpreting higher percentages of introduced species in island biotas is that no account is taken of failed introductions and amount of effort devoted to attempting to establish exotic species. Many islands (e.g., Hawaii, New Zealand, La Réunion, Mauritius) had active acclimatization societies that tried to redress the paucity of native bird life by introducing species from elsewhere. Thus, for example, ~70 species of perching birds and doves have been introduced to Hawaii (compared to only 40 in all of the continental United States), of which about half survive, a number that rivals the surviving native species in these groups (most of which are now endangered). Only 13 introduced species of perching birds and doves established populations in the continental United States. However, several acclimatization societies strove mightily to establish these birds in Hawaii, with large propagules and multiple introductions for many. It has been established that propagule pressure—the number of individuals introduced and number of attempts—is a key determinant of the likelihood that an introduced species will establish a population.

Though island species are not generically maladapted weaklings, there does appear to be an aspect of many island biotas that predisposes them to be particularly susceptible to certain kinds of introductions. This is the fact that many islands, especially oceanic islands, lack large vertebrate predators and social insects. Native species on these islands therefore did not evolve adaptations to such species, and an absence of such adaptations often led them to be devastated upon introduction of predators and ants. Many island birds, for example, nest on the ground, making them easy prey for the small Indian mongoose, rats, cats, pigs, and mustelids. Approximately 90% of bird extinctions since AD 1500 that can be ascribed to predation have occurred on islands. Similarly, Guam has no native snakes or predators that could have countered the introduced brown tree snake, so native birds (and other species) were easy prey for a new type of predator. Furthermore, continental birds introduced to islands had evolved in the presence of vertebrate predators, so they were preadapted to replace vulnerable native birds. Isolated islands also lack large mammalian grazers and browsers, so it is no wonder that the dominant native plants were devastated by introduced goats, reindeer, sheep, and cattle. By contrast, continental plants introduced to islands had evolved with large grazers and browsers, so it is unsurprising that they replaced many native island plants.

A second factor predisposing island communities to be vulnerable to certain kinds of invaders is that islands are smaller than continents, usually much smaller. Thus, it is far less likely on an island that there will be a refuge available to a native species that an invader cannot reach. Even hurricanes have far greater biodiversity consequences for this reason; a number of bird species and subspecies have been eliminated by hurricanes on islands (e.g., five on Kauai from Hurricane Iniki in 1992), but none on continents, and this is not because island birds are somehow weaker than continental ones; rather, their entire range can be exposed to a hurricane.

ERADICATING INTRODUCED SPECIES ON ISLANDS

A growing number of introduced species have been eradicated on islands around the world, and the size of such projects has increased substantially as eradication technology advances. Terrestrial island invaders are particularly tempting targets for eradication because islands are generally far less likely to be reinvaded than are continental regions, simply by virtue of the water barrier separating islands from sources.

The first island eradication of an introduced insect was the elimination of the tsetse fly (*Glossina* spp.) from Principe in the Gulf of Guinea. The flies had arrived in cargo from Africa in 1825, and sleeping sickness devastated the human population beginning in 1859. However, the fly and the disease were eradicated between 1911 and 1914. In 1956, a new tsetse invasion was noted on Principe, and the fly was again eradicated in a massive effort using traps, insecticides, extensive brush clearing, and hunting to reduce populations of pigs and wild dogs. Perhaps the most famous island eradication of an introduced insect entailed a demonstration in 1954–1955 on Curaçao. There, release of enormous numbers of sterile males eliminated an entire population (the screw-worm fly, *Cochliomyia hominivorax*) by keeping females from mating with fertile males. This technique has since been employed in several insect eradications on islands. Notable recent eradications of introduced insects from islands include the Oriental fruit fly (*Dacus dorsalis*) from Rota and Guam and the melon fly (*Bactrocera cucurbitae*) from the entire Ryukyu Archipelago, including Okinawa.

Many invasive introduced mammals have been eradicated from islands. The most widely publicized of such projects have eliminated rats. Black, Norway, and Pacific rats have been eradicated from at least 100 islands worldwide, usually with a goal of saving threatened bird species. The largest island cleared of rats to date is 113-km^2 Campbell Island, cleared in 2001 by the New Zealand government, which is in the intial stages of planning an attempt to eliminate rats from Great Barrier Island (274 km^2). Probably the best-known eradication of an invasive mammal was the successful campaign completed in 1986 to eradicate nutria (*Myocaster coypus*) from Great Britain. Among other introduced mammals eradicated from islands have been feral cats and dogs, house mice, rabbits, muskrats (*Ondatra zibethicus*), and burros. Very recently, large, longstanding populations of goats and pigs have been eradicated from Santiago (58,000 ha) and Isabella (400,000 ha) in the Galápagos. The New Zealand government is currently planning a campaign to eradicate Norway, black, and Pacific rats as well as rabbits, stoats, hedgehogs (*Erinaceus europaeus*), and feral cats simultaneously from Rangitoto and Motutapu Islands.

There have been far fewer attempted eradications of plants on islands and very few successes. However, the United States government recently succeeded in its 16-year campaign to rid Laysan Island of an introduced sandbur (*Cenchrus echinatus*), which had spread to replace native plants on 30% of the vegetated part of the island, threatening the endangered endemic Laysan finch (*Telespiza cantans*) and Laysan duck (*Anas laysanensis*).

Eradicating invaders from an island does not automatically restore the desired state, and unexpected results are common. Eradication of rats has, on several islands, been followed by explosions in density of house mice (*Mus musculus*), which may prove equally detrimental to the species targeted for restoration. Elimination of a predator may also increase herbivore populations, to the detriment of native plants, whereas eradicating an introduced herbivore can profit introduced weeds at the expense of native plants. Eradication of rabbits from Motunau Island (New Zealand) led to increased populations of introduced boxthorn (*Lycium ferocissimum*), whereas removal of grazing livestock from Santa Cruz Island (California) caused explosive increases in fennel (*Foeniculum vulgare*) and other introduced plants. Changes in plant community composition and structure after herbivore eradication can, in turn, affect animal populations. On Mana Island (New Zealand), removal of cattle decreased native lizard populations by modifying vegetation.

SEE ALSO THE FOLLOWING ARTICLES

Biological Control / Invasion Biology / Land Snails / Pigs and Goats / Rodents

FURTHER READING

D'Antonio, C. M., and T. L. Dudley. 1995. Biological invasions as agents of change on islands versus mainlands, in *Islands: biological diversity and ecosystem function*. P. M. Vitousek, L. L. Loope, and H. Adsersen, eds. Berlin: Springer-Verlag, 103–121.

Elton, C. S. 1958. *The ecology of invasions by animals and plants*. London: Methuen.

Lockwood, J. L., M. F. Hoopes, and M. P. Marchetti. 2007. *Invasion ecology*. Malden, MA: Blackwell.

Simberloff, D. 1995. Why do introduced species appear to devastate islands more than mainland areas? *Pacific Science* 49: 87–97.

Towns, D. R., I. A. E. Atkinson, and C. H. Daugherty. 2006. Have the harmful effects of introduced rats on islands been exaggerated? *Biological Invasions* 8: 863–891.

Veitch, C. R., and M. N. Clout, eds. 2002. *Turning the tide: the eradication of invasive species*. Gland, Switzerland: IUCN Species Survival Commission.

Williamson, M. 1996. *Biological invasions*. London: Chapman & Hall.

INVASION BIOLOGY

GEORGE RODERICK

University of California, Berkeley

PHILIPPE VERNON

University of Rennes 1, Paimpont, France

A biological invasion is the establishment and spread of a locally nonindigenous species. Invasive species have had significant ecological and economic impacts on islands, including displacement or extirpation of native species, changes in physical geography such as erosion and silting of streams and offshore habitats, and rising costs associated with management of urban and agricultural pests. Invasive species can colonize new areas on their own, inadvertently through actions of humans, or as a result of purposeful introductions with unexpected consequences. With the expansion of global trade and connectivity, the importance of invasive species is intensifying, and biological invasion is appropriately considered an agent of global change. Invasive species are a particular environmental concern for islands, especially those that have been isolated ecologically.

STATUS AND IMPACTS

The structural isolation of islands limits the number of species that colonize, resulting in communities comprising an unrepresentative collection of founding species. Adaptive radiation and other processes, which often scale with isolation, further accentuate the unrepresentative nature of species within island communities. It is for these reasons that new invasive species on islands can have such large impacts. The importance of invasive species has been recognized since the time of Darwin. However, with the recent increase in global trade and the accompanying homogenization of the world's flora and fauna, biological communities on islands are becoming less isolated and are experiencing increasing ecological and socioeconomic impacts associated with the arrival of many invasive species.

An invasive species is a locally nonindigenous species that colonizes a new geographical area and thus expands its range. Invasive species have a "pestlike" connotation, in that they typically directly or indirectly cause ecological or economic damage. Government entities often use the term invasive alien species (IAS), which, like the term exotic species, emphasizes that such species are nonindigenous. Knowledge of the ecological origins of a species (i.e., indigenous or nonindigenous) is crucial in understanding the biology of invasive species as well as in their management. Determining which species are endemic (i.e., native and found nowhere else) is usually not difficult. However, resolving which species are indigenous (i.e., native, but also found elsewhere) can be problematic—accordingly, many types of data are used to make this determination, including biogeography, comparative systematics, molecular genetic variation, and anthropological associations. Using these approaches, biodiversity surveys show that nonindigenous species can make up a large proportion of island biota. For example, in the Hawaiian archipelago, of the 2142 flowering plant species and 8151 insect species, 53% and 34%, respectively, are nonindigenous (Table 1). In the paucispecific cold oceanic islands, a dramatic increase in the number of insect species has been recently observed—on the Kerguelen Islands, for example, of 39 known insect species (Fig. 1), 16 (41%) are alien.

Severe ecological and environmental impacts have been attributed to invasive species on islands, such as displacement of native species, changes in demographic rates, gene flow modifications, and drastic perturbations in pristine ecosystems. For example, in the Hawaiian archipelago the introduction of the plant *Myrica faya* has displaced endemic plants, and because it also fixes nitrogen through a symbiotic relationship with an actinomycete bacteria, its presence alters the soil, facilitating colonization by a diversity of other alien species. Several features of islands and their biota may accentuate these effects: large edge effects (relative to area), large topological diversity within small geographical areas, demographic and genetic consequences related to small population sizes that characterize indigenous species on islands, and the nonrepresentative species composition and subsequent evolution in the absence of many mainland taxa. Although the role of these features requires more study, it is broadly understood that invasive species have a disproportionate impact in insular environments in general, whether oceanic islands, mountaintops, ponds and lakes, or even old trees in a field.

TABLE 1
Numbers of Hawaiian Species Tabulated by the Bishop Museum's Hawaii Biological Survey, Honolulu

Taxon	Total Species	Endemic Species	Indigenous Species	Nonindigenous Species
Algae	1118	104	?	?
Other protists	1229	?	?	?
Angiosperms	2142	896	107	1139
Other plants	639+	226		37+
Fungi and lichens	3185+	972		?
Cnidaria	108	32	31	45
Annelida	352	81	?	?
Mollusca	2073+	1096	848+	129
Crustacea	1407	?	?	?
Insecta[a]	8151	5245	124	2782
Araneae (spiders)	248	144	?	104
Acari (mites)	656	168	17	471
Other Arthropoda	97	42	14	41
Other invertebrates	2115	?	?	?
Echinodermata	309	150	159	0
Lower Chordata	77	?	?	?
Fish	1245	157	1033	55
Amphibians	7	0	0	7
Reptiles	29	0	3	26
Aves (birds)[b]	183	63	65	55
Mammalia	44	2	23	19
Totals	25,615	9378	2424	4910

SOURCE: Eldredge and Evenhuis (2003).
[a] Current estimates suggest there may be as many as 10,000 endemic insects and spiders in Hawaii, with many still undescribed.
[b] Many birds are known extinct, and there are thought to have been previously 200–300 species.

FIGURE 1 In the Kerguelen Islands, the subantarctic wingless fly *Anatalanta aptera* (shown here) has suffered great impacts from the predatory beetle *Oopterus soledadinus*, introduced in 1912 from the Falkland Islands. This fly also interacts with the blowfly *Calliphora vicina*, accidentally introduced in 1978. Photograph by M. Laparie.

One of the primary determinants that appears to dictate the impact of invasive species is isolation. Isolation is a complex concept, incorporating not only distance, but also island size and topological diversity, the nature of the matrix, and the dispersal capabilities of the taxa of interest. The timing of colonization of invasive species is also associated with isolation. For example, extremely isolated habitats have only relatively recently experienced the impact of global homogenization, accentuating the changes currently observed. Here, we consider invasive species of islands generally, and focus on the processes of invasion and establishment.

PROPAGULES

Propagules are individuals, or sets of individuals of a given species, that can give rise to new populations. Propagule pressure, or the rate at which propagules arrive or attempt to colonize, is an important predictor of the likelihood that a species will become established. Propagule pressure typically scales inversely with isolation, so that remote islands experience fewer propagules with only the most dispersive propagules being capable of reaching the most remote islands under natural conditions. For example, to account for the 8000 described indigenous terrestrial arthropod species in the Hawaiian Archipelago, it has been estimated that approximately 400 colonists must have arrived to initiate the various lineages, many of which diversified subsequently through adaptive radiation. If the age of the oldest high island, Kauai, of 4–6 million years is used as a reasonable older limit for most lineages, this number of colonists translates to a rate of successful establishment of approximately one every 10,000–15,000 years. Of course, many propagules that arrive are not successful, and thus it is difficult to estimate the magnitude of natural (historical) propagule pressure experienced by a given island.

Even using the roughest estimates of propagule pressure, it is clear that recent growth of traffic and transportation routes have resulted in a huge increase in the number of propagules reaching remote locations. For example, ballast water in ships is thought to account for the movements of 7,000–10,000 species simultaneously. In a study of ships arriving at 243 ports worldwide, expected invasion rates have been estimated at up to 2.94×10^{-4} species per km^2 per year, with large increases predicted in so-called hotspots. A study of 44,000 air transportation routes estimates that numbers of passengers transported will increase by 8% per year. International air travel also figures prominently in the movement of invasive species—for example, 73% of pest interceptions in the United States Port Information Network database were at international airports. Modeling efforts including both passenger and freight traffic suggest the greatest risks of invasion are between airports that are connected by numerous high-capacity routes and that have similar climates.

The relationship between establishment and propagule pressure can be important also for deciding how to manage invasive species. For example, greater propagule pressure may lead to a monotonic increase in probability of establishment, in which case any attempt to limit the number of propagules can limit further establishment. However, if the probability of establishment does not continue to increase with propagule pressure but levels off at some point, then additional effort to limit propagules after this point will have little effect. For ballast water, it has been suggested that moderate reductions in the per-ship-visit chance of introducing an invasive species can reduce the probability of invasion, and, by contrast, eliminating key port-to-port shipping connections will have negligible effects. This analysis suggests that reducing or eliminating organisms in ballast water will be effective, as for example in onshore treatment at particular ports or by onboard ballast water treatment.

Propagules are only a small sample of their source populations. Accordingly, colonists of islands typically show a reduction in genetic diversity, a genetic bottleneck, relative to their source. However, it is not obvious that genetic bottlenecks limit the spread of invasive species. One reason for this is that if a population can increase in size rapidly following a bottleneck, much genetic variation can be maintained among and within individuals, even though the rare alleles have been lost. Some loss of alleles may actually contribute to invasive success. For example, different species of invasive ants, such as the Argentine ant *Linepithema humile*, demonstrate a tendency toward larger multi-queen unicolonial structures that facilitate cooperation among otherwise competitive colonies. It has been hypothesized that bottlenecks have resulted in a loss of alleles that genetically code for diverse cuticular hydrocarbons, which are responsible for colonies recognizing each other. Hence colonies can grow largely unchecked in their new environment. In the red imported fire ant, *Solenopsis invicta*, the transition from monogyne form to a more damaging polygyne form appears to be associated with ecological constraints favoring cooperative breeding, but it is also controlled by genetic factors.

Multiple colonization events can mitigate, or even increase the genetic diversity of founding populations if the propagules stem from sources that are genetically different. For example, the brown anole, *Anolis sagrei*, has invaded Florida multiple times from separate, genetically differentiated locations in Cuba, resulting in invasive populations that are more genetically diverse than source populations. When invasive populations of the same species are subdivided, both genetic drift and independent selection in the subpopulations may preserve genetic diversity. Other species may show greater genetic diversity in invasive populations than in source populations. For example, in the common myna bird, *Acridotheres tristis*, populations introduced to New Zealand, South Africa, Hawaii, Australia, and Fiji show higher genetic variation than in native populations from India. In the house finch, *Carpodacus mexicanus*, populations introduced into Hawaii and eastern North America have greater variation in amplified fragment length polymorphism (AFLP) markers than populations in their native western North America range, differences that are consistent with morphological variation.

Invasive and colonizing often can adapt quickly to new environments, despite the effects of genetic bottlenecks. For example, in invasive Australian cane toads, *Bufo marinus*, populations at the edge of the expanding range have evolved longer legs and greater speed, suggesting that there is an advantage to founding a site first. One explanation for adaptive response to selection despite a recent genetic bottleneck is that novel interactions among genes in low-density and growing populations may enable populations to retain and even increase genetic diversity, upon which selection can then act. It has also been hypothesized that many invasive species have been the objects of selection over many generations for the ability to cope with low genetic diversity, and those that could not withstand low genetic diversity have been selected against. This hypothesis may help to explain why one of the best predictors for whether a plant species will be invasive is that it is invasive elsewhere.

One way that an invasive species may overcome a potential loss in the ability to adapt genetically is though phenotypic plasticity. If a species can naturally expand its ecological range upon colonization, adaptation through natural selection may not be necessary. Plasticity associated with colonization has been better studied in plants than animals, and there is some evidence to support two observations, both of which deserve more study in a greater variety of systems: (1) invasive species may be more plastic than noninvasive or native species, and (2) populations in the introduced range of an invasive species may evolve greater plasticity than populations in the native range. While the role of plasticity in invasions is likely more important than is currently appreciated, there are limits to the novelty of an environment that an organism is capable of exploiting. For example, for insects, climate-matching models do fairly well in predicting the geographic limits of species, suggesting that at least some organisms are already at their physiological limits, particularly with respect to temperature. The recent interest in predicting spread of invasive species associated with global temperature change will no doubt shed more light on this topic.

Multiple colonization events may also overcome demographic stochasticity—the loss of a population through chance events of survival and/or reproduction—in small invasive populations. In studies of conservation biology, it is thought that demographic stochasticity, coupled with Allee effects (see the following paragraph), are likely more important in determining the persistence of populations on an ecological time frame, and the same is likely true for invasive species in the early stages before populations have grown to sizable numbers.

A difficulty in understanding the relationship between establishment and propagule pressure, and in understanding the early phases of a biological invasion in general, is that individuals can be more difficult to detect at low population sizes. Thus, invasive species can be resident and established in an area for many years while escaping detection. One reason that invasive populations may remain at a low level in the initial stages and then suddenly appear may simply be a consequence of exponential growth. Additionally, small populations may display so called Allee effects, in which their growth rate may be reduced at low population sizes, as for example, when there are so few individuals that it is difficult for an individual to find mates. Both these phenomena make it difficult to use population numbers to estimate how long an invasive species has been established, making predictive models of invasions more difficult.

Islands may be the best places to study demographic processes associated with invasions. For example, because of the unrepresentative collection of species on islands and well-defined boundaries, invasions may be discovered at lower population sizes than might be possible in more complex ecological communities. For example, data on abundance and spread of the glassy-winged sharpshooter,

FIGURE 2 The glassy-winged sharpshooter, *Homalodisca vitripennis* (shown here on papaya), has been accidentally introduced into Hawaii, French Polynesia, and other Pacific Islands, where it is an urban nuisance because of the honeydew it produces, but is also an agricultural threat because of physical damage and its ability to vector a bacterial plant pathogen *Xylella fastidiosa*, which causes a variety of lethal scorchlike diseases in susceptible hosts. The Pacific invasions of the glassy-winged sharpshooter may have come from California; the insect is native to the southeastern United States and northern Mexico. Photograph by J. Grandgirard.

Homalodisca vitripennis (Fig 2), in Tahiti suggests that this leafhopper was discovered very quickly following invasion, prompting a successful program in biological control. It has also been possible to document the growth and spread of island populations of the cane toad, *Bufo marinus*, leading to the development of novel molecular genetic and statistical methods to investigate demographic history of invasive species.

VECTORS AND TRANSPORTATION

Many native species made use of vectors, or biological transportation agents, to reach islands naturally. For example, crab spiders, snails, and multiple plant lineages likely arrived on remote islands with birds. Likewise, the use of vectors and analogous transportation vehicles (noted in the preceding section) is characteristic of many invasive species, allowing them to spread to habitats and regions they would otherwise not be able to reach. Because the movement of the vector or other agent of transportation determines in large part, or completely, the movement of associated invasive species, management often focuses on the biology of the vector or mode of transportation.

Propagules and/or their vectors often follow well-defined pathways of introduction, though the pathways can take many forms. For example, invasive weeds are often found along routes used by human vectors, such as hiking trails. In the marine realm, ballast water is a well-known pathway for marine invasions worldwide (see the preceding section), as are ship hulls for fouling organisms. Similarly, the transport of fruit and vegetables is associated with movement of agricultural pests, as for example the introduction of tephritid flies, including the Mediterranean fruit fly, *Ceratitis capitata*, associated with citrus and other crops. Likewise, nursery or horticultural stock trade is a common mode of introduction of pests, including the glassy-winged sharpshooter, which mostly likely came to Tahiti from California as inconspicuous eggs on horticultural plants. Avian malaria, *Plasmodium relictum*, arrived in Hawaii with alien birds and vector mosquitoes around 1825, which has resulted in eliminating native birds at lower elevations, where mosquitoes and parasites persist.

Pathways associated with global trade, particularly regular shipping or airplane routes, can also be used to predict the arrival of potential invasive species. For example, invasive species managers in Hawaii are constantly vigilant for the brown tree snake, *Boiga irregularis*, which has been established in Guam for more than 50 years and has decimated the island's indigenous bird fauna. Risk modeling approaches are also used to predict routes of transportation. In a model that included volume of commodities transported, infestation rates of pests, efficacy of inspection, and the probability of establishment, transport of mosquitoes on airplanes has been identified as the most likely route for West Nile virus (an encephalitis) to reach the Galápagos, which would have devastating effects on endemic avifauna but also on reptiles.

FEATURES OF THE COMMUNITY

The presence or absence of other organisms is thought to play a significant role in the success of invasive species in novel habitats. Numerous studies have attributed the success of invasive species in new environments to having escaped many of their predators and parasites present in their indigenous range, although in some communities the invasive species may also pick up new predators and parasites. In a broad study of 26 host species of molluscs, crustacea, fish, birds, mammals, amphibia, and reptiles, the number of parasite species in native populations was double that in newly established populations, which also had lower rates of parasitism. A contributing factor may be that introduced populations are less likely to be infected upon arrival—propagules typically originate from small subsets of native populations and perhaps also from uninfected life-history stages. Invasive species have likely escaped pressures of direct competitors, though this effect is not as well studied. Over time, invasive species are likely to accumulate more predators, parasites, and competitors as a result of functional, numerical, or evolutionary responses. An increase in biological interactions over time may explain the downward slope of the often observed "hump" in population dynamics of invasive species, when, following an initial increase in population size, numbers eventually decline. One example is the glassy-winged sharpshooter in Hawaii, whose numbers declined without intervention, likely as a result of parasitoids introduced for the biological control of other sap-feeding species.

Although the foregoing examples indicate the importance of lack of elements of the biotic community in contributing to invasion success, the reverse can also be true. Accordingly, in some cases the establishment of one invasive species may facilitate the establishment and growth of a second (and third or more) invasive species, in which case the impact of the invasive species can be extreme and lead to a phenomenon termed *invasional meltdown*. For example, invasive ants have been shown to facilitate the invasion of sap-feeding insects, and invasive plants can facilitate invasive insect herbivores. In an example from Christmas Island, an invasion of the yellow crazy ant, *Anoplolepis gracilipes*, has decimated populations of a land crab, a keystone species in the native forest dynamics. The ants have also facilitated the increase of scale insects, whose honeydew causes sooty mold, reducing the health of the forest vegetation, eventually causing light gaps. The light gaps and a decrease in crab numbers have resulted in a decrease in other native species and invasions by weeds.

A factor that is often associated with invaded habitats is ecological disturbance and/or open space. For example, following Hurricane Iniki on the island of Kauai in 1992 there was a large increase in the abundance of invasive plants, which colonized open space. One might ask why the endemic species do not themselves fill open space caused by disturbance? One part of the answer may be that island endemics frequently have reduced dispersal abilities compared to invasive species that might occupy similar niches. This difference in dispersal between indigenous and invasive species, coupled with the unrepresentative assembly of island species, may also help to explain why invasive species are thought to have such a large ecological impact on islands.

Low species diversity may also encourage invasibility as a result of availability of open niches to a new spectrum on invasive species. Theory and empirical studies both suggest that historical accumulation of species in a community takes time, with time being dictated by isolation. Accordingly, particularly remote islands may take a long time to reach an equilibrium number of species (i.e., a number that stays consistent over time), and some islands may never reach such an equilibrium. Likewise, somewhat less isolated habitats that are relatively "young," such as San Francisco Bay in California, have been heavily impacted by invasive species, and this may also be because the natural community had not reached equilibrium in terms of species diversity.

The unrepresentative nature of taxa on islands also seems to be important in explaining the ability of invasive species to establish. For example, for plants, the current diversity in Hawaii is greater than what it was historically, and in New Zealand recent invasions have doubled the numbers of species from approximately 2000 to 4000. For freshwater fish in Hawaii the relative increase in numbers of species is even greater, with 40 alien species and only five indigenous species. This does not seem to be true for all taxa. For example, the total diversity of birds currently on the Hawaiian archipelago, including nonindigenous species, is approximately the same as it has been historically, despite the loss of bird diversity since human colonization.

Phylogenetic relationships between invaders and indigenous taxa may be used to predict not only which species can colonize but also the relative impact of invasive species. For grasses in California, for example, the most invasive species are not closely related to indigenous species in comparison to the less invasive exotics. A similar analysis of the association between invasiveness of species and their phylogenetic similarity to native species has not been conducted for remote islands, though the unrepresentative nature of species on remote islands may make such a comparison difficult.

THE FUTURE

As noted in other chapters, invasive and introduced species have large economic, environmental, and societal impacts on island communities. As a result, and because of the well-defined and compartmental nature of islands, invasive species on islands continue to be the object of basic and applied research and can provide much needed cross-fertilization among disciplines. Island systems can offer much to basic research in invasion biology. Several features are particularly worth noting. First, islands offer replicate systems in which to observe invasions and colonizations. Though invasions are natural experiments that are unplanned, observations of invasions can span a diversity of taxa under a range of conditions and time scales and thus complement more controlled manipulations. Second, island invasions are frequently noted very soon after arrival, so the initial stages can be studied. Thus, the early period of an invasion, which usually goes undetected, may be shorter on islands. Samples from the early stages are critical as benchmarks for studies of ecological and genetic change. Third, because of the clear boundaries and typically restricted size of islands it should be possible to obtain accurate, and replicated, climate data that can then be used to improve the parameterization of models used to predict the spread of species in the context of global climate change. Such models will both contribute to an understanding of responses to global change and also be used to manage threatened island ecosystems. In sum, while the threats of invasive species on islands are real and daunting, islands can offer a promise of better understanding of processes over a range of temporal and spatial scales.

SEE ALSO THE FOLLOWING ARTICLES

Adaptive Radiation / Ants / Biological Control / Dispersal / Introduced Species

FURTHER READING

Carlton, J. T. 1999. The scale and ecological consequences of biological invasions in the world's oceans, in *Invasive species and biodiversity management*. O. Sandlund, P. Schei, and Å. Viken, eds. Dordrecht, The Netherlands: Kluwer, 195–212.

D'Antonio, C. M., and T. L. Dudley. 1995. Biological invasions as agents of change on islands versus mainlands, in *Islands: biological diversity and ecosystem function*. Ecological Studies 115. P. M. Vitousek, L. L. Loupe, H. Anderson, eds. Berlin: Springer-Verlag, 103–121.

Eldredge, L.G., and N.L. Evenhuis. 2003. Hawaii's biodiversity: A detailed assessment of the numbers of species in the Hawaiian Islands. *Bishop Museum Occasional Papers* 76: 1–30.

Gillespie, R. G., E. M. Claridge, and G. K. Roderick. 2008. Biodiversity dynamics in isolated island communities: interaction between natural and human-mediated processes. *Molecular Ecology* 17: 45–57.

Lockwood, J. A., P. Cassey, and T. Blackburn. 2005. The role of propagule pressure in explaining species invasions. *Trends in Ecology and Evolution* 20: 223–228.

Pyšek, P., D. M. Richardson, J. Pergl, V. Jarošík, Z. Sixtová, and E. Weber. 2008. Geographical and taxonomic biases in invasion ecology. *Trends in Ecology and Evolution* 23: 237–244.

Reaser, J. K., L. A. Meyerson, Q. Cronk, M. De Poorter, L. G. Eldrege, E. Green, M. Kairo, P. Latasi, R. N. Mack, J. Mauremootoo, D. O'Dowd, W. Orapa, S. Sastroutomo, A. Saunders, C. Shine, S. Thrainsson, and L. Vaiutu. 2007. Ecological and socioeconomic impacts of invasive alien species in island ecosystems. *Environmental Conservation* 34: 98–111.

Richards, C. L., O. Bossdorf, N. Z. Muth, J. Gurevitch, and M. Pigliucci. 2006. Jack of all trades, master of some? On the role of phenotypic plasticity in plant invasions. *Ecology Letters* 9: 981–993.

Sax, D. F., J. J. Stachowicz, J. H. Brown, J. F. Bruno, M. N. Dawson, S. D. Gaines, R. K. Grosberg, A. Hastings, R. D. Holt, M. M. Mayfield, M. I. O'Connor, and W. R. Rice. 2007. Ecological and evolutionary insights from species invasions. *Trends in Ecology and Evolution* 22: 465–471.

Simberloff, D. 1995. Why do introduced species appear to devastate islands more than mainland areas? *Pacific Science* 49: 87–97.

Vernon, P., G. Vannier, and P. Trehen. 1998. A comparative approach to the entomological diversity of polar regions. *Acta Oecologica* 19: 303–308.

IRELAND

SEE BRITAIN AND IRELAND

ISLAND ARCS

RICHARD J. ARCULUS

Australian National University, Canberra

Island arcs are chains of concurrently or potentially active volcanic islands, consistently associated but displaced spatially more than 100 km from a deep-sea trench. Much of the eruptive activity is strongly explosive. Although some chains are strongly arcuate, others are linear. Adjacent to many island arcs in the western Pacific are back-arc basins floored by crustal spreading centers. Some arcs have associated nonvolcanic individual islands or island chains between the volcanic arc and trench, comprising uplifted, trench-accreted sediment.

CONTEXT

Oceanic lithosphere created at zones of plate divergence (mid-ocean ridges) is returned to the Earth's interior at sites of plate convergence, specifically at deep-sea trenches, which mark the surface trace of gently to steeply dipping subduction zones. During its interactive exposure at Earth's surface, oceanic lithosphere acquires an imprint of chemical exchange with the oceans and a superstrate of sediment derived predominantly from biological and hydrothermal activity in the oceans plus detritus transported from continents. This diverse rock and sediment package is variably recycled into the Earth's mantle, leading to a complex series of processes including energetic earthquake activity and fluid release accompanying metamorphic changes to the package; the fluids rise buoyantly into the wedge of mantle overlying the subducted lithosphere (or "slab"), triggering magma generation through partial melting. The magma rises towards the Earth's surface, eventually forming the chains of volcanoes known as island arcs. The world's greatest ocean deeps, biggest gravity anomalies, largest earthquakes, most explosive volcanic eruptions, and highest mountains all occur at zones of plate convergence and subduction. In the simplest terms, island arcs are the surface expression of a complex spectrum of magmatic and tectonic processes triggered by plate subduction at zones of convergence. In addition, the geochemical characteristics of island arc magmas most closely match those of the continental crust, unlike mid-ocean ridge and hotspot types; accordingly, a persistent stimulus for island arc research has been the potential link with genesis and evolution of the continents.

Island arcs differ in crucial ways from other chains and clusters of islands produced by volcanic activity, such as those of Hawaii or Galápagos. The magmas constructing the latter are the partial melt products of isolated, individual plumes of mantle material rising from thermal boundary layers within the Earth (e.g., the core–mantle or lower–upper mantle transition zones). Plumes (or "hotspots") are generally independent of zones of plate divergence or convergence, and consequently their magmatic products may be passively transported away from the active plume locus by ambient motion of the plate they happen to be erupted onto. In contrast, island arcs develop parallel to sites of plate subduction, and numerous loci of magmatic activity are concurrently formed over distances that can extend in individual arc segments for thousands of kilometers. Although portions of a previously active arc may be abandoned as a "remnant arc" by zones of crustal spreading, rifting an arc apart during "back-arc basin" formation, regular progressions from fringing reef to atoll are generally not well developed on the subsiding remnant arc edifices.

Historically, the development of arcuate chains of islands in some western Pacific arcs attracted attention: In 1903 W. J. Sollas demonstrated that such an arc could trace an outcropping planar fault on the Earth's spherical surface, and in 1931 P. Lake suggested these fault surfaces were defined by continentward-dipping zones of earthquake foci, documented the same year by K. Wadati for Japan. Note, however, that many so-called arcs, such as the Izu–Bonin, Solomons, New Hebrides, and Tonga–Kermadec, are linear geographically, presumably related to the local morphology of the subducted slab.

GEOGRAPHIC DISTRIBUTION

There is a gradation globally between island arcs apparently formed in intra-oceanic settings (Izu–Bonin–Mariana, New Hebrides, Tonga–Kermadec), those developed on fragments of continental crust that have migrated away from an adjacent continental land mass (Japan, Philippines, New Zealand), and those subduction-related volcanoes active at the margins of continents (Cascades in North America, Central America, and the Andes in South America) (Fig. 1). Specific gradational examples include the Aleutians, which extend onto the Alaskan Peninsula and Mainland; the Kurile chain, which extends between Kamchatka and Japan, a portion of continental crust that migrated eastward from the East Asian continental plate about 20 million years ago; and the Kermadec Arc, which strikes southward onto the continental crust of New Zealand and which migrated away from the margin of Antarctica some 80 to 55 million years ago.

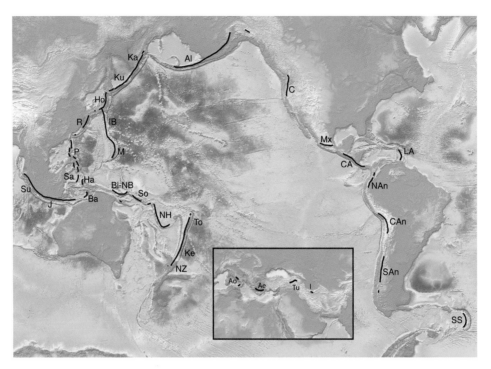

FIGURE 1 Global distribution of island arcs based on Global Topography base (http://topex.ucsd.edu/marine_topo/mar_topo.html); main panel centered on the Pacific with inset of Mediterranean and Middle East. Abbreviations: Ae, Aegean; Al, Aleutians; Ao, Aolian; Ba, Banda; Bi-NB, Bismarck–New Britain; C, Cascades; CA, Central America; CAn, Central Andes; Ha, Halmahera; Ho, Honshu; I, Iran; IB, Izu–Bonin; J, Java; Ka, Kamchatka; Ke, Kermadec; Ku, Kurile; LA, Lesser Antilles; M, Mariana; Mx, Mexico; NAn, Northern Andes; NH, New Hebrides; P, Philippines; R, Ryukyu; Sa, Sangihe; Su, Sumatra; So, Solomons; SAn, Southern Andes; SS, South Sandwich; To, Tonga; Tu, Turkey.

The Andean continental arcs are connected by transform faults to two island arcs in the western Atlantic: (1) in the north of the Andes to the Lesser Antilles Island arc and the subduction of the Atlantic portion of the North American plate beneath the Caribbean plate and (2) in the south of the Andes to the South Sandwich Island Arc, where the Atlantic portion of the South American plate is being subducted beneath the Scotia plate. Although the Sumatra–Java–Banda arc system appears superficially to comprise particularly large subaerial and emergent chains of volcanoes forming extensive islands, in fact these eruptive centers are developed on the margin of a mostly submerged continental crust extended southward from the Asian plate (Fig. 1).

Other arcs include the Aeolian and Aegean in the Mediterranean, involving subduction of the African beneath the Eurasian plate. Eastward, the further collision between the African and Eurasian plates is marked by a limited chain of continental arc-type volcanoes striking from Turkey southeastward to Iran. But eruptive activity in arcs generally ceases where prolonged collision between extensive continental portions of plates has occurred. For example, much of the subduction zone active between India and Asia lacks normal arc volcanism. Elsewhere, volcanic arc activity has also ceased where subduction of oceanic lithosphere is taking place at a relatively shallow dip, as is the case currently beneath much of Peru and parts of central Chile.

REPRESENTATIVE CROSS-SECTION

Although the structure of plate convergence zones along strikes (i.e., parallel to the deep-sea trench) is fundamentally important, as a first approach, a two-dimensional cross-section is useful for graphically illustrating the physical and chemical processes involved in arc genesis. A representative composite cross-section (Fig. 2) shows the following important features:

SUBDUCTING SLAB

The subducting slab (feature 1 in Fig. 2) at the trench comprises a superficial sediment layer (Layer 1; depending on specific oceanic setting, these sediments can include carbonate ooze, siliceous ooze, hydrothermal deposits, and continent-derived muds and sands) up to several hundred meters thick; Layer 2, formed by basaltic pillow lavas underlain by feeder dikes that are tapped from Layer 3, the gabbroic and olivine–clinopyroxene-dominated, intra-crustal magma chamber cumulative fractions. Depending on the magma production rate at a mid-ocean ridge, the igneous crustal thickness can range between 0 (at amagmatic transform fault–ridge intersections) and about 10 km.

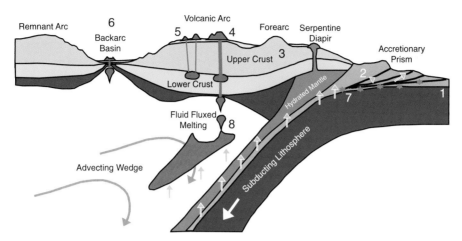

FIGURE 2 Schematic cross-section of an island arc from trench to back-arc basin. Numbers refer to the features described in the section titled "Representative Cross-Section." Stars indicate major thrust-mechanism earthquakes in the shallower (seismogenic) portion of the Wadati–Benioff zone of earthquakes.

Below the crustal layers is the residual upper mantle from which the basaltic portion of the ocean crust is derived. Alteration of the crust and underlying mantle by circulation of seawater produces hydrated minerals including clays, epidote and amphibole groups, and serpentine in the case of olivine-pyroxene-dominated lithologies.

ACCRETIONARY PRISM

An accretionary prism (feature 2 in Fig. 2) has developed in the fore-arc of only some systems; it consists of sediments and sporadic igneous crustal fragments, including major portions of oceanic seamounts. Packages of sediment are scraped off and sequentially underplated by thrust faulting; uplift of the prism as more and more sediment is underplated can develop to the point of emergence, forming islands such as Barbados between the southern Lesser Antilles Island arc and trench; Simelue, Nias, and Siberut between the Sumatra volcanic arc and adjacent trench; and Kodiak Island in Alaska.

FORE-ARC

In those arcs lacking an accretionary prism, the fore-arc (feature 3 in Fig. 2) is dominated by igneous rocks and volcaniclastic-rich sediments shed from the arc itself. The igneous rocks can include the crust upon which the active arc has been built and arc crust constructed during early stages of arc development. Emergent examples of the latter include the islands of Shikotan (Kurile), Chichi- and Hahajima (Bonin), Tinian–Saipan–Guam (Marianas), and Vavaʻu–Tongatapu–Eua (Tonga). Some nonaccretionary fore-arcs, such as those of the Izu–Bonin–Mariana system, are penetrated in diapiric manner by major seamounts dominated by serpentinized rocks and mud, derived by serpentinization of the shallow mantle wedge through fluid release from the subducted Pacific slab.

VOLCANIC FRONT

The most trenchward locus of eruptive volcanic activity (sometimes known as the "volcanic front," feature 4 in Fig. 2) comprises the volcanic arc *sensu stricto*; volcanic islands are the emergent tips of large, steeply sided submarine structures, typically of several thousand meters relative elevation between summits and local crustal basement. It is not clear why focused spacing of magma supply, leading to island formation rather than a continuous eruptive sheet is developed. Globally, there are major differences between the total number and proportions of emergent and submerged arc volcanoes in any given arc. In the past decade, detailed multi-beam sonar swath mapping accompanied by hydrothermal plume surveys and rock dredging has documented the true incidence of volcanic centers in a number of island arcs. For example, in the case of the Aleutians, the majority of individual volcanoes are emergent with an along-strike spacing of ~100 km. In contrast, in the Tonga–Kermadec system, there are only about 12 subaerial and more than 80 submerged volcanic centers with an average spacing of ~40 km. Some ephemeral islands in this region are instructive in terms of understanding the stabilization required of emergent submarine structures, particularly by blanketing lava flows. It is known that the crustal loads exerted by larger volcanic structures tend to capture more of the rising magma flux from the mantle—larger edifices grow at the expense of smaller vents. Thus, subaerial emergence is controlled by several major factors: the localized magmatic flux; the bathymetric depth of the plate surface overriding the subduction zone reflecting overall compressional, tensional, and gravitational force balance; the extent of collapse and degradation of the arc volcanoes; and the degree of tectonic subsidence or uplift of the arc accompanying formation of back-arc basins. Along the strike of the volcanic front in the Izu–Bonin Arc are also coupled variations in degree of emergence, crustal thickness, and dominant magma composition. For example, the islands in this arc tend to be dominated by basaltic eruptions overlying relatively thick (~20-km) crust, whereas the submerged structures include many rhyolite-dominated calderas overlying thin (~15-km) crust.

REAR-ARC CHAINS

Cross- and parallel rear-arc chains of volcanoes (feature 5 in Fig. 2) extending orthogonal to and along strike with the volcanic front. In a number of arcs such as the Izu–Bonin–Mariana system, there are subaerial (e.g., Niijima and Kozu-shima) to submarine chains of seamounts extending at a high angle away from the arc front. In the case of Izu–Bonin, the volcanoes are young toward the arc front and may relate to arc migration following back-arc basin formation. In other intraoceanic island arcs, there are volumetrically minor but volcanically active edifices, some of which form islands (e.g., Alaid, Kuriles) and others of which are submerged. Cross-chains can be quite localized in development, restricted to a small region of the arc, as is the case with the volcanically active Willaumez Peninsula and Witu Islands in the New Britain Arc of Papua New Guinea. In the case of major islands such as Honshu in Japan and Java in Indonesia, there are prominent volcanoes distributed irregularly further from the trench but parallel to the strike of the volcanic front, some of which have experienced major catastrophic eruptions in historical time, such as Tambora (AD 1815 eruption) on the island of Sumbawa. In general, the magma types forming cross- and arc-parallel rear-arc chains tend to be more alkali (Na + K)–rich at any given bulk SiO_2 content than those forming the volcanic front.

BACK-ARC BASINS

Although island arcs are located in zones of overall plate convergence, the most significant driving force behind plate motions is slab pull in subduction zones, and in many of these systems, no horizontal compressive stress is transmitted by the downgoing to the overriding plate. Slab roll-back is common in the western Pacific, resulting in advance of the trench toward the incoming plate, fore-arc collapse, and extension in arc and back-arc, leading in some instances to spreading and formation of a basin (feature 6 in Fig. 2). Many of the island arcs of the western Pacific have magmatically active zones of ocean floor spreading and new crustal creation, similar in many morphological respects to the major mid-ocean ridges. Actively spreading basins include the Mariana Trough, Manus Basin, Andaman Sea, Coriolis Troughs–North Fiji Basin (with several spreading loci), Lau Basin (with several spreading loci)–Havre Trough, and Scotia Sea. The only island in any of these basins is Barren Island in the Andaman Sea. The duration of spreading in back-arc basins appears to be restricted from a few to ~15 million years, and several other inactive basins in the western Pacific are known. These include the Kurile back-arc basin, Sea of Japan, West Philippine Basin, Shikoku–Parece Vela Basins, and South Fiji Basin. In a number of cases, rifting occurs within the volcanic arc, leading to development of a spreading system splitting a remnant arc from the sustainedly active volcanic front. Cessation of magmatic supply to the remnant arc generally leads to subsidence below sea level as with the cases of the remnant Kyushu–Palau Ridge–active Izu-Bonin-Mariana Arc and remnant Lau Ridge–Tofua Arc pairs. In the latter case, however, the Lau Group of islands at the northern end of the Lau Ridge are still emergent. In other cases, rifting takes place totally behind the arc, leading to migration of the entire arc structure and no remnant arc formation (e.g., New Hebrides). During episodes of back-arc basin initiation, volcanic front activity may temporarily cease; it appears that the back-arc basin rift captures the total mantle wedge–derived magmatic flux. This is currently the situation in the case of the volcanic front in the northern Tofua Arc, adjacent to the nascent spreading occurring in the Fonualei Rifts, where volcanic structures north of the volcanic island of Fonualei are planated ~30 m below current sea level and are carbonate capped. In the case of the Mariana Arc, the volcanic front became reestablished on newly created back-arc basin crust, requiring a longer period of time to emerge above sea level for a given arc magma flux than would have been the case if they were constructed on the bathymetrically shallower fore-arc.

DOWN DIP IN THE SUBDUCTED SLAB

Increases in pressure and temperature control a sequence of metamorphic dehydration reactions, accompanied at depths of less than ~30 km along the slab–overriding plate interface by large thrust earthquakes (feature 7 in Fig. 2). Some of these may propagate to the sea floor, triggering large tsunamis, as was the case with the December 2004 earthquake off the northwestern coast of Sumatra. The term Wadati–Benioff Zone is given to the continuum of a dipping, shallow-to-deep plane of earthquakes from the deep-sea trench, beneath the island arc and on to great depths (~600 to 700 km, transition zone in the mantle). The descending slab imposes an inverted temperature gradient in and forces advective motion of the wedge. Further down the slab dip, a complex and continuous series of dehydration reactions, involving breakdown of phases such as zoisite, epidote, lawsonite, amphibole, and mica, releases fluid into the overlying mantle wedge. Where sufficiently detailed studies can be made, double zones of earthquakes are found, one close to the slab–wedge interface and the other ~30 km deeper within the slab.

FOCUSED MAGMA CORNER

Motion of the advecting wedge results in focusing of the magma by porous flow to a "corner" (feature 8 in Fig. 2). It is an empirical fact that with some rare exceptions, the volcanic front is located on average about 100 km above the subjacent slab–wedge interface, and possibly above the corner of focused magma flow. Magmas do erupt above greater depths to the slab. In general, and unlike the situation at mid-ocean ridges and hotspots, where adiabatic decompression is the prime partial melting trigger, it is believed that fluid-fluxed partial melting of the mantle wedge is particularly important in arcs. A very wide variety of magma compositions is erupted globally, ranging at the basaltic end from low-K_2O olivine tholeiite through to high-K_2O feldspathoid-bearing compositions. Where cross-arc magmatism exists, a zonation from low- to high-K_2O types with increasing depth to slab is generally observed. Other restricted magma types include high-MgO, low-SiO_2 picrites. Rare high-MgO intermediate (~55–60 wt% SiO_2) types called boninite are particularly found accompanying arc inception and more rarely in active arcs developed above strongly basaltic-melt depleted mantle wedges, such as beneath northern Tonga. Arc basalts generally contain significantly more H_2O (~2 to 6 wt%) than is the case with mid-ocean ridge (less than 0.2 wt%) and hotspot (less than 1 wt%) basalts. With reduction in ambient pressure during ascent of the magma to the Earth's surface, exsolution of an H_2O-rich volatile phase drives the characteristically explosive volcanism of arcs. Return of H_2O to the Earth's atmosphere is part of a large-scale H_2O recycling process in the Earth from hydrothermal alteration at a mid-ocean ridge through fixation in and subduction of oceanic lithosphere, and fluid-fluxed melting of the mantle wedge to explosive submarine or atmospheric emissions. Other volatile compounds are exsolved in this gas phase including CO_2 and sulfur- and halogen-bearing molecules. Recognizing the importance of H_2O in the genesis of arc magmas and its plausible link with genesis of the continental crust, in 1983 I. H. Campbell and S. R. Taylor coined the aphorism "no water, no granites—no oceans, no continents." Numerous factors control the composition of the magma erupted in arcs: composition of the mantle wedge, particularly with respect to previous melting history; depth of melting of the wedge; composition of the fluid addition from the slab, which is a function of subducted sediment characteristics; type of alteration of igneous lithologies; extent of lithologic mixing in the surface layers of the slab (the fluids may range from H_2O-dominant through to H_2O-bearing silica-rich melts and possibly supercritical); ascent paths and degree of interaction between magma and host lithologies during ascent; extent and depth of fractional crystallization of the magma; and extent of magma mixing. The most important process generating the spectrum of basalt–andesite–dacite–rhyolite volcanic rock compositions is fractional crystallization of olivine–pyroxene–plagioclase–spinel (fom Cr- to Fe-rich) assemblages. These cumulative assemblages can be found as relic plutons in deeply exposed arc crustal sections. Although the low-MgO, intermediate-SiO_2 rock type called andesite is characteristic of many arcs, it is not volumetrically always the most abundant, particularly in the submarine realm of intraoceanic arcs where basalt is prominent.

INCEPTION, DEVELOPMENT, AND DEMISE

Several decades of research combining land-based studies with marine research voyages, deep-sea drilling, and submersible investigation have established the history of a number of western Pacific arcs in particular, from inception through to current stages of development. For example, it is known that the Izu–Bonin–Mariana and former Solomons–New Hebrides–Tonga–Kermadec ("Vitiaz") systems were initiated ~50 million years ago. Reconstruction of this episode requires closure of subsequently developed back-arc basins (e.g., Mariana Trough, Parece Vela, and Shikoku) juxtaposing the remanant arc of the Kyushu–Palau Ridge with the current amagmatic fore-arc. Paleomagnetic investigations have shown a clockwise rotation of ~90° of a formerly east-west trending system at equatorial latitudes for Izu–Bonin–Mariana since initiation. Inception through cannibalization of transform faults accompanying a change in plate motion has been a popular hypothesis, constituting an example of forced initiation. It is also recognized that gravitational instability of oceanic lithosphere more than ~10 million years old with respect to the underlying asthenosphere might plausibly lead to spontaneous subduction initiation.

Although the present location of systems such as Izu–Bonin–Mariana and Tonga–Kermadec appear to be remote (hence the term *intraoceanic*) from the nearest continents, it is important to bear in mind the evolution of these systems. For example, a series of continental fragments of eastern Australia (Lord Howe Rise, Norfolk Ridge) have been rifted off during the late Cretaceous through Tertiary periods, accompanied by a series of arc–back-arc developments along their eastern and northern margins. Similarly, the Kyushu–Palau Ridge is developed across a series of Cretaceous arc–back-arc systems that may have been proximal to Asia. The important point is that nominally intraoceanic arcs may have been initiated much closer to continental masses. We have few examples of subduction initiation in truly remote intraoceanic settings: the New Britain–New Ireland–Bougainville–Solomons–New Hebrides may be one such. Other island arcs are clearly developed on continental fragments such as Japan; during the Miocene epoch (20 million years ago), fragments of modern Japan migrated away from eastern Asia with the development of a back-arc basin (Sea of Japan). Some of this archipelago's most frequent earthquakes are on the Sea of Japan side of Honshu, marking the inception of a new subduction zone and closure of the Sea. Japan is destined to collide with its formerly rifted parent land mass.

More generally, despite the inherent gravitational instability of oceanic lithosphere, it is plausible that subduction initiation is propagated from elsewhere as a type of "tectonic infection," emphasizing the importance of the third dimension in our cross-sectional depiction of arc systems. We know that initial development of the Izu–Bonin–Mariana system was submarine in an extensional setting involving extensive pillow lava–dike (including boninite) emplacement. Ocean drilling has shown that subaerial, explosively eruptive conditions were temporarily reached a few million years after inception. Submarine growth stages are likely characteristic of many island arcs, as a growing body of survey data in the western Pacific has revealed. But we have sparse current examples of subduction inception: The Hjort Trench–Macquarie Ridge complex south of New Zealand is a possibility, but detailed swath mapping has failed to reveal volcanic activity, with the exception of a single active subaerial volcano (Solander).

The demise of arc systems includes the cessation of eruptive activity on remnant arcs and their general submergence. Changes in general plate tectonic frameworks can also lead to subduction and arc demise, as in the case of the Greater Antilles (Cuba–Hispaniola–Puerto Rico), where relatively large emergent islands remain. Another example is the South Shetlands off the northern coast of the Antarctic Peninsula, where sparse volcanic activity continues (e.g., Deception Island). Collision with a neighboring continent also leads to arc demise: Examples include multiple collisions of arc systems with western North America and the southern margin of central Asia during the Mesozoic. An ongoing example is the collision of the West Bismarck Arc with mainland Papua New Guinea. Volcanic activity is terminating progressively in this collision zone from northwest to southeast.

In conclusion, island arc systems are dynamic constructs with geologically ephemeral subaerial island development at varying distances from large land masses. Peripheral addition to continents and reworking of the constituent lithologies through metamorphism and igneous processes are important in continued continental evolution. The mobility of arc systems from initiation through migration and potential collision with continents is clearly significant from biogeographic perspectives, and there is much to learn from the combination of such studies with geological understanding.

SEE ALSO THE FOLLOWING ARTICLES

Antilles, Geology / Hydrothermal Vents / Island Formation / Japan's Islands, Geology / Kurile Islands / Oceanic Islands / Plate Tectonics

FURTHER READING

Abers, G. A., and P. E. van Keken. 2006. The thermal structure of subduction zones constrained by seismic imaging: Implications for slab dehydration and wedge flow. *Earth and Planetary Science Letters* 241: 387–397.
Arculus, R. J. 2004. Evolution of arc magmas and their volatiles. *Geophysical Monograph* 150: 95–108.
Grove, T. L., and R. J. Kinzler. 1986. Petrogenesis of andesites. *Annual Review of Earth and Planetary Sciences* 14: 417–454.
Stern, R. J. 2002. Subduction zones. *Reviews of Geophysics* 40: doi:10.1029/2001RG000108.
Tatsumi, Y., and S. M. Eggins. 1995. *Subduction zone magmatism*. Oxford, UK: Blackwell.

ISLAND BIOGEOGRAPHY, THEORY OF

JOSÉ MARÍA FERNÁNDEZ-PALACIOS

La Laguna University, Tenerife, Spain

The theory of island biogeography, developed by Robert H. MacArthur and Edward O. Wilson successively in 1963 and 1967, argues for the existence of a dynamic balance in species richness on islands, as a function of the addition of species through the immigration of propagules to the island, plus any speciation within it (dictated by the degree of isolation from the mainland), and of species extinction from the island (dictated by island area). The result of these opposing forces, given enough time, is a dynamic equilibrium in which the species number remains approximately constant through time but species composition is continually changing. As a result of its quantitative approach and predictive power, the theory has transformed island biogeography into a mature scientific discipline. It rapidly achieved paradigmatic status, with numerous studies setting out either to confirm or reject its predictions, focused both on a large variety of taxonomic groups (plants, vertebrates, invertebrates, protozoa) and island types (real and habitat islands). It has thus inspired a substantial advance in our knowledge of insular biotas and processes. Finally, it has provided the foundation of a substantial (if controversial) body of theoretical work on the design/implications of protected area networks.

FUNDAMENTAL PRINCIPLES OF THE THEORY

The theory of island biogeography was developed in large measure to account for apparently systematic variation in species–area relationships. Basically, the theory considers that there are two major processes shaping the species richness that a given island can carry. How they vary is described in a straightforward graphical model (Fig. 1), considered by some a key ingredient in the adoption of the theory. The first process is the immigration (arrival) of a propagule of a species that is new to the island, and its rate depends on the island's isolation (the distance from the mainland source). Nearer islands are easier to reach by mainland species than farther ones, therefore the immigration rate curve (I) tends both to flatten with increasing islands isolation, and to decline exponentially through time as initially empty islands fill up with species. Hypothetically, the immigration rate would decline to zero if the islands were near enough to the mainland to enable the whole mainland species pool (P) to disperse to them. However, because islands are smaller than mainland sources this point is never reached because of the counterbalancing second process, extinction. The theory of island biogeography posits that the extinction, or total disappearance, of a species from an island is dependent on the island area. So, larger islands can support larger populations than smaller ones, and as extinction risk is inversely related to population size, the extinction rate curve (E) tends both to flatten with increasing island area and to increase exponentially as initially empty islands (where $E = 0$) fill up with species. The increase of E with species richness is explained by the fact that the higher the number of species on a given island, the smaller the population sizes that can be maintained for a given species, hence the higher the extinction risks.

When the trajectories of the immigration and extinction rates (as the ordinates) for a given island with a specific area and isolation are plotted on a graph against species richness on the abscissa (Fig. 1), the projection to the abscissa from the point at which the two curves intersect defines the species richness that this island can carry, whereas the projection from this point to the ordinate defines the turnover rate (T), or number of species extinguished and replaced per unit time. Thus, species richness, as well as immigration, extinction, and turnover rates, are island-specific parameters that vary with the island's area and isolation in a dynamic equilibrium, where species richness tends to remain constant through time although species composition continues to turn over. Each combination of island area and isolation should produce a specific combination of species richness and turnover rate. Thus, the theory predicts that large, near islands tend to have higher species richness and lower species turnover than small, far islands, whereas small, near islands as well as far, large islands should have intermediate values (Fig. 1).

Finally, colonization, defined as the relatively lengthy persistence (e.g., through at least one life cycle) of an immigrant species on an island, can be plotted through time to obtain the colonization curve, or temporal change of numbers of species found together on an island (Fig. 2). The colonization curve starts from zero, when the island

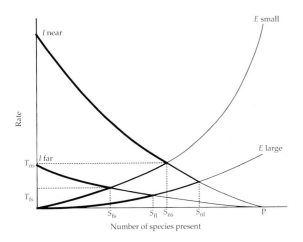

FIGURE 1 A version of MacArthur and Wilson's (1963, 1967) equilibrium model of island biogeography, showing how immigration rates are postulated to vary as a function of distance, and extinction rates as a function of island area. The model predicts different values for S (species number), which can be read off the ordinate and for turnover rate (T) (i.e., I or E, as they are identical at equilibrium). Each combination of island area and isolation should produce a unique combination of S and T. To prevent clutter, only two values for T are shown. Source: Whittaker and Fernández-Palacios (2007).

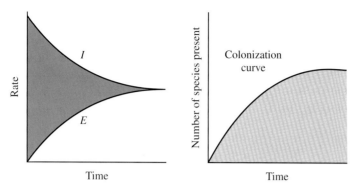

FIGURE 2 Integration of immigration and extinction curves (left) should theoretically produce the colonization curve as shown (right). Source: Whittaker and Fernández-Palacios (2007).

has not yet been colonized by any species, then rises rapidly as the island accumulates new species (notwithstanding the gradual rise in extinction events) until the curve flattens approaching asymptotically the island's species carrying capacity, as determined by its area and isolation.

OBJECTIONS AND EMBELLISHMENTS TO THE THEORY OF ISLAND BIOGEOGRAPHY

The theory of island biogeography assumes that all species are equal in their probabilities of immigrating onto the island or of going extinct once there. With the development of Hubbell's *Unified Neutral Theory of Biodiversity and Biogeography*, there has been a resurgence of theory regarding the importance of such neutral processes in dictating species composition in a given community. It appears that species do differ in their chances of colonizing and persisting on an island.

Although MacArthur and Wilson's monograph (1967) includes a chapter called "Evolutionary Changes Following Colonization," the mechanism through which speciation can substitute for immigration is not well developed, and there is general agreement that the theory is more readily applied to islands driven by "ecological" processes, where the frequency of immigration events precludes speciation, than for very remote islands, where species diversity is dictated mainly by "evolutionary" processes and in situ speciation is more frequent than immigration as the source of new species. Several approaches have been developed to include more information about speciation, the third main biogeographical process, together with immigration and extinction, within the framework of the theory of island biogeography (M. V. Lomolino, L. R. Heaney, etc.), in order to increase the generality of the theory.

Another limitation in the model is that isolation is considered to dictate rates of immigration only. However, island isolation has been shown also to play an important role in influencing extinction rates through a phenomenon known as the rescue effect (Fig. 3): the supplementary immigration from the mainland of propagules of species present on the island in small population sizes and that would otherwise go extinct (such supplementation does not count as immigration if the focal species is rescued prior to actually going extinct). This possibility was noted by MacArthur and Wilson but was not included in the model, yet the effect can be important on small islands near to the mainland, thus modifying the extinction rate, and equilibrium point. In addition, area is considered to dictate only extinction in the model, although MacArthur and Wilson noted the potential importance of island area in influencing the immigration rate as well. Through the latter phenomenon, known as the target effect (Fig. 4), larger islands provide easier targets for passively dispersing propagules, such as windborne and waterborne plants, thus increasing their immigration rates. These two effects undermine the predictive power of the simplified model in Fig. 1.

MacArthur and Wilson also recognized that biologists can rarely, if ever, be certain of recording all immigration and extinction events in real-world systems, especially if large islands are considered, and thus species turnover calculations for large islands can contain important

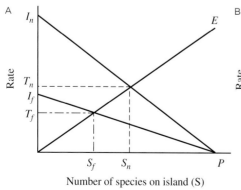
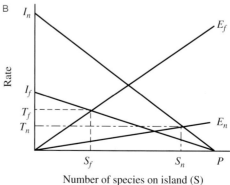

FIGURE 3 The rescue effect is the reduction in the extinction rate of near islands versus distant ones. Whereas the MacArthur and Wilson model predicts higher turnover on near islands (A), the rescue effect may increase turnover on more distant islands (B). T_n is the turnover rate on the near island; T_f is the turnover rate of the far island. Source: Gotelli (2001).

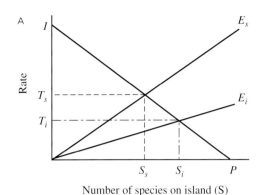 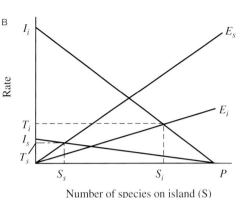

FIGURE 4 The target effect is the increase in the immigration rate on large islands versus small ones. Whereas the MacArthur and Wilson model predicts higher turnover on small islands (A), the target effect may increase the turnover on large islands (B). T_s is the turnover rate on the small island; T_l is the turnover rate on the large island. Source: Gotelli (2001).

biases. Two main problems arise when trying to calculate species turnover rates: cryptoturnover and pseudoturnover. Cryptoturnover is the real turnover of species not detected because of too great a gap between census intervals, as for instance, when a species undergoes extinction and later re-colonization, or *vice versa*, in the time interval occurring between two consecutive inventories. The effect leads to underestimation of the real turnover rate. Pseudoturnover, conversely, is the apparent disappearance and reimmigration of species in consecutive surveys when they were actually present throughout, or alternatively, were only ever present as vagrant individuals. Such incomplete surveying leads to overestimation of the real turnover rates. Attempts have been made to quantify both of these sources of error, but they are inherently hard to estimate precisely, especially for larger islands.

It has been argued that habitat diversity and not area per se is the true determinant of island species richness, because more habitats will offer more opportunities for the colonization of species differing in their ecological requirements. Although this is undoubtedly true, it is also the case that habitat diversity is generally correlated with island area. Indeed, it was partly for the lack of adequate data on habitat diversity (still a limitation today) that MacArthur and Wilson focused only on area in their monograph. Their specific mention of this point makes clear that they used area in the theory as a measure of the combined effects of area per se and habitat diversity.

Finally, it has been argued that some insular systems do not achieve equilibrium, even after extended periods of time. Three alternative states have been recognized. First, "static" nonequilibrium systems include those where species losses (such as those due to postglacial isolation of mountaintop systems preventing further immigration across arid lowlands) are occurring, but so slowly as to be effectively unmeasurable on ecological time scales. Second, "dynamic" nonequilibrium systems are those that are frequently impacted by extreme events, such as volcanic eruptions or hurricanes, which resets the community development iteratively, preventing the attainment of equilibrium. Third, "static" equilibrium systems are those archipelagoes that, over ecological time scales, have a clear species–area relationship (consistent with the theory of island biogeography) but with no turnover (not consistent with the theory), as for instance appears to apply to some oceanic island avifaunas.

These different scenarios have allowed fine-tuning of the theory of island biogeography, with an expanded framework.

THE THEORY OF ISLAND BIOGEOGRAPHY AND CONSERVATION

One of the main applications of the theory of island biogeography outside its academic context has been within the field of conservation biology. The increasing worldwide anthropogenic fragmentation experienced by continental ecosystems has transformed once continuous habitats into complex landscapes containing many patches of relict habitats, differing in size, shape, isolation, or degree of disturbance, surrounded by a more or less penetrable matrix of arable land, pasture, urbanized areas, and/or built infrastructure (e.g., roads). As the new geographical framework for the species of the relictual habitats increasingly resembles an archipelago more than a continent, many authors have attempted to apply the principles of the theory to generate guidelines for these new anthropogenic landscapes. Although there have been many theoretical developments (e.g., metapopulation scenarios, source–sink relationships, edge effects), at the core of this work is the use of the theory of island biogeography to predict the eventual species losses following fragmentation, as immigration rates decline as a result of increased isolation, and extinction rates rise as a result of reduction in contiguous area of habitat. The process of species losses is termed relaxation. At the point of habitat disruption, residual fragments may become supersaturated (i.e., have temporarily high species numbers), and may take time to reach their new equilibrium. The terms "lag effect" (duration of delay) and "extinction debt" (magnitude of the losses) are each used to describe the delay in species losses following fragmentation of habitat. Rather more controversially, the theory of island biogeography has also been invoked in the design of protected areas networks. It has been proposed that, in general, the theory favors deploying resources to protect fewer large reserves rather than many smaller ones (the "SLOSS Debate"), short rather than long inter-reserve distances, circular rather than elongated reserves (to minimize edge effects), and the use of corridors, when ever possible, to connect reserves and facilitate dispersal between them.

SEE ALSO THE FOLLOWING ARTICLES

Extinction / Fragmentation / Relaxation / Species–Area Relationship

FURTHER READING

Cody, M. 2006. *Plants on islands. Diversity and dynamics of a continental archipelago*. Berkeley: University of California Press.
Gotelli, N. J. 2001. *A primer of ecology*, 3rd ed. Sunderland, MA: Sinauer.
Lomolino, M. V. 2000. A call for a new paradigm of island biogeography. *Global Ecology and Biogeography* 9: 1–6.
Losos, J. B., and D. Schluter. 2000. Analysis of an evolutionary species-area relationship. *Nature* 408: 847–850.
MacArthur, R. 1972. *Geographical ecology. Patterns in the distribution of species*. Princeton, NJ: Princeton University Press.
MacArthur, R., and E. O. Wilson. 1963. An equilibrium theory of insular zoogeography. *Evolution* 17: 373–387.
MacArthur, R., and E. O. Wilson. 1967. *The theory of island biogeography*. Princeton, NJ: Princeton University Press.
Rosenzweig, M. L. 1995. *Species diversity in space and time*. Cambridge, UK: Cambridge University Press.
Simberloff, D. 1974. Equilibrium theory of island biogeography and ecology. *Annual Review of Ecology and Systematics* 5: 161–182.
Whittaker, R. J., and J. M. Fernández-Palacios. 2007. *Island biogeography. Ecology, evolution and conservation*, 2nd ed. Oxford: Oxford University Press.

ISLAND FORMATION

PATRICK D. NUNN

University of the South Pacific, Suva, Fiji

To understand how islands form, continental islands must be distinguished from oceanic islands, the former being pieces of continents with the connection submerged, the latter being younger islands that originated exclusively within the ocean basins. However they appear today—low or high, limestone or volcanic—all oceanic islands began life as ocean-floor volcanoes. Those that have not yet reached the ocean surface (and many never do so) are referred to as seamounts, whereas those that were once emergent but have since been submerged are often distinctively flat-topped and are called guyots.

OCEAN-FLOOR VOLCANOES: ORIGINS AND GROWTH

It comes as no surprise to learn that we are not very knowledgeable about ocean-floor volcanism because of the difficulties in actually observing it. Most ocean-floor volcanism occurs in the dark beneath 4 km of ocean water. It is not that the technological difficulties are insurmountable, just that it is difficult to be sure that researchers are getting an accurate picture of what is going on. In this regard, places where the ocean floor actually rises above the ocean surface are extremely valuable as observation sites. Second best are places where seamounts have been thrust up above sea level and have their insides exposed for scientists to see how they were built up.

The finest example of the first situation—where the ocean floor actually rises above the ocean surface—is the island of Iceland in the northern Atlantic Ocean. Iceland is part of the Mid-Atlantic Ridge (a divergent plate boundary) that lies at a plate triple junction and where eruptive activity has been unusually voluminous over the past few million years. The mid-ocean ridge—a common site of ocean-floor volcanism—actually passes through the center of Iceland. From studies of this, we learn that the earliest type of ocean-floor volcanism is commonly along fissures. As fissure eruptions continue, some parts of the fissure become blocked, and eruptions begin to occur at points. Point volcanism results in the build-up of the earliest types of seamounts.

Studies of emerged seamounts—which rise from ocean floor that has been thrust upward by tectonic forces—have also given us a lot of information about the undersea development of oceanic islands. In particular, it is clear that intrusion of igneous rocks is at least as important as extrusion is in building seamounts in many parts of the ocean basins.

Another important issue is the depth of overlying ocean water in places where seamount eruption occurs. In most places below about 600 m the weight of overlying water is so great that, however powerful the volcanic eruption, it will not be explosive, and the material produced will generally be pillow lava. At depths shallower than 600 m (the hydroexplosive zone), on the other hand, the weight of overlying water is not always sufficient to subdue explosive eruptions, and there is a reaction between the cold ocean water and the hot magma (liquid rock) that causes the latter to solidify rapidly as numerous small fragments. Some of this fragmental (clastic) material may reach the ocean surface, where it floats as pumice, but most of it sinks and becomes draped over the sides of the more solid seamount as a sediment apron. Given the fragmented nature of the eruptive products produced in the hydroexplosive zone, a growing seamount often takes far longer to build itself up through this part of the ocean than through deeper areas.

HOW ISLANDS RISE ABOVE SEA LEVEL

The three main ways in which a seamount can grow above the ocean surface are extrusion, intrusion, and uplift; each is discussed separately below. These categories should not be regarded as exclusive, for it is often a combination of these processes that actually leads to emergence.

Island Emergence by Extrusion

Underwater volcanoes that erupt within the hydroexplosive zone sometimes cause islands to form. These are often referred to as "jack-in-the-box" islands because they alternately appear (during eruptions) and then disappear (between eruptions because of wave erosion). Examples are comparatively common in southwestern Pacific island groups such as Tonga and Vanuatu, located close to convergent plate boundaries. The reason that these islands do not endure long above the ocean surface is because they are composed entirely of unconsolidated and uncemented rock fragments and are promptly destroyed by wave erosion after the island-forming eruption comes to an end.

For such an island to endure above the ocean surface, clearly it needs to be made of more resistant material. The way this happens has been recorded only once, with the exceptionally voluminous and lengthy eruption of an underwater volcano off the south coast of Iceland from 1964 to 1967, which produced the island Surtsey. In the case of this eruption, the amount of volcaniclastic (fragmented volcanic) material erupted created such a large island that at one point it isolated the volcano's vent from ocean water. Once this happened, there was no longer any explosive reaction between ocean water and magma, and as a result, volcaniclastic rocks were replaced by lava. The lava flowed out over the surface of the volcaniclastic island, armoring it against erosion and leading to the establishment of a permanently emerged island.

Island Emergence by Intrusion

One of the big unknowns in the emergence of oceanic islands is the precise role of intrusion—the emplacement and solidification of igneous rocks within (not outside) an existing island edifice. Studies of former oceanic islands, now uplifted far above sea level and partly denuded to allow glimpses of their anatomy, show that intrusions can comprise as much as 70–80% of the total mass of such islands. Whether or not this is typical is uncertain, but it does underline the importance of intrusion.

Intrusion begins to affect island growth from almost its beginnings, but it appears to be generally less important when an island is significantly emergent. Intrusion in early stages of island growth may be mostly through sill formation, although later, when the island edifice is sufficiently large to accommodate them, large intrusive bodies (batholiths, stocks) may come to dominate.

Island Emergence by Uplift

Uplift refers to the upward forcing of an island, irrespective of its extrusive or intrusive activity. Island uplift is most common at convergent plate boundaries, where arcs of non-volcanic islands are often found, produced by the movement of the overriding (upper) plate across the top of the downgoing (lower) plate. Examples include the limestone islands of Tonga (South Pacific) and those of the Mentawai group (Indonesia). At convergent plate boundaries, the downgoing plate is commonly flexed (bent) upward before it goes down into the Earth's interior, producing island emergence; examples include the Loyalty Islands (Southwest Pacific).

Island uplift can also occur in the middle of plates, where islands are carried on moving plates across hotspots or other ocean-floor irregularities such as intraplate swells. Some of the Tuamotu Islands (South Pacific) have emerged as a result of such a process.

More complex cases of island emergence result from a variety of causes. For example, there are various islands that are largely composed of pieces of ocean floors (ophiolites) that have been peeled off and pushed upward across a crustal irregularity. Examples include Crete (Mediterranean Sea) and La Grande Terre in New Caledonia (Southwest Pacific).

ROLE OF SEA-LEVEL CHANGE IN ISLAND FORMATION

It may seem paradoxical, but both sea-level rise and sea-level fall can cause islands to form.

Sea-level rise floods dry land and, in doing so, can transform a large island, for example, into a series of smaller islands. This is what happened as sea level rose after the last glacial maximum in the Channel Islands off the coast of California, where people about 12,000 years ago occu-

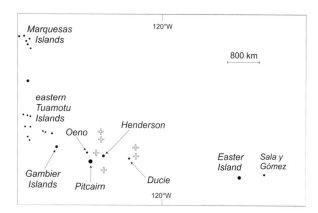

FIGURE 1 Islands of the east central and southeast parts of the Pacific Ocean. The biota of high Pitcairn Island shows affinities with the Gambier and eastern Tuamotu island groups. Henderson Island was uplifted during the late Quaternary (about 200,000 years ago) but was periodically submerged in earlier times. Ducie and Oeno atolls have been regularly submerged during high sea-level stages of the Quaternary; their terrestrial biota is derived from Pitcairn yet is consequently much less diverse. The existence of shallow-water (less than 130 m) submerged islands (marked by crosses on the map) that emerged during glacial low sea-level stages (most recently between 22,000 and 16,000 years ago) has undoubtedly aided the successive recolonization of Ducie and Oeno. The lack of any submerged islands between the Pitcairn group and Easter Island–Sala y Gómez accounts for the marked lack of biotic similarity between these island groups.

pied a large island (named Santarosae), which was subsequently broken up into several smaller ones, something that had discernible effects on societal evolution there. Yet in the same period of sea-level rise, some ten of the 26 islands that existed during the last glacial maximum in the Channel Islands were completely submerged.

A similar situation occurred in East Asia. During Quaternary glaciations, when sea level was low, the main islands of Japan often formed a single land mass (named Hondo), sometimes connected to the Asian mainland. Yet this connection was severed when sea level rose at the end of these glaciations, isolating Japan (and its occupants) from the continent and transforming one large island into a number of smaller ones. When Japan became a land of islands, people's diets changed focus from terrestrial- to marine-dominated.

For almost all of the past 2–3 million years, the ocean surface has lain tens of meters below its present level. Thus, for most of this time, the geography of islands has been quite different from the way it appears today. Many more islands were emergent at times of lower sea level, having emerged solely because of exposure as sea level fell to below their surface level. Many biogeographers have speculated that at such times the dispersal of plants and animals throughout the ocean basins was much easier than it would have been in a drowned world such as we occupy today. A good example is provided by the remote Pitcairn island group in the southeastern Pacific Ocean (Fig. 1).

SEE ALSO THE FOLLOWING ARTICLES

Continental Islands / Earthquakes / Motu / Oceanic Islands / Sea-Level Change / Seamounts, Geology / Surtsey

FURTHER READING

Kayanne, H., H. Yamano, and R. H. Randall. 2002. Holocene sea-level changes and barrier reef formation on an oceanic island, Palau Islands, Western Pacific. *Sedimentary Geology* 150: 47–60.
Menard, H. W. 1986. *Islands*. New York: Scientific American Books.
Nunn, P. D. 1994. *Oceanic islands*. Oxford: Blackwell.
Nunn, P. D. 2007. Holocene sea-level change and human response in Pacific Islands. Transactions of the Royal Society of Edinburgh: Earth and Environmental Sciences 98: 117–125.
Nunn, P. D. 2009. *Vanished islands and hidden continents of the Pacific*. Honolulu: University of Hawaii Press.
Sinton, J., E. Bergmanis, K. Rubin, R. Batiza, T. K. P. Gregg, K. Gronvold, K. C. Macdonald, and S. M. White. 2002. Volcanic eruptions on mid-ocean ridges: new evidence from the superfast spreading East Pacific Rise, 17°–19° S. *Journal of Geophysical Research* 107: doi:10.1029/2000JB000090.

ISLAND RULE

SHAI MEIRI

Imperial College London, United Kingdom

The island rule is a name given for the supposed tendency of small-bodied animals to evolve larger sizes on islands whereas large animals evolve toward relatively smaller body sizes. The evolutionary forces that drive the observed patterns and the circumstances under which the phenomenon is manifest are widely debated but are thought to include changes in resource abundance, lower interspecific competition, elevated levels of intraspecific competition, and reduced predation on islands.

SIZE EXTREMES ON ISLANDS

Anecdotal observation can lead to the conclusion that in many clades, island-dwelling species are characterized by extreme body sizes relative to their mainland counterparts, especially on large oceanic islands. Several animal groups have their largest or smallest representatives on islands. For example, the St. Helena earwig *Labidura herculeana* and the New Zealand wetas (*Deinacrida* spp.) may be the largest representatives of their clades. The world's largest bat is the Philippine-endemic golden-crowned flying fox (*Acerodon jubatus*). Brown bears on Kodiak (*Ursus arctos middendorffi*) are considered the largest land carnivores. At the other end of the scale, at roughly 100 kg, *Elephas falconeri* from the Middle Pleistocene of Sicily was by far the world's smallest elephant. The largest bird known is the Madagascar elephant bird, *Aepyornis maximus* (although *Dromornis stirtoni* from the late Miocene of Australia may have been even larger), and the smallest snake is probably the Lesser Antillean threadsnake, *Leptotyphlops bilineata*. Both the largest and smallest lizards (*Varanus komodoensis* and *Sphaerodactylus ariasae*, respectively) are insular endemics.

However impressive this array of examples may seem, they offer no more than anecdotal evidence that extreme-sized animals arise on islands. Even under an appropriate null model, the largest and/or smallest members of some clades are expected to be insular endemics by chance. The smallest member in three of 23 orders of terrestrial mammals, for example, is an insular endemic. This may not be surprising given that about 18% of the world's mammal species are insular endemics. Thus, although many insular dwarfs and giants are known from several taxa, it is not clear whether the numbers are really different from those expected by drawing species randomly from a global size distribution. This issue can be easily examined using randomization tests, where a same number of species equivalent to the known number of insular endemics are drawn from a global species pool and their masses compared to those actually observed. As yet, such tests have not been conducted, and the issue of whether islands really harbor an unusual number of giants and dwarfs is therefore unresolved.

COMPARATIVE STUDIES OF SIZE EVOLUTION ON ISLANDS

In the 1960s John Bristol Foster compared closely related mammals (usually conspecifics) on islands and mainlands. He found that the body size of most of the island rodent populations he studied (mainly of the deer mouse, *Peromyscus maniculatus*) was larger than on the adjacent mainland. Most of the carnivore, even-toed ungulate, and lagomorph (rabbits and hares) populations he examined, however, were characterized by smaller body size on islands. Island marsupial and shrew populations showed no general tendency toward either gigantism or dwarfism.

Lee Van Valen interpreted Foster's findings as a tendency of large animals to grow small on islands and of small animals to grow larger on islands. He named this phenomenon "the island rule." Such a pattern was later shown statistically in mammals in general by Mark Lomolino, who quantified the body size of insular mammals as a function of the body size of the mainland population. Statistically the island rule is shown when the slope of this relationship is significantly lower than 1 (Fig. 1).

Alternatively such a relationship can be quantified when island/mainland body size ratios are regressed against the body size of the mainland population. The island rule will be manifest in cases where the slope of this regression is negative (Fig. 2).

Although the latter method is easier to visualize, it is fraught with statistical difficulties, because a regression of a ratio against its denominator is likely to produce more negative than positive results even if island and mainland body sizes are chosen at random. An interesting property of the aforementioned regression equations is that it is possible to compute a value at which body size will be the same on both islands and mainlands. Under the island rule, insular mammals that are smaller than this value tend to be larger than their mainland relatives, and those that are larger than this value tend to be smaller than their mainland relatives. The threshold itself is sometimes viewed as an evolutionary attractor toward which

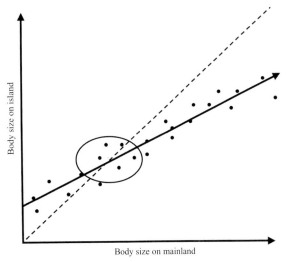

FIGURE 1 A regression of island body size on mainland body size. Each dot represents a pair of conspecific populations: one from an island, the other from a nearby mainland. Under the island rule, at small body sizes, dots will tend to fall above the line of equality (dashed line, where body size is the same on the island and the mainland) because small insular animals are larger than their mainland conspecifics. For large animals, dots will tend to fall below the line of equality (dotted line), because large animals get smaller on islands. The regression line (filled line) will thus have a slope of less than 1 (shallower than that of the line of equality). At body sizes where the regression line intersects the line of equality, little evolution is predicted: mainland and island animals will be roughly the same size (circle). This is the predicted "optimal" body size.

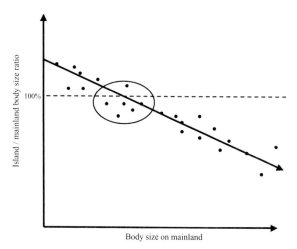

FIGURE 2 A regression of size ratio on mainland body size. Each dot represents a ratio: island body size/mainland body size (on the y axis) versus the body size of the mainland population. Under the island rule, at small body sizes, dots will tend to fall above the line of equality (dotted line, size ratio = 1): size ratios will be greater than 1 (island sizes larger). At large sizes, dots will tend to below the line of equality (size ratios of less than 1, large animals dwarf). The regression line (filled line) will thus have a negative slope. At body sizes where the regression line intersects the line of equality, little evolution is predicted: Mainland and island animals will be roughly the same size (circle). This is the predicted "optimal" body size.

size always tends to evolve (sometimes interpreted as an "optimal body size"; see below). Similar patterns to those described for mammals were shown for birds, turtles, snakes, and primates, but not for members of the mammalian order Carnivora or for lizards.

MECHANISMS OF SIZE EVOLUTION ON ISLANDS

The island rule is thought to emanate from a combination of evolutionary forces, most of which are related to the overall low species richness on islands. Because islands support fewer animal species than the adjacent mainland, insular animals face fewer potential competitor species on islands, fewer predator species and, for carnivores, fewer prey species as well. This in turn is taken to imply that interspecific competition and predation risk are also lower on islands. Thus, if predation and competition influence size evolution on the mainland, these selective forces are thought to be relaxed on islands. Insular animals are therefore thought to be evolving "back" toward the mean (or modal) body sizes of their group (the most common size classes within taxa are usually perceived to be closer to the evolutionary attractors discussed above).

Alternatively, if species occupying a certain niche in the ecological space on the mainland are missing from an island altogether, members of some other group may evolve to fill the niche. The existence of very large birds on oceanic islands (e.g., moas on New Zealand), for example, is often thought to result from the absence of large mammals. Thus, birds that perhaps cannot compete with large mammals on the mainland can evolve larger size only where mammals are absent.

A feature of islands that is thought to promote dwarfism is their small area, supposedly via an overall shortage of resources. However, it is not clear why this should be so, because the extent of dwarfism needed to alleviate a food scarcity problem is probably far beyond that actually seen in nature, even taking into account the fast evolution of island animals: Body size usually does not diminish drastically enough for overall resource consumption to be meaningfully lower, and population densities on islands are often higher than on the mainland. Furthermore, it is not clear why size reduction, even if it benefits the population as a whole, is advantageous to an individual: If large individuals have the upper hand in intraspecific confrontation, then they will be able to control more food resources and have better reproductive success even when the population as a whole is stressed for food. Sicily, home to the smallest of the insular elephants, the 100 kg *Elephas falconeri*, was shown to be large enough to maintain a healthy population of 15-ton elephants, yet *E. falconeri* dwarfed there to a mere 1% of the size of its mainland ancestor.

SELECTION PRESSURES AND IMPLICATIONS

The existence of an evolutionary attractor at medium body sizes assumes that both large and small sizes have their advantages. Large size is often associated with better ability to control territories, resources, and mates; larger prey size (for carnivores); better ability to fend off predators; higher number of offspring (in reptiles); better cold endurance; and a capacity to survive longer without food. Small size is correlated with earlier maturation and higher number of offspring (in mammals), shorter inter-birth intervals, better heat dissipation ability, lower overall resource requirements, and a higher chance of avoiding predation via the use of micro-habitat characteristics (e.g., burrows). Different species or groups of species were shown to respond differently to these selection pressures.

Being at the very end of a group's size spectrum is often problematic because of physiological and mechanical constraints. For example, very small homeotherm vertebrates have high surface-to-weight ratios. If resource acquisition scales isometrically with body size, then they must spend more time foraging than do larger homeotherms, and thus heat loss may preclude maintaining homeothermy below a size of one gram or so. A medium size is perceived as a compromise between the selective pressures favoring size extremes. The forces that are thought to maintain a wide spectrum of body sizes on mainlands are, therefore, predation and interspecific competition. If predation exerts a diverging force on body size (small animals benefit from being smaller still and large animals from being yet larger), and if the maintenance of these sizes is costly, then under relaxed predation pressure, as on islands, size will evolve toward medium. With competition, maintenance of a specific size need not even be costly, as long as by evolving a more similar size to a competitor, a species is able to exploit some resources used by that competitor (assuming resources are limiting). A small species is likely to have fewer large competitors and a large species fewer small competitors on islands by chance alone, and so reduced number of competitors can lead to medium size.

It has even been claimed repeatedly (but for different reasons) that large groups of animals (e.g., mammals) each have a single optimal size (the optima suggested for mammals differ by an order of magnitude: 100 g to 1 kg), and that the sole force maintaining the entire mammalian size distribution is interspecific competition. The island rule, with its regression toward intermediate sizes, was used as empirical evidence in support of all these theories. However, these models face some serious theoretical and empirical difficulties, not least from some patterns of insular size evolutions themselves: It has been shown empirically that different "optima" can be calculated for different mammalian subclades using island–mainland comparisons and that these perceived optima are probably artifacts of the specific data sets used rather than having a real biological meaning.

FUTURE PROSPECTS

Although the existence of insular giants and dwarfs among many clades is undeniable, much work needs to be done to understand better both the patterns and processes of insular size evolution. Merely recording extreme body sizes is not enough: the existence of 200 kg rodents and 450 kg birds on islands may be consistent with one formulation of the rule (Foster's) but not with others (Van Valen's and Lomolino's). Furthermore, it needs to be shown statistically that the numbers of dwarfs and giants on islands really deviate from a random draw of a similar number of species actually found on islands from an appropriate species pool. Otherwise, insularity need not be invoked to explain the existence of extreme body sizes on islands.

The terminology gets even more messy with attempts to invoke the island rule for phenomena that are unrelated in either time or space, such as susceptibility to extinction (e.g., Australian "time dwarfs," where only smaller relatives of large Australian mammals survived the wave of extinctions at the end of the Pleistocene), size evolution in mainland areas during the nineteenth and twentieth centuries, or size patterns in the deep sea. Paradoxically, we may understand the processes that shape size evolution on islands, because selection pressures affecting size can be demonstrated, but the resulting patterns are obscured by our inability to predict which of the prevailing evolutionary forces will be most influential under a given set of ecological circumstances.

More importantly, we have a limited theoretical framework that can be used to predict size evolution on islands. Mainland body size, which is assumed to be ancestral, is usually the sole factor used. Island area and island isolation feature less prominently and are often shown to have low explanatory power, or no effects on size evolution at all. This may be because patterns in nature are highly complex and may often be clade-specific, island-specific, and contingent on the colonization history of the island. The same animals may show strikingly different patterns of size evolution even within archipelagos in relation to the prevailing ecological conditions on different islands. Rather than viewing *Homo floresiensis*, the hotly debated "hobbit" of Flores, as an obvious case of insular dwarfism in large mammals, it may be more rewarding to ask why humans should dwarf on highly productive islands such as Flores, with its large predators (Komodo dragons) and potential prey (*Stegodon*). Features such as species-specific body masses and island areas are easy to quantify but may not be good predictors of the direction and magnitude of size evolution on islands. Quantifying aspects of the biotic and abiotic environments on different islands that will be relevant for different species is much more challenging, but it may offer our best chance of obtaining a better understanding of size evolution. The identity of community members is likely to prove more important than their numbers. To make good predictions we are likely to need data on the feeding habits, territoriality, reproductive strategy, guild membership, and numbers of smaller and larger guild members for the species in question. We will also need estimates of the island-mainland differences in the identity of predators and (for carnivorous animals) prey, and perhaps primary productivity and population density as well.

Understanding which animals will show which pattern of size evolution on different islands can offer a unique opportunity to decipher the mechanisms of size evolution in general, and better and more varied data may promise the best chance of achieving this.

SEE ALSO THE FOLLOWING ARTICLES

Dwarfism / Gigantism / Komodo Dragons / Moa / Rodents

FURTHER READING

Brown, J. H., P. A. Marquet, and M. L. Taper. 1993. Evolution of body size: consequences of an energetic definition of fitness. *American Naturalist* 142: 573–584.

Case, T. J. 1978. A general explanation for insular body size trends in terrestrial vertebrates. *Ecology* 59: 1–18.

Clegg, S. M., and I. P. F. 2002. The 'island rule' in birds: medium body size and its ecological explanation. *Proceedings of the Royal Society B* 269: 1359–1365.

Lomolino, M. V. 1985. Body size of mammals on islands: the island rule reexamined. *American Naturalist* 125: 310–316.

Meiri, S. 2007. Size evolution in island lizards. *Global Ecology and Biogeography* 16: 702–708.

Raia, P., and S. Meiri. 2006. The island rule in large mammals: paleontology meets ecology. *Evolution* 60: 1731–1742.

JAN MAYEN

SEE ARCTIC REGION

JAPAN'S ISLANDS, BIOLOGY

LÁZARO M. ECHENIQUE-DIAZ AND MASAKADO KAWATA
Tohoku University, Aoba-ku, Japan

JUN YOKOYAMA
Yamagata University, Japan

The Japanese Archipelago consists of 4 major islands (Hokkaido, Honshu, Kyushu, and Shikoku) and 6848 smaller islands, arranged across 3500 km parallel to the eastern coast of the Asian continent, and separated from it by the Sea of Japan. Overall, the Japanese Archipelago extends over a latitudinal range of 25° and a longitudinal range of 31°. Japan's rich biota is reflective of the complex geological history of the archipelago and of the great diversity of climates it encompasses. It lies within two major biogeographic regions: the Palearctic and the Indo-Malay in faunal categories and the Holarctic and the Paleotropic in floristic categories. The biogeographic boundaries for flora and fauna do not coincide, the latter being slightly shifted northward (Fig. 1).

BIOTA OF JAPAN: FLORA

The flora of Japan is characterized by its rich diversity. About 7100 vascular plant taxa are recognized, more than

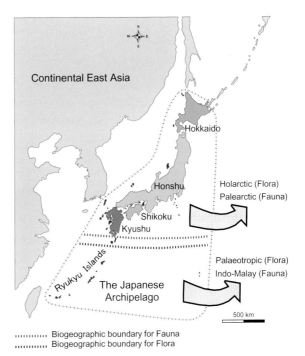

FIGURE 1 Geographic setting of the Japanese Archipelago, and the major biogeographic regions where it lies. The red dotted line marks the boundary between the Palearctic and the Indo-Malay biogeographic regions (defined for global patterns of animal distributions), and the black-dotted line marks the boundary for the Holarctic and the Paleotropic biogeographic regions (defined for global patterns of plants distribution). Smaller islands appear in red, and the four major islands appear differently colored.

1900 of them being endemic to the Archipelago. In addition, about 1000 species of bryophytes and more than 4500 algal species are recognized. This diversity is in part due to the different climate regimes that characterize the following distributional zones.

1. The subtropical zone, including the Ryukyu and Ogasawara island groups, and the lowland areas of

Kyushu and Shikoku islands, is characterized by the presence of evergreen broad-leaved trees (e.g., species of *Camellia*, *Machilus*, *Distylium*, *Lithocarpus*, and *Castanopsis*). In oceanic islands such as the Ogasawara group and a part of the Ryukyu group (Daito Islands), members of Fagaceae are absent as a result of isolation from mainland Japan. Several conifers are also present in the subtropical zone, such as *Pinus luchuensis* and *Podocarpus macrophyllus*.

2. The temperate zone ranges between the sea level to an altitude of about 1500 m in the mountainous areas of central and northeastern Honshu. In Hokkaido, it expands from the sea level to an altitude of about 600 m. In southern Honshu, Kyushu, and Shikoku, this zone is sparsely represented given the low occurrence of high mountains. It is characterized by the predominance of a large number of deciduous tree species (*Quercus*, *Acer*, *Kalopanax*, *Fagus*, *Aesculus*, *Juglans*, and other genera). Conifers such as *Abies firma* and *Tsuga sieboldi* are also typical. This zone is usually subdivided into three: the warm-temperature zone of broad-leaved evergreen forests dominated by evergreen *Quercus* species, covering most of southern Honshu, Shikoku, and Kyushu; the cool-temperature zone of broad-leaved deciduous forests dominated by *Fagus* and deciduous *Quercus* species, covering central and northern Honshu; and the southeastern part of Hokkaido (Oshima peninsula), and temperate broadleaf and mixed forests established in the transitional zone between the temperate and boreal zone. The third type of forests is dominated by *Betula*, deciduous *Quercus* species, and coniferous genera *Abies* and *Picea*.

3. The boreal zone covers areas above 1500 m on Honshu, above 600 m in central Hokkaido, and from the sea level in restricted areas in the easternmost part of Hokkaido. This zone is dominated by coniferous species of the genera *Abies*, *Tsuga*, *Picea*, and *Larix*. Deciduous broad-leaved trees such as *Betula ermani* also represent important members of this zone.

Most of the components of Japanese flora came from continental East Asia, and some families endemic to East Asia are also distributed in Japan (e.g., Cercidiphyllaceae, Eupteleaceae, Stachyuraceae, Trochodendraceae). Two monotypic families, Sciadopityaceae and Glaucidiaceae, and more than 20 genera (e.g., *Anemonopsis*, *Peltoboykinia*, *Pteridophyllum*, *Ranzania*) are endemic to Japan. About 90 genera of plants in Japan show a disjunct distribution in East Asia and North America. Distinctive genera include *Buckleya*, *Caulophyllum*, *Diphylleia*, *Stewartia*, *Schisandra*, *Illicium*, *Wisteria*, *Lespedeza*, *Penthorum*, *Itea*, *Astilbe*, *Menispermum*, and *Shortia*. The Ogasawara Islands and most of the Ryukyu Islands belong to the Paleotropic floristic kingdom, having common floral components with Polynesia (Ogasawara) and Southeast Asia (Ryukyus). For example, some of the endemic species of the Ogasawara Islands (e.g., *Clinostigma savoryana*, *Loberia boninensis*, *Metrosideros boninensis*, and two species of *Boninia*) are considered as having a Polynesian origin. Many plant genera such as *Archidendron*, *Arenga*, *Argostemma*, *Bischofia*, *Cananga*, *Dipteris*, *Erycibe*, *Epipremnum*, *Flagellaria*, *Grewia*, *Lepidagathis*, *Macaranga*, *Melicope*, *Murraya*, *Rhynchotechum*, *Schima*, or *Wendlandia*, are considered as South-East Asian elements of the Ryukyu Islands.

BIOTA OF JAPAN: TERRESTRIAL FAUNA

The Japanese fauna is relatively rich in comparison with those of other countries of similar size and given the fact that most of Japan belongs to the temperate climatic zone. This richness has a geographical foundation. Most islands of the Japanese Archipelago are continental islands and have a large basic stock of fauna that migrated from the Eurasian continent by different routes. Isolation and vicariant events due to glacial retreats that cut off connections with the mainland are the foundations of many cases of relict distributions. The Tokara Straits (Watase line) and the Tsugaru Strait (Blakiston line) (Fig. 2) represent connectedness gaps in the geological history of the archipelago and divide it into

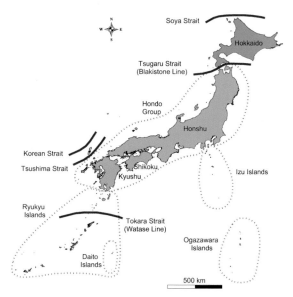

FIGURE 2 Major island groups of the Japanese Archipelago, and zoogeographical regions as determined by distributional patterns and historical connectedness gaps. Islands with the same color have greater faunistic affinities than islands with different colors.

three zoogeographic regions: Hokkaido; other main islands of the Archipelago (Honshu, Shikoku, and Kyushu) with their surrounding islands; and the Ryukyu and Ogasawara island groups. Each part has a specific fauna (more noticeable in birds and mammals), resulting in a very diverse and rich fauna for the whole archipelago (Fig. 2).

Japanese Mammals

Japanese Mammals are characterized by their endemism (40%), species abundance, and the scarcity of large species. Endemic and relict mammals as the result of isolation are found all across the Japanese Archipelago. On the Ryukyu Islands, separated from the continent in the late Tertiary, there are 14 endemic terrestrial species (48.3%), including two endemic genera (*Pentalagus* and *Tokudaia*). Honshu, Shikoku, Kyushu, and the surrounding islands (known collectively as Hondo) appeared in the Quaternary (around a million years ago), and together host 27 endemic species (46.5%). Most of the species in the genera *Apodemis*, *Eothenomys*, and *Aschizomys*, very common in Honshu, are endemic. On the other hand, Hokkaido shows a more monotonous fauna, with no endemics at the species level and many species in common with the Eurasian continent, such as *Ursus arctos*, *Sciurus vulgaris*, *Tamias sibiricus*, and *Lepus timidus*.

Japanese Birds

The prime characteristic of Japanese birds is regionality. Their distribution is largely affected by the Watase and Blakiston lines (Fig. 2), so that in the Ryukyu Islands, birds belonging to the Indo-Malay biogeographic region are more common, while Palearctic avifauna mainly from the Siberian region is distributed in Hokkaido. Southern Palearctic species from East Asia are more frequently found in Honshu, Kyushu, and Shikoku. In Hokkaido and the Ryukyu Islands, migratory birds represent more than 80% of the population, while in Honshu they account for 60%, and in Shikoku and Kyushu 40%. Many endemic birds are found across the Japanese Archipelago (e.g., *Phasianus soemmerringii*, *Phasianus versicolor*, *Synthliboramphus wumizusume*, *Picus awokera*), and some taxa are found only in remote oceanic islands. These include *Turdus celaenops* (Izu Islands), *Apalopteron familiare hahasima* (Bonin Islands), *Diomedea albatrus* (Torishima Islands), and *Otus scops interpositus* (Daito Islands) (Fig. 2).

Japanese Reptiles and Amphibians

About 87 taxa (species and subspecies) of reptiles and 60 of amphibians are known in Japan, symbolizing the evolutionary importance of the archipelago. Lizards and snakes all across the Japanese Archipelago are characterized by their high endemism (over 80%). The same occurs for amphibian species, showing over 75% endemism. Another characteristic of the Japanese herpetofauna is the abundance of salamander species. Particularly interesting examples are hynobiid salamanders, which had undergone a significant geographic separation in restricted areas of Japan, resulting in more species of this group in the Japanese Archipelago than on the continental mainland. The world's largest living amphibian, the Japanese giant salamander *Andrias japonicus*, a primitive species, is also an example of the evolutionary significance of the Japanese herpetofauna, where many relict species are found.

Japanese Invertebrates (Insects, Spiders, and Land Snails)

Most insects in Japan are common East Asian species, mainly derived from northeast Asia and from the South China–Himalayan region. They are characterized by a large number of species (e.g., over 8000 species of beetles and weevils, 310 of butterflies, 5000 of moths, 140 of cicadas and aquatic hemipterans, and 175 of dragonflies). Around 100,000 species are estimated for the whole archipelago, of which 30,000 are known). Some groups show high endemism (e.g., 70% in Carabinae ground beetles), and some species are representative of relict insect groups. Among these is an archaic dragonfly, *Tanypteryx pryeri*, a member of a group of which only ten species survive in the Pacific Rim, and subterranean species of the family Grylloblattidae (a group of wingless insects that live on top of mountains, sometimes called ice crawlers or icebugs).

Spiders in Japan comprise over 1400 species in 53 families. Endemic species can be found across the Archipelago and some oceanic islands. Conspicuous examples are *Doosia japonica*, *Desis japonica*, *Verpulus boninensis*, and *Alopecosa hokkaidensis*.

The characteristic for Japanese land snail fauna is a north-to-south cline in species richness, with more species found in southern parts of the archipelago than in northern areas. Endemism is also another feature of this group, with some oceanic islands showing the highest values (e.g., over 100 species in the Ogasawara Islands, 90% of which are endemic, including seven endemic genera such as *Ogasawarana*, *Mandarina*, *Hirasea*, *Hirasiella*, and *Boninena*).

Japanese Freshwater Fishes

The indigenous freshwater ichthyofauna of Japan is closely related to the fauna of continental East Asia and consists of about 214 native species, of which 24.3% are endemics. Strictly freshwater fish fauna is more abundant in central and western Honshu, where Lake Biwa is the main species reservoir. Of the peripheral freshwater fish fauna, northern species such as salmonids, sticklebacks, and sculpins are well represented in northeast Honshu and Hokkaido, and southern elements such as gobies occur mostly in western and southern areas.

JAPANESE MARINE BIOTA

Three major ecosystems with a diverse fauna can be found along the Japanese shores. Algal beds on rocky shores (the subtidal ecosystems) are dominated by *Sargassum* species, kelp beds are typically represented by *Laminaria*, *Eisenia*, and *Ecklonia*, while the sandy mud bottom is dominated by *Zostera marina* sea grass. In the Ryukyu Islands, the latter species is replaced by *Thalassia hemprichii*. The coral reef ecosystem in Japan covers an area of approximately 900 km^2, extending from the Ryukyu Islands to the southern areas of Kyushu, Shikoku, and the Kii Peninsula, with several gaps in between. It is characterized by a large number of genera and species, favored by the warm Kuroshio current. This region is considered a marine biodiversity hotspot, with 1315 fish species of which 26 are endemic (1.97%), ranking fourth among the world's top ten areas for endemic reef fishes. Conspicuous species include *Stegastes altus*, *Chromis albomaculata*, and *Chaetodon daedalma*. It also harbors a large number (400) of coral species, including the oldest and bigger community of blue coral (*Heliopora coerulea*) in the Northern Hemisphere. The tideland ecosystem is found in estuarine environments, supporting a rich fauna of marine invertebrates.

EXTINCTIONS IN THE JAPANESE BIOTA

Recent extinctions of plants and animals in Japan are well documented. Among the relatively recent extinctions are at least four species of mammals (two endemic bat species, the Japanese sea lion, and the Japanese wolf), while a fourth endemic species, the Japanese otter (*Lutra nippon*) is seemingly extinct. Similarly, 13 species of birds, 3 fish species, 2 insect species, and 41 species of plants (including eight extinct in the wild) are reportedly extinct. These extinctions are attributed to human influences, although natural causes cannot be excluded (e.g., impact of volcanic activity). Drastic changes in the natural lands of Japan began with the process of modernization and were especially severe after World War II, resulting in the decline and disappearance of wildlife habitats throughout the archipelago.

NATURE CONSERVATION IN THE JAPANESE ARCHIPELAGO

Although about 67.0% of the Japanese archipelago is forested, original primary vegetation only remains in few places (26% of the national land). The ratio of natural vegetation is regionally unbalanced, with Hokkaido and Okinawa showing the highest values (48.7% and 47.9%, respectively), while other regions had less than 20%. A large proportion of the natural vegetation lies within National Parks and other nature reserves, a network of 37,500 km^2 of legally enforced conservation areas.

SEE ALSO THE FOLLOWING ARTICLES

Climate on Islands / Endemism / Extinction / Freshwater Habitats / Japan's Islands, Geology

FURTHER READING

Abe, H. 1999. Mammals of Japan, their diversity and conservation, in *Recent advances in the biology of Japanese Insectivora: proceedings of the symposium on the biology of insectivores in Japan and on wildlife conservation*. Y. Yokohata, ed. Hiwa, Japan: Hiwa Museum for Natural History.

Goris, R., and N. Maeda. 2004. *Guide to the amphibians and reptiles of Japan*. Melbourne, FL: Krieger Publishing.

Japan Integrated Biodiversity Information System (J-IBIS). http://www.biodic.go.jp/english/J-IBIS.html.

Masuda, H., K. Amaoka, C. Araga, T. Uyeno, and T. Yoshino. 1984. *The fishes of the Japanese Archipelago*. Tokyo: Tokai University Press.

Ministry of the Environment, Government of Japan. *Nature and Parks*. http://www.env.go.jp/en/nature/.

Murata, J. 2000. Flora of Japan, in *The botany of biodiversity 1. Floristic Research*. K. Iwatsuki and M. Kato, eds. Tokyo: University of Tokyo Press, 24–47.

Otsuka, H., H. Ota, and M. Hotta, eds. 2000. International symposium: The Ryukyu Islands—arena for adaptive radiation and extinction of island fauna. *Tropics* 10.1: 1–241.

Wen, J. 1999. Evolution of eastern Asian and eastern North American disjunct pattern in flowering plants. *Annual Review of Ecology and Systematics* 30: 421–455.

JAPAN'S ISLANDS, GEOLOGY

S. MARUYAMA
Tokyo Institute of Technology, Japan

S. YANAI
Japan Geocommunications, Co., Ltd., Tokyo

Y. ISOZAKI
University of Tokyo, Japan

D. HIRATA
Kanagawa Prefectural Museum of Natural History, Odawara, Japan

The Japanese islands are a bow-shaped chain of islands extending over 3000 km along the eastern margin of Asia. Four arcs—from north to south, the Kurile, Japan, Izu–

Mariana, and Ryukyu arcs—are combined to form Japan. The Izu–Bonin arc extends south from the Japan arc (Fig. 1). The climate of Japanese islands is variable depending on latitude, ranging from snowy (from Hokkaido to northeastern Japan) to moderate (southwestern Japan) to subtropical (on the Ryukyu and Ogasawara arcs). One hundred and thirty million people live on the Japanese islands. Over 200 active volcanoes and their accompanying hot springs are present. Moreover, earthquakes along subduction zones and inland inner arcs pose additional geologic hazards.

EARLY GEOLOGIC STUDIES

The geology of the Japanese islands was first studied about 120 years ago by German geologist Edmund Nauman (1854–1927), who initially, at the age of 20, thought Japan was composed entirely of volcanic rocks, because they are located in the Pacific Ocean. Performing extensive geologic study during his 10-year stay in Japan, he recognized that high-grade regional metamorphic rocks in Japan were similar to those in continental areas, and therefore he realized that Japan had been a part of Asia before the opening of the Sea of Japan in the Tertiary and had rifted away and become isolated from Asia. Since 1950, modern geologic mapping has been carried out, and the evolution of the Japanese islands is now well defined, as outlined here. The history of the Japanese islands is now confirmed back to 1.9 billion years ago, as a part of the South China continent.

GEOLOGIC OUTLINE

Plate Boundaries around Japan

Japan faces four major plates, with the Pacific plate subducting at 10 cm/yr under Northeast Japan, where seismologically detected slabs reach to a depth of 660 km under Beijing, China. The Japan Trench is a topographically depressed region ~5–6 km deep, corresponding to the western end of the Pacific plate. The Philippine Sea plate subducts from southeast to northwest at 4 cm/yr under Southwest Japan. The boundary is called the Nankai Trough or Trench and can be traced on-land through Honshu to its eastern end, south of Tokyo. Northeast Japan (northern Honshu) and Hokkaido are part of the North American plate, which is underthrust from the west by the Eurasian plate. Two trench–trench–trench (TTT) triple junctions are present near Tokyo, one near Mt. Fuji and another to the south of Tokyo.

Northeast Japan has been colliding against Southwest Japan to form the Japanese Alps by uplift of the leading edge of the Eurasian plate since 0.5 million years ago. To

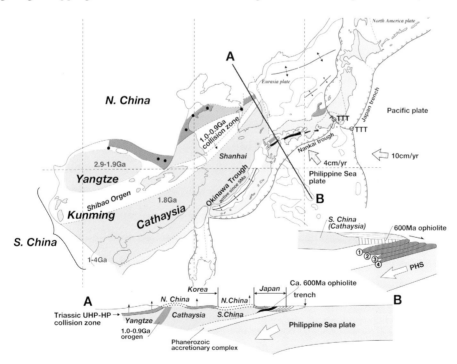

FIGURE 1 Geographic position of the Japanese islands. The island arc chains around Japan consist of the Kurile, Japan, Ryukyu, and Izu-Mariana arcs. The *Fossa Magna* separates the Japan arc into two segments; Northeast Japan on the North American plate and Southwest Japan on the Eurasian plate. The Pacific and Philippine Sea plates are juxtaposed with the Northeast Japan–Izu-Mariana arcs and the Japan–Ryukyu arc, respectively. The Sea of Japan behind the Japan arc is a rifted basin. The Okinawa trough behind the Ryukyu arc is currently active. Tectonic frameworks of the Japan and Ryukyu arcs are linked with the Cathaysia block, the southern part of the South China block.

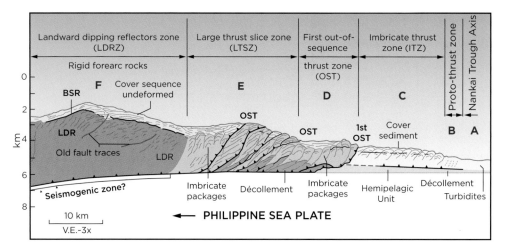

FIGURE 2 Seismic reflection profile of accretionary complex along Nankai Trough off Southwest Japan. The principal fabrics around the converging plate boundary are marked by development of decollement and associated imbrication pileups (duplexes). Source: Kuramoto et al. (2000).

the north of the Ryukyu arc, the Okinawa Trough marks where back-arc spreading has been on-going since 5 million years ago. The spreading began from the southern end and propagated northeastward to northern Kyushu.

Accretionary Front

One of the characteristic geologic phenomena is the formation of an accretionary complex on the leading edge of the Eurasian plate. The seismic reflection profile off Shikoku island (Fig. 2) shows thick turbidite units, right above the subducting Philippine Sea plate, that are highly folded and have thrust-bounded oceanward vergency. A series of high-angle reverse faults flatten at depth. Thrust-converged parts form a decollement zone between the underlying Philippine Sea plate and the accretionary wedge above, successively forming Zones A, B, C, D, E, and F. Zone A, the youngest one, is not yet deformed and originally derived from trench turbidites formed along the submarine canyons. The material and physical boundaries are somewhat different. Channelized fluid pipes pass along the decollement zone and branch to several active reverse faults and to the surface, where biological communities form colonies. Oceanic materials are often accreted together with trench turbidites.

Japan has grown by adding accretionary complexes through time since 500 million years ago, but also has eroded tectonically at trenches to lose already-formed accretionary complexes. The eroded materials either underplate the hanging wall at great depth or are removed down to the deep mantle. The near-trench zone in Northeast Japan is an example where only small amounts of post-Miocene accretionary complex are present.

Arc Volcanism and Underlying Mantle

Japan's arc volcanoes form as a result of secondary mantle convection triggered by cold slab subduction, and the arc runs parallel to the active trench. Volcanoes occur 100–120 km above the subducted slab, regardless of the angle of the subducting slab. If the angle is large, as in the case of the Mariana arc, the distance between volcano and trench becomes short.

An example from Northeast Japan is shown as a cross section (Fig. 3). High resolution P-wave tomographic images clearly document the high-velocity subducting cold Pacific slab (blue). The hot, low-velocity perturbation (orange-yellow) in the hanging wall corresponds to secondary mantle convection rising up to intersect the arc crust where the majority of volcanoes are formed (called the volcanic front). Minor volcanoes form behind the front, called the second chain of volcanoes on the Sea of Japan side.

On the cross sections, earthquakes are observed in the Pacific slab and the Japan arc. Earthquakes are absent along the top boundary of the slab. Double seismic zones occur within the Pacific slab, namely at the topmost slab mantle and in the center of the slab. These earthquakes

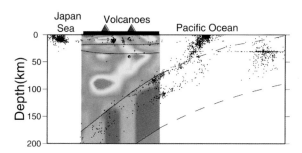

FIGURE 3 P-wave seismo-tomographic image of the Northeast Japan arc. The Pacific slab, blue-colored (high-P wave velocity), can be well discerned. The orange-yellow-colored low-velocity zone is developed in the hanging wall of the mantle wedge, probably driven as a counterflow of mantle convection triggered by the Pacific slab subduction. A rising counterflow of mantle intersects with arc crust to release arc magma by melting mantle as a result of high water content derived from the Pacific slab and decompression. Source: Hasegawa et al. (2008).

are triggered by dehydration embrittlement of the hydrous magnesium silicates. Water-rich fluids liberated from dehydration reactions lower the stress in the plate under extremely high-pressure conditions, triggering earthquakes.

Deep earthquakes in the mid-ocean ridge basalt (MORB) crust are abundant to 50–60 km depth, and limited to 100 km depth, by dehydration reactions in the hydrated MORB crust.

Earthquakes are also abundant in the upper half of the arc crust as a result of the brittle-ductile transformation of granitic materials. The boundary between absence and presence of earthquakes, corresponds to the 350 °C isotherm, above which granitic materials behave in a brittle fashion, hence forming earthquakes.

Geotectonic Divisions

Rocks of the Japanese islands are classified into four kinds: (1) accretionary complexes dominated by trench turbidites with enclosed oceanic materials, (2) igneous rocks derived from arc magma, (3) regionally metamorphosed rocks from 10–60 km depth above the subducting slab, and (4) normal sedimentary sequences. These four kinds can be grouped into one set of units, formed at the same time in different positions. For example, during the Cretaceous, ~120–80 million years ago, extensive volcano-plutonism occurred to form a huge batholith belt accompanied by acidic effusive rocks. Simultaneously, the deep-seated subduction zone complex was metamorphosed along the Benioff plane at 10–60 km depth and returned to the surface near the fore-arc region, a new accretionary complex formed oceanward, and active sedimentation took place in the fore-arc region. This group of geologic units is termed the Cretaceous orogenic unit. Japan is composed of six of these orogenic units formed 450, 320, 250, 120–80, and 60 million years ago and post-Tertiary. These are well-documented in Northeast and Southwest Japan, and presumably in the Ryukyu islands, but not in Izu-Ogasawara and eastern Hokkaido. These two regions originally formed above the ~100-million-year oceanic lithosphere by subduction of slabs to become arcs, and subsequently they collided to become parts of the Japanese islands.

The major parts of Japan are composed predominantly of accretionary complexes, including even regionally metamorphosed ones. The material incorporated into the accretionary complexes is a key to estimate the age of subducted slabs with ocean plate stratigraphy (Fig. 4). Thus, the

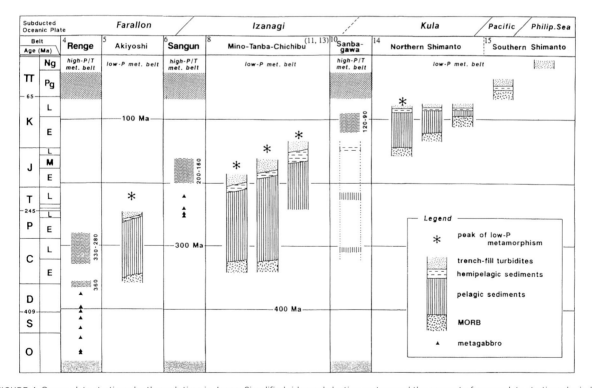

FIGURE 4 Ocean plate stratigraphy through time in Japan. Simplified ridge-subduction system and the concept of ocean plate stratigraphy indicate the age gap between two distinct horizons; the horizon between pillowed MORB and pelagic chert marking the oceanic plate; the horizon between hemipelagic mudstone and terrigenous clastics marking the arrival at the trench, which represents the total travel time of the subducting oceanic plate from mid-oceanic ridge to trench or, in other words, the age of the subducting oceanic plate at trench.

formation of accretionary complexes has been estimated to have occurred 500–450, 400–320, 300–250, 200–120, and 80–70 million years old, and the post-Tertiary.

The Japanese islands were initiated by continental rifting of the Cathaysia continent with 1.9–billion-year-old rocks. These old basement rocks are now exposed on small islands in the Sea of Japan, Oki-Dogo island. Further to the south, the structural top unit is the regionally metamorphosed unit formed 240 million years ago by the collision of South China against North China. The oldest basement is underneath this unit. The oldest ophiolite sequence (~600 million years old) is the oceanic lithosphere made by the opening of the paleo-Pacific ocean when the Rodinia supercontinent was formed and is equivalent to the oldest ophiolites along the western margin of North America, such as the Klamath ophiolite. After the conversion of the plate boundary from passive to active margin 450 million years ago, an accretionary complex formed and periodically exhumed the deep-seated accretionary complex to the surface, while the volume of arc crust increased rapidly by calc-alkaline volcano-plutonism.

Later tectonic modification, specifically the opening of the Sea of Japan and the collision of the Izu-Ogasawara arc and eastern Hokkaido, strongly modified pre-existing geologic units of Japan along the Itoigawa–Shizuoka Tectonic Line (originally called the *Fossa Magna* by E. Nauman), the Median Tectonic Line, the Tanakura Tectonic Line, and the Ishikari Tectonic Line and Kannawa thrust fault.

Oceanic plate stratigraphy and regional ultrahigh- and high-pressure metamorphic suites of accreting units of southwestern Japan are assigned dates from late Paleozoic to Neogene. The oceanward and tectonically downward-younging polarity is noted, as shown in the structural profile compiled schematically in Fig. 1.

Geologic History

In the following sections, the geotectonic evolution of Japan is illustrated by a series of paleogeographic maps.

RODINIA BREAKUP, ~750–600 MILLION YEARS AGO

During the late Proterozoic, 1.1–0.9 billion years ago, all fragmented continents were united through extensive collisions to form a supercontinent, Rodinia, cemented by several Grenvillian orogens. The Rodinia supercontinent was present in the Southern Hemisphere, where most continents (except for West Africa, Amazonia, or both) formed another supercontinent, were rifted, and drifted apart under the force a huge upwelling mantle plume, the Pacific superplume at the triple junction. South China (with proto-Japan) was juxtaposed with Laurentia (present-day western North America) and situated close to the divergent triple junction.

One of the important Grenville orogens is called the Shibao, which was a failed rift in South China separating the southern Cathaysia and northern Yangtze cratons. The Cathaysia was a part of North America, whereas Yangtze was a part of East Antarctica or Australia. Rodinia had broken up by ~600 million years ago, and more than 10 smaller continents had separated and migrated.

FORMATION OF GONDWANA AND ITS BREAKUP AND DISPERSION (540–300 MILLION YEARS AGO)

The smaller continents were amalgamated to form the semi-supercontinent Gondwana at 540 million years ago. North America subsequently docked to Baltica to form Caledonides and collided later against Africa to form Hercynides. On the other hand, Gondwana began fragmenting in Early Paleozoic, and the pieces were later amalgamated to form Laurasia during the late Paleozoic. The new supercontinent Pangaea formed near the end of Paleozoic to early Mesozoic.

Subduction of the Pacific seafloor occurred at almost all of the continental margins around the Pacific in the Early Ordovician, ~480 million years ago. Proto-Japan, near South China, had remained near the same corner of the Australian continent since the Cambrian, but it was now enrolled for the first time into a subduction regime.

According to the distribution of the continental blocks in the Early Devonian, ~400 million years ago, South China and Japan were close to Gondwana. Following the collision and closure of the seaway between Laurentia and Baltica that formed Laurasia (the Caledonian orogenic belt, colored in red), the remnant segments of the Iapetus Ocean were narrowed by successive subduction. South China moved to the north of Australia and became isolated in the Pacific. North China was also isolated from the other continental blocks and belonged to an independent faunal province.

FORMATION OF PALEOASIA (LAURASIA), ~200 MILLION YEARS AGO

In the Late Carboniferous ~300 million years ago, several continental blocks, including South China, North China, Tarim, and Indochina, moved northward from the Southern Hemisphere (Fig. 5) because a large-scale cold superplume (colored in purple in the figure) developed and dragged the blocks. In addition, the other continental blocks in modern Asia, such as Siberia and Kazakhstan, were also dragged in the same domain. On

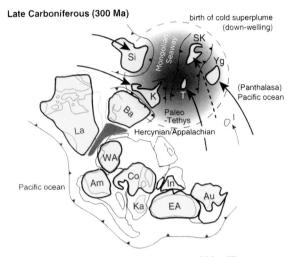

FIGURE 5 Birth of the Asian cold superplume ~300 million years ago. Am = South America; Au = Australia; Ba = Baltica; Co = Congo; EA = Eastern Antarctica; I = Indochina; In = India; K = Kazakhstan; Ka = Kalahari; La = Laurentia; Si = Siberia; SK = North China; T = Tarim; WA = West Africa; Yg = South China.

the other hand, the closure of the Iapetus ocean was completed along the Hercynian–Appalachian orogenic belts, and the merger of Gondwana and Laurasia consequently formed the supercontinent Pangaea.

As shown in Fig. 5, North China and South China continents moved northward to merge with Siberia, closing seaways among these blocks ~280 million years ago. Their mutual isolation resulted in the development of three distinct floristic provinces, namely the northern Cathaysia, southern Cathaysia, and Angara flora, respectively. Proto-Japan was on the southeastern continental margin of South China and has grown with the pasted materials of the Pacific Ocean, or Panthalassa. The Carboniferous to Permian Akiyoshi–Sawadani seamount chains, capped by a reef limestone, were approaching the proto-Japan margin of the South China continent, driven by subduction of the Farallon plate.

COLLISION OF SOUTH CHINA AGAINST NORTH CHINA 240 MILLION YEARS AGO

The Yangtze, or northern South China, started to collide against North China from Dabie Promontory during the Triassic period, ~240 million years ago, closing the paleo-Tethyan seaway. The collision suture developed along the Qinling–Dabie mountains in central China. The Mongolian seaway between the North China and Siberia continents also narrowed through double-vergent subduction on both sides. Along the southern margin of South China, the active subduction of the Farallon plate formed the Late Permian Akiyoshi accretionary complex (colored in light pink in Fig. 6), incorporating fragments

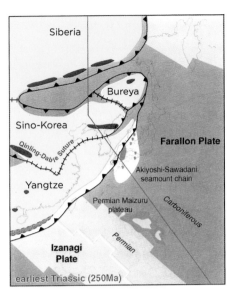

FIGURE 6 Collision of South China (Yangtze) against North China (Sino-Korea) during the Triassic period. The figure shows the modern coastlines of east Asia as dotted lines for reference, land as dark yellow, sea as blue, green seamount and oceanic plateau as green, granite batholith belt as red, and mid-oceanic ridges as light yellow.

from the subducted Akiyoshi–Sawadani seamount chain and the Permian Maizuru oceanic plateau.

After or around 210 million years ago, the Mongolian seaway mostly closed, and a delta formed in the eastern end of the suture. The Qinling–Dabie suture exhumed the ultrahigh-pressure metamorphic unit from the depth where diamonds form. Abundant terrigenous clastics were transported along the suture and finally brought to a deep-sea fan formed at the trench along the Pacific

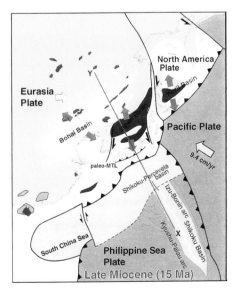

FIGURE 7 Separation of Japan from the Asian continental margin around 20 million years ago.

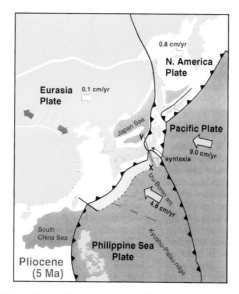

FIGURE 8 Collision of Izu–Bonin arc 5 million years ago.

ocean. Owing to the abundant supply of clasts from the suture, accretion along the Yangtze margin revived and constructed the Late Triassic–Early Cretaceous accretionary complex of the Mino–Tanba–Chichibu belt, Southwest Japan. As the triple junction of ridge–trench–trench moved northeastward, the young and buoyant Farallon and Izanagi Plates subducted, tectonically exhuming the high-pressure/low-temperature Sangun metamorphics and generating a granite batholith belt beneath the coeval volcanic arc.

OPENING OF THE SEA OF JAPAN, ~20–15 MILLION YEARS AGO

The Japanese islands became an island arc. The activity of the sub-Asia plume peaked, accelerating the rifting in major basins in East Asia, such as the Sea of Japan, the Baikal basin, and the Bohai basin in northern China. Bimodal (both basaltic and rhyolitic) volcanism characterized the initial phase of rift-related basin formation (Fig. 7).

The back-arc basins reached their full extent, generating several across-arc strike-slip faults. In contrast to the back-arc extension, the fore-arc of the Japanese islands suffered from a local contraction that tectonically juxtaposed the Ryoke granites above the Sanbagawa high-pressure/low-temperature metamorphics by the subhorizontal paleo–Median Tectonic Line in Southwest Japan. Another back-arc basin opened in the Philippine Sea plate, in which the proto–Izu-Bonin arc split the Kyushu–Palau ridge (arc) and created the Shikoku–Parece Vela basin. The subduction of the young and hot Philippine Sea plate generated slab melting along the subduction zone.

COLLISION OF IZU–BONIN ARC AGAINST HONSHU ARC SINCE 5 MILLION YEARS AGO

The back-arc basins that were opened during the Miocene, namely, the Sea of Japan and the South China Sea, had already been turned into a contraction regime by 5 million years ago. The subduction of the Philippine Sea Plate beneath Asia accompanied the collision/subduction of the Izu–Bonin arc against the Northeast Japan arc in central Japan (Fig. 8). A fore-arc sliver at the front of the Kurile arc was formed by the oblique subduction of the Pacific Plate.

The Japanese islands are currently located on four distinctive plates: the Eurasian, Okhotsk (North American), Philippine Sea, and Pacific Plates. Three active arc-trench systems developed in Southwest Japan–Ryukyu, Northwest Japan, and Izu–Bonin. The oblique subduction of the Philippine Sea plate generated fore-arc slivers along the Southwest Japan–Ryukyu arc. The linear neo–Median Tectonic Line cutting the low-angle paleo–Median Tectonic Line corresponds to the landward margin of the fore-arc sliver in Southwest Japan. Another back-arc basin with hydrothermal activity, the Okinawa trough, is opening in the Ryukyu.

SEE ALSO THE FOLLOWING ARTICLES

Earthquakes / Island Arcs / Japan's Islands, Biology / Kurile Islands / Pacific Region

FURTHER READING

Maruyama, S., Y. Isozaki, and G. Kimura. 1997. Paleogeographic maps of the Japanese islands: plate tectonic synthesis from 750 Ma to the present. *Island Arc* 6: 121–142.

Maruyama, S. 1997. Pacific-type orogeny revisited; Miyashiro-type orogeny proposed. *Island Arc* 6: 91–120.

Maruyama, S., M. Santosh, and D. Zhao. 2007. Superplume, supercontinent, and post-perovskite: mantle dynamics and anti-plate tectonics on the core-mantle boundary. *Gondwana Research* 11: 7–37.

Maruyama, S., D. A. Yuen, and S. Karato. 2007. Plumes and superplumes through Earth's history, in *Superplumes: beyond plate tectonics*. D. A. Yuen, S. Maruyama, S. Karato, and B. F. Windly, eds. Berlin: Springer-Verlag, 441–502.

REFERENCES

Hasegawa, A., J. Nakajima, S. Kita, Y. Tsuji, K. Nii, T. Okada, T. Matsuzawa, and D. Zhao. 2008. Transportation of H_2O in the NE Japan Subduction Zone as inferred from seismic observations: supply of H_2O from the slab to the arc crust. *Journal of Geography* (Tokyo Geographic Society) 117.1: 59–75.

Kuramoto, S., A. Taira, N. L. Bangs, T. H. Shipley, G. F. Moore, and EW99-07, 08 Scientific Parties. 2000. Seismogenic zone in the Nankai Accretionary Wedge: General summary of Japan–U.S. Collaborative 3-D Seismic Investigation. *Journal of Geography* (Tokyo Geographic Society) 109.4: 531–539.

JUAN FERNANDEZ ISLANDS

SIMON HABERLE
Australian National University, Canberra

The Juan Fernandez Islands are located in a warm temperate region of the far southeastern Pacific Ocean and consist of three large volcanic islands that harbor a flora of remarkably high endemism (about 67%). Historic human-induced changes to the island environments and their isolation from a continental landmass have contributed to degradation of the island biota, which is rapidly becoming one of the most threatened in the world.

GEOLOGY

The Juan Fernandez archipelago is made up of three large volcanic islands—Isla Robinson Crusoe (or Masatierra; 33°37′ S, 78°51′ W; area 47.9 km^2; elevation 915 m), Isla Alejandro Selkirk (or Masafuara; 33°45′ S, 80°45′ W; area 49.7 km^2; elevation 1380m), and Isla Santa Clara (33°41′ S, 79°00′ W; area 2.2 km^2, elevation 374 m)—and several small rocky islets that lie between 570 and 720 km west of the Chilean coast. Currently the islands are administered by the Chilean government and are inhabited by around 600 residents based in the present village of San Juan Bautista on Isla Robinson Crusoe and whose livelihoods are derived primarily from fishing (lobsters), tourism, and cattle.

The islands are volcanic in origin, forming over a hotspot that underlies the eastward drifting Nazca plate. These isolated volcanic islands range in age from around 4 to 1 million years old, with the oldest, Isla Robinson Crusoe, having a very rugged and deeply dissected topography compared to the younger Isla Alejandro Selkirk. The islands lie within the northern margins of the subantarctic region of the South Pacific Ocean and are under the influence of major ocean currents: the subantarctic Chile-Peru Current (Humboldt Current) and the subtropical Peru Oceanic Countercurrent. The climate data available for Isla Robinson Crusoe Island records a warm temperate climate with a wet winter and dry summer, a mean annual rainfall of around 1000 mm, and a mean annual temperature of 15.2 °C at sea level. The higher Isla Alejandro Selkirk, 187 km to the west, generates abundant orographic rain, primarily from southwesterly and southeasterly winds, in addition to precipitation delivered from the western storm track. The high altitudes are considered to be almost perpetually cloud covered, which along with occasional frosts and snow falls, contribute to the existence of a climatic forest limit at an elevation of around 700–750 m.

BIOGEOGRAPHY

The Juan Fernandez Islands have a very limited fauna, with no native mammals, reptiles, or amphibians. Around 30 landbird and seabird species breed on the islands. The islands have three endemic bird species, the most striking being the picaflor or Juan Fernandez firecrown, *Sephanoides fernandensis* (Fig. 1), the only endemic hummingbird known on oceanic islands. The present vegetation of the islands is well known with a total of 383 species of flowering plants, which includes 104 endemic, 52 native, and 227 introduced species. There are 51 species of ferns. Table

FIGURE 1 (A) The Juan Fernandez firecrown, *Sephanoides fernandensis*, is an endemic hummingbird on Isla Robinson Crusoe, seen here feeding on the nectar of *Dendroseris littoralis*. (B) *Robinsonia gayana*, an endemic Asteraceae on exposed southern cliff face, Robinson Crusoe Island.

TABLE 1
Diversity of Species on the Juan Fernandez Islands

	Endemic		Native		Introduced		Totals	
	Total	% before 1574	Total	% before 1574	Total	% after 1574	before 1574	after 1574
Flowering plants	104	67	52	33	227	59	156	383
Dicotyledons	90	58	29	19	195	50	119	314
Monocotyledons	14	9	23	14	32	8	37	69
Ferns (pteridophytes)	17	33	34	66	0	0	51	51
Mosses and liverworts	38	24	119	76	0	0	157	157
Terrestrial birds	3	42	4	58	7	50	7	14
Insects	440	72	170	28	77	11	687	687

1 shows the composition and origin of the flora with the pre-1574 flora consisting of around 67% endemic species. The origin of these plants is dominantly from the southern South American mainland (55%), with smaller proportions being widespread and derived from the Neotropics, Pacific, Australia, and New Zealand. The distinctive rosette-like trees, shrubs and herbs within the Asteraceae (five endemic genera including 26 endemic species derived predominantly from the Americas) have diversified most dramatically on the islands and occupy a wide range of habitats from lowland forest to alpine areas (Fig. 1B). Other genera that have numerous endemic species include *Gunnera*, *Peperomia*, *Wahlenbergia*, *Chenopodium*, and *Eryngium*.

The remaining forest cover on the islands falls into a subtropical montane rainforest classification that is dominated by trees of *Myrceugenia*, *Drimys*, and *Fagara*. The endemic palm species *Juania australis* forms an occasional emergent. At higher elevation, dwarf trees, shrubs (*Ugni*, *Pernettya*, and *Escallonia*), and tree ferns (*Blechnum* and *Dicksonia*) become dominant. Above the climatic tree line on Isla Alejandro Selkirk, the "alpine tundra" is distinguished by cushion plants (*Abrotanella*). In all these vegetation zones, invasive species have taken hold, mainly in the lowlands, resulting in more than 70% of endemic species being classified as threatened under IUCN threatened-species criteria.

Insects are also very diverse and have evolved extensive adaptations to island ecosystems. This is reflected in the high percentage of endemics (~70%), which is comparable to endemicity within the flowering plants of the Juan Fernandez Islands. The majority of these species are derived from southern Chile, with some Pacific and Indo-Malaysian elements present.

DISCOVERY AND EXPLORATION

The Juan Fernandez Islands were discovered in 1574, and a small colony of Spanish and South American Indians was established in 1591–1596. They introduced goats and pigs, cut firewood, grew vegetables, and caught and dried fish for the Spanish colonies in Chile. Short-lived settlement of the islands characterized much of the seventeenth and eighteenth centuries, which included groups of castaways or deserters inhabiting the islands for brief periods as British and Spanish interests competed for control. The most famous of these was marooned in 1704, when Alexander Selkirk took voluntary leave of Dampier's squadron and remained ashore for four and a half years—with his experiences later becoming the basis of Daniel Defoe's allegorical romance *Robinson Crusoe*. In 1749 the Spanish Viceroy ordered the formal colonization and protection of the Juan Fernandez Islands, with the construction of a substantial fort in Cumberland Bay and the arrival of 62 soldiers, 171 colonists, and 22 convicts, together with cows, sheep, mules, pigs, and poultry. By the 1790s, the settlement, essentially a penal colony, consisted of about 300 people, although this was abandoned by 1817. Permanent settlement began in 1877, and the islands were declared a Chilean national park in 1935; UNESCO declared the park a world biosphere reserve in 1977. As of 1976, the Chilean national park service, CONAF, has delivered administrative an environmental protection services to the islands.

CONSERVATION AND FUTURE CHALLENGES

Despite a relatively short history of permanent human settlement dating back only 130 years, the island flora and fauna are threatened by changes brought about by human activity including over-exploitation of forest and animal resources (e.g., indigenous sandalwood, *Santalum fernándezianum*, probably extinct since the beginning of the twentieth century, and earlier devastation of fur seal and elephant seal colonies—with 3 million skins being taken from Isla Alejandro Selkirk alone during 1797–1804), soil erosion, fire, introduced vascular plants (around 227

FIGURE 2 Comparison of landscape images in Cumberland Bay, Isla Robinson Crusoe by (A) George Anson, 1740; (B) Carl Skottsberg, 1918; (C) Simon Haberle, 2000. Notice the replacement of natural forest vegetation by degraded grassland and introduced tree species, including *Pinus* sp., *Eucalyptus* sp., *Aristotelia chilensis*, and *Ugni molinae*.

invasive species known in 1998), and introduced animals including cattle, goats, and the European rabbit. Descriptions of the islands by explorers such as George Anson dating to the mid-eighteenth century suggest that the lowland valleys were forested. Scientific exploration of the islands has been extensive with the work of the Swedish naturalist Carl Skottsberg in the early twentieth century providing a remarkable account of island environments at this time, suggesting that the vegetation became much more degraded after the mid-eighteenth century under the influence of increased human activity. Extensive botanical surveys and photographic comparisons spanning the twentieth century (Fig. 2) show that this process is ongoing and that the rapid invasion of exotic species will need to be halted if threatened and rare species are to survive to the end of the twenty-first century.

Since the islands were discovered in AD 1574 by the explorer Juan Fernandez, the conversion of natural vegetation into pastures, the occurrence of extensive fires, and the introduction of alien plant and animal species have had a profound impact on the composition and extent of natural biotic communities. Only one extinction has so far been observed (*Santalum fernandezianum*), but population sizes of many endemics have become small, with some having less than 25 known individuals left. Continued conservation efforts by CONAF, local residents, and international interests have the potential to ensure the ongoing preservation of this remarkable island environment.

SEE ALSO THE FOLLOWING ARTICLES

Deforestation / Exploration and Discovery / Insect Radiations / Invasion Biology / Pigs and Goats

FURTHER READING

Bernardello, G., G. J. Anderson, T. F. Stuessy, and D. J. Crawford. 2006. The angiosperm flora of the Archipelago Juan Fernandez (Chile): origin and dispersal. *Canadian Journal of Botany* 84: 1266–1281.

Castilla, J. C., ed. 1987. *Islas océanicas Chilenas: conociemento cientifico y necesidades de investigationes*. Santiago: Universidad Catolica de Chile.

Dirnböck, T., J. Greimler, P. Lopez, and T. F. Stuessy. 2003. Predicting future threats to the native vegetation of Robinson Crusoe Island, Juan Fernandez Archipelago, Chile. *Conservation Biology* 17: 1650–1659.

Haberle, S. G. 2003. Late Quaternary vegetation dynamics and human impact on Alexander Selkirk Island, Chile. *Journal of Biogeography* 30: 239–255.

Mueller-Dombois, D., and F. R. Fosberg. 1998. *Vegetation of the tropical Pacific Islands*. New York: Springer.

Skottsberg, C. 1953. The vegetation of the Juan Fernandez Islands, in *The natural history of Juan Fernandez and Easter Islands, vol. 2 (botany)*. C. Skottsberg, ed. Uppsala: Almquist & Wiksells, 793–960.

Woodward, R. L. 1969. *Robinson Crusoe's island: a history of the Juan Fernandez Islands*. Chapel Hill: University of North Carolina Press.

K

KARST ISLANDS
SEE MAKATEA ISLANDS

KERGUELEN
SEE INDIAN REGION

KERMADEC ISLANDS
SEE PACIFIC REGION

KICK 'EM JENNY

JAN LINDSAY
University of Auckland, New Zealand

Kick 'em Jenny is a submarine basaltic volcano located approximately 8 km north of Grenada in the eastern Caribbean (Fig. 1). It is the most frequently active volcano in the Lesser Antilles island arc and the only known submarine volcano in the region.

VOLCANO MORPHOLOGY AND GEOLOGY

Kick 'em Jenny is a conical-shaped volcano that rises 1300 m from the sea floor. It is asymmetric, as it abuts the Grenadines shelf to the east. It has a summit crater about 320 m in diameter, and the highest point on the crater rim is about 180 m below sea level. There is an actively degassing inner crater in the northwest of the essentially flat crater floor, which, at about 265 m below sea level, is the deepest point within the crater. Growth of a dome within the

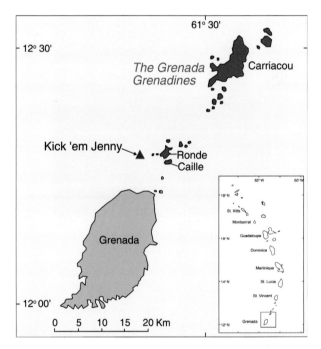

FIGURE 1 Location of Kick 'em Jenny in the Grenadines. Inset shows the islands of the Lesser Antilles.

crater between 1976 and 1978 brought the highest point on the volcano to within 160 m of the sea surface, but subsequent collapse during an eruption in 1988 destroyed the dome and breached the crater wall to the northeast.

The volcano lies within a large (5 km wide), horseshoe-shaped collapse scarp produced during a major sector collapse at some stage in Kick 'em Jenny's past. Debris avalanche deposits extending up to 30 km from the volcano into the Grenada Basin to the west represent the material carved off the volcano during this event (Fig. 2). The large volume of collapsed material (> 10 km^3) indicates that the summit of the ancestral Kick 'em Jenny volcano might have protruded above sea level before the collapse.

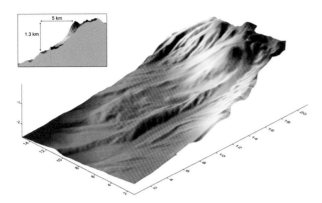

FIGURE 2 View of Kick 'em Jenny and debris avalanche deposits, with axes shown in km (approx. 2× vertical exaggeration). Inset shows side view at 5× vertical exaggeration.

Kick 'em Jenny eruptions are typically either nonexplosive (producing pillow lava) or explosive (producing tephra), depending on the rate and extent of magma degassing and magma–seawater interaction. The rocks are typically olivine basalt and basaltic andesite and contain unusually high abundances of large amphibole crystals that have been assimilated by the ascending magma from the surrounding crust.

The Kick 'em Jenny volcano lies on the western edge of a field of other shallow submarine volcanic features (Fig. 3), and nearby islands represent points where this bathymetric high rises above the sea surface. A sustained eruption from Kick 'em Jenny might cause the volcano to breach the sea surface and develop a small island, as was the case at Surtsey, Iceland, in 1963. With continuing eruptive activity over a long time period (thousands to millions of years), it may grow big enough to join with nearby islands to form one large island, thus illustrating the typical early stages of island growth in the Lesser Antilles arc.

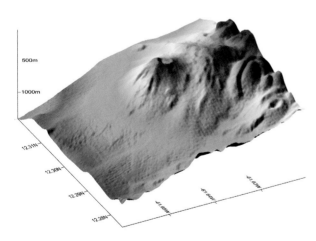

FIGURE 3 Seabeam image of Kick 'em Jenny and nearby submarine features (view from the southwest). Colors reflect depth and grade from red (shallow) to dark blue (deep).

HISTORICAL ERUPTIVE ACTIVITY AND HAZARDS

Throughout historical time Kick 'em Jenny has been the most frequently active volcano in the Lesser Antilles, erupting at least 12 times since 1939. The presence of the volcano was first revealed by a phreatomagmatic eruption in July 1939. The eruptions ejected ash-laden columns to heights of up to 300 m above the sea surface and generated water turbulence and numerous small earthquakes. Material was also ejected into the air during eruptions in 1974 and 1988; all other eruptions, including the most recent in December 2001, were entirely submarine.

Among the world's many submarine volcanoes, Kick 'em Jenny is one of the few that is in shallow water close to significant population centers. It lies directly beneath a major trade route frequented by local interisland traffic, large numbers of yachts, and cruise liners.

During eruptions, water disturbances and ejection of hot rocks pose a threat to boats in the vicinity of the crater. Continuous hydrothermal venting is occurring in the inner crater as well as on the southwestern flanks of the volcano, and periods of elevated magma degassing (both during and between eruptions) may reduce the water density (and thus the buoyancy of boats) above the volcano.

Although there has been significant and variable activity in the crater area of Kick 'em Jenny over the past 70 years, including at least one dome collapse event, to date there have been no confirmed accounts of any tsunami generated by this activity. At the current depth to the crater floor of about 265 m, hydrostatic pressures inhibit the occurrence of large explosive eruptions that could generate tsunamis, but this could change if the volcano grows closer to the surface. Collapse of the ancestral Kick 'em Jenny probably generated a large tsunami. Many similar large collapse scarps and deposits have been identified in other volcanic island settings around the world, including the Caribbean, and this appears to be a normal part of the evolution of these islands.

SEE ALSO THE FOLLOWING ARTICLES

Antilles, Geology / Hydrothermal Vents / Island Arcs / Tsunamis / Volcanic Islands

FURTHER READING

Devine, J. D., and H. Sigurdsson. 1995. Petrology and eruption styles at Kick'em Jenny submarine volcano, Lesser Antilles island arc. *Journal of Volcanology and Geothermal Research* 69: 35–58.

Koschinsky, A., R. Seifert, A. Knappe, K. Schmidt, and P. Halbach. 2007. Hydrothermal fluid emanations from the submarine Kick 'em Jenny volcano, Lesser Antilles island arc. *Marine Geology* 244: 129–141.

Lindsay, J. M., J. B. Shepherd, and D. Wilson. 2005. Volcanic and scientific activity at Kick 'em Jenny submarine volcano 2001–2002: implications for volcanic hazard in the southern Grenadines, Lesser Antilles. *Natural Hazards* 34: 1–24.

Lindsay, J. M., and J. B. Shepherd. 2005. Kick 'em Jenny and Ile de Caille, in *Volcanic Hazard Atlas of the Lesser Antilles*. J. M. Lindsay, R. Robertson, J. Shepherd, and S. Ali, eds. St. Augustine, Trinidad and Tobago: University of the West Indies, Seismic Research Unit, 108–126. .

Smith, M., and J. Shepherd. 1993. Preliminary investigations of the tsunami hazard of Kick 'em Jenny submarine volcano. *Natural Hazards* 7: 257–277.

Smith, M. and J. Shepherd. 1995. Potential Cauchy-Poisson waves generated by submarine eruptions of Kick 'em Jenny volcano. *Natural Hazards* 11: 75–94.

Watlington, R. A., W. D. Wilson, W. E. Johns, and C. Nelson. 2002. Updated bathymetric survey of Kick 'em Jenny submarine volcano. *Marine Geophysical Research* 23: 271–276.

KĪPUKA

AMY G. VANDERGAST
U.S. Geological Survey, San Diego

"Kīpuka" is one of several Hawaiian terms adopted by geologists to describe volcanic features, and is defined as a fragment of land surrounded by one or more younger lava flows (Fig. 1). Kīpuka are essentially habitat islands, and when present on islands themselves, kīpuka can be thought of as islands within islands. Kīpuka may play a unique role in shaping the ecology and evolutionary trajectories of species. Kīpuka undoubtedly act as refugia during flow events and provide source populations for colonists throughout ecosystem succession on new lava flows. As lava flows age, they are gradually recolonized by plant and animal communities until the distinction between kīpuka and surrounding lava flows becomes negligible. Thus, populations of plants and animals can become cyclically isolated and connected in this landscape, promoting population divergence and, potentially, speciation. Kīpuka systems also provide an ideal natural laboratory for examining invasions, fragmentation effects, metapopulation dynamics, and the state factors that regulate ecosystem development.

KĪPUKA AS REFUGIA

Kīpuka can remain ecologically distinct from surrounding lava flows for many decades (and even centuries) after isolation. For example, mature, closed-canopy forest kīpuka on the island of Hawai'i are more stable in temperature and humidity (generally cooler and wetter), supporting markedly different understory vegetation and invertebrate communities than surrounding 150-year-old flows. However, the main canopy tree, *Metrosideros polymorpha*, is one of the first colonizers of bare lava. As these trees mature and fill in, the closed-canopy forest regenerates, colonized from source populations in adjacent kīpuka. Kīpuka also provide refuges for native organisms from invasive species, as they seem to be less susceptible to invasions than surrounding lava flows. For example, invasive plant and spider species are largely restricted to younger surrounding lava flows in kīpuka systems on Hawai'i. This difference is presumably due to differences in niche availability, with younger and more open environments providing less competition for colonizing species.

KĪPUKA AS ACCELERANTS OF EVOLUTIONARY CHANGE

On volcanic archipelagoes, geologic processes play a central role in shaping patterns of biodiversity. On Mauna Loa and Kilauea volcanoes on Hawai'i, lava flows cover surfaces at rates of about 40–90% per 1000 years. This ongoing volcanic activity has created a shifting mosaic of habitats as large areas are destroyed or fragmented into kīpuka, and the new lava flows are subsequently recolonized. Populations of plant and animal species on the slopes of these volcanoes have been subject to repetitive extinctions, fragmentation, founder events, and population growth on time scales of hundreds to thousands of years. Based on studies of Hawaiian *Drosophila* flies, it has been hypothesized that a combination of metapopulation structure and founder effects can promote genetic differences among populations, sometimes leading to the evolution of new character states and species. Concordantly, recent work has shown that genetic differentiation increases among kīpuka in native *Tetragnatha* spiders. The potential importance of isolation of small populations due to lava flows has been invoked in explanations of diversification in several other Hawaiian lineages, as well as those from other systems

FIGURE 1 Kīpuka formed during the Pu'u 'O'o-Kupaianaha eruption on the east rift zone of Kilauea Volcano, Hawai'i. Photographed by J.D. Griggs on January 13, 1987, U.S. Geological Survey.

such as the Galápagos and Canary Islands and the Albertine Rift of central Africa.

KĪPUKA AS NATURAL LABORATORIES

Although kīpuka have been studied globally, ranging from sagebrush patches in the northwestern United States to tropical forest fragments in Africa, forested kīpuka systems in the Hawaiian Islands are understood particularly well. Kīpuka constitute a dominant landscape feature in these volcanic islands, with hundreds of these habitat fragments found on the Island of Hawai'i. The substrate ages of many of these kīpuka and their surrounding flows have been dated using historical records and geologic methods. The juxtaposition of older and younger lava substrates, ranging across gradients of elevation, temperature, soil type, and nutrient content, creates a unique system in which to study the state factors that regulate ecosystem development and function. Primary succession and ecosystem development can be studied across lava flows in a chronological sequence while other potentially confounding factors (such as altitude, substrate type, nutrient load, slope, and aspect) are held constant. Studies of vegetation communities in forest kīpuka have revealed that mature closed canopy forest can develop on new lava flows in as few as 300 or as many as 3000 years. These temporal differences depend on interactions among both biotic and abiotic factors (slope, aspect, altitude, lava type). Kīpuka systems can also be useful in investigations of the effects of habitat fragmentation over time and in comparisons of natural versus human-mediated fragmentation. Because habitat loss and fragmentation are major threats to global biodiversity, understanding the long term consequences of natural fragmentation processes may lead to a greater ability to protect biodiversity.

THREATS TO KĪPUKA SYSTEMS

Kīpuka on active volcanoes may not undergo the impact of human use and urbanization to the same degree as other, geologically more stable, environments. However, threats to these systems still exist. Threats from invasive species are prominent in nearly all island systems, even though kīpuka themselves may be more resistant to invasion than their surrounding younger lava flows. For example, invasions on Hawaiian lava flows by the nitrogen-fixing tree *Myrica faya* and the fountain grass *Pennisetum setaceum* drastically alter nutrient availability and fire regimes, decreasing the ability of native species to colonize these flows and disrupting the natural successional progression. Introduced ungulates are well established even in remote forest kīpuka on the island of Hawai'i. These animals physically alter the understory through feeding and rooting. Impacts to the forest understory in kīpuka may be severe, particularly in very small fragments, where large proportions of the understory can be affected. Perhaps even more importantly, introduced mammals transport seeds, pollen, and invertebrates as they move easily from patch to patch. These activities can homogenize once-distinct fragments, disrupting ecological and evolutionary processes associated with isolation. In some cases, active management, particularly control of non-native species, may be necessary to preserve the functionality of kīpuka systems.

SEE ALSO THE FOLLOWING ARTICLES

Fragmentation / Invasion Biology / Lava and Ash / Metapopulations / Refugia / Succession / Volcanic Islands

FURTHER READING

Carson, H. L., J. P. Lockwood, and E. M. Craddock. 1990. Extinction and recolonization of local populations on a growing shield volcano. *Proceedings of the National Academy of Sciences of the United States of America* 87: 7055–7057.

Vandergast, A. G., and R. G. Gillespie. 2004. Effects of natural forest fragmentation on a Hawaiian spider community. *Environmental Entomology* 33: 1296–1305.

Vandergast, A. G., R. G. Gillespie, and G. K. Roderick. 2004. Influence of volcanic activity on the population genetic structure of Hawaiian *Tetragnatha* spiders: fragmentation, rapid population growth and the potential for accelerated evolution. *Molecular Ecology* 13: 1729–1743.

Vitousek, P. M., G. H. Aplet, J. W. Raich, and J. P. Lockwood. 1995. Biological perspectives on Mauna Loa Volcano: a model system for ecological research, in *Mauna Loa revealed: structure, composition, history and hazards*. J. M. Rhodes and J. P. Lockwood, eds. Washington, DC: American Geophysical Union, 117–125.

KOMODO DRAGONS

TIM JESSOP

Zoos Victoria, Parkville, Australia

The Komodo dragon (*Varanus komodoensis*) is infamous for being the world's largest lizard, reaching a maximum length of 3 m and a maximum body mass of up to 87 kg (Fig. 1). As adults, this monitor lizard is capable of killing and consuming mammals, including water buffalo (*Bubalus bubalis*), Timor deer (*Cervus timorensis*), and wild pigs (*Sus scrofa*), which coexist on five rugged islands in eastern Indonesia. Monitor lizards are monophyletic and comprise three clades; the Komodo dragon is assigned to the Indo-Australian lineage. Its closest sister species, as inferred from a mitochondrial gene tree, is the Eastern Australian lace monitor (*Varanus*

FIGURE 1 Two adult male Komodo dragons engaged in combat during the winter mating period. Photograph by Achmad Ariefiandy.

varius). The Komodo dragon is estimated to have differentiated from Australasian ancestors ~4 million years ago, and historically, its range extended further east across to the island of Timor.

DIET AND DISTRIBUTION

Ungulate prey constitutes the majority of the diet of large Komodo dragons (i.e., > 15 kg in body mass), whereas smaller dragons consume small reptiles, birds, and rodents. Within Komodo National Park, where this species is best studied, Komodo dragons persist on four islands that vary in area from 10 km² up to 350 km². Across these islands, the availability and distribution of ungulate prey varies considerably, with deer being found on all islands but buffalo and pigs only on the two larger, and human-habited, islands. Differences in large-prey density and availability appear to be an important factor underpinning ecological and evolutionary processes influencing Komodo dragons.

PHENOTYPIC DIVERGENCE AMONG ISLANDS

Major differences in the morphology of the Komodo dragon appear to be associated with conspicuous island differences in large-prey availability. For instance, maximal body size among the four islands varies nearly fourfold, indicating that the Komodo dragon exhibits both dwarf and giant populations. Maximal body size is positively correlated with insular prey density, independent of genetic relatedness among lizard populations. Hence, even genetically similar island populations can exhibit large differences in body size associated with local prey density, suggesting considerable phenotypic plasticity in body size. Other ecological hypotheses that are implicated in morphological divergence among island populations, including character displacement due to competition or predation, are intuitively unlikely given that this lizard is the sole large predator persisting on these islands.

ISLAND POPULATION DEMOGRAPHY

Differences in island prey density appear to be associated with conspicuous differences in the population demography of Komodo dragons in Komodo National Park. For example, low densities of dwarf Komodo dragons persist on the two small islands, in contrast to high densities of large-bodied lizards occurring on the two big islands. Despite individuals exhibiting a much smaller body size, and in turn decreased energetic requirements, the two small-island dragon populations appear to be uncompensated by increased population density. Thus an inverse relationship between body size and population density, as predicted by the energetic equivalence rule, is not observed for Komodo dragons.

SPATIAL HABITS AND DISPERSAL

On emergence from their subterranean nests, hatchling Komodo dragons (~100 g) directly climb trees and exhibit an arboreal life-history stage. Limited natal dispersal is apparent, with juveniles moving slowly in a mostly linear direction, away from their nests. Arboreal living by juvenile Komodo dragons presumably reduces predation from larger terrestrial dragons, while enabling access to smaller arboreal prey. Once reaching several kilograms, juvenile lizards transition to predominantly terrestrial activity and develop home ranges of ~0.25 km². Spatial requirements increase considerably with body size, with the largest adult lizards occupying home ranges of ~7 km².

Long-distance dispersal of Komodo dragons within and among islands appears to be an ecologically rare event. Even within the large islands, the exchange of individuals (based on mark-recapture estimates and telemetry) between closely adjacent valleys is infrequent. Komodo dragons thus exhibit strong philopatry to their resident valleys (and to islands), a feature also generally supported by molecular data indicating significant genetic structure within, and among, island populations.

POPULATION GENETICS AND CONSERVATION

Neutral genetic estimates suggest varying population differentiation among extant populations. The populations of Rinca, Nusa Kode, and Western Flores are the most closely related, reflecting their proximity to one another and their higher rates of gene flow. In contrast, the more isolated populations are more differentiated. The Komodo island population exhibits the highest level of genetic divergence and allelic distinctiveness. In contrast, the small island of Gili Motang exhibits a low level of hetrozygosity, consistent with its small population size, suggesting that this island is most at risk from stochastic processes.

Demographic evidence from Komodo National Park suggests the two small island populations are declining in abundance, while the two large island populations appear relatively stable. The factors underpinning the declines on the two small islands are not clearly identified. Nevertheless, given the inherent vulnerability of island populations to extinction, and that Komodo dragons appear to exhibit limited dispersal among islands, ongoing and robust population monitoring is advocated to ensure that extant populations of Komodo dragons, both within and outside Komodo National Park, are conserved.

SEE ALSO THE FOLLOWING ARTICLES

Dispersal / Dwarfism / Gigantism / Indonesia, Biology / Island Rule / Lizard Radiations

FURTHER READING

Auffenberg, W. 1981. *The behavioral ecology of the Komodo monitor.* Gainesville: University Presses of Florida.

Ciofi, C., J. Puswati, D. Winana, M. E. De Boer, G. Chelazzi, and P. Sastrawan. 2007. Preliminary analysis of home range structure in the Komodo monitor, *Varanus komodoensis. Copeia* 2007: 462–470.

Imansyah, M. J., T. S. Jessop, C. Ciofi, and Z. Akbar. 2007. Ontogenetic differences in the spatial ecology of immature Komodo dragons. *Journal of Zoology* doi: 10.1111/j.1469-7998.2007.00368.x.

Jessop, T. S., T. Madsen, J. Sumner, H. Rudiharto, J. A. Phillips, and C. Ciofi. 2006. Maximum body size among insular Komodo dragon populations covaries with large prey density. *Oikos* 112: 422–429.

Jessop, T. S., T. Madsen, J. Sumner, H. Rudiharto, J. A. Phillips, and C. Ciofi. 2007. Differences in population size structure and body condition: conservation implications for Komodo dragons. *Biological Conservation* 135: 247–255.

Murphy, J. B., C. Ciofi, T. Walsh, and C. de la Panouse, 2002. *Komodo dragons: biology and conservation.* Washington DC: Smithsonian Institution Press.

KON-TIKI

ROBERT C. SUGGS

Boise, Idaho

In 1947, the Norwegian explorer Thor Heyerdahl made a drift voyage of 101 days from Peru to Polynesia on the balsa-log raft *Kon-Tiki.* Heyerdahl's stated purpose was to prove his theory that Polynesia was settled from South America by light-skinned followers of a foreign god.

WHO WAS "KON-TIKI?"

Heyerdahl named the *Kon-Tiki* after the god he identified as "Kon-Tiki Viracocha." This god was described as a bearded, white-skinned individual with reddish or blond hair and blue eyes who came to South America from across the Atlantic. Heyerdahl stated that this god remained in Peru for a time, before being driven out into the Pacific, where he was subsequently known to Polynesians as "Tiki" and worshipped as a sun god.

The name "Kon-Tiki" is a distorted rendering of the Quechua title for one of the many aspects of the sun god in Peru: *ápu qon téXsi wiraqúcha,* where the letter X represents a sound similar to that of *k* in *milk.* The word *teXsi* is therefore not a cognate of the Marquesan word *tiki.* The Polynesian Tiki was not a god but a Marquesan mythological character who created the first human by copulating with a heap of sand. No sun gods were present in the Polynesian pantheon; all Polynesian gods were deified ancestors.

The legends of Viracocha do not describe a migration from Peru into the Pacific by a white-skinned god and his followers. They describe the solar deity, with shining hair and beard, moving across the sky on his annual cyclical course between solstices. At the winter solstice, the sun, as seen from Peru, drops lower on the northwestern horizon, as though sinking into the Pacific. The supposed Caucasoid features of Viracocha appealed to Heyerdahl, who at one time was a correspondent of Hans Guenther, the author of Nazi racist anthropological texts that emphasized the superiority of the Nordic "race." The "red" beard and blue eyes of Viracocha are nowhere mentioned in Peruvian legends; these attributes were added by Heyerdahl to strengthen the Nordic connection.

THE *KON-TIKI* RAFT: CONSTRUCTION AND LOGISTICS

The *Kon-Tiki* raft was constructed from balsa logs. It generally resembled balsa log rafts seen by early European explorers of South America, such as Bartolomé Ruiz in 1526. There was never any serious question about the ability of balsa logs to survive long immersion in sea water: balsa rafts were a standard part of pre-European South American maritime technology. *Kon-Tiki* had a square-rigged sail, a large steering oar, and keel boards, *guaras,* to aid in tacking. A deckhouse sheltered Heyerdahl and his five companions, including a cameraman and a radioman.

The main food source was a large quantity of canned U.S. Army rations. Coconuts, bananas, and sweet potatoes were also taken aboard for symbolic value, but they spoiled within approximately three weeks. Primus stoves were used for cooking. The canned food was supplemented by 275 liters of water in cans. This quantity of water would have provided each man with approximately a half-liter per day, an amount insufficient for life under exposed conditions in subequatorial seas. Additional liquid was obtained from raw fish, from rain, or by mixing sea water with canned water, although solar stills supplemented the water supply. The essential modern life support supplies, including canned food and water, stove, lanterns, radio, electric generator, and solar water stills, invalidate the voyage as a test of pre-European Peruvian voyaging capability.

AT SEA

The *Kon-Tiki* was unable to cross the strong Humboldt Current, which flows northward along the Peruvian coast. The craft was therefore towed out to sea by the Peruvian Coast Guard and was released more than 50 miles from the coast. Once the raft was released from the tow, steering proved impossible. The raft drifted at the whim of winds and currents. Upon reaching the Tuamotu Archipelago, the *Kon-Tiki* passed near several atolls, finally washing ashore on the atoll of Raroia after 101 days adrift. The crew escaped without injury.

SIGNIFICANCE

Heyerdahl claimed that the voyage proved his theory according to which Caucasoid Peruvians first settled Polynesia and were later followed by an invasion of Native Americans from the Northwest Coast.

This single drift voyage provided no new facts or insights to challenge the data already available from the disciplines of archeology, ethnology, physical anthropology, and linguistics, all of which pointed to a Central-Western Pacific origin for the Polynesians. Additional archeological evidence against Heyerdahl's theory appeared in the decade following the *Kon-Tiki* voyage, and today, after 60 years of further archeological investigations in Polynesia, no evidence of South American contact has ever been discovered anywhere in Polynesia.

Although Peruvians never reached Polynesia, the Polynesians, descendants of an Asian maritime culture that originated on the coast of China and Taiwan approximately 4500 years BCE, are now known to have reached the New World in pre-Columbian times.

The *Kon-Tiki* raft voyage was a great financial and public relations success. Heyerdahl's book sold millions of copies, in many languages, and the film of the voyage won an Academy Award (Fig. 1). It also successfully proved that a group of physically fit personnel, on a crude balsa raft, with modern survival supplies and equipment, could survive an uncontrolled drift voyage from South America to Polynesia—once they were towed out beyond the Humboldt Current, which otherwise would have carried them up the South American coast to Panama.

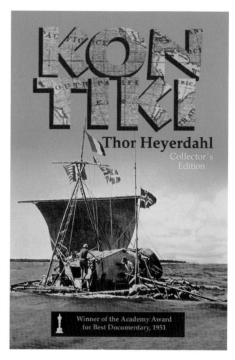

FIGURE 1 A photograph used in advertising the documentary film depicts the *Kon-Tiki* approaching Raroia, Tuamotu Archipelago. Image courtesy of Janson Media.

SEE ALSO THE FOLLOWING ARTICLES

Easter Island / Peopling the Pacific / Polynesian Voyaging / Popular Culture, Islands in

FURTHER READING

Bierbach, A., and H. Cain. 1996. *Religion and Language of Easter Island.* Baessler-Archiv, Neue Folge. Beiheft 9. Berlin: Dietrich Reimer Verlag.
Clark, R. 1990. Austronesian languages, in *The world's major languages.* B. Combe, ed. New York: Oxford University Press, 899–912.
Demarest, A. A. 1981. *Viracocha: the nature and antiquity of the Andean high god.* Peabody Museum Monographs, No. 6. Cambridge, MA: Harvard University.
Finney, B. 1994. *Voyage of rediscovery: a cultural odyssey through Polynesia.* Berkeley: University of California Press.
Heyerdahl, T. 1950. *Kon-Tiki: across the Pacific by raft.* New York: Rand McNally and Company.
Kirch, P. V. 2000. *On the road of the winds: an archeological history of the Pacific islands before European contact.* Berkeley: University of California Press.
Liping Jiang and Li Liu. 2005. The discovery of an 8000-year-old dugout canoe at Kuahuqiao in the Lower Yangzi River, China. *Antiquity* 79 (305).
Suggs, R. C. 1960. *Island civilizations of Polynesia.* New York: New American Library.
Storey, A. A., J. M. Ramirez, D. Quiroz, D. V. Burley, D. J. Addison, R. Walter, A. J. Anderson, T. L. Hunt, J. S. Athens, L. Huynen, and E. A. Matisoo-Smith. 2007. Radiocarbon and DNA evidence for a pre-Columbian introduction of Polynesian chickens to Chile. *Proceedings of the National Academy of Sciences of the United States of America* 104: 10335–10339.

KRAKATAU

ROBERT J. WHITTAKER
Oxford University, United Kingdom

The complete, or near-complete, sterilization of the Krakatau Islands in a devastating sequence of volcanic eruptions in 1883 provided a remarkable opportunity for natural scientists to monitor the processes and patterns of island recolonization and primary succession. Subsequent survey data, although intermittent in nature, extend to the present day and collectively (1) form a well-specified descriptive account of ecosystem development and (2) provide valuable opportunities for testing theories concerning the turnover and dynamics of insular systems. Most notably, Krakatau was the first case study system of colonization and turnover used by Robert H. MacArthur and Edward O. Wilson in evaluating their dynamic equilibrium model of island biogeography. Their findings, based on species data for the period 1883–1934, were equivocal, showing the apparent establishment of dynamic equilibrium for birds, but not for plants. Subsequent research suggested a slight upward drift in numbers of bird species and a continuing increase in plant species number and a variety of patterns for other taxa, consistent with strong successional structure in the patterns of recolonization. An overall dynamic equilibrium across different taxonomic and ecological groups had not been established by the start of the twenty-first century.

THE ENVIRONMENTAL CONTEXT AND HISTORY OF ERUPTIVE ACTIVITY

The Krakatau Islands (6° S, 105° E) are located in the Sunda Strait and are roughly equidistant between Java (approximately 40 km away) and Sumatra (30 km). The islands experience a tropical seasonal climate with a few dry months, classified as "Afa" in the Koeppen system. Because of their proximity to a point of lateral stress crossing a destructive plate margin, the Krakatau Islands have undergone repeated phases of volcanic activity. They experienced at least one caldera collapse event prior to 1883, when the group consisted of three islands, in order of diminishing area, Rakata (730 m maximum altitude, 17 km^2 area; formerly "Krakatau" or Pulau Rakata Besar), Sertung (180 m, 12 km^2; Verlaten Island), and Panjang (140 m, 3 km^2; Rakata Kecil, Lang Island).

Little is known about the islands prior to the events of 1883, although it is established that they were forest-covered and had been largely dormant from 1680 until May of 1883, when one of three volcanic cones (Perboewatan, Danan, and Rakata) on the largest island began a series of eruptions that ended on August 27, 1883, in exceptionally destructive eruptions. Two-thirds of that island, including the volcanoes Perboewatan and Danan and part of Rakata, were violently displaced, with vast quantities of ejecta thrown into the atmosphere, and pyroclastic surges crossing the sea to slam into mainland Sumatra. An estimated 36,000 people lost their lives in the coastal fringes of the Sunda Strait, mostly as a result of a series of tsunamis generated in the collapse. The advent of the telegraph shortly before the eruptions meant that news of the event traveled rapidly around the world, initiating an enduring scientific interest in the causes and consequences of the eruption, with the latter including pressure waves that encircled the world several times and a slight global cooling over the following 1–2 years.

ISLAND STERILIZATION

Although the main island (now known as Rakata) lost the majority of its land area, all three islands also gained extensive areas of new land resulting from the emplacement of pyroclastic deposits on to the preexisting foundations. These strata were some 60 to 80 m in thickness in the lowlands, and to this day the vast majority of the three islands remain mantled in these unconsolidated ashes, with little solid geology exposed at the surface. The islands can be taken to have been as near to completely sterilized in August 1883 as to make no practical difference, although this conclusion was bitterly contested by C. A. Backer in the early twentieth century. No evidence for any surviving plant or animal life was found by the scientific team led by the geologist R. D. M. Verbeek later in 1883, and in May 1884 the only life spotted by visiting scientists was a spider. The first signs of plant life, a "few blades of grass," were detected in September 1884.

ANAK KRAKATAU

The islands remained dormant until June 1927, when activity commenced once again from the sea bed in the center of the group. By August 12, 1931, a new island, Anak Krakatau (Child of Krakatau), had established a permanent presence, reaching nearly 50 m in height in little over a year. This island has remained highly active and now exceeds 300 m in elevation and 3.5 km^2 in area. As a result of its active volcanism, relatively little of this island has become vegetated, and Anak Krakatau has also caused repeated episodes of widespread accelerated mortality of trees within forests on Panjang and Sertung Islands but not, to date, on Rakata.

FAUNAL AND FLORISTIC SURVEYS

The fauna and flora of Krakatau have thus colonized since 1883 from an array of potential source areas, many of which were also impacted by the eruption; the closest of these, the island of Sebesi, is 12 km distant. The first botanical survey data are from 1886, with biological survey work being carried out intermittently since then (e.g., around 1897, 1905, 1919–1932, 1951, 1979, and 1983 onward). The 1886 survey recorded a few beach plants, and in the interiors it recorded mosses, blue-green algae, ferns, and a few higher plants (grasses, composites). Other life forms arrived quickly thereafter, and by 1897 Rakata supported young trees interspersed within tall, dense grasslands and an abundance of ferns. Since then, cumulative data from botanical surveys indicate a marked and rapid increase in the colonization of vascular plants on Krakatau, although few additional solely sea-dispersed plant species have colonized since 1930 (Fig. 1). By 1934, Rakata, Panjang, and Sertung islands collectively held nearly 300 plant species, and by 1983, between 423 and 456 species had been documented. The cumulative total of vascular plants now stands at approximately 540 species.

Colonization by animals was less well documented than that of the flora, but nonetheless, data that have been gathered provide valuable information for the analysis of trends in colonization, extinction, and turnover. The islands have been colonized by a wide variety of invertebrate species and at least 89 vertebrate species, including

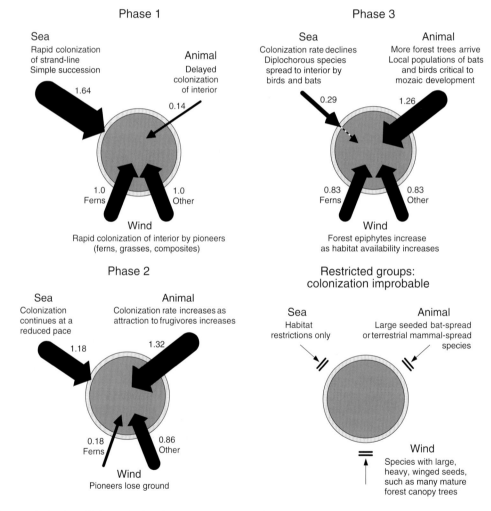

FIGURE 1 Plant recolonization of Rakata Island (Krakatau group) since sterilization in 1883. The three phases correspond with survey periods and represent convenient subdivisions of the successional process. Phase 1, 1883–1897; phase 2, 1898–1919; phase 3, 1920–1989 (subsequent survey data are not included in this analysis). Arrow widths are proportional to the increase in cumulative species number in species year[1] (these values are also given by each arrow). The flora is subdivided into the primary dispersal categories (i.e., the means by which each species is considered most likely to have colonized: animal-dispersed [zoochorous], wind-dispersed [anemochorous], and sea-dispersed [thalassochorous]). The model distinguishes between strandline (outer circle) and interior habitats, and in the fourth figure identifies constraints on further colonization. (From Whittaker and Jones, 1994b.)

54 birds, 11 microchiropterans (micro-bats), 8 pteropodid fruit bats, 11 reptiles, two snakes, two rats, and a pig (of uncertain origin and identity).

COLONIZATION, SUCCESSION, AND TURNOVER

Trends in colonization and loss of species have been analyzed by several authors within the framework of MacArthur and Wilson's equilibrium theory of island biogeography, largely to determine whether the data support a smooth (i.e., monotonic) progression of declining immigration and increasing extinction rates, approaching asymptotically the dynamic equilibrium condition predicted in their model. Based on data for the first 50 years, MacArthur and Wilson found support for this process in birds but not plants, which continued to increase steadily in number and showed no sign of approaching an asymptote. Subsequent analyses reported a further slight rise in bird numbers, a continued increase in plant species numbers, and a variety of patterns for other taxa (e.g., only two species of lizards have gone extinct, in both cases due to habitat loss, with no evidence of a dynamic equilibrium condition featuring continued turnover). Opinions have differed as to whether these various results and analyses can be incorporated within the framework of MacArthur and Wilson's theory: The findings clearly require at least some modification of their basic colonization model to recognize the successional structure inherent in the system (Fig. 1).

Summarizing a great deal of empirical detail, succession along the coastline was driven by the colonization of sea-dispersed plants, was rapid, and involved little compositional change and little turnover. The set of coastal plant species establishing early and on each island have undergone next to no species extinction, although some species (e.g., mangrove species) repeatedly reach the islands but fail to establish because of lack of habitat.

In the interiors, the first colonists were exclusively wind-dispersed species: Presumably the islands held no interest to passing frugivorous birds or bats at this stage. However, within the first decade of the recolonization commencing, a limited local source of fruit would have become available in the strand lines, as some of the early colonizing sea-dispersed species were diplochores, producing fruit that, in addition to their primary dispersal mode, provided a fruit resource for vertebrate frugivores. This inference is consistent with reports of the first restrictively animal-dispersed (zoochorous) plants being found in close association with the strandline vegetation. Some of these early colonizing strictly zoochorous plant species were fig (*Ficus*) species, subsequently shown to be bat dispersed. Indeed, fruit bats are now understood to have had a pivotal role in the early stages of forest development, introducing several ecologically important species and extending their distributions across the island interiors. However, frugivorous birds have introduced a much larger number of plant species, with fruit pigeons playing a particularly important role in the process.

The initially open fern-, herb-, and grass-dominated vegetation types (most species of which were wind-dispersed) gradually gave way to forest, in which most trees and shrubs were bat- or bird-spread. As the forest closed over during the 1920s, the rapid reduction in open habitat drove crashes in populations of open-habitat animal and plant species, leading to a measurable pulse of islandwide species extinctions. Coincidentally, the wide availability of new forest habitat was matched by the establishment of forest-specialist species, including many wind-dispersed epiphytic plants (e.g., ferns and orchids). The turnover pulse during forest closure is evidenced in the island colonization data for several taxa and is responsible, for example, for MacArthur and Wilson's premature claim for a dynamic equilibrium in bird species numbers.

ONGOING ACTIVITY

The emergence of Anak Krakatau around 1927–1931 has added a further dimension to the "natural experiment" of the Krakatau islands, both providing a further island for a rerun of recolonization and succession and impacting heavily on two of the other islands, Sertung and Panjang, which intermittently suffer significant disturbance from volcanic ejecta produced by Anak Krakatau. This is evidenced in deposits of between 1 and 2 m in depth of mostly fine-grained volcanic products across the great majority of each of these islands. To date, Rakata Island has not been directly affected in this way. Partly as a consequence, there are some differences in the succession of forest types evident across the different islands.

The Krakatau Islands remain of considerable interest to natural scientists. Geologists continue to debate the details of the 1883 events and to monitor Anak Krakatau's activity against the prospect of future significant hazard to the inhabitants of the Sunda Strait region. Recent biological work has included monitoring of forest dynamics, studies of bat–plant and fig wasp–fig mutualisms, and analyses of the genetic relationships of colonist plant species to populations from potential source areas.

In recognition of their special scientific interest, the islands enjoy protected status and have been largely uninhabited since 1883, although they are regularly visited by scientists, fishermen, tourists, illegal pumice gathering teams, and others.

SEE ALSO THE FOLLOWING ARTICLES

Eruptions: Laki and Tambora / Extinction / Island Biogeography, Theory of / Succession / Tsunamis

FURTHER READING

Compton, S. G., S. J. Ross, and I. W. B. Thornton. 1994. Pollinator limitation of fig tree reproduction on the island of Anak Krakatau (Indonesia). *Biotropica* 26: 180–186.

Shilton, L. A., J. D. Altringham, S. G. Compton, and R. J. Whittaker. 1999. Old World fruit bats can be long-distance seed dispersers through extended retention of viable seeds in the gut. *Proceedings of the Royal Society, London B* 266: 219–223.

Simkin, T., and R. S. Fiske, eds. 1983. *Krakatau 1883—the volcanic eruption and its effects.* Washington, DC: Smithsonian Institution Press.

Thornton, I. W. B. 1996. *Krakatau: the destruction and reassembly of an island ecosystem.* Cambridge, MA: Harvard University Press.

Whittaker, R. J., M. B. Bush, and K. Richards. 1989. Plant recolonization and vegetation succession on the Krakatau Islands, Indonesia. *Ecological Monographs* 59: 59–123.

Whittaker, R. J., R. Field, & T. Partomihardjo. 2000. How to go extinct: lessons from the lost plants of Krakatau. *Journal of Biogeography* 27: 1049–1064.

Whittaker, R. J., and S. H. Jones. 1994a. The role of frugivorous bats and birds in the rebuilding of a tropical forest ecosystem, Krakatau, Indonesia. *Journal of Biogeography* 21: 245–258.

Whittaker, R. J., and S. H. Jones. 1994b. Structure in re-building insular ecosystems: an empirically derived model. *Oikos* 69: 524–530.

Whittaker, R. J., S. H. Jones, and T. Partomihardjo. 1997. The re-building of an isolated rain forest assemblage: how disharmonic is the flora of Krakatau? *Biodiversity and Conservation* 6: 1671–1696.

KURILE ISLANDS

ALEXANDER BELOUSOV AND MARINA BELOUSOVA

Institute of Volcanology and Seismology, Petropavlovsk, Russia

THOMAS P. MILLER

U.S. Geological Survey, Anchorage, Alaska

The Kurile (or Kuril) Islands are one of the last blank spots on the world map, and their very remoteness results in a uniquely pristine environment. The biodiversity of the islands is remarkable, ranging from broad-leaved subtropical forests with magnolia, ligneous lianas, and Kurile bamboo in the south to subarctic moss tundra, alder shrubs, and stunted birches in the north. The landscapes are impressive, combining rocky capes, heavy fogs, surrealistic volcanic cones, boiling crater lakes, and almost impenetrable giant grasses. The Kurile Islands have often been compared to the nearby Aleutian Islands, and with good reason in terms of geology, remoteness, and notoriously bad weather. But the critical, though usually ignored, difference in orientation between the 1200-km-long, northeast-trending Kuriles and the 1800-km-long, east-west-trending Aleutians results in major differences in climate and accordingly in flora and fauna.

GEOGRAPHY

General Description

The Kurile Islands are located in the northwestern part of the Pacific Ocean, forming a 1200-km long island arc stretched over 8° of latitude from the Kamchatka Peninsula (Russia) southward to the island of Hokkaido (Japan) (Fig.1). This island arc separates the Sea of Okhotsk from the Pacific Ocean and represents an important geographical and geological boundary. The arc consists of 22 main islands and 30 smaller islets with a total area of 15,600 km^2. The largest islands—Iturup (3200 km^2), Paramushir (2000 km^2), Kunashir (1500 km^2), and Urup (1450 km^2)—are very narrow across the arc and extended along the arc, in contrast to the smaller islands, which tend to be oval or irregularly shaped.

The Kuriles are subdivided longitudinally into two approximately parallel island chains: the Greater Kuriles and the much shorter Lesser Kuriles, located in the southern part of the arc. The chains are separated by the 50-km-wide and 130-m-deep Southern Kurile Strait. The Lesser Kuriles consist of Shikotan Island and six small islands (the Habomai group). The Greater Kuriles include all of the remaining Kurile Islands, from Shumshu southward to Kunashir. Both island chains represent emerged summits of approximately parallel undersea ridges: the Greater Kurile Ridge, which connects to the Shiretoko Peninsula of Hokkaido, and the Lesser Kurile Ridge or Vityaz Ridge, which connects to the Nemuro Peninsula of Hokkaido. The oceanic slope of the Vityaz Ridge descends into the deep (10,542 m) Kurile-Kamchatka Trench, which lies along the entire length of the archipelago and represents the surface expression of subduction of the Pacific Plate under the Okhotsk Plate (formerly considered part of the North American Plate).

The island arc is subdivided transversely into three groups of islands separated by deep and wide straits. The Northern Kurile Islands (Shumshu to Shiashkotan) are separated from the Central Kurile Islands (Matua to Simushir) by the Kruzenstern Strait (1900 m deep and 80 km wide). The Central Kurile Islands are, in turn, separated from the Southern Kurile Islands (Chirpoy to Kunashir) by the Boussole Strait (2300 m deep and 67 km wide). Most of the Kurile Islands have mountainous relief punctuated by tall volcanoes, many of which are active. The highest volcanoes are Alaid (2339 m, Atlasova Island), Tyatya (1819 m, Kunashir Island), and

Chikurachki (1816 m, Paramushir Island). Tyatya is considered one of the most beautiful volcanic cones in the world (Fig. 2). Shumshu Island and islands of the Lesser Kuriles have no volcanoes and a low flat relief.

Although rivers and lakes are common on the larger Kurile Islands, several small islands have no sources of drinking water. Island rivers are commonly short and rapid (whitewater rivers) with many waterfalls. The 140-m high Ilia Murometz waterfall on Iturup is one of the highest in Russia. Many of the lakes are located in volcanic craters and calderas. The deepest (>264 m) and most beautiful is Kol'tsevoye (Circular) Lake (Fig. 3), which is located inside the Tao-Rusyr caldera at Onekotan Island. Some lakes and rivers located in hydrothermal areas have acid thermal waters, where special thermophilic microorganisms and algae flourish (e.g., Kipyasheye Lake on Kunashir, with a pH of 2.8, and Yur'eva river on Paramushir, with a pH of 1.6).

Climate

The Kuriles, in general, have a maritime monsoon climate influenced by sea currents (both cold and warm, Fig. 1) of the Pacific Ocean and the Sea of Okhotsk, as well as by air masses coming from eastern Asia or the Bering Sea region. The southern part of the Sea of Okhotsk is under the influence of the warm Soya sea current, whereas the cold Kamchatka current travels south along the Pacific coast. As a result, the climate of the western slopes of the largest southernmost islands, warmed by the Soya current and protected by high ridges from the cold Pacific, is close to subtropical. The climate of the eastern slopes is notably colder, resulting in strikingly different vegetation. The northernmost islands and the small islands of the Central Kuriles are surrounded by the cold sea and have a subarctic climate.

Precipitation is high throughout the year, from 700 to 1000 mm on the northern islands and 1000 to 1100 mm on the southern islands. Thus, the climate is rather humid, and the islands are almost continuously shrouded by cloud and fog; extended rain (drizzle) is common. In winter the precipitation occurs in the form of heavy snowfalls; snowstorms are frequent. By the end of winter, the Sea of Okhotsk is extensively choked by ice fields that can block the western coasts of the islands.

Population

Similar artifacts (pottery, stone tools, etc.) dated 14,000–11,000 years ago have been found in Japan, southern Kamchatka, and southern Alaska, indicating that in the past the Kurile Islands formed a migration route between Japan and Kamchatka that could have been involved in maritime migrations in and out of the North American continent.

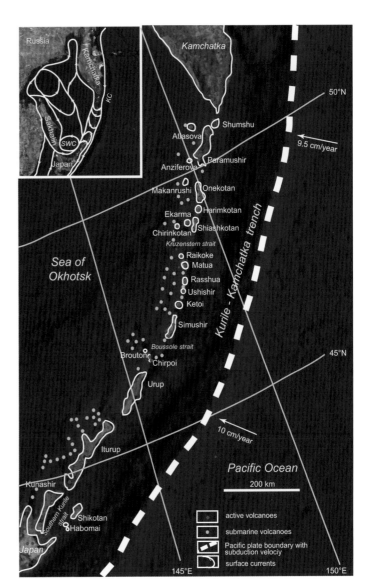

FIGURE 1 Schematic map showing the location of the Kurile Islands with inset showing their location in the Northwest Pacific. In the inset, the surface currents are shown with directions of circulation. SWC: Soya warm current; KC: Kamchatka Current.

Ancient settlements have been discovered on almost all the Kurile Islands. During the last 7000 years, the Kurile Islands were inhabited by several ethnically different groups of people that replaced one another. First known settlers were people of the Jomon (7000–2000 years ago), Epi-Jomon (2000–1300 years ago), and Okhotsk I and II (1300–700 years ago) cultures. Ainu people inhabited the islands after 700 years ago but were gradually displaced by Japanese and Russians in the eighteenth century.

Since the end of World War II in 1945, all the islands politically belong to the Sakhalinskaya oblast' (Sakhalin District) of Russia, although Japan claims the southernmost islands (Iturup, Kunashir, and the Lesser Kuriles).

FIGURE 2 Tyatya volcano, Kunashir Island. View from the southwest from the Pacific coast. Photograph by A. Belousov.

Only the largest islands (Paramushir, Iturup, Kunashir, and Shikotan) are presently inhabited, and the population of about 20,000, mostly Russian fishermen and coast guard personnel, is concentrated in several small towns.

GEOLOGY

Tectonic Setting

The Kurile Islands were formed by geological processes associated with subduction of the Pacific Plate under the Okhotsk Plate. The rate of subduction of the Pacific Plate is estimated at 95 mm/yr in the north (where the plate motion is normal to the Kurile Trench) and 100 mm/yr south of Boussole Strait (where the arc makes a sharp 22–23° turn to the west and plate subduction becomes oblique). Boussole Strait is considered to be a graben formed by northeast-southwest tension, caused by westward motion of the southern part of the arc due to oblique subduction of the Pacific plate since the Late Miocene (6–7 million years ago). The dip angle of the subduction plane is 48–55° in the northern part and 38–46° in the southern part of the arc. Crustal thickness is 25–36 km below the Northern Kuriles, 26–32 km below the Central Kuriles, and 25–44 km below the Southern Kuriles.

Overall, the Kurile Arc is seismically and volcanically much more active than the Izu-Bonin–Mariana and Ryukyu Arcs to the south, but less active than Kamchatka to the north. Six earthquakes with magnitude >8 were recorded in the twentieth century. Seismicity in the subducting slab occurs to the depth up to 650 km. The most intense seismicity is recorded in the southern sector of the Kurile Arc.

Some earthquakes and volcanic eruptions have generated tsunamis. The most deadly historic tsunami, up to 20 m high, occurred in 1952 in the northern part of the arc, when the town Severo-Kurilsk (Paramushir Island) along with multiple fishing settlements of the Pacific coast of the islands were demolished, with an estimated death toll of 5000. About 50 small to moderate-scale tsunamis have been recorded since 1952, and studies of paleotsunami deposits revealed multiple strong tsunamis throughout the Holocene in the Kuriles. Great earthquakes on November 16, 2006, and January 13, 2007, with magnitudes of 8.3 and 8.1, generated tsunamis more than 20 m high that affected unpopulated shores of the Central Kuriles.

At the present time, the Kurile Islands experience slow ground deformation between major local earthquakes and more rapid deformation (commonly in a reverse direction) during earthquakes. Long-term tide gauge data show that the west coast of Shikotan Island was uplifting at a rate of 12.6 mm/yr until the October 5, 1994, earthquake, when it experienced a 50-cm drop. Recent GPS measurements have shown horizontal motion of the south of Urup Island, with a rate of 18 mm/year in a direc-

FIGURE 3 Kol'tsevoye lake located in Tao-Rusyr caldera. Since the caldera formed 7500 years ago, Krenitzin Peak stratovolcano has formed in the central part of the caldera. The 1952 crater and the lava dome are visible on the slope and at the foot of the volcano. Photograph by A. Belousov.

tion coinciding with the direction of subduction. During the period of the winter 2006–2007 earthquakes, Ketoy Island experienced horizontal motion exceeding 60 cm in a direction opposite to the subduction direction.

Stratigraphy

The Kuriles are built of predominantly volcanic rocks (both volcaniclastic and effusive), and chemical and biochemical sedimentary rocks are rare. The geology of the Lesser and Greater Kuriles is notably different. The Lesser Kuriles are built of Late Cretaceous–Paleogene mafic volcaniclastic rocks intercalated with basalt and basaltic andesite lava flows. The lower part of the sequence (K/Ar ages 105–62.5 million years ago) was deposited in submarine conditions, while the upper (K/Ar ages 61–59 million years ago) formed subaerial shield volcanoes. No Neogene or younger rocks occur in the Lesser Kuriles.

The Greater Kuriles are built of a much wider spectrum of volcanic rocks of Late Miocene age and younger (<12 million years ago). Rock compositions range from basalt to rhyolite (basalts 20%; andesite 64%; dacite 13%, and rhyolite 3%), and the rocks were formed by diverse volcanic processes common for volcanic island arcs. The oldest rock sequence is commonly exposed along the lowermost part of an island sea cliff. It is represented by sub-aquatic tuff and breccia that formed when the islands emerged from the ocean. In the Northern and Central Kuriles, the sequence consists of predominantly basic-composition rocks (commonly palagonitized). In the Southern Kuriles, the sequence is represented by more evolved rocks. The uppermost, predominantly andesitic sequence comprises the volcanic formation processes of Quaternary age.

The Kurile Islands have a blocky structure: geologic formations are disrupted by multiple faults, forming horsts and grabens. Folds are rare and mostly associated with large faults. Intrusive bodies of various types and compositions are widespread.

Geologic History

Formation of the Kurile Archipelago began in the Late Cretaceous (approximately 100 million years ago) when subduction along the Siberian continent was blocked by a large terrain. A new subduction zone appeared in the ocean far southeast from the previous position, where the Kurile-Kamchatka Trench was formed and volcanism started above the subducting slab. The Lesser Kurile Ridge (Vityaz Ridge) was formed during Late Cretaceous. Volcanic activity and uplift in the region intensified during the Paleocene and Eocene (35–60 million years ago), when the islands of the Lesser Kuriles emerged from the sea. Then a Late Eocene–Middle Miocene volcanic hiatus followed.

Late in the Miocene (approximately 12 million years ago), volcanic activity resumed along the Kurile-Kamchatka Trench but was concentrated in a zone parallel to the Vityaz Ridge in the location of what is now the Greater Kurile Ridge. During the past 10 million years, the Greater Kurile Ridge has experienced intense volcanic activity and crustal uplift. Although the oldest rocks in the Greater Kurile Ridge are of Late Miocene age, sediment records indicate that main chain of the Kurile Islands probably did not emerge above the sea surface until the early Pliocene (5 million years ago). Some small islands, such as Atlasova, have been formed by volcanic activity in the Holocene (less than 10,000 years ago). The subduction process that led to the formation of the Kurile Archipelago is still in progress, and the elevation of islands continues to change with new eruptions and crustal movements.

Volcanoes

The Kurile Islands, together with the nearby Kamchatka Peninsula and Hokkaido Island, compose a single volcanic arc, a part of the so-called Pacific Ring of Fire. Because of a high subduction rate, the Kuriles are among the more active volcanic areas in the world. In the islands of the Greater Kuriles, 68 subaerial volcanic centers with a total of 200 Quaternary (less than 2 million years old) volcanoes were identified, and 32 volcanoes have been active in historic time. The list is not complete, since only fragmentary records of volcanic activity exist for the area beginning in 1711, and until the early twentieth century many eruptions, especially small ones, passed undocumented. During the historical period, 17 strong eruptions, with volcanic explosivity index (VEI) greater than 3, were reported. The most active volcanoes (with more than five recorded eruptions) are Alaid, Ebeko, Chikurachki, Severgin, Sarychev Peak, and Goryaschaya Sopka. The highest volcanic activity is observed north of the Boussole Strait, where the plate motion is normal to the Kurile Trench.

In addition to subaerial volcanoes, 96 submarine volcanic edifices were discovered in the course of bathymetric and geophysical surveys. They are situated mostly at the Sea of Okhotsk slope of the Greater Kurile Ridge. Depth to the subducting plate is 110–140 km under land volcanoes and 160–220 km under submarine volcanoes (Fig. 1).

Volcanism of the Kuriles has features typical of convergent plate boundaries of the Mariana type. Volcanic products are characterized by evolved compositions (andesite to rhyolite), whereas basalt is relatively rare. Volcanic eruptions of the Kurile arc are highly explosive and are commonly accompanied by pyroclastic flows and surges that, together with laharic deposits, accumulate in large quantities around volcanoes. Effusive eruptions of evolved, viscous lava, which commonly follow the explosive stage, produce steep-sided lava domes or thick lava flows with blocky surfaces.

Many volcanoes demonstrate complex eruptive history with formation of one or more collapse calderas. During the last 45,000 years, more than 15 such calderas with diameters up to 10 km were formed. Many of the calderas are partly submerged, such as 7 × 9 km Lvinaya Past (Lion's Jaw) caldera, formed 9400 years ago, and five underwater calderas are known in the arc.

The caldera-forming eruptions produced extensive sheets of ignimbrites (both welded and nonwelded), mostly of dacitic to rhyolitic composition. The youngest of the large calderas of the Kurile Islands, the 7-km-wide Tao-Rusyr caldera at Onekotan Island, was formed 7500 years ago (Fig. 3). Its formation was accompanied by deposition of extensive sheets of nonwelded ignimbrites of andesitic composition. The most recent (<2500 years ago) Zavaritsky caldera at Simushir has a diameter of 3 km.

Active volcanism in combination with a high precipitation rate results in strong development of hydrothermal activity in the form of solfataras and hot springs. Steam is extracted from wells at Kunashir and Iturup Islands to produce electricity in geothermal power plants. Wide areas of colorful hydrothermally altered rocks occupy summit areas of most volcanoes of the islands. Alteration weakens volcanic structures and, in combination with triggering effects of eruptions and earthquakes, leads to frequent large-scale flank collapses, the volumes of which reach several cubic kilometers. The collapsed mass transforms into fast-moving debris avalanches that generate tsunamis upon entering into the sea. Collapses also result in formation of broad horseshoe-shaped craters. The most recent collapse (volume 0.4 km²) associated with strong explosive eruption occurred at Severgin volcano (Harimkotan Island) in 1933, causing a tsunami up to 20 m high that killed two people on nearby Onekotan Island.

The long, complex history of many volcanoes of the Kurile Islands resulted in an equally complex morphology of their edifices. Steep-sided composite cones, frequently complicated by horseshoe-shaped craters and summit and/or flank lava domes, represent dominant type of volcanic structures in the Kuriles. In many cases, such cones are merged into volcanic ridges. Large calderas partly filled with younger cones are also common. Less common are symmetric stratovolcano cones, which belong to the youngest and most active volcanoes erupting basic magmas.

Minor gold deposits in hydrothermally altered volcanic rocks have been discovered at Kunashir and Iturup Islands. Sulfur was mined from many volcanic craters during Japanese times, but these deposits are now considered economically insignificant.

Sea Level and Glaciations

The presence of four sea terraces on the islands is evidence of sea transgressions in the past. Sea terraces with the following elevations and ages were distinguished: 3–7 m (climatic optimum of Holocene, ca. 6000 years ago); 20–40 m (Riss-Wurm interglacial, Late Pleistocene 67,000–128,000 years ago); 80–120 m (Middle Pleistocene 180,000–230,000 years ago); 200–250 m (Early Pleistocene 300,000–330,000 years ago).

During the Late Pleistocene (10,000–30,000 years ago), at least two significant sea-level regressions occurred, caused by global glaciations. Sea level dropped on the order of 100 m and possibly 200 to 300 m below the present elevations. The last of these major sea-level regressions occurred 18,000 yBP, when a broad underwater terrace with depths 120–140 m formed around many of the islands. During these periods, shallow straits separating Kunashir and the Lesser Kurile islands from Hokkaido became dry land. Similarly, the Shumshu and Paramushir islands became connected to the Kamchatka Peninsula. Thus, islands of the northern and southern tips of the archipelago have not been isolated from the mainland for a long time.

During the Pleistocene glaciations, glaciers covered the northern and central islands of the Kuriles. Glaciations probably did not extend any further south than central Iturup Island, as there are no signs of glaciation on Kunashir or on the Lesser Kurile islands.

BIOTIC LIFE

The flora and fauna of the Kuriles are not completely studied. Taxonomic diversity of biota of Kurile Islands is comparatively high; for example, among vascular plants 1194 species, 4550 genera, and 135 families were described. Endemic species compose less than 2% of all vascular plants; most of them are so-called neo-endemics (the differences from species of nearby lands are not significant, and not all botanists agree that they represent separate species). Among 300 known species of birds (either living on the islands or migrating), only one species (*Cepphus columba snowi*) is confirmed as endemic, indicating no prolonged isolation of the islands.

Flora

The flora of the Kurile Islands changes notably from north to south as well as with altitude. The main botanical boundary passes through the Boussole Strait between the islands Simushir and Chirpoy. The flora of the northern Kuriles is similar to that of the Kamchatka Peninsula: cedar and alder shrubs surrounded by tundra and meadow vegetation are widespread. From north to south along the arc the vegetation becomes more luxuriant and taller as the climate becomes milder. In the middle part of the Kurile Island Arc (Rasshua to north of Iturup), small-leaved forests (birch, alder, and poplar) dominate. Flora of two southern islands (Kunashir and southern Iturup) is similar to that of Hokkaido and Sakhalin Island: broad-leaved and coniferous forests are widespread, and dense shrubs (less than 3 m tall) of Kurile bamboo (*Sasa kurilensis*) are common (Ketoy Island is the northernmost limit of the bamboo distribution).

The total area covered by forest is estimated at about 80% at Iturup, 60% at Kunashir, and 20% at Shikotan. Some islands (Raikoke, Ushishir, and Brouton) have no forest at all. Grass vegetation is very rich on all the islands, and grass commonly grows unusually tall (up to 3 m). This phenomenon is so pronounced that it has a special name: "Far East gigantism." Dense vegetation of the islands, mountainous relief, and absence of roads make the internal parts of many islands very difficult to access by foot.

Flora of the Kurile Islands exhibits a clear vertical zoning because of the presence of high mountains. In the southern Kuriles (Iturup and Kunashir Islands), the belts of vegetation include (from sea level upward): (1) broad-leaved forests (oak, white elm, maple); (2) dark coniferous forests; (3) birch forests (*Betula ermanii*); (4) cedar and alder shrubs; (5) alpine tundra. At Shikotan an additional belt of juniper shrubs could be distinguished.

Altitude ranges of the belts depend on many factors (e.g., slope exposition, influence of oceanic currents and volcanic activity), and thus the belts are not continuous. From south to north along the archipelago, the lowermost belts gradually disappear, and on Paramushir Island only belts 4 and 5 are present.

Fauna

In total, terrestrial mammals on the Kuriles include six species of chiropterans (bats), nine species of rodents (squirrels, hamsters, and different kinds of rats and mice), nine species of carnivores (foxes, bears, ermines, weasels, minks, and sables), species of insectivores (shrews), and one lagomorph (hare). The largest mammal in the Kuriles is the brown bear, with a total population of about 700. Resident populations of brown bears exist on the largest islands (Kunashir, Iturup, and Paramushir); smaller islands do not support a resident bear population. The most widely distributed species of terrestrial mammals are red and blue arctic foxes, which were introduced on many islands by the Russian-American Trade Company in the nineteenth century. Similarly, American and European mink were introduced by the Japanese in the twentieth century.

Abundant marine mammals live in waters of the Sea of Okhotsk and the Pacific Ocean around the islands: 15 kinds of whales, including blue whales, white whales, sperm whales, and various dolphins. Steller's sea lion, sea otters, spotted seals, and harbor seals are common on rocky shores.

More than 300 species of birds are known in the Kuriles. Cormorants, seagulls, and diving-pigeons form more than 20 giant rookeries.

Seven species of amphibians and reptiles have been discovered in the Kuriles: one species of salamander, one of skink, three of snakes (all of them found only at Kunashir) and two species of frogs (at Kunashir, Shikotan, and at small southern islets). More than 3000 species of insects have been recorded on the islands, but the true number probably is much larger.

Convergence of warm and cold sea currents has made the Kuriles one of the richest fishing zones at the world. Fish from colder water include cod, mackerel, flounder, halibut, and five species of salmon. Subtropical species include saury, sardines, and tuna. Eighteen species of freshwater fishes occur on the islands. Other species, such as crab, shrimp, sea urchin, squid, scallops, and sea cucumbers, are also abundant.

The most significant biogeographical boundary within the Kurile Islands is the Boussole Strait. Of lesser importance are two other straits: the De Vries Strait (between Iturup and Urup Islands) and the Fouth Kurile Strait (between Onekotan and Paramushir Islands).

SEE ALSO THE FOLLOWING ARTICLES

Archaeology / Island Arcs / Japan's Islands, Geology / Pacific Region / Tsunamis

FURTHER READING

Avdeiko, G. P., O. N. Volynets, A. Yu. Antonov, and A. A. Tsvetkov. 1991. Kurile island-arc volcanism: structural and petrological aspects. *Tectonophysics* 199: 271–287.
Gorshkov, G. S. 1958. Kurile Islands, in *Catalog of active volcanoes of the world*. Vol. 7. Rome: IAVCEI, 1–99.
Gorshkov, G. S. 1970. *Volcanism and the upper mantle: investigations in the Kurile island arc*. New York and London: Plenum Press.
International Kuril Island Project (IKIP). http://artedi.fish.washington.edu/okhotskia/ikip/, http://depts.washington.edu/ikip/.
Pietsch, T. W., V. V. Bogatov, K. Amaoka, et al. Biodiversity and biogeography of the islands of the Kuril Archipelago. *Journal of Biogeography* 30: 1297–1310.
[*Vegetation and animals of Kurile Islands* (materials of Kuril Island Biocomplexity Project)]. 2002. Vladivostok: Dal'nauka (in Russian).

LAKES, AS ISLANDS

SHELLEY ARNOTT

Queen's University, Kingston, Canada

Lakes are like islands; they are islands of water surrounded by a sea of land (Fig. 1). They have discrete boundaries (i.e., the water–land interface), yet there is some flow of materials, energy, and organisms from one system to the other. For example, whale carcasses occasionally wash ashore on islands and provide food energy for terrestrial-based organisms. Similarly, insects, mammals, and birds fall into lakes and are eaten by aquatic organisms. As on islands, organisms residing in lakes are spatially separated from other lakes by an inhospitable matrix. Organisms do not readily move from island to island or lake to lake, although some dispersal certainly occurs. Isolation and the restriction of movement of organisms can influence the number and kinds of species found on islands and in lakes.

THE VALUE OF CONSIDERING LAKES AS ISLANDS

Over the past few decades, scientists have been investigating how the spatial arrangement of islands, lakes, and other habitat patches influences the movement of individuals among islands (and island-like habitats), and therefore the number and kinds of species that inhabit particular islands. The role of regional-level processes, such as the movement of organisms among islands and lakes, has been investigated both mathematically and experimentally. On islands and in lakes, local environmental conditions (e.g., nutrient levels, predation, and competition) play an

FIGURE 1 An aerial photo of the many ponds in Old Crow Flats in the Yukon Territory, Canada. Ponds are islands of water in a matrix of land that is uninhabitable by pond residents. Ponds have varying degrees of connectedness. Some are physically isolated whereas others are connected by streams. Photograph by Jon Sweetman, Parks Canada.

important role in determining community composition, but ecologists and evolutionists are becoming increasingly aware that the movement of organisms between habitats (and therefore the spatial arrangement of islands or lakes) can also influence the number and kinds of species present. The number of species present on an island or in a lake results both from colonization of species from other locations and from speciation processes such as adaptive radiation. Both of these processes are influenced by the movement of individuals; immigration increases colonization by new species but homogenizes genes and therefore reduces speciation.

PATTERNS AND PROCESSES ON ISLANDS AND IN LAKES

Some of the theories developed for predicting species richness and the maintenance of populations on islands have relevance for community dynamics in lakes. Some of the same processes that drive community composition on islands may also operate in lakes.

Island Biogeography

The theory of island biogeography, developed by MacArthur and Wilson in 1967, suggests that species composition on islands is dynamic through time, being influenced both by the arrival of new species (i.e., dispersal from a mainland source) and by local extinction. Extinction and immigration rates are influenced by the characteristics of the island and ultimately determine the equilibrium number of species present on the island.

Island size influences the extinction rates of species. Large islands can maintain more species than can small islands because of a greater number of habitat types and because extinction rates are lower because of higher population sizes. Island size may also influence immigration rates. Larger islands are bigger targets for organisms that disperse by wind, water currents, or animal vectors—and therefore large islands may have higher immigration rates than small islands do. Isolation also influences the number of species on an island. Islands that are far from the mainland (i.e., a source of potential colonists) have a lower probability of receiving a dispersing individual than do islands that are close to the mainland. Because of these relationships, small, isolated islands are expected to have fewer species than are large islands located near the mainland. These relationships have been well documented in lakes. Large lakes have more species of phytoplankton, zooplankton, snails, molluscs, fish, and so forth than do small lakes. Isolation-related factors, such as the number of stream connections, the size of the nearest neighboring lake, and the distance to the nearest road, also influence the number of species in a given lake.

Species-Area Relationships

One of the predictions of the theory of island biogeography is that the number of species increases with island size. This has been documented for numerous organisms for islands, island-like habitats on the mainland, and lakes. A recent meta-analysis of 794 species–area relationships (SAR) has revealed that SAR for mainland habitat patches is indistinguishable from SAR from islands. Surprisingly, lakes tend to have lower slopes for SAR than does the more contiguous ocean, suggesting that habitat isolation is not a primary factor driving SAR. Despite this, the slope of the SAR was related to species body size, with larger organisms having steeper slopes. This may be related to dispersal ability or relative isolation of habitats as larger species tend to be poorer dispersers than are smaller-bodied species. However, other factors such as sampling method and spatial distribution patterns of small and large organisms may also influence SAR slopes for both islands and lakes.

Metapopulations and Metacommunities

The idea that island habitats have dynamic species composition resulting from local extinctions and colonization from distant sources has been extended to metapopulation or metacommunity theory. Metapopulation theory suggests that populations can be maintained within a region by dispersal of individuals between habitat patches (or islands), despite occasional local extinctions resulting from stochastic demographic changes. In addition to being used to understand population and community dynamics in terrestrial systems, metapopulation and metacommunity theory has been used to study lake regions, particularly in areas where there are high densities of lakes with high potential for overland dispersal or dispersal along waterways (e.g., stream connections).

Ecologists have recognized four paradigms for metacommunities (the patch-dynamic view, species-sorting, mass effects, and the neutral view; Fig. 2), resulting from differences in competitive and dispersal ability of species within a community. The view that applies to a particular region of islands (or lakes) will depend on the habitat characteristics of the islands, the competitive abilities of the species, the dispersal ability of the species, and the connectedness of the island habitats.

Evolutionary Processes

Speciation is an important source of diversity in isolated habitats such as islands and lakes. Adaptive radiation is a mechanism responsible for the evolution of ecological diversity. It results from the differentiation of a single ancestor to multiple species that inhabit a variety of habitats. Each species has different traits that allow it to exploit those habitats. The adaptive radiation of species into different ecomorphs (i.e., the evolution of ecological diversity) is influenced by the degree of isolation of island or lake habitats. Speciation is higher on large islands because the opportunity for geographic isolation increases with area and the diversity of habitats probably increases with island area. Remote islands have higher speciation rates than do islands close to colonist sources. High immigration rates associated with more connected islands tend to homogenize genes and therefore prevent divergence among ecomorphs.

THE RELEVANCE OF ISLAND-BASED THEORIES FOR LAKES

For island-based theories to apply to lakes, there are several assumptions that must be met. Of primary importance is the ability of some of the organisms residing in lakes to be able to disperse from source lakes to other lakes, even

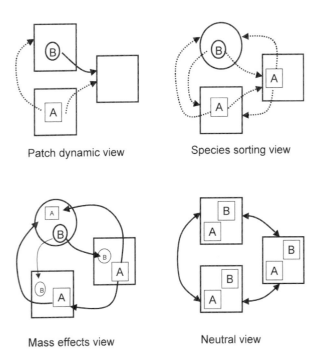

FIGURE 2 Illustration of the four metacommunity paradigms proposed by Leibold et al., 2004. Two competing species with populations A and B are considered. The competitive ability of each species is indicated by matching the shape of the species symbol (square or circle) with the shape of the habitat (square or circle). Species with a square symbol are competitive dominants when in a square habitat. Arrows indicate dispersal. Species with solid arrows are better dispersers than are species with dashed arrows. In the patch dynamic view, species A is a better competitor in all sites, but species B is able to persist because it is a better disperser and can more readily fill vacant niches. In the species sorting view, there is variation in habitat characteristic, and species are separated into spatial niches. Dispersal is low. In the mass effects view, dispersal is high. Populations of species not well suited for a particular habitat can be maintained through high dispersal in source habitats. Populations of species that are well suited to their habitat are sinks. In the neutral view, all species have equal dispersal abilities, and all habitat patches are equal in quality.

if at a low probability. Community composition must be dynamic: That is, extinction and immigration of species occurs. Dispersal of organisms, in addition to local environmental conditions (which contribute to extinction of species), must contribute to the composition of communities (i.e., the number and kinds of species) found in lakes. Dispersal will influence not only the colonization of species to new environments but also speciation or adaptive radiation on individual islands.

Dispersal

Dispersal of freshwater organisms among lakes, and of terrestrial organisms among islands, is hard to study, but seems important. The movement of organisms between lakes and islands is notoriously difficult to quantify, and much of the evidence is therefore based on indirect methods of assessment, such as collecting animals in traps, finding organisms attached externally to animals, and measuring viability of organisms that have passed through animal guts. Some studies have used observation of organisms in newly formed ponds or islands recently denuded of species to assess dispersal, although this method examines successful colonization rather than dispersal. Quantification of dispersal through stream connections or water currents is possible, but this method ignores the contribution of overland dispersal. The use of markers has been limited because marked individuals are difficult to find again, and it is uncertain how marking influences dispersal behavior. Analysis of genetic markers such as microsatellites and phylogenetic analyses are promising tools that have been used increasingly over the past decade. Despite limitations associated with estimating dispersal, multiple lines of evidence suggest that aquatic organisms do readily disperse, and for many taxa, community structure is influenced by dispersal of organisms from other lakes and islands.

MODES OF DISPERSAL OF AQUATIC ORGANISMS

There is a range in dispersal modes and abilities of aquatic organisms, depending on their size, their life history strategies, and their interaction with humans and other dispersal vectors. Larger organisms, such as fish, probably move between lakes through stream connections, flooding events that temporarily connect lakes, and tornadoes or waterspouts that lift fish from one lake and drop them into other lakes along their path. Over the past century, humans have probably become the most influential vectors associated with fish dispersal, purposefully stocking desired sport fish into lakes and accidentally releasing bait fish into popular fishing locations.

Zooplankton, bryozoans, and phytoplankton have several possible modes of dispersal. They can move between lakes through stream connections or flooding events, although, in contrast to fish, their movement tends to be passive rather than active. Some species are probably transported overland by wind or carried by waterfowl or other vertebrate vectors. Crossing the inhospitable land matrix is possible because plankton and bryozoans have desiccation-resistant resting stages, in which their metabolism is reduced and they are able to withstand a wide range of temperatures. The resting stages of many species involve shapes and structures such as spines that facilitate the transportation of these species between lakes via wind, duck feathers, or animal fur or inside the digestive tracts of animals. There is also evidence that humans transport zooplankton (including nonnative species) because they adhere to anchor ropes, fishing lines, and boat hulls and are then carried from one lake to another as humans move among lakes.

For many other aquatic organisms, such as snails and aquatic insects, modes of dispersal vary. Some aquatic insects have winged stages, in which they are able to fly distances up to several kilometers or more if carried in the wind. Snails and other molluscs probably have limited dispersal abilities and move short distances through stream connections or in some cases may be transported by waterfowl, flying insects, or amphibians. The spread of the invasive zebra mussel throughout North American waters suggests that human vectors may also contribute to mollusc dispersal.

DISPERSAL ABILITIES

Dispersal ability probably varies greatly among organisms, although much of our evidence of dispersal ability is anecdotal. Small organisms are generally thought to be better dispersers than are large organisms, although dispersal data for small propagules are sparse. Bacteria and algae can disperse overland via wind and through watercourses and are thought to have high dispersal abilities because of their ubiquitous distribution across continents and globally. Zooplankton are probably more variable in their dispersal abilities, although actual mechanisms and rates of dispersal are poorly understood. Evidence based on the arrival of species in traps or newly formed ponds suggests that rotifers, daphniids, and cyclopoid copepods tend to arrive within the first several weeks, suggesting that their dispersal rates are high. Despite the fast arrival of some species, the number of species that colonizes a newly formed pond or zooplankton trap tends to be much lower than the regional species pool, an indication that there is large variation in dispersal rates among species. Regional studies of species distributions indicate that some species, such as calanoid copepods and species that reside in the bottom of deep lakes, have limited, if any, dispersal among lakes. Fish probably have the most limited movement between lakes because they are primarily restricted to movement along watercourses. Human-mediated introductions of fish to new locations have increased the dispersal ability and geographical distribution of some fish species. The dispersal ability of organisms within a lake will partially determine the connectivity of the lakes in the landscape and will ultimately determine the community structure of the lakes.

Dynamic Communities

As suggested by the theory of island biogeography and metacommunity theory, empirical work has demonstrated that species composition of lakes is indeed dynamic, with species appearing and disappearing through time, even in relatively undisturbed systems (i.e., systems without large directional changes in environmental conditions). Fish species turnover in lakes is generally less than 1% per year, when turnover attributable to sampling error is removed. Zooplankton have high turnover rates—a study in south central Ontario indicated that they can be as high as 16% per year, although it is uncertain how much of this results from species abundances fluctuating around detection limits. For crustacean zooplankton, turnover rates decrease with increasing latitude but show no relationship to lake area. In temperate zone lakes, turnover rates decrease with increasing species richness, suggesting that more diverse communities are more resistant to colonization and extinction events.

Immigration and Extinction of Species

Aquatic ecologists have long recognized that local environmental conditions can influence extinction probabilities of species, and therefore the number and kinds of species that are present in a particular lake. These factors include the amount of primary production (often correlated with total phosphorus concentration), the acidity, the calcium concentration, and the presence of predators. Lake size and depth are also important determinants of community structure, likely because they are related to habitat heterogeneity. Large, deep lakes have more species because they have a diversity of habitats and food resources and therefore have lower extinction rates of species. In the same way, terrestrial islands with a diversity of habitats and much vertical structure (for example, forests) have high species richness and steep species-area relationships (i.e., the number of species increases rapidly with island size).

Over the past several decades, ecologists have gathered evidence that the spatial arrangement of lakes on the landscape can also be important in determining species composition and richness—probably because the spatial arrangement of lakes influences dispersal probabilities. Some of the earliest work examined the species composition of zooplankton communities in Quebec lakes. Space (the geographic arrangement of lakes in a region) primarily influenced community composition because it was associated with environmental gradients. Research conducted in Northern Wisconsin lakes drew similar conclusions for crayfish, snails, and fish. Crayfish abundance and fish species richness were related to lake order, which is a proxy for lake connectivity. Snail richness was higher in low-order, connected lakes than in high-order, isolated lakes, suggesting that dispersal may be limiting. However, lake order was also related to environmental conditions so it was not possible to separate out the effects of space and local conditions (i.e., lake order and local environmental conditions co-varied).

In 34 interconnected ponds in De Maten, Belgium, the spatial configuration of the ponds played an important role in determining zooplankton community composition. Ponds that are close to each other are more similar in zooplankton composition than are ponds that are far apart. In fact, the spatial configuration of the ponds (and presumably dispersal among them) accounted for 15% of the variation in zooplankton species composition, and local environmental conditions (primarily the presence of macrophytes and fish) accounted for approximately the same amount of variation, 17%. Similar results were found for zooplankton communities in lakes and ponds on Ellesmere Island. An analysis of 158 data sets that included lakes, streams, estuaries, marine areas, and terrestrial ecosystems for a variety of taxa, revealed that, on average, 22% of the variation in species composition was attributable to local environmental conditions and 10% of the variation was attributable to pure spatial configuration of the habitats. Although island habitats were not explicitly examined, results were similar for lakes and terrestrial ecosystems.

The relative importance of dispersal compared to local environmental conditions differs among aquatic taxonomic groups. In 18 Quebec lakes, bacteria and phytoplankton did not appear to be functioning as metacommunities because neither overland dispersal nor stream connections were important in determining community composition. In contrast, for zooplankton, overland dispersal and stream connections were equally important, together contributing approximately the same amount of variation in community composition as local environmental conditions. For fish, stream connections were the only form of dispersal influencing composition, and environmental factors did not play any role (in contrast to other studies that have shown acidity and lake size to be important). The results from these studies suggest that although metacommunity theory may not apply to all aquatic taxa, there is strong evidence to support the consideration of zooplankton communities in lakes and ponds as metacommunities (Fig. 3).

Zooplankton Metacommunities

There is increasing evidence that zooplankton consist of metacommunities in lake regions with high numbers of lakes. In the highly connected ponds in De Maten, where

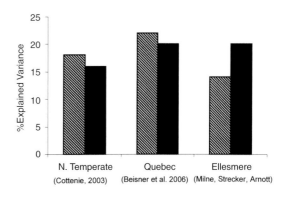

FIGURE 3 Results of variation partitioning of zooplankton communities in De Maten, Belgium, Quebec, and Ellesmere Island. Shaded bars indicate the variation in zooplankton communities that is explained by environmental factors, and the solid bar indicates the variation that is explained by space (i.e., the geographical arrangement of the lakes).

zooplankton move at an average rate of 3600 individuals per hour, dispersal provides a high number of species to each lake. Selective agents within each pond (e.g., the presence of macrophytes and fish) strongly shape the zooplankton community. This situation has close parallels to the species-sorting view of metacommunities, described in the previous section. However, some species are able to persist despite having characteristics unsuitable for a given environment because of high dispersal rates. This situation resembles the mass-effects view. It may be that several metacommunity paradigms operate within a lake region because of differences in environmental tolerances and dispersal abilities of individual species.

Work done in ponds in Michigan provided similar findings. Local zooplankton communities in experimental mesocosms were resistant to high dispersal of zooplankton from lakes in the surrounding area. The local communities were best suited for the local environment and prevented the establishment of immigrating species. However, when the abundance of the local species was reduced or when local conditions changed because of the addition of a predator, immigrating species were able to establish populations, a fact that is suggestive of the patch-dynamic and species-sorting views of metacommunities. It seems that the nature of the environmental change is important in determining the invasibility of the dispersing zooplankton. Jessica Forrest and I conducted an experiment in mesocosms in a lake in Killarney Park, Canada, where we investigated the influence of zooplankton dispersers on communities that were undergoing environmental change associated with a nutrient addition. We found that although there was an increase in zooplankton biomass in the mesocosms that received a nutrient addition, it was the result of increased abundances of the local species and not the establishment of the dispersing species. In a similar study in the same lake, however, Angela Strecker and I found that dispersers did establish in mesocosms where the environmental change was associated with the addition of an exotic invertebrate predator. These results strongly support the species-sorting view of metacommunities, where species composition tracks variation in local environmental conditions because high dispersal provides a variety of propagules with a diversity of character traits. The local environment selectively acts on these propagules, causing some to flourish and others to be driven to local extinction.

Speciation and Adaptive Radiation

There is strong evidence to suggest that similar processes exist in islands and lakes. For example, in freshwater lakes, there are often sympatric pairs of fish species: one benthic, consuming mostly invertebrates found in the vegetation and sediments, and one planktonic, consuming zooplankton in the open water. These two ecomorphs have been observed in a variety of freshwater fish, including smelt, whitefish, sticklebacks, and char. On islands, there is evidence of adaptive radiation in orb-weaving spiders on the Hawaiian Islands and lizards on the Greater Antilles. As with freshwater fish, spiders and lizards occupy several ecomorphs on each island, even though the species composition may be different among islands.

Isolation probably influences islands and lakes in a similar manner (i.e., there is a trade-off between the arrival of individuals [which also acts to homogenize genes] and the creation of ecological diversity through adaptive radiation). Speciation on Caribbean islands larger than 3000 km^2 is the dominant source of new species of *Anolis* lizards, whereas immigration is the dominant source of new species for islands smaller than 3000 km^2. In contrast, speciation of cichlids in Africa has occurred in all lakes, even the smallest ones. But as with terrestrial islands, speciation steeply increases in large lakes over 1000 km^2.

CONCLUSIONS

Experimental and observational data on organisms residing in lakes suggest that lakes are like islands and that theory developed for island habitats applies to lakes. Species dynamics in lakes are a function of both the local environment and the connectedness of lakes in the landscape. This approach is important because lakes are facing many human-induced changes resulting in the local loss of biodiversity. Some of the most important stressors (e.g., habitat loss and fragmentation, pollutants, climate change, and invasive species) result in changes to local environmental conditions and lake connectivity (i.e., the movement of energy and organisms among lakes). Ultimately, the connectedness of lakes on the landscape may determine how lake communities respond to environmental stressors. For example, climate change is expected to impact local populations and to change the range distributions of many organisms. Will organisms in lakes have the ability to colonize new habitats as local conditions become unfavorable with increasing changes in climate? Commercial transportation of goods between countries has increased the movement of species between continents. Currently, there are over 180 non-indigenous species in the Great Lakes, many of which are spreading to inland lakes resulting in negative impacts on native species (e.g., declines in diversity). Understanding how species move from one lake to another is critical for reducing the spread of exotic species throughout the thousands of inland lakes in North America. Understanding how species move between lake islands and what barriers to dispersal exist will improve our ability to predict and manage the impacts of these stressors. Explicitly considering lakes as islands will help us understand the full impact of these stressors on lake ecosystems.

SEE ALSO THE FOLLOWING ARTICLES

Cichlid Fish / Freshwater Habitats / Island Biogeography, Theory of / Marine Lakes / Metapopulations / Species–Area Relationship

FURTHER READING

Beisner, B. E., P. R. Peres-Neto, E. S. Lindstrom, A. Barnett, and M. L. Longhi. 2006. The role of environmental and spatial processes in structuring lake communities from bacteria to fish. *Ecology* 87: 2985–2991.

Bilton, D. T., J. R. Freeland, and B. Okamura. 2001. Dispersal in freshwater invertebrates. *Annual Review of Ecology and Systematics* 32: 159–181.

Cottenie, K., and L. De Meester. 2004. Metacommunity structure: synergy of biotic interactions as selective agents and dispersal as fuel. *Ecology* 85: 114–119.

Cottenie, K., E. Michels, N. Nuytten, and L. De Meester. 2004. Zooplankton metacommunity structure: regional vs. local processes in highly interconnected ponds. *Ecology* 84: 991–1000.

Forrest, J., and S. E. Arnott. 2006. Immigration and zooplankton community responses to nutrient enrichment: a mesocosm experiment. *Oecologia* 150: 119–131.

Leibold, M. A., M. Holyoak, N. Mouquet, P. Amarasekare, J. M. Chase, M. F. Hoopes, R. D. Holt, J. B. Shurin, R. Law, D. Tilman, M. Loreau, and A. Gonzalez. 2004. The metacommunity concept: a framework for multi-scale community ecology. *Ecology Letters* 7: 601–613.

Leibold, M. A., and J. Norberg. 2004. Biodiversity in metacommunities: plankton as complex adaptive systems? *Limnology and Oceanography* 49: 1278–1289.

MacArthur, R. H., and E. O. Wilson. 1967. *The theory of island biogeography*. Princeton, NJ: Princeton University Press.

Shurin, J. B. 2000. Dispersal limitation, invasion resistance, and the structure of pond zooplankton communities. *Ecology* 81: 3074–3086.

LAND CRABS ON CHRISTMAS ISLAND

PETER GREEN

La Trobe University, Bundoora, Australia

Christmas Island (105°40′ E, 10°30′ S) is a small (135 km^2), elevated (to 361 m above sea level) oceanic island lying 360 km south of Java in the northeastern Indian Ocean. It is an external territory of Australia, composed mostly of limestone and covered by rain forest where it has not been cleared for phosphate mining. Land crabs are ubiquitous components of the island's terrestrial fauna. One species, the red land crab *Gecarcoidea natalis,* determines the dynamics of seedling recruitment, a rare example of single-species dominance of this process in tropical rain forest.

SPECIES DIVERSITY AND DISTRIBUTION

Air-breathing crabs characterize the terrestrial fauna on many tropical, oceanic islands. Twenty species occur on Christmas Island (Table 1), including true land crabs (Brachyura: Grapsoidea: Gecarcinidae), a variety of grapsid crabs (Brachyura: Grapsoidea: Grapsidae), and two species of ghost crabs (Brachyura: Ocypodoidea: Ocypodidae). Also present are several species of hermit crabs (Anomura: Coenobitidae), including the coconut crab *Birgus latro,* the world's largest terrestrial crustacean (Fig. 1). Many species are widespread throughout the Indo-Pacific region, and only one (*Sesarma jacksoni*) is endemic to the island. The red crab *G. natalis* is widely regarded as an island endemic, but it also occurs in low numbers in the Cocos (Keeling) islands, 900 km further west. Most species have locally restricted distributions and occur in association with the island's few beaches, in freshwater streams and soaks, or in forest bordering the rugged island's coastline (Table 1). Two species, the coconut crab *B. latro* and the red crab *G. natalis,* are common throughout

TABLE 1
The Land Crabs of Christmas Island (Indian Ocean)

Common Name	Species	Habitat	Local Range	Abundance
Infraorder Anomura, family Coenobitidae				
Coconut crab/robber crab	*Birgus latro*	RF	W	A
Purple hermit crab	*Coenobita brevimanus*	ST, SL	R	C
Red hermit crab	*Coenobita perlatus*	SL	R	C
Tawny hermit crab	*Coenobita rugosus*	SL	R	R
Infraorder Brachyura, family Gecarcinidae				
Red crab	*Gecarcoidea natalis*	RF	W	A
Purple crab	*Gecarcoidea lalandii*	RF	W	R
Blue crab	*Discoplax hirtipes*	FS	R	C
Brown crab	*Epigrapsus politus*	SL	R	R
Infraorder Brachyura, family Grapsidae				
Little nipper	*Geograpsus grayi*	RF	W	C
Yellow nipper	*Geograpsus crinipes*	ST, SL	W	C
Red nipper	*Geograpsus stormi*	SL	W	R
White-stripe crab	*Labuanium rotundatum*	ST	R	R
Yellow-eyed crab	*Sesarma obtusifrons*	SL	W	C
Jackson's crab	*Sesarma jacksoni*	ST	R	R
Mottled crab	*Metasesarma rousseauxi*	SL	R	C
Freshwater crab	*Ptychognathus pusillus*	FS	R	R
Sandy-rubble crab	*Cyclograpsus integer*	SL	R	R
Grapsus	*Grapsus tenuicrustatus*	SL	W	A
Infraorder Brachyura, family Ocypodidae				
Horn-eyed ghost crab	*Ocypode ceratophthalmus*	SL	R	C
Smooth-handed ghost crab	*Ocypode cordimanus*	SL	R	C

NOTE: For habitat, FS = freshwater stream or seepage, RF = rain forest, ST = shore terrace (forest immediately fringing the coastal sea cliffs), SL = shoreline. For range, W = widespread, and R = restricted. For abundance, A = abundant, C = common, and R = rare. Many shoreline species have restricted ranges because they occur in association with the island's few beaches.

FIGURE 1 The coconut crab, *Birgus latro*, also known locally as the "robber" crab. This species is the world's largest terrestrial crustacean and was once common throughout islands in the Indian and Pacific oceans. The largest and most intact population is now on Christmas Island, where individuals regularly achieve enormous dimensions. They are omnivorous, eating fruits and seeds, or acting as predators of red crabs. Photograph copyright of Peter Green.

rain forest on the island, occurring right to the summit and several kilometers inland from the coast.

THE RED LAND CRAB, *GECARCOIDEA NATALIS*

The red crab is by far the most abundant land crab on the island, occurring at mean densities and biomasses from 0.4 to 1.8 crabs per m^{-2}, and 1.0 to 1.5 tons per ha. The largest crabs grow to more than 120 mm in carapace width and more than 500 g in weight and may live for more than 20 years. The total adult population is estimated at 40–45 million crabs. With the exception of some breeding-related activities, the crabs are completely terrestrial and live singly in dry, shallow burrows that do not reach the water table.

FIGURE 2 At the beginning of each wet season, the red land crab *Gecarcoidea natalis* migrates en masse to coastal areas to breed. Some crabs travel several kilometers over rough terrain and can cover more than 1 kilometer in a single day. Large numbers of crabs are killed each year as they cross the island's few roads, but in recent years this has been reduced through management intervention. Photograph copyright of Greg Miles.

Diel and seasonal patterns of activity are driven mostly by the need to avoid desiccation. The crabs are diurnally active but retreat to their burrows when the relative humidity falls below about 85%. They can remain below ground for weeks during the dry season, often with a plug of leaves in the entrance to their burrows to further reduce water loss. Approximately half the population migrates to the coast in spectacular, massed breeding migrations at the start of each wet season (Fig. 2), timing egg maturation to coincide with a high tide during the last quarter of the moon. Both sexes "dip" in the ocean on their arrival at the coast and then retreat into forest fringing the coast to mate. Eggs are released directly into the ocean and hatch immediately upon contact with seawater, taking from 17 to 30 days to complete several larval stages prior to the return of juveniles onto land. Recruitment is sporadic, with the massed emergence of juvenile crabs occurring once every few years.

Most land crabs are omnivorous and are widely regarded as phytophagous filters because many species include seeds and seedlings in catholic diets. Red land crabs have a

FIGURE 3 Red land crabs control key aspects of rain-forest ecology on Christmas Island, including litter processing, nutrient cycling, and the dynamics of seedling recruitment. The photographs show small experimental plots from which red crabs were either (A) allowed continued access, or (B) excluded for a period of four years. Photographs copyright of Peter Green.

profound effect on the seedling recruitment and rainforest dynamics on Christmas Island because these crabs are widespread, abundant, and eat a variety of seeds and seedlings. Small-scale exclusion experiments have demonstrated that red crabs cause a 20-fold reduction in seedling density and a tenfold reduction in species richness (Fig. 3). The seeds and seedlings of rain-forest plants are differentially susceptible to red crabs, and the seedling community is composed of a low diversity and low abundance of mostly non-palatable species. In this respect, the persistence of many canopy species, the seeds and seedlings of which are susceptible to red crabs, and which appear to rarely recruit seedlings, is puzzling.

THREATS AND CONSERVATION

Since 1989, more than 80% of the forest and 63% of the coastline have been included in the Christmas Island National Park, affording the island's land crabs substantial protection. However, a number of threats past and present have affected several species to varying degrees, some of which still require active management. Phosphate mining has been ongoing for more than 100 years and has reduced rain forest habitat for the widespread red crabs, coconut crabs, and little nippers by 25%. In addition, transport infrastructure has had an ongoing impact on migrating red crabs, with an estimated annual road toll of more than 1 million red crabs during much of the 1960s to 1980s. In recent years, the government agency Parks Australia North has made significant progress in managing the migration to reduce the road toll, including road closures and the use of portable drift fences that direct migrating crabs into strategically sited concrete tunnels (Fig. 4). Coconut crabs and blue crabs have been hunted in the past, with unknown impacts on their populations. Better legislative control, community education, and improved economic conditions have substantially reduced the level of hunting in recent years.

Since the late 1990s, the most serious threat to the land crabs of Christmas Island has been the invasion of rain forest by the introduced yellow crazy ant, *Anoplolepis gracilipes,* listed by the IUCN as one the world's 100 worst invasive species. Present on the island since at least 1930, this species began forming expansive (tens to hundreds

FIGURE 4 To reduce the annual road toll inflicted on red crabs, Parks Australia North uses (A) portable plastic drift fences (B) to direct migrating crabs into preformed concrete tunnels (C) through which the crabs walk to safety on the other side. There are many such tunnels at strategic crossing points around the island. Photographs copyright of Peter Green.

FIGURE 5 The invasion of rain forest by the yellow crazy ant *Anoplolepis gracilipes* has transformed the understory over large tracts of rain forest, because within supercolonies the ants extirpate red crabs. These photographs show the state of the understory in (A) areas uninvaded by yellow crazy ants, where local populations of red crabs are intact and there are few seedlings or leaf litter, and (B) areas three years after invasion by yellow crazy ants, from which red crabs have been extirpated and there are dense thickets of seedlings and saplings, with a persistent litter layer. The extirpation of red crabs by the yellow crazy ant has recapitulated at a landscape scale the effects of small-scale, experimental exclusion of red crabs (compare with Fig. 3). Photographs copyright of Peter Green.

of ha), high-density (~1000 ants per m^{-2}) supercolonies in undisturbed rain forest around 1997. Yellow crazy ants form mutualistic associations with a variety of sap-sucking scale insects, the honeydew excretions of which are used by the ants as a source of carbohydrate. Yellow crazy ants use formic acid to subdue their prey, and in supercolonies they extirpate local populations of red crabs and coconut crabs. So far, *Anoplolepis* has killed about one-third of the red crab population, or about 20 million individuals. The impact of yellow crazy ants on seedling recruitment and understory density over large areas of rain forest, via its impact on red crabs, has been dramatic (Fig. 5). In 2002 a helicopter was used to spread toxic bait that successfully knocked out *Anoplolepis* supercolonies over 2500 ha. Since then, Parks Australia North has maintained an extensive program of monitoring, surveillance, and suppression.

SEE ALSO THE FOLLOWING ARTICLES

Ants / Invasion Biology / Succession / Vegetation

FURTHER READING

Green, P. T., P. S. Lake, and D.J. O'Dowd. 1999. Monopolization of litter processing by a dominant land crab on a tropical oceanic island. *Oecologia* 119: 235–244.
Green, P. T., D. J. O'Dowd, and P. S. Lake. 1997. Control of seedling recruitment by land crabs in rain forest on a remote oceanic island. *Ecology* 78: 2474–2486.
Green, P. T., D. J. O'Dowd, and P. S. Lake. 2008. Recruitment dynamics in a rainforest seedling community: context-independent impact of a keystone consumer. *Oecologia* 156: 373–385.
Hicks, J., H. Rumpff, and H. Yorkston. 1990. *Christmas crabs,* 2nd ed. Christmas Island Natural History Association.
O'Dowd, D. J., P. T. Green, and P. S. Lake. 2003. Invasional 'meltdown' on an oceanic island. *Ecology Letters* 6: 812–817.
O'Dowd, D. J., and P. S. Lake. 1991. Red crabs in rain forest, Christmas Island: removal and fate of fruits and seeds. *Journal of Tropical Ecology* 7: 113–122.

LANDSLIDES

SIMON J. DAY

University College London, United Kingdom

Landslides take a wide variety of forms, but all involve the movement of rock, sediment, or soil under the influence of gravity. They may involve sliding of coherent rock or cohesive sediment masses on discrete surfaces or zones, but more commonly they involve rapid movement of fragmented material: The largest and most hazardous landslides are debris avalanches from volcanic islands.

NON-VOLCANIC LANDSLIDES

Mountainous islands at tectonic plate boundaries, such as Papua New Guinea and Taiwan, experience frequent landslides as a result of coastal erosion and cliff formation; rapid incision of rivers in response to tectonic uplift; and the proximity of the ocean base level. The landslides themselves may be triggered by high rainfall (particularly extreme rainfall events) or by frequent seismic shaking. Recent large, destructive landslides of this type include Kaiapit, Papua New Guinea (1988), and Tsao-Ling, Taiwan (1999). These frequent landslides are both an important hazard and a primary contributor to the high rates of erosion and sediment discharge to adjacent ocean basins characteristic of these islands, and of similar islands such as New Zealand and Japan (Fig. 1). Landslides also control development of the amphitheater-headed valleys that

FIGURE 1 Mountain valley partly filled with sediment released by the 1988 Kaiapit landslide, Finisterre Mountains, Papua New Guinea. Such landslides contribute greatly to the high erosion rates in, and sediment fluxes from, mountainous islands.

dominate many islands. Another frequent type of landslide at islands involves failure of the steep fronts of coral reefs, often removing raised reef and barrier islands on the reef top.

LANDSLIDES AT VOLCANIC ISLANDS

Large lateral collapse landslides at volcanic islands affect both the subaerial volcanic edifices and their flanks below sea level and may reduce islands to a fraction of their former size or even remove them altogether. Such landslides are usually but not always associated with volcanic eruptions or intrusions within the volcano. They range from small, thin landslides (for example, at Stromboli in 2002; Table 1) to giant lateral collapse landslides that remove sequences up to 3 km thick and generate debris avalanches and debris flows that travel for tens of kilometers over the adjacent ocean floors. The largest known example, the Nuuanu landslide deposit off Oahu (Hawaii), has a volume of ~5000 km^3, extends more than 200 km, and includes individual blocks up to 30 km long and 2 km thick. Giant lateral collapse landslides, with volumes of hundreds of cubic kilometers, affect most oceanic island volcanoes in their most active shield stage of growth: In addition to numerous Hawaiian examples, such volcanoes occur around the Canary Islands (Fig. 2), the Cape Verde Islands, and at Réunion Island, among other places. These

TABLE 1
Historical Lateral Collapses and Large Landslides at Island and Coastal Volcanoes

Date	Volcano	Landslide volume, if known	Maximum local tsunami run-up	Distant tsunami run-ups (examples)	Landslide type
1640	Komagatake, Hokkaido, Japan	0.3 km^3	> 8 m		Ocean entry by landslide from subaerial volcano lateral collapse
1741	Oshima-Oshima, Sea of Japan	2.4 km^3 (large block facies of deposit); 2.5 km^3 (collapse scar)	34 m?; certainly > 15 m (on coasts 60–80 km distant)	3–4 m on Korean coast, 1200 km distant	Partly submarine volcano lateral collapse
1792	Unzen, Japan	0.34 km^3 (collapse scar volume)	> 10 m	none: collapse into enclosed bay	Ocean entry by landslide from subaerial volcano lateral collapse
1883	Augustine, Alaska	~0.5 km^3	> 19 m?	6–8 m, ~100 km distant	Ocean entry by landslide from subaerial volcano lateral collapse
1888	Ritter Island, Papua New Guinea	~4 to 5 km^3 [a]	> 15 m (on coasts up to 50 km distant)	8 m at Hatzfeldhafen, 370 km distant; 4.5 m at Rabaul, 540 km distant	Partly submarine volcano lateral collapse
1928	Paluweh, Indonesia	Volume poorly constrained due to subsequent eruption	3 waves, from 5 to 10 m		Ocean entry by subaerial landslide at start of large explosive eruption
1933	Harimkotan, Kuriles	~1 km^3 (collapse scar volume)	20 m	Significant damage on adjacent islands	Ocean entry by landslide from subaerial volcano lateral collapse
1966	Tinakula, Solomon Islands	< 0.01 km^3 [b]	Small local waves only		Ocean entry by landslide from small failure near summit
1979	Ili Werung, Indonesia	0.05 km^3	9 m		Ocean entry by subaerial landslide
2002	Stromboli, Italy	0.02 km^3	10 m	2 m (140 km distant)	Two thin-slope parallel landslides, one subaerial and one submarine

[a] Volume estimate from collapse scar, not deposit volume.
[b] Tinakula is sometimes cited as an example of a volcano collapse without a significant tsunami. However, the main collapse scar was described some years before 1966, and examination of aerial photographs taken soon after the 1966 event show most of the collapse scar to be vegetated, with bare rock confined to a narrow chute from a small rockfall scar near the apex of the collapse scar: the latter is a prehistoric feature.

FIGURE 2 Cumbre Nueva giant lateral collapse scar, La Palma, Canary Islands. This collapse scar is at least 20 km wide. Although it has been partly filled by the younger Bejenado volcano (center of view) and the active Cumbre Vieja volcano (foreground), the collapse scar headwall cliff (right) is still about 1 km high.

giant ocean island volcano collapses are rare, with a worldwide frequency of known events of around 1 every 20,000 years. Lateral collapses at island arc volcanoes are smaller, with volumes usually in the range of 1–10 km^3, but they occur much more frequently. The most recent island arc volcano lateral collapses occurred at Ritter Island in 1888 and at Oshima-Oshima in 1741 (Table 1).

LANDSLIDE HAZARDS AND ECOSYSTEM EFFECTS DUE TO LANDSLIDES AT ISLANDS

In addition to destruction of areas covered by landslides themselves, large landslides at islands, especially volcano lateral collapse landslides, produce destructive tsunamis when they enter the ocean. Well-documented tsunamis produced by events such as the 1741 Oshima-Oshima and 1888 Ritter Island collapses provide key evidence to support computer models that predict catastrophic ocean-wide tsunamis as a result of giant ocean island volcano collapses. More locally, volcano collapses such as that of Ritter Island in 1888 can eliminate the pre-collapse island ecosystems, leading to instances of island recolonization.

SEE ALSO THE FOLLOWING ARTICLES

Canary Islands, Geology / Island Arcs / New Guinea, Geology / Taiwan, Geology / Tsunamis

FURTHER READING

Blong, R. J., and G. O. Eyles. 1989. Landslides: extent and economic significance in Australia, New Zealand and Papua New Guinea, in *Landslides: extent and economic significance*. E. E. Brabb and B. L. Harrod, eds. Rotterdam: Balkema, 343–355.
Moore, J. G., W. R. Normark, and R. T. Holcomb. 1994. Giant Hawaiian landslides. *Annual Reviews of Earth Planetary Science* 22: 119–144.

Siebert, L. 1996. Hazards of large volcanic debris avalanches and associated eruptive phenomena, in *Monitoring and mitigation of volcano hazards*. R. Scarpa and R. I. Tilling, eds. Berlin: Springer-Verlag, 541–572.

LAND SNAILS

BRENDEN S. HOLLAND

University of Hawaii, Manoa

Land snails are surprisingly adept at dispersing across vast stretches of open ocean, a fact supported by their presence on virtually all tropical and subtropical islands globally. Island snail radiations make fascinating subjects for the study of biogeography and diversification, as many archipelagoes have well-developed and diverse endemic snail faunas.

WHAT MAKES A SNAIL A SNAIL

Land snails are familiar molluscs with several characteristics that make them easily identifiable. They usually have paired eyes located at the tips of tentacles, a second pair of sensory tentacles, and a single, coiled shell into which the animals can generally withdraw their soft bodies for protection from predators and desiccation. Beneath the head is a mouth equipped with a radula, a highly specialized, elongated, rasping tongue-like organ used to scrape plant or fungal material, or in some cases to bore holes in the shells of other molluscs. Hard, calcified snail shells are secreted by a specialized layer of tissue called the mantle. Most snails have a flattened, muscular, tapering foot on which the animals glide.

Land snails typically dwell on the ground or in or under bushes, shrubs, or trees; feed on decaying organic matter; and deposit eggs on or in damp soil. Taxonomically, snails are members of the Gastropoda, a globally distributed class that contains more species than all of the other classes in the phylum combined.

The often drab coloration of familiar snails from temperate continental Asia, Europe, and North America, with the notable exception of a few taxa, such as the *Cepaea* species of Europe, contrasts dramatically with the often beautiful, brightly colored, and intricately patterned and banded shells of many island snails, such as tropical tree-dwelling groups like the Achatinellinae of the Hawaiian Islands, *Amphidromus* spp. of Indonesia, *Liguus* spp. and *Polymita* spp. of Cuba, *Papuina* spp. of Melanesia, and

FIGURE 1 Images of a number of conspicuous endemic island land snail shells from several Pacific diversity hotspot regions including Melanesia (New Guinea with over 1000 species) and Polynesia (Hawaiian Islands with over 750 species). A number of species represented are extinct, some are endangered, and the remainder are surviving but threatened. These images are shown at approximately accurate relative sizes and provide just a hint of the tremendous diversity and dazzling beauty of endemic island land snail faunas. Species identifications and geographic origins are as follows: (A) *Placostylus albersi*, New Caledonia; (B) *Papuina micans*, Solomons; (C) *Placostylus hargravesi*, Solomons; (D) *Achatinella decipiens*, Oahu; (E) *Placostylus strangei*, Solomons; (F) *Achatinella fulgens*, Oahu; (G) *Laminella sanguinea*, Oahu; (H) *Coxia* sp., New Guinea; (I) *Parahytida dictyodes*, New Caledonia; (J) *Papuina pulcherrima*, New Guinea; (K) *Trocomorpha* sp., Fiji; (L) *Carelia* sp., Kauai; (M) *Carelia* sp., Kauai; (N) *Amastra spirizona*, Oahu; (O) *Succinea kuhnsi*, Hawaii; (P) *Papuina mendana*, New Guinea; (Q) *Partulina proxima*, Molokai; (R) *Papustyla hindei*, New Guinea. All photographs by B. S. Holland. See also Fig. 2.

Partulidae of French Polynesia and the islands of the South Pacific (see Figs. 1 and 2 for selected examples).

DIVERSITY OF ISLAND SNAILS

Island snail diversity and biogeography have fascinated biologists since the time of Darwin, who wrote in a letter to A. R. Wallace in 1857, "One of the subjects on which I have been experimentising and which cost me much trouble, is the means of distribution of all organic beings found on oceanic islands and any facts on this subject would be most gratefully received: Land-Molluscs are a great perplexity to me." In spite of weak active dispersal, a relatively sedentary lifestyle, and minimal seawater tolerance, a number of land snail families have distributions that span ocean basins. In fact, land snails constitute some of the major terrestrial species radiations on oceanic islands, offering excellent opportunities for the study of historical biogeography, microevolution, and the diversification of insular lineages. Examples of island snail species diversity estimates include the Canaries (350), the Caribbean (1200), the Galápagos (100), the Hawaiian Islands (750), the Marquesas (300), Micronesia (400), New Guinea (1000), New Caledonia (400), the Ogasawaras (100), Pitcairn (30), Rapa (100), Rota (40), the Samoan Islands (100), and the Society Islands (200).

The broad distribution of snails on oceanic Pacific islands contributed to the development of early biogeographic theories including the "mid-Pacific continent"

FIGURE 2 More conspicuous endemic island land snail shells from Pacific hotspot regions. (A) *Placostylus koroensis*, Fiji; (B) *Amastra cylindrica*, Oahu; (C) *Achatinella byronii*, Oahu; (D) *Samoana stevensoniana*, Samoa; (E) *Samoana* sp., Marquesas; (F) *Partula affinis*, Tahiti; (G) *Crystallopsis tricolor*, Solomons; (H) *Placostylus fulguratus*, Fiji; (I) *Samoana ganymedes*, Marquesas; (J) *Achatinella livida*, Oahu; (K) *Achatinella pulcherrima*, Oahu; (L) *Partulina* sp., Hawaiian Islands; (M) *Corilla* sp., Samoa; (N) *Trochonanina rectangular*, Marquesas; (O) *Achatinella fulgens*, Oahu; (P) *Opiella pfeifferi*, Fiji. All photographs by B. S. Holland.

hypothesis, which proposed that a massive continent stretching across the South Pacific had once existed and had subsequently sunk beneath the ocean.

Origins of most island snail faunas are poorly understood. Multiple dispersal events from multiple geographic source regions, in which the nearer the potential geographic source, the higher the probability of colonization, is thought to be the predominant mode of origin for oceanic island snail faunas. Rafting attached to drifting tree trunks and other floating items, aerial dispersal including by tropical storms and hurricanes, and transport attached to birds have all been suggested as mechanisms of dispersal to and among oceanic islands. For long-distance dispersal, most malacologists agree that rafting is not likely to have been an important passive vector because of a general inability of land snails to tolerate saltwater. Small stones and pebbles of the mass of certain snails have been sampled in aerial plankton studies. Transport of land snails attached to the feet and feathers of migratory birds has been documented. To complicate the understanding of dispersal pathways and history further, founding propagules may not have originated on islands that are currently emergent but may have come from those that, because of erosion and or subsidence, no longer exist.

The dynamic geological and climatic processes that build up, tear down, and profoundly transform volcanic islands have no doubt influenced the diversity of land snail faunas that have persisted through long periods of island evolution. But vicariant processes such as island separation and coalescence, sea-level fluctuation, and lava flows do not hold the potential to generate novel lineages. Such processes fragment populations and have therefore played a role in enhancing allopatric speciation on a local scale, resulting in complexes of closely related sister taxa. Thus, although there is evidence that vicariance has impacted island snail species diversity, the results of island vicariance are phylogenetically relatively shallow (affecting only more recent tip clades), compared with the effects of long-distance dispersal and multiple colonizations by divergent, independent lineages, from various geographic sources. Deeper phylogenetic divisions, and therefore more ancient levels of diversity, are driven and shaped by passive long-distance dispersal. Therefore, both dispersal and vicariance have played important roles in shaping island snail faunas, but to differing degrees and at different phylogenetic levels.

Although no single compilation of the overall numbers of non-marine island snail species exists, species lists are available for various geographic regions, including most island groups, some recent, others more than 100 years old. Such lists are notorious for taxonomic inaccuracy, including the fact that many species remain undescribed, whereas others have been described multiple times as different species. Nevertheless, using these compilations, it is possible to arrive at rough estimates of diversity. One such recent estimate suggests that there are perhaps 24,000 described species of terrestrial snails,

FIGURE 4 More live oceanic island snails in their natural habitats. (A) Unidentified helicarionids, Ua Pou, Marquesas; (B) *Achatinella sowerbyana*, Oahu, Hawaiian Islands; (C) Unidentified helicarionid, Nuku Hiva, Marquesas; (D) *Partulina tappaniana*, Maui, Hawaiian Islands; (E) *Partulina redfieldi*, Molokai, Hawaiian Islands; (F) *Perdicella helena*, Molokai, Hawaiian Islands. All photographs by B. S. Holland.

FIGURE 3 Live oceanic island snails in their natural habitats. (A) *Lamprocystis* sp., Bora Bora, Society Islands; (B) *Partulina crocea*, Maui, Hawaiian Islands; (C) *Samoana* sp., Tahiti, Society Islands; (D) *Catinella* sp., Oahu, Hawaiian Islands; (E) *Succinea lumbalis*, Kauai, Hawaiian Islands; (F) *Succinea* sp., Tahiti, Society Islands. All photographs by B. S. Holland. See also Figs. 4 and 5.

although estimates as high as 80,000 species have been published. Likewise, a rough estimate of the total number of snails that have evolved on islands (excluding New Zealand, Papua New Guinea, and Madagascar) is about 6000, or 25% of the lowest total global estimate of snail species. In light of the minute fraction of global land area constituted by islands, this estimate shows the disproportionately important role played by islands in the generation of land snail biodiversity. For example, a comparison of the number of species in all of North America with the species diversity of the main Hawaiian Islands reveals that, in spite of the fact that the land area of North America exceeds that of the Hawaiian Islands by about a thousandfold, snail species diversity is roughly equivalent between the two regions.

The presence of diverse endemic assemblages of land snails on islands, including oceanic island groups that are thousands of kilometers from the nearest neighboring island or continent, is a testament not only to the extraordinary ability of these molluscs to passively disperse over long distances and establish new colonies, but also to their tendency to radiate into large numbers of species from relatively rare initial propagules, often in

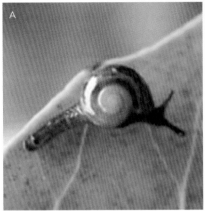 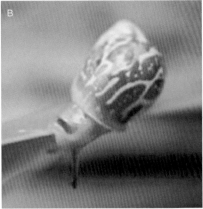 

FIGURE 5 More live oceanic island snails in their natural habitats. (A) Unidentified helicarionid, Hiva Oa, Marquesas; (B) *Samoana* sp., Tahiti, Society Islands; (C) *Achatinella sowerbyana,* Oahu, Hawaiian Islands. All photographs by B. S. Holland.

adaptive species complexes (see Fig. 3). Because of these characteristics, island land snail radiations are increasingly valued as informative systems in illuminating and advancing the general understanding of the processes and patterns of evolution and adaptive radiation on islands.

Factors considered most important in the generation of high species diversity include island area, latitude, age, elevation, rainfall, and plant community diversity. These factors are frequently interrelated, and probably influence species diversity on islands in concert, and in complex ways.

CONSERVATION CONCERNS

Non-marine molluscs comprise one of the most threatened groups of animals on Earth and include 99% of all molluscan extinctions. Habitat loss, degradation, and fragmentation and environmental stresses such as climate change, pollution, and the introduction of invasive species such as black rats (*Rattus rattus*), the wolf snail (*Euglandina rosea*), and the flatworm (*Platydemus manokwari*) all play important and often cascading synergistic roles in island land snail extinction. Although terrestrial vertebrate extinctions are well documented, invertebrate extinctions usually go unnoticed by the general public and undocumented by biologists and conservation agencies. Only a tiny fraction (less than 2%) of known molluscan species have had their conservation status scientifically assessed. Among the high-diversity island faunas, the Hawaiian land snails are relatively well studied, and species losses are daunting, estimated at 65–90%. Although most archipelagoes have not received the level of scientific scrutiny as have the Hawaiian Islands, similarly dire conditions probably exist on islands across all of the ocean basins.

Ecologists have begun to recognize important inputs to ecosystem function for a variety of invertebrate taxa, whether they be in the form of aeration of soils by earthworms, pollination of flowers by insects, decomposition of dead timber by termites, or other processes. Clearly, removal of such species has the potential to induce negative impacts on ecosystem balance, energy flow, health, and function. Although the precise roles of abundant, conspicuous terrestrial island snails in ecosystem function are not well understood, snails may play important roles in the maintenance of healthy forest ecological equilibrium (they may be involved in the calcium cycle, the carbon budget, and the breakdown of leaf litter and other organic material, and they may have a role as predator/prey in food webs and nutrient cycles).

Island snail assemblages are highly sensitive to habitat alteration and to the presence of invasive predatory species. On the majority of oceanic high islands, especially at elevations below about 300 m, native flora has been replaced by introduced weeds, ornamentals, and agricultural species. Most island land snails require native plants. We know from the notes of early twentieth-century expedition scientists that island floras were altered long ago; for example, Adamson in 1932 wrote of the Marquesas that "The native flora below 1000 ft has been replaced in large measure by immigrants, and to a considerable extent up to 2,500 feet." Thus, in general, the native snail fauna is presently restricted to upper elevations of high islands, because these areas harbor the only remaining native forest in many island environments.

The degree of island land snail imperilment is poorly documented and almost certainly underestimated. This view is supported by the continual discovery of undescribed snails, especially on tropical island archipelagoes throughout the world, many of which have been largely deforested and on which numerous harmful invasive species have become established. For example, a multiyear field survey of the terrestrial snails of the Papuan Peninsula and nearby islands by Florida Museum of Natural History biologists has recently uncovered dozens of undescribed species, suggesting that although the area has long been considered a biodiversity hotspot, the land snail fauna is far more diverse than was previously known. Surveys conducted in the southeastern United States focusing on small snails restricted to specialized habitats, including seeps and springs, are also uncovering previously undescribed species. Such surveys are also important in uncovering previously undocumented introduced species. Early detection of introduced species is crucial in the facilitation of efforts to control and prevent establishment, because after establishment, eradication is impossible.

CONCLUSION

This brief overview is intended to demonstrate that in terms of biodiversity hotspots, tropical and subtropical island land snails are important yet frequently overlooked components in terms of their contributions to unique island biodiversity. Ongoing efforts to preserve island land snails have had some success in recent years and have included captive rearing programs for rare Hawaiian achatinelline tree snails (University of Hawaii, Manoa) and French Polynesian partulid tree snails (mainly the London Zoological Society and the Jersey Zoo, but also the Detroit Zoological Park, the John G. Shedd Aquarium, and the San Diego Zoo); efforts to fence out, trap, or poison predators; habitat restoration; population translocation; and conservation genetics studies. However, the need persists for intensification of such efforts and for creative novel conservation strategies and solutions in the face of accelerating environmental change.

SEE ALSO THE FOLLOWING ARTICLES

Adaptive Radiation / Dispersal / Hawaiian Islands, Biology / Invasion Biology / Oceanic Islands / Vicariance

FURTHER READING

Barker, G. M. 2001. *The biology of terrestrial molluscs.* CAB International.
Cowie, R. H. 1996. Pacific Island land snails: relationships, origins and determinants of diversity, in *The origin and evolution of Pacific Island biotas, New Guinea to Eastern Polynesia: patterns and processes.* A. Keast and S. E. Miller, eds. Amsterdam: SPB Academic Publishing, 347–372.
Cowie, R. H., and B. S. Holland. 2006. Dispersal is fundamental to biogeography and the evolution of biodiversity on oceanic islands. *Journal of Biogeography* 33: 193–198.
Lydeard, C., S. A. Clark, K. E. Perez, R. H. Cowie, W. F. Ponder, A. E. Bogan, P. Bouchet, O. Gargominy, K. S. Cummings, T. J. Frest, D. G. Herbert, R. Hershler, B. Roth, M. Seddon, E. E. Strong, and F. G. Thompson. 2004. The global decline of nonmarine mollusks. *BioScience* 54: 321–330.

LAVA AND ASH

KATHARINE V. CASHMAN
University of Oregon, Eugene

Lava and ash are two different products of volcanic eruptions that cover the surfaces of volcanic islands. These two substrates have different volcanic origins and different physical properties, particularly with respect to their interaction with water (water transmission, storage, and susceptibility to erosion). The prevalence of one component or the other depends on the types of volcanic eruptions responsible for island formation. Eruption style, in turn, is determined primarily by the island's location with respect to tectonic plates.

LAVA FLOWS

Basaltic lava flows cover much of the surface area of volcanic islands that form over hotspots, such as Hawai'i. In these settings, the surfaces of active volcanoes are almost entirely composed of vast lava flow fields. J. D. Dana, an early geologist to visit Hawai'i on Charles Wilkes's U.S. Exploring Expedition, commented that "Areas, hundreds of square miles in extent, are covered with the refrigerated lava floods, over which the twistings and contortions of the sluggish stream as it flowed onward are everywhere apparent; other parts are desolate areas of ragged scoria." Dana also noted that degradation of lava-mantled surfaces was dependent

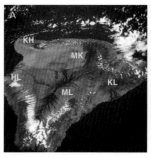 

FIGURE 1 Two NASA images of the island of Hawai'i, USA, available on the Earth from Space website (http://earth.jsc.nasa.gov/sseop/efs/). (A) Image STS61A-050-0057 (November 1985). A low-oblique, north-looking photograph that shows the five volcanoes that form the island. From youngest to oldest these are Kilauea (KL), Mauna Loa (ML), Hualalai (HL), Mauna Kea (MK), and Kohala (KH). Also evident in this image is the prominent rain shadow on the western and southern flanks of Mauna Kea and Kohala (seen as brown desert in contrast to the lush greenery of the northeastern slopes of the volcanoes). (B) Image STS051-102-083 (September 1993). Kohala volcano, the oldest on the island, shows the effect of topography-induced variations in rainfall on patterns of vegetation (green vs. brown in the image) and rainfall-induced erosion (seen as deep erosional gullies that have formed on the windward side of the volcano).

on both time and rainfall; in Hawai'i this is evidenced by the pronounced difference in vegetative cover and erosional dissection of slopes on the windward (high rainfall) and leeward (near-desert) sides of the island of Hawai'i (Fig. 1).

Basaltic lava flows are traditionally classified into two types on the basis of their surface morphology: pāhoehoe and 'a'ā. Pāhoehoe flows are characterized by smooth glassy surfaces that are often folded into ropy textures as a consequence of surface compression while they are still flowing (Fig. 2A). Pāhoehoe flow fields are fed through lava tubes, which develop within large lava sheets by protracted flow of lava at low to moderate rates. Crevices between surface folds may trap moisture and allow early colonization by plants with windblown spores, such as ferns (Fig. 2B). The surface crust of pāhoehoe flows is easily removed, and resulting shallow holes were used by native Hawaiians to plant upland taro and sweet potato. The thin glassy surface of pāhoehoe flows was also easily chipped away to create petroglyphs (Fig. 2C).

'A'ā flow surfaces are covered in rough clinkers that are typically 20–50 cm in diameter and form by shearing and tearing of partially crystalline lava along the flow margins and upper surface (Fig. 2D). 'A'ā flows form when eruption rates are high or in distal parts of flow fields (when the lava has had time to cool and crystallize). Because the rubbly character of 'a'ā flow surfaces makes them difficult to traverse, native Hawaiians constructed paths across these flows by breaking large clinkers into small pieces. The clinkery flow surfaces also make the boundaries between flows highly porous and permeable to water, despite less porous flow interiors. For this reason, active volcanic regions that are dominated by 'a'ā (or blocky) flows rarely have rivers or other manifestations of surface water, but instead are characterized by large aquifers and low-elevation springs.

ASH

Volcanic ash, in a strict sense, is a term used for solid fragments of volcanic material that are less than 2 mm in size. These fragments can include both frothy glass (quenched melt) and fragments of crystals (Fig. 3A). Ash is formed by explosive eruptions, where rapidly expanding gases within the ascending and cooling magma literally blow the magma apart, or by disintegration of growing lava domes during collapse. Ash-forming eruptions are most common along volcanic arcs that border subduction zones. Magma generated in this tectonic setting is rich in dissolved water, which powers highly explosive eruptions when magma ascends toward the Earth's surface. Like lava, ash is relatively impermeable

FIGURE 2 Photographs of Hawaiian basaltic lava flows. (A) An active pāhoehoe flow from Kīlauea Volcano, Hawai'i, slowly advances; surface wrinkles form by compression of the newly formed surface crust during flow. (B) Ferns growing on a young (1992) pāhoehoe flow, Kīlauea Volcano. (C) Petroglyph carved in an older pāhoehoe flow, Kīlauea Volcano. (D) Active 'a'ā flow from Kīlauea Volcano; flow surface is covered with clinkers that form as the flow surface is pulled apart during flow advance. These clinkers form a rubbly, and permeable, upper and lower surface on solidified flows.

to surface water, but unlike lava it is also highly mobile, easily eroded, and prone to airborne redistribution.

Explosive volcanic eruptions occur when water (or other magmatic gases such as carbon dioxide and sulfur) that is originally dissolved in the melt comes out of solution (vesiculates) as magma rises toward the surface (depressurizes), in a process similar to the release of carbon dioxide in a bottle of soda when pressure is released as the cap is removed. As bubbles form and expand within the melt, they both drive the bubble–melt mixture upward and blow the magma apart, ultimately creating violent eruptions that produce large volcanic plumes (Fig. 3B). Such plumes may carry fine ash fragments into the stratosphere, where they can then be transported, literally, around the globe. Most ash particles, however, fall out of the plume within tens to hundreds of kilometers from the volcano, along the trajectory of the wind.

More common are less violent eruptions that are associated with the growth of lava domes. Such eruptions may last for years, with ash posing persistent hazards to aircraft and covering fields and houses with thick deposits that are easily mobilized by heavy rainfall. A recent example of this type of activity can be found on the island of Montserrat, part of the Lesser Antilles island arc in the Caribbean. Here the Soufriere Hills volcano became active in 1995; by the end of 1997 the capital city of Plymouth had

FIGURE 3 (A) Scanning electron microscope image of ash particles from Mount St. Helens volcano, Washington. Seen here are ragged bubbly fragments of quenched melt and smooth-surfaced crystal fragments. Image is about 300 microns in width. (B) Eruption column from Mount Cleveland, Alaska, as photographed by J. N. Williams (NASA) from the International Space Station on May 23, 2006. Image available at http://antwrp.gsfc.nasa.gov/apod/image/0606/volcanoplume_iss_big.jpg. (C) Aerial view of ash-damaged buildings on Montserrat; both roof collapse and partial burial are the result of ash fall. (D) Summit region of Stromboli volcano, Italy. Summit was covered with large frothy lava "bombs" as well as ash during the eruption of April 5, 2003.

been largely covered by ash and pyroclastic debris. Activity persists today, and nearly 60% of this small island is currently uninhabitable (Fig. 3C).

Ash is also a component of basaltic Strombolian eruptions, named for the island of Stromboli (Italy), which has been erupting almost continuously for at least the past 1000 years, thus earning it the name "Lighthouse of the Mediterranean." Here several small eruptions occur every hour; less frequent, larger eruptions, such as one that occurred on April 5, 2003, cover the volcano's summit with frothy volcanic bombs as well as basaltic ash (Fig. 3D). More energetic basaltic eruptions produce more ash. Although not located on an island, an example of such an eruption is that of Parícutin volcano in Mexico, which was born in a cornfield on February 20, 1943. For the next nine years, strong explosions built up over 12 m of ash and scoria deposits in areas near the volcano, forcing local residents to move to more distant locations. Now, more than 50 years after the end of the eruption, much of the ash remains, but much has also been eroded and redistributed by annual monsoon rains.

SEE ALSO THE FOLLOWING ARTICLES

Eruptions: Laki and Tambora / Hawaiian Islands, Geology / Island Formation / Volcanic Islands

FURTHER READING

Cashman, K. V. 2004. Terrestrial volcanism, in *Volcanic worlds*. R. Lopes and T. Gregg, eds. Worthing, UK: Praxis Publishing Ltd., 5–42.
Druitt, T. H., and B. P. Kokelaar, eds. 2002. *The eruption of Soufrière Hills volcano, Montserrat, from 1995 to 1999*. Geological Society of London Memoir 21.
Macdonald, G. A. 1953. Pahoehoe, 'a'ā, and block lava. *American Journal of Science* 251: 169–191.
Sigurdsson, H., B. F. Houghton, S. McNutt, H. Rhymer, and J. Stix, eds. 2000. *Encyclopedia of volcanology*. San Diego, CA: Academic Press.
Soule, S. A., and K. V. Cashman. 2005. The shear rate dependence of the pahoehoe-to-'a'ā transition: analog experiments. *Geology* 33: 361–364.

LAVA TUBES

JIM KAUAHIKAUA
U.S. Geological Survey, Hawaii National Park

FRANK HOWARTH
Hawaii Biological Survey, Honolulu

KEN HON
University of Hawaii, Hilo

Lava tubes, originally called "pyroducts," form within lava flows as thermally insulated conduits through which lava is carried away from a volcanic vent and supplied to an active lava flow. After draining and cooling, these same lava tubes, along with the right mixture of water, in the form of trapped humid air, and food, in the form of organic debris and root systems, can provide a very specialized ecological niche that requires organism adaptation. The opportunity to colonize these forbidding environments is responsible for some of the fastest evolutionary adaptation currently known.

VOLCANIC ISLANDS AND LAVA TUBES

Volcanic islands, known to have lava tube caves, grow by repeated eruptions of basaltic lava. These islands form in two distinctly different settings—over hotspots or in island arcs. Most basaltic ocean islands are related to high rates of magma production either in the middle of tectonic plates or near spreading ridges. These are the "hotspot" volcanoes and include Hawai'i, Iceland, the Azores, Réunion, and many other island chains scattered throughout the world's oceans. The processes that lead to the formation of these islands are not well understood but seem to be related to varying contributions of deep mantle plumes. As these islands grow older or move away from their source, the erupted lavas often become richer in silica and alkalis (sodium and potassium).

Basaltic islands may also form along "primitive" sections of island arcs and are related to subduction of one ocean plate beneath another. Eruptions of tube-forming basalt are most common as arc island volcanoes emerge from the sea in places like the Izu–Bonin, Tonga–Kermadec, Banda Api, Aeolian, and South Sandwich subduction zones. As the arc ages and the islands grow, the magmas become much higher in SiO_2, and eruptions take on the explosive nature that characterizes most "Ring of Fire" subduction zone volcanoes. This change in lava chemistry diminishes the possibility that tubes will form.

Most basaltic volcanic islands begin as submarine seamounts that grow by eruptions of pillow lavas. As they near the surface and emerge from the sea, interaction with seawater produces pyroclastic eruptions, like those of Surtsey in Iceland, that mantle the seamount in glassy sand and ash. Ensuing subaerial eruptions of lava flows build shield and cinder cone complexes that give the islands their distinctive appearance.

FORMATION OF LAVA FLOWS

Effusive eruptions are the result of gentle gas expansion that drives magma from the ground in spectacular fissure eruptions and lava fountains. The erupted lava is very hot (typically 1100–1200 °C) and fluid, allowing the gases to escape easily and non-explosively. The resulting lava flows pour down the slopes of the volcanoes and form either pāhoehoe or ʻaʻā, depending on a combination of composition, temperature, water content, ground slope, and effusion rate.

Effusion rates in excess of about 5 m^3/s favor formation of ʻaʻā flows, whereas lower effusion rates favor pāhoehoe flows. In regions of low ground slope, pāhoehoe lava flows are favored, even during high effusion rate eruptions or at great distances from the vent. Effusion rates are generally highest at the beginning of an eruption and decrease as the eruption continues. Therefore, long-lived eruptions and flow fields typically become dominated by pāhoehoe flows even if they were initially ʻaʻā.

FORMATION OF LAVA TUBES

Lava tubes may form in both pāhoehoe and ʻaʻā flows. ʻAʻā channels may begin to crust over, allowing the transport of hotter, more fluid lava that favors pāhoehoe-morphology flows. Within a short period of time, the ʻaʻā flow is covered with pāhoehoe channel overflows, and the channel becomes roofed over, forming a lava tube.

Pāhoehoe channels may also crust over to form lava tubes. Pāhoehoe flows characteristically "self-seal" as the molten surface solidifies rapidly upon contact with air, forming an insulating crust that may inflate in response to continued lava supply. As the flow advances downslope, the lava eventually develops preferred pathways that become lava tubes. Open lava tube caves are less common in inflated flows because they occur most often on flat slopes that do not drain at the end of an eruption.

As both ʻaʻā and pāhoehoe flows mature, lava tubes develop. For broader flows, the tube network can develop a complicated, braided, or anastomosing network. With time, the network becomes consolidated, sometimes into a single tube. The development of tubes favors efficient transport of lava because solid basalt is a very good thermal insulator. Lava flowing through tubes on Kīlauea rarely loses more than a single degree C per kilometer of travel.

Lava tubes allow lava flows to extend great distances away from their vents. The Undara flows of Australia exceed 100 km in length, but on ocean islands, the length of the flows and lava tube systems are limited by the size of the island. Examples of long lava tube caves are the 42-km-long Kazumura lava tube cave on Hawaiʻi Island and a single sealed lava tube 13 km in length on Jeju Island, South Korea.

The concentration of all shallow tubes into a single tube begins the process of downcutting through a combination of melting and abrasion. Within a month, tubes can become deeper, creating a keyhole shape in cross section. A cross section through an entire mature flow will reveal the original anastomosing shallow caves and a "master tube" that is greatly enlarged and downcut by flowing lava. Any of these conduits can form caves when drained.

Localized collapse of the roof of active tubes creates skylights (Fig. 1) that frequently form over or near lava falls. Tiered lava tube systems may form around skylights

FIGURE 1 View through a skylight of lava flowing through one of the many lava tubes formed during the current eruption of Kīlauea volcano. The lava stream is about 1150 °C, about 2 m wide, and flowing away from the viewer at approximately 3 m/s. Note the incipient shelf formation over the stream surface.

by several processes. Cooling through a skylight can cause the lava channel to crust over, forming an internally roofed, two-tiered tube section. Overflows from skylights and ruptures in the tube may be caused by blockages of the tube due to collapse or by increased lava supply in the tube. The resulting overflows from the tube can form a new set of lava tubes and create a stacked lava tube system.

LAVA TUBE STRUCTURES

Internal ornamental structures that form while lava tubes are active include shark's tooth lava stalactites, hollow straw lava stalactites, anhydrite encrustations, driblet spire lava stalagmites, extruded lava buds, secondary lava flows, elephant skin textures, wall slumps, channel gutters, fluted walls, and floor volcanoes. Shark's tooth lava stalactites are drip structures that form during the rise and fall of lava against the tube roof. Growth rings inside these stalactites record individual coats of lava added during these events. Channel gutters and ledges on the walls of lava tubes record decreasing lava levels within the tubes due to either downcutting or declining lava supply. Many shelves are incomplete roofs downslope of skylights. Slumped and crumpled walls (elephant skin textures) result when partially molten tube walls begin to deform. Water-rich partial melts within the tube walls, roof, and floor may be expulsed to form hollow straw stalactites and cogenetic driblet spires (Fig. 2), secondary lava flows, and floor volcanoes. All of these features have distinctive chemical compositions that show that they were produced by remelting previously solidified tube roof, wall, and floor material.

Prolonged reaction of the remelted wall rock with the high-oxygen atmosphere within lava tubes creates a unique suite of high-temperature minerals. Most of the interior surfaces of lava tubes are coated with magnesioferrite, which gives many cave interiors a metallic gray appearance. Parts of the tube interior may also have an appearance of brownish ceramic-like glaze, where surfaces have not been exposed for sufficient time to develop crystalline magnesioferrite and related minerals. High-temperature anhydrite sublimates have also been found as drusy crystals coating the walls and roof of active lava tubes.

Many secondary minerals form during cooling of lava tubes after lava ceases to flow within the system. This process can take many months, and during this time the original high-temperature oxides and sulfates react and form lower-temperature iron oxides and hydroxides as well as a host of low-temperature sulfate minerals. Sections of lava tubes that cool near open skylights commonly develop bright red hematite glazes. Deep within the cave systems, the original magnesioferrite coatings are transformed into dull metallic to brownish iron oxides. Many of the sulfate minerals and speleothems form rapidly when the temperature of the cave drops below 100 °C, allowing water to infiltrate the tube. In wet climates, these deposits are washed away within months, whereas in arid climates they may last for years. Several lava tube caves on Jeju Island, South Korea, have well-developed calcite speleothems and flowstones due to leaching of water through calcareous sand dunes that overlie the tube systems.

LAVA TUBE CAVES

After a lava tube system has drained and cooled enough to allow human entry, each lava tube cave is defined by its traversibility. Therefore, a single lava tube system may result in several different lava tube caves, with each cave being defined as any length of tube passable by humans that includes at least one entrance. Although the lava tube cave systems may not have apparent physical connections, the extensive crack systems within lava flows allow small organisms ample mobility.

Besides lava tube caves (Fig. 3), basaltic lava flows produce abundant additional voids of varying sizes that may not be passable by humans. These are made by degassing,

FIGURE 2 Soda straw lava stalactites and stalagmites in a lava tube cave that last carried lava about two years before this photograph was taken.

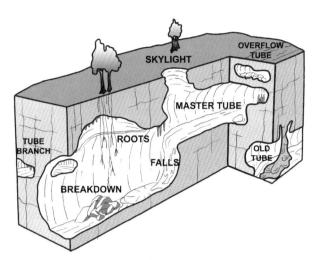

FIGURE 3 Schematic of a lava tube cave.

expanding flow lobes, gaps between new flows and the older substrate, buried 'a'ā rubble, and cracking of the cooling lava. These voids are interconnected by a vast system of cracks that radiate through the flow. The cracking is so pervasive that, after cooling, few lava tube caves can carry water except during times of high rainfall. The smaller, capillary-sized spaces created by gas vesicles are less important biologically because their small size limits the amount of food resources they can hold and transport. Voids larger than about 5 cm can transport large volumes of food into the underworld.

COLONIZATION OF THE LAVA FLOW SURFACE

The surfaces of basaltic lava flows are marked by numerous cracks, crevices, and broken lava tubes. Many of these surface openings connect with the system of subterranean voids. Any organic material falling on the surface is likely to fall or be washed into these voids and to subsequently sink underground. Thus, the surface of these flows can appear barren for decades to centuries, depending on the time it takes for the surface voids to fill and soil to form. The barren environment is an extreme one for life: The dark lava absorbs solar radiation and heats the surface to intolerable temperatures. Rain disappears underground almost as it falls, making the surface extremely dry (xeric). The changing heat and open environment generate strong surface winds. However, on each island, a group of specialized organisms soon colonize the surface.

The first to arrive are crickets, spiders, and other invertebrates, which colonize the flow within a month of the surface solidifying. Many of these are specifically adapted to live only on these barren lava flows. They hide deep in moist cracks during the day and emerge at night to feed on windborne organic matter concentrated in cracks on the flow. This aeolian, or wind-supported, ecosystem disappears shortly after plants arrive and the surface cracks become blocked. Within a year, cave-adapted animals begin colonizing the subterranean habitat, arriving from the voids in neighboring older flows. Predators and omnivores arrive first and feed on lost windblown animals and organic matter that washes or falls into the deeper moist voids.

The arrival of pioneering plants on the surface paves the way for colonization by the root- and litter-associated fauna. The plants can begin within a decade in wet climates but may take more than a century under desert conditions. Because of the lack of soil and surface water (due to the lava flow's high porosity and extensive cracking), colonizing plants must send their roots deep underground to obtain water and nutrients. In Hawai'i, a 1-m-tall pioneering *Metrosideros* tree on the surface requires the support of a branched root system that may extend 15 m or more underground (Fig. 4). In addition to plant roots, food resources in lava tube caves include chemoautotrophic bacteria and organic material brought in by roosting or lost animals washed in by rain or blown in by wind.

The young flow remains inhospitable for most soil and litter animals. The abundant underground food resources have allowed a few surface animals living on each volcanic island to adapt to live permanently underground on the same island.

FIGURE 4 *Metrosideros polymorpha* ('ōhi'a) roots in a lava tube cave within a lava flow erupted from Mauna Loa, Hawai'i, in 1881.

LAVA TUBE CAVE HABITATS

The cave habitat is dominated by water vapor and carbon dioxide. Water vapor is lighter than air and tends to rise out of caves via openings. However, long passages with convoluted shapes and smaller voids tend to trap water vapor, and the air remains saturated. Saturated air is above the equilibrium humidity of blood, and cold-blooded animals literally drown without special adaptations to cope with the excess water. Carbon dioxide, which is produced by respiration, is heavier than air and tends to concentrate in deeper and more isolated passages, especially within the intermediate-sized spaces in the lava. Uniquely high CO_2 concentrations may be found in lava tube caves located in active volcanic regions that are emitting the gas. High concentrations of CO_2 force animals to breathe more, intensifying the respiratory effect of excess water.

Cave-adapted animals prefer the stagnant, water-saturated atmosphere found in the innermost deep zones. Lava tube caves are three-dimensional, complex, maze-like passages with lethal or near-lethal gas concentrations; scattered, hard-to-find food resources; unforgiving rocky substrates; and wet, slippery, vertical surfaces. This perpetually dark and humid environment is, at least, stressful or outright inhospitable for most organisms. Few animal groups can exploit this habitat.

ADAPTATION BY CAVE ORGANISMS

In Hawai'i, both the lava flow crickets and big-eyed lava flow spiders (Fig. 5) have founded cave populations. Other colonizers have come from rainforest and marine littoral habitats. More than 75 species of cave arthropods have been discovered to date in the Hawaiian Islands, with 26 species known from Hawai'i Island, including beetles, crickets, planthoppers, spiders, moths, flies, earwigs, and true bugs.

FIGURE 5 *Adelocosa anops* (no-eyed big-eyed hunting spider) in a Kaua'i, Hawai'i, lava tube cave.

The ancestors of cave-adapted animals lived in damp, dark environments and could already cope with some of the stresses in the lava tube environment. The planthoppers are a good example. Nocturnal animals living in damp, wet-rock habitats, such as rocky sea coasts, stream margins, and lava flows, already possessed many of the necessary adaptations for subterranean life. The accumulating food resources provided the impetus to give up surface life and move underground permanently.

Cave adaptation has occurred independently on each island from surface ancestors living on the island. At least 12 separate groups have adapted independently to lava tubes at least twice in Hawai'i, indicating that cave adaptation is a natural process wherever the environment is suitable and animals are present. Because the cave animals move underground from older lava tubes to newer tubes, the cave species are nearly always older than the tubes in which they occur. Thus, the maximum time necessary for cave adaptation could be as old as the island itself.

The hallmark of cave-adapted species is the loss of conspicuous structures such as eyes, body color, protective armor, and for some insects, wings. These structures are useless underground but expensive for the body to make and maintain. Therefore, they can be lost quickly when selection is relaxed. A clue to how such losses might happen quickly is demonstrated by the cave-adapted planthoppers. The nymphs of surface relatives also feed on plant roots and have reduced eyes and body color, whereas the adults have big eyes, cryptic colors, and strong wings. The cave-adapted adult descendents maintain the nymphal eyes, color, and other characters into adulthood, a phenomenon known as neoteny.

Organisms adapting to live permanently underground also need to make changes in their physiology, form, and behavior to cope with the stressful environment, especially the high relative humidity and occasional episodes of high carbon dioxide concentrations. Because many of the normal cues used by surface animals behave abnormally or do not occur underground (e.g., wind, light-dark cycles, sound), the organisms also need to adapt to take advantage of new cues to find mates and food. Consider trying to follow a scent plume, whether from a potential mate or food resource, in a three-dimensional, dark maze; an animal would need to take an indirect path to the source. In addition, predators might take advantage of the same plume and be waiting for the unwary. Jumping or falling might land a hapless animal in a pitfall trap. Most lava tube arthropods have specialized elongated claws to hold on to the glassy wet rock surfaces. Many have elongated legs to step across cracks rather than having to descend and climb the other side. Small surface insects are too heavy or are unable to climb the meniscus at the edge of small pools and eventually drown. However, many cave-adapted insects have unique knobs near the base of each elongated claw that allow them to climb the meniscus and escape. Some of the latter species are predators or scavengers, who wait on pools for victims.

SEE ALSO THE FOLLOWING ARTICLES

Caves, as Islands / Crickets / Insect Radiations / Lava and Ash / Seamounts, Geology / Spiders

FURTHER READING

Calvari, S., and H. Pinkerton. 1998. Formation of lava tubes and extensive flow field during the 1991–1993 eruption of Mount Etna. *Journal of Geophysical Research*. 103: 27,291–27,301.

Cashman, K. 2004. Volcanoes on Earth: our basis for understanding volcanism, in *Volcanic worlds: exploring the solar system's volcanoes*. R. M. Lopes and T. P. Gregg, eds. Chichester, UK: Springer and Praxis Publishing, 5–42.

Culver, D. C., and W. B. White, ed. 2004. *The encyclopedia of caves*. Burlington, MA: Elsevier Academic Press.

Helz, R. T., C. Heliker, K. Hon, and M. Mangan. 2003. Thermal efficiency of lava tubes in the Puʻu ʻŌʻō-Kupaianaha eruption, in *The Puʻu ʻŌʻō-Kupaianaha eruption of Kīlauea volcano, Hawaiʻi: the first 20 years*, USGS Professional Paper 1676. C. Heliker, D. Swanson, and T. J. Takahashi, eds. Reston, VA: U.S. Geological Survey, 105–120.

Howarth, F. G. 1983. Ecology of cave arthropods. *Annual Review of Entomology* 28: 365–389.

Kauahikaua, J., K. Cashman, T. N. Mattox, C. C. Heliker, K. Hon, M. Mangan, and C. R. Thornber. 1998. Observations on basaltic lava streams in tubes from Kilauea volcano, island of Hawaiʻi. *Journal of Geophysical Research* 103: 27,303–27,323.

Sigurdsson, H., B. F. Haughton, S. R. McNutt, H. Rymer, and J. Stix, eds. 2000. *Encyclopedia of volcanoes*. San Diego, CA: Academic Press.

Wilkins, H., D. C. Culver, and W. F. Humphreys, eds. 2000. *Subterranean ecosystems, ecosystems of the world, 30*. Amsterdam: Elsevier Press.

LEMURS AND TARSIERS

ROBERT D. MARTIN

The Field Museum, Chicago, Illinois

Lemurs and tarsiers—two of the five main groups of living primates—are island inhabitants. Lemurs have undergone a major diversification on Madagascar, resulting in over 100 modern species (including 16 recently extinct representatives). Diversification of tarsiers on islands of Southeast Asia has been more modest, generating 17 modern species. However, 15 of those tarsier species are found on the relatively small island complex of Sulawesi, which is a "hotspot" for evolutionary divergence.

PRIMATES ON ISLANDS

The mammalian order Primates, to which we ourselves belong, contains five major groups: (1) lemurs; (2) lorisiforms (bushbabies and lorises); (3) tarsiers; (4) New World monkeys; and (5) Old World simians (monkeys, apes, and humans). Nonhuman primates are generally restricted to tropical and subtropical regions, and most are forest inhabitants. They are found on all mainland areas of the southern continents, except Australia. New World monkeys occur in South and Central America, whereas both lorisiforms and Old World simians are widely distributed throughout Africa and Asia. However, two of the five major groups of primates are restricted to island areas in the south: lemurs on Madagascar, and tarsiers on the island archipelago of Southeast Asia.

Perhaps because of their isolation on islands and the resulting lower levels of competition with other animals, both the lemurs and the tarsiers have remained relatively primitive in many features. Although tarsiers are actually quite advanced in certain respects and more closely related to higher primates (monkeys, apes, and humans), they have also retained many primitive features. At least in terms of the morphology of their cheek teeth, tarsiers have remained relatively unchanged over the past 40 million years or more. This has led some authors to describe them loosely as "living fossils."

LEMURS

Lemurs are found only on Madagascar, and they are the only primates to occur there. Molecular evidence has now confirmed previous indications from shared morphological features that lemurs are all derived from a single common ancestor. Within Madagascar, they have undergone a remarkable adaptive radiation. For example, dentitions are more diverse in lemurs than in any other living primate group, and the range of dietary adaptations is correspondingly extensive. The modern lemur fauna contains almost 100 species, ranging in size from mouse lemurs, weighing just 40 g, up to the indri, with a body mass of about 6.3 kg. In addition, there are 16 species that died out just a few thousand years ago, probably because of climatic change combined with human colonization of Madagascar. These subfossil lemurs, which are really part of the modern fauna, are generally very big and extend the body size range up to 100 kg or more. There is no true fossil record for lemurs. Relationships among lemurs must therefore be established from a combination of morphological and molecular evidence.

The remarkable diversification of Malagasy lemurs has occurred within a geographical area far smaller than that occupied by any other major group of primates. Lemurs account for around one-fifth of the genera and species of modern primates, and the ranges occupied by individual species are correspondingly small. The diversity of lemurs is due at least in part to limited competition from other mammals. Only three other land mammal groups are long-established inhabitants of Madagascar: tenrecs, rodents, and carnivores.

Lemurs show a number of general trends with increasing body size: from nocturnal to diurnal habits, from insectivory through frugivory to folivory, and from solitary foraging to gregarious social life, although most show strict seasonal breeding. Five distinct subgroups of lemurs are recognized: (1) mouse and dwarf lemurs (*Allocebus, Cheirogaleus, Microcebus, Mirza, Phaner*); (2) sportive lemurs (*Lepilemur*); (3) true lemurs and bamboo lemurs (*Eulemur, Hapalemur, Lemur, Prolemur, Varecia*); (4) the indri and its relatives (*Avahi, Indri, Propithecus*); and (5) the aye-aye (*Daubentonia*). Molecular and chromosomal evidence indicates that the aye-aye diverged first, but subsequent separations between the other four subgroups occurred in quite rapid succession.

A key question regarding the adaptive radiation of Madagascar lemurs concerns the process by which so many species could have arisen within the island. As a rule, a barrier of some kind between populations is required for speciation to occur. The combination of a cooler, elevated central plateau and several major rivers running down to the sea apparently accounts for effective isolation of a ring of forest regions around the coasts of Madagascar. Marked climatic zonation in Madagascar endows each region with a particular range of environmental conditions. Regarding rainfall, trade winds deposit most of their moisture along the east coast, leaving the western part of the island relatively dry. There is a latitudinal gradient for mean temperatures in the coldest month, which are higher in the north than in the south. The vegetation of any region isolated between two major rivers reflects the local climatic conditions, with a spectrum ranging from dense tropical rain forest in the northeast down to semi-arid, shrubby vegetation in the southwest. It thus seems highly likely that the animals in any coastal or lowland region are typically isolated and adapted to local conditions, including special features of both climate and plant life. Occasional migration of closely related species between lowland regions would lead to competition between sibling species, often resulting in divergent specialization.

This proposed model of lemur speciation in Madagascar has recently been subjected to a number of tests. Molecular evidence has shown that major rivers do, indeed, represent boundaries between extant sister species in many cases. Moreover, chromosomal and/or molecular evidence has allowed identification of many new species, in particular small-bodied nocturnal species ("cryptic species"), which correspond quite well to the major lowland forest regions recognized in the model. The most spectacular example for new cryptic species is the sportive lemur, *Lepilemur*. For many years, classifications recognized only a single species, *Lepilemur mustelinus*, with an extensive distribution throughout the forests of Madagascar. Eventually, this species was divided into two, with the red-tailed sportive lemur (*L. ruficaudatus*) inhabiting the humid forests in the east and the weasel mouse lemur (*Lepilemur mustelinus*) typically inhabiting the drier forests in the west. However, chromosomal and molecular evidence has progressively revealed the existence of at least 24 *Lepilemur* species. An equally striking case is that of the lesser mouse lemur, *Microcebus*, with the long-recognized island-wide species *M. murinus* initially divided into two: the brown mouse lemur (*Microcebus rufus*) in the east and the gray mouse lemur (*Microcebus murinus*; Fig. 1) in the west.

FIGURE 1 Gray lesser mouse lemur (*Microcebus murinus*), eating an insect. This small-bodied nocturnal species has an omnivorous diet of arthropods and various plant items. At one time, all lesser mouse lemurs were allocated to this single species, but molecular evidence has now revealed the existence of at least 16 species. Photograph by David Haring.

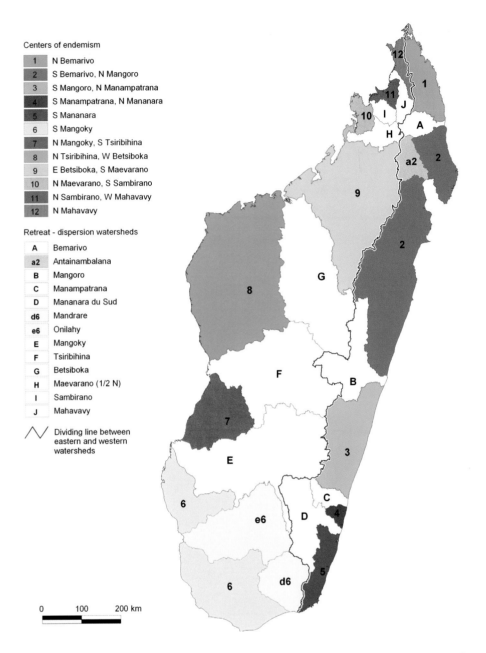

FIGURE 2 Topography of Madagascar. The elevated central plateau is the source of major river drainages that serve as routes of retreat into refuges and subsequent dispersion (white zones with letters). Centers of endemism (colored zones with numbers) are assemblages of smaller watersheds with sources at lower elevations isolated between retreat-dispersion watersheds. Recent distributions of lemurs at the level of species or subspecies are broadly concordant with the proposed centers of endemism. Illustration courtesy of Steven M. Goodman. Modified from Wilmé et al. (2006), Science 312: 1063–1065. Reprinted with permission from AAAS.

Subsequently, molecular evidence has gradually revealed the existence of at least 16 species of *Microcebus*, and it is likely that more remain to be discovered.

A more elaborate model has been developed to explain the general confinement of individual animal species to restricted geographical ranges (microendemism) in Madagascar using an analysis of watersheds in the context of Quaternary climatic shifts (Figs. 2–3). River catchments with sources at relatively low elevations were identified as zones of isolation that led to speciation of locally endemic taxa. By contrast, river catchments with sources at higher elevations were identified as zones of retreat and dispersion in which microendemism is correspondingly less pronounced. The resulting model provides a valuable framework for interpreting the process of explosive speciation on the island.

FIGURE 3 A female ringtail lemur (*Lemur catta*) with her ventrally carried infant. Ringtail lemurs, which are restricted to relatively dry regions of southwestern Madagascar (areas 6, d6, and e6 in Fig. 2), are day-active and live in troops containing around two dozen individuals. Photograph by the author.

FIGURE 4 Like lesser mouse lemurs, the Philippine tarsier (*Tarsius syrichta*), shown here with a lizard in its mouth, is a small-bodied nocturnal species, but it shows a number of special features. The very long hindlimbs reflect the specialized locomotor pattern of vertical-clinging and leaping between thin, vertical trunks. Relative to body size, the enormous eyes are bigger than those of any other mammal. Tarsiers are also the only primates that feed exclusively on animal prey. Photograph by David Haring.

TARSIERS

In contrast to lemurs, the tarsiers are members of a modest adaptive radiation, now confined to islands in the Southeast Asian archipelago. (Early fossil relatives have been documented from China and Thailand.) All species are allocated to the single genus *Tarsius*. One striking feature of tarsiers is the long hindlimb, notably including an elongated ankle region (tarsus). This feature, linked to their specialized locomotor pattern of vertical-clinging and leaping between thin, vertical trunks of saplings, has given rise to their common and scientific names. In connection with their nocturnal habits, they also have enormous eyes, which are by far

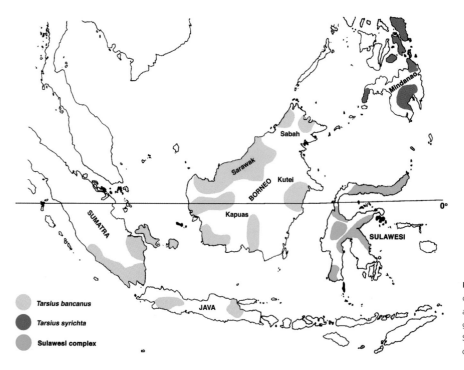

FIGURE 5 Distribution of tarsier species on islands of the Southeast Asian archipelago. Myron Shekelle has suggested that the species complex on Sulawesi may contain 15 tarsier species. Updated from Niemitz, 1994.

the largest (relative to body size) of any mammal. All tarsiers are quite small, with body masses ranging from 60 to 135 g. Uniquely among primates, tarsiers feed exclusively on animal prey: mainly arthropods, but also small vertebrates—including highly venomous snakes. They have never been seen to ingest plant food of any kind.

Until 1987 only three tarsier species were recognized: the western tarsier (*Tarsius bancanus*) on Borneo and Sumatra, the Philippine tarsier (*Tarsius syrichta*; Fig. 4), and the eastern tarsier (*Tarsius spectrum*) on Sulawesi. However, following a 1987 report on a pygmy species (*Tarsius pumilus*) discovered in museum collections, a fifth species (*Tarsius dianae*), also found on Sulawesi, was announced in 1991. By 2003, eight Sulawesi species had been recognized, and a study of vocalizations then revealed 15 distinct populations, all of which may be separate species (Fig. 5).

As with the lemurs of Madagascar, it must be asked how so many tarsier species could have evolved on Sulawesi. There are numerous other examples of animal groups, such as macaques and frogs, that have shown equally spectacular speciation on Sulawesi, the eleventh largest island in the world. In contrast with the model developed for Madagascar, geographical barriers such as major rivers are not prominent features of Sulawesi, there is little climatic zonation, and the original vegetation covering the island was relatively uniform tropical rainforest. On the other hand, Sulawesi has a peculiar geological history, and this seems to account for much of the zonation connected with subdivision between individual species. Reconstructions have shown that four major land fragments drifted together from different directions to form the island.

SEE ALSO THE FOLLOWING ARTICLES

Madagascar / Mammal Radiations

FURTHER READING

Alterman, L., G. A. Doyle, and M. K. Izard, eds. 1995. *Creatures of the dark: the nocturnal prosimians.* New York: Plenum Press.
Martin, R. D. 2000. Origins, diversity and relationships of lemurs. *International Journal of Primatology* 21: 1021–1049.
Pastorini, J., U. Thalmann, and R. D. Martin. 2003. A molecular approach to comparative phylogeography of extant Malagasy lemurs. *Proceedings of the National Academy of Sciences USA* 100: 5879–5884.
Tattersall, I. 1982. *The primates of Madagascar.* New York: Columbia University Press.
Wilmé, L., S. M. Goodman, and J. U. Ganzhorn. 2006. Biogeographic evolution of Madagascar's microendemic biota. *Science* 312: 1063–1065.
Wright, P. C., E. L. Simons, and S. Gursky. 2003. *Tarsiers: past, present, and future.* New Brunswick: Rutgers University Press.

LINE ISLANDS

CHRISTOPHER CHARLES AND STUART SANDIN

Scripps Institution of Oceanography, La Jolla, California

Oceanographers are often given to building transects, where contrasting properties can be observed within a relatively narrow geographic area. Such is the opportunity afforded by the Line Islands in the central tropical Pacific, an island chain that stretches from Johnston atoll in the north to the Tuamotu Islands in the south. The very name of the chain invites this transect approach, and, although the basic connotation of "The Line Islands" does not do justice to the complexity of its geological origin, the appellation does quite aptly describe the gradients in geographic properties manifested on the various islands.

A "NATURAL LABORATORY"

The atolls of the Line Island chain span several oceanographic and climatic zones. As a result, the terrestrial characteristics of the islands vary from lushly vegetated to nearly desert-like terrains, while the submerged coral reef fauna is subject to varying influences of the El Niño phenomenon. Furthermore, although the Line Islands archipelago is the most isolated collection of islands on the planet—no continent is closer than 5000 km from any of the Line Islands—the chain is now characterized by concentrated pockets of human disturbance; this concentration of human influence has important implications for the study of island resource management. All of these aspects of the Line Islands make the archipelago an essential natural laboratory for understanding the development, evolution, and governing processes of the tropical atoll environment.

ORIGIN AND GEOLOGICAL FEATURES

Many linear island chains are often assumed to have arisen from the passage of a crustal plate over a fixed "hotspot" or mantle plume. In the case of the Line Islands, it is clear that this hotspot model cannot explain the age, distribution, and origin of the chain. First, the Line Islands are not a simple chain; instead, they must be understood as part of a complex array of islands and seamounts. Given this geometry, reconstruction of plate movement, or plate "backtracking," does not lead to a convergence of all the islands and seamounts back to a

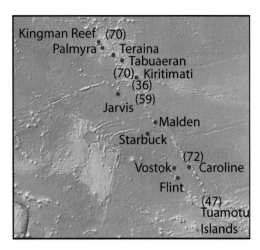

FIGURE 1 General bathymetric chart illustrating the geological context of the Line Islands. The numbers in parentheses are the ages (in millions of years) of sea-floor rock samples collected at the indicated locations.

single point source. Furthermore, there is no northward progression of age of the islands or seamounts, in contrast to what one might expect if a plate passed over a single hotspot or mantle plume. In fact, the crustal age of the northernmost islands is similar to that of their southernmost counterparts, and the ages of the ridges and platforms upon which the Line Islands rest fall into two main clusters with the formation of the northern Line Islands in the Late Cretaceous (70–80 million years ago) and the subsequent formation of some of the islands and neighboring seamounts in the Paleogene (35–50 million years ago) (Fig. 1). Finally, although regular age-depth relationships often characterize the sea floor away from spreading centers (because of the progressive cooling and sinking of the marine crust), the sea floor associated with the Line Islands deviates markedly from the expected trend with age—the northern Line Islands are more shallow than one would predict on the basis of the age of the surrounding marine crust. Taken together, the observations argue that the formation of the Line Islands was most likely an extension of the same general volcanic construction processes that created the Marquesas and the neighboring islands of the South Pacific—a phenomenon of intraplate volcanism termed the "South Pacific Superswell." Detailed study of the history of this volcanism and the age-depth relationships in the Line Island area may ultimately lead to a more complete picture of convection processes in the mantle.

The relative antiquity of the Line Islands implies that the present morphology of the islands must be largely a product of the complex interplay between (1) reef construction and sedimentation during sea level transgressions and (2) erosion and karstification of the emergent surface during sea-level regressions. The cycle of construction and erosion must have occurred repeatedly for all the Line Islands, given the history of sea-level fluctuations over the past 40 million years. As of yet, there have been no deep drilling campaigns that capture the entire sequence from island surface to basaltic crust, but one might nevertheless surmise the overriding importance of preexisting (antecedent) topography in determining the architecture of modern reef and land surfaces. This is especially true for the Line Islands, because there are no active tectonics in the region that otherwise might create significant structure.

This antecedent topography must surely account for much of the variability in the morphology of the islands. Only five of the islands have typical lagoons that are connected directly to the surrounding ocean water (Kingman, Palmyra, Tabuaeran [Fanning], Kiritimati [Christmas], and Karoraina [Millennium]). Flint and Vostok are small islands occupying most of the reef platform upon which they rest, whereas Karoraina is a crescent-shaped atoll. The margins of Starbuck and Malden consist of continuous land, whereas the interiors of these islands have salty inland lakes. Teraina (Washington), on the other hand, features an inland freshwater lake. Tabuaeran has discontinuous islands around its rim and a complex network of reefs within its lagoon. Kiritimati Island is by far the largest of the Line Islands, although at least 25% of its interior surface consists of an interconnected network of saline lakes of variable depths. In general, the Line Islands are characterized by uniformly low-lying topography, with elevations rarely exceeding 10 m above present mean sea level. Terrestrial substrates are typically reef sands and rubble, but one notable feature of many of the islands is that the coral rubble can reach boulder size (these coral boulders were presumably deposited by either past tsunamis or large storms).

CLIMATE

The characteristics of the Line Islands are shaped not only by their geological history but also by their location with respect to major oceanographic and climatic zones. The southern Line Islands (e.g., Flint, Vostok, Jarvis, Kiritimati) fall within the westward-flowing south equatorial current; the northern islands (Teraina, Palmyra, and to a lesser extent, Tabuaeran) lie in the realm of influence of the north equatorial countercurrent; and the northernmost islands (Kingman Reef, Johnston) are bathed by the westward-flowing north equatorial current. Forced by the winds, the surface ocean currents deliver differing heat and nutrient concentrations to the coral reef communities of the Line Islands. For example, Jarvis and Kiritimati lie on western edge of the equatorial "cold tongue," the region most directly influenced by the upwelling of colder, nutrient-rich subsurface water.

Across the Line Islands, average annual rainfall varies over tenfold, ranging from approximately 500 mm yr^{-1} in the most southern islands (Flint and Vostok) to over 5000 mm yr^{-1} in the north (Palmyra). These average rainfall patterns are related directly to the mean annual position of the intertropical convergence zone, which, while undergoing some seasonal north–south migration, nevertheless remains north of the equator throughout the year in the region of the Line Islands.

Annual averages of climate statistics do not provide a complete picture for the Line Islands because the interannual variability of temperature and especially rainfall in this part of the Pacific Ocean is highly significant. In fact, of all island chains across the tropics, the Line Islands are closest to the center of action for the El Niño–Southern Oscillation phenomenon (although the nature of the climatic influence does vary from island to island; for example, the temperature effect is maximized for the more equatorial islands) (Fig. 2). During strong El Niño years, the surface temperature in the region encompassing the Line Islands increases by as much as 3 °C during the December–March period, and the increase in rainfall may exceed historical averages by 200 to 300%. Kiritimati, for example, is subject to interannual sea-level variations of the order of 30–40 cm associated with El Niño warm events, and the greatly increased rainfall floods much of the atoll's interior, transforming the normally dry hypersaline flats into shallow brackish lakes. Thus, the fauna and flora must be capable of withstanding significant interannual extremes in climate. On a serendipitous note, the coral boulders exposed on the Line Island beaches offer one of the best available means to access the history of El Niño over the last few millennia (the chemistry of coral skeletons records a snapshot of ambient conditions at the time of coral growth).

BIOGEOGRAPHY

The extreme degree of isolation influences much of the islands' biology. Relative to less isolated sites, the islands have few species, either terrestrial or marine. The difficulty of dispersal across wide expanses of deep ocean limits the number of species that arrive to the Line Islands, and the small sizes of the islands provide only a limited number of habitats for arriving species to exploit. The species that are successful in surviving in the archipelago have created biological communities that efficiently exploit the available

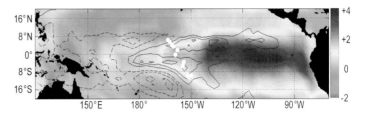

FIGURE 2 The Line Islands (white circles) are close to the center of the climatic disturbance associated with the El Niño–Southern Oscillation phenomenon. Shown here are the anomalies in temperature (colors, in °C) and rainfall (contours, in mm/day, negative is dashed) during the 1982–1983 El Niño event.

resources. A subset of species that use the Line Islands are temporary visitors. As islands in the middle of the deep central Pacific, the Line Islands provide rare shallow water and land resources to wide-ranging species such as whales, dolphins, pelagic sharks, and other fishes, sea turtles, and seabirds. The isolation of these islands also has limited the history and extent of human activities in the region.

Terrestrial Biota

Each of the Line Islands (except for Kingman) supports either forest or shrubland. The calcareous sand, low topography, and consequent lack of significant stored groundwater all imply that plants on the Line Islands cannot depend upon the soil to retain water between rainstorms. These conditions then create an intimate association between the floral assemblage and the amount of rainfall reaching the island. The most thickly vegetated islands are, not surprisingly, those that receive the most annual rainfall.

There is limited species diversity within the plant assemblage across the archipelago. Most species have large, buoyant seeds or small, light seeds capable of dispersing very long distances across the sea (for example, by flotation, by wind, or by attaching to feathers of seabirds). Because of the propensity for long-distance dispersal, there are few endemic plants and instead the Line Islands are populated by a host of wide-ranging, pan-Pacific species. One of the most notable species is *Pisonia grandis,* a tree species that historically dominated many Pacific coral atolls. *Pisonia* forests are renowned for their great stature and dense foliage, but they are becoming rare across the Line Islands because of a host of stressors. An introduced pest (a green scale insect, *Pulvinaria urbicola*) is believed to be causing mass mortality within the *Pisonia* population on Palmyra, whereas direct felling and replacement with coconut palms (*Cocos nucifera*) has hastened decline on other islands.

The flora of the islands is home to a limited, yet distinctive, collection of land animals. Most numerous on many of the Line Islands are the hermit crabs (principally

Coenobita brevimanus). Like their marine cousins, these hermit crabs use large snail shells (with openings up to 8 cm in diameter) to protect their bodies as they roam the forest floor for food. Also abundant is the larger land crab (*Cardisoma carnifex*), with a carapace width typically from 6 to 12 cm, which live in burrows and typically scavenge at night. Most impressive, however, is the coconut crab (*Birgus latro*), named for its ability to crack open coconuts with its claws. These crabs can grow to be 70 cm in thoracic length, have a leg span of 1 m, weigh a few kilograms (with some estimates over 4 kg), and live for decades. Because of their large size and their fabled taste, the coconut crab has been harvested extensively from the inhabited Line Islands. However, remnant populations of crabs thrive on the uninhabited islands (Palmyra, Millennium, Flint), with population estimates of hundreds of thousands to millions of crabs on the southern islands.

The land and vegetation of the Line Islands serve a critically important role as nesting areas for at least 19 species of seabirds. Because of the paucity of land in the central Pacific, the Line Islands attract millions of birds annually to complete their life cycles. The birds nest in the native vegetation, including the *Pisonia* forests, the shrubby beach naupaka (*Scaevola sericea*), and the tree heliotrope (*Tournefortia argentea*). Some of the more spectacular seabirds to visit these islands are the red-

FIGURE 3 General aspect of fore reef habitats at (A) Kingman and (B) Kiritimati, showing the degradation from a reef dominated by top predators and corals to a reef dominated by small planktivorous fishes and algae.

FIGURE 4 Representative 0.5-m^2 photos of the bottom at (A) Kingman and (B) Kiritimati, again showing reef degradation.

footed boobies, the great frigatebirds, and both the red- and white-tailed tropicbirds; enormous aggregations of sooty terns can fill the sky in pre-nesting displays. Some harvesting of seabirds and their eggs has reduced colony sizes on Kiritimati, Tabuaeran, and Teraina, but the uninhabited Line Islands continue to support a significant proportion of the seabird nesting for the central Pacific populations.

Marine Biota

The Line Islands are each surrounded by expansive fringing reefs. On the windward sides of the islands, the fringing reefs are exposed to frequent large Pacific swells and have thus been structured into sharp, steep slopes beginning near to shore. Leeward reefs typically include an extended shallow terrace (100–300 m wide and 10–20 m deep), dropping off to a steeper seaward reef slope. In contrast to much of French Polynesia, the majority of coral reef development in the Line Islands is fore-reef (habitats seaward of the reef crest).

The Line Islands are part of the central tropical Pacific bioregion, with marine species diversity showing much overlap with that from neighboring archipelagoes. Each of these island groups is fairly isolated from the rest of the shallow Pacific, and thus the total number of species in the bioregion is relatively low. Nevertheless, the Line Islands are still home to over 400 species of fishes, 200 species of algae, and at least 50 genera of reef-building corals. However, the taxonomic catalogs for the Line Islands are far from complete and are likely to be augmented during each new research survey.

Although not spectacular in their species diversity, the Line Islands are home to among the most healthy, ecologically intact coral reefs remaining on the planet. The reefs are characterized by very high fish biomass, dominated by predatory species such as snappers, groupers, and sharks, and a complex reef substrate, dominated by reef-building corals and coralline algae. Because of their isolation, the reefs of the Line Islands have remained greatly protected from the major stress caused by local human activities, predominately the stress of fishing. Many insights are being gained through the study of coral reef ecology in the Line Islands, in large part because of the opportunity to compare the structure and functioning of the community with and without local human influences (Figs. 3, 4).

The lagoons of the Line Islands, as in many other places, tend to serve as nursery habitats. For example, the lagoons on Kingman and Palmyra support abundant populations of juvenile reef sharks that find safety from predators that frequent the fore-reef. Much of the benthic fauna of these lagoons is distinctive from fore-reef habitats. In particular, many species of clams live in the reef matrix of these lagoons. Most impressive among these are the giant clams (mainly *Tridacna maxima*), a species that can grow to upward of 40 cm in length and has been reported at densities of over 20 adults per square meter at Kingman. The flora and fauna of the freshwater and hypersaline pools contain a number of unique and endemic species about which very little is known to date.

On the other side of the fore-reef habitat is the nearshore, pelagic environment. The Line Islands attract a large number of pelagic visitors, just as they attract seabirds above the sea's surface. Predatory fishes, such as yellowfin tuna, wahoo, and mahi mahi, are found frequently on the leeward sides of the islands. These species feed on smaller bait fishes that aggregate in areas of nearshore upwelling and on larval reef fishes produced and advected from the nearby reefs. Three species of dolphins are found commonly across the archipelago: bottlenose dolphins (*Tursiops truncates*), spinner dolphins (*Stenella longirostris*), and melon-headed whales (*Peponocephala electra*). Genetic analyses have revealed that these animals have significant site fidelity, with each island having its own resident population that mixes little, if at all, with populations from nearby islands. A number of other cetacean species are found in the area, including various beaked whales that have been found nowhere besides the Line Islands.

Humans and the Line Islands

At the time of European discovery in the late eighteenth century, there were no human inhabitants in the Line Islands. Archaeological remains found on a number of the islands suggest that the islands were visited sporadically by seafaring Pacific islanders. However, the lack of sufficient rainfall on many of the islands and adequate soil for agriculture would have limited the people's ability to develop long-term communities in the archipelago.

Since the beginning of the nineteenth century, the islands have been used for a number of purposes. In the late nineteenth century, a number of the Line Islands were colonized by small guano-mining operations. Additionally, a number of copra plantations were created on a number of islands by clearing native forests and planting coconut palms. Most importantly, however, three of the islands (Kiritimati, Tabuaeran, and Teraina) were populated by the I-Kiribati people. During the first part of the twentieth century, these populations grew slowly, but during the later part of the century, an active population relocation program began moving people to the Line Islands from the overpopulated Kiribati island of Tarawa. As of 2005, the I-Kiribati population in the northern Line Islands was estimated at 8500 people (5100 on Kiritimati, 2500 on Tabuaeran, and 900 on Teraina). Although fishing and limited agriculture supports a fraction of the food needs, cargo ships from Tarawa and Hawaii heavily subsidize the food and fuel demands of the population.

In recognition of the unique aspects of the Line Islands, a number of national and international efforts are in place to ensure the protection of the island chain. The three United States protectorates in the Line Islands (Kingman, Palmyra, and Jarvis) are designated as national wildlife refuges, managed by the U.S. Fish and Wildlife Service. A number of the Kiribati protectorates in the southern Line Islands also are protected as Kiribati wildlife sanctuaries. Finally, international efforts are under way to include the Line Islands as part of the nominated Central Pacific World Heritage Project, a program of the United Nations.

SEE ALSO THE FOLLOWING ARTICLES

Atolls / Climate Change / Coral / Pacific Region / Seamounts, Geology

FURTHER READING

Agardy, D. T. 2001. Scientific research opportunities at Palmyra atoll. A report submitted to The Nature Conservancy.
Cobb, K. M., C. D. Charles, H. Cheng, and R. L. Edwards. 2003. El Niño–Southern Oscillation and tropical Pacific climate during the last millennium. *Nature* 424: 271–276.
Davis, A. S., L. B. Gray, D. A. Clague, and J. R. Hein. 2002. The Line Islands revisited: New $^{40}Ar/^{39}Ar$ geochronologic evidence for episodes of volcanism due to lithospheric extension. *Geochemistry, Geophysics, Geosystems:* doi: 10.1029/2001GC000190.
Dawson, E. Y. 1959. Changes in Palmyra atoll and its vegetation through the activities of man 1913–1958. *Pacific Naturalist* 1: 51.
Dinsdale, E. A., O. Pantos, S. Smriga, R. A. Edwards, F. Angly, L. Wegley, M. Hatay, D. Hall, E. Brown, M. Haynes, L. Krause, E. Sala, S. A. Sandin, R. Vega Thurber, B. L. Willis, F. Azam, N. Knowlton, and F. Rohwer. 2008. Microbial ecology of four coral atolls in the northern Line Islands. *PLoS ONE* 3: e1584.
Handler, A. T., D. S. Gruner, W. P. Haines, M. W. Lange, and K. Y. Kaneshiro. 2007. Arthropod surveys on Palmyra atoll, Line Islands, and insights into the decline of the native tree *Pisonia grandis* (Nyctaginaceae). *Pacific Science* 61: 485–502.
Keating, B. H. 1992. Insular geology of the Line Islands, in *Geology and offshore mineral resources of the central Pacific basin,* Circum-Pacific Council for Energy and Mineral Resources Earth Science Series, vol. 14. B. H. Keating and B. R. Bolton, eds. New York: Springer-Verlag, 77–99.
Sandin, S. A., J. E. Smith, E. E. DeMartini, E. A. Dinsdale, S. D. Donner, A. M. Friedlander, T. Konotchick, M. Malay, J. E. Maragos, D. Obura, O. Pantos, G. Paulay, M. Richie, F. Rohwer, R. E. Schroeder, S. Walsh, J. B. C. Jackson, N. Knowlton, and E. Sala. 2008. Baselines and degradation of coral reefs in the northern Line Islands. *PLoS ONE* 3: e1548.
UNESCO. 2003. Central Pacific World Heritage Project, International Workshop Report.
Woodrofe, C. D., and R. F. McLean. 1998. Pleistocene morphology and Holocene emergence of Christmas (Kiritimati) Island, Pacific Ocean. *Coral Reefs* 17: 235–248.

LIZARD RADIATIONS

MIGUEL VENCES

Technical University of Braunschweig, Germany

Lizards belong to the clade Squamata, together with snakes, and among nonflying terrestrial vertebrates, they are the ones most commonly observed on islands. Lizards are characterized by a great facility in colonizing islands and adapting to novel ecological circumstances by changes in their morphology, physiology, and reproductive biology. They have consequently become an important model group for the inferential and experimental study of adaptive radiations.

LIZARDS ON ISLANDS

On major land-bridge islands with favorable climates (i.e., in the tropical, dry, and temperate zones) both liz-

FIGURE 1 Emblematic island lizards. (A) *Gallotia stehlini*, Gran Canaria. (B) *Gallotia galloti*, Tenerife. (C) *Chalcides sexlineatus*, Gran Canaria. (D) *Tarentola delalandii*, Tenerife. These species and their relatives have originated on the Canary Islands. Photographs by Miguel Vences.

ards and snakes are commonly encountered, with snake species richness often being similar to lizard species richness. On 14 major islands and island groups of the Mediterranean Sea, there are 152 occurrences of 30 species of lizards and 28 species of snakes. However, native extant snakes are missing on many smaller islands and on oceanic archipelagoes such as the Macaronesian Islands (Canary and Cape Verde Islands, Savage Islands, Madeira, and the Azores), where native species and even endemic radiations of lizards are present (Fig. 1). The most remote oceanic islands (e.g., Hawaii) are devoid of both native lizards and snakes.

Although within-island diversification is rare in snakes and is limited to very large islands such as Madagascar, lizards have diversified on medium-sized islands such as the Greater Antilles as well (see below). Of the currently known ~5000 species and 26 families of lizards, representatives of the Gekkonidae, Iguanidae, Lacertidae, and Scincidae are most commonly encountered on islands.

Continental islands, especially, may frequently act as an evolutionary reservoir by enabling the survival of remnants of lineages that became extinct or very rare on the mainland. Such is the case of the tuataras, two species of lizard-like reptiles which are the last extant representatives of the Sphenodontia (the sister group of squamates). At present, tuataras are confined to various small islands off New Zealand, although fossil remains demonstrate their past presence on the New Zealand mainland, and that of their relatives on other continents. On the Balearic Islands in the Mediterranean Sea, the lizard *Podarcis lilfordi* is present only on tiny offshore islands surrounding the larger islands of Mallorca and Menorca, where they are extinct. On Madagascar, the radiation of snakes in the subfamily Pseudoxyrhophiinae is very diverse, but

FIGURE 3 Both of the smallest lizards worldwide occur on islands. (A) The gecko *Sphaerodactylus ariasae* occurs on Isla Beata and adjacent areas of Hispaniola. Photograph by S. Blair Hedges. (B) Adult male Malagasy leaf chameleon of an undescribed species in the genus *Brookesia* from the extreme north of Madagascar. Photograph by Frank Glaw.

this lineage has only a few representatives in Africa, where it probably has been replaced by other snakes.

Both the largest and smallest extant lizards occur on islands: The largest is the Komodo dragon (*Varanus komodoensis*) with a maximum snout–vent length of over 1500 mm (Fig. 2); the smallest (Fig. 3) are two species of *Sphaerodactylus* geckos (*S. ariasae* and *S. parthenopion*) from the Caribbean, with adult snout–vent lengths of about 16 mm, and several species of Malagasy leaf chameleons (*Brookesia*) with adult snout–vent lengths of 14–19 mm. Lizards appear to show a trend of island gigantism and dwarfism opposite to what is generally considered as a rule: In lineages of small lizards, the island populations and species become even smaller, and in lineages of large forms, the island representatives become even larger, especially in carnivorous taxa. Snakes also show size changes in island populations and species, and snakes that evolved to become small on islands did so to a relatively greater degree than those that became large. The observed pattern suggests that snake body size is principally influenced by prey size, with large snakes mainly feeding on nesting seabirds and small snakes mainly feeding on lizards.

Many island lizards have adapted to resources that differ from those available on the nearby mainland. The most famous is the marine iguana from the Galápagos (*Amblyrhynchus cristatus*), the only lizard that feeds on algae while diving in the ocean. Many lizards of the family Lacertidae were originally insectivorous but became herbivorous on islands. In fact, herbivory in mainland lineages may be an important "preadaptation" that allows for successful colonization of island habitats.

A further intriguing difference between island and mainland populations of lizards is population density, which is generally one order of magnitude higher on islands. This phenomenon is likely driven by distinctly lower numbers of predators and competitors. These same factors may also have allowed island lizards to expand

FIGURE 2 The largest lizard worldwide, the Komodo dragon, *Varanus komodoensis*. Photograph by Thomas Ziegler.

their diet to include nectar, pollen, and fruit. Indeed, in several island ecosystems, lizards also occupy an important role as pollinators and seed dispersers.

Few studies have addressed changes in reproductive strategy in island populations of lizards, but in species of the family Lacertidae a trend of reduced clutch size and larger egg size on islands has been noted.

COLONIZATION OF ISLANDS BY LIZARDS

Recent years have seen a paradigm shift in our understanding of the occurrence of many taxa on islands. This has involved a shift from the dominance of vicariance explanations to hypotheses in which dispersal plays at least an equally important role. In general, the mode of reproduction of lizards and snakes, with internal fertilization, favors overseas dispersal because the arrival of a single gravid female to an island can be sufficient to give rise to a new population. For lizards, there is no doubt that their dispersal capacities are high and that they have on many occasions colonized islands over water from the mainland or from other islands. For green iguanas, direct evidence exists that after a hurricane in 1995, at least 15 individuals arrived on a mat of logs and uprooted trees on the eastern beaches of Anguilla and other islands in the Caribbean, and some specimens survived there for at least three years. Molecular genetic analyses have provided evidence for various events of long-distance dispersal between Africa and South America (e.g., in geckos of the genera *Tarentola* and *Hemidactylus*, and in skinks of the genus *Trachylepis*). For example, *Trachylepis atlantica* from the Fernando de Noronha Archipelago in the Atlantic, 350 km east of the Brazilian coast, belongs to this mainly African and Malagasy genus rather than to the related Neotropical genus *Mabuya*. Its ancestors presumably colonized by overseas dispersal from Africa rather than from nearby South America.

Native populations of lizards (and often endemic species) are found on many oceanic islands: on major archipelagos such as Macaronesia, the Galápagos, the Gulf of Guinea islands, the Comoros, and the Mascarenes, but also on many small and isolated islands. The Australian region, including small islands such as those of the Solomon and Bismarck archipelagoes, harbors a massive radiation of the scincid genus *Sphenomorphus*, and other skinks (genus *Emoia*) have radiated on most islands in the southwestern Pacific, including, among many others, the Fiji, New Caledonia, Solomon, and Bismarck archipelagoes. This further demonstrates the capacity for overseas dispersal of lizards.

FIGURE 4 An endemic species of chameleon from the Comoro island of Mayotte, *Furcifer polleni*. Photograph by Frank Glaw.

Inverse routes of colonization, from islands back to the mainland, have occurred as well. This appears to be the case for a Central and South American clade within the genus *Anolis*, which probably originated from a West Indian ancestor, and it is possibly also true for chameleons, which may have dispersed multiple times from Madagascar to mainland Africa, and which certainly have dispersed from Madagascar to Mayotte (Fig. 4).

On the Gulf of Guinea islands (São Tomé, Principe, and Annobon), a relatively high proportion of endemic burrowing species of lizards and snakes occur, indicating that the capacity of overseas dispersal also extends to species living in humid soil and leaf litter. A combination of ocean currents, floating islands, and reduced surface salinity caused by freshwater discharges from large rivers may be favorable to overseas dispersal events in general and may also enable such soil-dwelling species to colonize islands. Eggs of some lizards are known to be resistant to immersion in seawater. In the case of *Anolis sagrei*, this may explain the survival of populations of this lizard on small islands vulnerable to hurricanes, but it also may allow the overseas rafting of lizard eggs in tree holes or mats of vegetation.

In some cases, commensal species of lizards have been translocated by humans. Several species of geckos of the genus *Hemidactylus* have a transcontinental distribution that in some cases is due to natural colonization but often may reflect deliberate or, more probably, accidental introductions. *Lipinia noctua*, a scincid lizard that lives alongside humans on islands of the central and eastern Pacific, displays a phylogeographic pattern concordant with the "express train" hypothesis: Specimens may have been transported as stowaways on early Polynesian canoes during the rapid human colonization of Polynesian islands.

PATTERNS OF INSULAR LIZARD RADIATIONS

The process of speciation can be either (1) adaptive (i.e., the process of an ancestral population diverging and giving rise to two daughter lineages adapted to different niches) or (2) nonadaptive (e.g., the separation of the daughter species by geographic barriers or by differentiation of features that serve for species recognition).

Most lizard radiations on smaller islands probably belong to the category of nonadaptive and allopatric speciation on different islands. This same mode of speciation has also taken place within some islands of sufficient size. A few possible examples also exist for sympatric adaptive speciation within an island.

As an instance of nonadaptive speciation on different islands, the western Canary Islands are populated by small radiations of skinks and geckos (*Chalcides* and *Tarentola*), but on each island or group of islands, only one species of each genus occurs. The situation is slightly more complex in the Canarian lacertid lizards, genus *Gallotia*: Here an initial split is observed between large-sized and small-sized species, and sympatry occurs only between (ecologically strongly differentiated) representatives of either group (on Hierro, Gomera, Tenerife, and probably La Palma, if extant species and natural occurrences are considered). Day geckos of the genus *Phelsuma* have radiated on the Seychelles and Mascarenes, and on each of these two archipelagoes there is a monophyletic lineage of various species and subspecies. At least on the Seychelles, the available evidence favors allopatric speciation of the three endemic taxa on different islands, with secondary sympatry in some cases.

Crucial to test hypotheses of radiation on islands are robust phylogenies. However, critical data on the interplay of dispersal and vicariance can be provided by the geological age of an island or of its last connection to the mainland, and hence the age of evolutionary splits in the lineage under study. For example, the two Galápagos iguanas (the terrestrial genus *Conolophus* and the marine iguana *Amblyrhynchus*; Fig. 5) occur on the same islands and do form a monophyletic group. This could be interpreted as an example of speciation by ecological specialization under sympatric conditions. However, the age of the evolutionary divergence between these species predates the geological origin of the current Galápagos Islands. This indicates that either (1) they must have diverged on a previous, now submerged land mass, or (2) both species originated on the mainland, they colonized the Galápagos independently, and their mainland relatives subsequently went extinct. In general, the possibility of extinction must always be taken into account to under-

FIGURE 5 Galápagos iguanas. (A) The marine iguana, *Amblyrhynchus cristatus*. Photograph by Ylenia Chiari. (B) A terrestrial iguana, *Conolophus subcristatus*. Photograph by Scott Glaberman.

stand the biogeographic history of lizard populations on islands.

The best-studied case of an insular lizard radiation is that of the Caribbean genus *Anolis* (Iguanidae), the anoles, which are among the most common terrestrial vertebrates in the Caribbean and are found on almost every island in this region. There are over 400 species of anoles, of which nearly 150 are Caribbean. Their origin has been estimated at around 40 million years ago, and fossil specimens preserved in amber are known from the Oligocene to the Miocene of the Dominican Republic. The patterns of anole radiation have been intensively studied by Jonathan B. Losos and colleagues. Summaries are found in Losos (1998) and Losos and Thorpe (2004), from where much of the following information has been extracted.

Anoles are very good dispersers, evidenced by cases of related taxa occurring on islands of great geographic distance. However, by far the highest proportion of Caribbean anoles are endemic to single island banks (more than 85%). A few cases of natural hybridization are known, but in general, mismating among species of these lizards is prevented by the throat fans ("dewlaps") of males, which show specific colors and patterns used in species recognition. In fact, sympatric species of anoles always differ in the size, color, or patterning of their dewlaps. Up to 11 species of anoles can coexist at a single site, and such sympatric species almost always differ in terms of habitat use and morphology or physiology.

The number of anole species coexisting on a certain island is significantly correlated with island size. Considering only small islands (i.e., islands of a surface of 1500 km² or less), the species–area relationship is stronger for islands that were in the past connected by land bridges to other land masses than for isolated islands, highlighting the importance of historical effects: Land-bridge islands probably had a higher number of species at the time of isolation, and through subsequent extinctions, species numbers adjusted to the island-specific ecological carrying capacity. In contrast, isolated islands depend fully on over-water colonization as the source for species. Isolated islands mostly are populated by a single species of anole only, with a maximum of two species per island (which then differ in their ecology). Apparently, colonization of small isolated islands by anoles can be successful only if (1) the island does not yet harbor any anole population or (2) the island is populated by an anole species that differs in ecological requirements from the new colonizers.

Evolutionary diversification of anoles appears to occur on a single island when its size is above a certain threshold. In the Caribbean, within-island diversification has occurred on the Greater Antilles (Jamaica, Puerto Rico, Hispaniola, and Cuba). Each of these large islands harbors endemic divergent lineages, which contain various species and, hence, very probably originated on the island. Within-island speciation can be invoked for at least 70% of the Greater Antillean anoles. A few examples from smaller islands or island groups exist of co-occurrence of endemic taxa that could have arisen on the same island, but these cases are not compelling. Hence, a certain island area is necessary for within-island speciation, a conclusion that highlights the importance of geography for this process.

The *Anolis* radiations on the four Greater Antillean islands (although phylogenetically independent) show recurrent patterns. As was first pointed out by Ernest Williams, different types of habitat specialists (ecomorphs) occur on all or most of the Greater Antilles. These are usually represented by several species on each island (Figs. 6–8). Initially six ecomorphs were proposed, but others have since been distinguished. Interestingly, molecular

FIGURE 6 Ecomorphs of Caribbean *Anolis*. All species shown are from Hispaniola. Names roughly denote the preferred habitat of each ecomorph. (A) Crown giant: *Anolis baleatus*. (B) Trunk crown: *A. coelestinus*. Note that the photographs are not to scale; Crown Giants are much larger than all other ecomorphs. Photographs by S. Blair Hedges.

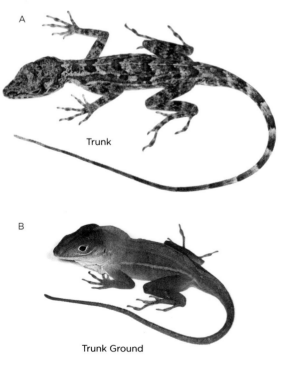

FIGURE 7 Ecomorphs of Caribbean *Anolis*, continued. (A) Trunk: *A. christophei*. (B) Trunk ground: *A. cybotes*. Photographs are not to scale. Photographs by S. Blair Hedges.

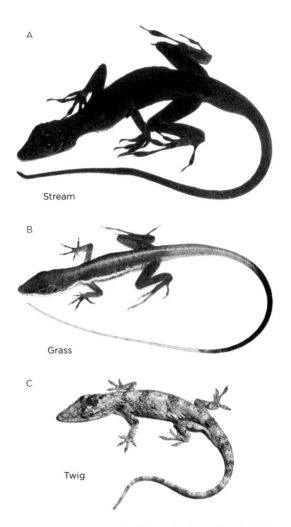

FIGURE 8 Ecomorphs of Caribbean *Anolis*, continued. (A) Stream: *A. eugenegrahami*. (B) Grass: *A. semilineatus*. (C) Twig: *A. placidus*. Photographs are not to scale. Photographs by S. Blair Hedges.

Cuba was fragmented during the Miocene. The *Anolis alutaceus* group, also on Cuba, contains 12 species with narrow distributions, mostly centered on different mountain ranges, a pattern that is also seen in other groups.

Which prevalent pattern of species formation gave rise to the current diversity of anoles? Adaptive speciation in sympatry or parapatry may occur in Caribbean anoles, but it is probably not the main driving force explaining their diversity. In many cases, populations became isolated on small land-bridge islands or reached isolated small islands by overseas dispersal. Geographically and thus genetically separated from other anole populations, they evolved different morphologies and dewlaps, probably largely because of adaptation to new ecological conditions. On the larger islands, species belonging to the main ecomorphs underwent allopatric speciation (e.g., on different mountain ranges or on parts of their island that were separated by water barriers in periods of rising sea levels). As summarized in the following section, many examples indicate that adaptation can occur in the absence of speciation in Caribbean anoles. But it is still uncertain how the initial differentiation of ecomorphs on each of the Greater Antillean islands took place.

PHYLOGEOGRAPHY AND EXPERIMENTAL TESTS OF SELECTION

Deciphering radiations is possible by looking at general patterns across a whole group or by examining in more detail the microevolutionary processes. Comparison of DNA sequences allows phylogeographical analyses where chiefly the geographical distribution of differentiated alleles (haplotypes) is mapped, and the phylogenetic relationships among these haplotypes is determined. The assumption is that haplotypes evolve through mutation, and different haplotypes get fixed in genetically isolated populations. In various studies on anoles and Canarian lizards, Roger S. Thorpe and colleagues have found evidence for discordance between historical and adaptive patterns. For example, in *Gallotia* lizards on the Canarian island of Tenerife, a historical boundary of mitochondrial haplotype lineages exists between western and northeastern areas, whereas within both groups, morphological differences were found between northern and southern populations, reflecting strong ecological differences between the humid north and arid south of the island. On Dominica, *Anolis oculatus* shows a complex phylogeographical structure that is not fully concordant with the phenotypic variability encountered.

These examples demonstrate that morphological adaptations to local conditions, especially in terms of col-

data show that, with two exceptions, the ecomorphs arose independently on the different islands: Different ancestors diversified independently and gave rise to the same ecological and morphological adaptations.

Species belonging to different ecomorphs usually occur sympatrically, but species belonging to the same ecomorph generally are geographically separated within an island (and have different dewlap colors or patterning). In addition to the six main ecomorphs, many islands harbor further habitat specialists, but these usually occur on a single island only.

In several cases, the different species of one ecomorph occur in geographically separated populations scattered across an island. In the *Anolis carolinensis* group, three evolutionary lineages can be distinguished and have ranges corresponding to three paleo-archipelagoes into which oration, can evolve very fast in island lizards. This is also witnessed by the large variability of lacertid lizard species inhabiting Mediterranean islands (e.g., Adriatic islands, satellite islands of the Balearics, or Tyrrhenic islands in Greece). From many of these archipelagoes, a plethora of subspecies have been described based on color patterns and partly on variation in scale numbers, but molecular studies have rarely found any significant differentiation between these populations, indicating that the external differences evolved extremely rapidly, on a geological timescale. Other work has yielded evidence that in *Anolis sagrei*, the number of body scales increases with increasing precipitation and with decreasing temperature in open arid habitats, and the variation in scale numbers is probably heritable. In further experiments, the effects of a potential predator (the ground-dwelling lizard *Leiocephalus carinatus*) on the behavior of *Anolis sagrei* was tested by introducing the potential predator on six small islands on the Bahamas and using six other predator-free islands as control sites. As a result, anoles altered their behavior by using the ground less often, but in addition, a strong selection took place: Surviving specimens on the experimental islands had larger body sizes and longer hindlimbs than those on control sites, probably reflecting their better capacities to escape.

Evidence for strong selection pressures acting on island lizards also comes from further experimental studies. The Dominican *Anolis oculatus* displays various ecomorphological variants related to different conditions between the east and west coasts and the montane regions of the island. In experiments, lizards were translocated to large lizard-proof enclosures in regions occupied by other habitat types than those in their source population. Morphology (coloration, scale counts, body proportions) of the translocated lizards were scored, and each lizard individually marked. Several months later, survivors were collected and identified. Morphological differences were found between survivors and non-survivors (e.g., of specimens of the montane population in enclosures of the relatively xeric west coast), and the intensity of selection was dependent on the magnitude of ecological change experienced by the specimens in the enclosures.

How these intraspecific processes of fast morphological variation relate to the actual process of species formation and adaptive radiation is not clear. Evidence of parapatric forms with restricted gene flow among them comes from the islands of Dominica and Martinique; on Martinique this may constitute evidence for adaptive (ecological) species formation because the forms are distinguished by current habitat and not by historical allopatry. It seems clear that these lizards have a strong potential to adapt to new ecological conditions by changes in morphology and coloration, and this may have favored adaptive speciation (mostly under allopatric conditions). This may also be a factor explaining the recurrent evolution of similar ecomorphs.

SEE ALSO THE FOLLOWING ARTICLES

Adaptive Radiation / Convergence / Dispersal / Komodo Dragons / Snakes

FURTHER READING

Losos, J. B. 1994. Integrative approaches to evolutionary ecology: *Anolis* lizards as model systems. *Annual Reviews of Ecology and Systematics* 25: 467–493.
Losos, J. B. 1998. Ecological and evolutionary determinants of the species-area relationship in Caribbean anoline lizards, in *Evolution on islands*. P. R. Grant, ed. Oxford: Oxford University Press, 210–224.
Losos, J. B., and R. S. Thorpe. 2004. Evolutionary diversification of Caribbean *Anolis* lizards, in *Adaptive speciation*. U. Dieckmann, M. Doebeli, J. A. J. Metz, and D. Tautz, eds. Cambridge: Cambridge University Press, 322–344.
Olesen, J. M., and A. Valido. 2003. Lizards as pollinators and seed dispersers: an island phenomenon. *Trends in Ecology and Evolution* 18: 177–181.
Williams, E. E. 1983. Ecomorphs, faunas, island size, and diverse end points in island radiations of *Anolis*, in *Lizard ecology*. R. B. Huey, E. R. Pianka, and T. W. Schoener, eds. Cambridge, MA: Harvard University Press, 326–370.

LOPHELIA OASES

SANDRA BROOKE

Marine Conservation Biology Institute, Bellevue, WA

The deep-water stony coral *Lophelia pertusa* (Linnaeus 1758) creates extensive and complex structures on hard-bottomed areas in the deep sea, including continental shelf bedrock, lithified sediment mounds, volcanic basalt, and (microbially mediated) authigenic carbonate. Large colonies of *L. pertusa* have abundant tangled branches that provide habitats for diverse and abundant associated communities. These long-lived and slow-growing coral ecosystems are currently under threat globally from negative human impact, and although some areas have been placed under protective legislation, continued international effort is needed to ensure the future of these valuable resources.

CORAL BIOLOGY

There are several species of "framework-building" deep-water corals (*Lophelia pertusa*, *Oculina varicosa*,

Enallopsammia profunda, *Solenosmilia variabilis*, and *Goniocorella dumosa*), all of which have similar characteristics. Individual colonies are complex branching structures that can be several meters high. These corals are broadcast-spawning species, releasing eggs and sperm into the water column for larval development. In the North Atlantic, *Lophelia* spawns during late February and early March, and each new coral colony is formed from the settlement of a single larva onto hard substrate; the resulting coral polyp divides asexually, and as the colony grows, the outer branches block the flow to the inner (older) parts of the colony, which eventually die. Subsequent invasion by boring and encrusting organisms weakens the inner core of the colony, and it falls apart, exposing the dead center to overgrowth by living coral. The standing dead coral provides hard substrate for a variety of other fauna and provides a micro-habitat for many small organisms. The living coral branches may come into contact with each other as the colony grows, and the braches commonly fuse with each other, which provides additional stability. Over time, this sequential growth, death, and overgrowth can create massive mounds composed of unconsolidated dead coral and sediment, with an outer layer of live coral (Fig. 1). These are referred to as bioherms and may be hundreds of meters deep. Deep-water corals do not have the algal symbionts common in shallow-water species, and they survive primarily on zooplankton. Corals require a moderate (and fairly continuous) current; water flow delivers food and oxygen, removes metabolic waste products, and reduces accumulation of sediment, which can suffocate the polyps. Growth rates for *Lophelia* have been estimated at 5–25 mm yr^{-1}. At this rate, it would take thousands of years to form the extensive coral ecosystems that have been discovered in recent years. Experimental and field observations have shown that the upper thermal limit for *Lophelia* survival is approximately 12 °C, and temperature is undoubtedly a controlling factor in the depth distribution of this coral. Deep coral communities may also provide information about climate change in the deep ocean. Because of their worldwide distribution and longevity, cold-water corals are an excellent proxy for reconstructing past changes in global climate and ocean conditions.

GLOBAL DISTRIBUTION OF *LOPHELIA PERTUSA* ECOSYSTEMS

Deep-water corals have been recorded from many topographic features throughout most of the world's oceans. The highest known density of deep-water coral ecosystems occurs in the North Atlantic, but this may be an artifact of the high level of research effort in that area. These communities have provided the basis for much of the current knowledge of deep-water corals. Norway has the largest known *Lophelia* systems (approximately 2000 km^2) particularly along the mid-Norwegian shelf at a 200–400 m depth. The shallowest known *Lophelia* systems (40 m in depth) occur in the fjords of Norway, where deep oceanic water intrudes into narrow channels and recreates deep-water conditions at shallow depths. At the other extreme, *Lophelia* has also been found on the Mid-Atlantic Ridge at more than 3000 m, but is most commonly found between 200 m and 1000 m.

The most northerly *Lophelia* structure in the western Atlantic covers a small, 1-km-long area at the mouth of the Laurentian Channel in Canada, but the continental margin between the Blake plateau off North Carolina and the Miami terrace in South Florida supports the most well developed and extensive *Lophelia* complexes off the North American coast. In contrast to the Atlantic coast, there are few well-developed coral complexes in the Gulf of Mexico. The limestone bedrock of the western Florida shelf supports a series of *Lophelia* mounds at 500 m depth, and elsewhere in the Gulf the soft sediment sea floor is interspersed with boulders of authigenic carbonate, which supports the development of hard-bottom communities. The most extensive *Lophelia* communities in the Gulf of Mexico have been found at depths of 420–530 m on Viosca Knoll, a large mound approximately 100 km south of Mobile Bay.

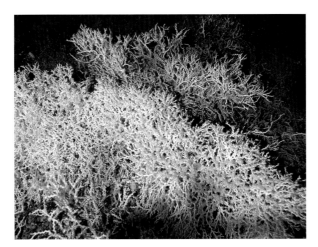

FIGURE 1 A well-developed thicket of *Lophelia* from deep-water bioherms along the southeast coast of Florida. The bright white branches are live *Lophelia*, with open polyps giving the colony a fuzzy appearance. The darker branches are dead skeleton, which is often colonized by other animals. Photograph courtesy of Brooke *et al.*, 2005, NOAA Office of Ocean Exploration (http://oceanexplorer.noaa.gov/explorations/05deepcorals).

Colonies of *Lophelia* have been observed in the eastern Pacific from California to Alaska. In 2006, *Lophelia* was discovered in the Olympic Coast National Marine Sanctuary off Washington State. Although there are currently no records of *Lophelia* elsewhere in the Pacific Ocean, it is probable that there are communities still to be discovered.

Information on deep-water coral distribution is particularly lacking from tropical and subtropical regions where countries do not have the funds or technology to conduct deep-water exploration.

Cold-water corals have also been documented on artificial substrates such as oil installations and wrecks. Hundreds of *Lophelia* colonies were observed on oil platforms in the North Sea. Large *Lophelia* colonies were also observed growing on a World War II wreck at 554 m in the northern Gulf of Mexico. With fossil fuel operations moving into deeper waters, oil and gas platforms may provide substrate for deep-water coral development in the Gulf of Mexico as they do in the North Sea.

CORAL-ASSOCIATED COMMUNITIES

The complex structure produced by the coral branches provides substrate for sessile benthic organisms, food and refuge for many small invertebrates and fish, and food for larger predators. Like shallow tropical reefs, deep-water corals also provide structures for spawning aggregations and nursery habitats for various fish species. Coral complexes form biodiversity hotspots in deep water, and although there have been few quantitative studies of coral-associated fauna, a great deal of census information has been collected over the past decade. A complex reef structure can be divided into three general areas: the live coral, the dead coral rubble underneath, and the rubble/soft sediment at the base of the coral mound.

FIGURE 2 View of *Lophelia* thicket showing the bright red squat lobster *Eumunida* sp., with claws outstretched, presumably waiting for prey. Small anemones and a pencil urchin are also visible among the branches. Photograph courtesy of Brooke et al., 2005, NOAA Office of Ocean Exploration (http://oceanexplorer.noaa.gov/explorations/05deepcorals).

FIGURE 3 Close-up view of a spectacular yellow glass sponge (family Hexactinellidae) nestled between dead *Lophelia* branches. A new species of amphipod was found living inside this species of sponge from the South Atlantic. Photograph courtesy of Brooke et al., 2005, NOAA Office of Ocean Exploration (http://oceanexplorer.noaa.gov/explorations/05deepcorals).

Relatively few organisms live on the live coral (polyps contain stinging cells that may repel larval settlement); however, the carnivorous polychaete (*Eunice* sp.) is frequently found within the live branches. The worm excretes a soft tube, which the coral then overlays with calcareous skeleton, thus strengthening the tube for greater protection. In return, these calcified tubes make the colony more robust. Mobile organisms are also observed among the live branches; one of the most common is a red squat lobster (*Eumunida* sp.), which sits in the coral with claws outstretched, apparently waiting for food to pass by (Fig. 2). The dead coral framework supports a much greater abundance and diversity of organisms than does the live coral. Hundreds of different species from many taxonomic groups live in this habitat, with gorgonians, black corals, anemones, and a diverse array of glass sponges (Fig. 3) being among the most common of the larger fauna. Some species bore into the skeleton, weakening it and causing collapse over time; some encrust the branches; and others simply use the dead structure as a refuge or hunting ground. In the northeast Atlantic, a census identified over 1300 species associated with *Lophelia* colonies. Deep coral communities have also been identified as habitats for hundreds of species of fish, many of which are commercially valuable.

ANTHROPOGENIC IMPACTS ON *LOPHELIA* ECOSYSTEMS

There are several potential threats to deep-water coral ecosystems, some more prevalent and potentially damaging than others. These include physical impacts such as commercial bottom fishing, hydrocarbon extraction, deployment of cables and pipelines, bio-prospecting, and coral harvesting. The greatest of these is bottom trawling, which drags weighted nets held open by heavy "doors" over the fragile corals. Large trawl nets can cover many square kilometers in a single fishing trip, and reccovery (if it occurs) may take hundreds of years. Another potential threat to deep water coral communities comes from impending changes in the ocean as a byproduct of burning fossil fuels. As carbon dioxide increases in the atmosphere, it diffuses into the ocean and upsets the balance of ocean chemistry, making the ocean more acidic. Coral skeleton is made from a calcium carbonate, which dissolves under acidic conditions, therefore as the pH drops, the corals will find it harder to make their supporting skeletons.

Fossil fuel exploration and extraction activity is a potential threat to deep-water corals if they are found in close proximity to large oil and gas deposits, particularly in the North Sea and the Gulf of Mexico. Live *Lophelia* has been observed close to drilling operations; however, more research is needed to determine the effects of fossil fuel extraction on deep-water coral communities. Laying gas pipelines and communication cables may also damage deep-water communities because large anchors are often used to stabilize the surface vessel during deployment. Other threats to deep-water corals exist, but their impact is relatively minor at present.

CONSERVATION AND MANAGEMENT OF DEEP-WATER CORALS

Many deep-water coral systems lie outside national exclusive economic zones (EEZ) and are therefore not covered by any legal jurisdiction; however, these vulnerable ecosystems are currently the focus of international efforts to create legal protection under the United Nations Convention on the Law of the Sea (UNCLOS). Over the past five years, several countries have enacted or are in the process of establishing regional measures to protect and manage their deep-water coral ecosystems. These measures vary greatly and range from requirements for environmental impact assessments prior to conducting activities around deep-water corals to designation of marine protected areas with specific regulations in place. Norway was the first country to implement protection of *Lophelia* reefs after it was estimated that 30–50% of the *Lophelia* had been damaged by bottom trawling. The Norwegian fisheries authorities established a regulation that prohibited intentional destruction of coral habitat and provided special protection to five selected areas by banning the use of bottom gear altogether. Norway is also in the process of implementing an additional series of marine protected areas. Off the coast of Scotland, a series of *Lophelia* bioherms called the Darwin Mounds was placed under special protection in 2004; bottom fishing was permanently prohibited over a 1300 km^2 area. Other countries such as Ireland, the Azores, the Canary Islands, and Madeira also implemented similar policies in 2004. In the United States, several large areas have been placed under protection from bottom-damaging activites in the past five years, and more protected areas are currently being proposed.

FUTURE CHALLENGES

As advances in technology allow humans to further exploit the deep ocean, the need for protective legislation is essential. Deep-water coral systems are usually far offshore and encompass large areas, so, even with protection in place within national EEZs, enforcement of regulations is extremely difficult and expensive, and protection from trawling on the high seas is yet a greater challenge. In the past decade, our understanding of deep-water corals has greatly increased due to funding from national and international sources and to the collaborative efforts of scientists and governmental institutions. However, these oases of biodiversity and abundance in the vastness of the deep sea are in urgent need of research, conservation, and protection from human exploitation.

SEE ALSO THE FOLLOWING ARTICLES

Coral / Fish Stocks/Overfishing / Marine Protected Areas

FURTHER READING

Continental Shelf Associates. 2007. Characterization of northern Gulf of Mexico deepwater hard bottom communities with emphasis on *Lophelia* coral. U.S. Department of the Interior, Minerals Management Service, Gulf of Mexico OCS Region, New Orleans, LA. OCS Study MMS 2007-044.

Freiwald, A., J. H. Fossa, A. Grehan, T. Koslow, and J. M. Roberts, 2004. Cold-water coral reefs: out of sight no longer out of mind. Biodiversity Series No. 22, Cambridge: UNEP-WCMC. Available online at http://www.unep-wcmc.org/resources/publications/UNEP_WCMC_bio_series/22htm

Freiwald, A., and J. M. Roberts, eds. 2005. *Cold-water corals and ecosystems*. Erlangen Earth Conference Series. Heidelberg, Germany: Springer Publishing House.

Gianni, M. 2004. *High seas bottom trawl fisheries and their impacts on biodiversity of vulnerable deep sea ecosystems: options for international action*. Gland, Switzerland: IUCN.

Lumsden, S. E., T. F. Hourigan, and A. W. Bruckner, eds. 2007. The state of deep coral ecosystems of the United States. NOAA Technical Memorandum NOS-CRCP-3. Silver Spring, MD.

Messing, C. G., J. K. Reed, S. D. Brooke, and S. W. Ross. 2008. Deep-water coral reefs of the United States, in *Coral Reefs of the USA*. B. Riegel and R. Dodge, eds. Heidelberg, Germany: Sprnger Publishing House, 763–787.

Rogers, A. D. 1999. The biology of *Lophelia pertusa* (Linnaeus 1758) and other deepwater reef-forming corals and impacts from human activities. *International Review of Hydrobiology* 84: 315–406.

LORD HOWE ISLAND

CAROLE S. HICKMAN
University of California, Berkeley

Lord Howe Island is the spectacular remnant of a large shield volcano that was active 7 million years ago in an isolated part of the southwestern Pacific Ocean between Australia and New Zealand. The high proportion of rare and endemic plants and animals, the rugged natural beauty of the landscape, and the occurrence of the southernmost coral reef in the world contribute to its status as a UNESCO World Heritage Site. The Permanent Park Preserve encompasses 75% of the island. Scientific studies dating from the 1850s have provided an unusually detailed documentation of natural history phenomena, making Lord Howe one of the most intensively studied sites in Australia. Because it is one of the last island biodiversity hotspots on the planet to be discovered and colonized by humans, the original composition of the biota is relatively well known. Successful management efforts to eliminate introduced species and site-based initiatives for conserving native habitats and threatened species place Lord Howe at the forefront of island conservation practice.

GEOGRAPHIC SETTING

Lord Howe Island is the largest emergent feature on the western flank of the Lord Howe Rise, at the southern end of a north–south chain of seamounts and guyots extending from southwest of New Caledonia to the Challenger Plateau, west of New Zealand. The island is 702 km northeast of Sydney, Australia, at a latitude of 31° S. The Lord Howe Rise is bounded by the Tasman Sea on the west and the New Caledonia Basin on the east. The rise is no more than 2000 m below sea level, whereas depths in the Tasman Sea between the rise and Australia exceed 4000 m. The rise is underlain by continental crust that separated from eastern Australia during the Cretaceous, moving eastward to its current position during the 30-million-year-old opening of the Tasman Basin. The Lord Howe World Heritage Site includes offshore islands, islets, and rocks: the Admiralty group to the northeast, Mutton Bird and Sail Rock to the east, Rabbit Island within the lagoon of the Lord Howe reef, Gower Island off the southern end, and Ball's Pyramid 25 km south of the island. The main island is crescent shaped, measuring 10 km from north to south and approximately 2 km in width.

FIGURE 1 Prominent volcanic features of the southern peaks of Lord Howe Island. (A) Mt. Lidgbird (left) and Mt. Gower, rising spectacularly out of the ocean to nearly 100 meters and capped by cloud. (B) Layers of horizontal basalt on Mt. Lidgbird, remnants of the final sequence of lavas that filled the ancient caldera after collapse of the Lord Howe volcano. (C) Oblique basalt dikes representing a final swarm of intrusions of magma into the horizontal layers of Mt. Gower.

PHYSICAL FEATURES

Topographically, Lord Howe is dominated by the southern basalt peaks of Mount Lidgbird (777 m) and Mount Gower (875 m), which rise nearly vertically out of the sea forming steep cliffs and slopes covered with dense subtropical rain forest (Fig. 1A). A narrow lowland isthmus separates the southern peaks from an older northern highland that forms the shear cliffs of the northern coastline. Only the lowland isthmus is habitable.

The two major rock types exposed on Lord Howe are basalt and calcarenite. The majority of the exposed rock is layered basalt, with individual flows ranging in thickness from a few centimeters to more than 30 m (Fig. 1B). The nearly horizontal basalt layers of Mount Lidgbird are part of a sequence that formed near the close of volcanic activity, filling an immense caldera that resulted from collapse of the summit of the original volcano. In addition to basalt, the older sequences of volcanic rock at the north end of the island and in the Admiralty Islands include tuff and volcanic breccia, consolidated fragmental deposits from a more explosive earlier phase of volcanism.

The basalt layers on Lord Howe are extensively crosscut by dikes. These features formed when younger basalt was intruded into fissures cutting obliquely across the older basalt layers (Fig. 1C).

The low-lying coastal strip of Lord Howe is dominated by calcarenite, a sedimentary rock that normally does not occur on high volcanic islands. Calcarenite consists of calcium carbonate sand, formed by mechanical breakdown of the skeletons of coral, calcareous algae, and shells. The calcarenites of Lord Howe represent major episodes of erosion of the coral reef during ice-age fluctuations of sea level and subsequent formation of beach sands and windblown dunes around the flanks of the old volcano. Cementatation of the deposits and their subsequent erosion has formed many interesting sedimentary features (Fig. 2A).

The calcarenite is also notable for its fossil content, which includes both marine and terrestrial biota. Fragments of the bones of seabirds are common (Fig. 2B) along with shells of endemic land snails and large quantities of bone of the extinct giant horned turtle *Meiolania platyceps*. The British Museum of Natural History and the Australian Museum both contain hundreds of specimens resulting from collections as early as the 1880s. The skull of a new species of extinct endemic bat, an extinct species of penguin, and bird eggs are included in the interesting fossil discoveries. Some of the best-preserved fossil invertebrates are in beach rock exposed at sea level when storms wash away the overlying beach sand (Fig. 2C).

GEOLOGIC HISTORY

When the emergent portion of Lord Howe Island formed, 7 million years ago during the Miocene epoch, its subaerial extent was 40 times greater than it is today. The subaerial phase of eruption and building of the shield volcano culminated in a collapse of a caldera estimated to be 900 m deep and 5 km by 2 km. Rapid filling of the caldera

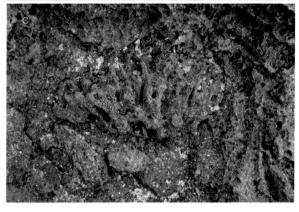

FIGURE 2 Sedimentary rocks and fossils. (A) Outcrop of the Neds Beach Calcarenite, a sequence of cross-bedded, windblown, calcium carbonate skeletal grains that formed during low stands of Pleistocene sea level and erosion of exposed reef. (B) A fragment of fossil bird bone in place in the Neds Beach Calcarenite. (C) Fossil coral imbedded in beachrock on Neds Beach.

with horizontal basalt layers ended the volcanic phase of island history 6.4 million years ago.

The modern landscape is predominantly a consequence of erosion of the original edifice. Very little is known of the larger submarine portion of the island, which is more than 25 km in diameter at its base. The Admiralty Islands and small rocks and islands adjacent to Lord Howe are part of the original shield volcano. Ball's Pyramid, the dramatic

551-m pinnacle located 25 km southeast of Lord Howe Island, is the remnant of a different shield volcano. It is separated from Lord Howe by depths of approximately 500 m, and the two were never connected by land.

Beach sand units and windblown dune units include fossil soil layers. The history and ages of the Lord Howe calcarenites have been studied in detail using five different dating techniques. The cyclical pattern of formation can be linked to periods of sea-level change during successive ice ages.

CLIMATE

Lord Howe Island has a humid subtropical climate, with mean summer temperatures of 23 °C and mean winter temperatures of 16 °C. Mean annual rainfall is approximately 165 cm, with higher rainfall during the winter. The peaks of Mt. Lidgbird and Mt. Gower must receive significantly higher precipitation throughout the year.

The marine climate is notable for its mix of temperate and tropical water. The tropical East Australia Current flows south along the Great Barrier Reef and into the northern Tasman Sea, but in some years there are strong incursions of cold subantarctic currents from the south.

BIOTA

Fauna

There are no large terrestrial vertebrates in the native fauna, which is restricted to a skink, a gecko, and a small bat. Of the 15 native land bird species at the time of discovery of Lord Howe, nine are now extinct. Two were eaten to extinction by sailors (white gallinule, white-throated pigeon) and a third was eliminated by early settlers because it was a crop pest (red-fronted parakeet). Five additional species were eliminated when the black rat reached the island in 1918. Two of the seven remaining native landbirds are endemic (Lord Howe woodhen, Lord Howe island silvereye). The most conspicuous members of the avifauna are seabirds (Fig. 3A). Fourteen species nest on Lord Howe and adjacent islets, including huge colonies of tens of thousands of individuals of flesh-footed shearwaters, sooty terns, and providence petrels.

The terrestrial invertebrate fauna is rich in unusual, rare, and endemic species. There are at least 85 endemic species of land snails and a remarkable evolutionary radiation of freshwater hydrobiid snails. More than 100 species of spiders have been identified, and 50% of these are believed to be endemic to the island (Fig. 3B). There is an endemic freshwater shrimp, an endemic freshwater crab, an endemic leech, ten endemic earthworm species, an endemic amphipod, 12 endemic species and one endemic genus of terrestrial isopod, and an endemic cicada. Genus-

FIGURE 3 Native flora and fauna of Lord Howe Island. (A) A nesting colony of masked boobies at Mutton Bird Point. (B) A colorful orb-weaving spider, part of the large and poorly known fauna of native spiders. (C) Native cloud forest on the slope of Mt. Lidgbird.

level endemism in the insects includes the hemipterin bug *Howeria* and the cricket *Howeta*.

One of the most unusual insects is the Lord Howe Island phasmid, *Dryococelus australis*, a giant stick insect reaching lengths of 15 cm. It was common until the arrival of the black rat in 1918 and extinct by 1920. Rediscovery of a small population of the "extinct" phasmid on Ball's Pyramid in 2001 is an example of the "Lazarus phenomenon." It has triggered a vigorous debate about conservation options for reappearing species.

The marine fauna of Lord Howe is remarkable for its mix of tropical and temperate species. The reef has attracted attention not only as the most southerly coral reef in the Pacific, but also because it has unusually high coral cover and high algal biomass. The 83 reported species of coral form some unique associations of tropical species. More than 400 species of fish have been reported. As with the rich marine invertebrate fauna, the fish are a unique mix of tropical and temperate species.

Flora

There are 241 native vascular plant species in the Lord Howe Island group, and 105 (44%) are endemic. The richness of the flora can be attributed in part to the variety of habitats. Twenty-five vegetation associations have been recognized. It is unusual for an island as small as Lord Howe to have altitudes supporting a true cloud forest (Fig. 3C). The moss forests at the summits of Mt. Gower and Mt. Lidgbird are rich in orchids as well as mosses.

The endemics are not all rare, high-elevation species. There are four endemic species of palm in three endemic genera. *Howea forsteriana* is the most notable, forming dense lowland forests.

Exotic species pose one of the greatest threats to the native flora. There are 230 introduced species, including 17 that are considered noxious weeds. Most of these are restricted to the settlement area, and most have not invaded the indigenous plant communities.

Biogeographic Affinities

The Lord Howe Island biota is a composite of organisms with different geographic affinities. Many endemic elements in the flora have their closest affinities with New Zealand, but there is a mix of tropical and temperate components. Insects show many different patterns. There are beetles whose closest relatives are on New Caledonia and Norfolk Island. The endemic species and genera of Lord Howe crickets also have their closest affinities with crickets on New Caledonia and Norfolk. Four species of caddis flies endemic to Lord Howe are in a genus endemic to Australia, and the endemic muscid flies of both Lord Howe and Norfolk have an Australian origin. The Lord Howe stick insect is closely related to a genus in New Guinea. The land snails also show several different biogeographic patterns. In one family there is a close affinity with Norfolk Island, whereas the Lord Howe *Placostylus* is closest to a species in New Zealand.

Reconstructing the deep history of island biotas of the western Pacific requires understanding 100 million years of plate tectonics events that both created and destroyed islands. These events were set in motion with separation of the immense continental crustal block containing the modern emergent islands of New Zealand, Lord Howe, New Caledonia, and Norfolk from Antarctica and Australia. The two main ridges on the block, the western Lord Howe Rise and the eastern Norfolk Ridge, separated 65 million years ago with the opening of the New Caledonia Basin. The probability of preexisting islands, connections, and extinct island biotas is strong, but it lacks preserved geologic evidence.

HUMAN HISTORY AND CULTURAL HERITAGE

There is no evidence of prehistoric habitation of Lord Howe Island. The earliest recorded sighting of the island was by Henry Lidgbird Ball, in command of HMS *Supply* bound from Sydney to Norfolk Island in 1788. He landed on his return trip to Sydney, claiming the island for Great Britain and naming several prominent features (Mt. Lidgbird, Ball's Pyramid) for himself. The island itself he named for the first lord of the British Admiralty, Lord Howe. Although ships in search of food and freshwater visited the island, it was not settled until 45 years after its discovery. Sparse early settlement in 1833 and 1844 was restricted to the lowland and supported by subsistence farming and supplying passing ships. The only "industry" in the latter part of the nineteenth century was the marketing of seeds of the *Howea* palm, an adaptable and hearty indoor plant that achieved great popularity during the Victorian era.

Tourism has been the only other "industry," initially by steamship service and small rustic guesthouses. Today the number of tourist beds on the island is controlled, and the natural history, beauty, and tranquility are the major attractions.

The island is under jurisdiction of the New South Wales Government and is administered locally by the Lord Howe Island Board. There is no private land ownership. Leaseholders must reside on the island, under a management plan for the settlement. The board also manages the Permanent Park Preserve, which encompasses 75% of the land on the island and has its own management plan. Listing of the island group as a world heritage property included 1455 hectares of the main island, offshore islets, and Ball's Pyramid. In 1988, the New South Wales government created the marine park that expanded the area to 145,000 hectares.

CONSERVATION

Although Lord Howe Island has not been drastically modified relative to most Pacific islands, the most obvious steps in conservation have focused on eradicating introduced feral animals. Feral pigs were eliminated in 1995. Feral cats have been eliminated, and goats may be totally eliminated. Rodent eradication assessments have been made, and eliminating rats appears feasible if external funding can be obtained to meet the relatively high costs.

Intensive efforts have been made on behalf of a number of endemic species listed as threatened or endangered. The Lord Howe Island woodhen, *Tricholimnas sylvistris*, had been reduced to three adult pairs by 1980. They were transferred to a captive breeding facility on the island. By the end of three breeding seasons, 57 individuals had been released on the island. By 1992, the population was estimated at 250–300 birds. Eradication of pigs and cats has been critical to the success of the captive breeding program.

There is a recovery plan for the Lord Howe *Placostylus*, a large, critically endangered land snail. It occurs in the fossil calcarenites of the island and is most often encountered today as empty shells (Fig. 4). The recovery plan emphasizes community involvement in conserving suitable habitat for the species.

Although Lord Howe Island has not been drastically modified relative to most Pacific islands, the "people pressure" is a constant threat. An Australian voluntary conservation movement views Lord Howe as "a paradise in peril," and has generated a "management strategy" to afford better protection to its world heritage values.

AUSTRALIAN ISLAND TERRITORIES

The Australian Commonwealth includes several island territories in addition to the islands such as Lord Howe that are under the jurisdiction of mainland territories. Off the east coast there are two island territories. The Coral Sea Islands Territory is a vast complex of uninhabited reefs and atolls northeast of Queensland and the Great Barrier Reef. Norfolk Island is a small, subtropical, volcanic island territory on the Norfolk Ridge, midway between New Zealand and New Caledonia, twice as far from Sydney as Lord Howe. The territory comprises three islands: Norfolk and the small adjacent Phillip and Nepean Islands. Norfolk Island is the only Australian territory with self-governance. Like Lord Howe, Norfolk is the remnant of a submarine volcano, but it differs from Lord Howe in its low topographic relief and considerably younger age (2–3 million years). In contrast to Lord Howe, there is archaeological evidence of late prehistoric Polynesian occupation, although there was no indigenous population at the time of its discovery in 1774 by Captain James Cook. It was more heavily colonized and disturbed following European discovery, and only 5% of the native forest remains intact. The origins of its endemic plants and animals have been of considerable interest to biogeographers.

SEE ALSO THE FOLLOWING ARTICLES

Endemism / Extinction / Fossil Birds / Island Formation / Land Snails / Spiders

FURTHER READING

Brooke, B. P., C. D. Woodroffe, C. V. Murray-Wallace, H. Heijnis, and B. G. Jones. 2003. Quaternary calcarenite stratigraphy on Lord Howe Island, southwestern Pacific Ocean and the record of coastal carbonate deposition. *Quaternary Science Reviews* 22: 859–880.

Francis, M. P. 1993. Checklist of the coastal fishes of Lord Howe, Norfolk, and Kermadec Islands, southwest Pacific Ocean. *Pacific Science* 47: 136–170.

Harriott, V. J., P. L. Harrison, and S. A. Banks. 1995. The coral communities of Lord Howe Island. *Marine and Freshwater Research* 46: 457–465.

Hutton, I. 1986. *Lord Howe Island*. Australian Capital Territory: Conservation Press.

Hutton, I., J. P. Parkes, and A. R. E. Sinclair. 2007. Reassembling island ecosystems: the case of Lord Howe Island. *Animal Conservation* 10: 22–29.

McDougall, I., B. J. J. Embleton, and D. B. Stone. 1981. Origin and evolution of Lord Howe Island, southwest Pacific Ocean. *Journal of the Geological Society of Australia* 28: 155–176.

Miller, B., and K. J. Mullette. 1985. Rehabilitation of an endangered Australian bird: the Lord Howe Island woodhen *Tricholimnas sylvestris* (schlater). *Biological Conservation* 34: 55–95.

Pickard, J. 1983. Vegetation of Lord Howe Island. *Cunninghamia* 1: 133–266.

Priddel, D., N. Carlile, M. Humphrey, S. Fellenberg, and D. Hiscox. 2003. Rediscovery of the 'extinct' Lord Howe Island stick-insect (*Dryococelus australis* (montrouzier)) (Phasmatoidea) and recommendations for its conservation. *Biodiversity and Conservation* 12: 1391–1403.

Standard, J. C. 1963. Geology of Lord Howe Island. *Proceedings of the Royal Society of New South Wales* 96: 107–121.

FIGURE 4 Empty shell of the endangered Lord Howe *Placostylus* (right) and an unidentified land snail (left).

MACARONESIA

SEE ATLANTIC REGION

MACQUARIE, BIOLOGY

JENNY SCOTT

University of Tasmania, Hobart, Australia

Macquarie Island (54°30′ S, 158°56′ E) is a remote subantarctic island 1500 km south-southeast of Tasmania in the southern Pacific Ocean (Fig. 1). It has luxuriant herbaceous vegetation but no trees or shrubs, and it supports huge concentrations of seabirds and seals. As with all subantarctic islands, the terrestrial ecosystem of Macquarie evolved without mammals. Their introduction by humans has resulted in significant ecological impacts.

BIOLOGICAL SETTING

Macquarie Island's position just north of the Antarctic Polar Frontal Zone gives it a uniformly cool, wet, and windy climate (3.3–7.0 °C, 920 mm precipitation) with no permanent ice or snow. Its remote location and wholly oceanic origin means that all its native terrestrial flora and fauna arrived by long-distance dispersal, either by wind or sea.

The marine environment is dominated by the Macquarie Ridge and associated trenches, with the westward-flowing Antarctic Circumpolar Current passing through gaps in the ridge north and south of the island. The island is 34 km long, up to 5.5 km wide, and

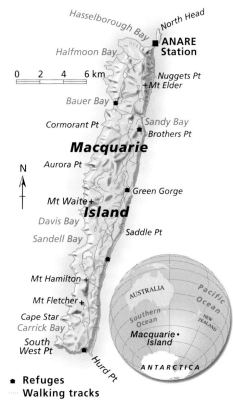

FIGURE 1 Map of Macquarie Island. Map reproduced with permission of *Australian Geographic Magazine*.

almost 13,000 hectares in area. Several small groups of islets lie to the north and south. The main island forms an elongated plateau, with a maximum elevation of 433 m, that is surrounded by steep coastal slopes 100–250 m in height (Fig. 2). The coastal environment is

FIGURE 2 Rugged tussock-covered slopes on the west coast, with rocky coastline and small sandy beaches. Photograph by J.J. Scott.

dominated by rocky shores with numerous small sandy beaches.

MARINE AND TERRESTRIAL LIFE

Vegetation

The rocky intertidal zone has a diverse and extensive benthic algal flora (seaweeds), dominated by spectacular growths of the giant Antarctic kelp *Durvillea antarctica*. At least 103 species of benthic algae have been recorded, over twice the number of terrestrial vascular (flowering) plants recorded (47 species, including four endemics and five alien species, two of which have been eradicated). The major plant communities are fellfield, herbfields, short grassland, mires, and tall tussock grassland. The most visually striking plants are the large tussock grass *Poa foliosa* (Fig. 3) and the megaherbs *Stilbocarpa polaris* (Macquarie Island cabbage) and *Pleurophyllum hookeri*. The latter two species have a greater abundance and vigor on Macquarie than anywhere else in their restricted global distribution. Bryophytes (mosses and liverworts) and lichens are abundant and diverse, with over 130 bryophytes and over 150 lichens recorded. More than 127 species of freshwater and terrestrial algae, and over 200 species of macro-fungi, have been reported. There are no known endemic bryophyte species, and endemic levels of the lichen, algal, and macro-fungal taxa are not yet known.

Fauna

The marine benthic and pelagic invertebrate and fish fauna around Macquarie Island is not well known, but it appears to be typical of other subantarctic islands. The coastal environment of Macquarie Island supports enormous concentrations of marine vertebrates during the breeding season (Figs. 3–4), with an estimated 80,000 seals, mainly southern elephant seals (*Mirounga leonina*) and small numbers of three fur seal species. At least 27 bird species breed on the island. Around 3.5–4 million seabirds, mainly four species of penguins, congregate along the coasts during summer. The endemic royal penguin (*Eudyptes schlegeli*) is the most abundant, with an estimated 850,000 breeding pairs (Fig. 4). Four species of albatross and at least 19 species of petrels, prions, and other birds also breed on the island. These include three alien bird species (a fourth species was eradicated). Less than 300 species of terrestrial invertebrates have been reported, mainly of unknown or cosmopolitan distribution. Approximately 10% are believed to be endemic, and there are at least nine alien species. Currently there are three species of alien mammals (European rabbit *Oryctolagus cuniculus,* ship rat *Rattus rattus*, and house mouse *Mus musculus*), and a fourth (cat, *Felis catus*) was recently eradicated.

FIGURE 3 The south coast during summer breeding season with black-browed albatross (*Thalassarche melanophris*) in the foreground, and rockhopper penguins (*Eudyptes chrysochome*), royal penguins (*E. schlegeli*), and the large tussock grass *Poa foliosa* in the background. Photograph by J.J. Scott.

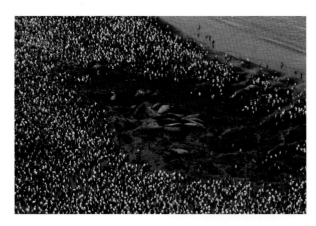

FIGURE 4 Royal penguins (*Eudyptes schlegeli*) and southern elephant seals (*Mirounga leonina*) crowd the beaches during summer. Photograph by J.J. Scott.

HUMAN IMPACTS AND CONSERVATION

Human presence on Macquarie Island initially involved exploitation of seals and penguins, mostly during the eighteenth and nineteenth centuries, followed thereafter by scientific studies, conservation protection, and a limited amount of commercial fishing and tourism. In recognition of its substantial natural values, the island is a UNESCO World Heritage Area and Biosphere Reserve, and a Tasmanian Nature Reserve. In recognition of the island's marine values and the interconnectedness of its marine and terrestrial environments, the World Heritage Area extends out to 22.2 km from low water mark and the Nature Reserve to 5.6 km, whereas the majority of the 370-km Exclusive Economic Zone around Macquarie Island has been declared a Commonwealth Marine Park covering around 16.2 million hectares.

Since the 1960s, natural resource conservation on the island itself has largely involved the management or eradication of mammals (cats, rabbits, and rodents) that were introduced during the nineteenth century and adapted to the subantarctic conditions, resulting in substantial impacts. In the first decade of the twenty-first century, a massive increase in rabbit numbers has caused widespread damage to vegetation through grazing and digging, with associated impacts to erosion and hydrological regimes and flow-through effects to habitats of threatened seabird species. This has prompted the preparation of a detailed eradication plan for all three remaining mammal pest species (rabbits, rats, and mice). Implementation is planned for 2010, after which substantial recovery of the terrestrial ecosystem is expected.

SEE ALSO THE FOLLOWING ARTICLES

Biological Control / Introduced Species / Macquarie, Geology / Marine Protected Areas / Seabirds

FURTHER READING

Banks, M. R., and S. J. Smith, eds. 1988. Macquarie Island Symposium, Hobart, May 1987. Symposium proceedings. *Papers and Proceedings of the Royal Society of Tasmania* 122: 1–318.
Environment Australia. 2001. *Macquarie Island Marine Park management plan 2001–2008.* Commonwealth of Australia. http://www.environment.gov.au/coasts/mpa/publications/pubs/macquarie-plan.pdf
Parks and Wildlife Service. 2006. *Macquarie Island Nature Reserve and World Heritage Area management plan.* Parks and Wildlife Service, Department of Tourism, Arts and the Environment, Hobart, Tasmania, Australia. http://www.parks.tas.gov.au/publications/tech/macquarie/macquarie.pdf
Parks and Wildlife Service and Biodiversity Conservation Branch. 2007. *Plan for the eradication of rabbits and rodents on subantarctic Macquarie Island.* Parks and Wildlife Service, Department of Tourism, Arts and the Environment, Tasmania, and Biodiversity Conservation Branch, Department of Primary Industries and Water, Tasmania, Australia, 1–30, http://www.environment.gov.au/heritage/publications/draft-macquarie-rabbit-eradication-plan.html

Selkirk, P. M., R. D. Seppelt, and D. R. Selkirk. 1990. *Subantarctic Macquarie Island: environment and biology.* Cambridge: Cambridge University Press.
Terauds, A., and F. Stewart. 2008. *Subantarctic wilderness: Macquarie Island.* Crows Nest, Australia: Allen and Unwin.

MACQUARIE, GEOLOGY

ARJAN DIJKSTRA
University of Neuchâtel, Switzerland

Macquarie Island is a unique island of great geological importance, because it is the only locality in the world where a complete section of young ocean crust formed at a spreading center is exposed above sea level. It has thus become a type-locality for ocean crust and has played a major role in the development of theories of seafloor spreading, one of the key processes of plate tectonics. The island was put on the UNESCO World Heritage List in 1997 for this reason. The island also provides a key geological record of a spreading ridge system that became a convergent plate boundary and that will probably develop into a subduction zone over time.

GEOLOGICAL SETTING

Macquarie Island is the subaerially exposed summit of a 1500-km-long submarine mountain chain, the Macquarie Ridge Complex (Fig. 1). The island measures 34 by 5.5 km, with a

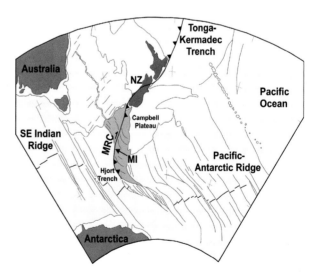

FIGURE 1 Location of Macquarie Island on the Australian-Pacific plate boundary. Continents in red, continental margins in tan, oceanic crust in green. Blue indicates ocean floor that was created at plate boundary prior to 5 million years ago, when it was a zone of seafloor spreading. Abbreviations: MI, Macquarie Island; MRC, Macquarie ridge complex; NZ, New Zealand. Diagram modified from Sutherland (1995).

maximum elevation of 433 m above sea level. Two groups of small islets also lie on this topographic ridge, namely the Judge and Clerk Islets 14 km to the north and the Bishop and Clerk Islets 34 km to the south of Macquarie Island. The Macquarie Ridge Complex marks the boundary between the Australian and Pacific tectonic plates. The adjacent sea floor lies at 4000 m below sea level, but trenches immediately next to the ridge system are locally as deep as 6500 m. At present, the Macquarie Ridge is a zone of oblique, right-lateral convergence. This plate boundary can be traced to the northeast into the Alpine fault zone of New Zealand and the Tonga–Kermadec subduction zone. The plate boundary at and around Macquarie Island is seismically very active, with several recorded earthquakes of a magnitude greater than 8. The absence of deep (greater than 35-km) earthquakes along the plate boundary southwest of New Zealand shows that subduction has not started yet in the region, although subduction may just be initiating in the Hjort Trench segment. Before 5 million years ago, the plate boundary was divergent (i.e., it was a zone of seafloor spreading). It was during this phase of sea-floor spreading that the rocks on Macquarie Island were formed; biostratigraphic and radiometric dating give ages for the formation of the rocks on the island between 10 and 6 million years ago.

ROCKS AND STRUCTURES RELATED TO SEAFLOOR SPREADING

Because of recent tilting and erosion of the rock units, a unique section of the oceanic crust and into the upper part of the mantle is exposed on the northern part of Macquarie Island. Here, a top-to-bottom sequence of basaltic pillow lavas, a sheeted-dike complex, various types of gabbros (Fig. 2), and other associated plutonic rocks is exposed

FIGURE 2 Outcrop of layered gabbros representing the lower levels of the oceanic crust exposed on the northern part of Macquarie Island (Elizabeth and Mary Point).

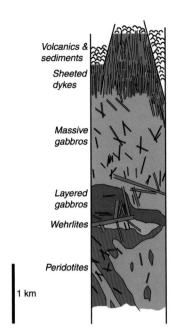

FIGURE 3 Schematic section through the rock sequence exposed on the northern part of Macquarie Island, restored to its assumed original orientation in which the sheeted dikes are vertical, and the layering in the lowermost gabbros is horizontal. Simplified from Dijkstra and Cawood (2004).

(Fig. 3). Finally, peridotites, rocks from the upper part of the mantle, are found at the base of the sequence. Such a sequence is characteristic for ophiolite complexes, and it is generally assumed that typical ocean floor consists of a similar rock sequence, as near-identical rocks are also found exposed at modern mid-ocean ridges. However, some volcanic rocks on the island are nearer in composition to volcanics typically found in seamounts formed at some distance from mid-ocean ridges. In contrast to the northern part of the island, the middle and southern parts of the island have only lavas and some sheeted-dike outcrops. Sedimentary rocks on the island include conglomerates, breccias, and sandstones, interlayered with volcanic rocks. These rocks were deposited at the base of early, sea-floor spreading–related fault scarps.

UPLIFT AND EROSION

Oblique convergence of the Pacific and Australian plates, starting 5 million years ago, has created a 50-km-wide zone of uplifted oceanic crust, the Macquarie Ridge Complex. Interestingly, the numerous recent fault scarps found on the island itself seem to be produced by normal rather than the reverse fault movements that are typical for convergent plate boundaries, with minor strike slip. These recent faults control the shapes and locations of most landforms, including the lakes on the island. Raised Pleistocene beach deposits at various levels clearly attest to recent uplift and marine erosion. Dating of paleobeaches has yielded an average uplift rate for the island of 0.8 mm per year and has shown that the island probably emerged above sea level 600,000–700,000 years ago. Since then, the island has been shaped by periglacial processes during the Pleistocene glacial cycles.

SEE ALSO THE FOLLOWING ARTICLES

Lava and Ash / Macquarie, Biology / Oceanic Islands / Plate Tectonics / Seamounts, Geology

FURTHER READING

Adamson, D. A., P. M. Selkirk, D. M. Price, N. Ward, and J. M. Selkirk. 1996. Pleistocene uplift and palaeoenvironments of Macquarie Island: evidence from palaeobeaches and sedimentary deposits. *Papers and Proceedings of the Royal Society of Tasmania* 130: 25–32.

Daczko, N. R., S. Mosher, M. F. Coffin, and T. A. Meckel. 2005. Tectonic implications of fault-scarp-derived volcaniclastic deposits on Macquarie Island: sedimentation at a fossil ridge-transform intersection. *Geological Society of America Bulletin* 117: 18–31.

Dijkstra, A. H., and P. A. Cawood. 2004. Base-up growth of ocean crust by multiple phases of magmatism: field evidence from Macquarie Island. *Journal of the Geological Society London* 161: 739–742.

Goscombe, B. D., and J. L. Everard. 2001. Tectonic evolution of Macquarie Island: extensional structures and block rotations in oceanic crust. *Journal of Structural Geology* 23: 639–673.

Varne, R., A. V. Brown, and T. Falloon. 2000. Macquarie Island: its geology, structural history, and the timing and tectonic setting of its N-MORB to E-MORB transition. *Geological Society of America Special Paper* 349: 301–320.

Wertz, K. L., S. Mosher, N. R. Daczko, and M. F. Coffin. 2003. Macquarie Island's Finch-Langdon fault: A ridge-transform insider-corner structure. *Geology* 31: 661–664.

MADAGASCAR

STEVEN M. GOODMAN

The Field Museum, Chicago, Illinois

As a result of its plant and animal endemism, nearly unparalleled in other portions of the world, and of the notable levels of threat associated with human activities, Madagascar has been designated as one of the priority biodiversity hotspots. This outstanding biological richness is a result of Madagascar's long isolation, having been separated from Gondwana some 160 million years ago and having had no subsequent land connection. Furthermore, the island has notable topographic and geological complexity, providing different mechanisms for speciation. One remarkable aspect of this island nation is the number of new taxa being described each year. As it is further explored, measures of species richness, already extraordinary, increase at a nearly exponential rate.

GEOGRAPHY OF THE TERRITORY

Madagascar is located in the western portion of the Indian Ocean and is separated from the African continent by 400 km of the Mozambique Channel (Fig. 1). Madagascar is the fourth largest island in the world, measuring 581,500 km^2, which makes it slightly larger than France or California. It is one of the oldest existing islands, with rocks dating back 3200 million years, and it was one of the last large island masses in the world to be colonized by humans—an event currently estimated by archaeological evidence to have taken place about 2300 years ago. This mini-continent is approximately 1600 km long and 600 km at its widest point, spanning the latitudinal range from 12° to 25.5° S; hence, the southern portion of the island falls outside of the Tropic of Capricorn. To the northwest are the Comoros Islands (Grande Comoro, Anjouan, Mohéli, and Mayotte), which are of volcanic origin and are no more than 7 million years old, and to the north is the western edge of the Seychelles archipelago (Aldabra group), which scatters across another 1000 km of ocean to the east. Approximately 850 km to the east of coastal eastern Madagascar are the isolated Mascarene Islands (La Réunion and Mauritius), in the Indian Ocean.

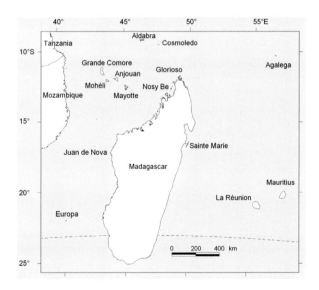

FIGURE 1 Map of Madagascar and surrounding islands. Aldabra and Cosmoledo form islands in the western portion of the Seychelles, and the balance of this archipelago continues off the map in an east-northeast direction. Map by Lucienne Wilmé.

Given its size, latitudinal breadth, and different oceanic currents, Madagascar shows remarkably different regional climatic regimes. Portions of the extreme northeast receive over 6.5 m of rainfall per year, whereas the extreme southwest, south of Toliara, generally receives less than 0.5 m per year. Furthermore, given the considerable elevation gradient of the island (Fig. 2A)—particularly the north–south oriented eastern mountain chain, which has numerous peaks surpassing 2000 m—overlaid on the latitudinal gradient, temperatures are highly variable. This includes nightly low temperatures well below freezing in the summital zone of Andringitra in the central south to daily highs reaching 45 °C in lowland areas of the west and southwest. As a generalization, for the humid eastern portions of the island, the rainy and warm season is from December to April, and the dry and colder season is from May to October or November. In contrast, the drier southwestern portions of the island have a short and generally unpredictable rainy season from December to February, and the balance of the year is dry. These different climatological patterns are closely associated with different natural vegetation cover (Fig. 2B).

Since the detachment of Madagascar from Gondwana 170–155 million years ago, there has been considerable geological activity on the island, including volcanic eruptions, mountain uplifting, and substantial formation of sedimentary rock. The island can be roughly divided into an eastern two-thirds composed of Precambrian rocks and a western one-third composed of Phanerozoic unmetamorphosed sedimentary rocks.

BIOLOGICAL CHARACTERISTICS

There is no other land mass in the world of equivalent size to Madagascar that surpasses its level of species richness and endemism at different taxonomic levels. Certain groups of organisms are considered Gondwana relicts, and others have colonized the island since its separation and undergone extensive speciation, forming distinctive adaptive radiations. These patterns can be found across different groups of plants and animals, and in some cases they rival the diversity of body forms classically cited for adaptive radiations, such as the Galápagos finches (Fig. 3).

The humid evergreen forest is located on the eastern side of Madagascar with an extension into the extreme

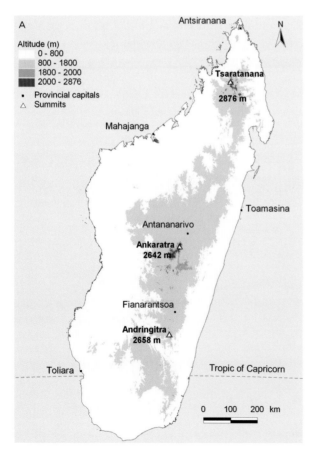
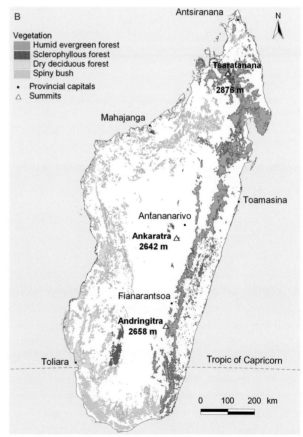

FIGURE 2 (A) Altitude and (B) vegetation maps of Madagascar. Little of the original natural vegetation of this biodiversity hotspot island nation remains. Maps by Lucienne Wilmé.

FIGURE 3 Line drawing of an extraordinary example of an adaptive radiation on Madagascar, here as illustrated by the vangas (family Vangidae), which are endemic to Madagascar and the Comoro Islands. Note the differing body sizes and bill shapes that function in a similar fashion to different tools (forceps, thin pliers, large pliers, etc.) and allow the different species of this monophyletic group to exploit different food resources. Drawing by John W. Fitzpatrick.

northwest. This zone has the least pronounced dry season on the island and, as a result, supports considerable floristic and associated faunistic diversity. Classically, the humid evergreen forest formation was divided into several vegetational types, from lowland (sea level to around 800 m) to mid-elevation (800 to 1800 m) to mountain (1800 to 2000 m). However, these different formations show a continuous gradient, and the precise divisions associated with this classification are oversimplified. Furthermore, variation in slope and aspect can have a dramatic influence on vegetational structure. In considerable portions of the zone formerly covered with humid evergreen forest, the natural vegetation has been removed by human actions, forming a secondary savanna or pseudo-steppe, where regenerating vegetation is dominated by introduced plants and animals (ants, earthworms, rodents, etc.).

On the higher mountains in the upper portion of the humid evergreen forest, overlapping with the mountain zone or above, there is a particular vegetational formation known as sclerophyllous forest. This vegetation type is dominated by ericaceous plants of the family Ericaceae, particularly the genera *Erica* (= ex *Philippia*), *Vaccinium*, and *Aguaria*, and is generally elfin (2–3-m tall). The sclerophyllous forest zones experience remarkable variation in weather patterns. For example, in mid-August (end of austral winter) in the summital zone of the Andringitra Massif (2600 m), daily temperatures vary from −11 to 32 °C. Given this temperature differential of over 40 °C and the fact that snow has been recorded on this massif, the local biota has numerous adaptations to withstand such extreme temperature vicissitudes.

Much of the central portion of Madagascar, generally referred to as the Central Highlands, representing about 40% of the island and defined as the zone above 800 m, has little remaining natural forest cover. The eastern side of the Central Highlands is well defined by an escarpment, whereas the western side is less so, and it generally descends gradually to the sea. The Central Highlands are currently covered by vast areas of open habitat, including a grassland formation dominated by introduced plants. Based on pollen cores from paleontological sites dating from the Late Quaternary, it has been shown that portions of the Central Highlands were not covered by dense forest formations before the recent large-scale habitat transformation of this zone by humans. Hence, the previous hypothesis that Madagascar was solid forest from coast to coast just before human colonization needs to be reconsidered. Rather, the Central Highlands were probably a mosaic of different forest types, including marshlands, dense humid evergreen forests, and open wooded savannas. The vast majority of the former marshlands of this zone were converted into agricultural lands, particularly rice paddies.

The vegetation formation along the western slope in the northern half of the island, below the Central Highlands, is dry deciduous forest. In the lowland portions of the extreme southwest and south, the natural vegetation is composed largely of spiny bush (also referred to as Didieraceae–*Euphorbia* thicket or xerophytic bush). The dry deciduous forest zone can experience annual dry seasons of up to 6 to 8 months, during which a considerable portion of the vegetation drops its leaves, although some plants remain green using an assortment of adaptations to retain moisture. In contrast, the spiny bush region can experience annual dry seasons of up to 10 months, and in some cases, several years can pass without any substantial precipitation. Here, the majority of the plants show adaptations to survive these long periods of drought, and they exhibit a remarkable level of micro-endemism.

Other vegetation communities on the island include mangrove (particularly on the western coast), aquatic freshwater zones along rivers and inland lakes, and a particular Central Highland vegetation known as *tapia* forest, dominated by the genus *Uapaca*. Botanists have been actively working on the flora of Madagascar for several hundred years, and even with this historical foundation, the estimate of the number of vascular plants on the island has increased from about 7400 in 1936 to 12,000 in 1971 to 14,000 in 2006. Madagascar possesses numerous endemic plant families.

Although the figures differ between groups from about 40% to 100%, the fauna of the island is largely endemic, including families not found elsewhere in the world. As with plants, a remarkable number of new animal taxa are described each year, running the gamut from invertebrates to mammals. To provide a few examples, the ant fauna of Madagascar contained about 300 species in the 1990s, and it is now estimated that 1000 species occur on the island, approaching 96% endemism. The amphibian fauna was estimated in 1984 to comprise 150 species; by 2008 about 230 have been described, and it is projected that 500 species occur on the island, with a level of endemism approaching 99%. Of the native mammal terrestrial fauna, 100% are endemic and composed of four lineages (carnivorans of the family Eupleridae, tenrecs of the family Tenrecidae, rodents of the family Nesomyidae, and lemurs of the five different families within the order Primata); a considerable number of the over 112 species have been named since 1984. Slightly over 50% of the island's bird fauna is endemic, including several families and subfamilies restricted to the Madagascar region (including the Comoro Islands).

ORIGINS OF BIODIVERSITY

As would be expected given Madagascar's size, geological history, and isolation from other land masses, the sequence of events that led to the evolution of its remarkably rich and unique biota is complicated, and a portion of these events have their origin deep in time. The island is a fragment of the former supercontinent Gondwana. During the earlier portions of the Mesozoic (the geological period spanning from 245 to 65 million years ago), this large land mass comprised areas now referred to as Africa, South America, Australia, Madagascar, India, Sri Lanka, Antarctica, and portions of the Seychelles. Gondwana started to break apart about 160 million years ago, and the eastern section of this massive land mass (composed of Madagascar, India, Australia, Antarctica, and portions of the Seychelles) started to drift toward the east. Approximately 140 million years ago, the landmass referred to as Indo-Madagascar, composed of Madagascar and the Indian subcontinent attached to its eastern flank, was completely separated from the African land mass, hence severing any direct means of plant or animal dispersal from Africa not passing over oceanic waters.

Several groups of plants and animals occurring on Madagascar presumably have their origins during the period before the breakup of Gondwana. Examples include the plant *Podocarpus*, found across portions of the Southern Hemisphere; different groups of invertebrates, such as trapdoor spiders (family Migidae), known from South America, Africa, Madagascar, Australia, New Zealand, and New Caledonia; and several different vertebrates, including the iguanid lizards (family Iguanidae) from the southern portion of the Americas and Madagascar. Hence, when Indo-Madagascar broke off from other portions of this former continent, these groups were already present on Madagascar. The severance of the former distributional range of these organisms, known as a vicariant event, associated with plate tectonics, led to distinct patterns of speciation on Madagascar. Subsequently, about 80 million years ago, India broke away from Madagascar and moved north toward Asia, carrying with it a number of Gondwana relicts.

After the Gondwana breakup, different groups, predominantly from Africa and to a lesser extent from Asia, colonized Madagascar by a variety of means. For plants, these means include wind-dispersed pollen, seeds dispersed in the guts of frugivores, buoyant seeds dispersed by sea currents, and sticky or spiny seeds dispersed in the feathers of birds. For numerous species of ferns, for example, spores can be transported in the stratosphere and dispersed over remarkable distances. In the case of animals, the mechanism of dispersal to Madagascar is often cited as rafting on vegetation, aestivating or outright hibernating in floating clumps of vegetation or holes in tree trunks, swimming, or flying directly (in the case of certain insects, bats, and birds). For larger land vertebrates, the process of colonization was logistically more complicated and rare than it was, for example, for certain groups of plants. Recent research on the native land mammals of the island indicate that all four living groups (see above) can be explained by merely four independent colonization events, underlining how exceptional successful colonization was and how such events led to subsequent diversification into different adaptive forms.

THREATS TO BIODIVERSITY

Other than New Zealand, Madagascar was the last large island mass in the world to be colonized by humans, who arrived from southeastern Asia. Based on current archeological information, people colonized Madagascar about 2300 years ago. There is no evidence of Neolithic or Paleolithic cultures on the island. Over the past few thousand years, Madagascar has experienced a considerable number of extinctions of large-bodied land vertebrates, including lemurs, tortoises, elephant birds, and a variety of other remarkable animals. For example, Bibymalagasia is an extinct order of mammals endemic to the island known only from subfossil bones dating from the Holocene. The period some of these animals disappeared roughly coincides with the first colonization of humans on the island. Explanations for the disappearance of these animals range from natural cycles of climatic change and associated shifts in resources, resulting in extinction events, to humans, through hunting or massive habitat modification that pushed these animals beyond the brink. Although this debate is partially unresolved, the best explanation is probably a combination of these two explanations. Radiocarbon dates from the bones of the extinct megafauna indicate that numerous species continued to exist until a few hundred years ago and existed contemporaneously with humans for over two millennia. This would indicate, at least for these animals, that the idea of massive extinction associated with direct human pressure or the introduction of a virulent hyperdisease associated with domestic animals are not tenable hypotheses to explain the disappearance of these animals.

However, over the past 1000 years, humans have colonized virtually all areas of Madagascar, and in their wake, there has been considerable habitat modification and destruction. This has reached a level such that less than 8% of the natural habitats existing before human colonization of the island remain today. Numerous forested areas have become isolated and fragmented, others show clear signs of extensive human usage and degradation, and hundreds of thousands of hectares have completely disappeared. Several decades of recent socioeconomic turmoil has considerably exacerbated this situation, giving rise to the "biological crisis" of the island. Major steps have been taken in past decades to conserve the remaining biotopes of the island and the organisms that they contain.

Madagascar was the first country in the African region to develop a national environmental action plan, dating from the early 1990s. It is one of the few regional examples of a country where biological data were applied to advance conservation programs. Madagascar has been at the forefront in conceptualizing and implementing numerous new policies into its national conservation programs to safeguard the unique biota of the island and, at the same time, allow human communities to retain their identities and advance in socioeconomic development. However, these often very innovative programs, even when properly implemented, have still produced few results at the needed level for changing the economic situation of people living near the forest edge or patterns of habitat disturbance.

The Malagasy Government has taken a series of more recent steps to rectify and ameliorate the protection of the island's unique natural patrimony. In September 2003, the President of Madagascar, M. Marc Ravolomanana, declared at the World Parks Congress held in Durban, South Africa, that over the subsequent five years the total protected areas system of Madagascar would be increased threefold, reaching 10% of the island's surface. Excellent progress is being made with regard to this commitment, and a series of new parks and reserves have and are about to be named; this process will help to protect aspects of the unique biota of Madagascar. Furthermore, several generations of Malagasy conservation biologists have been trained, and they have exceptional capacity to lead their nation. This, combined with programs to ameliorate the island's economic problems and improve its educational system, hold considerable promise for the future of the unique organisms and ecosystems found on Madagascar. Some of the critical aspects that need to be properly reinforced in the system of protected area management are the training of staff, the means of enforcing laws, and the revamping of the associated judicial system.

A variety of introduced organisms pose serious problems as invasive colonizers, particularly in naturally (e.g., cyclone damaged) or human disturbed habitats. These include plants such as *Lantana camera* and *Psidium cattleianum,* which form the principal vegetation in certain zones, and animals such as the ant *Technomyrmex,* the fish *Channa,* and the rodent *Rattus* that apparently outcompete or prey upon their endemic counterparts. However, the problems associated with invasive species on Madagascar, although a serious matter, are notably less pervasive than on neighboring islands such as the Comoros and the Mascarenes.

With the opening in recent years of Madagascar to foreign investors and companies, a new problem has developed with respect to the exploitation of the island's considerable mineral resources. There are current plans at various stages to commercially exploit a variety of different sites for ilmenite, nickel, cobalt, and different precious and semi-precious gems. Furthermore, there is currently considerable exploration of Madagascar's offshore and deep-sea oil reserves, with the intent of exploitation.

Unfortunately, from a conservation perspective, many of these sites are associated with important tracts of forest or with freshwater habitats, some within, some on the perimeter, and some lying considerable distances from protected areas. The Malagasy Government is currently formulating policies on how these proposed exploitation projects will unfold, and it is hoped that the conservation concerns will not be completely overridden by the means of important economic gain for the country.

SEE ALSO THE FOLLOWING ARTICLES

Archaeology / Comoros / Endemism / Lemurs and Tarsiers / Mammal Radiations / Mascarene Islands, Geology / Rafting

FURTHER READING

Battistini, R., and G. Richard-Vindard, eds. 1972. *Biogeography and ecology of Madagascar*. The Hague, Netherlands: W. Junk.
Burney, D. A., L. P. Burney, L. R. Godfrey, W. L. Jungers, S. M. Goodman, H. T. Wright, and A. J. T. Jull. 2004. A chronology for late Prehistoric Madagascar. *Journal of Human Evolution* 47: 25–63.
Dewar, R. E., and H. T. Wright. 1993. The culture history of Madagascar. *Journal of World Prehistory* 7: 417–466.
de Wit, M. J. 2003. Madagascar: heads it's a continent, tails it's an island. *Annual Review of Earth Planetary Science* 31: 213–248.
Donque, G. 1975. *Contribution géographique à l'étude du climat de Madagascar*. Tananarive: Nouvelle Imprimerie des Arts Graphiques.
Goodman, S. M., ed. 2008. *Paysages naturels et biodiversité de Madagascar*. Paris: Muséum national d'Histoire naturelle.
Goodman, S. M., and J. P. Benstead, eds. 2003. *The natural history of Madagascar*. Chicago: The University of Chicago Press.
Lowry, P. P., G. E. Schatz, and P. B. Phillipson. 1997. The classification of natural and anthropogenic vegetation in Madagascar, in *Natural change and human impact in Madagascar*. S. M. Goodman and B. D. Patterson, eds. Washington, D.C.: Smithsonian Institution Press, 93–132.
Paulian, R. 1961. *La zoogéographie de Madagascar et des îles voisines*. Volume 13 de *Faune de Madagascar*. Tananarive: l'Institut de Recherche Scientifique.

MADEIRA ARCHIPELAGO

DORA AGUIN-POMBO AND MIGUEL A. A. PINHEIRO DE CARVALHO

University of Madeira, Portugal

Madeira is a small archipelago of volcanic origin with a highly diversified flora and fauna. A substantial amount of this diversity is harbored in the evergreen laurel forest, a formerly widespread type of vegetation that covered southern Europe and North Africa during the Tertiary period.

LOCATION, ORIGIN, AND CLIMATE

The archipelago of Madeira is located in the Atlantic Ocean between 32 and 33° N and between 16 and 17° W, lying closer to Africa (~635 km) than to Europe (~794 km). Despite its size being less than 800 km^2, it comprises two inhabited islands, Madeira and Porto Santo; three islets of only 15 km^2 known as the Desertas Islands (Ilheu do Chão, Deserta Grande, and Deserta Pequena); and about ten offshore rocks. It is believed that this archipelago originated from a plume of the Earth's mantle in the Mid-Atlantic Ridge about 70 million years ago. Of all the islands, Porto Santo is the oldest (~14 million years), whereas Madeira and Desertas are more recent (from more than 4.6 to ~0.7 million years old). The composition of these islands is mainly basaltic, and there has not been any recent volcanic event in the last 6000 years.

The island of Madeira represents almost 93% of the archipelago extension and is rugged and steep, with about 90% of its surface above 500 m. It has a mountain ridge running east–west, reaching 1862 m at it highest point and much lower in its eastern part (below 200 m). The climate of the archipelago is Mediterranean, but because of differences in sun exposure, humidity, and annual mean temperature, on the island of Madeira there is a clear north–south differentiation. Here, the temperatures are mild year-round, between 15 and 22 °C at lower altitudes and between 5 and 15 °C at the highest altitudes. Humidity varies according to altitude, being greater at high and medium altitudes in the forest areas, where there is fog and a persistent cloud cover from 600–800 m up to 1600 m. In contrast, Desertas and Porto Santo are smaller and lower in altitude (less than 520 m) with temperatures similar to those in the warmer parts of Madeira, but they are much drier (less than 400 mm per year).

FLORA AND VEGETATION

On the basis of flora, the archipelago is biogeographically included in the Macaronesian region and shows affinities to the Mediterranean. In contrast to Desertas and Porto Santo, which are covered mainly with herbaceous vegetation, the flora of Madeira Island, as a result of its dimension, altitude, and orography, is more diverse, showing a marked altitudinal stratification, which is related mainly to temperature. There are four main types of vegetation: coastal vegetation, evergreen dry and wet forest, and upland vegetation. Coastal vegetation is below 300 m and includes a community of herbs and shrubs that have as dominant species *Euphorbia piscatoria*, *Echium nervosum*, and *Globularia salicina*, all endemic to Macaronesian archipelagoes (Fig. 1).

FIGURE 1 Vegetation of Madeira. Eastern part of Madeira (Ponta de São Lourenço) showing herbaceous vegetation and *Matthiola maderensis*, an endemic plant species occurring at low altitudes mainly in coastal rocks and cliffs. Photograph by T. Dellinger.

The dry evergreen forest has been much reduced and is typical of lower altitudes with high mean temperatures and low annual precipitation. *Apollonias barbujana*, *Laurus novocanariensis*, *Myrica faya*, and *Ilex canariensis* are the dominant canopy tree species. The evergreen wet laurel forest, which occupies 20% of the island, is the main type of vegetation. This luxuriant forest grows from 300–800 m to 1400 m in humid areas with mild temperatures, high precipitation, and frequent coastal fogs (Fig. 2) and contains many rare endemic species. Here are found hygrophilous tree species of Lauraceae exclusive to Macaronesia, such as *Laurus novocanariensis*, *Ocotea foetens*, and *Clethra arborea*. At higher altitudes, this forest is replaced mainly by herbaceous plants and shrubs, with *Erica arborea* being the dominant shrub species.

As occurs on other islands, the flora and fauna of this archipelago are species-poor but very interesting in terms of endemic species and taxonomically isolated groups. The vascular flora comprises about 1200 species including native and naturalized plants, of which 10% are endemic (Figs. 3, 4). Most of the endemics are found among trees and shrubs, and fewer among annuals. There are no endemic taxa above genus level in vascular plants, but there are five endemic genera, with three being monospecific, and about 18 genera out of 44 being exclusive to Macaronesia. Ferns and bryophytes are very diverse probably because of the high humidity; they are represented by approximately 75 and 512 species, respectively. Within the flowering plants, some genera such as *Argyranthemum* (four spp.), *Helichrysum* (four spp.), and *Sinapidendron* (five spp.) have diversified prolifically. One of the most remarkable features of vascular plants is the high number of woody endemic species with herbaceous relatives on the mainland, such as the genera *Euphorbia*

FIGURE 2 Vegetation of Madeira. Wet laurel forest with *Euphorbia mellifera*, an endemic woody species characteristic of moist and shady places on Madeira. Photograph by T. Dellinger.

(Fig. 2) and *Echium* (Fig. 4). Some genera show biogeographic disjunctions with taxa from far territories, such as the tree species *Apollonias barbujana*, with its closest relatives outside the islands being *A. arnottii* from southern India, or the genus *Picconia*, closely related to the genus *Notelaea* from Australia and Tasmania. In contrast to plants, lichens and fungi have been less studied.

FIGURE 3 *Andryala glandulosa*, a common endemic species on Madeira. Photograph by T. Dellinger.

FIGURE 4 *Echium candicans*, an endemic rare woody species found at high altitudes on cliffs in the laurel forest on Madeira. Photograph by T. Dellinger.

However, a checklist of the lichens and lichenicolous fungi of Madeira indicates more than 700 species already.

FAUNA

Among all animal species, about 25% are endemic, and all show great affinities to those of the Mediterranean and Europe, with about 4200 species having been reported thus far. Like other oceanic islands, Madeira has a characteristic disharmonic fauna, and this is especially visible in vertebrates. Native vertebrates include around 50 species, including one fish (an eel), five species of mammals (bats), one reptile (a lizard), and about 38 species of known resident birds, of which seabirds are the most diverse. However, there are only seven known endemic species of vertebrates: a lizard (*Teira dugessi*) (Fig. 5); three bird species including the Zino's petrel (*Pterodroma madeira*), the Trocaz pigeon (*Columba trocaz*), and the Madeira firecrest (*Regulus madeirensis*); and one bat and two bird species endemic to Madeira and the Canary Islands.

FIGURE 5 *Teira dugessi*, a very common endemic lizard species on Madeira. Photograph by T. Dellinger.

Although invertebrates are still insufficiently known, they are the most diverse, with land snails showing the most remarkable speciation rates and one of the highest rates of endemic speciation per square kilometer of all oceanic islands. About 70% of the approximately 250 recorded species are exclusive, and 18 genera are endemic, probably as the result of no more than 40 colonization events. High levels of single-island endemism and speciation have been predominantly a within-island phenomenon: In some extreme cases such as *Discula,* 15 endemic taxa occur on an island (Porto Santo and offshore rock) of only 42 km^2. Other genera, such as *Leiostyla* (30 spp.), *Actinella* (19 spp.), and *Amphorella* (12 spp.), have also undergone considerable radiation on Madeira.

Arthropods, especially millipedes, woodlice, and insects, represent about 87% of all known fauna. Of these, beetles stand out for the largest number of species, with approximately 900. There are 13 endemic genera of arthropods, most of which are monospecific. In contrast to this, some non-endemic genera—especially of beetles, millipedes, and moths—have diversified greatly. Particularly remarkable are the millipedes of the genus *Cylindroiulus,* which gave rise to the largest number of endemic species (30 spp.) within a single genus, and beetles of the genus *Laparocerus* (33 spp.).

In contrast to terrestrial fauna, marine biodiversity has been less studied and, although diverse, is not so remarkably rich in endemic species. However, the isolation of Madeira from the closest mainland and nearby islands by great depths (greater than 2000 m) and the warm Gulf Stream current that reaches this archipelago are both responsible for its unique fauna. Because of this, the marine fauna has its greatest affinities to the Atlantic Mediterranean areas and to the tropical and subtropical species of the eastern Atlantic, but there are also pantropical and amphiatlantic tropical species. These characteristics also explain the presence of Macaronesian and endemic species, among which are especially remarkable marine molluscs.

CURRENT STATUS AND CONSERVATION

After its discovery, the strategic position of the archipelago resulted in its extensive use by early sailors as a point of recharge for food on their way to America. Human population increased considerably after colonization in 1425, and today the number of inhabitants per square kilometer is about 330, the highest rate in Portugal and one of the largest in Europe. In addition to this large resident human population, Madeira receives about 1 million tourists per year. Thus, demographic pressure and tourism, together with exotic species, agriculture, forest degradation, and erosion, represent major challenges for conservation.

During and after colonization several non-native species of vertebrates such as rabbits, rats, mice, ferrets, goats, frogs, fishes, pigs, cats, and geckos were introduced, either purposely or accidentally, by humans, and many have become naturalized. Of these, mammals have been responsible for causing significant modifications to the native flora. Moreover, some, such as the house mouse (*Mus musculus domesticus*), have undergone considerable evolution during the short period of approximately 500 years since introduction. Indeed, this species represents an extreme case of rapid speciation, in which a single species has given rise to six different chromosomal forms probably because of geographical isolation and fragmentation.

The outstanding biodiversity of Madeira has been well known since the time of Wollaston and Darwin, yet much work remains in order to understand the extent and nature of this diversity. However, protected areas have been established for about 40% of the archipelago's surface in order to preserve the biodiversity of the island. The laurel forest, although it has been considerably reduced, still represents the largest surviving area of this kind in Macaronesia and, because of its size and conservation status, was recognized in 1999 by UNESCO as a World Natural Heritage Site. In addition, two marine reserves have been established on Madeira Island, whereas the entire Desertas Islands, both marine and terrestrial habitats, are considered natural reserves.

SEE ALSO THE FOLLOWING ARTICLES

Adaptive Radiation / Azores / Canary Islands, Biology / Cape Verde Islands / Endemism / Insect Radiations

FURTHER READING

Britton-Davidian, J., J. Catalan, M. G. Ramalhinho, G. Ganem, J.-C. Auffray, R. Capela, M. Biscoito, J. B. Searle, and M. L. Mathias. 2000. Environmental genetics: rapid chromosomal evolution in island mice. *Nature* 403: 158.
Carine, M. A., S. J. Russell, A. Santos-Guerra, and J. Francisco-Ortega. 2004. Relationships of the Macaronesian and Mediterranean floras: molecular evidence for multiple colonizations into Macaronesia and back-colonization of the continent in *Convolvulus* (Convolvulaceae). *American Journal of Botany* 91: 1070–1085.
Cook. L. M. 1996. Habitat, isolation and the evolution of Madeiran land-snails. *Biological Journal of the Linnean Society* 59: 457–470.
Gittenberger, E., D. S. J. Groenenberg, B. Kokshoorn, and R. C. Preece. 2006. Biogeography: molecular trails from hitch-hiking snails. *Nature* 439: 409–409.
Vanderpoorten, A., F. J. Rumsey, and M. A. Carine. 2007. Does Macaronesia exist? Conflicting signal in the bryophyte and pteridophyte floras. *American Journal of Botany* 94: 625–639.
Wetterer, J. K., X. Espadaler, A. L. Wetterer, D. Aguin-Pombo, and A. M. Franquinho-Aguiar. 2006. Long-term impact of exotic ants on the native ants of Madeira. *Ecological Entomology* 31: 358–368.

MAKATEA ISLANDS

LUCIEN F. MONTAGGIONI

University of Provence, Marseille, France

The term *makatea*, derived from Polynesian words (*maka*: slingstone; *tea*: white), relates to tropical Pacific islands possessing emergent (uplifted) limestones, mainly of coral reef origin and dissected by karst.

ORIGIN AND TECTONIC EVOLUTION

Makatea islands possess a volcanic basement built by a number of different mechanisms affecting the Earth's tectonic plates (hotspots, volcanism at or near divergent plate boundaries, arc volcanism at or near convergent plate boundaries). As the islands drowned due to crustal cooling, coral reefs developed around the volcanic cores. Locally, the volcanic pedestals were totally overtopped by reefal deposits. The time of deposition varies from site to site but was usually between the early Miocene and mid Quaternary.

The time of uplift also differs from island to island, but it was generally not before the early Quaternary. Mechanisms vary and reflect regional tectonics. On islands close to recent hotspot volcanoes (Makatea island in the Tuamotus), uplift was driven by the overload of the volcano (e.g., Tahiti), forming a bulge on the surrounding crust. By contrast, on Nauru (southwestern Pacific), emergence is attributed to the uplift of sea floor carried up onto a mantle bump by plate motion. Guam (in the Marianas) was uplifted at the crest of the frontal arc formed at the convergence of Pacific and Philippine plates.

GEOMORPHIC EVOLUTION

The major topographic features of the islands were acquired when the limestones were subaerially exposed as a result of uplift and/or global sea-level falls. At the time of emergence, two main island types existed: those with a volcanic core surrounded by an annular limestone platform and those high in elevation, composed of limestone capped by extensive plateaus (Fig. 1).

Landscapes resulted from differential karst erosion during periods of humid climate. This process promoted the

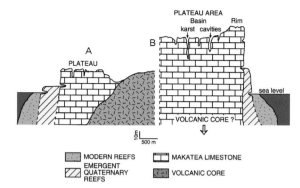

FIGURE 1 Idealized cross-sections of the borders of makatea island types. These islands are commonly fringed by fossil (mid- to late Quaternary) and modern coral reefs. Heights and distances (in meters) are given as indicative measures. (A) Island with a volcanic core surrounded by barrier of limestones. (B) High, atoll-like island, deeply dissected by karst. The volcanic pedestal lies at an unkown depth (presumably several hundreds of meters) below the carbonate pile.

formation of rim ramparts on the margins of the plateaus. Depressions developed coevally within plateau interiors. The final result was upland basin-and-rim structures and coastal cliffs.

SEE ALSO THE FOLLOWING ARTICLES

Atolls / French Polynesia, Geology / Pacific Region / Phosphate Islands / Sea-Level Change

FURTHER READING

Dickinson, W. R. 2004. Impacts of eustasy and hydro-isostasy on the evolution and landforms of Pacific atolls. *Palaeogeography, Palaeoclimatology, Palaeoecology* 213: 251–269.
Mylroie, J. E., J. W. Jenson, D. Taborosi, J. M. U. Jocson, D. T. Vann, and C. Wexel. 2001. Karst features of Guam in terms of a general model of carbonate island karst. *Journal of Cave and Karst Studies* 63: 9–22.
Nunn, P. D. 1994. *Oceanic islands*. Oxford: Blackwell Publishers.
Stoddart, D. R., C. D Woodroffe, and T. Spencer. 1990. Mauke, Mitiaro and Atiu: geomorphology of makatea islands in the Southern Cooks. *Atoll Research Bulletin* 341: 1–65.
Vacher, H. L., and T. M. Quinn. 1997. *Geology and hydrogeology of carbonate islands*. Amsterdam: Elsevier Science B.V.

MALDIVES

PAUL KENCH

The University of Auckland, New Zealand

The Republic of Maldives consists of 21 atolls and four reef platforms that straddle the equator in the northern Indian Ocean. Comprising 2041 reefs and 1190 reef islands, the archipelago is globally unique in the reef structures it possesses and their mode of evolution. The reef islands are low-lying and small in size, which has generated widespread concern as to their vulnerability to future sea-level and climatic change.

LOCATION AND GEOLOGIC STRUCTURE

Located in the northern Indian Ocean, the Maldives archipelago is an 868-km-long network of coral reefs that extends from the northern atoll of Ihavandhippolhu (6°57′ N) to Addu Atoll (0°34′ S), just south of the equator (Fig. 1). The Maldives constitute the central section

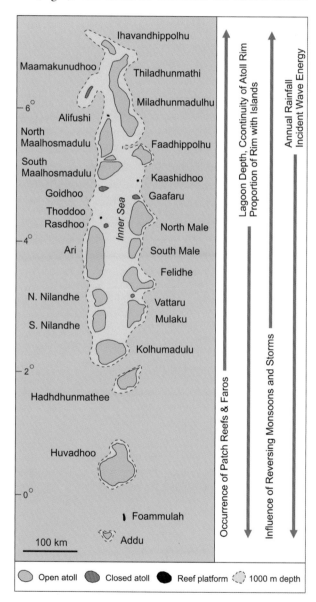

FIGURE 1 Configuration of the Maldives archipelago in the northern Indian Ocean showing latitudinal gradients in climate, oceanography, and physical atoll characteristics. Direction of arrow indicates an increase in each characteristic.

of a submarine ridge that stretches from the islands of the Lakshadweep (Laccadives) group in the north to the Chagos archipelago in the south.

The archipelago consists of a double chain of atolls on either side of an inner sea, tapering to single atolls to the north and south (Fig. 1). The basement rocks of the atoll chain are Eocene volcanics (55 million years old), which are capped with over 2000 m of Tertiary limestones. The atoll system has not formed according to the model of subsidence proposed by Charles Darwin. Rather, carbonate accumulation through the Tertiary has been dominated by lateral progradation from the outer twin atoll chain toward the inner sea. During the Quaternary, vertical reef growth has dominated over lateral progradation. This vertical growth is controlled by oscillations in sea level, with solutional lowering of reef platforms and lagoons during low sea-level stages and vertical accretion at times of higher sea level.

CORAL REEFS AND ATOLLS

The Maldives consists of 2041 individual reefs, with a total reef area of 4513 km^2. This complex network of reefs is organized into a number of distinct atoll and reef types. Open atolls are the dominant atoll type (16 in total) and are characterized by heavily dissected atoll rims, which in plan form appear as a sequence of individual reef platforms enclosing a central lagoon. Open atolls are large structures, ranging from 290 to 3790 km^2 in area, that contain numerous lagoonal reefs and collectively account for 99.5% of all Maldivian reefs. A striking feature of Maldivian open atolls is the presence of faros, ring-shaped coral reefs located within atoll lagoons. At a global scale, faros are scarce, yet they are abundant in the Maldives, where their formation remains a puzzle. There are five closed atolls in the archipelago, which are smaller in area than open atolls and have near-continuous reef platforms enclosing lagoons. Four oceanic reef platforms also occur. These reefs have no lagoon, and islands cover a large proportion of the reef surface.

REEF ISLANDS

The archipelago contains 1190 reef islands perched on top of reef surfaces, 200 of which are inhabited. The islands provide the only living space for the Maldivian population of approximately 330,000. Mid-Holocene in age, the reef islands formed during a major phase of deposition 5500–4000 years ago. Composed entirely of carbonate sands and gravels derived from the surrounding reefs, the islands are typically small and have a mean elevation of less than 1 m above sea level. They are dynamic landforms that exhibit rapid morphological adjustments in response to changing climatic (particularly wave-energy) and sea-level conditions.

LATITUDINAL GRADIENTS IN ATOLL CHARACTERISTICS AND PROCESSES

The physical characteristics of open atolls in the Maldives show marked spatial variations along the north–south gradient (Fig. 1). Northern atolls are characterized by a heavily dissected atoll reef rim, numerous lagoonal patch reefs and faros, and moderate lagoon depths (40–50 m). Reef islands are located on the peripheral and lagoonal patch reefs. Toward the south, atolls are characterized by more continuous atoll reef rims, a higher proportion of peripheral reef rim containing islands, deeper lagoons (70–80 m), and fewer lagoonal patch reefs.

Latitudinal variations in atoll morphology have been attributed to broad north–south gradients in climate and oceanographic conditions. Annual rainfall reduces from south to north along the archipelago (Fig. 1). Over the longer term, this rainfall gradient has influenced solutional lowering of lagoons during Quaternary glacial periods. The archipelago is subject to monsoon conditions that switch from west to northeast in a predictable fashion and influence wave and current patterns. The intensity of oscillating monsoon conditions increases to the north. In contrast, incident wave energy reduces in magnitude in the northerly direction. This energy gradient controls contemporary coral reef growth and island building processes.

SEE ALSO THE FOLLOWING ARTICLES

Atolls / Climate Change / Indian Region / Reef Ecology and Conservation / Sea-Level Change

FURTHER READING

Gardiner, J. S. 1903. *The fauna and geography of the Maldives and Laccadive archipelagoes.* Cambridge: Cambridge University Press.
Kench, P. S., and R. W. Brander. 2006. Response of reef island shorelines to seasonal climate oscillations: South Maalhosmadulu atoll, Maldives. *Journal of Geophysical Research* 111: F01001:1-12.
Kench, P. S., R. F. McLean, and S. L. Nichol. 2005. New model of reef-island evolution: Maldives, Indian Ocean. *Geology* 33: 145–148.
Nasser, A., and B. G. Hatcher. 2004. Inventory of the Maldives' coral reefs using morphometrics generated from Landsat ETM+ imagery. *Coral Reefs* 23: 161–168.
Purdy, E. G., and G. T. Bertram. 1993. *Atoll and carbonate platform development in the Maldives, Indian Ocean.* AAPG Studies in Geology No. 34. Tulsa, OK: American Association of Petroleum Geologists.

MAMMAL RADIATIONS

LAWRENCE R. HEANEY AND STEVEN M. GOODMAN

The Field Museum, Chicago, Illinois

Islands that have been isolated since their formation or for very long periods of time often have land mammal faunas that are made up largely or entirely of endemic species, and often these species are members of species-rich endemic clades. Though members of these endemic clades are each other's closest relatives, they typically show highly diverse body size, morphology, behavior, and ecology. These constitute classic cases of adaptive radiation, in which local speciation has produced spectacular diversity.

LOCATION AND EXTENT OF ISLAND MAMMAL RADIATIONS

On very isolated and small islands, such as those in the central Pacific and southern Atlantic Oceans, native mammals are either absent or very low in diversity. On progressively less isolated and larger islands, especially in warm seas, bats are present in increasing diversity, with moderately rich faunas often present before non-volant mammals make an appearance. Even in some archipelagoes, such as the Galápagos and the Canary Islands, which are relatively near to continents and have a dozen or more islands, non-flying mammals have low total diversity and show evidence of only limited diversification within the archipelago. However, on single large islands that once were parts of continents but have been isolated for tens of millions of years, such as Madagascar; in groups of large, geologically old islands with mixed continental and oceanic origins, such as the Greater Antilles and the islands west of New Guinea; and in large, complex oceanic archipelagoes, such as the Philippines, remarkably diverse endemic faunas are present, the great majority produced by local speciation.

Perhaps the best known of the island mammalian radiations occur on Madagascar, with a surface area of 587,000 km². Here the modern non-flying and native mammal fauna of well over 110 species, all of which are endemic, includes 26 species of rodents (family Nesomyidae), eight carnivorans (family Eupleridae), 31 tenrecs (family Tenrecidae), and over 50 lemurs (five different families). Within each of these different radiations, one can find a range of body forms and sizes, with striking convergence to other mammalian groups found elsewhere in the world. This is well exemplified by the tenrecs, which span three orders of magnitude in body mass from about 2.5 to over 2000 g (Fig. 1). Within one of the tenrec genera, *Microgale,* the most speciose on the island with 22 recognized species, are included, for example, small shrew- or mole-like animals, some of which are semi-fossorial; at least partially arboreal, moderately sized species with partially prehensile tails three times their head and body length; and primarily terrestrial, relatively large carnivorous or omnivorous species with robust dentitions. Furthermore, what is extraordinary about these four large Malagasy mammalian radiations is that this notable endemic diversity can be explained by only four successful and independent colonizations of the island by terrestrial mammals, which in turn indicates how rare successful colonization events were through geological time. Finally, recent biological exploration of Madagascar has revealed that a considerable proportion of its diversity was unknown to science, with, for example, over 30 land mammals being described as new since 1986. (These figures do not include descriptions of new lemur taxa based exclusively on genetic data.)

FIGURE 1 Malagasy tenrecs have diversified into a spectacular radiation. Here several forms are represented (clockwise from upper right-hand corner): *Setifer setosus, Geogale aurita, Echinops telfairi,* and *Hemicentetes semispinosus.* Reprinted from Goodman and Benstead (2003), with permission from Link Olson.

Less widely known but similarly diverse mammalian radiations have taken place in the Philippines, a group of 7000 islands totaling a land surface area of about 300,000 km². Most remarkable among these are the rodents in the family Muridae. Although members of this family are generally known as "rats" and "mice," most of the Philippine species look and behave little like the pest species familiar to most people. One endemic clade includes 15–20 species of cloud rats and their relatives. These nearly all live in trees; have long, furry, or hairy tails; and feed in the rainforest canopy on tender young leaves, fruits, and/or seeds. Another endemic clade includes at least 35 species of smaller, short-tailed animals that live mostly in cloud forest and feed on earthworms and other soft-bodied soil invertebrates (Fig. 2). As with Madagascar, recent field studies have documented many previously unknown species; 20 species have been described since 1986, and many others are currently under study.

THE ORIGINS OF DIVERSITY

In the cases of Madagascar, the oceanic Philippines, and many other islands, the non-flying mammals colonized from a nearby continent across ocean waters, probably floating on a mass of vegetation or in hollow tree trunks. In some of these cases (e.g., cloud rats and "vermivore mice" from the Philippines), species that are most closely related to one another occur in different parts of the archipelago, and the evidence suggests that much speciation takes places in populations that share a recent common ancestor but have become geographically isolated because of dispersal over water barriers.

However, we can also typically see not just one or two, but many species from the same clade living sympatrically on the large island of Luzon. For example, in the high, wet Central Cordillera of Luzon Island, up to five species of cloud rats may live together on one forested hillside; their weights are 15 g, 125 g, 200 g, 1.4 kg, and 2.6 kg, with the smaller members feeding on seeds and small fruit and the larger ones feeding on leaves and buds. Up to seven species of "vermivore mice" may occur on the same hillside, with two species that burrow through the thick layer of humus, two that forage in the leaf litter, one that patrols long trails over the moss-covered ground, and three that forage opportunistically on the ground surface and in the lower portions of trees and shrubs. Similarly, on Madagascar, up to 14 species of lemurs and 17 species of tenrecs may occur in close proximity in the same forest block, and in many cases, the sympatric species are each others' closest living relatives. In such cases, the speciation most likely took place within the island, with isolation caused by land features such as rivers, mountain ranges (for lowland species), or forest types that are not suitable habitat.

These examples illustrate several primary features of mammalian radiations. First, certain species within a given island or archipelago occurring in geographic isolation differentiate genetically, morphologically, and ecologically from one another. In some cases, with the passage of time, changing circumstances reduce or

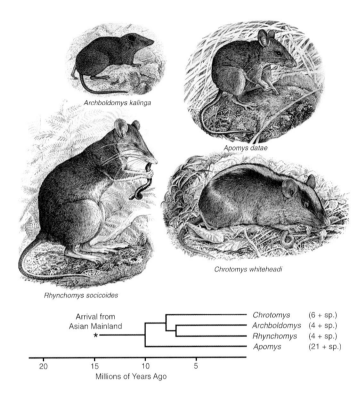

FIGURE 2 Four genera containing at least 35 species make up the endemic clade of vermivorous Philippine rodents, with a phylogeny showing the estimated times of divergence since their common ancestor arrived in the Philippines 12–15 million years ago from the Asian mainland. Phylogeny adapted from Jansa et al. (2006). Drawings by V. Simeonovski, copyright Field Museum of Natural History.

remove the geographic isolating barrier, and they are able to occur sympatrically. This process is repeated many times, usually over the passage of several hundred thousand or million years, so that species richness within communities increases over time. The members of a clade that live sympatrically—all descended from a single common ancestral species, and all derived by speciation within the island or archipelago—nearly always differ from each other in conspicuous ways, such as body size, pelage coloration, vocalizations, preferred food or foraging zones, time of day they are active, and so forth. They may continue to use the same kinds of resources as their shared ancestor, but they evolve in response to selection to exploit different portions of the resource base efficiently.

The net result is a diverse community of species, each using resources in a unique way (i.e., each with its own niche). Often these communities parallel the extent of mammalian diversity that we might see in a continental area, and the species on the islands superficially resemble those on continental landmasses, but they originated separately. For example, all tenrecs are related to one another, but certain species are shrewlike (small *Microgale*), hedgehog-like (several genera of the subfamily Tenrecinae), molelike (*Oryzorictes*), otter-like (*Limnogale*), and mouselike (*Geogale*). The carnivorans of Madagascar form an endemic radiation, superficially resembling other groups of African and Asian carnivores, such as genets and mongooses. The vermivorous shrew-mice of Luzon (*Archboldomys* spp.) are very shrewlike, as their English name implies, and the cloud rats of the Philippines are, in some respects, ecological equivalents of the leaf-eating monkeys of continental Asia and Africa, with some similar aspects to their anatomy (e.g., very large cecums for digesting tough leaves) and body size. Such comparisons have led to the concept of convergence, in which species in different places come to resemble one another because of similar selective pressures associated with use of similar resources. Although such comparisons are general and can rarely be made precisely, the concept does call attention to the fact that island mammal radiations produce a great many distinctive species from a single ancestor, and those species fill a wide range of niches, with overall patterns of resource use being similar to those in very distantly related groups of mammals.

CONSERVATION OF ISLAND MAMMAL RADIATIONS

A disproportionately large number of mammal species that have become extinct in the last 400 years were island endemics, leading to fears that island mammals may, in general, be vulnerable to extinction. Several lines of evidence indicate that the picture is complex, and no single pattern can be identified.

One of the clearest and most extreme cases for examining this issue involves the mammals of the West Indies, which have an exceptionally rich fossil and subfossil history. Of 76 species of Antillean non-flying mammals that existed 20,000 years ago (including insectivores, sloths, primates, and rodents) about 90% have become extinct, including 14% of the 59 species of bats (and 24% of those that still survive have undergone local but not complete extinctions). Thirty-six of the 67 extinct non-flying species (54%) disappeared prior to the arrival of humans on these islands, due to loss of open, grassy habitats (replaced by wetter and more densely forested habitats), reduction in island area (due to the 120 m rise in sea level at the close of the last glacial maximum), and flooding of coastal caves (especially in the case of bats). Hunting and massive habitat disturbance that came with the subsequent arrival of Amerindians about 4500 years ago, followed by Europeans with their attendant house rats and feral dogs, cats, goats, and pigs about 500 years ago, and mongooses more recently, were associated with the extinction of another 37 species of non-flying mammals, leaving only nine as survivors today.

The picture that emerges from studies in Madagascar is similar in some respects, but different in others. Rich deposits of subfossil lemur bones dating from the Late Quaternary include 17 species that have gone extinct, all larger in body mass than any living taxon. Three different species of dwarf hippos were recovered from these deposits, as well as the enigmatic genus *Plesiorycteropus,* which is placed in its own order of mammals (Bibymalagasia), and an assortment of other mammals, all of which have vanished. Some of these extinctions are associated with natural climatic change, including shifts in the west and south associated with aridification, and in other cases humans, who colonized the island some 2500 years ago, presumably delivered the *coup de grace.* The non-native rat *Rattus rattus* has colonized virtually all of the remaining humid forest zones of the island and has introduced bubonic plague to the native nesomyine rodents, which has had epidemic consequences for the latter. Overall, unlike the case of the Caribbean islands, a great diversity of Malagasy mammals is still extant.

Yet a different picture emerges from the Philippines. Several species of large mammals (an elephant, a rhinoceros, and a dwarf water buffalo) are known only as fossils or subfossils; their age and cause of extinction is unknown. The extant fauna is highly diverse, with about 200 land-living species. Recent studies have shown that alien pest species (primarily *Rattus*) are abundant in agricultural and residential areas but are entirely absent from areas with natural vegetation, where native small mammals are abundant. In disturbed natural vegetation, the alien *Rattus* are present but uncommon and are replaced by the native species as the natural forest regenerates. The paucity of fossils or subfossils makes the earlier history of extinction very uncertain, but there is no evidence that the extant native species are especially vulnerable to extinction; rather, they seem to persist well in the face of anything short of overwhelming habitat destruction and/or severe overhunting. Unfortunately, those conditions do occur in parts of the Philippines, and some extinction of mammals, though not yet recorded, seems inevitable.

Taken together, these examples indicate that there is a wide range in the susceptibility to extinction of endemic mammals on islands, just as there is on continents. Continued biological inventories of different islands around the world have documented far greater species diversity than previously thought and provide a wealth of new ecological data. This information provides new insights into understanding the evolutionary history of mammalian radiations, as well as new information on how we may conserve these remarkable fauna.

SEE ALSO THE FOLLOWING ARTICLES

Convergence / Endemism / Lemurs and Tarsiers / Madagascar / Philippines, Biology / Radiation Zone / Rodents

FURTHER READING

Burney, D. A., L. P. Burney, L. R. Godfrey, W. L. Jungers, S. M. Goodman, H. T. Wright, and A. J. T. Jull. 2004. A chronology for late Prehistoric Madagascar. *Journal of Human Evolution* 47: 25–63.
Goodman, S. M., ed. 2008. *Paysages naturels et biodiversité de Madagascar.* Paris: Muséum national d'Histoire Naturelle.
Goodman, S. M., and J. P. Benstead, eds. 2003. *The natural history of Madagascar.* Chicago: The University of Chicago Press.
Heaney, L. R. 2000. Dynamic disequilibrium: a long-term, large-scale perspective on the equilibrium model of island biogeography. *Global Ecology and Biogeography* 9: 59–74.
Jansa, S., K. Barker, and L. R. Heaney. 2006. Molecular phylogenetics and divergence time estimates for the endemic rodents of the Philippine Islands: evidence from mitochondrial and nuclear gene sequences. *Systematic Biology* 55: 73–88.
Steppan, S., C. Zawadski, and L. R. Heaney. 2003. Molecular phylogeny of the endemic Philippine rodent *Apomys* and the dynamics of diversification in an oceanic archipelago. *Biological Journal of the Linnean Society* 80: 699–715.
Woods, C. A., and Florence E. Sergile, eds. 2001. *Biogeography of the West Indies: patterns and perspectives,* 2nd ed. Boca Raton, FL: CRC Press.

MANGROVE ISLANDS

PETER SAENGER

Southern Cross University, Lismore, Australia

Mangrove islands are islands composed entirely or partially of mangrove vegetation and comprise microcosms of generally high productivity, high biodiversity, and structurally complex habitats in the nearshore environment. Such mangrove islands are largely confined to the tropics, with their latitudinal limits occurring in those regions where water temperatures never exceed 24 °C throughout the year. In comparison to shoreline mangroves, the vegetation of mangrove islands is subjected to much greater fluctuations in hydrological and meteorological conditions, resulting in its high dynamism and leading to its enhanced susceptibility to drastic change.

MANGROVE ISLAND TYPES

Mangrove islands can be subdivided into two broad types: islands consisting only of mangroves and islands where mangroves form a part of the island vegetation. Those mangrove islands consisting only of mangroves commonly occur where sediments have built up sufficiently to form an island at low tide, which has then been colonized by mangroves (Fig. 1). These are known as overwash mangroves as the islands are completely submerged at high tide, and only the mangroves are visible. Such overwash forests are commonly associated with deltas of rivers carrying high terrigenous silt loads (e.g., the Meghna–Brahmaputra delta of Bangladesh or the Fly River delta of Papua New Guinea) or in shallow waters where abundant sediments of biogenic carbonates accumulate (e.g., the overwash mangrove islands of the Florida Keys). Clearly, however, the mangrove vegetation will differ structurally and functionally in the deltaic, terrigenous setting from that of the carbonate setting.

FIGURE 1 Overwash mangroves on carbonate shoals in the Ras Mohammed National Park in the northern Red Sea, Egypt. These mangroves consist of only one species, *Avicennia marina,* whose pneumatophores form a 20–30-cm-high dense fringe to the stand.

The second type of mangrove islands, where mangroves form a significant but non-exclusive part of the island vegetation, are usually associated with coral reef islands. There mangroves may grow directly over a reef flat or a fossil reefal platform, or in the lee of shingle ramparts, which provide protection from trade winds. Mangrove islands on reef flats are common off the coast of Central America (e.g., Belize) and on many islands of the Pacific (e.g., Palau and Ponape), where low energy conditions and relatively low tidal ranges occur. Although such mangroves are generally more extensive on "high islands"—which are not included here as mangrove islands—they are also common on "low islands" (e.g., the atoll islands of Nui, Vaitupu, and Nanumanga of Tuvalu). On the Great Barrier Reef of Australia, persistent southeasterly trade winds occur during April to November, resulting in shingle ramparts on the windward, southeastern shores of these "low islands" and leeward sand cays with intervening mangroves in the lee of the shingle. These are referred to as "low wooded islands" and occur commonly in northeastern Queensland, north of 15° S. Low wooded islands display a complex range of features, which occupy around 30–50% of the reef flat. The windward shingle banks are often partially colonized by mangrove scrub (e.g., *Avicennia marina, Aegialitis annulata, Pemphis acidula*) and other halophytes (e.g., *Sarcocornia, Sesuvium, Suaeda*), and moats form to the leeward of the shingle ramparts, which retain water at low tide and allow micro-atoll growth (commonly of *Porites andrewsi*) to occur. The micro-atolls grow upward to mean low water level, and provide a convenient platform for the anchoring of drifting propagules and the subsequent growth of mangroves, which ultimately cover the entire moat with a closed canopy mangrove forest. On the Great Barrier Reef, the primary colonizing species is *Rhizophora stylosa*, although other species, such as *Avicennia marina* and *Lumnitzera racemosa*, quickly follow. In the northern Great Barrier Reef, closed canopy mangroves on low wooded islands consist of around 15 species. Because of the small catchment areas of these islands, freshwater availability is restricted, and mangrove communities on low wooded islands are restricted to regions where the annual rainfall exceeds 1200 mm. Low Isles, off Cooktown in North Queensland, is the most studied low wooded island on the Great Barrier Reef, being originally surveyed by the 1928–1929 Great Barrier Reef Expedition, which spent a year surveying the biota of this island. The island and its mangroves have been resurveyed in 1945, 1954, 1965, and 1973, and over that period, the *Rhizophora* woodland has increased markedly. Some, but not all, other low wooded islands also showed mangrove expansion, and it seems that reef-top topography and the extent of micro-atoll formation appear to be the major factors in the extent of mangrove development. Cyclones are an annual occurrence in this area, and destructive winds can arrest or even reverse mangrove expansion on low wooded islands for considerable periods.

ENVIRONMENTAL SETTINGS OF MANGROVE ISLANDS

As outlined above, mangrove islands occur in two contrasting settings: in river mouth deltas and on reefal platforms. Clearly, the deltaic setting is river-dominated, and the reefal setting is tide-dominated. These dominating factors result in significant differences in the structure and functioning of mangrove islands.

In the deltaic setting, mangrove islands are flooded by river water as well as by tides, so salinity remains moderate. Silt and nutrients are carried by the river water, but they flocculate and are deposited when mixed with seawater. As a result, nutrient-rich sediments are deposited around mangrove pneumatophores and stiltroots on mangrove islands, and luxuriant and diverse mangrove communities rapidly develop. Organic-matter production is generally washed out of the system (strongly outwelling) and replenished by the river-borne nutrient supply. In turn, the outwelling of organic matter trophically supports nearshore food webs, particularly for molluscs, penaeid shrimp, and juvenile fish.

By way of contrast, reefal settings bear the brunt of the tides, which are generally full-strength seawater. Only biogenic sediments are available, brought in by daily tides (bidirectional flux), and nutrients are generally deficient. As a consequence, mangroves are slowed in terms of their organic-matter production, nutrient coupling is usually tight, and organic matter tends to accumulate amongst the vegetation. Reefal mangrove islands are susceptible to erosion, particularly around their periphery, and tend to have less luxuriant mangrove vegetation, often wind-pruned and limited in species richness to those species capable of growing in full-strength seawater. The tendency toward erosion, however, may be offset in this setting, as root material does not break down as rapidly as in deltaic systems and may accumulate as mangrove peat, which facilitates vertical accretion and habitat stability. In Central America, in particular, wherever coring of reefal mangrove islands has been undertaken, peat layers of up to 10 m in depth have been found, and carbon dating suggests that the mangrove communities were initiated through rising sea levels around 8000 years ago and have accumulated peat at a rate that allowed them to keep pace with the rising sea levels of the late Holocene. On the Great Barrier Reef, the mangrove communities of low wooded islands, less reliant on peat accumulation than on micro-atoll growth, were initiated around 6000 years ago.

ECOLOGICAL SERVICES PROVIDED BY MANGROVE ISLANDS

Mangrove islands play an important role in stabilizing and protecting shoaling sand and mud flats. Their roots bind the sediments, and their stiltroots and pneumatophores reduce the water velocity around them, leading to the further deposition of sediments.

Mangrove islands, particularly those associated with deltas, tend to be strongly outwelling. Organic matter is broken down by various biotic and physical process into small particles, and this detritus supports a range of dependent

FIGURE 2 Brown pelicans and darters roost in the canopy of *Avicennia germinans* on reefal mangrove islands in Quintana Roo, Mexico. Although such roosting aggregations may provide significant nutrient enrichment, it does not offset the physical damage they do to the upper canopy.

nearshore species, including penaeid shrimp (prawns) and detritivorous fish, such as mullet (family Mugilidae) and bream (family Sparidae). Many commercial and subsistence fisheries are focused in and around such mangrove islands.

Finally, mangrove islands constitute structurally complex habitats that provide roosting and nesting sites, particularly for bats and seabirds. The surrounding waters, the mangrove vegetation, and the mangrove-associated invertebrates provide a diverse source of food for such nesting and roosting aggregations. On the Great Barrier Reef, around 30 species of seabirds nest on low wooded islands; one species, the Torres Strait pigeon (*Myristicivora spilorrhoa*), relies on such mangrove islands in its annual migrations between Australia and Papua New Guinea. Although such nesting aggregations in the reefal setting bring much needed nutrients, this comes at a physical cost in that dense aggregations of seabirds damage the mangroves (Fig. 2).

SEE ALSO THE FOLLOWING ARTICLES

Climate on Islands / Coral / Great Barrier Reef Islands, Biology / Hurricanes and Typhoons / Hydrology / Tides

FURTHER READING

Cintrón, G., A. E. Lugo, D. J. Pool, and G. Morris. 1978. Mangroves of arid environments in Puerto Rico and adjacent islands. *Biotropica* 10: 110–121.
Hopley, D. 1982. *The geomorphology of the Great Barrier Reef: Quaternary development of coral reefs.* New York: John Wiley & Sons Inc.
Macintyre, I. G., M. A. Toscano, R. G. Lighty, and G. B. Bond. 2004. Holocene history of the Mangrove Islands of Twin Cays, Belize, Central America. *Atoll Research Bulletin* 510: 1–16.
Saenger, P. 2002. *Mangrove ecology, silviculture and conservation.* Dordrecht, Netherlands: Kluwer Academic Publishers.

Stoddart, D. R. 1980. Mangroves as successional stages, inner reefs of the northern Great Barrier Reef. *Journal of Biogeography* 7: 269–284.
Woodroffe, C. D. 1987. Pacific Island mangroves: distribution and environmental settings. *Pacific Science* 41: 166–185.

MARIANAS, BIOLOGY

GORDON H. RODDA

U. S. Geological Survey, Fort Collins, Colorado

The Mariana archipelago is a line of small oceanic islands in the tropical northern Pacific Ocean about 1800 km east of the Philippine Islands. Because of its remoteness from Melanesian and Asian source areas, its small land area, its limited elevational range, and the sparse opportunities for faunal exchange among islands, the native biota of the Mariana Islands is relatively depauperate (species-poor), and native vertebrates are limited to species that can fly (birds, bats) or raft on floating vegetation and withstand contact with seawater (small lizards). Many of these vertebrates have been extirpated, and all but one of the Mariana endemic vertebrates have experienced range reductions consequent to the arrival of humans and human-introduced species.

BIOGEOGRAPHY

The Mariana island chain consists of 15 primary islands stretching 920 km from north to south and ranging in size from 541-km^2 Guam (13.5° N) to 2-km^2 Farallon de Pajaros (20.5° N). Guam is more than four times the size of the next larger Mariana island (Saipan, at 123 km^2) and by itself makes up over half the land area of the Marianas. Guam is also the largest island in Micronesia, the cluster of mostly tiny islands (hence the "micro" in Micronesia) stretching more than 5000 km east–west across the vast area of tropical northern Pacific Ocean west of the International Date Line.

The Mariana Islands arose from the depths of the ocean, in response to melting of the Pacific plate as it pushed under the Philippine plate. Accordingly, none of the islands has ever been connected to a continental land mass or to its neighboring islands. The southern arc of islands (Guam to Farallon de Medinilla) arose at about the same time, during a land-building period that occurred from 42 to 8 million years ago. Although seismically active, the southern arc has experienced no volcanism in the modern era.

The more northerly islands arose about 5 million years ago, with continued volcanism. The northern islands are relatively small (the largest is Pagan, 18.2° N, at 48 km²). Periodic eruptions throughout historic time have eliminated much of the vegetation on Pagan. In addition to being younger and smaller, the northern islands experience prevailing westerly winds, lower rainfall (Pagan receives about 1900 mm/y compared to Guam's 2400 mm/y), and greater seasonal variation in temperature and rainfall. In general, seasonal variation in rainfall and temperature increases from south to north, with Guam being the most equable (mean temperature: 26.3 °C; annual variation in monthly mean temperature: 1.2 °C; driest month: March [78 mm]; wettest month: September [411 mm]).

Unlike the low atolls of eastern Micronesia, which are mostly limited to seashore or strand vegetation, the Mariana Islands are considered "high" islands by virtue of their uplifted volcanic and limestone peaks and plateaus, most (11 of 15) of which have maximum elevations greater than 300 m (965 m on Agrihan, the highest point in the archipelago). Consequently they have extensive interior forests that benefit from high-elevation cloud cover and coolness. Thus, whereas the coastal plants and animals of the Mariana Islands are similar to those found throughout Micronesia, the interior species of the Mariana Islands are generally unique to high islands. However, the islands rarely attain enough elevation for independent life zones to have developed, as they have in Hawaii and many other large Pacific islands, and very few plant or animal species are limited to high elevation.

The Mariana Islands are relatively remote; no larger land masses are found within 1500 km. The nearest larger island to Guam is the slightly larger island of Manus, just south of the equator and north of New Guinea, 1740 km south of the Mariana Islands. Furthermore, the tropical source area for species climatically suitable for the Mariana Islands is to the south and west, whereas the prevailing winds and currents are from the northeast. Drift dispersal northward from the south is further inhibited by the powerful cross-currents supplied by the easterly equatorial current and the westerly equatorial countercurrent. Thus, the natural immigration rate of new species to the Marianas is very low, and biotic exchange among the islands is fairly limited. The southern islands are comparatively species-rich, as befits their greater age and larger area.

The Mariana island chain well illustrates the ecological factors affecting species colonization and diversification; few species have colonized, and little diversification has occurred. Clusters of remote islands, such as Hawaii, have proven to be incubators of evolutionary diversification if the islands are old enough, are close enough together, and have enough native habitat differences to promote, sustain, and exchange evolutionary novelties. However, most of the Marianas are too young, too far apart, and too limited in habitat diversity to support the dramatic evolutionary radiations seen in the Galápagos or Hawaii. Nonetheless, many Mariana Island populations have been isolated for such a long period that they have diverged into unique species; thus, the Mariana Islands are a hotspot of endemism.

ORIGINAL BIOTA OF THE MARIANA ISLANDS

With the exception of extant vertebrates, the biota of the Mariana Islands is not well documented. The earliest vertebrate inventory data come primarily from the subfossil excavations of David Steadman and colleagues on the islands of Rota, Tinian, and Aguijan. Steadman estimates that the prehuman vertebrate fauna of Rota consisted of 30–35 land birds, about ten lizards, one subterranean snake, and 2–3 bats. Originally all of the Mariana Islands probably supported breeding colonies of pelagic seabirds. Presumably, the primordial species richness would have been substantially greater on the much larger island of Guam, for which the prehuman fauna has yet to be excavated.

Guam now has about 325 native species of plants, of which about 25% are endemic to the Mariana Islands. About half of the 60 species of land molluscs (snails) are endemic, and an estimated 45% of the native insects are believed endemic. Endemism is lower in lizards (perhaps because of genetic swamping by repeated colonization) and higher in birds (as a result of a relatively high diversification rate). Near-shore marine fish species in the Marianas number around 900, but relatively few appear to be endemic to the Marianas, presumably because of a steady influx of planktonic larvae (which swamp out local adaptation). All of these taxa reached the Marianas primarily from the Caroline Islands to the south, but most Caroline Island lineages arose on the Asian mainland (fish) or in Australasia (other vertebrates and plants) and reached the Caroline Islands by island hopping via Indonesia (especially birds), the Philippine Islands (especially fish), or New Guinea and its environs (especially lizards).

HUMAN IMPACT

The Mariana Islands were first settled between 3500 and 4000 years ago, by people who originated in southeastern Asia, island hopped to Indonesia, and sailed eastward to the western Caroline and Mariana Islands. Their seclusion in the isolation of western Micronesia was abruptly terminated by the arrival in 1521 of Ferdinand Magellan and his starving crew. Over the ensuing several centuries (until 1896) the Mariana Islands were considered a possession of Spain, which forcibly converted the islanders to Catholicism through a series of small wars that—along with exotic diseases—largely depopulated the islands. In 1896 the Spanish administration of Guam was terminated by U.S. warships, which neglected to visit the more northerly Marianas, as the U.S. government erroneously believed them to be unpopulated. The U.S. took Guam as a spoil of the Spanish-American war, and Spain sold the Northern Marianas to Germany, which lost the islands to the Japanese during World War I. During World War II, the islands north of Guam were taken from the Japanese and later folded into the U.S. Strategic Trust Territory of the Pacific, a trusteeship sanctioned by the United Nations. The Strategic Trust Territory was dissolved in the 1980s following various plebiscites on self-government, which in the case of the Northern Marianas took the form of a request to be made part of the United States. The Mariana islands north of Guam are today recognized as the "Commonwealth of the Northern Mariana Islands" or CNMI. The CNMI and Guam are each territories of the United States. Mariana Island residents are U.S. citizens, but they cannot vote in U.S. elections, have no direct say in U.S. governance, and do not pay federal taxes.

The four large southern Mariana Islands have dense human populations, especially Guam (168,000) and Saipan (62,400). Since the cessation of Japanese colonization of the islands north of Saipan after World War II, the far northern islands have been generally uninhabited. Six of these (Guguan, Asuncion, the three islands of Maug, and Farallon de Pajaros) are permanently preserved as nature reserves under the terms of the CNMI constitution.

Most of the bird and mammal species of the Mariana Islands have been progressively eliminated by humans or by species introduced by humans. For example, Rota today has only ten land birds, eight native lizards, one native snake, and one bat. Guam is a more extreme example, having lost all but two of its native land birds. The circumstances of these extirpations warrant a closer look.

Prehistoric Extinctions and Introductions

About half of the bird species loss occurred during prehistoric times. Presumably, large, edible, flightless birds were the most vulnerable and disappeared first. For example, Steadman lists six species of undescribed rails eliminated from the southern Marianas. The proximate cause was probably direct human consumption, as their bones were found burned. On many oceanic islands, rats were an important early pressure on native birds, but this appears to have been less important in the Mariana Islands. The only rat present in the Marianas prior to the modern era was the small Pacific rat (*Rattus exulans*), which appeared relatively late in the prehistoric period (about 1000 years before present, or more than 2000 years after human colonization), and it was the only mammal introduced prior to Magellan's arrival in 1521.

Historic Introductions

Chickens, cattle, dogs, pigs, cats, deer, and goats were all intentionally introduced soon after Magellan's discovery. Of these, cats are likely to have taken the largest toll on the native fauna, especially flightless birds; ungulates (deer, pigs, cattle, goats) undoubtedly had the greatest impact on vegetation. Potentially even more destructive, however, were species introduced by accident: rats, ants, mosquitoes, lizards, snakes, and diseases. Early Western observers wrote with great awe about the irruption of rats in the Mariana Islands and the loss of food crops to the rodent plague. Mosquitoes appear to have arrived with Magellan; invertebrate introductions have continued to the present time, and the rate of new species' arrival has increased each decade. Plant introductions have probably followed the same temporal pattern, with nearly 600 species now recorded as having been introduced to Guam. Vertebrate introductions have been more carefully documented, with six frog species established as of 2007, as well as two or more turtle species, six lizards, one snake, eight birds, and 11 mammals.

Impacts of Historical Introductions

Every introduced species has the potential to transform the ecosystem that it invades, but generally only a few species have such a large impact. Introduced plants and herbivores often alter the structure and composition of the vegetation. Predatory species tend to depress the abundance of, or even eliminate, vulnerable prey. Prey species tend to subsidize the abundance of predators, which then become abundant and depress or eliminate native prey species. The Mariana Islands have notable examples of all three types of impacts.

ECOSYSTEM TRANSFORMERS

The key vegetation transformers in the Mariana Islands were an introduced legume tree (*Leucaena leucocephala*, known locally as tangantangan), introduced forest-covering vines (of diverse species), pigs (*Sus scrofa*), deer (*Cervus mariannus*), and goats (*Capra hircus*). The tangantangan tree readily colonizes disturbed sites, often forming dense monotypic stands that persist for decades or perhaps centuries, thereby preempting natural succession to native forest species. Introduced vines also play a major role in inhibiting natural succession, primarily by covering the forest canopy and preventing light penetration to native forest stands attempting to regrow after typhoons.

Regeneration is also inhibited by browsing of introduced ungulates. Curiously, the deer species found in the Marianas, although native to the Philippine Islands, was first described scientifically from a specimen taken in the Mariana Islands after the deer was introduced to Guam in 1771—hence the misleading species name *mariannus*. Regenerating plants are especially nutritive and defenseless, and are thus especially vulnerable to the unnatural browsing from introduced ungulates. Goats have been particularly damaging to native vegetation, especially on uninhabited islands, where the ungulates can sometimes reach densities enabling them to remove virtually all leaves within reach; plans are progressing for goat removal from selected northern islands.

INTRODUCED PREDATORS

Many predators have been introduced to the Mariana Islands, of which the most important were the rat (*Rattus rattus*), shrew (*Suncus murinus*), and brown tree snake (*Boiga irregularis*). Although *R. rattus* is primarily a herbivore, this species climbs trees very well and preys opportunistically on eggs and small animals. Worldwide, *R. rattus* has a strong negative influence on nesting birds, but data specific to the Mariana Islands are generally lacking. The large, primarily insectivorous shrew (20–45 g) arrived in the 1960s and can depress abundances of small vertebrates such as lizards.

The brown tree snake has had a dramatic impact on birds, mammals, and lizards, although fortunately it has colonized only Guam to date. Losses include three pelagic seabirds no longer nesting on Guam (a fourth disappeared for other reasons), and extirpations of ten of 12 species of native forest birds, one or more of the three native bats, and between one and five of the nine to 12 native lizards. Although introduced predators often depress the abundances of their prey, it is unusual for this depression to progress to outright extirpation. It is even more unusual for an introduced predator to extirpate species from more than one vertebrate class.

PREY AS SUBSIDIES FOR INTRODUCED PREDATORS

One reason for the snake's exceptionally severe impact is that native prey species in the Mariana Islands evolved without defenses against a nocturnal arboreal predator similar to the brown tree snake. An additional consideration is that several introduced prey species were present to subsidize the introduced predator when the predator's abundance would otherwise have declined in response to declining native prey populations.

The key introduced prey species were the rat, shrew, and lizard *Carlia ailanpalai*. All three species have reached spectacular densities in the Mariana Islands, and these abundances have sustained the snake when it would otherwise have declined in response to disappearing native prey species. For example, in spite of the brown tree snake's suppression of lizard populations on Guam, more than 180 million *Carlia ailanpalai* lizards are estimated to live on the islands of Guam, Saipan, and Tinian (mean density: 2364/ha). Thus, the primary ecological impact of *Carlia*'s introduction to the Mariana Islands is that this prey item subsidized populations of the introduced snake and made possible the extinction of several species of alternate prey.

CURRENT STATUS OF BIOTA

From the perspective of conserving global biodiversity, the greatest extinction risk is to species with very limited ranges: endemic species. Although the Mariana Islands are relatively depauperate, they are a hotspot of endemism. There are at least 15 endemic vertebrates, primarily birds, that were originally found nowhere beyond the Mariana Islands (Table 1). Endemic island species tend to be especially vulnerable to introduced predators and ecosystem transformers. Both factors have been detrimental to the Mariana Island endemics, most of which are now endangered or extinct. One unlisted endemic bird (the bridled white-eye) still occupies all of its original range. The other unlisted endemic species occupy from 17 to 45% of their original ranges; listed endemic species occupy from 0 to 17%.

The same introduced species have affected the non-endemic resident species, although perhaps to a lesser degree. The seabird colonies have been lost from most inhabited islands, with the exception of Rota. Thus, the Mariana Islands exemplify the challenges facing conservation of island biotas: loss of endemic and resident species, introduced predators, ecosystem transformation, and introduced prey as predator subsidies. These problems affect all archipelagoes to a degree, but have affected the Marianas more than most. Although there are various small nature

TABLE 1
Endemic Land Vertebrates of the Mariana Islands

Common Name	Species Name	Guam	Rota	Agui	Tini	Saip	Anat	Sari	Gugu	Alam	Paga	Agri	Asun	Percent Range Occupied	ESA Status
Reptiles															
Mariana skink	*Emoia slevini*	X	X	X	X		R	R	R	R	R	R	R	17	
Mammals															
Little Mariana fruit bat	*Pteropus tokudae*	X												0	Extinct
Birds															
Bridled white-eye	*Zosterops saypani*			R	R	R								100	
Golden white-eye	*Cleptornis marchei*		X	R	X	R								41	
Guam bridled white-eye	*Zosterops conspicillatus*	X												0	Extinct
Guam flycatcher	*Myiagra freycineti*	X	X											0	Extinct
Guam kingfisher	*Todiramphus cinnamominus*	X												0	Endangered[a]
Guam rail	*Gallirallus owstoni*	X	X											0	Endangered[a]
Mariana crow	*Corvus kubaryi*	X	D											13	Endangered
Mariana fruit dove	*Ptilinopus roseicapilla*	X	R	R	R	R								37	
Mariana mallard	*Anas oustaleti*	X			X	X								0	Extinct
Mariana swiftlet	*Aerodramus bartschi*	D	X	R	X	R								16	Endangered
Nightingale reed-warbler	*Acrocephalus luscinia*	X		?	X	R					R			17	Endangered
Rota white-eye	*Zosterops rotensis*		D											10	Endangered
Tinian monarch	*Monarcha takatsukasae*				R	X								45	

NOTE: Distribution, percent of original range still occupied, and status under the Endangered Species Act (ESA). Original range with reference to Steadman (2006); fossil-only taxa excluded. Major islands are represented by the initial four letters of their names and are ordered from south to north (see Fig. 1). R = resident islandwide in suitable habitat; X = extirpated from that island; D = diminished range on that island; ? = uncertain.

[a] Species persists in zoos or experimental populations.

preserves in the Mariana Islands, these have been powerless to protect the fauna from introduced species.

SEE ALSO THE FOLLOWING ARTICLES

Introduced Species / Invasion Biology / Land Snails / Marianas, Geology / Rodents / Snakes

FURTHER READING

Fritts, T. H., and G. H. Rodda. 1998. The role of introduced species in the degradation of island ecosystems: a case history of Guam. *Annual Review of Ecology and Systematics* 29: 113–140.

Furey, J., ed. 2006. *Island ecology and resource management: Commonwealth of the Northern Mariana Islands.* Saipan: Northern Marianas College Press.

Jaffe, M. 1994. *And no birds sing.* New York: Simon and Schuster.

Quammen, D. 1996. *The song of the dodo.* New York: Scribner's.

Rodda, G. H., T. H. Fritts, and D. Chiszar. 1997. The disappearance of Guam's wildlife: new insights for herpetology, evolutionary ecology, and conservation. *BioScience* 47: 565–574.

Rodda, G. H., Y. Sawai, D. Chiszar, and H. Tanaka, eds. 1999. *Problem snake management: the habu and the brown treesnake.* Ithaca, NY: Cornell University Press.

Steadman, D. W. 2006. *Extinction and biogeography of tropical Pacific birds.* Chicago: University of Chicago Press.

MARIANAS, GEOLOGY

FRANK A. TRUSDELL

U.S. Geological Survey, Hawaii National Park

The Mariana Islands are the summits of a large volcanic mountain range that stretches 650 km from Guam to Uracas (Farallon de Pajaros; Fig. 1). The subaerial islands are but a small fraction of the mass, estimated at 0.5 to 1.5% of the volcanoes that form the Mariana Arc.

GEOLOGIC SETTING AND STRUCTURE

The Northern Mariana volcanic islands form the upper 2–3 km of the East Mariana Ridge, which rises about 2–4 km above the ocean floor. To the east of the Mariana Ridge is the Mariana Trench, which is, at a depth of nearly 10 km, the deepest in the world. To the west of the Northern Mariana Islands is the Mariana Trough, which is partly filled with young lava flows and volcaniclastic sediment (Fig. 1). The Mariana Trench and the Northern Mariana Islands (East Mariana Ridge) overlie an active subduction zone, where the Pacific plate, moving northwest at about 11 cm/year, passes beneath the Philippine plate, moving west-northwest at 8.6 cm/year. Beneath the Northern Mariana Islands, earthquake hypocenters at depths of 50–250 km mark the location of the west-dipping subducting slab. Farther west, the slab becomes nearly vertical and extends to a 700-km depth. During the past century, more than 40 earthquakes of magnitude 6.5–8.1 have occurred along this subduction zone.

The Mariana Islands form two subparallel, concentric, concave-west arcs (Fig. 1). The southern islands make up the outer arc and extend north from Guam to Farallon de Medinilla. They consist of Eocene to Miocene volcanic rocks and uplifted Tertiary and Quaternary limestone. Converging plates cause uplift of the outer arc volcanoes. The nine northern volcanic islands extend from Anatahan to Uracas and form part of the active inner arc. This inner arc extends south from Anatahan, where some of the volcanoes form seamounts west of the older, outer arc (Fig. 1). Other volcanic seamounts of the active arc formed on top of the East Mariana Ridge. Six volcanoes (Uracas, Asuncion, Agrigan, Mount Pagan, Guguan, and Anatahan) and several seamounts (Ruby, Northwest Rota-1, Ahyi, Supply Reef, and Esmeralda) of the volcanic arc have erupted during the past century.

Rock Composition

The southern Mariana Islands comprise a basement of dacitic and andesitic rocks and a cap of marine limestone. The Northern Mariana Islands are made up of volcanic rocks ranging in composition from basalt to dacite. The suites of rocks of the Mariana Islands are transitional between tholeiites and cal-alkaline types.

Previous Investigations

Despite centuries of settlement, few detailed studies have been made of the geology and historic eruptive activity of most Mariana volcanoes. A few eruptions were reported as early as the 1600s by seafaring explorers. A large eruption was recorded from Pagan in 1872–1873, and relatively

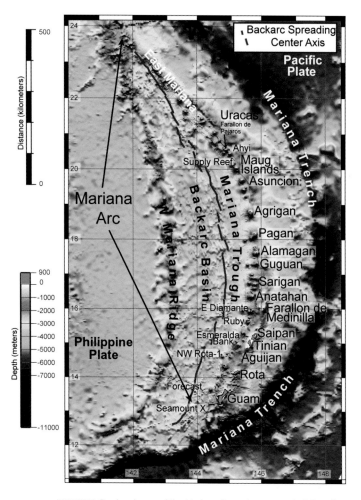

FIGURE 1 Regional map of the Mariana Arc volcanoes and of the adjacent Mariana Trench. The Commonwealth of the Northern Mariana Arc volcanoes extends from Guam in the south to Uracas in the north. Figure courtesy of Susan Merle, Oregon State University NOAA Vents Program.

minor activity was noted during the early part of the twentieth century (Fig. 2).

After World War II, the U.S. military commissioned geologic studies of Pagan, Saipan, Tinian, and Guam. More recently, the geology of Alamagan, North Pagan, and Anatahan were compiled.

The large Plinian eruption of Mount Pagan on May 15, 1981, resulted in the evacuation of Pagan residents. Intermittent ejection of mainly phreatic ash and other products continued for another 15 years.

The first historical eruption on Anatahan Island began on May 10, 2003, from the east crater of the volcano. Ash emissions continued until September 2005, and gas emissions and low levels of seismicity continue to the present. The eruption was not an immediate threat to human life, but volcanic ash posed a great hazard to aircraft.

CLIMATE

The Northern Mariana Islands are dominated by easterly trade winds, with seasonal variations, throughout the year. Winds blow from the east from July to November and from the northeast from December to March. In the summer, the winds are light and variable. The rainy season, from July through November, brings frequent heavy showers. Typhoons and tropical storms are common from July to December. Average annual rainfall is about 180–310 cm. In general, the climate of the Mariana Islands ranges from subtropical in the north to tropical in the south.

SOILS

Soils usually form from chemical and physical breakdown of parent material. Soils on the southern Mariana Islands are a direct result of those processes. Most of the soils are alkaline, due to the calcareous parental material. In the Northern Marianas, most soils are the products of volcanic eruptions and are therefore composed of volcanic ash and tephra, along with organic matter.

STRUCTURE AND MORPHOLOGY

Geomorphic Shape and Age

The young volcanoes have a nearly pristine conical shape and form, but older volcanoes are deeply incised with arroyos and stream valleys. The youngest volcanoes include Uracas, Asuncion, Mount Pagan, South Pagan (Butkan Paliat), North Guguan, and the east crater of Anatahan.

Over time, some of the southern Mariana volcanic islands subsided beneath the sea, and coral began

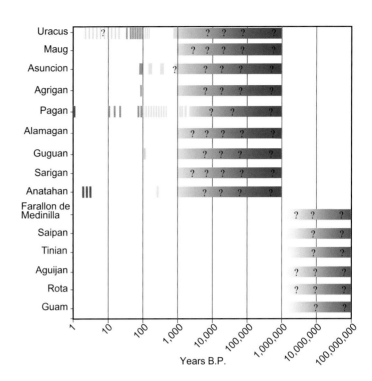

FIGURE 2 Generalized ages of Mariana Islands, modified after Simkin and Siebert (1994).

to cover the volcanic rocks with limestone. Currently, all the southern Mariana Islands are being uplifted as a result of the convergence of the Pacific and Philippine plates. Tinian has been uplifted by more than 100 m since the Pleistocene and by an additional 1.8 m during the Holocene. Rota has been uplifted ~500 m. In all cases, the southern Mariana Islands are now raised volcanic islands with caps of limestone. The limestone grew either as coral growth kept pace with the subsidence rate or during higher stands of the sea.

The Northern Marianas range in age from Holocene to Quaternary, whereas the southern islands range in age from 15 to 56 million years. Islands sizes are shown in Fig. 1.

Sector Collapse

Landslides and other large-scale mass movements are common on high volcanic islands, leading to the degradation of the volcanic edifice. In the Northern Mariana Islands, landslides are small and represent localized slope failure of unconsolidated volcanic pyroclasts. On most islands, they do not play as important a role in the deterioration of the Mariana volcanoes as in other ocean island environments, such as Hawai'i, Tenerife, and the Aleutians.

ISLANDS

Uracas

Uracas, or Farallon de Pajaros, is a single stratocone, mostly devoid of vegetation. The volcano stands 319 m above sea level, is ~2 by 1.5 km in size (north–south by east–west), with a surface area of 3.1 km^2. The volcano's subaerial volume is 0.25 km^3. The northeast flank has an embayment, semicircular in shape, suggesting flank failure. The lack of vegetation indicates recurring volcanic activity on a 1–3-year frequency. The island's remote location makes an absolute determination of eruptive frequency difficult. Pyroclastic flow deposits underlie the northern two-thirds of the island. The southern one-third is covered by ʻaʻā lava flows. On the southeast and southern coast are remnants of two older scarps of a former volcanic edifice, now destroyed. Resurgent activity (the current Uracas) has replaced the older edifice. The summit crater contains solfataras that fume continuously. Slopes greater than 30° are found on all flanks of the volcano, the northeast embayment, and the west coast.

Maug

Maug comprises three islets that were once part of a stratovolcano. Explosive activity destroyed the original edifice. The volcano stands 227 m above sea level and is ~3 by 3 km (north–south by east–west) in size, with a surface area of 4.2 km^2. The volcano's subaerial volume is 0.1 km^3. Bathymetric surveying by the National Oceanic and Atmospheric Administration (NOAA) in 2004 shows a resurgent cone within the lagoon between the islets. Maug is made up of stacked lava flows intercalated with minor pyroclastic deposits. Dikes are common in the interior walls. Slopes in excess of 30° are found on all flanks of the volcano except the southwest, where slopes of less than 5° are present. The coastline is mostly rocky, with one or two pocket beaches.

Asuncion

Asuncion is a conical volcano with slopes at the angle of repose. The volcano stands 857 m above sea level and is ~3.5 by 3 km (north–south by east–west) in size, with a surface area of 11.7 km^2. The volcano's subaerial volume is ~2.2 km^3. The summit of the volcano contains a shallow crater with a spatter cone from the 1906 eruption. Steaming, probably related to residual heat from the 1906 eruption, can be observed on clear days. The coastline is rocky, with no beaches. Slopes in excess of 30° are found on all flanks of the volcano; slopes of less than 5° are uncommon and are found only on the southwest flank of the volcano. Volcanics erupted on the island are andesite in composition.

Agrihan

Agrihan, the highest-standing stratovolcano and largest (by subaerial volume) in the Northern Mariana Islands, stands 882 m above sea level. The island is ~10 by 6.5 km (north–south by east–west) in size, with a surface area of 52.7 km^2. The volcano's subaerial volume is ~15.9 km^3. The summit contains a large depression, roughly 1.5 by 1.2 km in diameter, and 380 m deep. A spatter cone and flows from the 1917 eruption cover ~50% of the crater floor. This large crater implies a local edifice with shallow magma storage within the volcano. The flanks of the volcano are steep (more than 30°), with deep furrows extending radially away from the crater. To the north is a large canyon into which a recent, large ʻaʻā flow advanced to form a delta on the coast. Pyroclastic flow deposits mantle most of the interior of the island. Rocks erupted on the island range from basalt to andesite. The southwest coast has several beaches composed of mineral sands; otherwise, the coast is rocky.

Pagan

Pagan is an island made up of a string of volcanoes originating from three volcanic centers trending northeast to southwest, distinguishing the island from the rest of the Mariana Islands. A large caldera with a resurgent volcano at its center, Mount Pagan (579 m above sea level) is at the northern end of the island. The central part of the island is an isthmus composed of another volcanic center (the highest point is Togari Mountain at 579 m above sea level) containing a chain of deeply incised volcanoes representing the oldest rocks on the island. Another caldera occurs in the south and includes three volcanic cones. The largest, Bulikan Paliat, is 548 m above sea level and rises 248 m above the caldera floor, which is covered by recent ʻaʻā flows. Pagan Island is ~14 km long (north–south) and varies in width from 1.5 to 6 km (east–west), with a surface area of 64.1 km^2. The island's subaerial volume is ~7.1 km^3.

Mount Pagan is the second most active volcano in the Northern Mariana Islands, after Uracas (Fig. 2). The large Plinian eruption in 1981 caused the evacuation of residents, and intermittent ejection of mainly phreatic ash and other products continued for another 15 years. A few eruptions were reported as early as the 1600s; a large eruption was reported in 1872–1873, and relatively minor activity was noted during the 1920s. Mount Pagan most recently erupted on December 5, 2006, emitting fine ash for four days; elevated activity continued until December 19, 2006.

Rocks erupted on the island range from basalt to andesite. Eruptive products include scoria, ʻaʻā and pāhoehoe flows, tuffs, and other pyroclastic deposits. The roughly circular caldera of north Pagan is ~5.5 km in diameter and contains a prominent southern wall, with more subdued, partly buried northern and eastern scarps. Preliminary calculations of the volume of the volcanic edifice that occupied the region of the former caldera are between ~4.7 and 5.8 km^3 implying that the caldera-forming eruption was as large as volcano explosivity index (VEI) 5, (the same as the May 18, 1980, eruption of Mount St. Helens).

On the western flank of Mount Pagan is a maar, a ~2 by 2–km depression with a lake on its floor. Satellitic vents define a northwest to southeast trend on the flanks of Mount Pagan. To the northwest is a pair of cones, and on the south-southeast, four more cones are located on the southeast flank of Mount Pagan, with the largest standing 244 m high. The shoreline of north Pagan is rocky on all sides, except for a narrow beach 500 m long, flanked by a fringing reef, on the north coast. The Bandeera Peninsula, on the west coast, is a remnant caldera rim that forms two small bays with mineral sands. The central volcanic region has fringing reefs and organic (coral) sand beaches on the east side. The west-side beaches are a mixture of mineral and coral sands. The southernmost coastline is rocky, as a result of recent eruptions and wave erosion. The central volcanic deposits are deeply eroded and exhibit evidence of local mass wasting and/or landslide events. The oldest rocks on the island, located in this region, are probably Late Quaternary in age.

North Pagan has two lakes on its western flank. The inner, Laguna Sanhalom, at 10.8 ha, is 8 m deep and filled with brackish water. The southwest corner of the lake was the former site of a hot spring. Since the 1981 eruption, Laguna Sanhalom decreased in area by 40% because of infilling by lahars and debris flows. The outer lake, Laguna Sanhiyon, covers 12.8 ha, is 12 m deep, and also contains brackish water.

South Pagan has a caldera, containing three cones of relatively young age. The caldera measures 3 km (northeast–southwest) by 2 km (northwest–southeast), and ʻaʻā flows cover ~60% of the floor. To the northwest of Bulikan Paliat is another cone, Bulikan Bulifli. This cone is the oldest of the group and contains mud pots and a solfatara field. The southern tip of Pagan Island consists of tuffaceous deposits and lava flows.

Slopes of the northern half of Pagan Island are low to moderate (less than 20°), thus making this part of the island the flattest and most inhabitable of the Northern Marianas. Slopes in excess of 30° are common in the central and southern volcanic regions. The southern caldera is perched 200–300 m above sea level and has low slopes, excluding the volcanic cones. The outermost caldera flanks are steep (greater than 30°).

Alamagan

Alamagan stands 744 m above sea level, is ~4 by 3.5 km (north–south by east–west) in size, and has a surface area of 13.4 km^2. Its subaerial volume is ~2.5 km^3. Alamagan has had no eruptions during historical time. The island has a large crater just south of the topographic high, and its steep flanks are deeply furrowed. Near the summit are several steaming areas. The two most recent eruptions produced extensive pyroclastic flow deposits, dated by radiocarbon at 1077 ± 87 and 1410 ± 80 years ago. Pyroclastic flow deposits underlie about two-thirds of Alamagan. The northern coast is composed entirely of a single massive lava flow. Rocks erupted on the island range from basalt to andesite. The shoreline is rocky, but two small beaches can be found—on the northwest and southwest coasts. Both are rocky, composed of boulders, cobbles, and broken rocks.

Guguan

Guguan last erupted in 1883 and produced a tephra cone and ʻaʻā lava flows on the northern half of the island. The southern half of the island is made of an older edifice, eroded and faulted, composed of interbedded lava flows and pyroclastic deposits. The highest peak, found in the south, stands 287 m above sea level. Guguan is ~3.0 km by 2.5 km (north–south by east–west) in size, with a surface area of 16.8 km^2. The volcano's subaerial volume is ~0.6 km^3. Rocks erupted on the island range from basalt to andesite. The shoreline is rocky, composed of cobbles and boulders.

Sarigan

Sarigan, an old volcano with steep slopes (greater than 30°) as a result of mass wasting and/or erosion on the southeast and southwest flanks, has had no historic eruptions. The volcano stands 538 m above sea level and is ~3 km by 2 km (north–south by east–west) in size, with a surface area of 6.2 km^2. The volcano's subaerial volume is ~0.8 km^3. The summit of the volcano has a high plateau composed of dense spatter and blocky ʻaʻā lava flows. North of the summit is another flat mesa that was the site of a second eruptive center. These two regions are the flattest areas on the island. Sarigan consists of ~30%

lava flows and 70% pyroclastic material. The shoreline is rocky, composed of broken-down flows, cobbles, and boulders.

Anatahan

Anatahan measures 9 km (east–west) by 3.7 km (north–south), with a surface area of 46.2 km^2. The highest peak is 787 m above sea level, and the island has a subaerial volume of 9 km^3. The summit of the island is marked by an elongated caldera, 5 km by 2 km. Located in the eastern part of the caldera is a pit crater, 1.4 by 1.2 km and ~200 m deep. Pyroclastic flow deposits mantle about 80% of Anatahan. Prior to the eruption in 2003, Anatahan had active solfataras, springs, and mud pots on the floor of the east crater. Temperatures in the springs and mud pots ranged from a maximum of 98.8 °C to a minimum of 67.4 °C, and the pH ranged from 1.7 to 4.3.

The first historical eruption on Anatahan Island began on May 10, 2003, from the pit crater of the volcano. The eruption was preceded by several hours of seismicity. Two and a half hours before the outbreak, the number of earthquakes surged to more than 100 events per hour. Plume heights were 4500 to 13,000 m for the initial phases of the eruption. Prior to 2003, an earthquake swarm in April 1990 resulted in the evacuation of the island.

From submarine bathymetry, it appears that Anatahan is a single edifice (Chadwick *et al.*, 2005), as opposed to two coalesced volcanoes, as was previously reported. The flanks of the volcano are steep (greater than 30°), with deep furrows extending away from the highlands. Within the elongated caldera, the slopes are low (less than 10°). The shoreline of Anatahan is rocky, composed of sea cliffs, broken-down flows, blocks, and boulders. A small cobble beach is located on the west coast.

Saipan

Saipan measures 23 km (north–south) by 2.5–10 km (east–west), with a surface area of 141 km^2. The highest peak is 473 m above sea level, and the island has a subaerial volume of 10.3 km^3. The island is composed of mid-Eocene tuffs, lava flows, and volcanic breccias at its core, mantled by Miocene limestone and Plio-Pleistocene coralline algae and coral limestone. Ninety-five percent of the island is covered by limestone. The oldest rocks on the island have been dated to 41 million years ago. Fringing reefs are present on Saipan's east coast, and a barrier reef with a shallow lagoon and coral sand beaches is found along the west coast. Slopes in excess of 30° are rare and are found in mountainous regions; most slopes are moderate to low (less than 7°).

Tinian

Tinian measures 19 km (north–south) by 3–8 km (east–west) and has a surface area of 110 km^2. The highest peak is 187 m above sea level, and the island has a subaerial volume of 6.4 km^3. The island is composed of Eocene tuffs and volcanic breccias at its core, mantled by Miocene limestone and Plio-Pleistocene coralline algae and coral limestone, similar to Saipan. Holocene beach and raised reef deposits are also found on Tinian. Limestone covers 99% of the island, which is mostly flat, except for remnant fault scarps. The shoreline is rocky, composed of broken-down flows, limestone cobbles, and boulders.

Rota

Rota is 16 km (north–south) by 6 km (east–west) in size, with a surface area of 110 km^2. The highest peak is 187 m above sea level, and the island has a subaerial volume of 6.4 km^3. The geology of Rota has not been systematically studied but appears similar to that of adjacent islands. The island is composed of Eocene tuffs and volcanic breccias at its core, mantled by Miocene limestone and Plio-Pleistocene coralline algae and coral limestone. Limestone covers more than 90% of the island, which is mostly flat, except for remnant fault scarps. The island is surrounded by a fringing reef.

Guam

Guam is 48 km long (north–south) by 13 km wide (east–west), with a surface area of 546.2 km^2. The highest peak is 406 m above sea level, and the island has a subaerial volume of 53 km^3. Approximately 35% of the island is composed of exposed volcanic rocks, located mostly in the south. The northern portions of the island are exposed limestone of Neogene age.

SEE ALSO THE FOLLOWING ARTICLES

Earthquakes / Island Arcs / Lava and Ash / Marianas, Biology

FURTHER READING

Banks, N. G., R. Y. Koyanagi, J. M. Sinton, and K. T. Honma. 1984. The eruption of Mount Pagan volcano, Mariana Islands, 15 May 1981. *Journal of Volcanology and Geothermal Research* 22: 225–269.

Chadwick, W. W., R. W. Embley Jr., P. D. Johnson, S. G. Merle, S. Ristau, and A. Bobbitt. 2005. The submarine flanks of Anatahan volcano, Commonwealth of the Northern Mariana Island. *Journal of Volcanology and Geothermal Research* 146: 8–25.

Cloud, P. E., R. G. Schmidt Jr., and H. W. Burke. 1956. Geology of Saipan, Mariana Islands, part 1. General geology. U. S. Geological Survey Professional Paper 280-A.

Dickinson, W. R. 2000. Hydro-isostatic and tectonic influences on emergent Holocene paleoshorelines in the Mariana Islands, western Pacific Ocean. *Journal of Coastal Research* 16: 735–746.

Karig, D. E. 1971. Structural history of the Mariana Island arc system. *Geological Society of America Bulletin* 82: 323–344.

Kato, T., Y. Kotake, S. Nakao, J. Beavan, K. Hirahara, M. Okada, M. Hoshiba, O. Kamigaichi, R. B. Feir, P. H. Park, M. D. Gerasimenko, and M. Kasahara. 1998. Initial results from WING, the continuous GPS network in the western Pacific area. *Geophysical Research Letters* 25: 369–372.

Simkin, T., and L. Siebert. 1994. *Volcanoes of the world.* Tucson, AZ: Geoscience Press, Inc.

Simkin, T., R. I. Tilling, P. R. Vogt, S. H. Kirbey, P. Kimberly, and D. B. Stewart. 2006. This dynamic planet: world map of volcanoes, earthquakes, impact craters, and plate tectonics. U. S. Geological Survey Map I-2800, scale 1:12, 500.

Tanakadate, H. 1940. Volcanoes in the Mariana Islands in the Japanese mandated South Seas. *Bulletin Volcanologique* 18: 199–225.

Trusdell, F. A., R. B. Moore, and M. K. Sako. 2006. Preliminary geologic map of Mount Pagan volcano, Pagan Island, Commonwealth of the Northern Mariana Islands. U. S. Geological Survey Open-File Report 2006-1386.

MARINE LABORATORIES

SEE RESEARCH STATIONS

MARINE LAKES

MICHAEL N DAWSON AND LAURA E. MARTIN
University of California, Merced

LORI J. BELL AND SHARON PATRIS
Coral Reef Research Foundation, Koror, Palau

Marine lakes are bodies of seawater entirely surrounded by land. They come in a great variety of shapes, sizes, and distances from the "mainland" sea and can be described further in terms of their water-column characteristics and biotic complements, which may exhibit differences due, in part, to dissimilar physical connections with the sea. Marine lakes are "habitat islands" that exhibit the biogeographic, ecological, and evolutionary characteristics of "true islands" (with varying degrees of isolation), mainland fragments, or otherwise patchily distributed habitat.

DISTRIBUTION AND FORMATION

The distribution of marine lakes is poorly documented. They are mentioned, often incidentally, in literature for tourists or on wetlands, and they can be found most easily by searching high-resolution topographic maps, aerial photographs, and satellite images. Such sources, with some ground-truthing, clearly indicate over 200 marine lakes, concentrated in four regions worldwide (Fig. 1). These regions are characterized by coastal karst semisubmerged in the sea (Fig. 2). This association has two important implications. First, considering the distribution of karst in maritime areas globally, the true number of marine lakes, and regions with marine lakes, likely far exceeds our modest estimates. Marine lakes thus provide abundant natural experiments in ecology and evolution, which may be used to investigate patterns of biodiversity and biogeography globally. Second, all known marine lakes likely formed as depressions in porous, fissured, karst landscapes flooded by rising sea level after the last glacial maximum (LGM) around 18,000 years ago. Measurements of water depths of modern lakes in Berau, Palau, and Papua demonstrate that none is deeper than 60 m, and most are much shallower. Measurement of downward flux of sediment in a subset of lakes in Palau, using sediment traps and radioisotope analyses of cores, indicate sedimentation rates of 1–7 mm per year. Assuming even the slow compaction rates typical of highly flocculent rich organic sediments, all but the highest sedimentation rates applied to the deepest lakes indicate that the underlying karst is less than 100 m below current sea level and therefore must have been covered in dry valleys, not lakes, during the LGM. Indeed, the age of one lake in Palau has been corroborated at ~10,000 years by coring and radiocarbon analyses of sediments immediately overlaying the karst bedrock. The modern physical diversity of karstic marine lakes therefore formed during the last two decamillenia.

PHYSICAL CHARACTERISTICS

Modern marine lakes range from around 1.5 m to 60 m deep and from about 50 m to 2 km in their maximum horizontal dimension. They are a few to many hundreds of meters from the sea, surrounded by forest-covered ridges and peaks from several (although usually several tens) and occasionally several hundreds of meters high. All have a measurable tidal cycle, the amplitude and timing of which can differ greatly from those occurring in the surrounding ocean, indicating variously restricted physical connections. The lakes closest to the sea may have tunnels that are short enough and wide enough for a person to swim through; the lakes farthest from the sea have no such obvious, direct connections. Partly as a consequence of tidal mixing, lakes range from those

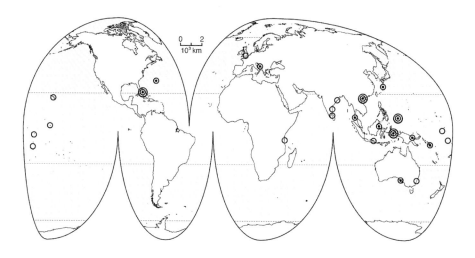

◉ 10 or more marine lakes in clusters
◉ fewer than 10 marine lakes
○ other enclosed and semi-enclosed marine ecosystems

FIGURE 1 Global distribution of marine lakes. The distribution and frequency of marine lakes are inferred from peer-reviewed scientific literature, natural history and dive magazines, satellite images, personal experience, and conservation management publications (including Hamner and Hamner 1998; Donachie et al. 1999; Porter et al. 2001; hsea.unep-wcmc.org/sites/wetlands/; oceancolor.gsfc.nasa.gov/cgi/landsat.pl; GoogleEarth). ◉, tens of marine lakes occur in clusters among drowned karst in at least four regions (Bahamas, Vietnam, Palau, Papua). ◉, marine lakes also occur in smaller numbers in additional locations. ○, other enclosed and semi-enclosed ecosystems with marine origins are widespread, numerous, and may share some of the ecological and evolutionary characteristics of marine lakes. A global total of approximately 200 marine lakes is a conservative minimum estimate due to extremely poor knowledge and underreporting of this class of marine ecosystem.

FIGURE 2 Oblique aerial photograph of Mecherchar Island, Palau, showing elevated karst ridges covered in tropical forest among which are nestled a handful of marine lakes (an additional four are obscured by the ridges). The largest, in the foreground, is approximately 2 km long. Ongeim'l Tketau, or "Jellyfish Lake," is in the background (top right). Photograph courtesy of Patrick L. Colin.

that are physically very similar with the surrounding ocean—"holomictic," isothermal and well-oxygenated to depth, and having a salinity of 34 practical salinity units (psu). Others, because of damped tidal flux, shelter from wind, high rainfall, high insolation, and the lack of distinct seasons, are physically and chemically very different—"meromictic," vertically stratified by temperature and salinity, anoxic at depth, and brackish. Thus, marine lakes are variously connected to the ocean, forming a continuum from lagoon-like to highly isolated, and they constitute a wide variety of habitats and corresponding assemblages. Such environmental heterogeneity yields a range of novel selective regimes with the potential to structure communities and modify species over time.

BIOLOGICAL CHARACTERISTICS

Marine lakes in Palau are famous for the *Mastigias* jellyfish that populate them. In one such jellyfish lake, Ongeim'l Tketau, millions of medusae, from the size of a pea to a small soccer ball, migrate a quarter-mile eastward in the morning, turn around at noon, and migrate the quarter-mile westward back to where they began. The results are spectacular aggregations of many hundreds of medusae per square meter during the late morning and late afternoon (Fig. 3). This migration, like the morphology of this jellyfish (Fig. 4), which is just one of five subspecies each restricted to a single lake in Palau, must have evolved in situ (in ≤ 10,000–15,000 years) after colonization from an ancestral population in the surrounding lagoon. Remarkably, that ancestral population, the most recent common ancestor to all marine lake populations, still exists in the lagoon in Palau, providing a situation rare in its clarity for studying the genotypic and phenotypic evolution of *Mastigias*.

What is true for *Mastigias* logically should also be true for many of the other tens to hundreds of species that inhabit marine lakes. However, with only preliminary biotic inventories completed in a subset of lakes in Palau, Berau, and West Papua, much remains unknown about these other taxa, including other cnidarians, a few fishes, bivalves and gastropods, echinoderms, green algae, bryozoans, copepods, ascidians, and many sponges. Some species are found in all regions, whereas others are regionalized; some are shared among lakes, and some are not; some are common in the surrounding sea, whereas others are known only from a single individual in a single lake. As many as 10–20% of species and subspecies in marine lakes may be new to science, either because they are endemic to marine lakes or

FIGURE 3 *Mastigias* swarm in Ongeim'l Tketau, Palau, viewed from above and below water. The perspectives of these two photographs are perpendicular to each other, with the forest to the right in the lower photograph.

because they are simply very rare elsewhere. Determining their taxonomic status is a necessary precursor to further biogeographic research.

Geographic variation in the physical structure and community composition of marine lakes is also accompanied by variation in population dynamics. Because marine lakes are much smaller, and isolated, bodies of seawater, they are expected to amplify the effects of changing weather patterns and climate, relative to larger, and open, marine environments that have greater thermal inertia and mixing. Marine lakes in Palau, at the far western edge of the Western Pacific Warm Pool, the region in which El Niño–Southern Oscillation (ENSO) events initiate, are indeed particularly sensitive to ENSO conditions in the eastern Pacific, extreme events of which may cause the *Mastigias* population in Ongeim'l Tketau to vary in size by seven orders of magnitude. Interestingly, given the high sedimentation rates in the marine lakes, such physical variation, linked population dynamics—as well as colonization and extirpation dynamics affecting community constitution—should be recorded in great detail.

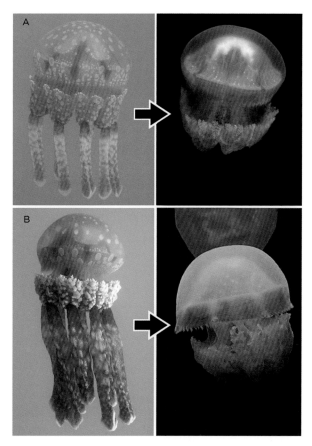

FIGURE 4 The evolution of *Mastigias* medusae in marine lakes occurs in parallel within species and archipelagoes, but also remarkably between species separated by many millions of years of evolution and thousands of kilometers of ocean (plus a few hundred meters of land). (A) Palau—the morphotype that occurs in coves within the lagoon (Ngerchaol Island) colonized marine lakes giving rise to, after about 10,000 years, the morphotype now found in Ongeim'l Tketau. (B) West Papua—the ocean morphotype from Gam Island and a derived marine lake morphotype from southeastern Misool (December 2007); in this image, the top of the "bell" is reflected by the water's surface. Photograph in right half of (B) courtesy of Precious Planet: Eric Battistoni.

RELEVANCE, THEORY

The analogies between marine lakes and other kinds of islands are remarkable because for much of the last three decades, the majority of marine science literature discounted the existence of marine islands. The example of marine lakes, which provide a particularly clear venue for studying patterns and processes in the ecology and evolution of marine taxa, demonstrate beyond doubt that marine species and marine communities can be influenced by the processes described by ecological and evolutionary theory regularly applied to freshwater and terrestrial systems. Concomitantly, over the last decade, publications reporting geographic isolation have blossomed, and papers on rapid evolution, species-area relationships, and "island rule" evolution have appeared. Such theory may be particularly important in understanding the ecological and evolutionary dynamics of marine resources on island archipelagoes, networks of marine protected areas, and species invasions.

Beyond basic tests of island theory such as species–area relationships, colonization–diversity curves, and "the island rule," the range of studies to which marine lakes are likely to contribute in the future is immense. Deep lakes can be cored to compare trajectories of community assembly, population dynamics (related to colors of environmental variation), colonization, and extirpation dynamics. By analogy, modern lakes of different depths may provide analogues of various stages in the flooding and formation of what are now deep lakes, illuminating biotic changes associated with transitions from holomictic to meromictic lakes. Because lakes are small and easy to sample comprehensively, and because they yield natural subsamples of regional species pools, variation between modern lakes' physical structures can be clearly related to α-, β-, and γ-diversity tied into comparisons of latitudinal and longitudinal trends in diversity. Thus, correlates (and causal factors) affecting physical connectivity, gene flow, and community similarity—and also those influencing physical heterogeneity, heterozygosity, and species diversity—can be studied, including in a community phylogenetics framework. Such analyses are likely to result in greater synthesis in studies of marine and terrestrial systems.

An important question, given this long list of similarities, is whether studies of marine lakes will provide anything novel? The combination of a very high-resolution sediment record of subfossils and associated climate is likely unparalleled in almost every freshwater and terrestrial system. In addition, the on-average larger range size of and strong stabilizing selection on marine taxa in ocean environments means that many marine lakes are colonized independently from the same ancestral population (or equivalent conspecific phenotypes), providing perhaps more possibilities to examine the combined stochastic and deterministic effects of genetic drift, genomic canalization or genetic lines of least resistance, and altered adaptive landscapes than any other system. Moreover, the pattern has been repeated in region after region (Fig. 4). In the case of marine lakes, the tape of life has been run again, and again, and again.

SEE ALSO THE FOLLOWING ARTICLES

Climate Change / Freshwater Habitats / Island Rule / Lakes as Islands / Palau / Sea-Level Change / Species–Area Relationship / Tides

FURTHER READING

Dawson, M. N, and W. M. Hamner. 2005. Rapid evolutionary radiation of marine zooplankton in peripheral environments. *Proceedings of the National Academy of Sciences of the USA* 102: 9235–9240.

Dawson, M. N, and W. M. Hamner. 2008. A biophysical perspective on dispersal and the geography of evolution in marine and terrestrial systems. *Journal of the Royal Society Interface* 5: 135–150.

Donachie, S. *et al.* 2004. The Hawaiian archipelago: a microbial diversity hotspot. *Microbial Ecology* 48: 509–520.

Hamner, W. M. 1982. Strange world of Palau's salt lakes. *National Geographic* 161: 264–282.

Hamner, W. M., and P. P. Hamner. 1998. Stratified marine lakes of Palau (Western Caroline Islands). *Physical Geography* 19: 175–220.

Martin, L. E., M. N Dawson, L. J. Bell, and P. L. Colin. 2005. Marine lake ecosystem dynamics illustrate ENSO variation in the tropical western Pacific. *Biology Letters* 2: 144–147.

Porter, J. S., P. E. J. Dyrynda, J. S. Ryland, and G. R. Carvalho. 2001. Morphological and genetic adaptation to a lagoon environment: a case study in the bryozoan genus *Alcyonidium*. *Marine Biology* 139: 575–585.

Tomascik, T., A. J. Mah, A. Nontji, and M. K. Moosa. 1997. *The ecology of Indonesian seas. Part I & II*. The Ecology of Indonesia Series, Vol. VII. Singapore: Periplus Editions.

MARINE PROTECTED AREAS

ALAN M. FRIEDLANDER

University of Hawaii, Honolulu

Marine protected areas (MPAs) are any intertidal or subtidal areas, together with their associated flora, fauna, and historical and cultural features, that have been set aside by law or other effective means to protect part or all of the designated environments. Marine reserves are a more restrictive subset of MPAs and are defined as areas permanently and completely protected from extractive harvest and other major human uses.

MPA PRINCIPLES AND THEORY

As a result of overfishing and overall degradation of marine ecosystems, marine protected areas (MPAs) have increasingly been proposed as an ecosystem-based management tool to conserve biodiversity and manage fisheries. Closing certain areas to harvest for periods of time has been practiced for centuries by Pacific Islanders to help sustain healthy populations of marine resources; area closure has more recently come into increased use because of the failure of more "modern" management methods. By protecting populations, habitats, and ecosystems within their borders, MPAs provide a spatial refuge for the entire ecological system they contain and provide a powerful buffer against human uncertainty and natural variability. In addition to resource management, MPAs also contribute to the long-term livelihoods of island people though the strong cultural and economic connections between islanders and the sea, as well as their interdependence on a healthy marine environment for survival and prosperity.

Theory and experience show that populations of exploited species, when protected within MPAs, respond by producing larger and more abundant individuals (Fig. 1). Larger individuals produce exponentially more, and healthier, offspring, and higher population densities improve the likelihood of reproductive success. By increasing reproductive output, MPAs can serve as a source area for larvae that can restock the protected area itself, as well as export larvae to adjacent areas open to fishing. These changes in population structure help to conserve fish stocks within MPA boundaries and provide fisheries benefits outside these protected areas through enhanced reproductive output and adult spillover (Fig. 2). Spillover occurs when high population densities within MPAs result in net movement of individuals into nearby areas (Fig. 3). This density-dependent emigration can enhance adjacent populations but may also diminish the reproductive potential of the MPA itself if the protected area is too small and the home ranges of the organisms extend out into fished areas where they can be caught.

MPAs can protect entire marine ecosystems by conserving multiple species and essential habitats such as spawning areas and nursery grounds. Island ecosystems often contain limited numbers of these critical habitats, which are interconnected by the movement of organisms at a variety of spatial and temporal scales. By protecting

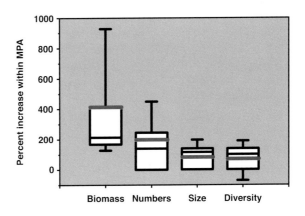

FIGURE 1 Species protected inside MPAs have been shown to increase in total biomass, numbers, size, and diversity over time. Average increases are most pronounced for biomass (413%) and number of individuals (200%). Box plot showing 25th, 50th, and 75th percentiles, with 10th and 90th percentiles as error bars and red bars as means. Adapted from Halpern (2003).

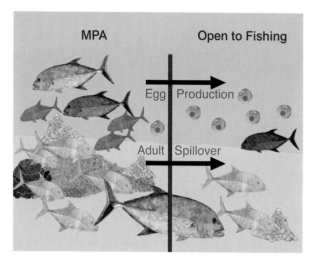

FIGURE 2 Larger-bodied fishes in higher densities within MPAs produce exponentially more and healthier offspring that can replenish stocks both inside and outside the protected area. Adults and juveniles may spillover from MPAs into adjacent areas as a result of density-dependent effects.

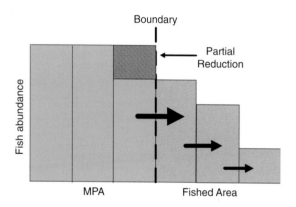

FIGURE 3 Spillover from MPAs to adjacent areas may result from emigration of fishes from a reserve due to increased competition for resources or other density-dependent mechanisms (black arrows). Shading represents partial reduction in abundance of fishes from inside the MPA that occurs when home ranges extend beyond the MPA boundaries and fish are caught. Adapted from Abesamis et al (2006).

the diversity of habitats and interactions necessary for proper ecosystem function, MPAs represent a holistic ecosystem-based approach to management.

BENEFITS OF MPAS

Fisheries Benefits

From a fisheries standpoint, MPAs may be used for a number of purposes. Goals include (1) protecting genetic diversity, size distributions, sex ratios, or other stock characteristics; (2) reducing bycatch impacts on vulnerable species in multispecies assemblages; (3) rebuilding overfished stocks; (4) maintaining habitat characteristics necessary for productive populations; (5) maintaining ecosystem processes necessary for productive populations; (6) providing a reference point to guide future management decisions; (7) increasing overall catches despite reducing the fishing area; and (8) ensuring future catches against management mistakes.

Non-Fisheries Benefits

MPAs also have many non-fisheries benefits, such as protecting biodiversity and ecosystem structure, serving as biological reference areas, providing nonconsumptive recreational activities, and maintaining other ecosystem services such as shoreline protection, nutrient cycling, climate control, and so forth. Nonconsumptive access to protected areas includes enhanced economic opportunities, diversified social activities, and increased public awareness. Because of the importance of nearshore ecosystems to islanders, MPAs can also provide a mechanism for cultural maintenance and revival through increased food and economic security.

With species loss in the sea accelerating, the irreplaceability of these species makes MPAs a powerful tool for marine conservation by protecting species and their associated habitats. The long-term decline in marine ecosystem health has led to the "shifting baseline syndrome," where there are no truly natural places left to compare against current conditions. The establishment of MPAs provides an unparalleled opportunity to study marine ecosystems and to better understand ecosystem function in the absence of fishing pressure and other major human impacts.

MPA DESIGN CRITERIA

Results suggest that in order to help sustain fisheries, MPAs should cover from 10 to 20% of an area, whereas 30 to 50% protection may be necessary to ensure high long-term fisheries catch levels. The size and shape of individual MPAs can have important effects on ecological and socioeconomic performance. Individual areas need to be large enough to contain the short-distance-dispersing larvae (~1 kilometer) and spaced far enough apart so that long-distance-dispersing larvae (tens to hundreds of kilometers) released from one MPA can settle in adjacent ones. Whether the goals are to enhance fishing or to conserve natural ecosystems, it is desirable to design MPAs so that most adults remain inside whereas some of their reproduction flows out. There is no ideal shape for protecting species and their associated habitats, although swaths stretching from shore into deep water are more likely to contain a diversity of habitats than reserves without as much depth range, and

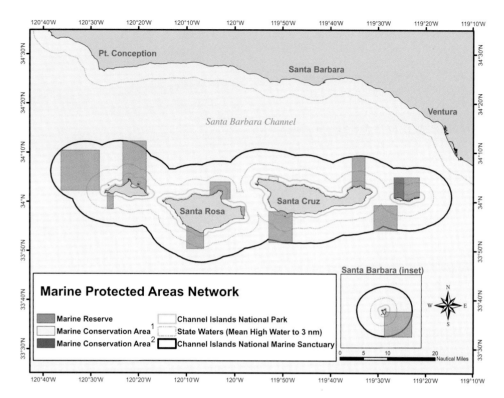

FIGURE 4 Example of a network of MPAs varying in size, use, and level of protection. Channel Islands National Marine Sanctuary boundaries showing the location of existing state and proposed federal marine reserves (no-take) and marine conservation areas that allow limited harvest of lobster and pelagic finfish (1, recreational harvest only; 2, recreational harvest of finfish and both recreational and commercial harvest of lobster). The total MPA network covers about 22% of the sanctuary. Image courtesy of NOAA/Channel Islands National Marine Sanctuary.

they may also encompass common natural migration pathways from shallower nearshore to deeper habitats.

Individual MPAs need to be networked in order to provide large-scale ecosystem benefits (Fig. 4). An MPA network consists of a series of protected areas that are connected by larval dispersal or juvenile and adult movement. MPA networks have the greatest chance of protecting all species, life stages, and ecological linkages if they encompass representative portions of all ecologically relevant habitat types in a replicated manner.

For fisheries purposes, many small reserves in a network may be preferred because of the higher rates of juvenile and adult spillover and more regional benefits through greater larval export than from fewer, larger areas. However, a smaller number of larger MPAs that limit fishing and preserve a greater amount of habitat will provide more benefits for biodiversity conservation. Enforcement and compliance will be greatly aided if reserve borders are straight lines or utilize other obvious navigational reference points.

SOCIOECONOMIC FACTORS

The traditional ecological knowledge held by many island peoples is critical to the development and design of MPAs. Traditional customary management systems have included various forms of area protection, and incorporating elements of these established and recognized practices into a contemporary framework can increase the legitimacy of decisions regarding MPAs, as well as aid in compliance with regulations. Locally managed marine areas that incorporate traditional concepts of customary marine tenure have been effective on many Pacific islands, and participatory community approaches in other parts of the world have also proven to be effective means by which to involve stakeholders in the process, and therefore achieve the intended benefits established for MPAs.

SEE ALSO THE FOLLOWING ARTICLES

Fish Stocks/Overfishing / Refugia / Sustainability

FURTHER READING

Abesamis, R. A., G. R. Russ, and A. C. Alcala. 2006. Gradients of abundance of fish across no-take marine reserve boundaries: evidence from Philippine coral reefs. *Aquatic Conservation: Marine and Freshwater Ecosystems* 16: 349–371.

Halpern, B. S. 2003. The impact of marine reserves: do reserves work and does reserve size matter? *Ecological Applications* 13: S117–S137.

Palumbi, S. R. 2004. Marine reserves and ocean neighborhoods: the spatial scale of marine populations and their management. *Annual Review of Environment and Resources* 29: 31–68.

Roberts, C. M. 2005. Marine protected areas and biodiversity conservation, in *Marine conservation biology: the science of maintaining the sea's biodiversity*. E. Norse and L. Crowder, eds. Washington, DC: Island Press, 265–279.

Russ, G. R. 2002. Yet another review of marine reserves as reef fisheries management tools, in *Coral reef fishes: dynamics and diversity in a complex ecosystem*. P. F. Sale, ed. San Diego, CA: Academic Press, 421–443.

Sladek Nowles, J., and A. M. Friedlander. 2005. Marine reserve design and function for fisheries management, in *Marine conservation biology: the science of maintaining the sea's biodiversity*. E. Norse and L. Crowder, eds. Washington, DC: Island Press, 280–301.

MARQUESAS ISLANDS

SEE PACIFIC REGION

MARSHALL ISLANDS

NANCY VANDER VELDE

Majuro, Marshall Islands

The loosely strung double archipelagoes of Ratak and Rālik, with their 29 atolls and five solitary coral islands, make up what is now known as the Marshall Islands. Marine life associated with these north central Pacific islands is rich and varied. Terrestrial environments range from lush forests and inland mangrove ponds to dry

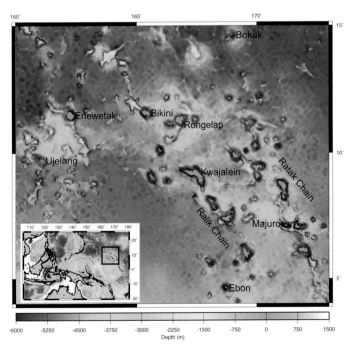

FIGURE 1 Map of the Marshall Islands, with locator inset. Satellite altimetry data: Geoware GMT Companion CD-R Vol. 1, Version 1.9, June 2006. Image prepared by Jenny Paduan and David Clague, MBARI.

FIGURE 2 Sandy beach along the ocean shore of Eneu, Bikini atoll.

shrublands, with a relatively limited diversity of plant and animal species. All the atolls and islands are low in elevation. Some of the northern atolls were used for nuclear tests, the impact of which is still being studied.

GEOLOGY

The atolls and islands of the Marshall Islands were formed as marine animals and plants continually built upon the foundation of submerging volcanoes. Sea level was 1 to 1.5 m higher 4000–6000 years ago, during which time the present Marshall Islands were probably just coral reef. After sea level dropped slightly, an estimated 2000 years ago, the atolls became inhabitable. Deep drilling in the Marshall Islands has shown evidence of volcanoes beneath generations of accumulated marine organisms.

The Marshalls' atolls, with their thousand-plus individual islets, are part of the geographical region called "Micronesia" ("small islands"). Total land is now about 180 km^2 (roughly the area of Santa Catalina Island, California) spread over 2 million km^2 of ocean, located from 14°37′ to 04°37′ N and 172°9′ to 160°55′ E. (Fig. 1). Although the land is relatively flat, rarely reaching even 4 or 5 m in elevation, marine topography is varied, with seamounts, guyots, and pinnacles scattered across oceanic plains.

The narrow, irregularly shaped rings of islets encircling lagoons of the Marshalls' atolls have many features of classic, idyllic paradises (Figs. 2 and 3). Nevertheless, there are harsh environmental conditions—desiccating winds, salt spray, and an annual dry season. Typhoons are uncommon but are devastating when they occur.

BIOLOGY

The equatorial countercurrent, equatorial currents, seabirds, and wind continually bring animals and plants.

Both above and below water, the biota displays a close relationship with the far western Pacific. Over 1000 fish species and 100 coral species have been reported; however, there are no native terrestrial mammals, less than three

FIGURE 3 Mangrove wetlands, Jaluit atoll.

dozen reptile species, and only about 100 native bird species (although isolated nesting islets can be home to spectacular numbers of individual birds) (Fig. 4). No one atoll is home to more than a few dozen native plants.

Endemics are limited and are primarily marine, with the most visible being the three-banded anemonefish (*Amphiprion tricinctus*). On land, there are a few endemics, including unique pseudoscorpions (i.e., *Garupus ornatus* of Bikini), insects (i.e., several species of longhorn beetles and sandflies), a landsnail (*Assiminea nitida marshallensis*), and lizards (i.e., the Arno skink, *Emoia arnoensis*). The only extant land bird, the Ratak Micronesian pigeon, is an endemic subspecies from the Marshalls' eastern chain. There are distinct horticultural varieties of pandanus (Fig. 5).

The array of the atolls' overall biodiversity is far from being homogenous, largely because of species' serendipitous arrival and survival. The atolls' relative isolation affords habitat to globally endangered migratory species, such as the hawksbill turtle and bristle-thighed curlew.

Although less impacted by invasives than other parts of the world, foreign introductions are of concern; for example,

there are probably less than 75 native species of land plants, but almost ten times that number of introduced plants can be found in urban areas. Concern has been expressed over the possible impacts of sea-level rise and climate change.

HUMAN HABITATION

The first inhabitants likely came from the south in outrigger canoes capable of heading into the wind, and the people continued to interact with distant places even after settlement. They imported valuable plants and animals, including giant swamp taro (*Cyrtosperma merkussi*), breadfruit (*Artocarpus* spp.), and red jungle fowl (*Gallus gallus*).

Modern Marshallese people are probably largely descendants of the original inhabitants and still largely speak their own language, also called Marshallese, a Micronesian language within the Austronesian family. English, although rapidly becoming widely known, remains the secondary language.

INFLUENCES BEYOND THE REGION

Spain first "discovered" the Marshall Islands in the sixteenth century but distant countries had little impact until the mid- to late nineteenth century, when Germany established the copra industry and Christendom's missionaries from the United States caused major changes. Forests were cleared for coconut plantations, and exotic species were introduced.

Change continued under foreign administration (Germany, Japan, the United States Navy, Trust Territory of the Pacific). In 1979, the Republic of the Marshall Islands came into being, and modernization continues

FIGURE 5 A horticultural variety of pandanus from Likiep atoll. The fruits of most varieties are used extensively for food. Other varieties serve other purposes.

FIGURE 4 Baby white tern from southwestern Bikini atoll.

apace. By the beginning of the twenty-first century, the atolls were home to over 51,000 people.

NUCLEAR TESTING

Some of the northwestern atolls were used for the first postwar nuclear testing ever conducted. From 1946 to 1958, 67 nuclear devices were detonated on Bikini and Enewetak atolls, with radioactive material that spread over the rest of the country with long-term ramifications that are still being investigated and debated.

Such is only part of what remains to be discovered about the Marshall Islands. The atolls' overall charming appearance and outwardly simple environment belies their true diversity, complexity, and importance.

SEE ALSO THE FOLLOWING ARTICLES

Atolls / Introduced Species / Nuclear Bomb Testing / Pacific Region

FURTHER READING

Amerson, A. B., Jr. 1969. Ornithology of the Marshall and Gilbert Islands. *Atoll Research Bulletin* 127: 1–216.
Crisostomo, Y. A. 2000. *Initial communication under the United Nations framework convention on climate change.* Majuro: Republic of the Marshall Islands Environmental Protection Authority.
Erdland, A. 1914. Die Marshall Insulanur. Leben und sitte, sinn und religion eines sudseevolkes. *Antropos Bibliothek* 2: 1–376.
National Biodiversity Team of the Republic of the Marshall Islands. 2000. *The Marshall Islands: living atolls amidst the living sea.* The National Biodiversity Report of the Marshall Islands. Majuro: RMI Biodiversity Project.
Neidenthal, J. 2001. *For the good of mankind: a history of the people of Bikini and their islands,* 2nd ed. Majuro, Marshall Islands: Bravo Publishers.
Republic of the Marshall Islands Biodiversity Clearing House Mechanism. http://www.biormi.org/.

MASCARENE ISLANDS, BIOLOGY

CHRISTOPHE THÉBAUD
Paul Sabatier University, Toulouse, France

BEN H. WARREN AND DOMINIQUE STRASBERG
University of La Réunion, Saint-Denis, Réunion.

ANTHONY CHEKE
Oxford, United Kingdom

The Mascarenes are an island group lying near the Tropic of Capricorn in the southwestern Indian Ocean ~700 km east of Madagascar. This archipelago comprises three high volcanic islands (Réunion, Mauritius, Rodrigues), scattered along a ~600 km west-east axis, and a group of small coralline islands (Cargados Carajos Shoals) ~400 km to the north of Mauritius, which sit upon a submarine bank of volcanic origin that extends a further 700 km or more to the northeast. The Mascarene Islands have an extraordinary status among islands: Mauritius was the former home of the dodo, the universal symbol of human-caused species extinction on islands. Although their recent history, since the first permanent human settlements in the seventeenth century, has been an endless series of ecological disasters and species extinctions, these islands still harbor up to 25% of their original forest cover and are extremely rich in species and habitats, with high degrees of endemism. Consequently, they are listed among the world's top biodiversity hotspots.

THE GEOGRAPHICAL CONTEXT FOR THE EVOLUTION OF BIODIVERSITY IN THE MASCARENE ISLANDS

Although less well-studied than the Hawaiian Islands, the Mascarene Islands, Rodrigues excluded, are generally believed to result from the same process of plate movement over a stationary hotspot. Today the Réunion hotspot is the source of frequent volcanism on the island of Réunion. The Réunion hotspot's activity, however, can be traced northeast along the Mascarene Plateau to India, where massive Deccan volcanism coincided with the Cretaceous–Tertiary (K/T) mass extinction event. Rodrigues, which sits next to the Central Indian ridge, is thought to have arisen in relation to the tectonic evolution of the Rodrigues triple junction, located 950 km to the southeast of the island. Typical of such archipelagoes, the Mascarene islands of today have have never been connected to larger land masses. Thus the biogeography and endemic biodiversity of these islands are the product of oceanic dispersal alone. The three main islands of today, Réunion, Mauritius, and Rodrigues, are very different in their size and current topography but are united by their relative geographic proximity and volcanic origin. Réunion, the largest (2512 km^2) and most southerly (21° S, 55.5° E), is nearest to Madagascar (665 km), whereas Mauritius, next in size (1865 km^2) is 164 km east-northeast of Réunion. Rodrigues, the smallest (104 km^2) and currently the most isolated, is located 574 km east of Mauritius.

The islands are separated from each other by fracture zones, and each island has developed independently. The most ancient dated lavas from Réunion, Mauritius, and Rodrigues are dated at 2.1, 7.8, and 1.5 million years ago, respectively. However, many exposed lavas in the Mascarenes are the result of recent reactivation, and new data suggests that Rodrigues, instead of being the youngest, is at least as old as Mauritius. Thus Mauritius and Rodrigues have been available for colonization by diverse biota for about 8–15 million years, while Réunion became habitable much later, about 2–3 million years ago. The extent to which Mauritius and Rodrigues have been isolated from larger land masses over the course of their history has been influenced by Pliocene and Quaternary sea level changes. Information from past sea level curves and current ocean floor bathymetry supports the existence of several large islands, as recently as 18,000–10,000 years ago, along the 115,000-km^2 Mascarene Plateau (currently under water with depths ranging from 8 to 150 m) between the granitic Seychelles and the Mascarenes. Drilling projects establish a volcanic origin for (or volcanic contribution to) these islands, with erosion and subsidence thereafter. It is likely that these islands, and also the Chagos and Maldives when fully above water, have played a role as a source of colonists for the present islands. In addition, chains of smaller islands would have reduced the distance for oceanic dispersal and could have served as steppingstones for dispersal between India and the Mascarenes.

As a consequence of erosion and subsidence on older islands and volcanic activity on younger islands, the greatest elevations above sea level are currently found on Réunion, with two main summits: Piton des Neiges (3070 m), which is the highest peak in the Indian Ocean, and Piton de La Fournaise (2631 m), one of the most active volcanoes in the world. The highest points of Mauritius (Black River Peak, 828 m) and Rodrigues (Mt. Limon, 398 m) are low in comparison. Réunion, like other young volcanic islands, has a very dramatic topography, being highly dissected into huge caldera-like valleys (cirques) caused by erosion under very high rainfall, with very narrow outlets to the sea through deep gorges. Mauritius, in spite of being an old island, has undergone dramatic geological transformation until recently. Volcanic eruptions have reshaped the island into a series of small, eroded, "geological" islands (age 7.5–5.1 million years) embedded in a matrix of recent lava flows (0.7–0.025 million years old). Thus, both Mauritius and Réunion show considerable spatial heterogeneity in their topography. While the significance of such heterogeneity for the evolution of colonist lineages is evident in the case of Réunion, the biological implications of the "islands within the island" structure of Mauritius, though obvious, have been overlooked by most biologists until very recently. As on the other islands, the main relief of Rodrigues is composed of basaltic lava, but Rodrigues also has an area of limestone plateau made of consolidated coral sands and punctuated with caves.

Owing to their geological history, geographic isolation, and current climate, the Mascarene Islands show more similarities to the Hawaiian Islands than to any other archipelago, even though these two island systems differ greatly in the numbers of islands currently present and the degree of isolation from the nearest other masses. Colonization of the Mascarene Islands by immigrating lineages has occurred relatively recently, but in spite of the simplicity of the present geographic setting, evolutionary diversification in the archipelago has been strongly influenced by a rather complex volcanic evolution combined with a regional geographic configuration that has greatly changed since the first island was formed.

THE ECOLOGICAL THEATER

The Mascarene Islands have a tropical climate; that is, temperatures are warm and show little seasonal variation. The climate is strongly influenced by the humid prevailing winds blowing from the southeast, with annual rainfall varying from 500 mm in the driest leeward areas to about 12 m in the wettest areas on the windward slopes of Réunion. Such climate generally promotes the development of forests. From early reports and ecological inference from what is left of the original vegetation, all three main islands were completely forested when discovered. Exceptions are the high-elevation environments above 1900 m on Réunion, where the forests give way to a subalpine scrub.

The Mascarene Islands share with other oceanic islands the habitat destruction and transformation associated with human activity. Low-altitude areas have been subject to a much higher impact than high-altitude areas. Although native vegetation remains, all the original forest covering Rodrigues has been destroyed, and a mere 2% of the original cover has been left in Mauritius. In contrast, about 25% of the estimated original extent of Réunion's habitats are still in a good state. As a result of deforestation and rugged topography, Réunion's forest remnants are severely fragmented, with large tracts found only above 500 m elevation, and with no more than 1% of lowland forest remaining. Lowland forest remnants are mostly located on the slopes of the active volcano, where they often take the form of forest islands embedded in a matrix of lava flows of various ages. Unfortunately, many of these forest islands were wiped out by a massive volcanic eruption in 2007. Descriptions of the vegetation zones, using historical accounts and subfossil record when necessary, have emphasized five natural plant formations arranged in broad moisture and altitudinal zones. Dry lowland forests dominated by palms (*Latania* spp., *Dictyosperma album*), screw-pines (*Pandanus* spp.), and trees such as *Terminalia bentzoe* (Combretaceae) were present from sea level to 200 m elevation in areas with less than 1000 mm average annual rainfall. This ecosystem was probably the habitat of some of the most spectacular endemic animals, in particular the now extinct giant tortoises (*Cylindraspis* spp., Testudinidae), but it no longer exists on the main islands. Some relicts may be found on a small islet (Round Island) off the northern tip of Mauritius and in a few places on Réunion.

Semi-dry sclerophyllous forests occurred between coastal areas and 360 m on all sides of Mauritius and Rodrigues, but were restricted to 200–750 m elevation on the western slopes of Réunion, where they still exist in small forest remnants. This ecosystem has an average annual rainfall of 1000–1500 mm and is characterized by ebonies (*Diospyros* spp., Ebenaceae) and other trees such as *Pleurostylia* spp. (Celastraceae), *Foetidia* spp. (Lecythidaceae), *Olea europea* subsp. *africana* (Oleaceae), *Cossinia pinnata* (Sapindaceae), *Dombeya* spp. (Sterculiaceae), and a variety of Sapotaceae species (*Sideroxylon boutonianum*, *Mimusops* spp.). The ecosystem is also home to several spectacular endemic species of *Hibiscus* (Malvaceae). Many species of this zone, such as *Zanthoxylum* spp. (Rutaceae), *Obetia ficifolia* (Urticaceae), and *Scolopia heterophylla* (Flacourtiaceae), exhibit developmental heterophylly, with juvenile leaves being more divided than those of adults. Such convergence may have evolved to deter herbivory by extinct giant tortoises.

Lowland rainforests occur above 360 m (on Mauritius) and all over the eastern lowlands from the coast to 800–900 m and, on the western side, from 750 to 1100 m (on Réunion) (average annual rainfall 1500–6000 mm). These forests have a canopy of tall trees up to 30 m high and represent the richest plant communities of the Mascarene Islands. Characteristic plants include trees in the plant family Sapotaceae (e.g. *Mimusops* spp., *Labourdonnaisia* spp., *Sideroxylon* spp.), Hernandiaceae (*Hernandia mascarenensis*), Clusiaceae (*Calophyllum* spp.), and Myrtaceae (*Syzygium* spp., *Eugenia* spp., *Monimiastrum* spp.); shrubs in the plant family Rubiaceae (*Gaertnera* spp., *Chassalia* spp., *Bertiera* spp., *Coffea* spp.); and numerous species of orchids (e.g., *Angraecum* spp., *Bulbophyllum* spp.) and ferns (e.g., *Asplenium* spp., *Hymenophyllum* spp., *Trichomanes* spp., *Elaphoglossum* spp.).

Dense cloud forests occur on Réunion between 800 and 1900 m on eastern slopes (average annual rainfall 2000–10,000 mm) and between 1100 to 2000 m on western slopes (average annual rainfall 2000–3000 mm) and are also restricted to a small area of Mauritius around Montagne Cocotte above 750 m on Mauritius (average annual rainfall 4500–5500 mm). On both islands these low forests, with a canopy of 6 to 10 m high, are rich in epiphytes (orchids, ferns, mosses, lichens), emergent tree ferns (*Cyathea* spp.), and, originally, palms (*Acanthophoenix rubra*), but these now survive only in areas of Réunion where poaching has not wiped them out. Untransformed cloud forests still cover large areas on Réunion (44,000 ha in 2005). These forests are characterized by trees such as *Dombeya* spp. (on Réunion only) and species in the plant family Monimiaceae (*Monimia* spp., *Tambourissa* spp.) as canopy species, with small trees and shrubs such as *Psiadia* spp. (Asteraceae) and *Melicope* spp. (Rutaceae) in the understory. They also include large areas of three monodominant plant communities, forests with *Acacia heterophylla* (Fabaceae) as canopy species that are very similar to *Acacia koa* forests in Hawaii, thickets dominated by *Erica reunionensis* (Ericaceae), or hyperhumid screw-pine forest (*Pandanus montanus*).

Finally, above the tree line, at elevations where frosts occur regularly in winter (1800–2000 m), is a unique subalpine scrub dominated by shrubs in the plant families of Ericaceae (*Erica* spp.), Asteraceae (*Hubertia* spp., *Psiadia* spp., *Stoebe passerinoides*), and Rhamnaceae (*Phylica nitida*), with some notable endemic species such as *Heterochaenia rivalsii* (Campanulaceae), *Eriotrix commersonii* (Asteraceae), and *Cynoglossum borbonicum* (Boraginaceae) (average annual rainfall 2000–6000 mm). The summits of the volcanoes are covered by large mineral areas with sparse grasslands rich in endemic grasses (Poaceae, e.g., *Festuca borbonica*, *Agrostis salaziensis*, *Pennisetum caffrum*) and orchids (Orchidaceae, e.g., *Disa borbonica*), ericoid thickets, or thickets of the small tree *Sophora denudata* (Fabaceae), depending on substrate texture and age.

The Mascarene Islands are surrounded by approximately 750 km² of coral reef. Rodrigues has nearly continuous fringing reefs bounding an extensive lagoon with deep channels, whereas Mauritius is surrounded by a discontinuous fringing reef and a small barrier reef. In contrast, Réunion has very short stretches of narrow fringing reefs along the western and southwestern coasts only. The islets of the Cargados Carajos Shoals, which have a very depauperate terrestrial biota owing to being so low-lying and swamped during cyclones, are bound to the east by an extensive arc of fringing reef, which accounts for ~30% of the reefs of the Mascarene Islands. Lagoon reefs and reef flats are dominated by scleractinian corals such as branching and tabular *Acropora*, *Porites* massives, foliaceous *Montipora* and *Pavona*, and sand consolidated with beds of seagrass such as *Halophila* spp. (Hydrocharitaceae). Among coral reef fishes, wrasses (Labridae), damselfish (Pomacentridae), carnivorous groupers (Serranidae), and surgeonfishes (Acanthuridae) are particularly well represented.

BIODIVERSITY AND ENDEMISM

Identifying species that are unique (endemic) to a group of islands or individual islands relies upon extensive biological inventories, including in-depth systematic investigations, and the ability to recognize cryptic species in lineages that lack substantial morphological differentiation across their range. Although there is currently much effort to fill the gaps, there are still many taxonomic groups in the Mascarene Islands that have not been thoroughly investigated and for which rigorous figures for endemism are not yet available. The Mascarenes therefore present an exciting relatively unchartered study system for island biologists. However, the Mascarene biota, like most other oceanic island biotas, has suffered many recent extinctions that are a source of information bias, particularly in groups of organisms that leave no subfossil materials or that were not recorded and described by the early travelers.

The Mascarene biota exhibits high levels of endemism in many groups of related taxa: about three-quarters of the approximately 960 native flowering plant species, ~65% of the Coleoptera (~1550 species), and 90% of the nonmarine molluscs (~200 species) are endemic (Table 1). These degrees of endemism are very close to those observed in similar groups in the Hawaiian Islands or New Caledonia. The Mascarene Islands had the richest oceanic island reptile fauna before the arrival of people. For nonmarine reptiles, the percentage of endemic species has been estimated to be 94%, but more than half of the 30 endemic species known to have occurred in the Mascarene Islands have gone extinct in the last four centuries, including five species of Indian Ocean giant tortoises. Of three endemic snakes, only one boa still exists (*Casarea dussumieri*). Apart from bats there are no terrestrial mammals, but all three species of fruit bats (*Pteropus* spp.; 1 extinct) and at least two (two newly recognized species of *Mormopterus*) of the four species of microbats are endemic. The Mascarene Islands once had a very rich avifauna, with an estimated 81 native species, 54 of which (67%) were endemic to the archipelago. Apart from three seabirds (*Pterodroma baraui*, *Pseudobulweria aterrima*, and an undescribed *Pterodroma* known only from subfossil bones), most endemic species were landbirds. A large fraction of these endemic birds, especially the larger ones, have now gone extinct, among which were found the legendary dodo (*Raphus cucullatus*, Columbidae, formerly Rhaphidae), its flightless relative the Rodrigues solitaire (*Pezophaps solitarius*), and the Réunion solitaire (*Threskiornis solitarius*, which was not related to the dodo but was an ibis, Threskiornithidae).

For the marine biota, there is a dearth of comprehensive systematic investigation at the scale of the archipelago. The average percentage of marine species in the Mascarene Islands that are endemic is apparently lower (2–15%) than in the terrestrial biota. However, the degree of endemism may vary considerably among taxonomic groups, and it is likely that cryptic species, having virtually indistinguishable morphologies, are more widespread in the sea than previously thought. Future broad-scale examination of groups with many wide-ranging species, such as marine molluscs (with at least 3000 species of gastropods occurring in the western Indian Ocean region), crustaceans (with a minimum total of 780 species for the western Indian Ocean), or bryozoans (with at least 500 species in the western Indian Ocean), using molecular taxonomy and new morphometric approaches, may reveal similar levels of endemism to those observed in some terrestrial groups.

There are many endemic genera of plants and animals in the terrestrial biota. For example, 32 genera of flowering plants (11% of the total number of genera) and 89 genera of Coleoptera (14% of the total number of genera) are restricted to the Mascarene Islands. Some of these genera provide spectacular examples of diversification within the archipelago, such as weevils (*Cratopus*, 86 species), leaf beetles (*Trichostola*, >25 species) and several shrubs (e.g., *Badula* [14 species], *Heterochaenia* [3 species], *Trochetia* [6 species]) (Fig. 1). However, the highest numbers of

TABLE 1
A Summary of Mascarene Biodiversity and Endemism for Different Taxonomic Groups with Adequate Systematic Knowledge

Taxonomic Group	Number of Species	Number of Endemic Species	Percent of Endemic Species
Flowering plants	959	691	72
Ferns and allies	265	58	22
Mosses and allies	~800	40–80	5–10
Nonmarine mammals	7	4	57
Reef fishes	923	42	5
Landbirds	60	51	85
Seabirds	21	3	14
Nonmarine reptiles	32	30	94
Nonmarine molluscs	200	180	90
Coleoptera	1538	979	64

FIGURE 1 *Badula borbonica*, or bois de savon (Myrsinaceae). *Badula* is a species-rich genus, endemic to the Mascarene Islands. Here is shown a large-leaved species that forms a medium-sized unbranched shrub in the dense cloud forests of Réunion. Photograph by Christophe Thébaud.

FIGURE 2 *Forgesia racemosa*, or bois de Laurent-Martin. This enigmatic and beautiful species, common in cloud forests, is in the Escalloniaceae family. Its closest known relatives live in the Andes, South America. Photograph by Christophe Thébaud.

endemics are often found in nonendemic genera (numbers of species in parenthesis), for example, *Gonospira* landsnails (28), *Phelsuma* geckos (9), *Diospyros* trees (14), *Dombeya* trees (13), *Gaertnera* shrubs (14), *Pandanus* screw pines (22), or daisy trees *Psiadia* (26). Such a pattern suggests that a number of very recent species radiations have been a significant factor in the buildup of endemic biodiversity in the Mascarene Islands.

PHYLOGEOGRAPHY, PROCESS OF SPECIES FORMATION, AND ADAPTIVE RADIATION

An important question for understanding endemic biodiversity is the geographical origin of colonizing lineages. As expected from current geography, Mascarene biota has close affinities with Madagascar and Africa. However, unexpectedly, many elements are related to more remote regions, notably Asia and the Indo-Pacific region. In flowering plants, about two-thirds of the genera are shared between the Mascarene Islands and Madagascar and Africa (e.g., *Angraecum, Diospyros, Dombeya, Psiadia*) whereas at least 20% are shared with Asia and the Indo-Pacific region (e.g., *Astelia, Ochrosia, Terminalia*) (Fig. 2). This pattern has been suggested for many other groups, including birds, insects, and even reptiles, including the endemic Mauritian boa family Bolyeridae. Recent phylogenetic analyses using DNA markers have confirmed either the western (e.g., the fruit fly *Drosophila mauritiana*, *Falco* kestrels, *Cylindrapsis* tortoises, *Phelsuma* geckos, *Angraecum* orchids, *Gaertnera* shrubs, *Phylica* shrubs, *Polyscias* trees, *Psiadia* daisy trees) or the eastern (e.g., *Mormopterus* free-tailed bats, *Leiolopisma* skinks, *Nactus* geckos, *Hypsipetes* bulbuls, *Aerodramus* swiftlets, *Psittacula* parakeets, *Alectroenas* pigeons, the climbing shrub *Roussea simplex*) origins of some groups (Fig. 3). They have also revealed the Asian origins in groups with poorly understood evolutionary history. The dodo and the Rodrigues solitaire, whose geographic origins have long been mysterious, appear to have dispersed from southeast Asia to the Mascarene Islands at some point in the past. The phylogenetic relationships of Indian Ocean white-eyes (*Zosterops*) point to an Asian origin for Mascarene species, contrary to intuition (Fig. 4). That a significant portion of colonist lineages comes from the east emphasizes the likely role played by now-submerged land masses between the Mascarene Islands and India and also sea currents and winds in drawing high numbers of colonists from Asia and the Indo-Pacific region into the southwestern Indian Ocean region.

FIGURE 3 The day gecko *Phelsuma cepediana*, one of the seven surviving Mascarene species, is currently the sole pollinator and seed disperser of *Roussea simplex*, a climbing shrub endemic to the mountains of Mauritius that was named after Jean-Jacques Rousseau, the Swiss philosopher of the Enlightenment. Photograph by Dennis Hansen.

FIGURE 4 *Zosterops mauritianus* belongs to Mascarene gray white-eyes, an anomalous group of warbler-like white-eyes with no "white-eye" with Asian affinities that appears to have undergone a cryptic adaptive radiation in the Mascarenes. Photograph by Charlie Moores.

How biodiversity builds up in an archipelago like the Mascarene Islands after the first island has appeared above sea level depends on patterns of dispersal and subsequent diversification of founding lineages within the nascent archipelago, including within-island speciation. Some taxa were never successful in colonizing the archipelago. For example, amphibians were absent from the original fauna. Among lineages that colonized the archipelago, some may have repeatedly colonized the archipelago or diversified more than others. Data on the diversification of the Mascarene biota are still too scanty to draw any generalization, but recent molecular studies provide good illustrations of the processes that have led to species diversity in this region. An example in which species diversity within the archipelago reflects multiple successful colonizations from source areas comes from fig trees (*Ficus*), fruit bats (*Pteropus*), and orchids (*Angraecum*). Recent phylogenetic hypotheses imply that the five species of figs found in the Mascarenes, three of them being endemic, have arisen from five separate colonization events. The three endemic fruit bat species apparently originated from three distinct colonizations. Concerning *Angraecum*, a genus represented by approximately 30 species in the Mascarene Islands, 21 of which are endemic, phylogenetic data implies at least 20 independent colonization events. In contrast, biodiversity in other groups appears to result from single or a few colonization events followed by the evolution of species radiations. The daisy trees (*Psiadia*), the second most species-rich genus of flowering plant in the Mascarene Islands (*Angraecum* is the first), display phylogenetic relationships that are consistent with a double archipelago colonization, followed by extensive species radiations within both Mauritius and Réunion. The phylogenetic hypothesis for day geckos (*Phelsuma*) is concordant with a single colonization of the archipelago, something that is also true for other endemic reptile taxa in the Mascarene Islands (e.g., *Cylindrapsis* tortoises, *Nactus* geckos). The nine Mascarene species of day geckos are best explained by a combination of inter-island dispersal and intra-island speciation events. Understanding how speciation proceeds within small islands in groups as diverse as flowering plants, reptiles, insects, and even birds is not well understood yet in the Mascarenes. Considerable morphological variation is found in diverse organisms on both Mauritius and Réunion. Many species of streptaxid land snails and several species of day geckos display high among-population morphological and/or genetic variation within the islands. The Mascarene gray white-eye (*Zosterops borbonicus*), a bird endemic to Réunion, shows spectacular variation in plumage traits that coincide with differences in the habitats occupied within this topographically and climatologically diverse island. Thus, it seems likely that natural selection is the key to diversification both among and within islands, although the detailed processes involved remain to be studied.

CONSERVATION ISSUES

Before the arrival of the Europeans in the sixteenth century, the Mascarene Islands had evaded discovery by seafarers. The early visitors released ungulates and, on Mauritius, rats and monkeys, but the islands were settled only in the the mid-seventeenth century, when commercial rivalry induced European trading nations to annex and settle the islands. The Mascarene Islands had not experienced major perturbations of the biota when early visitors started to describe what they found. Hence, nowhere in the world has the tragic loss of species and the alteration of pristine tropical island ecosystems been documented as thoroughly as in the Mascarenes.

Historical records demonstrate unambiguously that forest clearance, human hunting, and the introduction of nonindigenous predators have been the primary causes of species extinctions in these islands. In total, the Mascarenes have lost about 40% of their native vertebrate species, and most of these were already gone by the middle of the nineteenth century. Some iconic species such as the dodo, the Rodrigues solitaire, and the Réunion solitaire even disappeared from Mascarene landscapes by the late seventeenth or early to mid-eighteenth century. The last giant tortoises were seen in Rodrigues around 1800, and when Charles Darwin visited Mauritius aboard the *Beagle* in 1836, there had been no record of any tortoise on this island for more than a century. These extinctions and several others mostly predate the first spells of extensive human-caused habitat destruction. This fact implicates human hunting and predation by rats, cats, and pigs, rather than forest clearance, as the primary cause of vertebrate extinctions in the Mascarenes. Patterns of vertebrate species loss also show clearly that the impacts of human settlement and introductions of predators occurred very rapidly. They may have been exacerbated by the fact that many species possessed characteristics that increased their susceptibility to human hunting (large body size) or the impact of nonindigenous predators (lack of mammalian predator escape response, including, e.g., flightlessness, tameness).

As a consequence of growing human populations and colonial policies, the rate of forest destruction peaked during the nineteenth century. Forests were cleared for agriculture development to produce sugar cane in both Mauritius and Réunion and maize, coffee, and geranium oil in Réunion, and for slash-and-burn agriculture combined with free-range livestock production in Rodrigues. Such forest destruction likely caused many extinctions by wiping out entire habitats (e.g., semi-dry sclerophyllous forests). Another, almost inevitable, consequence of large-scale forest destruction was increased fragmentation of habitats coupled with invasion by a wide range of introduced plants, additional animals (e.g., *Herpestes* mongooses, *Calotes* agamid lizard, and *Lycodon* wolf snakes), and pathogens. Notable extinctions that occurred during this period include the hoopoe starling (*Fregilupus varius*, a bird species that was still common in forested areas of Réunion into the early 1850s but had vanished before 1860), the pigeon hollandais (*Alectroenas nitidissima*), the Mauritius lizard-owl (*Mascarenotus sauzieri*, last seen in the 1820s or 1830s), the slit-eared skinks (*Gongolymorphus* spp., which disappeared with the arrival of the wolf snake and survive only on islets offshore), and the anomalous hole-roosting flying-fox (*Pteropus subniger*, which vanished on Réunion around 1840).

During the twentieth century, destruction of forests continued (often for short-lived agricultural initiatives or even make-work programs), fragmentation of forest remnants increased, nonindigenous species continued to arrive in the Mascarenes, and the number of species on the verge of extinction rose steadily. Today, Mauritius and Rodrigues are so extensively altered that most of their native biotas are already extinct or severely threatened. In addition, Mauritius forest remnants are permeated with hordes of invasive mammal species such as monkeys, deer, pigs, and mongoose. By contrast, Réunion is relatively free of these animals. Hence, the survival of relatively intact Mascarene ecosystems largely depends on adequate conservation on the island of Réunion, although even here the best surviving lowland forest was mostly lost to ill-conceived forestry in the 1970s. From a network of reserves finally set up in the 1980s and onwards, approximately 1000 km^2 of Réunion (40% of the island area) were designated as a national park in 2007, with a further 35 km^2 of marine nature reserves. In Mauritius, nominally protected areas are much longer established (dating from 1951), and a national park covering 66 km^2 and much of the best remaining habitat was established in 1994. Total reserves cover a mere 75 km^2, but considerable effort has been devoted to conservation management work. Pioneering habitat restoration programs have been underway since the late 1960s, but until recently the programs have been focused on a handful of endangered bird species. Through captive-breeding and release programs, including eradication of nonindigenous predators, the Mauritius kestrel (*Falco punctatus*), pink pigeon (*Nesoenas mayeri*), and echo parakeet (*Psittacula eques*) were rescued from imminent extinction, while the future of Mauritius fody (*Foudia rubra*) and Mauritius olive-white-eye (*Zosterops chloronothus*) now also looks more secure. An extraordinary result is that the population of Mauritius kestrels recovered from a single wild breeding pair in 1974, when its prospects were considered to be hopeless, to over 900 individuals in the wild today. The populations of echo parakeets and pink pigeons have also bounced back from 10 (early 1980s) to 300 and 10 (1991) to 360 individuals in the wild today, respectively.

In the Mascarene Islands, eradication of nonindigenous predators has become a high conservation priority to prevent further extinctions and is often a prerequisite to ecosystem restoration work. On islets around Mauritius and Rodrigues such as Round Island, Ile aux Aigrettes, or Gunner's Quoin, eradication programs have succeeded in clearing these islands from feral goats, rabbits, rats, cats, and mice, although house shrews, agamid lizards, and wolf snakes have proved harder to remove. Removal of these predators has led to an increase in numbers of native plant and animal species living on the islets, and the now stabilized conditions on Round Island have allowed the palm forest to recover and the unique reptiles there to thrive, and one species, Telfair's skink (*Leiolopisma telfairii*), has been reintroduced onto islands it formerly inhabited, now again rat-free. However, complete eradication of all predators is not always possible, and eradication programs easily become a difficult conundrum for conservation. In Réunion, where cats and rats threaten endemic petrels that breed on mountain tops, conservation managers have to take into account possible mesopredator effects. While intuition is that cats should be eliminated first, this may not be the best strategy if a population explosion of rats might follow and hit the petrel populations even harder. However, cats mostly eat adult petrels, while rats only attack eggs and chicks. Thus, even if the eradication of cats leads to an increase in rat density, this might not necessarily mean a decline in the petrel populations. Removal of cats and rats in nesting areas is under way, with long-term monitoring projects to ensure that conservation action leads to increased population sizes, not to unwanted decreases due to mesopredator effects.

Nonindigenous plant invasions are widespread in the remnant native ecosystems of the Mascarenes. Most invaders colonize human-disturbed sites most successfully, with sizable forest remnants being still dominated by native species. In Mauritius and Rodrigues, and to a lesser extent in the lowlands of Réunion, forest remnants are small, disturbed, and heavily invaded. To improve the prospects of long-term survival of these fragments and the species that inhabit them, pioneering restoration programs began in the late 1960s in Mauritius, and have been much extended, with Rodrigues added, from the early 1980s. Conservation management areas were established in remnants of the major original habitat types and have demonstrated that habitat restoration can effectively reduce the rate of species loss if based on sound ecological knowledge. These areas were fenced to reduce access by introduced herbivores and regularly weeded to control populations of nonindigenous plants. They usually showed an improvement in natural regeneration of native plants in less than ten years, including the legendary tambalacoque tree (*Sideroxylon grandiflorum*), popularly supposed to be dependent on the extinct dodo. However, although some species that were not regenerating before management were now thriving, nearly half of the species were still not regenerating. This lack is likely related to alterations of plant-pollinator and plant-disperser interactions as a result of extreme habitat fragmentation and animal species extinctions. The loss of native mutualists can limit natural regeneration of native plants that were once dependent on them, through the shortage of pollinators or seed dispersers (Fig. 5). Pioneering work is currently under way to reconstruct missing elements of these critically endangered ecosystems by filling the gaps left by the lost species using analogues (e.g., related species surviving in other parts of the Indian Ocean). However, suggestions to reintroduce from Mauritius species lost

FIGURE 5 Giant Aldabra tortoises (*Aldabrachelys gigantea*), introduced into Iles aux Aigrettes (Mauritius), can be used as ecological analogue seed dispersers of *Syzygium mamillatum*, a rare endemic tree. Photograph by Dennis Hansen.

in Réunion (several birds and a fruit bat), and vice versa (Réunion harrier, *Circus maillardi*), have so far not been acted on.

SEE ALSO THE FOLLOWING ARTICLES

Biological Control / Coral / Deforestation / Dodo / Mascarene Islands, Geology

FURTHER READING

Atkinson, R., J. C. Sevathian, C. N. Kaiser, and D. M. Hansen. 2005. *A guide to the plants in Mauritius*. Vacoas, Mauritius: Mauritius Wildlife Foundation.

Austin, J. J., E. N. Arnold, and C. G. Jones. 2004. Reconstructing an island radiation using ancient and recent DNA: the extinct and living day geckos (*Phelsuma*) of the Mascarene islands. *Molecular Phylogenetics and Evolution* 31: 109–122.

Blanchard, F. 2000. *Guide des milieux naturels: La Réunion–Maurice–Rodrigues*. Paris: Editions Eugen Ulmer.

Bosser J., T. Cadet, J. Guého, and W. Marais. 1976–2005. *Flore des Mascareignes*. Paris: Editions de l'Institut de Recherche pour le Développement (IRD).

Cheke, A., and J. Hume. 2008. *Lost land of the Dodo: an ecological history of the Mascarene Islands*. London: T & AD Poyser.

Griffiths, O. L., and V. F. B. Florens. 2006. *A field guide to the non-marine molluscs of the Mascarene Islands (Mauritius, Rodrigues and Réunion) and the northern Dependencies of Mauritius*. Mauritius: Bioculture Press.

Motala, S. M., F.-T. Krell, Y. Mungroo, and S. E. Donovan. 2007. The terrestrial arthropods of Mauritius: a neglected conservation target. *Biodiversity and Conservation* 16: 2867–2881.

Probst, J. M. 1997. *Animaux de La Réunion: guide d'identification des oiseaux, mammifères, reptiles, et amphibiens*. Saint-Denis, Réunion: Editions Azalées.

Turner, J., and R. Klaus. 2005. Coral reefs of the Mascarenes, Western Indian Ocean. *Philosophical Transactions of the Royal Society A* 363: 229–250.

Warren, B. H., E. Bermingham, R. P. Prys-Jones, and C. Thébaud. 2006. Immigration, species radiation, and extinction in a highly diverse songbird lineage: white-eyes on Indian Ocean islands. *Molecular Ecology* 15: 3769–3786.

MASCARENE ISLANDS, GEOLOGY

ROBERT A. DUNCAN
Oregon State University, Corvallis

The west-central Indian Ocean volcanic islands of Reunion, Mauritius, and Rodrigues are collectively known as the Mascarene Islands. They are volcanoes related to an age-progressive trend of north-to-south volcanic activity that includes the coral-capped Mascarene plateau and the Chagos-Maldive-Laccadive ridge, extending northward to the Deccan flood basalts of western India (Fig. 1). The origin and distribution of these elevated features, rising from ocean floor depths of 4–5 km, are thought to be the result of plate motions over the Reunion hotspot, a persistent upper mantle melting anomaly maintained by focused mantle upwelling. The primary evidence for this idea is (1) common elements of volcano composition and evolution and (2) a northward increase in age of volcanic activity consistent with rapid northward motion of the Indian plate, followed by slower northeastward motion of the African plate over the last 65 million years.

TEMPORAL DISTRIBUTION OF VOLCANIC ACTIVITY

The volcanic islands of Reunion, Mauritius, and Rodrigues lie between 19° and 22° S and between 55° and 64° E (Fig. 1). Radiometric age determinations, by K–Ar total fusion and $^{40}Ar-^{39}Ar$ incremental heating methods, provide the time frame for volcanism on the three islands.

Reunion Island (21°12′ S, 55°32′ E) rises from an ocean floor depth of ~5 km to a maximum elevation of 3069 m above sea level (Fig. 2). Two volcanoes make up the island: Piton des Neiges is inactive and forms the northwest two-thirds of the island, whereas Piton de la Fournaise forms the southeastern part of the island and is one of the most productive volcanoes in the world. The centers of the two volcanoes are 30 km apart. The oldest lava flows found on the island are about 2 million years old. Piton des Neiges has not been active for 70,000 years, and the volcano is eroding into steep valleys and cirques. Active for 360,000 years, Piton de la Fournaise is being built on the flank of the older volcano in much the same way as Kilauea is forming on the flank of Mauna Loa, at the island of Hawaii. The island's west coast hosts an intermittent fringing coral reef.

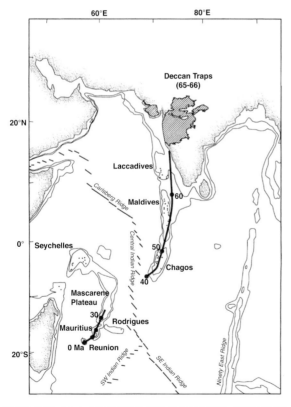

FIGURE 1 Regional plate tectonic map of the western Indian Ocean, with the volcanic trail of the Reunion hotspot. Black line shows predicted trail of the hotspot, in 10-million-year ticks of activity, modeled from an assumed stationary hotspot and plate motions over the last 65 million years.

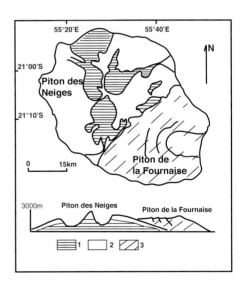

FIGURE 2 Geological map of Reunion Island, composed of two volcanoes: the older Piton des Neiges and the younger Piton de la Fournaise. The oldest shield lava flows of Piton des Neiges are exposed in steep-sided river canyons (unit 1), overlain by surface lava flows of more evolved composition (unit 2). The young Piton de la Fournaise volcano (unit 3) is growing on the southeast flank of the older volcano.

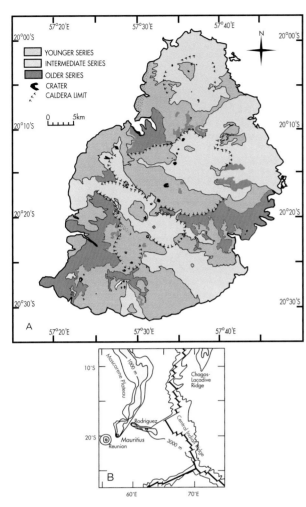

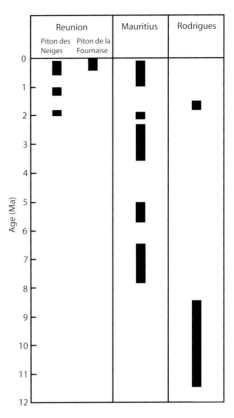

FIGURE 4 Time frame of volcanic construction at the Mascarene Islands, from radiometric age determinations.

FIGURE 3 Geological map of Mauritius Island, with distribution of three phases of volcanic activity.

Mauritius Island (20°20′ S, 57°30′ E) is inactive and rises from a coastal plain to a central plateau of between 275 and 730 m elevation. The island was built in at least three eruptive episodes (Fig. 3). The Older Series lavas comprise the erosional remnants of a single large volcano that was built up above sea level from ocean depths of 4.5 km, between 7.8 and 6.8 million years ago (Fig. 4). Waning activity persisted intermittently until about 5.5 million years ago. After a 2-million-year hiatus and extensive erosion of the Older Series rocks, smaller volumes of lavas were erupted between 3.5 and 1.9 million years ago and form the Intermediate Series. The Younger Series lavas are confined to localized vents along a system of NNW–SSE fissures that traverse the island. These were erupted between 1.0 and 0.1 million years ago. The island is surrounded by a fringing coral reef at a distance of several hundred meters to 5 km offshore.

Rodrigues Island (19°42′ S, 63°25′ E) lies about halfway between the Central Indian sea floor spreading ridge and Mauritius (570 km distant to the south-southwest), at the eastern end of the Rodrigues ridge. It rises only a few hundred meters above sea level as the erosional remnant of a volcano that ceased activity about 1.5 million years ago. The entire island sits within an extensive, reef-fringed lagoon of around 200 km². Rocks dredged from the Rodrigues ridge are 7.5 to 11.0 million years old, indicating that the island is the most recent, emergent part of an older volcanic feature.

COMPOSITIONAL VARIATIONS

As products of hotspot volcanic activity, the Mascarene Islands show many similarities in composition and evolution to the classic Hawaiian volcanoes. In both Reunion and Mauritius, construction of the main volcanic edifice ("shield") occurred with eruptions of rather homogeneous magmas derived from moderately high degrees (5–10%) of melting of the upper mantle, followed by low-pressure (i.e., crustal level) fractional crystallization (gravitational removal of crystals of olivine and pyroxene). Many of these early lavas contain significant accumulation of olivine crystals. At the termination of the shield-building phase at both islands, highly differentiated lavas were erupted, consistent with smaller degrees of partial melting of the mantle and more extended cooling in crustal magma chambers prior to eruption. Late-stage, shield-capping lavas in Hawaii are distinctly more alkaline than earlier ones, whereas this is not the case on Piton des Neiges. The Intermediate Series at Mauritius, consisting of highly silica-undersaturated lavas erupted in small volumes following a long hiatus and period of erosion, is analogous to the post-erosional series of Hawaii. The Younger Series lavas at Mauritius, which are less alkaline and were erupted in larger volumes than the Intermediate Series lavas, have no obvious compositional analog in Hawaii.

Geochemical data (major and trace element concentrations, and Sr, Nd, Pb isotopic compositions) from lava flows from Reunion, Mauritius, and Rodrigues islands support the proposed relationship to a common mantle plume. These data suggest that the plume composition is heterogeneous, however, with a more "fertile" (lower temperature and higher degree melt fraction) component dominating the early shield-building phase, and a less "fertile" (higher temperature and lower degree melt fraction) playing a more significant role in post-erosional phases (Intermediate and Younger Series at Mauritius). The protracted time scale of volcanic activity at Mauritius (7.8 to 0.1 million years ago) presents a challenge for the hotspot model, because even at relatively slow rates of African plate motion, eruptions have taken place over 200 km away from the hotspot (now beneath Reunion). Explanations involve entrainment of the plume "downstream" (in the direction of plate motion) from the hotspot along the underside of the African plate, together with flexing of the oceanic lithosphere caused by the load created by the growing island of Reunion (2 million years ago to present).

RELATIONSHIP TO REGIONAL GEOLOGICAL DEVELOPMENT

The location and timing of volcanic activity at Mauritius and the Rodrigues ridge imply that they are products of the focused mantle upwelling now beneath Reunion (the Reunion mantle "plume"), which apparently produced extraordinary melting at the time of the Deccan flood basalts in western India, 65 million years ago. Northward motion of the Indian plate away from the plume left a trail of volcanic islands and ridges now seen as the Laccadive-Maldive-Chagos island chains between 65 and 36 million years ago. The Central Indian spreading ridge then migrated across the plume, from southwest to northeast, causing the subsequent Mascarene plateau to Mascarene islands trend of volcanic activity, 36 million years ago to present, to be left on the African plate (Fig. 1).

SEE ALSO THE FOLLOWING ARTICLES

Indian Region / Mascarene Islands, Biology / Volcanic Islands

FURTHER READING

Duncan, R. A. 1981. Hotspots in the southern oceans—an absolute frame of reference for motion of the Gondwana continents. *Tectonophysics* 74: 29–42.

Duncan, R. A., and M. A. Richards. 1991. Hotspots, mantle plumes, flood basalts, and true polar wander. *Reviews of Geophysics* 29: 31–50.

Gillot, P.-Y., and P. Nativel. 1989. Eruptive history of the Piton de la Fournaise volcano, Reunion Island, Indian Ocean. *Journal of Volcanology and Geothermal Research* 36: 53–65.

Mahoney, J. J., R. A. Duncan, W. Khan, E. Gnos, and G. R. McCormick. 2002. Cretaceous volcanic rocks of the South Tethyan suture zone, Pakistan: implications for the Reunion hotspot and Deccan Traps. *Earth and Planetary Science Letters* 203: 295–310.

McDougall, I. 1971. The geochronology and evolution of the young oceanic island of Reunion, Indian Ocean. *Geochimica et Cosmochimica Acta* 35: 261–270.

McDougall, I., and F. H. Chamalaun. 1969. Isotopic dating and geomagnetic polarity studies on volcanic rocks from Mauritius, Indian Ocean. *Geological Society of America Bulletin* 80: 1419–1442.

Morgan, W. J. 1981. Hotspot tracks and the opening of the Atlantic and Indian Oceans, in *The Sea*, Vol. 7. C. Emiliani, ed. New York: Wiley, 443–487.

Paul, D., W. M. White, and J. Blichert-Toft. 2005. Geochemistry of Mauritius and the origin of rejuvenescent volcanism on oceanic island volcanoes. *Geochemistry, Geophysics, and Geosystems* 6, doi: 10.1029/2004GC000883.

White, W. M., M. M. Cheatham, and R. A. Duncan. 1990. Isotope geochemistry of Leg 115 basalts and inferences on the history of the Reunion mantle plume. *Proceedings of the Ocean Drilling Program Scientific Results* 115: 53–61.

MAURITIUS

SEE INDIAN REGION

MEDITERRANEAN REGION

JOHN WAINWRIGHT
University of Sheffield, United Kingdom

Islands make up a significant component of the Mediterranean coastline, with about 19,000 km of length compared to the total of some 46,000 km. However, there are only five large islands—Sicily, Sardinia, Cyprus, Corsica, and Crete, in decreasing order—but numerous groups of smaller islands (Fig. 1). This article describes the groups and individual islands following a clockwise direction from the northwest.

GEOGRAPHIC SETTING

The Mediterranean Sea has an area of ~2.54 million km^2, with an average water depth of about 1500 m and a total volume of about 3.7 million km^3. It stretches from about 45°42′ N in the northern Adriatic to 30°15′ N off the coast

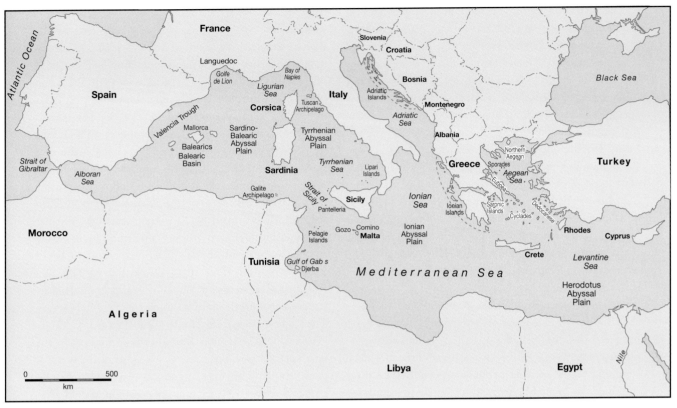

FIGURE 1 The Mediterranean Sea, showing the location of the main islands and groups of islands. Drawing by Paul Coles.

of Libya and from 35°57′ E off northern Lebanon. At its westernmost point, approximately 35°56′ N, 5°36′ W, it joins the Atlantic Ocean via the Strait of Gibraltar. Just over 14 km wide at its narrowest, the Strait is formed of a shallow sill no more than 400 m deep, which restricts inflow from the Atlantic to about 0.65–0.85 Sverdrup (i.e., 650,000–850,000 m³ per second). Freshwater inputs to the basin are dominated by a small number of major rivers fed from the mountains surrounding the sea and from the Nile and the Black Sea. The size of the Mediterranean and the dominance of the encircling mountains lead the region to have a characteristic "Mediterranean" climate, which is typified by hot, dry summers and relatively warm, wet winters with rainfall dominated by convective events. However, the broad definition of a Mediterranean climate conceals a great deal of variability, with both temperature and rainfall gradients from south to north and east to west. Within these broad gradients, west-facing land masses will also typically have greater rainfall because of the dominance of westerly air flows. As the Mediterranean is an enclosed basin with generally high temperatures and low inflows from both rainfall and runoff, evaporation rates are high. Circulation from west to east produces increasingly saline waters that eventually sink and return at depth in the currents of the Levantine Intermediate Water. Outflow occurs below the surface across the sill at the Strait of Gibraltar and amounts to about 0.6–0.8 Sverdrup (600,000–800,000 m³/s), reflecting the net evaporative loss across the basin.

Physiologically, the Mediterranean is divided into two large basins by the constriction of the Strait of Sicily. The position of islands and the irregular coastline further divides the area into more or less discrete units, with the Alboran Sea, Balearic Basin, the Ligurian Sea, and Tyrrhenian Sea in the west and the Adriatic, Ionian, Aegean, and Levantine Seas in the east. The western basin is the smaller, with a surface area of 0.87 million km², and descends to depths of 2806 m on the Sardino-Balearic Abyssal Plain and 3427 m on the Tyrrhenian Abyssal Plain. The continental shelf tends to fall off rapidly except for the area of the Alboran Sea, the Valencia Trough, Languedoc, and areas north and east of Corsica and south of Sardinia. The eastern basin has an area of 1.67 million m², and its deepest points are on the Ionian Abyssal Plain and the Herodotus Abyssal Plain, at 4140 m and 3219 m, respectively. There are large areas of much shallower water between Sicily and Tunisia and to the east of Tunisia, in the Adriatic, and parts of the Aegean. Elsewhere, the continental shelf again tends to fall off rapidly.

The formation of the Mediterranean is geologically complex and began around 260 million years ago with the breakup of the Pangaea supercontinent. This breakup produced at least 11 major crustal blocks as well as the Iberian subplate (which included the crust underlying Corsica and Sardinia) and the main African and Eurasian plates. Sea floor spreading produced a narrow ocean that originally linked the proto-Atlantic to the proto-Indian Oceans. This ocean has been called Tethys (after the woman in Greek mythology who was sister and wife of the god of the sea, Okeanus). Rotation of the African plate by about 65 million years ago produced a series of subduction zones that produced a contraction of the width of Tethys and collisional zones that produced most of the surrounding mountain ranges. The modern Mediterranean came into being around 9 million years ago with the final closure of the seaways to the east through what is now southeastern Turkey. A subsequent, relatively short-lived series of closures of the Strait of Gibraltar during the period from about 5.75 to 5.32 million years ago produced the so-called Messinian salinity crisis, in which large parts of the Mediterranean, and possibly the entire sea, dried up. Many Mediterranean rivers have overdeepened gorges as a result of this event, and the resulting evaporite deposits are now exposed in a number of areas, not least at the eponymous site on Sicily. The region is tectonically and volcanically active.

BALEARICS

The main island of the group is Mallorca, with an area of 3640 km^2 and a highest point of 1445 m. Menorca lies to the northeast while Ibiza to the southwest. There are a number of smaller islands in the group, the biggest of which are Formentera and Cabrera. Part of the Iberian subplate, which became sutured onto the rest of the Euroasian plate with the formation of the Pyrenees by 35 million years ago, the Balearics rotated clockwise away from the mainland by limited sea floor spreading in the Valencia Trough between around 23 and 10 million years ago.

The Serra de Tramuntana mountain range to the north and the Serres de Llevant in the east of Mallorca are dominated by limestones of Triassic to Cretaceous age. The central area is characterized by some Tertiary limestones as well as alluvial fan and other unconsolidated deposits. Menorca has a similar division between older limestones of the Tramuntana to the north and Tertiary sediments to the south. Ibiza similarly has a series of southwest-northeast-trending, uplifted limestone massifs with Tertiary basins between them.

CORSICA AND SARDINIA

Corsica and Sardinia were also originally part of the Iberian subplate. By about 35 million years ago, lateral movement along the Pyrenean fault zone had emplaced the area that is now Corsica to be adjacent to the Provençal coast of southeastern France and the present west coast of Sardinia to run across the Golfe du Lion. The spreading of the Valencia Trough from around 23 million years ago caused movement away from the French coast, and subsequent displacement of the spreading center to the east of the Balearics rotated Corsica and Sardinia away from these islands, forming the extensive, deep Sardino-Balearic Abyssal Plain by 10 million years ago. Further back-arc spreading over a similar time period created the Tyrrhenian Sea and Abyssal Plain, thus rotating Corsica and Sardinia away from what is now mainland Italy.

Corsica has an area of 8680 km^2 and is dominated by steeply sided, north-south-trending chains of mountains. The highest points are Monte Cinto (2706 m), Monte Rotondo (2622 m), and Monte Padro (2393 m) in the north-central part of the island. This part of the island is made up predominantly of granites as well as some gabbros and related volcanic rocks that date to around 280–250 million years ago. To the east are a series of schists that were formed during the early Tertiary, together with limited limestone outcrops. The narrow coastal plain on the eastern side of the island is composed of breccias and marls that date to around 18 million years ago.

At 23,813 km^2, Sardinia is the second largest of the Mediterranean islands. The eastern half of the island is composed of the same batholith structure that dominates western and central Corsica. The 11-km-wide Strait of Bonifacio separates the two islands. The highest points in Sardinia are in this eastern zone, reaching 1359 m at Punta Batestrieri in the north and 1834 m at Punta La Marmora in the center of the island. Granites and related igneous rocks are again intruded into an early Palaeozoic metamorphic basement. The southwest of the island is also made up of similar granites and metamorphic rocks, reaching a high point of 1017 m at Punta Maxia. Uplift and thrusting in this zone were the result of the Pyrenean faulting of the early Tertiary. The remainder of the eastern part of the island is made up of Jurassic-Cretaceous limestones overlain by Tertiary molassic sediments. Rifting in the southwest produced the Campidano Graben during the Late Oligocene to Early Miocene, with the opening of the Sardino-Balearic Basin. The graben contains carbonate deposits as well as subsequent conglomerates, and there are related Tertiary volcanics in the western part of the island.

ISLANDS OF THE TYRRHENIAN SEA

Islands in the Tyrrhenian can be divided into three groups. The Tuscan Archipelago is located between the northeast of Corsica and the Italian coast. The main island is Elba (224 km^2), which has a complex geology with multiple nappes relating to the Apennine development and intrusive and volcanic igneous activity dating to the later Tertiary. The island was known from prehistory for its copper and iron sources. The other islands of the archipelago—Giglio, Capraia, Pianosa, Gorgona, Montecristo, and Giannutri—have late Tertiary volcanic origins and are all less than 24 km^2 in area.

The Bay of Naples has a number of small islands, again of volcanic origin. Ischia is the largest, with an area of 46 km^2, and reaching a maximum height of 788 m at Mount Epomeo. The islands of Procida and Capri also belong to this group. This volcanic activity is ongoing on the nearby mainland at Vesuvius, which also relates to the subduction of the Ionian Abyssal Plain beneath southern Italy.

The Lipari or Aeolian Islands are also related to this subduction zone. It contains the eponymous island of Vulcano (from Vulcan, the Roman god of fire and metalwork), with its typical moderately explosive activity. Further to the north, Stromboli exhibits the typical, almost continuous, rhythmic eruptions of this type of volcano. Other islands in the group are Lipari (the largest at 38 km^2), Salina, Panarea, Alicudi, Filicudi, and Ustica. Volcanic activity is recent and ongoing.

SICILY

Sicily's 25,700 km^2 make it the largest of the Mediterranean islands. It rises to a maximum in the east of the island of 3323 m at the summit of Mount Etna, which has built up over the last 700,000 years in a series of eruptions producing subalkaline lavas. Major eruptions have continued to occur through the historical period, some of which have been highly explosive and caused widespread damage. The volcano is also an important global source of CO_2, producing on average 26 million metric tons per year, and of SO_2, with an average production of 0.6 million metric tons per year.

The north of the island is dominated by the Peloritani, Nebrodi, and Madonie mountains. The mountains are principally of limestone with some metamorphic rocks, and relate to Tertiary thrusting events as part of the development of the Apennine chains through Italy. The south-central and northeastern tip (around the type location of Messina) of the island contain extensive evaporitic deposits relating to the Messinian salinity crisis in the Mediterranean, described earlier in this article. These deposits are locally overlain by extensive Pliocene and Pleistocene sediments.

ISLANDS OF THE STRAIT OF SICILY

Malta (245 km^2), Gozo (65 km^2), and Comino (4 km^2) are located about 80 km to the south of Sicily, from which they are separated by the Malta Channel. Both are composed of more or less horizontally bedded coralline limestone overlain by Globigerina limestone and blue clays and topped by a further coralline limestone. Perched water tables in the upper coralline limestone have almost totally been exhausted by overpumping, and groundwater tables in the lower coralline limestone are becoming increasingly vulnerable to seawater intrusion.

Pantelleria, located midway between southwest Sicily and Tunisia, is a volcanic island of 83 km^2 composed of basalts, pantellerites (alkaline rhyolites), trachytes, and tuff; there have been a complex series of eruptions since ~320,000 years ago. Most volcanic rocks exposed date to eruptions over the last 50,000 years. Obsidian is found on the island and was widely distributed in the western Mediterranean in the Neolithic period.

The Pelagie Islands, made up of Lampedusa and the smaller islands of Linosa and Lampione, are also volcanic but sit on the edge of the African continental shelf. They are important sites for the endangered loggerhead turtle.

ADRIATIC ISLANDS

There are approximately 1250 islands in the Adriatic, mostly located along the eastern edge. They are predominantly located off the coast of Croatia, with smaller numbers off Bosnia-Herzegovina and Montenegro. Krk and Cres are both located towards the north of the chain and are approximately the same size at 405 km^2. Of the other main islands, Pag, Brač, Hvar, and Korčula are the biggest. Most of the islands have steep slopes and are deeply incised. They are composed of Jurassic and Cretaceous limestones and dolomites and exhibit extreme karstic features.

IONIAN ISLANDS

The Ionian Islands are mainly located off the west coast of the Greek mainland. Kerkira (Corfu) is the northernmost, 180 km^2, with Paxos (80 km^2), Levkas (360 km^2), Ithaka (120 km^2), Cephalonia (910 km^2), and Zakinthos (or Zante, 405 km^2) progressively further south. Many of the islands were extensively affected by a major earthquake occurring in 1953. Kithira (280 km^2) is separate from the main group of Ionian Islands; it is located off the southern Peleponnese and joined via a submarine ridge to Crete. The Ionian geotectonic zone also continues onto the Peleponnese mainland and consists of a series of Triassic evaporites, Jurassic-Cretaceous limestones in extensional terranes, and Tertiary flysch deposits, reflecting the compression of the Alpine phase.

CRETE

Crete is bounded to the south by the subduction zone of the Hellenic Trench, which forms the deepest part of the Mediterranean (at 4661 m) to the west of the island. The island covers an area of 8340 km² and is principally composed of east-west-trending high mountains. There are three principal massifs of crystalline limestone: the White Mountains (Leuka Ori) in the west, peaking at 2452 m; Mount Idhi or Psiloritis in the center (2456 m); and the Dikti Mountains in the east (2148 m). These Triassic, Jurassic, and Eocene limestone areas are made up of high plateaux, numerous caves and other karstic features, and deep gorges. Locally, there are upthrust phyllites and quartzites, as well as flysch deposits. Lower areas were infilled with clays, marls, and conglomerates in the Tertiary, and there are localized Quaternary fan and gravel deposits. Igneous rocks are sparsely distributed through the island, and the southeast and central parts of the island are made up of Jurassic ophiolites. There is a small area of gneiss near the southernmost tip of the island.

AEGEAN ISLANDS

Although there are approximately 1200 islands in the Aegean, most of them are very small (only 50 are larger than 40 km²), like the Adriatic islands, and are sparsely inhabited or uninhabited. The region as a whole is undergoing extensional tectonics as a result of indentation from the Turkish subplate, and there are extensive areas of normal faulting as a result. The islands can be considered in six main groups. The Cyclades are located in an arc approximately 110–250 km to the north of Crete. A number of these islands are volcanically active and are related to the subduction of oceanic crust at the Hellenic Trench to the south of Crete. The volcanic islands include Milos (another prehistoric source of obsidian), Antiparos, and Santorini. The latter surrounds an 85 km² caldera that is the result of the prehistoric megaeruption of Thera *ca.* 1629 BCE, which buried the Bronze Age city of Akrotiri on the island to a depth of 7 m; ejected an estimated 13–40 km³ of material, some of which reached as far as the Black Sea, eastern Turkey, and the Nile Delta; and possibly caused a tsunami that destroyed coastal settlement on Crete. Islands closer to the subduction zone such as Santorini and Milos tend to produce calc-alkaline volcanics, while Antiparos is made up of alkaline lavas. Some of the larger islands such as Naxos (430 km²) and Andros (380 km²) are made up of high-grade metamorphic rocks, such as blueschists and greenschists, that formed through the Cretaceous and Tertiary and have been uplifted as a result of back-arc extension. Naxos and Paros also have ophiolites exposed. Other parts of the group are made of limestone with karst morphology well developed.

The Saronic Islands are located in the Saronic Gulf between the Attic peninsula and the Peleponnese. There are two principal islands of the group. Aegina (85 km²) is an extension of the calc-alkaline volcanics in the Cycladic arc, while Salamis (96 km²) is essentially a drowned part of the mainland, located only 2 km from the port of Piraeus.

Evvoia (Euboea) is aligned on a similar southeast-northwest axis as the Cycladic islands of Andros, Tinos, and Mikonos from which it extends, and runs subparallel to the mainland Greek coastline of Attica, Boeotia, and Phthiotis. Evvoia has an area of 3685 km², and its highest point is Mount Dirphys at 1745 m. At the Euripus Strait, Evvoia is only 40 m away from the mainland, and the island would have been connected to the mainland during lower sea levels in the Pleistocene. The island has a complex geology, with thrusting juxtaposing shallow-water limestones of Jurassic age, with Triassic basalts and metamorphic rocks including ophiolites, and sediments.

The Sporades are composed of four main islands—Alonnisos (65 km²), Skiathos (50 km²), Skopelos (95 km²), and Skiros (210 km²)—and 20 smaller islands, to the north and east of Evvoia. They are dominated by calc-alkaline igneous rocks relating to Hellenic Trench subduction.

The Dodecanese occur to the east of the Cyclades and close to the Turkish mainland. Again, some of the northern part of the archipelago relates to volcanic activity from subduction at the Hellenic Trench. Active alkaline volcanics are found on Kos (290 km²), while potassic lavas dominate further south at Yali (20 km²) and Nisyros (42 km²). Rodhos (Rhodes) is the largest island at 1400 km², reaching a peak of 1216 m on Mount Attavyros. The uplands are dominated by Mesozoic limestones with karst morphology and deeply incised gorges, and there are Tertiary flysch and molasse deposits as well as ophiolites to the west of the island. These ophiolites as well as those on the neighboring island of Karpathos are late Cretaceous and thus more closely linked to the ophiolites of Turkey than to those of Crete, the Cyclades, and the Balkans. Lower-lying areas to the north of the island have been infilled with Plio-Pleistocene clastic sediments and shallow-water limestones. Karpathos also exhibits karst morphology in its limestone areas.

Northern Aegean islands include Samos (480 km²), with recent alkaline volcanics relating to Hellenic Trench subduction. Further north, Khios (Chios, 840 km²) preserves the remains of an earlier collisional belt, with turbidites and phyllites dating to the Early Carboniferous. Overlying these are Mesozoic limestones with developed karst; there are also middle Miocene andesites, basalts, and alkaline and calc-alkaline rhyolites in the southeast of the island. Lesvos (1630 km²) can be broadly divided into two parts. To the northwest are early Miocene rhyolites and pyroclastic deposits; to the southeast, ophiolites separate the younger volcanics from metamorphic rocks including marbles, schists, quartzites, and phyllites. Limnos (480 km²) also contains Miocene volcanics above localized Eocene marls and sandstones. There are basaltic, andesitic, and trachyandesitic lava flows and tuffs and small outcrops of quartz monzonite. Gökçeada (280 km², also known as Imbros or Imroz) is composed of an Eocene sedimentary sequence passing from turbidites to nearshore and fluvial sediments overlain by limestone. Locally there are Miocene potassic volcanics. Samothraki (180 km²) also has potassic volcanics, Miocene granites, and ophiolites. The furthest north of the Aegean islands, Thasos (380 km²), is 7 km from the mainland coast of Macedonia. It is dominated by metamorphic rocks including gneiss, marble, and migmatites. The metamorphism is Oligocene to Miocene in age and relates to the development of the Rhodope metamorphic core complex on the adjacent mainland. In antiquity, the island was well known for its gold mines.

CYPRUS

Cyprus is the third largest of the Mediterranean islands, with a surface area of 9250 km². Except for the small islands of Arwad in Syria and Tyre in Lebanon (joined to the mainland since Alexander the Great built a causeway after conquering the city), it is also the furthest east of the Mediterranean islands. There are two main mountain ranges: the Troödos to the center west and the Kyrenia range running along the northern edge of the island. Between these are the lower-lying plains of the Mesaoria. The summit of Mount Troödos is at 1952 m and is composed of an Upper Cretaceous ophiolite suite surrounded by pillow-lava basalts. The area is extensively mineralized and has some of the world's major sources of copper (the name for which comes from the Roman name for the island) as well as iron, nickel, cobalt, and chrome. To the west of Troödos, the Mamonia terrane contains Palaeozoic and Cretaceous metamorphic rocks and limestones. The lower parts of the south of the island are made up of Upper Cretaceous to mid-Tertiary limestones and marls.

The Kyrenia or Pentadaktylos Range of the north of the island is composed of Permian to mid-Cretaceous limestones. These rocks are steeply dipping following the compression that closed this part of the Tethys Ocean in the late Cretaceous and early Tertiary, thrusting up material from the former sea bed. Some scattered late Miocene evaporites are found in the west and south. The central part of the island has unconsolidated marls and clastic sedimentary deposits from the Pliocene and Quaternary.

ISLANDS OF THE NORTH AFRICAN COAST

The North African coast has relatively few islands compared to the rest of the basin, and none are of any real size. The Gulf of Gabès in eastern Tunisia is the location of Jerba and the Kerkennah Islands. Jerba has an area of 515 km² and is the only African Mediterranean island with any significant human population. It is just over 2 km from the mainland and has a maximum elevation of 53 m. The island is made up of Miocene and later sediments and is adjacent to the major offshore natural gas field of Ezzaouia. The Kerkenneh Islands are a small archipelago about 20 km from the mainland at Sfax. They are dominated by Quaternary sands and typically only a few meters above sea level. Along the north Tunisian coast there are the small island of Zembra in the entrance to the Bay of Tunis, and further west the Galite Archipelago, which is about 40 km from Cape Serrat. The six islands are made of granite and related volcanics.

Other small island groups are the Habibas off Oran (Algeria) and the Chararinas Islands off Morocco. Finally, Alborán, in the center of the sea of the same name, is located about 50 km off the Moroccan coast and 80 km from Spain.

GEOMORPHOLOGY

Surface features on the islands are dominated in nonlimestone terrains by intense surface erosion. The dominance of convective storm events, coupled with the relatively sparse vegetation cover, produces rapid surface runoff. Soil erosion accelerates rapidly where rills and gullies form. Coupled with the active tectonic regime over most of the region, these processes produce typically steep slopes, and the lack of significant shelf areas around most islands means that there is often little in the way of fan or alluvial plain buildup in the coastal zone. Wildfire is also an important process that acts to remove vegetation cover and thus increase the potential for erosion. The west-east trend of rainfall suggests that there should be an asymmetry in erosion rates, with the western side of most islands exhibiting more erosion, but this pattern is complicated not only by high geological and soil variability but also by the fact that enhanced water supply will tend to increase vegetation cover and thus reduce this effect. Many of the steeper slopes have been terrraced so that agriculture can be carried out without too high soil loss rates. Landslides are also a common feature, especially in unconsolidated Neogene sediments, and rockfalls are also common in limestone areas, especially in areas of oversteepened slopes. Local badlands also form (e.g., on Sardinia) as a result of intense erosion, usually in the areas of Neogene sediments. Rivers are typically steep and short, with gravel beds. In many areas the irregular rainfall and high temperatures mean that rivers are ephemeral and the only permanent water supplies are from groundwater.

In limestone areas, karst topography often dominates. Rapid infiltration produces little in the way of surface runoff, especially once soil cover is removed. Residual soils are often found in deep solutional hollows. On many of the islands of the eastern Mediterranean, but also on Sardinia and Corsica, dust deposited on the surface carried from the Sahara is often a significant soil-forming material. It has often been used to explain the distinctive bright red color of many *terra rossa* soils formed on limestone in the Mediterranean. Crete and some of the larger Adriatic islands have extensive upland plateaux with limestone pavement and poljes. Cave systems and deep-cut river gorges are typical of these landscapes, and rivers on the larger islands may be perennial here because they are fed from groundwater from deep springs, as evaporative loss is lower.

The more upland areas exhibit relict periglacial features such as rock debris slopes, from the cold periods of the Pleistocene, but none of the islands are sufficiently high to have been glaciated, unlike parts of the adjacent mainland. A more important consequence of Quaternary climate change is the sea-level change and its effect on joining present-day islands to the mainland, with important consequences for their biogeography as well as for oceanic circulation patterns. For example, Kerkira and Levkas in the Ionian Islands and Jerba would have been joined to the mainland in this way. Of the larger islands, only Sicily would have become joined to the mainland along a narrow isthmus, as the shallowest part of the Strait of Messina is approximately 70 m deep. Many of the Cyclades would have been joined into a single larger island under such conditions, although there was never a complete land bridge across the southern Aegean during the Pleistocene either by this route or via Crete. The Black Sea was cut off from the Mediterreanen during lowstands. Many islands exhibit wave-cut notches from higher sea levels both in the past interglacials and the mid-Holocene. Raised (and drowned) beaches and cliff lines are frequent because of neotectonic movements. Coastal erosion is reduced in many areas because of the very low amplitude of the tidal range in the Mediterranean.

ECOLOGY

The Mediterranean has a wide range of species that are considered to be characteristic, although close observation of their extents shows that most are more restricted by other conditions rather than a strict adaptation to drought and other features of the Mediterranean climate. The highly accentuated relief on most islands leads to a strong vertical distribution of plants. Of the arboreal species, Aleppo pine (*Pinus halepensis* or other pines such as *P. pinaster* or *P. pinea*) and holm oak (*Quercus ilex*) tend to dominate below 500–600 m above sea level. From about 500 m to 1500 m, they may be joined with white (deciduous) oak (*Q. pubescens*) and chestnut (*Castanea sativa*), then beech (*Fagus sylvatica*) above 1400 m, and finally Scots pine (*Pinus sylvestris*) and fir (*Abies* spp.). Some islands have species that are endemic, for example Corsican pine (*P. laricio*), or only found in Europe on certain islands; for example, Cretan pine (*P. brutia*) and *Zelkova* are restricted to Crete. Nonarboreal species are also commonly endemic—about 10% of the Cretan flora is endemic. Plants may also be highly specialized to local conditions even within the same group of islands. Within the Balearics, 40 endemic taxa are only found on a single island in the archipelago, for example. Even before extensive human disturbance, forest cover may not have been extensive in the Mediterranean because of frequent disturbance from drought and fire. At lower levels, a plagioclimax community tends to develop, with tree species occurring in shrubby forms or even with just shrub forms (for example, the kermes oak *Q. coccifera*). These two stages are usually characterized as *maquis* (or *macchia* in Corsican dialect) and *garrigue* after the French terms. In Spanish, the use is to call both *matorral,* while *phrygana* in Greek is close to *garrigue*. The extent to which garrigue replaces maquis is most likely related to the frequency of disturbance, including human-induced. Islands where the geology and water availability has limited soil production may be reduced to desert-like conditions with bare or steppe grassland surfaces.

Because of the isolation of many of the Mediterranean islands, animal species also exhibited a high degree of endemism, although many of these species have subsequently become extinct following human settlement. Cyprus was home to species of pygmy hippopotamus (*Hippopotamus minutis*) and elephant (*Elephas cypriotes*), which seem to have been hunted to extinction in the late Pleistocene or early Holocene. Dwarf forms of these animals (*H. melitensis* and *E. falconeri*) have also been found in Pleistocene sediments on Malta, Crete (*H. creutzbergi*), and Sicily (*H. pentlandi* and *E. falconeri*). *Myotragus balearicus* is a now-extinct dwarf antelope that inhabited the Balearic Islands. On Sardinia and Corsica, an endemic rabbit (*Prolagus sardus*) survived into the Holocene, and there are numerous smaller endemic species throughout the Mediterranean islands. Not all species that have been considered endemic have proved to be so. The mouflon of Corsica and Sardinia is now thought to descend from feral sheep following human settlement in the Neolithic.

HUMAN SETTLEMENT OF THE MEDITERRANEAN ISLANDS

Of the major islands, only Sicily seems to have been settled through the Pleistocene, although during the colder periods, there would have been a connection to the mainland. Seafaring is attested for the last few millennia at least of the Pleistocene, by the occurrence of obsidian from Milos on the Argolid mainland of Greece. Cyprus was first occupied around 12,300–11,200 BCE, and the island may have undergone several phases of settlement, as at present there are hiatuses in the early Neolithic archaeological record of the island. Early settlers from the Levantine mainland introduced sheep, cattle, pigs, and deer. Crete does not seem to have been settled until 7000–6200 BCE, and, as on many of the Mediterranean islands, the settlement seems related to the development of simple agriculture in the Neolithic. Sicily, Kerkira, and the Dalmatian islands seem to have been settled around the same time, and these settlements may reflect underlying seafaring activities of hunter-gatherer populations in the Mediterranean. Mesolithic populations are known from slightly earlier than this date on Corsica and Sardinia, and shortly afterward agriculture seems to have been practiced on these islands. Malta was settled by the start of the sixth millennium BCE, and Pantelleria was known as a source of obsidian in the middle Neolithic, indicating that even the smallest of islands were visited. Although there were also previous explorations of the Balearics, agriculture arrived fairly late on the island in the second half of the fourth millennium BCE. From the later Neolithic onward, settlement frequently intensified on the islands, and as elsewhere in the Mediterranean this led to increased disturbance of the vegetation and often the acceleration of erosion rates.

FURTHER READING

Allen, H. D. 2000. *Mediterranean ecogeography.* London: Prentice-Hall.
Blondel, J., and J. Aronson. 1999. *Biology and wildlife of the Mediterranean region.* Oxford: Oxford University Press.
Grove, A. T., and O. Rackham. 2001. *The nature of Mediterranean Europe: an ecological history.* New Haven, CT: Yale University Press.
Ricou, L. E. 1994. Tethys reconstructed: plates, continental fragments and their boundaries since 260 Ma from Central America to south-eastern Asia. *Geodinamica Acta* 7: 169–218.
Thompson, J. D. 2005. *Plant evolution in the Mediterranean.* Oxford: Oxford University Press.
Wainwright, J., and Thornes, J. B. 2003. *Environmental issues in the Mediterranean: processes and perspectives from the past and present.* London: Routledge.
Woodward, J. C., ed. 2009. *The physical geography of the Mediterranean.* Oxford: Oxford University Press.

METAPOPULATIONS

DAG ØYSTEIN HJERMANN

University of Oslo, Norway

Metapopulations are "populations of populations"—collections of island populations bound loosely together by occasional migration between the islands. Metapopulation theory has been applied to terrestrial environments (where the "islands" are patches of habitat in an "ocean" of unsuitable habitat) as well as to real island archipelagoes. The theory is especially useful in conservation biology.

THE CONCEPT

A basic concept of MacArthur and Wilson's theory of island biogeography was that the collection of species found on an island is dynamic. The metapopulation theory, first used by Richard Levins in 1970, shares this dynamic view of animal populations. In contrast to MacArthur and Wilson's theory, few people paid much attention to metapopulation theory before about 1990. It differs from the theory of island biogeography in two ways. First, it focuses typically on just one species (animal or plant). This species inhabits islands—which may be real islands in the ocean, or terrestrial or oceanic habitat islands (in an "ocean" of unsuitable habitat). Secondly, there is no mainland where the species is constantly present. Each island holds only a small population of the species, so the population of a specific island will sooner or later die out. However, from time to time, animals are able to migrate from an inhabited island to an uninhabited one and establish (or reestablish) a new population (Fig. 1). As in MacArthur and Wilson's theory, over time, the system will reach equilibrium. In the metapopulation case, the fraction of islands that are inhabited by the species is roughly constant over time; however, specifically *which* islands are inhabited by the species will change over time. The scale of a metapopulation can be anywhere from millimeters to thousands of kilometers, depending on the migration capability of the species.

MATHEMATICAL MODEL

Although it is not essential for understanding, Levins's original mathematical model of a metapopulation may shed further light on metapopulation theory. Let us say that there are 100 identical islands with suitable habitat for a given species. At a given moment, p islands are occupied by different populations of the species (whereas

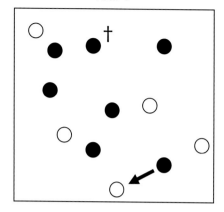
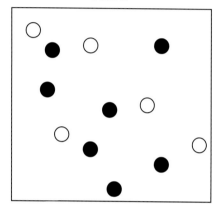

FIGURE 1 A schematic map of a metapopulation showing islands with (•) and without (○) a population of a certain species. In year 1, one population becomes locally extinct (marked with a †), but this is balanced by a colonization of an empty island from a neighboring populated island (shown with an arrow).

$100 - p$ are "empty"). For a single occupied island, the probability that the population becomes extinct during a time step (say, one year) is e. For the entire metapopulation, $p \times e$ extinctions occur every year (the number of islands multiplied by each island's extinction probability). The probability that animals from a specific occupied island will colonize a specific "empty" island is c. The number of colonization events in the entire metapopulation during one year is $p \times (100 - p) \times c$, depending on the number of "donor" and "recipient" islands as well as the colonization frequency. At equilibrium, the number of extinction events necessarily equals the number of colonization events: $p \times e = p \times (100 - p) \times c$. Solving this equation for p, we find that at equilibrium, the number of islands that are occupied is $p = 100 - e/c$. Thus, the entire metapopulation can be maintained only if the colonization rate c is high enough (in this case, $c > 0.01e$). Metapopulation models have, since 1980, been developed to be more realistic, for instance by allowing for differences in island size and isolation.

LOCAL EXTINCTION

For most metapopulation theories, it is assumed that the habitat is suitable at all times; that is, extinction events are stochastic (the model can be modified to include deterministic extinction, for instance if the vegetation over time changes to be unsuitable). Four types of processes can lead to stochastic extinction events: (1) demographic stochasticity (caused by inherent variation in, for example, litter size and sex ratio), (2) environmental stochasticity (e.g., variation in survival and recruitment as a result of weather conditions), (3) genetic stochasticity (the loss of genetic variance by genetic drift and inbreeding), and (4) catastrophes (e.g., catastrophic weather events). The risk of population extinction as a result of the first three extinction types increases strongly as the population becomes smaller, whereas the fourth is quite independent of population size. Also, immigration from other islands may rescue that population from extinction ("the rescue effect"). In practice, habitat quality may differ between islands, so some islands (sinks) have a tendency toward extinctions, which is balanced by inflow from islands with high habitat quality (sources).

PRACTICAL USE OF THE THEORY

Metapopulation theory has a number of important implications, especially in the field of conservation biology. Threatened species are often found as small, scattered populations on the brink of extinction, a population structure that typically is the result of fragmentation of the species habitat by human land use. Thus, the metapopulation concept is well suited to such species. One implication of metapopulation theory is that every habitat island counts if we want to save a threatened species—even habitat islands where the species currently does not exist. Because the species survive by recolonizing new habitat islands as fast as local populations go extinct, the entire collection of habitat islands is vital to avoid regional extinction in the long term. A second message of metapopulation theory is that destruction of habitat has a non-linear effect on species abundance and long-term survival: As habitat area is reduced, the species will become regionally extinct long before total habitat area reaches zero. A third message is that migration is important: Without migration, regional extinction is inevitable.

It is important to realize, however, that a species that lives in patchy environments does not necessarily function as a metapopulation. Species with good migration capabilities, such as many birds, easily travel between the islands of an oceanic archipelago. For such species, the entire archipelago functions as a single habitat for a single population.

SEE ALSO THE FOLLOWING ARTICLES

Dispersal / Extinction / Fragmentation / Island Biogeography, Theory of / Population Genetics, Island Models in

FURTHER READING

Fahrig, L. 2002. Effect of habitat fragmentation on the extinction threshold: A synthesis. *Ecological Applications* 12: 346–353.
Hanski, I. 1999. *Metapopulation ecology.* Oxford: Oxford University Press.
Hanski, I. A., and M. E. Gilpin, eds. 1997. *Metapopulation biology: ecology, genetics, and evolution.* San Diego, CA: Academic Press.
Harrison, S. 1991. Local extinction in a metapopulation context: an empirical evaluation. *Biological Journal of the Linnean Society* 42: 73–88.
Hastings, A., and S. Harrison. 1994. Metapopulation dynamics and genetics. *Annual Review of Ecology and Systematics* 25: 167–188.
McCullough, D. R., ed. 1996. *Metapopulations and wildlife conservation.* Washington, DC: Island Press.

MICRONESIA

SEE PACIFIC REGION

MIDWAY

ELIZABETH FLINT

U.S. Fish and Wildlife Service, Honolulu, Hawaii

Midway atoll (28°15′ N, 177°20′ W) consists of three sandy islets (Sand Island: 4.56 km², Eastern Island: 1.36 km², and Spit Island: 0.05 km²), for a total of 5.98 km² in terrestrial area, lying within a large, elliptical barrier reef measuring approximately 8 km in diameter (Fig. 1). Although geographically part of the Hawaiian archipelago, Midway is not part of the State of Hawaii and is an unincorporated territory of the United States.

CLIMATE

The climate of Midway is influenced by the marine tropical or marine Pacific air masses, depending upon the season. During the summer, the Pacific high pressure system becomes dominant with the ridge line extending across the Pacific north of Kure and Midway. This places the region under the influence of easterly winds, with marine tropical and trade winds prevailing. During the winter, especially from November through January, the Aleutian low moves southward over the North Pacific, displacing the Pacific high before it. The Kure-Midway region is then affected by either marine Pacific or marine tropical air, depending upon the intensity of the Aleutian low or the Pacific high pressure system.

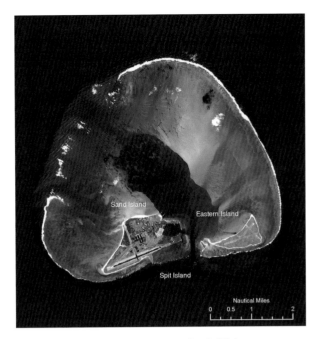

FIGURE 1 Midway atoll. Image courtesy of DigitalGlobe.

GEOLOGY

Nowhere else on the planet is the tropical island evolution process, with examples of every stage of development, illustrated so beautifully and linearly as in the northwestern Hawaiian Islands. The 1200-mile-long string of islands represents the longest, clearest, and oldest example of island formation and atoll evolution in the world. The ten islands and atolls extending northward from Kauai represent a classic geomorphological sequence, consisting of highly eroded high islands, near-atolls with volcanic pinnacles jutting from surrounding lagoons, true ring-shaped atolls with roughly circular rims and central lagoons, and secondarily raised atolls, one of which bears an interior hypersaline lake. These islands are also surrounded by over 30 submerged banks and seamounts. This geological progression along the Hawaiian ridge continues northwestward beyond the last emergent island northwest of Midway, Kure atoll, as a chain of submerged platforms. Numerous patch reefs formed by reef-building coralline algae and 16 species of corals provide habitat for a wide variety of coral reef species.

Midway is at the northern end of the Hawaiian archipelago. The atoll, which is 28.7 million years old, is surrounded by more than 356 km² of coral reefs.

FLORA AND FAUNA

The biota of Midway is a combination of native and introduced species that are a product of its land-use history. The islands boast enormous nesting colonies of Laysan albatrosses (*Phoebastria immutabilis*; ~450,000 breeding pairs) (Fig. 2) and black-footed albatrosses (*Phoebastria nigripes*; ~24,000 breeding pairs), making the atoll the largest single colony of albatrosses in the world. Another 14 species of tropical seabirds also breed in abundance at Midway. Introduced canaries breed among historic buildings that mark the beginning of cable communication across the Pacific in the early twentieth century.

FIGURE 2 Laysan albatrosses (*Phoebastria immutabilis*) incubating on Sand Island, Midway atoll. Photograph by U.S. Fish and Wildlife Service.

Currently, the land cover on all of the islands at Midway is approximately 30% paved or structures, 23% grass and forbs, 18% woodland, 7% sand and bare ground, 22% shrublands, and less than 0.23% wetland. Of the at least 354 species of plants that have ever been observed at Midway, only 14 are considered indigenous, and only three of those are endemic to the Hawaiian Islands. There are no plant species considered endemic to Midway atoll alone. A total of 508 species of terrestrial arthropods are listed from Midway. Only 41 of these species (8%) are endemic to Midway, and 50 additional species (10%) are indigenous to the tropical Pacific. The large number of alien species that have been introduced and the failure to re-collect three of the four native seed bugs and all three native moths in recent years suggests that the high rate of alien species introductions has reduced the numbers of native insects.

Hawaiian monk seals and green turtles forage in the waters offshore but come to the sandy beaches of the atoll to breed. There are 163 species of reef fish reported from Midway, and the shallow-reef fish community is remarkable in the abundance and size of fish in the highest trophic levels. Recent research efforts in the Hawaiian Islands and the Line archipelago have demonstrated that coral reef communities with low levels of human exploitation and disturbance are characterized by fish communities dominated by apex predators.

HUMAN HISTORY

There are no records yet discovered of Polynesian visits to Midway, but Captain N. C. Brooks of the *Gambia* first claimed it under the Guano Act of 1856 after discovering it on July 5, 1859. Midway atoll's central location in the Pacific made it a critical link in communications and transportation history in the Pacific in the early twentieth century. One of the most important battles of World War II in the Pacific, the Battle of Midway, was fought both at Midway atoll and in the waters beyond it. Midway continued to have a significant military role after World War II; it was an active Navy installation during the Cold War and served as an aircraft and ship refueling station during the Vietnam War.

CONSERVATION

Noteworthy is the role that Midway played in the early history of wildlife conservation in the United States. In 1903 President Theodore Roosevelt sent in the U.S. Marines to stop the slaughter of seabirds at Midway atoll by feather hunters.

In 1988 the atoll was designated an overlay national wildlife refuge, and in 1993 the Navy closed the naval air facility and embarked on a major environmental cleanup in which many buildings and underground fuel tanks were removed. In 1996 the Navy formally transferred Midway atoll and the ocean waters out to 19 km to the Department of Interior as Midway Atoll National Wildlife Refuge. In 1999, the U.S. Navy identified the Battle of Midway as one of the two most important events in its history, and in 2000, the U.S. government designated Midway as the Battle of Midway National Memorial. Papahānaumokuākea Marine National Monument was established by Presidential Proclamation 8031 on June 15, 2006. This resulted in bringing waters out to 80 km around Midway atoll, along with the rest of the northwestern Hawaiian Islands waters, into the largest fully protected marine conservation area in the world.

Midway's terrestrial native vegetation and insect communities have been greatly altered by more than a century of human occupation. The U.S. Navy, the U.S. Fish and Wildlife Service, and the U.S. Department of Agriculture successfully eradicated black rats (*Rattus rattus*), introduced in 1943, from all of Midway, and invasive ironwood trees (*Casuarina equisetifolia*) have been entirely removed from Eastern Island. An active program of invasive weed eradication and native plant propagation is ongoing to restore the plant community present prior to human occupation. This has improved living conditions for the 16 species of tropical seabirds that breed at the atoll as well as for migrant shorebirds such as bristle-thighed curlews (*Numenius tahitiensis*) and Pacific golden plovers (*Pluvialis fulva*) that winter there. A translocated population of Laysan ducks (*Anas laysanensis,* an endangered species whose sole remaining population was at Laysan Island until the translocation) is thriving as it forages on the introduced insect community at Midway. Tragically, the Laysan rail population that was moved to Midway prior to its extirpation at Laysan Island in the 1920s went extinct in 1943 when *Rattus rattus* arrived at Midway and before individuals could be returned to repopulate Laysan Island.

Today, Midway serves as a window to the rest of the Marine National Monument as the only atoll in the chain open for public visitation. It is the site of active ecological restoration efforts and research for conservation science.

SEE ALSO THE FOLLOWING ARTICLES

Atolls / Hawaiian Islands, Geology / Invasion Biology / Reef Ecology and Conservation / Seabirds

FURTHER READING

Amerson, A. B., Jr., R. B. Clapp, and W. O. Wirtz II. 1974. The natural history of Pearl and Hermes Reef, northwestern Hawaiian Islands. *Atoll Research Bulletin* 174. Washington, DC: Smithsonian Institution.

Clague, D. A. 1996. Growth and subsidence of the Hawaiian-Emperor volcanic chain, in *The origin and evolution of Pacific Island biotas, New Guinea to eastern Polynesia: patterns and processes*. A. Keast and S. E. Miller, eds. Amsterdam: SPB Academic Publishers, 35–50.

Healy, M. 1993. *Midway 1942, turning point in the Pacific*. Oxford, UK: Osprey Publishing.

Hinz, E. 1995. *Pacific Island battlegrounds of World War II: then and now*. Honolulu, HI: The Bess Press.

Nishida, G. 1998. *Midway terrestrial arthropod survey*. Final report prepared for the U.S. Fish and Wildlife Service. Honolulu, HI: Hawaii Biological Survey, Bishop Museum.

Rauzon, M. J. 2001. *Isles of refuge: wildlife and history of the northwestern Hawaiian Islands*. Honolulu: University of Hawaii Press.

MIGRATION

SEE DISPERSAL

MISSIONARIES, EFFECTS OF

ALAN I. KAPLAN
El Cerrito, California

VINCENT H. RESH
University of California, Berkeley

Missionaries have gone to islands to convert local populations away from indigenous religions for millennia. They have been successful because of geographical isolation and cultural aspects of island life, as well as internal and external political influences. The activities of missionaries on islands have resulted in improvements in health, agricultural development, and education but also have brought about drastic cultural changes. Missionaries often had positive influences on island peoples' lives but, having the range of human frailties, sometimes also did irreparable harm.

ISLANDS AND MISSIONS

Islands have been the recipients of a great deal of missionary activity relative to the size of the area they comprise worldwide, although missionary efforts have been far greater in China, Africa, India, and the Americas than on oceanic islands. Missionaries have been going to islands in their search for converts for millennia, proselytizing for Buddhism (the Indian Emperor Asoka sent Buddhist missionaries to Sri Lanka in the third century BC) and Islam (the ruler of Brunei invited Islamic missionaries there after his conversion from Hinduism around 1425) as well as for Christianity, as is most familiar today. As an early example of the latter, the legendary St. Patrick went to Ireland in the mid-fifth century AD to convert the population, ordain priests, and establish churches. In response, Ireland began to send missionaries into the rest of Europe shortly after the death of Patrick, and these represented one of the most vibrant Christian missionary movements in history. Ireland continues this influence today as the country that sends the highest number of missionaries per capita in the world.

There is evidence of some groups concentrating their early evangelizing efforts on islands. For example, Catholic missionaries went to Guam in 1668 to establish the first Pacific outpost of European civilization and religion. The Moravians sent the first Protestant missionaries of the modern era to the West Indies in 1732. Mormon missionaries who were originally sent to the British Isles in 1837 sent British converts in turn to Illinois. Mormons also came early to Polynesia in 1844, when Addison Pratt and others sailed there from Boston.

Geopolitical and historical contingencies also have played a role in bringing missionaries to islands. For example, Roman Catholic missionaries went to the Philippines with Spanish traders, Lutherans to New Guinea after it became a German protectorate in 1884, Roman Catholics to New Caledonia and Tahiti with French military intervention, Russian Orthodox to the Alaska islands with fur traders from Tsarist Russia, and American varieties of Protestantism to Haiti during the U.S. military occupation there from 1915 to 1934.

EVANGELIZATION AND CONVERSIONS ON ISLANDS

Conversions by missionaries on islands can be seen to occur at multiple levels: the option taken by an individual; a collective decision (e.g., an entire group makes this decision); or through the authority of a leader. For example, on New Guinea, missionaries withheld individual baptisms (i.e., the outward signs of personal conversions to Christianity) until there were large numbers of conversions to take place. Wesleyan Methodists converted Tongan royalty in 1830s, which greatly encouraged the conversion of others. Today, the Tongan Free Wesleyan Church (Methodist Church of the United States) is headed by the Tongan monarch, and about one-third of the population belongs to it. On Fiji, little progress was made by either the London Missionary Society or Methodist missionaries until the principal chief Thakombau was baptized in 1854. The conversion of the King and Queen of Huahine (a tributary of the Tahitian ruler King Pomare) resulted in almost the entire population of the island converting to Christianity.

There is a clear model of how missionary programs can be successful: the charismatic authority or leader who succeeds in making converts because of his or her personal qualities. The problem is that when that leader leaves or dies, the mission often fails.

Even with a charismatic leader, the isolation of missions and missionaries was often a problem. When a missionary died, a long time often passed before a replacement arrived. Isolation sometimes also caused missionaries to become "tropo" or "to go native." Related to this is the question of whether missionaries should dress like locals, or induce the locals to dress like them?

The background and training of missionaries varied greatly. For example, the London Missionary Society sent competent lay tradesmen (blacksmiths, carpenters) with religious zeal but little or no formal theological training. In contrast, other groups (e.g., Jesuits) emphasized religious training.

Sometimes, societies with long-established religious ties to one sect are very resistant to conversion to other sects. On Guam and Saipan, Underwood (2005) noted that the Catholic Church is "religion, family, culture, custom, and country," and the Chamorro culture there fiercely ostracizes converts who leave the Catholic religion. These converts to other religions often find themselves rejected by their immediate family as well as the greater society. As a consequence, converts in the Marianas are usually from the non-Chamorro population on those islands. On predominantly Catholic Guam, Mormon missionaries were chased with rocks, machetes, and even shotguns as recently as 1975 as they went about evangelizing.

Missionaries on islands often deal with small populations, which results in increased effort in translating the Bible and other holy texts into local languages. For example, speech-group size might be as small as a few hundred, and in Melanesia even a large speech group might only have a few thousand speakers. This contrasts with some Polynesian islands with many thousands, or continents with hundreds of thousands to millions, speaking the same language.

Missionaries often had to provide written languages for populations with largely oral traditions. The introduction of the alphabet and written languages (done by such organizations as the Wycliffe Society to aid in the translation of the Bible, but also done by Buddhists on some islands of Southeast Asia) provided a chance to record oral tradition more permanently. Moreover, translation of sacred texts into local languages raised the status of a language, because their written language now included the "Word of God." Arguably, it also often resulted in the loss of the oral tradition as biblical stories were used in place of local legends.

By providing a written language, missionaries could have great and lasting influence on islands. Wesleyan missionaries developed a written form of Fijian in 1850 by choosing one of the 300 available island dialects as the language for translation of the Bible; this dialect then became Standard Fijian. Fijians were quick to convert to Methodism because of the use of this Fijian Bible and the use of the Fijian language in their services. Today, 90% of Fijian Christians are Methodists.

The use of local legends by missionaries was widespread on islands. For example, the traditional creation story of the island of Moorea in the Society Islands featured an octopus. For this reason an octagonal-shaped building was constructed as the first Protestant church in French Polynesia (Fig. 1). It is widely accepted today that missions have been most successful when they incorporate elements of local beliefs into the religion they are promoting.

FIGURE 1 Octagonal-shaped church on the island of Moorea, French Polynesia, built over a *marae* (a ceremonial site) honoring an octopus deity. Photograph by Cheryl Resh.

THE IMPACT OF MISSIONARIES ON ISLANDS

A vivid, popular culture image of the influence that missionaries have had on island communities is the "fire and brimstone"-preaching Rev. Abner Hale, who is the protagonist in the novel and movie *Hawaii* by James Michener. Even the most ardent believer in evangelization and missionary activities cringes at the havoc Hale creates. In contrast, the presentation of the young mid-twentieth-century Mormon missionary John Groberg on Tonga in the autobiographical book *In the Eye of the Storm* and movie *The Other Side of Heaven* is extremely positive and even inspiring. These contrasting views reflect current perceptions about both the activities and the role that missionaries have played on islands.

Sociologists of religion are clearly divided or at least ambivalent in terms of evaluating the effects of missionary activities on native peoples. In part, this is because societies change with evangelization, and change can be viewed as either positive or negative depending on the perspective of the observer. It can be argued, however, that people choose to be missionaries because they believe that converting people to their own religious beliefs truly will benefit the converted (e.g., through achieving salvation) and thus make their lives better.

An individual missionary on an island, perhaps because of the small population usually present, can have great influence in the process of evangelization and on the people. On the Scottish coastal island of St. Kilda, two sides of this influence can be seen. The Church of Scotland's Reverend Neil MacKenzie arrived in 1830 and improved agriculture and education, introducing formal training in reading, writing, and arithmetic. However, a later-arriving missionary instituted day-long, mandatory church services on Sunday, forbade children from playing on that day, and required them to carry a Bible wherever they went. For 24 years his strictures seriously interfered with the practical necessities of maintaining island life.

Slavery was fought by missionaries in the South Pacific, who tried to protect native populations from "blackbirding," which included the capture of people for labor on sugar cane fields in Queensland or mines in Peru. But some missionaries actually engaged in slavery themselves. Jesuit priest Honoré Laval terrorized and enslaved inhabitants of the Polynesian island of Mangareva for 37 years. He destroyed statues of their gods and forced them to build a 1200-seat coral and mother-of-pearl cathedral on nearby Rikitea. He was later tried for murder in Tahiti, and not surprisingly found insane.

In contrast to Fr. Laval, William Knibb, sent to Jamaica by the Baptist Missionary Society, worked to end slavery. Although firmly instructed by his sending society not to interfere in civil or political affairs, Knibb was outspoken in his support of abolition, resulting in the burning of his chapel and school in Jamaica, his arrest there, and frequent libelous accusations that appeared about him in the press. However, through his efforts, slavery ended in the British colonies in 1838. There have been many other examples of missionaries intervening on behalf of local people. For example, Russian Orthodox priests protected Kodiak Island Alaskans from the warlike behavior of fur traders.

Modern missionaries on islands also build schools, orphanages, and hospitals (as well as churches), teach hygiene and public health, provide sanitation and clean water facilities, and educate. Father Damien de Veuster, and the story of his leper colony on Molokai in Hawaii, is still presented to children as an example of self-sacrifice for others. Missionaries, in fact, have been presented as the primary means of modernization in education, medicine, and technology on islands. For example, the introduction of metal to Pacific Islands for creation of longer-lasting tools has been attributed to missionaries.

Undoubtedly, some early missionaries were the inspiration for out-of-country aid activities (for example, the Peace Corps program in the United States) that involve volunteers "doing good to help others." In many cases, the activities of modern missionaries are indistinguishable from those of secular NGO (nongovernmental organization) participants. More recently, there has been an increase in faith-based foreign aid programs, which have been reported to increase prosperity of local economies, increase standard of living, and even contribute to greater global security through poverty alleviation.

Even critics of missionary influence would agree that some missionary activities have improved life on the islands. However, in doing so they have changed cultures dramatically. On islands, perhaps because of their isolation, the impact of missionaries is magnified compared to the effects on continents. A common criticism applied to some missionaries on oceanic islands is that "They came to do good and did well," implying personal economic and political gain resulting from their activities.

Throughout the islands of the world, missionaries stopped what Western culture would view as horrific practices such as infanticide (which was often done as a means of reducing overpopulation) and human sacrifices. On the Marquesas in the Society Islands, missionaries put a stop to death rituals performed by female relatives of deceased people that included harming themselves by cutting their hands and faces with sharks' teeth and other sharp objects. On Samoa, although instructed by the London Missionary Society to restrict their activities to the religious sphere, missionaries ended warfare, polygamy, abortion, and other activities, fundamentally restructuring Samoan society within a few years in the nineteenth century.

Some cultural effects of missionary activities are truly indicative of, at least, cultural insensitivity and would today be viewed as foolishness. For example, throughout the South Pacific, social mores such as tattoos and betel-nut chewing were outlawed and upper-body nudity replaced by full body-covering garments. Religious shrines were destroyed and, more unfortunately, oral histories were lost.

There has more recently been a response by some island people to preserve their past. Although in the Cook Islands and French Polynesia, tribal religions are now virtually without followers, elements of tribal religion clearly influence the expression of island Christianity. Likewise, Chamorros (Guamanian natives) today are nearly entirely Roman Catholic, but their belief practices reflect Filipino animism, Chamorro ancestor veneration, and religious icon worship. Similar practices can also be seen elsewhere in Micronesia. Tattooing, which was banned by Catholic missionaries on the Marquesas in 1867, was allowed again in 1985 because the Catholic bishop felt that the Marquesans needed these marks as a means of identifying themselves to the rest of the world.

Islands today are filled with contradictions because of the missionary influence. On Tahiti, the celebration of the Bible arriving there is a national holiday. However, it is now celebrated with festivals in which the islanders compete in dances that the missionaries tried to forbid! Ironically, missionaries came to islands, had the local people dress conservatively, and required them to abstain from activities on Sunday except churchgoing. Today, tourists from these same missionary-sending countries attend Sunday church for the island-cultural experience, but they often dress skimpily and then follow church services with the activities previously banned on Sunday!

Arguably, the advances in agriculture and animal husbandry brought by the missionaries were significant contributions to island economies. However, the subsistence levels of agriculture on Pacific islands probably were sufficient and did not necessarily require greater food production. These increased activities often were instigated because of the missionaries' view that hard work kept people from "idleness." Furthermore, the high incidence of diabetes among Pacific Islanders may have resulted in part from changes in diet.

Punishment for failing to obey religious laws could be severe. For example, banishment (which on a small island typically meant death) was sometimes a punishment for repeatedly disobeying church laws such as drinking alcohol and Sabbath nonobservance. On Rarotonga in the Cook Islands, missionaries created a "virtual religious police state" (one in six people worked for the police), and in the period 1835–1880, when missionaries were at the height of their powers there, the police rigidly enforced a variety of morality laws. For example, any man who walked with his arm around a woman after dark had to carry a torch in his other hand!

Although difficult to separate from modernization in general, higher divorce rates, less stable family lives, and loss of many traditional taboos have resulted after conversions because of the abandonment of the more rigid social structure. For example, in the Cook Islands, the nuclear family replaced communal structure, and living quarters were limited to the smaller units, which reduced society cohesiveness.

It should be noted that missionaries with good intentions were often duped by others with less noble motives. In 1863, four evangelical missionaries on Penryn in the Cook Islands, hoping to raise funds for a new church, were tricked by Peruvian slave traders into recruiting their congregation to work in Peru, with the missionaries serving as overseers. With most of the chiefs and men gone (and never to return), the line of leadership succession was eliminated and the social structure of the island was destroyed.

COMPETITION AND CONFLICTS AMONG MISSIONARIES

Evangelizing efforts on islands often take the form of a series of consecutive arrivals of competing sects. After the U.S. purchase of Alaska from Russia in 1867, Baptist and other Protestant missions came into the Alaska islands. They told the Alaskan natives that they had accepted the "wrong" form of Christianity (Russian Orthodoxy) and needed to convert over again. This competition among Christian sects caused entire villages to revert to shamanism.

Violence and warfare sometimes resulted from evangelization. The original missionaries' success on Tonga brought more missionaries to the island, and consequent religious wars between Christian and non-Christian Tongans were fought in 1826, 1837, 1840, and 1852.

Twentieth-century missionary conflicts have been less bloody but no less disruptive. In recent years, a great deal of competition among new religious groups on the outer atolls of the Marshall Islands has hurt these communities. Furthermore, competition among sects for reconversions can be seen in American Samoa, where Congregationalists have declined over a 40-year period because of losses to other churches, chiefly to Latter-day Saints (Mormon), Catholic, and Seventh Day Adventist churches.

Ironically, increased evangelization does not always result in increased conversions. On many predominantly Buddhist islands in Southeast Asia, a small portion of the population is Christian. Christian missionary activities typically fail to make new converts among any of the Buddhists but are successful in drawing previous converts from different Christian sects.

Even on nearby islands, differences in religion and conflicts may be pronounced. Tokelau (north of Samoa) consists of three clusters of islets that have a long history of fighting among each other. The ~1500 residents occupying the 11 km^2 of land comprised by the islands ascribe to different branches of Christianity. Two of the three clusters are predominantly Congregationalist, and the third is entirely Roman Catholic.

NATIVE MISSIONARIES

From the very earliest days of evangelization, there has been a zeal on the part of the recently converted to spread the message of their new faith, and religious organizations have used these native missionaries in their conversion activities. Both the Anglican Melanesian Mission and the London Missionary Society used native converts as missionaries in the Pacific Islands from the beginning of their efforts in the early and middle nineteenth century. From the Cook Islands, the Congregationalists sent 70 of their own native missionaries to Papua between 1872 and 1896, and over 200 have been sent since the practice began in 1830. Maori Mormon missionaries today are in the Cook Islands, working to convert (native) Congregationalists.

Fijians have sent out missionaries to other countries for more than 150 years, and in 2001 nearly 100 native Fijians served overseas. Tahitians were sent as missionaries throughout the Pacific in the nineteenth century, but this practice has declined in the last 40 years and likely will not be revived.

Besides sending native missionaries to other islands, some groups send them to continental countries that have large emigrant populations from those islands. For example, many of Jamaica's 60 independent churches have missionary programs to Great Britain because of the large number of Jamaicans who have migrated there. However, Jamaica still receives far more missionaries than it sends. Because it is a predominantly Christian nation today, these missionaries are competing among sects for reconversion to other ones.

The most recent trend in missionary evangelization has been the increased number of Protestant missionaries from countries that themselves were fields of missions. For example, in 2002 there were 44,000 missionaries from India, with 60% going on external missions. Today, missionary recruitment is greater from the Southern Hemisphere than from the West, and soon the majority of missionaries will be from Latin America, Africa, and Asia. Clearly, the pattern of South-to-South mission activity (rather than North-to-South) will be a major trend in the future on both islands and continents.

CONCLUSION

Missionaries have clearly had major effects on island culture and practice. This is not surprising: as agents of change, missionaries have been highly successful. The authors believe that much of this influence also resulted from the perceptions of the missionaries among themselves and their view of potential converts. The missionaries often viewed themselves in terms of their social positions, and the relationships among them were important. In contrast, their relationship toward the indigenous people was sometimes remote, superior, estranged, and often they viewed the local people as dangerous. The latter view clearly would evoke a need for changing the population's beliefs. Sometimes, however, native values helped missionaries. For example, the generosity and kindness of Tongans set an example for the behavior of the Mormon missionaries operating there.

In *The Voyage of the Beagle,* Charles Darwin wrote that a marooned sailor would be grateful if missionaries had previously arrived on an island where the unfortunate seaman might find himself. He said that this was because the missionaries would have tamed the "murderous habits" of the natives toward sailors. Unfortunately, the behavior of some missionaries toward natives on islands has often been less than honorable.

SEE ALSO THE FOLLOWING ARTICLES

Cook Islands / Easter Island / Popular Culture, Islands in / Tonga / Voyage of the *Beagle*

FURTHER READING

Barrett, D. B., G. T. Kurian, and T. M. Johnson. 2001. *World Christian encyclopedia,* 2nd ed. Oxford: Oxford University Press.
Hiney, T. 2000. *On the missionary trail: a journey through Polynesia, Asia and Africa with the London Missionary Society.* New York: Atlantic Monthly Press.
Johnston, A. 2003. *Missionary writing and empire, 1800–1860.* Cambridge, UK: Cambridge University Press.
Montgomery, R. L. 1999. *Introduction to the sociology of missions.* Westport, CT: Praeger.
Neill, S. C. 2005. Missions: Christian missions, in *Encyclopedia of religions,* 2nd ed. L. Jones, ed. Farmington Hills, MI: Thomson Gale.
Stackhouse, M. L. 2005. Missions: missionary activity, in *Encyclopedia of religions,* 2nd ed. L. Jones, ed. Farmington Hills, MI: Thomson Gale.
Underwood, G., ed. 2005. *Pioneers in the Pacific: memory, history, and cultural identity among the Latter-day Saints.* Provo, UT: Religious Studies Center, Brigham Young University.
Walters, J. S. 2005. Missions: Buddhist missions, in *Encyclopedia of religions,* 2nd ed. L. Jones, ed. Farmington Hills, MI: Thomson Gale.
Whiteman, D. L. 1993. Oceania, in *Toward the 21st century in Christian mission.* J. M. Phillips and R. T. Coote, eds. Grand Rapids, MI: Wm. B. Eerdmans Publishing Co.
Wright, L. B., and M. I. Fry. 1936. *Puritans in the South Seas.* New York: Henry Holt and Co.

MOA

ALLAN J. BAKER

Royal Ontario Museum, Toronto, Canada

Isolated island archipelagoes with ecologically diverse habitats can be veritable laboratories of evolution for many organisms. Some of the most spectacular examples involve colonizing species of birds, such as Darwin's finches in the Galápagos Islands and Hawaiian honeycreepers, both of which have undergone adaptive radiations primarily involving incredible differences in bill morphology. Another equally spectacular but less well-known adaptive radiation occurred in the now extinct moa on the two larger islands of New Zealand, the North and South Islands. Moa were giants of the bird world; species varied in body mass from 20 to 250 kg, with the largest genus (*Dinornis*) reaching up to about 3.5 m when the neck was extended, although recent studies suggest moa carried their heads much lower than this because of the articulation of the vertebral column at the back of the skull (Fig. 1).

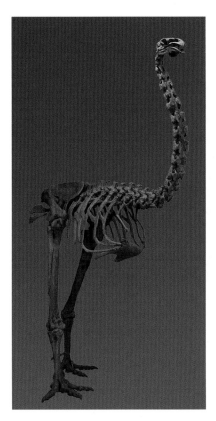

FIGURE 1 Cast of a skeleton of the South Island giant moa (*Dinornis robustus*), standing 2.5 m high in this mount. Image by Claiton Martins Ferreira and Daniel Baker.

DISCOVERY AND ORIGIN OF THE MOA

Moa were first discovered scientifically in the nineteenth century when European colonists shipped bones of the giant birds back to England. The renowned comparative anatomist Sir Richard Owen originally dismissed a partial bone fragment as coming from an ox but realized later on closer inspection that it represented an "unknown struthious bird," a discovery published in 1840 amid much skepticism. As more collections were received and studied, he described a number of new species from isolated bones. Part of the doubt about Owen's original deduction of a new type of flightless bird was that it seemed beyond belief that such a large bird, thought to be about the size of an ostrich, could have evolved solely on the relatively small islands of New Zealand. This issue has puzzled biogeographers ever since, leading to the formulation of two competing hypotheses. Many scientists argued that the large flightless birds called ratites—such as kiwis and moa of New Zealand, emus and cassowaries of Australia, rheas of South America, and ostriches and elephant birds of Africa and Madagascar—could not have dispersed to these widespread regions of the world. Thus, they could not have evolved from a common ancestor and do not form a monophyletic group. Others have argued that they do share a common ancestor and were rafted on drifting land masses following the fragmentation of the supercontinent Gondwana to their present geographic locations in the southern hemisphere. A third hypothesis based on the discovery of flighted fossil lithornithiforms in the northern hemisphere suggests that the ancestor of ratites may have flown to these locations and become secondarily flightless.

So how did moa originate and become confined to an isolated island archipelago like New Zealand? Studies of shared morphological characters produced a phylogenetic tree that placed ratites in a monophyletic group. Additionally, moa were most closely related to kiwis on a basal branch of the tree, inferring that this was an ancient divergence event consistent with the separation of New Zealand from Gondwana about 82 million years ago. With the discovery that DNA could be extracted from well-preserved moa bones retrieved from swamps or from mud in cave floors, it became possible to test the Gondwana hypothesis by amplifying and sequencing fragments of mitochondrial DNA (mtDNA) because of the high copy number of mitochondria in cells. Furthermore, these sequences could be used to build a tree of relationships from multiple genes, and two laboratories were even able to reconstruct the complete mtDNA genomes of three genera of moa. The approximately even rate of molecular evolution of the genes provided a molecular clock to date the divergence of moa about the time New Zealand separated from Gondwana. As in other DNA studies, kiwis were shown not to be closely related to moa at all, but instead to be close relatives of emus and cassowaries of Australia. The ancestor of moa must have originated separately in Gondwana, and there is no need to posit flying ancestors or island hopping as a way for moa to reach and proliferate in the isolation of New Zealand. Sadly, moa became extinct around 1300, about 100 years after humans arrived in New Zealand, as they apparently were tame and were an easy source of meat.

TIMING AND MODE OF ADAPTIVE RADIATION OF MOA

The diversity of moa species that evolved in New Zealand has been gradually whittled down from an earlier estimate of 64 species to a low of ten species. Once again, ancient DNA extracted from bones provided completely new insights about species diversity and showed why previous estimates based on bone morphology had been so difficult to interpret. Innovative studies that were able to amplify fragments of a nuclear gene that identified the sex of moa species came up with a really surprising result—several genera contained a mixture of smaller and larger individuals that had been classified into two species in each genus. The sexing test instead showed that the smaller individuals were males, the larger ones were females, and intermediate-sized birds could be of either sex. Consequently, this confirmed the earlier removal of a number of (but not all) putative species from the list, consistent with an earlier statistical analysis of morphometric variation suggesting that bimodal distributions represented sexual dimorphism within species. Genetic typing of bones from 125 specimens detected 14 major lineages of moa and indicated misidentifications of bones in collections. By combining the control region sequences with sequences from nine other mtDNA genes, a strongly supported phylogeny of these lineages was derived. If all these lineages represent species, then moa were about as diverse as Darwin's finches (13 species) and Hawaiian honeycreepers (14 surviving and eight extinct species).

To date the divergence times of these lineages, and thus to estimate the tempo of moa evolution, a method of molecular dating was employed that allowed the rate of evolutionary change to vary across the phylogeny. The beginning of the diversification of moa lineages was dated to 15–23 million years ago, coinciding with the Oligocene "drowning" of New Zealand, in which much of the land surface was thought to have been submerged under water. This probably erased the earlier history of moa, especially if the surviving ancestral stock was small. However, by dating the control region tree of 125 specimens, a cycle of lineage-splitting was estimated to have occurred about 4–6 million years ago, when the landmass was fragmented by tectonic plate movements and mountain-building

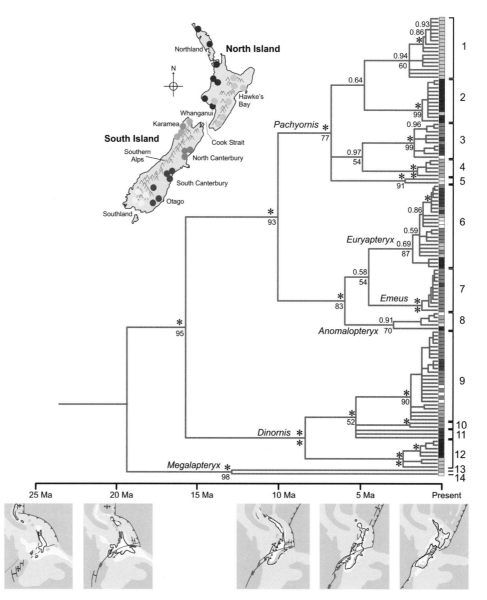

FIGURE 2 Bayesian tree constructed by using 658-bp control region sequences from 125 moa specimens. Numbers at the branch tips identify the 14 major lineages. Specimens are color-coded according to geographic locations plotted together with place names. Major mountain ranges are represented by peaks. Unfilled bars at the branch tips indicate specimens without locality data. Asterisks at the nodes indicate posterior probabilities (above the nodes) or maximum likelihood bootstrap values (below the nodes) of 1.0 and 100%, respectively. (Lower Insets) Extent of the New Zealand landmass and movement of tectonic plates from 25 million years ago to present. Faults, subduction zones, and sea-floor spreading centers are shown in red. Modified from Baker et al., 2005.

events, with ocean-level rises severing the link between the North and South Islands, and with general cooling of the climate occurring, including Pleistocene glacial cycles (Fig. 2). This resulted in the geographic isolation of lineages and ecological specialization in different habitats in the North and South Islands.

Evidence that moa exploited different ecological niches comes from subfossil remains and observations that no more than four species in different genera existed in the same biogeographic regions of New Zealand and ate different foods. Diets ranged primarily from coarse twigs (*Dinornis*), soft leaves and berries (*Euryapteryx* and *Emeus*), and tough leaves (*Pachyornis* and *Anomalopteryx*) to plants in forest edges and adjacent high altitude grasslands (*Megalapteryx*). Because moa did not have foregut fermentation, it has been speculated that they evolved long intestines to ferment their plant diets, and therefore have correlated large body sizes. These findings jointly provide strong evidence that the mode of evolution was by allopatric speciation, and the flowering of lineages in the last 6 million years was indeed a striking example of an adaptive radiation in the isolation of the islands of New Zealand.

SEE ALSO THE FOLLOWING ARTICLES

Bird Radiations / Flightlessness / Fossil Birds / Gigantism

FURTHER READING

Anderson, A. 1989. *Prodigious birds: moas and moa hunting in New Zealand.* Cambridge: Cambridge University Press.

Baker, A. J., L. Huynen, O. Haddrath, C. D. Millar, and D. M. Lambert. 2005. Reconstructing the tempo and mode of evolution in an extinct clade of birds with ancient DNA: the giant moa of New Zealand. *Proceedings of the National Academy of Sciences of the USA* 102: 8257–8262.

Haddrath, O., and A. J. Baker. 2001. Complete mitochondrial DNA genome sequences of extinct birds: ratite phylogenetics and the vicariance biogeography hypothesis. *Proceedings of the Royal Society of London, Series B* 268: 1–7.

Holdaway, R. N., and T. H. Worthy. 1991. Lost in time. *New Zealand Geographic* 12: 51–68.

Huynen, L., C. D. Millar, R. P. Scofield, and D. M. Lambert. 2003. Nuclear DNA sequences detect species limits in ancient moa. *Nature* 425: 175–178.

Paton, T., O. Haddrath, and A. J. Baker. 2002. Complete mtDNA genome sequences show that modern birds are not descended from transitional shorebirds. *Proceedings of the Royal Society of London, Series B* 269: 839–846.

Worthy, T. H., and R. N. Holdaway. 2002. *The lost world of the moa: prehistoric life in New Zealand.* Bloomington: Indiana University Press.

MOTU

FRANCIS J. MURPHY

University of California, Berkeley

A motu is a small island, made entirely of coral reef sediment, which forms on top of a barrier reef (Fig. 1). These islands are variably capable of supporting vegetation, seabird colonies, and human settlements, but because of their low elevation and small size, they are also highly susceptible to the effects of major storms and changes in sea level.

FIGURE 1 This small motu on the southeast corner of Raiatea has formed just a short distance from the reef crest. Photograph © DANEEHAZAMA.com.

TERMINOLOGY

Motu is a Polynesian-language term that means "small island" (the plural is also *motu*), and it is most commonly used for all reef islands associated with high islands, almost-atolls, and atolls. In the scientific classification of reef islands, a distinction is made between motu and cays. In this context, a "motu" is a coral reef island that is found in high-energy environments, has a seaward shingle ridge and a leeward sand deposit, and is relatively permanent. A "cay" is normally made of homogeneous sediment (either sand or shingle) and is ephemeral. Often cays do not support vegetation.

MOTU GEOMORPHOLOGY

Because motu are derived from coral reef material, they can be found only where coral reefs form (i.e., in tropical seas where the sea-surface temperatures range between 18 and 34 °C). Furthermore, motu require large storms for their development, so their distribution is only common in tropical areas that support coral reefs and experience cyclonic storms. The Pacific has the largest number of motu, which make up the bulk of the land on the atolls of the Federated States of Micronesia, the Marshall Islands, Tuvalu, the Cook Islands, and French Polynesia. Atolls, such as those of Kiribati, that lie close to the equator and are therefore out of the dominant storm tracks do have some motu but are mostly covered with sand cays.

Toward the end of the last glacial period, around 14,000 years ago, sea level was at a lowstand, at approximately 120 m below its present level. As sea level rose during the Holocene, old exposed coral reefs were once again inundated, and existing barrier reefs developed. Dates vary from island to island but generally in the Pacific, sea level approached present levels from 8000 to 6000 years ago, and coral reefs, lagging somewhat, reached present levels about 4000 years ago. Once sea level and reef crests reached a state where sedimentation on the reef flat was possible, then the stage was set for motu to form. There is evidence that in some places where motu occur there was a late-Holocene highstand, with relative sea levels up to 1 m above present that persisted until as recently as 2000 years ago. In this situation, motu could not have formed until relative sea levels dropped to close to present levels.

Motu have a pattern of development and maintenance whereby the two different sediment classes experience erosion and sedimentation at different times and rates: The large-scale sediment deposits quickly during occasional high-energy storms and erodes slowly during normal periods, and the small-scale sediment deposits slowly during normal times and erodes quickly during high-energy events. Large storms often create or add to steep berms on the seaward shore made of sediment that ranges from shingle to boulders. These then create a protected area directly leeward, where during calm periods finer sediment—sand and gravel—accumulates. These same large storms can, however, create waves that completely submerge the island. In these cases, small sediment is usually swept into the lagoon.

Motu are found both on reefs that show no sign of emergence (drop in sea level) and on reefs that are slightly emerged, such as those on the south shore of Rangiroa in the Tuamotu Archipelago. Where emergence has raised a shoreline to the point that wave-driven sedimentation can no longer occur, there are no motu—for instance, on Niau in the Tuamotu.

Motu are stabilized somewhat by the cementation of sediment into features such as conglomerate platforms and beachrock. Reef sediment that is wetted sufficiently (i.e., in or just above the intertidal zone) will be bonded together by calcareous cement that precipitates between sediment particles. On beaches, this creates beachrock, which has the slope and sediment characteristics of the original beach. Rubble banks created by large storms cement from the bottom up to create conglomerate platforms that have a characteristically level surface and a wide range of clast sizes. Both these features become exposed over time when the loose sediment covering them is removed by subsequent storm action.

MOTU LIFE

Today, motu and sand cays are often thought of as the ideal paradise islands, with iconic palm-shaded white-sand beaches adjacent to a turquoise lagoon. They are, in fact, great places to visit, but long-term settlement is difficult because they are harsh environments for both plants and animals. The native vegetation of motu is characterized by hardy plants that can withstand salt spray and survive with little freshwater and a very narrow spectrum of organic nutrients. These include shrubs such as *Pemphis acidula* and *Suriana maritima*, and trees such as *Hernandia sonora*, *Pandanus tectorius*, *Pisonia grandis*, and *Tournefortia argentea*. As plants become established, they stabilize and add nutrients to the motu sediment, and they create roosting areas for birds, which in turn also contribute nutrients to the motu.

Freshwater that falls on motu seeps into the porous coralline base of the island and sits on top of the heavier saltwater. The potential size of this reservoir is relative to the area and elevation of the motu, such that there can be 40 times the amount of water below sea level as there is above it. The actual volume is regulated by rainfall and utilization of the aquifer by plants and people. It is clear from this relationship that small changes in sea level can effect large changes in freshwater capacity.

The habitability and carrying capacity of motu for humans varies with the size of the motu and the amount of rainfall that it receives. People living on motu have to depend heavily on marine life for food because it is difficult to grow food crops in this harsh environment. Crops that can be cultivated on motu include coconuts, taro, bananas, and arrowroot. Because of these generally harsh conditions, high-density populations on motu are unusual, and the normal high end is around 500 people per square kilometer, such as in the villages of Avatoru and Tiputa on Rangiroa in the Tuamotu Archipelago. In extreme cases such as Ebeye Island on Kwajalein Atoll in the Marshall Islands there are more than 33,000 people per square kilometer. More common densities of inhabited motu, however, are below 50 people per square kilometer.

Radiocarbon dating of conglomerate platforms across the Pacific (which ranges from 4000 to 1500 years ago), and archaeological evidence for the same region, shows that the creation of motu only slightly predates the period over which most of the islands in the central and eastern Pacific were being settled, with atolls in general being settled later than the high islands. Even as human populations were moving out through Oceania, searching for new landfall, the combined effects of sea level, storms, reef growth, and plant dispersal and settlement, were producing new motu for them to live on.

Rising sea levels, an increase in frequency of major storms, and the lowering of pH in ocean surface water are all likely consequences of global climate change. For motu this may have grave consequences. The predicted sea-level rise of 30–80 cm over the next 100 years will erode motu shores and reduce the volume of the freshwater lens. Major storms will do the same, only more drastically, as well as directly removing flora and fauna. A lowering of pH will slow calcium carbonate production and eventually increase chemical erosion of reef sediments. Because of human-induced climate change, these landforms, and all of the biota and human populations associated with them, will increasingly become endangered and may in fact disappear completely within a few generations.

SEE ALSO THE FOLLOWING ARTICLES

Atolls / Coral / Global Warming / Oceanic Islands / Sea-Level Change

FURTHER READING

Alkire, W. H. 1978. *Coral islanders.* Arlington Heights, IL: AHM Publishing.
Guilcher, A. 1988. *Coral reef geomorphology.* New York: John Wiley and Sons.
Nunn, P. D. 1994. *Oceanic islands.* Oxford: Blackwell.
Stoddart, D. R. 1975. Almost-atoll of Aitutake: geomorphology or reefs and islands. *Atoll Research Bulletin* 190: 31–57.
Stoddart, D. R., and J. A. Steers. 1977. The nature and origin of coral reef islands, in *Biology and geology of coral reefs.* O. A. Jones and R. Endean, eds. New York: Academic Press, 59–105.

NEW CALEDONIA, BIOLOGY

JÉRÔME MURIENNE

Harvard University, Cambridge, Massachusetts

Because of its extremely diverse biology, New Caledonia is one of few islands to be designated a biodiversity hotspot. The long isolation of the territory (since the breakup of Gondwana 80 million years ago), in conjunction with its climatic stability, was often proposed as an explanation for its outstanding biodiversity. Recent evolutionary studies are more in accordance with submersion of the territory from 65 to 45 million years ago, establishing a new paradigm for the origin of biodiversity in this island.

GEOGRAPHY OF THE TERRITORY

New Caledonia is a Melanesian archipelago comprising two sets of islands (Fig. 1). The first includes the main island (Grande Terre, approximately 500 km long and 50 km wide), the Belep Islands to the north, and the Isle of Pines to the south. The second includes more recent volcanic islands (Ouvéa, Lifou, Tiga, and Maré), called the Loyalty Islands. The inclusion of other small dependencies (Chesterfield, Matthew, and Hunter) brings the total land area to 18,972 km².

The climate is subtropical, and the mean annual temperature varies between 22 and 24 °C. There are two major seasons: a hot season from mid-November to mid-April and a cool season from mid-May to mid-September. The Grande Terre harbors an asymmetric mountain chain. The eastern part is abrupt, whereas the western part ends in long plains. The culminating points are Mont Panié (1629 m) in the north and Mont Humboldt (1618 m) in the south. The mean annual rainfall is 1700 mm, but precipitation is also uneven, mainly because of the mountains. Eastern parts of the island can receive five times more rain than their western counterparts.

Different soils are found in New Caledonia. Among them, ultramafic soils cover one-third of the territory (mainly in the south). Their richness in metal and poverty in nutritive elements makes that area a very particular environment for the flora and its associated fauna.

BIOLOGICAL CHARACTERISTICS

Compared to other islands, New Caledonia, despite its small size, exhibits an extraordinary species richness and rate of regional endemism (Fig. 2). In addition, several groups are considered relictual. Local endemism to restricted areas is a general characteristic of groups such as terrestrial squamates or some insects and can also be found in some plants. New Caledonia has long been seen as a distinctive floristic and faunal entity and, as mentioned, is now recognized as a biodiversity hotspot.

The dense evergreen rain forest is the richest vegetation type. It is mainly located on the east coast of the island because of the high precipitation there. This forest is present on every kind of substrate, from 300 m to the highest altitudes.

The sclerophyll forest is the most endangered vegetation type. Including small-sized trees (10 to 15 m), this dry forest is located on the west coast between the altitudes of 0 and 400 m. Twenty-four percent of its plant species are locally endemic.

The scrubland, or *maquis,* because it is mostly associated with ultramafic soils, is often referred to as *maquis minier* ("mining maquis"). High-altitude maquis covers 100 km², and low- to mid-altitude maquis covers 4400 km².

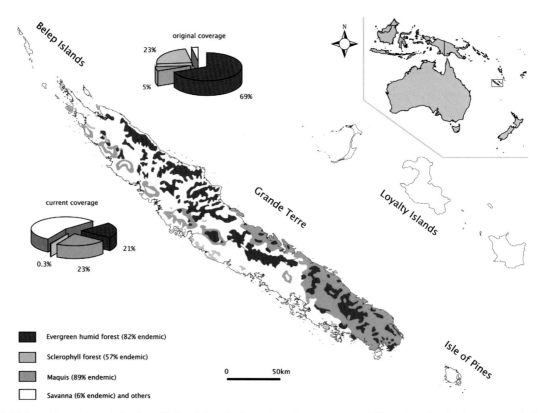

FIGURE 1 Principal types of vegetation found in New Caledonia Grande Terre (current coverage). Plant rates of endemism are noted in the legend. Diagrams represent the evolution from the original to the current coverage.

The savanna covers 6000 km², and its major tree component is the niaouli, *Melaleuca quinquenervia*. Other types of land include mangrove forests, swampland, and secondary shrubland.

New Caledonia presents a very distinctive flora and fauna. Thirteen endemic species of coniferous trees in the genus *Araucaria* are found in this territory, out of only 19 worldwide species. Illustrating the unique nature of the island, the endemic *Amborella trichopoda* is the sole member of the family Amborellaceae, which is a sister group to all the remaining flowering plants. Likewise, the kagu (*Rhynochetos jubatus*), an emblematic, almost flightless forest bird, is the sole representative of the endemic family Rhynochetidae, an additional lowland species endemic to the island having gone extinct in the Holocene.

ORIGIN OF THE BIODIVERSITY

New Caledonia has often been described as a "museum" in which its tremendous biodiversity has been explained by a long accumulation of species in ancient groups since Gondwanan times. The geological history of the region, however, contradicts this classical view. The total submersion of the territory between 65 to 45 million years ago (subsequent to separation from Gondwana) was either unrecognized or disregarded because of the presence of the "relictual" groups.

The recent use of molecular phylogenetic methods has enabled the investigation of the origin of the biodiversity in New Caledonia in an evolutionary framework. Most studies on plants, vertebrates, and invertebrates show that for supposedly Gondwanan groups, their origin in New Caledonia is never older than 40 million years ago and can involve multiple dispersal events from the neighboring regions. Additionally, their diversification can be extremely recent (a few million years), especially in insect groups. A new paradigm has now emerged that presents a more balanced view, with either single or multiple dispersal events and possibly different tempos of evolution inside New Caledonia. The potential existence of refugium islands during submersion times could explain the presence of "relictual" groups in New Caledonia.

On a local scale (inside New Caledonia) the high heterogeneity of the territory, rather than its long isolation, is certainly an important factor of diversification. The presence of ultramafic soils (toxic and poor in nutrients) could have also been involved in some process of adaptive radiation.

THREATS TO THE BIODIVERSITY

Fire is an important factor in the degradation and the transformation of the vegetation. Because of the fire resistance of the niaouli tree, the expansion of the savanna (to the detriment of the primary vegetation) has been significant. Even though fires are attested as a natural process before human arrival, recent fire is mainly due to human activities.

Invasive species constitute a major threat to the biodiversity. Among plants, around 1300 species are non-native, and 67 taxa are considered invasive. The rusa deer, *Cervus timorensis,* was introduced less than 150 years ago and retards the regeneration of the forest. With its fierce sting, the little fire ant, *Wasmannia auropunctata,* was introduced in 1960. In addition to harming humans, cattle, and cultures, it induces a serious ecological problem by competing with other arthropods (it is known to use its venom against other ant species) and potentially affecting the entire food web.

Even though open-pit mining is better managed than in the past, it is still a concern. The high international demand for nickel has led to the development of new projects that may threaten some relictual forests. The necessary economical development of the island (mining representing the major source of income) needs to better integrate the question of the preservation of New Caledonia's exceptional biodiversity.

Numerous protected areas exist, but their impact on the conservation of plant diversity has proven to be relatively low. Perhaps the most urgent and promising initiative has been the creation in 2001 of the "dry forest conservation program." This program groups ten different partners (including national and local agencies and international NGOs) that join together to put a stop to the erosion of the most endangered New Caledonian ecosystem.

SEE ALSO THE FOLLOWING ARTICLES

Ants / Endemism / Invasion Biology / New Caledonia, Geology / Vegetation

FURTHER READINGS

Bauer, A. M., and R. A. Sadlier. 2000. *The herpetofauna of New Caledonia.* Ithaca, NY: Society for the Study of Amphibians and Reptiles.

Beauvais, M.-L., A. Coléno, and H. Jourdan. 2006. *Invasive species in the New Caledonian archipelago.* Paris: Institut de Recherche pour le Développement.

Dry Forest Conservation Program. http://www.foretseche.nc [in French and English].

Jaffré, T., P. Bouchet, and J. M. Veillon. 1998. Threatened plants of New Caledonia: is the system of protected areas adequate? *Biodiversity and Conservation* 7: 109–135.

Jaffré, T., P. Morat, J.-M. Veillon, F. Rigault, and G. Dagnostini. 2004. *Composition and characterization of the native flora of New Caledonia.*

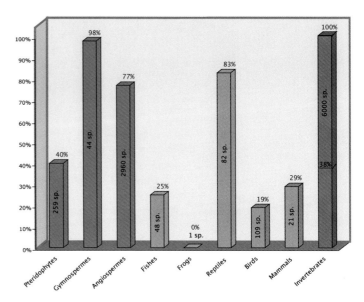

FIGURE 2 Rates of endemism in different groups. Common names have been used even though they do not always represent monophyletic groups. Numbers for plants include only native species. Endemism in invertebrates varies between 38% in the butterflies to 100% in the less mobile groups.

Documents Scientifiques et Techniques II 4. Nouméa, New Caledonia: Institute de Recherche pour la Développement.

Marquet, G., P. Keith, and E. Vigneux. 2003. *Atlas des poissons et des crustacés d'eau douce de Nouvelle-Calédonie.* Collections Patrimoines Naturels 58. Paris: Muséum National d'histoire Naturelle.

Morat, P. 1993. The terrestrial biota of New Caledonia. *Biodiversity Letters* 1: 69–71. [This special edition includes numerous review papers on the subject.]

NEW CALEDONIA, GEOLOGY

TIMOTHY J. RAWLING

University of Melbourne, Australia

The New Caledonia archipelago is a group of islands that form the emergent northern part of the Norfolk Ridge in the Southwest Pacific Ocean. The archipelago consists of six major islands as well as numerous smaller islands. The closest neighboring island group is Vanuatu, 500 km to the northeast, while New Zealand and Australia are situated 1500 km to the south and west, respectively. The New Caledonian territorial boundaries are approximately between latitude 18° and 23° S and longitude 158° and 172° E, and the land area of the archipelago is in the order of 18,600 km².

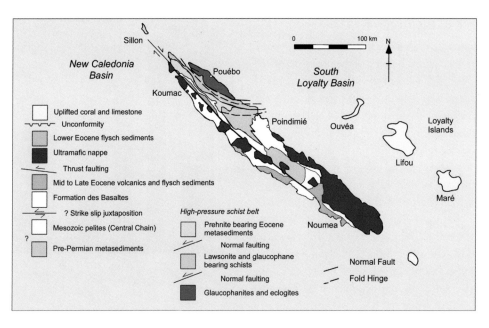

FIGURE 1 Regional geology map of New Caledonia including major faults and fold hinges.

GEOGRAPHIC SETTING

The largest island in the New Caledonia group, commonly referred to as "Grande Terre," is elongate (about 400 km long and 60 km wide; 16,350 km²) and strikes northwest–southeast. The geomorphology of Grande Terre is highly variable and includes a mountain chain that extends from the northeast coast through the central chain to the southern plateau, numerous evolved erosional river valleys, and a gently undulating west coast disrupted by several steep-sided, mesa-like plateaus. The highest peaks are Mt. Panié (1628 m), Mt. Colnett (1505 m), and Mt. Ignambé (1311 m) on the northeast coast and Mt. Humboldt (1618 m) in the south. The climate on Grande Terre is highly variable. New Caledonia is situated immediately north of the Tropic of Capricorn and thus is on the very edge of the tropics.

The large range of mountains that extend much of the way down the east coast act as a barrier to rain-bearing clouds approaching from the east or northeast. As a result, the much flatter west coast lies within a rain shadow and receives considerably less rain. Rainfall on the northeast coast is typically in excess of 4000 mm per year, whereas the west coast may receive as little as 1000 mm per year.

The other major islands of the New Caledonia archipelago include the Loyalty Islands, Maré, Lifou, and Ouvéa, as well as the smaller Île des Pines and Belep Island. The Loyalty Islands are situated on the Loyalty Ridge, which runs parallel to the northern Norfolk Ridge, about 100 km to the east. These islands are volcanic in origin and are covered by low-lying coral reef. Recent uplift has resulted in the emergence of the reefs and exposure at present levels. These islands have little relief and no developed drainage systems (Fig. 1).

GEOLOGY

The main island of New Caledonia is composed of a complex mixture of metamorphic, igneous, and sedimentary rocks with both continental and oceanic affinities. These rocks record a geological evolution that spans 300 million years and is punctuated by three major phases of tectonic activity (or deformation and heating): (1) late Paleozoic to early Mesozoic convergent-margin tectonism along the eastern margin of Gondwana, (2) mid- to late-Cretaceous extensional tectonism associated with the dismemberment of the eastern passive margin of Gondwana, (3) collisional orogenesis, peridotite nappe emplacement, and high-pressure metamorphism during the Tertiary.

The geology of New Caledonia can be divided into six geologically distinct tectonostratigraphic terranes: (1) the ultramafic nappe, (2) the central mountain chain, (3) the western coastal belt, (4) the eastern coastal belt, (5) the northern region, and (6) the Loyalty Islands.

The Ultramafic Nappe

The ultramafic nappe is an extensive ophiolite complex, or sheet of ocean crust, that was originally thrust over much of the island during the late Eocene during a major collisional tectonic event but has subsequently been disrupted and locally removed by faulting. The ultramafic rocks, primarily harzburgite and dunite, comprise over 40% of the rocks exposed in New Caledonia. The age of ophiolite emplacement has been established in the western part of the island, because the peridotite at the base of the sheet is in thrust contact with underlying Upper Cretaceous and Lower Tertiary sediments. In addition, there is a Late Eocene flysch (or marine sediment formed at the time of collision) that unconformably overlies part of the ophiolite sheet. This unit contains basalt pebbles derived from the basement, as well as serpentinized peridotite pebbles derived from the overthrust ophiolite sheet. During emplacement the ophiolite can be seen to have overpushed material to the southwest, indicating northeast to southwest–directed collision and thrusting.

The ultramafic rocks are thought to be relics of a once continuous nappe of Cretaceous oceanic crust and mantle, derived from the southern flank of a WNW-ESE oceanic ridge in the Loyalty Basin. Geophysical data confirms that the ultramafic rocks exposed in New Caledonia are continuous with crust-mantle structures in the Loyalty Basin.

Tropical weathering of the ophiolite material has formed widespread and deep laterites that are strongly enriched in nickel. These deposits account for about 25% of the world's nickel reserves, and the associated mining industry forms a critical part of the local economy.

The Central Chain

The oldest rocks in New Caledonia are the Permian-Jurassic quartzo-feldspathic metasediments and metavolcanics of the central chain. These rocks are interpreted to have been deposited along an active continental margin and are thought to have been sourced from emergent land to the southwest at that time. They were accreted to the eastern margin of Gondwana by the early Cretaceous. The igneous rocks of the central chain include mainly tholeiitic basalts, dolerites, and gabbros, indicative of an active continental margin setting. Extrusive material from this suite is overlain by Senonian-age sedimentary sequences. The rocks of the central chain have undergone varying amounts of metamorphism, or burial and heating, and metamorphic grade ranges from low-pressure/low-temperature prehnite-pumpellyite facies in the southwest to low-pressure/moderate-temperature greenschist facies with in the northeast. In this region some localized high-pressure/low-temperature metamorphic assemblages containing the minerals glaucophane and lawsonite have also been identified, which may be indicative of collision- or subduction-related tectonic activity. Metamorphism occurred between 152 and 132 million years ago and has been interpreted to be associated with either (1) southwest dipping subduction beneath the Gondwana margin or (2) uplift and emergence of the New Caledonian core region, which at this time is thought to have been attached to the eastern margin of Gondwana.

The Western Coastal Belt

The western coastal belt contains a diverse mixture of rock types. Deformed rocks of Jurassic and older age are overlain by Cretaceous sediments associated with deepening sea levels (or transgression), which were probably deposited contemporaneously with the opening of the Tasman Sea. The stratigraphy and palaeontology of these sequences closely resemble rocks of the similarly aged Murihiku terrane in New Zealand, and both are interpreted to be part of a marginal Gondwana sedimentary package. These in turn are overlain by Eocene conglomerates and shallow-water sequences, as well as Upper Eocene flysch sequences.

Basaltic igneous material overlies rhyolitic tuffs and is intercalated with marine sediments varying in age from Senonian to late Eocene. Dating of basalts from the western coastal belt indicates that they were erupted between 42 and 59 million years ago; however, the lower age limit is absolutely constrained by emplacement of the ophiolite in the late Eocene. Basalts of the Poya Formation are typically mid-ocean ridge type. The temporal relationship between these basalts to the rhyolitic tuffs is interpreted to indicate basalt eruption occurred shortly after the rift formed during the late Cretaceous and that rifting continued at least until the early Tertiary.

The Eastern Coastal Belt

The eastern coastal belt includes a thin strip of sediments and volcanics that separate rocks of the Central Chain from the east coast of Grande Terre. These rocks are Cretaceous to Eocene in age and are broadly similar to those of the western coastal belt. Based on age and stratigraphic affinities they are also generally considered to have formed in a similar tectonic environment. Also included in the eastern coastal belt is the root zone of the ophiolitic nappe complex described above.

The Northern Region

The northern region of Grande Terre is dominated by variably metamorphosed metasedimentary and metavolcanic sequences associated with an Eocene-aged high-pressure/low-temperature metamorphic belt. This region contains one of the world's largest, most spectacular and continuous exposures of high-pressure (blueschist- and eclogite-facies) metamorphic rocks in the world. The western part of the belt is dominated by fine-grained siliceous siltstones (phtanites) and limestones that are variably metamorphosed from prehnite-pumpellyite facies to lawsonite facies, increasing in grade to the northeast. The eastern part of the belt, or Pouébo terrane, contains a mixture of mafic material (derived from oceanic precursors) and more siliceous material (with continental origins) that was metamorphosed to blueschist or eclogite facies during the Eocene.

The western and eastern sectors of the northern region are geologically distinct; however, they do represent different levels of the same structural and metamorphic zone. The northern region is strongly dissected by numerous generations of brittle faults and shear zones that formed at various stages during the exhumation history of this once deeply buried package of rocks.

High-pressure/low-temperature metamorphic rocks are important because they typically form in subduction zones where cold oceanic crust is forced down to great depths (and pressures) as it is overridden by another crustal block. The eclogite facies rocks in northern New Caledonia are believed to have been metamorphosed at pressures of 20 kbar, which equates to burial to depths of 60 km or more. These rocks were then transported via some (still relatively poorly understood) tectonic process back to the Earth's surface, where they are now exposed in mountainous ridges high above sea level. This all happened in a geologically short period of time (on the order of 20–40 million years).

Loyalty Islands

The Loyalty Islands are a series of four emergent islands (Ouvéa, Lifou, Maré, and Walpole) that lie on the Loyalty Ridge, 100 km to the east of Grande Terre. The Loyalty Ridge probably represents an extinct volcanic arc associated with the Eocene subduction described above. The islands are probably volcanic, but, apart from a small outcrop of basalt on Maré, the basement rocks have not been identified because they are overlain by the thick coral and limestone sequences that have developed on top of the islands.

Evolution of the High-Pressure Schist Belt

Since plate tectonic theory was first applied to the Southwest Pacific, it has been recognized that a detailed understanding of the evolution of New Caledonia's high-pressure schist belt was critical to any tectonic interpretation or reconstruction of the region.

The most recent tectonic models for the evolution of the region suggest that a collision took place between two segments of oceanic crust, resulting in east-to-northeast-directed subduction beneath the Loyalty Basin. This subduction was interrupted by the arrival of a sliver of continental crust from the west. This material was partially subducted, resulting in stacking of continental thrust slices at depth and crustal thickening. Buoyancy of this material resulted in failure of the subduction zone and emplacement of the ophiolite from the northeast via obduction. Exhumation was accomplished by a combination of buoyancy-driven diapiric uplift, erosion, and tectonic unroofing. Subsequent to the collision, late-stage extensional tectonism resulted in the formation of a series of low-angle ductile shear zones that caused the unroofing and rapid exhumation of the region and, ultimately, the exposure of the blueschists and eclogite rocks.

New Caledonia's steep topography, its high elevation, and a number of geomorphologic clues (such as stranded rivers, elevated coral reefs, and weathering surfaces) indicate that tectonic processes are actively maintaining the relief of the island. This uplift does not appear to be related to any active faulting, because there is very little record of recent seismicity beneath the island. It is likely, however, that this relief is being maintained at least in part by the lithospheric bulge associated with the subduction of the Australian plate in the New Hebrides subduction zone several hundred kilometres to the east.

SEE ALSO THE FOLLOWING ARTICLES

New Caledonia, Biology / Pacific Region / Plate Tectonics / Vanuatu

ADDITIONAL READING

Aitchison, J. C., G. L. Clarke, D. Cluzel, and S. Meffre. 1995. Eocene arc–continent collision in New Caledonia and implications for regional SW Pacific tectonic evolution. *Geology* 23: 161–164.

Black, P. M., and R. N. Brothers. 1977. Blueschist ophiolites in the melange zone, Northern New Caledonia. *Contributions to Mineralogy and Petrology* 65: 69–78.

Clarke, G. L., J. C. Aitchison, and D. Cluzel. 1997. Eclogites and blueschists of the Pam Peninsula, NE New Caledonia: a reappraisal. *Journal of Petrology* 38: 843–876.

Cluzel, D., J. C. Aitchison, G. L. Clarke, S. Meffre, and C. Picard. 1994. Point de vue sur l'évolution tectonique et géodynamique de la Nouvelle-Calédonie. *Comptes Rendus de l'Academie des Sciences, Serie II* 319: 683–690.

Rawling, T. J., and G. S. Lister. 1999. Oscillating modes of orogeny in the Southwest Pacific and the tectonic evolution of New Caledonia, in *Exhumation processes: normal faulting, ductile flow, and erosion*. U. Ring, M. T. Brandon, G. S. Lister, S. D. Willett, eds. Special Publication 154. London: Geological Society of London, 101– 102.

Schellart, W. P., G. S. Lister, and V. G. Toy. 2006. A Late Cretaceous and Cenozoic reconstruction of the Southwest Pacific region: tectonics controlled by subduction and slab rollback processes. *Earth-Science Reviews* 76: 191–233.

NEWFOUNDLAND

HAROLD WILLIAMS
Memorial University, St. John's, Canada

The island of Newfoundland is a textbook example of a collisional geologic mountain belt. It formed through the opening and closing of an ancient Iapetus Ocean, which preceded the modern North Atlantic. The cycle of opening and closing lasted for about 300 million years. The major geologic divisions of Newfoundland represent the margins and vestiges of Iapetus.

GEOGRAPHIC SETTING

The island of Newfoundland forms the northeast extremity of the Appalachian Mountain Belt, or Appalachian Orogen, in eastern North America (Fig. 1). The northeast coastline of the island displays a complete cross section of the Appalachians in superb wave-washed cliff exposures. Since the wide acceptance of continental drift and plate tectonics, many of the concepts of geological mountain building, especially those of continental collisions, originated in Newfoundland. The island exhibits many of the best examples of plate tectonic processes in the remote geological past. Some of its rocks, such as those of Gros Morne National Park, are designated by UNESCO as World Heritage; others, such as those of Fortune Head and Green Point, are the world's type-examples of certain geological boundaries—in this case, the Precambrian–Cambrian and Cambrian–Ordovician, respectively. Many other rocks and fossil localities are of provincial heritage status.

MAJOR GEOLOGIC DIVISIONS

The major divisions of rocks of the Newfoundland Appalachians, from west to east, are the Humber, Dunnage, Gander, and Avalon Zones (Fig. 2). These geological divisions in Newfoundland represent the continental margins and vestiges of an ancient Iapetus Ocean that opened and closed between 600 and 300 million years ago. The

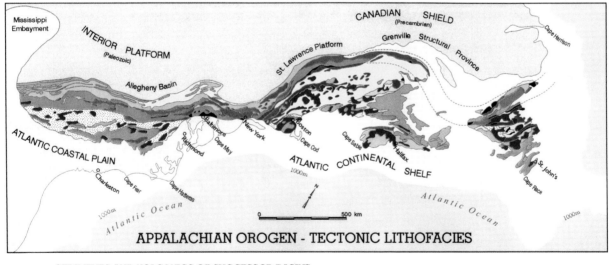

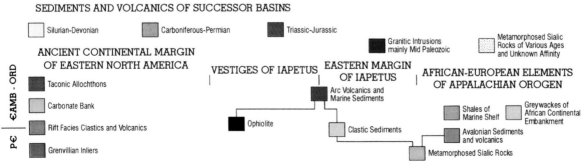

FIGURE 1 Extent of the Appalachian Mountain Belt in eastern North America and interpretation of its major geological divisions.

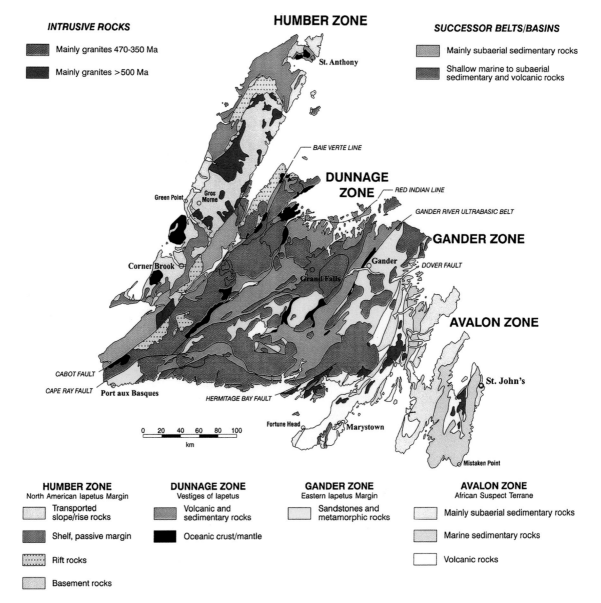

FIGURE 2 Generalized interpretive map of the Newfoundland Appalachians. Modified after a compilation by J. P. Hayes, 1987, and digitized by T. Paltanavage, 1994.

Newfoundland zones have been extended the full length of the Appalachian Mountain Belt from Newfoundland 3000 km southward to Alabama. The closure of Iapetus resulted in the assembly of a supercontinent, Pangea. The break-up of Pangea, beginning 250 million years ago, gave rise to the modern Atlantic Ocean. Opening of the Atlantic Ocean dispersed segments of the Appalachians and correlatives to the North Atlantic borderlands of northwestern Africa, Europe, Scandinavia, and eastern Greenland (Fig. 3).

HUMBER ZONE

The Humber Zone represents the ancient continental margin of eastern North America or the western margin of Iapetus. The Dunnage Zone represents vestiges of Iapetus. The Gander Zone represents the eastern margin of Iapetus, and the Avalon Zone originated somewhere east of Iapetus and is of African affinity.

Rocks and structures of the Humber Zone fit the model of an evolving continental margin and spreading Iapetus Ocean. This began with rifting of existing continental crust at about 600 to 550 million years ago. The rifting is evidenced by liquid injections that filled cracks in the older crust and fed volcanic eruptions. It also led to the deposition of coarse fragmental sedimentary rocks. This was followed by the development of a passive continental shelf with mainly limestone deposition, like that of the present Bahamas, with contemporary continental slope/rise deposits. After about 100 million years, this ended with the deposition of coarse sandstones that are a harbinger of forthcoming catastrophic events. Destruction of the Humber margin is marked by the transport of rocks from the compressed uplifted continental slope and rise landward above the former continental shelf. These transported slope/rise rocks are, in turn, structurally overlain by slabs of Iapetan oceanic crust and mantle, such as the Tablelands of Gros Morne National Park.

The western boundary of the Humber Zone is drawn where deformed rocks of the Appalachians pass into flat-lying rocks of the continental interior. The eastern boundary of the Humber Zone is a steep belt marked by discontinuous occurrences of Dunnage Zone ocean crust and underlying mantle.

DUNNAGE, GANDER, AND AVALON ZONES

The Dunnage Zone is recognized by its abundant volcanic rocks and oceanic crust and mantle rocks. It also contains chaotic mixtures of discrete resistant blocks surrounded by shales. Sedimentary rocks are all of deep marine deposition. The Dunnage Zone is the widest and best preserved in northeastern Newfoundland because of matching morphological embayments in the opposing margins of Iapetus. It is narrow or absent in southwestern Newfoundland, where matching promontories took the brunt of the collision. The boundary between the Dunnage and Gander Zones is also marked by the occurrence of Iapetan oceanic crust and mantle rocks.

The Gander Zone has a thick, monotonous sequence of sandstones, siltstones, and shales that grade eastward into deformed and altered rocks. Its analysis is far less sophisticated than that of the Humber Zone. Almost half of its rocks are granitic intrusions, and half of the remainder are deformed and altered beyond recognition.

The Avalon Zone is defined by its well-preserved sedimentary and volcanic rocks, dated at about 650 to 550 million years. Overlying shales contain a trilobite fauna completely different from trilobites in equivalent rocks of the Humber Zone. The Avalon Zone extends 600 km offshore, making it the broadest zone of the entire

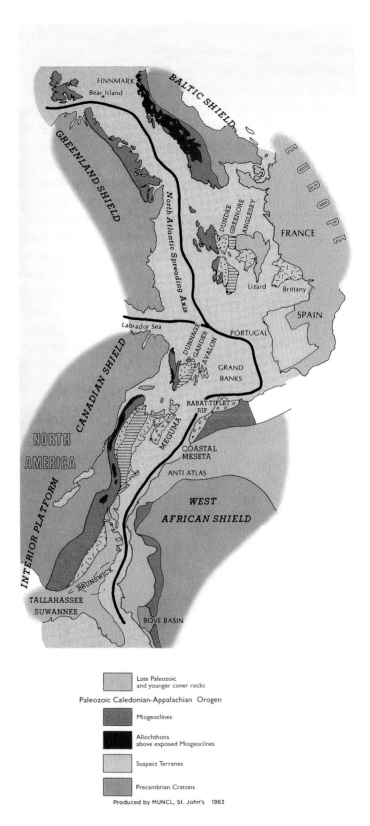

FIGURE 3 Restored North Atlantic region showing the spreading axis of the imminent North Atlantic Ocean and dispersed correlative rocks (blue and yellow colors) of the Appalachian mountain belt.

Appalachians, more than twice the combined width of all other zones. The western boundary of the Avalon Zone is a major fault.

LATER HISTORY

Rocks younger than about 400 million years overlie those of the fundamental zones. They are sedimentary and volcanic rocks that show an upward change from marine to terrestrial, with all rocks deformed together and cut by granites. These changes mark the final closing phases of Iapetus. After Iapetus closure, the youngest Appalachian rocks are everywhere the same, mainly subaerial red and gray sandstones, coal measures, shallow marine limestones, and salt deposits.

OTHER FEATURES

Rocks of the fundamental zones and overlying deposits record the full history of the Iapetus cycle. Apart from kinds of rocks and structures, zones are expressed also by geophysics, paleontology, metallogeny, plutonism, metamorphism, isotopic signatures, and other features. Younger rocks show less contrast because they were deposited during the dying phases of the Iapetus cycle or after its closure.

RELATIONSHIPS BETWEEN THE NORTH ATLANTIC AND IAPETUS

The North Atlantic Ocean and its margins provide an actualistic model for the Iapetus Ocean, the destruction of which led to the Appalachian Mountain Belt. Just as Atlantic rifting involved a broad area of several hundred kilometers, so too did Iapetan rifting extend well inland toward the continental interior. The transition from rifting to continental drifting at the Atlantic margin, defined by seismic reflection, deep drilling, and the age of adjacent oceanic crust, has an Iapetan counterpart in the Appalachian Humber Zone. The widths of the North Atlantic shelf/slope/rise and the thickness of modern sediments are comparable to restored widths of the Humber Zone and thicknesses of its sedimentary rocks. The offshore form of the North Atlantic margin at the Grand Banks mimics an Iapetan promontory in the Gulf of St. Lawrence and provides an explanation for the sinuosity of the Humber Zone along the Appalachian Orogen. The crust and mantle beneath the North Atlantic are analogous to Iapetan volcanic rocks and its oceanic crust and mantle rocks of the Dunnage Zone, and Atlantic micro-continents and oceanic volcanic islands and sea mounts are typical of some Iapetan unexplained terranes.

CONCLUDING REMARKS

Newfoundland has attracted the attention of geologists worldwide, as well as other prominent dignitaries. Visitors are encouraged to visit Gros Morne World Heritage Site as well as the World Heritage Viking Site at nearby L'Anse aux Meadows. The Johnson GeoCentre in the capital of St. John's is also a place of more than passing geological interest.

SEE ALSO THE FOLLOWING ARTICLES

Atlantic Region / Britain and Ireland

FURTHER READING

Williams, H., ed. 1995. Geology of the Appalachian-Caledonian Orogen in Canada and Greenland. Geological Survey of Canada, Geology of Canada, no. 6 (also Geological Society of America, The Geology of North America, v. F-1).

Williams, H., S. A. Dehler, A. C. Grant, and G. N. Oakey. 1999. Tectonics of Atlantic Canada. *Geoscience Canada* 26: 51–70.

NEW GUINEA, BIOLOGY

ALLEN ALLISON
Bishop Museum, Honolulu, Hawaii

New Guinea, the world's largest and highest tropical island, was one of the last parts of the globe to be explored, earning it the nickname "the last unknown." Although many of its species have yet to be scientifically named, we do know that is inhabited by an extraordinarily rich assemblage of plants and animals, derived from both Southeast Asia and Australia, with diversity exceeding that of the much larger Australian continent and rivaling that of the Amazon Basin. Overall it has ~8% of the world's biota. At least 70% of species are endemic.

GEOGRAPHICAL COVERAGE

In this treatment, New Guinea includes the main island and associated satellite islands, such as the Raja Ampat group, the islands in Cenderawasih Bay, and the Aru Islands in the west, and in the east the d'Entrecasteaux and Louisiade groups together with Woodlark Island and the Trobriand Islands. The western half of mainland New Guinea and satellite archipelagoes are politically part of Indonesia. The eastern half, together with archipelagoes to the north and east (which are not included in this treatment), are comprised in the sovereign state of Papua New Guinea.

GEOLOGICAL SETTING

Elements of New Guinea began forming in the late Cretaceous at the leading edge of the Australian tectonic plate. As this plate separated from Antarctica at the beginning of the Cenozoic, it began moving northward, colliding with a complex subduction system that included island arcs, oceanic plateaux, seamounts, and plate fragments. Today, New Guinea is a geological composite consisting of at least 32 separate terranes. The evolutionary history of the biota is linked to the accretion of these terranes to the Australian craton, and to the uplift, volcanism, and rifting that accompanied these tectonic events.

The island can be naturally divided into five main biogeographic provinces, based on geological origin of these regions (Fig. 1): (1) the Australian craton, (2) fold belt (leading edge of the Australian craton), (3) accreted terranes, (4) the Vogelkop composite terrane, and (5) the East Papuan composite terrane. The Aru Islands, located off the south coast of New Guinea, have important biotic differences from the New Guinea mainland, particularly for fishes, and are sometimes recognized as a separate biogeographic province, although part of the Australian craton. Similarly, some of the the Raja Ampat islands at the western tip of New Guinea and a sliver of the northern Vogelkop have a separate geological origin from the adjacent Vogelkop Peninsula and are sometimes treated as a separate province.

Although many genera are widespread, a conspicuous number of genera, or species groups within genera, are restricted to single biogeographic provinces. In addition, the fold belt, which comprises the mountainous spine of the island, has tended to promote speciation through the creation of extensive montane habitat and by dividing the north and south coast lowlands.

COMPOSITION AND RICHNESS OF THE BIOTA

New Guinea was, and arguably still is, one of the world's great biological unknowns. Although early collections were made from coastal areas beginning in the late 1700s, it was not until the mid-1870s that biologists first penetrated the mountainous interior. Large parts of the island still remain unexplored, particularly in the west. It is likely that upwards of half the biota remains unknown to science.

Plants

There are no comprehensive treatments of the flora of New Guinea, and the number of vascular plant species is the subject of considerable controversy, with estimates ranging from 11,000 to more than 30,000.

The composition of the New Guinea flora is generally similar to that of Southeast Asia, and the two areas are often floristically grouped together into a region called Malesia. However, the New Guinea flora tends to have fewer plant families and a much higher proportion of endemic genera and species than do other parts of Malesia. New Guinea is also noteworthy in having a high proportion of orchids and ferns; collectively these groups make up about a quarter of the named plant species. The orchids, which are represented by more than 3000 species, comprise the largest family of flowering plants in New Guinea. The madder family, Rubiaceae, is next, with about 1000 species, followed by the grasses (Poaceae), spurges (Euphorbiaceae [*sensu lato*]) and palms (Arecaceae).

Animals

The vertebrates are far better understood than invertebrates or plants, but many groups remain incompletely known. Although the New Guinea vertebrates form a rich assemblage, many lineages found in Southeast Asia are missing from New Guinea. For example, there are only four major groups, comprising six orders, of terrestrial mammals (monotremes, marsupials [three orders]), rodents, and bats) in New Guinea, in contrast to 10–11 orders in most of Southeast Asia. Only one of the three orders of amphibians (frogs) occurs in New Guinea. A number of bird families that are widespread in Asia (e.g., woodpeckers) do not occur in New Guinea.

The largest vertebrate group in New Guinea is the marine fishes, which make up ~62% of the New Guinea

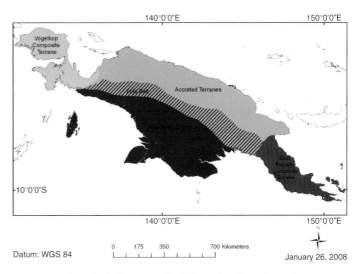

FIGURE 1 Main geological components of New Guinea (based on various sources, particularly Pigram and Davies 1987; Hill and Hall 2003).

vertebrate fauna. Freshwater and brackish fish species comprise nearly 8%. Birds make up nearly 15% of New Guinea vertebrates, followed by amphibians and reptiles at nearly 10%. Mammals account for only 5% of the total.

If we focus on land and freshwater vertebrates (i.e., exclude marine and brackish-water fishes as well as other exclusively marine vertebrates such as sea turtles and sea snakes), there are 1759 land and freshwater vertebrates known from New Guinea; 1247 (71%) of these are endemic. The level of endemism ranges from a low of 50% for crocodiles, which tend to have large geographic ranges, to a high of 93% for frogs, which have limited dispersal capacity and are represented by a high percentage of restricted-range species. Most native species that are not endemic to the island of New Guinea are shared with Australia and to a lesser extent with Maluku, Southeast Asia, or the Bismarck Archipelago and Solomon Islands. A few indigenous species also have extensive ranges in the Pacific Basin.

New Guinea vertebrates comprise 8% of currently recognized world vertebrate species and range from about 4% of the world total for snakes and lizards to a high of nearly 11% for fishes (Table 1). If we restrict the comparison to land and freshwater vertebrates, including fishes, the New Guinea proportion drops to about 4% of the world total; it is 6.5% if fishes are excluded.

The freshwater fish fauna is relatively small, with about 400 species known, representing less than 1% of world fishes. A few of these are primary-division fishes (species belonging to families that have always been intolerant of saltwater), but most are thought to have evolved from

FIGURE 2 *Choerophryne* sp. This small treefrog is about 25 mm long and inhabits mid-montane rain forest on the northeast coast of New Guinea.

FIGURE 3 *Varanus salvadorii*. This large lizard, which is widespread in the lowlands of western and central New Guinea, can exceed 3 m in total length.

TABLE 1
The Vertebrate Fauna of New Guinea in Comparison with That of the World

Taxon	New Guinea	World	New Guinea as % of World
Fishes	3,200	30,000	10.7
Frogs	314	5,500	5.7
Turtles	18	308	5.8
Crocodiles	2	23	8.7
Lizards	208	5,002	4.2
Snakes	113	3,113	3.6
Birds	553	9,704	5.7
Mammals	341	5,415	6.3
Total	4,749	59,065	8.0

NOTE: Figures for fishes are from Allen (1991, 2007) and Allen and Swainston (1992). Amphibian and reptile data for New Guinea are from Bishop Museum (2007), and for the world are from Frost (2007). Data for world reptiles are from J. Craig Venter Institute (2007). Data for birds are from Mack and Dumbacher (2007). Mammal data are from Helgen (2007). Where necessary, figures have been updated to include recently described species.

marine ancestors. About three-quarters of the freshwater fishes lack a marine larval stage, and of these more than 80% are endemic to New Guinea.

The marine fishes present quite a different pattern and include an estimated 2600 to 3000 species, which represent ~11% of the world total. About a third of the world's reef fishes occur in New Guinea, although endemism is low and many species occur widely in the Indo-Pacific region.

The frogs are probably the most poorly known group of New Guinea vertebrates. There are currently 314 species known from New Guinea, but this total will almost certainly double when all species have been scientifically named. Each major herpetofaunal survey during the past five years has turned up five to ten, and sometimes more, new species in a single drainage basin. All but 23 of New Guinea's frogs are endemic. The indigenous species are generally shared with Australia.

The lizards are somewhat better known than frogs, with 208 described New Guinea species. However, dozens of new species are known to reside in collections, and ongoing studies of several large genera are likely to result in the recognition and description of many additional species. The overall total is likely to reach around 300 species. Lizards tend to have a higher capacity for dispersal than do frogs, and only about two-thirds of them are endemic. The indigenous species include widespread Indo-Pacific species as well as species that are shared only with Australia or Maluku.

The New Guinea snake fauna is relatively depauperate, with 113 species. Only 87 of these are terrestrial; the rest are marine. About half the terrestrial species are endemic, while only two of the marine species are endemic. As a result of ongoing systematic studies and field surveys, the total number of New Guinea snake species is likely to reach about 140.

The New Guinea herpetofauna includes a number of charismatic species, such as the largest tree frog in the world (*Litoria infrafrenata*), some of the world's smallest frogs (genus *Oreophryne*), frogs with peculiar snouts (genus *Choerophryne*; Fig. 2), the world's largest crocodile (*Crocodylus porosus*), and the world's longest lizard (*Varanus salvadorii*; Fig. 3).

Birds are generally the most conspicuous elements of the vertebrate fauna and, because of their vagility, tend to have larger geographic ranges than do the more sedentary amphibians and reptiles. They also have a much larger popular following and for these reasons are far better studied than other vertebrate groups. New species are still occasionally discovered, but the cumulative number of species for New Guinea has reached a plateau, and the current total of 579 resident species, of which 325 (56%) are endemic, is not expected to increase by much with further surveys.

Perhaps the most famous group of birds in New Guinea are the birds of paradise (family Paradisaeidae; Fig. 4), which comprise 42 species, most of which are endemic. They have long fascinated naturalists because of the males have extremely colorful plumage and elaborate courtship displays. In addition, 11 of the world's 18 species of bowerbirds (family Ptilonorhynchidae) occur in New Guinea. Bowerbirds, which also occur in Australia, are renowned for constructing elaborate structures called "bowers" that are used for courtship displays. New Guinea is also home to three species of cassowaries—large flightless birds distantly related to ostriches—as well as the world's largest pigeon (*Goura victoria*) and, at the other extreme, the world's smallest parrots (*Micropsitta* spp.).

Another group of interesting birds in New Guinea are the pitohuis, six passerine bird species that contain neurotoxic alkaloids. These compounds, which are chemically related to batrachotoxin (found in Central and South American poison arrow frogs of the family Dendrobati-

FIGURE 4 *Paradisaea raggiana*. The raggiana bird of paradise is found throughout much of eastern New Guinea and is a national symbol of Papua New Guinea.

dae), are thought to ward off ectoparasites and predators. Several species are brightly colored, suggesting that this is aposematic or warning coloration. Geographic races of other pitohui species sometimes mimic these brightly colored species, a phenomenon known as Müllerian mimicry, in which the mimic and model are both toxic, but gain a mutual advantage by sharing, and therefore reinforcing, the same anti-predator warning coloration and pattern.

Mammals are much more poorly known than birds. More than half the ~340 mammal species known from New Guinea have been described since 1900. At least two-thirds of these species are endemic. Recent taxonomic work has revealed the existence of many cryptic species, and field surveys have turned up dozens of undescribed taxa. It is likely that more than 100 species remain to be named.

The monotremes, an ancient group of egg-laying mammals now restricted to Australia and New Guinea, are perhaps the most noteworthy elements of the vertebrate fauna (Fig. 5), followed by the marsupials, which include tree kangaroos and small rat-sized carnivores. There are at least two, and probably at least four, species of monotremes in New Guinea (one shared with Australia). There are also large radiations of bats and rodents.

The major groups of insects and other invertebrate groups inhabiting New Guinea are similar to those found

FIGURE 5 *Zaglossus bruijni*. One of three monotremes or egg-laying mammals in the world, this species is endemic to New Guinea and is found in montane regions throughout much of the island.

in Southeast Asia. Although there are no comprehensive checklists for most groups of invertebrates, an estimated 80,000 species are known. This probably represents less than a quarter of the actual total, and it is likely that the overall number exceeds 300,000 species. However, a few groups are relatively well known, and some trends can be identified. In the best-known group, butterflies, the descriptions of new species appears to be approaching an asymptote, suggesting that New Guinea has around 960 species or 5.5% of the world total. The Odonata, which includes dragonflies and damselflies, are also relatively well known at a world level. New Guinea has around 580 species, or 9.4% of the world total. Aquatic Heteroptera, which include nearly 5000 species worldwide, have been well surveyed in New Guinea and include around 350 species, or about 7% of the world total. Approximately 70–80% of New Guinea insects are endemic.

Recent ecological studies have demonstrated that trees in New Guinea and those in temperate regions support similar numbers of insect species, and that levels of host-specificity are similar in both areas. This suggests that the high diversity of insects found in New Guinea, and the tropics generally, may be related to much higher levels of plant diversity found in the tropics compared to temperate regions (a sevenfold difference). Interestingly, these recent studies have also demonstrated that herbivorous insects tend to feed on clades of closely-related species (e.g., genera) and are not, as had been commonly assumed, mostly restricted to individual tree species.

Some of the particularly noteworthy elements of the New Guinea insect fauna include the world's largest butterflies and moths (the Queen Alexandra's Birdwing, *Ornithoptera alexanderae*, and the atlas moth, *Attacus atlas*, respectively); stick insects reaching 20 cm or more in total length), and the bizarre stalk-eyed and moose-antlered flies in which the males joust for access to females for mating. There are also long-lived, high-elevation beetles (genus *Gymnopholus*) that have pits on their dorsum that support a growth of algae and other primitive plants, providing excellent camouflage for the host, which would otherwise be highly visible in its mossy forest habitat. Rotifers, protozoans, and other microorganisms live within the plants, a phenomenon termed episymbiosis.

DISTRIBUTION PATTERNS

Biogeographical Patterns

The biogeographic affinities of New Guinea vertebrates are complex. Repeated land connections during the late Cenozoic facilitated faunal exchange between New Guinea and Australia, resulting in many similarities of the vertebrate faunas of these two areas, particularly at the level of genus and higher. This is particularly true for mammals, for which marsupials and monotremes are such important and distinctive components of the fauna, but is also true for many groups of birds and other vertebrates. For example, of the 119 genera of amphibians and reptiles that occur in New Guinea, 56 (47%) also occur in Australia.

New Guinea also has a strong affinity with Southeast Asia, but oceanic barriers have always separated it from that region, as demarcated by the famous "Wallace's Line." These oceanic barriers, which prevented a wide array of vertebrate lineages from reaching New Guinea, have been less of a barrier to plants and invertebrates, which are largely of Southeast Asian origin.

The geological history of New Guinea has strongly influenced the distribution of plants and animals. Its many isolated mountain ranges and lowland basins have produced pockets of endemism, particularly beyond the northern margins of the Australian craton, where, for example, lizard species richness is greatest. In contrast, the central mountains have the richest assemblages of frogs.

The separation of northern and southern lowlands by the central mountains also strongly influences biotic distributions, even in relatively recently evolved groups such as the seven birds of paradise in the genus *Paradisaea*, which are probably are no more than 500,000 years old (Fig. 6).

Ecological Patterns

Although mean temperatures do not show much seasonality, temperatures are strongly influenced by elevation, ranging from a mean of around 26 °C in the lowlands down to 5–7 °C on the high peaks. These differences in temperature, together with considerable geographic variation in the total amounts and seasonal patterns of rainfall, have combined with the geological history to produce a rich array of habitats and ecological diversity.

Focusing on vegetation, which is closely linked to climate and substrate, much of the coast, particularly in the south, is fringed by mangroves, which make up about 4% of forested areas. Other broadleaf forests—mostly rain forests—cover about two-thirds of the island. These forests, which are floristically diverse and occur to around 3000 m, generally decrease in canopy height and tend to have fewer lianas and a lower incidence of tree buttressing with increasing elevation. A band from about 3000 to 3200 m is dominated by shrubs and small trees. There is a shrub-grassland mosaic from 3200 to 3400 m. Areas above 3400 m and extending to about 4200 m are dominated by grasslands, sometimes with pockets of shrubbery, particularly on slopes. Areas above 4200 m, which include a total of area of 5000 km^2, support a tundra-like vegetation or are covered in bare rock. The highest mountain in New Guinea, Puncak Jaya (4884 m), supports a permanent glacier.

Large parts of southern New Guinea, south of the Fly and Digul Rivers and covering about 15% of the island area, have strongly seasonal rainfall patterns and are covered in woodland savanna or seasonally deciduous monsoon forest. Similar forests are found along the southeast coast of Papua New Guinea, and patches extend nearly to the eastern tip of the island and comprise another 5% of the island's land area.

More than 31% of the land area of New Guinea is above 1000 m, and nearly 2% is above 3000 m elevation. Many of the mountains are the result of rapid uplift during the Tertiary period. This has tended to promote speciation along elevational gradients in which closely related species occupy contiguous and often mutually exclusive elevational ranges. This is seen, for example, in the lizard genus *Papuascincus*. One species, *P. morokanus*, occurs from about 900 m to 1800 m throughout much of montane New Guinea. A second species, currently unnamed, occurs from 1800 m to nearly 2600 m. A third species, *P. stanleyanus*, occurs from about 2000 m to nearly 2800 m. Similarly, three species of acanthizid mouse-warblers of the genus *Crateroscelis* occur throughout much of New Guinea. *Crateroscelis murina* occurs from sea level to about 1700 m. A montane species, *C. robusta*, generally occurs from 1750 to 3600 m. A third species, *C. nigrorufa*, which has an extremely patchy distribution, occupies an intermediate altitudinal band from 1300 to 2000 m.

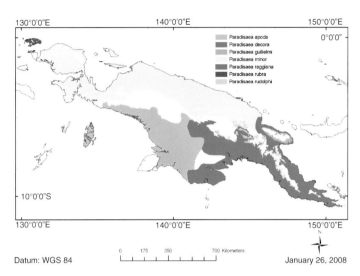

FIGURE 6 Distribution of the seven species of birds of paradise of the genus *Paradisaea* in New Guinea (data from Frith et al. 1998). The red bird of paradise, *P. rubra*, is endemic to islands of Waigeo and Batanta in the Raja Ampat group at the western trip of New Guinea. Adjacent areas of the New Guinea mainland, such as the Vogelkop Peninsula, are occupied by the lesser bird of paradise, *P. minor*, which also occurs throughout much of western New Guinea. The greater bird of paradise, *P. apoda*, occupies southwestern New Guinea including the Aru Islands. The Raggiana bird of paradise, *P. raggiana*, occupies much of eastern New Guinea. The blue bird of paradise, *P. rudolphi*, occupies montane regions of central New Guinea. The final two species have restricted ranges: the emperor bird of paradise, *P. guilielmi*, is endemic to the Huon Peninsula of northern Papua New Guinea; and Goldie's bird of paradise, *P. decora*, is endemic to Fergusson and Normanby islands in the Entrecasteaux Archipelago near the eastern tip of New Guinea. Both these areas were created by relatively recent geological uplift.

COMPARISON OF NEW GUINEA AND BORNEO

In order to put the biota of New Guinea into a regional perspective, it is useful to compare it with the island of Borneo. The second largest tropical island in the world, Borneo is moderately smaller than New Guinea (743,330 km^2 vs. a land area of 790,000 km^2 for New Guinea). Both islands have high mountains (>4000 m) and formed through complex processes of accretion of island arcs, oceanic crustal material, and other tectonic fragments onto Paleozoic continental cores. Borneo began forming during the Mesozoic and was connected by land to mainland Southeast Asia until at least the Eocene; it has subsequently been reconnected to that region for varying amounts of time coinciding with periods of lowered sea level during the Tertiary and Quaternary. It is today topographically far less complex than New Guinea, with wide expanses of lowlands surrounding the central mountains.

The occurrence of past land connections between Borneo and continental Asia, the uniformity of the climate throughout the region during the Tertiary, and the absence of much topographic relief has resulted in the fauna of Borneo being rather similar to that of other parts of SE Asia. New Guinea, by contrast, with its more complex geological history, its relative isolation from continental areas with similar climates, its greater topographic diversity, and the impact of these factors on the evolutionary diversification of species, has developed a richer and more endemic biota than has Borneo. However, Borneo is generally richer in higher taxa or lineages (e.g., families and orders).

The two islands are thought to have about the same number of plant species, approximately 14,000. Forests in Borneo have been shown to be exceedingly rich in numbers of species; by some accounts they are the richest forests in the world. New Guinea's forests are less rich but have a much higher proportion of endemic plant species than do Bornean forests. New Guinea is much more poorly known than Borneo and it is likely that with further collecting the total number of plants occurring there will ultimately exceed the total for Borneo.

There are 11 orders and at least 219 species of mammals from Borneo; 46 (21%) species are endemic. New Guinea has six orders of mammals (monotremes, three orders of marsupials [Dasyuromorphia, Peramelemorphia, and Diprotodontia], rodents, and bats) with 341 species, of which 252 (74%) are endemic. A total of 434 species of breeding birds are known from Borneo, but only 39 of these are endemic. New Guinea. by contrast. has 579 species of breeding birds, of which 325 species are endemic.

Borneo has a higher number of families and genera of amphibians and reptiles (30) than does New Guinea (23) and, in addition, has caecilians, an amphibian order that is absent from New Guinea. At the species level Borneo also has a higher number of snakes (157) than does New Guinea (113), reflecting the importance of this group in Southeast Asia. However, it has only 150 species of frogs and 117 species of lizards. In comparison New Guinea has 314 frogs and 208 lizards. Overall Borneo has only about two-thirds the number of species that are found in New Guinea.

Borneo and New Guinea both have a similar number of freshwater fishes, about 400 species, but around a quarter of the freshwater ichthyofaunas of both areas are composed of fishes with a marine larval stage. Borneo is dominated by primary-division fishes; about 40% of these species are endemic. In contrast, New Guinea essentially lacks primary-division fishes, and most species of its freshwater fishes are thought to have evolved from marine ancestors; approximately 180 (84%) of currently recognized New Guinea freshwater species lacking a marine larval stage are endemic. In general, species richness in Borneo tends to be higher within a drainage basin than in New Guinea.

For example, the large Kapuas River system in western Kalimantan has 320 species. By contrast, the Fly River drainage in New Guinea has fewer than 110 species.

The New Guinea region is estimated to harbor around 2600 to 3000 species of marine fishes, including ~30% of the world's reef fishes; Borneo likely has a comparable number of marine species. Many of the marine species are widespread in the Indo-Pacific region and are found in both Borneo and New Guinea. Species richness of the marine biota of the Indo-Australian region is among the highest is the world.

These differences tend to highlight the fact that while New Guinea has been colonized by fewer evolutionary lineages than has Borneo, the complex geological history of New Guinea and its topographic diversity has resulted in enormous adaptive radiations of species that have produced much higher levels of species richness in New Guinea than have been attained in Borneo.

HUMAN IMPACTS AND FUTURE CONSIDERATIONS

New Guinea has been inhabited by humans for 50,000 years. Agriculture first developed there some 6000 years ago, about the same time as food plants were first cultivated in the Fertile Crescent of Asia. Initial human population density in New Guinea was low and is still only around 10–15 people per square kilometer, less than half that of the United States. About a third of New Guinea's human population of about nine million people is concentrated along the coastal fringe, where slash and burn is the prevailing type of agriculture, and the other two-thirds are in highland areas, where large valleys are covered in grassland, likely the result of frequent burning to clear fields for agriculture. However, human impacts tend to be localized, and overall less than 1% of the island is covered in urban or built-up areas or agricultural systems; vast areas of the island are virtually uninhabited and seemingly pristine.

This has led to a persistent belief among biologists, conservation professionals, and others that New Guinea is in no immediate conservation peril. However, huge tracts of old-growth forests are being lost to timber harvesting and mining. Approximately half of forests suitable for logging (on slopes < 30%) have been logged, and logging concessions have been granted for much of the rest. Land conversion to cash-crop agriculture, particularly oil palm plantations, and to urban development are taking an increasing toll. The introduction of alien species, illegal collecting driven by the pet trade, and other factors are also endangering elements of New Guinea's biodiversity.

The 1992 Papua New Guinea Conservation Needs Assessment and the 1997 Irian Jaya Conservation Priority-Setting Workshop helped to focus attention on areas of endemism in New Guinea. Subsequent work has largely tended to confirm the importance of areas identified in these exercises and highlighted the need for additional field surveys. There is a compelling, indeed crucial, need for a comprehensive biological survey of the island to better document the biota and to guide and inform the designation of protected areas to help preserve some of the richest assemblages of biodiversity on the planet.

SEE ALSO THE FOLLOWING ARTICLES:

Borneo / New Guinea, Geology / Wallace's Line

FURTHER READING

Beehler, B. M., T. K. Pratt, and D. A. Zimmerman. 1986. *Birds of New Guinea*. Princeton, NJ: Princeton University Press.
d'Abrera, B. 1977. *Butterflies of the Australian region*. Melbourne: Lansdowne Press.
Flannery, T. F. 1995. *Mammals of New Guinea*. Ithaca, NY: Comstock/Cornell.
Gressitt, J. L. 1982. *Biogeography and ecology of New Guinea*. Monographiae Biologicae. The Hague: W. Junk.
Marshall, A. J., and B. M. Beehler, eds. 2007. *The ecology of Papua*. Singapore: Periplus Editions.
Novotny, V., P. Drozd, S. E. Miller, M. Kulfan, M. Janda, Y. Basset, and G. D. Weiblen. 2006. Why are there so many species of herbivorous insects in tropical rainforests? *Science* 313: 1115–1118.

REFERENCES

Allen, G. R. 1991. *Field guide to the freshwater fishes of New Guinea*. Madang, Papua New Guinea: Christensen Research Institute.
Allen, G. R. 2007. Fishes of Papua, in *The ecology of Papua: Part 1*. B. M. Beehler and A. J. Marshall, eds. Singapore: Periplus Editions, 637–653.
Allen, G. R., and R. Swainston. 1992. *Reef fishes of New Guinea: a field guide for divers, anglers and naturalists*. Madang, Papua New Guinea: Christensen Research Institute.
Bishop Museum. 2007. Papuan herpetofauna: a Bishop Museum project. http://www.bishopmuseum.org/research/pbs/papuanherps/.
Frith, C. B., B. M. Beehler, and W. T. Cooper. 1998. *The birds of paradise: Paradisaeidae*. Oxford: Oxford University Press.
Frost, D. R. 2007. Amphibian species of the world: an online reference. Version 5.1 (October 10, 2007). http://research.amnh.org/herpetology/amphibia/index.php.
Helgen, K. M. 2007. A taxonomic and geographic overview of the mammals of Papua, in *The ecology of Papua: Part 1*. B. M. Beehler and A. J. Marshall, eds. Singapore: Periplus Editions, 689–749.
Hill, K. C., and R. Hall. 2003. Mesozoic–Cenozoic evolution of Australia's New Guinea margin in a West Pacific context, in *Evolution and dynamics of the Australian plate*. R. R. Hillis and R. D. Müller, eds. Special Paper 372. Boulder, CO: Geological Society of America, 265–290.
J. Craig Venter Institute. 2007. The reptile database. http://www.tigr.org/reptiles/search.php.
Mack, A., and C. Dumbacher. 2007. Birds of Papua, in *The ecology of Papua: Part 1*. B. M. Beehler and A. J. Marshall, eds. Singapore: Periplus Editions, 654–688.
Pigram, C. J., and H. L. Davies. 1987. Terranes and the accretion history of the New Guinea orogen. *BMR Journal of Australian Geology* 10(3): 193–211.

NEW GUINEA, GEOLOGY

HUGH L. DAVIES

University of Papua New Guinea

The island of New Guinea is made up of elements of Australian and Pacific geology. It is in a dynamic part of the world and has been the subject of benchmark studies into plate tectonics, ophiolite, rifting of continents, Quaternary sea levels, and other fields. Because the geological history of the island is relatively short, it is more readily deciphered than for older regions of Earth's crust.

PHYSIOGRAPHY

New Guinea lies across the northern margin of Australia. It is the second largest island in the world, 2200 km long and up to 750 km wide, and one of the most mountainous, with peaks to almost 4900 m above sea level. A central mountain range runs the length of the island and is bounded to the north by lesser mountain ranges and plains and to the south by a broad plain. The Mamberamo and Sepik rivers drain the north side of the central range, and the Digul and Fly rivers drain the south (Fig. 1).

Beyond the southern plains a broad, shallow shelf extends to the Australian coast. Other shorelines are steeper, and some are bounded by deep-sea trenches (Fig. 2). Small ocean basins lie to the northeast and southeast and a great submarine plateau (Ontong Java Plateau) lies to the extreme northeast, beyond the islands of the Bismarck Archipelago. Smaller submarine plateaus lie south of the eastern part of New Guinea.

Politically, the island is divided between the independent state of Papua New Guinea (PNG) in the east and

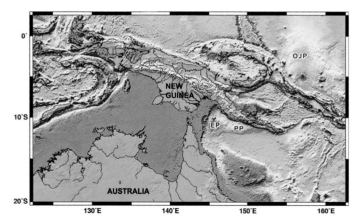

FIGURE 1 Physiographic map of New Guinea (seafloor topography from Smith and Sandwell 1997; Satellite Geodesy, Scripps Institute of Oceanography 2008). OJP, Ontong Java Plateau; EP, Eastern Plateau; PP, Papuan Platform.

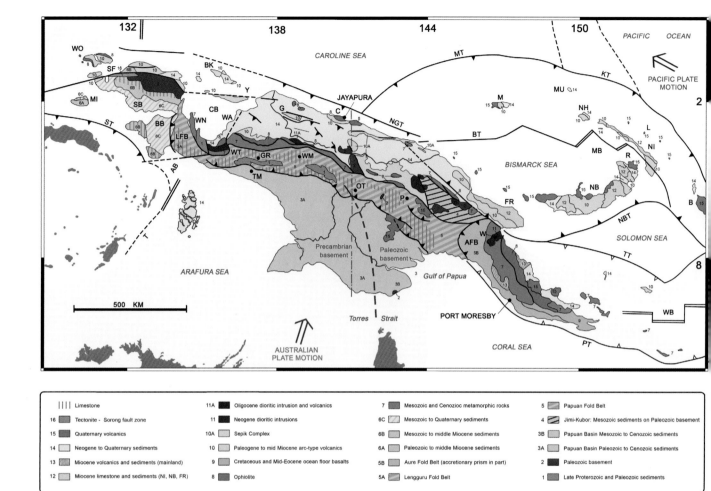

FIGURE 2 Geological map of New Guinea. AB, Aru Basin; AFB, Aure fold belt; B, Bougainville; BB, Bintuni Basin; BK, Biak; BT, Bismarck Sea Transform; C, Cyclops Mountains; CB, Cenderawasih Bay; FR, Finisterre Range; G, Gauttier Range; GR, Grasberg Mine; KT, Kilinailau Trench; L, Lihir Island (mine); LFB, Lengguru Fold Belt; M, Manus; MB, Manus Basin; MI, Misool; MT, Manus Trench; MU, Mussau; NB, New Britain; OT, Ok Tedi; P, Porgera; PT, Pocklington Trough; R, Rabaul; SB, Salawati Basin; SF, Sorong Fault; ST, Seram Trench; T, Timor Trough; TT, Trobriand Trough; W, Wau; WA, Waipona Basin; WB, Woodlark Basin; WM, Wamena; WN, Wandamen Peninsula; WO, Waigeo; WT, Weyland Thrust; Y, Yapen.

Indonesia in the west, with a boundary that coincides, for the most part, with the 141° E meridian. The western half was known as Irian Jaya and is now divided into two provinces known as Western Irian Jaya (the westernmost point of the island, commonly known as the Bird's Head and Neck peninsula) and Papua (the rest of the Indonesian part of New Guinea). The population of PNG is 6 million and is predominantly Melanesian. The population of the Indonesian provinces is 2.1 million and comprises 60–70% indigenous Melanesian and 30–40% migrants from Java, Sulawesi, and Ambon.

GEOLOGICAL SETTING

New Guinea is at the interface between the northward-moving Australian plate and the west-northwest-moving Pacific plate. The resultant motion is convergence at a rate of 110 mm/yr on an azimuth close to 070°.

Convergence has led to a succession of collisions of the Australian craton with the microcontinents and volcanic islands of the Pacific and with fragments of the craton that had been separated from the craton and then docked again. While the southern half of the island was always part of the Australian continent, the northern part has been added by successive collisions. In geological terms, the southern part is autochthonous and the northern part allochthonous, being made up of accreted terranes.

The boundary between the Pacific and Australian plates is marked by a number of microplates. Those offshore are bounded by spreading ridges, deep-sea trenches, and transform faults, and those onshore by thrust, extensional, and strike-slip faults and folds. Earthquakes are located on the microplate boundaries (Fig. 3). Volcanic activity is associated with the deep-sea trenches and spreading ridges.

THE PAPUAN BASIN

The Papuan Basin occupies all of autochthonous New Guinea. Sediments of the Papuan Basin underlie the southern plains and are exposed in the fold belt, where the strata have been folded and faulted.

The basin is underlain by Australian craton of Precambrian age in the west and of Paleozoic age in the east. The sedimentary section in the west is 16 km thick and includes Late Proterozoic and Paleozoic strata. The sedimentary section in the east is 4 km thick and is entirely Mesozoic and Cenozoic.

The Mesozoic sediments in the east are similar to those in the west (Kembelangen Group) but tend to be less mature and less well sorted. Quaternary volcanoes are present only in the east.

In both the east and the west, the sediments at the northern margin of the Papuan Basin fold belt have been metamorphosed to phyllitic and micaceous graphitic schists. In the east the transition from unmetamorphosed to metamorphosed sediments is gradational, and the metamorphic rocks are included in the area shown as fold belt on the map. In the west the contact is faulted and the metamorphosed sediments, along with other metamorphic rocks, have been picked out as a separate rock unit.

The evolution of the Papuan Basin after the Permian period took place in five stages (Figs. 3–4):

1. Triassic and Jurassic rifting of the northern margin of the Australian continent, accompanied by the development of rift-related volcanics and syn-rift pockets of sediment
2. Jurassic and Cretaceous postrift copious siliciclastic sedimentation on the subsiding rifted margin, with some volcanic activity in Early Cretaceous
3. End-Cretaceous uplift and erosion of Cretaceous sediments in the east, due to thermal uplift associated with the opening of the Coral Sea Basin
4. Paleocene and Eocene to mid-Miocene: Slow subsidence and deposition of limestone followed by calcareous shale, except in the extreme northeast, where there was rapid clastic sedimentation from an emerging part-volcanic mountain mass
5. Late Miocene (12–8 million years ago) to present: Development of fold belt, uplift of mountains, rapid erosion and deposition of coarse clastic sediments, accompanied by volcanic activity

Much of the fold belt comprises a broad, asymmetric, south-facing, thrust-bounded anticline upon which are superimposed lesser structures. East of 142° E the single anticline gives way to parallel thrust slices and valley-and-ridge topography.

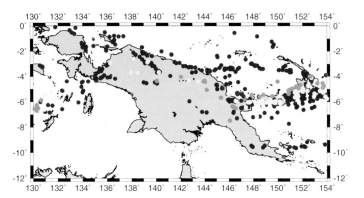

FIGURE 3 Earthquakes in the period 1963–2004, stronger than magnitude 6. Focal depths as follows: red < 50 km, yellow < 100 km, green < 200 km, blue < 300 km, purple < 400 km, brown < 500 km, gray > 500 km. Shallow earthquakes (< 50 km) mark active plate boundaries. Map by Emile Okal.

The strike of the fold belt changes at the international border. This may indicate that the border coincides with the contact between Paleozoic basement in the east and Precambrian in the west. However, the presence of Permo-Triassic metamorphic rocks further west, in the upper reaches of the Eilanden River at 140.2° E, suggests that the contact may be more complex.

JIMI-KUBOR TERRANE

The Mesozoic and Cenozoic sediments of the Kubor Range and the Jimi River drainage rest on Paleozoic basement. The sequence is broadly similar to the Papuan Basin sequence and is regarded by some as an *in situ* part of the Papuan Basin. However, there are sufficient differences to suggest that Jimi–Kubor is a terrane, probably one that separated from the Australian margin in the Triassic and redocked with the margin in the Paleocene or Early Eocene.

TERRANES OF THE CENTRAL RANGE, 136–141° E

In western New Guinea, along the north side of the central range, there is a belt of metamorphic rocks. This is bounded northward by ophiolite and, beyond the ophiolite, Cenozoic volcanic arc rocks.

The metamorphic rocks, for the most part, are graphitic schists formed from the black shales and siltstones of the Kembelangen Formation and are thus part of the Papuan Basin sequence. However, there are blueschist facies metabasites, eclogites, and high-temperature amphibolites toward the northern margin. Such rocks typically are associated with arc-continent collision and the emplacement of ophiolite.

Ultramafic and gabbroic rocks of the ophiolite, with minor basalt, are exposed for a length of 450 km on the north side of the central range. The ophiolite is thought to represent Late Cretaceous oceanic crust and mantle and to have been emplaced in the Late Cretaceous. Soils overlying ultramafic rocks are depleted in the lighter elements and enriched in iron and support vegetation poorly, with the result that the boundary of the ultramafic rocks can be mapped on vegetation pattern alone.

Arc-type volcanic rocks and associated dioritic intrusive rocks that adjoin the ophiolite in the northwest are Late Miocene, whereas those further east are Late Eocene to Oligocene.

ISOLATED TERRANES NORTH OF THE CENTRAL RANGE, 136–141° E

North of the central range is the Mamberamo basin, a vast area underlain by Late Miocene to Quaternary folded and faulted clastic sediments and some limestone. Within the basin are isolated blocks of basement rocks. The Gauttier mountain block comprises ultramafic rocks with some Cenozoic volcanic arc rocks. The Efar and Sidoas mountain blocks, adjacent to the north, are of ultramafic rocks.

The Cyclops Mountains, near Jayapura, comprise ultramafic and gabbroic rocks and moderate to high-grade metamorphic rocks: amphibolite, gneiss, and schist. Immediately adjacent are Miocene sediments and Cenozoic volcanic arc rocks.

Directly south of the Cyclops Mountains are the Border Mountains, made up of Paleozoic metasedimentary rocks intruded by Permo-Triassic granodiorite, diorite, and gabbro with some ultramafic rocks and andesitic porphyry. The metamorphic rocks include amphibolite, schist, quartzite, and garnet gneiss.

TERRANES OF EASTERN NEW GUINEA 141–146° E

The Sepik complex comprises faulted blocks of metamorphic, ophiolitic, Eocene volcanic arc, and sedimentary rocks, with Oligocene to Miocene dioritic intrusive rocks

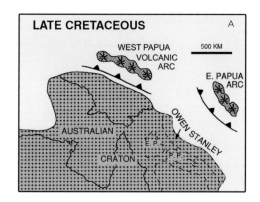

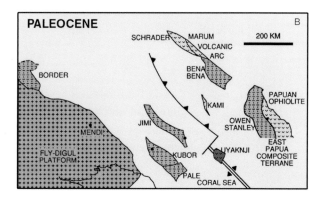

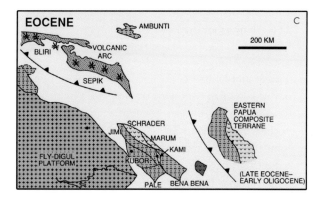

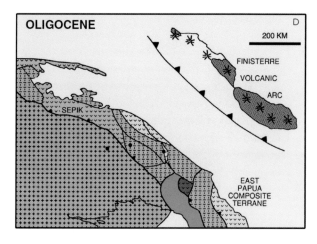

FIGURE 4 History of accretion of terranes. (A) Late Cretaceous: Arc-continent collision in Late Cretaceous in west emplaced Irian ophiolite. (B) Paleocene: Rifting of continent margin to produce Jimi–Kubor terrane, opening of Coral Sea; arc-continent collisions formed East Papua Composite Terrane (EPCT) and captured Marum ophiolite. (C) Eocene: Jimi-Kubor sutured to craton; Salumei volcanic arc develops. (D) Oligocene: Arc-continent collision formed the Sepik Complex; Finisterre volcanic arc developed; EPCT sutured to craton toward end of Oligocene.

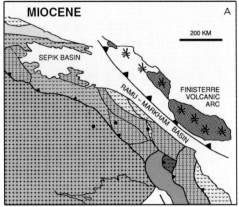
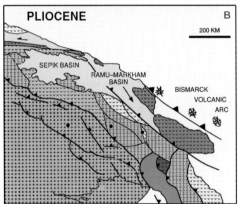
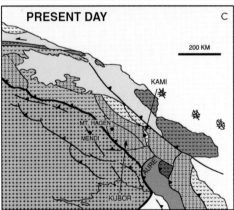
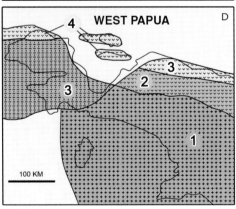

and Miocene-Quaternary sedimentary cover. The metamorphic rocks include blueschist facies metabasites and eclogite. The complex extends from the northern margin of the fold belt north to about the center line of the northern ranges and underlies the Miocene-Quaternary sediments of the Sepik basin. The complex includes Cretaceous high-grade metamorphic rocks associated with ultramafic rocks in the northeast.

Basement rocks of the northern slopes of the Bewani–Torricelli ranges are Oligocene to earliest Miocene arc volcanic rocks and associated sediments and may be related to the arc volcanic rocks of the Finisterre Range. Near the international border, Miocene limestone partly conceals ultramafic rocks that may be an extension of the ultramafic rocks of the Cyclops Mountains.

The Adelbert, Finisterre, and Saruwaged Ranges comprise Oligocene to Early Miocene arc volcanic rocks overlain by Miocene and younger limestone. The structure of the mountains is a south-facing thrust-based antiform marked by limestone dip-slopes on the north side and by rapidly eroding volcanic rocks in the south.

WESTERNMOST NEW GUINEA: THE BIRD'S HEAD AND NECK, 130–136° E

The area north of the Sorong Fault comprises terranes that are mostly but not entirely of oceanic affinity. South of the Sorong Fault are terranes of continental affinity.

Rocks north of the Sorong Fault include Permo-Triassic granite, Cenozoic volcanic arc rocks, and younger sediments; ophiolite is present in adjacent islands. South of the Sorong Fault a basement of Paleozoic low-grade metamorphics is exposed in the northern mountains and on Misool Island. This is overlain by a thick sequence of platform sediments of Permian to Middle Miocene age that comprise the Salawati and Bintuni sedimentary basins.

In the Bird's Neck area, south of Cendrawasih Bay, the Weyland Thrust has transported Paleozoic sediments, metamorphic rocks, and ophiolite slices and Miocene diorites southward over a terrane of continental affinity.

FIGURE 5 History of accretion of terranes. (A) Miocene: Finisterre volcanic arc sutured to continent, development of Ramu-Markham and Sepik basins, Maramuni volcanic arc igneous activity triggered by uplift rather than subduction? (B) Pliocene: Converging of continent and Bismarck volcanic arc caused thrust faults to develop in Finisterre terrane and in Papuan Basin fold belt. (C) Present day. (D) Bird's Head terrane history: 1. Australian Precambrian craton; 2. terranes that had docked by 25 million years ago; 3. terranes that had docked by 10 million years ago, those in Bird's Head have Paleozoic continental basement; 4. terranes that had docked by 2 million years ago.

SOUTHEASTERN NEW GUINEA, 146–154.4° E

The mountainous peninsula that forms southeastern New Guinea is sometimes referred to as the Papuan peninsula because it is part of what was formerly the Australian Territory of Papua.

At the heart of the peninsula and islands is the East Papua Composite Terrane (EPCT). This comprises ophiolite on the northeast side and metamorphic rocks on the southwest. The two are separated by a major fault, which allows crustal extension and uplift of the metamorphic rocks.

Overlying the ophiolite are Middle Eocene arc-type volcanics, Middle Miocene and younger volcanics, rapidly deposited clastic sedimentary rocks and some limestone, and Pliocene–Quaternary volcanic rocks, including those of the intermittently active major volcanoes Lamington and Victory.

The association of metamorphic rocks and ophiolite extends northeast to the D'Entrecasteaux Islands and east-southeast to Misima, Sudest, and Rossel islands. The Trobriand Islands are raised coral and may be underpinned by Pliocene–Quaternary volcanic rocks. Woodlark (Muyua) Island comprises Quaternary limestone cover on Early Cenozoic volcanic arc rocks.

The Aure Fold Belt is made up of a thick sequence of rapidly deposited Late Oligocene to Miocene and Pliocene mostly-clastic sediments, folded and faulted in response to westward movement of the EPCT. The Aure Fold Belt extends offshore as far east as 146.8° E.

East of 146.8° E the fold belt gives way to thrust-bounded strike ridges of Paleocene and Eocene fine siliceous sediments and minor Oligocene coarser sediments intruded by Oligocene gabbro. The sequence is interpreted to have formed as an accretionary prism above an Eocene–Oligocene northeast-dipping subduction system.

Late Cretaceous and Middle Eocene tholeiitic basalts with rare interbeds of pelagic limestone form much of the peninsula east of 148° E. The basalts are 3000–4000 m thick and represent former ocean crust. Scattered stocks of syenite and related alkali-rich rocks of Middle Miocene age intrude the basalts.

ISLANDS OF THE BISMARCK ARCHIPELAGO

The large islands of the Bismarck Archipelago have a common origin, as is indicated by the similarities in their geology. Each has a basement of volcanic arc rocks of Eocene or Oligocene age unconformably overlain by Miocene limestone, which is in turn unconformably overlain by Pliocene and Quaternary clastic sediments and volcanics. The sequence is similar to that seen in the Finisterre Range on the mainland.

The volcanoes of the Bismarck volcanic arc include the caldera collapse volcanoes Long Island, Dakataua, Witori, and Rabaul, each of which has been the source of devastating eruptions in the past. Ash emission from an eruption of Long Island in about 1665 AD was sufficiently voluminous to block out the light of the sun for several days, an event recalled in legend as a time of darkness.

ONTONG JAVA PLATEAU

The Ontong Java Plateau is a continent-scale mass that comprises a sequence of basaltic lava flows overlain by 1 km of pelagic sediments. The lavas were emplaced in one remarkable major magmatic event over a period of less than 7 million years around 122 million years ago (Early Cretaceous, Aptian). The plateau is entirely submerged except for isolated atolls.

SMALL OCEAN BASINS

The Coral Sea Basin opened in the Paleocene. Adjacent submarine plateaus are rifted fragments of Australian craton. The Solomon Sea basin, north of the Trobriand Trough, opened in the Late Eocene–Early Oligocene and has been subducted at both the New Britain Trench and the Trobriand Trough. The Manus and Woodlark basins both opened in the last 6 million years.

TECTONIC ACTIVITY

The rapid oblique convergence of the Australian and Pacific plates results in ongoing tectonic activity.

In northwestern New Guinea a series of east-west-trending faults links the Bismarck Transform in the east with the Sorong Fault in the west. Movement on the faults is complemented by subduction at the New Guinea Trench and transpressional folding and faulting in the fold belt and in the Mamberamo Basin. Repeated occupation of survey stations shows that the western side of Cendrawasih Bay is moving west-southwest at a rate of 93 mm/yr. This motion likely caused the opening of Cendrawasih Bay, the development of Waipona Basin, and the development of the Lengguru Fold Belt. The motion is almost the same motion as that of the Pacific Plate (110 mm/yr) and suggests that the western part of the New Guinea Trench is locked.

Further east, the New Guinea Trench is active. Seismic tomography suggests that a subducted slab dips at a gentle angle southward from the trench beneath the central ranges. The subducted slab may be be the trigger for igneous activity in the fold belt, such as the intrusive rocks at Grasberg and Ok Tedi, and it may explain the transfer of convergent stress from the New Guinea Trench to the southern front of the fold belt, a distance of 300 km.

In northeastern New Guinea, the ongoing collision of the Finisterre and Sarawaged ranges with the Bismarck volcanic arc causes uplift of the north coast of the Huon Peninsula at averaged rates of 1–3 mm/yr. Study of raised coral terraces on the peninsula has yielded a high-quality record of fluctuations in sea level during the Late Quaternary. At the same time the Finisterre mountain mass rides southward and causes the down-warping of the northern end of the Papuan peninsula, which is subsiding at rates of up to 5 mm/yr.

In eastern New Guinea, active sea floor spreading in the Woodlark Basin is advancing westward and causing north-south extension of the mainland and adjacent islands. One result is the emergence in the Pliocene of domes and half-domes of metamorphic rocks by low-angle extensional faulting. Another is the opening of small rift basins offshore. Spreading within the last 1.2 million years has caused the separation of Misima Island from a position adjacent to Woodlark (Muyua) Island (152.8° E).

ECONOMIC ASPECTS

Oil is produced from Miocene reefs in the Salawati Basin. A 17-trillion cubic foot reserve of gas beneath Bintuni Bay is being developed for export as liquefied natural gas (LNG). In Papua New Guinea oil and gas are produced from structures in the fold belt. Copper and gold are produced from major mines at Grasberg and Ok Tedi, and gold from Porgera and Lihir Island. In the eastern Bismarck Sea (Manus Basin), gold and base metal sulfide mineralization on the sea floor is associated with active spreading ridges.

SEE ALSO THE FOLLOWING ARTICLES

Earthquakes / New Guinea, Biology / Plate Tectonics / Pocket Basins and Deep-Sea Speciation

FURTHER READING

Cloos, M., B. Sapiie, A. Quarles van Ufford, R. J. Weiland, P. Q. Warren, and T. P. McMahon. 2005. *Collision delamination in New Guinea: the geotectonics of subducting slab breakoff.* Special Paper 400. Boulder, CO: Geological Society of America.
Dow, D. B., G. P. Robinson, U. Hartono, and N. Ratman. 1988. *Geology of Irian Jaya.* Indonesia: Geological Research and Development Centre.
Hill, K. C., and R. Hall. 2003. Mesozoic–Cenozoic evolution of Australia's New Guinea margin in a west Pacific context. , in *Evolution and dynamics of the Australian plate.* R. R. Hillis and R. D. Müller, eds. Special Paper 372. Boulder, CO: Geological Society of America, 265–290.
Hope, G. S., and K. P. Aplin. 2007. Paleontology of Papua, in *The ecology of Papua: Part 1.* B. M. Beehler and A. J. Marshall, eds. Singapore: Periplus Editions, 247–254.
Parris, K. 1996. Central Range Irian Jaya Geology Compilation 1:500,000 scale geological map. Jakarta: P. T. Freeport Indonesia.
Pieters, P. E., C. J. Pigram, D. S. Trail, D. B. Dow, N. Ratman, and R. Sukamto. 1983. The stratigraphy of western Irian Jaya. *Bulletin of the Geological Research and Development Centre, Bandung, Indonesia* 8:14–48.
Pigram, C. J., and H. L. Davies. 1987. Terrranes and the accretion history of the New Guinea orogen. *BMR Journal of Australian Geology and Geophysics* 10:193–211.
Visser, W. A., and J. J. Hermes. 1962. *Geological results of the exploration for oil in Netherlands New Guinea.* Verhandelingen, Geologische Serie 20. The Hague: Koninklijk Nederlands Geologisch Mijnbouwkundig Genootschap.

REFERENCES

Satellite Geodesy, Scripps Institute of Oceanography. 2008. *Global topography.* http://topex.ucsd.edu/marine_topo/mar_topo.html.
Smith, W. H. F., and D. T. Sandwell. 1997. Global seafloor topography from satellite altimetry and ship depth soundings. *Science* 277: 1957–1962.

NEW ZEALAND, BIOLOGY

STEVEN A. TREWICK AND MARY MORGAN-RICHARDS

Massey University, Palmerston North, New Zealand

New Zealand, spanning more than 1400 km of latitude on the southwest edge of the Pacific Ocean, supports a distinct assemblage of plant and animal groups. Species-level endemism in the wet temperate forests and alpine habitats is high; however, compared to many other oceanic islands, species diversity is not. New Zealand is a small part of a large continent, Zealandia, that sank beneath the surface of the sea after separation from Gondwanaland and is thus often considered a continental island. Whether any of the New Zealand biota originated in Zealandia is uncertain, but a number of animals that lack close living relatives elsewhere in the world may have arrived in that way. As with true oceanic islands, New Zealand biodiversity is dominated by speciation in relatively recent geological time, mostly from overseas colonists.

GEOLOGICAL HISTORY AND GEOGRAPHIC SETTING

New Zealand is composed of continental crust; a property it shares with just a few other islands, including New Caledonia and Madagascar. In fact, the geological histories of New Zealand and New Caledonia are closely linked, as both are small, emergent parts of an otherwise submerged continental fragment, Zealandia. The continent of Zealandia was somewhat larger than India when it rifted from Gondwana starting about 85 million years ago. During the following 60 million years the continental crust of Zealandia stretched, thinned, and sank, so that today 93% of its area is beneath the sea. Subsequently, parts of Zealandia have re-emerged as a result of local tectonic activity to form islands, and these include New Caledonia, the Chatham Islands, and most if not all of New Zealand.

The New Zealand archipelago lies in the southwest Pacific Ocean and has a total area of about 270,000 km². The nearest continent, Australia, is about 1,500 km to the west. New Zealand consists of three main islands: North Island and South Island separated by the Cook Strait, and the rather smaller Stewart Island separated by Foveaux Strait (Fig. 1). There are a number of other smaller inhabited islands: the Chatham Islands (963 km²) about 800 km east and the islands of Waiheke and Great Barrier in the Hauraki Gulf of North Island, plus numerous uninhabited islands near the mainland coast that share a recent biological history, because they were connected when sea level was lower in the Pleistocene (Fig. 2). In contrast, the more distant islands to the north (Poor Knights, Three Kings), south (sub-Antarctic islands), and east (Chatham Islands) have been isolated longer or were never linked to mainland New Zealand. Diversity and endemicity are not homogeneous even across the main islands; distinct zones of higher endemicity are particularly conspicuous among the flora (Fig. 2). The climate is predominantly temperate but ranges from cool temperate in the south (latitude 47°) to subtropical in the north (latitude 34°). A large proportion of New Zealand can be classified as mountain land (60% of South Island and 20% of North Island), and this contributes to habitat diversity. The majority of alpine habitat is on the Southern Alps, which run the length of South Island, but there are smaller ranges and a number of volcanic mountains in North Island (Fig. 1). The Southern Alps reach to ~3000 m above sea level (the highest, Mt. Aoraki/Cook, is 3753 m), but the treeline in New Zealand is relatively low (averaging 1300 m above sea level), so the alpine zone is a relatively large and important ecotone.

For students of island biogeography, New Zealand is an enigma. The biology has features reminiscent of both continental lands and oceanic islands. For example, the absence of native terrestrial mammals is typical of an island fauna, but the presence of an endemic order of reptiles (tuatara; Sphenodontia), two endemic orders of birds (moa and kiwi), and an endemic family of amphibians (frogs; Leiopelmatidae) is seen as more continental in nature. Recognition that New Zealand is a continental island, and that fragmentation of Gondwanaland played an important role in its geological formation, has strongly influenced interpretation of its biogeography. In contrast to dispersal origination of biota on true oceanic islands,

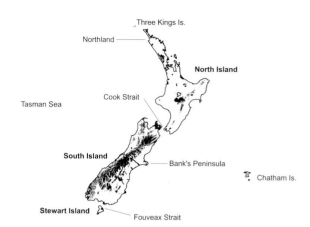

FIGURE 1 The islands, straits, and and other locations in modern New Zealand, with the distribution of montane land (1000 m) indicated (black), formed largely since the Pliocene (5 million years ago).

the biota of New Zealand have been widely treated as elements of an ancient, continental "ark." There has been an overemphasis on inferring biogeographic origins from distribution patterns rather than recognizing the role of speciation in New Zealand biology.

PATTERNS OF SPECIES DIVERSITY

Prior to human arrival the natural vegetation over about 85% of New Zealand was mixed temperate rain forest, with southern beech (*Nothofagus*), tree ferns (*Cyathea, Dicksonia*), and species of the Southern Hemisphere gymnosperm family Podocarpacea being prominent features of the forests. The biggest New Zealand tree is the endemic kauri (*Agathis australis*), the only representative of this genus in New Zealand. Kauri belongs to a group of ancient conifers Araucariaceae, which has most of its diversity in subtropical and tropical Oceania.

New Zealand plants tend not to have prominent, showy flowers; instead, a large proportion of angiosperms have small, often pale or white flowers pollinated by generalists, including flies, rather than specialist butterflies and bees. Among the few flamboyant flowering trees are kowhai (*Sophora*), with large yellow flowers, and pohutakawa or Christmas tree (*Meterosidros*), which produces abundant scarlet inflorescences in December; both have close relatives elsewhere through the southern hemisphere and Pacific.

Like the flora, New Zealand's birds are not showy; native forest birds tend to have cryptic plumage, many are nocturnal (21%), and there is also little sexual dimorphism in plumage. For example, New Zealand species of *Petroica* robins are monomorphic and monochrome, whereas male Australian *Petroica* robins have bright pink or scarlet chests. However, several New Zealand birds show traits indicative of resource partitioning, including pronounced beak dimorphism in the huia (*Heteralocha acutirostris,* extinct) and probably also, but to a lesser extent, the Chatham Island rail (*Gallirallus modestus,* extinct) and kaka parrot (*Nestor meridionalis*), and size dimorphism in some moa that was so extreme that bones from males and females were initially identified as belonging to different species.

The alpine habitat is rather youthful in New Zealand, probably largely developed during the last 5 million years as the mountain ranges formed and the climate cooled in the late Pliocene. Nevertheless there are many alpine specialists, including snow tussock grass (*Chionochloa*), cushion-forming herbs such as "vegetable sheep" (*Haastia* and *Raoulia*), buttercups (*Ranunculus*), alpine daisies (*Celmisia*), skinks, geckos, birds including the rock wren (*Xenicus gilviventris*), and the world's only alpine parrot, the kea (*Nestor notablis*). Specialist alpine invertebrates are numerous and include cicadas (*Maoricicada*), a black ringlet butterfly (*Percnodaimon*), short-horned grasshoppers (e.g., *Sigaus*), cockroaches (*Celatoblatta*), and weta. Some insects, including the alpine species of weta (*Hemideina* and *Deinacrida*), grasshoppers (*Sigaus Brachaspis*), and cockroaches (*Celatoblatta*), are tolerant of freezing and over-winter under snow and ice as adults and juveniles (Figs. 3–6). These alpine-adapted species all belong to recent radiations that include alpine and lowland representatives.

ASSEMBLY OF THE BIOTA

The New Zealand biota has been described as ill-balanced because of the variance in diversity and endemicity exhibited among different plant and animal groups, and this has frequently been attributed to New Zealand's supposed long isolation in the Pacific. However, the composition is broadly consistent with other island assemblages and indicative of long-standing (albeit intermittent) interactions with other biotas. Its geographic position, hundreds of kilometers from other land for some millions of years, has inevitably restricted successful migration of plants and animals, but not precluded it. As a result, a large proportion of species in many groups are endemic to New Zealand, but endemicity above this level is much lower. For example, New Zealand

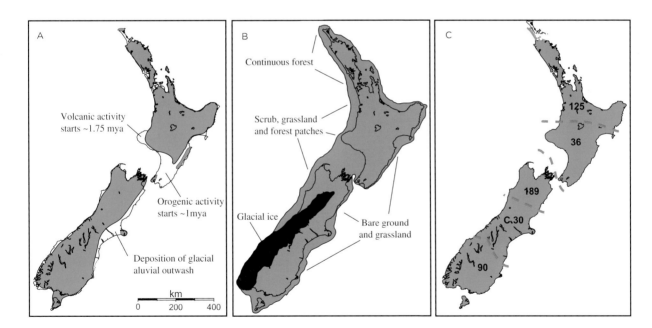

FIGURE 2 The changing shape of New Zealand and its biota. Changes in the distribution, extent and topography of land during the last 5 million years are thought to have influenced the distribution of endemism within New Zealand. (A) During the early Pleistocene (1.8 million years ago) there were two major islands separated more than today. A number of smaller islands were subsequently joined to the mainland (for example, Bank's Peninsular and the Northland archipelago, which still have endemic species). Estimated outline of land during the early Pleistocene (green) is superimposed on the present shoreline. (B) During glaciation sea level fell, and at the last glacial maximum (0.02 million years ago) a single major island would have existed (green). During this time glaciers extended across the Southern Alps (black) (McGlone 1985). (C) Five broad zones of plant endemicity have been identified among plants (dashed lines, values indicate numbers of endemic species) (Wardle 1963). Not surprisingly, the land areas with fewest endemics are those that are youngest or most disturbed during the late Pleistocene. The extent of forests was much reduced during glacials.

FIGURE 3 Biodiversity patterns in New Zealand, exemplified by anostostomatid weta. The Anostostomatidae are a relatively diverse group of orthopterans in New Zealand, expressing a range of biological features characteristic of the biota generally. The tusked weta (A) have their closest relatives in New Caledonia. The (B) giant (*Deinacrida*) and (C) tree (*Hemideina*) weta of New Zealand are unique among Anostostomatidae for eating leaves, fruit, and flowers. Secondary sexual characteristics such as enlarged heads and jaws in males have evolved in *Hemideina* tree weta, and a similar evolutionary path has resulted in the tusks of male tusked weta (A).

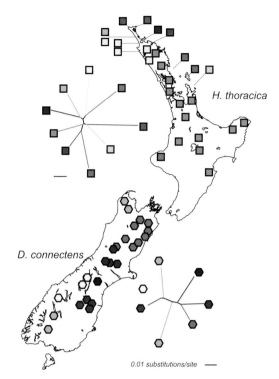

FIGURE 4 Biodiversity patterns in New Zealand, exemplified by anostostomatid weta. Genetic structure within two widespread species of weta in New Zealand. The Auckland tree weta (*Hemideina thoracica*) in North Island has greatest genetic diversity in the far North, where islands exisited during the Pliocene and where forests grew even during Pleistocene glacial cycles and after the volcanic eruptions of the central North Island. The giant scree weta (*Deinacrida connectens*) in South Island is fragmented on mountain peaks, but during glacial cycles its distribution would have been more continuous at a lower altitude. Genetic diversity within this species dates to uplift of the southern mountains.

has more than 20,000 endemic invertebrate species (95% endemism at species-level) but just five endemic invertebrate families, and 2000 endemic species of vascular plants but no endemic families.

New Zealand is subjected to a prevailing westerly wind and circumpolar oceanic current, and some animals are regular visitors. For instance, many seabirds regularly traverse the oceans but nest in New Zealand (e.g., sooty shearwater, *Puffinus griseus*); cuckoos (*Chrysococcyx lucidus, Eudynamys taitensis*) travel to and from islands in the Pacific Ocean; and godwits (*Limosa lapponica*) migrate to breed in Alaska.

For other taxa, migration is less frequent, but among the landbirds there are many post-human colonists (see the section "Human Contact"), and Australian species that arrived before recording began (e.g., the pukeko *Porphyrio porphyrio*, the harrier *Circus approximans*, the fantail *Rhipidura fuliginosa*, and the owl *Ninox novaezealandiae*), in addition to endemic species that have a close affinity (genus level) to

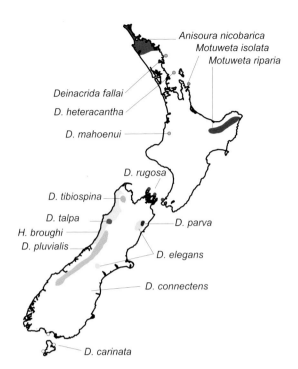

FIGURE 5 Biodiversity patterns in New Zealand, exemplified by anostostomatid weta. The distribution of tusked and giant weta in New Zealand. The greatest diversity of *Deinacrida* is in the habitat-diverse South Island mostly associated with the mountains. The three tusked weta species are restricted to northern North Island.

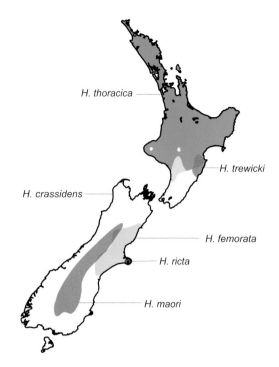

FIGURE 6 Biodiversity patterns in New Zealand, exemplified by anostostomatid weta. The distribution of tree weta reveals broad regional allopatry and evidence for range expansion following climate change. *H. thoracica* extend its range south probably after the last glacial maximum, excluding the cold-adapted *H. crassidens* from lowland forests of central North Island but leaving isolated populations of *H. crassidens* marooned in the subalpine zone of mountains in the region.

taxa elsewhere (e.g., *Petroica* robins in Australia). In many instances, New Zealand species have extraordinary forms compared to their nearest (frequently Australian) counterparts. For example the large, endemic flightless takahe (*Porphyrio hochstetteri*) shares a common ancestor with the smaller, flying purple swamphen or pukeko (*P. porphyio*), and the giant extinct eagle (*Harpagornis moorei*) shares a recent (Plio-Pleistonce) ancestor with the Australian little eagle (*Aquila morphnoides*). However, New Zealand also has taxa that have been classified as distinct at higher levels (e.g., bird families including wrens Acanthisittidae, kiwis Apterygidae).

In the freshwater realm, the fish are either themselves diadromous or are related to diadromous taxa. Eels are common (*Anguilla*) and have been an important food resource of people. Other native fish in New Zealand lakes and streams include species of *Galaxias, Neochanna,* and *Gobiomorphus,* most of which are endemic, but the genera also occur elsewhere, including Australia. Endemic freshwater invertebrates include two species of crayfish (*Paranephrops*) (related taxa exist among the more diverse Parastacidae fauna of Australia) and a number of insects with unusual life histories, such as caddis flies and dragonflies with semi-terrestrial larvae. Freshwater invertebrates tend to be highly distinctive (e.g., 20 of 21 genera of stoneflies, Plecoptera, are endemic), although there is a species of freshwater crab that also occurs in Australia. An analogous pattern is evident in the terrestrial flora too, with endemism high at species level (80%) but less so at higher taxonomic levels. The flora, and other elements of the biota, have been subjected to substantial change over time. There is evidence for bouts of diversification associated with geophysical events in New Zealand's prehistory, including significant changes in area, habitat diversification, and climate fluctuation. During and since the Miocene, changing diversity has resulted from extinction (e.g., *Eucalyptus* gum trees disappeared from New Zealand), colonization (e.g.; *Fuscospora* beech arrived), and speciation (e.g., *Coprosma*).

Many elements of the biota are described as "Gondwanan." Such taxa are simply those that have a distribution largely restricted to modern land areas that originated from the breakup of Gondwanaland: Africa (usually just southern Africa), Madagascar, South America, India, Australia, and New Caledonia. It is often assumed that a Gondwana distribution is evidence of a vicariant Gondwanan origin, and this concept has been a central tenet in New Zealand biogeography since the acceptance of continental drift in the early 1970s. A broad range of New Zealand taxa are attributed to this origin, including moa and kiwi (ratite birds), southern beech trees (*Nothofagus*), weta (anostostomatid crickets), land snails (Punctidae, Charopidae), and peripatus (Onychophora or velvet worms). Following a form of reciprocal illumination, some biogeographers have confounded the apparent evidence from Gondwana biological distributions with geological evidence for the Gondwanan origin of the continental crust from which New Zealand is formed and concluded that the biology of New Zealand is first and foremost Gondwanan. That is, that the biota of New Zealand has evolved in isolation since separation by continental drift from Gondwana some 62–80 million years ago. A stream of recent evidence from molecular studies in particular reveal that this is, in fact, far from true. A prime example comes from *Nothofagus* beech, an iconic Gondwanan taxon, long assumed to be incapable of transoceanic dispersal. Yet, recent research has revealed that extant New Zealand beech arrived after separation of Zealandia from Gondwana. In a similar vein, although the presence of peripatus (Onychophora) in New Zealand is consistent with a vicariant Gondwanan origin, this is not the only explanation. There are also peripatus in Jamaica (an emergent Miocene island more than 600 km from a continent). Similarly, weta (Anostostomatidae), which have a largely Gondwanan distribution (i.e., Southern Hemisphere), also occur in Japan, and speciose land snail families (Charopidae, Punctidae) are well represented on oceanic Pacific islands. It may remain useful to describe some of these as Gondwanan in terms of distribution (extant and fossil), but not necessarily in terms of the process that led to their current distribution. Gondwanan taxa are essentially southern hemisphere taxa and these are the most likely to arrive in southern lands.

For most plants and for almost all terrestrial animals, the fossil record in New Zealand is, as yet, too patchy to be highly informative about timing of origin and persistence in the biota. Birds and some insects and plants are represented by fossils in caves, middens, swamps, and sand dunes; such sites often have abundant material, but because they are of Holocene age, they are informative about the last few thousand of years of New Zealand's prehistory and not the preceding millions. Recent discoveries of rich Miocene fossil sites in southern South Island will no doubt open important windows into the past biology. Already a small mammal, a crocodile, and a community of now-extinct aquatic birds are known to have existed 18 million years ago and reveal much about the turnover of the New Zealand biota.

FEATURES OF THE BIOTA

Eclectic Mix

There are no living snakes, land turtles, or crocodiles in New Zealand, and lizards belong to just two groups (geckos and skinks), but the tuatara is the only living representative of the sphenodontid reptiles anywhere in the world. Among invertebrates, many orders are missing (e.g., Scorpionida, Embioptera, Zoraptera), and others are poorly represented (e.g., Mecoptera, one species; Formicidae, 12 species, compared to ~1300 in Australia). Similarly, within orders, representation is patchy even in comparison to neighboring land areas. For example, Orthoptera have high endemicity and diversity among the Rhaphidophoridae (18 genera, ~50 species), Anostostomatidae (five genera, ~70 species), and to a lesser extent Acrididae (four endemic and two native genera, ~16 species, mostly associated with alpine habitat), fewer still in Tettigonidae (two Australian katydids), Gryllidae (two endemic and one Australian species), and Gryllotalpidae (one endemic), while others are absent (e.g., Gryllacrididae).

Curious Endemics

Among the birds, the large parrot kakapo (*Strigops habaproptilus*) is exceptional in its combination of life history characteristics and is unique among parrots for any one of these characteristics: lek breeding, nocturnal, and flightless. The sporadic breeding cycle of the kakapo is linked to the masting of forest trees.

The New Zealand avifauna included some 16 ratites (all endemic genera), including five of the smallest living members of the group, the kiwi (*Apteryx*). Kiwi invest all their reproductive effort into a single, very large (up to 450 g) egg per nesting. They are predominantly nocturnal and are the only birds with nostrils at the tip of the bill. Olfactory cues are important to kiwi, which forage on the ground at night, but it was recently discovered that kiwi have vibration sensors in the bill tip like those of wading birds (Scolopacidae), which are used when probing for prey.

New Zealand amphibian diversity consists of just four extant frogs, belonging to an endemic family (Leiopelmatidae). New Zealand frogs lack ears, pass through the tadpole stage in their eggs, and have paternal care of froglets.

The only recent native mammals are three species of bat. The long-tailed bat (*Chalinolobus tuberculatus*) probably arrived from Australia in the Pleistocene. The two short-tailed bats (one now extinct) belong to an endemic family (Mystacinidae), and *Mystacina tuberculate* is unusual in having evolved the ability to move efficiently on the ground, where they forage on insects, fruit, pollen, and nectar. The largest known eagle in the world (*Harpagornis moorei*) hunted moa (as revealed by talon marks in Holocene fossil moa hip bones), had a wingspan of 2.5 m, and had evidently evolved to exploit this large prey. Two genera of New Zealand weta (*Deinacrida* and *Hemideina*) are unusual in having adopted a largely herbivorous diet (Fig. 3). They feed primarily on green leaves or fruit or flowers, and rarely on animals, which is the normal diet of other genera in this family of crickets (Anostostomatide) in New Zealand and elsewhere. The only significantly poisonous animal in New Zealand is the katipo spider (*Latrodectus katipo*), which, judging by its close relation to the infamous Australian red-back (*Latrodectus hasselti*), must have arrived on New Zealand beaches in recent geological time.

Gigantism

Gigantism has in the past been identified as a distinctive feature of the biota, but most of the animal groups usually identified as being represented by giant forms in New Zealand (e.g., earthworms, centipedes, land snails, flatworms, millipedes, slugs, stick insects, weta, longhorn beetle, a weevil, a moa, and an eagle) also have species as large or, in most cases, larger in other parts of the world. A weta (cricket) is credited in New Zealand as being the heaviest insect recorded anywhere; female *Deinacrida heteracantha* in the wild average about 32 g (a much higher, anomalous weight of 71 g is famously recorded from a captive egg-engorged female). On the whole, animal groups with large species in New Zealand also contain many (usually the majority) smaller taxa. This is true even for the famous moa; although *Dinornis giganteas* was very big (240 kg), a bigger ratite (*Aepyornis maximus*) is known to have existed in Madagascar, and other moa were considerably smaller (*Megalapteryx didinus* ~25 kg). There are a number of "megaherbs" with large, glossy leaves and large, colorful floral displays associated with the moist environments of some offshore islands, including Chatham Island forget-me-not (*Myosotidium hortensia*) and Campbell Island daisy (*Pleurophyllum speciosum*). The extinct New Zealand eagle (*Harpagornis moorei*) and the giraffe weevil (*Lasiorhynchus barbicornis*, which owes half of its length to a long, thin snout) are the largest of their respective kinds, but there is little evidence for a dominant evolutionary pattern across the biota.

Flightlessness

About a third of New Zealand land birds at the point of human contact were flightless; many are now extinct. All modern ratites are flightless, and New Zealand had diverse fauna of moa (11) and kiwi (5). Similarly, six species of penguin (Spheniscidae) breed in New Zealand. Other flightless species were members of volant groups. Rails have the greatest propensity to evolve flightless forms, as observed on many other islands, and 11 species (70%) are known from the New Zealand archipelago. *Porphyrio*, *Gallirallus*, *Gallinula*, and *Fullica* are each represented by one or more flightless species on New Zealand main or offshore islands; each is a product of a separate colonization by a flying ancestor. Other flightless taxa include a parrot (*Strigops*), the strange rail-like predator (*Aptornis*), ducks (*Anas*), geese (*Cnemiornis*), and wrens (*Xenicus*). Much, and in some cases everything, that is known of these birds has been gleaned from their Holocene bones preserved in sand dunes, swamps, and caves.

Floral Peculiarities

The New Zealand flora consists of some 2300 native species, of which 85% are endemic. A relatively large proportion of the angiosperms have separate sexes or some degree of sexual dimorphism (23% of genera). There is a predominance of white flowers and unspecialized pollination systems among the largely evergreen trees and shrubs of New Zealand. Many (e.g., *Nothofagus*, *Dacrydium*, *Chionocloa*) display a high variance in fruiting from one year to the next (masting), and this has a significant impact on breeding of endemic birds (in particular kakapo, *Strigops*) and introduced mammals (mice, *Mus*). There is a high frequency of small-leaved, tangle-branched shrubs (divaricating habit), a form that has evolved independently in 20 plant families (e.g., *Coprosma*, *Myrsine*, *Melicytus*, *Pseudopanax*, *Pittosporum*, *Olearia*). There are about 60 species with tiny leaves and interlacing branchlets, and about 14% of these represent juvenile stages of plants that later grow into adult leafy trees of normal habit and foliage (e.g., matai, *Prumnopitys taxifolia*; putaputaweta, *Corpodetus serratus*). Two competing hypotheses explain the unusual abundance of this plant growth form: moa browsing and climate. From subfossil remains we know that moa did indeed eat these divaricating plants. However, the probable recent origin and distribution of the divaricating species, added to the fact that they have not been replaced in the >200 years since moa went extinct, suggests that climate is the more likely selective force. Skinks and geckos appear to be important seed dispersers for many of these small/divaricating shrubs.

Honeydew is a sugar-rich secretion produced by sap-sucking insects. In New Zealand, beech trees (*Nothofagus*) are frequently infested with scale insects (*Ultracoelostoma*) that secrete honeydew through a long, hairlike, waxy tube that extends from the bark of host trees and is an important resource for native birds (e.g., kaka, tui, bellbird). Exotic bees and wasps (especially *Vespula*) compete for honeydew, and in some forests they take almost all of it, depriving local birds.

Group Diversity and Radiations

Many invertebrate groups have high diversity and endemism that are the product of species radiations, including some carabid beetles (e.g., *Mecodema*), alpine cicadas (*Maoricicada, Kikihia*), weta (Fig. 2), and land snails large (predatory taxa of Rhytidae including *Powelliphanta*) and small (including Punctidae and Charopidae). The latter show high levels of sympatric species diversity in some parts of New Zealand.

Species diversity within bird groups is less prominent; there are relatively few endemic species of birds in New Zealand (176 endemics of 245 species at human contact) but high representation of taxonomic diversity (20 of 27 orders). Among the more speciose are kiwi (5) and moa (11), and the *Cyanoramphus* parakeets represent a young radiation of some 10 species in the New Zealand archipelago. Seabirds are well represented, and even today albatross (*Diomedea epomophora*), gannet (*Sula bassana*), and two penguins (*Megadyptes antipodes, Eudyptula minor*) nest on the mainland, although many others are now restricted to offshore islands.

Notably speciose plant groups include *Coprosoma, Hebe, Ranunculus* (buttercup), *Celmisia* (daisy), and *Asplenium* (fern); the products of relatively young radiations.

HUMANS IN AOTEAROA

Human Contact

As with most oceanic islands, the arrival of humans in New Zealand had a major impact on the composition and structure of the flora and fauna. The first people (Maori) to colonize the New Zealand archipelago (Aotearoa), came from central Polynesia and made first contact with these islands 1000 to 600 years ago. They introduced, during a succession of exchanges, a number of commensal animals, including the Pacific rat (kiore, *Rattus exulans*) and dogs (the extinct kuri) and some plants for cultivation (e.g., kumara, *Ipomoea batatas*). Early Maori hunted birds that provided a ready and abundant resource of food and materials (e.g., feathers woven into cloaks known as *korowai*). Moa were probably hunted to near extinction within about 100 years of colonization, and it is likely that low fecundity and slow growth resulted in their subsequent extinction. Other large ground birds (the flightless goose *Cnemiornis*, the adzebill *Aptornis*) were also soon extinct, and other forest birds (the weka rail *Gallirallus* and the keruru pigeon *Hemiphaga*) were harvested using specialized techniques including snares and traps. Sea mammals including seals and sea lions were taken in great numbers at their coastal rookeries, and seabirds (titi petrels, *Puffinus*) were gathered at their nesting grounds that were often far inland. Maori also used fire to clear land as tribal structure developed.

The first confirmed contact with New Zealand by a European was made by Abel Tasman in 1642. James Cook and Jean François Marie de Surville reached the islands more than 100 years later in 1769, and subsequently whalers, traders, and missionaries began arriving. Colonization by European people did not start in earnest until 1840 but resulted in accelerated modification of the landscape and biology of New Zealand. In particular, land was cleared of trees as timber was harvested, and pasture farming developed. Today, about 22% of the prehuman vegetation/habitat remains in a relatively pristine condition.

Introduced wild or feral animals included mice, cats, pigs, rats (ship and Norway), deer, goats, hares, and rabbits, and early European settlers continued to add familiar species from "home" after settlement. Many "garden" birds (e.g., song thrush, sparrow, blackbird, several finches, pheasant) and hedgehogs were also introduced under the auspices of regional Acclimatization Societies. A total of 33 bird species and 34 terrestrial mammal species have established in New Zealand. Trout, salmon, and 18 other freshwater fish have been introduced, to the detriment of native freshwater fish and invertebrates. In an early (1882) misguided attempt at biocontrol, mustelids (stoats, weasels) were introduced to limit the burgeoning rabbit population. It soon became apparent that, as predicted by a few scientists of the time, the predators did not restrict their attentions to rabbits, and today mustelids remain major predators of native birds, lizards, and invertebrates. Despite early attempts in the late nineteenth century to protect notable endemic species, the impacts of introduced predators and natural history collecting continued to be felt. For instance, a species of wren (*Xenicus insularis*) was extinguished from Stephen's Island in the Cook Strait in 1895 through the attention of cats brought to the island by the lighthouse keepers. In the 1890s a pair of huia (*Heteralocha acutirostris*), captured for translocation to a reserve island, were in fact sold illegally to a collector in the expectation that more could be found and conserved, but this never happened and the species was lost. The brush-tailed possum (*Trichosurus vulpecula*) was introduced from Australia in 1837 with the aim of supplying the fur trade; this possum is now a major pest in New

Zealand. Eating leaves, flowers, and fruits of native trees, it competes with native birds, and it also directly preys on eggs and nestlings. Some 76 native bird species have thus become extinct since the arrival of people in New Zealand, including 41% of the 176 endemic species. A large number of plant species have also been introduced (about 1630 alien plant species have established), and many of these are now invasive weeds. For vascular plants, land mammals, land birds, and freshwater fish, over 40% of the species now found in New Zealand are exotics. Habitat modification also appears to have opened the way for a number of self-introductions; first records for birds include the silvereye *Zosterops* in 1832, the welcome swallow *Hirundo tahitica* in the 1950s, the spur-wing plover *Vanellus miles* in 1932, and the cattle egret *Bubulcus ibis* in 1963. Exotic insects include the monarch butterfly *Danaus plexippus,* which arrived in the 1880s. Although perhaps facilitated by human activities, this process very much continues a prior persistent feature of sporadic colonization (see the section on "Assembly of the Biota").

Conservation

Despite the ravages of habitat modification and exotic taxa, a number of distinctive bird species have been kept from extinction by the efforts of New Zealand conservationists, who have pioneered management techniques now applied worldwide. Prominent successes include the black robin (*Petroica traversi*), resurrected from just five birds (two females, three males) in 1980 using a combination of translocations, cross-fostering, and supplementary feeding. The intensely managed night-parrot, kakapo (*Strigops habroptilus*), which, despite an extreme male bias and intermittent breeding, has experienced a gradual improvement in its meager population, which today stands at 90.

Among the most valuable resources available to New Zealand conservation are the offshore islands, from which introduced predators are removed to provide vital reserves for protected species. Several endangered taxa owe their survival thus far to remnant populations on offshore islands (e.g., Hamilton's frog, *Leiopelma hamiltoni,* on Stephens Island; hihi, *Notiomystis cincta,* Little Barrier Island; saddleback or tieke, *Philesturnus carunculatus,* on Hen Island and Big South Cape Island). New Zealanders also pioneered mammal eradication techniques to clear islands of introduced pest species; the largest island to have been successfully cleared of rats is Campbell Island (114 km^2). A dozen species of exotic mammal have been eradicated from other islands in the region including mice, possums, cattle, pigs, goats, rabbits, and cats. The use of sophisticated fencing techniques and predator control programs have also allowed the development of "mainland islands" that protect dwindling biodiversity and provide a valuable point of contact between people and the natural environment. Conservation efforts are now supplemented by stringent biosecurity measures that strictly limit the importation of further exotic species to New Zealand.

CONCLUSION

New Zealand as a land mass has a long and complex biological history. Proximity to a large continent (Australia) and composition of continental crust have complicated inferences about the origins of the biota. The biota has assembled over many millions of years, but relatively few (conceivably none) of the lineages that must have been present when Zealandia broke from Gondwana survive today. Episodes of extinction and colonization have acted upon the biological assemblage, but, as with most islands, the key influence on New Zealand's biological character has been speciation.

SEE ALSO THE FOLLOWING ARTICLES

Bird Radiations / Flightlessness / Gigantism / Madagascar / New Caledonia, Biology / New Zealand, Geology / Vicariance

FURTHER READING

Campbell, H., and G. Hutching. 2007. *In search of ancient New Zealand.* Hong Kong: Penguin Books.
Gibbs, G. 2006. *Ghosts of Gondwana: the history of life in New Zealand.* Nelson, New Zealand: Craig Potton Publishing.
Goldberg, J., S. A. Trewick, and A. M. Paterson. 2008. Evolution of New Zealand's terrestrial fauna: a review of molecular evidence. *Philosophical Transactions of the Royal Society* B 363: 3319–3334.
McGlone, M. 1985. Plant biogeography and the late Cenozoic history of New Zealand. *New Zealand Journal of Botany* 23: 723–749.
Pole, M. S. 2001. Can long-distance dispersal be inferred from the New Zealand plant fossil record? *Australian Journal of Botany* 49: 357–366.
Trewick, S. A., A. M. Paterson, and H. J. Campbell. 2007. Hello New Zealand. *Journal of Biogeography* 34: 1–6.
Wardle, P. 1963. Evolution and distribution of the New Zealand flora, as affected by Quaternary climates. *New Zealand Journal of Botany* 1: 3–17.

NEW ZEALAND, GEOLOGY

HAMISH CAMPBELL
GNS Science, Lower Hutt, New Zealand

CHARLES A. LANDIS
University of Otago, Dunedin, New Zealand

In geological terms, New Zealand may be regarded as an emergent portion of a sunken continent. This is unusual globally; the Kerguelen Plateau may be the only other modern example of a large tract of submerged continental crust, or sunken continent. The New Zealand land area represents just 7% of a much larger area of submerged continental crust referred to as Zealandia. New Zealand owes its existence as land to tectonic collision along a northeast-southwest-oriented segment of the plate boundary separating the Australian Plate (to the west) from the Pacific Plate (to the east), a process that has been especially active since the onset of Miocene time 24 million years ago.

THE ISLANDS OF NEW ZEALAND

The New Zealand "mainland" is oriented more or less northeast-southwest, stretching some 1500 km between latitudes 34° and 47° S (Figs. 1, 2). It is largely the product of plate collision tectonism and consists of two large islands, North Island (150,437 km^2) and South Island (113,729 km^2); the smaller Stewart Island (1,680 km^2); and about 700 much smaller islands, including islets that are in close proximity to the coast. The largest of these include Great Barrier, D'Urville, Resolution, Waiheke, Secretary, Arapawa, Ruapuke, Codfish, Big South Cape, and Kapiti.

Beyond the "mainland" but considered part of New Zealand are the "offshore islands." These are White Island (Fig. 3), the Three Kings Islands, and Kermadec Islands to the north; the Chatham Islands (including Chatham, Pitt, Mangere, Southeast, Little Mangere, Forty Fours, Sisters,

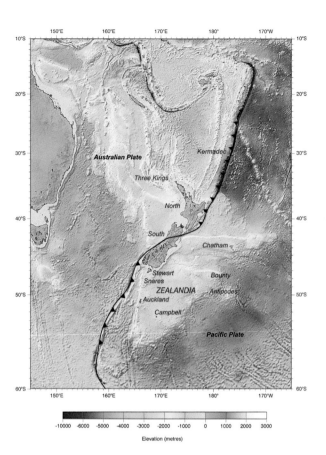

FIGURE 2 The major islands of New Zealand presented on a map of the "New Zealand Continent." This is a bathymetric map color-coded to show water depth. The 2,500-m isobath serves as a proxy for the boundary between oceanic crust and continental crust. New Zealand represents 7% of the area of Zealandia. In this context New Zealand can be regarded as an emergent part of a large sunken continent. Location of the active plate boundary between the Australian and Pacific plates is shown. The "teeth" indicate direction of down-going subducting slabs of oceanic crust. Diagram by GNS Science.

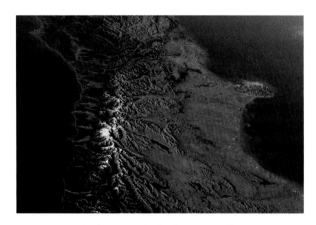

FIGURE 1 "Mainland" New Zealand is the product of active plate collision between the Pacific Plate (to the east) and the Australian Plate (to the west). This image, taken from the Space Shuttle (NASA), is looking north from above southernmost South Island. From space, the South Island appears as a giant welt or bruise within the Earth's crust. This is not surprising as it is the locus of continent-continent collision, giving rise to the Southern Alps. The plate boundary is the Alpine Fault, which appears as a remarkably straight line that runs northeast–southwest down the western side of the snow-capped high peaks of the Southern Alps. In a sense, the New Zealand land surface is being pushed up and held up against its will. If and when the tectonic forces responsible diminish or cease, the land will slowly subside. Photograph by NASA.

and Pyramid), Antipodes Islands, and Bounty Islands to the east; and Solander Island, the Snares Islands, and the Subantarctic Islands to the south, namely the Auckland Islands and Campbell Island.

Whereas "mainland" New Zealand consists mainly of diverse continental rocks (igneous, metamorphic, and sedimentary) of Paleozoic, Mesozoic, and Cenozoic age, a number of the smaller New Zealand islands are volcanic. Some owe their existence to subduction-related late Cenozoic (Miocene-Recent) arc volcanism including Great Barrier, Little Barrier, Great Mercury, Solander, White, and the Kermadec Islands, but of these only White Island and the Kermadec Islands (Raoul, Macauley, Curtis, Cheeseman, and L'Esperance Rock) are classed as active.

Conversely, two islands in the Bay of Plenty, Mayor and Motiti, relate to extensional tectonism associated with active rifting of the Taupo Volcanic Zone (TVZ).

FIGURE 3 White Island, New Zealand's most active volcano, located offshore some 50 km north of the Bay of Plenty coast of the northeastern North Island. This is an example of an active subduction-related volcano. Photograph by GNS Science/Lloyd Homer.

The TVZ is often referred to as a "back arc basin," but it is nevertheless a rift within continental crust and hence is the locus of new continental crust. "Fresh" granite magmatism at depth manifests itself at the surface as rhyolite volcanism accompanied by voluminous production of pumice, ash, and ignimbrite. This continental rift is responsible for the Y shape of the North Island, involving relative rotation of Northland (trending to the northwest) and East Cape (trending northeast) with respect to Mount Ruapehu (approximately). The TVZ is the main locus of rifting with a history of active normal faulting and east-west extension that has been measured at rates of about 10 mm per year.

Islands proximal to (Mercury Islands) and to the north of the Coromandel Peninsula (Great Barrier, Little Barrier) in the eastern part of the Hauraki Gulf may also be regarded as rift related but are of older Miocene–Pliocene age.

A number of other more substantial islands relate to extinct Cenozoic (Miocene) intraplate basalt volcanism, including the Auckland Islands and Campbell Island.

The Chatham Islands owe their existence to Neogene uplift, but their origins can be traced back to older Cretaceous intraplate basalt volcanism with subsequent lesser Cenozoic (Eocene and Pliocene) activity. New research suggests that some tectonic process (other than volcanism) must also be at work. Such a process is necessary to explain uplift within the past 3 million years during Late Pliocene to Pleistocene time. Most probably the uplift relates to localized perturbation (thermal inflation or upwelling) within the mantle.

Several islands in the western part of the Hauraki Gulf near Auckland, including Rangitoto and Ponui, relate to active intraplate basalt volcanism.

Most nonvolcanic islands around New Zealand relate to erosion resulting from Pleistocene glaciation and sea-level rise, especially in Fiordland (Resolution, Secretary, Anchor, Cooper, Chalky, Long, and Coal) and Stewart Island (Ruapuke, Codfish, and Big South Cape), but also the Bounty Islands and Antipodes Islands. Note that, of course, volcanic islands have also been subject to these processes, including the Auckland Islands and Campbell Island.

Some islands relate to downdrop (crustal sag) associated with subduction of the Pacific Plate below the Australian Plate, such as in Cook Strait between the North and South islands, and in the Marlborough Sounds (Kapiti, Mana, D'Urville, Arapawa, Blumine, Forsyth, Chetwode, Stephens). In this context, the Marlborough Sounds are indeed a drowned landscape, but not because sea level has risen; rather, the crust has sunk (Fig. 4). It is thought that continental crust on the Australian Plate has been drawn down by the descending "slab" of oceanic crust on the adjacent Pacific Plate.

In summary, the islands of New Zealand are of diverse origin. Most, including the largest islands (North, South, Stewart), are the product of tectonic uplift, localized tectonic downdrop or subduction-related volcanism as a result of collision between the Pacific Plate and the Australian Plate. Other islands relate to processes such as localized continental rifting, intraplate volcanism, and possibly localized mantle inflation. Yet others are the

FIGURE 4 The Marlborough Sounds, northeastern South Island, an example of island formation in a drowned landscape caused by tectonic downwarp. This view is looking north across Queen Charlotte Sound with part of D'Urville Island (top left) and Arapawa Island (right). Photograph by GNS Science/Lloyd Homer.

product of erosion relating to a combination of Pleistocene glaciation and/or sea level rise.

In order to explain this diversity, it is necessary to consider the geology of New Zealand. Let us start by considering aspects of New Zealand today.

YOUNG NEW ZEALAND

New Zealand is widely regarded as "young" geologically. Perhaps the main basis for this claim is the presence of active and very conspicuous volcanoes in the central North Island. These include subduction-related arc volcanoes that represent a southwestern extension of the Pacific "Ring of Fire": the Taranaki, Ruapehu, Tongariro (Ngauruhoe), Tarawera, and White Island volcanoes. Of these, only Taranaki is more than 100,000 years old. They each have long eruption histories that vary in frequency.

Other less conspicuous volcanoes, also classed as active but considered much more dangerous and productive, are the eight rift-related rhyolite calderas within the TVZ including Taupo and Okataina (Rotorua). The eruption products (tephras, ignimbrites) of these "supervolcanoes" are widespread and conspicuous throughout central North Island and are present in Auckland. These volcanoes are less than two million years old, and Taupo is less than 500,000 years old.

Other reasons for regarding New Zealand as geologically young relate to earthquake activity, the presence of active faults, and spectacular fault scarps. More than 16,000 earthquakes are recorded annually in New Zealand.

Auckland, New Zealand's largest city, is built on an active volcanic field of about 50 volcanic centers (cones, maars, craters). The age of eruption of each cone is imprecise, but the entire field is less than 250,000 years old, and the youngest eruption (Rangitoto Island) was about 600 years ago.

By contrast, the Canterbury Plains are composed of an extensive sheet of youthful alluvial gravels related primarily to active braided river development and redistribution of glacial outwash during Pleistocene time (the past 1.8 million years), mainly since the last ice age just 20,000 years ago. Correlative gravel accumulations, widespread throughout the South Island on both sides of the Southern Alps, relate to vigorous uplift and erosion of the Southern Alps, largely within the last five million years.

In summary, there is ample evidence of active geological processes in New Zealand. The effects of active volcanism, active deformation of the landscape, and mountain building are conspicuous. However, it cannot be claimed that New Zealand is necessarily younger than many other parts of the world, especially those that are also affected by active plate boundary collision, such as Japan, Indonesia, Alpine Europe, or Himalayan Asia.

To explain and understand these active geological processes better, it is first necessary to consider the present-day tectonic setting of New Zealand.

TECTONIC SETTING OF NEW ZEALAND

Modern New Zealand straddles an active plate boundary zone. The actual boundary can be drawn as a line that extends from the southern end of the Tonga–Kermadec Trench and runs down (heading from northeast to southwest) but entirely offshore of the eastern coast of the North Island, cuts across the upper half of the South Island along the Hope Fault, then connects with the Alpine Fault on the western side of the Southern Alps, and heads offshore at the entrance of Milford Sound (western Fiordland) and into the Puysegur Trench. In broad terms, the Pacific Plate is moving westwards and the Australian Plate is moving northwards. The relative rate of plate motion convergence (collision) has been measured using satellite GPS surveillance at about 4–5 cm per year.

In terms of the crust, the collision varies greatly along the New Zealand segment of the plate boundary, and it is this variation that dictates the diversity of topography and geological processes manifest in the New Zealand landscape. The entire North Island consists of continental crust of the Australian Plate and is subject to collision with oceanic crust on the subducting Pacific Plate. This is an example of conventional continent–ocean collision, whereby dense oceanic crust slides down or is drawn down below less dense continental crust (subduction). Hence, the active subduction-related volcanoes are restricted to the North Island.

The South Island is utterly different. It represents collision between continental crust of the Australian Plate and continental crust of the Pacific Plate (continent–continent collision). Hence, the Southern Alps are comparable to the European Alps and the Himalayas, and there is an absence of subduction-related volcanism.

At the southern end of the South Island the plate boundary reverts back to continent–oceanic collision, but here it is subduction of oceanic crust on the Australian Plate beneath continental crust of the Pacific Plate. This is the reverse of what is happening in the North Island. Solander Island is the only subduction-related arc volcano above sea level on this segment of the plate boundary and may rightly be regarded as part of the Pacific Ring of Fire.

Much of the interest and geological intrigue of New Zealand relates to the diversity of geology afforded by the present tectonic setting, all within one small part of the globe.

The nature and age of the actual rocks that form the substrate or basement of New Zealand are considered next in terms of the three major episodes in the geological history of the New Zealand land mass: Gondwanaland (545–83 million years ago), Zealandia (83–23 million years ago), and New Zealand (23 million years ago to the present).

GEOLOGICAL HISTORY OF NEW ZEALAND

Gondwanaland

The oldest rocks known from mainland New Zealand are of Early Cambrian age, about 505 million years old (Tasman Formation). These rocks are fossil-bearing limestone and subduction-related volcanic rocks exposed in northwest Nelson, South Island. However, older rocks are known from Campbell Island and are of latest Precambrian age, about 545 million years old (Complex Point Schist). In geological terms, these occurrences fall within the Takaka and Buller terranes (respectively) of the Western Province.

Modern geological interpretation (mapping) of the older basement rock of New Zealand recognizes tectonostratigraphic units referred to as terranes (Fig. 5). These are elongate belts of rock measurable in terms of crustal thickness (between 5 and 25 km thick), tens of kilometers wide, and hundreds of kilometres long. Each terrane is composed of variably metamorphosed sedimentary or volcanic rocks that are fault-bounded and share a common history. Series of related volcano–sedimentary terranes are grouped into "provinces." In New Zealand there are two provinces, Western and Eastern, separated by the predominantly igneous Paleozoic–Mesozoic Median Batholith (effectively a long-lived magmatic province) that is dominated by granite and gneiss. Two Paleozoic terranes are recognised in the Western Province and seven Paleozoic–Mesozoic terranes in the Eastern Province (from west to east): Brook Street, Murihiku, Dun Mountain–Maitai, Caples, Waipapa, Rakaia, and Pahau.

The dominant sedimentary rock type that is common to all nine terranes is greywacke and its metamorphic equivalent, schist. More accurately described perhaps as muddy sandstone or silty sandstone derived from erosion of either granite or volcanic rocks, greywacke, more than any other rock type, is characteristic of New Zealand. More importantly, it is characteristic of sediment accumulation in marine continental accretionary margin settings. Greywacke dominates and pervades the New Zealand landscape.

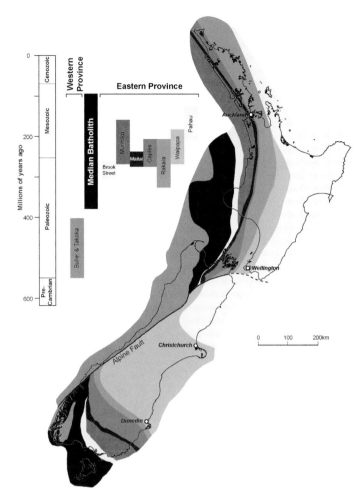

FIGURE 5 Distribution of basement rocks in New Zealand in terms of tectonostratigraphic units: nine terranes grouped into provinces. The orientation of these elongate belts of rock reflects systematic continental growth of eastern Gondwanaland by tectonic accretion during 459 million years of Paleozoic–Mesozoic time. The oldest continentward terranes are to the west, the youngest oceanward terranes are to the east. Diagram by GNS Science.

Systematic research on the provenance or source of the original sediments that make up the sedimentary (greywacke) rocks of all nine New Zealand terranes has established that they are all derived from unique yet closely related source areas. Broadly speaking, these terranes represent a series of more or less independent sedimentary basins. Each terrane represents sediment accumulation from distinctive source areas within eastern Gondwanaland, and mainly the Queensland sector of what is now Australia. Furthermore, the terranes are all allochthonous to varying degrees. Those of the Eastern Province are more so than those of the Western Province and indeed have been referred to as "exotic." This means that they have been tectonically removed (excised along major faults) and are now detached from both their original source rocks and their original geographic setting.

This prolonged period of accumulation and subsequent tectonic displacement spans 459 million years, from Cambrian to Cretaceous time—from about 542 to 83 million years ago—and represents the Gondwanan or Gondwanaland history of New Zealand.

Both provinces, along with the Median Batholith, were assembled into their present configuration with respect to each other within Early Cretaceous time. They were assembled during oblique plate boundary convergence (collision tectonism), similar to what is happening today in the North Island. It involved prolonged subduction of oceanic crust of the Panthalassa Ocean floor on the paleo–Pacific Plate beneath continental crust of eastern Gondwanaland.

In this regard, the basement rock of New Zealand may be regarded as a remnant of a Cretaceous land mass composed of continental crust that developed from sustained growth by continental accretion over a period of about 460 million years along a segment of the eastern margin of Gondwanaland. As would be expected from any orderly accretionary process, the terranes of New Zealand reflect systematic eastward growth from the continent outwards.

In summary, in Cretaceous time this part of the world, best described as proto–New Zealand, established itself in two significant respects: as continental crust and as land.

The basement rocks of New Zealand are largely composed of sediments that accumulated in the Panthalassa Ocean, deposited by rivers draining eastern Gondwanaland, and that were subsequently bulldozed (accreted) onto the eastern Gondwanaland continental margin by subduction. The "bulldozer" was the oceanic crust of the paleo–Pacific Plate, and it was at work for about 460 million years. Contrast this with the present "bulldozing" of the Pacific Plate, which has been at work for only about 25 million years.

Cretaceous time was seminal in the geological history of New Zealand, with comparative stability reigning for about 50 million years, until the rifting of Zealandia away from Gondwanaland.

Zealandia

Zealandia is the name of the large fragment of eastern Gondwanaland that broke away about 83 million years ago (Fig. 6). Almost half the size of Australia, this new continent drifted in a northeasterly direction in response to continental rifting and the formation of fresh oceanic crust in

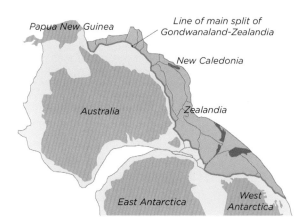

FIGURE 6 A reconstruction showing the configuration and extent of Zealandia about 90 million years ago, prior to rifting from eastern Gondwanaland about 83 million years ago and opening of the Tasman Sea. Note the location of "mainland" New Zealand (dark green shading) and also New Caledonia. Diagram by GNS Science.

what is now the Tasman Sea floor. The age and duration of this sea floor spreading process is based on paleomagnetic interpretation coupled with radiometric dating. The oldest sea floor is dated at 83 million years, and the youngest is about 60 million years old, so the Tasman Sea floor formed in less than 25 million years and has been more or less inactive since earliest Cenozoic (Paleocene) time.

As it drifted away, the continental crust of Zealandia was stretched, thinned, and cooled, and as a consequence the continent slowly sank. It did so over a sustained period of 60 million years until about 23 million years ago. The crust of Zealandia effectively lost buoyancy and sank up to 2500 meters with respect to sea level. Modern bathymetric mapping reveals the extent of continental crust of Zealandia as shallower than the 2500-meter isobath.

The Late Cretaceous to mid-Cenozoic (earliest Miocene) geology of New Zealand provides an excellent record of this process: the slow, steady submergence and drowning of Zealandia as a function of marine transgression—inundation by the sea. A transgressive sequence becomes increasingly more oceanic as the sea encroaches farther onto the land. A New Zealand–wide marine trangressive sequence accumulated during the time period from 83 to 23 million years ago. The geology suggests that sea level was rising, but what was really happening was tectonic sinking of the crust. Superimposed on this primary long-term process, involving sea level changes measurable in hundreds to thousands of meters, was second-order, much shorter-term sea level fluctuation involving tens of meters of rise and fall.

Recent analysis of the geological evidence for land in the New Zealand region of Zealandia in earliest Miocene time (24–22 million years ago) has raised the possibility of total submergence. Current geological evidence cannot conclusively confirm nor deny the existence of continuous land at that time. Furthermore, a strong geological argument can be made in favor of total submergence. Needless to say, this idea is controversial, because it has major implications for the origins and antiquity of the native terrestrial biota of New Zealand.

This does not mean to say that all of Zealandia was submerged 23 million years ago. New Caledonia has been emergent for at least 30 million years, although it was totally submerged prior to the onset of Oligocene time (34 million years ago). So, whereas the New Zealand region of Zealandia may have been completely submerged about 23 million years ago, New Caledonia was land at that time. This is an important consideration, because New Caledonia may have been a significant source of the ancestors of the native terrestrial biota of modern New Zealand, even though it lies more that 1200 km to the north of New Zealand.

Maximum submergence of the New Zealand region of Zealandia culminated in latest Oligocene and Early Miocene time, about 23 million years ago, with an abrupt change in regional tectonism. This heralded a new active configuration along a segment of the boundary between the Pacific and Australian plates that cut its way clean through continental crust of old Zealandia. Zealandia no longer represented a single tract of largely submerged continental crust. It was now split asunder, shared between two plates.

New Zealand

The present plate boundary configuration can be traced back to Eocene time, 45 million years ago. However, collision tectonism became vigorous with the onset of Miocene time. This process manifests itself in the geological record as a New Zealand–wide regressive sequence. In other words, the geology reflects retreat of the sea from the New Zealand region, commencing in Early Miocene time and continuing to the present day. The cause, however, is not sea-level change *per se*, but sustained tectonic uplift as a function of plate collision. It is this process that gave rise to the New Zealand land mass as we know it and it continues today. In this context, the geological history of New Zealand, strictly speaking, only relates to the last 23 million years.

DISCUSSION

In light of the preceding account of the geology of New Zealand, both "mainland" New Zealand and "off-shore islands," it is clear that New Zealand is very much part of a much bigger entity, namely the Zealandia continent, which rifted away from Gondwanaland about 83 million years ago and slowly submerged over a period of 60 million years. New Zealand has gradually become emergent in response to tectonism associated with a major change in plate configuration about 23 million years ago. New Zealand owes its location to the modern plate boundary. Without it, New Zealand would still be substantially underwater.

New Zealand can be regarded as part of a largely sunken continent. The Kerguelen Plateau in the southern Indian Ocean may be considered as another possible example of a sunken continent or large tract of submerged continental crust.

This modern understanding of New Zealand has been realized primarily from fresh insight into our understanding of the nature of the Earth's crust, how it works, and at what rates. This enables us to better explain the origin of the New Zealand islands, and yet some uncertainties and mysteries remain.

For instance, an explanation for the Pliocene–Recent uplift of the Chatham Islands is required. The Chatham Islands are too far removed to be affected by any uplift associated with the active plate boundary through "mainland" New Zealand, and late Cenozoic volcanism seems insufficient (volumetrically) to account for tectonic uplift that appears to involve the entire Chatham Islands area of a least 10,000 km². One possible explanation, a larger-scale mantle inflation effect, is being considered.

Like "mainland" New Zealand, the Chatham Islands and several other "offshore" islands are continental: the Snares Islands, Auckland Islands, Campbell Island, the Antipodes Islands, and Bounty Islands. They all have basement rocks of continental affinity: granite, greywacke, or schist. In addition, most of these are also the centers of basaltic intraplate volcanism. The presence of basement on these islands either at or above sea level is a function of chance preservation beneath a resistant cap of much younger basaltic rock. In a sense, the volcanoes act as storage vessels, harboring for scientific posterity a record of the rock they buried when they erupted. All of these islands appear to be Miocene or younger. They are variously subduction-related, associated with continental rifting in the TVZ, or intraplate basalt volcanoes. The latter are of particular interest because a satisfactory explanation of their location remains as yet uncertain. These include the Auckland Islands, Campbell Island, the Chatham Islands and the islands of the Auckland Volcanic Field.

The idea that the New Zealand 'mainland' may not have existed in earliest Miocene time, about 23 million years ago, is of considerable interest. Some of the key evidence is geomorphological, and in particular the origin or formation of regional planar surfaces in the New Zealand landscape. These features have been reinterpreted as remnants, not of a terrestrial "peneplain" but a regional wave-cut surface referred to as the Waipounamu Erosion Surface (WES). This surface formed as a result of marine planation between Cretaceous and Miocene time, commensurate with the slow sinking and submergence of Zealandia. Inevitably, the WES is commonly superimposed upon an older Cretaceous land surface. However, the Cretaceous land surface is easily recognized because it does not form planar geomorphic features in the landscape but exhibits demonstrable relief and is invariably overlain by fluvial sediments, whereas the younger WES is invariably overlain by marine sediments.

The geological origins of New Zealand's islands continue to fascinate the research world, and yet it is somewhat surprising that so much new insight has been gained in recent years. It is not easy work. Most difficult is determining the age of surfaces in the landscape. Conversely, it is relatively easy to determine the age of the rocks into which the surfaces are cut. Geomorphology, stratigraphy, radiometric dating, and a keen understanding of crustal processes remain the principal keys to further geological understanding of islands and their histories.

SEE ALSO THE FOLLOWING ARTICLES

Continental Islands / New Caledonia, Geology / New Zealand, Biology / Volcanic Islands

FURTHER READING

Campbell, H. J., and G. J. Hutching. 2007. *In search of ancient New Zealand*. Auckland, New Zealand: Penguin Books and GNS Science.
GNS Science Website. http://www.gns.cri.nz.
Graham, I. J. (ed.) 2008. *Continent on the move*. Geological Society of New Zealand Miscellaneous Publication 124. Wellington, New Zealand: Geological Society of New Zealand Inc.
Hicks, G., and H. J. Campbell. 1998. *Awesome forces: the natural hazards that threaten New Zealand*. Wellington, New Zealand: Te Papa Press.
Landis, C. A., H. J. Campbell, J. G. Begg, D. C. Mildenhall, A. M. Paterson, and S. A. Trewick. 2008. The Waipounamu erosion surface: questioning the antiquity of the New Zealand land surface and the terrestrial fauna and flora. *Geological Magazine* 145: 173–197.
LINZ Website (for inventory of New Zealand islands): http://www.linz.govt.nz.
Mortimer, N. 2004. New Zealand's geological foundations. *Gondwana Research* 7: 261–272.
Te Ara online encyclopedia of New Zealand Website. http://www.teara.govt.nz.

NORFOLK ISLAND

SEE PACIFIC REGION

NUCLEAR BOMB TESTING

EDWARD L. WINTERER

Scripps Institution of Oceanography, La Jolla, California

Nuclear bomb testing by the United States began just two years after the end of World War II and continued intermittently until late 1962, running up a total of 105 tests of fission (A-bomb) and fusion (H-bomb) weapons, all of them in the atmosphere or in shallow lagoon waters. The tests were conducted not only to evaluate successive bomb designs but also to observe the effects of nuclear explosions on ships and crews.

Islanders were evacuated from Bikini atoll (now Pikini) successively to a series of other atolls, where they fared poorly because of poor environments for fishing and growing crops. Their home island is still not safe. On Eniwetok atoll (now Enewetak), A-bomb tests began in 1948, followed by H-bomb tests in 1952, which later alternated between there and Bikini. A further 24 U.S. tests were made as air drops close to Christmas atoll, a British island in the Line Islands chain. Two very high-altitude rocket tests over Johnston atoll in 1962 closed out the American atoll testing. Meanwhile, the British, at Christmas and Malden, exploded nine devices, and the French at Mururoa conducted some 175 tests, including 134 underground tests in lagoon boreholes, finally ending all testing there in 1996.

EARLY BIKINI TESTS

Only a few months after the detonation of the atomic bombs over Hiroshima and Nagasaki in August 1945, U.S. military and political authorities began to plan for improvements in the U.S. atomic arsenal and to plan for testing of these new bombs at Bikini (now Pikini) atoll in the Marshall Islands in the western Pacific (Fig. 1). The Marshall Islands had been under Japanese rule under a League of Nations Mandate but were occupied by U.S. military forces during 1944 and 1945. From the war's end until the establishment of a UN trust territory in 1947, the islands were under U.S. military control. In February 1946, the U.S. military commander of the Marshall Islands obtained the permission of the chief of the group of 167 Bikini people to move all the people to another atoll (Rongerik, about 300 km east of Bikini) during the tests. By early March 1947, the Bikinians were on Rongerik, and Americans began arriving at Bikini

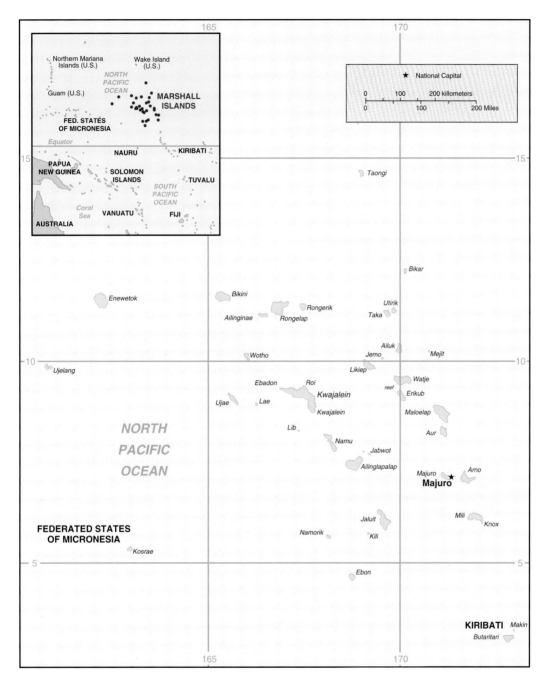

FIGURE 1 Marshall Islands. Image courtesy of the U.S. Central Intelligence Agency. (As a work of the United States Government, all images by the CIA are in the public domain, with the exception of classified information.)

atoll to prepare for the tests, termed Operation Crossroads, which were conducted in July. Besides the military and the bomb scientists and technicians, the arriving group included teams of scientists—oceanographers, biologists, and geologists—to map the atoll and its waters prior to any explosions.

One of the main purposes of the tests was to evaluate the effects of atomic bombs on ships, planes, and equipment, so a fleet of some 90 target vessels, along with some animals was assembled in the lagoon and on Bikini atoll. The first test, named Able, an air drop about 500 feet over the lagoon, caused little visible damage to the islands, whereas the second, Baker, submerged at a depth of about 30 m, threw up a huge cloud of water and irradiated debris from the floor of the lagoon (Fig. 2). The two test bombs were each 23 kilotons. Able damaged ships, but radioactivity faded enough for crews to board the vessels in two days. Baker sank eight ships, and the radiation persisted

FIGURE 2 Cloud of water vapor and dust from Bikini lagoonal bottom sediments resulting from Baker explosion, July 25, 1946. The dome evolved to a mushroom cloud as it rose. Image courtesy of the U.S. Government Defense Threat Reduction Agency. (As a Work of the United States Government, all images by the Defense Threat Reduction Agency are in the public domain.)

for weeks, preventing all but brief visits to remaining ships. The ships were towed to Kwajalein atoll for decontamination and offloading ammunition, work that continued into 1947. Eight ships were towed to the United States for inspection, and 12 were sailed there by their crews. All the rest were sunk. In 1947, scientists returned to Bikini atoll for a resurvey.

Meanwhile, the Bikini people were in worsening physical condition on Rongerik and pleaded to be relocated. Plans were made to move them to Ujelang atoll, about 60 km southwest of Eniwetok atoll, but instead, Eniwetok was evacuated and its people moved to Ujelang atoll. The Bikinians were moved first to one of the islands on Kwajelein atoll and finally, in November 1948, to Kili atoll, a tiny (0.75 km²) island about 800 km southeast of Bikini. There they stayed until October 1972, when a few families moved back to Bikini, which was declared safe by U.S. authorities. In 1975, a restudy discovered that the water and food available on Bikini were too radioactive for human consumption, but the population was not moved; rather, food was brought in by ship. In 1978, the remaining people on Bikini were moved back to Kili.

ENIWETOK TESTS

Test operations after Crossroads were moved to Eniwetok atoll (now Enewetak), where a series of three fission bombs were exploded from 60-m towers on three different islands in April and May 1948, during a 10,000-man operation named Sandstone. Yields were from 18 to 49 kilotons, and the tests (successful) were to evaluate a new, more efficient design than the ones for the bombs used over Japan and at Bikini. Most of the people of Eniwetok had already, in May of 1946, been evacuated to Kwajalein atoll, in preparation for the Bikini tests. In December 1947, the balance of the Eniwetok population was taken to Ujelang atoll.

After three more years, in April 1951, a new series of fission bomb tests, termed Operation Greenhouse, was conducted at Eniwetok. The series comprised at least four shots, with yields from 14 to 81 kilotons, and included an odd design (test George) that was intended to test the mechanism of radiation implosion, an element in the design of the H-bomb (Fig. 3).

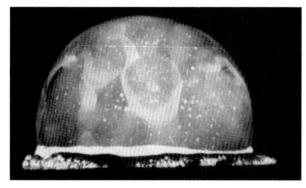

FIGURE 3 Fireball from test device George, 20 milliseconds after detonation. Eniwetok, 1951. (This work is in the public domain in the United States as a work of the U.S. Federal Government under the terms of Title 17, Chapter 1, Section 105 of the U.S. Code.)

In November 1952, there began the first tests of the thermonuclear hydrogen fusion bomb, first declared a goal by President Truman in 1950, following shortly after the first successful Soviet fission bomb test. In test Mike, the first H-bomb was detonated, with a yield of 10.4 megatons (versus the 15 kilotons of the Hiroshima fission-type bomb). The cylindrical device, called Sausage by its builders, was sheltered in a "cab" and weighed about 80 tons (Fig. 4). It comprised two stages: a fission bomb as the primary stage and a flask of liquid deuterium (a heavy isotope of hydrogen) with a central rod of plutonium and an enveloping 5-ton layer of uranium as the secondary. The device was encased in a cylinder of 30-cm-thick steel about 2 m across and 6 m long, and left a crater more than 1.5 km wide and 50 m deep. The fireball was over 5 km wide, and the mushroom cloud rose to 17 km in less than 90 seconds. Little more than a minute later it had reached 36 km in altitude, with the top eventually

FIGURE 4 "Sausage" device Mike (vertical cylinder) at Eniwetok. Seated man with mandolin gives scale. 1952. (This work is in the public domain in the United States as a work of the U.S. Federal Government under the terms of Title 17, Chapter 1, Section 105 of the U.S. Code.)

spreading out to a diameter of 160 km. The blast created a crater about 1800 m in diameter and almost 50 m deep where the isle of Elugelab had once stood; the blast generated 6-m water waves. Irradiated coral debris fell on ships stationed 50 km from the blast. Two weeks later, fission bomb King was airdropped, testing a 500-kiloton design, at about the limit for a fission bomb at that time and finishing the test series.

LATER TESTS AT BIKINI

Further testing of H-bombs, with the first test on March 1, 1954, took place on an isle at the northwestern corner of Bikini atoll. This test, named Bravo, was of a device (Shrimp) of an appearance similar to the huge Sausage, but weighing only about 24 tons. The novel feature was that about 40% of the fuel was of enriched lithium-6 deuteride. The case was of aluminum rather than heavy steel. The unexpectedly high yield of 15 megatons was a surprise and resulted from the supposedly inert lithium-7, which, when struck by high-energy neutrons, could fragment into a helium and a tritium atom. The tritium caused very energetic fusion that increased the expected 4–8-megaton yield. Bravo created a crater on the reef rim and adjacent part of the lagoon about 2 km wide and 75 m deep, and threw all the excavated debris into the sky. The cloud rose in only six minutes to a height of 40 km. After 8 minutes, even the cloud bottom had risen to about 15 km. A few weeks after this test, the new bomb design was weaponized, and a bomb was put into production (275 bombs, 4.5-megaton yield) during 1955 and 1956.

The fallout moved mainly east and northeast and blanketed atolls and ships as distant as 250 km with ash, exposing the (completely unwarned) people to dangerous levels of radiation (Fig. 5). A Japanese fishing vessel, the *Fifth Lucky Dragon,* working about 80 km northeast of Bikini, was blanketed by ash, and its crew received about 300 rad of radiation. One member of the crew died of this exposure. It took about two days for the military to send evacuation ships to the affected atolls of Rongelap, Rongerik, Utirik, and Ailinginae, where people received estimated doses of from 14 to 175 rad. Bikini atoll was declared too contaminated for people, and further tests, which used non-enriched lithium deuteride, were controlled by radio from a ship. The first device was detonated

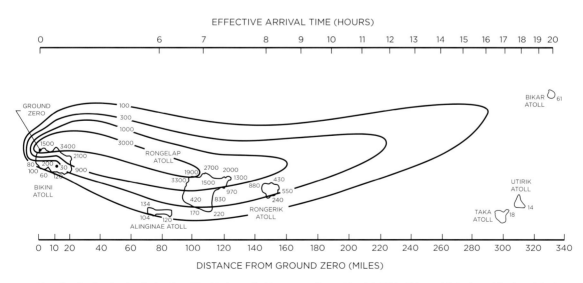

FIGURE 5 Map showing levels of radiation (in millirads) downwind from event Bravo, March 1, 1954. (This work is in the public domain in the United States as a work of the U.S. Federal Government under the terms of Title 17, Chapter 1, Section 105 of the U.S. Code.)

on a barge in the lagoon, rather than onshore like Bravo, and yielded about 11 megatons. There then followed four more tests, from 110 kilotons to 11 megatons, all from barges, ending on May 14, 1954. Most of these produced significant fallout at the atolls east of Bikini.

BACK TO ENIWETOK

Testing was resumed, this time back on Eniwetok, in May 1955, in which a small (50-cm diameter) warhead and very small (12-, 20-, and 28-cm diameter), lightweight systems were detonated. It was a test series for proving thermonuclear designs of actual weapons. Through July 21, 1955, 17 shots were detonated, of which seven were at Bikini and ten at Eniwetok, some from barges, some from towers, and one from the ground. Yields ranged from 1 kiloton to 5 megatons.

Further operations began at Eniwetok and Bikini in May 1958 and comprised 35 detonations. One shot was from a balloon at 25-km altitude, about 135 km northeast of Eniwetok, and another from about 150 m underwater in waters about 975 m deep. Yields ranged from to 1.7 kilotons to 9.3 megatons. The last test (OAK) of a thermonuclear device, with a yield of 8.9 megatons, left a large crater and debris field on the edge of the reef flat (Fig. 6). Two tests over Johnston atoll, at about 16°45′ N, 169°31′ W, were of a warhead designed for antiballistic missiles, which was fired from a rocket to altitudes of about 80 km. On August 18, 1958, the last explosion was detonated, completing all nuclear testing in the Marshall Islands. Total cost is estimated at about $2.5 billion.

BRITISH TESTS AT CHRISTMAS ATOLL AND MALDEN ISLAND

Christmas atoll (1°50′ N, 157°20′ W), now the island of Kiritimati in the Republic of Kiribati (formerly the Gilbert Islands), and Malden Island (4°1′ S, 154°56′ W), about 300 km south of Christmas, were used for a series of nuclear bomb tests beginning in 1956, with the arrival of British military personnel and civilian scientific people. Next, extensive construction was begun on Christmas atoll for wharves, airfields, and living quarters. Between May 15 and 19, 1957, three megaton-size bombs were dropped by airplanes and detonated at about 2500 m over the sea about 50 km south of Malden. Next, tests were conducted over Christmas atoll itself. Women and children had been evacuated to Fanning Island, about 400 km to the north, and the remaining plantation staff was put on a ship, below decks, during the early morning tests. The thermonuclear devices were dropped from airplanes off the island, one (1.8 megatons) on November 8, 1957, and three more (2–3 megatons) from April 28 to September 11, 1958. A bomb in the megaton range was detonated from a balloon on August 22, 1959, and another balloon-dropped device with a yield in the kiloton range on September 23.

AMERICAN TESTS AT CHRISTMAS ATOLL

Further tests at Christmas Atoll resumed in April 1962, this time by the United States. In all, 24 devices, ranging in yield from 3 kilotons to 7.6 megatons, were dropped from airplanes and detonated at altitudes of 750 to 4500 m in he vicinity of the atoll. The tests were concluded on July 11, 1962.

TESTS AT JOHNSTON ATOLL

To evaluate the effects of nuclear bombs to be used in the U.S. missile defense system, the United States, on July 9, 1962, at 0900 GMT, conducted a high-altitude nuclear test over Johnston atoll, using a rocket that carried the thermonuclear device to an altitude of about 400 km, where it detonated with a yield of 1.4 megatons. The result was a spectacular light show, visible from Kwajalein, some 2500 km to the southwest, as a white flash that quickly turned into a green ball projecting long white streaks arcing toward the poles, followed by a set of concentric rings that rose high in the sky. Then, a red glow grew to a broad red arc in the northeast, stretching high up toward

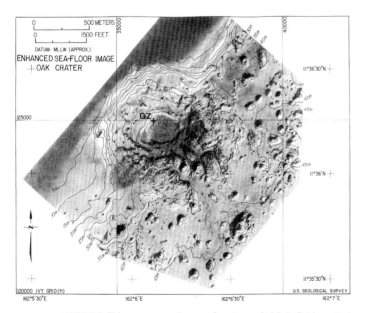

FIGURE 6 Side-scan sonar image of crater and debris field created during 8.9-megaton test OAK on the lagoon side of the west rim of Eniwetok atoll, June 28, 1958. (From Foulger et al., 1986; this work is in the public domain in the United States as a work of the U.S. Federal Government under the terms of Title 17, Chapter 1, Section 105 of the U.S. Code.)

the zenith. The Kwajalein show lasted for about 10 minutes. The show at Johnston was even more flamboyant: a red disk at the zenith and down to about 45 degrees from there. A yellowish-white streak grew from the zenith to the north along magnetic north–south. The auroral band widened quickly, then receded from the north and extended to the south. It was over in less than 10 minutes. In Hawaii, about 1600 km away, street lights went dark, TVs and alarms malfunctioned, and microwave phone links were shut down. Radiation belts generated by the explosion crippled some satellites and destroyed others. The experiment had produced a global belt of high-energy electrons, some persisting for five years.

Testing of eight more nuclear bombs at Johnston atoll continued to November 4, 1962. These were partly of bombs dropped from airplanes at altitudes of 3000 to 3600 m and missile airbursts at altitudes of 13 to 90 miles.

FRENCH TESTS AT MURUROA AND FANGATAUFA ATOLLS

Following a series of 17 nuclear tests in the Sahara from 1960 to 1966, France elected to move its testing to a remote site in the South Pacific, at Mururoa atoll (21°52′ S, 138°55′ W) and Fangataufa (22°14′ S, 138°45′ W) atoll, 65 km southeast of Mururoa, where testing began in July 1966. After 41 atmospheric tests of yields from 20 to 1000 kilotons ended in September 1974, there followed a series of 134 underground tests in boreholes, ranging from 5 to 150 kilotons in yield. That series ended on January 27, 1996. Shortly afterward, in March, France and Britain signed the Raratonga Treaty, which created a nuclear-free zone in the South Pacific, thereby foreclosing further tests in Polynesia. The United States signed the treaty, but the Senate never ratified it.

LESSONS LEARNED FROM TESTING AT PACIFIC ATOLLS

Without question, the military preparedness objectives of three nations—the United States, Great Britain, and France—were advanced. Many new designs of nuclear weapons were tested, and the results translated into the construction of both strategic and tactical weapons. The human costs were large: People died or were sickened by exposure to fallout, and people were deprived of their ancestral homelands and exiled—for some people, forever—to unsuitable substitute islands. The biota of the atolls was, at least locally, seriously impacted by some of the test explosions, but the long-term effects are probably very minor compared to more far-reaching environmental changes associated with, for example, global warming. Scientifically, the surveys and test borings of the test-range atolls, especially Bikini, Eniwetok, and Mururoa, have deepened our understanding of the origin of atolls and their physical, biological, and geological makeup. It is highly unlikely that this knowledge could ever have been achieved without the science being done in conjunction with the testing. It was a devil's bargain.

SEE ALSO THE FOLLOWING ARTICLES

Atolls / Marshall Islands

FURTHER READING

Foulger, D. 1954. Geology of Bikini and nearby atolls. U.S. Geological Survey Professional Paper 260-A.
Johnston, W. R. 2005. Nuclear tests—databases and other material. http://www.johnstonsarchive.net/nuclear/tests/index.html.
Micronesia Support Committee. 1981. *Marshall Islands: a chronology: 1944–1981.* Honolulu, HI:
Sublette, C. 1994–2007. The nuclear weapon archive. http://nuclearweaponarchive.org/.

OASES

SLAHEDDINE SELMI
Faculté des Sciences de Gabès, Tunisia

THIERRY BOULINIER
Centre d'Ecologie Fonctionnelle et Evolutive, Montpellier, France

The term "oasis" is often taken in its metaphorical and very broad sense: a spot of life within an inhospitable environment. In that way, it has repeatedly been used to designate patches of vegetation in less-vegetated and dry landscapes, isolated ice-free areas in Antarctica, isolated life-rich areas in marine ecosystems, and every other kind of isolated habitat. Even though these systems are comparable in that they are isolated and different from their surroundings, their structure, origin, and evolution, as well as the factors affecting their dynamics, are widely different. Oases are more classically defined as relatively more fertile areas in a desert or wasteland, made so by the presence of water. There are numerous so-defined oases in North Africa, eastern Asia, Australia, and the southwestern region of North America (Baja, in particular), and as such, they represent important models to investigate the role of isolation for the dynamics and conservation of biodiversity in arid landscapes. This article focuses on the original meaning of "oasis" (i.e., a date palm grove in a desert), and provides information on the ecology and dynamics of these particularly poorly known island-like systems.

OASES AS SEMI-NATURAL CONTINENTAL ISLANDS

The world distribution of date palm oases shows that they are mainly associated with Saharan regions in Arabia, eastern Asia, the Middle East, and North Africa, where water resources are localized and where the date palm trees *Phoenix dactylifera* have been cultivated for thousands of years (Figs. 1–3). The structure, evolution, and functioning of these particular continental island systems depends upon a great complexity of environmental, historical, and socioeconomic factors. Their faunas and floras are also greatly shaped by those factors.

The desert environment surrounding oases is characterized by harsh climatic conditions, with annual rainfalls rarely exceeding 100 to 200 mm and a summer temperature

FIGURE 1 Aerial image of Chebika oasis, southwestern Tunisia (34°19′00″N, 7°56′20″ E). Modified from Google Earth© views.

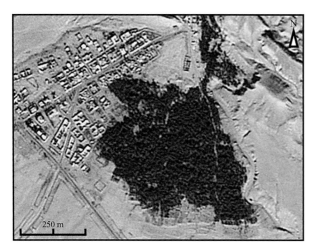

FIGURE 2 Aerial close-up of Chebika oasis. Modified from Google Earth© view.

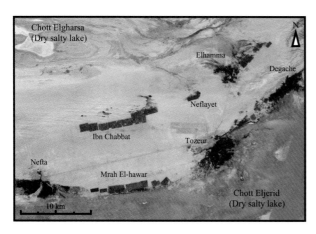

FIGURE 3 Aerial photography of the oasis archipelago of the Jérid area in southwestern Tunisia. The darker and more irregular patches correspond to traditional oases; the lighter and more regular patches correspond to modern plantations. Modified from Google Earth© view.

often exceeding 40 °C. In this environment trees, are mostly absent, and the vegetation is composed of sparse steppe shrubs. Nevertheless, the particular geological conditions of some areas, in particular the existence of major faults, permitted the emergence of fossil groundwater as springs. The use of this water for irrigation by local human populations has allowed the practice of agricultural activities and the development of a thick vegetation typically composed of three distinct layers (palm trees, fruit trees, and herbaceous plants), which has induced a local microclimate that strongly contrasts with the arid climate of the surroundings (Fig. 4). This so-called oasis effect is responsible for the insular character of desert oases.

Given the pronounced physical and climatic contrasts between an oasis interior and the surrounding environment, and because oases are directly dependent on the availability of water and on human activities for irrigation and maintenance, they can be considered as seminatural continental islands. Furthermore, given that the geographic location of an oasis is primarily constrained by the existence of a water spring, oases are not randomly distributed in the desert, but are generally concentrated in some areas where geological conditions have permitted the emergence of groundwater, leading to regional "archipelagoes" of oases.

FIGURE 4 Understory view within Teboulbou oasis (33°50′30″ N, 10°07′33″ E; southeastern Tunisia), showing the strong stratification of the vegetation. Photograph by S. Selmi.

INSULAR SPECIFICITIES OF OASIS COMMUNITIES

In spite of their originality and the current threats to which they are exposed, the flora and animal biodiversities of oases have not been much considered by biologists until recently. For instance, within the abundant ecological literature on the biodiversity of continental insular systems, there is very little information on the dynamics of animal and floral oasis communities. The main contributions on these aspects have dealt with oasis bird communities from southern Tunisia (work by the authors). These studies have shown that the occurrence of several non–desert adapted Palearctic bird species, such as the common blackbird (*Turdus merula*), blue tit (*Cyanistes caeruleus*), chaffinch (*Fringila coelebs*), serin (*Serinus serinus*), orphean warbler (*Sylvia hortensis*) and woodchat shrike (*Lanius senator*), in the arid land of southern Tunisia is strictly related to the oasis habitat. They also highlighted the very insular characteristics of oasis breeding-bird communities, notably the fact that their species richness is linked to oasis-area vegetation characteristics and degree of spatial isolation. In particular, it was shown that the number of breeding species in a given oasis is positively related to its area (Fig. 5). Such a typical species-area relationship is an important characteristic of island systems. Species richness of avian communities also depends on vegetation structure: Oases with a diversified structure (notably in terms of the presence of date palm trees, fruit trees, and herbaceous plants), and with a dense structure showing a sharp contrast with the surrounding desert, host richer bird communities than do more open oases where vegetation is less dense and diversified. In addition, it was shown that bird species richness and composition varied as a function of the degree of geographic isolation of the oasis. Oases close to each other host very similar communities, independently of their area and the quality of their habitat. This suggests that the dispersal of individuals and the exchange of species between neighboring oases play an important role in shaping oasis communities.

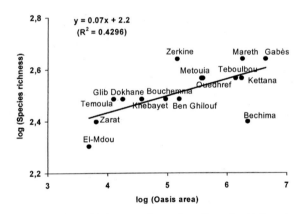

FIGURE 5 Relationship between the number of breeding-bird species and oasis area in a sample of oases of southeastern Tunisia.

Overall, the work on the breeding bird communities of Tunisian oases has shown that a combination of ecological, geographical, and historical, but also socioeconomic, factors seems to determine the diversity of oasis faunas. At the scale of an oasis archipelago (a geographic area regrouping a set of oases), the sets of species constituting the local communities would be subsets of a pool of potential colonists that would have reached the region. The size and the composition of the pool of potential colonists (regional richness) are determined by the geographic location of the region of interest with regard to sources of potential colonists (notably, the closest woodland and semi-woodland areas). Those characteristics of the regional communities will also depend on the capacities of the species to extend their area of distribution and to colonize new areas.

At the scale of an oasis, the physical characteristics, notably area and the suitability of habitat for the various species, will determine the richness and composition of the oasis's community, which will be a subset of the regional pool. The physical characteristics of oases are in fact the result of the interactions between various environmental factors, notably water availability, and socioeconomic factors linked to the agricultural system in place in the oasis. If the environmental factors are critical for determining the mere existence of the oasis and can strongly affect its area by limiting its possible extension, then the socioeconomic factors are directly responsible for the quality of the within-oasis habitat that will be available to the various species. Finally, colonization processes from nearby oases, as well as local extinction, seem to play a role. Such metapopulation processes could also largely be responsible for the presence of some species at the regional scale, and thus for the maintenance of regional richness. Comparable mechanisms are likely to be involved in the dynamics of other animal and plant communities, although rates of colonization and extinction may be very different.

THREATS TO OASIS BIODIVERSITY

For an oasis to exist within a desert matrix, two elements are needed: water and the exploitation of that water by humans for agricultural activities. This results in the creation of a green area. An oasis is thus a sort of semi-artificial continental island, but at the same time it is a very precarious and sensitive ecosystem. It can decay or actually disappear relatively quickly if water resources decline or if the oasis farming practices change, notably when oasis farmers change their way of life. Such factors are currently threatening the mere existence of numerous old oases, which can lead to the loss of the original and sensitive biodiversity that is living in oases.

Traditionally, the agricultural systems run by local oasis farmers fed a mainly self-sufficient local economy. Date palm and fruit trees, together with vegetables and food to feed animals, were grown using a stratified system of plantation that enabled the oasis farmers to optimize their use of water while producing a wide diversity of products. The traditional agricultural practice (notably based on complex local irrigation systems) was characterized by reliance on inter-family networks and by the use of traditional tools. It is those traditional agricultural activities that led oases to look like continental islands, and it is also those activities that seem necessary for the maintenance of local diversity of oasis animal and plant communities. This system, however, has seen and is currently seeing dramatic changes, directly linked with the socioeconomic changes occurring in some oasis societies. For instance, new types of palm plantations have been established in Tunisia since the middle of the twentieth century as a means of maximizing the production and exportation of dates of the well-known Deglet Nour variety. Those modern palm plantations are actual monocultures of date palm trees, which lack the vegetation structure of traditional oases, notably the stratification of the vegetation. The agricultural production is done by employees and uses modern techniques and tools. Those modern palm plantations do not lead to a sharp climatic contrast between the oasis and the surrounding desert, and they do not host a wild biodiversity as rich as that of traditional oases. Moreover, those new oases are in competition with the traditional ones for limited fossil water. Hence, several springs that were irrigating traditional oases have dried up, leading to severe drought problems for those oases. In addition to these water-availability problems, socioeconomic issues linked with the abandonment of traditional agricultural practices are affecting oasis vegetation structure. The largest of such socioeconomic problems are the non-profitability of traditional agricultural production in the current economic context and the tendency for farmers to focus on a monocultural approach, such as date production, or to switch to more rewarding activities such as tourism and industry; the fragmentation of real estate within the oases over generations; the concurrent tendency for young people to move toward cities and to migrate to Europe; and the fast urbanization of some oases.

Overall, it seems that if oases are created by humans, then their possible disappearance can also logically be caused by humans. The case of oases illustrates how the development of modern agricultural practices, done to maximize profit, can represent an important threat to the biodiversity of precarious agro-ecosystems. As with other island and island-like entities that host specific and sensitive ecosystems, special efforts should be devoted to the protection of traditional oases. In this respect, a sound knowledge of their animal and plant communities is needed to better understand the dynamics of biodiversity in such systems. Any action plan for the conservation of these very original sociohistorical and ecological islands will have to consider the human aspect of things—economic, social, and cultural issues—as well as the physical and biological aspects.

SEE ALSO THE FOLLOWING ARTICLES

Continental Islands / Hydrology / *Lophelia* Oases / Metapopulations / Species–Area Relationship / Vegetation

FURTHER READING

Kassah, A. 1996. *Les oasis Tunisiennes: aménagement hydro-agricole et développement en zone aride.* Tunis, Tunisia: Centre d'Etudes et de Recherches Economiques et Sociales.
Riou, C. 1990. Bioclimatologie des oasis. *Options Méditerranéennes* A(11): 207–220.
Rodríguez-Estrella, R., M. C. Blazquez, and J. M. Lobato. 2005. Avian communities of arroyos and desert oasis in Baja California Sur: implications for conservation, in *Biodiversity, ecosystems, and conservation in northern Mexico.* J.-L. Cartron and G. Ceballos, eds. Oxford: Oxford University Press, 334–356.
Selmi, S., and T. Boulinier. 2003. Breeding bird communities in southern Tunisian oases: the importance of traditional agricultural practices for bird diversity in a semi-natural system. *Biological Conservation* 110: 285–294.
Selmi, S., T. Boulinier, and R. Barbault. 2002. Richness and composition of oasis bird communities: spatial issues and species-area relationship. *The Auk* 119: 533–539.
Zaid, A. 2002. *Date palm cultivation.* Rome: Food and Agricultural Organization of the United Nations.

OCEANIC ISLANDS

PATRICK D. NUNN

University of the South Pacific, Suva, Fiji

Those of us who live close to the edges of the world's continents are likely to be familiar with islands, often as places for recreation or retreat. In the past, they were sometimes places of refuge for people or other biota escaping continental calamities ranging from warfare to ice advance. In a geological sense, such islands are commonly slivers of continent, their connections drowned by the high-sea-level conditions in which we live today. Oceanic islands are quite different, often smaller and more remote, and to find them, the continental dweller generally has to travel much farther offshore, into the hearts of the ocean basins.

DEFINING AND UNDERSTANDING OCEANIC ISLANDS

The crust of the Earth is divisible into two distinct types: continental and oceanic. Relative to the ocean surface, the less dense continental crust rests higher—and therefore forms Earth's largest contiguous land masses—than the denser oceanic crust, most of which is covered by ocean. Only rarely does the ocean crust push above the ocean surface and form oceanic islands. Two further differences between the continental and oceanic types of crust are also important to mention.

The first is their age. The continental crust is old—in places as much as 5000 million years old—but the oceanic crust is almost nowhere older than 120 million years in age. The reason for this astonishing difference has, of course, to do with plate tectonics, widely acknowledged as the driver of the evolution of the Earth's surface. The mid-ocean ridges, mostly under water, steadily create new oceanic crust along their axes, pushing it laterally outward. At the other end of this oceanic "conveyor belt," the old oceanic crust is pulled down into the Earth's interior along ocean trenches, eventually perhaps to be regurgitated along mid-ocean ridges. So the oceanic crust is being continually moved sideways: pushed from one end, and pulled from the other.

The second important difference between continental and oceanic crust is composition. Understandably, given that continents have been around so much longer, continental crust is far more diverse in terms of its rock types than is oceanic crust, which—at least at its surface—is rarely anything other than a stack of basalt.

Oceanic islands are bits of the oceanic crust that have somehow reached above the ocean surface. The obvious difficulty of this achievement, given the comparatively short time available (120 million years maximum) and the height involved (perhaps 4 km from ocean floor to ocean surface), explains why there are so few oceanic islands. This in turn explains why, for decades, while geologists explored the continents in minute detail, oceanic islands were marginalized, typically regarded as unremarkable adjuncts to continents or—even more pejoratively—as the detritus left in the wake of drifting continents.

Not surprisingly, then, the earliest ideas about the evolution of the Earth's surface all had a continental bias, and because they effectively ignored the other 73% of the Earth's surface (the ocean basins), they have since been proven largely wrong. Some of the earliest investigations of oceanic islands and, more generally, of the ocean floor can today be read as full of pointers to the critical importance of these features in understanding the formation of the Earth's surface, but it was not until after World War II, when the U.S. Navy (among others) turned some of its resources and expertise to gathering scientific information about the ocean floor, that this breakthrough in perspective occurred. The results of these investigations, which led eventually to the formulation of the theory of plate tectonics in 1967, also underlined the importance of knowing about oceanic islands.

OCEANIC ISLANDS AND PLATE TECTONICS

Plate tectonics envisages the Earth's crust as divided into huge chunks (plates) that are generally rigid, interlocking, and continually moving. Plates include both continental and oceanic crust, but it is only the latter that moves independently. To understand the variety of ways in which oceanic islands form, it is helpful to classify their origin, as in Table 1. At the highest level, this separates oceanic islands formed along plate boundaries from those—far fewer—that form in the middle of plates.

Divergent Plate-Boundary Islands

In the scheme of plate tectonics, a single plate may have a divergent plate boundary—typically marked by a mid-ocean ridge—along which (sea-floor) spreading takes place. This is therefore marked by divergence or extension, and all the world's ocean basins have one main divergent plate boundary, on either side of which ocean floor (and the islands that rise from it) increases in age with increasing distance from the ridge axis.

Islands occur in places where (part of) a mid-ocean ridge rises above the ocean surface, an unusual occurrence best exemplified by Iceland in the North Atlantic.

TABLE 1
Genetic Classification of Oceanic Islands

Level 1 Classification	Level 2 Classification	Examples
Plate-boundary islands	Divergent plate boundary (mid-ocean ridge)	Iceland (North Atlantic)
		Niuafo'ou (Tonga, South Pacific)
	Convergent plate boundary (island arc)	Lesser Antilles group (Caribbean Sea)
		Solomon Islands (western Pacific)
		Sunda arc (Sumatra-Java, eastern Indian Ocean)
	Transform plate boundary	Cikobia (Fiji, South Pacific)
Intraplate (mid-plate) islands	Linear island groups (hotspot island chains)	Hawaii group (northeastern Pacific)
		Tristan da Cunha–Walvis ridge (South Atlantic)
		Réunion-Laccadive (Indian Ocean)

Here there is a plate triple junction where high heat flow has elevated the ocean floor, causing part of it to emerge above sea level. Much of what has been learned about divergent plate boundaries comes from Iceland, but there are smaller divergent plate-boundary islands. Among these are the island Niuafo'ou (Tonga) in the Southwest Pacific, whose doughnut shape is a result of a stretched caldera that has become filled with water and forms Lake Vai Lahi.

Convergent Plate-Boundary Islands

As well as a divergent boundary, a plate may also have a convergent boundary, one type of which involves an oceanic plate pushing down into the Earth's interior beneath another oceanic plate. In terms of island formation, these are the most productive places in the ocean basins. This ocean–ocean convergent plate boundary is generally marked by an ocean trench with parallel lines of islands, sometimes along both sides. Ocean trenches are asymmetrical, with the more gently sloping side being that of the downgoing plate, and the steeper side—along which collapses often occur, generating tsunamis—being that of the overriding plate.

Along convergent plate boundaries, islands can form and emerge in one of three locations: along the volcanic island arc on the overriding plate, along the non-volcanic island arc on the overriding plate, or along a crustal flexure on the downgoing plate.

VOLCANIC ARC ISLANDS

At a convergent plate boundary, the downgoing plate is pulled down into the Earth's interior, where—because of the intense heat—it melts. The liquid rock (magma) is lighter than the solid rock, so it tries to rise back to the Earth's surface. Where it succeeds, it will erupt on the ocean floor and may eventually build a line of volcanoes (which may grow into volcanic islands) parallel to the associated ocean trench. The composition of the volcanic rocks in these islands can be linked to the type of material being pulled down into the trench, particularly whether or not this includes significant amounts of the sediments that accumulate in the bottoms of trenches.

Because ocean trenches are usually arcuate in plan, the lines of associated volcanic islands are likewise arc-shaped; hence, they are referred to as volcanic island arcs. Examples come from the Caribbean (Lesser Antilles) and western Pacific (Marianas–Izu). In youthful volcanic island arcs, there are typically many underwater islands; some of these occasionally erupt just beneath the ocean surface,

FIGURE 1 Oceanic islands forming. (A) The summit of Kavachi volcano in Solomon Islands lies 50–100 m below the ocean surface, but when it erupts, as here in October 2002, it forms a conspicuous sight. Kavachi occasionally forms islands, but these are short-lived, being eroded by waves when the eruption ends. (Photograph by Corey Howell). In 1453 the giant Kuwae volcano in Vanuatu blew itself to pieces in one of the largest eruptions by volume in the last 10,000 years. Today, an undersea caldera lies where Kuwae once stood, and occasionally a smaller volcano named Karua that has grown up from the caldera rim erupts. The eruption shown occurred on February 22, 1971, (B) and formed an island one day later (C). (Photographs by Don Mallick, used with permission of the Vanuatu Cultural Centre.)

making their presence manifest (Fig. 1A), and may form islands of unconsolidated pumice that are washed away when the eruption ends (Figs. 1B and 1C).

Under many volcanic island arcs, particularly when they have grown comparatively large, much of the rising magma may not reach the surface of the crust, so it solidifies below it and forms intrusive igneous rocks. Recent work, particularly in the Canary Islands, has demonstrated that the importance of intrusive rocks to the growth of oceanic islands in such locations is far greater than once suspected.

ISLANDS OF NON-VOLCANIC ARCS

It is clear that, along convergent plate boundaries in the ocean basins, not only is one plate being pulled down, but the other is being thrust upward over the top of it. This overriding plate is therefore being pushed not only sideways but also upward as it rides over the downgoing one, a process that is amplified when the surface of the downgoing plate is highly irregular.

The uplift of the overriding plate often causes its edge to emerge above sea level, typically producing a line of islands, parallel to both the volcanic island arc and the adjacent ocean trench (Fig. 2A). Although these islands have volcanic basements, when they emerge, the basements are draped with thick piles of ocean-floor sediments. If they emerge within the coral seas, the emergent islands will commonly exhibit a thick cover of coral reef, testimony to their journey through the uppermost layers of the ocean. Examples include parts of several larger Caribbean islands (Hispaniola, Jamaica, Puerto Rico), the Mentawai Islands of Indonesia, and islands such as Choiseul in the northern Solomon Islands in the western Pacific.

Such islands are generally not visibly volcanic, and therefore form a non-volcanic arc, distinct from its volcanic counterpart, which is farther away from the trench axis. Some of the most distinctive types of non-volcanic islands of this kind are those whose form is that of a staircase of broad limestone steps, each of which represents an emerged coral reef. The highest emerged coral reef is expected to be the oldest—the first to emerge above sea level—whereas the lowest is the most recent (Fig. 2B).

ISLANDS ALONG THE CRUSTAL FLEXURE ON A DOWNGOING PLATE

Oceanic crustal plates are stiff, 15-km thick slabs of solid rock that naturally resist being forced upward or downward at convergent plate boundaries in the ocean basins. One clear manifestation of this resistance to be found on the downgoing plate is the way in which it rises upward slightly before being thrust down. This upward rise produces a flexure (or bulge) in the ocean floor up which submerged islands rise and sometimes poke their heads above the ocean surface when they get close to the flexural crest. Thereafter, it is all downhill, with many formerly emergent islands being pulled down into the bottoms of the ocean trenches where they are eventually dismembered and destroyed. Examples associated with the Tonga trench in the southwestern Pacific include the emergent island Niue, which is rising up the flexure, and the underwater Capricorn seamount, which is presently on its way down the trench slope.

It is rare to have a line of islands form along the crustal flexure on a downgoing plate, because usually not very many seamounts (or guyots) are appropriately positioned, but some do occur. One of the best-studied examples is the Loyalty Islands of New Caledonia in the southwestern Pacific. Of these, the largest (Maré) is close to the flexural

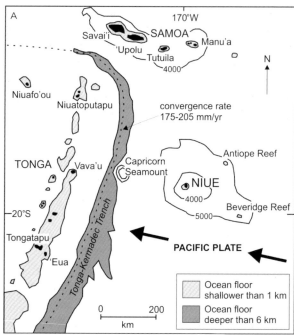

FIGURE 2 Islands close to convergent plate boundaries. (A) Map of part of the South Pacific showing the form of the ocean floor in the area where the Pacific plate in the east is converging with the plate in the west along the Tonga-Kermadec Trench. The non-volcanic arc is represented by a line of high uplifted reef-limestone islands from 'Eua in the south through Vava'u to Niuatoputapu in the north. The volcanic arc runs parallel 30–50 km to the west. To the east of the Tonga-Kermadec Trench, Niue Island is rising up the crustal bulge, whereas Capricorn seamount is on its way down into the trench. Niuafo'ou Island formed along a small divergent plate boundary, whereas the Samoa islands are a chain of hotspot islands. (B) View of the Talava Arches in northwest Niue Island. The flat top is formed by the emergence of a fringing reef since the last interglaciation about 120,000 years ago. (Photograph by the author.)

crest, whereas the two smaller ones (Ouvéa and Lifou) are on their way upward.

Transform Plate-Boundary Islands

If a rectangular plate has one divergent boundary and one convergent boundary, then its movement (from divergent to convergent) is fixed, which means that the other two boundaries can be neither of these. They are, in fact, places where one plate slides past an adjoining plate, theoretically with no net divergence or convergence: a type of boundary termed transform (or strike-slip). These are notorious as sites of large earthquakes—the San Andreas Fault is the best-studied—but they are not generally thought of as places where islands form.

Islands form along transform plate boundaries only where there are slight irregularities (kinks) in these that lead to localized convergence. The Fiji island of Cikobia may be an example of just such an island, formed at a kink in the Fiji fracture zone, a transform plate boundary in the southwestern Pacific.

Linear Groups of Intraplate Islands

Away from the edges of oceanic plates, in places where crustal quiescence rather than crustal activity is the norm, islands also form, although these are generally smaller and fewer and have less complex histories than their plate-boundary counterparts. Most such islands occur in approximately straight lines, something much remarked upon in early accounts of oceanic islands. Later work showed something even more remarkable: namely, that the age and size of these islands generally increased uniformly from one end to another of the island chain. And at the younger, larger end, there always seemed to be an active volcano.

The combination of these observations led to the formulation of the hotspot hypothesis, the idea that lines of intraplate islands were produced when an oceanic plate passed over a hotspot—a fixed place in the Earth's crust thin enough for underlying magma to punch its way through to the surface. The movement of the oceanic plate led to the volcano over the hotspot being gradually pulled away from it, eventually becoming extinct and being replaced by another volcano. In time, this process gives rise to a line of volcanic islands whose age increases with greater distance from the hotspot and that will slowly subside and thereby become smaller (Fig. 3).

Lines of hotspot islands are common in intraplate locations and include the Hawaii–Emperor island and seamount chain in the northern Pacific and the Samoa–Tuvalu island chain in the central Pacific. Réunion Island in the Indian Ocean and Tristan da Cunha Island in the South Atlantic are both volcanic islands built on top of active hotspots.

VARIETIES OF OCEANIC ISLANDS

The classification of oceanic islands given in Table 1 tells us about island origins but not necessarily about

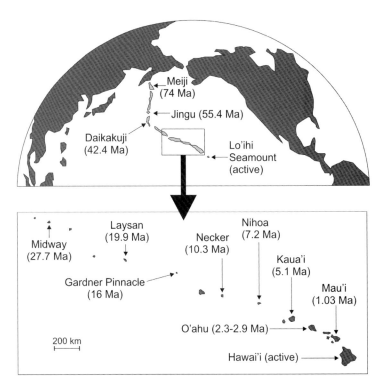

FIGURE 3 Ages of the most recent eruptions of island volcanoes along the Hawaii-Emperor island seamount chain (dates in millions of years ago). The upper map shows the location of the Hawaiian Ridge within the northern Pacific, and the oldest part of this hotspot chain (Meiji Seamount: 74 million years old). The lower map shows the younger, largely emerged part of the island chain from Midway Atoll (27.7 million years old) to still-active Hawai'i Island. Note the presence of newly active Lo'ihi Seamount, which is growing directly above the hotspot while Hawai'i moves away from it.

the way these islands look now. This is far less easy to generalize about systematically because various processes have caused islands to emerge or submerge relative to the ocean surface (which itself changes), irrespective of their origin. Thus, current appearance is an unhelpful guide to the origin of an island. But it is something worth knowing about, not least to help explain the nature of island biotas and—something highly topical at present—the vulnerability of particular oceanic islands to erosion, even erasure, by sea-level rise.

Every oceanic island began life as an ocean-floor volcano, but not all retain an immediately recognizable volcano form. When we look at the appearance of oceanic islands, three major groups can be identified: volcanic islands, high limestone islands, and atoll islands. There are other, more complex, types, often composites of volcanic rock and limestone, but their environments are generally reflective of the dominant rock types.

Volcanic islands vary in appearance depending largely on their lithology (rock type composition) and their age. Basalt volcanoes that form over very productive hotspots, for example, tend to involve huge volumes of moderately viscous material piled up in shield volcanoes—by far the most common type on Earth—in a series of comparatively low-energy, effusive eruptions. In contrast, andesite volcanoes tend to be much steeper sided—a reflection of the high viscosity of the eruptive material—and to erupt comparatively explosively. In a similar fashion, youthful—maybe even active—volcanoes form islands that generally betray that fact, whereas older (long extinct) volcanic islands may have been thoroughly disguised by post-eruptive denudation and flank collapse.

Limestone is a rock that forms only beneath the ocean surface, so a high limestone island has by definition emerged. For this reason, the form of limestone islands tends to broadly reflect the flat-topped form of undersea deposits. Sharp risers (slopes) to the next flat surface indicate successive periods of island emergence.

Atoll islands (motu) are distinguished by their comparative lowness—they usually rise no more than 3 m above sea level—and their transient nature. Part of this is because they are largely composed of unconsolidated sediments that are comparatively easy for the sea to remove.

Oceanic islands are more numerous in the warmer parts of the world's oceans than in the cooler parts. This is because in the former exist coral reefs, which can build up above a sunken volcanic island. It is a moot point whether living coral reefs actually constitute an oceanic island because they cannot generally grow above low-tide level, yet on almost every reef, there are accumulations of debris, swept up from below sea level by waves, that indeed reach above it—and in some parts of the world form islands large enough to be habitable by humans. Such islands—commonly called motu—are accumulations of largely unconsolidated sand and gravel, typically cemented by beachrock or phosphate rock along their fringes.

LIFE CYCLES OF OCEANIC ISLANDS

For anyone who is interested in explaining the distribution of islands in the world's ocean basins, it makes no sense to confine a survey to those islands that are currently emergent (above sea level): There are many "islands" whose summits lie below the ocean surface.

Some of these islands may have once been emergent but have since subsided (sunk) and/or been drowned as a result of sea-level rise. Traditionally these islands are classified as guyots, characterized by a flat top beveled by wave erosion as the island was slowly submerged.

Conversely, there are many islands that rise from the ocean floor but have not yet pushed their heads above the ocean surface. Some may do so eventually, and some not. Irrespective of whether they attain the surface, they should be included in any survey of oceanic islands. Such islands are generally referred to as seamounts, characterized by a conical form, attesting to their volcanic origins.

As noted above, in contrast to continental crust, oceanic islands are transient entities, never more than 120 million years in age and rarely emergent (above sea level) for even 50% of that time. Of course, there are exceptions, including oceanic islands that have become scraped off along continental margins and now lie far above the reach of the ocean, their insides exposed for all to see. Good examples are found in the accreted Wallowa terrane (Oregon) and others along the western side of the North American continental core.

There are three main external influences on the life cycles of oceanic islands: first, island tectonics—the rises and falls of the island itself; then islands and sea-level changes; and finally, island landscape evolution.

Oceanic-Island Tectonics

In any part of the world, the solid Earth's crust can rise and fall, but in the ocean basins these processes are more widespread and are a major cause of oceanic island emergence and submergence. Long-term uplift and subsidence also affect islands as a result of changing water loads on the ocean crust. Particularly following land-ice melt during deglaciation, rapid inputs of water into the oceans can cause the ocean floor to deform. Yet the principal cause of individual island emergence and submergence over shorter time periods is vertical tectonics—movements of the Earth's crust resulting from the accommodation of stresses associated with plate movements.

The process of crustal rise is known as uplift, and many oceanic islands have been uplifted, especially near convergent plate boundaries. Uplift can be continuous, in which case it is usually slow; Maré Island in the Loyalty Islands of New Caledonia in the southwestern Pacific has been climbing the crustal flexure (described previously) at rates as high as 1.9 mm/year during the last half million years or so. Uplift can also be sporadic, however. This uplift type is typified by rapid bursts of uplift during large-magnitude earthquakes, typically causing ground level to rise 1–2 m, and is termed coseismic uplift. Yet between these infrequent bursts of coseismic uplift, there is often slow subsidence, so the net uplift over long time periods may be comparatively slow. A recent example comes from Ranongga Island in the Solomon Islands in the southwestern Pacific where, early in the morning on April 2, 2007, a large earthquake raised up the entire island 2 m exposing the surface of its fringing reefs. A similar event happened during the December 26, 2004, earthquake in Indonesia (which caused the devastating tsunami) when Simeulue Island off the coast of Sumatra was raised 1.5 m in a few minutes.

Subsidence can also be rapid and abrupt, perhaps coseismic, but more often it is an expression of the gravity-induced collapse of an island's flanks. On Hawai'i Island, during the Kalapana Earthquake on November 29, 1975, a 60-km stretch of the south coast sank 3.5 m and moved seaward some 8 m, causing a 10-m-high tsunami. Yet far more common among the global population of oceanic islands is slow, continuous, monotonic subsidence, typically the outcome of an island being carried on a moving plate into deeper water. Thus, islands that move away from the mid-ocean ridges or from hotspots usually subside as the underlying oceanic crust cools and comes to lie at increasing depths below the ocean surface. Some of these rates of subsidence are minute but continue for extremely long periods of time; the atoll island Enewetak (Marshall Islands, northwestern Pacific) has been sinking at an average rate of 0.03 mm/year for 45 million years.

Oceanic Islands and Sea-Level Changes

By comparing the maps of islands in the southwestern Pacific 18,000 years ago, when sea level was around 120 m lower than it is today, with maps from the present (Fig. 4), it is possible to get a sense of just how important sea-level changes are in causing islands to alternately emerge and submerge. Over the past 2–3 million years, sea level has oscillated between glacial (ice-age) low stands and interglacial high stands every 100,000 years or so. Today, in the middle of the Holocene interglacial period, which began around 12,000 years ago, we live in a drowned world; the ocean surface is higher than it has been for around 95% of the past 150,000 years. Thus, islands are far rarer today than they were during the last glaciation, something that has implications for various types of biota (including humans) that have dispersed across the oceans, as well as for islands themselves.

An island that is submerged is immune from many of the processes of erosion that affect its subaerial counterparts. Conversely, it also ceases to be a viable habitat for terrestrial biota, meaning that it has to be recolonized if it emerges again. Moreover, the process of alternate submergence and emergence may affect the stability of an oceanic island through the successive application of pressure and then the abrupt release of that pressure, which can accelerate the large-scale collapse of steep island flanks.

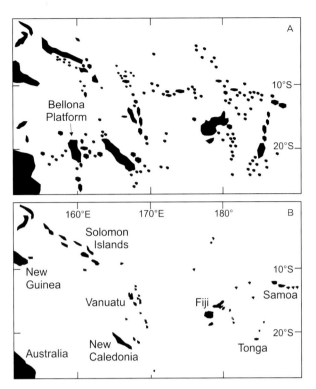

FIGURE 4 The changing geography of the southwest Pacific. (A) About 20,000 years ago during the lowest sea level (–120 m) of the last glaciation, much more land was exposed in this region. Note particularly the large island between New Caledonia and Australia, marked by the Bellona platform from which a few isolated reefs rise today. (B) The modern geography of the region for comparison.

Landscape Evolution on Oceanic Islands

Landscape evolution on oceanic islands is—as it is on continental landmasses—controlled largely by climate and lithology (rock type), and it is therefore difficult to generalize about. That said, it is clear that oceanic islands, largely because of their discrete nature (their boundedness) and their steep-sidedness—itself the outcome of their oceanic location—are susceptible to quite different processes of landscape evolution from those that operate on continents.

In terms of their discrete nature, it is the fact that most oceanic islands are comparatively small and not part of larger land masses that makes their landscapes evolve in isolation. Thus, for example, oceanic islands in the trade wind belts may have well-defined wet and dry sides, where different sets of geomorphic processes operate. Many oceanic islands are, on account of their comparative smallness, entirely coastal, which means that ocean-driven processes affect the entire island; thus, it may change far more rapidly as a result.

Next, there is the issue of the steep-sidedness of many oceanic islands, something that results from their having grown upward from the deep ocean floor. Like steep slopes anywhere, those that form the flanks of such oceanic islands are more prone to failure (collapse) than are gentler slopes, a process that is exacerbated for some islands by earthquake activity. The geological record is full of incidences of island flank collapse, ranging from the uncommon gigantic ones—such as the 5000-km^3 Nu'uanu Slide on Hawai'i Island 2.1 million years ago—to more frequent, yet smaller, ones. Contained in the sedimentary (underwater) apron that surrounds the Marquesas Islands of the central eastern Pacific, there is many times more volcanic material than there is in the modern islands, suggesting that earlier islands collapsed and rebuilt themselves several times in the past.

SEE ALSO THE FOLLOWING ARTICLES

Coral / Earthquakes / Island Arcs / Island Formation / Plate Tectonics / Sea-Level Change / Volcanic Islands

FURTHER READING

Menard, H. W. 1986. *Islands*. New York: Scientific American Books.
Nunn, P. D. 1994. *Oceanic islands*. Oxford: Blackwell.
Nunn, P. D. 2006. Island origins and environments, in *A world of islands: a physical and human approach*. G. Baldacchino, ed. Malta: Agenda, 5–37.
Nunn, P. D. 2009. *Vanished islands and hidden continents of the Pacific*. Honolulu: University of Hawaii Press.
Nunn, P. D., C. D. Ollier, G. S. Hope, P. Rodda, A. Omura, and W. R. Peltier. 2002. Late Quaternary sea-level and tectonic changes in northeast Fiji. *Marine Geology* 187: 299–311.
Sigmundsson, F. 2006. *Iceland geodynamics: crustal deformation and divergent plate tectonics*. Berlin, Germany: Springer.
Strahler, A. N. 1998. *Plate tectonics*. Cambridge, MA: Geo-Books.

ORCHIDS

DAVID L. ROBERTS AND
RICHARD M. BATEMAN

Royal Botanic Gardens, Kew, United Kingdom

Assisted by their dustlike seeds, orchids are among the first plant families to colonize islands, often speciating into the many unexploited niches on newly formed or newly disturbed islands. Reduced (or at least temporarily reduced) competition on some islands may allow more radical evolutionary shifts, as well as the establishment of new relationships between an orchid lineage and its necessary partners—animals for pollination and mycorrhizal fungi for germination and nutrition. Furthermore, the tendency of orchids to be pollinator-limited, and thus to occur as small populations, has resulted in orchids frequently evolving through founder effect and genetic drift.

SEED DISPERSAL

Orchids are well known for producing vast quantities of seeds, in some cases several million. This did not go unnoticed by Charles Darwin, who painstakingly recorded around 6200 seeds from a single capsule of *Orchis* (now *Dactylorhiza*) *maculata*. In his classic book *On the Various Contrivances by which British and Foreign Orchids are Fertilised by Insects, and on the Good Effects of Intercrossing* (1877) he stated whimsically that

> To give an idea what the above figures really mean, I will briefly show the possible rate of increase of *O. maculata*: an acre of land would hold 174,240 plants, each having a space of six inches square, and this would be just sufficient for their growth; so that, making the fair allowance of 400 bad seeds in each capsule, an acre would be thickly clothed by the progeny of a single plant. At the same rate of increase, the grandchildren would cover a space slightly exceeding the island of Anglesea; and the great grand-children of a single plant would nearly (in the ratio of 47 to 50) clothe with one uniform green carpet the entire surface of the land throughout the globe. But the number of seeds produced by one of our common British orchids is as nothing compared to that of some of the exotic kinds.

The exceptional dispersibility of dustlike orchid seeds in air currents over considerable distances has allowed successful colonization of islands hundreds or thousands of kilometers from the nearest seed source. Also, unusually for flowering plants, orchids have pollen that on average travels a shorter distance than the seed. However, most orchid seeds fall close to the mother plant. Orchids, particularly tropical species, are often pollinator-limited, resulting in low levels of fruiting success. Hence, tropical orchids have on average considerably lower fruiting success than do temperate species. In addition, many temperate species have several characteristics that maximize reproductive success under conditions of infrequent pollination.

COLONIZATION OF ISLANDS

Seed dispersal is only the first step in the successful colonization of an island. The seed must fortuitously land in a suitable place to germinate, on a surface that provides at least one compatible mycorrhizal fungal associate. An immigrant orchid seed may face not one fungally mediated barrier to colonization but two. There is an increasing body of evidence suggesting that, in many orchids, a member of one group of mycorrhizal fungi is necessary for successful germination of the microscopic seeds, whereas a member of a second group of fungi is needed to supply nutrition to the mature plant. There is now growing evidence that availability of suitable mycorrhizal partners is a key controlling factor of orchid populations. Recent research in Australia has suggested that rare orchids are associated with rare mycorrhizal partners. Thus, the main factor determining the success or failure of an immigrant orchid seed may be the happenstance presence, within the substrate on which it lands, of appropriate members of both cohorts of fungi.

The resulting seedling must then survive long enough to flower, and if that were not a sufficient challenge, it also needs to form a relationship with a pollinator that is sufficiently competent to remove its pollinia but not so competent that it refuses to visit another compatible flower—one that has opened in the vicinity of the first and at the same moment in time. Fortunately, the vastly improbable becomes probable when extended over a geological time scale.

BAKER'S LAW AND REPRODUCTIVE ASSURANCE

It is often stated, with some justification, that orchids are among the most specialized of all flowering plant families. Not surprisingly, therefore, they tend to evolve specialized pollination systems, with 60% of species supposedly being pollinated by a single species of animal. Whereas many flowering plants rely on various chemical inhibition systems to prevent self-pollination and promote outcrossing, most orchids are self-compatible. Rather than chemical compatibility barriers, they have evolved floral mechanisms that promote outcrossing through attracting, and influencing the behavior of, appropriate pollinators. This confers on orchids an advantage in colonizing new territory, as there is no longer an absolute need for a second compatible plant in order to generate viable seeds. Even pollinators, though undoubtedly helpful, may not be essential. Instead, the plant may be able to resort to vegetative reproduction or apomixis (production of viable seed without male intervention), although the latter process is rare among orchids. These advantages of self-compatibility for island colonization have become known as Baker's Law.

As a result, we find that orchid species on tropical oceanic islands exhibit either of two contrasting strategies to achieve successful reproduction. Many species emulate those of tropical rain forests in producing low frequencies of viable fruits. In contrast, the second group shows much higher levels of fruiting success through efficient self-pollination. For example, this approach to what is known as reproductive assurance is evident in the endemic species of *Jumellea* from the island of La Réunion in the Mascarene Island archipelago, which are pollinated by hawkmoths. In order to succeed, some orchids have reduced the size of their floral spurs to better fit the proboscides of the moths, whereas others have become self-pollinating.

THE CONSEQUENCES OF REPRODUCTIVE ASSURANCE FOR ISLAND ORCHIDS

Ultimately, reproductive assurance means that the orchid successfully passes on its genes to the next generation. However, reproductive assurance also means that characters such as floral display, scent production, and method of nectar presentation are no longer selected for by the relevant pollinator(s). What is the point in having complex, energy-consuming features that are supposed to elicit pollination if no pollinator is present to appreciate them? Thus, although the genus *Angraecum* is well known for producing a strong scent at dusk to attract hawkmoth pollinators, *A. borbonicum* (Fig. 1), endemic to La Réunion, has lost its scent and considerably reduced its nectar production. In its putative mainland ancestor, the ability to attract an effective pollinator must outweigh the considerable energetic cost of producing the scent. But if the species becomes self-pollinating, the pollinator is eliminated from the equation, and scent production thus becomes irrelevant to the long-term well-being of the orchid. Once selection has been relaxed, there is no longer pressure to purge from the population the mutations that prevent scent production. Furthermore, the energy the orchid saves by not producing a scent can usefully be reallocated toward higher-priority objectives such as producing healthy fruit.

There are, of course, good reasons why the majority of orchids cross-pollinate. Notably, self-pollination allows

FIGURE 1 *Angraecum borbonicum* from La Réunion, Mascarenes.

FIGURE 2 Two species of *Himantoglossum* showing speciation via directional change in a lineage through anagenesis: (A) *H. metlesicsianum* from Tenerife; (B) *H. robertianum* from the mainland, Macaronesia.

much easier perpetuation of mutationally induced morphological novelties, most of which are likely to prove competitively inferior to their parents. However, ecologically mediated competition is often lower on islands, increasing the probability that such "hopeful monsters" will be perpetuated to form new evolutionary lineages. Potential examples of such monsters from La Réunion include the endemic genus *Bonniera*, which probably evolved through saltational evolution from species of *Angraecum* when the labellum (lip) was replaced by a less differentiated, petal-like organ. This developmental shift has also resulted in the loss of the spur and therefore of nectar production. This transition occurred independently twice: *A. conchoglossum*, also found on La Réunion, gave rise to *B. corrugata* through cladogenesis, whereas *A. arachnites* from Madagascar gave rise to *B. appendiculata* through anagenesis. Another example of anagenesis is *Himantoglossum metlesicsianum* (Fig. 2A), a rare endemic of Tenerife that has a floral morphology distinctly divergent from that of its widespread mainland sister, *H. robertianum* (Fig. 2B). The two species also show substantial genetic differences, suggesting that the ancestor(s) of *H. metlesicsianum* migrated to Tenerife 1–2 million years ago. Of 13 orchid species currently occurring on the Macronesian islands, one may not merit species-level recognition, and all of the remaining 12 species could have arisen by anagenesis, demonstrating that not all orchids indulge in spectacular evolutionary radiations after colonizing an island.

ALTITUDINAL ASPECTS OF ISLAND HOPPING

It is widely recognized that species richness commonly peaks at intermediate elevations. If the range of a species centers on the mid-elevation level, then the species has a greater potential to occupy more of the island, as in theory it can expand to the top and bottom of the island. However, if the mid-point of a species's range occurs at the bottom or the top of the altitudinal range offered by the island, then that species can spread within the island in only one direction—up or down, respectively. Widespread species therefore have midpoints near the center of the elevational range, whereas localized species with a narrow altitudinal range are equally likely to be found anywhere across the elevational gradient. It follows that the greater is the proportion of widespread species in a particular orchid flora, the greater is the likelihood that species richness will peak toward the middle of the altitudinal range of the island. This phenomenon is termed the "mid-domain effect."

Let us once again consider the Mascarene Islands (Mauritius, La Réunion, and Rodrigues), which lie about 900 km east of Madagascar. Morphological and DNA-based studies suggest that the lineages that have diversified to give the Mascarene Islands their present orchid flora of about 150 species largely originated from among the 960 species of orchids that currently occupy Madagascar. The similarity in altitude of Madagascar (2876 m) and La Réunion (3069 m) is reflected in the behavior of their respective orchid flora components of La Réunion (i.e., those species that are shared between Madagascar and La Réunion, and those species endemic to La Réunion), whereas there is a strong contrast between the distributions of comparable orchids on La Réunion and nearby Mauritius, which rises to only 828 m. These observations suggest that a species has difficulty escaping from its original altitudinal range, even when it colonizes an island over three times the height of the source terrain on which it evolved (Fig. 3).

In contrast, no such pattern is evident when we compare the orchid floras of the Gulf of Guinea Islands (Annobon, Sao Tomé, Príncipe, and Bioko), 60–300 km off the West African coast. Like the Mascarene Islands, the Gulf of Guinea archipelago evolved from a volcanic hotspot. However, the much closer proximity of the Gulf of Guinea islands to the African mainland has permitted multiple colonization events by orchids. The effect of the proximity to major land masses can be seen in the percentage of endemic orchid species, which is 11% for Príncipe and 16% for Bioko. In contrast, the 68% endemism observed in the isolated Mascarene Islands suggests that there is little movement of orchids between the islands and the mainland but considerable movement among the islands.

SPECIATION ON ISLANDS

The most important factors determining the species richness of an island are its age, size, and distance from potential sources of immigrants. Altitudinal extent and diversity of geological substrates also strongly affect the ability of the island to differentiate habitats and thereby to multiply niches that could potentially be occupied by orchids. It has also been suggested that species diversity might itself help to drive speciation. Increasing the number of species present increases the number of likely ecological interactions, which in turn increases the number of potential niches that can then be filled by novel species—a classic positive feedback loop.

Such "niche-filling" and subsequent speciation has been documented on La Réunion, where the pollination of two of the endemic species of *Angraecum, A. bracteosum*, and *A. striatum* (from the endemic Section *Hadangis*), has been captured on film. Pollen transfer was effected not by the expected hawkmoths, but rather by birds—the endemic white-eyes, *Zosterop olivaceus* and *Z. borbonicus*, respectively. These orchids, and a further related species endemic to the Mascarene Islands, may have evolved to fill an available niche as a result of encountering the depauperate hawkmoth fauna of the island.

Why then are the orchid floras of islands not even more diverse? The factors that discourage a veritable explosion of species are (1) the failure to find an alternative coevolutionary partner on the island and (2) extinction. The latter can reflect the absence or precariousness of a particular habitat. For example, *Bulbophyllum variegatum* has been extirpated from Mauritius and is declining on Réunion, where it grows epiphytically on only one species of tree, *Agarista salicifolia*—a tree that has a regrettable predilection for an unpredictable substrate, specifically as a primary colonizer of recently extruded lava flows. However, extinction is typically caused by competition among species for finite resources such as nutrients or the attention of pollinators. In this context, Baker's Law has important implications for conservation, because it suggests that increased pollinator specialization and/or self-incompatibility will predispose a species to extinction. Self-pollinating species may therefore be better suited to survive in the current changing climate by means of reproductive assurance and the ability of some species to maintain a dual reproductive strategy, indulging in outcrossing when conditions permit but using self-pollination as a failsafe.

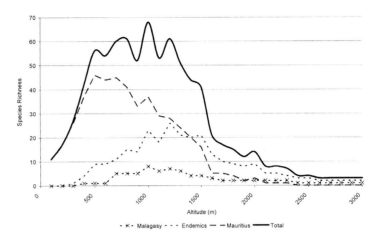

FIGURE 3 Altitudinal distribution of the orchid flora of La Réunion, Mascarenes.

SPECIATION BEYOND CONVENTIONAL ISLANDS

One of the main biological effects of island colonization is the founder effect and subsequent genetic drift. This scenario has been proposed not just for orchids on geological islands but also for the evolution of the family as a whole.

Orchids are well-known for being reproductively pollinator-limited and can be even further restricted by resource constraints. Consequently, only a small proportion of the population often gives rise to the next generation. This situation is further skewed by the fact that most orchid populations, particularly in the tropics, are small, occurring in an unusually fluid ecosystem rich in underexploited niches. This leads to frequent changes in appearance or behavior among generations and subsequent diversification and speciation. In other words, low reproductive success leads to a low proportion of reproducing individuals (technically termed a small effective population size, N_e). This often results in genetic drift being the initial cause of evolution, with Darwinian adaptation, driven by directional or disruptive selection, later imposing itself to better fit the new generation to the local pollinator population.

SEE ALSO THE FOLLOWING ARTICLES

Anagenesis / Dispersal / Founder Effects / Mascarene Islands, Biology / São Tomé, Príncipe, and Annobon

FURTHER READING

Arditti, J., and A. K. A. Ghami. 2000. Numerical and physical properties of orchid seeds and their biological implications. Tansley Review No. 110. *New Phytologist* 145: 367–421.

Bateman, R. M., and W. A. DiMichele. 2002. Generating and filtering major phenotypic novelties: neoGoldschmidtian saltation revisited, in *Developmental genetics and plant evolution*. Q. C. B. Cronk, R. M. Bateman, and J. A. Hawkins, eds. London: Taylor & Francis, 109–159.

Bateman, R. M., P. M. Hollingsworth, D. Devey, and D. L. Roberts. 2005–2006. When orchids challenge an island race 1–4. *Orchid Review* 113: 334–337; 114: 36–41; 114: 98–102; 115: 212–217.

Bateman, R. M., and P. J. Rudall. 2006. The good, the bad, and the ugly: using naturally occurring terata to distinguish the possible from the impossible in orchid floral evolution. *Aliso* 22: 481–496.

Colwell, R. K., and D. C. Lees. 2000. The mid-domain effect: geometric constraints on the geography of species richness. *Trends in Ecology and Evolution* 15: 70–76.

Dixon, K. W., S. P. Kell, R. L. Barrett, and P. J. Cribb. 2003. *Orchid Conservation*. Kota Kinabalu, Sabah: Natural History Publications.

Dressler, R. L. 1990. *The Orchids: natural history and classification*. Cambridge, MA: Harvard University Press.

Pridgeon, A. M., P. J. Cribb, M. W. Chase, and F. N. Rasmussen. 2001–2006. *Genera Orchidacearum, vols. 1–4*. Cambridge: Cambridge University Press.

Tremblay, R., J. D. Ackerman, J. K. Zimmerman, and R. N. Calvo. 2005. Variation in sexual reproduction in orchids and its evolutionary consequences: a spasmodic journey to diversification. *Biological Journal of the Linnean Society* 84: 1–54.

ORGANIC FALLS ON THE OCEAN FLOOR

CRAIG R. SMITH

University of Hawaii, Manoa

Organic falls are large parcels of organic matter (e.g., dead fish, marine mammal carcasses, wood and other vascular plant debris, masses of macroalgae) that sink largely intact to the sea floor. Because most of the sea floor underlies deep water and is food-limited (fed by a diffuse rain of organic material from surface waters), large organic falls on the deep-sea floor create food-rich islands in an energy-poor desert. Suites of deep-sea species rapidly consume a broad range of organic-fall types, causing these food-rich islands to be relatively ephemeral (i.e., lasting for days to decades).

NATURE OF ORGANIC FALLS

The rates and patterns of exploitation of large organic falls at the deep-sea floor depend on a number of factors. These include (1) parcel size (ranging from 10^1–10^5 kg), (2) lability or digestibility of the organic material (ranging from easily digested muscle protein in fish to recalcitrant cellulose and lignin in wood), and (3) the physical structure of the parcel (for example, vertebrate soft tissues, calcified whale bone impregnated with whale oil, or heavily bored wood). All three factors influence the succession of organisms utilizing a large organic fall and the persistence time of such food-rich islands at the sea floor.

WHALE FALLS

Sunken whale carcasses (or "whale falls"), ranging in mass from 10^3 to 10^5 kg, are the end-members in size and lability of large organic falls. A single 5×10^4-kg whale fall yields a massive pulse of labile proteins and lipids to the sea floor; in one moment the ocean floor underlying the whale carcass receives the equivalent of hundreds to thousands of years of background carbon flux. The assemblage of organisms exploiting a deep-sea whale fall can pass through at least three stages of community succession:

1. A mobile scavenger stage, during which aggregations of voracious, highly active scavengers (e.g., sleeper sharks, hagfish, rattail fish, and lysianassid amphipods) consume the whale's soft tissue over time scales of months to years
2. An enrichment opportunist stage, lasting months to years, during which organically enriched sediments and exposed bones are colonized by dense assemblages of opportunistic worms and crustaceans
3. A sulfophilic (or "sulfur-loving") stage which can last for decades, during which a large, species-rich assemblage lives on the skeleton as it emits sulfide from anaerobic breakdown of bone lipids; the sulfide effluxing from the bones supports a chemoautotrophic assemblage of animals deriving nutrition from sulfur-oxidizing bacteria (Fig. 1).

FIGURE 1 Whale skeleton on the sea floor at 1670-m depth off the California coast with white and red mats of sulfur-oxidizing bacteria, small white anemones, and an asteroid. This whale carcass has been on the sea floor for about 4.5 years. Photograph by Craig R. Smith.

Whale falls can harbor hundreds of species and include at least 33 species of "whale fall specialists" (i.e., animals that apparently require whale falls to maintain their populations). Whale-fall communities also contain a number of chemoautotrophically dependent generalists found at hydrothermal vents and cold seeps.

WOOD FALLS

Sunken wood and other vascular plant debris (e.g., coconut husks, palm fronds) also support specialized deep-sea communities (Fig. 2), with a total of over 200 animal species known to be associated with vascular plant material recovered from the deep-sea floor. Keystone species on vascular plant debris include "wood-eating" bivalves in the genus *Xylophaga* (meaning "wood eater") that bore into sunken wood and consume it from within, much as termites reduce wood in terrestrial habitats. These wood-eating clams contain endosymbiotic bacteria in their gills, which produce enzymes to break down the otherwise recalcitrant cellulose in the wood. *Xylaphaga* can colonize isolated wood parcels at the deep-sea floor within weeks, consuming most of the mass of 1-kg wood parcels within a year, but requiring many years to reduce large tree trunks or shipwrecks. These wood-eating clams are considered keystone species because their fecal material and biomass can support a substantial local food web of detritivores (including polychaete worms and crustaceans) and predators (e.g., nemertean worms and galatheid crabs). The tunnels formed by the boring clams, as well as other cracks and crevices in the vascular plant debris, also provide habitat for a diversity of benthic invertebrates seeking physical shelter. Still other faunal components on wood-fall islands include surface grazers such as limpets, which consume bacteria and fungi growing on the wood surface.

FIGURE 2 Wood parcel at 1670-m depth off California. The wood is covered with the whitish siphons of the wood-eating bivalve *Xylophaga* protruding from holes in the wood, and by the muddy tubes of polychaete worms. A rockfish, galatheid crabs, scale worms, and a brittle star are also visible living around the wood parcel. Photograph by Craig R. Smith.

MACROALGAL FALLS

Sunken masses of giant kelp and other macroalgae (typically $1–10^3$ kg in mass) provide yet another type of large organic fall. In the deep sea, kelp falls often are rapidly colonized by shrimp, galatheid crabs, limpets, and amphipods that (1) graze on the kelp and associated microbes, (2) prey on the biomass of the attracted kelp fauna, and (3) use the kelp (especially the rootlike holdfast) as an attachment substrate and as shelter. Very large kelp falls can undergo anaerobic decomposition and produce sulfide which can then foster a chemoautotrophic assemblage of bacteria and clams. Very large masses of dead kelp and seagrass accumulating in submarine canyons are known to support an extraordinary abundance of benthic invertebrates (e.g., more than 10^6 amphipods and other crustaceans per m^2) and sustain some of the highest rates of secondary production ever measured at the sea floor. Faunal diversity on macroalgal falls can also be high, with more than 50 species exploiting a 100-kg kelp parcel at any given time.

CONCLUDING REMARKS

The speed at which animals find and exploit organic falls is remarkable, especially considering the spatial rarity of food falls in the deep sea. Even in the richest deep-sea regions, research submersibles may travel many kilometers across the sea floor without encountering a dead fish, parcel of wood, or kelp fall. Nonetheless, nearly all large organic falls form oases of biomass and biodiversity, sustaining species rarely if ever encountered in the food-poor background deep sea. It is clear that these ephemeral food-rich islands have supported adaptive radiation in the deep sea, fostering specialized fauna that contribute significantly to biodiversity and evolutionary novelty in the ocean.

SEE ALSO THE FOLLOWING ARTICLES

Cold Seeps / Hydrothermal Vents / Succession / Whale Falls

FURTHER READING

Distel, D. L., and S. J. Roberts. 1997. Bacterial endosymbionts in the gills of the deep-sea wood-boring bivalve *Xylophaga atlantica* and *Xylophaga washingtona*. *Biological Bulletin* 192: 253–261.
Smith, C. R. 1985. Food for the deep sea: utilization, dispersal and flux of nekton falls at the Santa Catalina Basin floor. *Deep-Sea Research I* 32: 417–442.
Smith, C. R., and A. R. Baco. 2003. Ecology of whale falls at the deep-sea floor. *Oceanography and Marine Biology Annual Review* 41: 311–354.
Vetter, E. W. 1994. Hotspots of benthic production. *Nature* 372: 47.
Wolff, T. 1979. Macrofaunal utilization of plant remains in the deep sea. *Sarsia* 64: 117–136.

PACIFIC REGION

ANTHONY A. P. KOPPERS

Oregon State University, Corvallis

The Pacific Region is extraordinary in many aspects. It is the largest ocean on Earth, harbors the deepest trenches, has the highest abundance of islands, and underneath its sea surface it encompasses the largest tectonic plate on Earth. Its enormity is emphasized by the fact that Christmas Island in the center of the Pacific Ocean lies more than 8,000 km away from any continent. As a whole the islands in the Pacific Region are referred to as Oceania, the tenth continent on Earth. Inherent to their remoteness and because of the wide variety of island types, the Pacific Islands have developed unique social, biological, and geological characteristics.

ISLAND FORMATION IN THE PACIFIC

Little was known about the formation of islands in the Pacific until the voyages of Captain Cook in 1768–1780, and until the HMS *Beagle* and HMS *Challenger* visited the region in 1835 and 1875, almost a century later. Before these famous expeditions, knowledge about the island geography in the Pacific Region was limited to the first maps of the "Quiet Ocean" or *Maris Pacifici* by Abraham Ortelius in 1589, based on the sparse data collected during the Magellan expedition, the very first to circumnavigate the world (Fig. 1).

Today, the formation of most of the 25,000 islands in the Pacific can be explained by a singular geological phenomenon, described by the theory of plate tectonics. The first ideas for this theory date back to the early work of Alfred Wegener (1915) on continental drift, yet the fact that tectonic plates move over the Earth's surface with rates up to 100 mm/year did not get wide recognition until the late 1960s. Now it is well understood how the sometimes violent interactions at the boundaries of these moving plates generate most of the earthquakes and volcanism on Earth. In the Pacific Region the majority of these plate interactions are embodied by the so-called Ring of Fire, which is a 40,000-km-long stretch of subduction zones where the Pacific Plate is actively being destroyed (together with two smaller plates along its eastern edge) while causing many earthquakes and producing many volcanoes and volcanic islands (Fig. 2).

Whereas the subduction of an oceanic plate (such as the Pacific Plate) underneath a continent does not often generate islands (Fig. 2A), subduction underneath another oceanic plate normally results in a remarkable semicircle of volcanic islands, named an island arc (Fig. 2B). The Aleutian Islands located to the north of the similarly named Aleutian Trench, and all volcanic islands located to the west of the Kurile, Japan, Izu-Bonin, Marianas, Bougainville, Tonga, and Kermadec trenches (Fig. 2) are typical examples of island arcs. New island-building volcanism in these arcs has taken place as recently as August 9–11, 2006, near Home Reef in the Tonga Islands, resulting in a new volcanic island and massive pumice rafts visible by satellite for hundreds of miles in the Pacific Ocean. A few years earlier, in May 2000, a new phase of island building was witnessed for Kavachi, a submarine volcano in the Solomon Islands, which appeared above the sea surface for the first time in 1951 and which has been destroyed by wave erosion and rebuilt by volcanic eruptions several times since then. Myojinsho Island breached the sea surface for the first time in 1952–1953 along the Izu-Bonin island arc,

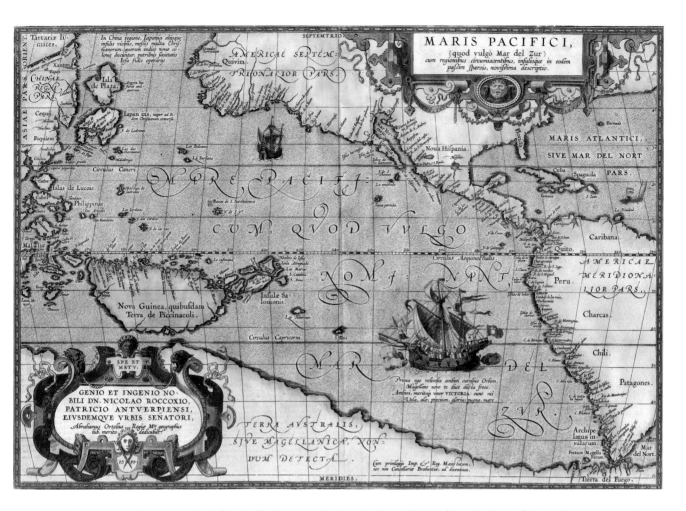

FIGURE 1 Map No. 12, called *Maris Pacifici* (of the Pacific Ocean) by Abraham Ortelius in 1589. This first printed map of the Pacific Region included "a very new description of the peaceful sea, commonly called the South Sea with the regions lying around it, and its islands, scattered everywhere" and was created following the first expedition by Magellan. Despite the small number of prior explorations of the Pacific Region at that time, this map contains references to New Guinea (*Nova Guinea*), the Japan Islands (*Iapan Ins.*), the Galapagos Islands (*Y. de Galopagos*), the Hawaiian Volcanic Islands (*Los Bolcanes* and *La Farfana*), and the Solomon Islands (*Insulae Salomonis*).

420 km south of Tokyo. During 12 months of continuous eruption, more than 1000 phreatomagmatic explosions were recorded at this newly born volcanic island. These examples emphasize the continuous and sometimes explosive volcanic activity that is characteristic of island arcs and the cause of unremitting production of new islands in the Pacific.

Only in three places do the boundaries comprising the Ring of Fire have a different character, and those are along the San Andreas fault in the western United States and Baja California, the Islands of New Zealand, and the Solomon Islands (Fig. 2). It follows that the converging motions of the tectonic plates do not everywhere result in the formation of a typical subduction zone. In some cases, convergence occurs under significant angles, causing lateral motions that form so-called transform faults or spreading centers in order to accommodate the non-orthogonal displacements. For example, horizontal shear motions between two plates have created the San Andreas transform fault and explain the high frequency of earthquakes observed along this active fault line. On the other hand, diverging motions caused Baja California to rift apart after a spreading center (partly) followed the Farallon Plate in its subduction underneath North America. In the case of New Zealand, the direction of subduction is reversing along an axis running through the North and South Islands, a reversal that can be accommodated only by the formation of a transform fault in between. The Solomon Islands are unusual because the Pacific plate has a hard time subducting beneath the Australian plate as a result of the 122-million-year-old and 30-km-thick Ontong Java Plateau (the biggest large igneous province on Earth) riding on top of it. This rare obstacle has been obstructing the

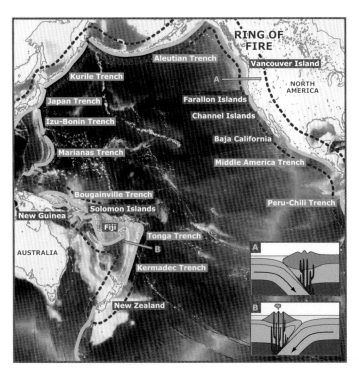

FIGURE 2 Ring of Fire and Island Arcs of the Pacific Region. Insets: (A) Ocean–continent subduction zone. (B) Ocean–ocean subduction zone. Black and white base map provided by Jasper Konter.

subduction of the Pacific plate for at least the last 6 million years and resulted in the upthrusting, or obduction, of this abnormally thick oceanic crust as the Solomon Islands were formed.

One of the most extraordinary aspects of the Pacific Region is that it is littered with volcanic islands that formed far away from tectonic plate boundaries and island arcs, in a so-called intraplate setting (Fig. 3). Some of these volcanoes are currently active, such as Kīlauea on the Big Island of Hawaii. Others experienced historical eruptions, such as Savai'i Island in the Samoan Archipelago and Macdonald Seamount in French Polynesia. Their existence in the middle of the wide Pacific Ocean instantly captured the imagination of Charles R. Darwin and James D. Dana in the nineteenth century. But despite Darwin's and Dana's detailed explorations around and on these active islands, they could not explain why this kind of volcanism and island formation occurred in these remote locations. Neither could twentieth-century scientists when they attempted to apply the plate tectonic theory to the problem, despite the large amounts of new oceanographic data collected in the decades following World War II. In the early 1970s, and immediately following the widespread acceptance of plate tectonics, J. Tuzo Wilson and W. Jason Morgan proposed the hotspot and mantle plume models (Fig. 3A). In these models, thermally "hot" anomalies are presumed present deep in the Earth's mantle, causing narrow upwellings of mantle material that eventually reach the lithosphere. Decompression in these mantle plumes as they rise to depths just below the lithosphere causes melting, and the magmas that are produced penetrate the lithosphere to form intraplate volcanoes. Because tectonic plates move vast distances over geological time, each volcanic center is moved away from the hotspot and becomes inactive. At its place a new hotspot volcano or volcanic island gets produced, in a process that has generated many linear chains of volcanoes in the Pacific Region, with the Hawaii–Emperor chain being the prime example.

However, today most of these extinct hotspot volcanoes exist below the sea surface as seamounts (Fig. 4). Over geological time the volcanic islands eroded, subsided, and sank several kilometers deep, because the Pacific Plate on which they ride cooled down and became more dense with age (Fig. 3A). This process typically has produced drowned flat-topped seamounts that we know as guyots. More than 50,000 of these seamounts and guyots are estimated to exist in the Pacific Region alone. They tell us of a complex history and volcanic evolution, involving many more ancient islands than we can observe above the sea surface today.

ISLAND ARCS

Most islands of the Pacific Region formed along the Pacific Rim as part of the Ring of Fire (Fig. 2). Subduction of the Pacific oceanic plate underneath the surrounding North American, Philippine, and Australian plates caused the formation of numerous island arcs, most of them currently active and not older than a few million or tens of million years. Because of their location on the fringes of the Pacific Ocean, these islands have a strong continental connection in a sociohistorical sense, in their biological evolution, and in their climate. These island arcs functioned as the springboard for the early Polynesian people in their explorations of the largest ocean on Earth in search of new island habitats. They also created a buffer zone during World War II, where many combat campaigns and bombardments took place outboard of the continents and on the Solomon Islands (Tulagi, Guadalcanal, Bougainville), the Marianas and Volcano Islands (Saipan, Truk, Tinian, Guam, Iwo Jima), and the Aleutian Islands (Attu, Adak). Their proximity close to the continental land masses makes arc islands less isolated from the import of new species, allowing for more biological diversity (and less endemism) despite the rather young age of many of these islands. And, finally, as

FIGURE 3 Volcanic Islands and Atolls of the Pacific Region, most of which are concentrated in the equatorial Pacific. (Inset A) Hotspot trail formation by mantle plumes, showing the age-progressive nature of the volcanoes formed over long periods of plate motion away from the hotspot and the subsequent "drowning" of each volcano as the oceanic crust cools down over geological time and sinks in the more fluid or plastically behaving asthenosphere. Black and white base map provided by Jasper Konter.

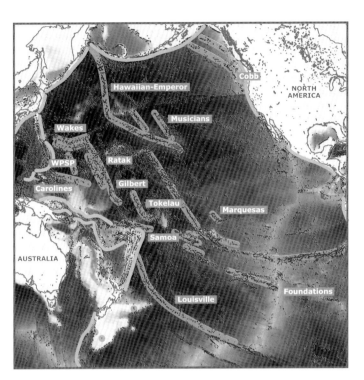

FIGURE 4 Seamounts of the Pacific Region. Only a selection of seamount trails are indicated by the light blue lines, whereas red dots indicate the entire distribution of seamounts, estimated to total ~50,000 in the Pacific Ocean alone. Note that WPSP is short for the West Pacific Seamount Province, a region mostly including Cretaceous seamounts. Black and white base map and seamount locations provided by Jasper Konter.

another consequence of their location nearby the continents, many Pacific island arcs are first in line to receive a full beating of incoming typhoons or tsunamis.

Aleutian Islands

This island arc in the northeast corner of the Pacific Ocean contains more than 300 volcanic islands and spans almost 2000 km between the peninsulas of Alaska and Kamchatka. Toward the west the Aleutian arc sharply bends southwest into the Kurile island arc segment. Fifty-seven historically active volcanoes are present on the islands, of which two (Bogoslof and Fire Island) surfaced less than two centuries ago. Its location above 52° N latitude makes this island arc most extreme in terms of its climate, compared to all other Pacific islands. With short growing seasons, generally low temperatures, and strong prevalent winds, trees remain remarkably short and only a small variety of vegetation and animal species are present. Most of the islands also may have been connected during the sea level lowstand of the last glacial period, effectively creating a land bridge between Eurasia and North America that provided a likely immigration route for the earliest occupation of the Americas. At that time the Aleutian Islands were covered by extensive ice caps and glaciers.

Kurile Islands

The Kurile Islands form a 1300-km-long island arc that connects Japan with the Aleutian Islands in the north. This island arc contains 56 islands with about 100 volcanoes, of which 40 are active. It lies in one of the world's most remote places and is characterized, like the Aleutians, by harsh climate. However, its northeast-southwest orientation provides a strong latitudinal difference that results in rather dissimilar subtropical weather conditions and wildlife toward the southern end of the island arc. Nonetheless, prevalent rains, snowstorms, and ice fields keep these islands in a strong weather grip most of the year. The earliest settlements date back to 7000 BC, but today no more than 17,000 inhabit the Kurile Islands.

Japan Islands

Recent volcanism in the Japan Islands has been caused by the dual subduction of the Philippine Sea Plate in the south and the Pacific Plate in the northeast. The entire island arc system has been active since the Permian (260–299 million years ago), at which time the closing of the Tethys Seaway initiated the westward subduction of the (now completely destroyed) Farallon Plate underneath the Asian continental margin. Overall, more than 3000 islands make up the Japanese Islands. Mount Fuji is Japan's highest peak and the most famous of its more than 200 active volcanoes. Japan's climate ranges from cold snowy peaks on the northeastern islands to subtropical conditions on its southern islands. The Japan Islands also are the most heavily populated in the Pacific Region, with more than 130 million inhabitants. This dense human population may be the cause of the (near) extinctions of four mammal species (two bat species, the Japanese wolf, and the Japanese otter), 15 bird and fish species, and more than 40 plants. Fighting in the Pacific Region during World War II ended after two nuclear weapons were detonated above the cities of Hiroshima and Nagasaki on August 6 and 9, 1945. More than 220,000 civilians perished from these bombings, the only nuclear attacks in human history.

Izu-Bonin and the Marianas

Both the Izu-Bonin and Mariana island groups form a combined island arc that stretches from Guam at 12° N to the beginning of the Japan Trench at 21° N. The current island arc is the most recent product of the ongoing subduction of the Pacific Plate to the west, which began around Eocene times (34–56 million years ago). Six new volcanic islands and a similar number of seamounts formed during the past century. Behind the strongly curved Marianas island arc, an inner arc runs parallel to the outer arc and is characterized by active volcanism as well. In recent years at least 50 earthquakes with magnitudes between 6.0 and 8.1 have occurred along these arc segments. Occupation of the Izu-Bonin and Mariana Islands started in prehistoric times, resulting in a major (50%) loss among the larger and edible flightless bird species on the islands. Their extinction continued during historic times, when colonists settled on the islands and introduced cats (and other predatory invasive species) to the further detriment of the indigenous birds. Even more dramatic was the involuntary introduction of the brown treesnake (*Boiga irregularis*) on Guam during World War II, which resulted in the disappearance of three seabirds, ten native forest birds, one native bat, and maybe as many as five native lizards from the island. A recent estimate places more than 2500 snakes per square kilometer on Guam, making it the island with the highest snake density in the world.

Bougainville and the Solomon Islands

Bougainville and the Solomon Islands are both part of a 1500-km-long island group in Melanesia containing about 900 islands. Most islands are volcanic in origin and fringed by coral reefs. However, many small islands exist that consist of uplifted reefs (up to 800 m above current reef levels) that formed by regional tectonic forces as a result of the (failed) subduction of the Ontong Java Plateau. In fact, the downward-going Pacific Plate is bulging upward at the Bougainville Trench, and Cretaceous volcanic rocks from the Ontong Java Plateau itself are obducted at this suture to form the islands of Maliata and St. Isabel. These volcanic islands are quite different (in composition and age) from the island arcs formed elsewhere in the Pacific Region. They are seismically active as well, with an 8.1 magnitude earthquake occurring on April 2, 2007, and triggering a tsunami with waves 5 to 10 m high on the Solomon Islands. Early settlers from Papua New Guinea started to arrive as early as 30,000 BC, yet it took until 1200 to 800 BC before the ancestors of the Polynesians arrived on the islands and started to explore the wider Pacific Region. Finally, the Battle of Guadalcanal between August 1942 and February 1943 was one of the most epic (and most costly) battles in the Pacific Region during World War II, turning the strategic table in favor of the Allied Forces.

Kermadec, Tonga, and Fiji

The Kermadec–Tonga island arc forms a group of small South Pacific islands (and many more seamounts) between the tropical islands of Samoa in the north and New Zealand in the south. At 29° S latitude, the two arc segments join where Osbourne Guyot (the oldest seamount in the Louisville seamount trail) is being subducted. A seismic gap appears at this location, in what seismically and volcanically otherwise is the most active plate boundary in the world. The latest volcanic episode was highlighted by the formation of a new volcanic island at Home Reef in August 2006, whereas the latest high-impact earthquake, with a magnitude of 7.9, occurred about 160 km northeast of Nuku'alofa on May 4, 2006. The Fiji Islands lie sub-parallel and to the west of the Tonga Islands and have a more prolonged history (starting around 40 million years ago), also related to the subduction of the Pacific Plate. However, over the last 3 million years volcanic rocks erupted on Fiji have a more intraplate (similar to those on, e.g., the Hawaiian or Samoan Islands) composition, reflecting its current location inbetween the Lau and Fiji back-arc basins and away from an active subduction zone. The Kermadec–Tonga–Fiji islands were first settled by Polynesians some

3000 years ago, yet with more than 100,000 inhabitants today, Fiji has become one of the most densely-populated island groups in the Pacific Region. As a result, significant environmental pressures have been exerted on the flora and fauna of the islands. Many species declined or disappeared because of the intense consumption by the large human population, whereas introduced alien mammals (rats and cats, in particular) decimated other forest plants and animals. Exporting for the international pet industry also decreased the numbers of (endemic) Fijian shining parrots and lorikeets, up to the level of critical endangerment or extinction. The Kingdom of Tonga is the only surviving monarchy existing in the Pacific Region. Fiji's Great Sea Reef (north of Vanualevu) is one of the largest barrier reefs in the world.

CONTINENTAL ISLANDS

Many continental islands formed following the last glacial period, after the Earth warmed up and sea level started to rise steadily. Sea level rise (including the effects caused by global warming) thus accounts for the formation of many of the islands along the continental margins of the Pacific Region. Land bridges that once existed during the last glacial lowstand were relinquished to the sea, truncating entire lineages of species and civilizations from the continental hinterland. Sea level rise, for instance, helped to form the Channel Islands off the California Coast and produced the configurations of multiple Japanese and Hawaiian islands as we know them today. Continental islands are often not volcanic in origin, like most other islands existing in the Pacific Region. Nearly all are part of terranes accreted to the continents over longer periods of subduction (e.g., microcontinents, ancient island arcs). Over time some eroded to elevations only slightly higher than sea level. Others separated from the main continental masses during periods of continental rifting, which in some cases were followed by the formation of new ocean basins through sea floor spreading. Under this scenario, the continental fragment of Papua New Guinea became an independent island once it separated from the Australian continent.

Channel and Farallon Islands

The Channel Islands lie just off the coast of California and, during the last ice age, were connected to each other because of the lower global sea level. These eight islands have a geology similar to the granitic Sierras that form most of Southern California. The one big island that formed during the Pleistocene is named Santarosae and provided free roaming space for many species, including a dwarf mammoth (now extinct) and many cypress and pine species. Even though not connected to the mainland, new species arrived at Santarosae on "debris rafts" carried across the narrow seaway by storms or shifting sea currents. The island fox (*Urocyon littoralis*) is the smallest North American canid that evolved as a special Channel Islands species since more than 10,000 years ago, after its predecessor, the gray fox, arrived by rafting or alternatively aided by the Chumash native people, for whom the fox was sacred. With the start of the interglacial period, the Channel Islands separated from each other, allowing the island fox (and many other species) to evolve into endemic island species. The introduction of the South African iceplant in the late nineteenth century to California and the Channel Islands has been devastating to native ecosystems. This plant now covers most of the islands with a thick cover and leaches high concentrations of salt into the native soil, making a saline environment inhospitable to most native plants.

New Zealand

The North and South Islands make up the majority of New Zealand and together are about 1600 km long. Because of the inclusion of the Chatham Islands, located 800 km to the east, the country has the seventh-largest Exclusive Economic Zone (EEZ) in the world, about 15 times its land area. Mount Cook in the Southern Alps is the highest peak at 3754 m, whereas the same area includes another 17 peaks taller than 3 km. Most volcanism occurs on the North Island, with Mount Ruapehu being the tallest and most active volcano. Despite its active volcanism, New Zealand is not considered to be a volcanic island, as is common in the Pacific Region. Instead, New Zealand primarily consists of continental rocks and (together with New Caledonia) forms the submerged continent Zealandia. This continental fragment separated from the supercontinent Gondwanaland around 80 million years ago, giving many of the New Zealand taxa a "Gondwanan" character that dates back to many lineages that previously evolved when Gondwana was a single landmass. The arrival of the Polynesian Maori happened only a millennium ago, yet their habituation of New Zealand changed the existing lineages in flora and fauna drastically. For example, when these Maori arrived, about 30% of the land birds were flightless, but almost none survive these days as a result of the consistent consumption by people. European settlers in the early nineteenth century in their turn altered the New Zealand landscape considerably by clearing all but 22% of the original trees and vegetation. With the colonial immigrants came

33 introduced bird species, 34 mammal species, and 20 freshwater fish, together causing a loss of 41% of the 176 endemic species. Despite these large changes in the flora and fauna, New Zealand retains a spectacular and widely varying landscape that notably inspired the filming of the *Lord of the Rings* trilogy there.

New Guinea

The island of New Guinea is the second largest island in the world and was formed when the Torres Strait flooded as sea levels rose following the last ice age. This effectively separated New Guinea from Australia, a transgression that is recorded in the emerged reef staircases on Huon Peninsula. This geological phenomenon provided the first (tectonically corrected) eustatic records that accurately determined past sea level elevations for the New Guinea region, with lowstands and highstands that varied between −17 and +10 m over the last 10,000 years compared to present-day sea levels. The highland peaks of Mount Wilhelm (4509 m) and Puncak Jaya (4884 m) are part of an east-west mountain range in New Guinea that is one of a few regions around the equator that receives snowfall and has permanent glaciers, now disappearing because of changing climate and global warming. Even though the first humans arrived on the island about 60,000 years ago, it took more than 50,000 years until the first plant domestications were successful in the highlands. At that time, the mountain population of hunters and gatherers started to cultivate many (indigenous) plants into garden crops, including sugar cane, bananas, yams, and taro. Today, much of New Guinea is still unexplored, leaving countless plant, insect, and animal species undiscovered. New Guinea also is infamous for its ritual cannibalism or "head-hunting" that was practiced by a few ethnic groups up to a few decades ago. A diet low in protein (owing to the small size of edible indigenous animals on the island) is often theorized to have sparked this morbid habit or ritual, even though pigs were introduced a few millennia ago.

Vancouver Island

This large island off Canada's Pacific coast formed as an accretionary terrane that is now part of the Western Cordillera in the Cascadia subduction system, a convergent plate boundary stretching south to northern California. This terrane, named Wrangellia, welded itself to the North American Plate about 100 million years ago, after it traversed from south of the equator and got entangled in an ancient version of the Cascadia subduction zone. Based on fossil records and paleomagnetic measurements, it is clear that during the Upper Cretaceous the Wrangellia terrane was located around a 25° N paleolatitude, equivalent to the location of today's Baja California. Most of the Vancouver Island landscape was formed by strong glaciations, as was the current outline of the island itself.

Because of a combination of severe winds and strong ocean currents, many ships have been stranded or were wrecked on the rocky coasts of Vancouver Island. With as many as one wreck per mile, this island is referred to as the Graveyard of the Pacific.

INTRAPLATE VOLCANIC ISLANDS AND ATOLLS

A noteworthy exception in plate tectonics is the formation of intraplate volcanic islands, such as the Islands of Hawaii, the French Polynesian Islands and the Samoan Archipelago, which each may require the presence of a mantle plume originating deep in the Earth's mantle. Overall there are more than 750 volcanic islands in the middle of the Pacific, many of them arranged in linear island chains. The highest concentration can be found in the equatorial Pacific, in particular, in the triangular area between Samoa, Easter Island, and the Marshall Islands (Fig. 3). This also explains the high proportion of fringing reefs and atolls that formed around and on top of these volcanic islands because the tropical conditions in this part of the Pacific Ocean were most favorable for prolonged coral growth. Because many of these intraplate volcanoes are located far from the tectonic plate boundaries, they also are remote from the North American and Eurasian continents on the neighboring tectonic plates. As a result, the intraplate volcanic islands and atolls of Midway, Eniwetok, Bikini, Christmas, Johnston, Mururoa, and Fangataufa gained critical strategic value in the naval campaigns during World War II and became the remote testing grounds for the development of nuclear weapons during the Cold War era.

From a biological standpoint, these volcanic islands are "hotspots" of biodiversity, characterized by species unique to only certain islands or island groups (high endemism) but with an otherwise limited variability (low species diversity) because of the small size of suitable habitats. The lineages of these species are typically short-lived since the intraplate volcanic islands do not survive for more than a few tens of million years in the plate tectonic environment. Some lineages may be prolonged through sequential colonization over the lifetime of entire island chains, whereby the species hop from island to island when older islands disappear underwater and new islands form farther up the chain by new volcanism. The extreme isolation of each island (or island chain) within the Pacific Region makes the survival of many of its unique species more difficult, particularly upon invasion by non-native predatory species, parasites, or competitors. These invading species typically have been brought onshore by the Polynesian settlers and (starting in the early eighteenth century) European immigrants and caused an increased pressure on the island species, effectively reducing enemy-free spaces on the islands. Without refuge many island species diminished quickly, many toward extinction. For example, from the 120 island birds that have become extinct since 1500 AD, it is estimated that more than 80% were caused by the introduction of predatory carnivores on the remote islands of the Pacific Region.

Caroline Islands

These islands are part of Micronesia in the West Pacific and include many small coral islands and a few larger volcanic islands. The larger islands are mostly located in the eastern part of the Caroline Islands and are considered erosional remnants of extinct hotspot volcanoes. Toward the west the volcanic edifices have mostly subsided and in some cases have only a limestone platform or coral top that surfaces above the Pacific Ocean. Hotspot volcanism started around 12 million years ago and formed the Caroline Ridge between Kosrae Island and Chuuk Atoll. These are the youngest intraplate volcanoes that lie just south of the otherwise Cretaceous atolls and seamounts in the West Pacific Seamount Province. All other islands in the Caroline archipelago have more complex histories, including the formation of ancient island arcs, transform faults, spreading centers, and intraplate volcanoes.

The Caroline Islands also are located close to an area of intense typhoon activity. While typhoons in the West Pacific can occur at any time, they mostly occur between February and April, with five severe typhoons per year that pass over or close by the Caroline Islands, through the so-called Typhoon Corridor.

Cook-Austral Islands

The Cook and Austral Islands are both part of Polynesia and form a complex system of short volcanic chains in the South Pacific between Samoa and the Society Islands. Fifteen islands make up the Cook Islands and range in size from the large volcanic island of Rarotonga to the smallest solitary atolls in the north. Only seven islands make up the Austral Islands, which actually lie on a continuation of the Southern Cook Islands that converges on the recently active Macdonald volcano. Although the islands generally are believed to have formed by hotspot volcanism, the age and distance systematics of this intraplate island cluster are rather complex. Alternative processes may be required to explain the complications, including tectonic control on the Pacific Plate, the presence of multiple plumes/plumelets in the mantle underneath, or even the upwelling of a 2000-km-wide hotline, rather than only a hotspot, of mantle material.

The Cook–Austral Islands lie just east of the International Date Line and were inhabited between 500 and 800 AD by Polynesians crossing the Pacific Ocean from the islands of Tonga, Samoa, Tahiti, Marquesas, and Society. The extreme isolation of these islands toward the southern boundary of Polynesia makes for a low biological diversity, yet there is a significant difference in flora and fauna between the Cook–Austral islands and atolls. About 4% of the angiosperms are endemic to the islands, yet the Miti'aro fan-palm and the Pacific fruit bat of Rarotonga and Mangaia are unique to this part of the world. Today most of the vegetation is dominated by species introduced by Polynesian and European settlers, including taro, yam, ferns, coconut palms, bananas, and papaya. Because the livable area of the combined 22 main islands is extremely small, solid waste management, land use practices, overfishing, and ecotourism are all taking a large toll on the islands' habitat and their delicate ecosystems. For example, wetlands often are used as waste disposal sites or filled in for construction, natural drainage and groundwater systems are disturbed by extensive land cultivation, and coastal landscapes have been altered by sand mining practices. Uncontrolled fishing in the numerous lagoons of the Cook–Austral islands have had destructive effects on fish populations and the health of the coral reefs, as have numerous toxins from sewage and pesticides leached into the lagoons.

Easter Island

This solitary volcanic island is one of the most remote islands in the Pacific Region, located more than 2000 km east of Pitcairn Island, the closest inhabited island in its vicinity. Its remoteness also explains its late discovery by the Marquesas or Mangareva people in 400–600 AD, making it the last inhabited island by the Polynesians. Easter Island is rather small (164 km^2) and forms only one of two isolated subaerial volcanoes along the 2700 km Easter–Sala y Gomez ridge on the Pacific Plate. Plate tectonic reconstructions show that Easter also is related to the formation of the Nazca and Tuamotu ridges on the Nazca Plate. These submarine ridges formed from the same Easter hotspot when it was located directly underneath the mid-ocean spreading center of the East Pacific Rise, simultaneously forming two hotspot trails on two tectonic plates (in a remarkable V-shaped pattern). However, Easter hotspot volcanics less than 3 million years old are rather unusual because of their continental-type igneous character. These "granitic" rocks confused many early geologists, including Alfred Wegener, who used this oddity in labeling Easter Island as a remnant of continental drift.

At a site called Rano Raraku these rocks exist in the form of volcanic ash and tuff deposits that are easily workable and thus were used to carve out most of the *moai,* the famous stone statues or "heads" of Easter Island. Almost 900 moai statues have been found on Easter Island, and generally they are thought to have been erected between 1000 and 1700 AD. By the time the last statue was erected, overpopulation of the island and profligate consumption of trees in the transport of the statues seem to have caused a sudden drop in crop productivity and the loss of all forests and their resident birds and animals. With that, the Easter population was decimated and did not stabilize until the Europeans settled on the island. The census of 2002 shows that currently only 3791 people reside on Easter Island, one of the least dense island populations in the Pacific Region.

Galápagos Islands

This archipelago of 13 islands, shown on sixteenth-century maps as *Insulae de los Galopegos* (Islands of the Tortoises), was formed starting about 10 million years ago by the Galápagos hotspot. This hotspot is located close to the equator (about 965 km off the coast of Ecuador) and is still active, with the latest volcanic eruptions taking place in 2005 on the islands of Isabela and Fernandina. Before 8 million years ago the Galápagos hotspot was located on the Galápagos Spreading Center and, like the Easter hotspot, formed two concurrent volcanic traces on two oceanic plates, namely the submarine Carnegie and Cocos Ridges. Both ridges delineate the northeastern or southeastern continuation of the Galápagos Islands, except that they are entirely composed of seamounts that once were emerged volcanic islands in the same archipelago. Today this hotspot resides about 200 km south of the Galápagos Spreading Center and only continues to form the Carnegie Ridge on the Nazca Plate. Because of its close proximity to this spreading center, the Galápagos Islands have been formed on relatively thin oceanic lithosphere with increased tectonic stresses. This may explain the complex clustering of the Galápagos volcanic islands, which is contrary to the classical linear alignment of volcanic islands for the Hawaiian hotspot.

The Galápagos Islands do not have any historical settlers, so these islands were not discovered until European explorers landed in 1535. By the late eighteenth century, whalers and sailors were using Galápagos as a base station and started to kill and capture many thousands of the Galápagos tortoises for their meat, rich in protein and fat, bringing many species to the brink of elimination and some to extinction. Charles Darwin visited the islands in 1835 onboard the HMS *Beagle,* and even though this young naturalist spent only one month on Galápagos, his discoveries and observations on the geology and biology of the islands formed the very basis of his renowned book *On the Origin of Species.* On Galápagos Darwin studied, in particular, the subspecies of finches and found that there are distinct differences in these species among the 13 islands. The Galápagos Islands also are home to the marine iguana, which according to recent DNA studies evolved from a common land ancestor not more than 20 million years ago. To explain this recent observation, the drowned seamounts in both the Carnegie and Cocos Ridges must have acted as steppingstones covering the entire evolution of this species, starting more than 20 million years ago.

Despite the fact that the Galápagos Islands have been occupied for only half a millennium, invasions by other (mammal) species have occurred again and again, and changes to the islands' habitat and species inventory have been significant. For example, goats introduced on Isla Pinta in 1959 reproduced extremely rapidly, reaching a population of 20,000 by 1971. These introduced animals prevented regeneration of trees because of their intense seedling browsing. As a result, the original vegetation on this Galápagos island survived only in a few areas inaccessible to the goats.

Hawaiian Islands

Formerly known as the Sandwich Islands, Hawaii appears as a linear group of 19 islands and atolls about 2600 km long and extending from the island of Hawaii (the Big Island) in the southeast to the atolls of Midway and Kure in the northwest. In this island chain the volcanoes become gradually older toward the northwest, where they are volcanically inactive, have almost no surface morphology, and are entirely capped with coral reefs. In the 1960s these observations were confirmed by radiometric age dating using the K/Ar method. These ages indeed showed a systematic increase from less than 0.6 million years on the Big Island to about 5.8 million years at Kauai (where the very last subaerial volcanic rocks could be sampled) to about 27 million years for rocks cored at Midway Atoll.

With these data in hand, geophysicists J. Tuzo Wilson and W. Jason Morgan theorized that a stationary and deep-seated mantle plume impinges on the overriding Pacific Plate, producing a trace of volcanic islands on top of that plate that becomes older toward the northwest and that directly reflects the direction and speed of past Pacific plate motions. In this model the Hawaiian hotspot must be located almost underneath the "zero-aged" Big Island, which explains the ongoing extrusion of magma at Kīlauea crater, where lava flows described as pāhoehoe (low-viscosity lava that is smooth and ropey when hardened) and ʻaʻā (high-viscosity lava that is rough when hardened) have been erupting from 1983 until today at a rate of ~0.1 km^3 per year. Seismic activity is not only related to the movement of magmas in the 3–7-km-deep magma chamber of Kīlauea. On October 15, 2006, an earthquake with a 6.7 magnitude was recorded off the northwest Kona coast. Other large earthquakes are triggered by large landslides and avalanches on the (submarine) flanks of the Hawaiian Islands and have also generated local tsunamis in the past. For example, tsunami deposits have been found as high as 170 m above current sea level on the leeward side of the islands of Kohala, Maui, Lanai, and East Molokai. Tsunamis also have been caused by earthquakes happening along the outer perimeter of the Pacific Region, in the Ring of Fire. In 1960, the Great Chilean earthquake (believed to have had a 9.5 magnitude) resulted in a large tsunami that devastated Hilo with waves reaching 11 meters. Despite the remoteness of the Hawaiian Islands from Chile, this tsunami was able to kill at least 61 people.

The isolation of these islands toward the north side of Polynesia has dictated the limited species inventory of Hawaii as well, which originally did not include ants, ginger plants, terrestrial reptiles, or amphibians, and only two mammal species. Most of these Hawaiian species have rather short lineages since they started to differentiate only about 6 million years ago, at the time when the volcanic island of Kauai was formed. A good example are the Hawaiian honeycreepers, representing a diverse group of 56 endemic species of cardueline finches, which evolved from a single ancestral species about 3 to 4 million years ago. More than 70% of these finch species are by now extinct, and most of the remaining species are currently endangered. Hunting, habitat modification, or the introduction of predators (such as the rat) by the early Polynesians may have caused these extinctions. Modern threats to the honeycreepers include their particular susceptibility to the avian poxvirus and malaria, spread around the oceanic islands by mosquitoes originating in Africa.

Line Islands

The Line Islands include a group of 11 atolls and low coral islands, located in extreme isolation in the central Pacific. Only three islands are inhabited, with a population of 8809 in 2005, up from 300 in the early twentieth century. Kiritimati, or Christmas Island, is the largest atoll in the world at 642 km^2 and was first discovered by Captain James Cook on Christmas Eve in 1777. The Line Islands, and in particular Palmyra Atoll and Kingman Reef, contain the most pristine coral reefs remaining in the Pacific Region and the world. Geologically speaking, the formation of this island group is undecided, because they are part of a linear 4,000-km-long chain of atolls, submarine ridges, and seamounts that is rather complex, with numerous *en echelon* and cross-cutting seamount chains. Based on recent radiometric age dating of basaltic lavas, this island chain does not possess an obvious linear age progression, unlike the Hawaiian Islands. This requires alternate explanations, including the possibility that more than one hotspot may have been in play during the formation of the Line Islands or that extension in the Pacific Plate may have created volcanic islands and seamounts during three or four narrow eruption periods. The United States and Great Britain used Christmas Island for testing hydrogen bombs during Operation Grapple in 1957 and Operation Dominic in 1962.

Marquesas, Pitcairn, and the Society Islands in French Polynesia

French Polynesia comprises several island chains over an area of 2.5 million km^2 in the equatorial Pacific, including the Marquesas, Pitcairn, Society, Tuamotu, and Austral islands. Their combined land area, including 118 islands and atolls, is rather small, yet with 84 atolls French Polynesia contains the majority of all atolls on Earth. All of these island chains are volcanic in origin and are thought to have been formed by hotspots, similar to the Hawaiian Islands. Most French Polynesian volcanic islands are younger than 16 million years, except Tuamotu, which is as old as 40 million years. As an archipelago, French Polynesia is the most isolated in the world; it has a rather limited flora and fauna of which 80 endemic species are documented to have become extinct, and because of its generally low relief it is one of the Pacific island groups that will endure the most immediate consequences of any future sea level rise driven by global warming. France used the atolls of Fangataufa and Moruroa for the (underground) testing of nuclear bombs between 1970 and 1997.

The Marquesas Islands are relatively young, with its oldest island dated at 3.8 million years. They contain 14 volcanic islands and one atoll about 9° south of the equator, with Mount Temetiu on Hiva Oa being the highest volcanic peak at 1190 m above sea level, and Nuku Hiva being the second largest island in French Polynesia. Only 8632 people were living on the islands as of the 2007 census, despite their early discovery by Polynesians around 150 BC and the fact that in the sixteenth century an estimated 100,000 Polynesians inhabited the islands. Diseases such as measles, syphilis, dysentery, smallpox, tuberculosis, malaria, and leprosy were brought in by the Europeans, and frequent epidemics decimated the population to their current level by the end of the nineteenth century. The age-distance relationships between the Marquesas Islands provide us with a 10.4 cm/year estimate for the past velocity of the Pacific Plate, which is similar to, and thus confirms, the observations for the Hawaiian hotspot. However, the 115° azimuth of the Marquesas island chain is ~20° different from that of Hawaii and the other volcanic chains in French Polynesia, which suggests that the formation of the Marquesas islands and the emplacement of its volcanoes may have been controlled by weaknesses in the Pacific Plate induced by the Marquesas Fracture zone.

The Duke of Gloucester–Moruroa–Gambier–Pitcairn island chain is 1650 km long and runs south of the Tuamotu plateau. The oldest volcanic island was formed around 12 million years ago, whereas Pitcairn itself formed around 0.45 million years ago, and active volcanism is happening 100 km southeast of Pitcairn at any one of 20 submerged seamounts in this chain. The velocity of the Pacific Plate at this location is estimated at 11.0 cm/year and thus is well attuned with other estimates in French Polynesia and the Hawaiian Islands. Pitcairn Island is not surrounded by a fringing reef. Pitcairn became uninhabited when the original Polynesian settlers died out in the cold climate interval of the fourteenth century. In 1790 it was settled again, but this time by mutineers from the HMS *Bounty*, a Royal Navy ship charged by King George III to sail to Tahiti to obtain breadfruit plants. The mutineers, a mixed party of nine British sailors, six Polynesian men, and 13 Polynesian women, sought refuge on the island and established a small society that today still occupies the island with less than 50 remaining descendants of the original eighteenth-century families.

The Society Islands are maximally 4.3 million years old, and the plate velocity measured based on the ages and distances between the islands is 10.9 cm/year. This island chain is composed of five atolls and nine islands, of which Tahiti is the largest island and has the highest volcanic peak at 2241 m above sea level. This island consists of two coalescent eruptive systems and was active prior to 0.87 million years ago. The youngest volcanic activity is concentrated around Mehetia, where volcanic eruptions started around 0.3 million years ago and new seamounts are forming 50 km southeast. The insect fauna species groups on the Society Islands (and French Polynesia) have adaptive radiation patterns that in many cases have resulted in 100% endemism. Good examples are 70 endemic species of *Mecyclothorax* on Tahiti, 35 species of *Rhyncogonus* in French Polynesia, and 29 endemic species of black flies *Simulium* in the Society Islands. Many endemic species also disappeared, such as 60 tree snails following the introduction of the carnivorous snail *Euglandina rosea*.

Marshall Islands

The Marshall Islands consist of the oldest surviving atolls (29) and coral islands (5) on Earth that formed on top a group of hotspot volcanoes that erupted in the French Polynesia area between 97 and 67 million years ago. Today the islands are organized in the sub-parallel Ratak and Rālik (sunrise and sunset) chains and were settled by Micronesians in the second millennium BC. British Captain John Marshall was the first European to visit the islands in 1788.

During World War II, the Marshall Islands were occupied by the United States, who also started to use the atolls for the testing of their nuclear weapons almost directly after the war. In total, the United States tested 66 nuclear weapons between 1946 and 1958, including the 1954 Castle Bravo, the largest nuclear detonation by the United States, with a yield of 15 megatons and the first ever test of a dry-fuel thermonuclear fusion bomb. Bikini Island (and many other atolls in the Marshall Islands) are still unsafe because of the lasting effects of radiation from the contaminated soils (mostly by wind driven fallout) and thus remain uninhabited following the evacuation of the Micronesian people.

The native flora and fauna are dominated by over 1000 fish species and 100 coral species, yet no endemic mammal species exist on the islands, and only one endemic bird subspecies, the Ratak Micronesian Pigeon.

Samoa and American Samoa

The Samoan archipelago is divided into Western Samoa (officially Independent Samoa or Samoa) and American Samoa (an unincorporated territory of the United States) to the east. Samoa and American Samoa have populations of 214,265 and 57,291 respectively (from censuses in 2000), of whom more than 90% are native Samoans. Geologically speaking, the archipelago is a volcanic hotspot chain of nine main islands and coral atolls located between 13° S and 15° S in the South Pacific. Savai'i (the fifth largest island in the tropical Pacific) is the oldest island in Samoa, which started to form around 5.0 million years ago, whereas Vailulu'u is the most recent volcano that still resides underwater and to the east of Ta'u Island. Hydrothermal and volcanic activity on Vailulu'u are ongoing, with the latest volcanic episode occurring in 2003, at which time a new 350-m-high volcanic cone (named Nafanua) formed inside the main crater, at approximately 1 km water depth. Rose Atoll, located farthest east in the Samoan archipelago, is much older, suggesting that its evolution began where French Polynesia is located today and from an entirely different hotspot in that area.

The first settlers in Samoa arrived from Malesia in the west more than three millennia ago, and the Manu'a Islands therefore have one of the oldest Polynesian histories. Because of the proximity of Samoa to New Guinea and the Australian continent, most biodiversity in the terrestrial and marine species is derived from these regions. For example, the native Samoan flora is the largest in Polynesia and includes 550 angiosperm species and 228 pteridophyte species, 32% of which are endemic. Modern hunting, however, has brought down the extant purple-capped fruitdove (*Ptilinopus porphyraceus*) and many-colored fruitdove *(Ptiliopus perousii)* to rare and declining subspecies of *manutagi*. On the other hand, the Pacific pigeon (*Ducula pacifica*), or *manume`a,* is still common on Savai'i and Upolu and is important to Samoan culture with its haunting vocalizations. Now the national bird of Samoa, the pigeon is threatened by hunting and the loss of forest habitat.

The recurrence of devastating cyclones in combination with increased agricultural deforestation have caused the once completely forested Samoan Islands to loose 60% of its rainforest cover in less than 40 years.

SEAMOUNTS AND GUYOTS

Seamounts are underwater volcanic mountains that are taller than 500 m and form isolated edifices (or ridges) on the ocean floor. The term *seamount* is typically used very broadly to describe all submarine features ranging from the smallest bathymetric disturbance to the largest volcanic structure that spans up to tens or hundreds of miles across its base. On the other hand, *guyots* are more specifically regarded as remnants of volcanic islands that eventually got planed off by wave erosion, resulting in the formation of typically large, flat-topped seamounts. These ancient volcanic islands disappeared from the Earth's sea surface because the tectonic plates they are riding on cool down over geological time, making the plates more dense and allowing them to sink into the upper mantle, taking the guyots with them into the deep of the oceans. However, once drowned, the seamounts do not survive far back into geological time, because the subduction zones surrounding the Pacific have been destroying all the older ones. The oldest dated seamount is Look Seamount, located to the west of the Marshall Islands, with an age of ~140 million years.

Compared to our knowledge about the volcanic islands in the Pacific Region, we know very little about seamounts and guyots and how they formed. Knowing how many are present in the Pacific Ocean is an even more difficult question, because our maps of the ocean floor (still) are rather incomplete and not even up to par to the detailed topographic maps we have of the Moon, Mars, or any other planet body in our solar system. Not only are the seamounts poorly mapped; they have hardly been surveyed or sampled. This makes deciphering their histories a challenge, and thus it remains unclear how many islands have existed (and perished) in the Pacific Region over the last hundred million years. We also understand very little about the role of ancient seamounts (once volcanic islands) in the island-hopping of certain species, thereby significantly prolonging their lineages in certain island groups through time.

Cobb Seamounts

The Cobb–Eickelberg seamount trail is one of two hotspot trails existing in the northeast Pacific, located west of the Oregon and Washington coast. This seamount chain stretches from the 33-million-year-old Patton and Murray seamounts near the Aleutian trench to Axial Seamount, the youngest volcano in the chain, with eruptions and a magnitude 4.7 earthquake occurring as recent as January 28, 1998. The flat-topped Cobb Seamount formed during the Oligocene, but drowned only recently (because of the global sea level rise following the last ice age) and thus its summit is submerged only 34 meters below the sea surface. These shallow depths sustain a rich marine environment with fish communities typical for the subarctic North Pacific and the hard rock substrate of this volcanic seamount. Rockfish (*Sebastes*) species seem to dominate and be isolated around Cobb, where their larvae do not spread out more than 30 km from the seamount. With its active volcanism Axial Seamount denotes the current location of the Cobb hotspot, which also intersects the Juan de Fuca Ridge at this point in time. The coincidence with this mid-ocean spreading center has a clear effect on the evolution of Axial Seamount, which has anomalous shallow bathymetry due to an oversupply of magma and a geochemistry with mixed hotspot and mid-ocean ridge signatures.

Foundation Seamounts

The Foundation chain is unlike the Hawaii–Emperor hotspot chain, because its volcanoes are volumetrically small and because no islands exist in this hotspot chain, only seamounts. The 1350-km-long seamount chain is located about 5 degrees to the southeast of Easter Island, it includes dozens of seamounts up to 21 million years in age, falling in a narrow band maximally 200 km wide, and overall the chain itself trends parallel to most other seamount chains in the Pacific Region at 70° west of north. A first-order age progression is evident and gives a plate velocity of 9.1 cm/year, but in detail the age systematics are more complex. For example, over the last 10 million years the Foundation hotspot has been slowly approaching the Pacific–Antarctic spreading center, which instilled many small-scale complexities in the resulting Foundation seamount chain. Secondary mechanisms such as the "reheating" of normal oceanic crust, local stress-induced fracturing of the lithosphere, the interaction with the growing Selkirk microplate, and the creation of "extra thin" lithosphere as a result of the combined hotspot and ridge melting, may all have played a role in the formation of the Foundation Seamounts and thus cause the observed disturbances in the age-distance relationships.

Hawaii-Emperor Seamounts

The Hawaii–Emperor seamount chain is 5800 km long and contains more than 80 submerged seamounts and guyots. It starts out with the Meiji Seamount, 83 million years old, located at the edge of the Aleutian Trench, and extends to Loihi Seamount, 35 km southeast of the Big Island of Hawaii. Loihi is volcanically and hydrothermally active, is a relatively small submarine volcano with a summit that lies still 980 m below the sea surface, and generally is considered to be situated directly above the Hawaiian hotspot. With that Loihi represents the "preshield" stage in the volcanic evolution of a typical Hawaiian volcano, a standard model adopted by many researchers studying hotspot volcanoes worldwide. The slightly older Big Island of Hawaii and the current eruptions at Kīlauea crater represent the "shield" stage, which follows after a few hundred thousand years, which is the main volcanic stage responsible for building up more than 90% of the hotspot volcano structure. One of the most engaging characteristics of the Hawaiian hotspot is that its volcanoes become gradually older toward the west and that beyond Kure Atoll the hotspot chain bends sharply to the northwest and continues underwater in the Emperor seamounts. Traditionally, these observations have been interpreted to reflect a sudden and pronounced 60° change in the motion of the Pacific Plate, now dated to have occurred about 50 million years ago. However, recent paleomagnetic measurements on ocean drilling samples from Detroit, Suiko, Nintoku, and Koko seamounts in the Emperors is casting some serious doubt on this interpretation. These measurements show that the paleolatitudes of these seamounts are up to 15° different from the latitude of the Hawaiian hotspot itself, which would be impossible if this hotspot and its associated mantle plume were to be fixed in the Pacific mantle over the last 80 million years. The Hawaiian hotspot is showing signs of significant motion prior to 50 million years ago instead, which may entirely or partly explain the bend in the Hawaii–Emperor seamount chain.

Louisville Seamounts

Formerly called the Louisville Ridge, this 4300-km-long seamount chain is the only true counterpart of the Hawaiian hotspot in the Pacific Region. In a similar fashion as Meiji, the oldest Hawaii–Emperor seamount, Osbourne Guyot (76–79 million years old) at the northwestern end of the Louisville chain is being consumed by the Tonga–Kermadec subduction zone. The likeliness to the Hawaiian hotspot continues, because the Louisville seamount chain also has a rather systematic age progression and because it has been active for more than 75 million years as well. No other seamount trail in the Pacific Region shows the same kind of prolonged volcanic history as the Hawaiian and Louisville hotspots. The outstanding question now is whether the Louisville hotspot shows the same kind of motion in its mantle plume before 50 million years ago.

Tokelau Seamounts

The Tokelau seamount chain mainly consists of ancient submarine volcanoes, yet the Howland and Baker Islands appear in the north of this chain, whereas three Tokelau atolls delineate the southern most end. The chain also shows a bend, equivalent to the bend in the Hawaii–Emperor chain, toward its southern end, yet the Tokelau seamounts dated with the $^{40}Ar/^{39}Ar$ method range in age between 58 and 72 million years, making this bend at least 8 million years older than the start of the Hawaii–Emperor bend. Beside the latest evidence for moving mantle plumes and hotspots, the asynchronous bends create another enigma for the classical hotspot hypothesis, which would predict that changes in Pacific Plate motion create bends in all "active" hotspot trails at the exact same time.

West Pacific Seamount Province

The ocean floor in the West Pacific is littered with seamounts, guyots, and atolls. Most of these already formed in the early Cretaceous, yet many still have surviving coral reefs, even tens of million of years after drowning. Paleomagnetic measurements show that these seamounts formed in the tropics between 10° S and 30° S, but despite their motion to the north over time due to Pacific Plate motions, these seamounts have remained within reach of the equatorial Pacific, explaining their sustained coral growth, which was able to keep track of the gradual subsidence of these aging volcanoes. The Western Pacific seamounts also are important fishing grounds for bottomfish and seamount groundfish. However, many seamounts experience overfishing, which is especially harmful because many lutjanid, serranid, and lethrinid species (and their larvae) only appear in isolated groups that differ per seamount without many interconnections. Overfishing thus easily disrupts the ecosystems on the seamounts, making reestablishment of some fish stands slow to impossible. Protection of these seamount fish stocks from environmentally destructive fishing therefore is paramount in keeping the seamount fish populations healthy.

FURTHER READING

Menard, H. W. 1986. *The ocean of truth: a personal history of global tectonics.* Princeton, NJ: Princeton University Press.
Menard, H. W., and H. S. Ladd. 1963. Oceanic islands, seamounts, guyots and atolls, in *The sea*, Vol. 3. M. N. Hill, ed. New York: Wiley & Sons, 365–387.
Nunn, P. D. 1994. *Oceanic islands.* Oxford: Blackwell Publishers.
Wegener, A. 1915. *Die Entstehung der Kontinente und Ozeane.* Braunschweig: Sammlung Vieweg Heft.

PALAU

ALAN R. OLSEN
Belau National Museum, Koror, Palau

Located in the western equatorial Pacific Ocean, Palau is the westernmost group of islands in Micronesia. The archipelago rests on the eastern edge of the continental shelf of the Philippine Plate, approximately 800 km east of Mindanao. Palau is a mixture of old volcanic islands, raised limestone islands, coralline platform islands, and atolls representing an exposed crest of the now-dormant southern section of the Palau–Kyushu Ridge. It is estimated that the volcanic islands emerged approximately 30 million years ago during the late Oligocene.

GEOGRAPHY

Palau, approximately 7°30′ N latitude and 134°35′ E longitude, extends northeast to southwest for 700 km, with over 500 islands and an estimated total land area of 450 km^2. One island, Babeldaob, accounts for 75% of the land area. The climate is wet tropical, with annual rainfall of 350–450 cm, temperatures of 22–32 °C, and relative humidity averaging 82%.

A coral barrier reef system surrounds five central islands—from north to south, Babeldaob, Koror, Arakabesan, Malakal, and Peleliu—in a lagoon that covers over 1,000 km^2. Hundreds of small raised coralline limestone islands are distributed throughout the southern lagoon. These are the famous (and photogenic) Rock Islands of Palau, which are a popular tourist attraction (Fig. 1).

FIGURE 1 One of Palau's famous rock islands. These raised limestone formations are found throughout the southern lagoon. Photograph by Milang Eberdong.

To the south, the platform island of Angaur is separated from the lagoon by a deep channel. Five small remote platform islands occur 300–600 km southwest of the lagoon: Fanna, Sonsorol, Merir, Pulo Anna, and Tobi. There are three atolls: Ngeruangel and Kayangel lie north of the lagoon and Helen Reef is south of Tobi Island. Helen Reef is a protected sanctuary for coral reefs, marine turtle nests, and seabird rookeries.

HISTORY AND CULTURE

According to archeological evidence, Palau was settled 1000–3000 years ago. The exact origins of modern Palauans are subject to debate. Palau's matrilineal culture retains its language and many ancient traditions, which are distinct from others in Micronesia. The people of the remote southwest islands are distinct from other Palauans, having cultural and linguistic affinities with the peoples of central Micronesia.

The Spaniard Ruiz Lopez de Villalobos made first European contact in 1543. The Englishman Henry Wilson was shipwrecked in Palau in 1783, leading the British Empire to claim Palau. In 1885, Pope Leo XIII restored the islands to Spain, which sold Palau to Germany in 1899. At the end of World War I, Palau was ceded to Japan and then to the United States of America following World War II. Palau became independent on October 1, 1994. Currently, there are 20,000 Palauans, the majority living on the central volcanic islands.

BIODIVERSITY

Palau has many types of marine habitats that support a diversity of life. In addition to 235 km of coral barrier reef, there are coral fringing reefs and mangrove forests associated with many islands. Over 50 marine lakes are located in the interiors of various rock islands. These unusual lakes are connected to the lagoon by fissures and tunnels in the limestone rock. Some of the lakes contain unusual fauna such as Palau's golden medusa (*Mastigias* sp.) jellyfish, which have algae incorporated into their tissues. The algae provide a portion of the medusas' nutritional needs through photosynthesis.

Palau's marine biodiversity is reflected in the 385 species of reef-building corals found there. Marine fauna includes an endemic nautilus (*Nautilus belauensis*), sea skaters (*Halobates*), 1450 species of fish, and breeding populations of green turtles (*Chelonia mydas*), hawksbill turtles (*Eretmochelys imbricata*), saltwater crocodiles (*Crocodylus porosus*), and endangered dugongs (*Dugong dugon*).

Terrestrial habitats include a freshwater lake, rivers, tropical rain forests, swamp forests, and savannas. Palau's terrestrial habitats are largely undisturbed, with little deforestation. For the most part, the terrestrial flora and fauna have affinities with the biota of the large land masses in the Philippines and Indonesia with minor influences from Australia, Papua New Guinea, and Polynesia.

Terrestrial biodiversity encompasses an estimated 1200 species of plants, an estimated 10,000 invertebrates, 38 reptiles, and the only endemic frog (*Platymantis pelewensis*) in Micronesia. The avifauna numbers 148 recognized species, with 11 endemic landbirds. Palau harbors populations of the Micronesian megapode (*Megapodius laperouse*), an endangered mound-building forest bird extinct throughout much of its original range in Micronesia. There are two bats: the Pacific sheath-tailed bat (*Emballonura semicaudata*) and the Micronesian flying fox (*Pteropus mariannus*). An endemic bat, *Pteropus pilosus*, is believed extinct.

Palau has an active conservation program that includes marine protected areas, a protected area network, ecosystem-based management initiatives, important bird areas and ecosystem management plans for the southern lagoon, and other important areas. There are active programs to monitor coral reefs, crocodiles, dugongs, forest birds, and sea turtles. The Micronesia Challenge, which was recently issued through the Convention on Biological Diversity in 2006, is a regional conservation initiative that originated in Palau.

FOLKLORE: AN UNUSUAL ICON

Interestingly, a spider is identified with a benevolent demigod who is central to the folklore of Palau's matrilineal society. Palauan folklore features a cycle of myths centered on an unusual demigod, Mengidabrudkoel, who transformed from spider into human form to introduce traditional rituals to Palau such as those involving childbirth. Among the many man-spider legends is a popular tale of a mythical island where Mengidabrudkoel planted a magical breadfruit tree that yielded a fish when a branch was broken off. Greedy villagers chopped down the tree expecting to reap a bounty of fish. Instead, a flood of ocean water poured from the tree stump to sink the entire island under the sea. Another legend in the cycle resembles the biblical account of Jonah and the whale.

Spider imagery is firmly embedded in modern Palauan society, with ubiquitous icons of spiders appearing in architecture, art, handicrafts, postal stamps, textiles, and logos of government agencies and athletic organizations. Palau's giant golden orb-weaver (*Nephila pilipes*) (Fig. 2),

FIGURE 2 Palau's giant golden orb-weaver, *Nephila pilipes*, is found throughout Palau except the southwest islands. Photograph by Alan R. Olsen.

a 15-cm spider, is the living symbol of the demigod and bears the vernacular name *mengidarudkoel*. The English name alludes to the golden hue of its 3-m web.

SEE ALSO THE FOLLOWING ARTICLES

Archaeology / Marine Lakes / Reef Ecology and Conservation

FURTHER READING

Colin, P. L., and A. C. Arneson. 1995. *Tropical Pacific invertebrates.* Beverly Hills, CA: Coral Reef Press.
Colin, P. L. 2007. *Marine environments of Palau.* Koror, Palau: Indo-Pacific Press.
Crombie, R. I., and G. K. Pregill. 1999. A checklist of the herpetofauna of the Palau Islands (Republic of Palau), Oceania. *Herpetological Monographs* 13: 29–80.
Force, R. W., and M. Force. 1972. *Just one house: a description and analysis of kinship in the Palau Islands.* Bernice P. Bishop Museum Bulletin 235.
Kayanne, H., ed. 2007. *Coral reefs of Palau.* Koror, Palau: Palau International Coral Reef Center.
McManus, Fr. E. G. 1977. *Palauan-English dictionary.* J. S. Josephs, ed. PALI Language Texts: Micronesia. Honolulu, HI: University Press of Hawaii.
Wiles, G. J. 2005. A checklist of the birds and mammals of Micronesia. *Micronesica* 38: 141–189.

PANTELLERIA

SEE MEDITERRANEAN REGION

PANTEPUI

VALENTÍ RULL

Botanic Institute of Barcelona, Spain

Pantepui (*pan*, Greek for "all," and *tepui*, South American indigenous name for "table mountains") is a discontinuous biogeographical entity shaped by the assemblage of the flat-topped summits of the Guayana (northern South America) table mountains, or Guayana Highlands (Figs. 1 and 2), above 1500 m in altitude. These summits are isolated from each other and from the surrounding lowlands by spectacular vertical cliffs, and they hold a singular biota with unique adaptations and amazing levels of biodiversity and endemism. The origin of such biotic patterns is a still-unresolved evolutionary enigma.

THE TEPUIS

The indigenous (Pemón) word *tepui*, meaning "stone bud," has been adopted as a physiographical term to name the table mountains of the Guayana Highlands (e.g., Auyán-tepui). A typical tepui is a tabular mountain made of sandstones and quartzites (with occasional intrusive rocks, mostly diabases), with a more-or-less flat summit limited by a rim, and isolated from the surrounding lowlands by vertical escarpments in the upper part and steep talus slopes in the foothills (Fig. 3). To understand the origin of the tepuis, it is necessary to go back to the Cretaceous (145 to 65 million years ago), when Africa and South America were joined in the Gondwana supercontinent. The separation began around 80–100 million years ago and determined the initial opening of the Atlantic Ocean, which led to the formation of the huge Amazon and Orinoco basins, among others. By that time, the Guayana region was covered by extensive erosional plains modeled on the Precambrian sandstones and quartzites of the Roraima

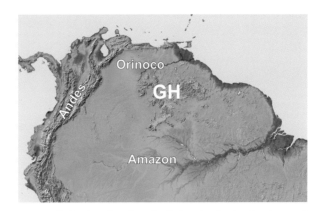

FIGURE 1 Radar view of northern South America showing the placement of the Guayana highlands (GH) region, with respect to the Orinoco and Amazon basins, and the Andean range. (Image courtesy of NASA/JPL Caltech.)

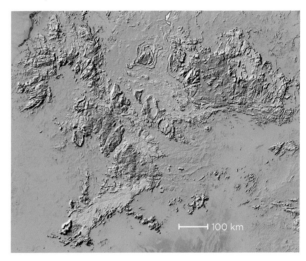

FIGURE 2 Close-up view of the Guayana highlands, showing this area's characteristic tabular topography, composed of several erosion surfaces and culminated by the tepuis. Lowlands (100–500 m altitude) are in green and yellow, whereas uplands and highlands (500–1500 and 1500–3000 m, respectively) are in light brown. (Image courtesy of NASA/JPL Caltech.)

FIGURE 3 View of the Upuigma-tepui, showing the typical flat summit (2100 m altitude and more than 1 km² in surface area), the vertical cliffs, the extensive forested slopes, and the basal level, which in this case is part of the so-called Gran Sabana, and is around 850 m in altitude. Note the characteristic savanna vegetation, spiked by clusters of "morichales" or gallery forests dominated by the palm *Mauritia flexuosa*. In the background, two other table mountains are present: the Angasima-tepui (left), at 2250 m in altitude, and the extensive Chimantá massif (right), composed of several tepuian summits between 2200 and 2700 m in altitude, attaining a total surface of more than 600 km². Photograph by the author.

group (1400–2300 million years ago). These sedimentary rocks began to be denuded by the incipient fluvial systems, in a process that is still ongoing. Weathering and erosion proceeded more easily on anticlines, where water penetration has been favored by an open fracture system. Synclines, however, are more resistant to erosion, and several of them have persisted as isolated topographical remnants: the tepuis.

More than 50 tepuis and tepuian massifs have been recognized as such, most of them in Venezuela—where they attain their maximum development—with a few representatives in Guyana and Brazil. The tepui summits are variable in both altitude and surface area, ranging from 1100 to 3014 m (typically 2000–2600 m) in maximum altitude and less than 1 to more than 1000 km² (typically 200–500 km²) in area. The total Pantepui surface is ~5000 km² in area. The degree of physical isolation of these summits is also variable. The surrounding lowlands are situated between 100 and 1200 m elevation (commonly 100 to 400 m), and the vertical difference between them and the tepui summits ranges from 200 to 2400 m (usually 800 to 1800 m). Despite these numbers and the visual impression of the tepuian landscape, only ~20% of the tepui summits are really isolated topographically; the others are connected to the lowlands by extensive river valleys, ridges, or eroded walls.

BIOTA

The climate atop the tepuis is mild (10–18 °C average annual temperature) and very humid (2500–3500 mm of total annual rainfall), which allows development of dense vegetation types, most of them unique and characteristic of Pantepui. Among forests, the more emblematic are the dense high-tepui forests dominated by *Bonnetia* (Theaceae), associated with diabase intrusions and watercourses. The more characteristic shrublands, exclusive of one single tepuian massif (the Chimantá), are organized around a few species of the endemic Asteraceae genus *Chimantaea* (Figs. 4–6). The typical tepui meadows are dominated by broad-leaved plants without gramineous morphology, such as *Stegolepis* (Rapateaceae) and *Xyris* (Xyridaceae). Grasses and sedges are minor elements atop the tepuis. Characteristic pioneer communities with cyanobacteria, fungi, and incrustant lichens grow on bare rock. Vascular plants, the best-known organisms of Pantepui, are commonly used to illustrate the biodiversity and endemism patterns of the tepui summits. So far, around 2500 species (630 genera and ~160 families) are known, of which 62% are endemic to the Guayana region, 42% are endemic to Pantepui, and 25% are local endemics (i.e., endemic to a single tepui or tepuian massif). Local endemism can reach up to 60% in some tepuis, which is comparable to the most isolated oceanic islands. There are 23 endemic genera (~4%) but not any endemic family. Most of the Pantepui vascular plant genera are of neotropical distribution (~70%), followed by paleotropical (~20%), cosmopolitan (~5%), and temperate (~5%) elements. Among neotropical affinities, 25% of the genera correspond to Guayana endemics, and 6% are shared with the tropical Andes; the rest are more widespread.

Among the animals, the most studied are birds, followed by frogs and reptiles, and then mammals. Around 100 species of birds have been described in Pantepui, of

FIGURE 4 Inflorescences of *Stegolepis ligulata* (Rapateaceae) from the summit of the Apakará-tepui (Chimantá massif), around 2200 m in altitude. This species is endemic to the summits of the Chimantá massif, where it dominates the broad-leaved meadows. Photograph by the author.

FIGURE 5 *Chimantaea mirabilis* (Asteraceae) from the Apakará-tepui, around 2200 m in altitude. This species dominates the unique and characteristic "paramoid" shrublands of the Chimantá massif, to which it is endemic. Photograph by the author.

which one-third are endemic to Guayana and 10–20% are endemic to the highlands. The diversity of herpetofauna (frogs and reptiles) is about half that of birds, but the level of endemism is higher, reaching 60% in some tepuis. Mammals, most of them small, are also represented by even lower number of species, mainly of bats, rodents, and some marsupials (opossums). So far, felids, monkeys, and medium to large herbivores have not been observed atop the tepuis. Insects have not been studied systematically, but recently, a new genus of damselflies was described (*Tepuibasis*), with all its seven species being endemic to Pantepui.

The origin of such biotic features has been debated for long time. Based on floristic data, earlier researchers (up to about 1970) explained the uniqueness of the Pantepui biota as the result of evolution in isolation since the Cretaceous. According to this hypothesis, present species would be very old in origin. However, ecological and paleoecological evidence favoring genetic interchange among summit biotas is increasing. The proposed mechanism is related to the Quaternary (the last 2.6 million years) glaciations. Glacial cooling promoted downward migration and lowland spreading of summit taxa, whereas interglacial warming favored upward migration and colonization of new summits. Molecular phylogenetic studies on key taxa, such as *Stegolepis* and *Myoborus,* a genus of redstarts, agree with this view and favor a recent origin, probably Plio-Pleistocene (the last 5 million years), for their species. Such evolutionary processes would have promoted adaptative radiation, favored by the elevated habitat heterogeneity and ecological diversity of the tepui summits. A good example can be found in *Brocchinia* (Bromeliaceae), a genus of both lowlands and highlands with known morphological and functional adaptations. The issue of the origin of the Pantepui biota is a fascinating subject, which is still open to new ideas and further research efforts.

CONSERVATION

Pantepui is still virtually pristine. Indigenous people do not visit the tepui summits because they consider them the homeland of gods and therefore to be sacred places. Activities such as cultivation, lodging, burning, mining, tourism, and so forth are prevented by several protection efforts, including national parks, natural monuments, biosphere reserves, and a World Heritage Site. In addition, Pantepui has been considered by the WWF/IUCN as one of the neotropical plant diversity centers (SA-2), as well as a critical ecoregion (ER-45) of the Global 2000 Project. Therefore, the tepui summits seem to be well protected against direct human intervention. However, the potential consequences of the ongoing and predicted future global warming have not been fully realized until very recently. Increasing temperatures will cause an upward displacement of suitable environmental conditions for mountain species, such that a number of them may lose their habitat. In the tepuis' summits, this effect will be enhanced by the flat topography, which prevents further upward displacement. Preliminary estimations show that roughly 200–400 endemic vascular plant species (~30–50% of the total) of Pantepui are threatened with habitat loss because of the projected 2–4 °C warming for the end of this century. Owing to the singularity of the Pantepui biota, this would be a serious danger for Guayanan, as well as for global, biodiversity.

FIGURE 6 The insectivorous plant *Heliamphora minor* (Sarraceniaceae) from the summit of the Eruoda-tepui (Chimantá massif), at about 2600 m in altitude. Photograph by the author.

SEE ALSO THE FOLLOWING ARTICLES

Adaptive Radiation / Continental Islands / Global Warming

FURTHER READING

Berry, P. E., and R. Riina. 2005. Insights into the diversity of the Pantepui flora and the biogeographic complexity of the Guayana shield. *Biologiske Skrifter* 55: 145–167.
Briceño, H. O., and C. Schubert. 1990. Geomorphology of the Gran Sabana, Guayana shield, southeastern Venezuela. *Geomorphology* 3: 125–141.
Huber, O., ed. 1992. *El macizo del Chimantá*. Caracas: Oscar Todtmann Ed.
Huber, O. 1988. Guayana lowlands versus Guayana highlands: a reappraisal. *Taxon* 37: 595–614.
Maguire, B. 1970. On the flora of the Guayana highland. *Biotropica* 2: 85–100.
Rull, V. 2004. Biogeography of the "Lost World": a palaeoecological perspective. *Earth-Science Reviews* 67: 125–137.
Rull, V., and T. Vegas-Vilarrúbia. 2006. Unexpected biodiversity loss under global warming in the neotropical Guayana highlands. *Global Change Biology* 12: 1–9.
Steyermark, J. A. 1986. Speciation and endemism in the flora of the Venezuelan tepuis, in *High-altitude tropical biogeography*. F. Vuilleumier and M. Monasterio, eds. Oxford: Oxford University Press, 317–373.
Steyermark, J. A., P. E. Berry, and B. K. Holst, eds. 1995–2005. *Flora of the Venezuelan Guayana*. St. Louis: Missouri Botanical Garden Press.

PAPUA NEW GUINEA

SEE NEW GUINEA

PEOPLING THE PACIFIC

PATRICK V. KIRCH

University of California, Berkeley

The islands of the Pacific Ocean were settled by humans in two major episodes. The earliest phase began in the Late Pleistocene, at least 40,000 years ago, and involved the movement of hunting-and-gathering populations into Near Oceania. The second major phase commenced about 4000 years ago and involved the diaspora of the Austronesian-language speakers into Remote Oceania, as well as into the Indian Ocean as far as Madagascar. The most isolated islands and archipelagoes of Remote Oceania, including Hawai'i, Easter Island (Rapa Nui), and New Zealand (Aotearoa), were settled by Polynesians between AD 800 and 1200.

GEOGRAPHIC BACKGROUND: NEAR AND REMOTE OCEANIA

The Pacific Islands, or Oceania, were classically subdivided into three main geographic regions, following the scheme of the French explorer Dumont d'Urville in the early nineteenth century: Melanesia, Micronesia, and Polynesia. Although these geographic terms continue to be widely used, except for "Polynesia," they have little cultural or historical basis. Only Polynesia stands out as a culturally and linguistically meaningful category. More recently, historical anthropologists and archaeologists stress the distinction between Near Oceania in the western Pacific (including New Guinea, the Bismarck Archipelago, and the Solomon Islands) and Remote Oceania (which includes all of island Melanesia southeast of the Solomons, along with Polynesia and Micronesia). Near Oceania, which was first settled by *Homo sapiens* in the late Pleistocene, is characterized by intervisible islands with a highly diverse biota, capable of supporting hunter-and-gatherer populations. The widely dispersed islands of Remote Oceania were discovered and settled only within the past 4000 years, by horticultural peoples who introduced food crops and domestic animals to these biotically more depauperate and resource-limited islands.

PLEISTOCENE SETTLEMENT OF NEAR OCEANIA

During periods of glaciation in the Late Pleistocene, lowered sea levels exposed the continental shelf joining New Guinea to Australia (and Tasmania to Australia in the south). This enlarged land mass is known to biogeographers as Sahul. Similarly, the Malaysian peninsula and much of Indonesia formed another exposed land mass called Sunda. The region between Sunda and Sahul, known as Wallacea, always had water gaps that acted as barriers to biotic dispersal. However, human entry into Sahul was facilitated when these water gaps were at their narrowest, and human expansion throughout Australasia occurred rapidly once people entered the region, at least 40,000 years ago. A number of occupation sites are now radiocarbon-dated to ~36,000 years ago, on the large island of New Guinea, and on New Britain and New Ireland in the adjacent Bismarck Archipelago. Late Pleistocene sites in the Admiralty Islands, New Ireland, and Buka (Solomons) all indicate the existence of open-ocean transport, thereby suggesting the presence of some form of early watercraft (possibly rafts, bark boats, or dugouts). There is no evidence, however, for human expansion beyond the eastern end of the main Solomon Islands chain until the Middle to Late Holocene.

The earliest human colonists in Near Oceania were hunters and gatherers, who exploited tropical rain forests but also made use of inshore marine resources. The presence of simple flake tools of obsidian (originating on the island of New Britain) at these sites provides evidence for long-distance communication and exchange between communities, often on separate islands. By the early Holocene (~8000 BC), archeobotanical evidence indicates that tree, root, and tuber crops (such as the *Canarium* almond and various aroids) were being domesticated within Near Oceania. The Kuk Swamp site in the highlands of New Guinea has produced structural evidence in the form of drains and ditches for water control and possible taro horticulture beginning as early as 7000 BC. Such archeobotanical evidence supports long-standing ethnobotanical claims that Near Oceania was an important region for the domestication of a number of tropical root, tuber, and tree crops.

Historical linguistic evidence suggests that the Pleistocene settlers of the Near Oceanic islands spoke a diversity of languages, all of which were non-Austronesian. These highly diverse languages (of which at least 900 are recorded) are often lumped together under the rubric *Papuan*, but in fact a number of distinct language families are involved.

AUSTRONESIAN ORIGINS

Beginning around 3000 BC, a major diaspora of peoples speaking languages belonging to the Austronesian language family began to expand from island Southeast Asia into the Near Oceanic region. The immediate homeland of the Austronesians has generally been regarded as including the island of Taiwan, as well as adjacent areas of mainland China (e.g., the Fujian coast). Linguistic reconstructions of Proto-Austronesian language include a number of terms for the outrigger sailing canoe and its components, and the ability of early Austronesians to disperse rapidly is doubtless due to their invention of this technology. Linguistic evidence likewise indicates that the Austronesians were horticulturalists who transported root, tuber, and tree crops including taro (*Colocasia esculenta*), the true yams (*Dioscorea alata* and other species), coconut, and many other crop species via their canoes. They also raised domestic pigs, dogs, and chickens. Other animals, such as the Pacific rat (*Rattus exulans*), may have been carried as stowaways, although possibly also for food.

Archaeologically, the Austronesian dispersal from Taiwan through the Philippines and onto other archipelagoes and islands of Southeast Asia and Oceania is marked by the presence of a variety of ceramic assemblages, along with related material culture including stone adzes and shell technology (fishhooks, ornaments, etc.). The oldest pottery associated with Austronesians is the Ta-p'en-k'eng culture of Taiwan (~4300–2500 BC), which is followed by several local varieties of red-slipped pottery. The dispersal of Austronesian speakers into coastal regions of Near Oceania and on into the islands of Remote Oceania is marked by sites of the Lapita cultural complex (see below).

The Austronesian diaspora rapidly encompassed the major archipelagoes of island Southeast Asia; one branch of Austronesian-speakers expanded along the north coast of New Guinea into the Bismarck Archipelago. This branch is known to linguists as Oceanic, and the Oceanic languages (numbering about 450 modern languages) include most of those spoken throughout the Pacific Islands.

THE LAPITA CULTURAL COMPLEX AND REMOTE OCEANIA

The movement of Austronesian speakers into the Bismarck Archipelago has been correlated with the presence of a distinctive style of pottery known as Lapita. Early Lapita pottery assemblages include vessels of various shapes, such as bowls supported with pedestal feet, flat-bottomed dishes, large carinated jars, and globular jars with restricted necks. Many of these vessels were covered with lime-infilled decorations, with motifs made largely through a technique of dentate stamping, although incising was also used. Some of the pottery, such as the globular jars, was not decorated. Aside from the characteristic pottery, Lapita sites yield a variety of other portable artifacts, such as *Tridacna*-shell adzes and *Trochus*-shell fishhooks, as well as a diversity of ornaments and exchange valuables. Obsidian from sources in the Bismarck and Admiralty Islands is also common at Lapita sites and is further evidence of extensive trade or exchange between Lapita communities. The earliest phase of the Lapita cultural complex dates to ~1500–1300 BC and is represented by sites in the Bismarck Archipelago, such as the waterlogged Eloaua Island sites of the Mussau group. Often, these early Lapita sites were hamlets or villages consisting of houses elevated on posts or stilts, situated over tidal reef flats or along shorelines (such as the large Talepakemalai site on Eloaua Island). Excavated plant and animal remains indicate a mixed economy with horticulture and marine exploitation.

It is very likely that the makers of Lapita pottery can be correlated with the Proto-Oceanic stage in the history of the Austronesian language family dispersal. Similarly, genetic evidence (such as mitochondrial DNA and hemoglobin markers) suggests that the Lapita population derived from an actual demic intrusion into the Bismarck Archipelago, deriving out of island Southeast Asia. However, it is also likely that the Proto-Oceanic speakers interacted extensively with the indigenous Papuan-speaking populations who were already in place in Near

Oceania. For these reasons, the Lapita phenomenon is now interpreted as the outgrowth of simultaneous cultural processes of intrusion, integration, and innovation (the so-called Triple-I model).

After the initial phase of Lapita development in Near Oceania, Lapita populations expanded eastward into Remote Oceania, first to the far eastern Solomon Islands, by at least ~1300 BC, and soon thereafter into Vanuatu, the Loyalty Islands, and New Caledonia. The 850-km water gap between the Eastern Solomons and Fiji may have been crossed by 1100 BC. The archipelagoes of Tonga and Samoa, along with the nearby islands of Futuna and ʻUvea, had Lapita settlements in place by 900 BC.

SETTLEMENT OF MICRONESIA

Archeological evidence for initial Austronesian dispersal into Micronesia comes from the western Micronesian archipelagoes of Palau and the Marianas. In these islands, the earliest sites contain a red-slipped, sand-tempered pottery, some of which is decorated with lime-filled, impressed designs. Radiocarbon dates from these sites suggest that humans settled the Mariana and Palau archipelagoes no later than 1500 BC, and possibly as early as 2000 BC; the immediate homeland of these voyagers may have been in the Philippines or Molucca Islands. Around 2000 years ago, Oceanic speakers who made plainware pottery (a late form of Lapita) and who used shell adzes, fishhooks, and other implements moved northward into central Micronesia and founded settlements on several volcanic islands, including Chuuk, Pohnpei, and Kosrae. The atolls of the Marshall Islands were also colonized around this time. This two-phase settlement of Micronesia, with initial Austronesians moving in from the Philippines ~2000–1500 BC, followed by later movement of late Lapita (Oceanic-speaking) colonists into central Micronesia, is reinforced by historical linguistic evidence. The Austronesian languages spoken in western Micronesia are distinct from those of central and eastern Micronesia, with the latter belonging to the Oceanic subgroup. However, later contact between island groups often resulted in considerable linguistic borrowing; the Yapese languages, for example, had several phases of interaction with both western and central Micronesian languages.

POLYNESIAN DISPERSALS

Ancestral Polynesian culture and Proto-Polynesian language developed in the Tonga–Samoa region between ~900 and 500 BC, directly out of the founding Lapita cultural complex. Thus, the Tonga–Samoa region, referred to as Western Polynesia, is considered the immediate homeland of Polynesian culture and language. A number of changes were involved in the transition from late eastern Lapita to Ancestral Polynesian culture, among them the gradual loss of the art of making pottery. A significant number of lexical changes also occurred in the language spoken by the dialect chain linking Tonga and Samoa, resulting in a distinctive Proto-Polynesian language.

The final stage in the human settlement of the Pacific Islands began in the first millennium AD, with the additional expansion of Polynesian-speaking peoples eastward out of Tonga and Samoa into the archipelagoes of central Eastern Polynesia. This last great phase in the expansion of Austronesians into the Pacific was enabled by the invention and development, probably in the Tonga-Samoa region, of the double-hulled voyaging canoe. These large canoes were capable of carrying 40 to 60 people and their cargo (including plants and domestic animals to reestablish their subsistence economy on new islands), for voyages lasting a month or more. There has been considerable debate among archaeologists regarding the exact chronology and sequence of the Eastern Polynesian dispersals. Recent radiocarbon dating of sites in the Society Islands, the Cook Islands, the Marquesas, and Mangareva suggest that initial settlements were in place by around AD 900, and possibly slightly earlier in the case of the Society Islands. Remote Easter Island is likely to have been discovered by AD 900–1000. Evidence from pollen cores on Oʻahu Island likewise suggests that Polynesians were clearing lowland forests there by AD 800–1000. The large, temperate islands of New Zealand were among the last to be colonized by Polynesians, after AD 1200.

As noted earlier, Polynesia is the only term of the original tripartite classification of Oceanic peoples (including Melanesia and Micronesia) that can properly be considered a monophyletic cultural and linguistic group. All of the ethnographically attested Polynesian societies can be demonstrated to have descended from a common Ancestral Polynesian culture and Proto-Polynesian language. For this reason, Polynesia is regarded as an ideal region for testing phylogenetic models of cultural differentiation from a common ancestor.

CONTACTS WITH THE AMERICAS

The question of contacts between Polynesia and the Americas has long interested scholars. The older view of Thor Heyerdahl that Polynesian populations originated in North and South America has not stood the test of archaeological, linguistic, or human biological evidence. However, it is certain that later Polynesian voyaging canoes had the technical capability to reach the Americas, and to return. That the Polynesians did indeed reach South America and return is strongly suggested by preserved remains of the sweet potato (*Ipomoea batatas*), a South American domesticate, recovered from several prehistoric Polynesian sites (such as the Tangatatau rockshelter on Mangaia, Cook Islands). Further evidence of contact has recently been provided by the bones of chickens (*Gallus gallus*, a Polynesian domesticate) from a pre-Columbian archeological site in Chile. More controversial has been the claim that Polynesians may have introduced the technology of plank-built canoes to the Chumash of the Channel Islands, off the coast of California.

SEE ALSO THE FOLLOWING ARTICLES

Archaeology / Cook Islands / Easter Island / Exploration and Discovery / Human Impacts, Pre-European / *Kon-Tiki* / Polynesian Voyaging / Tonga

FURTHER READING

Bellwood, P. 1985. *Prehistory of the Indo-Malaysian archipelago*, revised ed. Honolulu: University of Hawaii Press.
Irwin, G. 1992. *The prehistoric exploration and colonisation of the Pacific*. Cambridge: Cambridge University Press.
Kirch, P. V. 2000. *On the road of the winds: an archaeological history of the Pacific Islands before European contact*. Berkeley: University of California Press.
Kirch, P. V., and R. C. Green. 2001. *Hawaiki, Ancestral Polynesia: an essay in historical anthropology*. Cambridge: Cambridge University Press.
Lilley, I., ed. 2006. *Archaeology of Oceania: Australia and the Pacific Islands*. Oxford: Blackwell Publishing.
Spriggs, M. 1997. *The island Melanesians*. Oxford: Blackwell Publishing.

PHILIPPINES, BIOLOGY

RAFE M. BROWN
University of Kansas, Lawrence

ARVIN C. DIESMOS
National Museum of the Philippines, Manila

The Philippines (Fig. 1) is one of the Earth's most spectacular island archipelagoes. The country spans the Asian–Australian faunal zone interface at the sharpest biotic demarcation (Wallace's Line) on the planet. Although collectively comprising a land mass approximately the size of the U.S. state of Arizona, the Philippines is a complex archipelago with more than 7100 distinct islands. Geographically situated on the edge of multiple colliding tectonic plates, the Philippines has an ancient and complex geological history that has only recently come to light. Ancient land mass movements, environmental gradients along steep volcanic slopes, and sea level–induced alterations of connectivity between neighboring islands have all presumably fueled the in situ evolutionary process of diversification on a magnitude seen in few other island systems. The result is a spectacular array of biodiversity, unparalleled levels of vertebrate endemism, and some of the planet's most spectacular examples of life on islands. The Philippines is a global superpower of biodiversity that may possess one of the highest concentrations of vertebrate life on Earth.

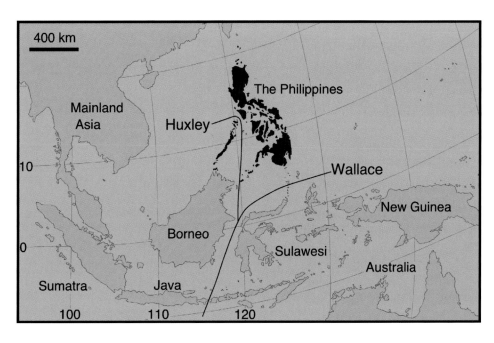

FIGURE 1 The position of the Philippines (darkly shaded islands) in relation to surrounding Southeast Asian and Australasian land masses. The positions of Wallace's and Huxley's lines are indicated for reference. Wallace's line marks the edge of the Sunda Shelf and the transition between Asian and Australian faunal regions. Huxley's line separates Palawan and associated land bridge islands from the truly oceanic portions of the Philippines.

FIGURE 2 Species exemplars from the extraordinarily diverse endemic Philippine rodent radiation: (A) Luzon hairy-tailed rat, *Batomys granti;* (B) Cordillera striped earth-mouse, *Chrotomys whiteheadi;* (C) Luzon needle-nosed shrew-rat, *Rynchomys* cf *soricoides;* (D) Kalinga shrew-mouse, *Archboldomys kalinga;* (E) silver earth-mouse, *Crotomys silaceus;* and (F) common Philippine forest rat, *Rattus everetti*. Photographs by Rafe M. Brown.

GEOGRAPHICAL AND GEOLOGICAL SETTING

The Philippines is situated south of Taiwan and north of Borneo and Sulawesi (between the equator and the Tropic of Cancer), separated from the Asian mainland by the South China Sea. Collectively the archipelago covers 2 million km² with a total land mass of approximately 300,000 km².

The western part of the archipelago (Palawan, Balabac, Busuanga, Coron, and smaller associated islands) has had a stable geologic configuration and has been intermittently connected (or nearly connected) to northern Borneo via a narrow land bridge during maximum exposure of the world's largest continental shelf (the Sunda Shelf). The remaining portions of the Philippines are classified as oceanic islands and are believed to have risen directly from the ocean floor as a result of volcanic activity associated with subduction of the Philippine Plate at the western boundary of the Pacific Ring of Fire.

BIODIVERSITY AND ENDEMISM

The Philippines is recognized internationally as a global stronghold of biodiversity. The shallow, warm seas surrounding the country support the richest coral reef communities on the planet and are literally the epicenter of Southeast Asian and southwestern Pacific marine diversity. Terrestrial ecosystems in the Philippines are similarly diverse, supporting a wealth of natural resources, habitat heterogeneity, and a rich array of species diversity. High levels of alpha-diversity (species richness) and beta-diversity (variation among adjacent regions) place the Philippines among the world's most biodiversity-rich countries.

Some of this diversity is truly astonishing. Highlights include dwarf water buffalos or "Tamaraw" (*Bubalus mindorensis*), the largest extant forest eagle in the world (*Pithecophaga jefferyi*), a radiation of the world's largest flowers (genus *Rafflesia*), the rarest batagurid turtle in the world (*Siebenrockiella leytensis*), an extensive and exceptionally diverse rodent radiation (including such novelties as raccoon-sized "cloud rats" and needle-nosed shrew-rats that feed entirely on worms; Fig. 2), four endemic genera of snakes, five endemic species of spectacularly maned wild pigs, and such "living fossils" as the primitive and relictual flat-headed frog (*Barbourula busuangenis*). The fossil record shows that many extraordinary, large-bodied animals have gone extinct with the arrival of modern humans, including dwarf buffalos, elephants, and giant land tortoises.

Terrestrial ecosystems are fantastically diverse. Recent summaries of birds recognized 593 species (32% endemic); mammal diversity currently is estimated at 175 native terrestrial mammals (65% endemic). Recent reviews of the classification of Philippine amphibians and reptiles recognized 105 species of amphibians (79% endemic) and 264 reptiles (68% endemic). These estimates emphasize conspicuous terrestrial vertebrates, but total country estimates (Table 1) are awe inspiring, with as many as 15,000 plants (and their relatives) and 38,000+ animals (vertebrates and invertebrates), for a startling total of 53,500 species. These numbers should be viewed as conservative approximations; numerous recent studies have shown that terrestrial biodiversity of the Philippines is substantially underestimated, in some cases grossly so. In poorly studied groups such as earthworms, more than a hundred

TABLE 1
A Summary of Recent Estimates of Total Countrywide Philippine Species Diversity

Taxonomic Group	Philippine Total
Vertebrates	3,308–3,325
Invertebrates	34,940–35,000
Plants	14,000–15,310
Others (algae, lichens, fungi, etc.)	6,100+
Total	53,500+

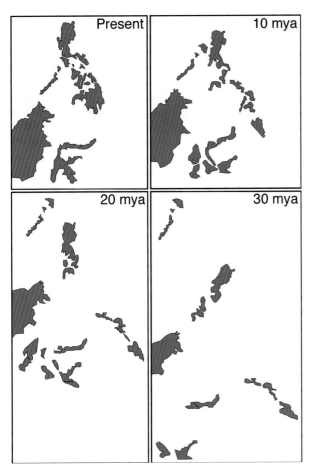

FIGURE 3 Estimated Cenozoic movements of Philippine land masses (modified from work of Hall 1998) 10, 20, and 30 million years ago.

new species have recently been discovered. Molecular phylogenetic studies have confirmed the exceptional species diversity of numerous Philippine radiations; many have uncovered suites of previously unrecognized, cryptic species. Several of these key studies of species boundaries have produced startling results and increased species diversity in selected groups by 30–50%. Virtually every molecular phylogenetic study that has been published in the last 10 years has included the discovery of new hidden species, including many groups of frogs, lizards, insects, worms, forest mice, bats, rats, and birds.

One example is the case of the Philippine forest frogs of the genus *Platymantis*. Past studies were based solely on morphology; species numbers grew slowly from 7 to 12 named species between 1950 and 1990. It was at this point that a group of herpetologists began to focus on bioacoustic characters (analysis of the mating calls of male frogs) and applied molecular techniques (DNA sequence data) to the problem of species boundaries in this group. By emphasizing different aspects of the phenotype, these workers provided fine-scale taxonomic partitioning of

FIGURE 4 Spectacular Philippine endemic species: (A) the endemic Philippine freshwater crocodile, *Crocodylus mindorensis* (photograph by Rafe M. Brown); (B) the newly discovered Calayan Island flightless rail, *Gallirallus calayanensis* (photograph by Marge Babon/CEAE); (C) the giant flower *Rafflesia manillana* (photograph by Arvin Diesmos); (D) and one of the world's only fruit-eating monitor lizards, *Varanus olivaceus* (photograph by Charles Linkem).

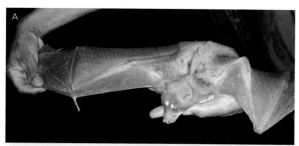

FIGURE 5 Spectacular Philippine endemic species: (A) the recently discovered Mindoro stripe-faced flying fox, *Styloctenium mindorensis* (photograph by Jake Esselstyn); (B) the rarest batagurid turtle in the world, the Philippine forest turtle, *Siebenrockiella leytensis* (photograph by Rafe M. Brown); (C) the endemic Philippine dwarf water buffalo or Tamaraw, *Bubalus mindorensis* (photograph courtesy of Department of the Environment and Natural Resources-Protected Areas and Wildlife Bureau Tamaraw Conservation Program); (D) the brightly colored Igorot frog, *Rana igorota* (photograph by Rafe M. Brown).

forest frogs and opened the way for a flood of discoveries and new species descriptions. It is clear that *Platymantis*, currently estimated at 27 described species, represents one of the Earth's major island radiations of frogs. Several dozen species remain to be described. When the same analytical tools are applied to other frog groups throughout the archipelago, it becomes clear that the species diversity of Philippine Amphibia has been underestimated by as much as 30–40%.

Despite the lack of a complete knowledge of the biodiversity of the Philippines, today's taxonomic estimates allow researchers to calculate the number of species per unit area. When this calculation is carried out for terrestrial species, given the available land mass (300,000 km^2), the end result is *the highest concentration of biodiversity on Earth*.

BIOGEOGRAPHY AND PROCESSES OF DIVERSIFICATION

Several generations of biologists have converged on a startling consensus: The extraordinary biodiversity of the Philippines has likely been produced by three major processes of the geographic template (i.e., geology + climate). We consider these process to be the major "generators" of diversity in the Philippines. The first involves the complex geological history of the archipelago. Briefly, the formation of the Philippines involves more than 50 million years of collisions of multiple separate plates of the earth's crust, resulting in a history characterized by constantly changing configurations of islands. The Philippines of 10, 20, or 30 million years ago looked almost nothing like the country today (Fig. 3). The inferred result of these ancient movements of land masses is that the distant ancestors of many of today's Philippine endemic species may have first invaded and become isolated on Philippine paleo-islands several to many tens of millions of years before present. These events appear to have given rise to deep phylogenetic diversity and the presence of numerous "old endemics," or highly divergent, taxonomically distinctive taxa (Figs. 4 and 5).

The second generator of biodiversity in the Philippines is the process of evolutionary differentiation along dramatic elevational, atmospheric, and environmental gradients from sea level to 2000+ meters (Fig. 6). The oceanic portions of the Philippines are islands produced primarily by volcanism. Many of the smaller islands (and many isolated peaks within larger islands) were formed as active volcanoes arose from the ocean floor over the last 100 million years. Steep elevational gradients (replicated

numerous times along dozens of isolated volcanic peaks) have given rise to atmospheric variation, microclimate variability, and forest community heterogeneity—resulting in biodiversity gradients along these sheer mountain slopes. As a result, the study of Philippine biodiversity is largely the study of species succession along environmental gradients. The last 50 years of biodiversity research in the Philippines have documented the enormous impact that elevation-associated atmospheric and habitat gradients that have played a prominent role in the evolutionary processes of diversification.

The third generator of Philippine biodiversity (and the one that is, for better or for worse, most often invoked) is the repeated "species pump" action of rising and falling sea levels between the mid- to late Pleistocene (350,000–12,000 years ago). Repeated cooling episodes during this recent time period resulted in the capture of the Earth's water at the expanding polar ice caps and a concomitant lowering of sea levels throughout the globe. In the Philippines the result of these sea level oscillations was repeated episodes of connectivity and isolation between islands separated by channels of 120–180 meters (Fig. 7). Biogeographers now recognize at least eight faunal provinces that correspond to these Pleistocene Aggregate Island Complexes (PAICs), which represent maximum exposure of expanded islands. Each PAIC is a major center of biodiversity and endemism, with distinctive flora and fauna. The tracing of submarine bathymetric contours to reveal Pleistocene exposure of land provides biologists with an estimate of former paleoisland connection and serves as a basis for predictions of taxonomic affinity (in the absence of phylogenetic data) for the species inhabiting those islands. This exercise has also identified several minor subcenters of biological diversity in the form of small islands. Finally, it is clear that both in situ evolutionary diversification and repeated colonization events have contributed to the accumulation of diversity in the archipelago.

The development of molecular phylogenetic methods has allowed for unprecedented study of both biodiversity in general and the historical and temporal framework for

FIGURE 6 Striking elevational gradients and volcanic landscapes of the Philippines: (A) Pagudpud area, northern Luzon Island; (B) Lake Danao, a high-elevation volcanic crater lake in northern Leyte Island; (C) the saw-toothed peaks of Mt. Guiting-guiting, Sibuyan Island; and (D) terraced fields along the slopes of the northern Cordillera Mountain range (between Bontok and Banaue). Photographs by Rafe M. Brown.

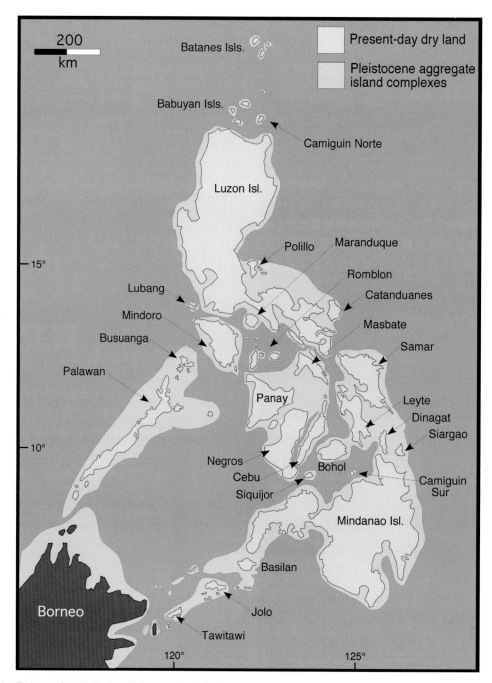

FIGURE 7 A detailed map of the Philippines (light green shaded islands) with Pleistocene Aggregate Island Complexes (PAICs) indicated by tracing of 120 m underwater bathymetric contour. Areas shaded purple indicate shallow seas that may have been exposed as many as ten times during the middle- to late Pleistocene. These land-positive connections (land bridges) may have allowed for past floral and faunal exchange between islands that today are isolated by marine channels.

the generation of Philippine biodiversity. Several modern phylogenetic studies have produced surprises that contradict earlier hypotheses and threaten to topple the prevailing view of diversification in the Philippines. The result is an emerging consensus suggesting that Philippine biodiversity is far more complex than previously thought.

Recent studies of Philippine flying lizards, forest geckos, spotted stream frogs, and forest mice suggest that for some taxa, periodic sea level oscillations have had a profound impact on the production and distribution of species diversity. These studies have all demonstrated a near complete adherence of species to PAIC boundaries and Pleistocene seashores. For much of the terrestrial vertebrate life in the Philippines, these historical events have played a predominant role in shaping species distributions—much more so, in fact, than the traditional biogeographic variables such as island size and distance from mainland. This dominant paradigm of Philippine biogeography is so prevalent, in fact, that the shared presence of species across PAIC boundaries is now viewed as an extreme outlier. For the most part, this has been a healthy development for the understanding of Philippine biodiversity. By heightened attention to the unique evolutionary histories of PAIC endemics, biogeographers are now beginning to arrive at an appreciation of Philippine biodiversity that is concordant with evolutionary history—one that recognizes distinct and cohesive lineage segments as evolutionary species. The end result is that a truly *evolutionary* classification of Philippine vertebrate life is now perceived as an obtainable goal.

Several recent studies have dramatically upset accepted scenarios concerning the predominant processes of evolutionary diversification in the Philippines. One study, a phylogenetic analysis of Asian "Fanged Frogs" conducted by Ben Evans and colleagues, revealed a pervasive and entirely unexpected zoogeographic link between Sulawesi and the Philippines. In that study, DNA sequence data revealed that multiple over-water dispersal events have allowed for successive waves of exchange between Sulawesi and the Philippines. As might be expected, the oldest invasions gave rise to proportionately more species and have spread across more PAICs than the younger arrivals. This Philippine–Sulawesi connection stands in opposition to all previous zoogeographic evidence provided by the last 25 years of studies of Philippine vertebrates, particularly mammals.

Additional studies have upset the land-bridge (Palawan) versus oceanic (remaining Philippines) dichotomy. Phylogenetic studies of fanged frogs, spiny rats, shrews, litter frogs, spotted stream frogs, slender stream frogs, emerald tree skinks, flying lizards, geckos, fresh water fish, river shrimp, and numerous groups of insects all demonstrate that a large portion of Palawan's endemics are most closely related to the truly oceanic portions of the Philippines, to the exclusion of the (expected) species from the islands of the Sunda Shelf. In fact, many Palawan endemics are nested within the truly Philippine radiation, suggesting recent dispersal to Palawan from the oceanic portions of the Philippines, and a lack of faunal exchange with Borneo. The simple depiction of Palawan as a faunal extension of northern Borneo (based largely on mammal and bird taxonomy) is a taxon-biased expectation that is not supported by the bulk of available phylogenetic evidence. In fact, Palawan is an exceptional amalgamation of greatly divergent ancient taxa, recent dispersal events from Borneo (e.g., mammals), and much older biogeographic elements that are nested within Philippine radiations and represent old dispersal events from the oceanic portions of the archipelago. In this sense, the slender island of Palawan represents a true biogeographic novelty, literally a crossroads spanning the most prominent biogeographic boundary in the world and one of the exceptions to expectations based on Wallace's and Huxley's Lines.

An additional class of exceptions to the PAIC-centric tradition of biogeography in the Philippines is the increasingly prevalent pattern of micro-endemism on small land bridge islands. Because small land bridge (or even deep-water) islands next to the larger land masses in the Philippines were expected to possess a nested subset of species diversity observed on large neighboring islands, they have often gone unsurveyed and unscrutinized by biologists. Many of these tiny, seemingly irrelevant islands are now known to harbor endemics, some of which are spectacularly divergent evolutionary novelties. Examples include Lubang, Camiguin Sur, Dinagat, Siargao, Sibuyan, and the Gigante Islands. All demonstrate extensive levels of endemism, commensurate with their status as deep-water minor subcenters of diversity. Recent work suggests that our collective knowledge of these minor subcenters of Philippine biodiversity is far from complete.

A final exception to the prevailing paradigm of PAIC-level partitioning of biodiversity of the Philippines is an emerging pattern of autochthonous speciation on the large islands. Numerous recent phylogenetic studies involving amphibians, reptiles, shrews, bats, rats, birds, plants, and insects have drawn attention to patterns of endemism *within* the large islands of Luzon, Palawan, Mindanao, Negros, Panay, and Samar-Leyte. These studies provide evidence of evolutionary and ecological processes (e.g., other than rising sea level vicariance) fueling diversification. Many of these are very interesting because they provide support for adaptive evolution, sympatric, and/or ecological speciation within larger islands. As such, these examples stand in contrast to the "passive" process of non-adaptive divergence following isolation inferred for the diversification of many other groups of species in the archipelago.

HUMAN HISTORY AND IMPACT

Colonizing from mainland Asia, humans first arrived in the Philippines many thousands of years ago. Before the arrival of the Spanish (sixteenth century), the Philippines was 90–95% forested, with only scattered settlements surrounding various coastal areas. Despite a growing population and a great demand for timber, the next 300 years probably resulted in the loss of only an additional 20% of forest cover. We know that at the turn of the twentieth century, Cebu Island (Ferdinand Magellan's capitol) was almost completely deforested and that nearby islands of Negros, Guimaras, and Panay were covered by less than half of their original forest area. But other islands maintained vast expanses of intact forest, and perhaps as much as 70% of the country's forest persisted. The industrial and agricultural revolution, periods of American and Japanese occupation, plus the period of economic expansion following World War II took an extremely heavy toll on Philippine forests. During this the last century, it is estimated that the Philippines lost between 50% and 70% of its original forest cover.

During this post-war period, many factors weighed heavily on Philippine biodiversity, including high human population growth associated with the industrial revolution (and a concomitant application of low-level pressure on forests; Fig. 8), a long tradition of colonial control and exploitation, shortcomings in the country's (U.S.-installed) education system, crippling poverty in many undeveloped portions of the country, and chronic government graft and corruption that have pandered to a wealthy elite and foreign hegemony.

However, by far the most destructive force has been unchecked environmental destruction associated with large-scale logging and mining. Large-scale logging and mining were introduced to the country during the American occupation. Over the last century, the Philippines has been the target of wholesale exploitation of natural resources combined with wide-scale plantation-style agricultural efforts (principally hemp, bananas, copra, and sugar cane), which have converted many of the country's most fertile natural ecosystems into monoculture and barren wasteland. The end result of this unchecked colonial rule and political instability has been systematic exploitation of Philippine natural resources, and catastrophic environmental destruction at rates exceeding those almost anywhere on the planet. Originally >90% forested, the Philippines now retain only 4–8% of its original old-growth forest.

CONSERVATION PROSPECTS AND CHALLENGES FOR THE FUTURE

The Philippines shares only with Madagascar the distinction of being both a megadiverse country and a Global Conservation Hotspot. Despite a greater understanding of Philippine biodiversity gained during the past 15 years, such knowledge is accumulating too slowly to conserve and stem the loss of the archipelago's spectacular biodiversity. There can be no doubt: the Philippines is of the planet's highest conservation priorities.

Part of the roots of the global conservation crisis currently unfolding in the Philippines is related to the so-called Linnaean shortfall: the disparity between where taxonomists have focused their attention over the last 400 years (e.g., predominantly vertebrates, selected insect groups, and angiosperms) and where the greatest proportions of undescrbed biodiversity remain to be studied (invertebrates, other plants, parasites, and soil microbes, among many other groups). The Linnaean shortfall represents a failure of taxonomy and at the same time a tremendous opportunity for future generations of researchers who choose to focus their attention on these lesser-known groups. Nevertheless, because a lack of knowledge of biodiversity in part contributes to its wanton exploitation and neglect, the end result of this ignorance is biodiversity loss and extinction.

Despite these bleak assessments, there is increasing cause for hope and renewed efforts towards fostering sustainable development, ecosystem restoration, and reliance on renewable natural resources (e.g., nontimber forest products such as rattan). The National Integrated Protected Areas System (NIPAS) Act has established a protocol for the establish-

FIGURE 8 Primary causes of deforestation in the Philippines: (A) Large-scale commercial logging was introduced to the Philippines by the Americans. In this image, loggers use refurbished World War II weapons carriers to skid mature hardwood trunks from lowland forest of Mt. Busa, South Cotobato Province, southern Mindanao Island. (B) Small-scale timber poaching and clearing of land for agricultural purposes. *Kaingineros* like this man from Mt. Malinao, Albay Province, Luzon Island, routinely clear small patches of forest, burn debris, and then plant a single crop of rapidly-growing vegetables for sustenance and sale at local markets. Photographs by Rafe M. Brown.

ment and support of nationally protected areas, and establishment of new parks continues to this day at a growing rate. Grassroots community environmentalism has sprung up throughout the country, and a growing protected-area system continues to spread coverage and jurisdiction over the remaining forested regions. The country's National Commission on Indigenous Peoples has taken a strong leadership role in protecting natural resources through management of tribal ancestral homelands, and an expanding base of young, energetic conservation biologists have enthusiastically taken up the battle to protect the country's natural heritage. There is great cause for hope in Philippine conservation, and unique Philippine flagship species (Fig. 9) play an increasingly important role in spreading environmental awareness to the Filipino public.

An understanding of Philippine biodiversity will be greatly enhanced by renewed, vigorous attention to four primary avenues of conservation-related research. First and foremost, these include a need for large-scale, countrywide, and faunistically comprehensive surveys of the remaining natural habitats of the Philippines. Great progress has been made by a handful of researchers over the past 20 years, but this progress needs to be increased by an order of magnitude in both scope and urgency. The next several decades must see a massive resurgence in biodiversity studies if the Philippine conservation crisis (and expected catastrophic extinction event) is to be averted. Second, thorough taxonomic revisionary studies will be required to avert the Linnaean shortfall in the Philippines and promote biodiversity conservation as a global priority. This field of study is surprisingly thankless and increasingly threatened by changing academic landscapes and emphasis on "high-impact" publications in science. Third, the future is very bright for molecular phylogenetic studies of endemic Philippine radiations. Many of the truly spectacular Philippine radiations have gone unstudied. A much-needed synthesis of geological evidence and time-calibrated phylogenetic studies shows tremendous promise for exposing the temporal framework for evolutionary diversification of Philippine biodiversity. Finally, for a variety of reasons, it is clear that the future of biodiversity research and conservation in the Philippines is an effort that must be led by Filipinos. For this to occur, public and government perception of biologists, societal values, and educational emphasis must all shift to endorse the preservation of natural resources and environmental quality. Aside from their inherent value, biodiversity, forested ecosystems, and environmental health all provide societal services such as clean water, food, renewable resources, and buffering from inclement weather. Filipinos have a rich cultural and historical legacy that is tightly linked to their natural heritage through these natural resources. These must all be celebrated and preserved for future generations.

FIGURE 9 Flagship species of Philippine biodiversity conservation and symbols of an emerging surge in environmentalism: (A) the poorly-known Philippine tarsier, *Tarsius syrichta;* (B) the largest eagle in the world, the Philippine "monkey-eating" eagle, *Pithecophaga jefferyi.* Photographs by Rafe M. Brown.

SEE ALSO THE FOLLOWING ARTICLES

Borneo / Frogs / Indonesia, Biology / Madagascar / Philippines, Geology / Sustainability / Wallace's Line

FURTHER READING

Brown, R. M. 2007. Introduction, in *Systematics and zoogeography of Philippine Amphibia*. R. F. Inger. Kota Kinabalu, Malaysia: Natural History Publications, 1–17.

Brown, R. M., A. C. Diesmos, and A. C. Alcala. 2002. The state of Philippine herpetology and the challenges for the next decade. *The Silliman Journal* 42: 18–87.

Catibog-Sinha, C. S., and L. R. Heaney. 2006. *Philippine biodiversity: principles and practice*. Quezon City, Philippines: Haribon Foundation for Conservation of Natural Resources.

Collar, N. J., N. A. D. Mallari, and B. Tabaranza, Jr. 1999. *Threatened birds of the Philippines*. Makati City, Philippines: Bookmark, Inc.

Department of Environmental Resources and United Nations Environment Programme. 1997. *Philippine biodiversity: as assessment and action plan*. Makati City, Philippines: Bookmark, Inc.

Diesmos, A. C., R. M. Brown, A. C. Alcala, R. V. Sison, L. E. Afuang, and G. V. A. Gee. 2002. Philippine amphibians and reptiles, in *Philippine Biodiversity Conservation Priorities: a Second Iteration of the National Biodiversity Strategy and Action Plan*. P. S. Ong, L. E. Afuang, and R. G. Rosell-Ambal, eds. Quezon City, Philippines: Department of the Environment and Natural Resources, 26–44.

Environmental Center of the Philippines Foundation. 1998. *Environment and natural resources atlas of the Philippines*. Manila, Philippines: ECRF.

Hall, R. 1998. The plate tectonics of Cenozoic SE Asia and the distribution of land and sea, in *Biogeography and geological evolution of southeast Asia*. R. Hall and J. D. Holloway, eds. Leiden: Brackhuys, 99–132.

Mallari, N. A. D., B. R. Tabaranza, Jr., and M. J. Crosby. 2001. *Key conservation sites in the Philippines*. Makati City, Philippines: Bookmark, Inc.

Posa, M. R. C., A. C. Diesmos, N. S. Sodhi, and T. M. Brooks. 2008. Hope for threatened tropical biodiversity: lessons from the Philippines. *BioScience* 58: 231–240.

PHILIPPINES, GEOLOGY

GRACIANO YUMUL, JR., AND CARLA DIMALANTA

University of the Philippines, Quezon City

KARLO QUEAÑO

Department of Environment and Natural Resources, Quezon City, Philippines

EDANJARLO MARQUEZ

University of the Philippines, Manila

Island arc systems such as the Philippines are produced through accretion brought about by collision of geologic blocks, resulting in volcanism and emplacement of crust and mantle fragments on land. The various igneous, sedimentary, and metamorphic rock types in island arcs reflect the complex processes involved in their generation and evolution. Island arc-related processes involve interactions of geological features (e.g., trenches, volcanoes, and faults) that result in specific tectonic evolution, geologic hazards, and mineral deposits.

GEOLOGIC AND TECTONIC SETTING OF AN ISLAND ARC SYSTEM

The Philippines, an arc system lying off the Asian land mass, is trapped at the margins of the Eurasian–Sundaland and the Philippine Sea Plates. The northwest-southeast oblique convergence between these plates is currently being absorbed by two oppositely dipping subduction zones, namely (1) the Manila–Negros–Sulu–Cotobato trench system along which the marginal basins (i.e., South China Sea, Sulu Sea, and the Celebes Sea) on the eastern edge of the Eurasian–Sundaland Plate are being subducted west of the Philippine arc, and (2) the East Luzon Trough–Philippine Trench system, along which the West Philippine Basin of the Philippine Sea Plate is being subducted east of the arc (Fig. 1).

The subduction zones extend approximately 1500 km, delineating an approximately 400-km-wide, seismically active deformed zone known as the Philippine Mobile Belt within the archipelago. Intense deformation also affects different parts of the Philippine Mobile Belt with the activity of the Philippine Fault Zone (Fig. 1). This major left-lateral strike-slip fault system transects the archipelago for more than 1200 km, from northwestern Luzon to eastern Mindanao. It has both transpressional and transtensional components and both horizontal and vertical displacements. Several major strike-slip faults branch out from the main trace of this major fault system. These faults and fault segments can be recognized on the basis of displacements in exposed rock sequences, fault scarps, sag ponds, and pressure ridges. Linear features on aerial photographs, satellite, and remote sensing images also serve to define these faults.

The Philippine Fault Zone, which formed during the Middle Miocene, accommodates the lateral component corresponding to the excess stress resulting from the oblique convergence between the Philippine Sea Plate and the Philippine Mobile Belt through shear partitioning mechanism.

Data obtained from the Global Positioning System networks within the Southeast Asian region have provided measurements of the convergence rate between the Sundaland–Eurasian margin and the Philippine Sea Plate. The Sunda block to the west of the Philippine archipelago is moving with respect to Eurasia at around 10 mm/year in the direction 78° east of south along its northern margin and ~6 mm/year toward 61° east of south along its southern portion. On the eastern side, the Philippine Sea Plate is moving northwestward at approximately 7 cm/year in the region northeast of Luzon and around 9 cm/year southeast of Mindanao (see inset in Fig. 1).

The Manila Trench on the western part of the archipelago continues on as collision zones in the central

Philippines (e.g., Mindoro and Panay islands) and connects to the Negros Trench in the Visayas region. It merges with the collision zone in southwestern Mindanao, where fragments of Eurasian–Sundaland affinity have been accreted to the Philippine Mobile Belt (see inset in Fig. 1). These continent-derived fragments, which were rifted from the southern portion of mainland Asia, form part of the Palawan Microcontinental Block, an aseismic block composed of Upper Paleozoic to Mesozoic sedimentary and igneous bodies.

A number of marginal basins surround the Philippine archipelago: the South China Sea, Sulu Sea, Celebes Sea, Molucca Sea, and West Philippine Basin (see Fig. 1). The Middle Oligocene to Miocene South China Sea oceanic crust is characterized by nearly east-west-trending magnetic lineations. This suggests a general north-south opening direction for this marginal basin. Several mechanisms that have been proposed to explain the formation of this marginal basin include an Andean-type subduction, extrusion tectonics, and large-scale shear-related strike-slip movements. The Early to Middle Miocene Sulu Sea can be divided into the northwest and southeast subbasins. The northwest subbasin is underlain by an island arc basement, whereas the southeast subbasin is made up of oceanic crust. East-west-trending (relative to present-day geographic location) magnetic lineations recognized in the Sulu Sea indicate a north-south opening. The Sulu Sea basin is postulated to have formed by back-arc spreading. The Celebes Sea basin, on the other hand, represents an oceanic crust of Early to Middle Eocene age that also opened in a north-south direction, based on observed east-west magnetic lineations. The Eocene West Philippine basin is one of the subbasins comprising the Philippine Sea Plate. This oceanic marginal basin formed from spreading along the Central Basin Spreading Center in a northeast-southwest direction (60–45 million years ago), followed by opening in a north-south direction (45–35 million years ago). Models to explain the formation of this large marginal basin include an entrapment model and a back-arc origin model. The Molucca Sea basin consists of oceanic crust that is Eocene in age. The closure of this oceanic basin has been scissor-type, with closure complete in the north and the southern portion undergoing basin closure at present.

Subduction along the trench systems surrounding the Philippine archipelago produced the volcanoes and volcanic plugs, which are distributed along linear or arcuate belts. Subduction of the South China Sea crust along the Early Miocene Manila Trench is marked by a volcanic arc that extends from Taiwan to western Luzon. The nearly 1200-km-long chain of stratovolcanoes and volcanic necks

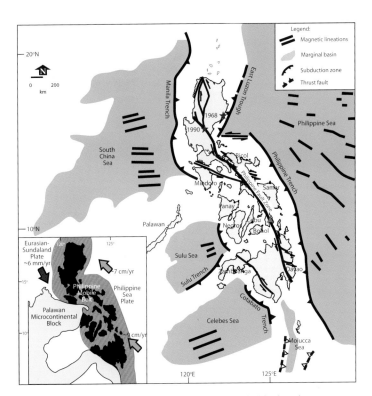

FIGURE 1 The Philippine archipelago is bounded on both sides by subduction systems: Manila-Negros-Sulu-Cotabato on the west and East Luzon Trough-Philippine Trench on the east. The entire archipelago is bisected by the left-lateral strike-slip Philippine Fault Zone. Marginal basins (e.g., South China Sea, Sulu Sea, Celebes Sea, Molucca Sea, and Philippine Sea) are being consumed along the bounding trench systems (Rangin 1990 and references therein). Red stars show the location of the epicenter (and year when the event happened: 1968 in Quezon; 1990 in Nueva Ecija) of two of the most destructive earthquakes that have been experienced in the Philippine archipelago (data from the Web site of the Philippine Institute of Volcanology and Seismology). The orange triangle shows the location of Guinsaugon, the site of a massive landslide February 17, 2006. (Inset) Continent-derived fragments are juxtaposed against arc rocks as a result of the collision between the Palawan Microcontinental Block and the Philippine Mobile Belt. Arrows indicate the direction of motion of the Sundaland-Eurasian (red arrow) and Philippine Sea Plates (green arrow).

includes the active Pinatubo and Taal volcanoes (Fig. 2). In the Visayas region, the Negros arc was produced by subduction of the Sulu Sea crust along the Middle Miocene Negros Trench. The only active volcano in this nearly north-south-trending linear arc of volcanoes in Negros island is Canlaon volcano (Fig. 2). The Cotabato arc, which includes Parker volcano, was produced by subduction of the Celebes Sea crust along the Cotabato Trench, which began in the Late Miocene. The volcanic arc extending from southeastern Luzon to Leyte in the Visayas region is attributed to the subduction of the Philippine Sea crust along the Philippine Trench. Subduction along this trench began during the Pliocene, approximately 3–5 million years ago. The segment of this arc in southeastern Luzon is made

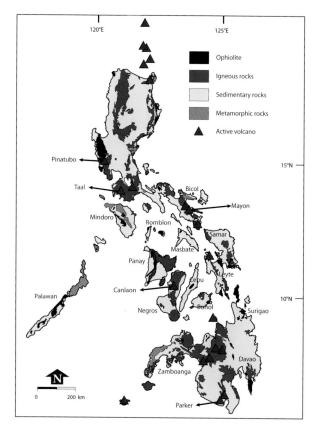

FIGURE 2 Map showing the distribution of igneous, sedimentary, metamorphic, and ophiolitic rocks. Red triangles show the location of the active volcanoes found in the Philippines, which include Pinatubo, Mayon, Taal, Canlaon, and Parker. Lithologies associated with the Palawan Microcontinental Block are continent derived, whereas for the Philippine Mobile Belt, the origin of the lithological suites are varied (map drawn using data from the Mines and Geosciences Bureau–Department of Environment and Natural Resources and data from the Web site of the Philippine Institute of Volcanology and Seismology).

eastern and central portions of the country. Ophiolites generated during the Eocene to Oligocene found along the western side of the archipelago have also been mapped (see Fig. 2). The majority of these ophiolite complexes are believed to have formed in subduction-related marginal basins. They are mostly classified as supra–subduction zone ophiolites. The whole rock geochemistry and mineral chemistry of most of these ophiolite complexes manifest island arc signatures. Chert with terrigenous sediments caps the oceanic lithospheric fragments, consistent with generation in land-bounded marginal basins.

Thick accumulations of Upper Oligocene to Holocene sedimentary units make up the onshore and offshore sedimentary basins found within the archipelago (Fig. 3). These sedimentary basins consist of shallow to deep

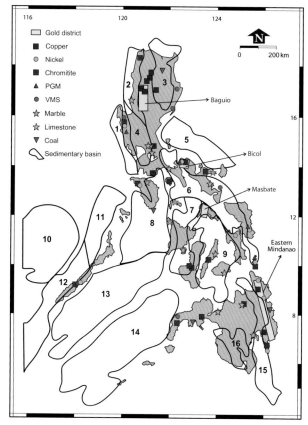

FIGURE 3 Some of the metallic and nonmetallic mineral deposits found in the archipelago include chromium, nickel, and the platinum group of metals. The major copper and gold deposits are found in northern Luzon (Baguio), southeastern Luzon (Bicol), and eastern Mindanao. There are 16 onshore and offshore sedimentary basins in the Philippine archipelago: (1) West Luzon; (2) Ilocos; (3) Cagayan; (4) Central Luzon; (5) Bicol Shelf; (6) Southeastern Luzon; (7) West Masbate–Iloilo; (8) Mindoro-Cuyo Platform; (9) Visayan; (10) Reed Bank; (11) Northwest Palawan; (12) Southwest Palawan; (13) East Palawan; (14) Sulu Sea; (15) Agusan-Davao; and (16) Cotabato (figure drawn using data from the Mines and Geosciences Bureau–Department of Environment and Natural Resources and the Department of Energy).

up of twelve large stratovolcanoes and smaller cones that include Mayon volcano (Fig. 2).

Some areas in the Philippine archipelago are floored by a metamorphic basement. Most of the metamorphic belts recognized in the Philippine archipelago are related to regional metamorphism brought about by large-scale geologic processes (e.g., collision, regional igneous activities). Some metamorphic rocks, mostly ophiolite-derived lithologies, are products of ocean floor metamorphism. The metamorphic rocks are found all throughout the country (Fig. 2).

Ophiolite complexes, representing oceanic crust–upper mantle sequences, have been mapped in different parts of the Philippine archipelago. A complete ophiolite sequence consists of (from bottom to top) peridotite, layered gabbro, massive gabbro, dike-sill complex, and volcanic rocks. Ophiolites recognized in the Philippine archipelago are mostly of Cretaceous age, especially those exposed in the marine, deltaic, and fluvial clastic and carbonate rocks. The younger Neogene section in these sedimentary basins is generally undeformed, with the older Paleogene section showing more intense deformation. Most of these basins have axes trending north-south, whereas those basins near Palawan have a northeast trend. Sedimentary basins near southeastern Luzon are characterized by a more northwesterly trend. These sedimentary basins have been explored for oil and gas accumulations.

The geologic features comprising the Philippine archipelago are defined by distinct geophysical anomalies. Low-gravity anomalies coincide with these deep-sea trenches (Fig. 4) because the trenches are filled with water or low-density sediments instead of rock or high-density materials. Onland, low-gravity anomalies generally characterize the sedimentary basins as a result of the low densities of sedimentary rocks. Distinct high-gravity anomalies are seen over areas underlain by ophiolitic and igneous units because these rocks have high densities. The Philippine Fault Zone and other major structural features are defined by linear contours and by displacements in the geophysical anomalies.

Gravity and seismic data were used to determine the thickness of the crust. Available data show that the Philippines is generally characterized by crust with a thickness varying from ~17 to 30 km. This is typical thickness for island arcs worldwide. The available data also suggest a thicker crust (greater than 30 km) in Central Luzon and in the Bicol–Panay–Central Mindanao area (Fig. 4). These regions experienced several episodes of arc magmatism, which served to thicken the crust.

In terms of paleomagnetic investigations, results show that during the Early Miocene, Mindoro and Marinduque rotated counterclockwise whereas Panay rotated clockwise (Fig. 4). The rotations are attributed to the collision of the Palawan Microcontinental Block with the Philippine Mobile Belt. Paleomagnetic data also showed that northern Luzon occupied subequatorial latitudes throughout a significant portion of the early Cenozoic history of the Philippine archipelago.

GEOLOGIC HISTORY OF A COMPOSITE TERRANE

The pre-collision history of the Palawan Microcontinental Block and the Philippine Mobile Belt is evident from its contrasting stratigraphy (Fig. 5). The Palawan Microcontinental Block consists of Paleozoic or older metamorphic rocks and a chert-clastic sequence of continental affinity. The lithologic and biostratigraphic correlation of the different rock units in the Palawan Microcontinental Block with those found in China, Russia, and Japan provides evidence for a similar origin for these units—that is, these were derived from the east Asian margin. Carbonate deposition was prevalent in the Palawan Microcontinental Block from Eocene to Oligocene, as indicated by the limestone deposits that cap the older units. On the other hand, the Philippine Mobile Belt evolved mainly as a Cenozoic arc developed on a Mesozoic to Paleogene ophiolitic or metamorphic basement. Palinspathic reconstruction for the early Cenozoic puts the Philippine Mobile Belt at the subequatorial region along the proto-Philippine Sea Plate and Indo-Australian plate margin.

The evolution of the Philippine archipelago began in the Mesozoic when a fragment of the Asian margin broke off to become the Palawan Microcontinental Block. The margin experienced lithospheric extension in the Late Cretaceous. This was followed by the formation of oceanic crust and opening of the South China Sea in the

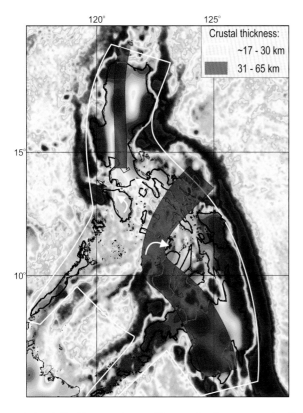

FIGURE 4 Gravity anomaly map (drawn using the map construction tool at http://www.serg.unicam.it/Gravity.htm) shows low-gravity anomalies (dark blue-purple colors) over the deep-sea trenches surrounding the Philippine archipelago. The white arrows show the counterclockwise rotation recorded for Mindoro and Marinduque and clockwise rotation for Panay during the Early Miocene. The area defined by the white solid line corresponds to crustal thickness from ~17 to 30 km. The brown-shaded regions have crustal thickness greater than 30 km.

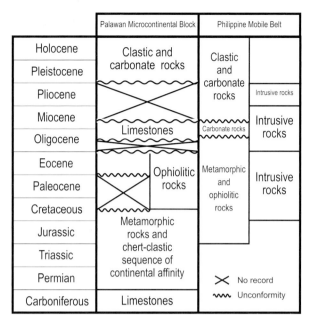

FIGURE 5 The difference between the Palawan Microcontinental Block and the Philippine Mobile Belt is evident in the contrasting lithologies observed in the two areas. The Palawan Microcontinental Block is made up of Paleozoic to Mesozoic strata of continent affinity overlain by younger carbonate rocks. The Philippine Mobile Belt is composed of metamorphic, ophiolitic, and intrusive units capped by clastic and carbonate sequences. There are three well-recognized episodes of magmatism (Cretaceous to Eocene, Oligocene to Miocene, and Pliocene), as revealed by the igneous rocks distributed in different parts of the Philippine Mobile Belt.

Late Oligocene to Early Miocene. Indentation of the oceanic leading edge of the Palawan Microcontinental Block with the Philippine Mobile Belt following its northward translation started during the Miocene, with the collision terminating by the Pliocene. The Philippine Mobile Belt, in itself, is an agglomeration of several terranes of varied origin (e.g., ophiolitic, island arc). With the collision of the Palawan Microcontinental Block with the Philippine Mobile Belt, the whole island arc system can be classified as a composite terrane.

Dioritic, alkalic, and andesitic rocks distributed within the Philippine Mobile Belt represent pulses of magmatism during the Cretaceous to Paleogene, Miocene, and Pliocene times (Fig. 5). Subduction in certain parts of the Philippine archipelago commenced during the Cretaceous. Sedimentary basins within the Philippine Mobile Belt began accumulating clastic and carbonate rocks during the Oligocene. A regional Middle Miocene unconformity, indicating rapid uplift or a transgressive event, is inferred from the sedimentary sequences in these basins. Episodes of oceanic floor generation, which began during the Cretaceous within marginal basins that have long been subducted, are evident from the ophiolites that are currently exposed on land (Fig. 2). The mechanisms responsible for the emplacement of these ophiolitic units include large-scale faulting, onramping, and subduction upwedging, among others.

PLATE INTERACTIONS AND THE CORRESPONDING MINERAL AND ENERGY RESOURCES

Late Mesozoic to Cenozoic intrusive rocks (e.g., diorite–quartz diorite, granodiorite, tonalite, monzonite, syenites) and Plio-Pleistocene volcanic rocks (e.g., dacite) (Fig. 3) that make up the island arc system are host to the gold–copper deposits in the Philippines. More than 50% of the country's gold production comes from epithermal veins (temperatures of 50 to 300 °C), and the remainder is derived mostly from porphyry copper–gold deposits. Volcanogenic massive sulfide deposits have been developed recently for the associated gold deposits. Most of the mineralization in the country can be grouped into two major episodes occurring during the Middle Miocene and Late Miocene–Pliocene periods. It should be noted that Cebu hosts the Atlas porphyry copper deposit, which is of Cretaceous age.

The ophiolitic rocks host chromitite, nickel, and in some instances, platinum and palladium (Fig. 2, 3). Together with iron and cobalt, nickel also occurs as a secondary product within laterites, which form as a result of the weathering of ultramafic rock units. As mentioned, volcanogenic massive sulfide deposits have been recognized to be viable sources not only of base metals (e.g., lead and zinc) but, more importantly, of gold (Fig. 3). Most of these deposits are hosted by the volcanic carapace units of ophiolite complexes or the overlying sedimentary sequences that have been metamorphosed (e.g., quartz–sericite schist). Potential offshore mineral resources in the Philippines comprise mainly placer deposits that include chromitite, magnetite, silica, and, to a limited extent, gold.

In terms of nonmetallic resources, Miocene and Plio-Pleistocene limestone deposits provide the raw materials for cement. The most significant deposits are found in northern and central Luzon, central Visayas, and central Mindanao (Fig. 3). Marble, the metamorphic counterpart of limestone, which is used in construction and building materials, is distributed in various parts of the archipelago. The most well-known marble deposits are found in the central Philippines (i.e., Romblon) (Fig. 2, 3).

The Philippine archipelago has 16 sedimentary basins over an area of 700,000 km^2, which have potential for hydrocarbon resources (Fig. 3). The most productive is the northwest Palawan Basin, which has, to date, produced 54 million barrels of oil, 4 million barrels of condensate, and 67 billion cubic feet of gas. Hydrocarbon deposits were also discovered in the Cagayan Valley Basin and in the Visayan Basin. Coal resources also abound in the Philippines, the most significant of which are found in Mindoro and in Zamboanga del Sur in western Mindanao (Figs. 2, 3).

GEOLOGIC HAZARDS AND ISLAND ARC DEVELOPMENT

The Philippines experiences, on the average, around 20 earthquakes daily, but most of these are hardly felt. The frequency with which earthquakes occur is due to the presence of earthquake generators such as the trenches on either side of the archipelago. Other earthquake generators are active faults, the most significant of which is the Philippine Fault Zone. The country has been devastated by 12 destructive earthquakes, with magnitudes between 6.2 and 7.9, during the period 1968–2003. This includes the earthquake that took place on August 2, 1968, with a magnitude of 7.3 and intensity VIII, reported in Casiguran, Quezon. A second devastating earthquake was the one that struck Luzon on July 16, 1990, with a magnitude of 7.9. Its epicenter was located near the town of Rizal in Nueva Ecija, but the earthquake was felt in north and central Luzon (see Fig. 1).

Aside from destructive earthquakes, the Philippines has also been ravaged by volcanic eruptions from the 22 active volcanoes throughout the country. The most active volcanoes are Mayon, Canlaon, Bulusan, Taal, Hibok-Hibok, and Pinatubo (Fig. 2). Hazards related to volcanic eruptions include lava flows, pyroclastic flows, ash fall, and lahars. The deadliest eruption of Mayon Volcano was on February 1, 1814, which was characterized as a Plinian eruption accompanied by pyroclastic flows and lahar. Another active volcano is Mount Pinatubo, which erupted on June 1991 after around 460 years of inactivity. The eruption column rose to a height of up to 30 km from the volcano's vent. Lahars resulted when typhoon Yunya combined with the ash cloud from Pinatubo and caused most of the deaths and destruction in western Luzon. With more than 5 km^3 of volcanic ash and rock fragments on the slopes of Pinatubo, lahar deposition went on for several years, filling up river channels and forming dammed lakes.

Another natural hazard with which Filipinos have to deal is mass wasting or landsliding, which is usually triggered by excessive rainfall. The most recent tragic landslide happened on February 17, 2006, and buried the town of Guinsaugon, Southern Leyte, Philippines (see Fig. 1 for location). More than 600 mm of rainfall, brought on by a La Niña event, was recorded in the weather station closest to Guinsaugon, 10 days prior to the actual landslide. The tragedy led to 154 deaths, with 973 persons still missing (and presumed dead). Considering the geographic location and geologic characteristics of the Philippines, an understanding of these natural hazards and how to mitigate the effects are critical in the development of this island arc system.

SUMMARY

The Philippine archipelago is an interesting laboratory that allows the observation of features related to past and present-day active tectonic processes. Its location at the convergence between the Sundaland–Eurasian and Philippine Sea plates has produced an array of tectonic features that shaped the evolution of this island arc system. A composite terrane made up of an aseismic Palawan Microcontinental Block and the seismically active Philippine Mobile Belt, this island arc system bears witness to various geological processes and their corresponding effects. These include, among others, collision resulting to island rotations, marginal basin closure leading to ophiolite emplacement, and large-scale strike-slip fault formation as a consequence of oblique subduction. As a result, metamorphic suites are formed, volcanic arc belts are generated, and sedimentary basins receive their fill.

The formation and evolution of the Philippine archipelago also saw the formation of both metallic and nonmetallic mineral resources. Gold, copper, nickel, chromium, platinum-group metals, volcanogenic massive sulfide, limestone, and marble are some of the more economically attractive resources being developed now. Oil and gas resource exploration activities are being undertaken in both onshore and offshore sedimentary basins of the country.

Because the country is a composite terrane in a seismically and volcanologically active region (the Pacific Ring of Fire), various geologic hazards are present. Earthquakes, volcanic eruptions, landslides, and their related hazards are some of the challenges with which people in the Philippine archipelago have to contend. These problems are exacerbated by the meteorological and hydrological hazards that hit the country. An understanding of these hazards and how to mitigate their negative effects is important to ensure the development of the Philippines.

SEE ALSO THE FOLLOWING ARTICLES

Earthquakes / Island Arcs / Landslides / Lava and Ash / Philippines, Biology

FURTHER READING

Balce, G. R., R. Y. Encina, A. Momongan, and E. Lara. 1980. Geology of the Baguio District and its implications on the tectonic development of the Luzon Central Cordillera. *Geology and Paleontology of Southeast Asia* 21: 265–288.

Dimalanta, C. B., and G. P. Yumul, Jr. 2004. Crustal thickening in an active margin setting (Philippines): the whys and the hows. *Episodes* 27: 260–264.

Queaño, K. L., J. R. Ali, J. Milsom, J. C. Aitchison, and M. Pubellier. 2007. North Luzon and the Philippine Sea Plate motion model: insights following paleomagnetic, structural and age-dating investigations. *Journal of Geophysical Research* 112: B05101, doi: 10.1029/2006JB004506.

Rangin, C. 1990. The Philippine Mobile Belt: a complex plate boundary. *Journal of Southeast Asian Earth Sciences* 6: 209–220.

Yumul, G. P. Jr., C. B. Dimalanta, R. A. Tamayo, Jr., and R. C. Maury. 2003. Collision, subduction and accretion events in the Philippines: New interpretations and implications. *The Island Arc* 12: 77–91.

PHOSPHATE ISLANDS

JAMES R. HEIN

U.S. Geological Survey, Menlo Park, California

Phosphate islands host deposits of phosphate rock (also called phosphorite) of sufficient quantity and quality to be economically mined. Most of the phosphate rock deposits were derived from bird droppings (guano). The phosphate rock was mined at various times in the past, but the only extant mine in the Pacific is on the island of Nauru in the west equatorial Pacific. The most common phosphate minerals in insular phosphate rocks are composed of calcium (Ca^{2+}) and phosphate (PO_4^{3-}), combined with charge-balancing anions such as fluoride (F^-), chloride (Cl^-), and hydroxide (OH^-). Calcium phosphates are used predominantly in agriculture as fertilizer, although they have many other industrial applications.

DISTRIBUTION OF PHOSPHATE ISLANDS

Phosphate islands are located mostly at low and middle latitudes in the Pacific Ocean, with a few exceptions such as Christmas Island in the eastern Indian Ocean and Navassa Island in the Caribbean (Fig. 1). Two types of phosphate islands exist, which are distinguished by elevation. Low islands are atolls or low-elevation (less than roughly 50 m) carbonate platforms; phosphate rock occurs on islets that rim the atoll, within the lagoon that is enclosed by the atoll reef, or on the carbonate platform. Most high islands (more than roughly 50-m elevations) that host phosphate rock deposits are carbonate platforms formed by the uplift of atolls, during which the lagoons were filled in with carbonate debris. A few insular phosphate rock deposits formed on volcanic high islands.

Phosphate deposits on the low islands were mostly small and were completely mined out during the late nineteenth century, with the exception of Starbuck Island in the central Pacific, which remained in production until 1927. Other examples of these low islands include

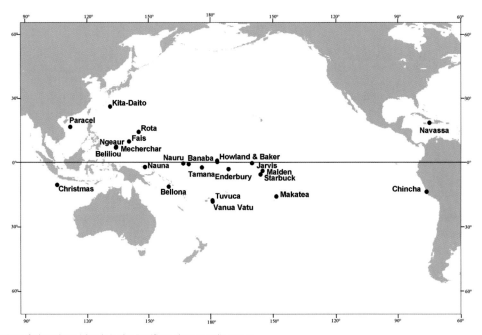

FIGURE 1 Location of phosphate islands in the Pacific and surrounding area.

Jarvis, Howland/Baker, and Malden islands (Fig. 1). The low-island phosphate deposits are the younger of the two types of deposit and include contemporary guano accumulations as well as somewhat older deposits formed by the interaction of guano with Holocene (last 10,000 years) carbonate sediments and rocks (see the section on origin of the phosphate deposits).

Phosphate deposits on some of the high islands were mined until relatively recently, with deposits on Makatea Island exhausted in 1966 and deposits on Banaba Island (formerly called Ocean Island) depleted in 1979. Since 1979, only Nauru Island in the Pacific has produced phosphate, which is exported mostly to Australia and New Zealand. Nauru, Makatea, and Banaba islands are known as the three great phosphate rock islands. Nauru had one of the highest per capita incomes in the world soon after it gained its independence in 1968 because of its phosphate rock export, but now it is in financial difficulties, and the island itself is nearly 80% uninhabitable because of the phosphate mining (Fig. 2). Mining on Nauru began in 1907, declined significantly since the 1980s, and came to a halt in 2003. With the help of Australia, the mining infrastructure has been rebuilt; exports started again near the end of 2006, and they are expected to continue until about 2010. Phosphate has also been mined on Christmas Island (Australia) in the eastern Indian Ocean for more than a hundred years, with a short hiatus in mining from 1987 to 1991. Several new mining operations have been recently proposed for Christmas Island. All these high-island deposits are generally older than the low-island deposits and consist of Neogene (24.1 to 1.8 million years ago) and Quaternary (1.8 to present) phosphate rocks commonly associated with karst topography.

Other high-island phosphate-rock deposits were exploited mainly during the first half of the twentieth century, especially during the two World Wars. Examples of those include many western Pacific island-arc-hosted deposits, such as Kita-Daito, Rota, Fais, Ngeaur (formerly spelled Angaur), Mecherchar (formerly known as Eil Malk), and Beliliou (formerly spelled Peleliu) islands (Fig. 1). A few phosphate islands exist outside the map area of Figure 1, such as Juan de Nova Island in the western Indian Ocean, where phosphate was mined from the early 1900s until 1970. Many other islands in the global ocean host small phosphate rock or guano deposits that are not now and never have been economically minable, so those islands cannot technically be classified as phosphate islands.

COMPOSITION

Guano is an organic-rich mixture composed of uric ($C_5H_4N_4O_3$), phosphoric (H_3PO_4), oxalic ($H_2C_2O_4$), and carbonic (H_2CO_3) acids as well as ammonia (NH_3), calcium, potassium, sodium, and magnesium, among other constituents. The most commonly occurring phosphate minerals derived from the interaction of guano with calcium carbonate (limestone) are composed of calcium (Ca^{2+}) and phosphate (PO_4^{3-}), combined with charge-balancing anions such as fluoride (F^-), chloride (Cl^-), and hydroxide (OH^-). These calcium phosphate minerals are predominantly varieties of apatite, including for example, carbonate apatite, carbonate hydroxyapatite, and carbonate fluorapatite. Other guano-derived phosphate minerals occur in insular phosphates but are rather rare; they include brushite, monetite, and whitlockite. Trace elements found in the apatites are derived from dissolution of the limestone or from seawater.

Modern guano is about 5% phosphorus, which is upgraded from about 1.7% phosphorus in fresh seabird droppings containing about 22% nitrogen. With time, chemical reactions (mostly leaching) decrease the nitrogen content and increase the phosphorus content to about 9–12% phosphorus. Further upgrading to as much

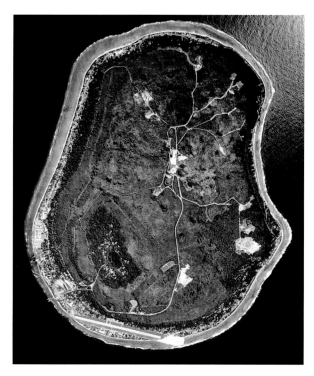

FIGURE 2 Aerial photograph of Nauru Island, which has an area of 21 square kilometers. Most of the interior of the island has been mined, with the exception of the area around the lake in the southwest quadrant. Courtesy of the U.S. Department of Energy's Atmospheric Radiation Measurement Program.

as 18% phosphorus can occur when the guano reacts with the carbonate rocks and sediments, thereby forming a phosphate-rock deposit of economic importance.

ORIGIN OF THE PHOSPHATE DEPOSITS

The seabirds most responsible for producing large guano deposits include boobies, terns, and frigates. Two characteristics are essential for the formation of guano-derived phosphate rock deposits. The first is high primary productivity in surface waters surrounding the islands. This is essential to produce the great amount of fish (mostly anchovies) required to support the vast numbers of seabirds that deposit guano on the islands, which can range to millions of birds for a single island. It is estimated that enriched guano on Chincha Island off Peru (Figure 1) was more than 45 m thick, with pits up to 125 m thick. Primary productivity in the surface waters is supported by a process called upwelling, wherein cold, nutrient-rich waters rise from deeper levels (a few hundred meters) to the sea surface, carrying the silicate, phosphate, and other nutrients required for the flourishing of plankton, which occupy the base of the food chain.

The second criterion is low rainfall, which allows for the preservation of the guano once it accumulates and also inhibits the growth of vegetation that would decrease the area available for ground-nesting birds. Further, high rainfall leaches the guano of its valuable nitrates, a highly valued component for fertilizer in addition to the phosphorus. These criteria for formation of phosphate rock—high primary productivity and low rainfall—may not describe the current conditions around some islands where older insular phosphate deposits occur, deposits that formed at a time in the past when the then extant climate was different from the modern climate.

Once guano is deposited, it interacts chemically with the carbonate rock and sediment (limestone) of the underlying reef, which requires limited rainfall. The leaching solution, strongly charged with phosphate, percolates into the limestone and may form a phosphate hardground or phosphate coating on grains within the vadose zone (zone above the water table where pores contain air as well as water). Islands commonly have a lens-shaped zone of saturated rock (where the pores are filled with water) that is freshwater above and saltwater below, with mixing at the boundary. Some scientists have proposed that large insular phosphate deposits formed within the seawater part of the lens, whereas others have suggested that they formed in the freshwater lens or in the vadose zone. Phosphate rock forms when phosphate in groundwater replaces the limestone via dissolution and immediate precipitation of apatite, and precipitation of apatite in void space in turn commonly cements carbonate grains.

It is clear that some chemical elements found in insular phosphates are not derived from guano—for example, the high fluorine and uranium contents—because those elements occur in very low quantities in guano. Rather, those elements must have been derived from the limestone via reactions with the phosphate-rich waters, from the dissolution of limestone during the formation of karst topography, or from seawater. Limestone and seawater are likely sources of these elements for different insular phosphate deposits; however, some deposits have never come in contact with seawater and must have acquired those elements from the dissolution of limestone.

PHOSPHATE COMMODITY

Phosphorus is an essential element for plant and animal nutrition. Mined phosphate rock is usually processed to produce phosphoric acid for fertilizer and elemental phosphorus for other applications. Insular phosphates with high nitrate contents are in especially high demand, because they fulfill the need for both phosphorus and nitrogen in agriculture. A third essential element needed to fertilize crops is potassium, which is also enriched in some phosphate rock deposits.

Examples of nonfertilizer applications of phosphate derivatives include food additives (such as phosphoric acid in soft drinks), detergents, herbicides, pesticides, water treatment, lubricants, matches, flares, and fireworks, to name just a few.

SEE ALSO THE FOLLOWING ARTICLES

Atolls / Island Formation / Makatea Islands / Pacific Region / Seabirds

FURTHER READING

Burnett, W. C., and A. I. N. Lee. 1980. The phosphate supply system in the Pacific Region. *GeoJournal* 4(5): 423–436.

Hein, J. R., B. R. McIntyre, and D. Z. Piper. 2005. *Marine mineral resources of Pacific Islands—a review of the exclusive economic zones of islands of U. S. Affiliation, excluding the state of Hawaii.* U. S. Geological Survey Circular 1286.

Hutchinson, G. E. 1950. The biogeochemistry of vertebrate excretion. *American Museum of Natural History Bulletin* 96.

Piper, D. Z., B. Loebner, and P. Aharon. 1990. Physical and chemical properties of the phosphate deposit on Nauru, Western Equatorial Pacific Ocean, in *Phosphate deposits of the world,* Vol. 3. W. C. Burnett and S. R. Riggs, eds. Cambridge, UK: Cambridge University Press, 177–194.

Stoddart, D. R., and T. P. Scoffin. 1983. Phosphate rock on coral reef islands, in *Chemical sediments and geomorphology: precipitates and residua in the near-surface environment.* A. S. Goudie and H. Pye, eds. London: Academic Press, 369–400.

PIGS AND GOATS

ELIZABETH MATISOO-SMITH
University of Auckland, New Zealand

Pigs (*Sus scrofa*) and goats (*Capra hircus*) were two important domesticated animals taken to islands by early agriculturalists. Later introductions of these animals by European explorers and sailors, who often left them on islands as provisions for passing ships or shipwreck survivors, further extended their island distributions. Unfortunately, the impact of these animals on island ecosystems has been significant. In recent years eradication measures have been undertaken to remove these invasive species and restore the native island habitats they destroyed.

ORIGINS, INITIAL DOMESTICATION, AND EARLY ISLAND INTRODUCTIONS

Pigs and goats belong to the order Artiodactyla. Both species are omnivores and particularly adaptable, which no doubt was one factor in their being two of the earliest domesticated forms of livestock. This adaptability would have also been a significant factor leading to their continued survival when introduced to many marginal or isolated islands around the world. Wild boar and goat have been important components of the human diet for tens of thousands of years, and recent molecular evidence suggests that both species were probably domesticated multiple times independently in several different locations.

Archaeological and genetic evidence on goats suggest at least three major domestication events beginning around 10,500 years ago. These domestication events have been linked to the Near East and Indus regions (the source populations of later European and African lineages) and to East Asia. The fact that they provide not only meat and hides but milk and fiber in a relatively compact package that was easy to transport by boat made goats ideal animals to take on early voyages to the various islands in the Mediterranean region. Archaeological remains of domesticated goats and sheep appear from as early as 8000 years ago on Crete and Cyprus and in later Neolithic and Bronze Age settlements on other large islands in the Mediterranean and Baltic regions.

The origin of *Sus scrofa* appears to be western island Southeast Asia. From there, populations dispersed into the Indian subcontinent and later radiated into East Asia, Eurasia, and Western Europe. Domestication appears to have occurred independently in all of these areas beginning around 9000 years ago. Domesticated pigs were also taken to the major islands of Europe with Neolithic farmers. Despite earlier suggestions for domestication of wild boar on some islands, such as Sardinia, genetic evidence suggests that all island populations studied are descendants of pigs domesticated on the mainland.

The East Asian domesticated pigs eventually were transported to the islands of the Pacific by prehistoric peoples. There is much debate about the date of pig introductions to New Guinea, which is also a center of early plant domestication. Some have argued that pigs may have been introduced to New Guinea during this early phase of agricultural development as early as 10,000 years ago. Both translocation and domestication of some of the other Southeast Asian *Sus* species, such as *Sus celebensis*, have also occurred in island Southeast Asia.

The spread of *Sus scrofa* throughout the rest of the Pacific clearly did not occur until around 3300 years ago, when they were transported as part of the Lapita cultural complex and introduced to the islands of the Bismarck Archipelago and the Solomon Islands and out into Remote Oceania as far as Samoa and Tonga. From there they were taken to most of the Polynesian islands with the initial colonists. Pigs were not, however, successfully introduced to New Zealand or Easter Island by Polynesian colonists. Interestingly, despite their important use either as protein or for the long-term storage of surplus carbohydrates, pig populations were not always maintained on Pacific islands. Archaeological evidence suggests that on some small Pacific islands, particularly atolls, pigs were often extirpated, presumably on account of their impact on and competition for fragile and limited resources. This may explain the lack of pigs in Micronesia at European contact.

Little is known for sure about the physical or behavioral characteristics of the first Pacific pigs. Early European accounts generally describe them as small, short, and dark with sharp backs, stiff bristles, and long snouts. These are indeed the characteristics of the wild pigs found in New Zealand today, often referred to as "Captain Cookers," which are believed to be the descendents of the first pigs introduced to New Zealand by Captain James Cook. Whether these were "native" Pacific pigs he picked up in the islands he visited before arriving in New Zealand, European pigs he carried with him, or both is not clear.

EUROPEAN INTRODUCTIONS

European exploration and colonialism beginning in the sixteenth and seventeenth centuries resulted in the introduction of many domesticated plants and animals to islands around the world. Pigs and goats were among the most successful introductions. Sailors, traders, whalers, and sealers would often release pigs and goats on islands, often uninhabited, so that they would establish natural feral populations, providing a guaranteed supply of fresh meat for future passing voyages and castaways. The first recorded introduction of goats to an island was in the Madeiras in 1458. When the Spanish arrived in the Canary Islands in the early 1500s, they recorded that goats and pigs were already present. These were most likely transported by the Guanches, the aboriginal peoples of Tenerife, who settled the islands from North Africa.

During his second and third Pacific voyages of 1773 and 1777, Captain James Cook released pigs and goats to most of the islands he visited, including Hawaii and New Zealand. The pigs brought by the Europeans often replaced the Polynesian pig populations on islands where they were present, as the European breeds were generally larger and fatter than the original introductions. La Pérouse introduced both pigs and goats to Easter Island in 1786. Trade ships from the East Indies and the Philippines were the main sources of pig and goat introductions to Micronesia. In 1790, two ships from the British East India Company introduced eight she-goats, two rams, five sows, and two boars to Palau. Missionaries introduced pigs and goats to many Pacific islands throughout the nineteenth century. Twentieth-century introductions of both pigs and goats to islands for game hunting, or by fishermen as in the case of Pinta in the Galápagos, resulted in the modern distribution, making these two species among the most broadly distributed mammals, after rats, worldwide.

ECOLOGICAL IMPACTS OF PIGS AND GOATS

Given that most oceanic island ecosystems have evolved in the absence of grazing and browsing ungulates, the introduction of both pigs and goats have a major negative impact on island floras. They are both considered to be in the top 100 of the world's worst invading species.

The grazing habits of goats result in significant loss of native plant species, particularly woody plants and shrubs. They remove leaves and young shoots and strip bark, resulting in plant death. This, in combination with trampling and other secondary impacts, leads to further devastation such as soil erosion and loss of dependent faunas. In some locations goats have been identified as the primary cause of island plant extinctions. The case of the introduction of three goats to Pinta Island in 1959 provides a dramatic example of the damage that can be done in a remarkably short period of time. Within 10 years of their initial introduction, the two females and one male released on the 60-km^2 island resulted in a population of 5000 to 10,000 goats. This resulted in the loss of four endemic shrubs from the lowland areas and the dramatically reduced distribution of five other species. Unfortunately, other islands to which goats have been introduced have had similar stories.

Pigs have an equally devastating impact on island ecosystems. Pig rooting not only digs up plants but significantly alters soil content, causing oxidation and leaching of key minerals. The destruction of leaf litter habitats by pigs often results in the loss of several native vertebrate and invertebrate species. In Hawaii, feral pigs destroy tree ferns, a favorite food, along with their associated epiphytes. As a secondary impact on native flora and fauna, pigs, like goats, often disperse the seeds of invasive species such as the strawberry guava, which has been particularly devastating in Hawaii. Pigs are known to eat the eggs and young of numerous ground-nesting birds and amphibians. They also feed on exposed reefs at low tide on many Pacific islands (Fig. 1), causing damage to the intertidal zone. Feral pigs transmit a number of diseases including pseudorabies, leptospirosis, and Japanese encephalitis, and they are known for carrying parasites that can be passed to humans and other animals. Pigs are known to be strong swimmers and thus can self-disperse across significant water gaps. The naturalist A. R. Wallace, in fact, described seeing pigs swimming from Singapore to the Malacca Peninsula.

FIGURE 1 Pacific pig feeding on a reef in Tonga. Photograph by Sean P. Connaughton.

ERADICATION AND POPULATION CONTROL

Given the devastating impact of goats and pigs on island ecosystems, not surprisingly, eradication and population control are major issues for many governments and conservation groups. Invasive mammals can be removed from islands using a range of approaches including hunting, poisoning, trapping, or biocontrol. Previous eradication programs for both feral goats and pigs have shown that the use of a combination of approaches is most effective, particularly if a sustained effort and population monitoring is maintained. In island locations where both goat and pig removal is being attempted, it has been shown that the eradication of pigs prior to that of goats dramatically reduces the costs of removal. If goats are removed before pigs, or if removal attempts are simultaneous, the rapid recovery of dense foliage in the absence of goats makes pig removal much more difficult and time consuming.

In New Zealand, pig eradication from small islands began in the mid-twentieth century and employed hunting as the primary means of removal; however, the combination of hunting and poisoning is now generally used on larger islands. The extreme mobility of both pigs and goats can make eradication efforts difficult and expensive, particularly on large islands. Once animals are removed from one area, they are quickly replaced by new populations. In Hawaii, the use of fencing has aided pig control in areas such as the Volcanoes National Park.

Goats have been successfully removed from over 120 islands ranging in size from 1 hectare (Marielas Sur, in the Galapagos) to 132,867 hectares (Flinders Island, Australia). Hunting is the most common method used for goat eradications, involving game hunting, hunting with dogs, and shooting from helicopters. The gregarious nature of goats has also lead to a particularly successful hunting aid: the use of Judas goats. Judas goats are goats that are captured, fitted with radio collars, and released. These goats then search out other goats, thus identifying their location, so that hunters can then locate populations that might otherwise evade discovery. Once the identified group has been located and removed, the Judas goat is rereleased to seek out new groups. The use of Judas goats has been most successfully employed in areas where goats may only be found at low densities (for example, after major eradication programs using other methods have reduced an originally dense population) or where topography makes tracking and access difficult.

The successful removal of goats and pigs from island ecosystems has been shown to have a dramatic and rapid effect on vegetation. Unfortunately, non-native plant species are often the most rapidly recovering plants after removal of pigs and goats. The use of pig and goat exclosures to determine how the various plant communities will react to the loss of these species is recommended prior to eradication so that an environmental management scheme can be developed. Studies in Hawaii have shown that in terms of native plant recovery, lowland grasslands are perhaps the least likely to recover once pigs and goats have been removed. Native woody species have been shown to recover relatively quickly in lowland areas after removal of goats in particular. The more upland areas and rainforests recover differently depending on the degree of goat and pig impact. Where the impact has not been great because of low density or short period of pig or goat presence, native plants recover relatively well. However, where the impact has been severe, recovery even 6 to 8 years after removal has been negligible.

Eradication of pigs or goats on large islands or on islands where reintroduction is likely must include some degree of community consultation, outreach, and education. Both pigs and goats have recreational and economic uses, and local communities may react negatively to eradication attempts. Unsuccessful eradications have occurred on several large islands such as Great Barrier Island in New Zealand and Lord Howe Island in Australia because of lack of local support or other political issues. Successful eradication of goats and pigs from islands requires the cooperation of biologists, ecologists, anthropologists or sociologists, educators, and local communities and government agencies.

SEE ALSO THE FOLLOWING ARTICLES

Biological Control / Hawaiian Islands, Biology / Human Impacts, Pre-European / Introduced Species / Invasion Biology / New Zealand, Biology

FURTHER READING

Campbell, K., and C. J. Donolan. 2005. Feral goat eradication on islands. *Conservation Biology* 19: 1362–1374.

Giovas, C. M. 2006. No Pig Atoll: island biogeography and the extirpation of a Polynesian domesticate. *Asian Perspectives* 45: 69–95.

Hide, R. 2003. *Pig husbandry in New Guinea: a literature review and bibliography.* ACIAR Monograph No. 108. Canberra: Australian Centre for International Agricultural Research.

Larson, G., K. Dobney, U. Albarella, M. Fang, E. Matisoo-Smith, J. Robins, S. Lowden, H. Finlayson, T. Brand, E. Willerslev, P. Rowley-Conwy, L. Andersson, and A. Cooper. 2005. Worldwide phylogeography of wild boar reveals multiple centres of pig domestication. *Science* 307: 1618–1621.

Luikart, G., L. Gielly, L. Excoffier, J.-D. Vigne, J. Bouvet, and P. Taberlet. 2001. Multiple maternal origins and weak phylogeographic structure in domestic goats. *Proceedings of the National Academy of Sciences of the United States of America* 98: 5927–5932.

PITCAIRN

NAOMI KINGSTON
National Parks and Wildlife Service, Dublin, Ireland

NOELEEN SMYTH
National Botanic Gardens, Dublin, Ireland

Pitcairn Island is one of four islands in the Pitcairn group, the most easterly island group in Polynesia: Pitcairn, a relatively young, high volcanic island; Henderson, an atoll uplifted by the eruption of Pitcairn; and two non-uplifted atolls, Ducie and Oeno. The group is extremely remote, separated from both New Zealand and South America by over 4500 km; from Easter Island, the nearest neighbor to the east, by 1570 km; and from the Gambier Islands to the west by 450 km. The isolation of Pitcairn makes it of equal interest to those studying the island's biota and its origins, those studying its geology, and also those studying the unique culture that has developed since its most recent phase of human settlement in 1790.

GEOGRAPHIC SETTING

Pitcairn Island (25° 04′ S, 130° 06′ W) is very small, being only 4 × 2 km, and with the highest point at 347 m. As the island is of volcanic origin, its terrain is very rugged, with the soil being derived from volcanic ashes, and the underlying rock types formed of consolidated ash. Four different volcanic episodes account for varying ash substrates across the island. The island is not protected by a fringing reef, so the coast is surrounded by cliffs with few small-boulder beaches. On the south of the island these cliffs and steep slopes reach almost to the highest point at 347 m.

From the highest point in the southwest across to the northeast, the terrain is more gradually sloping, and the cliffs below Adamstown are much lower (~50 m, and lower again at St. Paul's Point). The only area with low coastal cliffs (~30 m) on the south of the island is at Tautama. A flat area known as Aute Valley covers a considerable portion of the southeast corner of the island, at an altitude of approximately 180 m. Tedside is used as a general name for the area to the west and northwest of Big Ridge and Garnets Ridge, although it is actually composed of several small valleys. It slopes from High Point right down to a beach, which is also occasionally used as a harbor and may have been the location of the main Polynesian occupation.

Overall only about 10% of the island is flat land, an area of about 550 ha, the remainder being at a 20–45° slope, and much steeper in some of the remoter valleys. Another 160 ha are on slopes suitable for some cultivation, although the plots on such slopes are subject to erosion.

CLIMATE

The climate of Pitcairn is subtropical, with mean annual rainfall of about 1716 mm, but with considerable annual variation. Mean temperatures range in summer from 17 to 28 °C, and in winter from 13 to 23 °C, with winter being wetter and windier. There are no permanent springs on the island, but during wet periods streams run down the centers of several valleys. As for the rest of this region of the south Pacific, the trade winds blow from the southeast, and although Pitcairn is affected by complex climatic patterns that affect the whole of the region, it is out of the line of cyclones and only occasionally hit by them.

GEOLOGY AND SOILS

Pitcairn formed comparatively recently (~0.75–1 million years ago), and basaltic lava dating shows that the high islands of the Gambier group (5.2–7.2 million years) and the Tuamotu island of Mururoa (6.5–8.4 million years), to the west, formed at the same hotspot as Pitcairn. Farther east from Pitcairn along the same hotspot alignment there are over 20 underwater volcanic edifices (seamounts), the shallowest of which is only 60 m below sea level. The island is migrating along this alignment at a rate of 12.7 ± 5.5 cm per year. The young geological age of Pitcairn may account for the lack of fringing reef, a feature commonly associated with volcanic islands in southeast Pacific.

A detailed soil survey of the island, carried out in 1958, identified three main soil suites:

- Pulau suite, derived from tuffs, ash, and tuff agglomerates
- Adamstown suite, derived from basaltic lavas
- Taro Ground suite, derived from iron-rich basaltic lavas and agglomerates

All of the soils are deep and fertile, but with major erosion in parts.

FLORA, FAUNA, AND HABITATS

Pitcairn has a variety of habitats, which consist of rugged sea cliffs on the north and south coasts, scrub and eroding slopes, ridge vegetation, invaded *Syzygium jambos* forest, *Homalium taypau* (taypau) forest, *Pandanus tectorius* (thatch) forest, and *Meterosideros collina* (rata) forest. The flora consists of 81 native vascular plant species (including 11 endemic species) and a further 250 introduced species. Non-native species are found right across the island, but native species still dominate in the remoter valleys on the south of the island. Most of the plant communities contain high numbers of non-native species, and even in areas where there are a high number of native species, the dominant species is often an introduced taxon. Analyses have found that 63% of the native flora is threatened on Pitcairn, while 22% is globally threatened. The combination of taxa found in the Pitcairn group are most closely related to the taxa found in the Austral group of islands to the west of Pitcairn, which were in turn derived from the southeast Polynesian biogeographic region of the Pacific.

The fauna of Pitcairn is less well known than the flora. The avifauna falls into four categories: endemic breeding landbirds, migrant land- and shorebirds, the seabirds visiting the waters of the region, and the breeding seabirds. The islands are categorized as a high-priority endemic bird area by BirdLife International, with internationally significant populations of seabirds. Bird diversity is low but highly specialized; the only landbird found on Pitcairn Island is the endemic Pitcairn Reed-Warbler. The land snail fauna is also of immense interest with 16 species, eight of which are endemic. Little is known about other invertebrate groups, but they would undoubtedly show similar levels of endemicity.

The fish of the Pitcairn Islands show a low degree of endemicity, and green turtles and hawksbill turtles occur around the islands. Marine mammals are in need of assessment within the Pitcairn group, and it is probable that there are many cetacean species occurring in the surrounding waters.

SETTLEMENT AND GOVERNANCE

Pitcairn was settled by Polynesians from about the tenth century, at a time when world climates were warmer and calmer seas allowed easy exploration of the Pacific. There was certainly trade with other Polynesian islands, as timber and tools from Pitcairn have been found elsewhere in Polynesia, including Easter Island. Similarly, pearl shells from the Tuamotus and Gambier Islands have been found on Pitcairn Island. Contact was lost between Pitcairn and the more westerly Polynesian islands in about the fourteenth century, during a period of climatic instability, and the population probably died out soon after this. Settlement may have never been permanent, but rather the island may have been used as a stop-off point for long voyages, or as a quarry for obsidian.

Pitcairn Island was rediscovered on July 2, 1767, by Carteret for Britain, but was charted incorrectly. This is probably the reason why the *Bounty* mutineers sought out the island as their new home and hideaway and relocated to it in 1790. The story of the mutiny on the HMS *Bounty* is one of the most famous in seafaring lore. The *Bounty* was charged by King George III to sail to Tahiti and collect breadfruit plants (*Artocarpus altilis*), which were to be brought to the West Indies as a source of food for slaves. Having spent longer then was originally planned in Tahiti, the crew were not keen to go back to Britain, and they mutinied in 1789. Captain William Bligh and his supporters were cast adrift, and Fletcher Christian, reputed to be the ringleader of the mutineers, returned with the rest of the men to Tahiti. At Tahiti they took women and some native men and went to Tubuai to set up a colony. This was unsuccessful, and so, after only a few months they went to sea again, aiming for the Solomon Islands, but instead finding Pitcairn Island.

The initial settlers counted of nine mutineers, six Polynesian men, and 13 Polynesian women. Difficulties soon developed between the men on the island with disputes over land and women, the result of which was a period of unrest and murder. This meant that just 10 years after landing on the island only one man, John Adams, was left with 10 women and 23 children who had been born on the island. During this time an interesting culture had been developing on the island, with a mixture of Polynesian and European influences.

The Pitcairnese creole language, still in use, had begun to develop, although it was later to incorporate words learnt from American whalers and others who settled on the island. Food and cooking methods were predominantly Polynesian, while the homes were built like European dwellings, but from local timbers such as taypau (*Homalium taypau*), miro (*Thespesia populnea*), and huliandah (*Cerbera manghas*). Tools were similarly a mixture of those taken from the *Bounty* ship and from Polynesia by the women. Traditional Polynesian tapa cloth was made, as the source plant, aute (*Broussonettia papyrifera*), survived following the earlier period of Polynesian settlement. Fishing was performed from the rocks until in 1795 the first European style canoe was built.

The curious place names that have been given to all parts of the island give an insight into the history and culture of the Pitcairnese people (Fig. 1).

The islanders have always been astutely environmentally aware, as their lives have always depended on the fine balance between population size and resource availability. The Pitcairn laws through the nineteenth century reflect their concerns about environmental sustainability. A report in the 1850s noted regulations about cutting timber for enclosures, highlighting that in less than 100 years of settlement, timber resources were becoming scarce. Their complete dependence on island resources for food was apparent when a law stated "no coconuts were to be taken from T'otherside" (now Tedside) unless the collectors were accompanied by someone in authority. Erosion and drought problems were becoming evident, attributed to the loss of the island trees. In 1856, because of

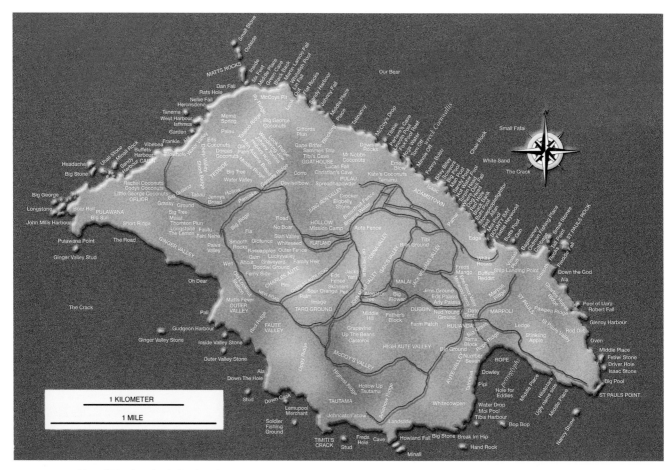

FIGURE 1 Map of Pitcairn Island, showing the intriguing place-names. Image courtesy of Pitcairn Island Administration, modified by Mark Vity.

overpopulation, the 194 islanders were evacuated to colonize Norfolk Island, by then abandoned as a penal colony.

However, in 1859, two families returned, followed in 1864 by another four families, increasing the island's population to 43. It was from this group that the extant population of Pitcairn is now largely descended. Even today close links are kept between the families on Norfolk and Pitcairn Island. In 1887 the islanders converted to Seventh-Day Adventism, a faith they retain to this day.

From the Pitcairners' first contact with the British Admiralty in 1814, the island has been under British Rule as an Overseas Dependent Territory. The Wellington-based British High Commissioner to New Zealand holds the office of Governor of the Pitcairn Islands, an appointment made by Her Majesty. The Governor holds formal powers "to make laws for the peace, order and good government of the islands," and all the laws are styled "Ordinances." Administration is via New Zealand through the Island Commissioner based at the Pitcairn Islands Office in Auckland. The Pitcairn Island Council is responsible for the local government and administration of internal affairs within the group, and the Island Council comprises the Island Mayor (elected every three years), the Island Secretary, Chairman of Internal Affairs Committee, four officers (elected annually), and two advisers, one appointed by the governor and one by the elected members.

CULTURE

The population of Pitcairn declined through the 1970s, 1980s and 1990s, and today the island population is about 50. However, over one-third of the native born Pitcairners live away from the island, with most of them resident in New Zealand. Contact with the outside world is via telephone and e-mail. Many islanders are also ham-radio enthusiasts. Until recently the small population of Pitcairn lived a subsistence existence, with extensive areas of the island in cultivation for a variety of food crops. Breadfruit is a particularly important carbohydrate food stable, although, in contrast to other Polynesian islands, the breadfruit trees on Pitcairn are generally not maintained or pruned, and the small fruits are often shot down from the trees with rifles.

Currently the main employment on Pitcairn is in local government and community services, such as conservation officer and postmaster. Supplementary income is provided by the sale of woodcarvings termed "curios" to passing cruise ships and, to a lesser extent, by mail order. The social and economic status of Pitcairn Island is one area in which dramatic changes are forecast for the coming years. In the past the islands' main income was obtained from philately. However, during the last ten years, income from this activity has been reduced by over 80%, and the islands entered into budgetary aid in 2004. Business plans for the island are currently being developed; the aim is to create a sustainable economy and self-sufficiency. The developments include a new trade link and memorandum of understanding between the Pitcairn Islands and French Polynesia, the installation of wind turbines to harness electricity, a new breakwater/harbor to encourage cruise ships to visit Pitcairn as well as to stop at the other islands in the group, and the development of international markets for the sale of local produce (carvings and honey) to provide a boost for the island's economy and development.

Supplies arrive by ship from New Zealand every 1–3 months, and a small co-op shop keeps surplus goods. Access to the island is extremely difficult, because large ships cannot dock in the small harbor. All supplies to the island currently have to be offloaded into longboats at sea and transported into Bounty Bay (Fig. 2). There is no airstrip, and plans to construct a runway on the flat part of the island have for now been shelved and a fast catamaran will be obtained, which will allow a quicker sea link to Pitcairn via Gambier Islands.

CONSERVATION

The main threats affecting the island biota are posed by habitat clearance, spread of invasive species, small population sizes or restricted distributions, loss of genetic diversity within species, erosion, and exploitation. The extinction of fruit-eating birds from the island is also preventing the successful dispersal and germination of many native plant species.

Conservation management programs are underway to address these threats through species-specific recovery plans and control of invasive species. Cleared areas are being replanted with native and economically important species, resulting in an environmentally and economically sustainable resource. Planting, in turn, controls erosion and reduces unchecked exploitation of native species. In addition, three potential reserve areas have been identified (at Tautama, Big Ridge, and Down Rope), which would set aside areas for nature conservation and prevent development in these areas.

An environmental management plan for Pitcairn has been produced; it sets out a series of sustainable solutions that will allow the island economy and infrastructure to develop for future generations, while maintaining the integrity of the island's unique ecosystems. The success of future conservation measures is reliant on the involvement of the local community, and so the plan has been drawn up in consultation with them and mindful of their current and future interests.

SEE ALSO THE FOLLOWING ARTICLES

Easter Island / Exploration and Discovery / Lava and Ash / Peopling the Pacific / Sustainability

FURTHER READING

Benton, T., and T. Spencer, eds. 1995. *The Pitcairn Islands: biogeography, ecology and prehistory.* London: Academic Press.
Göthesson, L-Å. 1997. *Plants of the Pitcairn Islands including local names and uses.* Sydney: University of New South Wales.
Kallgard, A. 1991. *Fut yoli noo bin laane aklen? a Pitcairn Island word list.* Sweden: University of Goteborg.
Kingston, N., and S. Waldren. 2003. The plant communities and environmental gradients of Pitcairn Island: the significance of invasive species and the need for conservation management. *Annals of Botany* 92(1): 31–40.
Kingston, N., and S. Waldren. 2004. A conservation appraisal of the rare and endemic vascular plants of Pitcairn Island. *Biodiversity and Conservation* 14: 781–800.
Nicolson, R. B. 1966. *The Pitcairners.* London: Angus & Robertson Ltd.
Paulay, G. 1989. Marine invertebrates of the Pitcairn Islands: species composition and biogeography of corals, molluscs and echinoderms. *Atoll Research Bulletin* 326: 1–28.

FIGURE 2 A view of the main landing point at Bounty Bay. Photograph by Noeleen Smyth.

PLANT DISEASE

ULLA CARLSSON-GRANÉR, LARS ERICSON, AND BARBARA E. GILES

Umeå University, Sweden

Diseases can increase mortality, decrease reproduction and growth of plants, and ultimately influence the sizes and genetic structures of populations and the species composition in plant communities. Plant populations situated on small and distant islands may more easily escape diseases than those on the mainland, where host and pathogen populations lie in close proximity. However, if pathogens spread to island populations that have previously evolved in absence of diseases, their effects may be severe. Studies of patterns of disease in insular systems have shown that the ages and sizes of plant populations and the distance between islands affect disease spread in archipelagoes.

DISEASES IN PLANTS

Plants serve as hosts to a wide range of parasites. These organisms, which can be fungi, oomycetes, nematodes, protozoa, bacteria, or viruses, live in or on their plant hosts, from which they derive their nutrients. If a parasite causes disease in its host plant and reduces the host's fitness, we call the parasite a plant pathogen. The ways in which pathogens affect host fitness varies widely among different groups of parasites. For example, necrotic generalist pathogens among oomycetes or fungi that can also live on dead host tissue often result in extensive mortality, particularly among weakened plants suffering from waterlogging, oxygen deficit, or other stresses. Biotrophic pathogens that infect flowers (e.g., anther smut fungi, Basidiomycota) and developing fruits also have a strong effect on host fitness by preventing seed production, but these pathogens rarely affect host survival. Foliar biotrophic pathogens, including fungi such as rust, attack plant leaves and may reduce photosynthesis, ultimately decreasing resources available for growth, reproduction, and survival. However, the strength of their negative effects on plant individuals vary substantially between years (e.g., with extreme weather conditions).

Due to growth of international trade and human travel around the globe, the spread of diseases that result in plant epidemics has increased over the years. This may pose a particular threat to island plant communities that may have evolved in the absence of disease or in isolation from particular diseases. Today, we have records of a number of plant pathogens that have been introduced and spread on islands. For example, an average of more than one new rust species has been found per year on New Zealand. These rusts have often been of northern temperate origin, but introductions from Australia also seem common. Increased numbers of introduced parasites causing disease have been noted on crops grown on the Maldive Islands. However, plant pathogens could also be used for controlling exotic invading species on islands. The invasion of ecosystems by alien species is actually one of the most important sources of biodiversity loss on islands.

Diseases affect plant populations and communities and spread between plants and populations. Because the effects of parasitic fungi and oomycetes are better known than the effects of other groups of parasites, all examples in this article are taken from these kinds of systems. The spatial structure of islands (i.e., the distances between populations and the sizes of host populations) is predicted to affect the dynamics of host-pathogen interactions. Long-term studies of diseases in an archipelago in Sweden have presented interesting results.

EFFECTS OF DISEASES ON POPULATIONS AND COMMUNITIES OF PLANTS

When a large proportion of plants are diseased and the fitness of infected individuals is reduced, the growth rate of populations may decrease, leading to reductions in the sizes and densities of plant populations. This can ultimately change the structure of plant communities, especially when large dominant keystone species are attacked.

The potential numerical effects that pathogens might have on their specific hosts have been used to control plants that become invasive when introduced to new geographical areas. Such programs have been successful in controlling some exotic invasive plants on the islands of Hawaii, e.g., introduction of the host-specific fungal pathogen *Septoria passiflorae* (asexual Ascomycota) to Hawaii has resulted in an up to 90% decrease in its host, the invasive banana poka (*Passiflora tarminianan*). However, there are also examples of limited success in using fungal pathogens as control agents. For example, release of the rust fungus *Puccinia chondrillina* to control the Mediterranean invasive rush skeletonweed (*Chondrillina juncea*) in Australia reduced the abundance of the susceptible and most widespread clone of the weed, whereas the two clones that were resistant to the pathogen increased in abundance. Introduction of the pathogen resulted in selection and increased the frequencies of resistance in the skeletonweed population, leading to decreased effectiveness of biocontrol. Similar evolutionary changes might explain the varied success of other biological control programs.

Evolutionary changes in host resistance can also mediate reciprocal changes in infectivity and aggressiveness in the pathogen population through frequency-dependent selection. In populations with a high frequency of resistant plants, more infective fungal isolates may be favored, and the pathogenicity of the pathogen population may increase. This, in turn, selects for rare resistant host genotypes, which then increase in frequency until a new pathogen strain manages to infect these plants, and so on.

The most obvious large-scale numerical effects of diseases have been seen when pathogens migrate over large distances to new geographical regions (usually by the means of humans) and encounter new hosts for the first time. On the Seychelles, for example, a new fungal disease has caused high death rates of the native takamaka tree (*Calophyllum inophyllum*). Spores of the fungus appear to be spread by a native bark beetle, but the origin of this pathogen is still unclear. It may represent an introduced pathogen or a pathogen that has evolved to infect a new host in its home range. That hybridization between two pathogen strains may lead to novel pathogenicity has been observed on several occasions. Clearly, knowledge of the factors that allow a parasite to establish within a host population is essential for understanding and controlling diseases on islands and other natural ecosystems.

DISPERSAL, INFECTION, AND ESTABLISHMENT OF DISEASES

To be able to colonize and parasitize a plant, a pathogen has to solve two problems. First, it must "find" a host plant; then it must infect that plant and grow and reproduce on or in it. Some parasites are directly transmitted by contact between susceptible hosts (e.g., from adult plants to seed offspring or between ramets of the plant). Others spread by means of wind, water, or animal vectors or through the soil. Although most pathogen spores usually land relatively close to the source of inoculum, some spores of wind-dispersed pathogens such as rusts can travel for great distances. Vector-transmitted pathogens, such as anther smut on caryophyllaceous hosts, which is spread by pollinators of the plant, spread more locally. The spread of soil-borne diseases such as *Phytophthora* spp. (Oomycota) is even more local, and in these systems the host usually disperses over longer distances than the pathogen.

Once a parasite has dispersed to a new plant, it must be able to overcome host defenses to infect, grow, and reproduce. First of all, at least some individuals of a potential host species must be susceptible toward the particular pathogen isolate. Plants may, however, actively defend themselves biochemically by producing secondary metabolites that are toxic or otherwise inhibitory to pathogen growth. Host resistance can also be based on passive physical factors that allow the plants to avoid and escape the disease. Even when a pathogen manages to spread, overcome a host plant's defenses, and cause an infection, the disease will not necessarily spread through the plant population. Theory predicts that the basic reproductive rate of the pathogen, R_o (the average number of new infections produced from a single infective individual), must exceed 1 to allow a pathogen to invade. It is commonly assumed that host density or population size must reach a critical threshold for a specific pathogen to persist. From agricultural situations, where plants are cultivated in homogenous stands, it is well known that increasing the density of plants can increase disease spread. In natural communities, plant populations vary from being dense, large, and continuous to being highly subdivided in small patches. Each plant population may in turn be distinctly isolated from other populations or be a part of a network where populations exchange seeds and pollen. This variation in spatial structuring of host populations can affect disease spread.

DISEASE SPREAD IN SPATIALLY STRUCTURED SYSTEMS

Theory predicts that increasing subdivision of host populations into smaller, more isolated patches slows the dispersal rate and increases the probability of local extinction of pathogens. It has been proposed that similar mechanisms could result in escape from diseases for plant species that are introduced to new, distant areas. Without diseases, these species may experience a demographic release with high recruitment, growth rates, or survival that may make them invasive in the new area (i.e., the enemy release hypothesis). When host patches are closer and more connected, pathogens may disperse frequently among populations, leading to long-term persistence of pathogen populations.

In insular systems, plant populations on isolated islands are predicted to escape diseases more easily than populations that are closer to other populations. However, small, isolated populations on islands may be inbred with low diversity in resistance genes, which may limit their evolutionary response if diseases are introduced, for example, by human trade. As shown in the case of the introduced chestnut blight (*Cryphonectria parasitica*, Ascomycota) in North America, where all plants are more or less susceptible, a new pathogen can rapidly spread. When particular host species become rare, the pathogen may also evolve ways for persisting in an isolated situation. For example, generalist pathogen strains that are able to infect more than one species may be favored (given that these traits are genetically variable in the pathogen), which can increase the overall impact of the disease in the community.

Disease dispersal and pathogen persistence are predicted to be higher in plant populations on islands found in archipelagoes. In these systems, interpopulation dispersal of both host and pathogen may occur often enough to rescue the pathogen from regional extinction, although local extinctions may occur. Theoretical studies have also shown that genetic variation in host and pathogen is easiest to maintain in such spatially structured situations. This can further increase pathogen persistence, because a few susceptible plants and a few infective pathogens are likely to occur in some sites in the system. At the same time, some host populations may escape the disease and the overall impact of a pathogen may be smaller than in large continuous populations or local isolated populations.

FIGURE 1 Islands of different ages and phases of primary succession in the Skeppsviks Archipelago. (A) A new island just emerged above sea level. Terrestrial plants have not yet colonized the islands. (B) A young small, open island (~50 years) with grasses and tall herbs (*Valeriana sambucifolia* and *Filipendula ulmaria*) growing in the middle part. Seedling establishment of the first tree, *Alnus incana*, occurs. Inundated under high-water periods. (C) A ~100-year-old island. The central part of the island has scattered *A. incana* and *Sorbus aucuparia* bushes. *Trientalis europaea* may establish, and after a few more years, also *Silene dioica*. Inundated during autumn storms. (D) *A. incana* trees form closed stands in the central part on islands that are ~150 years old. *V. sambucifolia* populations decline and occur only at the outer part of the *Alnus* border together with *F. ulmaria*. *T. europea* in dense populations and *S. dioica* populations begin to expand. (E) A ~250-year-old island with an *A. incana* border toward the shore succeeded by *S. aucuparia* and then *Betula pendula* and spruce (*Picea abies*). *T. europaea*, *S. dioica*, and *F. ulmaria* are relatively abundant but become increasingly restricted to the shoreward parts. (F) An old island (>300 years) fringed with a narrow *A. incana* border, succeeded by *P. abies* at the maximum high-water level. In the *Alnus* border, *T. europaea* is relatively abundant and *S. dioica* occurs in small patches or as single individuals. Photograph by B. E. Giles.

The theoretical framework for the importance of spatial structure in determining disease dynamics has grown alongside the development of the island biogeography and the metapopulation theories. However, detailed field studies of pathogen distributions and disease spread within and between island plant populations are restricted to a few long-term studies in Baltic archipelagoes in Sweden and Finland.

DISEASE PATTERNS IN ARCHIPELAGOES

The Skeppsviks Archipelago in northeastern Sweden has been the focus for several studies of natural plant-pathogen interactions. This area is particularly suitable for studying how various factors affect the dynamics of host-pathogen systems in island networks. The archipelago consists of about 100 islands of different ages and sizes within a 20-km^2 area, and primary succession is in strikingly different phases on these islands (Fig. 1). By studying the demography of plants and patterns of disease spread in relation to the age and size of host populations and colonization and extinction rates, the long-term consequences of host-pathogen interactions can be estimated. The islands, which are composed of moraine deposited in a north-south direction by land ice during the latest glaciation, were initially left under water when the ice melted 7700 years ago. Since then, there has been an isostatic land uplift in the area, and the land still rebounds at a rate of about 0.85 cm per year. New land is continually made available for colonization, either as new islands rising above sea level or as extensions of existing islands. The height of an island above sea level is correlated with island age and also with the time since it was first colonized by plants, and it is possible to estimate the maximum population ages of plants that colonize during the early succession stages.

Studies of the interactions between the plants *Valeriana sambucifolia* and the rust fungus *Uromyces valerianae*, between *Trientalis europaea* and the smut fungus *Urocystis trientalis*, and between *Silene dioica* and the anther smut *Microbotryum violaceum* in Skeppsviks Archipelago have shown that the percent of individuals infected in populations varies considerably among island populations of different ages. Since hosts must be present prior to their obligate pathogens, young populations generally show low disease levels. When island populations become older and larger, the probability of disease increases, and all intermediate-aged populations of the studied species, are diseased although the frequency of disease in different populations and host-parasite systems varies. Among old populations, fewer populations are diseased, and the percent of individuals infected is generally lower (Fig. 2). This effect is in part due to populations becoming more scattered on older islands as succession proceeds on the islands.

Long-term studies of the *M. violaceum*/*S. dioica* system confirm this pattern; that is, island populations of *S. dioica* in which the anther smut has colonized have been larger than populations that have remained healthy. Moreover, even though *M. violaceum* causes a systemic infection and can live for many years in diseased plants, the pathogen has been lost from some populations. These populations have been smaller on average than diseased populations where the pathogen has persisted. The rate of disease spread is

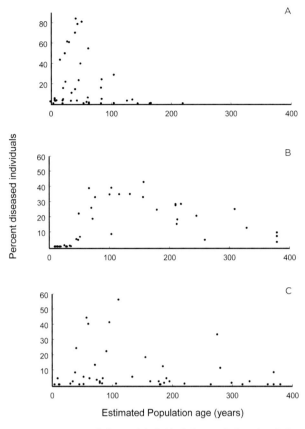

FIGURE 2 Percent of diseased individuals in populations in relation to estimated population age (years) for (A) *Valeriana sambucifolia-Uromyces valerianae*, (B) *Trientalis europaea-Urocystis trientalis*, and (C) *Silene dioica-Microbotryum violaceum* in Skeppsviks archipelago (from Carlsson et al. 1990).

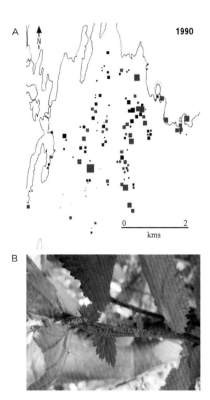

FIGURE 3 (A) Spatial patterns in the disease occurrence of the rust *Triphragmium ulmariae* among 129 populations of *Filipendula ulmaria* growing on islands (shown as squares) in Skeppsviks Archipelago. The size of squares is proportional to host population size; red squares represent diseased populations, and black squares represent healthy populations (from Burdon et al. 1995). (B) Spores of *T. ulmariae* on *F. ulmaria*.

also affected by the levels of resistance in *S. dioica* populations; that is, populations showing an increase in disease frequency tend to be more susceptible than populations where the disease has remained at low levels. Differences in resistances among populations are established during founding of populations; that is, by chance some populations are established by susceptible plants and others by more resistant plants. Although islands exchange seeds and pollen, populations remain differentiated because of limited gene flow between islands.

Data from another long-term study in the Skeppsviks Archipelago on the interaction between the rust *Triphragmium ulmariae* and its host *Filipendula ulmaria* have similarly shown that the disease is more often found in larger host populations but also in populations that are in close proximity to larger diseased populations (Fig. 3). *T. ulmaria* is a nonsystemic rust that causes local lesions on its host, and this system is characterized by drastic winter bottlenecks when plants die back to an underground rootstock during the winter. It therefore shows large fluctuations in disease frequencies, both among populations and among years.

Clearly, the results obtained in these studies have confirmed some of the predictions generated from spatial host-pathogen models: the importance of population size and distance and the dynamic nature of disease. Whether these predictions also fit with the dynamics of disease in more isolated insular systems and in oceanic islands remain to be tested. Moreover, the dynamics of disease may be very different in other types of pathogens than in the biotrophic host-specific pathogens studied in the Skeppsviks Archipelago.

SUMMARY: FROM HERE AND BEYOND

Today we have a better understanding about diseases in natural plant communities than we had a mere decade ago. In particular, we have become aware that isolation and the spatial structure of host and pathogen populations can greatly affect the dynamics of plant-pathogen interactions. It has also become evident that pathogens may play dual roles in the dynamics and evolution of natural plant communities. The prevailing view is that plant pathogens are destructive organisms that reduce fitness of individual plants and cause declines of populations of particular species. However, diseases also play an important role in maintaining genetic diversity within species and biodiversity at the community level. Their role is, however, difficult to assess without careful experimentation, but such strategies have so far been adopted only in a very limited number of studies. Further empirical work should also focus on improving our understanding about the importance of life history of hosts and pathogens for spread and the numerical and evolutionary dynamics of diseases. This knowledge is essential if we are to understand not only the risk and the challenges of plant diseases on islands and in mainland communities, but also how we conserve these often neglected organisms.

SEE ALSO THE FOLLOWING ARTICLES

Biological Control / Deforestation / Dispersal / Metapopulations / Vegetation

FURTHER READING

Burdon, J. J., L. Ericson, and W. J. Müller. 1995. Temporal and spatial changes in a metapopulation of the rust pathogen *Triphragmium ulmariae* and its host, *Filipendula ulmaria*. *Journal of Ecology* 83: 979–989.

Burdon, J. J, P. H. Thrall, and L. Ericson. 2006. The current and future dynamics of disease in plant communities. *Annual Review of Phytopathology* 44: 19–39.

Carlsson, U., T. Elmquist, A. Wennström, and L. Ericson. 1990. Infection by pathogens and population age of host plants. *Journal of Ecology* 78: 1094–1105.

Carlsson-Granér, U., and T. M. Pettersson. 2005. Patterns of host susceptibility and disease occurrence in a metapopulation of *Silene dioica*. *Evolutionary Ecology Research* 7: 353–369.

Carlsson-Granér, U., and P. H. Thrall. 2002. The spatial distribution of plant populations, disease dynamics and evolution of resistance. *Oikos* 97: 97–110.

Gilbert, G. S. 2002. Evolutionary ecology of plant diseases in natural ecosystems. *Annual Review of Phytopathology* 40: 13–43.

Hunter, D. G., and A. Shafia. 2000. Diseases of crops in the Maldives. *Australasian Plant Pathology* 29: 184–189.

McKenzie, E. H. C. 1998. Rust Fungi of New Zealand: an introduction, and list of recorded species. *New Zealand Journal of Botany* 36: 233–271.

Mill, M., D. Currie, and N. J. Shah. 2003. The impacts of vascular wilt disease of the takamaka tree *Calophyllum inophyllum* on conservation value of islands in the Granite Seychelles. *Biodiversity and Conservation* 12: 555–566.

Parker, I., and G. S. Gilbert. 2004. The evolutionary ecology of novel plant-pathogen interactions. *Annual Review of Ecology, Evolution and Systematics* 35: 675–700.

Trujillo, E. E. 2005. History and success of plant pathogens for biological control of introduced weeds in Hawaii. *Biological Control* 33: 113–122.

PLATE TECTONICS

ROGER C. SEARLE

Durham University, United Kingdom

Plate tectonics was developed in the 1960s and 1970s as the unifying, global theory of the Earth sciences. It assumes that the outer surface of the Earth is made up of thin, brittle tectonic plates, which have rigid interiors and interact only at their edges, where their relative motions produce earthquakes and volcanoes. By measuring the relative velocities of the plates in a finite number of places, their motions across the world and over the geological past can be computed using relatively simple geometric techniques. The assumption of rigid, undeformable plates breaks down to some extent in some continental areas but has proved to be a highly successful concept to describe the geology of the ocean basins.

CONTINENTS, OCEANS, AND TECTONIC PLATES

The Earth can be divided into a number of concentric regions based on its composition: the crust (up to 70 km thick), the mantle (2900 km thick), and the core (3500 km thick). The crust can also be subdivided into thicker (30–70 km) continental crust, made of relatively low-density rocks such as sandstone, limestone, and granite, and thinner (3–7 km) oceanic crust, consisting mainly of denser basalt and gabbro (Fig. 1A). Thus the thicker continental crust "floats" higher than the thin but dense oceanic crust, by an average of about 4.6 km, thus forming the ocean basins.

However, if we divide the Earth into layers on the basis of their mechanical strength instead of composition, we find an outer layer some 100 km thick, called the lithosphere, which constitutes the tectonic plates and comprises both crust and uppermost mantle. The lithosphere overlies a layer within the mantle that is still solid but somewhat weaker, called the asthenosphere (Fig. 1B).

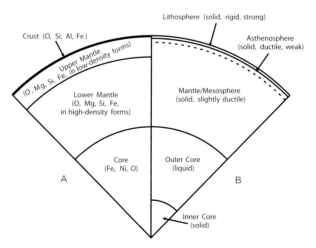

FIGURE 1 Internal structure of the Earth. (A) Compositional subdivisions of the Earth. The most common elements in each layer are given in descending order of abundance. (B) Mechanical subdivisions of the Earth.

It is this weak asthenosphere that, over millions of years, allows the lithosphere both to adjust its depth and also to move around as described by plate tectonics. The lithosphere itself is divided into some 12 large, independently moving tectonic plates (plus a larger number of small or "micro" plates). It is important to note that a tectonic plate can contain both oceanic and continental crust; for example, the North American Plate comprises both the continental area of north America and the western half of the north Atlantic Ocean (Fig. 2).

PLATES AND THEIR BOUNDARIES

The Earth's mantle, though solid, is actually slightly ductile (like very strong toffee), and is very slowly convecting as it transports heat from the deep interior towards the surface. The tectonic plates actually represent the tops of these convection cells, where the Earth has become cool enough to behave in a brittle, rather than ductile, manner. The plates are in contact with each other everywhere on the Earth's surface and are driven this way and that by the convection (like scum on boiling jam). Thus, the relative motions between plates may consist of one of three types of motion: divergence (at mid-ocean ridges), convergence (at subduction zones), and pure horizontal slip (at transform faults). Figure 2 shows the major plates with their boundary types and relative motions.

Mid-Ocean Ridges

At mid-ocean ridges the plates diverge, and the underlying asthenosphere is drawn up to fill the gap. As the asthenosphere material rises, the pressure drops and the melting point decreases, so that a small proportion melts to form liquid basaltic magma. This may rise to feed volcanic eruptions on the sea floor, with any residue being trapped below, where it cools to form diabase or gabbro and is incorporated into new lithosphere. Thus, new lithosphere is continually created at mid-ocean ridges. Approximately 3 km² of new lithosphere is created this way every year. As the plates pull apart, they become fractured, and this cracking is manifested in small earthquakes.

Because this new ridge is hot, it is quite buoyant, so the young lithosphere at ridges is relatively shallow. As the lithosphere moves away from the ridge, it ages, cools, becomes less buoyant, and sinks deeper, which is why divergent boundaries form ridges! Note, however, that these mid-ocean ridges are very broad features, extending the whole width of the ocean basins and deepening from less than 2900 m at their crests to over 5000 m on their flanks, which may be some 100 million years old. Most ridges are too dense to reach the sea surface, although in a few places, such as Iceland, the mantle is unusually buoyant and raises the ridges above sea level.

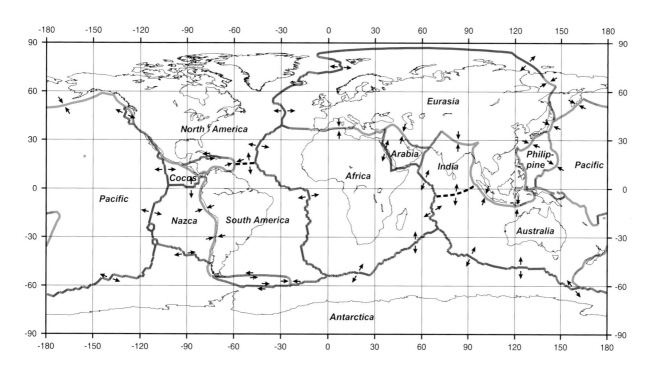

FIGURE 2 Map of the world with boundaries, boundary types, and names of the major plates. Red: divergent boundaries or ridges; green: convergent boundaries or trenches; blue: transform boundaries. Black dashed lines between North America and South America and between India and Australia indicate diffuse boundaries with very slow relative motion. Double arrows show relative motion directions across boundaries, but not relative speeds.

The erupted lavas contain significant amounts of the naturally magnetic mineral magnetite (lodestone), and as they cool, they acquire a magnetization that is proportional and parallel to the Earth's magnetic field. This field reverses its direction (i.e., north and south magnetic poles switch over) every few hundred thousand years, and the alternating field direction is recorded in the sea floor lavas. We can find the times at which the field reversed by dating rocks at the reversal boundaries, and can thus work out how rapidly the lithosphere moved away from the ridge axis—the so-called spreading rate—which helps us determine plate velocities.

Subduction Zones

We have good reason to believe that the Earth is not expanding, so if new lithosphere is created at mid-ocean ridges, an equivalent amount must be lost somewhere. This occurs at convergent plate boundaries or subduction zones. Here, two plates collide and one is pushed back into the Earth's interior; the boundary is marked by a deep trench, such as the 11-km-deep Mariana Trench. The motion of the descending plate causes earthquakes, including the largest, really destructive ones. As the plate descends, it heats up, and water that was trapped in the minerals making up the crust is released. This lowers the melting point of the mantle overlying the descending plate and causes it to start melting; the magma thus formed rises to feed volcanoes on the overriding plate. Because of the presence of the water and a somewhat different melting depth (pressure), the magma at subduction zones contains somewhat more silica than mid-ocean ridge basalts, and produces a rock type called andesite (so-called because it was first recognized in the Andes, in volcanoes formed above the Peru–Chile subduction zone).

When one of the plates consists of continental lithosphere, it is too buoyant to subduct, and andesite volcanoes form on it above sea level. Examples are the Cascade volcanoes in western North America. If both plates are oceanic, the andesite will initially be erupted on the sea floor but will gradually build up on top of the plate, and eventually volcanoes will emerge above sea level. Because the subducting plate, and therefore the line of melting, is usually curved, these volcanoes form an arc referred to as an island arc. A good example is the Lesser Antilles at the eastern end of the Caribbean Sea, where the North American Plate is subducting westward beneath the Caribbean Plate and forming volcanic islands such as Guadaloupe, Dominica, and Martinique. Andesitic magma is relatively viscous and also contains large amounts of gas, so these volcanoes tend to be explosive and steep-sided—the classic volcanic cone.

Transform Faults

Where plates slide past one another without growing or being consumed, the boundary is called a transform fault. Many of these faults, such as the San Andreas Fault (separating the North American and Pacific plates) lie entirely within continental lithosphere. The longest ridge–ridge transform is the 1000-km-long Romanche Transform in the Equatorial Atlantic. The Alpine Fault in New Zealand is also a transform fault cutting through a continental island.

Hotspots and Plumes

Although they are not plate boundaries, hotspots are an important cause of oceanic islands. In plate tectonic terms, a hotspot is a small geological region (as opposed to an extensive plate boundary) that is characterized by volcanic and seismic activity. They may lie on plate boundaries (Iceland is a good example), but often are far from the boundaries in mid-plate locations; the archetype of these is Hawaii. Hotspots are underlain by parts of the mantle that are unusually hot or, perhaps, have unusually low melting points so that they produce excess magma. It has been suggested that they may lie above upwelling plumes of hot mantle, but the depth from which they arise (whether a few hundred km down in the upper mantle, or perhaps 3000 km down at the boundary of the core) is still controversial.

At these hotspots, the mantle melts to produce basaltic magma, similar to that produced at mid-ocean ridges (though with some subtle differences in the amounts of trace elements). This magma is eventually erupted onto the sea floor, forming volcanic seamounts that may grow large enough to rise above sea level. The Big Island of Hawaii, measured from its submarine base to the summit of Mauna Kea, is the world's tallest volcano and is higher than Mount Everest. Basalt, being much less viscous than andesite and containing little gas, flows freely and produces shield-shaped volcanic domes with gentle slopes.

If the hotspot persists for millions of years and is fixed in the Earth's mantle, the lithospheric plate may drift over it so that the hotspot leaves a trail of volcanoes of steadily increasing age going away from it. As the plate ages and sinks, so will the volcanoes; if they were originally subaerial (i.e., above the surface) islands, they eventually sink to become underwater seamounts. This also explains the origin of atolls. Tropical volcanic islands typically develop fringing coral reefs; as the island sinks, its radius shrinks but the reef grows upwards, forming a barrier reef with a lagoon inside. Eventually, the island sinks completely, leaving just the reef as an atoll.

ISLANDS AND PLATE TECTONICS

From the above discussion it will be realized that islands may have a number of different plate tectonic and geological origins and settings: continental fragments of all sizes, arc volcanoes at subduction zones, and hotspot volcanoes near ridges or in ocean basins.

Continental Islands

Some islands are actually whole continents, such as Australia or Greenland. Their geology will generally be a complex combination of different types of rocks, generated and altered over hundreds of millions of years, and so displaying a great deal of variability.

Many continents have small islands just offshore, where a shallow waterway separates them from the mainland, with which they share a common geological structure (e.g., Long Island, New York; Sri Lanka; and countless smaller examples). Others are continental slivers that have been separated from their parent continents by plate motions and may have drifted far away from their origins. Examples are the Seychelles in the Indian Ocean (which, unusually for mid-ocean islands, display extensive outcrops of granite) and Madagascar, whose linear east coast reflects its past motion along a transform fault as it rifted away from Antarctica.

Arc Volcanoes

Many oceanic islands are arc volcanoes above subduction zones, and all these lie near and parallel to deep-sea trenches. There are numerous examples in the Lesser Antilles, the South Sandwich Islands, and the western Pacific. All of these are based on andesite volcanoes, so they typically have active or dormant volcanoes with steep sides, explosive eruptions, and associated earthquakes. Such volcanoes, especially in the tropics, can produce rich, volcanic soils, but they also pose considerable hazards from volcanic eruptions (lava and ash flows) and mud flows.

Hotspot Volcanoes

Most islands within the deep ocean basins are hotspot volcanoes. They may lie on mid-ocean ridges (e.g., Iceland), near ridges (the Azores, the Galápagos), or far from them in the plate interiors (e.g., the Hawaiian, Marquesas, and Friendly Islands in the Pacific, the Canaries and Tristan da Cunha in the North and South Atlantic, and Réunion and Mauritius in the Indian Ocean).

These volcanoes are basaltic, so they have low slopes and are shield-shaped. Islands with active volcanoes may have significant areas of recent lava flows with no vegetation, although vegetation soon establishes itself in a matter of years. As the volcanoes age, they become eroded and, especially in the tropics, may develop steep, dramatic ridges and valleys as a result. In many hotspot volcanoes the flanks eventually collapse and fall into the sea, producing dramatic sea cliffs (such as the north coast of Molokai in Hawaii), while the process of collapse may generate tsunamis.

SEE ALSO THE FOLLOWING ARTICLES

Continental Islands / Earthquakes / Island Formation / Oceanic Islands / Seamounts, Geology / Volcanic Islands

FURTHER READING

Cox, A., and R. B. Hart. 1986. *Plate tectonics—how it works.* Oxford: Blackwell Scientific Publications, Inc.

DeMets, C., R. G. Gordon, D. F. Argus, and S. Stein. 1990. Current plate motions. *Geophysical Journal International* 101: 425–478.

DeMets, C., R. G. Gordon, D. F. Argus, and S. Stein. 1994. Effect of recent revisions to the geomagnetic reversal timescale on estimates of current plate motions. *Geophysical Research Letters* 21(20): 2191–2194.

Isacks, B. L., J. Oliver, and L. R. Sykes. 1968. Seismology and the new global tectonics. *Journal of Geophysical Research* 73: 5855–5900.

Kearey, P., and F. J. Vine. 1996. *Global tectonics*, 2nd edition. Oxford: Blackwell Science.

Sella, G. F., T. H. Dixon, and A. Mao. 2002. REVEL: A model for recent plate velocities from space geodesy. *Journal of Geophysical Research* 107(B4): ETG 11, 1–32.

Tackley, P. 1996. Mantle convection and plate tectonics: toward an integrated physical and chemical theory. *Science* 288: 2002–2007.

Vine, F. J. 1966. Spreading of the ocean floor: new evidence. *Science* 154: 1405–1415.

POCKET BASINS AND DEEP-SEA SPECIATION

BRUCE H. ROBISON

Monterey Bay Aquarium Research Institute, Moss Landing, California

WILLIAM M. HAMNER

University of California, Los Angeles

Isolation is an important agent in the evolution of new species. When two populations of a single species become sufficiently isolated that there is no exchange of genetic material, then random genetic mutations and genetic drift over time will eventually render them distinct from each other. Thereafter, if the environmental conditions that affect these two populations begin to differ, genetic separation (speciation) can proceed even faster. Island chains at and near the sea surface and pocket basins in the deep seafloor provide the isolation necessary for speciation in the ocean.

NATURAL LABORATORIES

Scientists often rely on the manipulation of natural systems or processes in order to understand them, but the vast size of oceanic ecosystems makes it clearly impossible to conduct experimental research at this scale. Instead, we must look for naturally occurring manipulations, for natural laboratories where the local variability of environmental conditions provides us with the equivalent of experimental results. The most familiar examples of these processes occur on islands like the Galápagos, where populations of related species have evolved into different forms, shaped by the particular conditions on the individual islands they inhabit.

In the deep ocean, however, there are relatively few barriers to genetic exchange because the great basins of the Pacific, Atlantic, and Indian Oceans extend for thousands of kilometers and their deep-water environments are relatively homogeneous. Consistent with this global pattern is the substantial number of cosmopolitan deep-sea species, those found in deep-sea basins worldwide.

For shallow-water marine animals as well, speciation is clearly related to the isolation provided by islands. The area of the Indo-Pacific bounded by the Philippines, New Guinea, Indonesia, and Borneo has the highest density of islands on earth. Within this area there are more shallow-water marine species than anywhere else in the sea. These islands are correctly considered to be the "cradle of marine biodiversity."

POCKET BASINS BENEATH INDO-PACIFIC SEAS

For many kinds of deep-sea animals, the Indo-Pacific is also considered to be the center of origin and distribution. This is where species diversity is the highest, and again, speciation is clearly linked to the multitude of islands. In this case the islands and the submerged ridges between them provide barriers to the exchange of genetic material between species that inhabit deep water. Between many of the island groups there are pocket basins, areas of the seafloor that are thousands of meters deep yet isolated from each other by islands and submerged ridges. Typically, the ocean area above a pocket basin, circumscribed by islands and ridges, is called a sea. Familiar examples include the Sulu Sea of the Philippines and the Banda Sea of Indonesia.

While the bottom of a pocket basin may be several thousand meters deep, the sill depth, or the deepest part of the rim around a basin, has an important effect as a barrier to the exchange of genetic material between the deep-living animals that occupy adjacent basins. For example, the bottom depth of the Sulu Sea is 5200 m, but the deepest place along the rim of its basin is only 420 m. This means that unless a deep-living species can tolerate the decompression associated with depth changes of perhaps several thousands of meters, it cannot exchange genetic material with a related population in a neighboring basin.

DEEP-SEA LIFE CYCLES AND SPECIATION

Many deep-living species do have eggs that float upward so that their larvae can develop in the highly productive waters near the surface. But after initial development at shallow depths, the maturing juveniles move back down to the greater depth range of the adults. This ontogenetic or developmental vertical migration allows many deep-living species to broadcast their eggs widely, some of which are carried by surface currents to surrounding seas. Yet other species have eggs and larvae that remain at depth or have particular requirements that can be met only locally, and these species remain genetically isolated in their natal pocket basins.

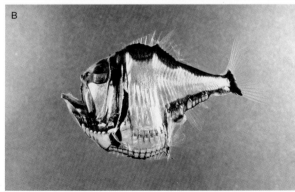

FIGURE 1 Deep-sea fishes: (A) lanternfish, *Myctophum*, with small, rounded bioluminescent organs on its flanks and along its ventral surface; (B) hatchetfish, *Argyropelecus*, with elaborate light organs along its underside, and upward-directed eyes.

For example, fishes of the family Myctophidae ("lanternfish" because of their light-producing organs, Fig. 1A), live in deep, dark waters during the day, hundreds of meters beneath the sea surface. At night they migrate upward to feed near the surface during the hours of darkness. Their bioluminescent organs are used to find prey, find mates, and avoid predators. Lanternfish are found worldwide, and they belong to one of the most speciose of all fish families. The center of their geographical distribution, and the place where more myctophid species have been recorded than anywhere else, is the Indo-Pacific region off Southeast Asia. Similar patterns of distribution and species richness exist for many other deep-sea fish families, including hatchetfish (Fig. 1B), dragonfish (Fig. 2A), and anglerfish (Fig. 2B). Although the fishes in each of these families have different feeding patterns, depth ranges, and life histories, they all appear to have originated in the Indo-Pacific and have speciated widely.

The Sulu and Banda Seas lie among the narrow maze of twisted trenches and pocket basins that link the great basins of the Indian and Pacific Oceans (Fig. 3). These interlocking depressions in the sea floor provide the only deep waters between the continental shelves of Australia

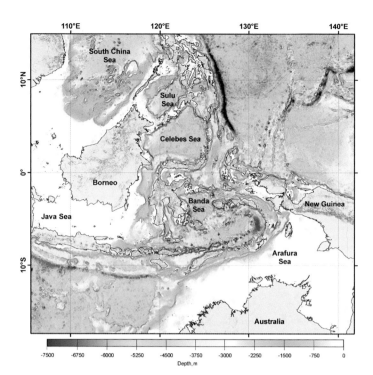

FIGURE 3 Bathymetry of the Indo-Pacific region. Three pocket basins beneath the Sulu, Celebes, and Banda Seas are circumscribed by islands and submerged ridges that isolate their deep-dwelling inhabitants. Satellite altimetry data: Geoware, GMT Companion CD-R Vol. 1 Version 1.9, June 2006. Image prepared by Jenny Paduan and David Clague, MBARI.

FIGURE 2 Deep-sea fishes: (A) dragonfish, *Idiacanthus*, with a luminous chin barbel for attracting prey; (B) anglerfish, *Caulophryne*, with a luminous lure, small eyes, and elongate fin rays for detecting movement in the dark waters it inhabits.

and Southeast Asia. The many seas of this region are filled with water from the western Pacific, but they are hydrographically distinct. The numerous island groups and sills that separate the seas also restrict the flow of subsurface waters between them, and these semi-isolated basins with shallow sills provide the reproductive isolation needed for the speciation of deep-sea animals.

In the complex topography of the seafloor in the Indo-Pacific region the many deep pocket basins are the inverse of islands at the sea surface. They isolate populations of related deep-sea animals from one another, thereby leading to evolutionary speciation.

SEE ALSO THE FOLLOWING ARTICLES

Cold Seeps / Hydrothermal Vents / Indonesia, Geology / Marine Lakes / Philippines, Geology

FURTHER READING

Briggs, J. C. 1974. *Marine zoogeography*. New York: McGraw-Hill.
Herring, P. J. 2002. *The biology of the deep ocean*. New York: Oxford University Press.
Koslow, J. A. 2007. *The silent deep*. Chicago: University of Chicago Press.
Robison, B. H. 1995. Light in the ocean's midwaters. *Scientific American* 273: 60–64.
Robison, B. H. 2004. Deep pelagic biology. *Journal of Experimental Marine Biology and Ecology* 300: 253–272.

POLYNESIA

SEE PACIFIC REGION

POLYNESIAN VOYAGING

ATHOLL ANDERSON

Australian National University, Canberra

Polynesia consists of Samoa and Tonga (West Polynesia), settled initially by Lapita voyagers about 3000 years ago, and the dispersed archipelagoes of East Polynesia, especially Hawaii, French Polynesia, Easter Island, Cook Islands, and New Zealand, colonized 1100–700 years ago (Fig. 1). The term "Polynesian voyaging" refers to the means by which island colonization was effected and the extent to which interaction occurred between distant islands. One extreme of opinion envisages exclusively accidental colonization by one-way voyaging that precluded development of long-range interaction, while at the other extreme, purposeful and navigated voyaging within a strategic system of colonization, multiple contact, and interaction is proposed. The evidence at issue consists of Polynesian traditions, early European observations, and linguistic, archaeological, and experimental data concerning the sequence, timing, and capabilities of Polynesian voyaging.

TRADITIONS AND HISTORICAL OBSERVATIONS

Polynesian traditions of origin are varied, but most refer to ultimate ancestry in mythical homelands, notably Hawaiki, which is reflected in island names such as Hawaii, Havaii (Raiatea in the Society Islands), and Savaii (in Samoa). Voyaging traditions are common in Polynesia, and where they have been studied in detail, notably in New Zealand, it is apparent that spare, enigmatic early nineteenth-century records had been elaborated subsequently by indigenous and European scholars. Stories about fleets of canoes, detailed descriptions of particular voyages, and precise navigational instructions, including methods, cannot be traced back to reliable early traditions. However, a general congruence of traditional genealogies among widely dispersed tribal populations in New Zealand and, to some extent, between marginal and central East Polynesia, does suggest that widespread colonization voyaging occurred in East Polynesia during a period 20–30 generations before AD 1800; that is, about AD 1100–1300.

There are no historical records (sixteenth to early nineteenth centuries) of Polynesian voyaging; that is, no voyaging canoes were observed far at sea between archipelagoes. The records consist, instead, of numerous observations of Polynesian canoes sailing and ashore, and some accounts of Polynesian geography. Of the latter, the map constructed for James Cook in 1769 according to information from the Raiatean scholar Tupaia is the most important. It contains islands between West Polynesia and the Marquesas, although Tupaia claimed no first-hand knowledge beyond the Societies and nearer Australs. His map and other contemporary information from Polynesians suggest that by the late prehistoric era there was frequent interaction within each archipelago, infrequent interaction between them, and no contact between the central (now French Polynesian) groups and the marginal groups of Hawaii, Easter Island, and New Zealand.

Historical observations record that "double canoes" were mostly used for inter-island travel. There was considerable diversity in their rigging. The oceanic lateen sail, probably derived from Micronesia and ultimately of Indian Ocean origin, was common in West Polynesia (Fig. 2). In the seventeenth century it was rigged in a rudimentary form, but by the late eighteenth century it conformed to the Micronesian style, in which, to put

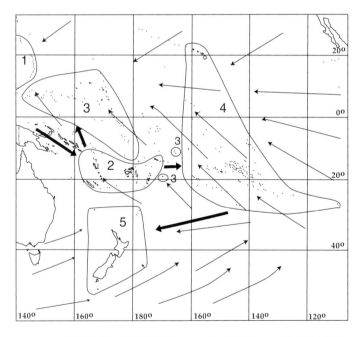

FIGURE 1 Colonization (thick arrows) in relation to prevailing wind directions (thin arrows): 1 = West Micronesia 1500 BC, probably from Philippines; 2 = Lapita colonization, from New Guinea islands, reaches West Polynesia 1000 BC; 3 = colonization of East Micronesia and West Polynesian marginal islands 200 BC from Tonga or Samoa; 4 = colonization of East Polynesia AD 1000 from West Polynesia; 5 = colonization of South Polynesia (New Zealand region) AD 1200.

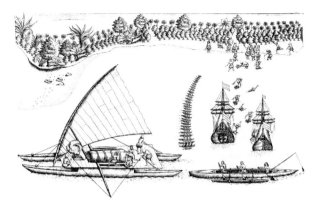

FIGURE 2 A Tongan double canoe with an oceanic lateen sail. Photo lithograph of ink drawing by Isaac Gilsemans, 1642, at Tongatapu Island, titled, in translation, "Our Ships at Anchor in the Roadstead." Published with permission from the Alexander Turnbull Library, Wellington, New Zealand, PUBL-0106-001.

FIGURE 3 A Hawaiian double canoe with an oceanic spritsail. Engraving of drawing by John Webber, 1788, at Hawaii, titled "Tereoboo, king of Owhyee, bringing presents to Captain Cook." Published with permission from the Alexander Turnbull Library, Wellington, New Zealand, C-131-061.

the vessel about, the mast was canted over and the tack point (or tack) of the sail moved from one end to the other, a maneuver called "shunting." In East Polynesia, lateen sails were rare, and the main type was an oceanic spritsail slung, point down, between two upright spars (Fig. 3). Generally, one spar was used as, or attached to, a mast, but in New Zealand and possibly in the Marquesas the earliest records, although not definitive, suggest that the spars were free-standing, held up by wind pressure against the sheets and used only before the wind. Navigation methods were not recorded in any detail, except to note that the moon and stars were used in some way. They probably enabled estimation of latitude, to which dead reckoning and use of land-finding indicators, such as swell patterns and homing seabirds, provided some additional aid.

LINGUISTIC, ARCHAEOLOGICAL, AND EXPERIMENTAL SEAFARING EVIDENCE

Comparison of Polynesian languages shows that proto-Polynesian split into three groups: proto-Tongic, proto–nuclear Polynesia (mostly in the Polynesian outliers), and proto-Ellicean, which comprised mainly the Samoan language and Eastern Polynesian. The latter, in turn, split into a central eastern group (Tahitian, western Australs and Tuamotu, Cooks, Maori, and, slightly different, Marquesan and Hawaiian) and a southeastern group comprising the languages of Mangareva, Easter Island, and the eastern Tuamotu and Australs. This sequence is mirrored approximately in the archaeological data of initial island colonization. These show a considerable "pause" in colonisation between Samoa–Tonga at 3000 years ago and marginal West Polynesian islands such as Niue and Tuvalu about 2200 years ago; another until the earliest settlement of the Societies and Cooks at 1100–1000 years ago; and then a briefer pause before colonization of the subtropical islands such as Rapa and Easter Island and of temperate New Zealand about 800 years ago. The overall pattern of island colonization is west to east, center to periphery, and clearly episodic rather than continuous.

Two canoe planks and a steering oar from Huahine island, Societies, dated to about AD 1000, constitute the best, and almost the only, direct archaeological evidence of possible voyaging canoes. Notably absent is any evidence of early voyaging sails and rigging, apart from a few depictions of Polynesian spritsails in rock art, especially in Hawaii. Historical linguistic reconstructions indicate that the double canoe came into existence in West Polynesia, its cargo-carrying capacity perhaps a critical prerequisite for further voyaging to the east. However, terms for "mast" and "standing rigging" are not known for early Polynesian languages. Chemical analysis of transported adzes to establish stone sources can provide useful information about prehistoric movement, but in some cases where long-distance movement is asserted, it is apparent that the adzes are not from prehistoric contexts. At present, stone sourcing studies confirm frequency of movement across the central Polynesian islands but the isolation of marginal archipelagoes such as Hawaii, New Zealand, and Easter Island.

One way of attempting to overcome the paucity of direct evidence for Polynesian voyaging is by experiment or simulation. This has taken two forms. The first consisted of building and sailing ocean-going vessels regarded as "performance-accurate" in terms of potential prehistoric types. The most famous of these was the *Kon-Tiki*, built in the form of a Peruvian sailing raft and sailed to Polynesia in

1947 by Thor Heyerdahl in support of his contention that South America represented a major source of Polynesian culture. Another well-known experimental vessel is the *Hokule'a,* a double canoe, inspired by Ben Finney's belief in the capabilities of prehistoric seafarers, which sailed from Hawaii to the most distant points of Polynesia.

The other form of experimental sailing, increasingly preferred since its inception in 1973 by Levison, Ward, and Webb, has been voyaging by computer simulation. The core idea is to map the frequency of wind directions across the Pacific, then put numerous virtual canoes with defined sailing characteristics to sea from different islands in order to determine which potential routes were more likely used, the relative rates of successful landfall, and the probable sequence of island discovery. Experiment and simulation have shown, in general, that simple drifting would probably not have been sufficient to people the remote Pacific islands, but that a sophisticated voyaging ability, including weatherly vessels and astral navigation, should have led to faster and more continuous colonization. Prehistoric voyaging capabilities lay somewhere in between.

MODELS OF POLYNESIAN VOYAGING

In the light of these various sources of evidence, it is broadly agreed that the overwhelming contribution to Polynesian populations and cultures came from the western Pacific. Claims for an American influence are not generally accepted, although close similarities in aspects of material culture between South America and Easter Island, and the existence of South American sweet potato in prehistoric Polynesia, suggest some level of contact. As large, capable sailing rafts existed in Ecuador at the time of Spanish arrival, it is quite possible that they were the agent of transport, rather than Polynesian canoes. Chicken bones discovered recently in a Chilean site are not of any distinctive Polynesian type, and as they date very close to the Spanish era, the evidence should be regarded cautiously. Claims of Polynesian voyaging to North America are implausible.

Within mainstream scholarship, the various kinds of evidence have been constructed into widely differing models. The most prominent from the late nineteenth to the mid-twentieth century was "traditionalism." Represented in the writings of Sir Peter Buck (Te Rangi Hiroa), for example, it accepted the expanded corpus of indigenous traditions at face value and assumed, as in traditional thought, that the abilities of the ancestors in seafaring exceeded those of their descendants. Advanced navigation techniques and frequent long-distance voyaging and interaction between remote islands were regarded as characterizing prehistoric voyaging. In reaction to this perspective, Andrew Sharp argued that many traditions were corrupted, Polynesian geographical knowledge was limited, advanced navigation was impossible, and Polynesian canoes suffered problems in sailing, seaworthiness, and seakeeping.

His incisive attack served, however, to inspire a revival of traditionalism, as in the work of Ben Finney. "Neotraditionalism" uses the traditionalist assumptions to argue that Polynesian voyaging technology declined once most of the islands were colonized and that historical evidence is thus only a pale reflection of earlier prehistoric ability. Therefore, the original Polynesian voyaging canoes must have been more advanced than any recorded historically. This maxim was put into practice in building and sailing experimental canoes such as *Hokule'a,* which combined the most advanced of widely dispersed Polynesian technologies; Micronesian navigation techniques; European buoyancy, fastenings, and rigging; and sails that were twice as large, relative to waterline length (a proxy for displacement, which cannot be estimated independently from historical data), than on eighteenth-century Polynesian double canoes of similar size. The experimental sailing data are therefore at the most liberal extreme of probability. Nevertheless, in the absence of alternatives, they have been used in voyaging simulation studies, notably by Geoffrey Irwin, to argue that such fast, weatherly vessels were used strategically in exploration; sailing first toward the prevailing wind, which allowed easier return, and only later, as uninhabited islands became few, in more difficult directions across and before the prevailing wind.

Recently, opinions have been changing again. Neotraditionalism remains the preferred model among indigenous scholars and in popular opinion, as shown in the recent Vaka Moana exhibition (Auckland Museum, New Zealand) and book, but some archaeologists have tried to rethink the problems of understanding Polynesian voyaging from more basic starting points. I have returned to the early historical descriptions to argue that they show not a supposed devolution of sailing technology following an age of exploration but rather a late evolution (for example in the introduction of the lateen sail, which encouraged greater use of fixed masts and standing rigging), and therefore that, contrary to traditionalist assumption, earlier technology might have been more rudimentary. If it was largely confined to sailing before the wind, then the episodic nature of Polynesian dispersal eastward could reflect approximately correlated periods of high El Niño frequency, notably about 3000 and 1000 years ago, in which easterly winds were reversed into tropical westerlies. Some simulation studies have also returned to a more basic model in which only rudimentary sailing ability (drifting, paddling) is assumed. This can be shown as sufficient to reach west Polynesia but inadequate for east Polynesian conditions. Even in allowing assumptions of greater ability (e.g., weatherly sailing), recent simulations show that voyaging in East Polynesia was difficult.

CONCLUSIONS

The difficulty of understanding Polynesian voyaging arises most directly from the virtual absence of direct evidence, such as archaeological remains of boats and rigging. Were these more abundant, and the form and capabilities of seafaring, which is the critical practice of voyaging, much better known, most of the current arguments would be resolved. As it is, the burden of debate is carried by experimental and simulated sailing and the comparison of results from those against archaeological evidence of island colonization patterns. At present, this method shows that Polynesian voyaging was more difficult than envisaged in traditionalist models, except by assuming the existence of very sophisticated watercraft of types used experimentally but not recorded historically. Most East Polynesian canoes were capable off the wind but unsuited to sailing long distances against it.

In that circumstance, the episodic nature of Polynesian dispersal, and the scarcity of evidence for frequent interaction over long distances, seem readily understandable. The larger questions about what actuated episodes of voyaging remain unanswered. Was it improvements in technology, periodic population growth and social pressure (as reflected in traditional references to exile), changing climate, or mere chance? These propositions need much research before it will be possible to write a more satisfactory account of Polynesian voyaging.

SEE ALSO THE FOLLOWING ARTICLES

Archaeology / Exploration and Discovery / Human Impacts, Pre-European / *Kon-Tiki* / Peopling the Pacific

FURTHER READING

Anderson, A. J. 2000. Slow boats from China: issues in the prehistory of Indo-Pacific seafaring, in *East of Wallace's Line*. S. O'Connor and P. Veth, eds: Leiden: Balkema, 13–50.
Anderson, A. J., J. Chappell, M. Gagan, and R. Grove. 2006. Prehistoric maritime migration in the Pacific Islands: an hypothesis of ENSO forcing. *The Holocene* 16: 1–6.
Buck, Sir P. 1954. *Vikings of the sunrise*. Christchurch: Whitcombe and Tombs.
Di Piazza, A., P. Di Piazza, and E. Pearthree. 2007. Sailing virtual canoes across Oceania: revisiting island accessability. *Journal of Archaeological Science* 34: 1219–1225.
Finney, B. R. 1979. *Hokule'a: the way to Tahiti*. New York: Dodd, Mead and Co.
Golson, J., ed. 1963. *Polynesian navigation: a symposium on Andrew Sharp's theory of accidental voyages*. Wellington: The Polynesian Society.
Howe, K. R., ed. 2006. *Vaka Moana: voyages of the ancestors. The discovery ansd settlement of the Pacific*. Auckland: David Bateman.
Irwin, G. J. 1992. *The prehistoric exploration and colonisation of the Pacific*. Cambridge, UK: Cambridge University Press.
Levison, M., R. G. Ward, and J. W. Webb. 1973. *The settlement of Polynesia: a computer simulation*. Minneapolis: University of Minnesota Press.
Sharp, A. 1956. *Ancient voyagers in the Pacific*. Harmondsworth: Penguin.

POPULAR CULTURE, ISLANDS IN

VINCENT H. RESH
University of California, Berkeley

JONATHAN P. RESH
Undaunted Design Co., Chicago, Illinois

Popular culture is the culture of the people. It includes the fashions, movies, television shows, advertising, and even video games that are easily accessible to individuals with a wide variety of social backgrounds. Emphasis in popular culture is on instant accessibility; no prior or profound knowledge is required. The use of islands in popular culture is an excellent example of how media can use strong images and perceptions to influence people's taste and behavior.

ISLAND IMAGES AND INFLUENCES

Island images are generally positive—they emphasize sensuality, escape, solitude, seduction, and self-sufficiency. However, negative images are there as well—islands also can be lonely, inhospitable, forbidden, or mysterious (Fig. 1). Islands can elicit stereotypical responses strictly on their location. For example, Arctic islands are typically depicted as inhospitable and isolated, while Caribbean islands are friendly, plentiful, and carefree. No islands, however, can evoke romance and allure as those of the South Seas do. Tahiti, like Timbuktu, is almost a "brand," producing instant responses in the human imagination.

Advertisers capitalize on the charms of tropical islands in the Caribbean and the South Pacific to promote travel getaways or casual clothing or just as a mechanism for escape and fantasy. The message in selling the island experience is clear—islands will do for you what continental life cannot, and people can come to islands either to find themselves or to lose themselves.

FIGURE 1 Artist's rendering of the dual nature of islands as depicted in popular culture. Drawing by Jonathan Resh.

Island images and influences are present in many aspects of Western popular culture, including the media, food, drinks, restaurants, and the arts.

TELEVISION

Clearly, the archetype of television shows featuring islands is *Gilligan's Island*. Universally ridiculed by critics, this 1960s television show was produced for only four years, but reruns of its 98 episodes continue to be seen, and cartoons and even movies have revived the story. A recent book (*Gilligan Unbound* by Paul A. Canto) has actually maintained that this television series mirrors the beliefs of 1960s America and is a "window" into the liberal democratic culture that was present at that time.

More recently, the first popular reality television show in the United States, *Survivor*, was first set on an uninhabited island off Borneo, and many islands, including the Marquesas, Pearl Islands, Cook Islands, Palau, Vanuatu, and Fiji, have been the location of subsequent installments. Concurrently, British television featured *Castaway 2000*, in which contestants spent months living on an uninhabited island off the Scottish coast. Unlike the (sometime) comedic message of *Gilligan's Island*, these shows emphasize the danger of islands, where contestants are forced to be resourceful in order to "survive" in the island wild. They also serve as a populist social experiment, showing viewers how a select society may react when confronted by the forced isolation of islands.

Even though the participants in these more recent shows choose to be stranded, these are not island experiences for the vacation-bound.

Islands have been locations for other popular television shows as well, including *Hawaii Five-O*, *Hawaiian Eye*, and *Magnum PI*, all based in Hawaii, with *Temptation Island* filmed in the Caribbean and *Fantasy Island*, *Man from Atlantis*, and *Lost* set on fictional islands. In all these shows, some of the benefits of island living were presented, although crime and human frailties were always present as an undercurrent. Television shows have also contributed catch-phrases to popular culture, sometimes becoming painful with their repetition but fortunately short-lived such as "Book 'em, Dano" from *Hawaii Five-O*, "The plane, the plane" from *Fantasy Island*, and "Voted off the island" from *Survivor*.

MOVIES

Well over a thousand movies have the word "island" appearing in their title. However, rather than drawing on real islands, many of these films draw on the metaphor of island isolation or loneliness. There are, however, hundreds of movies that do not have "island" in the title but take place on islands. These latter films evoke the idea of survival or the noble savage (*Robinson Crusoe*, *Castaway*), freedom from adults (*Blue Lagoon*, *Lord of the Flies*), an idyllic paradise interrupted by an outside event (the arrival of Europeans or missionaries as in *Mutiny on the Bounty* and *Hawaii*, respectively), and internal conflicts or disagreements (*Rapa Nui* and *Tabu*). Of course, World War II movies that recount great battles on islands of the Pacific (*Guadalcanal Diary* and *Tora! Tora! Tora!*), and that contrast beautiful island settings with the atrocities of war, have long been a staple of Hollywood.

Like other forms of popular culture, cinema has also explored the duality of island images, exploiting both the positive and the darker sides. Beyond loneliness, films set on islands have suggested the possibility of being eaten (*Cannibals of the South Seas*), visiting the lair of villains (*Dr. No*, *The Island of Dr. Moreau*), and even encounters with nonhuman monsters (*King Kong* and *Jurassic Park*). More recently, the *Pirates of the Caribbean* series has added a supernatural element to island adventures on film.

The 1958 movie *South Pacific* gave us one of the most romantic movie images ever created—the mythical Bali Hai—which like Shangri-la (set in the Himalayas) in *Lost Horizon* has become synonymous with paradise, love, and transformation. The key song in that musical evokes this feeling by suggesting that people live on a lonely island and long for another, special, island. The name Bali Hai has been applied to resorts throughout the Pacific, sometimes thousands of miles apart.

LITERATURE AND ART

Although movies and television are more recent additions to the popular culture canon, islands have figured prominently in literature for centuries (*The Odyssey, The Tempest*). The island novel that is foremost in many readers' minds is Daniel Defoe's *Robinson Crusoe*. Based on the marooning of a Scottish seaman on Màs a Tierra (400 miles from Chile), it changed the images of islands from fearful places (albeit lonely ones) to sites of redemption and freedom. Since then, several novels set on islands that draw on traditional island motifs of romance and danger, including Ballantyne's *The Coral Island*, Stevenson's *Treasure Island*, Wells' *The Island of Dr. Moreau*, Golding's *Lord of the Flies*, Huxley's *Island*, Charrière's *Papillon*, Verne's *The Mysterious Island*, and Garland's *The Beach*. Although these stories have taken place on islands located throughout the world (e.g., *The Beach* was based on one of the islands off the coast of Thailand), many of the great island novels are based on stories of the South Pacific.

The South Pacific has spawned a rich number of stories that have become popular lore. American writers over successive generations, from Melville, Stevenson, and London to Nordhoff and Hall to Michener, along with writers of other nationalities such as Burke (Australia) and Loti (France), have contributed to the island fantasy. A recent trend has been to revisit islands and sites from previous voyages (e.g., Horowitz's *Blue Latitudes* retraces Cook's voyages) or of those evoking strong popular culture images (Theroux's *Happy Isles of Oceania*). Unfortunately, some of this genre seems to be searching for a zoolike paradise, and comments about modern life on islands range from being condescending to mean-spirited when the authors discover that DVD players have joined traditional culture.

Paul Gauguin, probably the artist most closely linked to the image of island life, is also an example of the linkage between high culture and popular culture. Now recognized as one of the great impressionist painters, he went from France to Tahiti following a quest to find the "noble savage" and to live the free life he envisioned away from continental civilization. His last home on Hiva Oa in the Marquesas Islands, which he named "Le Maison de Jouir," matches the images of a hedonistic life that drew him to these islands. Depictions of his paintings and sculptures are instantly recognizable as "Gauguins," whether on a T-shirt or parodied in advertisements. The novel by Somerset Maugham and subsequent movie *Moon and Sixpence,* based very loosely on Gauguin's life, helped create the popular mystique of the islands as a source of creative muse for the artist, and also the "free love and free coconuts" lifestyle that can supposedly be found there.

FOOD, DRINK, AND RESTAURANTS

Just as sight and sound convey images of islands in television and movies, culinary tastes may reflect images of islands. Perhaps not unexpectedly, many of these tastes are actually continental creations. For example, from the 1930s to the present day, Polynesian-island themed bars (often referred to as "Tiki bars") and restaurants have appeared with names like "Tiki Village," "Don the Beachcomber's," and "Trader Vic's." By providing decorative escapism and an assortment of rum-based drinks with exotic names such as Black Widow, Zombie, Missionary's Downfall, Samoan Fog Cutter, and Mai Tai, they have captured the exotic spirit of island mythology. Foods served in Tiki bars, which traditionally have contained non-Polynesian ingredients such as cooking sherry and Worcestershire sauce, reportedly were the origin of the Chinese "take-out" food widely consumed today. Décor of the Tiki bars included (and still includes) fishing nets and floats, A-frame design, masks, and spears (Fig. 2).

While many of these Tiki bars have disappeared, reappeared, and disappeared again, others like the "Tonga Room" and "Trader Vic's" in San Francisco continue the tradition of the Tiki bar/restaurant. One of the interesting aspects of the design of many of these island-themed restaurants is that oftentimes patrons have to cross a bridge (sometimes over a gurgling stream) to get from the entrance to the restaurant interior. Some have suggested that this represents a "symbolic crossing" and escape from reality to fantasy.

FIGURE 2 Interior of an "ultimate" tiki restaurant in Emeryville, California. Photograph by Cheryl Resh.

MUSIC AND DANCE

The often idyllic and exotic styles of "island music" also have become part of popular culture. Arguably, Jamaican music has had a greater influence on worldwide popular culture than the music of any other island. Interest and appreciation of reggae music, as well as its precursors (e.g., rocksteady, ska) and its derivatives (e.g., dancehall, dub), make reggae one of the few aspects of island popular culture that have extended into both developed and developing countries. The word "reggae" is practically synonymous with Jamaica, and the icon for the reggae movement remains the late Bob Marley. His songs are as likely to play on classic rock radio stations as they are on TV commercials or as background in elevators, and his face is as common on T-shirts as it is on music covers. Marley and the Rastafarian lifestyle associated with reggae spread the music's popularity largely through its distinctively colorful, countercultural aspects (as portrayed by Jimmy Cliff in the cult film *The Harder They Come*). While the dreadlock hairstyle, open use of marijuana, and even tricolor flag—all mainstays of reggae style—had their roots in Jamaica's cultural upheaval, they have since been widely adopted by non-Jamaicans seeking to express a sense of acceptable nonconformity.

Other Caribbean islands have equally rich musical traditions, some of which developed from century's-old African and European influences. The beginnings of calypso in Trinidad purportedly were a musical means of communicating information among working slaves because conversations in the fields were banned. But by the 1950s, modern calypso, enhanced by the steel-pan drum, became a worldwide craze, largely from Harry Belafonte's "Banana Boat Song." A dance/parlor game, the limbo, also became associated with music from the West Indies. In terms of culture proceeding in the opposite direction, other islands altered Western music to create their own native sound, most notably Cuba, as popularized by the film *Buena Vista Social Club*.

Polynesian music, mostly emanating from the tourist areas of Hawaii, is also immediately recognizable. The first broad exposure to Hawaiian music was at the Panama-Pacific International Exhibition in San Francisco in 1915, introducing the slack-key guitar and the jaunty strumming of the ukulele. Hawaiian music, or at least Americanized versions of it, was an immediate hit. Hundreds of Hawaiian-type songs were penned by American musicians, with such titles as "Oh How She Could Yacki Hacki Wicki Wicki Wo," "On the Beach at Waikiki," and "Lovely Hula Hands." Popular singers like Bing Crosby and Jimmie Rodgers added Hawaiian sounds to their repertoire.

Hawaiian music had a particularly strong effect on country music, which incorporated the smooth tones onto steel guitar (best exemplified by masters such as Jerry Byrd and Chet Atkins), and still heard in the genre today. The ukulele was popularized in the United States by Bing Crosby and Arthur Godfrey, and it often had a pineapple shape that both reflected an island image and produced a "warmer sound." The height of the Hawaiian Islands' music popularity was arguably Don Ho's "Tiny Bubbles" in 1966. By any measure, the song was a diluted offering of true Hawaiian styles, but it brought attention to the island's sounds, with Ho himself serving as a figurehead for that style.

Closely associated with Hawaiian music is the hula, a traditional dance with both ancestral and derivative versions seen throughout the Pacific Islands but universally recognized as a distinct cultural attribute of the Hawaiian Islands. Popularized variations of the dance's image have been spread through depictions in movies, television, and tourist spectacles, as well as the ever popular hula hoop, hula costumes (e.g., grass skirt, leis), figurine dolls (often placed on car dashboards), and even hula girl tattoos.

Few other islands in the world have had as much pop-culture impact as the Caribbean and Pacific islands, with two notable exceptions: Ibiza and Ireland. Although the Mediterranean islands have long had an association with music (including the irresistible song of the Sirens from *The Odyssey*), the Spanish island of Ibiza became a tourist destination precisely because of its music. In the 1980s, with the rise of electronic dance culture and techno/raves, Ibiza started many trendy nightclubs. By the 1990s, its reputation as a hedonistic, sun-drenched, and music-saturated paradise gave Ibiza iconic status as a popular culture sanctuary for both partying and music (as depicted in *It's All Gone Pete Tong*).

Over the last quarter-century, Ireland has been the island with some of the most popular and successful musical groups. Many of these groups maintain ties to the country's older roots by drawing upon Celtic melodies and rhythms. For example, the band U2 has embedded a distinctly Celtic ethos and context in their music and has incorporated Irish symbols, which evoke a mystical, isolated romanticism of Ireland as an island. Meanwhile, the sounds of traditional jigs, reels, and drinking songs can be heard in more or less authentic renditions of Irish pubs in cities all over the world.

ADVERTISING

The allure of islands has widely been used by advertisers to attract consumers. They have learned that the top fantasy vacation for men is being marooned on a tropical island with several members of the opposite sex. Advertisers have played on sensual images of islands very effectively. Couples holding hands and sharing loving glances (often with the male considerably older than the female in magazines catering to men), sunsets, and deserted beaches have been the staple of tourist advertising to specific islands or collective, exotic island locations. A further staple of island vacation advertising is promoting the impossible combination of luxuriant tropical vegetation (which needs quite a bit of rain) and continual sunshine!

However, travel is not the only product that island advertising sells: skin-care products are named "Bali Orchid Body Lotion" and "Bora Bora Sand Scrub," rum commercials offer "slow down" images, and "Souper Star Hawaii soups" appear on television and in print advertisements. The successful clothing chain Tommy Bahama's slogan is "purveyor of island lifestyles" and has expanded from clothing to include even home furnishings. The message in using islands in advertising is clear: islands will do for you what continental life will not.

Islands have long been used as indications of personal status. The elite American families have gone to Nantucket and Martha's Vineyard for generations, private islands in the Thousand Islands (of the St. Lawrence River) have castles built on them, and Atlantis (the lost continent originally described by Plato) has become a luxurious island resort in the Bahamas. The inaugural issue of *Fortune* (1930) magazine suggested that "As a great symbol of possession, the privately owned island may yet supplant the steam yacht."

Every island that is seeking tourists is also advertised as having friendly and welcoming natives. Of course, like many island images, this can be far from reality. Reportedly, one crime-ridden island even has a "Visitors' Society" that provides clothing to tourists whose luggage has been stolen and has members visit victims of violent crime during their stays in local hospitals!

VIDEO GAMES, COMICS, AND GESTURES

Video games may take place on real (*Monkey Island* in the Caribbean or *Over the Edge* in the Mediterranean) or fictional islands (*Sonic the Hedgehog, Super Mario, Legend of Zelda, Myst, Yoshi's Island, Adventure Island,* and *Amazing Island*). The same is true for comics with *X-man* (Atlantic Islands), *Patrouille des Castors* (Caribbean), *Corto Maltese* (Pacific) appearing on actual islands and *Whiz Comics, Exciting X Patrol,* and *Teen Titans* appearing on fictional islands.

The "shaka" sign is a common gesture associated with Hawaii and surfing but now widely used by teenagers to denote a carefree, "hang loose" attitude. This gesture involves extending the thumb and pinkie while keeping the middle fingers curled and rotating the wrist. This gesture even has a popular culture origin. Supposedly, it originated with a Hawaiian sugar-cane worker who lost his three middle fingers in an accident. His hand gesture of "all clear," using only the thumb and pinkie, became the shaka!

CONCLUSION

Every day we see images to remind us of islands—people in Hawaiian shirts, dreadlocked travelers, advertising posters, and beauty products. A popular parlor game that has been played for decades is "If you were stranded on a desert island, what books (or music or movies) would you want to take with you?" The connotation here is one of isolation but also of potential contentment.

Through popular culture, islands provide an accessible view of the exotic—"Come to an island and your life will change, even if it's only while you are there." Perhaps we have made islands a romantic icon because we need something to help us to escape life's tedium and troubles. Maybe without such a dream of going to a "better place" (even if it's only through what we wear, taste, or see), the tedium of normal life would be overwhelming. But beware—isolation, loneliness, and especially boredom may await the island visitor as well.

SEE ALSO THE FOLLOWING ARTICLES

Kon-Tiki / Missionaries, Effects of / Prisons and Penal Settlements

ADDITIONAL READING

Briand, P. L. Jr. 1966. *In search of paradise: the Nordhoff and Hall story.* Honolulu: Mutual Publishing.

Cantor, P. A. 2001. *Gilligan unbound: pop culture in the age of globalization.* Lanham, MD: Roman and Littlefield.

Clarke, T. 2001. *Searching for paradise. a grand tour of the world's unspoiled islands.* New York: Ballantine Books.

Day, A. G. 1986. *The lure of Tahiti: an armchair companion.* Honolulu: Mutual Publishing.

Day, A. G. 1987. *Mad about islands: novelists of a vanished Pacific.* Honolulu: Mutual Publishing.

Kernahan, M. 1995. *White savages in the South Seas.* London: Verso.

Kirsten, S. 2004. *Tiki style.* Cologne: Taschen.

Marsh, T., and J. Sparks. 2002. *The magic of the Scottish Isles.* Newton Abbot, UK: David & Charles.

Michener, J. A. 1951. *Return to paradise.* New York: Fawcett Crest.

Young, L. B. 1999. *Islands: portraits of miniature worlds.* New York: W. H. Freeman and Company.

POPULATION GENETICS, ISLAND MODELS IN

JEFFREY D. LOZIER

University of Illinois, Urbana-Champaign

Species are rarely spatially continuous or homogenous in their distributions; more often they are geographically subdivided into local subpopulations, within which individuals frequently interact but between which interactions are less common. Such population structure is often complex and can alter the way in which evolutionary forces (gene flow, genetic drift, and natural selection) act in nature. Island models and their variants provide a useful framework in which to investigate genetic variation and its potential ecological and evolutionary consequences in subdivided populations.

POPULATION GENETICS OF STRUCTURED POPULATIONS

In many scientific disciplines, researchers are interested in measuring demographic parameters for the organisms under study, including numbers of individuals and dispersal rates among neighboring populations. For many organisms, direct counts and observations of movement are impractical, and modern molecular methods that make use of putatively neutral genetic markers (e.g., DNA sequencing or microsatellite genotyping) provide an indirect way to estimate such parameters. However, making inferences from genetic data requires models that predict how variation is distributed within and among individuals and populations, which in turn requires assumptions about how individuals and populations are spatially arranged in nature. A few such models are highlighted in this article.

"CLASSIC" MODELS OF POPULATION STRUCTURE

Wright's Island Model and F-statistics

To investigate evolutionary dynamics in a structured population, Sewall Wright developed his classic island model, which continues to be commonly used for making inferences from genetic data. The model consists of an array of discrete subpopulations each of size N that exchange genes each generation by contributing equal proportions of individuals (m) to a migrant pool, which are then redistributed randomly and equally among subpopulations (Fig. 1). Gene frequencies

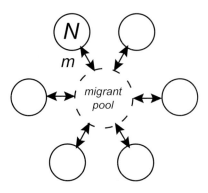

FIGURE 1 A schematic representation of Wright's island model consisting of six subpopulations of size N that contribute and receive an equal proportion of migrants (m) each generation.

in the subpopulations diverge over time as a result of random genetic drift, which is dependent on the population size, while at the same time are homogenized as a result of the movement of individuals. The degree of differentiation is expressed by the standard inbreeding coefficient F_{ST}, which can be thought of as the relatedness of a pair of genes sampled within a subpopulation relative to a pair sampled from the population as a whole. Theoretically, F_{ST} can range from 0 to 1 (though in practice this can vary), where 0 represents complete random mating and 1 represents complete isolation; high levels of gene flow thus tend to make subpopulations more similar, while low levels allow populations to diverge. With an infinite number of subpopulations at equilibrium between migration and drift, F_{ST} is related to the number of migrants per generation according to the equation $F_{ST} \approx 1/(4Nm + 1)$, which, theoretically, can be used to estimate migration rates. Implementing this relationship in practice can be problematic for a number of reasons. Of particular importance is that drift and migration are confounded in Nm, and F_{ST} estimated from genetic data can be highly influenced by recent demographic history as well as migration. Populations that are recently isolated, perhaps as a result of an island colonization event or habitat fragmentation, can exhibit low F_{ST} despite an absence of gene flow, because of their recent shared ancestry. Similarly, islands that receive migrants from a continental population but that never directly exchange migrants may also exhibit low F_{ST}. Interested readers should refer to the 1999 paper by Whitlock and McCauley, which critically examines the assumptions of the so-called Fantasy Island model. However, F_{ST} remains a useful measure of overall population structure, as long as researchers are cautious in its interpretation.

Steppingstone Models and Isolation-by-Distance

The steppingstone model (SSM) of population structure relaxes the particularly unrealistic assumption of equal probabilities of gene flow among all subpopulations; it is based on the empirical observation that individuals in most species are in some way dispersal limited. The SSM comprises an array of discrete subpopulations that can be arranged in one or two dimensions and is spatially explicit in that migration occurs most frequently between neighboring steppingstones (Fig. 2). This property gives rise to a positive correlation between genetic differentiation (usually estimated using pairwise F_{ST} or some similar estimator) and the geographic distance between the two subpopulations, and is known as isolation-by-distance. Isolation-by-distance can also arise in continuous populations as long as individuals are dispersal limited in some way. Tests for isolation-by-distance can be a useful way to get a general idea of the spatial scale of dispersal or to assess the equilibrium status of a population.

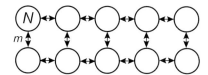

FIGURE 2 A schematic representation of a two-dimensional stepping-stone model.

"MODERN" MODELS OF POPULATION STRUCTURE

Several population genetic methods have made use of statistical advancements and increased availability of computational resources in recent years. Many of these methods use coalescent theory and realistic demographic models to infer parameters such as population sizes, divergence times, and migration rates. These methods relax a number of the assumptions of classical models and can estimate more complex parameters such as asymmetrical migration and populations that differ in size or change in size over time.

In some situations, researchers are interested in defining which sets of sampled individuals actually belong to discrete randomly mating populations, and there are several methods that make use of genetic data for this purpose. Some of these can also be used to estimate current migration events at the individual level, which is sometimes of greater interest than the long-term estimates provided by F_{ST}.

CONCLUSION

Overall, there is much insight to be gained about demographic history from the study of genetic variation in structured populations, as long as researchers are aware of the assumptions and limitations of the particular models being implemented. As methods with more realistic models are developed, many of the limitations of the more basic models will be overcome, allowing for better estimation of the parameters of interest in particular situations.

SEE ALSO THE FOLLOWING ARTICLES

Dispersal / Inbreeding / Island Biogeography, Theory of / Metapopulations

REFERENCES

Hey, J., and C. A. Machado. 2003. The study of structured populations—new hope for a difficult and divided science. *Nature Reviews Genetics* 4: 535–543.

Pearse, D. E., and K. A. Crandall. 2004. Beyond F_{ST}: Analysis of population genetic data for conservation. *Conservation Genetics* 5: 585–602.

Slatkin, M. 1993. Isolation by distance in equilibrium and non-equilibrium populations. *Evolution* 47: 264–279.

Whitlock, M. C., and D. E. McCauley. 1999. Indirect measures of gene flow and migration: $F_{ST} \neq 1/(4\ Nm+1)$. *Heredity* 82: 117–125.

Wright, S. 1951. The genetical structure of populations. *Annals of Eugenics* 15: 323–354.

PRINCE EDWARD ISLAND

SEE INDIAN REGION

PRISONS AND PENAL SETTLEMENTS

EPHRAIM COHEN

Hebrew University of Jerusalem, Rehovot, Israel

Prisons are institutions in which people are confined and deprived of a large range of liberties. The obvious parallel between islands and prisons resides in isolation, whether from mainland or from society. The more distant an island is from the mainland, and the deeper, colder, stormier, and more infested with predators (e.g., sharks) its surrounding waters are, the more effective it is as a prison. Opportunities to escape from such islands are drastically reduced or even completely eliminated. Whereas prisons on the mainland are usually cell prisons, on islands convicts were not always confined but rather held in camps. On certain islands, prisoners were merely slaves of the free settlers or were subjected to harsh work in the mining or timber industries. Some ex-convicts had to serve time following release; thus, although freed, they were not allowed to leave the islands. Under such conditions the islands were regarded as penal settlements that had considerable economic benefits for the mainland. Prison islands have nowadays become tourist attractions because famous criminals or political figures were detained there. Some of the islands, which hold natural beauty, became attractive targets for tourism (including ecotourism) and for mainland vacationers.

AUSTRALIAN ISLAND PRISONS

The practice of sending prisoners from England to its North American colonies stopped as the United States of America gained independence in 1783. Because of prison overcrowding, the British Crown devised the solution (initiated in 1786) of shipping convicts to the newly discovered Australia. The convicts, (both criminal and political) suffered a long and harsh 8-month journey under overcrowded conditions, being often chained and supplied with meager food rations.

Throughout the eighteenth and nineteenth centuries Tasmania and some offshore islands of Australia served as prisons for criminals and political convicts. Later, for economic reasons, some islands became penal settlements. From the beginning, although provided with basic supplies from England, prisoners were expected to become self-sufficient.

Cockatoo Island

Cockatoo Island, located in Sydney Harbor, New South Wales, was a prison during 1839–1869. The convicts were employed in constructing silos for storing the colony grain and in excavating rocks for construction projects in Sydney.

Norfolk Island

Norfolk Island, located in the Pacific Ocean between Australia, New Zealand, and New Caledonia, is part of the external territories of Australia. In order to claim the island before the French colonized it, in preparation for commercial development, England in 1788 sent 15 convicts and 17 free men there. More convicts were later sent to create a penal settlement that would be capable of providing farm supplies to Sydney. This first penal settlement ceased to exist in 1813. However, a second penal settlement was established in 1824 with the purpose of transporting there the worst detainees, those who had committed crimes after arrival as convicts in New South Wales. Norfolk Island was notorious for exposing convicts to harsh punishment, torture, constant flogging, and scarce food. The second penal settlement was closed in 1855 when the last prisoners were transferred to Tasmania.

Port Arthur

Port Arthur is located on the Tasman peninsula about 60 km southeast of the capital city, Hobart. The peninsula is separated from the main island of Tasmania by a narrow 30-m-long stretch of land called Eaglehawk Neck. Port Arthur served as a penal colony from 1830 until its closure in 1877. Escape of prisoners was almost impossible because the surrounding waters were shark-infested and the narrow isthmus was guarded by soldiers and dogs. Port Arthur was the preferred destination for the hardiest English convicts as well as for second offenders, those who had committed another offense after arriving in Australia. The place was also a destination for juvenile convicts, 9–18 years old, separated from adults in a special prison located at Point Puer.

The first 150 convicts in Port Arthur worked as slaves to establish the timber industry. By exploiting the hard work of the convicts, Port Arthur became Tasmania's industrial center for timber, shipbuilding, brickmaking, wheat growing, and shoe making. The convicts suffered cruel physical conditions and were punished by flogging. Later, according to new ideas of treating convicts developed in England and based on the design of Pentonville Prison in London, Port Arthur turned into a Model Prison in 1852. Physical punishments were replaced by psychological ones, using absolute silence and complete isolation in tiny cells (Fig. 1).

FIGURE 1 Interior of the Model Prison, Port Arthur, Tasmania, Australia. Photograph by Joern Brauns.

During 1830–1877 Port Arthur housed about 12,000 convicts (men and boys), and after closure it became a geriatric home and mental asylum for ex-convicts. Today Port Arthur is one of the main tourist attractions for visitors to Tasmania.

Rottnest Island

Rottnest Island, which is located 19 km offshore from Perth, became a prison for Western Australian Aborigines in 1838. Since the end of the nineteenth century the island has been a favorite holiday resort for Perth inhabitants.

St. Helena Island

St. Helena Island was the principal Queensland prison for men from 1867–1993. It is located 21 km east of Brisbane in Moreton Bay, a few kilometers from the mouth of Brisbane River. When an Aboriginal nicknamed Napoleon was exiled to this island, it received its present name from the better-known St. Helena in the South Atlantic, where Napoleon Bonaparte spent his final years. During the early days it was a secure prison, and the worst criminals were sent there. It was known as "the hellhole of the Pacific" and as "Queensland's Inferno" for its physically harsh conditions. St. Helena Island was initially meant to be a self-sufficient prison, even exporting products to the mainland. Today, as a national park, the island has become a tourist destination for visitors to the Brisbane area.

Sarah Island

Sarah Island is located in Macquarie Harbour, west of Tasmania. The penal colony was established in 1822 and closed 11 years later, in 1833. The worst convicts who had escaped other penal systems were sent there. The prison was notorious for very harsh treatment of inmates, who suffered crowded conditions, hard forced labor, malnutrition, scurvy, and dysentery in addition to widespread severe flogging as punishment. Convicts were employed in the shipbuilding industry and logging the Huon pine trees, while in chains. Topographical conditions such as access via a narrow gate ("Hell's Gate") of dangerous sea and rough mountains, which separate the penal colony from the other settlements on the island, made escape dangerous and almost impossible. Aspects of the horrible British penal system have been described in several books and a play. The ruins of the penal settlement are today on the Tasmanian Wilderness World Heritage list.

SOUTH AND CENTRAL AMERICA

Coiba Island

Coiba is a large island (15 × 50 km) off Panama's Pacific coast. Its penal colony was established in 1912 and closed down in 2004. The island is far from the mainland, and its surrounding waters are famous for strong currents and aggressive sharks, which make any escape quite difficult. The worst Panamanian criminals were imprisoned there, as were political opponents of the military regime of General Noriega. The prisoners worked long hours and were subjected to extreme torture. Today Coiba is a tourist attraction.

Dawson Island

Dawson Island is located in the Strait of Magellan, 100 km south of Punta Arenas city in Chile. The island became a prison for political detainees following the military coup in 1973 and was closed down in 1975.

Devil's Island

Devil's Island, one of three islands off the coast of French Guiana, was part of a French penal colony system. It was established as a prison by Napoleon the III in 1852 and closed down in 1952 by the French government. Throughout that century the prison was used by France for sending political convicts like the anarchist Clément Duval (1938) as well as thieves and murderers. The total number of prisoners spending their term on Devil's Island exceeded 80,000. At the timber camps on the island the convicts had to endure forced labor in malaria-infested areas. Only a few prisoners managed to escape through the surrounding dense jungle.

The French authorities attempted to convert the island into a penal settlement because, according to an 1854 law, convicts who had spent more than 8 years were forced to stay permanently in French Guiana. Convicted women were sent to the territory in order to marry freed convicts (as was done in Australia).

Devil's Island became famous in history because in 1895 the Jewish French army captain Alfred Dreyfus was convicted of treason and sent to the prison. The sentence evoked a heated political reaction (*J'Accuse* by Émile Zola), and the case, as well as the antisemitic atmosphere in France, had a great impact on the Jewish journalist Theodor Herzl. The Dreyfus case was eventually instrumental in convincing Herzl that Jews needed to create their own homeland, and thus in his becoming the father of modern Zionism.

The Devil's Island prison was a subject of books such as *Papillon* by ex-convict Henri Charrière, *Dry Guillotine* by Rene Belbenoit, and *Plan de Evasion* by Adolfo Bioy Casares, as well as a movie based on *Papillon* and folk songs.

Gorgona Island

Gorgona Island is located 50 km off the Colombian Pacific coast. The island was used as a high-security jail from the 1950s until 1984. Escape of prisoners was highly discouraged by the presence of poisonous snakes on the island as well as shark-infested waters that separated the island from the Colombian mainland. In 1985 the island became a National Natural Reservation Park, where its endemic species, subtropical forests, and coral reefs are strictly preserved.

Martin Garcia

Martin Garcia is located in Rio de la Plata, 3.5 km off the Uruguayan coast. In 1765, the Spaniards installed a military prison for deserters. For about 90 years (1870–1957) the island served as prison for both political and criminal detainees. The crewmen of the German battleship *Graf Spee* were detained there; the Nicaraguan poet Ruben Dario lived on the island (though not as a prisoner), as did the Argentinian dictator Juan Peron. The island was a notorious site for detention and torture of political opponents to the military dictatorship in Argentina during 1976–1983.

San Cristobal

On San Cristobal in the Galápagos Archipelago a penal colony was established during 1850–1860 for Ecuadorian prisoners. A second penal colony was established in 1946, turned essentially into a concentration camp, and was closed down in 1959.

SOUTH AFRICA: ROBBEN ISLAND

Robben Island is located 12 km offshore from Cape Town, South Africa. Since the end of the seventeenth century the island has been used as prison for political leaders from various Dutch colonies or for isolating lepers (1836–1931). Under apartheid, Robben Island became notorious for imprisoning black political leaders, notably Nelson Mandela. The Island was declared a World Heritage Site and has been transformed into a popular tourist attraction.

NORTH AMERICA

Alcatraz Island

Alcatraz Island is located in the middle of San Francisco Bay, California (Fig. 2). It served as a lighthouse (first erected in 1854), a military fortification, and a military prison. The island was transformed into a federal prison in 1934, was closed in 1963, and has since become a national recreation area. A few dozen escapes were recorded, yet scores were apparently successful. Escape attempts from the penitentiary were the subject of a motion picture (*Escape from Alcatraz*, 1979) and a TV series.

FIGURE 2 Alcatraz Island as seen from Coit Tower in San Francisco. Photograph by Jon Sullivan, PD Photo.org.

Johnson's Island

This prison was located in Sandusky Bay on Lake Erie, Ohio, and served for Confederate prisoners of war captured during the Civil War. The prison was opened in 1862 and was closed at the end of the war.

ASIA

Indonesia

Indonesia has a long history of using islands in its vast archipelago as prisons. For example, Buru Island in the Maluku region was a place for political prisoners after the attempted 1965 coup that led to the ouster of President Sukarno; Atauro Island, off the north coast of East Timor, was used as a penal colony for rebellious East Timorese.

Russia: Sakhalin

Sakhalin is a large island located north of Japan, in the entrance of the Sea of Okhotsk in the Russian Far East. Being a remote and harsh place, it served as a prison island and was the largest Russian penal colony under the tsars; political opponents were also exiled there. Following a visit to Sakhalin, the famous Russian writer Anton Chekhov described the hellish life in the island in his book *The Island: A Journey to Sakhalin*. Today Sakhalin has become a major energy supplier because of its vast oil reserves.

The Solovetsky Islands

This is a group of six islands located in the White Sea, Russia, in Onega Bay. They served as an exile location during the Russian Empire and became a prison in 1926 after the Russian Revolution. The islands were an integral part of the vast Gulag system where many prisoners were detained. Being close to the Finnish border, the prison was closed in 1939 at the beginning of World War II. The islands have become a tourist attraction, primarily because of the famous fifteenth-century monastery, and are on UNESCO's World Heritage List.

EUROPE

Chateau d'If

Chateau d'If was initially a fortress and a castle built (during the period 1516–1529) by the French king François I on a small rocky island in the Mediterranean Sea about 3 km off the city of Marseilles in southern France. In 1634 the chateau became a state prison for mostly religious (e.g., several thousand Huguenots) and political (e.g., Gaston Crémieux, a leader of the Paris Commune) detainees. The notorious prison was made famous by Alexandre Dumas's novel *The Count of Monte Cristo* (1844–1845). The prison was closed at the end of the nineteenth century and since 1890 has become a popular tourist attraction.

İmralı

Only in 1935 was a prison built on the Turkish island of İmralı located in the Sea of Marmara. A number of political detainees, notably the former Prime Minister Adnan Mederes, served terms there. Since 1999 the jail has become a maximum-security prison for its only inmate, Kurdish rebel leader Abdulla Ocalan.

INDIAN AND PACIFIC OCEANS

Andaman Islands

The Andaman Islands are a group of Indian islands located in the Bay of Bengal, separated from the Malay Peninsula by the Andaman Sea. The first British penal settlement established there, in 1789, was closed in 1796. Dangerous convicts who broke the law in the penal settlement were sent to Viper Island, where they were put in chains while being subjected to forced hard labor. A second penal settlement was established in 1858 for rebels against British colonial rule in India. A prison called the Cellular Jail was built there from 1896 to 1906 and was used first by the British and, during World War II, by the Japanese occupation. Today the Andaman Islands are a territory of India, and the Cellular Jail is a national memorial.

Mauritius

Mauritius is an island in the Indian Ocean, east of Madagascar. Primarily, the island can be regarded as a penal settlement for Indian convicts who were sent on demand, especially as a labor force for the sugar cane industry. The island was a penal settlement from 1790 until 1853, when the last convict was liberated.

New Caledonia

New Caledonia became a French colony in 1853 and, since 1986, has been included under the United Nations list of Non-Self-Governing Territories. The island is located in the Melanesia region of the southwestern Pacific. A penal colony was established on the island in 1864 by Napoleon III. Most prisoners were French criminals, but inmates also included political convicts, such as 4300 socialists following the insurrection of the Paris Commune in 1871. Following a revolt of Arab and Berber tribes in Algeria, many thousand prisoners were transported to New Caledonia. It is estimated that about 22,000 prisoners were kept in the island between 1864 and 1922, when the penal colony was closed down.

SEE ALSO THE FOLLOWING ARTICLES

Popular Culture, Islands in / Rottnest Island / Tasmania

FURTHER READING

Charrière, H. 1971. *Papillon*. New York: Pocket Books.
Hughes, R. 1988. *The fatal shore: the epic of Australia's founding*. New York: Vintage Books.

RADIATION ZONE

KOSTAS A. TRIANTIS AND
ROBERT J. WHITTAKER

Oxford University, United Kingdom

Species richness on island systems is a function of colonization, speciation, and extinction. In island faunas and floras, there exist radiation zones, in which phylogenesis (including by adaptive radiation) increases with distance from the major source region. Within-island speciation and within-archipelago speciation (sometimes termed simply archipelago radiation) can thus be major contributors to species richness.

THE THEORETICAL BACKGROUND

In their seminal book *The Theory of Island Biogeography*, MacArthur and Wilson (1967) wrote, "In equilibrial biotas . . . the following prediction is possible: adaptive radiation will increase with distance from the major source region and, after corrections for area and climate, reach a maximum on archipelagos and large islands located in a circular zone close to the outermost dispersal range of the taxon." They termed such peripheral areas the *radiation zone* (Fig. 1).

The biogeographical circumstances in which radiations take place are reasonably distinctive. Radiations are especially prevalent on large, high, and remote islands, lying well beyond a group's normal dispersal range. Here the low diversity of colonists, and the disharmony evident in the lack of a normal range of interacting taxa, facilitates in situ diversification for those few representatives of a group that do manage to reach the islands. For less dispersive taxa, their radiation zones may coincide with less remote archipelagoes, which have a greater degree of representation of interacting taxa than do the most remote archipelagoes. Thus, macroevolutionary aspects of island biogeography complement ecological aspects. That is: (1) within-island radiations should occur most often on distant, large islands of complex topography; (2) colonization should occur more often on near than distant islands; (3) speciation should occur more often on distant islands than on near islands. Hence, the circumstances for radiation reach their synergistic peak on the most remote islands, the epitome being the Hawaiian Island group (Fig. 2).

To illustrate how the low diversity of colonists, and the disharmony evident in the lack of a normal range of interacting taxa, facilitate in situ diversification, ants dominate arthropod communities across most of the world, but in Hawaii and southeastern Polynesia they are absent (or were, prior to human interference). In their place, there have been great radiations of carabid beetles and spiders, and even caterpillars have in a few cases evolved to occupy predatory niches.

NATURE'S EVOLUTIONARY RADIATION EXPERIMENTS

Examples that fit this idea of maximal radiation near the dispersal limit include, among others, birds on Hawaii and the Galápagos, frogs on the Seychelles, gekkonid lizards on New Caledonia, lemurs on Madagascar, ants on Fiji, and *Anolis* lizards in the Caribbean.

Although the Galápagos are renowned for other endemic groups, notably the tortoises and plants, the most famous group of endemics are Darwin's finches (Emberizinae). The radiation of the lineage has taken place in the context of a remote archipelago, presenting extensive "empty niche" space, in which the considerable (but not excessive) distances between the islands have led to phases of inter-island exchange—but only occasionally. Differing environments have apparently selected for different feeding niches both between and within islands during a radiation process of some 3 million years, involving both allopatric and sympatric phases. Thereafter, behavioral differences between forms maintain sufficient genetic distance between sympatric populations to enable their persistence as (to varying degrees) distinctive lineages.

The Hawaiian honeycreepers (or honeycreeper-finches) have shown an even greater radiation than Darwin's finches. They are a monophyletic endemic group perhaps best considered a subfamily, the Drepanidinae. Estimates of the number of species known historically range from 29 to 33, with another 14 having recently been described from subfossil remains (and it seems more are in the process of being described); most extinctions occurred between the colonization of the islands by the Polynesians and European contact.

The Hawaiian "Experiment"

The Hawaiian endemic crickets are thought to have derived from as few as four original colonizing species, each being flightless species arriving in the form of eggs carried by floating vegetation. The ancestral forms were a tree cricket and a sword-tail cricket from the Americas and two ground crickets from the western Pacific region. Three of the successful colonists have radiated extensively, and Hawaii now has at least twice as many cricket species as the continental United States. Much later, a further eight species have colonized, but these are considered to have been introduced by humans. The tree crickets (Oecanthinae) have been calculated on phylogenetic grounds to have colonized Hawaii about 2.5 million years ago, radiating into three genera and 54 species (43% of the world's known species), with the greatest diversification seen within the older islands, which were occupied earliest. They have radiated into habitats not occupied by their mainland relatives.

Drosophila and several closely related genera in its subfamily include about 2000 known species, of which Hawaiian drosophilids (the closely related genera *Drosophila* and *Scaptomyza*) account for some 600–700 species—although, with further taxonomic work, the eventual figure could be as great as 1000 species. The ancestors of the Hawaiian drosophilids (a single species, or at most two) probably arrived on one of the older, now submerged islands and have radiated within and between islands. In general, the older islands to the west contain species ancestral to those of the younger islands to the east; that is, most, but not quite all, of the inter-island colonizations have been from older to younger islands (this pattern is termed the progression rule). The general trend of colonization from old to young Hawaiian islands in *Drosophila* is matched by similar trends in the radiation of the silversword alliance

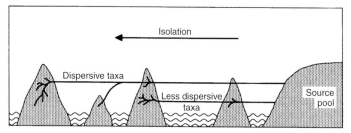

FIGURE 1 Dispersive taxa radiate best at or near to their effective range limits (the radiation zone), but only moderately, or not at all, on islands near to their mainland source pools. Less dispersive taxa show a similar pattern, but their range limits are reached on much less isolated islands. The increased disharmony of the most distant islands further enhances the likelihood of radiation for those taxa whose radiation zone happens to coincide with the availability of high island archipelagoes. From Whittaker and Fernández-Palacios (2007).

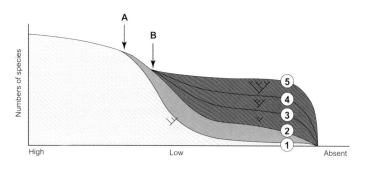
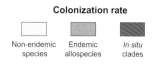

FIGURE 2 Conceptual model showing the development of species richness on large islands or archipelagoes that experience varying rates of colonization as a result of varying degrees of isolation. As explained by Heaney, on islands near a species-rich source that initially lack the study taxon, all species will be present through direct colonization because high rates of gene flow will swamp out any potential speciation; thus, no endemic species will be present, but many non-endemics will be present. As the average rate of gene flow drops below approximately the level of $Nm = 1$ (gene flow equal to one individual per generation) for the study taxon (point A), anagenesis will begin to take place, and endemic species will develop, although they will be outnumbered by non-endemic species. These endemic species (between lines 1 and 2) will have their sister taxon in the source area, not on the island/archipelago. As colonization becomes still less frequent, and as time passes, phylogenesis will produce endemic clades in which the endemic taxa have their sister taxon on the island/archipelago, not in the source area (species between lines 2 and 3). As more time passes, the oldest clades will become progressively more species-rich (radiation zone) (between lines 3 and 5). From Heaney (2000).

(tarweeds), which is a group of 30 species in three genera (*Dubautia*, *Argyroxiphium*, and *Wilkesia*) in the Asteraceae. The alliance appears to have descended from a colonization event by a bird-dispersed herbaceous colonist (itself a hybrid from the genera *Madia* and *Raillardiopsis*) about 5 million years ago, matching the age of the oldest large island, Kauai. They have diversified into a range of life forms, including monocarpic rosette plants, vines, mat-forming plants, and trees, and they occur in habitats ranging from desert-like to rain forest.

The Macaronesian "Experiment"

The Canarian invertebrates provide an outstanding example of radiation on rather less isolated islands than Hawaii and the Galápagos. There may be as many as 6000 native Canarian invertebrate species, of which half are considered to be endemic. In general, the Canarian invertebrate fauna is characterized by the absence of certain groups (e.g. scorpions, Cicadae, Solifugae, and dung beetles), but these absences have been compensated for through active speciation within colonist lineages, producing a high species/genus ratio. Additionally, 99 genera are considered exclusive to the archipelago, 58 of them being present in more than one island (multi-island genera) and 41 restricted to one island, of which 25 are found on the island of Tenerife. Radiation is particularly evident in land snails, spiders, and beetles, with at least 24 different genera producing 15 or more endemic species. The largest genera of beetles (*Laparocerus*), diplopod millipedes (*Dolichoiulus*), and spiders (*Dysdera*) are represented in all the major habitats, from the coast up to 3000 m, in arid and wet zones, in forested and open areas (even in lava tubes), all over the archipelago, a pattern strongly supportive of the label "adaptive radiation."

For Macaronesia, the estimated native vascular flora comprises some 3200 species, of which about 680 (~20%) are endemic. The Canaries are the largest in area, the highest, and the richest, with 570 endemics, about 40% of the native flora. Forty-four Macaronesian lineages are endemic at the generic level also, with exactly half of them restricted to the Canaries. Support for the use of the term "adaptive radiation" in several of the above Macaronesian genera comes from studies of features such as habit, leaf morphology, and habitat affinities. Evidence of similar morphology in species of the same and of different genera (termed parallel or convergent evolution) occurring in similar habitats, is supportive of a model of selection having favored particular adaptive outcomes. Within the Macaronesian flora as a whole, most genera are represented by only one or two species, and most of those with over four occur in the Canaries. In general, radiation of lineages has been greatest on the larger islands with the greatest diversity of habitats. The island with the greatest number of endemic plant species (320 species) is Tenerife, and it also has the largest numbers of *Aeonium* (12 species), *Sonchus* (11 species), and *Echium* (9 species). Most endemic plant species have a restricted distribution, 48% occurring on a single island, and a further 15% on two.

EVOLUTIONARY RADIATION NOT UNIVERSAL EVEN IN THE RADIATION ZONE

Although there seem to be numerous examples of taxa radiating at or near their outermost dispersal range, there are also plenty of exceptions: taxa that have not radiated much at remote outposts. Such cases include terrestrial mammals on the Solomon Islands and snakes and lizards on Fiji, and within the invertebrates of the Canary Islands, for example, as many as 57 of the endemic genera are monotypic. Similarly, nearly 50% of the ~1000 native flowering plant species of Hawaii are derived from fewer than 12% of the ~280 successful original colonists. Most of the remaining colonists are represented by single species. Thus, it is clear from even the most remote island archipelagoes that not all lineages within a single taxon have radiated to the same degree. Such differences may reflect the length of time over which a lineage has been present and evolving within an archipelago. As MacArthur and Wilson (1963) noted, "To say that the latter [nonradiating] taxa have only recently reached the islands in question, or that they are not in equilibrium, would be a premature if not facile explanation. But it is worth considering as a working hypothesis." Nonetheless, recent phylogenetic analysis has made it clear that some lineages persist for lengthy periods on oceanic islands without radiation. In short, although early colonization of remote, topographically complex archipelagoes may favor evolutionary radiations, these are not the only factors of significance, and such circumstances do not lead to radiations within all lineages.

SEE ALSO THE FOLLOWING ARTICLES

Adaptive Radiation / Crickets / *Drosophila* / Galápagos Finches / Honeycreepers, Hawaiian / Lemurs and Tarsiers / Silverswords / Taxon Cycle

FURTHER READING

Heaney, L. R. 2000. Dynamic disequilibrium: a long-term, large-scale perspective on the equilibrium model of island biogeography. *Global Ecology and Biogeography* 9: 59–74.

MacArthur, R. H., and E. O. Wilson. 1963. An equilibrium theory of insular zoogeography. *Evolution* 17: 373–387.

MacArthur, R. H., and E. O. Wilson. 1967. *The theory of island biogeography*. Princeton, NJ: Princeton University Press.

Ricklefs, R. E., and E. Bermingham. 2007. The West Indies as a laboratory of biogeography and evolution. *Philosophical Transactions of the Royal Society B: Biological Sciences* 363: 2393–2413.

Schluter, D. 2000. *The ecology of adaptive radiation.* Oxford: Oxford University Press.

Stuessy, T. F., and M. Ono, eds. 1998. *Evolution and speciation of island plants.* Cambridge, UK: Cambridge University Press.

Wagner, W. L., and V. L. Funk, eds. 2005. *Hawaiian biogeography: evolution on a hot spot archipelago.* Washington, DC: Smithsonian Institution Press.

Whittaker, R. J., and J. M. Fernández-Palacios. 2007. *Island biogeography: ecology, evolution, and conservation,* 2nd ed. Oxford: Oxford University Press.

RAFTING

CHRISTOPHE ABEGG

German Primate Centre, Padang, Indonesia

Natural rafting is the rare occurrence in which animals or plants of any kind succeed in crossing a sea strait using tree parts and vegetation. Although rafting has understandably been more often regarded as a means of dispersal for marine plants and animals, it also happens to terrestrial plants and non-volant animals. It is thought that animals are launched by chance from riversides (at any time for reptiles or small mammals but probably at night and during a flood for macaques; Fig. 1), or along the coast (for any animal whenever a tsunami strikes). Once the plant or animal is carried out to sea, another series of contingencies is needed for more than one individual (i.e., at least one male and one female) to reach an oceanic island and populate it. The chances of survival on a natural raft, once drifting on the sea, seem to differ according to species, judging by the terrestrial species that made their way to islands that were never connected to any land. Reptiles, for instance, are usually able to resist starvation for longer periods than are mammals.

FIGURE 1 After flooding, rivers carry tree trunks and branches, which can serve as rafts for a variety of plants and animals. The longtailed macaque, shown here, is a lightly built terminal branch feeder, whose ecological niche has raised its chances of dispersal by rafting. This species is even a good candidate for the title of world champion rafter among mammals.

OVER-SEA DISPERSAL BY RAFTING: SPECIES AND CONTEXT

What taxa tend to use natural rafting, with what vectors, and to get where? Long-distance dispersal of reef corals by rafting has been studied extensively. Rafting on algae, pumice, and other floating materials has also been proposed for marine invertebrates. In contrast to active dispersal and to passive dispersal by wind or water, rafting is a rare phenomenon. Some reptiles, lizards, and skinks appear to have had some success with the use of rafts, as their distribution on oceanic islands attests. The distribution and diversification of *Anolis* lizards in the Caribbean may be explained by infrequent rafting events; indeed, anoles have been observed rafting between islands in the Lesser Antilles after a hurricane. The *Neotarentola* geckos seem to have reached Cuba from North Africa by following north equatorial currents across the Atlantic Ocean, a journey of 6000 km. A frog endemic to the islands of Sao Tomé and Principe, 300 km off the west coast of Africa, is thought to have colonized its habitat using rafts and oceanic currents, possibly using rotten trunks. Rafting via floating mats of vegetation has been invoked to explain the frequent occurrence of land snails on remote oceanic islands, such as the Galápagos and Hawaii. South Pacific top shells have colonized many islands of the Pacific Ocean, most likely through rafting, possibly using buoyant rafts of kelp, from Australia and New Zealand up to South America. In the same way, certain insects appear to use rafting rather than flying as a means of dispersal. Weevils (*Rhyncogonus*) are widespread on the remote islands of the Pacific, and their distribution is attributed to their tendency to attach their eggs to leaf phyllodes, facilitating transport on vegetation mats. The occurrence of certain ant lineages in Fiji has been attributed to their ability to nest in plant cavities. Rafting may likewise allow terrestrial plant species to colonize remote islands.

Among medium-sized mammals, the genus *Macaca* (Mammalia: Cercopithecidae) is unique among non-human primates for the range of habitats it has colonized, from equatorial to temperate ecosystems, from evergreen primary forests to grassland or even habitats modified by humans, but also from continents to deepwater islands.

Eleven macaque species inhabit the Indonesian archipelago. Their ancestors colonized oceanic islands as did no other primates or even medium-sized mammals in the world, and they are found on a sufficient number of islands to allow significant assessment of dispersal and stocking determinants.

For terrestrial mammals, colonization by rafting has been inferred from their presence on offshore islands that were most likely never connected to the mainland during their geological history. However, for some animals that can float, an alternative to rafting, such as floating and swimming, can also be inferred; this dispersal mode was indeed suggested for the colonization of the Galápagos by tortoises. Rafting between islands that were often connected by land during the past few million years cannot be excluded, but the most plausible hypothesis remains terrestrial dispersal—for instance, the dispersal of plants and animals between peninsular Malaysia and Sumatra.

As a means of reaching most oceanic islands, rafting is an unpredictable event for terrestrial animals. However, in the Indonesian archipelago, the occurrence of rafting is connected with climate changes because the distances that had to be crossed to reach an oceanic island were shorter during cold periods, when straits' widths were reduced by lower sea levels. Indeed, there have been climatic fluctuations over much of Earth's history, but they intensified with polar ice accumulations from 5 to 1.6 million years ago (i.e., during the Pliocene); then the predominance of glacial intervals became characteristic of the last 1.6 million years (i.e., the Pleistocene). During the Holocene (from 10,000 years BP), the global climate entered an interglacial period and reached the present warm conditions. Interglacial periods are correlated with high sea levels, and glacial periods with low sea levels. Lands situated as low as 120 m below present sea level (BPL) were regularly exposed, and some early glacial periods might have lowered sea levels even more than 120 m, up to 200 m BPL. Quaternary climatic changes associated with periodic glaciations had a profound influence on the distribution of primates and other mammals. In such circumstances, the existence of refuges, the emergence of land bridges, and the possibility of sea rafting strongly affected the fate of populations.

Because glacially lowered sea levels provoked successive exposures of the southern end of the continental shelf of Asia (the Sunda Shelf), mammals, including a macaque progenitor, were further able to enter this region, which is today made up of islands. Because few terrestrial mammals are believed capable of dispersing over sea channels using natural rafting, past sea-level changes represent a crucial factor in understanding the present distribution of mammals in the region. Significantly, beyond those lands that have never been exposed during the last million years, there now exists an endemic mammalian fauna that is well differentiated from mainland Asian fauna, for instance on the Philippines and Sulawesi, but also on islands off West Sumatra.

The region crossed by the Wallace Line represented an effective barrier that has drastically filtered the dispersal of terrestrial mammals of Asian origin. Similarly, mammals' dispersal to the oceanic islands found off Sumatra's west coast was sharply restricted. At the eastern end of the Indonesian archipelago where the Australian region begins, the only Asian mammals to be found, apart from the rodents, were probably dispersed by humans. Off this 120- to 180-m sea-depth limit, where the sea floor has been regularly exposed in the past, at the periphery of the Sunda shelf, some islands are today nonetheless inhabited by terrestrial mammals of Asian origin. Judging by the high degree of endemism of these islands' faunas, their ancestors are consequently assumed to have crossed sea barriers early on, using either a land bridge connection that later disappeared, or natural rafts.

DISPERSAL BY RAFTING FOR MAMMALS: THE MACAQUE EXAMPLE

Among non-volant mammals, very few species could disperse by rafting over sea straits and populate islands successfully: only some rodents, shrews, civets, and a few macaques, each with varying degrees of success. Indeed, the only native terrestrial mammals to occur on the more isolated islands of the central Pacific are bats and seals. It has been suggested that lemurs, tenrecs, carnivores, and rodents have crossed the 400-km strait separating Madagascar from the African mainland by rafting. However, if ancient vicariance is improbable because Africa and Madagascar separated before those animals had evolved, then terrestrial migration via a transitory land bridge should not be excluded, as the gap to cross by rafting to reach Madagascar from Africa was exceptionally wide for medium-sized mammals. Among such mammals thought to have used natural rafts, the macaques are probably the biggest in size. Their ability to disperse by raft also differs according to species. For instance, in about a third of the pigtailed macaque's time, the long-tailed macaque colonized ten oceanic islands, more or less, whereas the pigtailed taxon is thought to have colonized only two oceanic islands. Let us follow chronologically how the pigtailed macaque and then, much more efficiently, the long-tailed macaque used natural rafts to colonize oceanic islands otherwise unreachable by any other medium-sized primate and by very few other mammals.

Separated from the continental shelf by trenches over 180 m deep, Sulawesi and the Mentawai Islands would represent refuges for the first descendants of the ancestral macaque coming from India, refuges that the long-tailed macaque was not able to reach in later times. From India, the first macaques to enter insular Southeast Asia dispersed by land, taking advantage of the glacially induced emergence of the Sunda Shelf. A macaque ancestral stock probably reached Sulawesi by island hopping and natural rafting, whereas it reached the Mentawai Islands over a transitory land bridge created by low sea levels during a maximal Plio-Pleistocene glaciation. The long-tailed macaque progenitor would have dispersed only long after.

Sulawesi macaque progenitors probably migrated during the Pleistocene by natural rafting from Borneo across a water gap of about 60 km (the Macassar strait), with only a few winners. The most convincing argument for this latter possibility lies in the absence of other primate taxa in Sulawesi, unlike in the Mentawai Islands. The strait separating the Mentawai from the Sumatran shelf is 200 m deep and 30 km wide. With the exception of the Mentawai macaques, which are closely related to the Sulawesi taxa suspected to have colonized their island by raft, the other three endemic primates of the Mentawai Islands do not have any Asian congeners on oceanic islands, suggesting that overwater dispersal had extremely low chances of success for these primates. The progenitors of today's five endemic Mentawai primates most probably arrived by land from the Sunda shelf at about the time or soon after macaques first colonized Sulawesi.

It can be hypothesized that after a particularly cold glacial episode, the ancestral macaques restricted their distribution to a few refuges such as Sulawesi and the Mentawai Islands and that only afterwards did new taxa disperse toward more central ranges. Two species were able to take advantage of climate recovery and recolonize mainland insular Southeast Asia, whereas nine remained in their refuges.

According to some authors, the pigtailed macaque's ancestor, now populating central ranges, may have originated in the Mentawai or Sulawesi Islands, which acted as refuges during cooling. The Mentawai archipelago is more plausible in this regard, as it is comparatively less isolated from the adjacent continental shelf than is Sulawesi. However, this would imply that a recolonization of the adjacent mainland, Sumatra, occurred from the Mentawai Islands either by rafting during an interglacial period or by land during a glacial maximum that exposed the 200-m-deep present sea floor. Dispersal from Siberut to the mainland via a land bridge is unlikely because if this had occurred, then the fauna of the Mentawai would not be as specific, with such a high degree of primate and mammal endemism, as it is today. More modern taxa would have taken advantage of such a land bridge and would be present on the islands now. Most likely, for the second time in their history after the colonization of Sulawesi, the ancestral pigtailed macaques dispersed by rafting, from Siberut to Sumatra by island hopping, and then succeeded in invading central ranges of Southeast Asia.

THE LONG-TAILED MACAQUE: A CHAMPION OF NATURAL SEA RAFTING

Why were long-tailed macaques so successful in reaching oceanic islands, in comparison with pigtailed macaques and all other medium-sized mammal taxa worldwide?

As the long-tailed macaque's present distribution suggests, its progenitor was probably able to use natural rafting over sea straits, from 17 up to 150 km wide during maximal sea regressions, to reach such islands as Maratua, Simeulue, or Nicobar, whereas only a handful of smaller terrestrial mammals did so. There are ten *M. fascicularis* subspecies from mainland Southeast Asia, going even beyond the Sunda Shelf. Such diversification of subspecies is best explained by a "two-wave" hypothesis. The first wave would have promoted the evolution of the present dark pelage subspecies found on oceanic islands off the Sunda Shelf (i.e., the most differentiated long-tailed macaque populations). This supposedly happened during a recent glacial maximum (~160,000 years), when the distance to be covered by rafting dispersal would have been greatly diminished. The case of the Philippines is interesting, as it represents a composite of oceanic islands separated by more than 180-m-deep sea floors. The two-wave hypothesis suggests that the only macaque present in the Philippines, the long-tailed macaque, may have dispersed in the course of its evolution on two occasions over straits between islands, by natural rafting from Borneo to the Philippines, presumably during two glacial maxima. As a result, only populations from the northern islands are of dark pelage (*M. fascicularis philippinensis*), whereas the ones in the southern islands show pale to intermediate pelage. The occurrence of an earlier dispersal also applies to the deepwater islands of Nicobar, Simeulue, and Lasia, situated off Sumatra's northwestern coast, as well as to Maratua Island off Borneo, where the remaining dark pelage subspecies are found.

The only other medium-sized primates that are thought to have dispersed by rafting in Asia to a deepwater island are the progenitor of Sulawesi macaques, and probably that of today's pigtailed macaques. Worldwide, the only other medium-sized primate thought to have used a raft to colonize its island habitat is the vervet monkey of Pemba Island, 50 km off the East African coast. The particular riverine habits of the long-tailed macaque would have favored its dispersal by sea rafting. After examining the long-tailed macaque's distribution and its exceptional ability to cross sea straits up to 150 km wide, one tends to favor the hypothesis of a land bridge at the origin of Madagascar's colonization by mammals, given that a 400-km sea strait had to be crossed, unless these mammals were able to hibernate.

Although long-tailed macaques rely mainly on riverine and coastal forests including mangroves, pigtailed macaques prefer inland primary forests. In Sumatra, long-tails are rarely observed more than 300 m from river banks: This placement is likely correlated with the species's morphological features—light weight and long tail—which are adequate features for a terminal branch feeder living in a riverine habitat. From these features, it is proposed that *M. fascicularis* was equipped to successfully disperse over water barriers. Indeed, natural rafts originate mainly from estuaries fed by relatively wide rivers that carry diverse pieces of wood and organic matter assemblages from inland, rather than from beaches or rock-bound coasts. Long-tailed macaques often feed and usually sleep in certain preferred trees above rivers, high above ground as a protection against predators. In all likelihood, this species, because of its habitat use, has the highest probability among primates and mammals of similar size to be found on a natural raft off the coast.

Because long-tailed macaques succeeded in dispersing to a higher number of deepwater islands than any other primate representatives, something other than a sea barrier must have stopped them from populating Sulawesi and the Mentawai archipelago. The absence of the long-tailed macaque in Sulawesi and the Mentawai Islands does not mean that some migrants did not reach these islands but rather that they could not establish themselves there as a distinct population because these islands were already inhabited by macaque species belonging to a more ancestral lineage. Compared with terrestrial colonization, migration by raft involves an extremely low number of migrants and an even lower number of successful ones. Two elimination mechanisms may have acted against such migrants. First, animals recovering from the state of near-starvation following long-distance rafting dispersal need to eat upon arrival. If a primate with a similar diet was already present in the new insular habitat, then competitive exclusion of newcomers for food resources must have diminished their chances of survival. Second, whenever newcomers were able to survive and gain access to food, they would have quickly intermingled with populations of a different taxon: Because species within the *Macaca* genus are inter-fertile (i.e., able to produce fertile offspring between species), it is likely that long-tailed macaques thereafter produced hybrids. As a consequence, they soon were absorbed by the considerably broader gene pool of the first settled taxon. Therefore, further survival and/or stocking opportunities for long-tailed macaque migrants were most likely held in check in Sulawesi and the Mentawai Islands. However, there were also many opportunities for long-tailed macaques to reach macaque-free oceanic islands because members of the first lineage, which preceded the newcomers, had considerably fewer chances to disperse over sea barriers. Because terrestrial colonization involves a relatively high number of migrants, resident species reach a demographic equilibrium with newcomers in a quite different way from that following overwater colonization.

In accordance with the view expressed here is the fact that fossil primates *Paralouatta* found in Cuba, Jamaica, and Hispaniola did not cross sea straits wider than 150 km, in line with long-tailed macaques' maximal performance in Southeast Asia.

SEE ALSO THE FOLLOWING ARTICLES

Dispersal / Ephemeral Islands, Biology / Refugia / Wallace's Line

FURTHER READING

Abegg, C., and B. Thierry. 2002. Macaque evolution and dispersal in insular Southeast Asia. *Biological Journal of the Linnean Society* 75: 555–576.

Ricklefs, R. E., and E. Bermingham. 2008. The West Indies as a laboratory of biogeography and evolution. *Philosophical Transactions of the Royal Society of London* 363: 2393–413.

Thiel, M., and L. Gutow. 2005. The ecology of rafting in the marine environment. I. The floating substrata. *Oceanography and Marine Biology: An Annual Review* 42: 181–265.

Thiel, M., and P. A. Haye. 2006. The ecology of rafting in the marine environment. III. Biogeographical and evolutionary. *Oceanography and Marine Biology: An Annual Review* 44: 323–429.

REEF ECOLOGY AND CONSERVATION

ROBERT H. RICHMOND
University of Hawaii, Manoa

WILLY KOSTKA
Micronesia Conservation Trust, Kolonia, Pohnpei

NOAH IDECHONG
Palau National Congress, Koror

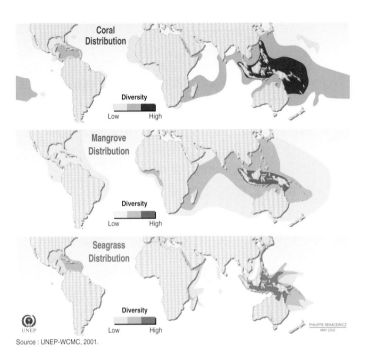

FIGURE 1 Map of global coral reef and associated mangrove and seagrass biodiversity. Illustration courtesy of United Nations Environmental Program (UNEP).

Coral reefs are the most biologically diverse marine ecosystems on Earth, rivaling terrestrial tropical rain forests. They are ecologically, economically, and culturally valuable resources that provide billions of dollars in goods and services to millions of people. An estimated 30% of the world's reefs have been substantially degraded during the past few decades, with predictions that nearly 60% are in jeopardy of being lost by the year 2050 if present trends continue. Anthropogenic stressors including overfishing, sedimentation, pollution, eutrophication, and warming tied to global climate change are responsible for losses and are reducing natural resilience and the potential for recovery. The bridging of research findings to the development and implementation of sound policies is critical if a legacy of intact reefs is to be left for future generations to enjoy.

CORAL REEF DISTRIBUTION AND CONSERVATION NEEDS

Stony corals are distributed worldwide, with reef development being predominantly restricted to the tropical band, approximately 25° north and south of the equator. Clear coral biodiversity patterns exist globally, with the greatest number of species being found in the Indonesia–New Guinea–Philippines area, known as the coral triangle, which houses nearly 600 species (Fig. 1). Moving from west to east, Palau has over 400 species of corals, Guam has nearly 300, Hawaii drops to the mid-60s, and the eastern Pacific of Panama has approximately 12. The Caribbean is comparable to Hawaii, with around 62 species. These observed patterns and gradients are attributed to several factors including (1) the geologic age of the provinces and the time elapsed since major catastrophic events (e.g., glacial cooling of shallow coastal waters in the Caribbean), (2) tectonic events including the movement of oceanic plates with their associated fauna and the rise of the Isthmus of Panama that separated the Caribbean from the Pacific Ocean, with subsequent speciation and extinction events (vicariance theory), and (3) the dispersal of larvae into and out of diversity centers via oceanic currents. The high diversity found in the Coral Triangle is theorized to be the result of long-term stability, hybridization during mass spawning events, and possibly the accumulation of larvae from other regions.

Scleractinian corals are found primarily in shallow tropical waters, but they can also be found in temperate areas, in the deep ocean, and across the world, including off the coast of Africa, in the Indian Ocean, and in the Red Sea. Although substantial biogeographical data are available on the distribution and abundance of coral species, little is known about genotypic diversity within species. The loss of sensitive genotypes, due to both anthropogenic and natural disturbances, has not been measured and is an increasing cause for concern for maintaining future coral populations and biodiversity.

Both the quantity and quality of corals reefs are declining worldwide, and the impacts are being felt ecologically, economically, and culturally, through the loss of critical resources and ecological services. The major anthropogenic causes of coral reef loss include overfishing leading to ecosystem-level shifts to fleshy algal domination when herbivorous grazers are depleted, runoff and sedimentation associated with poor land-use management, destructive fishing practices including the use of dynamite and cyanide, coastal eutrophication from human sewage and agricultural runoff, pollution from maritime activities as well as land-based sources, and temperature-related bleaching events tied to global climate change. Increases in atmospheric CO_2 are responsible for ocean acidification, which is expected to become more of a problem for corals and other calcifying organisms in the future.

Conservation efforts at local, regional, and global scales are needed to address the documented declines in coral reefs if these precious centers of biodiversity are to survive into the future. The establishment of community-driven marine protected areas (MPAs) in the Philippines and parts of Indonesia has increased fisheries yields outside of the boundaries and has reestablished trophic interactions supporting reef recovery. In addition to networks of MPAs, efforts are also under way to reduce land-based impacts on coastal reefs through the application of integrated watershed management practices in parts of the Coral Triangle, the Caribbean, and the Pacific Islands. Although the United States and other industrialized nations have invested financial, human, and institutional resources into coral reef conservation efforts, development and implementation of effective policies have lagged behind the available science, and the focus has often been on outputs (workshops, publications, and meetings) rather than on outcomes (improving metrics of coral reef health and resilience). Some of the best conservation efforts to date can be found in small tropical islands, where communities are actively engaged in conservation and management. Traditional conservation practices, often tied to reef tenure and resource ownership in the Pacific, provide examples of coral reef resource stewardship that serve as models for other nations aspiring to leave a legacy of functional coral reefs for future generations.

CORAL REEF BIOLOGY

Coral reefs are composed of a myriad of interacting invertebrate, vertebrate, algal, and bacterial species. Although corals may not be the most abundant or most taxonomically diverse elements of a coral reef, these organisms are the trophic foundation through their symbiosis with unicellular algae called zooxanthellae and are the main framework-builders providing essential habitat for other reef creatures. These diverse biological communities are often integral components of larger ecosystems including coastal mangroves, seagrass beds, sand flats, lagoons, and deep-water oceanic systems. Coral reefs provide important ecological services, are economically valuable, and provide for a variety of cultural activities central to the fabric of tropical island and coastal societies.

Corals are relatively simple creatures on the evolutionary hierarchy, with a tissue level of organization. The three types of tissues found in corals and related cnidarians include the inner gastrodermis (housing symbiotic dinoflagellate algae called zooxanthellae, intracellularly within the gastrodermal cells), an outer epidermis, and a layer of mesoglea in between. Reef-building corals, also referred to as hermatypic corals, are colonies made up of interconnected flower-like polyps. The combined animal–algal association is called the holobiont.

Corals receive the majority of their energy through photosynthetically derived metabolites released from their algal symbionts; hence, water clarity and associated light penetration are key determining factors affecting the state of coral reefs. The remaining energetic needs are met by the capture of small organisms, which is accomplished through the use of tentacles that are armed with adhesive or penetrating nematocycts. In order for heterotrophic feeding to occur, the surfaces of the coral must be relatively clean of sediment and debris.

The living portion of a coral is a relatively thin layer covering and slightly penetrating a non-living exoskeleton composed of calcium carbonate (aragonite). Coral taxonomy is primarily based on skeletal characteristics, including the arrangement of septa within the calices (cups) or troughs housing the individual polyps. Intraspecific variations in colony morphology are often a reflection of environmental parameters including wave energy, water clarity, and depth.

Coral reefs persist through the dual processes of reproduction, the formation of new individuals from prior stock, and recruitment, the process through which these new individuals become part of the reef population. Reproduction in corals can occur either asexually (e.g., through the fragmentation of colonies) or sexually through the formation and fusion of male and female gametes. Most corals are simultaneous hermaphrodites, producing both eggs and sperm within the same polyp. Larval development can occur internally, with resulting brooded planula larvae that are immediately competent to settle and metamorphose into the primary polyp of the new colony upon release. Most corals spawn their gametes, often in combined egg-sperm clusters, with external fertilization and development of embryos into the planula stage. The parental line of symbiotic zooxanthellae is usually transmitted to the larvae in brooding species, whereas many spawning species take up their zooxanthellae from surrounding waters following settlement and metamorphosis.

There are six chemically mediated steps involved in coral reproduction and recruitment: (1) synchronization among conspecifics in gamete development and release, (2) egg-sperm interactions leading to fertilization, (3) embryological development, (4) settlement, (5) metamorphic induction, and (6) acquisition of zooxanthellae if not transmitted via the parent. If any of these steps are disrupted, replenishment of populations will not occur. Freshwater runoff, coastal pollution, and sedimentation can all lead to reproduction and recruitment failure, and all have been documented to be problems on many coastal coral reefs.

THE VALUE OF CORAL REEFS

Coral reefs across Australia, the Indo-Pacific, Asia, the Caribbean, the Middle East, India, and Africa are ecologically, economically, and culturally important ecosystems. Damage to and destruction of reefs can affect the structure and function of adjacent mangrove, seagrass, and lagoonal communities. Ecological benefits of coral reefs include protection of shorelines and coastal communities from erosion and damage from storm generated waves, with an estimated value of billions of dollars. Coral reefs also generate revenue in the billions of dollars annually from diving tourism and related recreational activities. Coral reefs provide habitat for marine resources of economic and cultural value, and when these resources are depleted or destroyed, social, economic, and nutritional problems often arise.

Coral reefs are home to a variety of edible species, including fishes, crustaceans, holothurians, and molluscs. Some reef organisms are used for their chemical or medicinal properties (e.g., the sea cucumber *Holothuria atra* [Fig. 2], which contains a chemical called holothurin,

FIGURE 2 The sea cucumber *Holothuria atra*, used for the toxic chemicals in the skin for tidepool fishing and extracting octopuses from lairs.

used to kill and collect fish from tidepools or to tease octopuses from their lairs). The high biodiversity of coral reefs offers a diverse source of natural chemical products of interest to the fields of medicine and pharmaceutical research.

Whereas the food value of coral reef species is evident, the cultural value of traditional activities is often overlooked and is one of the most essential and important benefits of healthy and functional coral reefs to island and coastal communities. Communal fishing, sharing of resources, and the physical demands of reef fishing and gleaning are important to societies adjacent to coral reefs, and the value of these activities cannot be replaced by the provision of canned and imported foods alone.

Diving-related tourism has proven to be an important economic asset for destinations possessing coral reefs, and in this context, reefs can be viewed as the "geese laying the golden eggs." One famous dive site in Palau, the "Blue Corner," generates over $3 million annually on dives alone, not including hotel stays and associated restaurant revenues. Recognizing the museum value of their coral reefs, Palau introduced a dive permit program and a funding mechanism for supporting a protected area network, with revenues going to coral reef protection, including the placement of mooring buoys at popular dive sites and integrated watershed management activities. Similar examples can be found in the Caribbean and Atlantic (Cayman Islands, Bermuda, and Florida) where the economic benefits of corals reefs have been recognized and serve as the impetus for conservation-oriented efforts. Protecting specific reefs as museums for both the local residents and visitors is a sound management decision that provides ecological, economic, and cultural benefits.

NATURAL AND HUMAN-INDUCED DISTURBANCES AFFECTING CORAL REEFS

Modern scleractinian corals and coral reefs date back 65 million years, to the beginning of the Cenozoic Period. Over geological time, coral reefs undoubtedly have been subjected to a variety of natural disturbances, including volcanic eruptions, changes in sea level, wave events associated with hurricanes, outbreaks of corallivores including the crown-of-thorns starfish *Acanthaster planci*, freshwater runoff, sedimentation, and El Niño–Southern Oscillation–related warming/bleaching events. Disturbance is important to maintaining diversity on coral reefs, as corals can aggress against one another (e.g., by using sweeper tentacles and mesenterial filaments) or can simply overgrow and/or shade slower growing and less dominant species. Coral reefs have recovered from the multitude of natural disturbances over long periods of time and are particularly well adapted to rebound from acute disturbances.

Anthropogenic disturbance has been affecting coral reefs for a fraction of their time on earth, but such stressors tend to be more chronic in nature, preventing recovery from occurring. Overfishing, particularly of herbivorous species that keep fleshy and filamentous algae in check, runoff and sedimentation above natural levels as a result of poor land-use practices, coastal eutrophication from sewage and agricultural activities, disease outbreaks associated with human-induced stress, marine pollution, and the frequency and magnitude of bleaching events have all been increasing in the past several decades, with documented impacts on reef populations. Although some of these stressors are associated with population centers and urbanization, even the most remote Pacific atolls have been affected by roving fishing fleets and global climate change.

Runoff and sedimentation from development and poor land-use practices have had damaging consequences on coral reef communities. Turbidity reduces light penetration and affects the nutrition of corals by shading their photosynthetic zooxanthellae. Decreases in water and substratum quality interfere with critical chemically mediated events tied to coral spawning, larval development, and larval settlement. Source reefs, corridors of transport, and recipient reefs are all being affected by coastal pollution, and even the establishment of marine protected areas will not protect reefs unless integrated management practices are implemented within adjacent watersheds on land.

During the past two decades, regional bleaching events tied to global warming have been responsible for the loss of hundreds to thousands of square meters of coral reefs. Elevated seawater temperatures cause the breakdown of the coral–zooxanthellae symbiosis, and many species of corals cannot recover following such bleaching episodes. Additionally, elevated levels of atmospheric CO_2 are responsible for ocean acidification, which threatens the ability of corals and coralline algae to calcify. Some coral reefs damaged by regional bleaching events (e.g., in Palau following the 1998 event) have recovered over the subsequent decade, but affected reefs exposed to local anthropogenic stressors such as pollution, runoff, and sedimentation have not. This observation supports the critical value of local conservation and management efforts in supporting recovery following losses associated with global climate change, and the importance of enhanced efforts in response to what many scientists and managers consider the most serious threat to the future of coral reefs.

CORAL REEF CONSERVATION: EXAMPLES FROM THE PACIFIC ISLANDS

Marine Conservation Practices

There are irrefutable data demonstrating that coral reefs are in decline on a global scale and that conservation efforts are essential for addressing this problem. The United States and other developed countries have only a few decades of experience in addressing coral reef management concerns. Present regulations, enforcement capabilities, and the allocation of financial and human resources are insufficient to adequately protect coral reefs. Valuable lessons can be learned from Pacific islands that have centuries to millennia of experience dealing with resource conservation and utilization in the context of limited carrying capacities, and whose traditional leaders consider the needs of future generations in their actions.

Many Pacific islands still have traditional governance systems, and these lend themselves well to resource stewardship. Reef tenure or ownership is one of the most important and effective features for conservation in many of these islands, and it varies from complete and formal ownership of the reef and resources by villages to modified systems of local stewardship but central government jurisdiction.

Yap State in the Federated States of Micronesia and parts of Fiji and New Guinea are among the most traditional island groups in the Pacific and have complex and sophisticated resource management frameworks. Although these jurisdictions have elected governments, their traditional chiefs still hold positions of responsibility, and villages own their coastal reefs and associated resources. Permission of the village leaders is required to access the reefs, which are typically closed to fishing or collecting by outsiders.

Yap still maintains a type of caste system, which is reflected in a variety of cultural interactions including fishing techniques. Specific clans use unique tools for fishing, including breadfruit kites (Fig. 3), which employ loops of muscle bands from the pectoral/abdominal region of sharks instead of fish hooks. These bands are "danced" on the ocean's surface and ensnare the teeth of long-nosed needlefish that bite them. Unlike nets and traps, this technique is an example of a specific type of fishing gear that is highly selective for a desired species with no unintended by-catch.

On nearby Tam Tam Island, certain species of fish (e.g., the humphead parrotfish) can be eaten only by

FIGURE 3 Yapese fishing kite made from a breadfruit leaf, supported with *Pandanus* spines and flown with coconut sennet. Shark muscle bands are used instead of hooks for capturing specific species of long-nosed needlefish.

chiefs, reducing the fishing pressure on these ecologically important species. Large schools of these desirable fish, which are also important grazers keeping algae in check, can still be seen on many reefs where traditional leaders regulate fisheries, and community members comply. Although MPAs have gained popularity as a conservation tool in the United States and internationally during the past two decades, such restricted areas have been in place for centuries on some Pacific Islands.

Fishers from Satawal still use traditional sailing canoes to travel to the uninhabited island of West Fayu to fish as a way of reducing pressure on the reefs of their home island (Fig. 4). In island systems like this, where bartering still takes place, and food is shared among families, there are no market forces pushing toward over-exploitation of coral reef fisheries, which, although diverse, cannot withstand heavy export pressure.

FIGURE 4 Traditional navigators and fishers from Satawal building a new sailing canoe for fishing on West Fayu and for inter-island travel.

Palau provides one of the best examples of the value of traditional knowledge and its reapplication. Palau was part of the post–World War II Trust Territory of the Pacific Islands (TTPI) under U.S. jurisdiction. During the previous Japanese Mandate years through the TTPI period, internal governance and traditional practices within these islands were suspended, and in many cases, lost. Coral reef fisheries, which are not sustainable at high levels of exploitation, were found to be in serious decline by the 1980s as evidenced from data gathered by the Palau Marine Resources Division. Through a series of meetings with villagers and traditional leaders, the Marine Protection Act of 1994 was developed with the key provision of reestablishing the traditional system of *bul*, which allows traditional leaders to close sites and fisheries based on the location and timing of spawning and feeding aggregations. In addition, the commercial export of coral reef fishes was phased out. The combination of these traditionally based management approaches, now national law and part of the recently passed legislation on Palau's Protected Area Network, is a model from which the United States and other nations can learn, with key elements of ownership, stewardship, responsibility, and legacy.

American Samoa is a U.S. Territory that also has an intact traditional leadership system that approaches coral reef conservation using a village-based approach. The Matai system is one of extended family economic and political units, with the Matai as the leader. As with the other islands, modern American Samoan society is a hybrid of elected officials and traditional leaders. Through the Matai system, areas, species, and fisheries can be regulated and protected. American Samoa, using modern scientific data as well as traditional knowledge, has been proactive in protecting reef areas and large oceanic fishes. The American Samoa government is notable for recently establishing a population task force to address the key issue of island resource availability and carrying capacity as affected by immigration and local population growth.

Land–Sea Connections

Because of the relatively small size of Pacific Islands or the steeply sloping topographic relief that defines watersheds on larger land masses, land-based activities quickly and drastically affect adjacent coastal and oceanic ecosystems. Traditional land partitioning in Hawai'i is based on *Ahupua'a*, parcels that extend from the tops of ridges to the sea. Such "ridge to reef" systems are found in other Pacific Islands as well and reflect the understanding of the land–sea connection. Pacific Islanders are aware of and sensitive to upstream effects on downstream communities, as activities often affect members of the same village. Coral reef conservation begins on land and requires an integrated watershed management approach.

Village leaders in Enipein Village, Pohnpei, established a continuous protected area from the mountaintop rain forests through the coastal mangroves and out into the lagoon when studies found upland clearing for farming sakau (a plant whose root is a narcotizing agent and which is a major cash crop) was responsible for landslides and sedimentation stress on coastal coral reefs. Similarly, clearing and grading of mangroves for house lots in Airai Village, Palau, was halted when it was found that the loss of buffering mangroves was responsible for increased sedimentation impacts to adjacent coral reefs and associated fisheries. In both of these cases, data made available to traditional leaders resulted in policy development and implementation decisions within a period of weeks, rather than years as is often the case in Westernized democracies, which require legislation and in which compromise often limits the effectiveness of environmental protection measures.

STEWARDSHIP VERSUS THE TRAGEDY OF THE COMMONS

Coral reef conservation practices in many of the Pacific Islands reflect a stewardship ethic that stems from ownership of the resource, as well as responsibility for addressing problems and their solutions. This is in contrast to the Western "Tragedy of the Commons" scenario, where there is universal ownership and no direct lines of responsibility by stakeholders. The advantage of traditional leadership, which still exists in a variety of forms in the Pacific, is the speed at which conservation-based decisions can be put into practice. For example, it took nearly five years for conservation policies to be developed and enforced following studies demonstrating overexploitation of a sea cucumber fishery in the Galápagos Islands. In Yap, it took only one day to move from data presentation to village chiefs to a policy decision to close the fishery. A distinguishing feature of most traditional leadership systems is community compliance without the need for legislation, enforcement activities, and legal proceedings.

Recent efforts to cite "the right to fish" as part of traditional cultures to argue against the establishment of MPAs and other conservation measures are falsely based. Fishing, including the use of specific types of gear, access to particular areas, and the consumption of certain species, is a privilege granted by chiefs or master fishermen and was never a right. These privileges are granted by traditional leaders based on considerations including resource availability and sustainability and may be rescinded if conditions warrant.

The Future of Coral Reefs

Many challenges remain in the area of coral reef conservation, especially as a result of global climate change and regional bleaching events superimposed over other anthropogenic and natural stressors. If reefs in their present (degraded) state are the legacy left to future generations, then today's society will have failed as a steward of these remarkable ecosystems. Corals and coral reefs are resilient, and actions can be taken to reverse the present trend of increasing degradation and decline. Efforts at coral reef restoration are receiving heightened attention recently, but so far the best means of recovery is tied to restoring those conditions that allow natural recovery to occur.

Select case histories have demonstrated that elements of traditional management practices that support the responsibility of communities for resource stewardship can be effective. A shift to community-based management in the Philippines, Indonesia, and other areas in the Coral Triangle has resulted in improved compliance and resource availability. Science is an important tool for guiding conservation efforts and for developing tools and metrics to determine the efficacy of mitigation measures. Emerging technologies, including the use of molecular biomarkers of stress, enable the determination of cause-and-effect relationships between stressors and coral reef responses and can measure changes at the cellular and organismal level (for better or worse) within the timeframe of days and weeks, rather than the years needed using standard assessment and monitoring techniques focused on species diversity and abundance.

The social sciences and economics are also critical elements of conservation program development, as the reasonable goal is to change preventable human behaviors responsible for coral reef decline, rather than to change the behavior of coral reef organisms. Conservation successes emerging in the Caribbean, Africa, and Asia have required buy-in from communities that have a vested interest in resource sustainability, as they allow for present needs to be met without compromising the future state of coral reefs and related resources. The two-, four-, and six-year electoral cycles of Western democracies often make longer-term considerations irrelevant to politicians looking to stay in office, yet it is the legacy concerns of traditional leaders and societies that provide the context for effective conservation-based policies and their implementation. Enhanced awareness is needed to generate the political will required for policy development and implementation to insure the future of coral reefs. Bridging science to policy is a key step in moving from outputs to outcomes.

SEE ALSO THE FOLLOWING ARTICLES

Coral / Fish Stocks/Overfishing / Global Warming / Marine Protected Areas

FURTHER READING

Cinner J., M. Marnane, T. Clark, T. McClanahan, J. Ben, and R. Yamuna. 2005. Trade, tenure, and tradition: influence of sociocultural factors on resource use in Melanesia. *Conservation Biology* 19: 1469–1477.

Downs, C. A., C. M. Woodley, R. H. Richmond, L. L. Lanning, and R. Owen. 2005. Shifting the paradigm for coral-reef 'health' assessment. *Marine Pollution Bulletin* 51: 486–494.

Hughes T. P., and J. Connell. 1999. Multiple stressors on coral reefs: a long-term perspective. *Limnology and Oceanography* 44: 932–940.

Jackson, J. B. C., M. X. Kirby, W. H. Berger, K. A. Bjorndal, L. W. Botsford, B. J. Bourque, R. H. Bradbury, R. Cooke, J. Erlandson, J. A. Estes, T. P. Hughes, S. Kidwell, C. B. Lange, H. S. Lenihan, J. M. Pandolfi, C. H. Peterson, R. S. Steneck, M. J. Tegner, and R. R. Warner. 2001. Historical overfishing and the recent collapse of coastal ecosystems. *Science* 293: 629–638.

Johannes, R. E. 1978. Traditional marine conservation methods in Oceania and their demise. *Annual Review of Ecology and Systematics* 9: 349–364.

Johannes, R. E. 1981. *Words of the lagoon: fishing and marine lore in the Palau district of Micronesia.* Berkeley: University of California Press.

Johannes, R. E. 1997. Traditional coral-reef fisheries management, in *Life and death of coral reefs.* C. Birkeland, ed. New York: Chapman and Hall, 380–385.

Pandolfi, J.M., R. H. Bradbury, E. Sala, T. P. Hughes, K. A. Bjorndal, R. G. Cooke, D. McArdle, L. McClenachan, M. J. Newman, G. Paredes, R. R. Warner, and J. B. Jackson. 2003. Global trajectories of the long-term decline of coral reef ecosystems. *Science* 301: 955–958.

Richmond, R. H., T. Rongo, Y. Golbuu, S. Victor, N. Idechong, G. Davis, W. Kostka, L. Neth, M. Hamnett, and E. Wolanski. 2007. Watersheds and coral reefs: conservation science, policy and implementation. *BioScience* 57: 598–607.

Wolanski, E., R. H. Richmond, and L. McCook. 2004. A model of the effects of land based human activities on the health of coral reefs in the Great Barrier Reef and in Fouha Bay, Guam, Micronesia. *Journal of Marine Systems* 46: 133–144.

REFUGIA

ANGUS DAVISON

University of Nottingham, United Kingdom

A refugium is a geographic area in which organisms survive during adverse conditions. Although the term is most frequently applied to the glacial–interglacial cycles of the Pleistocene, species that are endangered as a result of human actions or ongoing climate change are restricted to modern-day "refugia." One common feature of refugia is that they may contain the greatest of relictual biodiversity, meaning that they merit special conservation attention.

REFUGIA DURING GLACIAL–INTERGLACIAL CYCLES OF THE PLEISTOCENE

The impact of the Pleistocene glacial–interglacial cycles on the flora and fauna of high latitudes has been well characterized, especially in North America and Europe. In these regions, large tracts of land were covered by ice caps during the glacial periods (or were under permafrost), so temperate species were restricted to southern refugia. In contrast, the impact of the glacial cycles at low latitudes has largely been overlooked, especially on oceanic islands. The form and extent of refugia in these regions and their role in shaping biodiversity are poorly understood, yet also controversial, because of their potential impact as "drivers" of speciation. This lack of knowledge is unfortunate in light of the traditional and continuing contribution of island species (e.g., Darwin's finches, *Partula* snails, Lord Howe Island palms) to understanding speciation theory.

Changes in sea level are generally considered to be one of the most important causes for islands appearing and disappearing, so one main impact of high-latitude glaciations was to increase globally the area of land above sea level. At the greatest extent of glaciation, when the sea level was about 100–130 m below that of the present day, many islands both were larger and had a greater range of elevations. Other islands became incorporated into nearby mainland (e.g., Britain), and archipelagoes became a single island (Figure 1, 2). Some generalist species may therefore have reached their greatest extent and population size during the glaciations—meaning, in effect, that their present-day distribution is refugial.

A second main impact of the high-latitude glaciations was a global change in climate. At high latitudes, islands had extensive ice caps, so many species must have gone extinct, subsequently recolonizing from southerly mainland refugia. At low latitudes, it is more difficult to generalize, and the data are sparse, because climatic adjustments were more local in their action, especially on oceanic islands. In consequence, the nature of potential refugia remains undetermined for the vast majority of low-latitude island species. Because island species are often uniquely adapted to specific habitats, having undergone an adaptive radiation, then perhaps their most likely response during glaciations was to track available habi-

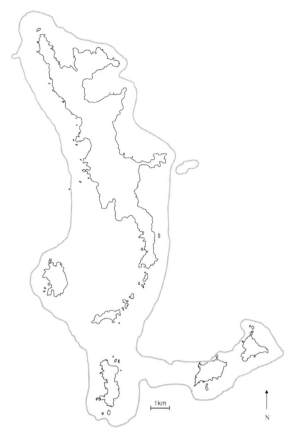

FIGURE 1 The Hahajima islands south of Japan are presently an archipelago. During the Pleistocene glaciations, they were a single island with a much greater geographic extent (gray line).

FIGURE 2 Mt. Chubu on Hahajima is 463 m above sea level, with moist forest at the summit. At the height of glaciations, the mountain would have been ~570 m above sea level, so the habitat diversity was likely greater. Genetic data indicate that snails in the region of this mountain did not undergo bottlenecks, unlike populations sampled from the low-lying peninsula in the foreground.

tat. If so, the geographic extant of refugia on islands was likely dependent upon the range of available elevations and consequent habitat zones. More generally, populations may sometimes become geographically restricted to refugia, so that they speciate in allopatry (e.g., in caves or in islands of forest surrounded by lava).

Several methods can be used to discover places that historically served as refugia, with molecular genetic and paleobotanical studies perhaps the most powerful. In the Neotropics, one explanation for a correspondence between paleo-pollen diversity and changes in global temperature is that fluctuating climate forced plants into refugial habitat islands, directly enabling the diversification of species in allopatry, and so explaining the extraordinary diversity that is found there. In the rather few tropical island species that have been investigated, the overall impact of the Pleistocene on biodiversity is unclear. The genetic diversity of the damselfly *Megalagrion xanthomelas* is greatest on the Big Island of Hawaii, where habitat availability has been most constant during glacial cycling; similarly, genetic data indicate that *Mandarina* snails in the low lying islands of the Japanese Bonin Islands (Ogasawara) underwent much more severe bottlenecks than those in the central highlands. At the opposite climatic extreme, it is claimed that some species survived the glaciations on islands in northerly refugia that happened to be free of ice. Other species may have clung on on mountain nunataks.

MODERN-DAY REFUGIA

As centers of endemism, islands harbor a high proportion of species that are found nowhere else. Island species are also particularly vulnerable to extinction and have been especially affected by introduced exotics. Nonetheless, islands are also frequently the last remaining refuges for many species, either by accident or as a consequence of their remoteness; species that were long since made extinct on the mainland, linger on remote islands. Prominent examples include the Lord Howe stick insect *Dryococelus australis,* recently rediscovered after being considered extinct for 70 years, and the tuatara *Sphenodon*, confined to a handful of offshore islands and mainland locations of New Zealand. Several Hawaiian forest birds survive in high-elevation refugia, where they cannot be reached by mosquitoes carrying avian malaria.

Islands are also sometimes used as natural holding grounds (refugia) for seminatural breeding; introduced predators can be extirpated from offshore islands, and then endangered species (re)introduced.

SEE ALSO THE FOLLOWING ARTICLES

Adaptive Radiation / Climate Change / Galápagos Finches / Land Snails / Lord Howe Island / Sea-Level Change / Vicariance

FURTHER READING

Davison, A., and S. Chiba. 2006. The recent history and population structure of five *Mandarina* snail species from sub-tropical Ogasawara (Bonin Islands, Japan). *Molecular Ecology* 15: 2905–2919.

Hewitt, G. M. 2004. Genetic consequences of climatic oscillations in the Quaternary. *Philosophical Transactions of the Royal Society of London Series B: Biological Sciences* 359: 183–195.

Jaramillo, C., M. J. Rueda, and G. Mora. 2006. Cenozoic plant diversity in the Neotropics. *Science* 311: 1893–1896.

Priddel, D., N. Carlile, M. Humphrey, S. Fellenberg, and D. Hiscox. 2003. Rediscovery of the 'extinct' Lord Howe Island stick-insect (*Dryococelus australis* (Montrouzier)) (Phasmatodea) and recommendations for its conservation. *Biodiversity and Conservation* 12: 1391–1403.

RELAXATION

KENNETH J. FEELEY
Wake Forest University, Winston-Salem, North Carolina

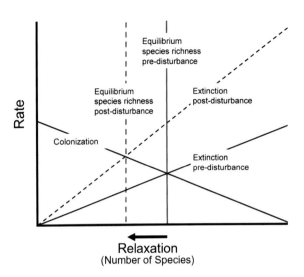

FIGURE 1 A hypothetical example of how an increase in local extinction rates following a disturbance that reduces the area of an island (e.g., a rise in sea level) will decrease the equilibrium number of species that can be sustained. The resulting loss of species richness over time is referred to as relaxation.

Relaxation is the process by which species are lost from an island following a disturbance event that increases the rate of local extinction, decreases the rate of colonization, or both. The disturbance is typically a decrease in area or increase in isolation, but it may alternatively be a change in habitat quality or any other disturbance event that decreases the number of species that the island can support at equilibrium (Fig. 1). Relaxation is most commonly referred to in the setting of continental islands or mainland habitat fragments (e.g., sky islands). In both of these settings, habitat that was originally connected to, or embedded in, a larger "mainland" becomes isolated (e.g., in the case of continental islands, by rising sea levels). Following isolation, the newly formed island will generally be "supersaturated," meaning that the number of species occurring on the island is greater than can be supported as based on the new conditions. In a supersaturated community, extinctions will exceed colonizations and the number of species will decrease through time until the appropriate equilibrium number is achieved.

PATTERNS OF RELAXATION AMONG ISLANDS

Studies tracking the process of relaxation have observed that the rate of species loss generally mimics an exponential decay process: species loss is rapid at first and then decelerates through time until the new equilibrium rate of extinction is reached. Furthermore, it has been noted that the rate of relaxation is not constant between islands but may depend on several factors, including area and degree of isolation. Large islands (or habitat fragments) typically lose species slower than do smaller islands.

The reason for the relationship between area and rate of relaxation is poorly understood. One potential explanation is that if extinctions are primarily a passive process resulting from stochastic population fluctuations, then the reduced populations occurring on smaller islands will have elevated risks of extinction, resulting in accelerated rates of relaxation. Alternatively, the loss of area on smaller islands may lead to synergistic effects, such as changes in resource availability or predation pressures, that can in turn hasten species loss. For example, numerous studies have shown that the increased relative edge length (perimeter/area) on small islands may lead to deleterious "edge effects," including changes in abiotic conditions (such as temperature, humidity, or wind speed), as well as changes in the biotic environment such as greater exposure to edge-frequenting predators. Conversely, a few studies have demonstrated "positive" edge effects that may actually act to decrease relaxation rates on small islands. Examples of positive edge effects include the allochthonous input of marine nutrients via seabird guano or ocean detritus.

PATTERNS OF RELAXATION AMONG SPECIES

The order in which species go locally extinct during relaxation is generally not random. Rather, certain species or groups of species appear to be particularly sensitive to changes in area and hence tend to be lost first. Extinction vulnerability has been variably related to a large number of species characteristics, depending on the system and/or group of species studied. Examples of species attributes that have been related to extinction vulnerability include population size, trophic position, dispersal capability, degree of resource specialization, body size, and reproductive rate. As a result of selective extinction processes, newly formed archipelagoes or networks of habitat fragments often exhibit "nested" patterns of species distributions, such that species occurring on species-poor islands also occur on all of the more diverse islands.

CONSERVATION IMPLICATIONS

As a result of relaxation, isolated reserves will not sustain the same number of species as they did pre-isolation. By understanding the processes and patterns of relaxation it may be possible to predict the rates at which species will be lost as well as which species are at the greatest risk of extinction. This information can in turn help improve reserve design and management.

SEE ALSO THE FOLLOWING ARTICLES

Continental Islands / Extinction / Fragmentation / Island Biogeography, Theory of / Sky Islands / Species–Area Relationship

BIBLIOGRAPHY

Brown, J. H. 1971. Mammals on mountaintops: nonequilibrium insular biogeography. *The American Naturalist* 105(945): 467–478.
Diamond, J. M. 1972. Biogeographic kinetics: estimation of relaxation times for avifaunas of Southwest Pacific. *Proceedings of the National Academy of Sciences of the United States of America* 69(11): 3199–3203.
Karr, J. R. 1982. Population variability and extinction in the avifauna of a tropical land bridge island. *Ecology* 63 (6): 1975–1978.
MacArthur, R. H., and E. O. Wilson. 1967. *The theory of island biogeography.*, Princeton, NJ: Princeton University Press.
Patterson, B. D. 1987. The principle of nested subsets and its implications for biological conservation. *Conservation Biology* 1(4): 323–334.
Pimm, S. L., H. L. Jones, and J. M. Diamond. 1988. On the risk of extinction. *The American Naturalist* 132(6): 757–785.
Terborgh, J. W. 1974. Preservation of natural diversity: the problem of extinction prone species. *BioScience* 24(12): 715–722.
Wilcox, B. A. 1978. Supersaturated island faunas: a species-age relationship for lizards on post-Pleistocene land-bridge islands. *Science* 199(4332): 996–998.

RESEARCH STATIONS

NEIL DAVIES

University of California, Berkeley

Most islands have limited indigenous resources (human, financial, and physical) to support research, and they rely on importing the people (skills) and equipment needed for scientific studies. Some island research stations are little more than specialized guest houses, but others rival the most advanced mainland laboratories. Where a station falls along this spectrum depends on a number of factors, including (1) the scientific interest of the location, (2) its socioeconomic context, and (3) its potential to support educational and public service activities. All of these are being radically transformed by (4) new information and communication technologies.

SCIENTIFIC POTENTIAL

Studies at research stations investigate natural processes (including the ecological role of humans) through experiment and observation. Sites with a field station or marine laboratory represent parts of the world where we can increase the precision of biophysical observations. A research station is thus a significant investment in a geographic area of particular scientific interest. Far from research stations, the ecological matrix is fuzzy, but it snaps into focus closer to them. Researchers consider archipelagoes to be natural laboratories, providing systems of discrete replicates facilitating the design of scientific studies. Such natural systems are particularly important for scientists who, for ethical and/or practical reasons, cannot easily manipulate their study subjects. For example, one cannot put a species (humans in the case of anthropology) in separate cages, each differing in some key environmental factor, and leave them for a million years to see what evolves. Natural experiments are messier than laboratory studies, to be sure, but researchers often find it easier to separate the signal from noise in island systems. Another advantage of smaller islands for science is that they are complete but relatively simple ecosystems and thus are scientifically more tractable for holistic study than are larger more complex regions. Major scientific breakthroughs often result from the focus on such model systems. For example, cell and molecular biology advanced tremendously by concentrating efforts on a few model species such as mice, fruit flies (*Drosophila*), and nematodes (*Caenorhabditis elegans*). Similarly, rapid progress in ecology and evolution has already come from studies of island model ecosystems, but much more potential remains to be explored. Nor is it sufficient to investigate only one model: Integration across a network of ecosystems is critical for understanding regional or global drivers of local change. Comparisons of biologically complex islands (e.g., those in the western Pacific) with simpler ones (e.g., those in eastern Polynesia) can provide insights into the role of biodiversity in ecosystem structure, function, and resilience to perturbations such as climate change and globalization (e.g., invasive species). These represent some of the most important issues facing science and society in the twenty-first century.

The fundamental scientific questions addressed on islands often invoke a long time scale. Research stations provide permanent facilities in relatively remote areas and represent a source of programmatic continuity as institutions that persist after individual researchers leave or particular projects end. Research stations host long-term monitoring programs as well as shorter-term, process-oriented studies. Monitoring programs do not always address an explicit research question, and mechanistic studies are sometimes a series of quasi-independent projects connected more by their common geography than by an integrated research program. Exceptions exist, however, such as the U.S. National Science Foundation's (NSF) Long-Term Ecological Research (LTER) program, which combines empirical studies with the long-term data acquisition needed to address specific questions over appropriate ecological timescales. The NSF-funded LTER network includes a number of sites based at island research stations, from the Pacific (Gump Station, Moorea) to the Caribbean (El Verde Field Station, Puerto Rico) and the Antarctic (Palmer Station, Anvers Island). The International LTER (ILTER) initiative had 32 member networks in May 2006, and as these develop, they are likely to include more island sites and associated research stations. The LTER-Europe was founded in June 2007 and could be particularly influential given the global distribution of islands associated with the European Union (Fig. 1).

Just as they significantly influenced science in the nineteenth and twentieth centuries (e.g., inspiring Darwin), islands and their research stations are particularly well placed to advance the frontiers of science in the twenty-first century. For example, biocomplexity addresses the causes of environmental, biological, and social changes and the interaction among them. Sustainable development relies on ecointelligence—understanding biocomplexity and managing human society's relationship with the natural world rationally. The interacting processes underlying biocomplexity occur at multiple spatial scales, and island research stations enable the simultaneous study of local drivers (island- and archipelago-scale) as well as regional drivers such as climate change or globalization (oceanic/continental-scale).

Island research stations therefore address issues of general concern to human society as well as more applied projects of primarily local importance. As the new knowledge generated by fundamental research is shared beyond the local community, so is the potential funding base commensurately larger. To succeed in attracting (national or international) support, however, island research stations must overcome significant barriers. Although all islands benefit from local investment in applied science, only some are able to attract the external resources necessary to sustain long-term fundamental research.

SOCIOECONOMIC CONTEXT

Field research, whether marine or terrestrial, refers to those scientific studies that take place outside the confines of

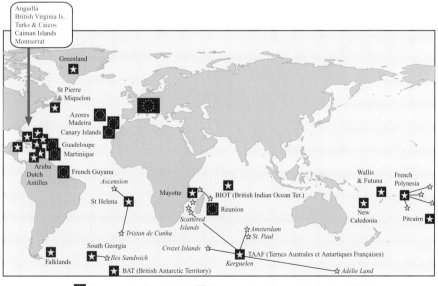

FIGURE 1 The European Union's seven Overseas Regions, and 20 Countries and Territories. Reprinted with permission from IUCN.

the laboratory. In other words, it describes the investigation of the natural world in vivo. In theory, field research can be conducted anywhere in the world. For practical economic and cultural reasons, however, it tends to be concentrated in the environs of large cities in the developed world within easy reach of major museums and universities. Field research is more challenging further from these core centers, if only because of the travel costs and the need to carry equipment to a makeshift base camp (which, in rural or wilderness settings, can lack even the most essential utilities such as power and water). Such forays are described as "expeditions" or "field trips." In some (rare) cases, the base camps become permanent, and a research station is established.

Given their isolation and small size, all research stations are "islands" in a socioeconomic sense, lacking the economies of scale of a large campus. The decision to invest in such permanent infrastructure is thus a trade-off between the scientific productivity of a locality and the increased costs of conducting research there. Although the decision to establish or maintain a research station might be made after a rigorous assessment of the scientific return on investment, many island research stations were not established on the basis of a carefully formulated strategic plan. Nevertheless, the subsequent development of research stations depends on their continued ability to attract researchers and funds. A process of natural selection, therefore, eventually determines whether island research stations survive or perish: Schumpeterian, or Darwinian, creative destruction! A non-exhaustive list of existing island research stations is shown in Table 1.

Of course, the need to attract researchers is not unique to island research stations; mainland institutions must also struggle to recruit the best scientists and students. Research stations, however, tend to *host* rather than *employ* the majority of their scientists, and so they face challenges that are perhaps more familiar to the tourism industry than to academia. This is perhaps a fortuitous parallel, as it aligns island research stations with a major source of external revenue for island economies. Research stations are part of the tourism sector but with an important difference: Their clients, visiting researchers, not only spend hard currency in the local supermarket but also contribute to the island's heritage capital. By increasing the knowledge base of the island, researchers help to improve the management of its natural and cultural resources (a prerequisite for sustainable tourism), as well as providing even more direct benefits for ecotourism.

EDUCATION AND PUBLIC SERVICE

If research stations can be seen as part of an island's tourism sector, then students are the ultimate ecotourists. Like any visitor, students might appreciate the beaches or nightlife of an island, but their primary purpose is to learn. Other tourists might share this motivation, but it is not usually their sole reason to travel. Many research stations are financially viable thanks to the income they generate from teaching field courses. This is not just a money-making activity, however, as it also fulfils a core mission of many research stations, particularly those operated by universities. Combining research with advanced training programs provides increased economies of scale, helping to overcome some of the problems of being a small and remote site. It also offers richer experiences for researchers and students alike.

Many research stations provide education and training for the local community in addition to their programs for visiting students. In this case, the outreach provides another crucial function, reinforcing the social contract between the research station and the local community. Research stations are a valuable tool for sustainable development and a source of conservation know-how, and such public service is another core part of their mission. Indeed, research stations sometimes represent the only scientific expertise available in some locations and represent an important link for island communities to the international network of scientific knowledge and resources. Public service can also include applied research projects commissioned (and funded) at the local level to solve local problems. This is a win-win exchange that provides the research station with more knowledge about the local ecosystem (as well as revenue toward its operations overhead) while the island receives high-quality services at lower cost thanks to the capital (human and material) already invested in the station (often from external sources).

TABLE 1
Partial Listings of Island Research Stations

World Register of Field Centers, Royal Geographical Society	http://www.rgs.org/
National Association of Marine Labs	http://www.naml.org/
European Marine Research Stations Network	http://www.marsnetwork.org/
Organization of Biological Field Stations	http://www.obfs.org/
Association of Marine Laboratories of the Caribbean	http://www.amlc-carib.org/

The educational and outreach products of island research stations can have impacts far beyond the local community. Islands make particularly charismatic showcases illustrating the challenges facing modern (and prehistoric) societies. Many of these are most evident in insular systems and are being documented by today's island research stations. For example, rising sea levels and elevated ocean temperatures are submerging tropical islands and increasing the bleaching of their coral reefs. Similar impacts will be experienced in mainland systems, but islands are the "canaries in the mine" warning us of the impending global crisis. Island research stations are not only documenting the present and modeling the future: They are also reconstructing the past. Some traditional island societies succeeded in establishing sustainable economies despite the finite natural resources available to them. Others failed, and overexploitation led to collapse with terrible consequences for the ecosystem and the well being of its human population. Such lessons provide a vital service reminding humanity that the Earth is our common "island" home.

TECHNOLOGY

Advances in information and communication technologies have revolutionized the potential of islands in many economic, social, and cultural arenas, including scientific research. These technologies are particularly disruptive for island economies because they attack the root of an island's comparative disadvantage, its isolation. It is now possible for a knowledge worker to live in the mountains of California, or the islands of Polynesia, and be (virtually) as connected to his or her colleagues and clients as if he or she were in a major metropolitan area. Interestingly, this pulls in two directions: It is becoming less necessary for scientists to physically visit a remote site in order to study it, but at the same time, it is now possible to live permanently at a remote site without being entirely disconnected from the intellectual environment of a large city. These two trends will transform research stations (and much else) on islands.

Research stations occupy a key position at the beginning of the digital supply chain that delivers ecological knowledge to the global scientific community. Historically, they catered to the most basic human needs (shelter, food, etc.), but modern research stations will need to support advanced sensor technologies embedded in their field sites. Increasingly, research stations will host engineers and computer scientists as well as their more traditional clientele. Furthermore, the technologies associated with genomics and molecular biology are also being applied outside of the laboratory, enabling whole new worlds to be studied in the field (e.g., microbial communities and their ecological role).

One of the most basic requirements for scientific investigation is a detailed description of the study system. Thus, the core infrastructure of a research station is not just its buildings, equipment, and staff but also a digital rendition (preferably in real time) of the island's physical, biological, and social characteristics. This "living encyclopedia" must be readily accessible and motivates the development of online climatic databases (weather records), maps (GIS), and biotic guides (species identification), as well as a host of sophisticated informatics tools. Just as model species such as the fruit fly or mouse were described in ever-increasing detail as technology advanced (e.g., whole-genome sequencing), island model ecosystems are also being "sequenced" (e.g., using short genetic "bar codes" that enable rapid species identification). Most research stations already maintain databases of meteorological data (technologically the most simple to collect and archive), but soon they will also participate in the acquisition and management of massive amounts of information (environmental, biological, and social).

As they become more technology intensive, research stations are increasingly being compared to astronomical observatories. For the moment, however, there are no known systems in the universe as complex as those on Earth, and none more immediately important for us to understand. Ecological observatories on the "blue planet" will require (and will probably receive) unprecedented levels of investment as society wakes up to the twin crises of climate change and mass extinction. Islands and their research stations will be on the front line of the epic struggle for sustainable development.

SEE ALSO THE FOLLOWING ARTICLES

Global Warming / Marine Protected Areas / Reef Ecology and Conservation / Sea-Level Change / Sustainability

FURTHER READING

Check, E. 2006. Treasure island: pinning down a model ecosystem. *Nature* 439: 378–379.
Diamond, J. 2005 *Collapse: how societies choose to fail or succeed.* New York: Viking, Penguin Group.
ILTER (International Long-Term Ecological Research Network). http://www.ilternet.edu/
LTER-EU (Long-Term Ecosystem Research and Monitoring, European Union). http://www.lter-europe.ceh.ac.uk/
LTER-US (Long-Term Ecological Research Network, United States) http://www.lternet.edu/
Michener, W. K., T. J. Baerwald, P. Firth, M. A. Palmer, J. L. Rosenberg, E. A. Sandlin, and H. Zimmerman. 2001. Defining and unraveling biocomplexity. *Bioscience* 51: 1018–1023.

Vitousek, P. M. 2002. Oceanic islands as model systems for ecological studies. *Journal of Biogeography* 29: 573–582.

RÉUNION

SEE INDIAN REGION

RODENTS

DAVID TOWNS

New Zealand Department of Conservation, Newton

Islands throughout the world have been modified by introduced rodents. A few highly destructive species have been deliberately released to establish a fur industry. Mice and four species of rats have reached islands as passengers during exploration, warfare, and commerce. These have become the perfect invasive species, able to spread over wide distances and with significant ecological effects wherever they colonize.

DISTRIBUTION OF RODENTS ON ISLANDS

There are more species of rodents (around 2000) than of any other group of mammals. Isolation and extraordinary dispersal ability have led to considerable evolutionary radiation on less remote islands and continental fragments, with distinctive faunas in Madagascar, Sri Lanka, the Philippines, Australia, and even the Galápagos Islands. Madagascar alone has 22 species of native rodents. However, three species of rats and one of mice are distinctive—not for their evolutionary significance, but for a long history of association with people, for their ability to stow away on all kinds of seagoing craft, and for their spread across enormous geographic distances. They now occupy all continents other than Antarctica and most islands from the subantarctic Southern Hemisphere to the arctic Northern Hemisphere. In order of distances covered, these species are the Pacific

FIGURE 1 The four rodent species most widely spread by people. (A) A wild house mouse (*Mus musculus*) and offspring. (B) Norway rat (*Rattus norvegicus*). Photographs courtesy of New Zealand Department of Conservation—Rod Morris. (C) Ship rat (*R. rattus*) attacking a native forest bird at its nest in New Zealand. Photograph by David Mudge. (D) Pacific rat (*R. exulans*) in New Zealand forest. Photograph courtesy of New Zealand Department of Conservation—Dick Veitch.

rat or kiore (*Rattus exulans*), the Norway or brown rat (*R. norvegicus*), the ship or black rat (*R. rattus*), and the house mouse (*Mus musculus*) (Fig. 1). The origin of the house mouse is unclear, although current evidence points to the Indian subcontinent. They have since been spread to all inhabited parts of the world and may be the most widely distributed mammal other than people. Similarly, Norway rats have become so widespread that their center of origin is unclear, but it may be China or Siberia. They now live from South Georgia in the South Atlantic Ocean to the northernmost islands of Europe and North America. Ship rats probably originated in India, but they have since reached islands from the southern Indian Ocean to Scotland. Pacific rats are derived from Southeast Asia but have been spread across the Pacific basin as far east as Easter Island and south to New Zealand. In combination, mice and these three rats have reached at least 80% of the world's island groups.

SPREAD TO ISLANDS

Origins

Rodents have been spread to islands in four ways: in cargo, by abandoning ships in port or after shipwrecks, through natural dispersal either by swimming or floating on debris, and through deliberate spreading by people. As examples, ship rats escaped onto Midway Island from military stores offloaded during World War II. The same species invaded Big South Cape Island off southern New Zealand along the mooring lines of fishing boats in about 1962. Norway rats reached Raoul Island in northern New Zealand after the wreck of the *Columbia River* in 1921. In 2004, a male Norway rat swam at least 400 m between islands in New Zealand while carrying a radio transmitter. In contrast, ship rats are reluctant or unable to swim more than 200 m, and Pacific rats failed to invade some islands separated by only 50 m. Therefore, the distribution of mice and of the more recently arrived rats is most likely to have been accidental through rodents abandoning vessels or being hidden in stores. However, some anthropologists have claimed that Polynesians deliberately introduced Pacific rats to islands as a source of food. Other analyses have found that Pacific rats were more likely to be present on islands with easy access by canoe than on islands where landings were more difficult. The pathways for the spread of rodents are now being determined by genetic analyses using mitochondrial DNA. This can be used to trace the origins of species spread accidentally by boats or transported by people many centuries ago.

Additional species have been deliberately spread in attempts to establish fur industries. For example, American beavers (*Castor canadensis*) were introduced to the island of Tierra del Fuego in South America, and also to islands off western Canada.

Speed of Invasion

The invasion biology of rodents on islands has only recently been studied. Some invasions have certainly originated from a single pregnant female rat or mouse. This raises the question of how they avoid the effects of inbreeding. In New Zealand, pregnant female Norway rats were found carrying the embryos of multiple fathers; one litter of rats can include the genetic information from several male lines. If invading rodents become established, populations can expand with great speed. For example, the ship rats that invaded Big South Cape Island (900 ha) off southern New Zealand, and the single known female Norway rat that in 1995 reached Frégate Island (210 ha) in the Seychelles, each produced sufficient offspring to spread throughout the entire island within two years of arrival.

Population Densities

Once established, rodent populations on islands may periodically reach very high densities. For example, mice on Gough Island reached densities of 224 per ha, and in pasture on Mana Island, New Zealand, they became so abundant that improvised bucket traps caught over 200 per night. Elsewhere on New Zealand islands, Norway rats reached densities of 13 per ha, ship rats were recorded at densities of 50 per ha, and Pacific rats in grasslands were noted at densities of greater than 100 per ha. On many islands, the populations go through wide fluctuations in abundance. All three species decline during the cool winter in New Zealand, and Pacific rats similarly decline during the dry season on some islands in the tropical Pacific.

EFFECTS ON ISLAND FAUNA AND FLORA

Species That Modify Environments

Introductions of beavers and coypu (*Myocastor coypus*) in failed attempts to establish a fur industry have led to radically transformed landscapes. Beavers have modified watercourses by constructing ponds, and in South America they leave extensive areas of dead southern beech (*Nothofagus*) forest. Similarly, coypu undermined the banks of watercourses and damaged wetlands. Gray squirrels (*Sciurus carolinensis*) introduced to islands off Canada stripped the bark from sensitive plant species and damaged the acorns of native oaks. As a result, forest composition has changed on islands where alien squirrel populations established.

Species That Modify Biodiversity

EFFECTS OF MICE

On islands, mice primarily feed on seeds and small invertebrates such as insect larvae. On Marion Island in the southeastern Atlantic Ocean, mice were found to have significant effects on the caterpillars of an endemic species of moth. On Gough Island, landbirds became confined to cliffs inaccessible to the mice. The mice also fed on the living chicks of albatrosses until the birds died of their wounds. A 90% decline of Tristan Albatross may be attributable to predation by mice. On Mana Island, populations of giant flightless crickets, geckos, and large nocturnal skinks greatly increased after mice were removed, which indicates that mice were affecting species equal to or greater than their own body weight. However, despite their widespread distribution, in general the effects of mice on native species are poorly known.

EFFECTS OF RATS

Information on the effects of rats, although better documented than for mice, is still scattered, with little comparison between different species that occupy similar biogeographic regions. Much of the data available has been obtained on islands in New Zealand, but even there, the studies have been selective. Like mice, rats are omnivores and carrion feeders. Unlike mice, the diet of rats is extremely wide and includes plants, invertebrates, amphibians and reptiles, birds, and other mammals.

Effects on Plants All three invasive species of rats affect forest plants largely through predation of seeds, but in rare cases, also through browsing seedlings. They may also climb trees and feed on flowers and fruit. For example, the tree-climbing abilities of ship rats led to them becoming a threat to macadamia nut production in the Hawaiian Islands. In New Zealand, Pacific rats severely suppress at least two species of forest canopy trees, which on some islands have become rare. They also suppress at least another nine and perhaps up to 17 species of forest plants, one of which is a native palm. Norway rats can have similar effects and were responsible for recruitment failure of southern beech on Breaksea Island. In the Canary Islands off northwestern Africa, ship rats consume about half of the fleshy-fruited tree species of the laurel forest and may be partly responsible for changes in forest structure. Particularly devastating effects are attributed to rats in the tropical Pacific. Heavy seed predation by Pacific rats on Rapa Nui (Easter Island) appears to have been instrumental in the loss of palm forests. Pacific rats have also been associated with extreme forest modification on the Hawaiian Islands.

Rats also have indirect effects on plants. In the Canary Islands, ship rats compete with the Canary robin for the fruits of *Vibernum,* reducing the distance that seeds are dispersed. In the Balearic Islands, rats affect fruit-eating lacertid lizards, which are the main seed dispersers for species of endemic plants.

Effects on Invertebrates Rats feed on a very wide range of invertebrates. Numerous species of invertebrates are eaten by rats in the Hawaiian Islands, where entomologists have even discovered unknown species in rat stomachs. Some species of rats feed at husking sites where invertebrate fragments can be collected and identified. Husking sites of Pacific rats analyzed in New Zealand revealed the remains of over 60 species of native invertebrates. Such predation may not have serious effects on some species, but others have disappeared from islands invaded by the rats. Particularly vulnerable species are those that are nocturnal, large, and flightless, including one species of large (miturgid) spider, darkling (tenebrionid) beetles, and large flightless (anostostomatid) crickets. The invasion of ship rats on Big South Cape Island led to local extinction of large *Hadramphus* weevils, either through predation or because of heavy damage to the weevil's host plants. Evidence of heavy predation on subfossil deposits of large *Placostylus* land snails in New Zealand proved useful in estimating the time of invasion by the rats. In some locations, the snails became extinct after the rats arrived. Other invertebrates affected by rats include *Gecarcinus* crabs in the Caribbean.

Effects on Vertebrates The three invasive rat species have been associated with declines in all major groups of vertebrates, ranging from amphibians to mammals. In New Zealand, Pacific rats appear responsible for the loss of three species of endemic frogs, for range reductions of geckos, for local extinctions of at least six species of nocturnal skinks, and for breeding failure of at least four species of small seabirds. When Pacific rats were removed from islands around New Zealand, juvenile recruitment increased for the large endemic iguana-like tuatara (*Sphenodon*) and for three species of skinks and two species of geckos. Pacific rats also probably eliminated five species of small ground-dwelling birds in New Zealand and are also likely to have directly or indirectly instigated the demise of some species in Hawaii.

Norway rats have been found to affect a wider range of vertebrates than have Pacific rats, including a range of larger ground-dwelling birds such as endemic wrens in the Falkland Islands and seabirds in New Zealand that weigh up to 750 g. Norway rats appear to have wiped out the endemic Canary Island lark and are implicated in the declines of seabird populations in numerous island groups. Along with ship rats, Norway rats have heavily suppressed populations of burrowing seabirds around New Zealand.

The most damaging species of rat is the ship rat because of its climbing abilities. In the Caribbean, ship rats almost wiped out an endemic species of racer snake. In the Canary Islands, they destroyed up to 90% of the nests of endemic pigeons through egg predation. In New Zealand, sudden irruptions of ship rats are associated with dramatic declines of a forest bird, the yellowhead, and the orange-fronted parakeet. When ship rats invaded Big South Cape Island, three species of terrestrial birds and one bat became extinct. Likewise, when ship rats escaped a wrecked ship, five species of endemic forest birds became extinct on Lord Howe Island. Groups affected by ship rats range from bats to other species of rodents, including two species of endemic rodents in the Pacific, and perhaps another in the Balearic Islands, that went extinct.

Effects on Island Ecosystems In New Zealand, comparisons of islands invaded by rats with those uninvaded revealed that islands without rats had greater seabird burrow densities, higher soil fertility, greater abundance of primary consumers (herbivorous nematodes and land snails), and larger numbers of secondary consumers (enchytraeids, microbe-feeding nematodes, rotifers, and collembolans). Furthermore, plants in soil from rat-free islands grew more rapidly than did those inhabited by rats. The study revealed complex interactions between belowground food webs and aboveground plant nutrient levels and biomass related to changes in soil fertility and disturbance as an indirect result of invasions by rats.

Effects on People Pacific rats were widespread throughout Polynesia by the thirteenth century, and in some parts they became a valued item of food for local people, although in New Zealand they were also a pest because they ate stored food, and Maori people built elaborate storage structures to exclude them. The spread of ship rats presented other problems. When they reached Madagascar in the fourteenth century, they carried bubonic plague, with serious effects on the local inhabitants. Disease carried by rats remains a problem in Madagascar.

REMOVAL FROM ISLANDS

Eradication programs against invasive rodents on islands have become increasingly successful, and their targets include mice and all three species of rats. From such programs' beginnings on small islands in the early 1960s, rats have been removed from over 300 islands around the world. The largest eradication was of Norway rats from 11,000-ha Campbell Island off southern New Zealand. These large eradications have been made possible by the combined use of helicopters with purpose-built bait spreaders, modern GPS systems to ensure accuracy of bait spread, and second generation toxicants that are highly effective. The eradications can be controversial, especially if it is unclear how rats have affected native species or if there are native species that might themselves be sensitive to the toxicants. One such example was Anacapa Island off California, where the eradication of ship rats in order to protect seabirds could have threatened a local subspecies of native *Peromyscus* deer mouse. The solution was to remove large numbers of deer mice and hold them in captivity during the eradication. Once returned to the island, *Peromyscus* became far more abundant than it was while the rats were present, indicating that the rats had suppressed this species, as well as the seabirds.

Controversy would be reduced if the benefits of eradications were better known. Rapid responses have been reported after eradications. When Pacific rats were removed, the fledging success of Cook's petrels immediately increased in New Zealand. After Norway rats were removed, native shrews became more abundant in islands off France, and a rare snipe recolonized Campbell Island off New Zealand. After ship rats were removed, rare seabirds successfully nested on Anacapa Island, and burrowing seabirds recolonized St. Paul Island in the South Indian Ocean.

For these benefits to persist, it is necessary to prevent reinvasion, which is possible if islands are regularly checked, if there are restrictions of movement of bulk materials, and if care is taken with all luggage transported ashore. Numerous devices are now available that can detect invading rodents. Studies in New Zealand indicate that the most effective detection is with specially trained rat dogs. Unfortunately, the effects of invasions are most often only realized well after the rodents have established and unmistakeable changes to the island plants and animals have occurred. The speed and range of spread of these four species of rodents demonstrate that there is only one way to avoid the loss of additional island species: constant vigilance.

SEE ALSO THE FOLLOWING ARTICLES

Biological Control / Dispersal / Invasion Biology / Madagascar / New Zealand, Biology

FURTHER READING

Atkinson, I. A. E. 1985. The spread of commensal species of *Rattus* to oceanic islands and their effects on island avifaunas, in *Conservation of island birds*. P. J. Moors, ed. ICBP Technical Publication Number 3, 35–81.

Fukami, T., D. A. Wardle, P. A. Bellingham, C. P. H. Mulder, D. R. Towns, G. W. Yeates, K. I. Bonner, M. S. Durrett, M. N. Grant-Hoffman, W. M. Williamson. 2006. Above- and below-ground impacts of introduced predators in seabird-dominated island ecosystems. *Ecology Letters* 9: 1299–1307.

King, C. M. 2005. *The handbook of New Zealand mammals.* Melbourne, Australia: Oxford University Press.

Russell, J. C., D. R. Towns, S. H. Anderson, and M. N. Clout. 2005. Intercepting the first rat ashore. *Nature* 437: 1107.

Towns, D. R., I. A. E. Atkinson, and C. H. Daugherty. 2006. Have the harmful effects of introduced rats on islands been exaggerated? *Biological Invasions* 8: 863–891.

Wanless, R. M., A. Angel, R. J. Cuthbert, G. M. Hilton, and P. G. Ryan. 2007. Can predation by invasive mice drive seabird extinctions? *Biology Letters* online: 1–4.

ROTTNEST ISLAND

ANNE BREARLEY

University of Western Australia, Crawley

FIGURE 1 Longreach Bay, Rottnest Island, 2008, with view of Bathurst Point and the smaller of the Rottnest lighthouses, built in 1900. An eroded foredune area with *Spinifex* and *Acacia* lines the beach, with dark green tea trees on higher ground. Norfolk Island pines, which have been planted along most beaches in Australia, tower above the native vegetation. Reef platforms and small islets typical of the bays lie to the left of the lighthouse. Buildings of the city of Perth on the mainland are visible in the background behind the small islets. The city lies on the Swan coastal plain, and the 400-m-high hills of Darling Scarp form the eastern boundary of the Perth sedimentary basin. The clear waters are host to many different seaweeds and seagrasses. Boats at anchor illustrate the pressures of modern life on the natural environment. Photograph by O. Paterson.

Rottnest Island, 1900 ha in area and located 17 km off the coast of Perth, is an iconic holiday destination for many Western Australians and popular tourist venue for 500,000 visitors each year. Limestone buildings from the colonial period in the 1800s still form the heart of the settlement. Motor vehicles are used only for maintaining services, with visitors riding bicycles, walking, or using buses to explore the island. The scenic bays with clear turquoise waters are ideal for swimming, and the colorful diverse marine life is of great fascination to all (Fig. 1).

HISTORICAL PERSPECTIVE

Willem de Vlamingh, a member of the Dutch East India Company, in 1696 provided the island with its name, a derivation of "Rottenest" or "rat's nest," referring to the small rat-like marsupial quokka *Setonix brachyurus* found on the island (Fig. 2). Vlamingh's charts also showed the mainland coast and the Swan River, named for its black swans, which was to become the focal point of British settlement of the west coast in 1829 with the foundation of Perth and Fremantle. Settlement on Rottnest commenced the following year, but farming on the island was a failure, and with growing conflict between the colonists and the aboriginal people a decision was made in 1838 to use the island as a prison for offenders of the new laws. The island was used as a prison for about 70 years but also served as a boys' reformatory and a vice-regal summer retreat until the decision was made to convert it to a holiday resort. During World War I, the island was again used as a prison for internees and prisoners of war. Throughout the Second World War, the island, with its sweeping ocean views, was the first line of defense for protecting Fremantle Harbour, which, following the attack on Pearl

FIGURE 2 The quokka among leaves and seed pods shed by the overhanging native *Acacia*, with clumps of green onion–like *Trachyandra*. Photograph by H. Lambers.

Harbor and the Japanese advance through the Pacific, was a major base for Allied Forces. With the return of peace, the island resumed its holiday mode, although the Kingstown Barracks area was not relinquished until 1984, when it was handed back to the State, and the former army buildings converted to the Environmental Education Centre to foster public interest in the islands unique natural and historical heritage. The island was declared an A-class Conservation Reserve primarily for public recreation in 1917. The flora and fauna of the area have been well documented, and a research station is managed through University of Western Australia. The island is a major tourist attraction for day visitors and also features some short-term cottage and resort accommodations.

CLIMATE

The island experiences a Mediterranean-type climate characterized by cool, wet winters and hot, dry summers. In winter, the westward passage of systems of low pressure bring winds from the northwest to west. In summer, winds are predominantly from the east in the morning and from the southwest in the afternoon. These sea breezes frequently exceed 50 km per hour. The tides are diurnal, occurring only once per day, and are of small amplitude, with a daily range of 0.4–1.1 m. Sea-level changes are, however, also affected by air pressure, winds, oceanic swells, and currents.

In contrast to other west-facing coastlines, the marine waters are warm because of the southward-flowing Leeuwin Current, which brings warm, less saline, nutrient-poor water from the tropics in autumn and winter. As a result, the waters around the island are up to 3 °C higher than along the coast.

THE ISLAND AND ITS SETTING

Rottnest, 11 km long and 5 km at the widest point, with the highest point 45 m above sea level, is the largest of a chain of small islands and reefs representing one of a series of dune lines on the coastal plain and continental shelf, which were separated from the mainland 5000–7000 years ago. The island is composed of aeolian or dune limestone (the Tamala or Coastal Limestone) of Late Pleistocene or Early Holocene age, with an intercalated Late Pleistocene coral reef (the Rottnest Limestone) exposed at Fairbridge Bluff, overlain dunes, and weakly lithified Holocene shell beds (Herschell Limestone) around the lakes, along with swamp deposits and dune sands. The coastline of rocky headlands and bays with wide sandy beaches backed by sand dunes is fringed by shallow shoreline platforms cut in the Tamala Limestone.

The shoreline platforms are undercut with overhanging visors or sloping ramps. A narrow storm bench at 3 m above the notch and visor are commonly formed below the limestone cliffs. The outer edge of the reef platform often has a raised rim, covered with coralline algae, which commonly covers much of the exposed rock surface. On some reefs, the surface is divided into a mosaic or into polygons formed by lines of brown algae. These do not appear to be associated with features in the reef surface, and their origin has been attributed to fish grazing. In other areas, mobile nodules of coralline algae or rhodoliths are abundant. Terraces up to 70 cm higher than the main platform also occur near the outer edge of some platforms.

The reef platforms are covered with turfing and foliose algae and with a diversity of invertebrates including the tropical echinoid *Echinometra mathei* that occupy holes excavated into the reef surface. The large turban shell *Turbo intercostalis,* a temperate species, is also common, and broken shells and opercula deposited by Pacific gulls are found on some headlands.

Salt lakes, the deepest of which is about 8.5 m, cover about 10% of the island. These are believed to overlie dolines formed during periods of lower sea level. Water levels rise with the winter rainfall and fall again in summer, when salinities may exceed 150,000 mg/L. Some of the lakes dry out completely and were a source of salt production in the late 1800s to mid-1950s. The lake waters are often pink in color because of pigments in the tissues of microscopic algae. Brine shrimp *Artemia,* which may have been introduced with salt harvesting equipment, provide food for a variety of birds, including transequatorial migratory species that congregate here in summer. Algal mats and stromatolites–thrombolites cover the lake bottoms. Brackish groundwater with a thin freshwater lens underlies some parts of the island, and there are a number of small fresh- and brackish-water swamps. However, a number of these are now hypersaline following excavation in the 1970s for road-building material.

TERRESTRIAL VEGETATION AND FAUNA

Descriptions of the island prior to European settlement record extensive areas of Rottnest Island pine *Callistris preissii,* tea tree *Melaleuca lanceolata,* and wattle *Acacia rostelliferia,* forming a low woodland that probably covered about 65% of the island. By 1941 the area of forest was reduced to about 23%, and it currently covers less that 5% of the island, with an additional 6% of indigenous and exotic species. In 2003, 196 plant species were recorded, and a low heath dominated by the shrub prickle lily *Acanthocarpus presissii,* the tussock grass *Austrostipa flavescens,* and the introduced geophyte *Trachyandra divaricata* now covers most of the island. Changes in the vegetation are considered to be the result of human activity, initially by bushfires and woodcutting, but more recently through grazing by the marsupial quokka.

The island is home to limited number of vertebrates. The wallaby-like quokka, numbering about 10,000, was at the time of European settlement also common on the mainland where it is now restricted to a few isolated areas of uncleared swamps. The dugite, a venomous snake, is smaller and darker in color than in populations on the mainland and is considered to be a separate subspecies. Lizards, geckos, and about 50 species of bird are also present on the island. Red-capped robins, golden whistlers, and singing honeyeaters inhabit the woodland and heath, and Australian shelducks, banded stilts, and migrant waders including stints, sandpipers, and turnstones are common around the lakes. Terns and gulls are also common along the shore, and ospreys nest on islets around the island.

MARINE ENVIRONMENT

The 12,000-km coastline of western Australia spans three marine biogeographical regions: the northern tropical, which is continuous with the Indo–western Pacific; the southern warm temperate; and the western coast, an area of overlap with tropical and warm temperate species and a small number of endemics. Rottnest, at 32° S, lies within the overlap zone and influence of the Leeuwin Current, and the waters are host to a range of temperate and tropical species (Fig. 3).

About 350 species of algae, including 170 southern, 52 tropical, 50 temperate, and 59 west coast endemics, are found around the island. The largest are the kelp *Ecklonia radiata* and *Sargassum*. Nine species of seagrass are also found around the island. These include the large southern seagrasses that form extensive meadows in the shallow, clear water: the ribbon weeds *Posidonia* (three species), wireweeds *Amphibolis* (two species), and *Thallassodendron*. Smaller species (*Halophila* spp., *Syringodium*, and *Zostera* spp.) grow at the edge or within meadows and are often regarded as colonizers.

Thirty species of coral have been recorded. Generally, these grow as isolated colonies; however, there are some extensive areas of *Pocillopora damicornis* in shallow water on the southern coast. About 360 species of fish have also been recorded. Most are temperate species, but others are typically more common among tropical corals. In response to escalating recreational fishing, four small Sanctuary Zones have been formed, and their effect

FIGURE 3 Typical shallow-water habitat Parker Point, southern coast of Rottnest Island. The tropical hermatypic, or reef-building, coral *Pocillopora damicornis* and alga *Sargassum* sp. growing on limestone reef among areas of sand, and the temperate seagrass *Posidonia sinuosa* with the western buffalo bream *Kyphosis cornelii*. Photograph by G. Kendrick 2003.

is currently being investigated. Sea lions and migrating whales are also seen around the island, and the populations appear to be recovering following the nineteenth-century closure of the sealing and whaling industries.

SEE ALSO THE FOLLOWING ARTICLES

Marine Lakes / Prisons and Penal Settlements / Research Stations / Whales and Whaling

FURTHER READING

Bradshaw, S. D., ed. 1983. Research on Rottnest Island. *Journal of the Royal Society of Western Australia* 66: 1–55.
Joske, P., C. Jeffery, and L. Hoffman. 1995. *Rottnest Island: a documentary history*. Centre for Migration and Development Studies, University of Western Australia.
Playford, P. E. 1988. *Guidebook to the geology of Rottnest Island*. Geological Society of Australia, Western Australian Division, Excursion Guidebook 2.
Rippey, E., M. C. Hislop, and J. Dodd. 2003. Reassessment of the vascular flora of Rottnest Island. *Journal of the Royal Society of Western Australia* 86: 7–23.
Rottnest Island Authority. 2003. *Rottnest Island Management Plan 2003–2008*.
Saunders, D., and P. de Rebeira. 1985. *The birdlife of Rottnest Island*. Western Australia: DAS & CpdeR.
Wells, F. E., D. I. Walker, H. Kirkman, and R. Lethbridge, eds. 1993. *Proceedings of the Fifth International Marine Biological Workshop: the marine flora and fauna of Rottnest Island Western Australia 1991* (2 vols.). Perth: Western Australian Museum.

SAMOA, BIOLOGY

A. C. MEDEIROS

U.S. Geological Survey, Makawao, Hawaii

The Samoan archipelago is a volcanic chain of nine main inhabited islands, high islets, and low coral islands located about 13°–15° south of the equator in the central South Pacific (though disjunct Swains Island is 11° S). The archipelago is divided into two political entities: Western Samoa (officially called Independent Samoa or Samoa) to the west and American Samoa (an unincorporated territory of the United States) to the east. Western Samoa consists of Savai'i (1820 km² area; 1860 m elevation) and 'Upolu (1110 km²; 1100 m), the fifth and eighth largest islands of the tropical Pacific, and a number of sizable volcanic islets (Aleipata Islands, Manono, and Apolima). The Aleipata islands have great potential as restoration sites for native birds, some of which are now endangered on the larger islands. American Samoa consists of five high islands—Tutuila (124 km²; 653 m), 'Aunu'u (2.6 km²), Ofu (5 km²; 495 m), Olosega (4 km²; 640 m), and Ta' (39 km²; 945 m)—and two remote coral islands—Rose Atoll (2.6 km²) and Swains Island (1.5 km²). Mount Silisili, the summit of Savai'i, is the highest point in Samoa, the tallest peak of central Polynesia, and the sixth tallest peak of the tropical Pacific.

CLIMATE AND TOPOGRAPHY

Located well within the tropics, the Samoan islands are hot, humid, and often rainy throughout the year. There is a wet summer season (October–May) and a shorter, slightly cooler and drier winter season (June–September).

FIGURE 1 Samoan rain forest, northern Tutuila island. *Cyathea* tree fern in center. Photograph by the author.

Total rainfall varies throughout the islands because of orographic effects associated with topography and ranges from an average of 318 cm/yr on the leeward coastal plain of Tutuila to more than 500 cm/yr in the mountains. Samoa's position in the southern trade wind zone would suggest wetter windward and drier leeward exposures, as occurs strongly in the Hawaiian archipelago. However, that is not the case, as rain forest vegetation dominates both windward and leeward exposures, presumably because of the lower elevation of Samoan mountain summits and the archipelago's orientation, roughly parallel to southeasterly tradewinds. Larger islands are dissected by numerous short stream drainages, most flowing only intermittently.

TERRESTRIAL BIOLOGY

Except for the two coral islands, the Samoan islands are volcanic, created by the westward movement of the Pacific Plate over a near-stationary hotspot beginning some 2 million years ago. The Samoan biota has been derived through colonization across 500 km of ocean and through in situ evolution and speciation. Like other island groups in the South Pacific, much of the ancestral terrestrial biota of Samoa arrived from Malesia in the west. Terrestrial and marine species diversity in the South Pacific is highest near New Guinea and generally declines eastward through Melanesia and Polynesia.

Vegetation

The high island vegetation of Samoa has been ascribed to seven plant communities: littoral, wetland, lowland rain forest, montane rain forest, cloud forest and scrub, vegetation on recent volcanic surfaces, and modified vegetation. Rain forests are the predominant vegetation form and still dominate much of Samoa from just above the shoreline to mountain summits. Samoan rain forests are nearly as diverse as Melanesian rain forests and comprise some of the most diverse lowland rain forest in the Pacific. The native Samoan flora is the largest in Polynesia except for Hawaii, with 550 angiosperm species in 300 genera and 228 pteridophyte species. About 30% of the flowering plants of Samoa are endemic, with a single endemic plant genus, *Sarcopygme*.

Fauna

The native fauna of Samoa consists of 24 land and freshwater birds, 20 seabirds, three mammals, seven skinks, four geckos, two marine turtles, one snake, and numerous invertebrates. Of particular interest in the latter group are 94 species of native land snails, 63% unique to the archipelago. U.S. federally listed endangered species include the humpback whale and two marine turtles. Four other vertebrate species are known as species of concern (Polynesian sheath-tailed bat, many-colored fruit dove, friendly ground dove, spotless crake).

The nectar-feeding honeyeaters include one of the most common and one of the rarest forest birds of the archipelago. The common and indigenous ʻiao, or wattled honeyeater, (*Foulehaio carunculata*), though plainly colored, is an energetic and boisterous addition to native forest and village garden. The sweet singing segasegamauʻu, or cardinal honeyeater (*Myzomela cardinalis*), common around villages and gardens, is dimorphic with scarlet males and gray-olive females. One of the rarest birds of Oceania, is endemic to the Samoan archipelago: the mao, or maʻomaʻo (*Gymnomyza samoensis*), one of the largest honeyeaters of the tropical Pacific, endemic to the Samoan islands of Savaiʻi, ʻUpolu, and formerly Tutuila (last recorded in 1977). Described as common in certain areas of ʻUpolu in the mid-1980s, this nectarivore has declined markedly and is now a Red List Endangered species of the World Conservation Union; current population is estimated at 1000–2500 birds. In contrast, its closest living relative, the giant forest honeyeater (*Gymnomyza viridis*) of Fiji, is still relatively common in some areas.

Of the two native starlings of Samoa, the larger endemic fuia, or Samoan starling (*Aplonis atrifusca*), is the more adaptable species. The wide-ranging miti vao, or Polynesian starling (*A. tabuensis*), is generally restricted to forested surroundings.

Five species of gallinules occur in Samoa. The most easily seen include the common veʻa, or banded rail (*Rallus philippensis*), and the manu aliʻi, or purple swamphen (*Porphyrio porphyrio*). The rare, shy punaʻe, or Samoan woodhen (*Gallinula pacifica*), is thought to be restricted to primary rain forests of Savaiʻi. Old accounts suggest that the burrowing punaʻe was once common and was hunted with dogs and nets; it was last seen in 1908.

Samoan pigeons are the subject of special cultural interest. Prior to modern hunting techniques and cyclone impacts, the manutagi, or purple-capped fruit-dove (*Ptilinopus porphyraceus*); the manuma, or many-colored fruit-dove (*Ptiliopus perousii*); and the tuʻaimeo, or friendly ground dove (*Gallicolumba stairi*), were once common but are now infrequent to rare and declining. The still relatively common lupe, or Pacific pigeon (*Ducula pacifica*), is important to Samoan culture, and in prehistory its capture influenced the construction of the large, stone platforms called tia seulupe (star-mounds), in part for pigeon catching. When first discovered in 1839, the secretive manumeʻa, or tooth-billed pigeon (*Didunculus strigirostris*), caused quite a stir, as it was initially thought to be related to the extinct dodo of Mauritius. In fact, the genus name *Didunculus* means "little dodo." Now recognized as the national bird of Samoa, the manumeʻa is restricted to forests of Savaiʻi and ʻUpolu. This thick-bodied, short-tailed pigeon, with its strongly hooked upper bill and notched or "toothed" lower bill, is the only extant member of the subfamily Didunculinae, or tooth-billed pigeons. The bill is used in feeding primarily on the tough fibrous fruits of the rain forest tree maota (*Dysoxylum* spp.). The only other representative of the genus is the extinct Tongan tooth-billed pigeon (*Didunculus placopedetes*), known only from subfossils from ʻEua island. Losing forest habitat through agricultural deforestation as well as hunting, populations of tooth-billed pigeon were halved to an estimated 2500 birds by the 1990s and continue to decrease. The tooth-billed pigeon currently is listed as a Red List Endangered species by the World Conservation Union.

Samoa has one native snake: the nonpoisonous gata, or Pacific boa (*Candoia bibroni*), found locally on Savaiʻi,

'Upolu, and Ta'u islands. The only native mammals of Samoa are three bat species, including a small insectivore and two fruit bat species. The pe'ape'avai, or Polynesian sheath-tailed bat (*Emballonura semicaudata*), was once common throughout Polynesia and Micronesia but now is either extinct or nearly so in Samoa. The extremely wide-ranging pe'a fanua, or Tongan fruit bat (*Pteropus tonganus*), is found from New Guinea to the central Pacific (Cook Islands). The diurnal pe'a vao, or Samoan fruit bat (*P. samoensis*), still occurs in Samoa and Fiji but is extinct on Tonga. Native Samoan frugivores, especially fruit bats, pigeons and doves, and starlings, play a critical role in the islands' ecology by dispersing seeds of many native plant species.

SAMOAN MARINE BIOLOGY

Samoa has 890 species of coral reef fishes and over 200 coral species. The rate of endemism in this rich native marine biota is low, apparently because of its proximity to other South Pacific islands and the high mobility of many marine species during at least some life stages.

Samoa has two marine turtle species, both of which are protected by the U.S. Endangered Species Act. Both sea turtles are declining throughout the Pacific, primarily as a result of harvesting for food and shell, incidental mortality in commercial fisheries, and disturbance of nesting beaches. Though widely distributed, the laumei uga, or hawksbill (*Eretmochelys imbricata*), nests in low numbers throughout Samoa and does not often migrate long distances. In contrast, the green turtle (*Chelonia mydas*) is highly migratory and wide-ranging (> 1000 miles); it nests primarily on Rose Atoll. Excessive harvesting of this species, the so-called "buffalo of the sea," for its shell, cartilage, and meat has made it among the most exploited of turtle species worldwide.

THREATS TO SAMOAN BIODIVERSITY

Human impacts (overfishing, pollution and accelerated erosion) and crown-of-thorns starfish (*Acanthaster plana*) threaten the sustainability of coral reef communities in Samoa. Although Polynesians have lived on the islands for over 3000 years, accelerated population growth and economic development in the past 50 years is leading to extensive modification of coral reef and nearshore ecosystems. The current human population of American Samoa (65,000) is increasing at a rate of 2% per annum, the vast majority living within several hundred meters of the shoreline. Climate change and consequent sea level rise obviously have potential to bring about serious damage to Samoan biodiversity.

Deforestation (for timber in Western Samoa and expanding agriculture generally) and replacement by non-native invasive plant species seem to pose the greatest threat to long-term survival of native rain forest biota of Samoa. Prior to human settlement, Samoa was almost entirely forested. Agriculture, timber harvesting, and invasion by non-native plant and animal species have reduced forest cover. In 1954, an estimated 74% of Savai'i and 'Upolu (93% of Samoa's total land mass) was covered in native rain forest. That figure had declined to 40% by 1990, and forest reduction continues.

Invasive species are notorious as degraders of Pacific island biodiversity. Among the most damaging of vertebrate invaders in Samoa are feral pigs and rats, including the Norway rat on Tutuila. Introduced *Euglandina rosea* snails prey on complexes of endemic snail species. Increasing numbers of non-native plant species are becoming established in Samoa. Currently, two of the most damaging invasive trees of Samoan rain forests are silkrubber (*Funtumia elastica*) and Moluccan albizia (*Falcataria moluccana*). *Funtumia*, brought from Africa and planted widely on 'Upolu and Savai'i, has now spread beyond plantations and is invading intact rain forest throughout the country up to 700 m elevation. *Funtumia* may be the worst forest invader of Western Samoa (it currently does not occur in American Samoa) and may be one of the most potentially damaging invaders of rain forests of Pacific islands.

Falcataria moluccana is a large, quick-growing tree species native to rain forests of the Moluccas, New Guinea, New Britain, and the Solomon Islands. Introduced as a forestry resource, *Falcataria* now occurs in about 35% of Tutuila's forests. Nitrogen-fixing abilities by *Falcataria* may disrupt succession in native Samoan forests by promoting establishment of non-native plant species and by deterring recruitment of natives. Other notable threats to Samoan rainforest include two Neotropical species: the shrub *Clidemia hirta*, widespread in Samoa, and the small tree strawberry guava (*Psidium cattleianum*), reported only from small naturalized populations on Tutuila and 'Upolu.

Land tracts set aside for biological and cultural diversity include several national parks in Western Samoa (Tafua and Falealupo on Savai'i and 'O le Pupū-Pu'e on 'Upolu) and the 3600-ha National Park of American Samoa (NPSA), which includes sections of the four main islands of American Samoa. The persisting culture and tradition (fa'a Samoa) of native Samoans can be a powerful force supportive of the management of natural ecosystems. In 2005 the NPSA joined with local villages and NGOs in eliminating *Falcataria* trees from more than 250 hectares on Tutuila, with over 3000 large trees killed (by girdling) as of 2007. Cooperative community-based efforts like this may serve as a model for future land management in Samoa.

SEE ALSO THE FOLLOWING ARTICLES

Deforestation / Fish Stocks/Overfishing / Reef Ecology and Conservation / Samoa, Geology / Sustainability

FURTHER READING

Craig, P. 2002. *Natural history guide to American Samoa. A collection of articles.* Pago Pago: National Park of American Samoa and American Samoa Department of Marine and Wildlife Resources.
Mueller-Dombois, D., and F. R. Fosberg. 1998. *Vegetation of the tropical Pacific islands.* New York: Springer-Verlag.
Watling, D. 2001. *A guide to the birds of Fiji and Western Polynesia.* Fiji: Environmental Consultants Ltd.
Whistler, W. A. 2002. *The Samoan rainforest: A guide to the vegetation of the Samoan Archipelago.* Honolulu: Isle Botanica.
Whistler, W. A. 2004. *Rainforest trees of Samoa.* Honolulu: Isle Botanica.

SAMOA, GEOLOGY

JAMES H. NATLAND

University of Miami, Florida

The Samoan Islands are at the eastern end of a chain of volcanoes, most of them submerged, and very near the southwestern edge of the main Pacific Basin at the Tonga Trench (Fig. 1). The islands proper span a distance of about 400 km along a trend of about 290°, but the entire chain, in the form of shallow banks and submarine volcanic pinnacles, extends to the west for about another 1200 km. The islands resemble the Hawaiian Islands in many respects, but some features of them are distinct. The principal difference results from being near the Tonga Trench rather than in the middle of the basin. In plate tectonic terms, the Samoan Islands are at the edge of the Pacific plate in a most unusual spot, just to the north of where the trench turns to the west. South of there, the Pacific plate is disappearing into the trench, and the path of its subduction can be traced seismically by a dipping belt of earthquakes that extends to depths of about 700 km beneath the Tonga arc. However, because the trench changes trend just south of Samoa, the islands manage to ride the plate west past the trench and are not subducting. But it is a near thing, and the effect on Samoan volcanism is profound. The chain may even exist because of this peculiar relationship.

AGE PROGRESSION

From east to west, the main islands are Ta'u and Ofu-Olosega (two islands separated by a narrow channel, denoted here by the hyphen) in the Manu'a group of American Samoa; a larger island, Tutuila, also in American Samoa; and the two largest islands, Upolu and Savai'i, in the independent country of Samoa, which was called

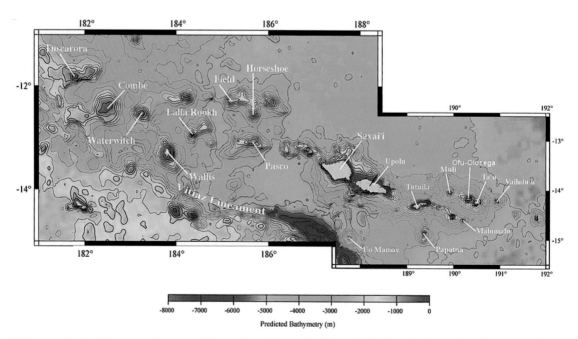

FIGURE 1 Bathymetry and location of the principal Samoan Islands (gray—Savai'i, Upolu, Tutuila and Ta'u are labeled), banks, and seamounts (Muli, Malumalu, and Vailulu'u, eastern Samoa; Papatua and Ua Mamae—also called Macias—north of the northwesterly-trending portion of the Tonga Trench. The Vitiaz lineament is shown as a continuation of the northern termination of the Tonga Trench, and is here interpreted as a trench-trench transform fault.

Western Samoa until 1997. A number of smaller islands scatter about these, and several substantial seamounts rise from the seafloor near these main islands. One of them, Vailulu'u, at the eastern end of the chain, is an active volcano. To the west are submarine banks, the nearest being Pasco and the farthest Tuscarora, which is a shallow submerged reef platform. This article will sometimes refer to the collection of islands and seamounts as Samoa, with no geopolitical connotation.

American explorer James Dwight Dana first described the geology of the Samoan Islands in 1849. Dana noted many similarities to Hawaii, and in particular that the largest island, Savai'i, has a youthful, uneroded, domed shape and is studded with many small and obviously recently erupted volcanic cones, much like Mauna Kea volcano on the island of Hawaii. The islands to the east appeared to him to be successively more dissected by erosion, and they have wide rather than narrow offshore reefs, indicating they are older. In these respects, they are similar to Oahu and Kauai in the Hawaiian chain. But whereas those older Hawaiian Islands lie to the west, in Samoa the islands most similar to these are to the east. Dana wrote, "It is hence evident that the fires were soonest extinct to the east, and burnt longest and to the latest period on the western island, Savai'i" (Dana 1849: 335).

This comparison to Hawaiian volcanism was pivotal, in my view, to Dana's later development of the doctrine of fixity of continents and ocean basins, which, more than 60 years hence, was one among the array of ideas used in criticism of Alfred Wegener's theory of continental drift. Dana construed island chain volcanism in the Pacific to occur along fractures and largely simultaneously along the entire lengths of the ridges at first, but then dying out toward one end or the other—to the west at Hawaii, to the east at Samoa. The Pacific crust beneath, consequently, had to be fixed in place. Nevertheless, even after plate tectonics was demonstrated and linear island chains were considered to form in consistent directions as plates passed over fixed hotspots, Samoa remained an anomaly. However, subsequent, more precise comparisons with the stages of volcanism at Hawaii cast the matter in a different light. The work of several investigators showed that Savai'i and its nearest neighbor, Upolu, are both experiencing a second major phase of volcanism that has buried older volcanoes that should be considered separately for the purpose of establishing the age progression along the chain. Comparison to Hawaii dictated at the time that these rocks be termed "post-erosional" with respect to the eroded volcanoes they surmounted (Fig. 2). Thus, potassium–argon dating of rocks sampled in light

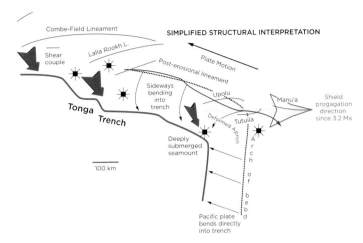

FIGURE 2 Generalized structural relationships of the principal Samoan Islands, including lineaments linking central volcanoes in red and the great Samoan post-erosional volcanic rift system in blue. Arrows are drawn to indicate plate motion, subduction direction, propagation directions, and inferred plate bending and shear orientation near the transform system.

of this geological distinction shows that there is indeed a Hawaiian-like age progression to the underpinnings of the Samoan chain (Fig. 3). The older phase of volcanism at Savai'i began at about 5 million years ago, at Upolu about 3.5 million years ago, and at Tutuila at about 1.5 million years ago. Ta'u has had some barely prehistoric eruptions, and Vailulu'u is active at the present day. The age progression along the islands is thus perfectly normal for the Pacific plate and is consistent with the hotspot model; indeed it has been extended to the submerged banks to the west as far away as Combe Bank, which is 14.1 million years old.

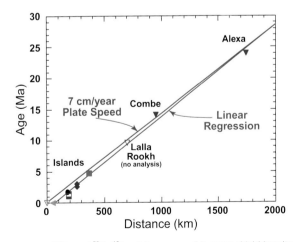

FIGURE 3 K/Ar and $^{39}Ar/^{40}Ar$ plateau ages of Samoan shield basalts, plus an unanalyzed sample from Lalla Rookh bank. The linear regression is nearly identical to the estimated 7 cm/year rate of motion over a hotspot at Vailulu'u fixed with respect to an assumed fixed Hawaiian hotspot. Many more alkaline basalts and other dated samples fall below the regression (not shown), including all dated post-erosional samples.

STAGES OF VOLCANISM: A MISFIT TO HAWAII

The key to understanding the age progression turned out to be distinguishing the stages of volcanism at Samoa that are comparable to those of Hawaii: namely, preshield, shield, postshield, and rejuvenated. However, aspects of this sequence, and especially the terms, are somewhat misleading when considering the Samoan case. One can, however, see provisionally how they apply to Samoa, consider the contrasts between Samoa and Hawaii, and then try to understand those differences in evaluating the causes of Samoan volcanism.

The preshield stage at Hawaii is that of nascent volcanism in which alkalic basaltic lava erupts to form a seamount at the youthful end of the chain; this is presently the active submarine volcano of Loihi, offshore of Kilauea volcano. At Loihi, volcanism has shifted through time to tholeiitic compositions, some of which erupted in the past few years. The Samoan equivalent to Loihi is Vailuluʻu, the summit lavas of which are indeed alkalic basalt; as discussed below, however, no Hawaiian-type tholeiite has been found there. The volcano has a summit crater about 2 km in diameter.

Among the Hawaiian Islands, the next stage of volcanism is construction of large shield volcanoes made of tholeiitic basalt. The structures are huge, centered on principal conduit systems beneath the tallest parts of the volcanoes, but they also include enormous rift zones that are supplied laterally by magma pumping from staging areas just beneath the summit regions. Deep exposures of old rifts reveal densely spaced dikes, all with nearly the same orientation. Some of the rifts reach offshore as far as 200 km from their central conduits. The shield volcanoes overlap and structurally influence each other to the extent that rift systems nest younger against older, and dike swarms conforming to the resulting stress field are parallel. Then comes the postshield stage of waning volcanism, formation of large calderas, and eruption of alkalic basalt and affiliated differentiates, typically hawaiite, mugearite, and trachyte. Mapping reveals that most of this volcanism is centered on or at least near the older tholeiitic summits, but it is infrequent enough for some of the lavas to collect in valleys carved by rivers.

At Samoa, no shield volcano is nearly as large as the typical Hawaiian shield, and they do not coalesce or combine in the same way, or even at all. No rift zone nesting occurs. Again, Hawaiian-type tholeiite has not been found on any of them. The islands of Taʻu and Ofu–Olosega conform pretty much to the postshield Hawaiian equivalent; Taʻu has a large caldera and has erupted alkalic basalt and some hawaiite (Fig. 4A). These are relatively youthful structures with much of their original uneroded surfaces intact, and magmatic lineages have not produced the light-colored felsic differentiates mugearite and trachyte. The older islands of Tutuila and Upolu have substantial and now fairly deeply eroded shield volcanoes, with well-developed magmatic lineages that include alkalic olivine basalt, hawaiite, mugearite, and trachyte (Figs. 4B and C). On Tutuila the trachyte occurs as prominent plugs that intruded an elliptical caldera ring fault that has principal axial lengths of 5 km × 10 km; this now nearly surrounds Pago Pago Bay. A few similar plugs are just outside the crater. Another large, caldera-like ring fault also bounds most of Fagaloa Bay on Upolu. Trachyte cobbles have been sampled from beaches near this fault, but actual trachyte plugs there have not been mapped. Varieties of tholeiitic basalt have been obtained from deep exposures or outlying districts at both Tutuila and Upolu, but the classification of these rocks as tholeiites, and their place in the morphological development of those volcanoes, is not simple. Again, this topic is considered separately below.

Structurally, the volcanoes from which these lavas erupted are not strictly comparable to the shield volcanoes of Hawaii. Thus, four separate volcanoes along the 40 km of the length of the island were mapped at Tutuila, and a fifth older one lies partly buried and without physiographic expression. However, some of the details of that early mapping amidst the jungle are now problematic, and where the mapping is still sound, the question now is whether at least two of these volcanoes may simply have been satellitic summits about a principal vent centered on what is now the caldera at Pago Pago Bay. Thus the separate summits are very closely spaced, and the total line of them would easily fit along less than half the length of the Kilauea eastern rift from the summit to the shoreline. Further, potassium–argon ages indicate that, if there were indeed separate volcanoes, four of them erupted basaltic lava simultaneously from about 1.5 to 1.2 million years ago; then mainly differentiates to about 1.1 million years ago. Late-stage trachytes are nearly completely restricted to the area around the ring fault at Pago Pago Bay and are 1.03 million years old. The most reasonable interpretation now is that there was one principal conduit system that split into separate upper-level conduits that were alternately supplied differentiated lava as the rate of magma supply diminished, and that as volcanism waned further, it became more constricted to above the vicinity of the main conduit.

The final Hawaiian stage of volcanism is eruption of basaltic lava with low to very low SiO_2 contents, namely alkalic olivine basalt, basanite, olivine nephelinite, and

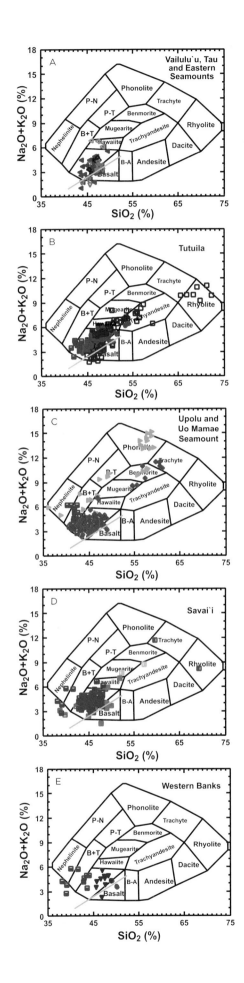

olivine melilitite (Figs. 4B–E). On the islands of Oahu and Kauai, these erupted chiefly as volcanic scoria at cones such as Diamond Head, but with some lava, along short fracture systems that trend across the older shield rift systems. The lavas carry numerous ultramafic xenoliths.

Following eruption of trachytes on both Tutuila and Upolu, volcanism shut down at both places for nearly a million years. Then, in the last few thousand years, renewed alkalic basaltic volcanism has taken place at both islands in the form of cinder cones and lava flows. The lavas are alkalic olivine basalt, basanite, and olivine nephelinite, but not olivine melilitite. The lavas carry ultramafic xenoliths. On Tutuila, the volume of this volcanism has been small enough that it was at first strongly likened to Hawaiian rejuvenescent volcanism. But on Upolu, the extent of this volcanism was much greater, and the eruptions there are also clearly related structurally to even more substantial Holocene to historic volcanism on the western island of Savai'i, where some of the lava flows, including two historic ones, are extensive. Almost all of this renewed volcanism has occurred along a single prominent volcanic rift system that spans the length of Savai'i and Upolu and that only recently has reached Tutuila (see Fig. 2). At Samoa, the scale of this volcanism has been much greater that at Hawaii, especially at Savai'i, where accumulations of the younger lavas are thousands of meters thick, and there are no exposures of older shield volcanoes on the island; those are represented only by offshore rocks obtained by dredging.

Thus the archetypal Hawaiian sequence of stages does not truly fit Samoa. The concept was useful when it came to establishing the age progression. However, in detail the misfit is substantial. The nearest equivalents to shield volcanoes are much smaller than those of Hawaii and do

FIGURE 4 Diagrams of total alkalis ($Na_2O + K_2O$) versus SiO_2 for different portions of the Samoan chain from literature and unpublished data sources. The green lines divide Hawaiian shield tholeiitic (below) from alkalic basalt (above). All chemical analyses from 1902 to 2008 are combined in these diagrams. On all diagrams, analyses of posterosional samples are given by half-filled blue squares. (A) Vailulu'u, Ta'u, and eastern Samoan seamounts are all given different symbols; (B) Tutuila—shield and postshield samples are divided into red symbols—Masafau volcano; and open squares—the coeval volcanism of Taputapu, Pago, Pago Intracaldera, Olomoana, and Asofau volcanoes of Stearns (1944), here combined into the Greater Pago volcano. (C) red diamonds = the Fagaloa volcano of 'Upolu; green triangles = lavas from Uo Mamae (Machias) Seamount. (D) Savai'i – various colors besides blue for offshore seamounts and trachyte cobbles from the Vanu River; (E) samples from western Samoan banks analyzed by Johnson et al (1986); red circles are Alexa Bank, inverted triangles are Combe, Lalla Rookh, and Field Banks, plus Wallis Island. Distinctive magmatic lineages leading to soda rhyolite (Greater Pago of Tutuila), trachyte (Fagaloa of Upolu), and phonolite (Uo Mamae Seamount) occupy islands of the central part of the Samoan chain.

not overlap or structurally interfere with each other. The postshield summit structure of Tutuila is different, having broken into four smaller centers. Hawaiian-type tholeiites are not found anywhere, and much of the shield-building stage of volcanism is alkalic, not tholeiitic, in character; and the final stage of volcanism is so prominent on two of the islands that the term "rejuvenescent" clearly is inadequate to describe it. The formerly used term "post-erosional" for this stage of volcanism at Samoa is not very good either. Later, we shall consider the substantial geochemical contrasts between Hawaii and Samoa. But given all of this, how should Samoan volcanism be described?

SAMOA CONSIDERED PHYSIOGRAPHICALLY

Consider that the second major phase of basaltic volcanism, which spans the length of Upolu and Savai'i and reaches Tutuila, is a single straight volcanic lineament superimposed on older features (see Fig. 2). What does the rest of the Samoan chain look like? We must look at the offshore geology. It does little good just to consider island exposures, which only amount to a few percent of the entire volume of the volcanic superstructure of the chain.

Bathymetry derived from satellite altimetry (see Fig. 1) combined with subaerial topography in the islands reveals several ridges along the Samoan chain that are 100–250 km long (see scale bar in Fig. 2). These are curving volcanic ridges that all probably combine several tall volcanoes. This is most obvious at the eastern end of the chain. Thus the easternmost of these combines, from west to east, Mali seamount, Ofu–Olosega, Ta'u, and Vailulu'u. A second links the several closely spaced summits of Tutuila. At its western end, this ridge curves toward the southwest and the westerly trending portion of the Tonga Trench; at its eastern end, the ridge curves back toward the southeast, reaching Malumalu seamount. West of Savai'i, some of these ridges can clearly be linked end to end as longer curving ridges that are broken by short gaps. Some of the curving ridges approach each other, and their general concavity is also toward the Trench (Fig. 2). The longest curving ridge extends from Combe Bank to Field Bank. Ages for these banks are shown in Fig. 3. Two shorter curving ridges, one including Lalla Rookh Bank, are both closer to and even more strongly inclined toward the westerly trending portion of the Tonga trench.

The general impression given by these trends and the radiometric ages is that the Samoan chain consists of a number of curving ridges, most of them diverging from the westerly trending part of the Tonga Trench, and with an overall progression that is younger toward the east.

The general tendency for the ridges to curve toward the Tonga Trench is notable because this sort of curvature does not occur anywhere else in the Pacific, and certainly not along the Hawaiian chain. Along the eastern portion of the chain, a major younger volcanic lineament is superimposed. It is probably longer than just the combined lengths of Savai'i and Upolu and reaches all the way to Tutuila. Hypothetically, if it additionally spans the lengths of the two ridges that align with it west of Savai'i, this feature could be more than 500 km long. This is the great Samoan post-erosional volcanic rift zone.

THE PROBLEM OF SAMOAN THOLEIITE

Although a number of classification schemes have been used to distinguish tholeiitic from alkalic basalt, the simplest one that has been applied to Hawaiian rocks is represented as a single straight line on a plot of total alkalis ($Na_2O + K_2O$) versus silica (SiO_2), separating basalts of the principal shield stages of several of the volcanoes, which all had the characteristics of tholeiitic basalt based on other classification schemes, and the alkalic basalts and differentiates of the postshield alkalic cappings. When it came to Tutuila, however, no tholeiites of the Hawaiian type were found even among what were obviously basalts of the shield stage as described for Tutuila.

Presently, with many more analyses in hand, this conclusion still holds, even though a few analyzed specimens, especially from Tutuila, fall below the Hawaiian dividing line (Fig. 4). Some of these are fairly obviously lavas leached of their K_2O by flow of groundwater. Several others are picritic, being highly charged with olivine phenocrysts, which serves to dilute total alkalis in a bulk-rock analysis, drawing compositions below the line. One major difference lies in the proportion of K_2O in a bulk rock analysis, which is rarely below 0.8% and often more than 1% among least altered Samoan basalt of shield volcanoes not charged with olivine crystals; it is typically less than 0.5%, and in some cases not even half of that, in the most nearly similar Hawaiian tholeiite. Samoan basalts have concomitantly higher concentrations of elements with geochemical behavior similar to potassium, and most have a different isotopic signature, described by geochemists as more enriched. Put in the most general terms, although some scattered lavas might fall into the tholeiitic field on a plot of alkalis versus silica, a *principal* tholeiitic stage of volcanism has not been found anywhere in Samoa, whether considered looking downward into the deepest accessible stratigraphy left by erosion on the islands or upward from the structures of seamounts. Indeed, basalt at Vailulu'u seamount is passably similar to basalt of any part of Ta'u, and basalt there in turn resembles much of the basalt at Tutuila. Furthermore, some portions of the shield systems of Tutuila and Upolu, and at least two of the offshore volcanoes, are distinctly more alkalic (that is, further removed from tholeiite, with higher K_2O and total alkalis; e.g., Ua Mamae, Fig. 4C) than Ta'u or Vailulu'u in character.

Even more difficult to understand is that this geochemical quality called "enrichment," which is usually considered in terms of isotopic ratios such as $^{87}Sr/^{86}Sr$ and $^{206}Pb/^{204}Pb$, is inconsistent from one Samoan volcano to the next. Thus the Fagaloa shield volcano of Upolu becomes *more* enriched through an upward (younging) succession of basalts, whereas at Hawaii the opposite consistently occurs on almost every volcano. At Tutuila, the extent of enrichment is greatest in the oldest exposed lavas; it then starts to scatter as one proceeds upward in the succession into younger lavas. Then abruptly the geochemical bottom drops out, and everything erupted since about 1.28 million years ago is least enriched on this island and similar in $^{87}Sr/^{86}Sr$ to lavas of Ta'u and Vailulu'u. Yet each shield volcano also has different $^{206}Pb/^{204}Pb$ regardless of $^{87}Sr/^{86}Sr$, and all of the second-phase alkalic basaltic lavas in turn are different isotopically from every older age-progressive shield volcano that they happen to overlie (Fig. 5).

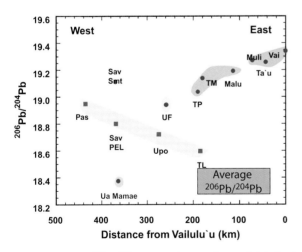

FIGURE 5 Average $^{206}Pb/^{204}Pb$ versus distance from Vailulu'u for the Samoan Islands compiled from all literature sources. Coding for central volcanoes is Vai = Vailulu'u; Ta'u seamount and island; Muli = Muli volcano; Malu = Malumalu volcano, TM = Tutuila Masefau volcano; TP = Tutuila Greater Pago volcano; UF = Upolo Fagaloa volcano; SavSmt = offshore Savai'i; and Ua Mamae volcano. Coding for post-erosional lavas is TL = Tutuila Leone lavas; Upo = Upolu, Samoa; Sav PEL = Savai'i post-erosional lavas; and Pas = ANTP 239 from Pasco Bank. From Upolu eastward, each volcano has a distinctive average $^{206}Pb/^{204}Pb$ increasing toward the east, and post-erosional lavas (blue band) are isotopically different from them.

THE MECHANISM OF SAMOAN VOLCANISM

Samoan volcanism has been attributed to volcanism along a series of sea floor fractures or to a mantle plume, with the plume perhaps being either generated or distorted by stresses associated with the nearby subduction of the Pacific plate, or to thermal/convective disturbances triggered by subduction in the mantle above about 700 km, which is the depth of deepest seismicity in the nearby Tonga Trench.

The strongest case for a fracture is the second major phase of basaltic volcanism, that most usually termed post-erosional volcanism, which occurs along a fissure system of closely spaced volcanic cones spanning the axes of Savai'i and Upolu and reaching Tutuila. The fissure system is too long and narrow to be a plume, its lavas are chemically distinct from everything else, and there is clear evidence for distortion of the Pacific plate, probably by bending of the plate southward between the islands of Savai'i and Upolu toward the westward-trending portion of the Tonga Trench. In this scenario, the axis of southward plate bending is directly beneath the two islands, and the Pacific plate is moving, by being either pushed or pulled, directly along the length of the bend. The base of the bending material lies within the region of partial melting in the mantle, and here it is also in compression. A combination of ponding of relatively small partial melts of enriched source material and squeezing aggregates those partial melts and focuses their eruption along the narrow fissure system. The melts probably pond at the rheological base of the lithosphere until a combination of buoyancy forces and regional stresses acts to release them to the surface, sometimes in large volume.

The older, age-progressive volcanoes are not so readily interpreted in terms of a fracture mechanism; thus, most recent workers have favored an origin for them in a mantle plume. Both the age progression and the general form of the volcanoes, which resemble those of Hawaii in so many ways (Fig. 3), are consistent with a plume model. But the plume would still have to be viewed as very circumstantially located just at this unusual tectonic juncture in the Pacific plate; the plume itself has produced no geochemical consistency from one volcano to the next or even within the history of a single island (Fig. 5); and the pattern of arcuate ridges mainly directed toward the westerly trending portion of the Tonga Trench (Fig. 2) is anomalous in the context of other volcanic chains in the Pacific.

The idea that shield volcanoes are caused by thermal-convective disturbances triggered by subduction into the nearby trench, allows the chain to have mainly Hawaiian-like structural attributes and an appropriate age progression. Yet it lets portions of the chain be distorted by shear stresses so that the curving volcanic ridges tend to curve toward the trench. And, finally, it is indifferent to geochemical complexity in the mantle except to the extent that source heterogeneity is restricted to the upper 700 km of the mantle. The enriched sources sampled by the volcanoes would lie in the relatively shallow mantle. Why they are so enriched in the first place has to do with the likely ancient prehistory of mantle sources in this region of the Pacific rather than with the specific manner of their arrival in the shallow mantle beneath Samoa, where partial melting has occurred.

SEE ALSO THE FOLLOWING ARTICLES

Hawaiian Islands, Geology / Lava and Ash / Pacific Region / Samoa, Biology / Volcanic Islands

FURTHER READING

Daly, R. A. 1924. The geology of American Samoa. *Carnegie Institution of Washington Publication* 340: 95–145.

Dana, J. D. 1849. U.S. exploring expedition during the years 1838–1842 under the command of Charles Wilkes, U.S.N. *Geology* 10: 307–336.

Hart, S.R., H. Staudigel, J. Blusztajn, E. T. Baker, R. Workman, M. Jackson, E. Hauri, M. Kurz, K. Sims, D. Fornari, A. Saal, and S. Lyons. 2000. Vailulu'u undersea volcano: the new Samoa. *Geochemistry, Geophysics, Geosystems* 1: 1–13, 2000GC000108.

Kear, D., and B. L. Wood. 1959. The geology and hydrology of Western Samoa. *New Zealand Geological Survey Bulletin* 63.

Macdonald, G. A. 1944. Petrography of the Samoan Islands. *Geological Society of America Bulletin* 56: 861–872.

Natland, J. H. 1980. The progression of volcanism in the Samoan linear volcanic chain. *American Journal of Science* 280A: 709–735.

Natland, J. H. 2003. The Samoan chain: a shallow lithospheric fracture system. www.mantleplumes.org.

Stearns, H. T. 1944. Geology of the Samoan Islands. *Geological Society of America Bulletin* 55: 1279–1332.

Workman, R. K., S. R. Hart, M. Jackson, M. Regulous, K. A. Farley, J. Blustahn, M. Kurz, and H. Staudigel. 2004. Recycled metasomatized lithosphere as the origin of the Enriched Mantle II (EM2) end-member: Evidence from the Samoan volcanic chain. *Geochemistry, Geophysics, Geosystems* 5.4. doi 10.1029/2003GC00623.

SÃO TOMÉ, PRÍNCIPE, AND ANNOBON

D. JAMES HARRIS

University of Porto, Vila do Conde, Portugal

The islands of the Gulf of Guinea are part of one of the world's biodiversity hotspots. Part of a volcanic chain that includes Mount Cameroon on the continent, the islands are Bioko, São Tomé, Príncipe, and Annobon. Although Bioko was connected to the continent during the last interglacial period, São Tomé, Príncipe, and Annobon are all surrounded by deep-sea trenches and are thus true "oceanic islands." These islands are striking centers of endemism. In contrast to most oceanic islands, where habitat loss is probably the chief conservation concern, in these islands control of introduced species may play a more critical role.

GEOGRAPHICAL AND GEOLOGICAL BACKGROUND

These islands are formed by shield volcanoes along the 1600-km-long Cameroon line, a linear rift zone that extends from Cameroon into the Atlantic. The volcanic chain was formed during the middle to late Tertiary and includes Mount Cameroon and the islands of Bioko, São Tomé, Príncipe, and Annobon (Fig. 1). Bioko, formerly Fernando Po, is the largest and closest to Africa, lying about 32 km from Cameroon. Príncipe, 225 km off the northwest coast of Gabon, has an area of 136 km^2 and reaches 948 m at Pico de Príncipe. Several smaller islets surround Príncipe, including Ilhéu Bom Bom, Ilhéu Caroço, Tinhosa Grande, and Tinhosa Pequena. The oldest geological date of Príncipe is 31 million years ago. When sea levels were lower in the last glacial periods, Príncipe would have been consid-

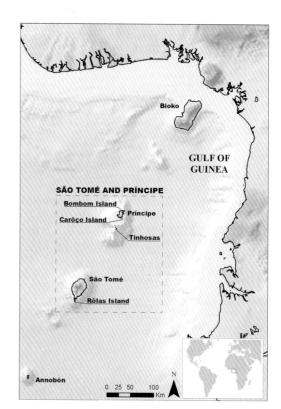

FIGURE 1 Map of the Gulf of Guinea islands.

erably larger. São Tomé, 140 km southwest of Príncipe, is approximately 845 km^2, about 48 km north–south and 32 km east–west. The southern tip of the island is just 2 km north of the equator, which passes through the small islet of Rolas. The highest point is 2024 m at Pico de São Tomé. The oldest dated volcanic rocks of São Tomé are 15.7 million years old, although most are basaltic lava less than 1 million years old, especially in the north, where pyroclastic cones are common. The central and southern parts of the island were exposed by a volcanic period around 3 to 8 million years ago, whereas the most southerly tip of the island and the Rolas islet are composed of pyroclastic and lava cones less than 400,000 years old. The most isolated island of the group is Annobon, about 160 km southwest of São Tomé and 350 km from the west coast of Africa. It has an area of approximately 17 km^2, rising to 598 m at its peak. It is composed of a single extinct volcano with a central crater lake. Its oldest dated rocks are approximately 4.8 million years old.

All the islands are within the wet tropical belt. Annual rainfall on São Tomé ranges from 1000 mm in the northeast to over 4000 mm in the southwest. Average annual temperatures range from 18 to 21 °C minimums to 30 to 33 °C maximums. The rainy season runs from October to May. Príncipe is very similar to São Tomé, whereas Annobon is notably drier. All are generally very steep, with lowlands at the base of the volcanoes constituting the only (relatively) flat land. The volcanic soils of basalts and phonolites are relatively fertile. There are currently estimated to be over 40 km^2 of primary forest on Príncipe and 240 km^2 on São Tomé.

ORIGINS OF ISLAND FAUNA

Sea levels between Bioko and the continent are shallow; thus, this island would have been connected to the mainland by sea-level fluctuations during the last glacial periods. This is not the case for the islands of São Tomé, Príncipe, and Annobon, each of which is surrounded by deep sea trenches and is thus a true "oceanic island," having never been connected to the mainland or to its neighbors. Thus, although the fauna of Bioko is essentially continental in nature, the other islands have fewer species but far more endemism. In many other Atlantic archipelagoes, such as the Canary Islands or the Cape Verde Islands, there is a strong trend for island radiations, where a colonizer reaches the islands and then radiates on to the other islands of the group. However, the islands of the Gulf of Guinea are fewer and further apart, and phylogenetic studies indicate that in several cases islands were colonized independently from the continent rather than taking a more classic "stepping-stone" model, in which the fauna first colonize the island closest to the continent (e.g., Príncipe) and then travel successively on to the more distant islands. Many non-volant species may have reached the islands by rafting: Currents are favorable in carrying material from the continent toward the islands, and major rivers such as the Niger and Congo discharge into the area. During wet periods and flooding, sea-surface salinity can drop substantially, and this may have been critical for successful colonization by some species, such as the endemic amphibians.

ENDEMISM

The Gulf of Guinea islands are a striking center of endemism. Between them, they include 29 endemic bird species, about one-third of all the endemic birds of the overall Guinea forest biodiversity hotspot and more than the number found across the famously diverse Galápagos archipelago. There are four endemic genera on São Tomé and another on Príncipe. In a global review of priority areas for bird conservation, both São Tomé and Príncipe were considered critically important because of the high number of restricted range species occurring there. Even the tiny island of Annobon has endemic birds such as the Annobon white-eye, *Zosterops griseovirescens,* and the Annobon Paradise flycatcher *Terpsiphone smithii.* As is typical in oceanic islands, most endemic mammals are bats. São Tomé has two endemic species, *Myonycteris brachycephala* and *Chaerephon tomensis,* the latter of which was only recently discovered, along with other widespread bat species. Príncipe and Annobon both have endemic subspecies. São Tomé also has an endemic shrew, *Crocidura thomensis,* whereas Príncipe has an endemic subspecies of *Crocidura poensis.* Unlike most oceanic islands, which generally have few endemic amphibians, São Tomé has four endemic frog species, and Príncipe has three. On São Tomé there is also an endemic caecilian, *Schistometopum thomense.* There are no amphibians on Annobon. There are large numbers of endemic reptiles to the islands, including three scolecophidian snakes, a legless skink, and multiple gecko species. Recent analyses of genetic diversity show that several island forms, currently assigned to widespread mainland species, may also deserve recognition as distinct island endemic species. This is true for the widespread skinks *Mabuya* and *Afroblepharus,* and it may also be true of the green tree snakes. Fish are essentially marine, occurring in estuarine habitats.

Rates of endemism are also exceptional in invertebrate groups. For example, 75% of terrestrial gastropods are endemic to the islands, with several endemic genera and a monospecific endemic family, *Thyrophorella thomensis*. Similarly, within the Geometridae (Lepidoptera), Príncipe and São Tomé have 24 and 30 species, respectively, with 15 and 26 respectively being endemics. Most of the endemics are found on single islands, with few shared across islands. Regarding plants, the forests of São Tomé and Annobon have the highest fern diversity and density in Africa. Annobon alone holds 208 species of vascular plants, of which about 15% are endemic. Approximately 601 and 314 species of vascular plants can be found on São Tomé and Príncipe, respectively, again with high levels of endemism (14% and 8%). In particular, the Rubiaceae, Orchidaceae, and Euphorbiaceae are characteristic of the islands' flora and have large numbers of endemics. The islands are also designated as "centers of plant diversity" by the World Wildlife Fund (WWF) and the International Union for Conservation of Nature (IUCN). Only 16 of the regional endemic plants are shared by more than one island, emphasizing the extreme isolation under which each island community has evolved and further supporting the hypothesis that each island was generally colonized independently from the mainland.

HUMAN HISTORY, GEOGRAPHY, AND INFLUENCE

São Tomé and Príncipe were uninhabited prior to their discovery by the Portuguese in 1470–1471, as was Annobon, which was discovered on January 1, 1473. São Tomé and Príncipe were quickly settled, with sugar cane being the dominant crop. By the mid-seventeenth century, São Tomé was an important transit point for ships engaged in the slave trade. In the nineteenth, coffee and cocoa were introduced, and extensive plantations of these crops were developed: By the early 1900s São Tomé was the world's largest producer of cocoa. The dominant export remains cocoa, which accounts for 95% of all exports, followed by coffee, copra, and palm products. São Tomé and Príncipe became independent from Portugal in 1975. In 1778 Annobon and Bioko passed to Spanish control as part of a larger land swap, and these islands now form part of Equatorial Guinea. Annobon has no major exports. Currently, numbers of inhabitants are approximately 137,500; 6000; and 5000 for São Tomé, Príncipe, and Annobon, respectively.

Following independence in São Tomé and Príncipe, many plantations were abandoned, and there has been some regeneration to secondary forest. The relatively large area of primary montane forest remaining, coupled with a low level of exploitation of this forest, means that it is a relatively secure habitat. However, many endemic species are associated with particular areas, such as the lowland forests, which have been widely cleared for agriculture. Ongoing forest destruction of the remaining patches remains a threat to diversity. Hunting pressures, although important for some endemic bird species, are relatively light: Birds, wild pigs, monkeys, and the common, non-endemic bat *Eidolon helvum* are the primary targets. Medicinal plant use is similarly almost exclusively of non-endemic species. Trapping birds for the cage-bird trade may be damaging for some species, such as lovebirds and gray parrots. It is not at all clear whether current levels of exploitation are sustainable. Small-scale agricultural practices on Annobon have generally been less damaging to terrestrial biodiversity than they have been on São Tomé and Príncipe.

On both islands, large areas of secondary forest are regenerating on old plantations. On Annobon, much of the lowland forest has been replaced by savanna grasslands and banana plantations, with the exceptions of the high peaks of Santa Mira and Quioveo. Primary forests of all types cover approximately 28.5% of the islands. However, lowland forest is now restricted to a few areas. Many bird species, such as the dwarf olive ibis, *Bostrychia bocagei,* are restricted to these areas, and further major losses of this habitat would almost certainly cause multiple species to go extinct. Montane primary forest constitutes the majority of the primary forest, with large areas on the center of São Tomé. Mist forest is limited to the Pico de São Tomé. Mature secondary and shade forest cover 32.4% of the islands. Although not as species-rich as primary forest, these areas still support substantial diversity. Along the coasts of São Tomé and Príncipe, there are small patches of mangroves.

All of the islands have the usual introductions associated with humans, such as rats, cats, dogs, pigs, and so forth. Civets, weasels, and monkeys have been introduced to São Tomé. Most of these have been introduced for more than 100 years, and although there are no records of their effect on the endemic fauna rats, weasels and civets are very likely to have a deleterious effect, especially on nesting birds. Common house geckos, *Hemidactylus mabouia,* have also been introduced. The islands have no endemic venomous snakes, but the cobra, *Naja,* has been introduced to São Tomé. Many more unreported introductions are also likely and are continuing unabated. For example, despite its extreme isolation—with the lack of an airport, a port, or any other major facility—recent assessments of the herpetofauna of Annobon reported two recently introduced species, the house gecko and the blind snake *Ramphotyphlops braminus,* the presence of which increases the reptile fauna from five species to seven. Similar recent surveys of every Atlantic island group, such as the Canary Islands and the Cape Verde Islands, indicate similar levels of ongoing new introductions.

Probably the greatest ongoing threat to the diversity of the islands remains introduced species. It is now impossible to assess the damage caused by the various mammals, both domestic and wild, introduced over the centuries. Reports of recent introductions ranging from terrestrial gastropods to snakes indicate that extremely high levels of introduction continue. It is unlikely that all the endemic species will be able to survive in the face of this constant alien tide.

SEE ALSO THE FOLLOWING ARTICLES

Atlantic Region / Exploration and Discovery / Introduced Species / Oceanic Islands / Orchids / Rafting

FURTHER READING

Measey, G. J., M. Vences, R. C. Drewes, Y. Chiari, M. Melo, and B. Bourles. 2007. Freshwater pathways across the ocean: molecular phylogeny of the frog *Ptychaden newtoni* gives insights into amphibian colonization of oceanic islands. *Journal of Biogeography* 34: 7–20.

SARDINIA

SEE MEDITERRANEAN REGION

SEABIRDS

MARK J. RAUZON

Marine Endeavours, Oakland, California

SHEILA CONANT

University of Hawaii, Honolulu

From the nearshore waters to the open ocean, seabirds are the most conspicuous component of marine systems. Soaring on brisk winds, floating buoyantly amidst cresting waves or flying underwater, this varied group of birds has adapted to the demanding life at sea, utilizing the marine environment to feed and returning to land, primarily islands, to breed.

SEABIRD SPECIES

There are approximately 350 species of seabirds in seven orders. Here, we highlight four orders most characteristic of insular systems: Sphenisciformes (penguins), comprising 16–18 species; Procellariiformes (albatross, shearwaters [Fig. 1], petrels, and storm petrels), with over 100 species; Pelicaniformes (pelicans, cormorants, boobies [Fig. 2], frigate birds, and tropicbirds), with about 55 species; and Charadriiformes (gulls [Fig. 3], terns [Fig. 4], auks, jaegers, and phalaropes), with about 125 species. Ornithologists define "seabird" broadly, including sea ducks (Anseriformes), grebes (Gaviiformes), and loons (Podicipediformes), which are primarily marine but are largely nearshore feeders. The latter three orders and phalaropes are not included in this article.

FIGURE 1 Wedge-tailed shearwaters (*Puffinus pacificus*). Photograph by Mark Rauzon.

FIGURE 2 Brown boobies (*Sula leucogaster*), Pearl and Hermes Reef, Hawaiian Islands National Wildlife Refuge. Photograph by Sheila Conant.

FIGURE 3 Red-legged kittiwakes (*Rissa brevirostris*), endemic to the Bering Sea. Photograph by Mark Rauzon.

FIGURE 4 Brown noddy (*Anous stolidus*) and chick, Manana Island, Hawaii. Photograph by Sheila Conant.

TROPHIC RELATIONSHIPS

All seabirds rely on the productivity of the ocean to one degree or another. For most, the food web begins at the edge of the continental shelf, where strong upwelling currents convey submerged nutrients to the sunlit surface. Primary productivity of phytoplankton fuels a marine food web that changes seasonally and along clines of salinity, temperature, and nutrient availability. Water regimes, major currents, and slipstreams create feeding habitats that tend to be patchy and ephemeral. Sustained marine productivity centers are important bioregions for seabird evolution, such as the Bering Sea or the Humboldt Current off South America. Water masses define how species evolve; some are actually endemic to discrete currents or to other water masses.

Some tropical seabirds such as sooty terns and brown noddies (Fig. 4) depend on a commensal relationship with dolphins and, especially, tuna. Schools of tuna drive small fish and squid to the surface, where plunge-diving seabirds and surface seizers have access to prey. Without the driving force of tuna, many tropical seabirds would have less available food. Food availability forges behavioral life history strategies, energetic costs of foraging behaviors, colony visitation schedules, clutch size and offspring development, and population and colony size.

Resource partitioning of patchy food supplies has shaped seabirds' foraging ecologies, which have evolved several methods to capture food in the water column. Surface seizing, pursuit diving, and plunge diving are the main techniques that various members of seabird families may share. For example, plunge diving is well developed among the Pelicaniformes. Tropicbirds plunge from a hundred meters over the sea, and pelicans drop from tens of meters above the sea, tucking in the wings and then hitting the water and opening their large throat pouches to net their prey. These birds dive as deep as 6.3 m, and may spend up to 6.7 seconds underwater. The surface-seizing gulls are among the most successful because they are opportunists. Western gulls in the San Francisco Bay eat french fries and hot dogs as well as common murre eggs, fish offal, carrion, and invertebrates.

SEABIRD ADAPTATIONS TO LIFE AT SEA AND ON LAND

There are myriad adaptations to marine life, including webbed feet and aerodynamic wings designed for soaring and floating on ephemeral winds. Seabird wings may be supported by hollow bones reinforced with internal struts similar to models for airplane wing design, or hydrodynamic wings and flippers for flying underwater, or glands for processing the toxic salt levels that confront them, or glands whose oil waterproofs plumage. Exceptions to adaptations exist in almost every family; for example, frigate birds produce very little oil, because they rarely are in contact with saltwater. With the highest wing surface to body weight ratios, they are masterful fliers and depend on flight maneuverability and speed to secure food by kleptoparasitism (theft from other birds) or outmaneuvering flying fish in the air. They cannot afford to get wet and cannot take off again if they accidentally hit the sea surface. Galápagos flightless cormorants also produce very little oil and lack functional wings because they must remain underwater and overcome buoyancy to pursuit-dive effectively after subsurface prey. On land, Pelicaniformes use their gular (throat) pouches to thermoregulate. The gular flutter is a conspicuous behavior of boobies, gannets, and cormorants, an adaptation to shed excessive heat, especially in the tropics.

Seabirds must deal with harsh contingencies of living on the ocean. Albatross, fulmars, shearwaters, storm petrels, and petrels have special glands at the base of their beaks that maintain the body's salt balance. These glands extract salt from the blood that passes through them and excrete drops of highly concentrated brine through tubes mounted on the culmen. This latter structure is the basis for referring to the order as tubenoses, which are also known as the most accomplished of fliers. The tubenoses—albatross, shearwaters, petrels, and relatives—have long, narrow, saber-like wings, and they bank and glide easily in gale-force winds. These birds spend their early adolescence—usually several years—at sea before returning to the natal colony. Other species, such as the white tern (Fig. 5) return to land during their second year, although they may not breed until several years later.

FIGURE 5 White tern (*Gygis alba*) incubating an egg. Photograph by Mark Rauzon.

BEHAVIOR AND ECOLOGY OF SEABIRD GROUPS

Albatross, the largest of all seabirds, use the energy-saving flight known as *dynamic soaring*. The albatross glides into the wind with one wing pointed to the water, and dips its wing tip down, slicing a wake in the sea surface. Again and again, the albatross repeats the looping circles of flight without flapping its wings so long as the wind blows across the ocean. One albatross was measured flying at 55 miles per hour; another flew for six hours without flapping its wings. They fly for days without landing and may cover several million miles in their lifetimes. To test their *homing* ability, several Laysan albatross were taken from their nests on Midway, flown by airplane to distant airports, and then released. One albatross was taken to Washington State; it returned to Midway in 12 days, traveling over 3,000 miles—averaging over 300 miles per day.

Recent advances in telemetry suggest such large-scale movements are the norm rather than the exception for many tubenoses. Hawaiian petrels regularly visit the sub-Arctic on two- to three-week foraging excursions while provisioning chicks in the nest burrow in the subtropics. Sooty shearwaters annually migrate over 20,000 nautical miles from the Austral breeding ground to their boreal foraging grounds in the north Pacific and Atlantic Oceans to feed on seasonally abundant euphausiids, or krill. The diminutive Leach's storm petrel covers thousands of miles from its sub-Arctic nesting colonies as far south as equatorial waters looking for food. Like their namesake, St. Peter, who was reputed to walk on water, storm petrels use their tiny, paddle-like webbed feet to patter the ocean surface, picking up the minute arthropods this attracts, including the water striders (*Halobates*), the only insect that lives on the open sea. The tropical storm petrels have long legs and short rounded wings whereas northern storm petrels tend to have shorter legs and more pointed wings. Despite many species' affinities to particular water regimes, seabirds as a group are free to wander the ocean. Indeed, some are called gadfly petrels.

No other seabirds have mastered the marine realm as have the penguins of the Southern Hemisphere. The 18 species of penguins are masters of underwater flight. Propelled by wings that have evolved into flippers, and steered by feet and tail, they fly underwater after squid and fish. On land the upright stance of the penguin and comical waddle combined with the black-and-white plumage attire has helped make them one of the most popular of animals.

Alcids are the northern counterparts to penguins. The family Alcidae is comprised of 23 species including puffins, murres, murrelets, auklets, and guillemots, all of which are pursuit-divers, capable of both traditional and underwater flight. But with versatility comes a price: alcids must flap continuously to stay aloft, unlike tubenoses, and they are limited to nesting on sea cliffs, where they can free-fall to gain flight momentum. The great auk was flightless and swam in pursuit of prey, propelled by its small wings. Because of its flightlessness, the great auk was quickly over-hunted and became one of the first North American birds to become extinct at the hands of man in 1844.

Other members of the Chadriiformes include seven species of jaegers and skuas, powerfully built hawklike gulls that out maneuver other seabirds to rob them of their catch. On land they eat rodents, birds, and bird eggs. Pomarine, parasitic, and long-tailed jaegers are boreal breeders that migrate into the southern hemisphere to spend the winter, while south polar skuas migrate into the Northern Hemisphere, avoiding the Austral winter. In summer, skuas prey on penguin chicks and scavenge at seal rookeries.

Sooty terns are one of the most abundant seabirds in the world. Occurring in all tropical oceans, colonies often number in the hundreds of thousands. Sooty terns are built like miniature frigate birds and can spend as long as nine months on the wing. Scientists speculate sooty terns can sleep on the wing by resting half of the brain—one hemisphere at a time. Coincidentally, their nickname, "wide-awake" tern, refers to their incessant cries at the colony.

DISTRIBUTION

Although there are numerous seabird rookeries on the coastlines of the Earth's continents, today most are found on islands. Islands in the Southern oceans, especially the Southern Pacific Ocean, have a great diversity of breeding seabirds (Table 1). A few other remote island groups,

TABLE 1
Numbers of Breeding Seabirds on Some of the World's Islands

Region	Breeding Species
Southern Ocean Archipelagoes	
Crozet Archipelago (IO)[a]	34
Kerguelen Islands (IO)	31
Auckland Islands and Campbell Island, NZ (PO)	29
Prince Edward Islands (AO)	28
South Georgia (AO)	27
Tristan da Cunha (AO)	22
Seychelles (IO)	19
Galápagos Islands (PO)	19
Falkland Islands (AO)	16
Juan Fernandez Islands (PO)	16
Mascarene Islands (IO)	8
Northern Ocean Archipelagoes	
Britain	24
Hawaiian Islands	22
Aleutian Islands	26
Seabird-Poor Islands	
Wallacea, Indonesian Islands, and Philippines	13
Andaman Islands	4
South of Java (except Christmas Island)	Virtually none
Bay of Bengal	Virtually none
Oceans/Seas	
North Pacific/Bering Sea	35
Tropical Central Pacific (160° E to 140° W)	29
Indian Ocean	23
Arctic	20
North Atlantic	21
Tropical Atlantic	14

SOURCES: Croxall *et al.* 1984, Gaston 2004, and references therein.
[a] Abbreviations: AO = Atlantic Ocean, PO = Pacific Ocean, IO = Indian Ocean.

such as the Galápagos Islands, the Hawaiian Islands, and the Juan Fernandez Islands, have high seabird diversity. Island groups with surprisingly low numbers of seabird species are the islands of Wallacea, the Indonesian islands, and the Philippine Islands, which comprise hundreds of islands that support only about a dozen seabird species. In the Northern Oceans, the Aleutian Islands support breeding colonies of 26 seabird species, while Britain has 24.

The Pacific Ocean seabird fauna is by far the most diverse, with the tropical Pacific supporting 29 breeding species and the North Pacific 35. In contrast, the North Atlantic has 21 breeding species and the Tropical Atlantic 14. The Arctic and Indian Oceans have similar numbers: 20 and 23, respectively. Among the world's largest seabird colonies are Kirimati Island in the Pacific Ocean, where there used to be as many as 4 to 6 million nesting birds of 18 different species, and Isla Guafo, where an estimated 4 million sooty shearwaters breed.

HUMAN IMPACTS

Unfortunately the activities of humans have resulted in dramatic declines in seabird populations and distributions. For example, the apparent higher level of endemism observed in the shearwaters and petrels of the tropics is largely an artifact of local extinctions as the islands were settled 1000 to 2000 years ago. Species with restricted ranges today such as the Tahiti petrel (*Pterodroma rostrata*) and Abbott's booby (*Sula abbotti*) were much more widespread prior to human colonization of the Pacific. Virtually all islands for which paleontological surveys exist experienced dramatic reductions in numbers of species of both land birds and seabirds after humans arrived. Perhaps the most spectacular example is Easter Island, where only one species, the sooty tern, of the original 22 seabirds known from the prehistoric record still breeds. Ironically, among other things, vast flocks of seabirds helped exploring Polynesians find the islands they colonized in the first place.

Climate changes in the past millennium have altered species dynamics and distribution. Recently discovered subfossil bones on Bermuda indicate that a colony of short-tailed albatross existed there 450,000 year ago. A recent eastern range expansion and increase in number of Laysan albatross in Mexico may be a result of greater food availability through ocean regime shifts. El Niño–Southern Oscillation (ENSO) events cause unpredictable changes to the seabirds' world. The recently elucidated Pacific decadal oscillation, which increases ocean mixing approximately every other decade, can add a positive pulse to the trickle-down economy in which seabirds live. Annual hurricanes and typhoons can set seabird recovery back by causing reproductive failure, while the vagaries of anthropogenic forces such as chronic oil leaks and spills, plastics in the ocean, pesticides in the environment, and other human disturbances limit the birds' capacity to recover from natural events.

Commercial fishing poses one of the greatest threats to seabirds. It is estimated that over 100,000 seabirds are hooked as bycatch and drown each year in the long-line fishing industry. A major ecological shift observed in the trophic structure of the Southern Ocean was originally thought to be caused by climatic change but in fact was due to depletion of fish stocks. Industrial fishing removes adult fish with K-selected (long life span, late age of first reproduction, limited production of young) life histories. The resulting declines in breeding fish stocks, as well as of annually reproducing forage fish, has contributed to declines in piscivorious seabirds such as macaroni, Gentoo penguins, and Imperial shags, which seem unlikely to recover.

CONSERVATION

Seabird conservation has made great strides devising ways to limit fishery bycatch through the use of scaring devices, setting lines at night, and other techniques. Great strides have been made in eliminating introduced predators to seabird islands (Fig. 6). By removing introduced Arctic foxes, feral cats, rodents, and various other mammals and weeds, some seabird colonies have recovered to their historic population levels. Eradication programs also provide new ecological insights on a geographic scale. Introduced foxes had altered the vegetative ecology of Aleutian Islands by eating the seabirds that fertilized the islands with guano. With decreased guano, the islands' grasslands succeeded into scrubby tundra.

FIGURE 6 Dead wedge-tailed shearwaters, killed by dogs at Ka'ena Pt. on O'ahu, Hawaii. Photograph by Lindsay Young.

Seabirds are more vulnerable than ever before. During the last 20 years, the world's human population has increased 34%, and trade is almost three times greater, thus increasing the number of seagoing vessels and consequently increasing the risk of spreading invasive species to more islands (UN 2007). Looming on the horizon is global warming, which threatens penguins in Antarctica and tropical nesting islands and brings about shifting of water masses and increased storm intensity. Seabirds will need continued and expanded conservation efforts to ensure their survival for the future.

SEE ALSO THE FOLLOWING ARTICLES

Bird Disease / Bird Radiations / Climate Change / Fish Stocks/Overfishing / Midway

FURTHER READING

Croll, D. A., J. L. Maron, J. A. Estes, E. M. Danner, G. V. Byrd. 2006. Introduced predators transform subarctic islands from grassland to tundra. *Science* 307: 1959–1961.
Croxall, J. P., P. G. H. Evans, and R. W. Schreiber, eds. 1984. *Status and conservation of the world's seabirds*. Tech. Rep. 2. Cambridge, UK: International Council for Bird Preservation.
Dickinson, E. C., R. S. Kennedy, and K. C. Parkes. 1991. *The birds of the Philippines*. London: British Ornithologists' Union.
Gaston, A. J. 2004. *Seabirds: a natural history*. New Haven, CT: Yale University Press.
Harrison, P. 1983. *Seabirds: an identification guide*. Boston: Houghton Mifflin.
Lloyd, E. G., M. L. Tasker, and K. Partridge. 1991. *The status of seabirds in Britain and Ireland*. London: Poyser.
Rauzon, M. J. 2001. *Isles of refuge: the history and wildlife of the northwestern Hawaiian islands*. Honolulu: University of Hawaii Press.
Steadman, D. W. 2007. *Extinction and biogeography of tropical Pacific birds*. London: Oxford University Press.
Stoddart, D. R. 1984. Breeding seabirds of the Seychelles and adjacent islands, in *Biogeography and ecology of the Seychelles Islands*. D. R. Stoddart, ed. The Hague: W. Junk, 575–592.
UNEP (United Nations Environmental Programme). 2007. *Global environmental outlook*. Malta: Progress Press Ltd.
White, C. M. N., and M. D. Bruce. 1986. *The birds of Wallacea*. London: British Ornithologists' Union.

SEA-LEVEL CHANGE

W. H. BERGER

University of California, San Diego

Sea level has been rising around the world for the last hundred years. For the second half of the last century, the overall rate of the global rise is between 1.5 and 2 mm per year, according to various compilations. When projecting this same rise forward in time, the total rise by the end of the century turns out to be between 15 cm and 20 cm, an estimate that agrees with the latest assessment offered by the Intergovernmental Panel on Climate Change (IPCC) of the World Meteorological Organisation (WMO) and the United Nations Environment Programme (UNEP). It is distinctly lower than various estimates for the total rise by the end of the century offered since 1990 by scientists considering the issue (typically 0.5 to 1 m). There are good reasons why all such predictions are subject to a high level of uncertainty.

SOURCES OF UNCERTAINTY FOR ASSESSING THE PRESENT RISE

Two major problems arise when assessing present and future rise of sea level. The first is that it is difficult to establish the historical rise of global sea level because the records of tide gauges (which are at the base of the assessment) are influenced by many regional factors that must be considered in addition to global change. The second is that the mix of factors responsible for the global rise in the last century is unlikely to persist for the current or the next century.

Some of the various factors contributing to the tide gauge records are as follows: (1) changes in the effects from tides; (2) changes in the average wind field; (3) changes in the velocity or direction of offshore current flow (which results in dynamic adjustments of sea level from the changing geostrophic balance); (4) uplift or sinking of the coast; (5) local settling or shifting of the underground from mass movements (stimulated by earthquakes or changes in rainfall); (6) changes in the contribution of runoff into a harbor bearing the tide gauge (salinity effects on density); (7) global sea-level rise. It is only the last of these that is of interest. However, the changes induced by the other factors (especially dynamic adjustments to currents and winds) are very large with respect to the signal sought.

To extract the signal from a plethora of noisy data, long records must be studied, and many of them must be combined from different regions on the globe. Tide gauges tend to be clumped in highly populated regions, some of which are close to formerly glaciated regions in the north, so that tectonic adjustments to the unloading of a formerly ice-covered Canada and Scandinavia are important. Tide gauges of tropical islands have other problems; for example, some are heavily influenced by El Niño–Southern Oscillation (ENSO)-related dynamic changes of sea level (Fig. 1). The weighting of the tide gauge records, depending on their geography, will influence results, naturally.

SOURCES OF THE GLOBAL COMPONENT OF THE RISE

The mix of factors determining global sea-level rise includes expansion of the water column from warming, the addition of freshwater from the melting of mountain glaciers, and a possible contribution from polar ice caps, especially

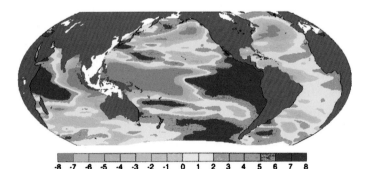

FIGURE 1 Anomalous sea-level positions during El Niño conditions, winter 1997/1998. Source: Topex/Poseidon, NASA and NOAA, as given in Künzi (2002), modified.

from Greenland. The first two of these factors are amenable to estimation. The temperature change can be measured and the resulting expansion can be calculated. The fate of mountain glaciers can be assessed from images taken from spacecraft. But the present and future contribution from melting in polar regions is an open question. The reason is that warming can produce increased snowfall on large ice fields. As long as the regional climate is cold enough to retain the additional snow, there is then a buildup of ice compensating for the melting at lower elevations. With further warming, melting undoubtedly wins the contest, but the point of transition is not readily determined. In the most recent Intergovernmental Panel on Climate Change (IPCC) assessment, the assumption is made that the ice on land in Antarctica and in Greenland will retain its mass for the current century, reflecting the uncertainties mentioned. It is important to realize that this assumption simply reflects a preference for a conservative approach to the problem of rising sea level, rather than the result of reliable assessment. An assessment of an increasing contribution from polar regions will be possible only after such a contribution has materialized for some time.

We note, from the foregoing discussion, that the derivation of a trend in sea-level change depends on the availability of well-spaced data points over a sufficiently long time interval and on the mathematical treatment accorded such data. When contemplating the significance of observed trends, it must also be remembered that there are hardly any measurements before the late nineteenth century; that is, there are no good data for the time before the onset of the trend. This means that the timing of the onset of the modern rise of sea level is obscure. Furthermore, given the uncertainties in the determination of the average rate of sea-level rise, an increase in the rate of rise cannot be demonstrated for the twentieth century, although it is not unlikely.

Available data suggest that thermal expansion of the ocean's water column accounts for about one-half of the observed rise. The expansion comprises the upper layer of the ocean, roughly the upper 1000 m, where the warming is most noticeable. The second most important factor is thought to be addition of water from the melting of mountain glaciers.

Mountain glaciers hold sufficient ice to raise sea level by 0.5 to 1 m upon melting, and they have been retreating practically everywhere since the middle of the nineteenth century. Their retreat is especially obvious in western North America and is readily recognized from the exposure of fresh rock with no soil and but little growth of lichen on such rock (Fig. 2). Records of glacier conditions

FIGURE 2 Remnant of a mountain glacier surrounded by freshly exposed rock and moraine debris. Glacier National Park, Montana, late summer 2006. Photograph by the author.

are commonly kept in terms of the glacier length, and height of a glacier's surface within its valley, which allows the estimation of changes in volume and hence change of volume over the period of observation.

OPEN QUESTIONS REGARDING CAUSES FOR THE OBSERVED RISE

One important problem is that the values attributed to the main factors (expansion, melting of glaciers other than polar ice masses) do not add up to the sea-level rise observed for the twentieth century. Calling on melting of polar ice could solve the problem, but this hypothesis has implications for changes in the rotation of the Earth that have not been observed. Thus, it is possible that the general rise of sea level has been overestimated for the twentieth century, perhaps as a result of changes in the shape of the sea surface as it adjusted to changes in circulation. In California, for example, a slowing of the California Current produces a rise in sea level along the coast.

Of course, the fact that the apparent rise of sea level for the last century is not fully explained implies that predictions of future sea-level rise are quite unreliable. A demonstration of lack of knowledge, however, does not necessarily support complacency about future developments: uncertainty works in both directions. As mentioned, the Fourth Assessment of the IPCC (whose results have been available since early in 2007) does not attempt to specify likely contributions from the melting of polar ice masses, since the dynamics of polar ice are poorly understood.

LESSONS FROM POSTGLACIAL COLLAPSE OF ICE SHEETS

While a study of the more distant past (i.e., the last several ice age cycles) cannot fill the gap in such knowledge, it can deliver some idea about the range of possible rates of sea-level rise, during the collapse of great ice masses.

Large-scale collapse of ice sheets presumably helped cause the high rates of sea-level rise indicated in recent reconstructions of deglaciation; that is, the period between 16,000 and 9,000 years ago, when the Canadian and Scandinavian ice shields disintegrated. It is generally agree that the disintegration proceeded in pulses, the two major steps of deglaciation being separated by a cold spell with static sea level known as the Younger Dryas (YD) episode. The YD cold period lasted somewhat longer than one millennium. It had almost fully glacial climate conditions and it kept sea level at roughly one half of its total rise (120 m) at an intermediate position (60 m), halfway through deglaciation. The overall rate of the rise was between one and two meters per century, for a thousand years at a time. The crucial information allowing assignment of such rates is absolute dating of corals growing near sea level.

Uncertainties remain because corals have a depth *range* for habitat, rather than a depth level. However, a rise of 2 to 3 m per century is plausible, for intervals spanning a few centuries at a time, given the rate estimates obtained. The estimates of contributions from the various ice shield sources vary; roughly two thirds of the rise is thought to be from North America and Greenland, the rest being shared by the Fennoscandian shield and Antarctica. Calculations regarding relative contributions are based on isostatic rebound statistics and have large error bars.

An important question in the context of expectations for future sea-level rise concerns the vulnerability of ice sheets that remain after deglaciation to additional warming. At present, the ice remaining in Antarctica, if melted, could raise sea level by about 61 m, while the ice in Greenland could provide for a rise in sea level of 7 m. The question is difficult to decide on the basis of ice physics, because of the mass-wasting aspect of ice sheet disintegration. However, clues to answers are contained in the behavior of ice sheets and sea level over the past million years or so.

PAST SEA-LEVEL RISE AS SEEN IN THE DEEP-SEA RECORD

The relevant data are oxygen isotope ratios within the shells of microscopic organisms called foraminifers, a minor but important part of the plankton of the ocean. They also have representatives living on the sea floor; that is, benthic foraminifers. The isotopic composition of their shells reflects the isotopic composition of seawater at the time of growth; records are recovered by coring the sediments on the deep seafloor.

The isotopic composition of the benthic shells, in particular, is largely controlled by changes in the mass of polar ice sheets. The water used in making ice, which is extracted from the sea, is depleted in the heavier of the two main oxygen isotopes (^{18}O). Thus, when ice grows, the seawater is enriched in this isotope. The range of change over the last million years, in benthic foraminifers, is roughly 1 permil (one tenth of one percent of change in the deviation from a standard). This change is readily measurable (to about 5% precision). The corresponding change in sea level can be estimated if it is assumed that the temperature effect on the composition is tightly linked to the ice effect. Also, the isotopic composition of the ice is taken as constant through time. The errors stemming from these assumptions are thought to be modest, relative to the estimates of sea-level change (on the order of 10% or less).

When studying the patterns of change in the isotope values with the goal of reconstructing rates of sea-level change, one finds that rates of sea-level rise of more than 1 m per century are not highly unusual over the last million years. A similar pattern of rates (although based on few data) is obtained for values drawn from the fifth percentile of sea-level positions, on the high end, which would seem to indicate that ice sheets left over after deglaciation (the present situation) remain vulnerable to removal up to a +10-m limit. Thus, the available information from the history of sea-level change through the last million years suggests that the present sea level is not privileged but can readily move upward by 5 m or more, and that such rise could proceed at a rate of 1.5 m per century given sufficient motivation (that is, sufficient warming in high latitudes).

SUMMARY

In the twentieth century, sea level has been rising between 1 mm and 2 mm each year, for a total rise of about 15 cm for that century. Estimates for the rise in recent years are between 2 and 3 mm per year, suggesting an acceleration of the rise, which is expected but not demonstrated. Precise assignment of sea-level rise is not possible, because observations are not globally distributed and are greatly influenced by local conditions. Most of the inferred rise is attributed to thermal expansion of the water column as the ocean has warmed. Some of the rise is attributed to the melting of mountain glaciers, whose ice mass is shrinking quite rapidly (which is readily verified by satellite surveys and from historical records). Forward extrapolation of observed and inferred sea-level rise yields estimates for the end of the present century with values that vary between less than 30 cm and distinctly less than 1 m. These estimates assume little or no contribution from polar ice masses. In the recent geologic past, such ice masses have disintegrated at rates equivalent to a sea-level change of around 1 to 2 m per century when stimulated by summer warming. There is no evidence that such stimulation will not produce similar rates in the near future. Instead, the available record of ice behavior for the last million years suggests that at present there is enough vulnerable ice in existence to raise sea level by between 5 and 10 m, within a few centuries.

CONCLUSIONS

The uncertainties for assessment for present and future sea-level rise are rather large. Data concerning past sea-level behavior suggest that, once started, disintegration tends to persist for some time, on the scale of centuries, which points to internal positive feedback. However, the resolution (1000 years) of the observational series for the last million years or so is insufficient to belabor the issue for time scales below that resolution.

SEE ALSO THE FOLLOWING ARTICLES

Climate Change / Coral / Surf in the Tropics / Tides

FURTHER READING

Berger, W. H. 2008. Sea level in the Late Quaternary: Patterns of variation and implications. *International Journal of Earth Sciences* 97: 1143–1150

Church, J. A., J. M. Gregory, P. Huybrechts, M. Kuhn, K. Lambeck, M. T. Nhuan, D. Qin, P. L. Woodworth, *et al*. 2001. Changes in sea level, in *Climate change 2001: the scientific basis. Contribution of Working Group I to the Third Assessment Report of the Intergovernmental Panel on Climate Change*. J. T. Houghton, Y. Ding, D. J. Griggs, M. Noguer, P. J. van der Linden, X. Dai, K. Maskell, and C. A. Johnson, *et al*., eds. Cambridge, UK: Cambridge University Press, 639–693.

Künzi, K. 2002. Ozeane aus der Ferne gesehen [Oceans looked at from afar], in *Der Ozean—Lebensraum und Klimasteuerung* [The ocean—life habitat and climate control]. G. Hempel and F. Hinrichsen, eds. Bremen: Hausschild Verlag, 63.

Munk, W. 2002. Twentieth century sea level: an enigma. *Proceedings of the National Academy of Sciences of the United States of America* 99.10: 6550–6555.

Revelle, R. R., ed. 1990. *Sea-level change*. Studies in Geophysics. Washington, DC: National Academy Press.

SEAMOUNTS, BIOLOGY

MALCOLM CLARK

National Institute of Water and Atmospheric Research, Kilbirnie, New Zealand

Seamounts occur in all oceans of the world, from the tropics to the poles, and cover depth ranges from near the surface to the abyss. They provide a wide variety of habitat types for a huge range of animals and often feature high levels of biodiversity and abundance. This can make them important components of oceanic ecosystems, yet also the target of commercial exploitation.

THE SEAMOUNT ENVIRONMENT

Seamounts have three important characteristics that distinguish them from the surrounding deep-sea habitat. First, they are "islands" of shallow sea floor, surrounded by the deep ocean, and they provide a range of depths for different communities. Second, the majority of the ocean sea floor is covered with fine, unconsolidated sediments. In contrast, the often steep slopes of seamounts, combined with accelerated currents over them, can keep sediment from depositing, and bare rock surfaces can be common. Third, the physical structure of some seamounts enable the formation of hydrographic features and current flows that can restrict the dispersal of larvae and plankton and keep species and production processes concentrated over the seamount rather than dispersing them into the wider ocean system.

SEAMOUNT FAUNA

The existence of seamounts has been known for a long time, but their biology received little attention until the late 1950s. In a review in 1987, it was reported that about 96 seamounts had been sampled, but recent estimates by the Census of Marine Life program on seamounts (CenSeam: http://censeam.niwa.co.nz) have the number now up to about 300 (Fig. 1). But of this total, most have been sampled opportunistically, with few specific or comprehensive biological studies having been undertaken. Geographically, the most comprehensive data on seamounts come from the northeastern Atlantic, the southwestern Pacific, and the southeastern Pacific. Other scattered seamount groups have also received study, but major gaps exist in global coverage, which makes robust estimation of biodiversity and interpretation of biogeographical patterns difficult.

Nevertheless, the benthic communities of species found on seamounts are rich and varied. This is partly related to the wide variety of habitat types that seamounts can offer animals. On seamounts where currents are slow and there are sandy or muddy sediments, we can find a predominance of deposit-feeding species that utilize the matter sinking down from mid-water layers. These macrofaunal communities include polychaetes, echinoderms (ophiuroids, holothurians, echinoids), various crustacean groups, sipunculids, nemertean worms, molluscs, sponges, and nematodes. However, where faster water

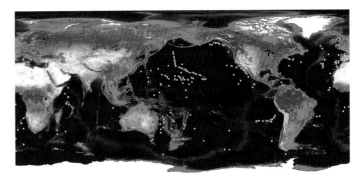

FIGURE 1 The location of seamounts where biological samples have been taken and data are available in the Seamounts Online database (as of October 2007).

flows occur over seamounts (often caused by the size and shape of the seamounts themselves), there can be greater regions of hard-bottom habitat. These rocky areas can be dominated by suspension feeders, which eat material suspended in the water by the enhanced current flow. These animal groups include corals, crinoids, hydroids, ophiuroids, and sponges. The large corals and sponges can form extensive and complex reef-like structures, which themselves provide a habitat for smaller mobile fauna.

SEAMOUNT BIODIVERSITY

It is not known how many species occur on seamounts. Accounts of seamount biodiversity show a rapid increase in number of known species with research in the last few decades. A global review in 1987 recorded 449 species of fish and 596 species of invertebrates from 100 seamounts. Research in the southwestern Pacific around New Caledonia has found 730 species from just 18 seamounts at present, and 192 species of invertebrates and 171 species of fish have been recorded from a single seamount chain off South America. Worldwide, almost 800 species of fish have been recorded from seamounts. Data on seamount biodiversity is growing daily, through both national research programs and through the efforts of CenSeam, and the data are being input to a public database, Seamounts Online (http://seamounts.sdsc.edu). This database in 2007 has over 3000 taxa (although not all to species) from 250 seamounts. This number will continue to increase as more seamounts around the world, at different depths and with different ages and geological histories, are sampled. However, selectivity of sampling gear, and generally a small number of samples per seamount, means that not all species present are caught. Typically, biodiversity is not well described even after a large number of samples have been taken from a group of seamounts. It seems certain that biodiversity is not fully known for any seamount.

Seamounts can be hotspots of biodiversity. Sessile animals such as sponges and corals can attach to the hard substrate. Corals, in particular, can form large reeflike structures that provide a home for hundreds of other species. Deep-sea, or "cold-water," corals do not need light to exist (as tropical shallow corals do) but feed on matter swept up by the currents on seamounts. These corals can be hundreds of years old and very slow growing.

Despite sampling limitations, comparisons of species richness have been made between some seamounts. Results comparing seamounts with adjacent continental slopes and with deep sea floor areas have been mixed, with no consistent trend of elevated or depressed biodiversity. It is likely that various seamounts and regions differ in their comparative biodiversity levels, with some being high and others not. In many terrestrial and marine habitats, there is a relationship between habitat size and the number of species present. Hence, because of the small size of seamounts, fewer species are expected. This is likely to be the case, but the biodiversity of seamounts is, nevertheless, relatively high for their size. It is possible that seamount habitat is relatively stable over evolutionary time, which means that species may accumulate.

A common feature of seamount biodiversity is low levels of species evenness, where the fauna can be dominated by a few very abundant species. The biomass of fishes and other planktivorous animals is often unusually high over seamounts. Upwelling of currents can occur, but is generally not strong enough, or permanent enough, to increase the growth of local zooplankton populations. Trophic enrichment of seamounts can occur but is caused by processes such as the entrapment of vertically migrating plankton by the sea floor, whereby plankton descending in the morning strike the seamount summit or flanks and accumulate, thus providing a rich food source for fishes. Enhanced horizontal fluxes of suspended food also occur, caused by increased current flow over and around a seamount. Hence, the biological "enrichment" of seamount fauna may not be due to increased primary production but to a bottom-up pathway that provides a greater food supply for carnivores.

SEAMOUNT ISOLATION

Seamounts close together can have markedly different faunal communities. The degree of isolation is affected by whether recruitment occurs from animals already living on the seamount, or whether recruits come in from wider areas. For large fish species and pelagic animals, there is evidence that seamounts are not isolated populations. Tuna, turtles, marine mammals, and seabirds are known to migrate between seamounts and between non-seamount regions. It is not as clear with smaller species or invertebrates. Most studies to date (on fish) have not supported the hypothesis that seamounts are highly isolated, and this is in conflict with the frequent observations of endemic species. However, benthic invertebrates may have different characteristics. For some groups, it is highly likely that distance between seamounts can drive isolation, as animals need to disperse over areas of ocean and locate the relatively small surface area of a seamount. Alternatively, reproductive mechanisms need to be adapted so that eggs and larvae remain local in the vicinity of the seamount where they were spawned. Oceanographic conditions may also play a role in retaining small animals and in promoting self-recruitment to the same seamount.

Seamount communities often have faunal affiliations with the nearest continental margin. Despite instances of high endemism and high variability of some seamount faunas, they tend generally to comprise the same broad biogeographic regional fauna as found on the continental slope.

Seamounts can also occasionally yield taxa that have been thought extinct. These "relicts" or living fossils may exist on offshore seamounts long after environmental changes have occurred in continental shelf or slope waters.

SEAMOUNT ENDEMISM

A feature of seamount fauna that is often referred to is the occurrence of highly endemic faunas. Studies have recorded variable levels of endemism, from 5 to 10% for fishes on some seamount groups up to levels as high as 50% for invertebrates. Despite being highly variable, endemism levels are possibly lower in fishes than invertebrates, given the wider distribution of fish ranges compared to many invertebrate groups, and also given generally broad reproductive outputs. A recent review of levels of endemism on seamounts found an overall average of about 20%. True rates of endemism are rarely known, especially on seamounts, where sampling has been so limited. Published rates of endemism are best regarded as indicative: They could increase as more cryptic species are differentiated or decrease as species ranges are better described.

HUMAN EXPLOITATION

Dense aggregations of animals can occur on seamounts. These include fishes, among which are a number of commercial species that are the target of trawl or line fisheries: for example, orange roughy, slender armourhead, and alfonsino. Seamounts are also of increasing interest for mining for deposits of polymetallic sulfides and cobalt-rich ferromanganese crusts.

However, seamount communities may be severely affected by exploitation. Deep-water coral–dominated communities are highly vulnerable to physical disturbance, with the coral matrix being fragile to impact. As well as direct removal of species and structural habitat, there can be indirect effects caused by sediment resuspension. Furthermore, changes in community structure can occur with selective removal of species, and the interactions between pelagic and benthic components of seamount ecosystems may change. Given the slow growth rates of many seamount benthic species, recovery rates may be slow, if in fact disturbed systems ever return to their original state. Natural changes are also occurring, and long-term ocean acidification will have important consequences for seamount coral species, which are affected by the chemical composition of the water at depth. However, in the short term, direct human impacts are of most concern, and careful management of seamount resources is required to balance exploitation and conservation of the habitat.

SEE ALSO THE FOLLOWING ARTICLES

Climate Change / Reef Ecology and Conservation / Seamounts, Geology

FURTHER READING

Pitcher, T. J., T. Morato, P. J. B. Hart, M. R. Clark, N. Haggan, and R. S. Santos, eds. *Seamounts: ecology, fisheries, and conservation*. Blackwell Fisheries and Aquatic Resources Series 12. Oxford, UK: Blackwell Publishing.

Richer de Forges, B., J. A. Koslow, and G. C. Poore. 2000. Diversity and endemism of the benthic seamount fauna in the southwest Pacific. *Nature* 405: 944–947.

Rogers, A. D. 1994. The biology of seamounts. *Advances in Marine Biology* 30: 305–350.

Stocks, K., and P. J. B. Hart. 2007. Biogeography and biodiversity of seamounts, in *Seamounts: ecology, fisheries, and conservation*. T. J. Pitcher, T. Morato, P. J. B. Hart, M. R. Clark, N. Haggan, and R. S. Santos, eds. Blackwell Fisheries and Aquatic Resources Series 12. Oxford, UK: Blackwell Publishing, 255–281.

Wilson, R. R., and R. S. Kaufmann. 1987. Seamount biota and biogeography, in *Seamounts, islands and atolls*. B. H. Keating, P. Fryer, R. Batiza, and G. W. Boehlert, eds. *Geophysical Monograph* 43, 355–377.

SEAMOUNTS, GEOLOGY

PAUL WESSEL

University of Hawaii, Manoa

Seamounts are traditionally defined as undersea mountains whose summits rise more than 1000 m above the sea floor; however, modern studies describe seamounts down to several tens of meters in height. They generally exhibit a conical shape with a circular, elliptical, or more elongated base. Seamounts are some of the most ubiquitous landforms on Earth and are present in uneven proportions in all ocean basins. Being volcanic in nature, seamounts are mostly found on oceanic crust and to a much lesser extent on extended continental crust. They are generated near mid-ocean spreading ridges, in plate interiors over upwelling plumes (hotspots), and in island-arc convergent settings. Oceanic islands form a small subset of large seamounts that have breached sea level.

THE VOLCANIC ORIGIN OF SEAMOUNTS AND ISLANDS

Most seamounts and islands are constructional aggregates of basalt, reflecting their volcanic origin. Seamounts are typically formed in one of three distinct tectonic settings, each imparting unique tectonic characteristics to its offspring.

Intraplate Seamounts

The majority of larger seamounts found in the ocean basins were formed in an intraplate setting. Because of their frequent alignment into linear, subparallel chains that correlate with the direction of past plate motions, the consensus origin of such seamounts is given by the hotspot hypothesis, which states that these seamounts formed above more or less stationary mantle plumes (or hotspots) in the Earth's mantle. As the plates move, the seamounts thus formed are carried away from the source of magma and cease volcanic activity, building a line of extinct volcanoes that exhibits a monotonic age progression reflecting the plate motion history. Numerous hotspots have been proposed for sites of unusual volcanic activity, yet conclusive imaging of mantle plumes using seismic tomography remains elusive. Although the simple age progressions predicted by the hotspot hypothesis have been confirmed for several seamount chains, others show complex age patterns, which cast doubt on the hotspot theory as the only explanation.

Seamounts formed by hotspot volcanism may grow quite large (Fig. 1A). In particular, hotspot seamounts formed on old (and hence thicker and stronger) oceanic lithosphere can in some cases reach almost 10 km (measured from their base), making Mauna Kea (one of five volcanoes that form the Big Island of Hawaii) the tallest mountain on Earth. Seamounts formed on oceanic crust must reach at least 2.5 km in height just to match the typical mid-ocean ridge depth; however, most larger seamounts were formed in even deeper water; hence, only truly large seamounts will

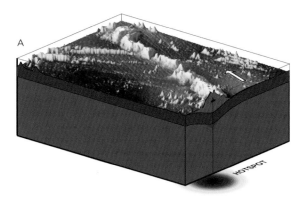

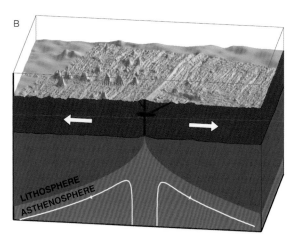

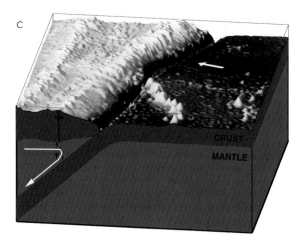

FIGURE 1 (A) Intraplate seamount formation over the Hawaii hotspot. On thick lithosphere, seamounts can grow very tall and even breach sea level to form islands. The volcano deforms the lithosphere, which responds by flexure. The hotspot feeds the active volcanoes by a network of feeder dikes; magma may pond beneath the crust as well. As plate motion carries the volcanoes away, they cease to be active and form a linear seamount chain. (B) Seamount formation near the East Pacific Rise. A thin plate cannot sustain large volcanoes, and typically only smaller cones are found. Excess magma is diverted into feeder dikes that reach the surface on the ridge flank, forming small volcanoes. (C) Island arc formation behind the Kermadec Trench. The subducting Pacific plate and its sediments will induce melt at depth, eventually erupting to create a volcanic arc that parallels the subduction zone. Note that the oldest part of the Louisville seamount chain (another intraplate chain) is currently being subducted.

become islands or have a shallow-water presence. Because large seamounts often penetrate the euphotic zone, they have been the main focus of ecological studies, despite being a small subset of all seamounts globally.

Mid-Ocean Ridge Seamounts

Most seamounts are small and were formed near a divergent plate boundary. Here, excess amounts of magma percolate through the thin, fractured crust to form small, sub-circular seamounts—often just a few tens to hundreds of meters tall. Occasionally, larger seamounts can be formed (Fig. 1B). It is likely that most small seamounts formed in this near-ridge environment as the thickness of the lithosphere rapidly increases away from ridges, making the ascent of small amounts of magma from an increasingly deeper source less likely. Consequently, seamount production rates decrease with increasing crustal age and lithospheric thickness, being highest close to the ridge axis. At fast-spreading ridges (e.g., the East Pacific Rise), small seamounts form on the flanks of the ridge where the crust is just 0.2–0.3 million years old, and their abundance correlates with spreading rate. At slow-spreading ridges (e.g., the Mid-Atlantic Ridge), small seamounts are produced almost exclusively within the median valley. Many new seamounts undergo extensive tectonic deformation by normal faulting, which reduces their original heights considerably. Because of increased sediment coverage on older sea floor, the smallest and most numerous seamounts, with heights less than 100 m, are likely to be buried after a few tens of millions of years.

Island Arc Seamounts

Island arc seamounts form at subduction zones where one oceanic plate is being forced to subduct beneath the other. As plates descend into the mantle, the higher pressure and friction and the increasing temperatures and water content eventually cause decompressional melting that produces an ascending basaltic melt of a different magmatic composition than the basalt available at spreading centers (Fig. 1C). The magma may be more volatile, thus increasing the chance of explosive eruptions. The distribution of island arc seamounts and islands reflects the trend of the convergent plate boundaries, and the overall plate tectonic geometry places strong constraints on the evolution of such seamounts. These island arc seamounts are found in the relatively narrow collision zones between the converging tectonic plates, thus occupying a small area of the total sea floor. Like hotspot-produced seamounts, island arc seamounts can reach considerable height and often form islands. Unlike hotspot-produced seamounts, the volcanic activity along an active arc is essentially simultaneous, geologically speaking, with older seamounts constantly being overprinted by younger ones.

MORPHOLOGY AND THE EVOLUTION OF SEAMOUNTS

Seamounts are born kilometers below the sea surface. Following a pathway of preexisting cracks or weaknesses, buoyant magma finds its way to the ocean floor. Here, emerging seamounts may be exposed to overburden pressures of 25–50 MPa. Consequently, volcanic gases within the magma cannot expand, and extrusive flows are effusive. The cooling effect of seawater affects the shape of the volcano, allowing construction of steeper flanks (greater than 10°) than would generally be possible once the volcano builds up above sea level (less than 10°). At first, the seamount is fed from a central vent, yielding an almost circular feature, and some develop summit craters. Many seamounts do not develop beyond this stage. However, if adequate magma supply is available, and the seamount is allowed to grow taller, then gravitational stresses in the flanks of the seamount, possibly enhanced by flexural stresses transmitted from the increasingly deformed subsurface, will favor the development of rift zones. These break the circular symmetry and promote construction of long ridges from fissure eruptions. As the summit of the seamount approaches sea level, water pressure can no longer keep gases locked up in the magma, and explosive eruptions become common. The extrusive products tend to be finer-grained, more vesicular, and structurally less resistant to erosion, which begins to shape the islands, augmented by catastrophic submarine landslides. The combined effect of rift zones, erosion, and landslides is to modify the basic circular form of seamounts into stellate forms.

Once the island is well established, the volcano enters the shield-building stage, during which large flows of ʻaʻā and pāhoehoe lava are extruded. When active construction finally wanes, the island no longer regenerates to keep up with the destructive forces of erosion, which combine with long-term thermal subsidence of the sea floor and isostatic adjustments to bring the summit area back to sea level, where wave erosion forms a flat-topped guyot. Coral growth may keep up with the subsidence rate, capping many volcanic islands with a thick coral reef layer before subsidence eventually drowns the seamount. Complex interplays between eustatic sea-level changes, vertical isostatic adjustments, and latitude changes caused by plate motion result in a wide variety of seamounts, some with fringing reefs, others with lagoons with calcareous sediments, and others that never developed a coral cap and may have drowned long ago.

THE DISTRIBUTION OF SEAMOUNTS

Seamounts are distributed both in space (geographically) and time (temporally), and studies of these variations have provided key insights into several factors that control the formation of seamounts.

Spatial Distribution

The number of seamounts varies considerably between ocean basins (Fig. 2). Seamounts form both linear and

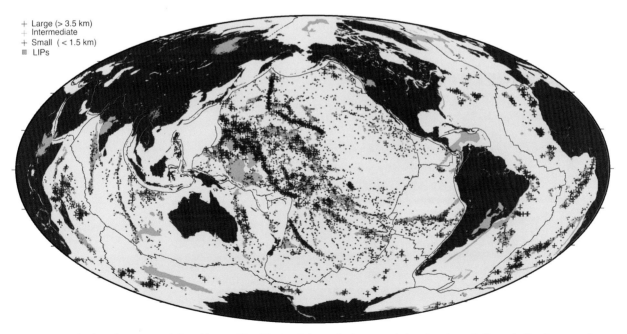

FIGURE 2 Distribution of seamounts inferred by satellite altimetry. Colors reflect seamount sizes (see legend). The majority of seamounts can be found in the Pacific basin, with the remainder divided between the Atlantic and Indian Oceans. Large igneous provinces (LIPs) are outlined in orange; these are often associated with seamount provinces.

random constellations, and their sizes and distributions provide invaluable information about their origins. Studies have found that the distribution of seamounts over a wide range of sizes is well approximated by an exponential or power-law model. Such models reflect the observation that most seamounts are small, and by extrapolating the power-law, there may be perhaps as many as 100,000–200,000 seamounts reaching heights of 1 km or more (Fig. 3A). Extrapolation further down to the smallest seamount sizes observed (a few tens of meters) would predict seamount populations reaching into the millions, but the majority of such small seamounts will likely be buried, given the typical thickness (100–200 m) of sediment in the world's ocean basins. Consequently, the smallest seamounts are observed only on young sea floor with modest sediment cover. Seamount summit depths are normally distributed around a mean depth of ~3 km. Notably, the shallow end seems to have an additional number of shallow seamounts and islands, possibly reflecting the ability of coral reef growth to keep up with subsidence for long periods of time (Fig. 3B).

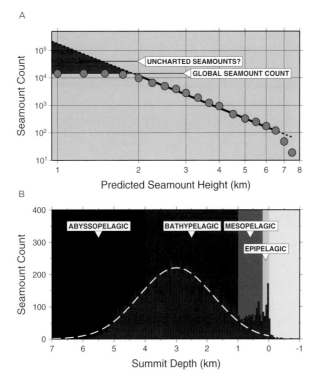

FIGURE 3 (A) Seamount size–frequency distribution. The green circles indicate the number of seamounts taller than a given size. For seamounts greater than 2 km tall, the data are well explained by a scaling rule (solid line). For heights less than 2 km, the trend levels off because numerous smaller seamounts fall below the resolution of altimetry data. Only ~15,000 out of a potential of ~200,000 seamounts greater than 1 km in height may have been mapped. (B) Summit depths follow a normal distribution centered on a ~3-km depth, with large and shallow seamounts appearing as outliers, possibly as a result of coral growth.

The abundance of seamounts has been shown to vary considerably among the ocean basins. The Pacific basin is host to nearly half of the seamounts that are large enough (greater than ~2 km) to be mapped by satellite altimetry. The Atlantic and Indian Oceans combine to contain most of the remaining seamounts, with considerably fewer seamounts appearing on plate segments located at high latitudes (e.g., northern Atlantic on either the North American or Eurasian plates) or on relatively small plates (e.g., Cocos, Philippine Sea).

It is not clear what causes seamount abundances to vary spatially. One factor may be the underlying distribution of mantle plumes, which are found in higher numbers beneath plates with numerous seamounts. However, one would still expect excess magma at the spreading center to produce the smaller and more numerous seamounts. Another factor may be systematic variations in plate stresses, with smaller plates possibly being in a compressional stress state, which would not favor the intrusion of magma. Smaller plates are also less likely to have a directional regional stress dominating the state of stress. In contrast, the large Pacific plate, in particular the equatorial region, appears to be under tension from distant slab pull forces, as evidenced by widespread extensional volcanism associated with neither hotspots nor mid-ocean ridges. Finally, plates that move the fastest appear to have the highest seamount abundances, provided they share at least one spreading plate boundary.

Island arcs aside, the distribution of seamounts appears as a superpositioning of two separate processes: Divergent plate boundaries produce a near-steady stream of new, small seamounts, most of which exhibit no particular clustering pattern, whereas mantle plumes or hotspots generally create both small and large seamounts, which are often organized in linear groups by plate motions. Frequency-size analysis (Fig. 3A) of the combined seamount populations does not immediately separate out the two modes of production, but this possibly reflects the inability of satellite altimetry to detect smaller seamounts (less than 1 km) and the lack of significant spatial coverage of small-size seamount provinces using multibeam techniques.

Temporal Distribution

Seamounts are among the youngest geologic features on Earth, reflecting the youthfulness of the oceans and the regenerative processes of plate tectonics. Only a few seamounts are currently volcanically active, and they tend to be restricted to (1) the very youngest volcanoes of hotspot island chains (such as Hawaii, Samoa, Réunion, and others), (2) various places along active island arcs, and (3) newly formed smaller seamounts associated with mid-ocean ridges. Many volcanic islands, but only a few seamounts, have been dated using radiometric techniques, yet the sparse age data, the underlying sea-floor age, and the size of seamounts imply that the production of seamounts is not steady-state. During the Cretaceous (146–65 million years ago) the Pacific seamount production was almost twice as high, resulting in numerous large seamounts now residing in the western Pacific. This period also saw the formation of several large oceanic plateaus, such as the Ontong Java, the Manahiki, the Shatsky, and the Mid-Pacific Mountains; hence, plateau and seamount formation appear correlated.

THE IMPACT OF SEAMOUNTS

Geologic Impact

Seamounts are windows into the mantle that allow scientists to study the nature of erupting magma. Minor changes in the chemical and isotopic composition of basaltic lavas can be used to make inferences about magma source depth and composition. Seamounts represent a significant fraction of the entire crust production, perhaps as much as 5–10%, and variations in this intraplate volcanic budget shed light on plate tectonics and how Earth gets rid of excess heat. The alignment of seamount chains provides a means to decode the motion of tectonic plates over long geologic intervals, enabling an understanding of the climatic changes experienced at islands that simply follow from latitudinal migration of plates carrying seamount provinces on their backs. Many seamounts have active hydrothermal convection systems that may have a significant effect on element cycles involving seawater, and they also participate in the dissipation of residual heat from the formation of both seamount and sea floor. Finally, seamounts and islands act as measuring sticks for relative sea-level variations, which can have both eustatic and tectonic components.

Oceanographic Impact

Bathymetry influences ocean circulation in several ways. The first-order features such as ridges and plateaus steer currents and, in places, act as barriers that prevent deep waters from mixing with warmer, shallower waters. Smaller-scale bathymetry, such as seamounts, may play a largely overlooked role in the turbulent mixing of the oceans. Measurements suggest that mixing around a shallow seamount is many orders of magnitude more vigorous than in areas far from seamounts. Understanding how the climate will evolve depends on how quickly heat and carbon dioxide can penetrate into the deep oceans, and assumed rates of vertical mixing can considerably affect model predictions.

Mineral Resources Impact

Older seamounts may accumulate a ferromanganese oxide crust enriched in the elements cobalt, copper, manganese, and sulfur, typically occurring at depths exceeding 3 km. The total cumulative amounts of such marine mineral resources might exceed the amounts currently available on land. So far, the cost of harvesting deep ocean nodules and crusts has been prohibitive. However, rising prices associated with depletion of terrestrial resources will likely make deep ocean resources more attractive, especially because the bulk of these are in international waters.

SEE ALSO THE FOLLOWING ARTICLES

Island Arcs / Plate Tectonics / Sea-Level Change / Seamounts, Biology

FURTHER READING

Batiza, R. 2001. Seamounts and off-ridge volcanism, in *Encyclopedia of ocean sciences*. J. H. Steele, S. A. Thorpe, and K. K. Turekian, eds. San Diego, CA: Academic Press, 2696–2708.
Keating, B. H. 1987. *Seamounts, islands, and atolls*. Washington, DC: AGU.
Macdonald, G. A. 1986. *Volcanoes in the sea*, 2nd ed. Honolulu: University of Hawaii Press.
Schmidt, R., and H.-U. Schmincke. 2000. Seamounts and island building, in *Encyclopedia of volcanoes*. H. Sigurdsson, ed. San Diego, CA: Academic Press, 383–402.
Wessel, P. 2001. Global distribution of seamounts inferred from gridded Geosat/ERS-1 altimetry. *Journal of Geophysical Research* 106: 19, 431–419, 441.
White, S. M. 2005. Seamounts, in *Encyclopedia of geology*. R. C. Selley, et al., eds. London: Elsevier, 475–484.

SEXUAL SELECTION

KENNETH Y. KANESHIRO
University of Hawaii, Manoa

RICHARD T. LAPOINT
University of California, Berkeley

Sexual selection is defined as the differential mating success among individuals of the same sex, males in most cases. Sexual selection is viewed as a dynamic, frequency-dependent process, a driver for evolutionary change, and a synergist for the formation of new species.

ISLANDS AND ISLAND BIOTA

Because of their isolated nature, islands are among the best places for investigating evolutionary processes, and research on specific groups of organisms that evolved on islands, or island-like habitats, has begun to shed light on processes by which species diversify and originate, processes such as sexual selection. For example, the Hawaiian Archipelago is often considered to be the world's most outstanding living laboratory for the study of evolutionary biology, and in the most diverse lineages (e.g., *Drosophila* and crickets), sexual selection is hypothesized to drive evolution in combination with other factors. Likewise, the African Rift lakes have provided an outstanding context for studying the role of sexual selection in driving proliferation of cichlid fish. At the same time, in other radiations (e.g., *Anolis* lizards in the Caribbean and *Tetragnatha* spiders in Hawaii), sexual selection appears to play a secondary role to natural selection.

MODELS OF SEXUAL SELECTION

Though known to be important at the population level, sexual selection is also believed to be a major force in divergence evolution. In his text on sexual selection, Darwin (1871) discussed how extreme secondary sexual characters often characterize diverse species groups. A number of models have been proposed for explaining how sexual selection might lead to rapid speciation. Runaway sexual selection was initially alluded to by Darwin and later was mathematically explored by R. A. Fisher (1930). The general concept is that one sex (usually female) develops a preference for a specific trait, or ornament, in the other sex. Quantitative genes for this preference become genetically correlated with the quantitative genes for the ornament. Over time this creates a strong positive feedback loop, where stronger preference drives a more exaggerated ornament. Very rapidly the ornament becomes more embellished to the point where the fitness benefits provided by enhanced mating success is outweighed by the selective disadvantage such an extreme trait confers. Runaway sexual selection requires the female preference to evolve for a male trait present in the population and that this trait be quantitative and heritable.

Several authors, such as Russell Lande and Mary Jane West-Eberhard, have provided models showing that sexual selection, in conjunction with drift or ecological specialization, could cause speciation. After a large population with diverse mating behaviors or morphology becomes isolated, different morphologies, and subsequently female preferences, could become fixed in allopatry. When these two lineages come into contact, differences in mating behavior prevent gene flow.

In the specific case of Hawaiian *Drosophila*, studies by Hampton Carson and Kenneth Kaneshiro have shown that there is a range of mating types segregating in both sexes within a population, some males being more successful in mating than others. Similarly, there are females that are highly discriminant in mate choice and others that are not as choosy. In this situation, it appears that sexual selection occurs via a form of density-dependent selection.

Many of the most diverse groups of island organisms are characterized by extreme secondary sexual characteristics and with sexual selection appearing to play a key role in diversification. Hawaiian *Drosophila* are a prime example, with many species that are closely related but clearly distinguishable based on exaggerated secondary sexual characters such as intricate wing patterns or modified tarsal segments and complex mating behaviors. *Laupala*, or Hawaiian crickets, are similarly a diverse group characterized mainly by mating calls. In the same way, New Guinea's birds of paradise are strikingly diverse in their mating displays and plumage. In the insular habitat of the African Rift Lakes, cichlid fish have evolved a large variety of color morphs used in mate recognition. In *Laupala* at least, a number of attributes of the system have lead researchers to conclude that speciation is the result of forces acting on secondary sexual traits. This is because closely related species are morphologically cryptic and can only be distinguished by the pulse rate of the male courtship song, a secondary sexual trait (females prefer males with pulse rates of their own species); moreover, different species are often syntopic and synchronic. A similar argument has been used to suggest that the spectacular diversification in African cichlid fish is driven by sexual selection on male coloration and perception thereof.

THE FOUNDER PRINCIPLE AND SEXUAL SELECTION

Since islands are defined by isolation, it is assumed that any new population is most likely to have been founded by a small number of individuals. Mayr first proposed and further developed the general concept of the founder principle. He argued that the reduced genetic diversity that accompanies founder events provides the mechanism by which a "genetic revolution" or new coadapted genetic system can arise and speciation could occur. Hampton Carson proposed a slightly modified version of the founder principle, suggesting that drift served to rearrange the coadapted genetic system of founding populations. His model hypothesized that the founder event is followed by rapid population growth as a result of relaxed selection pressures in the new environment. This was referred to as the "founder-flush" model. In Carson's model, then, the reduction in genetic variability is minimized as a result of rapid increase in population size immediately following the founder event, resulting in increased genetic variability.

A paradox emerges under models of sexual selection on populations simultaneously affected by founder effects. In certain instances, if Carson is correct, the loss of variability due to founder effects can be mitigated. Sexual selection, under the classical models (e.g., runaway selection), reduces genetic variability, especially in males. Also, unless secondary sexual characters are linked to or enhance other components of fitness, these characters are energetically costly to produce and maintain in the population, and individuals possessing such traits would be strongly selected against by predation or other environmental pressures.

Based on the results of mating studies of Hawaiian *Drosophila* species, Kenneth Kaneshiro proposed an alternative view of the sexual selection process on islands. The tremendous morphological diversity observed in the Hawaiian drosophilids, especially in the secondary sexual characters found in the males, gave a clue to early researchers that sexual selection may have played a significant role in the evolution and explosive speciation in this island fauna. Indeed, the complex courtship displays observed in the Hawaiian drosophilid fauna provided an opportunity to test some of the classical concepts of sexual selection. Kaneshiro observed that a frequent outcome of mate preference experiments among closely related species was that of asymmetrical sexual isolation between reciprocal crosses between two populations. That is, the females of, say derived population A, may readily accept the courtship displays of males from ancestral population B, resulting in successful matings. However, in the reciprocal combination, females of population B were less likely to mate with males of population A. Initially, Kaneshiro suggested that such shifts in the behavior of derived populations were the result of the severe drift conditions and the genetic revolution that accompanied founder events. Subsequently, it was suggested that the sexual selection process strongly influenced the shift in behavior in the most derived populations, which resulted in the observed asymmetrical sexual isolation between the two reciprocal crosses.

It was suggested that during founder events, when population size is small, there is strong selection for females that are less choosy. Highly discriminant females are strongly selected against because they may never even encounter males that are able to satisfy their courtship requirements. If the population bottleneck condition persists over a few generations, there will be a shift in the distribution of mating types in the population toward an increase in frequency of less choosy females. Such a shift in the mating distribution may result in a corresponding change in gene frequencies in the population, further resulting in the destabilization of the coadapted genetic systems that had evolved in the population as it adapted to the environmental conditions of the habitat in which it lives. The destabilized genetic environment of the population provides the opportunity for genetic change conducive to speciation. The breakup of coadapted gene complexes allows for novel genetic recombinants to be generated, some of which may be better adapted to the environmental conditions that led to the drastic reduction in population size. Especially when such novel genetic recombinants are linked to or correlated with the genotypes of less choosy females, it is easy to visualize how such preadaptive gene complexes can spread quickly in subsequent generations, providing an evolutionary mechanism by which populations can regenerate genetic variability that had been lost during reduced population size. Clearly, then, at least during the initial stages of colonization following the founder event, sexual selection, and especially the dynamics of sexual selection, may be playing an important role in providing a genetic environment that is most conducive to the formation of new species.

In the same way, for African cichlid fish, as in the Hawaiian *Drosophila*, much of the sustained rapid speciation has also been explained by an interaction between drift and sexual selection in subdivided populations.

SEXUAL SELECTION AND GENETIC VARIABILITY IN ISLAND POPULATIONS

The dynamics of the sexual selection process as described here indeed provide an evolutionary mechanism by which founder populations can, under small-population conditions, generate novel genetic recombinants that can be selected for adaptation to the new habitat. In fact, when viewed as a dynamic, frequency-dependent system, sexual selection can play a significant role in influencing, maintaining, and in some cases enhancing levels of genetic variability in natural populations. When the population is large and healthy, sexual selection may serve as a stabilizing force in maintaining a balanced, polymorphic mating system that maintains an appropriate level of genetic variability required to survive in a particular habitat. However, if the population undergoes a bottleneck due to some kind of environmental stress, the shifts in the distribution of mating types as described above play an important role in replenishing reduced levels of genetic variability by the generation of novel recombinants (i.e., not by increasing the mutation rate). The shift toward an increased frequency of less choosy females and corresponding shift in gene frequencies toward the genotypes of the less choosy females result in the destabilized genetic environment and the breakup of coadapted gene complexes, allowing for the new recombinants to be generated. Those recombinants that are better adapted to the environmental stress conditions will be strongly selected and spread quickly throughout the population, especially if linked or correlated with the genotypes of the less choosy females. Thus, sexual selection can play an important role in mitigating the fragility of island populations that can be easily perturbed by changing environmental conditions. Sexual selection provides a mechanism by which genetic variability can be restored within relatively few generations following a bottleneck event.

SEXUAL SELECTION IN CONJUNCTION WITH ECOLOGICAL ADAPTATIONS

In many well-known adaptive radiations, secondary sexual characters are remarkably diverse. These secondary characters provide no real adaptive benefit and yet are nearly ubiquitous. Sexual selection has been inferred to be a more potent force in adaptive radiations than previously thought. In cases such as the Hawaiian *Drosophila*, and more so *Laupala*, there is sometimes little ecological differentiation between the most closely related species. Sexual selection may actually increase the rate of diversification beyond that of ecological adaptation alone. Studies on cichlid fish show that while niche adaptation is an important component to this rapid radiation, sexual selection may be even more important in some cases.

Intraspecific sexual selection may also increase the adaptive ability within these radiations. On islands in the West Indies, sexual dimorphism in *Anolis* lizards allows increased niche exploitation, as size differences allow for increases in the number of ecomorphs present in each species. This may speed up the already rapid filling of niche space for this group. On the Galápagos Islands, females of a few species of Darwin's finches as well as those of marine iguanas were found to prefer larger males, a trait that is correlated to environmental exploitation in this case. Sexual selection may operate very quickly on an adaptive trait.

THE ROLE OF SEXUAL SELECTION IN CONSERVATION BIOLOGY

Although insular systems are often recognized for nature's creativity and their high levels of endemism, they are also known for their fragility and the relatively large number of rare and endangered species. Sexual selection and recognition have been implicated directly in conservation issues in African cichlids. Here, mate recognition is critical to the dynamics of the system. However, eutrophication of the lakes has led to a breakdown of the recognition system, with resulting extensive hybridization.

Clearly, understanding the nature of mate recognition and the biology of small populations is crucial for addressing conservation issues, since rare and threatened species are faced with extinction, primarily as a result of drastically reduced population size. These demographic conditions are no different from those occurring during early stages of colonization following a founder event. The sexual selection model described in this article permits the generation of novel genetic recombinants, some of which are better adapted to the new habitat or to the stress conditions that caused the population to decline. What is inferred here, then, is that if the habitat of those species that are faced with small population size can be sustained by the removal of the threats that impact on the habitat, these populations have the capacity to replenish the genetic variability that may have been lost due to drift. Furthermore, the sexual selection system may facilitate natural hybridization with related sympatric species, which can result in the introgression of genetic elements that are less susceptible to the environmental stress conditions that resulted in the population crash. Thus, when considering the management of rare and endangered species, it is important to consider the interaction with other related species with overlapping distributions, from which genetic material may be "leaking" across species barriers to maintain adequate levels of genetic variability in the population.

CONCLUDING REMARKS

For some of the most diverse species found on islands, founder event speciation along with sexual selection is considered to be a likely model for the formation of new species and the explanation for the spectacular adaptive radiation among island biota. Most evolutionary biologists acknowledge the fact that natural selection can play a dominant role in directing the course of evolution within species. However, the recent research on sexual selection theory and its role in the evolutionary process suggest that the evolution of novelty is enhanced when populations are subjected to bottlenecks and that the dynamics of sexual selection are extremely important, especially during the initial stages of species formation.

SEE ALSO THE FOLLOWING ARTICLES

Adaptive Radiation / Cichlid Fish / Crickets / *Drosophila* / Founder Effects

FURTHER READING

Carson, H. L. 1986. Sexual selection and speciation, in *Evolutionary processes and theory.* S. Karlin and E. Nevo, eds. London: Academic Press, 391–409.
Darwin, C. 1871. *The descent of man, and selection in relation to sex.* New York: Penguin Putnam, Inc.
Fisher, R. 1930. *The genetical theory of natural selection.* Oxford: Clarendon Press.
Kaneshiro, K. Y. 1989. The dynamics of sexual selection and founder effects in species formation, in *Genetics, speciation, and the founder principle.* L. V. Giddings, K. Y. Kaneshiro, and W. W. Anderson, eds. New York: Oxford University Press, 279-296.
Lande, R. 1981. Models of speciation by sexual selection on polygenic traits. *Proceedings of the National Academy of Sciences of the United States of America* 78: 3721–3725.
Mayr, E. 1982. Processes of speciation of animals, in *Mechanisms of Speciation.* C. Barigozzi, ed. New York: Alan R. Liss, 1–19.
Mendelson, T. C., and K. L. Shaw. 2005. Rapid speciation in an arthropod. *Science* 433: 375–376.
O'Donald, P. 1980. *Genetical models of sexual selection.* Cambridge, UK: Cambridge University Press.
Seehausen, O. 2004. Hybridization and adaptive radiation. *Trends in Ecology and Evolution* 19: 198–207.
West-Eberhard, M. J. 1983. Sexual selection, social competition and speciation. *Quarterly Review of Biology* 58: 155–183.

SEYCHELLES

JUSTIN GERLACH

Nature Protection Trust of Seychelles, Cambridge, United Kingdom

The Republic of Seychelles comprises 115 islands spread over 1.3 million km² of the western Indian Ocean. The majority of these are small and uninhabited. Of the Seychelles' 80,000 human inhabitants, 90% live on the largest island, Mahé (153 km²), and 7% on the next largest, Praslin (28 km²). The islands can be divided into two groups: the northern islands (the granitic islands and the coral cays of Bird and Denis) and the southern coral islands (Fig. 1).

GEOGRAPHY AND GEOLOGY

The granitic islands are mostly composed of Precambrian granite (750 million years old) and represent fragments of the ancient supercontinent of Gondwana. The breakup of

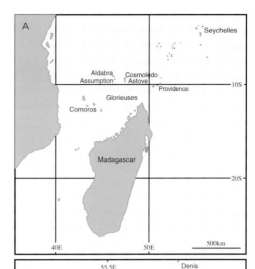

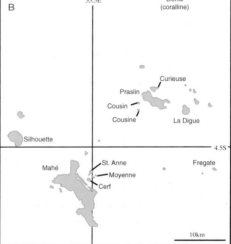

FIGURE 1 Map of the Seychelles islands.

Gondwana over hundreds of millions of years finally ended with the separation of India and the Seychelles 65 million years ago. The resultant Seychelles microcontinent has gradually submerged and eroded, until today only islands representing the granitic microcontinent's mountain tops remain. The islands are very rocky and steep, reaching a maximum height (at Morne Seychellois) of 905 m.

Periods of volcanic activity and tectonic movement led to the formation of other islands, some granitic (Silhouette and North) and others volcanic. The latter have long since subsided and eroded, leaving only raised coral reefs at or near the surface. These are now represented by coral cays and atolls. These coral islands rise no more than 8 m above sea level, and most are less than 1-m high (Fig. 2).

CLIMATE

The islands have an equatorial climate, with relatively small variations in temperature (24–32 °C at sea level) and constantly high humidity (over 70%). The low-lying

FIGURE 2 Cosmoledo, a low-lying coral island.

FIGURE 3 Silhouette, the most natural of the high granitic islands.

FIGURE 4 The Seychelles pitcher plant *Nepenthes pervillei*.

coral islands have lower rainfall and, consequently, lower humidity. On the high islands, there may be significant local variation in climate, with high forest areas being cooler (18–24 °C) and wetter than lowlands.

BIODIVERSITY AND ECOLOGY

The islands have a diverse fauna and flora and are well known for remarkable species such as giant tortoises (*Dipsochelys* species) and the coco-de-mer palm, or double coconut, *Lodoicea maldivica*. Two distinct species assemblages are found, corresponding to the origin of the islands. The coral islands, which represent land raised up from under the sea, have been colonized by organisms that have crossed large expanses of ocean, some from Asia to the north and east, some from Africa to the west, but most from the closest land mass, Madagascar, to the south. These small, flat islands have relatively few species, and almost all of those they do have are widespread outside of the Seychelles. The exception is Aldabra atoll, which supports some 1500 species, 60% of which are believed to be endemic to the atoll.

The high granitic islands of the Seychelles support a diverse flora and fauna with many interesting affinities to both Asia and Madagascar (Fig. 3). The plants include some 300 species, of which 40% are endemic. These include the only member of the Dipterocarpaceae found outside of Southeast Asia (*Vateriopsis seychellarum*) and a whole family restricted to the islands (the monotypic family Medusagynaceae, represented by *Medusagyne oppositifolia*). There are populations of an endemic carnivorous pitcher plant, *Nepenthes pervillei* (a genus restricted to Southeast Asia, the Seychelles, and Madagascar) (Fig. 4). A major component of the natural vegetation is the palm family; in addition to the widespread tropical coconut *Cocos nucifera*, there are six endemic genera of palm: the coco-de-mer, *Deckenia nobilis*, *Nephrosperma vanhouetteana*, *Phoenicophorium borsigianum*, *Roscheria melanochaetes*, and *Verschaffeltia splendida*. Most of these palms have prominent spines on their stems and leaf bases, a feature generally thought to be a defense against the giant tortoises that used to roam free in the islands.

Animal life is mainly small and inconspicuous, with a large proportion of the fauna being specialized to live in high forest leaf litter or in the narrow spaces between palm leaves. Unusual animal groups include a wide range of carnivorous snails, specialized carrion-feeding caddis flies, and land-dwelling diving beetles. More conspicuous are the birds, reptiles, and amphibians. The Seychelles are well known as a bird-watching destination because, although they support only a small number of species, most of these are endemic, such as the Seychelles magpie robin *Copsychus seychellarum*, the Seychelles black paradise flycatcher *Terpsiphone corvina*, and the Seychelles kestrel *Falco araea*. All of the endemic birds are close relatives of species found in Madagascar. The reptiles are far more diverse and are the dominant vertebrates of the islands. Most notable are the giant tortoises, which were historically the largest herbivores in the islands. Four species have been described from the Seychelles, of which the

Aldabra tortoise *Dipsochelys dussumieri* is still abundant in the wild, Daudin's tortoise *Dipsochelys daudinii* is extinct (known from only two specimens), and the Seychelles giant tortoise *D. hololissa* and Arnold's tortoise *D. arnoldi* were thought to be extinct. Recently (Fig. 5), individuals thought to be the few survivors of these latter two species were found and are being bred in captivity. Arnold's tortoises were reintroduced to the wild on Silhouette Island in 2006. The islands also support declining populations of *Pelusios* terrapins and large breeding populations of the hawksbill turtle *Eretmochelys imbricata* and the green turtle *Chelonia mydas*. Other reptiles comprise the little-known chameleon *Calumma tigris*, geckos, skinks, and snakes. The geckos are easily visible, with abundant *Phelsuma* day geckos in coastal habitats and introduced house geckos (*Gehyra mutilata* and *Hemidactylus frenatus*) in buildings. There are also more specialized species, such as the *Ailuronyx* bronze geckos in palm woodlands and the strange sucker-tailed gecko *Urocotyledon inexpectata*, which lives in caves, often occupying potter wasp nests.

FIGURE 5 Arnold's tortoise *Dipsochelys arnoldi*.

Amphibians are among the most important of groups from a biogeographical perspective, as they tend to reflect ancient land connections because their permeable skins make it very difficult for them to cross open sea. The Seychelles support four families of amphibian with differing biogeographical affinities. The most widespread is Ranidae, represented by the Mascarene frog *Ptychadena mascariensis*. This species was probably introduced from Mauritius. The Hyperoliidae is represented by the Seychelles tree frog *Tachycnemis seychellensis*, which is closely related to Malagasy species and is thought to be descended from frogs that were carried to the islands on rafts of vegetation. The last frog family in the islands is the Sooglossidae, an endemic family related to an obscure Indian species. This is an ancient family of very specialized frogs,

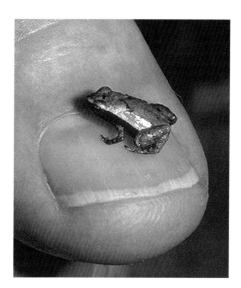

FIGURE 6 The minute Gardiner's frog *Sooglossus gardineri*.

none of which have free-living tadpoles. The mountain streams of the Seychelles are seasonal and very fast flowing, so frogs avoid living in the water and become terrestrial or arboreal instead, either carrying their tadpoles or developing their young entirely in the eggs. This latter strategy is adopted by Gardiner's frog *Sooglossus gardineri*, which is one of the smallest frogs in the world (adult size 8–11 mm), the eggs of which hatch into frogs 3-mm long (Fig. 6). Strangest of all the Seychelles amphibians are the caecilians (Caeciliidae). The Seychelles have six species of these unusual and very poorly known legless amphibians. This is another group that originated on the ancient continent of Gondwana. Almost all these amphibians form part of an adaptive radiation, with ten of the endemic species being found in five endemic genera and only one being an apparently recent arrival.

Insect diversity is high, with approximately 6,000 species recorded. A high proportion of these are endemic: between 50% (Lepidoptera and Diptera) and 80% (Orthoptera). Although mollusc diversity is lower (only 80 species), 88% are endemic. These include remnants of the ancient Gondwana fauna, for example the large *Stylodonta* snails of the sothern African, Madagascan, Sri Lankan, and Australian family Acavidae and the *Pachnodus* tree snails related to species from the Congo. *Pachnodus* is a typical island radiation, having diverged into 12 distinct species through a combination of geographical isolation on different islands and habitat specialization. A similar radiation is found in the carnivorous snail family Strepaxidae, where ten species form an endemic radiation of six distinct genera. The remaining three species are also endemics but belong to African and Asian genera.

The islands are fringed by coral reefs, which are diverse and productive environments. Over 1200 marine fish have been recorded (compared to only eight freshwater fish). Despite several decades of interest in the marine biology of the region, very little is known of the invertebrate communities of the reefs of Seychelles. These reefs were very badly damaged by the coral bleaching event of 1998, which resulted in the death of over 80% of corals. The corals are recovering in some areas, although the process is extremely slow.

HUMAN HISTORY AND DEVELOPMENT

The Seychelles islands were uninhabited when discovered by European explorers in the fifteenth and seventeenth centuries. There have been suggestions that Melanesian sailors settled the islands on their way to colonize Madagascar, but there is no evidence of this. Similarly, suggestions that Arab traders occupied Aldabra atoll remain speculative. The first recorded sighting of the islands was by Vasco da Gama's expedition of 1498, but the first landing did not occur until 1609 (on North Island). The first settlement of the islands was the French colonization of St. Anne in 1770. This led to the settlement of most of the granitic islands by the mid-1800s, originally by French plantation owners and their slaves. Britain took possession of the islands in 1810 but did little to change the organization of the colony beyond using the islands as a place to release slaves from captured French slave traders. Small numbers of Indian and Chinese merchants also settled the islands at this time. These people make up the ancestry of the racially mixed population of the present day.

In the nineteenth century, most of the larger islands were settled and the lowland forests cleared for agriculture, especially for coconut plantations. On Mahé Island, large plantations of cinnamon *Cinnamomum verum* were established, and further deforestation occurred as wood was cut to supply cinnamon oil distilleries. Human impacts on the environment were high at this time, with the exploitation of giant tortoises and turtles as food sources and the deliberate extermination of the crocodile *Crocodylus porosus* and the Seychelles green parakeet *Psittacula wardi*, the latter of which was perceived as a pest of fruit plantations. By 1900 all the islands were extensively modified, with the exception of Aldabra atoll, which remained inhospitable and difficult to reach. A small settlement had been established on the atoll, but the human impact remained minimal.

Economically the islands relied on agriculture throughout the nineteenth century and well into the twentieth century. Copra from the coconut plantations was the main export, with relatively small levels of production of whale oil, cinnamon oil, vanilla, turtle meat, seabird eggs, and fish. Periodic attempts were made to increase the value of fish exports, but this remained insignificant until the arrival of large international tuna fishing fleets in the Indian Ocean in the 1980s. Collapse of the market for copra and cinnamon oil in the mid-1900s left the Seychelles economically dependent on other sources of income, particularly tourism after the opening of the international airport in 1971. Tourism dominates the economy of the islands today, with some 120,000 visitors a year. Infrastructure development has increased dramatically in recent years, and now most coastal properties are built upon, with plans for the development of 60 new hotels in the next ten years. There are concerns that the planned rate of development is unsustainable.

From 1810 the islands were governed by the British administration in Mauritius, not receiving their own governor until 1903. In 1976 the islands became an independent republic. The first president, James Mancham, led a coalition government headed by the Democratic Party for one year, after which point he was deposed by a military coup, in 1977. Power was assumed by the Seychelles People's Progressive Front, led by Albert René, under a one-party state system until 1993. In that year, international pressure led to the adoption of a new democratic constitution. René was returned to office twice, in 1998 and 2001, and then resigned in favor of his vice president, James Michel, in 2004. Michel won his own mandate in 2006. The SPPF has dominated Seychelles politics since 1977 and retains 68% of the seats in the National Assembly, although it held only 54% of the presidential vote in the 2006 election.

CONSERVATION

The islands were heavily exploited from their settlement in 1772 until the mid-1900s. Lowland forests were cleared, and all usable timber trees felled; large animals were eaten or traded. Despite this, there have been few recorded extinctions from those times: Those species that did go extinct are the crocodiles, the green parakeet, chestnut-flanked white-eye *Zosterops semiflava,* the giant tortoise *Dipsochelys daudinii,* the terrapin *Pelusios seychellensis,* and the snails *Pachnodus ladiguensis* and *P. curiosus*. Although one of the first administrators of the islands was concerned over the rate of forest loss and established a park for the tortoises in 1787, potential problems received little attention until 1874, when eminent British scientists, led by Charles Darwin, appealed to the governor of Mauritius to take measures to protect the declining Aldabra tortoise population. This action prompted the protection of Aldabra, and from 1891 commercial users of the atoll were required to protect the tortoises. The tortoise population on the atoll probably numbered fewer than 1000 animals in 1900 but recovered rapidly, to some 120,000 in 1974. Today, the giant tortoise is secure from exploitation, and steps are being taken to conserve the remnant populations of granitic island tortoises.

Although parts of Aldabra were effectively managed as a reserve from 1955, the atoll did not receive formal protection until 1979. Other reserves were established in the late 1970s, primarily to protect water catchment areas. Today, 43% of the land area of the islands is legally protected, as are a further 228 km^2 of marine habitat. These areas are managed mainly by the Seychelles government, although the most significant reserves are managed by non-governmental organizations. Management of these reserves has largely focused on conserving threatened bird populations, particularly through translocation. This has led to population recovery for the Seychelles magpie robin, the Seychelles warbler *Acrocephalus sechellensis,* and the Seychelles white-eye *Zosterops modesta.* Management of other species based on scientific research is a more recent development. This is largely at the stage of population assessment for reptiles, amphibians, invertebrates, and most plants. Science-based population management is now in place for the critically endangered Seychelles sheath-tailed bat *Coleura seychellensis* and for plant species such as *Impatiens gordonii.*

The main threats to be biodiversity of the Seychelles islands have changed over the years. In the early years of human settlement, habitat loss and exploitation were the main threats, but for much of the twentieth century, invasive species have been considered the primary problem. These include predators such as cats and rats, as well as invertebrates such as the crazy ant *Anoplolepis gracilipes,* which affect some species. Far more widespread in effects have been the invasive plants, especially cinnamon, Chinese guava *Psidium cattleianum,* and *Clidemia hirta.* Since 2000 there has been a great expansion of development for housing and for the tourism industry. This is leading to new developments in coastal areas, and potentially to the construction of roads to remote areas. Such roads would open up new areas to invasive species and to exploitation and would also directly impact some of the few remaining fragments of primary habitat.

In the future, there will be additional pressures from climate change. For small islands, sea-level rise is an important issue. Some of the coral islands will be badly affected by any rise in sea level, although the main islands are relatively high. Despite this, the impacts of sea-level rise may be notable because most of the human population lives in coastal areas at risk from flooding. Changes in climate are expected to make the weather patterns less stable, with an increase in extreme storms on the granitic islands and a reduction in rainfall on the coral islands. Increased length and frequency of dry periods has already caused the extinction of one species, with the disappearance of the Aldabra banded snail *Rachistia aldabrae* in 1997.

SEE ALSO THE FOLLOWING ARTICLES

Frogs / Granitic Islands / Lizard Radiations / Madagascar / Tortoises

FURTHER READING

Amin M., D. Willetts, and A. Skerrett, eds. 1994. *Aldabra—world heritage site.* Nairobi: Camerapix Publishers International.
Gerlach, J. 2004. *Giant tortoises of the Indian Ocean.* Frankfurt: Edition Chimaira.
Gerlach, J., ed. 2007. *Terrestrial and freshwater vertebrates of Seychelles.* Leiden, Netherlands: Backhuys Publishers.
McAteer, W. 1991. *Rivals in eden.* London: The Book Guild.

SHIPWRECKS

JAMES HAYWARD

University of California, Berkeley

Island life and exploration have always necessitated the use of ships and other vessels, be it for transportation, for exploration, for commerce, or for recreation. From rafts and canoes, to sailing ships and steamers, to modern cruise ships and tankers, there have always been some of these vessels that failed to make it safely to their destination. Every shipwreck, whether one unnamed and unknown or an icon in world history such as the *Titanic,* has the potential to make an impact on island life. While each shipwreck may have a variety of consequences, the most common and most influential include cultural, political, or biological impacts.

ANATOMY OF A SHIPWRECK

Shipwrecks occur for a variety of reasons, inclement weather and navigation errors being among the most common. Some ships were intentionally grounded, including the famous *HMS Bounty,* which was grounded on Pitcairn Island, stripped, and then burned in order to avoid sighting by British ships, which might have then arrested the mutineers. In October 1942, after striking two mines entering the Espiritu Santo Harbor in Vanuatu, the SS *President Coolidge*, a troop transport ship, was driven aground in order to enable all but two of the 5440 men aboard to abandon ship safely, before she settled back into the harbor waters.

What we know about shipwrecks depends primarily on their causes and where they end up. Many ships break up rapidly in the powerful and dynamic environments that lead to their sinking. Some ships may settle mostly intact to the bottom, only to be covered up by sediment over many decades. In some cases rapid sedimentation can provide anoxic conditions where ship, crew, and cargo can be preserved for hundreds of years. Shipwrecks in deep freshwater, such as the Great Lakes, have been found in relatively pristine condition.

CULTURAL IMPACTS

When a shipwreck occurs, the survivors who are able to find a safe harbor can bring cultural changes with them. In 1841 the slave ship *Trouvador* was wrecked off the Turks and Caicos in the Caribbean. While a disaster for those who died, the wreck was a mixed blessing for the slaves on board, who found themselves in an emancipated British colony instead of the slave market in Cuba, their original destination. Ironically, though free, the majority of the Africans were required to work for a year in the island salt ponds to pay for their rescue. On the other side of the world, birth rates are known to have spiked on Rapa Nui (Easter Island) after several shipwrecks in the 1800s. In some cases, cultural differences are a disadvantage to the shipwreck survivors. In 1840 the *Maria* was wrecked in Western Australia. After making it safely to shore, several of the crew attempted to walk back to civilization with the help of the local Aboriginals, who reportedly killed them for refusing to respect the Aboriginal culture and repeatedly harassing the Aboriginal women.

POLITICAL AND FINANCIAL IMPACTS

In some cases, the greatest impact a shipwreck may have may not be felt locally or might not be appreciated for years to come. Such is the case of the *Dunotter Castle*, a British vessel en route from Australia to California with a load of coal in 1886. After she ran aground on uninhabited Kure Atoll in the northern Hawaiian Islands, there was nearly an international incident when British government representatives in Hawaii launched a rescue mission that could have paved the way to claiming the island for Great Britain. A shipwreck that did ultimately change the political future of an island was the wreck of the *Sea Venture* in Bermuda in 1609. While leading a fleet with 600 settlers and supplies to the foundering Jamestown settlement, the *Sea Venture* wrecked on the coast of Bermuda during a hurricane. Although many of the survivors were able to make their way to Jamestown, a small band remained on the island until 1612, when the first ship of intended settlers arrived from England.

BIOLOGICAL IMPACTS

Shipwrecks can have long-term impacts on island biology. In 1780 a Japanese shipwreck delivered Norway rats, a very destructive species, to islands off the coast of Alaska. These rats are still a problem today, feeding on the eggs of nesting sea birds. In more modern times, shipwrecks have included oil tankers such as the *Exxon Valdez*, which coated 2100 km of shoreline in oil after running aground on Bligh Reef, between Bligh Island and Glacier Island in the Prince William Sound area of Alaska. Fortunately, in some cases, shipwrecks may actually have a positive impact on the environment. As discussed below, the presence of shipwrecks in some environments can become the basis for an entire ecosystem. Much like a volcanic island breaking the surface of the sea for the first time, the sudden arrival of a shipwreck on the ocean floor will often lead to colonization and productivity. In some situations, the initial impact of a shipwreck may be negative, but over a number of years, decades or centuries, the shipwreck environment may recover and eventually surpass the original.

SHIPWRECKS AS ISLANDS

When a vessel comes to rest on a sandy sea floor, it may offer a new fixed substrate that allows for coral settlement, sea grass bed development, or even kelp forest formation. These habitats then attract additional species of invertebrates and fish until a thriving ecosystem has taken hold in what may have been a previously barren environment (Fig. 1). Chuuk lagoon in Micronesia, the site of two major battles between the United States and Japan during World War II, now holds the remains of over 50 vessels and many more aircraft, which provide homes to a highly diverse marine ecosystem. On all coasts of the United States, and in many locations around the world, there are very successful examples of decommissioned vessels being purposely sunk to provide recreational opportunities for SCUBA divers and anglers, while giving the local marine species extra reef space to call home.

FIGURE 1 Marine life takes over the hull of the *Benwood*, in the Florida Keys.

SEE ALSO THE FOLLOWING ARTICLES

Archaeology / Easter Island / Introduced Species / Reef Ecology and Conservation

FURTHER READING

Advisory Council on Underwater Archaeology. http://www.acuaonline.org.
Australasian Institute for Maritime Archaeology. http://www.aima.iinet.net.au.
Ballard, R. D., L. E. Stager, D. Master, D. Yoerger, D. Mindell, L. Whitcomb, H. Singh, and D. Piechota. 2002. Iron Age shipwrecks in deep water off Ashkelon, Israel. *American Journal of Archaeology* 106: 151–168.
Bass, G. F. 1975. *Archaeology beneath the sea.* New York: Walker Publishing Co. Inc.
Bathurst, B. 2005. *The wreckers.* New York: HarperCollins.
Jurisic, M. 2000. *Ancient shipwrecks of the Adriatic: maritime transport during the first and second centuries AD.* Oxford: Archaeopress.

SICILY

SEE MEDITERRANEAN REGION

SILVERSWORDS

BRUCE G. BALDWIN

University of California, Berkeley

Silverswords belong to an ecologically diverse lineage of endemic Hawaiian woody and semi-woody members of the sunflower family (Compositae) collectively known as the Hawaiian silversword alliance (31 species in *Argyroxiphium, Dubautia,* and *Wilkesia*). True silverswords and greenswords (*Argyroxiphium;* five species), named for their narrow, sword-shaped leaves (silvery-hairy in silverswords), are spectacular rosette plants of alpine cinder slopes, forest edges, mesic scrub, and bogs on Maui and Hawai'i (Fig. 1). These famous, young-island endemics are closely related to the bizarre, fibrous-leaved rosette plants in *Wilkesia* (two species), found on generally dry or exposed slopes of western Kaua'i, and to trees, shrubs, mat-plants, cushion-plants, and lianas in *Dubautia* (24 species), found in wet, dry, and bog habitats across the six major high islands of the Hawaiian archipelago (Fig. 2). The silversword alliance is an extreme, well-documented example of insular adaptive radiation following long-distance dispersal in plants.

ADAPTIVE RADIATION OF THE HAWAIIAN SILVERSWORD ALLIANCE

The silversword alliance exemplifies adaptive radiation by exhibiting a high rate of diversification accompanied by major ecological shifts and associated morphological change. Unlike some (but not most) other prominent plant and animal examples of adaptive radiation in the Hawaiian Islands, such as the lobelioids and drosophilids, the silversword alliance radiation is evidently contemporary with the modern high islands and does not appear to date back to islands further northwest that have been reduced by erosion and subsidence to atolls, islets, or submarine seamounts. The estimated maximum age of the most recent common

FIGURE 1 Rosette plants of the Hawaiian silversword alliance. (A) The Haleakala silversword (*Argyroxiphium sandwicense* subsp. *macrocephalum*), a semelparous, thick-leaved plant endemic to dry, alpine cinder on Haleakala, East Maui. Photograph by Donald W. Kyhos. (B) The iliau (*Wilkesia gymnoxiphium*), a semelparous fibrous-leaved plant endemic to dry or open sites on western Kaua'i. Photograph by Gerald D. Carr.

ancestor of the silversword alliance based on molecular data (5.2 ± 0.8 million years) approximates the age of the oldest high Hawaiian island, Kaua'i (5.1 ± 0.2 million years). The resulting minimum diversification rate estimated for the silversword alliance (0.56 ± 0.17 species per million years) falls within the range of rates estimated for other prominent examples of adaptive radiation in plants and animals.

Major ecological change associated with diversification is evident in part from comparative analysis of habitats and life forms within lineages of the silversword alliance endemic to particular islands of the Hawaiian chain. On the oldest high island, Kaua'i, for example, one endemic lineage includes rosette plants of mostly dry, open sites (*Wilkesia*), sprawling shrubs of wet forests (*Dubautia raillardioides*), and erect, relatively compact shrubs of open bogs (*D. paleata*). Similarly, lineages of *Dubautia* section *Railliardia*, restricted to younger islands or volcanoes (east of Kaua'i), generally differ markedly in habitat (e.g., moisture availability) and often in life form; for example, an endemic Maui Nui lineage includes a wet-forest tree (*D. reticulata*), up to 8-m tall, and a shrub of dry, open scrub or barrens (*D. menziesii*) that occur on a single volcano, Haleakala.

Studies of plant structure and function by S. Carlquist and by R. H. Robichaux and colleagues have shown that differences in leaf and wood anatomical traits among species of the silversword alliance are generally consistent with expectations based on water availability in the habitats occupied, except in species from bogs, wherein features otherwise associated with dry habitats may reflect either physiological drought (compromised root function in bogs) or retention of characteristics from ancestors of dry environments. For example, leaves of species from dry habitats, such as *Dubautia menziesii*, often contain copious hydrophilic extracellular "mucilage," which confers resistance to wilting under drought stress; leaves from species of wet habitats, such as the closely related *D. reticulata*, contain lower or undetected amounts of these polysaccharides and wilt at levels of drought stress that are tolerated by species of dry situations.

HISTORICAL BIOGEOGRAPHY AND ECOLOGY

Molecular phylogenetic data indicate that the silversword alliance in general follows the "progression rule"

FIGURE 2 Ecological diversity in *Dubautia*. (A) *D. waialealae*, a cushion plant from bogs of Kaua'i. Photograph by Kenneth Wood. (B) *D. latifolia*, a liana or woody vine from mesic forests of Kaua'i. Photograph by Bruce G. Baldwin. (C) *D. scabra* subsp. *scabra*, an often mat-forming, soft shrub from young lava of Hawai'i and East Maui. Photograph by Bruce G. Baldwin. (D) *D. menziesii*, a shrub from high, dry cinder barrens and scrub of East Maui. Photograph by Susan J. Bainbridge.

of Hawaiian biogeography; namely, dispersal evidently has been mostly from older to younger islands (northwest to southeast). Dispersal between islands has been uncommon; all but five species are known only from one island. In the most widespread genus, *Dubautia,* all species endemic to one island can be explained by diversification on that island from a single founder species.

Most single-island endemic species of the silversword alliance (14 of 25) occur on the oldest modern high island, Kaua'i, where *Dubautia* and *Wilkesia* evidently originated, based on molecular phylogenetic data and levels of genetic variation within and among species there compared to such variation in the younger-island lineages. Absence of true silverswords and greenswords (*Argyroxiphium*) from Kaua'i and O'ahu has been puzzling because, based on molecular analyses, divergence of the *Argyroxiphium* lineage preceded the origin of Maui Nui and Hawai'i, where the true swords are endemic. Prehistoric extinction of *Argyroxiphium* on islands older than Maui Nui is consistent with loss of high-elevation habitat on those islands through erosion and subsidence and the predominant occurrence of most species of *Argyroxiphium* at elevations that exceed those of the highest summits of the older islands.

HYBRIDIZATION AND GENE FLOW

Experimental hybridization studies by G. D. Carr and D. W. Kyhos have shown that crosses between species of the silversword alliance reliably yield vigorous hybrids of low to high fertility, even when made between members of the most evolutionarily divergent lineages. Documentation of natural hybrids representing 41 different species combinations from various environments throughout the archipelago indicates that prezygotic reproductive isolation has not reached a level that is sufficient to prevent gene flow between species.

Chromosomal structural mutations (specifically, one to three whole-arm reciprocal translocations) account for reduction in interfertility between some species of the silversword alliance. These chromosomal rearrangements alone do not result in strong post-zygotic reproductive barriers between species; experimental studies have shown that even the least interfertile members of the silversword alliance are capable of gene exchange under ecologically permissive conditions. For example, Carr found that natural hybrids between the closely sympatric Haleakala silversword (*Argyroxiphium sandwicense* subsp. *macrocephalum;* Fig. 1) and Haleakala kupaoa (*Dubautia menziesii;* Fig. 2) produce progeny (~5% seed set) by backcrossing to either of the parent species, despite highly reduced fertility (~10%) of first-generation hybrids from extensive chromosomal structural heterozygosity (the parent species differ by three whole-arm chromosomal interchanges). Resulting backcross plants sometimes reach maturity in nature and can be highly fertile and vigorous. One additional generation of backcrossing to the same species can yield fully fertile plants that could be mistaken for that species. In summary, hybridization and gene flow remain possible between sympatric members of the silversword alliance and appear to be limited in part by post-dispersal selection against hybrid or backcross phenotypes in various natural settings (i.e., extrinsic or environment-mediated post-zygotic reproductive isolation). Differences in peak flowering time between some species (e.g., *Dubautia paleata* and *D. raillardioides*) may also contribute to reproductive isolation.

The possibility that hybridization may play an important evolutionary role in the silversword alliance must be taken seriously given the likely importance of environmental factors in restricting gene flow or recombination between species of the group. During periods of major environmental change, selection favoring hybrid or backcross phenotypes could drive evolution of new, stable lineages. Phylogenetic evidence for a lasting impact of hybridization on the genetic constitution of a limited number of species in the silversword alliance has been obtained from comparison of phylogenetic data from different genes and genomes. Conversely, some completely interfertile species that occur in proximity and are known to hybridize in nature do not show convincing molecular evidence of gene exchange.

ORIGIN OF THE SILVERSWORD ALLIANCE

The silversword alliance belongs to the helianthoid subtribe Madiinae, a monophyletic group that also includes ~90 species of mostly annual herbs commonly known as tarweeds (so named because of their sticky glandular exudates). All species of continental tarweeds occur in western North America, and all but two are native to the California Floristic Province, with one (*Madia sativa*) also indigenous to southern South America (Chile and Argentina). Multiple lines of phylogenetic data indicate that the silversword alliance belongs to one of four major lineages of the tarweed subtribe, the "Madia" lineage (including *Anisocarpus, Carlquistia, Harmonia, Hemizonella, Jensia, Kyhosia,* and *Madia,* in addition to the three Hawaiian genera), and arose well after the onset of tarweed diversification in western North America. Although tarweeds and silverswords are strikingly different in general appearance, genetic similarity between the continental and Hawaiian members of the Madia lineage remains sufficient to allow for production of vigorous (though highly sterile) tarweed × silversword-alliance hybrids. Such successful crosses have been made between arborescent or shrubby species of *Dubautia,* on one hand, and perennial herbaceous species or hybrids of *Anisocarpus*, *Carlquistia*, and/or *Kyhosia*, on the other.

Multiple lines of evidence for descent of the silversword alliance from western North American tarweeds indicate that the ancestor of the Hawaiian group must have dispersed across more than 3500 km of open ocean to reach the Hawaiian archipelago. No land bridge or intervening islands would have allowed for shorter dispersal intervals. External bird dispersal appears to be the most likely means by which the ancestor of the silversword alliance reached the Hawaiian Islands; the dry, single-seeded fruits of tarweeds of the Madia lineage and silverswords have features such as sticky enveloping bracts or persistent bristles (pappus) and fruit hairs that promote adhesion to animals. Birds have also been inferred to be the probable dispersal vectors for other Hawaiian plants of temperate North American origin, such as the endemic Hawaiian sanicles (*Sanicula*) and mints (*Haplostachys*, *Phyllostegia*, and *Stenogyne*).

Understanding establishment of the original colonizing ancestor of the silversword alliance in the Hawaiian Islands requires consideration of sporophytic self-incompatibility (SI), which is inferred to have been ancestral in the group based on its patterns of occurrence among continental and Hawaiian members of the Madia lineage and the theoretical difficulty of reconstructing SI once lost. Inability of self-incompatible plants to set seed from self-pollination may mean that multiple seeds of the original colonist were dispersed together. Original introduction of multiple individuals can be inferred for various other indigenous plants (and animals) of remote oceanic islands, such as ancestrally dioecious species, which bear only pollen or ovules (not both) on individual plants. Among plants of the most remote oceanic islands, the silversword alliance has been regarded as a rare exception to Baker's Rule (that colonizing plant species tend to be self-compatible), although examples of SI plants on remote oceanic islands continue to be documented. SI is absent and presumed to have been lost in some members of the silversword alliance (e.g., *Dubautia scabra*) that pioneer new habitats produced by recent lava flows, in keeping with Baker's Rule.

Enforced outcrossing (from SI) and ancestral allopolyploidy (= hybrid polyploidy) of the original colonist species from North America may have helped to counter potentially deleterious consequences of excessive inbreeding and contribute to adaptive radiation of the silversword alliance. Although no polyploids with the same genomic constitution as the Hawaiian species are known among the continental tarweeds, both continental lineages that contributed a genome to the tetraploid ancestor of the silversword alliance are still extant. Vigorous hybrids have been produced between those two continental tarweed lineages (i.e., between *Anisocarpus scabridus* and *Carlquistia muirii*), and the resulting plants produced some viable, diploid pollen—the raw material for producing allopolyploid progeny—thereby verifying the biological potential for allopolyploidization between the inferred ancestral diploid lineages.

Other inferences about ancestral characteristics of the silversword alliance are aided by comparison of the group to the continental tarweed members of the Madia lineage. For example, all of the continental species in the Madia lineage are herbaceous, unlike the woody or semi-woody members of the Hawaiian group; evidently, a switch from herbaceousness to woodiness and an associated increase in plant stature accompanied colonization of the islands, as has been often inferred for island plants.

Invasive non-native plants and animals, habitat loss, and other human-related factors endanger diversity in the silversword alliance. Feral mammals, such as mouflon, goats, and pigs, consume or uproot plants and have been especially destructive to true silverswords and greenswords. The Haleakala greensword, *Argyroxiphium virescens,* was evidently driven to extinction by such impacts. Extensive fencing of habitats to exclude ungulates and an intensive, genetically informed breeding and outplanting program have been highly effective at reversing population declines of silverswords. An emerging threat is destruction of the main pollinators of silverswords—native yellow-faced bees (*Hylaeus*)—by introduced Argentine ants (*Linepithema humile*). Strategies similar to those used for silversword recovery, in addition to aggressive efforts to control invasive plants, may be necessary to prevent extinction of endangered species of *Dubautia* and *Wilkesia* as well, some of which are known from fewer than 50 surviving individuals.

SEE ALSO THE FOLLOWING ARTICLES

Adaptive Radiation / Dispersal / Hawaiian Islands, Biology / Inbreeding / Invasion Biology / Pigs and Goats

FURTHER READING

Baldwin, B. G. 2006. Contrasting patterns and processes of evolutionary change in the tarweed-silversword lineage: revisiting Clausen, Keck, and Hiesey's findings. *Annals of the Missouri Botanical Garden* 93: 64–93.

Carlquist, S., B. G. Baldwin, and G. D. Carr, eds. 2003. *Tarweeds & silverswords: evolution of the Madiinae (Asteraceae)*. St. Louis: Missouri Botanical Garden Press.

Carr, G. D. 1985. Monograph of the Hawaiian Madiinae (Asteraceae): *Argyroxiphium*, *Dubautia*, and *Wilkesia*. *Allertonia* 4: 1–123.

Forsyth, S. A. 2003. Density-dependent seed set in the Haleakala silversword: evidence for an Allee effect. *Oecologia* (Berlin) 136: 551–557.

Friar, E. A., L. M. Prince, E. H. Roalson, M. E. McGlaughlin, J. M. Cruse-Sanders, S. J. DeGroot, and J. M. Porter. 2006. Ecological speciation in the East Maui-endemic *Dubautia* (Asteraceae) species. *Evolution* 60: 1777–1792.

Lawton-Rauh, A., R. H. Robichaux, and M. D. Purugganan. 2003. Patterns of nucleotide variation in homoeologous regulatory genes in the allotetraploid Hawaiian silversword alliance (Asteraceae). *Molecular Ecology* 12: 1301–1313.

Purugganan, M. D., and R. H. Robichaux. 2005. Adaptive radiation and regulatory gene evolution in the Hawaiian silversword alliance (Asteraceae). *Annals of the Missouri Botanical Garden* 92: 28–35.

Robichaux, R. H., G. D. Carr, M. Liebman, and R. W. Pearcy. 1990. Adaptive radiation of the Hawaiian silversword alliance (Compositae—Madiinae): ecological, morphological, and physiological diversity. *Annals of the Missouri Botanical Garden* 77: 64–72.

SKY ISLANDS

JOHN E. MCCORMACK, HUATENG HUANG, AND L. LACEY KNOWLES

University of Michigan, Ann Arbor

Sky islands are high-elevation habitats that are geographically subdivided and isolated among different mountain ranges (Fig. 1). Because of differences in climatic history and dispersal dynamics, the ecological and evolutionary processes and patterns characterizing sky islands may not always parallel those for traditional oceanic archipelago systems. Nevertheless, like their oceanic counterparts, sky islands are generators of diversity over multiple spatial and temporal scales and offer considerable potential for investigating how different evolutionary processes such as natural selection and genetic drift lead to species formation.

BIOGEOGRAPHY OF SKY ISLANDS

The term "sky islands" was originally coined by Weldon Heald in his writings on the mountains of southeastern Arizona, which harbor isolated montane forest or woodland surrounded by a sea of desert (Fig. 2). Analogous to the water of oceanic archipelagoes, low-elevation habitat acts as a barrier to dispersal on sky islands, facilitating divergence of isolated populations. The general concept of sky islands has been expanded to include a variety of settings in which high-elevation habitats are separated by inhospitable lowlands, such as plateaus, ridges, páramos, and alpine meadows.

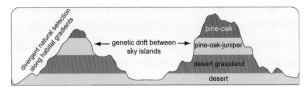

FIGURE 2 Schematic of the altitudinal distribution of habitats and examples of the evolutionary forces acting within and between typical Madrean sky islands of the southwestern United States and northern Mexico.

There are roughly 20 sky island complexes in the world (Table 1), each with a distinct origin, spatial arrangement, age, and climate history. As with oceanic archipelagoes, the geologic processes forming sky islands, which include volcanic activity, erosion, and mountain uplift, provide a temporal reference for determining the age of the islands. These differences have important consequences for evolution in sky islands. For example, ancient sky islands, such as those in the Pantepui region of Brazil, Guyana, and Venezuela, might be expected to show higher levels of divergence and species diversity than would younger sky islands. The Pantepui region formed through erosion of a formerly more widespread plateau that covered a large portion of the Guiana Shield. Rivers gouge this plateau, creating isolated, high-elevation mesas (tepuis), which harbor cool, humid subtropical forests among the surrounding tropical lowlands. Many of the endemic species and subspecies found in this region are thought to have evolved in situ (i.e., they are not immigrants from non–sky island habitats) through isolation among tepuis. Similarly, the sky islands of the Cameroon Line in western Africa, which harbor many endemic species, were uplifted from lowland tropical forest during the Cretaceous

FIGURE 1 Map of the sky islands from the Rocky Mountains of northwestern North America. The distribution of the most likely habitat for montane grasshoppers in the genus *Melanoplus* is shown in white, as determined by ecological niche-modeling based on sampled populations (blue circles).

TABLE 1
Sky Island Complexes Discussed in the Text

Name	Region	Description
Madrean archipelago	Southwestern United States (Arizona, New Mexico, and western Texas); Northern Mexico (Sonora, Chihuahua, and Coahuila)	Complex of 30–40 forested mountain ranges in the desert lowlands between the Rocky Mountains and Mexican Sierra Madres
Great Basin archipelago	Southwestern United States (Nevada and Western Utah)	Complex of ~200 mountain ranges arranged in close proximity surrounded by Great Basin desert
Rocky Mountains	Western United States (Colorado, Wyoming, Idaho, and Montana); Western Canada (British Columbia and Alberta)	Hundreds of isolated high-elevation meadows within a matrix of lower-elevation boreal forest
Ethiopian Highlands	Northeastern Africa (Ethiopia and Western Eritrea)	Mountain massif sundered by a rift valley with satellite forested mountains isolated by lowland desert and savanna
East African Arc	Eastern Africa (mainly Tanzania)	Circular complex of ~20 mountains (some of volcanic origin) harboring rain forest surrounded by arid woodland
Cameroon Line	Western Africa (Cameroon)	Linear chain of ~10 mountain ranges of volcanic and tectonic origin that also includes oceanic islands in the Gulf of Guinea
Annamite Range	Southeastern Asia (Vietnam and Laos)	Large cordillera and nearby satellite ranges of tropical rain forest and higher-elevation evergreen forest
Western Ghats	Southern India	Extremely isolated chain of ~5 mountain peaks harboring humid rain forest surrounded by arid lowlands
Pantepui region	Venezuela, Guyana, and northern Brazil	> 25 relictual plateaus of cool subtropical forest bisected by rivers and surrounded by tropical lowlands

NOTE: Many other sky island systems exist worldwide, as do other fragmented high-elevation habitats with characteristics similar to those of sky islands.

through tectonic and volcanic activity, and their large diversity of endemic species have apparently accumulated over this long history by in-situ speciation. Interestingly, oceanic islands in the nearby Gulf of Guinea formed concurrently through the same processes. The mountain isolates of the East African Arc, the Western Ghats in India, and the Annamite Range of Vietnam and Laos are also relatively old sky islands where differentiation among sky islands has promoted diversification. In contrast, the sky islands of North America, especially those in the Rocky Mountains, are relatively young geologically, on the order of 40–70 million years. Despite their short history, they have still promoted species diversification and have accumulated an impressive level of diversity.

Species in many sky island complexes, especially those in temperate regions, have experienced multiple and frequent climate-induced distributional shifts in association with the Pleistocene glacial cycles. This dynamic history contrasts with oceanic island systems, where inter-island dispersal is fairly rare and the geographic configuration of species across islands remains relatively constant over geologic time. Glacial cycles occurred regularly throughout the Pleistocene and intensified in the last 700,000 years, causing dramatic latitudinal and elevational shifts in species' distributions. During past glacial periods, displacement of sky island habitats to lower elevation meant that formerly isolated populations came into contact. This retreat into refugia was then followed by recolonization of the isolated mountaintops as climate warmed during relatively short interglacial periods. Tropical sky islands—for example those in the Cameroon Line and the Ethiopian highlands—were also affected by Pleistocene glacial cycles through global cooling, although the effect was likely moderate compared to that occurring at higher latitudes. A dynamic history of climate change could potentially explain why the sky islands of the relatively high-latitude Madrean archipelago in the southwestern United States and northern Mexico have low levels of endemism despite harboring high diversity by virtue of lying at a crossroads of several biogeographic regions. In contrast, the climate regime of the sky islands of the endemic-rich East African Arc is thought to have been quite stable throughout the Pleistocene, thanks to the ameliorating influence of Indian Ocean currents.

In contrast to oceanic islands, colonization of sky islands can also occur through niche shifts from neighboring low-elevation habitats or through short-distance migration, in addition to long-distance colonization from other sky islands. For this reason, the dynamics of species assemblage in sky islands differ in many ways from the equilibrium theory of island biogeography set forth by MacArthur and Wilson in 1967, which postulates that island species diversity is a balance between colonization (distance between habitat patches) and extinction (size of patch). Research by Brown and colleagues showed that species composition of mountaintop mammals in the

Great Basin was largely governed by extinction and that the type of habitat that potential colonizers had to traverse to reach a new island, rather than geographic distance per se, determined when colonization by immigration was likely. Studies on birds were more consistent with predictions of the theory of island biogeography, showing an effect of habitat area and degree of isolation on species diversity and endemism. One particularly extreme example is the Sierra del Carmen of northern Coahuila, Mexico, an isolated sky island that shows strong effects of insularity typical of oceanic islands, including reduced species diversity and resulting ecological niche expansion, and augmented density of several resident bird species.

SKY ISLANDS: WINDOWS INTO EVOLUTIONARY PROCESS

Sky islands, like oceanic islands, are ideal settings for evolutionary study. The replicated spatial arrangement of similar, isolated habitats provides a natural experiment for determining what ecological and geographic features facilitate diversification. Selection and genetic drift are both particularly powerful in island systems: The isolation between islands is conducive to divergence by genetic drift, whereas opportunities for selection may occur because of differences in ecology or factors relevant to sexual selection among islands.

Additionally, divergence by natural selection might occur across altitudinal gradients within individual sky islands. For example, the Madrean sky islands are distributed among 30–40 mountain ranges of various sizes and orientations that link the Sierra Madre massifs with the Rocky Mountains in stepping-stone fashion. The present configuration of habitats within these sky islands is postglacial (less than 20,000 years ago). Whereas relatively contiguous woodlands covered the region during glacial periods, there is greater habitat contrast in the Madrean sky islands today: arid oak woodlands in low-elevation canyon, mid-elevation pine-oak forest, and boreal forest in some of the highest-elevation sky islands (Fig. 3). These ecological transitions are comparable to changes manifest at much larger geographic distances (e.g., over hundreds of kilometers in latitude), yet in sky islands they occur over just a few kilometers—and within the dispersal capacity of most species. Plant and animal populations that span these ecological transitions experience drastically different environments, both abiotic (e.g., rainfall and temperature) and biotic (e.g., food resources, competitors, and predators), which can result in divergent selective pressures. If strong enough, selection can generate observable differences between populations that are connected by gene flow. This "gradient model" of divergence has been validated by empirical studies showing elevational diversification of ecologically divergent sibling species as well as population-level divergence in adaptive morphology.

FIGURE 3 The Chisos Mountains, a sky island in Big Bend National Park in southwestern Texas and part of the Madrean archipelago. Chihuahuan desert gives way to desert grassland at mid-elevation and pine forest at high elevation. The nearest sky island is the Sierra del Carmen in northern Coahuila, Mexico, approximately 60 km away.

SKY ISLANDS AS GENERATORS OF DIVERSITY

Much debate surrounds how species diversification has been promoted in sky islands. The likely effect of isolation among sky islands on the formation of new species is not controversial; however, the effect of the Pleistocene glacial cycles on species divergence is a contentious issue. Because of the relatively short amount of time between glacial cycles, there are disagreements over whether diversification would have been inhibited or promoted. Although divergence is expected within sky islands from the divergent selective pressures generated from altitudinal habitat differences, as well as from opportunities for differentiation among isolated sky islands by genetic drift and/or selection, differentiation among populations that accumulated during one glacial cycle may have been lost during subsequent distributional shifts. Alternatively, if sufficient reproductive isolation evolved in any one cycle, speciation could actually have been promoted by the frequent distributional shifts, as hypothesized by the "species pump" model of diversification.

Molecular data play a critical role in distinguishing among evolutionary scenarios, as neutral genetic markers provide a means for estimating the timing of divergence and reconstructing the history of divergence across the sky island landscape. In the Madrean archipelago, fossilized plant material from packrat (*Neotoma* spp.) nests has allowed for a thorough reconstruction of paleohabitats through the last glacial maximum. According to these data, fragmentation of sky islands occurred approximately 10,000 years ago, offering a hypothesis that can be tested with intraspecific genetic data from sky island species. So far, molecular studies in this region have produced conflicting results on the timing of divergence, with some species showing no evidence for postglacial divergence among sky islands and other species showing evidence for divergence much older than the last glaciation. This discord between paleoecological and genetic data can be attributed in part to the difficulty of detecting and timing recent genetic divergence. Large sample sizes and thorough regional sampling will undoubtedly be necessary to thoroughly test this hypothesis.

What has emerged from the many molecular phylogenetic and phylogeographic studies on sky islands is that some species show strong patterns of differentiation that are coincident with the glacial cycles, whereas divergence predates the Pleistocene in other taxa. For example, in the northern Rocky Mountains of North America, some insect groups, such as flightless montane grasshoppers in the genus *Melanoplus*, radiated during the Pleistocene. Genetic data indicate that divergence among multiple glacial refugia, as well as differentiation during the recolonization of sky islands, was facilitated by genetic drift. However, sexual selection has also been implicated in this diversification because of the rapid divergence in male genitalia, which can play a role in reproductive isolation. Sexual selection has also likely been important to incipient speciation among populations of a jumping spider (*Habronattus pugillis*) in the Madrean archipelago. Populations on different mountains have divergent phenotypes, seismic songs, and courtship displays, which result in mating incompatibilities between forms. Also in the Madrean archipelago, pollinator-mediated natural selection seems to be influencing morphological differentiation among populations of a perennial plant (*Macromeria viridiflora*). Thus, in North American sky island systems, natural selection, sexual selection, and genetic drift all appear to be important evolutionary processes driving Pleistocene diversification. However, the timing of speciation in other taxa from the Rocky Mountains—most notably the mammals—predates the Pleistocene, suggesting that divergence in these species may have been inhibited by the frequent displacements from their sky island habitats.

A pattern of pre-Pleistocene origination also has been found in a number of tropical bird and mammal species, casting doubt on the hypothesized "species pump" model of diversification among tropical sky islands. Nevertheless, diversification of tropical taxa across sky islands, albeit pre-Pleistocene, demonstrates how this geographic setting is conducive to divergence. More examples of divergence among sky islands in tropical regions will undoubtedly be uncovered with further study and will be important for confirming that tropical sky islands do not conform to the "species pump" model of diversification.

Regardless of the timing of diversification, sky islands are clearly promoters of diversification from the level of the population to the species. Despite the fact that our most thorough descriptions of genetic and phenotypic divergence in sky islands come from temperate North America, this sampling bias should not be taken as evidence that tropical sky islands are not important centers of diversification. In the Pantepui region of South America, for example, passerine birds (genus *Myioborus* and *Atlapetes*) have differentiated at both the subspecies and species level among the isolated plateaus. Similarly, populations of poison dart frogs inhabiting sky peninsulas in central Peru have diverged rapidly in coloration despite close proximity. In a particularly thorough study of the starred robin (*Pogonocichla stellata*) on sky islands in the East African Arc, Rauri Bowie and colleagues showed how considerable genetic diversification has accumulated among phenotypically divergent sky island populations. Further studies promise to reveal more about patterns and timing of diversification in tropical sky islands, which have been little explored compared to their temperate counterparts. Lastly, such information will also provide an interesting counterpart to the considerable attention given to the diversification across isolated tropical habitat fragments, such as the remnants of previously widespread wet-tropical forest fragments found in North Queensland, Australia.

IMPACT OF HUMAN-INDUCED CLIMATE CHANGE ON SKY ISLANDS

Climate change poses a unique threat to sky islands. Temperature increases of as little as a few degrees could push sky island habitats to higher elevations, reducing their area and potentially causing local extinction of endemic taxa and divergent populations harboring unique genetic and phenotypic diversity. Sky islands in North America and Mexico are already being affected by climate change, with increases in drought, fire, and outbreaks of invasive insects. Although these resilient systems have endured large-scale shifts in climate throughout the Pleistocene, the pace of human-induced climate change may represent an insurmountable challenge for sky islands, with potentially devastating consequences to their biodiversity and evolutionary potential.

SEE ALSO THE FOLLOWING ARTICLES

Fragmentation / Island Biogeography, Theory of / Pantepui / Refugia / Sexual Selection

FURTHER READING

Bowie, R. C. K., J. Fjeldså, S. J. Hackett, J. M. Bates, and T. M. Crowe. 2006. Coalescent models reveal the relative roles of ancestral polymorphism, vicariance, and dispersal shaping phylogeographic structure of an African montane forest robin. *Molecular Phylogenetics and Evolution* 38: 171–188.

Brown, J. H. 1978. The theory of insular biogeography and the distribution of boreal birds and mammals. *Great Basin Naturalist Memoirs* 2: 209–228.

Heald, W. F. 1951. Sky islands of Arizona. *Natural History* 60: 56–63, 95–96.

Lovett, J. C., and S. K. Wasser. 1993. *Biogeography and ecology of the rain forests of eastern Africa*. Cambridge: Cambridge University Press.

Mayr, E., and W. H. Phelps Jr. 1967. The origin of the bird fauna of the south Venezuelan highlands. *Bulletin of the American Museum of Natural History* 136: 273–327.

Pielou, E. C. 1991. *After the Ice Age: the return of life to glaciated North America*. Chicago: University of Chicago Press.

Smith, T. B., K. Holder, D. Girman, K. O'Keefe, B. Larison, and Y. Chan. 2000. Comparative avian phylogeography of Cameroon and Equatorial Guinea: implications for conservation. *Molecular Ecology* 9: 1505–1516.

Warshall, P. 1995. The Madrean sky island archipelago: a planetary overview, in *Biodiversity and management of the Madrean archipelago: the sky islands of the southwestern United States and northwestern Mexico*. L. DeBano, P. Ffolliott, A. Ortega-Rubio, G. Gottfried, R. Hamre, and C. Edminster, eds. Fort Collins, CO: USDA Forest Service Rocky Mountain Forest and Range Experiment Station, 6–18.

SNAILS

SEE LAND SNAILS

SNAKES

GORDON H. RODDA

U. S. Geological Survey, Fort Collins, Colorado

Snakes (3000+ spp.) are a highly specialized and successful limbless form of lizard. Their low metabolic rate combined with jaw anatomy that accommodates the ingestion of relatively prodigious meals allows snakes the energetic option of long fasts between large meals, permitting them to rely on infrequent food sources such as annually nesting birds. Avian and mammalian predators require more continuous food sources and therefore cannot survive on islands lacking year-round prey. Accordingly, snakes are the top predator on many islands, especially land-bridge islands that experience pulses of visiting birds. However, snakes are rare on most oceanic islands because they are not well suited for dispersing across saltwater. Snakes may dramatically rearrange the vertebrate ecology of formerly snake-free oceanic islands when they are introduced by humans.

EVOLUTION

Snakes were the last major group of reptiles to appear, evolving from a varanid lizard–like progenitor at least 135 million years ago. Because the eyes of snakes are radically different from those of their immediate lizard ancestors, proto-snakes are believed to have passed through an evolutionary transition in which their eyes degenerated as a result of living underground or in murky water. Returning to the aboveground environs suited them well and set off an adaptive radiation yielding over 3000 modern species, covering all habitats except high arctic and cold marine environments.

What are the ingredients for this evolutionary success? One is a superbly developed ability to smell prey, perhaps a benefit retained from the time spent underground or in murky water. Another is the loss of legs. Leglessness facilitates locomotion underground, in vegetative thickets, and in water. Another is their ectothermic heritage, which saved them the expense of having to feed continuously to generate heat. The low metabolic cost of ectothermy allows a predatory style that emphasizes stealth, patience, and surprise over speed and endurance. Mammals and birds are incessantly in motion, searching insatiably for each next bite of food; snakes are well suited to the complementary lifestyle: patiently conserving energy while waiting for a frenetic endotherm to wander within striking range. A less obvious asset of snakes is their very light and supple jaws, which arose in the course of ophidian evolution to permit the ingestion of extraordinarily large meals (at maximum, more than 100% of their body mass). These lightweight jaws are unsuited for biting prey into submission, so snakes developed constriction and venom (or perhaps their possession of venom and constriction enabled the concurrent evolution of lightweight jaws).

Thus, snakes are often successful by virtue of visual- and olfactory-targeted sit-and-wait predation from the security and concealment of water, a burrow, or a thicket. Sit-and-wait snakes generally subdue their prey by venom or constriction. Other snakes are successful by patrolling for relatively defenseless prey, such as eggs, that are large but rarely found: Having low metabolic needs, the snake can afford to wait for the rare but highly rewarding (i.e., large) prey item.

THE ROLE OF SNAKES ON ISLANDS

Snakes have limited aerobic capacity, are poor long-distance swimmers, and, of course, cannot fly, so they are not particularly well suited for dispersal to islands that have never been connected to land (i.e., oceanic islands). A snake drifting across the surface of the ocean without concealment in vegetation would also be highly vulnerable to marine predators. Most reptiles on oceanic islands are presumed to have reached their destination by rafting on flood-swept vegetation, usually in the form of eggs buried in earth or tree hollows. Should an adult succeed in making the overwater trip, a single gravid or pregnant female could start a new population. One snake species (the flowerpot snake, *Ramphotyphlops braminus*) is parthenogenetic, allowing each individual to clone itself on arrival to a new island; this species is pan-tropical and is found on many islands. However, most snakes are oviparous and lay eggs with leathery (semi-permeable) shells; these rarely withstand long contact with seawater. Thus, there are many remote island chains that have no native snakes. For example, only the Pacific tree boa (*Candoia bibroni*) colonized the central Pacific Ocean islands, and it did not manage to disperse eastward from Samoa (it may have reached even Samoa only with assistance from prehistoric humans).

Should they reach an island, snakes are well suited for island living, especially if the island has suitable prey that are available infrequently. Many Mediterranean, East Indian, and West Indian islands, for example, experience annual or semi-annual pulses of migratory bird visitation. Seabirds usually visit their breeding islands at a particular time of year. Before the seabird nestlings are fledged, the chicks are often superabundant and relatively defenseless against snakes. Medium-sized mammals are often a predatory threat to snakes, but mammals disperse poorly to islands and are particularly poorly suited to survival on islands with infrequent pulses of food. Thus, snakes can often do very well on islands, if they can get there.

NATIVE SNAKES ON ISLANDS

A few sea snakes take refuge on small islets to rest and breed. Most sea snakes are live bearers and have no need to ever crawl onto land, but the half-dozen species of the Australasian sea krait genus *Laticauda* are less fully evolved for a marine existence and—like sea turtles—must crawl up on land once a year to lay eggs. They also may come ashore to mate and rest. Sea kraits can be impressively abundant on the islands where they mate, but they are of little energetic consequence to the island's terrestrial ecosystem because they forage exclusively offshore.

Native terrestrial snakes have been studied on a wide variety of islands with simple food chains. On large land-bridge islands with complex food webs, snakes have an energetic role much as they do on adjacent continental land masses. However, on small or oceanic islands, which often have simplified food chains, snakes may take advantage of their unique adaptive advantages to attain extraordinary densities. Conditions that benefit a population may not be favorable for the constituent individuals, however: At the high densities achieved, the competition for food is intense, and each individual snake may have difficulty acquiring enough energy for rapid growth. This may lead to insular dwarfism. The proximate cause for small size on an island may be a shortage of food (phenotypic response) or a genetic disposition for reduced growth (which improves survival in the face of limited food availability).

Conversely, the food that is available on an island may be of a size requiring a larger body size (e.g., if large birds such as boobies are the main food source). In this case, insular gigantism may occur. Both dwarfism and gigantism have been widely noted of snakes on islands, but in only a few cases is it known whether the observed size differences (in comparison to the nearest mainland population) are due to phenotypic or genotypic adaptation. In the best studied case, there was a strong interaction: The island snakes had a genetic disposition to allow them to take maximal advantage of large prey sizes, but they achieved large size phenotypically in response to elevated intake. Snake species tend to diversify on islands by evolving adaptations to the local environment, but those locally adapted species rarely colonize adjacent islands, as is characteristic of adaptive radiations of other taxa such as island birds or island lizards.

SNAKES INTRODUCED TO ISLANDS BY HUMANS

Most human-introduced snake populations on islands probably arose accidentally. For example, the flowerpot snake has been accidentally introduced around the world in association with potted plants and soil. Because this diminutive, termite-eating subterranean snake is rarely seen—and because termites do not have many defenders—this introduction is rarely noted and less often lamented. Horticultural shipments may be an important pathway for the introduction of snakes, as plant products are probably responsible for colonization of several introduced snakes in the West Indies. Corn snakes, *Pantherophis* (= *Elaphe*) *guttatus,* probably concealed in ornamental plants shipped from Florida, have become established on Anguilla, Bonaire, Curaçao, and the Cayman Islands.

One of the strangest snake introductions occurred around 1971 on the island of Cozumel, Mexico, following the filming of a movie. Non-native *Boa constrictor* were brought to the island for the filming and were subsequently released. Twenty years later, the residents noted the extraordinary abundance of the new boas and the disappearance of a number of Cozumel's unique wildlife species. More than a dozen endemic subspecies are believed to be at risk, including birds, mammals, and lizards. Boas have subsequently been released and are presumed established on Curaçao and Aruba in the Dutch West Indies.

Perhaps the frequent traveler award for snakes should go to the rear-fanged wolf snake, *Lycodon aulicus*, named for its anterior teeth, which resemble those of a wolf. Introduced to several places in the Mascarene Islands during the nineteenth century and to Australia's Christmas Island around 1987, it is also suspected of having been introduced at an unknown date to the Philippine Islands, and possibly to Java, Borneo, and Sumatra. The wolf snake is variable in size and diet, but on Christmas Island, it eats primarily lizards (of which Christmas Island has five vulnerable endemic species, as well as an endemic shrew). The wolf snake is suspected of causing multiple lizard extirpations throughout the Mascarene Islands, especially on Mauritius.

The oldest reported snake introductions are of snakes apparently placed on Mediterranean islands by the Romans or Carthaginians. Scientific attention has focused on the threats to native wildlife in Spain's Balearic Islands (Mallorca, Menorca, etc.). Two snake species are involved: *Macroproton cucullatus*, the false smooth snake, and *Natrix maura*, the misleadingly named viperine snake (a nonvenomous water snake unrelated to vipers). The false smooth snake is associated with the endangerment of an endemic lizard, Lilford's wall lizard, *Podarcis lilfordi*, and perhaps other species. The viperine snake is best known as the threat to the continued survival of the endemic Mallorcan midwife toad, *Alytes muletensis*. The wall lizard is listed by the International Union for Conservation of Nature (IUCN) as endangered, and the toad is judged vulnerable. Both prey species persist only in highly restricted ranges in the Balearic Islands, suggesting that their ecological fates hang in the balance. This uncertainty is truly extraordinary given that the exotic predators were introduced about 2000 years ago. Presumably other prey species were eliminated more quickly. The tenuous persistence of the toad and lizard suggest that the time course of island extinctions may be vastly longer than the usual duration of scientific studies, a cautionary tale for scientists wishing to conclude that a given introduction is harmless as no negative impacts have yet been demonstrated.

Perhaps the earliest example of ecoterrorism may be manifest in the occurrence of *Vipera aspis*, the asp viper, on the island of Sicily. Some believe that the viper was introduced to Sicily by the Carthaginians during their conquest of the island in the years 398–368 BC. The basis for the speculation is that the Carthaginians were known to load a small boat with a collection of venomous snakes and push the boat toward enemy ships as a means of terrorizing their opponents prior to combat. In the case of an island to be conquered, the victors would then be rewarded with acquiring their new territory infested with an introduced venomous snake, a rather questionable achievement and perhaps an early example of tactical cleverness untempered by consideration of the strategic consequences.

The Brown Tree Snake

The best documented snake introduction comes from the island of Guam, in the Mariana Islands of the western Pacific, and it too involves a snake translocated by naval forces (but this time inadvertently). The brown tree snake, *Boiga irregularis*, was accidentally introduced to Guam in war materiel salvaged from the New Guinea area immediately after World War II. For at least 35 years, the snake population growing in the southern part of the previously snake-free island was a subject of interest but not concern. The concurrent disappearance of virtually all bird life from the southern portion of the island attracted little attention until the 1980s, by which time the imminent extinction of Guam's native forest birds set off a frantic search for the disease that was assumed to be killing the birds.

A comprehensive search turned up no diseases that could account for the bird disappearances. Consideration then turned to pesticide contamination and other possible causes, again to no avail. Only when all other reasonable hypotheses were rejected did ecologists begin to accept the notion that an introduced snake could be responsible. Resistance to the snake hypothesis was based on the presumption that snakes were too rare and ecologically inconsequential to extirpate species. In this particular case, it turned out that not only was the brown tree snake responsible for the demise of virtually the entire forest avifauna of Guam (ten of 12 species have been extirpated to date, with the two other species teetering on the brink), but it was also responsible for extirpating mammals (three bats were lost,

FIGURE 1 Brown tree snakes reach high densities in their introduced population on Guam. These were collected along 0.3 km of road edge in 1989. Photograph by G. H. Rodda, USGS.

though it is still unclear what caused the first two species to disappear) and many of the island's lizards (between one and five of the island's nine to 12 native lizards disappeared in association with the snake).

Ecologists are well aware that introduced mammals frequently wipe out island populations of birds; why then was it so hard for ecologists to accept similar culpability on the part of a snake? This is a problem for psychologists, but ecology played a role. Recall that the evolutionary success of snakes hinges on their cryptic and stealthy predatory mode. Most prey individuals never saw the snake that bit them (until it bit them). Humans are no better at spotting snakes; some humans are ophiophobic because snakes are so invisibly omnipresent. As a consequence, ecologists frequently underestimate the abundance of snakes in an area. This was the case in Guam (Fig. 1), where the brown tree snake was later found to have reached extraordinary population densities at its peak (in excess of 100/ha). This was more than enough to wipe out the island's birds (estimated to have totaled about 30/ha for all species combined in optimal habitat), but even at their peak density the snakes were largely invisible, as they are active only at night and usually in heavily vegetated areas where the very slow-moving, vinelike, drab brown snakes are very difficult to see. Despite high snake densities persisting on the island, Guam's visitors and residents rarely see brown tree snakes.

The example of the brown tree snake on Guam revolutionized ecologists' understanding of the potential importance of snakes in island ecosystems. All island ecosystems are at risk from the introduction of new snake species, but especially vulnerable are ecosystems such as Guam that evolved without any snakes present. Unfortunately, the difficulty of detecting snakes at low densities continues to challenge managers keen on preventing colonizations of snakes on islands. The evolutionary innovation that has made snakes so successful—their invisibility and tolerance of long periods without food—is also the feature that makes them one of the most difficult invasive species to prevent or manage.

SEE ALSO THE FOLLOWING ARTICLES

Cozumel / Introduced Species / Lizard Radiations / Mascarene Islands, Biology

FURTHER READING

Fritts, T. H., and G. H. Rodda. 1998. The role of introduced species in the degradation of island ecosystems: a case history of Guam. *Annual Review of Ecology and Systematics* 29: 113–140.
Greene, H. W., M. Fogden, and P. Fogden. 1997. *Snakes: the evolution of mystery in nature.* Berkeley: University of California Press.
Jaffe, M. 1994. *And no birds sing.* New York: Simon and Schuster.
Quammen, D. 1996. *The song of the dodo.* New York: Scribner's.
Rodda, G. H., T. H. Fritts, and D. Chiszar. 1997. The disappearance of Guam's wildlife: new insights for herpetology, evolutionary ecology, and conservation. *BioScience* 47: 565–574.
Rodda, G. H., Y. Sawai, D. Chiszar, and H. Tanaka, eds. 1999. *Problem snake management: the habu and the brown treesnake.* Ithaca, NY: Cornell University Press.
Shine, R. 1991. *Australian snakes: a natural history.* Ithaca, NY: Cornell University Press.
Zug, G. R., L. J. Vitt, and J. P. Caldwell. 2001. *Herpetology: an introductory biology of amphibians and reptiles,* 2nd ed. San Diego, CA: Academic Press.

SOCIETY ISLANDS

SEE PACIFIC REGION

SOCOTRA ARCHIPELAGO

KAY VAN DAMME

Ghent University, Belgium

The Socotra Archipelago (Yemen), situated in the Arabian Sea, consists of four main islands on an ancient microplate. Socotra, the largest island, is known as the "Galápagos of the Indian Ocean" because of its high biodiversity in both terrestrial and marine realms. A long period of isolation from the Afro-Arabian mainland and a significant geological diversity have resulted in a high endemism with remarkable relicts.

"JEWEL OF THE ARABIAN SEA"

The Socotra Archipelago is located in the Arabian Sea, situated 380 km southeast from the coast of Yemen, to which it politically belongs, and about 100 km east from the Horn of Africa (Cape Guardafui, Somalia). It consists of one major island in the west, Socotra, about 130 km long and 40 km wide; three smaller islands, Samhah, Darsa ("The Brothers"), and Abd al Kuri, to the east; and a few rocky limestone outcrops inhabited only by birds. The largest island is populated by approximately 43,000 inhabitants (census 2004) with highest concentration in two major towns, the capital Hadiboh and the coastal town Qalaansiyah. The population is rapidly increasing from immigration, resulting in strong urban expansion. Socotra's high cultural and natural diversity and the complexity of conservation attract a global interest. Listed as a UNESCO Man and Biosphere Reserve (2003), a Global 200 Ecoregion by the World Wildlife Fund (WWF), and a Centre of Plant Diversity by Plantlife International, Socotra has become Yemen's national pride, its "Jewel of the Arabian Sea." The country recognizes Socotra's importance in biodiversity. For example, Yemen designated the Detwah Lagoon on Socotra as the country's first Ramsar site (a wetland of international importance) in 2007. Since July 2008, the Socotra Archipelago is officially a UNESCO World Heritage site, a historical event. Seventy-five percent of the land surface is indicated as terrestrial core area and consists mainly of elevated areas, the rest is terrestrial buffer zone. The marine core area is smaller, 7.6% of the total marine area, and aims to protect mainly the coral reefs of the Archipelago.

GEOLOGY OF SOCOTRA

Islands of the Socotra Archipelago (Fig. 1) lie together on a submerged Precambrian basement, the Socotran Platform. This granite microplate of continental origin was formed 700–800 million years ago as part of the Afro-Arabian continent, close to Eastern Oman. At this time, Socotra was situated relatively close to India, until a series of tectonic events in the Mesozoic leading to the breakup of Eastern Gondwana. The latter events caused structural changes in the Socotra Platform and formed the onset of rifting of the Gulf of Aden, which would lead to a final separation of Socotra from the Afro-Arabian mainland. Geological and tectonic evidence suggest that the Socotran Platform remained close to Southern Arabia until oceanic rifting of the Gulf of Aden in the Oligocene–Miocene, with an estimated timing of separation about 20–18 million years ago. Sea floor spreading increased the distance between Socotra and the Arabian mainland since 18 million years ago. The region was subjected to a general uplift since this period, and the Archipelago gradually rose.

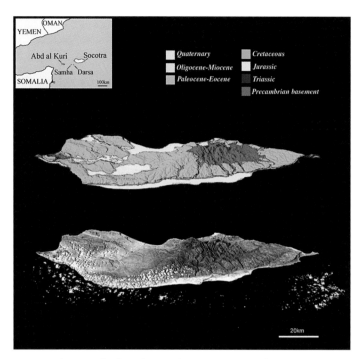

FIGURE 1 Socotra Island, geology and topography, view from the south. Geology shows a Precambrian basement in the central eastern half of the island, surrounded by the dominant Palaeocene-Eocene limestone plateaus. The center (Zahr Plain) and coastal areas consist of lowlands, and two plateaus crop out in the east. Inset: Location of the Socotra Archipelago. Image courtesy of Kay Van Damme, satellite image based on Landsat 7 ETM (2001) draped over USGS DEM.

These tectonic events, together with several large sea transgressions, shaped the islands. Socotra consists of a granite center, cropping out in the Haggeher Mountains (1600 m), which formed as part of the Precambrian basement complex. Although there is no direct geological evidence, several authors believe that these mountains have remained above sea level since their formation and thus provided a refuge for relict taxa. Bordering the Haggeher are a series of limestone plateaus reaching up to 1000 m in altitude (Diksam Plateau), deposited during several marine transgressions in the Cretaceous and Palaeocene-Eocene. These limestone plateaus from the Paleocene–Eocene transgression make up the largest surface of Socotra Island, in most areas strongly karstified and with extensive cave systems. In the center (Zahr Basin), south (Noged plain), and north of the island, these plateaus are bordered by coastal lowlands consisting of Quaternary sands and elevated coral reefs. At periods of lower sea level during the last glacial maximum, most of the Socotra Platform emerged, connecting the main island to Samha and Darsa by land bridges.

Thus there is a marked variety in topography and geological features on Socotra: Precambrian granite outcrops, Paleocene–Eocene limestone plateaus with steep cliffs and large cave systems and Quaternary coastal sandy regions. The smaller islands (Abd al Kuri and The Brothers) are built up mainly of limestone cliffs.

CLIMATE

Because Socotra is situated at the margins of the subequatorial and northern tropical climate belts, its climate is governed by the Intertropical Convergence Zone and related monsoon cycles. Two seasons, the northeast monsoon (January) and the southwest monsoon (July), are most marked, and the intensity of their winds has kept the island isolated during these periods even up to the last decades. Both are quite different; the January monsoon blows dry air on land and creates upwelling of cooler waters in the sea, whereas the July monsoon brings moisture, but parts of the island (the north) are in a rain shadow and remain dry. Annual temperatures fluctuate strongly depending on region because of the differences in topography, ranging between 8 and 31 °C in the mountains (Skand) and from 28 to 43 °C at the coast (Hadiboh). Also, air humidity and precipitation strongly vary with location, some areas depending completely on permanent fog for precipitation. Overall, climate on Socotra depends strongly on place and time, and the island shows both tropical and arid features.

RESEARCH HISTORY

Biological research on Socotra Island started at the end of the nineteenth century, with main expeditions by British botanist I. B. Balfour of the Royal Botanic Gardens of Edinburgh (RBGE) and the zoologists H. Forbes and W. R. Ogilvie-Grant of the Liverpool Museums. Little coordinated research was carried out in the twentieth century until political stability increased accessibility to the island again in the late 1990s. Major expeditions, such as a multidisciplinary expedition by the RBGE and marine surveys by Senckenberg Museum in Germany since 1999. Building on the earlier works, these recent expeditions form the basis for our current faunistic and floristic knowledge of the island and provide an important framework for conservation.

ENDEMISM IN TERRESTRIAL FLORA

Geological history, topography, and microclimates on Socotra, combined with its long period of isolation, lie at the base of a high biodiversity and endemism in terrestrial biota. Of 825 terrestrial plant species recorded, 37% are endemic to the archipelago, with 15 unique genera. In comparison, Mauritius has 31–35% endemic plant species and the Galápagos about 42%, ranking Socotra high in island biodiversity. Distributions of species are relatively limited, with local hotspots of endemism such as the granite mountains, which count over a hundred endemic species. The mainly xeric vegetation has been classified into seven types: coastal mosaic, croton shrubland, succulent shrubland, woody-based herb communities, semi-evergreen woodland, submontane shrubland, and montane mosaic (Fig. 2). Extensive research has been done on the plants, in particular on the ethnobotany. The endemic flora contains several relicts from formerly widespread taxa. The Socotran dragon's blood tree (*Dracaena cinnabari*) is such an example, a local representative of a genus that was more widespread in surrounding regions during the Miocene. *Dracaena* is considered a genus of Tethyan origin based on fossil evidence, now with fragmented distribution because of climatic shifts. Remaining species of *Dracaena* have survived in refugia on Socotra, parts of the Arabian Peninsula, northeastern Africa, and northwestern Africa. Known since antiquity for its red resin, the Socotran dragon's blood tree has become a symbol for the island. Other well-known floristic elements of Socotra are the frankincense trees *Boswellia* (a genus that radiated on Socotra), the medicinal *Aloe,* and the cucumber tree *Dendrosicyos socotrana*, the only arborescent member of the Cucurbitaceae. Most plant species show clear xeromorphic adaptations to the dry and windy conditions governing these islands, others are restricted to wet refugia.

FIGURE 2 Montane shrubland vegetation (altitude 1000 m) at Rewged, Socotra Island, with the typical umbrella-shaped Socotran dragon's blood tree (*Dracaena cinnabari*). The mountaineous areas of Socotra Island form a local hotspot of endemism.

TERRESTRIAL FAUNA

Endemism is high in terrestrial fauna, reaching up to 100% in some groups. Endemism (at species level) for spiders is ~60%, and for isopods 73%. Large groups of insects such as butterflies, dipterans, hymenopterans, grasshoppers, and beetles need taxonomic revision, and at least a third to half of the known species are endemic to the archipelago. Scorpions and Amblypygi are all endemic. Endemism and diversity in the terrestrial molluscs (Fig. 3) is exceptionally high,

FIGURE 3 *"Tropidophora" socotrana* on *Dracaena* wood. About 95% of the land snail species in Socotra are endemic. They evolved and radiated on the archipelago, most likely after separation of Socotra from the Arabian mainland. This particular operculate species of the Pomatiidae lives in close association with the Socotran dragon's blood tree.

with 95% of ~100 species currently described, endemic to the archipelago; 75% of the land snail genera on the archipelago are unique. A large portion was only discovered in the last decade, and more species are being described. High endemism of the Socotran molluscs provided an important argument for its World Heritage nomination.

Dominating the freshwater habitats (wadis) are endemic, semiterrestrial freshwater crabs of the genus *Socotrapotamon*. The dominant species, *Socotrapotamon socotrensis*, is common on the main island. A second terrestrial species, *Socotra pseudocardisoma*, is restricted to rock holes in the high limestone plateaus. A separate evolution can be noted for cave invertebrates, of which seven species are unique to the island and, in several cases, to a particular cave system. The cave systems on Socotra are extensive, reaching 13 km in length (Ghiniba Cave), and harbor endemic species that have found refuge here during dry periods. Among them, a typical group of Gondwana origin, are the whip spiders (Amblypygi), with four endemic species of the genera *Phrynichus* and *Charinus*. Among the stygobionts, several new freshwater crustaceans were discovered from Socotra in the last decade, including the endemic genera *Paradoniscus* and *Dioscoridillo*.

As to vertebrates, primary freshwater fish and amphibians are absent. Their absence may be the result from extinctions during long periods of drought in Socotra's history. In reptiles, about 90% of the 34 terrestrial species are endemic, with all six snake species endemic, and a radiation in the geckos (e.g., *Hemidactylus* and *Pristurus*). Indigenous mammals are absent, though the insectivores (shrews and bats) deserve closer study. For birds, the global importance of Socotra was recognized by Birdlife International, which lists 22 important Bird Areas here. Six species of birds (e.g., the Socotra sunbird) are endemic to Socotra. Several subspecies (e.g., the Socotra buzzard) are under taxonomic research and may prove to be additional true endemic species in the future. For other species, such as the Egyptian vulture and several seabirds, Socotra harbors globally significant populations. In total, Socotra counted 192 bird species in 2006.

NOTE ON AFFINITIES OF THE BIOTA

The origin of the Socotran endemics depends on the group in question. Both vicariance and dispersal hypotheses have been postulated. Although for many groups a link with the Arabian mainland and Africa is clear, Oriental, Afrotropical, and Palearctic elements may also be present. The biogeography of Socotra, because of its ancient continental origin, is a complex matter and should be carefully examined jointly with dispersal capacities of each group. Only few studies have focused thoroughly on this aspect of Socotra, and relatively more work has been done on the floral links. The following are a few examples. In isopods, species composition on Socotra shows interesting biogeographical links with Afrotropical and Oriental regions, whereas important Afro-Arabian groups are missing. A radiation of Oriental isopod species (*Serendibia* and four genera of the Trachelipodidae), groups that are not present in Africa and Arabia, indicate a long evolutionary history and isolation of the isopod fauna of the Socotra Archipelago. Species of *Serendibia* seem to derive from a single common ancestor, and these "oriental" elements are considered survivors of relict taxa, no longer present on the surrounding mainland. The terrestrial molluscan fauna, strongly limited in dispersal, is suggested as ancient as well, of central Gondwanan origin, with virtually no species overlap in distribution ranges between the different islands. A similar (Gondwana) age is suggested for the Amblypygi. Several radiations of the latter groups have taken place on the main island. A relatively more recent (Oligocene–Miocene) age was suggested using molecular clocks for two endemic Socotran snakes belonging to the Colubridae, likely invading from Africa.

The endemics of Socotra contain a mix of biogeographical histories. Dispersal is likely in the majority of terrestrial plants, with closest relationships to the adjacent Afro-Arabian mainland (Eritreo-Arabian floristic subregion). Unraveling the connections is not simple. Relationships of the Socotran *Exacum,* a genus with a vicariance-like pattern, is suggested now to result from long-distance dispersal. Molecular studies confirm vicariance for a few groups. In the cucumber tree *Dendrosicyos,* for example, age of lineage is estimated at ~40 million years (Oligocene), predating separation of the island.

Information on biogeography of Socotra is scattered in phylogenetic studies, not all equally focused, and different scenarios exist for different groups. The number of true relicts versus recent colonizers depends strongly on the group and needs further investigation. In general, true Socotran relicts are considered relatively old (i.e., Miocene or older). Furthermore, a comprehensive geological framework is lacking, with the age of separation of the island only recently becoming clear. Little has been done on the biogeographical connections between the islands of the Socotra Archipelago. Scientists are currently working jointly, across disciplines, on compiling the biogeographical data and investigating main patterns.

ALOES, FRANKINCENSE, AND DRAGON'S BLOOD

Socotra is known for several products that made the island a popular trading center in antiquity. The most common products were tree resins such as frankincense and dragon's blood (for pigment), and sap from aloe. These products are still used. In addition, the island was famous for important components in the perfume industry, such as ambergris (from whales) and musk (from the lesser civet, introduced from India). Nearly all historical sources (e.g., Marco Polo), mention one or another of these products in relation with Socotra, in particular the high-quality incense and aloes. In addition to the use of several plants for trade, Socotri traditionally maintain a vast practical etnobotanical knowledge.

EXTINCTIONS

In contrast to many islands, Socotra has remained in a relatively pristine state until the last decades. The island has been strongly isolated for both political and climatological reasons (strength of the monsoons), and only since 2001 has a long period of isolation been breached. Four plant species have not been recorded since Balfour's collections in the last century and are considered extinct. For the fauna, there is only circumstantial evidence for extinctions. A single historical source, a detailed account from the first century AD (the Periplus), mentions the presence of crocodiles, large lizards, and three species of land turtle on Socotra. Use of the large lizards and one "giant" mountain tortoise are described, the latter being used for making Roman tableware. There is no paleontological evidence on Socotra for the presence of these vertebrates, and the evidence is, as yet, inconclusive as to whether they actually existed here. However, as the Periplus is known for being a quite accurate source, we cannot exclude the possibility that these species, including a giant land turtle, may have gone extinct as a result of human interference. Major risks for extinctions on the island have arisen only in the last decade.

THREATS

Major threats to the island are many. Several that have arisen in the last decade can be expected to have a major impact on biodiversity on Socotra.

Traditional practices, land-use systems, and local laws concerning the environment and use of resources are being lost. Traditionally, fisheries and livestock management of the local communities (fishermen and herders) were aimed at sustaining natural resources. Through centuries of experience, Socotri have become careful stewards of their island, well aware of the meaning of sustainable use. Disruption of traditional ways negatively impacts biodiversity through overconsumption, such as firewood collection, overgrazing, and overfishing. Overgrazing by goats is a major problem, causing severe soil erosion through loss of vegetation. Regeneration is low, in combination with the dry climate, weakening the ecosystem. Some plants such as the cucumber tree may also suffer, being used as fodder in times of scarcity. The breakdown of traditions has a direct effect on overexploitation of the Socotran resources, on which many people depend.

Invasive species are another threat. Several aliens, such as the Indian house crow (*Corvus splendens*), the Norway rat (*Rattus norvegicus*), and four species of plants (e.g., *Argemone mexicana, Prosopis juliflora*) reproduce quickly. The number of invasive species may be relatively low but all need closer monitoring. We expect an increase in rodent populations in lowland areas following expansion of human settlements. Impacts on the terrestrial fauna of Socotra, which evolved in the absence of mammals, are likely to increase. An attempt to introduce rabbits recently was fortunately avoided by the local authorities.

Pollution is caused by improper waste management in the major cities, and the use of chemicals in agriculture. With the increase of population and imports of goods, chemical pollution rapidly increases. Agricultural crops have also been imported recently. Their maintenance leads to the use of pesticides, which until now have stayed clear of Socotra.

Uncontrolled development, clearing, and habitat destruction are a major threat for the island. Infrastructure projects, particularly roads, far exceed local needs and have a direct negative impact on the environment, especially in the coastal regions, where exceedingly wide roads take up a large portion of the surface, even crossing through protected areas. The result of uncontrolled planning, road projects especially should be coordinated to minimize negative impacts on biodiversity.

The number of tourists on Socotra is increasing exponentially, reaching beyond local capacities. Although all effort is put into increasing visitor awareness and attracting ecotourism, the island still attracts a significant proportion of "beach" tourists. Socotra is not that type of destination, but is rather an island with a natural and cultural heritage, globally recognized for its value and importance.

A synergy of threats, together with climate effects, may cause a fast switch in the Socotra ecosystems. The recent nomination as UNESCO World Heritage site is a positive step toward long term conservation and sustainable development of the archipelago and provides an opportunity to address main biodiversity threats.

SEE ALSO THE FOLLOWING ARTICLES

Caves, as Islands / Dispersal / Endemism / Ethnobotany / Sustainability / Vicariance

FURTHER READING

Cheung, C., and L. Devantier. 2006. *Socotra: a natural history of the islands and their people*. K. Van Damme, ed. Hong Kong: Odyssey Books.
Miller, A. G., and M. Morris. 2004. *Ethnoflora of the Soqotra archipelago*. Edinburgh, UK: The Royal Botanic Gardens.
Sohlman, E. 2004. A bid to save the Galápagos of the Indian Ocean. *Science* 303: 1753.
Thiv, M., T. Mats, K. Norbert, and L. H. Peter. 2006. Eritreo-Arabian affinities of the Socotran flora as revealed from the molecular phylogeny of *Aerva* (Amaranthaceae). *Systematic Botany* 3: 560–570.

SOLOMON ISLANDS, BIOLOGY

ORLO C. STEELE

University of Hawaii, Hilo

Within the South Pacific Island nation of the Solomon Islands there are hundreds of islands, ranging from large high volcanic islands to low atolls, with a total of 4023 km of coastline and an economic exclusive zone of 1,340,000 km^2. The biodiversity of the Solomon Islands appears to be the richest among the Pacific Island nations, with the exception of Papua New Guinea. However, the collection of specimens and field observations has been limited, and new species are discovered with each new inventory.

GEOGRAPHIC SETTING

The Solomon Islands extend 1450 km in a southeast direction from 5° S and 152° E to 12° S and 170° E. Most of the 28,785 km^2 of land area is found on six large high volcanic islands, the largest being Guadalcanal with 5,310 km^2. These six high islands are arranged in a double chain that extends 850 km southeastward from the Papua New Guinea island of Bougainville. The northern chain includes (from west to east) the islands of Choiseul, Santa Isabel, and Malaita, while the southern chain is composed of the New Georgia group in the west and the larger islands of Guadalcanal and Makira (San Cristobal) in the east. The maximum elevations of these islands range from 2331 m on Guadacanal to 795 m on Vella Lavella in the New Georgia group.

The Santa Cruz Islands form a second group of medium-sized islands in the southeast part of the country, which are geologically and biologically more related to the Vanuatu islands (formerly known as the New Hebrides). This group is approximately 375 km east of Makira, with its major islands being Nendo, Tinakula, Utupua, and Vanikoro. In addition, the raised limestone islands of Bellona and Rennell are about 250 km to the south of Guadacanal, and the atoll of Otong Java lies approximately 450 km to the north.

TERRESTRIAL BIOLOGY

Vegetation and Flora

The major vegetation types of the large islands include coastal strand, mangrove forests, freshwater swamp forests and herbaceous wetlands, lowland rain forest, seasonally dry forests and grasslands, and montane rain forest. In addition, cloud forests may form along windward slopes at higher elevations. The dominant vegetation on most of the high islands is lowland rain forest, with grasslands occurring on the northern plains and foothills of Guadalcanal.

The flora is primarily composed of Southeast Asian elements and pertains to the same floristic province as Bougainville in Papua New Guinea, with the exception of the Santa Cruz Islands, on which the flora is more closely related to the flora of Vanuatu. There have been approximately 3571 species of flowering plants recorded for the country, with 3.5% (125 species) being endemic. The origin of the Solomon Islands flora has been mostly from short-distance migrations from westward islands, which has not fostered high levels of speciation; however, the genus *Ficus* is an exception, with 35% of the 63 species being endemic. Compared to the islands of Bougainville, the rain forests are relatively species-poor, with the notable absence of the Dipterocarpaceae and Fagaceae, which are important tree families further west. The lowland rainforests are typically dominated by the genera *Pometia, Dillenia, Elaeocarpus, Endospermum, Campnosperma, Calophyllum, Terminalia, Canarium, Agathis, Metrosideros,* and *Sararanga*, all of which are found on mainland New Guinea. Montane rain forests can occur as low as 700 m on wet windward slopes but more typically occur above 1000 m. Common tree genera of the the montane forest include *Syzygium, Metrosideros, Ardisa, Psychotria, Schefflera, Rhododendron,* and *Ficus*. These mountain trees are typically covered with epiphytes, primarily orchids, with an understory largely composed of pandans, gingers, and bamboo species. In freshwater swamp forests, single tree species such as *Campnosperma brevipetiolata, Inocarpus fagifer,* and *Terminalia brassi* may dominate. In contrast to the diversity of upland forests, mangrove forests are relatively rich, with 26 species (43% of the world's mangroves), which are commonly represented by the genera *Rhizophora, Bruguiera,* and *Lumnitzera* (Fig. 1).

FIGURE 1 Kolombangara Island, 1981. View of mangrove forests in the foreground and lowland rainforest ascending the slopes of this cone-shaped volcano. Photograph by Dieter Mueller-Dombois.

Fauna

Among the terrestrial fauna of the Solomon Islands, the birds have been the most well studied, with approximately 173 species of resident birds and 50 species of seabirds recorded. This represents the highest avifauna diversity of all South Pacific nations. The Solomon Islands show very high levels of bird speciation, with 44% of the species and 38% of the subspecies being endemic. Three notable endemics are the fearful owl (*Nesasio solomonensis*), the Solomons sea-eagle (*Haliaeetus sanfordi*), and the megapode (*Megapodius freycinet*).

The native terrestrial mammals of the Solomon Islands are composed entirely of bats and rats. There are eight species of rats, most of which are large (up to 1 kg) and arboreal. Of the 44 species of bats, 26 are flying foxes (*Pteropus* sp.) and 18 are smaller insectivorous species. This represents one of the highest diversities of bats and rats in the world, with 26, or 50%, of these species being endemic.

There are 17 species of amphibians in the Solomon Islands, all of which are frogs, with three genera and 7 species that are endemic to the country. There are 61 species of native terrestrial reptiles (lizards and snakes); three genera and 25 species are endemic. One notable endemic is the Solomon Islands' prehensile-tailed skink (*Corucia zebrata*), which is a very large, arboreal skink that feeds primarily on the leaves of epiphytes.

The terrestrial invertebrate fauna is poorly known, with the butterflies being the best described among the arthropods. There are reported to be 130 species of butterflies in the country with 35% of the species endemic. Many of these species are spectacular bird wings (*Ornithoptera*), which are farmed to supply butterfly collectors. Two of the more prominent and spectacular bird wing species are *Ornithoptera allotae* and *O. victoriae*, which depend upon specialized food plants for their reproduction. In addition, the blue emperor swallowtail (*Papilio ulysses*) is also found in the Solomons. Finally there are reported to be 200–270 species of land snails in the country, but this group requires considerable more study.

FRESHWATER BIOLOGY

The freshwater streams of the Solomon Islands have a diverse assemblage of catadromous fish, molluscs (Neritidae), and crustaceans (Atyidae and Palaemonidae). The icthyofauna include several species of gobies (Gobiidae), sleeping gobies (Eleotridae), river mullet (Mugilidae), and freshwater eels (Anguillidae). Because these organisms are of marine origin, they are capable of tolerating a wide range of saltwater concentrations, with the number of species decreasing further inland. In addition, there are a large number of aquatic insects that live in streams for all or part of their life cycle. Notably abundant are the dragonflies and damselflies (Odonata), caddisflies (Trichoptera), and black flies (Simuliidae).

MARINE BIOLOGY

Corals and Reef Fish

The Solomon Islands have a very high diversity of coral species and are within the "Coral Triangle" of the Indo-Pacific, which has the highest diversity of coral species worldwide. Based on recent field surveys, there are at least 485 species of corals in 76 genera found in the Solomon

Islands which is second only to Indonesia. This high diversity is thought to be due to a wide range of bathometric and current regimes.

There have been 1019 nearshore reef fish species reported from 2–60-m depth in the Solomon Islands in 348 genera and 82 families. This diversity is very high and places the Solomon Islands in the top ten most diverse reef fish areas of the world. Single reef sites may contain over 200 species of reef fish, with the highest recorded on the island of Gizo in the New Georgia Province. Sixty percent of the reef fish are found in ten families, with gobies, damselfishes, and wrasses being the richest and most abundant groups. Because of the broad dispersal capacities of these fish, there are only two endemic species of reef fish in the Solomon Islands (Fig. 2).

FIGURE 2 White-bonnet clownfish, *Amphiprion leucokranos*, 2004. This species is restricted to the western central Pacific and is associated with anemones. Photograph by Gary Allen.

Reptiles

There are five species of sea turtles in the Solomon Islands, namely the hawksbill, green, leatherback, loggerhead, and Olive Ridely. Five species of sea snakes are indigenous to the country, and there is one endemic species of sea krait (*Laticaudata crockeri*), found only in Lake Te-Nggano on Rennell Island. There is also one species of saltwater crocodile (*Crocodylus porosus*), which also inhabits brackish streams.

Mammals

A total of eleven species of cetaceans in nine genera and four families have been observed in the waters of the Solomon Islands. Most common of these are the spinner dolphin, the pan-tropical dolphin, and the common bottlenose dolphin. Other dolphins also reported are the Indo-Pacific bottlenose, Risso's, and rough-toothed dolphins and the orca. In addition the following whales have been reported: short-finned pilot whale, *Mesoplodon* beaked whale, blue whale, and sperm whale. The sea cow (*Dugong dugong*) is also found in the shallow waters of lagoons and estuaries, where it feeds on sea grasses.

Macroinvertebrates

As with their terrestrial counterparts, not much is known of the marine invertebrates, with the exception of those of commercial value. There are 19 species of sea cucumbers (holothurans), several of which are marketed as "bêche-de-mer." Pearl oysters are also important commercially, of which there are three species: blacklip (*Pinctada margaritifera*), goldlip (*P. maxima*), and brownlip (*Pteria penguine*). There are six species of giant clams, with *Tridacna maxima, T. squamosa,* and *T. crocea* being the most abundant. Other commercially important molluscs are the green marine snail (*Turbo marmoratus*) and trochus shell (*Trochus niloticus*).

The spiny Pacific lobster (*Panulirus penicillatus*) and painted crayfish (*P. versicolor*) are harvested for food locally, along with a number of other reef and estuary crustacean species.

HUMAN IMPACTS

The larger Solomon Islands have been occupied by Melanesians for approximately 30,000 years, with the smaller outlying islands of Ontong Java, Bellona, Rennell, and Tikopia populated by Polynesians approximately 3500 years ago. Both of these groups brought with them their traditional crops and food animals and probably had relatively minor environmental impact as they lived in small villages and practiced subsistence agriculture and fishing (Fig. 3). In the late eighteenth century European and American whalers arrived, followed by other foreign traders in sandalwood, bêche de -mer, trochus and pearl shell. These new industries certainly affected local populations but had limited negative impacts on entire ecosystems within the archipelago.

FIGURE 3 Typical coastal village in the Solomon Islands, 2004. Houses are made from locally available materials, with subsistence agriculture and fishing practiced nearby. Photograph by Emre Turak.

Adverse human impacts on the environment have increased dramatically within the last 50 years. The major threats to biodiversity are habitat destruction and fragmentation, overexploitation, and introduced species. Although 77.6% of the Solomon Islands land area is reported to still be under forest cover (FAO 2005), extensive areas of lowland rain forests have been clear-cut for timber export. These cleared areas are then often converted into monoculture oil palm and copra plantations. This has resulted in contamination of freshwater and marine environments from increased sedimentation and chemical runoff. In addition, there has been increased pressure placed on lowland forests from expanding village gardens. The population of the Solomon Islands in 2007 was reported to be 566,842 and is growing at annual rate of 3.5%, which is one of the highest growth rates worldwide. Several species are currently overexploited for foreign markets such as crocodiles, sea turtles, pearl oysters, trochus shell, bêche de mer, and green marine snails. Finally, introduced invasive species such as pigs, cats, rats, and little red fire ants (*Wasmannia auropunctata*) have had devastating impacts in lowland rain forests. Feral cats, in particular, have nearly wiped out several species of native rats on Guadalcanal, and little red fire ants have locally reduced arthropod biodiversity. Because of these adverse impacts, 16 higher plants, 20 mammals, 23 resident birds, four reptiles, and two fish are threatened with extinction (IUCN 2002).

SEE ALSO THE FOLLOWING ARTICLES

Coral / Freshwater Habitats / New Guinea, Biology / Solomon Islands, Geology / Vanuatu

FURTHER READING

Green, A., P. Lokani, W. Atu, P. Ramohia, P. Thomas, and J. Almany, eds. 2006. Solomon Islands Marine Assessment: technical report of the survey conducted May 13 to June 17, 2004. *TNC Pacific Islands Country Report* No. 1/06.
Keast, A., and S. E. Miller, eds. 1996. *The origin and evolution of Pacific Island biotas, New Guinea to eastern Polynesia: patterns and processes.* Amsterdam: SPB Academic.
Leary, T. 1991. *Survey of wildlife management in the Solomon Islands.* SPREP Project PA 17. Report prepared from Solomon Islands Government, South Pacific Regional Environment Programme and TRAFFIC Oceania.
Mayer, E., and J. M. Diamond. 2001. *The birds of northern Melanesia.* Cambridge, MA: Harvard University Press.
McCoy, M. 2006. *Reptiles of the Solomon Islands.* Sofia and Moscow: Pensoft Publishers.
Mueller-Dombois, D., and F. R. Fosberg. 1998. *Vegetation of the tropical Pacific islands.* New York: Springer-Verlag.
Peake, T. F. 1969. Patterns in the distribution of Melanesian land Mollusca. *Philosophical Transactions of the Royal Society of London B* 255: 235–306.
Randall, J. E., G. R. Allen, and R. C. Steene. 1990. *Fishes of the Great Barrier Reef and the Coral Sea.* Bathurst, Australia: Crawford House Press.
Whitmore, T. C. 1969. The land flora: geography of the flowering plants. *Philosophical Transactions of the Royal Society of London B* 255: 5499–5566.

SOLOMON ISLANDS, GEOLOGY

HUGH L. DAVIES

University of Papua New Guinea

The Solomon Islands are part of the Outer Melanesian Arc, a discontinuous chain of islands that stretches from the Bismarck Archipelago in the northwest to Fiji and Tonga in the southeast. The islands lie within latitudes 5–12° S.

PHYSIOGRAPHY

The Solomon Islands comprise more than 1000 generally mountainous islands that are distributed in two geographic areas. Most islands are in the western and central area, where they form a 1200-km-long northwest-trending double chain, founded upon a basement ridge (Fig. 1). These islands include Bougainville and Buka, which are politically a part of Papua New Guinea. Islands of the Santa Cruz Group and the Outer Eastern Islands are located 300 km to the east, at the northern end of the Vanuatu island chain.

The western and central islands and the basement ridge are bounded on both sides by deep-sea trenches. The southwestern trench (Makira or South Solomons Trench, Fig. 2) partly adjoins the young ocean crust of

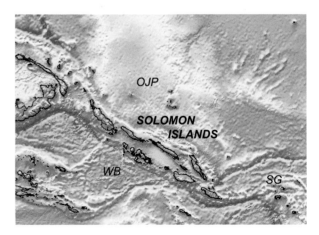

FIGURE 1 Relief map of the Solomon Islands (sea floor topography from Smith and Sandwell (1997); http://topex.ucsd.edu/marine_topo/mar_topo.html). Spreading ridges and transforms shown in the Woodlark Basin are from B. Taylor, University of Hawaii, unpublished data. OJP = Ontong Java Plateau, SG = Santa Cruz Group, WB = Woodlark Basin.

the Woodlark Basin; in this sector the trench is poorly defined. The northeastern trench (Kiliniailau and North Solomons trenches) adjoins the Ontong Java Plateau, a vast submarine plateau that stands 2500 m above the surrounding sea floor. Atolls occur atop volcanic peaks that rise from the plateau.

The climate of the Solomon Islands is wet tropical and dominated by a northwest monsoon in December to March and southeast trade winds in May to October. The capital city of the independent state of Solomon Islands is Honiara on Guadalcanal Island. The population of the islands is approximately 600,000 and is mostly Melanesian.

GEOLOGY

The geology of the islands has been described in terms of three provinces: a northeastern province of Cretaceous–Paleocene oceanic basalts and pelagic sediments; a central province of more complex geology that includes metamorphic and ultramafic rocks and Eocene to Early Miocene arc-type volcanics; and a southwestern province of Late Miocene to Quaternary volcanic islands.

The islands of the northeastern province are Malaita, Ulawa, a northeastern part of Makira, and the northeastern flank of Santa Isabel. On these islands Cretaceous and Paleocene submarine basalts with intercalated pelagic limestone and mudstone are overlain by younger sediments that include terrigenous material. The Cretaceous basalts on Malaita are 3.5 km thick. Those on Makira are metamorphosed. The basalts and pelagic sediments originated as part of the Ontong Java submarine plateau.

On Malaita an intrusion of alnoite (a silica-poor mafic rock) is of interest because it has brought to the surface fragments of unusual mantle rocks, including garnet peridotite. The age of the intrusion is 34 million years (Oligocene).

The islands of the central province are Choiseul, the southwestern part of Santa Isabel, the Florida Islands, Guadalcanal, and Makira (San Cristobal). Basement on these islands comprises Cretaceous basalt, greenschist- and amphibolite-facies mafic schists, and ultramafic rocks; the metamorphic rocks have radiometric ages of around 50 million years, Early to Middle Eocene. Basement is intruded by volcano-related dioritic stocks and is unconformably overlain by arc-type volcanic rocks of Oligocene and Early Miocene age and some possibly as old as Late Eocene. A thick sequence of Miocene to Pliocene clastic sediments covers much of Guadalcanal Island.

The southwestern province comprises Late Miocene to Holocene arc-type volcanic rocks and associated sediments and intrusive rocks. This association extends from Bougainville and the Shortland Islands in the northwest to the New Georgia Group and the near end of Guadalcanal Island in the southeast. There is an outlier of Pliocene volcanic and intrusive rocks in central Guadalcanal, at Gold Ridge.

Bougainville Island combines characteristics of the central and southwestern provinces, though with minor differences in age. A basement of Middle to Late Eocene and Middle to Late Miocene arc-type volcanic rocks is partly overlain by Miocene limestone and Plio-Quaternary volcanoes and is intruded by Pliocene dioritic stocks.

The islands of the Santa Cruz Group, 300 km east of Makira, comprise Miocene to Holocene volcanic rocks and sediments. They include an active volcano on Tinakula Island.

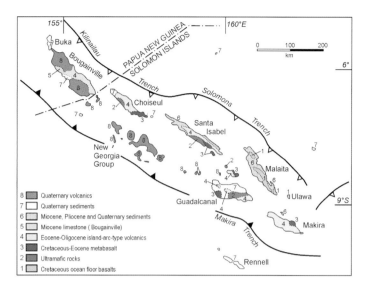

FIGURE 2 Geological map of the central and western Solomon Islands. Courtesy of Davies et al. (2005).

ONTONG JAVA PLATEAU

The Ontong Java Plateau comprises $4-5 \times 10^7$ km^3 of basaltic lava flows. Whereas normal ocean crust is 7–10 km thick, the crust beneath the plateau is more than 30 km thick. The basaltic lavas appear to have been emplaced in one major magmatic event over a period of less than 7 million years at around 122 million years ago (Early Cretaceous, Aptian) and are chemically identical to the basalts exposed on Malaita Island. They are overlain by 1 km of pelagic sediments.

PLATE BOUNDARIES AND SEISMICITY

The islands are located on the boundary between the west-northwest-moving Pacific Plate and the north-northeast-moving Woodlark and Australian plates. The net motion at the boundary is a convergence of the order of 110 mm/yr in a northeast–southwest direction. Most convergence is

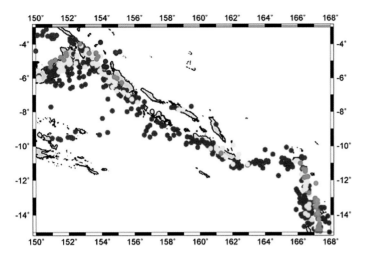

FIGURE 3 Earthquakes of magnitude 6 or greater, 1963–2007. Colors indicate depth below Earth's surface as follows: red, less than 50 km depth; yellow, 50–100 km; green, 100–200 km; pale blue, 200–300 km; dark blue, 300–400 km; purple, 400–500 km. The shallow earthquakes coincide with plate boundaries at the earth's surface. The deeper earthquakes are generated by movement of the subducted slabs. Map by Emile Okal.

accommodated on the Makira Trench, but some is taken up by slow subduction at the North Solomons Trench, and some probably by deformation of crustal rocks.

Shallow earthquakes coincide with the surface trace of the Makira Trench and the Woodlark Basin spreading ridges and transforms (Fig. 3). Intermediate and deep earthquakes are aligned southeastward from beneath Bougainville and indicate the existence of a steeply dipping subducted slab beneath Bougainville and beneath the open water between the two chains of islands.

On April 2, 2007, a magn. 8.1 shallow earthquake located near Gizo in the New Georgia Group caused significant damage. A tsunami that followed 5 minutes after the earthquake damaged coastal villages in the New Georgia Group and as far away as the south coast of Choiseul Island. The earthquake and tsunami caused the loss of 52 lives.

GEOLOGICAL EVOLUTION

The Solomon Islands have been constructed by sea floor volcanism with no influence or input from continental crustal material. The earliest volcanism was the prolific outpouring of great volumes of basalt on to the ocean floor at the time of the development of the Ontong Java Plateau, at around 122 Ma. The volcanic activity is thought to have been associated with a mantle plume. The plateau developed at a great distance to the southeast of its present location. The plateau is thought to have remained submerged throughout the time of its development and subsequently.

The next activity was arc-type volcanism in the Eocene as a result of southwestward subduction of Pacific plate oceanic lithosphere at the Kilinailau Trench. Volcanic material accumulated on the seafloor to the point where volcanic islands were formed.

Metamorphosed basalts from the Eocene subduction system and ultramafic rocks from the earth's mantle were then exhumed (brought to the earth's surface), where they became the platform upon which further arc-type volcanic rocks and sediments were deposited, in the Oligocene. We do not know the reason for exhumation but can speculate that it was caused by a change in plate motion resulting in an interval of crustal extension. The Oligocene and Early Miocene volcanics were generated by continuing southwestward subduction at the Kilinailau Trench.

In the Middle Miocene, continuing subduction of the Pacific plate brought the Ontong Java Plateau to the Kilinailau Trench. The thick lithosphere of the plateau could not be subducted. The result was delamination and westward thrusting of the upper part of the plateau. Basalts from the upper part of the plateau were carried on thrust faults southwestward to the present location of Malaita and the other islands of the northeastern province. As a result of the emplacement of the thrust sheets, the trace of the Kilinailau Trench stepped northeastward to its present location at the North Solomons Trench. The remaining lower layers of Ontong Java Plateau continued to be slowly and steeply subducted.

Because the subduction of Pacific plate was halted, or almost so, the convergence between the Pacific and Australian plates was taken up by development of a new subduction system on the southwestern flank of the islands. This required northeastward subduction of the Australian plate and the development of the Makira Trench. Volcanism associated with the Makira Trench began 8 or 10 million years ago.

The Woodlark Basin developed by seafloor spreading beginning at 6 million years ago or possibly earlier.

ECONOMIC ASPECTS

A major porphyry copper–gold mine was opened at Panguna on Bougainville Island in 1972 but was closed in 1989 on account of civil unrest. A smaller gold mine at Gold Ridge on Guadalcanal opened in 1997 and closed in 2000 due to civil unrest elsewhere on the island. Alluvial gold is mined on a small scale on both Bougainville and Guadalcanal islands.

SEE ALSO THE FOLLOWING ARTICLES

Earthquakes / Lava and Ash / Solomon Islands, Biology / Tsunamis / Volcanic Islands

FURTHER READING

Coleman, P. J. 1965. Stratigraphical and structural notes on the British Solomon Islands with reference to the first geological map. *British Solomon Islands Geological Record (1959–1962)* 2: 17–31.
Hughes, G. W., P. M. Craig, and R.A. Dennis. 1981. Geology of the outer Eastern Islands. *Geological Survey Division Solomon Islands Bulletin* 4.
Kroenke, L. W. 1984. Solomon Islands: San Cristobal to Bougainville and Buka, in *Cenozoic tectonic development of the southwest Pacific*. L. W. Kroenke, ed. Technical Bulletin 6. Suva, Fiji: Committee for Co-ordination of Joint Prospecting for Mineral Resources in South Pacific Offshore Areas, ch 4.
Mahoney, J., G. Fitton, P. Wallace, and the Leg 192 Scientific Party. 2001. ODP Leg 192: basement drilling on the Ontong Java Plateau. *JOIDES Journal* 27.2: 2–6 and covers.
Reagan, A. J., and H. M. Griffin. 2005. *Bougainville before the conflict*. Canberra: Pandanus Books/ANU.
Vedder, J. G., and T. S. Bruns. 1989. *The geology and offshore resources of Pacific island arcs: Solomon Islands and Bougainville*. Earth Science Series 12. Houston: Circum-Pacific Council for Energy and Mineral Resources.
Vedder, J. G., K. S. Pound, and S. Q. Boundy, eds. 1986. *The geology and offshore resources of Pacific island arcs: central and western Solomon Islands*. Earth Science Series 4. Houston: Circum-Pacific Council for Energy and Mineral Resources.

REFERENCES

Davies, H. L., et al. 2005. Geology of Oceania, in *Encyclopedia of Geology, vol. 4*. R. C. Selley, L. R. M. Cocks, and I. R. Plimer, eds. Oxford: Elsevier, 109–123.
Smith, W. H. F., and D. T. Sandwell. 1997. Global seafloor topography from satellite altimetry and ship depth soundings. *Science* 277: 1957–1962.

SOUTH GEORGIA

SEE ATLANTIC REGION

SOUTH SANDWICH ISLANDS

SEE ATLANTIC REGION

SPECIES–AREA RELATIONSHIP

DAVID A. SPILLER AND THOMAS W. SCHOENER
University of California, Davis

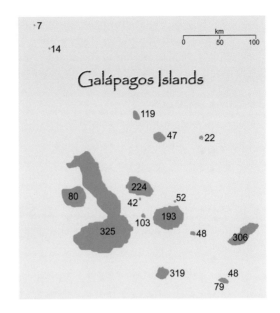

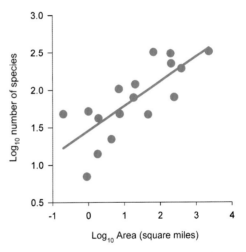

FIGURE 1 A good example of the species–area relationship is the number of land plant species on the Galápagos Islands. Top: Map of the Galápagos Islands showing the number of land plant species. Note that the largest island contains the most species and that numbers tend to be higher on large islands than on small islands. However, several deviations from this pattern exist, such as the "unexpected" high number of species (319) on the medium-sized island at the southern end of the archipelago, suggesting that factors other than area may sometimes be important in determining the number of species on islands. Bottom: Relationship between the number of species and island area. Regression line is the least-squares estimate: log(number of species) = 1.46 + 0.33 log(area). Data taken from Hamilton et al. (1964).

For over a century, ecologists have been captivated by the tendency for number of species within a taxonomic group to increase with island area. This "species–area relationship" has been found for a broad range of organisms in numerous archipelagoes around the world. A partial list of studies demonstrating the relationship includes land plants on the Galápagos (Fig. 1) and Aleutian Islands; insects on the Tuscan Islands and on subantarctic islands; reptiles on islands in the Gulf of California and in the West Indies; birds on the Canary, Solomon, and Aegean Islands; and mammals on islands in the Philippines and North American Great Lakes. Although most studies are of higher organisms, even protozoans and diatoms have shown the relationship. To explain the occurrence of the species–area relationship, ecologists have proposed several hypotheses on causal mechanisms, which shed light on the processes that structure biological communities. Conservation biologists have applied the relationship to the design of nature reserves.

STATISTICAL MODELS

The species-area relationship can be estimated statistically using the power equation

$$S = kA^z \tag{1}$$

where S = number of species, A = area of island, and k and z are fitted parameters. Logarithmic transformation of both sides of the power equation yields the linear equation

$$\log S = \log k + z \log A \tag{2}$$

in which z is the slope of the estimated species–area relationship (i.e., the rate of increase of log S with log A). Because the scales are logarithmic, the z-value is independent of scale, allowing comparisons between different studies even when the units for area are different. A second equation, with S instead of log S in Equation 2, is often used instead.

Log-transformed data on number of land plant species and area for the Galápagos Islands (Fig. 1, bottom) show a positive linear relationship with a slope (z-value) of 0.33. Similarly, many other studies have found that the data fit the log-log model with z-values usually ranging from 0.20 to 0.40. Preston developed a mathematical explanation for the prevalence of the linear log-log relationship, with specific assumptions about the distribution of species abundances and other biological processes, in which the z-value is expected to be approximately 0.27. However, a detailed analysis of 100 studies by Connor and McCoy showed that the data often fit other statistical models just as well or better, suggesting that biological interpretation of parameters in statistical models should be made cautiously.

WHY DOES THE NUMBER OF SPECIES INCREASE WITH ISLAND AREA?

Although the species–area relationship is one of the surest generalizations in ecology, there has been much debate on the causal factors and processes. There are six major hypotheses, as follows.

Random Sampling

Assuming that the individuals on a given island are a random sample from a nearby mainland or other source containing all species, and larger islands contain more individuals, then the number of species should increase with island area. An analogy would be a bowl of colored marbles with ten marbles of each of ten different colors. The larger the handful of marbles you take, the more different colors you will get on average. This simple explanation may serve as a null model to compare with the other five hypotheses, all of which incorporate biological processes.

Larger Populations and Less Extinction on Larger Islands

The hypothesis that populations are larger on larger islands, implying lower extinction rates, is an integral part of the equilibrium theory of island biogeography developed by MacArthur and Wilson in 1964. According to this theory, the number of species present on an island is determined by the dynamic equilibrium between the rate of immigration of species not already on the island from a source pool and the rate of extinction of species already on the island (Fig. 2A). Because larger islands

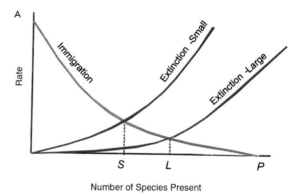

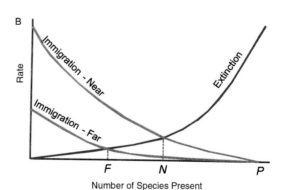

FIGURE 2 Graphical representations of the MacArthur-Wilson equilibrium model of island biogeography. (A) The effect of island area; S = equilibrium number of species on a small island, L = equilibrium number of species on a large island, P = number of species in the pool on the mainland (source). (B) The effect of distance of the island from the source of immigration; N = equilibrium number of species on a near island, F = equilibrium number of species on a far island.

contain more individuals (larger population size) in each species, and extinction rate is inversely proportional to population size, extinction rate of species on small islands is higher than on large islands, so the equilibrium number of species is positively related to island area. In 1976 Simberloff provided experimental evidence for this hypothesis by showing that the number of species on mangrove islets decreased when he reduced the area of islets. In addition to the area effect, the equilibrium theory contains a distance effect; immigration rates are higher for islands near the source pool than for those farther away (Fig. 2B).

Larger Interception Area and More Immigration on Larger Islands

In addition to the lower extinction rate in the equilibrium model, a larger island may also be a bigger "target" for colonists; this causes the immigration curve of a large island to be higher than that of a small island, thereby increasing the equilibrium for the large island.

Higher Habitat Diversity on Larger Islands

Because the number of different types of habitats may increase with island area and different species often occur in different habitats, more species occur on larger islands. Hence, the species-area relationship may be indirect via the positive effect of area on habitat diversity. This hypothesis is most popular among ecologists familiar with the natural histories of the species in their studies. Early on, Watson found that habitat diversity was a better predictor of Aegean bird species number than by area. Very recently, Morrison found that the number of ant species on islets in the Bahamas was predicted by number of land plants species better than was area; in this case, different plant species may serve as different types of habitats or provide different types of food resources for ants. Returning to Galápagos land plants, Hamilton and collaborators showed in 1964 that elevation (a proxy for habitat diversity) was a better predictor of number of species than was area.

Lower Abiotic Disturbance on Larger Islands

The impact of abiotic disturbances, such as hurricanes and tsunamis, may be more devastating on small islands than on large islands, exterminating species mostly on the former and thereby causing the positive species–area relationship. Such an effect could go well beyond the effect of population size on extinction, as equivalently sized populations could be more vulnerable on smaller islands. Whittaker pointed out that such disturbances can affect island communities for decades or even centuries, during which time species would not be at a dynamic equilibrium as in the MacArthur–Wilson model. In addition to the effect of episodic major disturbances, chronic non-catastrophic disturbances, such as wind and salt spray, may affect small islands more than large ones because of the greater perimeter-area ratio on small islands. Only species that can tolerate these harsh conditions could persist on small islands.

Greater Speciation on Larger Islands

The larger the island, the more likely that geographic barriers exist that cause isolation between viable subpopulations of a species; this can lead to allopatric speciation. The effect may be geometric, as the opportunity for speciation is itself proportional to species number, causing an especially high species–area slope.

Lomolino's Combined Model

Lomolino proposed in 2000 a general model for the species–area relationship that integrates disturbance and equilibrium dynamics along with speciation on islands. In this model the species–area relationship has three regions (Fig. 3): (1) On small islands, the stochastic effects of disturbances are predominant, making the relationship between number of species and island area variable and unpredictable (an example is given in the later discussion of spiders in the Bahamas). (2) On medium to large islands, the deterministic effects of area on extinction/immigration and habitat diversity predominate, and number of species increases at a decreasing rate as the number of species in the source pool is approached. (3) The largest islands contain barriers isolating species populations, leading to allopatric speciation. Losos and Schluter assessed the speciation component for *Anolis* lizards of the West Indies. For these relatively poor dispersers, the sharp increase in the species–area slope due to speciation obliterates the flattening portion (region 2) of Lomolino's model.

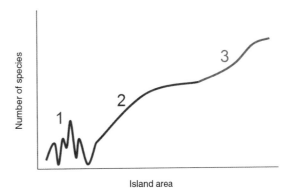

FIGURE 3 Lomolino's general species-area model containing three different regions (see text for explanation). Modified from Lomolino and Weisen (2001).

ECOLOGICAL CORRELATES OF VARIATION IN THE SPECIES–AREA SLOPE

Comparisons among different kinds of organisms have revealed substantial variation in the slope of the species–area relationship. Wright showed that on islands in the West Indies, the slope was greater for non-flying mammals than for bats or birds, suggesting that dispersal ability influences the relationship. A plausible explanation is that islands are more isolated for species with low dispersal ability, so most of them do not occur on small islands, whereas species with high dispersal ability are frequently found on small islands because their immigration rates are high even though they may not persist for long. Schoener suggested that z-values should be greater for species occurring at low density (e.g., territorial carnivores) than for those occurring at high density because the low-density species can only persist on larger islands. Holt and collaborators developed a mathematical model predicting that the z-value for species at the top of the food chain is higher than for species in lower trophic levels.

SPIDERS ON BAHAMIAN ISLANDS

Studies of web spiders occurring on islands in the Bahamas by Schoener and Spiller illustrate the species–area relationship and address several of the issues discussed above (Fig. 4). Complete censuses of all web spiders occurring over the entire areas of 64 islands were conducted annually over a ten-year period. Number of species (mean over time) was positively correlated with island area for four reasons. First, larger islands tended to have larger populations, which in turn had lower extinction rates than did small populations, increasing their number of species. Second, larger islands have a higher immigration rate: the "target effect." Third, some species are habitat generalists (e.g., *Argiope argentata*), occurring in areas with high or low vegetation, whereas others are more specialized (e.g., *Gasteracantha cancriformis*), living in only high vegetation (Fig. 5). Small islands tend to have only low vegetation, whereas large islands contain areas with both low and high vegetation. Therefore, only

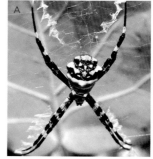 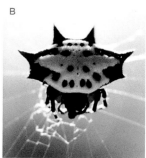

FIGURE 5 (A) *Argiope argentata*. (B) *Gasteracantha cancriformis*. Photographs by D. Spiller.

generalists occur on small islands, whereas both generalists and specialists occur on large islands with more types of habitats. Fourth, tropical storms can affect smaller islands more drastically: During several recent hurricanes, smaller Bahamian islands, which are lower, were completely inundated by high water, apparently killing all spiders. Note that variation in the numbers of spider species on small islands is relatively high as in Lomolino's model. Another factor that can affect number of species is the presence of lizards, which are major predators of spiders and which occurred on about half of the study islands: Islands with lizards tended to have fewer spider species (Fig. 6). This "lizard effect" is more apparent for larger islands than for smaller islands, and the slope of the species-area relationship is greater for islands without lizards (Fig. 7). Hence, the z-value is greater for islands without lizards, on which spiders occupy a higher level in the food web, as in Holt's model.

FIGURE 4 Aerial photograph showing a portion of the Bahamian spider study area.

FIGURE 6 Illustration of a lizard eating a spider by G. Dan.

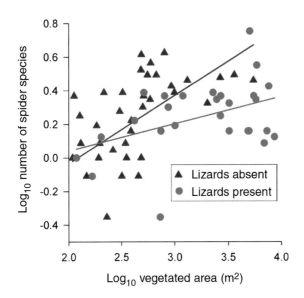

FIGURE 7 Species–area relationship for Bahamian orb spiders on 37 islands without lizards and 27 islands with lizards. Regression lines for islands without and with lizards are respectively: log(number of species) = -0.85 + 0.41log(area), log(number of species) = -0.28 + 0.16log(area). Unpublished data (T. Schoener and D. Spiller).

THE DESIGN OF NATURE RESERVES

In addition to "true islands" (bodies of land surrounded by water), the species–area relationship has been well documented for many types of "island analogues" (patches of suitable habitat for species isolated by unsuitable habitat) on mainlands. Knowledge of the factors that shape the species–area relationship can be used to evaluate different strategies for designing nature reserves that maintain the highest number of species. Of course, the ideal strategy would be to have many huge preserves, but this may not be feasible. Hence, the problem that conservation biologists have pondered is whether one large reserve or several small reserves with the same total area is better. The answer to this question, sometimes referred to as the SLOSS (single large or several small) debate, is not obvious: A single large reserve will have a lower per-species extinction rate than will any smaller reserve, but the more reserves there are, the less chance there will be for a species to disappear simultaneously from all of them. In a review of land plants, insects, and vertebrates on islands, Quinn and Harrison showed that total numbers of species on groups of several small islands were higher than on a comparable area consisting of only one or a few large islands. Among other explanations, they suggest that the several small islands were distributed over a larger region and contained more types of habitats, enabling more species to exist on the entire set of islands. Similarly, genetic diversity within a species, which reduces extinction likelihood, may be increased with a certain amount of reserve-area fragmentation. In conclusion, several small preserves distributed over a large or varied region may sometimes be the better conservation strategy.

SEE ALSO THE FOLLOWING ARTICLES

Extinction / Fragmentation / Galápagos Islands, Biology / Island Biogeography, Theory of / Spiders

FURTHER READING

Connor, E. F., and E. D. McCoy. 1979. The statistics and biology of the species-area relationship. *American Naturalist* 113: 791–833.
Hamilton, T. H., R. H. Barth, and I. Rubinoff. 1964. The environmental control of insular variation in bird species abundance. *Proceedings of the National Academy of Sciences of the United States of America* 52: 132–140.
Lomolino, M. V., and M. D. Weisen. 2001. Toward a more general species-area relationship. *Journal of Biogeography* 28: 431–445.
MacArthur, R. H., and E. O. Wilson. 1967. *The theory of island biogeography*. Princeton, NJ: Princeton University Press.
Quinn, J. P., and S. P. Harrison. 1988. Effects of habitat fragmentation and isolation on species richness: evidence from biogeographic patterns. *Oecologia* 75: 132–140.
Whittaker, R. J. 1995. Disturbed island ecology. *Trends in Ecology and Evolution* 10: 421–425.

SPIDERS

MIQUEL A. ARNEDO

University of Barcelona, Spain

The ability to produce silk is a distinctive feature of spiders. Silk-mediated airborne dispersal has allowed spiders to colonize even the most remote archipelagoes. Spiders on islands have served as models for the study of the evolutionary and ecological underpinnings of biodiversity. Because of their generalist predatory habits, introduced spiders may pose a serious threat to islands' native fauna.

DEFINING A SPIDER

Spiders (order Araneae) comprise a megadiverse group of arthropods that includes close to 40,000 species distributed in 109 families. The origin of spiders can be traced back to the Devonian, about 400 million years ago, representing some of the earliest evidence of terrestrial life on Earth. As the dominant non-vertebrate predators in most terrestrial ecosystems, spiders have enormous ecological importance.

ACCESSING ISLANDS

Dispersal capabilities and generalist predatory habits make spiders formidable pioneers. Spiders have a unique method of dispersal called ballooning. During suitable meteorological conditions (high turbulence, vertical updrafts, and relatively weak horizontal winds), spiders climb to a height from which they will hang suspended from a silk thread to facilitate takeoff. Some spiders have evolved a more specialized adaptation, "tiptoeing" or "tipping behavior," which consists of releasing a silk thread into the air while stretching their legs and raising their abdomens (Fig. 1). Ballooning behavior, which occurs mostly in spiderlings, is triggered by innate responses to food shortage and environmental conditions (i.e., temperature). Aerial dispersal allows spiders to be among the first inhabitants of devastated or newly formed areas, such as volcanic islands. The first known colonist of Krakatau (Indonesia), nine months after the eruption that exterminated all life on the island, was a spider. The 1980 eruption of Mount St. Helens created a large, barren area denuded of any resident arthropods; from 1981 to 1986, ballooning spiders represented over 23% of the windblown arthropod fallout.

Native spider faunas on remote islands are generally impoverished and disharmonic because of geographic, mechanical, and ecological constraints upon dispersal. The chances of arriving on an island will ultimately depend on the distance from the source of colonization (Table 1). Only 15 out of the 109 spider families are represented in the native biota of the Hawaiian Islands, located more than 3800 km from the closest mainland. In contrast, the spider fauna of Bioko, a volcanic island situated just 32 km from the African continent, is rich, well-balanced, and shows close affinities to the nearby mainland of Cameroon. Unsurprisingly, spiders with well-known ballooning capabilities such

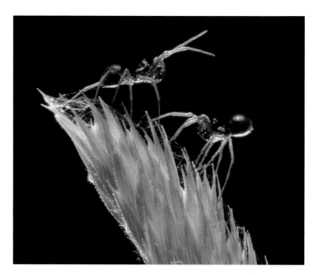

FIGURE 1 Two male *Erigone* spiders on a grass seed head preparing for aerial take-off. The lower one is in a pre-ballooning posture, known as the "tip-toe" position, ready to disperse. Photograph by Andy Reynolds, Dave Bohan, and James Bell of Rothamsted Research.

as the orb-weaving spiders and their relatives (Araneidae, Tetragnathidae) and some non–web-building families (Lycosidae, Salticidae, and Thomisidae) are usually overrepresented among the spider faunas of oceanic islands.

Medium-sized and large ground-dwelling spiders have most likely colonized oceanic islands by rafting on logs or other vegetal debris washed by rivers and waves. This seems to be especially true for spiders associated with littoral or near-tidal areas, such as the mygalomorph genus *Nihoa*, which includes species endemic to several Pacific islands from New Guinea to the Hawaiian Islands. Similarly, it is suggested that the woodlouse hunter spider *Dysdera*, usually found in damp and warm ground habitats,

TABLE 1
Total Diversity and Approximate Levels of Endemism for Spiders on Different Island Groups

Island Group	Island Type	Ocean	Distance to Continent (km)	Area (km²)	Isolation Index (minimum)	Altitude (m)	% Endemism	Total Species
Ascension	Oceanic	Atlantic	1700	97	119	859	9	43
Azores	Oceanic	Atlantic	1300	2,333	74	2,351	19	121
Balearic Islands	Continental	Atlantic	85	5,015	28	1,445	10	278
Bioko	Oceanic	Atlantic	32	2,017	17	3,008	36	50
Canary Islands	Oceanic	Atlantic	100	2,007.8	22	3,718	63	473
Cape Verde	Oceanic	Atlantic	620	4,033	53	2,829	45	105
Galapagos Islands	Oceanic	Pacific	965	7,845	60	1,707	60	146
Hawaiian Islands	Oceanic	Pacific	3000	16,636	119	4,169	60	262
Kuril Islands	Oceanic	Pacific	5	10,291.7	7	2,339	0	427
Madagascar	Continental	Indic	370	587,713.3	58	2,876	85	459
Madeira	Oceanic	Atlantic	560	749.4	66	1,861	37	150
New Caledonia	Continental	Pacific	1200	16,648.4	88	1,618	90	226
New Guinea	Continental	Pacific	155	785,753	37	5,030	60	618
Principe	Oceanic	Atlantic	220	148.5	39	948	47	30
Sao Tome	Oceanic	Atlantic	225	854.8	39	2,024	60	56
Selvages	Oceanic	Atlantic	375	2.73	-	163	10	33
Tasmania	Continental	Pacific	200	65,021.8	35	1,617	60	215

has colonized the Canary Islands from the neighboring northeastern African coast by transporting itself on floating islands.

In contrast to oceanic islands, the presence of spiders on continental or fragment islands can be attributed to vicariance in response to a changing geography rather than chance dispersal. Plate tectonics and eustatic sea-level changes have isolated spider communities formerly continuously distributed along large continental areas. For instance, the occurrence of the tree trap-door spiders of the family Migidae in South America, Africa, Australia, Madagascar, New Zealand, and New Caledonia illustrates the former connection of these land masses in the Gondwana supercontinent, which started gradual sundering about 165 million years ago. The major islands of the western Mediterranean—Corsica, Sardinia, and the Balearic Islands—are continental terranes that drifted toward their present-day location following retreat from their original position on the eastern Iberian Peninsula, about 30 million years ago. The temporal sequence of species formation in the genus *Parachtes* (Dysderidae), endemic to the region, closely follows the geological sequence of separation of the main terranes, suggesting that their present distribution was determined by the disjunction of the islands.

SPIDERS ON OTHER ISOLATED SYSTEMS

Mountaintops and caves are among the continental areas effectively isolated by ecological barriers that show a more characteristic and richer spider fauna. Montane populations of the jumping spider *Habronattus pugillis* (Salticidae) became isolated on the "sky islands" of southeastern Arizona as a result of climatic changes, displaying striking amounts of phenotypic divergence between mountaintops. Similarly, species of the woodlouse-hunter genus *Harpactocrates* (Dysderidae) show non-overlapping distributions across major cordilleras in the Iberian Peninsula and the Alps.

Many spiders have evolved striking features as a result of the adaptation to caves, including eye reduction, depigmentation, and appendage elongation. Spider families with the largest representation of troglobitic species include Pholcidae and Telemidae, worldwide; Agelenidae, Dysderidae (Fig. 2), Leptonetidae, Linyphiidae, and Nesticidae in the Holarctic region; and Ochyroceratidae in the Southern Hemisphere.

EVOLUTION, ECOLOGICAL ADAPTATIONS, AND COMMUNITY ASSEMBLY

Spiders that have undergone adaptive radiations on islands have served as models for examining the ecological and evolutionary processes underpinning island bio-

FIGURE 2 *Dysdera unguimmanis*, a troglobitic spider from the lava tubes of the island of Tenerife, Canary Islands, illustrating some of the adaptations to cave life in spiders: eye reduction, depigmentation, and appendage elongation. Photograph by Pedro Oromí.

diversity. The Hawaiian *Tetragnatha*, with 37 described and over 55 estimated species, and the Canarian woodlouse-hunter spiders *Dysdera*, with 49 endemic species, rank at the top of the most species-rich spider lineages on islands. Other examples of island genera with a dozen or more closely related endemic species include the crab spiders *Mecaphesa* (Thomisidae); the cobweb spiders *Theridion* (Theridiidae); the *Orsonwelles* (Linyphiidae) and *Lycosa* wolf spiders (Lycosidae) in the Hawaiian Islands; the genera *Spermophorides* and *Pholcus* (Pholcidae); the wolf-spider *Alopecosa*; the wall spider *Oecobius* (Oecobiidae); and the ground-spider *Scotognapha* (Gnaphosidae) in the Canary Islands. Ecological adaptation seems to have played a marginal role in the diversification of some of these lineages. For example, the major force driving speciation in the Hawaiian linyphiid genus *Orsonwelles* has been population isolation associated with both island hopping and vicariance caused by erosion of volcanic ridges. Similarly, the diversification history of the Hawaiian jumping spiders *Havaika* has been shaped mostly by the successive and independent colonization down the island chain of two lineages that diverged in size and genitalic features early in the evolution of the group. Additionally, inter-island colonization within the archipelago has also play a key role in the diversification of the six-eyed pholcid *Spermophorides* in the Canary Islands, where endemic species from the same island have non-overlapping distributions.

In other groups, however, local diversification has likely been triggered by ecological factors. Endemic Hawaiian *Tetragnatha* (Tetragnathidae), living in the same sites, are represented by orb weavers that construct webs with different architectures together with cursorial spiny-leg species that differ in color and size. Similarly, *Dysdera* species co-occurring in the Canary Islands show remarkable differences in body size and chelicera shape. In the three former examples, phylogenetic analyses show that coexisting species with divergent ecological traits are frequently each other's closest relatives. Although this observation may suggest sympatric speciation, this may not be necessarily the case: Morphological differentiation may have evolved as a result of intraspecific competition following secondary contact of reproductively isolated allopatric populations. In this context, phylogeographic studies have revealed that populations of two species of Hawaiian *Tetragnatha*, as well as those of the Canarian species *Dysdera lancerotensis,* are subjected to

FIGURE 4 *Orsonwelles macbeth* from Molokai, Hawaiian Islands: a remarkable example of island gigantisms in spiders. This genus belongs to the family Linyphiidae also known as dwarf spiders. Photograph by Gustavo Hormiga.

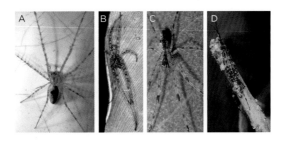

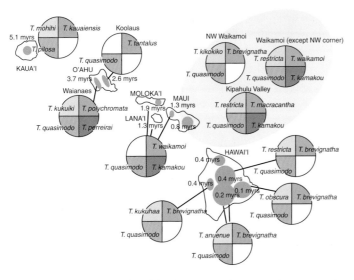

FIGURE 3 Examples of the four ecomorphs of spiny-leg *Tetragnatha* species endemic to the Hawaiian Islands: (A) green, *T. waikamoi,* (B) maroon, *T. kamakou,* (C) large brown, *T. kukuhaa,* and (D) small brown, *T. quasimodo.* Distribution of the four ecomorphs across the main Hawaiian volcanoes (gray circles), with age indicated in millions of years (myrs). Each section of a pie represents a different ecomorph (color codes as shown in the top pictures) whenever a morph is present at a site. Never are two species that share the same ecomorph found in the same locality. Phylogenetic evidence indicates that ecomorphs have evolved largely independently within islands. From Gillespie (2004). Reprinted with permission from the American Association for the Advancement of Science.

increased genetic subdivision as a result of fragmentation by lava flows. This may ultimately drive allopatric speciation.

The adaptive radiation of habitat-associated, polychromatic Hawaiian spiny-leg *Tetragnatha,* coupled with the snapshots of the evolutionary process provided by the chronological arrangement of the islands, has illustrated the deterministic nature of community assembly (Fig. 3). That is, at any given locality, similar sets of ecomorphs arise. This leads to a dynamic species assembly: a maximum number of species occurs in communities of intermediate age, whereas first stages are characterized by ecological release probably as a response to reduced interspecific competition. Ecological release following island colonization has been reported in the ray spider *Wendilgarda galapagensis* (Theridiosomatidae), endemic to the island of Cocos, which shows greater variation in terms of habitat selection and web design and construction on islands than its continental counterparts. Ecological release through decreasing levels of interspecific competition may also explain the evolution of gigantism in linyphiid spiders of the genera *Orsonwelles* (Fig. 4) in the Hawaiian Islands or *Laminacauda* in the Juan Fernández Islands.

Unlike the former examples, daddy-longlegs spiders of the genus *Pholcus,* endemic to Madeira and the Canary Islands, do not show any somatic or ecological differentiation. In these taxa, diagnostic characters are confined to structures involved in copulation, suggesting that diversification in this group has been driven by sexual selection.

The main ongoing evolutionary processes associated with continental islands are relictualization and formation of paleoendemics. These processes are evident in several spider lineages occurring in Madagascar, such as the family Migidae. In this case, the three Malagasy genera of the family form a lineage, the closest relatives of which are taxa occurring in Australia, New Zealand, South America, and Africa. This conforms to the biogeographic predictions for the dismantling of the Gondwana supercontinent.

CONSERVATION ISSUES

Exotic pest introductions pose a major threat to unique island faunas, including spiders. The biota of the Hawaiian Islands and southeastern Polynesia evolved without ants or other social hymenopterans. Over the last century, more than 40 species of ants have been introduced to Hawaii. Native spiders lack effective defense mechanisms against direct predation or competition from these aggressive predators. Three species have been shown to be particularly damaging in Hawaii: the argentine ant *Linepithema humile,* the big-headed ant *Pheidole megacephala,* and the long-legged ant *Anoplolepis longipes.* The little fire ant *Wasmannia auropunctata* threatens the spider biota on many of the islands of the central and southern Pacific. Spiders themselves are generalist predators and hence may become aggressive invaders. As an example, in Hawaii, the introduced araneid *Gasteracantha mammosa* is considered a nuisance to farmers, to residences, and potentially to endangered Hawaiian arthropods.

Unfortunately, a complete assessment of the conservation status of many island native spiders is hampered by a lack of taxonomic knowledge. As a result, the only island endemic spider currently listed in the International Union for Conservation of Nature (IUCN) Red List is the Kauai cave wolf spider (*Adelocosa anops*), and its inclusion was motivated by the deterioration of its cave habitat.

SEE ALSO THE FOLLOWING ARTICLES

Adaptive Radiation / Ants / Caves, as Islands / Continental Islands / Dispersal / Rafting / Sky Islands / Vicariance

FURTHER READING

Australasian Arachnological Society. http://www.australasian-arachnology.org/
Gillespie, R. G. 2004. Community assembly through adaptive radiation in Hawaiian spiders. *Science* 303: 356–359.
Gillespie, R. G. 2005. The ecology and evolution of Hawaiian spider communities. *American Scientist* 93: 122–131.
Gillespie, R. G., and G. K. Roderick. 2002. Arthropods on islands: colonization, speciation and conservation. *Annual Review of Ecology and Systematics* 47: 595–632.
Griswold, C. E. 2003. Araneae, spiders, in *The natural history of Madagascar.* S. Goodman and J. Benstead, eds. Chicago: University of Chicago Press, 579–587.
Jocqué, R., and A. S. Dipepenaar-Schoeman. 2006. *Spider families of the world.* Tervuren, Belgium: Royal Museum for Central Africa.

SPITSBERGEN

MARIA WŁODARSKA-KOWALCZUK

Institute of Oceanology, Polish Academy of Sciences, Sopot, Poland

Spitsbergen is the largest island (38,000 km^2) of the Svalbard archipelago located at the northeastern edge of the Barents Sea shelf.

CLIMATE

Despite its high Arctic location (74 to 81º N), the island experiences relatively mild weather conditions, with average air temperatures ranging from –12 ºC in February to 5 ºC in July. Heat is transported to these high latitudes by the Atlantic waters of the west Spitsbergen current, a distant branch of the Gulf Stream. The Greenland Sea waters off the west coast are ice-free throughout most of the year. The Barents Sea polar water masses occurring north and east of the island are colder; here the sea is more often covered with ice, and air temperatures on the east coast are usually a few degrees lower than those on west Spitsbergen. Its location at the front between the cold Arctic and warm Atlantic water masses renders the island susceptible to early signs of global climate change.

LANDSCAPE

The Spitsbergen landscape was shaped during repeated Quaternary glaciations; it is dominated by steep mountains with sharply pointed peaks ("Spitsbergen" means "jagged peaks") and large fjord systems. The two largest fjords—Isfjorden on the west and Wijdefjorden on the north—nearly cut the island in two. The broad flat plains, which measure up to 10 km in width, are partially covered by marine deposits and were formed during interglacial periods; these are referred to as "strandflats." There are also raised marine terraces that occur commonly along the coast. Northeastern Spitsbergen is covered by large ice caps, whereas in the western and southeastern parts of the island, numerous valley glaciers flow from elevated ice fields and often terminate in the sea. Once every several decades the glaciers surge; this means that they make a massive downward movement that can advance the glacier front by several kilometers. Episodic surge events and short-term local fluctuations do not affect the general trend of Spitsbergen glacier retreat that has been observed throughout the last century. This is linked to global climate change—namely, to increases in air temperatures.

FLORA AND FAUNA

Permafrost (ground that is permanently frozen except in summer, when a layer just a few decimeters deep thaws), low temperatures, and a very short growing season (6 to 10 weeks) limit the island's vegetation to about 170 species of vascular plants that form Arctic tundra communities. In summer, the island hosts huge populations of birds; the fjord cliffs are colonized by hundreds of thousands of auks (Brunnich's gullemot, *Oria lomvia*; black guillemot, *Cepphus grille*; little auk, *Alle alle*) and gulls (kittiwake, *Rissa tridactyla*). Although only three species of terrestrial mammals occur on the island—the arctic fox (*Alopex lagopolus*), the Svalbard reindeer (*Rangifer tarandus platyrhynchus*), and the polar bear (*Ursus maritimus*)—there are a number of sea mammals, including walruses (*Odobenus rosmarus*), ring seals (*Phoca hispida*), bearded seals (*Erignathus barbatus*, Fig. 1), and white beluga whales (*Delphinapterus leucas*). Wildlife is protected in national parks, nature reserves, and bird or plant sanctuaries that together comprise more than 60% of the area of the Svalbard archipelago.

FIGURE 1 Bearded seals resting on an iceberg in front of a tidal glacier in Kongsfjorden, one of west Spitsbergen's fjords. Photograph by Wojtek Moskal.

HISTORY

The island was discovered in 1596 by Willem Barentsz, a Dutch Explorer who was leading an expedition in search of the northeast passage to southeast Asia. For the next two centuries, the island and the surrounding Greenland Sea waters were the site of intensive whaling activities. The hunting, led by Dutch and British companies, targeted primarily the Greenland right whale (*Balaena mysticetus*), but other marine mammals, including walruses and seals as well as cetaceans, were also taken in large numbers. Ultimately, this led to catastrophic reductions in the natural populations. The main economic activity on the island in the twentieth century was coal mining. The Svalbard Treaty, which was ratified by the international community in 1920, established Norwegian sovereignty throughout the archipelago but also guaranteed rights for all signatory countries to conduct commercial and scientific activities. At present, about 2500 inhabitants reside year-round in a few settlements located on the west coast of the island, with most of the population in the "capital" town of Longyearbyen. Since the end of the twentieth century, the economic significance of mining has gradually declined in favor of tourism and scientific research.

SEE ALSO THE FOLLOWING ARTICLES

Arctic Region / Climate Change / Whales and Whaling

FURTHER READING

Conway, M. 1906. *No man's land*. Cambridge: Cambridge University Press.
Hisdal, V. 1998. *Svalbard: nature and history*. Norsk Polarinstitutt Polarhandbok Number 12. Oslo.
Mehlum, F. 1989. *Birds and mammals of Svalbard*. Norsk Polarinstitutt Polarhandbok Number 5. Oslo.
Rønning, O. I. 1996. *The flora of Svalbard*. Norsk Polarinstitutt Polarhandbok Number 10. Oslo.

SRI LANKA

COLIN GROVES
Australian National University, Canberra

KELUM MANAMENDRA-ARACHCHI
Wildlife Heritage Trust, Colombo, Sri Lanka

Sri Lanka is, for its size, the most biologically diverse of all islands. It has three distinctive climatic zones, each with its own characteristic fauna, two of which show only the most distant biological affinities with nearby India. A tropical (5°55′–9°51′ N) island nation, Sri Lanka is classed as part of South Asia and is separated from southern India by the narrow (20 km wide) Palk Strait. It is teardrop shaped, with a north–south length of 432 km and a maximum east–west width of 224km. The total area is 65,610 km^2.

GEOLOGY AND GEOGRAPHY

Sri Lanka is geologically part of the Indian subcontinent, and as such, it was part of the supercontinent of Gondwana (along with South America, Africa, Australia, and Antarctica) during much of the Mesozoic ("Age of Dinosaurs"). During the latter part of this period, Gondwana began to break up, and India, with Sri Lanka, moved northward until it collided with Eurasia. After this, fauna of Eurasian origin spread into India and Sri Lanka during successive dispersal events and then differentiated in the variety of habitats found there.

The highlands in the center of Sri Lanka rise to 2524 m (at Pidurutalagala); the highlands grade downward into the wide flat coastal plane in the west and southwest, but elsewhere drop down in a series of escarpments. The largest river, the Mahaweli, rises in the central highlands and initially flows west, then circles to the northeast and finally flows east.

THE WET ZONE

The Wet Zone is located in the southwestern quarter of the island and extends into the highlands where lowland rain forest grades into lower montane forest (Fig. 1). Temperatures in this zone vary monthly from about 22 to 33 °C, and rainfall is high (2500–5000 mm per annum) and not strongly seasonal. The climax vegetation is evergreen tropical rain forest, dominated by Dipterocarpaceae, whose diversity here is second only to that found in Southeast Asia.

FIGURE 2 Typical vegetation of the dry zone, inside Yala National Park, extreme southeastern Sri Lanka. The dry zone is in most respects an extension of peninsular India, and the spotted deer (*Axis axis*), of which a stag is seen in this photo, is one of many species that occur on both sides of the Palk Strait. Photograph by Colin Groves.

FIGURE 1 Remnant rain forest in the wet zone, at the archaeological site of Batadomba Lena, near Ratnapura in southwestern Sri Lanka. Very little rain forest vegetation now remains. Photograph by Daniel Rayner.

THE DRY ZONE

This is by far the largest climatic zone, covering the entire northern, eastern, and southeastern lowlands (Fig. 2). The vegetation varies from arid to dry forest formations. Rainfall here is much lower (less than 2000 mm per annum) and is distinctly seasonal, but the temperatures are comparable. The border between dry and wet zones is ill-defined in the lowlands, and there is quite a wide belt of intermediate climate and mosaic vegetation.

THE CLOUD FOREST (OR UPPER MONTANE FOREST) ZONE

In this zone, mostly above 1500 m in altitude though as low as 800 m in places, the temperature fluctuates seasonally between 7 °C and about 26 °C. Rainfall is as little as 1500–2000 mm, but humidity is enhanced because the low temperatures reduce evaporation, so the landscape is wreathed in moisture-bearing cloud on a daily basis. Low trees (under 10 m, often only 3–4 m high) are interspersed with tussock grassland (Fig. 3), and there are abundant

FIGURE 3 A view of forest and tussock grassland on Horton Plains, about 1800 m, central Sri Lanka. Photograph by Colin Groves.

lichens, mosses, ferns, climbers, and epiphytic orchids. In the forest, Dipterocarps are replaced by *Rhododendron, Prunus, Ilex,* and *Berberis,* whose closest relatives are in the Western Ghats, and by species of *Syzygium* and *Eugenia* related to plants in the wet zone.

The largest cloud forest area, a vaguely triangular zone about 50 km from east to west and extending about the same distance to the north, is in the Central Highlands; this is separated (by the Mahaweli Valley, at only 500 m) from a smaller zone, a 30-km-long north–south trending ridge, on the Dumbara or Knuckles range to the north. The Sinharaja forest to the south of the Central Highlands incorporates the high Rakwana Hills, on the summit of which is a further cloud forest zone; there is a fourth, tiny area of cloud forest at Namunukula just to the east of the Central Highlands.

The boundaries of these three zones have fluctuated in the recent past, in phase with the strengthening and weakening of the southwest monsoon. The Horton Plains, in the cloud forest zone of the Central Highlands, were a species-poor, even semiarid, environment from at least 24,000 to 18,500 years ago, becoming semi-humid as the monsoon became stronger, and then eventually becoming hyperhumid about 8700 years ago. Humidity rapidly decreased again until 3600 years ago, when the present extent of the cloud forest became established. There were, however, two short humidity events about 600 and 150 years ago.

UNIQUE DIVERSITY IN A SMALL ISLAND

Sri Lanka is only slightly larger than Tasmania, somewhat smaller than Banks Island in the Northwest Territories of Canada, and smaller by at least 1000 km² than Hispaniola in the Caribbean, Sakhalin in the Russian Far East, Hokkaido in Japan, and Ireland. None of these other islands is even remotely comparable to Sri Lanka in terms of biodiversity or endemism. Yet the Palk Strait is only 10 m deep, and during times of low sea level, most recently during the time of the last glaciation of the temperate zones (approximately 28,000–12,000 years ago), a land bridge 140 km wide joined Sri Lanka to India.

The comparatively unproductive dry zone, which lies opposite the Indian mainland and was continuous with the dry country of southeastern India, forms a semiarid barrier protecting the wet zone and cloud forest, the centers of the biological hotspot, from incursions from the South Asian mainland. The wet zone and cloud forest appear to have long been isolated from comparable regions elsewhere. Moreover, the wet zone is dissected by large rivers, notably the Kalu, north and south of which many groups have diverged into different species, and some of the wet zone forests, such as the Kanneliya forest, may have been partially isolated from the rest of the tropical rain forest belt over a considerable period of time. The hill forests of the wet zone change with altitude, and the ridges, slopes, and valleys have distinct vegetational assemblages. The existence of three separate cloud forest areas (excluding the tiny, and poorly known, Namunukula cloud forest) further enhances biodiversity.

The flora and fauna reflect the variety of dispersal routes by which Sri Lanka has been populated. Probably the oldest stratum survives in the wet zone, the dipterocarp-rich forests of which reflect an affinity with Southeast Asia. The flora of the highlands, especially the cloud forest, has its predominant affinities with the Western Ghats of India, but there are also apparent survivors is from Gondwanan time, such as the Jade Vine, *Strongylodon* (Fabaceae), which is found also in northern Australia, New Guinea, New Caledonia, and Madagascar, and the spider *Diallomus,* whose closest relatives are found in Madagascar and the highlands of South America. This article concentrates on a few vertebrate groups, whose affinities are less ancient, but which still illustrate the amazing diversity of this small island.

Amphibians

The diversity of amphibians in Sri Lanka is extraordinary (106 species recorded so far); although Sri Lankan species tend to have their closest relatives in the Western Ghats of India, they form distinct clades and have been long separated from their Indian counterparts by the intervening dry zone. Some examples will illustrate the nature and age of this amazing amphibian biodiversity and the ways in which it is divided up zonally.

The species of the frog genus *Philautus* (Rhacophoridae), known as flying frogs, live in eastern Asia and India, but nine species groups (a greater known level of diversity than in any other part of the range of the genus) occur in the cloud forests of Sri Lanka, with most of them being entirely restricted to the cloud forests. Their greatest diversity is in the Central Highlands; some have representatives in the Knuckles Range, and fewer in the Rakwana Hills. Molecular clock dates put the separations between related species in the three areas at anything from 4–6 to 8–12 million years old; the extreme age of these clades is part of a general picture. Another rhacophorid, *Polypedates,* known as the whipping frog, has two species widespread in the wet zone, a third that is also from the wet zone but is restricted to the Sinharaja forests, a fourth in the cloud forest of the Central Hills, and a fifth—a dry zone species—that is shared with India; the other species of the genus live in Southeast Asia.

The tree-hole frog *Ramanella* (family Microhylidae) is restricted to Sri Lanka and India. It has four species in Sri Lanka, one widespread in the wet zone up to 1200 m, one in the Central Highlands at about 2000 m, and one restricted to the little-known Kanneliya Forest Reserve in the southeastern part of the wet zone; the fourth is a dry zone species that is shared with India.

Finally, there are three complete genera of amphibians that are endemic to the wet and cloud zones of Sri Lanka: the dwarf toad *Adenomus*, the fang-bearing frog *Lankanectes*, and the true frog *Nannophrys*.

Reptiles

The biodiversity of reptiles in Sri Lanka is hardly less than that of amphibians. The gecko genus *Cyrtodactylus* has seven species in Sri Lanka, all in the wet zone and reaching well into the lower montane forests but only regionally into the cloud forest (in the Knuckles Range and Namunukula). Oddly, the genus is not represented in peninsular India; the closest relative of the Sri Lankan group is in far northeastern India, and the genus extends through Southeast Asia to the Solomon Islands.

The agamid genus *Otocryptis* has a species in the dry zone as well as one in the wet zone. Interestingly, these two are sister species, and a South Indian species is sister to the Sri Lankan clade.

Two agamid genera, *Cophotis* and *Lyriocephalus*, are endemic to Sri Lanka. *Cophotis* has two species, both restricted to the cloud forest: *C. ceylanica* in the Central Highlands and the recently described *C. dumbarae* in the Knuckles Range. *Lyriocephalus*, the sister genus to *Cophotis*, has populations in the Knuckles and in the wet zone, the relationships between which have not been fully studied.

Finally, 98 snake species are found in Sri Lanka, of which 45 are endemic, including five endemic genera. The majority of endemic species in this case are restricted to the lowland wet zone.

Mammals

Early knowledge of the mammals of Sri Lanka was due to Kelaart in the mid-nineteenth century, Phillips in the early twentieth century, and most especially Deraniyagala in the middle years of the twentieth century. These authors were well aware of the zonation of the mammals, which they generally assessed at subspecific level. For example, Phillips distinguished at minimum one subspecies each in the wet zone, dry zone, and cloud forest for the primates *Loris tardigradus* (slender loris) and (what is now referred to as) *Trachypithecus vetulus* (purple-faced leaf monkey), whereas Deraniyagala did the same for the giant squirrel *Ratufa macroura*. More recent studies have concluded that the distinctions between the taxa in different zones have been underestimated. Groves and his colleagues, for example, separated the small red lorises of the wet zone at specific level from the larger gray dry zone loris—which Groves deemed conspecific with the loris of southern India—and the wet zone and dry zone mouse deer (*Moschiola*) from one another, though neither is identical to the mouse deer of India.

The biodiversity of mammals goes deeper than this. It is likely that most or all of the wet zone mammals are specifically distinct from their relatives, if any, in the dry zone, but the cloud forest contains many endemic genera, including the shrew *Solisorex* and the murid rodent *Srilankamys* (both of these appear to have their closest living relatives in Southeast Asia). So far, just one of the endemic genera, *Feroculus*, the long-clawed shrew, appears to be shared with the corresponding zone in the Western Ghats.

FOSSIL HISTORY

Much remains to be discovered about the fossil history of the Sri Lankan fauna, which in the main has been recovered in piecemeal fashion during excavations for gems, especially at Ratnapura. Deraniyagala divided the Pleistocene fossils into an older *Hippopotamus* stage and a younger *Elephas maximus* stage, but could not date them given the absence of much stratigraphic control. A Pleistocene lion fossil has long been known, and it has recently been shown that there was also a fossil tiger. The two were apparently not contemporary; the tiger remains date from 16,500 years ago, whereas the lion may have become extinct before (or possibly as a consequence of) the arrival of modern humans and dates to 37,000 years ago. Two species of rhinoceros, *Rhinoceros unicornis* (Indian rhinoceros, at present confined to the monsoon forests of northern India and Nepal) and *R. sondaicus* (Javan rhinoceros, a Southeast Asian rain forest species), are known from the Pleistocene, but whether they were contemporaries is unclear; remains of the latter date to 80,000 years ago.

Another species known from the Ratnapura deposits is the gaur, *Bos gaurus*, a large wild ox that still occurs in mainland South and Southeast Asia. This is of special interest because of literary evidence that it survived in Sri Lanka until the seventeenth century.

CONSERVATION

The conservation status of the Sri Lankan fauna and flora is alarming. Of 34 anuran species confirmed to be extinct worldwide, 21 were endemic to Sri Lanka. Some 95% of the wet zone forest has been lost, because this is the area of highest human population density; only 750 km² remains, and not a single national park or other protected area is located there. Many of the animal and plant species that survive there, most of them endemic for reasons already stated here, are endangered. The largest remaining block of lowland rain forest is the Sinharaja World Heritage Area, to the south of the Central Highlands. Matters are little better in the cloud forest; much of the forest has been cleared for tea plantations, and although there are some protected areas, the whole zone is at the mercy of climate change. Over the period 1869–1995, the average annual temperature at Nuwara Eliya (1800 m in altitude) has increased by 1.3 °C, and the average annual rainfall has decreased by 20%. Only the dry zone is reasonably well served by protected areas, including the Yala National Park in the far southeast.

SEE ALSO THE FOLLOWING ARTICLES

Climate on Islands / Frogs / Indian Region / Lizard Radiations / Rodents

FURTHER READING

Bambaradeniya, C. N. B., ed. 2006. *The fauna of Sri Lanka: status of taxonomy, research and conservation.* IUCN Publications.
Bossuyt, F., M. Meegaskumbura, N. Beenaerts, D. J. Gower, R. Pethiyagoda, K. Roelants, A. Mannaert, M. Wilkinson, M. M. Bahir, K. Manamendra-Arachchi, P. K. L. Ng, C. J. Schneider, O. V. Oommen, and M. C. Milinkovitch. 2004. Local endemism within the Western Ghats—Sri Lanka biodiversity hotspot. *Science* 306: 479–481.
Pethiyagoda, R., ed. 2005. *Raffles Bulletin of Zoology,* Supplement 12: special issue on Sri Lanka.
Phillips, W. W. A. 1980. *Manual of the mammals of Ceylon,* 2nd ed. Colombo, Sri Lanka: Wildlife and Nature Protection Society.
Werner, W. 2001. *Sri Lanka's magnificent cloud forests.* Colombo, Sri Lanka: WHT Publications (Private) Limited.

ST. HELENA

PHILIP ASHMOLE AND MYRTLE ASHMOLE

Peebles, Scotland, United Kingdom

St. Helena—one of the most isolated inhabited islands in the world—lies at latitude 15°58′S and longitude 5°43′W, about 800 km east of the Mid-Atlantic Ridge in the southern Atlantic Ocean. The island played a significant part in the development of biological concepts such as endemism, extinction, and the origins of insular biota. Less than a decade after publication of Charles Darwin's *The Origin of Species,* the botanist Joseph Hooker used his firsthand knowledge of St. Helena in a seminal lecture on insular floras to the British Association for the Advancement of Science, and the chapter on St. Helena in Alfred Russel Wallace's *Island Life* (1892) includes an elegant analysis of key factors in the colonization of oceanic islands and the origin and evolution of their distinctive endemic species. In relation to current thinking, the isolated islands of St. Helena and its distant neighbor Ascension provide insights into evolutionary processes distinct from those offered by archipelagoes such as the Galápagos and Hawaiian Islands.

SITUATION AND GENERAL CHARACTERISTICS

Africa is the nearest continent to St. Helena, with the coast of Angola lying 1800 km to the east and South America resting 3260 km to the west. Ascension Island, the closest land, lies 1300 km to the northwest. The island lies in a region of low marine productivity, although it is influenced by the Benguela current flowing to the northwest from the coast of southern Africa. The richer waters close inshore probably once supported seals, and there are still numerous dolphins and moderately diverse marine life, although coral reefs are absent. St. Helena has a subtropical oceanic climate with relatively slight seasonal change. The southeast tradewinds, dominant throughout the year, generate much condensation on the mountains. Annual rainfall is about 1 m on the central ridge, but less than a fifth of this in the driest areas.

St. Helena is about 17 km long and 10 km wide, with an area of 121.7 km². It is the eroded summit of a 5000-m conical volcanic pile rising from the floor of the deep ocean. The island emerged into the air some 12–14 million years ago, eventually forming a mountain much higher than the modern central ridge, which still reaches 820 m above sea level. The rocks were laid down by successive activity from two volcanic centers producing long series of eruptions, mainly of relatively liquid basaltic lavas from fissures now preserved as striking dike swarms. The northeastern volcano was active during the emergence of the island and for up to another 3 million years. Much of it has now disappeared, but the north of the island is made up of rocks generated during this phase, comprising basal breccias and the predominant mass of subaerial basaltic lavas and pyroclastics.

The southwestern volcano created the greater part of the modern island between 11 and 7 million years ago. Its initial phase produced the lower shield, composed mainly of pyroclastics, the erosion of which formed the great amphitheater of Sandy Bay, open to the southeast.

The main shield, made up primarily of basalt and trachybasalt lava flows, forms most of the southern and western parts of the island. Recent analysis by Ian Baker indicates that about 9 million years ago a huge landslide removed several cubic kilometers of the eastern flank of the older (northeastern) volcano. The resultant depression was rapidly infilled by thick trachybasalt and trachyandesite flows erupting from near the modern peaks. These eastern flows (previously termed the upper shield) formed most of the relatively level areas in the northeast of St. Helena.

A late intrusive phase, which occurred about 7.5 million years ago, injected massive intrusions of trachyte and phonolite into the southwestern volcano; some of these are now exposed, forming a series of landmarks. There has been no later volcanic activity, so marine and terrestrial erosion has shaped the modern island. It is ringed by sea cliffs up to 400 m high, interrupted by steep-sided valleys, and is surrounded by a broad and irregular shelf created by wave action. This was largely exposed during glacial episodes in the Pleistocene, sometimes doubling the size of the island.

HISTORY

St. Helena was untouched by humans until its discovery by João de Nova in 1502. For one and a half centuries it remained free of resident people but was frequently visited, initially by Portuguese mariners and later also by the Dutch and English. Settlement was organized in 1659 by the English East India Company, and the island became an important staging point on return voyages from Asia. St. Helena came to international prominence in 1815 when Napoleon Bonaparte was exiled there, where he remained until his death in 1821. Later in the nineteenth century, the island was used as a base by the British navy for suppression of the slave trade, and since that time it has been governed from Britain. The resident population had origins in England, West Africa, Malaysia, China, the Maldive Islands, and many other places. Natural resources are few, and the economy of the island is heavily subsidized. Access is only by sea, although an airport is now planned.

PLANTS AND ANIMALS

The extreme isolation of St. Helena has ensured that few kinds of land plants and animals reach it naturally; it prevents colonization by poor dispersers such as freshwater fish, amphibians, and terrestrial reptiles. However, the considerable age of the island has led to high rates of endemism among successfully colonizing taxa and to some splitting of lineages after arrival. The affinities of the indigenous biota are overwhelmingly with southern Africa.

Of the 37 endemic flowering plant species (six of which are now extinct), the four species of *Commidendrum* and the one species of *Melanodendron* (both Asteraceae) probably arose from a single dispersal event, and the three modern species of *Trochetiopsis* (Sterculiaceae) derive from one other colonization. Additional stocks are represented by the seven monotypic genera *Trimeris* (Campanulaceae); *Lachanodes*, *Petrobium*, and *Pladaroxylon* (all Asteraceae); *Nesiota* (Rhamnaceae); *Nesohedyotis* (Rubiaceae); and *Mellissia* (Solanaceae) and by several other groups with one or more endemic species. Half a dozen non-endemic flowering plants are coastal halophytes and are probably native, but almost all of the roughly 250 additional species have been introduced, from many parts of the world. St. Helena also has 30 species of native ferns (about 13 of them endemic) and a rich bryophyte flora.

Fossil evidence indicates that ancestors of the modern *Trochetiopsis* and *Lachanodes* species, and also the tree fern *Dicksonia* and other fern genera, were already present some 9 million years ago. Also represented among the fossils are plant groups not found in the modern flora, including two lineages of palms (including *Voamniola*) and the genus *Gunnera* (Haloragaceae), which probably became extinct as a result of major volcanism more than 7 million years ago.

The only land vertebrates that reached St. Helena naturally were birds. Deposits of fossil bird bones show that the island once supported large colonies of seabirds, which are now reduced to remnants. At least eight species of seabird have been lost to the island, and three of them are now extinct: *Bulweria bifax*, *Pterodroma rupinarum*, and *Puffinus pacificoides* (earlier extinction). Endemic landbirds known only from fossils are the two rails *Aphanocrex* (ex *Atlantisia*) *podarces* and *Porzana astrictocarpus*, a cuckoo *Nannococcyx psix*, a hoopoe *Upupa antaios*, and a dove *Dysmoropelia dekarchiskos*. Songbirds are not represented in the fossil record. The only surviving endemic bird is the wirebird *Charadrius sanctaehelenae*. Introduced landbirds comprise two gamebirds, two pigeons, and five passerines.

Much of the zoological interest of the island stems from its diverse endemic invertebrates. Around 80 genera and 400 species (~38% of the fauna) of invertebrates are currently recognized as endemic and many species—as in the plants—are highly distinctive descendants from ancient colonizations. Beetles are by far the most diverse group, with about 150 endemic species (~58%) and 32 endemic genera. There have been spectacular radiations in the weevil family Curculionidae (77 endemic species) and the related Anthribidae (27 endemics). Speciation has also been striking in the Lepidoptera, with some 50 endemic moth species including at least 20 in the genus *Opogona* (Tineidae). Spiders also have about 50 endemic species (~48%), but these are spread among many families. Other groups showing significant radiations include snails of the family Charopidae, in which the only survivor is the ammonite snail *Helenoconcha relicta*; bugs of the subfamily Phylinae (family Miridae) with at least ten species; and the homopteran family Cicadellidae with 13 endemic species probably derived from only a few colonizing stocks. The best known endemic invertebrate is the St. Helena giant earwig *Labidura herculeana* (Labiduridae). This is the world's largest dermapteran, with a maximum length of over 80 mm, but it has not been seen alive for half a century.

ECOLOGY, CONSERVATION, AND ECOLOGICAL RESTORATION

The original habitats of St. Helena were diverse. The fringes of the island were either semi-desert or were occupied by scrubwood formed by several endemic shrubs. Inland from these were ebony gumwood thicket and dry gumwood woodland, with the latter grading upward into moist gumwood woodland and then into cabbage tree woodland containing half a dozen tree species. The highest part of the central ridge (above 700 m) was once covered by a cloud forest of tree fern thicket with the most drought-intolerant endemic trees and shrubs. The moist mountain habitats still support a high proportion of the rich fern and bryophyte flora of the island.

Invertebrate diversity is also highest in the forests on the central ridge, but surviving fragments of dry gumwood woodland and the semi-desert regions also have many endemic species. Of particular interest is an arid area in the east comprising Prosperous Bay Plain and its immediate surroundings, which unfortunately is also the most appropriate site for an airport. The area has six genera and around 40 species of invertebrates that have been recorded only in this part of the island and nowhere else in the world. The fauna of the plain includes a remarkable array of endemic nocturnal wolf spiders (Lycosidae).

Discovery of St. Helena in 1502 was followed by introduction of a range of herbivores and predators. Pigs, dogs, cats, and rats decimated seabird colonies and drove vulnerable endemic land birds to extinction. Goats, released within a decade or so of 1502, were the prime destroyers of the native vegetation, although rabbits, doubtless, also played a part. Tree felling for timber and firewood, along with burning of wood in the making of lime, became major factors after settlement in 1659. By the early nineteenth century, much of the island had been denuded of native trees. Near the end of the same century, cultivation of the New Zealand flax *Phormium tenax* was started and soon led to massive destruction of surviving native forest on the central ridge. Invasion of forest remnants by introduced plants (especially flax) continued through the second half of the twentieth century. Drier parts of the island also have many invasive plants, including the creeper *Carpobrotus edulis*, the prickly pear *Opuntia* spp., and *Lantana camara*.

Within a few decades of settlement, some enlightened administrators of St. Helena instituted measures to prevent extinction of an endangered tree species (St. Helena redwood *Trochetiopsis erythroxylon*) and to conserve the native forests. Failure to ensure continuity of effort resulted in reduction of the original forests that once covered the greater part of the island to about 1 ha of gumwood woodland and 16 ha of cabbage tree woodland and tree fern thicket along the central ridge. The ecological devastation of St. Helena through human agency, recorded in the correspondence of the East India Company, was one of the examples that influenced the development—during the late seventeenth to early nineteenth centuries—of western philosophical and scientific ideas about the relationship of humans with the natural world.

In recent years, serious efforts have been made to halt the loss of native forest and to care for relict stands of endemic plants. Establishment of the Diana's Peak National Park in 1996 focused attention on a key area. Educational efforts have raised awareness of endemic species, and the endangered St. Helena wirebird is the subject of a special conservation initiative. There have been important conservation gains, including rediscovery and protection of the St. Helena ebony *Trochetiopsis ebenus*

FIGURE 1 St. Helenian conservation volunteers caring for the endemic boxwood *Mellissia begoniifolia* in scree formed of trachyte/phonolite boulders (bottom left) derived from Lot's Wife pinnacle. Note the wind screen erected to protect the plants. The boxwood was thought to have become extinct in the nineteenth century, but this small stand was found in 1998.

FIGURE 2 St. Helena olive *Nesiota elliptica* (Rhamnaceae). The death of the last individual of this species (the only member of an endemic genus) was witnessed by the authors in 2003.

and boxwood *Mellissia begoniifolia* (Fig. 1), both of which had been thought to be extinct. However, the recent death of the last specimen of the St. Helena olive *Nesiota elliptica* (Fig. 2), a generic endemic, is a poignant reminder that in attempting to preserve extremely rare species, success can only be assured in the short term, whereas failure is forever.

Recent plans for construction of an airport and the development of tourism will bring new threats to the native habitats and species on the island. Conservationists on the island are continually overstretched, and resources are limited. Nonetheless, some initiatives in ecological restoration have been undertaken. These aim to re-create large areas of native forest and have sometimes involved participation by a large proportion of the population. Experience suggests that the future of the natural environment and biodiversity of St. Helena will depend on the presence of individuals who care and have relevant expertise, on funding and support by the authorities even in the face of commercial pressures, and on the maintenance of protection and restoration measures over the long term.

SEE ALSO THE FOLLOWING ARTICLES

Ascension / Deforestation / Fossil Birds / Insect Radiations / Introduced Species / Wallace, Alfred Russel

FURTHER READING

Ashmole, P., and M. J. Ashmole. 2000. *St Helena and Ascension Island: a natural history*. Oswestry, UK: Anthony Nelson (current distributor: www.kidstonmill.org.uk).
Baker, I. 2004. *St Helena—one man's island*. Windsor, UK: Wilton 65.
Cronk, Q. C. B. 2000. *The endemic flora of St Helena*. Oswestry, UK: Anthony Nelson.
Edwards, A. 1990. *Fish and fisheries of Saint Helena Island*. Newcastle upon Tyne, UK: Government of Saint Helena and the University of Newcastle upon Tyne.
Grove, R. H. 1995. *Green imperialism: colonial expansion, tropical island edens and the origins of environmentalism, 1600–1860*. Cambridge: Cambridge University Press.
Rowlands, B. W., T. Trueman, S. L. Olson, N. McCulloch, and R. K. Brooke. 1998. *The birds of St Helena*. Tring, UK: British Ornithologists' Union Checklist Number 16. British Ornithologists' Union.
Weaver, B. 1999. *A guide to the geology of Ascension Island and St Helena*. School of Geology and Geophysics, University of Oklahoma, Norman, Oklahoma 73019, USA.

STICKLEBACKS

MICHAEL A. BELL

Stony Brook University, New York

The stickleback fish family (Gasterosteidae) comprises five major subgroups (genera), three of which are primitively marine but commonly colonize freshwater (Fig. 1). The habitats of these freshwater colonists contrast sharply with the marine environment from which they came, and their descendants rapidly evolve behavioral, physiological, and morphological traits that adapt them to diverse freshwater environments. Glaciation and isostatic depression eliminated freshwater fishes over wide areas of the Holarctic.

FIGURE 1 Stickleback species that commonly colonize freshwater habitats on islands from the ocean: (A) threespine (male with red and blue reproductive coloration), (B) fourspine, and (C) ninespine stickleback. Photographs by Joseph Ross/PhotoQuery.

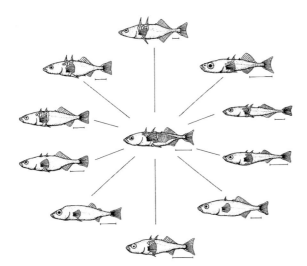

FIGURE 2 Variation in the shape, size, and armor of the threespine stickleback. The fish in the middle is a highly armored anadromous specimen, and those around the periphery indicate the range of armor and body shape among freshwater populations from western North America. All specimens were drawn at the same size, and the scale bars represent 1 cm. Important forms of armor variation include the distribution of lateral plates along the flanks, the number and length of dorsal spines, and the size and presence of the pelvis. Reprinted with permission from Bell and Foster (1994).

When those areas became exposed, sticklebacks and other fishes that move readily through the ocean but tolerate freshwater quickly colonized these newly formed, fish-free, freshwater habitats. Consequently, freshwater stickleback populations are common on islands within their Holarctic ranges. The threespine stickleback (*Gasterosteus aculeatus*) is ubiquitous in postglacial freshwater habitats, and isolated freshwater populations have formed spectacular adaptive radiations in Eurasia and North America (Fig. 2). Two other stickleback groups, the ninespine and fourspine sticklebacks, colonize freshwater habitats on islands. Thus, sticklebacks are abundant, ecologically important, and scientifically interesting members of insular freshwater ecosystems.

ISLAND STICKLEBACKS

Threespine (*Gasterosteus aculeatus* species complex), fourspine (*Apeltes quadracus*), and ninespine (*Pungitius* spp.) sticklebacks are commonly anadromous (sea-run) or can pass through the ocean to enter freshwater. The fifteenspine or sea stickleback (*Spinachia spinachia*) of northwestern Europe never colonizes freshwater, and the North American brook stickleback (*Culaea inconstans*) does not occur in the ocean. The threespine stickleback is the most widespread and abundant freshwater stickleback. The ninespine stickleback comprises several species, which together are widely distributed and locally abundant in freshwater. The fourspine stickleback is limited to northeastern North America.

ISLANDS ON WHICH FRESHWATER STICKLEBACKS OCCUR

Threespine Stickleback

The threespine stickleback species complex is widespread on islands around the north Pacific, Atlantic, and Arctic basins, including Japan, islands of the Russian Far East, the Aleutians, the islands of the Bering Sea, Kodiak Island, the Alexander archipelago, Haida Gwai (Queen Charlotte Islands), and Vancouver Island. They are also widespread on islands in the eastern Canadian Arctic and extend eastward into the Atlantic basin on Greenland and Iceland. They have been reported as far south as Chesapeake Bay, and in northwestern Europe, they are common in the British Isles and islands of Scandanavia and western Arctic Russia. The threespine stickleback is the only Icelandic stickleback.

Ninespine Stickleback

Species of ninespine sticklebacks are also widespread. They occur throughout Japan, the Russian Far East, the Aleutian Islands, islands of the Bering Sea, and Kodiak Island in Alaska. They are widespread around the Arctic Ocean and occur in the Atlantic south to Newfoundland and Prince Edward Island, Canada; Long Island, New York; Greenland; northwestern Europe; and the British Isles. They are conspicuously absent in Iceland.

Fourspine Stickleback

This species has the most limited distribution, occurring only in the northeastern United States and the adjacent Canadian Maritimes. Its insular freshwater distribution includes islands around the Gulf of St. Lawrence in Canada and Long Island, New York.

CHARACTERIZATION AND IDENTIFICATION OF FRESHWATER ISLAND SPECIES

Sticklebacks are easy to identify even without capture. Threespine, fourspine, and ninespine sticklebacks rarely exceed 8 cm total in length and are usually smaller. Free dorsal spines precede the dorsal fin on the back, and each spine is followed by a membrane. The pelvis is robust and has a large spine on either side. Sticklebacks often erect these spines when they feel threatened, making identification easy. They routinely swim by sculling with the pectoral fins, holding the body straight or curving it to turn. They swim in short bursts, between which they hover to strike at food or search the bottom for prey. Thus, sticklebacks can be distinguished from other fishes by observation of the tempo of their swimming through the surface of the water or can be caught and identified.

Threespine Stickleback

The threespine stickleback (Fig. 1A) is a complex of morphologically diverse populations, some of which represent separate biological species. Members of this species complex usually have two large and one small dorsal spine, and a series of large, bony lateral plates in a single row on each side. Most anadromous and marine stickleback and some freshwater populations have a continuous row of about 33 plates running from head to tail (complete plate morph). However, the plates may form separate rows near the head and tail with an unplated gap in between (partial morph) or may number fewer than ten plates restricted to the front of the body (low morph). Some freshwater populations contain all three plate morphs. Threespine stickleback range in color from silvery in open waters (including the ocean) to drab beige with dull green to brown vertical bars in shallower, more heavily vegetated habitats. During the breeding season, males typically develop red throats, but the red color may cover much of the lower body (Fig. 1A). Threespine stickleback often can be seen in the open near shore in schools or small groups. They may be mistaken for small trout or salmon, but the tempo of their swimming—stopping, darting, and swimming again—differs strikingly from the more continuous undulation of trout.

Ninespine Stickleback

The ninespine stickleback (Fig. 1B) is more elongated but generally similar in shape to the threespine stickleback. Although it too may have a complete row of plates on the flanks, these plates are smaller and usually form a short row on the sides in front of the tail fin. Ninespine sticklebacks have seven to 12 dorsal spines, which are smaller than those of the threespine stickleback and which tilt alternately to the sides. Ninespines tend to be more brownish to gray in color than threespines, and reproductive males are partly or completely black with contrasting white pelvic spines. They are secretive and rarely seen in the field.

Fourspine Stickleback

The fourspine stickleback (Fig. 1C) is usually less than 5 cm long, relatively deep bodied compared to the threespine, and lacks lateral plates. It usually has more than three but fewer than seven dorsal spines, and they tilt to the right and left, distinguishing it from the other sticklebacks. The body is generally brown and beige, and reproductive males have red pelvic spines.

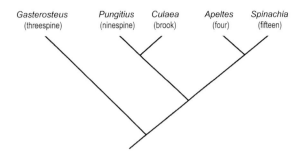

FIGURE 3 Cladogram depicting the sequence of branching during evolution of the genera in the stickleback family (Gasterosteidae).

DIVERSITY AND RELATIONSHIPS

The sticklebacks constitute the small family Gasterosteidae with five distinctive genera (Fig. 3). Relationships of the stickleback genera have been inferred from behavior, morphology, chromosome structure, and DNA sequences. Aside from the typical morphological features already noted, sticklebacks and a few close relatives are distinguished from most other fishes by the nests males build from plant fibers that they glue together with a kidney secretion. The cladogram (family tree) in Fig. 3 depicts the current understanding of relationships among types of sticklebacks.

Only a single species of the fifteenspine, fourspine, and brook stickleback are presently recognized. Several geographically isolated species of ninespine sticklebacks are usually recognized. The threespine stickleback has one close but distinctive relative, the black-spotted stickleback (*Gasterosteus wheatlandi*) of northeastern North America. However, freshwater populations of both threespine and ninespine sticklebacks exhibit fascinating morphological variation, and some of the morphologically differentiated populations clearly represent unnamed biological species that are typically restricted to a single lake. Thus, although there is only one named species of threespine stickleback and a small number of named ninespine stickleback species, named species of both *Gasterosetus* and *Pungitius* may include a great deal of biodiversity that is easily underestimated.

STICKLEBACK BREEDING BEHAVIOR

The elaborate breeding ritual of the threespine stickleback has been studied intensively for more than 50 years, and it can readily be observed in the field and aquaria. In the spring, a male will establish a territory that he defends against other males. He digs a pit within the territory, often near cover, brings plant fibers to the pit, and glues them together with his kidney secretion. The male probes and pulls on the fibers to shape them into a nest. When the pile of fibers is ready, he creeps through it to form a chamber. Now he is ready to court gravid (ripe) females.

Gravid females are easily recognized by their bulging bellies, and males prefer to spawn with larger females. When a male sees a gravid female, he swims in her general direction using short bursts alternating to the right and left to perform the classic zig-zag dance. If the female is receptive, she will tip her head up at about a 30° angle when the zig-zagging male approaches to signal her readiness to spawn with him. The male may lead her back to the nest immediately or stick her with his dorsal spines (dorsal pricking), but he usually returns to the nest to perform nest-oriented behaviors, including gluing, fanning, or creeping through the nest, before he zig-zags back to her again. After a final bout of male zig-zagging and a female heads-up response, he leads her back to the nest, lies on his side and points his snout into the nest entrance (showing) where he wants her to enter. Now she may enter, steal a mouthful of eggs and flee, or break off the courtship. If she enters the nest, he will vibrate his snout against the base of her tail fin, which usually stimulates her to release her eggs. He may bite her tail to induce her to leave the nest after she spawns or has failed to do so, after which he creeps through the nest to release sperm and fertilize her eggs. Having achieved his goal, he chases her out of the territory. Another male (a sneaker) may observe this ritual, dash into the nest, and fertilize the eggs before the owner of the nest.

The male may repeat this courtship ritual and spawn with several females before he enters the parental phase, during which his major activity is fanning water over the eggs with his pectoral fins, while he swims forward furiously with his tail fin to hold his place above the nest. His other major activities are removing dead eggs and chasing off predators, including other stickleback that may cannibalize his eggs or fry. The eggs hatch about a week after spawning, and the male tends the fry for another week, after which they become unmanageable and swim away. Only a minority of stickleback clutches survive. The male may build a new nest and repeat this process or die of exhaustion.

Ninespine and fourspine sticklebacks deviate from these courtship habits but share territoriality, elaborate courtship, nest building, male parental care, and the use of glue to build the nest. However, they nest off the bottom in vegetation, and male fourspine stickleback build a separate nest for the clutch of each female and suck water through the nest instead of fanning.

OTHER ECOLOGICAL PROPERTIES

Sticklebacks are used by a wide range of pathogens, multicellular parasites, and predators. They are important prey for a variety of fish, birds, aquatic insects (e.g., dragonfly larvae), and other predators. Most sticklebacks eat any animal material they can capture, but some threespine stickleback populations specialize on plankton or benthic prey. They usually live one or two years before breeding one or more times during a single season and dying. Females ripen eggs in batches, which they spawn as a single clutch at about weekly intervals. Clutch size depends strongly on body size and ranges from a few tens of eggs to several hundred.

ADAPTIVE RADIATION OF THREESPINE STICKLEBACK

There is relatively little variation in the marine or anadromous populations of any stickleback species. However, there is extensive variation within and among freshwater stickleback populations, and it is most conspicuous in the threespine stickleback. This variation was poorly understood until the late 1960s, when it became clear that many stickleback traits are adaptations to local conditions. Subsequent studies showed that variation among freshwater stickleback populations can evolve within decades, and may produce striking differences between populations only a few meters apart. For example, populations in a lake and its tributary or outlet streams may look and behave very differently. Diet (plankton versus bottom prey) and predation regime (the mix of insect, fish, and bird predators) are the major factors for stickleback adaptive radiation, but dissolved salt concentration also influences evolution of armor. Because so many Holarctic lakes contain threespine stickleback, similar conditions in multiple lakes have caused the independent evolution of similar traits in numerous geographically distant freshwater populations, enabling the detection of associations between specific environmental conditions and the evolution of specific stickleback traits. Ninespine and brook sticklebacks exhibit similar population differentiation in lakes, but they have not been studied as intensively.

IMPORTANCE OF THE THREESPINE STICKLEBACK IN BIOLOGICAL RESEARCH

The threespine stickleback was among the first species named by Linnaeus in 1758. Niko Tinbergen's research in the middle of the twentieth century established it as a classic subject for research in ethology. Research in the late 1960s led to the discovery of exceptional variation in behavior, life cycles, color patterns, skeletal traits, and other properties. Subsequent research on the evolution of the threespine stickleback revealed high levels of skeletal variation within and among populations around the Holarctic. This variation and other favorable properties of the threespine stickleback for genetic analysis (e.g., short generation time, small body size) attracted the interest of geneticists who are beginning to use threespine stickleback to study how variation in DNA sequences affects development and phenotypic variation. In 2006, the threespine stickleback became one of the first fish to have its genome sequenced. Thus, research originally motivated by questions about adaptation and species formation has led to the development of the threespine stickleback for use in biomedical research.

SEE ALSO THE FOLLOWING ARTICLES

Adaptive Radiation / Arctic Islands, Biology / Freshwater Habitats / Lakes as Islands

FURTHER READING

Bell, M. A., and S. A. Foster. 1994. *The evolutionary biology of the threespine stickleback.* Oxford: Oxford University Press.
McKinnon, J. S., and H. D. Rundle. 2002. Speciation in nature: the threespine stickleback model systems. *Trends in Ecology and Evolution* 17: 480–488.
Östlund-Nilsson, S., I. Mayer, and F. A. Huntingford. 2007. *Biology of the threespined stickleback.* Boca Raton, FL: CRC Press.
Wootton R. J. 1976. *The biology of the sticklebacks.* London, UK: Academic Press.
Wootton, R. J. 1984. *A functional biology of sticklebacks.* Berkeley: University of California Press.

SUCCESSION

BEATRIJS BOSSUYT

University of Ghent, Belgium

Succession consists of the often-predictable series of changes in an ecological community over time after a disturbance. These changes occur through colonization and extinction of species. Primary succession involves the assembly of a plant community on a newly formed substrate, whereas secondary succession indicates community changes after a disturbance that has destroyed the vegetation but left the soil to a large extent intact. During primary succession, species accumulate in the plant community through dispersal, environmental selection, and biotic interactions, resulting in an increase of species richness with time. On islands, both dispersal limitations and environmental selection differ from those on the mainland. The isolated position of the island and the large distance to source populations strongly hamper seed dispersal processes, retarding the arrival of species. The newly formed substrate provides a very harsh and stressful environment for the colonizing organisms. Much as in mainland succession situations, biotic interactions, such as facilitation and competition, impact the successional pathway.

DISPERSAL LIMITATION

The degree of isolation of the island (i.e., the distance to source populations on the mainland or on other islands that may act as stepping stones) is a first filter on the potential local species pool. The first pioneers on islands are in most cases good dispersing species. These species may establish on the coastlines, although this may be hampered by the presence of cays. Islands contain fewer species, and in particular the subset of the best dispersing species, compared to the vegetation on the mainland; this results in a disharmonic flora.

Because seed dispersal is to a large extent a stochastic process, and because the occurrence and sequence of seed arrival are unpredictable, the successional pathway during the first stages on islands is largely non-deterministic. The arrival of particular species on the island often depends on chance events (e.g., the establishment of a bird colony on the island, which may result in an accelerated establishment of plant species). The importance of chance events results in a divergent succession pattern, meaning that the successional pathway and the resulting species composition can strongly differ between islands with a similar area and degree of isolation.

The same applies for continental islands, isolated habitat patches within a matrix of habitat that is unsuitable for the colonizing species. Also in this case, seeds of species arrive in the habitat patch from source populations at larger distances, and the stochastic nature of the probability and sequence of arrival may result in a divergent succession pattern.

ENVIRONMENTAL SELECTION AND BIOTIC INTERACTIONS

New islands often arise after volcanic eruptions or because of a lowering of the lake water table. In the first stages, this substrate is very harsh and stress-imposing, unstable and with an extremely low nutrient availability, in particular of nitrogen, and a high water deficit. Cyanobacteria are known to be among the first organisms to stabilize the substrate. Some microorganisms may slightly increase the nutrient availability by nitrogen fixation. The colonization of vascular plant species and bryophytes is a slow process, because only a limited number of the stress-tolerant species that can cope with the low nutrient and water availability are able to establish at the most favorable microsites. During the early stages, bryophytes are the most important colonizers, establishing on the most favorable areas and loosening the substrate. During the next stages, vascular plant species, in particular dwarf shrubs, establish in the moss carpet, often in association with scattered forbs and graminoids. Symbiosis with mycorrhizae, facilitating the uptake of nutrients by the plant, probably plays an important role in the first successional stages. Moreover, the small number of species present in pioneer populations lead to a high degree of inbreeding and a decreased sexual reproduction.

In the early stages, positive interactions between neighbors (i.e., facilitation) are generally more important than negative interactions, such as competition. Facilitation by nitrogen-fixing species, which add nitrogen to the soil, can be considered an important step in the successional process. In this way, pioneer species facilitate the establishment of later successional species. However, nitrogen fixing is energy demanding. Although phosphate is released during the soil weathering process, nitrogen fixing may be hampered by phosphate shortage, such that primary succession can often be considered as nitrogen and phosphate co-limited. Some species may also act as nursery plants, protecting seedlings of later arriving species from extreme microclimate conditions.

In the course of the primary succession, when the number and abundance of species gradually increase, environmental conditions change. The substrate stabilizes, and there is an increasing amount of organic matter and nutrients and a decrease in pH because of the building up and the decomposition of organic matter and because of symbiotic nitrogen fixation. An ameliorated soil structure increases the soil's water-holding capacity.

There is often a trade-off between a species's level of stress tolerance and its dispersal capacity, which may retard succession in the case that stress-tolerant species are not able to reach the island and good dispersing species do not manage to establish a viable population. If stress-tolerant species are hampered by dispersal barriers, a long period of slow habitat change must occur before species accumulate. As conditions ameliorate, the rate of succession gradually accelerates. Species become more abundant, and total cover increases, such that competitive ability becomes a more important condition for establishment than stress tolerance. Species that can more efficiently use nutrients will outcompete less dominant species. The sequence of arrival of different species determines the outcome of these competitive interactions. When succession further continues, tree and shrub species colonize, and the canopy closes. At that point, competition for nutrients shifts toward competition for light. Early successional species will disappear from the vegetation because they can not cope with the increased shade conditions. For this reason, there is often a species richness peak in the middle successional stages because of the temporal co-occurrence of early successional and late successional, shade tolerant species.

Similar to isolation, habitat conditions and competitive interactions restrict the species composition to some subset of the potential species pool. In contrast to dispersal limitations, the effect of habitat conditions is to a larger extent predictable. A similar plant community can be expected on islands with equal ecological conditions, resulting in community convergence. If habit conditions and biotic interactions are the prevailing factors in determining the course of succession, then a deterministic successional pathway will result.

If large-scale disturbances, such as hurricanes, destroy late successional vegetation, then a secondary succession starts. Secondary succession follows a totally different pathway than primary succession and is in most cases a much faster process because propagules are already present, and soil nutrient conditions and water availability are not restricting plant community assembly. In contrast, species can often profit from the accelerated decomposition of organic matter and the release of nutrients after a disturbance. Competition will play a much more important role in structuring plant communities during secondary succession.

Parallel with the process of the colonization by plant species, animal species colonize the island. Much less is known, however, about animal successional pathways. As is the case with plant species, both distance to source populations and dispersal capacity, along with environmental factors, such as climate and soil conditions and the structure and composition of plant communities, will determine the occurrence of animal species. The first colonizers during primary succession are often almost exclusively predators, whereas herbivores and decomposers appear only in later stages. Animal species may affect the successional pathway of plant species through soil organic matter decomposition, seed dispersal, and herbivory.

DIFFERENCES BETWEEN ISLANDS

Because of the importance of dispersal limitation, differences in the degree of isolation of islands are to a large extent responsible for differences in their successional pathways. Distant islands reach the equilibrium much more slowly—and with a balance between immigration and extinction—than do islands near the mainland. In addition to isolation, area determines the number of species accumulating. Large islands contain larger and more differentiated habitats, so more species can co-occur, and species accumulation is much faster. Overall, species

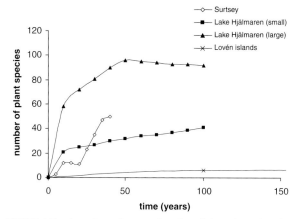

FIGURE 1 Number of species as a function of time since the island appeared, for Surtsey (Iceland), small and large islands in Lake Hjälmaren (average values) (Sweden), and the Lovén Islands (Norway).

accumulation and community assembly will be slower on more isolated and smaller islands, in comparison to large islands at short distances from the mainland.

General ecological conditions on an island at the start of the succession (e.g., climate and initial soil conditions) have been found to strongly determine the rate of succession. Increasing temperatures and precipitation enhance rates of organic matter decomposition and soil formation. Succession and species accumulation rates on boreal islands will hence be slower than on islands at tropical latitudes, because processes of weathering and soil formation are slower. Moreover, the potential species pool is smaller at higher latitudes. Additionally, characteristics of the substrate will determine the successional rate. Parent material with a fine texture will allow a faster succession because it is likely to weather more rapidly, increasing phosphate availability and the formation of an appropriate soil structure.

Fig. 1 illustrates the species accumulation rates of three island examples in the boreonemoral and boreal climate zone of the northern hemisphere. The 40 islands of Lake Hjälmaren (482 km^2, south-central Sweden, 60° N) appeared after a lowering of the water table in 1886, and ranges in size between 20 m^2 and 1 km^2. The islands can be divided into large (greater than 0.3-ha) and small (less than 0.3-ha) islands. As can be seen, large islands reach equilibrium at a time scale of about 50 years, after which species richness slightly decreases, whereas small islands still continue to accumulate species, and the equilibrium may last for more than a century. The number of species at small islands is at every point in time always smaller compared to that of large islands. The Lovén islands (78° N), situated in Spitsbergen, Norway, were successively released from beneath the ice during the regression of a valley glacier. The islands are about 3 km from the mainland and separated by at least 1 km of open water. Because these islands are situated in the boreal climate zone, species accumulation is very slow, and only six species were established on the island after a period of 100 years. Surtsey (63° N) covers 1.4 km^2 and is situated at the southern coast of Iceland; it appeared after a volcanic eruption in 1967. The slower species accumulation rate on the Lovén islands and Surtsey, in comparison with the islands in Lake Hjälmaren, despite the similar latitudinal position, may be attributed to the much higher degree of isolation of this oceanic island in comparison with the lake islands. Plant species richness increased for the first 15 years and then stabilized. The steep increase of species richness after 20 years is due to the establishment of a seagull colony on the islands. Seeds are transported by the birds, and there is a spatially concentrated input of nutrients, allowing a larger subset of species to establish. This clearly illustrates the importance of chance events in early successional stages on islands.

SEE ALSO THE FOLLOWING ARTICLES

Continental Islands / Convergence / Dispersal / Species–Area Relationship / Surtsey / Vegetation

FURTHER READING

Baldursson, S., and A. Ingadóttir. 2006. *Nomination of Surtsey for the UNESCO World Heritage List.* Reykjavík: Icelandic Institute of Natural History.
Burns, K. C. 2005. A multiscale test for dispersal filters in an island plant community. *Ecography* 28: 552–560.
Kadmon, R., and H. R. Pulliam. 1993. Island biogeography: effect of geographical isolation on species composition. *Ecology* 74: 977–981.
Rydin H., and S.-O. Borgegård. 1988. Plant species richness on islands over a century of primary succession: Lake Hjälmaren. *Ecology* 69: 916–927.
Thornton, I. W. B. 1996. *Krakatau: the destruction and reassembly of an island ecosystem.* Cambridge, MA: Harvard University Press.
Thornton, I. W. B. 2007. *Island colonization: the origin and development of island communities.* Cambridge: Cambridge University Press.
Whittaker, R. J., and S. H. Jones. 1997. The rebuilding of an isolated rain forest assemblage: how disharmonic is the flora of Krakatau? *Biodiversity and Conservation* 6: 1671–1696.

SURF IN THE TROPICS

GRAHAM SYMONDS

CSIRO Marine and Atmospheric Research, Wembley, Australia

THOMAS C. LIPPMANN

Ohio State University, Columbus

The wave climate around islands is highly variable in space and time as a result of the level of exposure to incident waves and the frequency of local and distant storm events. The magnitude of the incident waves affects the distribution of benthic communities such as algae, corals, and fish assemblages, as well as the distribution of sediment and the accretion and erosion of shorelines. Topographic effects often provide waves highly sought by surfers. Waves breaking on coral reefs around islands can have a significant impact on sea level and mean currents in an otherwise sheltered lagoon, flushing water, nutrients, and pollutants through the system and helping to maintain a healthy marine environment.

LINEAR WAVE THEORY

Field research carried out on tropical islands over the past several decades has led to fairly robust understanding of wave transformation, water levels, and circulation on fringing reef systems. The basics of the wave physics are the same as for open coast sandy beaches found worldwide; however, the importance of bathymetric irregularities and characteristic geometries of fringing reef systems on wave and current dynamics renders the circulation significantly different. The close link between the circulation and the health of the living reef ecosystem is recognized. Linking sediment and solute transport to predictive models of wave transformation and wave-driven flow with wind and tidal forcing is a difficult problem, but much progress has been made. The following sections discuss basic elements of wave physics and the dynamic consequences of waves impinging on a typical fringing reef system.

Wave Characteristics

Linear wave theory adequately describes many of the observed features of wave propagation and transformation, and the reader is referred to the list of references following this article. A schematic of an idealized wave is shown in Fig. 1, where wave height is the distance between wave crest and trough, wave period is the time taken for a wave crest at position B to reach position A, and wavelength is the distance between consecutive wave crests. Wave periods are typically in the range of 5 to 20 seconds, and the corresponding range of wavelengths in 20 m depth is 39 m to 270 m.

Waves in the open ocean are generated by winds. In general, wave growth depends on the strength of the wind, the duration over which the wind blows, and the region (fetch) that the storm covers. As waves propagate outside their generation region, they begin to disperse, with longer-period waves traveling faster than shorter-period waves. As the waves sort themselves out, they form wave packets, or groups. In a typical group, there are anywhere from 5 to 10 wave crests that vary in height from small amplitudes at the front and rear of the group to larger amplitudes in the middle. The group pattern travels at one-half the speed of the individual waves in deep water; thus, the individual waves pass through the groups as the group pattern propagates shoreward. The energy of the ensemble of all waves in the wave field propagates with the group. In the absence of wave breaking or bottom drag, the transmission of wave energy, called the energy flux, is conserved for a given length of wave crest.

It is important to recognize that waves represent a propagation of energy, not mass, and that the energy propagates with the wave group rather than the individual waves. Beneath the surface, individual water particles describe almost closed ellipses (or circles in deep water) as shown in Fig. 1. As the wave crest passes, the particles at the top of the elliptical orbit are moving in the direction of wave motion, and as the trough passes, the particles at the bottom of the elliptical orbit are moving opposite to the direction of wave motion. Fig. 1 also illustrates a number of changes that occur as waves enter shallow water. Through the shoaling zone, the speed of the wave crest decreases with depth, and because the wave period must remain the same (to conserve the number of waves in a given time span), the wavelength must also decrease. In order to conserve energy flux, there must be a corresponding increase in wave height, thereby increasing its steepness. Eventually the wave crest overturns, or breaks, onto the front face, creating turbulent vortices and dissipating energy, and thus the wave height diminishes through the surf zone, approaching zero at the shoreline. Waves with residual height are reflected off the beach face and head back out to sea.

Wave Refraction

The dependence of the speed of the wave crests on the water depth leads to the process known as refraction.

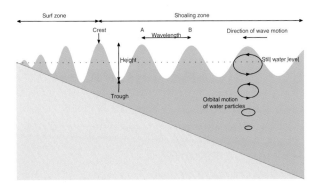

FIGURE 1 Schematic of the transformation of ocean surface waves approaching a shoreline. Through the shoaling zone, waves slow down, their wavelength decreases, and their height increases as they approach the surf zone. Through the surf zone, wave height is attenuated by dissipation processes and decreases to near zero at the shoreline.

Consider a very long crested wave approaching an island. That part of the wave crest that first encounters the shallow water around the island will slow down, whereas the part of the crest out in deeper water will be moving faster and will thus move ahead of the slower-moving part of the wave. This process is called refraction, and it causes the wave crests to bend toward the island, wrapping into what would otherwise be a shadow region behind the island. However, as the waves refract, they are effectively being stretched along the crest, and conservation of energy causes the wave height to decrease such that the wave energy flux (proportional to wave height squared times the group velocity) per unit length of wave crest remains constant. The greater the angle between the direction of the refracted wave and the corresponding deep water direction, the greater the reduction in wave height. State-of-the-art numerical models do a reasonably good job of simulating wave refraction over complex bathymetry, although care must be taken to properly represent intersecting waves refracting around opposite sides of an island.

Wave Momentum

Consider a column of water with a horizontal area of 1 m² extending from the sea surface to the bottom. Under the wave crest, this water column has a greater mass than it has under a trough. Momentum, given by mass times velocity, is thus greater under the crest than under the trough, and thus, when averaged over a wavelength, results in a flux of momentum in the direction of wave motion. This momentum flux caused by the presence of surface waves is known as radiation stress. As waves approach a shoreline or a sufficiently shallow reef, they eventually break, dissipating energy and transferring momentum to the water column. To conserve momentum, the gradient in wave momentum translates to a force that can drive a mean current in the direction of the waves or that is balanced by an equal and opposite force, such as a pressure gradient resulting from water piling up against a shoreline. On coral reefs around tropical islands, these wave-driven flows can be quite strong. A schematic representation of wave-driven flow over a submerged reef is shown in Fig. 2. Through the surf zone, the gradient in wave momentum drives a current in the direction of wave motion. However, when the waves reach the top of the reef slope, they cease to break, the gradient in wave momentum vanishes and is thus unable to provide a force to drive the flow over the reef top. In this one-dimensional case, assuming the current is uniform throughout the water column, conservation of mass requires the current to increase as

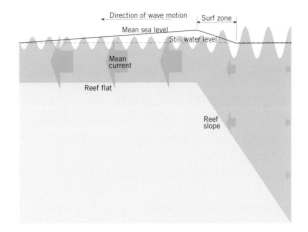

FIGURE 2 Schematic of wave-driven mean flow (large arrows) and corresponding mean sea level (solid blue line) over a submerged reef. Through the surf zone, the mean current is forced by radiation stress gradients and an opposing pressure gradient associated with the sea surface slope (blue line). Seaward and shoreward of the surf zone, the current is forced only by the pressure gradient caused by the sea surface slope.

the depth decreases, as illustrated in Fig. 2 by the larger arrows representing stronger currents. A pressure gradient is established by increasing mean sea level through the surf zone to a maximum at the top of the reef slope. From there, the mean sea surface slopes down to the level of water in the lagoon, providing a pressure gradient sufficient to drive the flow over the reef crest. Through the surf zone, the slope of the mean sea surface opposes the cross-reef flow such that increasing the mean sea surface elevation at the top of the reef slope increases the flow across the reef flat while decreasing the flow through the surf zone until the mass transport over the reef flat is equal to the transport through the surf zone. The magnitude of the cross-reef flow depends on the magnitude of the incident waves offshore, on the width of the reef flat, and on the depth of water over the reef flat. In the idealized case shown in Fig. 2, the mean sea level decreases to the still water level at the downstream, or lagoon, side of the reef flat. Observations have shown that in the presence of waves breaking on a reef, the mean sea level in the lagoon may also increase above the still water level, effectively reducing the cross-reef pressure gradient and cross-reef flow. It is possible for this wave-induced setup to raise the lagoon sea level sufficiently to prevent any cross-reef flow. Because of the typically narrow continental shelf around islands, wave setup can cause significantly larger sea-level fluctuations than can wind-induced storm surges that occur along continental margins.

As waves wrap around an island, refraction does not always orient the wave crests entirely parallel with the bathymetry, particularly on islands where the shelf is very steep. In this case, obliquely incident waves will also drive mean flows along the reef parallel to the depth contours.

WAVES ON REEFS

Natural coral reefs are significantly more complex than the idealized case shown in Fig. 2, with highly variable bathymetry across and along the reef and occasional deep passages cutting through the reef. Lord Howe Island (31°34′ S, 159°05′ E), shown in Fig. 3, is situated 630 km off the east coast of Australia, and is approximately 10 km long and up to 2.6 km wide. It is roughly crescent shaped, with a lagoon facing southwest and bounded by a fringing reef about 6 km long; it is the southernmost coral reef in the world. The lagoon and reef are shown in Fig. 3, with depth contours in gray; the white regions associated with wave breaking indicate the location of the reef crest. The reef is cut by three deep passages connecting the lagoon and open ocean, one at the northern end of the reef (North Passage) and two at the southern end (Erscotts Passage and South Passage). Mean currents, obtained by averaging 2-Hz data over five minutes, were recorded continuously for a three-week period at a number of sites indicated on Fig. 3. The current data shown in Fig. 3 have been smoothed and decimated to 12 hourly vectors, removing most of the semi-diurnal tidal component. These data confirm the presence of a mean inflow over the reef crest as predicted by the idealized one-dimensional model described above. However, in the two-dimensional case, once the flow enters the lagoon, it may be deflected alongshore toward the deeper passages, where the absence of wave breaking allows the water to exit back out to sea. The current pattern shown in Fig. 3 indicates a persistent outflow in the passages. Current speeds in the lagoon are generally weaker, with occasional short pulses of strong currents. On a natural reef, current magnitude is also limited by bottom friction governed by the roughness of the bottom. Observations from Kaneohe Bay, Oahu, Hawaii, have shown that the dissipation of wave energy by bottom friction is similar to depth-induced breaking over a very wide reef flat.

Clearly, there is variability in current speed and direction not always consistent with wave forcing, and it may reflect contributions that are the result of other processes such as wind and tidal forcing, which will certainly dominate during periods of low waves. Application of a state-of-the-art numerical model reveals the complex spatial variability of the wave-driven currents, as shown in Fig. 4. In this case, the model is forced by a constant wave height and direction at the offshore boundary with no tidal or wind forcing; thus, the spatial variability in the modeled flow field is mainly determined by spatial variability in the bathymetry. Although the model shows inflow over the reef and outflow in the major channels, qualitatively similar to the observations shown in Fig. 3, bathymetric low points along the reef crest are also sufficient to cause outflow.

The magnitude of the cross-reef flows and the relative shallowness of many coral reef lagoons leads to effective flushing of the lagoon with time scales on the order of the semi-diurnal tidal period or less. Tropical waters surrounding coral reefs are generally nutrient-poor, but the rapid flushing maintains a high through-flow of low-nutrient water sufficient to maintain the highly diverse coral reef ecosystem. Nutrient uptake by some corals and benthic communities has also been found to increase with increasing flow rate. Wave-driven flows also wash away fine suspended sediment from terrestrial runoff, which can have a detrimental effect on corals by reducing

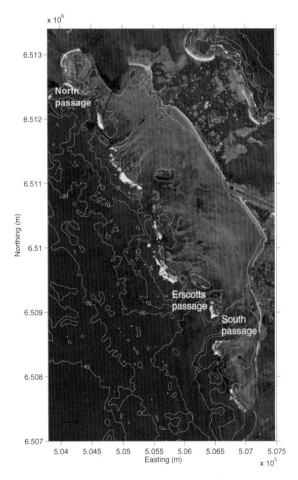

FIGURE 3 Mean current vectors measured along the reef and in the lagoon at Lord Howe Island, November 27 to December 14, 2004.

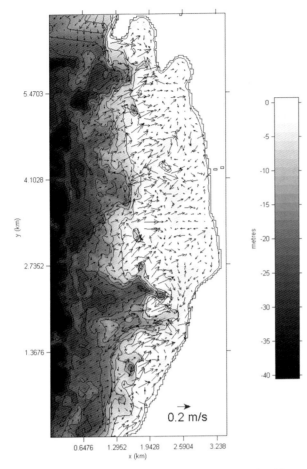

FIGURE 4 Numerical simulation of wave-driven currents at Lord Howe Island using a depth-averaged two-dimensional hydrodynamic numerical model (2DBeach, http://www.asrltd.co.nz). Background shading from light to dark indicates increasing depth according to the scale on the right.

ambient light levels and smothering benthic communities. Terrestrial runoff can also contain pollutants such as fertilizers, and rapid flushing of these pollutants helps prevent toxic algal blooms from developing in the lagoon. Most fish species have a larval stage, during which time dispersal and subsequent settlement are largely governed by local and large-scale current patterns.

SEE ALSO THE FOLLOWING ARTICLES

Beaches / Lord Howe Island / Reef Ecology and Conservation / Sea-Level Change / Tides / Tsunamis

FURTHER READING

Brown, J., A. Colling, D. Park, J. Phillips, D. Rothery, and J. Wright. 1993. *Waves, tides and shallow water processes,* G. Bearman, ed. Open University, Pergamon Press.
Dean, R. G., and R. A. Dalrymple. 2001. *Coastal processes with engineering applications.* Cambridge: Cambridge University Press.
Komar, P. D. 1998. *Beach processes and sedimentation,* 2nd ed. Upper Saddle River, NJ: Prentice Hall.
Kraines, S. B., T. Yanagi, M. Isobe, and H. Komiyama. 1998. Wind-wave driven circulation on the coral reef at Bora Bay, Miyako Island. *Coral Reefs* 17: 133–143.
Longuet-Higgins, M. S., and R. W. Stewart. 1964. Radiation stresses in water waves; a physical discussion, with applications. *Deep Sea Research* 11: 529–562.
Monismith, S. G. 2007. Hydrodynamics of coral reefs. *Annual Review of Fluid Mechanics* 39: 37–55.
Storlazzi, C. D., A. S. Ogston, M. H. Bothner, M. E. Field, and M. K. Presto. 2004. Wave- and tidally-driven flow and sediment flux across a fringing coral reef: Southern Molokai, Hawaii. *Continental Shelf Research* 24: 1397–1419.
Symonds, G., K. P. Black, and I. R. Young. 1995. Wave driven flow over shallow reefs. *Journal of Geophysical Research* 100(C2): 2639–2648.

SURTSEY

STURLA FRIDRIKSSON

Reykjavik, Iceland

Off the southern shore of Iceland in the North Atlantic Ocean is a group of 14 islands and a number of skerries called Vestmannaeyjar or the Westman Islands. The largest of these is Heimaey at 11.6 km^2, the only populated member in the archipelago. The youngest and the outermost island is Surtsey. It was formed during a volcanic eruption that started on November 14, 1963.

AN ISLAND CREATED

That morning a fishing boat was situated some 20 km southwest off Heimaey. At dawn the captain saw a column of smoke coming out of the sea and realized that a volcanic eruption had started.

The pillar of smoke rose ever higher with pulsating explosions, constantly ejecting cinders and ashes, while lava bombs were being flung up and splashed into the sea. The ash particles blown into the atmosphere were positively charged, and shortly after an explosion from the volcano, they triggered great lightning displays that were spectacular to watch and were seen for miles around.

It took only 24 hours of volcanic activity for an island to take form, built up from the ocean floor of 130 m in depth (Fig. 1). This island was later named Surtsey after the fire giant from Nordic mythology, Surtur the Black—a name that suited the island's fiery origins and its black basalt lava formations.

SURTSEY SECURED FOR SCIENCE

Quickly recognizing the scientific importance of the new island, the Icelandic authorities decided to protect

FIGURE 1 A volcanic island emerges out of the sea off the southern coast of Iceland in November 1963. Photograph by Sturla Fridriksson.

it, making it accessible only to scientists. The legislation was passed despite the enormous tourist interest in the eruption. Thus, the island became an unique laboratory for scientific studies and was secured as a typical isolate, where natural forces alone could act without any interference from humans.

Surtsey has continued to awaken world attention, first and foremost because of the legal isolation of the island, but also because of the research that has been conducted on the island under the protection of law. Surtsey was an unusual geological phenomenon, and at the same time, it created an exceptional opportunity for biological research, even including the origin of life. An experiment was carried out to demonstrate that organic material could be created by the actions of volcanic eruptions.

Surtsey is a small replica of Iceland, where the same elements are active and can be studied on a condensed scale. For all these reasons, Surtsey has been nominated for the World Heritage List.

Those scientists who were interested in Surtsey and its development realized the need to organize their research and founded the Surtsey Research Society in 1965 to promote research in geological and biological sciences on the island. It publishes the journal *Surtsey Research* (www.surtsey.is).

GEOLOGICAL DEVELOPMENT

The eruption experienced several distinct phases. In the first, called the explosive phase, a crater formed in the center of the newly created island. The brim of the crater was low, and the seawater would gush into the volcanic shaft. The ocean constantly came in contact with the upwelling magma, rapidly cooling it off and fracturing it into ash particles and cinder, which burst into the air with great explosions of steam.

The eruption continued from the island's crater until early February 1964, when it shifted to a new opening, farther to the west of the island. Ashes from the two craters piled up to form two crescent-shaped cones, Austurbunki and Vesturbunki.

The explosions from the new crater continued until April 4, 1964, when the volcanic shaft developed a watertight lining, and the ocean no longer flooded over the crater wall. When this occurred, floating lava emerged, and a lava fountain was formed, spouting 50- to 100-m high columns of glowing lava that bubbled and splashed up from the red-hot lake. This marked the beginning of the effusive phase.

Sometimes the lava overflowed the crater rim, and floods of thin, fiery streams swept down from the crater over the south part of the island, gushing toward the sea with speeds of up to 70 km/h. The glowing lava often flowed through closed veins, retaining a temperature of 1100 °C and ending its journey by cascading off the cliffs into the ocean (Fig. 2). When the scorching lava came into contact with the cool ocean, steam whipped up, forming a white fringe around the edge of the island. The lava solidified, forming a hard dome of a lava surface that sloped gently to the sea on the eastern and southern sides of the island.

During the eruption, new veins with outlets were formed from the main volcanic shaft. A few small craters opened up on the island, and two satellite islands formed off the shore of Surtsey. They consisted only of cinder cones and, lacking a lava crust, were washed away when the eruptions ceased.

When the main eruption ended on June 5, 1967, it had lasted over three and a half years. Surtsey had increased in size to 2.7 km² in area and reached a height of 175 m above

FIGURE 2 Lava flows over the southern part of Surtsey. Photograph by Sturla Fridriksson.

ocean level. Because the depth of the sea was about 130 m, the total height of the island was approximately 300 m.

GEOLOGICAL STUDIES

An island had formed and was secured with a cover of lava that coated the tuff cones like frosting on a layer cake. It demonstrated to geologists that the so-called table mountains or tuya of the Icelandic highlands had been formed in a similar way, as islands in glacial lakes, during the Ice Age.

The total volume of material produced during the eruption was about 1.1 km³. Of this material, 70% was tephra and 30% was lava. The central part of Surtsey was made of tuff breccias, which have gradually transformed into palagonite. This process has been followed carefully by geologists on Surtsey.

On the southern part of the island, two kinds of lava were formed: ropy lava, or pāhoehoe, and rough 'a'ā lava. Gradually, the southwestern edge of the island was eroded by breaking waves, created by the prevailing southwest wind. In 40 years, the island has lost approximately 3 ha a year and is at present, in 2008, only 1.4 km² in area and 155 m high. Some of the eroded material has been carried toward the leeward side at the northern edge, where it forms a spit of sand and boulder rims. The various changes in the landscape and the gradual wind and water erosion of the island are constantly being followed, and geomorphologic evolution on Surtsey has been well recorded. The climate of Surtsey is maritime, windy and rainy with relatively warm winters (average 2 °C) and cool summers (average 10 °C) (Fig. 3).

BIOLOGICAL INVESTIGATIONS

Biologists used the opportunity afforded by the formation of Surtsey to study various ecological phenomena and to understand how life would gradually colonize the island. They had to consider the location of the territory and investigate the source of available species for dispersal of life forms. A survey was made of the means of dispersal of plants and animals. The landing facilities were observed and living conditions on the island studied, as was the multiplication of various colonists. Records were made of every new introduction of plant and animal species and its performance on the island.

Diaspores could be carried by air and ocean. Seeds drift in by ocean currents, and those surviving the saltwater would sometimes be able to establish themselves on the arid island. Grass knobs would fall from the nearby islands and were carried to Surtsey, taking with them insects and other small biota.

Even fish aided in such transports. A number of sea purses, capsulated eggs of skate, drifted to Surtsey. A number of seeds attached themselves to the outer coat of the sea purses' capsules when the purses lay on a distant shore. The purses acted as cargo boats, carrying their loads to this new land.

Similarly, many diaspores, such as spores of molds, moss, lichens, and ferns, have been carried by air from the mainland of Iceland. Light seeds with feathers, such as cotton grass *Eriophorum* spp. and groundsel *Senecio vulgaris,* which are adapted to wind dispersal, have been found on the island, although these two species have not yet succeeded in growing there. Other species that have established themselves on Surtsey, such as dandelion and three types of willows, have probably grown from windborne seeds.

Birds carry seeds attached to their feet or feathers or in their alimentary tracts. Migrating birds coming to Iceland in the spring were found to carry viable seeds long distances over the ocean. In this way the seed of the lady's thumb plant, *Persicaria,* was brought by a snow bunting with the seed lodged in its gizzards. This snow bunting was of a race that lives in Scotland over the winter and migrates to Greenland in the spring. By analyzing the grit in the gizzards, it could be concluded that the minerals were of Scottish origin and that the seed had simultaneously been

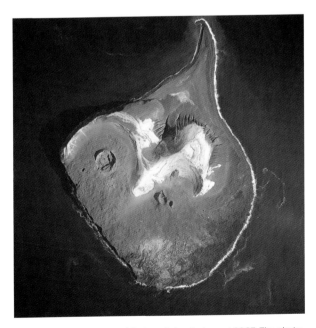

FIGURE 3 An aerial view of Surtsey, taken in August 2007. The photograph shows the gravel and sandy spit in the north, eroded water gullies (small, black furrows), the two cinder cones (light colored), the two main craters, the wind and ocean eroded edge of the island (at left), and the green spot of plant colonization. Photograph by Loftmyndir EHF, Reykjavik, Iceland.

picked up in the Scottish Highlands. This was proof of a long-distance transport of seed by birds.

Seagulls also carry vegetative material for their nest building, and they bring some vegetables to their young with their fish dinner, further adding to Surtsey's ecosystem.

Some insects may fly to Surtsey on their own wing support, but invertebrates are also carried by birds or can float to Surtsey on driftwood. Moths and butterflies from the continent of Europe are carried by favorable winds to Surtsey, and spiders come to Surtsey gliding on cobweb threads.

Birds fly easily to Surtsey, and seals come there by swimming. Thus, a complex ecosystem has slowly evolved on the once barren island.

MARINE LIFE

Monitoring of the marine life around Surtsey began in 1964 and has continued ever since.

Benthic Vegetation

Marine algae were among the first plants to drift to Surtsey. Knotted wrack, *Ascophyllum nodosum,* was occasionally washed ashore during the first summer of the island's existence. Some marine bacteria and diatoms were pioneers of the marine benthic vegetation of Surtsey.

In the summer of 1965, some filamentous green algae were added to the marine flora of the island. The rather slow process of colonization of the marine algae on Surtsey is due to the relative isolation of the island and the severe environmental conditions. In 1966, the rate of colonization increased markedly, and a noticeable zonation was formed. By 1970, the marine vegetation at Surtsey had increased to 35 species of algae and 11 of diatoms. In 1997 altogether 76 species had been recorded, omitting diatoms.

The marine vegetation on Surtsey has apparently not yet reached the climax community. The kelp forests are slow in developing. However, many of the most common marine vegetation species of the adjacent areas have colonized the submarine pedestal of Surtsey.

Pelagic Biota

The volcanic eruption had little effect on the pelagic biota of the ocean around the island. These waters are relatively rich in marine life, and the banks around the Vestmannaeyjar are some of the best fishing grounds in Iceland.

Benthic Fauna

The benthic animals were destroyed during the eruption. While the island was still under formation, the colonization of benthic animals began, and the pumice was invaded by numerous larvae.

Already by November 1964, eight animals were found by a scraper from a depth of 70 m. Among them was the tube-dwelling trumpet worm *Pectinaria koreni.* In 1966 some bottom samples were collected at a 100-m depth. The species found indicated the presence of the most common animals, such as 17 species of marine bristle worms (Polychaeta).

This study has continued ever since. The marine fauna has developed normally, zonation is slowly forming, and gradually the communities on Surtsey are reaching the same balance as those of the adjacent islands.

TERRESTRIAL LIFE

The terrestrial life on Surtsey has been intensely investigated. In the first years of the island's existence, the presence of bacteria and molds was observed, and there already existed organic sources of energy on the island to support various living organisms, in the form of bird excrement as well as seaweed, kelp, and all sorts of driftwood.

Mosses

Although mosses easily occupy barren ground in Iceland, it was not until 1967 that the first moss colonies were discovered on Surtsey. The plants were all of the same species, common cord moss *Funaria hygrometrica*. The same year, a second location of mosses was observed at the edge of the central lava crater, where the colonies consisted of two species: cord moss and the silver moss *Bryum argenteum.* Since then, moss has gradually invaded the lava fields.

Lichens

Lichens are common on lava fields in Iceland. Surprisingly, it was not until the summer of 1970 that the first lichen was found on Surtsey. A lava crust, northeast of the central crater, was then found occupied by small specimens of fructicose lichen *Stereocaulon vesuvianum,* a common pioneer lichen on Icelandic lava flows. The other habitat was situated on the outer side of the northern slope of the crater. Two species, disk lichen *Trapelia coarctata* and bullseye lichen *Placopsis gelida,* had settled this area. Lichens have since gradually occupied the lava fields.

Vascular Plants

Vascular plants were earlier colonizers than most of the lower plants. A number of seeds were collected on the shore in 1964. They proved viable and capable of germinating. The first higher plant to colonize the island was found on the northern shore in 1965. These were small seedlings of the sea rocket *Cakile edentula.* This showed that living seeds could be carried by sea at least 20 km, which is the distance from Heimaey to Surtsey.

The second attempt of higher plants to invade Surtsey was in 1966, when four seedlings of sea lyme grass *Leymus arenarius* were found, and in 1967 a third species, the sea sandwort *Honckenya peploides,* started to grow on the island. The sea sandwort has since spread out widely over the sand-filled lava fields of Surtsey and has become the dominating plant of the island.

Vascular plants are gradually colonizing Surtsey. During the first 43 years, 69 species of higher plants have been discovered on the island. There was a relatively constant addition of one new species a year in the first ten-year period, but then a ten-year stagnant period followed. The formation of a gull nesting colony from 1981 to 1985 became a turning point in the rate of arrival of new vascular plants, with a rapid increase of colonizers appearing in the following years. The species found have had a rather high percentage of survival, as 90% of these are recorded annually.

Plant Communities

Since 1990 a number of permanent plots 10 × 10 m in area have been set up for studying the development of vegetation communities and soil in the various habitats on the island. At the same time, GPS instruments have been used to record locations of rare individuals discovered on Surtsey. The vegetation may be classified into four groups: (1) coastal community, dominated by the sea sandwort and lyme grass *Leymus arenarius,* (2) gravel flat community, which is more variable than the coastal community and has developed on a fine-textured substrate at the upper part of the lava, (3) skerry community on sand-filled lava, dominated by saltmarsh grass *Puccinellia coarctata* as well as scurvy grass *Cochlearia officinalis,* and (4) grassland community, fertilized by the colony of gulls since 1985. This community has smooth meadow grass *Poa pratensis* and red fescue *Festuca rubra* as dominant species, but it is also rich in various forbs. The succession of these communities has been followed intensely.

Invertebrate Fauna

The first terrestrial invertebrate to be found on Surtsey was the midge *Diamesa zernyi* recorded in May 1964. A number of species were brought in by wind and birds or on various driftage during the first years of the island's existence. In 1966 a small spider was found, which probably had glided from the mainland on its spinning thread. But there was a lack of food for the sustenance of these first settlers. With increased vegetation, the invertebrate fauna benefited, and in the first ten years, a total of 170 species had been identified on Surtsey. When the gull colony became established in 1985, there developed communities of invertebrates with their complex food chains. In 1993 the first earthworm was discovered, and the number of mites had climbed to 62 species. A total of 335 invertebrate species had been recorded on Surtsey in 2006.

Bird Invasion

The bird life on Surtsey has been followed intensely. Seagulls were seen landing on the island two weeks after it was formed. The island is the southernmost landing place in Iceland for migrating birds. The place was therefore used to investigate migrants and to record the landing of stragglers. The entire island was monitored, and records were kept of visiting birds. A total of 89 species of birds have been seen on Surtsey. The black guillemot and the northern fulmar were the first birds to breed on Surtsey in 1970. Since then, 13 species of birds have been found nesting on the island.

Of all these birds, the gulls have had the greatest impact on Surtsey. The lesser black-backed gulls started to nest in 1981. They began with only a couple of nests but ultimately formed a colony with 120 nests within four years. Their colony is the focus of pioneering lower life forms, worms, and insects, and to their breeding area the birds annually carry new plant material and fertilize diverse communities of plants with their droppings and offal.

Mammals on Surtsey

Seals are the only mammals found on Surtsey. Since 1983 both common and gray seals have annually bred on the island. They find the surrounding ocean abundant in food, as do some whales that are often seen off the coast of Surtsey.

THE FATE OF THE ISLAND

Surtsey offers more varieties of habitats than do the other outer islands. On the southeastern shore, coastal plant communities have developed. As sand blows into the vegetated area, dunes have been formed reaching a height of 1.5 m. The most conspicuous spot of vegetation on Surtsey is on the southern lava apron, where the lesser black-backed gull and the herring gull breed and fertilize the area heavily. There, the vegetation developed, and the plant community increased annually in size and density. In 20 years it covered an area of up to 10 ha. Almost 8% of Surtsey had turned green.

Gradually red fescue will probably dominate the area and will eventually cover the top of Surtsey as it does on all the other islands in the archipelago. Now, as the island is

FIGURE 4 Surtsey seen from the northern spit. The photograph shows rounded boulders and sand that are partly overgrown by the sea sandwort. Water gullies are in the slope of the cinder cone, Austurbunki. The island already looks old. Photograph by Borgthor Magnusson.

fast diminishing in area, these habitats will start to become more uniform and to resemble those of the other outer islands that have lost all of their flatlands and only retain the hardened core. Some of the other islands of the archipelago are 6000 years old; by that measure, it can be assumed that Surtsey will last for thousands of years (Fig. 4).

SEE ALSO THE FOLLOWING ARTICLES

Iceland / Island Formation / Lava and Ash / Seabirds / Vegetation / Volcanic Islands

FURTHER READING

Einarsson, T. 1966. *Gosið í Surtsey*. Reykjavik: Heimskringla.
Fridriksson, S. 1975. *Surtsey: evolution of life on a volcanic island*. London: Butterworths.
Fridriksson, S. 2005. *Surtsey: ecosystems formed*. Reykjavik: Vardi, the Surtsey Research Society.
Nomination of Surtsey to the UNESCO World Heritage List 2007. Reykjavik: Institute of Natural History.
Surtsey Research Progress Reports 1. 1965 to 12. 2006. Reykjavik: Surtsey Research Society.
Thorarinsson, S. 1967. *Surtsey: the new island in the North Atlantic*. New York: Viking Press Inc.

SUSTAINABILITY

R. R. THAMAN

University of the South Pacific, Suva, Fiji

Sustainability, or "sustainable development," is the ability of island nations or communities to acquire the income needed to purchase material and non-material goods from the modern cash economy that are needed to make life healthier, safer, more productive, and more enjoyable, but, at the same time, doing so without destroying the local natural and cultural capital needed for the material and cultural survival of future generations. Sustainability is about balancing these two, often conflicting, objectives: achieving the right balance between cash and subsistence economies, between self-reliance and dependency. For many traditional and indigenous peoples, sustainability is not so much about "production," but rather about the "reproduction" and survival of the cultures and communities that have proved relatively sustainable for millennia. It is suggested that, for islands, their biodiversity and ethnobiodiversity constitute the most important natural and cultural capital, which must be managed as the living bank account and foundation for sustainability. Such diversity has been the foundation for "subsistence affluence" and self-reliance for millennia, a foundation that is now seriously threatened.

EROSION OF BIODIVERSITY-BASED SUSTAINABILITY ON ISLANDS AND THE CBD

The accelerating erosion of the Earth's biodiversity, along with climate change, population growth, poverty, and overconsumption (which are closely linked), constitute perhaps the most serious obstacles to sustainable human occupation of the Earth. The recognition of this mainly human-induced biodiversity crisis resulted in the ratification by most nations, including most island nations, of the United Nations Convention on Biological Diversity (CBD), launched at the United Nations Conference on Environment and Development (UNCED), the "Earth Summit," held in Rio de Janeiro in 1992. Islands, particularly small islands, with their limited and fragile biodiversity inheritances, are among the most seriously affected. Although most island nations are signatories to the CBD, the evidence is that most of the island "developmental elite," when asked what "biodiversity" is, neither understand nor can define it in any detail, nor do they recognize that we have a "biodiversity crisis" that threatens the very sustainability of the "development" they promote. They also do not seem to understand or appreciate the extreme uniqueness and fragility of island biodiversity, which has become the focus of the most recent work program of the CBD, the Work Program on Island Biodiversity, approved in Curitiba, Brazil, in April 2006.

For most small island nations and communities, there are few options for modern, market-driven industrial or export-oriented development, and the conservation, sustainable use, and equitable sharing of the benefits from biodiversity, the three main objectives of the CBD, constitute the most important foundation for sustainable development. For most small island developing states (SIDS), the sustainability of their island cultures and economic well being is dependent on their island biodiversity inheritances. Although the main examples used here are the islands of the tropical Pacific Ocean, where biodiversity inheritances are relatively intact, the importance of biodiversity as a foundation for sustainability holds true for most islands. Sadly, despite the current apparent state of well-being and the absence of real poverty and associated social breakdown in the Pacific Islands, relative to other areas of the "developing" world, there is clearly a "biodiversity crisis" of unprecedented proportions" that undermines the sustainability of island life.

There are three simple messages: (1) The conservation and sustainable use of biodiversity and the protection of what has been referred to as "subsistence affluence" constitute, perhaps, the single most important foundation for short- and long-term poverty alleviation and sustainability on islands; (2) there is clearly a "biodiversity crisis of unprecedented proportions," both on land and in the seas surrounding islands, which undermines the sustainability and the integrity of island cultures; and (3) if not addressed as an integral component of all development thinking, the ultimate result of the erosion of island biodiversity will be the abject poverty that we have come to associate with continental societies that have pillaged their biodiversity, leaving behind degraded and life-depleted deserts, scrublands, polluted lakes and rivers, dying reefs, squatter settlements, and associated communities that are now among the poorest of the poor.

To achieve sustainability on islands, there is a need for rethinking the allocation of aid and funding for education, research, business development, and so forth toward the realization of a much more environmentally and culturally benign form of development, a development that builds on existing biocultural foundations rather than eroding or destroying these foundations. If such a strategy were to be adopted, it might just be possible to avoid the inevitable tightening of the vicious circle of resource depletion, environmental degradation, loss of biodiversity, and resultant economic and cultural breakdown that has doomed so many developing and "developed" (read "biodiversity destroyed") countries to a future of abject poverty and has marginalized their time-tested biodiversity-use traditions.

ISLAND BIODIVERSITY: A DEFINITION

Island biodiversity is defined as including (1) diversity of island types, (2) diversity of island and associated marine ecosystems and habitats, (3) species and taxonomic diversity within these ecosystems, (4) genetic diversity within and between species populations, and (5) ethnobiodiversity.

Island diversity includes the almost unbelievable diversity of island types found in the tropical Pacific Ocean and elsewhere on Earth. The diversity among islands affects the richness of respective biodiversity inheritances and the ability of governments, nongovernmental organizations (NGOs), and local communities to mount systematic programs for conservation and sustainable use.

Island ecosystem diversity includes all natural and "cultural" terrestrial, freshwater, and marine ecosystems and habitat types found on or around these islands and the "ecosystem services" they provide. These include terrestrial and freshwater ecosystems such as tropical lowland and upland forests; swamp and riverine forests; mangrove and coastal beach forests; woodlands and savannas; meadows; scrublands; deserts; marshes; rivers, streams, and lakes; marine ecosystems such as algal and seagrass beds; beaches; a range of reef and lagoon types; estuaries; offshore slopes, terraces, shelves, canyons, sea mounts and abyssal plains; and subsets of these, such as seabird rookeries, sea turtle nesting areas, and upwelling systems in the ocean. Also included are cultural or humanized ecosystems, such as secondary and fallow forests, shifting and permanent agricultural lands, grazing lands and livestock production systems, fish ponds and reservoirs, urban gardens, and towns. Table 1 is a simplified classification of Pacific Island ecosystems that can be used at the community or landowner, national, and regional levels, and in schools, to promote biodiversity conservation and serve as a basis for systematically gathering traditional knowledge (ethnobiodiversity) about these ecosystems.

Species and taxonomic diversity includes all species and taxonomic groupings (e.g., vascular and non-vascular plants, vertebrates and invertebrates, and microorganisms) found in island ecosystems and their surrounding seas. Table 2 attempts to summarize some of the main taxonomic groups from both scientific and indigenous Pacific Island perspectives. It is stressed that almost all categories are known to, and have recognized economic, cultural, and ecological utility for, Pacific Island societies.

Genetic diversity includes all subspecies, genetic types, breeds, cultivars, or varieties of wild and domesticated or cultivated plants and animals found in these ecosystems (e.g., cultivars of yams, taros, sweet potatoes, sugarcane, coconuts, breadfruit, mangoes, pandanus, etc., as well as breeds of pigs, chickens, dogs, and other animals), and, for the purpose of protecting intellectual property rights, all chemical extracts

TABLE 1
Terrestrial, Freshwater, and Marine Ecosystems of the Pacific Islands

Terrestrial/freshwater ecosystems

    Lowland native forest
    Upland or montane rain forest
    Mature fallow forest
    Plantation forest
    Grassland/woodland
    Scrubland/scrub-fernlands
    Shifting agricultural land
    Permanent/semi-permanent agricultural land
    Plantations
    Pasture
    Houseyard/urban gardens
    Intensive livestock holdings
    Ruderal sites
    Wetlands/swamps
    Rivers/streams/lakes/ponds
    Fish ponds/aquaculture
    Mangroves[a]
    Coastal strand vegetation
    Beaches and dunes
    Bare rock
    Caves
    Built/urban areas

Marine ecosystems

    Mangroves[a]
    Estuaries
    Intertidal zone
    Lagoons/bays
    Fish ponds/maricultural areas
    Coral reefs
    Island shelf/reef platform/ocean floor
    Open ocean

NOTE: Ecosystems listed are those that (1) constitute major resource-use zones and (2) could serve as the focus for community-based, national, and regional biodiversity conservation in the Pacific Islands. Adapted from Thaman 1994a.
[a] Mangroves are listed as both terrestrial and marine ecosystems.

from these organisms. In most traditional polycultural Pacific Island shifting agricultural systems, for example, there are invariably many distinct cultivars of staple food plants, all of which have differential utilities and levels of resistance to drought, tropical cyclones, diseases, floods, and so forth. Genetic diversity also includes human lineages.

"Ethnobiodiversity" is defined as the knowledge, uses, beliefs, resource-use systems, management systems and conservation practices, taxonomies, and language that a given society (or ethnic group), including the modern scientific community, has for their islands, ecosystems, species, taxa, and genetic diversity. It is stressed that this final category or "level" of biodiversity is central to the definition of biodiversity itself because, in the Pacific Islands and in most other traditional or indigenous island societies, people and their knowledge, traditions, and spirituality are seen as inseparable from their terrestrial, freshwater, and marine ecosystems (e.g., in western Melanesia as embodied in the Melanesia pidgin concepts of *kastom*/custom or *ples*/place; in Fiji in the concepts of *vanua*/land and *iqoliqoli*/fisheries; in Polynesia under the all-encompassing pan-Polynesian concept of land/*fonua, fanua, fenua, whenua, henua* or *'enua*, depending on where you are; or the concepts of *te aba* and *bwirej*, in Kiribati and the Marshall Islands, respectively).

ISLANDS AS BIODIVERSITY HOTSPOTS AND COOLSPOTS

Although biodiversity is the foundation for sustainability on most islands, there is great diversity among and disparity between biological inheritances. The larger continental islands (e.g., the islands of New Guinea, New Caledonia, and Madagascar) and the isolated high volcanic islands (Hawai'i and the Galápagos) have among the richest biodiversity inheritances and highest levels of endemism on Earth and are considered global "biodiversity hotspots." Many, such as Hawai'i and the Galápagos, have historically served as laboratories of evolution and extinction. Many of these "hotspots" also have considerable natural resource endowments and potential for modern economic development. Other islands with reasonably high levels of endemism and relatively rich biodiversity, but with less modern development potential, include the Solomon Islands, Vanuatu, and Fiji in Melanesia; Samoa, Tahiti, and some of the other islands of French Polynesia; and Palau and Pohnpei in Micronesia.

Conversely, many small, low-lying islands, particularly atolls, have among the poorest or least diverse terrestrial and freshwater floras and faunas on Earth, with virtually no endemic species. They, in effect, constitute the Earth's "biodiversity coolspots." Moreover, their terrestrial floras and faunas are among the most highly degraded and threatened on Earth. However, this fragile, very limited biodiversity inheritance is, for most isolated small island communities and nations, their only foundation for survival!

In terms of marine biodiversity, the disparity between large and small islands is not so great, although there is still an attenuation or dropoff in the number of species from west to east as we move from the center of diversity in Malesia (the tropical island world between Asia and Australia including Indonesia, the islands of Malaysia, the Philippines, New Guinea, and Taiwan) and the Indo–western Pacific to the small, more isolated, islands of the central Pacific.

In summary, although the larger "hotspot" high islands have far greater diversity of ecosystems than the smaller, low-lying "coolspot" islands, all islands have terrestrial and marine ecosystems, and in the case of larger islands, mon-

TABLE 2
Biological Resources of Community-Level Ecosystems in the Pacific Islands

Class	Subclass	Specific Type	Utility
Lower life forms		Bacteria	E,s,c
		Viruses	E,s,c
Plants	Indigenous	Phytoplankton	E,s,c
	Aboriginal introductions	Algae	E,S,C
	Recent introductions	Fungi	E,s,c
	Wild plants	Mosses	E,s
	Domesticated plants	Other lower plants	E,s,c
	Food plants	Ferns	E,S,C
	Non-food plants	Herbs/forbs	E,S,C
	Terrestrial	Grasses/sedges	E,S,C
	Freshwater	Vines	E,S,C
	Marine	Shrubs	E,S,C
	Trees		E,C,C
Animals	Indigenous	Protozoa	E,s,c
	Aboriginal introductions	Zooplankton	E,s,c
	Recent introductions	Sponges	E,s,c
	Wild animals	Corals	E,S,c
	Domesticated animals	Jellyfish	e,s,c
	Food species	Worms	E,s,c
	Non-food species	Molluscs	E,S,C
	Terrestrial	Insects	E,s,c
	Freshwater	Crustaceans	E,S,C
	Marine	Echinoderms	E,S,C
		Other invertebrates	E,s,c
		Finfish	E,S,C
		Amphibians	E,s,c
		Reptiles	E,S,C
		Birds	E,S,C
		Non-human mammals	E,S,C
		Humans	E,S,C

NOTE: Classes, subclasses, specfic types, and the utility of terrestrial, freshwater, and marine resources that constitute the pool of ecologically important and functionally useful biological resources of community-level ecosystems in the Pacific islands. Under "Utility," E, S, and C = direct major ecological, subsistence, or commercial/export utility to people at the community and national level in Melanesia, Polynesia, or Micronesia, and e, s, and c = minor or indirect ecological, subsistence, or commercial/export importance (e.g., amphibians are found on very few islands and plankton is of indirect importance to commercial tuna fishing in terms of its importance in marine food chains). It must be stressed that taxa in some categories may also be harmful or have a negative impact on sustainable development (e.g., pathogenic viruses or bacteria, mosquitoes, etc.).

tane freshwater ecosystems, that provide for most of the subsistence and cash needs of their resident rural communities. It is stressed that, regardless of island type, size, or degree of isolation, all islands have ecosystems, plants, animals and microorganisms that are critical to the continuing health and survival of each island's biodiversity itself and to the human communities that depend on the islands for their ecological, economic, and cultural survival. For many of the smaller, more isolated islands, their dependence on biodiversity and associated ethnobiodiversity is an "obligate dependence" because there are no other options. The "pyramid of sustainable development" (Fig. 1) based on the conservation of biodiversity is an attempt to illustrate the critical importance of biodiversity and associated ethnobiodiversity as factors that underpin all forms of development, including subsistence production (the main protection against poverty) and production for local sale and export, both of which provide an economic, cultural, and material base for sustainability.

CURRENT STATUS OF ISLAND BIODIVERSITY

Although all island societies have critically important terrestrial, freshwater, and marine biodiversity inheritances, including associated ethnobiodiversity, as we begin the twenty-first century, there are frightening signs of the loss or endangerment of this living inheritance that has supported island societies for millennia. These include a wide range of terrestrial, freshwater, and marine ecosystems and species that are now rare or endangered. This is not a new phenomenon, but rather a phenomenon that began long before European contact with the islands, when the early Pacific Islanders severely deforested many islands,

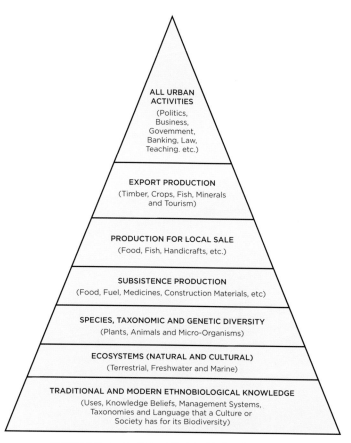

FIGURE 1 Pyramid of Island Sustainable Development illustrating the critical importance of (1) biodiversity as a foundation for most forms of development, including subsistence production and production for local sale and export, which provide the economic and material base for sustainability, and (2) "ethnobiodiversity"—the knowledge that underpins the conservation and sustainable use of limited island biodiversity inheritances.

including islands in the Marquesas, the Austral and Hawaiian Islands, and Easter Island (Rapa Nui). They also brought many bird species to extinction or extirpation (local extinction), extinguished much of the lowland biota throughout much of Polynesia, and brought to extirpation some species of giant clams in Fiji, Tonga, and Samoa. The process has, however, intensified, and the identification of the ecosystems, species, and genetic resources that are rare, endangered, or in need of protection or re-establishment is critical to the success of biodiversity conservation efforts and for the maintenance of cash and non-cash incomes (subsistence affluence), at *all* levels (regional, national, and local). An attempt to do this for the entire Pacific is shown in Table 3.

As can be seen from Table 3, there is a wide range of ecosystems and organisms that are in need of protection. Although the need for the protection of these organisms is slightly less in the larger "biodiversity hotspot" islands of Melanesia, where there is greater ecosystem and biotic diversity, larger areas of strategic ecosystems, larger populations of individual organisms, and lower human population densities, there is, nevertheless, a serious need to protect designated ecosystems and specific endangered organisms throughout the Pacific. Even in the larger island countries, there exist many small outer island communities with limited resources and high population densities that experience the same trends of degradation and loss of biodiversity as the communities in the smaller "coolspot" islands of the eastern Pacific. This assessment is based on over 15 years' field observation and in-depth interviews with local communities.

THREATS TO ISLAND BIODIVERSITY

Table 4 is an attempt to classify the most significant threats to the conservation and sustainable use of island biodiversity. They include (1) direct threats, which seriously degrade and upset the stability of natural and cultural ecosystems and their biodiversity and (2) social, institutional, or infrastructural activities or phenomena that indirectly threaten or undermine capacity to conserve or sustainably use island biodiversity. Most of these threats are of both global and local concern and can be addressed, in some way, at all levels. Some, however, such as global warming, the depletion of stratospheric ozone, the international trade of endangered or potentially invasive plants and animals, and pest and disease infestations and epidemics, are best dealt with regionally or internationally, whereas the protection of endangered or threatened species and ecosystems are, perhaps, best dealt with at the national or community levels. Similarly, many of these threats overlap or feed into each other and, if not addressed in some way, could lead to a dangerous negative synergistic effect and to the collapse of entire island ecosystems or biological communities and the countries and cultures that depend on them. Conversely, if a number of threats are addressed simultaneously, the result could be very positively synergistic and could lead to significant gains in the mainstreaming of conservation and sustainable use of biodiversity in the Pacific Islands.

It is beyond the scope of this article to discuss each of these threats. Suffice it to say that all of these threats, many of which are interrelated, are of serious concern in varying degrees throughout the Pacific Islands and other islands globally. If efforts to promote island biodiversity conservation are to be successful, these threats need to be systematically addressed at a range of different levels (government, school systems, local communities), by different agencies (e.g., NGOs, private enterprises, funding/aid agencies), and in a range of different international forums.

TABLE 3
Pacific Islands Taxa in Need of Protection

Category	Melanesia	Polynesia	Micronesia
Ecosystem			
Uninhabited islands	++	+++	+++
Coastal littoral and mangrove forests	++	+++	+++
Lowland forests	++	+++	+++
Montane/cloud forests	++	++	++
Rivers and lakes	+++	+++	+++
Wetlands/swamps	++	+++	+++
Shifting agroforestry lands and agroforests	++	+++	+++
Semi-permanent/intensive agricultural areas	++	++	+++
Houseyard and village gardens	++	++	+++
Selected productive reefs	+++	+++	+++
Intertidal zone and seagrass beds	++	+++	+++
Reef passages	++	+++	+++
Coral reefs	+++	+++	+++
Terrestrial organisms			
Native coastal and mangrove plants	++	+++	+++
Native inland trees and plants	++	+++	+++
Cultivated trees and plants	++	+++	+++
Plant cultivars/varieties	++	+++	+++
Native insects/arthropods	++	+++	+++
Land crabs	++	+++	+++
Native molluscs	++	+++	+++
Other native invertebrates	++	+++	+++
Native amphibians	+++	NP	NP
Native reptiles	+++	++	++
Native birds	+++	+++	+++
Native mammals	+++	+++	+++
Humans (ethnobiological knowledge)	+++	+++	+++
Freshwater organisms			
Freshwater plants	++	+++	+++
Crustaceans	++	+++	+++
Shellfish	++	++	++?
Insects	+++	+++	+++
Finfish/eels	++	+++	+++
Amphibians	++	NP	NP
Reptiles	+++	+	+
Marine organisms			
Seaweeds (marine macro-algae)	++	++	++
Sea grasses	+++	+++	+++
Stony reef-forming corals	+++	+++	+++
Shellfish (giant clams, trochus, turban snail, pearl oyster, triton)	+++	+++	+++
Bêche-de-mer/holothurians	+++	+++	+++
Crabs, lobsters, mantis shrimp	++	+++	+++
Reef and lagoon fish	++	++	++
Eels (conger, moray)	++	++	++
Large demersal finfish (rockcods, wrasses, parrotfish)	+++	+++	+++
Sharks and rays	++	+++	+++
Billfish	+++	+++	+++
Turtles	+++	+++	+++
Crocodiles	+++	–	+
Seabirds	+++	+++	+++
Mammals (whales, dolphins, dugongs)	+++	+++	+++

NOTE: Ecosystems and groups or taxa of terrestrial, freshwater, and marine plants and animals that are rare, endangered, or in short supply and in need of protection in the Pacific Islands (+++ = of serious widespread concern and in need of immediate protection; ++ = of some widespread concern or of serious concern in specific areas; + = of limited or localized concern; – = of no concern; NP = not present).

TABLE 4
Threats to Biodiversity in the Pacific Islands

Direct Threats to Biodiversity

1. High frequency of extreme events/natural disasters
2. Global warming/eustatic sea-level rise
3. Stratospheric ozone depletion and increasing UV-B radiation
4. Breakdown and simplification of the species composition and trophic structure of terrestrial, freshwater, and marine ecosystems and ecosystem functions
5. Degradation of uninhabited islands
6. Upland and inland deforestation and forest degradation
7. Coastal and mangrove deforestation and degradation
8. Degradation of freshwater resources and ecosystems
9. Agricultural simplification and degradation, agrodeforestation, and the loss of biodiversity in agricultural systems
10. Overgrazing and degradation of biodiversity by domestic livestock
11. Destruction caused by feral animals
12. Alien invasive plants and animals
13. Pest and disease infestations and epidemics
14. Soil degradation and accelerated soil erosion
15. Fire
16. Destruction and degradation of productive marine ecosystems and disruption or change in the dynamics of marine ecosystems
17. Overuse/overexploitation/unsustainable use of terrestrial plant and animal resources
18. Overfishing/overexploitation/unsustainable use of marine resources
19. Use of destructive fishing technologies
20. Illegal fishing
21. Pollution of freshwater resources
22. Air pollution
23. Marine pollution
24. Indiscriminate and increasing use of pesticides
25. Hazardous/toxic waste disposal
26. Nuclear/radioactive pollution and contamination

Social, Institutional, and Infrastructural Threats

1. Uncontrolled population growth
2. Loss of traditional ethnobiological/ethnobiological knowledge
3. Breakdown in traditional diversified subsistence economy
4. Inadequate modern scientific baseline knowledge of the nature and status of biodiversity
5. Inadequate systems of marine and terrestrial conservation areas
6. Inadequate capacity to deal with terrestrial, freshwater, and marine invasive species
7. Inadequate legislation/legal instruments
8. Inadequate infrastructure/capacity for biodiversity conservation
9. Inappropriate modern education and curricula
10. Rapid and uncontrolled urbanization
11. Unforeseen large-scale developments
12. Free trade/globalization and increasing international free trade in biodiversity
13. Poverty and economic deterioration
14. Gender inequity in the control, use, and management of biodiversity
15. Political instability and political ignorance or lack of political will to commit to conservation

NOTE: Table lists significant reasons for the loss of biodiversity or threats to biodiversity and biodiversity conservation in the Pacific Islands that can be addressed in the mainstreaming of biodiversity conservation at the regional, national, and local community levels.

One threat that should be singled out, however, is the loss or lack of ethnobiodiversity, because the loss of traditional and indigenous knowledge about the uses, beliefs, management systems, taxonomies, and language, and the lack of modern scientific knowledge related to biodiversity, could be among the most serious obstacles to successful biodiversity conservation and sustainability on islands. As suggested diagrammatically in the "Pyramid of Sustainable Island Development," good knowledge, both traditional and modern, about biodiversity is the basic requirement for all island biodiversity conservation and sustainability. At the local level, site-based biodiversity conservation will be problematic if local people cannot marry traditional conservation strategies with modern scientific models and findings as part of co-management systems. If local people no longer know the local names, uses, and management systems for their biodiversity, chances are that they will *not* place a priority on its preservation. At the same time the modern scientific community is bemoaning the lack of financial support for the training of good modern taxonomists to replace those who pass away, many of the best traditional Pacific Island men and women scientists and taxonomists are dying and are not being replaced by a younger generation that is less interested in the natural world. Along with them dies traditional ethnobiodiversity that has been accumulated over thousands of years in close contact with the island environment.

MAINSTREAMING BIODIVERSITY CONSERVATION AS A FOUNDATION FOR SUSTAINABLE ISLAND LIFE

The conservation and sustainable use of island biodiversity as a foundation for sustainability is, perhaps, the single most important obligation we have, as the current generation, to future generations of islanders. The challenge before us is clearly to get this message out to all island stakeholders, at all levels, and from all walks of life, and to get them to understand its real meaning. Hopefully, as a result, they will also take on board, as individuals and as groups or institutions, the protection and wise use of this living inheritance as central to their philosophies of life and theories of island development. This should be the objective of "mainstreaming" island biodiversity conservation.

Fortunately, there are an increasing number of local, national, and international initiatives that are spreading this message and implementing programs, particularly community-based programs, to promote the conservation and sustainable use of biodiversity in the Pacific Islands. Many of these initiatives are led by consortia or networks of NGOs, regional organizations, national and local government agencies, the private sector, and local landowners and resource users and are funded by an increasing number of entities including the MacArthur and Packard Foundations and Australian, New Zealand, French, American, and Japanese funding agencies. Prominent among these include international conservation NGOs, such as the WWF—World Wide Fund for Nature (known in the United States as the World Wildlife Fund), The Nature Conservancy (TNC), Conservation International (CI), the Wildlife Conservation Society (WCS), Birdlife International, the Foundation for the Peoples of the South Pacific International (FSPI) and its national counterparts, the World Conservation Union (IUCN), the United Nations Education, Social and Cultural Organization (UNESCO), as well as Pacific Island regional organizations, such as the Secretariat of the Pacific Regional Environment Programme (SPREP), the Secretariat of the Pacific Community (SPC), the Secretariat for the South Pacific Applied Geoscience Commission (SOPAC), the University of the South Pacific (USP), the University of Hawaii, the University of Guam, the Bishop Museum, and the Institute for Research and Development (IRD) in the French Territories. A number of local NGOs are also participating, such as Nature Fiji–Mareqeti Viti and Micronesians in Island Conservation (MIC). Both alone and together, with funding from bilateral and multilateral sources and foundations, these entities have, over the last decade, established regional, national, and local community–based networks of terrestrial and marine protected areas, produced educational materials, assessed the status of biodiversity, and championed programs and projects designed to address the major threats to biodiversity conservation and sustainable use and to create alternative sources of revenue to take pressure off threatened biodiversity. In an increasing number of cases, governments have sanctioned and become integral partners in these initiatives, and in the many successful cases the initiatives have focused on the full involvement of local communities and resources users and owners in the planning, implementation, and monitoring of the initiatives.

Some of the more notable initiatives include the Fiji and Asia-Pacific Locally Managed Marine Areas Networks (FLMMA and APLMMA), the Pacific-Asia Biodiversity Transect Network (PABITRA), the Marine Aquarium Council (MAC), the South Pacific Regional Initiative on Forest Genetic Resources (SPRIG), the Micronesia Conservation Trust (MCT), the Micronesians in Island Conservation Network (MIC), the Micronesian Challenge, the accession to the Ramsar Wetlands Convention by five Pacific countries, the Pacific Invasives Initiative (PII), and the Pacific Invasives Learning Network (PILN).

Under the FLMMA and wider APLMMA, over the past decade, some 300 locally managed marine areas with associated management plans have been established, over 200 of which are in Fiji. The main partners in this initiative are Fiji, Solomon Islands, the Federated States of Micronesia, Palau and Papua New Guinea in the Pacific, and the Philippines and Indonesia in Asia. Results indicate that fisheries stocks have increased in most areas and that local communities are taking biodiversity conservation into their own hands. Much of the success has been based on the sharing of success stories and methodologies for community-based conservation. Also in Fiji, prohibited fishing zones have been set aside in Fiji's Great Sea Reef (known locally as Cakau Levu) to conserve Fiji's most extensive area of coral reef, which covers more than 200,000 km^2 and is home to thousands of marine species, including marine turtles, dolphins, sharks, and 43 new hard coral species. The reef is also an important fishing ground for local communities.

For over a decade, PABITRA has promoted comparative studies and built local research capacity for the study of island ecosystems across the Pacific, with a focus on local people's perceptions and multiple ecosystem-use systems along the lines of the Hawaiian *ahu pua'a*, "summit to sea" or "ridge to reef" integrated ecosystem and land-use approach. Under PABITRA, stress has also been placed on the importance of the protection, recording, and application of indigenous knowledge about island biodiversity and ethnobiodiversity.

The Marine Aquarium Council (MAC) has, over the past ten years, established a program to certify entities involved in the trade of aquarium fish, coral, and other marine organisms. The program has been very successful in a number of areas of the Pacific in promoting best practices in this area. There are similar, but so far less successful, initiatives addressing certification of timber exploitation from indigenous rain forests in some Pacific countries, and the SPRIG program has had, as its main focus, the protection, collection, and propagation of forest genetic diversity as a basis for the development of sustainable plantation forestry and agroforestry systems in the Pacific Islands.

The Micronesian Challenge, which was launched in 2006 by the president of Palau, involves committing at least 30% of nearshore marine and 20% of forest resources across Micronesia to conservation by 2020. Similarly, in 2006, the Kiribati government designated the atolls and marine area of the Phoenix Islands as the world's third largest marine protected area, and uninhabited Ant (Ahnd) Atoll in Pohnpei State of the Federated States of Micronesia has been recently designated as a UNESCO Biosphere Reserve, as has the Lake Tenggano area of East Rennell, in the Solomon Islands.

Also, as of 2008, five Pacific countries (Papua New Guinea, Fiji, Samoa, Palau, and the Marshall Islands) have ratified the Ramsar Wetland Convention and designated "Wetlands of International Importance" that are now under some form of conservation.

The Pacific Invasives Initiative (PII) and the Pacific Invasives Learning Network (PILN) are focused on reducing the impacts of invasive species on island economies and ecosystems through prevention and management of invasive species and through building capacity at the national level to deal with invasive species.

To mainstream biodiversity conservation on islands is to ensure that all individual and institutional stakeholders involved in island development (local communities, private enterprises, governments, NGOs, the aid community, and other international agencies) include the conservation and sustainable use of biodiversity as a priority concern, if not as the most important precondition, for our own individual or institutional well being. To make this vision a reality, there is an urgent need for all stakeholders to clearly understand (1) what island biodiversity really is, its current status, and why biodiversity conservation should be a priority concern as a basis for sustainable development and the well being of all and (2) what we can do as individuals and institutions to promote its conservation. Once this level of awareness of the issues is achieved (and there are, as outlined above, a number of very positive developments in this direction), it is hoped that the incentives and reward systems will serve to motivate all stakeholders to address the increasingly serious threats to island biodiversity, even if it requires sometimes drastic changes in the way we conduct, and with whom we conduct, international and local business.

If efforts to mainstream the conservation of this "living foundation" are successful, then those who live on islands, their children, and their children's children will hopefully be able to walk along island mountain trails, streams, shores, and reefs and through island towns, and continue to marvel at the cowries, fishes, trees, flowers, birds, and other island plants and animals that, for thousands of millennia, have provided the cultural, economic, and ecological foundation for the rich (not poor) biodiverse island cultures of today. If these efforts are not successful, then widespread impoverishment of future generations and their environments and nature-based cultures and widespread unsustainability will be the result.

SEE ALSO THE FOLLOWING ARTICLES

Ethnobotany / Marine Protected Areas / Reef Ecology and Conservation

FURTHER READING

Clarke, W. C., and R. R. Thaman, eds. 1993. *Pacific Island agroforestry: systems for sustainability*. Tokyo: United Nations University Press.

Kay, E. A. 1999. Biogeography, in *The Pacific Islands: environment and society*. M. Rapaport, ed. Honolulu, HI: The Bess Press, 353–365.

Steadman, D. W. 1995. Prehistoric extinctions of Pacific Island birds: biodiversity meets zooarchaeology. *Science* 267: 1123–1131.

Stoddart, D. R. 1992. Biogeography of the tropical Pacific. *Pacific Science* 46: 276–293.

Thaman, R. R. 1999. Pacific Island biodiversity on the eve of the 21st century: current status and challenges for its conservation and sustainable use. *Pacific Science Association Information Bulletin* 51: 1–37.

Thaman, R. R. 2004. Sustaining culture and biodiversity in Pacific Islands with local and indigenous knowledge. *Pacific Ecologist* Autumn–Winter: 43–48.

Thaman, R. R. 2005. Biodiversity is the key to food security. *Spore* 117 (June): 1–3.

SVALBARD

SEE ARCTIC REGION

TAIWAN, BIOLOGY

MAN-MIAO YANG

National Chung Hsing University, Taichung, Taiwan

KUANG-YING HUANG

Yangmingshan National Park, Taipei, Taiwan

Taiwan, an island of approximately 36,000 km², is situated in the western Pacific Ocean and separated from China by the Taiwan Strait. It is also known as Formosa (Fig. 1), meaning "the beautiful island," a name given to it by Portuguese mariners who encountered it in the sixteenth century. Around the main island of Taiwan, there are more than 80 smaller islands, including the granite-origin continental islands, volcanic oceanic islands, cubic basalt-formed Pescadores (Penghu archipelago) (Fig. 2A), and coral-reef based islands (e.g., Pratas atoll [Fig. 2B]). This article focuses on the main island of Taiwan.

LANDSCAPE AND CLIMATE

Lying directly in the center of the East Asian islands, Taiwan belongs to the Pacific Ring of Fire, an area of dense volcanic action. The island of Taiwan is located at the place where the Philippine plate is subducting under the Eurasian plate. Consequently, the elevated Central mountain range, stretching from north to south in the middle of Taiwan, forms the spine of the island and is a natural watershed for the rivers. In addition, there are four additional parallel mountain ranges: to the west is the Hsuehshan (Snow Mountain) range in the north, the Yushan mountain range in the middle, and the Alisan

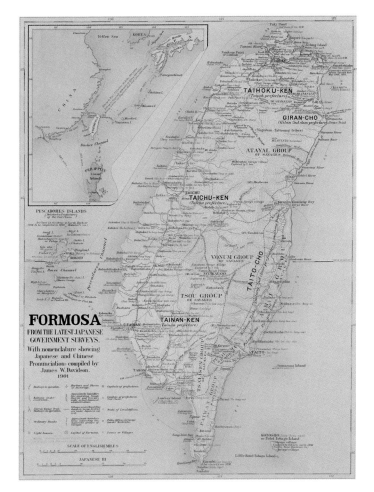

FIGURE 1 There are more than 80 islands in the area of Taiwan, and they have diverse origins. This map of the main island Taiwan (Formosa) in 1901 clearly shows its relative position in East Asia and the distribution of various indigenous people living on it. Courtesy of SMC Publishing Inc., after James W. Davidson (1901).

TABLE 1
Altitudinal Vegetation Zones and Their Temperature Ranges in Central Taiwan

Altitudinal zone	Vegetation zone	Elevation (m)	Mean annual temperature (°C)
Alpine	*Juniperus–Rhododendron* zone	3600–4000	< 5
Subalpine	*Abies* zone	3100–3600	5–8
Upper montane	*Tsuga–Picea* zone	2500–3100	8–11
Montane	*Quercus* zone	1500–2500	11–17
Submontane	*Machilus–Castanopsis* zone	500–1500	17–23
Foothill	*Ficus–Machilus* zone	0–500	> 23

SOURCE: Su (1984).

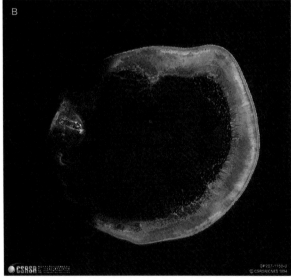

FIGURE 2 (A) Aerial photograph of Jishan Islet of the Pescadores archipelago, which reveals the typical geology of andesite volcanism with plateau basalt forming flat-topped tablelands. Photograph by Po-Lin Chi. (B) Satellite photograph of Pratas Atoll with its main island situated on the west of the reef ring. The 350,000-ha area was newly established as Dongsha Marian National Park in 2007. Courtesy of ©CSRSR/CNES 1994.

mountain range in the southeast; to the east is the lower Coastal mountain range. More than 200 mountain peaks in Taiwan exceed an altitude of 3000 m, with Yushan, or Mt. Jade, the highest peak in East Asia, reaching 3952 m. Some vestiges of the glacial age may be seen in the alpine areas of these mountains, such as a glacier cirque on Snow Mountain. Short and rushing rivers, including 19 primaries, 32 tributaries, and 100 minors, cut through these steep mountains, with the longest one, Chuo-Shuei River, flowing 186 km into the Taiwan Strait. Although two-thirds of the island is covered by mountains, the remaining area, which mainly lies along the western coast, consists of hills, tablelands, coastal plains, and basins. The length of coastal line around the main island is 1250 km, and the southern part features rich coral reefs.

Lying on the Tropic of Cancer and extending from 21°45′25″ to 25°56′20″N latitude, Taiwan features a climate that is mainly subtropical, with the southern part being tropical; areas at higher elevations are considered temperate. Snow may be seen on high mountain caps, but the lowland rarely has temperatures below 10 °C. The average temperature is 28 °C in summer and 14 °C in winter, and annual precipitation is higher than 2000 mm in most areas. In general, Taiwan has an oceanic and subtropical monsoon climate, evidently affected by its topography. Spring monsoon is also known as "plum rain," and the length varies from year to year. Although typhoons often attack Taiwan in the summer and autumn, causing floods and damage, they also bring plentiful rainfall to fill the reservoirs for supporting farming and other livelihoods.

FLORA

The nearly 4000 recorded vascular plants denote the high floristic diversity of Taiwan, considering its relatively small area. As a consequence of the high annual precipitation and the lofty mountains, forests are the dominant natural vegetation types, and various forest vegetation zones are characterized by different dominant plant groups (Table 1), with temperature gradients changing with elevation. In general, lowland broadleaf forests are dominated by Moraceae, Euphorbiaceae, and some Lauraceae. Between altitudes of 500 and 2500 m, Lauraceae and Fagaceae are the dominant groups. Above 2500 m, conifer forests are found in the region with various zones being dominated by different evergreen trees; the ecotone may be obvious in high mountain areas.

The recorded numbers of families for ferns, gymnosperms, and angiosperms in Taiwan are 38, 8, and 228, respectively, representing more than 50% of world flora in each category. The flora of Taiwan is not only abundant for its high diversity but is also remarkable for its high endemism (Fig. 3A). There are about 4000 species of indigenous vascular plants in Taiwan, and nearly a quarter of the plant species are endemic. Additionally, because of its transitional position biogeographically, Taiwan has become the home of many relic species, such as the Taiwan red cypress (*Chamaecyparis formosensis*), Taiwan cypress (*Chamaecyparis obtusa* var. *formosana*), the Formosan redwood or Taiwan cedar (*Taiwania cryptomerioides*), the wheelstamen tree (*Trochodendron aralioides*), the Chinese sweet gum (*Liquidambar formosana*), Taiwan Pieris (*Pieris taiwanensis*), the fern *Dipteris conjugata* (Fig. 3B), and so forth.

FAUNA

The known fauna of Taiwan includes more than 32,000 species and many more surely remain to be explored. Recorded species (TaiBNET 2008) include 113 species of mammals, 461 birds, 99 reptiles, 36 amphibians, nearly 3000 freshwater and marine fishes, and about 20,000 insects. Located on the boundary of the tropics and subtropics, Taiwan's transitional position is also depicted by its separation from mainland China by the Taiwan Strait on the west and its division from its southeast island neighbor Lanyu by Kano's extension of Wallace's Line on the east. These factors, along with Taiwan's diverse landscapes, have created a variety of ecological habitats for wildlife and have nurtured many endemic species, such as the Formosan macaque (*Macaca cyclopis*), Swinhoe's pheasant (*Lophura swinhoii*), the Formosan blue magpie (*Urocissa caerulea*) (Fig. 4A), the Taipei tree frog (*Rhacophorus taipeianus*), the Formosan greater horseshoe bat (*Rhinolophus formosae*) (Fig. 4B), the snake *Rhabdophis swinhonis* (Fig. 5A), and the highland red-belly swallowtail butterfly (*Atrophaneura horishana*) (Fig. 5B), as well as many endemic subspecies, such as the Formosan mole (*Mogera insularis insularis*), the Formosan hare (*Lepus sinensis formosus*), the crested serpent eagle (*Spilornis cheela hoya*) (Fig. 6A), the Formosan land-locked salmon (*Oncorhynchus masou formosanus*), and the Formosan long-armed scarab (*Cheirotonus macleayi formosanus*) (Fig. 6B). The high mountain ranges in Taiwan play an important role as a natural barrier for

FIGURE 3 Endemic plants and animals of Taiwan. (A) The Taiwan lily, *Lilium formosanum*, a widely distributed endemic plant among four indigenous lilies of Taiwan; (B) *Dipteris conjugata*, a relic fern that survived through ice ages and has lived on the Earth since the Jurassic Period. Photographs by Kuang-Ying Huang.

FIGURE 4 Endemic animals of Taiwan. (A) The Formosan blue magpie, *Urocissa caerulea*; (B) the Formosan greater horseshoe bat, *Rhinolophus formosae*. Photographs by Kuang-Ying Huang.

FIGURE 5 Endemic animals of Taiwan. (A) *Rhabdophis swinhonis*, an endemic snake named after Robert Swinhoe, a British consul on Taiwan in the nineteenth century who made great contributions to the natural history of Taiwan; (B) the highland red-belly swallowtail butterfly, *Atrophaneura horishana*. Photographs by Kuang-Ying Huang.

many species. Species assemblages in the west and the east of the island may differ, as is the case with some freshwater fishes and insects. The endemism is as high as 64% in mammals, 31% in reptiles, 27% in amphibians, and 18% in birds (TaiBNet 2007). The Estimate of insect endemism in Taiwan is more than 60%.

The unique location of Taiwan also provides roosting places for migratory birds. The population of the endangered black-faced spoonbill, *Platalea minor* (Fig. 7), numbers only slightly over a thousand worldwide, and Taiwan has served as an overwinter area for more than two-thirds of this population. The gray-faced buzzard *Butastur indicus* is a spring and fall migrant in Taiwan. Magnificent views can be seen when large numbers of these buzzards utilize rising air currents to gain lift and then soar into the sky.

BIOGEOGRAPHIC AFFINITIES

The flora and fauna of Taiwan originated in the Northern Hemisphere and dispersed to the mid-high altitude of Taiwan during several ice ages, the latest of which was between 7000 and 10,000 years ago. During these periods, a land bridge connected Taiwan and mainland Asia, which allow organisms to disperse between the two land masses. Organisms arrived in Taiwan via two routes: one from the west, through the Himalayas, and the other from the northern continental region. Evidence is found in many birds, insects, and other organisms that affinity species distributed in a disjointed manner from their center of origin to Taiwan. Moreover, the montane cloud forests, often covered by cloud or mist, nurture magnificent virgin cypress forests in Taiwan (Fig. 8), accounting for the largest known mass of old growth conifers in the subtropical regions.

HUMAN HISTORY AND IMPACTS

Located in a biogeographical transitional zone, Taiwan encompasses a remarkable biodiversity in both biota and culture. There are more than 13 indigenous Austronesian tribes, who have been lived on the island for more than 5000 years; additionally, many immigrants from southeastern China have settled on the island in the last 400 years. Linguistic and some other archeological evidence support Taiwan as the center of origin of the Austronesians, who later dispersed between Easter Island and Madagascar.

FIGURE 6 Endemic animals of Taiwan. (A) the crested serpent eagle (*Spilornis cheela hoya*), a snake predator commonly seen in the lowland area; (B) the Formosan long-armed scarab (*Cheirotonus macleayi formosanus*), an endemic subspecies of large-sized beetle that is on the protected species list. Photographs by Kuang-Ying Huang.

The heavy population pressure in Taiwan, which currently houses 23 million people, has resulted in immense forest clearing in the lowlands for several centuries. No virgin forest is left on the plain area, and only a few patches remain on the foothills. Many temperate old growth forests in the montane zone, at 1500–2500 m, faced extensive logging over the last 100 years. Excessive farming in this area during the last several decades has also caused severe erosion. The change of habitats certainly impacts the biota

FIGURE 7 Two-thirds of the world's population of the endangered black-faced spoonbill, *Platalea minor*, overwinter in the estuarine wetlands of southwestern Taiwan. Photograph by Chieh-Te Liang.

and fauna. Some native species, such as the Formosan clouded leopard (*Neolelis nebulosa*) and the Formosan sika deer (*Cervus Nippon taiouanus*), were reported to be extinct in the wild because of habitat loss in recent centuries.

As the twentieth century ends and the twenty-first begins, both government and non-government organizations have made concerted efforts to promote nature conservation. Pertinent laws have been enacted, such as the Cultural Heritage Preservation Act in 1981 and the Wildlife Conservation Act in 1989. To protect the island's diverse biological resources and ecosystems, approximately 20% of the land is now officially designated as protected areas, including six terrestrial national parks, one marine national park, 19 nature reserves, eight forest reserves, 15 wildlife refuges, and 29 major wildlife habi-

FIGURE 8 The fog forests of the cypress oldgrowth in the Magao area of Taiwan. Cloud fall and fog forest are typical phenomenon seen on the mountains in the altitude range between 1500 and -2500 m. Photograph by Kuang-Ying Huang.

FIGURE 9 A wildlife underpass, part of the corridor system in Yangmingshan National Park, reconnects habitats divided by a road to prevent roadkills. (A) Guiding net; (B) an endemic Formosan gem-faced civet, *Paguma larvata taivana*, emerging from the underpass (B). Photographs by Kuang-Ying Huang.

tats. Many long-term biodiversity research projects and hotspot surveys have been initiated. Furthermore, some mitigation projects to reduce human impact on wildlife have been implemented, such as the wildlife underpass system, which has successfully reduced 40% of roadkills in Yangmingshan National Park (Fig. 9).

SEE ALSO THE FOLLOWING ARTICLES

Hurricanes and Typhoons / Island Formation / Pacific Region / Taiwan, Geology / Wallace's Line

FURTHER READING

Blundell, D. 2000. *Austronesian Taiwan: linguistics, history, ethnology, prehistory.* Phoebe A. Hearst Museum of Anthropology and Shung Ye Museum of Formosan Aborigines.
Kuo, C. M., ed. 2002. Discovery of green Taiwan. Taiwan Forestry Bureau and BCSD-Taiwan, R. O. C.
Lai, Y. M., ed. 2000. Vanishing dancers. The special issue on the conservation of rare and endangered animals in Taiwan. Council of Agriculture and BCSD-Taiwan, R. O. C.
Peng, C. I., ed. 1992. The biological resources of Taiwan: a status report. Institute of Botany, Academia Sinica.
Su, H. J. 1984. Studies on the climate and vegetation types of the natural forests in Taiwan. (II). Altitudinal vegetation zones in relation to temperature gradient. *Quarterly Journal of Chinese Forestry* 17: 57–73.
Su, H. J. 1994. Species diversity of forest plants in Taiwan, in *Biodiversity and terrestrial ecosystems.* C. I. Peng and C. H. Chou, eds. Institute of Botany, Academia Sinica. Monograph Series 14: 87–98.
Taiwan Biodiversity International Network (TaiBNET). 2008. http://taibnet.sinica.edu.tw/ajaxtree/allkingdom.php.

TAIWAN, GEOLOGY

YUE-GAU CHEN

National Taiwan University, Taipei

Taiwan, located in the western margin of the Pacific with a center at 24° N, 121° E, is an olive-shaped island with a length and width of ~400 km and 200 km, respectively. The highest mountain peak, Jade Mountain, rises ~4000 m above sea level. Approximately two-thirds of the island's area is characterized by mountains, hills, and tablelands. Only the remaining one-third is settlement-suitable plain, an area that is mainly distributed along the western coast.

MOUNTAINOUS ISLAND AND ARC-CONTINENT COLLISION

Taiwan is dominated by mountain ranges because of its tectonic setting. Based on geologic and seismologic lines of evidence, Taiwan has been defined as one of the modern examples representing an arc–continent collision.

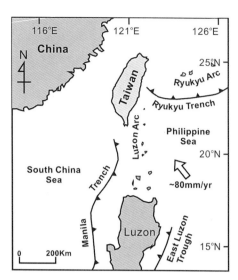

FIGURE 1 Map showing the tectonic environment of Taiwan and its adjacent region. The Philippine Sea plate is moving northwestward and subducting under the Eurasian plate along the Ryukyu trench. Accordingly, the Luzon arc, generated by the subduction of South China Sea crust under the Philippine Sea plate, collides with the continental margin to create the island of Taiwan.

The "arc" is a volcanic arc (i.e., the Luzon arc) formed by subduction of South China Sea oceanic crust beneath the Philippine Sea plate (Fig. 1). However, the Philippine plate moves northwesterly toward the Eurasian plate and dives under it along the Ryukyu trench, conveying the Luzon arc into the subduction zone once it approaches the trench. The Luzon arc is a huge mass sitting on the top of the plate, so during subduction, it inevitably causes a collision between the arc and the margin of the "continent." In response to such collisional compression, the crustal materials are thickened to form an accretionary prism, which is the reason for the island of Taiwan's existence.

WESTWARD DEVELOPMENTS OF NORTH-SOUTH-DISTRIBUTED MOUNTAIN RANGES

Under such a geologic situation, the rocks exposed on land consist of materials from both the arc and continent sides, which are divided by the longitudinal valley (LV) in eastern Taiwan (Figs. 2 and 3). To the east of the LV, a parallel mountain range is composed of a rock association related to the arc system (i.e., andesitic volcanics, limestones, and deep-sea sedimentary turbidite sequences). To the west of the LV, a suite of mountain ranges are distributed westward all the way to the western coastal plain (Fig. 3), which has been regarded as the deformed continental margin. Additionally, the metamorphic grade of the rocks decreases from the LV westward. The eastern flank of the Central Range, located immediately next to the LV in the west,

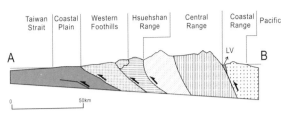

FIGURE 3 A cross-island section showing west-vergent development of the fold-and-thrust belt of Taiwan. The easternmost mountain range (i.e., the Coastal range) is situated as a back-stopper to westerly plow the continental margin and to incrementally form mountains by developing the major thrust faults toward the western coast.

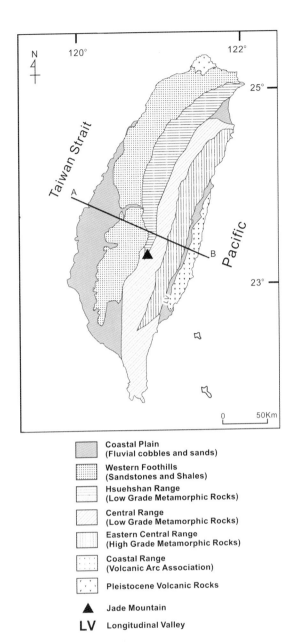

FIGURE 2 Simplified geological map of Taiwan showing north-northeast–south-southwest trending geological entities in response to the stress caused by the tectonic collision.

is mainly constituted of schists, gneisses, and marbles, which are believed to be the preexisting internal materials of the continental crust. The western flank of the Central Range and the Hsuehshan Range, which is connected with the Central Range in the west, are dominated by low-grade metamorphic rocks including slates and meta-sandstones, which were originally sedimentary rocks that were deposited on the continental shelf prior to the tectonic collision mentioned above, but that were then transformed and deformed as a result of their deep burial during the collision. On the west of the Hsuehshan Range is a hilly zone with a width of ~30 km, called the Western Foothills, which are mainly composed of non-metamorphic sedimentary rocks. These have a similar origin as the Hsuehshan Range but have not undergone deep burial. In the western margin of these foothills and further west in the coastal plain, terraces and tablelands are widely distributed and are mainly composed of fluvial sands and cobbles, indicating the early stage of the mountain front depositional systems. However, they soon became the early stage of mountain building because the tips of the deformation belt migrated westward into the plain area. The entire mountain-building history and topographic geometry have been studied for several decades, and a model of thin-skinned critical wedge taper first proposed in the early 1980s can still explain most of what we observe.

HAZARDS UNDER HIGH PRECIPITATION, HIGH MOUNTAINS, AND HIGH SEISMIC OCCURRENCE RATES

Because the neighboring subduction and mountain building associated with arc-continent collision are still occurring, Taiwan and its surrounding regions are seismically active. Seismic hazards are certainly one of the threats to people living on this island. Despite the numerous earthquakes generated by subduction processes offshore, onshore large earthquakes are in fact the major seismic hazard source. This is because when the epicenter is located on land, the shaking damage will be obvious, and surface ruptures may further extend the degree of the hazards. Theoretically, land earthquakes are caused by active fault systems, which are major structural elements of the accretionary prism and are widely distributed on land in Taiwan. In the past century, five large earthquakes (1906, 1935, 1946, 1951, and 1999) occurred with surface ruptures along the surface trace of active faults. Each time, the earthquake event seriously damaged vital systems where the epicenter and rupture were located.

On the other hand, the genetic mountain-building process produces a generally high rate of uplift in most of the Taiwan region, which induces an increase of the mass of debris on the mountain slopes. Unfortunately, for six months per year (during typhoon season), Taiwan undergoes tropical storms because of its geographic location. The abrupt high-mountain relief results in extremely high precipitation occurring within a day or less. Hence, landslides, floods, and debris often strike the colonial or other urban areas. Such hazards are significantly enhanced in the event that a typhoon quickly follows a large earthquake. Being better prepared to mitigate natural hazards requires having better knowledge about all these geologic processes. Such mitigation is urgently needed because Taiwan is one of the few places in world where the rates of geologic processes are extraordinarily high.

SEE ALSO THE FOLLOWING ARTICLES

Earthquakes / Landslides / Plate Tectonics / Taiwan, Biology / Volcanic Islands

FURTHER READING

Angelier, J. 1986. Geodynamics of the Eurasia-Philippine sea plate boundary: preface. *Tectonophysics* 125, ix–x.
Davis, D., J. Suppe, and F. A. Dahlen. 1983. Mechanics of fold-and-thrust belts and accretionary wedges. *Journal of Geophysical Research* 88: 1153–1172.
Ho, C. S. 1986. A synthesis of the geologic evolution of Taiwan. *Tectonophysics* 125: 1–16.
Suppe, J. 1984. Kinematics of arc-continent collision, flipping of subduction, and back-arc spreading near Taiwan. *Memoir of the Geological Society of China* 6: 21–33.
Teng, L. S. 1986. Late Cenozoic arc-continent collision in Taiwan. *Tectonophysics* 183(1990): 57–76.
Tsai, Y. B. 1986. Seismotectonics of Taiwan. *Tectonophysics* 125: 17–38.

TARSIERS

SEE LEMURS AND TARSIERS

TASMANIA

ALASTAIR M. M. RICHARDSON
University of Tasmania, Hobart, Australia

Tasmania is a medium-sized continental island lying southeast of the Australian mainland. It is topographically diverse, largely forested, and supports a number of relictual and endemic species. It has been occupied by aboriginal people for at least 30,000 years and by Europeans for just over 200 years.

LOCATION

Tasmania is a continental island of about 68,400 km^2, lying between latitudes 40 and 43° S, about 200 km off the coast of southeastern Australia, from which it is separated by Bass Strait. Several substantial offshore islands are associated with the mainland of Tasmania, notably Bruny and Maria Islands off the east coast, King Island halfway between Tasmania's northwest tip and the Australian mainland, and the Furneaux Group (Cape Barren, Flinders, etc.), which forms a chain linking the northeast tip with Wilson's Promontory, the southernmost point of the Australian mainland. Tasmania is a state of the Commonwealth of Australia, with a human population of about 500,000.

GEOLOGICAL AND POSTGLACIAL HISTORY

As part of the Australian tectonic plate, Tasmania's history can be traced back to the supercontinent of Gondwana, and it shared Australia's long isolation following the separation from Antarctica (80 million years ago) that lasted until the collision with southeast Asian plates about 14 million years ago. As the most southern part of the Australian land mass, Tasmania has always been farthest from the source of animals and plants colonizing from the north.

Its southern location and relatively high topography meant that Tasmania felt the effects of the Pleistocene glaciations more strongly than anywhere else in Australia. At the glacial maximum, almost half the land mass was covered by ice, and periglacial influences extended to sea level in some places. The proximity of the continental shelf edge to the east and west coast allowed only a small increase in land area as sea levels fell, but Bass Strait disappeared and was re-flooded several times as the glaciers advanced and retreated. During the glacial maxima, the Bassian plain was probably dry, cold, and treeless, providing quite a strong filter to potential colonizers. After the last glacial maximum, about 18,000 years ago, sea levels rose for the last time, and by about 10,000 years ago Tasmania was separated from the mainland.

GEOLOGY AND TOPOGRAPHY

In many ways, Tasmania is an island of two halves: the cool, wet west and the drier east. This pattern results from the interaction between the topography and the westerly weather patterns of the "roaring forties" latitudes. Western Tasmania is formed of old, hard, folded, and sometimes mineralized rocks, whereas the east is mostly composed of softer rocks of Permian and Triassic age, into which were injected large sills of dolerite during the Jurassic. Extensive faulting, accompanied by weathering of the softer rocks to expose the dolerite surfaces, has produced in the east a landscape of plateaus and grabens, whereas the folded rocks of the west give a much more rugged topography. There are some exceptions to this general pattern, notably the granites of the northeast, but the east–west divide is striking. The mountainous nature of the west, with its high rainfall and low fertility, has largely protected the region from European settlement, with the result that it is now mostly in the Tasmanian Wilderness World Heritage Area.

CLIMATE

Westerly winds blowing in from the southern ocean rise over the western mountains depositing over 2000 mm of rain annually in some places, but leaving a rain shadow to the east, where annual rainfall may be below 500 mm per year. Annual temperatures also increase from west to east, with mean January maxima and July minima of around 23 °C and 5 °C, respectively, at sea level. Although Tasmania's climate can generally be described as cool maritime, its location to the south of a large dry continent, which extends into low latitudes, allows hot, dry air to affect the island when northerly winds blow. Tasmania's weather is largely the product of a series of high and low pressure systems that move across the island from west to east. As an area of high pressure moves to the east of the island, winds become northerly or northwesterly; in summer, they bring hot, dry air, and temperatures can exceed 35 °C but usually for only a short period. This pattern may be sharply terminated by a cold front, after which winds turn southerly or southwesterly, bearing squalls of rain, or snow at higher altitudes. Above 300 m, frosts may be experienced in any month, but snow does not lie for long periods, even at the highest elevations.

This typical pattern has been modified in recent years by the increasing strength and frequency of El Niño events. During strong El Niños, anticyclones pass south of Tasmania, breaking the westerly pattern and increasing rainfall in the east, as warm, moist easterly or northeasterly winds blow in from the Tasman Sea.

VEGETATION AND SOILS

The east–west divide in topography and climate is reflected in the vegetation and soils. The climax vegetation is forest over most of the island below 1000 m; in the west the climax is rain forest, dominated by southern beech, *Nothofagus cunninghamii,* in areas where the annual rainfall exceeds 50 mm per month. In the east, the forests are dominated by various species of *Eucalyptus:* very tall closed forest in the wetter areas, grading to lower open savanna forest in the driest areas. These forest climaxes are often diverted by edaphic conditions or by fire. Large areas of western Tasmania are covered by a tussock sedgeland growing on acidic peat soils, often dominated by buttongrass, *Gymnoschoenus sphaerocephalus,* and maintained by a fire frequency of 20–30 years. At the eastern boundary of the range of southern beech, the forest is often in the form of "mixed forest," (i.e., an overstory of tall, old eucalyptus and an understory of rain forest species, including southern beech). In the absence of fire, these forests become rain forest when the eucalyptus trees, unable to regenerate without fire, reach the end of their lives at 350 or more years; fire in this forest type is catastrophic, although very infrequent, and it results in large areas of even-aged regeneration of eucalyptus from seed shed from standing trees after the fire.

Non-forest vegetation is found at altitudes above 1000 m, although endemic pines may form patchy woodlands; the alpine vegetation is highly sensitive to fire. Deciduous beech, *Nothofagus gunni,* the only native deciduous tree, is found in subalpine regions. Around the coast, wind and salt deposition allow the development of a florally rich coastal heath of shrubs and herbs. *Poa* grasslands are also found at low altitudes in dry or frost-hollow situations.

FAUNA

The Tasmanian fauna shows the characteristics that might be expected of a continental island in its particular location: The fauna is depauperate compared to that of the adjacent mainland, but it includes several groups with relictual members and species that have, at least until recently, found a refuge there from introduced species on the Australian mainland. The list of breeding landbirds, for example, includes about 104 species, compared to 176 in equivalent habitats in Victoria, the adjacent mainland state. This number fits quite well the species-area curve for island avifaunas, and the number of endemic species (12) is in the same range as that of other large continental islands close to the source. The presence of one endemic flightless bird, the Tasmanian native hen (*Gallinula mortierii*) (Fig. 1A), reflects the island status.

Tasmania is perhaps best known for two members of its marsupial fauna, although one of these (the thylacine, *Thylacinus cynocephalus*) is extinct, and the other (the Tasmanian devil, *Sarcophilus harrisii*) (Fig. 1B) is rapidly declining as a result of a highly unusual communicable

FIGURE 1 Endemic animals from Tasmania. (A) The flightless Tasmanian native hen (*Gallinula mortieri*) (photograph by Dave Watts). (B) The largest extant marsupial carnivore, the Tasmanian devil (*Sarcophilus harrisii*) (photograph by Menna Jones); (C) Mountain shrimps (*Anaspides tasmaniae*), syncarid crustaceans (photograph by Niall Doran); the largest specimen is about 60 mm long. (D) The giant freshwater crayfish, or *tayatea* (*Astacopsis gouldi*), the world's largest non-marine invertebrate; this specimen is about 40-cm long and weighs about 2 kg (photograph by Niall Doran).

facial tumor. The condition was first observed in 1996, and the tumors are spread directly, as cancerous cells, when the animals bite each other—for example during competition when feeding at carcasses. The island is also a refuge for two other large dasyurids, the eastern quoll (*Dasyurus viverrinus*) and the spotted-tailed quoll (*D. maculatus*), and supports good populations of the bettong (*Bettongia gaimardi*), pademelon (*Thylogale billardierii*), long-nosed potoroo (*Potorous tridactylus*), and barred bandicoot (*Perameles gunni*), which are extinct or threatened on the Australian mainland. Other common marsupials include the brush-tail possum (*Trichosurus vulpecula*) and Bennett's wallaby (*Macropus rufogriseus*). Wombats (*Vombatus ursinus*) are widely distributed and are common in sedgelands, heaths, and grasslands.

Two monotremes, the echidna (*Tachyglossus aculeatus*) and platypus (*Ornithorhynchus anatinus*), are widespread and common, but there is concern about an outbreak of a necrotic fungal disease, first observed in 1982, in the platypus populations in some northern rivers.

The reptile fauna is modest by Australian standards, comprising 17 skinks, one agamid, and three snakes; however, several of the skinks are short-range endemics, with highly restricted populations on isolated mountain ranges and, in the case of one (the pedra branca skink, *Niveoscincus palfreymani*), on a small off-shore island. The frog fauna (11 spp.) is also limited compared with the rest of Australia but includes three endemics. Chytrid fungus, which attacks cartilage and is implicated in worldwide frog declines, has been identified in Tasmania, and there is evidence that the populations of several frog species have decreased.

The native fish fauna (25 native species) is dominated by members of the Galaxiidae (15 spp.), and once again levels of endemism are high (12 spp.). The galaxiids are all small species and generally suffer from predation by the introduced salmonids, principally brown and rainbow trout.

It is difficult to do justice to Tasmania's invertebrate fauna, but the non-marine crustaceans serve as an example of the relictual nature of the fauna. Most widely known, at least to specialists, are the Syncarida, or mountain shrimps, primitive malacostracans with a body plan that suggests the morphology of the earliest members of that group. *Anaspides tasmaniae* (Fig. 1C), which can exceed 50 mm in length, is common in mountain streams and lakes where trout are absent or where there are good refuges. *Paranaspides lacustris* is another large syncarid, an inhabitant of weed beds, and is convergent with decapod shrimps. The lakes and streams of Tasmania also support diverse faunas of amphipods and phreatoicoid isopods. On land, the leaf litter in the wetter forest is often dominated by very large numbers of talitrid amphipods, sometimes exceeding 5000 animals per square meter. At least 15 species of these relatives of the familiar coastal beachfleas are found on the island.

The most spectacular member of the crustacean fauna is the giant freshwater lobster, *Astacopsis gouldi* (Fig. 1D), known to aboriginal people as *tayatea*. It can exceed 1 m and a weight of 4 kg, making it the world's largest freshwater invertebrate. Very large animals are now rare as a result of fishing pressure, and although fishing is now illegal, poaching and habitat deterioration from land clearance remain problems for this species. This giant is one of over 30 species of freshwater crayfish found in Tasmania; these include species living in the acid peaty sedgelands, and others that have become highly terrestrial, living in burrows in clay soils of the rain forests, where their water supply is collected from the surface runoff and stored in underground chambers.

The crustacean fauna is not the only invertebrate group that shows special features. Many insect orders include primitive endemics, such as the Gondwanan dragonfly *Archipetalia auriculata*. Other endemic arthropods include the enigmatic centipede *Craterostigmus tasmanianus* and the Tasmanian cave spider *Hickmania troglodytes*, one of a very small number of hypochilid spiders found in the Southern Hemisphere that represent the basal form of the araneomorph spiders.

ENDEMISM AND GIGANTISM

Endemism in the Tasmanian fauna can be related directly to the vagility of each group (Table 1); almost all the groups with very low powers of dispersal show 80% or more endemism. Gigantism, either globally or locally to Australia, is a feature of several invertebrate groups: for example, syncarids, parastacid crayfish, collembolans, and stoneflies. Among the birds, some species are larger than their mainland counterparts (e.g., the masked owl *Tyto novaehollandiae*, the wedge-tailed eagle *Aquila audax*, the superb blue wren *Malurus cyanaeus*), whereas others are smaller (e.g., the yellow-tailed black cockatoo *Calyptorhynchus funereus*, the tawny frogmouth *Podargus strigoides*, the eastern spinebill, *Acanthorhynchus tenuirostris*); several show colors darker than those of equivalent mainland species.

ABORIGINAL COLONIZATION

Aboriginal people have lived in Tasmania for at least 35,000 years. The earliest ^{14}C dates for human occupation, obtained from fire hearths, are around 33,850 years old, but estimated sea levels suggest that the Bass Strait was dry and would have offered a passage to Tasmania as

TABLE 1
Examples of Various Animal Groups Occurring in Tasmania: The Relationship between Vagility and Level of Endemism

Vagility	Group	Species in Tasmania	Endemic species	Endemism (%)
High to medium	Land mammals	26	3	12
	Landbirds	104	14	13
	Reptiles	21	7	33
	Frogs	11	3	27
	Water beetles (Dytiscidae)	26	1	4
	Dragonflies	28	6	21
	Water bugs	37	4	11
Low	Caddisflies	157	116	74
	Stoneflies	46	40	87
	Terrestrial snails	49	33	67
Very low	Freshwater crayfish	33	31	94
	Terrestrial amphipods	19	18	95
	Mountain shrimps	8	7	88
	Torrent midges	6	6	100
	Trechine beetles	63	62	98

long as 130,000 years ago. Pollen records from sediment cores suggest an increase in fire frequency at about that time, which may have been the result of aboriginal burning. Although debate continues about the date of aboriginal colonization, cave deposits provide good evidence that the island was occupied during the last glacial maximum. At the time of European colonization, the aboriginal population was probably between 4000 and 6000. It is likely that the fire regime in Tasmania changed with the establishment of aboriginal people, and fires became more frequent throughout the island.

After Europeans arrived, the aboriginal population declined sharply as a result of disease, displacement, and direct conflict with the colonists. Attempts to establish aboriginal settlements on Flinders Island and later at Oyster Cove in the southeast did nothing to prevent the decline, and the woman considered to be the last full-blooded aboriginal, Truganinni, died in 1876. However, the aboriginal community in Tasmania remains strong and has recently been involved in a vigorous worldwide campaign to repatriate aboriginal skeletal material held in museums.

EUROPEAN COLONIZATION

There is no evidence to suggest that coastal vessels from Southeast Asia, or Arab traders, penetrated as far south as Tasmania, and the first non-aboriginal group to discover the island was almost certainly Abel Tasman and the crew of the ships *Heemskerck* and *Zeehaen* in 1642. French and English explorers, including Du Fresne, D'Entrecasteaux, and Bligh, visited the island, but the first European settlement was not established until 1803. European sealers, however, had already discovered the seal colonies in Bass Strait, and they established informal settlements on the islands, taking aboriginal women as partners.

After the sealers, the main economic use of the island was as a penal colony, and prisons were set up in various locations, with the largest at Port Arthur. Free settlers began a small pastoral industry, but economic growth began in earnest toward the end of the nineteenth century with the discovery of metal ores, especially copper, in the western mountains. Extraction and smelting of zinc, aluminum, and other metals continues today, and these metals still lead the list of the state's exports, but forestry, both for lumber and paper production, follows closely.

Charles Darwin spent ten days in Hobart during the voyage of the HMS *Beagle* in 1836; he climbed Mt. Wellington and collected fossils near the town but made few other observations. Scientific observations and collections flourished in the nineteenth century, and the Royal Society of Tasmania was the first of its type in Australia when it was established in 1843. The University of Tasmania was founded 1890, making it the fourth oldest in Australia.

TABLE 2
Summary of Species Listed under the Tasmanian Threatened Species Protection Act 1995

	Vertebrates	Invertebrates
Extinct	5	4
Endangered	35	17
Vulnerable	19	16
Rare	13	89
Total	72[a]	126[b]

NOTE: A further five species are listed under the Commonwealth of Australia's Environment Protection and Biodiversity Conservation Act 1999.
[a]Includes 31 species from oceanic or Antarctic waters, or Macquarie Island.
[b]Includes 43 species of hydrobiid freshwater snails with highly localized distributions.

CONSERVATION ISSUES

Over 40% of the Tasmanian land mass is protected in national parks and other reserves, including the Tasmanian Wilderness World Heritage Area established in 1982, but the eastern half of the island is underrepresented in reserves, and land clearance or conversion to non-native vegetation continues. Just three vertebrates (the thylacine, the Tasmanian emu *Dromaius novaehollandiae diemenensis,* and the King Island emu *D. minor*) are known to have become extinct since European colonization; four invertebrates (two caddisflies, a beetle, and a spider) are listed as extinct under the Tasmanian Threatened Species Protection Act 1995, but the ranges of many more have been severely fragmented (Table 2).

A number of European plants, fish, birds, and mammals were deliberately introduced following European settlement, although Tasmania did not suffer as badly in this regard as New Zealand. More recently, several species have invaded the island with human help, all of which have potentially devastating effects. European carp, *Cyprinus carpio,* have appeared in two lakes, and the arrival of the European red fox, *Vulpes vulpes,* may result in the elimination of the mid-weight range species of marsupials, for which Tasmania has been such a valuable refuge. Vigorous efforts are being made to eradicate both of these invaders.

Several birds have colonized Tasmania naturally in the last 50 years: The kelp gull, *Larus dominicanus,* and the cattle egret, *Ardea ibis,* are two examples. Other Australian native birds have become established, probably as a result of deliberate introductions or aviary escapes: Examples include the kookaburra (*Dacelo novaeguineae*), the superb lyrebird (*Menura novaehollandiae*), the galah (*Cacatua roseicapilla*), the little and long-billed corellas (*C. sanguinea* and *C. tenuirostris*), and the rainbow lorikeet (*Trichoglossus haemotodus*).

SEE ALSO THE FOLLOWING ARTICLES

Climate on Islands / Gigantism / Prisons and Penal Settlements / Vegetation / Voyage of the *Beagle*

FURTHER READING

Doran, N., A. M. M. Richardson, and R. Swain. 2001. The reproductive behaviour of *Hickmania troglodytes*, the Tasmanian cave spider (Araneae, Austrochilidae). *Journal of Zoology* 253: 405–418.

Hamr, P. 1990. Rare and endangered: Tasmanian giant freshwater lobster. *Australian Natural History* 23: 362.

Parks and Wildlife Service Tasmania: Nature of Tasmania website. http://www.parks.tas.gov.au/nature.html

Reid, J. B., R. S. Hill, M. J. Brown, and M. J. Hovenden. 1999. *Vegetation of Tasmania*. Canberra: Australian Biological Resources Study.

Smith, S. J., and M. R. Banks, eds. 1993. *Tasmanian wilderness—world heritage values*. Hobart: Royal Society of Tasmania.

Williams, W. D. 1974. *Biogeography and ecology in Tasmania*. The Hague: Dr. W. Junk.

TATOOSH

EGBERT GILES LEIGH, JR.
Smithsonian Tropical Research Institute, Balboa, Panama

ROBERT T. PAINE
University of Washington, Seattle

Tatoosh (Fig. 1) is a set of islets covering 17 to 18 ha at 48°24′ N, 124°44′ W, 0.6 km off Cape Flattery, the northwest tip of the Olympic Peninsula. A lighthouse on the largest islet marks the southern lip of the strait of Juan de Fuca, which leads from the Pacific Ocean to Seattle and Vancouver. These islets' rocky shores support a luxuriant community of intertidal organisms. Research there by Robert Paine and his students and colleagues has revealed many of the factors that govern what lives where on rocky shores, and has shed light on other, very different ecosystems.

THE SITE

Native Americans almost certainly had occupied Tatoosh for millennia. They were progressively displaced after its lighthouse was built in 1857, and they rarely visited after the early 1900s. The U.S. Coast Guard automated the lighthouse and abandoned Tatoosh in 1976, and ownership of Tatoosh has been restored to the Makah Nation. Its intertidal zone has been relatively free of human disturbance for a century.

Tatoosh is at a strait's mouth. The continental margin (the 100-fathom line) is only 2.5 km distant, so wave action is heavy. Its different shores experience very different conditions. The western shores face the ocean, with its many long processions of great swells. These "weather shores" support a particularly lush intertidal community, whose barnacle zone can extend over 5 m above mean lower low water. In a relatively sheltered area between two islets, the barnacle zone extends less than half as high. These differences in wave exposure govern the distribution and abundance of many organisms.

Tatoosh has gloomy, rainy winters and drier summers. Hard freezes are relatively rare. The main intertidal herbivores and predators are slow-moving invertebrates—sea urchins, chitons, snails, starfish—which crawl over the rock. Tatoosh also has a large gull colony. Like the recently returned sea otters, gulls eat a wide variety of invertebrates. The ecological roles of many species of animals and plants have been assessed by using cages, strips of poisonous paint, or frequent manual removal to exclude consumers or competing plants, and quantifying the impacts of the exclusion.

An intertidal zone is alternately submerged by the sea and exposed to the air. This zone separates the marine world from the land. At Tatoosh, nearly vertical walls rise from lush stands of kelps exposed only at the lowest tides. Above is a barer zone with brightly colored crusts, low-statured invertebrates, and scattered algae, followed by dense mussel beds; sometimes a zone of short, springy algae; a barnacle zone; and, finally, lichen-encrusted rock (Fig. 2). More gently sloping surfaces reveal subtle variations on the same theme: a few meters' walk upward from the low tide level leads through lush stands of several kinds of kelp to a broad band of mussels, followed by a barnacle zone, sometimes separated by a zone of springy algae, *Mazzaella*. These "zones," some dominated by plants, others

FIGURE 1 An aerial view of Tatoosh, facing roughly south, at or near low tide. The lighthouse is visible on the largest of this set of islets. The western side of Cape Flattery appears at the upper left. Photograph by Alan Trimble.

FIGURE 2 Zonation along an east-facing wall of a surge channel at Tatoosh. From the top, the greenish tinge of the green alga *Prasiola*, then a broad band of the barnacle *Balanus glandula*, separated from the bed of mussels, *Mytilus californianus*, by a thin band of the red alga *Endocladia*. Below the mussels is a mixture of the barnacle *Semibalanus cariosus* and the gooseneck barnacle *Pollicipes*. The band closest to the water is occupied by a mixture of sea anemones, *Anthopleura*; sponges; kelps; and the occasional bright orange hydrocoral, *Allopora*. Photograph by Robert T. Paine.

by sessile animals, differ more fundamentally than the zones of montane forest, cloud forest, elfin forest, and paramo separating lowland rain forest from the bare rock or perpetual snow of a high tropical mountain. A moderate tidal range (maximum of 3.8 m) and an abundance of waves and swells make the zonation at Tatoosh particularly clear. The factors maintaining these striking patterns, and the variation in distribution and abundance of this community's species, have been our primary focus. Discussed below are four themes of research at Tatoosh, some of which were initiated on nearby, ecologically similar, mainland weather shores.

INTERTIDAL KELPS AND OTHER SEAWEEDS

A visitor cannot help being impressed by the variety of form among Tatoosh's hundreds of species of seaweeds: sheets, tubes, straps on stalks, miniature "palms," and crusts that are more limestone than living plant. High in the intertidal are extensive carpets of the springy, 2–3-cm high *Mazzaella cornucopiae*. Paine installed marker screws in various places for recording how this carpet's sharp upper limit varied through time. Surprisingly, this limit has receded 30 to 40 cm since 1978. *Mazzaella* supports about 7 m^2 of fronds per square meter of rock.

Lower down, in gaps within the most wave-beaten mussel beds, grow sea palms, *Postelsia palmaeformis*, which support a thatch of fronds atop a stout, hollow, 50-cm stalk. Sea palms need waves to sweep away adjacent, encroaching mussels and protect the palms from grazers.

Near the low-water mark is the laminarian zone, populated by large kelps and an abundance of grazers. The canopy of these kelps shelters a diverse flora of understory seaweeds and many species of algal crust. On most shores *Laminaria*, with an upright stalk 45–50 cm tall carrying a long, smooth broad frond, dominates the lowermost intertidal. Where they are present, sea urchins, *Strongylocentrotus purpuratus*, eliminate *Laminaria*: it returns if the sea urchins are removed. These kelps grow where waves restrict urchin activity or natural enemies keep urchins out.

On gentle, less exposed shores, another perennial kelp, *Hedophyllum*, flourishes above the *Laminaria*. *Hedophyllum*'s broad, heavy fronds grow from a flat holdfast. *Hedophyllum* grows where chitons, *Katharina*, eat their competitors. The faster-growing annual kelp *Alaria marginata* occupies plots cleared of both *Hedophyllum* and chitons, but *Hedophyllum* recolonizes plots with chitons. Settings sheltered enough to allow sea urchins to aggregate, however, lack *Hedophyllum*. Instead, these "urchin barrens" favor coralline crusts, which are hard to eat because of their limestone content. Experiments show that the fastest-growing coralline algae are most susceptible to grazers. This trade-off between fast growth and effective defense helps maintain the diversity of corallines, and probably of many other groups. The influence of herbivores on what plants grow where is especially clear on these lower shores, where herbivores so clearly influence zonation, species composition, and productivity.

WHO EATS WHOM AND WHY IT MATTERS

Another theme is who eats how much of whom, where, and how these eating habits organize the intertidal community. To answer this question one must observe who eats whom, and remove consumers to see how they affect the species they eat. In 1963, Paine removed the mussel-eating starfish, *Pisaster ochraceus* from a plot on a weather coast 10 km south of Tatoosh. Removing this starfish allowed mussels to spread lower in the intertidal, overgrowing the scattered kelps, coralline crusts, sponges, and the like that grew where starfish had kept the mussels out. This experiment, since repeated twice at Tatoosh, spotlighted *Pisaster* as a "keystone species" whose impact on the community was far greater than its numbers suggested. Similar studies by Paul Dayton showed how limpets affected the algae they ate, how some limpets influence barnacle species composition by "bulldozing" young barnacles of some species, and how large starfish, *Pycnopodia*, that live from the low tide downward protect

Laminaria by eating, and chasing off, the sea urchins that would destroy them. Large green sea anemones, which eat many of the sea urchins fleeing from *Pycnopodia,* also aggregate where they can catch mussels dislodged by foraging *Pisaster*. In 1974 Kenneth Sebens learned that the normally sessile anemones can move—slowly—to where they can catch more food. Timothy Wootton, Jennifer Ruesink, and Paine have used a variety of experimental methods to assess who eats how much of whom, who replaces whom, and how fast they do it. This knowledge is essential for understanding how the ways different species affect each other organize the community to which they belong.

MUSSEL BEDS

Mussels, *Mytilus californianus,* form a conspicuous zone on gently sloping, rather wave-beaten shores. In 1978 Thomas Suchanek showed that a thick, old mussel bed may shelter over 300 macroscopic species. Mussels grow faster lower down, but starfish, *Pisaster ochraceus,* limit how far down they spread. Where waves pound hardest, limiting the activity of these starfish, or where starfish are regularly removed, the mussels spread downward. When mussels spread among *Hedophyllum* or a shrubby, many-fronded kelp of more exposed shores, *Lessoniopsis,* the mussels replace the kelps because waves shred the fronds of the kelps against the many sharp edges of the mussel bed. Moreover, long-term data on the upper limit of the mussel bed has shown how it has tracked the 18.5-year cycle in gravitational influence on tides.

Waves unravel mussel beds that have grown too thick and multilayered, clearing gaps of several square meters or more, exposing bare rock. How often do gaps open? How long does it take gaps of given size to close again? In prime mussel territory, waves strip mussels from a given point every 7–10 years on the average. Gaps of various ages provide a dynamic patchwork of variety within the mussel bed. The centers of big gaps, beyond the reach of grazers sheltering under the mussels, are colonized by fast-growing kelps—*Postelsia palmaeformis* on the most wave-beaten angles, *Alaria nana* elsewhere. Mussel beds, with their rich array of associated species, are the biological centerpiece of these exposed shores.

THE IMPORTANCE OF WAVES

A fourth research theme is how exposure to waves influences intertidal life. The intertidal zone is wider and more luxuriant on more exposed shores, where sea spray reaches higher. Starfish, sea urchins, limpets, and snails cannot forage where waves are breaking too vigorously, so they are less able to control their food species. As we have seen, a shore's zonation reflects its degree of wave exposure.

Where their surrounding water moves faster, plants and filter-feeding animals extract nutrients and food from it more easily, and the wastes they shed disperse more quickly. Kelps of wave-beaten shores, where wave damage is the primary threat, have smooth, sleek fronds. Sheltered waters house other species whose crinkled or ruffled fronds are designed to excite turbulence in the surrounding water, helping nutrients to penetrate and wastes to leave the boundary layer of still water around the frond. On sheltered coasts, *Hedophyllum* is cabbage-like, with short, wrinkled upstanding fronds quite unlike the broad smooth *Hedophyllum* fronds of more exposed shores.

Starting in 1979, Mark Denny measured the force of breaking waves on Tatoosh. He calculated the force needed to dislodge various sea urchins, limpets, and snails when they were foraging, and showed how the threat of breakage or dislodgement by waves limited the size of intertidal organisms, especially those of weather shores.

On the most exposed angles waves allow sea palms, *Postelsia,* to support over 15 m^2 fronds/m^2 rock by stirring their fronds enough to bring light to each. Sea palms can produce as much dry matter in five months as wheat carefully cultivated for maximum production in a mock spaceship. In mature rain forests, tall trees shade their neighbors: canopy leaves are flooded with light, while forest floor herbs receive barely enough light to survive. Unequal sharing of light means that a hectare of mature rainforest carries less than ten hectares of leaves. By limiting maximum height, thereby enforcing the Roman maxim *debellare superbos, parcere subjectis* (beat down the proud, spare the meek), waves allow a sufficiently even distribution of light among fronds to enable *Postelsia*'s extraordinary productivity.

LESSONS FROM TATOOSH

What lessons emerge from this story? First, the research of a group of biologists attracted to a site of singular beauty and fascination, each working on a project of his or her own design, provides a coherent picture of how that site's community functions. "Unity of place" focuses the research, enabling each project to provide context for the others, raise new questions, and provide a basis for answering them.

Second, we cannot infer process from pattern. Organisms do not always live where they grow best, because predators or competitors may keep them out.

Third, certain keystone species exert an impact utterly disproportionate to their numbers. For example, by restricting the downward spread of mussels, the starfish *Pisaster* makes room for a diversity of kelps, sponges, coralline algae, and their consumers.

Fourth, we learn how the degree of wave action modulates the processes that organize the community. Kelps that require ceaseless waves to stir their many fronds are restricted to the most wave-beaten shores; sea urchins are more active, and more devastating, where they are safer from waves.

Finally, Tatoosh represents an extreme of interdependence. Any weather coast collects energy from swells stirred anywhere upwind in the vast expanses of ocean they face. At Tatoosh, these waves enhance intertidal productivity. The ocean near Tatoosh is also a rich source of food. Its abundant plankton allows filter feeders to compete on even terms with primary producers. This ocean feeds planktonic larvae; how dispersal of these larvae enhances the genetic variation of their species or restricts their capacity for local adaptation is an open question. This ocean's fish feeds the gulls and puffins, cormorants and auklets, that abound on Tatoosh. Excrement of these birds fertilizes plankton in the surrounding water. The gulls supported by this ocean eat the sea urchins they can reach, helping to protect kelp. Compared to most rain forests, Tatoosh is an emporium that depends utterly on the wealth it collects from a remarkably wide area.

SEE ALSO THE FOLLOWING ARTICLES

Barro Colorado / Succession / Tides

FURTHER READING

Dayton, P. K. 1971. Competition, disturbance and community organization: the provision and subsequent utilization of space in a rocky intertidal community. *Ecological Monographs* 41: 351–389.

Denny, M. W. 1987. Life in the maelstrom: the biomechanics of wave-swept rocky shores. *Trends in Ecology and Evolution* 2: 61–66.

Leigh, E. G., Jr., R. T. Paine, J. F. Quinn, and T. H. Suchanek. 1987. Wave energy and intertidal productivity. *Proceedings of the National Academy of Sciences of the United States of America* 84: 1314–1318.

Paine, R. T. 1974. Intertidal community structure: experimental studies on the relationship between a dominant competitor and its principal predator. *Oecologia* 15: 93–120.

Paine, R. T. 2002. Trophic control of production in a rocky intertidal community. *Science* 296: 736–739.

Paine, R. T., and S. A. Levin. 1981. Intertidal landscapes: disturbance and the dynamics of pattern. *Ecological Monographs* 51: 145–178.

Rosenfeld, A. W., and R. T. Paine. 2002. *The intertidal wilderness*. Berkeley: University of California Press.

Wootton, J. T. 1993. Size-dependent competition: effects on the dynamics vs. the end point of mussel bed succession. *Ecology* 74: 195–206.

TAXON CYCLE

MANDY L. HEDDLE

Center for Environmental Education, Ahmedabad, India

The taxon cycle was originally conceived to explain species distributions on islands where, within a particular species complex, some species are widespread, abundant, and found in coastal habitats, whereas others are fragmented, rare, and occupy montane habitats. The mechanism for the cycle is driven by new colonists (early-stage species) that establish populations on coastal, marginal habitats, outcompeting established populations, which experience shifts in ecological adaptations as they are forced toward more interior habitats. As more colonists arrive, older taxa (late-stage species) become further specialized for particular ecological conditions and occupy increasingly interior habitats, such that their populations become fragmented and more susceptible to extinction. These species finally go extinct, completing the cycle (Fig. 1). The taxon cycle is controversial as a theory to explain species distribution.

DEFINING THE TAXON CYCLE

The taxon cycle was first defined by E. O. Wilson in 1959 to describe island populations of Melanesian ants. Although the concept of species proceeding through predictive stages, similar to the life cycle of an individual, had been proposed by several authors prior to Wilson, the taxon cycle specifically describes the characteristics of these stages by their geographical distribution and ecological adaptations.

REFINEMENTS OF THE TAXON CYCLE

Several authors have proposed modifications and refinements to the taxon cycle. In 1972, Ricklefs and Cox explored the concept of competition as it pertains to the taxon cycle proposing that competitive ability will change at different stages of the cycle. For example, habitat shifts could result in changes in coevolutionary relationships with predators, and as species become marginalized and rare, they escape density-dependent predation. Erwin (1985) proposed an alternate, non-cyclical model of progressive specialization that he termed a "taxon pulse," in which taxa make adaptive shifts along deterministic pathways depending on the taxon. As with the refinements offered by Ricklefs and Cox (1972), competition and

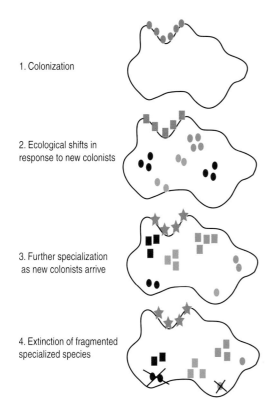

FIGURE 1 Stages of the taxon cycle.

habitat shift are predicted as primary mechanisms driving the shifts. Roughgarden and Pacala (1989) have applied the taxon cycle and competitive displacement concepts to body size of lizards on islands rather than to ecological specialization.

PROPOSED MECHANISMS FOR THE TAXON CYCLE

Various mechanisms by which the taxon cycle operates have been theorized, but all stem from the impact of competition between established and colonizing species both directly and indirectly. However, critics of the taxon cycle theory have contended that competition between colonizing species and established species is unlikely to result in the successful establishment of the newly arrived species. Roughgarden and Pacala's (1989) model of competitive displacement based on body size has resulted in a range of outcomes, only some of which are consistent with a taxon cycle scenario.

PHYLOGENETIC EVIDENCE FOR THE TAXON CYCLE

Another criticism of the taxon cycle is whether it exists at all as an evolutionary process or is instead a pattern resulting from extrinsic ecological events such as natural disturbances or climate change. If the taxon cycle is an intrinsic evolutionary process, then species lineages will reflect a history consistent with the cycle. The taxon cycle predicts that clades will form within islands and contain representatives of diverse ecological phenotypes, whereas an extrinsic process predicts simultaneous evolutionary events across multiple lineages. Phylogenetic studies permit testing of these hypotheses but have produced variable results. Although some researchers have found that periods of expansion or contraction for particular species do not necessarily correspond to extrinsic environmental events, and that the stages of the taxon cycle do show a historical progression, with correlation between increased specialization and geographical distribution, other studies have uncovered evolutionary histories that show that species occupying similar ecological niches on multiple islands are more closely related to each other than are species within an island. The taxon cycle therefore remains a controversial theory for explaining species distribution and abundance on islands.

SEE ALSO THE FOLLOWING ARTICLES

Ants / Extinction / Fragmentation / Lizard Radiations

FURTHER READING

Erwin, T. L. 1985. The taxon pulse: a general pattern of lineage radiation and extinction among carabid beetles, in *Taxonomy, Phylogeny and Zoogeography of Beetles and Ants*. G. E. Ball, ed. The Hague: Dr. W. Junk, 437–472.
Liebherr, J. K., and A. E. Hajek. 1990. A cladistic test of the taxon cycle and taxon pulse hypotheses. *Cladistics* 6: 39–59.
Pregill, G. K., and S. L. Olson. 1981. Zoogeography of West Indian vertebrates in relation to Pleistocene climatic cycles. *Annual Review of Ecology and Systematics* 12: 75–98.
Ricklefs, R. E., and E. Bermingham. 2002. The concept of the taxon cycle in biogeography. *Global Ecology and Biogeography* 11: 353–361.
Ricklefs, R. E., and G. W. Cox. Taxon cycles in the West Indian avifauna. *American Naturalist* 106: 195–361.
Roughgarden, J., and S. Pacala. 1989. Taxon cycles among *Anolis* lizard populations: review of the evidence, in *Speciation and its Consequences*. D. Otte and J. A. Endler, eds. Sunderland, MA: Sinauer Associates, 403–432.
Wilson, E. O. 1959. Adaptive shift and dispersal in a tropical ant fauna. *Evolution* 13: 122–144.

TEPUI

SEE PANTEPUI

TIDAL WAVES

SEE TSUNAMIS

TIDES

MARLENE NOBLE
U.S. Geological Survey, Menlo Park, California

Tidal sea-level fluctuations, which are fundamentally caused by the gravitational attractions between the rotating Earth and the moving positions of the sun and the moon, not only cause the sea level to regularly cover and expose beaches over most of the Earth but also transport water, suspended material, nutrients, larvae, and debris onto or off of the adjacent landmass. Because of their regularity, these tidal fluctuations are well known at any particular location not only by humans, who predict the depth of water in a harbor in order to dock a ship, but also by the many living organisms in the sea that make use of tidal currents to transport a regular flux of nutrients or suspended food particles past them in order to feed. The regular cycle of exposure and inundation of beaches, tidal pools, and tidal flats is essential for the survival of many oceanic life forms.

SURFACE AND INTERNAL TIDES: GENERAL CHARACTERISTICS

Tidal sea-level fluctuations and tidal currents are generated by rhythmic interactions among the gravitational forces between the Earth, the moon, and the sun. Because the moon is much closer to the Earth than the sun is, and because the gravitational attraction force scales as one over the distance squared between objects, the strength of the moon's gravitational attraction on the Earth is about twice that of the sun, even though the sun is much larger and more massive. Hence the largest tides are caused by the gravitational attraction between the Earth and the moon. One might think that because (1) the moon's force on the Earth is largest on the side of the Earth that is closest to the moon, and (2) the moon's position with respect to the Earth remains relatively constant in a 24-hour period, the ocean on an entirely water-covered Earth would bulge out toward the moon on the side of the Earth that is closest to the moon (Fig. 1A). Because the Earth spins on its axis once a day, this would lead to a daily (diurnal) tidal cycle in water-level fluctuations as a particular location on the Earth moves under, then out from under, the tidal bulge in sea level.

However, because the Earth and moon jointly revolve around a common center of mass once every 27.3 days, the centrifugal force on the Earth caused by that rotation is directed approximately opposite to the gravitational force from the moon. The total centrifugal force in the Earth–moon system balances the gravitational forces at the center of their joint rotation. But these forces are not everywhere the same. The centrifugal force is less than the gravitational force on the side of the Earth facing the moon; it is larger than the gravitational force on the side of the Earth farthest from the moon. Hence, sea level on a water-covered Earth has two bulges, one toward and one away from the moon (Fig. 1B). Therefore, a particular point on a rotating Earth generally experiences two highs and two lows in sea level each day (i.e., a semidiurnal tide). The semidiurnal tides would be of similar magnitudes if the moon revolved directly around the Earth's equator. However, because the moon's orbit is inclined to the equator, the tidal bulge is not located on the axis of the Earth's rotation (Fig. 1C). A

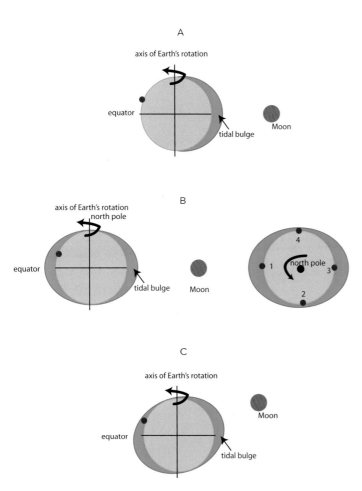

FIGURE 1 Equilibrium tide on a water-covered Earth, (A) if the Earth and moon were not rotating around a common center of mass and (B) if the Earth and moon are rotating around a common center of mass as viewed from the side and from above the north pole of the Earth. The red dot denotes a specific location on the rotating Earth. Note that in a single day, two high (and low) tides are experienced as the red dot moves through positions 1 and 3 (2 and 4). (C) The equilibrium tide when the moon's orbit is inclined to the Earth's equator.

particular spot on the Earth usually experiences two high tides a day, but often the amplitude of one tide is much larger than that of the other.

Both diurnal and semidiurnal tidal cycles would be observed on a water-covered Earth-Moon system. Given that the Earth–sun system has analogous gravitational forces, although different rotational cycles, the interactions of all these planetary bodies cause some of the complexity in the tidal cycles seen on the Earth. Based on the strength of the astronomical forcing, the largest tidal constituent is the semidiurnal principal lunar tide, M_2, which has a period of 12.42 hours. The other large semidiurnal constituent is the principal solar tide, named S_2, which has a period of 12.00 hours. The diurnal tides, which are the other dominant astronomical tidal constituents, are the principal lunar diurnal tide, O_1, with a period of 25.82 hours, and the luni-solar diurnal tide, K_1, with a period of 23.93 hours.

An additional layer of complexity is added when one realizes that the Earth is not a water-covered sphere. The Earth's ocean basins are separated by large continental land masses that prevent the tidal bulges in sea level from moving freely around the Earth. In addition, the combination of basin topography and the influence of Coriolis force on the ocean currents causes the tides in ocean basins to travel around amphidromic points within each basin. Tidal elevations are nearly zero at the amphidromic points and increase with distance from that point. Hence tidal fluctuations at islands in the same ocean basin may be zero if the island is close to an amphidromic point, such as Tahiti or Madagascar, and be much larger if the island is farther from that point.

The amplitudes of tidal sea-level fluctuations at the major frequencies are typically 0.5 to 2 m. At any one location, the tidal range for any particular constituent, such as M_2, is fairly constant because the causes of the tidal fluctuations are linked to the movements of the Earth, sun, and moon and the large-scale topography of that site. However, because there are several different tidal constituents in the major tidal bands, such as M_2 and S_2 or O_1 and K_1, the individual tidal constituents move in and out of phase with each other, adding or subtracting to the amplitude of the observed sea-level fluctuations. In the semidiurnal band, the principal lunar tide, M_2, usually beats against the principal solar tide, S_2, moving in and out of phase with each other every 14.8 days (Fig. 2). This is denoted the Spring/Neap cycle. The principal diurnal tides beat against each other every 13.6 days. Only twice a year are the two principal diurnal and the two principal semidiurnal tides in phase. This is typically when one observes the highest, and lowest, tides of the year.

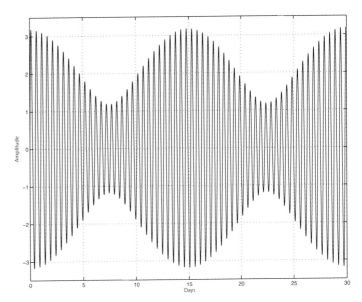

FIGURE 2 The Spring/Neap cycle in tidal heights created by using simulated semidiurnal M_2 (12.42 hour) and S_2 (12.00 hour) tidal constituents. In this example, the amplitude of M_2 is twice as large as S_2.

The currents intrinsically caused by the tidal forcing tend to be classified into two types of flows: barotropic and baroclinic. The barotropic, or surface, tidal current is linked to the tidal fluctuations in sea level. They are oscillating flows at tidal periods that, theoretically, have uniform amplitude with depth and time. Barotropic tidal currents are the only tidal current present in a body of water that has a constant density. Baroclinic, or internal, tidal currents exist because there are vertical gradients in the temperature and salinity, and therefore in density, in the ocean. They are usually generated when barotropic tidal currents flow over submerged topographic features in the ocean. Baroclinic tidal currents account for most of the vertical structure in observed tidal currents and are not connected to the tidal sea-level fluctuations. The amplitude of the internal tides varies with time, partly because local density gradients vary with time.

TIDES, BEACHES, AND BIOTA

In the shallow regions that fringe an island's beaches, the most obvious effect of tidal sea-level fluctuations is that they regularly raise and lower sea level nearly simultaneously around the entire island. The beaches can be covered at high tide (Fig. 3A), whereas tidal flats are exposed at low tide (Fig. 3B). A rising, or flooding, tide will fill lagoons adjacent to the beach with salty ocean water that may be nutrient-rich and contain suspended material such as plankton. If the tidal lagoon is river-dominated, the seawater will mix with the freshwater. The strength and extent of mixing will depend on the amplitude of the

FIGURE 3 The (A) rising and (B) falling tide on the south coast of Molokai, Hawaii. Figures courtesy of Mike E. Field, U. S. Geological Survey.

flooding tide and the strength of the river flow. Several hours later, the falling, or ebb, tide will remove much of this mixed water from the lagoon. This regular cycle of flooding and ebbing tidal currents is essential for the health of the lagoon, for it tends to enhance the exchange of larvae, plankton, and other suspended materials between the lagoon and the coastal ocean.

The raising and lowering of sea level just off the island beaches, or over the fringing coral reefs, also changes the physical processes in the coastal ocean that affect these regions. During low tide, surface waves propagating toward the shore from the surrounding ocean feel the seabed at some distance from the shoreline. Typically a surface wave will feel the bed and begin to shoal or break when the water depth is less than one-half the wavelength of the wave. On a rising tide, the water depth increases, and deep-water surface waves can travel farther toward the shore. Larger, locally generated, wind-driven waves can also develop in this deeper water. These waves shoal, or break, closer to the shoreline on a rising tide, thus moving the location where the energy from those waves can resuspend fine sediment closer to the beach. Some of this resuspended sediment is then carried alongshore by the prevailing currents or offshore on the subsequent ebbing of the tide. This moving zone where sediment is resuspended, and then transported on a tidal cycle, can be critical to the development of shallow coral reefs because coral prefers to grow in clear water. The corals can expand their habitat because the flooding tide allows waves to carry sediment deposited in these shallow waters out of the region.

Wind-driven shore-parallel currents, normally reduced in the shallow water off the beach by friction at low tide, can also be enhanced at high tide. These shore-parallel currents move larvae, suspended materials, and pollutants over the shallow water just off the beach.

Although tidal fluctuations obviously affect the shallow regions around island margins, they can also alter the current flow in narrow straits between adjacent islands. For example, because sea-level fluctuations do not rise and fall synchronously on both ends of the strait between Maui and the Big Island in Hawaii, an along-strait pressure gradient is generated that drives strong tidal currents up and down this strait. These strong currents can enhance local wind-generated waves, delay ships sailing against the current, and enhance the transport of water and other materials along the islands.

TIDES AND BARRIER ISLANDS

Barrier islands are low-relief islands that lie just offshore of continental land masses, usually where one or more large rivers supply sediment to the coastal ocean. The barrier islands are typically grouped as a series of small landmasses elongated parallel to a continental shoreline. A typical set of barrier islands is found in the Gulf of Mexico, near the mouth of the Mississippi River. Barrier islands react to tidal fluctuations similarly to other islands, but because they lie close to a continental shoreline, they are even more sensitive to interactions between tidal fluctuations and other processes that change sea-level height. The local alongshore winds can be strong enough to raise sea level at the continental and barrier island shoreline by tens of centimeters. When this wind-driven increase in sea level is associated with a rising, rather than a falling, tide, the potential for flooding on that barrier island is substantially increased.

TIDES AND SEAMOUNTS

Seamounts, which are found throughout the ocean basins, are essentially submerged islands. Their tops can lie anywhere from 100 to 2000 m below the sea surface, so they are not affected by the relatively small rise and fall of the surface tides. Nor do the currents associated with the surface tide, which have amplitudes of 2–4 cm/s in the deep ocean, directly affect them. However, when surface tidal currents flow over the very gently sloping summit of a large seamount, they can generate much larger internal tidal currents, with speeds of 10 cm/s and higher. Depending on the location of the seamount, these internal tidal currents either propagate away from the seamount or are trapped to that seamount. The frequencies of both the freely propagating and trapped internal tidal currents depend on the shape of the seamount, the density profile of the surrounding ocean water, and the latitude of the seamount.

The amplification of tidal currents around seamounts is locally important in that the enhanced internal tidal currents tend both to increase the supply of particulate food to benthic organisms and to disperse their larvae over and possibly off the seamount. It is clear that the internal tidal currents do reach speeds strong enough to sweep detritus and fine sediment off the seamount, thus allowing ferromanganese crusts to precipitate out of the cold ambient seawater onto the hard rock substrate of that seamount. These crusts, which occur over vast areas of the sea floor, are rich in economically important minerals, such as cobalt and platinum.

SEE ALSO THE FOLLOWING ARTICLES

Barrier Islands / Beaches / Sea-Level Change / Seamounts, Geology

FURTHER READING

Grant, S. B., B. F. Sanders, A. B. Boehm, J. A. Redman, J. H. Kim, R. D. Mrse, A. K. Chu, M. Gouldin, C. D. McGee, N. A. Gardiner, B. H. Jones, J. Svejkovsky, G. V. Leipzig, and A. Brown. 2001. Generation of *Enterococci* bacteria in a coastal saltwater marsh and its impact on surf zone water quality. *Environmental Science and Technology* 35: 2407–2416.

Holloway, P. E., and M. A. Merrifield. 1999. Internal tide generation by seamounts, ridges, and islands. *Journal of Geophysical Research* 104: 25,937–25,951.

Noble, M., D. A. Cacchione, and W. C. Schwab. 1988. Observations of strong mid-Pacific internal tides above Horizon Guyot. *Journal of Physical Oceanography* 18: 11,300–11,306.

The Open University. 2006. *Waves, tides and shallow-water processes*, 2nd ed. Jointly published by the Open University, Milton Keynes, and Butterworth Heinemann.

Storlazzi, C. D., E. K. Brown, and M. E. Field. 2006. The application of acoustic doppler current profilers to measure timing and pattern of larval dispersal. *Coral Reef* 25: 369–381.

Storlazzi, C. D., A. S. Ogston, M. H. Bothner, M. E. Field, and M. K. Presto. 2004. Wave- and tidally-driven flow and sediment flux across a fringing coral reef: southeastern Molokai, Hawaii. *Continental Shelf Research* 24: 1397–1419.

TIERRA DEL FUEGO

MATTHEW J. JAMES
Sonoma State University, Rohnert Park, California

JOHN M. WORAM
Rockville Centre, New York

Tierra del Fuego is the extensive archipelago of large and small islands at the southern tip of South America, separated from the mainland by the Strait of Magellan. Its total area is 73,746 km², two-thirds of which is owned by Chile, one-third by Argentina. The largest island within the archipelago is Isla Grande de Tierra del Fuego (Fig. 1).

GEOGRAPHIC SETTING

The archipelago is further divided by the Beagle Channel running along the southern coast of Isla Grande. Along the Argentine territory, it forms the border with the Chilean islands to the South. The Chilean territory contains Cape Horn and False Cape Horn (both located on islands). Cape Horn is located on Isla Hornos in the Hermite Islands group, a small archipelago at the very southern extent of Tierra del Fuego. The cape was not named for its shape, but rather for the city of Hoorn in the Netherlands, the birthplace of Dutch navigator Willem Cornelisz Schouten, who first sailed around the cape in 1616. Although Cape Horn is considered the south-

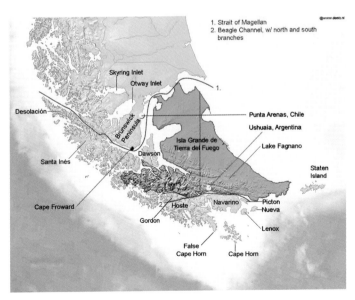

FIGURE 1 Isla Grande and other islands constituting the Tierra del Fuego archipelago.

ern tip of South America, Islas Diego Ramirez are farther south, with Islote Aguila being the southernmost in that group. False Cape Horn is a headland at the southern tip of Isla Hoste and is the southernmost point of one of the large islands that constitute Tierra del Fuego, but it is quite often mistaken for Cape Horn itself, especially by sailors approaching from the west.

Also in Chilean territory, Cape Froward is the southernmost extremity of the South American continental land mass, located on the Brunswick Peninsula on the north shore of the Strait of Magellan, south of Punta Arenas.

FLORA AND FAUNA

The region supports an attenuated southern flora and fauna, as well as dominant introduced species such as the North American beaver, the European rabbit, and sheep, as well as the native guanaco.

Forests within the Tierra del Fuego National Park contain tree species of *Lenga, Guindo,* and *Ñire.* In his *Journal and Remarks* (later, *Voyage of the Beagle*), Charles Darwin mentions "vegetation thriving most luxuriantly, and large woody stemmed trees of Fuchsia and Veronica in full flower" that was noted on a previous expedition. Darwin himself claims to "have seen parrots feeding on the seeds of the winter's bark, south of latitude 55°."

HUMAN HISTORY

Events that took place in Tierra del Fuego during the first of three surveying voyages of HMS *Beagle* include the suicide of Royal Navy Captain Pringle Stokes in 1828, the appointment of Flag Lieutenant Robert FitzRoy as his replacement, and FitzRoy's subsequent kidnapping of four Fuegian Indians in 1830, which ultimately led to Charles Darwin's participation in the second *Beagle* circumnavigation-surveying voyage in 1831–1836. It was on this voyage that Captain FitzRoy returned three of the Fuegian Indians—a fourth had died of smallpox in England—to their home on Isla Navarino, on the Chilean side of the Beagle Channel.

Tierra del Fuego in general is known for its harsh weather, and Cape Horn in particular is renowned in the history of sailing for the difficulty experienced by vessels and their crews when "rounding the Horn." Buccaneer William Ambrosia Cowley described his experience in 1684: "The weather in the lat. of 60 deg. was so extream cold that we could bear drinking 3 quarts of Brandy in 24 hours each Man, and be not all the worse for it." With perhaps less brandy on hand, Captain David Porter of the U. S. frigate *Essex* offered this recommendation after his own passage in February 1813: "I would advise those bound into the Pacific, never to attempt the passage of Cape Horn, if they can get there by any other route." Adding to the negative experiences from tempestuous passages around Cape Horn were the remarks of Philo White (1789–1883), who rounded the Horn in 1841 in the U.S. sloop of war *Dale* and wrote: "It is now going on four weeks, that we have been off Cape Horn! buffeting strong gales of contrary winds,—wearing and tacking ship, in endeavors to make progress against tremendous head swells, and strong adverse currents,—amidst furious snow squalls, and hail and sleet storms: There is consequently much suffering amongst the crew."

Tierra del Fuego has the world's southernmost city (Ushuaia), national park (Parque Nacional Cabo de Hornos), highway (Argentina's RN 3), and brewery (Cervecería Fueguina).

SEE ALSO THE FOLLOWING ARTICLES

Continental Islands / Juan Fernandez Islands / Voyage of the *Beagle*

FURTHER READING

Armstrong. P. 2004. *Darwin's other islands*. New York: Continuum International Publishing Group.
Olivero, E. B., and D. R. Martinioni. 2001. A review of the geology of the Argentinian Fuegian Andes. *Journal of South American Earth Sciences* 14: 175–188.

TOKELAU ISLANDS

SEE PACIFIC REGION

TONGA

DONALD R. DRAKE

University of Hawaii, Manoa

Tonga is an archipelago of small, tropical, Pacific islands consisting of limestone or volcanic rock. The natural vegetation is tropical rain forest, and the native vertebrate fauna includes bats, birds, and lizards. The terrestrial environment has been strongly modified by humans, resulting in deforestation, animal extinctions, and invasion by alien species, although significant remnants of the native biota still exist.

GEOGRAPHY AND PHYSICAL ENVIRONMENT

The Kingdom of Tonga is a South Pacific nation consisting of about 700 km² of land divided among 170 scattered islands (Fig. 1). The archipelago lies just west of where the Pacific tectonic plate is being subducted beneath the Indian-Australian plate. This geological activity has produced the 10,000-m-deep Tonga Trench in the east and

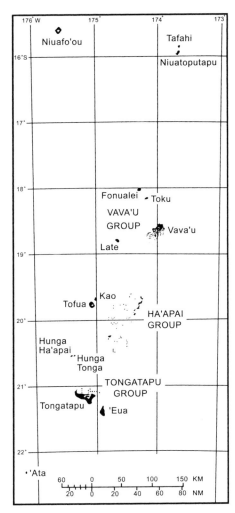

FIGURE 1 The Kingdom of Tonga. From Steadman (2006), courtesy of University of Chicago Press.

the Tongan Islands in the west. The islands form two north–south lines paralleling the trench: raised limestone islands in the east and volcanoes in the west. The limestone islands range from small cays to high, terraced landforms (Fig. 2). The limestone islands are covered with 1–4 m of fertile soil derived from volcanic ash. The volcanic islands are all either active or dormant and, in many places, have relatively young surfaces with poorly developed soil. Although there are no permanent streams, several islands have freshwater lakes, ponds, or marshes.

The climate is tropical to subtropical. The prevailing winds are the southeast trade winds. Mean annual rainfall is approximately 1900–2300 mm, with November to April being slightly warmer and wetter. Cyclones occasionally strike the islands during the wet season.

HUMAN HISTORY

Tonga was originally colonized by the ancestors of modern Polynesians 2800–3000 years ago and first made significant contact with Europeans through Dutch explorer Abel Tasman in 1643. The current population of approximately 110,000, two-thirds of whom live on Tongatapu, is spread across 45 islands. The combination of long human occupation, high population density, and small island size has resulted in significant environmental modification, including extensive deforestation and extinction.

Tonga has always been an independent nation with indigenous governance, and for much of its history it was ruled by a paramount chief. In 1875, the ruler established a constitutional monarchy and became the first in a hereditary line of five kings and queens that extends to the present. Tonga is the last Polynesian country to be ruled by a hereditary monarch. Since 1875, the government has largely been controlled by members of the royal family and the traditional nobles. In recent years, however, a pro-democracy movement has been advocating political change.

Tonga's economy is somewhat constrained by the country's small land area, which limits its natural resources. Land is widely distributed among the population, and most families cultivate a diverse range of crops for home consumption or sale. Many people are subsidized by remittances from relatives living overseas. There is a modest tourist industry.

BIOTA

Biogeography

Because Tonga has never been connected to a continent, its native biota is limited to those organisms that were able to colonize via long-distance dispersal. Hence, the biota is less diverse than that of Australasia, but more diverse than that of the distant island groups of eastern Polynesia. Tonga's biota is largely a subset of that of its larger neighbors, Fiji and Samoa. Although few species are endemic to Tonga, many are endemic to the Tonga–Fiji–Samoa biogeographic

FIGURE 2 Small, raised limestone islands in the Vava'u Group. Note the extensive fringing reef around Taunga (left) and the terraced topography on 'Euaiki (right).

FIGURE 3 Diverse rain forest covers the upper slopes of the national park on 'Eua, a 312-m-high island consisting of a volcanic core covered by limestone terraces.

region. In addition, this region represents the furthest natural penetration into the Pacific for many groups of organisms, including gymnosperms, mangroves (Rhizophoraceae), snakes, and honeyeaters (Aves: Meliphagidae).

Terrestrial Flora and Vegetation

Tonga's native flora has about 340 species of angiosperms, two species of gymnosperms, and 83 species of ferns and fern allies. Only eight angiosperms, one gymnosperm, and one fern are endemic. The natural vegetation for nearly all of Tonga is tropical rain forest (Fig. 3). Mangrove vegetation occurs in some lagoons and other sheltered, coastal sites. Forests on coral-sand beaches are dominated by coastal trees common throughout the tropical South Pacific, whereas inland rain forests are dominated by species with narrower geographic ranges. Today, most variation in rain forest composition on the limestone islands is determined by a site's history of disturbance by humans or cyclones rather than by variation in the physical environment. Only the highest islands exhibit significant altitudinal variation in species composition. Vegetation on the sparsely inhabited volcanic islands is poorly known.

Terrestrial Fauna

Tonga's native vertebrate fauna is limited to those groups of animals most capable of long-distance dispersal across seawater: bats, birds, and reptiles. The invertebrate fauna is less well known.

Tonga once supported two species of large, fruit-eating bats (flying-foxes) and one small, insectivorous bat (*Emballonura semicaudata*). One of the flying foxes (probably *Pteropus samoensis*, based on subfossil remains) is locally extinct but still occurs in Fiji and Samoa. The other, *P. tonganus*, is a widespread South Pacific species that is still common in Tonga. With the extinction of many of Tonga's large, native birds, *P. tonganus* plays a key role in forest dynamics as one of the few seed dispersers of large-seeded trees.

Birds are Tonga's most diverse native vertebrates. There are 21 species of land birds in 20 genera today; 26 more species are known from the subfossil record but disappeared following human colonization. The only extant endemic birds are the Tongan whistler (*Pachycephala jacquinoti*) of Vava'u and the Niuafo'ou megapode (*Megapodius pritchardii*) of Niuafo'ou. Pigeons and fruit doves (Columbidae) appear to have been major components of the prehistoric bird fauna, but many are extinct. In addition, about 20 species of seabirds occur in and around the islands, though not all breed there.

Tonga has many species of reptiles but no amphibians. Although sea turtles and sea snakes use coastal habitats, the only truly terrestrial reptiles in Tonga today are lizards, including four genera of geckos and five of skinks. Tonga also has extinct and extant iguanas (*Brachylophus* spp.), a tropical American group of reptiles otherwise known in the South Pacific only from Fiji. The subfossil remains of the Pacific boa (*Candoia bibronii*) suggest this snake was once native to Tonga.

Marine Biota

Tonga's limestone islands are surrounded by extensive coral reefs that support a diverse marine flora and fauna. During the winter, humpback whales (*Megaptera novaeangliae*) migrate from Antarctica to Tonga to give birth and mate.

CONSERVATION

Tonga's natural environment has been strongly impacted by the direct and indirect effects of humans and the hundreds of alien plant and animal species they have introduced. Today, most of the land is used for villages or agriculture, and few islands retain more than 10% of their original forest cover. Most remaining forest is on steep slopes or sparsely inhabited volcanic islands. However, the government has also set aside several areas, such as 'Eua National Park (Fig. 3), in an effort to conserve Tonga's natural heritage. Alien mammals, such as rats (*Rattus* spp.) and cats (*Felis catus*) pose significant threats to plants and animals that evolved in the absence of mammalian predators.

SEE ALSO THE FOLLOWING ARTICLES

Fiji, Biology / Fossil Birds / Makatea Islands / Peopling the Pacific / Samoa, Biology

FURTHER READING

Fall, P. 2005. Vegetation change in the coastal-lowland rain forest at Avai'o'vuna Swamp, Vava'u, Kingdom of Tonga. *Quaternary Research* 64: 451–459.
Flannery, T. 1995. *Mammals of the south-west Pacific and Moluccan Islands.* Chatswood, Australia: Reed Books.
Franklin, J., S. K. Wiser, D. R. Drake, L. E. Burrows, and W. R. Sykes. 2006. Environment, disturbance history and rain forest composition across the islands of Tonga, Western Polynesia. *Journal of Vegetation Science* 17: 233–244.
McConkey, K. R., and D. R. Drake. 2006. Flying foxes cease to function as seed dispersers long before they become rare. *Ecology* 87: 271–276.
McConkey, K. R., D. R. Drake, H. J. Meehan, and N. Parsons. 2003. Husking stations provide evidence of seed predation by introduced rodents in Tongan rain forests. *Biological Conservation* 109: 221–225.
Meehan, H. J., K. R. McConkey, and D. R. Drake. 2002. Potential disruptions to seed dispersal mutualisms in Tonga, Western Polynesia. *Journal of Biogeography* 29: 695–712.
Mueller-Dombois, D., and F. R. Fosberg. 1998. *Vegetation of the tropical Pacific Islands.* New York: Springer-Verlag.
Nunn, P. D. 1994. *Oceanic islands.* Oxford, UK: Blackwell Publishers.
Steadman, D. W. 2006. *Extinction & Biogeography of Tropical Pacific Birds.* Chicago: University of Chicago Press.
Wiser, S. K., D. R. Drake, L. E. Burrows, and W. R. Sykes. 2002. The potential for long-term persistence of forest fragments on a large island in western Polynesia. *Journal of Biogeography* 29: 767–787.

TORTOISES

CHARLES R. CRUMLY

University of California, Berkeley

Size change is one of the most common patterns of evolution on islands—dwarfism in some cases and gigantism in others. Land tortoises become giants. Three independently evolved lineages of giant tortoises survived until historic times (Figs. 1–4). The Galápagos Islands tortoise populations include a dozen or so species, some extinct and others with small but recovering populations. Two other lineages of giants evolved on different island groups of the Indian Ocean. Only one species of these two lineages survives, mostly on the isolated atoll of Aldabra. Land tortoises represent, in the public conscience, both the pattern of insular gigantism and a vivid example of the process of evolution.

THE TESTUDINIDAE

All land tortoises belong to the Testudinidae, a monophyletic lineage of turtles that includes approximately 50 living species (including those extinct within historic times) and appears first in the fossil record around 60 million years ago. Tortoises are found on all continents except Australia and Antarctica. Giant continental forms disappeared from mainland ecosystems at the close of the Pleistocene, perhaps coincident with the spread of humanity out of Africa. Most land tortoises possess a suite of characteristics that distinguish them from all other turtles, such as a domed carapace, elephantine feet with two phalanges or fewer in each digit, and several skull characters. A xeric lifestyle, common among tortoises, may be one reason why tortoises survived and became gigantic on island ecosystems prone to climatic and resource variability and lengthy drought.

Phylogenetic studies of comparative anatomy and more recent studies of molecular systematics have confirmed both the monophyly (shared common ancestry) of tortoises and the independent origin (polyphyly) of insular giant tortoises.

FIGURE 1 *Chelonoidis nigra porteri* from Santa Cruz. Photograph by Adalgisa Caccone and gratefully used here with permission.

FIGURE 2 *Chelonoidis nigra hoodensis* from Española. Photograph by Adalgisa Caccone and gratefully used here with permission.

FIGURE 3 A captive *Aldabrachelys gigantea* from the Ménagerie du Jardin des Plantes, Paris.

FIGURE 4 *Cylindraspis vosmaeri* from the Muséum National d'Histoire Naturelle (a female specimen, number AC-A5222).

THE PATTERN IN VERTEBRATE EVOLUTION

Size change is common in vertebrate evolution, especially on islands. Some lineages are well known for including species much larger, whereas others include lineages much smaller than close mainland relatives. Although it often seems obvious that a particular radiation of organisms evolved by changing size, it can be difficult to accumulate evidence sufficient to support a hypothesis regarding the cause of size increase or decrease. Even determining the size of the ancestor can be complicated by the absence of a fossil record, a poorly studied phylogenetic history, and an inadequate number of specimens of intermediate sizes. Thus, evolutionary size change both is perceived as common and, paradoxically, is difficult to corroborate because affirming evidence is scant. This was certainly true for land tortoises until the recent publication of reliable and corroborated hypotheses of tortoise phylogeny.

Whether island animals become giants or dwarfs seems partly influenced by physiology. Those animals with body temperatures that do not vary with ambient temperatures, referred to as endotherms, can become gigantic or can become dwarfs, such as the smaller elephants, rhinoceroses, and hippopotami. In contrast, vertebrates such as tortoises, whose body temperatures match the ambient temperature of the environment (ectotherms), tend to become gigantic. Examples of this pattern include several lineages of lizards as well as land tortoises. Many, but not all, of these groups are herbivores and capable of living through lengthy periods of drought and starvation. Insular ectothermic giants often possess long life spans, show sporadic and unpredictable levels of recruitment, and inhabit islands usually free of predators and competitors (Table 1).

INDEPENDENT EVOLUTION

In land tortoises, there have been many instances of size change; dwarfism on continents is just as common as gigantism on islands. Indeed, because the phylogenetic history is relatively well documented, it is easier to confirm that island lineages are giants. It is not as easy to determine whether the first tortoises to reach these islands were as large as they are today. Indeed, it is likely that all of these cases involve size increase and that they evolved independently. And in every case, the tortoises reached their island refuges by over-water dispersal. In fact, there are many reports of tortoises found at sea after being pitched overboard accidentally or intentionally during storms or conflicts, or naturally in the central lagoon of Aldabra. Because giant tortoises can survive lengthy periods of over-water transport, dispersal is acknowledged as the likely means of colonization.

Galápagos

Tortoises of the Galápagos are the most diverse surviving lineage of insular giant tortoises, and they are most closely related to the tortoises of South America. As many as 13 taxa are classified as subspecies of *Chelonoidis nigra*. Five subspecies share Isabela, each restricted to a volcanic cone separated from neighboring cones by impassable lava fields. Pinzon, San Cristóbal, Santa Cruz, and Santiago each harbor a separate unique species. A single adult male ("Lonesome George") is the last of the Pinta population, and he now lives at the Charles Darwin Research Station on Santa Cruz. Three species—two named (one on Fernandina, another on Santa Maria) and one unnamed (on Santa Fe)—are extinct. In addition, genetic studies have revealed unnamed populations that might warrant taxonomic recognition.

TABLE 1
Giant Tortoises on Islands Listed by Fritz and Havas (2007)

Taxon	Distribution	Estimated Population Size
Galápagos		
Chelonoidis nigra abingdonii	Pinta	1 (extinct in the wild)
Chelonoidis nigra becki	Isabela, Volcán Wolf	1000–2000
Chelonoidis nigra chathamensis	San Cristóbal (introduced onto Rábida)	500–700
Chelonoidis nigra darwini	San Salvador (or Santiago)	500–700
Chelonoidis nigra ducanensis	Pinzón	150–200
Chelonoidis nigra guntheri[a]	Isabela–Sierra Negra	100–300
Chelonoidis nigra hoodensis	Española	15 (native)
Chelonoidis nigra microphyes[a]	Isabela–Volcán Darwin	500–1000
Chelonoidis nigra nigra	Santa María (or Floreana)	Extinct
Chelonoidis nigra phantastica	Fernandina	Extinct
Chelonoidis nigra porteri	Santa Cruz	2000–3000
Chelonoidis nigra vandenburghi[a]	Isabela–Volcán Alcedo	3000–5000
Chelonoidis nigra vicina	Isabela–Cerro Azul	400–600
Aldabra, the Seychelles, and Madagascar		
Aldabrachelys abrupta	Western and central Madagascar	Extinct
Aldabrachelys gigantea	Aldabra and Granitic Seychelles	85,000
Aldabrachelys grandidieri	Southwestern Madagascar	Extinct
Mascarenes		
Cylindraspis indica	Réunion	Extinct
Cylindraspis inepta	Mauritius	Extinct
Cylindraspis peltastes	Rodrigues and Île aux Aigrettes	Extinct
Cylindraspis trisserata	Mauritius	Extinct
Cylindraspis vosmaeri	Rodrigues	Extinct

NOTE: Galápagos estimated population sizes from Caccone *et al.* 2002.

[a] Some experts have suggested that *guntheri*, *microphyes*, and *vandenburghi* are conspecific with *vicina*. Genetic studies support a close relationship among all the taxa on Isabela (see Fig. 7), but these populations are geographically isolated from one another.

These are among the largest of the living land tortoises, approaching 300 kg and 1.2 m long. Although the maximum age in nature is not known, it is estimated that Galápagos tortoises often live to be at least 200 years old.

Galápagos tortoises are informally divided on the basis of shell shape (Fig. 5). Some have domed carapaces similar to those of familiar mainland tortoises. But some possess a carapace whose anterior perimeter is raised, creating a shape like a medieval Spanish saddle. Tortoises with saddlebacked carapaces are smaller than their domed relatives and usually occur on lower, drier islands. A raised anterior carapacial edge may confer a competitive advantage. Thomas Fritts reported observations of male/male contests; the individual whose head is raised highest wins. In his observations of mixed captive herds, large domed forms lose contests with smaller saddlebacked opponents, because the raised anterior edge of the shell permits the smaller form to raise its head further. Furthermore, gaining access to higher branches of trees and shrubs is an advantage conferred by a saddlebacked carapace. Examples of the saddlebacked form include forms from Pinzon, Española, and Pinta (Fig. 6). Domed forms are found on Santa Cruz and on Isabela. Intermediate forms also occur.

During the first decade of the twenty-first century a team of researchers at Yale University, including Adalgisa Caccone, Jeffrey Powell, and Michael Russello, studied the genetic relationships among and between Galápagos tortoises. This team was able to corroborate the earlier hypothesized relationship between Galápagos tortoises and the Chilean tortoise *Chelonoidis chilensis*. They also proposed the first explicit hypothesis of the phylogeny of Galápagos tortoises (Fig. 7) and discovered that some populations are related to nearby populations—as would be expected. For example, all isolated populations on Isabela are more closely related to each other than to any tortoise population found elsewhere in the archipelago; in fact, there is so little genetic variation that some believe nearly all the Isabela populations are the same form. But Caccone and co-workers also discovered unexpectedly close relationships between geographically distant popu-

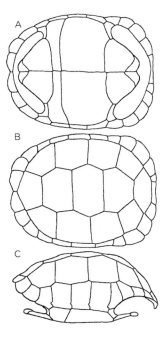

FIGURE 5 The shells of Galápagos tortoises can be domed, like those of most other tortoises, or saddlebacked. Here are three views of the domed condition of the Santa María tortoise *Chelonoidis nigra nigra* with the front of the shell to the left: (A) ventral view of the plastron; (B) dorsal view of the carapace; (C) lateral view of the shell. This is Plate 18 from Garman (1917), drawn from Museum of Comparative Zoology specimen number 4479.

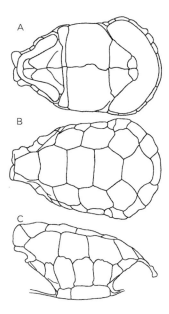

FIGURE 6 Three views of the saddlebacked condition of Galápagos tortoise shells as exemplified by the Pinta tortoise *Chelonoidis nigra abingdonii*: (A) ventral view of the plastron; (B) dorsal view of the carapace; (C) lateral view of the shell. This is Plate 40 from Garman (1917) drawn from Günther's (1877) plate 40 and 41. It is the type specimen of *C. n. abingdonii* and is stored in the Natural History Museum in London.

lations (e.g., *Chelonoidis nigra hoodensis* from Española in the far southeast of the archipelago share genetic markers with *C. n. abingdonii* from Pinta in the far north). Finally, by interpreting mitochondrial DNA evidence, the researchers were able to hypothesize instances of humans transporting tortoises from one island to another.

Seychelles and Aldabra

Once present on several islands of the Indian Ocean (Madagascar and islands of the Seychelles group), this lineage is now represented by a single species, mostly restricted to the atoll of Aldabra. The ecosystem on Aldabra is dominated by tortoises – one of the rare cases wherein a herbivore is the dominant taxon. Indeed, the ground vegetation on Aldabra is referred to as "tortoise turf." Individuals of this species (*Aldabrachelys gigantea*) have also been introduced elsewhere and captive populations are maintained, partly against the event of the catastrophic loss of the only remaining native populations on Aldabra. Although several different scientific names have been proposed for living tortoises in this lineage, unambiguous genetic evidence documents variation typical of a single living species that is most closely related to tortoises of Madagascar.

Composed of several smaller islets (Grande Terre, Malabar, and Picard), the atoll of Aldabra supports three separate populations, each with slight differences in size, abundance and certain demographic factors such as fecundity. About 90% of Aldabra tortoises occupy Grande Terre, the largest part of the atoll. These tortoises lay about a third the number of eggs laid by Malabar tortoises and about a fourth the number of eggs laid by Picard tortoises. The average egg size and the typical maximum body size also vary from islet to islet. The largest individuals occur on Picard, whose tortoises lay the greatest number of the smallest eggs.

All members of this lineage, extinct as well as the one living species, possess a feature unique among all turtles: a vertically expanded external narial opening. This feature is coincident with an unusual drinking behavior. On Aldabra infrequent rainfall gathers in shallow depressions. In most tortoises, drinking is done by submerging the head in water and oscillating the gular region of the neck to pump water into the esophagus. This behavior requires water sources deep enough to permit full emergence. But the puddles on Aldabra are too shallow to allow this. Instead, Aldabra tortoises dip their nostrils into puddles and suck water up through their nasal passages. Specialized internal soft tissues prevent the fouling of the olfactory epithelium.

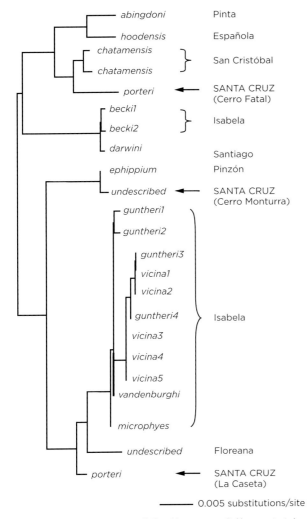

FIGURE 7 The evolutionary relationships among Galápagos tortoises (modified slightly from Russello et al. 2005) confirm that they are very closely related. Other studies (Caccone et al. 2002) propose a shared common ancestor from around 2–3 million years ago. Most parts of this diagram are well supported by evidence from mitochondrial DNA. The relationships among very closely related Isabela populations are not as well supported. The numbers after species names are used to document populations of the same taxon with almost identical and yet still distinctive types of mitochondrial DNA.

Mascarenes

The most poorly known of the giant island tortoises are members of the *Cylindraspis* lineage, which includes five named species, all of which are extinct. They disappeared before significant museum collecting in the mid-1700s to early 1800s. Those specimens that did make it into museum collections were often fragmentary or without reliable locality data. Thus, there are too few specimens with too little information on provenance to allow for detailed systematic analysis. One feature that unifies these tortoises is the tendency toward shell ankylosis when approaching adult size. In addition, some *Cylindraspis* possess a single, rather than paired, gular scute and, unlike *Aldabrachelys*, lack a nuchal scute.

A CONFUSION OF NAMES

There are many names associated with giant island tortoises and where they have lived. Early studies of Galápagos tortoises used English names and, for some islands, there are multiple Spanish names. Thus, linking specimens to localities can be difficult and is sometimes impossible. Human transport of specimens, without any record or reliable documentation, further confounds efforts to name and classify tortoises. This is true for both Galápagos tortoises and the tortoises of the Indian Ocean.

Limited samples of Galápagos tortoises with poor locality data contributed to the same species being named more than once. An inadequate appreciation of variability also generated invalid names. Studies of the phylogeny of testudinids have required changes in the classification of all tortoises including Galápagos species. Once most tortoises of the neotropics were allocated to *Testudo*, then to *Geochelone*, and are now referred to as *Chelonoidis*. For technical nomenclatural reasons, the species epithet *elephantopus* was replaced first by *nigrita* and then by *nigra* despite many years of consistent usage of *elephantopus*. The application of different species concepts generated both species and subspecies names for the Galápagos forms. Most races are essentially isolated reproductively, a prerequisite for species recognition using the biological species concept. But genetic evidence suggests that, in some races, interbreeding has occurred naturally as well as resulting from human transport of tortoises from one population to another. The capacity to interbreed is, under the biological species concept, an indication that individuals are the same species. Thus, recognition of the species versus subspecies status of Galápagos tortoises depends upon the importance one places on observations that can be interpreted differently.

The name used for the Aldabra tortoise is even more confused. Stable and often used for nearly a century, the epithet *gigantea* was rejected by some well-meaning but misguided experts based on disputable or ambiguous interpretation in favor of the strict adherence to nonbiological codes of nomenclature. Over the past 20 years, questionable nomenclatural choices have compounded this confusion. And in some instances, definitive evidence has been ignored or discounted in order to cling to falsified hypotheses. Thus, in the most recent checklist of turtles compiled by Uwe Fritz and Peter Havaš, nearly 40 names, employed during the last 30 years, were included in the synonymy of *Aldabrachelys gigantea*.

HUMANS AND INTRODUCED SPECIES

All insular giant tortoise populations evolved and survived on islands free of predators and relatively free of competition until human exploration and invasive species began processes of decline and extinction. Rats escaped sailing ships and ate eggs and hatching tortoises. Goats, brought on ships for meat and milk, became feral and ate vegetation that might have sustained tortoises. And, of course, sailors loaded their ships with tortoises, whose flesh sustained crews during long ocean voyages.

The tortoises of the Indian Ocean were driven extinct by the combined impact of all these factors. And these same factors have been in operation in the Galápagos; why have the tortoises of Galápagos fared better? The difference seems partly due to when disturbances began. In the case of Indian Ocean tortoises, sailing ship visitations, colonization, hunting, and introduced animals occurred several hundred years earlier than in Galápagos. In addition, human visits to the Mascerenes and Seychelles were more frequent than to the Galápagos. Furthermore, permanent settlements were established on Indian Ocean islands long before any settlements were established in Galápagos. Aldabra tortoises avoided extinction because the atoll is remote and outside regular sailing routes, and there is no permanent water.

CONSERVATION

The success of most conservation programs depends on timing and effort. For *Cylindraspis*, no effort was made because conservation was not a priority in the eighteenth and nineteenth centuries. *Aldabrachelys* on Madagascar became extinct between 750 and 1250 years before present, well after the first appearance of humans on the island.

Today, active conservation efforts are ongoing in Galápagos and on Aldabra. The Charles Darwin Research Station was built by the Charles Darwin Foundation and inaugurated in 1964. The Station is headquartered on Santa Cruz and manages continuing efforts to help in recovery of Galápagos tortoises. For races rare in their native range, the Station raises hatchlings until they are large enough to be released back to their native habitat. Other activities include the eradication of goats and rats. Financial support comes from organizations and institutions, as well as individuals. The government of Ecuador, which exercises sovereignty over the archipelago, established the Galápagos National Park Service and deserves special praise for the commitment made to preserve Galápagos biodiversity. The work of the Station is one of the success stories in conservation of biodiversity and habitat restoration.

Aldabra has been vigorously protected through a variety of programs and with considerable international participation. Many of the same institutions and organizations that support the activities of the Charles Darwin Research Station also support conservation efforts on Aldabra.

SEE ALSO THE FOLLOWING ARTICLES

Adaptive Radiation / Galápagos Islands, Biology / Gigantism / Madagascar / Seychelles

FURTHER READING

Austin, J. J., E. N. Arnold, and R. Bour. 2003. Was there a second adaptive radiation of giant tortoises in the Indian Ocean? Using mitochondrial DAN to investigate speciation and biogeography of *Aldabrachelys* (Reptilia, Testudinidae). *Molecular Ecology* 12: 1415–1424.
Caccone, A., G. Gentile, J. P. Gibbs, T. H. Snell, H. L. Snell, J. Betts, and J. R. Powell. 2002. Phylogeography and History of Giant Galápagos Tortoises. *Evolution* 56.10: 2052–2066.
Fritts, T. H. 1984. Evolutionary divergences of giant tortoises of Galapagos. *Biological Journal of the Linnean Society* 21: 165–176.
Fritz, U., and P. Havaš. 2007. Checklist of the Chelonians of the world. *Vertebrate Zoology* 57: 149–368.
Garman, S. 1917. The Galapagos tortoises. *Memoirs of Museum of Comparative Zoology* 30.4: 261–296.
Günther, A. 1877. *Gigantic land-tortoises (living and extinct) in the collections of the British Museum.* London: Taylor and Francis.
Le, M., C. J. Raxworthy, W. P. McCord, and L. Mertz. 2006. A molecular phylogeny of tortoises (Testudines: Testudinidae) based on mitochondrial and nuclear genes. *Molecular Phylogenetics and Evolution* 40: 517–531.
Pritchard, P. C. H. 1996. The Galápagos tortoises: nomenclatural and survival status. *Chelonian Research Monographs* 1: 1–85.
Russello, M. A., S. Glaberman, J. P. Gibbs, C. Marquez, J. R. Powell, and A. Caccone. 2005. A cryptic taxon of Galápagos tortoise in conservation peril. *Biology Letters* 1.3: 287–290.
Van Denburgh, J. 1914. The gigantic land tortoises of the Galapagos archipelago. *Proceedings of the California Academy of Sciences* 4th Ser., 2.1: 203–374.

TRADE WINDS

SEE CLIMATE ON ISLANDS

TRINIDAD AND TOBAGO

CHRISTOPHER K. STARR

University of the West Indies, St. Augustine, Trinidad and Tobago

Trinidad and Tobago are two small islands with a combined land area of about 5100 km², lying just off the northeast edge of the South American continent (Fig. 1) at 10°02′–11°21′ N and 60°31′–61°55′ W. Southwest Trinidad is separated from the mainland by an 11-km strait, whereas in the northwest there are steppingstone islands between Trinidad and the mainland. Tobago is separated from Trinidad by a 36-km strait. Trinidad's Northern

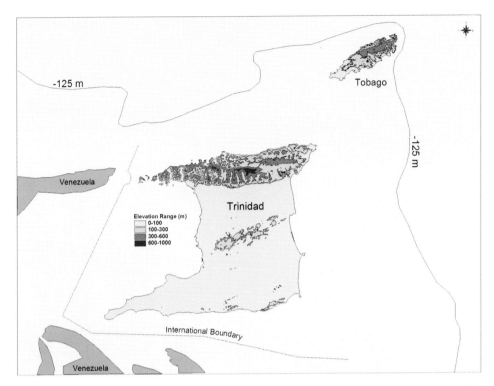

FIGURE 1 Trinidad and Tobago, position and topography. The present -125-m line approximates the coastline at the height of the most recent glaciation about 20,000 years ago. Map by Bheshem Ramlal.

Range and Tobago are eastern extensions of Venezuela's long Coastal Range.

CLIMATE AND TOPOGRAPHY

The islands are characterized by moderate topography—maximum elevation 940 m for Trinidad, 576 m for Tobago—and by a climate typical of their tropical latitude. Mean annual rainfall varies from about 125 to about 325 cm, according to locality, with a moderately distinct dry season from about mid-January to late May. Mean daily temperature fluctuation is estimated at 10.4 °C, with very little seasonal difference. Trinidad and Tobago lie south of the usual path of Atlantic hurricanes and have not been significantly affected by them in most decades.

PEOPLE AND GOVERNMENT

The two islands, together with various associated islets, form the Republic of Trinidad and Tobago. About 80% of the populace of 1.3 million is of Indian and African descent in equal proportions, with small minorities of people of other races and of mixed descent. English is the language of public affairs, and no other language is spoken by large numbers. The government of this former British colony, independent since 1962, is a parliamentary democracy in the British model. The economy is semi-industrialized and is heavily reliant on the petroleum industry (Trinidad) and on tourism (Tobago) and hardly at all on agriculture. Per-capita GDP is variously estimated at US$15,500–17,500. Life expectancy at birth is 74 years for women, 68 years for men.

ENVIRONMENT

The tectonic history of Trinidad and Tobago is complex and controversial. However, they appear to have undergone no significant movement or other gross disturbance since the Tertiary. Although it is difficult to plot Quaternary sea-level changes, they are thought to have caused several cycles of isolation and reunification with the mainland. The age of present isolation is generally estimated at 10,000 years for Trinidad and 14,000 years for Tobago, although a minority view holds that a land bridge connected Trinidad to the mainland at least intermittently until much more recently.

These fluctuations in land area were presumably accompanied by cyclical changes in gross habitat type, as throughout northern South America. The greatest extent of savanna, relative to forest, occurred during glacial maxima (most recently about 20,000 years ago), and it is estimated that seasonal evergreen forest came to cover about 75% of the land surface by 10,000 years ago and to remain at about that figure through pre-Columbian times. Forest cover is now reduced to about 20–30%, depending on definition, although the decline of agriculture over about the last century has slowed the pace of deforestation.

The predominant natural land habitat is evergreen seasonal forest, found in wetter areas up to about 250 m. Other habitats of note include swamp forest (most notably on the east coast of Trinidad), mangrove (on the east and west coasts of Trinidad and in southwest Tobago), savanna (in central and southwest Trinidad), and lower montane forest (above about 250 m on both islands), with some elements of montane forest in the highest parts of Trinidad's Northern Range. Coastal habitats include many sand beaches, the major Buccoo Reef at the southwest end of Tobago, and several lesser coral reefs in Tobago and northeast Trinidad. Each island has a great many streams, but no significant rivers or natural lakes.

BIOTA

Whereas the rest of the West Indies—the Antilles—are oceanic islands, Trinidad and Tobago are typical continental islands. That is, they show only slight endemism, and they closely resemble comparable nearby mainland habitats in their (harmonic) biotic composition and diversity (Fig. 2). In addition, they are relatively resistant to invasive species and their effects. Endemism among the approximately 6600 species of seed plants, for example, is estimated at 2.1%. To cite some other well-studied examples, the corresponding figure for land vertebrates is 2 of 521 species (0.4%) (Fig. 3), for butterflies (*sensu stricto*, excluding Hesperiidae) is 5 of 387 species (1.3%), and none of the 42 known species of freshwater fishes is endemic. In line with this trend, no family of plants or animals with strong representation in the Guianas or eastern Venezuela appears to be absent from Trinidad and Tobago. As rough estimates, these islands harbour about 3% of the world's land and freshwater animal species and about 2% of plant species.

It is expected that over an extended period of time, a continental island will increasingly partake of the biotic features of an oceanic island: decreased diversity, increased disharmony, and increased endemism. We can refer to these outcomes col-

FIGURE 2 Like much of Trinidad and Tobago's biota, the social wasp *Mischocyttarus alfkeni* is very broadly distributed in South America. It nests in a variety of lowland habitats on many different substrates. Photograph by Allan W. Hook.

FIGURE 3 The golden treefrog, *Phyllodytes auratus* (A), one of Trinidad and Tobago's very few putative endemic species, known only from the upper reaches of Trinidad's two highest peaks. It breeds in the water that accumulates among the bracts of *Glomeropitcairnia erectiflora* (B). This tank bromeliad, although not rare, is known only from high elevations in Trinidad and nearby parts of Venezuela. Photographs by Daniel G. Thornham.

lectively as the "island effect." The earliest of these features to appear is likely to be the first, a lowering of diversity as a result of uncompensated local extinction, or "relaxation," which may be the engine of the island effect as a whole.

To what extent is an island effect manifest in Trinidad and Tobago? This question is only now coming to be addressed, by way of floristic and faunistic comparisons between Trinidad's Northern Range and similar habitat in Venezuela's Paria Peninsula. After some 10,000 years of separation, it is predicted that the magnitude of Trinidad's island effect will vary in a meaningful way among taxa. Preliminary results suggest, for example, that the diversity of social wasps (Polistinae) is much the same in Trinidad as in comparable habitats on the mainland, whereas that of stingless bees (Meliponini) is markedly lower.

CONSERVATION ISSUES

Trinidad and Tobago are a signatory of several international agreements relating conservation and the environment, including CITES, the Convention on Wetlands (Ramsar), the Convention on Biological Diversity, and the Cartagena Convention. Furthermore, a relatively high proportion of land area is under public ownership, and much of this remains in a natural or semi-natural state. A contributing factor here is undoubtedly the heavy dependence of the national economy on petroleum and, to a lesser extent, tourism, which limits pressure on the land for agricultural purposes.

At the same time, legal protection remains weak. Much of the country's conservation policy and infrastructure dates back to colonial times. There is still no formal system of national parks and protected areas that meets today's international standards, and the few designated conservation areas enjoy little real protection. Even in these areas, poaching and logging are relatively unchecked.

However, the growth of ecotourism, together with the presence of a number of active conservation-related NGOs federated under a national umbrella body, are promising signs. Allied with this latter factor is a perceptible, ongoing shift in government toward an increased local participation in management of the natural environment.

The most striking conservation success story of recent times is the rise of community-based patrolling of sea-turtle nesting beaches in both Trinidad and Tobago. This earns substantial revenue from both domestic and foreign ecotourism and has reduced poaching of adult turtles and eggs to a fraction of its former level. Another promising development is a move toward formal designation of a well-preserved, 90-km^2 forested area in northeastern Trinidad as the Matura National Park, again with community involvement.

SEE ALSO THE FOLLOWING ARTICLES

Antilles, Biology / Endemism / Island Biogeography, Theory of / Relaxation / Sea-Level Change

FURTHER READING

Brereton, B. 1981. *A history of modern Trinidad, 1783–1962.* Kingston, Jamaica: Heinemann.
Living World, journal of the Trinidad and Tobago Field Naturalists' Club. http://livingworldjournal.googlepages.com/home.
Murphy, J. C. 1997. *Amphibians and reptiles of Trinidad and Tobago.* Malabar, FL: Krieger.
Woods, C. A., and F. E. Sergile, eds. 2001. *Biogeography of the West Indies: patterns and perspectives*, 2nd ed. Boca Raton, FL: CRC.

TRISTAN DA CUNHA AND GOUGH ISLAND

PETER G. RYAN

University of Cape Town, South Africa

Renowned for supporting the most remote human community, the Tristan archipelago and Gough Island are small, cool-temperate, volcanic islands in the central South Atlantic. The islands range in age from 0.2 to 18 million years, resulting in a wide diversity of topography. Their isolation has led to high levels of endemism among the biota. Despite being discovered more than 500 years ago, Tristan was settled only in the early 1800s and is the only permanently inhabited island. The other islands have been relatively little impacted by humans. Currently, the main threats to native species are introduced rodents as well as a suite of introduced plants.

LOCATION AND PHYSICAL STRUCTURE

Tristan da Cunha is an archipelago of three main islands located almost midway across the Atlantic Ocean between the southern tip of Africa and South America. Gough Island lies 350 km south-southeast of Tristan (Fig. 1). They are the only cool-temperate islands in the South Atlantic; the nearest other islands are St. Helena to the north, and frigid Bouvet Island to the south. The islands are volcanic in origin, rising up steeply from the abyssal plain. Despite being only 20–30 km apart, they are separated by deep trenches, with water that is more than 500 m deep between Inaccessible and Nightingale, and more than 2000 m deep between these islands and Tristan. The islands differ greatly in age and, as a consequence, in size and height (Table 1).

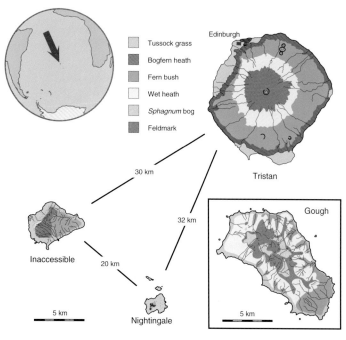

FIGURE 1 Map of the Tristan archipelago and Gough Island showing main vegetation types. Scale bars apply to the islands only.

The youngest and largest island, Tristan, is still an active volcano, whereas the oldest and smallest island, Nightingale, has been eroded until only the core of tough trachyte rocks remains. Marine erosion outstrips fluvial erosion, resulting in steep sea cliffs and hanging valleys. Soils are generally shallow and poorly developed, with a mantle of peat in many areas. Peat slips are frequent, especially at Gough Island. There are numerous perennial streams on Gough and Inaccessible, but not on the other two islands. On Tristan, rain rapidly soaks into the porous lava flows, emerging as springs around the base of the island. The low elevation of Nightingale results in lower rainfall, but swampy ponds occur in depressions on the island's summit (Fig. 2).

CLIMATE

The climate is cool-temperate and oceanic, with relatively little seasonal variation in temperature or rainfall. Gough Island is distinctly colder, wetter, and windier than the Tristan group, lying as it does on the edge of the roaring forties. The average temperature near sea level at Gough is 12 °C (−3 to 25 °C), compared with 15 °C (2 to 25 °C) at Tristan. Rain falls year round, usually associated with the passage of cold fronts, averaging 1670 mm per year at Tristan but closer to 3000 mm at Gough. Temperatures decrease and rainfall increases with elevation. Snow is regular on the peak of Tristan and on the highlands of Gough in winter, and orographic clouds are often found over the islands. The prevailing winds are from the west but veer to the northeast prior to the passage of a cold front, then back steadily to the south or southwest as the front moves through. Average wind speed is 36 km/h at Tristan and 44 km/h at Gough, with a tendency for stronger winds in winter.

Tristan and Gough were not glaciated during the last ice age, and analysis of pollen cores suggests that the vegetation (and hence the climate) has remained fairly constant for at least the last 20,000 years. However, over the last 40 years, average air temperatures have increased by 0.6 °C, and climate change models predict further increases of 1–5 °C over the next century. A warmer climate is likely to favor alien, introduced species that outcompete the native biota.

BIOTA AND TERRESTRIAL HABITATS

Tristan and Gough have never been connected to a continent, so all terrestrial animals and plants, and the shallow-water marine biota, have had to disperse across the ocean. Most immigrants arrived from South America and adjacent islands, thanks to the prevailing westerly winds and currents. However, some species have arrived from southern Africa, and some species are shared with Amsterdam and St. Paul, temperate islands in the central Indian Ocean. Once they reached the islands, many evolved to adapt to their new island home, resulting in endemic species. These include all seven landbirds, four of 22 breeding seabirds, 27 of 50 native flowering plants, 16 of 35 ferns, and close to 100 invertebrates (total diversity unknown). Levels of endemism are high among some marine groups too, notably seaweeds and bivalves.

In some cases, the new colonists underwent adaptive radiations to exploit the many vacant niches. At Tristan, the best-studied example is among the endemic

TABLE 1
The Area, Height, and Age of the Tristan da Cunha and Gough Islands

	Area (km²)	Height (m)	Age (million years)	Most recent eruption
Tristan	96	2060	0.2	1961
Inaccessible	14	600	3–4	50,000 years ago
Nightingale	4	350	18	200,000 years ago
Gough	65	910	3–5	100,000 years ago

FIGURE 2 Swampy "ponds" on the summit of Nightingale Island, with Middle Island and Tristan in the background.

Nesospiza buntings, with small-billed dietary generalists and large-billed specialists evolving to exploit the woody fruits of *Phylica* trees. This has resulted in two well-segregated species at Nightingale Island and three only partly segregated ecomorphs at Inaccessible Island. Other radiations have occurred among *Scaptomyza* flies (with two flightless, strap-winged species), *Tristanodes* weevils (11 species), *Balea* land snails (nine species), *Agrostis* grasses (seven species) and *Nertera* chicken berries (three species). Unfortunately, the relative paucity of competitors and predators on the islands renders their biota highly susceptible to extinction when new species are introduced.

Terrestrial habitats tend to be segregated by altitude. Coastal areas are dominated by tussock grassland. *Spartina arundinacea* dominates the coastal lowlands and cliffs up to 500 m at the Tristan islands, and is joined by the smaller *Paridochloa flabellata* at Gough Island. Fern bush is a diverse community found above coastal tussock, up to around 800 m at Tristan and 500 m at Gough. It is characterized by two large and distinctive species: the cycad-like fern *Blechnum palmiforme* and the island trees *Phylica arborea*. Wet heath is a fairly short, transitional vegetation type, containing elements of other vegetation types. It occurs from the upper limit of fern bush to above 800 m and contains fewer ferns than fern bush, with a higher proportion of mosses, grasses, sedges, and other flowering plants. At even higher elevations and on more exposed ridges, wet heath gives way to feldmark, an assemblage of dwarf, cushion-forming plants. Bogs are widespread at the islands, forming in hollows where drainage is impeded. There are two main types: Some support floating mats of big bog grass *Scirpus sulcatus*, whereas others are dominated by *Sphagnum* moss.

HUMAN HISTORY

The islands were first discovered by Portuguese explorers pioneering a sailing route around Africa: Gough in 1505 and Tristan in 1506. Despite plentiful water, fish, seals, and seabirds, the islands remained uninhabited because they lacked safe anchorages. A proposal to establish a British penal colony on Tristan was rejected in favor of Australia, and it was only when commercial sealing started in the late eighteenth century that protracted visits were made to the islands. Gangs of sealers were put ashore from 1790, killing thousands of seals for their skins and oil. Vegetable gardens were established at Tristan, and goats, pigs, and poultry were introduced. The first attempt to settle the islands was led by Jonathan Lambert, a Yankee whaler, in 1810, but this foundered when Lambert drowned in 1813.

The islands were annexed by Britain in 1816, when a garrison was stationed at Tristan to prevent the French from using the islands as a base from which to free Napoleon from his exile on St. Helena. When the garrison withdrew in 1817, William Glass was given permission to remain at Tristan. The fledgling community, augmented by castaways and crew from passing ships, flourished until the 1870s, thanks to the many vessels, especially whalers, calling to trade for fresh produce. But thereafter the number of ships dwindled, thanks to the switch to steamships, the opening of the Suez Canal, and collapsing whale stocks.

Tristan's isolation was greatest during the early twentieth century, resulting in an increasing reliance on seabird populations for food and guano. Links with the outside world increased during the Second World War, when a naval garrison was stationed at Tristan. The island was evacuated in 1961, when a volcano erupted next to the settlement, causing the entire community to flee to the United Kingdom. Most residents returned to the island in 1963. Today, the islands form the U.K. Overseas Territory of Tristan da Cunha, led by an Administrator and elected Island Council. The community of some 270 people is largely self-sufficient, generating income from fishing and the sale of stamps, and meeting most of their food needs from farming sheep, cows, poultry, potatoes, and other crops (Fig. 3). Careful stewardship of marine resources is crucial to Tristan's economy.

HUMAN IMPACTS AND CONSERVATION

Early visitors and colonists exploited natural resources, leading to the near extinction of seals, whales, and some seabird species. Subantarctic fur seals (*Arctocephalus tropicalis*) found a refuge on the remote western coast of Gough Island, and their numbers have recovered since the cessation of sealing. However, southern elephant seals (*Mir-*

FIGURE 3 Potato patches on the settlement plain at Tristan, backed by the steep cliffs that lead to the Base. The *Spartina* tussock that once covered the plain has been replaced by a mix of introduced pasture plants, and of the native landbirds, only a few Tristan thrushes persist in the deep gulches.

ounga leonina) have not recovered, and only a small, relict population persists on Gough's northeast coast. Human impacts on seabirds were most severe at the main island of Tristan, where Tristan albatrosses (*Diomedea dabbenena*) and southern giant petrels (*Macronectes giganteus*) disappeared, and populations of other seabirds were greatly reduced, exacerbated by predation by rats, which arrived on a shipwreck in 1882. Fortunately, the other islands have remained free of rats, but house mice (*Mus musculus*) were introduced by sealers to Gough Island in the 1800s and now pose a serious threat to the chicks of Gough buntings (*Rowettia goughensis*) and winter breeding seabirds, as well as to many native invertebrates (Fig. 4). At Inaccessible Island, feral pigs almost wiped out spectacled petrels and reduced the last population of Tristan albatrosses breeding at the northern islands to just a few pairs. Luckily, the pigs died out before they managed to finish the job.

Introduced plants outnumber native species at Tristan, and several species have reached the other islands. At Tristan the native vegetation has been almost entirely replaced by introduced pasture species on the coastal plains, and pasture species are also widespread on the island's plateau or Base. There are also many introduced invertebrates at the islands, including representatives of many groups that had not reached the islands prior to the arrival of humans. Their impacts are not well known, but some native species have become quite rare.

Fortunately, the need for environmental protection was recognized. In 1976, a conservation ordinance proclaimed Gough Island a nature reserve and provided some protection for seabirds at Tristan. Inaccessible Island subsequently was made a reserve, and together with Gough forms one of only two British Natural World Heritage Sites. The traditional harvesting of seabirds is now confined to Nightingale Island, where only great shearwaters and rockhopper penguin eggs may be collected. Tristan has protected almost half of its land area and has taken active steps to conserve its marine heritage. The community recently adopted a biodiversity action plan and appointed a conservation officer. Controls on imported goods are being tightened to reduce the risk of further accidental introductions to the islands, and eradication programs for invasive alien species are under way, targeting New Zealand flax (*Phormium tenax*) at Inaccessible and Nightingale and procumbent pearlwort (*Sagina procumbens*) at Gough Island. Plans are also being drawn up to eradicate rats and mice from Tristan and mice from Gough.

SEE ALSO THE FOLLOWING ARTICLES

Atlantic Region / Biological Control / Introduced Species / Rodents / Seabirds / St. Helena

FURTHER READING

Baker, P. E., I. G. Gass, P. G. Harris, and R. W. le Maitre. 1964. The volcanological report of the Royal Society Expedition to Tristan da Cunha, 1962. *Philosophical Transactions of the Royal Society of London A* 256: 439–578.

Crawford, A. B. 1982. *Tristan da Cunha and the roaring forties.* Edinburgh: Charles Skilton.

Munch, P. A. 1945. Sociology of Tristan da Cunha. *Results of the Norwegian Scientific Expedition to Tristan da Cunha 1937–1938* 13: 1–331.

Preece, R. C., K. D. Bennett, and J. R. Carter. 1986. The Quaternary palaeobotany of Inaccessible Island (Tristan da Cunha group). *Journal of Biogeography* 13: 1–33.

Ryan, P. G., ed. 2007. Field guide to the animals and plants of Tristan da Cunha and Gough Island. Newbury: Pisces Publications.

Wace, N. M., and M. W. Holdgate. 1976. Man and nature in the Tristan da Cunha Islands. *International Union for the Conservation of Nature and Natural Resources Monograph* 6: 1–114.

FIGURE 4 The Gough bunting *Rowettia goughensis* is confined to Gough Island. Its population has decreased alarmingly in recent years, apparently as a result of predation on eggs and chicks by house mice that were inadvertently introduced to the island by early sealing expeditions.

TSUNAMIS

EMILE A. OKAL
Northwestern University, Evanston, Illinois

Tsunamis are gravitational oscillations of the entire body of water of an ocean basin, following a disruption in the bottom (or exceptionally the surface) of the ocean. They differ from more conventional swells by their much longer periods (typically from 10 min to 1 hr) and relatively faster speeds over deep ocean basins (typically 220 m/s, or the speed of a modern jetliner). Tsunamis are capable of exporting death and destruction across entire ocean basins, their propagation being limited only by continental masses.

TSUNAMI SOURCES

Although tsunamis were once called "tidal waves," they are not caused by tides. Most tsunamis are generated when very strong earthquakes deform the ocean floor. Such a mechanism can move extremely large amounts of water (the fault rupture reached 1200 km in the 2004 Sumatra event), but only over relatively short distances (at most 20 m in the largest earthquakes), resulting in long waves of considerable energy (> 10^{22} erg for the Sumatra tsunami). Secondary, less frequent, tsunami sources include landslides, which can be either submarine (e.g., Papua New Guinea, 1998; Storrega, Norway, 6000 BC) or aerial, falling into the water (e.g., Stromboli, 2002; Aysen, Chile, 2007). Finally, catastrophic volcanic eruptions in the marine environment (e.g., Santorini, 1630 BC; Krakatau, AD 1883) and bolide impacts at sea (Yucatan, 65 million years ago) can also give rise to major tsunamis. However, non-earthquake sources obey different source scaling laws, which in lay terms means that they displace water over considerable distances (up to hundreds of km), but involve more contained volumes (rarely exceeding 50 km in linear dimensions). As a result, their tsunamis, which can be devastating in the near field (less than 1000 km from the source), feature shorter wavelengths and experience more efficient dispersion while propagating over large distances, resulting in generally benign amplitudes in the far field.

As the floor of the ocean has finite rigidity, it reacts to the passage of the tsunami by deforming elastically, resulting in a small, but significant, coupling of the tsunami to the solid Earth. Conversely, an earthquake source embedded in the solid Earth can excite a tsunami in an overlying ocean, but because the coupling is weak, appreciable transoceanic tsunamis are generated only by truly great earthquakes (of moments $\geq 5 \times 10^{28}$ dyn cm or so-called "moment magnitude" ≥ 8.7). The resulting tsunami wave remains in all cases relatively small on the high seas: Even for the catastrophic 2004 Sumatra tsunami, satellite altimetry provided a direct measurement of only 70 cm zero-to-peak in the Southern Bay of Bengal. Such low amplitudes are also spread over considerable wavelengths (~300 km), giving the tsunami a flat aspect ratio and rendering the wave undetectable on the open sea by classical (visual or optical) means.

THE INTERACTION OF TSUNAMI WAVES WITH COASTLINES

When approaching a shoreline, a tsunami undergoes shoaling; that is, the wave slows down in the shallower water, while its amplitude increases considerably, with run-up heights at the shoreline having reached, during the Sumatra event, 32 m in the near field and up to 12 m in the far field. If the structure of the wave remains stable at the shoreline, it can continue to propagate over initially dry land and inundate the coastal areas in the form of a progressively rising swell over distances having reached, again in 2004, 10 km in the near field and 3 km in the far field. Otherwise, the wave breaks like surf and hits the shore as a wall of water or "bore," particularly destructive but unable to propagate far inland.

In very general terms, tsunamis with shoreline run-up of decimetric amplitudes (< 1 m) are generally benign; amplitudes of a few meters will result in significant destruction of individual structures and in instances of loss of life; and dekametric amplitudes (≥ 10 m) in total eradication of infrastructure and population.

The exact run-up amplitude at a given shore location is a very complex function of the shape of the coastline and of the small-scale bathymetry and topography at the receiving beach. In particular, bays, harbors, and coves can resonate at typical tsunami frequencies, and their nonlinear responses can locally increase the wave amplitude. While such effects can be successfully modeled, they must be addressed on a case-by-case basis. In this general context, the following properties have been regularly observed, and justified theoretically.

Near-Field Scaling

In the near field, the maximum run-up observed from a nearby seismic source along a smooth, linear coastline featuring no substantial indentation does not exceed twice the amplitude of the seismic slip on the fault. This simple rule of thumb (known as "Plafker's law") expresses a general scaling of the near field to the seismic source, verified by numerical simulations and in the field during the 2004 Sumatra tsunami: its local run-up, exclusive of splashes on cliffs, reached 32 m for a seismic slip estimated at 15 to 20 m. Any departure from the Plafker law is a proxy for the presence of an ancillary source, such as a submarine landslide, as was the case in Papua New Guinea (1998), Riangkroko, Flores, Indonesia (1992) or Unimak, Aleutian Islands (1946).

Effect of Island Geometry

The response of an island to a distant tsunami depends crucially on the geometry of its structure, from the ocean floor up. In this respect, atolls, characterized by small dimensions, and steep underwater structures (with slopes reaching 40°) offer an overall smaller cross-section to the onslaught of the tsunami than do traditional islands sloping more gently to the ocean floor. As a result, all other parameters being equal, run-up on atolls has generally been smaller than on high islands. In practical terms, the tsunami is able to flow unimpeded around the structure, largely ignoring it, while the gently dipping, and necessarily larger structure of a high island would provide a physical barrier against which the wave has to abut, leading to a substantial transfer of momentum. This property was observed during the 2004 Sumatra tsunami in the Maldives, where the run-up was relatively contained (≈ 2 m), while it reached 9 m along the coast of Somalia, essentially in the same azimuth but at double the distance along the same ray paths. These results are a scaled-up expression of the well-known value of pillared structures (houses, etc.) in tsunami mitigation: the water flows effortlessly around the pillars whereas it would take down a continuous wall at the same location.

Effect of Fringing Reefs

Coral structures fringing high islands provide some degree of protection to the shorelines. Although a tsunami can penetrate a lagoon and reach the shore of a reefed island, it propagates very inefficiently over the irregular and extremely shallow topography of the lagoon, resulting in a significant loss of energy before it reaches the high ground. As a result, the type of island most vulnerable to a tsunami is the unreefed high island. Numerous examples exist of this difference in vulnerability; for example, in Polynesia the Marquesas Islands, which are young, reefless volcanic high islands, have traditionally suffered much larger run-ups from distant tsunamis (Aleutian, 1946; Chile, 1960) than the nearby Society or Austral Islands, of comparable geological structure and age, but protected by substantial reef systems. Similarly, Mauritius (reefed) was much less affected by the 2004 Sumatra tsunami than its sister island of Réunion (unreefed).

Effect of Small-Scale Topography

Small-scale island topography also plays a significant role in controlling the run-up of a tsunami wave at a coastline. In this respect, river beds and gulches are known to function as efficient channels of tsunami flow, often doubling or trebling the local amplitude of run-up. For example, the river valleys in the Marquesas Islands recorded up to 20 m of run-up during the 1946 tsunami, while nearby overland locations were typically 5 to 7 m. Similarly, during the 1993 Japan tsunami, run-up at a gulch on the West Coast of Okushiri Island reached 32 m, in rough numbers double its values along most of the nearby coastline. Numerical modeling at Okushiri has explained such high run-up as a result of the concentration of tsunami energy in the cove formed by the estuary of the gulch. Unfortunately, human settlement usually favors estuaries, which provide freshwater and communication routes to the hinterland, as well as locales featuring gaps in coral reefs, which provide easy access to the high seas, but considerably restrict the mitigating effect of the reef.

Refraction Around Islands

Circular island structures with dimensions comparable to tsunami wavelengths can result in refraction of the wave along the island and focusing on the lee side of the island. This situation, which may go against common-sense intuition, was demonstrated dramatically at Babi Island during the 1992 Flores, Indonesia, tsunami. Even though this local tsunami approached the island from the north, the southern shore of the island suffered more devastation. This "Babi Island effect" was later both reproduced in the laboratory and explained theoretically.

Wave Energy Spectrum

Even though the most perceptible energy in a tsunami is usually in the millihertz (mHz) frequency range (typical periods from 10 minutes to one hour), large earthquake sources contribute tsunami energy throughout a broad spectrum. Higher frequencies (typically in the 5–15 mHz range) have shorter wavelengths and no longer qualify as shallow-water waves. As a result, they are considerably dispersed; that is, their velocities across the ocean basins can be reduced to as little as 70 m/s, and this portion of the tsunami can reach distant coastlines as much as five hours after the main components of the tsunami. This "tail" to the tsunami wave was identified for the first time on hydrophone and seismic records of the 2004 Sumatra tsunami. While it carries comparatively less energy than the more traditional (and more rapidly propagating) components at longer periods, it can trigger oscillations of large amplitudes in specific harbors when the tsunami spectrum matches their resonance frequencies. During the 2004 Sumatra tsunami, this has led to incidents in which large vessels broke their moorings in harbors of the Western Indian Ocean (Réunion, Madagascar) *several hours* after the passage of the more traditional tsunami waves. Although these phenomena can be simulated numerically given an adequate model of the harbor, they raise very sensitive issues regarding the duration of tsunami alerts and, in particular, the issuance of an "all clear" message to harbor communities.

TSUNAMI WARNING

Because tsunamis propagate much more slowly than seismic waves (typically 220 m/s rather than 3–10 km/s), it is possible, at least in principle, to issue a warning to coastal communities, based on the interpretation of seismic waves in terms of earthquake source, and on the evaluation of the potential of the source for tsunami genesis. The full description of tsunami warning procedures transcends the scope of this article; however, the remaining challenges in this field are fundamentally of two kinds.

First, it remains difficult to accurately quantify truly gigantic earthquakes in real time. In the case of the 2004 Sumatra event, the true size of the earthquake took about six weeks to assert, through a study of the free oscillations of the Earth. The major problem in this respect is that all real-time evaluation algorithms were by necessity (and to a large extent, continue to be) designed, implemented, and tested on earthquakes of lower magnitude, and the adjustment of their parameters to mega-events is far from a trivial task. Note, however, that the triggering of a tsunami alert is fundamentally a matter of overcoming a threshold, beyond which the exact size of the earthquake source is not crucial. In this respect, the Sumatra earthquake had been evaluated as having widespread tsunami potential within about 40 minutes of its source; the failure to issue adequate warnings for the far field had more to do with communications than with pure science.

A second challenge is that of anomalous earthquakes disobeying scaling laws, whose rupture proceeds more slowly than in conventional sources, resulting in a significant deficiency of seismic release at the high frequencies (1 Hz) typical of shaking and damage to property (and to some extent at the intermediate frequencies (0.1 to 0.01 Hz) traditionally recorded on seismometers), while low-frequency waves such as tsunamis are vigorously excited. The geological context in which these so-called "tsunami earthquakes" can take place remains obscure, and while we are making progress toward identifying their anomalous character in real time from their seismic waves, they remain a formidable challenge, notably because they are poorly felt by the local population, which may then not be receptive to the issuance of a tsunami alert. Such a dramatic scenario occurred in Java on July 17, 2006, where waves ran up to 20 m and 700 people were killed despite a warning issued by the Pacific Tsunami Warning Center in Hawaii, which remained largely ignored by authorities along a section of shoreline where the earthquake, distant only 200 km, had hardly been felt.

TSUNAMI MITIGATION

Efforts to minimize the effect of tsunamis can take several forms. Passive mitigation relies on the building of structures designed to absorb or reflect the wave's momentum before it can reach coastal infrastructure, houses, or individuals. Among them, tsunami walls have long been used, as in Japan, and are now engineered to optimize the reflection of the wave back towards the sea. However, they remain only as good as their height in relation to that of the incoming wave. The role of vegetation (mangroves or forest) has also been researched both in situ and through scaled experiments in the laboratory. The relocation of critical facilities (hospitals, schools, fire houses) is also necessary in low-lying areas prone to inundation during a tsunami. Finally, building construction can lessen tsunami damage, in particular through the use of stilts or pillars that provide a free flow through an empty first floor offering no cross section to the wave's momentum.

Active mitigation by individuals consists essentially of taking refuge at a combination of altitude and distance from the shore that the wave is not expected to reach. This evacuation depends crucially on the amount of time available before the arrival of the tsunami. In the near field, this may be as short as a few minutes, and the value of a centralized warning becomes marginal; the responsibility for evacuation away from the shore must be borne by the individual, as soon as shaking is felt or an anomalous behavior of the sea, most notably a regression exposing a normally submerged beach, is observed. (Note, however, that automatic systems not requiring human intervention, including closing sluices and stopping trains, can be successfully implemented in the near field.) A generally appropriate rule of thumb is to evacuate to an altitude of 15 m and to remain there at least three hours after anomalous wave activity has ceased. Motorized vehicles should be avoided in the near field, as they will almost certainly contribute to traffic jams. In low-lying areas providing no adequate relief, vertical evacuation must be used. During the 1998 Papua New Guinea tsunami (2200 deaths), some villagers survived by quickly climbing trees; high-rise buildings can serve (and are occasionally built for) the same purpose in developed communities, but the use of elevators during evacuation should be avoided. Evacuation platforms standing on pillars have been built in Japanese ports to provide harbor workers with a means of vertical evacuation at the workplace. Tsunami evacuation drills are regularly conducted in countries at risk, such as Japan and Peru.

In the far field, tsunami alerts may benefit from several hours' advance notice, which can be used for a more profound level of orderly evacuation over greater distances. In both fields, a critical aspect of a successful evacuation is some advance knowledge of the geometry of the expected flooding. This is achieved by running, before the fact, numerical simulations of the extent of flooding for a given community under various scenarios of local or distal tsunamis. These simulations use models of expectable sources, and computer codes solving the equations of hydrodynamics under the relevant initial conditions to map the inundation of the wave down to the scale of a city block, based on available small-scale bathymetry and topography. Their output is made available to civil defense and law enforcement officials, who can then review zoning, optimize evacuation procedures, and conduct drills. For example, the entire west coast of the United States is presently undergoing a systematic program of inundation mapping for all coastal communities.

THE VALUE OF EDUCATION

Above all, a number of recent occurrences have repeatedly shown that tsunami fatalities can be significantly reduced among a population educated to this kind of hazard. For example, following the Papua New Guinea tsunami of 1998, an informative video was developed, translated into many local languages, and shown in neighboring countries, including on battery-operated televisions in remote villages. When a large earthquake hit the island of Pentecost in Vanuatu just a few months later in 1999, the village chief immediately ordered its evacuation. Minutes later, the village was destroyed by the tsunami, with all but a handful of residents unharmed.

In addition to formal education of children in the classroom, tsunami awareness can come as part of a community's cultural heritage through parental or ancestral education in regions regularly affected by tsunamis. For example, fishing communities in Southern Peru suffered no casualties during the 2001 tsunami, as the villagers took to the hills as soon as they felt the earthquake and noticed a down-draw of the sea. By contrast, farm workers hired from the hinterland shrugged off the earthquake and were swept by the waves as they tended to crops in the delta of the Camana River. Similarly, a number of tsunami-aware tourists, mostly from Japan, but also an 11-year-old British girl who had been taught about tsunamis at school, escaped the catastrophic 2004 tsunami on Thai beaches by recognizing anomalous down-draws as harbingers of disaster and immediately evacuating the beaches.

Unfortunately, tsunami awareness inherited from ancestral tradition will fade after an estimated four or five generations in the absence of a recurring event; it is estimated that the recurrence time of an event of the size of the 2004 one is at least 400 years in Sumatra, and thus the local populations were not educated to this hazard (with the possible exception of the Moken people of the Surin, Andaman Islands, who live in complete isolation and may have been able to preserve their heritage longer). In this context, it is crucial to emphasize both the value of, and the need for, permanent education of populations at risk. The fundamental messages are simple: (1) Tsunamis are a natural phenomenon associated with the dynamic nature of the Earth, as opposed to supernatural occurrences; hence they must and will recur. (2) Upon feeling any kind of shaking along a shore line, or noticing an anomalous behavior of the sea, and in particular a strong down-draw, one should immediately evacuate to higher ground. Such simple precautions have repeatedly been proved to save lives.

SEE ALSO THE FOLLOWING ARTICLES

Earthquakes / Eruptions / Landslides / Surf in the Tropics

FURTHER READING

Geist, E. L., V. V. Titov, and C. E. Synolakis. 2005. Tsunami: wave of change. *Scientific American* 294: 56–63.

Okal, E. A. 2008. The excitation of tsunamis by earthquakes, in *Tsunamis*. E. N. Bernard and A. R. Robinson, eds. The Sea 15. Cambridge, MA: Harvard University Press, 137–177.

Okal, E. A., and C. E. Synolakis. 2004. Source discriminants for near-field tsunamis. *Geophysical Journal International* 158: 899–912.

Synolakis, C. E., E. A. Okal, and E. N. Bernard. 2005. The mega-tsunami of December 26, 2004. *The Bridge* 35.2: 26–35.

TUAMOTU ISLANDS

SEE PACIFIC REGION

TYPHOONS

SEE HURRICANES AND TYPHOONS

VANCOUVER

MARTIN L. CODY

University of California, Los Angeles

Vancouver Island, 48–51° N latitude, is the largest island off the Pacific coast of North America, part of the western Canadian province of British Columbia, and the location of its provincial capital Victoria. The island is renowned for its soaring mountains, abundant lakes and waterfalls, spectacular coastal scenery, and imposing coniferous forests especially along the cooler and wetter western coastlines. It measures 450 km on its long axis (Cape Scott in the northwest to Victoria in the southeast), about four times its maximum width, with an area of 32,000 km^2; the 2200-m Mt. Golden Hinde, in Strathcona Provincial Park, is at the highest point along the island's rugged backbone.

ORIGINS: GEOGRAPHY AND GEOLOGY

Vancouver Island is a continental shelf island isolated from the mainland by shallow straits: the Juan de Fuca Strait in the south separates the island from Washington State's Olympic Peninsula, whereas mainland British Columbia lies east across the Georgia Strait, which narrows abruptly northward into the Johnstone Strait before opening into the Queen Charlotte Strait. The first recorded circumnavigation of the island in 1792 by Captain George Vancouver on the British Navy's ships *Discovery* and *Chatham* undoubtedly drew on navigational talents honed on his earlier voyage to the region under Captain James Cook. The eponymous commander's mission: to counter Spanish influence in the region, secure the fur trade (in sea otter pelts), and settle the question of a northwest passage.

Most of Vancouver Island, like other segments of the continent's western coastline, is part of a tectonic microplate or "wandering terrane" termed Wrangellia, which dates from Devonian times (400 million years ago) and originated in the southern paleo–Pacific Ocean 10,000 km from its present position. Volcanics, marine carbonates, and intruded granites are mostly souvenirs of the terrane's northward drift. Permian and Mesozoic fossil corals and ammonites preserved in the old sediments reveal the historical legacy in their close affinity with ammonites in southern Asia rather than to those on the North American plate. Accretion of Wrangellia to the North American plate occurred in the early Cretaceous (130 million years ago).

Present-day island topography, resulting from platelet deformation, subduction, mountain uplift, deposition of new sediments, and most recently glaciation, evolved long since Vancouver Island docked with the continent; it was almost completely ice-covered during the successive glacial episodes of the Pleistocene, when contiguous ice sheets extended from the mainland over the Straits, the island, and 50–60 km beyond the present western coast. The last glacial recession (after ~18,000 years ago) signaled a reconstituted island status and recolonization via northern advances of forest, woodland, and coastal vegetation and associated fauna from southern refuges, including the ice-free southern half of the Olympic Peninsula.

VEGETATION

The current diversity of habitats on Vancouver Island owes much to the climate gradient from outer to inner coast, determined largely by the mountainous interior. Points on the outer, western coast receive rainfall in excess of 3 m per year, about three times that of corresponding inner-coast stations; the west has cooler summer temperatures (by ± 4 °C, with 30 rather than 80 days a year with daily maximum temperatures over 20 °C and less sunshine); winters are cooler on the inner coast. This climate shift corresponds to a turnover from cool and damp coniferous forests dominated by cedar *Thuja plicata,* hemlock *Tsuga heterophylla,* and spruce *Picea sitchensis* on the outer coast to drier, more open forest and woodlands of madrone *Arbutus menziesii,* Douglas fir *Pseudotsuga menziesii,* and Garry oak *Quercus garryana* on the inner coast. Overall, most of the southeastern and mainland-facing parts of the island enjoy moderate and benign climates unknown elsewhere in Canada, and the profusion of formal gardens and gardeners around the provincial capital attests to this.

Some 2717 plant species are listed for British Columbia, 22.6% of them aliens; of these, 1604 are known to occur on Vancouver Island, and 150 or more are absent from the mainland. The dominant trees of the island's magnificent coniferous forests range from northern California to Alaska, but many bog, alpine, or shoreline plants have much broader distributions (e.g., Circumboreal, Palearctic, or Nearctic). There are no plant endemics, although Vancouver Island is the only provincial locale for several British Columbian plant species whose ranges extend further north or further south. Two threatened near-endemics are Macoun's meadowfoam *Limnanthes macounii* (Limnanthaceae) and Vancouver Island beggarticks *Bidens amplissima* (Asteraceae) of southeastern coastal areas, both now known to occur in a few sites off the island, with the former having recently been discovered (1998) in California. The Brooks Peninsula, a provincial park located on the northwestern coast of Vancouver Island, remained ice-free at the last glacial maximum and represents a refugium for alpine plants shared with the Haida Gwai (Queen Charlotte Islands) and Alaskan mountains to the north.

FAUNA

With few exceptions, the terrestrial vertebrates of mainland and island are very similar. The island's coasts abound in marine mammals (ten species of whales, dolphins, porpoises, sea lions, and seals) and bald eagles (*Haliaeetus leucocephalus*), and support diverse seabird colonies of gulls, cormorants, and alcids on the outlying rocks. The threatened marbled murrelet (*Brachyramphus marmoratus*) is an inland-breeding seabird still common around inlets that have not yet been logged. There is one spectacular mammal endemic, a large (up to 7 kg) ground squirrel, the Vancouver Island marmot (*Marmotus vancouverensis*), which is related to the similarly restricted Olympic marmot (*M. olympus*), and to the more widespread Hoary marmot (*M. caligula*). Although listed as "endangered" since 1979, a captive breeding program has been successful; reintroductions to the wild have tripled the population size from its 1998 low of around 70 animals.

HUMAN HISTORY

With the last glacial recession, ice-free coastal areas likely permitted the continent's first humans, crossing from northeast Asia to North America via the Bering Strait land bridge, to expand south rapidly, settling the productive island coasts by as early as perhaps 12,000 years ago. When European interest and activity in the region intensified in the latter half of the eighteenth century, explorers found the coastal regions densely populated by First Nations peoples, such as the Nuu-chah-nulth (Nootka) on the west coast, living in complex organized and sophisticated societies. They had found ingenious ways to utilize a wide variety of marine and forest resources, especially salmon and cedar. Energetic territorial defense necessitated a near-constant state of internecine warfare, the lethality of which European arms, along with European diseases, enhanced.

CONSERVATION ISSUES

The old-growth forests are now largely harvested out (about three-quarters gone, and only 10% left in the valley bottoms where the tallest trees occur); some of the remainder is preserved in the few parks and reserves (6% of the island area); the rest is being harvested at levels much above those sustainable over the longer term. The old growth is replaced by managed timber that cannot support the previously diverse canopy ecosystem (e.g., habitat-specific insects, spiders, and orobatid mites; voles; long-eared bats; hole-breeding owls and swifts; nesting murrelets) and sub-canopy biota (e.g., ferns, slugs, lichens, the many endemic forest floor rove beetles, northern red-legged frogs [*Rana aurora*], and several salamander species), various components of which are considered endangered, threatened, or at risk. Forest streams and their denizens survive logging operations poorly; the salmon runs become sparse and sporadic. There is a strong popular voice for banning raw log exports and for conserving the remaining sparse acreage of old-growth forest. In that event, tourism based on the spectacular landscape, seascape, natural history, and more persistent natural resources will offer a better, more economically sound future.

SEE ALSO THE FOLLOWING ARTICLES

Climate on Islands / Deforestation / Frogs / Vegetation

FURTHER READING

Cannings, R., and S. Cannings. 1996. *British Columbia: a natural history.* Vancouver, BC: Greystone Books, Douglas & McIntyre.

Cody, M. L. 2006. *Plants on islands: diversity and dynamics on a continental archipelago.* Berkeley: University of California Press.

Cody, M. L., and J. McC. Overton. 1996. Short-term evolution of reduced dispersal in island plant populations. *Journal of Ecology* 84: 53–61.

Douglas, G. W., G. B. Straley, D. E. Meidinger, and J. Pojar. 1998–2002. *Illustrated flora of British Columbia.* 8 vols. Victoria, BC: Ministry of Environment, Lands and Parks, Ministry of Forests, British Columbia Provincial Government.

Duff, W. 1997. *The Indian history of British Columbia: the impact of the white man.* Victoria, BC: Royal British Columbia Museum.

Krajina, V. J. 1973. *Biogeoclimatic zones of British Columbia.* Victoria, BC: British Columbia Ecological Reserves Committee.

Ludvigson, R., and G. Beard. 1994. *West Coast fossils: a guide to the ancient life of Vancouver Island.* Vancouver, BC: Whitecap Books.

Pojar, J. 1980. Brooks Peninsula: possible Pleistocene refugium on northwestern Vancouver Island. *Botanical Society of American Miscellaneous Series Publication* 158: 89. Vancouver: British Columbia Ministry of Forests and Lone Pine Publishing.

Winchester, N. N. 1998. Severing the web: changing biodiversity in converted northern temperate ancient coastal rainforests, in *Structure, process, and diversity in successional forests of coastal British Columbia.* J. A. Trofymow and A. MacKinnon, eds. *Northwest Science* 72 (special issue #2), Washington State University Press.

Yorath, C. J., and H. W. Nasmith. 1995. *The geology of southern Vancouver Island: a field guide.* Victoria, BC: Orca Book Publishers.

VANUATU

JÉRÔME MUNZINGER
Institut de Recherche pour le Développement, Nouméa, New Caledonia

Vanuatu, an archipelago in the southwestern Pacific, is famous for its active volcanoes, both emergent and under the ocean's surface. Largely because of its relatively modest terrestrial biodiversity, Vanuatu was recently included in the newly expanded East Melanesian Islands hotspot. The archipelago's recent volcanic origin, the result of interaction between the subducting Australian plate and the Pacific plate, is often proposed as an explanation of its relative low level of distinctiveness. However, Vanuatu has also been poorly investigated, and some islands have never been explored; thus, it may provide surprises in the future.

GEOGRAPHY OF THE TERRITORY

Vanuatu, officially the Republic of Vanuatu, is a Melanesian archipelago in the southwestern Pacific Ocean, comprising 83 islands and islets, between 13–21° S and 166–170° E (Fig. 1), with a total land area of ~12,220 km^2

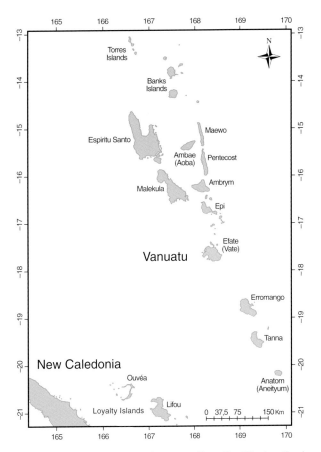

FIGURE 1 Map of the position of Vanuatu. (From the Millenium Coral Reef Mapping Project.)

and an exclusive economic zone (EEZ) of 680,000 km^2. The archipelago forms a Y shape, with the longest branch extending over 900 km, with a northwest–southeast orientation, and located some 1150 km east of the northeastern Australian coast. The base of the Y (Anatom Island) is only 220 km northeast of the Loyalty Islands (New Caledonia). Fourteen islands have surface areas exceeding 150 km^2, with the largest island, Espiritu Santo, reaching ~3900 km^2 and culminating in Vanuatu's highest mountain, Tabwemasana, at 1879 m (Fig. 2). Several islands are quite high, and eight of them bear active volcanoes, such as Yasur on Tanna Island and Marum and Benbow on Ambrym Island.

The archipelago's climate is generally hot and humid, with oceanic characteristics, but there are some differences between the islands, ranging from hot, very humid, and with little seasonality in the north to warm, humid, and with well-marked wet and dry seasons in the south. The climate variation follows three gradients: latitudinal, from south toward the equator; altitudinal, from sea level to mountain summits; and longitudinal, from east to west. The latter gradient results from the effects of the trade

FIGURE 2 Tabwemasana Mountain range. Photograph courtesy of Y. Pillon (IRD, Nouméa).

winds ("Alizés") and their interaction with relief, creating a windward and a leeward side of each island.

Mean annual temperature is greater than 23 °C; temperature varies less than 5 °C between seasons; average annual rainfall is greater than 2000 mm (from 2000 mm in the south to 4000 mm in the north); and usually there is no month with a rainfall deficit.

Most of the soils in Vanuatu are derived from volcanic rocks, of both submarine and aerial origin, and from sediments derived from the latter. Some calcareous substrates of reef origin are also found. The oldest rocks occur in the western (Malekula, Espiritu Santo, Torres) and eastern arcs (Pentecost, Maewo), where some Oligocene and Miocene substrates occur, without recent volcanic forms. Some ultramafic rocks have been reported on Pentecost Island.

Vanuatu's population is modest in size with only 220,000 habitants for the entire country, although the annual rate of increase is about 2.8%. The capital, and largest town, Port-Vila, on Efate (Vate) Island, has nearby 40,000 inhabitants.

BIOLOGICAL CHARACTERISTICS

Vanuatu's flora was recently included in a broad East Melanesian Islands hotspot, which also encompass the Bismarck Archipelago, the Solomon Islands, and the Santa Cruz Islands (Temotu).

The vegetation of Vanuatu can be divided into six main categories: (1) lowland rain forest, (2) montane cloud forest and related vegetation, (3) seasonal forest, scrub, and grassland, (4) vegetation on new volcanic surfaces, (5) coastal vegetation, including mangroves, and (6) secondary and cultivated woody vegetation. The first of these is subdivided into six variants, such as the remarkable *Agathis–Calophyllum* forest or mixed-species forests lacking gymnosperms and *Calophyllum*. Montane cloud forests are typically dominated by species of the genera *Metrosideros, Syzygium, Weinmannia, Geissois, Quintinia,* and *Ascarina*. In humid summit areas, the trees are covered with epiphytes, mostly filmy ferns and liverworts, but also with epiphytic shrubs such as *Vaccinium*. The seasonal forest, scrub, and grassland can also be divided into three variants, including the *Acacia spirorbis* forest, locally referred to as "gaiac forest," where sandalwood can grow; in drier locations, introduced shrubs are dominant, such as *Leucaena leucocephala, Acacia farnesiana,* and *Psidium guajava*.

Intact forest landscapes would cover ~710 km^2, or about 6% of the country. The Torres Islands appear to be particularly well preserved, without damage from logging and with few introduced species.

Compared to its neighbors such New Caledonia or Fiji, the biological diversity of Vanuatu is modest, with 870 native vascular plant species in 534 genera, and local species endemism that is quite low (~17%); a single genus is endemic to Vanuatu (the monotypic palm *Carpoxylon*), and only four species of gymnosperms are recorded, of which one, the kauri (*Agathis macrophylla*), is an emergent forest tree that is imposing in appearance and provides a special physiognomy to vegetation where it occurs. With exception of bats (including flying foxes), no native mammals are present, although rats, mice, pigs, cows, and horses have been introduced. About 121 bird species have been recorded in Vanuatu, nine of which are endemic, and one, the buff-bellied monarch (*Neolalage banksiana*), belongs to a monotypic endemic genus; ten introduced bird species are known. Amphibians are represented by a single species, introduced during the 1970s, whereas reptiles are slightly more diverse, with only two snakes known (including just the Pacific boa as indigenous), but two families of lizards (Gekkonidae and Scincidae) being present and having several species each, including some endemics. A Fijian iguanid was also recently introduced to Efate Island only.

Vanuatu remains poorly studied, with the exception of the work of a few individual researchers (such as Kajewski or Guillaumin), a large international botanical expedition led by the Royal Society of London in 1971, a Japanese botanical expedition on Espiritu Santo in 1996–1997, and more recently, a large multidisciplinary expedition conducted in 2006, focusing just on Espiritu Santo but studying both marine (lagoon) and terrestrial habitats. All of these endeavors yielded many discoveries in various groups such as plants (including large trees), molluscs, and arthropods.

BIOLOGICAL AFFINITIES

Although the closest territory to Vanuatu is New Caledonia, especially the Loyalty Islands, this is not correlated with the dominant biogeographic relations. Several studies of species richness in various plant groups, such as palms, figs, and ferns, have indicated closer relationships with the Solomons and Fiji. Molecular studies of kauri trees also show the species present in Vanuatu are more closely related to taxa in Fiji, Malaysia, and Australia than to species of New Caledonia. Exceptions do, however, exist, such as the genus *Tinadendron* (Rubiaceae), only known from calcareous substrates on New Caledonia and also on Erromango and Anatom Islands in Vanuatu, or *Megastylis gigas* (Orchidaceae), recorded only on the main island in New Caledonia and on Anatom. Nevertheless, as a whole, Vanuatu's flora is considered to represent an extension of that of Malaysia, through Bougainville and the Solomons. Based on information from several zoological groups, the fauna of Vanuatu is typical of an oceanic island and does not require land bridges or former continents to explain its origin. Closer affinities are observed with Fiji than with New Caledonia, just as with plants, although there are also exceptions to this, such as the genus *Emoia* (scincid lizards), which has two species in New Caledonia, which are restricted to the Loyalty Islands and are presumed to be the result of an introduction from Vanuatu. Finally, the insect fauna is especially poorly known, and several papers dealing with biogeography of insects in the Pacific lack samples from Vanuatu.

The first humans arrived in Vanuatu about 3300 years ago; they are identified with the culture called "Lapita" by the very distinctive pottery that they left behind them and that marks out their migration routes. Their impact was direct, by changing natural space for agricultural use, multiplying the incidence of fires, and introducing invasive species (voluntarily and accidentally).

THREATS TO THE BIODIVERSITY

Logging appears to be an important threat to the forests of Vanuatu, with special pressure being particularly high on kauri (Agathis) and kohu (Intsia) species. Sandalwood (Santalum) has been also strongly used in the past and was cut for decades; it is still under exploitation today. Invasive species also represent a major threat to biodiversity, and invasives occur in every group. A notable invasive plant is the liana Merremia peltata, one of the worst pests of the world, which is said to have been introduced during the Second World War to hide military infrastructures (a story whose status as myth or reality is unknown), and now cover large areas of secondary forest, killing the remaining trees and preventing regeneration (Fig. 3). For

FIGURE 3 Slopes of the volcano of Vanua Lava Island (Banks), with forest completely covered by introduced invasive liana. Photograph courtesy of Michel Lardy (IRD, Nouméa).

insects, the recent discovery (in 2006) of the tiny fire ant, *Wasmannia auropunctata*, in Luganville, is terrible news for the biodiversity in Santo Island, given the impact of this insect in New Caledonia, where it destroys the native entomofaune and has a heavy impact on reptiles as well.

SEE ALSO THE FOLLOWING ARTICLES

Ants / Deforestation / Fiji, Biology / Invasion Biology / New Caledonia, Biology

FURTHER READING

Bouchet, P., H. Le Guyader, and O. Pascal, eds. 2007. Santo2006 Expedition Progress Report. Santo2006/Gamma, Paris.
Corner, E. J. H., and K. E. Lee. 1975. A discussion on the results of the 1971 Royal Society–Percy Sladen expedition to the New Hebrides. *Philosophical Transactions of the Royal Society of London B* 272: 267–486. [This special edition includes numerous review papers on the subject].
Iwashina, T., T. Hashimoto, and E. Bani, eds. 1998. Contributions to the flora of Vanuatu: scientific results of the botanical expedition to Vanuatu and adjacent countries in 1996 and 1997. Tsukuba Botanical Garden, National Science Museum, Tsukuba.
Mueller-Dombois, D., and R. F. Fosberg. 1998. *Vegetation of the tropical Pacific Islands.* New York: Springer-Verlag.
Quantin, P. 1992. Les sols de l'archipel volcanique des Nouvelles-Hébrides (Vanuatu). Editions de l'ORSTOM, Paris.
Tardieu, V., and L. Barnéoud. 2007. Santo: les explorateurs de l'île-planète. Belin.

VEGETATION

DIETER MUELLER-DOMBOIS
University of Hawaii, Manoa

Vegetation is, simply, the plant cover of landscapes. It can be subdivided into plant communities on the basis of differences in structure (such as forest, shrubland, savanna, open forest, closed forest, summer deciduous forest, etc.), species dominance and composition (for example, eucalyptus forest, pine forest, softwood/hardwood forest, etc.). Vegetation is the most obvious and important biological component of terrestrial ecosystems and provides basic environmental services: Through its role as an absorber of carbon dioxide, an oxygen emitter, and a primary producer, it provides for life in terrestrial environments; it acts as an air filter or air conditioner in the Earth's biosphere; and it filters, cleans, and regulates the flow of water in and on the soil. The forest recycles rainwater by evapotranspiration into the atmosphere from layers deeper in the soil than can be reached by evaporative power alone. The "wick action" of a forest can result in greater circulation of water from evapotranspiration per ground area than from an open water surface.

VEGETATION CONCEPTS

Vegetation can be a natural plant cover, such as a rain forest. It can also be an artificial (i.e., under human control) plant cover, such as a field of sugar cane. Vegetation is also a hierarchical concept. There are such broad categories as island versus continental vegetation, and there are progressively narrower categories. For example, within island vegetation, there are biomes or ecosystems such as montane rain forests, lowland dry forests, freshwater swamp forests, mangroves, and seagrass beds (Fig. 1). Such biomes are also found in tropical continental areas and along their coasts.

A major building block of vegetation is the flora of an area. The flora usually differs greatly among plant covers on islands versus on continents, and it differs also among island areas. Thus, although different islands may support the same biome, such as a tropical rain forest, these rain forests differ from island area to island area in their structure and floristic composition. We may distinguish "dominance type" forests from "multi-species type" forests on the basis of structure and floristic composition of their canopies. Below these categories, we can usually recognize more narrowly defined community types by differences in forest undergrowth patterns of recurring species groups.

VEGETATION DEVELOPMENT

Although the flora of a region is the major structural component of vegetation, it is not the only component. The flora is the broad matrix of species growing in an area. From this matrix, only certain species form assemblages at a specific geoposition or site. Floras are usually treated in taxonomic books or in checklists with descriptions, keys, diagrams, and pictures for identifying the species.

Vegetation, in contrast, is built from a spectrum of environmental and biotic factors. These are summarized as a function of six factors with two overriding dimensions as follows:

$$\text{Vegetation development} = f\,(cl,\,g,\,d,\,fl,\,ac,\,e)$$

where cl = climate (regional, local, and microclimate); g = geoposition (geographic position, geology, geomorphology, and ground = soil); d = disturbances including human interventions; fl = flora of the region, ac = access potential of a species to the geoposition in question; e = ecological/evolutionary properties of species coming together in a vegetation or plant community (such as growth form and function, e.g., shade intolerance versus tolerance).

The overriding dimensions of time and space determine the state of vegetation development after disturbance. In harsh geopositions, vegetation development

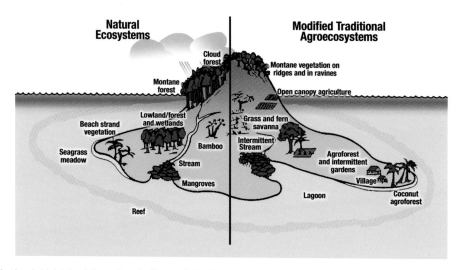

FIGURE 1 Typical volcanic high island. Reproduced with permission from Imamura and Towle (1987).

takes much longer than in ameliorated sites, and large disturbed areas are generally much slower to recover in vegetation than are small disturbed areas.

THE ISOLATION FACTOR WITH HABITAT RESTRICTIONS

The study of island ecology was highly stimulated by the island biogeography theory of MacArthur and Wilson. These authors developed a model on the natural assemblage of island biota based on island size and distance from biotic source areas (Fig. 2). The model makes two predictions. First, it predicts that the rate of invasion of species in islands near biotic source areas will be faster than in similarly sized islands further removed. Secondly, it predicts that the rate of species extinction will be greater on small islands than on larger ones.

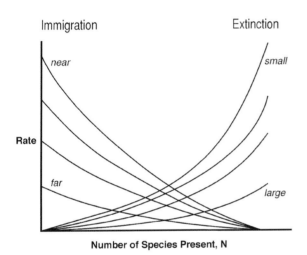

FIGURE 2 The island biogeography model of MacArthur and Wilson (1967), slightly modified. Reproduced with permission from Princeton University Press.

Although the first prediction is generally true, the second prediction assumes native species extinction to be related to the invasion of new native species at certain saturation points. These species saturation points are assumed to be lower in remote versus near-source islands. This is an unrealistic prognosis, because there is no known state of species saturation. Even if such were to exist, there is no reason to assume remote islands to have a lesser capacity for species diversity than islands near biotic source areas.

Moreover, competition is not the main cause of species extinction. Species extinction is primarily due to loss of habitat. In this respect, smaller islands are more prone to species loss than larger islands when subjected to natural as well as human disturbances. This is a relationship of disturbance intensity, frequency, and size to habitat size and diversity. Certainly, smaller islands have more restricted habitats. With regard to species extinction through habitat loss, smaller islands can be interpreted as habitat-reduced larger islands. This also means that they offer smaller target areas for invasion.

Generally, there is much greater species packing on islands than there is species extinction. This is particularly true because the isolation factor has been broken by introduction of species by humans. Naturally invaded and successfully established species prior to human interference are the native species, known as indigenous. In islands they are the progenitors of the other group of native species known as endemics. The isolation factor, prior to being broken by humans, had the distinct effect of facilitating the evolution of additional species, the endemics.

FLORISTIC IMBALANCES

A small archipelago, such as Palau, which is near a biotic source area (the Philippines), may have the same number of native plant species (~1000) as a large archipelago far removed from any biotic source area, such as the Hawaiian Islands. But among Palau's native plant species, only 10% may be endemic, versus 90% in Hawai'i.

However, the fraction of endemics among an archipelago's native species is not simply a function of isolation in terms of distance and time. It is also a function of the propensity among indigenous species to evolve into different species. For example, in the Hawaiian Islands, only 10% of the approximately 280 indigenous species gave rise to the native flora of about 1000 species. The isolation factor had a distinct effect on primary exclusion and secondary enrichment of taxa.

Primary Exclusion

Certain families of plants present on other tropical Pacific islands never reached Hawai'i naturally. They include the terminalia family, the cunonia family, the mahogany family, the melastome family, and others. In contrast to the southwestern Pacific islands, Hawai'i never received gymnosperms naturally. Likewise, mangrove stands, which form important shore stabilization communities in Melanesia and Micronesia, did not reach Hawai'i naturally. Thus, there is an inherent primary impoverishment of taxonomic groups on oceanic islands.

Secondary Enrichment

The initial impoverishment is taxonomically counterbalanced by a most remarkable secondary enrichment of species through endemism. In Hawai'i for example, the pantropical African violet family has only one genus,

Cyrtandra. But this single genus has proliferated into 51 endemic species. The cosmopolitan bellflower family, which is considered by some botanists to be the "crown jewel" of Hawai'i, has five endemic genera. The bellflower genus *Cyanea* is the most prolific with 52 endemic species. The only palm genus in Hawai'i, *Pritchardia*, has speciated into 19 endemic species. This is more than half of all species recognized in this palm genus. Because of this primary exclusion and secondary enrichment, island biotas have been characterized as "disharmonic" by some. Others have argued that island forests are not in balance because they lack certain important life forms, such as mammalian herbivores and their predators.

No doubt, disharmony applies to island floras and biota in general when compared to continental floras. But does this floristic disharmony also result in disharmonic or imbalanced island vegetation? This question will be further elucidated in the following sections.

THE BIODIVERSITY FACTOR WITH FUNCTIONAL SURPRISES

In spite of floristic disharmonies or imbalances, island forests are functionally sound. Healthy forests are found on all Pacific islands, where climate and soil are amenable for tree growth. In this respect, there is no difference between island and continental forests. Most island forests have also continuously renewed themselves in spite of their great distances from continental source areas.

Limitations in self-maintenance or sustainability may have occurred in the initial stages of island forest development until trees arrived that were able to become established through generational turnover. Indications for inefficiencies in establishing new forests are present in Hawai'i today. For example, the swamp mahogany (*Eucalyptus robusta*) from northeastern Australia has been planted in wet areas as replacement cover for native 'ōhi'a lehua (*Metrosideros polymorpha*) forest; yet stands of swamp mahogany do not seem to be self-sustainable in spite of seemingly favorable environmental conditions. Similarly, fig tree species, whose seed was spread in abundance on stumps in the early twentieth century, largely failed to become established in the target areas. Instead, the alien paperbark tree (*Melaleuca quinquenervia*) established itself after planting and now is in the process of forming self-maintaining forests. The three species (*E. robusta, M. qinquenervia, M. polymorpha*) are members of the same family (Myrtaceae). Their tiny seeds are emitted from dry capsules and are distributed by wind.

In spite of this similarity in seed size and seed dispersal mechanism, these three species have quite different ecological properties. This has become abundantly clear from an operational experiment to fix what has become known in Hawai'i as the "Maui Forest Trouble." A native *Metrosideros* rain forest on the lower windward slope of East Maui was noticed to deteriorate rapidly in the early twentieth century. Initially, this was thought to be the result of a new forest disease. Yet after a decade of research, no disease agent was found. Instead, the research conclusion was that the native *Metrosideros* rain forest could not persist because it was a pioneer vegetation unable to adapt to aging soils. This reasoning argued for the introduction of non-native trees to save the Hawaiian watersheds. The idea was put into practice in the Depression years of the 1930s. Plantations of *Eucalyptus robusta* and *Melaleuca quinquenervia* were established in half of the area with deteriorating native forest. Both these alien tree species grow well on this wet and swampy soil. *Metrosideros* is still present and reproducing on that site, but it grows only in the form of dwarf trees or shrubs. *Melaleuca quinquenervia* trees, on the other hand, are now invading this area in large numbers (Fig. 3).

FIGURE 3 Photograph of alien paperbark trees invading the "Maui Forest Trouble" area. Photograph by Dieter Mueller-Dombois, 2006.

This example shows that the native rain forest has floristic-functional and structural limitations: The tree-to-shrub size reduction in the recovering *Metrosideros* population is an adaptation of this species to bog formation. Indeed, in more advanced bogs, *Metrosideros* has formed new dwarf varieties or ecotypes. There are no native trees available to grow tall on hydromorphic or swampy soils, except perhaps the hala tree (*Pandanus tectorius*). But this tree lacks an efficient seed disperser in terrestrial environments, as do many other native tree species, which can be interpreted as another limitation in biodiversity function.

The process of forest deterioration on East Maui was not simply a response to soil aging but also a response to a complete landscape change from normally drained wet rain forest soil to incipient bog and stream formation. On that long time scale of geomorphologic aging, involving breakdown of the volcanic shield, we now see little justification for introducing alien trees to slow down the natural process of bog formation. Bogs may be just as efficient in watershed protection as are forests.

THE SIMPLIFICATION FACTOR WITH ADDED COMPLEXITIES

Metrosideros polymorpha still rules today as the dominant tree species in the wet rain forests from the youngest island (Hawai'i) to the oldest high island (Kaua'i). This can be attributed to a floristic-functional simplification. Vitousek made great use of this simplification by emphasizing Hawai'i as a model for ecosystem research. In most primary chronosequences associated with soil aging, a change in dominant forest trees is expected in continental environments. The simplification in structure and species composition of forests as evolved in isolation has eluded their functional interpretation in the past. This is best shown by the example of forest dieback and succession as researched in Hawai'i.

Forest Dieback

During the early 1970s a major decline of native *Metrosideros* rain forest in the form of canopy dieback was discovered on the island of Hawai'i. As in the "Maui Forest Trouble," the general assumption of research foresters was that a new disease was killing the forest, and the same prognosis was made that the native Hawaiian rain forest was doomed. After a decade of intensive disease and insect pest research, no biotic disease agent could be established as causing the dieback. Causal research was continued for environmental stresses in soil, climate variability, and extreme weather events. All of these factors were found to be involved in the canopy dieback, but the answer for the principal cause remained elusive until the early 1980s. Eventually, the principal cause was found in the *Metrosideros* forest itself, in its population structure and landscape mosaic. Individual stands of *Metrosideros* forest display a simplified uniform cohort structure. There is a canopy tree cohort and a cohort of small seedlings present in mature stands. Saplings are lacking or are very scarce. This indicates a common generational origin. *Metrosideros* stands become reestablished as generational or cohort stands after catastrophic events that destroy an existing forest (such as volcanic explosions) or after a clearing of the former canopy (such as resulting from canopy dieback). Once a new cohort stand is formed, it develops to maturity more or less synchronously like a stand of planted trees. If not cut, burned, or otherwise destroyed, the cohort stand will eventually reach senescence and break down in the form of canopy dieback. Synchrony is inherent in such a stand because the trees are mostly of the same generation and are similarly stressed through habitat constraints and old age. A tropical storm can be the dieback trigger in an aging or otherwise stressed cohort forest when such a storm has removed a large proportion of canopy foliage. Senescing stands lack the reserves of non-structural starch to replace their canopy. Biotic agents then can hasten tree death, with the result that new openings are created that allow for stand rejuvenation. A new cycle of cohort stand development may follow because canopy dieback itself is a major disturbance in an otherwise evergreen forest. A model developed for the theory of cohort senescence with subsequent rejuvenation is shown in Fig. 4.

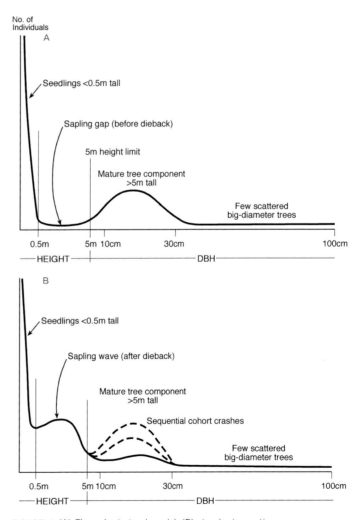

FIGURE 4 (A) The cohort stand model; (B) stand rejuvenation upon canopy dieback (Mueller-Dombois 1987). Reprinted with permission from *BioScience* 37, by American Institute of Biological Science. DBH = diameter at breast height.

Canopy dieback that fits the cohort senescence theory has been found on other islands, such as New Zealand, New Guinea, Lord Howe Island, Norfolk Island, and the Galápagos. Cohort senescence, rather than single-tree senescence, can be considered a consequence of structural and functional simplification in island forests. Permanent plot research has given evidence of stand recovery after dieback, with *Metrosideros* being the leading species again after 30 years in most cases. Invasion of new species can change this pattern.

Succession

Another related simplification is the peculiar succession of island forests. Many islands lack successional trees. Forest turnover is generally a simple process known as auto-succession or direct succession, whereby the same species resume dominance after some initial recovery with less tall species such as grasses and shrubs. Auto-succession differs from what has been termed obligatory or normal succession, where different tree species dominate one another in succession. The latter is typical in continental temperate environments. The succession model of Hawaiian rain forest (Fig. 5) illustrates the concept of auto-succession along a long-term chronosequence with dieback-induced secondary successions.

After the soil fertility peak, the height, diameter, and biomass of the leading canopy tree, *Metrosideros polymorpha*, declines with decreasing soil fertility. This regression

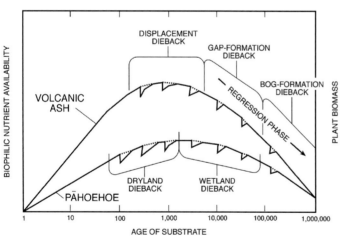

FIGURE 5 The Hawaiian rain forest succession model with five *Metrosideros* dieback types on the two main volcanic substrates. Checkmark-like symbols indicate episodic stand breakdowns with secondary successions (Mueller-Dombois 1986). Displacement dieback refers to suppression of *Metrosideros* regeneration from competition by other plants, such as tree ferns (*Cibotium* spp.); gap formation dieback refers to small groups of trees dying in the form of patches; bog formation dieback is a form of stand reduction relating to landscape change through geomorphic aging; dryland dieback is related to well-drained substrates, wetland dieback to poorly drained and firm substrates. Reprinted with permission from the *Annual Review of Ecology and Systematics* 17 © 1986 by Annual Review www.annualreviews.org.

phase may not be surprising, but in the continental tropics, tree diversity evolved in such a way that even soils with low nutrient contents support relatively high biomass. Speciation in the island flora has rarely resulted in successional replacers of the dominants. Most endemic trees other than the leading dominants are distributed spatially in different locations, often separated by topographic barriers that prevent efficient gene flow. Some of them are associated with the leading trees, but they do not take over as dominants.

Invasive Species, an Added Complexity

The virtual absence of native successional species provides resources for two types of tree life forms not found among the native trees. First, there are the quick responders, the species with r-type strategies, which include many of the cosmopolitan second-growth species that make up the secondary forests in the continental tropics. Competitive replacement occurs when alien species overtop the native shade-intolerant pioneer trees. A second group of successional invaders providing threats of competitive displacement are slow-growing shade-tolerant trees that grow taller than the native forest trees, the so-called K strategists. Because they usually require specific animal seed dispersers, they have not yet been generally observed as threats in native Hawaiian rain forests. But some of them have been dispersed by humans, thereby causing displacement of native tree species.

FUTURE OUTLOOK

The demise of the native Hawaiian rain forest has been predicted twice in earlier scientific publications, first by Lyon (1918) and thereafter by Petteys et al. (1975). In both cases the argument was that the native *Metrosideros* rain forest could not sustain itself naturally. The same argument could be made today, but for another reason. Invasive species have become the modern threat. There is now a new, unstoppable, natural dynamic on account of human-introduced invasive species. Because this problem is caused by human interference, it also now requires human interference to steer the problem into a more amenable direction. It is therefore necessary to apply silvicultural measures based on proper ecological knowledge. This means that the new dynamic trajectories are carefully assessed, and control measures are taken that divert them into more desirable directions. This requires willingness to compromise, because one cannot expect original conditions to be maintained, given that vegetation and plant communities are living systems that change with changing environmental constraints. A more general problem in island vegetation that applies to all Pacific islands is loss of indigenous forests. As already discussed, this is the major cause of native species extinction. Measures to counteract it will come only through ecological research, education, and training for mutual capacity building in cooperation with the Pacific islanders. Mutual capacity building is needed to approach traditional island cultures with respect and for bridging indigenous knowledge together with modern scientific understanding of the dynamics of local ecosystems and landscapes. To be successful, such approaches must also be integrated with conservation management and decision making for environmental policy.

SEE ALSO THE FOLLOWING ARTICLES

Hawaiian Islands, Biology / Island Biogeography, Theory of / Mangrove Islands / Succession / Sustainability

FURTHER READING

Cuddihy, L. W., and C. P. Stone. 1990. *Alteration of native Hawaiian Vegetation: effects of humans, their activities and introductions.* Honolulu: University of Hawaii Press.
Holt, R. A. 1983. The Maui Forest Trouble: a literature review and proposal for research. *Hawaii Botanical Science Paper* No. 42. www.botany.hawaii.edu/pabitra
Huettl, R. F., and D. Mueller-Dombois, eds. 1993. *Forest decline in the Atlantic and Pacific regions.* New York: Springer-Verlag.
Imamura, C. K., and E. Towle. 1987. *Integrated renewable resource management in U.S. insular areas (Island study 1987).* Paper commissioned for the U.S. Office of Technology (OTA), Pacific Basin Council Research Institute, Honolulu.
Lyon, H. L. 1918. The forests of Hawaii. *Hawaiian Planter's Record* 20: 276–281.
MacArthur, R., and E. O. Wilson. 1967. *The theory of island biogeography.* Princeton, NJ: Princeton University Press.
Mueller-Dombois, D. 1987. Natural dieback in forests. *BioScience* 37.8: 575–583.
Mueller-Dombois, D. 2006. Long-term rain forest succession and landscape change in Hawai'i: The "Maui Forest Trouble" revisited. *Journal of Vegetation Science* 17: 685–692.
Mueller-Dombois, D., and F. R. Fosberg. 1998. *Vegetation of the tropical Pacific Islands.* NY: Springer-Verlag.
Petteys, E. Q. P., R. E. Burgan, and R. E. Nelson. 1975. *Ohia forest decline: its spread and severity in Hawaii.* USDA Forest Service Research Paper PSW-105. Pacific SW Forest and Range Experiment Station, Berkeley, California.
Vitousek, P. 2004. *Nutrient cycling and limitation: Hawai'i as a model system.* Princeton, NJ: Princeton University Press.

VICARIANCE

MICHAEL HEADS

Ngaio, Wellington, New Zealand

In plants and animals, closely related species and groups of species often occur in different areas, and this pattern is called vicariance. Its origin can be explained by a process, also termed vicariance, in which a widespread common ancestor differentiates and breaks up, more or less in situ, into related descendants. For example, consider a variety of plant or animal that is found only on (i.e., is endemic to) a certain island. It may have its closest relative on a nearby mainland. Isolation and differentiation of the island population from the mainland one could have arisen by vicariance if the island and its life were separated from the mainland by a geological process, such as erosion or subsidence. In an alternative explanation, dispersal theory, the island is colonized after it forms by long distance, overwater dispersal of an individual or seed from the mainland. Colonization then stops, and the new population differentiates to become an island endemic. The well-known "equilibrium theory" of island biogeography is based on dispersal theory and suggests that immigration from a mainland center of origin and extinction together determine biogeography.

DISPERSAL AND VICARIANCE

In dispersal theory, a related group, or taxon, evolves in a restricted area—its "center of origin"—and attains its distribution by dispersing out from there. Each taxon comes into existence on its own, not with others. In contrast, during vicariance of a widespread ancestor, the population of each different sector evolves into a new taxon where it is, often over a wide area. There is no center of origin and no "dispersal," although subsequent range expansion or contraction may take place. In dispersal theory, a "founder" population is established through normal dispersal, but at some point migration stops, and the founder becomes isolated from its parent population. It has never been explained why exactly dispersal would stop, but this is a critical question. In dispersal theory the end of dispersal is attributed to chance rather than to geological or climatic change, as in vicariance.

Normal, "ecological" dispersal, the physical movement of plants and animals, is observed every day. It does not involve the differentiation of new taxa, unlike long-distance, chance dispersal, which is a theory of speciation. All organisms move, and given the amount of geological time available, they should be able to migrate to areas of suitable habitat around the world. However, in fact, most groups show marked local and regional endemism. This creates an apparent paradox: The movement that exists everywhere does not seem related to the most fundamental aspect of distribution. In addition, distribution patterns are usually shared by many taxa with different ecology and means of dispersal. These observations conflict with the dispersal theory of speciation and biogeography and have led to ongoing debate about biogeographic processes. Normal "ecological" dispersal and vicariance are accepted by all workers. The center of origin/founder dispersal mode of speciation is much more controversial.

DATING TAXA

Attempts have been made to distinguish between origin by vicariance or by dispersal through dating taxa, because dispersal theory generally proposes a geologically younger age for a taxon than does vicariance. There are difficulties with this approach, however. Fossils provide only minimum ages for groups, and most groups are much older than their oldest known fossil. Fossil-based molecular clock calculations also give minimum ages for taxa, but these are often misrepresented as maximum dates and used to rule out earlier vicariance events as irrelevant. Simplistic correlations with paleogeography, such as calibrating all Pacific-Atlantic divergence with the rise of the Isthmus of Panama at 3.5 million years ago, can also give ages for taxa that are much too young.

Dating studies usually assume that evolutionary differentiation is more or less continuous over time and so roughly clock-like. Degree of differentiation would then be more or less proportional to the time since divergence. However, evolution probably proceeds in distinct phases with long periods of stasis, and so taxa may show little or no differentiation despite having been separated for many millions of years.

CENTERS OF ORIGIN

The center of origin for a particular group has often been deduced from the arrangement of taxa in a cladogram or phylogenetic tree. However, the "basal" branch or clade is not a primitive, ancestral group occurring at the center of origin—just a small sister-group, and no more primitive, just less diverse, than the main clade. A series of vicariant taxa that branch off sequentially in a cladogram represents a geographic sequence of differentiation in a widespread ancestor, not a series of chance dispersal events.

VICARIANCE IN THE PACIFIC

The Dispersal Model

Dispersal and vicariance models of Pacific biogeography were vigorously debated by the Victorian naturalists, but after the First World War, dispersal theory became almost universally accepted. According to the theory, all Pacific taxa derive from ancestors that originated in Asia or America and migrated into the Pacific. The dispersalists recognized the existence of prior land in the Pacific, as they knew that many volcanic islands had sunk there, each leaving only an atoll as a trace. Nevertheless, they interpreted distributions among the islands of the Pacific as the result of dispersal among extant, rather than former, islands. For example, the distinctive Pacific clade of *Cyrtandra* shrubs is endemic to rain forest on islands between the Carolines and southeastern Polynesia. The very wide range has been taken as proof that the plants are highly vagile and have a remarkable capacity for long-distance dispersal. However, this proposal overlooks the many single-island endemic species in the group, the clear-cut vicariance between the group and its relatives (which are absent from the central Pacific), and the volcanism that has been widespread and continuous in the Pacific basin since its formation.

Volcanism and Metapopulations

Dispersal theory for islands involves random volcanism, a center of origin, and long distance dispersal, whereas vicariance emphasizes recurrent volcanism, normal migration among unstable local populations any of which may go extinct, and regional persistence of taxa despite changing geography. The dispersal model assumes that the Pacific region was originally devoid of islands and island life, but this is very unlikely. Volcanism does not occur at random, but takes place around particular sectors over periods that are much longer than the age of the individual volcanoes. Oceanic volcanic islands form at subduction zones, spreading centers, hotspots, and propagated fissures, which have always been active in and around the Pacific. The individual islands are relatively short-lived, but new islands are constantly being formed in the vicinity. These new islands are colonized by "ordinary dispersal" from nearby islands, not by "long-distance dispersal," and there is no speciation involved. The taxon originates and survives in the region as a population of populations, a metapopulation, and whether or not the current islands have ever been joined to a mainland or not (the distinction between "oceanic" and "continental" islands) is not relevant for their biogeography. Populations on volcanic islands and atolls, and their reefs, survive and evolve in the same way that they do on any other recurring habitat "islands," such as termite mounds or forest gaps.

Establishing the age of an individual volcanic island is not straightforward and involves more than just dating the exposed strata. An island may be composed of very young volcanics or limestone but could be very old as an island, if new material is added as old material is removed by subsidence, erosion, or burial, which is usually the case. In practice, the age of an island's rocks is not nearly as important for biogeographic analysis as the age and history of the associated subduction zone, fissure, or hotspot that has been generating the volcanism. In Hawaii and the Galápagos, for example, biologists have suggested that endemic taxa are much older than the rocks that currently form the islands. The populations have survived in the region by constantly dispersing from older, now largely eroded islands to nearby younger islands. In one model, island chains form by plate movement over hotspots, and so archipelagoes may eventually join with others. This process might explain the Hawaii–southeastern Polynesia connections seen in many marine and terrestrial taxa. However, some geologists now suggest that linear island chains are formed not by hotspots but by propagated fissures resulting from plate tectonics processes. The fissures, like subduction zones, may be much older than the individual islands.

Vicariance Model

Recently, several authors have discussed a model for the Pacific which assumes that plants and animals have always occurred there. The fact that the central Pacific is a large, well-marked center of endemism that includes smaller areas of endemism is not well accounted for in traditional biogeographic theory. If the Pacific had been populated by dispersal from a western center of origin, a simple dropping out of individual species across the Pacific from west to east might be predicted. But regional areas of endemism occur throughout the Pacific and often involve parts of one archipelago and parts of another. Across the Pacific, there is a west-to-east drop in total diversity of terrestrial and shallow-water marine species, and this is sometimes cited as evidence for dispersal. However, the decline is simply due to the islands of Indonesia and Melanesia being larger than the islands of Polynesia.

The rain forest trees of *Metrosideros* (Myrtaceae) are a typical example of central Pacific endemism. The distributions of the five main clades are mostly vicariant (Fig. 1), with overlap only in Vanua Levu, Fiji (although clade 3 is only known there from one mountain), and North Island, New Zealand (where clade 1 has a very restricted range). Each of the *Metrosideros* clades occupies an area of endemism also held by many other, very different taxa. Southeastern Polynesia, an important biogeographic sector that is often overlooked, is illustrated here by the land snail *Tubuaia* (Fig. 1). *Metrosideros* has close relatives endemic to New Guinea, the Philippines, South Africa, and Chile, indicating that it originated by vicariance, as the central Pacific representative in a South Africa–Pacific–Chile group. This group, in turn, has vicariant relatives in Australia.

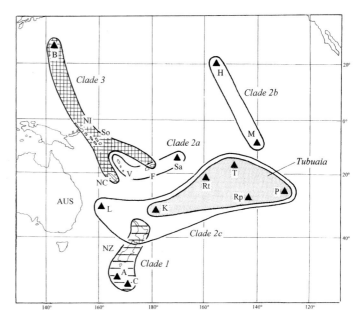

FIGURE 1 Distribution of *Metrosideros* (Myrtaceae), endemic to the central Pacific. The five clades (1, 2a, 2b, 2c, and 3) are mainly vicariant. Distribution of the land snail genus *Tubuaia* (Achatinellidae) is also shown (stippled). A = Auckland Is., AUS = Australia, B = Bonin Is., C = Campbell I., F = Fiji, H = Hawaii, M = Marquesas Is., NZ = New Zealand, Rp = Rapa I. (Austral or Tubuai Is.), Rt = Rarotonga (Cook Is.), Sa = Samoa, K = Kermadec Is., L = Lord Howe I., NI = New Ireland, P = Pitcairn I. (Tuamotu ridge), So = Solomon Is., T = Tahiti (Society Is.), V = Vanuatu.

Atolls, Extinction, and Survival

The atoll zone of the central Pacific comprises a vast area from the Carolines to southeastern Polynesia—an area that has undergone subsidence and massive extinction. Land snails characteristic of high, forested islands occur as fossils on the atolls of the Marshall Islands and Midway Island, west of Hawaii. These atolls are the remains of former high islands, which subsided at least 1500 m over the Cenozoic. As the high islands collapsed, the peaks no longer caught the mists, and this wiped out the moisture-dependent snails and other forest taxa, perhaps including *Metrosideros*. On some Pacific islands, relic species in groups with a high survival coefficient, such as flies and certain birds, still survive on the young limestone covering the sunken volcanics.

PLATE TECTONICS

Rifting of the Earth's crust is a well-known mode of vicariance. Geologists have proposed that the Solomons, Vanuatu, Fiji, and Tonga once formed a continuous island arc, which was rifted apart into separate chains, and zoologists at the Smithsonian Institution have interpreted disjunct distributions in reef fishes and terrestrial lizards as a direct result of this.

A geological terrane is a fault-bounded block of the Earth's crust that has had its own independent history.

Accretion, or the docking and fusion of terranes at convergent plate margins, can also modify and create vicariance patterns. New Guinea, New Caledonia, and New Zealand are larger than the other Pacific islands and include older, continental-type rock. They are often described as "fragments of Gondwana" but are in fact geological and biological composites (like the Greater Antilles in the Caribbean). Each comprises an older, Gondwanic part plus many other terranes, including island arcs, which have accreted to the Gondwana fragments after arriving from the Pacific side with the encroaching Pacific plate. The regions of accreted terranes, like others in Indonesia and the Philippines, show high diversity in many groups and indicate evolution by juxtaposition rather than by radiation from a center of origin. Geological and biological accretion has also occurred in the archipelagoes of Indonesia and the Philippines, and accreted island arc complexes now form vast areas of western North, Central, and South America.

In the central Pacific, very large igneous plateaus formed in the Cretaceous and then moved south, west, and east, colliding with countries like the Solomons, New Zealand, and Colombia. The plateaus are now largely submarine but include many formerly emergent seamounts up to 24-km across, as well as sediment layers with fossil wood.

In summary, the plants and animals of the Pacific islands are probably the result not of Neogene founder dispersal from Asia or America but of long-term survival and evolution in the Pacific basin since its Jurassic origin. Despite enormous extinction, metapopulations in different parts of the Pacific have preserved patterns of endemism and vicariance that reflect plate tectonic rifting and convergence.

OTHER ISLANDS

Outside the Pacific, fossil-calibrated dating studies have inferred dispersal from mainland Africa to Madagascar, but fossils of many groups are scarce there and Mesozoic–Early Cenozoic vicariance better explains the close biogeographic relationship among eastern Africa, the southwestern Indian Ocean islands, and India/Sri Lanka. The Galápagos–West Indies connection was discussed in some early vicariance analyses and illustrates the predictive power of the method. Although molecular dating is generally misleading, molecular cladogram topologies are more reliable and are revealing a vast amount of previously hidden vicariance in all large clades. Recent studies show that Galápagos finches (*Geospiza*, etc.) and Galápagos mockingbirds (*Nesomimus*) each have their sister group in the Caribbean, not on nearby mainland South America. Some geologists now trace the origin of the Caribbean plate to the Galápagos hotspot, and this would provide an explanation for the distribution pattern.

SEE ALSO THE FOLLOWING ARTICLES

Convergence / Dispersal / Endemism / Island Biogeography, Theory of / Metapopulations / Plate Tectonics

FURTHER READING

Arbogast, B. S., S. V. Drovetski, R. L. Curry, P. T. Boag, G. Suetin, P. R. Grant, B. R. Grant, and D. J. Anderson. 2006. The origin and diversification of Galapagos mockingbirds. *Evolution* 60: 370–382.

Craw, R. C., J. R. Grehan, and M. J. Heads. 1999. *Panbiogeography: tracking the history of life.* New York: Oxford University Press.

Fitton, J. G., J. J. Mahoney, P. J. Wallace, A. D. Saunders, eds. 2004. Origin and evolution of the Ontong Java plateau. *Geological Society of London, Special Publication* 229: 1–369.

Foulger, G. R., and D. M. Jurdy, eds. 2007. Plates, plumes, and planetary processes. *Geological Society of America, Special Paper* 430: 1–998.

Heads, M. 2006. Seed plants of Fiji: an ecological analysis. *Biological Journal of the Linnean Society* 89: 407–431.

VOLCANIC ISLANDS

JOHN M. SINTON

University of Hawaii, Honolulu

In the broadest sense, volcanic islands include all islands that form by volcanic processes, even those islands that represent ancient volcanic terranes that have been tectonically exposed, such as Gorgona Island offshore of Colombia or Macquarie Island in the southwestern Pacific. It also should be noted that most atolls represent a calcareous carapace surmounting oceanic volcanoes, the volcanic part of which has subsided below sea level, a concept originally proposed by Charles Darwin and subsequently confirmed by drilling through shallow carbonate atolls into the underlying volcanoes of the Hawaiian and Tuamotu island chains. However, attention here will be restricted to those islands primarily formed by volcanic processes more or less in situ, and where portions of the original volcano are still exposed above sea level. This restriction still leaves a subject that encompasses hundreds of islands and island groups scattered around the world's oceans.

ISLAND ARCS AND OCEANIC ISLANDS

Volcanic islands can be broadly divided into two principal categories, reflecting fundamentally different geologic processes responsible for the volcanism. One class of islands forms in regions of plate convergence above subduction zones, whereas the second class is associated with anomalous upwelling zones of upper mantle, not associated with subduction zones. Differences in mantle sources and melting processes in these two environments result in contrasting chemical and mineralogical compositions of the volcanic rocks that form. The recognition of two categories of oceanic islands predates modern concepts of plate tectonics. For example, in 1912 Patrick Marshall distinguished the high-SiO_2 basalts, andesites, and rhyolites of islands of the easternmost South Pacific from the dominantly basaltic lavas of the Pacific basin lying to the west of what he called the Andesite Line. The development of the plate tectonic concept more than 50 years later provided a framework for the reinterpretation of the Andesite Line to coincide with subduction zones associated with plate convergence in the western Pacific. The rocks that make up the two main classes of volcanic islands are now generally known as island arc volcanics (IAV) and ocean island basalts (OIB). IAV are dominated by SiO_2-rich lavas, whereas OIB span a considerable range in silica and alkali contents, although generally basaltic compositions are by far the most common.

MELTING PROCESSES

The geologic processes responsible for the formation of volcanic islands depend on their tectonic setting. Melting beneath island arcs is initiated in the mantle wedge overlying subducted lithosphere as a consequence of the addition of water and other volatiles derived by dehydration of the downgoing plate. This melting is commonly referred to as flux melting, because the addition of water reduces the melting temperature of the overlying mantle in this region. Mantle melting in this environment produces basaltic magmas that carry the imprint of relatively shallow, hydrous melting that is manifest in generally higher silica and volatile contents compared to basaltic magmas elsewhere. In addition, the common presence of very high-silica magmas (dacites and rhyolites) in some arc islands is generally ascribed to direct melting of the lower crust, which can be heated above the melting point when voluminous basaltic magma ponds above the crust–mantle interface.

The formation of oceanic islands away from subduction zones signifies the presence of melting anomalies in the ambient mantle. In this case, melting is a direct consequence of mantle upwelling, commonly referred to as decompression melting. There is considerable debate about the origin of these melting anomalies or hotspots, mainly centered around the relative contributions from thermal and compositional anomalies in the underlying mantle. Several oceanic island provinces appear to be associated with thermal upwelling from the deep mantle that can contain the chemical signature of recycled oceanic lithosphere. Other island provinces may be associated with chemical heterogeneities in the mantle that require neither an origin in the deeper mantle nor unusually high mantle potential temperatures. Although the presence of melting anomalies is independent of plate boundary processes, some melting anomalies coincide with divergent plate boundaries or mid-ocean ridges, most notably in Iceland, the Azores, and the Galápagos. This coincidence probably reflects the weakening of the lithosphere in the presence of melting anomalies, such that spreading ridges tend to localize around them, rather than the other way around.

Whether or not melting anomalies are spatially fixed with respect to one another is another issue of considerable debate. For example, the geometry of some linear island chains have been used to define the motion of lithospheric plates over the deeper mantle. Such "absolute" plate motion models, using a relatively fixed hotspot reference frame, have been used successfully to predict the locations and azimuths of linear island provinces and other major bathymetric anomalies of the ocean basins, such as the Hawaiian–Emperor chain (Hawaiian hotspot), the Iceland–Faeroe Ridge (Iceland), the Walvis Ridge and Rio Grande Rise (Tristan da Cunha hotspot), and the Carnegie and Cocos ridges (Galápagos hotspot). However, the fixity of deep-mantle melting anomalies has been challenged by numerical modeling constraints on the stability of thermal plumes in convecting mantle and also by paleomagnetic studies of island and seamount lavas that suggest significant variation in latitude of formation.

The evidence for crustal melting and the presence of rhyolitic magmas in ocean islands is much more limited that in island arcs, although notable examples have been reported from Iceland, Easter Island (Rapa Nui), Ascension Island, and Pitcairn. Quartz-normative lavas with SiO_2 contents greater than 55 wt% are extremely rare in the Hawaiian Islands, with the notable exception of the Waiʻanae volcano on Oʻahu.

ERUPTIVE ACTIVITY AND VOLCANIC PRODUCTS

The nature of volcanic activity is directly related to the composition of magmas in different environments. The generally high volatile content of arc magmas promotes relatively explosive eruptions in that environment. The solubility of volatiles is inversely related to pressure so that as volatile-rich magmas ascend beneath arc volcanoes, volatile components exsolve as gases, which can lead to large increases in pressure under confining conduit conditions. High conduit pressures promote explosive eruptions and lava fragmentation, two characteristics that commonly distinguish arc volcanic processes from those in oceanic islands. Although there are many exceptions to this generality, typical island arc volcanoes contain a much higher proportion of tephra (ash and other ejecta) to lava, compared to typical ocean island volcanoes.

In contrast, most ocean island volcanic eruptions are relatively quiet, generally effusive, and dominated by lava rather than tephra, although again there are notable exceptions. Some of these exceptions can be related to the interaction of shallow magma with meteoric water in the edifice substructure, but others may be related to high primary magma volatile contents, such as some of the eruptions in the Azores and Iceland provinces.

VOLCANO STRUCTURES

Volcanoes that produce significant amounts of tephra tend to build composite volcanoes with relatively steep slopes and an internal stratification consisting of mixed layers of lava and tephra; such volcanoes are sometimes referred to as stratovolcanoes. In contrast, volcanic islands dominated by lava eruptions tend to have much gentler slopes and can generally be described as shield volcanoes. Arc islands are almost exclusively composite volcanoes, whereas oceanic islands tend to be dominated by shield volcanoes. Individual islands can be composed of single volcanoes or multiple, coalesced volcanoes in both environments (Fig. 1). However, the number of coalesced volcanoes in arc islands rarely exceeds two, whereas single oceanic islands can comprise five or more individual volcanoes, such as the Island of Hawai'i in the Hawaiian archipelago or Isabella in the Galápagos archipelago. Iceland is by far the largest and most complex volcanic island, consisting of tens of volcanic systems and hundreds of individual volcanoes.

The largest individual volcano on Earth is Mauna Loa on the island of Hawai'i, with an estimated total volume of ~40,000 km^3. Although this volcano clearly is atypical, it is certainly true that most ocean island volcanoes tend to be significantly larger that most arc volcanoes. The diameters of ocean island volcanoes commonly exceed 10 km, whereas most arc volcanoes have diameters less than 5 km.

Although most ocean island volcanoes can generally be described as shield volcanoes, there can be considerable variation in the morphology and structure of them, particularly with respect to the development of rift zones, the overall shape, and the nature of calderas (Fig. 1). Hawaiian volcanoes commonly develop rift zones that radiate away from a summit region characterized by the presence of a caldera during the main growth phase of activity (see below). Eruptions can occur either from the summit region or from along the rift zone. Rift zones of Hawaiian volcanoes can be more than 50-km long, in some cases with a submarine extension that is longer than the subaerial rift. The Hawaiian example appears to be exceptional, however, as many other oceanic island volcanoes are much more circular in plan form, and evidence for highly elongated rift zones is much less common. For example, most eruptions on Fernandina volcano in the Galápagos archipelago occur from circumferential fissures near the summit region, whereas eruptions on Mauna Loa can occur far down the

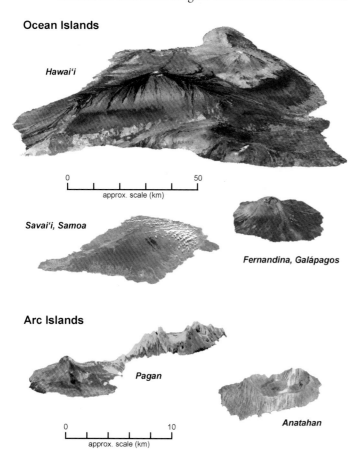

FIGURE 1 Perspective views of selected ocean islands (top) and arc islands (bottom); note different scales. Oceanic islands tend to be much larger than typical arc islands. Hawai'i Island comprises five separate shield volcanoes; Pagan Island comprises two separate volcanoes. Lava flows on Hawai'i emanate from along linear rift zones, radiating away from the summits, whereas Fernandina lacks prominent rift zones, and lava flows mainly emanate from near the summit region. The caldera of Fernandina is several times larger in diameter and deeper than that on Mauna Loa. The diameter of the caldera of Anatahan is more than 50% of the entire island.

rift zones away from the summit (Fig. 1). Highly elongated rift zones are rare in arc volcanoes, which mainly have a nearly conical shape with eruptions that are strongly concentrated near the central summit region.

Summit depressions, or calderas, characterize most volcanic islands while they are active. However, there can be considerable variation in the dimensions of calderas and probably also in the processes that are responsible for their formation. In arc volcanoes, cataclysmic explosive eruptions can result in calderas with dimensions up to ~80% of the total volcano diameter. Perhaps the most well-known such eruption is the AD 1883 eruption of the Indonesian volcano Krakatau. The 1883 eruption, however, was only the latest giant eruption of this region, occurring within a 7-km-wide caldera that had formed during a previous large event around AD 416. Such cataclysmic processes are much rarer in ocean island volcanoes. Calderas that characterize the summit regions of active island volcanoes tend to be smaller features formed by collapse or down-sagging above relatively shallow, quasi-steady-state magma reservoirs within the volcanic edifice. Even within ocean island volcanoes, dramatic differences in size, shape, and depth of calderas can be present, as exemplified by the differences in Galápagos and Hawaiian volcanoes (Figs. 1, 2).

ISLAND GROWTH

The initiation of volcanic islands begins below sea level. Deep submarine volcanism mainly produces pillow lavas where water pressures are sufficient to inhibit lava fragmentation. Below ~1000 m, H_2O, which is by far the most abundant volatile component of most magmas, remains dissolved in the liquid magma. At lower pressures and shallower depths of eruption, volatile exsolution increases the amount of fragmentation, and most volcanic islands probably have variable amounts of hyaloclastite (literally, broken glass) within the volcanic edifice. As the volcano emerges from the sea, violent explosions are characteristic, but once the magmatic conduits are insulated from interaction with seawater, the nature of volcanic eruptions becomes progressively less explosive and more effusive.

Hawaiian volcanoes are known to proceed through a series of volcanic stages that reflect variations in the composition of magma and the nature of volcanic eruptions. Subaerial Hawaiian volcanoes are dominated by a shield stage of evolution, characterized by the eruption of basalts relatively low in alkali elements. This stage can last for a million years or more. Many Hawaiian volcanoes are known to have evolved into a later stage of activity, characterized by less frequent eruption of lavas that form by lower extents of melting of the underlying mantle, and by evolution in

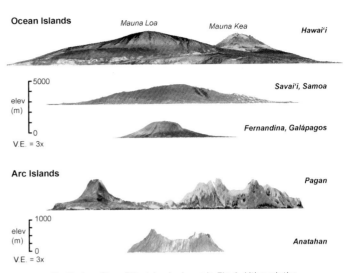

FIGURE 2 Vertical profiles of the islands shown in Fig. 1. Although the two classes of islands are shown at different scales, both are displayed with 3D vertical exaggeration. It is apparent that ocean island shield volcanoes tend to be much larger and have lower slopes than do typical arc volcanoes.

much deeper magma chambers within the volcanic edifice. This later stage is called the postshield stage of volcanism. In most Hawaiian volcanoes, this stage is known to last approximately 200,000 years, although there are examples where it was much longer lived, and yet others in which this stage of activity is entirely lacking.

The details of other oceanic island provinces are less well known than in Hawai'i, but the concepts of shield and postshield volcanic stages have been extended to the geological evolution of several volcanoes in the Samoan, Society, Marquesas, and Austral island chains, although the characteristic variations in lava chemistry are not always identical to those in Hawai'i.

One of the most intriguing and least understood aspects of many oceanic islands is the presence of rejuvenation volcanism (i.e., volcanic activity that recurs after a period of quiescence lasting up to a million years or more). During the quiescent period, erosion, subsidence, and reef growth dominate the evolution of the island. Although rejuvenation volcanism was first recognized in the Hawaiian Islands, geological and geochronological investigations have confirmed that it also characterizes some Samoan and Marquesan islands, as well as most of those in the Cook-Austral chain, and the island of Wallis, north of Fiji.

ISLAND EVOLUTION BY EROSION, MASS WASTING, AND SUBSIDENCE

Important processes that affect the morphology of volcanic islands include those that modify the original landscapes. Among these are erosion, mass wasting, and subsidence. Erosion occurs throughout the history of a

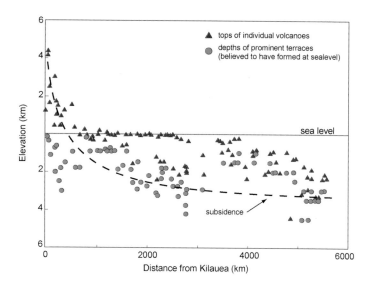

FIGURE 3 Variation in elevation along the Hawaiian-Emperor volcanic chain, illustrating the importance of subsidence in island evolution. The effect of subsidence is shown schematically by the dashed line. The uniform elevation of volcanoes near sea level ~700-2500 km from Kilauea indicates the region of prominent reef growth. Most islands in this province have subsided at least 2 km since formation. Figure modified from Moore (1987).

volcanic island, but it has the greatest effect at times of relative volcanic inactivity. Evidence for erosion of oceanic volcanoes mainly takes the form of stream valleys incised into the volcanic flanks and sea cliffs cut into the volcano during relative high stands of the sea.

One of the most important processes affecting island morphology is mass wasting, which can occur at a variety of scales. Mass wasting in the form of rock falls, landslides, and debris avalanches is the principal process by which valleys widen. However, the growth of volcanic edifices surrounded by the sea produces edifices that are inherently unstable. Large-scale flank collapses can have a dramatic effect on volcanic islands. Evidence from the Hawaiian chain indicates that these large collapse events can occur at any stage of volcanic activity, although they are most likely to occur when the volcano is actively growing and the edifice is least stable.

Mass wasting of volcanic flanks of island arc volcanoes is less well known than in ocean island volcanoes but almost certainly also occurs in such environments. High-resolution mapping of the submarine flanks of some arc volcanic islands shows large sedimentary aprons that probably represent debris flows associated with mass wasting events. Several historical tsunamis originating in various island arcs are thought to be related to submarine landslides in those regions.

The important role of subsidence in affecting island topography was fully appreciated only after the advent of detailed studies of the submarine flanks and deep drilling of some islands. These studies documented the transition from subaerial to submarine lava sequences and other evidence for ancient shorelines lying up to several kilometers below present sea level. Studies of submerged shorelines around oceanic islands have contributed to this understanding by determining the rates of subsidence on some islands. It is now clear that the dominant reason why the Hawaiian Islands have become gradually lower in elevation with age has more to do with subsidence than with erosion (Fig. 3).

Islands subside because the underlying lithosphere is not rigid but behaves somewhat elastically. Thus, subsidence is inevitable following loading during volcanic growth. However, the resistance of the underlying lithosphere to flexure and subsidence depends on its age and on the amount of reheating that has taken place during volcanic growth. In general, subsidence should be much greater where ocean islands are forming on young lithosphere, such as the Galápagos Islands and Iceland, than for those forming on older and stronger oceanic lithosphere.

SEE ALSO THE FOLLOWING ARTICLES

Darwin and Geologic History / Hawaiian Islands, Geology / Island Formation / Krakatau / Lava and Ash / Oceanic Islands

FURTHER READING

Clague, D. A., and G. B. Dalrymple. 1987. The Hawaiian Emperor volcanic chain, part I. Geologic evolution. *U. S. Geological Survey Professional Paper* 1350: 5–54.
Darwin, C. 1837. On certain areas of elevation and subsidence in the Pacific and Indian Oceans, as deduced from the study of coral formations. *Proceedings of the Geological Society of London* 2: 552–554.
Marshall, P. 1912. The structural boundary of the Pacific basin. *Australasian Association for the Advancement of Science* 13: 90–99.
Moore, J. G. 1987. Subsidence of the Hawaiian ridge. *U.S. Geological Survey Professional Paper* 1350: 85–100.
Morgan, W. J. 1972. Plate motions and deep mantle convection. *Geological Society of America Memoir* 132: 7–22.
Volcano World website. http://volcano.und.edu/.

VOYAGE OF THE *BEAGLE*

JERE H. LIPPS

University of California, Berkeley

HMS *Beagle*, with Capt. Robert FitzRoy and young Charles Darwin aboard, left the island of Britain for hydrographic surveying, particularly in the Southern

Hemisphere, on December 27, 1831, on a voyage spanning nearly five years, to October 2, 1836. Over 40 islands were visited or closely approached, including the Galápagos Islands, which Darwin later made famous. Darwin observed and recorded information about the geology, biology, and people of the many places he visited, which was then used in his later books and papers.

CAPTAIN FITZROY, HMS *BEAGLE*, AND CHARLES DARWIN

Robert FitzRoy had been an officer on the earlier voyage of the *Beagle* (1828–1831) to southern South America, and he was made captain of the ship after Captain Pringle Stokes committed suicide. Continuing the hydrographic survey of Tierra del Fuego, FitzRoy held four Fuegian natives hostage in exchange for a boat that had been stolen and then decided to take them back to England, and he made anomalous and confusing magnetic measurements, which he attributed to the possible presence of magnetic minerals in the mountains nearby. He determined then that he should return to Tierra del Fuego to return the Fuegians and to take a geologist with him on his next voyage to investigate the magnetism problem. In mid-1831, the British Admiralty agreed to his request to return the Fuegians, but also provided him with voluminous instructions on additional surveying in South America and the acquisition of navigational, magnetic, and astronomical measurements at meridians around the world. *Beagle* was recommissioned on July 4, 1831, and FitzRoy began planning, including a search for a naturalist to accompany him.

Charles Darwin had the good fortune to do geological fieldwork during the summer of 1831 with Reverend Adam Sedgwick, his professor of geology at Cambridge University. Chiefly for this reason, he was chosen as naturalist on board HMS *Beagle*. After responding to an inquiry from the Reverend John Henslow, Darwin was interviewed by FitzRoy. At this time Darwin thought of himself more as a geologist than a biologist. Indeed, although he often called himself a "naturalist"; the only time he ever referred to himself as something else, he chose "geologist." Thus Darwin was invited aboard *Beagle* by FitzRoy as an intellectual companion as well as a naturalist. His father initially opposed the idea but later relented. FitzRoy received the approval of the Lords of the Admiralty to include Darwin on the voyage. FitzRoy's mission and needs dictated where Darwin would go and what he would see on the voyage of the *Beagle*. He would take good advantage of this arrangement.

Darwin, like many people, was fascinated by islands and even had plans to visit the Canary Islands before he was offered the position on *Beagle*. Islands attracted Darwin because they are isolated, often contain different or unusual biotas, are generally simple in ecological structure, and can be neatly circumscribed and thus appear to be more readily understood than the contiguous geology and biology of much larger continents. So Darwin looked forward to the voyage with great anticipation.

ISLANDS VISITED BY THE *BEAGLE*

Beagle left England from Devonport (Fig. 1, no. 1) and headed south through the Atlantic Ocean (Fig. 1) with

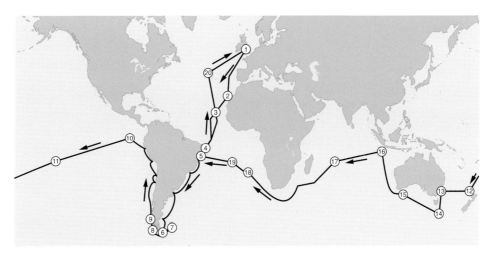

FIGURE 1 Route of the HMS *Beagle*, December 27, 1831, to October 2, 1836, showing the major islands visited by the ship and Darwin. Points visited, in the order of visit, were (1) Britain; (2) Canary Islands; (3) Cape Verde Islands; (4) St. Paul's Rocks; (5) Fernando Noronha Island; (6) Tierra del Fuego, Straits of Magellan, and Beagle Channel; (7) East Falkland Island; (8) Chonos Archipelago; (9) Chiloé; (10) Galápagos Islands; (11) Tahiti; (12) Bay of Islands, New Zealand; (13) Sydney, Australia; (14) Hobart, Tasmania; (15) King George's Sound, Western Australia; (16) Cocos (Keeling) Atoll; (17) Mauritius; (18) St. Helena; (19) Ascension Island; (3) Cape Verde Islands again; (20) Azores.

FitzRoy intending to put in at various islands to make detailed measurements for the Admiralty. After ten miserable days at sea, during which time *Beagle* passed Madeira and Piton Rock from a distance and searched in vain for the Eight Stones, Darwin was excited to approach Tenerife in the Canary Islands (Fig. 1, no. 2). He had read Humboldt's description of the island and its biota, and made much about the book and its descriptions. He thus was anxious to see Humboldt's island for himself. But *Beagle* had been blocked from making that visit by port authorities, although Darwin could clearly see the brightly colored houses of Santa Cruz as *Beagle* moved to within a half mile of the island. *Beagle* had been quarantined because the authorities feared she might harbor cholera, and so the ship would have to lie at anchor for 12 days to demonstrate she was free of the disease. FitzRoy would have none of that, however, and set sail immediately south along Africa for the Cape Verde Islands, passing through the Canary Islands close to Gran Canaria. Darwin could see from the ship the volcanic nature of the islands. Modern geology considers these islands to be hotspot volcanoes, although not with certainty. Although greatly disappointed by not visiting Tenerife, Darwin was still excited by his imagination of what that island and the disappearing Eight Stones islands held.

Heading south along the African coast, Darwin was disappointed by the Cape Verde Islands and St. Jago (now Santiago) Island especially, also now best considered hotspot volcanoes, because they were so desolate and miserable. On St. Jago and adjacent tiny Quail Island, he began to examine and compare the volcanic rocks to those at home, the larger corals living there with the small solitary forms near Edinburgh, and the various sea levels he could recognize on the islands. He recalled his geological instruction from Sedgwick and began to understand the geology of these islands. The tropical plants impressed him, and he devoted much space in his notes to the people living there.

After three weeks in the Cape Verde group (Fig. 1, no. 3), *Beagle* sailed southwesterly to the middle of the Atlantic Ocean nearly at the equator and found tiny St. Paul's Rocks (Fig. 1, no. 4). These rocks, not large and not high, were uninhabited except for a huge number of birds, as they are today, and Darwin's stay lasted only a few hours. FitzRoy described his crew throwing stones at the birds to kill them for food, and noted that Darwin successfully used his geological pick in a similar fashion. Later Darwin used it appropriately to collect samples of green and black rock types; these are now known to be pieces of an intrusion from deep in the earth's mantle

on the Mid-Atlantic Ridge upon which St. Paul's Rocks lie. Although Darwin did not know this origin then, his description of the rock types was accurate. He also was impressed by the simplicity of the biota—two birds (a booby and a noddy) plus associated insects. *Beagle* continued southwesterly and put in for a day at another small island in Fernando de Noronha group (Fig. 1, no. 5), about 370 km off the Brazilian coast. These islands were volcanic as well, forested and contained an array of insects, all noted by Darwin in just a few hours. After stopping for two weeks at Bahia on mainland Brazil, FitzRoy took the ship to the Abrolhos Islands lying in shallow waters 32 km off the coast to commence lead-line bathymetric surveys. Here Darwin observed a "saurian" under every rock, vivid green vegetation consisting of just a few species, and many spiders and rats. He also observed the relationship of these volcanic islands to their coral development and recalled the ones he had previously seen.

From these islands, *Beagle* continued south to the tip of South America, landing along the way at various places on the east coast. At last, on December 16, 1832, the ship was off Tierra del Fuego and the crew could see fires on the shore lit by the people there. The southern part of South America includes a large number of islands, many of different sizes, and all pieces of continental South America isolated during the past glaciations and sea level rise.

Darwin spent more time in these islands (Fig. 1, no. 6) than anywhere else on the voyage, for *Beagle* and FitzRoy had particular aims to accomplish. FitzRoy was determined to return the three Fuegians (one had died in England) he had taken on his previous voyage to their homelands and of course to finish surveying. Foul weather interfered with these objectives. Nevertheless *Beagle* or its whaleboats and yawl sailed among this archipelago for 75 days (Fig. 2), putting ashore in various places. The area, of course, is huge,

FIGURE 2 HMS *Beagle* at Tierra del Fuego (painted by Conrad Martens). From Leakey (1979).

and Darwin could see only glimpses of any of it, and he had little time to study the people, geology and biology of any particular place. In *Beagle's* smaller boats, FitzRoy, Darwin and members of the crew traversed the length of the Beagle Channel, named by FitzRoy on his previous journey to this region, twice, once in each direction.

Since Darwin's actual views of the region were incomplete, he once again drew general conclusions, but they have remained more or less valid up to the modern times. He observed that the western and southern parts of the group were made largely of igneous and metamorphic rock with trends like those of the Andes and that the northeastern part had sedimentary rocks identical to those in Patagonia. Darwin was piecing together a much bigger picture than he had observed. His biological observations, focused on birds, insects, marine invertebrates, and the *Nothofagus* forests, dealt with a broader range of conclusions, including some about the behaviors and associations of the animals and of the plants. His observations of the penguins and ostriches (rheas) led him to make comparations of the life styles of the two birds that he retained even in the *Origin of Species*. He was not impressed with the people, whom he characterized as "miserable creatures, stunted in their growth, their hideous faces daubed with white paint & quite naked"; and later he would refer them to the lowest rung of all the peoples he met on *Beagle's* voyage, a rather characteristic Eurocentric view at the time, but one no longer accepted.

Upon reaching the *Beagle,* they sailed northward to the Falkland Islands and Patagonia for surveying work and replenishment and repair of the ship. The Falkland Islands, in the southwestern Atlantic east of Tierra del Fuego (Fig. 1, no. 7), were visited twice, once from December 1832 to February 1833 and again from March to April 1834. In between these times, the ship returned to the east coast of South America to continue surveying and to sail south down the Straits of Magellan before turning back to exit the Strait, and then to explore the eastern and southern (again) coast of Tierra del Fuego. Both times at the Falklands, *Beagle* spent her time in Berkeley Sound anchored near the English settlement of Port Luis on the eastern extremity of East Falkland Island. For 10 weeks, Darwin was able to study a relatively small area longer than any other place on the voyage, while FitzRoy was compelled to deal with shipwrecked French sailors, American whalers' destruction of British property, sovereignty issues about the islands, and unrest among the inhabitants resulting in mistrust and murder, as well as his chief mission of surveying the area. Although not happy there, Darwin kept busy documenting the geology, marine life and domesticated animals of Berkeley Sound. He made detailed geologic notes with cross sections of the folded strata he encountered, wondered about the "stone runs" of angular rocks, ranging from a few centimeters to meters in size, that formed long, narrow lineaments in the valleys, and noted that fossils he found suggested a different environment in much earlier times. Darwin was confronted by evidence of catastrophe while still holding his gradualistic, Lyellian views of geology. He collected more animals and plants, marine and terrestrial, continuing his comparative approaches by noting that the extremely abundant kelp forests were similar to those of Tierra del Fuego and to the tropical rain forests in abundance he had seen earlier in South America.

On leaving the Falklands in April 1834, more than halfway through her voyage, *Beagle* made for and then passed through the Straits of Magellan to the myriad of islands just offshore of southwestern South America. *Beagle* struggled north along the islands, passing the Archipélago de los Chonos, which she would later visit in the next summer. As winter was upon it, *Beagle's* passage was rough, and all on board were relieved when the ship landed first on June 28, 1834, at San Carlos on the north end of Chiloé Island (Fig. 1, no. 9), southwest of Puerto Montt, Chile. FitzRoy stayed only two weeks at San Carlos, while Darwin walked inland and along the coast. Then the ship headed north to Valparaiso to present papers and gain permission from authorities in Santiago to survey coastal Chile, as well as for resupply, refitting, and surveying in the region. At Valparaiso, Darwin climbed the Andes and found fossil plants and clams. Returning to Chiloé for November and December 1834, Darwin, along with the ship's crew, explored the island. During January 1835, *Beagle* went further south to survey the Chonos Archipelago, before going back in February and March to Chiloé for a third time to complete surveying there. At Chonos, Darwin collected barnacles, perhaps part of the incentive to monograph them later. Darwin spent much of time on Chiloé commiserating with the crew about the overcast and depressing weather, although he considered Chiloé a "fine island" in terms of its geology and temperate rain forests. On the western islands, Darwin saw evidence of sea level changes he attributed to the rise of the land and of active volcanism in the Andes that could be seen from the islands. Using the small boats and horses, Darwin explored Chiloé and the nature of the biota and human populations. His impressions from Chiloé lasted the rest of his life, perhaps more so than those from the Galápagos, and he incorporated them into his books and correspondence repeatedly.

After finishing his surveys of the intricate coastal areas, FitzRoy headed north along Chile and Peru. On September 15, 1835, the ship began survey work in the Galápagos Islands (Fig. 1, no. 10) for five weeks. Darwin, however, was mainly restricted to *Beagle* as it moved among the islands, and he spent only 19 days or parts of a day ashore on only four of the islands (Fig. 3) although he passed closely by eight others. Indeed he did not like the Galápagos because they were too hot and barren, nor did he achieve any great insights at the time of his visit. He was awed by the evidence of the very recent volcanic activity, and wondered about the giant tortoises and their future given the intense predation on them by ships' crews, including that of *Beagle*. Although Darwin was impressed by the volcanism and low diversity of the fauna and flora, he still made large collections that served him and others well later. These included examples of the finches and other birds, tortoise shells, molluscs, insects, fish, plants, and rocks. Many of these were studied by others, such as John Gould, who later corresponded with Darwin about their significance.

FIGURE 4 Darwin saw Moorea as a picture in a frame from Tahiti. The barrier reef with waves crashing on it formed the frame, the rather small and narrow lagoon was the mat around the picture, and the island itself was the picture. Image of the reef, lagoon, and island taken by J. H. Lipps at Viare, Moorea, 2002.

FIGURE 3 Darwin's view of the Galápagos Islands. Darwin saw the Galápagos as a series of young volcanoes on which unusual varieties of organisms had later developed; among them, the land iguana, which he called "ugly animals" with a "singularly stupid appearance" but that "when cooked, yield a white meat." Image taken by J. H. Lipps on South Plazas Island, Galápagos Islands, 1982.

From the Galápagos, *Beagle* sailed directly to the Society Islands, passing by the eastern low islands and finally anchoring in Matavai Bay, Tahiti (Fig. 1, no. 11), on September 15. The sequence of islands that Darwin saw from the ship intrigued him because he was formulating his own theory of reef formation. FitzRoy set about making magnetic and navigational measurements at Point Venus, where Cook had done so in 1769; this gave Darwin time to climb the adjacent mountain high enough (above 800 m) to see clearly the nearby island of Moorea, then known as Eimeo. He saw this view of Moorea like a picture, the reef being the frame, the smooth lagoon the mat, and the island itself the picture (Fig. 4). This view, together with his observations of the other Society Islands, came together to confirm his ideas, first formulated off South America, about the formation of atolls. He carefully studied the details of the reefs of Tahiti while wading and canoeing over them. He determined that the corals grew best in the upper 35 m of the ocean and not as well below. The seaward edge of the reefs, he discovered, fell away sharply into much deeper water. His descriptions again were perfect, and his conclusions well drawn. He placed every coral island, whether low or high, that he encountered into his own grand picture of the process of fringing, barrier, and atoll reef formation. So sure of his new idea, he wrote a manuscript proposing it while on board *Beagle* in the month after he left Tahiti. Darwin also noted that the volcanic rocks of Tahiti were well weathered, and from that observation he concluded that the island had been active at some time in the distant past, unlike the rocks at the Galápagos. He was impressed by the Tahitians, who he found happy and generally quite religious, although he said little about the lack of dress of the women. After just 11 days, *Beagle* left Tahiti bound for New Zealand, where FitzRoy would take more measurements at the Bay of Islands (Fig. 1, no. 12).

As the voyage extended well beyond the two years it was planned for, FitzRoy and the crew began to move more quickly and do less extensive surveying work. Thus, *Beagle* found anchorage on the northern tip of North Island in the Bay of Islands for only seven days. This was the only place Darwin visited in New Zealand, and he was not impressed because of the accumulation of whalers and seamen from other places. He was interested in the Maoris and what he regarded as their lowly life style, adding yet another rung to his ladder of human advancement. The missionaries, he thought, were doing fine work among them by raising their expectations, however.

The ship then proceeded to Australia, arriving 12 days later for restocking of food and supplies. Darwin would spend a total of 37 days at Port Jackson (now Sydney), New South Wales; Hobart Town, Tasmania; and King George's Sound, Western Australia (Fig. 1, no. 13, 14, 15). At Port Jackson, Darwin undertook an excursion inland. He became very much aware of the dissimilarity of the fauna and flora of Australia relative to the previous places he had visited. He remarked on the nature of eucalyptus forests, and on marsupials and the platypus. As impressed by those differences as he was, he did find similarity between an ant lion he watched and those of England. His biogeography was developing into a tool for understanding evolution, although he was a long time from enunciating it. Even in 1836, Darwin recorded that fire was a prevalent feature of the landscape. He also noted the major geological features of the area.

Leaving Port Jackson 18 days after arriving, *Beagle* set sail to Tasmania. There Darwin engaged in his usual studies of geology around Hobart, also making observations on the native people, climbing Mt. Wellington to view the spectacular large eucalyptus trees, and collecting organisms. Again he dwelt on the geology, in particular his discovery of some "Devonian or late Carboniferous" fossils, and the evidence of possible former sea levels and uplift of the land. He compared the local geology to that he saw in New South Wales, using his comparative approach to try to understand the rock types and sequences he described. He was also concerned about the conflict between the Aboriginals and the Europeans, concluding that the native population would become extinct unless they were removed to an area remote from the Europeans.

After 10 days, *Beagle* left Hobart for Port Williams on Australia's southwestern coast. He felt that his stay at Port Williams was the most "dull and uninteresting time" of the entire voyage. Yet Darwin expressed interest in the sparse plants, such as the grass tree, and he was fascinated by smooth domes of granite penetrated by numerous veins. A group of Aborigines called the "White Cockatoo men" visited and were persuaded to put on a dance that evening. Darwin described the dancing, often imitating emus or kangaroos, in negative terms: "all moving in hideous harmony, formed a perfect display of a festival amongst the lowest barbarians." Overall, Darwin was not impressed, writing, "he who thinks with me will never wish to walk again in so uninviting a country." Indeed, he did not find Australia in general appealing and thought the idea of being served by convicts repulsive. "Farewell, Australia! You are a rising child, and doubtless some day will reign a great princess in the South: but you are too great and ambitious for affection, yet not great enough for respect. I leave your shores without sorrow or regret."

Although Darwin had finished his essay on the formation of atolls after leaving Tahiti, he had never actually set foot on one. Cocos Keeling Islands, lying nearly 2000 km northwest of Perth, Australia, had originally been suggested to FitzRoy as a landing place by the Admiralty to determine their exact position. However, *Beagle* was now far to the south off southeastern Australia, and FitzRoy wrote a letter to the Admiralty that he would take the ship directly across the Indian Ocean, then to England. Darwin, however, may have influenced FitzRoy to visit the islands, as originally suggested, because of his own interest in atolls. In any case, FitzRoy set sails for those islands, much to Darwin's delight.

The Cocos Keeling Islands (Fig. 1, no. 16) consist of two atolls with many small islets on the coral rims. The *Beagle* anchored off the southernmost of the two where the only population of Europeans and Malays lived. Darwin visited at least five of the islands, examining the vegetation, beaches with pumice and plant debris tossed up by waves, the spiders, insects, and large coconut crabs (*Birgus latro*), the lagoon with its variety of marine animals including many giant clams (*Tridacna*), and the communities of people. He collected much from the rather depauperate flora and fauna, and he thought he had representatives of every plant (20) save two, which existed as single trees whose seed were likely tossed ashore by the waves. The first tree to occupy newly formed land in the atolls, he noted, was *Pemphis acidula*. Darwin was probably satisfied that his atoll theory seemed reasonable in view of what he saw, and he might have wondered how far below his feet was the volcano that supported this atoll. He was quite happy about his visit and seemed to enjoy his stay.

Continuing the homeward trip, *Beagle* next anchored at Port Louis on the northern side of Mauritius (Fig. 1, no. 17). Although Darwin spent much time walking about town and watching the society at work, he also examined the relationship of the coral reefs to the volcanic rocks of the island, the apparent recent elevation of the land, and the nature of volcanic structures observed in the interior; he was able to place this island into his theory of reefs through comparisons to the other islands he had visited. Darwin did not collect much, but he did find a frog, which puzzled him since amphibians were nonexistent on islands, presumably because they could not cross saltwater. Later, the frog was discovered to have been introduced.

From Mauritius, FitzRoy steered around Cape Horn into the Atlantic with stops at St. Helena, Ascension, and, finally, Terceira in the Azores (Fig. 1, nos. 18–20). The first two are hotspot volcanoes close to the Mid-Atlantic Ridge of basaltic composition, younger than 15 million years, and well eroded by the sea. Darwin thought these islands and St. Jago, which *Beagle* had visited early in its voyage, had rims raised by volcanic forces that surrounded the interiors of the islands. Each was experiencing volcanism, uplift, and denudation eroding them lower. On Ascension, he saw that the lava flows were far fresher than on St. Helena, concluding that it was also far younger. In fact, later geologic studies have confirmed this. The *Beagle* arrived at Terceira, after it had revisited Bahia in Brazil, to complete FitzRoy's measurements on the meridians, on September 19. Darwin seemed less than enthusiastic about his short stay, but in spite of that he made three excursions on the island, one to see fumaroles. From these, he concluded that the lava flowed from the central part of the island outwards toward the shores.

On October 2, 1836, after nearly five years, the *Beagle*, FitzRoy, and Darwin arrived at Falmouth on their home island. Darwin never again left the British Isles, the last islands he would ever see and the islands of his birth, until his death April 19, 1882.

WHAT DARWIN LEARNED ON THE VOYAGE OF THE *BEAGLE*

Darwin's scientific observations came from a vast experience on the islands and two continents (South America and Australia) on the voyage and provided him with a strong grasp on both his geological and biological hypotheses. He developed his comparative research style, and focused more on geology of these places than on their biology. Overall, he absorbed much about both fields and integrated them into his thinking. Islands contributed a great deal to his books, articles, and correspondence. He saw volcanic islands, continental islands, very isolated islands, tropical and subpolar islands, and those in close proximity to their neighbors. These differences in environment, Darwin realized, accounted for the islands' differing biotas.

Darwin's interest in geology comes through in his notes and theory development while on the *Beagle*. He wrote 1383 pages of geological notes, only 368 pages of zoological ones, and none on botany. He developed the theory of atoll formation as he sailed, noting that the coral rings could easily form by subsidence of the volcano and growth of the corals. He explained each stage from corals growing closely together and against the shore of an island to barrier reefs and finally atolls. His theory, widely adopted by geologists and biologists everywhere, did not include a mechanism for subsidence. Darwin then surmised, based on his observations of submergence of islands in the middle of oceans, the Pacific in particular, and of continents such as South America rising, that when some part of the Earth's surface submerges, another must rise to balance it. Only since the 1960s and the development of plate tectonics did a proper mechanism became available for the submergence of volcanoes as they are carried away from hotspots of rising magma by the moving plates on which they lie. Darwin, of course, had no idea of this, but his theory of atolls remains valid.

Darwin wrote of conservation through his observations on many of the islands he visited. He was concerned, for example, that the harvesting of Galápagos tortoises by ships' crews for food might lead to their extinction. Likewise, in the far more complex situation of the Tasmanian Aborigines, whose conflicts with the European population Darwin well understood from his other island observations, led him to suggest that the only way to preserve them as a people was to remove them from Tasmania and place them in a locale far from Europeans. Without this move, he predicted that they would go extinct as a pure race, which indeed they did less than 40 years after his statement.

Darwin was also concerned that the introduction of alien species to islands caused destruction and extinction of native species. At the Bay of Islands on New Zealand, he admired the many fine English plants that grew around the houses, but he saw that many weeds and the common black rat had also been introduced. These he regarded as a threat to the native species, especially the rat, which Darwin knew had already eliminated the Polynesian rat. At St. Helena, he noted the same situation—a large number of introduced plants crowding out the natives. There he also noted that the introduced goats had demolished the trees and much of the other vegetation, and that this resulted in a cascade of extinction of a number of species of land snails and probably impacted the insects too. On Ascension, he saw many feral cats, which he condemned as a plague on the land. He even took note of the impact by humans through removal of the trees on the formerly well-forested island.

While recording his observations, Darwin commonly referred to "struggles" between various natural elements. This concept developed early in the voyage and he returned to it often. He referred to the struggle of native species to resist those recently introduced, of corals to resist waves, of one species in competition with another, and of waves against the shore. Darwin had seen enough, chiefly on islands, during the voyage of the *Beagle* to write many books and articles that changed biology and geology forever.

SEE ALSO THE FOLLOWING ARTICLES

Atolls / Cape Verde Islands / French Polynesia, Geology / Galápagos Islands, Geology / St. Helena / Tasmania / Tierra del Fuego / Tortoises

FURTHER READING

Armstrong, P. 2004. *Darwin's other islands.* London: Continuum.

Barlow, N., ed. 1958. *The autobiography of Charles Darwin, 1809–1882, with original omissions restored.* London: Collins.

Browne, J. 1996. *Charles Darwin voyaging.* Princeton, NJ: Princeton University Press.

Darwin, C. 1842. *The geology of the voyage of the* Beagle. *Part 1. The structure and distribution of coral reefs.* London: Smith, Elder and Co.

Darwin, C. 1844. *The geology of the voyage of the* Beagle. *Part 2. Geological observations on volcanic islands.* London: Smith, Elder and Co.

Darwin, C. 1846. *The geology of the voyage of the* Beagle. *Part 3. Geological observations on South America.* London: Smith, Elder and Co.

Darwin, C. 1859. *On the origin of species by means of natural selection, or the preservation of favoured races in the struggle for life.* London: John Murray.

Darwin, C. 1899. *The descent of man and selection in relation to sex,* 2nd ed. London, United Kingdom: John Murray.

FitzRoy, R. 1839. *Narrative of the surveying voyages of His Majesty's Ships Adventure and Beagle between the years 1826 and 1836, describing their examination of the southern shores of South America, and the Beagle's circumnavigation of the globe. Proceedings of the second expedition, 1831–36, under the command of Captain Robert Fitz-Roy, R.N.* London: Henry Colburn.

Herbert, S. 2005. *Charles Darwin, geologist.* Ithaca, NY: Cornell University Press.

Leakey, R., abr., illus. 1979. *The illustrated Origin of Species by Charles Darwin.* London: Faber and Faber.

WALLACE, ALFRED RUSSEL

ELIN CLARIDGE
University of California, Berkeley

Alfred Russel Wallace (1823–1913) is best known as the co-author of the theory of evolution by natural selection, with Charles Darwin. However, he was also an influential British scientist and social thinker of his time. As a young man, he traveled extensively in the Amazon River basin, and then in the Malay Archipelago, working as a professional collector, naturalist, and explorer. Besides his contributions to the development of evolutionary theory, Wallace is considered a founding father of the field of biogeography. He published articles and books covering a wide array of topics in both the natural and social sciences.

WALLACE'S THEORY OF EVOLUTION

Wallace developed his theory of evolution by natural selection independently of Darwin. In his autobiography, Wallace recounts that the idea of natural selection came to him as he was lying in bed delirious with a malarial fever, pondering Malthus's ideas about checks on populations. At the time he was traveling in the Malay Archipelago (present-day Malaysia and Indonesia). After recovering from his fever, he wrote a short essay outlining his ideas and sent a copy to Darwin, an esteemed colleague, without realizing that Darwin himself had been developing these same ideas over 20 years prior to Wallace's discovery but had not yet published on the subject because of its radical implications. Wallace's essay was an unpleasant shock for Darwin but provided the impetus he needed to start writing up his own research.

WALLACE'S EARLY LIFE

Wallace was an unconventional and somewhat controversial scientist with broad interests. Unlike most of the gentleman scientists of his time, he came from a modest background and although he received some formal schooling, he was primarily self-educated. He was also a radical socialist who championed unpopular causes, without consideration of the impact it might have on his reputation or career. He was one of the first scientists to raise concerns about the environmental impacts of human activities. He was highly critical of the economic and social injustices that he saw in nineteenth century Britain. His advocacy of spiritualism and his belief that natural selection could not explain the origin of consciousness or human intellect put him in direct conflict with many of the other proponents of evolutionary theory.

Wallace was born in Wales in the village of Llanbadoc, near Usk in Monmouthshire. He was the eighth of nine children. Wallace's father had a law degree but never actually practiced law because he had inherited some property that generated sufficient income to support the family; however, he, like his son, had a knack for making bad investments, and a series of failed business ventures led the family into financial decline. The Wallace family moved to Hertford, north of London, when Alfred was five years old, and there he attended Hertford Grammar School until, at the age of 13, financial difficulties meant that he had to withdraw. He initially moved to London to work with his brother John, who was an apprentice builder. In 1837, he left London to work for his older brother William, who was a surveyor. Over the following years, he became an experienced surveyor working primarily in the west of England and Wales. Much of the surveying work that they undertook was associated with the Tithe Act and General Enclosure Act, which modified land rights, in restricting the small farmers' rights to graze the common lands. Wallace was outraged by the social injustice. He considered that these Acts were nothing more than the persecution of poor farmers by rich landowners and lawmakers. This sense of social injustice never left him; he was a socialist at heart, who believed that everyone had a fundamental right to share the world's resources.

When in 1843 his brother was unable to find sufficient work for them both, Wallace left the surveying business and found a position as a schoolmaster at the Collegiate School in Leicester, teaching technical drawing and surveying. In Leicester, he continued to read widely and frequented the library, where he read Malthus's pivotal work on populations. He also met the young entomologist Henry Walter Bates. They became good friends, and Wallace was inspired by Bates to start collecting insects. They shared a common passion for natural history and insect collecting, and they exchanged frequent written correspondence, including lists of insects they had collected and discussions of their own scientific ideas and the most recent literature. Wallace also became interested in mesmerism (hypnotism) and phrenology while he was in Leicester. He even discovered that he was able to mesmerize some of his students.

In 1845 when his brother William died, Wallace left his teaching position in order to deal with his brother's business affairs in Neath. He was unable to re-launch the business, but instead found work as a civil engineer, working on a survey for a proposed railway line in the Vale of Neath. The railway line was never built, but Wallace's surveying work, hiking through the remote Welsh countryside, fed his passion for insect collecting and natural history.

In early 1846 Wallace was able to persuade his brother John to join him in starting another architecture and civil engineering firm in Neath. They rented a small cottage, and Wallace's mother and sister came to live with them. Their firm carried out a number of projects including designing a building for the Mechanics Institute of Neath. William Jevons, the founder of that institute, was a family friend and persuaded Wallace to lecture on science there for two winters. During this period Wallace was still reading avidly and exchanging letters with Bates; in particular, they were discussing the anonymous evolutionary treatise *Vestiges of the Natural History of Creation*, which propounded the radical idea of species transmutation; Charles Darwin's journal and remarks from his adventures on the *Beagle*; and Charles Lyell's *Principles of Geology*.

TRAVEL AND EXPLORATION

When Bates came to visit Wallace in Wales in 1847, inspired by the travel chronicles of earlier naturalists, they hatched a plan to embark on their own big adventures. Having read William Edwards's "Voyage Up the Amazon," they decided that they too would travel to South America, where they could work as professional collectors in the Amazon rain forest, sending exotic and unusual specimens back to England for sale. They would also be able to indulge their own passion for natural history and might even find evidence for the transmutation of species. In 1848, the companions left for Brazil on the sailing barge *Mischief*. Wallace and Bates spent most of their first year collecting near Belém do Pará, before deciding to part company and explore the more remote reaches of the rain forest separately. Wallace spent a total of almost four years charting the Rio Negro, collecting specimens for sale and for his personal collection, as well as writing prodigious notes on the natural history, people, and places he visited. In 1849, he met up with another young explorer, who would become a lifelong friend, the botanist Richard Spruce, as well as Wallace's younger brother Herbert, who planned to work as Wallace's assistant. After some time together, Herbert decided that he was not cut out to be a collector, so he and Wallace parted company. Sadly, Herbert died of yellow fever while he was in Pará waiting for his passage back to Liverpool. Wallace probably felt partly responsible for this family tragedy, and it seems likely that this incident inspired, at least in some small part, his later interest in spiritualism.

In July 1852, Wallace decided to return home on the brig *Helen*. This was a fateful decision, because the brig's cargo of balsam caught fire three weeks away from Brazil. The vessel eventually sank, and Wallace lost all his personal collections and most of his notes, not to mention some of the live animals he was hoping to sell to collectors upon his return. However, he escaped with his life and was even able to save part of his diary and some sketches. The journey was an ordeal: Wallace and the crew spent ten days at sea in an open life boat before being picked up by another brig, the *Jordeson*, which was in a sad state of disrepair and running low on rations and water.

Fortunately, his agent Samuel Stevens had insured Wallace's collections, on his behalf, and Wallace spent the next 18 months in London living off the insurance payment and the profits he had made by selling the collections that he had sent back during his stay in the Amazon. During this time, he wrote six academic papers and two books describing his experiences in the Amazon. One of the papers dealt with the monkey species found in the Amazon. He had noticed that closely related monkey species tended to occur in close proximity, often found on neighboring sides of a river, and he was still actively thinking about species transmutation.

Despite the rigors of his Amazon expedition, Wallace decided, almost as soon as he returned, to plan another expedition, this time to the even less explored Malay Archipelago. He was able to get his passage to Singapore arranged by the Royal Geographic Society, so in 1854 he left England once more. From Singapore he traveled to Borneo; it was the start of an eight-year-long journey. During that time, he collected more than 125,000 specimens (more than 80,000 beetles alone); of these, more than a thousand were species new to science.

WALLACE'S GREAT DISCOVERY

Wallace's travels around the many islands of the Malay archipelago also got him thinking about geographic distributions. He was struck by the zoological differences between the islands and island groups that he visited. He was amazed by the faunal discontinuity he found across the narrow Lombock Strait, dividing Bali and Lombock. He later proposed this as a zoogeographical boundary, which is now known as Wallace's Line, that divides the Australasian and Asian fauna. He also had ample opportunity to refine his thoughts on transmutation. In September 1855, he published a paper in the *Annals and Magazine of Natural History* "On the Law Which Has Regulated the Introduction of Species." In this paper, he did not discuss the mechanistic aspect of species transmutation, but he concluded that all species have come into existence coincident, in both space and time, with a closely allied species.

His next paper went much further. In February 1858, Wallace sent a copy of his essay "On the Tendency of Varieties to Depart Indefinitely from the Original Type" to Darwin, with a request that he review it and pass it on to Charles Lyell, if he thought it worthwhile. This may sound like a strange request, as Darwin and Wallace had met only once and then only very briefly, but during his time in the Malay Archipelago, Wallace had begun corresponding with Darwin, and he clearly respected Darwin's opinions.

Wallace's essay outlined the means by which species could diverge as a result of environmental pressures, although he did not use Darwin's term "natural selection." Darwin was shocked to receive such a concise synthesis of a theory that was so very similar to the one he had been working on for 20 years but had not yet published. Darwin asked for advice from Charles Lyell and Joseph Hooker, both respected scientists and close friends of Darwin's. It was decided that Wallace's essay would be published in a joint presentation, together with some unpublished writings that would highlight Darwin's priority. Wallace's essay was presented to the Linnean Society of London on July 1, 1858. The reading of the essay generated remarkably little reaction. It was not until the publication of Darwin's *On the Origin of Species* in 1859 that the radical implications of this theory generated the public scrutiny Darwin had so much feared.

Wallace had little choice but to accept the arrangement after the fact; he may even have been grateful to receive any credit at all. Wallace had neither the social nor the scientific status that Darwin had, and it is unlikely that his views on evolution and the transmutation of species would have received much attention without being linked to Darwin's name. When Wallace returned to England in 1862, he found that he now had access to elite British scientific circles. Much to Darwin's delight, Wallace turned out to be a staunch defender of natural selection. During the 1860s, he wrote several outspoken and eloquent articles, demolishing the arguments put forward by opponents of the theory.

WALLACE'S VS. DARWIN'S THEORY OF EVOLUTION

Wallace and Darwin did not always share the same ideas about natural selection; in fact, there are fundamental differences in the emphasis of Darwin's and Wallace's theories. Wallace was always struck by geographic differences between neighboring islands or habitat islands, and he emphasized the role of the physical environment in driving adaptation and evolution, whereas Darwin emphasized the role of competition between individuals of the same species, and in particular the role of sexual selection. This fundamental difference in emphasis accounts for the frequent scholarly disagreements they had.

WALLACE'S LIFE AFTER HIS TRAVELS

Back in England, Wallace moved in with his sister Fanny Sims and her husband Thomas, who owned a photographic company. Wallace occupied himself giving lectures, writing articles about his travels and scientific findings, and dealing with both his personal and family business affairs, which took a lot of his time and energy.

In 1864 Wallace became engaged to a young woman named Marion Leslie, whom he had been courting for some time. However, she soon broke off the engagement, for unclear reasons, leaving Wallace heartbroken and unable to concentrate on his scientific work. The return of his good friend Richard Spruce from his South American travels was the perfect distraction. Through Spruce, Wallace met William Mitten, who was a pharmacist by trade and bryologist by passion. Mitten had four daughters; the eldest, Annie, became a good friend of Wallace's, and in 1866 they were married.

SPIRITUALISM

During the summer of 1865, about the same time that he first met Annie Mitten, Wallace began to investigate spiritualism. His sister Fanny Sims had been involved in it for some time. Wallace reviewed the literature on the topic and attended séances; he became convinced that at least some of the phenomena he saw were real. He was not alone; spiritualism was very popular at the time. Many educated Victorians found that strict religious doctrine was at odds with their new understanding of the world, but they were not ready to adopt the completely materialistic and mechanical views that were increasingly emerging from nineteenth-century science, in no small part associated with Darwinian philosophy. Unfortunately, Wallace's public advocacy of spiritualism seriously damaged his scientific reputation and strained his friendships with Henry Walter Bates, Thomas Huxley, and Darwin. Other scientists were more publicly hostile.

At about the same time that Wallace began to become interested in spiritualism, he also began to vocally maintain that natural selection could not account for human intellect and artistic talents. Darwin disagreed vehemently, arguing that sexual selection could easily explain such apparently non-adaptive mental phenomena. Wallace's views were at odds with the fundamental Darwinian idea that evolution is neither teleological nor anthropocentric.

THE MALAY ARCHIPELAGO

In 1869, Wallace finally published a popular account of his travels and observations in *The Malay Archipelago*. His book became a bestseller; it was kept in continuous print by its original publisher for over 40 years. The book brought in some much needed income, particularly as Annie was now pregnant with their second child, Violet. Despite having accumulated a considerable amount of money from the sale of specimens while he was away traveling, Wallace had squandered most of this money on bad business investments.

FINANCIAL STRUGGLES

Financial difficulties were a recurring theme throughout Wallace's life. Despite his scientific prowess, Wallace was never able to secure a permanent salaried position, possibly because of some of his more controversial and outspoken opinions. He relied almost entirely on income generated by writing to support his family. He graded government exam papers, wrote articles for commission, edited scientific works, and gave lectures. Finally in 1881, Darwin was able to secure a government pension of £200 per year for Wallace, in recognition of his contributions to natural history and science, but it required much hard lobbying on Darwin's part. The modest income that the pension provided helped Wallace's uncertain financial situation considerably.

ZOOGEOGRAPHY

In 1874, tragedy struck the Wallace family: Wallace's eldest son Herbert (named after his deceased uncle) died, after a long period of sickness. Wallace was devastated, and he threw himself headlong into his new project on the geographic distributions of animals.

The Geographical Distribution of Animals was a monumental work that was published in 1876; in it, Wallace dealt with the patterns and causes of the global distribution of animal species. He emphasized the role that both historical and physical factors played in shaping the observed distributions of animals, and his exhaustive treatment of the patterns he observed provided a basis for defining the zoogeographic regions that are still recognized today. This two-volume work was quickly adopted as a classic text for students and was widely used for over 80 years after its initial publication.

His next book, *Island Life* (1880), also addressed the processes shaping the geographic distributions of animals, but focused exclusively on islands. Wallace classified islands into three different types. Oceanic islands, such as the Hawaiian Islands and Galápagos, were categorized as those that had been formed in the middle of the ocean and had never been part of any large continent. As such, their biota are the result of transoceanic dispersal and subsequent in situ diversification. Such islands are notable for the almost complete absence of terrestrial mammals or amphibians. Continental islands, those that had once been part of a larger landmass, were divided into two separate classes depending on whether the connection was recent or ancient. Islands that have been isolated for a long time retain elements of ancient continental faunas that may no longer be abundant or even extant on the continents themselves. Wallace discussed the roles that isolation and climate change would have on island biota. He devoted a significant portion of the book to a discussion of the causes and impacts of glacial cycles on species distributions. It was an innovative and interesting publication, which received wide public acclaim.

SOCIAL AND POLITICAL INTERESTS

From the late 1870s until his death, Wallace became increasingly interested in social and political issues, writing on a wide array of topics from environmental issues, to land reform, to women's rights, to a critique of militarism, to opposing obligatory smallpox vaccination. His broad interests and talent for writing concise, clear prose meant that he was constantly being asked to contribute articles to popular journals and newspapers.

THE AMERICAN TOUR AND DARWINISM

In 1886 Wallace was invited to give a lecture series at the Lowell Institute in Boston, and given that he was also in need of a financial boost, he decided to make the invited lecture series part of a ten-month lecture tour of the United States. He was a gifted speaker and had an undeniable knack for making complex scientific ideas easily accessible. Most of the lectures he gave were on Darwinism (evolution and natural selection), but he also lectured on biogeography, spiritualism, and social/economic reform. During the trip he traveled to California, to Oakland, to meet up with his brother John, who had emigrated there more than 30 years previously. He also got the opportunity to meet John Muir and visited the mighty redwood forests; he was much impressed by their grandeur and dismayed by humanity's thoughtless destruction of these majestic trees. Before he left America, he made sure to spend a week exploring the Rocky Mountains with the American botanist Alice Eastwood as his guide. He had developed a passion for alpine floras. While in the States, he also met many other prominent American naturalists and viewed their natural history collections. Despite all the new and exciting places and people, Wallace was very happy to arrive back in Liverpool in August 1887. He had been troubled with minor ailments throughout the trip and was beginning to feel he was too old for these kinds of things.

When he returned home, he set to work adapting the information he had compiled for the lecture series and other evidence he had accumulated during his American trip for his 1889 book *Darwinism*. It was one of Wallace's most scholarly books, in which he explained and defended evolution by natural selection and developed carefully argued discussions of evolutionary phenomena. In this book, he presented a mechanism by which natural selection could drive reproductive isolation and thus result in the formation of new species. He argued that once two populations had become adapted to their own particular environments, hybridization could be disadvantageous, if the hybrid offspring had intermediate characters that were less adapted to their environment than the parental forms. Natural selection would select against these hybrids, whereas mechanisms that prevented hybridization from occurring would be favored, reinforcing the reproductive isolation of these two species. This idea is sometimes known as the Wallace Effect. Wallace also wrote at length about the adaptive significance of animal coloration, in particular the evolution of warning coloration in unpalatable insects. He and Darwin had corresponded at length about this topic. Darwin considered that sexual selection could explain most bright color patterns and ornamentation in the animal kingdom; Wallace disagreed, and even suggested that there might be alternative hypotheses to explain some of the examples of sexual selection Darwin had proposed. He also went on at some length about his ideas about humanity and evolution, once more reiterating his belief that human intellect was beyond the realms of natural selection.

A LIFETIME OF ACHIEVEMENT HONORED

Throughout the 1890s, Wallace continued to write prolifically, tackling projects such as *The Wonderful Century*, reviewing the scientific achievements and failures of the nineteenth century. His lifetime of achievement was recognized and awarded; he was offered an honorary doctorate from Oxford University in 1889, and in 1890 he was awarded the Darwin Medal by the Royal Society. In 1892 he was awarded a founder's medal by the Royal Geographic Society and was given the Royal Medal at the Linnean Society. Always a modest and unassuming character, Wallace was rather embarrassed by all of this recognition. Even his autobiography, *My Life*, which he published in 1905, is extremely self-effacing.

The year 1908 marked the 50th anniversary of the joint-reading of Darwin's and Wallace's papers, and the Linnean Society produced a medal in honor of the event. He also received the prestigious Copley Medal that same year and much to his surprise and amusement, he was presented with an Order of Merit by the British Sovereignty. This was quite extraordinary, given that he was a self-acclaimed radical socialist who had openly opposed the Boer Wars and had campaigned for land nationalization.

WALLACE'S DEATH

Toward the end of his life, Wallace became more reclusive and spent most of his time at home with his family and his beloved garden. He died at home, peacefully, on November 7, 1913. His death was widely reported in the press. Some of Wallace's colleagues even suggested that he be buried in Westminster Abbey, like Darwin; however, following his personal wishes, his family had him buried in the small cemetery at Broadstone, Dorset, near their home. In lieu of this honor, a committee of prominent scientists lobbied to have a medallion of Wallace placed in Westminster Abbey, near Darwin's burial site. The medallion was uncovered in 1915.

CONCLUDING REMARKS

During his lifetime, Wallace published at least 747 papers and 22 books, on a wide variety of subjects. He made significant contributions to evolutionary theory and biogeography and was deservingly honored as an exceptional scientist. His early exploration of both the Amazon and the islands of the Malay archipelago defined him as one of the foremost explorers and naturalists of his time. Despite all of his achievements, after his death, Wallace received relatively little recognition for his role in the development of evolutionary theory, at least in comparison with Darwin. This may be due to his vocal support of ideas that cannot be rationalized by evolutionary thinking, or it may be because of his modest and unassuming manner: He had always been willing to stand in Darwin's shadow.

SEE ALSO THE FOLLOWING ARTICLES

Continental Islands / Oceanic Islands / Sexual Selection / Voyage of the *Beagle* / Wallace's Lines

FURTHER READING

Fichman, M. 2004. *An elusive Victorian: the evolution of Alfred Russel Wallace.* Chicago: University of Chicago Press.
Raby, P. 2002. *Alfred Russel Wallace: a life.* Princeton, NJ: Princeton University Press.
Severin, T. 1997. *The Spice Islands voyage: the quest for Alfred Wallace, the man who shared Darwin's discovery of evolution.* New York: Carroll and Graf.
Slotten, R. A. 2004. *The heretic in Darwin's court: the life of Alfred Russel Wallace.* New York: Columbia University Press.
Wallace, A. R. 1905. *My life.* Google Books.

WALLACE'S LINE AND OTHER BIOGEOGRAPHIC BOUNDARIES

JEREMY D. HOLLOWAY

The Natural History Museum, London, United Kingdom

Wallace's Line represents the first attempt to establish a boundary between two major biotic regions of the world within an entirely archipelagic context. A prerequisite is the recognition of such biotic regions: gross areas that support distinctive but generally distributed assemblages of plants and animals. These regions are all continent-based, with climate being a secondary factor in the Northern Hemisphere: temperate versus tropical. A regional boundary must represent the most significant transition between two regions in terms of losses and gains of components. This transition should also be relatively abrupt: a discontinuity.

HISTORICAL BACKGROUND

The eighteenth century saw a significant increase in exploration of the world by European navigators, traders, and scientists, and an increasing realization that other continents supported plants and animals of form and diversity that often contrasted very strongly with the European experience. The impact of the biotas of Australia and New Zealand, coming soon after the Linnaean revolution in taxonomy, were of particular force relative to the better-known ones of the "East Indies." Concepts of distinct biogeographic regions were promoted and, inevitably with this, a perceived need emerged to establish the boundaries between them. Only the boundary between the Oriental and Australian regions would be likely to be discovered amid a geography that was entirely archipelagic. Alfred Russel Wallace provided the initial stimulus for such a boundary. Debate over its position and nature continued for the next century, as did further biological exploration, yielding an ever more detailed picture of plant and animal distributions. The very weight of this information resource led in the latter half of the twentieth century to a much more analytical approach to the classification of the organisms themselves, through cladistics, and of their distributions, through several distinctive and often antagonistic approaches to biogeography. These new approaches coincided with better understandings of a much more dynamic Earth surface than had previously been agreed upon and included the recognition of the dynamism brought about by the processes of plate tectonics, processes that have shaped the Indonesian Archipelago and its relationship with major continents in a most dramatic manner.

WALLACE AND HIS LINE

By the time Wallace undertook his voyages in the "Malay Archipelago" in the mid-nineteenth century, much biological information and material had already been brought to Europe by the disciples of Linnaeus and by Dutch, British, and French explorers, including expedition scientists and traders. Wallace was aware of this work and so went out with at least some background knowledge. The stage was effectively set for his particular contribution: to make zoological collections intensively on a transect from west to the east through the archipelago, visiting the major islands of Sundaland and many of the Lesser Sundas, the extremities of Sulawesi, many of the Moluccas, and various places and offshore islands in western New Guinea. In this he had the support of two similarly indefatigable assistants, Charles Allen and Ali, recruited as teenagers, the latter from Sarawak in Borneo. Wallace's combination of toughness, enthusiasm, curiosity, and perspicacity led him to accumulate and realize the significance of an immense amount of data on the distribution of vertebrates and insects through the archipelago, and, from his background knowledge and later research project, to recognize elements that were Australian in more westerly localities, particularly among the better known bird fauna.

Wallace had a clear concept of what he considered to be the Indo-Malayan fauna, grouping the Philippines with this fauna, so he was particularly struck by the disappearance of many key elements of this fauna, especially bird groups, coupled with the appearance of elements he considered to be Australian, when he traveled from Bali to Lombok. The "Australian" taxa were not noted on his visits to Sulawesi, where he found a high level of peculiarities to the island, together with similar impoverishment of Indo-Malayan elements. These were the major influences in his first publication on the zoogeography of the archipelago, a work that appeared in 1860 (with an ornithological note in 1859) when he was still on his travels, as related in his more extensive account of them published in 1869, some years after his return. In 1860 he stated that the Strait of Lombok marked the limits and abrupt separation of two of the great zoological regions of the globe, but in 1869 his views appeared more measured on reflection, as he implied that it was only the Indo-Malayan fauna proper that terminated at a line running between Bali and Lombok (the Strait of Lombok), north between Borneo and Sulawesi (the Makassar Strait), and thence northeast between Sulawesi and the Philippines. Areas to the east were either distinctive like Sulawesi or had an increasingly strong Australasian influence as shown by the Lesser Sundas (his Timor group of islands) and lands from the Moluccas eastward. By this time, T. H. Huxley had termed this boundary Wallace's Line, but he proposed that its northern course should run to the west of the main Philippine archipelago, but to the east of Palawan.

FURTHER LINES

Although Wallace's Line initially met with general acceptance, the accumulation of data in the decades following led to alternative proposals (Fig. 1), the most significant of which was Weber's Line of Faunal Balance, sited further to the east. This proposal was premised on the grounds that Weber's Line marked the transition between islands where the fauna is predominantly Oriental and those where the fauna is predominantly Australasian. This line runs between the Sulawesi group of islands and the Lesser Sundas on one hand and the Moluccas on the other, with its southern extremity running to the east of the Tanimbar group. Wallace's Line was seen increasingly as the eastern boundary of the Oriental region, with an equivalent boundary, Lydekker's Line, marking the western limits of the Australo-Papuan region, which extends along the continental shelf of Australia and New Guinea.

Geological information on the area was also becoming more detailed, and it became apparent that Huxley's version of Wallace's Line marked the extent of lands that had been connected to the Asian mainland during periods of low sea level during the glaciations. This included Bali but not Lombok, and also Palawan but not the rest of the Philippines. The intervening areas, particularly Sulawesi, the Moluccas, and the Philippines, were regarded more as a zone of transition, often termed Wallacea.

The value of such lines in themselves has also come under scrutiny, a process that commenced in the middle of the last century, as exemplified in an essay by Ernst Mayr in 1944 from a zoological viewpoint, and in a review by Scrivenor and a group of zoologists and botanists in 1943; botanical and zoological views were found to differ considerably.

THE IMPACT OF PLATE TECTONICS AND NEW BIOGEOGRAPHIC METHODS

The last half-century has seen major advances in two areas. The first is in our understanding of plate tectonics,

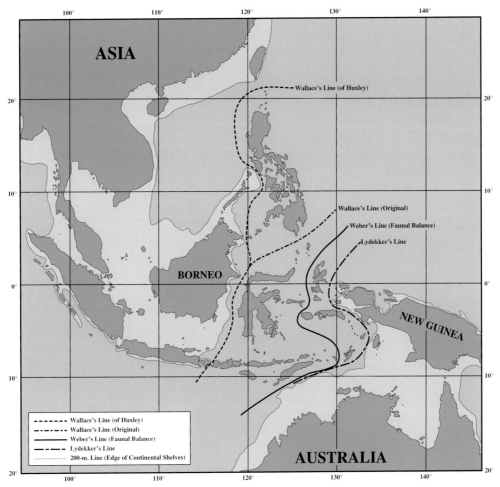

FIGURE 1 Zoogeographic boundaries proposed between the Oriental and Australian regions. The shaded areas are the continental shelves; these would have been mostly exposed in periods of low sea level during the glaciations. Adapted from Mayr (1944).

a process leading to movement of the major continents relative to each other, and to extensive island arc formation and uplift in zones of plate convergence and subduction. A glance at a map of Indonesia and the western Pacific reveals the influence and complexity of such processes in the area. The movement of Australia northward on the Indian Ocean plate, and the shearing influence of the Pacific plate moving from east to west across the north of this, can be witnessed in the swirls and whorls of the emergent lands, the massive mountains in central New Guinea, the arcs of volcanoes, and the deep ocean basins and trenches. Thus, land areas have emerged, submerged, and moved relative to each other, leading to separation or convergence, often with fusion. Many of today's islands are composites formed from different island systems, some incorporating continental fragments detached from Asia or Australia; Sulawesi incorporates all of these. This process has been operative incrementally through most of the Tertiary and has most recently been augmented by the effects of sea-level fluctuations during the Quaternary glaciations.

It is often said that Earth and life have evolved together. It is probable, therefore, that the mere drawing of lines on maps to explain the complexity of life in the archipelago between Asia and Australia is inevitably overly simplistic, given that the evolution of both Earth and life has involved significant mobility.

The second major advance has been in the methodology used to study the distribution of plants and animals on a geographic scale, although this has involved some dichotomy of purpose and approach that reflects the evolutionary dichotomy just mentioned as well as a perceived need to separate the analysis of pattern from the interpretation of that pattern in terms of process. The basic data to hand are much more comprehensive today, and the taxonomy of many groups has been modernized by application of cladistic methodology to suites of morphological and molecular data to provide strong phylogenetic hypotheses for the groups concerned, including some still controversial attempts to date events in the phylogeny, particularly for groups with poor fossil records.

The phylogenetic hypotheses can be combined with distributional data, especially where most individual taxa are endemic to islands or parts of islands, to provide hypotheses of the interrelationships of islands as an indication of Earth's history (cladistic biogeography, also known as vicariance biogeography). The more extreme proponents of this approach have yet to attract more than fleeting attention from geologists, particularly given the difficulties of dating just mentioned and the difficulties of separating components of the patterns revealed, which owe their origins primarily to geological mobility rather than to the mobility of the organisms themselves.

The potential for organisms to disperse across water gaps between islands and thus to reach remote and recently emergent islands, such as in the Hawaiian archipelago, is essential if such islands are to develop biotas with a degree of endemism but problematic if the preferred methodology postulates that patterns derived from dispersal are uninformative. The very nature of the Indo-Australian archipelago and what we know of its geological evolution makes it likely that such patterns will predominate. The alternative biogeographic approach in such a situation is to place more reliance on geological hypotheses and dating and to use biogeographic patterns to explore processes of dispersal, colonization, and speciation, while still allowing for reciprocal illumination between geology and biogeography. But, whereas geologists identify the lack of dating of biological events as a deficiency, biologists can also be frustrated by the reluctance of geologists to identify the extent and timing of their structures as dry land.

LINES AS DISTRIBUTIONAL DISCONTINUITIES

So do Wallace's Line or any of the others proposed have any reality? Even in the absence of modern taxonomic treatments, the distributions of species and higher taxa can be used to identify discontinuities in such distributions: major changes between islands in terms of components of their fauna or flora. Such analyses, usually involving some sort of "phenetic" clustering methodology applied to similarity coefficients, identify several major discontinuities in the archipelago coincident with all or parts of the lines of Weber and Wallace. That of the former is usually recaptured in entirety, but that of the latter is confirmed between Sulawesi and Borneo, and to some extent on either side of the Philippines; however, varying links between the Philippines and Sulawesi are not uncommon and may predominate over those between the former and Borneo once the effects of faunal size disparities have been factored out.

The Lesser Sundas from Lombok to Timor can show some integrity as a group, but usually they are revealed as an attenuation of the Javan fauna, with an increasing Australian element as one passes eastward. Bali was united with Java and the rest of Sundaland during Pleistocene sea-level falls, with the Strait of Lombok persisting throughout. Hence, the striking contrast between the two islands noted in the bird faunas by Wallace and others was mostly due to a sharp decline in the diversity of Indo-Malayan elements, which rendered the small number of Australian elements more conspicuous. There is an even greater contrast in less dispersive vertebrate groups such as the mammals and freshwater fish, although the diversity of the former was greater in the recent past, as revealed by fossil faunas of Indo-Malayan character in the Lesser Sundas and southwest Sulawesi.

The major discontinuities show variation in their importance from group to group, with Weber's Line being emphasized for butterflies and birds, but the northerly part of Wallace's Line being stronger for flowering plants and hawkmoths. In most such analyses, Sulawesi occupies a relatively isolated position, and it is often unclear whether its closest link is with the Philippines, the southern Moluccas, or even Java and the Lesser Sundas.

THE POSSIBILITY OF TWO AUSTRALASIAN BIOTAS

Whereas the Oriental region is predominantly tropical in character, a significant part of the area and biota of Australasia can be considered temperate. During its northward drift, the original uniform Gondwanan biota of the Australian continent became more polarized into tropical and temperate elements, combined with increasing aridity in the interior. The tropical archipelago that developed to the north of Australia into the major island of New Guinea, with the Moluccas to the west and the Bismarck and Solomon groups to the east, now has a flora and a fauna that in some groups are more Indo-Malayan in character than Australian. This is particularly true of the insects, where the Indo-Malayan influence also extends strongly down through Queensland to northern New South Wales, leading J. L. Gressitt to suggest that the boundary between the Oriental and Australian regions for most insect groups occurred within Australia, rather than at Wallace's Line! There is a contested possibility also that all the Australian butterflies were derived from the north by dispersal. It has also been suggested, from fossil pollen evidence, that India contributed significant Gondwanan elements to the Indo-Malayan flora when it made contact with Asia in the Early Tertiary, and that these have also spread eastward, probably via Sulawesi.

The distributional discontinuities are evident in all these patterns, but their role in them is variable. For example, the two major lineages of cicadas in the Australasian tropics have an Indo-Malayan origin rather than originating within the temperate Australian fauna, yet they barely extend west of Wallace's Line. The monitor lizard genus *Varanus* has an Australian lineage (including the Komodo dragon) that extends to Wallace's Line through the Lesser Sundas, but also has a separate tropical Australasian lineage that is related to an Indo-Malayan lineage, which it replaces east of Wallace's Line.

MONTANE FLORAS AND FAUNAS

The previous sections have focused on the very rich lowland biotas of the archipelago, but it must not be forgotten that there are also distinctive ecosystems at higher altitudes on many of the islands, forming an archipelago of montane habitats within the greater archipelago. The recency of tectonic uplift on the majority of islands, except, perhaps, for parts of Sundaland (especially Borneo), means that distribution patterns of montane plants and animals can be very different from those of the lowlands, transgressing the discontinuities observed in the lowland flora and fauna. The oak-laurel forests of middle altitudes can support a diversity of animal life that is as rich as, or richer than, that of the lowlands, and the zones above this, with conifers and Ericaceae dominating and elements of a herbaceous alpine flora sometimes present, can also show significant diversity.

North-temperate elements extend both from the Himalayan regions down though Southeast Asia to the mountains of Sumatra, Java, and the Lesser Sunda chain of islands, and from Taiwan to the Philippines, northern Borneo, and Sulawesi. But the major central archipelago of montane habitat in Borneo, Sulawesi, the Moluccas, and New Guinea shows strong representation of south-temperate taxa as well as some intrinsic elements that have radiated particularly within New Guinea. Thus, on these central mountains, there is an intriguing mixture of southern coniferous and myrtaceous shrubs, Himalayan ericaceous shrubs, and general alpine herbs in genera such as *Ranunculus* and *Euphrasia*.

SEE ALSO THE FOLLOWING ARTICLES

Borneo / Indonesia, Geology / Island Arcs / New Guinea, Geology / Philippines, Geology / Plate Tectonics / Vicariance / Wallace, Alfred Russel

FURTHER READING

Gressitt, J. L. 1974. Insect biogeography. *Annual Review of Entomology* 19: 293–321.
Hall, R., and J. D. Holloway, eds. 1998. *Biogeography and geological evolution of South East Asia*. Leiden, Netherlands: Backhuys.
Holloway, J. D., and N. Jardine. 1968. Two approaches to zoogeography: a study based on the distribution of butterflies, birds and bats in the Indo-Australian area. *Proceedings of the Linnean Society of London* 179: 153–188.
Mayr, E. 1944. Wallace's Line in the light of recent zoogeographic studies. *Quarterly Review of Biology* 19: 1–14.
Scrivenor, J. B., T. H. Burkill, M. A. Smith, A. S. Corbet, H. K. Airy Shaw, P. W. Richards, and F. E. Zeuner. 1943. A discussion of the biogeographic division of the Indo-Australian archipelago, with criticism of the Wallace and Weber Lines and of any other dividing lines and with an attempt to obtain uniformity in the names used for the divisions. *Proceedings of the Linnean Society of London* 154: 120–165.
van Steenis, C. G. G. J. 1965. Plant geography of the mountain flora of Mt. Kinabalu. *Proceedings of the Royal Society of London B* 161: 7–38.
Wallace, A. R. 1860. On the zoological geography of the Malay archipelago. *Proceedings of the Linnean Society of London* 4: 172–184.
Wallace, A. R. 1869. *The Malay archipelago*. New York: Dover Edition (1962).
Whitmore, T. C., ed. 1981. *Wallace's Line and plate tectonics*. Oxford, UK: Clarendon Press.
Whitmore, T.C., ed. 1987. *Biogeographical evolution of the Malay archipelago*. Oxford, UK: Clarendon Press.

WARMING ISLAND

KURT M. CUFFEY
University of California, Berkeley

Warming Island, situated at the north end of Liverpool Land in eastern Greenland (71°29' N; 21°51' W), is not particularly significant as a physical feature. Instead this island is noteworthy because of its connection to two major themes: the formation of new islands and the impacts of climate warming on glacial landscapes. Warming Island became an island only recently—between 2002 and 2005—when the glacial isthmus connecting it to the mainland was destroyed by melt and disintegration of ice, a consequence of climate warming (Fig. 1).

CONFIGURATION, FORMATION, AND DISCOVERY

Quaternary glaciation has sculpted the island into three narrow parallel ridges, whose sides plunge steeply to footings below sea level. The highest elevation is about 520 m. The island is about 7 km in length and is separated from the mainland by a narrow strait. This strait is becoming wider as the glaciers emanating from the mainland continue to retreat. The terrain is mostly barren rock, with several small glaciers filling the heads of valleys. The bed-

FIGURE 1 Aerial view of Warming Island, showing the newly formed strait separating it from the glaciers flowing off the mainland, seen at left. Photograph by Jeff Shea, taken during the 2006 expedition led by Dennis Schmitt.

rock consists of very old (Proterozoic) metamorphosed sedimentary sequences that were deformed in a tectonic collision with Eurasia around 400–450 million years ago. Igneous intrusions are also exposed in this region.

Warming Island does not appear as a separate landmass in the first detailed aerial surveys of the east Greenland coast, in the mid-twentieth century. A partial map of the region from 1957 seems to show Warming Island separated from the mainland, but this map is missing numerous features of the region—including a whole island and a large part of Liverpool Land—and cannot be regarded as complete or accurate. Satellite images show that in the mid-1980s the island was still connected to the mainland by glaciers flowing from the highlands of northern Liverpool Land. At this time, the ice surface close to the island appears to have been nearly flat, suggesting that this was a floating ice shelf. Like many ice masses in eastern Greenland, this ice shelf subsequently thinned, and its margin retreated; by 2002, the ice connecting Warming Island to the mainland was a narrow neck, no more than 1 km in width. In September 2005, a group of explorers led by Dennis Schmitt of California discovered that the ice neck had vanished. The explorers sailed through the open waters of the new strait. Schmitt named the newly isolated island Uunartoq Qeqertoq, an Inuit-language phrase for "the warming island." This discovery coincided with new scientific reports of increased melt and ice flow on the nearby Greenland Ice Sheet, and with increased general concern about global warming. Warming Island thus became an object of public interest, a symbol of Earth's transformation in a warming climate. It was featured in media reports, including an article in *The New York Times*. Editors of the *Oxford Atlas of the World* designated Warming Island as the "Place of the Year" in 2007.

THE ISLAND AS A SYMBOL OF GLOBAL WARMING

Does it make sense to use Warming Island as a symbol of the transformative effects of climate change—and of human-caused warming in particular? There is essentially no specific information available about the glacier flow and the climate history in the immediate vicinity of Warming Island. And it is inherently problematic to use events at a single place (or a few places) to represent a complex, diverse, and global-scale phenomenon. But diminishing glacial ice is without doubt a central issue of climate warming. Diminishing glaciers provide visually captivating evidence of the warming itself and, because they contribute to sea-level rise, are one of the warming's most important consequences. The majority of glaciers in the world's high mountain ranges are retreating because of climate warming. This is occurring, for example, in the Himalayas, the Andes, the Canadian Rockies, and the ranges of southern Alaska. Retreat in eastern Greenland is thus part of a global phenomenon. In many locations, retreat began in the late nineteenth century, before human-caused warming was significant. Brightening of the sun was likely the most important cause of this initial warming. By the first decade of the twenty-first century, however, most of the cumulative warming—including essentially all of the warming after 1970—was due to human-caused increases of atmospheric greenhouse gases. Thus ongoing worldwide glacier retreat can be attributed, in large part, to human causation.

In Greenland, warming of air and ocean waters has increased melt of glaciers and thinning of floating ice shelves. On the main Greenland Ice Sheet, warming in the 1990s and the 2000s expanded the zone of melting. The ice sheet shrank as a result of both increased melt and increased glacier flow to marine margins where icebergs form. The glaciers of Liverpool Land and Warming Island are not part of this main ice sheet, but are subject to the same climatic and oceanic influences. Measurements of air temperature show that eastern Greenland is now about 2 °C warmer than at the start of the twentieth century. Similarly warm temperatures prevailed in the 1930s. The intervening decades were cooler, but not as cold as the first years of the twentieth century. Given this history, it is likely that the glacier retreat in eastern Greenland began in the first warm period (the 1930s) and was rejuvenated by the more recent warming. It is thus reasonable to regard the glacier retreat here as partly natural and partly human-caused. Over the next century, the human-caused warming is expected to become very much larger than natural variations.

Glacier retreat around Warming Island is not documented nearly as well as glacier retreat in some other locations, such as the European Alps. Nor has the retreat at Warming Island been as extensive and dramatic as that in places like coastal Alaska and Patagonia. But the Warming Island case is culturally compelling. Not only did retreat here lead to the formation of a new coastal island, but it also occurred at 71° latitude, in the heart of the Arctic, adjacent to one of the world's great ice sheets. The attention given to Warming Island as a symbol of climate change seems fully justified, provided it is recognized that glacier retreat in this particular location is caused by local climate changes whose relation to global-scale warming is complex and not fully elucidated.

ISLAND FORMATION BY GLACIER RETREAT

Warming Island is also interesting as an example of the formation of a new island, an event that is rarely witnessed. The best known new islands, like Surtsey in Iceland, owe their birth to volcanic eruptions. But the majority of new islands formed in the last 15,000 years are the result of glacier retreat. Glaciers can scour bedrock to well below sea level. They scoured deep valley systems near the edges of the Quaternary ice sheets in Scandinavia, Patagonia, and several regions in North America (the Pacific Northwest, Canadian Arctic, and North Atlantic). Retreat then produced numerous islands as ocean water replaced ice. The formation of Warming Island is a good modern example of this process.

Two other processes form new islands as ice sheets retreat. First, retreat causes uplift of the Earth's crust; removing the weight of the ice leaves a low-pressure zone in the underlying mantle, which is gradually filled by lateral flow of mantle rock. This causes the surface above to rise. This process—isostatic rebound—creates islands when topographic summits emerge from the surface of the ocean or a large lake. Second, when large ice sheets melt, the resulting rise of sea level drowns continental watersheds and converts topographic summits into islands. Many islands were formed by this process as sea level rose by approximately 130 m at the end of the last ice age.

SEE ALSO THE FOLLOWING ARTICLES

Global Warming / Island Formation / Sea-Level Change

FURTHER READING

Box, J. E., D. Bromwich, B. A. Veenhuis, L.-S. Bai, J. C. Stroeve, J. C. Rogers, K. Steffen, T. Haran, and S. Wang. 2006. Greenland Ice Sheet surface mass balance variability (1988–2004) from calibrated polar MM5 output. *Journal of Climate* 19: 2783–2800.

Henriksen, N., A. K. Higgins, F. Kalsbeek, and T. C. R. Pulvertaft. 2000. *Greenland from Archaean to Quaternary. Descriptive text to the geological map of Greenland, 1:2,500,000. Geology of Greenland Survey Bulletin* 185. (Geological Survey of Denmark and Greenland.)

Kaser, G., J. G. Cogley, M. B. Dyurgerov, M. F. Meier, and A. Ohmura. 2006. Mass balance of glaciers and ice caps: consensus estimates for 1961–2004. *Geophysical Research Letters* 33: L19501.

Oerlemans, J. 2005. Extracting a climate signal from 169 glacier records. *Science* 308: 675–677.

Rignot, E., and P. Kanagaratnam. 2006. Changes in the velocity structure of the Greenland Ice Sheet. *Science* 311: 986–990.

Rudolf, J. C. 2007. A peninsula long thought to be part of Greenland's mainland turned out to be an island when a glacier retreated. *The New York Times*, January 16, 2007.

Solomon, S., *et al.* Technical summary, in *Climate change 2007: the physical science basis*. Contribution of Working Group 1 to the Fourth Assessment Report of the Intergovernmental Panel on Climate Change. Cambridge: Cambridge University Press.

Steffen, K., S. V. Nghiem, R. Huff, and G. Neumann. 2004. The melt anomaly of 2002 on the Greenland Ice Sheet from active and passive microwave satellite observations. *Geophysical Research Letters* 31: L20402.

WATER RESOURCES

SEE HYDROLOGY

WHALE FALLS

AMY BACO

Associated Scientists at Woods Hole, Massachusetts

"Whale fall" is the general term for a sunken whale carcass resting on the sea floor. Whale falls pass through a series of successional stages ranging from scavenging of soft tissue through to a highly diverse chemoautotrophic assemblage, fueled by sulfides from the anaerobic breakdown of lipids contained in the whale bones. Whale falls have species overlap with hydrothermal vents and cold seeps and thus have been hypothesized to have played a role in the dispersal and evolution of vent and seep fauna. In addition to these shared species, whale fall communities have components of endemic and specialized fauna that are specific to whale falls.

HISTORY

Scientists have theorized that sunken whale carcasses might be important to deep-sea fauna since the 1930s, and whale bones with attached fauna had been brought up in fisheries' trawls for over 150 years. The first observed whale fall was discovered accidentally off southern California in 1987, during other deep-sea studies with the submersible *Alvin*. A second whale fall was discovered nearby during military mapping work in 1995. In addition to these two natural whale falls, a number of whale carcasses, from animals that died of natural causes, have been sunk for scientific study off southern California, central California, Japan, Sweden, and New Zealand. Fossilized whale bones with associated fauna have also been found, dating to 30–40 million years ago.

SUCCESSIONAL STAGES

When a fresh whale fall reaches the sea floor, it represents a food bonanza to an otherwise food-poor deep-sea fauna. The soft tissue of the carcass will be rapidly consumed by sharks, hagfish, crabs, and amphipods, at rates as high as 50 kg/day. This "mobile scavenger stage" slowly transitions to an "organic enrichment stage" in which millions of smaller animals and bacteria, in the sediments and on the bone surfaces, continue to break down and consume the putrid remnants of the whale soft tissues, eventually leaving the bones exposed (Fig. 1).

FIGURE 1 A whale fall that has been on the sea floor for 18 months. Hagfish swarm among the bones, looking for leftover bits of soft tissue. Photograph courtesy of Craig Smith and Mike DeGruy.

In addition to the highly labile carbon provided by the whale blubber, organs, and other soft tissues, the bones of whales contain large volumes of lipids (fats and oils), up to 60% by volume, which are slowly broken down by heterotrophic bacteria with inorganic sulfate as oxidizer, producing sulfides. These sulfides in turn fuel chemosynthetic production by various free-living microbes on the bones and endosymbiotic bacteria hosted by various invertebrate taxa living on or near the bones. This stage is termed the "sulfophilic stage" and is characterized by a highly diverse suite of fauna with a complex food web.

COMPARISON TO OTHER CHEMOSYNTHETIC ECOSYSTEMS

Whale falls, like other types of deep-sea chemosynthetic ecosystems (hydrothermal vents, cold seeps, sunken wood), are a type of deep-sea habitat island. Throughout each of the successional stages, the fauna found on the whale falls are very distinct in species composition and abundance from that of the surrounding sea floor. Abundant taxa in the sulfophilic stage include vesicomyid clams, bathymodiolin mussels, and siboglinid worms. These are the same taxa that characterize many vent and seep habitats. In fact, much of the fauna found at whale falls is similar to what is found at vents and seeps, with many of the same families and genera, and even some of the same species, being represented.

Unlike vents and seeps, whale falls are not restricted to certain types of geological features. They are also potentially abundant on the sea floor, estimated at one whale fall every 12–36 km, based on abundance and mortality of the nine largest whale species. Whale falls from the largest species of whales are estimated to harbor chemosynthetic fauna for up to 100 years. These factors, combined with their faunal similarities to vents and seeps, led to the hypothesis that whale falls may act as dispersal stepping stones for vent and seep fauna. There has been DNA sequence–based confirmation of species overlap in several taxa between whale falls, vents, and seeps, supporting this hypothesis. An interesting spinoff of these studies is that whale falls, along with other types of organic remains such as sunken wood, may have also played some role in the evolution of vent and seep fauna.

Although whale falls share species with vents, seeps, and sunken wood, they have much higher levels of biodiversity on a local scale than do vent and seep habitats, and also more than any other type of deep-sea hard-substrate habitat, with nearly 200 species being found on a single skeleton in the sulfophilic stage. Large whales have existed for sufficient time (30–40 million years) to plausibly have evolved a whale-fall specialist or endemic fauna, which may contribute to higher diversity levels. In fact there are close to 40 species (approximately 10% of the known species from whale falls) that have so far been found only on whale falls and in no other type of habitat. Sulfide, which fuels the whale fall communities, is also toxic to most animals. Whale falls have lower levels of sulfide than do vents and seeps, which may allow more of the background fauna to colonize, again contributing to the high levels of biodiversity.

SEE ALSO THE FOLLOWING ARTICLES

Cold Seeps / Hydrothermal Vents / Organic Falls on the Ocean Floor / Succession / Whales and Whaling

FURTHER READING

Baco, A. R., and C. R. Smith. 2003. High biodiversity levels on a deep-sea whale skeleton. *Marine Ecology Progress Series* 260: 109–114.

Baco, A. R., C. R. Smith, A. S. Peek, G. K. Roderick, and R. C. Vrijenhoek. 1999. The phylogenetic relationships of whale-fall vesicomyid clams based on mitochondrial COI DNA sequences. *Marine Ecology Progress Series* 182: 137–147.

Distel, D. L., A. R. Baco, E. Chuang, W. Morrill, C. M. Cavanaugh, and C. R. Smith. 2000. Do mussels take wooden steps to deep-sea vents? *Nature* 403: 725–726.

Rouse, G. W., S. K. Goffredi, and R. C. Vrijenhoek. 2004. *Osedax*: bone-eating marine worms with dwarf males. *Science* 305: 668–671.

Smith, C. R. 1992. Whale falls. *Oceanus* 35: 74–78.

Smith, C. R. 2007. Bigger is better: the role of whales as detritus in marine ecosystems, in *Whales, Whaling and Ocean Ecosystems*. J. Estes, ed. Berkeley: University of California Press.

Smith, C. R., and A. R. Baco. 2003. Ecology of whale falls at the deep-sea floor. *Oceanography and Marine Biology Annual Review* 41: 311–354.

Smith, C. R., H. Kukert, R. A. Wheatcroft, P. A. Jumars, and J. W. Deming. 1989. Vent fauna on whale remains. *Nature* 341: 27–28.

WHALES AND WHALING

JOE ROMAN

University of Vermont, Burlington

Whaling has impacted all species of great whales. It has also changed island ecosystems and cultures, spread invasive species and disease, and brought some island species, such as the Galápagos tortoise, to the brink of extinction. As hunting has declined, whale watching has risen as an important economic activity on many islands.

WHALE DISTRIBUTION

There are currently about 80 recognized species of cetaceans, divided into two suborders, the Mysticeti, or filter-feeding baleen whales, and the Odontoceti, or toothed whales, a category that includes dolphins, porpoises, and sperm whales. The great whales, a group based on the cultural history of whaling rather than systematics, are mostly composed of mysticetes—the gray whale (Esrichtiidae), right whales (Balaenidae), and rorquals (Balaenopteridae) such as humpbacks and blue whales—and one odontocete, the sperm whale *Physeter macrocephalus*.

The distribution of whales is linked to environmental features such as temperature and underwater topography. Many species of cetaceans are attracted to the coastal margins of islands and continents to feed, where highly productive habitats are found. Physical processes such as transverse circulation and tidal mixing can bring nutrients into the euphotic zone in these areas, enhancing primary and secondary productivity. Such productivity concentrates zooplankton, fish, and squid on shelf fronts, attracting whales, seals, and seabirds. Whales may also exploit island wakes, feeding on predictable aggregations of plankton and nekton along localized upwellings and fronts.

Islands provide shallow breeding areas for whales. For coastal species such as humpback and right whales, such waters are especially important. Many northern Pacific humpback whales, for example, migrate from high-latitude feeding grounds to the Hawaiian Islands in the central Pacific, where they gather to breed. In the North Atlantic, humpbacks travel from summer feeding areas in northeast North America, northern Europe, and Iceland to breeding grounds in the West Indies and Cape Verde Islands. Many species exhibit site fidelity to their nursery grounds, indicating that these are important, and perhaps limited, areas for reproduction. Such fidelity to feeding and calving areas has made whales vulnerable to human hunting and disturbance.

TRADITIONAL AND ABORIGINAL WHALING

Humans have hunted large cetaceans for millennia, and whaling has played an important role in the culture and ecology of several island groups around the globe. Archaeological evidence indicates that whaling cultures arose about 2000 years ago along the Bering and Chukchi Seas and on the coasts of Korea and Norway. In the North Pacific, Arctic and temperate aboriginal whalers used dugout boats or *umiaks* (large, open boats covered with skin), and harpoons and lances. Whaling for bowhead and gray whales changed societies around the Pacific Rim, supporting the establishment of permanent villages and trade networks. The hunt continued for centuries along the islands around the rim of the North Pacific, from Japan to Vancouver Island.

Some islanders developed their own traditions of hunting. Poison darts were used to hunt right whales off the Aleutians, and flax nets were employed by Japanese whalers to capture humpbacks, right whales, and other species. In the tropics, aboriginal hunting of humpbacks and sperm whales began in the Philippines and Indonesia around the 1600s, employing open boats and handheld harpoons or hooks.

In at least one case, prehistoric hunting continues to affect island ecology. Thule Inuit whalers in the High Arctic thrived on whale meat and built the structures of their winter settlements out of bowhead-whale bones (Fig. 1). Somerset Island, in what is now Nunavut, Canada, has the greatest concentration of these structures, which were occupied between AD 1200 and 1600. The towing of whales ashore, where they were flensed (stripped of their blubber), and the concentration of human activity changed the water quality and planktonic assemblages of the island's ponds. Although whalers abandoned the area more than four centuries ago, the legacy of these human disturbances is still evident in the present-day limnology of the island's ponds, characterized by elevated nutrient concentrations and atypical biota.

Subsistence hunting continues on some of these islands: Bowheads are hunted in Saint Lawrence Island in the North Pacific, humpbacks in the West Indies and in Tonga, and Bryde's whales in the Philippines. A small hunt for sperm whales and other odontocetes continues, with wooden boats and hand-delivered harpoons, on the Indonesian islands of Solor and Lembata.

COMMERCIAL WHALING

The development of whaling as a commercial industry is generally credited to the Basques, with the earliest known records from around AD 1000. The reach of whaling in the centuries that followed would touch almost every island: tropical and polar, continental and oceanic, those whose inhabitants had a history of whaling, and those who had only seen a cetacean as a distant blow or a windfall of meat and when it cast ashore. Svalbard, the Azores, Cape Verde, Saint Helena, South Georgia, Nantucket, the West Indies, the Galápagos, the Hawaiian Islands, New Zealand, and the Seychelles are a few of the islands central to the whaling trade.

In Basque-style whaling, open boats were deployed from shore or coastal ships, typically manned by six oarsmen, a captain, and a harpooner. Whalers hunted with hand-held harpoons and lances, and much of the processing of the oil was typically done onshore. Basque whalers were hired by other European nations, and their technique was employed around the rim of the Atlantic. Mysticetes were the first commercially exploited whales, first the North Atlantic right and then extensively the "common whale" or bowhead. One of the most productive whaling grounds in history was the region surrounding the Arctic island of Spitsbergen in the Svalbard archipelago. In 1599, the Dutch expedition of Willem Barents reported the presence of numerous whales in the area. Within a decade, the Basques and the British, Dutch, and Danes, often with Basques onboard, were competing for the bowhead whale, attractive for its long and valuable baleen ("whalebone"), and high oil yields. On Spitsbergen, as on islands before and after, shorelines became littered with the debris of the fishery—piles of blubber, bones, and guts—and lined with buildings to support the shore-based effort. Before the hunt for bowheads began, there may have been 500,000 of these polar whales in the Arctic. There are now probably less than 10,000, with numbers in the tens around Svalbard. The removal of this species from the marine ecosystem altered the food web in the area, from plankton to cod and seabirds.

The development of onboard tryworks in the eighteenth century enabled American and European whalers to process and cask oil at sea, opening up whaling grounds in the Pacific. They still used hand-thrown harpoons and petal-shaped lances or "killing irons" to finish off their quarry, but the primary target of American-style pelagic whaling was the sperm whale, with its clean-burning oil and spermaceti, which was made into candles. Pelagic hunting soon brought commercial hunters to remote islands around the world in search of provisions, shore leave, and a chance for some unhappy crewmembers to desert a luckless voyage or overbearing captain. Whalers collected firewood and seabirds and their eggs on isolated

FIGURE 1 The direct and indirect effects of whaling have altered the ecology of many islands. This house on Somerset Island, Nunavut, built from the bones of bowhead whales, dates back at least 400 years. The island's ponds retain high nutrient levels and unusual planktonic assemblages as a result of the centuries-old hunt of Thule Inuit whalers. Photograph by John P. Smol, Queen's University.

atolls in the central Pacific. They also helped spread infectious disease. Epidemics of tuberculosis, typhoid, influenza, and smallpox followed colonization and commercial shipping, including whaling, throughout Polynesia. Dysentery hit Tahiti in 1807 after the passage of the whale ship *Britannia*. As a result of these diseases, many Polynesian islands lost more than 80% of their population.

The Hawaiian Islands were an important stopover for nineteenth century whalers, providing freshwater, fruits, and vegetables. The presence of *kanaka* (Hawaiian men) willing to fill empty berths, and brothels full of welcoming *wahine*, made stops at Honolulu and Lahaina attractive to captain and crew alike. Further south, American, British, and French whalers set up hundreds of stations along the bays and open coasts of New Zealand to hunt for right whales.

The most famous whaling grounds in the Pacific in the nineteenth century were the Offshore Grounds (5–10° S, 105–125° W), west of the Galápagos Islands. Discovered in 1818, the grounds were visited by whalers in all seasons. The Galápagos were attractive for the provision of food and freshwater, but they were no tropical paradise. Herman Melville remarked on the "emphatic uninhabitableness" of the islands. "No voice, no low, now howl, is heard. The chief sound of life here is a hiss." There were some inhabitants: iguanas, birds, and, most prized to whalers, turtles. Galápagos tortoises were easy prey, supplied much-needed fresh meat, and could be stowed aboard ship for months. In 1848, the *Daniel Lincoln* took 273 tortoises from Chatham Island in just five days. With tens of thousands of tortoises removed by whalers, many islands were soon emptied of adults. Rats, introduced accidentally, removed the remaining eggs. Goats were intentionally introduced to the Galápagos, probably by New England whalers, to provide a larder for ships when they provisioned on the islands. Long after American whaling declined in the late nineteenth century, goats remain a problem, though eradication efforts are under way.

In the twentieth century, commercial whalers employed diesel-powered catcher boats and deck-mounted cannons that could kill even the swiftest whale. One of the first whaling stations used to exploit the Southern Ocean was founded in Grytviken on South Georgia in 1904. The island had already been emptied of seals in the eighteenth century, but Carl Anton Larsen, a famous Norwegian mariner and whaler, saw the potential as a whaling station. Not only did large populations of whales surround the island, but its shores also remained ice-free in the winter. The coastal humpback was soon depleted; fin whales and blues followed. To make the island more like home, Larsen had reindeer released on the islands, and several herds persist there to this day. When he visited the island in 1950, British surgeon R. B. Robertson described the station as "the most sordid, unsanitary habitation of white men to be found the world over." The abandoned station still causes environmental and health hazards, with asbestos and petroleum remaining among the ruins. After the rise of factory ships and the near depletion of many coastal species, whaling became almost entirely pelagic in the mid-twentieth century, with stern slipways allowing the carcasses to be hauled quickly onboard for processing.

We see the ramifications of whaling in the oceans today. Centuries of exploitation have greatly reduced cetacean populations in the Northern Hemisphere. Four hundred years after the Svalbard fishery began, the horizon lines off these Arctic islands are bereft of the once-abundant V-shaped blows of bowhead whales. Right whales in both the North Atlantic and North Pacific number just a few hundred individuals. Gray whales in the east Pacific now number more that 25,000, yet only a remnant population of about 100 whales survives along the western rim of the ocean. In the Southern Hemisphere, more than 2 million whales were killed in the twentieth century. Humpbacks and southern right whales remain rare, and blue whales show little sign of recovery.

We also see the impacts of whaling on islands, in the extinction of endemic species such as turtles, in the persistence of intentionally and unintentionally introduced species, and in the declining health of Hawaii's birds. The crew of the British whaling ship *Wellington* has frequently been blamed for introducing the first *Culex* mosquitoes to Hawaii, in drinking water from Mexico in 1826. The subsequent epidemics of avian malaria and poxviruses that broke out among native birds may have been initiated with this ship. Although several authors challenge the specificity of the event, there is no doubt that the frequent trafficking between islands by whale ships inevitably spread species and disease. American whalers have also been implicated in the spread of mosquitoes to Australia and many other Pacific islands. (In one case, whales played a minor role in the *eradication* of an invasive species. When the Asian citrus blackfly, *Aleurocanthus woglumi*, was found in Key West, Florida, in 1934, an emulsion of whale-oil soap, paraffin oil, and water was sprayed on infected trees. The fly never reached the other keys.)

Industrialized whaling also had impacts on the ocean ecosystems themselves. One of the most highly productive areas in the Northern Hemisphere was the Aleutian arc, an oceanographic hotspot that attracted whales and later whalers. More than 60,000 fin, sei, and sperm whales were killed in this area in the 1940s through 1960s, while they were on their summer feeding grounds. (Right whales, bowheads, humpbacks, gray whales, and blues had already been depleted decades earlier.) The near extirpation of most whale species may have prompted a dietary shift in the largest mammal-eating odontocete, the killer whale. As killer whales switched from a diet of offshore whales to sea otters and Steller sea lions, they may have caused the decline of both nearshore species in the Aleutians.

Since the moratorium on commercial whaling was imposed in 1986, direct impacts from hunting have been greatly reduced. Only a few island nations continue industrial whaling. Iceland resumed commercial whaling in 2006, despite protests from Great Britain, the United States, and other nations. Japan uses a loophole in the International Whaling Commission's bylaws, selling meat that has been collected for what it claims to be a scientific hunt. Both countries have a long tradition of whaling, dating back at least a thousand years in surrounding waters. Japan's contemporary, and controversial, whale hunt is now largely conducted in the Southern Hemisphere.

Once viewed as goods, whales are widely recognized for the ecosystem services they provide, largely through tourist operations. Hawaii, many islands in the West Indies, the Canary Islands, the Galápagos, and Iceland all have thriving whale-watching industries. Despite the shift away from commercial whaling, humans continue to impact cetacean populations. Whales are killed in collisions with ships and after being entangled in fishing gear. Seismic air guns and sonar can cause internal hemorrhages and gas-bubble disease, or the bends. Persistent anthropogenic noise from ships and active sonar reduces the ability of whales to communicate. Even whale watching itself can change whale behavior, feeding patterns, and mating activity.

WHALE ISLANDS

One persistent and fascinating connection between whales and islands is the mythology of whales *as* islands. The whale-island motif dates back at least 2000 years and can be found among the folklore of maritime cultures around the world. Two of the most famous tales involve St. Brendan the Navigator and Sinbad. On his epic voyage in search of Tir na nÓg (translated as the Enchanted Isles or the Land of Promise), Brendan and his fellow monks set foot on a barren island on Easter Sunday to say Mass. As he prayed at the altar, the monks lit a fire to cook breakfast, and the ground stirred beneath them. When they reached their boat, the island swam away. The deceptive island was in fact an enormous whale, who would later offer his back to help the monks on their Atlantic journey (Fig. 2). On Sinbad's first voyage, he and his fellow merchants also land

FIGURE 2 Stories of whale islands date back at least 2000 years. During St. Brendan's legendary Atlantic voyage in the sixth century, he and his fellow monks stopped at a barren island to say Mass. After they lit a fire, the island—actually a whale—awoke, tossing the pilgrims into the sea. Image courtesy of the New Bedford Whaling Museum.

on a small island and light a fire. But for Sinbad, who loses his treasure when the waking whale plunges into the ocean, the elusive cetacean is an obstacle to his pursuit of wealth. Sinbad's mistrusting view, and the whaler's dogged pursuit of cetaceans, would dominate the relationship between humans and whales for centuries. By the late twentieth century, however, as perceptions of cetaceans changed, Brendan would become the patron saint of whales.

Such tales are more fancy than fact, of course. Although right whales can be seen logging at the surface in calm days at sea, it is hard to imagine anyone mistaking the blubbery slab for an actual island. Yet there are real whale islands. After death, whale falls in the deep sea attract scavenger assemblages that recycle soft tissue over a short time. These carcasses serve as habitats for numerous endemic species, including polychaetes, molluscs, and other chemoautotrophic organisms. They also provide nutrients for scavenging crustaceans, fish, and sharks. A large whale can support successional communities for at least 80 years before it decomposes completely. Just as whaling changed island ecosystems, it may have endangered these communities before they were even discovered.

SEE ALSO THE FOLLOWING ARTICLES

Bird Disease / Galápagos Islands, Biology / Iceland / Spitsbergen / Whale Falls

FURTHER READING

Ellis, R. 1991. *Men and whales.* New York: Knopf.
Estes, J. A., D. P. Demaster, D. F. Doak, T. M. Williams, and R. L. Brownell Jr., eds. 2006. *Whales, whaling, and ocean ecosystems.* Berkeley: University of California Press.

Perrin, W. F., B. Würsig, and J. G. M. Thewissen, eds. *Encyclopedia of marine mammals*. 2002. San Diego, CA: Academic Press.

Reeves, R. R. 2002. The origins and character of 'aboriginal subsistence' whaling: a global review. *Mammal Rev*iew 32: 71–106.

Roman, Joe. 2006. *Whale*. London: Reaktion.

Tønnessen, J. N, and A. O. Johnsen. 1982. *The history of modern whaling*. R.I. Christophersen, trans. Berkeley: University of California Press.

WIZARD ISLAND

DAVID W. RAMSEY

U.S. Geological Survey, Vancouver, Washington

Rising steeply above the water like a sorcerer's pointed hat, Wizard Island is the most prominent and recognizable feature in Crater Lake, Crater Lake National Park, Oregon. Crater Lake, the deepest lake in the United States and the seventh deepest lake in the world (594 m depth relative to the shoreline), partially fills the caldera that formed approximately 7700 years ago by the eruption and subsequent collapse of an approximately 3700-m volcano called Mount Mazama. Since the climactic eruption of Mount Mazama, there have been several less violent, smaller post-caldera eruptions within the caldera itself. Wizard Island is one of four known volcanic vents within the caldera and is the only postcaldera volcano visible above the surface of Crater Lake.

MOUNT MAZAMA AND CRATER LAKE

Mount Mazama, a major, ~3700-m andesite-dacite stratovolcano in the Cascade Range, collapsed during a climactic eruption approximately 7700 years ago, leaving an 8 × 10-km caldera, which is now partially filled by Crater Lake (Fig. 1). Prior to the climactic event, Mount Mazama had a 400,000 year history of cone-building activity comparable to that of other long-lived Cascade volcanoes such as Mount Shasta. Since the climactic eruption, volcanism has been confined to the caldera, where most of the products of post-caldera eruptions are obscured beneath Crater Lake's surface. Exploration of the lake floor by remotely operated vehicles (ROV) and the manned submersible *Deep Rover,* dredged and cored samples, and a recent high-resolution multibeam bathymetric survey of the entire lake nearly to its shoreline have revealed the geology, geomorphology, and post-caldera eruptive history of Crater Lake. These surveys also discovered evidence that Crater Lake filled rapidly, has remained at or near its present level for an extended period of time, and maintains this level through a balance of precipitation, evaporation, and leakage into the subsurface, especially through permeable glacial sands and gravels found along the waterline in the northeast caldera wall.

GEOLOGY AND FORMATION OF WIZARD ISLAND

Wizard Island is one of four known volcanic vents within the caldera and is the only post-caldera volcano visible above the surface of Crater Lake. However, the 1.6-km² visible cinder cone and blocky lava flows of andesitic Wizard Island volcano represent only 2.4% of the volume of the total edifice, which rises around 750 m above the lake floor, has an areal extent of 9.0 km², and has a volume of at least 2.6 km³ (Fig. 2). Based on recent modeling of the history of lake filling, it is believed that Wizard Island began erupting within a few decades of caldera collapse and ceased erupting about 200 to 500 years later.

The subaqueous flanks of Wizard Island consist of relatively flat slopes (2–10°) of lava that transform abruptly to steep talus slopes (29–36°) of broken lava fragments, representing former shorelines where subaerial lava flows entered water, chilled, and fractured, forming slopes of subaqueous breccia (talus) (Fig. 2). The transition from

FIGURE 1 Panoramic photograph of Crater Lake and Wizard Island taken from the visitor overlook on The Watchman by Peter Dartnell. Digital photographic processing by Eleanore Ramsey. View is to the east.

FIGURE 2 Digital perspective view of generalized geologic map of the lake floor draped over shaded-relief image of 2-m bathymetry (after Ramsey et al., 2003). Light greens in foreground highlight lava flows and breccia slopes of Wizard Island volcano that are obscured beneath Crater Lake.

subaerial lava to subaqueous breccia is called a passage zone. The presence of passage zones on Wizard Island volcano indicates that the edifice was actively growing as Crater Lake was filling. Preservation of successive passage zones on the flanks of Wizard Island that would otherwise have been overridden by younger lava implies changes in the source vent location for the edifice or a decrease in eruption rate in comparison to lake-level rise.

ECOLOGY OF WIZARD ISLAND

Wizard Island is separated from the crater walls on its west side by narrow Skell Channel, only about 300 m wide and less than 100 m deep (Fig. 3). The channel is easily breached by avian and aquatic fauna as well as the seeds of vascular plants that are blown to Wizard Island by the wind or carried there by seed-eating birds. Yet only around 100 species of the nearly 600 vascular plant species found in the rest of Crater Lake National Park are found on Wizard Island. Unstable or youthful substrate development on a relatively young volcano, temperature extremes, a short growing season, and soil moisture availability limit plant development and stratify plant communities on the sparsely vegetated island.

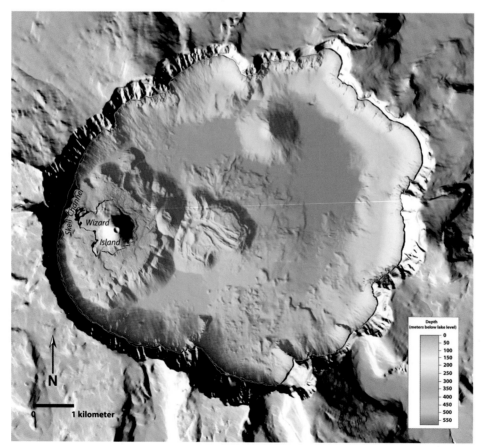

FIGURE 3 Shaded-relief image of the lake floor color-coded by depth (after Gardner et al., 2001). Narrow Skell Channel separates Wizard Island from the caldera walls.

Limitations on the vascular plant community are reflected in the wildlife found on the island, especially the birds. Absent or scarce are birds that feed in meadows, such as juncos, robins, and sparrows, because of the lack of substrate suitable for the development of meadow flora. Common are forest-inhabiting species such as Clark's nutcracker, Stellar's jay, nuthatches, and chickadees. The island is forested by eight conifers: mountain hemlock, Shasta red fir, lodgepole pine, western white pine, ponderosa pine, sugar pine, subalpine fir, and whitebark pine. The hemlocks and firs dominate the lower flanks of the island, where soil moisture is greatest, whereas the crater rim is dominated by whitebark pine. Harsh conditions at the crater rim contribute to high mortality rates in whitebark pine and may leave the stressed trees susceptible to attack by the mountain pine beetle, dwarf mistletoe, and white pine blister rust, which have reduced whitebark pines on Wizard Island in recent years. Herbs and shrubs on the island exhibit similar stratification, with Davis's knotweed being the most abundant plant on the unstable cinder slopes, but being conspicuously absent from the lower flanks, where Crater Lake currant thrives. The flowering vegetation on the island feeds nectar-eating rufous hummingbirds.

Wizard Island is home to many insects, including ants, bees, butterflies, dragonflies, and spiders. Frogs and toads also inhabit the island, as do garter snakes. Mice, pikas, minks, golden-mantled ground squirrels, two species of chipmunk, and bats are the only mammals living on Wizard Island. Deer and bears have been reported to have visited the island. Non-winged mammals most likely arrived on the island by swimming across Skell Channel or by crossing to the island by land during a time of lower lake level when Wizard Island was erupting subaerial lava flows across Skell Channel.

SEE ALSO THE FOLLOWING ARTICLES

Eruptions / Frogs / Insect Radiations / Lakes, as Islands / Lava and Ash

FURTHER READING

Applegate, E. I. 1934. The flora of Wizard Island. *Crater Lake Nature Notes* 7: 7–8.

Bacon, C. R., J. V. Gardner, L. A. Mayer, M. W. Buktenica, P. Dartnell, D. W. Ramsey, and J. E. Robinson. 2002. Morphology, volcanism, and mass wasting in Crater Lake, Oregon. *Geological Society of America Bulletin* 114: 675–692.

Bacon, C. R., and M. A. Lanphere. 2006. Eruptive history and geochronology of Mount Mazama and the Crater Lake region, Oregon. *Geological Society of America Bulletin* 118: 1331–1359.

Campbell, B. 1934. The birds of Wizard Island. *Crater Lake Nature Notes* 7: 6.

Count, E. W. 1932. Wizard Island. *Crater Lake Nature Notes* 5: 4–5.

Gardner, J. V., P. Dartnell, L. Hellequin, C. R. Bacon, L. A. Mayer, M. W. Buktenica, and J. C. Stone. 2001. Bathymetry and selected perspective views of Crater Lake, OR. U.S. Geological Survey Water Resources Investigation Report 01-4046.

Huestis, R. R. 1937. Mammals of Wizard Island. *Crater Lake Nature Notes* 10: 35–36.

Jackson, M. T., and A. Faller. 1973. Structural analysis and dynamics of the plant communities of Wizard Island, Crater Lake National Park. *Ecological Monographs* 43: 441–461.

Jones, J. G., and P. H. H. Nelson. 1970. The flow of basalt lava from air into water: its structural expression and stratigraphic significance. *Geological Magazine* 107: 13–19.

Nathenson, M., C. R. Bacon, and D. W. Ramsey. 2007. Subaqueous geology and a filling model for Crater Lake, Oregon. *Hydrobiologia* 574: 13–27.

Ramsey, D. W., P. Dartnell, C. R. Bacon, J. E. Robinson, and J. V. Gardner. 2003. *Crater Lake revealed*. U.S. Geological Survey Geologic Investigations Series I-2790.

Z

ZANZIBAR

N. D. BURGESS
University of Cambridge, United Kingdom

R. A. D. BURGESS
Cambridge Regional College, United Kingdom

The tropical archipelago of Zanzibar comprises two large islands and 53 smaller ones. It is located in the Indian Ocean at longitude 39° E and latitude 6° S. The larger island has traditionally been referred to as Zanzibar Island but is known locally as Unguja. It is 35 km offshore of the Tanzanian mainland, from which it is separated by the Zanzibar Channel, and it is distinct from its sister island of Pemba, which is located further north across the deep-water Pemba Channel (Fig. 1). These islands are a part of a global center of species diversity in the terrestrial and marine realms, with significant local endemism. This diversity reaches its peak in the forests on land and in the coral reefs in the ocean. Species diversity is strongly influenced by the process of island formation and a long history of human habitation, land holding, development, and use. These islands' historical importance as an entrepôt and their contemporary engagement in tourism defines their remaining habitats, endemic species, and natural resources and will influence their survival.

BACKGROUND

Land coverage of Unguja (hereinafter referred to as Zanzibar) is 1666 km²; the main settlement area (Zanzibar Stone Town), developed around a sheltered harbor on the west, is now a World Heritage site. The main language spoken is Kiswahili, with English used in commerce and business

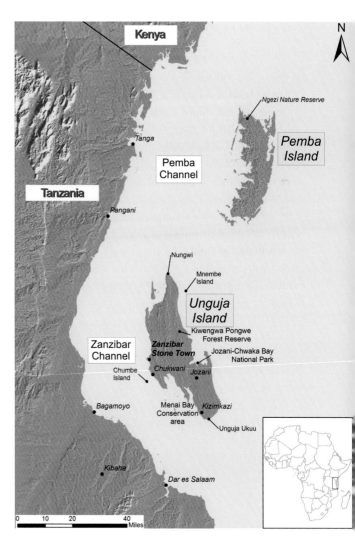

FIGURE 1 Map indicating the position of Unguja (Zanzibar) and Pemba Islands off the coast of mainland Tanzania and showing the location of the key sites for conservation and human history on these islands.

organizations. The origin of the name is disputed but could well have come from the Persians (Zanzibar meaning the "coast land of the black"; Zenj or Zangh meaning Negro), although Omani Arabs state that Zayn Zal Barr ("Fair is the Island") is more apt. The name could also be a reference to the production of ginger (genus *Zingiber*) traditionally grown on this "spice" island. There is no doubt of Zanzibar islands' importance as a main trading port throughout history on the East African equatorial coast. The official human population (2002 census) is 981,754; of these, 97% are Muslim (predominantly Sunni), and the remaining 3% are Christian, Hindu, or Sikh. Zanzibar's people, with their distinct features, owe much to Arab, Indian, and African influences. The Swahili language is primarily a blend of African tribal languages, especially Bantu, but around one-third of its words are derived from Arabic.

GEOLOGICAL HISTORY

Zanzibar is composed of uplifted sedimentary rocks, including coral reef limestone, and marine and fluviatile sediments. The uplifting that formed Zanzibar Island is associated with the rifting of East Africa, which started in the Middle Tertiary, some 30 million years ago. The sea level of eastern Africa has fluctuated considerably since then, with low sea levels at the end of the Oligocene around 26 million years ago and in the past 1 million years, when Zanzibar was connected to the mainland as recently as 100,000 to 10,000 years before present. Since the last period of uplifting, the coral limestone has been eroded by rainfall, creating karst features, such as underground caverns, sinkholes, and jagged rocky surfaces.

CLIMATE

Zanzibar has a tropical oceanic climate with little seasonal variation in day length (11–13 hours per day), temperature (between 15 and 39 °C, with a mean of 27 °C), or annual average humidity (71%), but with significant variation in rainfall, due to the northern and southern movements of the inter-tropical convergence zone (ITCZ), bringing rain in November and December (short and lighter rains) and March to June (long and heavier rains). Annual rainfall is between 800 and 1600 mm per annum, and although most falls during the rainy seasons, tropical storms can occur in any month.

BIOGEOGRAPHY AND BIOLOGICAL IMPORTANCE

Terrestrial

The islands offshore of the Tanzanian mainland, including Unguja and Pemba, are part of the so-called Zanzibar–Inhambane regional mosaic, which extends from southern Somalia through the coastal regions of Kenya and Tanzania into Mozambique. This distinct biogeographical region has been defined scientifically as a center of global endemism for plants and animals and has thus been prioritized for conservation attention by international and national conservation organizations. Some species in this biogeographical region have affinities to those further west in the Congo basin and in West Africa; others have affinities with other Indian Ocean Islands, but there are also many local endemics. Species with affinities to those in the western and central African rain forests indicate the former existence of a tropical forest belt across the entire African continent; this belt has now disappeared in East Africa as the forest has largely been replaced by savanna–woodland habitats except along the moister coast. Species with affinities to those of other Indian Ocean islands are often mobile, such as fruit bats or birds, or are plants that have been dispersed either by the ocean or perhaps by people.

Zanzibar would have originally been covered by various types of forest habitat. On the deeper and better soils, higher-canopy eastern African coastal forest would have predominated, with stunted scrub (coral rag) forests being found on the outcropping limestone and along the exposed eastern seaboard, and swamp forests occurring in freshwater wetlands. In sheltered marine bays, mangroves would have dominated.

The most biologically important habitats are the forests. However, the coastal forest habitats of Zanzibar have fairly low biological values when compared to the similar habitats on the Tanzanian mainland, or those of Pemba Island. The flora is rather impoverished, with four island endemic plants and few regional endemics. The island supports around 2500 individuals of the unique Zanzibar red colobus monkey (*Procolobus kirkii*) and under 1000 individuals of the dwarf antelope Aders's duiker (*Cephalophus adersi*), which is also known from coastal Kenya. Three butterflies are also confined to this island, as is the newly described amphibian *Kassina jozani,* and there are four endemic subspecies of birds (three shared with other islands) and mammals (including the Zanzibar leopard, which has not been seen in recent years).

Marine

The marine habitats of Zanzibar are a part of an eastern African marine region recently named the East African Marine Ecoregion, which extends from Somalia through Kenya, Tanzania, Mozambique, and northern South Africa. It is the most species-rich marine region of Africa and one of the most diverse marine areas in the world, but there are only a few locally endemic marine species. The main oceanographic influence on Zanzibar is the South Equatorial Current, which hits the East African mainland in northern Mozambique and then flows north as the East Africa Coastal Current and south as the Mozambique Current.

The seas around Zanzibar and Pemba contain a variety of habitats. These include expansive areas of coral reef and seagrass habitats, interspersed with areas of sand and mud. These marine habitats are among the most diverse in Africa and are part of one of the global centers of coral reef diversity. The waters around Zanzibar also support populations of green turtle (*Chelonia mydas*), hawksbill turtle (*Eretmocheles imbricata*), loggerhead turtle (*Caretta caretta*), olive ridley turtle (*Lepidochelys olivacea*), and leatherback turtle (*Dermochelys coriacea*). There are also healthy numbers of sailfish, blue and black marlin, dolphins, and migrating humpback whales.

HUMAN HISTORY

Archaeological excavations in Unguja Ukuu, in the south of Zanzibar island during the early 1990s, date evidence of settlement occupation (glass beads, pottery, worked metal objects) from between AD 400 and 600. Seen in the wider context of East Africa, Zanzibar Island may have been settled for millennia by the mainland African tribes especially the Hadimu and Tumbatu. It is known that the West Central African tribe, the Nyamwezi, pioneered routes to the Zanzibar coast and were very important carriers and traders, as documented early in the nineteenth century. By the mid-nineteenth century, Zanzibar Arabs financed by Indian traders had monopolized the trade in ivory and slaves and exchanged guns, small arms, cloth, and manufactured goods to areas beyond the shores of Lake Tanganyika and Lake Victoria.

Zanzibar's location was the prime reason for its increasing importance as a trading center. It is claimed that the Zanzibar Island was known to the classical world; the Greeks referred to it, and Roman merchant seaman sailing under Arab captains are thought to have provisioned and exchanged goods circa AD 80. The Omani Arabs from Muscat are documented as having arrived in the eighth century, and there is a tradition that relates how Abi Ben Sultan Hazan of Shiraz in Persia sailed from Bushehr in the Persian Gulf with seven dhows and landed unexpectedly on Zanzibar in 975.

The Portuguese (Vasco de Gama, 1499) visited Zanzibar on the journey from the Cape of Good Hope and were escorted by Arabs on the trade route to India. It is documented how they were astounded at the wealth of the Zanzibaris; from then on, the Portuguese began to dominate the trade routes by constructing forts along the mainland coast and on Zanzibar during the sixteenth and seventeenth centuries. It is also apparent that the Chinese traded with Zanzibar from well before this time. Later, the East Indian Trading Company capitalized on the knowledge of trade in this area.

Philanthropic and Christian efforts to stop the slave trade led to legislation by 1822, but this remained ineffective until the end of the nineteenth century, when this trade was finally halted. Contemporary evidence of this trade, which centered on the market in Stone Town, is now an integral part of cultural tourism of the area.

Historically, goods were traded to and from Zanzibar by sailing boats that took advantage of the seasonal wind patterns (trade winds or monsoons; Swahili: *kazkazi*), using the island as an emporium for various commodities and goods. From November to March, the hot northeast monsoon blows down the East African coast from Asia, a weather pattern that assisted sailing passage for oceangoing dhows from Aden, Hadhramaut, Muscat (Persia), and India laden with barter goods. At the end of March, the winds change to the southwest, and this assisted ships bound back to these ports laden with ivory, timber, charcoal, building poles, gum copal, spices, grain, and slaves. This trade, and the need for agricultural lands, decimated the forests of Zanzibar and also resulted in the depletion of elephant populations and forest lands well into the Tanzanian mainland. Even today, Zanzibar is a major transit port for logs cut from the coastal forests and woodlands of the mainland, for charcoal burned from the same forests and woodlands, and for building poles cut from mangroves.

The explosion of interest in natural sciences and exploration of mainland Africa by colonial powers in Europe and America (particularly Johann Krapf, David Livingstone, Richard Burton, John Speke, and Henry Morton Stanley) meant that in the late nineteenth century, Zanzibar was unrivalled in its importance as a starting and finishing point for expeditions into East and East Central Africa. A symbiosis developed between the Sultans of Zanzibar, the traders such as Tippo Tip (with their caravans, porters, and the knowledge of local tribes and provisioning skills), and the explorers. This cooperation led to the international advances in knowledge of the natural science of Zanzibar and mainland Africa and the economic resources of this region, and it also furthered the work of missionary societies in converting mainland Africans to the Christian religion.

Archaeological evidence from existing buildings illustrates many facets of Zanzibar's existence and development: the Shirazi Dimbani Mosque ruins at Kizimkazi dated 1107, the Old Slave Market and Anglican cathedral, Livingstone House, Kirk House, Guliani Bridge, Dhow Harbour, the Beit el Ajaid (House of Wonders), Beir al Sahel (People's Palace), Jamhuri Gardens, Mbweni Palace ruins, Portuguese Mvuleni ruins, Hamamni Persian baths at Kidichi, and the Bububu Railway. All point to the importance and wealth of this area, which was the greatest exporter of slaves, ivory, and cloves.

POLITICAL HISTORY AND LEGAL FRAMEWORK

The political history of this island has influenced the development of protected areas for the conservation of the biodiversity (marine and terrestrial, especially forests) of Zanzibar.

The dominance of the sultanate, the importance of trade, and the emergence of the main European powers in the late nineteenth century (French, German, and British) with the scramble for economic resources from Africa led to Zanzibar's status of British protectorate in the 1890s. The existence of the protectorate ensured that the British resident of Zanzibar was responsible to the London-based Colonial Secretary, a system that lasted until Zanzibar obtained independence in 1963. During this period, no protected areas were established on Zanzibar, whereas many were declared on mainland Tanganyika.

On December 9, 1961, Julius Nyerere obtained independence for Tanganyika (the mainland), and there was a growing political awareness on Zanzibar. On January 12, 1964, the government was overthrown, and this resulted in the flight of the Sultan (Jamshid ibn Abd Allah). By April 26, 1964, the Revolutionary Government of Zanzibar signed an Act of Union with Tanganyika to form a united republic; this was later renamed the United Republic of Tanzania, and the Revolutionary Government of Zanzibar is an integral part. However, the Revolutionary Government of Zanzibar has a distinct and separate legal system with its own assemblies, and it shares only the Court of Appeal of the United Republic with mainland Tanzania. Because of longstanding agreements, it is represented by an entitlement of 30% of the seats at the Union Assembly. Zanzibar's high court, zadhis courts, and magistrates courts serve the five district areas, which are further subdivided into ten administrative districts, including Pemba. The president of Zanzibar is also the executive of the United Republic of Tanzania and has several distinct political and statutory roles.

Up until the early 1990s, Zanzibar's separate status ensured that mainland Tanzanians could visit only with a passport and could not own property; now there is a greater influence from the Tanzanian mainland. However, Zanzibar is entitled to keep all its foreign exchange earnings, and so it remains, in part, economically independent. It is only in the past decade that conservation has become a focal area of work for the Zanzibar government, with a network of protected areas being developed to cover marine and terrestrial habitats and key endemic or severely threatened species.

CURRENT ECONOMY

The economy of Zanzibar is dominated by agricultural-related activities, but tourism has developed rapidly in recent years. Local people rely on subsistence agriculture and on harvesting natural resources for survival: wildlife, forestry, and fishing. Commercial crops include spices (cloves, vanilla, cinnamon, peppers, nutmeg, ginger) and tropical fruit (especially coconuts). There is some small-scale manufacturing, especially for the tourist trade and for trade with the mainland.

Zanzibar's exotic image, location, beaches, ocean, wildlife, and history have led to its growing importance for international tourism (particularly ecotourism) and associated services in retail and hospitality, located mainly in Stone Town and the coastal fringes. Current estimates (2006) place the number of visitors in excess of 100,000 per year, of which more than 20% visit the Jozani–Chwaka Bay National Park. Tourism activities include deep-sea fishing (especially for tuna, marlin, and shark), wildlife watching (dolphins, Zanzibar red colobus, and birds), and excursions to neighboring islands. Zanzibar is developing its tourism experience to include visiting its protected areas, globally important habitats, and unique species. This strategy aims to give value to these areas and thus ensure their conservation in the long term.

HUMAN IMPACTS ON HABITATS AND SPECIES

Over the past 2000 years of human expansion, natural habitats have been reduced to fragments and replaced by areas cultivated for food production and spice export. Zanzibar and Pemba Islands are famous for their plantations of spices such as cloves, cinnamon, nutmeg, pepper, ginger, and cardamom and tree crops such as coconut, jackfruit, banana, and citrus. Today, the largest patches of natural habitat on Zanzibar are found in the Jozani–Chwaka Bay region (southwest), in the Kiwengwa forest (northeast), and in areas of coral rag thicket as these cannot be cultivated. Because of the small remaining areas of habitat and the high levels of hunting, which have affected some species, populations of several of the key species on Zanzibar are either threatened by extinction or declining.

CONSERVATION

Forest conservation on Zanzibar is the responsibility of the Department of Commercial Crops and Forestry, and marine reserves are managed by the Department of Fisheries and Marine Resources. The protected area network is newly established and still expanding. It includes two marine reserves (Menai Bay, 476 km^2, and Mnemba Island, 0.6 km^2), one national park (Jozani–Chwaka Bay, 50 km^2), one forest reserve (Kiwengwa–Pongwe, 32 km^2), and one private island/marine reserve (Chumbe, 0.3 km^2). There have also been efforts to conserve the remaining populations of Zanzibar red colobus and Aders's duiker that are found in farmland areas unprotected by official reserves.

SEE ALSO THE FOLLOWING ARTICLES

Archaeology / Coral / Exploration and Discovery / Indian Region / Marine Protected Areas / Missionaries, Effects of

FURTHER READING

Burgess, N. D., and G. P. Clarke. 2000. *The coastal forests of eastern Africa.* Cambridge, UK: IUCN Forest Conservation Programme.
Burgess, N. D., J. D'Amico Hales, E. Underwood, E. Dinerstein, D. Olson, I. Itoua, J. Schipper, T. Ricketts, and K. Newman. 2004. *Terrestrial ecoregions of Africa and Madagascar: a continental assessment.* Washington, DC: Island Press.
Dale, G. 1969. *The peoples of Zanzibar: their customs and religious beliefs.* New York: Negro Universities Press.
Horton, M., and K. Clark. 1985. Archaeological survey of Zanzibar. *Azania: The Journal of the British Institute in Eastern Africa* 20: 167–171.
Ingrams, W. H. 1967. *Zanzibar, its history and its people.* London: Frank Cass.
Pakenham, T. 1991. *The scramble for Africa.* London: Abacus Books.
Sheriff, A. 1987. *Slaves, spices and ivory in Zanzibar.* London: James Currey.

GLOSSARY

The glossary that follows defines over 900 specialized terms that appear in the text of this encyclopedia. Included are a number of terms that may be familiar to the lay reader in their common sense but that have a distinctive meaning within these fields of study. Definitions have been provided by the encyclopedia authors so that these terms can be understood in the context of the articles in which they appear.

'a'ā Basaltic lava flow of typically higher viscosity, which cools to produce a very rough surface. (A term originating from the Hawaiian language.)

abiotic disturbance A non-living cause, such as wind or earthquake, that results in a change in the environment (perturbation).

aboriginal introduction A plant or animal intentionally or unintentionally introduced to an area by the original colonizers (as in the case of Oceania, by the Micronesians, Melanesians, and Polynesians).

accretionary wedge/prism/complex A mass of sediments that is scraped off the subducting plate at a convergent plate boundary and accreted or stuck to the non-subducting plate to form a mound-shaped mass.

adaptive radiation Evolution of ecological diversity within a rapidly multiplying lineage.

adaptive trait A characteristic of an organism that allows it to maximize its fitness.

adventive species Species in a particular location that arrived from elsewhere.

agamids A family of lizards that is spread worldwide and occurs especially in tropical and subtropical regions; mostly characterized by their very long legs and fast movement.

aggressiveness Of plant pathogens, the ability to invade and establish within the host fitness.

alien species See NON-INDIGENOUS SPECIES.

alkaline magmas Magmas that are relatively poor in silicon and rich in sodium and potassium.

allele One of two or more alternative forms of a gene.

allelic drift See GENETIC DRIFT.

allochthonous Formed elsewhere than its present place.

allopatric Occurring in separate, non-overlapping geographic areas.

allopatric speciation/allopatry The formation of new species that results when an extrinsic geographic barrier (e.g., a mountain or a sea strait between islands) prevents the interbreeding of two or more populations of a species.

allopolyploidy The condition of having two or more sets of chromosomes as a result of combining genomes from evolutionarily distinct lineages.

allospecies A pair of species that are each other's closest relatives occupying non-overlapping (i.e., allopatric) distributions.

alpine tundra An ecozone that does not contain trees because the climate (generally with no month having an average temperature in excess of 10 °C) limits tree growth.

altimetry Satellite-borne radars measuring small, permanent sea-surface topography caused by the gravitational attraction of the seafloor relief. These undulations may be inverted to yield information about seamount shapes.

ambergris A fatty substance secreted in sperm whale intestines. After long exposure to seawater and sunlight, it develops a dark gray color and pleasant odor and may wash up on island shores. Because it retains fragrance molecules, ambergris has been used since antiquity as a basis for perfumes.

ammonoids A group of extinct (since the end of the Cretaceous) cephalopod molluscs with an external shell divided internally into chambers that control flotation; related to the nautilus.

amphidromic point A location within a tidal system where the tidal range is near zero. Amphidromic points occur because the Coriolis effect alters current paths on a rotating Earth and because the topography of oceanic basins interferes with the free propagation of the tidal wave. The tidal wave pattern within an oceanic basin or bay rotating around an amphidromic point tends to have the appearance of spokes in a wheel.

AMS radiocarbon dates A method of radiocarbon dating in which an accelerator-based mass spectrometer is used to count all radiocarbon atoms found in a sample, thereby reducing the amount of material needed for a radiocarbon date.

amygdule A secondary deposit of minerals found in spherical, elongated, or almond-shaped cavities (vesicles) in igneous rock. The cavities are created by the expansion of gas bubbles or steam within lava.

anadromous Describing or referring to a life cycle in which fish breed in freshwater, and the progeny run to the sea where they live most of their life before returning to freshwater to breed. Also, SEA-RUN.

anagenesis Speciation by gradual genetic and morphological divergence due largely to genetic drift.

anastomosing Connecting in a network pattern, as, for example, branches of rivers, leaf veins, or blood vessels.

anatexis The partial melting of rock in the Earth's crust.

anchialine Describing or referring to marine or brackish water bodies that lack surface connections to the sea but with subsurface hydrologic connections; influenced by both marine and terrestrial ecosystems.

andesite A gray to black volcanic rock with between about 52 and 63% silica (SiO_2); produced at subduction zones by melting of wet mantle.

angiosperms Flowering plants.

anhydrobiosis See DESICCATION TOLERANCE.

ankaramite A basalt containing abundant (more than 10%) olivine and clinopyroxene.

Anomura An infraorder within the suborder Pleocyemata of the order Decapoda (crabs and shrimps).

anoxic Lacking oxygen.

Antarctic polar frontal zone/Antarctic convergence A circumpolar oceanic front in the Southern Ocean where cold, dense, north-flowing Antarctic waters sink beneath the relatively warm, less dense subantarctic waters. The zone is up to 100 km wide and variable in latitude.

anthropogenic Derived from human activities, as opposed to effects or processes occurring in natural environments without human influences.

anticline A convex fold in rock, the central part of which contains the oldest section of rock.

anurans The group of amphibians that includes frogs and toads.

aphotic Lacking light.

apomictic Of plants, producing seeds without pollination.

apozooxanthellate Of corals, ordinarily possessing zooxanthellae but having lost them as a result of environmental stress.

apterous Lacking wings.

arboreal Living in trees.

arc-continent collision A condition in which two tectonic plates, one of continental lithosphere and the other of volcanic arc lithosphere, converge and, because neither can be subducted, collision ensues.

archaeology The study of past lifeways by analysis of material remains.

archaeophyte A plant species that is not native to a region but that was introduced prior to AD 1500.

archipelago A group of geographically or geologically related islands.

Arctic The area north of the northern tree-line or north of the 10 °C July isotherm.

arc-type volcanic rocks Volcanic rocks that have the physical and chemical features characteristic of arc volcanoes.

arc volcanism A process creating arcuate island chains at the boundaries of convergent tectonic plates.

Area of Special Scientific Interest (ASSI) See SITE OF SPECIAL SCIENTIFIC INTEREST (SSSI).

Armorica A Roman name for the Brittany region of northwestern France.

arthropods The most diverse animal phylum (Arthropoda), which includes segmented animals with a chitinous exoskeleton and a series of paired articulated appendages.

artifact Any object made by humans, such as a tool or piece of pottery.

Artiodactyla Even-toed ungulates (hoofed animals). Members of this order, also known as the cloven-hoofed animals, have two hooves on each foot. Examples are goats, pigs, deer, and cows.

ash In volcanology, particles of volcanic glass, crystals, and rock that are less than 2 mm in size and are either ejected from a volcanic vent or formed by secondary fragmentation of volcanic material.

asthenosphere A weak layer of solid but ductile rock underlying the lithosphere, and on top of which the Earth's lithospheric tectonic plates move. It extends from the bottom of the lithosphere to depths of several hundred kilometers into the Earth's upper mantle.

atoll A ring-shaped oceanic reef formation surrounding a lagoon, caused by coral reef growth around a volcanic island that subsides into the ocean.

Australasian Pertaining to Australasia, the area including Indonesia, Papua New Guinea, and Australia, where Asian faunas may mix with Australian faunas across Wallace's Line.

Australo-Melanesians A diverse group of peoples with shared physical traits presently found in the Andaman Islands, in parts of Southeast Asia, in Australia, and in Melanesia. They include Australian aborigines, Melanesians, and Negrito populations scattered across the Philippines, the Malay Peninsula, and the Andaman Islands and are believed to be the descendants of the first wave of modern humans to populate the region, arriving from Africa via the southern coast of Asia around 65,000 years ago.

Austronesian 1. The family of languages (including more than 1200 modern languages) that extends from Madagascar to Easter Island and includes all of the Polynesian, Micronesian, island Melanesian, and Indonesian languages; believed to have originated in Taiwan. **2.** A person who speaks any of the Austronesian languages.

authigenic carbonate Carbonate rock created in situ through chemosynthetic activity associated with cold seeps.

autochthonous Formed in the place where it is currently found. *Autochthonous species* do not result from dispersal, but rather, have evolved in situ where they exist today.

avifauna The birds of a given region.

azooxanthellate Of corals, naturally without zooxanthellae; non-photosynthetic.

back-arc basin A small oceanic basin on the side of an island arc opposite to the trench and subduction zone; commonly a site for minor seafloor spreading. Also, RETRO-ARC BASIN.

backcross A cross between a hybrid and one of the parental types from which it was formed.

Baker's Law The theory that it is more likely for self-compatible than self-incompatible species to establish sexually reproducing colonies after long-distance dispersal. (Formulated by American evolutionist G. Ledyard Stebbins and named for British ecologist Herbert G. Baker.)

balsa 1. A Spanish colloquial term for a species of tree (*Ochroma lagopus*) with extremely light wood. **2.** A raft made from balsa wood. **3.** A pond or pool.

barrier reef A reef that is separated from the shore by an open-water lagoon.

basalt A common gray to black volcanic rock with less than about 52% silica (SiO_2); produced when magma erupts into either air or water.

basanite Dark-colored lava containing the minerals olivine, pyroxene, calcium-feldspar, and nepheline; a type of alkaline basalt.

basement Underlying or deeper rocks; a term used to distinguish cover rock sequences from underlying rocks. Typically, basement rocks are igneous and metamorphic rocks found beneath a sedimentary cover.

basin A feature of seafloor topography; a low part of the lithosphere lying between continental masses.

bathymetry 1. The measurement of underwater depth or topography. **2.** Terrain, or relief, of the seafloor.

beach nourishment An engineering technique used to slow beach erosion that involves placing a significant volume of sand derived from elsewhere on the eroding beach.

beach profile A cross section of the beach surface, commonly used to compare different beach types. Repeated beach profile surveys are one method used to measure changes in beach morphology.

beach ridge A relict inland shore ridge of a similar alignment to the modern beach. Beach ridge complexes are indicative of prograding coastlines.

beachrock A sedimentary rock formed in the intertidal zone through cementation of calcareous grains and skeletal fragments.

Benioff (or Benioff-Wadati) zone A deep active seismic area in a subduction zone. Differential motion along the zone produces deep-seated earthquakes, the foci of which may be as deep as ~700 km. They develop beneath volcanic island arcs and continental margins above active subduction zones. Also, WADATI-BENIOFF ZONE.

benthic Of or relating to the sea floor, including organisms living in or on the seabed.

Bergmann's Rule An ecogeographic phenomenon predicting that within species of homeothermic vertebrates (e.g., birds and mammals), large-bodied individuals will inhabit higher latitudes, or colder areas, whereas smaller conspecifics will inhabit lower latitudes, or warmer areas. (Formulated in 1847 by German biologist Christian Bergmann.)

berm A shore-parallel ridge or platform on the upper beach that marks the break in slope between the foreshore and backshore.

biodiversity The variety of living things in an area, measurable from genetic to ecosystem levels.

biodiversity coolspot An area (such as an atoll or other small island) that has very limited biodiversity inheritances, little or no endemism, and often with high cultural dependence on this limited biodiversity.

biodiversity hotspot An area with exceptional endemism of plants and/or animals as well as significant levels of habitat loss, as first defined by British ecologist Norman Myers in 1988 and 1990. Thirty-four hotspots are now recognized worldwide.

biogeographic disjunction The discontinuous distribution of closely related taxa.

biogeography The study of the geographic distribution of organisms.

bioherm A carbonate rock formation consisting of the fossilized remains of marine organisms such as coral, algae, and molluscs.

biological control/biocontrol The use of living organisms to control pest species or diseases.

biological species Groups of interbreeding populations that are reproductively isolated from other such groups, allowing them to evolve independently and accumulate trait differences.

bioluminescence The production of light by an organism through a chemical reaction, usually involving a protein (luciferin) and a catalyst (luciferase).

biomass The total mass of living matter, such as fish in a stock, within a given area; a biological measurement used to establish the importance of certain groups of living things in an ecosystem, as opposed to numbers of individuals.

biosecurity The protection and maintenance of ecological integrity.

biosphere reserve A UNESCO designation for a dedicated conservation area.

biota The entire assemblage of plants, animals, and other living organisms in a given region.

biotic disharmony A condition in which major high-order groups of plants or animals are absent from a particular biota; observed frequently in islands.

biotroph An organism that feeds on living hosts.

blackbirding Recruitment of laborers through kidnapping and trickery, especially for work on sugar plantations in Queensland and Fiji and in Peruvian mines; a practice that occurred primarily in the last half of the nineteenth century.

bleaching Loss of color in corals as they expel their symbiotic algae (zooxanthellae) under stress from increased water temperatures or changes in salinity, light availability, or sedimentation.

blueschist Metamorphic rock containing the mineral glaucophane; produced at low temperatures and high pressures when basalt or gabbro is subducted into the Earth's mantle.

bottleneck See GENETIC BOTTLENECK.

bovids Cloven-hoofed mammals of the family Bovidae, which includes cattle, sheep, goats, and buffaloes.

brachypterous Having small, nonfunctional, straplike wings.

Brachyura An infraorder within the suborder Pleocyemata of the order Decapoda (crabs and shrimps).

breccia A coarse-grained rock composed of angular rock fragments held together by a fine-grained matrix or cement.

breeding migration Migration by terrestrial crabs en masse over several kilometers from the land to the ocean for breeding, usually triggered by the first rains of the wet season.

broadcast spawning The release of gametes (eggs and sperm) into the water column.

Bronze Age A period in human cultural development when the most advanced metalworking led to a bronze alloy by melting copper and tin together, and casting them into bronze artifacts. In Greece and China the Bronze Age began before 3000 BC, in Britain around 1900 BC.

bycatch The nontarget species (e.g., unwanted marine species, juveniles of target species, and other marine wildlife, including seabirds) caught in fishing gear.

calcarenite A sedimentary rock formed on land by cementation of sand-sized fragments of calcium carbonate, derived by weathering of shells, calcareous algae, and coral.

caldera A wide, basin-shaped structure that forms abruptly at a volcano's summit as a result of collapse following a major eruption of magma.

California Current The southward flow of cool water from British Columbia to Mexico, extending offshore up to about 500 km; the current is the eastern edge of the North Pacific Gyre, the clockwise-circulating pool of water occupying most of the northern Pacific Ocean.

California Floristic Province A region of Mediterranean climate (winter-wet, summer-dry) and high floristic endemism in western North America (western California and southwestern Oregon in the United States and northwestern Baja California in Mexico).

canoe plant A plant, generally a useful one, that was carried by indigenous peoples, usually in a canoe (in the case of Oceania and the Caribbean), for planting in new areas that were being colonized.

canopy dieback See FOREST DIEBACK.

caracara A type of scavenging hawk found in South and Central America.

carrying capacity The maximum number of individuals that an ecosystem can support. Stocks above carrying capacity decrease in abundance, whereas stocks below it increase in abundance.

Caste War A rebellion of the Maya of the Yucatán Peninsula against the economically and politically dominant European-descended Yucatecos. The war lasted from 1847 to 1901, with skirmishes continuing until 1933. As a result, an independent Maya state developed in the southeastern part of the peninsula.

catadromous Of fish, living in freshwater and breeding in the sea.

catchment An area that collects and drains rain water.

Cathaysia The continental region assembled as a single landmass in tropical latitudes during the Carboniferous and Permian. The term was originally used to indicate an area of Sino-Malaya with a distinctive flora that includes *Gigantopteris*, but is now also used to identify a particular East Asian tectonic block that includes South China.

cay A small, low island composed primarily of coral or sand. (Called a "key" in American English.)

cay sandstone Lithified reef island sediments cemented by phosphate-rich cements, commonly associated with bird droppings (guano).

cenote A natural freshwater-filled sinkhole or dolina of karst origin. These geomorphologic structures are typical of limestone platforms in the Yucatán and are ritually significant to the Maya, representing the passage to the underworld. (From the Maya *dzonot*, meaning "sacred well.")

Cenozoic The most recent geological era, extending from ~65 million years ago to present; includes the Tertiary and Quaternary epochs.

centrifugal force A pseudo- or "fictitious" force that appears when a rotating reference frame is used for analysis of motion. The (true) frame acceleration is substituted by a (fictitious) centrifugal force that is exerted on all objects and directed away from the axis of rotation.

Chamorros The native Pacific island people of Guam.

chaparral The often dense evergreen plant community of drought-adapted shrubs and scrub oaks typical of Mediterranean-type climates in California and northern Baja. Similar growth forms occur in all Mediterranean-type climates.

character displacement An effect of competition or predation that results in morphological change within animals.

chelicerae (*singular* **chelicera**) The anteriormost pair of appendages in a spider, each one comprising a large basal part and a fang used to inoculate venom produced by modified salivary glands.

chemoautotroph An organism that uses chemosynthesis to make organic matter.

chemosynthesis The biological conversion of carbon dioxide to organic matter, as in photosynthesis, but using sulfide or some other inorganic molecule as the energy source, rather than sunlight.

chicle The natural sap of the chicozapote or sapodilla tree (*Manilkara zapota*; Sapotaceae), obtained in tropical rain forests of Mexico, Belize, and Guatemala by repeated tapping of trees. Chicle was the original source of chewing gum, replaced now by synthetic gum.

chuckawalla A large, herbivorous desert lizard found in the southwestern United States and in Baja California.

cinder cone A conical hill formed by accumulation of solidified bubble-rich droplets and clots of lava that fall around the vent during a single eruption.

circulation In the ocean, the movement of water masses over a reef system.

circumboreal Referring to plant or animal distributions at high, subarctic latitudes across both North America and Eurasia.

cirque A deep, steep-walled, half-bowl-like recess, located high on the side of a mountain.

clade A group of organisms descended from a common ancestor (monophyletic).

cladistics The classification of species into a hierarchy where groups are characterized by shared characteristics derived from a common ancestor (homologies), optimizing the recapturing of phylogeny (genealogy).

cladogenesis Speciation by splitting of the original gene pool and subsequent divergence, as commonly occurs during adaptive radiation in islands.

cladogram A branching diagram or family tree that groups species accordingly to recently evolved, shared traits. Only the branching pattern is represented; branch lengths do not represent time.

clast A single sediment unit in sedimentary rock.

climax For vegetation, the community representing the endpoint of succession in a particular climatic zone.

cloud forest A tropical or subtropical habitat type that occurs within a relatively narrow altitudinal zone, generally between 2000 and 3500 m on continents, though reduced to about 1000 m on some islands (e.g., Hawai'i), or even lower on small steep islands (e.g., Kosrae and Pohnpei in Micronesia), and is usually covered in mist and fog. The high moisture level and cool temperatures promote the growth of abundant mosses covering both ground and vegetation. Other characteristics include reduced tree stature and very high endemism. Also, FOG FOREST, MIST FOREST, MOSSY FOREST.

Cnidaria A phylum of predominantly marine invertebrate animals, all of which are carnivores; formerly termed Coelenterata.

coadapted genetic system The system of tightly linked genes that interact to allow high fitness within a local population.

coalescence time The total number of generations back to the common ancestor of gene copies in a present-day sample.

coalescent theory or **process** An approach that models the genealogical history of a set of sampled genes backward through time. Coalescent theory can be applied to many demographic models, including structured populations with migration or populations that change size over time, and can be used to make inferences about the historical processes giving rise to observed data.

coastal plain Low-lying and gently seaward-sloping plain along the coast, extending from the sea to elevated land; commonly depositional in nature and formed by shoreline progradation under relatively stable tectonic conditions.

coccolithophores Microscopic one-celled marine plants with calcium carbonate plates that live in the plankton.

coevolution Evolution of multiple species caused by interactions between the species.

cohort A generation of individuals having a statistical factor (such as age or size-class membership) in common in a demographic study.

cold seep A seafloor habitat where high concentrations of hydrocarbons are seeping out of the ocean floor; characterized by unusual communities of organisms dependent on chemoautotrophic production. As opposed to hydrothermal vents, cold seeps do not have a temperature anomaly compared to the surrounding water.

collective evolution A mechanism through which moderate levels of gene flow between differentiating species may increase the rate of divergence between them.

collisional orogenesis The process by which the Earth's crust is deformed and thickened, often resulting in the formation of mountains, as a result of the collision of continental and/or oceanic crustal material in a subduction zone.

colonization The successful establishment of a reproducing population of a species in a new environment.

colonization curve The change through time of numbers of species found together on an island.

commensal 1. Describing or referring to a relationship between two living organisms in which one benefits and the other is not significantly harmed or helped. 2. A member of such a relationship.

competitive exclusion An ecological theory stating that two species competing for the same resources cannot stably coexist, and one will therefore exclude the other.

composite volcano A steep volcano built by both lava flows and pyroclastic eruptions. Also, STRATOVOLCANO.

Conservation International The international environmental non-governmental organization that has identified 34 regions worldwide as biodiversity hotspots, covering 2.3% of the Earth's surface.

conservation of mass A law of physics that describes the balance of mass through convergences and divergences of water movements.

conspecific Belonging to the same species.

constraint In evolution, any impairment in the anticipated course of evolution as a result of the phylogenetic history of a lineage. In fact, species inherit from their ancestors developmental pathways that predetermine most of their design and functioning in a strict sense.

continental crust The basement rock for all continental land areas, normally between 35 and 50 km thick and characterized by granite, an igneous rock that is relatively enriched in the light elements silicon (Si) and aluminum (Al), and is therefore less dense than oceanic crust.

continental island An island on the continental shelf that is geologically part of the continent but, as a result of changes in sea level, is surrounded by water; if sea level drops, a continental island can become reconnected to the continental land mass.

continental plate One of the basic elements of the plate tectonics hypothesis. These plates underlie continents, whereas oceanic plates underlie major ocean basins. Because the material making up the continental plate contains light elements, it tends to ride up over a converging, heavier oceanic plate, creating young mountain ranges along the line of collision (its leading edge). The trailing edge of the plate is usually sinking slightly, which allows wide deltaic and coastal plains to form along that margin of the plate.

continental rocks The mixture of sedimentary, igneous, and metamorphic rocks that make up the continental crust.

continental shelf The region of submerged topography surrounding a continental island and leading to deep water.

convergent boundary/plate boundary/margin An actively deforming region where two tectonic plates move toward each other, usually involving one being thrust down into the Earth's interior and the other being thrust upward. In the oceans, the oldest (and densest) plate will subduct beneath the other plate, resulting in the formation of an island arc.

convergent evolution The development of similar adaptations by distantly related species living in different locations under similar environmental conditions.

Cope's Rule A theory that lineages of animals evolve to larger body sizes over their evolutionary histories; hence, descendant species will usually be larger than the species they evolved from. (Named after American paleontologist Edward Drinker Cope.)

coral rag Uplifted and eroded coral that supports desiccation-resistant thicket vegetation.

Coriolis force In motions observed in a rotating, rather than fixed, frame of reference, an apparent force that causes moving objects on the rotating Earth to be deflected from a straight path. Hence, freely moving objects on the surface of the Earth veer to the right in the Northern Hemisphere and to the left in the Southern Hemisphere. The Coriolis force does not appear when the motion is analyzed in an inertial frame of reference.

Cretaceous The geologic time span extending from about 145 million to 65 million years ago.

crown-of-thorns A sea star (Echinodermata) that preys on living coral, causing high levels of destruction of coral reefs when in outbreak phase.

crust The solid outer shell of the Earth, divisible into oceanic crust and continental crust.

cryoprotectors Substances synthesized within the animal body, such as glycerol, that have a protective function at low temperatures. They form stable hydrogen bonds with water molecules and thereby decrease the freezing point of any solution in which they are included.

cryptoturnover Turnover of species (extinction followed by immigration) that remains undetected because it occurred between two surveys.

cultural inheritance Traits that are learned from other organisms in the course of development and are passed on to future generations by learning without genetic encoding.

cumulate An igneous rock formed by the accumulation of crystals from a magma.

customary marine tenure The traditional rights of communities to regulate and manage access and use of marine resources.

dacite Light-colored volcanic rock made essentially of plagioclase and lesser quartz plus minor hornblende or biotite, and containing about 63 to 68% silica (SiO_2). Dacite generally erupts at temperatures between 800 and 1000 °C.

Darwinian fitness See FITNESS.

debris avalanche A sudden and rapid movement of rock and other debris, such as vegetation, driven by gravity; a fast-moving debris flow. May result from the collapse of the side of an oversteepened volcano or gravitational collapse of unconsolidated sediments.

debris flow A landslide of soil, sediment, or fragmented rock, saturated with water, that moves rapidly and over long distances with pervasive internal deformation and development of flow structures.

demersal Found on or near the bottom of an ocean or lake.

demographic stochasticity The variation in population growth rates arising from random differences among individuals in survival and reproduction.

density compensation On an island, compensation for the absence of mainland species by niche expansions and higher abundances such that the total population density of individuals of all species on islands equals the total mainland densities.

density dependence An effect (e.g., per capita birth rate or death rate) that increases in intensity as population density increases.

depauperate Having limited biodiversity.

depositional coasts Shorelines dominated by deltaic and coastal plains; the location of a majority of the major barrier islands of the world.

desiccation tolerance The ability of certain organisms to survive long periods of drought and to rehydrate fully when conditions improve. Also, ANHYDROBIOSIS.

deterministic extinction Extinction that occurs when the birth rate of a species becomes less than the mortality rate. Typically, the habitat is destroyed (by humans) or changes (by natural succession), the number of predators or competing species increases, or the species' food or prey decreases.

Devensian The last glacial stage within the Pleistocene epoch of the British Quaternary from about 110,000 to about 10,000 years ago. Equivalent to the North American Wisconsin.

diabase A coarse-grained intrusive rock of basaltic composition.

diadromous Migrating between freshwater and marine environments during a fish's life cycle.

diamictite Fragmental rock with angular clasts in a mud matrix interpreted to be formed by rocks dropped from melting ice into marine muds.

diaspores Seeds, spores, or buds that can disperse and form new plants.

differentiated lavas Lavas formed by partial crystallization of basaltic magma. Segregation of the crystallizing minerals yields a residual (i.e., differentiated) magma. Typically, residual melts have higher SiO_2, Na_2O, and K_2O, but lower MgO than does the parental basalt. Examples of differentiated lavas are hawaiite, phonolite, and trachyte.

differentiation series Igneous rocks that are closely linked in space and time and are related by cooling and partial crystallization of a basaltic magma.

dike A vertical or steeply inclined sheet of intrusive rock.

dipterocarp Any member of the tree family Dipterocarpaceae, characterized by seeds having two "wings," which enable them to be dispersed by air currents.

direct development In frogs, abbreviated or truncated larval development such that embryos develop (usually terrestrially) entirely in the egg capsule and emerge as fully formed froglets (thus omitting the aquatic larval stage or "tadpole" stage of development).

disharmony The different balance of species composition on islands when compared to similar patches of mainland. Thus, *disharmonic biota*, *disharmonic fauna*, and *disharmonic flora* are included in this comparison.

disjunction Geographically outlying populations, separated from the majority of populations within a species range.

dispersal Movement of species away from an existing population, such that the process either maintains or expands the species' distribution. In many cases organisms have evolved adaptations for dispersal that take advantage of various forms of environmental kinetic energy, such as water flow in rivers or ocean currents, or wind.

dispersal ability The ability of a species to diffuse over new habitats.

divaricating habit A distinctive, tangled plant growth form consisting of many flexible, interlacing branchlets and small leaves. Some biologists interpret this growth form as a response to moa browsing, although the consensus of evidence points to a climatic reason.

divergence The accumulation of differences (in genes and in traits) among isolated lineages resulting from distinct evolutionary trajectories.

divergent plate boundary A boundary where two tectonic plates move away from each other. In oceans, such zones are generally marked by spreading ridges where new ocean floor is created.

diversification rate The rate of origin of new, evolutionarily divergent lineages (speciation) minus the rate of extinction of such lineages.

diversity The relative degree of abundance of wildlife species, communities, or habitats per given area.

DNA (deoxyribonucleic acid) sequence The order of the four nucleotide bases A (adenine), C (cytosine), G (guanine), and T (thymine) within a given strand of DNA.

DNA sequencing Analysis of the order of bases of a particular region in order to reconstruct evolutionary relationships and history of the species or group of species.

dome Lava extruded as a dome-like feature.

dominance 1. In ecology, the ranked order in terms of relative abundance (number or dry weight per unit area, usually in a square meter of habitat) of each species in a community. This ranking is often related to the importance of each species' role in an ecosystem process. 2. In genetics, a relationship in which one of a pair of alleles suppresses the expression, or dominates the effects, of the other (recessive) allele.

Drepanidiinae The subfamily name for the Hawaiian honeycreepers, a single group or clade of evolutionarily related bird species living in the Hawaiian Islands that are derived from cardueline finch ancestors, family Fringillidae. Some authors still prefer to categorize this group as a family, Drepanididae.

drift See GENETIC DRIFT.

drip pool A small pool underneath dripping water.

drusy crystals Minute crystals that form a continuous coating over a surface.

dugong A marine mammal closely related to manatees (order Sirenia).

dunite A coarse-grained rock consisting primarily of the mineral olivine $(Mg,Fe)_2SiO_4$.

dynamic equilibrium 1. A condition of balance between opposing forces. 2. In the theory of island biogeography (developed by Robert H. MacArthur and Edward O. Wilson in 1967), the balance between rates of immigration (plus speciation) on the one hand and extinction on the other. Where these rates precisely balance, species richness remains constant through time although species composition is continually varying.

early-stage species Species newly colonizing an island, and therefore in the early stages of the taxon cycle.

Defined as having a widespread distribution in marginal, coastal habitats.

eclogite A type of rock consisting mostly of garnet and pyroxene, formed from basaltic rocks when they are subducted into the Earth's mantle.

ecological niche See NICHE.

ecological release Expansion of range, habitat, and/or resource usage by an organism when it reaches a community from which competitors, predators, and/or parasites may be lacking.

ecological sorting Determination of the membership of an ecological community through immigration, interaction, and extinction of species.

ecomorphology The study or classification of species on the basis of their ecological preferences and overall morphology. Repeated evolution of similar but unrelated ecomorphs on islands is a common theme in literature pertaining to the evolutionary biology of island life.

ecomorphs Populations or species whose appearance is determined by ecological factors.

ecosystem-based management A process that integrates biological, social, and economic factors into a comprehensive approach to ensure the sustainability, diversity, and productivity of an ecosystem.

ecotone A boundary between ecological zones.

ecotourism Specialized tourism where the primary attraction is the species or habitats of the area to be visited.

ectothermy Reliance by certain animals on the external environment for temperature control, as in most invertebrates, fish, amphibians, and reptiles.

edaphic Relating to the soil conditions.

edge In ecology, the amount of patch habitat that is in contact with contrasting environments. At the forest edge, there is an effect of physical (e.g., wind, sun) and biological conditions (e.g., exposure to predators and parasites) as compared to those found in the interior of the habitat patch.

edge effects Changes in abiotic and biotic conditions along habitat edges.

effective population size The number of breeding individuals in a typical population, usually smaller than the actual or census population size; an important statistic in population genetic studies.

effectively neutral Not showing an effect from the deterministic nature of natural selection (with beneficial alleles spreading and deleterious alleles declining) because natural selection is swamped by the effects of random sampling.

effusive eruption A volcanic eruption that occurs when magma reaches the Earth's surface and erupts passively, producing lava flows and lava domes; generally occurs when the gas content of the magma is low. Rocks formed during such an eruption are called *effusive rocks*.

El Niño A sea-surface temperature anomaly of greater than 0.5 °C across the southern tropical Pacific Ocean, occurring at irregular intervals of two to seven years and typically lasting one to two years. As warm water moves from the Indo-Pacific to the eastern Pacific, rainfall and storms increase along the eastern Pacific Ocean margin, and droughts become more prevalent in the western Pacific. (Originally named in the late 1890s by Peruvian fishermen who observed the changes in currents and weather associated with a small, warm coastal current that intensified during Christmas time and so named it "The Little Boy" for the Christ Child.)

El Niño Southern Oscillation (ENSO) The combination of the periodic warming of the sea surface in the central and eastern Pacific Ocean, El Niño, and the coincident negatively correlated variations in the sea-level pressures between the eastern South Pacific and the western equatorial Pacific, called the Southern Oscillation.

emergence In geology, the exposure of a coastline because of a relative drop in sea level.

emerging disease A disease circulating in a natural ecosystem that transfers to humans when the environment is perturbed.

endangered species Species that are considered in danger of imminent extinction and are usually formally listed as such by the U.S. Fish and Wildlife Service or other government agencies.

endemic Native to, and restricted to, a particular area, such as a mountain, island, or continent; found only there.

endemism The fraction of an area's biota that is found nowhere outside that area. Also, LOCAL ENDEMISM, MICROENDEMISM, REGIONAL ENDEMISM, SHORT-RANGE ENDEMISM.

endolithic Living inside rock.

endosymbiont Any organism that lives within the body or cells of another organism.

endosymbiotic Living inside the cells of another organism, typically in a mutually beneficial relationship.

endothermy The ability of some animals, such as mammals, birds, and some insects, to generate their own body heat from internal metabolic reactions.

enemy release hypothesis A hypothesis that attributes the success of introduced invasive species to the absence of natural enemies in the introduced range.

energetic equivalence rule The principle that smaller organisms, with reduced energetic needs, exhibit increased density relative to larger organisms with greater energetic needs.

enrichment opportunist An animal adapted to utilize conditions of organic enrichment, typically at the sea floor. Enrichment opportunists typically have high rates of colonization and production and are physiologically tolerant to habitat conditions associated with organic enrichment (e.g., low oxygen concentrations).

environmental stochasticity Variation in population growth rates arising from the influence of factors external to the population, such as weather, predation, disease, and competition.

Eocene The geologic epoch in the Paleocene period from about 56 million to 34 million years ago.

epicenter The point on the Earth's surface directly above the hypocenter of an earthquake.

epidemic disease A disease that develops and spreads rapidly over a short period of time.

epikarst The interface zone between soil and rock, characterized by fractures and small solution pockets; an ecotone between surface and subterranean water in karst.

epilithic Growing on stone.

epiphyte A plant that grows upon another plant, using it for support.

epithermal Referring to shallow depths (from surface down to 2 km) in the Earth, characterized by temperatures varying from 50 to 300 °C.

epithermal mineralization Mineralization associated with volcanic rocks usually forming within 1–2 km of the surface and typically dominated by gold in quartz vein systems.

epizootoic A disease outbreak among wild animals.

eradication In pest management, elimination from a site of all individuals of a species.

estuary A flooded river valley (lowstand valley) where freshwater and saltwater mix under the influence of rising and falling tides.

ethnobiodiversity The knowledge, uses, beliefs, management systems, classification systems (taxonomies), and language that a given culture, including modern scientific culture, has for biodiversity.

ethnobotany The study of the relationship between plants and people, including how plants are used and managed as food, medicine, housing materials, cordage, textiles, cosmetics, dyes, and other artifacts and practices that are a part of all cultures.

ethology The study of behavior in its natural context.

euphotic zone The upper layer of the ocean, where sufficient sunlight penetrates to drive photosynthesis.

eustatic Referring to global changes in sea level attributed to the ice ages.

evolution In biology, the change of inherited traits in successive generations, resulting from natural selection.

evolutionary attractor A value toward which things tend to evolve. If there is, for example, an evolutionary attractor at a body mass of 1 kg, then successive generations will be closer to 1 kg than their ancestors were.

exotic species See NON-INDIGENOUS SPECIES.

explosive eruption A volcanic eruption that involves the rapid expansion of gas, causing the surrounding rock or magma to fragment explosively. There are three types of explosive eruptions: magmatic, phreatomagmatic, and phreatic.

extensional tectonism The formation of structures associated with the stretching and thinning of the Earth's crust.

extinction In conservation biology, the total disappearance of all individuals throughout the range of a species.

extinction debt A situation in which, following habitat loss, conditions have become insufficient for survival of some species; these species are destined for extinction, yet are still extant because of the time delay in their response to the habitat loss.

extinction filter A reduction in the vulnerability of an assemblage to further species extinction as a consequence of previous exposure of that assemblage to the same threatening process.

extinction rate The number of species in a given area that become extinct per unit time.

extirpation Complete loss of a population of a species at a specific location, with the species continuing to survive elsewhere (as opposed to extinction, in which all populations of a species are lost).

extratropical cyclone A cold-core cyclonic storm of middle and high latitudes. Such storms include attendant frontal systems and are much larger than tropical cyclones.

extrusion The passive eruption of magma on the Earth's surface, either in the air or under water. Rocks and magmas formed by this process are termed *extrusive*.

facilitation In ecology, the enhancement of a process by other species or conditions.

facultative Able to exist with or without a given set of conditions.

fault A planar fracture in bedrock resulting from brittle failure under stress. A *strike-slip fault* is one in which the dominant displacement is horizontal. A *thrust fault* is low-angle and is the result of compression, allowing rocks to override one another. A *normal fault* is a fault in which the dominant displacement is vertical or high-angle.

fellfield A plant community characteristic of harsh cold climatic conditions where vegetation has less than 50% ground cover. Vegetation is sparse, low, and mainly composed of bryophytes, lichens, and small herbs.

feral Describing or referring to an individual of a domestic species that has shifted into a wild state and is no longer under the control of humans.

fernbrake Thick vegetation dominated by one or more fern species.

fetch The spatial region over which the wind blows when generating wave motion in the ocean.

fish In the context of fish stock management, not only vertebrate fish (finfish) but also molluscs, crustaceans, and all other forms of marine animal and even plant life.

fitness An individual's genetic contribution to the gene pool of the next generation relative to the average for the population. Also, DARWINIAN FITNESS.

floating islands Large portions of soil and vegetation removed and dragged away by the action of rivers and subsequently carried to neighboring islands by marine currents.

flocculation The process by which individual, minute, suspended particles are bound together to form aggregations, such as the settling out of suspension of clay particles in salt water.

flying fox A large diurnal bat of the family Pteropidae.

flysch Sediment deposited in deep water at a subduction zone.

focus See HYPOCENTER.

fog forest See CLOUD FOREST.

folivory Subsistence on a diet dominated by leaves.

foraminifera Amoeboid protists with a calcium carbonate shell. They may be either planktonic or benthic and are typically about 1 mm in diameter.

forb A broad-leafed non-woody plant.

forearc or **forearc basin** A depression on the sea floor located between the subduction zone and its associated volcanic or island arc. It is typically filled with sediments from the adjacent landmass and island arc in addition to trapped oceanic crustal material.

forest dieback or **decline** Stand-level loss of canopy foliage out of season; a vegetation process involving a structural dynamic change conditioned by chronic stress caused by aging and/or habitat constraints, and initiated by a trigger that usually remains elusive. Also, CANOPY DIEBACK.

Fossa Magna A rupture zone traversing the middle part of the Japanese islands and separating the Southwest Japan and Northeast Japan arcs.

founder effect Loss of genetic variation due to a colonizing population carrying only a small fraction of the total genetic variation of the parental population. Can lead to intense genetic drift and, in extreme cases, to speciation (founder effect speciation).

founder event Formation of a new population by one or a few individuals that are genetically, geographically, or behaviorally isolated from a source population; effective population size is much smaller than source population. May subsequently lead to genetic changes, yielding a founder effect.

founder flush A type of founder effect speciation that proposes a reduction in intraspecific competition and an increase in population size following a bottleneck.

fracture zone Linear fractures that extend lateral to transform faults on the ocean floor.

frankincense An aromatic resin derived from plants, used for creating a pleasant scent when burned. The frankincense tree, *Boswellia*, produces high-quality frankincense and has several endemic species on Socotra.

frequency-dependent selection The principle that a phenotype will experience variable selection pressure (e.g., predation), with fitness declining (negative frequency-dependent selection) or increasing (positive frequency-dependent selection) as the relative frequency of that phenotype increases.

fringing reef A reef system attached to a mainland or continental island shoreline.

frugivory Subsistence on a diet dominated by fruits.

fundamental niche The entire range of biological and physical environmental conditions under which a species can reproduce and survive.

gabbro A coarse-grained rock formed by intrusion of basaltic magma.

gene flow The movement of genes among populations as a consequence of organismic dispersal.

genetic bottleneck An event associated with a large decrease in the numbers of breeding individuals of a population. Also, BOTTLENECK, POPULATION BOTTLENECK.

genetic differentiation The accumulation of genetic differences between populations.

genetic drift Random changes in the frequency of alleles in a gene pool as a result of chance rather than natural selection; usually occurs with greater force in small populations. Also, DRIFT, ALLELIC DRIFT, RANDOM GENETIC DRIFT.

genetic recombinants New genetic combinations generated by the process of recombination via reassortment of chromosomes during meiosis and crossing-over.

genetic revolution A condition, generally associated with founder event speciation, in which coadapted sets of genes are broken apart and allowed to rearrange as a result of a change in the genetic environment of the population.

genetic transilience A type of founder effect speciation where reshuffling of alleles within an outbred founding population can lead to a shift in selective forces.

genetic variance The variance among individuals in a population resulting from the presence of different genotypes.

genotype The internally coded, heritable information of an organism.

ghost of interaction past A trait that is exhibited by an extant species and that evolved or was evolutionarily maintained by the interaction of that species with one or more other species that are now extinct.

gigantism An evolutionary trend toward increasing body size; common on islands where it is thought to be a result of reduced competition.

glacial An interval of colder temperatures and glacier advances (glaciation), in which the passage of large masses of slow-moving ice (glaciers) causes changes in the Earth's surface by erosion or deposition.

Gondwana A supercontinental configuration of Southern Hemisphere land masses that existed between Permian and Cretaceous times. All the continents of the hemisphere are fragments of the Gondwana continent that have been separated from each other by continental drift. (Named after the common occurrence of the distinctive "Gondwana sequence" in parts of India, Southeast Asia, Australia, Antarctica, South America, and Africa.)

graben A depressed block of land between two parallel faults.

gradient In climatology, the rate of change of a variable with distance; mathematically expressed as the rate of decrease.

granite Silica-rich intrusive igneous rock consisting mainly of alkali-feldspar and quartz.

granulite A metamorphic mineral assemblage formed when crustal rocks are metamorphosed at temperatures greater than 650 °C.

gravity anomaly A change in the strength of the Earth's gravitational field resulting from lateral differences in the density of materials.

greenhouse gases Atmospheric gases that contribute to global warming; they are, in order of abundance, water vapor, carbon dioxide, methane, nitrous oxide, and ozone.

Gruiformes A bird order (lit. "cranelike") containing the coots, cranes, and rails.

guano Droppings from birds, bats, or other vertebrates, rich in nitrogen and phosphorus.

Guiana shield A more-than-1-billion-year-old geological formation that underlies much of northeastern South America.

guyot An undersea mountain (seamount) with a flat top, considered to be an extinct volcano. (Named after Swiss-born U.S. geologist Arnold Henri Guyot.)

gymnosperms Non-flowering seed plants, including conifers and cycads.

habitat A place with a suitable environment (vegetation, food availability, and the like) for a particular species.

habitat island A distinct area of suitable habitat, surrounded by areas unsuitable for the species in question. Examples include forest patches in an agricultural landscape, mountaintops (for montane species), or coral reefs.

halieutic Of or referring to fishing.

halophilous Describing or referring to organisms that live in areas of high salt concentration; the organisms have special adaptations to permit them to survive in these environments.

halophytes Plants that are capable of surviving and growing in high salt concentrations.

hardground A limestone that has been cemented at a sea bed or lake bed, forming a hard and commonly irregular surface.

Hardy-Weinberg ratios The genotypic ratios (expressed in terms of the allele frequencies) expected in a large, isolated population given random mating, no mutation, and an equal fitness of all genotypes.

harmonic biota An island biota that has much the same composition as the nearest continental area.

harvest rate The fraction of a fish stock that is harvested in a given time period.

hawaiite A volcanic rock, a differentiated type of basalt with moderate sodium and potassium.

head Poorly sorted, poorly stratified deposits of locally derived angular rock fragments produced by solifluxion (i.e., downslope movement of soil material in which water acts as a lubricant rather than as an agent of transportation) under periglacial conditions; these deposits may mantle high ground or may occur on slopes or in valley bottoms.

henequen 1. A plant (*Agave fourcroydes*; Agavaceae), native to the Yucatán, whose leaves produce a fiber used primarily in cordage (ropes, cords, and twine). Synthetic fibers have now largely replaced its use. 2. The natural fiber derived from agave plants.

hermatypic A descriptive term for reef-building corals.

herpetofauna The reptiles and amphibians of a given region.

heterochrony Evolutionary changes in size and appearance caused by the differential timing of development of features.

Highland Clearances Forced displacements of people from the Scottish Highlands in the eighteenth century.

high-pressure/low-temperature metamorphics The metamorphic rocks formed in unusually low geothermal gradients, characteristic of metamorphism in subduction zones.

hinge rollback Movement of the subduction zone as subducted lithosphere descends into the mantle under the force of gravity. Subduction is often viewed as the result of plate convergence but can equally be considered as the result of one slab falling into the mantle with the upper plate extending to fill the space created.

Holarctic The biogeographic realm comprising arctic, boreal, and temperate regions of Eurasia and North America.

Holocene The geologic time span from about 11,500 years ago to the present.

holomictic lake A lake that is mixed completely from top to bottom.

holotype The specimen or specimens designated as representative of a newly described species.

hominin Any member of the tribe Hominini in the subfamily Homininae, which comprises all creatures believed to be modern humans or human ancestors; includes all species within the *Homo* and *Australopithecus* genera.

Homo erectus Latin for "upright man"; an extinct member of the human lineage that first evolved in Africa around 2 million years ago and lasted until about 400,000 years ago.

horizon A distinct spatial pattern in material culture.

horst A block pushed upward between faults.

host In biology, an animal or plant on or in which a parasite (including those that cause disease) or commensal lives.

hotspot 1. In geology, a small region of increased volcanic activity, thought to be caused by locally increased mantle melting due to increased temperature or lowered melting point, which produces a chain of islands (hotspot archipelago). 2. See BIODIVERSITY HOTSPOT.

Hoxnian Interglacial An interglacial period between 300,000 and 200,000 years ago.

Humboldt Current The cool current flowing northward along the west coast of South America; first scientifically described by German naturalist and geographer Alexander von Humboldt.

husking sites Sheltered sites used by rodents to strip inedible material (husks) from seeds and other collected food such as invertebrates.

hyaloclastite Rock made of fragmented volcanic glass.

hybridization Interbreeding between two different species (or populations), resulting in the production of offspring (a hybrid); may be either artificial or natural.

hydroexplosive zone The upper 600 m or so of the ocean, within which overlying water pressure is insufficient to prevent underwater volcanic eruptions from being explosive.

hydrology The scientific study of the behavior, distribution, and movement of global water (both liquid and solid) in the atmosphere and on and under the Earth's surface.

hydrothermal mineralization Mineral deposits created by the circulation of hot, watery fluids through the uppermost part of the Earth's crust.

hydrothermal vent A hot-water vent on the sea floor, occurring near volcanically active places and characterized by unusual, high-biomass communities dependent on chemoautotrophic production.

hypocenter The point at which a sudden breakage of rocks within the Earth starts, leading to an earthquake. Also, FOCUS.

Iapetus Ocean An ancient ocean, the opening and closing of which preceded the opening of the Atlantic Ocean. (Named for Iapetus, the mythical Greek father of Atlantis.)

ichthyofauna The fishes of a given region.

igneous rock Rock produced by solidification of magma on or beneath the Earth's surface.

ignimbrite A deposit formed as a result of an explosive eruption and subsequent deposition of hot clastic materials, such as ash.

immigration In the theory of island biogeography, the process of arrival of a propagule on an island. The fact of an immigration implies nothing concerning the subsequent duration of the propagule or its descendants.

immigration rate The number of new species arriving on an island per unit time.

impoverishment On an island, the reduction in the number of species or higher taxonomic ranks per unit area compared to the mainland.

indigenous Describing or referring to species occurring naturally within a specific geographic area, though they may also occur elsewhere. Also, NATIVE.

insectivory Subsistence on a diet dominated by arthropods.

intensity In seismology, a measure of the effect of an earthquake based on its macroseismic or "felt" impact; commonly reported according to the modified Mercalli intensity scale. A single earthquake can be associated with a wide range of effects and intensities that vary primarily as a function of distance away from the earthquake.

interfertility The capability to produce fertile progeny through interbreeding; often expressed at different levels depending on the proportion of viable spores or gametes produced.

interglacial An interval between glacial periods when climates are warmer, sea level is higher, and tree lines are higher on mountains.

internecine Relating to mutual violence within a larger group.

Intertropical Convergence Zone The boundary (convergence zone) between the northeast and southeast trade winds, seasonally shifting 10° north to south of the Equator.

intraplate Situated within the interior of a crustal tectonic plate.

intrinsic growth rate The rate at which a [fish] stock changes in size over a given time period.

introduced species See NON-INDIGENOUS SPECIES.

introgression The introduction of genes of one species into the gene pool of another through hybridization.

intrusion The emplacement of bodies of molten rock within the Earth's crust. Rocks and magmas formed by this process are termed *intrusive*.

invasional meltdown A process whereby two or more introduced species enhance one another's probability of establishment or impact on native species.

invasive Describing or referring to a species, usually non-indigenous, that expands its range, often causing negative environmental, economic, and/or human health impacts.

inversion An atmospheric layer in which temperature increases with increasing height; an exception to the normal decrease (or lapse) of atmospheric temperature with height.

inverted barometer The response of sea level to changes in atmospheric pressure (10 mb of pressure decrease corresponds to 0.1 m of sea level increase); a component of storm surge.

Iron Age A period in human cultural development, preceded by the Bronze Age, and marked by the development of tools and weapons made of iron. The adoption of this material coincided with other changes in some past societies, often including differing agricultural practices, religious beliefs, and artistic styles. In the ancient Near East and Greece, the Iron Age began in the twelfth century BC; in Great Britain it lasted from about the seventh century BC until the Roman conquest.

irruption Rapid and irregular increases in a species' abundance, which are often subsequently followed by rapid declines in their populations.

island arc An arcuate group of islands and seamounts formed by magma produced during the subduction of an oceanic plate beneath another oceanic plate. The subducted slab adds water, and decompressional melting generates basaltic magma that penetrates the overriding plate and forms volcanoes.

island biogeography A field of biogeography that attempts to establish and explain the factors that affect species richness in archipelagoes. The field began in the 1960s and was established by the ecologists Robert MacArthur and E. O. Wilson, whose theory of island biogeography attempted to predict the number of species that would exist on a newly created island. More recently, this theory has been extended to terrestrial fragmented systems.

islet Any small island, including any of many small islands comprising an atoll.

isometry A relationship in which one variable changes with another one while keeping the same proportion between the two (e.g., if heart weight doubles when body weight is doubled, then the relationship between them is isometric; but when body length doubles, surface area increases four-fold, and body weight increases eight-fold, the relationship between these variables is allometric rather than isometric).

isostasy The state of equilibrium of vertical forces acting on the Earth's crust.

isostatic rise or **subsidence** Changes in the elevation of a tectonic plate as its thickness or density changes (e.g., as by erosion or cooling) so as to maintain the overall gravitational equilibrium between the Earth's lithosphere and asthenosphere.

isotopic age An age expressed in years and calculated from the quantitative determination of radioactive elements and their decay products and known rates of decay.

IUCN The World Conservation Union, formerly the International Union for the Conservation of Nature and Natural Resources.

karst Carbonate rock terrain with complex erosional features, such as sinkholes and caves, formed through dissolution by water.

karstification The process by which an area of irregular limestone is eroded, producing fissures, sinkholes, underground streams, and caverns.

key innovation In evolution, any modification in structure or function that permits a lineage to exploit the environment in a more efficient or novel way and thereby leads to a comparatively rapid diversification of the lineage.

keystone species A species that usually plays a large role in community or ecosystem function relative to other species.

kīpuka An island-like area of older substrate, often supporting forest, that is surrounded by younger lava flows.

kīpuka systems Kīpuka, their surrounding lava flows, and their inhabitants.

lagoon A shallow body of water, separated from the sea by coral reefs or sandbars.

lahar An Indonesian term for water combined with volcanic rock debris deposited on the slope of a volcano that flows rapidly downslope.

lamprophyre Dark-colored igneous rock with large crystals of minerals such as hornblende, biotite, and alkali feldspar.

land-bridge island An island once connected to the adjacent land mass but later isolated by rising sea levels.

Landnám A Norse term (lit. "taking land") usually applied to the process of preparing the landscape for farming, such as forest clearance.

landscape A distinct area of land that may be occupied or unoccupied by natural vegetation; a geological landform with built-up environmental or human-constructed features; the geographical equivalent of ecosystems or interconnected ecosystems.

Lapita An archaeological complex marked by a distinctive style of earthen-made pottery decorated with dentate-stamped designs and found from New Guinea to Samoa during ~1500–500 BC. Makers of Lapita pottery are thought to have been the speakers of proto-Oceanic language and to have been direct ancestors of the Polynesians.

large organic fall A common term for a large parcel of organic matter, such as a dead whale, log, or detrital kelp mass, that has sunken to the sea floor.

last glacial maximum The maximum extent of ice sheet coverage during the most recent glaciation. Names for this period differ according to geographic distributions, the most intensively studied being the Wisconsin in North America, generally thought to have occurred 70,000–20,000 years ago, and the Devensian in the British Isles.

lateral collapse A very large landslide that removes a sector of the flank of a volcano to a depth of 0.5–3 km below the surface, and often also removes the summit.

lateral plates Modified scales that occur as a single row along the flanks of most sticklebacks. The plates mechanically link the spines on the back and pelvis of the threespine stickleback, producing a formidable defense structure.

late-stage species Species that have passed through several stages of the taxon cycle; they tend to have a narrow and fragmented distribution in interior or montane habitats.

laurel forest/laurisilva A previously widespread forest type that, in the Tertiary, occupied the area that is presently Europe and southwestern Africa.

lava A term for magma that has passively erupted onto the Earth's surface.

lava domes Roughly circular mounds of lava formed by the eruption of viscous (sticky) magma.

Lazarus phenomenon The reappearance of a species thought to be extinct, usually as a very small population, and sometimes in a previously undocumented place or habitat.

lecithotrophic Feeding on yolk supplied by an egg.

leptospirosis A disease caused by the bacterium *Leptospira* that can infect humans as well as wild and domestic animals, causing flulike symptoms. It is generally spread through contamination of soil or water supplies by the urine of infected animals.

Levallois The archaeological name for a type of flint knapping developed by humans during the Paleolithic.

limestone A sedimentary rock composed primarily of calcium carbonate ($CaCO_3$), usually as the mineral calcite.

Linnaeus, Carl The scientist who laid the foundations for binomial nomenclature (genus name followed by species name) and modern taxonomy in the mid-eighteenth century.

lithified Describing or referring to sediment that has been changed (over time and under certain conditions) into rock.

lithology The types and characteristics of the rocks of a region.

Lithornithiformes Early flying birds with a paleognathous palate, generally similar to tinamous.

lithosphere The outer, rigid part of the Earth, including the crust and part of the mantle to depths of about 100 km, forming the tectonic plates.

living fossils Species or species groups most closely related to a now totally extinct archaic lineage. Groups of living fossils usually have low taxonomic diversity.

local endemism See ENDEMISM.

loess A widespread, homogeneous, commonly nonstratified, porous, friable, slightly coherent, usually highly calcareous, fine-grained blanket deposit consisting predominantly of silt with secondary grain sizes ranging from clay to fine sand; formed as a windblown dust of Pleistocene age, and carried from land surfaces unprotected by a vegetative cover, such as glacial or glaciofluvial deposits uncovered by glacial recession.

lowstand valley A valley carved across coastal plains and the continental shelf when sea level was lower during the ice ages (Pleistocene epoch).

Macaronesia The biogeographic region comprising the archipelagoes of Azores, Canaries, Madeira, Selvagens, and Cape Verde and some parts of Morocco and Mauritania.

machair The land between a beach and the area where sand encroaches on peat bogs; former beach.

macrotidal Experiencing tides with a vertical range over 2 m.

mafic Composed chiefly of one or more ferromagnesian minerals such as olivine and pyroxene.

magma Molten rock that may include crystals and exsolved gases (bubbles); when magma erupts at the Earth's surface, it loses most of its gas and is called *lava*.

magmatic eruption An explosive volcanic eruption that occurs when dissolved gases in rising magma expand to form gas bubbles that then burst as the magma nears the Earth's surface, leading to explosive fragmentation of the magma.

magnetic anomaly A change in the strength of the Earth's geomagnetic field resulting from lateral differences in the magnetic mineral content of rocks.

magnetic lineation A linear pattern of magnetic anomaly that is preserved in rocks on the ocean floor.

magnitude A measure of the size of an earthquake, derived from seismic signals recorded on calibrated seismic instruments that are corrected for the distances between the earthquake and the recording instruments; the magnitude is a property of the earthquake rather than of its effects or intensities.

mainland species pool The whole set of species inhabiting a mainland that can potentially colonize a given island.

makatea 1. Fossilized coral, referring to a raised coral reef that encircles an island. 2. A type of island consisting of a volcanic center surrounded by an emerged (or uplifted) fringing coral reef. 3. A composite island (volcanic and limestone).

malacology The biological subdiscipline that deals with molluscs.

Malesia The geographic area that includes the countries of Indonesia, Malaysia, the Philippines, Papua New Guinea, Singapore, and Brunei Darussalam.

mammal-niche species Species other than mammals that occupy a similar ecological niche in their absence.

mangrove The tropical or subtropical vegetation that grows in the intertidal zone; often used more narrowly to refer to species in the family Rhizophoraceae, which frequently dominate mangrove vegetation.

mantle The ~2900-km thick part of the Earth between the crust and the core. It is divided on the basis of seismic wave velocities into upper mantle and lower mantle, separated by a transition zone.

mantle plume An upwelling of abnormally hot rock within Earth's mantle. Where such plumes interact with the lithosphere, they form volcanic provinces called hotspots.

mantle wedge Part of the Earth's convecting upper mantle (asthenosphere) overlying a subducted slab, composed of peridotite (olivine-pyroxene rock) but likely with heterogeneities established by prior melt loss and ingress.

maquis Evergreen xerophyllous and sclerophyllous shrubs in the Mediterranean region (analogous to chaparral in California, fynbos in South Africa, and mallee in Australia).

marginal basin A small oceanic basin that is adjacent to a continent and is separated from larger oceans by an island arc.

marine advection The dominant atmospheric process on islands because of their oceanic context. When the atmospheric circulation is constant all around the year, this process can lead to a drastic contrast between the wet windward part and the dry leeward part of a mountainous island.

marine carbonates Rock derived from calcium carbonate fixed by marine animals, of submarine origin.

marine protected area (MPA) Any intertidal or subtidal area, together with its associated flora, fauna, and historical and cultural features, that has been set aside by law or other effective means to protect part or all of the designated environment.

Maritime Continent The region encompassing Indonesia, Southeast Asia, and the southwestern Pacific. The numerous large islands influence the atmosphere in a manner similar to the equatorial continents of Africa and South America.

marl A soft, impure limestone that commonly makes fertile soils.

massif A massive topographic and structural feature, usually formed of rocks more rigid than its surroundings.

massive sulfide mineralization Mineral deposits with a high percentage of sulfide minerals, typically containing copper, lead, and zinc; commonly interbedded with marine volcanic horizons, and forming generally through exhalation of hydrothermal fluids through vents on the sea floor ("black smokers").

masting A group phenomenon in which plants within a population correlate their reproductive activity in both time and size of crop.

mating asymmetry Mating behavior in which one gender of a population or species is more likely (relative to the opposite gender) to mate with members of another population or species.

matrix The most extensive and connected land cover type in a landscape, which therefore plays the dominant role in the functioning of the landscape.

mean sea level The average sea level (including waves, currents, and tides) determined over time.

Mediterranean-type ecosystem (MTE) An ecosystem defined by a climate that resembles those of the lands bordering the Mediterranean Sea, with wet winters and dry summers. MTEs are found in five widely-separated regions of the world, namely Australia, California, Central Chile, South Africa, and the Mediterranean Sea.

megaherbs Spectacular herbaceous perennial flowering plants found on subantarctic and cool temperate islands in the Pacific and Indian Ocean regions. The plants are characterized by very large leaves and large, unusually colored flowers, and are well adapted to the harsh climatic conditions.

melanephelinite Dark-colored low-SiO_2 basaltic lava with large crystals of the minerals olivine and pyroxene, and also the mineral nepheline.

Melanesian Referring or belonging to Melanesia, those islands from New Guinea south to New Caledonia and east to Fiji.

melange A mappable unit composed of a heterogenous mixture of deformed rocks, typically with a pervasively sheared muddy matrix.

meromictic lake A lake that is not mixed seasonally from top to bottom but remains stratified for long periods of time because of density.

mesocosm An experimental container that is used to simulate the natural environment while allowing researchers to manipulate the local conditions under replicated conditions.

Mesolithic period A period in the development of human technology starting about 11,500 years ago (toward the end of the Pleistocene) and ending with the introduction of farming (around 5000 BC in northern Europe). Sometimes referred to as the Middle Stone Age (between the Paleolithic or Old Stone Age and the Neolithic or New Stone Age).

mesotidal Experiencing tides with a vertical range of 2–4 m.

Mesozoic The geological era in the Phanerozoic eon from ~250 million to 65 million years ago that includes the Triassic, Jurassic, and Cretaceous periods.

Messinian Salinity Crisis The period when the Mediterranean Sea evaporated partly or completely during the Messinian period of the Miocene epoch, approximately 6 million years ago.

metabasite Metamorphosed mafic rock, such as metamorphosed basalt.

metacommunity A set of local communities that are linked by dispersal of multiple species.

metallogeny Origin of mineral deposits.

metamorphism A change of mineral assemblage by heat and pressure. Deep in the Earth's crust, this action produces *metamorphic rock* (formed from sedimentary and other kinds of rocks). The resulting rock types are considered *metamorphosed* (e.g., marble is metamorphosed limestone, schist is metamorphosed shale).

metapopulation A collection of populations connected through immigration that undergo periodic extinction and recolonization.

metasediment A metamorphosed sedimentary rock.

microarray A microscopic array of single-stranded pieces of DNA fixed on a piece of glass or plastic used to screen a biological sample for the presence of specific genetic sequences. Fluorescent tags reveal actively transcribing genes that contain the sequence being probed. In this way, relative expression levels can be measured for a very large number of different genes.

micro-endemism See ENDEMISM.

microfossils Preserved remains such as foraminifera shells, diatoms (photosynthetic protists with silica skeletons) frustules, and plant pollen grains, which are too small to see with the naked eye but provide detailed information to scientists who study past climates.

Micronesia A group of islands in the Pacific Ocean that includes the Republic of Palau, the Mariana Islands (Guam and the Commonwealth of the Northern Mariana Islands), the Federated States of Micronesia (Yap, Chuuk, Pohnpei, and Kosrae states), and the Republic of the Marshall Islands.

microsatellites Highly variable, putatively neutral, dominant allelic markers located in the nuclear genome that consist of variable numbers of short, repeated base pair sequences.

microtidal Experiencing tides with a vertical range less than 2 m.

mid-ocean ridge A divergent plate boundary, where new crust is created from basaltic magma by volcanic and plutonic processes.

Miocene The geologic epoch in the Neogene period extending from about 23 million to 5 million years ago.

miogeocline Continental margin.

mist forest See CLOUD FOREST.

modern introduction A species that has been intentionally or accidentally relocated by humans in the Modern Era, or since approximately 1500 AD.

molasse Sediment deposited in shallow water at a subduction zone.

molecular markers Specific segments of the genome that are screened for mutations and used to gauge relatedness or associations with phenotypic traits.

molecular phylogenetics The use of macromolecular (usually DNA) data to infer relationships among evolutionarily divergent lineages.

molluscs A phylum of animals that comprise a monophyletic lineage originating early in the explosive Cambrian metazoan diversification more than 500 million years ago, including invertebrates such as snails, slugs, oysters, clams, octopuses, and squids, and characterized by bilateral symmetry, a well-developed foot, a mantle, a mantle cavity, a complete gut, and well-developed kidneys (metanephridia).

monophyletic Derived from a common ancestor.

monophyletic group An exclusive group containing all of the descendants from a unique common ancestor.

monotypic family A single species that is the sole member of the family to which it belongs.

monotypic genus A genus in which there is only one species.

monsoon A seasonal prevailing wind. The term is often applied to rainy seasons, but not all prevailing seasonal winds are wet (e.g., the winter monsoon of East Asia).

monsoon climate A climate characterized by an annual cycle of prevailing winds, typically featuring a rainy season and a dry season.

Moravian A denomination that was the first large-scale Protestant missionary movement. It has roots in the late fourteenth-century Catholic Church reform movement of Jan Hus and was revived in the eighteenth century in what is today Germany (but takes its name from a part of the Czech Republic).

MORB Mid-ocean ridge basalt, erupted at a seafloor spreading axis.

morphology The study of the structure and form of organisms and their parts.

mossy forest See CLOUD FOREST.

MPA network A series of marine protected areas (MPAs) connected by larval dispersal or juvenile and adult migration.

MTE Mediterranean-type ecosystem.

multibeam bathymetry A technique in which high-resolution bathymetry data are collected by ship echosounders that transmit multiple beams of sound simultaneously to collect a swath of data.

murids Members of the Muridae (order Rodentia), the largest family of mammals; the group of rodents that includes rats and mice.

mycorrhiza A mutual symbiosis between a fungus and the roots of a plant, in which the fungus extracts carbohydrates from the plant, and the plant uses the fungal mycelium to more efficiently extract nutrients from the soil.

mysticete A member of the whale suborder Mysticeti (order Cetacea), the baleen, or filter-feeding, whales, including right whales, rorquals, and grey whales.

mythicomyiid Belonging to the family Mythicomyiidae (order Diptera), a group of tiny humpbacked flies.

myxomatosis A ravaging disease of European rabbits (*Oryctolagus cuniculus*), caused by the myxoma virus and spread by fleas and mosquitos. Purposely introduced to Australia to control rabbit populations, it reached the United Kingdom in 1953.

nappe A sheet of rock that has moved a considerable distance as a result of tectonic activity and has been emplaced on or adjacent to rocks of different character.

National Wildlife Refuge System Lands and waters managed by the U.S. Fish and Wildlife Service, Department of Interior, for the conservation, management, and (where appropriate) restoration of wildlife (plants and animals) and their habitats for the benefit of present and future generations of Americans.

native See INDIGENOUS.

natural enemy An organism that feeds on another; a term used in biological control.

naturalized species An intentionally or unintentionally introduced species that has adapted to and reproduces successfully in its new environment without human help.

natural selection The process by which favorable heritable traits become more common in successive generations, because individuals bearing these favorable traits are more likely to survive and reproduce than those bearing less favorable traits.

nautilus A primitive cephalopod with an external spiral shell.

Nazca plate An oceanic tectonic plate in the eastern Pacific Ocean basin off the west coast of South America. (Named after the Nazca region of southern Peru.)

Neanderthal *Homo neanderthalensis*, a species of humankind extinct by about 30,000 years ago, distinguished from modern *Homo sapiens* by features of the skull and postcranial skeleton, and typically associated with a Mousterian stone-tool culture and Eurasian geographic range.

Nearctic The high arctic latitude zone of North America.

Near Oceania The part of Oceania lying west of Santa Ana Island; includes New Guinea, the Bismarck Archipelago, and the Solomon Islands. Parts of this region were settled by humans as early as ~40,000 years ago.

necrotroph An organism that kills the host and then lives as a saprobe on dead host tissue.

nematocyst A microscopic capsule unique to Cnidaria that is synthesized by a single cell (the nematocyte) and can be used just once; there are about 30 kinds (although animals of a given species typically possess no more than four to five kinds), some of which function offensively in prey capture, some defensively, and, in a few cases, for locomotion.

neo-endemic A species endemic to an area by reason of having evolved in situ, originating from an ancestor with origin on another island or on the mainland.

Neogene The geologic period in the Cenozoic era from about 23 million years ago to present that includes the Miocene, Pliocene, Pleistocene, and Quaternary epochs.

Neognathae The suborder of birds (lit. "new jaws") (order Aves) containing the vast majority of extant species.

Neolithic period A period that ran in northern Europe from around 5000 BC to 1700 BC (the beginning of the Bronze Age).

neoteny The retention of juvenile body characters in the adult state as a result of maturation of the reproductive system in a larval form that fails to undergo metamorphosis.

neotropics An ecozone that includes more tropical rainforest (tropical and subtropical moist broadleaf forests) than any other ecozone, extending from southern Mexico through Central America and northern South America to southern Brazil, including the vast Amazon rainforest.

nested distribution/nestedness A distribution of species in which the species occurring within species-poor communities are subsets of those occurring in richer communities.

niche 1. The particular environmental habitat to which a species is adapted. 2. The functional role of a species in an ecosystem, including the habitat in which it lives and the resources it uses. Also, ECOLOGICAL NICHE.

nitrogen fixation The process by which plant species interact with bacteria to incorporate atmospheric nitrogen (N_2) into compounds such as ammonia and nitrate that are useful for various chemical processes, including physiological processes of plants.

non-adaptive radiation Species proliferation that has not been attended by diversification of ecological roles.

nonequilibrium A condition of instability in ecological communities that change in structure and composition in response to the environment.

non-indigenous or **non-native species** A species found in an area in which it does not occur naturally. Also, ALIEN SPECIES, EXOTIC SPECIES, INTRODUCED SPECIES.

non-volant Unable to fly; among mammals, only bats are volant.

Norse Describing or referring to the people and cultural traditions of communities living in, or migrating out from, Norway, mostly beginning around 700 AD.

Notogean Realm The zoogeographical division of the Earth's surface that includes Australia, New Guinea, New Zealand, and the southwestern Pacific islands and which is based on shared geographical distributions of distinctive faunal elements (e.g., four of the six orders of marsupials).

null hypothesis A hypothesis that is set to be tested against the theory being developed. It usually involves the distribution of values to be obtained if only random processes were operating.

nunatak A mountain peak that protrudes above a glacier.

obligate Restricted to a particular way of life.

Oceania A large group of islands in the south Pacific including Melanesia, Micronesia, and Polynesia (and sometimes Australasia and the Malay Archipelago).

oceanic crust The mafic (magnesium- and iron-rich) basement rock beneath all oceans, normally about 7 km thick; it is thinner and denser than continental crust.

oceanic island An island (usually of volcanic origin) that begins its above-water existence in isolation from the mainland, without a prior land connection.

odontocete A member of the whale suborder Odontoceti (order Cetacea), toothed whales, including porpoises, dolphins, and sperm whales.

Oligocene The geologic epoch in the Paleogene period that extends from about 34 million to 23 million years ago.

omnivorous Eating both plants and animals.

ontogenetic Referring to the developmental stages that occur in the sequential transformation of a fertilized egg into an adult organism.

oolites Spherical-shaped sand grains composed mostly of calcium carbonate precipitated out of seawater in the form of spherical layers around a nucleus (commonly a small shell fragment).

ophidian 1. Pertaining to snakes. 2. A member of the suborder Ophidia, generally considered an alternate name for Serpentes.

ophiolite An assemblage of ultramafic and mafic rocks, widely thought to represent oceanic crust.

ophiolite complex or **suite** A fragment of ancient ocean floor emplaced onto a continent. The idealized ophiolite sequence consists of deep-sea sediments, underlain by basaltic (pillow) lavas, next to a sheeted-dike complex, and next to coarse gabbroic plutonic rocks, which together form the oceanic crust. Ophiolites also often contain slivers of the underlying mantle.

optimal body size A body size that serves as an evolutionary attractor under some theories of size evolution.

orogen A geological mountain belt of deformed, intruded, altered rocks.

orthopteran An insect of the order Orthoptera, a group with incomplete metamorphosis that includes grasshoppers, crickets, and locusts.

osteology Characterization of skeletal elements, often used to identify and classify species.

outcrossing The movement of genetic material between two separate individuals.

outrigger A section of a canoe's rigging that helps to stabilize the craft, especially when turning.

overriding See SUBDUCTION.

Pacific decadal oscillation (PDO) An oscillation phenomenon in the Pacific Ocean on a time scale of 20 to 30 years, in contrast to 6 to 18 months for the El Niño–Southern Oscillation (ENSO). The climatic fingerprints of the PDO are most visible in the North Pacific/North American sector, whereas secondary signatures exist in the tropics. In contrast, ENSO signatures are global.

paedomorphosis A condition in which the body shape of a descendant adult species resembles the juvenile shape of its ancestor species; one of a class of phenomena relating to phyletic alteration of growth trajectories known as heterochrony.

pāhoehoe Basaltic lava flow of typically lower viscosity, which cools to produce a smooth and ropy surface. (A term originating from the Hawaiian language.)

palagonite Clays resulting from hydration of volcanic glass, a line of evidence that a lava flow may have quenched in contact with water or ice to form the volcanic glass.

Palearctic The high, arctic latitude zone of Eurasia.

paleobotany The study of fossil remains of plants.

paleoclimatology The study of climates of the past through indirect methods such as the study of sedimentary records.

paleoecology The study of past environments through examination of the fossil records of organisms living in the environment at that time.

paleo-endemic A species endemic to an area by reason of other populations or close relatives elsewhere becoming extinct; has changed little after colonization.

Paleocene The geologic epoch in the Paleogene period from about 65 million to 56 million years ago

Paleogene The geologic period in the Cenozoic era from about 65 million to 23 million years ago, which includes the Paleocene, Eocene, and Oligocene epochs.

Paleognathae The suborder of birds (lit. "old jaws") containing the primitive tinamous and ratites.

Paleolithic The Old Stone Age, a period in human technological development during which stone tools were introduced and the climate changed frequently. The Lower Paleolithic (or Palaeolithic) ran from about 2.5 million years to 100,000 years ago; the Middle Paleolithic, 300,000 to 30,000 years ago; and the Upper Paleolithic 40,000 and 10,000 years ago. The period was followed by the Mesolithic.

paleomagnetism The direction and intensity of the Earth's magnetic field as recorded in igneous rocks when they cool.

paleontology The study of fossils.

Paleotropical Realm The zoogeographical division of the Earth's surface that includes Africa, Madagascar, India, and Indo-Malaya through the Lesser Sundas and Moluccas, and which is based on shared geographical distributions of distinctive faunal elements (e.g., elephants, pangolins, rhinoceroses, civets, and mongooses).

Paleozoic The geologic era in the Phanerozoic eon from about 540 million to 250 million years ago and including the Cambrian, Ordovician, Silurian, Devonian, Carboniferous, and Permian periods.

pandanus Tree and/or fruit of the monocotyledonous species *Pandanus tectorius,* sometimes referred to as screw pine.

Pangea The ancient supercontinent that included North America, Scandinavia, Europe, and Africa; produced by the closing of the Iapetus Ocean.

panmictic Describing or referring to a population in which all individuals are potential partners; there are no restrictions to mating.

pantepui A biogeographic region in Venezuela and elsewhere in the north central Guyana shield that includes all the isolated tepui summits. The region is noted for the distinctive biota adapted to the typically cold, wet, rocky, and windy conditions on these summits.

Panthalassa Ocean A name given to the vast ocean configuration that existed during the time of the Gondwanaland and Laurasia supercontinents (much of Paleozoic and Mesozoic time).

Papuan The large group of 900 or more non-Austronesian languages (including many different and ill-defined language families) centered on the island of New Guinea and neighboring islands.

paramo A high-elevation habitat found only in South America, associated with the Andes but also with some outlying ranges, dominated by grasses and herbs (notably bromeliads).

parasitoid An arthropod that completes its development attached to or inside a single host organism, which it ultimately kills (and often consumes) in the process.

Parks Australia North An agency of the Australian Government's Department of Environment and Water Resources, locally responsible for natural resource management on Christmas Island.

passage zone The surface that separates subaerial lava from underlying subaqueous breccia.

pathogen An agent, such as a bacterium or fungus, that causes disease.

pathogenicity The ability to cause disease.

pathway In invasion biology, a route or method along which an invasive species spreads.

pelagic In or on the open ocean, far from land.

peridotite An olivine plus pyroxene rock that is abundant in the upper mantle; it partially melts to form basaltic magma.

periglacial Around the margin of a glacier; said of the processes, conditions, areas, climates, and topographic features near the margins of former and existing glaciers and ice sheets influenced by the cold temperature of the ice.

permafrost Earth materials (soil, sediment, or rock) that maintain a temperature at or below 0 °C for a period of at least two years.

phenetic clustering 1. A statistical method for grouping objects. 2. In taxonomy, grouping of objects based on overall similarity, without consideration of whether characters are unique to the group or also occurred in the ancestral species; for analysis of evolutionary relationships, this method has largely been replaced by cladistics. 3. In biogeography, grouping of objects based on shared attributes (e.g., islands in terms of shared species).

phenotype The outward physical appearance of an organism.

phenotypic adaptation Change that allows an organism to become better suited to the environment; can occur through either *developmental plasticity* (frequently involving parental effects) or *adaptive plasticity*, allowing directional selection to act and thereby enhance fitness. Such plasticity allows adaptive evolution to occur on an ecological time scale and has been characterized colloquially as "nurture versus nature."

phenotypic plasticity Under different environmental conditions, the ability of organisms of a given genotype to either change their phenotype or produce different phenotypes as a result of developmental plasticity.

philopatry The tendency of an animal to return to its birthplace to breed (natal philopatry) and to return to the same nesting site in successive years (breeding philopatry).

phonolite A gray, fine-grained differentiated lava with high silica content (~60%) and consisting of alkali feldspars and nepheline. The name relates to its resonance when struck.

phreatic eruption or **explosion** An explosive volcanic eruption that occurs when confined, sub-surface geothermal waters are heated to temperatures above their boiling point and flash to steam, thereby expanding to form an explosion. No molten rock is involved.

phreatomagmatic eruption An explosive eruption that occurs when magma comes into contact with water.

phreatophytes Desert trees and shrubs with deep tap roots to the water table.

phylogenetics The classification of organisms based on their degree of evolutionary relatedness.

phylogenetic tree A branching diagram or family tree showing the evolutionary relationships among species or other entities relative to a common ancestor. Usually represented as a phylogram or cladogram.

phylogeny The pattern of relationships of species, usually represented in the form of a tree or hierarchy indicating commonality of descent.

phylogeography The study of the geographic distributions of genealogical lineages (populations or species) across the geographic landscape, and of the processes that have shaped these patterns.

phylogram A phylogenetic tree in which branch lengths represent time.

phytophagous Feeding on plants, including shrubs and trees; said especially of certain insects.

phytophagous filter An agent that affects the recruitment of plant species by eating some of them. (For example, land crabs can be phytophagous filters because, by differentially consuming seeds or seedlings of many rain-forest species, they can cause the community of established seedlings to be a nonrandom draw of all seeds falling to the forest floor.)

phytoplankton Algae that occur as single cells or in colonies in the open water of lakes, streams, and oceans.

picrite Basaltic rock rich in the mineral olivine.

Picts Ancient people of northern Britain, undefeated by the Romans. The Picts joined with the Scots from Ireland to form a kingdom (later to become Scotland) in the ninth century AD.

Pikermi fauna Fossil vertebrate fauna of Upper Miocene (Turolian) age, intermediate in form between European, Asiatic, and African types, and first found at Pikermi (Attiki, Greece) by Jean Albert Gaudry between 1855 and 1860. Representatives of this fauna have since been found in many other sites around the Mediterranean. *Hipparion* is a typical fossil genus of this fauna.

pillow lavas Elongated lava mounds formed by repeated oozing and quenching of hot basalt lava extruding underwater, and typically having a glassy crust.

pinnacle A pointed formation arising from the sea floor.

pinniped A marine mammal of the order Carnivora, including seals (Phocidae) and sea lions (Otariidae).

plankton Passively floating, drifting, or weakly motile organisms in a body of water, primarily comprising microscopic algae and protozoa.

planktotrophic Feeding on materials from the plankton.

plant macrofossils Large preserved plant remains that have formerly grown in a locality.

plate A rigid section of the Earth's lithosphere, usually containing a proportion of oceanic crust and of continental crust.

plate tectonics The study of the dynamics of the surface of the Earth; the movement and marginal interaction of the rigid tectonic plates that form the lithosphere.

platform reef An uplifted reef consisting of a flat-topped bench with or without a very shallow lagoon.

Pleistocene The geologic epoch in the Neogene period that began about 1.8 million years ago and ended about 11,500 years ago. The period was marked by the repeated formation of extensive ice sheets and other glaciers because of the cooling of the climate.

Pleistocene Aggregate Island Complexes (PAICs) Groups of islands separated by shallow seas (depths less than 120 m) that were connected during Pleistocene glaciations when the sea level was lower than it present is.

Pliocene The geologic epoch in the Neogene period from about 5 million to 1.8 million years ago.

plug Lava intruded as a cylinder-like body.

plutonism Intrusion of magmatic rocks such as granite.

pneumatophores Small roots arising from the cable root system of some mangroves (e.g., in genera such as *Avicennia, Sonneratia, Heritiera,* and *Xylocarpus*), which extend upward into the air as small conical projections that facilitate the movement of oxygen into the root system of mangroves growing in anoxic conditions.

polar frontal zone The area between two high-velocity regions of the Antarctic Circumpolar Current: the sub-Antarctic Front to the north and the Antarctic Polar Front to the south.

polje A valley formed by the collapse of the roof of a limestone cave tunnel, through which a stream flows from a cave mouth at one end to a cave mouth at the other.

polyp The body form of corals, which is a cylinder with only one body opening—the mouth—that is at one end and is surrounded by tentacles; in some cnidarians, the life cycle alternates between polyp and a jellyfish-like medusa.

population Individuals of the same species in a given area that interbreed.

population bottleneck See GENETIC BOTTLENECK.

porphyry copper-gold Low-grade copper and gold mineralization disseminated through a large mass of granitic to dioritic intrusive rock that typically shows porphyry texture (larger crystals in a finer groundmass).

porphyry mineralization Mineralization usually associated with subvolcanic intrusive rocks with a porphyritic texture, forming at depths of around 3–5 km and typically dominated by copper sulphide minerals in an intense network of quartz veins (stockwork).

portmanteau biota The assemblage of plants and animals transported by humans, both purposively and inadvertently, during human population dispersals and expansions. Also, SYNANTHROPIC BIOTA.

post-caldera Occurring after caldera formation.

post-zygotic reproductive isolation Lack of gene flow between evolutionarily distinct lineages in which mating and fertilization have occurred, because of intrinsic factors (hybrids are either developmentally unsound or sterile) or extrinsic factors (hybrids are sound and fertile but unfit in the current environment, especially in comparison with parental phenotypes).

Precambrian The geologic supereon from formation of the Earth 4.567×10^9 years ago to 540 million years ago; includes 90% of geologic time and the Hadean, Archean, and Proterozoic eons.

pressure gradient The slope of isobaric surfaces; most commonly associated with mean sea levels.

pre-zygotic reproductive isolation Lack of gene flow between evolutionarily distinct lineages because of factors that either prevent mating or prevent fertilization after mating.

primary productivity In marine systems, a measure of the amount of phytoplankton (chlorophyll-bearing plankton, "plant plankton") produced in surface waters, which occupy the base of the food chain; usually quantified as grams of carbon produced per square meter per unit time.

progradation The seaward or lakeward development of a shoreline by accumulation or deposition of sediments.

progression rule Colonization and divergence of a clade within an archipelago, occurring in a stepwise manner, from oldest to youngest island.

propagule 1. An individual or group of individuals that arrive at a site. 2. In the theory of island biogeography, the minimal number of individuals of a species required to achieve colonization of a habitable island.

propagule pressure A composite measure of the number and rate of arrival of propagules at a site.

protists A diverse group of largely unicellular eukaryotic organisms formerly considered to comprise the kingdom Protista.

pseudorabies A highly contagious swine disease caused by a herpes virus that can also affect horses, cattle, sheep, goats, dogs, and cats. The pseudorabies virus (PRV) causes reproductive problems, including miscarriages and stillbirths, and can lead to death. PRV is generally spread through animal-to-animal contact.

pseudoturnover The apparent loss and gain of particular species from an island resulting from incomplete census data, when they have actually been residents throughout or, alternatively, never properly colonized.

pumice A highly vesicular, light-colored fragment or its aggregate produced by volcanic eruptions with low bulk density.

pyroclastic Formed from fragments of volcanic origin. (From Greek *pyro*, meaning "fire," and *clastic*, meaning "broken.") Volcanic rock layers formed in this way are termed *pyroclastic deposits* (or *pyroclastics*). Examples of *pyroclastic rocks* are volcanic ash, pumice, and ignimbrites.

pyroclastic density current A current composed of hot fragments and gases produced by volcanic eruption and driven by gravity or its bulk density.

quartzite A metamorphic rock derived from pure sandstone and consisting primarily of quartz.

Quaternary The geologic period in the Cenozoic era that extends from about 1.8 million years ago to the present; includes the Pleistocene and Holocene epochs.

Quechua The native Peruvian language spoken by the Incas; spoken widely in Peru, Bolivia, and neighboring countries.

radiation In evolutionary biology, an increase in biodiversity by divergence, especially involving the formation of new species.

radiometric dating The dating of igneous rocks utilizing the constant decay of isotopic series elements such as uranium–lead (U–Pb series) and potassium–argon (K–Ar series).

rafting The passive transport of floating objects and their associated fauna to distant habitats.

rain shadow An area of reduced rainfall on the leeward side of a mountain range, caused by the cooling-induced precipitation as air is forced up over the range.

Ramsar List of Wetlands of International Importance An inventory of sites selected within the scope of the Ramsar Convention by the presence of representative, rare, or unique wetland types of international importance for conserving biological diversity. (Formerly the emphasis of the Ramsar Convention was the conservation and wise use of wetlands primarily as habitat for water birds, but over the years the Convention has broadened its scope of implementation to cover all aspects of wetland conservation and sustainable use.)

Ramsar site A wetland included in the Ramsar List of Wetlands of International Importance of the Convention on Wetlands, an intergovernmental treaty signed in Ramsar, Iran, in 1971.

random genetic drift See GENETIC DRIFT.

realized niche A narrower subset of the fundamental niche to which a species is confined as a result of interactions with other organisms.

recent introduction Another term for modern introduction, but it typically refers to very recent times.

reciprocal translocation Chromosomal mutation resulting in exchange of chromosomal segments between non-homologous chromosomes.

recruitment The addition of new individuals to a habitat, generally as a result of reproduction and growth of young to the adult population.

Red List An inventory of the conservation status of species in a particular region, according to criteria defined by the International Union for the Conservation of Nature and Natural Resources (IUCN). Categories of conservation status include: extinct, extinct in the wild, critically endangered, endangered, vulnerable, near threatened, least concern, data deficient, and not evaluated. The IUCN Red List Program aims to identify and document those species most in need of conservation attention if global extinction rates are to be reduced and to provide a global index of the state of degeneration of biodiversity.

reef A strip of coral that rises close to the surface of the water.

refugia (singular refugium) Locations of relictual populations of species that were once more widespread.

regional endemism See ENDEMISM.

reinforcement In evolutionary biology, the development of prezygotic barriers in response to selection against interspecific mating.

relaxation A decrease in species diversity over time on a continental island as a result of local species extinction without compensating immigration, associated with the decrease in area subsequent to island formation.

relic A human artifact or tradition that remains from a past era.

relict A surviving member of a lineage or other natural phenomena. A relict endemic is equivalent to a paleoendemic.

relictual Describing or referring to those organisms originally occurring in a location prior to fragmentation, which subsequently form part of the fragment biota; one of three classes of taxa inhabiting patchy landscapes.

relictual series A mixture of endemics within an island biota with differing degrees of phylogenetic divergence from non-island relatives as a result of different colonization times and lineage histories.

Remote Oceania The islands of Oceania east of Santa Ana in the eastern Solomon Islands, including Vanuatu, New Caledonia, Fiji, Micronesia, and Polynesia.

rescue effect The prevention of extinction of a small insular population (of a species) by the occasional influx of individuals from another (e.g., mainland) area.

research station A scientific facility to support research and educational programs focusing on the natural world and human interactions with it. Unlike institutions on the mainland and larger islands (e.g., universities or museums) that have permanent scientific staff, research stations (field stations and marine laboratories) provide permanent infrastructure typically for visiting researchers and students.

resistance The ability to be unaffected by something; for example, a host plant may resist the activity of a pathogen.

resource depression An effect of heavy predation pressure on one or more natural resource species, typically indicated by decreases in the abundance of larger or more desirable prey, or by changes in the demographic structure of prey populations (e.g., decreased number of older adults).

retro-arc basin See BACK-ARC BASIN.

rhyolite A light-colored, typically glassy, differentiated volcanic rock made of alkali feldspar and quartz with few iron-bearing minerals; erupted at temperatures of 700 to 850 °C.

Rim of Fire The ring of volcanoes around the Pacific Ocean, which is also the scene of much earthquake activity.

Rodinia supercontinent The oldest known supercontinent, which contained most or all of Earth's land mass at the time. It was formed during the Precambrian supereon between 880 and 1100 million years ago through an event known as the Grenville orogeny and began to rift apart no later than 750 million years ago.

runaway sexual selection An idea first proposed by the British geneticist R. A. Fisher to explain the evolution of extreme traits, generally in a single sex (e.g., the tail of a male peacock); these traits may result from sexual preference that thus confers reproductive success.

sacbé A raised causeway built by the ancient Maya civilization. The limestone used in the construction of these roads gave them the white appearance for which they were named. (From the Maya *sac*, meaning "white," and *bé*, meaning "road.")

Saladoids The earliest ceramic-making horticulturalists to settle islands in the Caribbean. Known by their distinctively styled pottery, with occupations that lasted from about 400 BC to AD 600.

saltation In genetics, a modification that is expressed as a profound phenotypic change across a single generation and results in a potentially independent evolutionary lineage.

sandstone A consolidated sedimentary rock composed of sand particles.

saprobe An organism that derives its nourishment from decomposing organic matter.

scarp A steep erosional feature in the beach face, resulting from rapid beach retreat during high-wave activity.

scleractinian A stony coral (order Scleratinia).

sclerites Needle-like calcium carbonate skeletal elements secreted by the tissues of octocoral polyps and tissues connecting the polyps of a colony; scattered in the tissues or fused to form a solid structure around or under the polyps and connecting tissues, they serve for protection and/or support.

sclerophyll forest Woodland that develops in regions that receive less than 1100 mm of precipitation per year and where the dry season can last for several months in a row. Also, TROPICAL DRY FOREST.

sclerophyllous Of plants, having hard leaves and short distances between leaves along the stem.

scoria cones Volcanic cones formed by explosive eruptions.

SCUBA Acronym for *s*elf-*c*ontained *u*nderwater *b*reathing *a*pparatus.

seabird Any of several groups of birds specifically evolved to life at sea, including penguins, albatrosses, shearwaters, petrels, storm petrels, pelicans, cormorants, boobies, gannets, frigate birds, tropicbirds, gulls, terns, auks/alcids (including auks, auklets, dovekies, guillemots, murres, murrelets, puffins, and razorbills), and jaegers.

seafloor spreading Gradual divergent movement of tectonic plates at mid-ocean ridges resulting in the formation of new oceanic crust through volcanic activity.

sea ice Ice created when the surface of the ocean freezes. Through the process of brine rejection, sea ice is largely freshwater ice, although the initial freezing of saltwater requires temperatures at or below −1.8 °C.

seamount An underwater mountain, generally of volcanic origin, that arises from the sea bottom but does not reach the ocean surface; after sustained growth, may produce an oceanic island volcano.

sea-run See ANADROMOUS.

sea skater A pelagic marine insect, family Gerridae.

sector collapse The collapse of a portion of a volcano, typically resulting from an earthquake, rising magma, or abundant precipitation in combination with high relief, steep slopes, and unstable or altered rock.

sedimentary breccia Poorly sorted rock dominated by angular fragments not greatly abraded during their deposition (as opposed to conglomerate, which has rounded or subrounded fragments).

sedimentary rock Rock that forms either as a result of erosion of pre-existing landscapes and deposition of the resulting detritus in a range of settings, or from precipitates and organic remains in seas and lakes.

seed dispersal The scattering of seeds away from each other and from the parent plant.

seedling recruitment In plant ecology, a population-level process beginning with seed germination and continuing through an early period of high seedling mortality, to the establishment of survivors with a relatively high probability of survival.

seismic tomography The use of seismic waves to identify velocity variations and, hence, three-dimensional mantle structures within the Earth.

semelparous Having only one set of progeny in a lifetime; in angiosperms, flowering once.

semi-fossorial Describing or referring to a species that forages by burrowing but that also spends some time on the surface of the ground.

senescence 1. The condition or process of aging. 2. In plants, the growth phase from full maturity to death that is characterized by accumulation of metabolic products, increase in respiratory rate, and loss of tree foliage out of season.

sessile Permanently attached to the substrate.

sexual dimorphism Any differences in morphology between the male and the female of the same species.

sexually dimorphic characters Morphological, behavioral, or other differences between males and females of animal species.

shamanism An animistic religion of northern Asia in which the shaman mediates between the visible and spiritual worlds. (From a Siberian word for "a person possessed by spirits who has mastered them.")

shear See WIND SHEAR.

sheeted-dike complex A rock unit that consists almost entirely of near-vertical, parallel dikes (i.e., vertical planar fissures up to several meters wide filled with igneous rock, typically diabase). Sheeted-dike complexes are locally found in ocean crust and are very characteristic of seafloor spreading environments.

shield volcano A gently sloping volcano with the shape of a flattened dome, built almost exclusively of lava flows.

shifting baseline In conservation, the concept that each human generation believes that current populations or ecosystems are at a natural state despite declines that have occurred over many generations.

shingle A type of sediment that is common to coral reef islands and is made up mostly of fragments of branching corals.

short-range endemism See ENDEMISM.

sill A flat sheet of igneous rock injected between existing rocky strata.

sister species Distinct species that descended from a single common ancestor.

Site of Special Scientific Interest (SSSI) A nationally designated site providing statutory protection for flora, fauna, or geological or physiographical features. In Ireland, such a site is called an Area of Special Scientific Interest (ASSI).

skerry A rocky islet or reef (Scottish).

skylight A hole in the roof of a lava tube or lava tube cave, usually formed by collapse.

SLOSS debate A debate in conservation biology as to whether it is preferable, for the conservation of biodiversity in a fragmented landscape, to protect a *s*ingle *l*arge *o*r *s*everal *s*mall (SLOSS) reserves.

socioecosystems Managed landscapes that have become inextricably linked with the human societies that inhabit them, through dynamically coupled human-natural processes.

spawning The often synchronous release of eggs and sperm by numerous coral species during specific lunar phases and seasons.

Special Protected Area (SPA) A site designated in accordance with the Birds Directive set up by the European Union in 1979, which aimed to conserve all species of naturally occurring birds and their habitats.

speciation The evolutionary process through which new species originate by descent and divergence from an ancestral species.

species-area relationship The relationship of area to number of species. As the area of an island or other potential (isolated) habitat increases, the number of species present also increases.

species richness The number of species in an ecological community; one of several important measures of biodiversity. Other measures of biodiversity include genetic diversity and individual behavioral attributes.

species turnover The replacement over space or time of one species or species assemblage by another.

speciose In evolutionary ecology, describing or referring to species-rich taxa.

speleothem A natural formation inside a cave.

spermaceti A wax from the barrel-shaped organ in a sperm whale's head, used to make fine candles that were the standard for the candlepower, a measure of artificial light.

spillover The emigration of adult or juvenile fishes or other marine organisms outside a marine protected area as a result of density-dependent effects such as competition or aggression.

spiritualism A religious movement that was popular in the 1840s–1920s, which centered around the belief that the spirits of the dead can be contacted by mediums and can provide guidance to those inhabiting the living world.

sporophytic self-incompatibility (SI) The inability to form progeny from self-fertilization or from mating with plants that share either allele at the SI locus (called S-alleles), regardless of which allele is carried by the pollen.

stable isotopes Versions of elements that are not radioactive but have differing numbers of neutrons and therefore slightly different weights; studies of the ratios of these elemental forms, such as oxygen, nitrogen, and carbon, provide useful measures of variations in past climates.

stalactite A cylindrical or conical protrusion from the roof of a cave.

stalagmite A cylindrical or conical protrusion from the floor of a cave.

state factors Factors that control the characteristics and dynamics of terrestrial ecosystems, such as time, substrate parent material, substrate age, relief, topography, and regional flora and fauna.

statistical model A model containing parameters that are varied to fit a set of observations.

still water level The level of the water in the absence of wave motions.

stiltroots Prominent arching, generally branched roots growing from the trunk of some mangroves into the substrate, facilitating the movement of oxygen into the root system.

stochastic extinction Extinction of a species because of random events; the opposite of deterministic extinction.

stock In conservation management, a species, subspecies, or geographical grouping that can be managed as a unit.

storm surge The abnormal elevation of sea level caused by the pressure fall and winds associated with a tropical cyclone.

stratocone A volcanic cone consisting of both lava and pyroclastic rocks.

stratovolcano See COMPOSITE VOLCANO.

structure In geology, the arrangement and disposition of rocks in the Earth's crust related both to Earth movements and to intrinsic morphological features such as joints.

stygobiont Of aquatic species, inhabiting underground or cave waters and adapted to life in permanent darkness.

subaerial In volcanology, erupted on the land surface, as opposed to subaqueous.

Subantarctic A biogeographic region in the Southern Ocean, north of the Antarctic region and south of the cool temperate region, characterized by large expanses of ocean with scattered small islands and strong connections between marine and terrestrial ecosystems. Its terrestrial vegetation has no trees or shrubs.

subaqueous Formed underwater.

subduction The process in which one of the plates that make up the Earth's crust is carried down into the Earth's interior beneath another plate with which it is converging. Disturbance to the rocks of the mantle by the release of volatiles or melt from the subducted slab may cause partial melting of the mantle and may trigger volcanic activity. Also, OVERRIDING.

subduction complex A mappable terrane composed of deformed sediments and associated volcanic rocks scraped off a downgoing oceanic plate, accreted to the overlying plate, and metamorphosed under relatively high pressures and low temperatures.

subduction zone An area where two tectonic plates meet and one slides beneath the other, carrying materials from the Earth's surface deep into its interior.

subfossil Bones or other materials from organisms that have been dead for some time but whose fossilization process is not complete; usually of relatively recent geological origin.

subsidence Sinking of crust, typically caused by thermal contraction or crustal thinning.

subsidized population A population that is artificially increased either directly or indirectly through human agency (e.g., as from feeding on an introduced species or by consuming garbage from a landfill).

substrate The base on which an organism lives.

succession Predictable changes in a community over time. Development of complex communities on new substrates is termed *primary succession*, whereas *secondary succession* is the process of ecosystem development on previously vegetated sites that have been cleared.

sulfide A general term for chemical compounds containing sulfur in it lowest oxidation state (e.g., hydrogen sulfide [H_2S]).

sulfophilic "Sulfur-loving" (i.e., thriving in habitats containing high concentrations of sulfide).

Sundaland The area of land formed by sea-level falls during the glaciations, uniting Borneo, Palawan, Sumatra, Java, and Bali with the Malay Peninsula and mainland Asia.

Sunda Shelf The southeasternmost extension of the Eurasian continental shelf, the eastern boundary of which is marked by Wallace's Line.

sunderbans The largest mangrove forest in the world, situated at the mouth of the Ganges River in India. It is crisscrossed by networks of tidal waterways that drain mudflats and create islands within the forest.

supercolony In ants, a group where aggression between worker offspring of conspecific queens is absent, contributing to high worker density over very large areas.

superplume A large-scale cold downwelling or a hot upwelling flow in the mantle.

supersaturated Having a larger number of species than can be maintained at equilibrium. Supersaturation generally occurs prior to relaxation.

suspect terrane A geologic entity of unknown origin.

sustainable development Development that generates sufficient income to purchase material and non-material goods from the modern cash economy to make life healthier, safer, more productive, and more enjoyable, without destroying the local natural and cultural capital needed for the material and cultural survival of future generations.

sverdrup A unit of river and ocean current flow, equal to 1 million m^3 of water per second.

swath mapping Multibeam sonar mapping of the bathymetry and acoustic backscatter of the ocean floor.

sweepstakes route A rare, chance colonization after a long-distance dispersal. Only a small number of individuals arrive and successfully establish on isolated islands. Their success is often followed by adaptive radiation.

syenite A coarse-grained rock with abundant feldspar; formed by intrusion of trachyte magmas.

sympatric 1. Having broadly overlapping geographic ranges. 2. Specifically, living in the same local community and able to interact.

sympatric speciation/sympatry Species formation occurring when populations diverge within the same geographic region.

synanthropic biota See PORTMANTEAU BIOTA.

syncline A concave fold in rock, the central part of which contains the youngest section of rock.

tacking A maneuver in sailing in which the bow of the boat is turned into the wind, causing it to fall off on the opposite site, which then allows the craft to change course.

target area effect The effect of island area on the rate of immigration, by which larger islands provide easier targets for passively dispersing propagules, thus enhancing colonization.

taxon (*plural* **taxa**) A clade, usually a named one.

taxon cycle Sequential phases of expansion and contraction of the ranges of species, associated generally with shifts in ecological distribution and differentiation. Stages range from expanding (Stage I), to differentiating (Stage II), to fragmenting (Stage III), and to endemic (Stage IV).

tectonic earthquake A rupture in the stiff, outermost part of the Earth (lithosphere). Tectonic earthquakes are triggered by the movement of tectonic plates relative to one another.

tectonic plates Pieces of the Earth's lithosphere (crust and upper mantle) that move in relation to one another on the Earth's surface. Earthquakes and volcanoes commonly occur along the boundaries between plates.

tectonic processes Large-scale movements of the Earth's crust, during which rocks are folded and faulted.

tectonics The study of how movements of the Earth's surface relative to its center shape the forms observable on the surface.

tectonostratigraphic terrane A group of rocks that have a common history in terms of both their tectonic and their sedimentary evolution.

teleseism A distant earthquake, as opposed to a "local" earthquake.

tephra Fragments of volcanic rock and lava of any size expelled from a volcano (e.g., ash, bombs, cinders).

tephrochronology A method of using ash layers (tephra) of known age as marker horizons in soil or sediment sections.

tepui A steep-sided mesa or table-shaped mountain, characteristic of Venezuela and elsewhere in the north central Guyana shield.

terrane A part of the Earth's crust that is bounded by faults on all sides and that differs from adjacent parts of the Earth's crust by the nature of its geology and geological history.

terrigenous Derived from the land; in most cases, terrigenous sediment is siliciclastic (i.e., noncarbonate).

Tertiary The geologic subera extending from about 65 million to 1.75 million years ago, including the Paleocene to Pliocene epochs.

Tethyan Pertaining to the age when the Tethys Ocean existed between Laurasia and Gondwana during the Mesozoic era until its closure in the Tertiary (Miocene). In relation to terrestrial vegetation, the term is used with reference to a wide "Madrean–Tethyan belt" of sclerophyllic taxa that, based on fossil evidence, occupied a large subhumid region across the Holarctic in this period (mainly Eocene–Miocene); a hypothesis often used to explain strong disjunctions.

tholeiitic magmas Magmas that are relatively rich in silicon and poor in sodium and potassium, usually generated by larger extents of melting of the mantle. *Tholeiite* or *tholeiitic basalt* is the rock name.

thrashers Large, noisy, mostly-terrestrial birds, family Mimidae, with a chiefly southwestern North American species radiation.

threatened species A species that faces a risk of extinction in the near future. The best-known international list of threatened species is the IUCN Red List.

tillite The rock equivalent of till, a sediment laid down by glaciers.

trachyandesite A fine-grained differentiated volcanic rock with relatively high silica content and consisting mainly of sodium-rich plagioclase and alkali feldspars.

trachybasalt A dark, fine-grained volcanic rock with higher silica content than basalt, which consists of calcium-rich plagioclase and alkali feldspars in equal proportions.

trachyte/trachytic rock A gray, fine-grained volcanic rock with high silica content (~60%) and consisting largely of alkali feldspars. It usually has a rough and gritty surface.

trade wind inversion The inversion layer associated with the trade winds, characterized by increasing temperature and sharply decreasing relative humidity; it represents the boundary between heated surface air and sinking air aloft.

trait Any distinct or quantifiable feature (structure, behavior, or physiological or developmental process) of an organism.

transform fault/transform plate boundary A plate boundary at which two plates slide past one another horizontally.

transmutation The alteration of one species into another.

trench A deep depression on the ocean floor, produced when a tectonic plate sinks upon colliding with another tectonic plate.

trilobite An extinct fossil marine arthropod characterized by an exoskeleton consisting of three sections.

troglobionts Terrestrial species that are obligate subterranean dwellers.

troglomorphic Pertaining to the morphological, behavioral, and physiological characteristics that are convergent in subterranean species.

trophic structure The ratio of the numbers of species in different dietary categories, such as herbivores, scavengers, nectar feeders, predators, parasites, and so on.

tropical dry forest See SCLEROPHYLL FOREST.

trough A feature of seafloor topography; a narrow depression that is less steep than a trench.

tryworks The equipment used to render oil from the blubber of whales.

tsunami A series of waves that travel across the surface of the ocean, generated by impulsive disturbances of the sea floor. Tsunamis are most often associated with submarine earthquakes or landslides.

tuff A consolidated rock composed of pyroclastic fragments and fine ash.

turbidite A sediment or rock deposited from a turbidity current, a bottom-flowing density current of suspended sediment moving quickly down a subaqueous slope and spreading out at the base of the slope, with deposits characterized by a fining-up grain-size distribution and changes in cross-stratification type that indicate an upward waning of current velocity.

turnover The product of opposing rates of species immigration to, and species extinction from, an island. Where these rates precisely balance, a dynamic equilibrium is formed, with species richness remaining constant through time, but with composition continually changing.

turnover rate The number of species eliminated and replaced per unit time.

tussock Grass that grows in clumps, not evenly.

tuya A flat-topped, steep-sided volcano formed when lava erupts through a thick glacier or water.

ultrahigh-pressure metamorphism (UHPM) A type of metamorphism observed at many plate collision zones. Metamorphics contain relic grains of coesite or microdiamond, indicating formation at depths as great as 100 km, commonly found in mountains uplifted by continent–continent collision.

ultramafic Composed entirely of mafic (magnesium- and iron-rich) minerals.

ultramafic soil A metalliferous soil resulting from the weathering of ultramafic rocks and characterized by low nutrient and (on New Caledonia) high nickel concentration.

underground water Accumulation of water underground, due to fast percolation of rainfall and surface water into porous soils and bedrocks. Occurs on islands where sandy deposits, limestone rocks (including coral), or volcanic layers are dominant.

UNESCO The United Nations Educational, Scientific and Cultural Organization, whose mission is to encourage the identification, protection, and preservation of cultural and natural heritage around the world considered to be of outstanding value to humanity.

ungulates Hooved animals such as pigs, deer, or cattle.

uplift The rise of part of the Earth's crust relative to its center.

upwelling A process by which warm, less dense surface water is drawn away from a shoreline by offshore currents and replaced by cold, denser water brought up from the subsurface.

vagile Highly mobile with good dispersal potential; the opposite of sessile.

vagility The capacity of animals to disperse, either under their own power or passively (e.g., as eggs or cysts).

varanid A lizard of the family Varanidae, the monitor lizards. The Komodo dragon is the world's largest varanid.

vascular plants Ferns, flowering plants, and other plants that circulate water and nutrients in conductive (vascular) tissue called phloem and xylem.

vector 1. An organism that transports another organism from one place to another. 2. A species such as a mosquito that transmits a disease between hosts.

vegetable sheep A prostrate, mat-forming perennial New Zealand herb with woolly, hoary leaves.

vegetative reproduction Asexual plant reproduction in which new individuals (ramets) arise without the production of seeds.

VEI See VOLCANIC EXPLOSIVITY INDEX.

vein In geology, a fissure, crack, or channel in rock or ice that has been filled with a mixture of minerals.

velamen radicum In desiccation-tolerant vascular plants, a specialized outer layer of mostly aerial roots, consisting primarily of dead cells that rapidly absorb water.

vent The opening at the Earth's surface through which volcanic materials are ejected.

vesiculation Exsolution of dissolved volatile components (water, carbon dioxide, etc.) from magma to form

bubbles; the energy provided by bubble expansion provides much of the energy of volcanic eruptions.

vicariance The separation of closely related organisms by a natural barrier or a vicariant event, such as a body of water or the rise of a mountain range isolating drainages, resulting in differentiation of the original group into new varieties, species, or other taxa.

volant Capable of flight.

volcanic arc A row of volcanoes formed where one of the plates that make up the Earth's outer shell is subducted below another. The cause of the volcanic activity is the release of volatiles or melt from the subducted slab. This causes partial melting in the adjacent mantle, and the melt rises to the earth's surface. The volcanoes typically lie along an arc when seen in plan view.

volcanic bombs Molten rock fragments larger than 65 mm in diameter that are ejected during a volcanic eruption; bombs may acquire aerodynamic shapes during transport through the air.

volcanic dome A rounded, steep-sided mound formed of very viscous magma, usually either dacite or rhyolite. Such magmas are typically too viscous to move far from the vent before cooling and crystallizing.

volcanic earthquake An earthquake characterized by high-frequency seismic signals thought to be generated by the fracturing of rock in response to the intrusion and migration of magma. Volcanic earthquakes often occur in swarms and usually precede the onset of volcanic activity, although they do not always culminate in a volcanic eruption.

volcanic explosivity index (VEI) A method for measuring the scale of an eruption by means of a 0 to 8 index of increasing explosivity.

Wadati-Benioff Zone See BENIOFF ZONE.

Wallace's Line A biogeographic boundary that separates Asian from Australasian species. It runs through the Malay archipelago between Borneo and Sulawesi with various northern extensions, but its precise location has been modified a number of times. (First proposed by Alfred Russel Wallace when he observed the sudden changes in bird families between the islands of Bali and Lombok.)

washover fan A fan-shaped sheet of sand deposited on top of a barrier island during an unusually high tide (either astronomically- or storm-induced).

washover terrace A continuous sheet of sand formed where several washover fans merge together.

wave The profile of the sea surface elevation between two successive downward zero-crossings of the mean sea surface.

wave breaking The process by which organized kinetic and potential wave energy is converted to turbulence and dissipated.

wave-dominated coast A depositional coast where waves are the dominant process shaping the geomorphology of the coast (e.g., long barrier islands, numerous parallel beach ridges on arcuate-shaped deltas, or continuous sandy shores), usually in areas with small tides.

wave-driven flows Currents that are driven by gradients in wave momentum, usually associated with wave dissipation by breaking or bottom friction.

wave-formed bar A large, ridgelike bedform that develops in the surf zone seaward of sand beaches.

wave group The association of several to many individual waves that vary in height from small to large to small again.

wave group velocity The speed and direction at which a wave group propagates.

wave refraction The bending of a wave as it moves into shallow water so that wave crests become parallel to bottom contours.

wave transformation The process by which waves change as they move from deep to intermediate to shallow water.

weatherly Capable of sailing upwind and relatively close to the direction of the wind.

West Indies The Antilles island arc in the Caribbean Sea extending from Cuba south and east as far as, but excluding, Trinidad and Tobago.

weta Any of a number of species of orthopterans belonging to the Anostostomatidae (king crickets) and Rhaphidophoridae (camel or cave crickets), most endemic to New Zealand.

wetlands Areas with sufficient water to support vegetation adapted for life in saturated soils, including swamps, marshes, bogs, and similar areas.

whale fall A common term for a whale carcass resting on the ocean floor.

wildlife translocation A wildlife management technique in which a population of plants or animals is moved to a site where it is not present currently.

wind shear Variation of the wind vector over a distance. Also, SHEAR.

World Conservation Union See IUCN.

World Heritage Site A site (such as a forest, mountain, or lake) that is on the list maintained by the international World Heritage Program administered by the UNESCO World Heritage Committee.

wrack Seaweed floating in the sea or growing on the shoreline.

xenolith In geology, an exotic fragment, typically coarse-grained, included in a volcanic rock (e.g., an upper mantle peridotite included in basalt).

xeric Of, marked by, or adapted to, a very dry habitat.

xeromorphic Adapted to drought and/or arid conditions, using strategies to store water in the leaves or stem (e.g., succulents).

xerophytic Describing or referring to plants adapted to cope with limited water supply.

Zealandia The continental crust of which New Zealand and some other islands form the modern emergent parts. Some 93% of Zealandia is today under the sea.

zonation Living in zones; the restriction of particular species to one of the life zones (e.g., wet zone, dry zone, cloud forest) instead of being found all over an island.

zooarchaeology The study of animal remains (bones, shell, teeth, etc.) to understand human diet, subsistence practices, and other phenomena of ancient times.

zoogeography A branch of biogeography that studies the distribution of animal species over space and time.

zoonosis A disease of wild animals that can be transmitted to humans.

zooplankton Small, free-swimming animals that drift in lake or ocean currents. Considered a pivotal component of food webs because they graze on algae and are a source of food for higher trophic levels such as aquatic insects and fish.

zooxanthellae Unicellular dinoflagellate algae that live intracellularly within the gastrodermal cells of corals as symbionts and which provide much of the daily energy needs for the host coral through the translocation of photosynthetically derived metabolites.

zooxanthellate Of corals, possessing zooxanthellae.

INDEX

Boldface indicates main articles. Italics indicate illustrations and tables. For information about specific islands, see groups or regions to which they belong, and vice versa. For information about specific taxa, see other taxa to which they belong, and common names.

'a'ā, 543, 545, 711, 823
aboriginal introductions, 272, 273, 274, 275, 891
aboriginal peoples
   Australia, 114, 768
   French Polynesia, 332
   Rottnest Island, 796
   Tasmania, 904, 907–908, 958, 959, 960
   Tenerife, 742
   whaling and, 975–976
Abrolhos Islands, 956
acclimatization societies, 473, 672
accretion
   Antarctic, 18
   Antilles, 29, 30, 31
   barrier islands and, 86
   Borneo, 657
   Canadian Shield and, 77
   faults and, 903
   Iceland, 429, 430
   island arcs and, 483
   Japan, 502, 503, 504, 505
   Maldives and, 587
   mangrove islands and, 592
   New Guinea and, 653, 657, 661, 662, 664
   New Zealand, 677, 678
   Philippines, 738
   Taiwan and, 902, 903
   Vancouver Island, 708, 937
   vicariance and, 950
   waves and, 880
achenes, *227*
acidification, ocean, 70, 379, 780, 782, 821. *See also* pH
acoustical signaling
   birds, 106, 110, 111, 355, 399, 413, 713
   crickets, 6, 207, 208, 209, 210, 211, 463, 465, 826
   fishes, 826

flies, 6, 465
frogs, 27, 725
humans, 762, 764
mammals, 589
spiders, 842
tarsiers, 553
Acrididae, 670
Adamson, 541
adaptation, local, 52, 224, 226, 353, 355, 594, 911
adaptive convergence. *See* convergence, evolutionary
adaptive radiation, **1–7**. *See also* convergence, evolutionary; insular radiation; *specific islands; specific organisms*
   anagenesis vs., 8
   autochthonous, 157, 465
   competition and, 89
   endemism and, 257
   environmental factors and, 353
   evidence for, 251–253, 257
   genes and, 465, 477, 526
   geology and, 221
   gigantism and, 373
   inbreeding and, 437
   invasive species and, 475
   isolation and, 13
   lakes and, 527–528, 531, 606
   sexual selection and, 828
   taxon cycles and, 912
   zones, 772–774
Aders's duiker, 983, 986
Admiralty Islands, 349, 471, 569, 720, 721
Adriatic islands, 471, 564, 622, 623, 625, 626, 628
advertising, 765
adzes, 42, 43, 45, 46, 47, 247, 721, 722, 759
Aegean islands, 482, 623, 626–627, 628, 857, 859

Aegean Sea, 628
Aeonium, 4, 128, 129, 774
Africa. *See also* cichlid fish; North Africa; South Africa; *specific locations*
   ants and, 36, 38, 40
   barrier islands, 83
   bird divergence, 108–109
   birds, 122
   Canary Islands and, 129, 131
   Comoros and, 178, 179, 180
   conservation, 784
   crickets, 208
   dispersals, 560
   Great Rift Valley lakes, 7, 165
   mammals, 121
   missionaries and, 637
   monocotyledons, 467
   Newfoundland and, 650
   rabbits, 121
   rock sequences, 66
   spiders, 863
   succulents, 468
   vicariance, 950
African plate, 64, 65, 71, 142, 144, 215, 388, 392, 394–395, 482, 620, 622, 624, *753*
African violets, 943–944
agamids, 180, 618, 869, 906
age of islands. *See also* progression rule
   ants and, 36, 37, 39
   Azores, 71
   beaches and, 94
   Cooks, 191
   cricket diversity, 209
   deforestation and, 222
   distribution of species and, 181
   diversity and, 73, 320, 840
   drowned islands and, 370
   endemic taxa and, 948–949
   endemism and, 254

1019

age of islands (continued)
    fossils and, 319
    fragmentation and, 186–187, 330
    French Polynesia, 342
    freshwater species and, 343–344, 345
    Gough Island, 930
    granitic vs. oceanic, 381
    Hawaii and, 398, 406, 407, 408
    hotspot theory and, 339
    insects and, 463, 464
    landslides and, 340
    Madagascar, 577
    mantle plumes and, 368
    motu, 642
    New Zealand, 676
    plant disease and, 750
    pre-European impacts and, 417
    radiation zone and, 773–774
    rate of succession and, 879
    Samoa, 802–803
    speciation rates and, 210
    spiders and, 864
    Surtsey Island, 888
    timing of species formation and, 412
    Tristan da Cunha, 929
age of taxa, 319
agoutis, 90
agriculture. See also domestication; ethnobotany
    Britain and Ireland and, 117–118
    Canary Islands, 130, 132
    Cape Verde, 144, 146
    Caroline Islands and, 148
    Channel Islands (British), 155
    Comoros, 180
    conservation and, 224
    Cook Islands, 195, 196
    coral reefs and, 784
    crickets and, 209
    deforestation and, 222
    Faroe (Faeroe) Islands, 291, 292
    fragmentation and, 330
    French Polynesia, 336
    Greek Islands, 391, 393
    Gulf of Guinea, 810
    lava and, 543
    Mediterranean, 629
    Mesoamerica and, 184
    missionaries and, 636
    motu and, 642
    New Guinea, 658
    New Zealand and, 672
    oases and, 688–689
    overviews, 942
    Phosphate Islands and, 740
    plant disease and, 749
    pre-European, 417
    Samoa, 801
    Seychelles, 832
    Socotra, 850
    Solomons and, 853, 854
    St. Helena, 873
    sustainability and, 889–890
    Taiwan, 901
    Tonga, 919

    Tristan da Cunha/Gough, 931, 932
    Vanuatu, 941
    Zanzibar, 984, 985
Agrihan, 594, *600*
Aguijan, 594
Ailsa Craig, 125
Ainu people, 522
air flow, 172–173
air transportation, 477, 479
Aitutaki Island, 192, 193, 194, 195, 196, 342
Alaid, 484
Alamagan, *598*, 599, 601
Alaska
    ash, 543
    global warming and, 973
    landslides, 536
    missionaries and, 634, 637
    plants, 938
    sea levels and, 48
    shipwrecks, 834
albatrosses, 10, 12, 16, 104, 358, 574, 632, 672, 794, 811, 812, 813, 814, 932
Albertine Rift, 513
Alcatraz, 770
Alcids, 813
Aldabra, 179, *577*, 830, 831, 832, 833, 921, 922, 924, 926
Alderney, 154
Aleipata, 799
Aleutian Islands, 52, 481, 483, 702, 705, 814, 874, 934, 977–978
Alexander archipelago, 52
Alexander Island, 11, 17, 18, 19, 874
algae
    Antarctic, 12
    arctic, 50, 55, 56
    atolls and, 68
    Bermuda, 96
    Canary Islands, 131
    Cape Verde, 146
    Cook Islands, 194
    coralline, 199
    ephemeral islands, as, 259
    Farallon Islands, 296
    French Polynesia invasive, 337
    Japan, 497
    jellyfish and, 716
    Macquarie, 574
    New Guinean beetles and, 656
    organic falls of, 701
    overfishing and, 310
    Philippines, 724
    Rottnest Island, 797, 798
    Surtsey Island, 886
    sustainability and, 891
    Tatoosh, 909, 910, 911
    waves and, 880
algal ridges, 68, 205, 206
algarrobo trees, 23, 27
Ali, 968
alien species. See also exotic species; introduced species; invasive species
    Antilles, 23, 26
    Ascension, 62
    Azores and, 71, 74

    Bermuda and, 97, 98
    Britain and Ireland and, 118
    climate change and, 15
    Darwin on, 959
    dispersal and, 228
    freshwater species and, 346
    Hawaii and, 36, 944
    lakes and, 531
    Macquarie, 574
    sub-Antarctic, 14–15
    Tonga, 920
    Vancouver, 938
alkalinity, 176–177
Allee effects, 478
Allen, Charles, 968
allopatric speciation. See also isolation
    caves, 151
    climate change and, 786
    fish, 166
    Galápagos, 773
    Indonesia, 452
    insect, 81, 123, 462
    lizards, 561, 563, 564
    moa, 640
    neotropical, 786
    refugia and, 786
    sexual selection and, 826
    size of island and, 859
    snails, 539
    spiders, 864
    Wallacea, 452
    weta, 669
allopolyploid speciation, 255, 838
de Almeida, Fernando, 298
aloes, 132, 850
Alpine Fault, *674*, 754
altitude. See elevation
altruism, 225, 226
aluminum ore, 149
Amanu, 334
Amazon, 964
amber, 23, 27, 38, 561
Amblypygi, 849
Ambon, 448, 660
Amborellaceae, 644
American Samoa, 104, 378, 637, 712–713, 783, 799, 801, 802. See also Samoa
Amerindians, 23, 24, 25
Amirante Island, 177
ammonites, 937
Ampère seamount, 64
amphibians
    Antilles, 27
    Bermuda, 98
    Borneo/New Guinea and, 658
    Britain and Ireland and, 122
    Channel Islands (California), 159
    Comoros, 179
    Fijian, 301
    Gulf of Guinea, 809
    Indonesia, 449, 450
    Japanese, 499
    Kurile Islands, 525
    Madagascar, 580
    New Guinea, 653, 654, 656

New Zealand, 667
overviews, 347
Philippines, 724, 729
pigs and goats and, 742
Seychelles, 830, 831, 833
sustainability and, 891
Taiwan, 899
Vanuatu, 940
Wallacea, 452
Zanzibar, 983
amphipods, 151, 701, 907, *907*
Amsterdam Island, 320, 441, *789*
Amundsen-Scott station, South Pole, 15
Anacapa Island, *156,* 157, 158, 161, 162, 163, 795
anagenesis, **8–10,** 698, *773*
Anak Krakatau, 206
analogues, 619
Anatahan Island, 598, *598, 599, 599,* 602, *952*
Anatolian plate, 388, 392
Andaman Islands, 114, 439, 771
Anderson, Edgar, 415
Andes, 217
Andesite Line, 951
andesites, 31, 694, 754, 755
Adriatic islands, 471
anemones, *700, 853, 909,* 910
anerobic oxidation of methane (AOM), 176–177
Ángel de la Guarda Island, 80, 81, 82
angiosperms, 145, 146, 193, 360, 466, *644,* 667, 671, 709, 713, 800, 899, 920
Anglesey, 125
Anguilla, 22, 24, 29, 31
Anjouan Island, 177, 178, 179, 180
Annamite range, 840
Annobón Island, 65, **808–811**
Anolis (anoles)
    amber, in, 23
    colonizations, 560
    convergence, 189, 190
    dwarfism, 236
    insular radiation, 3, 5, 7, 27, 531, 561–564, 773
    invasions, 477
    isolation, 859
    rafting, 775
    sexual selection and, 826, 828
Anson, George, 509
Antarctic
    adaptive radiation, 13
    biology, 10–16
    bird disease and, 105
    endangered species, 908
    flightless insects, 227
    flightlessness and, 317
    geology, 17–20, 504
    global warming and, 815
    island arcs and, 481
    research stations, 789
    rock sequences, 66
antelope, 51, *120,* 983
anthropogenic disturbances. *See* human impacts
Antigua, 23, 24, 31, 32, 35, 278
Antilles
    adaptive radiation, 24, 27

biology, 20–29, 36, 190
colonization, 278–279
convergent evolution, 7
geology, 29–35
Antipodes, 11, 16, 675
ants, **35–41**
    adaptive radiation, 37–38
    Antilles, 23
    Barro Colorado, 88, 91
    Britain and Ireland and, 122, 123
    crickets and, 211
    Fiji, 36, 38, 300, 775
    Hawaii and, 865
    invasive, 39, 40, 472, 473, 477, 479, 708
    Madagascar, 580
    maximal radiation, 773
    Melanesia, 38, 253, 912
    New Caledonia, 645
    radiation zone and, 772
    silverswords and, 838
    Vanuatu, 941
Anvers Island, 17, 18, 19
Anzhu Island, 51
aphids, 49
apomictically reproducing plants, 118
apomixis, 697
Appalachians, *152,* 649, 650, 652
aquaculturists, 346, 347
aquatic species. *See also* fish; marine life and environments; seabirds
    cave, 150, 151
    dispersal, 528
    invetebrates, 131
    New Guinea, 656
    plants, 101
    vertebrates, 22
Arabia, 104, 146, 215
Arabian plate, 753
Arabs, 832, 984
arachnids, 130, 146, 300–301. *See also* spiders
Arago sea mount, 342–343
Arakamchechen Island, 51
Arapawa Island, 675
arboriculture, 224
archaeology, **41–47.** *See also* fossils
    ants and, 40
    Bermuda and, 97
    Britain and Ireland and, 119, 120, 121, 124, 126
    Caribbean, 279
    Caroline Islands, 149
    Channel Islands (California), 160, 161
    Cozumel, 204
    Europe, in, 117
    Faroe (Faeroe) Islands, 287–288
    languages and, 759
    Micronesia, 722
    Oceania, 42
    pigs and, 741
    Polynesian voyaging and, 759
    pre-European impacts and, 417
    Sarawak caves, 114
    seafaring and, 276–277
    Sri Lanka, 867
    whaling and, 975
    Zanzibar, 984, 985

archeophytes, 119
archipelagoes. *See also specific archipelagoes*
    adaptive radiation and, 3
    anagenesis and, 9
    biodiversity and, 617
    caves and, 153
    glaciations and, 785
    insect radiations and, 460, 462–463
    mammals and, 588
    plant disease and, 749
    refugia and, 786
architecture
    Faroe (Faeroe) Islands, 292
    Polynesian, 46–47
    Rapanui, 246
    Solomons, 853
    whaling and, 976
arcs. *See* island arcs
Arctic
    fishes, 874
    seabirds, 814
    whales, 976
Arctic islands
    stereotypes, 761
arctic islands, biology, **47–54**
arctic islands, geology, **55–59**
Arctic Ocean, 55, 58, 86
arctic region, **59–61**
    birds, 126
arctic vegetation
    Britain and Ireland and, 124
area factors. *See also* size of islands; species-area relationship (SAR)
    equilibrium and, 487
    island rule and, 495
    oases and, 688
Argentina, ants and, 40
Argyranthemum, 4, 129, 583
Arizona, 839
armor, 874, 876
Armorican Massif, 154
Arran Island, 116, 118, 125, 255
art, 763
arthropods, 3, 13, 907
    ants and, 39, 40
    arctic, 51, 53
    Azorean, 71, 73, 74
    Borneo, 115
    Canary Islands, 130, 132
    Cape Verde, 145
    cave, 548–549
    Fijian, 300
    French Polynesian endemics, 335
    gigantism, 375
    Hawaii, 36
    lava cave, 548
    Madeira, 584
    matrix-derived taxa, 182
    Midway Island, 632
    New Caledonia, 645
    Solomons, 854
    sub-Antarctic, 15
    Vanuatu, 940
    volcanism and, 862
Aruba, 29

INDEX 1021

Aru Islands, 448, 653
Ascarina, 940
Ascension Island, **61–63**, 66, 218, 320, *789*
   geology, 951
   introduced species and, 471
   invasive species, 959
   research stations, 789
   spiders, 862
   volcanism, 959
asexual reproduction, 285
ash, volcanic, 543–544, 664, 744, 884
Asia
   ants and, 36, 38
   arctic and, 51
   biota, 114
   birds, 122
   conservation, 784
   crickets, 208
   frogs, 7
   mammals, 121
   missionaries and, 637
   prisons, 770–771
Asian plate, 482
Asians, 149
asteroids, *700*
asthenosphere, 752–753
Asuncion, 595, *598*, *599*, *600*
Atiu Island, 192, 193, 194, 196, 197
Atkinson, Carter, 414
Atlantic Ocean, 63, 288, 379, 426, 482, 650
Atlantic region, **63–67**
   bird radiation, 106
   Canary Islands and, 131
   coral reefs, 200, 781
   fishes, 874
   flightless birds, 314, 316–317
   mammals and, 588
   overfishing and, 311
   research stations, 789
   seamounts, 819, 823, 824
   vents, 424
atmospheric features, 172
atoll lagoons, 219
atolls, **67–70**
   Caroline Islands, 149
   Cook Islands, 193, 194
   Darwin and, 220
   Darwin on, 219, 958
   flora, 276
   French Polynesian, 711
   geology, 754
   Hawaiian, 631
   Indian Ocean region, 442
   Maldives, 587
   Marshall Islands, 712
   motu and, 641
   overviews, 694
   Pacific region, 704, 708–709
   snails, 949
   Solomons, 855
   tsunamis and, 934
   volcanism and, 950
   western Pacific, 715
Auckland Islands, 11, 12, 14, 16
   geology, 675, 679

weevils, 13
auklets, 813
   Channel Islands (California), 159
   Farallon Islands, 297
auks, 124, 811, 813
   Britain and Ireland and, 124
Aure Fold Belt, 664
aurocks, 120–121
   Britain and Ireland and, 120
Australasia, 971
   Gondwana and, 669
Austral Fracture Zone, 340, 343
Australia
   aborigines, 114
   barrier reef, 200
   beetles, 7
   biocontrol and, 102
   biotas, 114
   bird disease and, 105
   birds, 6, 304, 312
   crickets, 208
   Darwin and, 958
   extinctions, 170
   flies, 571
   frogs, 348
   geology, 447, 448, 458, 504, 755
   global warming and, 380
   Indonesia and, 453
   inselbergs diversity, 469
   introduced species, 470, 477
   island arcs and, 485
   island rule and, 493, 495
   lava, 545
   monitor lizards, 513–514
   New Guinea and, 656
   New Zealand and, 677
   orchids, 697
   pigs and goats and, 743
   plant disease and, 748
   prisons, 769
   seafaring and, 277
   shipwrecks, 834
   snakes, 844
   spiders, 863
   Wallace and, 968
   Wallace's Line and, 970
   wet forest, 842
Australian Island Territories, 572
Australian plate, 448, *753*
   Indonesia and, 459
   Macquarie Island and, 575, 576
   New Caledonia and, 648
   New Guinea and, 653, 660, 664
   New Zealand and, 674, 675, 676, 679
   overviews, 705
   Solomons and, 703
   Tasmania and, 904
Australian region
   lizards, 560
Austral Islands, 709
   architecture, 47
   Cook Islands and, 192
   endemism, 336
   geology, 342–343
   languages, 759

   overviews, 332
   volcanism, 953
Australo-Melanesians, 114
Austronesians, 42, 43
   Borneo and, 114, 115
   Near Oceania and, 721
   seafaring and, 277
   Taiwan and, 900
authigenic carbonates, 176–177
autochthonous, 660
autochthonous differentiation, 255
autochthonous speciation, 729
Avars, 54
Aves Ridge, 21, 30
avian influenza, 105
avian malaria and pox, 103–104
avifauna. *See* birds
Axel Heiberg Island, 50, 58, *59*
Axial Seamount, 713
Azores archipelago, 64, **70–74**, 127
   birds, 106
   coral conservation, 567
   Darwin and, 220
   Drosophila, 234
   eruptions, 952
   geology, 755, 951
   invasive species, 346
   spiders, 862
   volcanism, 959
   whaling, 976
Azores-Gibraltar fracture zone, 64, *133*
Azores Plateau, *424*, 426

Babeldaob Island, 149
back-arc basins
   Antarctic, 19
   Antilles, 30
   Atlantic region, 66
   Caroline Islands, 149
   Fiji, 306, 307
   Indonesia, 459
   Japan, 486, 502, 506
   Mediterranean, 626
   Pacific region, 481, 483, 484, 485, 486, 706
   Philippines, 733
   vents and, 424
back-colonizations, 254, 464, 585
Backer, C. A., 517
bacteria, 974
   inselbergs and, 467
   lakes and, 530
   natural selection and, 6
   organic falls and, 700
   Pantepui, 718
   plant disease and, 748
   succession and, 877
   Surtsey Island, 886
   sustainability and, 891
badgers, 120, 375
Baeropsis, Asteraceae, 78
Baffin Bay, 55, 58
Baffin Island, *48,* 50, 54, 58, 60, **76–79**
Bahamas
   ants, 859
   climate change and, 170

geology, 651
global warming and, 378
hydroclimate, 423
rodents, 23
spiders, 860
watercraft, 279
Bahamian Bank, 22
Baja California, **78–82**, 81, 157, 162, 703
Baker, Ian, 871
Baker's Law, 697, 699, 838
balance of species richness, 486–487
    Hawaiian, 944
Balcones Escarpment, *152*
Balearic Islands, 559, 624, 628, 629
    rats and, 794, 795
    spiders, 862
Balfour, I. B., 848
Bali, 450–451, 970
Ballard, R., 424
Balleny Islands, 11, 17, 20, *423*
Ball, Henry Lidgbird, 571
Baltica, 56, 504
Baltic archipelagoes, 750–751
Banda arc, 459
Banda Sea, 756, 757
bandicoots, 906
Bangladesh, 591
Banks Island, 50, 51
Banks, Sir Joseph, 219
Barbados, 24, 29, 31, 35, 483
Barbuda, 23, 24, 31, 35
Barentsz, Willem, *866*
Barlavento group, 65
barnacles, 426, 909, 910, 957
Barnes Ice Cap, 77
Barombi Mbo (Cameroon), 166
Barra, 123–124
barrage lakes, 331
Barren Island, 484
barrier islands, **82–87**, 92, 93, 211, 420, 474, 536, 673, 675, 743, 915, 916
barrier-reef islands, 333
barrier reefs, 69, 200
    Cook Islands, 194, 195
    Fijian, 299, 707
    overviews, 219
    western Pacific, 715, 716
barriers, insect diversity and, 463
Barro Colorado Island, **88–91**, *181*
Barton, C., 416
basalts
    adzes, 45
    Antarctic, 19, 20
    Antilles, 31
    arctic, 55, 58
    Atlantic, 64, 65, 66, 67
    Baffin Island, 76
    Baja California, 80
    beaches and, 93
    Borneo, 459
    Canary Islands, 135, 136, 137, 138, 139, 140
    Cape Verde, 144
    Caroline Islands, 149
    Channel Islands (British), 154
    Channel Islands (California), 163, 164

Comoros, 177
Cook Island, 193
coral and, 200
Cyprus, 213, 214
Darwin and, 218, 220
Faroe (Faeroe) Islands, 288
Fernando de Naronha, 298
Fiji, 306, 307, 308
French Polynesia, 338, 339, 340, 342
Galápagos, 370, 371
Hawaii, 404, 406, 409, 807
Iceland, 428, 429, 430, 431, 432, 433, 434
Indian region, 437, 438, 439, 440, 441, 442, 443, 444, 445
Isla de Guadalupe, 377
island arcs and, 483, 484, 485
Japan, 503, 505
Kick 'em Jenny, 510, 511
Kuriles, 522, 523
Laki vs. Tambora, 270
lava flows, 542, 543, 544, 545, 546, 547
Line Islands, 554, 711
Lord Howe Island, 568–569
Macquarie Island, 576
Madeira, 582
Mariana Islands, 598, 600, 601
marine species and, 378
Mascarenes, 613, 620, 622
Mediterranean Islands, 625, 626, 627
Mid-Atlantic Ridge islands, 959, 981
New Caledonia, 647, 648
New Guinea, 663, 664
New Zealand, 675, 679
North Atlantic, 288–289
oceanic islands and, 380, 381, 690, 693, 951
overviews, 690, 754
perched beaches and, 93
Pitcairn, 744
plate tectonics and, 752, 753, 754, 755, 822–823
Samoa, 803, 803, 804, 805, 806–807
São Tomé Island, 809
seamount, 821, 822, 825
Solomons, 855, 856
St. Helena, 870, 871
Surtsey Island, 883
Taiwan, 897, 898
volcanism and, 339, 693–694, 951, 953
basins. *See also* back-arc basins; *specific basins*
    East Asian, 505–506
    overviews, 752, 756
    Philippines, 735
    seamounts and, 824
    Solomons, 854, 856
Basque people, 976
Bates, Henry Walter, 963
bathymetry
    anomalous, 951
    arctic islands, 55
    Azores, 71
    Canary and Madeira Islands, 133
    Canary and Selvagen Islands, 141
    Caroline Islands, 148
    Channel Islands (California), 156
    Crater Lake, 979, 980

French Polynesia, 339, 341
Galápagos, 367
Indonesia, 447, 454, 455
New Zealand, 674, 678
ocean circulation and, 825
Philippines, 727, 728
Samoa, 802, 806
seamounts, 714
Sunda shelf, 112
tsunamis and, 933
waves and, 882
bathyspheres, 97
bats, 79, 81
    Antilles, 23, 26
    Azorean, 72
    Barro Colorado, 89, 90, 91
    Britain and Ireland and, 126
    Canary Islands, 129, 130–131
    Cape Verde, 144
    cave/islands, 152–153
    Channel Islands (California), 158
    Comoros, 179
    Cook Islands, 194
    dispersal, 158
    dwarfism and, 236
    extinctions, 845
    Farallon Islands, 295
    Gulf of Guinea, 809
    island rule and, 492
    Krakatau and, 519
    Lord Howe Island, 569, 570
    mangroves and, 593
    Mascarene Islands, 615, 617
    New Guinea, 655
    New Zealand, 670–671
    overviews, 588
    Pacific region, 776
    Palau, 716
    Philippines, 726
    Rarotonga/Mangaia, 709
    Samoa, 800, 801
    Seychelles, 833
    Socotra, 849
    Solomons, 852
    Taiwan, 899, 900
    Tonga, 920
    Vanuatu, 940
    Wizard Island, 981
    Zanzibar, 983
bauxite, 149
Bay of Bengal, 420
beaches, **91–94**. *See also* dunes
    Bermuda, 97
    Britain and Ireland and, 126
    Canary Islands, 127
    Cape Verde, 146, 147
    Channel Islands (California), 154
    Cook Islands, 194
    Cook Islands forests, 193
    geology, 642
    Great Barrier Reef, 385
    Hawaiian Island, 400
    Midway Island, 632
    seawalls and, 261
    tides and, 914, 915

beachrock, 86
Beagle, HMS, 702, 918, **954–961**
beaks
    character displacement and, 106
    divergence and, 108
    finches, 352, 355–356
    isolation and, 108
    kiwi, 670
    Madagascar, 579
    New Zealand, 667
    reproductive isolation and, 110–111
    Samoan bird, 800
bears, 981
    Baffin Island, 78
    Britain and Ireland and, 119, 120
    island rule and, 492
    Kurile Islands brown, 525
    polar, 54
Beaufort Island, 17
beavers, 471, 793, 918
    Britain and Ireland and, 120
Beebe, William, 97
Beeches Bay formation, 163
beech trees, 472, 670, 905
    New Zealand, 672
bees, 3, 23, 53, 472, 667, 672, 838, 929, 981
    introduced species, 472
    New Zealand, 672
beetles, 462, *902*
    ants and, 39
    arctic, 51, 52, 53
    Azorean, 72, 73
    Borneo, 114, 964
    Britain and Ireland and, 122–123
    Canary Islands, 128–129, 130, 774
    Cape Verde, 145, 146
    Comoros, 179
    Cratopus, 2, 615
    endangered species, 285
    Fijian, 300
    flightless, 318
    Galápagos, 361
    Hawaiian, 772
    Miocalles, 3, 335
    New Guinea, 656
    New Zealand, 672
    Polynesian, 772
    Rynchogonus, 3
    scarabs, 7, 460, 899, 902
    single-island radiations, 462
    St. Helena, 871
    Tasmania, 907
    taxon cycles, 253
    Wallacea, 452
behavior
    adaptive radiation and, 5
    Barro Colorado research, 88–89
    dispersal and, 225
    diversification and, 462
    diversity of, 166
    ecology and, 412–413
    extinction and, 286
    Hawaiian *Drosophila* and, 233
    learned, 109
Belau, 148

Beliliou Island, 149, *738, 739*
bellflowers, 944
Bellinghausen, 337
Bellona Platform, *695*
Bellwood, Peter, 45
Benioff zone, 30, 32
Bequ Island, 308
Berau, marine lakes and, 605
Beringian islands, 48, 50, 51
Bering Sea
    fishes, 874
    seabirds and, 812
Berkner Island, 17
Bermuda, **95–98**. *See also* oceanic islands
    ants and, 40
    shipwrecks, 834
Bermuda Islands, 65
Bermuda Rise, 65
bet-hedging strategies
    dispersal and, 226
bettongs, 906
Big Ben volcano, 440–441
Big Island, Hawaii
    crickets, 211
Big South Cape Island, 673, 674, 675, 793, 794, 795
Bikini (Pikini) atoll, 69, *610, 611,* 680–681, 683, 684, 712
Billefjorden, Svalbard, *58*
bills. *See* beaks
biocontrol. *See* biological control
biodiversity
    biogeography and, 107
    fragmentation and, 329
    influences on, 29
    kīpuka and, 512
    overviews, 888, 889–890
    threats to, 894
biodiversity hotspots
    Borneo, 112, 113
    coral, 566
    Florida snails, 542
    French Polynesia, 335
    Gulf of Guinea, 808, 809
    Indian region, 2
    Japan, 500
    Lord Howe Island, 568
    Madagascar, 577, 578
    Mascarene Islands, 612
    New Caledonia, 643
    New Guinea, 652
    overviews, 890, 892
    Pacific region, 538
    seamounts, 820
    Socotra, 848
    Sri Lanka, 868
    Sundaland/Wallacea, 447, 449, 450, 452
    Taiwan, 902
    Vanuatu, 939, 940
biogeography, **486–490**. *See also* Equilibrium Theory of Island Biogeography
    Antarctic, 12–13
    Antilles, 27, 28
    Baja California Islands, 82
    Barro Colorado and, 91

    cladistic, 970
    Darwin and, 220–221
    distribution of species and, 947–948
    diversity and, 107
    endemism and, 254
    fragmentation and, 328–329, 330
    French Polynesia, 332, 336
    geology and, 968–969, 970
    Greek Islands, 389
    overviews, 329–330, 527, 943
    pre-European impacts and, 417
    snails and, 538
    Socotra, 849–850
    vicariance and, 949
    Wallace and, 962
biogeomorphic agents, 197
bioherms, 565
Bioko Island, 65, 808
    spiders, 862, *862*
biological control, **99–102**
    Bermuda, 98
    freshwater habitat and, 346–347
    Hawaiian insects, 479
    invasive species, 478
    New Zealand, 672
    plant disease and, 748
The Biology of the Amphibia (Noble), 348
biota, Greek Islands, **389**
biotic environment. *See* communities
biotic stressors, 787
biotic variables, freshwater species and, 343
bird fossils, **318–326**
birds, 3, 189, 236. *See also* bird fossils; flightlessness; landbirds; migrants; seabirds; songbirds; *specific birds*
    adaptive radiations, 105–111
    Antarctic, 12
    Antilles, 21, 23, 26
    ants and, 39, 40
    arctic, 50, 51, 52, 53–54
    Ascension, 62
    Azorean, 72
    Baffin Island, 77, 78
    Baja California Islands, 78, 79, 80, 81
    Barro Colorado, 90, 91
    Barro Colorado biogeography and, 91
    Bermuda, 96, 98, *98*
    biogeography and, 841
    Borneo, 114
    breeding colonies/Fernando de Noronha archipelago, 298
    Britain and Ireland and, 118, 120, 122, 123–124
    Canary Islands, 131, 132
    Canary Islands flightless, 131
    Cape Verde, 144, 146, 147
    Channel Islands (California), 159
    Comoros, 178, 179, 180
    convergence of, 189
    Cook Islands invasives, 195
    deforestation and, 222
    disease, 103–105
    dispersal and, 227, 539
    dwarfism and, 238
    dwarfs, 237

endangered Javan, 451
endemism and, 256
extinctions, 228–231, 282, 283, 892, 920 (*see also* flightlessness)
Faroe (Faeroe) Islands, 288–289
food and, 843
fragmentation and, 186
French Polynesia, 335
freshwater species and, 345
Galápagos, 368 (*see also* Galápagos finches)
gigantism and, 373–374
Gould on, 220
Guam, 845
Gulf of Guinea, 809, 810
Hawaiian, 480, 786, 977
Hawaiian extinctions, 283
humans and, 44
Indonesia, 449, 450
introduced species and, 470, 471, 472, 473
island rule and, 493, 494
Japanese, 499
Japanese extinctions, 500
Krakatau and, 519
Kurile Islands, 525
Lord Howe Island, 569
mammals and, 321–326
Marianas, 594
Mascarene Islands, 615
Mauritius, 229
maximal radiation, 773
New Caledonia, 644
New Guinea, 653, 654–655
New Zealand, 99–100, 183, 638–641, 667, 669, 670, 671, 672, 673
oases, 687, 688
Pantepui, 718–719
Philippines, 724, 729
pigs and goats and, 742
Pitcairn, 745
Pitcairn extinctions, 747
radiations, 105–111
Rapanui climate change and, 245
Rottnest Island, 798
Samoa, 713, 800, 801
Seychelles, 830, 833
Seychelles extinctions, 832
sky islands, 842
Solomons, 852
species-area relationship and, 857
Spitsbergen, 866
St. Helena, 873
St. Paul's Rocks, 956
Surtsey Island, 885, 886, 887
sustainability and, 891
Taiwan, 899, 900
Tasmania, 907, 908
taxon cycles, 253
Tonga, 920
Vanuatu, 940
Wallacea, 452
Weber's Line and, 970
Wizard Island, 981
Zanzibar, 983
birds of paradise, 6, 303, 453, 655, 656, *657*, 826
New Guinea, 655, 655, 656, 657

sexual selection and, 826
Biscoe Islands, 17
Bismarck Archipelago
frogs, 349
geolgoy, 664
human colonization, 720, 721
humans and, 42, 43
lizards, 560
pigs, 741
bison, 51
bitterns, 122
Bjørnoøya, 59
Black and White Islands, 17
blackbirds, *122*
blackbutt forest, 331
black corals, 198
Black Hills, *152*
Black Sea, 82
Blake plateau, 565
Blanca formation, 162–163
Blanco Transform Fault, 425
bleaching (coral), 202, 336, 378, *379*
human impacts, 782
Seychelles, 832
blocks, crustal, 457
blue tits, *122*
bluffs, and insects, *318*
boars, Britain and Ireland and, *120*
boats. *See* watercraft
Boatswain Bird Island, 62
Boavista Island, 143, 144, 145, 146, 147
body size. *See also* dwarfism; gigantism; island rule
evolution and, 922
extinction and, 285, 788
food and, 640
Galápagos, 361
Komodo dragons, 514
lemurs, 550
spiders, 864
sticklebacks, 874, 877
body temperatures, 922
bogs, 836, 930
alien trees and, 945
Hawaiian, 401
Hawaiian trees and, 944
Holocene and, 433
succesion and, 946
Vancouver, 938
Bonaparte, Napoleon, 871, 931
Bonin, *277*, 307, 481–486, 499, 501, 506, 522, 545, 702
Bonnetia, 718
boobies, 197, 556, 812
Great Barrier Reef, 383, 384
human impacts, 814
Lord Howe Island, 570
overviews, 811
St. Paul's Rock, 956
Bora Bora, 333
snails, 540
volcanism and, 340
boreal forests, 185
fragmented, 185
Borneo, 36, **111–116**

climate, 971
climate change and, 171
frogs, 348, 349, 350
geology, 448, 457, 458–459
Indonesia, 451–452
Indonesian, 451–452
macaques, 778
mammals, 553
New Guinea and, 657–658
overviews, 447
snakes, 845
Bornhardt, Wilhelm, 466
bottlenecks
bird disease and, 103
birds and, 107
Galápagos, 372
invasive biology and, 477
refugia and, 786
sexual selection and, 827–828
Bougainville, 485, 706, 854, 855, 856
Boulenger, G.A., 348
boundaries, biogeographic, **967–971**. *See also* Wallace's Line
Bounty, HMS, 11, 16, 712, 745, 833
Bouvet Island, 67
Bouvetøya, 11
bowerbirds, 655
Bowie, Raurie, 842
Bowman, Bob, 361
boxwood, *872*, 874
Brabant Island, 17, 19
brachiopods, vent, 426
bracken, 124
Bransfield Strait, 19
Brava Island, 65, 143, 144
Brazil, 963
caves, 151
inselbergs, 466, 467, 469
introduced species from, 101, 195
sustainability, 888
tepuis, 718, 839, 840
Brazilian islands, 62, 65–66, 297, 298, 560, 956
Brecquou, 154
breeding islands, 320
whales and, 975
breezes, 173
Bridgeman Island, 19
brids, Socotra, 849
Britain and Ireland, **116–126**
age of, 428
birds, 120
Channel Islands (British), 154–155
fishes, 874
fishing, 310
glaciations and, 785
introduced species and, 472–473
missionaries and, 634, 637
overviews, 183
Pitcairn and, 746
rat eradication, 474
seabirds, 814
British Columbia, Canada, 4
British people, and Borneo, 115
Brocchinia, 719

Broken Ridge, 439–440
bromeliads, 152
Bronze Age, Mediterranean and, 741
Brook, B.W., 282
Brown, James H., 185, 840–841
brown noddies, 812
brown rats. *See* Norway rats
Brunei, 112
    frogs, 349
    overviews, 115
Brunet, A.K., 152
bryophytes, 12, 13, 14
    arctic, 51, 52, 54
    Atlantic, 72
    Azorean, 73, 74
    Britain and Ireland and, 124
    Cape Verde, 146
    Japan, 497
    Macquarie, 574
    Madeira, 583
    succession and, 877–878
    temperatures and, 48
Buck, Peter, 760
buckwheat, 130, 157
buffalos, 724, *726*
Bugis people, 115
bugs
    arctic, 53
    biocontrol and, 101
    Hawaiian, 462
    St. Helena, 872
    Tasmania, 907
Buka, 720, 854
bulls, Baffin Island, 78
Bunger Hills, Antarctica, 10
buntings, 108, 930–931, *932*
buoyancy, 258, 425
    island arcs and, 481
    mantle plumes, of, 440
    plate boundaries and, 753
Burton, Richard, 984
bustards, 122
Bute, 125
butterflies, 122
    Antilles, 23, 28
    arctic, 53
    Australia, 970
    Barro Colorado, 91
    biocontrol and, 102
    Britain and Ireland and, 118, 119, 122, 126
    Channel Islands (California), 159
    introduced species and, 473
    New Caledonia, 644
    New Guinea, 656
    New Zealand, 667
    novel adaptive zones and, 2
    Solomons, 852
    Surtsey Island, 886
    Taiwan, 899, 900
    Trinidad and Tobago, 928
    Wallacea, 452
    Weber's Line and, 970
    Zanzibar, 983
buttongrass, 905

cabbages, 125
Caccone, Adalgisa, 923, *923*–924
cacti, 81
    Galápagos, 359, 361, 362
Cadomian Orogeny, 154
caecilians, 831
cahows, 96, 97
Caicos Islands, 22
Cairn Peak, Antarctica, *10*
Caithness, 292
calcareous animals and plants, 70
calcium carbonate
    coral reefs and, 202
    overviews, 203
calcium, coral and, 199
calderas, 952, 953. *See also specific volcanoes*
    Kurile Islands, 523–524
Caledonia orogeny, 55, *56*, 57
California, avian diseases, 104
California Channel Islands, 5
California Continental Borderland, 162
California tarweeds, 4–5
Callitris forests, 331
Cambrian-Ordovician boundary, 649
Cameroon, 839
    inselbergs, 467
Cameroon line, 65, *840*
Campbell, I.H., 485
Campbell Island, 11, 16
    eradications, 673
    geology, 675, 677, 679
    gigantism, 671
    rat eradication, 474, 795
Canada
    coral, 565
    fishes, 874
    Greenland and, 50
    ponds, 526
    rodents and, 793–794
Canadian Arctic Archipelago (CAA), 47, 48, 50–51
    diversity of, 49
    human impacts, 54
    sedimentation and, 57–58
Canadian Shield, 77, *651*
Canary Islands
    adaptive radiations, 4, 774
    anagenesis, 9
    Beagle and, 956
    biology, 72, 127–133
    birds, 104, 106, 132
    Cape Verde and, 146
    coral, 567
    Drosophila, 234
    endemic species, 73
    flora, 146
    geology, 63, 64–65, 133–142, 691
    invertebrates, 462, 774
    kīpuka and, 513
    landslides, 536
    lizards, 557, 561, 563
    mammals and, 588
    pigs and goats and, 742
    rats, 794, 795

snails, 538
species-area relationship and, 857
spiders, 862, 863, 864
volcanism, 220
whales, 978
de Candolle, Augustin Pyramus, 254
canoe plants, 272, 275, 276
canoes, 758–759, 760
Canterbury Plains, 676
Canton Island, 173–174
Cape Horn, 918
capercaillie, 118, 122
Cape Romain, South Carolina, *84*
Cape Verde archipelago, 64, 65, 127, **143–147**
    bird divergence, 108–109
    Darwin and, 956
    Drosophila, 234
    gigantism, 373
    landslides, 536
    spiders, 862
    whales, 975, 976
Capian Sea, 83
Capricorn seamount, 692
carbon
    Antilles, 32
    coral and, 198, 199
    diversity and, 13
carbon dioxide, 70
    coral reefs and, 202
Caribbean, 36
    adaptive radiation of lizards, 531
    adaptive radiations, 3
    Anolis lizards, 775
    ants and, 37, 40
    Bermuda and, 96
    climate change and, 779
    conservation, 784
    coral reefs, 200, 378–379, 780, 781
    crickets, 209, 210
    Darwin's finches and, 359
    Drosophila, 233–234
    dwarfism and, 236
    ecological release and, 253
    ethnobotony and, 271, 273
    extinctions, 170
    flightless birds, 313, 316
    geckos, 559
    introduced species, 471
    invasive species, 347
    lizards, 5, 7, 561–562, 561–563, 773
    popular culture and, 764
    pre-Columbian seafaring, 278–279
    primates, 778
    rats and, 795
    reefs, 98, 779
    research stations, 789
    shipwrecks, 834
    snails, 538
    snakes, 845
    stereotypes, 761
    vicariance, 950
Caribbean plate, 950
caribou, 51, 78
Carlquist, Sherwin, 78, 836

Carmen Island, 80, 81
Carnegie ridge, 370, 951
carnivores
    extinct dwarfs, 237
    gigantism and, 375
    introduced species, 470
    island rule and, 494
    Madagascar, 588, 776
    New Guinea, 655
carnivorous plants, 830
Caroline Islands, **148–149**
    dispersal and, 594
    geology, 709
Caroline Ridge, 149
carp, 908
Carpenter, C. R., 88
Carr, G. D., 837
Carson, Hampton, 233, 327, 826
Carteret, 745
Carthaginians, 845
cascades of extinction, 284
Cascadia subduction system, 708
Caspian Sea, 82
cassowaries, 639, 655
casuarinas, 98, 633
Catalina schist, 162
catastrophes, 630, 957
caterpillars, 2, 15, 398, 472, 773, 794
Cathaysia, 504, 505
cats
    Antilles, 23, 24, 26
    Ascension, 61, 62
    Baja California Islands, 79
    biocontrol and, 102
    Britain and Ireland and, 120, 125
    Canary Islands, 132
    Channel Islands (California), 159, 161
    extinction and, 283
    Southern Ocean island, 15
    Tonga, 920
cattle
    Britain and Ireland and, 124, 126
    Hawaiian crickets and, 211
Caucasus, 121
causeways, 261
cave art, 119
caves, **150–153**. *See also* karst features
    Antilles, 24
    Azorean, 72, 74
    Bermuda, 95, 96
    Canary Islands, 130–131
    Cook Islands, 192
    crickets, 207, 210
    freshwater species and, 344
    insect convergence and, 463
    islands, as, 181
    lava tube, 545, 546–548
    Mascarene Islands, 613
    matrix-derived taxa, 182
    Mediterranean, 628
    overviews, 182
    refugia, as, 786
    Sarawak, 114
    Seychelles geckos, 831

Socotra, 847, 849
spiders, 863, 865
Caygill, David, 301
Cayman Islands, *21, 23,* 25, 27, 28, 235, 278, 374, 781, 845
cays (keys, quays). *See also* islets
    Antilles, 21, 22
    Cook Islands, 193
    Great Barrier Reef, 383, 384–385, 385–386, 387, 388
    islets and, 200
    motu vs., 641
    sand, 93
    Tonga, 918
cedars, 95, 97, 98, 899
centipedes, 385, *390, 399,* 671, 907
Central America, 23, 24, 769
Central Indian Ridge, *424*
central Pacific, 693, 949, 950, 977
cephalopods, 145
Cerralvo Island, 81, 82
cetaceans, 745, 866
Chadriiformes, 811, 813
chaffinches, *106,* 129
Chagos Archipelago, 442, 613
Challenger, 702
Challenger plateau, 568
chameleons, 178–179, 236, *559, 560,* 831
Chamorros people, 634, 636
chance (stochastic) events
    dispersals and, 35, 36, 775, 863, 877, 879, 947, 948
    extinctions and, 187, 284, 285, 329, 478, 527, 630, 787
    founder effects, 6, 326, 327, 437, 751
    immigrations, 107, 206, 386
    invasive species and, 478
    Komodo dragons and, 514
    New Zealand geology and, 679
    Polynesian dispersal and, 761
    species-area relationship and, 859
    succession and, 879
Channel Islands (British Isles), **154–155**
Channel Islands (California)
    adaptive radiation, 5
    biology, 155–161
    dwarfs, 236, 237
    formation of, 492
    geology, 154–155, 161–164, 707
    introduced species, 472
chaparral shrubs, 78
    Channel Islands (California), 157
character displacement, 106, 108
charcoal, 33, 44, 132, 221, *291,* 416, 984
Charriére, Henri, 769
Chateau d'If, 771
Chatham, 937
Chatham Islands, 322, 324, *324,* 461, 666, 667, 674, 675, 679, 707
    whaling and, 977
Chekhov, Anton, 770
chemosynthetic biological communities (CBC), 174, 175, 176
cherry trees, 157

Chiapas, Mexico, *328*
chickens, 43, 103, 104, 180, 223, 246, 337, 414, 415, 595, 721, 723, 760, 889
Chile, 957
    geology, 482
    tsunamis, 934
Chiloé, *423,* 957
China
    biocontrol and, 100
    Borneo and, 115
    climates, 423
    earthquakes, 242
    eruptions and, 266, 269
    geology, 504, 505
    patches, 466
    seafarers, 277
    tarsiers, 552
    watercraft, 277
    Zanzibar and, 984
Chincha Island, *738,* 740
Chinese, 832
Chinese sweet gum tree, 899
Chios, 395
chitons, 910
Christianity, 293. *See also* missionaries
Christman, M. C., 151
Christmas Atoll, nuclear tests and, 684
Christmas Island, *738, 738,* 739
    geology, 439, 711
    invasive ants, 39, 40, 472, 479, 708
    isolation of, 702
    lagoons, 554
    land crabs, 235, 532–535
    parallel evolution, 235
chuckawallas, 82
Chukchi Plateau, 55
Chumash people, 158, 160, 707
    Polynesians and, 723
Chuuk, FSM, *94,* 149
cicadas, 101, 300, 465, *465,* 499, 570, 667, 672, 774, 872 971
    Fijian, 300
cichlid fish, **165–169**
    Africa, 3, 6, 7, 165–169
    ecomorphs, 7
    isolation and, 531
    sexual selection, 6, 826, 827, 828
    size of lakes and, 531
cichlids, adaptive radiation, 2, 3, 166, 463
Cierva Point, Antarctica, 14
Cikobia, 693
civets, 179, 850, *902*
Civil War, 770
cladogenesis, 8, 9
clams, 174, *176,* 194, 424, 426, 557, 701, 853, 892, *893,* 957, 959, 974
Clarence Islands group, 17, 18
Clarión Fracture Zone, 80
Clarión Island, 80
clastic rocks, Philippines, *736*
clay, 149
Clermontia, 2, 3, 4, 399
cliffs, 261
    erosion and, 262

cliffs *(continued)*
    Faroe (Faeroe) Islands, 288
    French Polynesia, 334
    geology, 755
    Hawaiian Island, 400–401
    Pitcairn, 744
    Rottnest Island, 797
    seabirds and, 813
climate, **171–174**. *See also* rain; temperatures; weather
    Antarctic plants and, 13
    barrier islands and, 85–86
    dispersal and, 210
    endemism and, 257
    extinction and, 282
    flightlessness and, 312
    hydrology and, 421–422, 423
    invasive biology and, 477
    perhumid, 112, 113
climate change, **169–170**. *See also* glacial-interglacial cycles; global warming; ice ages; sea levels
    Antarctic, 15
    Antilles, 23, 26
    archaeology and, 44
    arctic, 54, 58
    atmosphere and oceans, 174
    Baffin Island and, 76
    beaches and, 263
    Britain and Ireland and, 119, 120
    Channel Islands (British) and, 154
    coral and, 565
    defined, 172
    erosion and, 261
    extinction and, 282–283
    Faroe (Faeroe) Islands and, 292
    fragmentation and, 182
    French Polynesia, 332
    Galápagos and, 365–366
    Gough Island, 930
    Greek Islands and, 389
    lakes and, 531
    Mediterranean and, 628
    Mesoamerican forests and, 186
    montane remnants and, 184–185
    motu and, 642
    neotropics and, 786
    rafting and, 776
    Rapanui vegetation and, 245
    reefs and, 779
    refugia and, 785
    reproductive strategies and, 699
    research and, 41
    Samoa and, 801
    seabirds and, 814
    seabirds research, 295
    seamounts and, 825
    Seychelles and, 833
    sky islands and, 842
    Spitsbergen and, 865
    Sri Lanka and, 870
    Tristan da Cunha, 930
    vegetation change and, 185
    Wallace on, 966
climate effects, volcanos and, 270

climate fluctuations, Iceland and, **435**
climax vegetation, Tasmania, 905
clinkers, *543*
cloth, 274, 745
cloud forests
    Ascension, 61, 62
    climate change and, 171
    Fijian, 299
    French Polynesia, 334, 335, 337, 338
    Lord Howe Island, 570, 571
    Mascarene Islands, 614, 616
    Mesoamerican, 182
    overviews, 184, 312
    Philippines, 589
    Polynesian, 194
    relictual taxa, 182
    Samoa, 800
    Solomons, 851
    Sri Lanka, 867–868
    Taiwan, 900, 902
    Vanuatu, 940
clouds, 172–173
    navigation and, 280
clownfish, *853*
CO_2 (carbon dioxide), 376, 379, 548, 625
    coral reefs and, 780, 782
    forests and, 942
    geology and, 485
coal, 58, 866
    Borneo, 112
    Caroline Islands, 149
    Greek Islands, 393
    Philippines, 737
coastal communities
    Surtsey Island, 887
    Trinidad and Tobago, 926
    Vanuatu, 940
coastal plains, Sri Lanka, 867
coastal processes
    arctic, 61
    cichlids, in, 166
    coral reefs and, 781
    erosion, 261–263
    Krakatau, 519
    overviews, 695
    tides and, 915–916
coastal shrub, 331
coastal vegetation, 335
    Madeira, 582
coastlines
    Britain and Ireland and, 124, 125
    climate change and, 171
    Cook Islands, 193
    Mediterranean, 215
coasts. *See also* beaches; shorelines
    Bermuda, 97
    Canary Islands, 128–129, 132
    Channel Islands (California), 155, 157
    Cook-Australs, 709
    French Polynesian, 336
    Hawaiian Island, 400–401
    Macquarie, 574
    Madeira, 583
    Pacific Islands, 94
    Singapore, 263

    Socotra, 848
    tsunamis and, 933–934
coatis, 90
Cobb-Eikelberg seamount trail, 713
Cockatoo Island, 768
cockatoos, 907
cockroaches, 97, 667
coconuts, 68, *273*
    crickets and, 208
    Seychelles, 830
Cocos Island
    bird radiation, 106
    finches, 352
    spiders, 864
Cocos-Keeling Islands, 439, 958
Cocos plate, 357, *753*
    Galápagos, 367–368, 370
Cocos Ridge, 370, 951
cod, 976
coevolution
    Barro Colorado, on, 88
    dispersal and, 226
    Drosphila, 235
    New Zealand birds, 325
    rabbits and virus, 121–122
cohort stand model, 945–946
Coiba Island, 769
cold seeps, **174–177**
cold temperate zones
    northern hemisphere, 14
Coleoptera, 113
    Cape Verde, 145
Coley, Phyllis, 90
collapses, 954
    overviews, 696
    sea levels and, 695
collapses, volcanoes
    Kurile Islands, 524
collective evolution, 6
colonialism, Borneo and, 115
colonization. *See also* community assembly; founder effects and events; propagules; rafting
    adaptive radiation and, 5
    ants and, 41
    Azores and, 72, 73–74
    Bermuda, 96
    biogeography and, 487–488
    bird, 81, 107
    crickets, by, 206
    dispersal and, 227
    diversity and, 617
    endemism and, 254
    Faroe (Faeroe) Islands, 292–293
    Galápagos, 357
    Galápagos age and, 372
    Gulf of Guinea, of, 809, 810
    Hawaiian Islands, 398
    human, 42
    human of Pacific Islands, 44–47
    insects, by, 460–461
    lizard, 560–561
    Madagascar, 580
    Marianas and, 594
    Mascarene Islands, 613

New Zealand, 669
oases, 688
orchids, 696–697
radiation zone and, 772
coloration
frog, 842
insect, 966
sexual selection and, 826
Colquhoun, D. J., 87
Columbian Exchange, 280
Columbus, Christopher, 139, 280
comics, 765
Commander Islands, 52
commercial activity, Cook Islands, 195
communities (biotic environment)
convergence and, 188–189, 191
coral, 119–120
dwarfism and, 236
community assembly, 6–7
Comoros Archipelago, 107, **177–180,** 444–445
frogs, 348
Comoros, lizards, 560
competition
adaptive radiation and, 6
ants and, 38
archaeology and, 47
Baja California Islands, 82
body size and, 514
Channel Islands (California), 160
character displacement and, 106–107
co-evolution and, 189
coral reefs and, 781–782
cricket morphology and, 211
dispersal and, 225, 226, 227, 228
divergence and, 108
diversity and, 168, 550
dwarfism and, 236, 238
ecological release and, 251, 252
extinction and, 282, 283–284, 943
gigantism and, 376
groups and, 88
Hawaiian Drosophila and, 233
insect speciation and, 461
introduced species and, 472
invasive species and, 346, 479
island rule and, 494, 495
macaques and, 778
metacommunity paradigms and, 528
place and, 911
spiders, 864
succession and, 877, 946
taxon cycles and, 912, 913
competitive exclusion, 89
condition-dependent dispersal, 225
"conies," 23, 24
coniferous trees, 644, 937, 971
Borneo, 114
Kurile Islands, 525
Taiwan, 900
Wizard Island, 981
conservation, 28–29. *See also* biological control; eradication programs; reserves, nature
African cichlids and, 169
Antarctic, 16
Antilles, 27

arctic, 54
Ascension and, 61, 63
Azores and, 71, 74
beach erosion and, 261, 262–263
Bermuda, 97–98
biogeography and, 489–490
Borneo, 115
Britain and Ireland and, 118, 119, 124, 125
Canary Islands, 132
Cape Verde, 146–147
Channel Islands (California), 158, 161
Christmas Island crabs, 534
Cook Islands, 196–197
coral, 567
Cozumel, 203
crabs, 534
Cuba, 22
Darwin on, 959
Darwin's finches, 356
Farallon Islands sharks, 296
Fijian, 305
forests, of, 224
fragmentation and, 187
Frazer Island, 332
French Polynesia, 332, 337–338
frogs, 351
Galápagos, 364–365
Greek Islands, 390, 391
Hawaiian, 403–404, 413–414
inbreeding management, 437
Indonesia, 452
introduced species and, 474
Japan, 500
Juan Fernandez islands, 508–509
Komodo dragons, 514–515
kīpuka and, 513
Krakatau, 519
Line Islands, 557
Lord Howe Island, 568, 571–572
Madagascar, 581–582
Madeira, 585
mammal radiations and, 590–591
Marianas, 596–597
Mascarene Islands, 617–618
metapopulations and, 629, 630–631
Midway Island, 632
New Caledonia, 645
New Guinea, 658
New Zealand, 673
organizations and initiatives, 895
overfishing and, 310, 311
Pacific Islands, 102
Palau, 716
Pantepui, 719
pest control and, 99
Philippines, 730–731
Pitcairn and, 747
reefs, 779–785
relaxation and, 788
research and, 41
seabird, 295–296, 815
sexual selection and, 828
Seychelles, 832
snails, 540–542
solenodons, of, 25

spiders, 865
Sri Lanka, 869–870
St. Helena, 872–873
Taiwan, 901–902
Tasmania and, 908
Tonga, 920
tortoises, 926
Trinidad and Tobago, 928–929
Vancouver, 938
Zanzibar, 985, 986
Zanzibar and Pemba, 982, 983
continental biota, 168
crickets, 207, 210
Hawaii and, 234
seamounts and, 820
continental collision, 55–56
continental islands, **180–185**. *See also* granitic islands; oases
biogeography and, 948
biota, 271
freshwater species and, 344, 345
geology, 707, 755
global warming and, 376, 377
Great Barrier Reef, 383, 386
lizards, 559
New Zealand, 679
oases, 688
overviews, 490, 665, 707–708
South America, 956
spiders, 863, 864
succession and, 877
Taiwan, 897
Trinidad and Tobago, 926–927
Wallace on, 965–966
continental lithosphere, 754
continental shelf islands, 9, **180–187**
arctic, 76
barrier islands, 87, 93
convergence, 191
endemism and, 254
Sunda shelf, 112
Vancouver, 937
continental shelves, 289
continents, 690
climatic effects, 172
collisions with, 902
ephemeral islands and, 260
geology, 481
island arcs and, 481, 485
continuous differentiation, 948
Convention on Biological Diversity (CBD), 888
convergence, evolutionary, **188–191,** 774, 876
caves and, 151
continental islands and, 186
flightlessness and, 315
gigantism and, 375
Hispaniola/Mauritius, 109
honeycreepers and, 412, 413
insects and, 463
Macaronesia and, 774
mammals and, 588, 590
overviews, 1, 7
succession and, 877, 878
Tasmania and, 906

convergence, plate. *See* plate convergence
convergent-plate boundaries, 691
Cook, James, 42, 195
    Christmas Island discovery, 711
    Frazer Island and, 332
    Vancouver, G. and, 937
    Hawaii and, 403, 759
    New Zealand and, 672
    overviews, 280, 702
    pigs and, 741
    pigs and goats and, 742
    Polynesian voyaging and, 758
    popular culture and, 763
    Rapanui and, 250
    Tahitians and, 222
Cook-Austral chain, 953
Cook Islands, **191–197**
    freshwater species, 345
    geology, 342, 343
    human colonization, 44, 759
    missionaries and, 636–637
    motu, 641
    overviews, 709
coots, 374
copepod fauna, 151, 152, 153, *153*
Cope's Rule, 239
copra, 68
coral, **197–203**. *See also* coral islands; coral reefs
    arctic, 52
    Atlantic, 65
    atolls and, 69
    Bermuda, 96, 98
    boulders, 554, 555
    Cape Verde, 145, 147
    Caroline Islands, 149
    Darwin on, 217
    dispersal, 779
    French Polynesian, 336, 337
    global warming and, 378, 379–380
    human impacts, 821
    hydrology and, 421
    Lesser Antilles, 22
    Mascarene Islands, 614
    overfishing and, 310
    overviews, 565
    Pacific region, 708
    Rapanui, 247
    Rottnest Island, 798
    sea level and, 817
    seamounts and, 819, 820
    shipwrecks and, 834
    Solomons, 852–853
    sustainability and, 891
    Tatoosh, 909, 910
    Vancouver, 937
    volcanism and, 956
    waves and, 880
coral islands, 641–642, 712, 799, 830, 833
coralline algal microatolls, 205
coral reefs. *See also* atolls; fringing reefs; reefs
    atolls, 68–69
        biology, 780–781
        Caribbean, 200–201
        conservation, 895

Cook-Australs, 709
Cook Islands, 192, 193, 195–196, 196–197
    Darwin on, 219, 958, 959
    ecology and conservation, 779–785
    French Polynesia, 332
    geology, 692, 754
    global warming and, 70, 377, 378–379
    Hawaiian Islands, 400, 408
    Japan, 500
    landslides, 536
    Line Islands, 555, 556, 557, 711
    Lord Howe Island, 571
    Maldives, 587
    mangroves and, 591
    Mascarene Islands, 614–615, 620
    Midway Island, 631
    motu, 641–642
    northernmost, 95
    overviews, 200–202, 566, 694
    rafting and, 775
    Rottnest Island, 797
    Samoa, 801
    seamounts and, 823, 824
    Seychelles, 831
    Solomons, 853
    Taiwan, 897, 898
    temperature and, 68
    tides and, 916
    Tonga, 920
    waves and, 882
    western Pacific, 715, 716
    west Pacific, 715
    Zanzibar, 984
Coral Sea Basin, 664
cordage plants, 273, 274
core, Earth's, 752
corellas, 908
Corliss, J., 424
cormorants, 140, 159, 294, 295, 296, *313*, *359*, 374, 525, 811, 812, 912 938
corn, *184*
Corsica, 624, 629
Corvo Island, 64, 71
cotton, *357*, 359
cottontails, 79, 81
Coulman Island, 17
The Count of Monte Cristo (Dumas), 771
cowbirds, 329
Cowley, William Ambrosia, 918
Cox, G. W., 912
coyotes, 81
coypu, *121*, 474, 793
Cozumel Island, **203–205**, 236, 845
crabs
    ants and, 39, 40
    Ascension, 62
    Cape Verde, 145
    Christmas Island, 479
    Cocos Keeling Islands, 958
    cold seeps and, 174
    dispersal, 425
    drosophilids parallel evolution, 235
    Line Islands, 556
    New Zealand, 669
    Socotra, 849

crakes, 320, 326
cranes, 122, 313, 321
Crater Lake, 979
crayfish, 346
Crémieux, Gaston, 771
Cretaceous-Tertiary mass extinction, 612
Crete
    biota, 237, 390, 391, 626, 741
    geography, 623
    geology, 389, 392, 393, 394–395, 492, 626, 628
    human colonization, 629
crickets, **206–212**
    Britain and Ireland and, 123
    Channel Islands (California), 159
    Hawaiian, 3, 6, 461, 462–463, 465, 548, 773
    Lord Howe Island, 569, 571
    mice and, 794
    New Zealand weta, 318
    sexual selection and, 826
crinoids, 819
crocodiles, 27, 304, 325, 654, 670, 716, *726*, 832, 850, 853, 854, *893*
Cromwell current, 366
Crosby, Alfred, 280
Crosby, Andrew, 415
crossbills, 26
cross breeding, 108
Cross Island, Alaska, *93*
Crozet Archipelago, 11, 12, 16, 444, *789*
Crozet plateau, 444
crustaceans. *See also specific crustaceans*
    Cape Verde, 145, 146
    Cook Islands, 194
    seamounts and, 819
    Socotra, 849
    Solomons, 852, 853
    sustainability and, 891
    Tasmania, 907
crustal processes. *See also* blocks, crustal; granitic islands; plate tectonics; subduction; *specific plates*
    arctic, 55, 58
    Atlantic, 65, 66, 67
    Borneo and, 113
    Canary Islands, 133
    Caroline Islands, 149
    Cook Islands, 191–193
    Cyprus and, 215
    drowned islands and, 370
    Fiji and, 306
    French Polynesia and, 339
    Galápagos, 371
    Greek Islands and, 395
    Hawaiian, 408
    Iceland, 429–430, 434
    Indian Ocean region, 445
    Kurile Islands, 523
    New Caledonia, 647
    New Zealand and, 675
    Pacific, 484
    sea level and, 261
    seamounts and, 714
crust, continental. *See also* crustal processes; *specific plates*

island arcs and, 482
Kerguelen Plateau and, 679
New Zealand, 676, 678
overviews, 689–690, 752, 754
seamounts and, 821
Taiwan and, 903
water and, 485
crust, oceanic. *See also* crustal processes; *specific plates*
Borneo/New Guinea and, 657
French Polynesia, 340–341, 342
island arcs and, 482–483
Kuriles and, 522
Newfoundland, 650, 651
New Guinea, 664
New Zealand, 676, 678
North Atlantic and, 652
oceanic islands and, 951
overviews, 689–690, 692, 694, 752, 754
Philippines and, 733, 735
Samoa, 803
seamounts and, 821, 825
Solomons, 855
thickness of, 735
Cruz, Eliecer, 366
cryptic species, 725, 820
cryptogams, 13, 14, 467
Cuba
ants, 36, 37, 38
bats, 26
biogeography, 20, 21–22, 28–29
birds, 26, 321
cave deposits, 26
colonization, 278
crickets, 210
Drosophila parallel evolution, 235
fish, 28
geckos, 775
land mammals, 25
lizard ecomorphs, 562
reptiles/amphibians, 27
rodents, 23, 24
sloths, 24
snails, 537
cuckoos, 109, 322, 450, 669, 871
Culebra Island, 22
Culver, D.C, 151
Curaçao, 474
curlews, 611, 633
currents, 61. *See also* rafting; *specific currents*
colonization and, 357
coral and, 565
dispersal and, 425
fish and, 525
freshwater species and, 344
global warming and, 376, 379
Juan Fernandez islands and, 507
Line Islands and, 555
sea levels and, 817
tides and, 915, 916, 917
cuscus, 453
Cyclades, 394, 626
cyclones, 178, 299, **418, 420**. *See also* hurricanes; typhoons
birds and, 800

mangroves and, 592
Samoa and, 713
cypress, 899
Taiwan, 900
Cyprus, **212–216**
geology, 627
human colonizations, 629

daisies, 157
Dale, 918
damselflies, 74, 88, *123*, 300, 402, 656, 719, 786, 852
Dana, James D., 200, 704, 803
dance, 764, 958
dandelions, 50
Daniel Lincoln, 977
Daphne Major, 106–107, 108, *355*
Dario, Ruben, 770
darters, 593
Darwin, Charles, 2–3, **217–221**. *See also* Beagle, HMS
Ascension and, 61, 62
atolls and, 68–69, 950
coral reefs, on, 200
endemism and, 254
Fernando de Noronha archipelago, 297
flightlessness, on, 227, 311–312
Galápagos and, 352, 710
geology and, 367, 587, 704
missionaries, on, 638
molluscs, on, 538
orchids, on, 696
sexual selection, on, 826
Tahiti, on, 334
Tasmania and, 908
Tierra del Fuego and, 918
tortoise conservation and, 833
Darwinian islands, 228
Darwinism (Wallace), 966
Darwin point, 201
Darwin's finches (Galápagos finches), 2–3, 106–107, 108, 110–111, **352–356**, 365, 461, 773, 828, 950
Dawson Island, 769
Dayak people, 112, 115
Dayton, Paul, 910
decapods, 145, 146, 345, 906, 345
Deccan Plateau, 441, 442, 443, 444, 445, 612, 620, 622
Deception Island, 19, 485
deciduous habitat
Channel Islands (California), 157
Indonesia, 453
Madagascar, 578, 579
New Guinea, 657
deciduous woodland, 155
deep corals, 198–199, 820
deep-sea speciation, **755–757**
deer, 901, 981
Baja California Islands, 79, 81
Britain and Ireland and, 118, 120, 121, 124, 125
Channel Islands (British Isles), 155
Channel Islands (California), 158, 160
dodo and, 230

dwarfism and, 236, 237, 238
Marianas, 596
New Caledonia, 645
Sri Lanka, 869
defense mechanism, loss of, 361
Defoe, Daniel, 508, 763
deforestation, **221–224**. *See also* timber
Borneo, of, 115
Britain and Ireland and, 117–118, 123
Channel Islands (California), 155
climate change and, 170
Comoros, 177, 180
Cook Islands, 195
Indonesia, 448, 450
introduced species and, 471
Line Islands, 557
Pacific region, 891–892
Philippines, 730
pre-European impacts, 416, 417
Samoa, 713, 800, 801
Seychelles, 832
Solomons, 854
St. Helena, 873
Taiwan, 901
Tonga, 920
deglaciation, 4
atolls and, 68
Faroe (Faeroe) Islands, 291
overviews, 694
sea levels and, 817
Deinandra, 5, 79, 157
De-Longa Island, 51
Denny, Mark, 911
de novo formation, 180–181, 183, 187, 460, 462
density, extinction and, 284–285
Deperet, Charles, 239
depth. *See* bathymetry
Desertas Islands, 64
deserts, 686–687
development. *See also* sustainability
Bermuda, 97, 98
biodiversity hotspots and, 890
Canary Islands, 132
Cook Islands, 196
coral reefs and, 782
erosion and, 261
hydrology and, 423
Madagascar, 581
New Guinea, 658
Pitcairn, 747
Seychelles, 832, 833
Socotra, 851
St. Helena, 874
Devil's Island, 769
Devon, 50
diagenetic processes, 176–177
Diamond, *106*
diatoms, 857
Surtsey Island, 886
dicotyledons, Fijian, 299
Diego Alvarez Island, 66
Diksam Plateau, 847
dimorphism
birds, 667
dispersal and, 227

INDEX 1031

dimorphism *(continued)*
    Hawaiian *Drosophila*, 232
    moa, 639
    New Zealand, 671
    Samoa, 800
dinosaurs
    dwarf, 237, 239
    New Zealand, 322
Diomede Island, 48, 51
diploids, dispersal and, 226
Diptera, Fijian, 300
Dipterocarpaceae, 830, 867
    Sri Lanka, 868
dipterocarp forest, 114
Disco, *423*
discovery, **276–281**. *See also individual explorers*
Discovery, 937
diseases. *See also specific diseases*
    Channel Islands (California), 158
    coral reefs and, 782
    Cozumel, 205
    extinction and, 282, 283
    Galápagos, 356
    Hawaiian bird, 414, 977
    honeycreepers and, 711
    introduced species and, 472, 473, 474
    invasive species and, 478
    Marquesas and, 712
    missionaries and, 636
    pigs and goats and, 742
    plant, 747–752
    rats and, 795
    refugia and, 786
    Tasmania, 905, 906
    Tierra del Fuego and, 918
    whaling and, 977
disharmony, 944
    radiation zone and, 772
dispersal, **224–228**. *See also specific methods; specific organisms*
    alien species and, 228
    altruism and, 225, 226
    aquatic species, 528
    behavior and, 225
    bet-hedging strategies and, 226
    biogeography and, 970
    chance events and, 35, 36, 761, 775, 863, 877, 879, 947, 948
    climate, 210 and
    coevolution and, 226
    colonization and, 227
    competition and, 225, 226, 227, 228
    condition-dependent, 225
    conservation and, 619
    currents and, 425
    dimorphism and, 227
    diploid, 226
    distribution of species and, 947–948
    diversity and, 167
    dormancy and, 226
    ecological succession and, 228
    endemism and, 360, 452, 907
    equatorial forests and, 210
    Equilibrium Theory of Island Biogeography and, 947

    extinction and, 225, 226, 284
    faulting and, 425
    flightlessness and, 460–461
    fragmentation and, 330
    fragmentation (patches) and, 226, 227–228
    freshwater species and, 345–346
    generalist species and, 225
    global warming and, 380
    habitat factors and, 224–225
    Holocene period, 776
    human impacts, 528, 529
    inbreeding and, 225, 226, 227
    inter-patch, 182
    introduced species and, 472
    invasive species, 347, 478, 742
    lakes and, 527, 528–530, 531
    local extinctions and, 225, 226
    longevity and, 226
    metacommunity paradigms and, 528
    mortality rates and, 227–228
    new islands and, 228
    nutrient availability and, 531
    ocean, 367–368
    plant disease and, 749
    Pliocene period, 776
    pollen and, 226–227
    predation and, 225, 531
    Quaternary period, 776
    radiation zone and, 772
    random, 225
    sexual difference and, 226
    size of islands and, 228, 527
    space, ecological and, 225, 226, 228
    specialist species and, 225
    species-area relationship and, 860
    stages of life and, 226
    storms and, 350
    succession and, 877
    time and, 225
    vicariance vs., 560, 561, 947–948
    volcanism and, 948
    wings and, 227
dispersal ability
    ants, 37
    Azores and, 72–73
    biogeography and, 12
    diversification and, 5
    endemism and, 256
    extinction and, 286, 788
    fragmentation and, 282–283
    freshwater species, 344
    frogs, 348, 349–350
    Hawaiian Islands and, 398
    immigration and, 488
    isolation and, 463
    lakes and, 527–528, 529
    loss of, 361
    open space and, 479
    succession and, 877
dissolution theory, 69
distance, 13, 14, 38, 209, 859
distribution of species, 181, 947–948, 969, 970, 971
disturbances, 942–943. *See also fires; storms*
divaricating species, 671

divergence, evolutionary, 826. *See also diversification; diversity*
    bird, 105, 108–109
    cricket morphology, 211
    cricket songs and, 209
    ecological release and, 252
    sky islands and, 841–842
    stochastic influences, 877
divergence, plate. *See plate divergence (spreading)*
divergent natural selection, 4
divergent plate boundaries, 690–691
divergent selection, 166
divers (birds), Britain and Ireland and, 123
diversification. *See also anagenesis; divergence; speciation; species-area relationship (SAR)*
    adaptive radiation and, 6
    ants and, 36
    Channel Islands (California), 157
    cichlids, of, 166
    constraints on, 462
    ecological release and, 252
    fragmentation and, 182
    insects, 461, 462
    overviews, 220
    Philippines, 729
    plasticity and, 6
    sky islands and, 841
    snails and, 538
diversity, 4, 23. *See also divergence*
    adaptive radiation and, 5
    ant, 37–38
    Antarctic, 11–12
    ants and, 40
    arctic, 52
    Azores and, 73
    Barro Colorado, 89–90
    Borneo, 113–114
    Britain and Ireland and, 117–118, 122, 124, 125–126
    Canary Islands, 128–129
    Cape Verde, 145
    Channel Islands (California), 157, 160
    cichlid fish, of, 165
    colonization/extinction and, 529
    Comoros, 179
    convergence and, 189
    Cook Islands low, 193
    cricket, 209
    endemism and, 254, 256
    French Polynesian, 335
    genetic, 178, 889–890
    human, 42
    influences on, 29
    inselbergs hotspots, 468–469
    island arcs and, 705
    morphological, 8
    Oceanian, 42
    overviews, 708, 889–890
    Philippines, 724–725, 726–727
    plant disease and, 751–752
    Samoan, 800
    temperatures and, 48–49
    variations over time, 9–10

divisions (fragments), 220–221
DNA, Hawaiian birds and, 412, 413
Dodecanese Islands, 395–396, 626
dodos, **228–231**, 316, 619
dogs, 43
   Antilles, 26
   Baja California Islands, 79
   Canary Islands, 132
dolphins, 812, 853, 975
   Britain and Ireland and, 120, 125
   Canary Islands, 132
   Cape Verde, 145
   Channel Islands (California), 159
   Line Islands, 557
   river, 113
   St. Helena, 870
   Vancouver, 938
domestication. *See also specific animals and plants*
   Britain and Ireland and, 120–121
   Faroe (Faeroe) Islands, 292
   Mediterranean, 629
   Near Oceania, 721
   New Guinea, 741
   Polynesians and, 723
Dominica, 29, 32, 35
   lizards, 563
Dominican Republic, 22, 23
   amphibians, 27
   biodiversity of, 23
   birds, 26, 27
   lizards, 561
   protection in, 29
   shrews, 25
Domnica
   lizards, 564
dormancy, 224
   dispersal and, 226
dormice, 126
   Britain and Ireland and, 120
Dorset peoples, 76
doves
   Baja California Islands, 80
   bird disease and, 105
   Fijian, 303
   Samoa, 800
   Tonga, 920
dragonflies, 852, 907, *907*
   Japan, 499
   New Guinea, 656
dragon's blood trees, 848, *849*, 850
Drake, Francis, 294
Dreyfus, Alfred, 769
drift, genetic
   adaptive radiation and, 6
   bird colonization and, 107
   bird disease and, 103
   conservation and, 828
   endemism and, 256
   founder effects and, 110, 327
   isolation and, 108
   orchids, 696, 699
   overviews, 766
   sexual selection and, 826–827
   sky islands and, 841
   speciation and, 166

Driftless Area, *152*
dripping water, 152–153
drip pools, 152
Drosophila, **232–235**
   adaptive radiation, 3, 6, 461, 462, 463, 616
   colonization, 465, 616
   island age and, 464, 773–774
   isolation and, 402, 512
   science and, 788
   sexual selection, 826, 827
droughts
   freshwater species and, 345
   Great Barrier Reef, 385–386
   Madagascar, 579
   oases and, 689
   Pitcairn, 745
   silversword adaptations to, 836
drowned islands, 678
   Galápagos, 370–371, 372
   Makatea, 585
   overviews, 483
   seamounts, 823
drowned reefs, 201
Ducie Atoll, 45
ducks, 671, 811
   Britain and Ireland and, 122, 124
   Galápagos, 359
   Hawaiian, 326
   introduced species and, 472
   Midway Island, 633
Dumas, Alexandre, 771
dunes
   barrier islands and, 83, 85, 86, 87
   Bermuda, 95
   Britain and Ireland and, 125
   Cape Verde, 144, 147
   Channel Islands (California), 157
   Frazer Island and, 331
   isolation and, 187
   Surtsey Island, 887
dung beetles, 7
Dunotter Castle, 834
D'Urville Island, *675*
Dutrou-Bournier, Jean-Baptiste Onesime, 250
Dutson, Buy, 302
dwarfism, 151, **235–239**. *See also* body size; island rule
   Channel Islands (California), 158, 160
   Cozumel, 205
   Galápagos, 361
   Hawaiian, 944
   island area and, 494
   Juan Fernandez islands, 508
   lemurs, 550
   lizard, 559
   mammoths, 707
   Mediterranean, 628
   Philippines, 724, 726
   snakes, 844
   succession and, 878
   tarsiers, 553
   tortoises, 922
   Zanzibar, 983
dye plants, 274–275
Dystaenia, 8

eagles, 122, 671
   Britain and Ireland and, 124, 125
   Channel Islands (California), 158, 159, 161
   gigantism, 374
   Philippines, 724, 731
   Solomons, 852
   Taiwan, 899, 902
   Tasmania, 907
   Vancouver, 938
earthquakes (seismic activity), **240–244**
   Antilles, 30, 32, 34–35
   arctic, 55
   Azores and, 72
   Chile, 711
   Darwin on, 217
   faults and, 481
   Greek Islands and, 392, 394, 395, 396
   Hawaiian, 711
   Iceland, 430–431, 435
   Indonesia, 454–455, 455–456
   Japan, 485, 502–503
   Kurile Islands, 522
   Macquarie Island and, 576
   Mediterranean, 625
   New Guinea, 661, 662, 664–665
   New Zealand, 676
   overviews, 481, 483, 484, 485, 693, 703, 753, 754
   Pacific region, 706, 802
   Philippines, 732, 737
   seamount, 713
   Solomons, 706, 856
   Taiwan, 903
   tsunamis and, 933, 934, 935
   uplift and, 694–695
Earth Summit, 888
earwigs, 872
East African arc, 840, 842
East African beetles, 7
East African islands, 778
East Asian plate, 481
East Azores Fracture Zone, 64
Easter Island (Rapanui), **244–251**
   archaeology, 47
   bird diversity, 320
   geology, 951
   human colonization, 41, 720, 722
   introduced species, 471
   overviews, 709–710
   rodents, 794
   seabirds, 814
   settlement of, 760
   shipwrecks and, 834
Eastern Plateau, *659*
East Malaya-Indochina block, *456*
east Pacific, 892, 977
East Pacific Rise, 78, 164, 425
Eberhard, Mary-Jane West, 5
ebony, 873
echidnas, 906
echinoderms, *194,* 379, *476,* 605, 797, 819, *891*
ecological factors, 8, 462, 488, 564, 729, 772. *See also* adaptive radiation; convergence, evolutionary; *specific factors*

ecological release, 5, 211, **251–253**, 375, 864
ecological specialization, 233–234
ecological succession, 228
ecology and behavior, 412–413
ecomorphs, 7, *190*, 191, 349, 531, 562–563, 931
economies. *See also* sustainability
    Cook Islands, 195
    coral reefs and, 781
    Cozumel, 205
    Pitcairn, 746–747
    research stations, 789–790
    Seychelles, 832
    Solomons, 856
    sustainability and, 891
    Tonga, 919
    Trinidad and Tobago, 927, 928
    Vancouver, 938
    Vanuatu, 939
    Zanzibar, 985
ecoregions, Indonesia, 448, 451
ecosystem diversity, 889–890, *893*
    vegetation, 942
ecoterrorism, 845
ecotherms, 922
ecotourism, 790
    bird disease and, 105
    Cook-Australs, 709
    Cook Islands, 197
    Socotra, 851
    Trinidad and Tobago, 929
    Zanzibar, 985
edge effects, 787
Edwards Plateau, *152*
Edwards, William, 963
eels, 124, 159, 669
    Cook Islands, 194
egg colors, 109, 110
egging, 294
Egg Island, Alaska, *86*
egrets, 908
Egypt, *591*
Ehrlich, Paul, 1
Eiao Island, 223–224, 337, 341
eiders, 125
Eigg Island, 124
Eil Malk Island, 149
elastic rebound theory, 240
Eldgja eruption, 434
elephant birds, 639
Elephant Islands group, 17, 18
elephants, 628, 922, 984
    Borneo pygmy, 113
    dwarf, 236, 237, 238
    island rule and, 493, 494
elephant seals, Farallon Islands, 296
Elephas, 869
elevation. *See also* inselbergs; mountains; plateaux, mountain
    anagenesis and, 9
    Antarctic, 11, 13–14
    Antilles, 22
    ants and, 38
    arctic diversity and, 51
    Borneo, 112, 113–114
    Borneo forests and, 114
    Canarian insects and, 130
    Canary Islands, 128, 132
    Cape Verde, 143, 146, 147
    Caroline Islands, 149
    climate change and, 15
    Comoros, 178
    deforestation and, 222
    divergence and, 841
    diversity and, 10, 186
    endemism and, 257, 551
    extinctions and, 171
    finches and, 352
    flightlessness and, 312, 317
    fragments and, 186
    freshwater species and, 344
    global warming and, 377, 378
    Gough Island, 929
    Great Barrier Reef, 384, 386
    groundwater and, 420–421
    Hawaiian birds and, 414
    Hawaiian colonization and, 398
    hydrology and, 421
    insect diversity and, 463
    inselbergs and, 466
    island-like systems and, 235
    Kurile Islands flora, 525
    Madeira and, 582, 583
    Marianas and, 594
    Mascarene Islands, 613
    montane remnants, 184
    New Caledonia, 643, 645, 648
    Philippines, 727
    Samoa, 799
    silverswords and, 837
    sky islands and, 839, 840
    snails and, 541
    species richness and, 698
    storms and, 419–420
    Taiwan, 898
    timber and, 274
    Tristan da Cunha, 929, 930
    Vanuatu, 939
    vegetation and, 578, 579
    wind and, 172
elks, Britain and Ireland and, *120*
Ellef Ringnes Island, 51
Ellesmere Island, 50, 51, *53*, *54*, 57
    zooplankton, 530
Ellice Islands, 69
El Niño, 81, 82
    climate change and, 174
    Cook Islands reefs and, 196
    Fiji and, 299
    Galápagos and, 365
    global warming and, 379–380
    Line Islands and, 553, 554
    Polynesian voyaging and, 760
    Tasmania and, 905
El Niño-Southern Oscillation (ENSO), 93, 171, 173–174, 345
    coral reefs and, 781
    corals and, 378
    Galápagos and, 355
    hydrology and, 423
    Line Islands and, 555
    Palau marine lakes and, 605
    seabirds and, 814
    storms and, 419
    tides and, 816
emergent islands, 692, 694, 695
Emperor seamounts, 201
emus, 639, 908
endangered and threatened species. *See also* conservation; introduced species; reserves, nature
    amphibians, 27
    Antilles, 25, 26
    arctic, 54
    birds, 845
    body size and, 285
    Britain and Ireland and, 122
    Cape Verde, 145, 146–147
    Channel Islands (California), 161
    Comoros, 179
    Cook Islands birds, 194
    Cozumel, 205
    Fijian, 299, 302, 304, 707
    finches, 356
    Frazer Island, 331
    French Polynesian, 332, 335, 337
    frogs, 351
    Galápagos, 365
    Greek Islands, 390–391
    Hawaiian, 282, 403, 413, 414, 711, 838
    Indonesia, 450, 451, 452
    introduced species and, 473
    Juan Fernandez islands, 508–509
    lizards, 845
    Lord Howe Island, 572
    New Zealand, 325, 673
    overviews, 891–892
    Pitcairn, 745
    Puerto Rican, 472
    reefs, 779
    Samoa, 799, 800, 801
    Solomons, 854
    spiders, 865
    Sri Lanka, 870
    Taiwan, 902
    Tasmania, 908
    Vancouver, 938
    vegetation, 643
    Zanzibar, 986
Endemic Bird Areas of the World, 147
endemic diseases, 104
endemism, **253–258**. *See also* biodiversity hotspots; biogeography
    Antilles, 20, 25, 26, 27, 28, 234
    arctic, 51
    Azorean, 72, 73, 74
    Bermuda, 96, 98
    Borneo, 114
    Britain and Ireland, 118
    Canadian Arctic Archipelago, 49
    Canary Islands, 129, 130
    Cape Verde, 145, 147
    caves, 151
    Channel Islands (California), 155, 157, 159, 161
    climate change and, 171
    Comoros, 177, 178, 179

Cook Islands, 194
Cozumel, 204, 205
extinction and, 319–320
Farallon Islands, 296
Fijian, 299, 301–304
French Polynesia, 332, 711
Greek Islands, 390
Gulf of Guinea, 809–810
Hawaii, 232, 233, 234
Hawaiian Island, 400
Indian Ocean, 107
Indonesia, 448
isolation and, 320, 335, 359, 773
Madagascar, 580
Marianas, 596, 597
Marianas (hotspot), 594
marine lakes and, 605
Marshall Islands, 712
Mascarene Islands, 612, 615–616
New Zealand, 183–184, 666, 669
oceanic islands, 808
overviews, 9, 708, 943
Pantepui, 718–719
Philippines, 724–725, 729
Rapanui, 245
Samoa, 713, 800
seamount, 820
Society Islands, 712
Southern Ocean Islands and, 13
spiders, 862
Tasmania, 907
Vanuatu, 940
vulnerability of, 35, 39
whale falls and, 974–975
Endler, 209
endotherms, 922
energetic equivalence rule, 514
energy, species richness and, 13
Enewetak (Eniwetok) atoll, 69, 680, 682, 684
    coral and, 200
    geology, 219
    subsidence, 695
environmental factors. *See also* convergence, evolutionary; habitat factors
    adaptive radiation and, 353
    Darwin's finches, 355
    lakes and, 529, 530
    reproductive isolation and, 327–328
    succession and, 877–878
environmental protection, 16
    Antilles diversity and, 29
    ants and, 40
environments, novel, 1
Eocene Jolla Vieja formation, 162
Eocene Poway conglomerate, 162
ephemeral islands, **258–260**, 700
    geology, 483–484
    organic falls, 700–701
ephemeral vegetation, inselbergs and, 468
Epi-Jomon people, 522
epikarst, 150–151, 152, 153
episymbiosis, 656
equatorial forests, dispersal and, 210
equatorial Pacific, 708, 754
equilibrium
    overviews, 947
    relaxation and, 787
    size of islands and, 879
    species-area relationship and, 858
    Surtsey Island and, 886
Equilibrium Theory of Island Biogeography
    conservation and, 28
    continental islands and, 185
    dispersal and, 947
    evidence, 384, 517, 519
    Krakatau and, 451, 517
    overviews, 858, 943
    sky islands and, 840–841
    species-area relationship and, 858–859, 943
eradication programs, 618
    Aleutian foxes, 815
    Galápagos, 926
    pigs and goats, 741, 743
    refugia and, 786
    rodent, 795
    Tasmania, 908
    Tristan da Cunha/Gough, 932
Ericaceae, 971
ericaceious plants, 579
ermine, 52
erosion. *See also* landslides
    atolls and, 68
    barrier islands and, 83, 85, 87
    beaches and, 92
    Borneo and, 112, 113
    Britain and Ireland and, 125
    Canary Islands, 134, 135, 142
    Cape Verde, 144
    Channel Islands (California), 161
    climate and, 85–86
    climate change and, 642
    coastal, 261–263
    Comoros, 180
    coral reefs and, 782
    Cyprus, 216
    deforestation and, 221–222
    exotic species and, 223
    Fernando de Noronha archipelago, 297–298
    Frazer Island, 331
    freshwater species and, 343, 344
    granitic islands and, 381
    Great Barrier Reef, 385
    Greek Islands, 393
    guyots and, 713
    Hawaiian Islands, 397, 398
    Indonesia and, 458
    Japan and, 502
    landslides and, 535
    Lord Howe Island, 569
    mangroves and, 592
    Marquesas, 223–224
    Mascarene Islands, 613
    Mediterranean, 627–628, 629
    motu and, 641–642
    New Caledonia, 646
    New Zealand, 675, 676
    pigs and goats and, 742
    Pitcairn, 744, 745, 747
    Samoa, 803
    seamounts and, 823
    St. Helena, 871
    tepuis and, 717, 718
    Tristan da Cunha, 930
    Uniformitarianism and, 218
    volcanic islands and, 953–954
    volcanoes and, 755
    waves and, 880
eruptions. *See also* lava; lava pioneer communities; magma; volcanism; *specific eruptions*
    Greek Islands, 394, 396
    Iceland, 434
    magma and, 952
    Marianas, 600–602
    Mascarene Islands, 613
    overviews, 544, 545, 755
    rifting and, 952–953
    seamount, 713
    tsunamis and, 933
escalation/diversification hypothesis, 1–2
Española, *921*
Espíritu Santo Island, 80, 81
Essex, 918
Estanque Island, 283
Ethiopian Highlands, *840*
ethnobiodiversity, 890, *892*, 894–895
ethnobotany, **271–276**
Eucalyptus, 905, 944, 958
Eurasia
    birds, 122
    squirrels, 119
Eurasian plate, 71, 392, 448, 482, 624, *753*
    Cyprus and, 215
    Indonesia and, 454
    Japan and, 501, 502, 506
    seamounts and, 824
    Taiwan and, 897, 902
Eurasian-Sundaland plate, Philippines and, 732
Euraud, Eugène, 249
Europe
    Azores and, 73
    barrier islands and, 83
    birds, 122, 126
    Borneo and, 115
    Britain and Ireland and, 126
    Canary Islands and, 131
    Channel Islands (California) and, 155
    fishes, 874
    glacial-interglacial cycles, 785
    Greenland and, 50
    Newfoundland and, 650
    prisons, 771
    research and, 789
    tectonism and, 55, 56, 57
    volcanism and, 58
Europeans. *See also* missionaries; *individual Europeans*
    Caroline Islands and, 148
    Cook-Australs and, 709
    explorers, 280
    Marianas and, 595
    Marshall Islands and, 611
    New Zealand and, 672
    overviews, 709

Europeans (continued)
    pigs and goats and, 741–742
    Tasmania and, 907–908
    Vancouver and, 938
eustatic processes, 261. *See also* sea levels
Evans, Ben, 729
evergreen and semi-evergreen forests, 128
    Canary Islands, 132
    Cozumel, 205
    Java, 451
    Madagascar, 578, 578–579
    Madeira, 582, 583
    New Zealand, 671
    Socotra, 848
    Trinidad and Tobago, 930
everwet (evergreen) forests, 113, 114
evolution
    Azores and, 73–74
    caves and, 151
evolutionary radiation, 253–254
evolutionary sorting, 189, 191
evolutionary trees, *412*
    chance and, 948
    New Zealand, 640
    Philippine rodents, 589
    tortoises, 924
Evvoia, 393, 626
Exacum, 850
exotic species. *See also* alien species; introduced species; invasive species
    birds and, 230
    Canary Islands, 132
    Cape Verde, 145, 146
    caves and, 151
    Channel Islands (California), 156, 160
    Comoros, 177, 179–180
    Cook Islands, 195
    Cozumel, 205
    deforestation and, 222–223
    eradication of, 618
    ethnobotany and, 272
    Farallon Islands, 295
    freshwater, 343
    Galápagos, 365
    Lord Howe Island, 571
    New Zealand, 673
    Rapanui, 245–246
expeditions
    Caroline Islands and, 148
    overviews, 790
    Socotra and, 848
    Vanuatu, 940
    Zanzibar and, 984
exploration, **276–281**. *See also individual explorers; specific explorers*
    boundaries and, 967
    Farallon Islands and, 294
    Juan Fernandez islands, 508
    Seychelles and, 832
    tortoises and, 926
    Tristan da Cunha/Gough and, 931
    Zanzibar and, 984
Explorer ridge, 426
explorers, 203

extinctions, **281–286**. *See also* bird fossils; endangered and threatened species; local extinctions; metapopulations; relaxation
    adaptive radiation and, 2
    Africa, 129
    African cichlids and, 169
    Antilles, 23, 24, 25
    Antilles birds, 26
    anuran (amphibian), 869–870
    Ascension, 62
    Australia, 905
    Baja California Islands, 78, 79, 80
    Barro Colorado, 91
    Bermuda, 96, 98
    biocontrol and, 101–102
    biogeography and, 487, 488
    birds, 103, 228–231, 314, 324, 871, 920 (see also bird fossils)
    Borneo and, 115
    Britain and Ireland and, 118, 119, 120, 120, 122, 124
    Canary Islands, 128, 131, 132
    Cape Verde, 146
    caves and, 151, 152, 153
    Channel Islands (California), 158, 160
    climate change and, 170, 171
    continental islands and, 183, 254
    Cozumel, 205
    dwarfism and, 236–237
    Fijian, 302, 304
    flowering plants, 871
    forests, 947
    fragmentation and, 187, 329, 330
    fragments and, 185
    French Polynesia, 332, 335, 337, 711
    Galápagos, 364–365
    Galápagos finches, 773
    Greek Islands, 389–390, 391
    Hawaiian, 403, 404, 413–414, 711
    Hawaiian birds, 410, 411
    honeycreeper, 773
    human-caused, 24, 25, 44, 107
    inbreeding and, 437
    Indonesia, 450, 451
    Indonesian biota, 448–449
    insect radiations and, 461
    inselbergs and, 466
    intraplate islands, 709
    introduced species and, 470, 471, 473
    invasive species and, 347
    island rule and, 495
    Japan, 500, 706
    Juan Fernandez islands, 508, 509
    lakes and, 529–530
    lizard, 559, 561
    logical, 630
    Lord Howe Island, 569, 570
    Madagascan lemurs, 549
    Madagascar, 581
    mammal endemics, 590
    Marianas, 706
    Marquesan birds, 223
    Mascarene Islands, 615, 617, 618
    Mauritanian ants and, 39

    Mediterranean, 628
    metapopulations and, 630
    Micronesian, 716
    Midway Island, 633
    New Caledonia, 644–645
    New Caledonian plants, 255
    New Zealand, 669, 671, 672, 673
    New Zealand bats, 671
    New Zealand birds, 314
    oases and, 688
    pigs and goats and, 742
    Pitcairn birds, 747
    pollination and, 699
    prehistoric, 595
    Rapanui, 245–246
    rats and, 795
    Samoa, 801
    seabirds, 814
    seamounts and, 820
    Seychelles, 832
    Seychelle tortoises, 831
    size of islands and, 859, 860
    sky islands, 842
    snakes and, 845
    Socotra, 849, 850
    species-area relationship and, 858, 943
    species richness variation and, 13
    St. Helena, 959
    Taiwan, 901
    Tasmania, 959
    Tonga, 920
    tortoises, 921, 922, 924, 926
    trees, 874
    Tristan da Cunha/Gough, 931
    vent, 426
    weevils, 794
    whaling and, 977
extinctions, local, dispersal and, 225, 226
extinctions, overviews, 220
extrusion, overviews, 491
extrusion tectonics, Philippines and, 733
Exxon Valdez, 834
eyes, 151
    caves and, 548
    Hawaiian crickets, 461
    lemur, 552
    snake, 843
    spider, 863, *863*
Eysturoy Island, 288, 290, 291, 292

Faroe (Faeroe) Islands, 50, 63–64, 104, **287–293,** 291, *310*
Faial Island, 64, 71, 72
falcons, 159
Falkland Islands, 11, 12, 66. *See also* continental islands
    Beagle and, 957
    continent and, 381–382
    rats and, 795
    research stations, 789
Falkland Plateau, 382
Fangataufa atoll, 336, *339*, 341, 685, 708, 711
Faraday/Vernadsky station, Antarctica, 15
Farallon Islands, **293–297,** 377

geology, 707
Farallon plate, 78, 164
    Baja California and, 703
    Japan and, 503, 505, 706
Farallon Ridge, 426
Farne Islands, *116*, 125
faros, 68, 587
Farquhar Island, 177
Fatu Hiva, 342
faulting. *See also* fracture zones; *specific faults*
    arctic, 56, 57
    Asian strike-slip, 506
    Baja California Islands, 80, 81
    beaches and, 92
    Canary Islands, 136, 142
    Caroline Islands and, 149
    CBCs and, 175
    Channel Islands (California) and, 161–162, 164
    Cyprus, 214, 215
    dispersal and, 425
    earthquakes and, 242
    Fiji and, 308
    Galápagos and, 370
    Greek Islands and, 392–393, 393–394, 395
    Hawaiian, 408
    Iceland earthquakes and, 435
    Indian region, 444
    Indonesia, 458, 459, 460
    island arcs and, 481
    Japan and, 504
    Kurile Islands, 523
    Macquarie Island, 577
    Mediterranean, 626
    New Caledonia, 647, 648
    New Guinea, 663, 664, 665
    New Zealand, 676
    overviews, 240
    Pyrenean, 624
    ridges and, 425
    Samoa, 807
    seamounts and, 822
    Solomons, 856
    strike-slip, 454, 457
    Taiwan, 903
    Tasmania and, 905
    vicariance and, 949
fauna
    arctic, 57
    Atlantic, 67
    Baja California Islands, 80
    Britain and Ireland and, 119–122
    Channel Islands (California), 155, 157–158
    Comoros, 178
    competition and, 284
    Great Barrier Reef, 385
    lava flows and, 547
    Samoan, 800–801
    Socotra, 848
    succession and, 878
    Vanuatu, 941
fecundity, 102, 225, 226, 285, 672, 924
Federated States of Micronesia (FSM), *94*, 148, 641, *681*, 770, 782, 895, 896

fellfield vegetation, *11*, 14, 289, 574
female choice, 6, *106*, 110–111, 233, 656, 826, 827, 828
Fernandez, Juan, 509
Fernandina Island, *313*, 356, *357*, *359*, 364, 366, 367, *368*, 369, 370, 371, 710, 952, *952*
Fernando de Noronha archipelago, 65, 66, **297–298,** 560, *955,* 956
Fernando Nerhona, 218
Fernando Po Island, 65
fernbrakes, 14
ferns
    Bermuda, 98
    biocontrol and, 101
    Britain and Ireland and, 118
    deforestation and, 221
    Fijian, 299
    Gough Island, 931
    Gulf of Guinea, 810
    inselbergs and, 468, 469
    Juan Fernandez islands, 507
    Madeira, 583
    St. Helena, 871
    sustainability and, 891
    Taiwan, 899, 899
    Tonga, 920
    Tristan da Cunha, 931
    Vanuatu, 940
Ficus, 852
fig trees, 88–89, 617, 944
Fiji
    amphibia, 305
    ants, 36, 38, 773, 775
    bats, 194, 801
    biocontrol and, 101–102, 305
    biology, 298–305
    birds, 302–304, 305, 325–326
    conservation, 305, 782, 895, 896
    crickets, 208
    endangered species, 285
    extinctions, 304–305
    fishes, 301, 305
    frogs, 348, 349
    geology, 305–309, 690, 693, 706
    humans and, 42
    introduced species, 305, 471, 477
    mammals, 304
    missionaries and, 634, 637
    reptiles, 301–302
    skinks, 560
    snails, 539
    traditional practices, 890
Fiji Fracture Zone, 307, 309, 693
finches, *106*, 131, 159, 326, *412*, 477. *See also* Galápagos finches; honeycreepers
Finney, Ben, 760
fires
    Borneo forests and, 115
    Channel Islands (California), 159
    deforestation and, 222
    extinctions and, 171
    introduced species and, 471
    Juan Fernandez islands, 509
    New Caledonia, 645

    New Zealand and, 672
    Tasmania, 905, 907
    Vanuatu, 941
Firth of Clyde, 125
fish. *See also* fish, freshwater; marine life and environments; shellfish
    arctic, 52, 57
    Baffin Island, 77
    Bermuda, 97
    Britain and Ireland and, 124
    Canary Islands, 131
    Cape Verde, 145
    Channel Islands (California), 159
    Comoros, 179, 180
    conservation, 783
    Cook Islands, 194, 197
    deep-sea, 756–757
    Faero Islands, 289
    French Polynesian endemics, 335, 336
    Gulf of Guinea, 809
    humans and, 43
    insular, 310–311
    introduced species and, 470
    Mascarene Islands, 615
    Midway Island, 632
    migratory, 311
    New Caledonia, 644
    New Guinea, 653–654
    Pitcairn, 745
    rifting and, 949
    Rottnest Island, 798
    seamounts and, 819, 820
    Seychelles, 832
    shipwrecks and, 834
    Solomons, 853
    Surtsey Island, 885
    sustainability and, 891
    Taiwan, 899
    Tasmania, 906
    waves and, 880, 884
    Zanzibar, 984
Fisher, John, 87
Fisher, R.A., 826
fish, freshwater
    Antilles, 28
    Borneo, 114
    Borneo/New Guinea and, 658
    Japan, 499–500
    lakes and, 530
    New Guinea, 654
    New Zealand, 669, 673
    Philippines, 729
    Seychelles, 832
    Solomons, 852
    sticklebacks, 873–875, 876
    Taiwan, 899
    Trinidad and Tobago, 927
fishing, **310–311**
    Bermuda and, 98
    Cape Verde, 147
    Comoros, 180
    conservation and, 782, 784, 895
    Cook Islands, 196
    coral and, 564, 567, 781

fishing *(continued)*
    Farallon Islands, 296
    Galápagos, 366
    Kurile Islands, 525
    mangroves and, 593
    marine protected areas and, 607, 608
    poison for, 275
    Rapanui, 245
    Rottnest Island, 798
    seabirds and, 814
    seamounts and, 820
    Solomons and, 853
    Tasmania and, 907
    Zanzibar, 985
fission, *37*
fissures, 949
fitness, 108, 225, 226, 437, 748, 751, 826, 827.
    See also natural selection
FitzRoy, Robert, 918, 954, 955
fjords, *54, 56, 57, 58,* 60, *77*
    Icelandic, 432, 436
    Indian Ocean region, 440
    Spitsbergen, 865
flamingoes, *358, 359*
flatworms, 124
fleas, 121, 906
flies
    alien species, 475, 478–479
    biocontrol and, 101–102
    Britain and Ireland and, 125
    Cape Verde, 145
    eradication of, 474
    French Polynesian endemics, 335
    New Guinea, 656
    Tasmania, 907
    whales and, 977
flight, 813
flightlessness, **311–318**
    Channel Islands (California), 160
    dispersals, 460–461
    Drosophila, 234
    extinction and, 313
    Fiji, 325
    fossils and, 319
    Galápagos, 359, 361, 812
    gigantism and, 375, 376
    global warming and, 378
    Hawaiian Islands, 326, 398–399, 773
    mammals and, 322, 323, 325
    Marianas, 595, 706
    New Caledonia, 644
    New Guinea, 655
    New Zealand, 184, 324, 671, 672, 707
    pre-European impacts, 416
    Tasmania, 905, 906
    Tristan da Cunha/Gough Island insects, 931
Flinders, Matthew, 332
floating objects, 258–259
flooding
    Taiwan, 904
    tides and, 916
    tsunamis and, 936
floor mats, 273–274
Flores Island, 64, 71, 236, 237, *267*, 276, 375,
    439, *446*, 448, 452, *454*, 459, 495, 514, 934

Florida, *152*, 175, 477, 542, 565, 781, 855, 977
Florida Keys, 236, 378, *419*, 420, 591, *835*, 855,
    977
flowering plants. *See also* ornamental plants
    Borneo, 113
    Britain and Ireland and, 118, 124, 126
    Cook Islands, 194
    endemism and, 255
    Galápagos, 365
    Gough Island, 930
    Hawaiian, 774
    Indonesia, 450
    Philippines, 724, 726
    Solomons, 851–852
    St. Helena, 871
    Tristan da Cunha, 930
    Wallace's Line and, 970
flycatchers, *106*, 159, 194, 197, *204*, 303, 304,
    335, 336, 337, 413, *597*, 809, 830
flysch, 212, 213, 214, 625, 626, 647
fodder, 146, 147
Fogo Island, 65, 143, 147
folding. *See also specific fold belts*
    Falklands, 957
    Fiji and, 308
    New Guinea, 663, 664, 665
    Tasmania and, 905
Fonualei, 484
food (nutrients). *See also* ethnobotany; nutrient
        availability; resource exploitation;
        resource limitation; trophic structure
    coral and, 781
    intertidal habitats, 910, 912
    island rule and, 494
    moa and, 640
    popular culture and, 763
    seabirds and, 812
    seamounts and, 820
    snakes and, 843, 844
    succession and, 877
    tides and, 914, 915, 916
    waves and, 911
foothills, *898, 903*
foraminifers, 817–818
Forbes, H., 848
forbs, 12, 81, 146 147, 632, 878, 887, *891*
fore-arc, 483
fore-reefs, 556, 557
forests. *See also* deforestation; trees; *specific*
        kinds of forests
    Antilles pine, 27
    arctic, 52, 58
    Ascension, 61
    Atlantic island, 72
    Baja California Islands, 78
    Barro Colorado, 89, 89–90
    Bermuda, 95, 98
    Borneo's, 115
    Britain and Ireland and, 117–118, 124
    Canary Islands, 128, 130, 132
    Channel Islands (California), 155, 157
    climate and, 613
    climate change and, 170, 171
    Comoros, 178, 179
    Cook Islands, 194

    dry, 335
    fragmentation, 182, 328, 329
    Gulf of Guinea, 809, 810
    humans and, 44
    Juan Fernandez islands, 509
    kīpuka and, 513
    Krakatau, 519
    Kurile Islands, 525
    macaques and, 778
    Madagascar, 582
    maritime, 83
    Mediterranean, 628
    New Caledonia, 645
    New Guinea, 658
    New Zealand, 666, 666–667
    overviews, 941–942
    Pacific region, 944
    Pacific region extinctions, 946–947
    Panamanian, 88
    Pitcairn and, 744
    Rapanui, 245
    refugia, as, 786
    rodents and, 793–794
    Rottnest Island, 797
    Samoan, 713
    snails and, 541
    Solomons, 851
    St. Helena, 872–873
    Taiwan, 898
    Tierra del Fuego, 918
    Tonga, 920
    Trinidad and Tobago, 930
    Zanzibar, 984
formation of islands, **490–492**
Formentera, 237, 321, 624
Forrest, Jessica, 530
fossil fuels. *See also* hydrocarbons
    arctic, 54, 58
    Borneo, 112, 115
    Channel Islands (California), 161
    coral and, 566, 567
    Iceland, 433
    Madagascar and, 582
    New Guinea, 665
    Philippines, 737
    Sakhalin, 770
    seabirds and, 814
    shipwrecks and, 834
    South Georgie, 977
    Trinidad and Tobago and, 929
fossils
    Antarctic, 19
    Antilles, 20, 25
    Antilles bat, 26
    arctic tree, 58
    Ascension bird, 62
    Bermuda, 96
    Canary Islands, 127, 131
    Cape Verde, 144, 145
    Channel Islands (California), 157, 158, 160
    cold seep, 177
    Darwin and, 221
    dodo, 230–231
    dwarf, 237, 238
    endemism and, 256

Greek Islands, 395
Hawaiian bird, 411
Lord Howe Island, 569
Newfoundland, 649
New Zealand, 670
Philippines, 724
primates, 778
Sri Lanka, 869
St. Helena, 871
Tasmania, 908
taxon age and, 948
whale, 974
Foster, John Briston, 493
Foster, Robin, 89–90
Foundation seamount chain, 714
founder effects and events, 5–6, **326–328**
    African lakes, in, 165
    Antarctic, 13
    bird, 107
    cricket, 209
    DNA and, 110
    Hawaiian Drosophila and, 233
    inbreeding and, 437
    ontogeny and, 9
    orchids, 696, 699
    plant disease and, 751
    sexual selection and, 826–827
    silverswords and, 836
foxes
    arctic, 49–50, 51, 52, 54
    Baffin Island, 78
    Britain and Ireland and, 120
    Channel Islands (California), 158, 160, 161, 707
    island rule and, 492
    Kurile Islands and, 525
    Tasmania and, 908
fracture zones. *See also* faulting; *specific fracture zones*
    Antilles, 20
    Baja California, 79
    Canary Islands, 142
    Cook Islands, 192
    Galápagos, 368
    hydrothermal vents and, 425
    Mascarene Islands and, 612
    Mid-Ocean Ridges and, 753, 822
    Pantepui and, 718
    Samoa, 803, 805, 807
fragmentation (patches), **328–330**. *See also* island-like systems; metapopulations
    biogeography and, 489–490
    dispersal and, 226, 227–228
    Drosophila, 235
    endemism and, 254
    ephemeral islands, 259
    extinction and, 282, 788
    Mascarene Islands, 618, 619
    speciation and, 585
    spiders, 864
    Sulawesi and, 553
    Tasmania, 908
    Zanzibar, 985–986
France
    bird disease and, 104

Channel Islands (California) and, 155
Comoros and, 180
François I, 771
frankincense, 850
Franklin Island, 17
Franz Josef Land, 47–48, 50, 51, 52
Fraser Island, **330–332**
freezing strategies, 14, 16
French Caribbean, 421, *422*
French Polynesia, **332–343**
    adaptive radiation, 3, 332, 712
    biocontrol and, 101
    Cook Islands and, 192
    geology, 708
    invasive species, 478
    missionaries and, 635
    motu, 641
    overviews, 711–712
    Polynesian voyaging and, 758
    research stations, 789
    snails, 537, 542
    volcanism, 704
freshwater habitats and organisms, **343–347**. *See also* aquatic species; fish, freshwater; hydrology; lakes
    Greek Islands, 391
    Hawaii, 402
    Line Islands, 557
    Madagascar, 580, 582
    motu, 642
    Rottnest, 797
    Socotra, 849
    Solomons, 851, 852, 854
    Tasmania, 906, 907–908
    threatened, 893, 894
frigatebirds, 62, 556, 811
fringing reefs
    Cook Islands, 192, 193, 194, 195
    Cyprus, 215
    Darwin on, 69, 219, 823
    French Polynesia, 334
    Hawaii, 420
    island arcs and, 481
    Marianas, 601, 602
    Mascarenes, 614
    oceanic islands and, 692, 695
    Palau, 149, 716
    Pitcairn, 744
    seamounts and, 823
    Tonga, 919
    tsunamis and, 934
    waves and, 880, 882
Fritz, Uwe, 925
Frobisher, Martin, 77
frogmouths, 907
frogs, **347–351**
    Antilles, 27
    Baja California Islands, 78
    Borneo, 114
    Borneo/New Guinea and, 658
    Comoros, 179
    Indonesia, 453
    introduced species and, 471
    Mauritius, 958
    maximal radiation, 773

New Caledonia, 644
New Guinea, 654, 656
New Zealand, 670
Palau, 716
Pantepui, 718–719
Philippines, 724, 725, 726, 727–728, 729
ranid, 7
Seychelles, 831
sky islands, 842
Solomons, 852
South American, 655
Sri Lanka, 868–869
Taiwan, 899
Tasmania, 906, 907
Wizard Island, 981
Fuerteventura, 64, 127, 128, 129, 132, 135, 140, *141*
fulmars, 50, 104, 124, 289, 812, 887
fumaroles, 67, 960
Funafuti atoll, 69, 420
fungi
    arctic, 54
    Barro Colorado, 89
    Britain and Ireland, 124
    Canary Islands, 130
    Drosophila and, 235
    Macquarie, 574
    Madeira, 583–584
    orchids and, 696–697
    Pantepui, 718
    Philippines, 724
    plant disease and, 748–749
    sustainability and, 891
    Swedish archipelagoes and, 750
    Tasmania, 906

GAARlandia theory, 21
gabbros
    Arctic, 58
    Atlantic region, 67
    Canary Islands, 140
    Channel Islands (British Isles), 154
    Channel Islands (California), 162
    Cyprus, 213
    Fiji, 306, 307, 308
    Iceland, 429, 430, 431
    Indian region, 440, 443, 445
    island arcs and, 482
    Macquarie Island, 576
    New Caledonia, 647
    New Guinea, 663, 664
    overviews, 576, 752, 753
    Philippines, 734
    Tahiti, 340
Gabrielino people, 158, 160
Gadow, Hans, 410
gadfly
galah, 908
Galápagos. *See also* Galápagos finches
    adaptive radiation, 361–364
    biology, 357–366
    bird disease and, 104, 105
    crickets, 206, 210
    Darwin and, 218, 220, 955, 958
    dispersal methods, 776
    endemism and, 257

Galápagos (continued)
    extinctions, 170
    flightlessness and, 318
    geology, 220, 357, 367–372, 710, 755, 951
    human impacts, 710
    invasive species, 474, 479
    kīpuka and, 513
    lizards, 560, 561
    mammals and, 588
    penal colony, 770
    pigs and goats and, 742, 743
    plants, 848, 859
    seabirds, 814
    snails, 538, 775
    species-area relationship and, 857
    spiders, 862
    subsidence and, 954
    tortoises, 921–926, 922, 923, 925, 926
    vicariance, 950
    volcanism, 953
    whaling and, 975, 976, 977, 978
Galápagos finches, 2–3, 106–107, 108, 110–111, **352–356,** 365, 461, 773, 828, 950
Galápagos Rift, 425
Galápagos Spreading Center, 424
gallinules, 184, 800
Galveston Hurricane, 420
da Gama, Vasco, 832, 984
Gambiers, 333, 341
game species, 117, 120, 126, 132, 160, 289, 295, 403, 742, 743. *See also* hunting
gannets, 122, 124, 125, 383, 672, 812
Gardiner's frog, 831
garnet peridotite, 855
gas, natural, 54, 58, 737. *See also* fossil fuels
gastropods
    Azorean, 73
    Bahamas, 378
    Canary Islands, 130
    Cape Verde, 146
    cold seeps and, 174
    Fijian, 300
    French Polynesia, 337
    freshwater, 344, 345
    human impacts, 416
    marine lakes, 605
    overviews, 537
    São Tomé Island, 809, 811
    vents, 426
Gatun Lake, Panama, *181*
Gauanches, 742
Gaugin, Paul, 763
Gau Island, 308
geckos
    Antilles, 23, 27
    Baja California Islands, 82
    Canary Islands, 131
    Caribbean, 559
    Comoros, 179
    dwarf, 27
    Gulf of Guinea, 809, 810
    introduced species, 472
    Mascarene Islands, 616, 617
    maximal radiation, 773
    mice and, 794
    New Zealand, 671
    nonadaptive radiations, 561
    Philippines, 727–728
    rafting and, 775
    Rottnest Island, 798
    Samoan, 800
    Seychelles, 831
    Socotra, 849
    Tonga, 920
geese, 49, 51, 78, 122, 671, 672
Geissois, 940
gene flow, 6, 766–767, *773*, 826, 827, 828, 837, 841, 946
gene pools, 6, 108, 778
generalist species
    diseases and, 748
    dispersal and, 225
    extinction and, 283, 285
    glaciations and, 785
    overviews, 461
    species-area relationship and, 860
genetic revolution model, 6, 235, 826
genotypic responses, 844
gentian, 126
geochemistry, 135–136, 141
The Geographical Distribution of Animals (Wallace), 965
geography. *See* biogeography; spatial considerations; *specific geographical factors*
geology. *See also specific islands, processes and structures*
    biogeography and, 970
    Darwin and, 217
    global warming and, 376
    overviews, 421
    radiations and, 2, 464
    time scales, 56
    turnover and, 38
geophytes, 798
Georgia coast, 83
geothermal energy, 435, 524
Gerald Island, 51
Germany, 148, 150
gestrues, 765
Ghyben-Herzberg lens, 421
gigantism, **372–376**. *See also* body size; island rule; tortoises
    barrier islands, 82
    birds and, 237
    Canary Islands lizards, 131
    Caribbean, 24
    Channel Islands (California), 160
    community composition and, 236
    Comoros, 178
    Cozumel, 205
    cricket, 211
    dwarfism and, 238–239
    Galápagos, 361
    Japan, 499
    Kurile Islands, 525
    Line Islands clams, 557
    lizard, 559
    Mascarene Islands, 619
    Mascarene Islands tortoises, 614
    New Zealand, 638–641, 671
    Philippines, 724
    Seychelles, 830–831
    snakes, 844
    spiders, 864
    Tasmania, 907
    tortoise, 921, 922
Gilsemans, Isaac, *759*
Gjalp eruption, 432
glacial-interglacial cycles. *See also* deglaciation; glaciers; ice sheets
    Borneo and, 113
    Channel Islands (California) foxes and, 707
    climate change and, 169
    diversification and, 841
    formation of islands and, 492
    Greek Islands and, 390
    Gulf of Guinea and, 809
    Iceland, 432, 885
    Indonesia and, 448
    Kurile Islands and, 524
    macaques and, 777
    marine lakes and, 603
    motu and, 641
    refugia and, 785
    sea levels and, 695
    sky islands and, 840, 841–842
    Spitsbergen and, 865
    Taiwan and, 898
    Tasmania and, 907
    Trinidad and Tobago and, *927*, 927
    Vancouver and, 937
    Wallace on, 966
glaciations. *See also* glacial-interglacial cycles; glaciers; ice ages; ice sheets
    Antarctic, 19
    arctic, 47–48, 49, 51, 52, 58–59
    arctic diversity and, 50
    arctic evidence, 56–57
    Atlantic, 65, 66
    atolls and, 69
    Baffin Island, 77
    Baltic archipelagoes and, 750
    birch and, 433
    Borneo and, 113
    Britain and Ireland and, 117–118, 122–123, 125
    coasts and, 83
    Faroe (Faeroe) Islands and, 287, 288
    Greenland, 58
    Hawaiian, 409
    Iceland, 431, 434
    New Zealand, 322, 666
    Pantepui and, 719
    South America and, 956
    southwest Pacific, 969
    Spitsbergen and, 865
    springtails and, 13
    Sri Lanka and, 868
    sticklebacks and, 873
    Tasmania and, 904
    Vancouver and, 708, 937
glaciers. *See also* glacial-interglacial cycles; ice sheets
    arctic, 60

Atlantic, 64, 67
Baffin Island, 77, 78
climate change and, 15, 171, 971–973
formation of islands and, 973
Iceland, 432, 434–435
Indian Ocean region, 444
Montana, 817
New Guinea, 657, 708
sea level and, 816–817
Spitsbergen, 865
succession and, 879
Glass, William, 931
glassy-winged sharpshooter, 478, 479
global warming, **376–380**. *See also* climate change
arctic and, 54
atolls and, 69
barrier islands, 85, 85–86
Britain and Ireland and, 118
coral reefs and, 202, 782
Europe and, 117
French Polynesia and, 336, 711
Galápagos, 356
hurricanes and, 419
New Guinea, 708
Pantepui and, 719
seabirds and, 815
sustainability and, 892
vent fauna and, 426
Warming Island and, 971–973
gneiss, 56, 66, 67, 154, 395, 466, 626, 627, 663, 677, 903
goats, **741–743**
Baja California Islands, 78
Canary Islands, 132
Cape Verde, 146
Channel Islands (California), 155
Galápagos, 365, 926
Marquesas, 223
Socotra, 850
St. Helena and, 873, 959
tortoises and, 926
whalers and, 977
gold, 149, 459, 524, 856
Golden whistlers, 106
Golson, Jack, 43
La Gomera Island, 64, 128, 130, 132, 134, 135, 136–137, *138*
Gonave Island, 22
Gondwana
biota, 325
birds, 312
Falklands and, 66
fauna, 831
flora, 868
fragments and, 187, 299
geology, 677
Indian region and, 445
Indonesia and, 447, 453, 457
moa and, 639
New Caledonia and, 644, 645, 646, 647
New Zealand and, 322, 665, 666, 669–670, 678, 707
overviews, 504, 580
Seychelles and, 829

Socotra and, 847, 849
South America and, 717
southwestern Pacific islands and, 950
spiders, 863, 865
Sri Lanka and, 866–867
Tasmania and, 904
tectonic activity and, 113
Gorda Ridge, 426
Gorgona Island, 769–770, 950
Gough archipelago, 11, 12, 13, 15, 16, 234, 316
Gough Island, 66, 793, 794, **929–932**
Gould, John, 220, 352, 361, 958
groundwater, Cook Islands, 193
Graciosa Island, 71
Graf Spee, 770
Gran Canaria, 132, 134, 135, *138*, 139–140, 557
Grand Comoro Island, 177, 178, 179
Grande Comoro, 107
Grande Terre, 646
granitic islands, **380–382**
Farallon, 377
Seychelles, 438, 829, 830, 832, 833, 922
Socotra, 847
Taiwan, 897
granitic rock
Aegean, 393, 394, 395
arctic intrusions, 57, 58
Atlantic region, 66, 67
Australia, 959
Baffin Island, 77, 78
Britain and Ireland, 125, 126
Channel Islands (British), 155
Channel Islands (California), 293, 707
continental islands, 755
earthquakes and, 503
Easter Island, 710
Fiji, 306, 307, 308, 309
Indian region, 338, 440, 445
Indonesia, 456, 457, 459
inselberg, 466
Japan, 505, 506
Mediterranean, 624, 627
Newfoundland intrusions, 650, 651, 652
New Guinea, 663
New Zealand, 675, 677, 679
overviews, 485
Tasmanian intrusions, 905
Vancouver intrusions, 937
Grant, Rosemary and Peter, 3, 106, 108, 353, 361
grasses. *See also* grasslands; sea grasses; tussocks
Antarctic, 12, 14, 15
arctic, 51
Ascension, 62
Baffin Island, 78
Baja California Islands, 80
Canary Islands, 132
Cape Verde, 146
Channel Islands (California), 157, 160
conservation and, 590
convergence, 189
Cook Islands, 194
deforestation and, 221, 222
diseases, 750
dispersal and, 862, 885

Easter Island, 221
ecological release and, 252
Faroe (Faeroe) Islands, 289, 290
Fiji, 330
grazing and, 265, 743
Great Barrier Reed, 384
Gulf of Guinea, 810
Hawaii, 401, 471, 513
invasive, 470, 471, 480, 513
Krakatau, 517, 518
Kurile Islands, 520, 525
Macquarie, 574
Madagascar, 579
Mascarene Islands, 614
matrix-derived taxa, 182
Midway Island, 632
New Guinea, 653, 657
Pantepui, 718
pigs and goats and, 743
Rapanui, 245, 246
Rottnest Island, 798
Solomons, 851
spiders and, 862
Sri Lanka, 867
succession and, 946
Surtsey Island, 887
sustainability and, 891
Tasmania, 905
Tristan da Cunha, 931
Vanuatu, 940
grasshoppers, 130, 206, 361, 452, 460, 464, 667, *839*, 842, 848
grasslands. *See also* grasses
Aleutians, 815
Annobon, 810
Britain and Ireland, 117, 118, 123, 124, 126
Cape Verde, 146
Channel Islands (British), 155
Channel Islands (California), 157
convergence, 189
Cook Islands, 194
Easter Island, 245
Faroe (Faeroe) Island, 210, 290
fire and, 222
Krakatau, 518, 519
Macquarie Island, 574
Madagascar, 579
mammals and, 781
Mascarenes, 614
Mediterranean, 628
moa and, 640
New Guinea, 657, 658, 666
patchy, 182, 183, 184
rats and, 793
sky island, 841
Solomon Islands, 851
Surtsey Island, 887
Tasmania, 905
Vanuatu, 940
gravel flats, 887
gravity, 30, 735, 911, 914, 932
grazing
Canary Islands, 132
Channel Island (California), 472
intertidal habitats, 910, 911, 912

INDEX 1041

grazing *(continued)*
    introduced species, 470, 471, 474
    isolated islands and, 473
    Marquesas and, 222–224
    Socotra, 850
Great Barrier Reef Islands (GBR), **382–388**, 474, 592
Great Basin, North America, 185–186, 840, 841
Greater Antilles
    adaptive radiation, 531
    amphibians, 27
    ants, 38
    biogeography, 20–22
    fish, 28
    geology, 485
    lizards, 3, 7, 190, 562
    mammals and, 588
Greater Kurile Ridge, 523
Greater Timor, 448
Great Lakes, North America, 104
Great Rift Valley lakes, 165
Great Sea Reef, 707
grebes, 811
Greek Islands, **388–396**
greenhouse gases, 126, 376, 426, 435, 972
Greenland
    age of, 428
    climate effects, 47
    diversity of, 49
    fishes, 874
    geology, 58, 755
    glaciation, 59
    global warming and, 972, 973
    human impacts, 54
    insects and plants, 51
    Newfoundland and, 650
    plants, 50, 51
    research stations, 789
    size of, 59, 111
    tectonism, 55, 56, 57
Greenland Ice Sheet, 60, 972, 973
Green Mountain, Ascension, 61, 62
Green, Roger, 42, 44
greenswords, 401, 835, 837, 838
Greenwich Island, 19
Grenada, 31, 32, 35
Grenada basin, 29
Grenada trough, 30
Grenadines, 26, 35, *510*
Grenvillian orogens, 504
Gressitt, J.L., 970
Gressitt Line, Antarctica, 12
de Grijalva, Hernando, 80
groundwater
    arctic, 61
    atoll, 68
    Channel Islands (British), 155
    coral and, 201
    Frasier Island, 332
    Hawaiian, 409
    magma and, 33
    overviews, 420–421
grouse, 51
groynes, 261, 262, 263
Gryllidae, 670

Gryllotalpidae, 670
Guadaloupe, 24, 31, 35, *422*
Guadalupe Island, Mexico, *5*, 157
Guadalupe Mountains, *152*
Guadeloupe, 23, 27, 31, 32
Guadeloupe archipelago, 421
Guam
    biology, 94
    extinctions, 706
    geology, 585
    human impacts, 595
    introduced species, 471, 473, 474, 479, 595, 845–846
    missionaries and, 633, 634
    overviews, 148, 593, 594
    species richness, 594
guanacos, 918
guano, 104, 222, 557, 739–740, 912
Guayana, 717, 718, 719
Guernsey, 154, 155
Guguan, 595, *598*, 599, 601
Guide to the Birds of Fiji & Western Polynesia (Watling), 302
Guillaumin, 940
guillemots, 125, 158, 289, 294, 295, 297, 813, 866, 887
Gulags, 770–771
Gulf Coast, U.S., 87, 916
Gulf of California, 164, 857
Gulf of Guinea Islands, 65, 560, 699, **808–811**, 840
Gulf of Mexico, 175, 565–566, 567
Gulf Stream, 71, 95, 96, 116, 124–125
gulls
    Britain and Ireland, 122
    Channel Islands (California), 159
    food and, 812
    Galápagos, 360
    New Zealand, 322
    succession and, 879
    success of, 812
    Surtsey Island, 866, 887
    Tasmania, 908
    Tatoosh, 909, 912
guyots, 134, 142, 201, 219, 339, 381, 490, 568, 610, 692, 694, 704, 706, 713–715, **713–715**, 823, 917
gymnosperms, 299, *644*, 667, 899, 920, 940

habitat factors. *See also* competition; predation
    anagenesis and, 9
    biogeography and, 489
    conservation and, 619, 630
    dispersal and, 224–225
    diversity and, 320
    endemism and, 256–257
    extinctions and, 282–283, 943
    Fijian, 299
    Indonesia, 448
    metacommunity paradigms and, 528
    size and, 187
    specialist species and, 285
    species-area relationship and, 859
    succession and, 877
    temporal, 330

Hahajima, *786*
Hainan, *423*
Haiti, *21*, 23, 24, 25, 26, 27, 28, 29, 108, 313, 634
Hale, Marie, 119
Halmahera Islands, *446*, 448, 453, 458, 459, 460, *482*
H na Ridge, 405
Hoa, 334
haplo-diploidy, 226
Hardy, Alister, 61
hares, 52, *53*, 78, *121*, 238, 289, 493, 672, 899
Harmattan (wind), 144
Hatuta'a, 224
Hatutu, 337, 338
Hava , Peter, 925
Hawaiian-Emperor chain, 951, *954*
Hawaiian Islands. *See also specific volcanoes*
    adaptive radiations, 399
    age of, 71, 464
    alien species, 36, 99
    anagenesis and, 9
    ants and, 36, 39
    arthropods, 36
    beaches, 93
    bees, 3
    beetles, 7
    biocontrol and, 100, 101, 102, 748
    biology, 397–404
    bird disease and, 103, 104, 283, 977
    bird extinctions, 283, 315–316, 324, 326
    birds, 106, 322, 480, 773, 813
    climate, 172, 173
    climate change and, 170
    coastline, 94
    collapses, 696, 954
    conservation, 865
    Cook Islands and, 193
    coral reefs, 783
    crickets, 206, 208, 209, 210, 211, 773
    Drosophila, 3, 6, 773
    earthquakes and, 242, 243, 244
    ecological release and, 252
    elevation and, 594
    endemism, 254, 255–256, 257, 943–944
    ethnobotony and, 272, 273
    extinct/endangered species, 282
    fauna, 320
    flightlessness and, 313, 314–315, 317
    flora, 3, 4, 78, 274, 275
    fragmentation and, 329
    freshwater species, 346
    geology, 201, 404–418, 621–622, 690, 695, 704, 707, 708, 710, 754, 950, 951, 954
    gigantism, 374
    honeycreepers, 2, 3, 4
    humans and, 41, 42, 46, 416, 417–418, 632, 720
    hurricanes and, 345, 420
    introduced species, 470, 471, 472, 473, 477
    invasive species, 346, 347, 478, 480
    kīpuka and, 512, 513
    lava, 542–543, 547
    lava caves, 548
    lava tubes, 545
    marine endemism, 96

name of, 758
popular culture and, 762, 764, 765
radiation zone, 772, 773
rats, 794
refugia and, 786
seabirds, 811, 812, 814
seamounts, 822, 824–825
shipwrecks, 834
silversword alliance, 4–5, 835
snails, 537, 538, 540, 541, 542, 775
source biotas, 461
species overviews, 475
spiders, 5, 7, 188, 531, 826, 862, 863–864
tides and, 916, 916
volcanism, 134, 135, 140, 704, 755, 804–805, 952, 953
whales, 975, 976, 977, 978
Hawaii Division Line, 434
Hawaii-Emperor chain, 693, 714
hawkmoths, 970
hawkweeds, 50
head hunting, 708
Heald, Weldon, 839
Heaney, L.R., 773
Heard Island, 11, 13, 16, 440–441
heat dissipation, 825
heat generation, 843
heather, 49, 119, 178, 290
heath forests, 114, 157
Hebrides, 116, 120, 123–124, 292
hedgehogs, 120, 132
Hekla, 433
Helen, 963
Hellenic Arc, 394
Henderson Island, 44, 45, 320, 490, 744
Henslow, John, 955
herbfields, 14
herbivores, 453, 462, 472, 559, 878. See also grazing
herbs, 891, 905, 981
Hercynian-Appalachian orogenies, 505
Herm, 154
Heron Island, 374
herons, 62, 147, 256, 320
Herre, Allen, 88
Herzl, Theodor, 769
Hesperelaea, Oleaceae, 78
heterochrony, 239, 312
heterosis, 225
Heyerdahl, Thor, 250, 515, 722, 760
El Hierro Island, 132, 133, 134, 135, 136
high limestone islands, 694
Hilo, 207
Hilton Head Island, South Carolina, 85
Hindus, 115, 445, 633, 983
hippopotami, 236, 237, 321, 322, 628, 869, 922
Hirta, island of, 121
Hispaniola
adaptive radiations, 24
ants, 36, 37, 38
bats, 26
biogeography, 20, 22, 23
birds, 26, 108–109, 110
butterfly diversity, 28
colonization, 278

crickets, 210
fish, 28
geology, 692
introduced species, 471
invasive species, 346
land mammals, 24, 25
lizards, 559, 562
reptiles/amphibians, 27
rodents, 23
shrews, 25
size of, 29
HMS *Challenger*, 97
hoa, 68
hobbits, 237, 495
Hokkaido, 498, 499, 500, 501, 503, 504, 520, 523, 524, 525, 536, 868
Hokule'a (watercraft), 760
Holarctic, 497, 863, 873, 874, 876. See also North America
holothurians, 174
Holt, 860
Holy Island, 125
Homo erectus, 237, 276
Hondo, 492, 499
honeycreepers
adaptive radiation, 2
Hawaii, 401, 402, 410–414, 711, 773
overviews, 639
radiation, 106
radiation zone, 773
honeydew, 472, 478, 479, 535, 672
honeyeaters, 800, 920
Honshu, Japan, 423, 482, 484, 486, 497, 498, 499, 500, 501, 506
Hooker, Joseph Dalton, 61, 870, 964
hoopoes, 316, 871
horses, 51, 120, 155
hotspot islands. See also mantle plumes
Ascension/ St. Helena, 66, 959
Canary Islands, 133, 134, 956
Caroline Islands, 709
Easter Island, 709–710
French Polynesia, 333, 335, 339, 341–342, 342–343
geology, 553
Iceland, 428
insects, 462, 463, 464
Juan Fernandez islands, 507
Mascarene Islands, 620–622
overviews, 339, 340, 342, 690, 693, 704
Pacific region, 704, 709–713
Samoa, 799–800, 803
hotspots, biodiversity. See biodiversity hotspots
hotspots, geological
Atlantic, 66
Caroline Islands, 149
Comoros, 177
Cook Islands and, 191–192
French Polynesia, 341
Galápagos, 367, 950
Hawaiian, 404
Hawaiian/Louisville Seamount, 714
Indian Ocean region, 438, 439, 441–442, 444, 445

Mascarene Islands, 612
Mediterranean region, 147
oceanic islands and, 951
overviews, 201, 714–715, 754
Pitcairn, 744
seamounts and, 821, 824
hotspot volcanoes, 755
house mice, 793
Hual lai volcano, 405
Hubbell, Stephen, 89–90
huia, 672
human colonizers. See also Austronesians; Europeans; Lapita; Polynesians
Bermuda, 834
Kurile Islands, 705
Madagascar, 577, 579, 580–581
Marianas, 594–595
Marshall Islands, 611
Mediterranean, 626, 629
motu, 642
New Guinea, 658, 708
Pacific region, 720–723
Palau (western Pacific), 715–716
Pitcairn, 745–746
Seychelles, 832
Solomons, 706
St. Helena, 871
Tasmania, 907
Tatoosh, 909
Tristan da Cunha, 929
West Indies, 590
Zanzibar, 984
human dwarfs, 237
human impacts. See also archaeology; dodos; ethnobotany; human colonizers; introduced species; *specific activities*; *specific peoples*
Antarctic, 14–15
ants and, 36, 40, 41
aquatic species dispersal, 528, 529
arctic, 54, 59
Ascension, 62, 959
Ascension Island, 61, 62
atolls, on, 68
Azores and, 71, 72, 74
Baja California Islands, 79–80, 82
barrier islands and, 83
beaches and, 94
Bermuda and, 96–98
biodiversity, on, 29, 107
birch, on, 433
bird disease and, 104
birds and, 103, 230, 319, 320
body size and, 285
Borneo and, 115
Britain and Ireland and, 117–118, 120, 121, 123, 125, 126
Canary Islands, 128, 131, 132
Cape Verde, 145, 146
Caroline Islands, 149
Channel Islands (California), 155, 157, 158, 159, 160–161
Christmas Island crabs, 534
Comoros, 177, 180
Cook Islands, 195, 196

human impacts *(continued)*
   coral reefs, on, 202–203, 566–567, 779–780
   coral reefs and, 782
   Cozumel, 205
   crickets and, 208, 209
   Darwin and, 220
   diversity, on, 169
   Easter Island, 710
   ephemeral islands and, 258, 259
   erosion, 261, 262
   Europe, 117
   extinction and, 170, 171, 282, 284, 286, 773
   Faroe (Faeroe) Islands, 289, 291–292
   Farallon Islands, 294–295, 297
   Fernando de Noronha archipelago, 297
   Fiji, 304–305, 707
   flightless birds, 314, 316, 322
   fragmentation and, 187, 328, 330
   Frazer Island, 332
   French Polynesia, 332, 336, 337, 338
   freshwater species, 343
   freshwater species and, 346
   Galápagos, 356, 364, 773, 958
   global warming and, 972
   Great Barrier Reef, 385
   Greek Islands, 388, 391
   Gulf of Guinea, 810
   Hawaii, on, 403, 711
   Hawaiian birds, 326
   Indian Ocean region, 442
   Indonesia, 451
   inselbergs and, 466–467
   invasive species vectors, 478
   isolated islands and, 99–100
   Japan, 500, 706
   Juan Fernandez islands, 508, 509
   Kurile Islands, 522, 525
   lakes and, 531
   Line Islands, 557
   lizards and, 560
   Lord Howe Island, 570, 571, 572
   Macquarie, 575
   Madagascar, 549, 579
   Madeira, 584
   mangroves and, 377–378
   Marianas, 706
   Marshall Islands, 611–612
   Mascarene Islands, 613, 617–618
   Mediterranean, 628
   Midway Island, 632
   New Caledonia, 325, 645
   New Guinea, 658–659
   New Zealand, 323, 639, 672–673, 707–708
   Panama, 88
   plant disease, 748
   predation and, 283
   rain forests and, 946
   Rapanui, 245, 248
   rats and, 793
   reefs and, 779
   refugia and, 785
   richness variation and, 13
   Rottnest Island, 797, 798
   Samoa, 713, 801
   Sarawak caves, 114
   seabirds and, 814
   Seychelles, 833
   Socotra, 850–851
   Solomons, 853–854
   St. Helena, 873
   sustainability and, 888, 894
   Taiwan, 900–901
   Tierra del Fuego, 918
   Tonga, 918, 919, 920
   tortoises, 924, 926
   Tristan da Cunha/Gough, 931–932
   vent communities and, 427
   Wallace on, 962, 966
   whales, 978
   Zanzibar, 985
human impacts, pre-European, **414–418**
Humboldt Current, 357, 368
   seabirds and, 812
humidity
   crabs and, 533
   crickets and, 210
   endemism and, 257
   frogs and, 301
   lemurs and, 550
   Madeira and, 582, 583
   Mediterranean climates and, 389
   overviews, 421
   Sri Lanka and, 868
hummingbirds, 91, 159, 236, 303, 507, 981
hunting, 115, 118, 743, 800. *See also* game species; whaling
Hurricane Allen, 202
Hurricane Hugo, *84*
Hurricane Iniki, *418,* 420, 474
hurricanes, **418–420**. *See also specific hurricanes*
   Bahaman spiders and, 860
   barrier islands and, 87
   Bermuda and, 95
   bird eradication and, 474
   Cozumel, 205
   crickets and, 211
   freshwater species and, 345
   seabirds and, 814
   succession and, 877
Huxley's Line, *723,* 729
Huxley, T.H., 968
hybridization
   adaptive radiation and, 6
   African lakes, 165
   Britain and Ireland, 118, 120, 125
   cichlid fish, 828
   coral, 779
   Darwin's finches, 354–355
   extinction and, 285
   Greek Islands, 390
   introduced species and, 472
   lizards, 561–562
   macaques, 778
   plant disease and, 749
   reproductive isolation and, 110
   selection and, 108
   silverswords and, 837
   Wallace on, 966
hydrocarbons, 174, 175, 458, 736–737. *See also* fossil fuels
hydrocorals, 199
hydroids, 819
hydrology, **420–424**
hydrothermal areas, 521, 524
hydrothermal vents, **424–427,** 425, 700, 974
hyenas, 119
Hymenoptera, 145

Iapetus Ocean, 57, 504, 505, 649, 650, 651, 652
Iberia, 118, 863
Iberian subplate, 624
Ibiza, 237, 321, 764
ice, 13, 49, 50, 59, 60, 77–78, *93*
ice ages, 23, 56–57, 58–59, 78, 170, 248
icebergs, 60, 181
Iceland, **428–436**. *See also* Surtsey Island biology; *specific eruptions*
   bird disease and, 104
   dispersal and, 50
   eruptions, 952
   fishes, 874
   fishing, 310
   formation of, 490
   geology, 55, 58, 690, 690–691, 753, 754, 755, 885, 951
   human impacts, 291
   hydroclimate, 423
   overviews, 63–64
   subsidence and, 954
   volcanism, 952
   whaling, 978
Iceland-Faroe (Faeroe) Ridge, 951
Iceland Plateau, 63, 428
ice sheets, 17, 60, 254, 255, 817–818. *See also* glaciation; glaciers
igneous rocks. *See also* granitic rock
   Antilles, 30
   arctic, 55, 56, 58
   Atlantic region, 65
   Canary Islands, 133, 135
   Cape Verde, 144
   central Pacific, 950
   Channel Islands (British), 154
   Channel Islands (California), 162, 163
   chemistry, 433
   Cyprus, 214
   Darwin on, 220, 957
   Easter Island, 710
   Fernando de Noronha, 298
   Galápagos, 371
   Iceland, 431
   Indian region, 437, 439, 441, 442
   island arcs and, 482, 483, 485, 486
   island formation and, 491, 492
   Japan, 503
   Mediterranean region, 624, 625, 626
   New Caledonia, 646, 647
   New Guinea, 662
   New Zealand, 675, 677
   oceanic islands and, 691
   overviews, 381, 691
   Pacific region, 703, 710
   Philippines, 732, 733, 734, 734, 735, 736, 736
   plutonic vs., 220
   seamounts and, 823

iguanas
- Tierra del Fuego, 957
- volcanic islands and, 950
- Warming Island, 972

iguanas
- Antilles, 27
- Comoros, 179
- Fijian, 301–302
- Galápagos, 359, 360, 365, 372, 559, 710, 828, 958
- Galápagos endemism and, 257
- Tonga, 920
- Vanuatu, 940

iguanids, 179
Ikhotsk plate, 522
I-Kiribati people, 557
immigration, 489, 526, 527, 529–530, 860. *See also* colonization; introduced species; propagules
- species-area relationship (SAR) and, 859

immune systems, 103
Important Birds Areas in Fiji (Masibalavu and Dutson), 302
mrali, 771
Inaccessible Island, 66, 108
inbreeding, **436–437**
- coefficient of, 766
- dispersal and, 225, 226, 227
- extinction and, 284
- rats and, 793
- silverswords and, 838

India
- Borneo and, 115
- drift, 177
- frogs, 348–349
- geology, 458
- Gondwana and, 580, 669
- mammals, 121
- missionaries and, 637
- snakes, 844

India-Australian plate, 454, 918
Indian Ocean plate, 439, 969
Indian Ocean region
- ants, 37, 39
- atolls, 67
- biogeography and, 107
- climate, 173
- continental effects, 172
- crickets, 208, 209
- currents, 840
- earthquakes, 242
- fishing, 832
- flightless birds, 313, 316
- geology, 437–446
- human colonization, 720
- islands of, 177
- prisons, 771
- research stations, 789
- seabirds, 814
- seamounts, 823, 824
- tortoises, 924
- vents, 426
- volcanism, 755
- weevils, 13

Indian plate, 440, 441, 454, 620, 753
Indians, 832

indigenous species, 15, 475, 943
Indo-Australian archipelago, 970
Indo-Australian plate, 306
Indochina-East Malaya block, 457
Indo-Malaysia, 274
Indonesia. *See also* Maritime Continent; Sunda; *specific eruptions*
- biology, 446–453
- Borneo and, 115
- conservation, 895
- coral reefs, 200, 780
- dwarfs, 237
- fishing, 310
- forests, 452
- frogs, 348, 349
- geology, 437, 454, 536, 695
- global warming and, 380
- invasive species, 347
- mammals, 776
- snails, 537
- tsunamis, 934
- whaling, 976

Indonesian Borneo, 451–452
Indonesian volcanic arc, **439**
Indo-Pacific region, 200, 277, 756, 757
influenza, 104–105
infrared radiation, 258
Inner Hebrides, 116, 124–125, 292
insectivores, 23
insects. *See also* flightlessness; *specific types of insects*
- adaptive radiation and, 4
- Antarctic, 12, 15
- Antilles, 28
- ants and, 40, 941
- arctic, 50, 53
- Australia, 970
- Azorean, 73, 74
- Baja California Islands, 78
- Barro Colorado coevolution, 90
- Bermuda, 98
- biocontrol and, 100
- Britain and Ireland and, 122–123
- Canary Islands, 130, 131
- Cape Verde, 145
- Channel Islands (California), 159
- dispersal, 227, 460
- endangered, 285
- extinctions, 500
- fecundity, 226
- introduced species and, 473
- Japan, 499
- Lord Howe Island, 569, 571, 786
- Marianas, 594
- Midway Island, 632–633
- New Caledonia, 645
- New Guinea, 655–656
- New Zealand, 667, 669, 672
- open sea, 813
- Philippines, 729
- radiations, 460–465
- rafting and, 775
- Seychelles, 831
- sky islands, 842
- Society Islands, 712

Socotra, 848
Solomons, 852
species-area relationship and, 857
Surtsey Island, 886
sustainability and, 891
Taiwan, 899, 900
Tasmania, 907
temperatures and, 14
trade-offs and, 89
Wizard Island, 981

inselbergs, 182, 186, **466–469**
in-situ speciation, 839–840, 947
in-situ vocanoes, 950
insular communities, 321, 328, 384, 687–688, 749
insular dwarfism. *See* dwarfism
insular radiation, 3–4, 326, 348, 561–562
integration, 44
Interior Low Plateau (U.S.), 152
inter-patch dispersers, 182
intertidal habitats, 160, 909, 912
Intertropical Convergence Zone (ITCZ), 144
intraplate islands
- Antarctic, 19–20
- Fiji, 706
- Indian region, 437, 439, 444, 445
- Line Islands, 554
- New Zealand, 675
- oceanic islands, 690
- overviews, 693, 708–710
- Pacific region, 704
- seamounts, 821–822

introduced species, **469–475**
- Channel Islands (California), 707
- Fiji, 707
- Galápagos, 356, 365
- Gulf of Guinea, 810
- Hawaiian, 414, 773, 865
- honeycreepers and, 413
- Juan Fernandez islands, 508–509
- Kurile Islands, 525
- Line Islands, 556
- Madagascar, 579
- Madeira, 585
- Marianas, 595–596
- Marshall Islands, 611
- Midway Island, 632
- New Guinea, 658–659
- New Zealand, 671, 672, 708
- Pitcairn, 744–745
- Rottnest Island, 797
- seabirds and, 815
- snakes, 843, 844–846
- St. Helena, 873
- Tasmania, 906, 908
- Tierra del Fuego, 918
- Tristan da Cunha/Gough, 929, 932
- Vanuatu, 940
- whaling and, 977

intrusive rocks
- arctic, 57, 58
- Newfoundland, 650, 651, 652
- New Guinea, 664, 665
- overviews, 491, 691
- Philippines, 736

intrusive rocks *(continued)*
   Solomons, 855
   St. Helena, 871
   St. Paul's Rocks, 956
   Tasmanian, 905
   Vancouver, 937
Inuit people, 54, 76, 77, 266, 976
invasional meltdown, 472, 479
invasive species, **475–480**. *See also* alien species; exotic species; human impacts; human impacts, pre-European; introduced species; Polynesians; *specific invasives*
   ants, 39, 40, 472, 479, 708
   dispersal of, 742
   ecological release and, 253
   French Polynesia, 336
   freshwater species and, 346–347
   Galápagos, 710
   Hawaiian, 403
   Juan Fernandez islands, 508, 509
   kīpuka and, 512, 513
   Lord Howe Island, 571–572
   Madagascar, 581
   Marianas, 706
   Mascarene Islands, 618, 619
   Midway Island, 633
   New Zealand, 673
   niaouli, 645
   overviews, 742, 748
   plant disease and, 749
   Samoa, 801
   Seychelles, 833
   silverswords and, 838
   size of islands and, 943
   snails and, 541–542
   Socotra, 850
   Solomons, 854
   succession and, 946
   sustainability and, 896
   Tasmania and, 908
   tortoises and, 926
   Vanuatu, 941
invertebrates
   Antarctic, 14, 19
   ants and, 41
   arctic, 52–53, 54
   Ascension, 62
   Azorean, 73
   biocontrol and, 100
   Britain and Ireland and, 122–123
   Canary Islands, 130–132, 774
   Cape Verde, 144–145
   Comoros, 179, 180
   extinctions, 540–541, 908
   Gough/Tristan da Cunha, 930
   Great Barrier Reef, 383
   Gulf of Guinea, 809–810
   introduced species and, 470
   Japan, 499
   lava flows and, 547
   Lord Howe Island, 569
   Macquarie, 574
   Madagascar, 580
   Madeira, 584

   mangroves and, 593
   New Caledonia, 644
   New Guinea, 655–656
   New Zealand, 667, 669, 670, 672
   Palau, 716
   Philippines, 724
   rafting and, 775
   rats and, 794
   reefs and, 201–202
   Rottnest Island, 797
   Samoa, 800
   seamounts, 819, 820
   Seychelles, 833
   shipwrecks and, 834
   Solomons, 852, 853
   St. Helena, 871–872
   Surtsey Island, 886, 887
   sustainability and, 891
   Tasmania, 906, 907
   Tatoosh, 909
   Tristan da Cunha/Gough, 932
Ionian Islands, 393–394, 625
Ireland, 292, 567, 764. *See also* Britain and Ireland
Irian Jaya, 660
irrigation, 224
Irwin, Geoffrey, 42, 760
Isla de Cedros, 79
Isla de Guadalupe, *377*
Isla de Mona, Puerto Rico, *93*
Isla Grande, 471, 472
Isla Guadalupe, 78
Isla Guafo, 814
Isla Isabela (Galápagos), 356, *357, 358, 359, 361, 363, 364*, 365, 366, 369, 370, 474, 710, *923, 952*
Islam, 115, 633
The Island. A Journey to Sakhalin (Chekhov), 770
island arcs, **481–486**. *See also specific arcs*
   Antilles, 29–31
   Baffin Island and, 76
   Borneo and, 112, 113
   ephemerals and, 260
   Fiji and, 306–308
   geology, 950–951
   Greek Islands, 388
   Indian Ocean region, 438, 439
   Japan, 500–501
   Java, 450–451
   lava tubes and, 545
   Lesser Sunda, 452
   Lyell and, 220
   mass wasting, 954
   overviews, 691, 705, 754
   seamounts and, 821, 822–823, 825
   volcanism, 951, 952
island effect, 167–168, 172, 928–929
island hopping, 107, 594, 639, 698, 713, 777, 863
Island Life (Wallace), 870, 965–966
island-like systems, 248, 258–260, 460, 613. *See also* caves; continental islands; fragmentation (patches); hydrothermal vents; kīpuka; marine lakes; oases;

organic falls; Pantepui; patches; seamounts; shipwrecks; sky islands; whale falls
island rule, 237, 238, 374, 375, **492–496**, 606
Isla Pinta, *360, 362, 363, 364*
Isla Porcia, *423*
Isla San Benedicto, 80, 204
Isla San Pedro Mártir, 81, 82
Isla Santa Cruz, *367*
Isla Santiago, *367, 368*
Islas Revillagigedo, 80, 82
Isle of Anglesey/Arran/Islay/Man/Skye/Wight/Scilly, *116*, 124–126
Isle of Youth, 21, 27
islets. *See also* cays (keys, quays)
   Aldabra, 924
   Antilles, 22
   Atlantic region, 65, 66
   Bermuda, 95
   Canary Islands, 132
   Cape Verde, 143, 147
   Caroline Islands, 149
   Cocos, 959
   Cook Islands, 192, 193
   Fernando de Naronha, 297, 298
   French Polynesia, 333, 334
   Greek, 388, 390, 391
   Gulf of Guinea, 808
   Hawaiian, 835
   Indian region, 441, 442, 445
   Juan Fernandez, 507
   Kuriles, 521, 525
   Lord Howe Islands, 568, 570, 571
   Macquarie Island, 573, 576
   Madeira, 582
   Marianas, 600
   Marshalls, 610, 611
   Mascarenes, 614, 618
   Midway, 631
   New Zealand, 674
   North Farallones, 293, 296–297
   Phosphate Islands, 738
   Rottnest Island, 796, 798
   Samoa, 637, 799
   snakes and, 844
   South America, 926
   species-area relationship and, 859
   Tatoosh, 909
   Vanuatu, 939
isolation. *See also* allopatric speciation; ecological isolation; fragmentation (patches); pocket basins; reproductive isolation
   adaptive radiation and, 352–353, 531
   alien species and, 99
   Azores and, 72
   Baja California Islands, 81
   biogeography and, 107, 488
   birds and, 103
   Borneo and, 113
   caves and, 150, 151, 152, 153
   Channel Islands (California) fauna and, 159
   characterized, 3
   continental islands and, 184, 185
   distance, by, 767

diversity and, 186, 272, 329
ecological, 210–211
endemism and, 52, 254, 256, 257, 320, 335, 359
evolution and, 2, 124, 234
extinction and, 787, 943
finch speciation and, 355
fragmentation and, 329
French Polynesia, 333
freshwater species and, 343, 344, 345, 346
frogs and, 350
Greek Islands and, 391
habitat loss vs., 330
Hawaii and, 209, 398, 400, 711
human, 43, 44–45
index of, 243
insect radiations and, 461, 463–464
introduced/invasive species and, 473, 475, 479–480, 513, 709
lakes and, 166, 169, 529–530
lemurs and, 550
Line Islands and, 555
Madagascar and, 577
mammals and, 588
Marianas and, 594
moa and, 640
New Guinea, 663
oases and, 688
plant disease and, 749
seamount, 820
snails and, 540
Socotra and, 850
speciation and, 292, 527, 585, 773
species-area relationship and, 859
St. Helena and, 871
succession and, 877, 878, 879, 945
traditional practices and, 936
vent biota and, 425
Wallace on, 966
isopods, 390, 399, 570, 848, 849, 907
isostatic processes, 68, 70, 78, 254, 261, 308, 381, 432, 434, 436, 603, 817, 823, 873, 973
Isthmus of Panama, 426, 948
Italy, 151, 188, 237, 392, *423*, *536*, *543*, 544, *623*, 624, 625
Iturralde-Vinent, Manuel, 21
ivory, 52
Izanagi plate, *503*, 505
Izu, 484, 501, 503, 504, 506, 522, 702, 704, 706
Izu-Bonin arc, 702, 706

jackrabbits, 81
J'Acuse (Zola), 769
jaegers, 811, 813
Jamaica
ants, 38
bats, 26
birds, 26, 106
caves, 151
coral reefs, 202
geology, 22, 692
introduced species, 471
land mammals, 24
lizards, 190
missionaries and, 635, 637

Onychophora, 670
reptiles, 27
rodents, 23, 24
size of, 29
James, Helen, 411
James Ross Island group, 17, 19
Jamshid ibn Abd Allah, 985
Jan Mayen, 50, 55, 58, 59, *423*
Janzen, D.H., 210
Japan archipelago, **497–506**. *See also* Old World tropics
Caroline Islands and, 148, 149
extinct dwarfs, 237
fishes, 874
fishing, 310
formation of, 492
geology, 484, 485, 500–506, 702–703, 705–706, 707
gigantism, 373
introduced species, 471, 474
invasive species, 346
landslides, 535, 536
mountain floras, 114
tsunamis, 934, 9365
weta, 670
whaling, 978
Japanese, 522, 525, 936
Japan Sea, 8
Japan Trench, 501
Jaussen, Tepano, 249
Java
biodiversity, 113, 451
Borneo and, 115
climate, 971
Drosophila, 234
geology, 112, 448, 450, 457, 458, 484
hydroclimate, 423
New Guinea and, 660
snakes, 845
tsunamis and, 935
Java Plateau, 306
jaws, 843
jays, 159, 160
Jean de Fuca Ridge, 713–714
jellyfish, 199, *604, 605, 606*, 716, *891*
Jersey, 154, 155
Jevons, William, 963
João de Castro seamount, 64
Johnson's Island, 770
Johnston atoll, 680, 684–685
Joinville Island group, 17, 18
jökulhlaups, 64, 433, 434
Jomon people, 522
Jordeson, 963
Juan de Fuca Ridge, 425, 426
Juan de Nova, 739
Juan Fernandez Islands, **507–509**, *704*, 814
juniper, 78, 128
Jura Island, *116,* 124–125
"jutias," 23–24

Kadavu, 302, 303, 308, 309
kagu, 644
Kaiapit landslide, *536*
Kajewski, 940

kakapos, *314*
Kalapana earthquake, 695
Kalimantan, 115, 451–452, 457, 458, 459
Kalogrea-Ardana Flysch, 212–213
Kamchatka current, 521
Kaneshiro, K.Y., 233, 826, 827
kangaroos, 331, 373, 374, 453, 655, 959
Kapingamarangi Atoll, 148
Kapiti Island, New Zealand, *184*
Karpathos, 237, 391, 626
karst features. *See also* caves
Bermuda, 95, 96
Cook Islands, 192
endemism and, 151
marine lakes and, 603
Mediterranean, 625, 626, 628
Phosphate Islands, 739
Socotra, 847
theory, 69
Zanzibar and, 983
Kaua'i
age of, 710
birds, 326, 413
cliffs, 400
coastline, 94
crickets, 208, 211
evolution and, 397
forests, 401
freshwater habitats, 402
geology, 405, 406
hurricanes and, 474
lava tube cave spider, 548
mesic forests, 401
plants, 774, 835, 836, 837
rainfall, 173
snails, 539
volcanoes, 408
kauri, 667, 941
Kavachi volcano, 702
Kayangel, 149
Keeling Atoll, 218, 219
kelps
Channel Islands (California), 156, 159
Falklands, 957
Macquarie, 574
organic falls, 701
shipwrecks and, 834
Surtsey Island, 886
Tatoosh, 909, 910, 911, 912
Kerguelen archipelago, 11, 12–13, 15, 16, 440, 471, 475, *789*
Kerguelen hotspot, 439, 440, 441
Kerguelen Plateau, 11, 440, 443, 673, 679
Kermadec, 483, 485
Kermadec-Tonga arc, 481, 706–707
kestrels, 618, 830
Ketoy Island, 522
key innovations, 1–2, 169, 252
keys. *See* cays (keys, quays)
keystone species, 15, 36, 479, 535, 701, 748, 910, 912
Kiawah Island, South Carolina, 83, 85, 87
Kick 'em Jenny, 32, 34, **510–511**
Kikihia cicadas, *465*
Kili atoll, 682

King George Island, 14, 17, 19
King Island, 51
kin selection theory, 225, 226
Kiribati, 380, 641, 896
Kirimati Island, 814
Kiritimati Island, 335, 554, *555, 556,* 558, 684, 711, 814
kites (birds), 122, 132
kittiwakes, *811*
kiwis, 314, 323, 639, 670, 671
Kīlauea, *207,* 243, 402, 405, 406, *407,* 512, *542, 543,* 545, 704, 711, 714
Knibb, William, 635
knotweed, 124, 981
Kodiak Island, *423,* 483, 492, 635, 874
Kohala, 408
kohu trees, 941
kokako, 472
Kolbeinsey Ridge, 428
Kolguyev Island, 51, 54
Kolombangara Island, *852*
Kol'tsevoye lake, *521*
Kolumadulu Atoll, *67*
Komodo dragons, 373, 452, **513–515**, 559, 971
Kon-Tiki, **515–516**, 759, *759*–760
kookaburra, 908
Koro, 309
Kosrae, FSM, *94*
kīpuka, 402, 404, **512–513**
Krafla Fires, 434
Krakatau, 206, 439, 451, 454–455, **517–519**, 862, 953
Krapf, Johann, 984
Kula plate, *503*
Kurile Islands, 481, 483, 484, 501, **520–525**, *536,* 705, 862
Kuroshio current, 500
Kwajalein, 642, 682, 685
Kyhos, D.W., 837
Kythrea Flysch, 213, 214
Kyushu, 149, 484, 485, 486, *497,* 498, 499, 500, 502, 506, 715

Laccadive archipelago, 442, *690*
Lack, David, 105, 361
lagoon reefs, 614
lagoons
    atoll, 219
    barrier islands and, 82, 85, 86, 87
    Bermuda, 95, 96
    Cook-Australs, 709
    Cook Islands, 193, 194, 195, 196, 197
    Cozumel, 205
    French Polynesia, 334, 336, 338
    geology, 68, 219, 754
    Hawaiia, 631
    Line Islands, 557
    Maldives, 587
    Mascarene Islands, 621
    nuclear tests and, 680, 681, 682, 683, 684
    pre-European impacts, 417–418
    seamounts and, 823
    tides and, 915
    Tonga, 920
    tsunamis and, 934
    Vanuatu, 940
    waves and, 880, 881, 882
Lake Malawi (Africa), 3, 165, 166, *167*
Lake, P., 481
lakes. *See also specific lakes*
    Britain and Ireland, 124, 125
    Comoros, 179
    Cook Islands, 194
    deglaciated, 4
    fishes, 876
    Frazer Island, 331
    French Polynesia, 335
    freshwater species and, 345–346
    islands, as, 165, 169, 181, 526–531
    Kurile Islands, 521
    Marianas, 601
    marine, 603–606
    Palau, 716
    Philippines, 727
    rate of succession, 879
    Rottnest Island, 797
    Solomons snakes and, 853
Lake Tanganyika, 165, *167*
Lake Victoria (Africa), 3, 165, 166, 169
Laki eruption, **263–266, 270,** 434
Lakshadweep, 442
Lambert, Jonathan, 931
Lambir, Sarawak, 90
Lāna'i, *94,* 374, 408–409, 711, *864*
landbirds. *See also* birds
    Antarctic, 13
    Cook Islands, 194
    diversity of, 318
    Farallon Islands, 295
    Gough Island, 932
    Great Barrier Reef, 383
    New Zealand, 673
    Palau, 716
    St. Helena, 871
    Tasmania, 905, 907
    Tristan da Cunha, 932
    Tristan da Cunha/Gough, 932
land-bridge islands
    Aleutians, 705
    Baja California Islands, 80, 81
    Borneo, 657–658
    lizards and snakes, 557–558, 561, 562
    Mediterranean, 628
    Pacific region, 707
    Philippines, 723, 724, 727
    predators, 843
    snakes and, 844
land bridges
    Antilles, 22
    Britain and Ireland and, 117, 118
    frogs and, 350
    human colonization and, 720
    Indian Ocean region, 440
    Indonesia and, 448
    Kurile Islands and, 524
    Madagascar and, 776, 778
    mammals and, 776
    Philippines, 728
    Socotra, 847
    Sunda shelf islands, 113
    Taiwan and, 900
    Tasmania, 904, 907
    Trinidad and Tobago and, 927
land crabs, **532–535**
Lande, Russell, 826
landscape evolution, 695–696
landslides, **535–537**
    age of islands and, 340
    diversification and, 464
    French Polynesia and, 341
    Hawaiian, 405, 408, 409, 711
    Indian Ocean region, 439, 441
    Marianas and, 599
    Mediterranean, 627–628
    Philippines, 732, 737
    seamounts and, 823
    St. Helena, 871
    Taiwan, 904
    tsunamis and, 933
land snails, **537–542**
land-use practices, 123, 196, 329, 416, 446, 473, 632, 779, 782, 850, 896
languages
    Borneo's, 115
    Caroline Islands and, 148–149
    missionaries and, 634
    Oceania, 721–722
    Pitcairn, 745
    Polynesian, 759
    Rapanui, 251
langurs, *867*
La Niña, 355, 737
Lanzarote, 65, 127, 128, 129, 131, 132, 134, 135, 140, *141,* 423
Lapita, 43–44, 277, 415, 721–722, 741, 758, 941
Lapithos Formation, 212
large igneous provices (LIPs), 11, 441, 703, *823,* 950
larks
    bird disease and, 104
    Cape Verde, 144
    Channel Islands (California), 159
Larsen, Carl Anton, 977
Lasia, 778
Latin America, 637
latitudinal effects, 14
    flightlessness, 317
    Maldives, 586, 587
    sky islands, 840
    Vanuatu, 939
Laupala, 3, 6, 207, 209, 211, 461, 463, 465, 826, 828
Laurasia, 187, 504–505
laurels, 72, 98, 128, 274, 582, 583, 585, 794, 971
Laurentian shield, 56
laurisilva, 72
lava, **542–544**. *See also* kīpuka; volcanism; *specific eruptions*
    Cape Verdes, 218
    crickets and, 206, 207, 208, 210
    Cyprus, 214–215
    Fernando de Noronha archipelago, 298
    fragmentation and, 329
    Galápagos, 370
    Mascarene Islands, 612–613

oceanic islands and, 952
spiders and, 864
Surtsey Island, 884, 885
Laval, Honoré, 635
lava pioneer communities, *402*, 512, 886
lava tubes, 72, 140, *141*, 151, 207, 208, 209, 410, 411, 461, 463, 464, 543, **544–549**, 774
Lawlor, Tim, 185
Laysan fever, 104
Laysan Island, 474
Leeuwin Current, 797, 798
Leeward Islands, 22, 143
leeward regions
　Cape Verde, 146
　erosion and, 93
　Fiji, 299
　freshwater species and, 344
　hydrology and, 421
　lava and, 543
　predators and, 557
　Rapanui, 245
　reefs, 556
　Samoa, 799
　Tikehau, 68
　Vanuatu, 940
　weather, 172
leglessness, 831, 843
legumes, 50, 130
Leibold, M.A., *528*
lemmings, 51, 52, 53, 78, *120*
Lemnos and Imroz, 395
lemurs, **549–553**
　adaptive radiation, 2
　Comoros, 180
　gigantism, 375
　Madagascar, 580, 588, 589, 776
　maximal radiation, 773
　parallel radiation of, 462
Lengguru Fold Belt, 664
leopards, 451, 901, 983
Lepidoptera, 113, *114*, 124, 126, 300, 317, 336, *390*, 810, 831, 871–872
Lesser Antilles
　biology, 7, 24, 26, 27, 28
　colonization, 278
　earthquakes, 32, 35
　geochemistry, 31, 32
　geology, 22–23, 29, 690, 691, 754, 755
　island rule and, 493
　tectonism, 30
　volcanism, 32–33, 34–35
lesser sheathbills, 12
Lesser Sundas, 448, 452, 970, 971. *See also* Wallacea
Lesvos Island, 391, 393, 395, 627
Levallois people, 117
Levins, Richard, 629
Lewis and Harris, *116*, 123–124
liana, *941*
lice, 23, 462
lichens
　Antarctic, 12
　arctic, 51, 52
　Atlantic, 67
　Baffin Island, 78, *78*

Canary Islands, 130
Cape Verde, 146
　inselbergs and, 467
　Macquarie, 574
　Madeira, 583–584
　Pantepui, 718
　Philippines, 724
　Surtsey Island, 886
　Tatoosh, 909
liestone, Rottnest Island and, 797
light, 200, 202, 884, 911
light houses, 294
lignite, 149
L'ihi, 404–405
lilies, *899*
limestone
　Cape Verdes, 218
　Cyprus, 212, 214, 215
　New Guinea, 663
　New Zealand, 677
　Philippines, 735, 736
　phosphates and, 740
　reefs, 798
　Tonga, 918, 919
　western Pacific, 715
Limestone Caribbees, 29, 31
limpets, 425, 910, 911
Lindisfarnre, 125
linear island groups (hotspot island chains), *690*, 693, 708
　biogeography and, 949
Line Islands, 173–174, **553–558**, 632, 680, 711
Linnaen shortfall, 730
Linnaeus, 876
lions, 52, 119, 869
literature, 763
Lithic/Archaic peoples, 278
lithosphere. *See also* subduction
　Galápagos and, 710
　Hawaii, 407
　Indonesia and, 456
　New Caledonia, 648
　oceanic islands and, 951
　overviews, 481, 485, 752–753, 754
　Philippines, 734
　seamounts and, 822
　Solomons and, 856
　subsidence and, 954
Lítla Dímun Island, *288*
Little Cumbrae, 125
Little Ice Age, 433
　Iceland and, 435
Little Swan Island, 23
littoral regions
　Samoa, 800
liverworts, 12, 940
　Canary Islands, 130
livestock
　Cape Verde, 146
　Channel Islands (California), 156, 160
Livingstone, David, 984
Livingston Island, 17, 19
lizards, 3, **557–564**
　adaptive radiation, 561
　Antilles, 27

ants and, 40
Baja California Islands, 80, 81, 82
Bermuda, 96
body size, 913
Borneo, 114
Borneo/New Guinea and, 658
Canary Islands, 131, 132
Cape Verde, 144
Caribbean Anolis, 5, 7, 531
Caribbean convergence, 189
Channel Islands (California), 159, 161
Comoros, 179
convergence, 190
dwarfism and, 236, 238
endangered, 845
extinctions, 846
Fiji, 774
Galápagos, 359, 362, 363
gigantism, 922
Greater Antilles, 531
Indonesia, 452
island rule and, 493, 494
Madagascar, 580
Madeira, 584
Marianas, 594, 596
maximal radiation, 773
monitor, 513–514
New Caledonia, 773
New Guinea, 654, 655, 656, 657
New Zealand, 670
Philippines, 726, 727–728
rafting and, 775
rifting and, 949
Rottnest Island, 798
Socotra, 850
species-area relationship and, 859
spiders and, 860–861
Tonga, 920
Vanuatu, 940, 941
Wallace's line and, 971
Lāna'i, *94*, 374, 408–409, 711, *864*
lobelioids, *2*, *3*, *4*, 400, 401, 835
lobsters, 907
　Cape Verde, 145
　Channel Islands (California), 159
　coral andanemones, 566
local adaptation, 52, 224, 226, 353, 355, 594, 911
local extinctions. *See also* relaxation
　Channel Islands (California) eagles, 158
　climate change and, 842
　continental islands and, 183
　defined, 281
　dispersal and, 225, 226
　ecological sorting and, 189, 531
　endemism and, 814
　human impacts and, 814, 892
　island effect and, 928–929
　isolation and, 282, 285, 466
　metapopulations and, 630
　oases and, 688
　overviews, 527
　pathogens, of, 749
　Polynesians and, 892
　skinks, 794
　Tonga, 920

local participation, 929–930. *See also* traditional practices
locusts, 206
Lofoten Islands, 59
logging, 115, 332, 769, 941. *See also* timber
Lohachara Island, 380
Lōʻihi, 404, 408, 693, 714, 804, *864*
Lōʻihi Seamount, 714
Lombok, 447
Lomolino, Mark, 493
Lomolino, Mark V., 859
Lomolino's Combined Model, 859, 860
longevity, dispersal and, 226
Long Island, 345, 755
    fishes, 874
longitudinal effects, 939–940
Look Seamount, 713
loons, 811
Lophlia, **564–567**
lorakeets, 908
Lord Howe Island, 210, **568–572**
    endangered species, 285
    insects, 786
    pigs and goats and, 743
    rats and, 795
    waves and, 484, 882
lorikeets, 98, 194, 303, 305, 335, 337, 707
Losos, Johathan, et al., 7, 859
Losos, Johathan B., 561
Louisville Seamounts, 714
low reef islands, 386, 387, 388
Loyalty Islands, 491, 643, 646, 647, 648, 692, 694, 722, *939*
Lunday Island, 255
Lusitania, 118
Luzon arc, 902
Lyakhovsky Island, 51
Lydekker's Line, 447, 968
Lyell, Charles, 218, 219, 220, 221, 964
Lyme disease, 104
lynx, 120
Lyon, H. L., 946
lyrebirds, 908

Ma'anyan speakers, 115
macaques, 775–776, 776–778, 899
Macaronesian Islands
    biogeography, 72–73, 127
    endemic species, 130
    fauna, 74
    geology, 64–65
    lizards, 560
    orchids, 698
    plants, 129
    radiation zone, 774
MacArthur, Robert H., 271, 629, 772, 774. *See also* Equilibrium Theory of Island Biogeography
Macdonald, Gordon, 342
Macdonald hotspot, 343
Macdonald seamount, 342
Macdonald volcano, 709
MacKenzie, Neil, 635
MacPhee, Ross, 21

Macquarie Island, 11, 15, 16, **573–577**, 908, 950
macroalgae, 701
macroevolutionary factors, 772
macropods, 453
Madagascar, **577–582**. *See also* Gondwana
    adaptive radiation, 578–579
    age of, 319
    ants, 36, 37, 38, 39
    birds, 106, 319, 321
    chameleons, 559
    climate change and, 170
    Comoros and, 178, 179, 180
    deforestation and, 170
    drift, 177
    dwarfs, 237
    freshwater species, 345
    frogs, 7, 348, 349
    geology, 445, 446, 755
    gigantism, 374
    Gondwana and, 669
    humans, 115, 720
    inselbergs, 469
    island rule and, 493
    lemurs, 773
    lizards, 559, 560
    mammals, 2, 322, 549–550, 588, 590
    monocotyledons, 467
    orchids, 698
    overviews, 183
    rodents, 792, 795
    snakes, 179, 559
    South American dispersals, 560
    spiders, 862, 863, 864–865
    succulents, 468
    tides, 915
    topography, 551
    tortoises, 922, 926
    tsunamis and, 935
Madagascar plateau, 444
madder, 157
Madeira archipelago, **582–585**
    Cape Verde and, 146
    conservation, 567
    coral conservation, 567
    fauna, 72, 73, 106, 129, 130, 559, 862, 864
    flightlessness and, 311–312, 317
    flora, 145, 146
    geology, 64, 127, 141, 142
    map, 133
    pigs and goats and, 742
    research stations, 789
    spiders, 862, 864
Madrean Archipelago, 235, 839–842, *840*, *841*, 841–842
Mafia Island, 179
Magellan, Ferdinand, 280, 595, 702, *703*
magma
    corners, 485
    eruptions and, 951–952
    Galápagos, 371
    Hawaii and, 405
    island arcs and, 481
    melting, 951

    overviews, 543, 691, 753, 754
    Philippines and, 736
    seamounts and, 823, 825
    subduction and, 822
    water and, 953
magmatic rocks, 19
magnetic fields, 754, 955
magpies, 899, *900*
mahogany, 157, 944
Maiao, 333
mainland effects, 123
Maio Island, 65, 143, 144, 145, 147
Majuro, RMI, *94*
Makah Nation, 909
Makatea, 333, 334, 336, *339*, 739
makatea islands, 192, 193, 196, 333, 335, **585–586**
Makira, 855
Makira Trench, 856
Malagasy songbirds, *106*
Malaita, 855
malaria, 103, 104, 414
Malay archipelago, 446, 964, 965
The Malay Archipelago (Wallace), 446, 965
Malay/Muslim peoples, 115, 276
Malaysia, 112, 113, 114, 115, 208, *310*, 347, 349. *See also* Sunda
Malden, 680, 684, *738*, 739
Maldive Islands, **586–587**
    climate change and, 171
    crickets, 208, 209
    geology, 442, 613
    hydrology and, 423
    photo, 67
    plant disease, 748
    wind and, 92–93
Male islet, 423
Malesians, 713
Malesian subkingdom, 447
mallards, 472, *572*
Mallorca, 321, 559, 623, 624, 845
Malta, 629
Malthus, 962
mammals, **588–593**
    Antilles, 21, 22, 23, 24, 26
    ants and, 40
    arctic, 51, 52, 53, 53–54
    Asia, 121
    Baffin Island, 76
    Baja California Islands, 80, 81
    Barro Colorado, 90
    Bermuda, 98
    bird fossils and, 320–326
    boreal forests fragments, 185–186
    Borneo/New Guinea, 114, 658
    Britain and Ireland, 119–120, 124
    Canary Islands, 129, 131
    Cape Verde, 144, 145
    Channel Islands (California), 155, 159–160
    convergence and, 188
    Cook Islands, 194–195
    dwarfism/gigantism, 236, 238, 375
    extinctions, 282, 330, 500, 845–846
    Fiji, 305

food and, 843
Great Barrier Reef, 383
Hawaii, 470
Indonesia, 447, 449, 450, 451, 452, 453, 513
island rule and, 493
Japanese, 499
kiwis and, 670
Komodo dragons and, 513, 514
Kurile Islands, 525
Macquarie, 574, 575
Madagascar, 580
Marianas, 595
Mascarenes, 618
Mediterranean, 629
New Caledonia, 644
New Guinea, 653, 654, 655
New Zealand, 99–100, 472, 667, 670, 671, 672, 673
overviews, 112, 184, 793
Pantepui, 719
Philippines, 724
rafting and, 775–776
Samoa, 800, 801
seamounts and, 820
silverswords and, 838
sky islands, 840–841, 842
snakes and, 844
Socotra, 849
Solomons, 774, 853
species-area relationship and, 857, 860
Spitsbergen, 866
Sri Lanka, 869
sustainability and, 891
Taiwan, 899
Tasmania, 907
Tonga, 920
Vancouver, 938
Wizard Island, 869, 981
mammoths, 51, 52, *120*, 158, 160, 237, 707
Mamonia Terrane, 214
Mana Island, 794
Mancham, James, 832
Mandela, Nelson, 770
Mangaia Island, 44, 192, 193, 194, 196, 342, 417
Mangareva, 45–46, 250, 417, 635
Mangareva people, 709
mangrove islands, **591–593**
mangroves
  Bermuda, 95, 96, 97
  Borneo, 114
  Comoros, 178
  crickets and, 208
  dispersals, 367, 368
  erosion and, 261–262
  ethnobotony and, 274–275
  Frazer Island, 331
  Gulf of Guinea, 810
  Hawaiian, 943
  Madagascar, 579
  Marshall Islands, 611
  New Caledonia, 644
  New Guinea, 657
  sea levels and, 377–378
  Solomons, 851, 852

South Atlantic, 298
Tonga, 920
Trinidad and Tobago, 928
Vanuatu, 940
Zanzibar, 983, 984
Manihiki Plateau, 191, 195
mantle. *See also* intrusive rocks
  Cook Islands and, 709
  hotspots and, 754
  Newfoundland, 651
  New Zealand and, 675, 679
  North Atlantic and, 652
  overviews, 752
  Samoa and, 808
  seamounts and, 825
  Solomons and, 855
  volcanic islands and, 951
mantle plumes
  anomalies and, 951
  Antarctic, 17, 20
  Antilles, 32
  Arctic, 63, 64
  Atlantic, 63, 64, 66, 67
  Canary Islands, 135, 142
  Cape Verde, 144
  Cook Islands and, 709
  French Polynesia, 340
  Galápagos, 367, 368–369, 370, 371
  Hawaiian, 406, 407, 711, 714
  Iceland and, 428
  Indian Ocean region, 439, 440, 445
  intraplate islands and, 708
  Laurasia and, 504–505
  line islands and, 553
  Madeira and, 582
  Marquesas, 342
  Mascarene Islands, 622
  moving, 714–715
  overviews, 481, 544, 704, 754
  Pacific region, 704
  Samoa and, 807
  seamounts and, 821, 824
  Solomons, 856
mantle rocks, 66
Manuae Island, 192, 194, 195, 341
manuring, 14
Manus basin, 664
Manx breeds, 125
mao (ma'oma'o), 800
Maore Island, 177
Maori people, 273, 470, 672, 707, 795, 958
Map Island, 149
Maratua, 777, 778
marble, 393, 394, 395, 736
Maré, 694
Mare-aux-Songes marsh, 230–231
Margarita, 23
Maria, 834
Maria Cleofas Island, 79
Maria Madre Island, 79
Maria Magdalena Island, 79
Mariana arc, 502, 691, 706
Marianas, 148, 483, 484, 485, **593–603**, 705
Mariana Trench, 149, 754

Marie Galante, 24, 31
marine lakes, **603–606**
marine life and environments. *See also specific types of marine life*
  archaeology and, 43, 44
  arctic, 54, 57
  Baja California Islands, 78, 81
  Bermuda, 96
  Borneo/New Guinea and, 658
  Britain and Ireland and, 125
  Canary Islands, 131–132, 140
  Cape Verde, 145, 146
  Channel Islands (California), 155, 159–160, 161
  Cook Islands, 194, 195
  eastern African, 983
  endangered, 893
  Faroe (Faeroe) Island, 288
  French Polynesia, 336, 338
  freshwater species and, 346
  Galápagos, 368, 710
  Galápagos endemism, 359–360
  global warming and, 378
  Gough Island, 930
  Gulf of Guinea, 809
  Japan, 500
  Kurile Islands, 525
  Lord Howe Island, 569, 571
  Macquarie, 574, 575
  Madeira, 584
  mangroves and, 592–593
  Marianas, 594
  marine lakes and, 606
  Mascarene Islands, 615
  New Guinea, 654
  organic falls and, 700–701
  Palau, 716
  Pitcairn, 745
  pre-European impacts, 417
  rafting and, 775
  Rapanui, 245
  Rottnest Island, 798
  Samoa, 801
  seamounts, 713
  Seychelles, 832
  shipwrecks and, 834
  South American, 69
  speciation of, 756
  St. Helena, 870
  Surtsey Island, 886
  sustainability and, 896
  Tonga, 920
  Tristan da Cunha/Gough, 932
  Vancouver, 938
  Zanzibar, 983–984, 986
marine protected areas (MPAs), 98, 196, 403, 427, 567, **607–609**, 716, 780, 783, 895, 896
marine reserves, Socotra, 847
Marion Island, 13, 14, 15, 16, 444, 471, 794
Marion mantle plume, 445
Maritime Continent, 172, 173. *See also* Indonesia
Marlborough Sounds, 675

marmots, 938
Marotiri, 342
Marquesa Fracture Zone, 339, 340, 341, 342, 712
Marquesan people, 195, 709
Marquesas
　deforestation, 222
　Drosophila, 234
　endemism, 335, 336
　freshwater species, 345
　geology, 333, 341–342, 696, 711–712
　grazing and, 223
　landscapes, 334
　marine life, 336
　missionaries and, 636
　snails, 538, 539, 540, 541
　tsunamis and, 934
　volcanism, 755, 953
Marshall Islands, **610–612**
　conservation, 896
　motu, 641, 642
　overviews, 69, 148, 680, 681, 712
Marshall, John, 712
Marshall, Patrick, 951
marshes, *97*, 194, 335
marsupials, 43, *188*, 453, 655, *796*, 798, 905, *906*, 908, 958
martens, 120
Martin Garcia, 770
Martinique, 27, 28, 31, 32, 35, 108–109, 564
Martinique Passage, 23
Martín Vaz Island, 65, 66
Mascarene Plateau, 612, 613, 620, 622
Mascarenes
　biogeography and, 107
　biology, 612–619
　birds, 179, 319, 320, 322
　extinctions, 170, 922
　flightless birds and, 316
　geology, 442–443, 620–622
　introduced species, 472
　lizards, 560, 561
　orchids, 698, 699
　snakes, 845
　tortoises, 922, 925
　weevils, 2
Masibalavu, Vilikesa, 302
mass effects, 527, *528,* 530
mass extinction event, 612
Massif de la Hotte, 24, 27
massifs, 182
mass wasting, 134, 953–954. *See also* landslides
Matai people, 783
material plants, 273–274
mating behaviors, 233, 413, 656. *See also specific organism*
mating signals, 108, 110, 465, 548. *See also* acoustical signaling
matrix contrast, 181
matrix-derived taxa, 182
Maug, 595, *598, 599,* 600
Maui
　birds, 413
　climate, 397, 398
　climate change and, 171

　coastline, 94
　crickets, 211
　dry forests, 401
　forests, 944–945
　silverswords, 835
　snails, 540
　tides and, 916
　volcanoes, 408
Maui Nui islands, 408
Mauke Island, 192, 193, 194, 196, 342
Mauna Kea, 134, *172,* 398, 402, *405,* 406, 407, 409, *542,* 754, 803, 821
Mauna Loa, 134, *172,* 173, 405, 406, *407,* 408, 409, *542, 547,* 620, 952
Maupihaa, 341
Maupiti, 333, 341
Mauritius. *See also* dodos
　ants, 37, 39
　beetles, 2
　biodiversity, 617
　biogeography and, 107
　bird divergence, 108–109
　conservation, 618
　Darwin and, 218, 958
　ecology, 614
　geology, 442, 443–444, 612, 613, 621–622
　gigantism, 373
　introduced species, 471
　orchids, 698, 699
　phylogeography, 616
　plants, 848
　prisons, 771
Maya people, 203
mayflies, Fijian, 300
Mayotte Island, 177, 178, 180, *560*
Mayr, *106,* 826
Mayr, Ernst, 327, 968
McCauley, D.E., 766
McDonald Islands, 11, 16, 441
McMurdo Dry Valleys, Antarctica, 10, 13
meadows, 72, 78, 471, 524, 652, 718, 798, 839, *840,* 887, 889, 938, 981. *See also* grasses
mealybugs, 39
Mecherchar Island, 149, *604, 738,* 739
Medellín, R.A., 152
Mederes, Adnan, 771
medicines, 90, 91, 146, 275, 781, 810, 848
Medioeuropean region, 127
Mediterranean plate, 392
Mediterranean region, **622–629**
　birds, 321
　Canary Islands and, 131
　Cape Verde and, 145, 146
　dwarfs, 237
　extinct dwarfs, 236–237
　extinctions, 170
　island arcs, 482
　lizards, 564
　Macaronesia and, 127
　mammals, 321
　pigs and goats and, 741
　plants, 128, 129
　popular culture and, 764
　prisons, 771
　seafaring and, 277–278

　snakes, 844, 845
　spiders, 863
Mediterranean Sea, 83
Medusagynaceae, 830
megafauna, 581
megaherbs, 574, 671
megapodes, 304, *314,* 325, 374, 416, 716, 920
Mehetia, 333, 341, 712
melaleuca, 944
Melanesia
　ants, 38, 253, 912
　bird radiation, 106
　conservation, 892, 893
　diversity, 800
　ethnobotony and, 272, 274, 275
　frogs, 348, 349
　humans and, 114, 277
　missionaries and, 634
　overviews, 42, 722, 890, 891
　snails, 537, 538
　sustainability and, 892
　traditional practices, 890
Melanesian arc, 300, 306, 309
Melanesians, 180, 634, 637, 660, 832, 853, 855
melting glaciers, 816
melting of mantle, 951, 953
Melville, Herman, 977
memertean worms, 819
Mendelson, T.C., 209
Mendocino Fracture Zone, *425*
Menorca, 321, 559, 624, 845
Mentawai Islands, *446,* 451, 491, 692, 777, 778
Mercalli scale, 241
Mesaoria Plain, 214–215
mesas, 182, 601, 646, 839
mesic forests, 335, 401, *836*
Mesoamerica, 182, 186, 205
Mesopotamia, 121
mesosphere, *752*
Messinian salinity crisis, 624, 625
metacommunities, 527–531
metamorphic rocks
　Antarctic, 18
　Antilles, 31
　arctic, 55, 56, 57
　Channel Islands (British), 154
　Channel Islands (California), 162
　Darwin and, 957
　Greek Islands, 393, 394, 395
　Indian region, 445
　Indonesia, 456, 457
　island arcs and, 481, 484
　Japan, 501, 504, 505, 506
　Mediterranean region, 624, 625, 626, 627
　New Caledonia, 646, 647–648
　Newfoundland, 649
　New Guinea, 661, 662, 663, 664, 665
　New Zealand, 675, 677
　Philippines, 732, 733, 734, 735, 736
　Solomons, 855, 856
　Taiwan, 902–903
　Warming Island, 972
metapopulations, 227, 527, **629–631,** 688, 948
methane, 175–176
Metrosideros, 940, 949

Mexican Caribbean Sea, 203
Mexico, 38, 168, 544, *593*, 840, 841
mice
    Antarctic, 15
    Baja California Islands, 79, 81
    biodiversity and, 794
    Britain and Ireland and, 120
    Channel Islands (California), 158, 161
    endemism and, 255
    extinctions, 283
    Farallon Islands, 295
    Gough Island, 932, *932*
    Philippines, 589, 727–728
    rats and, 474
    sub-Antarctic, 15
    wide-spread species, 792
    Wizard Island, 981
Michel, James, 832
microallopatric speciation, 166
microbes, 11–12, 89
microevolution, 108, 538, 563
Micronesia
    climate change and, 171
    human colonization, 722
    humans and, 42
    motu, 641
    pigs and, 741
    pigs and goats and, 742
    shipwrecks and, 834
    snails, 538
    sustainability and, 896
Micronesians, 712
microorganisms, 54, 62, 118, 174
Mid-Atlantic Ridge, 58, 63, 67, 426, 428, 435, 490, 565, 822, 956
Middle East, 121, *482*
mid-domain effect, 698
midges, 12, 13, 14–15, 23, 887, *907*
mid-ocean ridges. *See also specific ridges*
    Antarctic islands and, 18, 19
    CBCs and, 175
    Galápagos and, 368–369
    hotspot volcanoes and, 755
    melting anomalies, 951
    overviews, 690–691, 753–754
    seamounts and, 714, 822, 825
mid-Pacific continent, 538
Midway Island, 104, 105, 206, **631–633**, 710, 793
migmatites, *56*
migrants. *See also* dispersal; human colonizers
    birds, 26, 50, 78, 122, 124, 125, 126
    Canary Islands, 132
    Cape Verde, 144, 147
    cave, 152–153
    Christmas Island crab, 533
    Comoros, 179
    conservation and, 630
    endemism and, 254
    Farallon Islands, 295
    fishes, 311
    freshwater species, 344–345
    gene flow and, 766, 767
    Greek Islands, 391
    jellyfish, 605

Kurile Islands and, 522
mangroves and, 593
Marshall Islands, 611
metapopulations and, 629
New Zealand, 669
Samoan, 801
seabirds, 813
Surtsey Island, 887
Tonga, 920
vertical, 756, 757
Milankovitch Theory, 376
millipedes, 300
mimicry, *6*, 258, 655
mineral deposits. *See also* mining; *specific minerals*
    Fiji, 308
    Indonesia, 457, 460
    lava tubes, 546
    Mediterranean, 625, 627
    New Caledonia, 647
    Philippines, 734, 736
    seamounts and, 825
    subduction and, 483
    tides and, 917
mining. *See also* mineral deposits; Phosphate Islands
    arctic islands, 54
    Baffin Island, 77
    Borneo, 112, 115
    Caroline Islands, 149
    Christmas Island, 534
    Cook-Australs, 709
    Fiji and, 309
    Frazer Island, 332
    Greek Islands, 395
    Madagascar, 581–582
    New Caledonia, 645, 647
    New Guinea, 665
    Philippines, 730
    seamounts and, 820
    Solomons, 856
    Spitsbergen, 866
    Tasmania and, 908
    vent fauna and, 427
minks, 124, 525
Minoan eruption, *394*
Miogeoclines, *651*
mires, 14
Mischief, 963
Misima Island, 665
missionaries, **633–638**, 742, 958, 984
mist forest, 810
mites, 12, 13, 53, 100, 130, 887
Mitiaro Island, 192, 193, 194
Mittermeier, Russell, 449
mixed forests, 905
Mnaihiki, 195
moa-nalos, 324, 326
Moana, Vaka, 760
moas, *106*, 184, 315, 319, 322, 323, 324, 325, 373, 416, **638–641**, 667, 670, 671, 672
mockingbirds, 80, 359, 950
Moheli Island, 177, 178, 179, 180
Mohotane Island, 223
Mohotani, 337

Mojave Desert taxa, 81
molds, 44, 479, 885, 886
molecular research, 7, 8
moles, 899
molluscs
    Azores, 72, 73
    Bermuda, 98
    Canary Islands, 130, 132
    Cape Verde, 145, 146
    conservation, 540–541
    Fijian, 299–300
    French Polynesia, 336–337
    Greek Islands and, 393
    Madeira, 584
    Seychelles, 831
    Socotra, 848–849
    Solomons, 852, 853
    sustainability and, 891
    Vanuatu, 940
    vent, 426
Moloka'i, 46–47, *94*, **400**, 408–409, *540*, 635, *864*, *915*
Moluccas, 448, 453, 459, 968, 971. *See also* Wallacea
mongooses, 23, 24, 25, 26, 179, 305, 471, 473
monkeys, 21, 25, 88, 90, 113, 144, 236, 778, 983
monocotyledons, 467, 468, 469
monocultures, 99
monogamy, 413
monophylism, 1, 26, 52, 73, 165, 234, 411, 463, *464*, 513, 561, 579, 639, 722, 773, 837, 921
monotremes, 655, *656*
Monserrat, 80
monsoons
    Australasian and African, 172
    beaches and, 92
    Borneo, 112
    Cape Verde, 144
    climate change and, 174
    Comoros, 178
    forests, 452
    Kurile Islands, 521
    latitudinal effects, 587
    New Guinea, 657
    Socotra, 848
    Sri Lanka, 868
    Taiwan, 898
    Zanzibar, 984
Montagne Pelée, Martinique, 32, 35
Monteray shale, 163
Monterey Bay, *176*
Montserrat, 23, 24, 32, 35, *543*, 544
Morgan, W. Jason, 704, 711
Morocco, 104, 130
morphological evolution, 211, 233–234, 353–354, 355–356, 563–564, 923
Morrison, 859
mortality rates, 227–228
Moruruo, 336, 341
Moslow, Tom, 87
mosquitoes, 88, 103, 104, 105, 283, 472, 711, 977
mosses
    Antarctic, 12
    arctic, 50

INDEX 1053

mosses (continued)
  Atlantic, 67
  Bermuda, 98
  Gough Island, 931
  Lord Howe Island, 571
  Surtsey Island, 886
  sustainability and, 891
  Tristan da Cunha, 931
Mother Lode, 152
moths. See also caterpillars
  Antarctic, 15
  Antilles, 23, 28
  arctic, 49, 51, 53
  biocontrol (Fiji) and, 101–102
  Borneo, 114
  Britain and Ireland and, 123
  cave, 548
  Channel Islands (California), 159
  Fiji, 101–102, 300
  French Polynesia, 336
  Galápagos Galagete, 362–363
  Hawaii, 398, 400, 548
  Japan, 336
  Madeira, 584
  Midway, 632
  New Guinea, 656
  New Zealand, 318
  orchids and, 697
  St. Helena, 871–872
  Surtsey Island, 886
  Wallace's Line and, 970
motu, 67, 68, **641–642**, 694
Motu One, 333, 337, 341
mountains. See also elevation; sky islands
  arctic, 61
  Borneo's, 113, 114
  breezes, 173
  Britain and Ireland, 116, 125
  Canary Islands and, 127
  climate and, 172
  Cook Islands, 193
  Cyprus geology, 212
  endemism and, 257
  Guayana, 717
  Indonesia, 460
  Japan, 501
  New Zealand, 666–667
  North Atlantic island, 287
  Solomons, 852
  Tahiti, 334
  Taiwan, 897, 899, 902–903
  Tasmania, 9054
  Vancouver, 937–938
  Vanuatu, 940
Mount Cameroon, 808
Mount Cook, 707–708
Mount Desert Island, *380*
Mount Etna, 220, 549, 625
Mount Fuji, 706
Mount Kinabalu, 112, 113, 114
Mount Mazama eruption, 979
Mount Pagan, 600–601
Mount St. Helens, *543*, 862
mourning doves, 80
movies, 762–763

Moynihan, Martin, 88
Mozambique Channel, 177
Mt. Chubu, *786*
Mt. Erebus, 19–20
Mt. Etna, 220
Mt. Haddington, 19
Mt. Siple, 20
Mt. Vesuvius, 220
Muck Island, 124
mulberry, 274, 275
Mull, *116*, 124, 125
Müllerian mimicry, *6*, 655
mullets, 124
Mull Island, 124–125
multiple colonization, 477, 478
murrelets, 158, 161, 813, 938
murres, 813
  Baffin Island, 78
  Farallon Islands, 294, 296–297
Mururoa atoll, 69, 680, 685, 744
music, 764
musk oxen, 51
mussels, 425, 426, 529, 909, 911, 974
mustelids, 672
mutations, 6, 103, 125, 151, 225, 238, 355, 437, 563, 697, 698, 755, 828, 837
mutualism, 89, 91, 189, 519, 535, 619, 655, 920
Mwali Island, 177
My Life (Wallace), 966
Myojinsho Island, 702–703
Myrdalsjökull, 435
myxoma virus, 121–122

The Naked Island (Shindo), 423
Nankai Trough, 501, *502*
Nansen-Gakkel Ridge, 58
Nantucket, 976
Nason, John, 89
national parks. See also marine protected areas (MPAs); reserves, nature
  Antilles, 29
  Borneo, 115
  Canary Islands, 129, 132
  Channel Islands (California), 161
  Christmas Island, 534
  Cook Islands, 197
  Egypt, 591
  Fraser Island, 330, 332
  Galápagos, 365, 366, 926
  Greek Islands, 391
  Komodo, 514
  Mascarene Islands, 618
  Midway Island, 632
  Newfoundland, 649, 651
  prisons, 769, 771
  Samoa, 801
  Sri Lanka, 867
  St. Helena, 873
  Taiwan, 898, 901–902, *902*
  Tierra del Fuego, 917, 918
  Tonga, 919, 920
  Trinidad and Tobago, 929
  U.S., 979, 980
  Vancouver, 937, 938
  Zanzibar, 985

Native Americans, 909
natural enemies. See predation
natural gas, 112, 627
natural selection. See also fitness
  divergence and, 108
  drift and, 327
  Mascarene Islands, 617
  plasticity and, 5, 6
  speciation and, 166
  Wallace on, 220, 964, 965, 966
nature reserves. See reserves, nature
Nauman, Edmund, 501
Nauru Island, 585, 738, *738*, 739, *739*
Navidad Bank, 22
navigation, 204, 279–281, *280*, *281*, 759, 760
Nazca plate, 357, 367–368, *753*
Ndzuani Island, 177
Near Oceania, 720
needlefish, *783*
Negrito populations, 114
Neilson bank, 342
nematocysts, 199
nematodes, 10, 12, 14, 748, 788, 795, 819
Nenets, 54
neoendemics, 73, 74, 183, 184, 185, 253, 255, 256, 524
neoteny, 548
neotropical regions, *7*, 131, 234, 467, 468, 508, 560, 718, 786, 801, 925
nesting
  Line Islands, 556
  mangroves and, 593
  New Zealand, 669
  seabirds and, 813–814
  snakes and, 843, 844
  stickleback, 875–876
  Surtsey Island, 887
  turtle, 801, 929
Netherlands Antilles, 23
neuston, 258
neutral view, *528*
Nevis Island, 23, 34
New Amsterdam archipelago, 11
New Britain, 485
New Caledonia, **643–648**. See also Gondwana
  age of, 319
  biota, 324
  birds, 303, 304, 322
  crickets, 208
  endemism and, 254
  forests, 645
  freshwater species and, 345
  geology, 492, 644, 666, 679, 692, 707
  Gondwana and, 669
  humans and, 42
  lizards, 773
  Lord Howe Island and, 571
  missionaries and, 634
  overviews, 183
  paleoendemics, 255
  pre-European impacts, 416, 417
  prisons, 771
  research stations, 789
  seamounts, 819
  skinks, 560

snails, 538
spiders, 862, 863
Newcastle disease, 105
Newfoundland, **649–652**, 874
New Guinea
    ants, 36
    bananas, 272
    biology, 652–659
    bird radiation, 106
    Borneo biota and, 114
    climate, 971
    climate change and, 170
    coral reefs, 200
    diversity, 450
    ecoregions, 448
    extinctions, 170
    frogs, 348, 349, 350–351
    geology, 447, 448, 460, 656, 659–665, 708
    human colonization, 720, 721
    humans and, 42, 43
    insects, 571
    mammals and, 588
    missionaries and, 634
    mountain floras, 114
    overviews, 453
    pigs, 741
    pitohui birds, 5
    seafaring and, 277
    shape and climate, 172
    size of, 111
    snails, 538
    spiders, 862, 862
    Wallace and, 968
New Guinea Trench, 664–665
New Hebrides, 306, 309, 481, *482*, 485, 648
New Ireland, 485, 486, 720, *949*
new islands, 228
New, T.R., 206
newts, 122
New World, 7, 40, 52
New Zealand. *See also* Gondwana
    age of, 319
    alien species and, 99–100
    architecture, 47
    avian diseases, 104
    biocontrol and, 102
    biology, 665–673
    biota, 322–325
    bird disease and, 105
    birds, 106, 304, 314, 322, 323–324, 638–641
    conservation, 474
    Cook Islands and, 195
    Darwin and, 958
    deforestation and, 170, 221
    diversity and, 320
    ethnobotany and, 272
    Fijian skinks from, 302
    fishing, 310
    flightlessness and, 314, 317–318
    frogs, 348
    geology, 647, 665–666, 669, 673–680, 703, 707–708, 754
    gigantism, 373, 374, 375
    global warming and, 380
    humans and, 42, 707–708, 720, 722

insects, 465
introduced species, 470, 471, 472, 477, 959
invasive species, 474, 480
island rule and, 492, 494
landslides and, 535
lizards, 559
Lord Howe Island and, 571
mammals, 322–323, 672
mice, 793
millipedes, 300
overviews, 183
pigs and goats, 741, 742, 743
plant disease, 748
Polynesians and, 758
pre-European impacts, 416
rats, 792, 793, 794, 795
settlement of, 41
spiders, 863
sub-Antarctic islands, 11, 16
threatened species, 285
tuatara, 786
weed costs, 99
whaling, 976, 977
Ngeaur Island, 149, 739
niaouli tree, 644, 645
Nias, 483
Niau Island, 333, 337, 642
niches
    competition and, 106
    divergence and, 110
    empty, 233
    extinction and, 285
    novel, 252
    speciation and, 699
nickel, 647
Nicobar Islands, 439, 777, 778
Nightingale Island, 66, 108
Nihoa Island, 208, 211, 375, 411, 412, 862
Ni'ihau Island, 208
Ni'ihau shield, 406
Ninety East Ridge, 439–440
Nisyros, *396*
nitrogen, 13, 14, 470–471, 472, 475, 801, 877, 887, 888
Niue (Tonga), 692
Noble, G.K., 348
noddies, 956
non-adaptive radiations, 1, 461, 561
nonequilibrium, 489
non-native species. *See* alien species
nonradiating taxa, 774
nonscientific projects, 102
nontarget effects, 102
non-volcanic arcs, 691–692
Nordvestfjord, Greenland, 56
Norfolk Island, 210, 571, 572, 768
Normandy, 155
Normans, 121
North Aegean fault zone, 392
North Africa, 627, 686
North America. *See also specific locations*
    Antilles and, 21, 25
    arctic and, 51
    barrier islands, 83, 85
    Bermuda and, 96

birds, 26
Britain and Ireland and, 117, 126
extinctions, 813
fish, 28
fishes, 874, 875
geology, 504, 694, 754
glacial-interglacial cycles, 785
human colonization, 722–723
introduced species, 473
island formation, 973
plant disease, 749
Polynesian voyaging and, 760
prisons, 770
raccoons, 26
silverswords and, 837
sky islands, 839, 840
species-area relationship and, 857
tectonism, 56, 57
volcanism and, 58
North American Plate. *See also* Okhotsk plate
    Antilles and, 22, 30
    Azores and, 64, 71
    Baja California and, 78
    Bermuda and, 95
    Channel Islands (California) and, 161, 164
    Iceland and, 429
    Japan and, 501
    Kuriles and, 522
    overviews, 482, 705, 753, 754
    seamounts and, 824
    Vancouver/Wrangellia and, 708, 937
    vents, 425, 426
North Andros Island, 211
North Atlantic region. *See also* Iapetus Ocean
    coral, 565
    plate system, 429
    seabirds, 814, *814*
    tectonics, 428, 651
    volcanism, 755
    whales, 975, 976, 977
North Carolina coast, 87, 565
northeast Pacific region, 705, 713
North Equatorial Current, 172
northern hemisphere, 14, 977
Northern Ocean Archipelagoes, *814*
North Moluccas, 459
North Pacific, *55*, 158, 418, 419, 713, 813, 814, *814*, 874, 975, 976, 977
North Pole, 59
North Sea, 117, *258*, 266, 287–288, 289, 566, 567
North Slope, Alaska, 58, 85–86
Norway, 54, 289, 292, 423, 565, 567, 866, 879, 933, 975
Norway rats, 471, 472, 792, *792*, 793, 794, 795, 801, 834, 850
Norwegian-Greenland Sea, 55, 58
Nostostomatidae, 670
Notogean Realm, 446, 447, 448
de Nova, João, 871
Novaya Zemlya, 47, 48, 51, 52, 54, *55*, 57, 59, 60
novel environments, 1, **2–4**
novel niches, 252
Novosibirskiye Ostrova Islands, 48, 51, 59

nuclear weapons, 336, 341, 612, **680–685**, 706, 708, 711, 712
Nukuoro Atoll, 148
nunataks, 10, 12, 13, 17, 19, *48*, 433, 786
Nunn, Patrick, 248
nutria, 474
nutrient availability
    caves and, 547, 548
    dispersal and, 531
    diversity and, 13
    edge effects, 787
    introduced species and, 471, 513
    island rule and, 494
    rats and, 795
    seabirds and, 812
    succession and, 877–878
    waves and, 882
nutrients. *See* food (nutrients)
Nuu-chah-nulth people, 938
Nyerere, Julius, 985

O'ahu
    adaptive radiation and, 412
    beaches, 92
    climate, 172
    coastline, 94
    geology, 406, 951
    human colonization, 722
    insects, 317
    introduced species, 471
    landslides, 536
    plants, 3
    silverswords and, 837
    snails, 539, 540, 541
    waves and, 882
oak forests, 157, 182, 184, 186, 389, 525, 628, 793, 841, 938
oak-laurel forests, 971
oaks, 129
oases, 174, **686–689**
Oaxaca Valley, *184*, 186, 187
obsidian, 43, 46, 247, 277, 278, 394, 625, 626, 629, 720, 721, 745
Ocalan, Abdulla, 771
ocean circulation, 172, 288. *See also* Gulf Stream
ocean floor, 174, 175, *176*, 177, 576, 933. *See also* organic falls; vents; whale falls
Oceania, 41–43, 313–314, 720. *See also* Lapita
oceanic islands, **689–696**. *See also* de novo formation
    age of, 319
    anagenesis and, 9
    basalts, 951
    biogeography and, 948
    Comoros, 178
    diversity and, 10
    dwarfs, 238
    endemism, 129, 254, 257, 870
    erosion and, 954
    flightlessness and, 312
    frogs and, 350
    geology, 951
    global warming and, 376–377
    insects, 130
    introduced species and, 473
    overviews, 490
    seamounts and, 821
    snakes, 843, 844
    speciation on, 8
    spiders, 862
    subsidence and, 954
    Taiwan, 897
    volcanism, 950–951, 952, 953
    Wallace on, 965
oceans. *See* acidification, ocean; *specific oceans*
octopuses, 145, 160, *176*, 635, 781
Oeno Atoll, 45
Ogasawara Islands, 234, 498, 499, 504, 538
Ogilvie-Grant, W.R., 848
ohia trees, 402
oil. *See* fossil fuels
oil palm, 115
Okhotsk cultures, 522
Okhotsk plate, 506, 522
Okinawa, 474
Okinawa Trough, 502
Old Crow Flats, *526*
old growth, 88, 901, 938. *See also* timber
Old World flora and fauna, 52
Old World tropics, 7
olfactory cues, 210, 235, 670, 843, 924
olive trees, 874
olivewood bark, 95
Olson, Storrs, 411
One Tree Island, 385–386
"On the Law Which Has Regulated the Introduction of Species" (Wallace), 964
On the Origin of Species (Darwin), 710, 964
"On the Tendency of Varieties to Depart Indefinitely from the Original Type" (Wallace), 964
On the Various Contrivances by which British and Foreign Orchids are Fertilised by Insects, and on the Good Effects of Intercrossing (Darwin), 696
ontogeny, 9–10, 239, 515
Ontong Java Plateau, 306, *659*, 660, 664, 703–704, 706, 825, *854*, 855, 856
Onychophora, 670
oomycetes, 748
oopu, 402
ophiolites, 113
    Cyprus, 214, 215
    New Caledonia, 646, 647
    Newfoundland, 649
    New Guinea, 663, 664
    Philippines, 734, 735, 736
ophiuroids, 819
opportunists, 700
Öræfajökull Zone, 429, 433, 434
orangutans, 113
orbit, Earth's, 376
orchids, 467, 469, 571, 653, **696–699**
Oregon, 694
organic falls, **700–701**. *See also* whale falls
Orient, 114
Origin of Species (Darwin), 2–3
Orkney Islands, *116*, 120, 123, 292
ornamental plants, 275

ornithosis, 104
orogenies, 55–56, 57, 77, 112, 154, 308, 646
orographic effects, 172, *422*
Ortelius, Abraham, 702, *703*
Orthoptera, 670
Osborn, 1
Osbourne Guyot, 706, 714
osprey, 122, 798
ostriches, 639, 957
otters, 120, 159, 160, 294, 296 500, 525, 909, 937, 978
"Our Ships at Anchor in the Roadstead" (Gilsemans), *759*
Outer Banks, North Carolina, 87
Outer Hebrides (Western Isles), *116*, 120, 123–124, 292
outrigger canoes, 277
outwelling, 592–593
Ovalau, 302, 305, 308
overfishing, 709, 715, 779, 782
overwash forests, 591
Owen, Richard, 230–231, 638
owls, 26, 179, 295, 374, 852
oxen, 155, 869
oystercatchers, 122, 132, 295
oysters, 195, 336, 853, 854, 893
Ozarks, *152*
ozone, 892, 894

Pacala, S., 913
pacas, 90, 491
Pacific Ocean, 55, 172, 173, 566
Pacific-Philippine Sea plate, 454
Pacific plate
    Bougainville and, 706
    Channel Islands (California) and, 164
    Cook Islands and, 191, 192, 709
    Easter Island and, 709
    Farallon Islands and, 287
    Fiji and, 306
    French Polynesia and, 339, 340, 341, 342, 712
    Galápagos and, 357
    Guam and, 585
    Hawaii and, 404, 406, 711, 714
    Japan and, 503, 506, 705
    Kurile Islands and, 522
    Line Islands and, 711
    Macquarie Island and, 575, 576
    Marianas and, 593, 598, 706
    New Guinea and, 660, 664
    New Zealand and, 674, 675, 676, 678, 679
    overviews, 692, 702, 704, 705, 753, 754, 969
    Palau and, 149
    Samoa and, 802, 807
    seamounts and, 715
    Solomons and, 703–704, 856
    Tonga and, 918
    vicariance and, 950
Pacific rats, 415, 416, 470, 471, 472, 672, 721, 792, 793, 794
Pacific region (Oceania), **702–715**. *See also* Line Islands
    archaeology, 41
    bird diversity, 320–321

coral conservation, 782, 784
coral reefs, 783–784
crickets, 208–209
earthquakes, 240
ethnobotany and, 271–276
extinctions, 286
flightlessness, 312, 317
forests, 946–947
formation of, 490
frogs, 348
introduced species, 472
invasive species, 478
mammals and, 588
overfishing and, 311
pre-European impacts, 416, 891–892
rafting and, 775
rats and, 794, 795
reefs, 779
research stations, 789
seafaring, 276–277, 280
seamounts, 819, 823
snails, 537
snakes, 844
sustainability, 890–896
trees, 667
vents, 424, 425, 426
whaling, 976
Pacific ridge system, 426
Pacific slab, 502
pademelons, 905
Pagalu Island, 65
Pagan, 594, 598–599, 600–601, *952*
pāhoehoe, 543, 545, 711, 823
Paine, Robert, 909, 910
Palaearctic region, 120
Palau Islands, **715–716**
    archaeology and, 44
    architecture, 47
    conservation, 895, 896
    coral reefs, 781, 782, 783, 784
    frogs, 349
    marine lakes, 603, 604, 605, 606
    overviews, 148, 149
    pigs and goats and, 742
    vegetation and, 943
Palau Ridge, 484, 485
Palawan Island, *349*, 729
Palawan Microcontinental Block, *736*
Palearctic origination, 72–73, 144
Paleoasia (Laurasia), 504–505
paleoendemics, 73, 183, 184, 253, 255, 256, 390, 864. See also relictual taxa
paleofauna, 158
paleorelicts, 157
Paleotropical Realm, 447
La Palma Island, 64–65, 128, 131, 132, 134, 135, 136, *537*
Palmerston, 192, 194, 195
palmettoes, 95
palms
    Canary Islands, 130
    Cook Islands, 193
    date, 686, 689
    extinctions, 245
    Fijian endemics, 299

    Hawaiian, 944
    intraplate islands and, 709
    introduced species and, 471
    Juan Fernandez islands, 508
    Lord Howe Island, 571
    Mare aux Songes, 231
    Mascarene Islands, 613
    New Guinea, 653
    Rapanui, 247
    rats and, 794
    Seychelles, 830
    St. Helena, 871
    Vanuatu, 940
Palms of the Fiji Islands (Watling), 299
palm worms, 425
Panama, 88
Panama current, 357
pandanus, *611*
Pangaea, 63, 504, 505, 624
Pangea, 650
Pantepui, **717–719**, 839, *840*, 842
Panthalassa, 505
paperbark trees, 944
Papillon (Charriére), 769
Papuan Basin, 661
Papua New Guinea. See also Gondwana
    biocontrol and, 101
    bird plasticity, 6
    conservation, 895, 896
    ecology, 657
    freshwater species, 345
    geology, 485, 662, 707
    humans and, 42
    island arcs, 484
    landslides, 535, 536
    marine lakes and, 605, 606
    missionaries and, 637
    oil, 665
    overviews, 854
    tsunamis, 935–936, *936*
    vent communities, 427
parakeets, 618, 832
para-littoral forests, 335
parallel evolution. See coevolution
parapatric speciation, 166, 564
parasitism, 109, 472, 479, 742, 748. See also specific parasites
parasitoids, 101, 102
parrotfish, 782–783
parrots
    colonization by, 320–321
    Fijian, 707
    introduced species, 472
    New Guinea, 655
    New Zealand, 184, 322, 667, 670, 671
    Tierra del Fuego, 917
parthenogenesis, 74, 844
Partida Norte Island, 80
Partida Sur Island, 80
partitioning, 106
Patagonia, 973
patches. See fragmentation (patches); island-like systems
pathogens, 252, 748
PCR (polymerase chain reaction), 412

pea family, 157
Pearl Archipelago, 251
peat, *97*, 114, 123, 124, 290, 291, 592, 905, 931
peccaries, 91
pelagic animals, 820
pelagic birds, 12, 13, 557, 594
pelicans, 81, 158, *159*, 161, *593*, 811, 812
Pemba Island, 778
penal colonies, **767–771**, 908
Penan people, 115
Penguin Bank, 408
Penguin Island, 19
penguins
    Antarctic, 12
    Atlantic, 67
    bird disease and, 104
    Darwin on, 957
    diseases and, 104
    Galápagos, 357, 359, 361
    global warming and, 815
    human impacts, 814
    Lord Howe Island, 569
    Macquarie, 574, 575
    New Zealand, 322
    overviews, 811, 813
    Tristan da Cunha/Gough, 932
Peninsular Ranges, Mexico, 162, 163
Penrhyn, 195
Pentadaktylos Range, 212
peppers, 98
Perapedhi Formation, 214
perch, 169
perched lakes, 331
Periplus, 850
perissodactyls, 22
Perkins, R.C.L., 411
permafrost, 60, 86, 866
Peron, Juan, 770
La Pérouse, 742
Persian Gulf, 180, 421
persisters, 81
Peru, 104, 482, 936
Peru-Chile subduction zone, 754
Pescadores, 897, *898*
pesticides, 100, 161, 709, 850
pests, 89, 90, **99–102**, 121, 160, 556. See also invasive species
Peter I Island, 11, 17, 20
Peter the Great, *423*
petrels
    Bermuda, 96
    Britain and Ireland, 124
    Channel Islands (California), 158
    human impacts, 814
    introduced species and, 471
    Mascarenes, 619
    overviews, 811, 812, 813
    rats and, 795
    Tristan da Cunha/Gough, 932
petroglyphs, 543
petroleum. See fossil fuels
pet stores, 346
Petteys, 946
pH, 379, 521, 602, 642, 878. See also acidification, ocean

phalaropes, 811
Phanerozoic revolutions, 2
phasmids, 570
phenotypic responses, 844
pheromones, 210. *See also* olfactory cues
Philippine Fault Zone, *732, 733,* 735, *737*
Philippine Mobile Belt, *736*
Philippine plate, 149, 585, 593, 598, 705, 715, 724, *753,* 897
Philippines
    adaptive radiations, 3
    biology, 723–731
    Borneo and, 114–115
    climate, 971
    conservation, 895
    coral conservation, 780
    corals, 379
    deforestation, 221
    dwarfs, 237
    fishing, 310
    frogs, 348, 349
    geology, 725, 732–738
    humans and, 43
    invasive species, 346, 347
    island rule and, 492
    macaques, 777
    mammals, 553, 588–589, 590, 776
    missionaries and, 634
    snakes, 845
    species-area relationship and, 857
    whaling, 976
Philippine Sea, 149
Philippine Sea plate
    Indonesia and, 454, 458
    Japan and, 501, 502, 503, 506, 705
    Philippines and, 732, 733, 735
    Taiwan and, 902
pāhoehoe, 543, 545, 711, 823
phosphate, 149, 336, 387, 439, 532, 534, 694, 877, 878. *See also* Phosphate Islands
Phosphate Islands, **738–74**
phreatophytes, 81
phylogenesis, *773*
phylogenetic analysis
    ants, 3
    birds, 108, 183, 353, 412, 616
    Canary Islands, 129
    cichlids, 165, 168
    dispersal/vicariance and, 528, 539, 561, 948
    distribution of species and, 969–970
    Drosophila, 232, 234
    endemism and, 255
    flightlessness, 312, 317
    frogs, 348, 350
    gigantism, 376
    Gulf of Guinea, 809
    invasion biology and, 480
    lizards, 562, 563
    Mascarenes, 617
    New Caledonia, 645
    New Zealand, 639
    overviews, 464, –465
    Pantepui, 719
    Philippines, 725, 727, 729, 731
    planthoppers, 252

Polynesia, 722
    radiation zones and, 772, 773, 774
    rodents, 589
    silverswords, 836, 837
    snails, 363
    Socotra, 850
    spiders, 864
    taxon cycles and, 913
    tortoises, 359, 921, 922, 923, 925
    vent organisms, 424, 426
phylogenetic trees, *412, 464*
phylogeography, 13, 616
phytoplankton, 289, 527, 528, 812, *891*
picaflor, 507
Pico de Teide, Tenerife, 65
Pico Duarte, 22
Pico Island, 64, 71, 72
piculet, Antilles, 21, 23
pigeon guillemots, 158
pigeons
    Canary Islands, 131, 132
    colonization by, 320–321
    diversity and, 319
    dodos, 231
    Mascarene Islands, 618
    New Guinea, 655
    New Zealand, 672
    Samoa, 800
    Tonga, 920
pigments, 151, 234, 258, 461, 468, 797, *863*
pigments, dye, 274–275
pigs, **741–743**
    archaeology and, 43
    Baja California Islands, 80
    Canary Islands, 132
    Channel Islands (California), 155, 158, 160–161
    Philippines wild, 724
    pre-European impacts, 417
    Samoa, 801
    Tristan da Cunha/Gough, 932
Pikas, 120
Pikini (Bikini) atoll, 69, *610, 611,* 680–681, 683, 684, 712
pillow lava, 491, 953
Pimm, S.L., 282
pine forests, 27, 79, 128, 129, 130, 132, 182, 614, 841, 905, 942
pine-oak forests, 182, 184, 186, 841
pine trees, 27, 78, 128, 157, 288, 296, 331, 395, 450, 614, 628, 707, 769, 797, 981
pingos, 60
pinnipeds, 79, 159, 161, 294, 296, 297, 378
pioneer communities
    adaptive radiation and, 361
    ants, 36–37
    dispersal and, 877
    gigantism and, 372
    Great Barrier Reef, 384, 385
    invasive species and, 946
    Kerguelen, 15
    Krakatau, 518
    lava, 402
    lava tubes, 547
    Pantepui, 718

    silverswords, 838
    soils and, 944
    spiders, 861
    succession and, 877, 878
    Surtsey Island, 887
pirates, 203, 762, 918
Pisonia grandis, 556
Pitcairn, 538, 712, **744–747,** *789,* 834, 951
Pitcairn-Gambier, 341
Pitcairn Group, 44–46, 492
pitcher plant, 830
pitohui birds, 5, *6,* 655
Piton de la Fournaise, 443, 613, 620, 622
Piton des Neiges, 443, 613, 620, 622
plankton. *See also* zooplankton
    dispersal and, 357, 528
    Faroe (Faeroe) Islands, 288, 289, 290
    Galápagos, 360
    global warming and, 379, 817–818
    lake, 527, 528
    Marianas, 594
    Phosphate islands, 740
    seabirds and, 812
    seamounts and, 819, 820
    snails and, 539
    sticklebacks and, 876
    sustainability and, 891
    Tatoosh, 911, 912
    tides and, 916
    whales/whaling and, 975, 976
plant diseases, **747–752,** 749, *751*
planthoppers, 252, 463, 548
plants, seed, 129
plants, vascular. *See also* flowering plants
    alien species, Antarctic, 15
    Antarctic, 12, 14, 15
    arctic, 50, 51, 52
    Atlantic, 72
    Azorean, 71, 73, 74
    Canary Islands, 129, 132
    Fijian, 299
    global warming and, 719
    Gulf of Guinea, 810
    Hawaiian, 774
    Indonesia, 448, 449, 450
    inselbergs and, 467, 468, 469
    Juan Fernandez islands, 508
    Kurile Islands, 524
    Lord Howe Island, 571
    Macquarie, 574
    Madagascar, 580
    Madeira, 583
    Mascarene Islands, 616, 617
    New Guinea, 653
    New Zealand, 673
    organic falls and, 701
    Pitcairn, 744–745
    Southern Ocean Islands and, 13
    succession and, 877
    Surtsey Island, 886–887
    Taiwan, 898, 899
    temperatures and, 48
    Vanuatu, 940
    Wallacea, 452
    Wizard Island, 980–981

plants (flora). *See also* ethnobotany; flowering
    plants; plants, vascular; pollination;
    seeds; succession
  adaptive radiation and, 2, 3
  age of, 750
  Antilles, 27
  ants and, 40
  arctic, 52, 57
  Ascension, 61, 62
  Bahamas, 859
  Baja California Islands, 80, 81
  Bermuda, 98, 98
  Borneo/New Guinea and, 658
  Britain and Ireland and, 118, 119, 124
  Canary Islands, 128–130
  Cape Verde, 145–146
  Channel Islands (California), 155, 156–157
  competition and, 283
  convergence, 189
  dispersal, 556, 775, 776
  dodo and, 231
  endemism and, 256
  extinctions, 500
  Faroe (Faeroe) Islands, 291
  French Polynesian endemics, 336
  Galápagos, 848
  Great Barrier Reef, 384–385
  Gulf of Guinea, 810
  Hawaiian Islands, 2, 399–400
  intertidal, 911
  introduced species, 44, 470, 673
  invasive, 253, 474
  lava flows and, 547
  Mauritius, 848
  Mediterranean, 628, 629
  Midway Island, 632
  New Guinea, 653, 656
  New Zealand, 322, 667, 671, 672
  non-vascular, 130, 889
  ornamental, 44, 195, 275, 541, 844
  Philippines, 729
  rats and, 794, 795
  Samoa, 801
  Seychelles, 830, 833
  sky islands, 842
  snakes and, 844–845
  species-area relationship and, 857
  sub-Antarctic, 15
  Surtsey Island, 885
  sustainability and, 891
  Tenerife, 774
  Trinidad and Tobago, 927–928
  Tristan da Cunha/Gough, 932
  Vanuatu, 941
  vegetation vs., 942
  waves and, 911
  Zanzibar, 983
plasticity, 5, *6*, 15, 477–478, 480, 514
plateaus, mountain. *See also specific plateaus*
  Africa, 165
  Antilles, 20
  Arctic, 58
  Atlantic, 66
  Baja California, 78
  Britain and Ireland, 124, 154
  Cook Islands, 192, 193
  deforested, 223
  Faroe (Faeroe) Islands, 287, 288, 289
  Fernando de Noronha, 297, 298
  French Polynesia, 334, 335, 339
  Galápagos, 369
  glaciers, 434, 435
  Haiti, 24, 25
  Kaui, 208
  Macquarie Island, 573
  Madagascar, 469, 550, 551
  makatea, 585, 586
  Marianas, 594, 601
  Mauritius, 621
  Mediterranean, 626, 628
  New Caledonia, 646
  Pantepui, 839, 840, 842
  Socotra, 849
  Tasmania, 905
  Tristan da Cunha, 931
  U.S., 152
plateaux, oceanic. *See also specific plateaus*
  Atlantic, 64
  Azores, 71
  basalts, 431, 898
  French Polynesia, 339, 340
  Japan, 505
  New Guinea, 653, 654
  Pacific region, 950
  seamounts and, 825
plate-boundary islands, *690*, 692–693
plate convergence. *See also* uplifts
  extrusion and, 491
  Indian region, 437–438
  Indonesia and, 454
  Kurile Islands and, 523
  Macquarie Island and, 576
  Marianas and, 598
  New Caledonia, 646
  New Guinea and, 660
  New Zealand, 676, 678
  overviews, 481, 690, 691–693, 703, 753, 754
  Philippines, 733
  seamounts and, 822
  Vancouver Island and, 708
  vicariance and, 950
plate divergence (spreading)
  Baja California and, 703
  Galápagos, 371, 424
  Iceland and, 428, 429, 490
  Indian region, 437
  Macquarie Island, 575, 576
  melting anomalies and, 951
  overviews, 481, 484, 690, 690–691, 753
  seamounts and, 824
  Seychelles and, 382
  vents and, 424
plate tectonics, **752–755**. *See also* earthquakes;
    plate convergence; plate divergence
    (spreading); rifting; volcanism; *specific
    plates*
  age of crusts and, 690
  Antarctic, 18
  Antilles, 22, 29–30, 32
  arctic, 55–56, 57, 59
  Atlantic, 63, 64, 66, 71–72
  Baffin Island, 77
  Baja California Islands, 78, 79
  barrier islands and, 82
  beaches and, 92
  biogeography and, 949, 968–969
  Borneo/New Guinea and, 112, 657
  Channel Islands (British) and, 154
  Comoros, 178
  coral reefs and, 201
  Cyprus, 214
  earthquakes and, 240
  evolution and, 220
  fragments and, 187
  French Polynesia, 339
  freshwater species and, 343
  frogs and, 348–349
  Galápagos, 367–369
  granitic islands and, 381
  Greek Islands, 388, 396
  Iceland, 428–430, 429, 435
  Indian Ocean region, 437–446
  Indonesia and, 447
  Japan and, 501–502
  Kurile Islands and, 522
  Lord Howe Island and, 571
  Madagascar, 580
  Makatea Islands, 585
  Mediterranean, 625, 626, 627, 628
  moa and, 639
  New Caledonia, 646, 647, 648
  Newfoundland, 649
  New Guinea, 664–665
  New Zealand, 674, 676–677
  overviews, 690–693, 694, 702, 713
  reefs and, 779
  Samoa, 803
  seamounts and, 714, 821–822, 824, 825
  Seychelles and, 829
  sky islands and, 840
  Socotra and, 847
  Solomons and, 855–856
  spiders and, 863
  Vancouver and, 937
  Warming Island and, 972
platform islands, 715
platforms, 642, 738, 797, 803
platypuses, 906, 958
Pleistocene Aggregate Island Complexes
    (PAICs), 727, 728, 729
plovers, 78, 106, 159
plumes. *See* cold seeps; mantle plumes
plumes, volcanic, 544
plutonic rock
  Antarctic, 19
  Antilles, 30, 31
  Canary Islands, 138, 140, 141
  Channel Islands (British), 154, 162, 163
  Cyprus, 213
  Darwin on, 220
  Fiji, 306, 307, 308, 309
  French Polynesia, 339
  Iceland, 430, 433
  Japan, 503, 504
  Kerguelen, 440

plutonic rock *(continued)*
    Macquarie Island, 576
    Newfoundland, 652
pocket basins, **755–757**
Podocarpus, 580
Pohnpei, FSM, *94*, 211, 784
Poinar, George and Roberta, 23
Point Bennett formation, 163
polar deserts, 14, 48, 52, 78
polarity reversals, 459
polar regions, 85–86, 104, 171, 421–422, 816
polecats, 120
pollen, 44, 226–227, 291, 838, 907, 970–971
pollination
    arctic, 52
    conservation and, 619
    evolution of, 227
    extinction and, 284, 286
    orchid, 696, 697, 699
    selection and, 842
pollution, 54, 266, 782, 884
Polo, Marco, 850
polychaetes, 426, 819
Polynesia
    ants and, 36
    diseases, 977
    ethnobotony and, 273
    gods, 515
    humans and, 42, 43
    Japan and, 498
    overviews, 758
    pigs, 741, 742
    popular culture and, 764
    radiation zone, 772
    spider conservation, 865
    timber trees, 274
    trade and, 44–45
    traditional practices, 890
    tsunamis and, 934
Polynesians, **758–761**. *See also* Lapita; Maori people
    Americas and, 722–723
    Comoros and, 180
    Cook-Australs and, 709
    Cook Islands, 195
    deforestation and, 221, 222
    Easter Island and, 709
    extinctions and, 282
    flightless birds and, 316
    Hawaii and, 403, 413–414, 711
    homeland, 722
    impacts, 891–892
    lizards and, 560
    Marquesas and, 712
    New Zealand and, 672
    Norfolk Island and, 572
    overviews, 705, 706–707, 709
    Pacific region and, 720
    Pitcairn and, 745
    pre-European impacts, 415–416, 516
    rats and, 793, 795
    Samoa and, 713, 801
    seabirds and, 814
    Tonga and, 920
ponds, 400, *526*

pools, 467–468
poppies, 132, 157
popular culture, **761–765,** 769, 770, 771
population density, 417, 514, 559–560, 793, 826, 846, 912
population genetics, 766–767
population size, 788
porpoises, 120, 125, 159, 938, 975
Port Arthur, 768
Porter, David, 918
Porter's hypothesis, 359
portmanteau biota, 415
Porto Santo Island, 64
Portuguese, 180, 984
position of islands, 377
positive gravity anomalies, 339
Possession Island, 12, 15
possums, 673
potoroos, 905
pottery, 43, 721, 722, 941
poverty, 888, 889, 891, 896
Poway formation, 163
Powell, Jeffry, 923
Powell Island, 18
Pratas Atoll, *898*
praying mantis, 129, 130
Precambrian-Cambrian boundary, 649
precipitation. *See also* rain
    beaches and, 93
    elevation and, 173
    flowering and, 90
    lagoons and, 69
predation
    body size and, 514
    Channel Islands (California), 159
    co-evolution and, 189
    cricket morphology and, 211
    dispersal and, 225
    diversity and, 168–169
    dwarfism and, 236, 238
    ecological release and, 251, 252
    extinction and, 169, 282, 283
    flightlessness and, 317
    Great Barrier Reef, 383
    Indonesia, 451
    introduced species and, 471–472
    invasive species and, 479
    island rule and, 494, 495
    lizard adaptations and, 564
    place and, 911
    radiations and, 206
predators
    birds, 325, 326
    coloration and, 655
    dispersal and, 531
    eradication of, 474
    flightlessness and, 312
    food and, 843
    gigantism and, 376
    Hawaiian Islands and, 398–399
    introduced, 596, 672
    introduced species and, 473
    Mascarene Islands, 618
    succession and, 878
    taxon cycles and, 912

preserves, nature. *See* reserves, nature
President Coolidge, SS, 834
Président Thiers bank, 342
Preston, Chris, 118
prey, 596. *See also* predation; predators
Price, Trevor, 108
prickly pear, 132
primates, 23, 25, 238, 375, 494, 549, 775–776, 778. *See also specific primates*
Prince Edward Islands, 11, 12, 16, 444
Principe, 65, 474, 775, **808–811,** *862*
Principle of Geology (Lyell), 963
prisons, **767–771,** 796, 908
progression of life, 385
progression rule
    Azores, 73–74
    Drosophila and, 773
    endemism and, 254
    Hawaiian biota and, 233, 255
    Hawaiian Islands, 399
    insects and, 464
    silverswords and, 835–837
propagules, 475–479, 487, 531
protozoa, 62, 857, *891*
pseudoscorpions, 62, 130
ptarmigans, 50, 53
pteridophytes, 146, *644*, 713
Puercos Island, 251, 252
Puerto Rican Bank, 22, 472
Puerto Rico
    amphibians, 27
    ants, 38
    beaches, 93
    birds, 26
    fish, 28
    geology, 22, 29, 692
    hydroclimate, 423
    invasive species, 346
    rodents, 24
    shrews, 25
    sloths, 24
puffins, 124, 125, 158, 813
Pukapuka, 193
Pulo Anna Island, 149
pumas, 91
pumice rafts, 260
purging, 437

quail, 132
Quammen, David, 220
quays. *See* cays
Queen Charlotte Islands, 874, 938
Queen Charlotte Sound, *675*
Queen Maud Land, Antarctica, 66
Quintinia, 940
quokka, *796*, 798
quolls, 905

rabbits
    Antarctic, 15
    Baja California Islands, 81
    Britain and Ireland and, 121
    Canary Islands, 132
    Cape Verde, 146
    Channel Islands (California), 159

dwarfism and, 238
Farallon Islands and, 295
introduced species and, 473
Macquarie, 575
Mediterranean, 628
Tierra del Fuego, 917
raccoons, 26, 79, 236
radiation, nuclear test, 683
radiations. See also adaptive radiation
African, 168
chance and, 206
Comoros, 179
Hawaiian crickets, 208
overviews, 461
preludes to, 108
radiation zone, **772–774**
Rae Craton, 77, 78
rafting, **775–778**
Azores and, 72
Bermuda and, 96
crickets, by, 206
ephemeral islands and, 259
Fijian iguanas, of, 302
Galápagos and, 368
Gulf of Guinea islands and, 809
lizard, 560
pumice, 260
snails, 539
ragwort, 124
Raiatea, *641*, 758
rails
Ascension, 62
colonization by, 320–321
diversity and, 319
extinctions, 286, 320
flightless, 312, 313, 314, *314*, 316, 321, 325
gigantism, 374
Great Barrier Reef, 383
introduced species and, 471
Madagascar extinctions, 322
Midway Island, 633
New Zealand, 184, 322, 667, 671
Philippines, 726
Samoa, 800
St. Helena, 871
rain. See also precipitation
arctic, 51
atolls and, 68
climate and, 172
coral reefs and, 201
deforestation and, 222
dunes and, 95
estimation of, 421
freshwater species and, 345
guano and, 740
Gulf Stream and, 116
heat and, 172–173
mountains and, 172
prehumid climates, 112
storms and, 420
vegetation shifts and, 185
rain forests
Borneo, 111–112, 114, 115
Christmas Island, 532, 533–534
climate change and, 171

Cook Islands, 193
diversity and, 106
endemism, 949
Falklands, 957
Fijian, 299
fragments, 181
Frazer Island, 331
French Polynesia, 335
Hawaiian, 945
Indonesia, 451, 453
Mascarene Islands, 614
Maui, 944
New Caledonia, 643
New Guinea, 657
New Zealand, 667
overviews, 942
Samoa, 800, 801
Solomons, 851, 852
Sri Lanka, 867, 868, 870
succession and, 945–946
Tasmania, 905
Tonga, 918, 920
Vanuatu, 940
Zanzibar and, 983
Raivavae Island, 338, 342
Raja Ampat, 653
ranching, 157, 160
Rand, Stanley, 90
random dispersal, 225
range size, 284, 285
Rangiroa, 333, 334, 642
Rangitoto eruption, 676
ranid frogs, 7
Ranongga Island (Solomons), 694–695
Rapa, 342, 462, 538
Rapa Island, 3, 335–336
raptors, 124, 182
Raratonga Treaty, 685
rarity and restriction, 284–285
Rarotonga, 191–193, 194, 195, 196, 197, 335, *339*, 342, 416, 636, 709, *949*
Rasa Island, 80, 81
ratites, 639, 670, 671
rats
adaptive radiations, 3
Antarctic, 15
Antilles, 25
Ascension, 62
Baja California Islands, 79, 81
Bermuda, 97
biodiversity and, 794–795
Britain and Ireland and, 120, 121, 124
Canary Islands, 131, 132
Cape Verde, 146
Channel Islands (California), 161
Comoros, 179
dispersal, 793
eradication of, 474
Galápagos, 926
Gulf of Guinea, 809
Hawaiian birds and, 326
Indonesia, 453
introduced species, 470, 471
Lord Howe Island, 570
mammals and, 590

Marianas, 595, 596
New Zealand, 673, 959
Norway, 792
overviews, 742
Philippines, 588–589, 729
Rapanui, 245, 248
Samoa, 801
Solomons, 852
sub-Antarctic, 15
Tonga, 920
tortoises and, 925, 977
Tristan da Cunha, 932
Raven, Peter, 1
Ravolomanana, M. Marc, 581
rays, *194*
razorbills, 125
Reao, 334, *339*
rear-arc chains, 484
recolonization, 465, 517–519, 777, 840
red colobus, 986
Red Sea, 421
red-tailed tropic birds, 103–104
redwood, 899
reef ecology and conservation, **779–785**
reef fishes, 310, 500, 615, 632, 654, 658, 781, 783, 801, 852–853, 949
reef flats, 614
reef islands, 587
reefless islands, 934
reefs. See also coral reefs
age of, 803
beaches and, 93, 94
Bermuda, 95, 98
Caroline Islands and, 149
Cook Islands, 192–193, 194
Cyprus, 214, 215
French Polynesia, 334, 338
mangroves and, 592
pigs and goats and, 742
pre-European impacts, 417–418
Trinidad and Tobago, 927
waves and, 881
reefs, non-coral, 819
refugia, *182*, 463, 632, 645, **785–786**, 795, 847, 849, 905, 937, 938
regoliths, 433
reindeer, 15, 49, 51, 52, 53, 54, *120*, 236, 470, 471, 472, 473, 866, 977
Reitoru atoll, *334*
rejuvenation volcanism, 953
relaxation, 548, 697, **787–788**, 929
relaxation models, 151, 183, 185, 312, 489–490
relictual taxa
endemism and, 255
examples, 157, 181, 183
Fiji, 299
fragments and, 181, 185, 187
frogs, 348
Japan, 499
Madagascar, 578
Mascarene Islands, 614
New Caledonia, 643, 645
overviews, 182, 184
Pacific region, 949
Socotra, 847, 850

relictual taxa *(continued)*
    spiders, 864
    Taiwan, 899
    Tasmania, 905, 906
religious architecture, 46–47
The Reluctant Mr. Darwin (Quammen), 220
remnants, 329. *See also* fragmentation (patches)
Remote Oceania, 720, 721, 741
René, Albert, 832
reproduction
    corals, 780–781
    effort, 226, 238, 239
    extinctions and, 788
    insect radiations and, 465
    pressures and dwarfism, 236
    seamounts and, 820
    sticklebacks, 875–876
    strategies, 285, 560, 697, 831
    success, 283, 285, 361, 607
reproductive assurance, 697–698, 699
reproductive isolation
    birds, 105, 106, 108–111
    deep-sea animals, 757
    fish, 169
    founder effects and, 327
    plants, 837
    sky islands, 841, 842
    spiders, 864
    tortoises, 925
    Wallace on, 966
reptiles
    Antilles, 27
    Baja California Islands, 78, 80, 82
    Bermuda, 98, 98
    Borneo/New Guinea and, 658
    Britain and Ireland and, 122
    Canary Islands, 131
    Cape Verde, 144, 146
    Channel Islands (California), 155, 159
    Comoros, 178, 179, 180
    Cook Islands, 194
    dwarfism and, 236, 238
    dwarfs, 237
    fire ants and, 941
    Great Barrier Reef, 383
    Gulf of Guinea, 809, 811
    Indonesia, 449, 450
    Japanese, 499
    Kurile Islands, 525
    Mascarene Islands, 615
    New Caledonia, 644
    New Guinea, 654, 656
    New Zealand, 184, 667, 670
    overviews, 843
    Palau, 716
    Pantepui, 718–719
    Philippines, 724, 729
    rafting and, 774
    Seychelles, 830, 833
    Socotra, 849
    Solomons, 852, 853
    species-area relationship and, 857
    Sri Lanka, 869
    sustainability and, 891

Taiwan, 899
Tasmania, 906, 907
Tonga, 920
Vanuatu, 940
rescue effect, *488*
research. *See also* scientific operations
    ants and, 39, 41
    biocontrol and, 102
    Farallon Islands, 293, 294–295, 296–297
    Fijian, 299
    forest, 91
    frogs, 350–351
    invasive species, on, 480
    Krakatau and, 519
    overviews, 756
    Panama and, 88
    red deer, 124
    Socotra, 848
    Spitsbergen, 866
    Tatoosh, 909
research stations, 125, **788–791**, 884, 926
resemblance. *See* convergence, evolutionary
reserves, nature. *See also* marine protected areas (MPA's); national parks
    Antarctic, 16
    Antilles, 29
    ants and, 39
    arctic, 54
    Atlantic, 64, 66, 67
    Azorean, 74
    Baja California Islands, 78, 81
    barrier islands, 83
    beaches, 93
    Bermuda, 97–98
    Borneo, 112
    Britain and Ireland, 124, 125, 126
    Canary Islands, 132
    Cape Verde, 147
    Channel Islands (California), 161
    Cook Islands, 196–197
    coral reefs, 780, 782
    Cozumel, 205
    Farallon Islands, 293–294
    Fernando de Noronha archipelago, 297, 298
    Frazer Island, 330
    French Polynesian, 337–338
    Great Barrier Reef, 386
    Haiti, 27
    Hawaiian, 403
    Indian Ocean region, 440, 441
    Indonesia, 449, 450
    Juan Fernandez islands, 508
    Lord Howe Island, 568, 571
    Macquarie, 575
    Madeira, 585
    Micronesia, 896
    Norway, 567
    Panamanian, 88
    Pitcairn, 747
    Rottnest Island, 797, 798
    Seychelles, 832–833
    species-area relationship and, 861
    Spitsbergen, 866
    Taiwan, 901–902

Tasmania, 908
Tristan da Cunha/Gough, 932
vent communities, 427
resource exploitation. *See also* fishing; mining; oil; *specific activities*
    archaeology and, 44
    coral reefs, 783
    Gulf of Guinea, 810
    Juan Fernandez islands, 508
    Madagascar, 581–582
    Philippines, 730
    research on, 790
    Samoan turtles, 801
    Seychelles, 832
    Solomons, 854
    sustainability and, 889, 891
    Tristan da Cunha/Gough, 931
    Vancouver, 938
    Wallace on, 963
resource limitation, 236, 237–238
resource management, 417
restaurants, 763
restriction and rarity, 284–285
Resurrection plants, 469
La Réunion Island
    ants, 41, 755
    biogeography and, 107
    birds, 108–109, 229, 313, 316, 615, 617
    conservation, 618–619
    ecology, 614
    geology, 441, 442–443, 444, 445, 536, 612, 613, 620, 621, 622, 690, 693
    gigantism, 373, 374
    orchids, 697, 698
    research, 789
    research stations, 789
    seamounts, 824–825
    storms, 420
    tortoises, 922
    tsunamis and, 935
    weevils, 2
Reykjanes Ridge, 428
rheas, 639
rhinoceroses, 22, 51, 113, 285, 389, 395, 447, 450, 869, 922
Rhodes, 237, 388, *393*, 396, *623,* 626
rhododendron, 124, 852, 868, *898*
Rhyncocephalia, 184
Rhynochetidae, 644
rhyolites
    Antarctic, 19
    Antilles, 31
    Atlantic, 64, 65, 66, 67
    Canary Islands, 139
    Channel Islands (California), 162, 163, 164
    Iceland, 432, 433
    island arcs, 484, 485
    Mediterranean, 625, 627
    New Zealand, 676
    Pacific islands, 309, 523, 805, 951
Richter scale, 241
Ricklefs, R.E., 912
ridge-hotspots, 71, **441**
ridges, 424–425. *See also specific ridges*
ridge spreading, 425, 441

rifting
    continental islands and, 707
    East Asian, 505–506
    Gulf of Guinea, 808
    Hawaii, 804
    island arcs and, 484
    Japan and, 485, 504
    Mediterraean, 624
    New Guinea, 665
    New Zealand and, 675, 678
    North Atlantic and, 652
    Pacific region vicariance and, 949–950
    Samoa, 806
    Socotra and, 847
    volcanism and, 952–953
rift zones
    Canary Islands, 136
    Fiji, 308, 309
    Galápagos, 370
    Hawaii, 405
    Iceland, 429, 434
    Indonesia, 458, 460
    seamounts and, 823
    volcanic, 64
Rimatara Island, 335, 342
Ring of Fire, 240, 523, 545, 676, 702–703, 704, 711, 897
ringtail lemurs, 552
Rio Grande Rise, 951
riparian forests, 157, 182, 335, 345
Ritter Island, 537
rivers. *See also* granitic islands
    barrier islands and, 916
    Comoros, 180
    isolation and, 187
    Kurile Islands, 521
    mangroves and, 592
    rafting and, 775
    Taiwan, 898
    tides and, 915
    tsunamis and, 934
Robertskollen nunatak group, *10*
Robertson, R.B., 977
Robichaux, R.H., 836
robins, *122*, 667, 830, 833, 842
Robinson Crusoe (Defoe), 508, 763
rockfish, 159, 701, 713
rock outcrops, 466
rock pools, 467–468
Rocky Mountains, *840*, 841, 842
rodents, **792–795**. *See also specific rodents*
    Antilles, 23–24
    arctic, 49, 51
    extinctions, 205
    Galápagos, 365
    gigantism, 374–375
    Greater Antilles, 21
    Madagascar, 588, 776
    New Guinea, 655
    Philippines, 724
    sigmodontine, 24
    Sri Lanka, 869
Rodinia supercontinent, 504
Rodrigues Island, 107, 231, *313*, 316, 373, 374, 444

birds, 615
ecology, 614
geology, 442, 444, 612, 613, 621, 622
phylogeography, 616
Roggeveen, Jakob, 249–250
Romanche Transform, 754
Rongelap, 683
Rongerik, 680, 682, 683
Roosevelt Island, 17
Roosevelt, Theodore, 632
Rosario group, 163
Ross Island, 17
Ross Sea region, 20
Rota, 474, 538, 594, 595, *597, 598,* 599, 602, *738,* 739
rotation
    African plate, 624
    Canary Islands, 142
    Channel Islands (California), 162, 164
    Cyprus, 215
    Iceland, 435
    overviews, 485
    Pacific islands, 113, 308, 309, 458, 675, 735, 737
    Philippines, 735
rotation, Earth's, 817
Rothera station, Antarctica, 14
Roth, Louise, 238
Rothschild, Walter, 410
rotifers, 12, 298
Rottnest Island, 768, **796–798**
Roughgarden, J., 913
Roussel, Hippolyte, 249, 250
Royal Society Fiord, 77
rubber, 115
Rubiaceae, 653
Rubinoff, Ira, 88
Ruesink, Jennifer, 910
Rum, island of, 120
Rurutu, 333, 342, 343
Russello, Michael, 923
Russian Arctic islands, 49, 51–52, 54, 59, 874
Russians, 522
rust fungus, 748, 750, 751, 752
Ryan, Peter, 108
Ryukyu Islands, 474, *482,* 497, *498,* 499, 500, 501, 502, 503, 506, 522, 902

Saba, 26, 32
Sabah, 115
Sahara desert, 72, 95, 140, 144, 146, 208
Sahul, 720
sailing, 42, 43, 103, 276, 918, 984. *See also* discovery; exploration; seafaring
sailors, 228–229, 570, 710, 741, 742
Saint Cuthbert, 125
Saint Peter and Paul Rocks, 66
Saipan, CNMI, *94,* 483, 593, 595, 596, 597, 598, 599, 602, 603, 634, 705
Sakhalin, 770
salamanders, 159, 161, 182, 499
salinity, 202, 344, 345, 346, 347
salinity crisis, 624, 625
Sal Island, 65, 143, 144, 147
Salix herbacea, 290

salmon, 124, 899
Salsipuedes Island, 80
salt, 14, 797, 812
salt marshes, 331
Samoa, **799–808**
    bats, 194
    birds, 303, 304
    conservation, 896
    Cook Islands and, 195
    crickets, 208
    ethnobotony and, 274
    geology, 708, 712–713
    Lapita and, 43
    medicinal plants, 275
    missionaries and, 636
    seamounts, 824–825
    snails, 539
    snakes, 844
    volcanism, 704, 953
Samoa-Tuvalu island chain, 693
Samos, 395
Samothrace, 395
San Andreas Fault, 78, 161, 164, 693, 703
San Clemente Island, *156,* 157, 159, 162, 163
San Cristobal (Galápagos), *360, 361, 363,* 365, 770, *922,* 923, 925
Sand Adreas Fault, 754
sandalwood, 941
sand and bare ground, 632
Sanday Island, 124
San Diego, 162, 163
San Diego Island, 80, 82
Sandoy Island, 292
sandpipers, 78, 106
sandstones, *650, 651,* 677
Sandwich Islands, *423,* 755
San Estéban Island, 80, 82
San Francisco Island, 80
Sangihe arc, 459
San José Island, 80, 81
San Juanito Island, 79
San Lorenzo Island, 80
San Marcos Island, 80
San Miguel Island, *156,* 157, 158, 160, *160,* 161, 162, 163
San Nicolas Island, *156,* 159, 162, 163
San Onofre breccia, 162, 163
Santa Barbara Island, *156,* 158, 159, 161, 162, 163
Santa Catalina Island, 81, *156,* 157, 159, 162, 163
Santa Cruz Island, California, 80, 82, *156,* 157, 158, 159, 161, 162, 163, 257, 474
Santa Cruz Island, Galápagos, *111, 359,* 924
Santa Cruz Island fault, 162
Santa Cruz Islands, Solomons, 851, 854, 855, *921*
Santa Luzia Island, 143
Santa Maria Island, 64, 71, 73, 74, *923*
Santarosae, 158, 160, 492, 707
Santa Rosa Island, *156,* 157, 158, 159, 160, 162, 163
Santiago (St. Jago) Island, 143, 144, 146, 147, 217, 218, 474
Santo Antão Island, 65, 143, 144
São Jorge Island, 64, 71, 72
São José Islet, 65

São Miguel Island, 64, 71
São Nicolau Island, 143, 147
São Tomé Island, 65, 108–109, 775, 808–811, *862*
São Vicente Island, 65, 143, 147
Sarah Island, 769
Sarawak, 90, 115
sardines, 296, 525
Sardinia, 624, 629, 741
Sarigan, *598, 599*, 601
Sark, 154
Saronic Islands, 626
Satawal, 783
Sator, 82
saturation, 384
Sauromalus, 82
Savai'i, 758, *952*
savannas, 644, 657, *718,* 927, 983
scales, 98, 472
Scandinavia, 117, 121, 650, 874, 973
scarab beetles, 7
scarabs, 899, *902*
scavengers, 700, 974
scent, 697
Schluter, Dolph, 1, 4, 859
Schmitt, Dennis, 972
Schnierla, T.C., 88
Schoener, T., 860
Schouten, Willem Cornelisz, 917
science, 228–230, 784–785. *See also* research; research stations
scientific expeditions, 427, 542. *See also* Beagle, HMS
scientific operations
 alien species and, 16
 Bermuda and, 97
 biocontrol and, 101
 Britain and Ireland, 124
 Channel Islands (California), 161
 cold seeps and, 175
 Indian Ocean region, 440, 441, 443, 444
 outdated, 186
Scilly Isles, 116, *121,* 126, 337, *338*
scleractinian coral, 96, 195, 198, 199, 200, 201, 202, 614, 779, 781
sclerophyllous forests, 72, 128, 132, 156, 299, *578, 579,* 614, 618, 643
Scotia Arc, Antarctica, 10, 66
Scotia plate, 482
Scotia Sea formation, 18
Scotia Trench, 67
Scotland, 567
Scott Island, 17, 20
scribbly gum/wallum banksia forest, 331
Scrivenor, J.B., 968
scrub
 Baja California Islands, 81
 Cape Verde, 146
 Cook Islands, 193
 Maui, 836
 Mesoamerican, 184
 New Caledonia, 643
 Rapanui, 245
 Samoa, 800
 Vanuatu, 940

Zanzibar, 983
seabirds, **811–815**. *See also* birds; nesting
 Antarctic, 12, 13, 14
 arctic, 50
 Ascension tropical, 62
 Atlantic, 67
 Baja California Islands, 80, 81
 Bermuda, 96
 bird disease and, 105
 Britain and Ireland and, 122, 124, 125, 126
 Cape Verde, 144, 147
 Channel Islands (California), 158–159, 161
 Cook Islands, 194, 197
 extinctions, 931
 Farallon Islands, 293–294, 295–296, 296–297
 global warming and, 379
 Gough Island, 930
 Great Barrier Reef, 383, 384–385
 humans and, 43
 Indonesia, 453
 introduced species and, 471
 Line Islands, 556
 Lord Howe Island, 570
 Macquarie, 574
 Midway Island, 632, 633
 navigation and, 280
 New Zealand, 672
 phosphate and, 740
 pre-European impacts, 416
 rats and, 795
 Rottnest Island, 798
 seamounts and, 820
 Seychelles, 832
 Socotra, 849
 Solomons, 852
 St. Helena, 871
 Tatoosh, 911
 Tonga, 920
 Tristan da Cunha, 930
 Tristan da Cunha/Gough, 931
 Vancouver, 938
 whaling and, 976
sea cows, 22, 853
sea cucumbers, *781,* 853
sea depths, 491
seafaring, **276–281**, 292, 629. *See also* sailing
sea floor, 847, *854, 856. See also* deep-sea speciation; vents
 seamounts and, 825
sea grasses
 Bermuda, 96
 coral and, 780, 781
 Fiji, 302
 global distribution, 779
 Japan, 500
 Mascarene Islands, 615
 organic falls, 701
 patchy, 182
 Raratonga, 194
 Rottnest Island, 796, 798
 shipwrecks and, 834
 Solomons, 853
 threatened species, 893
 Zanzibar, 984

sealers, 742, 908, 931
sea levels, **815–818**. *See also* intertidal habitats; land bridges; tides
 barrier islands and, 83, 87
 Bermuda and, 95, 96
 biodiversity and, 113
 Borneo and, 112
 Britain and Ireland and, 117, 125, 126
 Canary Islands and, 127
 Channel Islands (British) and, 154
 Chatham Islands and, 323
 climate change and, 170, 171
 coasts and, 83
 coral and, 201
 coral reefs and, 219
 Darwin on, 957
 dispersals and, 776
 diversity and, 320
 erosion and, 261
 Fernando de Noronha archipelago and, 297
 Frazer Island and, 331
 French Polynesia and, 336, 711
 freshwater species and, 345
 glaciations and, 524, 785
 global warming and, 377–378
 granitic islands and, 381
 Great Barrier Reef and, 388
 Greek Islands and, 388, 389
 Gulf of Guinea, 808–809
 Indian Ocean, 178
 Indonesia and, 447, 448
 island formation and, 491–492, 973
 Line Islands and, 554
 Lord Howe Island and, 569, 570
 Maldives and, 587
 mangroves and, 592
 marine lakes and, 603
 Mascarene Islands and, 613
 Mediterranean, 628
 moa and, 640, 642
 motu and, 641, 642
 New Guinea and, 665, 708
 New Zealand, 666, 666, 675, 678, 679
 oceanic islands and, 695–696
 overviews, 693
 Pacific region, 707
 Philippines, 727, 728
 prediction of, 262
 seamounts and, 825
 Seychelles and, 833
 Socotra and, 847
 South America, 956
 southwest Pacific and, 969
 Tasmania, 904, 958
 Trinidad and Tobago and, 927
 volcanoes and, 268
 Wallace's Line and, 968
 waves and, 880, 881
sealing, 15, 54
sea lions, 79, 159, 161, 296, 358, 359, *368,* 525, 672, 798, 938, 978
Seal Nunataks, 17, 19
seals
 Antarctic, 12, 13, 14
 Atlantic, 67

Baffin Island, 76, 78
Baja California Islands, 78
Britain and Ireland and, 125, 126
Canary Islands, 132
Channel Islands (California), 159, 160, 161
currents and, 357
Farallon Islands, 294, 296
Galápagos, 368
human impacts, 14
Juan Fernandez islands, 508
Kurile Islands, 525
Macquarie, 574, 575
Midway Island, 632
Pacific region, 776
predators, 50
South Georgia, 977
Spitsbergen, 866
St. Helena, 870
Surtsey Island, 886, 887
Tristan da Cunha/Gough, 931–932
Vancouver, 938
seamounts, **818–825**. *See also specific seamounts*
age of, 824–825
Atlantic, 63, 64, 65
Bermuda, 95
Canary Islands, 133, 136, 140–141, 142
Caroline Islands, 149
central Pacific, 950
Galápagos, 371
Galápagos, 710
geology, 951
Line Island chains, 711
North Atlantic, 652
old, 706
overviews, 490, 491, 694, 704–705, 706, 713, 754
Pitcairn, 744
Samoa, 802, 803
Society Islands, 712
tides and, 916–917
Sea of Cortés Islands, 80–82
Sea of Japan, 536, 505–506, 537
sea palms, 910, 911
Searle, Jeremy, 124
seas, defined, 756
seasonal forest, 940
sea urchins, 566, 910, 911, 912
Sea Venture, 834
seawalls, 261, 262
seaweeds
Cape Verde, 146
cold seeps and, 175
Cook Islands, 194
ephemeral islands, as, 258, 259
Macquarie Island, 574
Rottnest Island, 796
Surtsey Island, 886
Tatoosh, 910
threatened, 893
Tristan da Cunha/Gough, 930
Sebens, Kenneth, 910
secondary sexual characteristics, 826, 827
sedges, 53, 78, 905, 930
Sedgwick, Adam, 955

sedimentary rocks. *See also* sedimentation
Antarctica, 19
New Caledonia, 646
Newfoundland, 650, 651, 652
New Zealand, 677
Philippines, 734
South American, 718
Taiwan, 903
Tierra del Fuego, 957
Warming Island, 972
Zanzibar, 983
sedimentation. *See also* flysch; sedimentary rocks
Antarctic, 18–19
Antilles, 30, 31
archaeology and, 44
arctic, 55, 56–57, 57–58
Atlantic, 67
atolls and, 68
Baffin Island, 77
barrier islands and, 82, 83, 84–85, 86, 86, 87, 93
beaches and, 91, 92, 93, 94
Borneo and, 113
Channel Islands (British), 154
Channel Islands (California), 162, 163, 164
Cook Islands, 194
coral and, 200
Cyprus, 212, 214, 214–215
deforestation and, 222
erosion and, 261
Fernando de Noronha archipelago, 298
Fiji and, 308
Frazer Island and, 331
freshwater species and, 344
global warming and, 379–380
Iceland, 436
Indonesia and, 456, 457, 458, 459
island arc volcanoes and, 954
landslides and, 535
Lord Howe Island, 569
Macquarie Island and, 576
Madagascar and, 578
Maldives and, 587
marine lakes and, 606
methane and, 175
motu and, 641
Newfoundland, 649
New Guinea, 661, 663, 664
New Zealand, 680
overviews, 481
seamounts and, 491, 819, 822, 824
subduction and, 482
tundra and, 60
Vancouver and, 937
volcanic arc islands and, 691
waves and, 880, 882–883
seeds, 225, 226–227, 299, 877, 885–886, 887, 920, 944
seeps, 426, 974
Segeberger Höhle, 150
seismic activity. *See* earthquakes (seismic activity)
selection pressures
dwarfs and, 237–238

ecological release and, 252
island rule and, 494–495
New Zealand, 671
orchids and, 699
sky islands, 841
self-incompatibility, 838
self-pollination, 697–698
Selkirk, Alexander, 508
Selvagen Islands, 64, 127, 133, 140–142, *862*
semi-deciduous forests, 193
Senckenberg Museum, 848
sensu stricto, 483
Seri Indians, 82
Sespe formation, 163
Severgin volcano, 524
Severnaya Zemla Ostrova, 47, 51, 52, 59
sewage, 196, 197
sexual difference, 226
sexual dimorphism, 232
sexual selection, **825–829**
adaptive radiation and, 3
diversification and, 6
Drosophila, 232
Hawaiian crickets, 463
mental phenomena and, 965
sky islands and, 842
Wallace on, 966
Seychelles, **829–833**
ants, 39, 41
bats, 179
bird disease and, 104
continent and, 381–382
endemic amphibians, 348
frogs, 348, 773
geckos, 179
geology, 177, 445–446, 755
introduced species, 472
lizards, 561
plant disease and, 748–749
rats and, 793
tortoises, 922, 924
whaling, 976
Seychelles Plateau, 177
Seymour Island, 17, 19
shags, 814
shales, 67, 214, 393, 439, 457, 651, 663
shape of islands, 172
Shark Bay, Australia, 56
sharks, 78, 145, *194*, 296, 301, 557
Sharp, Andrew, 760
Shaw, K.L., 209
shearwaters, 132, 811, 812, 813, 815, 932
sheep
Britain and Ireland and, 118, 121, 124, 125, 126
Canary Islands, 132
Channel Islands (California), 160–161
Faroe (Faeroe) Islands, 289, 290
Marquesas, 223
Mediterranean, 629
Rapanui, 245–246, 250
Tierra del Fuego, 917
sheeted-dyke complexes, 213, *214*, 576
Shekelle, Myron, 552
shellfish, *194*, 196

INDEX 1065

Shetland Islands, *116,* 123, 292
shields/shield volcanoes
    Antarctic, 19
    Arctic, 55, 56
    Atlantic, 64, 65
    Baja California and, 78
    Canadian, 77
    Canary Islands, 134, 135, 136, 137–139, 140, 141, 142
    French Polynesia, 339, 341
    Galápagos, 369–370
    Gulf of Guinea, 808
    Hawaiian stages, 404–409, 953
    Iceland, 432
    Iceland Holocene, 434
    Indian Ocean region, 444, 445
    inselbergs and, 466
    Kurile Islands and, 522
    landslides and, 536
    Lord Howe Island, 568, 569–570
    Mascarene Islands, 621–622
    overviews, 694, 754, 755, 952, 953
    Samoa, 803, 804, 805–806, 807–808
    seamounts and, 823
    St. Helena, 870–871
Shikoku, 484, 485, 497, 498, 499, 500, 502, 506
Shindo, Kaneto, 423
ship rats, *792,* 793, 795
ships. *See* trade; watercraft
shipwrecks, **833–835**
shorelines. *See also* beaches; coasts
    Bermuda, 98
    birds, 106
    colonization and, 206
    subsidence and, 954
shrews
    Antilles, 21, 23, 25
    Baja California Islands, 81
    Britain and Ireland and, 126
    Canary Islands, 129, 131, 132
    Greek Islands, 390
    Philippines, 729
    rats and, 795
    Socotra, 849
    Sri Lanka, 869
shrikes, 159
shrimps, 425, 426, 729, 797, *906,* 907
shrubs
    Baja California Islands, 81
    Channel Islands (California), 156, 157
    Hawaiian, 836, 944
    introduced species, 470–471
    Mediterranean, 628
    Midway Island, 632
    New Caledonia, 644
    New Guinea, 657
    Pacific region, 891, 948
    Pantepui, 718, 719
    pigs and goats and, 742
    Rottnest Island, 797–798
    Samoa, 801
    Socotra, 848
    St. Helena, 873
    succession and, 878
    Tasmania, 905

Vanuatu, 940
Wizard Island, 981
Siamese people, 115
Siberian Arctic, 51, 58
Siberut, 483, *774*
Sibumasu block, *456,* 457
Sicily, 183, 238, *423,* 492, 494, 625, 629
Signy Island, 14–15
Silver Bank, 22
silverswords, 4, 401, *402,* 773–774, **835–838**. *See also* tarweeds
Simberloff, D., 859
Simeulue, 483, 777, 778
Sinbad, 978
Sindia, 146
Singapore, 263, 282
single-island radiations, 462, 463, 464
sinkholes, 20, 24, 25, 150, 394, 983
sinks, *528,* 630
SiO_2, *406,* 804–805, 806, 951
Siple Island, 17, 20
sipunculids, 819
sister taxa, 359
size of islands. *See also* area factors; space, ecological; species-area relationship (SAR)
    adaptive radiation and, 531
    Antilles, 28
    ants and, 36
    biodiversity and, 107
    biogeography and, 488–489
    deforestation and, 222
    dispersal and, 228
    distribution and, 949
    diversity and, 186, 320, 329
    erosion and, 262
    evolution and, 124
    extinction and, 283, 285, 329
    fragments and, 187
    Galápagos biodiversity and, 371–372
    Gough Island, 930
    groundwater and, 420–421
    introduced species and, 473–474
    lakes and, 529
    Line Islands, 555
    overviews, 9
    research and, 788
    saturation and, 384
    seamounts, 820, 824
    snails and, 540
    species-area relationship and, 858–859
    species richness and, 259, 527
    succession and, 878–879
    Tristan da Cunha, 930
size of lakes, 166
size of population, 750
size responses. *See* dwarfism; gigantism
skerry communities, 887
skinks
    Bermuda, 96
    colonizations, 560
    Comoros, 179
    Gulf of Guinea, 809
    Mascarene Islands, 618
    mice and, 794

    New Zealand, 671
    nonadaptive speciation, 561
    Philippines, 729
    Samoan, 800
    Seychelles, 831
    Solomons, 852
    Tasmania, 906
    Tonga, 920
Skokholm Island, 255
Skottsberg, Carl, 509
skuas, 813
skunks, 158
Skúvoy Island, *289*
sky islands, **839–842**, **839–843**. *See also* Pantepui
    Hispaniola, 22
    insects and, 460
    relaxation and, 787
    spiders, 863
    U.S., 6, 235, 460, 863
skylark, 126
slabs, 484–485. *See also* subduction
slash and burn, 416
slip, plate, 753
sloths, 21, 23, 24–25, 90, 236, 321, 375, 590
Slovenia, 150, 151, 152, *153,* 623
slugs, 118, 145, 671, 938
small-island effect, 384
Small Isles, *116,* 124
small populations, 284, 285
Smith, A.C., 299
Smith Island, 17, 18
Smithsonian Institution, 88, 89, 411
smut fungus, 748, 749, 750
Snæfellsjökull, 434
Snæfellsnes Zone, 429, 433
snails
    adaptive radiation, 540
    Azores, 73
    Baja California Islands, 78
    Bermuda, 95, 96, 98
    biocontrol and, 100, 101
    Canary Islands, 130, 774
    conservation, 540–541
    dispersal, 539
    endangered species, 908
    extinctions, 283
    French Polynesian, 335–336, 337
    Galápagos, 363–364, 365
    Gondwana and, 670
    introduced species, 470, 473
    invasive species, 346–347
    Japan, 499, 786
    lakes and, 529–530
    Lord Howe Island, 569, 571, 572
    Marianas, 594
    Mascarene Islands, 617
    New Zealand, 672
    Pacific region, 670
    Pitcairn, 745
    Polynesian, 949
    rafting and, 775
    rats and, 794, 795
    refugia and, 786
    Samoa, 800, 801

Seychelles, 830, 831
Society Islands, 712
Socotra, 849
Solomons, 852, 853
St. Helena, 872
Tasmania, 907
Tristan da Cunha, 930
waves and, 911
snakes, **843–846**
Antilles, 27
Baja California Islands, 80, 82
body size, 373, 559
Borneo, 114
Borneo/New Guinea and, 658
Britain and Ireland, 122
Channel Islands (California), 159
colonization and, 560
Comoros, 179, 180
dispersal, 844
dwarfism and, 238
Fijian, 302, 774
Guam, 596, 706
Gulf of Guinea, 809, 810
introduced species and, 471
invasive species, 479
island rule and, 493, 494
land bridge islands and, 558–559
New Guinea, 654
Philippines, 724
Rottnest Island, 798
Samoa, 800–801
Seychelles, 831
Socotra, 849
Solomons, 853
Sri Lanka, 869
Taiwan, 899, 900
Tasmania, 906
Tonga, 920
Vanuatu, 940
Wizard Island, 981
Snares, 16
Snares Island, 11, 16
snipe, 795
Snowball Earth, 57
Snow Hill Island, 19
social facilitation, 285
societies, human, 41–42, 46–47
Society Islands
adaptive radiation, 712
biocontrol and, 101
birds, 335
Cook Islands and, 195
ethnobotony and, 273
freshwater species, 345
geology, 340–341, 712
human colonization, 759
insects, 712
marine life, 336
missionaries and, 634
snail, 283
snails, 538, 540
volcanism, 953
socioecosystems, 416
Socotra, 421, *423*, 847
Socotra archipelago, **846–851**

soft corals, 198–199
soils
aging, 944, 945
alien species and, 99
Borneo, 114
Britain and Ireland and, 123, 124
Britain's, 117
Cape Verde, 144, 146
Cook Islands, 193
coral reefs, on, 201
dwarfism and, 238
Hawaiian Islands, 398
hydrology and, 421
introduced species and, 471
lava flows and, 547
Line Islands, 555–556
Mediterranean, 628
movement, 13
New Caledonia, 643, 645
pigs and goats and, 742
Pitcairn, 744
plant disease and, 749
Rapanui, 251
rats and, 795
succession and, 877, 879, 946
Tasmania, 905
temperature and, 172
Tonga, 919
Tristan da Cunha, 930
ultramafic rocks and, 663
Vanuatu, 940
vegetation and, 942
volcanic, 755
Solander Island, 676
solar radiation, 109
solenodons, 21, 25
solid waste management, 196, 197
Sollas, W.J., 481
Solomons, **851–857**
conservation, 895
crickets and, 208
forests, 851
frogs, 395
geology, 306, 485, 690, 691, 702, 703
human colonization, 720
humans and, 42, 43
landslides, 536
lizards, 560
mammals, 774
overviews, 705, 706
pigs, 741
snails, 538, 539
species-area relationship and, 857
Solomon Sea basic, 664
Solovetsky Islands, 770–771
Somalia, 934
Somali plate, 178, 445
Somerset Island, *976*
songbirds
Channel Islands (California), 159
introduced species, 472
mating, 413
radiation, 106
reproductive isolation and, 110
St. Helena, 871

song, cricket, 210, 211
songes, *891*
songs, bird, 110–111, 413
songs, cricket, 209
Sonoran taxa, 80, 81, 82
Sonsorol Island, 149
soricomorphs, 21, 25
Sorong Fault, 454, 459, 663
Sorong fault, 459
sorting, 189, 191, 220, *528*
sorting, species, 527, *528*, 530, 531
Sotavento group, 65
Soufrière, Montserrat, 31, 32, 33–34, 35
South Aegean volcanic arc, 396
South Africa, 7, 16, 104, 477, 669, 770
South America. *See also specific regions*
African dispersals, 560
Antilles and, 21, 23
ants from, 36
beavers and, 793
Darwin on, 69, 217
fish, 28
Galápagos and, 358–359, 360, 361
Gondwana and, 669
human colonization, 722–723, 760
Juan Fernandez Islands and, 508
land mammals, 24
prisons and, 769
rats and, 793
seamounts, 819
spiders, 863
tectonism, 69
South American plate, 30, *31*, 67, 482, *753*
South Atlantic, 755
South Carolina coast, 83, 84, 87
South China, 113, 501, 504, 505, 506
South China Sea, *446*, 458, 506, 724, 732, 733, 735, 902
Southeast Asia
frogs, 348, 349
human colonization, 721
macaques, 777
mammals, 552
missionaries and, 634, 637
New Guinea and, 656
pigs, 741
primates, 549
threatened species, 285
southern hemisphere, 14
Southern Hemisphere whales, 977, 978
Southern Ocean islands, 10, 11, 12, 13, 15, 16, *814*
Southern volcanic zone, *429*, 433, 601
South Georgia, 10–11, 13–14, 16, 66–67, 104, 471, 472, *789*, 793, *814*, 976, 977
South Iceland Seismic Zone (SISZ), 428, *429*, 435
South Korea, 545, 546
South Orkney Islands, Antarctica, 10, 11, 17, 18, 66–67
South Pacific
geology, 951
literature and, 763
missionaries and, 635, 636
South Pacific Superswell, 554

INDEX 1067

South Sandwich Islands, Antarctica, 10, 66–67
South Sea Islands, 42, 761
South Shetland Islands, Antarctica, 10, 11, 17, 18, 485
southwest Pacific, 695
Soya current, 521
space, ecological. *See also* isolation
    adaptive radiation and, 1
    caves and, 152
    continental islands and, 183
    convergence and, 191
    coral and, 200
    dispersal and, 225, 226, 228
    diversity and, 13
    rivers vs. lakes, 169
space, open, 479
Spain, 121, 148
Spanish explorers, 203, 729
sparrows, 144, 159
spatial considerations, 529–530, 749, 751, 766, 809, 820, 824. *See also* isolation; size of islands; species-area relationship (SAR)
spawning areas, 607
specialist species. *See also* ecomorphs
    Barro Colorado insects, 90
    dispersal and, 225
    ecological release and, 252, 253
    extinction and, 283, 285, 788
    Galápagos birds, 110
    insects, 461
    inselbergs and, 469
    Krakatau, 519
    New Zealand alpine, 667
    orchids, 697
    overviews, 461
    Seychelles, 831
    whale falls and, 700
speciation. *See also* adaptive radiation; anagenesis
    adaptive radiation and, 5, 7
    Azorean, 73
    Cape Verde, 146
    caves and, 151, 153
    cichlids, of, 165–166
    distance and, 772
    founder effects and, 327
    freshwater species, 347
    genetic propensities to, 168–169
    Greek Islands, 390
    Hawaiian Islands, 399
    isolation and, 108
    mechanisms of, 166–167
    New Zealand, 669
    overviews, 755
    rapid, 3
    sexual selection and, 826, 827
    temporal separation and, 209
species-area relationship (SAR), 73, 527, 562, 688, **857–861**, 943
species boundaries, 725
species pump, 747, 841, 842
species richness
    biogeography and, 486–487
    elevation and, 698
    equilibrium theory and, 384
    extinction and, 284
    Hawaiian, 711
    island rule and, 494
    lakes and, 526
    Madagascar, 577, 578
    oases and, 688
    orchids, 699
    Philippines, 724
    seamounts and, 820
    succession and, 877
species-sorting, 527, *528,* 530, 531
Speke, John, 984
spiders, **861–864**
    arctic, 50, 51, 52, 53
    Australian, 671
    Barro Colorado, 90
    Bermuda, 98
    Canary Islands, 130, 131, 774
    Cape Verde, 145
    Channel Islands (California), 159
    convergence and, 188, 191
    dispersal, 861–862
    Hawaiian, 5, 7, 257, 531, 548
    Hawaiian convergence, 189
    Hawaiian Drosophila and, 233
    kīpuka and, 512
    Lord Howe Island, 570
    Madagascar, 580
    New Zealand, 671
    Palau, 716–717
    predators, 235
    sexual selection and, 6, 826
    sky islands, 842
    Socotra, 848, 849
    species-area relationship and, 860–861
    Sri Lanka, 868
    St. Helena, 872, 873
    Surtsey Island, 886, 887
    Tasmania, 907
spiderworts, 132
Spiller, 860
spinebills, 907
spiny bush, *578,* 579
spiritualism, 965
spit elongation, 87
Spitsbergen Island, 58, 172, *423,* **865–866**, 879, 976
sponges, 52, 819, 820, *909,* 911
spoonbills, *902*
Sporades, 626
spots, 110
spreading. *See also* plate divergence (spreading)
    continental islands and, 707
    Easter Island and, 709
    Galápagos and, 710
    Mediterraean, 624
    New Guinea, 665
    New Zealand and, 678
    overviews, 703
    seamounts and, 714, 821, 822
spreading, ridge, *854,* 856
spreading, sea floor, 847, 856
springtails
    Antarctic, 12, 13
    arctic, 48, 51, 52, 53
    climate change and, 15
    temperatures and, 14
Spruce, Richard, 963
spurges, 653
squirrels, 91, 119, 121, 125, 126, 132, 472, 793–794, 981
Sri Lanka, 346, 349, 445, 755, **866–869**
St. Anne, 832
St. Barthelemy, 23, 31
St. Brendan the Navigator, 978
St. Christopher, 23
St. Croix, U.S. Virgin Islands, *92*
St. Eustatius, 23, 24, 31, 32
St. Helena, *789,* **870–873**
    Drosophila, 234
    endangered species, 285
    endemism and, 256
    flightless birds, 316
    forests, 873
    introduced species and, 470, 959
    island rule and, 492
    rainfall, 173
    research stations, 789
    volcanism, 959
    whaling, 976
St. Helena (Australia), 769
St. Helena Island, 66
St. Helens, Mount, *543*
St. Jago (Santiago), 217, 218
    Darwin and, 956
    volcanism, 959
St. John, 395
St. Kilda, 121, 123–124, 126, 635
    map, 116
St. Kitts, 24, 31, 32, 34
St. Kitts Bank, 23
St. Lawrence Island, 48, 51
St. Lucia, 24, 31, 32, 35
St. Martin, 35
St. Martin Bank, 22, 24, 31
St. Paul Fracture Zone, 66
St. Paul Island, 441, *789,* 795
St. Paul/New Amsterdam archipelago, 11
St. Paul plateau, 441
St. Paul's Rocks, 218, 956
St. Vincent, 24, 31, 32, 35
stages of life, 226
Standley, Paul, 89–90
Stanley, Henry Morton, 984
starfish, 910–911
starlings, 800
static equilibrium, 489
Steadman, David, 314, 320, 594, 595
steep-sidedness, 695–696, 754, 755, 783, 823, 871, 952, *953*
Stephens Island, 283, 673
stepping-stone islands, 384, 767
stereotypes, 761
Stevens, Samual, 963
stickleback fish, 4, 5, 124, **873–877**
stochasticity. *See* chance (stochastic) events
Stokes, Pringle, 917, 955
stonechats, 129, 131
stonecrop, 157
Stóra Dimun Island, *288*

storms. *See also specific types of storms*
    climate change and, 171
    Comoros, 178
    coral and, 200
    erosion and, 262, 263
    freshwater species and, 344, 345
    frog dispersal and, 350
    global warming and, 379–380
    Great Barrier Reef, 385
    island arcs and, 705
    motu and, 641, 642
    Tierra del Fuego, 918
    waves and, 779–880
Strait of Sicily islands, 625
stratigraphy, 522–523
streams, 94
Strecker, Angela, 531
stress tolerance, 877
Streymoy Island, 292
strike-slip faults (transform convergence), 454, 661, 693, *732, 733*
stromatolites, 56, 797
stygobionts, 151, *152, 153*
sub-Antarctic, 15
subantarctic region, 14, 359
subduction. *See also specific plates*
    Caroline Islands and, 149
    Channel Islands (California) and, 161, 164, 162
    cold seeps and, 175
    Cyprus, 215, 216
    Fiji and, 306
    Indian Ocean region, 437
    Indonesia and, 454, 456, 457, 458, 459
    initiation, 484, 485
    island arcs and, 702
    Japan and, 502, 503, 504
    Kurile Islands and, 523
    lava tubes and, 545
    Macquarie Island and, 576
    Marianas and, 598
    Mediterraean, 625, 626
    New Caledonia, 648
    New Caledonia metamorphic rock and, 648
    New Guinea, 664–665
    New Zealand, 676, 678
    overviews, 424, 481–486, 691–692, 754
    Philippines and, 733
    seamounts and, 713
    Solomons, 856
    Vancouver and, 937
subduction-related islands, 17–19, 22, 30, *31*, 112, 113. *See also* island arcs
subpopulations, 766
subsidence
    atolls and, 69
    Darwin and, 220
    Darwin on, 219
    Enewetak and, 219
    erosion and, 261
    granitic islands and, 381
    Hawaiian Islands, 397, 408
    Indian Ocean region, 440
    Maldives and, 587
    overviews, 484, 694, 695

sea levels and, 695
Uniformitarianism and, 218
volcanic islands and, 953–954
subterranean species. *See* caves
succession, **877–879**
    Krakatau, 518, 519
    Mediterranean forests, 628
    organic falls and, 700
    plant disease and, 750, *750*
    rain forests and, 945–946
    Surtsey Island, 886, 887
succession, geologic, *650*
succulents, 468, 469
Sudan, 146
Sulawesi. *See also* Wallacea
    Borneo and, 115
    climate, 971
    distribution and, 970
    fauna, 553
    freshwater species, 346
    frogs, 349
    geology, 457, 458, 459, 969
    macaques, 777, 778
    mammals, 776
    New Guinea and, 660
    overviews, 114, 446, 452
    Philippines and, 729
    shape and climate, 172
    tarsiers, 549
    Wallace and, 968
sulfophilics, 700
sulfur, 176, 524
Sulu Sea, 756, 757
Sumatra
    biodiversity, 113
    Borneo biota and, 114
    climate, 971
    earthquakes, 242
    endangered species, 450
    endemism, 450
    geology, 439, 448, 456–457, 458
    Krakatau and, 517
    macaques, 778
    mammals, 553, 776
    mountain floras, 114
    overviews, 112
    snakes, 845
    tsunami, 933, 935
Sumatran fault, 454
Sumbawa island, 270, 454, 484
Sunda. *See also* Wallacea
    age of, 690
    diversity, 450
    geology, 455–456, 457, 458, 459
    human colonization and, 720
    volcanism, 439
    Wallace and, 968
Sundaland plate, 732
Sunda Shelf, 112–113, 448, 455, 724, 729, 776, 777
Sunderbans, 380
sunflowers, 132, 157, 359, 362, *363*, 835
sunflower star fish, 159
sunken continents, 673–674, 679
surface water, 421

surf in the tropics, **879–883**
Surtsey eruption, 63–64, 429, 432, 491, 511, 545, **883–885**
Surtsey Island biology, 879, **885–888**
sustainability, **888–896**. *See also* conservation
    overviews, 789
    Pitcairn and, 745
    rain forests and, 946
    Seychelles, 832
    successes, 784
    vegetation and, 944
Suwarrow, 195
Svalbard archipelago, 48–49, 50, 51, 53, 54, 55, 56, 57, 58, 59, 60, 976
Swahili, 983
swans, 122, 123, 796
swash bars, 85, *86*
Sweden, 122, 750–751, 879
sweepstakes routes, 36
sweet potatoes, 246, 723, 760
swells, topographic, 340
swiftlets, 182
swifts, 144
Swinhoe, Robert, *900*
Sylvianoris, 325
sympatric speciation
    amphibians, 27
    birds, 6
    Galápagos, 352, 355, 773
    insect, 461, 462
    lakes and, 166, 169, 531
    lizards, 561, 562, 563
    mammals and, 589
    Philippines, 729
    rats, 589
    sexual selection and, 828
    silverswords, 837
    snails, 672
    spiders, 864
Syowa station, East Antarctica, 14
Syzygium, 940

Tachigali, 89
Tahiti
    adaptive radiation, insects, 712
    Banks and, 219
    biocontrol and, 101
    Cook Islands and, 195
    Darwin and, 218, 958
    diseases, 977
    endemics, 335, 337
    forests, 222
    freshwater species, 345
    geology, 341, *341*, 585, 712
    humans and, 42
    insect diversity, 463
    invasive species, 478, 479
    missionaries and, 634, 636, 637
    ornamental plants, 275
    popular culture and, 763
    population of, 337
    Rapanui and, 250
    snails, 539, 540, 541
    tides, 915
    volcanoes, 333

Taiaro, 334, 337
Taiwan, **897–904**. *See also* continental islands; Old World tropics
 Borneo and, 114–115
 climate, 971
 fishing, 310
 human colonization, 721
 humans and, 43
 invasive species, 346, 347
 landslides, 535
 seafaring and, 277
takahes, *184*
Takapoto, 334
Takutea Island, 192, 193, 197
Tamang Bank, 149
tambalacoque trees, 619
Tambora eruption, 263, **267–270**, 439, 454, 484
Tam Tam Island, 782–783
tanagers, 26–27
tardigrade (water bear) species, 12, 14
target effect, 488, *489*
tarsiers, **549–553**, *731*
tarweeds, 78, 157, 837, 838. *See also* silverswords
Tasman, Abel, 672, 908, 919
Tasmania, **904–909**
 Darwin and, 958, 959
 freshwater species and, 345
 hydroclimate, 423
 overviews, 181, 183
 spiders, 862
Tasmanians, 114
Tasmanian Sea, 678
Tatoosh, **909–912**
Taupo Volcanic Zone (TVZ), 675, 676
Taveuni Island, 300, 301, 302, 309
taxon cycles, 5, 38, 253, **912–913**
taxonomic diversity, 889–890
Taylor, S.R., 485
Taymyr Peninsula, 51
tea trees, 797
technology, 791
tectonism. *See* plate tectonics
television, 762
temperatures. *See also* climate change; global warming
 caterpillars and, 15
 coral and, 200, 202, 565
 diversity and, 13, 14
 Drosophila and, 235
 elevation and, 173
 endemism and, 257
 extinctions and, 171
 extreme, 578, 579, 656–657
 flightlessness and, 317
 fluctuations, 555
 Holocene, 433
 insects and, 227
 overviews, 173, 421
 seabirds and, 81
Templeton, Alan, 233, 327
temporal separation, 209
Tenerife
 adaptive radiation, 774
 birds, 314

 lizards, 558, 563
 orchids, 698
 overviews, 127, 138
 pigs and goats and, 742
 species, 129, 130, 131, 132
 spiders, 863
 volcanism, 65, 134, 135, 137–139
tenrecs, 179, 550, 580, 588, 589–590, 776
tephra, 80, 139, 222, *263, 264,* 265, 288, 432, 433, 511, 599, 601, 676, 885, 952. *See also* ash, volcanic
 Surtsey Island, 885
tephrochronology, 433
tepuis, 182, 717, *718*
Terceira Island, 71, 72, 959
Terceira Rift, 64
"Tereoboo, king of Owhyee, bringing presents to Captian Cook" (Webber), 759
terns, 62, 81, 125, 248, 556, 611, 811, 812, 813, 814
terrapins, 831
Testudinidae, 921
Tethys Ocean, 215, 392, 426, 624, 627
Tettigonidae, 670
Texas coast, 83, 87
Thailand, 347
Thasos, 395
thatch, 273–274
The Theory of Island Biogeography (Macarthur and Wilson), 772
Thera, 394
Thera eruption, 626
thermal expansion, 816
thistles, *227*
tholeiites, *803,* 804, 806–807
Thomas, Jeremy, 118
Thompson Glacier, *59*
Thorarinsson, S., 433
Thornton, W.B., 206
Thorpe, Roger S, 563
thrashers, 81, 472
threatened species. *See* endangered and threatened species
thrombolites, 797
thrushes, 26, 104, *122, 931*
Thule people, 76–77, 976
Thurston Island, 17
Tiburón Island, 80, 81
ticks, 23, 104
tidepools, 181
tides, **914–917**
 arctic and, 61
 barrier islands and, 83, 84, 85, 86, 87
 beaches and, 91
 Bermuda, 95
 Britain and Ireland and, 125
 Channel Islands (California), 155
 marine lakes and, 603, 605
 Mediterranean, 628
 mussels and, 911
 sea levels and, 815–816
Tierra del Fuego, **917–918**, 955, 956–957
tigers, 869
Tikehau atoll, *68*
Tikopia, 417

timber. *See also* logging
 Borneo and, 115
 Canary Islands, 130
 ethnobotony and, 274
 Pitcairn, 745
 Vancouver, 938
time, dispersal and, 225
time dwarfs, 495
Timor, 459
Tinbergen, Niko, 876
Tinian, 594, 596, 597, *598,* 599, 602, *952*
Tip, Tippo, 984
Tiputa, 642
Tiree, *116*
Tjörnes Fracture Zone (TFZ), 428, *429,* 435
toads, 27, 118, 122, 477, 478, 845, 981
Toba eruptions, 455
Tobago, 23, 35
Tobi Island, 149
Tofua, 484
Tokelau, *256, 273,* 333, 637, *705,* 714–715
Tokelau seamount chain, 714
tomatoes, 125
Tonga, **918–921**
 bats, 194
 birds, 303, 304
 Cook Islands and, 195
 ethnobotony and, 274
 extinctions, 801
 Fiji and, 299
 geology, 306, 483, 485, 491, 690, 692, 702
 Lapita and, 43
 missionaries and, 634, 635, 637, 638
 overviews, 707
 pigs, 742
 plants, 273
 plate tectonics, 691
 watercraft, 759
 whaling, 976
Tonga-Kermadec subduction zone, 714
Tonga-Kermadec Trench, *692*
tortoises, **920–926**
 Bermuda, 96
 Canary Islands, 131
 dispersal, 922
 dispersal methods, 776
 extinct, 231
 Galápagos, 359, 361–362, 364–365, 372, 710, 958
 Galápagos endemism and, 257
 Mascarene Islands, 614, 615, 617, 619
 Seychelles, 830–831, 832–833
Tortuga Island, 82
tourism. *See also* ecotourism
 advertising and, 765
 Antarctic, 14
 arctic, 54
 atolls and, 68
 Azorean, 74
 Baja California Islands, 78
 beach renourishment and, 262
 Bermuda and, 97
 Channel Islands (California), 155
 Cook Islands, 195, 196
 coral reefs and, 781

Cozumel, 205
Frazer Island, 332
Galápagos, 356
Line Islands and, 555
Lord Howe Island, 571
Madeira, 584
Midway Island, 633
Pitcairn, 746
prisons and, 768, 769, 771
Rapanui, 250, 251
Rottnest Island, 796
Seychelles, 832, 833
Spitsbergen, 866
Tonga, 919
Vancouver, 938
western Pacific, 715
whales and, 978
towhees, 159
Townsend's shearwater, 80
trachytics, 66
trade
   ancient, 44–45
   Bermuda and, 97
   bird disease and, 104, 105
   Borneo and, 115
   Cook Islands and, 195
   dodo and, 229
   Farallon Islands and, 294
   Gulf of Guinea and, 810
   invasive biology and, 476–477, 815
   invasive species and, 346, 478–479
   Kurile Islands and, 525
   lakes and, 531
   pigs and goats and, 742
   plant disease and, 748
   Rapanui, 249–250
   Seychelles and, 832
   St. Helena and, 871
   sustainability and, 892
   Tristan da Cunha/Gough and, 931
   Zanzibar and, 984
trade winds
   Canary Islands and, 128, 131
   Cape Verde and, 144, 146
   climate and, 172
   climate change and, 174
   clouds and, 172
   Cook Islands and, 193
   Hawaii and, 173
   heat and, 173
   landscape evolution and, 695
   Vanuatu and, 939–940
traditional practices, 609, 782–783, 784, 850, 890, 894–895, 936, 947. *See also* ethnobiodiversity
Tragedy of the Commons, 784
Traill, Greenland, 60
transform faults, 753, 754, 755, *802*
transform plate boundaries, 703
transform plate-boundary islands, *690*, 692–693
Trans-Hudson Orogeny, 77
transilience, 233
translocation, 120
transported landscapes, 415–416

tree ferns, 13
treefrogs, 654, *655*
trees. *See also* conifers; deforestation; forests; *specific trees*
   Antarctic, 13
   Antilles, 27
   ants and, 40
   arctic, 58
   Ascension, 62
   Atlantic island, 72
   Baja California Islands, 78, 80, 81
   Barro Colorado, 91
   Barro Colorado coevolution, 88–90
   Britain and Ireland and, 118, 125, 126
   Cape Verde, 145
   Comoros, 178
   Cook Islands, 194
   Dominican Republic, of, 23
   Gough Island, 931
   New Guinea insects and, 656
   New Zealand, 667
   Pitcairn, 745
   Rapanui, 245
   Rottnest Island, 796
   Samoa invasives, 801
   Socotra, 848, 850
   Solomons, 852
   succession and, 877
   Tristan da Cunha, 931
   Vancouver, 938
   Vanuatu, 941
Trematolobelia, 2, 3, 4, 400
trenches, 691, 702, *704*, 854. *See also specific trenches*
Las Tres Marías, 79
trilobites, 651
Trindad Island, 23, 26, 35, 36, 65–66, 278, 298
Trinidad and Tobago, **926–929**
Trinity Peninsula group, 18
Tristan da Cuhna archipelago, **929–932**
   anagenesis, 9
   biotas, 11
   birds, 108
   Drosophila and, 234
   geology, 66, 690, 693, 951
   research stations, 789
Tristan/Gough plume, 66
Tristan (volcano), 66
Trobriand Islands, 664
troglobionts, 151, *152*, 863
Troodos Range, 213–214, 216
trophic structure, 385–386
tropical Atlantic, 814
tropical Pacific, 814, 889
Tropical Pacific Decadal Variability (TPDV), 345
tropical wilderness areas, 453
tropicbirds, 556, 811, 812
trout, 124
Trouvador, 834
Truman, Harry, 682
tsetse flies, 474
tsunamis, **933–936**
   2004, 439
   earthquakes and, 240–244, 695

   geology, 755
   Hawaiian, 408–409, 711
   Kick 'em Jenny and, 511
   Kurile Islands and, 522, 524
   landslides and, 537, 954
   Mediterranean, 626
   overviews, 484, 691
   Solomon Islands, 706, 856
Tuamotu Archipelago, 46, *68*, 333, 335, 336, 337, 340, 492, 553, 642, 711, 950
Tuamotu-Gambier archipelago, 334
Tuamotu plateau, *341*, 712
tuataras, 184, 670, 786
tube worms, 174, *176*, 424, 425, 426
tuna, 311, 338, 812, 820, 832, 891
tundra, *48, 50,* 51, 52, 53, 60, *77,* 78, 155, 291, 524, 657, 866
Tunisia, *687*, 688
Tupaia, 758
Turkish subplate, 626
turnover of species
   ants, 38, 41
   Krakatau, 519
   New Zealand, 670
   rates, 488, 489
   vegetation, 944
Turtle Island, 22
turtles
   Ascension, 62
   Baja California Islands, 80
   beaches and, 91, 92
   beach renourishment and, 262
   Bermuda, 96, 97
   Borneo, 114
   Canary Islands, 131–132
   Cape Verde, 145, 146, 147
   Cook Islands, 194, 197
   Fernando de Noronha archipelago, 298
   Great Barrier Reef, 383, 385
   Indonesia, 452
   island rule and, 493
   Marshall Islands, 611
   Mauritius, 229
   Midway Island, 632
   Pelagie Islands endangered, 625
   Philippines, 724, 726
   Pitcairn, 745
   Samoa, 800, 801
   seamounts and, 820
   Seychelles, 831
   Socotra, 850
   Solomons, 853
   Tonga, 920
   Trinidad and Tobago, 929
   whaling and, 977
   Zanzibar, 984
Tuscan Islands, *623,* 625, 857
tussocks, 14, 52, 108, 184, 401, 471, 574, 667, 798, 867, 905, *929,* 931, *932*
Tuvalu, 256, 333, 420, 591, 641, *681,* 693, 759
typhoons, **418–420,** 709, 814, 898, 904
Tyrrhenian Sea islands, 625

Uist, *116,* 123–124
Ullung Island, Korea, 8, *9*

ultraviolet radiation, 14, 15, 151, 258
ungulates, 132. *See also specific ungulates*
Unified Neutral Theory of Biodiversity and Biogeography (Hubbell), 488
Uniformitarianism, 218
United States
    Caroline Islands and, 148
    caves, 150, 151, 152
    coral conservation, 567, 780, 782
    coral reefs and, 783
    introduced species, 473
    mosquitoes and, 977
    shipwrecks and, 834
    snails, 542
uplifts
    coseismic, 694
    Darwin and, 220
    glaciers and, 973
    Indian Ocean region, 439
    Kurile Islands, 522
    Macquarie Island and, 576–577
    Madagascar and, 578
    Makatea Islands and, 585
    Marianas and, 594, 599
    Mediterranean, 624
    New Caledonia, 648
    Newfoundland, 651
    New Guinea, 657, 665
    New Zealand, 675, 676
    overviews, 491–492, 691–692, 694
    Philippines, 736
    sky islands and, 839
    Socotra, 847
    Solomons, 692, 704, 706
    Taiwan, 903
    Tasmania, 958
    Vancouver, 937
    Zanzibar and, 983
Uracas (Farallon de Pajaros), *599*, 600
Urals, 52, 57
Urup Island, 522
d'Urville, Dumont, 42, 720
Utirik, 683

vagrants (birds), 12, 122, 124, 126, 489
Vaigach Island, 51
Vancouver, 333, 375, *423*, *704*, 708, 874, **937–938**
Vancouver, George, 937
vangas, *579*
van Riper, Charles, 414
Vanua Levu, 299, 300, 301, 302, 304, 305–306, 308, 309
Vanuatu, 306, 380, 491, 834, 854, 936, **939–941**
Van Valen, Lee, 493
Vatnajökull glacier, 64, 435
vectors, 103, 478, 527, 749
vegetation, **941–947**. *See also* forests; plants
    Antarctic, 13–14
    arctic, 49
    Atlantic, 72
    barrier islands and, 83, 85
    beaches and, 91
    climate and, 85–86
    climate change and, 44
    Cook Islands, 193
    diversity and, 186
    endangered, 643
    extinction and, 282
    Gough Island, 930–931
    Japan, 500
    Midway Island, 632–633
    motu and, 642
    New Guinea, 657
    New Zealand, 707
    oases, 687
    overviews, 171
    pigs and goats and, 743
    Samoan, 800
    Spitsbergen, 866
    Surtsey Island, 887
    Taiwan, 898
    Tonga, 918
    Tristan da Cunha, 930–931
    volcanism and, 940
vegetative reproduction, 697
Venezuela, *21*, *30*, *35*, *36*, *71*, *839*, *840*, 926–927
vents, hydrothermal, **424–427**, *425*, 700, 974
Verbeek, R.D.M., 517
vertebrates
    Antarctic fossils, 19
    Antilles fossils, 20, 22
    ants and, 41
    Azorean, 72
    biocontrol and, 100
    Canary Islands, 129, 131
    Cape Verde, 144
    Cozumel, 205
    distribution, 970
    dwarfism and, 236
    evolution of, 922
    extinction and, 284
    extinctions, 581, 908
    food-limited populations, 121
    founder events and, 6
    humans and, 3
    Indonesia, 448, 453
    Madeira, 584
    Marianas and, 594
    New Guinea, 653–654, 656
    Philippines, 724, 729
    rats and, 794–795
    Rottnest Island, 798
    Socotra, 849
    Tonga, 918, 920
    Trinidad and Tobago, 927
    Vancouver, 938
    world, 654
Vespucci, Amerigo, 297
Vestfold Hills, Antarctica, 10
Vestiges of the Natural History of Creation, (anonymous), 963
de Veuster, Damien, 635
vicariance, **947–950**
    Antilles, 28
    dispersal vs., 560, 561
    diversity and, 182
    fragments and, 187
    Gondwana and, 669, 670
    Madagascar, 580
    overviews, 970
    reefs and, 779
    snails and, 539
    Socotra, 849
Victoria Island, 50
Victoria Land, Antarctica, 13
video games, 765
Vidoy Island, *290*
Vieques Island, 22
Vikings, 291–293
village weaverbirds, 108–109
de Villalobos, Ruiz Lopez, 716
vireos, 96
Virgin Islands, 22, *92*
viruses, 121–122, *891*
visual signaling, 211
Viti Levu, 299, 300, 301, 302, 303, 304, 305, *305–306*, 307, 308, 319, 322, 325–326
Vitousek, 945
de Vlamingh, Willem, 796
vocalizations. *See* acoustical signaling
volcanic arcs. *See also* island arcs
    collisions with continents, 902, 903
    geology, 755
    magma and, 952
    overviews, 953
Volcanic Caribbees, 29, 31
volcanic islands, **950–954**. *See also* island arcs; volcanic arcs
    age of, 825
    geology, 951
    Gough Island, 929
    growth of, 953
    landslides, 536–537
    lava tubes and, 544–545
    overviews, 693–694
    Pacific region, 704, 708
    Samoa, 799–800
    Seychelles, 829
    shield volcanoes and, 952
    Taiwan, 897
    Tristand da Cunha, 929
volcanism. *See also* ash, volcanic; hotspot islands; island arcs; kīpuka; lava; mantle plumes; seamounts; *specific eruptions*; *specific volcanoes*
    Antarctic, 11, 17, 19–20
    Antilles, 22, 29, 30, 31, 32–35
    ants and, 38
    arctic, 50, 55, 56, 58
    Ascension and, 61
    Atlantic, 58, 63–66, 67, 71–72
    Azorean, 74
    Baffin Island, 77
    Baja California Islands, 78, 79, 80, 81
    beaches and, 92, 93, 93, 94
    Bermuda and, 95
    Borneo, 112
    Britain and Ireland, 125
    Canary Islands, 127, 133–142
    Cape Verdes, 144, 218
    Caroline Islands, 149
    Channel Islands (British), 154
    Channel Islands (California), 162, 163, 164

climate effects, 266–267
Comoros, 178
Cook Islands, 191–193
coral and, 200, 956
Cyprus, 214
Darwin and, 217, 218, 219–220, 958, 959
deforestation and, 222
dispersal/vicariance and, 948
diversification and, 464
earthquakes and, 240–241
endemism and, 254
ephemeral islands and, 260
extinctions and, 500
Faroe (Faeroe) Islands and, 288
Fernando de Noronha archipelago, 297, 298
Fiji, 305
French Polynesia, 338–343
freshwater species and, 344, 345
geothermics and, 435
global warming and, 376
Greek Islands and, 388, 389, 393, 394, 395–396
Gulf of Guinea, 808, 809
Hawaiian Islands, 397, 404–408
Iceland, 430–434, 433, 434, 435
Indian Ocean region, 437–446, 441, 442–444
Indonesia, 454–455, 458, 459
isolation and, 187
Japan, 501, 502
Kurile Islands, 521, 522, 523–524
Lesser Antilles, 510
Lord Howe Island, 568, 569
Macquarie Island and, 576
Madagascar, 177, 578
Makatea Islands, 585
Marianas and, 593–594, 598–599, 706
Mascarene Islands, 612, 620–622
Mediterranean, 624, 625, 626, 627
navigation and, 280
New Caledonia, 648
Newfoundland, 649, 650, 651, 652
New Guinea, 661, 663, 664
New Zealand, 666, 674, 675, 676, 677, 679
North Atlantic, 652
northeast Pacific, 705
ocean floor, 490–491
overviews, 219, 483, 753, 754, 755
Pacific region, 706, 707
Philippines, 725, 727, 727, 733–734, 734, 737
Rapanui and, 243
Samoa, 802, 803, 804–808
seamounts and, 821, 822–823, 824–825
sky islands and, 840
Solomons, 855, 856
spiders and, 862
St. Helena, 870–871
Surtsey Island and, 484
Taiwan, 898
tectonism and, 339
Tonga, 919
Tristan da Cunha eruptions, 931
Uniformitarianism and, 218
Vancouver and, 937
Vanuatu, 939
voles, 49, 52, 53, *120, 126*

"Voyage Up the Amazon" (Edwards), 963
Vulcão de Paredão, 65
vultures, *181*

Wadati-Benioff zone, 32, 484–485
Wadati, K., 481
Waipounamu Erosion Surface (WES), 680
walking stick insects, 6
wallabies, 182
Wallace, Alfred Russel, 114, 220, 446, 742, 870, **962–967**. *See also* Wallace's Line
Wallacea, 114, 446, 447, 448, *450*, 452, 453, 720, 814, 968
Wallace effect, 966
Wallace's Line, 114, 447, 458, 656, 723, 729, 776, 899, 964
Wallis Island, 953
walruses, 76, 78, 866
Walvis Ridge, 951
warblers
    Antilles, 26
    Canary Islands, 131
    Cape Verde, 144
    Channel Islands (California), 159
    continental islands, 186
    Cook Islands, 194
    Cozumel, 204
    Fiji, 303
    French Polynesia, 335
    Galápagos, 352, 353, 354, 360
    Madagascar, 106
    Marianas, 597
    New Guinea, 657
    oases, 687
    Pitcairn, 745
    Rodrigues, 107
    Seychelles, 833
Warming Island, **971–973**
Warner, Richard, 414
washover fans/terraces, 83, *84*, 85
wasps, 23, 88–89, 90, 100, 102, *123*, 672, *927*, 929
water, 172, 409, 749. *See also* groundwater; tides; waves
water availability. *See also* hydrology
    aliens species and, 99
    arctic, 51
    Ascension, 62
    atolls, on, 68
    Cape Verde, 146
    Channel Islands (British), 155
    climate change and, 15
    Cook Islands, 192, 193
    coral reefs and, 782
    Cozumel, 205
    diversity and, 13, 14
    factors in, 201
    forests and, 128
    freshwater species and, 343
    geology, 485
    Hawaii, 409
    inselbergs and, 466, 467–468
    Kurile Islands, 521
    lava and, 543, 547, 548
    Madagascar, 579

oases, 687, 688, 689
seabirds and, 812
silverswords and, 836
weaverbirds and, 109–110
watercraft. *See also* Kon-Tiki; shipwrecks
    ancient, 277, 278, 279–280
    Austronesian, 721
    human colonization and, 720
    Polynesian, 722, 723, 758–759
    Satawal traditional, 783
    whaling, 975, 976, 977
water density, 421
waterfowl, 124, 323, 325, 326. *See also* seabirds
water movement, 13, 14, 131, 200. *See also* cold seeps
Watling, Dick, 299, 302, 304
Watson, 859
wattle, 797
waves. *See also* storms; surf in the tropics; tsunamis
    barrier islands and, 83, 85, 87
    beaches and, 91, 93, 262
    climate and, 86
    ephemeral islands and, 259
    erosion and, 261, 491, 885
    freshwater species and, 345
    Hawaii and, 409
    intertidal habitats and, 909, 910, 912
    Krakatau and, 517
    Maldives and, 587
    navigation and, 280
    reefs and, 386
    sea levels and, 916
    shorelines and, 93
    tsunami, 933–934
weakness of islands, 473
weather, 11, 92, 93, 193. *See also* climate; rain; temperatures
weaverbirds, 108–109, 110
Webber, John, 759
Weber's Line, 968, 970
weeds, 100, 330, 479
weevils
    adaptive radiation and, 2, 3
    arctic, 51
    Barro Colorado, 89
    Bermuda, 97
    biocontrol and, 101
    Canary Islands, 130
    French Polynesian endemics, 335
    Galápagos, 361, 363, 368
    Gough Island, 931
    New Zealand, 671
    rafting and, 775
    Rapa, 461
    single-island radiations, 462
    Southern Ocean Islands and, 12–13
    St. Helena, 871
    Tristan da Cunha, 931
Wegener, Alfred, 702, 710
Weichselian glaciation, 59
Weinmannia, 940
Weisler, M.I., 45
wekiu bugs, 402
Wellington, 977

west Africa, *651*
West Bismarck Arc, 485
West-Eberhard, Mary Jane, 826
Western Ghats, 840
Western Isles (Outer Hebrides), *116*, 120, 123–124
western Pacific, 755, 825
West Indies
　ants and, 40
　bird radiation, 106
　birds, 321
　frogs, 348, 349
　introduced species, 471
　lizards, 828
　mammals, 321, 590
　snakes, 844, 845
　species-area relationship and, 857, 860
　vicariance, 950
　whales, 975, 976, 978
West Nile virus, 105, 479
west Pacific, 715
West Pacific Seamount Province (WPSP), *705*, 715
West Virginia caves, 151
weta, 211, 285, 318, 667, *668, 669*, 670, 671, 672
wet forests and bogs, 331, 401–402, 842, 870, 907
wetlands
　Britain and Ireland and, 125
　Cape Verde, 147
　conservation, 896
　French Polynesian, 335
　global warming and, 377–378
　Greek Islands, 391
　Midway Island, 632
　Samoa, 800
　Solomons, 851
　Zanzibar, 983
whale falls, **973–975**, 978
whales, **975–978**
　Antarctic, 195
　Baffin Island, 77, 78
　Bermuda, 97
　Britain and Ireland and, 125
　Canary Islands, 132
　Cape Verde, 145
　Channel Islands (California), 159–160
　cold seeps and, 175
　falls, 700
　Farallon Islands, 296
　human impacts, 14
　Kurile Islands, 525
　Line Islands, 557
　Rottnest Island, 798
　Samoa, 800
　Socotra, 850
　Solomons, 853
　Surtsey Island, 887
　Tonga, 920
　Vancouver, 938
　wintering areas, 22
　Zanzibar, 984
whaling, **975–978**
　Antarctic, 15, 54
　Beagle and, 956
　Bermuda, 97
　Farallon Islands, 296
　Galápagos and, 710
　pigs and goats and, 742
　Pitcairn, 745
　Rottnest Island, 798
　Seychelles, 832
　Solomons and, 853
　Spitsbergen and, 866
　Tristan da Cunha, 931
Whataroa virus, 104–105
wheelstamen tree, 899
white-eyes, *617*
White Island, *423*, 674
White, Philo, 918
White, Thomas, 124
Whitlock, M.C., 766
Whitsunday Islands, *387*
wild cats, 120
Williams, Ernst, 562
Wilson, Edward O., 38, 253, 271, 629, 772, 774, 912. *See also* Equilibrium Theory of Island Biogeography
Wilson, Henry, 716
Wilson, J. Tuzo, 704, 711
wind
　ants and, 36, 37
　arctic, 49, 52, 61
　Ascension and, 61
　Azores and, 72
　Bermuda and, 95
　Borneo biota and, 114
　Canary Islands and, 127
　climate and, 172
　climate change and, 15
　dispersal and, 227, 368
　diversity and, 13
　erosion, 885
　flightlessness and, 317
　freshwater species and, 344, 345
　Line Islands and, 555
　nutrients and, 295
　plant disease and, 749
　reefs and, 68
　seafaring and, 279
　shorelines and, 92–93
　size of island and, 859
　waves and, 880
window lakes, 331–332
Windward Islands, 23, 143
windward regions
　Fiji, 299
　freshwater species and, 344
　hydrology and, 421
　lava and, 543
　Rapanui, 245
　reefs, 556
　Samoa, 799
　Vanuatu, 940
winglessness, 37, 130, 211, 318, 373, 375, *476*, 499, 548
wings, 227, 812
wireweeds, 97
Wisconsin glaciation, 59, 83
within-island radiations, 772

Wizard Island, **979–981**
wolverines, *120*
wolves, 52, 78, 120
The Wonderful Century (Wallace), 966
wood-eating clams, 701
woodhens, 572
Woodlark Basin, 664, 665
woodpeckers, 21
Wootton, Timorth, 910
World War II
　intraplate islands and, 708
　Marshall Islands and, 712
　movies, 762
　nuclear weapons, 706
　Philippines and, 730
　Phosphate Islands and, 739
　prisons and, 771
　rats and, 793
　Rottnest Island and, 796–797
　shipwrecks, 834
　snakes and, 845
　Tristan da Cunha/Gough and, 931
　Vanuatu and, 941
worms
　Antarctic, 15
　Antarctic nematode, 12
　cold seeps and, 174
　Cook Islands, 194
　flat, 124
　palm, 425
　Philippines, 724–725
　Surtsey Island, 886, 887
　sustainability and, 891
　tube, 176
　whale falls and, 974
Wrangel Island, 48, *49*, 51, 59, *423*
Wrangellia, 708, 937
wrens, 159, 322, 671, 672, 907
Wright, 209, 860
Wright, S. Joseph, 90
Wright, Sewall, 436, 766
Wright's island model, 766

xenoliths, 31, 805

Yap, FSM, 784
Yap Islands, *94*, 148, 149, *277*, 722, 782, *783*, 784
yellow crazy ants, 39, 472, 479, 534–535
Yemen, 100
Yttygran Island, 51

"zagoutis," 23, 24
Zambia, 146
Zanzibar, 39, 445, **982–986**
Zealandia, 322, 323, 325, 665, 666, 670, 673, 674, 677–680, 707
Zemlya Frantsa Iosifa, 59
Zetek, James, 88
Zola, Émile, 769
zoogeography, 965–966, *969*
zooplankton, 295, 527, 528–529, 530, 531, 565, 820, *891*
zooxanthellae, 780, 781, 782
zooxanthellate corals, 198, 199, 201, 202, 378

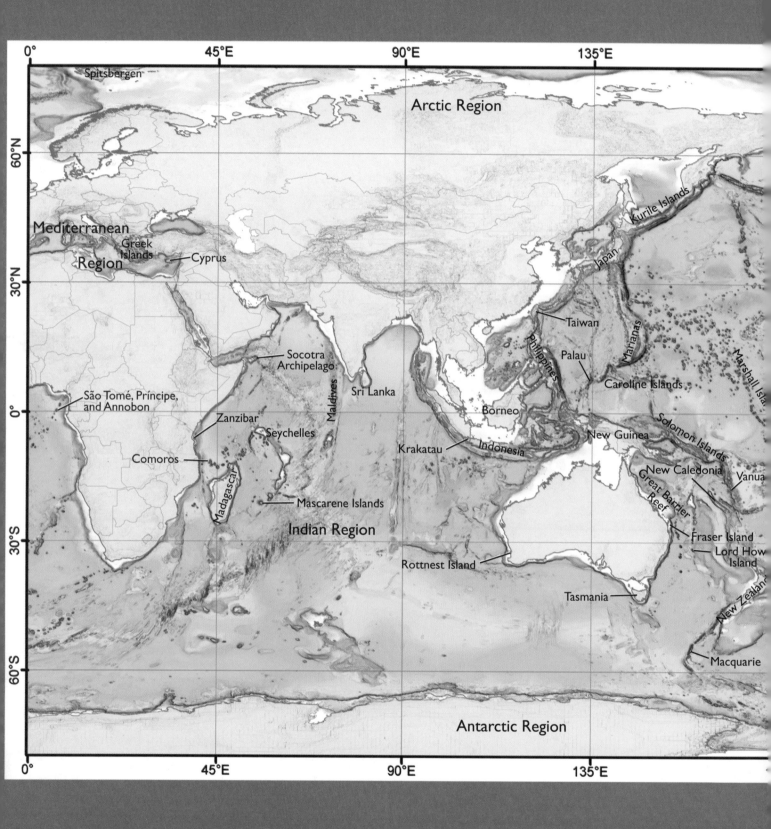